Principles of Bone Biology

Principles of Bone Biology

Fourth Edition

Volume 1

Edited by

John P. Bilezikian
Division of Endocrinology, Department of Medicine
College of Physicians and Surgeons
Columbia University, New York, NY, United States

T. John Martin
St. Vincent's Institute of Medical Research;
Department of Medicine at St. Vincent's Hospital
The University of Melbourne
Melbourne, Australia

Thomas L. Clemens
Department of Orthopaedic Surgery
Johns Hopkins University School of Medicine;
Baltimore Veterans Administration Medical Center
Baltimore, MD, United States

Clifford J. Rosen
Maine Medical Center Research Institute
Scarborugh, ME, United States

Academic Press is an imprint of Elsevier
125 London Wall, London EC2Y 5AS, United Kingdom
525 B Street, Suite 1650, San Diego, CA 92101, United States
50 Hampshire Street, 5th Floor, Cambridge, MA 02139, United States
The Boulevard, Langford Lane, Kidlington, Oxford OX5 1GB, United Kingdom

Library of Congress Cataloging-in-Publication Data
A catalog record for this book is available from the Library of Congress

British Library Cataloguing-in-Publication Data
A catalogue record for this book is available from the British Library

SET ISBN: 978-0-12-814841-9
Volume 1 ISBN: 978-0-12-819932-9
Volume 2 ISBN: 978-0-12-819933-6

For information on all Academic Press publications visit our website at
https://www.elsevier.com/books-and-journals

Publisher: Stacy Masucci
Acquisition Editor: Tari Broderick
Editorial Project Manager: Megan Ashdown
Production Project Manager: Poulouse Joseph
Cover designer: Greg Harris

Typeset by TNQ Technologies

Dedication of Fourth Edition to Lawrence G. Raisz

By the end of the 1970s, when the bone research community felt that it was ready for its own scientific society, Larry Raisz was one of the leaders of the group that founded the American Society for Bone and Mineral Research (ASBMR). The ASBMR had its first annual conference in 1979, with Larry serving as its second president. As the first editor of the *Journal of Bone and Mineral Research*, for a decade, Larry set the highest scientific standards for quality and integrity. That standard remains untarnished today.

Larry's knowledge of the facts in our field was prodigious. His expertise and experience in basic elements of bone biology were exceptional. He had great understanding and wisdom in interpretation of the clinical implications of basic bone biology. But he always wanted to know more. At ASBMR and other annual meetings, it was always Larry who rose to the microphone after a presentation to ask, not only the first question, but typically the best one! Remarkably, Larry could translate basic bone biology to the clinical arena. Few in our field then or now could so smoothly integrate clinical aspects of metabolic bone diseases with the burgeoning knowledge of underlying pathophysiological mechanisms. Adding to these talents was a collegiality and an exuberant enthusiasm that pervaded all venues of Larry Raisz's world. As osteoporosis became more widely recognized to be a medical scourge, then and now, Larry quickly grasped the need to speak about the burden of the disease and contributed to the international dialogue, raising awareness among us all. This awareness was a major factor in the recognition among countries that we are dealing with a disease that needs greater understanding at all levels. And, indeed, at all levels, Larry contributed so much.

These qualities made Larry Raisz a wonderfully effective coeditor of the first three editions of *Principles of Bone Biology*. Much more than that, though, he was a pleasure to work with as a colleague and friend, exceptionally efficient and with unfailing humor and optimism when faced with any adversity. Larry would share the highs and lows with you, but the lows were rare and short lived.

We remember him constantly and dedicate to Lawrence G. Raisz, MD, this fourth edition of what he called "Big Gray."

John P. Bilezikian
T. John Martin
Thomas L. Clemens
Clifford J. Rosen

Dedication to Gideon A. Rodan, reprinted from the Third Edition

We pay special tribute in this third edition of *Principles of Bone Biology* to one of the original three editors, Gideon Rodan, who passed away after a long illness on January 1, 2006. Gideon was a wonderful scientist who made outstanding contributions to our understanding of bone cell biology and to the treatment of metabolic bone diseases. His quiet but highly effective leadership style, superb intellect, and major scientific achievements brought together bone and mineral investigators from all over the world. He was a beloved friend whose insight, empathy, and sense of humor enriched our lives. Gideon's wisdom and breadth of knowledge were invaluable in selecting and evaluating the contributions to the first two editions of this book.

Gideon's education in mathematics and basic sciences in Israel, and his PhD at the Weitzman Institute in the physicochemical aspects of mineral metabolism, provided the fuel for a career of sustained achievement and scholarship. He began his academic career at the University of Connecticut Dental School, rapidly became chairman of the Department of Oral Biology, and built a program of research that brought that school to great prominence. He was a mentor supreme, with a large number of students, postdoctoral trainees, and close colleagues who went on to have successful careers. They remained intensely loyal to him. A former president of the American Society of Bone and Mineral Research (ASBMR), Gideon was also the first recipient of the ASBMR Excellence in Mentorship award, an award that has been named for him in perpetuity. After moving to the pharmaceutical industry in 1984 to lead research and development in bone biology and osteoporosis at Merck, Gideon fulfilled one of his obligations to that position many times over by selecting and then developing alendronate as a treatment for osteoporosis. This achievement set the bar for all future drug development programs in osteoporosis. Most remarkably at Merck, however, Gideon retained and developed even further the rigorous academic approach to bone biology that had always characterized him, wherever he was. In his never-ending quest to teach, to train, and to learn, Gideon was helped enormously by his wife, Sevgi, also his lifelong coworker.

We all owe much to the innovative thought that Gideon brought to all levels of bone and mineral research. His great contributions directed our thinking and our concepts for an entire generation that has followed him. Gideon would have contributed as much to this third edition as he did to the first two editions. It is with the greatest admiration and respect that we dedicate this third edition of *Principles of Bone Biology* to his memory.

T. John Martin
Lawrence G. Raisz
John P. Bilezikian

Contents

List of Contributors xiii
Preface to the Fourth Edition xxi

Volume 1

Part I
Basic principles

Section A
Cell biology

1. Molecular and cellular regulation of intramembranous and endochondral bone formation during embryogenesis 5
Christine Hartmann and Yingzi Yang

2. Skeletal stem cells: tissue-specific stem/progenitor cells of cartilage, bone, stroma, and marrow adipocytes 45
Pamela Gehron Robey and Mara Riminucci

3. Bone marrow and the hematopoietic stem cell niche 73
Laura M. Calvi

4. The osteoblast lineage: its actions and communication mechanisms 89
Natalie A. Sims and T. John Martin

5. Osteoclasts 111
Naoyuki Takahashi, Yasuhiro Kobayashi and Nobuyuki Udagawa

6. The osteocyte 133
J. Klein-Nulend and L.F. Bonewald

7. Transcriptional control of osteoblast differentiation and function 163
Gérard Karsenty

8. Wnt signaling and bone cell activity 177
Bart O. Williams and Mark L. Johnson

9. Vascular and nerve interactions 205
Ryan E. Tomlinson, Thomas L. Clemens and Christa Maes

10. Coupling of bone formation and resorption 219
Natalie A. Sims and T. John Martin

11. Modeling and remodeling: the cellular machinery responsible for bone's material and structural strength during growth, aging, and drug therapy 245
Ego Seeman

12. Aging and bone 275
Maria Almeida and Stavros Manolagas

Section B
Biochemistry

13. Type I collagen structure, synthesis, and regulation 295
George Bou-Gharios, David Abraham and Benoit de Crombrugghe

14. Collagen cross-linking and bone pathobiology 339

David M. Hudson, MaryAnn Weis and David R. Eyre

15. Secreted noncollagenous proteins of bone 359

Jeffrey P. Gorski and Kurt D. Hankenson

16. Bone proteinases 379

Teruyo Nakatani and Nicola C. Partridge

17. Integrins and other cell surface attachment molecules of bone cells 401

Pierre J. Marie and Anna Teti

18. Intercellular junctions and cell–cell communication in the skeletal system 423

Joseph P. Stains, Francesca Fontana and Roberto Civitelli

Section C
Bone remodeling and mineral homeostasis

19. Histomorphometric analysis of bone remodeling 445

Carolina A. Moreira and David W. Dempster

20. Phosphorus homeostasis and related disorders 469

Thomas O. Carpenter, Clemens Bergwitz and Karl L. Insogna

21. Magnesium homeostasis 509

Karl P. Schlingmann and Martin Konrad

22. Metal ion toxicity in the skeleton: lead and aluminum 527

J. Edward Puzas and Brendan F. Boyce

23. Biology of the extracellular calcium-sensing receptor 539

Chia-Ling Tu, Wenhan Chang and Dolores M. Shoback

Section D
Endocrine and paracrine regulation of bone

24. Parathyroid hormone molecular biology 575

Tally Naveh-Many, Justin Silver and Henry M. Kronenberg

25. Paracrine parathyroid hormone–related protein in bone: physiology and pharmacology 595

T. John Martin and Natalie A. Sims

26. Cardiovascular actions of parathyroid hormone/parathyroid hormone–related protein signaling 623

Sasan Mirfakhraee and Dwight A. Towler

27. Parathyroid hormone and parathyroid hormone–related protein actions on bone and kidney 645

Alessandro Bisello and Peter A. Friedman

28. Receptors for parathyroid hormone and parathyroid hormone–related protein 691

Thomas J. Gardella, Harald Jüppner and John T. Potts, Jr.

29. Structure and function of the vitamin D-binding proteins 713

Daniel D. Bikle

30. Vitamin D gene regulation 739

Sylvia Christakos and J. Wesley Pike

31. Nonskeletal effects of vitamin D: current status and potential paths forward 757

Neil Binkley, Daniel D. Bikle, Bess Dawson-Hughes, Lori Plum, Chris Sempos and Hector F. DeLuca

32. Cellular actions of parathyroid hormone on bone 775

Elena Ambrogini and Robert L. Jilka

33. Calcitonin peptides 789

Dorit Naot, David S. Musson and Jillian Cornish

34. Regulation of bone remodeling by central and peripheral nervous signals 809

Patricia Ducy

Section E
Other systemic hormones that influence bone metabolism

35. Estrogens and progestins 827

David G. Monroe and Sundeep Khosla

36. Physiological actions of parathyroid hormone-related protein in epidermal, mammary, reproductive, and pancreatic tissues 839

Christopher S. Kovacs

37. The pharmacology of selective estrogen receptor modulators: past and present 863

Jasna Markovac and Robert Marcus

38. Thyroid hormone and bone 895

Peter A. Lakatos, Bence Bakos, Istvan Takacs and Paula H. Stern

39. Basic and clinical aspects of glucocorticoid action in bone 915

Hong Zhou, Mark S. Cooper and Markus J. Seibel

40. Diabetes and bone 941

Caterina Conte, Roger Bouillon and Nicola Napoli

41. Androgen receptor expression and steroid action in bone 971

Venkatesh Krishnan

Section F
Local regulators

42. Growth hormone, insulin-like growth factors, and IGF binding proteins 985

Clifford J. Rosen and Shoshana Yakar

Index for Volumes 1 and 2 1017

Volume 2

43. The periodontium 1061

Stephen E. Harris, Audrey Rakian, Brian L. Foster, Yong-Hee Patricia Chun and Rubie Rakian

44. Notch and its ligands 1083

Stefano Zanotti and Ernesto Canalis

45. Fibroblast growth factor (FGF) and FGF receptor families in bone 1113

Pierre J. Marie, Marja Hurley and David M. Ornitz

46. Vascular endothelial growth factor and bone–vascular interactions 1141

Steve Stegen and Geert Carmeliet

47. Transforming growth factor-β and skeletal homeostasis 1153

Xin Xu and Xu Cao

48. Bone morphogenetic proteins 1189

David E. Maridas, Marina Feigenson, Nora E. Renthal, Shek Man Chim, Laura W. Gamer and Vicki Rosen

49. Extraskeletal effects of RANK ligand 1199

Andy Göbel, Tilman D. Rachner and Lorenz C. Hofbauer

50. Local regulators of bone 1205

Natalie A. Sims and Joseph A. Lorenzo

51. Prostaglandins and bone metabolism 1247

Shilpa Choudhary and Carol Pilbeam

Part II
Molecular mechanisms of metabolic bone disease

52. The molecular actions of parathyroid hormone/parathyroid hormone–related protein receptor type 1 and their implications 1273

Michael Mannstadt and Marc N. Wein

53. Multiple endocrine neoplasia type 1 1293

Francesca Giusti, Francesca Marini, Francesco Tonelli and Maria Luisa Brandi

54. Parathyroid hormone-related peptide and other mediators of skeletal manifestations of malignancy 1307

Richard Kremer and David Goltzman

55. Localized osteolysis 1335

Julie A. Rhoades (Sterling), Rachelle W. Johnson and Conor C. Lynch

56. Genetic regulation of parathyroid gland development 1355

Fadil M. Hannan and Rajesh V. Thakker

57. Genetic disorders caused by mutations in the parathyroid hormone/parathyroid hormone–related peptide receptor, its ligands, and downstream effector molecules 1379

Caroline Silve and Harald Jüppner

58. Molecular basis of parathyroid hormone overexpression 1405

Geoffrey N. Hendy and Andrew Arnold

59. Diseases resulting from defects in the G protein $G_s\alpha$ 1431

Lee S. Weinstein and Michael T. Collins

60. Renal osteodystrophy and chronic kidney disease–mineral bone disorder 1463

Sharon M. Moe and Thomas L. Nickolas

61. Osteogenesis imperfecta 1489

David W. Rowe

62. Hereditary deficiencies in vitamin D action 1507

Uri A. Liberman

63. Fibroblast growth factor 23 1529

Seiji Fukumoto

64. Tumor-induced osteomalacia 1539

Michael T. Collins, Iris R. Hartley and Pablo Florenzano

65. Osteopetrosis 1553

Antonio Maurizi and Anna Teti

66. Hypophosphatasia: nature's window on alkaline phosphatase function in humans 1569

Michael P. Whyte

67. Paget's disease of bone 1601

Frederick R. Singer and G. David Roodman

68. Genetic determinants of bone mass and osteoporotic fracture 1615

Yi-Hsiang Hsu, Charles R. Farber and Douglas P. Kiel

Part III
Pharmacological mechanisms of therapeutics

69. Pharmacologic mechanisms of therapeutics: parathyroid hormone 1633

Donovan Tay, Gaia Tabacco, Serge Cremers and John P. Bilezikian

70. Calcium 1643

Connie M. Weaver and Robert P. Heaney

71. Drugs acting on the calcium receptor: calcimimetics and calcilytics 1657

Cristiana Cipriani, Edward F. Nemeth and John P. Bilezikian

72. Clinical and translational pharmacology of bisphosphonates 1671

Serge Cremers, Matthew T. Drake, Frank H. Ebetino, Michael J. Rogers, John P. Bilezikian and R. Graham G Russell

73. Pharmacological mechanisms of therapeutics: receptor activator of nuclear factor–kappa B ligand inhibition 1689

Elena Tsourdi, Michael S. Ominsky, Tilman D. Rachner, Lorenz C. Hofbauer and Paul J. Kostenuik

74. Pharmacologic basis of sclerostin inhibition 1711

Hua Zhu Ke, Scott J. Roberts and Gill Holdsworth

75. Vitamin D and its analogs 1733

Glenville Jones and J. Wesley Pike

76. Mechanisms of exercise effects on bone quantity and quality 1759

Vihitaben S. Patel, Stefan Judex, Janet Rubin and Clinton T. Rubin

Part IV
Methods in bone research

77. Application of genetically modified animals in bone research 1787

Matthew J. Hilton and Karen M. Lyons

78. Bone turnover markers 1801

Patrick Garnero and Serge Cremers

79. Microimaging 1833

Steven Boyd and Ralph Müller

80. Macroimaging 1857

Klaus Engelke, Harry K. Genant and James Griffith

81. Methods in lineage tracing 1887

Brya G. Matthews, Noriaki Ono and Ivo Kalajzic

82. Bone histomorphometry in rodents 1899

Y. Linda Ma, David B. Burr and Reinhold G. Erben

83. Bone strength testing in rodents 1923

Mary L. Bouxsein and Frank C. Ko

84. Regulation of energy metabolism by bone-derived hormones 1931

Mathieu Ferron and Gérard Karsenty

Index for Volumes 1 and 2 1943

List of Contributors

David Abraham Centre for Rheumatology and Connective Tissue Diseases, University College London, London, United Kingdom

Maria Almeida Division of Endocrinology and Metabolism, Center for Osteoporosis and Metabolic Bone Diseases, University of Arkansas for Medical Sciences, Little Rock, AR, United States; The Central Arkansas Veterans Healthcare System, Little Rock, AR, United States

Elena Ambrogini Center for Osteoporosis and Metabolic Bone Diseases, University of Arkansas for Medical Sciences Division of Endocrinology and Metabolism, Little Rock, AR, United States; Central Arkansas Veterans Healthcare System, Little Rock, AR, United States

Andrew Arnold Center for Molecular Oncology and Division of Endocrinology and Metabolism, University of Connecticut School of Medicine, Farmington, CT, United States

Bence Bakos 1st Department of Medicine, Semmelweis University Medical School, Budapest, Hungary

Clemens Bergwitz Departments of Pediatrics and Internal Medicine, Yale University School of Medicine, New Haven, CT, United States

Daniel D. Bikle VA Medical Center and University of California San Francisco, San Francisco, California, United States

John P. Bilezikian Division of Endocrinology, Department of Medicine, College of Physicians and Surgeons, Columbia University, New York, NY, United States

Neil Binkley University of Wisconsin School of Medicine and Public Health, Madison, Wisconsin, United States

Alessandro Bisello Department of Pharmacology and Chemical Biology, Laboratory for GPCR Biology, University of Pittsburgh School of Medicine, Pittsburgh, PA, United States

L.F. Bonewald Indiana Center for Musculoskeletal Health, Departments of Anatomy and Cell Biology and Orthopaedic Surgery, Indiana University, Indianapolis, IN, USA

George Bou-Gharios Institute of Ageing and Chronic Disease, University of Liverpool, Liverpool, United Kingdom

Roger Bouillon Laboratory of Clinical and Experimental Endocrinology, Department of Chronic Diseases, Metabolism and Aging, KU Leuven, Belgium

Mary L. Bouxsein Center for Advanced Orthopaedic Studies, Beth Israel Deaconess Medical Center, Boston, MA, United States; Department of Orthopaedic Surgery, Harvard Medical School, Boston, MA, United States; Endocrine Unit, Department of Medicine, Massachusetts General Hospital, Boston, MA, United States

Brendan F. Boyce Department of Pathology and Laboratory Medicine, University of Rochester School of Medicine and Dentistry, Rochester, NY, United States

Steven Boyd McCaig Institute for Bone and Joint Health, The University of Calgary, Calgary, AB, Canada

Maria Luisa Brandi Department of Experimental and Clinical Biomedical Sciences, University of Florence, Florence, Italy

David B. Burr Department of Anatomy and Cell Biology, Indiana Center for Musculoskeletal Health, Indiana University School of Medicine, Indianapolis, IN, United States

Laura M. Calvi Department of Medicine and Wilmot Cancer Center, University of Rochester Medical Center, Rochester, NY, United States

Ernesto Canalis Departments of Orthopaedic Surgery and Medicine, and the UConn Musculoskeletal Institute, UConn Health, Farmington, CT, United States

Xu Cao Department of Orthopedic Surgery, Johns Hopkins University School of Medicine, Baltimore, MD, United States

Geert Carmeliet Laboratory of Clinical and Experimental Endocrinology, Department of Chronic Diseases, Metabolism and Ageing, KU Leuven, Leuven, Belgium; Prometheus, Division of Skeletal Tissue Engineering, KU Leuven, Leuven, Belgium

Thomas O. Carpenter Departments of Pediatrics and Internal Medicine, Yale University School of Medicine, New Haven, CT, United States

Wenhan Chang Endocrine Research Unit, Department of Veterans Affairs Medical Center, Department of Medicine, University of California, San Francisco, CA, United States

Shek Man Chim Regeneron Pharmaceuticals, Inc. Tarrytown, NY, United States

Shilpa Choudhary Department of Medicine and Musculoskeletal Institute, UConn Health, Farmington, CT, United States

Sylvia Christakos Department of Microbiology, Biochemistry and Molecular Genetics, Rutgers, New Jersey Medical School, Newark, NJ, United States

Yong-Hee Patricia Chun Department of Periodontics, University of Texas Health Science Center at San Antonio, San Antonio, TX, United States

Cristiana Cipriani Department of Internal Medicine and Medical Disciplines, Sapienza University of Rome, Italy

Roberto Civitelli Washington University in St. Louis, Department of Medicine, Division of Bone and Mineral Diseases, St. Louis, MO, United States

Thomas L. Clemens Department of Orthopaedic Surgery, Johns Hopkins University School of Medicine, Baltimore, MD, United States; Baltimore Veterans Administration Medical Center, Baltimore, MD, United States

Michael T. Collins Skeletal Disorders and Mineral Homeostasis Section, National Institute of Dental and Craniofacial Research, National Institutes of Health, Bethesda, MD, United States

Caterina Conte Vita-Salute San Raffaele University, Milan, Italy; Division of Immunology, Transplantation and Infectious Diseases, IRCCS San Raffaele Scientific Institute, Milan, Italy

Mark S. Cooper The University of Sydney, ANZAC Research Institute and Department of Endocrinology & Metabolism, Concord Hospital, Sydney, NSW, Australia

Jillian Cornish Department of Medicine, University of Auckland, Auckland, New Zealand

Serge Cremers Department of Pathology & Cell Biology and Department of Medicine, Vagelos College of Physicians and Surgeons, Columbia University Irving Medical Center, United States

Bess Dawson-Hughes Jean Mayer USDA Human Nutrition Research Center on Aging at Tufts University, Boston, Massachusetts, United States

Benoit de Crombrugghe The University of Texas M.D. Anderson Cancer Center, Houston, TX, United States

Hector F. DeLuca Department of Biochemistry, University of Wisconsin–Madison, Madison, Wisconsin, United States

David W. Dempster Regional Bone Center, Helen Hayes Hospital, West Haverstraw, NY, United States; Department of Pathology and Cell Biology, College of Physicians and Surgeons, Columbia University, New York, NY, United States

Matthew T. Drake Department of Endocrinology and Kogod Center of Aging, Mayo Clinic College of Medicine, Rochester, MN, United States

Patricia Ducy Department of Pathology & Cell Biology, Columbia University, College of Physicians & Surgeons, New York, NY, United States

Frank H. Ebetino Department of Chemistry, University of Rochester, Rochester, NY, United States; Mellanby Centre for Bone Research, Medical School, University of Sheffield, United Kingdom

Klaus Engelke Department of Medicine, FAU University Erlangen-Nürnberg and Universitätsklinikum Erlangen, Erlangen, Germany; Bioclinica, Hamburg, Germany

Reinhold G. Erben Department of Biomedical Research, University of Veterinary Medicine Vienna, Vienna, Austria

David R. Eyre Department of Orthopaedics and Sports Medicine, University of Washington, Seattle, WA, United States

Charles R. Farber Center for Public Health Genomics, Departments of Public Health Sciences and Biochemistry and Molecular Genetics, University of Virginia School of Medicine, Charlottesville, VA, United States

Marina Feigenson Department of Developmental Biology, Harvard School of Dental Medicine, Boston, MA, United States

Mathieu Ferron Institut de Recherches Cliniques de Montréal, Montréal, QC, Canada

Pablo Florenzano Endocrine Department, School of Medicine, Pontificia Universidad Católica de Chile, Santiago, Chile

Francesca Fontana Washington University in St. Louis, Department of Medicine, Division of Bone and Mineral Diseases, St. Louis, MO, United States

Brian L. Foster Biosciences Division at College of Dentistry at Ohio State University, Columbus, OH, United States

Peter A. Friedman Department of Pharmacology and Chemical Biology, Laboratory for GPCR Biology, University of Pittsburgh School of Medicine, Pittsburgh, PA, United States

Seiji Fukumoto Fujii Memorial Institute of Medical Sciences, Institute of Advanced Medical Sciences, Tokushima University, Tokushima, Japan

Laura W. Gamer Department of Developmental Biology, Harvard School of Dental Medicine, Boston, MA, United States

Thomas J. Gardella Endocrine Unit, Department of Medicine and Pediatric Nephrology, MassGeneral Hospital for Children, Massachusetts General Hospital and Harvard Medical School, Boston, MA, United States

Patrick Garnero INSERM Research Unit 1033-Lyos, Lyon, France

Harry K. Genant Departments of Radiology and Medicine, University of California, San Francisco, CA, United States

Francesca Giusti Department of Experimental and Clinical Biomedical Sciences, University of Florence, Florence, Italy

Andy Göbel Department of Medicine III, Technische Universität Dresden, Dresden, Germany; German Cancer Consortium (DKTK), Partner site Dresden and German Cancer Research Center (DKFZ), Heidelberg, Germany

David Goltzman Calcium Research Laboratories and Department of Medicine, McGill University and McGill University Health Centre, Montreal, QC, Canada

Jeffrey P. Gorski Department of Oral and Craniofacial Sciences, School of Dentistry, and Center for Excellence in Mineralized and Dental Tissues, University of Missouri–Kansas City, Kansas City, MO, United States

James Griffith Department of Imaging and Interventional Radiology, The Chinese University of Hong Kong, Hong Kong, China

R. Graham G Russell Mellanby Centre for Bone Research, Medical School, University of Sheffield, United Kingdom; Nuffield Department of Orthopaedics, Rheumatology and Musculoskeletal Sciences, The Oxford University Institute of Musculoskeletal Sciences, The Botnar Research Centre, Nuffield Orthopaedic Centre, Oxford, United Kingdom

Kurt D. Hankenson Department of Orthopaedic Surgery, University of Michigan Medical School, Ann Arbor, MI, United States

Fadil M. Hannan Department of Musculoskeletal Biology, Institute of Ageing and Chronic Disease, Faculty of Health & Life Sciences, University of Liverpool, Liverpool, United Kingdom; Academic Endocrine Unit, Radcliffe Department of Medicine, University of Oxford, Oxford Centre for Diabetes, Endocrinology and Metabolism (OCDEM), Churchill Hospital, Oxford, United Kingdom

Stephen E. Harris Department of Periodontics, University of Texas Health Science Center at San Antonio, San Antonio, TX, United States

Iris R. Hartley Interinstitute Endocrine Training Program, Eunice Kennedy Shriver National Institute of Child Health and Human Development, National Institutes of Health, Bethesda, MD, United States

Christine Hartmann Institute of Musculoskeletal Medicine, Department of Bone and Skeletal Research, Medical Faculty of the University of Münster, Münster, Germany

Robert P. Heaney Creighton University, Omaha, NE, United States

Geoffrey N. Hendy Metabolic Disorders and Complications, McGill University Health Center Research Institute, and Departments of Medicine, Physiology and Human Genetics, McGill University, Montreal, QC, Canada

Matthew J. Hilton Department of Orthopaedic Surgery, Department of Cell Biology, Duke University School of Medicine, Durham, NC, United States

Lorenz C. Hofbauer Center for Regenerative Therapies Dresden, Center for Healthy Aging and Division of Endocrinology, Diabetes, and Bone Diseases, Department of Medicine III, Technische Universität Dresden, Dresden, Germany

Gill Holdsworth Bone Therapeutic Area, UCB Pharma, Slough, United Kingdom

Yi-Hsiang Hsu Department of Medicine, Beth Israel Deaconess Medical Center and Harvard Medical School, Harvard School of Public Health, Hinda and Arthur Marcus Institute for Aging Research, Hebrew SeniorLife, Boston, MA, United States

David M. Hudson Department of Orthopaedics and Sports Medicine, University of Washington, Seattle, WA, United States

Marja Hurley Department of Medicine, University of Connecticut School of Medicine, UConn Health, Farmington, CT, United States

Karl L. Insogna Departments of Pediatrics and Internal Medicine, Yale University School of Medicine, New Haven, CT, United States

Robert L. Jilka Center for Osteoporosis and Metabolic Bone Diseases, University of Arkansas for Medical Sciences Division of Endocrinology and Metabolism, Little Rock, AR, United States; Central Arkansas Veterans Healthcare System, Little Rock, AR, United States

Mark L. Johnson Department of Oral and Craniofacial Sciences, UMKC School of Dentistry, Kansas City, MO, United States

Rachelle W. Johnson Vanderbilt Center for Bone Biology, Department of Medicine, Division of Clinical Pharmacology, Nashville, TN, United States

Glenville Jones Department of Biomedical and Molecular Science, Queen's University, Kingston, ON, Canada

Stefan Judex Department of Biomedical Engineering, Bioengineering Building, State University of New York at Stony Brook, Stony Brook, NY, United States

Harald Jüppner Endocrine Unit, Department of Medicine and Pediatric Nephrology, MassGeneral Hospital for Children, Massachusetts General Hospital and Harvard Medical School, Boston, MA, United States

Ivo Kalajzic Department of Reconstructive Sciences, UConn Health, Farmington, CT, United States

Gérard Karsenty Department of Genetics and Development, Columbia University Medical Center, New York, NY, United States

Hua Zhu Ke Angitia Biopharmaceuticals Limited, Guangzhou, China

Sundeep Khosla Department of Medicine, Division of Endocrinology, Mayo Clinic College of Medicine, Rochester, MN, United States; The Robert and Arlene Kogod Center on Aging, Rochester, MN, United States

Douglas P. Kiel Department of Medicine, Beth Israel Deaconess Medical Center and Harvard Medical School, Harvard School of Public Health, Hinda and Arthur Marcus Institute for Aging Research, Hebrew SeniorLife, Boston, MA, United States

J. Klein-Nulend Department of Oral Cell Biology, Academic Centre for Dentistry Amsterdam (ACTA), University of Amsterdam and Vrije Universiteit Amsterdam, Amsterdam Movement Sciences, Amsterdam, The Netherlands

Frank C. Ko Center for Advanced Orthopaedic Studies, Beth Israel Deaconess Medical Center, Boston, MA, United States

Yasuhiro Kobayashi Institute for Oral Science, Matsumoto Dental University, Nagano, Japan

Martin Konrad Department of General Pediatrics, University Children's Hospital Münster, Münster, Germany

Paul J. Kostenuik Phylon Pharma Services, Newbury Park, CA, United States; School of Dentistry, University of Michigan, Ann Arbor, MI, United States

Christopher S. Kovacs Faculty of Medicine, Memorial University of Newfoundland, St. John's, NL, Canada

Richard Kremer Calcium Research Laboratories and Department of Medicine, McGill University and McGill University Health Centre, Montreal, QC, Canada

Venkatesh Krishnan Lilly Research Laboratories, Eli Lilly & Company, Lilly Corporate Center, Indianapolis, United States

Henry M. Kronenberg Endocrine Unit, Massachusetts General Hospital, Harvard Medical School, Boston, MA, United States

Peter A. Lakatos 1st Department of Medicine, Semmelweis University Medical School, Budapest, Hungary

Uri A. Liberman Department of Physiology and Pharmacology, Sackler Faculty of Medicine, Tel Aviv University, Tel-Aviv, Israel

Joseph A. Lorenzo The Departments of Medicine and Orthopaedics, UConn Health, Farmington, CT, United States

Conor C. Lynch Department of Tumor Biology, Moffitt Cancer Center, Tampa, FL, United States

Karen M. Lyons Department of Orthopaedic Surgery/Orthopaedic Hospital, University of California, Los Angeles, CA, United States; Department of Molecular, Cell, & Developmental Biology, University of California, Los Angeles, CA, United States

Y. Linda Ma Biotechnology and Autoimmunity Research, Eli Lilly and Company, Indianapolis, IN, United States

Christa Maes Laboratory of Skeletal Cell Biology and Physiology (SCEBP), Skeletal Biology and Engineering Research Center (SBE), KU Leuven, Leuven, Belgium

Michael Mannstadt Endocrine Unit, Massachusetts General Hospital, Harvard Medical School, Boston, MA, United States

Stavros Manolagas Division of Endocrinology and Metabolism, Center for Osteoporosis and Metabolic Bone Diseases, University of Arkansas for Medical Sciences, Little Rock, AR, United States; The Central Arkansas Veterans Healthcare System, Little Rock, AR, United States

Robert Marcus Stanford University, Stanford, CA, United States

David E. Maridas Department of Developmental Biology, Harvard School of Dental Medicine, Boston, MA, United States

Pierre J. Marie UMR-1132 Inserm (Institut national de la Santé et de la Recherche Médicale) and University Paris Diderot, Sorbonne Paris Cité, Paris, France

Francesca Marini Department of Experimental and Clinical Biomedical Sciences, University of Florence, Florence, Italy

Jasna Markovac California Institute of Technology, Pasadena, CA, United States

T. John Martin St. Vincent's Institute of Medical Research, Melbourne, Australia; Department of Medicine at St. Vincent's Hospital, The University of Melbourne, Melbourne, Australia

Brya G. Matthews Department of Molecular Medicine and Pathology, University of Auckland, Auckland, New Zealand

Antonio Maurizi Department of Biotechnological and Applied Clinical Sciences, University of L'Aquila, L'Aquila, Italy

Sasan Mirfakhraee The University of Texas Southwestern Medical Center, Department of Internal Medicine, Endocrine Division, Dallas, TX, United Sates

Sharon M. Moe Division of Nephrology, Indiana University School of Medicine, Rodebush Veterans Administration Medical Center, Indianapolis, IN, United States

David G. Monroe Department of Medicine, Division of Endocrinology, Mayo Clinic College of Medicine, Rochester, MN, United States; The Robert and Arlene Kogod Center on Aging, Rochester, MN, United States

Carolina A. Moreira Bone Unit of Endocrine Division of Federal University of Parana, Laboratory PRO, Section of Bone Histomorphometry, Pro Renal Foundation, Curitiba, Parana, Brazil

Ralph Müller Institute for Biomechanics, ETH Zurich, Zurich, Switzerland

David S. Musson Department of Medicine, University of Auckland, Auckland, New Zealand

Teruyo Nakatani Department of Basic Science and Craniofacial Biology, New York University College of Dentistry, New York, NY, United States

Dorit Naot Department of Medicine, University of Auckland, Auckland, New Zealand

Nicola Napoli Unit of Endocrinology and Diabetes, University Campus Bio-Medico, Rome, Italy; Division of Bone and Mineral Diseases, Washington University in St Louis, St Louis, MO, United States

Tally Naveh-Many Minerva Center for Calcium and Bone Metabolism, Nephrology Services, Hadassah University Hospital, Hebrew University School of Medicine, Jerusalem, Israel

Edward F. Nemeth MetisMedica, Toronto, ON, Canada

Thomas L. Nickolas Division of Nephrology, Department of Medicine, Columbia University Medical Center, New York, NY, United States

Michael S. Ominsky Radius Health Inc., Waltham, MA, United States

Noriaki Ono Department of Orthodontics and Pediatric Dentistry, University of Michigan School of Dentistry, Ann Arbor, MI, United States

David M. Ornitz Department of Developmental Biology, Washington University School of Medicine, St. Louis, MO, United States

Nicola C. Partridge Department of Basic Science and Craniofacial Biology, New York University College of Dentistry, New York, NY, United States

Vihitaben S. Patel Department of Biomedical Engineering, Bioengineering Building, State University of New York at Stony Brook, Stony Brook, NY, United States

J. Wesley Pike Department of Biochemistry, University of Wisconsin–Madison, Madison, WI, United States

Carol Pilbeam Department of Medicine and Musculoskeletal Institute, UConn Health, Farmington, CT, United States

Lori Plum Department of Biochemistry, University of Wisconsin–Madison, Madison, Wisconsin, United States

John T. Potts, Jr. Endocrine Unit, Department of Medicine and Pediatric Nephrology, MassGeneral Hospital for Children, Massachusetts General Hospital and Harvard Medical School, Boston, MA, United States

J. Edward Puzas Department of Orthopaedics and Rehabilitation, University of Rochester School of Medicine and Dentistry, Rochester, NY, United States

Tilman D. Rachner Department of Medicine III, Technische Universität Dresden, Dresden, Germany; German Cancer Consortium (DKTK), Partner site Dresden and German Cancer Research Center (DKFZ), Heidelberg, Germany; Center for Healthy Aging and Division of

Endocrinology, Diabetes, and Bone Diseases, Technische Universität Dresden, Dresden, Germany

Audrey Rakian Department of Applied Oral Sciences, The Forsyth Institute, Cambridge, MA, United States; Department of Oral Medicine, Infection, Immunity, Harvard School of Dental Medicine, Boston, MA, United States

Rubie Rakian Department of Applied Oral Sciences, The Forsyth Institute, Cambridge, MA, United States; Department of Oral Medicine, Infection, Immunity, Harvard School of Dental Medicine, Boston, MA, United States

Nora E. Renthal Division of Endocrinology, Children's Hospital Boston, Boston, MA, United States

Julie A. Rhoades (Sterling) Department of Veterans Affairs, Nashville, TN, United States; Vanderbilt Center for Bone Biology, Department of Medicine, Division of Clinical Pharmacology, Nashville, TN, United States; Department of Biomedical Engineering, Nashville, TN, United States

Mara Riminucci Department of Molecular Medicine, Sapienza University of Rome, Rome, Italy

Scott J. Roberts Bone Therapeutic Area, UCB Pharma, Slough, United Kingdom

Pamela Gehron Robey National Institute of Dental and Craniofacial Research, National Institutes of Health, Department of Health and Human Services, Bethesda, MD, United States

Michael J. Rogers Garvan Institute of Medical Research and St Vincent's Clinical School; University of New South Wales, Sydney, Australia

G. David Roodman Department of Medicine, Division of Hematology and Oncology, Indiana University School of Medicine, and Roudebush VA Medical Center, Indianapolis, IN, United States

Clifford J. Rosen Maine Medical Center Research Institute, Scarborugh, ME, United States

Vicki Rosen Department of Developmental Biology, Harvard School of Dental Medicine, Boston, MA, United States

David W. Rowe Center for Regenerative Medicine and Skeletal Development, Department of Reconstructive Sciences, Biomaterials and Skeletal Development, School of Dental Medicine, University of Connecticut Health Center, Farmington, CT, United States

Janet Rubin Endocrine Division, Department of Medicine, University of North Carolina, Chapel Hill, NC, United States

Clinton T. Rubin Department of Biomedical Engineering, Bioengineering Building, State University of New York at Stony Brook, Stony Brook, NY, United States

Karl P. Schlingmann Department of General Pediatrics, University Children's Hospital Münster, Münster, Germany

Ego Seeman Department of Endocrinology and Medicine, Austin Health, University of Melbourne, Melbourne, VIC, Australia; Mary MacKillop Institute for Health Research, Australian Catholic University, Melbourne, VIC, Australia

Markus J. Seibel The University of Sydney, ANZAC Research Institute and Department of Endocrinology & Metabolism, Concord Hospital, Sydney, NSW, Australia

Chris Sempos Vitamin D Standardization Program, Havre de Grace, MD, United States

Dolores M. Shoback Endocrine Research Unit, Department of Veterans Affairs Medical Center, Department of Medicine, University of California, San Francisco, CA, United States

Caroline Silve Hôpital Bicêtre, Paris, France; Centre de Référence des Maladies rares du Calcium et du Phosphore and Filière de Santé Maladies Rares OSCAR, AP-HP, Paris, France; Service de Biochimie et Génétique Moléculaires, Hôpital Cochin, AP-HP, Paris, France

Justin Silver Minerva Center for Calcium and Bone Metabolism, Nephrology Services, Hadassah University Hospital, Hebrew University School of Medicine, Jerusalem, Israel

Natalie A. Sims St. Vincent's Institute of Medical Research, Melbourne, Australia; Department of Medicine at St. Vincent's Hospital, The University of Melbourne, Melbourne, Australia

Frederick R. Singer John Wayne Cancer Institute, Saint John's Health Center, Santa Monica, CA, United States

Joseph P. Stains Department of Orthopaedics, University of Maryland School of Medicine, Baltimore, MD, United States

Steve Stegen Laboratory of Clinical and Experimental Endocrinology, Department of Chronic Diseases, Metabolism and Ageing, KU Leuven, Leuven, Belgium; Prometheus, Division of Skeletal Tissue Engineering, KU Leuven, Leuven, Belgium

Paula H. Stern Department of Pharmacology, Northwestern University Feinberg School of Medicine, Chicago, IL, United States

Gaia Tabacco Division of Endocrinology, Department of Medicine, College of Physicians and Surgeons, Columbia University, New York, NY, United States; Endocrinology and Diabetes Unit, Department of Medicine, Campus Bio-Medico University of Rome, Rome, Italy

Istvan Takacs 1st Department of Medicine, Semmelweis University Medical School, Budapest, Hungary

Naoyuki Takahashi Institute for Oral Science, Matsumoto Dental University, Nagano, Japan

Donovan Tay Division of Endocrinology, Department of Medicine, College of Physicians and Surgeons, Columbia University, New York, NY, United States; Department of Medicine, Sengkang General Hospital, Singhealth, Singapore

Anna Teti Department of Biotechnological and Applied Clinical Sciences, University of L'Aquila, L'Aquila, Italy

Rajesh V. Thakker Academic Endocrine Unit, Radcliffe Department of Medicine, University of Oxford, Oxford Centre for Diabetes, Endocrinology and Metabolism (OCDEM), Churchill Hospital, Oxford, United Kingdom

Ryan E. Tomlinson Department of Orthopaedic Surgery, Thomas Jefferson University, Philadelphia, PA, United States

Francesco Tonelli Department of Experimental and Clinical Biomedical Sciences, University of Florence, Florence, Italy

Dwight A. Towler The University of Texas Southwestern Medical Center, Department of Internal Medicine, Endocrine Division, Dallas, TX, United Sates

Elena Tsourdi Department of Medicine III, Technische Universität Dresden, Dresden, Germany; Center for Healthy Aging, Technische Universität Dresden, Dresden, Germany

Chia-Ling Tu Endocrine Research Unit, Department of Veterans Affairs Medical Center, Department of Medicine, University of California, San Francisco, CA, United States

Nobuyuki Udagawa Department of Biochemistry, Matsumoto Dental University, Nagano, Japan

Connie M. Weaver Purdue University, West Lafayette, IN, United States

Marc N. Wein Endocrine Unit, Massachusetts General Hospital, Harvard Medical School, Boston, MA, United States

Lee S. Weinstein Metabolic Diseases Branch, National Institute of Diabetes, Digestive, and Kidney Diseases, Bethesda, MD, United States

MaryAnn Weis Department of Orthopaedics and Sports Medicine, University of Washington, Seattle, WA, United States

Michael P. Whyte Center for Metabolic Bone Disease and Molecular Research, Shriners Hospitals for Children - St. Louis, St. Louis, MO, United States; Division of Bone and Mineral Diseases, Department of Internal Medicine, Washington University School of Medicine at Barnes-Jewish Hospital, St. Louis, MO, United States

Bart O. Williams Program for Skeletal Disease and Center for Cancer and Cell Biology, Van Andel Research Institute, Grand Rapids, MI, United States

Xin Xu State Key Laboratory of Oral Diseases & National Clinical Research Center for Oral Diseases, Department of Cariology and Endodontics, West China Hospital of Stomatology, Sichuan University, Chengdu, PR China

Shoshana Yakar David B. Kriser Dental Center, Department of Basic Science and Craniofacial Biology, New York University College of Dentistry New York, New York, NY, United States

Yingzi Yang Department of Developmental Biology, Harvard School of Dental Medicine, Boston, MA, United States

Stefano Zanotti Departments of Orthopaedic Surgery and Medicine, and the UConn Musculoskeletal Institute, UConn Health, Farmington, CT, United States

Hong Zhou The University of Sydney, ANZAC Research Institute, Sydney, NSW, Australia

Preface to the Fourth Edition

The first edition of *Principles of Bone Biology* was published about 24 years ago, in 1996. Our field was ready then for a compendium of the latest concepts in bone biology. Now, several decades and two editions later, we are pleased to welcome you to the fourth edition of "Big Gray." Since the third edition was published in 2008, our field has continued to undergo sea changes of knowledge and insights. As a result of these advances since then, all chapters have undergone major revisions. In addition, areas not previously covered in depth are featured, such as vascular and nerve interactions with bone, interorgan communicants of bone, hematopoietic–bone cell interactions, nonskeletal aspects of vitamin D and RANK ligand, newly recognized signaling molecules and systems, and advances in methodological aspects of skeletal research. Illustrative of the vibrancy of our field, over 50% of the authors in this edition are new to it. Our returning and new authors are the very best.

We remember Larry Raisz. He, along with Gideon Rodan, constituted the triumvirate of coeditors for the first and second editions. We dedicated the third edition to the memory of Gideon. We dedicate *Principles of Bone Biology*, fourth edition, to the memory of Larry. In these front pages, we reprint our dedication to Gideon and remember Larry with a separate dedication for this edition. We miss them both very much.

We want to acknowledge Jasna Markovac, who has served as our liaison to our authors and our publisher. Given the nature of the times, this book would not have been completed without her dedication, perseverance, and single-minded purpose not to let anything disrupt our publishing goals. She accomplished this feat with an even handedness and a professionalism that was remarkable and remarkably effective. We are grateful to you, Jasna.

Finally, we are grateful to our authors, who have made this book what it is, namely a repository of knowledge and concepts in bone biology and a resource for us all in the years to come.

John P. Bilezikian
T. John Martin
Thomas L. Clemens
Clifford J. Rosen

Preface to the [illegible] Edition

Part I

Basic principles

Section A

Cell biology

Chapter 1

Molecular and cellular regulation of intramembranous and endochondral bone formation during embryogenesis

Christine Hartmann[1] **and Yingzi Yang**[2]

[1]*Institute of Musculoskeletal Medicine, Department of Bone and Skeletal Research, Medical Faculty of the University of Münster, Münster, Germany;* [2]*Department of Developmental Biology, Harvard School of Dental Medicine, Boston, MA, United States*

Chapter outline

Introduction 5
Intramembranous ossification 6
The axial skeleton 9
Somitogenesis 9
Sclerotome differentiation 11
The limb skeleton 11
Overview of limb development 11
Proximal–distal axis 12
Anterior–posterior axis 14
Dorsal–ventral axis 14
Mesenchymal condensation and patterning of the skeleton 14
Endochondral bone formation 15
Overview 15
The growth plate 18
Mediators of skeleton formation 19
Systemic mediators 19
Local mediators 19
Growth factor signaling pathways 19
Transforming growth factor β and bone morphogenetic proteins 19
Parathyroid hormone-related protein and Indian hedgehog 21
WNTs and β-catenin 23
Fibroblast growth factors and their receptors 24
C-type natriuretic peptide 25
Notch signaling 25
Transcription factors 25
Epigenetic factors and microRNAs 26
The functional roles of the vasculature in endochondral bone formation 27
References 27

Introduction

The skeletal system performs vital functions: support, movement, protection, blood cell production, calcium storage, and endocrine regulation. Skeletal formation is also a hallmark that distinguishes vertebrate animals from invertebrates. In higher vertebrates (i.e., birds and mammals), the skeletal system contains mainly bones and cartilage, as well as a network of tendons and ligaments that connects them. During embryonic development, bones and cartilage are formed by osteoblasts and chondrocytes, respectively, both of which are derived from common mesenchymal progenitor cells called osteochondral progenitors. Skeletal development starts from mesenchymal condensation, during which mesenchymal progenitor cells aggregate at future skeletal locations. As mesenchymal cells in different parts of the embryo are derived from different cell lineages, the locations of initial skeletal formation determine which of the three mesenchymal cell lineages contribute to the future skeleton. Neural crest cells from the branchial arches contribute to the craniofacial bone, the sclerotome compartment of the somites gives rise to most of the axial skeleton, and lateral plate mesoderm forms the limb mesenchyme, from which limb skeletons are derived.

Principles of Bone Biology. https://doi.org/10.1016/B978-0-12-814841-9.00001-4

How osteoblast cells are induced during bone development is a central question for understanding the organizational principles underpinning a functional skeletal system. Abnormal osteoblast differentiation leads to a broad range of devastating skeletal diseases. Therefore, it is imperative to understand the cellular and molecular mechanisms underlying temporal and spatial controls of bone formation. Bone formation occurs by two essential processes: intramembranous ossification and endochondral ossification during embryonic development. Osteochondral progenitors differentiate into osteoblasts directly to form the membranous bone during intramembranous ossification, whereas during endochondral ossification, they differentiate into chondrocytes instead to form a cartilage template of the future bone. Both ossification processes are essential during the natural healing of bone fractures. In this chapter, we focus on current understanding of the molecular regulation of endochondral and intramembranous bone formation and its implication in diseases.

Intramembranous ossification

Intramembranous ossification mainly occurs during formation of the flat bones of the skull, mandible, maxilla, and clavicles. The mammalian cranium, or neurocranium, is the upper and back part of the skull. It protects the brain and supports the sensory organs, such as the ear, and the viscerocranium, which supports the face. The neurocranium can be divided into calvarium and chondrocranium, which grow to be the cranial vault that surrounds the brain and the skull base, respectively. The calvarium is composed of flat bones: frontal bones, parietal bones, the interparietal part of the occipital bone, and the squamous parts of the temporal bone (Jin et al., 2016). In mice, the calvarium consists of frontal bones, parietal bones, interparietal bone, and squamous parts of the temporal bone, all going through intramembranous ossification (Ishii et al., 2015). By lineage analysis in mouse models, frontal bones show a major contribution from neural crest and a small contribution from head mesoderm, while parietal bones entirely originate from head mesoderm (Jiang et al., 2002; Yoshida et al., 2008; Deckelbaum et al., 2012). Neural crest–derived and head mesoderm–derived cells coalesce to form calvarial bone primordia (Jiang et al., 2002; Yoshida et al., 2008). The mandible and maxilla are derived from the neural crest cells originating in the mid- and hindbrain regions of the neural folds that migrate ventrally, while the clavicles are formed from mesoderm.

The process starts from mesenchymal condensation and progresses through formation of the ossification center, ossification expansion, trabecula formation, and compact bone formation and the development of the periosteum (Fig. 1.1).

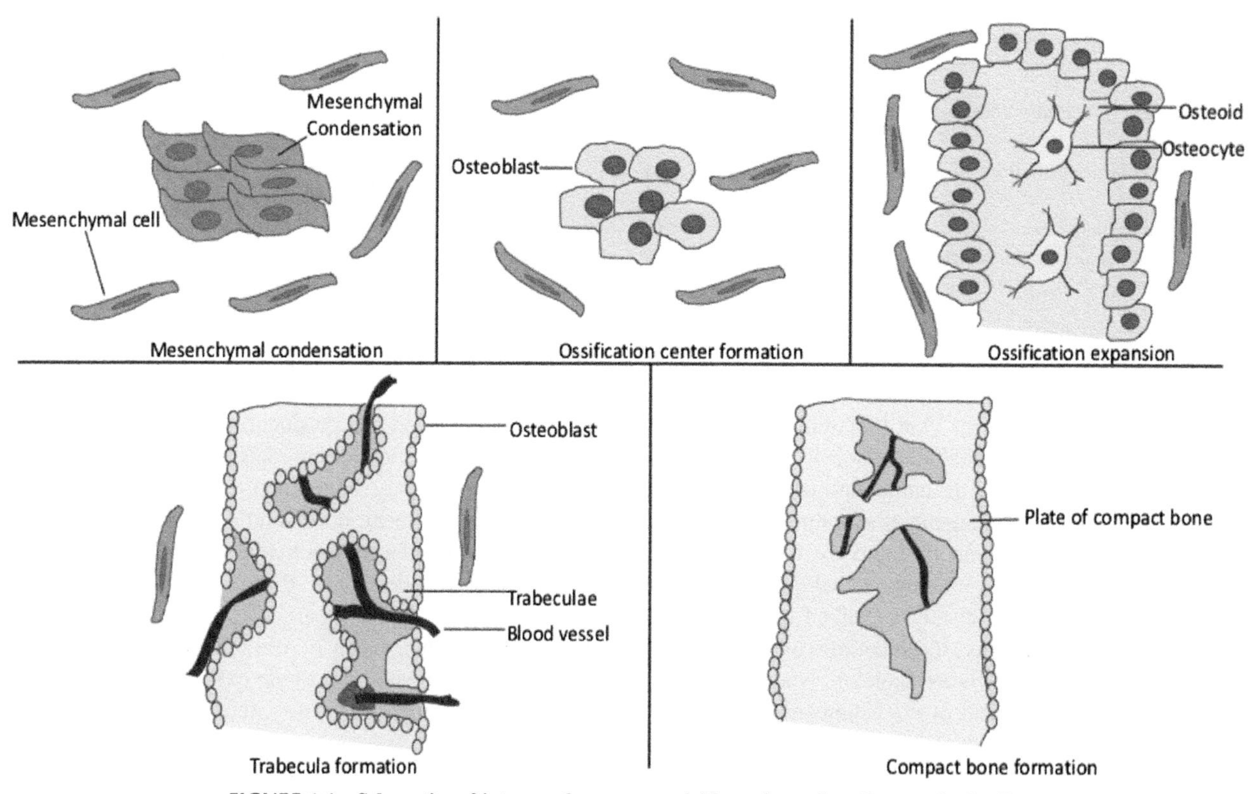

FIGURE 1.1 **Schematics of intramembranous cranial bone formation.** See text for details.

Condensation of mesenchymal progenitor cells is the first step for both intramembranous and endochondral ossification. During intramembranous ossification, mesenchymal progenitor cells differentiate into osteoblasts instead of chondrocytes as occurs during endochondral ossification. The osteoblasts that appear first in the condensation secrete bone matrix and form the ossification center. The early osteoblasts secrete osteoid, uncalcified matrix, which calcifies soon after, while the osteoblasts mature and terminally differentiate into osteocytes that are entrapped in the osteoid. As osteoblasts differentiate into osteocytes, more mesenchymal progenitors surrounding the osteoid differentiate into new osteoblast cells at the osteoid surface to expand the calcification center. Osteoid expansion around the capillaries results in a trabecular matrix of the spongy bone, while osteoblasts on the superficial layer become the periosteum. The periosteum is a layer that also contains mesenchymal progenitor cells, osteoblast differentiation of which contributes to the formation of a protective layer of compact bone. The blood vessels along with other cells between the trabecular bone eventually form the red marrow. Intramembranous ossification begins in utero during fetal development and continues on into adolescence. At birth, the skull and clavicles are not fully ossified. Sutures and fontanelles are unossified cranial regions that allow the skull to deform during passage through the birth canal. Sutures are joints between craniofacial bones, which are composed of two osteogenic fronts with suture mesenchyme between them (Fig. 1.2). Fontanelles are the space between the skull bones where the sutures intersect and are covered by tough membranes that protect the underlying soft tissues and brain. In humans, cranial sutures normally fuse between 20 and 30 years of age and facial sutures fuse after 50 years of age (Badve et al., 2013; Senarath-Yapa et al., 2012). Most sutures in mice remain patent throughout the animal's lifetime. Sutures and fontanelles allow the craniofacial bones to expand evenly as the brain grows, resulting in a symmetrically shaped head. However, if any of the sutures close too early (fuse prematurely), in the condition called craniosynostosis, there may be no growth in that area. This may force growth to occur in another area or direction, resulting in an abnormal head shape.

Apart from craniofacial bone development, intramembranous ossification also controls bone formation in the perichondral and periosteal regions of the long bone, where osteoblasts directly differentiate from mesenchymal progenitor cells. Yet, this requires a signal from the cartilaginous element. Furthermore, intramembranous ossification is an essential mechanism underlying bone repair and regeneration in the following processes: fracture healing with rigid fixation; distraction osteogenesis, a bone-regenerative process in which osteotomy followed by gradual distraction yields two vascularized bone surfaces from which new bone is formed (Ai-Aql et al., 2008); and blastemic bone creation, which occurs in children with amputations (Fernando et al., 2011).

Intramembranous ossification is tightly regulated at both molecular and cellular levels. Cranial malformations are often progressive and irreversible, and some of them need aggressive surgical management to prevent or mitigate severe impairment such as misshapen head or abnormal brain growth (Bronfin, 2001). For instance, craniosynostosis is a common congenital disorder that affects 1 in 2500 live births. It is characterized by premature cranial suture fusion, which may result in severe conditions such as increased intracranial pressure, craniofacial dysmorphism, disrupted brain development, and mental retardation. Craniosynostosis is generally considered a developmental disorder resulting from a disrupted balance of cellular proliferation, differentiation, and apoptosis within the suture (Senarath-Yapa et al., 2012; Levi et al., 2012; Slater et al., 2008; Lattanzi et al., 2012; Ciurea and Toader, 2009). Surgical correction followed by reshaping of the calvarial bones remains the only treatment available for craniosynostosis patients (Martou and Antonyshyn, 2011; Posnick et al., 2010; Hankinson et al., 2010). In contrast to craniosynostosis, cleidocranial dysplasia (CCD) is caused by reduced intramembranous bone formation, underdeveloped or absent clavicles (collarbones) as well as delayed maturation of the skull, manifested by delayed suture closure and larger than normal fontanelles that are noticeable as "soft spots" on the heads of infants (Farrow et al., 2018). Severe cases of CCD require surgical intervention. Identifying molecular pathways that control intramembranous ossification is critically important in the mechanistic understanding of craniofacial bone diseases and their targeted therapeutic development.

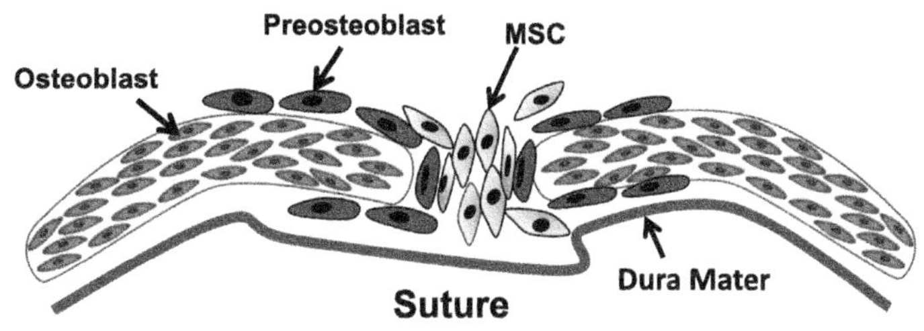

FIGURE 1.2 Schematics of cellular composition of the suture. In the suture, mesenchymal stem cells (MSCs) are located in the middle. They may first become committed preosteoblasts and then finally mature osteoblasts.

Studies of both developmental biology and rare genetic diseases have led to the identification of critical regulators of intramembranous ossification. Transcriptional regulation of the osteoblast lineage is considered in detail in Chapter 7. The runt-related transcription factor 2, RUNX2 (also known as CBFA1), and a zinc finger transcription factor, Osterix (OSX), are osteoblast lineage–determining factors required for both intramembranous and endochondral ossifications. *Runx2* is expressed in osteogenic progenitor cells and required for osteoblast cell fate determination by driving osteoblast-specific gene expression (Ducy et al., 1997; Otto et al., 1997). *Runx2* loss-of-function mutations are found in both mice and humans and cause CCD (Otto et al., 1997; Mundlos et al., 1997; Lee et al., 1997). RUNX2 induces the expression of *Osx*, which is required for osteoblast cell fate commitment, as loss of *Osx* leads to conversion from osteoblasts to chondrocytes (Nakashima et al., 2002). Under the control of RUNX2 and OSX, osteoblast cells produce osteoblast-specific collagen I together with a variety of noncollagenous, extracellular matrix (ECM) proteins that are deposited along with an inorganic mineral phase. The mineral is in the form of hydroxyapatite, a crystalline lattice composed primarily of calcium and phosphate ions.

Cell–cell communication that coordinates cell proliferation and differentiation also plays a critical role in intramembranous ossification. The WNT and Hedgehog (HH) signaling activities are required for cell fate determination of osteoblasts by controlling the expression of *Runx2*. Active WNT/β-catenin signaling is detected in the developing calvarium and perichondrium, where osteoblasts differentiate through intramembranous ossification. Indeed, enhanced WNT/β-catenin signaling enhances bone formation and *Runx2* expression, but inhibits chondrocyte differentiation and *Sox9* expression (Hartmann and Tabin, 2000; Guo et al., 2004; Day et al., 2005). *Sox9* is a master transcription factor that determines chondrocyte cell fate (Bi et al., 1999; Akiyama et al., 2002). Conversely, removal of *β-catenin* in osteochondral progenitor cells resulted in ectopic chondrocyte differentiation at the expense of osteoblasts during both intramembranous and endochondral ossification (Hill et al., 2005; Hu et al., 2005; Day et al., 2005). Therefore, during intramembranous ossification, WNT/β-catenin signaling levels in the mesenchymal condensation are higher, which promotes osteoblast differentiation while inhibiting chondrocyte differentiation. In addition, upregulated WNT/β-catenin signaling in the perichondrium also promoted osteoblast differentiation. In contrast to the WNT/β-catenin signaling, Indian hedgehog (IHH) signaling is not required for osteoblast differentiation of intramembranous bones in the skull (St-Jacques et al., 1999). It is still not clear what controls *Ihh*-independent *Runx2* expression during intramembranous ossification and it is important to understand further the differential regulation of intramembranous versus endochondral ossification by cell signaling. As removing Smoothened, which mediates all HH ligand-dependent signaling, does not abolish intramembranous ossification either (Jeong et al., 2004), HH signaling is likely to be activated in a ligand-independent manner in the developing calvarium. Indeed, it has been found that in the rare human genetic disease progressive osseous heteroplasia, which is caused by null mutations in *Gnas*, which encodes $G\alpha_s$, HH signaling is upregulated. Such activation of HH signaling is independent of HH ligands and is both necessary and sufficient to induce ectopic osteoblast cell differentiation in soft tissues (Regard et al., 2013). Importantly, *Gnas* gain-of-function mutations upregulate WNT/β-catenin signaling in osteoblast progenitor cells, resulting in their defective differentiation and in fibrous dysplasia that also affects intramembranous ossification (Regard et al., 2011). Therefore, $G\alpha_s$ is a key regulator of proper osteoblast differentiation through its maintenance of a balance between the WNT/β-catenin and the HH pathways. The critical role of WNT and HH signaling in intramembranous ossification is also shown in the suture. Mesenchymal stem cells that give rise to the cranial bone and regulate cranial bone repair in adult mice have been identified in the suture. These cells are either $GLI1^+$ or $AXIN2^+$ (Zhao et al., 2015; Maruyama et al., 2016), which marks cells that receive HH or WNT signaling, respectively (Bai et al., 2002; Leung et al., 2002; Jho et al., 2002).

Other signaling pathways, including those mediated by transforming growth factor (TGF) superfamily members, Notch, and fibroblast growth factors (FGFs), are also important in intramembranous ossification. Mutations in the FGF receptors FGFR1, FGFR2, and FGFR3 cause craniosynostosis. The craniosynostosis syndromes involving FGFR1, FGFR2, and FGFR3 mutations include Apert syndrome (OMIM 101200), Beare–Stevenson cutis gyrata (OMIM 123790), Crouzon syndrome (OMIM 123500), Pfeiffer syndrome (OMIM 101600), Jackson–Weiss syndrome (OMIM 123150), Muenke syndrome (OMIM 602849), crouzonodermoskeletal syndrome (OMIM 134934), and osteoglophonic dysplasia (OMIM 166250), a disease characterized by craniosynostosis, prominent supraorbital ridge, and depressed nasal bridge, as well as rhizomelic dwarfism and nonossifying bone lesions. All these mutations are autosomal dominant and many of them are activating mutations of FGF receptors. FGF signaling can promote or inhibit osteoblast proliferation and differentiation depending on the cell context. It does so either directly or through interactions with the WNT and bone morphogenetic protein (BMP) signaling pathways.

Apart from RUNX2 and OSX, other transcription factors are also important, as mutations in them cause human diseases with defects in intramembranous ossification. Mutations in the human *TWIST1* gene cause Saethre–Chotzen syndrome (OMIM 101400), one of the most commonly inherited craniosynostosis conditions. In addition, mutations in the homeobox genes *MSX1*, *MSX2*, and *DLX* are also associated with human craniofacial disorders (Cohen, 2000; Kraus and Lufkin, 2006). *MSX2* haploinsufficiency decreases proliferation and accelerates the differentiation of calvarial preosteoblasts, resulting in delayed suture closure, whereas its "overexpression" results in enhanced proliferation, favoring early suture closure (Dodig and Raos, 1999). It is likely that *MSX2* normally prevents differentiation and stimulates proliferation of preosteoblastic cells at the osteogenic fronts of the calvariae, facilitating expansion of the skull and closure of the suture. It would be critical to understand further how these transcription factors interact with one another and the signaling pathways to regulate intramembranous bone formation, maintenance, and repair.

The axial skeleton

The axial skeleton consists of the occipital skull bones, the elements of the vertebral column, and the rib cage (ribs and sternum). With the exception of the sternum, the axial skeleton is derived from the paraxial mesoderm, which is segmented into somites during early embryonic development. The occipital skull bones are generated from the fused sclerotomes of the cranial-most 4.5 somites (Goodrich, 1930). The bilateral anlagen of the sternum originate from the lateral plate mesoderm and fuse at the ventral midline in the course of the formation of the rib cage (Chen, 1952).

Somitogenesis

The basic body plan of vertebrates is defined by the metameric segmentation of the musculoskeletal and neuromuscular systems, which originates during embryogenesis from the segmentation of the paraxial mesoderm (for reviews see Winslow et al., 2007; Pourquie, 2000). The paraxial mesoderm is laid down during gastrulation, appearing as bilateral strips of unsegmented tissue (referred to as segmental plate in the avian embryo and presomitic mesoderm in the mouse). It flanks the centrally located neural tube and notochord and gives rise to the axial skeleton (head and trunk skeleton) and all trunk and limb skeletal muscles, as well as the dermis, connective tissue, and vasculature of the trunk. During development, the paraxial mesoderm is segmented through a series of molecular and cellular events in an anterior to posterior (craniocaudal) sequence along the body axis, the anterior-most somites being the more mature ones. The posterior, unsegmented part of the paraxial mesoderm is also referred to as the presomitic mesoderm (PSM), and the sequentially arising, paired tissue blocks are called somites. The PSM is a loose mesenchymal tissue. The cells reaching the anterior border of the PSM progressively undergo a mesenchymal-to-epithelial transition (Christ et al., 2007). Newly formed somites are epithelial balls with a mesenchymal core. As the somites mature, accompanied by the commitment of the cells to the different lineages, this organization changes. In response to signals from the notochord and the ventral floor plate of the neural tube (Sonic Hedgehog [SHH] and the BMP antagonist Noggin), cells on the ventral margin undergo an epithelial–mesenchymal transition, scatter, and move toward the notochord (Christ et al., 2004; Cairns et al., 2008; Yusuf and Brand-Saberi, 2006). These cells will express the transcription factors PAX1, NKX3.1, and NKX3.2 and form the sclerotome, giving rise to the vertebrae and ribs. The dermomyotome is specified by WNT ligands secreted from the dorsal neural tube and the ectoderm covering the dorsal somite. Low levels of SHH signaling are, in combination with WNT signaling, required to maintain the expression of dermomyotomal and myotomal markers (Cairns et al., 2008). The dermomyotome remains epithelial and eventually gives rise to the epaxial muscles of the back and vertebrae, the hypaxial muscles of the body wall and limb, the dermis underneath the skin of the trunk, and the brown adipose tissue (Scaal and Christ, 2004; Atit et al., 2006). Tendons and ligaments of the trunk arise from the fourth somitic compartment, the syndetome, which is induced by the newly formed sclerotome and dermomyotome (Brent et al., 2003; Dubrulle and Pourquie, 2003).

The molecular mechanism driving somitogenesis at the anterior end of the PSM is intrinsic to the PSM, while new cells are continuously added to the PSM from a posteriorly located progenitor pool (Martin, 2016). The so-called segmentation clock, a molecular oscillator coordinating the rhythmic activation of several signaling pathways and the oscillatory expression of a subset of genes in the PSM, is thought to be at the molecular heart of somite formation (Hubaud and Pourquie, 2014). One of the main signaling pathways with oscillatory gene expression is the Notch/Delta/DELTA pathway. This pathway also synchronizes the oscillations between the individual cells (Hubaud and Pourquie, 2014). Also, members of the WNT/β-catenin and the FGF signaling pathway display cyclic gene expression (Aulehla and Pourquie, 2008). The oscillatory expression of these genes appears to go like a wave from the caudal end, sweeping anteriorly through the PSM (Fig. 1.3A). Another molecular system involved in somite formation is the wave front, which is defined by opposing signaling gradients in the PSM (Fig. 1.3B). Here, a posterior–anterior gradient of FGF8 and nuclear β-catenin is opposed by an anterior–posterior gradient of retinoic acid (RA) activity (Mallo, 2016). Despite the fact that the existence of an RA gradient is debated, there is clear genetic evidence that a gradient of WNT signaling activity interacts with the

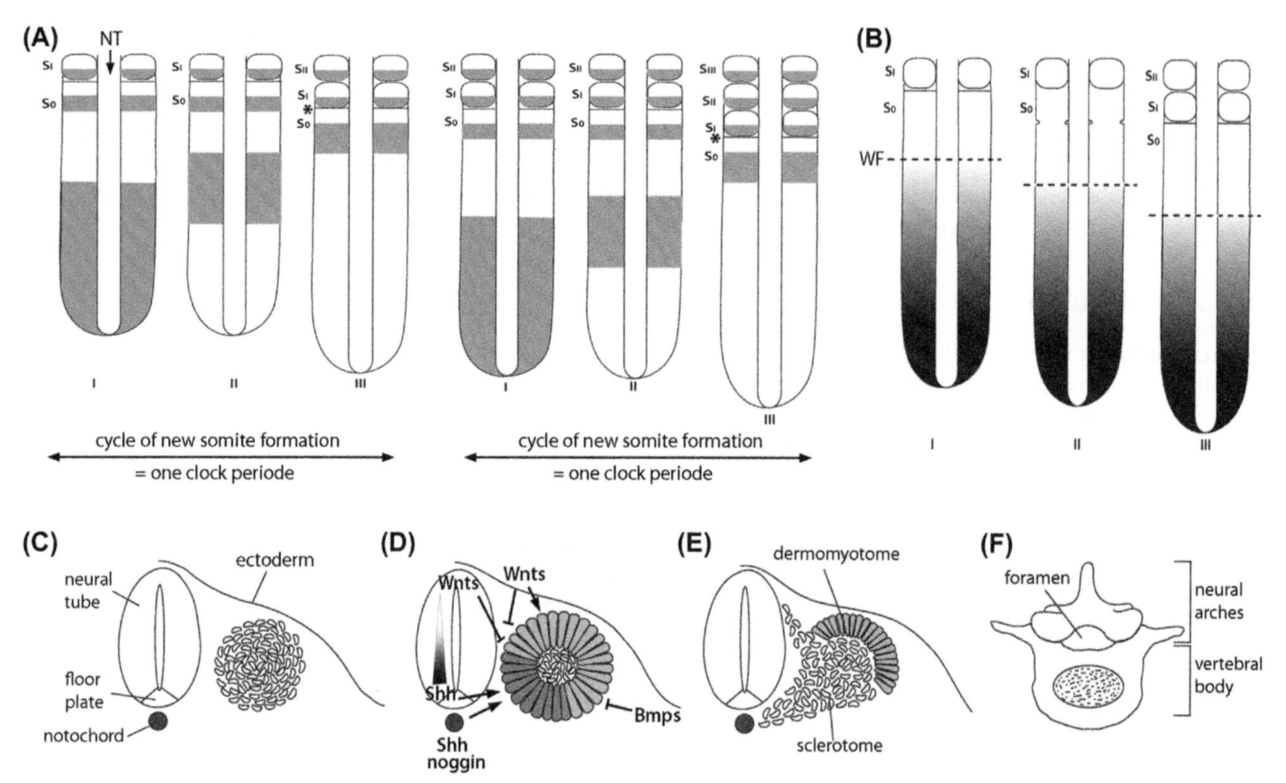

FIGURE 1.3 Somite formation and differentiation. (A) Cyclic gene expression during somite formation. The *asterisk* marks the position of new boundary formation. *NT*, neural tube; S_0, somite stage 0; S_I, somite stage I; S_{II}, somite stage II; S_{III}, somite stage III. (B) Signal gradient within the presomitic mesoderm (PSM), with the *dashed line* marking the position of the wave front (*WF*). (C–E) Schematic representations of the different somite stages. (C) Loose mesenchymal PSM, (D) epithelial ball stage (the ventral darker colored region marks the PAX1-positive sclerotomal region) and factors involved in the somite compartmentalization, (E) sclerotome differentiation. (F) Superior view of a vertebral element derived from the posterior and anterior sclerotomal compartments of two adjacent somites.

segmentation clock to determine the posterior border of a newly forming somite (Mallo, 2016). The morphological changes that eventually lead to the formation of a new somite at the anterior end of the PSM are triggered by Notch activity in combination with the T-box transcription factor, TBX6, and start with the expression of the basic helix–loop–helix transcription factor mesoderm posterior 2 (MESP2) (Saga, 2007; Sasaki et al., 2011). In cells posterior to the determination front, *Mesp2* is repressed by FGF signaling (Sasaki et al., 2011). In addition, *Mesp2* expression becomes restricted to the anterior half of the newly formed somite, as TBX6-mediated transcription of *Mesp2* is suppressed by the RIPPLY1/2 proteins expressed in the posterior part of the somite (Morimoto et al., 2007; Takahashi et al., 2007). MESP2 activity is essential for establishing somite polarity, which is in turn vital for the later formation of the vertebral bodies from the caudal/posterior part of one somite and the rostral/anterior part of the neighboring somite (Christ et al., 2007).

The positional identity of a somite defines the type of vertebral element (occipital, cervical, thoracic, lumbar, or sacral) it will eventually contribute to, and this is controlled, in part, by the regional code of *Hox* genes along the rostral–caudal body axis (for review see Wellik, 2007). Humans and all other bilateral animals have multiple *Hox* genes, encoding transcription factors with a homeobox DNA-binding domain, which are clustered together (Krumlauf, 1992). Through duplication events, the ancestral cluster of originally eight *Hox* genes has been multiplied to four gene clusters (*HoxA*, *HoxB*, *HoxC*, and *HoxD*) of 13 paralogous *Hox* genes in vertebrates. A particular feature of *Hox* gene expression from one cluster is that they are expressed in a temporal and spatial order that reflects their order on the chromosome, with the most 3′ *Hox* gene being expressed first and in the most anterior region. It is thought that the *Hox* genes provide a sort of positional code through their overlapping expression domains, which are characterized by a relatively sharp anterior border. For example, the expression of the *Hox5* paralogs (*HoxA5*, *HoxB5*, and *HoxC5*) correlates in different species such as mouse and chicken, always with the position of the last cervical vertebra, while the anterior domains of the *Hox6* paralogs lie close to the boundary between cervical and thoracic vertebrae (Burke et al., 1995; Burke, 2000). Yet, this correlation is not maintained at the levels of the somites, as mouse and chicken differ in their numbers of cervical elements. Changes in the HOX code can lead to homeotic transformation, which reflects a shift in the regional borders and axial identities.

Members of the polycomb family (*Bmi* and *Eed*) and the TALE class of homeodomain transcription factors are involved in further refining the positional identity provided by the *Hox* code. BMI and EED are transcriptional repressors limiting the rostral (anterior) transcription boundary of individual *Hox* genes (Kim et al., 2006). The TALE proteins, encoded by the *Pbx* and *Meis* genes, further modify the transcriptional activity of the Hox proteins through heterodimerization (Moens and Selleri, 2006).

Sclerotome differentiation

The earliest sclerotomal markers are the transcription factors *Pax1*, *Nkx3.1*, and *Nkx3.2/Bapx1*, which become expressed under the influence of SHH and Noggin signaling in the ventral somite region (Kos et al., 1998; Ebensperger et al., 1995; Murtaugh et al., 2001). *Pax9* expression appears slightly later in the sclerotome and overlaps in part with *Pax1* (Muller et al., 1996). Both genes act redundantly in the ventromedial region of the sclerotome, as in the *Pax1/Pax9* double-mutant mice the development of the ventral vertebra is strongly affected (Peters et al., 1999). NKX3.2 appears to act downstream of *Pax1/Pax9* and can be ectopically induced by PAX1 (Tribioli and Lufkin, 1999; Rodrigo et al., 2003). Although the initial *Pax1* expression is not affected by the loss of *Nkx3.2*, the vertebral differentiation also depends on the function of NKX3.2 (Tribioli and Lufkin, 1999). *Nkx3.1* mutant mice, on the other hand, do not display any skeletal defects (Schneider et al., 2000). As PAX1 is able to activate the expression of early chondroblast markers in vitro, it has been suggested that the activation of PAX1 is the key event that triggers sclerotome formation (Monsoro-Burq, 2005).

After their induction, the sclerotomal cells undergo epithelial–mesenchymal transition and migrate toward the notochord, around the neural tube, and in the thoracic segments also laterally, and then condense to form the vertebral bodies and the intervertebral discs, neural arches, and proximal part of the ribs, respectively (Fig. 1.3C–F). Some notochordal cells surrounded by sclerotomal cells die, while others become part of the intervertebral disc and form the nucleus pulposus (McCann and Seguin, 2016). The neural arches and spinous processes are derived from the mediolateral regions of the sclerotomes and from sclerotomal cells that migrated dorsally. The activity of PAX1/PAX9 is not required for these two compartments (Peters et al., 1999). The dorsally migrating sclerotomal cells contributing to the dorsal part of the neural arches and spinous processes do not express *Pax1* but another set of transcription factors, *Msx1* and Msx2 (reviewed in Monsoro-Burq, 2005; Rawls and Fischer, 2010). Other transcription factors, such as the winged-helix factor, MFH1 (FOXC2), are possibly required for the clonal expansion of cells taking place within the individual sclerotome-derived populations, as they migrate ventrally, laterally, and medially and then condense (Winnier et al., 1997). In addition, the homeodomain transcription factors *Meox1* and *Meox2* have been implicated in vertebral development and may even act upstream of PAX1/PAX9 (Mankoo et al., 2003; Skuntz et al., 2009). Within the individual sclerotomal condensations the chondrogenic and osteogenic programs are then initiated to eventually form the vertebral elements.

The limb skeleton

Overview of limb development

The mesenchymal cells contributing to the skeleton of the appendages (limbs) originate from the bilaterally located lateral plate mesoderm. The lateral plate mesoderm is separated from the somitic mesoderm by the intermediate mesoderm, which gives rise to the kidney and genital ducts. Our knowledge about limb development during embryogenesis is primarily based on two experimental model systems, chick and mouse. In all tetrapods, forelimb development precedes hindlimb development. The axial position of the prospective limb field is in register with the expression of a specific set of *Hox* genes within the somites (Burke et al., 1995). The limb fields are demarcated by the expression of two T-box transcription factors, *Tbx5* in the forelimb and *Tbx4* in the hindlimb field (Petit et al., 2017; Duboc and Logan, 2011). Yet, the identity of the limb is conveyed by the activity of another transcription factor, PITX1, which is expressed specifically in the hindlimb region and specifies hindlimb identity (Logan and Tabin, 1999; Minguillon et al., 2005). In mouse, the forelimb bud starts to develop around embryonic day (E) 9 and the hindlimb around E10. In chick, forelimb development starts on day 2½ (Hamburger Hamilton stage 16) with a thickened bulge (Hamburger and Hamilton, 1992). In humans, the forelimb is visible at day 24 of gestation. Experimental evidence from the chick suggests that WNT signaling induces FGF10 expression and the FGF-dependent initiation of the limb outgrowth (Kawakami et al., 2001). For continuous limb outgrowth the expression of *Fgfs* in the mesenchyme and in an epithelial ridge called the apical ectodermal ridge (AER) is essential (Benazet and Zeller, 2009; Martin, 2001) (Fig. 1.4A). Patterning of the outgrowing limb occurs along all three axes, the proximal–distal, the anterior–posterior, and the dorsal–ventral (Niswander, 2003). For example, in the human arm, the proximal–distal axis runs from the shoulder to the fingertips and can be subdivided into the stylopod (humerus),

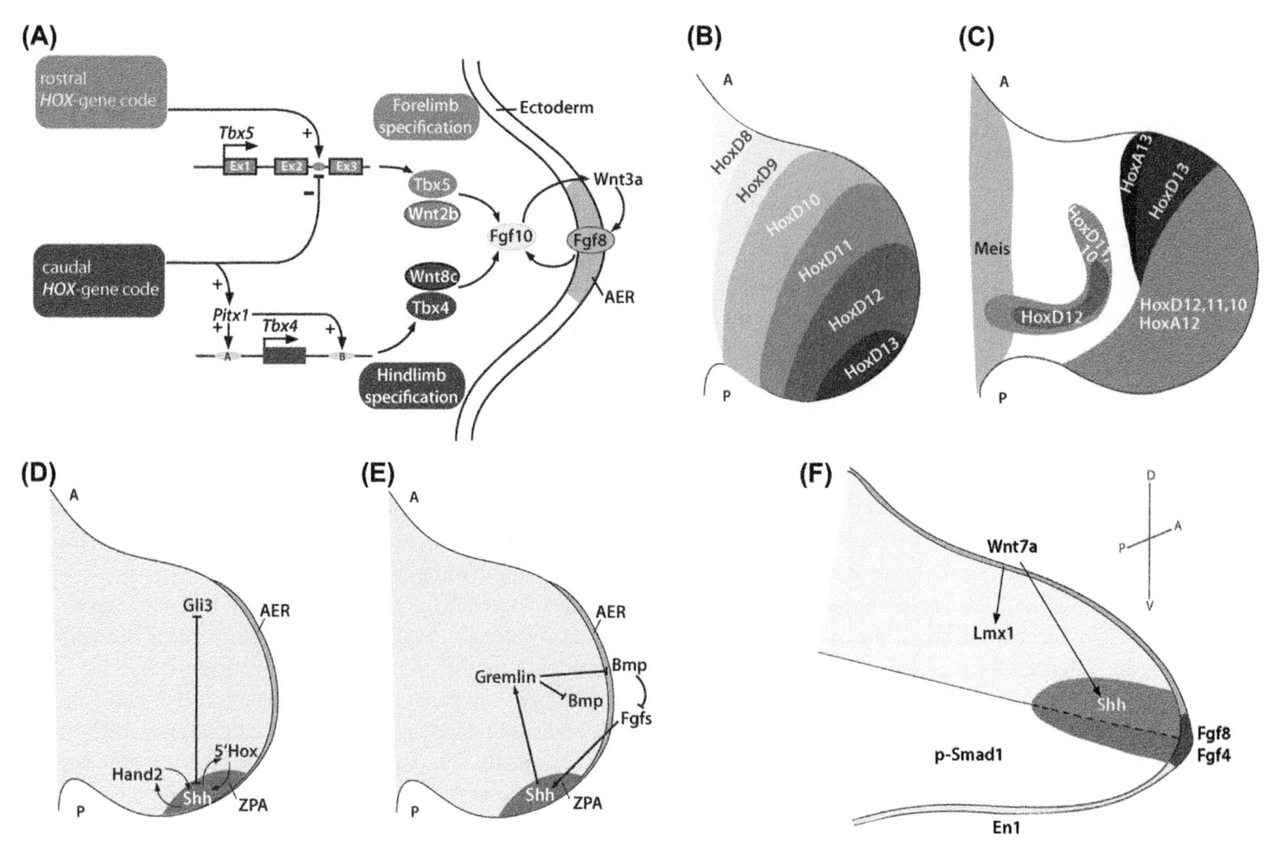

FIGURE 1.4 Limb development overview. (A) Early events in limb bud development: factors involved in the establishment of the limb identity and signals required for the initiation of limb outgrowth. *Hox* genes in the lateral plate mesoderm define the positions where the limbs will develop and activate or repress via specific enhancers the expression of *Pitx1* and the *Tbx4/5* genes. Together with the activity of limb field–specific WNTs an FGF10/WNT3a/FGF8 loop is established, which drives proximal–distal limb outgrowth. *AER*, apical ectodermal ridge. (B) Early nested expression of the HOXD cluster in the limb. *A*, anterior; *P*, posterior. (C) Late expression of the *HoxA* and *HoxD* genes in the autopod stage and expression of the proximal determinant *Meis1*. (D) Factors involved in anterior–posterior patterning of the limb, with *Shh* expressed in the zone of polarizing activity (*ZPA*) under the positive control of the transcription factors HAND2 and the 5′HOX proteins, while its activity in the anterior is opposed by the repressor GLI3. (E) Molecules involved in the interregulation of the anterior–posterior and proximal–distal axes. (F). Molecules involved in the specification of the dorsal–ventral axis: *Wnt7a* expressed in the dorsal ectoderm activates *Lmx1* expression in the dorsal mesenchyme specifying dorsal fate, while EN1 in the ventral ectoderm and phospho-SMAD1 in the ventral mesenchyme specify ventral fate. WNT7a also positively enforces the expression of *Shh*. *(A) Adapted from Fig. 1.2, Petit, F., Sears, K.E., Ahituv, N., 2017. Limb development: a paradigm of gene regulation. Nat. Rev. Genet. 18, 245–258.*

zeugopod (radius and ulna), and autopod regions (wrist and digits of the hand) (Fig. 1.5A). The anterior–posterior axis runs from the thumb to the little finger and the dorsal–ventral axis extends from the back of the arm/hand to the underside of the arm/palm. These three axes are established very early in development, and specific signaling centers, which will be briefly discussed in the following, coordinate the outgrowth and patterning of the limb.

Proximal–distal axis

As already mentioned, during the initiation stage, a positive FGF feedback loop is established between the *Fgfs* expressed in the mesenchyme (*Fgf10*) and the *Fgfs* in the AER (*Fgf8, Fgf4, Fgf9, Fgf17*). Mesenchymal FGF10 activity is essential for the formation of the AER (Sekine et al., 1999). In the positive feedback loop, FGF10 induces *Fgf8* expression in the AER, which is probably mediated by a *Wnt* gene's expression (*Wnt3a* in chick and *Wnt3* in mouse) (Kawakami et al., 2001; Kengaku et al., 1998; Barrow et al., 2003). The AER plays a critical role in the limb outgrowth. Removal of the AER at different time points of development leads to successive truncation of the limb (Saunders, 1948; Summerbell, 1974; Rowe and Fallon, 1982). The *Fgf* genes expressed in the AER confer proliferative and antiapoptotic activity on the distal mesenchyme and maintain the cells in an undifferentiated state (Niswander et al., 1994; Niswander et al., 1993; Fallon et al., 1994; Ten Berge et al., 2008). This is further supported by genetic studies showing that FGF4 and FGF8 are both required for the maintenance of the AER (Boulet et al., 2004; Sun et al., 2002). The most proximal part

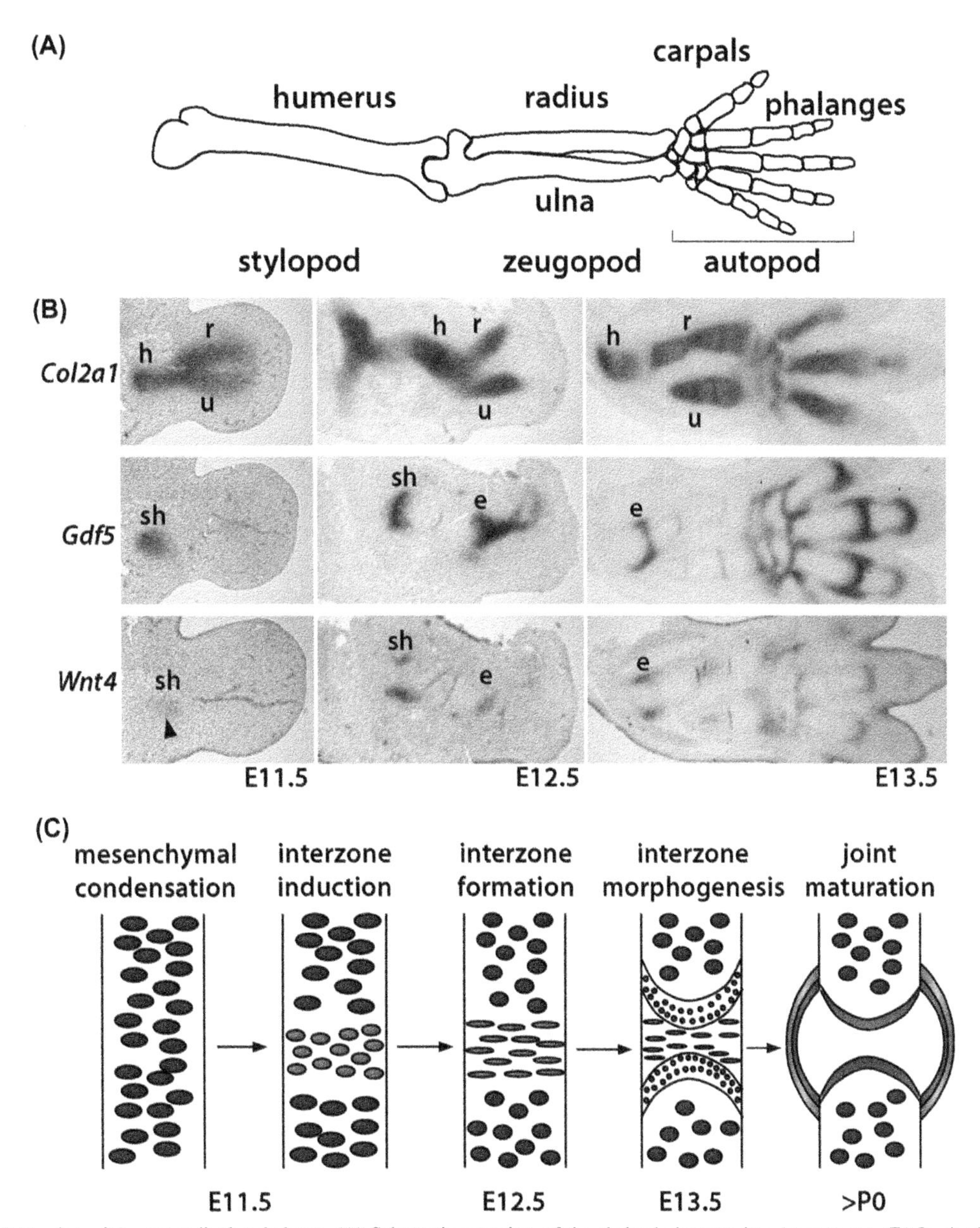

FIGURE 1.5 Patterning of the appendicular skeleton. (A) Schematic overview of the skeletal elements in a human arm. (B) In situ hybridizations on adjacent sections of a mouse forelimb (embryonic stages E11.5, E12.5, and E13.5), showing the branched structure of an early cartilaginous template (*Col2a1* expressing) consisting of the humerus (*h*), radius (*r*), and ulna (*u*). Note that at E11.5 markers of the joint interzone (*Gdf5* and *Wnt4*) are expressed in cells that also express the chondrogenic marker *Col2a1*. At E12.5, during interzone formation, *Col2a1* becomes downregulated in the shoulder (*sh*) and elbow (*e*) region, while the expression patterns of *Gdf5* and *Wnt4* undergo refinement. At E13.5, *Col2a1* is no longer expressed in the joint areas and the expression domains of *Gdf5* and *Wnt4* become distinct. (C) Schematic representation of the major steps during synovial joint formation.

of the limb expresses the TALE homeobox transcription factor MEIS1 under the control of opposing RA and FGF signaling (Mercader et al., 2000). MEIS1 alone is sufficient to proximalize the limb in the chick and mouse systems (Mercader et al., 1999, 2009). Along the proximal–distal axis, the 5′*Hox* genes, which are expressed early in a nested pattern (see Fig. 1.4B), are thought to provide positional cues for growth. As such, members of the group 11 paralogs (HOXA11 and D11 in the forelimb and HOXA11, C11, and D11 in the hindlimb) are required for the growth of the zeugopod, while the autopod establishment depends on the function of group 13 paralogs (Zakany and Duboule, 2007). *Hox* genes are also involved in connective tissue patterning in the limb (Pineault and Wellik, 2014). In addition to their role with regard to the proximal–distal axis, *Hox* genes also play an important role in establishing the signaling center within the limb bud regulating the anterior–posterior axis.

Anterior–posterior axis

Classical embryologic transplantation experiments uncovered the existence of a region present in the posterior limb bud conveying patterning information along the anterior–posterior axis (Saunder and Gasseling, 1968). Transplantation studies also revealed that this region, which was referred to as the zone of polarizing activity (ZPA), must contain some kind of positional information in the form of a secreted morphogen that specifies digit identity along the anterior–posterior axis (Tickle, 1981; Tickle et al., 1975; Wolpert, 1969). The molecular identity of this morphogen was uncovered only in 1993 with the cloning of a vertebrate homolog of the *Drosophila hh* gene, called *Shh*. *Shh* expression overlaps with the ZPA, and Shh-producing cells transplanted into the anterior mesoderm of the limb bud could reproduce mirror-image duplications of ZPA grafts (Riddle et al., 1993). Genetic experiments confirmed that *Shh* is required to establish posterior structures of the limb (Chiang et al., 1996). The *Shh* expression domain is established by the activity of positive and negative regulators. The transcription factor HAND2 (dHAND) is expressed in a posterior domain preceding and encompassing the *Shh* domain and acts as a positive regulator of SHH, which feeds back positively on the expression of HAND2 (Charite et al., 2000; Fernandez-Teran et al., 2000). Early in limb development, *Hand2* is expressed complementary to the transcription factor *Gli3* and GLI3 represses *Hand2* in the anterior (Wang et al., 2000). HAND2, on the other hand, represses *Gli3* in the posterior (Te Welscher et al., 2002). SHH signaling in the posterior prevents the cleavage of the full-length activator GLI3 into the GLI3 repressor (GLI3R) form. Hence, the GLI3R form is restricted to the anterior of the limb bud. The *5′Hox* genes and SHH signaling are also connected by a positive feed-forward regulatory loop (Tarchini et al., 2006; Ros et al., 2003), which may also involve FGF signaling (Rodrigues et al., 2017) (Fig. 1.4D). There is also an interconnection between the anterior–posterior and the proximal–distal axis: SHH signaling upregulates the BMP antagonist Gremlin in the posterior half of the limb. Gremlin antagonism of BMP signaling is required to maintain the expression of *Fgf4*, *Fgf9*, and *Fgf17* in the AER, and FGF signaling feeds positively onto *Shh* (Khokha et al., 2003; Laufer et al., 1994) (Fig. 1.4E).

Dorsal–ventral axis

The third axis that needs to be established is the dorsal–ventral axis. Here, the WNT ligand WNT7a is expressed in the dorsal ectoderm and regulates the expression of the LIM homeobox transcription factor LMX1 (LMX1B in the mouse) in the dorsal mesenchyme (Riddle et al., 1995; Vogel et al., 1995). LMX1B is required to maintain the dorsal identity of structures such as tendons and muscles in the limb (Chen et al., 1998). The ventral counterplayer is the transcription factor Engrailed 1 (EN1), which is expressed in the ventral ectoderm and the ventral half of the AER, and is essential for the formation of ventral structures (Davis et al., 1991; Gardner and Barald, 1992; Cygan et al., 1997; Loomis et al., 1996). BMP signaling appears also to be required for establishment of the dorsal–ventral axis, as the activated downstream component, phospho-SMAD1, is detected throughout the ventral ectoderm and mesenchyme (Ahn et al., 2001) (Fig. 1.4F). Deletion of a BMP receptor gene, *Bmpr1a*, from the limb bud ectoderm results in an expansion of *Wnt7a* and *Lmx1b* into ventral territories, an almost complete loss of *En1*, and severe malformation of the limbs missing the ventral flexor tendons (Ahn et al., 2001).

Mesenchymal condensation and patterning of the skeleton

Patterning of the somitic tissue and the limbs along the different axes is a prerequisite for the mesenchymal condensations to take place. In the craniofacial skeleton, epithelial–mesenchymal interactions occur during the precondensation phase (Hall and Miyake, 1995). Mesenchymal condensations are pivotal for intramembranous and endochondral bone formation. They define the positions and the basic shapes of the future skeletal elements. They can be visualized in the sclerotome, developing skull, and limbs in vivo and in micromass cell cultures in vitro by the presence of cell surface molecules that bind peanut agglutinin (Stringa and Tuan, 1996; Milaire, 1991; Hall and Miyake, 1992). During the prechondrogenic and preosteogenic condensation phase ECM molecules, such as the glycoproteins Fibronectin, Versican, and Tenascin; cell–cell adhesion molecules, such as N-CAM and N-cadherin; the gap-junction molecule Connexin43 (CX43); and Syndecans (type I transmembrane heparan sulfate proteoglycan) become upregulated, but their expression often changes dynamically during the subsequent differentiation process (for review see Hall and Miyake, 2000; DeLise et al., 2000). Cell adhesion and ECM proteins promote the formation of the condensations by establishing cell–cell contacts and cell–matrix interactions. Yet, through genetic studies, their functional requirement for the condensation process has not been demonstrated so far. For the cell–matrix interactions, integrins also play an important role as they act as receptors for Fibronectin ($\alpha5\beta1$; $\alpha V\beta3$), types II and VI collagen ($\alpha1\beta1$, $\alpha2\beta1$, $\alpha10\beta1$), Laminin ($\alpha6\beta1$), Tenascin ($\alpha9\beta1$, $\alpha V\beta3$, $\alpha8\beta1$, $\alpha V\beta6$), and Osteopontin (OPN) ($\alpha V\beta1$; $\alpha V\beta3$; $\alpha V\beta5$; $\alpha8\beta\beta1$) (Loeser, 2000, 2002; Tucker and Chiquet-Ehrismann, 2015; Docheva et al., 2014).

Various growth factors, such as members of the TGFβ superfamily, regulate the condensation process (reviewed in Moses and Serra, 1996). This has also been elegantly demonstrated in vitro for a subclass of this superfamily of growth factors, the BMP family (Barna and Niswander, 2007). For the proximal elements (femur, tibia, and fibula) in the hindlimb, genetics revealed a dual requirement for the zinc finger transcription factors GLI3 and PLZF to establish the correct temporal and spatial distribution of chondrocyte progenitors (Barna et al., 2005).

Mesenchymal cells within the condensations can differentiate into either osteoblasts (intramembranous ossification) or chondrocytes (endochondral ossification). WNT/β-catenin signaling is essential for the differentiation of osteoblasts, as no osteoblasts develop in conditional mouse mutants in which the β-catenin-encoding gene *Ctnnb1* was deleted in mesenchymal precursor cells of the limb and/or skull (Hu et al., 2005; Hill et al., 2005; Day et al., 2005). Instead, the precursor cells differentiate into chondrocytes (Day et al., 2005; Hill et al., 2005). Hence, β-catenin activity is not essential for chondrogenesis. WNT/β-catenin signaling is most likely acting as a permissive pathway at this early step of differentiation, as too high levels of WNT/β-catenin signaling block osteoblast as well as chondrocyte differentiation (Hill et al., 2005). WNT/β-catenin signaling in perichondrial cells is amplified by SOXC protein family members to further secure the nonchondrogenic fate of these cells (Bhattaram et al., 2014). For osteoblast differentiation to occur, the transcription factor RUNX2 needs to be upregulated within the preosteogenic condensations, while the HMG-box transcription factor SOX9 is required for the further differentiation of cells within the condensations along the chondrocyte lineage and probably also for the condensation process itself (Bi et al., 1999; Akiyama et al., 2002; Karsenty, 2001; Lian and Stein, 2003). The latter aspect has been challenged by the results of in vitro experiments by Barna and Niswander (2007) showing that *Sox9*-deficient mesenchymal cells compact and initially form condensations, yet the cells within the condensations do not differentiate into chondroblasts (Barna and Niswander, 2007).

The skeletal elements in the limbs, which are formed by the process of endochondral ossification, develop in part as continuous, sometimes bifurcated (pre)chondrogenic structures, such as, e.g., the humerus branching into the radius and ulna in the forelimb (Fig. 1.5B), being subsequently segmented by the process of joint formation (Shubin and Alberch, 1986; Hinchliffe and Johnson, 1980; Oster et al., 1988). Furthermore, studies have shown that the cartilage morphogenesis of the developing long bones also occurs in a modular way, with two distinct pools of progenitor cells contributing to the primary structures and the bone eminences (Blitz et al., 2013; Sugimoto et al., 2013). Cells within the bifurcated, SOX9$^+$ primary structures express the gene *Col2a1*, characteristic of chondroblasts/chondrocytes. Although they appear during early limb development (E11.5) to be morphologically uninterrupted, the region where a joint (here the shoulder joint) will be formed can be visualized using molecular joint markers, such as *Gdf5* (growth differentiation factor 5) or *Wnt4* (see Fig. 1.5B). Interestingly, the cartilage matrix protein Matrilin-1 is never expressed in the interzone region, nor in the adjacent chondrogenic region, which possibly gives rise to the articular cartilage (Hyde et al., 2007). How the position of joint initiation within the limb is determined is not completely understood as of this writing. A limb molecular clock operating in the distal region may be involved in this process. It has been proposed that two oscillation cycles of the gene *Hairy 2* (*Hes2*) are required to make one skeletal element in the zeugopod and stylopod region of the limb (Sheeba et al., 2016). As the joints develop sequentially along the proximal–distal axis at a certain distance from each other, secreted factors produced by the joint itself may provide some kind of self-organizing mechanism (Hartmann and Tabin, 2001; Hiscock et al., 2017). WNT/β-catenin signaling is also required for joint formation (Hartmann and Tabin, 2001; Guo et al., 2004; Spater et al., 2006a, 2006b). Yet, again, it may act in this process also as a permissive pathway, repressing the chondrogenic potential of the joint interzone cells. However, as WNT/β-catenin signaling also induces the expression of *Gdf5*, it may also play an active role in joint induction by inducing cellular and molecular changes required for joint formation. The AP1-transcription factor family member c-JUN acts upstream of WNT signaling in joint development regulating the expression of *Wnt9a* and *Wnt16*, which are both expressed in the early joint interzone (Kan and Tabin, 2013). Numerous other genes, including *Noggin*, *Hif1α*, *Gdf5*, *Gdf6*, *Gli3*, *Ihh*, *PTH/PTHrPR1*, *Tgfβ*, *Mcp5*, and *Crux1*, have been implicated in a variety of cellular processes during joint formation based on genetic or misexpression experiments (Brunet et al., 1998; Amano et al., 2016; Spagnoli et al., 2007; Longobardi et al., 2012), for review see (Archer et al., 2003; Pacifici et al., 2006).

Endochondral bone formation

Overview

The axial and appendicular skeletal elements are formed by the process of endochondral bone formation starting with a cartilaginous template (Fig. 1.6A–E). This process starts with the condensation of mesenchymal cells at the site of the future skeleton. As mentioned already, this involves alterations in cell–cell adhesion properties and changes in the ECM

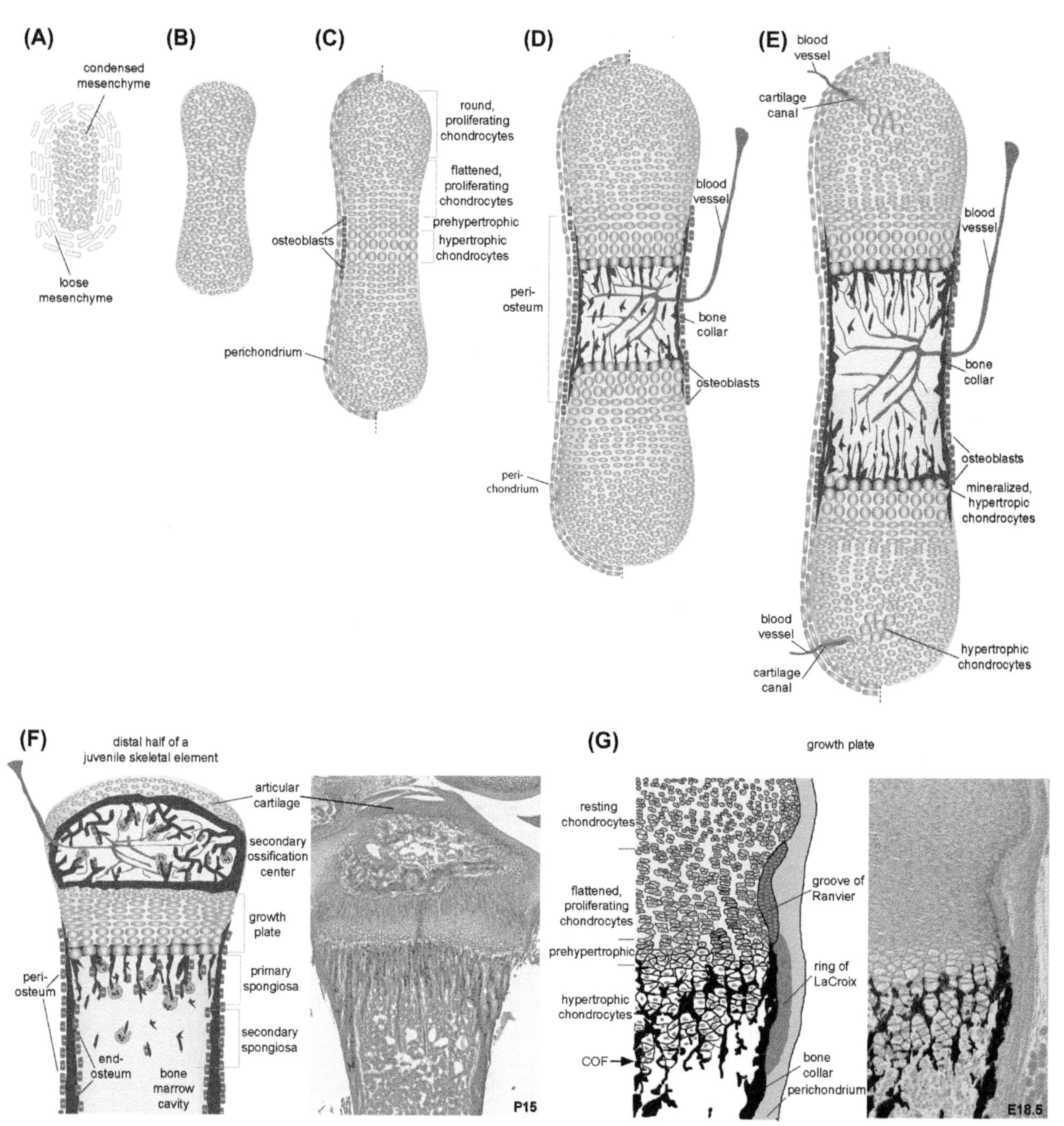

FIGURE 1.6 Schematic representation of the formation and growth of long bones by endochondral ossification. (A) Mesenchymal condensation with surrounding loose mesenchymal cells. (B) Cartilaginous template prefiguring the future skeletal element. (C) Chondrocyte differentiation within the cartilaginous template and differentiation of osteoblasts within a region of the perichondrium, which is then referred to as the periosteum. (D) Blood vessel invasion and onset of bone marrow cavity formation. (E) Onset of the formation of the secondary ossification center with differentiation of hypertrophic chondrocytes in the central region of the epiphysis and blood vessel invasion from the perichondrium through the cartilage canals. (F) Schematic representation on the left and corresponding Alcian blue/eosin−stained image of the proximal end of a postnatal day 15 (P15) mouse tibia on the right. (G) Schematic representation of the different features of a mouse growth plate based on the von Kossa/Alcian blue−stained proximal end of a mouse humerus at embryonic day 18.5 (E18.5). *COF*, chondro-osseous front.

(DeLise and Tuan, 2002a; Delise and Tuan, 2002b; Hall and Miyake, 1995; Bhat et al., 2011). Mesenchymal cells within the condensations start to express chondro-osteogenic markers, such as the transcription factors *Sox9* and *Runx2* (Hill et al., 2005; Akiyama et al., 2005; Wright et al., 1995). Next, the prechondrogenic precursor population of chondroblasts differentiates into chondrocytes, which produce an ECM rich in the proteoglycan aggrecan and fibrillar collagen of type II.

Cartilaginous template formation prefigures the future skeletal element and is surrounded by the so-called perichondrium, a layer of mesenchymal cells. As cartilage is avascular, limb vasculature regression needs to occur where cartilaginous structures form (Hallmann et al., 1987). Yet, interestingly, the chondrogenic condensation does express vascular endothelial growth factor (VEGF) (Eshkar-Oren et al., 2009). The outgrowth of vertebrate limbs occurs progressively along the proximal–distal axis (Newman et al., 2018; Zeller et al., 2009). Concomitantly, the skeletal elements develop in a proximodistal sequence, with the anlagen of the proximal elements (humerus in the forelimb and femur in the hindlimb) forming first, branching into more distal elements, and then being segmented into individual elements as the limb grows (Hinchliffe, 1994). The cartilaginous template increases in size by appositional and interstitial growth (Johnson, 1986). Interstitial growth by dividing chondrocytes allows the cartilage to grow rapidly along the longitudinal axis. The width of the cartilage element is controlled by appositional growth, whereby the perichondrium surrounding the cartilage template serves as the primary source of chondroblasts. Early on, all chondrocytes are still proliferating. As development progresses, the chondrocytes distant to the articulations in the central diaphysis will start to undergo a differentiation program. First, they flatten and rearrange into proliferative stacks of chondrocytes forming the zone of columnar proliferating chondrocytes. The elongation of these columns occurs internally through oriented cell division followed by intercalation movements of the daughters (Ahrens et al., 2009; Li and Dudley, 2009). A 2014 study showed that the daughter cells maintain intimate contact after cell division, preserving cadherin-mediated cell–cell interaction until the end of the rotational movement (Romereim et al., 2014). Interfering with cadherin-mediated cell–cell adhesion stalls the rotation process in vitro (Romereim et al., 2014). A similar rotation defect was observed in mice lacking integrin β1 (Aszodi et al., 2003). Chondrocytes at the lower end of the columns will then exit the cell cycle and become prehypertrophic; a stage that is not morphologically distinct but can be visualized using molecular markers such as the expression of the genes *Ihh* and parathyroid hormone/parathyroid hormone-like peptide receptor 1 (*Pthr1*). Next, the prehypertrophic chondrocytes increase dramatically in volume and become hypertrophic (Cooper et al., 2013; Hunziker et al., 1987). The almost 10-fold increase in volume occurs in parts by true cellular hypertrophy and swelling and significantly contributes to the longitudinal expansion of the skeletal elements as the cells are laterally restricted by matrix channels (Cooper et al., 2013). Hypertrophic chondrocytes (HCCs) are distinct in their ECM producing type X instead of type II collagen. Furthermore, they produce VEGF, which in this context attracts blood vessels to the diaphysis region (Gerber et al., 1999). The ECM of mature HCCs mineralizes and the cells produce matrix metalloproteinase 13 (MMP13) as well as OPN/SSP1. MMP13 (collagenase 3) breaks up the matrix of HCCs for the subsequent removal by osteoclasts (Inada et al., 2004; Stickens et al., 2004), while SSP1 has multiple functions; it regulates mineralization, serves as a chemoattractant for osteoclasts, and is functionally required for their activity (Franzen et al., 2008; Rittling et al., 1998; Boskey et al., 2002; Chellaiah et al., 2003). The final fate of HCCs has long been believed to be apoptotic cell death (Shapiro et al., 2005). Yet, ex vivo and in vitro experiments already hinted at an alternative fate, with HCCs transdifferentiating into osteoblasts (Shapiro et al., 2005). Lineage tracing experiments have confirmed this alternative fate, proposing a model of dual osteoblast origin (Zhou et al., 2014; Yang et al., 2014a, 2014b; Park et al., 2015). At least during embryonic development, about 20% of osteoblasts are chondrocyte derived and about 80% are derived from the perichondrium/periosteum. The latter population migrates into the bone marrow cavity along the invading blood vessels (Maes et al., 2010). This invasion originates from the periosteal collar, the area of the perichondrium in which osteoblasts differentiate and the bone collar is being formed (Colnot et al., 2004). In addition, monocytic osteoclast precursors as well as macrophages, both of which are of hematopoietic origin, enter the remodeling zone via the vascular system, which is attracted by VEGF (Henriksen et al., 2003; Engsig et al., 2000). Blood vessels have additional roles during trabecular bone formation in the primary spongiosa, which will be further discussed in the following. Endothelial cells, chondroclasts, and osteoclasts act together to erode the bone marrow cavity by removing HCC remnants. Interestingly, a bone marrow cavity can form in mouse mutants lacking osteoclasts or even macrophages and osteoclasts (Ortega et al., 2010). In these mutants, MMP9-positive cells are still present at the chondro-osseous junction and may be in part responsible for bone marrow cavity formation (Ortega et al., 2010). With the formation of the marrow cavity in the diaphysis, the two growth plates become separated from each other. The growth plates serve as a continual source of cartilage being converted into bone at the chondro-osseous front during the late stages of development and postnatally. In most species, a second ossification center appears during postnatal development within the epiphyseal cartilage. The onset differs between species for the individual bones and even within one bone for the two epiphyses (Adair and Scammin, 1921; Shapiro, 2001; Zoetis et al., 2003). Here, cartilage canals containing mesenchymal cells and blood vessels enter from the surrounding perichondrium, reaching eventually the hypertrophic center of the epiphysis (Blumer et al., 2008; Alvarez et al., 2005). After the formation of the secondary ossification center, the epiphyseal articular cartilage becomes distinct and the metaphyseal growth plate is sandwiched between the epiphyseal secondary ossification center and the primary ossification center in the diaphysis (Fig. 1.6F).

The growth plate

The cellular organization within the growth plate (schematically depicted in Fig. 1.6F) of a juvenile bone resembles the different zones in embryonic skeletal elements (Fig. 1.6G). There is a zone of small round chondrocytes, some of which are mitotically inert, that is often referred to as the resting zone. Stemlike or progenitor cells are thought to reside in this zone and require the activity of β-catenin for their maintenance (Candela et al., 2014). Concomitant with the growth plate closure that occurs in most vertebrates, with the exception of rodents, these progenitor cells eventually become senescent at the end of puberty and lose their proliferative potential, putting an end to long bone growth (Nilsson and Baron, 2004). The zone next to the resting zone contains flattened, stacked chondrocytes, which are mitotically active and form fairly regular columns. Eventually, the chondrocytes at the lower end of the zone will begin to enlarge, becoming first prehypertrophic and then HCCs (Ballock and O'Keefe, 2003). As already mentioned, some of the HCCs will undergo apoptosis (programmed cell death), while others survive and eventually differentiate into osteoblasts or other cells of the bone marrow cavity (Farnum and Wilsman, 1987; Shapiro et al., 2005; Tsang et al., 2015). The exact cellular and molecular mechanism of the transdifferentiation process of the surviving HCCs is not understood as of this writing. Earlier experiments suggested that this involves asymmetric cell division (Roach et al., 1995). According to the lineage tracing experiments, the transdifferentiating cells express at one point the gene *Col10a1*, encoding the α chain of type X collagen, but were they truly hypertrophic cells? If so, how was their cellular volume adjusted? Or alternatively, is there a pool of "stem cells" residing within the hypertrophic zone? So far, expression of stem cell markers has not been reported in HCCs of a normal growth plate. Yet, cells originating from the hypertrophic zone expressing the lineage tracer also express stem cell markers such as *Sca1* and *Sox2* in vitro (Park et al., 2015). Furthermore, a 2017 publication reported that during fracture healing HCCs express the stem cell markers *Sox2*, *Nanog*, and *Oct4* and that this is triggered by the invading vasculature (Hu et al., 2017). Other experiments such as one in rabbits, in which transdifferentiation was observed after physically preventing vascular invasion at the lower hypertrophic zone, suggest that the vasculature is not required for the transdifferentiation process to occur (Enishi et al., 2014). So far there are only a few molecules known to be required for the chondrocyte-derived differentiation of osteoblasts. One of them is β-catenin (Houben et al., 2016) and the other one SHP2, a protein tyrosine phosphatase (Wang et al., 2017). Mice lacking SHP2 activity in HCCs display a slight reduction in chondrocyte-to-osteoblast differentiation, and the mechanism behind this blockade is the persistence and/or upregulation of SOX9 protein in HCCs (Wang et al., 2017). Mice lacking β-catenin activity in HCCs display an even more severe reduction of chondrocytes differentiating into osteoblasts and its absence also affects in part the transdifferentiation of chondrocytes into other cell types (Houben et al., 2016). The mechanism by which β-catenin affects this transdifferentiation process is unknown as of this writing. Unlike what has been shown in perichondrial osteoblast precursors or in the case of SHP2, persistence of SOX9 protein was not observed (Houben et al., 2016). Furthermore, the loss of β-catenin activity in HCCs affects indirectly the differentiation of perichondrial-derived osteoblast precursors (Houben et al., 2016). HCCs also produce receptor activator of NF-κB ligand (RANKL) and its decoy receptor Osteoprotegerin, which positively and negatively, respectively, influence the differentiation of monocytes into osteoclasts at the chondro-osseous front (Usui et al., 2008; Silvestrini et al., 2005; Kishimoto et al., 2006). The expression of *Rankl* in HCCs is negatively controlled by β-catenin, leading to increased osteoclastogenesis and reduced trabecular bone formation in conditional *Ctnnb1* mice (Houben et al., 2016; Golovchenko et al., 2013; Wang et al., 2014a). As mentioned already, the matrix of the lower rows of HCCs mineralizes. HCCs utilize matrix vesicles to produce large amounts of microcrystalline, Ca^{2+}-deficient, acid-phosphate-rich apatite deposits in the collagen-rich matrix (Wuthier and Lipscomb, 2011). Matrix vesicle release occurs in a polarized fashion from the lateral edges of the growth plate HCCs, resulting in the mineralization of the longitudinal septae, while transverse septae remain unmineralized (Anderson et al., 2005a). The matrix vesicles then release the apatite crystals, which self-nucleate and grow to form spherical mineralized clusters in the calcified zone of the HCCs. Mitochondria may serve as storage containers for Ca^{2+}, with the mitochondria in HCCs reaching the highest Ca^{2+} concentrations and serving as the Ca^{2+} supply for matrix vesicles. The mitochondria loaded with Ca^{2+} can no longer produce sufficient amounts of ATP and the cells undergo a physiological energy crisis. As a consequence, the mitochondria produce increased amounts of reactive oxygen species (ROS) (Wuthier and Lipscomb, 2011). Increased ROS levels feed back on the chondrocytes, inducing them to hypertrophy (Morita et al., 2007).

Through knockout studies in mouse, numerous genes were identified that are involved in the regulation of the mineralization process, such as matrix Gla protein and tissue nonspecific alkaline phosphatase (encoded by the *Akp2* gene), ectonucleotide pyrophosphatase/phosphodiesterase type 1, progressive ankylosis gene, phosphoethanolamine/phosphocholine phosphatase, membrane-anchored metalloproteinase ADAM17, and, as already mentioned, OPN (Anderson et al., 2004, 2005b; Fedde et al., 1999; Hessle et al., 2002; Zaka and Williams, 2006; Harmey et al., 2004; Hall et al., 2013).

After the removal of HCCs, the mineralized longitudinal septae remain and are used by osteoblasts as a scaffold for the deposition of osteoid that calcifies into woven bone.

At the periphery, the growth plate is surrounded by a fibrous structure that consists of the wedge-shaped groove of Ranvier and the perichondrial ring of LaCroix (see Fig. 1.6G) (Brighton, 1978; Langenskiold, 1998). The groove of Ranvier serves as a reservoir for chondro-osteoprogenitor cells and fibroblasts, while the perichondrial ring of LaCroix may serve as a reservoir of precartilaginous cells (Fenichel et al., 2006; Shapiro et al., 1977). Interestingly, the two growth plates within a skeletal element have different activities leading to the differential growth of the distal and proximal parts (Pritchett, 1991, 1992; Farnum, 1994). Curiously, there seems to exist a temporal and local correlation between the appearance of the secondary ossification center and the activity of the nearby growth plate. For instance, in the humerus the secondary ossification center appears first in the proximal epiphysis and here, the proximal growth plate is more active than the distal one.

Mediators of skeleton formation

Accurate skeletogenesis, as well as postnatal growth and repair of the skeleton, depends on the precise orchestration of cellular processes such as coordinated proliferation and differentiation in time and space. Several signaling pathways impinge on the differentiation of the mesenchymal precursors as well as on the subsequent differentiation of chondrocytes and regulate the growth of the skeletal elements. Growth factor signaling is also partly controlled by the ECM and integrins (Munger and Sheppard, 2011; Ivaska and Heino, 2011). Cell-type-specific differentiation is under the control of distinct transcription factors with their activity being modulated by epigenetic factors and microRNAs. In addition to systemic and local factors, oxygen levels and metabolism also influence endochondral bone formation.

Systemic mediators

Longitudinal bone growth after birth is under the influence of various hormones, such as growth hormone (GH), insulin-like growth factors (IGFs), thyroid hormones, estrogen and androgens, glucocorticoids, vitamin D, and leptin. The importance of these hormones in skeletal growth has been demonstrated by genetic studies in animals and by "natural experiments" in humans (for reviews see Nilsson et al., 2005; Wit and Camacho-Hubner, 2011). Many of these systemic mediators interact with one another during linear growth of the juvenile skeleton and are differentially controlled by the nutritional status (Robson et al., 2002; Lui and Baron, 2011; Gat-Yablonski et al., 2008). Yet, only IGF signaling plays a role in endochondral ossification prior to birth.

Mice deficient for either *Igf1* or *Igf2* or the *Igf1r* gene display prenatal as well as postnatal growth defects, suggesting that IGFs act independent of GH on linear growth (Baker et al., 1993; Liu et al., 1993; Powell-Braxton et al., 1993). IGF1 was thought to affect chondrocyte proliferation, yet, a study on longitudinal bone growth in the *Igf1*-null mouse revealed no change in growth plate chondrocyte proliferation or cell numbers, despite the observed 35% reduction in the rate of long bone growth that was attributed to the 30% reduction in the linear dimension of HCCs (Wang et al., 1999). For more detailed information on the activities of GH and IGF signaling see reviews by Giustina et al. (2008), Kawai and Rosen (2012), Svensson et al. (2001), and Lindsey and Mohan (2016).

Local mediators

The various local mediators of endochondral and intramembranous ossification, which will be briefly discussed in the following, interact at multiple levels. Because of space constraints not all of these interactions can be mentioned.

Growth factor signaling pathways

Transforming growth factor β and bone morphogenetic proteins

The TGFβ superfamily is a large family of secreted polypeptides that can be divided into two subfamilies based on the utilization of the downstream signaling mediators, the regulatory SMADs (R-SMADs). The first one, encompassing TGFβ1−β3, activins, inhibins, nodal, and myostatin (GDF8), transduces the canonical signal through the R-SMADs 2 and 3. The second one consists of the BMPs 2 and 4−10 and most GDFs, transducing the canonical signal through R-SMADs 1, 5, and 8. The cofactor SMAD4 is utilized by both groups, forming a complex with the different activated R-SMADs. The receptor complexes are heterodimers consisting of serine/threonine kinase types I (ALKs 1−7) and II (TβRII, ActRII, ActRIIb, BMPRII, and MISRII) receptors. Ligand binding activates the type II receptor, leading to transphosphorylation of the type I receptor. In addition to the SMAD-dependent canonical signaling, TGFβ/BMPs can signal through numerous SMAD-independent noncanonical signaling pathways (reviewed in Wang et al., 2014c; Wu et al., 2016).

Many of the TGFβ and BMP signaling molecules are involved in endochondral bone formation. In the mouse, all three *Tgfβ* isoforms are expressed in mesenchymal condensations, perichondrium/periosteum, and appendicular growth plates (Pelton et al., 1990, 1991; Schmid et al., 1991). Despite the numerous in vitro reports indicating a role for TGFβ molecules promoting mesenchymal condensation and the onset of chondrocyte differentiation, none of the individual *Tgfβ* knockouts supports such an early role in vivo. It has been proposed that transient activation of TGFβ and/or activin signaling primes mesenchymal cells to become chondroprogenitors (Karamboulas et al., 2010). Of the individual *Tgfβ* knockouts, only the *Tgfβ2*$^{-/-}$ mutants displayed defects in intramembranous and endochondral bone formation (Sanford et al., 1997), some of which may be secondary due to defects in tendon formation (Pryce et al., 2009). The conditional ablation of the primary receptor for all three TGFβs and *Alk5*, in mesenchymal cells using the *Dermo1*-Cre line, resulted also in skeletal defects affecting intramembranous and endochondral bones (Matsunobu et al., 2009). The endochondral bone elements in the *Alk5*$^{-/-}$ animals were smaller and malformed, with ectopic cartilaginous protrusions present in the hindlimb. Conditional deletion of the *Tgfbr2* gene, encoding the TβRII receptor, in the limb mesenchyme with the *Prx1*-Cre line results in the absence of interphalangeal joints, probably due to a defect in downregulation of the chemokine MCP-5 in the joint interzone cells (Spagnoli et al., 2007; Longobardi et al., 2012). The appendicular skeletal elements of the *Tgfbr2*;*Prx1*-Cre embryos are also shorter, associated with altered chondrocyte proliferation and an enlarged HCC zone (Seo and Serra, 2007). This phenotype was also observed upon the expression of a dominant-negative form of the TβRII receptor or by expressing a dominant-negative TβRI (*Alk5*) construct in chondrocytes (Serra et al., 1997; Keller et al., 2011). Surprisingly, deletion of *Tgfbr2* in *Col2a1*-expressing cells resulted in defects only in the axial skeleton and not in the appendicular skeleton (Baffi et al., 2004). Nevertheless, the long bones of the *Tgfbr2;Col2a1*-Cre newborn mice were consistently shorter, but the difference was not significant. Sueyoshi and colleagues reported that deletion of *Tgfbr2* in HCCs results in a minor delay in chondrocyte differentiation around E14.5/15.5. Yet, at birth, no differences regarding the length of the long bones were observed, suggesting that this is a transient effect (Sueyoshi et al., 2012). Deletion of *Tgfbr2* in *Osx*-Cre-positive pre-HCCs and osteoblast precursors in the perichondrium led to postnatal alteration in the growth plate and affected osteoblastogenesis (Peters et al., 2017). This is probably associated with a loss of *TGFβ1* signaling (Tang et al., 2009). Nevertheless, inactivation of *Tgfbr2* may not be sufficient to eliminate all Tgfβ signaling, as TGFβ ligands were still capable of eliciting signals in the *Tgfbr2*$^{-/-}$ mice (Iwata et al., 2012). Furthermore, TGFβs can activate the canonical BMP/SMAD1/5/8 pathway through engagement of ALK1 (Goumans et al., 2002). TGFβ proproteins are sequestered by the ECM and can then be released and activated through, for instance, the activity of ECM degrading enzymes (Hildebrand et al., 1994; Pedrozo et al., 1998; Annes et al., 2003). For further information, in particular on the involvement of noncanonical TGFβ pathways in chondrogenesis and skeletogenesis and the implications of TGFβ signaling in osteoarthritis, see reviews by van der Kraan et al. (2009), Wang et al. (2014c), and Wu et al. (2016).

The cofactor SMAD4 is thought to mediate canonical signaling downstream of TGFβ and BMP signaling. Yet surprisingly, conditional mutants lacking *Smad4* in *Col2a1*-expressing cells are viable and display only mild phenotypic changes in the growth plate (Zhang et al., 2005; Whitaker et al., 2017). However, the prechondrogenic condensations do not form in mice lacking SMAD4 in the limb mesenchyme, supporting an essential role for TGFβ/BMP signaling in the early steps of chondrocyte differentiation, which appears to be independent of SOX9 (Lim et al., 2015; Benazet et al., 2012). Mice lacking either R-SMAD1/5 in *Col2a1*-expressing cells or all three R-SMADs (SMAD1, 5, and 8) acting downstream of BMP signaling are not viable and display a nearly identical severe chondrodysplasia phenotype (Retting et al., 2009). The axial skeleton is severely compromised, with vertebral bodies replaced by fibroblasts and loose mesenchymal tissue. This suggests that SMAD8 plays only a minor role in chondrogenesis. Furthermore, these results challenge the dogma that SMAD4 is required to mediate SMAD-dependent signaling downstream of BMPs and TGFβs.

Based on the analyses of gene knockout animals, the *Bmp*/*Gdf* family members *Bmp8*, *Bmp9/Gdf2*, *Bmp10*, and *Gdf10* appear to play no role in embryonic skeletogenesis (Zhao et al., 1996, 1999; Chen et al., 2004; Levet et al., 2013). The *short-ear* mouse is mutant for *Bmp5* and displays defects in skeletal morphogenesis and has weaker bones (Kingsley et al., 1992; Mikic et al., 1995). *Bmp6* mutants have sternal defects (Solloway et al., 1998). Mice mutant for *Bmp7* display skeletal patterning defects restricted to the rib cage, skull, and hindlimbs (Luo et al., 1995; Jena et al., 1997). In addition to *Bmp7*, *Bmp2* and *Bmp4* are expressed in the early limb bud. Conditional deletion of *Bmp2* and *Bmp4* in the limb mesenchyme results in an abnormal patterning of the appendicular skeleton with a loss of posterior elements in the zeugopod and autopod region probably due to a failure of chondrogenic differentiation of the mesenchymal cells caused by insufficient levels of BMP signaling (Bandyopadhyay et al., 2006). In addition, the skeletal elements that form are shorter and thinner. Chondrocyte differentiation within the skeletal elements is delayed but otherwise normal. Concomitantly, the endochondral ossification process is also delayed and bone formation is severely compromised in these mice (Bandyopadhyay et al., 2006). Yet, of the two BMPs, BMP2 appears to be the crucial regulator of chondrocyte proliferation and maturation (Shu et al., 2011). GDF11/BMP11 is required for axial skeleton patterning and acts upstream of the *Hox* genes (McPherron et al., 1999; Oh et al., 2002). Postnatally, GDF11 acts on bone homeostasis by stimulating osteoclastogenesis and inhibiting osteoblast differentiation (Liu et al., 2016). Mutations in human *GDF5 (BMP14, CDMP1)* or *BMPR1B* (*ALK6*) cause brachydactyly type C (OMIM 113100) and A2 (OMIM 112600), respectively (Lehmann et al., 2003; Polinkovsky et al., 1997). *Gdf5* and *Bmpr1b* mutant mice also display a brachydactyly phenotype (Storm et al., 1994; Baur et al., 2000; Yi et al., 2000). Closer examination of the *Gdf5* mutant brachypodism mouse revealed that the absence of the joint separating phalangeal elements 1 and 2 is due to the loss of the cartilaginous anlage and subsequent formation of the skeletal element by intramembranous instead of endochondral bone formation (Storm and Kingsley, 1999). The related family members *Gdf5*, *Gdf6*, and *Gdf7* are expressed in the interzone of different subsets of joints. *Gdf6* mutants display fusions of carpal and tarsal joints, and double mutants for *Gdf5/6* show additional skeletal defects (Settle et al., 2003). Interestingly, postnatally, GDF5 and GDF7 modulate the rate of endochondral tibial growth by altering the duration of the hypertrophic phase in the more active growth plate in opposite ways (Mikic et al., 2004, 2008). Bead-implant experiments in chicken and mouse embryos as well as various in vitro experiments revealed a prochondrogenic activity of BMP2, BMP4, or GDF5 protein, which can be antagonized by the secreted molecule Noggin (Zimmerman et al., 1996; Merino et al., 1999; Wijgerde et al., 2005). Consistent with this, Noggin-knockout mice display appendicular skeletal overgrowth and lack synovial joints (Brunet et al., 1998). Yet, surprisingly the caudal axial skeleton does not develop in the *Noggin* mutants. The vertebral phenotype can in part be reverted by the loss of one functional *Bmp4* allele, supporting the notion that too high levels of BMP4 signaling in the axial mesoderm may actually inhibit the differentiation of sclerotomal cells to chondrocytes. Instead, these cells take on a lateral mesodermal fate (Wijgerde et al., 2005; Murtaugh et al., 1999; Hirsinger et al., 1997). Double knockout of the BMP receptors *Bmpr1a* (*Alk3*) and *Bmpr1b* (*Alk6*) revealed a functional redundancy of these two receptors in endochondral ossification. Chondrocyte differentiation in the axial and appendicular skeleton is severely compromised in the mice lacking both receptors (Yoon et al., 2005). Conditional mutants for activin receptor type IA (*Alk2*) display only mild axial phenotypes. Double mutant analysis revealed a functional redundancy with *Bmpr1a* and *Bmpr1b* in endochondral skeletogenesis (Rigueur et al., 2015). Conditional postnatal deletion of *Bmpr1a* revealed a role for BMP signaling in the maintenance of the chondrogenic cell fate in the growth plate (Jing et al., 2013). Constitutively activating mutations in *ALK2* are found in patients with fibrodysplasia ossificans progressiva (OMIM 156400), a rare disorder in which the connective tissue progressively ossifies after traumatic injury (Shore et al., 2006). For further reading see reviews by Rosen (2006), Pogue and Lyons (2006), Wu et al. (2007, 2016), and Wang et al. (2014b).

Parathyroid hormone-related protein and Indian hedgehog

The paracrine hormone parathyroid hormone-related protein (PTHrP) and its receptor PTH1R are part of a crucial regulatory node, also referred to as the IHH/PTHrP feedback loop, coordinating chondrocyte proliferation with maturation in endochondral bone formation (Fig. 1.7). PTHrP is also required for normal intramembranous ossification (Suda et al., 2001). In the appendicular skeletal elements, PTHrP is expressed locally at high levels in the periarticular cells and at lower levels in the proliferating chondrocytes. Its receptor is expressed also at low levels in proliferating and at higher levels in pre-HCCs (Lee et al., 1996; Vortkamp et al., 1996; St-Jacques et al., 1999). *PTHrP* and *Pthr1* mutant mice display similar, but not identical phenotypes, with numerous skeletal abnormalities, including severely shortened long bones (Karaplis et al., 1994; Lanske et al., 1996). In both, the shortening of the long bones is associated with reduced chondrocyte proliferation and accelerated HCC maturation and bone formation (Amizuka et al., 1996; Lee et al., 1996; Lanske et al., 1998). Chimeric mice with *Pth1r*$^{-/-}$ clones in their growth plates revealed that the effects on chondrocyte maturation were direct but influenced by positional cues, as these clones expressed either *Ihh* or *Col10a1* ectopically dependent on their location within the proliferative zone (Chung et al., 1998). Concomitantly, mice overexpressing either PTHrP or a constitutively active form of PTH1R in chondrocytes show a delay in HCC maturation early and a prolonged persistence of HCCs associated with a delay in blood vessel invasion at later stages of development (Weir et al., 1996; Schipani et al., 1997b). PTH1R is a seven-transmembrane receptor coupled to heterotrimeric G proteins, consisting of α, β, and γ subunits. Its activation by PTHrP results in signaling via either the $G_s(\alpha)$/cAMP or the $G_q(\alpha)$/inositol-3-phosphate-dependent pathway. The two downstream pathways have opposing effects on chondrocyte hypertrophy with $G_q(\alpha)$/inositol-3-phosphate-dependent signaling cell-autonomously accelerating hypertrophic differentiation, while $G_s(\alpha)$/cAMP-signaling delays it (Guo et al., 2002; Bastepe et al., 2004). The intracellular mediator of the canonical WNT signaling pathway, β-catenin, interacts with the PTH1R and may modulate the switch from $G_s(\alpha)$ to the $G_q(\alpha)$ signaling (Yano et al., 2013; Yang and Wang, 2015). The $G_s(\alpha)$/cAMP signaling pathway is also involved in the maintenance of the pool of round proliferating

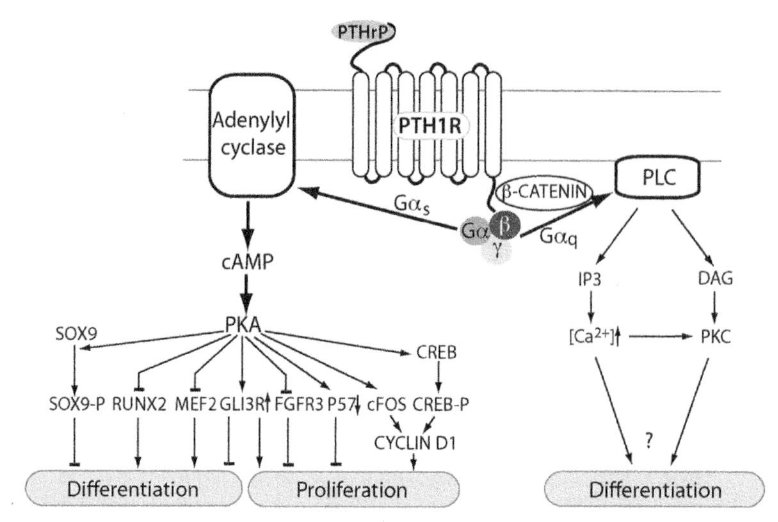

FIGURE 1.7 PTHrP/PTH1R signaling pathways and their functional consequences on chondrocyte differentiation and proliferation. The molecular mechanism underlying the differentiation-promoting effect of the PLC signaling branch is not yet understood. *PKA*, protein kinase A; *PLC*, phospholipase C; *PTHrP*, parathyroid hormone-related protein; *PTH1R*, PTHrP receptor.

chondrocytes (Chagin and Kronenberg, 2014). The inhibitory effect on chondrocyte hypertrophy is mediated through the activation of protein kinase A (PKA) downstream of $G_S(\alpha)$/cAMP signaling. This, in turn, promotes the following response: translocation of histone deacetylase 4 (HDAC4) into the nucleus where it binds to and inhibits the transcriptional activity of MEF2 transcription factors (Kozhemyakina et al., 2009). Furthermore, PTHrP signaling increases the expression of the transcription factor ZFP521, which negatively influences the transcriptional activity of RUNX2, again through recruitment of HDAC4 (Correa et al., 2010). In addition, PTHrP can decrease RUNX2 production and enhance its degradation specifically in chondrocytes (Guo et al., 2006; Zhang et al., 2009, 2010). MEF2 and RUNX2 are both positive regulators of chondrocyte hypertrophy (see later). PKA also phosphorylates SOX9, enhancing its DNA-binding activity, and stimulates GLI3 processing into its repressor fragment, thereby potentially interfering with chondrocyte maturation (Huang et al., 2000; Wang et al., 2000; Mau et al., 2007). PTH1R signal via PKA also inhibits the transcription of FGFR3 (McEwen et al., 1999). Furthermore, it leads to a downregulation of the cell-cycle-dependent inhibitor P57, a negative regulator of chondrocyte proliferation (Yan et al., 1997; MacLean et al., 2004). Last but not least, PTHrP signaling may stimulate proliferation through AP1/CREB dependent activation of cyclin D1 (Ionescu et al., 2001).

The findings in the different *Pthrp/Pth1r* mouse models can be correlated with activating mutations in the human PTH1R that lead to ligand-independent cAMP accumulation in patients with Jansen-type metaphyseal dysplasia (OMIM 156400) (Schipani et al., 1995, 1996, 1997a, 1999). On the other hand, the loss-of-function mutants correlate with the Blomstrand chondrodysplasia disorder (OMIM 215045) associated with the absence of a functional PTH1R (Karaplis et al., 1998; Zhang et al., 1998; Jobert et al., 1998). Interestingly, in the recessive Eiken skeletal dysplasia syndrome (OMIM 600002), a mutation leading to a C-terminal truncation of PTH1R has been identified that results in a phenotype opposite to that of Blomstrand chondrodysplasia and resembles a transgenic mouse model in which PTH1R signal transduction via the phospholipase C/inositol-3-phosphate-dependent pathway is compromised (Guo et al., 2002; Duchatelet et al., 2005).

Ihh, encoding a secreted molecule of the HH family, is expressed in pre-HCCs and has been shown to regulate the expression of *PTHrP* (Vortkamp et al., 1996; St-Jacques et al., 1999). This regulation is probably mediated by TGFβ2 signaling (Alvarez et al., 2002). *Ihh*-knockout mice display defects in endochondral and intramembranous bone formation (St-Jacques et al., 1999; Abzhanov et al., 2007; Lenton et al., 2011). In endochondral bone formation, IHH has multiple functions; it regulates proliferation and chondrocyte hypertrophy and is essential for osteoblastogenesis in the perichondrium. The last function of IHH apparently requires additional effectors other than RUNX2 (Tu et al., 2012). Conditional deletion of *Ihh* in *Col2a1*-CRE-expressing cells recapitulates the total knockout phenotype, including the multiple synostosis phenotype, a severe form of synchondrosis (Razzaque et al., 2005). In humans, *IHH* mutations are associated with brachydactyly type A1 (OMIM 112500), while copy number variations including the *IHH* locus are associated with syndactyly and craniosynostosis (Gao et al., 2009; Klopocki et al., 2011). The effects of *Ihh* on chondrocyte hypertrophy are PTHrP dependent as well as independent, while those on proliferation, osteoblastogenesis, and joint formation are PTHrP independent (Karp et al., 2000; Long et al., 2001, 2004; Kobayashi et al., 2005; Amano et al., 2016; Mak et al., 2008).

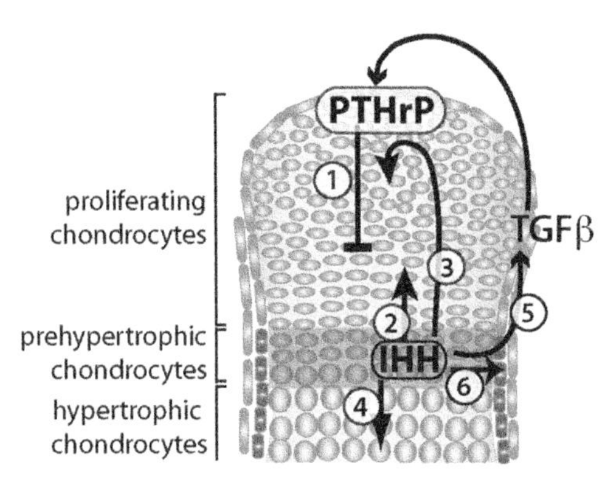

FIGURE 1.8 Parathyroid hormone-related protein (*PTHrP*) and Indian Hedgehog (*IHH*) interactions and functions in the growth plate. IHH and PTHrP participate in a negative feedback loop to regulate chondrocyte proliferation and differentiation. PTHrP is expressed from the perichondrial cells at the articular region and at low levels in round proliferative chondrocytes. It acts on proliferating chondrocytes, keeping them in a proliferative state and preventing their differentiation to prehypertrophic and hypertrophic chondrocytes (**1**). When the PTHrP concentration is sufficiently low enough, chondrocytes drop out of the cell cycle and differentiate into IHH-producing prehypertrophic chondrocytes. IHH, in turn, stimulates the proliferation of the adjacent flattened proliferating chondrocytes (**2**) and accelerates the progression of round to flattened proliferating chondrocytes (**3**) as well as the differentiation of prehypertrophic to hypertrophic chondrocytes (**4**). IHH also stimulates, probably mediated by transforming growth factor β (*TGFβ*) signaling, PTHrP production at the articular ends of the skeletal element (**5**) and acts on perichondrial cells, stimulating their differentiation into osteoblasts (**6**).

As mentioned earlier, the transcription factor GLI3 acts downstream of HH signaling, whereby HH signaling prevents the proteolytic conversion of GLI3 into the repressor form GLI3R. Mutations in *GLI3* are associated with Greig cephalopolysyndactyly (OMIM 175700) and Pallister–Hall syndrome (OMIM 146510) (Demurger et al., 2015). The mouse mutant extra-toes (Xt), a model for Greig cephalopolysyndactyly syndrome, has a deletion in the *Gli3* gene and displays numerous skeletal abnormalities, such as polydactyly, shortened long bones, split sternum, and craniofacial defects (Hui and Joyner, 1993; Vortkamp et al., 1992; Mo et al., 1997). Craniofacial abnormalities and shortened appendicular long bones are also reported in *Gli2* mutants (Mo et al., 1997). Interestingly, in double mutants for *Ihh* and *Gli3* the proliferation defect observed in the *Ihh* mutants is restored and the accelerated HCC differentiation, observed in $Ihh^{-/-}$ specimens, reverted (Hilton et al., 2005; Koziel et al., 2005). In contrast, the defects in osteoblastogenesis and cartilage vascularization are only partially rescued by the loss of *Gli3* (Hilton et al., 2005). Based on the observations in $Ihh^{-/-}$;$Gli3^{-/-}$ double mutants, Koziel and colleagues proposed a model whereby the IHH/GLI3 system regulates two distinct steps in chondrocyte differentiation: first, the transition from distal, round chondrocytes to the columnar chondrocytes, which appears to occur in a PTHrP-independent fashion, and second, the transition from proliferating to HCCs occurring in a PTHrP-dependent fashion (Koziel et al., 2005). Yet, Mak and colleagues proposed that Ihh also promotes chondrocyte hypertrophy in a PTHrP-independent way (Mak et al., 2008) (Fig. 1.8). In addition, Ihh activity is required for the maturation of the perichondrium and, in a cell-autonomous fashion, for the maintenance of endothelial cell fate (Colnot et al., 2005).

WNTs and β-catenin

As mentioned earlier β-catenin-mediated WNT signaling plays an important role as a permissive signal in the early steps of endochondral bone formation, enabling the differentiation of osteoblasts and cells contributing to the joint by repressing the chondrogenic potential within the respective precursor populations. The critical role of WNT/β-catenin signaling in osteoblastogenesis is first shown by the findings that human mutations in the WNT receptor LRP5 cause osteoporosis–pseudoglioma syndrome (OMIM 259770) (Gong et al., 2001; Lara-Castillo and Johnson, 2015). Mutations in the *WNT1* gene are causative for osteogenesis imperfecta, type XV, and an autosomal-dominant form of susceptibility to early onset of osteoporosis (OMIM 615220, 615221) (Keupp et al., 2013; Laine et al., 2013; Pyott et al., 2013). Numerous WNT-pathway molecules have been identified in genome-wide association studies related to skeletal phenotypes (Hsu and Kiel, 2012).

Stabilization of β-catenin in limb mesenchymal cells interferes with the initiation process of endochondral ossification (Hill et al., 2005). In contrast, expression of a constitutively active form of the downstream transcription factor LEF1 in *Col2a1*-expressing cells inhibits further maturation of chondrocytes and interferes with the formation of joints (Tamamura et al., 2005). Later during chondrocyte differentiation, WNT/β-catenin signaling regulates chondrocyte maturation in a positive manner (Hartmann and Tabin, 2001; Enomoto-Iwamoto et al., 2002; Akiyama et al., 2004; Day et al., 2005; Hill et al., 2005; Hu et al., 2005; Spater et al., 2006b; Joeng et al., 2011; Dao et al., 2012). This is mediated in multiple ways, via direct regulation of *Ihh*, through interference with SOX9, and in a RUNX2-dependent fashion (Akiyama et al., 2004; Yano et al., 2005; Spater et al., 2006b; Dong et al., 2006; Dao et al., 2012; Mak et al., 2008). In HCCs, β-catenin signaling downregulates the expression of *Rankl*, thereby locally regulating the differentiation of osteoclasts at the chondro-osseous border (Golovchenko et al., 2013; Wang et al., 2014a; Houben et al., 2016). Based on overexpression of an intracellular inhibitor of β-catenin, ICAT, it has been proposed that β-catenin positively regulates VEGF and MMP13 (Chen et al., 2008). Yet, this has not been confirmed in conditional β-catenin mutants. Transient activation of β-catenin during early postnatal development leads to abnormal growth plate closure and promotes secondary ossification center formation (Yuasa et al., 2009; Dao et al., 2012). For further information see reviews by Baron and Kneissel (2013), Wang et al. (2014d), and Usami et al. (2016).

In addition to WNT/β-catenin-mediated signaling, a number of additional WNT signaling pathways are important within the growth plate. One that is highly relevant is the planar cell polarity pathway and its components, WNT5a and receptor tyrosine kinase orphan receptor (ROR2). Mutations in WNT5a and ROR2 are associated with Robinow syndrome (OMIM 268310; 164975) and brachydactyly type B1 (OMIM 113000) (Patton and Afzal, 2002; Person et al., 2010; Roifman et al., 2015). In mice, loss-of-function mutations in *Wnt5a*, *Ror2*, *Vangl2*, *Prickle1*, and *Ryk* result in skeletal dysplasias resembling those associated with Robinow syndrome (DeChiara et al., 2000; Takeuchi et al., 2000; Wang et al., 2011; Andre et al., 2012; Macheda et al., 2012; Gao et al., 2011; Yang et al., 2013b; Liu et al., 2014). In mice, *Wnt5a* and its related family member *Wnt5b* are both expressed in pre-HCCs (Yamaguchi et al., 1999; Yang et al., 2003; Witte et al., 2009). WNT5a promotes chondrocyte proliferation, as such mice lacking *Wnt5a* develop shorter skeletal elements due to a reduction in chondrocyte proliferation in zone II of the proliferating chondrocytes, encompassing the flattened proliferating chondrocytes (Yamaguchi et al., 1999; Yang et al., 2003). Furthermore, chondrocyte differentiation of HCCs is severely delayed in $Wnt5a^{-/-}$ mice as it is in *Ror2* mutants (DeChiara et al., 2000; Takeuchi et al., 2000; Oishi et al., 2003). Overexpression of either *Wnt5a* or *Wnt5b* primarily in chondrocytes also delays chondrocyte differentiation, yet, the two WNT ligands act on different chondrocyte subsets (Yang et al., 2003). The intracellular pathways underlying these effects are not known as of this writing. Compromised differentiation of HCCs may be associated with the capacity of WNT5a to induce the proteolytic cleavage of the transcription factor NKX3.2, which inhibits chondrocyte hypertrophy (Provot et al., 2006). In vitro, WNT5a and WNT5b can both activate calcium-dependent signaling leading to nuclear localization of nuclear factor of activated T cells (NFAT), as well as NF-κB signaling, and the kinase JNK (Oishi et al., 2003; Bradley and Drissi, 2010, 2011). The two pathways have differential effects on chondrogenesis (Bradley and Drissi, 2010). WNT5a signaling has also been shown to downregulate WNT/β-catenin signaling (Topol et al., 2003; Mikels and Nusse, 2006). However, experiments suggest that it can also enhance WNT/β-catenin signaling during osteoblastogenesis (Okamoto et al., 2014). Which pathway is preferentially activated may be decided at the level of the coreceptors (Grumolato et al., 2010).

Fibroblast growth factors and their receptors

FGF signaling also plays a critical role in the growth plate. Mutations in all three human *FGFRs* cause skeletal malformations, such as craniosynostosis syndromes (see also intramembranous ossification) and chondrodysplasia. Constitutively activating mutations in *FGFR3* are associated with hypochondrodysplasia (OMIM 146000), achondrodysplasia (OMIM 100800), and thanatophoric dysplasias type I (OMIM 187600) and type II (OMIM 187601). For reviews see Robin et al. (1993) and Ornitz and Marie (2015).

In the murine growth plate, *Fgfr2* is expressed at low levels in the round proliferating zone, also referred to as the resting zone. Proliferating and pre-HCCs express high levels of *Fgfr3*, and HCCs express high levels of *Fgfr1* (Ornitz and Marie, 2015). The growth retardation in conditionally deleted *Fgfr2* mice is attributed to alterations at the chondro-osseous junction (Yu et al., 2003). Yet, chondrocyte proliferation was unaffected in these mice. An increase in the zone of proliferating chondrocytes as well as HCCs was observed upon loss of *Fgfr3* (Colvin et al., 1996; Deng et al., 1996). In contrast, mice carrying an *Fgfr3* gene with human achondroplasia mutations display the opposite phenotype, a decrease in chondrocyte proliferation and a reduced zone of HCCs (Chen et al., 1999; Li et al., 1999). Of the different ligands, FGF9 and FGF18 have been identified based on their mutant phenotypes to be relevant in endochondral bone formation. Both are expressed in the perichondrium and periosteum. FGF9 and FGF18 are both required for chondrocyte maturation, as the onset of hypertrophy is delayed in $Fgf9^{-/-}$ and $Fgf18^{-/-}$ embryos (Hung et al., 2007; Liu et al., 2007). Yet, in the $Fgf9^{-/-}$ mutant only the stylopod elements are affected (Hung et al., 2007). Due to the delay in chondrocyte maturation vascular invasion is also delayed in both mutants. However, there is evidence that FGF18 may directly stimulate the expression of VEGF (Liu et al., 2007). In addition, FGF18 is required for chondrocyte proliferation. A 2016 allelic series study of *Fgf9*/*Fgf18* mutant embryos revealed unique and redundant roles of the two ligands in endochondral ossification (Hung et al., 2016).

C-type natriuretic peptide

In the growth plate, C-type natriuretic peptide (CNP) and its receptor GC-B are primarily expressed in proliferative and pre-HCCs (Chusho et al., 2001). In humans, homozygous loss-of-function mutations in the receptor cause acromesomelic dysplasia Maroteaux type (OMIM 602875), while heterozygous mutations are associated with short stature (Bartels et al., 2004; Olney et al., 2006; Vasques et al., 2013). Yet, the CNP/GC-B system is widely distributed in the body and as such, it was unclear whether it acts systemically or locally on endochondral ossification. Evidence for the latter is based on conditional knockouts of *Cnp* or *Gc-b* in *Col2a1*-Cre-expressing cells that recapitulate the dwarfism phenotypes of the respective full knockouts (Chusho et al., 2001; Tamura et al., 2004; Nakao et al., 2015). Dwarfism is associated with a decrease in the proliferative zone and in the number and size of HCCs. In contrast, loss of the clearance receptor NPR-C results in skeletal overgrowth similar to that in mice overexpressing the related molecule BNP (Suda et al., 1998; Jaubert et al., 1999). Skeletal growth can also be stimulated in a dose-dependent fashion by interfering with the clearance of CNP by overexpressing osteocrin, a natural NPR-C ligand (Kanai et al., 2017). In humans, overexpression of CNP is also associated with skeletal overgrowth (Bocciardi et al., 2007). Craniofacial studies in mice suggest that CNP/GC-B signaling primarily stimulates endochondral ossification (Nakao et al., 2013). Downstream signaling involves cyclic GMP-dependent kinase II but also interferes with the activation of the mitogen-activated protein kinase cascade downstream of FGF signaling (Miyazawa et al., 2002; Ozasa et al., 2005). For further reading see the review by Peake et al. (2014).

Notch signaling

Mutations in the Notch signaling components cause at least two human disorders with vertebral column defects, spondylocostal dysostosis (OMIM 277300, 608681, and 609813) and Alagille syndrome (OMIM 118450 and 610205) (Baldridge et al., 2010). Gain-of-function mutations in *NOTCH2* are found in Hajdu–Cheney syndrome, a rare skeletal disorder characterized by osteoporosis (OMIM 102500) (Majewski et al., 2011; Isidor et al., 2011). These diseases highlight, among others, the critical role of the segmentation clock in human axial skeletal development.

In chick and mouse, the Notch receptors 1–4 and the ligands, Delta1 and Jagged1/2, are expressed in a dynamic way within the developing limb skeleton, and inhibition of Notch signaling disrupts chondrocyte differentiation (Williams et al., 2009; Dong et al., 2010). Misexpression of the ligand Delta1 in chick inhibits the transition from pre-HCC to HCC (Crowe et al., 1999). A similar phenotype is observed upon conditional expression of the active Notch intracellular domain (NICD) in chondrocytes within the long bones, while a loss of skeletal elements due to impaired chondrogenesis is observed in the axial skeleton (Mead and Yutzey, 2009). The latter is associated with a downregulation of *Sox9* and, as shown in additional studies, with an enhanced proliferation of the mesenchymal progenitor cells, which is dependent on the activity of the transcriptional cofactor RBPjκ (recombination signal binding protein for immunoglobulin κ J region), which interacts with the NICD in the nucleus (Dong et al., 2010; Chen et al., 2013). Consistent with the osteoporosis phenotype in humans, the gain of Notch signaling in mice affects osteoblastogenesis of endochondral and membranous bones (Hilton et al., 2008; Mead and Yutzey, 2009; Dong et al., 2010). In contrast, interference with the Notch pathway by conditional deletion of *Presenilin 1/2*, encoding proteins required for the NICD release, or the *Notch1/2* receptors in the limb mesenchyme results initially in a delay of the onset of chondrocyte maturation and later in a delay of terminal differentiation leading to an elongated hypertrophic zone (Hilton et al., 2008). Conditional loss of the Notch effector RBPjκ results in a similar phenotype (Kohn et al., 2012). RBPjκ-independent Notch signaling, in contrast, affects the morphology of all growth plate chondrocytes and enhances osteoblast maturation (Kohn et al., 2012). In the articular chondrocytes, Notch signaling may be required for the maintenance of a chondroprogenitor population (Sassi et al., 2011).

Transcription factors

SOX9 and RUNX2 are master transcription factors that determine chondrocyte and osteoblast cell fates, respectively. It is not surprising that genetic defects in chondrocyte or osteoblast cell fate determination cause severe skeletal defects. Haploinsufficiency of SOX9 protein in humans causes campomelic dysplasia (OMIM 114290) with cartilage hypoplasia and a perinatal lethal osteochondrodysplasia (Meyer et al., 1997). Mutations in human *RUNX2* cause CCD (OMIM 119600), an autosomal-dominant condition characterized by hypoplasia/aplasia of clavicles, patent fontanelles, supernumerary teeth, short stature, and other changes in skeletal patterning and growth (Mundlos et al., 1997). The transcription factor OSX/SP7 acts downstream of RUNX2 within the osteoblast lineage (Nakashima et al., 2002; Nishio et al., 2006). Mutations in the human SP7 gene may be associated with osteogenesis imperfecta type XII (OMIM 613849) (Lapunzina et al., 2010).

Genetic studies in mice revealed that SOX9 plays numerous roles in skeletogenesis, from the initial differentiation of mesenchymal cells to chondrocytes to the maintenance of chondrogenic phenotype, survival, and the control of chondrocyte maturation (reviewed in Lefebvre and Dvir-Ginzberg, 2017). Its necessity for chondrocyte differentiation was first demonstrated by chimeric studies showing that *Sox9*-deficient cells are excluded from the cartilage (Bi et al., 1999). SOX9 activates the expression of two related family members, *Sox5* and *Sox6*, and cooperates with them, establishing and maintaining chondrocyte identity (Smits et al., 2001; Akiyama et al., 2002). SOX9 also interacts directly with and blocks the activity of the transcription factor RUNX2 at target promoters (Zhou et al., 2006). *Runx2* is expressed in pre-HCCs, HCCs, and osteoblast precursors and is important for HCC and osteoblast differentiation (Komori et al., 1997; Otto et al., 1997; Inada et al., 1999; Kim et al., 1999). Thus, this interaction maintains chondrocytes in a proliferative state and blocks their differentiation into HCCs and transdifferentiation into osteoblasts (Dy et al., 2012). RUNX2 acts partially redundantly with the related RUNX-family member RUNX3 on HCC maturation (Yoshida et al., 2004). *Mef2c* and *Mef2d*, members of the myocyte enhancer factor 2 family of transcription factors, are also expressed in pre-HCCs/HCCs. In contrast to *Mef2d*-knockout mice, which have no reported skeletal phenotype, *Mef2c*-deficient mice have shorter long bones associated with a delay in chondrocyte hypertrophy and downregulation of *Runx2* expression (Arnold et al., 2007; Kim et al., 2008). A constitutively active form of MEF2C upregulates *Runx2* and promotes chondrocyte hypertrophy, suggesting that MEF2C acts upstream of RUNX2 (Arnold et al., 2007). The activity of both transcription factors, MEF2C and RUNX2, is modulated by the histone deacetylase HDAC4 (see later). RUNX2 activity in HCCs is probably also modulated by interactions with other transcription factors such as *Dlx5/6*, which both physically interact with RUNX2 (Roca et al., 2005; Chin et al., 2007). Two members of the forkhead family of transcription factors, *Foxa2* and *Foxa3*, also play a role in HCCs. Both are expressed in HCCs and the loss of *Foxa2* results in decreased expression of hypertrophic markers, such as *Col10a1* and *Mmp13*, which is aggravated by the additional loss of *Foxa3* (Ionescu et al., 2012). The *SoxC* genes, *Sox4*, *Sox11*, and *Sox12*, are initially expressed in the mesenchymal progenitors of endochondral and intramembranous bone and become restricted to the perichondrium and joint as the chondrocytes differentiate (reviewed in Lefebvre and Bhattaram, 2016). In the progenitors, SOXC proteins are required for cell survival (Bhattaram et al., 2010). Later, during endochondral ossification, they are required for growth plate formation in part by promoting noncanonical WNT5a signaling (Kato et al., 2015). Other transcription factors, such as *Prrx1/Mhox* in combination with *Prrx2*, *Msx2*, and the *AP1* family member *Fra2*, also play roles in endochondral ossification (Martin et al., 1995; Lu et al., 1999; Karreth et al., 2004; Satokata et al., 2000). These can be acting locally restricted as is the case for *Prrx1/2* (Lu et al., 1999). For further information see reviews by Hartmann (2009), Karsenty (2008), and Nishimura et al. (2018).

The hypoxia-inducible transcription factor HIF consists of an α subunit that is regulated by oxygen and a β subunit that is constitutively expressed (Semenza, 2012; Ratcliffe, 2013). In growth plate chondrocytes, which are hypoxic, the subunit protein HIF-1α is stabilized and, on one hand, induces the expression of VEGF in HCCs and, on the other hand, regulates the oxygen consumption of chondrocytes through stimulation of anaerobic metabolism or glycolysis. Both downstream mechanisms are necessary for chondrocyte survival (Maes et al., 2012; Schipani et al., 2001, 2015; Cramer et al., 2004; Zelzer et al., 2004). The delayed differentiation observed in *Hif1a* mutants is probably a consequence of the initial delay in the initiation of chondrogenesis earlier in development (Provot et al., 2007; Amarilio et al., 2007). In contrast, mutation in the related α-subunit-encoding gene *Hif2a* results in only a transient and modest delay in endochondral ossification (Araldi et al., 2011). Yet, HIF2a appears to play a more prominent role postnatally in articular chondrocyte homeostasis (Pi et al., 2015; Yang et al., 2010).

Epigenetic factors and microRNAs

Since 2009, novel regulators of chondrogenesis and osteoblastogenesis have emerged, including epigenetic factors (reviewed in Furumatsu and Ozaki, 2010; Bradley et al., 2015). Among them is the histone deacetylase HDAC4, which plays a prominent role in HCC differentiation (Vega et al., 2004). HDAC4 binds to and inhibits the activity of two transcription factors that promote HCC differentiation, RUNX2 and MEF2C (Vega et al., 2004; Arnold et al., 2007). Histone-acetyl transferases such as P300 are important cofactors for BMP/SMAD1- and TGFβ/SMAD3-dependent signaling (Furumatsu et al., 2005; Pan et al., 2009; Sun et al., 2009). P300 also acts as a cofactor within the WNT/β-catenin pathway (Levy et al., 2004) and interacts with SOX9 (Furumatsu et al., 2005). SOX9 is also acetylated, which reduces its transcriptional activity, and this can be modulated by the NAD-dependent class III protein deacetylase Sirtuin (SIRT1) (Buhrmann et al., 2014; Bar Oz et al., 2016). SIRT1 and the histone methyltransferases SET7/SET9 also interact with P300 on the type II collagen promoter, promoting transcription (Oppenheimer et al., 2014). Conditional mouse mutants for the histone methyltransferase *Eset* have severely shortened limbs, a split sternum, and a widening of the sagittal suture of the skull (Yang et al., 2013a). The growth plates of *Eset* conditional knockout mice are disorganized, and HCC differentiation appears to be accelerated. ESET interacts with HDAC4 to repress RUNX2 activity, thereby delaying hypertrophic differentiation (Yang et al., 2013a). Overall changes in the chromatin acetylation status in chondrocytes are induced through the interaction of the transcription factor TRPS1 with HDAC1 and HDAC4 (Wuelling et al., 2013).

Conditional deletion of *Dicer*, an enzyme that is required for the biogenesis of microRNAs, in chondrocytes revealed a functional role for microRNAs in chondrocyte proliferation and differentiation (Kobayashi et al., 2008). The latter is associated with a widened hypertrophic zone. Some of the specific microRNAs involved in these phenotypes are let-7 and miR-140 (Miyaki et al., 2010; Nakamura et al., 2011; Papaioannou et al., 2013). The noncoding RNA Dnm3os, a precursor for the microRNAs miR-199a, miR-199a*, and miR-214, is required for normal growth and skeletal development (Watanabe et al., 2008). In vitro, numerous microRNAs are differentially regulated during chondrogenesis and in osteoarthritis (Swingler et al., 2012; Crowe et al., 2016). For additional information on the role of microRNAs in skeletal development and homeostasis see Hong and Reddi (2012), Mirzamohammadi et al. (2014), and Fang et al. (2015).

Another class of RNA molecules with emerging functions in skeletal development are the long noncoding RNAs (lncRNAs). Mutations in the lncRNA *DA125942*, which interacts with PTHrP, result in brachydactyly type E (OMIM 613382) (Maass et al., 2012). The lncRNA *DANCR* promotes the chondrogenic differentiation of human synovial stem cell–like cells and is involved in osteoblastogenesis (reviewed in Huynh et al., 2017).

The functional roles of the vasculature in endochondral bone formation

Cartilage is an avascular and hypoxic tissue, yet, the ossification process and the remodeling of the cartilage template into cancellous bone require blood vessel invasion. Proliferating chondrocytes express numerous antiangiogenic factors, such as Chondromodulin I, Tenomodulin, Tissue-localized inhibitors of MMPs, and others (Maes, 2013). HCCs, in contrast, express VEGF, which is required to attract blood vessels to the perichondrium flanking the hypertrophic zone, as exemplified by mutant mice in which *Vegf* was deleted in cartilage or which lacked specifically the diffusible splice isoforms VEGF120 and VEGF164 (Zelzer et al., 2004; Maes et al., 2004, 2012). *Vegf* expression in HCCs is controlled by RUNX2 and, as mentioned earlier, by HIF1 (Zelzer et al., 2001). The invasion of blood vessels probably play an important role in the formation of the bone marrow cavity during endochondral ossification. Evidence for this is based on blocking VEGF signaling, which affects cartilage resorption, resulting in the elongation of the zone of HCCs (Gerber et al., 1999). Yet, as the monocytes, which are precursors for chondroclasts and osteoclasts, enter the bone marrow cavity via blood vessels, it is difficult to unambiguously distinguish between the functional requirements of the two components for the formation of the bone marrow cavity. Chondroclasts and osteoclasts produce matrix-degrading enzymes. Yet, the mineral dissolution function of osteoclasts is dispensable for the degradation of HCCs during long bone growth (Touaitahuata et al., 2014). Blood vessel endothelial cells also produce and secrete, among others, MMP9/Gelatinase B under proangiogenic conditions and may, therefore, be actively involved in the degradation of the cartilage matrix (Taraboletti et al., 2002). Blood vessels are, furthermore, important for trabecular bone formation during endochondral ossification. As mentioned previously, osteoblast precursors migrate into the forming bone marrow cavity along the blood vessels (Maes et al., 2010). In addition, it has been shown that the bone marrow cavity contains at least two types of blood vessels. In the embryo, an E and an L type can be distinguished, whereof the E type strongly supports osteoblast lineage cells (Langen et al., 2017). In the adult, the H-type vessels are the ones supporting osteoblast maturation (Kusumbe et al., 2014). Blood vessels also play a role as a structural component in trabecular bone formation. In addition to the mineralized cartilage matrix remnants, the vessels serve as structures for osteoid deposition (Ben Shoham et al., 2016).

References

Abzhanov, A., Rodda, S.J., Mcmahon, A.P., Tabin, C.J., 2007. Regulation of skeletogenic differentiation in cranial dermal bone. Development 134, 3133–3144.

Adair, F.L., Scammin, R.E., 1921. A study of the ossification centers of the wrist, knee and ankle at birth, with particular reference to the physical development and maturity of the newborn. Am. J. Obstet. Gynecol. 2, 35–60.

Ahn, K., Mishina, Y., Hanks, M.C., Behringer, R.R., Crenshaw 3rd, E.B., 2001. BMPR-IA signaling is required for the formation of the apical ectodermal ridge and dorsal-ventral patterning of the limb. Development 128, 4449–4461.

Ahrens, M.J., Li, Y., Jiang, H., Dudley, A.T., 2009. Convergent extension movements in growth plate chondrocytes require gpi-anchored cell surface proteins. Development 136, 3463–3474.

Al-Aql, Z.S., Alagl, A.S., Graves, D.T., Gerstenfeld, L.C., Einhorn, T.A., 2008. Molecular mechanisms controlling bone formation during fracture healing and distraction osteogenesis. J. Dent. Res. 87, 107–118.

Akiyama, H., Chaboissier, M.C., Martin, J.F., Schedl, A., De Crombrugghe, B., 2002. The transcription factor Sox9 has essential roles in successive steps of the chondrocyte differentiation pathway and is required for expression of Sox5 and Sox6. Genes Dev. 16, 2813–2828.

Akiyama, H., Kim, J.E., Nakashima, K., Balmes, G., Iwai, N., Deng, J.M., Zhang, Z., Martin, J.F., Behringer, R.R., Nakamura, T., De Crombrugghe, B., 2005. Osteo-chondroprogenitor cells are derived from Sox9 expressing precursors. Proc. Natl. Acad. Sci. U.S.A. 102, 14665–14670.

Akiyama, H., Lyons, J.P., Mori-Akiyama, Y., Yang, X., Zhang, R., Zhang, Z., Deng, J.M., Taketo, M.M., Nakamura, T., Behringer, R.R., Mccrea, P.D., DE Crombrugghe, B., 2004. Interactions between Sox9 and beta-catenin control chondrocyte differentiation. Genes Dev. 18, 1072–1087.

Alvarez, J., Costales, L., Lopez-Muniz, A., Lopez, J.M., 2005. Chondrocytes are released as viable cells during cartilage resorption associated with the formation of intrachondral canals in the rat tibial epiphysis. Cell Tissue Res. 320, 501–507.

Alvarez, J., Sohn, P., Zeng, X., Doetschman, T., Robbins, D.J., Serra, R., 2002. TGFbeta2 mediates the effects of hedgehog on hypertrophic differentiation and PTHrP expression. Development 129, 1913–1924.

Amano, K., Densmore, M., Fan, Y., Lanske, B., 2016. Ihh and PTH1R signaling in limb mesenchyme is required for proper segmentation and subsequent formation and growth of digit bones. Bone 83, 256–266.

Amarilio, R., Viukov, S.V., Sharir, A., Eshkar-Oren, I., Johnson, R.S., Zelzer, E., 2007. HIF1alpha regulation of Sox9 is necessary to maintain differentiation of hypoxic prechondrogenic cells during early skeletogenesis. Development 134, 3917–3928.

Amizuka, N., Henderson, J.E., Hoshi, K., Warshawsky, H., Ozawa, H., Goltzman, D., Karaplis, A.C., 1996. Programmed cell death of chondrocytes and aberrant chondrogenesis in mice homozygous for parathyroid hormone-related peptide gene deletion. Endocrinology 137, 5055–5067.

Anderson, H.C., Garimella, R., Tague, S.E., 2005a. The role of matrix vesicles in growth plate development and biomineralization. Front. Biosci. 10, 822–837.

Anderson, H.C., Harmey, D., Camacho, N.P., Garimella, R., Sipe, J.B., Tague, S., Bi, X., Johnson, K., Terkeltaub, R., Millan, J.L., 2005b. Sustained osteomalacia of long bones despite major improvement in other hypophosphatasia-related mineral deficits in tissue nonspecific alkaline phosphatase/nucleotide pyrophosphatase phosphodiesterase 1 double-deficient mice. Am. J. Pathol. 166, 1711–1720.

Anderson, H.C., Sipe, J.B., Hessle, L., Dhanyamraju, R., Atti, E., Camacho, N.P., Millan, J.L., Dhamyamraju, R., 2004. Impaired calcification around matrix vesicles of growth plate and bone in alkaline phosphatase-deficient mice. Am. J. Pathol. 164, 841–847.

Andre, P., Wang, Q., Wang, N., Gao, B., Schilit, A., Halford, M.M., Stacker, S.A., Zhang, X., Yang, Y., 2012. The Wnt coreceptor Ryk regulates Wnt/planar cell polarity by modulating the degradation of the core planar cell polarity component Vangl2. J. Biol. Chem. 287, 44518–44525.

Annes, J.P., Munger, J.S., Rifkin, D.B., 2003. Making sense of latent TGFbeta activation. J. Cell Sci. 116, 217–224.

Araldi, E., Khatri, R., Giaccia, A.J., Simon, M.C., Schipani, E., 2011. Lack of HIF-2alpha in limb bud mesenchyme causes a modest and transient delay of endochondral bone development. Nat. Med. 17, 25–26 (author reply 27-9).

Archer, C.W., Dowthwaite, G.P., Francis-West, P., 2003. Development of synovial joints. Birth Defects Res. C Embryo Today 69, 144–155.

Arnold, M.A., Kim, Y., Czubryt, M.P., Phan, D., Mcanally, J., Qi, X., Shelton, J.M., Richardson, J.A., Bassel-Duby, R., Olson, E.N., 2007. MEF2C transcription factor controls chondrocyte hypertrophy and bone development. Dev. Cell 12, 377–389.

Aszodi, A., Hunziker, E.B., Brakebusch, C., Fassler, R., 2003. Beta1 integrins regulate chondrocyte rotation, G1 progression, and cytokinesis. Genes Dev. 17, 2465–2479.

Atit, R., Sgaier, S.K., Mohamed, O.A., Taketo, M.M., Dufort, D., Joyner, A.L., Niswander, L., Conlon, R.A., 2006. Beta-catenin activation is necessary and sufficient to specify the dorsal dermal fate in the mouse. Dev. Biol. 296, 164–176.

Aulehla, A., Pourquie, O., 2008. Oscillating signaling pathways during embryonic development. Curr. Opin. Cell Biol. 20, 632–637.

Badve, C.A., K, M.M., Iyer, R.S., Ishak, G.E., Khanna, P.C., 2013. Craniosynostosis: imaging review and primer on computed tomography. Pediatr. Radiol. 43, 728–742 (quiz 725-7).

Baffi, M.O., Slattery, E., Sohn, P., Moses, H.L., Chytil, A., Serra, R., 2004. Conditional deletion of the TGF-beta type II receptor in Col2a expressing cells results in defects in the axial skeleton without alterations in chondrocyte differentiation or embryonic development of long bones. Dev. Biol. 276, 124–142.

Bai, C.B., Auerbach, W., Lee, J.S., Stephen, D., Joyner, A.L., 2002. Gli2, but not Gli1, is required for initial Shh signaling and ectopic activation of the Shh pathway. Development 129, 4753–4761.

Baker, J., Liu, J.P., Robertson, E.J., Efstratiadis, A., 1993. Role of insulin-like growth factors in embryonic and postnatal growth. Cell 75, 73–82.

Baldridge, D., Shchelochkov, O., Kelley, B., Lee, B., 2010. Signaling pathways in human skeletal dysplasias. Annu. Rev. Genom. Hum. Genet. 11, 189–217.

Ballock, R.T., O'keefe, R.J., 2003. The biology of the growth plate. J. Bone Joint Surg. Am. 85-A, 715–726.

Bandyopadhyay, A., Tsuji, K., Cox, K., Harfe, B.D., Rosen, V., Tabin, C.J., 2006. Genetic analysis of the roles of BMP2, BMP4, and BMP7 in limb patterning and skeletogenesis. PLoS Genet. 2, e216.

Bar oz, M., Kumar, A., Elayyan, J., Reich, E., Binyamin, M., Kandel, L., Liebergall, M., Steinmeyer, J., Lefebvre, V., Dvir-Ginzberg, M., 2016. Acetylation reduces SOX9 nuclear entry and ACAN gene transactivation in human chondrocytes. Aging Cell 15, 499–508.

Barna, M., Niswander, L., 2007. Visualization of cartilage formation: insight into cellular properties of skeletal progenitors and chondrodysplasia syndromes. Dev. Cell 12, 931–941.

Barna, M., Pandolfi, P.P., Niswander, L., 2005. Gli3 and Plzf cooperate in proximal limb patterning at early stages of limb development. Nature 436, 277–281.

Baron, R., Kneissel, M., 2013. WNT signaling in bone homeostasis and disease: from human mutations to treatments. Nat. Med. 19, 179–192.

Barrow, J.R., Thomas, K.R., Boussadia-Zahui, O., Moore, R., Kemler, R., Capecchi, M.R., Mcmahon, A.P., 2003. Ectodermal Wnt3/beta-catenin signaling is required for the establishment and maintenance of the apical ectodermal ridge. Genes Dev. 17, 394–409.

Bartels, C.F., Bukulmez, H., Padayatti, P., Rhee, D.K., Van Ravenswaaij-Arts, C., Pauli, R.M., Mundlos, S., Chitayat, D., Shih, L.Y., Al-Gazali, L.I., Kant, S., Cole, T., Morton, J., Cormier-Daire, V., Faivre, L., Lees, M., Kirk, J., Mortier, G.R., Leroy, J., Zabel, B., Kim, C.A., Crow, Y., Braverman, N.E., Van Den Akker, F., Warman, M.L., 2004. Mutations in the transmembrane natriuretic peptide receptor NPR-B impair skeletal growth and cause acromesomelic dysplasia, type Maroteaux. Am. J. Hum. Genet. 75, 27–34.

Bastepe, M., Weinstein, L.S., Ogata, N., Kawaguchi, H., Juppner, H., Kronenberg, H.M., Chung, U.I., 2004. Stimulatory G protein directly regulates hypertrophic differentiation of growth plate cartilage in vivo. Proc. Natl. Acad. Sci. U.S.A. 101, 14794–14799.

Baur, S.T., Mai, J.J., Dymecki, S.M., 2000. Combinatorial signaling through BMP receptor IB and GDF5: shaping of the distal mouse limb and the genetics of distal limb diversity. Development 127, 605–619.

Ben Shoham, A., Rot, C., Stern, T., Krief, S., Akiva, A., Dadosh, T., Sabany, H., Lu, Y., Kadler, K.E., Zelzer, E., 2016. Deposition of collagen type I onto skeletal endothelium reveals a new role for blood vessels in regulating bone morphology. Development 143, 3933–3943.

Benazet, J.D., Pignatti, E., Nugent, A., Unal, E., Laurent, F., Zeller, R., 2012. Smad4 is required to induce digit ray primordia and to initiate the aggregation and differentiation of chondrogenic progenitors in mouse limb buds. Development 139, 4250–4260.

Benazet, J.D., Zeller, R., 2009. Vertebrate limb development: moving from classical morphogen gradients to an integrated 4-dimensional patterning system. Cold Spring Harb. Perspect. Biol. 1 a001339.

Bhat, R., Lerea, K.M., Peng, H., Kaltner, H., Gabius, H.J., Newman, S.A., 2011. A regulatory network of two galectins mediates the earliest steps of avian limb skeletal morphogenesis. BMC Dev. Biol. 11, 6.

Bhattaram, P., Penzo-Mendez, A., Kato, K., Bandyopadhyay, K., Gadi, A., Taketo, M.M., Lefebvre, V., 2014. SOXC proteins amplify canonical WNT signaling to secure nonchondrocytic fates in skeletogenesis. J. Cell Biol. 207, 657–671.

Bhattaram, P., Penzo-Mendez, A., Sock, E., Colmenares, C., Kaneko, K.J., Vassilev, A., Depamphilis, M.L., Wegner, M., Lefebvre, V., 2010. Organogenesis relies on SoxC transcription factors for the survival of neural and mesenchymal progenitors. Nat. Commun. 1, 9.

Bi, W., Deng, J.M., Zhang, Z., Behringer, R.R., De Crombrugghe, B., 1999. Sox9 is required for cartilage formation. Nat. Genet. 22, 85–89.

Blitz, E., Sharir, A., Akiyama, H., Zelzer, E., 2013. Tendon-bone attachment unit is formed modularly by a distinct pool of Scx- and Sox9-positive progenitors. Development 140, 2680–2690.

Blumer, M.J., Longato, S., Fritsch, H., 2008. Structure, formation and role of cartilage canals in the developing bone. Ann. Anat. 190, 305–315.

Bocciardi, R., Giorda, R., Buttgereit, J., Gimelli, S., Divizia, M.T., Beri, S., Garofalo, S., Tavella, S., Lerone, M., Zuffardi, O., Bader, M., Ravazzolo, R., Gimelli, G., 2007. Overexpression of the C-type natriuretic peptide (CNP) is associated with overgrowth and bone anomalies in an individual with balanced t(2;7) translocation. Hum. Mutat. 28, 724–731.

Boskey, A.L., Spevak, L., Paschalis, E., Doty, S.B., Mckee, M.D., 2002. Osteopontin deficiency increases mineral content and mineral crystallinity in mouse bone. Calcif. Tissue Int. 71, 145–154.

Boulet, A.M., Moon, A.M., Arenkiel, B.R., Capecchi, M.R., 2004. The roles of Fgf4 and Fgf8 in limb bud initiation and outgrowth. Dev. Biol. 273, 361–372.

Bradley, E.W., Carpio, L.R., Van Wijnen, A.J., Mcgee-Lawrence, M.E., Westendorf, J.J., 2015. Histone deacetylases in bone development and skeletal disorders. Physiol. Rev. 95, 1359–1381.

Bradley, E.W., Drissi, M.H., 2010. WNT5A regulates chondrocyte differentiation through differential use of the CaN/NFAT and IKK/NF-kappaB pathways. Mol. Endocrinol. 24, 1581–1593.

Bradley, E.W., Drissi, M.H., 2011. Wnt5b regulates mesenchymal cell aggregation and chondrocyte differentiation through the planar cell polarity pathway. J. Cell. Physiol. 226, 1683–1693.

Brent, A.E., Schweitzer, R., Tabin, C.J., 2003. A somitic compartment of tendon progenitors. Cell 113, 235–248.

Brighton, C.T., 1978. Structure and function of the growth plate. Clin. Orthop. Relat. Res. 22–32.

Bronfin, D.R., 2001. Misshapen heads in babies: position or pathology? Ochsner J. 3, 191–199.

Brunet, L.J., Mcmahon, J.A., Mcmahon, A.P., Harland, R.M., 1998. Noggin, cartilage morphogenesis, and joint formation in the mammalian skeleton. Science 280, 1455–1457.

Buhrmann, C., Busch, F., Shayan, P., Shakibaei, M., 2014. Sirtuin-1 (SIRT1) is required for promoting chondrogenic differentiation of mesenchymal stem cells. J. Biol. Chem. 289, 22048–22062.

Burke, A.C., 2000. Hox genes and the global patterning of the somitic mesoderm. Curr. Top. Dev. Biol. 47, 155–181.

Burke, A.C., Nelson, C.E., Morgan, B.A., Tabin, C., 1995. Hox genes and the evolution of vertebrate axial morphology. Development 121, 333–346.

Cairns, D.M., Sato, M.E., Lee, P.G., Lassar, A.B., Zeng, L., 2008. A gradient of Shh establishes mutually repressing somitic cell fates induced by Nkx3.2 and Pax3. Dev. Biol. 323, 152–165.

Candela, M.E., Cantley, L., Yasuaha, R., Iwamoto, M., Pacifici, M., Enomoto-Iwamoto, M., 2014. Distribution of slow-cycling cells in epiphyseal cartilage and requirement of beta-catenin signaling for their maintenance in growth plate. J. Orthop. Res. 32, 661–668.

Chagin, A.S., Kronenberg, H.M., 2014. Role of G-proteins in the differentiation of epiphyseal chondrocytes. J. Mol. Endocrinol. 53, R39–R45.

Charite, J., Mcfadden, D.G., Olson, E.N., 2000. The bHLH transcription factor dHAND controls Sonic hedgehog expression and establishment of the zone of polarizing activity during limb development. Development 127, 2461–2470.

Chellaiah, M.A., Kizer, N., Biswas, R., Alvarez, U., Strauss-Schoenberger, J., Rifas, L., Rittling, S.R., Denhardt, D.T., Hruska, K.A., 2003. Osteopontin deficiency produces osteoclast dysfunction due to reduced CD44 surface expression. Mol. Biol. Cell 14, 173–189.

Chen, H., Lun, Y., Ovchinnikov, D., Kokubo, H., Oberg, K.C., Pepicelli, C.V., Gan, L., Lee, B., Johnson, R.L., 1998. Limb and kidney defects in Lmx1b mutant mice suggest an involvement of LMX1B in human nail patella syndrome. Nat. Genet. 19, 51–55.

Chen, H., Shi, S., Acosta, L., Li, W., Lu, J., Bao, S., Chen, Z., Yang, Z., Schneider, M.D., Chien, K.R., Conway, S.J., Yoder, M.C., Haneline, L.S., Franco, D., Shou, W., 2004. BMP10 is essential for maintaining cardiac growth during murine cardiogenesis. Development 131, 2219–2231.

Chen, J.M., 1952. Studies on the morphogenesis of the mouse sternum. I. Normal embryonic development. J. Anat. 86, 373–386.

Chen, L., Adar, R., Yang, X., Monsonego, E.O., Li, C., Hauschka, P.V., Yayon, A., Deng, C.X., 1999. Gly369Cys mutation in mouse FGFR3 causes achondroplasia by affecting both chondrogenesis and osteogenesis. J. Clin. Invest. 104, 1517–1525.

Chen, M., Zhu, M., Awad, H., Li, T.F., Sheu, T.J., Boyce, B.F., Chen, D., O'keefe, R.J., 2008. Inhibition of beta-catenin signaling causes defects in postnatal cartilage development. J. Cell Sci. 121, 1455–1465.

Chen, S., Tao, J., Bae, Y., Jiang, M.M., Bertin, T., Chen, Y., Yang, T., Lee, B., 2013. Notch gain of function inhibits chondrocyte differentiation via Rbpj-dependent suppression of Sox9. J. Bone Miner. Res. 28, 649–659.

Chiang, C., Litingtung, Y., Lee, E., Young, K.E., Corden, J.L., Westphal, H., Beachy, P.A., 1996. Cyclopia and defective axial patterning in mice lacking Sonic hedgehog gene function. Nature 383, 407–413.

Chin, H.J., Fisher, M.C., Li, Y., Ferrari, D., Wang, C.K., Lichtler, A.C., Dealy, C.N., Kosher, R.A., 2007. Studies on the role of Dlx5 in regulation of chondrocyte differentiation during endochondral ossification in the developing mouse limb. Dev. Growth Differ. 49, 515–521.

Christ, B., Huang, R., Scaal, M., 2004. Formation and differentiation of the avian sclerotome. Anat. Embryol. 208, 333–350.

Christ, B., Huang, R., Scaal, M., 2007. Amniote somite derivatives. Dev. Dynam. 236, 2382–2396.

Chung, U.I., Lanske, B., Lee, K., Li, E., Kronenberg, H., 1998. The parathyroid hormone/parathyroid hormone-related peptide receptor coordinates endochondral bone development by directly controlling chondrocyte differentiation. Proc. Natl. Acad. Sci. U.S.A. 95, 13030–13035.

Chusho, H., Tamura, N., Ogawa, Y., Yasoda, A., Suda, M., Miyazawa, T., Nakamura, K., Nakao, K., Kurihara, T., Komatsu, Y., Itoh, H., Tanaka, K., Saito, Y., Katsuki, M., Nakao, K., 2001. Dwarfism and early death in mice lacking C-type natriuretic peptide. Proc. Natl. Acad. Sci. U.S.A. 98, 4016–4021.

Ciurea, A.V., Toader, C., 2009. Genetics of craniosynostosis: review of the literature. J. Med. Life 2, 5–17.

Cohen Jr., M.M., 2000. Craniofacial disorders caused by mutations in homeobox genes MSX1 and MSX2. J. Craniofac. Genet. Dev. Biol. 20, 19–25.

Colnot, C., De La Fuente, L., Huang, S., Hu, D., Lu, C., St-Jacques, B., Helms, J.A., 2005. Indian hedgehog synchronizes skeletal angiogenesis and perichondrial maturation with cartilage development. Development 132, 1057–1067.

Colnot, C., Lu, C., Hu, D., Helms, J.A., 2004. Distinguishing the contributions of the perichondrium, cartilage, and vascular endothelium to skeletal development. Dev. Biol. 269, 55–69.

Colvin, J.S., Bohne, B.A., Harding, G.W., Mcewen, D.G., Ornitz, D.M., 1996. Skeletal overgrowth and deafness in mice lacking fibroblast growth factor receptor 3. Nat. Genet. 12, 390–397.

Cooper, K.L., Oh, S., Sung, Y., Dasari, R.R., Kirschner, M.W., Tabin, C.J., 2013. Multiple phases of chondrocyte enlargement underlie differences in skeletal proportions. Nature 495, 375–378.

Correa, D., Hesse, E., Seriwatanachai, D., Kiviranta, R., Saito, H., Yamana, K., Neff, L., Atfi, A., Coillard, L., Sitara, D., Maeda, Y., Warming, S., Jenkins, N.A., Copeland, N.G., Horne, W.C., Lanske, B., Baron, R., 2010. Zfp521 is a target gene and key effector of parathyroid hormone-related peptide signaling in growth plate chondrocytes. Dev. Cell 19, 533–546.

Cramer, T., Schipani, E., Johnson, R.S., Swoboda, B., Pfander, D., 2004. Expression of VEGF isoforms by epiphyseal chondrocytes during low-oxygen tension is HIF-1 alpha dependent. Osteoarthritis Cartilage 12, 433–439.

Crowe, N., Swingler, T.E., Le, L.T., Barter, M.J., Wheeler, G., Pais, H., Donell, S.T., Young, D.A., Dalmay, T., Clark, I.M., 2016. Detecting new microRNAs in human osteoarthritic chondrocytes identifies miR-3085 as a human, chondrocyte-selective, microRNA. Osteoarthritis Cartilage 24, 534–543.

Crowe, R., Zikherman, J., Niswander, L., 1999. Delta-1 negatively regulates the transition from prehypertrophic to hypertrophic chondrocytes during cartilage formation. Development 126, 987–998.

Cygan, J.A., Johnson, R.L., Mcmahon, A.P., 1997. Novel regulatory interactions revealed by studies of murine limb pattern in Wnt-7a and En-1 mutants. Development 124, 5021–5032.

Dao, D.Y., Jonason, J.H., Zhang, Y., Hsu, W., Chen, D., Hilton, M.J., O'keefe, R.J., 2012. Cartilage-specific beta-catenin signaling regulates chondrocyte maturation, generation of ossification centers, and perichondrial bone formation during skeletal development. J. Bone Miner. Res. 27, 1680–1694.

Davis, C.A., Holmyard, D.P., Millen, K.J., Joyner, A.L., 1991. Examining pattern formation in mouse, chicken and frog embryos with an En-specific antiserum. Development 111, 287–298.

Day, T.F., Guo, X., Garrett-Beal, L., Yang, Y., 2005. Wnt/beta-catenin signaling in mesenchymal progenitors controls osteoblast and chondrocyte differentiation during vertebrate skeletogenesis. Dev. Cell 8, 739–750.

Dechiara, T.M., Kimble, R.B., Poueymirou, W.T., Rojas, J., Masiakowski, P., Valenzuela, D.M., Yancopoulos, G.D., 2000. Ror2, encoding a receptor-like tyrosine kinase, is required for cartilage and growth plate development. Nat. Genet. 24, 271–274.

Deckelbaum, R.A., Holmes, G., Zhao, Z., Tong, C., Basilico, C., Loomis, C.A., 2012. Regulation of cranial morphogenesis and cell fate at the neural crest-mesoderm boundary by engrailed 1. Development 139, 1346–1358.

Delise, A.M., Fischer, L., Tuan, R.S., 2000. Cellular interactions and signaling in cartilage development. Osteoarthritis Cartilage 8, 309–334.

Delise, A.M., Tuan, R.S., 2002a. Alterations in the spatiotemporal expression pattern and function of N-cadherin inhibit cellular condensation and chondrogenesis of limb mesenchymal cells in vitro. J. Cell. Biochem. 87, 342–359.

Delise, A.M., Tuan, R.S., 2002b. Analysis of N-cadherin function in limb mesenchymal chondrogenesis in vitro. Dev. Dynam. 225, 195–204.

Demurger, F., Ichkou, A., Mougou-Zerelli, S., Le Merrer, M., Goudefroye, G., Delezoide, A.L., Quelin, C., Manouvrier, S., Baujat, G., Fradin, M., Pasquier, L., Megarbane, A., Faivre, L., Baumann, C., Nampoothiri, S., Roume, J., Isidor, B., Lacombe, D., Delrue, M.A., Mercier, S., Philip, N., Schaefer, E., Holder, M., Krause, A., Laffargue, F., Sinico, M., Amram, D., Andre, G., Liquier, A., Rossi, M., Amiel, J., Giuliano, F., Boute, O., Dieux-Coeslier, A., Jacquemont, M.L., Afenjar, A., Van Maldergem, L., Lackmy-Port-Lis, M., Vincent-Delorme, C., Chauvet, M.L., Cormier-Daire, V., Devisme, L., Genevieve, D., Munnich, A., Viot, G., Raoul, O., Romana, S., Gonzales, M., Encha-Razavi, F., Odent, S., Vekemans, M., Attie-Bitach, T., 2015. New insights into genotype-phenotype correlation for GLI3 mutations. Eur. J. Hum. Genet. 23, 92–102.

Deng, C., Wynshaw-Boris, A., Zhou, F., Kuo, A., Leder, P., 1996. Fibroblast growth factor receptor 3 is a negative regulator of bone growth. Cell 84, 911–921.

Docheva, D., Popov, C., Alberton, P., Aszodi, A., 2014. Integrin signaling in skeletal development and function. Birth Defects Res. C Embryo Today 102, 13–36.

Dodig, S., Raos, M., 1999. [Relation between month of birth and the manifestation of atopic diseases in children and adolescents]. Lijec. Vjesn. 121, 333–338.

Dong, Y., Jesse, A.M., Kohn, A., Gunnell, L.M., Honjo, T., Zuscik, M.J., O'keefe, R.J., Hilton, M.J., 2010. RBPjkappa-dependent Notch signaling regulates mesenchymal progenitor cell proliferation and differentiation during skeletal development. Development 137, 1461–1471.

Dong, Y.F., Soung Do, Y., Schwarz, E.M., O'keefe, R.J., Drissi, H., 2006. Wnt induction of chondrocyte hypertrophy through the Runx2 transcription factor. J. Cell. Physiol. 208, 77–86.

Duboc, V., Logan, M.P., 2011. Regulation of limb bud initiation and limb-type morphology. Dev. Dynam. 240, 1017–1027.

Dubrulle, J., Pourquie, O., 2003. Welcome to syndetome: a new somitic compartment. Dev. Cell 4, 611–612.

Duchatelet, S., Ostergaard, E., Cortes, D., Lemainque, A., Julier, C., 2005. Recessive mutations in PTHR1 cause contrasting skeletal dysplasias in Eiken and Blomstrand syndromes. Hum. Mol. Genet. 14, 1–5.

Ducy, P., Zhang, R., Geoffroy, V., Ridall, A.L., Karsenty, G., 1997. Osf2/Cbfa1: a transcriptional activator of osteoblast differentiation. Cell 89, 747–754.

Dy, P., Wang, W., Bhattaram, P., Wang, Q., Wang, L., Ballock, R.T., Lefebvre, V., 2012. Sox9 directs hypertrophic maturation and blocks osteoblast differentiation of growth plate chondrocytes. Dev. Cell 22, 597–609.

Ebensperger, C., Wilting, J., Brand-Saberi, B., Mizutani, Y., Christ, B., Balling, R., Koseki, H., 1995. Pax-1, a regulator of sclerotome development is induced by notochord and floor plate signals in avian embryos. Anat. Embryol. 191, 297–310.

Engsig, M.T., Chen, Q.J., Vu, T.H., Pedersen, A.C., Therkidsen, B., Lund, L.R., Henriksen, K., Lenhard, T., Foged, N.T., Werb, Z., Delaisse, J.M., 2000. Matrix metalloproteinase 9 and vascular endothelial growth factor are essential for osteoclast recruitment into developing long bones. J. Cell Biol. 151, 879–889.

Enishi, T., Yukata, K., Takahashi, M., Sato, R., Sairyo, K., Yasui, N., 2014. Hypertrophic chondrocytes in the rabbit growth plate can proliferate and differentiate into osteogenic cells when capillary invasion is interposed by a membrane filter. PLoS One 9 e104638.

Enomoto-Iwamoto, M., Kitagaki, J., Koyama, E., Tamamura, Y., Wu, C., Kanatani, N., Koike, T., Okada, H., Komori, T., Yoneda, T., Church, V., Francis-West, P.H., Kurisu, K., Nohno, T., Pacifici, M., Iwamoto, M., 2002. The Wnt antagonist Frzb-1 regulates chondrocyte maturation and long bone development during limb skeletogenesis. Dev. Biol. 251, 142–156.

Eshkar-Oren, I., Viukov, S.V., Salameh, S., Krief, S., Oh, C.D., Akiyama, H., Gerber, H.P., Ferrara, N., Zelzer, E., 2009. The forming limb skeleton serves as a signaling center for limb vasculature patterning via regulation of Vegf. Development 136, 1263–1272.

Fallon, J.F., Lopez, A., Ros, M.A., Savage, M.P., Olwin, B.B., Simandl, B.K., 1994. FGF-2: apical ectodermal ridge growth signal for chick limb development. Science 264, 104–107.

Fang, S., Deng, Y., Gu, P., Fan, X., 2015. MicroRNAs regulate bone development and regeneration. Int. J. Mol. Sci. 16, 8227–8253.

Farnum, C.E., 1994. Differential growth rates of long bones. In: Hall, B.K. (Ed.), Bone: Mechanisms of Bone Development and Growth. CRC Press, Boca Raton.

Farnum, C.E., Wilsman, N.J., 1987. Morphologic stages of the terminal hypertrophic chondrocyte of growth plate cartilage. Anat. Rec. 219, 221–232.

Farrow, E., Nicot, R., Wiss, A., Laborde, A., Ferri, J., 2018. Cleidocranial dysplasia: a review of clinical, radiological, genetic implications and a guidelines proposal. J. Craniofac. Surg. 29, 382–389.

Fedde, K.N., Blair, L., Silverstein, J., Coburn, S.P., Ryan, L.M., Weinstein, R.S., Waymire, K., Narisawa, S., Millan, J.L., Macgregor, G.R., Whyte, M.P., 1999. Alkaline phosphatase knock-out mice recapitulate the metabolic and skeletal defects of infantile hypophosphatasia. J. Bone Miner. Res. 14, 2015–2026.

Fenichel, I., Evron, Z., Nevo, Z., 2006. The perichondrial ring as a reservoir for precartilaginous cells. In vivo model in young chicks' epiphysis. Int. Orthop. 30, 353–356.

Fernandez-Teran, M., Piedra, M.E., Kathiriya, I.S., Srivastava, D., Rodriguez-Rey, J.C., Ros, M.A., 2000. Role of dHAND in the anterior-posterior polarization of the limb bud: implications for the Sonic hedgehog pathway. Development 127, 2133–2142.

Fernando, W.A., Leininger, E., Simkin, J., Li, N., Malcom, C.A., Sathyamoorthi, S., Han, M., Muneoka, K., 2011. Wound healing and blastema formation in regenerating digit tips of adult mice. Dev. Biol. 350, 301–310.

Franzen, A., Hultenby, K., Reinholt, F.P., Onnerfjord, P., Heinegard, D., 2008. Altered osteoclast development and function in osteopontin deficient mice. J. Orthop. Res. 26, 721–728.

Furumatsu, T., Ozaki, T., 2010. Epigenetic regulation in chondrogenesis. Acta. Med. Okayama 64, 155–161.

Furumatsu, T., Tsuda, M., Taniguchi, N., Tajima, Y., Asahara, H., 2005. Smad3 induces chondrogenesis through the activation of SOX9 via CREB-binding protein/p300 recruitment. J. Biol. Chem. 280, 8343–8350.

Gao, B., Hu, J., Stricker, S., Cheung, M., Ma, G., Law, K.F., Witte, F., Briscoe, J., Mundlos, S., He, L., Cheah, K.S., Chan, D., 2009. A mutation in Ihh that causes digit abnormalities alters its signalling capacity and range. Nature 458, 1196–1200.

Gao, B., Song, H., Bishop, K., Elliot, G., Garrett, L., English, M.A., Andre, P., Robinson, J., Sood, R., Minami, Y., Economides, A.N., Yang, Y., 2011. Wnt signaling gradients establish planar cell polarity by inducing Vangl2 phosphorylation through Ror2. Dev. Cell 20, 163–176.

Gardner, C.A., Barald, K.F., 1992. Expression patterns of engrailed-like proteins in the chick embryo. Dev. Dynam. 193, 370–388.

Gat-Yablonski, G., Shtaif, B., Abraham, E., Phillip, M., 2008. Nutrition-induced catch-up growth at the growth plate. J. Pediatr. Endocrinol. Metab. 21, 879–893.

Gerber, H.P., Vu, T.H., Ryan, A.M., Kowalski, J., Werb, Z., Ferrara, N., 1999. VEGF couples hypertrophic cartilage remodeling, ossification and angiogenesis during endochondral bone formation. Nat. Med. 5, 623–628.

Giustina, A., Mazziotti, G., Canalis, E., 2008. Growth hormone, insulin-like growth factors, and the skeleton. Endocr. Rev. 29, 535–559.

Golovchenko, S., Hattori, T., Hartmann, C., Gebhardt, M., Gebhard, S., Hess, A., Pausch, F., Schlund, B., Von Der Mark, K., 2013. Deletion of beta catenin in hypertrophic growth plate chondrocytes impairs trabecular bone formation. Bone 55, 102–112.

Gong, Y., Slee, R.B., Fukai, N., Rawadi, G., Roman-Roman, S., Reginato, A.M., Wang, H., Cundy, T., Glorieux, F.H., Lev, D., Zacharin, M., Oexle, K., Marcelino, J., Suwairi, W., Heeger, S., Sabatakos, G., Apte, S., Adkins, W.N., Allgrove, J., Arslan-Kirchner, M., Batch, J.A., Beighton, P., Black, G.C., Boles, R.G., Boon, L.M., Borrone, C., Brunner, H.G., Carle, G.F., Dallapiccola, B., De Paepe, A., Floege, B., Halfhide, M.L., Hall, B., Hennekam, R.C., Hirose, T., Jans, A., Juppner, H., Kim, C.A., Keppler-Noreuil, K., Kohlschuetter, A., Lacombe, D., Lambert, M., Lemyre, E., Letteboer, T., Peltonen, L., Ramesar, R.S., Romanengo, M., Somer, H., Steichen-Gersdorf, E., Steinmann, B., Sullivan, B., Superti-Furga, A., Swoboda, W., Van Den Boogaard, M.J., Van Hul, W., Vikkula, M., Votruba, M., Zabel, B., Garcia, T., Baron, R., Olsen, B.R., Warman, M.L., Osteoporosis-Pseudoglioma Syndrome Collaborative, G., 2001. LDL receptor-related protein 5 (LRP5) affects bone accrual and eye development. Cell 107, 513–523.

Goodrich, E.S., 1930. Studies on the Structure and Development of Vertebrates. Macmillan, London.

Goumans, M.J., Valdimarsdottir, G., Itoh, S., Rosendahl, A., Sideras, P., TEN Dijke, P., 2002. Balancing the activation state of the endothelium via two distinct TGF-beta type I receptors. EMBO J. 21, 1743–1753.

Grumolato, L., Liu, G., Mong, P., Mudbhary, R., Biswas, R., Arroyave, R., Vijayakumar, S., Economides, A.N., Aaronson, S.A., 2010. Canonical and noncanonical Wnts use a common mechanism to activate completely unrelated coreceptors. Genes Dev. 24, 2517–2530.

Guo, J., Chung, U.I., Kondo, H., Bringhurst, F.R., Kronenberg, H.M., 2002. The PTH/PTHrP receptor can delay chondrocyte hypertrophy in vivo without activating phospholipase C. Dev. Cell 3, 183–194.

Guo, J., Chung, U.I., Yang, D., Karsenty, G., Bringhurst, F.R., Kronenberg, H.M., 2006. PTH/PTHrP receptor delays chondrocyte hypertrophy via both Runx2-dependent and -independent pathways. Dev. Biol. 292, 116–128.

Guo, X., Day, T.F., Jiang, X., Garrett-Beal, L., Topol, L., Yang, Y., 2004. Wnt/beta-catenin signaling is sufficient and necessary for synovial joint formation. Genes Dev. 18, 2404–2417.

Hall, B.K., Miyake, T., 1992. The membranous skeleton: the role of cell condensations in vertebrate skeletogenesis. Anat. Embryol. 186, 107–124.

Hall, B.K., Miyake, T., 1995. Divide, accumulate, differentiate: cell condensation in skeletal development revisited. Int. J. Dev. Biol. 39, 881–893.

Hall, B.K., Miyake, T., 2000. All for one and one for all: condensations and the initiation of skeletal development. Bioessays 22, 138–147.

Hall, K.C., Hill, D., Otero, M., Plumb, D.A., Froemel, D., Dragomir, C.L., Maretzky, T., Boskey, A., Crawford, H.C., Selleri, L., Goldring, M.B., Blobel, C.P., 2013. ADAM17 controls endochondral ossification by regulating terminal differentiation of chondrocytes. Mol. Cell Biol. 33, 3077–3090.

Hallmann, R., Feinberg, R.N., Latker, C.H., Sasse, J., Risau, W., 1987. Regression of blood vessels precedes cartilage differentiation during chick limb development. Differentiation 34, 98–105.

Hamburger, V., Hamilton, H.L., 1992. A series of normal stages in the development of the chick embryo. 1951. Dev. Dynam. 195, 231–272.

Hankinson, T.C., Fontana, E.J., Anderson, R.C., Feldstein, N.A., 2010. Surgical treatment of single-suture craniosynostosis: an argument for quantitative methods to evaluate cosmetic outcomes. J. Neurosurg. Pediatr. 6, 193–197.

Harmey, D., Hessle, L., Narisawa, S., Johnson, K.A., Terkeltaub, R., Millan, J.L., 2004. Concerted regulation of inorganic pyrophosphate and osteopontin by akp2, enpp1, and ank: an integrated model of the pathogenesis of mineralization disorders. Am. J. Pathol. 164, 1199–1209.

Hartmann, C., 2009. Transcriptional networks controlling skeletal development. Curr. Opin. Genet. Dev. 19, 437–443.

Hartmann, C., Tabin, C.J., 2000. Dual roles of Wnt signaling during chondrogenesis in the chicken limb. Development 127, 3141–3159.

Hartmann, C., Tabin, C.J., 2001. Wnt-14 plays a pivotal role in inducing synovial joint formation in the developing appendicular skeleton. Cell 104, 341–351.

Henriksen, K., Karsdal, M., Delaisse, J.M., Engsig, M.T., 2003. RANKL and vascular endothelial growth factor (VEGF) induce osteoclast chemotaxis through an ERK1/2-dependent mechanism. J. Biol. Chem. 278, 48745–48753.

Hessle, L., Johnson, K.A., Anderson, H.C., Narisawa, S., Sali, A., Goding, J.W., Terkeltaub, R., Millan, J.L., 2002. Tissue-nonspecific alkaline phosphatase and plasma cell membrane glycoprotein-1 are central antagonistic regulators of bone mineralization. Proc. Natl. Acad. Sci. U.S.A. 99, 9445–9449.

Hildebrand, A., Romaris, M., Rasmussen, L.M., Heinegard, D., Twardzik, D.R., Border, W.A., Ruoslahti, E., 1994. Interaction of the small interstitial proteoglycans biglycan, decorin and fibromodulin with transforming growth factor beta. Biochem. J. 302 (Pt 2), 527–534.

Hill, T.P., Spater, D., Taketo, M.M., Birchmeier, W., Hartmann, C., 2005. Canonical Wnt/beta-catenin signaling prevents osteoblasts from differentiating into chondrocytes. Dev. Cell 8, 727–738.

Hilton, M.J., Tu, X., Cook, J., Hu, H., Long, F., 2005. Ihh controls cartilage development by antagonizing Gli3, but requires additional effectors to regulate osteoblast and vascular development. Development 132, 4339–4351.

Hilton, M.J., Tu, X., Wu, X., Bai, S., Zhao, H., Kobayashi, T., Kronenberg, H.M., Teitelbaum, S.L., Ross, F.P., Kopan, R., Long, F., 2008. Notch signaling maintains bone marrow mesenchymal progenitors by suppressing osteoblast differentiation. Nat. Med. 14, 306–314.

Hinchliffe, J.R., 1994. Evolutionary developmental biology of the tetrapod limb. Dev. Suppl. 163–168.

Hinchliffe, J.R., Johnson, D.R., 1980. The Development of the Vertebrate Limb. Clarendon Press, Oxford.

Hirsinger, E., Duprez, D., Jouve, C., Malapert, P., Cooke, J., Pourquie, O., 1997. Noggin acts downstream of Wnt and Sonic Hedgehog to antagonize BMP4 in avian somite patterning. Development 124, 4605–4614.

Hiscock, T.W., Tschopp, P., Tabin, C.J., 2017. On the formation of digits and joints during limb development. Dev. Cell 41, 459–465.

Hong, E., Reddi, A.H., 2012. MicroRNAs in chondrogenesis, articular cartilage, and osteoarthritis: implications for tissue engineering. Tissue Eng. B Rev. 18, 445–453.

Houben, A., Kostanova-Poliakova, D., Weissenbock, M., Graf, J., Teufel, S., Von Der Mark, K., Hartmann, C., 2016. beta-catenin activity in late hypertrophic chondrocytes locally orchestrates osteoblastogenesis and osteoclastogenesis. Development 143, 3826–3838.

Hsu, Y.H., Kiel, D.P., 2012. Clinical review: genome-wide association studies of skeletal phenotypes: what we have learned and where we are headed. J. Clin. Endocrinol. Metab. 97, E1958–E1977.

Hu, D.P., Ferro, F., Yang, F., Taylor, A.J., Chang, W., Miclau, T., Marcucio, R.S., Bahney, C.S., 2017. Cartilage to bone transformation during fracture healing is coordinated by the invading vasculature and induction of the core pluripotency genes. Development 144, 221–234.

Hu, H., Hilton, M.J., Tu, X., Yu, K., Ornitz, D.M., Long, F., 2005. Sequential roles of Hedgehog and Wnt signaling in osteoblast development. Development 132, 49–60.

Huang, W., Zhou, X., Lefebvre, V., De Crombrugghe, B., 2000. Phosphorylation of SOX9 by cyclic AMP-dependent protein kinase A enhances SOX9's ability to transactivate a Col2a1 chondrocyte-specific enhancer. Mol. Cell Biol. 20, 4149–4158.

Hubaud, A., Pourquie, O., 2014. Signalling dynamics in vertebrate segmentation. Nat. Rev. Mol. Cell Biol. 15, 709–721.

Hui, C.C., Joyner, A.L., 1993. A mouse model of greig cephalopolysyndactyly syndrome: the extra-toesJ mutation contains an intragenic deletion of the Gli3 gene. Nat. Genet. 3, 241–246.

Hung, I.H., Schoenwolf, G.C., Lewandoski, M., Ornitz, D.M., 2016. A combined series of Fgf9 and Fgf18 mutant alleles identifies unique and redundant roles in skeletal development. Dev. Biol. 411, 72–84.

Hung, I.H., Yu, K., Lavine, K.J., Ornitz, D.M., 2007. FGF9 regulates early hypertrophic chondrocyte differentiation and skeletal vascularization in the developing stylopod. Dev. Biol. 307, 300–313.

Hunziker, E.B., Schenk, R.K., Cruz-Orive, L.M., 1987. Quantitation of chondrocyte performance in growth-plate cartilage during longitudinal bone growth. J. Bone Joint Surg. Am. 69, 162–173.

Huynh, N.P., Anderson, B.A., Guilak, F., Mcalinden, A., 2017. Emerging roles for long noncoding RNAs in skeletal biology and disease. Connect. Tissue Res. 58, 116–141.

Hyde, G., Dover, S., Aszodi, A., Wallis, G.A., Boot-Handford, R.P., 2007. Lineage tracing using matrilin-1 gene expression reveals that articular chondrocytes exist as the joint interzone forms. Dev. Biol. 304, 825–833.

Inada, M., Wang, Y., Byrne, M.H., Rahman, M.U., Miyaura, C., Lopez-Otin, C., Krane, S.M., 2004. Critical roles for collagenase-3 (Mmp13) in development of growth plate cartilage and in endochondral ossification. Proc. Natl. Acad. Sci. U.S.A. 101, 17192–17197.

Inada, M., Yasui, T., Nomura, S., Miyake, S., Deguchi, K., Himeno, M., Sato, M., Yamagiwa, H., Kimura, T., Yasui, N., Ochi, T., Endo, N., Kitamura, Y., Kishimoto, T., Komori, T., 1999. Maturational disturbance of chondrocytes in Cbfa1-deficient mice. Dev. Dynam. 214, 279–290.

Ionescu, A., Kozhemyakina, E., Nicolae, C., Kaestner, K.H., Olsen, B.R., Lassar, A.B., 2012. FoxA family members are crucial regulators of the hypertrophic chondrocyte differentiation program. Dev. Cell 22, 927–939.

Ionescu, A.M., Schwarz, E.M., Vinson, C., Puzas, J.E., Rosier, R., Reynolds, P.R., O'keefe, R.J., 2001. PTHrP modulates chondrocyte differentiation through AP-1 and CREB signaling. J. Biol. Chem. 276, 11639–11647.

Ishii, M., Sun, J., Ting, M.C., Maxson, R.E., 2015. The development of the calvarial bones and sutures and the pathophysiology of craniosynostosis. Curr. Top. Dev. Biol. 115, 131–156.

Isidor, B., Lindenbaum, P., Pichon, O., Bezieau, S., Dina, C., Jacquemont, S., Martin-Coignard, D., Thauvin-Robinet, C., Le Merrer, M., Mandel, J.L., David, A., Faivre, L., Cormier-Daire, V., Redon, R., Le Caignec, C., 2011. Truncating mutations in the last exon of NOTCH2 cause a rare skeletal disorder with osteoporosis. Nat. Genet. 43, 306–308.

Ivaska, J., Heino, J., 2011. Cooperation between integrins and growth factor receptors in signaling and endocytosis. Annu. Rev. Cell Dev. Biol. 27, 291–320.

Iwata, J., Hacia, J.G., Suzuki, A., Sanchez-Lara, P.A., Urata, M., Chai, Y., 2012. Modulation of noncanonical TGF-beta signaling prevents cleft palate in Tgfbr2 mutant mice. J. Clin. Invest. 122, 873–885.

Jaubert, J., Jaubert, F., Martin, N., Washburn, L.L., Lee, B.K., Eicher, E.M., Guenet, J.L., 1999. Three new allelic mouse mutations that cause skeletal overgrowth involve the natriuretic peptide receptor C gene (Npr3). Proc. Natl. Acad. Sci. U.S.A. 96, 10278–10283.

Jena, N., Martin-Seisdedos, C., Mccue, P., Croce, C.M., 1997. BMP7 null mutation in mice: developmental defects in skeleton, kidney, and eye. Exp. Cell Res. 230, 28–37.

Jeong, J., Mao, J., Tenzen, T., Kottmann, A.H., Mcmahon, A.P., 2004. Hedgehog signaling in the neural crest cells regulates the patterning and growth of facial primordia. Genes Dev. 18, 937–951.

Jho, E.H., Zhang, T., Domon, C., Joo, C.K., Freund, J.N., Costantini, F., 2002. Wnt/beta-catenin/Tcf signaling induces the transcription of Axin2, a negative regulator of the signaling pathway. Mol. Cell Biol. 22, 1172–1183.

Jiang, X., Iseki, S., Maxson, R.E., Sucov, H.M., Morriss-Kay, G.M., 2002. Tissue origins and interactions in the mammalian skull vault. Dev. Biol. 241, 106–116.

Jin, S.W., Sim, K.B., Kim, S.D., 2016. Development and growth of the normal cranial vault : an embryologic review. J. Korean Neurosurg. Soc. 59, 192–196.

Jing, J., Ren, Y., Zong, Z., Liu, C., Kamiya, N., Mishina, Y., Liu, Y., Zhou, X., Feng, J.Q., 2013. BMP receptor 1A determines the cell fate of the postnatal growth plate. Int. J. Biol. Sci. 9, 895–906.

Jobert, A.S., Zhang, P., Couvineau, A., Bonaventure, J., Roume, J., LE Merrer, M., Silve, C., 1998. Absence of functional receptors for parathyroid hormone and parathyroid hormone-related peptide in Blomstrand chondrodysplasia. J. Clin. Invest. 102, 34–40.

Joeng, K.S., Schumacher, C.A., Zylstra-Diegel, C.R., Long, F., Williams, B.O., 2011. Lrp5 and Lrp6 redundantly control skeletal development in the mouse embryo. Dev. Biol. 359, 222–229.

Johnson, D.R., 1986. The cartilaginous skeleton. In: The Genetics of the Skeleton. Oxford University Press, New York.

Kan, A., Tabin, C.J., 2013. c-Jun is required for the specification of joint cell fates. Genes Dev. 27, 514–524.

Kanai, Y., Yasoda, A., Mori, K.P., Watanabe-Takano, H., Nagai-Okatani, C., Yamashita, Y., Hirota, K., Ueda, Y., Yamauchi, I., Kondo, E., Yamanaka, S., Sakane, Y., Nakao, K., Fujii, T., Yokoi, H., Minamino, N., Mukoyama, M., Mochizuki, N., Inagaki, N., 2017. Circulating osteocrin stimulates bone growth by limiting C-type natriuretic peptide clearance. J. Clin. Invest. 127, 4136–4147.

Karamboulas, K., Dranse, H.J., Underhill, T.M., 2010. Regulation of BMP-dependent chondrogenesis in early limb mesenchyme by TGFbeta signals. J. Cell Sci. 123, 2068–2076.

Karaplis, A.C., He, B., Nguyen, M.T., Young, I.D., Semeraro, D., Ozawa, H., Amizuka, N., 1998. Inactivating mutation in the human parathyroid hormone receptor type 1 gene in Blomstrand chondrodysplasia. Endocrinology 139, 5255–5258.

Karaplis, A.C., Luz, A., Glowacki, J., Bronson, R.T., Tybulewicz, V.L., Kronenberg, H.M., Mulligan, R.C., 1994. Lethal skeletal dysplasia from targeted disruption of the parathyroid hormone-related peptide gene. Genes Dev. 8, 277–289.

Karp, S.J., Schipani, E., St-Jacques, B., Hunzelman, J., Kronenberg, H., Mcmahon, A.P., 2000. Indian hedgehog coordinates endochondral bone growth and morphogenesis via parathyroid hormone related-protein-dependent and -independent pathways. Development 127, 543–548.

Karreth, F., Hoebertz, A., Scheuch, H., Eferl, R., Wagner, E.F., 2004. The AP1 transcription factor Fra2 is required for efficient cartilage development. Development 131, 5717–5725.

Karsenty, G., 2001. Minireview: transcriptional control of osteoblast differentiation. Endocrinology 142, 2731–2733.

Karsenty, G., 2008. Transcriptional control of skeletogenesis. Annu. Rev. Genom. Hum. Genet. 9, 183–196.

Kato, K., Bhattaram, P., Penzo-Mendez, A., Gadi, A., Lefebvre, V., 2015. SOXC transcription factors induce cartilage growth plate formation in mouse embryos by promoting noncanonical WNT signaling. J. Bone Miner. Res. 30, 1560–1571.

Kawai, M., Rosen, C.J., 2012. The insulin-like growth factor system in bone: basic and clinical implications. Endocrinol Metab. Clin. North Am. 41, 323–333 (vi).

Kawakami, Y., Capdevila, J., Buscher, D., Itoh, T., Rodriguez Esteban, C., Izpisua Belmonte, J.C., 2001. WNT signals control FGF-dependent limb initiation and AER induction in the chick embryo. Cell 104, 891–900.

Keller, B., Yang, T., Chen, Y., Munivez, E., Bertin, T., Zabel, B., Lee, B., 2011. Interaction of TGFbeta and BMP signaling pathways during chondrogenesis. PLoS One 6 e16421.

Kengaku, M., Capdevila, J., Rodriguez-Esteban, C., De La Pena, J., Johnson, R.L., Izpisua Belmonte, J.C., Tabin, C.J., 1998. Distinct WNT pathways regulating AER formation and dorsoventral polarity in the chick limb bud. Science 280, 1274–1277.

Keupp, K., Beleggia, F., Kayserili, H., Barnes, A.M., Steiner, M., Semler, O., Fischer, B., Yigit, G., Janda, C.Y., Becker, J., Breer, S., Altunoglu, U., Grunhagen, J., Krawitz, P., Hecht, J., Schinke, T., Makareeva, E., Lausch, E., Cankaya, T., Caparros-Martin, J.A., Lapunzina, P., Temtamy, S., Aglan, M., Zabel, B., Eysel, P., Koerber, F., Leikin, S., Garcia, K.C., Netzer, C., Schonau, E., Ruiz-Perez, V.L., Mundlos, S., Amling, M., Kornak, U., Marini, J., Wollnik, B., 2013. Mutations in WNT1 cause different forms of bone fragility. Am. J. Hum. Genet. 92, 565–574.

Khokha, M.K., Hsu, D., Brunet, L.J., Dionne, M.S., Harland, R.M., 2003. Gremlin is the BMP antagonist required for maintenance of Shh and Fgf signals during limb patterning. Nat. Genet. 34, 303–307.

Kim, I.S., Otto, F., Zabel, B., Mundlos, S., 1999. Regulation of chondrocyte differentiation by Cbfa1. Mech. Dev. 80, 159–170.

Kim, S.Y., Paylor, S.W., Magnuson, T., Schumacher, A., 2006. Juxtaposed Polycomb complexes co-regulate vertebral identity. Development 133, 4957–4968.

Kim, Y., Phan, D., Van Rooij, E., Wang, D.Z., Mcanally, J., Qi, X., Richardson, J.A., Hill, J.A., Bassel-Duby, R., Olson, E.N., 2008. The MEF2D transcription factor mediates stress-dependent cardiac remodeling in mice. J. Clin. Invest. 118, 124–132.

Kingsley, D.M., Bland, A.E., Grubber, J.M., Marker, P.C., Russell, L.B., Copeland, N.G., Jenkins, N.A., 1992. The mouse short ear skeletal morphogenesis locus is associated with defects in a bone morphogenetic member of the TGF beta superfamily. Cell 71, 399–410.

Kishimoto, K., Kitazawa, R., Kurosaka, M., Maeda, S., Kitazawa, S., 2006. Expression profile of genes related to osteoclastogenesis in mouse growth plate and articular cartilage. Histochem. Cell Biol. 125, 593–602.

Klopocki, E., Lohan, S., Brancati, F., Koll, R., Brehm, A., Seemann, P., Dathe, K., Stricker, S., Hecht, J., Bosse, K., Betz, R.C., Garaci, F.G., Dallapiccola, B., Jain, M., Muenke, M., Ng, V.C., Chan, W., Chan, D., Mundlos, S., 2011. Copy-number variations involving the IHH locus are associated with syndactyly and craniosynostosis. Am. J. Hum. Genet. 88, 70–75.

Kobayashi, T., Lu, J., Cobb, B.S., Rodda, S.J., Mcmahon, A.P., Schipani, E., Merkenschlager, M., Kronenberg, H.M., 2008. Dicer-dependent pathways regulate chondrocyte proliferation and differentiation. Proc. Natl. Acad. Sci. U.S.A. 105, 1949–1954.

Kobayashi, T., Soegiarto, D.W., Yang, Y., Lanske, B., Schipani, E., Mcmahon, A.P., Kronenberg, H.M., 2005. Indian hedgehog stimulates periarticular chondrocyte differentiation to regulate growth plate length independently of PTHrP. J. Clin. Invest. 115, 1734–1742.

Kohn, A., Dong, Y., Mirando, A.J., Jesse, A.M., Honjo, T., Zuscik, M.J., O'keefe, R.J., Hilton, M.J., 2012. Cartilage-specific RBPjkappa-dependent and -independent Notch signals regulate cartilage and bone development. Development 139, 1198–1212.

Komori, T., Yagi, H., Nomura, S., Yamaguchi, A., Sasaki, K., Deguchi, K., Shimizu, Y., Bronson, R.T., Gao, Y.H., Inada, M., Sato, M., Okamoto, R., Kitamura, Y., Yoshiki, S., Kishimoto, T., 1997. Targeted disruption of Cbfa1 results in a complete lack of bone formation owing to maturational arrest of osteoblasts. Cell 89, 755–764.

Kos, L., Chiang, C., Mahon, K.A., 1998. Mediolateral patterning of somites: multiple axial signals, including Sonic hedgehog, regulate Nkx-3.1 expression. Mech. Dev. 70, 25–34.

Kozhemyakina, E., Cohen, T., Yao, T.P., Lassar, A.B., 2009. Parathyroid hormone-related peptide represses chondrocyte hypertrophy through a protein phosphatase 2A/histone deacetylase 4/MEF2 pathway. Mol. Cell Biol. 29, 5751–5762.

Koziel, L., Wuelling, M., Schneider, S., Vortkamp, A., 2005. Gli3 acts as a repressor downstream of Ihh in regulating two distinct steps of chondrocyte differentiation. Development 132, 5249–5260.

Kraus, P., Lufkin, T., 2006. Dlx homeobox gene control of mammalian limb and craniofacial development. Am. J. Med. Genet. 140, 1366–1374.

Krumlauf, R., 1992. Evolution of the vertebrate Hox homeobox genes. Bioessays 14, 245–252.

Kusumbe, A.P., Ramasamy, S.K., Adams, R.H., 2014. Coupling of angiogenesis and osteogenesis by a specific vessel subtype in bone. Nature 507, 323–328.

Laine, C.M., Joeng, K.S., Campeau, P.M., Kiviranta, R., Tarkkonen, K., Grover, M., Lu, J.T., Pekkinen, M., Wessman, M., Heino, T.J., Nieminen-Pihala, V., Aronen, M., Laine, T., Kroger, H., Cole, W.G., Lehesjoki, A.E., Nevarez, L., Krakow, D., Curry, C.J., Cohn, D.H., Gibbs, R.A., Lee, B.H., Makitie, O., 2013. WNT1 mutations in early-onset osteoporosis and osteogenesis imperfecta. N. Engl. J. Med. 368, 1809–1816.

Langen, U.H., Pitulescu, M.E., Kim, J.M., Enriquez-Gasca, R., Sivaraj, K.K., Kusumbe, A.P., Singh, A., DI Russo, J., Bixel, M.G., Zhou, B., Sorokin, L., Vaquerizas, J.M., Adams, R.H., 2017. Cell-matrix signals specify bone endothelial cells during developmental osteogenesis. Nat. Cell Biol. 19, 189–201.

Langenskiold, A., 1998. Role of the ossification groove of Ranvier in normal and pathologic bone growth: a review. J. Pediatr. Orthop. 18, 173–177.

Lanske, B., Divieti, P., Kovacs, C.S., Pirro, A., Landis, W.J., Krane, S.M., Bringhurst, F.R., Kronenberg, H.M., 1998. The parathyroid hormone (PTH)/PTH-related peptide receptor mediates actions of both ligands in murine bone. Endocrinology 139, 5194–5204.

Lanske, B., Karaplis, A.C., Lee, K., Luz, A., Vortkamp, A., Pirro, A., Karperien, M., Defize, L.H., Ho, C., Mulligan, R.C., Abou-Samra, A.B., Juppner, H., Segre, G.V., Kronenberg, H.M., 1996. PTH/PTHrP receptor in early development and Indian hedgehog-regulated bone growth. Science 273, 663–666.

Lapunzina, P., Aglan, M., Temtamy, S., Caparros-Martin, J.A., Valencia, M., Leton, R., Martinez-Glez, V., Elhossini, R., Amr, K., Vilaboa, N., Ruiz-Perez, V.L., 2010. Identification of a frameshift mutation in Osterix in a patient with recessive osteogenesis imperfecta. Am. J. Hum. Genet. 87, 110–114.

Lara-Castillo, N., Johnson, M.L., 2015. LRP receptor family member associated bone disease. Rev. Endocr. Metab. Disord. 16, 141–148.

Lattanzi, W., Bukvic, N., Barba, M., Tamburrini, G., Bernardini, C., Michetti, F., Di Rocco, C., 2012. Genetic basis of single-suture synostoses: genes, chromosomes and clinical implications. Childs Nerv. Syst. 28, 1301–1310.

Laufer, E., Nelson, C.E., Johnson, R.L., Morgan, B.A., Tabin, C., 1994. Sonic hedgehog and Fgf-4 act through a signaling cascade and feedback loop to integrate growth and patterning of the developing limb bud. Cell 79, 993–1003.

Lee, B., Thirunavukkarasu, K., Zhou, L., Pastore, L., Baldini, A., Hecht, J., Geoffroy, V., Ducy, P., Karsenty, G., 1997. Missense mutations abolishing DNA binding of the osteoblast-specific transcription factor OSF2/CBFA1 in cleidocranial dysplasia. Nat. Genet. 16, 307–310.

Lee, K., Lanske, B., Karaplis, A.C., Deeds, J.D., Kohno, H., Nissenson, R.A., Kronenberg, H.M., Segre, G.V., 1996. Parathyroid hormone-related peptide delays terminal differentiation of chondrocytes during endochondral bone development. Endocrinology 137, 5109–5118.

Lefebvre, V., Bhattaram, P., 2016. SOXC genes and the control of skeletogenesis. Curr. Osteoporos. Rep. 14, 32–38.

Lefebvre, V., Dvir-Ginzberg, M., 2017. SOX9 and the many facets of its regulation in the chondrocyte lineage. Connect. Tissue Res. 58, 2–14.

Lehmann, K., Seemann, P., Stricker, S., Sammar, M., Meyer, B., Suring, K., Majewski, F., Tinschert, S., Grzeschik, K.H., Muller, D., Knaus, P., Nurnberg, P., Mundlos, S., 2003. Mutations in bone morphogenetic protein receptor 1B cause brachydactyly type A2. Proc. Natl. Acad. Sci. U.S.A. 100, 12277–12282.

Lenton, K., James, A.W., Manu, A., Brugmann, S.A., Birker, D., Nelson, E.R., Leucht, P., Helms, J.A., Longaker, M.T., 2011. Indian hedgehog positively regulates calvarial ossification and modulates bone morphogenetic protein signaling. Genesis 49, 784–796.

Leung, J.Y., Kolligs, F.T., Wu, R., Zhai, Y., Kuick, R., Hanash, S., Cho, K.R., Fearon, E.R., 2002. Activation of AXIN2 expression by beta-catenin-T cell factor. A feedback repressor pathway regulating Wnt signaling. J. Biol. Chem. 277, 21657–21665.

Levet, S., Ciais, D., Merdzhanova, G., Mallet, C., Zimmers, T.A., Lee, S.J., Navarro, F.P., Texier, I., Feige, J.J., Bailly, S., Vittet, D., 2013. Bone morphogenetic protein 9 (BMP9) controls lymphatic vessel maturation and valve formation. Blood 122, 598–607.

Levi, B., Wan, D.C., Wong, V.W., Nelson, E., Hyun, J., Longaker, M.T., 2012. Cranial suture biology: from pathways to patient care. J. Craniofac. Surg. 23, 13–19.

Levy, L., Wei, Y., Labalette, C., Wu, Y., Renard, C.A., Buendia, M.A., Neuveut, C., 2004. Acetylation of beta-catenin by p300 regulates beta-catenin-Tcf4 interaction. Mol. Cell Biol. 24, 3404–3414.

Li, C., Chen, L., Iwata, T., Kitagawa, M., Fu, X.Y., Deng, C.X., 1999. A Lys644Glu substitution in fibroblast growth factor receptor 3 (FGFR3) causes dwarfism in mice by activation of STATs and ink4 cell cycle inhibitors. Hum. Mol. Genet. 8, 35–44.

Li, Y., Dudley, A.T., 2009. Noncanonical frizzled signaling regulates cell polarity of growth plate chondrocytes. Development 136, 1083–1092.

Lian, J.B., Stein, G.S., 2003. Runx2/Cbfa1: a multifunctional regulator of bone formation. Curr. Pharmaceut. Des. 9, 2677–2685.

Lim, J., Tu, X., Choi, K., Akiyama, H., Mishina, Y., Long, F., 2015. BMP-Smad4 signaling is required for precartilaginous mesenchymal condensation independent of Sox9 in the mouse. Dev. Biol. 400, 132–138.

Lindsey, R.C., Mohan, S., 2016. Skeletal effects of growth hormone and insulin-like growth factor-I therapy. Mol. Cell. Endocrinol. 432, 44–55.

Liu, C., Lin, C., Gao, C., May-Simera, H., Swaroop, A., Li, T., 2014. Null and hypomorph Prickle1 alleles in mice phenocopy human Robinow syndrome and disrupt signaling downstream of Wnt5a. Biol. Open 3, 861–870.

Liu, J.P., Baker, J., Perkins, A.S., Robertson, E.J., Efstratiadis, A., 1993. Mice carrying null mutations of the genes encoding insulin-like growth factor I (Igf-1) and type 1 IGF receptor (Igf1r). Cell 75, 59–72.

Liu, W., Zhou, L., Zhou, C., Zhang, S., Jing, J., Xie, L., Sun, N., Duan, X., Jing, W., Liang, X., Zhao, H., Ye, L., Chen, Q., Yuan, Q., 2016. GDF11 decreases bone mass by stimulating osteoclastogenesis and inhibiting osteoblast differentiation. Nat. Commun. 7, 12794.

Liu, Z., Lavine, K.J., Hung, I.H., Ornitz, D.M., 2007. FGF18 is required for early chondrocyte proliferation, hypertrophy and vascular invasion of the growth plate. Dev. Biol. 302, 80–91.

Loeser, R.F., 2000. Chondrocyte integrin expression and function. Biorheology 37, 109–116.

Loeser, R.F., 2002. Integrins and cell signaling in chondrocytes. Biorheology 39, 119–124.

Logan, M., Tabin, C.J., 1999. Role of Pitx1 upstream of Tbx4 in specification of hindlimb identity. Science 283, 1736–1739.

Long, F., Chung, U.I., Ohba, S., Mcmahon, J., Kronenberg, H.M., Mcmahon, A.P., 2004. Ihh signaling is directly required for the osteoblast lineage in the endochondral skeleton. Development 131, 1309–1318.

Long, F., Zhang, X.M., Karp, S., Yang, Y., Mcmahon, A.P., 2001. Genetic manipulation of hedgehog signaling in the endochondral skeleton reveals a direct role in the regulation of chondrocyte proliferation. Development 128, 5099–5108.

Longobardi, L., Li, T., Myers, T.J., O'rear, L., Ozkan, H., Li, Y., Contaldo, C., Spagnoli, A., 2012. TGF-beta type II receptor/MCP-5 axis: at the crossroad between joint and growth plate development. Dev. Cell 23, 71–81.

Loomis, C.A., Harris, E., Michaud, J., Wurst, W., Hanks, M., Joyner, A.L., 1996. The mouse Engrailed-1 gene and ventral limb patterning. Nature 382, 360–363.

Lu, M.F., Cheng, H.T., Lacy, A.R., Kern, M.J., Argao, E.A., Potter, S.S., Olson, E.N., Martin, J.F., 1999. Paired-related homeobox genes cooperate in handplate and hindlimb zeugopod morphogenesis. Dev. Biol. 205, 145–157.

Lui, J.C., Baron, J., 2011. Effects of glucocorticoids on the growth plate. Endocr. Dev. 20, 187–193.

Luo, G., Hofmann, C., Bronckers, A.L., Sohocki, M., Bradley, A., Karsenty, G., 1995. BMP-7 is an inducer of nephrogenesis, and is also required for eye development and skeletal patterning. Genes Dev. 9, 2808–2820.

Maass, P.G., Rump, A., Schulz, H., Stricker, S., Schulze, L., Platzer, K., Aydin, A., Tinschert, S., Goldring, M.B., Luft, F.C., Bahring, S., 2012. A misplaced lncRNA causes brachydactyly in humans. J. Clin. Invest. 122, 3990–4002.

Macheda, M.L., Sun, W.W., Kugathasan, K., Hogan, B.M., Bower, N.I., Halford, M.M., Zhang, Y.F., Jacques, B.E., Lieschke, G.J., Dabdoub, A., Stacker, S.A., 2012. The Wnt receptor Ryk plays a role in mammalian planar cell polarity signaling. J. Biol. Chem. 287, 29312–29323.

Maclean, H.E., Guo, J., Knight, M.C., Zhang, P., Cobrinik, D., Kronenberg, H.M., 2004. The cyclin-dependent kinase inhibitor p57(Kip2) mediates proliferative actions of PTHrP in chondrocytes. J. Clin. Invest. 113, 1334–1343.

Maes, C., 2013. Role and regulation of vascularization processes in endochondral bones. Calcif. Tissue Int. 92, 307–323.

Maes, C., Araldi, E., Haigh, K., Khatri, R., Van Looveren, R., Giaccia, A.J., Haigh, J.J., Carmeliet, G., Schipani, E., 2012. VEGF-independent cell-autonomous functions of HIF-1alpha regulating oxygen consumption in fetal cartilage are critical for chondrocyte survival. J. Bone Miner. Res. 27, 596–609.

Maes, C., Kobayashi, T., Selig, M.K., Torrekens, S., Roth, S.I., Mackem, S., Carmeliet, G., Kronenberg, H.M., 2010. Osteoblast precursors, but not mature osteoblasts, move into developing and fractured bones along with invading blood vessels. Dev. Cell 19, 329–344.

Maes, C., Stockmans, I., Moermans, K., Van Looveren, R., Smets, N., Carmeliet, P., Bouillon, R., Carmeliet, G., 2004. Soluble VEGF isoforms are essential for establishing epiphyseal vascularization and regulating chondrocyte development and survival. J. Clin. Invest. 113, 188–199.

Majewski, J., Schwartzentruber, J.A., Caqueret, A., Patry, L., Marcadier, J., Fryns, J.P., Boycott, K.M., Ste-Marie, L.G., Mckiernan, F.E., Marik, I., Van Esch, H., Consortium, F.C., Michaud, J.L., Samuels, M.E., 2011. Mutations in NOTCH2 in families with Hajdu-Cheney syndrome. Hum. Mutat. 32, 1114–1117.

Mak, K.K., Kronenberg, H.M., Chuang, P.T., Mackem, S., Yang, Y., 2008. Indian hedgehog signals independently of PTHrP to promote chondrocyte hypertrophy. Development 135, 1947–1956.

Mallo, M., 2016. Revisiting the involvement of signaling gradients in somitogenesis. FEBS J. 283, 1430–1437.

Mankoo, B.S., Skuntz, S., Harrigan, I., Grigorieva, E., Candia, A., Wright, C.V., Arnheiter, H., Pachnis, V., 2003. The concerted action of Meox homeobox genes is required upstream of genetic pathways essential for the formation, patterning and differentiation of somites. Development 130, 4655–4664.

Martin, B.L., 2016. Factors that coordinate mesoderm specification from neuromesodermal progenitors with segmentation during vertebrate axial extension. Semin. Cell Dev. Biol. 49, 59–67.

Martin, G., 2001. Making a vertebrate limb: new players enter from the wings. Bioessays 23, 865–868.

Martin, J.F., Bradley, A., Olson, E.N., 1995. The paired-like homeo box gene MHox is required for early events of skeletogenesis in multiple lineages. Genes Dev. 9, 1237–1249.

Martou, G., Antonyshyn, O.M., 2011. Advances in surgical approaches to the upper facial skeleton. Curr. Opin. Otolaryngol. Head Neck Surg. 19, 242–247.

Maruyama, T., Jeong, J., Sheu, T.J., Hsu, W., 2016. Stem cells of the suture mesenchyme in craniofacial bone development, repair and regeneration. Nat. Commun. 7, 10526.

Matsunobu, T., Torigoe, K., Ishikawa, M., De Vega, S., Kulkarni, A.B., Iwamoto, Y., Yamada, Y., 2009. Critical roles of the TGF-beta type I receptor ALK5 in perichondrial formation and function, cartilage integrity, and osteoblast differentiation during growth plate development. Dev. Biol. 332, 325–338.

Mau, E., Whetstone, H., Yu, C., Hopyan, S., Wunder, J.S., Alman, B.A., 2007. PTHrP regulates growth plate chondrocyte differentiation and proliferation in a Gli3 dependent manner utilizing hedgehog ligand dependent and independent mechanisms. Dev. Biol. 305, 28–39.

Mccann, M.R., Seguin, C.A., 2016. Notochord cells in intervertebral disc development and degeneration. J. Dev. Biol. 4.

Mcewen, D.G., Green, R.P., Naski, M.C., Towler, D.A., Ornitz, D.M., 1999. Fibroblast growth factor receptor 3 gene transcription is suppressed by cyclic adenosine 3',5'-monophosphate. Identification of a chondrocytic regulatory element. J. Biol. Chem. 274, 30934–30942.

Mcpherron, A.C., Lawler, A.M., Lee, S.J., 1999. Regulation of anterior/posterior patterning of the axial skeleton by growth/differentiation factor 11. Nat. Genet. 22, 260–264.

Mead, T.J., Yutzey, K.E., 2009. Notch pathway regulation of chondrocyte differentiation and proliferation during appendicular and axial skeleton development. Proc. Natl. Acad. Sci. U.S.A. 106, 14420–14425.

Mercader, N., Leonardo, E., Azpiazu, N., Serrano, A., Morata, G., Martinez, C., Torres, M., 1999. Conserved regulation of proximodistal limb axis development by Meis1/Hth. Nature 402, 425–429.

Mercader, N., Leonardo, E., Piedra, M.E., Martinez, A.C., Ros, M.A., Torres, M., 2000. Opposing RA and FGF signals control proximodistal vertebrate limb development through regulation of Meis genes. Development 127, 3961–3970.

Mercader, N., Selleri, L., Criado, L.M., Pallares, P., Parras, C., Cleary, M.L., Torres, M., 2009. Ectopic Meis1 expression in the mouse limb bud alters P-D patterning in a Pbx1-independent manner. Int. J. Dev. Biol. 53, 1483–1494.

Merino, R., Macias, D., Ganan, Y., Economides, A.N., Wang, X., Wu, Q., Stahl, N., Sampath, K.T., Varona, P., Hurle, J.M., 1999. Expression and function of Gdf-5 during digit skeletogenesis in the embryonic chick leg bud. Dev. Biol. 206, 33–45.

Meyer, J., Sudbeck, P., Held, M., Wagner, T., Schmitz, M.L., Bricarelli, F.D., Eggermont, E., Friedrich, U., Haas, O.A., Kobelt, A., Leroy, J.G., Van Maldergem, L., Michel, E., Mitulla, B., Pfeiffer, R.A., Schinzel, A., Schmidt, H., Scherer, G., 1997. Mutational analysis of the SOX9 gene in campomelic dysplasia and autosomal sex reversal: lack of genotype/phenotype correlations. Hum. Mol. Genet. 6, 91–98.

Mikels, A.J., Nusse, R., 2006. Purified Wnt5a protein activates or inhibits beta-catenin-TCF signaling depending on receptor context. PLoS Biol. 4, e115.

Mikic, B., Clark, R.T., Battaglia, T.C., Gaschen, V., Hunziker, E.B., 2004. Altered hypertrophic chondrocyte kinetics in GDF-5 deficient murine tibial growth plates. J. Orthop. Res. 22, 552–556.

Mikic, B., Ferreira, M.P., Battaglia, T.C., Hunziker, E.B., 2008. Accelerated hypertrophic chondrocyte kinetics in GDF-7 deficient murine tibial growth plates. J. Orthop. Res. 26, 986–990.

Mikic, B., Van Der Meulen, M.C., Kingsley, D.M., Carter, D.R., 1995. Long bone geometry and strength in adult BMP-5 deficient mice. Bone 16, 445–454.

Milaire, J., 1991. Lectin binding sites in developing mouse limb buds. Anat. Embryol. 184, 479–488.

Minguillon, C., Del Buono, J., Logan, M.P., 2005. Tbx5 and Tbx4 are not sufficient to determine limb-specific morphologies but have common roles in initiating limb outgrowth. Dev. Cell 8, 75–84.

Mirzamohammadi, F., Papaioannou, G., Kobayashi, T., 2014. MicroRNAs in cartilage development, homeostasis, and disease. Curr. Osteoporos. Rep. 12, 410–419.

Miyaki, S., Sato, T., Inoue, A., Otsuki, S., Ito, Y., Yokoyama, S., Kato, Y., Takemoto, F., Nakasa, T., Yamashita, S., Takada, S., Lotz, M.K., Ueno-Kudo, H., Asahara, H., 2010. MicroRNA-140 plays dual roles in both cartilage development and homeostasis. Genes Dev. 24, 1173–1185.

Miyazawa, T., Ogawa, Y., Chusho, H., Yasoda, A., Tamura, N., Komatsu, Y., Pfeifer, A., Hofmann, F., Nakao, K., 2002. Cyclic GMP-dependent protein kinase II plays a critical role in C-type natriuretic peptide-mediated endochondral ossification. Endocrinology 143, 3604–3610.

Mo, R., Freer, A.M., Zinyk, D.L., Crackower, M.A., Michaud, J., Heng, H.H., Chik, K.W., Shi, X.M., Tsui, L.C., Cheng, S.H., Joyner, A.L., Hui, C., 1997. Specific and redundant functions of Gli2 and Gli3 zinc finger genes in skeletal patterning and development. Development 124, 113–123.

Moens, C.B., Selleri, L., 2006. Hox cofactors in vertebrate development. Dev. Biol. 291, 193–206.

Monsoro-Burq, A.H., 2005. Sclerotome development and morphogenesis: when experimental embryology meets genetics. Int. J. Dev. Biol. 49, 301–308.

Morimoto, M., Sasaki, N., Oginuma, M., Kiso, M., Igarashi, K., Aizaki, K., Kanno, J., Saga, Y., 2007. The negative regulation of Mesp2 by mouse Ripply2 is required to establish the rostro-caudal patterning within a somite. Development 134, 1561–1569.

Morita, K., Miyamoto, T., Fujita, N., Kubota, Y., Ito, K., Takubo, K., Miyamoto, K., Ninomiya, K., Suzuki, T., Iwasaki, R., Yagi, M., Takaishi, H., Toyama, Y., Suda, T., 2007. Reactive oxygen species induce chondrocyte hypertrophy in endochondral ossification. J. Exp. Med. 204, 1613–1623.

Moses, H.L., Serra, R., 1996. Regulation of differentiation by TGF-beta. Curr. Opin. Genet. Dev. 6, 581–586.

Muller, T.S., Ebensperger, C., Neubuser, A., Koseki, H., Balling, R., Christ, B., Wilting, J., 1996. Expression of avian Pax1 and Pax9 is intrinsically regulated in the pharyngeal endoderm, but depends on environmental influences in the paraxial mesoderm. Dev. Biol. 178, 403–417.

Mundlos, S., Otto, F., Mundlos, C., Mulliken, J.B., Aylsworth, A.S., Albright, S., Lindhout, D., Cole, W.G., Henn, W., Knoll, J.H., Owen, M.J., Mertelsmann, R., Zabel, B.U., Olsen, B.R., 1997. Mutations involving the transcription factor CBFA1 cause cleidocranial dysplasia. Cell 89, 773–779.

Munger, J.S., Sheppard, D., 2011. Cross talk among TGF-beta signaling pathways, integrins, and the extracellular matrix. Cold Spring Harb. Perspect. Biol. 3 a005017.

Murtaugh, L.C., Chyung, J.H., Lassar, A.B., 1999. Sonic hedgehog promotes somitic chondrogenesis by altering the cellular response to BMP signaling. Genes Dev. 13, 225–237.

Murtaugh, L.C., Zeng, L., Chyung, J.H., Lassar, A.B., 2001. The chick transcriptional repressor Nkx3.2 acts downstream of Shh to promote BMP-dependent axial chondrogenesis. Dev. Cell 1, 411–422.

Nakamura, Y., Inloes, J.B., Katagiri, T., Kobayashi, T., 2011. Chondrocyte-specific microRNA-140 regulates endochondral bone development and targets Dnpep to modulate bone morphogenetic protein signaling. Mol. Cell Biol. 31, 3019–3028.

Nakao, K., Okubo, Y., Yasoda, A., Koyama, N., Osawa, K., Isobe, Y., Kondo, E., Fujii, T., Miura, M., Nakao, K., Bessho, K., 2013. The effects of C-type natriuretic peptide on craniofacial skeletogenesis. J. Dent. Res. 92, 58–64.

Nakao, K., Osawa, K., Yasoda, A., Yamanaka, S., Fujii, T., Kondo, E., Koyama, N., Kanamoto, N., Miura, M., Kuwahara, K., Akiyama, H., Bessho, K., Nakao, K., 2015. The Local CNP/GC-B system in growth plate is responsible for physiological endochondral bone growth. Sci. Rep. 5, 10554.

Nakashima, K., Zhou, X., Kunkel, G., Zhang, Z., Deng, J.M., Behringer, R.R., DE Crombrugghe, B., 2002. The novel zinc finger-containing transcription factor osterix is required for osteoblast differentiation and bone formation. Cell 108, 17–29.

Newman, S.A., Glimm, T., Bhat, R., 2018. The vertebrate limb: an evolving complex of self-organizing systems. Prog. Biophys. Mol. Biol. 137, 12–24.

Nilsson, O., Baron, J., 2004. Fundamental limits on longitudinal bone growth: growth plate senescence and epiphyseal fusion. Trends Endocrinol. Metabol. 15, 370–374.

Nilsson, O., Marino, R., DE Luca, F., Phillip, M., Baron, J., 2005. Endocrine regulation of the growth plate. Horm. Res. 64, 157–165.

Nishimura, R., Hata, K., Nakamura, E., Murakami, T., Takahata, Y., 2018. Transcriptional network systems in cartilage development and disease. Histochem. Cell Biol. 149, 353–363.

Nishio, Y., Dong, Y., Paris, M., O'keefe, R.J., Schwarz, E.M., Drissi, H., 2006. Runx2-mediated regulation of the zinc finger Osterix/Sp7 gene. Gene 372, 62–70.

Niswander, L., 2003. Pattern formation: old models out on a limb. Nat. Rev. Genet. 4, 133–143.

Niswander, L., Jeffrey, S., Martin, G.R., Tickle, C., 1994. A positive feedback loop coordinates growth and patterning in the vertebrate limb. Nature 371, 609–612.

Niswander, L., Tickle, C., Vogel, A., Booth, I., Martin, G.R., 1993. FGF-4 replaces the apical ectodermal ridge and directs outgrowth and patterning of the limb. Cell 75, 579–587.

Oh, S.P., Yeo, C.Y., Lee, Y., Schrewe, H., Whitman, M., Li, E., 2002. Activin type IIA and IIB receptors mediate Gdf11 signaling in axial vertebral patterning. Genes Dev. 16, 2749–2754.

Oishi, I., Suzuki, H., Onishi, N., Takada, R., Kani, S., Ohkawara, B., Koshida, I., Suzuki, K., Yamada, G., Schwabe, G.C., Mundlos, S., Shibuya, H., Takada, S., Minami, Y., 2003. The receptor tyrosine kinase Ror2 is involved in non-canonical Wnt5a/JNK signalling pathway. Genes Cells 8, 645–654.

Okamoto, M., Udagawa, N., Uehara, S., Maeda, K., Yamashita, T., Nakamichi, Y., Kato, H., Saito, N., Minami, Y., Takahashi, N., Kobayashi, Y., 2014. Noncanonical Wnt5a enhances Wnt/beta-catenin signaling during osteoblastogenesis. Sci. Rep. 4, 4493.

Olney, R.C., Bukulmez, H., Bartels, C.F., Prickett, T.C., Espiner, E.A., Potter, L.R., Warman, M.L., 2006. Heterozygous mutations in natriuretic peptide receptor-B (NPR2) are associated with short stature. J. Clin. Endocrinol. Metab. 91, 1229–1232.

Oppenheimer, H., Kumar, A., Meir, H., Schwartz, I., Zini, A., Haze, A., Kandel, L., Mattan, Y., Liebergall, M., Dvir-Ginzberg, M., 2014. Set7/9 impacts COL2A1 expression through binding and repression of SirT1 histone deacetylation. J. Bone Miner. Res. 29, 348–360.

Ornitz, D.M., Marie, P.J., 2015. Fibroblast growth factor signaling in skeletal development and disease. Genes Dev. 29, 1463–1486.

Ortega, N., Wang, K., Ferrara, N., Werb, Z., Vu, T.H., 2010. Complementary interplay between matrix metalloproteinase-9, vascular endothelial growth factor and osteoclast function drives endochondral bone formation. Dis. Model Mech. 3, 224–235.

Oster, G.F., Shubin, N., Murray, J.D., Alberch, P., 1988. Evolution and morphogenetic rules: the shape of the vertebrate limb in ontogeny and phylogeny. Evolution 42, 862–884.

Otto, F., Thornell, A.P., Crompton, T., Denzel, A., Gilmour, K.C., Rosewell, I.R., Stamp, G.W., Beddington, R.S., Mundlos, S., Olsen, B.R., Selby, P.B., Owen, M.J., 1997. Cbfa1, a candidate gene for cleidocranial dysplasia syndrome, is essential for osteoblast differentiation and bone development. Cell 89, 765–771.

Ozasa, A., Komatsu, Y., Yasoda, A., Miura, M., Sakuma, Y., Nakatsuru, Y., Arai, H., Itoh, N., Nakao, K., 2005. Complementary antagonistic actions between C-type natriuretic peptide and the MAPK pathway through FGFR-3 in ATDC5 cells. Bone 36, 1056–1064.

Pacifici, M., Koyama, E., Shibukawa, Y., Wu, C., Tamamura, Y., Enomoto-Iwamoto, M., Iwamoto, M., 2006. Cellular and molecular mechanisms of synovial joint and articular cartilage formation. Ann. N.Y. Acad. Sci. 1068, 74–86.

Pan, Q., Wu, Y., Lin, T., Yao, H., Yang, Z., Gao, G., Song, E., Shen, H., 2009. Bone morphogenetic protein-2 induces chromatin remodeling and modification at the proximal promoter of Sox9 gene. Biochem. Biophys. Res. Commun. 379, 356–361.

Papaioannou, G., Inloes, J.B., Nakamura, Y., Paltrinieri, E., Kobayashi, T., 2013. let-7 and miR-140 microRNAs coordinately regulate skeletal development. Proc. Natl. Acad. Sci. U.S.A. 110, E3291–E3300.

Park, J., Gebhardt, M., Golovchenko, S., Branguli, F., Hattori, T., Hartmann, C., Zhou, X., De Crombrugghe, B., Stock, M., Schneider, H., Von Der Mark, K., 2015. Dual pathways to endochondral osteoblasts: a novel chondrocyte-derived osteoprogenitor cell identified in hypertrophic cartilage. Biol. Open 4, 608–621.

Patton, M.A., Afzal, A.R., 2002. Robinow syndrome. J. Med. Genet. 39, 305–310.

Peake, N.J., Hobbs, A.J., Pingguan-Murphy, B., Salter, D.M., Berenbaum, F., Chowdhury, T.T., 2014. Role of C-type natriuretic peptide signalling in maintaining cartilage and bone function. Osteoarthritis Cartilage 22, 1800–1807.

Pedrozo, H.A., Schwartz, Z., Gomez, R., Ornoy, A., Xin-Sheng, W., Dallas, S.L., Bonewald, L.F., Dean, D.D., Boyan, B.D., 1998. Growth plate chondrocytes store latent transforming growth factor (TGF)-beta 1 in their matrix through latent TGF-beta 1 binding protein-1. J. Cell. Physiol. 177, 343–354.

Pelton, R.W., Dickinson, M.E., Moses, H.L., Hogan, B.L., 1990. In situ hybridization analysis of TGF beta 3 RNA expression during mouse development: comparative studies with TGF beta 1 and beta 2. Development 110, 609–620.

Pelton, R.W., Saxena, B., Jones, M., Moses, H.L., Gold, L.I., 1991. Immunohistochemical localization of TGF beta 1, TGF beta 2, and TGF beta 3 in the mouse embryo: expression patterns suggest multiple roles during embryonic development. J. Cell Biol. 115, 1091–1105.

Person, A.D., Beiraghi, S., Sieben, C.M., Hermanson, S., Neumann, A.N., Robu, M.E., Schleiffarth, J.R., Billington JR., C.J., VAN Bokhoven, H., Hoogeboom, J.M., Mazzeu, J.F., Petryk, A., Schimmenti, L.A., Brunner, H.G., Ekker, S.C., Lohr, J.L., 2010. WNT5A mutations in patients with autosomal dominant Robinow syndrome. Dev. Dynam. 239, 327–337.

Peters, H., Wilm, B., Sakai, N., Imai, K., Maas, R., Balling, R., 1999. Pax1 and Pax9 synergistically regulate vertebral column development. Development 126, 5399–5408.

Peters, S.B., Wang, Y., Serra, R., 2017. Tgfbr2 is required in osterix expressing cells for postnatal skeletal development. Bone 97, 54–64.

Petit, F., Sears, K.E., Ahituv, N., 2017. Limb development: a paradigm of gene regulation. Nat. Rev. Genet. 18, 245–258.

Pi, Y., Zhang, X., Shao, Z., Zhao, F., Hu, X., Ao, Y., 2015. Intra-articular delivery of anti-Hif-2alpha siRNA by chondrocyte-homing nanoparticles to prevent cartilage degeneration in arthritic mice. Gene Ther. 22, 439–448.

Pineault, K.M., Wellik, D.M., 2014. Hox genes and limb musculoskeletal development. Curr. Osteoporos. Rep. 12, 420–427.

Pogue, R., Lyons, K., 2006. BMP signaling in the cartilage growth plate. Curr. Top. Dev. Biol. 76, 1–48.

Polinkovsky, A., Robin, N.H., Thomas, J.T., Irons, M., Lynn, A., Goodman, F.R., Reardon, W., Kant, S.G., Brunner, H.G., Van Der Burgt, I., Chitayat, D., Mcgaughran, J., Donnai, D., Luyten, F.P., Warman, M.L., 1997. Mutations in CDMP1 cause autosomal dominant brachydactyly type C. Nat. Genet. 17, 18–19.

Posnick, J.C., Tiwana, P.S., Ruiz, R.L., 2010. Craniofacial dysostosis syndromes: evaluation and staged reconstructive approach. Atlas Oral Maxillofac. Surg. Clin. North Am. 18, 109–128.

Pourquie, O., 2000. Segmentation of the paraxial mesoderm and vertebrate somitogenesis. Curr. Top. Dev. Biol. 47, 81–105.

Powell-Braxton, L., Hollingshead, P., Warburton, C., Dowd, M., Pitts-Meek, S., Dalton, D., Gillett, N., Stewart, T.A., 1993. IGF-I is required for normal embryonic growth in mice. Genes Dev. 7, 2609–2617.

Pritchett, J.W., 1991. Growth plate activity in the upper extremity. Clin. Orthop. Relat. Res. 235–242.

Pritchett, J.W., 1992. Longitudinal growth and growth-plate activity in the lower extremity. Clin. Orthop. Relat. Res. 274–279.

Provot, S., Kempf, H., Murtaugh, L.C., Chung, U.I., Kim, D.W., Chyung, J., Kronenberg, H.M., Lassar, A.B., 2006. Nkx3.2/Bapx1 acts as a negative regulator of chondrocyte maturation. Development 133, 651–662.

Provot, S., Zinyk, D., Gunes, Y., Kathri, R., Le, Q., Kronenberg, H.M., Johnson, R.S., Longaker, M.T., Giaccia, A.J., Schipani, E., 2007. Hif-1alpha regulates differentiation of limb bud mesenchyme and joint development. J. Cell Biol. 177, 451–464.

Pryce, B.A., Watson, S.S., Murchison, N.D., Staverosky, J.A., Dunker, N., Schweitzer, R., 2009. Recruitment and maintenance of tendon progenitors by TGFbeta signaling are essential for tendon formation. Development 136, 1351–1361.

Pyott, S.M., Tran, T.T., Leistritz, D.F., Pepin, M.G., Mendelsohn, N.J., Temme, R.T., Fernandez, B.A., Elsayed, S.M., Elsobky, E., Verma, I., Nair, S., Turner, E.H., Smith, J.D., Jarvik, G.P., Byers, P.H., 2013. WNT1 mutations in families affected by moderately severe and progressive recessive osteogenesis imperfecta. Am. J. Hum. Genet. 92, 590–597.

Ratcliffe, P.J., 2013. Oxygen sensing and hypoxia signalling pathways in animals: the implications of physiology for cancer. J. Physiol. 591, 2027–2042.

Rawls, A., Fischer, R.E., 2010. Development and functional anatomy of the spine. In: Kusumi, K., Dunwoodie, S.L. (Eds.), The Genetics and Development of Scoliosis. Springer.

Razzaque, M.S., Soegiarto, D.W., Chang, D., Long, F., Lanske, B., 2005. Conditional deletion of Indian hedgehog from collagen type 2alpha1-expressing cells results in abnormal endochondral bone formation. J. Pathol. 207, 453–461.

Regard, J.B., Cherman, N., Palmer, D., Kuznetsov, S.A., Celi, F.S., Guettier, J.M., Chen, M., Bhattacharyya, N., Wess, J., Coughlin, S.R., Weinstein, L.S., Collins, M.T., Robey, P.G., Yang, Y., 2011. Wnt/beta-catenin signaling is differentially regulated by Galpha proteins and contributes to fibrous dysplasia. Proc. Natl. Acad. Sci. U.S.A. 108, 20101–20106.

Regard, J.B., Malhotra, D., Gvozdenovic-Jeremic, J., Josey, M., Chen, M., Weinstein, L.S., Lu, J., Shore, E.M., Kaplan, F.S., Yang, Y., 2013. Activation of Hedgehog signaling by loss of GNAS causes heterotopic ossification. Nat. Med. 19, 1505–1512.

Retting, K.N., Song, B., Yoon, B.S., Lyons, K.M., 2009. BMP canonical Smad signaling through Smad1 and Smad5 is required for endochondral bone formation. Development 136, 1093–1104.

Riddle, R.D., Ensini, M., Nelson, C., Tsuchida, T., Jessell, T.M., Tabin, C., 1995. Induction of the LIM homeobox gene Lmx1 by WNT7a establishes dorsoventral pattern in the vertebrate limb. Cell 83, 631–640.

Riddle, R.D., Johnson, R.L., Laufer, E., Tabin, C., 1993. Sonic hedgehog mediates the polarizing activity of the ZPA. Cell 75, 1401–1416.

Rigueur, D., Brugger, S., Anbarchian, T., Kim, J.K., Lee, Y., Lyons, K.M., 2015. The type I BMP receptor ACVR1/ALK2 is required for chondrogenesis during development. J. Bone Miner. Res. 30, 733–741.

Rittling, S.R., Matsumoto, H.N., Mckee, M.D., Nanci, A., An, X.R., Novick, K.E., Kowalski, A.J., Noda, M., Denhardt, D.T., 1998. Mice lacking osteopontin show normal development and bone structure but display altered osteoclast formation in vitro. J. Bone Miner. Res. 13, 1101–1111.

Roach, H.I., Erenpreisa, J., Aigner, T., 1995. Osteogenic differentiation of hypertrophic chondrocytes involves asymmetric cell divisions and apoptosis. J. Cell Biol. 131, 483–494.

Robin, N.H., Falk, M.J., Haldeman-Englert, C.R., 1993. FGFR-related craniosynostosis syndromes. In: Adam, M.P., Ardinger, H.H., Pagon, R.A., Wallace, S.E., Bean, L.J.H., Stephens, K., Amemiya, A. (Eds.), GeneReviews. Seattle (WA).

Robson, H., Siebler, T., Shalet, S.M., Williams, G.R., 2002. Interactions between Gh, Igf-I, glucocorticoids, and thyroid hormones during skeletal growth. Pediatr. Res. 52, 137–147.

Roca, H., Phimphilai, M., Gopalakrishnan, R., Xiao, G., Franceschi, R.T., 2005. Cooperative interactions between RUNX2 and homeodomain protein-binding sites are critical for the osteoblast-specific expression of the bone sialoprotein gene. J. Biol. Chem. 280, 30845–30855.

Rodrigo, I., Hill, R.E., Balling, R., Munsterberg, A., Imai, K., 2003. Pax1 and Pax9 activate Bapx1 to induce chondrogenic differentiation in the sclerotome. Development 130, 473–482.

Rodrigues, A.R., Yakushiji-Kaminatsui, N., Atsuta, Y., Andrey, G., Schorderet, P., Duboule, D., Tabin, C.J., 2017. Integration of Shh and Fgf signaling in controlling Hox gene expression in cultured limb cells. Proc. Natl. Acad. Sci. U.S.A. 114, 3139–3144.

Roifman, M., Brunner, H.G., Lohr, J.L., Mazzeu, J.F., Chitayat, D., 2015. Autosomal dominant Robinow syndrome. In: Adam, M.P., Ardinger, H.H., Pagon, R.A., Wallace, S.E., Bean, L.J.H., Stephens, K., Amemiya, A. (Eds.), GeneReviews. Seattle (WA).

Romereim, S.M., Conoan, N.H., Chen, B., Dudley, A.T., 2014. A dynamic cell adhesion surface regulates tissue architecture in growth plate cartilage. Development 141, 2085–2095.

Ros, M.A., Dahn, R.D., Fernandez-Teran, M., Rashka, K., Caruccio, N.C., Hasso, S.M., Bitgood, J.J., Lancman, J.J., Fallon, J.F., 2003. The chick oligozeugodactyly (ozd) mutant lacks sonic hedgehog function in the limb. Development 130, 527–537.

Rosen, V., 2006. BMP and BMP inhibitors in bone. Ann. N.Y. Acad. Sci. 1068, 19–25.

Rowe, D.A., Fallon, J.F., 1982. The proximodistal determination of skeletal parts in the developing chick leg. J. Embryol. Exp. Morphol. 68, 1–7.

Saga, Y., 2007. Segmental border is defined by the key transcription factor Mesp2, by means of the suppression of Notch activity. Dev. Dynam. 236, 1450–1455.

Sanford, L.P., Ormsby, I., Gittenberger-DE Groot, A.C., Sariola, H., Friedman, R., Boivin, G.P., Cardell, E.L., Doetschman, T., 1997. TGFbeta2 knockout mice have multiple developmental defects that are non-overlapping with other TGFbeta knockout phenotypes. Development 124, 2659–2670.

Sasaki, N., Kiso, M., Kitagawa, M., Saga, Y., 2011. The repression of Notch signaling occurs via the destabilization of mastermind-like 1 by Mesp2 and is essential for somitogenesis. Development 138, 55–64.

Sassi, N., Laadhar, L., Driss, M., Kallel-Sellami, M., Sellami, S., Makni, S., 2011. The role of the Notch pathway in healthy and osteoarthritic articular cartilage: from experimental models to ex vivo studies. Arthritis Res. Ther. 13, 208.

Satokata, I., Ma, L., Ohshima, H., Bei, M., Woo, I., Nishizawa, K., Maeda, T., Takano, Y., Uchiyama, M., Heaney, S., Peters, H., Tang, Z., Maxson, R., Maas, R., 2000. Msx2 deficiency in mice causes pleiotropic defects in bone growth and ectodermal organ formation. Nat. Genet. 24, 391–395.

Saunder, J.W., Gasseling, M.T. (Eds.), 1968. Ecotdermal and Mesenchymal Interactions in the Origin of Limb Symmetry. Williams and Wilkins, Baltimore.

Saunders JR., J.W., 1948. The proximo-distal sequence of origin of the parts of the chick wing and the role of the ectoderm. J. Exp. Zool. 108, 363–403.

Scaal, M., Christ, B., 2004. Formation and differentiation of the avian dermomyotome. Anat. Embryol. 208, 411–424.

Schipani, E., Jensen, G.S., Pincus, J., Nissenson, R.A., Gardella, T.J., Juppner, H., 1997a. Constitutive activation of the cyclic adenosine 3',5'-monophosphate signaling pathway by parathyroid hormone (PTH)/PTH-related peptide receptors mutated at the two loci for Jansen's metaphyseal chondrodysplasia. Mol. Endocrinol. 11, 851–858.

Schipani, E., Kruse, K., Juppner, H., 1995. A constitutively active mutant PTH-PTHrP receptor in Jansen-type metaphyseal chondrodysplasia. Science 268, 98–100.

Schipani, E., Langman, C., Hunzelman, J., LE Merrer, M., Loke, K.Y., Dillon, M.J., Silve, C., Juppner, H., 1999. A novel parathyroid hormone (PTH)/ PTH-related peptide receptor mutation in Jansen's metaphyseal chondrodysplasia. J. Clin. Endocrinol. Metab. 84, 3052–3057.

Schipani, E., Langman, C.B., Parfitt, A.M., Jensen, G.S., Kikuchi, S., Kooh, S.W., Cole, W.G., Juppner, H., 1996. Constitutively activated receptors for parathyroid hormone and parathyroid hormone-related peptide in Jansen's metaphyseal chondrodysplasia. N. Engl. J. Med. 335, 708–714.

Schipani, E., Lanske, B., Hunzelman, J., Luz, A., Kovacs, C.S., Lee, K., Pirro, A., Kronenberg, H.M., Juppner, H., 1997b. Targeted expression of constitutively active receptors for parathyroid hormone and parathyroid hormone-related peptide delays endochondral bone formation and rescues mice that lack parathyroid hormone-related peptide. Proc. Natl. Acad. Sci. U.S.A. 94, 13689–13694.

Schipani, E., Mangiavini, L., Merceron, C., 2015. HIF-1alpha and growth plate development: what we really know. Bonekey Rep. 4, 730.

Schipani, E., Ryan, H.E., Didrickson, S., Kobayashi, T., Knight, M., Johnson, R.S., 2001. Hypoxia in cartilage: HIF-1alpha is essential for chondrocyte growth arrest and survival. Genes Dev. 15, 2865–2876.

Schmid, P., Cox, D., Bilbe, G., Maier, R., Mcmaster, G.K., 1991. Differential expression of TGF beta 1, beta 2 and beta 3 genes during mouse embryogenesis. Development 111, 117–130.

Schneider, A., Brand, T., Zweigerdt, R., Arnold, H., 2000. Targeted disruption of the Nkx3.1 gene in mice results in morphogenetic defects of minor salivary glands: parallels to glandular duct morphogenesis in prostate. Mech. Dev. 95, 163–174.

Sekine, K., Ohuchi, H., Fujiwara, M., Yamasaki, M., Yoshizawa, T., Sato, T., Yagishita, N., Matsui, D., Koga, Y., Itoh, N., Kato, S., 1999. Fgf10 is essential for limb and lung formation. Nat. Genet. 21, 138–141.

Semenza, G.L., 2012. Hypoxia-inducible factors in physiology and medicine. Cell 148, 399–408.

Senarath-Yapa, K., Chung, M.T., Mcardle, A., Wong, V.W., Quarto, N., Longaker, M.T., Wan, D.C., 2012. Craniosynostosis: molecular pathways and future pharmacologic therapy. Organogenesis 8, 103–113.

Seo, H.S., Serra, R., 2007. Deletion of Tgfbr2 in Prx1-cre expressing mesenchyme results in defects in development of the long bones and joints. Dev. Biol. 310, 304–316.

Serra, R., Johnson, M., Filvaroff, E.H., Laborde, J., Sheehan, D.M., Derynck, R., Moses, H.L., 1997. Expression of a truncated, kinase-defective TGF-beta type II receptor in mouse skeletal tissue promotes terminal chondrocyte differentiation and osteoarthritis. J. Cell Biol. 139, 541–552.

Settle JR., S.H., Rountree, R.B., Sinha, A., Thacker, A., Higgins, K., Kingsley, D.M., 2003. Multiple joint and skeletal patterning defects caused by single and double mutations in the mouse Gdf6 and Gdf5 genes. Dev. Biol. 254, 116–130.

Shapiro, F., 2001. Developmental bone biology. In: Pediatric Orthopedic Deformities - Basic Science, Diagnosis, and Treatment. Academic Press, San Diego, San Francisco, New York, Boston, London, Sydney, Tokyo.

Shapiro, F., Holtrop, M.E., Glimcher, M.J., 1977. Organization and cellular biology of the perichondrial ossification groove of ranvier: a morphological study in rabbits. J. Bone Joint Surg. Am. 59, 703–723.

Shapiro, I.M., Adams, C.S., Freeman, T., Srinivas, V., 2005. Fate of the hypertrophic chondrocyte: microenvironmental perspectives on apoptosis and survival in the epiphyseal growth plate. Birth Defects Res. C Embryo Today 75, 330–339.

Sheeba, C.J., Andrade, R.P., Palmeirim, I., 2016. Mechanisms of vertebrate embryo segmentation: common themes in trunk and limb development. Semin. Cell Dev. Biol. 49, 125–134.

Shore, E.M., Xu, M., Feldman, G.J., Fenstermacher, D.A., Cho, T.J., Choi, I.H., Connor, J.M., Delai, P., Glaser, D.L., Lemerrer, M., Morhart, R., Rogers, J.G., Smith, R., Triffitt, J.T., Urtizberea, J.A., Zasloff, M., Brown, M.A., Kaplan, F.S., 2006. A recurrent mutation in the BMP type I receptor ACVR1 causes inherited and sporadic fibrodysplasia ossificans progressiva. Nat. Genet. 38, 525–527.

Shu, B., Zhang, M., Xie, R., Wang, M., Jin, H., Hou, W., Tang, D., Harris, S.E., Mishina, Y., O'keefe, R.J., Hilton, M.J., Wang, Y., Chen, D., 2011. BMP2, but not BMP4, is crucial for chondrocyte proliferation and maturation during endochondral bone development. J. Cell Sci. 124, 3428–3440.

Shubin, N., Alberch, P., 1986. A morphogenic approach to the origin and basic organization of the tetrapod limb. Evol. Biol. 20, 319–387.

Silvestrini, G., Ballanti, P., Patacchioli, F., Leopizzi, M., Gualtieri, N., Monnazzi, P., Tremante, E., Sardella, D., Bonucci, E., 2005. Detection of osteoprotegerin (OPG) and its ligand (RANKL) mRNA and protein in femur and tibia of the rat. J. Mol. Histol. 36, 59–67.

Skuntz, S., Mankoo, B., Nguyen, M.T., Hustert, E., Nakayama, A., Tournier-Lasserve, E., Wright, C.V., Pachnis, V., Bharti, K., Arnheiter, H., 2009. Lack of the mesodermal homeodomain protein MEOX1 disrupts sclerotome polarity and leads to a remodeling of the cranio-cervical joints of the axial skeleton. Dev. Biol. 332, 383–395.

Slater, B.J., Lenton, K.A., Kwan, M.D., Gupta, D.M., Wan, D.C., Longaker, M.T., 2008. Cranial sutures: a brief review. Plast. Reconstr. Surg. 121, 170e-8e.

Smits, P., Li, P., Mandel, J., Zhang, Z., Deng, J.M., Behringer, R.R., De Crombrugghe, B., Lefebvre, V., 2001. The transcription factors L-Sox5 and Sox6 are essential for cartilage formation. Dev. Cell 1, 277–290.

Solloway, M.J., Dudley, A.T., Bikoff, E.K., Lyons, K.M., Hogan, B.L., Robertson, E.J., 1998. Mice lacking Bmp6 function. Dev. Genet. 22, 321–339.

Spagnoli, A., O'rear, L., Chandler, R.L., Granero-Molto, F., Mortlock, D.P., Gorska, A.E., Weis, J.A., Longobardi, L., Chytil, A., Shimer, K., Moses, H.L., 2007. TGF-beta signaling is essential for joint morphogenesis. J. Cell Biol. 177, 1105–1117.

Spater, D., Hill, T.P., Gruber, M., Hartmann, C., 2006a. Role of canonical Wnt-signalling in joint formation. Eur. Cells Mater. 12, 71–80.

Spater, D., Hill, T.P., O'sullivan, R.,J., Gruber, M., Conner, D.A., Hartmann, C., 2006b. Wnt9a signaling is required for joint integrity and regulation of Ihh during chondrogenesis. Development 133, 3039–3049.

St-Jacques, B., Hammerschmidt, M., Mcmahon, A.P., 1999. Indian hedgehog signaling regulates proliferation and differentiation of chondrocytes and is essential for bone formation. Genes Dev. 13, 2072–2086.

Stickens, D., Behonick, D.J., Ortega, N., Heyer, B., Hartenstein, B., Yu, Y., Fosang, A.J., Schorpp-Kistner, M., Angel, P., Werb, Z., 2004. Altered endochondral bone development in matrix metalloproteinase 13-deficient mice. Development 131, 5883–5895.

Storm, E.E., Huynh, T.V., Copeland, N.G., Jenkins, N.A., Kingsley, D.M., Lee, S.J., 1994. Limb alterations in brachypodism mice due to mutations in a new member of the TGF beta-superfamily. Nature 368, 639–643.

Storm, E.E., Kingsley, D.M., 1999. GDF5 coordinates bone and joint formation during digit development. Dev. Biol. 209, 11–27.

Stringa, E., Tuan, R.S., 1996. Chondrogenic cell subpopulation of chick embryonic calvarium: isolation by peanut agglutinin affinity chromatography and in vitro characterization. Anat. Embryol. 194, 427–437.

Suda, M., Ogawa, Y., Tanaka, K., Tamura, N., Yasoda, A., Takigawa, T., Uehira, M., Nishimoto, H., Itoh, H., Saito, Y., Shiota, K., Nakao, K., 1998. Skeletal overgrowth in transgenic mice that overexpress brain natriuretic peptide. Proc. Natl. Acad. Sci. U.S.A. 95, 2337–2342.

Suda, N., Baba, O., Udagawa, N., Terashima, T., Kitahara, Y., Takano, Y., Kuroda, T., Senior, P.V., Beck, F., Hammond, V.E., 2001. Parathyroid hormone-related protein is required for normal intramembranous bone development. J. Bone Miner. Res. 16, 2182–2191.

Sueyoshi, T., Yamamoto, K., Akiyama, H., 2012. Conditional deletion of Tgfbr2 in hypertrophic chondrocytes delays terminal chondrocyte differentiation. Matrix Biol. 31, 352–359.

Sugimoto, Y., Takimoto, A., Akiyama, H., Kist, R., Scherer, G., Nakamura, T., Hiraki, Y., Shukunami, C., 2013. Scx+/Sox9+ progenitors contribute to the establishment of the junction between cartilage and tendon/ligament. Development 140, 2280–2288.

Summerbell, D., 1974. A quantitative analysis of the effect of excision of the AER from the chick limb-bud. J. Embryol. Exp. Morphol. 32, 651–660.

Sun, F., Chen, Q., Yang, S., Pan, Q., Ma, J., Wan, Y., Chang, C.H., Hong, A., 2009. Remodeling of chromatin structure within the promoter is important for bmp-2-induced fgfr3 expression. Nucleic Acids Res. 37, 3897–3911.

Sun, X., Mariani, F.V., Martin, G.R., 2002. Functions of FGF signalling from the apical ectodermal ridge in limb development. Nature 418, 501–508.

Svensson, J., Lall, S., Dickson, S.L., Bengtsson, B.A., Romer, J., Ahnfelt-Ronne, I., Ohlsson, C., Jansson, J.O., 2001. Effects of growth hormone and its secretagogues on bone. Endocrine 14, 63–66.

Swingler, T.E., Wheeler, G., Carmont, V., Elliott, H.R., Barter, M.J., Abu-Elmagd, M., Donell, S.T., Boot-Handford, R.P., Hajihosseini, M.K., Munsterberg, A., Dalmay, T., Young, D.A., Clark, I.M., 2012. The expression and function of microRNAs in chondrogenesis and osteoarthritis. Arthritis Rheum. 64, 1909–1919.

Takahashi, Y., Yasuhiko, Y., Kitajima, S., Kanno, J., Saga, Y., 2007. Appropriate suppression of Notch signaling by Mesp factors is essential for stripe pattern formation leading to segment boundary formation. Dev. Biol. 304, 593–603.

Takeuchi, S., Takeda, K., Oishi, I., Nomi, M., Ikeya, M., Itoh, K., Tamura, S., Ueda, T., Hatta, T., Otani, H., Terashima, T., Takada, S., Yamamura, H., Akira, S., Minami, Y., 2000. Mouse Ror2 receptor tyrosine kinase is required for the heart development and limb formation. Genes Cells 5, 71–78.

Tamamura, Y., Otani, T., Kanatani, N., Koyama, E., Kitagaki, J., Komori, T., Yamada, Y., Costantini, F., Wakisaka, S., Pacifici, M., Iwamoto, M., Enomoto-Iwamoto, M., 2005. Developmental regulation of Wnt/beta-catenin signals is required for growth plate assembly, cartilage integrity, and endochondral ossification. J. Biol. Chem. 280, 19185–19195.

Tamura, N., Doolittle, L.K., Hammer, R.E., Shelton, J.M., Richardson, J.A., Garbers, D.L., 2004. Critical roles of the guanylyl cyclase B receptor in endochondral ossification and development of female reproductive organs. Proc. Natl. Acad. Sci. U.S.A. 101, 17300–17305.

Tang, Y., Wu, X., Lei, W., Pang, L., Wan, C., Shi, Z., Zhao, L., Nagy, T.R., Peng, X., Hu, J., Feng, X., Van Hul, W., Wan, M., Cao, X., 2009. TGF-beta1-induced migration of bone mesenchymal stem cells couples bone resorption with formation. Nat. Med. 15, 757–765.

Taraboletti, G., D'ascenzo, S., Borsotti, P., Giavazzi, R., Pavan, A., Dolo, V., 2002. Shedding of the matrix metalloproteinases MMP-2, MMP-9, and MT1-MMP as membrane vesicle-associated components by endothelial cells. Am. J. Pathol. 160, 673–680.

Tarchini, B., Duboule, D., Kmita, M., 2006. Regulatory constraints in the evolution of the tetrapod limb anterior-posterior polarity. Nature 443, 985–988.

Te Welscher, P., Fernandez-Teran, M., Ros, M.A., Zeller, R., 2002. Mutual genetic antagonism involving GLI3 and dHAND prepatterns the vertebrate limb bud mesenchyme prior to SHH signaling. Genes Dev. 16, 421–426.

Ten Berge, D., Brugmann, S.A., Helms, J.A., Nusse, R., 2008. Wnt and FGF signals interact to coordinate growth with cell fate specification during limb development. Development 135, 3247–3257.

Tickle, C., 1981. The number of polarizing region cells required to specify additional digits in the developing chick wing. Nature 289, 295–298.

Tickle, C., Summerbell, D., Wolpert, L., 1975. Positional signalling and specification of digits in chick limb morphogenesis. Nature 254, 199–202.

Topol, L., Jiang, X., Choi, H., Garrett-Beal, L., Carolan, P.J., Yang, Y., 2003. Wnt-5a inhibits the canonical Wnt pathway by promoting GSK-3-independent beta-catenin degradation. J. Cell Biol. 162, 899–908.

Touaitahuata, H., Cres, G., De Rossi, S., Vives, V., Blangy, A., 2014. The mineral dissolution function of osteoclasts is dispensable for hypertrophic cartilage degradation during long bone development and growth. Dev. Biol. 393, 57–70.

Tribioli, C., Lufkin, T., 1999. The murine Bapx1 homeobox gene plays a critical role in embryonic development of the axial skeleton and spleen. Development 126, 5699–5711.

Tsang, K.Y., Chan, D., Cheah, K.S., 2015. Fate of growth plate hypertrophic chondrocytes: death or lineage extension? Dev. Growth Differ. 57, 179–192.

Tu, X., Joeng, K.S., Long, F., 2012. Indian hedgehog requires additional effectors besides Runx2 to induce osteoblast differentiation. Dev. Biol. 362, 76–82.

Tucker, R.P., Chiquet-Ehrismann, R., 2015. Tenascin-C: its functions as an integrin ligand. Int. J. Biochem. Cell Biol. 65, 165–168.

Usami, Y., Gunawardena, A.T., Iwamoto, M., Enomoto-Iwamoto, M., 2016. Wnt signaling in cartilage development and diseases: lessons from animal studies. Lab. Invest. 96, 186–196.

Usui, M., Xing, L., Drissi, H., Zuscik, M., O'keefe, R., Chen, D., Boyce, B.F., 2008. Murine and chicken chondrocytes regulate osteoclastogenesis by producing RANKL in response to BMP2. J. Bone Miner. Res. 23, 314–325.

Van Der Kraan, P.M., Blaney Davidson, E.N., Blom, A., Van Den Berg, W.B., 2009. TGF-beta signaling in chondrocyte terminal differentiation and osteoarthritis: modulation and integration of signaling pathways through receptor-Smads. Osteoarthritis Cartilage 17, 1539–1545.

Vasques, G.A., Amano, N., Docko, A.J., Funari, M.F., Quedas, E.P., Nishi, M.Y., Arnhold, I.J., Hasegawa, T., Jorge, A.A., 2013. Heterozygous mutations in natriuretic peptide receptor-B (NPR2) gene as a cause of short stature in patients initially classified as idiopathic short stature. J. Clin. Endocrinol. Metab. 98, E1636–E1644.

Vega, R.B., Matsuda, K., Oh, J., Barbosa, A.C., Yang, X., Meadows, E., Mcanally, J., Pomajzl, C., Shelton, J.M., Richardson, J.A., Karsenty, G., Olson, E.N., 2004. Histone deacetylase 4 controls chondrocyte hypertrophy during skeletogenesis. Cell 119, 555–566.

Vogel, A., Rodriguez, C., Warnken, W., Izpisua Belmonte, J.C., 1995. Dorsal cell fate specified by chick Lmx1 during vertebrate limb development. Nature 378, 716–720.

Vortkamp, A., Franz, T., Gessler, M., Grzeschik, K.H., 1992. Deletion of GLI3 supports the homology of the human Greig cephalopolysyndactyly syndrome (GCPS) and the mouse mutant extra toes (Xt). Mamm. Genome 3, 461–463.

Vortkamp, A., Lee, K., Lanske, B., Segre, G.V., Kronenberg, H.M., Tabin, C.J., 1996. Regulation of rate of cartilage differentiation by Indian hedgehog and PTH-related protein. Science 273, 613–622.

Wang, B., Fallon, J.F., Beachy, P.A., 2000. Hedgehog-regulated processing of Gli3 produces an anterior/posterior repressor gradient in the developing vertebrate limb. Cell 100, 423–434.

Wang, B., Jin, H., Zhu, M., Li, J., Zhao, L., Zhang, Y., Tang, D., Xiao, G., Xing, L., Boyce, B.F., Chen, D., 2014a. Chondrocyte beta-catenin signaling regulates postnatal bone remodeling through modulation of osteoclast formation in a murine model. Arthritis Rheum. 66, 107–120.

Wang, B., Sinha, T., Jiao, K., Serra, R., Wang, J., 2011. Disruption of PCP signaling causes limb morphogenesis and skeletal defects and may underlie Robinow syndrome and brachydactyly type B. Hum. Mol. Genet. 20, 271–285.

Wang, J., Zhou, J., Bondy, C.A., 1999. Igf1 promotes longitudinal bone growth by insulin-like actions augmenting chondrocyte hypertrophy. FASEB J. 13, 1985–1990.

Wang, L., Huang, J., Moore, D.C., Zuo, C., Wu, Q., Xie, L., Von Der Mark, K., Yuan, X., Chen, D., Warman, M.L., Ehrlich, M.G., Yang, W., 2017. SHP2 regulates the osteogenic fate of growth plate hypertrophic chondrocytes. Sci. Rep. 7, 12699.

Wang, R.N., Green, J., Wang, Z., Deng, Y., Qiao, M., Peabody, M., Zhang, Q., Ye, J., Yan, Z., Denduluri, S., Idowu, O., Li, M., Shen, C., Hu, A., Haydon, R.C., Kang, R., Mok, J., Lee, M.J., Luu, H.L., Shi, L.L., 2014b. Bone Morphogenetic Protein (BMP) signaling in development and human diseases. Genes Dis. 1, 87–105.

Wang, W., Rigueur, D., Lyons, K.M., 2014c. TGFbeta signaling in cartilage development and maintenance. Birth Defects Res. C Embryo Today 102, 37–51.

Wang, Y., Li, Y.P., Paulson, C., Shao, J.Z., Zhang, X., Wu, M., Chen, W., 2014d. Wnt and the Wnt signaling pathway in bone development and disease. Front. Biosci. 19, 379–407.

Watanabe, T., Sato, T., Amano, T., Kawamura, Y., Kawamura, N., Kawaguchi, H., Yamashita, N., Kurihara, H., Nakaoka, T., 2008. Dnm3os, a non-coding RNA, is required for normal growth and skeletal development in mice. Dev. Dynam. 237, 3738–3748.

Weir, E.C., Philbrick, W.M., Amling, M., Neff, L.A., Baron, R., Broadus, A.E., 1996. Targeted overexpression of parathyroid hormone-related peptide in chondrocytes causes chondrodysplasia and delayed endochondral bone formation. Proc. Natl. Acad. Sci. U.S.A. 93, 10240–10245.

Wellik, D.M., 2007. Hox patterning of the vertebrate axial skeleton. Dev. Dynam. 236, 2454–2463.

Whitaker, A.T., Berthet, E., Cantu, A., Laird, D.J., Alliston, T., 2017. Smad4 regulates growth plate matrix production and chondrocyte polarity. Biol. Open 6, 358–364.

Wijgerde, M., Karp, S., Mcmahon, J., Mcmahon, A.P., 2005. Noggin antagonism of BMP4 signaling controls development of the axial skeleton in the mouse. Dev. Biol. 286, 149–157.

Williams, R., Nelson, L., Dowthwaite, G.P., Evans, D.J., Archer, C.W., 2009. Notch receptor and Notch ligand expression in developing avian cartilage. J. Anat. 215, 159–169.

Winnier, G.E., Hargett, L., Hogan, B.L., 1997. The winged helix transcription factor MFH1 is required for proliferation and patterning of paraxial mesoderm in the mouse embryo. Genes Dev. 11, 926–940.

Winslow, B.B., Takimoto-Kimura, R., Burke, A.C., 2007. Global patterning of the vertebrate mesoderm. Dev. Dynam. 236, 2371–2381.

Wit, J.M., Camacho-Hubner, C., 2011. Endocrine regulation of longitudinal bone growth. Endocr. Dev. 21, 30–41.

Witte, F., Dokas, J., Neuendorf, F., Mundlos, S., Stricker, S., 2009. Comprehensive expression analysis of all Wnt genes and their major secreted antagonists during mouse limb development and cartilage differentiation. Gene Expr. Patterns 9, 215–223.

Wolpert, L., 1969. Positional information and the spatial pattern of cellular differentiation. J. Theor. Biol. 25, 1–47.

Wright, E., Hargrave, M.R., Christiansen, J., Cooper, L., Kun, J., Evans, T., Gangadharan, U., Greenfield, A., Koopman, P., 1995. The Sry-related gene Sox9 is expressed during chondrogenesis in mouse embryos. Nat. Genet. 9, 15–20.

Wu, M., Chen, G., Li, Y.P., 2016. TGF-beta and BMP signaling in osteoblast, skeletal development, and bone formation, homeostasis and disease. Bone Res. 4, 16009.

Wu, X., Shi, W., Cao, X., 2007. Multiplicity of BMP signaling in skeletal development. Ann. N.Y. Acad. Sci. 1116, 29–49.

Wuelling, M., Pasdziernik, M., Moll, C.N., Thiesen, A.M., Schneider, S., Johannes, C., Vortkamp, A., 2013. The multi zinc-finger protein Trps1 acts as a regulator of histone deacetylation during mitosis. Cell Cycle 12, 2219–2232.

Wuthier, R.E., Lipscomb, G.F., 2011. Matrix vesicles: structure, composition, formation and function in calcification. Front. Biosci. 16, 2812–2902.

Yamaguchi, T.P., Bradley, A., Mcmahon, A.P., Jones, S., 1999. A Wnt5a pathway underlies outgrowth of multiple structures in the vertebrate embryo. Development 126, 1211–1223.

Yan, Y., Frisen, J., Lee, M.H., Massague, J., Barbacid, M., 1997. Ablation of the CDK inhibitor p57Kip2 results in increased apoptosis and delayed differentiation during mouse development. Genes Dev. 11, 973–983.

Yang, G., Zhu, L., Hou, N., Lan, Y., Wu, X.M., Zhou, B., Teng, Y., Yang, X., 2014a. Osteogenic fate of hypertrophic chondrocytes. Cell Res. 24, 1266–1269.

Yang, L., Lawson, K.A., Teteak, C.J., Zou, J., Hacquebord, J., Patterson, D., Ghatan, A.C., Mei, Q., Zielinska-Kwiatkowska, A., Bain, S.D., Fernandes, R.J., Chansky, H.A., 2013a. ESET histone methyltransferase is essential to hypertrophic differentiation of growth plate chondrocytes and formation of epiphyseal plates. Dev. Biol. 380, 99–110.

Yang, L., Tsang, K.Y., Tang, H.C., Chan, D., Cheah, K.S., 2014b. Hypertrophic chondrocytes can become osteoblasts and osteocytes in endochondral bone formation. Proc. Natl. Acad. Sci. U.S.A. 111, 12097–12102.

Yang, S., Kim, J., Ryu, J.H., Oh, H., Chun, C.H., Kim, B.J., Min, B.H., Chun, J.S., 2010. Hypoxia-inducible factor-2alpha is a catabolic regulator of osteoarthritic cartilage destruction. Nat. Med. 16, 687–693.

Yang, T., Bassuk, A.G., Fritzsch, B., 2013b. Prickle1 stunts limb growth through alteration of cell polarity and gene expression. Dev. Dynam. 242, 1293–1306.

Yang, Y., Topol, L., Lee, H., Wu, J., 2003. Wnt5a and Wnt5b exhibit distinct activities in coordinating chondrocyte proliferation and differentiation. Development 130, 1003–1015.

Yang, Y., Wang, B., 2015. Disruption of beta-catenin binding to parathyroid hormone (PTH) receptor inhibits PTH-stimulated ERK1/2 activation. Biochem. Biophys. Res. Commun. 464, 27–32.

Yano, F., Kugimiya, F., Ohba, S., Ikeda, T., Chikuda, H., Ogasawara, T., Ogata, N., Takato, T., Nakamura, K., Kawaguchi, H., Chung, U.I., 2005. The canonical Wnt signaling pathway promotes chondrocyte differentiation in a Sox9-dependent manner. Biochem. Biophys. Res. Commun. 333, 1300–1308.

Yano, F., Saito, T., Ogata, N., Yamazawa, T., Iino, M., Chung, U.I., Kawaguchi, H., 2013. beta-catenin regulates parathyroid hormone/parathyroid hormone-related protein receptor signals and chondrocyte hypertrophy through binding to the intracellular C-terminal region of the receptor. Arthritis Rheum. 65, 429–435.

Yi, S.E., Daluiski, A., Pederson, R., Rosen, V., Lyons, K.M., 2000. The type I BMP receptor BMPRIB is required for chondrogenesis in the mouse limb. Development 127, 621–630.

Yoon, B.S., Ovchinnikov, D.A., Yoshii, I., Mishina, Y., Behringer, R.R., Lyons, K.M., 2005. Bmpr1a and Bmpr1b have overlapping functions and are essential for chondrogenesis in vivo. Proc. Natl. Acad. Sci. U.S.A. 102, 5062–5067.

Yoshida, C.A., Yamamoto, H., Fujita, T., Furuichi, T., Ito, K., Inoue, K., Yamana, K., Zanma, A., Takada, K., Ito, Y., Komori, T., 2004. Runx2 and Runx3 are essential for chondrocyte maturation, and Runx2 regulates limb growth through induction of Indian hedgehog. Genes Dev. 18, 952–963.

Yoshida, T., Vivatbutsiri, P., Morriss-Kay, G., Saga, Y., Iseki, S., 2008. Cell lineage in mammalian craniofacial mesenchyme. Mech. Dev. 125, 797–808.

Yu, K., Xu, J., Liu, Z., Sosic, D., Shao, J., Olson, E.N., Towler, D.A., Ornitz, D.M., 2003. Conditional inactivation of FGF receptor 2 reveals an essential role for FGF signaling in the regulation of osteoblast function and bone growth. Development 130, 3063–3074.

Yuasa, T., Kondo, N., Yasuhara, R., Shimono, K., Mackem, S., Pacifici, M., Iwamoto, M., Enomoto-Iwamoto, M., 2009. Transient activation of Wnt/{beta}-catenin signaling induces abnormal growth plate closure and articular cartilage thickening in postnatal mice. Am. J. Pathol. 175, 1993–2003.

Yusuf, F., Brand-Saberi, B., 2006. The eventful somite: patterning, fate determination and cell division in the somite. Anat. Embryol. 211 (Suppl. 1), 21–30.

Zaka, R., Williams, C.J., 2006. Role of the progressive ankylosis gene in cartilage mineralization. Curr. Opin. Rheumatol. 18, 181–186.

Zakany, J., Duboule, D., 2007. The role of Hox genes during vertebrate limb development. Curr. Opin. Genet. Dev. 17, 359–366.

Zeller, R., Lopez-Rios, J., Zuniga, A., 2009. Vertebrate limb bud development: moving towards integrative analysis of organogenesis. Nat. Rev. Genet. 10, 845–858.

Zelzer, E., Glotzer, D.J., Hartmann, C., Thomas, D., Fukai, N., Soker, S., Olsen, B.R., 2001. Tissue specific regulation of VEGF expression during bone development requires Cbfa1/Runx2. Mech. Dev. 106, 97–106.

Zelzer, E., Mamluk, R., Ferrara, N., Johnson, R.S., Schipani, E., Olsen, B.R., 2004. VEGFA is necessary for chondrocyte survival during bone development. Development 131, 2161–2171.

Zhang, J., Tan, X., Li, W., Wang, Y., Wang, J., Cheng, X., Yang, X., 2005. Smad4 is required for the normal organization of the cartilage growth plate. Dev. Biol. 284, 311–322.

Zhang, M., Xie, R., Hou, W., Wang, B., Shen, R., Wang, X., Wang, Q., Zhu, T., Jonason, J.H., Chen, D., 2009. PTHrP prevents chondrocyte premature hypertrophy by inducing cyclin-D1-dependent Runx2 and Runx3 phosphorylation, ubiquitylation and proteasomal degradation. J. Cell Sci. 122, 1382–1389.

Zhang, P., Jobert, A.S., Couvineau, A., Silve, C., 1998. A homozygous inactivating mutation in the parathyroid hormone/parathyroid hormone-related peptide receptor causing Blomstrand chondrodysplasia. J. Clin. Endocrinol. Metab. 83, 3365–3368.

Zhang, Y., Ma, B., Fan, Q., 2010. Mechanisms of breast cancer bone metastasis. Cancer Lett. 292, 1–7.

Zhao, G.Q., Deng, K., Labosky, P.A., Liaw, L., Hogan, B.L., 1996. The gene encoding bone morphogenetic protein 8B is required for the initiation and maintenance of spermatogenesis in the mouse. Genes Dev. 10, 1657–1669.

Zhao, H., Feng, J., Ho, T.V., Grimes, W., Urata, M., Chai, Y., 2015. The suture provides a niche for mesenchymal stem cells of craniofacial bones. Nat. Cell Biol. 17, 386–396.

Zhao, R., Lawler, A.M., Lee, S.J., 1999. Characterization of GDF-10 expression patterns and null mice. Dev. Biol. 212, 68–79.

Zhou, G., Zheng, Q., Engin, F., Munivez, E., Chen, Y., Sebald, E., Krakow, D., Lee, B., 2006. Dominance of SOX9 function over RUNX2 during skeletogenesis. Proc. Natl. Acad. Sci. U.S.A. 103, 19004–19009.

Zhou, X., Von Der Mark, K., Henry, S., Norton, W., Adams, H., De Crombrugghe, B., 2014. Chondrocytes transdifferentiate into osteoblasts in endochondral bone during development, postnatal growth and fracture healing in mice. PLoS Genet. 10 e1004820.

Zimmerman, L.B., De Jesus-Escobar, J.M., Harland, R.M., 1996. The Spemann organizer signal noggin binds and inactivates bone morphogenetic protein 4. Cell 86, 599–606.

Zoetis, T., Tassinari, M.S., Bagi, C., Walthall, K., Hurtt, M.E., 2003. Species comparison of postnatal bone growth and development. Birth Defects Res. B Dev. Reprod. Toxicol. 68, 86–110.

Chapter 2

Skeletal stem cells: tissue-specific stem/progenitor cells of cartilage, bone, stroma, and marrow adipocytes

Pamela Gehron Robey[1] and Mara Riminucci[2]
[1]National Institute of Dental and Craniofacial Research, National Institutes of Health, Department of Health and Human Services, Bethesda, MD, United States; [2]Department of Molecular Medicine, Sapienza University of Rome, Rome, Italy

Chapter outline

Introduction 45
Developmental origins of bone and skeletal stem cells 47
Germ-layer specifications 47
Patterns of bone formation and development of pericytes/skeletal stem cells 47
The skeletal lineage 48
Regulation of SSC/BMSC fate 51
Hormonal regulation 51
Signaling pathways and transcription factors 52
Epigenetic controls 52
MicroRNAs 53
Cell–cell and cell–substrate interactions, cell shape, and mechanical forces 53
Isolation of SSCs/BMSCs 54
Characterization of SSCs/BMSCs 54
Potency 54
Markers 56
Determination of skeletal stem cell self-renewal 56
The role of SSCs/BMSCs in postnatal bone turnover and remodeling 57
Skeletal stem cells in disease 58
Fibrous dysplasia of bone and the McCune–Albright syndrome 58
Inherited forms of bone marrow failure 60
Role of SSCs/BMSCs in acquired inflammation 62
Skeletal stem cells in tissue engineering 63
Cell sources 64
Scaffolds 64
Skeletal stem cells and regenerative medicine 65
Stem cell and non–stem cell functions of skeletal stem cells 65
Summary 66
Acknowledgments 66
References 66

Introduction

Stem cells are often thought of as a relatively new discovery, yet the existence of some sort of a stem/progenitor cell in bone from either the periosteum or the marrow was suggested as early as the 1860s in studies on bone regeneration by Ollier and Goujon (Goujon, 1869; Ollier, 1867). The concept of a stem cell was further crystalized in the late 1800s through the early 1900s by a group of astute biologists from diverse fields of interest (e.g., Boveri, Häcker, Regaud, Weidenreich, Dantschakoff, Maximow, and others, reviewed in Ramalho-Santos and Willenbring, 2007; Robey, 2000). They hypothesized that the ability of a tissue to maintain itself throughout the lifetime of an organism is based on the existence of a "stem" cell ("Stammzelle," a term first coined in German by Haeckel; Haeckel, 1868) that would remain primarily quiescent and undifferentiated, but also be able to regenerate the functional parenchyma of its tissue of origin following injury or the need for tissue rejuvenation (tissue turnover). Originally thought to exist only in tissues with high rates of turnover, such as blood, skin, and the gastrointestinal tract, evidence suggested that stem cells do exist in tissues with substantially lower rates of turnover, such as bone, with the ability to repair injury. Even tissues never thought to turn

Principles of Bone Biology. https://doi.org/10.1016/B978-0-12-814841-9.00002-6

over or repair, such as brain, do in fact contain stem cells (reviewed in Robey, 2000; see https://stemcells.nih.gov/info/basics/1.htm for more information). It was subsequently thought that virtually every tissue in the body contains some type of a stem/progenitor cell, although the existence of cardiomyocyte stem cells has been called into question (Kretzschmar et al., 2018). Nonetheless, the very notion of stem cells has had a major impact on both biological and medical sciences since the early 1990s, in terms of understanding the dynamics of tissue homeostasis in health and disease and the thought that they could be useful in tissue regeneration.

Following upon the early work of Ollier and Goujon, the hematologists Tavassoli and Crosby, aiming to better understand the support of hematopoiesis, took bone-free fragments of hematopoietic marrow and transplanted them under the kidney capsule. Remarkably, they found that marrow could be regenerated, but only after the formation of bone (Tavassoli and Crosby, 1968). It was Friedenstein et al., later in collaboration with Owen et al., who determined that the origin of bone in those marrow fragments was in fact a subpopulation of nonhematopoietic cells of the stroma upon which hematopoiesis occurs. When single-cell suspensions of bone marrow were plated into tissue culture plastic dishes at low density, a small proportion of single fibroblastic cells rapidly attached and began to proliferate to form a colony in a density-independent fashion. Friedenstein termed the initial adherent single cells as colony-forming units–fibroblasts (CFU-Fs), being cognizant of the fact that many cells could attach, but could not form a colony in a density-independent fashion. Subsequently, when the progeny of CFU-Fs (bone marrow stromal cells, BMSCs) were transplanted in vivo inside diffusion chambers that prevent vascularization (closed system), cartilage was formed in the relatively anaerobic interior of the chamber, and bone formed at the periphery, which was in close proximity to (but not in direct contact with) blood vessels. When single colonies were transplanted in conjunction with a collagen sponge under the kidney capsule with complete access to the vasculature (an open system), some colonies not only were able to form bone, stroma, and marrow adipocytes of donor origin, but also supported the formation of hematopoietic tissue of recipient origin (an ectopic bone/marrow organ) (Friedenstein et al., 1974). From this clonal analysis (and notice the emphasis on clones; i.e., the progeny of a single cell), Friedenstein and Owen proposed the existence of the first multipotent tissue-specific stem/progenitor cell from a solid connective tissue, which they termed a bone marrow stromal stem cell, able to differentiate into four different phenotypes: chondrogenic, osteoblastic, stromogenic, and adipogenic (Fig. 2.1) (Owen and Friedenstein, 1988). Later, the term "skeletal stem cell" (SSC) was adopted based on this ability to re-create all of the cells found in skeletal tissue proper (reviewed in Bianco and Robey, 2004, 2015; Owen and Friedenstein, 1988). Of note, this groundbreaking work clearly demonstrated that the stroma created by these stem/progenitor cells transfers the hematopoietic microenvironment (Friedenstein et al., 1974), contributing to the concept of a "niche" as later formulated by Schofield (1978).

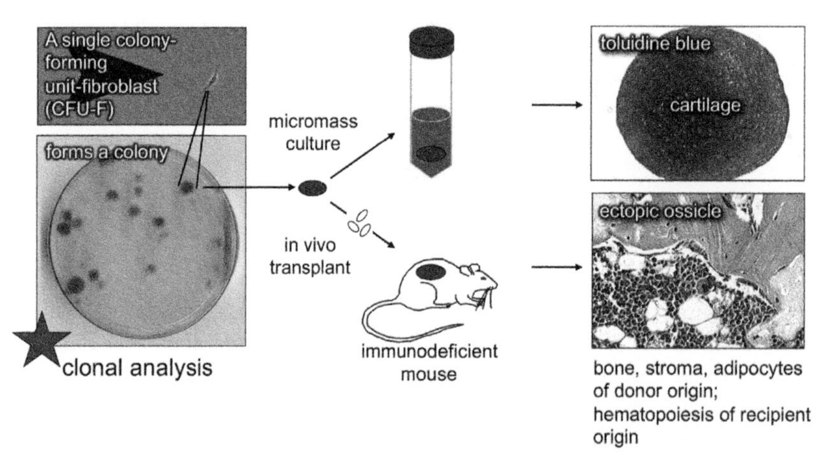

FIGURE 2.1 Proof of the existence of a postnatal multipotent skeletal stem cell. When single-cell suspensions of bone marrow are plated at low density, a single cell (colony-forming unit–fibroblast) rapidly attaches and proliferates to form a colony of bone marrow stromal cells. Clonal analysis is performed by placing clonally derived strains: (1) into micromass cultures (anaerobic) to observe the formation of cartilage or (2) onto appropriate scaffolds followed by subcutaneous transplantation into immunocompromised mice to observe the formation of a bone/marrow organ. This ectopic ossicle is composed of bone, hematopoiesis-supportive stroma, and marrow adipocytes of donor origin and hematopoiesis of recipient origin. *Modified from Bianco, P., Kuznetsov, S., Riminucci, M., Robey, P.G., 2006. Postnatal skeletal stem cells. Meth Enzymol 419, 117–148 and Bianco, P., Robey, P.G., Penessi, G., 2008a. Cell source. In: de Boer, J., van Blitterswijk, C., Thomsen, P., Hubbell, J., Cancedda, R., de Bruijn, J.D., Lindahl, A., Sohler, J., Williams, D.F. (Eds.), Tissue Engineering. Elsevier, Amsterdam, pp. 279–306.*

Developmental origins of bone and skeletal stem cells

Germ-layer specifications

During embryonic development, the body pattern is first laid out by gastrulation of the inner cell mass of the blastocyst (the source of embryonic stem cells) to form germ layers for definitive endoderm, mesoderm, and ectoderm (which includes neural crest). Bone is formed by three different specifications of mesoderm and ectoderm (Olsen et al., 2000). Temporary embryonic cartilages in the cranium are formed from paraxial mesoderm by a process of cartilage regression followed by intramembranous bone formation (Holmbeck et al., 2003). Paraxial mesoderm also forms part of the dorsal cranial vault through primarily intramembranous bone formation, and the axial skeleton through primarily endochondral bone formation. Somatic lateral plate mesoderm forms the appendicular skeleton through intramembranous and endochondral bone formation, and last, neural crest forms the frontal cranium and facial bones through both intramembranous and endochondral pathways (Berendsen and Olsen, 2015; Olsen et al., 2000). It has also been suggested that the dorsal root of the aorta gives rise to skeletal cells through a process whereby specialized angioblasts (mesoangioblasts) bud cells into the extravascular space that go on to form connective tissues, including skeletal tissue (Bianco and Cossu, 1999), analogous to hemangioblasts budding cells into the vessel lumen to form definitive hematopoietic stem cells (HSCs) (Dzierzak and Medvinsky, 2008). Consequently, there are at least three (and possibly four) embryonic sources of skeletal tissue; that is, there is no single common skeletal stem/progenitor cell during embryonic development (Fig. 2.2A and B).

During development of the axial skeleton, the primitive streak (the precursor of mesoderm and endoderm) migrates caudally to form mesendoderm. Cephalic mesendoderm further specifies into paraxial and lateral plate mesoderm. Paraxial mesoderm segments into somites with definitive borders (Pourquie, 2001). Somitomeres with less definitive borders are formed in the head region (Bothe et al., 2011). Somites further differentiate into the dermatome that will form the dermis upon which ectoderm will form skin, the myotome that will form skeletal muscle, the syndetome that will form tendons and ligaments, and the most ventral part of the somite, the sclerotome, which will form the axial skeleton. Lateral plate mesoderm undergoes a splitting process to form the intraembryonic coelom (future peritoneum) and the splanchnic and somatic lateral plate mesoderm, the latter of which goes on to form the appendicular skeleton along with part of the body wall and dermis. Splanchnic mesoderm forms viscera, the heart, and associated blood vessels, including the dorsal root of the aorta (Onimaru et al., 2011). Last, following delamination from the border of the neural plate, neural crest cells migrate into the branchial arches and form cartilage and bone of the head, all epidermal pigment cells, and most of the peripheral nervous system in the trunk (Bronner, 2015) (Fig. 2.2A). It is generally thought that neural crest cells do not contribute to skeletal tissues below the neck; however, based on the expression of the neural crest marker, Wnt1, and expression of Nestin, this concept is currently being questioned (Danielian et al., 1998; Isern et al., 2014).

Patterns of bone formation and development of pericytes/skeletal stem cells

The different patterns of skeletal tissue formation differ from one another based on the involvement (or lack thereof) of cartilage (Fig. 2.2B). During intramembranous bone formation, there is no cartilage template. Embryonic mesenchymal condensations organize into two opposing layers of committed osteoprogenitors that begin to secrete and deposit bone matrix proteins into a sequestered area between the two layers of cells. Due to the activity of alkaline phosphatase on the surface of the committed osteogenic cells, pyrophosphate (a mineralization inhibitor) is cleaved to provide free phosphate for precipitation of a carbonate-rich apatite. In endochondral bone formation, embryonic mesenchyme condenses and differentiates into a cartilage template, the interior portion of which undergoes hypertrophy and calcification, while the outer surface (perichondrium) becomes committed to osteogenesis (periosteum). In this case, bone formation occurs in two ways: (1) between a fully committed osteogenic layer of cells and a layer of hypertrophic chondrocytes, in the bony collar, and (2) through cartilage hypertrophy and subsequent replacement by bone (Hall, 2015). In another pattern of skeletal tissue formation, best exemplified in the cranium, embryonic cartilage templates undergo regression rather than hypertrophy. These cartilages serve as a pathway to guide the formation of new bone by an intramembranous process, and ultimately disappear (Holmbeck et al., 2003) (Fig. 2.2B).

Irrespective of the pattern of formation, newly formed capillaries invade the developing bone, but lack pericytes, a cell type that has numerous cell processes that wrap around developing blood vessels to provide stability (Armulik et al., 2011). As the "naked" blood vessels begin to invade the newly formed woven bone and hypertrophic cartilage (in the case of endochondral bone formation), they form an association with committed osteoprogenitors (and possibly with hypertrophic chondrocytes) based on their expression of fibroblastic cell surface proteins. Because the osteogenic cells are no longer opposing a cell layer that is secreting bone matrix proteins, they undergo a shape change and downregulate expression of bone matrix proteins due to their association with endothelial cells, and become pericytes, cells that wrap around blood

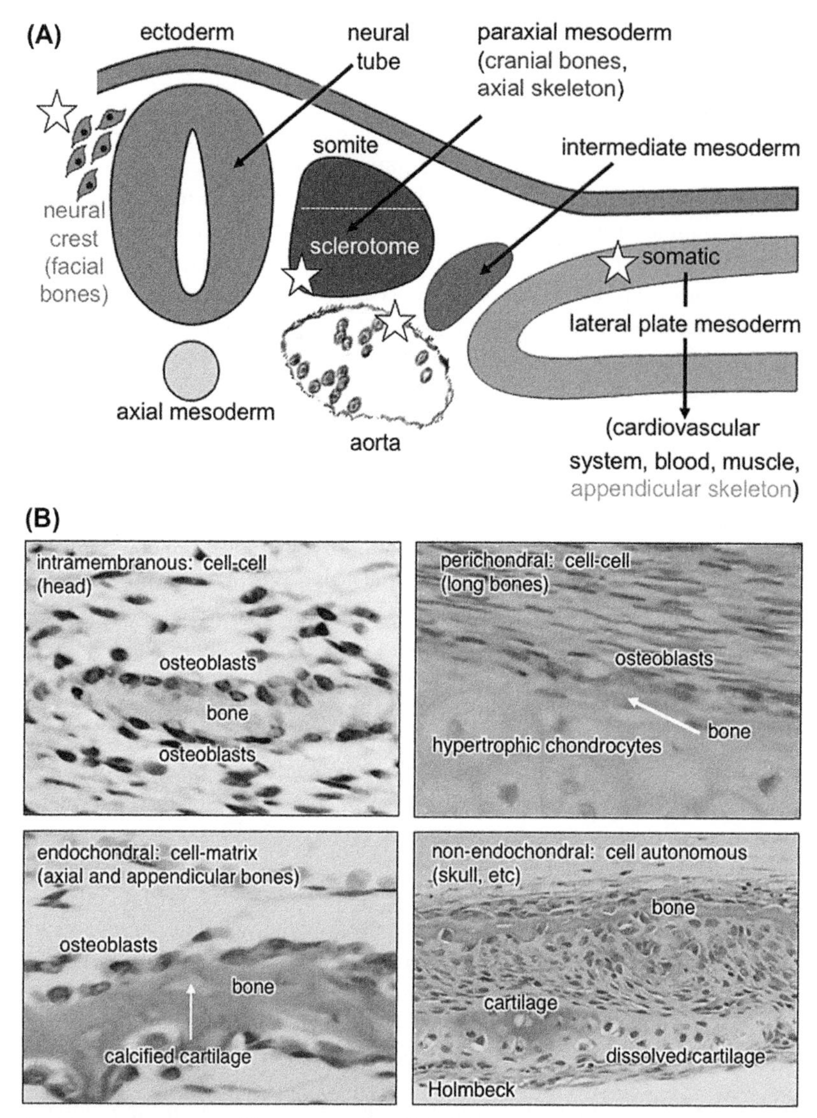

FIGURE 2.2 Development origins and patterns of bone formation. (A) During embryonic development, bone is formed by neural crest cells (ectoderm in origin) and two different specifications of mesoderm (paraxial mesoderm and somatic lateral plate mesoderm). It is also thought that mesangioblasts in the dorsal root of the developing embryo shed cells into the extravascular spaces to give rise to embryonic mesoderm. (B) Bone is formed through several different processes: (1) intramembranous (between two layers of osteoblastic cells), (2) perichondral (between a layer of osteoblastic and hypertrophic cells), (3) endochondral (calcified cartilage is replaced by woven bone), and (4) a nonendochondral replacement process (cartilage regresses, leading the way for intramembranous bone). *(A) Modified from Bianco, P., Robey, P.G., 2015. Skeletal stem cells. Development 142, 1023–1027.*

vessels to provide stability (Fig. 2.3). These pericytic cells proliferate along with the blood vessels into the interior of the bone, with osteoclasts leading the way to carve out the future marrow spaces. The pericytic cells proliferate further into the newly liberated space to form the prehematopoietic bone marrow stroma. Once sufficient bone has formed on the exterior of the rudiment to provide a protected environment, HSCs migrate through the circulation from the thymus, liver, and spleen to the bone, and escape from the circulation to establish definitive hematopoiesis in the bone marrow, the final home of the HSCs under normal conditions (Bianco et al., 1999, 2008b).

The skeletal lineage

As described earlier, embryonic skeletogenic mesenchyme, which contains the most primitive SSC, forms cartilage, bone, bone marrow stroma, and marrow adipocytes in a sequential fashion in time and space. Thus, during fetal development, chondrocytes, osteoblasts, SSCs/BMSCs, and marrow adipocytes can be seen as different stages of maturation rather than

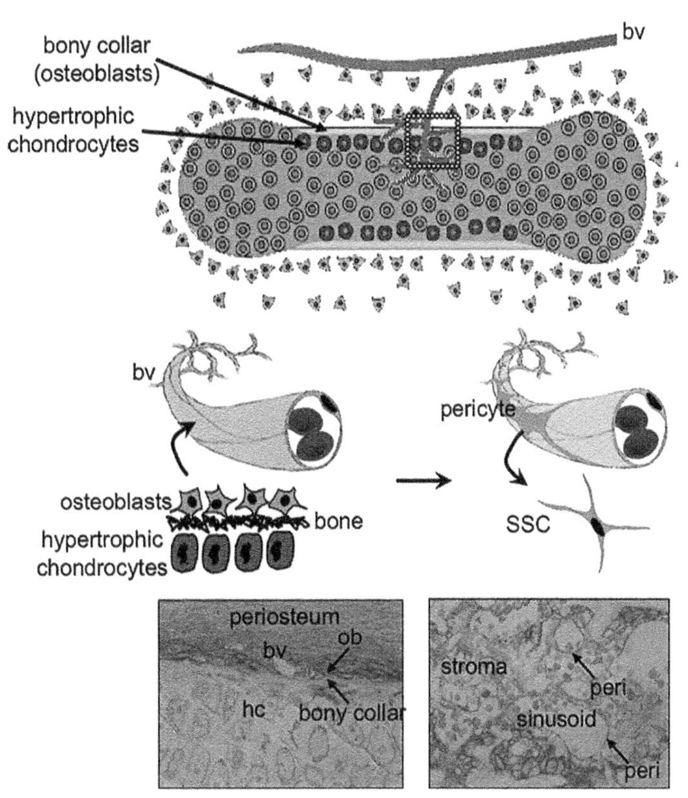

FIGURE 2.3 **Development of pericytes/skeletal stem cells.** As bone develops (in a long bone, as an example), blood vessels that are devoid of pericytes (cells surrounding blood vessels to give them stability) invade the developing rudiment, passing through the bony collar and hypertrophic cartilage. In doing so, ALP^+ osteogenic cells (colored blue in the cartoons and photomicrographs) from the perichondrium are relocated into the forming marrow cavity and become primitive stromal cells that coat the forming sinusoids as pericytes, as can be seen in rodents at embryonic day 18. The pericytes (a subset of which are skeletal stem cells) remain quiescent on the blood vessel surfaces until released by injury or the need for bone turnover. *bv*, blood vessel; *hc*, hypertrophic cartilage; *ob*, osteoblast; *peri*, pericyte; *SSC*, skeletal stem cell. *Modified from Bianco, P., Robey, P.G., 2015. Skeletal stem cells. Development 142, 1023–1027.*

as separate lineages (Bianco and Robey, 2015) (Fig. 2.4A). Chondrogenesis, osteogenesis, and adipogenesis are controlled by three master transcription factors: Sox 9 (chondrogenic) (Bi et al., 1999), Runx2 (osteogenic and, to a lesser extent, chondrogenic) (Komori et al., 1997), and peroxisome proliferator-activated receptor γ2 (PPARγ2) (adipogenic) (reviewed in Muruganandan et al., 2009). In the case of cartilage, there are two potential fates: to undergo hypertrophy and be replaced by bone or to remain as hyaline cartilage on the articular surfaces of long bones in joints and in the ears and nose. The factors that maintain cartilage in a hyaline state are not well delineated as of this writing. Knowledge of this key point of regulation is of high interest based on the need to make new cartilage from an appropriate cell type that will not undergo hypertrophy, such as periosteal cells or SSCs/BMSCs.

In the context of embryonic development, a subset of committed osteogenic cells is recruited by blood vessels to form pericytes, which subsequently form stroma that supports hematopoiesis (Bianco et al., 1999, 2008b; Maes et al., 2010). As longitudinal bone growth slows and establishment of hematopoiesis is at a sufficient level to support the organism, stromal cells convert into marrow adipocytes to form yellow marrow, primarily in the distal regions of long bones. During postnatal growth (modeling), homeostasis, and aging (remodeling), cells in bone marrow stroma give rise to bone, hematopoiesis-supportive stroma, and marrow adipocytes (but rarely cartilage under normal conditions), and these tissues are thought to emanate from a single common primordial precursor, the SSC (Fig. 2.4B). Consequently, whether it be during embryonic development, postnatal growth, or adult homeostasis, there is a lineage that gives rise to skeletal tissues (chondrocytes, osteogenic cells, stroma, and marrow adipocytes), not multiple different precursor cells that give rise to different phenotypes. This has implications for the maintenance of skeletal tissues and their adaptations. Cells in the skeletal lineage are flexible; i.e., they can convert from one cell phenotype to another based on changes in the microenvironment. For example, it is well known that with aging, there is an increase in marrow adipocytes (yellow marrow) at the expense of osteogenic differentiation and support of hematopoiesis (red marrow) (Bianco and Riminucci, 1998).

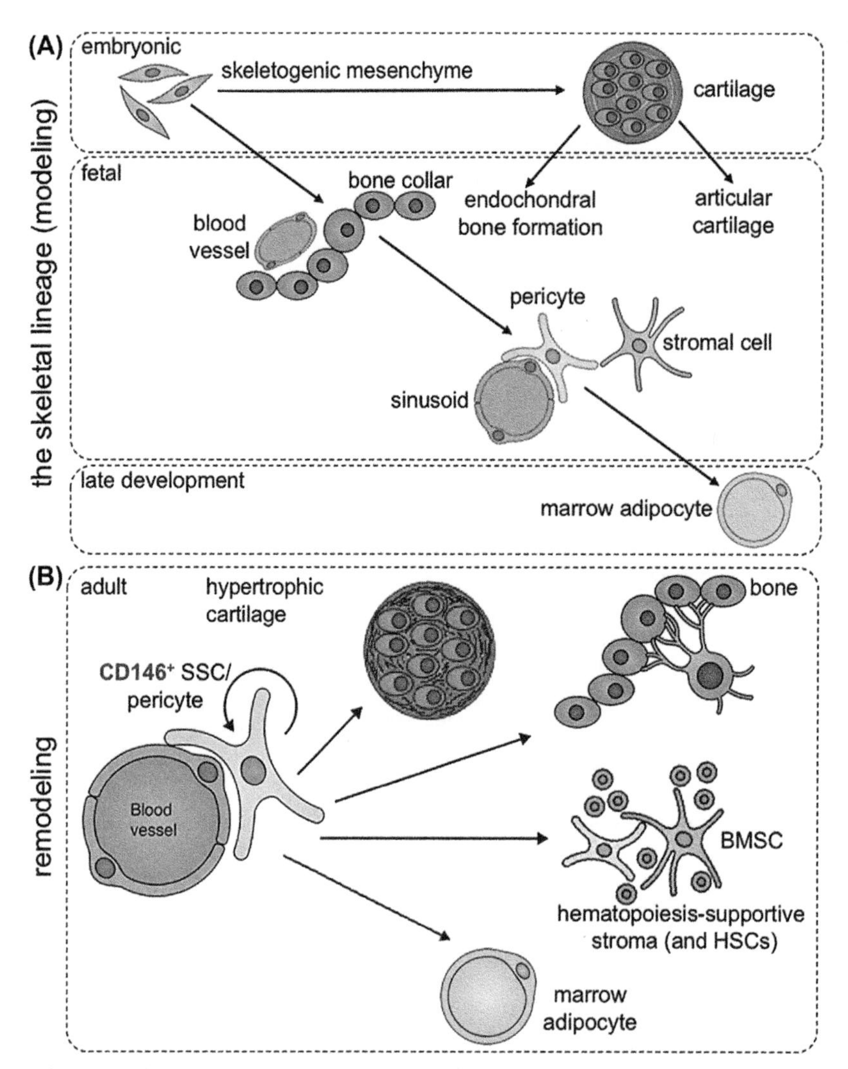

FIGURE 2.4 The skeletal lineage during embryonic development (modeling) and postnatal homeostasis (remodeling). (A) During embryonic development and postnatal maturation (modeling), cells emerging from skeletogenic mesenchyme sequentially form cartilage or bone. Cartilage either undergoes hypertrophy and is replaced by bone or remains as articular cartilage in joints. Osteogenic cells are recruited by blood vessels to form pericytes and stromal cells, which support hematopoiesis. At late stages of postnatal maturation, stromal cells accumulate fat and become marrow adipocytes. (B) In the mature postnatal organism, a subset of pericytes are skeletal stem cells that are capable of forming cartilage (rarely), bone, hematopoiesis-supportive stroma, or marrow adipocytes during skeletal remodeling. *BMSC*, bone marrow stromal cell; *HSC*, hematopoietic stem cell; *SSC*, skeletal stem cell. *Modified from Bianco, P., Robey, P.G., 2015. Skeletal stem cells. Development 142, 1023–1027.*

Cartilage is not normally seen in healthy bone marrow, and during fracture repair, the callus emanates from cells in the periosteum rather than from bone marrow stroma (O'Driscoll and Fitzsimmons, 2001). These periosteal cells are similar to, but not identical with, SSCs from bone marrow (Duchamp de Lageneste et al., 2018; Sacchetti et al., 2016). The role of marrow SSCs/BMSCs is probably to mediate remodeling of the bone in the fracture callus to restore the medullary cavity (reviewed in Einhorn and Gerstenfeld, 2015). This calls into question why SSCs/BMSCs would have the ability to make cartilage at all. The answer may be that during bone development, as blood vessels invade through the bony collar, they capture not only committed osteogenic cells, but hypertrophic chondrocytes as well, and orchestrate their conversion into pericytes. The association with endothelial cells may prevent their demise, but they may also retain a chondrogenic memory, to be recalled when released from the endothelial cell surface. Of note, it is most often reported that cartilage formed by SSCs/BMSCs undergoes hypertrophy and is converted into bone. This notion is supported by reports that not all hypertrophic chondrocytes die, and that some become osteogenic (Holmbeck et al., 2003; Muraglia et al., 2003; Roach et al., 1995; Scotti et al., 2013; Serafini et al., 2014), as has been rediscovered by Yang et al. (2014).

Regulation of SSC/BMSC fate

In postnatal life, SSCs are quiescent due to their association with blood vessels, until liberated by the need for bone turnover or injury. Upon liberation from blood vessel walls during bone turnover or following local injury, SSCs have the possibility of forming osteogenic cells, adipogenic cells, or more hematopoiesis-supportive stroma (Fig. 2.4B). The balance between osteogenic and adipogenic differentiation has received a large amount of attention due to its relevance to bone fragility and osteoporosis (Chen et al., 2016; Devlin and Rosen, 2015; James, 2013; Meyer et al., 2016; Teven et al., 2011). On the other hand, there is little information on how the SSC is directed into becoming a hematopoiesis-supporting stromal cell that is not a stem/progenitor cell. There are clear examples of the expansion of hematopoiesis, which relies on an increase in stroma, such as in hemolytic diseases where bone is lost to accommodate increased blood formation (Bianco and Riminucci, 1998). However, it has not been determined if the increase in hematopoiesis-supportive BMSCs is matched by an increase in SSCs. It has been shown that excess production of erythropoietin by platelet-derived growth factor subunit B-positive (PDGFB^{+}) cells (which include SSCs/BMSCs) causes expansion of red marrow (and thus, expansion of stroma) at the expense of bone and marrow adipocytes. However, the colony-forming-efficiency assay, the closest approximation of the number of SSCs available as of this writing, did not show a change in the number of colonies in these mice (Suresh et al., 2019). This suggests that expansion of the stromogenic pool does not necessarily equate with expansion of the SSC population.

Regulation of cell fate is complex, and is controlled by many factors, including hormones, cytokines, growth factors, and their downstream signaling pathways, all of which have an impact on the epigenetic pattern of the genome and subsequent transcriptional activity. Further control arises from posttranscriptional modifications (splice variants and isoforms), translational variation (small noncoding RNAs), and posttranslational modifications (reviewed in Chen et al., 2016). Mechanical forces and cell—matrix and cell—cell interactions are also influential. What follows is a very brief summary of regulatory factors influencing SSC/BMSC fate. However, it must be noted that there is some confusion in adipogenic fate choices based on the fact that it is considered by many that adipocytes from white or brown fat are equivalent to those in marrow, and they are not. Furthermore, in some cases, it is assumed that factors that induce osteogenic differentiation of SSCs/BMSCs will induce osteogenic differentiation of adipose-derived "mesenchymal stem cells" (and other nonskeletal "MSCs"). These nonskeletal "MSCs" are reported to become osteogenic based on an overreliance on artifactual in vitro assays after heavy treatment with chemical modifiers, and/or molecular engineering, but these nonskeletal "MSCs" are not inherently osteogenic.

Hormonal regulation

SSCs/BMSCs express receptors for many of the hormones that control skeletal growth and homeostasis, such as the parathyroid hormone (PTH) receptor, vitamin D receptor, glucocorticoid receptors, sex hormone receptors, thyroid hormone receptor, etc. If and how the corresponding ligands of these receptors control SSC/BMSC fate have not been closely examined in many cases. However, there are a few instances in which hormonal control or changes in their downstream signaling pathways (briefly described in the following) do have an impact on the fate of SSCs/BMSCs. For example, in hyperparathyroidism, it is thought that the ensuing endosteal fibrosis is due to the enhanced proliferation of stromal progenitor cells that accumulate and partially differentiate into osteoblastic cells (Bianco and Bonucci, 1991). Along this same line, excess PTH signaling due to a constitutively activating mutation in the PTH/PTH-related protein receptor (H223R, the causative mutation in Jansen's metaphyseal chondrodysplasia; Schipani et al., 1995), expressed specifically in mature osteoblasts in mice, led to the formation of excessive medullary bone due to the proliferation and maturation of SSCs/BMSCs, at the expense of marrow stromal cells and adipocytes (Kuznetsov et al., 2004). Interestingly, this phenotype resolved with age, with the subsequent removal of excess bone and establishment of a marrow cavity composed of stromal cells that were not multipotent SSCs based on in vivo transplantation assays. On the other side of the coin, it has long been known that estrogen deficiency leads to expansion of adipocytes at the expense of hematopoiesis-supportive stroma and bone, especially in females (e.g., Rosen and Bouxsein, 2006), which is thought to reflect a commitment of SSCs to adipogenesis at the expense of osteogenesis (e.g., Georgiou et al., 2012). However, it is also possible that direct conversion of stromal cells (rather than directing the fate of the SSC) into adipocytes contributes to this phenomenon. Likewise, glucocorticoid treatment in vivo leads to conversion of red marrow to yellow marrow, again thought to be due to a switch in fate of SSCs/BMSCs (e.g., Li et al., 2013). Paradoxically, a glucocorticoid, dexamethasone, is commonly used in medium used to osteogenically differentiate SSCs/BMSCs (Robey et al., 2014), implying that other factors are at play in vivo that drive glucocorticoid-mediated adipogenic differentiation of SSCs/BMSCs.

Signaling pathways and transcription factors

The levels of expression of Runx2, the master regulator of osteogenesis, and PPARγ, the master regulator of adipogenesis, are controlled by a number of signaling pathways activated by Wnts, members of the transforming growth factor β (TGFβ)/bone morphogenetic protein (BMP) superfamily, Notch, Hedgehogs, and fibroblast growth factors (FGFs) (and others), with extensive cross talk between these pathways (reviewed in Chen et al., 2012, 2016; Cook and Genever, 2013; Lin and Hankenson, 2011). Generally speaking, these pathways increase Runx2 expression with a concomitant decrease in PPARγ, or vice versa, during commitment and differentiation of SSCs. Runx2 is the first upregulated gene during osteogenic commitment and, in turn, upregulates Osterix, which along with Runx2 is essential for osteogenesis. Adipogenesis is initiated by upregulation of CCAAT/enhancer binding protein β (C/EBPβ) and C/EBPδ, which in turn induces PPARγ2. PPARγ2 upregulates C/EBPα, which maintains PPARγ2 expression via a positive feedback loop (reviewed in Cook and Genever, 2013).

Perhaps one of the most important signaling pathways that promotes osteogenesis of SSCs at the expense of adipogenesis is the Wnt/β-catenin pathway, which stimulates Runx2 expression and inhibits C/EBPα, thereby inhibiting adipogenesis (Monroe et al., 2012; Taipaleenmaki et al., 2011; reviewed in Cook and Genever, 2013). Members of the TGFβ superfamily also exert major influences mediated by pSmads 2/3, and have multiple effects on differentiation of SSCs/BMSCs. While TGFβs stimulate proliferation, they inhibit osteogenic differentiation (Alliston et al., 2001) and adipogenic differentiation (Kumar et al., 2012). In addition, BMPs play a major role in osteogenic commitment via pSmads 1/5/8 and Msx/Dlx homeoproteins to increase expression of Runx2 (reviewed in Cook and Genever, 2013). However, BMPs also appear to play a role in early adipogenic commitment, although the mechanisms are not clear. Adipogenic differentiation may be related to the type of BMP and its concentration and/or the type of receptors that are present on the cells at different stages of commitment (reviewed in Chen et al., 2012, 2016). It is generally reported that Notch signaling mediated by Delta/Jagged suppresses osteogenic differentiation (Zanotti and Canalis, 2016) and maintains SSCs in a primordial state (Hilton et al., 2008; Tu et al., 2012); however, positive effects on osteoblastogenesis have been reported (reviewed in Chen et al., 2012; Lin and Hankenson, 2011). Notch signaling also appears to have negative and positive effects during adipogenic commitment and differentiation (reviewed in Muruganandan et al., 2009). Indian hedgehog signaling, mediated by the Kinesin family protein Kif7, and the main intracellular Hedgehog pathway regulator, SUFU, is typically thought of in the context of growth plate dynamics, where it promotes osteogenic differentiation (Jemtland et al., 2003). However, in the context of the bone marrow microenvironment, Hedgehog signaling decreases with adipogenic differentiation of SSCs/BMSCs, suggesting that it is a negative regulator of adipogenesis (Fontaine et al., 2008). Hedgehogs may also interact with BMP signaling to promote osteogenic commitment (reviewed in Chen et al., 2016). FGFs can also be osteogenic (FGF2, FGF4, FGF3, FGF9, and FGF19) and adipogenic (FGF1, FGF2, and FGF10) and signal via a number of pathways that include extracellular signal-regulated kinase 1/2 (ERK1/2), p38 mitogen-activated protein kinase, Jun N-terminal kinase (JNK), protein kinase C, and phosphatidylinositol 3-kinase (PI3K) (Ling et al., 2006; Neubauer et al., 2004; reviewed in Chen et al., 2016).

As briefly listed, these are the major players that have been implicated in controlling the fate of SSCs/BMSCs, but there are other pathways initiated by factors such as insulin-like growth factor 1 (IGF-1) and PDGF (liberated from bone matrix during bone resorption), epidermal growth factor, and the transcription factor YAZ (reviewed in Chen et al., 2012, 2016; Cook and Genever, 2013; Lin and Hankenson, 2011). There are likely to be more regulators identified in the future.

Epigenetic controls

It is thought that epigenetic changes induced by signaling pathways operating in SSCs/BMSCs play a major role in fate decisions and stabilization of the osteogenic or adipogenic phenotype (Meyer et al., 2016). Modification of the conformation of chromatin (defined as DNA with bound protein) has an impact on transcription, with open chromatin (euchromatin) being accessible and condensed chromatin (heterochromatin) being inaccessible to transcriptional machinery, transcription factors, and other cofactors. Changes in chromatin architecture are brought about by DNA and histone modification. DNA methylation at CpG dinucleotides and CpG islands by methyltransferases is usually suppressive in nature. In addition to DNA methylation, chromatin structure is also influenced by modifications made to histones bound to DNA, such as methylation and acetylation (and there are others), that occur posttranslationally. Histones 3 and 4 are commonly methylated at specific sites by histone methyltransferases (HMTs). H3K4me3 is usually associated with euchromatin and gene activation, whereas H3K9me3 and H3K27me3 are usually associated with heterochromatin and gene repression. These methyl groups can be removed by histone demethylases (HDMs), and the changes in methylation are very dynamic, promoting flexibility. As an example, it has been demonstrated that formation of H3K27me3 by the

HMT EZH2 promotes adipogenesis of SSCs, whereas its removal by the HDM KDM6A promotes osteogenesis (Hemming et al., 2014). Histone acetylation by histone acetyltransferases and deacetylation by deacetylases (HDACs) is also dynamic. H3K9ac and HRK16ac are generally found in euchromatin and are indicative of transcriptional activation, whereas deacetylation leads to gene inactivation. While use of the HDAC inhibitor trichostatin A was thought to improve osteogenic differentiation of SSCs, it was determined that it profoundly decreased cell proliferation, and did not improve their osteogenic capacity upon in vivo transplantation (de Boer et al., 2006). Further work is needed to demonstrate what role histone acetylation and deacetylation plays in SSC/BMSC cell fate.

MicroRNAs

There are three forms of small noncoding RNAs: short-hairpin RNAs (shRNAs), microRNAs (miRs), and piwi RNAs. shRNAs are generally synthetic, and introduced exogenously (or created by processing of foreign dsRNAs), and piwi RNAs are primarily expressed in germ-line cells (reviewed in Carthew and Sontheimer, 2009; Ha and Kim, 2014); consequently, the role of miRs in controlling cell fate will be briefly summarized. miRs represent a way in which protein expression can be rapidly modified by either blocking translation or inducing the degradation of target mRNAs (reviewed in Jonas and Izaurralde, 2015). Interestingly, regulators of skeletal homeostasis such as BMPs and TGFβs are known to regulate miR expression and processing (reviewed in Lian et al., 2012). Numerous studies have been published indicating that miRs can control cell fate by downregulating the protein level of Runx2 or PPARγ, thereby enhancing adipogenesis or osteogenesis, respectively. Members of the miR-320 family were found to increase during adipogenic differentiation of SSCs/BMSCs by decreasing levels of Runx2 (among other genes) (Hamam et al., 2014). Conversely, miR-20a promotes osteogenesis by targeting PPARγ, as well as inhibitors of the BMP signaling pathway (Zhang et al., 2011). In addition, miR26a was found to decrease in SSCs during rapid bone loss in mice induced by ovariectomy. Overexpression of miR-26a was found to increase osteogenesis by SSCs by reducing levels of Tob1, which is a negative regulator of the BMP/Smad signaling pathway (Li et al., 2015). These are just a few examples, and others can be found in a 2014 review (Clark et al., 2014).

Cell–cell and cell–substrate interactions, cell shape, and mechanical forces

SSCs/BMSCs are not alone in the bone/marrow organ, and their interaction with other cell types undoubtedly influences fate. As already described, interaction with endothelial cells keeps them in a primordial state, and a study has shown that this is due to repression of Osterix (Meury et al., 2006). They also interact with HSCs; however, how HSCs and other hematopoietic cells influence SSC fate is not well known at this time, owing to the rather recent recognition by hematologists that SSCs are members of the HSC niche (reviewed in Ugarte and Forsberg, 2013).

Cell–matrix interactions occur primarily through: (1) integrins (which bind to many of the extracellular matrix components of bone), (2) discoidin domain receptors and urokinase plasminogen activator receptors, which bind to collagens, and (3) other cell surface molecules such as CD44, which binds to osteopontin (a major component of bone matrix) and hyaluronan. Of note, bone marrow is characterized by a preponderance of cell–cell interactions rather than cell–matrix interactions. Marrow contains thin reticular fibers composed primarily of type III collagen, and high levels of hyaluronan, which is gellike in nature (reviewed in Kuter et al., 2007). However, there is no doubt that after isolation from marrow, substrate has a profound influence on cell fate. For example, binding of αVβ1 to osteopontin inhibited adipogenic differentiation and stimulated osteogenic differentiation of murine SSCs, via downregulation of C/EBP expression (Chen et al., 2014). Modulation of cell shape also has an impact on fate. In a study where cell shape was controlled by plating at different densities, it was found that if cells were plated at low density and allowed to spread, the cells were directed toward osteogenesis, whereas plating at high densities, such that the cells remained rounded, induced adipogenesis. These phenomena were found to be mediated by the small GTPase RhoA, which is a regulator of the cytoskeleton (McBeath et al., 2004). There are a number of different methods by which substrates can be patterned, not only in shape, but in their surface character (D'Arcangelo and McGuigan, 2015). Patterns that promoted cytoskeletal contraction via JNK and ERK1/2 activity induced osteogenic differentiation, whereas when cytoskeletal contraction was not supported by a particular pattern, adipogenesis predominated (Kilian et al., 2010). Substrate elasticity and stiffness and the resulting mechanical forces sensed by SSCs/BMSCs control fate as well, with stiffer substrates generally favoring osteogenesis and softer substrates favoring adipogenesis. How substrate interactions control cell fate is extensively reviewed by Guilak et al. (2009). However, again the reader is cautioned that many of these studies purport to show differentiation of SSCs/BMSCs into nonskeletal phenotypes based on morphology and expression of a few markers, without demonstration of functionality. Last, matrix remodeling has a profound effect on the fate of SSCs. A dramatic example is found in the global deletion of the membrane-type matrix metalloproteinase MT1-MMP (which is a true collagenase), in mice. Although mice appear normal at birth, there is a rapid decrease in growth, and by 40 days, they exhibit severe dwarfism and multiple skeletal defects. Transplantation of SSCs/BMSCs from the mutant mice revealed an almost complete inability to re-form a bone/marrow organ, indicating that MT1-MMP activity is essential for SSC self-renewal (Holmbeck et al., 1999). This mouse model is a direct phenocopy of the human vanishing bone disease Winchester syndrome, now known to be caused by an inactivating mutation in MT1-MMP (Evans et al., 2012).

Isolation of SSCs/BMSCs

While the multipotent stem cell found in marrow is most properly called a bone marrow stromal stem cell or an SSC (reviewed in Bianco and Robey, 2004, 2015; Robey, 2017), it has also gone by the name of "mesenchymal stem cell," based on its ability to re-create bone tissues that originate primarily from embryonic mesenchyme (reviewed in Caplan, 2005). However, not all embryonic mesenchyme forms bone. Furthermore, mesenchyme is an embryonic tissue that forms not only connective tissues, but also blood and blood vessels (MacCord, 2012). No postnatal cell, including any kind of "MSC," has been found to form all three of these tissues. But by applying some of the techniques developed for the isolation and characterization of SSCs/BMSCs, numerous studies purported to identify "MSCs" with the ability to form cartilage, bone, and fat in virtually all connective tissues and beyond. However, the true differentiation capacity and ability to self-renew of non−bone marrow "MSCs" have often not been determined using rigorous assays, leading one to believe that "MSCs" from any tissue are equivalent to SSCs/BMSCs. Furthermore, while some non−bone marrow "MSCs" may form bone, usually after treatment with a BMP, or genetic modification, most, if not all, lack the ability to support the formation of marrow. Few publications on nonmarrow "MSCs" comment on the ability to support hematopoiesis; however, examination of the histological results of in vivo transplantation assays highlight the fact that "MSCs" from adipose tissue (Hicok et al., 2004), dental pulp (Gronthos et al., 2000), periodontal ligament (Miura et al., 2003), and muscle (Sacchetti et al., 2016), as examples, often do not form bone based on rigorous criteria as presented in Phillips et al. (Phillips et al., 2014), and do not support the formation of hematopoietic marrow. In evaluating the results of in vivo transplantation with appropriate scaffolds, it is critical to determine that bone with identifiable osteocytes and osteoblasts of donor origin is formed, rather than dystrophic calcification, which can arise in pathological conditions such as some forms of heterotopic ossification (reviewed in Xu et al., 2018).

SSCs/BMSCs can be isolated from iliac crest aspirates, core biopsies, and surgical waste (Robey, 2011; Robey et al., 2014). While aspirates are less invasive than core biopsies, aspirates are contaminated with large amounts of peripheral blood when large volumes are aspirated, even with frequent repositioning of the aspiration needle, and excess peripheral blood can have a negative impact on the growth of BMSCs (Kharlamova, 1975). When core biopsies and surgical waste are available, repeated washing releases high numbers of cells (reviewed in Bianco et al., 2006). Cells from periosteal tissue, generated via enzymatic treatment or by explant cultures (for examples, see O'Driscoll and Fitzsimmons, 2001; Duchamp de Lageneste et al., 2018), also contain SSCs, but they are not identical to those found in bone marrow.

Characterization of SSCs/BMSCs

The definition of a stem cell is often variable, based on properties that are specific to the tissue from which a stem cell has been isolated. However, the two critical definitions of a stem cell are that the progeny of a single cell are able to re-create the functional parenchyma of a tissue (potency) and that the single cell can self-renew (reviewed in Ramalho-Santos and Willenbring, 2007); i.e., that its divisions are controlled in such a way that the stem cell is maintained within the tissue (Fig. 2.5).

Potency

Rigorous potency assays are critical in defining the differentiation potential of a stem cell and are based on clonal analyses rather than the BMSC population as a whole (Fig. 2.1). Studies by a number of groups (Friedenstein et al., 1974; Gronthos et al., 1994; Kuznetsov et al., 1997; Sacchetti et al., 2007; Sworder et al., 2015) characterized the differentiation capacity of clonal strains by combination of single-colony-derived strains with an appropriate scaffold and transplantation into immunocompromised mice in vivo. From these studies, it was determined that only ~10% of the CFU-F-derived clonal strains were able to form a complete ossicle composed of bone, stroma, and marrow adipocytes of donor origin and hematopoiesis of recipient origin (multipotent clones), whereas ~50% of the clones formed only bone and fibrous tissue of donor origin and did not support hematopoiesis (osteogenic clones), and the remainder formed only fibrous tissue

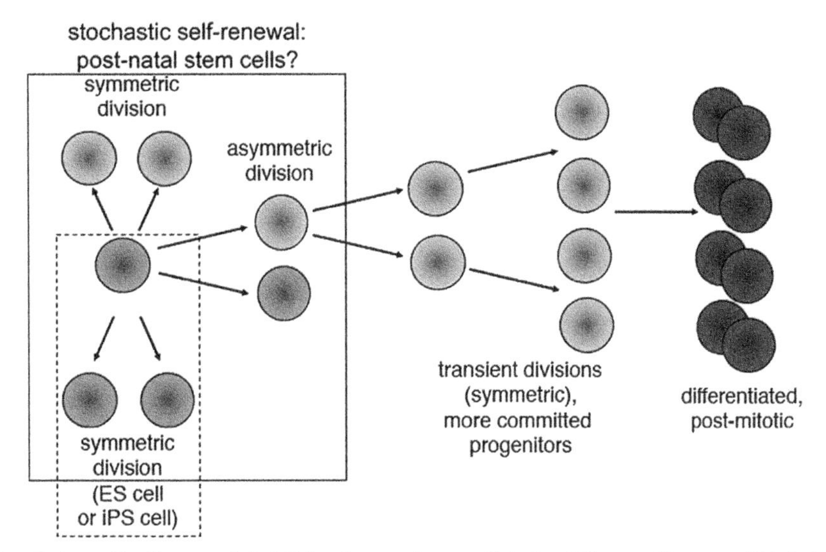

FIGURE 2.5 The kinetics of stem cell self-renewal. A defining feature of stem cells is the ability to self-renew, which is dependent on the type of cell division. Three patterns of stem cell division are possible: (1) asymmetric division in which one daughter cell remains a stem cell and the other is a committed transiently amplifying cell that ultimately differentiates and is postmitotic, (2) symmetric division in which both daughter cells remain as stem cells (as is exemplified by embryonic stem cells and pluripotent stem cells), and (3) symmetric division in which both daughter cells are committed transiently amplifying cells. The exact method by which postnatal stem cells are maintained within a given tissue is not well known, and it has been postulated that postnatal stem cells toggle from one type of division to another, based on the microenvironment and the needs of the tissue (Watt and Hogan, 2000). *ES*, embryonic stem; *iPS*, induced pluripotent stem. *Modified from Bianco, P., Robey, P.G., Penessi, G., 2008a. Cell source. In: de Boer, J., van Blitterswijk, C., Thomsen, P., Hubbell, J., Cancedda, R., de Bruijn, J.D., Lindahl, A., Sohler, J., Williams, D.F. (Eds.), Tissue Engineering. Elsevier, Amsterdam, pp. 279–306.*

(fibroblastic clones). These results indicate that while all of the multipotent clones originated from a single CFU-F, not all CFU-Fs are in fact multipotent. Based on the combination of clonal analysis and in vivo transplantation, one in five of the original CFU-Fs is multipotent, which is the closest approximation to date of the number of SSCs in the total BMSC population.

It must be noted that in vivo transplantation is the gold standard by which to evaluate osteogenic, stromagenic. and adipogenic capacity. Although in vitro assays are often used, they are highly prone to artifact. For the osteogenesis assay, alizarin red S cannot distinguish between dystrophic calcification induced by dead and dying cells versus matrix mineralization. In addition, if the cells make the enzyme alkaline phosphatase, the enzyme cleaves β-glycerophosphate that is in the osteogenic differentiation medium, and when the phosphate concentration in the medium becomes high enough, calcium phosphate precipitates, and it too stains with alizarin red S, but it is not hydroxyapatite. Of interest, there are a number of immortalized cell lines that are used to study the support of hematopoiesis in vitro. However, in vivo transplantation of several of these lines did not result in bone formation, or support hematopoiesis (Chang and Robey, unpublished results). In the adipogenic assay, many cells take up lipid from the serum in the medium and do not synthesize lipids de novo. From these findings, it is clear that in vivo transplantation with an appropriate scaffold is the gold standard by which to assess osteogenic, stromogenic, and adipogenic differentiation.

Conversely, the current gold standard for cartilage formation is the in vivo pellet culture as developed by Johnstone et al. (1998) and Barry et al. (Barron et al., 2015). However, the results of this assay are often misinterpreted. For the chondrogenic pellet culture, one must see bona fide chondrocytes lying in lacunae, surrounded by extracellular matrix that stains purple with toluidine blue (metachromasia). Alcian blue is not specific enough (osteoid stains lightly with Alcian blue), and many reports show pellets of dead cells that stain lightly with Alcian blue. With continued culture of cartilage pellets, the outer surface becomes osteogenic in nature, reminiscent of endochondral bone formation (Muraglia et al., 2003). If cartilage pellets are transplanted in vivo, without any additional scaffold, they remodel into bone either through a hypertrophy pathway (Scotti et al., 2013) or directly into a bone/marrow organ (Serafini et al., 2014). These results harken back to what is seen during development (modeling) and postnatal homeostasis (remodeling): under normal circumstances, SSCs/BMSCs do not form stable cartilage, although they may arise by conversion of a hypertrophic chondrocyte into a pericyte. However, it has been shown that SSCs/BMSCs may be maintained in vivo with an appropriate scaffold, perhaps due to a reawakening of the chondrogenic memory (Kuznetsov et al., 2019).

Markers

Based on their remarkable ability to form cartilage (at least temporarily), and a bone marrow/organ, there have been extensive studies on the characterization of the cell surface markers of human BMSCs, in hopes of identifying a signature that would be useful in isolating SSCs away from more committed BMSCs (reviewed in Bianco et al., 2006; Boxall and Jones, 2012). The consensus is that these cells are uniformly negative for hematopoietic markers such as CD14/LPS receptor, CD34/hematopoietic progenitor cell antigen, and CD45/leukocyte common antigen, and are devoid of many (CD31/PECAM1), but not all, endothelial markers. They are positive for a long list of markers, including (but not limited to) CD13/aminopeptidase N, CD29/ITGB1, CD44/phagocytic glycoprotein 1, CD49a/ITGA1, CD73/ecto-5′-nucleotidase, CD90/Thy-1, CD105/endoglin (also an endothelial marker), CD106/VCAM-2, CD166/ALCAM, Stro1/heat shock protein cognate 70 (a pericyte marker; Fitter et al., 2017; Simmons and Torok-Storb, 1991), and CD146/MCAM (also an endothelial marker) and CD271/LNGFR (Tormin et al., 2011), and there are others. However, these markers are not specific for SSCs/BMSCs and are found on virtually all fibroblastic and/or pericytic cells from many tissues, with some variation depending on the tissue source (Bianco et al., 2008b, 2013). As of this writing, there is no single cell surface marker that can be used to unequivocally identify an SSC, although combinations of markers such as CD146 (Sacchetti et al., 2007) and CD271 (Tormin et al., 2011) may be useful in enriching for SSCs (Robey et al., unpublished results). In addition, cell sorting is best applied to freshly isolated cells, rather than ex vivo-expanded cells. It is known that the expression of many markers changes with time and culture. For example, Stro-1 is known to rapidly decrease with time in culture (Bianco et al., 2008b, 2013).

The aforementioned markers pertain to human cells and in some cases to murine cells. However, there are some significant differences. For example, selection of $CD34^-/CD45^-/CD146^+$ human cells isolates all of the CFU-Fs, whereas murine cells with this phenotype are not CFU-Fs. Selection for CD106/VCAM-1 isolates murine CFU-Fs, but CD146 does not (Chou et al., 2012). With the ability to generate various reporter mice for proper lineage tracing, a number of studies have suggested that Nestin-GFP, Mx1, Lepr, PDGFRα, gremlin (reviewed in Chen et al., 2017, 2018) and CD164/podoplanin (Chan et al., 2018) are potentially relevant markers of murine SSCs.

Determination of skeletal stem cell self-renewal

Friedenstein and Owen, and others that followed, clearly demonstrated that BMSCs fulfilled the first criterion of a stem cell, e.g., they are able to re-create a functional parenchyma (a bone/marrow organ). However, determination of the precise origin and identity came later, along with evidence of self-renewal with the adaptation of a transplantation assay, as has been used by hematologists to determine the presence of HSCs (Sacchetti et al., 2007).

During mitosis, there are three fates for a stem cell: (1) the stem cell divides asymmetrically, and one daughter is maintained as a stem cell, the other as a transiently amplifying cell; (2) the stem cell divides symmetrically, and both daughters remain as stem cells, as in proliferation of embryonic and induced pluripotent stem cells in vitro; or (3) the stem cell divides symmetrically, but both daughters are transiently amplifying cells, with ultimate loss of the stem cell if this type of symmetrical division persists (Fig. 2.5). The allocation of stem cells among these types of divisions during embryonic development, postnatal growth, and homeostasis in adulthood is not well understood. It may be that all three types of divisions are operative under different circumstances, i.e., toggling from one type of division to another (Watt and Hogan, 2000).

Based on these types of cell divisions, it is clear that extensive proliferation is not evidence of self-renewal, as is often claimed in the literature. Using clonal analysis and the transplantation assay (which is key for the demonstration of self-renewal in the hematological system), evidence of self-renewal is linked with: (1) the isolation of a single cell with an identifiable phenotype whose progeny re-create a functional tissue and (2) reisolation of a single cell from the re-created functional tissue with the same identifiable phenotype. By comparing populations of bone-forming cells that were not able to support hematopoiesis to SSCs/BMSCs by fluorescence-activated cell sorting (FACS) for various different markers, it was determined that CD146 (also known as MCAM) was highly expressed by BMSCs, but not by normal human trabecular bone cells. Immunohistochemistry of human marrow identified the $CD146^+$ cells as pericytes. Clones established from freshly isolated human marrow were homogeneously positive for CD146, and when clonal strains were transplanted along with scaffolds into mice, again, ~10% were multipotent and able to form a bone/marrow organ. In those transplants, using an antibody specific for human CD146, the only human $CD146^+$ cells were pericytes on marrow sinusoids. Human osteocytes and osteoblasts, as identified by staining with antibody specific for human mitochondria, were not, indicating that ex vivo-expanded human $CD146^+$ cells were not only able to differentiate into osteogenic cells, but also able to self-renew into pericytes. Furthermore, human cells isolated by collagenase digestion of the transplants by

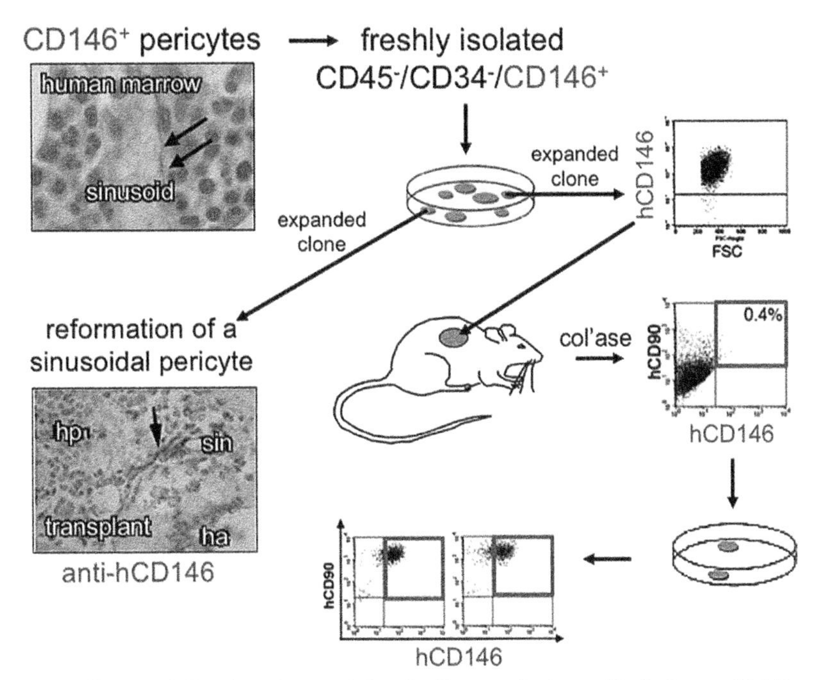

FIGURE 2.6 Skeletal stem cell self-renewal. Based on the use of cloned cells, an antibody specific for human CD146, and in vivo transplantation, it was determined that the multipotent subset of bone marrow stromal cells (BMSCs) does self-renew and contains bona fide skeletal stem cells (SSCs). It was first determined that CD146 is a marker of BMSCs that are able to support hematopoiesis (a defining feature of SSCs) and that these cells are sinusoidal pericytes. Freshly isolated human bone marrow $CD45^-/CD34^-/CD146^+$ cells are clonogenic. When a subset of individual $CD146^+$ clones (~1:5) is transplanted, they once again form a bone/marrow organ (ectopic ossicle) and human $CD146^+$ pericytes. Human $CD146^+$ cells can be reisolated from the ossicle as $CD146^+$ CFU-Fs, providing definitive evidence of self-renewal. *ha*, hydroxyapatite/tricalcium phosphate; *hp*, hematopoiesis; *sin*, sinusoid. *Modified from Sacchetti, B., Funari, A., Michienzi, S., Di Cesare, S., Piersanti, S., Saggio, I., Tagliafico, E., Ferrari, S., Robey, P.G., Riminucci, M., Bianco, P., 2007. Self-renewing osteoprogenitors in bone marrow sinusoids can organize a hematopoietic microenvironment. Cell 131, 324–336.*

way of magnetic cell sorting with anti-human-specific CD90 were clonogenic, and $CD146^+$, indicative of self-renewal of the CFU-F (Sacchetti et al., 2007) (Fig. 2.6).

The role of SSCs/BMSCs in postnatal bone turnover and remodeling

Bone turnover/remodeling is initiated by many factors, such as PTH, which is released from the parathyroid gland when the calcium-sensing receptor detects a decrease in serum calcium, and when there is a need to replace bone that has become microdamaged through mechanisms that are not yet clear. Formation of receptor activator of NF-κB (RANK)-expressing osteoclast precursors of the monocyte/macrophage series, as well as T cells that influence osteoclast formation (Li et al., 2011), is supported by SSCs/BMSCs by providing a "bed" (στρώμα (Greek)—*strṓma*) upon which they are formed. SSCs/BMSCs express a long list of hematopoiesis-associated cytokines and growth factors (Gene Ontology categories "hematological system development and function" and "hematopoiesis" are highly overrepresented in multipotent SSC/BMSC clonal lines; data are in GEO GSE64789; Sworder et al., 2015). SSCs/BMSCs also express high levels of macrophage colony-stimulating factor and RANK ligand (RANKL), both of which are essential for osteoclast formation, and osteoprotegerin (OPG), which serves as a decoy receptor for RANKL, preventing osteoclast formation. Osteoclast formation and bone resorption are dependent on the balance of RANKL and OPG expressed by BMSCs and more mature osteoblastic cells, and BMSCs have long been thought to play a major role in osteoclast formation (reviewed in Boyce and Xing, 2007). Based on mouse models with conditional or cell/tissue-specific deletion of RANKL, it has been proposed that osteocytes also control osteoclast formation based on their expression of RANKL. However, it appears that the $RANKL^+$ cell type that controls osteoclastogenesis may be site specific. For example, mice with RANKL deletion in mature osteoblasts and osteocytes using a DMP1-Cre driver maintain tooth eruption, which requires bone resorption to occur, whereas the long bones exhibit an osteopetrotic-like phenotype (reviewed in Sims and Martin, 2014; Xiong and O'Brien, 2012). Further studies are needed to determine the contributions of SSCs/BMSCs, more mature osteoblastic cells, and osteocytes to initiating osteoclast formation throughout the skeleton.

Subsequent to the formation of osteoclasts and dissolution of mineralized matrix, a plethora of growth factors, buried within bone due to their affinity for carbonate-rich apatite, is liberated, including BMPs, PDGF, IGF-1 and IGF-2, and TGFβs. Consequently, PTH, the released factors, as well as those synthesized by local cells (Wnt10b by $CD8^+$ cells, FGF and vascular endothelial growth factor by BMSCs), mediate the reversal stage, which is marked by a cessation in bone resorption and creation of a microenvironment conducive for bone formation, thereby coupling bone resorption with bone formation (reviewed in Crane and Cao, 2014; Sims and Martin, 2014). It is thought that through osteoclastic action, active TGFβ is released and establishes a gradient from the resorption site through the marrow to the pericytes located on a nearby blood vessel. This gradient has two effects. Due to the high concentration of TGFβ at the site of resorption, it inhibits migration of osteoclastic precursors into the area; however, the gradient initiates migration of SSCs/BMSCs into the area in need of restoration, but does not induce differentiation (Tang et al., 2009). Upon arrival in the resorption bay, BMSCs are induced to osteogenic differentiation by the interplay of the TGFβ/BMP, Wnt, and IGF-1 signaling pathways. Interestingly, PTH, in addition to initiating bone resorption, also plays a role in bone formation. PTH bound to its receptor, PTHR1, complexes with Lrp6 (Wnt coreceptor), which leads to stabilization of β-catenin and pSmad1 signaling, along with stimulating cAMP-mediated signaling. The PTH—PTHR1 complex can also bind to the TGFβRII receptor, leading to pSmad2/3 signaling (reviewed in Crane and Cao, 2014). IGF-1 signaling activates mTOR via the PI3K—Akt pathway and induces osteogenic differentiation of BMSCs. Alterations in any of these pathways due to mutation, or alterations in levels of expression due to changes in the microenvironment, can have significant consequences on the balance of bone resorption and formation necessary for skeletal homeostasis (reviewed in Crane and Cao, 2014; Sims and Martin, 2014).

Skeletal stem cells in disease

Based on the fact that SSCs/BMSCs are essential for new bone formation following injury, and the fact that they, in part, control osteoclast formation, these cells are important mediators of skeletal homeostasis. As such, it was hypothesized that any genetic (intrinsic) change or change in their microenvironment (extrinsic) that has an impact on their normal biological activity would result in a skeletal disorder. Furthermore, due to their key role in supporting hematopoiesis and as a component of the HSC's niche (Ugarte and Forsberg, 2013), it was also theorized that there may be some hematological diseases or disorders that are caused, or worsened, by dysfunction of SSCs/BMSCs. Support for these notions came from both genetic (intrinsic) and acquired (extrinsic) diseases that affect bone and hematopoiesis.

Fibrous dysplasia of bone and the McCune—Albright syndrome

The genetic disease, McCune—Albright syndrome (MAS; OMIM: 174800) is characterized by the triad of skin hyperpigmentation (café au lait spots with the "coast of Maine" profile), hyperactive endocrinopathies (renal phosphate wasting, precocious puberty, growth hormone excess, hyperthyroidism, hyperparathyroidism, etc.), and fibrous dysplasia of bone (FD). Patients with FD/MAS are somatic mosaics (the mutation occurs after fertilization) with activating missense mutations of *GNAS*, which codes for Gsα, leading to overproduction of cAMP. FD lesions are characterized by the replacement of normal bone and marrow with woven, poorly organized, and undermineralized bone and replacement of marrow (hematopoiesis and marrow adipocytes) with a fibrotic tissue (reviewed in Boyce et al., 1993) (Fig. 2.7). This fibrotic tissue was found to be composed of a high proportion of mutant BMSCs (Riminucci et al., 1997). The initial effects of the mutation are to dramatically increase their proliferation and to inhibit their ability to form hematopoiesis-supporting stroma and marrow adipocytes. Furthermore, their differentiation into mature osteoblastic cells is severely compromised, as characterized by their retracted morphology and abnormal synthesis and organization of bone matrix, resulting in the formation of Sharpey's fibers (Bianco et al., 2000; Riminucci et al., 1999). The abnormal bone matrix is osteomalacic, due to hypophosphatemia caused by excess production of the phosphate-regulating hormone FGF-23 by the mutated osteogenic cells (Riminucci et al., 2003a). The abnormal composition of the matrix may also play a role in the undermineralization noted in these lesions.

Studies utilizing the fibrotic marrow from FD lesional bone showed that there was a higher colony-forming efficiency compared with normal bone marrow (Fig. 2.7). The presence of transiently amplifying cells present in the FD sample most likely explains the increase in colony-forming efficiency (Kuznetsov et al., 2008).

When FD BMSCs were transplanted along with hydroxyapatite/tricalcium phosphate ceramic particles into immunocompromised mice, they formed a fibrous dysplastic ossicle that completely recapitulated the nature of the FD

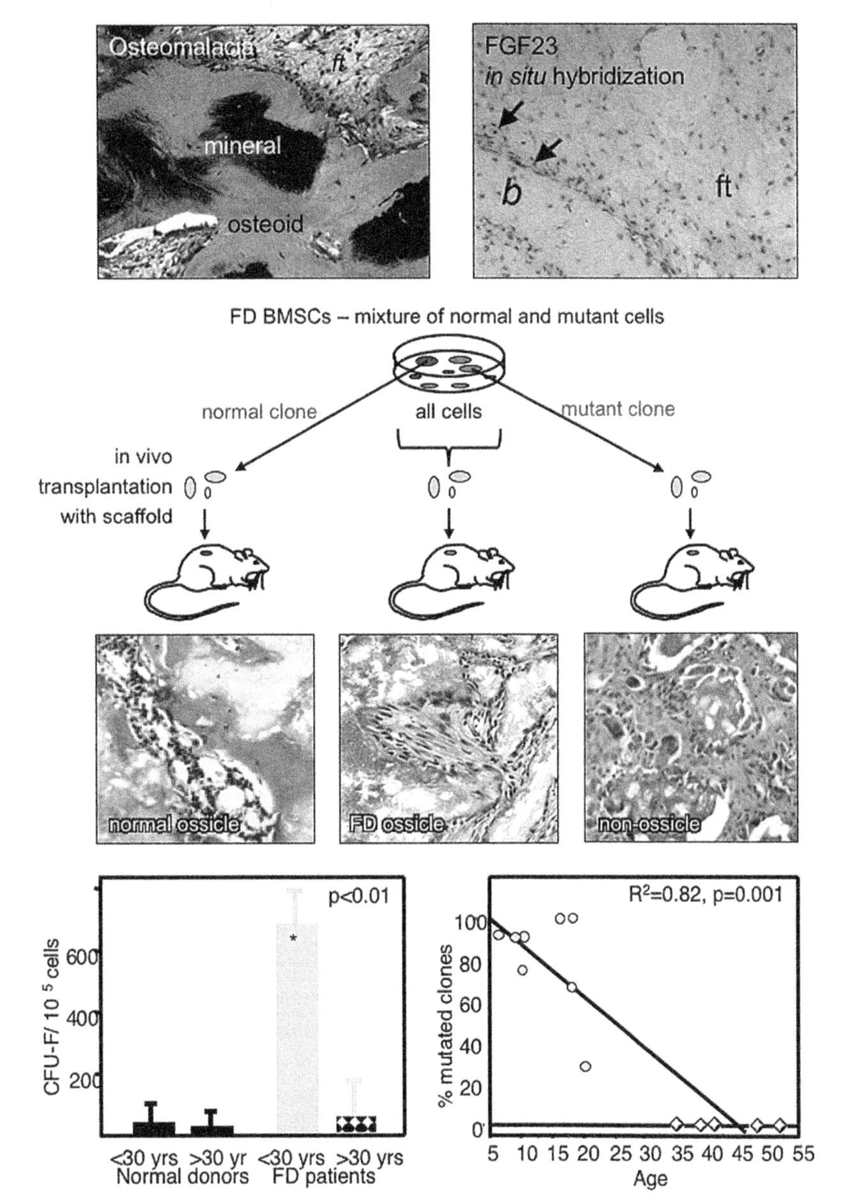

FIGURE 2.7 Fibrous dysplasia of bone/McCune–Albright syndrome (FD/MAS): a disease of skeletal stem cells. FD/MAS is a somatic mosaic disease caused by activating mutations of Gsα. Focal lesions are characterized by replacement of normal bone and marrow by highly disorganized and undermineralized bone and a fibrotic marrow devoid of hematopoiesis and marrow adipocytes. The fibrotic marrow is composed of a mixture of normal stromal cells, which are capable of forming normal bone in in vivo transplants, and mutated stromal cells, which on their own are not capable of surviving. It is only when normal and mutant cells are transplanted together that an FD tissue is formed. Colony-forming efficiency shows that in young patients with FD/MAS, there is a large increase compared with normal donors; however, with age colony-forming efficiency decreases dramatically, suggesting that colony-forming unit–fibroblasts (*CFU-Fs*) from young patients are, in fact, transiently amplifying cells rather than stem cells. Determination of the number of mutant CFU-Fs as a function of age also indicates that there is a dramatic decrease, leading to "normalization" due to the loss of mutated skeletal stem cells/transiently amplifying cells. *b*, bone; *BMSC*, bone marrow stromal cell; *FGF23*, fibroblast growth factor 23; *ft*, fibrous tissue. Arrows point out high levels of FGF23 expression by disfunction osteogenic cells on the bone surface. *Modified from Riminucci, M., Collins, M.T., Fedarko, N.S., Cherman, N., Corsi, A., White, K.E., Waguespack, S., Gupta, A., Hannon, T., Econs, M.J., Bianco, P., Gehron Robey, P., 2003a. FGF-23 in fibrous dysplasia of bone and its relationship to renal phosphate wasting. J. Clin. Invest. 112, 683–692; Riminucci, M., Robey, P.G., Saggio, I., Bianco, P., 2010. Skeletal progenitors and the GNAS gene: fibrous dysplasia of bone read through stem cells. J. Mol. Endocrinol. 45, 355–364; and Robey, P.G., Kuznetsov, S., Riminucci, M., Bianco, P., 2007. The role of stem cells in fibrous dysplasia of bone and the Mccune-Albright syndrome. Pediatr. Endocrinol. Rev. 4 (Suppl. 4), 386–394.*

lesion: abnormal woven and undermineralized bone, Sharpey's fibers, and lack of marrow formation (Bianco et al., 1998) (Fig. 2.7). In addition to transplantation of the total FD BMSC population, individual colonies were also evaluated. Clones were genotyped prior to transplantation, and as expected, clones with a normal genotype made a normal bone/marrow organ. Interestingly, mutant clones failed to survive transplantation; no bone was formed and human cells were not present (Bianco et al., 1998) (Fig. 2.7). This finding supports Happle's hypothesis that the mutation is embryonic lethal (Happle, 1986); that is, mutant cells would survive only when in combination with wild-type cells. Of note, there is no documented case in which FD/MAS was inherited; all cases are examples of de novo mutations after fertilization.

Evidence that FD is a "stem cell" disease (Riminucci et al., 2006) comes from the observation that as FD patients age, they "normalize" (Kuznetsov et al., 2008). Bone biopsies from patients of increasing age showed very high levels of apoptosis, up until approximately 30 years of age. Biopsies of older patients showed decreasing amounts of osteoid and increasing amounts of apparently normal bone and marrow. Marrow samples also showed a decrease in not only colony-forming efficiency, moving toward the normal level with increasing age, but also in the percentage of mutant cells in the total BMSC population, and a near-complete absence of mutant clones (Fig. 2.7). Transplants of BMSCs from older patients also showed increasingly normal ossicles, roughly correlated with age. From these observations, it was hypothesized that mutated SSCs were not able to undergo self-renewing types of division, and with time, transiently amplifying mutant cells differentiated and were ultimately cleared by apoptosis, leaving wild-type SSCs to undergo normal bone formation, albeit on an abnormal preexisting structure. Consequently, it can be reasoned that the impact of the mutation occurs at several levels in the skeletal lineage: (1) the mutation inhibits the self-renewal of mutant SSCs, (2) it causes extensive proliferation of mutant SSCs/BMSCs leading to premature apoptosis, and (3) it causes the derangement of BMSC differentiation into osteogenic, stromogenic, and adipogenic lineages.

Inherited forms of bone marrow failure

Inherited diseases of bone marrow failure are characterized by the inability to make normal numbers of all or specific types of blood cells (pancytopenia, or anemia, leukopenia, and thrombocytopenia) (Khincha and Savage, 2013). There are over 40 genes that have been reported to be mutated in association with these diseases (Ballew and Savage, 2013). Interestingly, many of these genes are not specific for hematopoietic cells based on the fact that patients with these diseases often have multiorgan manifestations. Dyskeratosis congenita (DC) is one such disease that is characterized by the triad of oral leukoplakia, nail dystrophy, and abnormal skin pigmentation, and also severe aplastic anemia, which is the main cause of death. These patients often have pulmonary fibrosis; stenosis of the esophagus, urethra, and/or lacrimal ducts; liver disease; premature graying of hair; and osteopenia (Ballew and Savage, 2013). Mutations have been detected in subunits of telomerase (*DKC1*, *TERT*, *TERC*, *NOP10*, *NHP2*), and in other genes involved in telomere biology (*WRAP53*, *TINF2*, *CTC1*, *RTEL1*). DC can be inherited in an X-linked recessive, autosomal dominant, or autosomal recessive fashion (Ballew and Savage, 2013; Nelson and Bertuch, 2012; Walne et al., 2013). In addition, a small proportion of patients with acquired aplastic anemia but without the other symptoms of DC have been found to have germ-line mutations in genes related to the telomerase complex (Alter et al., 2012; Yamaguchi et al., 2005). Based on these findings, patients with these various diseases and disorders have been grouped together as having telomere biology disorders (TBDs), and present with a very broad clinical spectrum ranging from limited bone marrow aplasia to a severe multiorgan phenotype as in DC.

Bone marrow failure caused by mutations in telomere biology-related genes is almost certainly caused, in part, by an intrinsic dysfunction of HSCs, due to short telomeres and a decrease in their ability to proliferate. However, the HSC's microenvironment (the niche) also controls the normal balance between quiescence and proliferation, self-renewal and differentiation (Schofield, 1978). As shown by Friedenstein et al., BMSCs are able to re-create the hematopoietic microenvironment upon in vivo transplantation into mice (Friedenstein et al., 1974; Owen and Friedenstein, 1988). Mature osteoblastic cells (Calvi et al., 2003), endothelial cells (Kiel et al., 2005), and BMSCs expressing CXCL12 (Omatsu et al., 2010; Sacchetti et al., 2007, 2016; Sugiyama et al., 2006), CD146 (Sacchetti et al., 2007), Nestin-GFP (Mendez-Ferrer et al., 2010), LEPR (Ding and Morrison, 2013), and others (reviewed in Ugarte and Forsberg, 2013) have been suggested to be constituents of the hematopoietic niche. In the quest for an identifiable niche-maintaining cell, current research is directed toward perivascular SSCs/BMSCs (Mendez-Ferrer et al., 2010; Omatsu et al., 2010; Pinho et al., 2013; Raaijmakers et al., 2010; Sacchetti et al., 2007, 2016). Thus, understanding the biological properties of SSCs/BMSCs has important implications not only for skeletal physiology and disease, but also for understanding the regulation of HSC physiology and dysregulation of hematopoiesis in blood disorders.

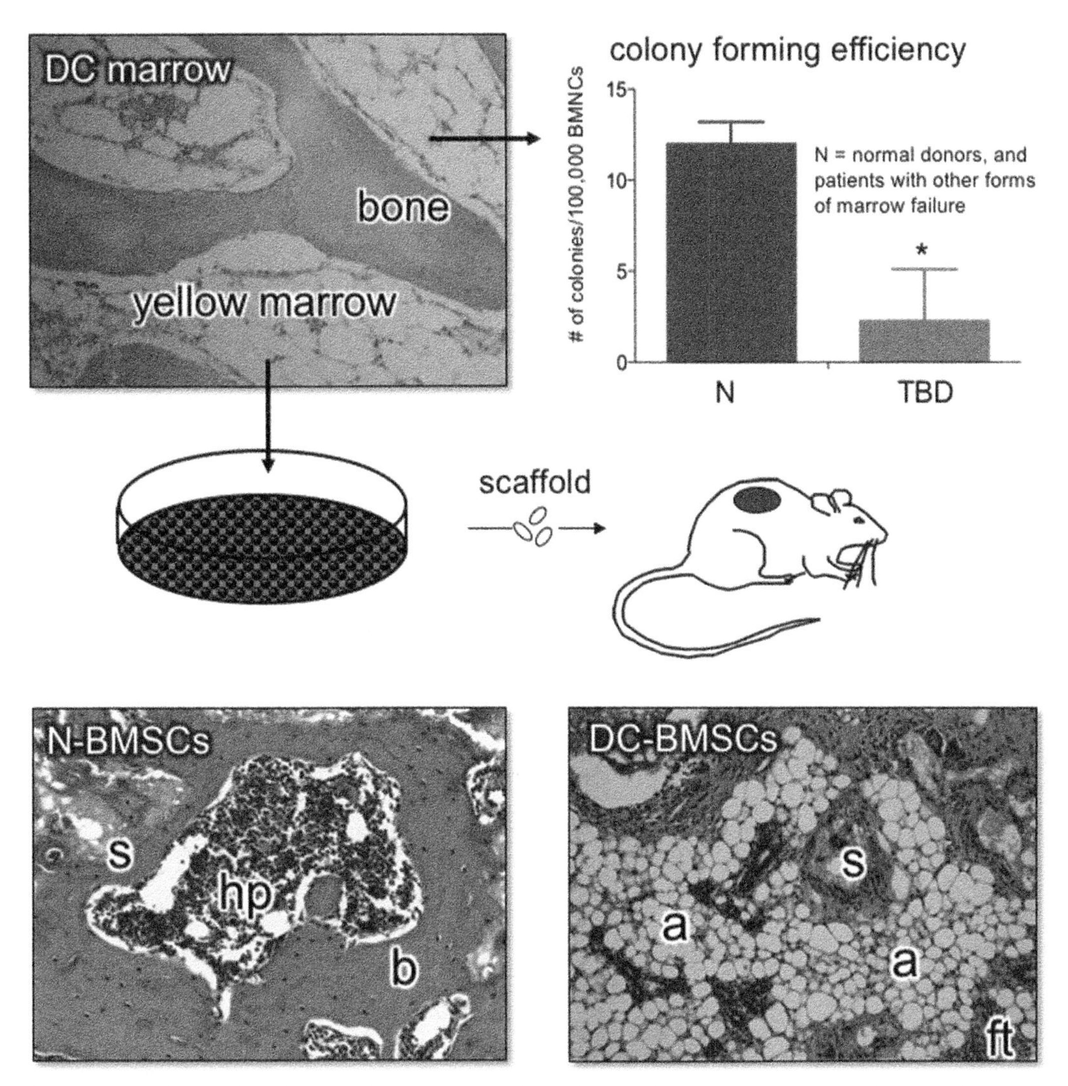

FIGURE 2.8 Bone marrow failure: the role of skeletal stem cells. A form of genetically acquired bone marrow failure syndromes is associated with mutations in genes that regulate telomere length, and they have been collectively termed telomere biology disorders (*TBDs*). Dyskeratosis congenita (*DC*) is one such TBD. Iliac crest biopsies from DC patients show predominately yellow marrow and demonstrate a dramatic decrease in colony-forming efficiency compared with normal donors. When DC bone marrow stromal cells (*BMSCs*) are expanded and transplanted, ectopic ossicles show a remarkable similarity to iliac crest biopsies: formation of a yellow marrow, completely devoid of hematopoiesis. *a*, adipocytes; *b*, bone; *BMNCs, bone marrow mono nucleated cells; ft*, fibrous tissue; *hp*, hematopoiesis; *s*, scaffold. *Modified from Balakumaran, A., Mishra, P.J., Pawelczyk, E., Yoshizawa, S., Sworder, B.J., Cherman, N., Kuznetsov, S.A., Bianco, P., Giri, N., Savage, S.A., Merlino, G., Dumitriu, B., Dunbar, C.E., Young, N.S., Alter, B.P., Robey, P.G., 2015. Bone marrow skeletal stem/progenitor cell defects in dyskeratosis congenita and telomere biology disorders. Blood 125, 793–802.*

By using BMSCs isolated from TBD patients to study the biological properties of TBD BMSCs, coupled with in vivo transplantation assay, it was found that SSCs contribute to the hematological phenotype in patients with TBD (Balakumaran et al., 2015) (Fig. 2.8). TBD BMSCs exhibited a drastically reduced colony-forming efficiency, as well as reduced telomerase activity. TBD BMSCs spontaneously differentiated into adipocytes and fibrotic cells even when cultured in osteogenic medium. In addition, they displayed increased senescence in vitro. Unlike normal BMSCs, upon in vivo transplantation into mice, TBD BMSCs failed to form bone or to support hematopoiesis. Knocking down *TERC* (a TBD-associated gene) in normal BMSCs using si*TERC*-RNA recapitulated the TBD BMSC phenotype as exemplified by reduced secondary colony-forming efficiency and growth rate, and accelerated senescence in vitro. Microarray profiles of control and si*TERC* BMSCs showed decreased hematopoietic factors at the mRNA level and decreased secretion of factors at the protein level. These findings are consistent with the notion that defects in SSCs/BMSCs, along with defects in HSCs, contribute to bone marrow failure in TBD patients. SSCs/BMSCs may be a target for drug treatment in some hematological diseases and disorders.

Role of SSCs/BMSCs in acquired inflammation

Changes noted in the ability of SSCs/BMSCs to support hematopoiesis as a function of an intrinsic (genetic) change in TBDs also led to the question whether extrinsic changes would influence the ability of SSCs/BMSCs to support hematopoiesis. Inflammation alters hematopoiesis, often by decreasing erythropoiesis and enhancing myeloid output. The mechanisms behind these changes and how the bone marrow stroma contributes to this process are not well known. These questions were studied in the setting of murine *Toxoplasma gondii* infection (Chou et al., 2012). The data revealed that infection alters early myeloerythroid differentiation, blocking erythroid development beyond the pre-megakaryocyte/erythroblast stage (erythroid crash), while expanding the granulocyte/monocyte precursor population (Fig. 2.9). The effects of infection in animal models deficient in cytokines known to regulate erythropoiesis were examined. The results indicated that in mice deficient in interleukin-6 (IL-6), the decrease in erythropoiesis was partially rescued, and it was also noted that in normal infected mice, serum IL-6 was elevated. From these results, IL-6 was found to be a critical mediator of the erythropoiesis, independent of hepcidin-induced iron restriction. Comparing bone marrow with the spleen showed that the hematopoietic response to infection was driven by the local bone marrow microenvironment. By using bone marrow transplantation to create bone marrow chimeras (IL-6-deficient marrow into normal recipients and vice versa), it was

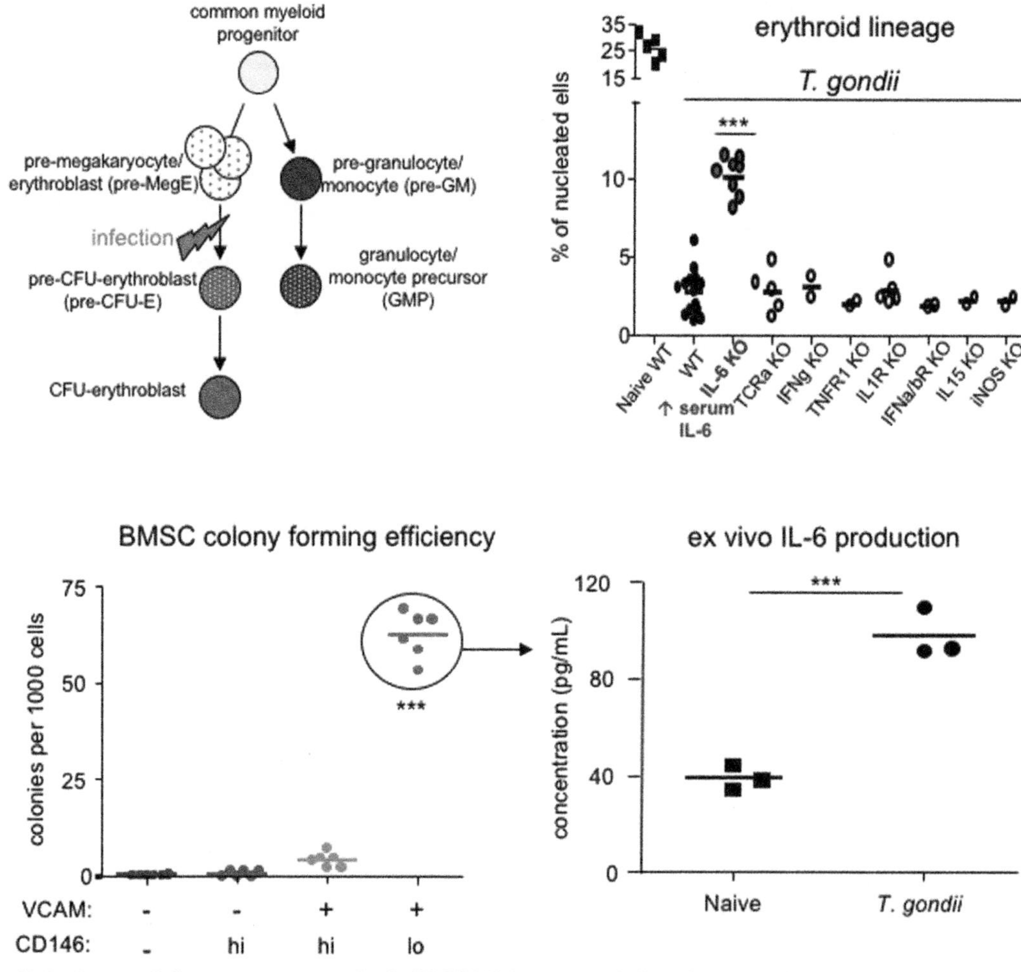

FIGURE 2.9 Skeletal stem cells/bone marrow stromal cells (SSC/BMSCs) in the sculpting of hematopoiesis. *Toxoplasmosis gondii* infection causes a crash in erythropoiesis due the blockade in differentiation of pre-megakaryocytes/erythroblasts into more differentiated cell types and a redirection of the common myeloid progenitor into the lymphoid series. Using mouse models deficient in various cytokines known to modulate erythropoiesis, it was determined that interleukin-6 (*IL-6*) deficiency partially rescued the erythroid crash and that IL-6 was increased in wild-type (*WT*) mice infected with *T. gondii*. Colony-forming unit (*CFU*)—fibroblasts ($CD45^-/Ter119^-/VCAM^+$) were isolated from naïve (uninfected) mice and infected mice, and were found to be the source of IL-6 induced by *T. gondii*. These results suggest that SSCs/BMSCs respond to external stimuli to modulate hematopoietic output and composition. *KO*, knockout. *Modified from Chou, D.B., Sworder, B., Bouladoux, N., Roy, C.N., Uchida, A.M., Grigg, M., Robey, P.G., Belkaid, Y., 2012. Stromal-derived IL-6 alters the balance of myeloerythroid progenitors during Toxoplasma gondii infection. J. Leukoc. Biol. 92, 123–131.*

demonstrated that radioresistant cells (BMSCs) were the relevant source of IL-6 in vivo. Finally, direct ex vivo sorting revealed that CD106/VCAM$^+$ colony-forming BMSCs significantly increased IL-6 secretion after infection (Fig. 2.9). These data suggest that SSCs/BMSCs regulate the hematopoietic changes during inflammation via IL-6 (Chou et al., 2012) and support the notion that extrinsic factors have an impact on stromal cell activity to sculpt hematopoietic output in response to microenvironmental changes.

Skeletal stem cells in tissue engineering

Based on their remarkable ability to form a complete bone/marrow organ, it is no wonder that many have pursued the use of SSCs/BMSCs in tissue engineering. Tissue engineering is typically described as the use of cells, growth factors, and/or cytokines and scaffolds, either alone or in various combinations, to restore tissue function lost due to trauma, tumor removal, and genetic and acquired diseases (Fig. 2.10). For skeletal tissue engineering, there are a number of parameters that affect outcome: the source of cells; the ex vivo expansion conditions, including the use of exogenous growth factors

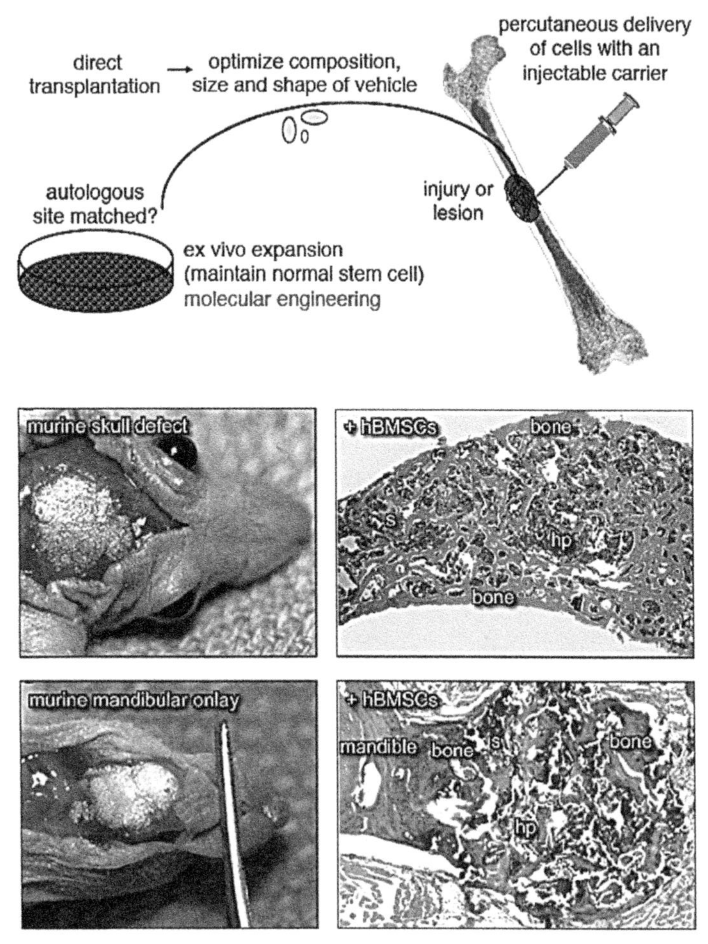

FIGURE 2.10 Skeletal stem cells in bone tissue engineering. In using skeletal stem cells/bone marrow stromal cells (SSCs/BMSCs), appropriate cell sources (e.g., autologous skeletal-derived cells such as bone marrow stromal cells) are needed. Ex vivo expansion conditions must maintain the SSC within the population to ensure that bone turnover can occur as needed. Culture of SSCs/BMSCs also offers the opportunity for molecular engineering to correct genetic defects (Piersanti et al., 2010). Ex vivo-expanded cells must be combined with an appropriate scaffold that will not only be osteoconductive, but also maintain the existence of SSCs. It may be possible to inject lesions directly, but an injectable carrier is needed to prevent their loss to the circulation and to hold them in place. Preclinical studies have been performed in mice to re-create jawbone and cranial bone (and also in dogs; Mankani et al., 2006a). *hp*, hematopoiesis; *s*, scaffold. *Modified from Bianco, P., Robey, P.G., Penessi, G., 2008a. Cell source. In: de Boer, J., van Blitterswijk, C., Thomsen, P., Hubbell, J., Cancedda, R., de Bruijn, J.D., Lindahl, A., Sohler, J., Williams, D.F. (Eds.), Tissue Engineering. Elsevier, Amsterdam, pp. 279–306; Mankani, M.H., Kuznetsov, S.A., Wolfe, R.M., Marshall, G.W., Robey, P.G., 2006b. In vivo bone formation by human bone marrow stromal cells: reconstruction of the mouse calvarium and mandible. Stem Cell. 24, 2140–2149; and Mankani, M.H., Kuznetsov, S.A., Marshall, G.W., Robey, P.G., 2008. Creation of new bone by the percutaneous injection of human bone marrow stromal cell and HA/TCP suspensions. Tissue Eng. Part A 14, 1949–1958.*

and/or cytokines; and the nature of the scaffold. It also must be mentioned that the microenvironment (e.g., gender and age of the recipient, presence or absence of bacterial infection, inflammation, etc.) in which such constructs are placed will have an impact on the outcome of tissue engineering approaches.

Cell sources

By far and away, BMSCs have been the most studied cell population for the use of skeletal tissue regeneration. As shown by clonal analyses, BMSCs, if properly expanded, have a subpopulation of SSCs that are multipotent and able to form a bone/marrow organ. Based on the fact that the SSC is a pericyte on the abluminal side of marrow sinusoids, the presence of marrow in a transplant of BMSCs is indicative of the presence of SSCs. However, that is not to say that more mature populations of osteogenic cells are not useful. There are instances in which building new bone will fit that bill, and mature osteogenic cells isolated from trabecular bone and BMSCs grown in basic FGF make copious amounts of bone, but do not support hematopoiesis (Sacchetti et al., 2007). However, based on the fact that more committed osteogenic cells will eventually disappear and will need to be replaced during bone turnover, it is clear that an SSC is required. Furthermore, based on the fact that differentiated osteogenic cells begin to express histocompatibility antigens, cells used for skeletal tissue regeneration must be autologous and would ultimately be rejected if allogeneic unless immunosuppression is employed, which is not without serious side effects (Robey, 2011).

Finally, it is known that bone arises from three different embryonic specifications during development: paraxial mesoderm, lateral plate mesoderm, and neural crest. It is known that there are subtle histological differences and growth differences of bone from different origins (Akintoye et al., 2006), and studies suggest that there are also differences in the patterns of gene expression (Kidwai et al., unpublished results). Consequently, it is not yet known if SSCs from one embryonic source can substitute for those from another embryonic source. Studies in which iliac crest (appendicular lateral plate mesoderm) was used to build up alveolar bone (neural crest) showed rapid resorption of the iliac crest, suggesting a potential incompatibility (Burnette, 1972). Study of pluripotent stem cells (human embryonic stem cells [ESCs] and human induced pluripotent stem cells [iPSCs]) provides a way to dissect out the regulatory pathways that govern the discrete stages of human bone development operative in the three distinct embryonic sources. Many studies have aimed at differentiating ESCs or iPSCs into SSCs (see Phillips et al., 2014, for examples). However, this and other studies did not attempt to determine from which embryonic specification the cells originated. Studies are under way that are designed to take pluripotent stem cells stepwise through specific stages of bone development (primitive streak, mesendoderm, paraxial and lateral mesoderm, and neural crest) (Kidwai et al., unpublished results).

Scaffolds

There have been a large number of preclinical studies, and a number of small human trials, showing the efficacy of using ex vivo-expanded SSCs/BMSCs in conjunction with an appropriate scaffold for direct orthotopic delivery into large segmental defects (reviewed in Chatterjea et al., 2010; Winkler et al., 2018) (Fig. 2.10). In some approaches, cells are precultured on a scaffold prior to transplantation. However, it is not clear that this is an advantage, especially if osteogenesis is induced, which may preclude the cell population's ability to maintain the SSC. In addition, the nature of the scaffold plays a major role in the performance of the cell population. While many synthetic and natural scaffolds have been fabricated, as of this writing, 3D scaffolds that contain ceramics (usually hydroxyapatite/tricalcium phosphate) as part of their formulation (reviewed in Oh et al., 2006) appear to be the most reliable with respect to the formation of bone and the support of hematopoiesis when seeded with SSCs/BMSCs. However, many of these scaffolds are resorbed very poorly and can persist for long periods of time in vivo. Consequently, scaffolds composed of polymers such as PLGA (poly(lactic-*co*-glycolic acid)) and PCL (poly(ε-caprolactone)), with and without calcium phosphate components or further functionalization, have been developed (reviewed in Kretlow and Mikos, 2007; Rezwan et al., 2006). However, it is not clear that these scaffolds also maintain the SSC/BMSC population's ability to support hematopoiesis. Of note, many studies to identify new scaffolds rely on in vitro assays to evaluate the performance of newly developed scaffolds; however, these assays do not adequately determine osteogenesis or support of hematopoiesis, and in vivo transplantation is required.

In addition to the use of 3D scaffolds, it would also be of benefit to develop techniques for the use of injectable scaffolds that would hold cells in place and support their differentiation, and thereby avoid the need for open surgery (Mankani et al., 2008) (Fig. 2.10). Injection of concentrated bone marrow aspirates has been shown to be beneficial in the treatment of avascular necrosis (Hernigou et al., 2005), but it would be of interest to determine if their efficacy could be enhanced through the use of an injectable carrier. It could also be envisioned that this approach would be useful for the treatment of unicameral bone cysts and nonunions.

Skeletal stem cells and regenerative medicine

It is thought that either systemic infusion of SSCs/BMSCs into the circulation or direct injection into a diseased area has a beneficial effect in treating not only skeletal diseases and disorders, but also a long list of diseases (e.g., graft vs. host disease, cardiovascular disease, osteoarthritis, multiple sclerosis, etc.) (reviewed in Galipeau and Sensebe, 2018). Initially it was thought that SSCs/BMSCs could transdifferentiate into cells outside of the skeletal lineage based on expression of a limited set of markers characteristic of a nonskeletal tissue. However, more rigorous assays have shown that SSCs/BMSCs do not functionally differentiate into nonskeletal tissues without extensive treatment with chemicals or molecular engineering to express tissue-specific transcription factors (reviewed in Bianco et al., 2013). Furthermore, it is known that cells are rapidly cleared from the circulation when injected systemically, and are even rapidly cleared upon direct injection (when they are not attached to a scaffold or carrier) (Bianco and Cossu, 1999; Bianco et al., 2008b). It has been proposed that even though the cells do not seem to home to sites of disease or injury, or to survive long term, they secrete growth factors, cytokines, and/or extracellular vesicles that exert paracrine, immunomodulatory, and/or immunosuppressive effects on recipient cells that initiate an endogenous repair process (reviewed in Galipeau and Sensebe, 2018). While some studies, such as the treatment of certain symptoms of acute graft versus host disease, show that there may be some beneficial effect (Le Blanc and Ringden, 2007), the results of the vast majority of studies have been equivocal (reviewed in Galipeau and Sensebe, 2018). However, it has not been determined that these putative effects are mediated by the subpopulation of SSCs within the BMSC population, and this is unlikely to be the case, based on the fact that SSCs are very rare. Any potential effect is mediated by the BMSC population as a whole, and cannot rightly be called a stem cell therapy at the time of writing. Furthermore, their rapid disappearance in the lungs and the distance of their demise from the site of disease or injury make a mechanism of action obscure.

Stem cell and non–stem cell functions of skeletal stem cells

As described earlier, the regeneration of bone and maintenance of a stem cell on the abluminal side of the marrow sinusoid is of utmost importance in maintaining skeletal homeostasis throughout life (Robey, 2011). However, SSCs/BMSCs also perform other functions that do not rely on their stem/progenitor cell status, that is, not on their potency or their ability to self-renew. For example, based on their pericytic nature, they produce factors that guide the formation of microvessels in conjunction with endothelial cells. In vitro, a mixture of SSCs/BMSCs and endothelial cells plated on Matrigel forms networks with branches, with endothelial cells on the interior, surrounded by SSCs/BMSCs on the exterior. Likewise, when SSCs/BMSCs are cotransplanted in vivo along with human umbilical vein endothelial cells in Matrigel (conditions that are not conducive for bone formation), they again form an extensive capillary-like network after about 3 weeks, and after 8 weeks, a mature thick-walled structure forms with endothelial cells surrounded by pericytes derived from the SSCs/BMSCs. These capillary-like structures are functional based on the presence of recipient blood cells inside their lumens due to anastomosis of the nascent structures with the recipient vasculature (Sacchetti et al., 2016). Furthermore, SSCs/BMSCs support hematopoiesis, one of their defining features, and are thought to be a part of the HSC niche that controls the activities of the HSC (Bianco, 2011). In addition, there is evidence that SSCs/BMSCs play a major role in shaping the composition of hematopoiesis, as presented earlier (Chou et al., 2012). As such, SSCs/BMSCs can be targets for therapeutics designed to control the quantitative and qualitative aspects of hematopoietic cells.

Last, SSCs/BMSCs provide a tool by which to study human skeletal disease. There are many instances in which a mutation associated with a human disease does not re-create the same phenotype when introduced into a mouse. For example, Meunke syndrome is caused by autosomal dominant activating mutations of FGFR3 and is characterized by craniosynostosis, ocular and hearing abnormalities, and fusion or missing bones in the hands and/or feet (Doherty et al., 2007). However, the same mutation in heterozygous mice does not exhibit Meunke syndrome, and only when the mice are homozygous for the mutation do they exhibit some, but not all, of the phenotypic features (Nah et al., 2012). The use of SSCs/BMSCs from human patients with monogenetic disease, or the introduction of mutation with the use of CRISP/Cas9 or other types of genetic engineering, can provide valuable tools with which to elucidate the pathogenetic mechanisms at play by using appropriate in vitro assays and in vivo transplantation assays as described earlier. For example, by studying normal BMSCs transduced with the R201C activating Gsα mutation, it was noted that mutant BMSCs highly overexpress RANKL, an essential factor for osteoclast formation (Piersanti et al., 2010). This finding provided an explanation for the florid increase in osteoclastogenesis during the establishment of expansile lesions (Riminucci et al., 2003b). Furthermore, it suggested the use of denosumab (a humanized anti-RANKL antibody) to treat expansile disease, as has been reported (Boyce et al., 2012). Alternatively, using mutant SSCs/BMSCs to create an ectopic bone/marrow organ provides a model for testing new therapeutics to prevent the formation of diseased bone. The ectopic bone/marrow organ may be useful in other diseases for which an appropriate animal model does not exist. Finally, the use of SSCs/BMSCs to create cartilage pellets may be helpful in screening small-molecule libraries for factors that prevent hypertrophy.

Summary

It is now clear that the bone/marrow organ contains a multipotent SSC that is able to re-create cartilage, bone, hematopoiesis-supportive stroma, and marrow adipocytes, and is able to self-renew. As such, these cells are central mediators of skeletal homeostasis. In the postnatal organism, fate choices of SSCs into chondrogenic (rarely), osteogenic, stromagenic, and adipogenic progeny are highly influenced by changes in the microenvironment in which they reside, and are mediated by numerous signaling pathways and genomic and epigenetic processes. Mutations (intrinsic changes) that affect their normal biological functions can have a profound effect on the skeleton and even on hematopoiesis based on their presence and activity in the HSC niche. In addition, changes in the microenvironment (extrinsic) that have an impact on SSC/BMSC function can also lead to skeletal and hematological diseases and disorders. Last, SSCs are an essential ingredient for any process aimed at enduring bone regeneration by the cells themselves due to their ability to mediate bone turnover.

Acknowledgments

The authors would like to acknowledge all of the genuinely seminal contributions of Professor Paolo Bianco to the evolution of the field of skeletal stem cell biology. He was truly a pioneer. The authors would also like to acknowledge all of the members, past and present, of the Skeletal Biology Section, NIDCR, and the Stem Cell Lab, Department of Molecular Medicine, Sapienza University of Rome.

The authors declare that they have no conflicts of interest. This work was supported by the DIR, NIDCR, of the IRP, NIH, DHHS, to P.G.R., and by Telethon (grant GGP15198), the EU (PluriMes consortium, FP7-HEALTH-2013-INNOVATION-1—G.A. 602423), and Sapienza University of Rome to M.R.

References

Akintoye, S.O., Lam, T., Shi, S., Brahim, J., Collins, M.T., Robey, P.G., 2006. Skeletal site-specific characterization of orofacial and iliac crest human bone marrow stromal cells in same individuals. Bone 38, 758—768.

Alliston, T., Choy, L., Ducy, P., Karsenty, G., Derynck, R., 2001. TGF-beta-induced repression of CBFA1 by Smad3 decreases cbfa1 and osteocalcin expression and inhibits osteoblast differentiation. EMBO J. 20, 2254—2272.

Alter, B.P., Rosenberg, P.S., Giri, N., Baerlocher, G.M., Lansdorp, P.M., Savage, S.A., 2012. Telomere length is associated with disease severity and declines with age in dyskeratosis congenita. Haematologica 97, 353—359.

Armulik, A., Genove, G., Betsholtz, C., 2011. Pericytes: developmental, physiological, and pathological perspectives, problems, and promises. Dev. Cell 21, 193—215.

Balakumaran, A., Mishra, P.J., Pawelczyk, E., Yoshizawa, S., Sworder, B.J., Cherman, N., Kuznetsov, S.A., Bianco, P., Giri, N., Savage, S.A., Merlino, G., Dumitriu, B., Dunbar, C.E., Young, N.S., Alter, B.P., Robey, P.G., 2015. Bone marrow skeletal stem/progenitor cell defects in dyskeratosis congenita and telomere biology disorders. Blood 125, 793—802.

Ballew, B.J., Savage, S.A., 2013. Updates on the biology and management of dyskeratosis congenita and related telomere biology disorders. Expert Rev. Hematol. 6, 327—337.

Barron, V., Merghani, K., Shaw, G., Coleman, C.M., Hayes, J.S., Ansboro, S., Manian, A., O'Malley, G., Connolly, E., Nandakumar, A., van Blitterswijk, C.A., Habibovic, P., Moroni, L., Shannon, F., Murphy, J.M., Barry, F., 2015. Evaluation of cartilage repair by mesenchymal stem cells seeded on a PEOT/PBT scaffold in an osteochondral defect. Ann. Biomed. Eng. 43, 2069—2082.

Berendsen, A.D., Olsen, B.R., 2015. Bone development. Bone 80, 14—18.

Bi, W., Deng, J.M., Zhang, Z., Behringer, R.R., de Crombrugghe, B., 1999. Sox9 is required for cartilage formation. Nat. Genet. 22, 85—89.

Bianco, P., 2011. Bone and the hematopoietic niche: a tale of two stem cells. Blood 117, 5281—5288.

Bianco, P., Bonucci, E., 1991. Endosteal surfaces in hyperparathyroidism: an enzyme cytochemical study on low-temperature-processed, glycolmethacrylate-embedded bone biopsies. Virchows Arch. A Pathol. Anat. Histopathol. 419, 425—431.

Bianco, P., Cao, X., Frenette, P.S., Mao, J.J., Robey, P.G., Simmons, P.J., Wang, C.Y., 2013. The meaning, the sense and the significance: translating the science of mesenchymal stem cells into medicine. Nat. Med. 19, 35—42.

Bianco, P., Cossu, G., 1999. Uno, nessuno e centomila: searching for the identity of mesodermal progenitors. Exp. Cell Res. 251, 257—263.

Bianco, P., Kuznetsov, S., Riminucci, M., Robey, P.G., 2006. Postnatal skeletal stem cells. Methods Enzymol. 419, 117—148.

Bianco, P., Kuznetsov, S.A., Riminucci, M., Fisher, L.W., Spiegel, A.M., Robey, P.G., 1998. Reproduction of human fibrous dysplasia of bone in immunocompromised mice by transplanted mosaics of normal and Gsalpha-mutated skeletal progenitor cells. J. Clin. Invest. 101, 1737—1744.

Bianco, P., Riminucci, M., 1998. The bone marrow stroma in vivo: ontogeny, structure, cellular composition and changes in disease. In: Beresfore, J., Owen, M.E. (Eds.), Marrow Stromal Cell Culture. Cambridge University Press, Cambridge, UK, pp. 10—25.

Bianco, P., Riminucci, M., Kuznetsov, S., Robey, P.G., 1999. Multipotential cells in the bone marrow stroma: regulation in the context of organ physiology. Crit. Rev. Eukaryot. Gene Expr. 9, 159—173.

Bianco, P., Riminucci, M., Majolagbe, A., Kuznetsov, S.A., Collins, M.T., Mankani, M.H., Corsi, A., Bone, H.G., Wientroub, S., Spiegel, A.M., Fisher, L.W., Robey, P.G., 2000. Mutations of the GNAS1 gene, stromal cell dysfunction, and osteomalacic changes in non-McCune-Albright fibrous dysplasia of bone. J. Bone Miner. Res. 15, 120–128.

Bianco, P., Robey, P.G., 2004. Skeletal stem cells. In: Lanza, R.P. (Ed.), Handbook of Adult and Fetal Stem Cells. Academic Press, San Diego, pp. 415–424.

Bianco, P., Robey, P.G., 2015. Skeletal stem cells. Development 142, 1023–1027.

Bianco, P., Robey, P.G., Penessi, G., 2008a. Cell source. In: de Boer, J., van Blitterswijk, C., Thomsen, P., Hubbell, J., Cancedda, R., de Bruijn, J.D., Lindahl, A., Sohler, J., Williams, D.F. (Eds.), Tissue Engineering. Elsevier, Amsterdam, pp. 279–306.

Bianco, P., Robey, P.G., Simmons, P.J., 2008b. Mesenchymal stem cells: revisiting history, concepts, and assays. Cell Stem Cell 2, 313–319.

Bothe, I., Tenin, G., Oseni, A., Dietrich, S., 2011. Dynamic control of head mesoderm patterning. Development 138, 2807–2821.

Boxall, S.A., Jones, E., 2012. Markers for characterization of bone marrow multipotential stromal cells. Stem Cell. Int. 2012, 975871.

Boyce, A.M., Chong, W.H., Yao, J., Gafni, R.I., Kelly, M.H., Chamberlain, C.E., Bassim, C., Cherman, N., Ellsworth, M., Kasa-Vubu, J.Z., Farley, F.A., Molinolo, A.A., Bhattacharyya, N., Collins, M.T., 2012. Denosumab treatment for fibrous dysplasia. J. Bone Miner. Res. 27, 1462–1470.

Boyce, A.M., Florenzano, P., de Castro, L.F., Collins, M.T., 1993. Fibrous dysplasia/McCune-Albright syndrome. In: Adam, M.P., Ardinger, H.H., Pagon, R.A., Wallace, S.E., Bean, L.J.H., Stephens, K., Amemiya, A. (Eds.), GeneReviews. Seattle, WA.

Boyce, B.F., Xing, L., 2007. The RANKL/RANK/OPG pathway. Curr. Osteoporos. Rep. 5, 98–104.

Bronner, M.E., 2015. Evolution: on the crest of becoming vertebrate. Nature 527, 311–312.

Burnette Jr., E.W., 1972. Fate of an iliac crest graft. J. Periodontol. 43, 88–90.

Calvi, L.M., Adams, G.B., Weibrecht, K.W., Weber, J.M., Olson, D.P., Knight, M.C., Martin, R.P., Schipani, E., Divieti, P., Bringhurst, F.R., Milner, L.A., Kronenberg, H.M., Scadden, D.T., 2003. Osteoblastic cells regulate the haematopoietic stem cell niche. Nature 425, 841–846.

Caplan, A.I., 2005. Review: mesenchymal stem cells: cell-based reconstructive therapy in orthopedics. Tissue Eng. 11, 1198–1211.

Carthew, R.W., Sontheimer, E.J., 2009. Origins and Mechanisms of miRNAs and siRNAs. Cell 136, 642–655.

Chan, C.K.F., Gulati, G.S., Sinha, R., Tompkins, J.V., Lopez, M., Carter, A.C., Ransom, R.C., Reinisch, A., Wearda, T., Murphy, M., Brewer, R.E., Koepke, L.S., Marecic, O., Manjunath, A., Seo, E.Y., Leavitt, T., Lu, W.J., Nguyen, A., Conley, S.D., Salhotra, A., Ambrosi, T.H., Borrelli, M.R., Siebel, T., Chan, K., Schallmoser, K., Seita, J., Sahoo, D., Goodnough, H., Bishop, J., Gardner, M., Majeti, R., Wan, D.C., Goodman, S., Weissman, I.L., Chang, H.Y., Longaker, M.T., 2018. Identification of the human skeletal stem cell. Cell 175, 43–56 e21.

Chatterjea, A., Meijer, G., van Blitterswijk, C., de Boer, J., 2010. Clinical application of human mesenchymal stromal cells for bone tissue engineering. Stem Cell. Int. 2010, 215625.

Chen, G., Deng, C., Li, Y.P., 2012. TGF-beta and BMP signaling in osteoblast differentiation and bone formation. Int. J. Biol. Sci. 8, 272–288.

Chen, K.G., Johnson, K.R., McKay, R.D.G., Robey, P.G., 2018. Concise review: conceptualizing paralogous stem-cell niches and unfolding bone marrow progenitor cell identities. Stem Cell. 36, 11–21.

Chen, K.G., Johnson, K.R., Robey, P.G., 2017. Mouse genetic analysis of bone marrow stem cell niches: technological pitfalls, challenges, and translational considerations. Stem Cell Reports 9, 1343–1358.

Chen, Q., Shou, P., Zhang, L., Xu, C., Zheng, C., Han, Y., Li, W., Huang, Y., Zhang, X., Shao, C., Roberts, A.I., Rabson, A.B., Ren, G., Zhang, Y., Wang, Y., Denhardt, D.T., Shi, Y., 2014. An osteopontin-integrin interaction plays a critical role in directing adipogenesis and osteogenesis by mesenchymal stem cells. Stem Cell. 32, 327–337.

Chen, Q., Shou, P., Zheng, C., Jiang, M., Cao, G., Yang, Q., Cao, J., Xie, N., Velletri, T., Zhang, X., Xu, C., Zhang, L., Yang, H., Hou, J., Wang, Y., Shi, Y., 2016. Fate decision of mesenchymal stem cells: adipocytes or osteoblasts? Cell Death Differ. 23, 1128–1139.

Chou, D.B., Sworder, B., Bouladoux, N., Roy, C.N., Uchida, A.M., Grigg, M., Robey, P.G., Belkaid, Y., 2012. Stromal-derived IL-6 alters the balance of myeloerythroid progenitors during Toxoplasma gondii infection. J. Leukoc. Biol. 92, 123–131.

Clark, E.A., Kalomoiris, S., Nolta, J.A., Fierro, F.A., 2014. Concise review: MicroRNA function in multipotent mesenchymal stromal cells. Stem Cell. 32, 1074–1082.

Cook, D., Genever, P., 2013. Regulation of mesenchymal stem cell differentiation. Adv. Exp. Med. Biol. 786, 213–229.

Crane, J.L., Cao, X., 2014. Bone marrow mesenchymal stem cells and TGF-beta signaling in bone remodeling. J. Clin. Invest. 124, 466–472.

D'Arcangelo, E., McGuigan, A.P., 2015. Micropatterning strategies to engineer controlled cell and tissue architecture in vitro. Biotechniques 58, 13–23.

Danielian, P.S., Muccino, D., Rowitch, D.H., Michael, S.K., McMahon, A.P., 1998. Modification of gene activity in mouse embryos in utero by a tamoxifen-inducible form of Cre recombinase. Curr. Biol. 8, 1323–1326.

de Boer, J., Licht, R., Bongers, M., van der Klundert, T., Arends, R., van Blitterswijk, C., 2006. Inhibition of histone acetylation as a tool in bone tissue engineering. Tissue Eng. 12, 2927–2937.

Devlin, M.J., Rosen, C.J., 2015. The bone-fat interface: basic and clinical implications of marrow adiposity. Lancet Diabetes Endocrinol 3, 141–147.

Ding, L., Morrison, S.J., 2013. Haematopoietic stem cells and early lymphoid progenitors occupy distinct bone marrow niches. Nature 495, 231–235.

Doherty, E.S., Lacbawan, F., Hadley, D.W., Brewer, C., Zalewski, C., Kim, H.J., Solomon, B., Rosenbaum, K., Domingo, D.L., Hart, T.C., Brooks, B.P., Immken, L., Lowry, R.B., Kimonis, V., Shanske, A.L., Jehee, F.S., Bueno, M.R., Knightly, C., McDonald-McGinn, D., Zackai, E.H., Muenke, M., 2007. Muenke syndrome (FGFR3-related craniosynostosis): expansion of the phenotype and review of the literature. Am. J. Med. Genet. 143A, 3204–3215.

Duchamp de Lageneste, O., Julien, A., Abou-Khalil, R., Frangi, G., Carvalho, C., Cagnard, N., Cordier, C., Conway, S.J., Colnot, C., 2018. Periosteum contains skeletal stem cells with high bone regenerative potential controlled by Periostin. Nat. Commun. 9, 773.

Dzierzak, E., Medvinsky, A., 2008. The discovery of a source of adult hematopoietic cells in the embryo. Development 135, 2343–2346.

Einhorn, T.A., Gerstenfeld, L.C., 2015. Fracture healing: mechanisms and interventions. Nat. Rev. Rheumatol. 11, 45–54.

Evans, B.R., Mosig, R.A., Lobl, M., Martignetti, C.R., Camacho, C., Grum-Tokars, V., Glucksman, M.J., Martignetti, J.A., 2012. Mutation of membrane type-1 metalloproteinase, MT1-MMP, causes the multicentric osteolysis and arthritis disease Winchester syndrome. Am. J. Hum. Genet. 91, 572–576.

Fitter, S., Gronthos, S., Ooi, S.S., Zannettino, A.C., 2017. The mesenchymal precursor cell marker antibody STRO-1 binds to cell surface heat shock cognate 70. Stem Cell. 35, 940–951.

Fontaine, C., Cousin, W., Plaisant, M., Dani, C., Peraldi, P., 2008. Hedgehog signaling alters adipocyte maturation of human mesenchymal stem cells. Stem Cell. 26, 1037–1046.

Friedenstein, A.J., Chailakhyan, R.K., Latsinik, N.V., Panasyuk, A.F., Keiliss-Borok, I.V., 1974. Stromal cells responsible for transferring the microenvironment of the hemopoietic tissues. Cloning in vitro and retransplantation in vivo. Transplantation 17, 331–340.

Galipeau, J., Sensebe, L., 2018. Mesenchymal stromal cells: clinical challenges and therapeutic opportunities. Cell Stem Cell 22, 824–833.

Georgiou, K.R., Hui, S.K., Xian, C.J., 2012. Regulatory pathways associated with bone loss and bone marrow adiposity caused by aging, chemotherapy, glucocorticoid therapy and radiotherapy. Am. J. Stem Cells 1, 205–224.

Goujon, E., 1869. Recherches experimentales sur les proprietes physiologiques de la moelle des os. J. de L'Anat et de La Physiol. 6, 399–412.

Gronthos, S., Graves, S.E., Ohta, S., Simmons, P.J., 1994. The STRO-1+ fraction of adult human bone marrow contains the osteogenic precursors. Blood 84, 4164–4173.

Gronthos, S., Mankani, M., Brahim, J., Robey, P.G., Shi, S., 2000. Postnatal human dental pulp stem cells (DPSCs) in vitro and in vivo. Proc. Natl. Acad. Sci. U.S.A. 97, 13625–13630.

Guilak, F., Cohen, D.M., Estes, B.T., Gimble, J.M., Liedtke, W., Chen, C.S., 2009. Control of stem cell fate by physical interactions with the extracellular matrix. Cell Stem Cell 5, 17–26.

Ha, M., Kim, V.N., 2014. Regulation of microRNA biogenesis. Nat. Rev. Mol. Cell Biol. 15, 509–524.

Haeckel, E. (Ed.), Natürliche Schöpfung-Geschichte, 1868, Druck und Verlag von Georg Reimer, Berlin, p. 832.

Hall, B.K., 2015. Bones and Cartilage, second ed. Academic Press, Waltham, MA.

Hamam, D., Ali, D., Vishnubalaji, R., Hamam, R., Al-Nbaheen, M., Chen, L., Kassem, M., Aldahmash, A., Alajez, N.M., 2014. microRNA-320/RUNX2 axis regulates adipocytic differentiation of human mesenchymal (skeletal) stem cells. Cell Death Dis. 5, e1499.

Happle, R., 1986. The McCune-Albright syndrome: a lethal gene surviving by mosaicism. Clin. Genet. 29, 321–324.

Hemming, S., Cakouros, D., Isenmann, S., Cooper, L., Menicanin, D., Zannettino, A., Gronthos, S., 2014. EZH2 and KDM6A act as an epigenetic switch to regulate mesenchymal stem cell lineage specification. Stem Cell. 32, 802–815.

Hernigou, P., Poignard, A., Manicom, O., Mathieu, G., Rouard, H., 2005. The use of percutaneous autologous bone marrow transplantation in nonunion and avascular necrosis of bone. J Bone Joint Surg Br 87, 896–902.

Hicok, K.C., Du Laney, T.V., Zhou, Y.S., Halvorsen, Y.D., Hitt, D.C., Cooper, L.F., Gimble, J.M., 2004. Human adipose-derived adult stem cells produce osteoid in vivo. Tissue Eng. 10, 371–380.

Hilton, M.J., Tu, X., Wu, X., Bai, S., Zhao, H., Kobayashi, T., Kronenberg, H.M., Teitelbaum, S.L., Ross, F.P., Kopan, R., Long, F., 2008. Notch signaling maintains bone marrow mesenchymal progenitors by suppressing osteoblast differentiation. Nat. Med. 14, 306–314.

Holmbeck, K., Bianco, P., Caterina, J., Yamada, S., Kromer, M., Kuznetsov, S.A., Mankani, M., Robey, P.G., Poole, A.R., Pidoux, I., Ward, J.M., Birkedal-Hansen, H., 1999. MT1-MMP-deficient mice develop dwarfism, osteopenia, arthritis, and connective tissue disease due to inadequate collagen turnover. Cell 99, 81–92.

Holmbeck, K., Bianco, P., Chrysovergis, K., Yamada, S., Birkedal-Hansen, H., 2003. MT1-MMP-dependent, apoptotic remodeling of unmineralized cartilage: a critical process in skeletal growth. J. Cell Biol. 163, 661–671.

Isern, J., Garcia-Garcia, A., Martin, A.M., Arranz, L., Martin-Perez, D., Torroja, C., Sanchez-Cabo, F., Mendez-Ferrer, S., 2014. The neural crest is a source of mesenchymal stem cells with specialized hematopoietic stem cell niche function. Elife 3 e03696.

James, A.W., 2013. Review of signaling pathways governing MSC osteogenic and adipogenic differentiation. Scientifica (Cairo) 2013, 684736.

Jemtland, R., Divieti, P., Lee, K., Segre, G.V., 2003. Hedgehog promotes primary osteoblast differentiation and increases PTHrP mRNA expression and iPTHrP secretion. Bone 32, 611–620.

Johnstone, B., Hering, T.M., Caplan, A.I., Goldberg, V.M., Yoo, J.U., 1998. In vitro chondrogenesis of bone marrow-derived mesenchymal progenitor cells. Exp. Cell Res. 238, 265–272.

Jonas, S., Izaurralde, E., 2015. Towards a molecular understanding of microRNA-mediated gene silencing. Nat. Rev. Genet. 16, 421–433.

Kharlamova, L.A., 1975. Colony formation inhibition in human bone marrow stromal cells exposed to a factor formed in vitro by peripheral blood leukocytes. Biull Eksp Biol Med 80, 89–91.

Khincha, P.P., Savage, S.A., 2013. Genomic characterization of the inherited bone marrow failure syndromes. Semin. Hematol. 50, 333–347.

Kiel, M.J., Yilmaz, O.H., Iwashita, T., Yilmaz, O.H., Terhorst, C., Morrison, S.J., 2005. SLAM family receptors distinguish hematopoietic stem and progenitor cells and reveal endothelial niches for stem cells. Cell 121, 1109–1121.

Kilian, K.A., Bugarija, B., Lahn, B.T., Mrksich, M., 2010. Geometric cues for directing the differentiation of mesenchymal stem cells. Proc. Natl. Acad. Sci. U.S.A. 107, 4872–4877.

Komori, T., Yagi, H., Nomura, S., Yamaguchi, A., Sasaki, K., Deguchi, K., Shimizu, Y., Bronson, R.T., Gao, Y.H., Inada, M., Sato, M., Okamoto, R., Kitamura, Y., Yoshiki, S., Kishimoto, T., 1997. Targeted disruption of Cbfa1 results in a complete lack of bone formation owing to maturational arrest of osteoblasts. Cell 89, 755–764.

Kretlow, J.D., Mikos, A.G., 2007. Review: mineralization of synthetic polymer scaffolds for bone tissue engineering. Tissue Eng. 13, 927–938.

Kretzschmar, K., Post, Y., Bannier-Helaouet, M., Mattiotti, A., Drost, J., Basak, O., Li, V.S.W., van den Born, M., Gunst, Q.D., Versteeg, D., Kooijman, L., van der Elst, S., van Es, J.H., van Rooij, E., van den Hoff, M.J.B., Clevers, H., 2018. Profiling proliferative cells and their progeny in damaged murine hearts. Proc. Natl. Acad. Sci. U.S.A. 115, E12245–E12254.

Kumar, A., Ruan, M., Clifton, K., Syed, F., Khosla, S., Oursler, M.J., 2012. TGF-beta mediates suppression of adipogenesis by estradiol through connective tissue growth factor induction. Endocrinology 153, 254–263.

Kuter, D.J., Bain, B., Mufti, G., Bagg, A., Hasserjian, R.P., 2007. Bone marrow fibrosis: pathophysiology and clinical significance of increased bone marrow stromal fibres. Br. J. Haematol. 139, 351–362.

Kuznetsov, S.A., Cherman, N., Riminucci, M., Collins, M.T., Robey, P.G., Bianco, P., 2008. Age-dependent demise of GNAS-mutated skeletal stem cells and "normalization" of fibrous dysplasia of bone. J. Bone Miner. Res. 23, 1731–1740.

Kuznetsov, S.A., Hailu-Lazmi, A., Cherman, N., de Castro, L.F., Robey, P.G., Gorodetsky, R., 2019. In vivo formation of stable hyaline cartilage by naive human bone marrow stromal cells with modified fibrin microbeads. Stem Cells Transl Med.

Kuznetsov, S.A., Krebsbach, P.H., Satomura, K., Kerr, J., Riminucci, M., Benayahu, D., Robey, P.G., 1997. Single-colony derived strains of human marrow stromal fibroblasts form bone after transplantation in vivo. J. Bone Miner. Res. 12, 1335–1347.

Kuznetsov, S.A., Riminucci, M., Ziran, N., Tsutsui, T.W., Corsi, A., Calvi, L., Kronenberg, H.M., Schipani, E., Robey, P.G., Bianco, P., 2004. The interplay of osteogenesis and hematopoiesis: expression of a constitutively active PTH/PTHrP receptor in osteogenic cells perturbs the establishment of hematopoiesis in bone and of skeletal stem cells in the bone marrow. J. Cell Biol. 167, 1113–1122.

Le Blanc, K., Ringden, O., 2007. Immunomodulation by mesenchymal stem cells and clinical experience. J. Intern. Med. 262, 509–525.

Li, J., Zhang, N., Huang, X., Xu, J., Fernandes, J.C., Dai, K., Zhang, X., 2013. Dexamethasone shifts bone marrow stromal cells from osteoblasts to adipocytes by C/EBPalpha promoter methylation. Cell Death Dis. 4, e832.

Li, J.Y., Tawfeek, H., Bedi, B., Yang, X., Adams, J., Gao, K.Y., Zayzafoon, M., Weitzmann, M.N., Pacifici, R., 2011. Ovariectomy disregulates osteoblast and osteoclast formation through the T-cell receptor CD40 ligand. Proc. Natl. Acad. Sci. U.S.A. 108, 768–773.

Li, Y., Fan, L., Hu, J., Zhang, L., Liao, L., Liu, S., Wu, D., Yang, P., Shen, L., Chen, J., Jin, Y., 2015. MiR-26a rescues bone regeneration deficiency of mesenchymal stem cells derived from osteoporotic mice. Mol. Ther. 23, 1349–1357.

Lian, J.B., Stein, G.S., van Wijnen, A.J., Stein, J.L., Hassan, M.Q., Gaur, T., Zhang, Y., 2012. MicroRNA control of bone formation and homeostasis. Nat. Rev. Endocrinol. 8, 212–227.

Lin, G.L., Hankenson, K.D., 2011. Integration of BMP, Wnt, and notch signaling pathways in osteoblast differentiation. J. Cell. Biochem. 112, 3491–3501.

Ling, L., Murali, S., Dombrowski, C., Haupt, L.M., Stein, G.S., van Wijnen, A.J., Nurcombe, V., Cool, S.M., 2006. Sulfated glycosaminoglycans mediate the effects of FGF2 on the osteogenic potential of rat calvarial osteoprogenitor cells. J. Cell. Physiol. 209, 811–825.

MacCord, K., 2012. Mesenchyme. In: Embryo Project Encyclopedia. Arizona State University. https://embryo.asu.edu/pages/mesenchyme.

Maes, C., Kobayashi, T., Selig, M.K., Torrekens, S., Roth, S.I., Mackem, S., Carmeliet, G., Kronenberg, H.M., 2010. Osteoblast precursors, but not mature osteoblasts, move into developing and fractured bones along with invading blood vessels. Dev. Cell 19, 329–344.

Mankani, M.H., Kuznetsov, S.A., Marshall, G.W., Robey, P.G., 2008. Creation of new bone by the percutaneous injection of human bone marrow stromal cell and HA/TCP suspensions. Tissue Eng. Part A 14, 1949–1958.

Mankani, M.H., Kuznetsov, S.A., Shannon, B., Nalla, R.K., Ritchie, R.O., Qin, Y., Robey, P.G., 2006a. Canine cranial reconstruction using autologous bone marrow stromal cells. Am. J. Pathol. 168, 542–550.

Mankani, M.H., Kuznetsov, S.A., Wolfe, R.M., Marshall, G.W., Robey, P.G., 2006b. In vivo bone formation by human bone marrow stromal cells: reconstruction of the mouse calvarium and mandible. Stem Cell. 24, 2140–2149.

McBeath, R., Pirone, D.M., Nelson, C.M., Bhadriraju, K., Chen, C.S., 2004. Cell shape, cytoskeletal tension, and RhoA regulate stem cell lineage commitment. Dev. Cell 6, 483–495.

Mendez-Ferrer, S., Michurina, T.V., Ferraro, F., Mazloom, A.R., Macarthur, B.D., Lira, S.A., Scadden, D.T., Ma'ayan, A., Enikolopov, G.N., Frenette, P.S., 2010. Mesenchymal and haematopoietic stem cells form a unique bone marrow niche. Nature 466, 829–834.

Meury, T., Verrier, S., Alini, M., 2006. Human endothelial cells inhibit BMSC differentiation into mature osteoblasts in vitro by interfering with osterix expression. J. Cell. Biochem. 98, 992–1006.

Meyer, M.B., Benkusky, N.A., Sen, B., Rubin, J., Pike, J.W., 2016. Epigenetic plasticity drives adipogenic and osteogenic differentiation of marrow-derived mesenchymal stem cells. J. Biol. Chem. 291, 17829–17847.

Miura, M., Gronthos, S., Zhao, M., Lu, B., Fisher, L.W., Robey, P.G., Shi, S., 2003. SHED: stem cells from human exfoliated deciduous teeth. Proc. Natl. Acad. Sci. U.S.A. 100, 5807–5812.

Monroe, D.G., McGee-Lawrence, M.E., Oursler, M.J., Westendorf, J.J., 2012. Update on Wnt signaling in bone cell biology and bone disease. Gene 492, 1–18.

Muraglia, A., Corsi, A., Riminucci, M., Mastrogiacomo, M., Cancedda, R., Bianco, P., Quarto, R., 2003. Formation of a chondro-osseous rudiment in micromass cultures of human bone-marrow stromal cells. J. Cell Sci. 116, 2949–2955.

Muruganandan, S., Roman, A.A., Sinal, C.J., 2009. Adipocyte differentiation of bone marrow-derived mesenchymal stem cells: cross talk with the osteoblastogenic program. Cell. Mol. Life Sci. 66, 236–253.

Nah, H.D., Koyama, E., Agochukwu, N.B., Bartlett, S.P., Muenke, M., 2012. Phenotype profile of a genetic mouse model for Muenke syndrome. Childs Nerv. Syst. 28, 1483–1493.

Nelson, N.D., Bertuch, A.A., 2012. Dyskeratosis congenita as a disorder of telomere maintenance. Mutat. Res. 730, 43–51.

Neubauer, M., Fischbach, C., Bauer-Kreisel, P., Lieb, E., Hacker, M., Tessmar, J., Schulz, M.B., Goepferich, A., Blunk, T., 2004. Basic fibroblast growth factor enhances PPARgamma ligand-induced adipogenesis of mesenchymal stem cells. FEBS Lett. 577, 277–283.

O'Driscoll, S.W., Fitzsimmons, J.S., 2001. The role of periosteum in cartilage repair. Clin. Orthop. Relat. Res. S190–S207.

Oh, S., Oh, N., Appleford, M., Ong, J.L., 2006. Bioceramics for tissue engineering applications – a review. Am. J. Biochem. Biotechnol. 2, 49–56.

Ollier, L., 1867. Traite experimentale et clinique de ea regeneration des os et de la production artificielle du tissu osseux. Victor Masson et fils, Paris, France.

Olsen, B.R., Reginato, A.M., Wang, W., 2000. Bone development. Annu. Rev. Cell Dev. Biol. 16, 191–220.

Omatsu, Y., Sugiyama, T., Kohara, H., Kondoh, G., Fujii, N., Kohno, K., Nagasawa, T., 2010. The essential functions of adipo-osteogenic progenitors as the hematopoietic stem and progenitor cell niche. Immunity 33, 387–399.

Onimaru, K., Shoguchi, E., Kuratani, S., Tanaka, M., 2011. Development and evolution of the lateral plate mesoderm: comparative analysis of amphioxus and lamprey with implications for the acquisition of paired fins. Dev. Biol. 359, 124–136.

Owen, M., Friedenstein, A.J., 1988. Stromal stem cells: marrow-derived osteogenic precursors. Ciba Found. Symp. 136, 42–60.

Phillips, M.D., Kuznetsov, S.A., Cherman, N., Park, K., Chen, K.G., McClendon, B.N., Hamilton, R.S., McKay, R.D., Chenoweth, J.G., Mallon, B.S., Robey, P.G., 2014. Directed differentiation of human induced pluripotent stem cells toward bone and cartilage: in vitro versus in vivo assays. Stem Cells Transl. Med. 3, 867–878.

Piersanti, S., Remoli, C., Saggio, I., Funari, A., Michienzi, S., Sacchetti, B., Robey, P.G., Riminucci, M., Bianco, P., 2010. Transfer, analysis, and reversion of the fibrous dysplasia cellular phenotype in human skeletal progenitors. J. Bone Miner. Res. 25, 1103–1116.

Pinho, S., Lacombe, J., Hanoun, M., Mizoguchi, T., Bruns, I., Kunisaki, Y., Frenette, P.S., 2013. PDGFRalpha and CD51 mark human nestin+ sphere-forming mesenchymal stem cells capable of hematopoietic progenitor cell expansion. J. Exp. Med. 210, 1351–1367.

Pourquie, O., 2001. Vertebrate somitogenesis. Annu. Rev. Cell Dev. Biol. 17, 311–350.

Raaijmakers, M.H., Mukherjee, S., Guo, S., Zhang, S., Kobayashi, T., Schoonmaker, J.A., Ebert, B.L., Al-Shahrour, F., Hasserjian, R.P., Scadden, E.O., Aung, Z., Matza, M., Merkenschlager, M., Lin, C., Rommens, J.M., Scadden, D.T., 2010. Bone progenitor dysfunction induces myelodysplasia and secondary leukaemia. Nature 464, 852–857.

Ramalho-Santos, M., Willenbring, H., 2007. On the origin of the term "stem cell". Cell Stem Cell 1, 35–38.

Rezwan, K., Chen, Q.Z., Blaker, J.J., Boccaccini, A.R., 2006. Biodegradable and bioactive porous polymer/inorganic composite scaffolds for bone tissue engineering. Biomaterials 27, 3413–3431.

Riminucci, M., Collins, M.T., Fedarko, N.S., Cherman, N., Corsi, A., White, K.E., Waguespack, S., Gupta, A., Hannon, T., Econs, M.J., Bianco, P., Gehron Robey, P., 2003a. FGF-23 in fibrous dysplasia of bone and its relationship to renal phosphate wasting. J. Clin. Invest. 112, 683–692.

Riminucci, M., Fisher, L.W., Shenker, A., Spiegel, A.M., Bianco, P., Gehron Robey, P., 1997. Fibrous dysplasia of bone in the McCune-Albright syndrome: abnormalities in bone formation. Am. J. Pathol. 151, 1587–1600.

Riminucci, M., Kuznetsov, S.A., Cherman, N., Corsi, A., Bianco, P., Gehron Robey, P., 2003b. Osteoclastogenesis in fibrous dysplasia of bone: in situ and in vitro analysis of IL-6 expression. Bone 33, 434–442.

Riminucci, M., Liu, B., Corsi, A., Shenker, A., Spiegel, A.M., Robey, P.G., Bianco, P., 1999. The histopathology of fibrous dysplasia of bone in patients with activating mutations of the Gs alpha gene: site-specific patterns and recurrent histological hallmarks. J. Pathol. 187, 249–258.

Riminucci, M., Robey, P.G., Saggio, I., Bianco, P., 2010. Skeletal progenitors and the GNAS gene: fibrous dysplasia of bone read through stem cells. J. Mol. Endocrinol. 45, 355–364.

Riminucci, M., Saggio, I., Robey, P.G., Bianco, P., 2006. Fibrous dysplasia as a stem cell disease. J. Bone Miner. Res. 21 (Suppl. 2), P125–P131.

Roach, H.I., Erenpreisa, J., Aigner, T., 1995. Osteogenic differentiation of hypertrophic chondrocytes involves asymmetric cell divisions and apoptosis. J. Cell Biol. 131, 483–494.

Robey, P., 2017. "Mesenchymal stem cells": fact or fiction, and implications in their therapeutic use. F1000Res 6.

Robey, P.G., 2000. Stem cells near the century mark. J. Clin. Invest. 105, 1489–1491.

Robey, P.G., 2011. Cell sources for bone regeneration: the good, the bad, and the ugly (but promising). Tissue Eng. B Rev. 17, 423–430.

Robey, P.G., Kuznetsov, S., Riminucci, M., Bianco, P., 2007. The role of stem cells in fibrous dysplasia of bone and the Mccune-Albright syndrome. Pediatr. Endocrinol. Rev. 4 (Suppl. 4), 386–394.

Robey, P.G., Kuznetsov, S.A., Riminucci, M., Bianco, P., 2014. Bone marrow stromal cell assays: in vitro and in vivo. Methods Mol. Biol. 1130, 279–293.

Rosen, C.J., Bouxsein, M.L., 2006. Mechanisms of disease: is osteoporosis the obesity of bone? Nat. Clin. Pract. Rheumatol. 2, 35–43.

Sacchetti, B., Funari, A., Michienzi, S., Di Cesare, S., Piersanti, S., Saggio, I., Tagliafico, E., Ferrari, S., Robey, P.G., Riminucci, M., Bianco, P., 2007. Self-renewing osteoprogenitors in bone marrow sinusoids can organize a hematopoietic microenvironment. Cell 131, 324–336.

Sacchetti, B., Funari, A., Remoli, C., Giannicola, G., Kogler, G., Liedtke, S., Cossu, G., Serafini, M., Sampaolesi, M., Tagliafico, E., Tenedini, E., Saggio, I., Robey, P.G., Riminucci, M., Bianco, P., 2016. No identical "mesenchymal stem cells" at different times and sites: human committed progenitors of distinct origin and differentiation potential are incorporated as adventitial cells in microvessels. Stem Cell Reports 6, 897–913.

Schipani, E., Kruse, K., Juppner, H., 1995. A constitutively active mutant PTH-PTHrP receptor in Jansen-type metaphyseal chondrodysplasia. Science 268, 98–100.

Schofield, R., 1978. The relationship between the spleen colony-forming cell and the haemopoietic stem cell. Blood Cells 4, 7–25.

Scotti, C., Piccinini, E., Takizawa, H., Todorov, A., Bourgine, P., Papadimitropoulos, A., Barbero, A., Manz, M.G., Martin, I., 2013. Engineering of a functional bone organ through endochondral ossification. Proc. Natl. Acad. Sci. U.S.A. 110, 3997–4002.

Serafini, M., Sacchetti, B., Pievani, A., Redaelli, D., Remoli, C., Biondi, A., Riminucci, M., Bianco, P., 2014. Establishment of bone marrow and hematopoietic niches in vivo by reversion of chondrocyte differentiation of human bone marrow stromal cells. Stem Cell Res. 12, 659–672.

Simmons, P.J., Torok-Storb, B., 1991. Identification of stromal cell precursors in human bone marrow by a novel monoclonal antibody, STRO-1. Blood 78, 55–62.

Sims, N.A., Martin, T.J., 2014. Coupling the activities of bone formation and resorption: a multitude of signals within the basic multicellular unit. Bonekey Rep. 3, 481.

Sugiyama, T., Kohara, H., Noda, M., Nagasawa, T., 2006. Maintenance of the hematopoietic stem cell pool by CXCL12-CXCR4 chemokine signaling in bone marrow stromal cell niches. Immunity 25, 977–988.

Suresh, S., de Castro, L.F., Dey, S., Robey, P.G., Noguchi, C.T., 2019. Erythropoietin modulates bone marrow stromal cell differentiation. Bone Res (in press).

Sworder, B.J., Yoshizawa, S., Mishra, P.J., Cherman, N., Kuznetsov, S.A., Merlino, G., Balakumaran, A., Robey, P.G., 2015. Molecular profile of clonal strains of human skeletal stem/progenitor cells with different potencies. Stem Cell Res. 14, 297–306.

Taipaleenmaki, H., Abdallah, B.M., AlDahmash, A., Saamanen, A.M., Kassem, M., 2011. Wnt signalling mediates the cross-talk between bone marrow derived pre-adipocytic and pre-osteoblastic cell populations. Exp. Cell Res. 317, 745–756.

Tang, Y., Wu, X., Lei, W., Pang, L., Wan, C., Shi, Z., Zhao, L., Nagy, T.R., Peng, X., Hu, J., Feng, X., Van Hul, W., Wan, M., Cao, X., 2009. TGF-beta1-induced migration of bone mesenchymal stem cells couples bone resorption with formation. Nat. Med. 15, 757–765.

Tavassoli, M., Crosby, W.H., 1968. Transplantation of marrow to extramedullary sites. Science 161, 54–56.

Teven, C.M., Liu, X., Hu, N., Tang, N., Kim, S.H., Huang, E., Yang, K., Li, M., Gao, J.L., Liu, H., Natale, R.B., Luther, G., Luo, Q., Wang, L., Rames, R., Bi, Y., Luo, J., Luu, H.H., Haydon, R.C., Reid, R.R., He, T.C., 2011. Epigenetic regulation of mesenchymal stem cells: a focus on osteogenic and adipogenic differentiation. Stem Cell. Int. 2011, 201371.

Tormin, A., Li, O., Brune, J.C., Walsh, S., Schutz, B., Ehinger, M., Ditzel, N., Kassem, M., Scheding, S., 2011. CD146 expression on primary nonhematopoietic bone marrow stem cells is correlated with in situ localization. Blood 117, 5067–5077.

Tu, X., Chen, J., Lim, J., Karner, C.M., Lee, S.Y., Heisig, J., Wiese, C., Surendran, K., Kopan, R., Gessler, M., Long, F., 2012. Physiological notch signaling maintains bone homeostasis via RBPjk and Hey upstream of NFATc1. PLoS Genet. 8 e1002577.

Ugarte, F., Forsberg, E.C., 2013. Haematopoietic stem cell niches: new insights inspire new questions. EMBO J. 32, 2535–2547.

Walne, A.J., Bhagat, T., Kirwan, M., Gitiaux, C., Desguerre, I., Leonard, N., Nogales, E., Vulliamy, T., Dokal, I.S., 2013. Mutations in the telomere capping complex in bone marrow failure and related syndromes. Haematologica 98, 334–338.

Watt, F.M., Hogan, B.L., 2000. Out of Eden: stem cells and their niches. Science 287, 1427–1430.

Winkler, T., Sass, F.A., Duda, G.N., Schmidt-Bleek, K., 2018. A review of biomaterials in bone defect healing, remaining shortcomings and future opportunities for bone tissue engineering: the unsolved challenge. Bone Joint Res 7, 232–243.

Xiong, J., O'Brien, C.A., 2012. Osteocyte RANKL: new insights into the control of bone remodeling. J. Bone Miner. Res. 27, 499–505.

Xu, R., Hu, J., Zhou, X., Yang, Y., 2018. Heterotopic ossification: mechanistic insights and clinical challenges. Bone 109, 134–142.

Yamaguchi, H., Calado, R.T., Ly, H., Kajigaya, S., Baerlocher, G.M., Chanock, S.J., Lansdorp, P.M., Young, N.S., 2005. Mutations in TERT, the gene for telomerase reverse transcriptase, in aplastic anemia. N. Engl. J. Med. 352, 1413–1424.

Yang, L., Tsang, K.Y., Tang, H.C., Chan, D., Cheah, K.S., 2014. Hypertrophic chondrocytes can become osteoblasts and osteocytes in endochondral bone formation. Proc. Natl. Acad. Sci. U.S.A. 111, 12097–12102.

Zanotti, S., Canalis, E., 2016. Notch signaling and the skeleton. Endocr. Rev. 37, 223–253.

Zhang, J.F., Fu, W.M., He, M.L., Xie, W.D., Lv, Q., Wan, G., Li, G., Wang, H., Lu, G., Hu, X., Jiang, S., Li, J.N., Lin, M.C., Zhang, Y.O., Kung, H.F., 2011. MiRNA-20a promotes osteogenic differentiation of human mesenchymal stem cells by co-regulating BMP signaling. RNA Biol. 8, 829–838.

Chapter 3

Bone marrow and the hematopoietic stem cell niche

Laura M. Calvi
Department of Medicine and Wilmot Cancer Center, University of Rochester Medical Center, Rochester, NY, United States

Chapter outline

Introduction 73
The niche concept: a historical prospective 74
Hematopoietic stem cell microenvironments in the embryo and perinatal period 74
The adult bone marrow niche 75
Mesenchymal stromal/stem cell populations 75
Adipocytes 75
Osteoblastic cells 76
Endothelial cells 77
Hematopoietic cells 77
Megakaryocytes 78
Macrophages 78
Neutrophils 78
T cells 79
Osteoclasts 79
Neuronal regulation of the hematopoietic stem cell niche 79
Hormonal regulation of the hematopoietic stem cell niche 79
Parathyroid hormone 79
Insulin-like growth factor 1 80
Niche heterogeneity for heterogeneous hematopoietic stem and progenitor cells 80
Conclusions 81
Acknowledgments 81
References 81

Introduction

In the adults of land-dwelling vertebrates, the skeleton is the obligate site for normal hematopoiesis during homeostasis. Throughout the life of an organism, hematopoiesis maintains the blood components necessary for oxygenation (erythropoiesis), clotting (megakaryopoiesis/thrombopoiesis), and immunity (myelopoiesis and lymphopoiesis). Since differentiated cells have varying life spans, from hours (in the case of neutrophils) to years (in the case of memory T cells and hematopoietic stem cells [HSCs]), a large component of the hematopoietic system is continually replaced by intermediate committed precursors derived from progenitors that may give rise to more than one differentiated lineage. Progenitors and precursors have limited self-renewal capacity and derive from HSCs, tissue-specific stem cells that persist throughout the life of an organism. While these stem cells are primarily quiescent, they are recruited when there is increased demand for blood products. Given their ability to regenerate the entire hematopoietic system and their relative accessibility, HSCs have been extensively studied as a model system for stem cell behavior. Clinically, HSCs are commonly used in the setting of stem cell transplantation for the treatment of hematologic malignancies and other disorders; therefore, extensive research has focused on understanding their regulation for regenerative purposes. In addition to cell-autonomous programs, a highly complex and dynamic microenvironment in the bone marrow contributes to the support and regulation of HSCs. The components of this system are likely to represent important potential therapeutic targets if their regulation and impact on HSCs are understood. Moreover, since bone and marrow are components of the same organ, dysfunction in bone may contribute to hematopoietic failure and vice versa. Here we present a historical prospective on the concept of the bone marrow microenvironment as a regulatory component for hematopoiesis, and describe the data supporting the role of niche constituents in the regulation of HSCs.

Principles of Bone Biology. https://doi.org/10.1016/B978-0-12-814841-9.00003-8

The niche concept: a historical prospective

In the 1960s and early 1970s the ability to transplant bone marrow cells that could regenerate the hematopoietic system (McCulloch and Till, 1960) and the development of in vivo and cell culture methods that could detect and study hematologic progenitor cells as colony-forming unit cells (Becker et al., 1963; Senn et al., 1967) initiated studies aimed at defining immature cells responsible for sustaining hematopoiesis and their regulation. Based on divergent observations of spleen colony-forming cells compared with in vivo HSCs, the stem cell niche concept was initially proposed in 1978 by Schofield (Schofield, 1978). This investigator single-handedly postulated that the "virtual immortality" of HSCs in the bone marrow may be conferred by association with other cells capable of preventing their maturation, maintaining their reconstitution capacity (Schofield, 1978). At the same time, studies suggested that the distribution of HSCs followed a hierarchical organization, with the most primitive cell populations being found at the endosteal surface of bone (Gong, 1978). Analogous to the niche in ecology, niches conceptually represent the microenvironmental conditions that support specific stem cell populations. Schofield proposed that niche cells would be found in close proximity to supported stem cells, so as to form a relatively "fixed" unit, and that they would provide signals that could modify stem cell fate choices, including bestowing stem cell potential to more differentiated daughter cells (Schofield, 1978). While this innovative idea was stimulated by observations in hematopoiesis, the niche was first demonstrated in the *Drosophila* gonad, where elegant studies showed that niche occupancy could alter cell fate and induce stem cell behavior in progenitor populations (Kiger et al., 2000; Tran et al., 2000; Xie and Spradling, 2000). Since these initial studies, niches have been defined in colonial chordates (Rosental et al., 2018), *Caenorhabditis elegans* (Austin and Kimble, 1987), zebrafish (Tamplin et al., 2015), and murine models for many tissue-based stem cells (Tumbar et al., 2004; Bjerknes and Cheng, 2001), including, most recently, the hematopoietic system. In the bone marrow, many challenges initially delayed identification of niches and critical populations within them. These challenges included the technical difficulties involved in the examination of the boundary between the hard bone tissue and the soft bone marrow, difficulty in direct visualization of the marrow, apparent anatomic disorganization of marrow populations, rarity of HSCs, lack of histologic distinguishing features of these populations, and limited definition of mesenchymal cell populations. However, the regulatory niche, if comprehensively defined, offers the potential for indirect stem cell manipulation and could therefore be a very powerful therapeutic target. In addition, understanding of the critical component of the niche is necessary to design adequate supportive conditions for HSCs ex vivo. For these reasons, definition of the HSC niche has been the focus of intense research efforts since the beginning of the 21st century.

Hematopoietic stem cell microenvironments in the embryo and perinatal period

While in the adult HSCs are found primarily in bone, stem cells with the ability to generate all hematopoietic lineages initially arise in the hemogenic endothelium within the aorta–gonad–mesonephros (AGM) (Ivanovs et al., 2011; Muller et al., 1994; Medvinsky and Dzierzak, 1996; Medvinsky et al., 2011; de Bruijn et al., 2000) as a result of endothelial-to-hematopoietic cell transition (Jaffredo et al., 1998; Bertrand et al., 2010; Boisset et al., 2010; Kissa and Herbomel, 2010; de Bruijn et al., 2002). Additional studies have shown that HSCs may develop not only in the AGM but also in the placenta (Gekas et al., 2005; Ottersbach and Dzierzak, 2005; Rhodes et al., 2008). The role of the placenta as a supportive microenvironment for HSCs has also been demonstrated in humans (Robin et al., 2009). From the AGM and the placenta, HSCs migrate to the fetal liver (murine embryonic day 1, E11), where they are supported by portal vessels (Khan et al., 2016). In the liver, HSCs dramatically expand prior to seeding the bone marrow (Morrison et al., 1995; Bowie et al., 2007; Ema and Nakauchi, 2000). HSC activity is first detected in the bone marrow at E16.5 (Coskun et al., 2014). Contemporary analyses of skeletal and hematopoietic development have shown that, in the mouse, this transition to the bone marrow corresponds to fetal long bone vascularization and incipient osteogenesis (Coskun et al., 2014). Notably, at E16.5, in the long bones osteoblasts are primarily found at the bone collar/periosteum, suggesting that at this developmental stage marrow vascular or mesenchymal populations may drive HSC homing. However, mice with genetic deletion of the transcription factor Osterix ($Sp7^{-/-}$ mice), which lack osteoprogenitors and their progeny, while their marrow vasculature is intact, have bone marrow HSCs that cannot properly home and engraft (Coskun et al., 2014), highlighting the importance of the osteogenic niche in the perinatal time. HSCs continue to migrate to the bone marrow through the first postnatal week, at a time when liver portal vessels are remodeled (Khan et al., 2016), while osteoblastic cell populations are rapidly expanding (Coskun et al., 2014). Once in the bone marrow, during the first 3 weeks of life, HSCs proliferate extensively while bone and marrow grow (Bowie et al., 2006). Subsequently, HSCs become primarily quiescent (Wilson et al., 2008), although the degree of their contribution to daily hematopoiesis remains disputed (Sun et al., 2014; Sawai et al., 2016; Busch et al., 2015).

The adult bone marrow niche

The discovery of cell surface markers that could more easily enrich for and prospectively isolate cells with HSC function using few stains (Kiel et al., 2005; Oguro et al., 2013), combined with advances in tissue imaging, reviewed in Tjin et al. (2018), and the availability of genetic models that could target cellular subsets, reviewed in Kfoury and Scadden (2015), is revolutionizing our understanding of the bone marrow microenvironment. We next summarize bone marrow cell populations that have been shown to contribute to the regulation and support of HSCs.

Mesenchymal stromal/stem cell populations

The definition and identification of mesenchymal stem/stromal cells (MSCs) remain a disputed topic in the bone field; however, niche studies have highlighted specific subsets of these populations that are able to support and modulate HSCs (Mendez-Ferrer et al., 2010). These studies have been performed primarily in the mouse, while the definition of MSCs and their role as HSC niches in the human bone marrow remain poorly understood. By using genetic models in which a key supportive molecule, stromal cell-derived factor 1 (SDF1, also known as C-X-C motif chemokine 12, or CXCL12) or stem cell factor (SCF) (Greenbaum et al., 2013; Ding and Morrison, 2013; Zhou et al., 2017), is genetically deleted in different mesenchymal subsets, and then quantifying HSCs, a number of marrow MSC populations have been reported as participating in HSC support. HSC-supportive cells have also been described, based on their degree of CXCL12 expression, as CXCL12-abundant reticular (or CAR) cells (Sugiyama et al., 2006). A number of populations identified based on either genetic labeling or flow cytometric expression have been shown to support HSCs: nestin—green fluorescent protein (GFP)-positive cells (Mendez-Ferrer et al., 2010), leptin receptor-positive cells (Zhou et al., 2017), paired related homeobox 1 (Prx1)-targeted cells (Greenbaum et al., 2013), and cells identified by their expression of integrin αV (CD51), the platelet-derived growth factor receptor α (Pinho et al., 2013), or CD51 and the cell surface marker Sca1 (Morikawa et al., 2009). Notably, there is evidence that some cell populations, for example, nestin—GFP$^+$ cells, are still heterogeneous (Kunisaki et al., 2013) and that they do not represent multipotent MSCs, since they cannot differentiate to adipocytes (Zhou et al., 2014; Itkin et al., 2016). In addition, there is probably a hierarchy in the organization of these populations. For example, Prx1$^+$ MSCs are able to generate leptin receptor-positive and CAR cells (Zhou et al., 2014; Omatsu et al., 2010), and deletion of the leptin receptor in Prx1$^+$ cells expands osteolineage populations while inhibiting adipocytes (Yue et al., 2016).

More recent data have suggested that the location of supporting cell populations may also have an impact on their ability to functionally influence HSCs. Studies have shown that perivascular cell populations in fact express the highest level of CXCL12 (Ding and Morrison, 2013). Some studies show that nestin—GFP$^+$ cells are distributed along arterioles and sinusoidal structures within the bone marrow (Kunisaki et al., 2013), and at these different locations they provide differential support to HSCs (Asada et al., 2017). Nestin—GFP$^+$ cells at arteriole sites express the pericyte marker neural/glial antigen 2 (NG2), and appear to play a role in the maintenance of HSCs (Pinho et al., 2018). These findings have been debated, as another study has suggested that niches at sinusoidal sites provide a quiescent niche for HSCs (Acar et al., 2015). As we will discuss later, these discrepancies are probably due not only to different experimental models (Joseph et al., 2013), but also to heterogeneity of HSC populations.

Adipocytes

Marrow adipocytes are differentiated from MSC populations (Horowitz et al., 2017). They are rare populations in the juvenile marrow but increase during aging and especially in the setting of myeloablation and anorexia (Calvo et al., 1976; Abella et al., 2002). Studies have suggested that marrow adipocytes have different characteristics compared with adipocytes in other depots and may be heterogeneous (Scheller et al., 2015); however, the lack of a genetic approach to target marrow compared with other adipocyte depots has as of this writing represented a challenge to support these claims. The contribution of marrow adipocytes to the support of HSCs is an area of debate, as is their impact on bone (Doucette et al., 2015; Ambrosi et al., 2017). Initial studies using genetic murine models and pharmacologic tools suggested that adipocytic populations may inhibit HSCs (Naveiras et al., 2009). More recently, Ambrosi et al. presented data that correlate age- and diet-dependent increases in adipocytes with hematopoietic and skeletal stem cell dysfunction (Ambrosi et al., 2017), while Zhou et al. found that adipocytes secrete SCF and promote the regeneration of HSCs (Zhou et al., 2017). While some of these discrepancies have been attributed to differential effects in different bones (Zhou et al., 2017), these data may also tap into the reciprocal relationship of osteolineage and adipocytic cells, long shown in vitro (Ge et al., 2016), and also into the emerging concept that adipocytes may represent a local energy depot to support the metabolically demanding daily hematopoiesis (Tabe et al., 2017). The ability to target specifically marrow adipocytes genetically will no doubt provide clarity to this important area of investigation, especially as it may modulate skeletal and hematopoietic regeneration, as well as contributing to a supportive microenvironment in the setting of leukemia (Tabe et al., 2017).

Osteoblastic cells

Since bone is the required site for hematopoiesis in adulthood, cells in the osteoblastic lineage were the first candidates as HSC-supportive niche cells. Initial in vitro studies demonstrated that cells of osteoblastic lineage, derived from mice or humans, could provide support to hematopoietic stem and progenitor cells (Taichman and Emerson, 1994; Taichman et al., 1996). Subsequently, data showed that genetic manipulation of osteoblastic populations could expand HSCs, providing the first evidence for the existence of a regulatory hematopoietic microenvironment (Calvi et al., 2003; Zhang et al., 2003). Additional experiments in which osteoblastic populations were ablated in adulthood also demonstrated loss of hematopoietic populations (Visnjic et al., 2004), although lymphocytes were lost prior to HSCs. Moreover, studies using bone marrow samples from patients showed that HSCs are found proximal to endosteal surfaces (Guezguez et al., 2013). In addition, direct imaging of the niche in the calvarial space demonstrated homing of HSCs preferentially to the endosteum (Lo Celso et al., 2009; Xie et al., 2009), even in the absence of myeloablative conditioning (Ellis et al., 2011), with hierarchical organization placing HSCs proximal to the endosteum compared with progenitor cells (Lo Celso et al., 2009). In a genetic model, depletion of osteocalcin-expressing osteolineage cells inhibits mobilization of HSCs (Asada et al., 2013; Ferraro et al., 2011). Finally, when *Sp7*, the gene encoding the transcription factor Osterix, is conditionally deleted, with resulting loss of osteoblastic differentiation, hematopoiesis is almost completely eliminated from the metaphyseal region (Zhou et al., 2010). However, the importance of osteoblastic populations in support of HSCs has been questioned by studies showing that, while the majority of hematopoietic cells reside in the metaphyseal regions, enriched with trabecular bone (Nombela-Arrieta et al., 2013; Guezguez et al., 2013), only a small subset of HSCs are found in direct proximity to osteoblasts (Lo Celso et al., 2009; Kiel et al., 2005, 2009; Sugiyama et al., 2006; Acar et al., 2015), pointing instead to perivascular sites as more frequent niches for HSCs. Later studies clearly demonstrated the close proximity of important endothelial structures to the endosteum, and their interaction with osteoblastic cells, suggesting that these populations collaborate in the support and regulation of HSCs (Kusumbe et al., 2014, 2016; Ramasamy et al., 2014; Xu et al., 2018). Skepticism regarding the role of osteolineage cells in HSC regulation was also supported by studies targeting deletions of two key pro-HSC signals, the chemokine CXCL2 (also known as SDF1) and the growth factor SCF, which demonstrated regulation of HSC numbers and function when removed from mesenchymal stem and progenitor cells but not from osteoblastic populations (Ding and Morrison, 2013; Greenbaum et al., 2013). However, these models reflect primarily the impact of these two factors on HSC numbers and function, and therefore limit the scope of the evaluation of the niche.

Numerous studies provide interesting evidence that osteoblastic cell populations support specifically the common lymphoid progenitor and/or the lymphoid lineage within the bone marrow (Zhu et al., 2007; Ding and Morrison, 2013; Wu et al., 2008), supporting the concept that there may be multiple niches within the bone marrow to support not just HSCs, but also other progenitor populations that are more specialized. For example, data have shown that thymus-seeding progenitors are supported by the Notch ligand Delta-like 4, produced by osteocalcin-expressing osteoblastic cells (Yu et al., 2015).

Within the osteoblastic lineage, there is also evidence that differentiation stage may contribute differentially to HSC support. Data from in vitro studies in which the differentiation stage of osteoblastic populations was controlled showed decreasing support of HSCs with osteoblastic maturation (Cheng et al., 2011; Nakamura et al., 2010). These findings are also supported by proximity-based isolation of endosteal osteoblastic populations found close to HSCs in vivo (Silberstein et al., 2016). Data have shown that genetic disruption of osteoprogenitor cells initiates bone marrow failure and pre-leukemia, while the same defect targeted to osteocalcin-positive populations does not (Raaijmakers et al., 2010). Moreover, while targeting of a constitutively active parathyroid hormone (PTH) receptor to osteoblastic cells expanded HSCs (Calvi et al., 2003), targeting the same construct to osteocytes did not (Calvi et al., 2012), despite similar degrees of osteoblastic expansion. A key PTH-regulated molecule that is produced by osteocytes is sclerostin, the protein encoded by the *Sost* gene, which is known to be downregulated by PTH. Notably, consistent with the data from mice with targeting of the PTH receptor to osteocytes, *Sost*$^{-/-}$ mice also demonstrate the bone anabolic effect of PTH but not the expansion in HSCs (Cain et al., 2012). Therefore, expansion of osteoblasts alone is not sufficient to expand HSCs, as also shown by treatment with strontium (Lymperi et al., 2008), and osteocyte activation alone does not increase HSCs. However, when osteocytes were depleted in adult mice, lymphopoiesis was disrupted, and mobilization of HSCs induced by granulocyte colony−stimulating factor (G-CSF) was inhibited (Asada et al., 2013). These data suggest that these populations of terminally differentiated osteoblasts that are embedded in matrix contribute to some aspects of HSC regulation. Interestingly, targeted deletion of $G_s\alpha$ in osteocytes induces a myeloproliferative disorder and splenomegaly through the microenvironment, in a G-CSF-dependent manner (Fulzele et al., 2013). Together these data suggest that cells in the osteoblastic lineage contribute to hematopoietic regulation in a differentiation-dependent manner. One limitation of these data is the unexpected targeting of CAR cells by the DMP-1 promoter (Zhang and Link, 2016). Further data are required to define mechanistically the contributions of these osteolineage populations to the regulation of HSCs and their progeny.

While the importance of osteoblastic cells in normal HSCs remains a hotly debated issue, the role of osteoblastic cells in hematologic malignancies is supported by extensive experimental data. In a murine model of myeloproliferative neoplasia there was an expansion of osteolineage populations that were, however, defective in their ability to support normal HSCs, enhancing clonal hematopoietic populations (Schepers et al., 2013). Numerous models of leukemia have shown its ability to disrupt osteoblastic populations, for example, in a model of blast crisis acute myeloid leukemia (Frisch et al., 2012) and in the MML-AF9 model of acute myeloid leukemia (Silva et al., 2010; Duarte et al., 2018). Loss of osteoblasts accelerated leukemia progression (Silva et al., 2011), and osteoblasts could inhibit leukemia cell engraftment and disease progression (Krevvata et al., 2014). In some cases, activation of the normal niche was able to decrease transformation of preleukemic clones to leukemia (Krause et al., 2013), pointing to the potentially protective role of normal niches.

Osteoblastic cells are also important for the ability of HSCs to recover from radiation injury (Dominici et al., 2009). In this setting, rapid expansion of endosteal osteoblasts required insulin-like growth factor 1 (IGF1) (Caselli et al., 2013). In elegant studies using proximity-based single-cell profiling in the setting of radiation conditioning, Silberstein et al. were able to define characteristics of endosteal osteoblastic populations proximal to HSCs, and identified secreted signals that regulate HSCs, including the secreted RNase angiogenin, interleukin 18, and the adhesion molecule Embigin (Silberstein et al., 2016). Interestingly, these studies were performed in neonatal calvaria, to avoid decalcification, and therefore whether these cell populations contribute to the support and regulation of adult HSCs remains untested. However, the novel signals identified by this approach were found to be important for hematopoietic stem and progenitor cell regulation, strongly implicating osteolineage populations in hematopoietic recovery after myeloablation (Silberstein et al., 2016). Notably, in the case of angiogenin, the same signal had differential effects on stem cell populations and progenitors, indicating a role for osteolineage populations in a coordinated response of the hematopoietic system to acute injury, in which angiogenin maintains HSC quiescence while enhancing progenitor cell proliferation (Goncalves et al., 2016).

Lessons learned from the dissection of mesenchymal osteoblastic components and their matrix were instrumental in attempts at recapitulating a supportive HSC niche ex vivo, as was shown in initial studies describing a biomimetic microenvironment using human cell populations (Bourgine et al., 2018).

Endothelial cells

Initial studies identifying HSCs as lineage$^-$, CD150$^+$, CD48$^-$, and CD41$^-$ cells showed that 60% of these cells were found in proximity to sinusoidal endothelial structures in the adult marrow at homeostasis (Kiel et al., 2005, 2007), suggesting that endothelial cells may contribute to HSC regulation. However, deletion of *CXCL12*, *Jagged1*, or *SCF* in endothelial populations did not have an impact on HSC maintenance although deletion of *Jagged1* in endothelial cells may impact HSF self-renewal (Ding et al., 2012; Greenbaum et al., 2013; Poulos et al., 2013; Ding and Morrison, 2013). However, the endothelium is clearly critical for recovery of HSCs after myeloablation (Hooper et al., 2009; Kobayashi et al., 2010; Butler et al., 2010; Winkler et al., 2012). Just as in the mesenchymal osteoblastic lineage, the heterogeneity of marrow endothelial structures is emerging (Kusumbe et al., 2014; Ramasamy et al., 2014). Vascular heterogeneity clearly imparts differential effects on hematopoiesis (Itkin et al., 2016), to some extent as a result of differential permeability of sinusoids compared with arteriolar structures. Moreover, coupling of vascular with mesenchymal osteolineage populations is being uncovered (Xu et al., 2018; Kusumbe et al., 2016). These data strongly suggest that different niche components are interdependent in their regulation of HSCs.

Hematopoietic cells

The progeny of HSCs are also recognized for their ability to regulate HSCs. In particular, megakaryocytes, marrow macrophages, senescent neutrophils, osteoclasts, and regulatory T cells have been demonstrated to regulate HSCs, either directly or through their ability to modulate the behavior of other niche cells.

Megakaryocytes

Megakaryocytes are the hematologic progenitors that give rise to platelets in the bone marrow when associated with endothelial structures. Several groups have shown that megakaryocytes are direct regulators of HSC quiescence. In the bone marrow, HSCs are often found in close proximity to megakaryocytes (Bruns et al., 2014). Moreover, signals derived from megakaryocytes, including the chemokine CXCL4 (also known as platelet factor 4), Thrombopoietin, and transforming growth factor β1 (TGFβ1), have been shown to mediate the quiescence of HSCs, at least at homeostasis (Bruns et al., 2014; Nakamura-Ishizu et al., 2014, 2015; Zhao et al., 2014). Megakaryocytes also promote murine osteoblastic expansion after radioablation (Olson et al., 2013), and provide a supportive niche preferentially for platelet- and myeloid-biased HSCs (Pinho et al., 2018). The study shows that cells contributing to HSC support can also have an impact on their fate, since deletion of megakaryocytes could induce myeloid- and platelet-biased HSCs to give balanced lineage contribution upon transplantation.

Macrophages

Macrophages are key regulators of both innate and adaptive immunity, and are critical for tissue homeostasis through their capacity to engulf debris, pathogens, and apoptotic cells. Macrophages are highly heterogeneous within different tissues and are derived both from the yolk sac and from the differentiation of monocytes derived from bone marrow (Heideveld and van den Akker, 2017). In the bone marrow, specialized macrophage activities include the engulfment of senescent neutrophils (Casanova-Acebes et al., 2013) and erythropoiesis through their contribution to erythroblastic islands (Bessis, 1958; Chow et al., 2013). Supporting a role for macrophages in the bone marrow niche, HSC dormancy is regulated through the interaction of CD234/Duffy antigen receptor for chemokines expressed on macrophages with CD82/KAI1 expressed on HSCs (Hur et al., 2016). Several lines of evidence demonstrate that marrow macrophages play an important role in trafficking of HSCs to and from the bone marrow. Macrophages regulate the retention of hematopoietic stem and progenitor cells in the bone marrow (Chow et al., 2011; Winkler et al., 2010), and specifically the release of HSCs from the niche, also known as mobilization, induced by G-CSF (Christopher et al., 2011). Interestingly, some of these actions appear to be through their ability to modulate mesenchymal and osteoblastic cell populations. For example, a subset of $CD169^{+}$ macrophages was found to maintain HSCs in the marrow by regulating MSC production of CXCL12 (Chow et al., 2011). In addition, a population of marrow macrophages marked by smooth muscle actin expresses cyclooxygenase 2 and produces prostaglandin E2 (Ludin et al., 2012). This macrophage population supports HSCs and is expanded by radiation (Ludin et al., 2012) and, on a daily basis, by circadian bursts of darkness-induced norepinephrine that induces melatonin through tumor necrosis factor (Golan et al., 2018). More recently, studies have begun to show that macrophages are also important for homing and engraftment of HSCs. For example, in the zebrafish caudal hematopoietic tissue (equivalent to the mammalian fetal liver), vascular cell adhesion molecule-positive macrophages were found to patrol the venous plexus, interact with HSCs through integrin α4, and regulate HSC retention (Li et al., 2018). In the mammalian bone marrow, a population of $CD169^{+}$ macrophages located within perivascular and endosteal regions was resistant to conditioning radiation and was found to promote long-term engraftment of HSCs (Kaur et al., 2018). A subset of marrow macrophages found in close proximity to the endosteum, also known as osteomacs, has been shown to regulate osteoblastic function (Chang et al., 2008) and also contribute to support of HSCs through their synergy with osteoblasts and megakaryocytes (Mohamad et al., 2017). Finally, a recent study demonstrated that marrow macrophages from aged mice become defective, with increased inflammatory signals and decreased phagocytic capacity, and that they determine HSC megakaryocytic skewing likely through Interleukin 1 β (Frisch et al 2019 JCI Insight. 2019 Apr 18;5. pii: 124213. doi: 10.1172/jci.insight.124213.PMID: 30998506).

Neutrophils

Neutrophils are hematopoietic cells characterized by segmented nuclei and the presence of granules critical for innate immunity. Neutrophils are very short lived and highly mobile, and represent the "first responder" population in immune and injury responses (Nathan, 2006). They are produced in the bone marrow, where they are preferentially found in close proximity to CAR cells rather than endothelial cells (Evrard et al., 2018). Mature and immature neutrophils exit the bone marrow to the circulation and tissues and senesce rapidly (Evrard et al., 2018). Following a circadian rhythm, a large number of senescent neutrophils expressing the CXCL12 receptor CXCR4 home daily to the bone marrow (Casanova-Acebes et al., 2013). In the bone marrow, senescent neutrophils are phagocytosed by macrophages and indirectly modulate microenvironmental CXCL12 levels (Casanova-Acebes et al., 2013), influencing signals that not only guide senescent neutrophils but also support HSCs.

T cells

High-resolution imaging of the HSC niche after transplantation showed colocalization of hematopoietic stem and progenitor cells with regulatory FoxP3 T cells (T regs) at the endosteum, and demonstrated that depletion of T regs resulted in the loss of HSCs (Fujisaki et al., 2011). These data support the hypothesis that T regs are necessary for the maintenance of allogeneic HSCs after transplantation. Follow-up studies identified a subset of CD150high T regs in the niche that maintain HSC quiescence (Hirata et al., 2018). Transcriptional analysis of these cells found increases in CD39 and CD73, cell surface ectoenzymes that generate extracellular adenosine (Hirata et al., 2018). The authors went on to demonstrate that adenosine is required to mediate the beneficial effects of CD150high T regs, and that adaptive transfer of these cells improves allogeneic engraftment of HSCs (Hirata et al., 2018). In addition to demonstrating the participation of T regs in the regulation of adult HSCs, these data provide strong evidence that definition and targeting of the niche can have a significant impact under clinically relevant conditions.

Osteoclasts

Osteoclasts, key specialized cells that resorb bone, are hematopoietically derived and play a crucial role in the maintenance of skeletal homeostasis. Osteoclasts, as key cells that resorb bone, are necessary for the formation of the bone marrow cavity in the embryo (Mansour et al., 2012). In addition, the ability of osteoclasts to resorb bone may also regulate HSCs through the release of matrix components and minerals. Data using HSCs from mice that lack the calcium sensor have suggested that HSCs preferentially home to regions of high calcium in the marrow (Adams et al., 2006). Similarly, TGFβ is a known matrix factor that can be released by osteoclastic bone resorption that also has been shown to induce quiescence in stem cell populations (Yamazaki et al., 2009).

Neuronal regulation of the hematopoietic stem cell niche

With improvements in the ability to prospectively isolate HSCs, experimentalists observed that the frequency of marrow HSCs is regulated by a circadian rhythm. Studies had uncovered how neuronal innervation of the marrow by the sympathetic nervous system (SNS) plays a role in mobilization of HSCs (Katayama et al., 2006). The same laboratory later discovered that the SNS is responsible for this circadian mobilization of HSCs through β-adrenergic signals (Mendez-Ferrer et al., 2008). Unexpectedly, β-adrenergic receptor agonists induced expression of the vitamin D receptor (VDR) in osteoblasts, and mice lacking the VDR had impairment in G-CSF-mediated osteoblastic inhibition and mobilization of HSCs, suggesting that the VDR is a modulator of HSC trafficking (Kawamori et al., 2010). Neuronal regulation of the hematopoietic stem cell niche is modulated by aging (Maryanovich et al., 2018), chemotherapy (Hanoun et al., 2014), and myeloproliferative syndromes (Arranz et al., 2014), although the mechanisms by which these conditions alter the SNS remain poorly understood. In addition to these direct effects, nonmyelinating marrow Schwann cells have been shown to induce HSC quiescence through their production of TGFβ (Yamazaki et al., 2011). Therefore, neuronal signals are clearly an important target for the manipulation of niches and the HSCs they support.

Hormonal regulation of the hematopoietic stem cell niche

The impact of hormones on the niche indicates that the niche is a dynamic system that is physiologically regulated and therefore represents a homeostatic mechanism to regulate and maintain HSCs. It is likely that these are simply examples of many systemic signals that can regulate HSCs through their niche.

Parathyroid hormone

Studies have demonstrated how constitutive activation of the PTH receptor 1 in osteolineage cells could expand HSCs, and how pharmacologic treatment with anabolic doses of PTH expands HSCs (Calvi et al., 2003; Bromberg et al., 2012; Adams et al., 2007; Itkin et al., 2012). Taking advantage of the use of teriparatide (recombinant PTH1-34) for the treatment of osteoporosis, Yu et al. collected blood from patients with osteoporosis undergoing treatment with teriparatide and quantified circulating HSCs (Yu et al., 2014). This study showed an increase in circulating HSCs in patients receiving anabolic treatment, a first demonstration of the impact of PTH on HSC regulation in humans. Since the bone marrow was not tested in this patient population, it is not known whether the increase in circulating HSCs was due to increases in their marrow populations or to mobilization to the periphery. Notably, the physiologic role of PTH or PTH-related protein acting through PTH receptor 1 in the niche is poorly understood.

Insulin-like growth factor 1

IGF1 is produced in the liver and by osteoblasts, and its increase in osteoblastic cells after radiation injury appears to have a beneficial impact on the recovery of the hematopoietic systems (Caselli et al., 2013). Since expression of IGF1 is increased with PTH treatment, it is possible that some of the PTH effects on the niche may be mediated by IGF1.

Niche heterogeneity for heterogeneous hematopoietic stem and progenitor cells

From the studies presented thus far, it is clear that, rather than a single niche cell population, the bone marrow microenvironment is heterogeneous and dynamic (Fig. 3.1). Developments in the understanding of hematopoietic stem and progenitor cells now suggest that rather than omnipotent undifferentiated stem cells, there may be functional heterogeneity of HSCs as well. Single-cell transplantation studies have shown that individual HSCs are biased toward myeloid or lymphoid differentiation (Dykstra et al., 2007; Muller-Sieburg et al., 2002; Wilson et al., 2015). This heterogeneity changes with development and with aging (Benz et al., 2012), and appears to be epigenetically imposed (Yu et al., 2017). Evidence of megakaryocytic-biased HSCs has also been shown by numerous laboratories (Sanjuan-Pla et al., 2013; Gekas and Graf, 2013; Shin et al., 2014). This heterogeneity may explain the conflicting data regarding localization of hematopoietic stem and progenitor cells in association with endosteal, periarteriolar, or perisinusoidal sites, or in association with megakaryocytes (as discussed earlier). Initial data to support this contention were provided by the Frenette laboratory, showing that lineage-biased HSCs are regulated by distinct niches, with platelet- and myeloid-biased HSCs

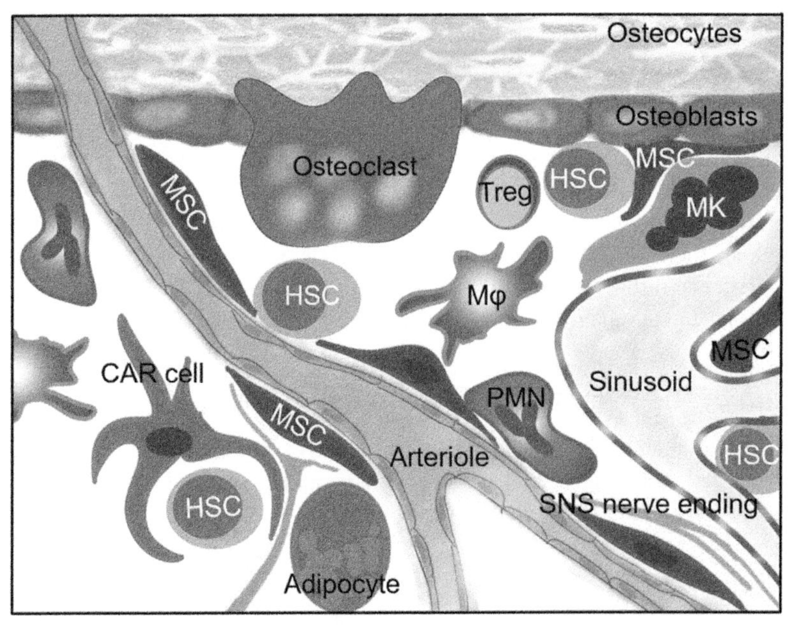

FIGURE 3.1 Heterogeneity of the hematopoietic stem cell niche in the bone marrow. Hematopoietic stem cells (HSCs) are found in close anatomic proximity to multiple regulatory populations in the bone marrow. A diagrammatic representation of the interactions of mesenchymal stroma cells (MSCs), HSCs, and other components of the bone marrow microenvironment is shown. Mesenchymal populations identified as HSC-regulating niche cells through genetic studies include subsets of MSCs (shown in *brown*), found in endosteal and more commonly perivascular locations, with CXCL12-abundant reticular (CAR) cells and leptin receptor-positive cells found primarily in perisinusoidal locations, while nestin–GFP^{+} cells are found primarily in periarteriolar sites. Multiple hematopoietic cells, including macrophages (Mφs), neutrophils (PMNs), megakaryocytes (MKs), osteoclasts, and regulatory T cells (T regs) contribute to HSC regulation. Sympathetic nervous system innervation modulates the circadian mobilization of HSCs in the circulation through its action on the niche.

found in association with and regulated with megakaryocytes, while $NG2^+$ arteriolar niche cells selectively regulate lymphoid-biased HSCs (Pinho et al., 2018).

Conclusions

As the data reviewed here show, our understanding of the HSC and progenitor microenvironment has evolved as our tool kit for their analysis has expanded, and it is likely to continue to increase as our ability to identify rare hematopoietic and mesenchymal populations in both murine models and humans improves. Progress in the definition of the HSC niche has already provided targets that may accelerate hematopoietic recovery and positively influence skeletal disease, especially if there are reciprocal interactions between skeletal stem cells and HSCs and/or their progenies. Moreover, given the potential for microenvironmental contribution to the support of either normal or malignant hematopoiesis, signals regulating interactions between bone microenvironmental populations and cancer cells could provide novel targets for the treatment of metastatic malignancy or primary malignancies in the bone and bone marrow.

Acknowledgments

The author wishes to thank the members of the Calvi, Becker, and Liesveld laboratories for helpful suggestions. This work was supported in part by the National Institutes of Health, National Cancer Institute, and National Institute on Aging awards CA166280 and AG046293; the University of Rochester Core Center for Musculoskeletal Biology and Medicine award AR061307 from the National Institute of Arthritis and Musculoskeletal and Skin Diseases; the Department of Defense (Award W81XWH1810485 to LMC) and the University of Rochester CTSA award UL1 TR002001 from the National Center for Advancing Translational Sciences.

References

Abella, E., Feliu, E., Granada, I., Milla, F., Oriol, A., Ribera, J.M., Sanchez-Planell, L., Berga, L.I., Reverter, J.C., Rozman, C., 2002. Bone marrow changes in anorexia nervosa are correlated with the amount of weight loss and not with other clinical findings. Am. J. Clin. Pathol. 118, 582–588.

Acar, M., Kocherlakota, K.S., Murphy, M.M., Peyer, J.G., Oguro, H., Inra, C.N., Jaiyeola, C., Zhao, Z., Luby-Phelps, K., Morrison, S.J., 2015. Deep imaging of bone marrow shows non-dividing stem cells are mainly perisinusoidal. Nature 526, 126–130.

Adams, G.B., Chabner, K.T., Alley, I.R., Olson, D.P., Szczepiorkowski, Z.M., Poznansky, M.C., Kos, C.H., Pollak, M.R., Brown, E.M., Scadden, D.T., 2006. Stem cell engraftment at the endosteal niche is specified by the calcium-sensing receptor. Nature 439, 599–603.

Adams, G.B., Martin, R.P., Alley, I.R., Chabner, K.T., Cohen, K.S., Calvi, L.M., Kronenberg, H.M., Scadden, D.T., 2007. Therapeutic targeting of a stem cell niche. Nat. Biotechnol. 25, 238–243.

Ambrosi, T.H., Scialdone, A., Graja, A., Gohlke, S., Jank, A.M., Bocian, C., Woelk, L., Fan, H., Logan, D.W., Schurmann, A., Saraiva, L.R., Schulz, T.J., 2017. Adipocyte accumulation in the bone marrow during obesity and aging impairs stem cell-based hematopoietic and bone regeneration. Cell Stem Cell 20, 771–784 e6.

Arranz, L., Sanchez-Aguilera, A., Martin-Perez, D., Isern, J., Langa, X., Tzankov, A., Lundberg, P., Muntion, S., Tzeng, Y.S., Lai, D.M., Schwaller, J., Skoda, R.C., Mendez-Ferrer, S., 2014. Neuropathy of haematopoietic stem cell niche is essential for myeloproliferative neoplasms. Nature 512, 78–81.

Asada, N., Katayama, Y., Sato, M., Minagawa, K., Wakahashi, K., Kawano, H., Kawano, Y., Sada, A., Ikeda, K., Matsui, T., Tanimoto, M., 2013. Matrix-embedded osteocytes regulate mobilization of hematopoietic stem/progenitor cells. Cell Stem Cell 12, 737–747.

Asada, N., Kunisaki, Y., Pierce, H., Wang, Z., Fernandez, N.F., Birbrair, A., MA'ayan, A., Frenette, P.S., 2017. Differential cytokine contributions of perivascular haematopoietic stem cell niches. Nat. Cell Biol. 19, 214–223.

Austin, J., Kimble, J., 1987. glp-1 is required in the germ line for regulation of the decision between mitosis and meiosis in *C. elegans*. Cell 51, 589–599.

Becker, A.J., Mc, C.E., Till, J.E., 1963. Cytological demonstration of the clonal nature of spleen colonies derived from transplanted mouse marrow cells. Nature 197, 452–454.

Benz, C., Copley, M.R., Kent, D.G., Wohrer, S., Cortes, A., Aghaeepour, N., Ma, E., Mader, H., Rowe, K., Day, C., Treloar, D., Brinkman, R.R., Eaves, C.J., 2012. Hematopoietic stem cell subtypes expand differentially during development and display distinct lymphopoietic programs. Cell Stem Cell 10, 273–283.

Bertrand, J.Y., Chi, N.C., Santoso, B., Teng, S., Stainier, D.Y., Traver, D., 2010. Haematopoietic stem cells derive directly from aortic endothelium during development. Nature 464, 108–111.

Bessis, M., 1958. Erythroblastic island, functional unity of bone marrow. Rev. Hematol. 13, 8–11.

Bjerknes, M., Cheng, H., 2001. Modulation of specific intestinal epithelial progenitors by enteric neurons. Proc. Natl. Acad. Sci. U.S.A. 98, 12497–12502.

Boisset, J.C., Van Cappellen, W., Andrieu-Soler, C., Galjart, N., Dzierzak, E., Robin, C., 2010. In vivo imaging of haematopoietic cells emerging from the mouse aortic endothelium. Nature 464, 116–120.

Bourgine, P.E., Klein, T., Paczulla, A.M., Shimizu, T., Kunz, L., Kokkaliaris, K.D., Coutu, D.L., Lengerke, C., Skoda, R., Schroeder, T., Martin, I., 2018. In vitro biomimetic engineering of a human hematopoietic niche with functional properties. Proc. Natl. Acad. Sci. U.S.A. 115, E5688–E5695.

Bowie, M.B., Kent, D.G., Dykstra, B., Mcknight, K.D., Mccaffrey, L., Hoodless, P.A., Eaves, C.J., 2007. Identification of a new intrinsically timed developmental checkpoint that reprograms key hematopoietic stem cell properties. Proc. Natl. Acad. Sci. U.S.A. 104, 5878–5882.

Bowie, M.B., Mcknight, K.D., Kent, D.G., Mccaffrey, L., Hoodless, P.A., Eaves, C.J., 2006. Hematopoietic stem cells proliferate until after birth and show a reversible phase-specific engraftment defect. J. Clin. Invest. 116, 2808–2816.

Bromberg, O., Frisch, B.J., Weber, J.M., Porter, R.L., Civitelli, R., Calvi, L.M., 2012. Osteoblastic N-cadherin is not required for microenvironmental support and regulation of hematopoietic stem and progenitor cells. Blood 120, 303–313.

Bruns, I., Lucas, D., Pinho, S., Ahmed, J., Lambert, M.P., Kunisaki, Y., Scheiermann, C., Schiff, L., Poncz, M., Bergman, A., Frenette, P.S., 2014. Megakaryocytes regulate hematopoietic stem cell quiescence through CXCL4 secretion. Nat. Med. 20, 1315–1320.

Busch, K., Klapproth, K., Barile, M., Flossdorf, M., Holland-Letz, T., Schlenner, S.M., Reth, M., Hofer, T., Rodewald, H.R., 2015. Fundamental properties of unperturbed haematopoiesis from stem cells in vivo. Nature 518, 542–546.

Butler, J.M., Nolan, D.J., Vertes, E.L., Varnum-Finney, B., Kobayashi, H., Hooper, A.T., Seandel, M., Shido, K., White, I.A., Kobayashi, M., Witte, L., May, C., Shawber, C., Kimura, Y., Kitajewski, J., Rosenwaks, Z., Bernstein, I.D., Rafii, S., 2010. Endothelial cells are essential for the self-renewal and repopulation of Notch-dependent hematopoietic stem cells. Cell Stem Cell 6, 251–264.

Cain, C.J., Rueda, R., Mclelland, B., Collette, N.M., Loots, G.G., Manilay, J.O., 2012. Absence of sclerostin adversely affects B-cell survival. J. Bone Miner. Res. 27, 1451–1461.

Calvi, L.M., Adams, G.B., Weibrecht, K.W., Weber, J.M., Olson, D.P., Knight, M.C., Martin, R.P., Schipani, E., Divieti, P., Bringhurst, F.R., Milner, L.A., Kronenberg, H.M., Scadden, D.T., 2003. Osteoblastic cells regulate the haematopoietic stem cell niche. Nature 425, 841–846.

Calvi, L.M., Bromberg, O., Rhee, Y., Weber, J.M., Smith, J.N., Basil, M.J., Frisch, B.J., Bellido, T., 2012. Osteoblastic expansion induced by parathyroid hormone receptor signaling in murine osteocytes is not sufficient to increase hematopoietic stem cells. Blood 119, 2489–2499.

Calvo, W., Fliedner, T.M., Herbst, E., Hugl, E., Bruch, C., 1976. Regeneration of blood-forming organs after autologous leukocyte transfusion in lethally irradiated dogs. II. Distribution and cellularity of the marrow in irradiated and transfused animals. Blood 47, 593–601.

Casanova-Acebes, M., Pitaval, C., Weiss, L.A., Nombela-Arrieta, C., Chevre, R., N, A.G., Kunisaki, Y., Zhang, D., Van Rooijen, N., Silberstein, L.E., Weber, C., Nagasawa, T., Frenette, P.S., Castrillo, A., Hidalgo, A., 2013. Rhythmic modulation of the hematopoietic niche through neutrophil clearance. Cell 153, 1025–1035.

Caselli, A., Olson, T.S., Otsuru, S., Chen, X., Hofmann, T.J., Nah, H.D., Grisendi, G., Paolucci, P., Dominici, M., Horwitz, E.M., 2013. IGF-1-mediated osteoblastic niche expansion enhances long-term hematopoietic stem cell engraftment after murine bone marrow transplantation. Stem Cell 31, 2193–2204.

Chang, M.K., Raggatt, L.J., Alexander, K.A., Kuliwaba, J.S., Fazzalari, N.L., Schroder, K., Maylin, E.R., Ripoll, V.M., Hume, D.A., Pettit, A.R., 2008. Osteal tissue macrophages are intercalated throughout human and mouse bone lining tissues and regulate osteoblast function in vitro and in vivo. J. Immunol. 181, 1232–1244.

Cheng, Y.H., Chitteti, B.R., Streicher, D.A., Morgan, J.A., Rodriguez-Rodriguez, S., Carlesso, N., Srour, E.F., Kacena, M.A., 2011. Impact of maturational status on the ability of osteoblasts to enhance the hematopoietic function of stem and progenitor cells. J. Bone Miner. Res. 26, 1111–1121.

Chow, A., Huggins, M., Ahmed, J., Hashimoto, D., Lucas, D., Kunisaki, Y., Pinho, S., Leboeuf, M., Noizat, C., Van Rooijen, N., Tanaka, M., Zhao, Z.J., Bergman, A., Merad, M., Frenette, P.S., 2013. CD169(+) macrophages provide a niche promoting erythropoiesis under homeostasis and stress. Nat. Med. 19, 429–436.

Chow, A., Lucas, D., Hidalgo, A., Mendez-Ferrer, S., Hashimoto, D., Scheiermann, C., Battista, M., Leboeuf, M., Prophete, C., Van Rooijen, N., Tanaka, M., Merad, M., Frenette, P.S., 2011. Bone marrow CD169+ macrophages promote the retention of hematopoietic stem and progenitor cells in the mesenchymal stem cell niche. J. Exp. Med. 208, 261–271.

Christopher, M.J., Rao, M., Liu, F., Woloszynek, J.R., Link, D.C., 2011. Expression of the G-CSF receptor in monocytic cells is sufficient to mediate hematopoietic progenitor mobilization by G-CSF in mice. J. Exp. Med. 208, 251–260.

Coskun, S., Chao, H., Vasavada, H., Heydari, K., Gonzales, N., Zhou, X., De Crombrugghe, B., Hirschi, K.K., 2014. Development of the fetal bone marrow niche and regulation of HSC quiescence and homing ability by emerging osteolineage cells. Cell Rep. 9, 581–590.

De Bruijn, M.F., Ma, X., Robin, C., Ottersbach, K., Sanchez, M.J., Dzierzak, E., 2002. Hematopoietic stem cells localize to the endothelial cell layer in the midgestation mouse aorta. Immunity 16, 673–683.

De Bruijn, M.F., Speck, N.A., Peeters, M.C., Dzierzak, E., 2000. Definitive hematopoietic stem cells first develop within the major arterial regions of the mouse embryo. EMBO J. 19, 2465–2474.

Ding, L., Morrison, S.J., 2013. Haematopoietic stem cells and early lymphoid progenitors occupy distinct bone marrow niches. Nature 495, 231–235.

Ding, L., Saunders, T.L., Enikolopov, G., Morrison, S.J., 2012. Endothelial and perivascular cells maintain haematopoietic stem cells. Nature 481, 457–462.

Dominici, M., Rasini, V., Bussolari, R., Chen, X., Hofmann, T.J., Spano, C., Bernabei, D., Veronesi, E., Bertoni, F., Paolucci, P., Conte, P., Horwitz, E.M., 2009. Restoration and reversible expansion of the osteoblastic hematopoietic stem cell niche after marrow radioablation. Blood 114, 2333–2343.

Doucette, C.R., Horowitz, M.C., Berry, R., Macdougald, O.A., Anunciado-Koza, R., Koza, R.A., Rosen, C.J., 2015. A high fat diet increases bone marrow adipose tissue (MAT) but does not alter trabecular or cortical bone mass in C57BL/6J mice. J. Cell. Physiol. 230, 2032–2037.

Duarte, D., Hawkins, E.D., Akinduro, O., Ang, H., De Filippo, K., Kong, I.Y., Haltalli, M., Ruivo, N., Straszkowski, L., Vervoort, S.J., Mclean, C., Weber, T.S., Khorshed, R., Pirillo, C., Wei, A., Ramasamy, S.K., Kusumbe, A.P., Duffy, K., Adams, R.H., Purton, L.E., Carlin, L.M., Lo Celso, C., 2018. Inhibition of endosteal vascular niche remodeling rescues hematopoietic stem cell loss in AML. Cell Stem Cell 22, 64–77 e6.

Dykstra, B., Kent, D., Bowie, M., Mccaffrey, L., Hamilton, M., Lyons, K., Lee, S.J., Brinkman, R., Eaves, C., 2007. Long-term propagation of distinct hematopoietic differentiation programs in vivo. Cell Stem Cell 1, 218–229.

Ellis, S.L., Grassinger, J., Jones, A., Borg, J., Camenisch, T., Haylock, D., Bertoncello, I., Nilsson, S.K., 2011. The relationship between bone, hemopoietic stem cells, and vasculature. Blood 118, 1516–1524.

Ema, H., Nakauchi, H., 2000. Expansion of hematopoietic stem cells in the developing liver of a mouse embryo. Blood 95, 2284–2288.

Evrard, M., Kwok, I.W.H., Chong, S.Z., Teng, K.W.W., Becht, E., Chen, J., Sieow, J.L., Penny, H.L., Ching, G.C., Devi, S., Adrover, J.M., Li, J.L.Y., Liong, K.H., Tan, L., Poon, Z., Foo, S., Chua, J.W., Su, I.H., Balabanian, K., Bachelerie, F., Biswas, S.K., Larbi, A., Hwang, W.Y.K., Madan, V., Koeffler, H.P., Wong, S.C., Newell, E.W., Hidalgo, A., Ginhoux, F., Ng, L.G., 2018. Developmental analysis of bone marrow neutrophils reveals populations specialized in expansion, trafficking, and effector functions. Immunity 48, 364–379 e8.

Ferraro, F., Lymperi, S., Mendez-Ferrer, S., Saez, B., Spencer, J.A., Yeap, B.Y., Masselli, E., Graiani, G., Prezioso, L., Rizzini, E.L., Mangoni, M., Rizzoli, V., Sykes, S.M., Lin, C.P., Frenette, P.S., Quaini, F., Scadden, D.T., 2011. Diabetes impairs hematopoietic stem cell mobilization by altering niche function. Sci. Transl. Med. 3, 104ra101.

Frisch, B.J., Ashton, J.M., Xing, L.P., Becker, M.W., Jordan, C.T., Calvi, L.M., 2012. Functional inhibition of osteoblastic cells in an in vivo mouse model of myeloid leukemia. Blood 119, 540–550.

Fujisaki, J., Wu, J., Carlson, A.L., Silberstein, L., Putheti, P., Larocca, R., Gao, W., Saito, T.I., Lo Celso, C., Tsuyuzaki, H., Sato, T., Cote, D., Sykes, M., Strom, T.B., Scadden, D.T., Lin, C.P., 2011. In vivo imaging of Treg cells providing immune privilege to the haematopoietic stem-cell niche. Nature 474, 216–219.

Fulzele, K., Krause, D.S., Panaroni, C., Saini, V., Barry, K.J., Liu, X., Lotinun, S., Baron, R., Bonewald, L., Feng, J.Q., Chen, M., Weinstein, L.S., Wu, J.Y., Kronenberg, H.M., Scadden, D.T., Divieti Pajevic, P., 2013. Myelopoiesis is regulated by osteocytes through Gsalpha-dependent signaling. Blood 121, 930–939.

Ge, C., Cawthorn, W.P., Li, Y., Zhao, G., Macdougald, O.A., Franceschi, R.T., 2016. Reciprocal control of osteogenic and adipogenic differentiation by ERK/MAP kinase phosphorylation of Runx2 and PPARgamma transcription factors. J. Cell. Physiol. 231, 587–596.

Gekas, C., Dieterlen-Lievre, F., Orkin, S.H., Mikkola, H.K., 2005. The placenta is a niche for hematopoietic stem cells. Dev. Cell 8, 365–375.

Gekas, C., Graf, T., 2013. CD41 expression marks myeloid-biased adult hematopoietic stem cells and increases with age. Blood 121, 4463–4472.

Golan, K., Kumari, A., Kollet, O., Khatib-Massalha, E., Subramaniam, M.D., Ferreira, Z.S., Avemaria, F., Rzeszotek, S., Garcia-Garcia, A., Xie, S., Flores-Figueroa, E., Gur-Cohen, S., Itkin, T., Ludin-Tal, A., Massalha, H., Bernshtein, B., Ciechanowicz, A.K., Brandis, A., Mehlman, T., Bhattacharya, S., Bertagna, M., Cheng, H., Petrovich-Kopitman, E., Janus, T., Kaushansky, N., Cheng, T., Sagi, I., Ratajczak, M.Z., Mendez-Ferrer, S., Dick, J.E., Markus, R.P., Lapidot, T., 2018. Daily onset of light and darkness differentially controls hematopoietic stem cell differentiation and maintenance. Cell Stem Cell 23, 572–585 e7.

Goncalves, K.A., Silberstein, L., Li, S., Severe, N., Hu, M.G., Yang, H., Scadden, D.T., Hu, G.F., 2016. Angiogenin promotes hematopoietic regeneration by dichotomously regulating quiescence of stem and progenitor cells. Cell 166, 894–906.

Gong, J.K., 1978. Endosteal marrow: a rich source of hematopoietic stem cells. Science 199, 1443–1445.

Greenbaum, A., Hsu, Y.M., Day, R.B., Schuettpelz, L.G., Christopher, M.J., Borgerding, J.N., Nagasawa, T., Link, D.C., 2013. CXCL12 in early mesenchymal progenitors is required for haematopoietic stem-cell maintenance. Nature 495, 227–230.

Guezguez, B., Campbell, C.J., Boyd, A.L., Karanu, F., Casado, F.L., DI Cresce, C., Collins, T.J., Shapovalova, Z., Xenocostas, A., Bhatia, M., 2013. Regional localization within the bone marrow influences the functional capacity of human HSCs. Cell Stem Cell 13, 175–189.

Hanoun, M., Zhang, D., Mizoguchi, T., Pinho, S., Pierce, H., Kunisaki, Y., Lacombe, J., Armstrong, S.A., Duhrsen, U., Frenette, P.S., 2014. Acute myelogenous leukemia-induced sympathetic neuropathy promotes malignancy in an altered hematopoietic stem cell niche. Cell Stem Cell 15, 365–375.

Heideveld, E., Van Den Akker, E., 2017. Digesting the role of bone marrow macrophages on hematopoiesis. Immunobiology 222, 814–822.

Hirata, Y., Furuhashi, K., Ishii, H., Li, H.W., Pinho, S., Ding, L., Robson, S.C., Frenette, P.S., Fujisaki, J., 2018. CD150(high) bone marrow tregs maintain hematopoietic stem cell quiescence and immune privilege via adenosine. Cell Stem Cell 22, 445–453 e5.

Hooper, A.T., Butler, J.M., Nolan, D.J., Kranz, A., Iida, K., Kobayashi, M., Kopp, H.G., Shido, K., Petit, I., Yanger, K., James, D., Witte, L., Zhu, Z., Wu, Y., Pytowski, B., Rosenwaks, Z., Mittal, V., Sato, T.N., Rafii, S., 2009. Engraftment and reconstitution of hematopoiesis is dependent on VEGFR2-mediated regeneration of sinusoidal endothelial cells. Cell Stem Cell 4, 263–274.

Horowitz, M.C., Berry, R., Holtrup, B., Sebo, Z., Nelson, T., Fretz, J.A., Lindskog, D., Kaplan, J.L., Ables, G., Rodeheffer, M.S., Rosen, C.J., 2017. Bone marrow adipocytes. Adipocyte 6, 193–204.

Hur, J., Choi, J.I., Lee, H., Nham, P., Kim, T.W., Chae, C.W., Yun, J.Y., Kang, J.A., Kang, J., Lee, S.E., Yoon, C.H., Boo, K., Ham, S., Roh, T.Y., Jun, J.K., Lee, H., Baek, S.H., Kim, H.S., 2016. CD82/KAI1 maintains the dormancy of long-term hematopoietic stem cells through interaction with DARC-expressing macrophages. Cell Stem Cell 18, 508–521.

Itkin, T., Gur-Cohen, S., Spencer, J.A., Schajnovitz, A., Ramasamy, S.K., Kusumbe, A.P., Ledergor, G., Jung, Y., Milo, I., Poulos, M.G., Kalinkovich, A., Ludin, A., Kollet, O., Shakhar, G., Butler, J.M., Rafii, S., Adams, R.H., Scadden, D.T., Lin, C.P., Lapidot, T., 2016. Distinct bone marrow blood vessels differentially regulate haematopoiesis. Nature 532 (7599), 323–328.

Itkin, T., Ludin, A., Gradus, B., Gur-Cohen, S., Kalinkovich, A., Schajnovitz, A., Ovadya, Y., Kollet, O., Canaani, J., Shezen, E., Coffin, D.J., Enikolopov, G.N., Berg, T., Piacibello, W., Hornstein, E., Lapidot, T., 2012. FGF-2 expands murine hematopoietic stem and progenitor cells via proliferation of stromal cells, c-Kit activation, and CXCL12 down-regulation. Blood 120, 1843–1855.

Ivanovs, A., Rybtsov, S., Welch, L., Anderson, R.A., Turner, M.L., Medvinsky, A., 2011. Highly potent human hematopoietic stem cells first emerge in the intraembryonic aorta-gonad-mesonephros region. J. Exp. Med. 208, 2417–2427.

Jaffredo, T., Gautier, R., Eichmann, A., Dieterlen-Lievre, F., 1998. Intraaortic hemopoietic cells are derived from endothelial cells during ontogeny. Development 125, 4575–4583.

Joseph, C., Quach, J.M., Walkley, C.R., Lane, S.W., Lo Celso, C., Purton, L.E., 2013. Deciphering hematopoietic stem cells in their niches: a critical appraisal of genetic models, lineage tracing, and imaging strategies. Cell Stem Cell 13, 520–533.

Katayama, Y., Battista, M., Kao, W.M., Hidalgo, A., Peired, A.J., Thomas, S.A., Frenette, P.S., 2006. Signals from the sympathetic nervous system regulate hematopoietic stem cell egress from bone marrow. Cell 124, 407–421.

Kaur, S., Raggatt, L.J., Millard, S.M., Wu, A.C., Batoon, L., Jacobsen, R.N., Winkler, I.G., Macdonald, K.P., Perkins, A.C., Hume, D.A., Levesque, J.P., Pettit, A.R., 2018. Self-repopulating recipient bone marrow resident macrophages promote long-term hematopoietic stem cell engraftment. Blood 132, 735–749.

Kawamori, Y., Katayama, Y., Asada, N., Minagawa, K., Sato, M., Okamura, A., Shimoyama, M., Nakagawa, K., Okano, T., Tanimoto, M., Kato, S., Matsui, T., 2010. Role for vitamin D receptor in the neuronal control of the hematopoietic stem cell niche. Blood 116, 5528–5535.

Kfoury, Y., Scadden, D.T., 2015. Mesenchymal cell contributions to the stem cell niche. Cell Stem Cell 16, 239–253.

Khan, J.A., Mendelson, A., Kunisaki, Y., Birbrair, A., Kou, Y., Arnal-Estape, A., Pinho, S., Ciero, P., Nakahara, F., MA'ayan, A., Bergman, A., Merad, M., Frenette, P.S., 2016. Fetal liver hematopoietic stem cell niches associate with portal vessels. Science 351, 176–180.

Kiel, M.J., Acar, M., Radice, G.L., Morrison, S.J., 2009. Hematopoietic stem cells do not depend on N-cadherin to regulate their maintenance. Cell Stem Cell 4, 170–179.

Kiel, M.J., Radice, G.L., Morrison, S.J., 2007. Lack of evidence that hematopoietic stem cells depend on N-cadherin-mediated adhesion to osteoblasts for their maintenance. Cell Stem Cell 1, 204–217.

Kiel, M.J., Yilmaz, O.H., Iwashita, T., Yilmaz, O.H., Terhorst, C., Morrison, S.J., 2005. SLAM family receptors distinguish hematopoietic stem and progenitor cells and reveal endothelial niches for stem cells. Cell 121, 1109–1121.

Kiger, A.A., White-Cooper, H., Fuller, M.T., 2000. Somatic support cells restrict germline stem cell self-renewal and promote differentiation. Nature 407, 750–754.

Kissa, K., Herbomel, P., 2010. Blood stem cells emerge from aortic endothelium by a novel type of cell transition. Nature 464, 112–115.

Kobayashi, H., Butler, J.M., O'donnell, R., Kobayashi, M., Ding, B.S., Bonner, B., Chiu, V.K., Nolan, D.J., Shido, K., Benjamin, L., Rafii, S., 2010. Angiocrine factors from Akt-activated endothelial cells balance self-renewal and differentiation of haematopoietic stem cells. Nat. Cell Biol. 12, 1046–1056.

Krause, D.S., Fulzele, K., Catic, A., Sun, C.C., Dombkowski, D., Hurley, M.P., Lezeau, S., Attar, E., Wu, J.Y., Lin, H.Y., Divieti-Pajevic, P., Hasserjian, R.P., Schipani, E., Van Etten, R.A., Scadden, D.T., 2013. Differential regulation of myeloid leukemias by the bone marrow microenvironment. Nat. Med. 19, 1513–1517.

Krevvata, M., Silva, B.C., Manavalan, J.S., Galan-Diez, M., Kode, A., Matthews, B.G., Park, D., Zhang, C.A., Galili, N., Nickolas, T.L., Dempster, D.W., Dougall, W., Teruya-Feldstein, J., Economides, A.N., Kalajzic, I., Raza, A., Berman, E., Mukherjee, S., Bhagat, G., Kousteni, S., 2014. Inhibition of leukemia cell engraftment and disease progression in mice by osteoblasts. Blood 124, 2834–2846.

Kunisaki, Y., Bruns, I., Scheiermann, C., Ahmed, J., Pinho, S., Zhang, D., Mizoguchi, T., Wei, Q., Lucas, D., Ito, K., Mar, J.C., Bergman, A., Frenette, P.S., 2013. Arteriolar niches maintain haematopoietic stem cell quiescence. Nature 502, 637–643.

Kusumbe, A.P., Ramasamy, S.K., Adams, R.H., 2014. Coupling of angiogenesis and osteogenesis by a specific vessel subtype in bone. Nature 507, 323–328.

Kusumbe, A.P., Ramasamy, S.K., Itkin, T., Mae, M.A., Langen, U.H., Betsholtz, C., Lapidot, T., Adams, R.H., 2016. Age-dependent modulation of vascular niches for haematopoietic stem cells. Nature 532, 380–384.

Li, D., Xue, W., Li, M., Dong, M., Wang, J., Wang, X., Li, X., Chen, K., Zhang, W., Wu, S., Zhang, Y., Gao, L., Chen, Y., Chen, J., Zhou, B.O., Zhou, Y., Yao, X., Li, L., Wu, D., Pan, W., 2018. VCAM-1(+) macrophages guide the homing of HSPCs to a vascular niche. Nature 564, 119–124.

Lo Celso, C., Fleming, H.E., Wu, J.W., Zhao, C.X., Miake-Lye, S., Fujisaki, J., Cote, D., Rowe, D.W., Lin, C.P., Scadden, D.T., 2009. Live-animal tracking of individual haematopoietic stem/progenitor cells in their niche. Nature 457, 92–96.

Ludin, A., Itkin, T., Gur-Cohen, S., Mildner, A., Shezen, E., Golan, K., Kollet, O., Kalinkovich, A., Porat, Z., D'uva, G., Schajnovitz, A., Voronov, E., Brenner, D.A., Apte, R.N., Jung, S., Lapidot, T., 2012. Monocytes-macrophages that express alpha-smooth muscle actin preserve primitive hematopoietic cells in the bone marrow. Nat. Immunol. 13 (11), 1072–1082.

Lymperi, S., Horwood, N., Marley, S., Gordon, M.Y., Cope, A.P., Dazzi, F., 2008. Strontium can increase some osteoblasts without increasing hematopoietic stem cells. Blood 111, 1173–1181.

Mansour, A., Abou-Ezzi, G., Sitnicka, E., Jacobsen, S.E., Wakkach, A., Blin-Wakkach, C., 2012. Osteoclasts promote the formation of hematopoietic stem cell niches in the bone marrow. J. Exp. Med. 209, 537–549.

Maryanovich, M., Zahalka, A.H., Pierce, H., Pinho, S., Nakahara, F., Asada, N., Wei, Q., Wang, X., Ciero, P., Xu, J., Leftin, A., Frenette, P.S., 2018. Adrenergic nerve degeneration in bone marrow drives aging of the hematopoietic stem cell niche. Nat. Med. 24 (6).

Mcculloch, E.A., Till, J.E., 1960. The radiation sensitivity of normal mouse bone marrow cells, determined by quantitative marrow transplantation into irradiated mice. Radiat. Res. 13, 115–125.

Medvinsky, A., Dzierzak, E., 1996. Definitive hematopoiesis is autonomously initiated by the AGM region. Cell 86, 897–906.

Medvinsky, A., Rybtsov, S., Taoudi, S., 2011. Embryonic origin of the adult hematopoietic system: advances and questions. Development 138, 1017–1031.

Mendez-Ferrer, S., Lucas, D., Battista, M., Frenette, P.S., 2008. Haematopoietic stem cell release is regulated by circadian oscillations. Nature 452, 442–447.

Mendez-Ferrer, S., Michurina, T.V., Ferraro, F., Mazloom, A.R., Macarthur, B.D., Lira, S.A., Scadden, D.T., MA'ayan, A., Enikolopov, G.N., Frenette, P.S., 2010. Mesenchymal and haematopoietic stem cells form a unique bone marrow niche. Nature 466, 829–834.

Mohamad, S.F., Xu, L., Ghosh, J., Childress, P.J., Abeysekera, I., Himes, E.R., Wu, H., Alvarez, M.B., Davis, K.M., Aguilar-Perez, A., Hong, J.M., Bruzzaniti, A., Kacena, M.A., Srour, E.F., 2017. Osteomacs interact with megakaryocytes and osteoblasts to regulate murine hematopoietic stem cell function. Blood Adv. 1, 2520–2528.

Morikawa, S., Mabuchi, Y., Kubota, Y., Nagai, Y., Niibe, K., Hiratsu, E., Suzuki, S., Miyauchi-Hara, C., Nagoshi, N., Sunabori, T., Shimmura, S., Miyawaki, A., Nakagawa, T., Suda, T., Okano, H., Matsuzaki, Y., 2009. Prospective identification, isolation, and systemic transplantation of multipotent mesenchymal stem cells in murine bone marrow. J. Exp. Med. 206, 2483–2496.

Morrison, S.J., Hemmati, H.D., Wandycz, A.M., Weissman, I.L., 1995. The purification and characterization of fetal liver hematopoietic stem cells. Proc. Natl. Acad. Sci. U.S.A. 92, 10302–10306.

Muller-Sieburg, C.E., Cho, R.H., Thoman, M., Adkins, B., Sieburg, H.B., 2002. Deterministic regulation of hematopoietic stem cell self-renewal and differentiation. Blood 100, 1302–1309.

Muller, A.M., Medvinsky, A., Strouboulis, J., Grosveld, F., Dzierzak, E., 1994. Development of hematopoietic stem cell activity in the mouse embryo. Immunity 1, 291–301.

Nakamura-Ishizu, A., Takubo, K., Fujioka, M., Suda, T., 2014. Megakaryocytes are essential for HSC quiescence through the production of thrombopoietin. Biochem. Biophys. Res. Commun. 454, 353–357.

Nakamura-Ishizu, A., Takubo, K., Kobayashi, H., SUZUKI-Inoue, K., Suda, T., 2015. CLEC-2 in megakaryocytes is critical for maintenance of hematopoietic stem cells in the bone marrow. J. Exp. Med. 212, 2133–2146.

Nakamura, Y., Arai, F., Iwasaki, H., Hosokawa, K., Kobayashi, I., Gomei, Y., Matsumoto, Y., Yoshihara, H., Suda, T., 2010. Isolation and characterization of endosteal niche cell populations that regulate hematopoietic stem cells. Blood 116 (9), 1422–1432.

Nathan, C., 2006. Neutrophils and immunity: challenges and opportunities. Nat. Rev. Immunol. 6, 173–182.

Naveiras, O., Nardi, V., Wenzel, P.L., Hauschka, P.V., Fahey, F., Daley, G.Q., 2009. Bone-marrow adipocytes as negative regulators of the haematopoietic microenvironment. Nature 460, 259–263.

Nombela-Arrieta, C., Pivarnik, G., Winkel, B., Canty, K.J., Harley, B., Mahoney, J.E., Park, S.Y., Lu, J., Protopopov, A., Silberstein, L.E., 2013. Quantitative imaging of haematopoietic stem and progenitor cell localization and hypoxic status in the bone marrow microenvironment. Nat. Cell Biol. 15, 533–543.

Oguro, H., Ding, L., Morrison, S.J., 2013. SLAM family markers resolve functionally distinct subpopulations of hematopoietic stem cells and multipotent progenitors. Cell Stem Cell 13, 102–116.

Olson, T.S., Caselli, A., Otsuru, S., Hofmann, T.J., Williams, R., Paolucci, P., Dominici, M., Horwitz, E.M., 2013. Megakaryocytes promote murine osteoblastic HSC niche expansion and stem cell engraftment after radioablative conditioning. Blood 121, 5238–5249.

Omatsu, Y., Sugiyama, T., Kohara, H., Kondoh, G., Fujii, N., Kohno, K., Nagasawa, T., 2010. The essential functions of adipo-osteogenic progenitors as the hematopoietic stem and progenitor cell niche. Immunity 33, 387–399.

Ottersbach, K., Dzierzak, E., 2005. The murine placenta contains hematopoietic stem cells within the vascular labyrinth region. Dev. Cell 8, 377–387.

Pinho, S., Lacombe, J., Hanoun, M., Mizoguchi, T., Bruns, I., Kunisaki, Y., Frenette, P.S., 2013. PDGFRalpha and CD51 mark human Nestin+ sphere-forming mesenchymal stem cells capable of hematopoietic progenitor cell expansion. J. Exp. Med. 210, 1351–1367.

Pinho, S., Marchand, T., Yang, E., Wei, Q., Nerlov, C., Frenette, P.S., 2018. Lineage-biased hematopoietic stem cells are regulated by distinct niches. Dev. Cell 44, 634–641 e4.

Poulos, M.G., Guo, P., Kofler, N.M., Pinho, S., Gutkin, M.C., Tikhonova, A., Aifantis, I., Frenette, P.S., Kitajewski, J., Rafii, S., Butler, J.M., 2013. Endothelial Jagged-1 is necessary for homeostatic and regenerative hematopoiesis. Cell Rep. 4, 1022–1034.

Raaijmakers, M.H., Mukherjee, S., Guo, S., Zhang, S., Kobayashi, T., Schoonmaker, J.A., Ebert, B.L., AL-Shahrour, F., Hasserjian, R.P., Scadden, E.O., Aung, Z., Matza, M., Merkenschlager, M., Lin, C., Rommens, J.M., Scadden, D.T., 2010. Bone progenitor dysfunction induces myelodysplasia and secondary leukaemia. Nature 464, 852–857.

Ramasamy, S.K., Kusumbe, A.P., Wang, L., Adams, R.H., 2014. Endothelial Notch activity promotes angiogenesis and osteogenesis in bone. Nature 507, 376–380.

Rhodes, K.E., Gekas, C., Wang, Y., Lux, C.T., Francis, C.S., Chan, D.N., Conway, S., Orkin, S.H., Yoder, M.C., Mikkola, H.K., 2008. The emergence of hematopoietic stem cells is initiated in the placental vasculature in the absence of circulation. Cell Stem Cell 2, 252–263.

Robin, C., Bollerot, K., Mendes, S., Haak, E., Crisan, M., Cerisoli, F., Lauw, I., Kaimakis, P., Jorna, R., Vermeulen, M., Kayser, M., Van Der Linden, R., Imanirad, P., Verstegen, M., Nawaz-Yousaf, H., Papazian, N., Steegers, E., Cupedo, T., Dzierzak, E., 2009. Human placenta is a potent hematopoietic niche containing hematopoietic stem and progenitor cells throughout development. Cell Stem Cell 5, 385–395.

Rosental, B., Kowarsky, M., Seita, J., Corey, D.M., Ishizuka, K.J., Palmeri, K.J., Chen, S.Y., Sinha, R., Okamoto, J., Mantalas, G., Manni, L., Raveh, T., Clarke, D.N., Tsai, J.M., Newman, A.M., Neff, N.F., Nolan, G.P., Quake, S.R., Weissman, I.L., Voskoboynik, A., 2018. Complex mammalian-like haematopoietic system found in a colonial chordate. Nature 564 (7736), 425–429.

Sanjuan-Pla, A., Macaulay, I.C., Jensen, C.T., Woll, P.S., Luis, T.C., Mead, A., Moore, S., Carella, C., Matsuoka, S., Bouriez Jones, T., Chowdhury, O., Stenson, L., Lutteropp, M., Green, J.C., Facchini, R., Boukarabila, H., Grover, A., Gambardella, A., Thongjuea, S., Carrelha, J., Tarrant, P., Atkinson, D., Clark, S.A., Nerlov, C., Jacobsen, S.E., 2013. Platelet-biased stem cells reside at the apex of the haematopoietic stem-cell hierarchy. Nature 502, 232–236.

Sawai, C.M., Babovic, S., Upadhaya, S., Knapp, D., Lavin, Y., Lau, C.M., Goloborodko, A., Feng, J., Fujisaki, J., Ding, L., Mirny, L.A., Merad, M., Eaves, C.J., Reizis, B., 2016. Hematopoietic stem cells are the major source of multilineage hematopoiesis in adult animals. Immunity 45, 597–609.

Scheller, E.L., Doucette, C.R., Learman, B.S., Cawthorn, W.P., Khandaker, S., Schell, B., Wu, B., Ding, S.Y., Bredella, M.A., Fazeli, P.K., Khoury, B., Jepsen, K.J., Pilch, P.F., Klibanski, A., Rosen, C.J., Macdougald, O.A., 2015. Region-specific variation in the properties of skeletal adipocytes reveals regulated and constitutive marrow adipose tissues. Nat. Commun. 6, 7808.

Schepers, K., Pietras, E.M., Reynaud, D., Flach, J., Binnewies, M., Garg, T., Wagers, A.J., Hsiao, E.C., Passegue, E., 2013. Myeloproliferative neoplasia remodels the endosteal bone marrow niche into a self-reinforcing leukemic niche. Cell Stem Cell 13 (3), 285–299.

Schofield, R., 1978. The relationship between the spleen colony-forming cell and the haemopoietic stem cell. Blood Cells 4, 7–25.

Senn, J.S., Mcculloch, E.A., Till, J.E., 1967. Comparison of colony-forming ability of normal and leukaemic human marrow in cell culture. Lancet 2, 597–598.

Shin, J.Y., Hu, W., Naramura, M., Park, C.Y., 2014. High c-Kit expression identifies hematopoietic stem cells with impaired self-renewal and megakaryocytic bias. J. Exp. Med. 211, 217–231.

Silberstein, L., Goncalves, K.A., Kharchenko, P.V., Turcotte, R., Kfoury, Y., Mercier, F., Baryawno, N., Severe, N., Bachand, J., Spencer, J.A., Papazian, A., Lee, D., Chitteti, B.R., Srour, E.F., Hoggatt, J., Tate, T., LO Celso, C., Ono, N., Nutt, S., Heino, J., Sipila, K., Shioda, T., Osawa, M., Lin, C.P., Hu, G.F., Scadden, D.T., 2016. Proximity-based differential single-cell analysis of the niche to identify stem/progenitor cell regulators. Cell Stem Cell 19, 530–543.

Silva, B., Krevvata, M., Manavalan, J.S., Zhang, C., Brentjens, R., Economides, A., Berman, E., Kousteni, S., 2011. Leukemia Progression Depends on the Presence of Osteoblasts. American Society for Bone and Mineral Research, San Diego, Ca, Usa, p. 1231.

Silva, B., Yoshikawa, Y., Johnson, L., Manavalan, J., Berman, E., Kousteni, S., 2010. Leukemia blasts compromise osteoblast function in a mouse model of acute myelogenous leukemia. In: American Society for Bone and Mineral Research Annual Meeting, October 2010. Toronto, Canada.

Sugiyama, T., Kohara, H., Noda, M., Nagasawa, T., 2006. Maintenance of the hematopoietic stem cell pool by CXCL12-CXCR4 chemokine signaling in bone marrow stromal cell niches. Immunity 25, 977–988.

Sun, J., Ramos, A., Chapman, B., Johnnidis, J.B., Le, L., Ho, Y.J., Klein, A., Hofmann, O., Camargo, F.D., 2014. Clonal dynamics of native haematopoiesis. Nature 514, 322–327.

Tabe, Y., Yamamoto, S., Saitoh, K., Sekihara, K., Monma, N., Ikeo, K., Mogushi, K., Shikami, M., Ruvolo, V., Ishizawa, J., Hail JR., N., Kazuno, S., Igarashi, M., Matsushita, H., Yamanaka, Y., Arai, H., Nagaoka, I., Miida, T., Hayashizaki, Y., Konopleva, M., Andreeff, M., 2017. Bone marrow adipocytes facilitate fatty acid oxidation activating AMPK and a transcriptional network supporting survival of acute monocytic leukemia cells. Cancer Res. 77, 1453–1464.

Taichman, R.S., Emerson, S.G., 1994. Human osteoblasts support hematopoiesis through the production of granulocyte colony-stimulating factor. J. Exp. Med. 179, 1677–1682.

Taichman, R.S., Reilly, M.J., Emerson, S.G., 1996. Human osteoblasts support human hematopoietic progenitor cells in vitro bone marrow cultures. Blood 87, 518–524.

Tamplin, O.J., Durand, E.M., Carr, L.A., Childs, S.J., Hagedorn, E.J., Li, P., Yzaguirre, A.D., Speck, N.A., Zon, L.I., 2015. Hematopoietic stem cell arrival triggers dynamic remodeling of the perivascular niche. Cell 160, 241–252.

Tjin, G., Flores-Figueroa, E., Duarte, D., Straszkowski, L., Scott, M., Khorshed, R.A., Purton, L.E., LO Celso, C., 2019 feb. Imaging methods used to study mouse and human HSC niches: current and emerging technologies. Bone 119, 19–35. https://doi.org/10.1016/j.bone.2018.04.022. Epub 2018 Apr 25. PMID:29704697.

Tran, J., Brenner, T.J., Dinardo, S., 2000. Somatic control over the germline stem cell lineage during Drosophila spermatogenesis. Nature 407, 754–757.

Tumbar, T., Guasch, G., Greco, V., Blanpain, C., Lowry, W.E., Rendl, M., Fuchs, E., 2004. Defining the epithelial stem cell niche in skin. Science 303, 359–363.

Visnjic, D., Kalajzic, Z., Rowe, D.W., Katavic, V., Lorenzo, J., Aguila, H.L., 2004. Hematopoiesis is severely altered in mice with an induced osteoblast deficiency. Blood 103, 3258–3264.

Wilson, A., Laurenti, E., Oser, G., Van Der Wath, R.C., Blanco-Bose, W., Jaworski, M., Offner, S., Dunant, C.F., Eshkind, L., Bockamp, E., Lio, P., Macdonald, H.R., Trumpp, A., 2008. Hematopoietic stem cells reversibly switch from dormancy to self-renewal during homeostasis and repair. Cell 135, 1118–1129.

Wilson, N.K., Kent, D.G., Buettner, F., Shehata, M., Macaulay, I.C., Calero-Nieto, F.J., Sanchez Castillo, M., Oedekoven, C.A., Diamanti, E., Schulte, R., Ponting, C.P., Voet, T., Caldas, C., Stingl, J., Green, A.R., Theis, F.J., Gottgens, B., 2015. Combined single-cell functional and gene expression analysis resolves heterogeneity within stem cell populations. Cell Stem Cell 16, 712–724.

Winkler, I.G., Barbier, V., Nowlan, B., Jacobsen, R.N., Forristal, C.E., Patton, J.T., Magnani, J.L., Levesque, J.P., 2012. Vascular niche E-selectin regulates hematopoietic stem cell dormancy, self renewal and chemoresistance. Nat. Med. 18, 1651–1657.

Winkler, I.G., Sims, N.A., Pettit, A.R., Barbier, V., Nowlan, B., Helwani, F., Poulton, I.J., Van Rooijen, N., Alexander, K.A., Raggatt, L.J., Levesque, J.P., 2010. Bone marrow macrophages maintain hematopoietic stem cell (HSC) niches and their depletion mobilizes HSCs. Blood 116, 4815–4828.

Wu, J.Y., Purton, L.E., Rodda, S.J., Chen, M., Weinstein, L.S., Mcmahon, A.P., Scadden, D.T., Kronenberg, H.M., 2008. Osteoblastic regulation of B lymphopoiesis is mediated by Gs{alpha}-dependent signaling pathways. Proc. Natl. Acad. Sci. U.S.A. 105, 16976–16981.

Xie, T., Spradling, A.C., 2000. A niche maintaining germ line stem cells in the Drosophila ovary. Science 290, 328–330.

Xie, Y., Yin, T., Wiegraebe, W., He, X.C., Miller, D., Stark, D., Perko, K., Alexander, R., Schwartz, J., Grindley, J.C., Park, J., Haug, J.S., Wunderlich, J.P., Li, H., Zhang, S., Johnson, T., Feldman, R.A., Li, L., 2009. Detection of functional haematopoietic stem cell niche using real-time imaging. Nature 457, 97–101.

Xu, R., Yallowitz, A., Qin, A., Wu, Z., Shin, D.Y., Kim, J.M., Debnath, S., Ji, G., Bostrom, M.P., Yang, X., Zhang, C., Dong, H., Kermani, P., Lalani, S., Li, N., Liu, Y., Poulos, M.G., Wach, A., Zhang, Y., Inoue, K., DI Lorenzo, A., Zhao, B., Butler, J.M., Shim, J.H., Glimcher, L.H., Greenblatt, M.B., 2018. Targeting skeletal endothelium to ameliorate bone loss. Nat. Med. 24, 823–833.

Yamazaki, S., Ema, H., Karlsson, G., Yamaguchi, T., Miyoshi, H., Shioda, S., Taketo, M.M., Karlsson, S., Iwama, A., Nakauchi, H., 2011. Non-myelinating Schwann cells maintain hematopoietic stem cell hibernation in the bone marrow niche. Cell 147, 1146–1158.

Yamazaki, S., Iwama, A., Takayanagi, S., Eto, K., Ema, H., Nakauchi, H., 2009. TGF-beta as a candidate bone marrow niche signal to induce hematopoietic stem cell hibernation. Blood 113, 1250–1256.

Yu, E.W., Kumbhani, R., Siwila-Sackman, E., Delelys, M., Preffer, F.I., Leder, B.Z., Wu, J.Y., 2014. Teriparatide (PTH 1-34) treatment increases peripheral hematopoietic stem cells in postmenopausal women. J. Bone Miner. Res. 29, 1380–1386.

Yu, V.W., Saez, B., Cook, C., Lotinun, S., Pardo-Saganta, A., Wang, Y.H., Lymperi, S., Ferraro, F., Raaijmakers, M.H., Wu, J.Y., Zhou, L., Rajagopal, J., Kronenberg, H.M., Baron, R., Scadden, D.T., 2015. Specific bone cells produce DLL4 to generate thymus-seeding progenitors from bone marrow. J. Exp. Med. 212, 759–774.

Yu, V.W.C., Yusuf, R.Z., Oki, T., Wu, J., Saez, B., Wang, X., Cook, C., Baryawno, N., Ziller, M.J., Lee, E., Gu, H., Meissner, A., Lin, C.P., Kharchenko, P.V., Scadden, D.T., 2017. Epigenetic memory underlies cell-autonomous heterogeneous behavior of hematopoietic stem cells. Cell 168, 944–945.

Yue, R., Zhou, B.O., Shimada, I.S., Zhao, Z., Morrison, S.J., 2016. Leptin receptor promotes adipogenesis and reduces osteogenesis by regulating mesenchymal stromal cells in adult bone marrow. Cell Stem Cell 18, 782–796.

Zhang, J., Link, D.C., 2016. Targeting of mesenchymal stromal cells by cre-recombinase transgenes commonly used to target osteoblast lineage cells. J. Bone Miner. Res. 31, 2001–2007.

Zhang, J., Niu, C., Ye, L., Huang, H., He, X., Tong, W.G., Ross, J., Haug, J., Johnson, T., Feng, J.Q., Harris, S., Wiedemann, L.M., Mishina, Y., Li, L., 2003. Identification of the haematopoietic stem cell niche and control of the niche size. Nature 425, 836–841.

Zhao, M., Perry, J.M., Marshall, H., Venkatraman, A., Qian, P., He, X.C., Ahamed, J., Li, L., 2014. Megakaryocytes maintain homeostatic quiescence and promote post-injury regeneration of hematopoietic stem cells. Nat. Med. 20, 1321–1326.

Zhou, B.O., Yu, H., Yue, R., Zhao, Z., Rios, J.J., Naveiras, O., Morrison, S.J., 2017. Bone marrow adipocytes promote the regeneration of stem cells and haematopoiesis by secreting SCF. Nat. Cell Biol. 19, 891–903.

Zhou, B.O., Yue, R., Murphy, M.M., Peyer, J.G., Morrison, S.J., 2014. Leptin-receptor-expressing mesenchymal stromal cells represent the main source of bone formed by adult bone marrow. Cell Stem Cell 15, 154–168.

Zhou, X., Zhang, Z., Feng, J.Q., Dusevich, V.M., Sinha, K., Zhang, H., Darnay, B.G., DE Crombrugghe, B., 2010. Multiple functions of Osterix are required for bone growth and homeostasis in postnatal mice. Proc. Natl. Acad. Sci. U.S.A. 107, 12919–12924.

Zhu, J., Garrett, R., Jung, Y., Zhang, Y., Kim, N., Wang, J., Joe, G.J., Hexner, E., Choi, Y., Taichman, R.S., Emerson, S.G., 2007. Osteoblasts support B-lymphocyte commitment and differentiation from hematopoietic stem cells. Blood 109, 3706–3712.

Chapter 4

The osteoblast lineage: its actions and communication mechanisms

Natalie A. Sims and T. John Martin

St. Vincent's Institute of Medical Research, Melbourne, Australia; Department of Medicine at St. Vincent's Hospital, The University of Melbourne, Melbourne, Australia

Chapter outline

Introduction **89**
The stages of the osteoblast lineage **90**
Mesenchymal precursors 90
Commitment of osteoblast progenitors (preosteoblasts) 91
Mature "bone-forming" osteoblasts 93
Bone-lining cells 94
Osteocytes 95
The process of osteoblast lineage differentiation **95**
At their various stages of development, cells of the osteoblast lineage signal to one another **96**
An example of contact-dependent communication: EphrinB2/EphB4 96
Communication between different stages of differentiation: IL-6 cytokines 97
Communication at different stages of differentiation: PTHrP/PTHR1 97
Physical sensing and signaling by osteoblasts and osteocytes 98
How does the osteoblast lineage promote osteoclast formation? **98**
Actions of the osteoblast lineage during the bone remodeling sequence **99**
Lessons in osteoblast biology from the Wnt signaling pathway **99**
From paracrinology to endocrinology in bone: the secretory osteoblast lineage **101**
References **102**

Introduction

The skeleton is a metabolically active organ in which the organizational pattern of its mineral and organic components determines its successful mechanical function. This is achieved by a combination of dense, compact (cortical) bone and cancellous (trabecular) bone, reinforced at points of stress (Chapter 11). Bone itself is a heterogeneous compound material. The mineral phase of bone, in the form of modified hydroxyapatite crystals, contributes about two-thirds of its weight. The remaining organic matrix consists largely of type I collagen (~90%), with small amounts of lipid (~2%), noncollagenous proteins (~5%), and water. Noncollagenous proteins within the bone matrix include signaling molecules (such as transforming growth factor β and insulin-like growth factor I) and regulators of mineralization (such as osteocalcin and dentin matrix protein 1 [DMP1]).

The word "osteoblast" has been used traditionally to describe those cells in bone responsible for bone collagen matrix production. However, we now know there are multiple stages within the osteoblast lineage, and they perform a much wider range of functions; in fact, these cell types within the lineage, when fully delineated, will merit naming in ways to reflect those functions. In addition to regulating mineralization and osteoclastogenesis directly, the lineage also produces paracrine and autocrine factors (cytokines, growth factors, prostanoids, proteinases), thereby forming communication systems profoundly influencing not only bone formation, but also bone resorption and hematopoiesis (see Chapter 3). To this is added the discoveries revealing bone as an endocrine organ, with cells of the osteoblast lineage being sources of circulating hormones such as fibroblast growth factor 23 (FGF23) and osteocalcin. These two hormones will be discussed in detail in Chapters 64 and 25, respectively.

Principles of Bone Biology. https://doi.org/10.1016/B978-0-12-814841-9.00004-X

Our approach in this chapter is to recognize the mixture of cells comprising the osteoblast lineage. We will consider their range of functions influencing bone structure and strength, without confining the discussion to only those particular cells in the lineage traditionally called osteoblasts because of their ability to produce the bone matrix.

The stages of the osteoblast lineage

The osteoblast lineage consists of a number of stages of differentiation through the mesenchymal lineage. These functions are identifiable by the morphology and location of the cells as well as their gene expression and synthetic capabilities (Fig. 4.1). We will describe each stage of the lineage in turn: mesenchymal precursors, committed preosteoblasts, mature matrix-producing osteoblasts, osteocytes, and bone-lining cells.

Mesenchymal precursors

A detailed discussion of the stem cells of bone will be provided in Chapter 2, but some matters most directly relevant to osteoblasts will be discussed briefly here.

Osteoblasts are derived from stromal mesenchymal cells present in the bone marrow. In vitro, these multipotent precursors are capable of differentiating into osteoblasts, chondrocytes, adipocytes, or myocytes (reviewed in Bianco et al., 2008; Bianco et al., 2010). In vivo their location in the marrow is required for their ability to differentiate into osteoblasts (Sacchetti et al., 2016). Our understanding of these pathways owes much to the work of Alexander Friedenstein and Maureen Owen (Friedenstein, 1976). Their seminal work identified the existence of osteogenic stem cells within the bone marrow stroma. This was achieved when intraperitoneal transplantation of diffusion chambers containing bone marrow cells in rabbits gave rise to a mixture of tissues, including bone and cartilage (Friedenstein, 1976). This diversity of differentiation led Bianco et al. (Bianco et al., 2008) to propose renaming their "osteogenic stem cells" to "skeletal stem cells" (SSCs). The ability of these cells to form bone depended on their location in the marrow: those bone marrow cells taken from close to the endocortical surface of the femora were more osteogenic than cells taken from the central marrow or the intermediate region (Ashton et al., 1984; Bab et al., 1984). Furthermore, when marrow cell distribution was correlated with the colony-forming efficiency of the cells in vitro, stromal fibroblasts from the endocortical surface formed four times as many colonies as the same number of core cells. These observations served to establish Friedenstein's concept of the CFU-F (colony-forming unit–fibroblastic), which he proposed was a self-renewing multipotential stem cell population. An elegant extension of this involved establishing single stromal cell colonies, transplanted under the renal capsule of mice, where they formed plaques containing bone, with the host cells establishing hemopoiesis and excavating a marrow cavity (reviewed in Owen and Friedenstein, 1988). This experiment also provided the first evidence that stromal or osteoblast lineage cells could promote osteoclast recruitment from circulating precursors. Direct evidence for stromal or osteoblast lineage regulation of osteoclast formation was obtained some years later by coculture of osteoblast lineage cells with hemopoietic cells (Takahashi et al., 1988; Udagawa et al., 1989), and is reviewed in Chapter 5.

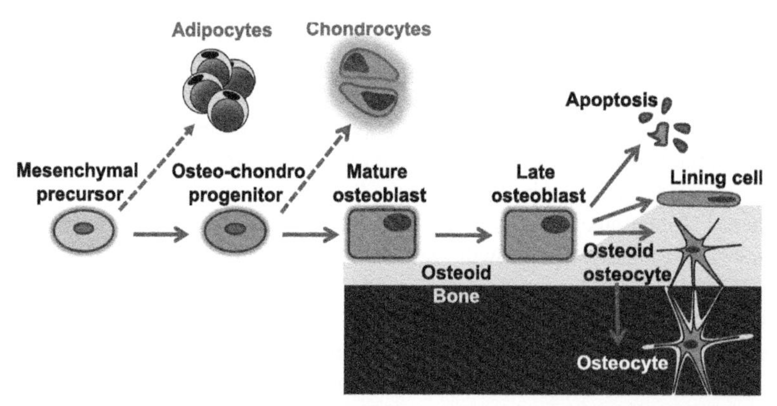

FIGURE 4.1 Stages of differentiation of the osteoblast lineage. Multipotent, replicating mesenchymal precursors, capable also of adipocyte differentiation, become more committed to be replicating osteo-chondro progenitors (osteoblast progenitors). These cells are capable of either chondrocytic or osteoblastic differentiation. When they reach the mature osteoblast stage, they reside on the bone surface above a layer of newly produced collagen-containing osteoid, and continue to differentiate. When they have completed the process of osteoid production, osteoblasts undergo one of three possible fates: (1) cell death through apoptosis; (2) remaining on the bone surface as flattened bone-lining cells, not actively producing osteoid; or (3) becoming embedded in the osteoid as "osteoid osteocytes," gradually becoming encased in mineralized bone, forming a network throughout the bone matrix, and remaining there as osteocytes.

These and subsequent studies using rigorous differentiation assays and CFU-F cells to show multipotency identified an SSC subset capable of forming skeletal tissue within bone marrow stroma. SSCs are rare in the marrow stromal cell population and need to be defined by rigorous clonal and differentiation assays. As an alternative some laboratories turned to the nonhemopoietic adherent cells of the bone marrow, which became known commonly as "mesenchymal stem cells" (MSCs). MSCs could be isolated from many tissues (da Silva Meirelles et al., 2006) and have been claimed to also be capable of forming osteoblasts in vitro. However, such studies have to be viewed with some reservation. The commonly used "osteogenesis assay," often used to confirm osteoblast differentiation, assesses mineral deposition by cells cultured in ascorbic acid and β-glycerophosphate. However, mineral stains rarely distinguish between collagen-containing nodule formation and diffuse mineral deposited by dead or dying cells (Orriss et al., 2014). It should be noted that any cell rich in alkaline phosphatase, including non–osteoblast lineage cells, like chondrocytes, can generate phosphorus by cleaving β-glycerophosphate, resulting in calcium phosphate precipitates, without expressing osteoblast marker genes or producing collagen-rich matrix. The International Society of Cell and Gene Therapy now recommends the term "mesenchymal stromal cells," which reflects only their in vitro properties, without clonal analyses or in vivo studies (Dominici et al., 2006). These caveats were summarized succinctly in a 2017 review (Robey, 2017).

A critical question remains as to whether other stromal cells derived from other sources can generate bone organoids (ossicles) supporting hemopoiesis in vivo. A comparison of stromal cells from bone marrow, white adipose tissue, umbilical cord, and skin led to the conclusion that only bone marrow–derived MSCs have this property (Reinisch et al., 2015). However, a different conclusion was reached when cord blood-borne fibroblasts (CB-BFs) were isolated. These CB-BFs, a rare population of cells, could generate complete ossicles in vivo with a functional hematopoietic stem cell niche, and did so through an endochondral bone formation program (Pievani et al., 2017). This outcome required as a starting point the CB-BF population rather than unselected MSCs from cord blood.

Further advances in understanding the osteoblast progenitor have come through cell lineage tracing methods. These studies localized osteoblast progenitors to vascular structures in the marrow, and suggested the same precursors may also give rise to cells forming the blood vessel and pluripotent perivascular cells (Doherty et al., 1998; Modder and Khosla, 2008; Otsuru et al., 2008). Support for this comes from mice with the smooth muscle α-actin (SMAA) promoter used to direct green fluorescent protein (GFP) to smooth muscle and pericytes; strong osteogenic differentiation was evident in cells positive for SMAA–GFP (Kalajzic et al., 2008). This indicates the capacity of SMAA to mark an osteoblast progenitor population, and reinforced earlier studies proposing the pericyte as an osteoblast progenitor during bone remodeling (Brighton et al., 1992; Doherty et al., 1998). It should be noted that pericytes appear to behave in an organ-specific manner, dictated by their anatomy and position (Bianco et al., 2008). For example, pericytes isolated from muscle generate myocytes in vitro (Dellavalle et al., 2007). Notably, only marrow-residing pericytes appear capable of becoming osteoblasts under normal conditions (Sacchetti et al., 2016). In pathological conditions, there are clearly osteoblast precursor populations resident in other tissues capable of forming bone, including the heterotopic ossifications observed after spinal cord injury (Genet et al., 2015). The generation of osteoblasts from bone marrow–specific pericytes illustrates the importance of the microenvironment in determining differentiation, probably by the generation and influence of local factors. Lineage tracing studies also suggest that bone-lining cells can form a population of osteoblast precursors (see "Bone-lining cells") (Matic et al., 2016).

Commitment of osteoblast progenitors (preosteoblasts)

For many years the "osteoblast phenotype" was studied in rodent cell culture systems, including stable and transformed cell lines, osteogenic sarcoma cell lines, and organ cultures. The limited characterization possible when cells were first grown from rodent bone fragments (Peck et al., 1964) was extended when methods were established to culture cells obtained from newborn rodent calvariae by enzymatic digestion (Luben et al., 1976). At the same time, cells were cultured from osteogenic sarcomata; these, and the clonal cell lines derived from them, were enriched in a number of osteoblastic properties (Majeska et al., 1980; Partridge et al., 1983). The concepts of the osteoblast phenotype developed in these rodent culture systems were extrapolated to adult bone in vivo. Limitations were not always realized, in particular the heterogeneity of primary rodent cultures and the identity of osteosarcoma cells, not as true osteoblasts, but as tumor cells enriched in some osteoblastic features. Nevertheless, these systems have provided useful foundational knowledge, particularly in studying osteoblast lineage intracellular signaling responses to hormones, growth factors, and cytokines (Crawford et al., 1978; Partridge et al. 1981, 1982; Livesey et al., 1982).

Stromal lineage commitment depends on the expression of key transcription factors that, on induction, initiate a cascade of events culminating in differentiation. These are discussed in depth in Chapter 7. Among the transcription factors regulated, osteoblast differentiation requires expression of Runx2 (Komori et al., 1997; Otto et al., 1997) and Osterix

(Nakashima et al., 2002) to commit progenitors to preosteoblasts. Other transcription factors, including ATF4 (Yang et al., 2004), AP-1 (Sabatakos et al., 2000), C/EBPβ, and C/EBPδ (Gutierrez et al., 2002), promote their transition to functional osteoblasts. Alternatively, commitment toward adipocytic differentiation requires expression of other transcription factors, including peroxisome proliferator-activated receptor (PPARγ) (Barak et al., 1999) and C/EBPα (Tanaka et al., 1997).

Because osteoblasts and adipocytes are derived from common progenitors, lineage commitment of precursor cells to osteoblasts results in a proportional decrease in adipogenesis. This inverse relationship is observed in cell culture (Walker et al., 2010; Poulton et al., 2012), and has been described in genetically altered mouse models, both where high osteoblast activity is associated with low marrow adipocyte volume (Sabatakos et al., 2000) and where low osteoblast numbers are associated with high marrow adipocyte volume (Sims et al., 2000; Walker et al., 2010; Poulton et al., 2012). Such reciprocal regulation is also observed clinically: high marrow adiposity is associated with age-related osteoporosis (Justesen et al., 2001) and observed in preclinical models of induced bone loss, such as ovariectomy (Martin et al., 1990) and immobilization (Ahdjoudj et al., 2002). Understanding the relationship between osteoblasts and adipocytes and how its dysregulation contributes to bone loss will provide key information required to improve treatments for skeletal disorders in adult bone metabolism.

The effects of pharmacological parathyroid hormone (PTH) provide another example of the reciprocal regulation of osteoblast/adipocyte commitment. Intermittent administration of amino-terminal preparations of PTH increases bone mass in part by promoting the differentiation of committed osteoblast precursors (Dobnig and Turner, 1995), decreasing osteoblast apoptosis (Manolagas, 2000), and suppressing production of sclerostin by osteocytes (Keller and Kneissel, 2005). Intermittent PTH treatment is also associated with reduced marrow adipocyte numbers in vivo (Sato et al., 2004). Similar observations of reduced osteoprogenitor recruitment and increased marrow adiposity have been noted in mice haploinsufficient for PTH-related protein (PTHrP) (Amizuka et al., 1996; Miao et al., 2005), and increased formation of adipocytes has been noted after cessation of PTH treatment in mice (Balani and Kronenberg, 2018). What is particularly interesting is recent evidence obtained through lineage tracing in mice indicating direct actions of PTH(1−34) through the PTH receptor (PTHR1) on very early osteoblast precursors, favoring osteoblast commitment at the expense of adipocytes (Balani et al., 2017); this is surprising because earlier studies of receptor activation in vitro indicated that responsiveness to PTH required more mature osteoblasts (Allan et al., 2008). The new data suggest that PTHR1 responsiveness is active earlier in the osteoblast lineage. This is consistent with very early work showing suppression of preadipocyte differentiation in vitro by PTHrP (Chan et al., 2001). In support of this, mice with PTHR1 deletion targeted to the osteoblast lineage, including early precursors (using Prx1-Cre), exhibited low bone formation and high marrow fat (Fan et al., 2017). This was associated with altered expression of ZFP467, a transcriptional cofactor, previously described to enhance adipocyte and blunt osteoblast differentiation, which is downregulated in osteoblast lineage cells by PTH or interleukin-6 (IL-6) family cytokine treatment (Quach et al., 2011). Transfection of ZFP467 into mouse calvarial cells substantially increased their transformation into adipocytes when transplanted into marrow (Quach et al., 2011).

A further link between the osteoblast and adipocyte lineages is provided by the finding that knockout of the leptin receptor (LepR) from limb bone marrow stromal cells by crossing *Prrx1.Cre* and $LepR^{fl/fl}$ mice revealed that the LepR also mediates local actions on very early SSC precursors to promote adipogenesis and inhibit bone formation (Yue et al., 2016). Such a role for the LepR is opposite to that described above for PTHR1 (Balani et al., 2017; Fan et al., 2017); whether there is any physiological connection between the two is not known.

Osteoblasts and chondrocytes are also derived from a common precursor (sometimes termed an "osteo-chondro progenitor"), and while this does not need to be considered in adult bone remodeling, the commitment of precursors to either of these lineages is important in developmental biology, fracture healing, and approaches being developed for joint cartilage repair. For example, the osteoblast commitment genes *Osterix* and *Runx2* both promote the final stage of chondrocyte differentiation (hypertrophy) preceding vascular invasion in endochondral ossification. This progression through to hypertrophy is blocked in both *Runx2*- (Inada et al., 1999; Kim et al., 1999) and *Osterix*-null mice (Nishimura et al., 2012). These two transcription factors appear to interact directly with each other to promote chondrocyte hypertrophy (Nishimura et al., 2012). In addition, overexpression of *Runx2* promotes chondrocyte hypertrophy, while knockdown reduces it and promotes adipogenesis (Enomoto et al., 2000; Takeda et al., 2001; Ueta et al., 2001). Although reciprocal regulation of chondrogenesis and osteoblastogenesis from the same common precursor has been suggested (Komori, 2018), mechanisms controlling this have not yet been identified. So, too, the question of whether hypertrophic chondrocytes also represent partially committed osteoblast precursors capable of transdifferentiating into osteoblasts continues to be investigated (Yang et al., 2014).

Mature "bone-forming" osteoblasts

Mature osteoblasts synthesize the organic components of bone and contribute to the events resulting in its mineralization. This is discussed in additional detail in Chapter 11.

Mature matrix-producing osteoblasts are recognized as groups of plump, cuboidal mononuclear cells lying on the unmineralized matrix (osteoid) they have synthesized. These cells do not operate in isolation, and in vivo are rarely seen even in small groups of two or three cells. Rather, active bone-forming surfaces are lined by a seam of osteoid, on the surface of which resides a team of osteoblasts with similar morphologic characteristics, including similar nuclear–cytoplasmic alignment, and extensive sites of contact between team members (Fig. 4.2). Formation of mineralized nodules in vitro also depends on a critical mass of differentiated osteoblasts, which form a cobblestone layer before matrix deposition occurs (Ecarot-Charrier et al., 1983; Abe et al., 1993). Matrix-producing osteoblasts communicate with one another, with adjacent bone-lining cells (Doty, 1981), and with osteocytes below the surface through gap junctions and direct cell–cell contact. This requirement of cell–cell contact between matrix-producing osteoblasts emphasizes the importance of juxtacrine and paracrine control mechanisms for osteoblast differentiation and bone formation.

Osteoblasts do not produce "bone" per se, but synthesize a collagen-rich osteoid matrix, which becomes gradually mineralized over time. This process of mineralization is controlled by noncollagenous proteins produced by late-stage osteoblasts and osteocytes. The ability of osteoblasts to produce large quantities of protein is clearly shown by their dense endoplasmic reticulum (Fig. 4.2).

When osteoid is deposited by osteoblasts, it has two potential forms depending on its collagen orientation and speed of production. Within the osteoblast Golgi, procollagen molecules that make up osteoid are biochemically modified to form collagen triple helices, which are released by exocytosis into the extracellular space (Leblond, 1989). During bone

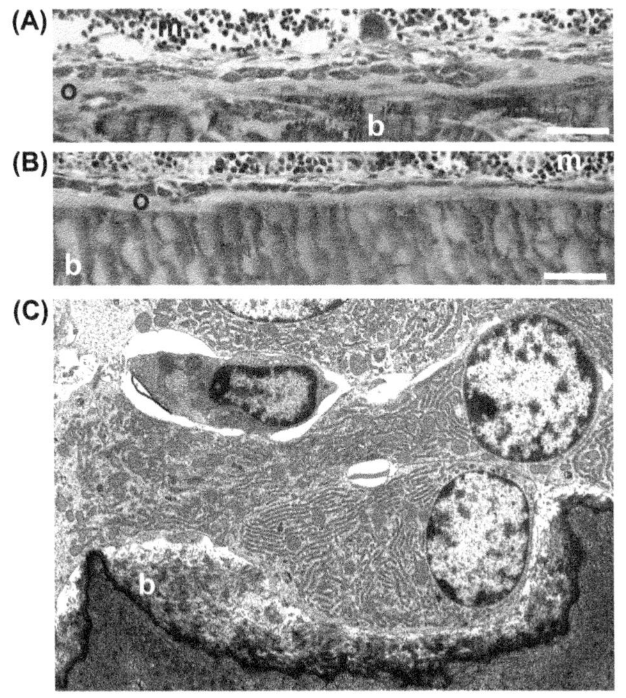

FIGURE 4.2 The appearance of matrix-producing osteoblasts by (A and B) light microscopy and (C) transmission electron microscopy. (A and B) Murine endocortical osteoblasts on newly formed metaphyseal (A) and diaphyseal cortical bone; osteoid (*o*) is the paler substance between the darker mineralized bone (*b*) and the line of cuboidal osteoblasts separating it from the marrow (*m*). Toluidine blue–stained undecalcified tibial sections from 6-week-old female C57BL/6 mice. Scale bar = 50 μm. (C) Osteoblasts on the surface of newly deposited osteoid above calcified cartilage in the primary spongiosa in 6-week-old female mice. Note the extensive contact between the osteoblasts and the extensive endoplasmic reticulum within the cells. *Image (C) courtesy Liliana Tatarczuch and Eleanor Mackie, The University of Melbourne, reproduced from Tonna, S., Takyar, F.M., Vrahnas C., Crimeen-Irwin, B., Ho, P.W., Poulton, I.J., Brennan, H.J., McGregor, N.E., Allan, E.H., Nguyen, H., Forwood, M.R., Tatarczuch, L., Mackie, E.J., Martin, T.J., Sims, N.A., 2014. EphrinB2 signaling in osteoblasts promotes bone mineralization by preventing apoptosis. FASEB J. 28 (10), 4482–4496.*

development and fracture healing, woven bone is deposited rapidly: this contains disordered collagen fibers. During adult bone remodeling, the more slowly deposited, and mechanically stronger, lamellar bone has collagen fibers oriented in perpendicular planes in adjacent lamellae (Giraud-Guille, 1988). How osteoblasts are instructed to form either of these two substances is not known. One clue comes from ultrahigh-voltage electron microscopy studies showing initial sparse and random deposition of collagen fibers, followed, with increasing distance from the surface osteoblasts, by thickening and a shift to being oriented parallel to the direction of growth (Hosaki-Takamiya et al., 2016).

After collagen is deposited it becomes progressively mineralized by the accumulation of hydroxyapatite crystals. This mineralization process has two phases. Within ~5–10 days, osteoid undergoes rapid primary mineralization, and over subsequent weeks, months, and years, secondary mineralization occurs (Glimcher, 1998). Usually, the tissue reaches approximately 50%–70% of its final mineral content during primary mineralization (Boivin and Meunier, 2002; Ruffoni et al., 2007). During secondary mineralization, mineral continues to accumulate at a slower rate (Fuchs et al., 2008), crystals become larger (Glimcher, 1998), and carbonate is substituted for phosphate groups within the matrix (Vrahnas et al. 2016, 2018). As mineral is deposited, the surrounding collagen fibers of bone also change, becoming more cross-linked (Paschalis et al., 2004) and more compact (Vrahnas et al. 2016, 2018).

The processes of mineralization initiation, mineral accrual, and crystal maturation are controlled by a number of noncollagenous proteins expressed by osteoblasts and osteocytes (see Table 4.1). Some proteins promote mineralization, such as the many phosphate-regulating and SIBLING proteins including alkaline phosphatase; phosphate-regulating neutral endopeptidase, X-linked; DMP1; matrix extracellular phosphoglycoprotein; and bone sialoprotein/integrin-binding sialoprotein. The importance of these proteins is clearly illustrated by defective bone mineralization in genetic insufficiencies of these proteins in both humans and mice (Whyte, 1994; Quarles, 2003; Holm et al., 2015). Mineralization inhibitors, such as osteopontin/secreted phosphoprotein-1 (Addison et al., 2010), are also expressed by osteoblasts and osteocytes. The final level of mineralization achieved in the bone substance varies locally within the bone matrix, and depends on the species, sex, age, and anatomical location of the bone (Boskey, 2013).

Bone-lining cells

Bone-lining cells are the abundant flattened cells lining both endocortical and trabecular bone surfaces (Miller and Jee, 1987). These cells have much less synthetic function, little cytoplasm or endoplasmic reticulum, and somewhat less cytoplasmic basophilia and alkaline phosphatase activity. They nevertheless possess gap junctions and may communicate both with one another and, probably through the canaliculi, with osteocytes residing in the bone matrix.

Although long regarded as a "resting" or "quiescent" population, bone-lining cells express receptors for endocrine and paracrine agents in common with mature osteoblasts. Their contraction from the bone surface in response to PTH (Jones and Boyde, 1976) was suggested to be a means of allowing osteoclasts access to the bone surface (Rodan and Martin,

TABLE 4.1 Selected proteins expressed at specific stages of osteoblast differentiation

Committed osteoblast progenitors	Early osteoblasts	Intermediate osteoblasts	Late osteoblasts	Early osteocytes	Late osteocytes
Runx2 (Komori et al., 1997) Osterix (Nakashima et al., 2002)	Pro-a(1)I collagen (Rodan and Noda, 1991) Fibronectin (Stein et al., 1990)	Alkaline phosphatase (Aubin, 1998) Osteonectin (Aubin, 1998) Matrix gla-protein (Stein et al., 1990). Bone sialoprotein Chen et al. (1992) PTH receptor (Allan et al., 2008)	Osteocalcin (Stein et al., 1990) Osteopontin (Stein et al., 1990)	Matrix extracellular phosphoglycoprotein (Igarashi et al., 2002). Dentin matrix protein 1 (Toyosawa et al., 2001) Phosphate-regulating neutral endopeptidase, X-linked (Westbroek et al., 2002)	Sclerostin (van Bezooijen et al., 2004) Fibroblast growth factor 23 (Yoshiko et al., 2007).

Shown is the stage of differentiation at which the marker is most highly expressed. Note: these proteins are expressed at lower levels at other stages of differentiation.

1981). The emergence of bone-lining cells from quiescence has been highlighted much more recently in other ways. This includes reactivation to resume their bone-forming ability by contributing to the anabolic effects of PTH (Kim et al., 2012) and anti-sclerostin treatment (Kim et al., 2017). The ability of bone-lining cells to form active osteoblasts is not restricted to "reactivation." After osteoblast ablation, bone-lining cells have been shown to proliferate and differentiate into mature osteoblasts; marrow-derived precursors were not capable of this (Matic et al., 2016). Bone-lining cells may therefore also provide a source of osteoblast precursors capable of proliferation during adulthood.

Another suggested function for bone-lining cells is their ability to lift from the bone surface to generate a canopy over the basic multicellular unit (BMU) to enclose its activities on the bone surface (Hauge et al., 2001), as discussed in Chapter 10. Bone-lining cells have also been shown to express receptor activator of NF-κB ligand (RANKL) and M-CSF (CSF1), and may therefore also participate in stimulating osteoclastogenesis (Matic et al., 2016). This has been suggested to be particularly important in estrogen deficiency (Streicher et al., 2017).

Osteocytes

Osteocytes, the matrix-embedded cells of the osteoblast lineage, are considered major "controlling cells" or "orchestrators" of bone due to their extensive networks and ability to signal to other cells (Bonewald, 2011; Schaffler et al., 2014). At this stage of the lineage, osteocytes exhibit a distinct pattern of gene expression (Table 4.1), and a distinct phenotype and location. Osteocytes are osteoblasts that became entrapped within the bone matrix during bone formation. As they are trapped, they form extensive dendritic processes through a fluid-filled network of communicating channels, allowing them to sense and respond to mechanical strain and microdamage to bone. They are the most abundant cell in bone by far, forming a highly complex cellular communication network through the bone matrix with a total of ~3.7 trillion connections throughout the adult skeleton (Buenzli and Sims, 2015). The nature and functions of osteocytes are considered in detail in Chapter 6.

One major protein involved in the communication network of osteocytes to osteoblasts is sclerostin, an inhibitor of Wnt-stimulated osteoblast differentiation (Bonewald and Johnson, 2008); the influence of this pathway on bone formation will be considered later. Sclerostin is likely to be important in limiting the amount of bone laid down in modeling, and its production in response to loading (Galea et al., 2011) or to local factors such as prostaglandin E_2 (Genetos et al., 2011), IL-6 family cytokines (Walker et al., 2010), or PTHrP (Ansari et al., 2018) also regulates bone formation.

The process of osteoblast lineage differentiation

Although we have summarized the main subsets of osteoblast lineage cells (precursors, matrix-producing osteoblasts, bone-lining cells, and osteocytes), differentiation is likely to proceed subtly even within each subset, with a range of phenotypic properties likely within any group (i.e., they are probably not homogeneous).

The first approach used to study osteoblast differentiation in vitro was to use primary cultured cells derived from stromal precursors or cell lines such as MC3T3-E1 (Sudo et al., 1983) and Kusa subclones (Allan et al., 2003). These have been used extensively to define pathways of osteoblast differentiation and equate each stage of differentiation and its gene expression profiles with some function in bone. The functions carried out by any member of the lineage depend on its stage of differentiation, its pattern of gene expression, its location in bone, and the influence of local and humoral factors. Thus, it is not appropriate to use the presence or absence of a single gene to define whether a cell is, or is not, an osteoblast. Rather, the expression of multiple genes may be used to define approximately the particular stages of osteoblast lineage differentiation, as illustrated in Table 4.1. While these gene products are often considered osteoblast markers, many are not exclusively expressed by osteoblasts and can be found in other cell types in the body. As single-cell sequencing methods are applied to this question, more will be learned about specific stages of differentiation and potential "subpopulations" of the osteoblast lineage.

When it comes to analyzing mutant mice, or mice treated with agents modifying osteoblast function, attempts to identify changes in the distribution of cells throughout the lineage (i.e., whether there are more cells at the precursor or collagen-producing stage of differentiation) still require morphologic and functional assessment. For example, our work using an antagonist to the EphrinB2/EphB4 interaction in cultured osteoblast lineage cells resulted in low mRNA levels specifically for genes associated with late osteoblasts and osteocytes (Takyar et al., 2013); when cell lineage–specific blockade was assessed in vivo, the midpoint blockade of osteoblast differentiation was reflected in elevated early-stage markers of osteoblast differentiation and reductions in late-stage markers (Tonna et al., 2014), suggesting an EphrinB2/EphB4 checkpoint midway through the differentiation process. Additional histomorphometric data confirmed this: osteoid production was maintained, while the initiation of bone mineralization (which requires expression of late osteoblast and osteocyte markers) was delayed, leading to a reduction in bone stiffness (Tonna et al., 2014). Thus, stage-specific changes in osteoblast differentiation can lead to changes in the process of bone mineralization, ultimately modifying bone strength.

Defining gene expression patterns at specific stages of osteoblast differentiation has been aided by the use of genetically modified mice with reporter genes under the control of specific promoters. These reporter genes allow sorting of specific populations by fluorescence-activated cell sorting (FACS), a method enabling microarray studies to define specific gene sets expressed by, for example, osteoblasts compared with osteocytes (Kalajzic et al., 2005; Paic et al., 2009). Such studies have led to the identification of gene products associated with bone dissolution expressed by osteocytes (Chia et al., 2015), and allowed purification of cells to confirm gene knockdown in genetically altered mice (Ansari et al., 2018).

Although cell sorting might be helpful for identifying the stages of osteoblasts present in cells isolated from rodents, there remain only very limited cell surface markers available to identify the stages of osteoblast differentiation in non–genetically modified bone. Efforts are being made to develop antibodies specific for SSCs and their progeny, but have to date been unsuccessful. In mice, when cells must be derived from bone by prolonged enzymatic digestion, cell surface markers are inevitably damaged; this hampers FACS analysis. This disadvantage does not apply to the analysis of hemopoietic cells, since they do not require enzymatic digestion; such analyses have been carried out successfully for many years. Despite this limitation, freshly isolated murine bone osteoblast progenitors, which are perhaps less firmly attached to the bone surface, can be separated broadly into early (Sca-1^{+}CD51^{-}) and relatively mature (Sca-1^{-}CD51^{+}) osteoblast progenitors, each of which consist of heterogeneous populations (e.g., Lundberg et al., 2007).

Histology and histomorphometry remain the gold standard for identifying and quantifying mature osteoblasts, osteocytes, and bone-lining cells, and rely on their distinct morphologies and locations; in this way increased osteoblast numbers or changes in the osteocyte network can be identified and compared with changes in osteoid deposition and the rate at which mineralization is initiated. It should be noted that cells very early in the lineage arising from SSCs cannot be recognized morphologically in bone without the use of specific markers. Identifying specific stages of differentiation has been challenging, and requires the use of immunohistochemistry and in situ hybridization. Such methods have been used to identify the presence of Runx2-positive precursors within the marrow space in pathological conditions (for example, Walsh et al., 2009). Needed above all else are markers of osteoblast development and methods of visualization for in situ applications to complement morphologic assessment. The method of laser scanning cytometry, in which image analysis, somewhat like cell sorting, can be carried out on immunostained tissue sections, is a promising approach (Fujisaki et al., 2011).

At their various stages of development, cells of the osteoblast lineage signal to one another

Osteoblast lineage precursors begin their differentiation in the BMU, during which they are likely to be in contact with (or at least in close proximity to) other osteoblast precursors and mature osteoblasts on the bone surface. Bone-lining cells and matrix-producing osteoblasts on the bone surface are also in contact, not only with one another, but also with the highly differentiated osteocytes within the bone matrix. Not surprisingly, therefore, the cells of the lineage exert their influence through communicating signals among its member cells. These may be through direct contact by gap junctions or by paracrine and autocrine signaling. Some influential signaling processes within the osteoblast lineage will be discussed as examples of interactions within the lineage contributing to differentiation and their abilities to form bone, and program the generation of osteoclasts.

An example of contact-dependent communication: EphrinB2/EphB4

We have drawn attention to the role of the contact-dependent interaction between the receptor tyrosine kinase EphB4 and its ligand EphrinB2 as an important checkpoint prior to the late stages of osteoblast differentiation (see earlier) (Takyar et al., 2013; Tonna et al., 2014). EphrinB2 is produced at constant levels throughout osteoblast differentiation in vitro (Allan et al., 2008). In contrast, EphrinB2, but not its receptor, was rapidly increased when cells were exposed in vitro to PTHrP or PTH (Allan et al., 2008). Acute treatment of mice or rats with PTH also results in a rapid increase in EphrinB2 mRNA production in bone (Allan et al., 2008). Osteoblast-specific deletion of EphrinB2 (*OsxCre.Efnb2$^{f/f}$* mice) caused a mild osteomalacia with compromised bone strength due to impaired late-stage osteoblast differentiation and high levels of osteoblast apoptosis (Tonna et al., 2014). The impairment in osteoblast differentiation was interesting because the osteoblasts expressed all early markers of osteoblast commitment at normal levels, but showed low levels of osteoblast marker expression after the stage of PTHR1 expression. This led to normal osteoid production, but delayed mineralization and diminished bone strength, pinpointing a failure of osteoblasts to pass through the EphrinB2/EphB4 checkpoint to become late-stage osteoblasts and osteocytes. A new insight into such contact-dependent communication processes came with deletion of EphrinB2 at a later stage of differentiation, in osteocytes. These *Dmp1Cre.Efnb2*$^{f/f}$ mice had brittle bones in which the maturing matrix incorporated mineral and carbonate more rapidly than controls, and strikingly, the osteocytes in the mutant mice exhibited a higher level of cellular autophagy (Vrahnas et al., submitted for publication). Thus EphrinB2-directed actions with differing outcomes can be recognized in these two stages of the osteoblast lineage.

Communication between different stages of differentiation: IL-6 cytokines

The IL-6 family cytokines have many roles in bone physiology mediated by the osteoblast lineage. The first function identified for these cytokines in bone was their ability to promote osteoclast formation in vitro. However, osteoclast formation by IL-6, IL-11, leukemia inhibitory factor (LIF), cardiotrophin 1 (CT-1), and oncostatin M (OSM) in vitro depends on the presence of osteoblast-lineage cells (Tamura et al., 1993), by virtue of their ability to produce RANKL in response to these cytokines (Richards et al., 2000; Palmqvist et al., 2002).

These cytokines also promote bone formation. IL-6, IL-11, CT-1, and OSM all promote osteoblast differentiation in vitro (Bellido et al., 1997; Song et al., 2007; Walker et al., 2008). OSM, CT-1, and LIF, all of which are produced by osteoblasts, stimulate bone formation in vivo (Cornish et al., 1993; Walker et al. 2008, 2010). OSM, CT-1, LIF, and IL-11 also inhibit adipocyte differentiation (Sims et al., 2005; Song et al., 2007; Walker et al. 2008, 2010; Poulton et al., 2012) and stimulate the transcription factors involved in osteoblast commitment described earlier (C/EBPβ and C/EBPδ), while inhibiting the adipogenic transcription factor PPARγ, indicating an influence of these cytokines on early osteoblast/adipocyte commitment; this is confirmed by the low bone formation, high marrow adipose phenotypes of the IL-11 receptor-, CT-1-, LIF-, and OSM receptor-null mice (Sims et al., 2005; Walker et al. 2008, 2010; Poulton et al., 2012). In addition, these cytokines, which are expressed at all stages of osteoblast differentiation, all strongly inhibit expression of sclerostin, confirming a direct action in osteocytes (Walker et al., 2010). Thus, they promote bone formation by direct actions at multiple stages of osteoblast differentiation.

Stage-specific roles of IL-6 family cytokines in the osteoblast lineage were studied further using two mouse models in which the common receptor subunit for these cytokines (gp130) was conditionally deleted either from the entire osteoblast lineage (*Osx1Cre*) or later in osteoblast/osteocyte differentiation (*Dmp1Cre*). Given the role of these cytokines in stimulating osteoclast formation, it was a surprise that no changes in osteoclast formation or RANKL expression were observed. Instead, the mice exhibited a low level of bone formation (Johnson et al., 2014). This indicated that the major physiological effect of these cytokines in the osteoblast lineage is not to support osteoclast formation, but to maintain bone formation (Johnson et al., 2014). Although these cytokines stimulate RANKL expression in the osteoblast lineage, that function is not required for the maintenance of normal bone mass, but may be required to generate osteoclasts specifically in pathological conditions; alternatively, the gp130-dependent support of osteoclast formation within the osteoblast lineage may occur prior to osteoblast commitment. Consistent with differing influences at specific stages of the osteoblast lineage, there was no effect of deletion of gp130 within the osteoblast lineage on marrow adipogenesis (Johnson et al., 2014), suggesting these cytokines also influence osteoblast–adipocyte commitment before the stage of *Osterix* expression. Ultimately, both mice showed the same structural phenotype, indicating that role of gp130 in the osteoblast lineage controlling bone structure is mediated by the osteocyte (Johnson et al., 2014).

Communication at different stages of differentiation: PTHrP/PTHR1

Another communication pathway among the osteoblast lineage is mediated by PTHR1, the G-protein-coupled receptor shared between the hormone PTH and the cytokine PTHrP.

The paracrine action of osteoblast lineage–derived PTHrP to promote bone formation and stimulate osteoclastogenesis has been demonstrated in mice globally deficient in PTHrP and in mice with PTHrP knockdown directed to mature osteoblasts and osteocytes using *ColI(2.3)Cre* (Miao et al., 2005). This revealed a mechanism analogous to the anabolic action of intermittent PTH treatment in osteoporosis (i.e., that local PTHrP retains trabecular bone mass by promoting bone formation). The communication by PTHrP within the lineage has been extended to osteocytes, since osteocyte-specific knockout of PTHrP also results in reduced bone formation and loss of bone mass and strength (Ansari et al., 2018), with the conclusion that osteocyte-derived PTHrP acts in a paracrine/autocrine manner on osteocytes and osteoblasts. Notably, deletion later in the lineage (with *Dmp1Cre*) did not result in reduced osteoclast formation (Ansari et al., 2018), indicating that PTHrP promotes bone formation through actions later in the osteoblast lineage than its actions to stimulate osteoclast formation, analogous to the aforementioned observations in gp130-deficient mice. This topic is considered in detail in Chapter 25.

Physical sensing and signaling by osteoblasts and osteocytes

Another possible signaling mechanism within the osteoblast lineage is their ability to "sense" changes on the bone surface. When rat calvarial cells were provided in vitro to bone slices with crevices made by osteoclasts, or mechanically excavated grooves, the cells made bone in those defects, filling them exactly to a flat surface (Gray et al., 1996). In this way, in bone remodeling, osteoblast filling of a BMU follows the dictates of the size of the resorption cavity made by the osteoclasts, and once the formation process is established, the participating cells themselves sense the spatial limits and fill the space. This may be achieved by the ability of osteoblast precursors to respond to changes in surface topography when the change is either much larger than the cell itself, as in the aforementioned study, or very much smaller than the cell. For example, in response to altered nanotopography, cells adhere to the surface and one another and proceed to differentiate (Dalby et al., 2006). This filling may then occur through chemical communication, possibly involving gap junctions both among matrix-producing osteoblasts and between osteoblasts and osteocytes (Doty, 1981). The changes brought about on the bone surface by cells in the reversal phase may also change the topography and thereby determine osteoblast function. The in vitro study of Gray used bone that did not contain osteocytes, so they are not necessary for osteoblasts to respond to topographic clues, at least in vitro.

Osteocytes also sense the need for new bone formation, in their case, by sensing mechanical strain within the lacuna–canalicular network. Osteocytes respond to these changes by modifying their production of a range of signaling proteins (Mantila Roosa et al., 2011). The most well defined of these modifications is their decreased production of the bone formation inhibitor sclerostin (Robling et al., 2006). This releases a brake on matrix-producing osteoblasts on the bone surface (Robling and Turner, 2009) such that bone formation is increased in response to load in regions under greatest strain, i.e., where new bone is needed. The properties of osteocytes and their actions in mechanotransduction are discussed in detail in Chapter 6; here we use it as an example of how different stages of the osteoblast lineage communicate with one another.

How does the osteoblast lineage promote osteoclast formation?

The first intercellular communication function ascribed to the osteoblast lineage arose when it was proposed that the osteoblast lineage controlled the formation of osteoclasts (Rodan and Martin, 1981). Cell culture methods established this to be the case (Takahashi et al., 1988) by showing that osteoclastogenic cytokines and hormones, such as IL-6, PTH, 1,25-dihydroxyvitamin D_3 ($1,25(OH)_2D$), and OSM, act first on osteoblast lineage cells to promote their production of a membrane-bound regulator of osteoclastogenesis. The essential product from the osteoblast lineage proved to be RANKL, a member of the tumor necrosis factor ligand family that acts upon its receptor RANK in the hematopoietic lineage (Yasuda et al., 1998). The interaction is restricted by a decoy soluble receptor, osteoprotegerin (OPG), also a product of the osteoblast lineage (Simonet et al., 1997). The essential physiological roles of these factors in osteoclastogenesis were established through genetic and pharmacological studies (Boyle et al., 2003).

It continues to be unclear which stage of the osteoblast lineage is most influential in providing RANKL to the osteoclast precursor population. Early studies indicated that RANKL was probably derived from cells relatively early in the osteoblast lineage (Udagawa et al., 1989; Kartsogiannis et al., 1999), which would be more likely to be in contact with the appropriate osteoclast precursor populations. Consistent with a role for pluripotent precursors as a source of RANKL, adipocytes have also been reported to support osteoclastogenesis (Kelly et al., 1998; Hozumi et al., 2009; Quach et al., 2011). It has also been suggested that bone-lining cells may make direct contact with osteoclast precursors (Streicher et al., 2017); this may particularly come into play when they lift from the bone surface to form the canopy prior to bone remodeling. Of all the cells of the osteoblast lineage, probably the least likely to play a part in the regulation of osteoclast formation are mature, bone-synthesizing osteoblasts.

It has also been proposed that osteocytes are a major source of RANKL for osteoclast precursors. This follows the findings that genetic deletion of RANKL causes osteoclast deficiency and osteopetrosis both when the deletion is targeted to the entire osteoblast lineage and (albeit less so) when it is targeted specifically to osteocytes (Nakashima et al., 2011; Xiong et al., 2011). Although genetic deletion of RANKL throughout the osteoblast lineage led to profound osteopetrosis, when genetic deletion of RANKL in that lineage was delayed until adulthood, a reduction of RANKL in the entire osteoblast lineage did not lead to osteopetrosis, leading the authors to suggest that it is only the osteocyte that provides RANKL for osteoclast formation. This finding was not reproduced by Fumoto et al., who achieved a similar level of delayed RANKL knockdown, albeit in younger mice, that resulted in osteopetrosis of the same severity as that in mice lacking RANKL in osteoblast lineage throughout life (Fumoto et al., 2013). Notable in the latter work, in direct contrast to the findings of Nakashima et al. (Nakashima et al., 2011), RANKL mRNA levels were higher in osteoblast-rich cell preparations compared with osteocyte-rich preparations. In both those studies the cell preparations being compared were impure, and would have contained hematopoietic cells, such as RANKL-expressing T cells, and natural killer cells, which are also capable of promoting osteoclast formation (Horwood et al., 1999; Soderstrom et al., 2010), although the proportion of contamination in each population is unknown. When highly purified populations of osteoblasts and osteocytes were prepared, in which those hematopoietic cells were removed, the level of RANKL expression in non-osteocytic mesenchymal cells (osteoblasts and their precursors) was approximately double that of purified osteocytes (Chia et al., 2015). This is consistent with early in situ hybridization studies, in which few osteocytes were reported to express RANKL (Kartsogiannis et al., 1999). It also parallels two other studies using transgenic expression of marker genes to isolate specific populations of the osteoblast lineage, in which osteoclastogenic support of osteoprogenitors was compared with a mixed population of mature osteoblasts and osteocytes (Li et al., 2010), with purified hematopoietic-cell-free osteocytes (Chia et al., 2015), or with an osteocyte-like cell line (McGregor et al., 2019). In all three works, while osteoblast progenitors supported $1,25(OH)_2D_3$-induced osteoclastogenesis, no osteoclasts were formed when the more mature osteocyte populations were used.

One important concept to consider in the role of osteoblast lineage cells in supporting osteoclastogenesis is that direct contact between the RANKL-expressing osteoblast lineage cells and the RANK-expressing hemopoietic osteoclast precursors was absolutely required for osteoclast formation in vitro (Takahashi et al., 1988; Suda et al., 1992). While recombinant soluble RANKL certainly promotes osteoclast formation from precursors in vitro (Quinn et al., 2001), and in vivo (Tomimori et al., 2009), there remains no evidence for shedding of soluble RANKL in the interaction between osteoblast and hemopoietic lineages, nor any convincing evidence of a physiological role for circulating RANKL. This means it is important to consider the location of the osteoblast lineage cells most likely to support osteoclast formation; cells within or in direct contact with the marrow, such as osteoblast precursors and bone-lining cells, rather than embedded osteocytes, are more likely to come into contact with osteoclast precursors and, therefore, are more likely to support osteoclast formation in normal remodeling. It has been difficult to understand how osteocytes, from within the matrix, could control RANKL availability to osteoclast precursors in the bloodstream through a contact-dependent mechanism, but some mechanisms have been proposed. Early confocal laser scanning microscope images have shown that osteocytic processes extend to the vascular-facing surface of the osteoblast (Kamioka et al., 2001). It has also been suggested that osteocyte-derived exosomes may participate: their release induced by apoptosis can stimulate osteoclastogenesis (Kogianni et al., 2008), and live-cell imaging showed release of microvesicles into the vasculature (Kamel-ElSayed et al., 2015). However, such a mechanism does not overcome the requirement for direct contact between RANKL-producing stromal cells and osteoclast precursors in vitro.

Actions of the osteoblast lineage during the bone remodeling sequence

Osteoblast differentiation is promoted by signals from a range of cell types within the BMU during remodeling. From the perspective of this chapter, it is important to consider the stages of osteoblast differentiation that participate in the remodeling sequence. Of course, their major known function in the bone remodeling sequence is the formation of the appropriate quantity of bone, in response to signals from the bone-resorbing osteoclast, including factors released from the bone matrix or factors produced by the osteoclasts themselves, often termed "coupling factors." Among the many possibilities are CT-1, sphingosine-1-phosphate, Wnt 10b, BMP6, and others (Pederson et al., 2008; Walker et al., 2008; Lotinun et al., 2013; Weske et al., 2018). Osteoclast-derived inhibitors of osteoblast differentiation might also contribute to the outcome. Examples of such inhibitors are Semaphorin 4D (Negishi-Koga et al., 2011), which acts on its receptor Plexin B1 in osteoblastic cells. These factors influence osteoblast precursors, mature matrix-producing osteoblasts, and osteocytes. They are discussed in Chapter 10 and have been described in reviews (Sims and Martin 2014, 2015).

There are additional functions of the osteoblast lineage in remodeling that should also be considered; these are also described in Chapter 10 but are noted briefly here. It has been proposed that the initiation of remodeling may stem from two activities at two different stages of the osteoblast lineage: (1) a signal from osteocytes in response to microdamage within the bone matrix (Schaffler et al., 2014) and (2) the lifting of bone-lining cells to form a canopy over the BMU (Hauge et al., 2001). It has also been proposed that differentiating osteoblast lineage cells on the bone surface are responsible for the duration of the reversal phase (Abdelgawad et al., 2016).

Lessons in osteoblast biology from the Wnt signaling pathway

A major pathway regulating osteoblast differentiation and bone formation identified and defined since 2001 is the Wnt signaling pathway, which modifies bone formation by altering, among other steps, the transcriptional control of osteoblasts. This new information has had a very great impact on the understanding of control mechanisms within the osteoblast lineage and the development of pharmacological approaches for skeletal fragility. These landmark findings arising out of human genetics are discussed and put into context in detail in Chapters 7 and 8. The implications of these findings for understanding the osteoblast lineage are discussed here.

Wnt signaling is specifically and powerfully inhibited by the action of sclerostin, a protein secreted by osteocytes and encoded by the *SOST* gene. The rare syndromes of skeletal enlargement, sclerosteosis, and van Buchem's disease were discovered to be caused by inactivating mutations in the *SOST* gene (Balemans et al. 2001, 2002). Genetically manipulated mouse models recapitulated the high bone mass observed in the human mutation syndromes (Loots et al., 2005). Within the cell, activation of canonical Wnt signaling leads to stabilization of β-catenin in the cytoplasm through inhibition of glycogen synthase kinase (GSK)-3β-mediated phosphorylation, resulting in accumulation of cytoplasmic β-catenin followed by its translocation to the nucleus and transcription of specific gene targets. Such activation in mesenchymal cells promotes osteoblast activity (Rawadi et al., 2003). Oral delivery of either a small-molecule inhibitor of GSK-3β (Kulkarni et al., 2006) or a less potent inhibitor, lithium chloride (Clement-Lacroix et al., 2005), each enhanced osteoblast differentiation in vitro and increased bone formation, bone mass, and strength in vivo.

It soon became clear that this pathway to control osteoblast differentiation offered appealing therapeutic prospects for the development of anabolic agents. The many possible targets used to increase Wnt/β-catenin signaling included extracellular agonists, inhibition of any of the several extracellular antagonists, and inhibition of GSK-3β. Many of these have been attempted in preclinical experiments, but the chosen approach that is now advanced in clinical development is blockade of sclerostin by treatment with neutralizing antibody.

The effects of the Wnt signaling pathway on bone are not restricted to direct actions on bone matrix production by osteoblasts, however. Mice prepared with a constitutively active β-catenin—and therefore constitutively active Wnt signaling—in the osteoblast lineage surprisingly showed no significant alteration in mature osteoblast numbers; instead the mice had a severe form of osteopetrosis, including failed tooth eruption (Glass et al., 2005). This phenotype appeared to be due to failure of osteoclast formation, caused by increased OPG production by osteoblast lineage cells in which active β-catenin was expressed. A similar osteopetrotic syndrome due to failed osteoclast development and increased OPG was observed in mice lacking the adenomatous polyposis coli protein (APC) in osteoblasts (Holmen et al., 2005). APC acts in a complex of proteins with GSK-3β to maintain the normal degradation of β-catenin. Its absence leads to accumulation of β-catenin, resulting in cell-autonomous activation of Wnt signaling in the osteoblast lineage.

As was predicted, inhibition of production or action of sclerostin resulting in enhanced Wnt canonical signaling led to increased bone mass in preclinical studies (Li et al., 2009; Kramer et al., 2010). In rodents and nonhuman primates, the tissue level mechanism by which anti-sclerostin increases bone is predominantly to promote bone formation on quiescent surfaces—thus a modeling effect. On preresorbed surfaces (remodeling) the amount of new bone formed is greater than that resorbed; this includes bone laid down over quiescent surfaces adjacent to remodeling sites. In the phase II and phase III clinical studies carried out with the humanized monoclonal antibody to sclerostin (romosozumab) (McClung et al., 2014; Cosman et al., 2016), romosozumab recapitulated the rapid increase in bone mineral density that had been seen in preclinical studies. In both clinical studies romosozumab treatment was associated with a transitory increase in the bone formation marker P1NP and a moderate but more sustained decrease in bone resorption markers. The latter effect remains unexplained but could be related to a change in the distribution of cells in the osteoblast lineage population. The profound increase in bone formation may shift the population such that there is a greater proportion of matrix-producing osteoblasts, and the push toward differentiation may lead to a lesser availability of less mature cells that stimulate RANKL production and therefore support osteoclast formation. Alternative explanations could be that Wnt signaling results in sufficient β-catenin activation to promote OPG production (*vide supra*), or that WISP1, induced by Wnt signaling (Holdsworth et al., 2018), inhibits osteoclast formation (Maeda et al., 2015).

The reason for the decrease in bone formation markers after an initial rise poses interesting questions concerning osteocyte/osteoblast interactions. Multiple studies using two different sclerostin antibodies in rats or mice identified a rapid increase in modeling-based bone formation and a transient increase in mineral apposition rate in remodeling sites that was not sustained with continued treatment (Ominsky et al., 2014; Stolina et al., 2014; Nioi et al., 2015; Taylor et al., 2016; Ma et al., 2017). Progressive increases were seen in mRNA levels of Wnt signaling antagonists *Sost* and *Dkk1* in osteoblastic cells, tibiae, and vertebrae. Supporting evidence was recently provided by Holdsworth et al. (Holdsworth et al., 2018), who showed that treatment of mice with anti-sclerostin increased expression in bone of transcripts for antagonists of Wnt signaling, including Sost, Dkk1, Dkk2, WIF1, SFRP2, SFRP4, SFRP5, and FRZB, as well as increased expression of WISP1, which inhibits osteoclast formation (Maeda et al., 2015). Thus, the intriguing possibility has been raised that the transience of anabolic action through sclerostin blockade is the result of self-regulation within the Wnt pathway.

Modulation of the program of osteoblast differentiation can also influence sclerostin expression by the osteoblast lineage. The transcription factor Osterix is absolutely required for osteoblast differentiation and is expressed at early stages of osteoblast commitment (Nakashima et al., 2002). Since sclerostin is expressed by osteoblasts that have reached the osteocytic stage, it was surprising to find that this early osteoblastic transcription factor also binds to the sclerostin promoter, thereby directly stimulating sclerostin expression (Yang et al., 2010). Whether this mechanism plays an important role in continued differentiation of the osteoblast lineage remains to be understood.

These aspects of anti-sclerostin action reveal a complexity of regulation that will undoubtedly be the focus of future attention. Ultimately, activation of Wnt signaling, by whatever means, induces osteoblast differentiation and osteoid formation, but the extent of these effects can be limited by built-in regulation, carried out by later cells of the lineage that produce inhibitors to block further Wnt signaling. They illustrate very well what crucial roles are played in the regulation of bone by cells of the osteoblast lineage; many are new roles revealed in the last 10 years. They draw attention to the fact that the mechanisms involved in sclerostin blockade are new to us, and they illustrate the lack of precision in using the word "osteoblast" as though it is being applied to a group of cells with common properties.

From paracrinology to endocrinology in bone: the secretory osteoblast lineage

The many constituent cells of the osteoblast lineage not only form bone, but also engage in paracrine and juxtacrine signaling among its members, as discussed earlier, as well as with hematopoietic cells in the regulation of osteoclast formation. This makes for a lineage of cells that carry out many functions that extend far beyond the matrix-producing role traditionally ascribed to osteoblasts. These additional functions of osteoblasts include support in physiology of the hematopoietic and immune systems (Calvi et al., 2003). In pathology, the osteoblast lineage also directs how cancer cells are housed, whether having their malignant progression enhanced, as in breast and prostate cancer (Sterling et al., 2011), or inhibited, as in multiple myeloma (Lawson et al., 2015). The osteoblast lineage can also play a central part providing an environment that favors cancer cell dormancy (reviewed in Croucher et al., 2016), in which the large population of lining cells could favor dormancy until changes in local events reawaken malignant cells.

A real surprise came with the realization that the osteoblast lineage also serves as a site of production and secretion of hormones. The first osteoblast lineage endocrine factor identified was FGF23, discovered as responsible for tumor-induced osteomalacia (Yamashita et al., 2000), and with missense mutations causing autosomal dominant hypophosphatemic rickets (Consortium, 2000). Its bone cell of origin was found to be the osteocyte (Liu et al., 2003). Then its hormonal nature was shown: circulating FGF23 acts on the kidney to promote phosphorus excretion and reduce 25(OH)D-1α-hydroxylase expression and hence circulating $1,25(OH)_2D_3$ (Shimada et al., 2004). FGF23 is reviewed in detail in Chapter 64.

Recognition of bone as an endocrine organ (Fukumoto and Martin, 2009) brings with it questions that relate to long-held views of regulatory mechanisms of hormone secretion from endocrine "glands." Usually a specific stimulus, e.g., low calcium for PTH, low glucocorticoid for adrenocorticotropic hormone, results in rapid response secretion of the appropriate hormone. In the case of FGF23, in which the cell source, predominantly osteocytes, is so widely distributed, there must be new endocrine mechanisms by which a need for FGF23 secretion by osteocytes is communicated. A high-phosphate diet increased FGF23 in rats and mice (Perwad et al., 2005), but the mechanism by which this increase occurred is not clear. The substantial increase in FGF23 mRNA and protein production following pharmacological administration of PTH(1−34) was blocked in mice with *Pthr1*-null bones (Fan et al., 2016). Given the evidence for the roles of paracrine PTHrP in osteoblast and osteocyte biology, this would support a view that locally generated PTHrP might regulate FGF23 secretion.

The questions become even more complex in the case of the second bone hormone discovered relatively recently, osteocalcin. Its endocrine role will be considered in detail in Chapter 86. Its direct relevance to this chapter is that it also is a product of the osteoblast lineage, specifically late osteoblasts and early osteocytes, secreted and stored as an abundant component of the bone matrix and with readily assayed circulating levels that can reflect both bone formation and bone resorption.

Osteocalcin is carboxylated at three glutamic acid residues through a vitamin K-dependent process, and is decarboxylated in the acid pH of the bone resorption lacuna to convert GlaOCN (osteocalcin) to GluOCN (undercarboxylated osteocalcin). The latter undercarboxylated and uncarboxylated forms are considered to be the hormonal forms capable of enhancing glucose uptake in muscle, insulin production and cell mass in the pancreatic β-cell, and insulin sensitivity in liver and adipose tissue (Lee et al., 2007; Ferron et al., 2010; Wei et al., 2014; Karsenty and Olson, 2016). These actions are thought to be mediated through the G-protein-coupled receptor family C group 6 member A, GPCR6A (Pi et al., 2011).

Mice with high osteoclast activity levels developed high circulating GluOCN and improved glucose tolerance and insulin sensitivity, while osteoclast suppression was associated with low GluOCN (Lacombe et al., 2013).

The effects of osteocalcin on energy production and utilization were not noted when osteocalcin (*Bglap*)-null mice were first prepared; at that time it was simply surprising to find that there was no striking bone phenotype (Ducy et al. 1996, 1997), although later studies revealed changed bone composition and strength (Boskey et al., 1998). The osteocalcin-null mice were later noted to have larger fat pads and mild hyperglycemia (Lee and Karsenty, 2008). The mice were also noted to breed poorly, and GluOCN was found to enhance male fertility by increasing testosterone production by Leydig cells (Oury et al., 2011). The reproductive role of osteocalcin will also be considered in Chapter 86.

Another feature of the osteocalcin-null mice (Ducy et al., 1997) was a noticeable passivity of behavior that translated into anxiety-like behavior and deficits in memory and learning that could be demonstrated experimentally (Oury et al., 2013). Although GPCR6A was found to be the receptor for osteocalcin in β-cells and myoblasts, the brain actions required a different receptor, since *Gpcr6a*-null mice had none of the behavioral abnormalities of osteocalcin deficiency, and *Gpcr6a* could not be detected at sites of action of osteocalcin in the brain (Khrimian et al., 2017). Rather, genetic experiments indicated that the receptor for osteocalcin in the brain is GPR158. Mice rendered deficient for *Gpr158* exhibited severe cognitive effects that could not be treated by undercarboxylated osteocalcin (Khrimian et al., 2017; Obri et al., 2018).

At the time of writing, these diverse effects of osteocalcin deficiency need confirmation in other species. Complete knockout of osteocalcin in the rat using CRISPR/Cas9 yielded rats with high trabecular thickness and trabecular bone volume, a phenotype not observed in the murine knockout. In addition, although a statistically significant increase in insulin sensitivity was noted in the osteocalcin-null rat, there were no changes in gonadal fat pad weight or body composition (Lambert et al., 2016).

These many effects of uncarboxylated osteocalcin and FGF23 as hormones are fascinating in pointing to an endocrine role for the osteoblast lineage that could never have been predicted. In the case of FGF23 the predominant cell of production is the osteocyte. It is not known whether there are osteocyte subpopulations that preferentially produce FGF23. Whether or not that is the case, FGF23 production is based on a large number of cells that are very widely distributed in the body. To serve as a hormone, mechanisms would be needed to ensure that FGF23 is secreted physiologically in response to signals that come from other organs (endocrine) or from neighboring cells in the same organ (paracrine).

Hormonal osteocalcin presents these same questions and others. Osteocalcin production in the osteoblast lineage is high at later stages of differentiation (e.g., Allan et al., 2003), and is generally regarded as a marker of "late" osteoblasts (Table 4.1). Expression of osteocalcin is also observed in the osteocyte cell line Ocy454 (Spatz et al., 2015), possibly at a lower level than in mature osteoblasts (Ansari et al., 2018). The few studies of osteocalcin localization by in situ hybridization and immunohistochemistry seem to point to either less production in osteocytes than in mature osteoblasts or production restricted to early osteocytes (Weinreb et al., 1990; Ikeda et al., 1992; Zhou et al., 1994). As with FGF23, questions about regulation are raised, such as, what are the immediate signaling mechanisms indicating the need for GluOCN secretion to facilitate energy utilization, reproductive activity, or cognitive activity? If acidification in resorption spaces is an important contributing mechanism, how could the levels be sufficiently tightly regulated for a hormone? Anabolic activity within the osteoblast lineage is associated with increased osteocalcin in the circulation, hence its role as a formation marker as well as a resorption marker, yet how does this modify glucose metabolism and brain function? Answers to these questions could uncover the osteoblast lineage at the center of a new endocrine system and illustrate an even more remarkable diversity in the influence of these cells.

References

Abdelgawad, M.E., Delaisse, J.M., Hinge, M., Jensen, P.R., Alnaimi, R.W., Rolighed, L., Engelholm, L.H., Marcussen, N., Andersen, T.L., 2016. Early reversal cells in adult human bone remodeling: osteoblastic nature, catabolic functions and interactions with osteoclasts. Histochem. Cell Biol. 145 (6), 603–615.

Abe, Y., Akamine, A., Aida, Y., Maeda, K., 1993. Differentiation and mineralization in osteogenic precursor cells derived from fetal rat mandibular bone. Calcif. Tissue Int. 52 (5), 365–371.

Addison, W.N., Masica, D.L., Gray, J.J., McKee, M.D., 2010. Phosphorylation-dependent inhibition of mineralization by osteopontin ASARM peptides is regulated by PHEX cleavage. J. Bone Miner. Res. 25 (4), 695–705.

Ahdjoudj, S., Lasmoles, F., Holy, X., Zerath, E., Marie, P.J., 2002. Transforming growth factor beta2 inhibits adipocyte differentiation induced by skeletal unloading in rat bone marrow stroma. J. Bone Miner. Res. 17 (4), 668–677.

Allan, E.H., Hausler, K.D., Wei, T., Gooi, J.H., Quinn, J.M., Crimeen-Irwin, B., Pompolo, S., Sims, N.A., Gillespie, M.T., Onyia, J.E., Martin, T.J., 2008. EphrinB2 regulation by PTH and PTHrP revealed by molecular profiling in differentiating osteoblasts. J. Bone Miner. Res. 23 (8), 1170–1181.

Allan, E.H., Ho, P.W., Umezawa, A., Hata, J., Makishima, F., Gillespie, M.T., Martin, T.J., 2003. Differentiation potential of a mouse bone marrow stromal cell line. J. Cell. Biochem. 90 (1), 158–169.

Amizuka, N., Karaplis, A.C., Henderson, J.E., Warshawsky, H., Lipman, M.L., Matsuki, Y., Ejiri, S., Tanaka, M., Izumi, N., Ozawa, H., Goltzman, D., 1996. Haploinsufficiency of parathyroid hormone-related peptide (PTHrP) results in abnormal postnatal bone development. Dev. Biol. 175 (1), 166–176.

Ansari, N., Ho, P.W., Crimeen-Irwin, B., Poulton, I.J., Brunt, A.R., Forwood, M.R., Divieti Pajevic, P., Gooi, J.H., Martin, T.J., Sims, N.A., 2018. Autocrine and paracrine regulation of the murine skeleton by osteocyte-derived parathyroid hormone-related protein. J. Bone Miner. Res. 33 (1), 137–153.

Ashton, B.A., Eaglesom, C.C., Bab, I., Owen, M.E., 1984. Distribution of fibroblastic colony-forming cells in rabbit bone marrow and assay of their osteogenic potential by an in vivo diffusion chamber method. Calcif. Tissue Int. 36 (1), 83–86.

Aubin, J.E., 1998. Advances in the osteoblast lineage. Biochem. Cell Biol. 76 (6), 899–910.

Bab, I., Ashton, B.A., Syftestad, G.T., Owen, M.E., 1984. Assessment of an in vivo diffusion chamber method as a quantitative assay for osteogenesis. Calcif. Tissue Int. 36 (1), 77–82.

Balani, D.H., Kronenberg, H.M., 2019. Withdrawal of parathyroid hormone after prolonged administration leads to adipogenic differentiation of mesenchymal precursors in vivo. Bone 118, 16–19.

Balani, D.H., Ono, N., Kronenberg, H.M., 2017. Parathyroid hormone regulates fates of murine osteoblast precursors in vivo. J. Clin. Invest. 127 (9), 3327–3338.

Balemans, W., Ebeling, M., Patel, N., Van Hul, E., Olson, P., Dioszegi, M., Lacza, C., Wuyts, W., Van Den Ende, J., Willems, P., Paes-Alves, A.F., Hill, S., Bueno, M., Ramos, F.J., Tacconi, P., Dikkers, F.G., Stratakis, C., Lindpaintner, K., Vickery, B., Foernzler, D., Van Hul, W., 2001. Increased bone density in sclerosteosis is due to the deficiency of a novel secreted protein (SOST). Hum. Mol. Genet. 10 (5), 537–543.

Balemans, W., Patel, N., Ebeling, M., Van Hul, E., Wuyts, W., Lacza, C., Dioszegi, M., Dikkers, F.G., Hildering, P., Willems, P.J., Verheij, J.B., Lindpaintner, K., Vickery, B., Foernzler, D., Van Hul, W., 2002. Identification of a 52 kb deletion downstream of the SOST gene in patients with van Buchem disease. J. Med. Genet. 39 (2), 91–97.

Barak, Y., Nelson, M.C., Ong, E.S., Jones, Y.Z., Ruiz-Lozano, P., Chien, K.R., Koder, A., Evans, R.M., 1999. PPAR gamma is required for placental, cardiac, and adipose tissue development. Mol. Cell. 4 (4), 585–595.

Bellido, T., Borba, V.Z., Roberson, P., Manolagas, S.C., 1997. Activation of the Janus kinase/STAT (signal transducer and activator of transcription) signal transduction pathway by interleukin-6-type cytokines promotes osteoblast differentiation. Endocrinology 138 (9), 3666–3676.

Bianco, P., Robey, P.G., Saggio, I., Riminucci, M., 2010. Mesenchymal" stem cells in human bone marrow (skeletal stem cells): a critical discussion of their nature, identity, and significance in incurable skeletal disease. Hum. Gene Ther. 21 (9), 1057–1066.

Bianco, P., Robey, P.G., Simmons, P.J., 2008. Mesenchymal stem cells: revisiting history, concepts, and assays. Cell Stem Cell 2 (4), 313–319.

Boivin, G., Meunier, P.J., 2002. The degree of mineralization of bone tissue measured by computerized quantitative contact microradiography. Calcif. Tissue Int. 70 (6), 503–511.

Bonewald, L.F., 2011. The amazing osteocyte. J. Bone Miner. Res. 26 (2), 229–238.

Bonewald, L.F., Johnson, M.L., 2008. Osteocytes, mechanosensing and Wnt signaling. Bone 42 (4), 606–615.

Boskey, A.L., 2013. Bone composition: relationship to bone fragility and antiosteoporotic drug effects. Bonekey Rep. 2, 447.

Boskey, A.L., Gadaleta, S., Gundberg, C., Doty, S.B., Ducy, P., Karsenty, G., 1998. Fourier transform infrared microspectroscopic analysis of bones of osteocalcin-deficient mice provides insight into the function of osteocalcin. Bone 23 (3), 187–196.

Boyle, W.J., Simonet, W.S., Lacey, D.L., 2003. Osteoclast differentiation and activation. Nature 423 (6937), 337–342.

Brighton, C.T., Lorich, D.G., Kupcha, R., Reilly, T.M., Jones, A.R., Woodbury 2nd, R.A., 1992. The pericyte as a possible osteoblast progenitor cell. Clin. Orthop. Relat. Res. 275, 287–299.

Buenzli, P.R., Sims, N.A., 2015. Quantifying the osteocyte network in the human skeleton. Bone 75, 144–150.

Calvi, L.M., Adams, G.B., Weibrecht, K.W., Weber, J.M., Olson, D.P., Knight, M.C., Martin, R.P., Schipani, E., Divieti, P., Bringhurst, F.R., Milner, L.A., Kronenberg, H.M., Scadden, D.T., 2003. Osteoblastic cells regulate the haematopoietic stem cell niche. Nature 425 (6960), 841–846.

Chan, G.K., Deckelbaum, R.A., Bolivar, I., Goltzman, D., Karaplis, A.C., 2001. PTHrP inhibits adipocyte differentiation by down-regulating PPAR gamma activity via a MAPK-dependent pathway. Endocrinology 142 (11), 4900–4909.

Chen, J., Shapiro, H.S., Sodek, J., 1992. Development expression of bone sialoprotein mRNA in rat mineralized connective tissues. J. Bone Miner. Res. 7 (8), 987–997.

Chia, L.Y., Walsh, N.C., Martin, T.J., Sims, N.A., 2015. Isolation and gene expression of haematopoietic-cell-free preparations of highly purified murine osteocytes. Bone 72, 34–42.

Clement-Lacroix, P., Ai, M., Morvan, F., Roman-Roman, S., Vayssiere, B., Belleville, C., Estrera, K., Warman, M.L., Baron, R., Rawadi, G., 2005. Lrp5-independent activation of Wnt signaling by lithium chloride increases bone formation and bone mass in mice. Proc. Natl. Acad. Sci. U. S. A. 102 (48), 17406–17411.

Consortium, A., 2000. Autosomal dominant hypophosphataemic rickets is associated with mutations in FGF23. Nat. Genet. 26 (3), 345–348.

Cornish, J., Callon, K., King, A., Edgar, S., Reid, I.R., 1993. The effect of leukemia inhibitory factor on bone in vivo. Endocrinology 132 (3), 1359–1366.

Cosman, F., Crittenden, D., Adachi, J., Binkley, N., Czerwinski, E., Ferrari, S., Hofbauer, L., Lau, E., Lewiecki, E., Miyauchi, A., Zerbin, C., Milmont, C., Chen, L., Maddox, J., PD, M., Libanati, C., Grauer, A., 2016. Romosozumab treatment in postmenopausal women with osteoporosis. N. Engl. J. Med. https://doi.org/10.1056/NEJMoa1607948.

Crawford, A., Atkins, D., Martin, T.J., 1978. Rat osteogenic sarcoma cells: comparison of the effects of prostaglandins E1, E2, I2 (prostacyclin), 6-keto F1alpha and thromboxane B2 on cyclic AMP production and adenylate cyclase activity. Biochem. Biophys. Res. Commun. 82 (4), 1195–1201.

Croucher, P.I., McDonald, M.M., Martin, T.J., 2016. Bone metastasis: the importance of the neighbourhood. Nat. Rev. Canc. 16 (6), 373–386.

da Silva Meirelles, L., Chagastelles, P.C., Nardi, N.B., 2006. Mesenchymal stem cells reside in virtually all post-natal organs and tissues. J. Cell Sci. 119 (Pt 11), 2204–2213.

Dalby, M.J., McCloy, D., Robertson, M., Wilkinson, C.D., Oreffo, R.O., 2006. Osteoprogenitor response to defined topographies with nanoscale depths. Biomaterials 27 (8), 1306–1315.

Dellavalle, A., Sampaolesi, M., Tonlorenzi, R., Tagliafico, E., Sacchetti, B., Perani, L., Innocenzi, A., Galvez, B.G., Messina, G., Morosetti, R., Li, S., Belicchi, M., Peretti, G., Chamberlain, J.S., Wright, W.E., Torrente, Y., Ferrari, S., Bianco, P., Cossu, G., 2007. Pericytes of human skeletal muscle are myogenic precursors distinct from satellite cells. Nat. Cell Biol. 9 (3), 255–267.

Dobnig, H., Turner, R.T., 1995. Evidence that intermittent treatment with parathyroid hormone increases bone formation in adult rats by activation of bone lining cells. Endocrinology 136 (8), 3632–3638.

Doherty, M.J., Ashton, B.A., Walsh, S., Beresford, J.N., Grant, M.E., Canfield, A.E., 1998. Vascular pericytes express osteogenic potential in vitro and in vivo. J. Bone Miner. Res. 13 (5), 828–838.

Dominici, M., Le Blanc, K., Mueller, I., Slaper-Cortenbach, I., Marini, F., Krause, D., Deans, R., Keating, A., Prockop, D., Horwitz, E., 2006. Minimal criteria for defining multipotent mesenchymal stromal cells. The International Society for Cellular Therapy position statement. Cytotherapy 8 (4), 315–317.

Doty, S.B., 1981. Morphological evidence of gap junctions between bone cells. Calcif. Tissue Int. 33 (5), 509–512.

Ducy, P., Desbois, C., Boyce, B., Pinero, G., Story, B., Dunstan, C., Smith, E., Bonadio, J., Goldstein, S., Gundberg, C., Bradley, A., Karsenty, G., 1996. Increased bone formation in osteocalcin-deficient mice. Nature 382 (6590), 448–452.

Ducy, P., Zhang, R., Geoffroy, V., Ridall, A.L., Karsenty, G., 1997. Osf2/Cbfa1: a transcriptional activator of osteoblast differentiation. Cell 89 (5), 747–754.

Ecarot-Charrier, B., Glorieux, F.H., van der Rest, M., Pereira, G., 1983. Osteoblasts isolated from mouse calvaria initiate matrix mineralization in culture. J. Cell Biol. 96 (3), 639–643.

Enomoto, H., Enomoto-Iwamoto, M., Iwamoto, M., Nomura, S., Himeno, M., Kitamura, Y., Kishimoto, T., Komori, T., 2000. Cbfa1 is a positive regulatory factor in chondrocyte maturation. J. Biol. Chem. 275 (12), 8695–8702.

Fan, Y., Bi, R., Densmore, M.J., Sato, T., Kobayashi, T., Yuan, Q., Zhou, X., Erben, R.G., Lanske, B., 2016. Parathyroid hormone 1 receptor is essential to induce FGF23 production and maintain systemic mineral ion homeostasis. FASEB J. 30 (1), 428–440.

Fan, Y., Hanai, J.-i., Le, P.T., Bi, R., Maridas, D., DeMambro, V., Figueroa, C.A., Kir, S., Zhou, X., Mannstadt, M., Baron, R., Bronson, R.T., Horowitz, M.C., Wu, J.Y., Bilezikian, J.P., Dempster, D.W., Rosen, C.J., Lanske, B., 2017. Parathyroid hormone directs bone marrow mesenchymal cell fate. Cell Metabol. 25 (3), 661–672.

Ferron, M., Wei, J., Yoshizawa, T., Del Fattore, A., DePinho, R.A., Teti, A., Ducy, P., Karsenty, G., 2010. Insulin signaling in osteoblasts integrates bone remodeling and energy metabolism. Cell 142 (2), 296–308.

Friedenstein, A.J., 1976. Precursor cells of mechanocytes. Int. Rev. Cytol. 47, 327–359.

Fuchs, R.K., Allen, M.R., Ruppel, M.E., Diab, T., Phipps, R.J., Miller, L.M., Burr, D.B., 2008. In situ examination of the time-course for secondary mineralization of Haversian bone using synchrotron Fourier transform infrared microspectroscopy. Matrix Biol. 27 (1), 34–41.

Fujisaki, J., Wu, J., Carlson, A.L., Silberstein, L., Putheti, P., Larocca, R., Gao, W., Saito, T.I., Lo Celso, C., Tsuyuzaki, H., Sato, T., Cote, D., Sykes, M., Strom, T.B., Scadden, D.T., Lin, C.P., 2011. In vivo imaging of Treg cells providing immune privilege to the haematopoietic stem-cell niche. Nature 474 (7350), 216–219.

Fukumoto, S., Martin, T.J., 2009. Bone as an endocrine organ. Trends Endocrinol. Metabol. 20 (5), 230–236.

Fumoto, T., Takeshita, S., Ito, M., Ikeda, K., 2013. Physiological functions of osteoblast lineage and T cell-derived RANKL in bone homeostasis. J. Bone Miner. Res.

Galea, G.L., Sunters, A., Meakin, L.B., Zaman, G., Sugiyama, T., Lanyon, L.E., Price, J.S., 2011. Sost down-regulation by mechanical strain in human osteoblastic cells involves PGE2 signaling via EP4. FEBS Lett. 585 (15), 2450–2454.

Genet, F., Kulina, I., Vaquette, C., Torossian, F., Millard, S., Pettit, A.R., Sims, N.A., Anginot, A., Guerton, B., Winkler, I.G., Barbier, V., Lataillade, J.J., Le Bousse-Kerdiles, M.C., Hutmacher, D.W., Levesque, J.P., 2015. Neurological heterotopic ossification following spinal cord injury is triggered by macrophage-mediated inflammation in muscle. J. Pathol. 236 (2), 229–240.

Genetos, D.C., Yellowley, C.E., Loots, G.G., 2011. Prostaglandin E2 signals through PTGER2 to regulate sclerostin expression. PLoS One 6 (3), e17772.

Giraud-Guille, M.M., 1988. Twisted plywood architecture of collagen fibrils in human compact bone osteons. Calcif. Tissue Int. 42 (3), 167–180.

Glass 2nd, D.A., Bialek, P., Ahn, J.D., Starbuck, M., Patel, M.S., Clevers, H., Taketo, M.M., Long, F., McMahon, A.P., Lang, R.A., Karsenty, G., 2005. Canonical Wnt signaling in differentiated osteoblasts controls osteoclast differentiation. Dev. Cell 8 (5), 751–764.

Glimcher, M.G., 1998. The Nature of the Mineral Phase in Bone: Biological and Clinical Implications. Academic Press, San Diego, CA.

Gray, C., Boyde, A., Jones, S.J., 1996. Topographically induced bone formation in vitro: implications for bone implants and bone grafts. Bone 18 (2), 115–123.

Gutierrez, S., Javed, A., Tennant, D.K., van Rees, M., Montecino, M., Stein, G.S., Stein, J.L., Lian, J.B., 2002. CCAAT/enhancer-binding proteins (C/EBP) beta and delta activate osteocalcin gene transcription and synergize with Runx2 at the C/EBP element to regulate bone-specific expression. J. Biol. Chem. 277 (2), 1316–1323.

Hauge, E.M., Qvesel, D., Eriksen, E.F., Mosekilde, L., Melsen, F., 2001. Cancellous bone remodeling occurs in specialized compartments lined by cells expressing osteoblastic markers. J. Bone Miner. Res. 16 (9), 1575–1582.

Holdsworth, G., Greenslade, K., Jose, J., Stencel, Z., Kirby, H., Moore, A., Ke, H.Z., Robinson, M.K., 2018. Dampening of the bone formation response following repeat dosing with sclerostin antibody in mice is associated with up-regulation of Wnt antagonists. Bone 107, 93–103.

Holm, E., Aubin, J.E., Hunter, G.K., Beier, F., Goldberg, H.A., 2015. Loss of bone sialoprotein leads to impaired endochondral bone development and mineralization. Bone 71, 145–154.

Holmen, S.L., Zylstra, C.R., Mukherjee, A., Sigler, R.E., Faugere, M.C., Bouxsein, M.L., Deng, L., Clemens, T.L., Williams, B.O., 2005. Essential role of beta-catenin in postnatal bone acquisition. J. Biol. Chem. 280 (22), 21162–21168.

Horwood, N.J., Kartsogiannis, V., Quinn, J.M., Romas, E., Martin, T.J., Gillespie, M.T., 1999. Activated T lymphocytes support osteoclast formation in vitro. Biochem. Biophys. Res. Commun. 265 (1), 144–150.

Hosaki-Takamiya, R., Hashimoto, M., Imai, Y., Nishida, T., Yamada, N., Mori, H., Tanaka, T., Kawanabe, N., Yamashiro, T., Kamioka, H., 2016. Collagen production of osteoblasts revealed by ultra-high voltage electron microscopy. J. Bone Miner. Metabol. 34 (5), 491–499.

Hozumi, A., Osaki, M., Goto, H., Sakamoto, K., Inokuchi, S., Shindo, H., 2009. Bone marrow adipocytes support dexamethasone-induced osteoclast differentiation. Biochem. Biophys. Res. Commun. 382 (4), 780–784.

Igarashi, M., Kamiya, N., Ito, K., Takagi, M., 2002. In situ localization and in vitro expression of osteoblast/osteocyte factor 45 mRNA during bone cell differentiation. Histochem. J. 34 (5), 255–263.

Ikeda, T., Nomura, S., Yamaguchi, A., Suda, T., Yoshiki, S., 1992. In situ hybridization of bone matrix proteins in undecalcified adult rat bone sections. J. Histochem. Cytochem. 40 (8), 1079–1088.

Inada, M., Yasui, T., Nomura, S., Miyake, S., Deguchi, K., Himeno, M., Sato, M., Yamagiwa, H., Kimura, T., Yasui, N., Ochi, T., Endo, N., Kitamura, Y., Kishimoto, T., Komori, T., 1999. Maturational disturbance of chondrocytes in Cbfa1-deficient mice. Dev. Dynam. 214 (4), 279–290.

Johnson, R.W., Brennan, H.J., Vrahnas, C., Poulton, I.J., McGregor, N.E., Standal, T., Walker, E.C., Koh, T.T., Nguyen, H., Walsh, N.C., Forwood, M.R., Martin, T.J., Sims, N.A., 2014. The primary function of gp130 signaling in osteoblasts is to maintain bone formation and strength, rather than promote osteoclast formation. J. Bone Miner. Res. 29 (6), 1492–1505.

Jones, S.J., Boyde, A., 1976. Experimental study of changes in osteoblastic shape induced by calcitonin and parathyroid extract in an organ culture system. Cell Tissue Res. 169 (4), 499-465.

Justesen, J., Stenderup, K., Ebbesen, E.N., Mosekilde, L., Steiniche, T., Kassem, M., 2001. Adipocyte tissue volume in bone marrow is increased with aging and in patients with osteoporosis. Biogerontology 2 (3), 165–171.

Kalajzic, I., Staal, A., Yang, W.P., Wu, Y., Johnson, S.E., Feyen, J.H., Krueger, W., Maye, P., Yu, F., Zhao, Y., Kuo, L., Gupta, R.R., Achenie, L.E., Wang, H.W., Shin, D.G., Rowe, D.W., 2005. Expression profile of osteoblast lineage at defined stages of differentiation. J. Biol. Chem. 280 (26), 24618–24626.

Kalajzic, Z., Li, H., Wang, L.P., Jiang, X., Lamothe, K., Adams, D.J., Aguila, H.L., Rowe, D.W., Kalajzic, I., 2008. Use of an alpha-smooth muscle actin GFP reporter to identify an osteoprogenitor population. Bone 43 (3), 501–510.

Kamel-ElSayed, S.A., Tiede-Lewis, L.M., Lu, Y., Veno, P.A., Dallas, S.L., 2015. Novel approaches for two and three dimensional multiplexed imaging of osteocytes. Bone 76, 129–140.

Kamioka, H., Honjo, T., Takano-Yamamoto, T., 2001. A three-dimensional distribution of osteocyte processes revealed by the combination of confocal laser scanning microscopy and differential interference contrast microscopy. Bone 28 (2), 145–149.

Karsenty, G., Olson, E.N., 2016. Bone and muscle endocrine functions: unexpected paradigms of inter-organ communication. Cell 164 (6), 1248–1256.

Kartsogiannis, V., Zhou, H., Horwood, N.J., Thomas, R.J., Hards, D.K., Quinn, J.M., Niforas, P., Ng, K.W., Martin, T.J., Gillespie, M.T., 1999. Localization of RANKL (receptor activator of NF kappa B ligand) mRNA and protein in skeletal and extraskeletal tissues. Bone 25 (5), 525–534.

Keller, H., Kneissel, M., 2005. SOST is a target gene for PTH in bone. Bone 37 (2), 148–158.

Kelly, K.A., Tanaka, S., Baron, R., Gimble, J.M., 1998. Murine bone marrow stromally derived BMS2 adipocytes support differentiation and function of osteoclast-like cells in vitro. Endocrinology 139 (4), 2092–2101.

Khrimian, L., Obri, A., Ramos-Brossier, M., Rousseaud, A., Moriceau, S., Nicot, A.S., Mera, P., Kosmidis, S., Karnavas, T., Saudou, F., Gao, X.B., Oury, F., Kandel, E., Karsenty, G., 2017. Gpr158 mediates osteocalcin's regulation of cognition. J. Exp. Med. 214 (10), 2859–2873.

Kim, I.S., Otto, F., Zabel, B., Mundlos, S., 1999. Regulation of chondrocyte differentiation by Cbfa1. Mech. Dev. 80 (2), 159–170.

Kim, S.W., Lu, Y., Williams, E.A., Lai, F., Lee, J.Y., Enishi, T., Balani, D.H., Ominsky, M.S., Ke, H.Z., Kronenberg, H.M., Wein, M.N., 2017. Sclerostin antibody administration converts bone lining cells into active osteoblasts. J. Bone Miner. Res. 32 (5), 892–901.

Kim, S.W., Pajevic, P.D., Selig, M., Barry, K.J., Yang, J.Y., Shin, C.S., Baek, W.Y., Kim, J.E., Kronenberg, H.M., 2012. Intermittent parathyroid hormone administration converts quiescent lining cells to active osteoblasts. J. Bone Miner. Res. 27 (10), 2075–2084.

Kogianni, G., Mann, V., Noble, B.S., 2008. Apoptotic bodies convey activity capable of initiating osteoclastogenesis and localized bone destruction. J. Bone Miner. Res. 23 (6), 915–927.

Komori, T., 2018. Runx2, an inducer of osteoblast and chondrocyte differentiation. Histochem. Cell Biol. 149 (4), 313–323.

Komori, T., Yagi, H., Nomura, S., Yamaguchi, A., Sasaki, K., Deguchi, K., Shimizu, Y., Bronson, R.T., Gao, Y.H., Inada, M., Sato, M., Okamoto, R., Kitamura, Y., Yoshiki, S., Kishimoto, T., 1997. Targeted disruption of Cbfa1 results in a complete lack of bone formation owing to maturational arrest of osteoblasts. Cell 89 (5), 755–764.

Kramer, I., Halleux, C., Keller, H., Pegurri, M., Gooi, J.H., Weber, P.B., Feng, J.Q., Bonewald, L.F., Kneissel, M., 2010. Osteocyte Wnt/beta-catenin signaling is required for normal bone homeostasis. Mol. Cell Biol. 30 (12), 3071–3085.

Kulkarni, N.H., Onyia, J.E., Zeng, Q., Tian, X., Liu, M., Halladay, D.L., Frolik, C.A., Engler, T., Wei, T., Kriauciunas, A., Martin, T.J., Sato, M., Bryant, H.U., Ma, Y.L., 2006. Orally bioavailable GSK-3alpha/beta dual inhibitor increases markers of cellular differentiation in vitro and bone mass in vivo. J. Bone Miner. Res. 21 (6), 910–920.

Lacombe, J., Karsenty, G., Ferron, M., 2013. In vivo analysis of the contribution of bone resorption to the control of glucose metabolism in mice. Mol Metab 2 (4), 498–504.

Lambert, L.J., Challa, A.K., Niu, A., Zhou, L., Tucholski, J., Johnson, M.S., Nagy, T.R., Eberhardt, A.W., Estep, P.N., Kesterson, R.A., Grams, J.M., 2016. Increased trabecular bone and improved biomechanics in an osteocalcin-null rat model created by CRISPR/Cas9 technology. Dis Model Mech 9 (10), 1169–1179.

Lawson, M.A., McDonald, M.M., Kovacic, N., Hua Khoo, W., Terry, R.L., Down, J., Kaplan, W., Paton-Hough, J., Fellows, C., Pettitt, J.A., Neil Dear, T., Van Valckenborgh, E., Baldock, P.A., Rogers, M.J., Eaton, C.L., Vanderkerken, K., Pettit, A.R., Quinn, J.M., Zannettino, A.C., Phan, T.G., Croucher, P.I., 2015. Osteoclasts control reactivation of dormant myeloma cells by remodelling the endosteal niche. Nat. Commun. 6, 8983.

Leblond, C.P., 1989. Synthesis and secretion of collagen by cells of connective tissue, bone, and dentin. Anat. Rec. 224 (2), 123–138.

Lee, N.K., Karsenty, G., 2008. Reciprocal regulation of bone and energy metabolism. Trends Endocrinol. Metabol. 19 (5), 161–166.

Lee, N.K., Sowa, H., Hinoi, E., Ferron, M., Ahn, J.D., Confavreux, C., Dacquin, R., Mee, P.J., McKee, M.D., Jung, D.Y., Zhang, Z., Kim, J.K., Mauvais-Jarvis, F., Ducy, P., Karsenty, G., 2007. Endocrine regulation of energy metabolism by the skeleton. Cell 130 (3), 456–469.

Li, H., Jiang, X., Delaney, J., Franceschetti, T., Bilic-Curcic, I., Kalinovsky, J., Lorenzo, J.A., Grcevic, D., Rowe, D.W., Kalajzic, I., 2010. Immature osteoblast lineage cells increase osteoclastogenesis in osteogenesis imperfecta murine. Am. J. Pathol. 176 (5), 2405–2413.

Li, X., Ominsky, M.S., Warmington, K.S., Morony, S., Gong, J., Cao, J., Gao, Y., Shalhoub, V., Tipton, B., Haldankar, R., Chen, Q., Winters, A., Boone, T., Geng, Z., Niu, Q.T., Ke, H.Z., Kostenuik, P.J., Simonet, W.S., Lacey, D.L., Paszty, C., 2009. Sclerostin antibody treatment increases bone formation, bone mass, and bone strength in a rat model of postmenopausal osteoporosis. J. Bone Miner. Res. 24 (4), 578–588.

Liu, S., Guo, R., Simpson, L.G., Xiao, Z.S., Burnham, C.E., Quarles, L.D., 2003. Regulation of fibroblastic growth factor 23 expression but not degradation by PHEX. J. Biol. Chem. 278 (39), 37419–37426.

Livesey, S.A., Kemp, B.E., Re, C.A., Partridge, N.C., Martin, T.J., 1982. Selective hormonal activation of cyclic AMP-dependent protein kinase isoenzymes in normal and malignant osteoblasts. J. Biol. Chem. 257 (24), 14983–14987.

Loots, G.G., Kneissel, M., Keller, H., Baptist, M., Chang, J., Collette, N.M., Ovcharenko, D., Plajzer-Frick, I., Rubin, E.M., 2005. Genomic deletion of a long-range bone enhancer misregulates sclerostin in Van Buchem disease. Genome Res. 15 (7), 928–935.

Lotinun, S., Kiviranta, R., Matsubara, T., Alzate, J.A., Neff, L., Luth, A., Koskivirta, I., Kleuser, B., Vacher, J., Vuorio, E., Horne, W.C., Baron, R., 2013. Osteoclast-specific cathepsin K deletion stimulates S1P-dependent bone formation. J. Clin. Invest. 123 (2), 666–681.

Luben, R.A., Wong, G.L., Cohn, D.V., 1976. Biochemical characterization with parathormone and calcitonin of isolated bone cells: provisional identification of osteoclasts and osteoblasts. Endocrinology 99 (2), 526–534.

Lundberg, P., Allison, S.J., Lee, N.J., Baldock, P.A., Brouard, N., Rost, S., Enriquez, R.F., Sainsbury, A., Lamghari, M., Simmons, P., Eisman, J.A., Gardiner, E.M., Herzog, H., 2007. Greater bone formation of Y2 knockout mice is associated with increased osteoprogenitor numbers and altered Y1 receptor expression. J. Biol. Chem. 282 (26), 19082–19091.

Ma, Y.L., Hamang, M., Lucchesi, J., Bivi, N., Zeng, Q., Adrian, M.D., Raines, S.E., Li, J., Kuhstoss, S.A., Obungu, V., Bryant, H.U., Krishnan, V., 2017. Time course of disassociation of bone formation signals with bone mass and bone strength in sclerostin antibody treated ovariectomized rats. Bone 97, 20–28.

Maeda, A., Ono, M., Holmbeck, K., Li, L., Kilts, T.M., Kram, V., Noonan, M.L., Yoshioka, Y., McNerny, E.M., Tantillo, M.A., Kohn, D.H., Lyons, K.M., Robey, P.G., Young, M.F., 2015. WNT1-induced secreted protein-1 (WISP1), a novel regulator of bone turnover and Wnt signaling. J. Biol. Chem. 290 (22), 14004–14018.

Majeska, R.J., Rodan, S.B., Rodan, G.A., 1980. Parathyroid hormone-responsive clonal cell lines from rat osteosarcoma. Endocrinology 107 (5), 1494–1503.

Manolagas, S.C., 2000. Birth and death of bone cells: basic regulatory mechanisms and implications for the pathogenesis and treatment of osteoporosis. Endocr. Rev. 21 (2), 115–137.

Mantila Roosa, S.M., Liu, Y., Turner, C.H., 2011. Gene expression patterns in bone following mechanical loading. J. Bone Miner. Res. 26 (1), 100–112.

Martin, R.B., Chow, B.D., Lucas, P.A., 1990. Bone marrow fat content in relation to bone remodeling and serum chemistry in intact and ovariectomized dogs. Calcif. Tissue Int. 46 (3), 189–194.

Matic, I., Matthews, B.G., Wang, X., Dyment, N.A., Worthley, D.L., Rowe, D.W., Grcevic, D., Kalajzic, I., 2016. Quiescent bone lining cells are a major source of osteoblasts during adulthood. Stem Cell. 34 (12), 2930–2942.

McClung, M.R., Grauer, A., Boonen, S., Bolognese, M.A., Brown, J.P., Diez-Perez, A., Langdahl, B.L., Reginster, J.Y., Zanchetta, J.R., Wasserman, S.M., Katz, L., Maddox, J., Yang, Y.C., Libanati, C., Bone, H.G., 2014. Romosozumab in postmenopausal women with low bone mineral density. N. Engl. J. Med. 370 (5), 412–420.

McGregor, N.E., Murat, M., Elango, J., Poulton, I.J., Walker, E.C., Crimeen-Irwin, B., Ho, P.W.M., Gooi, J.H., Martin, T.J., Sims, N.A., 2019. IL-6 exhibits both *cis* and *trans* signaling in osteocytes and osteoblasts, but only *trans* signaling promotes bone formation and osteoclastogenesis. J. Biol. Chem. In Press. 10.1074/jbc.RA119.008074.

Miao, D., He, B., Jiang, Y., Kobayashi, T., Soroceanu, M.A., Zhao, J., Su, H., Tong, X., Amizuka, N., Gupta, A., Genant, H.K., Kronenberg, H.M., Goltzman, D., Karaplis, A.C., 2005. Osteoblast-derived PTHrP is a potent endogenous bone anabolic agent that modifies the therapeutic efficacy of administered PTH 1-34. J. Clin. Invest. 115 (9), 2402–2411.

Miller, S.C., Jee, W.S., 1987. The bone lining cell: a distinct phenotype? Calcif. Tissue Int. 41 (1), 1–5.

Modder, U.I., Khosla, S., 2008. Skeletal stem/osteoprogenitor cells: current concepts, alternate hypotheses, and relationship to the bone remodeling compartment. J. Cell. Biochem. 103 (2), 393–400.

Nakashima, K., Zhou, X., Kunkel, G., Zhang, Z., Deng, J.M., Behringer, R.R., de Crombrugghe, B., 2002. The novel zinc finger-containing transcription factor osterix is required for osteoblast differentiation and bone formation. Cell 108 (1), 17–29.

Nakashima, T., Hayashi, M., Fukunaga, T., Kurata, K., Oh-Hora, M., Feng, J.Q., Bonewald, L.F., Kodama, T., Wutz, A., Wagner, E.F., Penninger, J.M., Takayanagi, H., 2011. Evidence for osteocyte regulation of bone homeostasis through RANKL expression. Nat. Med. 17 (10), 1231–1234.

Negishi-Koga, T., Shinohara, M., Komatsu, N., Bito, H., Kodama, T., Friedel, R.H., Takayanagi, H., 2011. Suppression of bone formation by osteoclastic expression of semaphorin 4D. Nat. Med. 17 (11), 1473–1480.

Nioi, P., Taylor, S., Hu, R., Pacheco, E., He, Y.D., Hamadeh, H., Paszty, C., Pyrah, I., Ominsky, M.S., Boyce, R.W., 2015. Transcriptional profiling of laser capture microdissected subpopulations of the osteoblast lineage provides insight into the early response to sclerostin antibody in rats. J. Bone Miner. Res. 30 (8), 1457–1467.

Nishimura, R., Wakabayashi, M., Hata, K., Matsubara, T., Honma, S., Wakisaka, S., Kiyonari, H., Shioi, G., Yamaguchi, A., Tsumaki, N., Akiyama, H., Yoneda, T., 2012. Osterix regulates calcification and degradation of chondrogenic matrices through matrix metalloproteinase 13 (MMP13) expression in association with transcription factor Runx2 during endochondral ossification. J. Biol. Chem. 287 (40), 33179–33190.

Obri, A., Khrimian, L., Karsenty, G., Oury, F., 2018. Osteocalcin in the brain: from embryonic development to age-related decline in cognition. Nat. Rev. Endocrinol. 14 (3), 174–182.

Ominsky, M.S., Niu, Q.T., Li, C., Li, X., Ke, H.Z., 2014. Tissue-level mechanisms responsible for the increase in bone formation and bone volume by sclerostin antibody. J. Bone Miner. Res. 29 (6), 1424–1430.

Orriss, I.R., Hajjawi, M.O., Huesa, C., MacRae, V.E., Arnett, T.R., 2014. Optimisation of the differing conditions required for bone formation in vitro by primary osteoblasts from mice and rats. Int. J. Mol. Med. 34 (5), 1201–1208.

Otsuru, S., Tamai, K., Yamazaki, T., Yoshikawa, H., Kaneda, Y., 2008. Circulating bone marrow-derived osteoblast progenitor cells are recruited to the bone-forming site by the CXCR4/stromal cell-derived factor-1 pathway. Stem Cell. 26 (1), 223–234.

Otto, F., Thornell, A.P., Crompton, T., Denzel, A., Gilmour, K.C., Rosewell, I.R., Stamp, G.W., Beddington, R.S., Mundlos, S., Olsen, B.R., Selby, P.B., Owen, M.J., 1997. Cbfa1, a candidate gene for cleidocranial dysplasia syndrome, is essential for osteoblast differentiation and bone development. Cell 89 (5), 765–771.

Oury, F., Khrimian, L., Denny, C.A., Gardin, A., Chamouni, A., Goeden, N., Huang, Y.Y., Lee, H., Srinivas, P., Gao, X.B., Suyama, S., Langer, T., Mann, J.J., Horvath, T.L., Bonnin, A., Karsenty, G., 2013. Maternal and offspring pools of osteocalcin influence brain development and functions. Cell 155 (1), 228–241.

Oury, F., Sumara, G., Sumara, O., Ferron, M., Chang, H., Smith, C.E., Hermo, L., Suarez, S., Roth, B.L., Ducy, P., Karsenty, G., 2011. Endocrine regulation of male fertility by the skeleton. Cell 144 (5), 796–809.

Owen, M., Friedenstein, A.J., 1988. Stromal stem cells: marrow-derived osteogenic precursors. Ciba Found. Symp. 136, 42–60.

Paic, F., Igwe, J.C., Nori, R., Kronenberg, M.S., Franceschetti, T., Harrington, P., Kuo, L., Shin, D.G., Rowe, D.W., Harris, S.E., Kalajzic, I., 2009. Identification of differentially expressed genes between osteoblasts and osteocytes. Bone 45 (4), 682–692.

Palmqvist, P., Persson, E., Conaway, H.H., Lerner, U.H., 2002. IL-6, leukemia inhibitory factor, and oncostatin M stimulate bone resorption and regulate the expression of receptor activator of NF-kappa B ligand, osteoprotegerin, and receptor activator of NF-kappa B in mouse calvariae. J. Immunol. 169 (6), 3353–3362.

Partridge, N.C., Alcorn, D., Michelangeli, V.P., Ryan, G., Martin, T.J., 1983. Morphological and biochemical characterization of four clonal osteogenic sarcoma cell lines of rat origin. Cancer Res. 43 (9), 4308–4314.

Partridge, N.C., Kemp, B.E., Livesey, S.A., Martin, T.J., 1982. Activity ratio measurements reflect intracellular activation of adenosine 3',5'-monophosphate-dependent protein kinase in osteoblasts. Endocrinology 111 (1), 178–183.

Partridge, N.C., Kemp, B.E., Veroni, M.C., Martin, T.J., 1981. Activation of adenosine 3',5'-monophosphate-dependent protein kinase in normal and malignant bone cells by parathyroid hormone, prostaglandin E2, and prostacyclin. Endocrinology 108 (1), 220–225.

Paschalis, E.P., Shane, E., Lyritis, G., Skarantavos, G., Mendelsohn, R., Boskey, A.L., 2004. Bone fragility and collagen cross-links. J. Bone Miner. Res. 19 (12), 2000–2004.

Peck, W.A., Birge Jr., S.J., Fedak, S.A., 1964. Bone cells: biochemical and biological studies after enzymatic isolation. Science 146, 1476–1477.

Pederson, L., Ruan, M., Westendorf, J.J., Khosla, S., Oursler, M.J., 2008. Regulation of bone formation by osteoclasts involves Wnt/BMP signaling and the chemokine sphingosine-1-phosphate. Proc. Natl. Acad. Sci. U. S. A. 105 (52), 20764–20769.

Perwad, F., Azam, N., Zhang, M.Y., Yamashita, T., Tenenhouse, H.S., Portale, A.A., 2005. Dietary and serum phosphorus regulate fibroblast growth factor 23 expression and 1,25-dihydroxyvitamin D metabolism in mice. Endocrinology 146 (12), 5358–5364.

Pi, M., Wu, Y., Quarles, L.D., 2011. GPRC6A mediates responses to osteocalcin in beta-cells in vitro and pancreas in vivo. J. Bone Miner. Res. 26 (7), 1680–1683.

Pievani, A., Sacchetti, B., Corsi, A., Rambaldi, B., Donsante, S., Scagliotti, V., Vergani, P., Remoli, C., Biondi, A., Robey, P.G., Riminucci, M., Serafini, M., 2017. Human umbilical cord blood-borne fibroblasts contain marrow niche precursors that form a bone/marrow organoid in vivo. Development 144 (6), 1035–1044.

Poulton, I.J., McGregor, N.E., Pompolo, S., Walker, E.C., Sims, N.A., 2012. Contrasting roles of leukemia inhibitory factor in murine bone development and remodeling involve region-specific changes in vascularization. J. Bone Miner. Res. 27 (3), 586–595.

Quach, J.M., Walker, E.C., Allan, E., Solano, M., Yokoyama, A., Kato, S., Sims, N.A., Gillespie, M.T., Martin, T.J., 2011. Zinc finger protein 467 is a novel regulator of osteoblast and adipocyte commitment. J. Biol. Chem. 286 (6), 4186–4198.

Quarles, L.D., 2003. FGF23, PHEX, and MEPE regulation of phosphate homeostasis and skeletal mineralization. Am. J. Physiol. Endocrinol. Metab. 285 (1), E1–E9.

Quinn, J.M., Itoh, K., Udagawa, N., Hausler, K., Yasuda, H., Shima, N., Mizuno, A., Higashio, K., Takahashi, N., Suda, T., Martin, T.J., Gillespie, M.T., 2001. Transforming growth factor beta affects osteoclast differentiation via direct and indirect actions. J. Bone Miner. Res. 16 (10), 1787–1794.

Rawadi, G., Vayssiere, B., Dunn, F., Baron, R., Roman-Roman, S., 2003. BMP-2 controls alkaline phosphatase expression and osteoblast mineralization by a Wnt autocrine loop. J. Bone Miner. Res. 18 (10), 1842–1853.

Reinisch, A., Etchart, N., Thomas, D., Hofmann, N.A., Fruehwirth, M., Sinha, S., Chan, C.K., Senarath-Yapa, K., Seo, E.Y., Wearda, T., Hartwig, U.F., Beham-Schmid, C., Trajanoski, S., Lin, Q., Wagner, W., Dullin, C., Alves, F., Andreeff, M., Weissman, I.L., Longaker, M.T., Schallmoser, K., Majeti, R., Strunk, D., 2015. Epigenetic and in vivo comparison of diverse MSC sources reveals an endochondral signature for human hematopoietic niche formation. Blood 125 (2), 249–260.

Richards, C.D., Langdon, C., Deschamps, P., Pennica, D., Shaughnessy, S.G., 2000. Stimulation of osteoclast differentiation in vitro by mouse oncostatin M, leukaemia inhibitory factor, cardiotrophin-1 and interleukin 6: synergy with dexamethasone. Cytokine 12 (6), 613–621.

Robey, P., 2017. Mesenchymal stem cells": fact or fiction, and implications in their therapeutic use. F1000Res 6.

Robling, A.G., Bellido, T., Turner, C.H., 2006. Mechanical stimulation in vivo reduces osteocyte expression of sclerostin. J. Musculoskelet. Neuronal Interact. 6 (4), 354.

Robling, A.G., Turner, C.H., 2009. Mechanical signaling for bone modeling and remodeling. Crit. Rev. Eukaryot. Gene Expr. 19 (4), 319–338.

Rodan, G.A., Martin, T.J., 1981. Role of osteoblasts in hormonal control of bone resorption–a hypothesis. Calcif. Tissue Int. 33 (4), 349–351.

Rodan, G.A., Noda, M., 1991. Gene expression in osteoblastic cells. Crit. Rev. Eukaryot. Gene Expr. 1 (2), 85–98.

Ruffoni, D., Fratzl, P., Roschger, P., Klaushofer, K., Weinkamer, R., 2007. The bone mineralization density distribution as a fingerprint of the mineralization process. Bone 40 (5), 1308–1319.

Sabatakos, G., Sims, N.A., Chen, J., Aoki, K., Kelz, M.B., Amling, M., Bouali, Y., Mukhopadhyay, K., Ford, K., Nestler, E.J., Baron, R., 2000. Overexpression of DeltaFosB transcription factor(s) increases bone formation and inhibits adipogenesis. Nat. Med. 6 (9), 985–990.

Sacchetti, B., Funari, A., Remoli, C., Giannicola, G., Kogler, G., Liedtke, S., Cossu, G., Serafini, M., Sampaolesi, M., Tagliafico, E., Tenedini, E., Saggio, I., Robey, P.G., Riminucci, M., Bianco, P., 2016. No identical "mesenchymal stem cells" at different times and sites: human committed progenitors of distinct origin and differentiation potential are incorporated as adventitial cells in microvessels. Stem Cell Reports 6 (6), 897–913.

Sato, M., Westmore, M., Ma, Y.L., Schmidt, A., Zeng, Q.Q., Glass, E.V., Vahle, J., Brommage, R., Jerome, C.P., Turner, C.H., 2004. Teriparatide [PTH(1–34)] strengthens the proximal femur of ovariectomized nonhuman primates despite increasing porosity. J. Bone Miner. Res. 19 (4), 623–629.

Schaffler, M.B., Cheung, W.Y., Majeska, R., Kennedy, O., 2014. Osteocytes: master orchestrators of bone. Calcif. Tissue Int. 94 (1), 5–24.

Shimada, T., Kakitani, M., Yamazaki, Y., Hasegawa, H., Takeuchi, Y., Fujita, T., Fukumoto, S., Tomizuka, K., Yamashita, T., 2004. Targeted ablation of Fgf23 demonstrates an essential physiological role of FGF23 in phosphate and vitamin D metabolism. J. Clin. Invest. 113 (4), 561–568.

Simonet, W.S., Lacey, D.L., Dunstan, C.R., Kelley, M., Chang, M.S., Luthy, R., Nguyen, H.Q., Wooden, S., Bennett, L., Boone, T., Shimamoto, G., DeRose, M., Elliott, R., Colombero, A., Tan, H.L., Trail, G., Sullivan, J., Davy, E., Bucay, N., Renshaw-Gegg, L., Hughes, T.M., Hill, D., Pattison, W., Campbell, P., Sander, S., Van, G., Tarpley, J., Derby, P., Lee, R., Boyle, W.J., 1997. Osteoprotegerin: a novel secreted protein involved in the regulation of bone density. Cell 89 (2), 309–319.

Sims, N.A., Clement-Lacroix, P., Da Ponte, F., Bouali, Y., Binart, N., Moriggl, R., Goffin, V., Coschigano, K., Gaillard-Kelly, M., Kopchick, J., Baron, R., Kelly, P.A., 2000. Bone homeostasis in growth hormone receptor-null mice is restored by IGF-I but independent of Stat5. J. Clin. Invest. 106 (9), 1095–1103.

Sims, N.A., Jenkins, B.J., Nakamura, A., Quinn, J.M., Li, R., Gillespie, M.T., Ernst, M., Robb, L., Martin, T.J., 2005. Interleukin-11 receptor signaling is required for normal bone remodeling. J. Bone Miner. Res. 20 (7), 1093–1102.

Sims, N.A., Martin, T.J., 2014. Coupling the activities of bone formation and resorption: a multitude of signals within the basic multicellular unit. Bonekey Rep. 3. Article number 481.

Sims, N.A., Martin, T.J., 2015. Coupling signals between the osteoclast and osteoblast: how are messages transmitted between these temporary visitors to the bone surface? Front. Endocrinol. 6, 41.

Soderstrom, K., Stein, E., Colmenero, P., Purath, U., Muller-Ladner, U., de Matos, C.T., Tarner, I.H., Robinson, W.H., Engleman, E.G., 2010. Natural killer cells trigger osteoclastogenesis and bone destruction in arthritis. Proc. Natl. Acad. Sci. U. S. A. 107 (29), 13028–13033.

Song, H.Y., Jeon, E.S., Kim, J.I., Jung, J.S., Kim, J.H., 2007. Oncostatin M promotes osteogenesis and suppresses adipogenic differentiation of human adipose tissue-derived mesenchymal stem cells. J. Cell. Biochem. 101 (5), 1238–1251.

Spatz, J.M., Wein, M.N., Gooi, J.H., Qu, Y., Garr, J.L., Liu, S., Barry, K.J., Uda, Y., Lai, F., Dedic, C., Balcells-Camps, M., Kronenberg, H.M., Babij, P., Pajevic, P.D., 2015. The Wnt inhibitor sclerostin is up-regulated by mechanical unloading in osteocytes in vitro. J. Biol. Chem. 290 (27), 16744–16758.

Stein, G.S., Lian, J.B., Owen, T.A., 1990. Relationship of cell growth to the regulation of tissue-specific gene expression during osteoblast differentiation. FASEB J. 4 (13), 3111–3123.

Sterling, J.A., Edwards, J.R., Martin, T.J., Mundy, G.R., 2011. Advances in the biology of bone metastasis: how the skeleton affects tumor behavior. Bone 48 (1), 6–15.

Stolina, M., Dwyer, D., Niu, Q.T., Villasenor, K.S., Kurimoto, P., Grisanti, M., Han, C.Y., Liu, M., Li, X., Ominsky, M.S., Ke, H.Z., Kostenuik, P.J., 2014. Temporal changes in systemic and local expression of bone turnover markers during six months of sclerostin antibody administration to ovariectomized rats. Bone 67C, 305–313.

Streicher, C., Heyny, A., Andrukhova, O., Haigl, B., Slavic, S., Schuler, C., Kollmann, K., Kantner, I., Sexl, V., Kleiter, M., Hofbauer, L.C., Kostenuik, P.J., Erben, R.G., 2017. Estrogen regulates bone turnover by targeting RANKL expression in bone lining cells. Sci. Rep. 7 (1), 6460.

Suda, T., Takahashi, N., Martin, T.J., 1992. Modulation of osteoclast differentiation. Endocr. Rev. 13 (1), 66–80.

Sudo, H., Kodama, H.A., Amagai, Y., Yamamoto, S., Kasai, S., 1983. In vitro differentiation and calcification in a new clonal osteogenic cell line derived from newborn mouse calvaria. J. Cell Biol. 96 (1), 191–198.

Takahashi, N., Akatsu, T., Udagawa, N., Sasaki, T., Yamaguchi, A., Moseley, J.M., Martin, T.J., Suda, T., 1988. Osteoblastic cells are involved in osteoclast formation. Endocrinology 123 (5), 2600–2602.

Takeda, S., Bonnamy, J.P., Owen, M.J., Ducy, P., Karsenty, G., 2001. Continuous expression of Cbfa1 in nonhypertrophic chondrocytes uncovers its ability to induce hypertrophic chondrocyte differentiation and partially rescues Cbfa1-deficient mice. Genes Dev. 15 (4), 467–481.

Takyar, F.M., Tonna, S., Ho, P.W., Crimeen-Irwin, B., Baker, E.K., Martin, T.J., Sims, N.A., 2013. EphrinB2/EphB4 inhibition in the osteoblast lineage modifies the anabolic response to parathyroid hormone. J. Bone Miner. Res. 28 (4), 912–925.

Tamura, T., Udagawa, N., Takahashi, N., Miyaura, C., Tanaka, S., Yamada, Y., Koishihara, Y., Ohsugi, Y., Kumaki, K., Taga, T., et al., 1993. Soluble interleukin-6 receptor triggers osteoclast formation by interleukin 6. Proc. Natl. Acad. Sci. U. S. A. 90 (24), 11924–11928.

Tanaka, T., Yoshida, N., Kishimoto, T., Akira, S., 1997. Defective adipocyte differentiation in mice lacking the C/EBPbeta and/or C/EBPdelta gene. EMBO J. 16 (24), 7432–7443.

Taylor, S., Ominsky, M.S., Hu, R., Pacheco, E., He, Y.D., Brown, D.L., Aguirre, J.I., Wronski, T.J., Buntich, S., Afshari, C.A., Pyrah, I., Nioi, P., Boyce, R.W., 2016. Time-dependent cellular and transcriptional changes in the osteoblast lineage associated with sclerostin antibody treatment in ovariectomized rats. Bone 84, 148–159.

Tomimori, Y., Mori, K., Koide, M., Nakamichi, Y., Ninomiya, T., Udagawa, N., Yasuda, H., 2009. Evaluation of pharmaceuticals with a novel 50-hour animal model of bone loss. J. Bone Miner. Res. 24 (7), 1194–1205.

Tonna, S., Takyar, F.M., Vrahnas, C., Crimeen-Irwin, B., Ho, P.W., Poulton, I.J., Brennan, H.J., McGregor, N.E., Allan, E.H., Nguyen, H., Forwood, M.R., Tatarczuch, L., Mackie, E.J., Martin, T.J., Sims, N.A., 2014. EphrinB2 signaling in osteoblasts promotes bone mineralization by preventing apoptosis. FASEB J. 28 (10), 4482–4496.

Toyosawa, S., Shintani, S., Fujiwara, T., Ooshima, T., Sato, A., Ijuhin, N., Komori, T., 2001. Dentin matrix protein 1 is predominantly expressed in chicken and rat osteocytes but not in osteoblasts. J. Bone Miner. Res. 16 (11), 2017–2026.

Udagawa, N., Takahashi, N., Akatsu, T., Sasaki, T., Yamaguchi, A., Kodama, H., Martin, T.J., Suda, T., 1989. The bone marrow-derived stromal cell lines MC3T3-G2/PA6 and ST2 support osteoclast-like cell differentiation in cocultures with mouse spleen cells. Endocrinology 125 (4), 1805–1813.

Ueta, C., Iwamoto, M., Kanatani, N., Yoshida, C., Liu, Y., Enomoto-Iwamoto, M., Ohmori, T., Enomoto, H., Nakata, K., Takada, K., Kurisu, K., Komori, T., 2001. Skeletal malformations caused by overexpression of Cbfa1 or its dominant negative form in chondrocytes. J. Cell Biol. 153 (1), 87–100.

van Bezooijen, R.L., Roelen, B.A., Visser, A., van der Wee-Pals, L., de Wilt, E., Karperien, M., Hamersma, H., Papapoulos, S.E., ten Dijke, P., Lowik, C.W., 2004. Sclerostin is an osteocyte-expressed negative regulator of bone formation, but not a classical BMP antagonist. J. Exp. Med. 199 (6), 805–814.

Vrahnas, C., Buenzli, P.R., Pearcson, T.A., Pennypacker, B.L., Tobin, M.J., Bambery, K.R., Duong, L.T., Sims, N.A., 2018. Differing effects of parathyroid hormone, alendronate and odanacatib on bone formation and on the mineralisation process in intracortical and endocortical bone of ovariectomized rabbits. Calcif. Tissue Int. 103 (6), 625–637.

Vrahnas, C., Pearson, T.A., Brunt, A.R., Forwood, M.R., Bambery, K.R., Tobin, M.J., Martin, T.J., Sims, N.A., 2016. Anabolic action of parathyroid hormone (PTH) does not compromise bone matrix mineral composition or maturation. Bone 93, 146–154.

Walker, E., McGregor, N., Poulton, I., Pompolo, S., Allan, E., Quinn, J., Gillespie, M., Martin, T., Sims, N.A., 2008. Cardiotrophin-1 is an osteoclast-derived stimulus of bone formation required for normal bone remodeling. J. Bone Miner. Res. 23, 2025–2032.

Walker, E.C., McGregor, N.E., Poulton, I.J., Solano, M., Pompolo, S., Fernandes, T.J., Constable, M.J., Nicholson, G.C., Zhang, J.G., Nicola, N.A., Gillespie, M.T., Martin, T.J., Sims, N.A., 2010. Oncostatin M promotes bone formation independently of resorption when signaling through leukemia inhibitory factor receptor in mice. J. Clin. Invest. 120 (2), 582–592.

Walsh, N.C., Reinwald, S., Manning, C.A., Condon, K.W., Iwata, K., Burr, D.B., Gravallese, E.M., 2009. Osteoblast function is compromised at sites of focal bone erosion in inflammatory arthritis. J. Bone Miner. Res. 24 (9), 1572–1585.

Wei, J., Hanna, T., Suda, N., Karsenty, G., Ducy, P., 2014. Osteocalcin promotes beta-cell proliferation during development and adulthood through Gprc6a. Diabetes 63 (3), 1021–1031.

Weinreb, M., Shinar, D., Rodan, G.A., 1990. Different pattern of alkaline phosphatase, osteopontin, and osteocalcin expression in developing rat bone visualized by in situ hybridization. J. Bone Miner. Res. 5 (8), 831–842.

Weske, S., Vaidya, M., Reese, A., von Wnuck Lipinski, K., Keul, P., Bayer, J.K., Fischer, J.W., Flogel, U., Nelsen, J., Epple, M., Scatena, M., Schwedhelm, E., Dorr, M., Volzke, H., Moritz, E., Hannemann, A., Rauch, B.H., Graler, M.H., Heusch, G., Levkau, B., 2018. Targeting sphingosine-1-phosphate lyase as an anabolic therapy for bone loss. Nat. Med. 24 (5), 667–678.

Westbroek, I., De Rooij, K.E., Nijweide, P.J., 2002. Osteocyte-specific monoclonal antibody MAb OB7.3 is directed against Phex protein. J. Bone Miner. Res. 17 (5), 845–853.

Whyte, M.P., 1994. Hypophosphatasia and the role of alkaline phosphatase in skeletal mineralization. Endocr. Rev. 15 (4), 439–461.

Xiong, J., Onal, M., Jilka, R.L., Weinstein, R.S., Manolagas, S.C., O'Brien, C.A., 2011. Matrix-embedded cells control osteoclast formation. Nat. Med. 17 (10), 1235–1241.

Yamashita, T., Yoshioka, M., Itoh, N., 2000. Identification of a novel fibroblast growth factor, FGF-23, preferentially expressed in the ventrolateral thalamic nucleus of the brain. Biochem. Biophys. Res. Commun. 277 (2), 494–498.

Yang, F., Tang, W., So, S., de Crombrugghe, B., Zhang, C., 2010. Sclerostin is a direct target of osteoblast-specific transcription factor osterix. Biochem. Biophys. Res. Commun. 400 (4), 684–688.

Yang, L., Tsang, K.Y., Tang, H.C., Chan, D., Cheah, K.S., 2014. Hypertrophic chondrocytes can become osteoblasts and osteocytes in endochondral bone formation. Proc. Natl. Acad. Sci. U. S. A. 111 (33), 12097–12102.

Yang, X., Matsuda, K., Bialek, P., Jacquot, S., Masuoka, H.C., Schinke, T., Li, L., Brancorsini, S., Sassone-Corsi, P., Townes, T.M., Hanauer, A., Karsenty, G., 2004. ATF4 is a substrate of RSK2 and an essential regulator of osteoblast biology; implication for Coffin-Lowry Syndrome. Cell 117 (3), 387–398.

Yasuda, H., Shima, N., Nakagawa, N., Yamaguchi, K., Kinosaki, M., Mochizuki, S., Tomoyasu, A., Yano, K., Goto, M., Murakami, A., Tsuda, E., Morinaga, T., Higashio, K., Udagawa, N., Takahashi, N., Suda, T., 1998. Osteoclast differentiation factor is a ligand for osteoprotegerin/osteoclastogenesis-inhibitory factor and is identical to TRANCE/RANKL. Proc. Natl. Acad. Sci. U. S. A. 95 (7), 3597–3602.

Yoshiko, Y., Wang, H., Minamizaki, T., Ijuin, C., Yamamoto, R., Suemune, S., Kozai, K., Tanne, K., Aubin, J.E., Maeda, N., 2007. Mineralized tissue cells are a principal source of FGF23. Bone 40 (6), 1565–1573.

Yue, R., Zhou, B.O., Shimada, I.S., Zhao, Z., Morrison, S.J., 2016. Leptin receptor promotes adipogenesis and reduces osteogenesis by regulating mesenchymal stromal cells in adult bone marrow. Cell Stem Cell 18 (6), 782–796.

Zhou, H., Choong, P., McCarthy, R., Chou, S.T., Martin, T.J., Ng, K.W., 1994. In situ hybridization to show sequential expression of osteoblast gene markers during bone formation in vivo. J. Bone Miner. Res. 9 (9), 1489–1499.

Chapter 5

Osteoclasts

Naoyuki Takahashi[1], Yasuhiro Kobayashi[1] and Nobuyuki Udagawa[2]

[1]*Institute for Oral Science, Matsumoto Dental University, Nagano, Japan;* [2]*Department of Biochemistry, Matsumoto Dental University, Nagano, Japan*

Chapter outline

Introduction **111**
Function of osteoclasts **112**
Morphological features of osteoclasts 112
Mechanism of bone resorption 113
DC-STAMP/OC-STAMP 113
Ruffled border formation 114
Role of osteoblastic cells in osteoclastogenesis **114**
Coculture system 114
Macrophage colony-stimulating factor 114
Osteoprotegerin and RANKL 115
Osteoclastogenesis supported by RANKL 116
Signal transduction in osteoclastogenesis **116**
The M-CSF receptor FMS 116
RANK 118
Tumor necrosis factor receptors 118
ITAM costimulatory signals **119**
Calcium signals 119
SIGLEC-15 and FcγR 119
WNT signals **121**
Canonical WNT signals 121
Noncanonical WNT signals 122
Induction of osteoclast function **122**
Adhesion signals 122
Cytokine signals 124
Characteristics of osteoclast precursors in vivo **124**
Conclusion and perspective **124**
References **125**

Introduction

We contributed a chapter with the title "Osteoclast generation" to *Principles of Bone Biology*, third edition, in 2008. Since then, our understanding of osteoclast differentiation and activation has greatly advanced. The previous chapter focused on the regulation of osteoclast differentiation by the receptor activator of NF-κB ligand/receptor activator of NF-κB/osteoprotegerin (RANKL/RANK/OPG) system. In this new chapter, we have added a section called "Function of osteoclasts" in addition to the regulation of osteoclastogenesis. In retrospect, the role of tumor necrosis factor receptor-associated factors (TRAFs) in osteoclast differentiation and activation was not adequately described in the previous chapter. Findings obtained since its publication have clarified the role of TRAFs in osteoclastogenesis. Studies have shown that the removal of endogenous inhibitory molecules from osteoclast precursors is particularly important for inducing osteoclast differentiation and function. We have attempted to summarize the regulatory mechanisms of osteoclastogenesis elucidated in the past decade in this chapter. In a PubMed search, the number of papers hit with the word "osteoclast" was 436 in 1997, when OPG was discovered, and this number increased to 1408 in 2017. Therefore, the identification of important studies has become more challenging. We did not describe the role of osteoclasts in functional coupling between bone resorption and bone formation; this is described in Chapter 10 (Sims and Martin). Since not all important findings on osteoclasts are introduced in this chapter, each reader will obtain a better understanding of osteoclasts by adding the findings that each considers important. We also describe our working hypothesis on osteoclast precursors in vivo in the final section of this chapter. We hope that the contents of this chapter will contribute to the reader's understanding of osteoclasts.

Principles of Bone Biology. https://doi.org/10.1016/B978-0-12-814841-9.00005-1

Function of osteoclasts

Morphological features of osteoclasts

Osteoclasts are multinucleated cells with a size of 20–100 μm, and are formed by the fusion of mononuclear pre-osteoclasts. Osteoclasts have a large number of pleomorphic mitochondria and a well-developed Golgi apparatus with a rough endoplasmic reticulum, indicating that active energy metabolism and protein synthesis are performed in these cells. Many lysosomes are present in osteoclasts, suggesting that the function of osteoclasts is related to protein degradation. The most characteristic feature of bone-resorbing osteoclasts is the presence of ruffled borders and clear zones (also called the sealing zones). Osteoclasts create resorption lacunae (also called Howship's lacunae) under ruffled borders. Resorption lacunae are acidic to promote the dissolution of bone minerals and proteins. The morphological features of a functioning osteoclast and the mechanisms underlying bone resorption are summarized in Fig. 5.1.

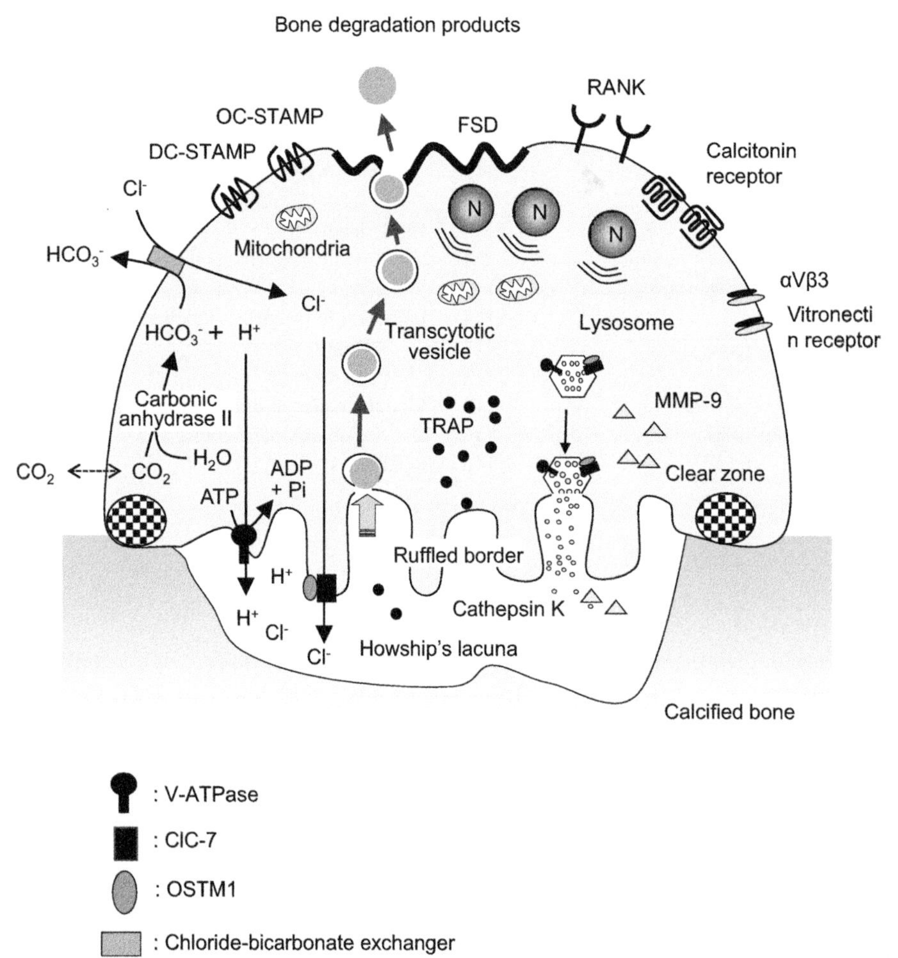

FIGURE 5.1 Schematic representation of a functioning osteoclast. Osteoclasts have several unique ultrastructural characteristics, such as multiple nuclei, abundant mitochondria, and numerous vacuoles and lysosomes. Functioning osteoclasts form ruffled borders and clear zones. The clear zone is recognized as a thick band of actin, isolating resorption lacunae from the surroundings. The resorbing area under the ruffled border is acidic. Vacuolar proton ATPase (*V-ATPase*) exists in the ruffled border and lysosomal membranes. Chloride channel 7 (*ClC-7*) and osteopetrosis-transmembrane protein 1 (*OSTM1*) colocalize in the ruffled border and lysosomal membranes. Matrix metalloproteinase-9 (*MMP-9*) and lysosomal enzymes containing cathepsin K are also secreted into resorption lacunae. Matrix degradation products are endocytosed from a ruffled border region into transcytotic vesicles and then secreted from the functional secretion domain (*FSD*) in the basolateral membrane. Protons are generated by carbonic anhydrase II. Passive chloride–bicarbonate exchangers expressed in the basolateral membrane simultaneously remove HCO_3^- and incorporate Cl^-. Osteoclasts also express large numbers of calcitonin receptors and αVβ3 vitronectin receptors. Dendritic cell-specific transmembrane protein (*DC-STAMP*) and osteoclast-specific transmembrane protein (*OC-STAMP*) are expressed in the osteoclast plasma membrane and play a role in the cell fusion of osteoclasts. Osteoclasts abundantly express calcitonin receptors and receptor activator of NF-κB (*RANK*). *TRAP*, tartrate-resistant acid phosphatase.

The cell membranes of bone-resorbing osteoclasts are divided into four distinct regions: the clear zone, ruffled border, and basolateral and functional secretion domain (FSD) regions (Väänänen and Laitala-Leinonen, 2008). The clear zone is named after its appearance as a bright area under an electron microscope, and this structure is also defined by a thick band of actin called actin rings. Clear zones are involved in the attachment of osteoclasts to bone surfaces and the isolation of resorption lacunae from the surroundings. The ruffled border, which is a foldlike structure formed by the invasion of the cell membrane into the cytoplasm, is the main functional site of bone resorption. Protons and lysosomal enzymes are secreted from ruffled borders into resorption lacunae for the degradation of minerals and matrix proteins. The degradation products of bone are taken up from the specific area of ruffled borders into transcytotic vesicles (Nesbitt and Horton, 1997; Salo et al., 1997; Mulari et al., 2003), are further digested in these vesicles, and are then released from the FSD by vesicular exocytosis.

Mechanism of bone resorption

Osteoclasts possess abundant amounts of carbonic anhydrase II, which is responsible for proton production (Sly et al., 1985). Vacuolar proton ATPase (V-ATPase) localizes to the ruffled border and secretes protons into resorption lacunae in an ATP hydrolysis-dependent manner (Blair et al., 1989; Li et al., 1999; Blair and Athanasou, 2004) (Fig. 5.1). V-ATPase is also detected in lysosomal membranes, and acidifies the inside of lysosomes. Cl^- is transported through chloride channel 7 (ClC-7) to the resorption lacunae to maintain an ionic equilibrium (Kornak et al., 2001). Osteopetrosis-transmembrane protein 1 (OSTM1) is a molecule that is essential for the localization and function of ClC-7 in lysosomes and ruffled borders (Ramírez et al., 2004). OSTM1 acts as the β subunit of ClC-7 (Lange et al., 2006). A mutation in the *OSTM1* gene has been reported in *gl/gl* (gray-lethal) osteopetrotic mice and in patients with hereditary osteopetrosis (Ramírez et al., 2004).

Resorption lacunae are maintained in an acidic environment of approximately pH 4, which favors the dissolution of bone minerals. Proton secretion into resorption lacunae leaves HCO_3^- inside osteoclasts. Passive chloride–bicarbonate exchangers are expressed in the basolateral membrane of osteoclasts and simultaneously remove and incorporate HCO_3^- and Cl^-, respectively (Teti et al., 1989; Wu et al., 2008). The degradation of type I collagen is also promoted under acidic conditions by cathepsin K, a lysosomal enzyme. Cathepsin K has been identified as an enzyme that is specifically expressed in osteoclasts (Tezuka et al., 1994b), and was confirmed to cleave type I collagen at acidic optimum pH (Bossard et al., 1996; Saftig et al., 1998). Matrix metalloproteinase-9 (MMP-9) is highly expressed in osteoclasts (Reponen et al., 1994; Tezuka et al., 1994a). MMP-9 is also secreted into resorption lacunae to digest bone matrix proteins. As described earlier, the degradation products of bone are taken up from the ruffled borders and are ultimately exocytosed to the intercellular space from the FSD. Thus, osteoclasts secrete acids and enzymes and simultaneously absorb degradation products. The flow of vesicles toward the ruffled borders is accompanied by microtubule- and microfilament-mediated vesicle transport (Abu-Amer et al., 1997; Teitelbaum and Ross, 2003; Jurdic et al., 2006). Small GTPases such as RAB7 and RAC are proposed to be involved in vesicle transport in osteoclasts (Zhao et al., 2001; Mulari et al., 2003; Teitelbaum, 2011).

Osteoclasts express a large amount of tartrate-resistant acid phosphatase (TRAP), and release it into blood when they resorb bone (Minkin, 1982). Therefore, TRAP staining is widely used to identify osteoclasts in bone tissues, and serum TRAP activity is measured to evaluate osteoclast number and function in vivo. However, the role of TRAP in bone resorption has not been fully elucidated. Osteoclasts abundantly express the vitronectin receptor, $\alpha V\beta 3$ (Horton et al., 1985). This integrin is considered to be involved in the early adhesion of osteoclasts to bone (McHugh et al., 2000). Osteoclasts also abundantly express the calcitonin receptor, a GTP-binding-protein-coupled receptor (Chambers and Magnus, 1982; Nicholson et al., 1986). Calcitonin is secreted from parafollicular cells (C cells) in the thyroids of mammals. Calcitonin activates cAMP/protein kinase A and Ca^{2+}/protein kinase C signaling via a heterotrimeric GTP-binding protein in osteoclasts (Suzuki et al., 1996; Sexton et al., 1999). Calcitonin-induced signals inhibit the bone-resorbing activity of osteoclasts by disrupting the cytoskeletal organization of osteoclasts (Tanaka et al., 2006; Martin and Sims, 2015). RANK is also abundantly expressed on the membrane of osteoclasts (described later).

DC-STAMP/OC-STAMP

Dendritic cell-specific transmembrane protein (DC-STAMP), a seven-transmembrane protein, was discovered as a membrane protein expressed by dendritic cells. Osteoclasts also abundantly express DC-STAMP (Kukita et al., 2004; Yagi et al., 2005). Mononuclear preosteoclasts, but not multinucleated osteoclasts, have been detected in DC-STAMP-deficient mice. Macrophage-derived foreign-body giant cells are also absent in DC-STAMP-deficient mice. Thus, DC-STAMP is a molecule that is needed for the cell fusion of macrophage lineage cells, including osteoclasts and foreign-body giant cells.

Mononuclear preosteoclasts in DC-STAMP-deficient mice express all osteoclast markers and form clear zones and ruffled borders on bone slices. These findings suggest that osteoclastic differentiation is nearly completed by the differentiation of precursor cells into mononuclear preosteoclasts.

Another seven-transmembrane protein, osteoclast-specific transmembrane protein (OC-STAMP), has been identified (Miyamoto et al., 2012b). OC-STAMP-deficient mice showed features very similar to those of DC-STAMP-deficient mice. OC-STAMP-deficient mice express DC-STAMP, whereas no multinucleated osteoclasts are observed in OC-STAMP-deficient mice. Many mononuclear preosteoclasts are also found in OC-STAMP-deficient mice. These findings suggest that DC-STAMP and OC-STAMP both play an important role in the fusion of osteoclasts, but cannot compensate for each other. DC-STAMP-deficient mice and OC-STAMP-deficient mice both develop weak osteopetrosis with age. Thus, DC-STAMP and OC-STAMP are essential factors for the cell fusion of osteoclasts and macrophages, and the multinucleation of osteoclasts appears to be necessary for increasing bone-resorbing capacity. Studies have also shown that the fusion of osteoclasts and macrophages is different from each other in some way. Signal transducer and activator of transcription 6 signals induced by interleukin 4 (IL-4) were required for the cell–cell fusion of macrophages but not osteoclasts (Miyamoto et al., 2012a). DC-STAMP is proposed to play an imperative role in bone homeostasis (Chiu and Ritchlin, 2016). Further studies will unravel the role of this interesting molecule in bone biology.

Ruffled border formation

The fusion of exocytotic vesicles to membranes appears to be mediated by the soluble *N*-ethylmaleimide-sensitive factor attachment protein receptor complex assembly. The synaptotagmin family is also involved in the fusion of exocytotic vesicles. Synaptotagmin I was initially identified as a Ca^{2+} sensor that is abundantly expressed on synaptic vesicles and regulates the fusion step of synaptic vesicle exocytosis. In osteoclasts, synaptotagmin VII plays an essential role in the fusion of lysosomal vesicles to secrete lysosomal enzymes into resorption lacunae (Zao et al., 2008). The degradation of type I collagen by cathepsin K was markedly decreased in synaptotagmin VII-deficient osteoclasts.

Autophagy-related proteins such as autophagy-related protein 5 (ATG5), ATG7, ATG4B, and light-chain protein 3, are also involved in the ruffled border formation of osteoclasts (DeSelm et al., 2011). These findings provide suggestions on how to make ruffled borders in osteoclasts. Osteoclasts possess numerous acidified lysosomes bearing V-ATPase and ClC-7 in the cytoplasm. Bone-resorbing osteoclasts transfer and fuse acidified lysosomes to the side of the plasma membrane facing the bone surface (Mulari et al., 2003; Teitelbaum, 2007). The insertion of vesicles into the plasma membrane markedly increases the surface area of cells and delivers V-ATPase and ClC-7 to the plasma membrane. Osteoclasts form ruffled borders, making extracellular secondary lysosomes and resorbing bone.

Role of osteoblastic cells in osteoclastogenesis

Coculture system

The differentiation and activation of osteoclasts are strictly regulated by the mesenchymal cells of osteoblast lineage cells (referred to as osteoblastic cells), such as osteoblasts, osteocytes, and bone marrow–derived stromal cells. In 1981, Rodan and Martin (1981) proposed a concept that bone-resorbing factors act on osteoblastic cells, but not osteoclasts, to induce osteoclastic bone resorption based on the observation that osteoblastic cells express the receptors for bone-resorbing factors, including parathyroid hormone (PTH) and prostaglandin E_2 (PGE_2). By incorporating this hypothesis, a murine coculture system of osteoblastic cells and hematopoietic cells was established in 1988 (Takahashi et al., 1988). Osteoclasts were formed in a coculture treated with bone-resorbing factors. Osteoblastic cells were shown to strictly regulate osteoclast differentiation and activation in the coculture. Cell-to-cell contact between hematopoietic cells and osteoblastic cells was required to induce osteoclast formation and activation. Experiments using this coculture system established the concept that osteoblastic cells express osteoclast differentiation factor (ODF) as a membrane-associated factor for the induction of osteoclast differentiation (Suda et al., 1992). It was also established that osteoclasts differentiate from various types of monocyte–macrophage lineage precursors in the coculture (Udagawa et al., 1990). The role of osteoblastic cells in the differentiation and function of osteoclasts is summarized in Fig. 5.2.

Macrophage colony-stimulating factor

Macrophage colony-stimulating factor (M-CSF; also called colony-stimulating factor 1) was originally identified as a growth factor that specifically stimulates the development of macrophage colonies. The critical role of M-CSF in osteoclast

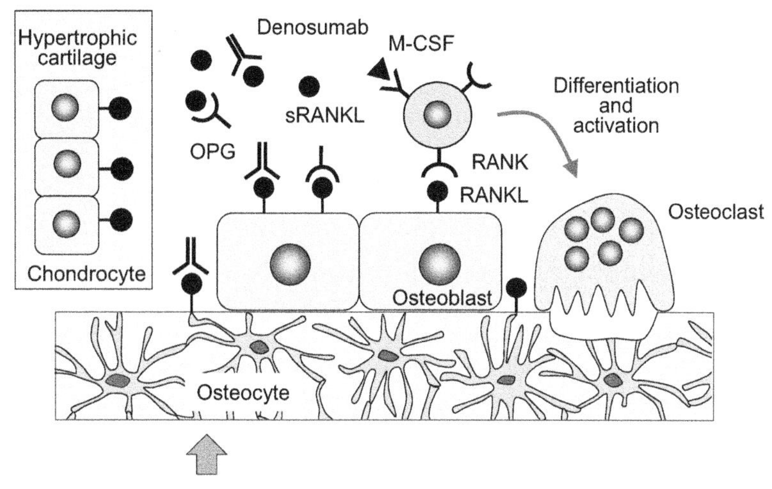

FIGURE 5.2 Schematic representation of osteoclast differentiation and activation. Receptor activator of NF-κB ligand (*RANKL*) expressed by osteoblasts, osteocytes, and hypertrophic chondrocytes supports osteoclast differentiation in the presence of macrophage colony-stimulating factor (*M-CSF*). Osteotropic factors such as 1α,25-dihydroxyvitamin D_3 (*1α,25(OH)$_2$D$_3$*), parathyroid hormone (*PTH*), prostaglandin E_2 (*PGE$_2$*), and interleukin 11 (*IL-11*) stimulate the expression of RANKL in the aforementioned cells. Osteoclast precursors express M-CSF receptor and RANK, and differentiate into osteoclasts in response to both cytokines. The membrane-associated and soluble (s) forms of RANKL support osteoclastogenesis. Osteoprotegerin (*OPG*) secreted by osteoblastic cells and other cells, and denosumab, an anti-RANKL antibody, inhibit osteoclast differentiation and activation by suppressing the RANKL–RANK interaction.

biology was established in a series of experiments on *op/op* (osteopetrosis) mice. Impaired bone resorption in *op/op* mice is associated with a marked reduction in the number of osteoclasts in skeletal tissues (Wiktor-Jedrzejczak et al., 1982). A point mutation in the coding region of the *M-CSF* gene was found in *op/op* mice, leading to the stop codon TGA, 21 bp downstream (Yoshida et al., 1990; Wiktor-Jedrzejczak et al., 1990). The administration of recombinant M-CSF to animals cured their osteopetrotic phenotype, confirming the essential role of this cytokine in the differentiation of osteoclasts (Felix et al., 1990; Kodama et al., 1991b). Hematopoietic cells derived from *op/op* mice differentiated into osteoclasts in a coculture with normal mouse-derived osteoblastic cells. On the other hand, *op/op* mouse-derived osteoblastic cells failed to support osteoclast formation in the coculture with normal hematopoietic cells (Kodama et al., 1991a; Takahashi et al., 1991). These findings further confirmed that M-CSF produced by osteoblastic cells is essentially involved in osteoclastogenesis.

Osteoprotegerin and RANKL

RANKL is a member of the tumor necrosis factor (TNF) superfamily (encoded by TNFSF11) (Anderson et al., 1997), which was originally identified as the T-cell-derived immunomodulatory cytokine, TNF-related activation-induced cytokine (TRANCE), in 1997 (Wong et al., 1997). RANKL is expressed in activated T cells and promotes the survival of dendritic cells by binding to RANK (encoded by *TNFRSF11a*), resulting in enhanced dendritic cell–mediated T cell proliferation. The discovery that RANKL is an essential factor for osteoclastogenesis was made by two independent groups at similar times. In 1997, Simonet et al. cloned a new member of the TNF receptor (TNFR) superfamily (encoded by *TNFRSF11b*) (Simonet et al., 1997). This protein lacked a transmembrane domain, indicating that it is a secreted member of the TNFR family. The hepatic expression of this protein in transgenic mice resulted in severe osteopetrosis. This protein was named "osteoprotegerin," reflecting its protection of bone. Tsuda et al. (1997) independently isolated a novel protein termed "osteoclastogenesis inhibitory factor" (OCIF) from the conditioned medium of a human fibroblast culture. The amino acid sequence of OCIF was identical to that of OPG (Tsuda et al., 1997; Yasuda et al., 1998a). OPG strongly inhibited osteoclast formation induced by 1α,25-dihydroxyvitamin D_3 (1α,25(OH)$_2$D$_3$), PTH, PGE$_2$, or IL-11 in the coculture. Using OPG as a probe, two groups simultaneously isolated a cDNA with an open reading frame encoding 316 amino acid residues from murine cell lines, and ODF and OPG ligand were found to be identical to RANKL (Yasuda et al., 1998b; Lacey et al., 1998). The expression of RANKL was induced in osteoblastic cells in response to 1α,25(OH)$_2$D$_3$, PTH, PGE$_2$, and IL-11. The combined treatment of hematopoietic cells with M-CSF and the soluble form of RANKL (sRANKL) induced osteoclast differentiation in vitro in the absence of supporting osteoblastic cells. Moreover, the targeted disruption of RANKL or RANK induced severe osteopetrosis in mice due to a defect in osteoclast differentiation (Kong et al., 1999; Dougall et al., 1999). RANKL was also shown to activate osteoclast function. Loss-of-function mutations in the genes encoding *RANK*, *OPG*, and *RANKL*, and gain-of-function mutations in the *RANK* gene, were detected in humans (Kobayashi et al., 2009; Crockett et al., 2011). These findings confirmed the essential role of the RANKL–RANK pathway in osteoclast development and function both in vitro and in vivo.

Osteoclastogenesis supported by RANKL

Osteocytes are osteoblast-derived cells embedded in the bone matrix. Osteocytes are connected to one another by gap junctions via cellular processes. Studies have shown that osteocytes also regulate osteoclast formation through the production of RANKL (Nakashima et al., 2011; Xiong et al., 2011). The deletion of RANKL specifically in osteoblastic cells using *osterix*-Cre and *osteocalcin*-Cre resulted in osteopetrosis with impaired tooth eruption, a typical symptom of severe osteopetrosis (Xiong et al., 2011). On the other hand, when *RANKL* was deleted in an osteocyte-specific manner using *dentin matrix protein 1*-Cre, bone resorption was suppressed 6 months after birth and bone mass increased. These findings indicated that RANKL expressed by osteocytes plays an important role in the process of bone remodeling. In addition, hypertrophic chondrocyte-derived RANKL was shown to be important for bone development. An osteopetrotic feature was developed when RANKL was disrupted in hypertrophic chondrocytes using *type X collagen*-Cre mice (Xiong et al., 2011). RANKL was shown to be strongly expressed by T lymphocytes. However, osteopetrosis does not develop with a T-cell-specific *RANKL* deficiency. Based on these findings, it was concluded that RANKL expressed by osteoblasts, osteocytes, and hypertrophic chondrocytes is important for the regulation of bone resorption in mice.

RANKL is produced as an integral membrane protein, and is also cleaved to produce sRANKL (Lacey et al., 1998). Previous studies using the coculture system suggested that the membrane-bound form of RANKL is required for osteoclast formation. However, transgenic mice overexpressing sRANKL or mice injected with sRANKL exhibited increased resorption. To establish the relative importance of membrane-bound versus sRANKL, mice expressing only membrane-bound form of RANKL were produced. The membrane-bound form of RANKL was shown to exert almost all of the functions of RANKL in bone (Xiong et al., 2017).

The anti-RANKL antibody denosumab was developed to suppress excessive bone resorption. Denosumab is a fully humanized monoclonal antibody with affinity and specificity for human RANKL (Lacey et al., 2012). Denosumab blocks the binding of RANKL to RANK expressed on osteoclast precursors and osteoclasts, causing the suppression of osteoclast differentiation and activation. Denosumab is used in the treatment of patients at high risk of bone fractures, including women and men with osteoporosis, men with prostate cancer, and women with breast cancer. Furthermore, denosumab is used to prevent skeletal-related events in patients with bone metastases originating from solid tumors. The roles of RANKL, OPG, and denosumab in osteoclast differentiation and activation are summarized in Fig. 5.2.

RANKL is expressed not only in bone, but also in mammary glands, the lymph nodes, the thymus, and intestinal Peyer's patches (Lacey et al., 1998). Subsequent studies demonstrated that the RANKL–RANK signal is essentially involved in mammary gland development (Fata et al., 2000), lymph node formation (Kong et al., 1999; Yoshida et al., 2002), the elimination of self-reactive T cells in the thymus (Rossi et al., 2007; Akiyama et al., 2008), and microfold cell differentiation in Peyer's patches (Knoop et al., 2009; Kanaya et al., 2012). It was shown in 2017 that the membrane-bound form of RANKL plays an essential role in microfold cell differentiation in intestinal Peyer's patches (Nagashima et al., 2017).

Signal transduction in osteoclastogenesis

The M-CSF receptor FMS

Osteoclasts develop from monocytic precursors from the hematopoietic lineage. PU.1 induces the expression of the M-CSF receptor FMS and is critical for monocyte–macrophage lineage commitment (Tondravi et al., 1997; Zaidi et al., 2003). Signals via FMS are essential for the induction of osteoclast differentiation (Hamilton, 1997; Feng and Teitelbaum, 2013; Stanley and Chitu, 2014). The receptor FMS has a tyrosine kinase domain in the intracellular region. Signals mediated by FMS are required not only for the proliferation of osteoclast progenitors, but also for the differentiation of postmitotic osteoclast precursors (Tanaka et al., 1993). The binding of M-CSF to FMS induces the autophosphorylation of the receptor at several tyrosine residues within the cytoplasmic domain (Feng et al., 2002; Stanley and Chitu, 2014). This phosphorylation of tyrosine residues recruits signaling molecules such as growth factor receptor–bound protein 2 (GRB2) (Insogna et al., 1997) and phosphatidylinositol 3-kinase (PI3K) (Mandal et al., 2009) during osteoclastogenesis. M-CSF mainly activates two signaling pathways: extracellular signal-regulated kinase (ERK) through GRB2 and a serine/

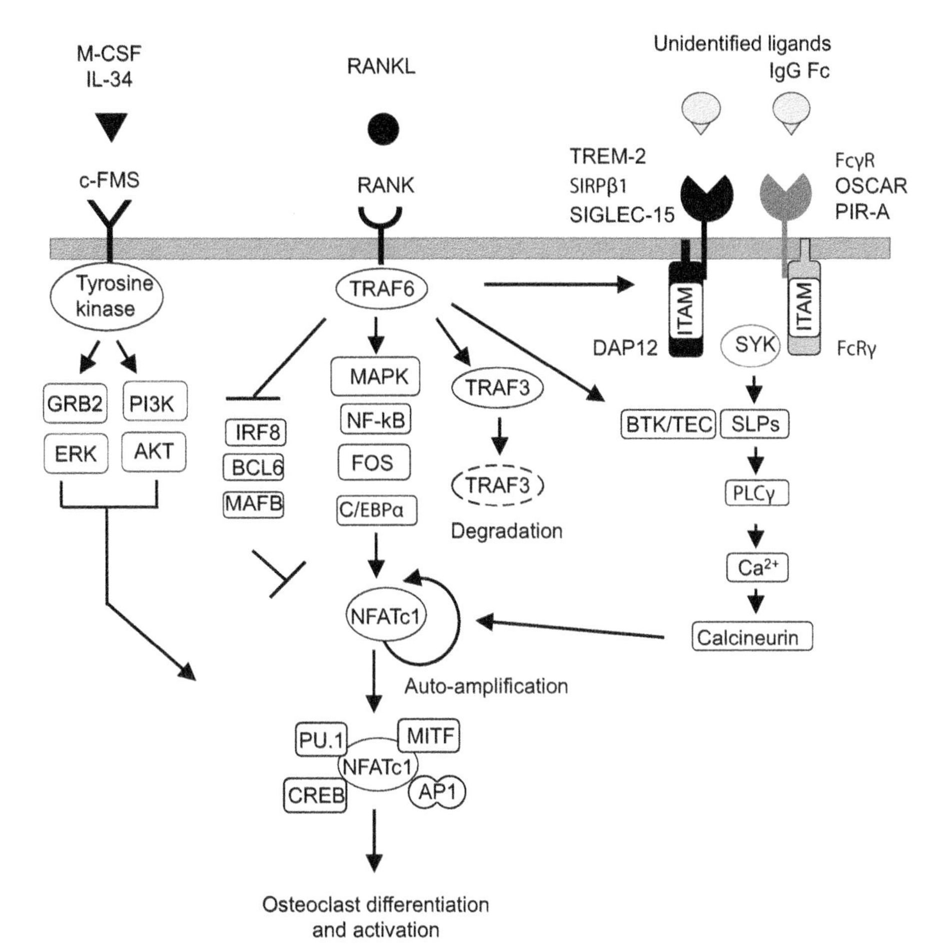

FIGURE 5.3 Signal transduction for osteoclast differentiation and activation. When M-CSF binds to the receptor FMS, the signaling pathways of GRB2–ERK and PI3K–AKT are activated. IL-34 is the second ligand for FMS and induces the same signals as M-CSF. TRAF6 acts as the main adaptor molecule of RANK signaling. TRAF6-mediated signals activate downstream targets such as MAPK, NF-κB, JNK, FOS, and C/EBPα, and ultimately induce the autoamplification of NFATc1. In addition, RANK signals suppress the expression of inhibitory molecules such as IRF8, BCL6, and MAFB. ITAM signaling via immunoglobulin-like receptors is an additional signal that regulates osteoclast differentiation and function. The adaptor proteins DAP12 and FcRγ possess the ITAM sequence and induce ITAM signals. RANK signals also activate ITAM signals. ITAM signals activate SYK, which, in turn, phosphorylates the tyrosine residues of SLPs. RANK signaling activates BTK and TEC. A signaling complex of BTK/TEC and SLPs is formed and activates PLCγ. Activated PLCγ induces the mobilization of intracellular Ca^{2+}, which activates calcineurin to induce the dephosphorylation and autoamplification of NFATc1. Amplified NFATc1 induces the transcription of osteoclast-associated genes in cooperation with PU.1, CREB, and MITF. These signaling cascades induce the differentiation and activation of osteoclasts. *AKT*, serine/threonine kinase; *BCL6*, B cell leukemia/lymphoma 6; *C/EBPα*, CCAAT/enhancer binding protein α; *c-FMS*, M-CSF receptor; *CREB*, cAMP-responsive element binding protein; *DAP12*, DNAX activation protein 12; *ERK*, extracellular signal-regulated kinase; *FcγR*, Fcγ receptor; *FcRγ*, Fc receptor common γ subunit; *GRB2*, growth factor receptor–bound protein 2; *IgG Fc*, immunoglobulin G Fc region; *IL-34*, interleukin-34; *IRF8*, interferon regulatory factor 8; *ITAM*, immunoreceptor tyrosine-based activation motif; *MAPK*, mitogen-activated protein kinase; *M-CSF*, macrophage colony-stimulating factor; *MITF*, microphthalmia-associated transcription factor; *NF-κB*, nuclear factor κB; *NFATc1*, nuclear factor of activated T cells 1; *OSCAR*, osteoclast-associated receptor; *PI3K*, phosphatidylinositol 3-kinase; *PIR-A*, paired immunoglobulin-like receptor A; *PLCγ*, phospholipase Cγ; *RANK*, receptor activator of NF-κB; *RANKL*, receptor activator of NF-κB ligand; *SIGLEC-15*, sialic acid-binding immunoglobulin-like lectin 15; *SIRPβ1*, signal-regulatory protein β1; *SLP*, SRC homology 2 domain-containing leukocyte protein; *SYK*, spleen tyrosine kinase; *TRAF*, tumor necrosis factor receptor-associated factor; *TREM-2*, triggering receptor expressed on myeloid cells 2.

threonine kinase, AKT, through PI3K (Feng and Teitelbaum, 2013; Stanley and Chitu, 2014). Both signaling pathways are considered to be involved in M-CSF-supported osteoclastogenesis. Signals mediated by GRB2 and PI3K are involved in the regulation of cytoskeletal organization in osteoclasts as well as osteoclast differentiation. Signal transduction that induces osteoclast differentiation and activation is summarized in Fig. 5.3.

M-CSF-deficient *op/op* mice exhibit monocytopenia and osteopetrosis (Wiktor-Jedrzejczak et al., 1982). However, several unusual phenomena have been observed in *op/op* mice. Osteoclasts are completely absent in young *op/op* mice, but appear in aged mice (Begg et al., 1993). The osteopetrotic characteristics of FMS-deficient mice are more severe than those of *op/op* mice (Dai et al., 2002). F4/80-positive macrophages exist in the splenic red pulp in *op/op* mice as well as in wild-type mice, and their number is regulated by a mechanism independent of M-CSF (Cecchini et al., 1994; Yamamoto et al., 2008). These phenomena may be explained by the discovery of IL-34, a second ligand for FMS (Lin et al., 2008; Nakamichi et al., 2013; Stanley and Chitu, 2014). The amino acid sequence of IL-34 is different from that of M-CSF; however, IL-34 binds to FMS and induces the same effects as M-CSF. IL-34 cannot substitute for M-CSF in vivo because its localization differs from that of M-CSF. IL-34 is strongly expressed in the spleen, but not in bone. On the other hand, M-CSF is strongly expressed in bone and the spleen. Osteoclast precursors were previously shown to be present in the spleen, but not in bone in *op/op* mice (Nakamichi et al., 2012). Splenectomy blocked M-CSF-induced osteoclastogenesis in *op/op* mice. Osteoclasts appeared in aged *op/op* mice with the upregulation of IL-34 expression in bone. Splenectomy in *op/op* mice also blocked the age-associated appearance of osteoclasts. These findings suggest that IL-34 plays a pivotal role in maintaining the splenic reservoir of osteoclast precursors, which are transferred from the spleen to bone via the bloodstream in response to diverse stimuli, in *op/op* mice. The administration of FYT720, a sphingosine-1-phosphate agonist, has also been shown to promote the egress of osteoclast precursors from hematopoietic tissues into the bloodstream (Ishii et al., 2009). These findings suggest that some osteoclast precursors are circulating in the bloodstream.

RANK

RANK-mediated signals are also essential for the induction of osteoclast differentiation and function (Fig. 5.3). When RANKL binds to RANK, RANK recruits TRAFs, which are adapter molecules that induce osteoclastogenesis (Wong et al., 1998). TRAFs 1, 2, 3, 5, and 6 are recruited to the cytoplasmic tail of RANK. Among them, the TRAF6-mediated signal is considered to induce osteoclast differentiation and function because TRAF6-deficient mice show severe osteopetrosis with a decreased number of osteoclasts (Lomaga et al., 1999; Naito et al., 1999; Walsh et al., 2015). Osteoclast precursors from TRAF6-deficient mice fail to differentiate into osteoclasts even in the presence of RANKL (Gohda et al., 2005). RANK-mediated signals activate downstream signals such as p38 mitogen-activated protein kinase (MAPK), Jun N-terminal kinase (JNK), and FOS (Wagner and Eferl, 2005; Takayanagi, 2007). These signals ultimately induce the expression of the transcription factor nuclear factor of activated T cells 1 (NFATc1), a master transcription factor that plays a pivotal role in osteoclastogenesis (Ishida et al., 2002; Takayanagi et al., 2002). This robust induction of NFATc1 is based on an autoamplifying mechanism affected through the persistent calcium signal−mediated activation of NFATc1. NFATc1 binds to NFAT-binding sites on its own promoter, constituting a positive feedback loop. In the nucleus, NFATc1 cooperates with other transcription factors, such as PU.1, cAMP-responsive element binding protein, and microphthalmia-associated transcription factor, to induce various osteoclast-specific genes (Sato et al., 2006; Okamoto et al., 2017). *CCAAT/enhancer* binding protein α (C/EBPα) is a key molecular determinant in myeloid lineage commitment. C/EBPα is strongly expressed in osteoclasts and has been identified as a critical *cis*-regulatory element-binding protein in the cathepsin K promoter (Chen et al., 2013). C/EBPα-deficient mice exhibit severely blocked osteoclastogenesis, suggesting that C/EBPα is a key regulator of osteoclast lineage commitment. Peroxisome proliferator-activated receptor γ is also reported to be important for osteoclast differentiation (Okazaki et al., 1999; Wan et al., 2007).

On the other hand, it is important for osteoclastogenesis to exclude factors that inhibit osteoclast differentiation. Interferon regulatory factor 8 (IRF8), a transcription factor expressed in immune cells, is a key inhibitory molecule for osteoclastogenesis (Zhao et al., 2009). IRF8 was found to inhibit osteoclast formation under physiological and pathological conditions, and was downregulated in response to a RANKL stimulation. RANKL also upregulates the expression of the transcription factor B-lymphocyte-induced maturation protein-1 (BLIMP1) (Nishikawa et al., 2010). BLIMP1 then suppresses the expression of other inhibitory transcription factors, such as B cell leukemia/lymphoma 6 and MAFB (Miyauchi et al., 2010). Thus, RANK signaling induces the downregulation of negative regulators together with the upregulation of positive regulators for osteoclastogenesis (Fig. 5.3).

Tumor necrosis factor receptors

TNFα together with M-CSF has been shown to induce osteoclastogenesis in the absence and presence of RANKL (Kobayashi et al., 2000; Zhang et al., 2001). Osteoclast formation induced by TNFα was inhibited by the addition of respective antibodies against TNFR type I (TNFRI, p55) and TNFR type II (TNFRII, p75), but not by OPG. This finding suggests that TNFα stimulates osteoclastogenesis via TNFRs. Osteoclast precursors obtained from RANK- or TRAF6-deficient mice differentiate into osteoclasts when stimulated with TNFα in the presence of cofactors such as transforming growth factor β (Kim et al., 2005). However, the osteoclastogenesis-inducing ability of TNFα is markedly weaker than that of RANKL. Studies have revealed the mechanisms by which TNF induces the differentiation of osteoclasts

(Boyce et al., 2015). When osteoclast precursors prepared from IRF8-deficient mice were used in an osteoclast formation assay, TNFα induced osteoclastogenesis as strongly as RANKL. Recombinant recognition sequence binding protein at the Jκ site (RBP-J) is a DNA-binding protein that functions as a transcriptional repressor or activator depending on the partner proteins. RBP-J strongly suppressed TNFα-induced osteoclastogenesis and inflammatory bone resorption (Zhao et al., 2012). In the absence of RBP-J, TNFα effectively induced osteoclastogenesis and bone resorption in RANK-deficient mice (Yao et al., 2009). In contrast to TRAF6, TRAF3 inhibits osteoclast formation by moderating nuclear factor κB (NF-κB) signaling (Xiu et al., 2014). TRAF3 suppresses the conversion of NF-κB p100 (inhibitory form) into NF-κB p52 (active form) through the degradation of NF-κB-inducing kinase, suggesting that TRAF3 signals induce the accumulation of inhibitory NF-κB p100. RANKL induced the degradation of inhibitory TRAF3 and NF-κB p100 in osteoclast precursors. In contrast, TNFα increased the expression of TRAF3 in osteoclast precursors (Yao et al., 2017). TNFα induced osteoclast formation to the same level as RANKL when NF-κB p100 or TRAF3 was deleted in RANK-deficient osteoclast progenitors. A TNFα injection strongly induced osteoclastogenesis in vivo in mice deficient in RANK together with NF-κB p100 (Yao et al., 2009; Boyce et al., 2015). These findings suggest that TNFα induces osteoclast formation at a similar level compared with RANKL through TNFR under specific conditions, and also that RANKL efficiently eliminates endogenous inhibitory molecules in osteoclast precursors.

ITAM costimulatory signals

Calcium signals

Immunoreceptor tyrosine-based activation motif (ITAM) signaling via immunoglobulin-like receptors is an additional signal regulating osteoclast differentiation and function (Koga et al., 2004; Mocsai et al., 2004; Humphrey and Nakamura, 2016, Okamoto et al., 2017). The adaptor proteins DNAX activation protein 12 (DAP12) and Fc receptor common γ subunit (FcRγ) possess the ITAM sequence (Fig. 5.3). DAP12 is coupled with triggering receptor expressed on myeloid cells 2, signal-regulatory protein β1, and sialic acid-binding immunoglobulin-like lectin 15 (SIGLEC-15), while FcRγ is coupled with osteoclast-associated receptor (Kim et al., 2002; Barrow et al., 2011) and paired immunoglobulin-like receptor A as well as Fcγ receptor (FcγR). DAP12- and FcRγ-mediated ITAM signals activate spleen tyrosine kinase (SYK). RANK-mediated signals also activate ITAM signals as well as MAPK, NF-κB, and FOS signaling. Activated SYK then phosphorylates the tyrosine residues of SRC homology 2 domain-containing leukocyte protein (SLP) adapter molecules such as B cell linker and SLP-76 (Shinohara et al., 2008; Lee et al., 2008). TEC and BTK are tyrosine kinases that belong to the TEC family. RANK signaling activates BTK and TEC by tyrosine phosphorylation. Mice lacking both BTK and TEC showed severe osteopetrosis caused by a defect in bone resorption. ITAM and RANK signaling results in the formation of a signaling complex of BTK/TEC and SLPs. This complex activates phospholipase Cγ (PLCγ), thereby increasing the mobilization of intracellular Ca^{2+}. This increase in intracellular Ca^{2+} activates calcineurin, a Ca^{2+}-dependent serine/threonine phosphatase. Calcineurin induces the dephosphorylation of NFATc1 and promotes its nuclear translocation to enhance the autoamplification of NFATc1. Amplified NFATc1 then induces the transcription of osteoclast-associated genes. Thus, the ITAM signal cooperates with the RANK-mediated signal to induce NFATc1.

Mice lacking both FcRγ and DAP12 exhibit severe osteopetrosis due to impaired osteoclast differentiation, confirming that ITAM signaling is an essential signal promoting osteoclast differentiation. Patients with Nasu−Hakola disease, a rare recessive genetic disease found in Japanese and Finnish people, have cyst formation at the limb epiphysis and exhibit multiple pathological fractures (Humphrey and Nakamura, 2016). Point mutations and defects in the *DAP12* gene have been reported in this disease.

Osteoclast differentiation is also induced from osteoclast precursors deficient in inositol 1,4,5-trisphosphate receptor type 2, which is essential for calcium mobilization. Therefore, there appears to be an osteoclast differentiation pathway that is not dependent on calcium ions (Kuroda et al., 2008; Oikawa et al., 2013). Several common pathways of bone metabolism and the immune system have been elucidated, and the term "osteoimmunology" has been proposed and widely used to obtain a clearer understanding of the regulation of osteoclastogenesis.

SIGLEC-15 and FcγR

Studies have revealed that SIGLEC-15 and FcγR play important roles in osteoclastic bone resorption (Fig. 5.4). SIGLEC family members are sialic acid-binding receptors that are primarily expressed by immune cells. SIGLEC-15 is the only SIGLEC highly conserved throughout vertebrate evolution (Angata et al., 2007). SIGLEC-15 was identified as a gene that is strongly expressed in human osteoclastoma (giant cell tumor of bone) (Hiruma et al., 2011) and also as a gene product

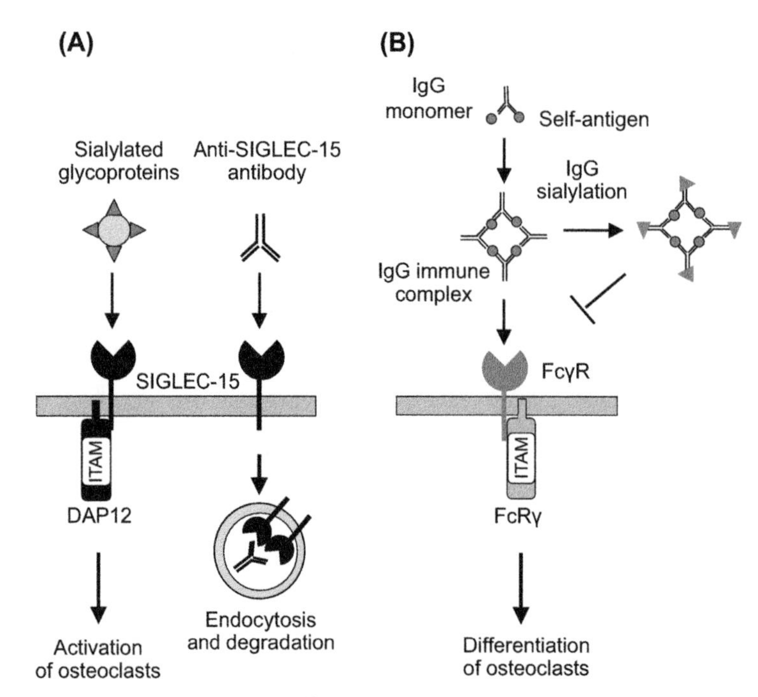

FIGURE 5.4 SIGLEC-15- and FcγR-mediated signaling. (A) Sialic acid-binding immunoglobulin-like lectin 15 (*SIGLEC-15*)-mediated signals. SIGLEC-15 is a sialic acid-binding receptor associated with DNAX activation protein 12 (*DAP12*). It activates spleen tyrosine kinase via DAP12 and promotes the bone resorptive function of osteoclasts. Osteoclasts lacking SIGLEC-15 cannot form actin rings. When an anti-SIGLEC-15 antibody is administered to mice, bone mass increases due to the suppression of bone resorption. Binding of the antibody to SIGLEC-15 induces the dimerization of SIGLEC-15, rapid internalization, and lysosomal degradation. **(B) Fcγ receptor (*FcγR*)-mediated signals.** Although immunoglobulin G (*IgG*) antibodies generally exist as monomers, IgG immune complexes (ICs) are formed in inflammation and autoimmune diseases. IgG ICs promote osteoclast formation via FcγR. The Fc portion of the IgG antibody undergoes sugar chain modifications, and sialic acid is added to a part of its end. Sialylated IgG ICs attenuate its ability to bind to FcγR. *FcRγ*, Fc receptor common γ subunit; *ITAM*, immunoreceptor tyrosine-based activation motif.

induced by NFATc1 (Ishida-Kitagawa et al., 2012). SIGLEC-15 associates with DAP12. SIGLEC-15-deficient mice exhibited osteopetrosis with the suppression of osteoclastic bone resorption. TRAP-positive osteoclasts were detected in SIGLEC-15-deficient mice at levels similar to those in wild-type mice, suggesting that SIGLEC-15 plays a role in osteoclast function (Hiruma et al., 2013; Kameda et al., 2013, 2015). Osteoclasts derived from SIGLEC-15-deficient mice could not form actin rings. In an antigen-induced arthritis model, the degree of periarticular bone loss in the proximal tibia is significantly lower in SIGLEC-15-deficient mice than in wild-type mice (Shimizu et al., 2015). When an anti-SIGLEC-15 neutralizing antibody was administered to mice, bone mass increased due to the suppression of bone resorption (Stuible et al., 2014; Fukuda et al., 2017). The binding of a neutralizing antibody to SIGLEC-15 induces the dimerization of the receptor, rapid internalization, and lysosomal degradation. SIGLEC-15-deficient mice also show resistance to bone loss by ovariectomy. SIGLEC-15-mediated signals activate PI3K and AKT together with PLCγ in osteoclasts. The ligand of SIGLEC-15 is proposed to be a glycoprotein having sialic acid that has not yet been identified. SIGLEC-15 may be a major target for the treatment of bone diseases with excessive bone resorption.

Autoantibody production and IgG immune complex (IC) formation are frequently observed in autoimmune diseases associated with bone loss. Studies have demonstrated that, under pathological conditions such as autoimmune diseases associated with hypergammaglobulinemia, IgG ICs enhance osteoclast formation through FcγR and promote bone resorption (Fig. 5.4) (Negishi-Koga et al., 2015; Harre et al., 2015). The Fc portion of the IgG antibody undergoes various sugar chain modifications. Sialic acid is added to the sugar chain terminal of the Fc portion. This sialylated IgG attenuates its ability to bind to FcγR. The desialylated, but not sialylated, human IgG IC promoted osteoclast formation in vitro. The relationship between the sialylation status of IgG and bone mass was examined in patients with rheumatism. Bone mass was found to decrease as the sialylation rate of IgG in these patients became lower. The administration of the sialic acid precursor *N*-acetyl-D-mannosamine to animals promotes the sialylation of IgG ICs, and suppresses IgG IC-enhanced osteoclast formation. These results demonstrate that ITAM costimulatory signals are important for regulating osteoclast differentiation and function.

WNT signals

Canonical WNT signals

WNT proteins activate β-catenin-dependent canonical and β-catenin-independent noncanonical signaling pathways. In the absence of WNT ligands, β-catenin is phosphorylated by a degradation complex composed of axin, adenomatous polyposis coli, glycogen synthase kinase 3β, and casein kinase 1. Phosphorylated β-catenin is degraded by the ubiquitin proteasome, and cytosolic β-catenin is then maintained at a low level. When the WNT ligand binds to a receptor complex composed of frizzled receptor and low-density lipoprotein receptor-related protein 5/6, the activity of the degradation complex is suppressed, and β-catenin then accumulates in the cytoplasm. Accumulated β-catenin translocates to the nucleus. Nuclear β-catenin together with the transcription factors T cell factor and lymphoid enhancer factor (LEF) induces the transcription of the target gene.

Glass et al. (2005) showed the importance of WNT/β-catenin signals in bone resorption in vivo using genetic approaches. Mice expressing the constitutively active form of β-catenin in mature osteoblasts had an increased bone mass with impaired osteoclast formation. The expression of OPG was increased in the long bones of these mice. The LEF-1 binding sequence is found in the promoter region of the *OPG* gene, and the LEF-1/β-catenin complex induces the expression of OPG. Thus, the activation of WNT/β-catenin signals induces the expression of OPG in mature osteoblasts and markedly inhibits bone resorption. The roles of β-catenin-dependent canonical WNT signaling and β-catenin-independent noncanonical WNT signals that regulate osteoclast differentiation and activation are summarized in Fig. 5.5 (Kobayashi et al., 2016).

WNT3A, a typical canonical WNT ligand, inhibited $1\alpha,25(OH)_2D_3$-induced osteoclast formation in a coculture, but not RANKL-induced osteoclast formation in osteoclast precursor cultures (Yamane et al., 2001). This finding indicates that WNT3A suppresses osteoclastogenesis through the expression of OPG in osteoblasts. The concept that WNT/β-catenin signals inhibit osteoclastogenesis through the expression of OPG was confirmed in a study using OPG-deficient mice (Kobayashi et al., 2015). WNT3A did not inhibit $1\alpha,25(OH)_2D_3$-induced osteoclast formation in a coculture of normal bone marrow cells with OPG-deficient osteoblastic cells.

Moverare-Skrtic et al. (2014) reported important roles for WNT16 in bone mass. WNT16 mRNA was strongly expressed in cortical bone among the various tissues tested, including trabecular bone, suggesting that WNT16 regulates bone formation and resorption in cortical bone. WNT16-deficient mice showed a marked decrease in cortical bone, but not in cancellous bone. WNT16 induced the expression of OPG through the activation of WNT/β-catenin signaling in MC3T3-E1 osteoblastic cells. Furthermore, osteoclast formation was enhanced in cocultures of WNT16-deficient osteoblastic cells with bone marrow cells, suggesting that WNT16 induced OPG through the activation of WNT/β-catenin signals in osteoblastic cells. WNT16 has also been shown to affect osteoclast precursors by activating noncanonical WNT signals. This topic is discussed later.

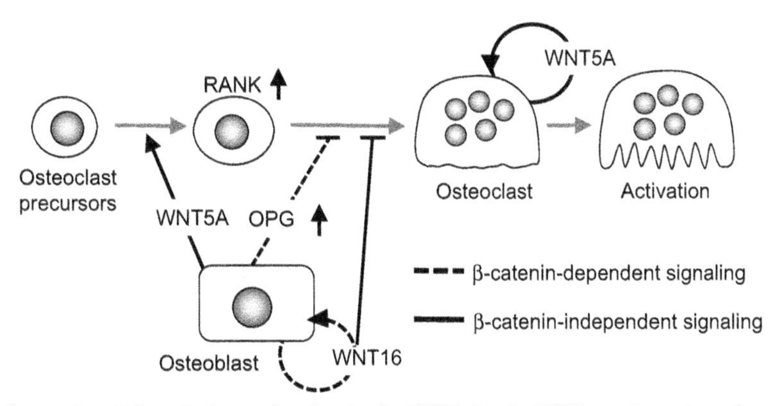

FIGURE 5.5 Regulation of osteoclast differentiation and activation by WNT signals. WNT proteins activate β-catenin-dependent canonical and β-catenin-independent noncanonical signaling pathways. Mice expressing constitutively active β-catenin in mature osteoblasts exhibit an increased bone mass with impaired osteoclast formation. Canonical WNT signals in osteoblastic cells enhance the expression of osteoprotegerin (*OPG*) and inhibit bone resorption. WNT16 is strongly expressed in cortical bone. WNT16 suppresses osteoclast differentiation through the β-catenin-dependent canonical pathway, which enhances OPG production, in osteoblastic cells. WNT16 also acts on osteoclast precursors and inhibits osteoclastic differentiation through noncanonical WNT signaling. WNT5A is a typical ligand for noncanonical WNT signals. WNT5A secreted by osteoblastic cells promotes receptor activator of NF-κB (*RANK*) expression in osteoclast precursors through receptor tyrosine kinase orphan receptor 2 and promotes their osteoclastic differentiation. WNT5A expressed by osteoclasts also induces osteoclast resorptive activity.

The roles of WNT/β-catenin signals in osteoclast precursors have also analyzed. Wei et al. (2011) developed mice expressing a constitutively active form of β-catenin (CA β-cat) or deleted the *β-catenin* gene (Δβ-cat) in osteoclast lineage cells. Both mice with CA β-cat and Δβ-cat osteoclast lineage cells exhibited osteopetrotic phenotypes due to decreased bone resorption. This study showed that β-catenin signals are required for the proliferation of osteoclast precursors, while the sustained activation of β-catenin signals inhibits the differentiation of precursors into osteoclasts. These findings suggest that finely balanced β-catenin signals in osteoclast precursors regulate osteoclastogenesis (Fig. 5.5).

Noncanonical WNT signals

WNT5A is a typical ligand that activates β-catenin-independent noncanonical WNT signals (Mikels et al., 2009) (Fig. 5.5). Receptor tyrosine kinase orphan receptor 1/2 (ROR1/2), coreceptors of WNT proteins, mainly mediate the action of WNT5A. WNT5A secreted by osteoblastic cells promotes osteoclast formation through ROR2, but not ROR1 (Maeda et al., 2012). The conditional deletion of ROR2 in osteoclast precursors and that of WNT5A in osteoblastic cells resulted in impaired osteoclast formation, suggesting that WNT5A−ROR2 signals between osteoblastic cells and osteoclast precursors promote osteoclast differentiation. WNT5A upregulated the expression of RANK in osteoclast precursors through the activation of JNK, thereby enhancing RANKL-induced osteoclast formation.

WNT16 suppressed osteoclast differentiation through the activation of β-catenin-dependent canonical pathways in osteoblastic cells. In contrast, the treatment of osteoclast precursors with WNT16 did not activate WNT/β-catenin signals, but suppressed RANKL-induced NF-κB activation and NFATc1 expression in osteoclast precursors (Moverare-Skrtic et al., 2014; Gori et al., 2015). Furthermore, WNT4 was also reported to inhibit RANKL-induced activation of NF-κB, which in turn suppressed osteoclast formation in a β-catenin-independent manner (Yu et al., 2014). WNT3A activated protein kinase A signals and suppressed the nuclear translocation of NFATc1 in osteoclast precursors, thereby inhibiting osteoclast formation (Weivoda et al., 2016). Thus, WNT16, WNT4, and WNT3A inhibit osteoclast formation through β-catenin-independent noncanonical WNT signals.

The expression of WNT5A and ROR2 was increased in osteoclast precursors during osteoclastic differentiation, suggesting that WNT5A regulates osteoclast function as well as osteoclast differentiation (Uehara et al., 2017). Osteoclast-specific *ROR2* conditional knockout (ROR2 cKO) mice exhibited impaired bone resorption with defects in actin ring formation, but with normal osteoclast formation. The overexpression of the constitutively active form of RHOA, but not that of RAC1, a small GTPase of the RHO family, rescued the impaired bone-resorbing activity of *ROR2* cKO osteoclasts. Osteoclasts strongly expressed protein kinase N3 (PKN3), a RHO effector. *PKN3*-deficient mice exhibited the high bone mass phenotype with the impaired bone-resorbing activity of osteoclasts. SRC activity was weaker in *ROR2* cKO and PKN3-deficient osteoclasts than in wild-type osteoclasts. These findings indicate that WNT5A promotes the bone-resorbing activity of osteoclasts through the ROR2−RHOA−PKN3−SRC signaling axis. The WNT5A−ROR2 signal was also found to promote the invasion of osteosarcoma cells through the activation of SRC and expression of MMP13. These findings suggest that the WNT5A−ROR2 signal plays a role in the degradation of extracellular matrices by activated osteoclasts (Fig. 5.5). Thus, the WNT5A−ROR2 signal in osteosarcomas as well as osteoclasts is critical for degrading extracellular matrices.

Induction of osteoclast function

Adhesion signals

For osteoclasts to become polarized, adhesion signals and cytokine signals need to be simultaneously transmitted into osteoclasts. The induction of osteoclast function begins with the adhesion of osteoclasts to the bone matrix via vitronectin receptors (αVβ3). Adhesion signals transmitted into osteoclasts are summarized in Fig. 5.6. Vitronectin receptors recognize the RGD sequence of extracellular matrix proteins in bone. The targeted disruption of SRC in mice induced osteopetrosis (Soriano et al., 1991). It was also revealed that osteoclasts express high levels of SRC and that SRC-deficient osteoclasts have defects in ruffled border formation (Tanaka et al., 1992; Boyce et al., 1992). Proline-rich tyrosine kinase 2 (PYK2) was identified as a major adhesion-dependent tyrosine kinase in osteoclasts. In osteoclasts derived from SRC-deficient mice, the tyrosine phosphorylation and kinase activity of PYK2 were markedly reduced (Duong et al., 1998). The adaptor molecule, p130 CRK-associated substrate ($p130^{CAS}$), is highly phosphorylated on tyrosine residues and is stably associated with the oncogene products CRK and SRC. Similar to PYK2, $p130^{CAS}$ is highly tyrosine phosphorylated upon osteoclast adhesion to extracellular matrix proteins (Nakamura et al., 1998; Lakkakorpi et al., 1999). $p130^{CAS}$ is not tyrosine

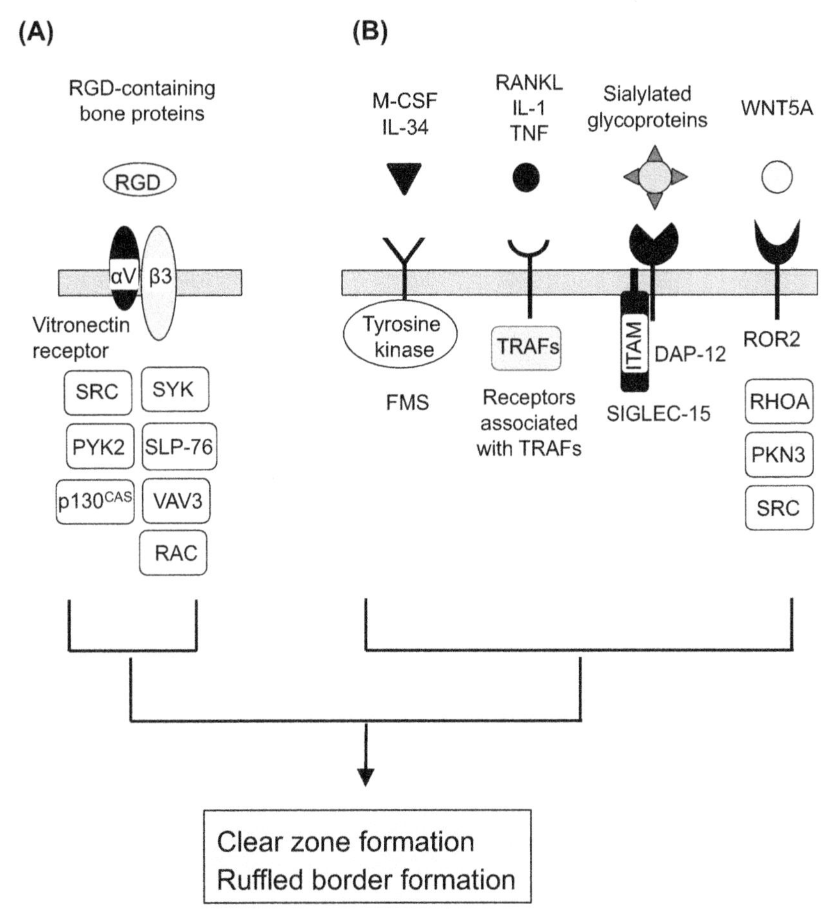

FIGURE 5.6 Signals that induce osteoclast activation. (A) Adhesion signals. The induction of osteoclast function begins with adhesion to the bone matrix via vitronectin receptors ($\alpha V\beta3$). Vitronectin receptors recognize the RGD sequence of extracellular matrix proteins in bone. Adhesion signals induce the formation of a signaling complex of SRC, PYK2, and $p130^{CAS}$. This signaling complex induces clear zone formation. Adhesion signals also recruit SLP-76, which functions as an adaptor of VAV3. VAV3 then activates RAC, which participates in αVβ3-mediated cytoskeletal organization in osteoclasts. These adhesion-induced signals are also involved in the induction of the bone-resorbing activity of osteoclasts. **(B) Cytokine signals.** In addition to adhesion signals, cytokine signals need to be transmitted into osteoclasts to induce bone-resorbing activity. M-CSF plays a role in the regulation of bone resorption in collaboration with αVβ3-mediated signals. IL-1 as well as RANKL activates osteoclast function. IL-1 and RANKL use TRAF6 as a signal transducing adaptor. TNFα, which mainly uses TRAF2 as an adaptor, may induce the activation of osteoclasts if each of the endogenous inhibitory molecules is decreased. SIGLEC-15-mediated signals induce the bone-resorbing activity of osteoclasts. WNT5A also promotes the bone-resorbing activity of osteoclasts through the ROR2−RHOA−PKN3−SRC signaling axis. Therefore, adhesion signals and cytokine signals need to be simultaneously transmitted to osteoclasts to induce their bone-resorbing activation. *FMS*, M-CSF receptor; *DAP-12*, DNAX activation protein 12; *IL*, interleukin; *ITAM*, immunoreceptor tyrosine-based activation motif; *M-CSF*, macrophage colony-stimulating factor; *RANKL*, receptor activator of NF-κB ligand; *$p130^{CAS}$*, p130 CRK-associated substrate; *PKN3*, protein kinase N3; *PYK2*, proline-rich tyrosine kinase 2; *ROR2*, receptor tyrosine kinase orphan receptor 2; *SIGLEC-15*, sialic acid-binding immunoglobulin-like lectin 15; *SLP*, SRC homology 2 domain-containing leukocyte protein; *SYK*, spleen tyrosine kinase; *TNF*, tumor necrosis factor; *TRAF*, tumor necrosis factor receptor-associated factor.

phosphorylated in SRC-deficient osteoclasts, indicating that $p130^{CAS}$ is a downstream molecule of SRC. Thus, the three molecules SRC, PYK2, and $p130^{CAS}$ make a complex that plays a role in osteoclast activation (Nakamura et al., 2012).

β3-integrin-null osteoclasts has been shown to be dysfunctional in vivo and in vitro (McHugh et al., 2000; Faccio et al., 2003; Teitelbaum, 2011). VAV3 is a Rho family guanine nucleotide exchange factor. VAV3-deficient osteoclasts showed defective actin cytoskeleton organization and polarization in response to αVβ3 integrin and FMS activation (Faccio et al., 2005). SYK is an upstream regulator of VAV3 in osteoclasts. SYK exerts its effects via intermediaries including SLP adaptor molecules. SLP of 76 kDa (SLP-76) is involved in the SYK-induced activation of VAV3 in osteoclasts (Reeve et al., 2009; Teitelbaum, 2011). Activated VAV3 converts the inactive GDP-bound form of RAC to the active GTP-bound form. RAC then participates in the αVβ3-mediated organization of the osteoclast cytoskeleton. VAV3-deficient mice showed an increased bone mass and were protected from bone loss induced by PTH and RANKL.

Cytokine signals

M-CSF enhances the survival of osteoclasts and also plays a role in the regulation of bone resorption in collaboration with αVβ3-mediated signals (Feng and Teitelbaum, 2013) (Fig. 5.6). The IL-1 receptor (IL-1R) as well as RANK uses TRAF6 as a signal transducing adaptor. IL-1 does not induce osteoclast differentiation, but strongly activates osteoclast function (Thomson et al., 1986; Jimi et al., 1999). Therefore, we considered TRAF6-mediated signals to induce the activation of osteoclasts. However, RANKL stimulation was found to induce the pit-forming activity of TRAF6-deficient osteoclasts when cultured on bone slices (Yao et al., 2017). This finding suggests that a TRAF(s) other than TRAF6 induces the bone-resorbing activity of osteoclasts treated with RANKL. As described previously, TNFα induces the resorptive activity of RANK-deficient osteoclasts in the absence of inhibitory molecules such as NF-κB p100, TRAF3, and RBP-J (Boyce et al., 2015; Zhao et al., 2012). TNFRI and TNFRII mainly use TRAF2 as a signal transducing adaptor. These findings suggest that the TRAF2-mediated signal induces osteoclast activation in the absence of endogenous inhibitory molecules. SIGLEC-15-mediated signals and ROR2-mediated signals in osteoclasts also induce the bone-resorbing activity of osteoclasts. Thus, osteoclast activation is induced through several ligand—receptor systems. These signals appear to cross talk with each other. Importantly, adhesion signals and cytokine signals must be simultaneously transmitted into osteoclasts to induce their resorptive activity (Fig. 5.6).

Characteristics of osteoclast precursors in vivo

RANKL has been considered to influence the site of osteoclast formation because it is expressed by osteoblastic cells. However, when sRANKL is administered to RANKL-deficient mice, osteoclasts are accurately induced in bone (Yamamoto et al., 2006; Takahashi et al., 2010). This is also the case for the administration of M-CSF to *op/op* mice: when M-CSF is administered to M-CSF-deficient *op/op* mice, osteoclasts appear only in bone (Nakamichi et al., 2012). We analyzed the mechanism by which osteoclasts are formed in bone only. Our findings revealed that the lineage-committed precursors that are determined to differentiate into osteoclasts exist in hematopoietic tissues such as bone marrow and the spleen (Mizoguchi et al., 2009; Arai et al., 2012). Lineage-committed precursors are nondividing cells: therefore, we named them "cell cycle-arrested quiescent osteoclast precursors," or qOPs (Muto et al., 2011). As described earlier, qOPs moved from the spleen to bone in *op/op* mice in response to an M-CSF injection. Some qOPs circulate in the bloodstream and settle in bone. qOPs present in bone strongly express RANK. We showed that M-CSF and WNT5A produced by osteoblastic cells induced RANK expression in osteoclast precursors. We named these osteoclast precursors located in bone tissue as "responding osteoclast precursors" (rOPs). rOPs quickly differentiate into osteoclasts without cell cycle progression in response to RANKL. Our hypothesis on osteoclast precursors in vivo is shown in Fig. 5.7.

When TNFα was injected into mice deficient in both RANK and NF-κB p100, osteoclasts formed only in bone. Similarly, a TNFα injection effectively induced osteoclast formation in bone in RBP-J-deficient mice. These findings suggest that if endogenous inhibitory molecules are removed, rOPs differentiate into osteoclasts in response to TNFα. The essential task of RANKL appears to be to effectively eliminate inhibitory molecules in rOPs. We assume that osteoblastic cells play a role in the conversion of qOPs to rOPs. We consider the distribution of rOPs to influence the accurate site of osteoclast formation in vivo.

Conclusion and perspective

Osteoclasts are special cells that secrete hydrochloric acid and proteolytic enzymes from ruffled borders into resorption lacunae and also take up degradation products from ruffled borders into transcytotic vesicles. Degradation products are exocytosed from the FSD. The ruffled border is formed by the fusion of acidified lysosomes to the side of the plasma membrane facing bone. The differentiation of osteoclasts is regulated by M-CSF and RANKL. RANKL expressed by osteoblasts, osteocytes, and hypertrophic chondrocytes, but not by T cells, may support physiological osteoclast formation. In addition to FMS and RANK-mediated signals, ITAM-mediated costimulatory signals are essentially involved in the induction of NFATc1, a master transcription factor for osteoclastogenesis. Canonical and noncanonical WNT signals are important players in the regulation of osteoclast differentiation and function. To induce the resorptive activity of osteoclasts, adhesion signals and cytokine signals must both be simultaneously transmitted into osteoclasts. When RANKL is injected into RANKL-deficient mice, osteoclasts are induced in bone, but not in other tissues. We demonstrate that certain cells such as rOPs are exclusively present in bone. Findings obtained over the past decade have deepened our understanding of osteoclasts. Still, important questions concerning osteoclasts remain as future studies. How do osteoclasts cease bone resorption? Is there a new signaling pathway that controls osteoclast differentiation and function? Is it possible to

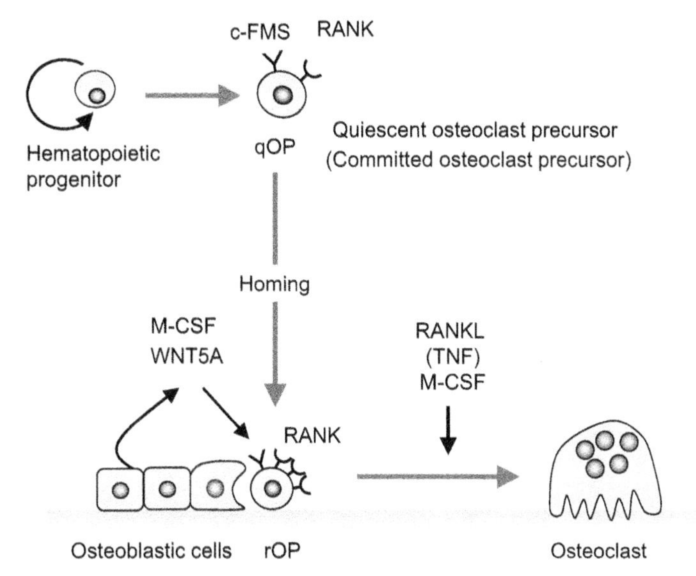

FIGURE 5.7 A hypothetical model of osteoclast differentiation. Osteoclasts are formed from cell cycle–arrested precursors in vivo. These precursor cells are named quiescent osteoclast precursors (*qOPs*). qOPs are the osteoclast lineage–committed precursors derived from hematopoietic tissues. Some qOPs circulate in the bloodstream and settle in bone. qOPs then differentiate into responding osteoclast precursors (*rOPs*). Macrophage colony-stimulating factor (*M-CSF*) and WNT5A produced by osteoblastic cells induce receptor activator of NF-κB (*RANK*) expression in qOPs, suggesting that osteoblastic cells play a role in the appearance of rOPs in bone. rOPs quickly differentiate into osteoclasts in response to receptor activator of NF-κB ligand (*RANKL*). When endogenous inhibitory molecules are removed, rOPs differentiate into osteoclasts in response to tumor necrosis factor α (*TNFα*). When the soluble form of RANKL is administered to RANKL-deficient mice, osteoclasts are induced in bone, but not in other tissues. By assuming rOPs, we may explain why osteoclasts form only in bone. *c-FMS*, M-CSF receptor.

prevent bone loss by adjusting the ITAM and WNT signals? What are the mechanisms integrating adhesion signals and cytokine signals to induce osteoclast activation? How do osteoblastic cells prepare rOPs in bone? These issues are expected to be clarified by future studies.

References

Abu-Amer, Y., Ross, F.P., Schlesinger, P., Tondravi, M.M., Teitelbaum, S.L., 1997. Substrate recognition by osteoclast precursors induces c-src/microtubule association. J. Cell Biol. 137, 247–258.

Akiyama, T., Shimo, Y., Yanai, H., Qin, J., Ohshima, D., Maruyama, Y., Asaumi, Y., Kitazawa, J., Takayanagi, H., Penninger, J.M., Matsumoto, M., Nitta, T., Takahama, Y., Inoue, J., 2008. The tumor necrosis factor family receptors RANK and CD40 cooperatively establish the thymic medullary microenvironment and self-tolerance. Immunity 29, 423–437.

Anderson, D.M., Maraskovsky, E., Billingsley, W.L., Dougall, W.C., Tometsko, M.E., Roux, E.R., Teepe, M.C., DuBose, R.F., Cosman, D., Galibert, L., 1997. A homologue of the TNF receptor and its ligand enhance T-cell growth and dendritic-cell function. Nature 390, 175–179.

Angata, T., Tabuchi, Y., Nakamura, K., Nakamura, M., 2007. Siglec-15: an immune system Siglec conserved throughout vertebrate evolution. Glycobiology 17, 838–846.

Arai, A., Mizoguchi, T., Harada, S., Kobayashi, Y., Nakamichi, Y., Yasuda, H., Penninger, J.M., Yamada, K., Udagawa, N., Takahashi, N., 2012. Fos plays an essential role in the upregulation of RANK expression in osteoclast precursors within the bone microenvironment. J. Cell Sci. 125, 2910–2917.

Barrow, A.D., Raynal, N., Andersen, T.L., Slatter, D.A., Bihan, D., Pugh, N., Cella, M., Kim, T., Rho, J., Negishi-Koga, T., et al., 2011. OSCAR is a collagen receptor that costimulates osteoclastogenesis in DAP12-deficient humans and mice. J. Clin. Invest. 121, 3505–3516.

Begg, S.K., Radley, J.M., Pollard, J.W., Chisholm, O.T., Stanley, E.R., Bertoncello, I., 1993. Delayed hematopoietic development in osteopetrotic (*op/op*) mice. J. Exp. Med. 177, 237–242.

Blair, H.C., Athanasou, N.A., 2004. Recent advances in osteoclast biology and pathological bone resorption. Histol. Histopathol. 19, 189–199.

Blair, H.C., Teitelbaum, S.L., Ghiselli, R., Gluck, S., 1989. Osteoclastic bone resorption by a polarized vacuolar proton pump. Science 245, 855–857.

Bossard, M.J., Tomaszek, T.A., Thompson, S.K., Amegadzie, B.Y., Hanning, C.R., Jones, C., Kurdyla, J.T., McNulty, D.E., Drake, F.H., Gowen, M., Levy, M.A., 1996. Proteolytic activity of human osteoclast cathepsin K. Expression, purification, activation, and substrate identification. J. Biol. Chem. 271, 12517–12524.

Boyce, B.F., Xiu, Y., Li, J., Xing, L., Yao, Z., 2015. NF-kappaB-Mediated regulation of osteoclastogenesis. Endocrinol. Metab. 30, 35–44.

Boyce, B.F., Yoneda, T., Lowe, C., Soriano, P., Mundy, G.R., 1992. Requirement of pp60c-src expression for osteoclasts to form ruffled borders and resorb bone in mice. J. Clin. Invest. 90, 1622–1627.

Cecchini, M.G., Hofstetter, W., Halasy, J., Wetterwald, A., Felix, R., 1994. Role of CSF-1 in bone and bone marrow development. Mol. Reprod. Dev. 46, 75–84.

Chambers, T.J., Magnus, C.J., 1982. Calcitonin alters behaviour of isolated osteoclasts. J. Pathol. 136, 27–39.

Chen, W., Zhu, G., Hao, L., Wu, M., Ci, H., Li, Y.P., 2013. C/EBPalpha regulates osteoclast lineage commitment. Proc. Natl. Acad. Sci. U.S.A. 110, 7294–7299.

Chiu, Y.H., Ritchlin, C.T., 2016. DC-stamp: a key regulator in osteoclast differentiation. J. Cell. Physiol. 231, 2402–2407.

Crockett, J.C., Mellis, D.J., Scott, D.I., Helfrich, M.H., 2011. New knowledge on critical osteoclast formation and activation pathways from study of rare genetic diseases of osteoclasts: focus on the RANK/RANKL axis. Osteoporos. Int. 22, 1–20.

Dai, X.M., Ryan, G.R., Hapel, A.J., Dominguez, M.G., Russell, R.G., Kapp, S., Sylvestre, V., Stanley, E.R., 2002. Targeted disruption of the mouse colony-stimulating factor 1 receptor gene results in osteopetrosis, mononuclear phagocyte deficiency, increased primitive progenitor cell frequencies, and reproductive defects. Blood 99, 111–120.

DeSelm, C.J., Miller, B.C., Zou, W., Beatty, W.L., van Meel, E., Takahata, Y., Klumperman, J., Tooze, S.A., Teitelbaum, S.L., Virgin, H.W., 2011. Autophagy proteins regulate the secretory component of osteoclastic bone resorption. Dev. Cell 21, 966–974.

Dougall, W.C., Glaccum, M., Charrier, K., Rohrbach, K., Brasel, K., De Smedt, T., Daro, E., Smith, J., Tometsko, M.E., Maliszewski, C.R., Armstrong, A., Shen, V., Bain, S., Cosman, D., Anderson, D., Morrissey, P.J., Peschon, J.J., Schuh, J., 1999. RANK is essential for osteoclast and lymph node development. Genes Dev. 13, 2412–2424.

Duong, L.T., Lakkakorpi, P.T., Nakamura, I., Machwate, M., Nagy, R.M., Rodan, G.A., 1998. PYK2 in osteoclasts is an adhesion kinase, localized in the sealing zone, activated by ligation of alpha(v)beta3 integrin, and phosphorylated by src kinase. J. Clin. Invest. 102, 881–892.

Faccio, R., Takeshita, S., Zallone, A., Ross, F.P., Teitelbaum, S.L., 2003. c-Fms and the alphavbeta3 integrin collaborate during osteoclast differentiation. J. Clin. Invest. 111, 749–758.

Faccio, R., Teitelbaum, S.L., Fujikawa, K., Chappel, J., Zallone, A., Tybulewicz, V.L., Ross, F.P., Swat, W., 2005. Vav3 regulates osteoclast function and bone mass. Nat. Med. 11, 284–290.

Fata, J.E., Kong, Y.Y., Li, J., Sasaki, T., Irie-Sasaki, J., Moorehead, R.A., Elliott, R., Scully, S., Voura, E.B., Lacey, D.L., Boyle, W.J., Khokha, R., Penninger, J.M., 2000. The osteoclast differentiation factor osteoprotegerin-ligand is essential for mammary gland development. Cell 103, 41–50.

Felix, R., Cecchini, M.G., Fleisch, H., 1990. Macrophage colony stimulating factor restores *in vivo* bone resorption in the op/op osteopetrotic mouse. Endocrinology 127, 2592–2594.

Feng, X., Takeshita, S., Namba, N., Wei, S., Teitelbaum, S.L., Ross, F.P., 2002. Tyrosines 559 and 807 in the cytoplasmic tail of the macrophage colony-stimulating factor receptor play distinct roles in osteoclast differentiation and function. Endocrinology 143, 4868–4874.

Feng, X., Teitelbaum, S.L., 2013. Osteoclasts: new insights. Bone Res 1, 11–26.

Fukuda, C., Tsuda, E., Okada, A., Amizuka, N., Hasegawa, T., Karibe, T., Hiruma, Y., Takagi, N., Kumakura, S., 2017. Anti-Siglec-15 antibody reduces bone resorption while maintaining bone formation in ovariectomized (OVX) rats and monkeys. In: 2017 ASBMR Annual Meeting, Abstract S112.

Glass 2nd, D.A., Bialek, P., Ahn, J.D., Starbuck, M., Patel, M.S., Clevers, H., Taketo, M.M., Long, F., McMahon, A.P., Lang, R.A., Karsenty, G., 2005. Canonical Wnt signaling in differentiated osteoblasts controls osteoclast differentiation. Dev. Cell 8, 751–764.

Gohda, J., Akiyama, T., Koga, T., Takayanagi, H., Tanaka, S., Inoue, J., 2005. RANK-mediated amplification of TRAF6 signaling leads to NFATc1 induction during osteoclastogenesis. EMBO J. 24, 790–799.

Gori, F., Lerner, U., Ohlsson, C., Baron, R., 2015. A new WNT on the bone: WNT16, cortical bone thickness, porosity and fractures. BoneKEy Rep. 4, 669.

Hamilton, J.A., 1997. CSF-1 signal transduction. J. Leukoc. Biol. 62, 145–155.

Harre, U., Lang, S.C., Pfeifle, R., Rombouts, Y., Fruhbeisser, S., Amara, K., Bang, H., Lux, A., Koeleman, C.A., Baum, W., Dietel, K., Gröhn, F., Malmström, V., Klareskog, L., Krönke, G., Kocijan, R., Nimmerjahn, F., Toes, R.E., Herrmann, M., Scherer, H.U., Schett, G., 2015. Glycosylation of immunoglobulin G determines osteoclast differentiation and bone loss. Nat. Commun. 6, 6651.

Hiruma, Y., Hirai, T., Tsuda, E., 2011. Siglec-15, a member of the sialic acid-binding lectin, is a novel regulator for osteoclast differentiation. Biochem. Biophys. Res. Commun. 409, 424–429.

Hiruma, Y., Tsuda, E., Maeda, N., Okada, A., Kabasawa, N., Miyamoto, M., Hattori, H., Fukuda, C., 2013. Impaired osteoclast differentiation and function and mild osteopetrosis development in Siglec-15-deficient mice. Bone 53, 87–93.

Horton, M.A., Lewis, D., McNulty, K., Pringle, J.A., Chambers, T.J., 1985. Monoclonal antibodies to osteoclastomas (giant cell bone tumors): definition of osteoclast-specific cellular antigens. Cancer Res. 45, 5663–5669.

Humphrey, M.B., Nakamura, M.C., 2016. A comprehensive review of immunoreceptor regulation of osteoclasts. Clin. Rev. Allergy Immunol. 51, 48–58.

Insogna, K.L., Sahni, M., Grey, A.B., Tanaka, S., Horne, W.C., Neff, L., Mitnick, M., Levy, J.B., Baron, R., 1997. Colony-stimulating factor-1 induces cytoskeletal reorganization and c-src-dependent tyrosine phosphorylation of selected cellular proteins in rodent osteoclasts. J. Clin. Investig. 100, 2476–2485.

Ishida-Kitagawa, N., Tanaka, K., Bao, X., Kimura, T., Miura, T., Kitaoka, Y., Hayashi, K., Sato, M., Maruoka, M., Ogawa, T., Miyoshi, J., Takeya, T., 2012. Siglec-15 protein regulates formation of functional osteoclasts in concert with DNAX-activating protein of 12 kDa (DAP12). J. Biol. Chem. 287, 17493–17502.

Ishida, N., Hayashi, K., Hoshijima, M., Ogawa, T., Koga, S., Miyatake, Y., Kumegawa, M., Kimura, T., Takeya, T., 2002. Large scale gene expression analysis of osteoclastogenesis *in vitro* and elucidation of NFAT2 as a key regulator. J. Biol. Chem. 277, 41147–41156.

Ishii, M., Egen, J.G., Klauschen, F., Meier-Schellersheim, M., Saeki, Y., Vacher, J., Proia, R.L., Germain, R.N., 2009. Sphingosine-1-phosphate mobilizes osteoclast precursors and regulates bone homeostasis. Nature 458, 524–528.

Jimi, E., Nakamura, I., Duong, L.T., Ikebe, T., Takahashi, N., Rodan, G.A., Suda, T., 1999. Interleukin 1 induces multinucleation and bone-resorbing activity of osteoclasts in the absence of osteoblasts/stromal cells. Exp. Cell Res. 247, 84–93.

Jurdic, P., Saltel, F., Chabadel, A., Destaing, O., 2006. Podosome and sealing zone: specificity of the osteoclast model. Eur. J. Cell Biol. 85, 195–202.

Kameda, Y., Takahata, M., Komatsu, M., Mikuni, S., Hatakeyama, S., Shimizu, T., Angata, T., Kinjo, M., Minami, A., Iwasaki, N., 2013. Siglec-15 regulates osteoclast differentiation by modulating RANKL-induced phosphatidylinositol 3-kinase/Akt and Erk pathways in association with signaling Adaptor DAP12. J. Bone Miner. Res. 28, 2463–2475.

Kameda, Y., Takahata, M., Mikuni, S., Shimizu, T., Hamano, H., Angata, T., Hatakeyama, S., Kinjo, M., Iwasaki, N., 2015. Siglec-15 is a potential therapeutic target for postmenopausal osteoporosis. Bone 71, 217–226.

Kanaya, T., Hase, K., Takahashi, D., Fukuda, S., Hoshino, K., Sasaki, I., Hemmi, H., Knoop, K.A., Kumar, N., Sato, M., Katsuno, T., Yokosuka, O., Toyooka, K., Nakai, K., Sakamoto, A., Kitahara, Y., Jinnohara, T., McSorley, S.J., Kaisho, T., Williams, I.R., Ohno, H., 2012. The Ets transcription factor Spi-B is essential for the differentiation of intestinal microfold cells. Nat. Immunol. 13, 729–736.

Kim, N., Kadono, Y., Takami, M., Lee, J., Lee, S.H., Okada, F., Kim, J.H., Kobayashi, T., Odgren, P.R., Nakano, H., Yeh, W.C., Lee, S.K., Lorenzo, J.A., Choi, Y., 2005. Osteoclast differentiation independent of the TRANCE-RANK-TRAF6 axis. J. Exp. Med. 202, 589–595.

Kim, N., Takami, M., Rho, J., Josien, R., Choi, Y., 2002. A novel member of the leukocyte receptor complex regulates osteoclast differentiation. J. Exp. Med. 195, 201–209.

Knoop, K.A., Kumar, N., Butler, B.R., Sakthivel, S.K., Taylor, R.T., Nochi, T., Akiba, H., Yagita, H., Kiyono, H., Williams, I.R., 2009. RANKL is necessary and sufficient to initiate development of antigen-sampling M cells in the intestinal epithelium. J. Immunol. 183, 5738–5747.

Kobayashi, K., Takahashi, N., Jimi, E., Udagawa, N., Takami, M., Kotake, S., Nakagawa, N., Kinosaki, M., Yamaguchi, K., Shima, N., Yasuda, H., Morinaga, T., Higashio, K., Martin, T.J., Suda, T., 2000. Tumor necrosis factor alpha stimulates osteoclast differentiation by a mechanism independent of the ODF/RANKL-RANK interaction. J. Exp. Med. 191, 275–286.

Kobayashi, Y., Thirukonda, G.J., Nakamura, Y., Koide, M., Yamashita, T., Uehara, S., Kato, H., Udagawa, N., Takahashi, N., 2015. Wnt16 regulates osteoclast differentiation in conjunction with Wnt5a. Biochem. Biophys. Res. Commun. 463, 1278–1283.

Kobayashi, Y., Udagawa, N., Takahashi, N., 2009. Action of RANKL and OPG for osteoclastogenesis. Crit. Rev. Eukaryot. Gene Expr. 19, 61–72.

Kobayashi, Y., Uehara, S., Udagawa, N., Takahashi, N., 2016. Regulation of bone metabolism by Wnt signals. J. Biochem. 159, 387–392.

Kodama, H., Nose, M., Niida, S., Yamasaki, A., 1991a. Essential role of macrophage colony-stimulating factor in the osteoclast differentiation supported by stromal cells. J. Exp. Med. 173, 1291–1294.

Kodama, H., Yamasaki, A., Nose, M., Niida, S., Ohgame, Y., Abe, M., Kumegawa, M., Suda, T., 1991b. Congenital osteoclast deficiency in osteopetrotic (op/op) mice is cured by injections of macrophage colony-stimulating factor. J. Exp. Med. 173, 269–272.

Koga, T., Inui, M., Inoue, K., Kim, S., Suematsu, A., Kobayashi, E., Iwata, T., Ohnishi, H., Matozaki, T., Kodama, T., Taniguchi, T., Takayanagi, H., Takai, T., 2004. Costimulatory signals mediated by the ITAM motif cooperate with RANKL for bone homeostasis. Nature 428, 758–763.

Kong, Y.Y., Yoshida, H., Sarosi, I., Tan, H.L., Timms, E., Capparelli, C., Morony, S., Oliveira-dos-Santos, A.J., Van, G., Itie, A., Khoo, W., Wakeham, A., Dunstan, C.R., Lacey, D.L., Mak, T.W., Boyle, W.J., Penninger, J.M., 1999. OPGL is a key regulator of osteoclastogenesis, lymphocyte development and lymph-node organogenesis. Nature 397, 315–323.

Kornak, U., Kasper, D., Bosl, M.R., Kaiser, E., Schweizer, M., Schulz, A., Friedrich, W., Delling, G., Jentsch, T.J., 2001. Loss of the ClC-7 chloride channel leads to osteopetrosis in mice and man. Cell 104, 205–215.

Kukita, T., Wada, N., Kukita, A., Kakimoto, T., Sandra, F., Toh, K., Nagata, K., Iijima, T., Horiuchi, M., Matsusaki, H., Hieshima, K., Yoshie, O., Nomiyama, H., 2004. RANKL-induced DC-STAMP is essential for osteoclastogenesis. J. Exp. Med. 200, 941–946.

Kuroda, Y., Hisatsune, C., Nakamura, T., Matsuo, K., Mikoshiba, K., 2008. Osteoblasts induce Ca^{2+} oscillation-independent NFATc1 activation during osteoclastogenesis. Proc. Natl. Acad. Sci. U.S.A. 105, 8643–8648.

Lacey, D.L., Boyle, W.J., Simonet, W.S., Kostenuik, P.J., Dougall, W.C., Sullivan, J.K., San Martin, J., Dansey, R., 2012. Bench to bedside: elucidation of the OPG-RANK-RANKL pathway and the development of denosumab. Nat. Rev. Drug Discov. 11, 401–419.

Lacey, D.L., Timms, E., Tan, H.L., Kelley, M.J., Dunstan, C.R., Burgess, T., Elliott, R., Colombero, A., Elliott, G., Scully, S., Hsu, H., Sullivan, J., Hawkins, N., Davy, E., Capparelli, C., Eli, A., Qian, Y.X., Kaufman, S., Sarosi, I., Shalhoub, V., Senaldi, G., Guo, J., Delaney, J., Boyle, W.J., 1998. Osteoprotegerin ligand is a cytokine that regulates osteoclast differentiation and activation. Cell 93, 165–176.

Lakkakorpi, P.T., Nakamura, I., Nagy, R.M., Parsons, J.T., Rodan, G.A., Duong, L.T., 1999. Stable association of PYK2 and p130(Cas) in osteoclasts and their co-localization in the sealing zone. J. Biol. Chem. 274, 4900–4907.

Lange, P.F., Wartosch, L., Jentsch, T.J., Fuhrmann, J.C., 2006. ClC-7 requires Ostm1 as a beta-subunit to support bone resorption and lysosomal function. Nature 440, 220–223.

Lee, S.H., Kim, T., Jeong, D., Kim, N., Choi, Y., 2008. The tec family tyrosine kinase Btk Regulates RANKL-induced osteoclast maturation. J. Biol. Chem. 283, 11526–11534.

Li, Y.P., Chen, W., Liang, Y., Li, E., Stashenko, P., 1999. Atp6i-deficient mice exhibit severe osteopetrosis due to loss of osteoclast-mediated extracellular acidification. Nat. Genet. 23, 447–451.

Lin, H., Lee, E., Hestir, K., Leo, C., Huang, M., Bosch, E., Halenbeck, R., Wu, G., Zhou, A., Behrens, D., Hollenbaugh, D., Linnemann, T., Qin, M., Wong, J., Chu, K., Doberstein, S.K., Williams, L.T., 2008. Discovery of a cytokine and its receptor by functional screening of the extracellular proteome. Science 320, 807–811.

Lomaga, M.A., Yeh, W.C., Sarosi, I., Duncan, G.S., Furlonger, C., Ho, A., Morony, S., Capparelli, C., Van, G., Kaufman, S., van der Heiden, A., Itie, A., Wakeham, A., Khoo, W., Sasaki, T., Cao, Z., Penninger, J.M., Paige, C.J., Lacey, D.L., Dunstan, C.R., Boyle, W.J., Goeddel, D.V., Mak, T.W., 1999. TRAF6 deficiency results in osteopetrosis and defective interleukin-1, CD40, and LPS signaling. Genes Dev. 13, 1015–1024.

Maeda, K., Kobayashi, Y., Udagawa, N., Uehara, S., Ishihara, A., Mizoguchi, T., Kikuchi, Y., Takada, I., Kato, S., Kani, S., Nishita, M., Marumo, K., Martin, T.J., Minami, Y., Takahashi, N., 2012. Wnt5a-Ror2 signaling between osteoblast-lineage cells and osteoclast precursors enhances osteoclastogenesis. Nat. Med. 18, 405–412.

Mandal, C.C., Ghosh Choudhury, G., Ghosh-Choudhury, N., 2009. Phosphatidylinositol 3 kinase/Akt signal relay cooperates with smad in bone morphogenetic protein-2-induced colony stimulating factor-1 (CSF-1) expression and osteoclast differentiation. Endocrinology 150, 4989–4998.

Martin, T.J., Sims, N.A., 2015. Calcitonin physiology, saved by a lysophospholipid. J. Bone Miner. Res. 30, 212–215.

McHugh, K.P., Hodivala-Dilke, K., Zheng, M.H., Namba, N., Lam, J., Novack, D., Feng, X., Ross, F.P., Hynes, R.O., Teitelbaum, S.L., 2000. Mice lacking beta3 integrins are osteosclerotic because of dysfunctional osteoclasts. J. Clin. Investig. 105, 433–440.

Mikels, A., Minami, Y., Nusse, R., 2009. Ror2 receptor requires tyrosine kinase activity to mediate Wnt5A signaling. J. Biol. Chem. 284, 30167–30176.

Minkin, C., 1982. Bone acid phosphatase: tartrate-resistant acid phosphatase as a marker of osteoclast function. Calcif. Tissue Int. 34, 285–290.

Miyamoto, H., Katsuyama, E., Miyauchi, Y., Hoshi, H., Miyamoto, K., Sato, Y., Kobayashi, T., Iwasaki, R., Yoshida, S., Mori, T., Kanagawa, H., Fujie, A., Hao, W., Morioka, H., Matsumoto, M., Toyama, Y., Miyamoto, T., 2012a. An essential role for STAT6-STAT1 protein signaling in promoting macrophage cell-cell fusion. J. Biol. Chem. 287, 32479–32484.

Miyamoto, H., Suzuki, T., Miyauchi, Y., Iwasaki, R., Kobayashi, T., Sato, Y., Miyamoto, K., Hoshi, H., Hashimoto, K., Yoshida, S., Hao, W., Mori, T., Kanagawa, H., Katsuyama, E., Fujie, A., Morioka, H., Matsumoto, M., Chiba, K., Takeya, M., Toyama, Y., Miyamoto, T., 2012b. Osteoclast stimulatory transmembrane protein and dendritic cell-specific transmembrane protein cooperatively modulate cell-cell fusion to form osteoclasts and foreign body giant cells. J. Bone Miner. Res. 27, 1289–1297.

Miyauchi, Y., Ninomiya, K., Miyamoto, H., Sakamoto, A., Iwasaki, R., Hoshi, H., Miyamoto, K., Hao, W., Yoshida, S., Morioka, H., Chiba, K., Kato, S., Tokuhisa, T., Saitou, M., Toyama, Y., Suda, T., Miyamoto, T., 2010. The Blimp1-Bcl6 axis is critical to regulate osteoclast differentiation and bone homeostasis. J. Exp. Med. 207, 751–762.

Mizoguchi, T., Muto, A., Udagawa, N., Arai, A., Yamashita, T., Hosoya, A., Ninomiya, T., Nakamura, H., Yamamoto, Y., Kinugawa, S., Oda, K., Tanaka, H., Tagaya, M., Penninger, J.M., Ito, M., Takahashi, N., 2009. Identification of cell cycle-arrested quiescent osteoclast precursors *in vivo*. J. Cell Biol. 184, 541–554.

Mocsai, A., Humphrey, M.B., Van Ziffle, J.A., Hu, Y., Burghardt, A., Spusta, S.C., Majumdar, S., Lanier, L.L., Lowell, C.A., Nakamura, M.C., 2004. The immunomodulatory adapter proteins DAP12 and Fc receptor gamma-chain (FcRgamma) regulate development of functional osteoclasts through the Syk tyrosine kinase. Proc. Natl. Acad. Sci. U.S.A. 101, 6158–6163.

Moverare-Skrtic, S., Henning, P., Liu, X., Nagano, K., Saito, H., Borjesson, A.E., Sjogren, K., Windahl, S.H., Farman, H., Kindlund, B., Engdahl, C., Koskela, A., Zhang, F.P., Eriksson, E.E., Zaman, F., Hammarstedt, A., Isaksson, H., Bally, M., Kassem, A., Lindholm, C., Sandberg, O., Aspenberg, P., Sävendahl, L., Feng, J.Q., Tuckermann, J., Tuukkanen, J., Poutanen, M., Baron, R., Lerner, U.H., Gori, F., Ohlsson, C., 2014. Osteoblast-derived WNT16 represses osteoclastogenesis and prevents cortical bone fragility fractures. Nat. Med. 20, 1279–1288.

Mulari, M.T., Zhao, H., Lakkakorpi, P.T., Vaananen, H.K., 2003. Osteoclast ruffled border has distinct subdomains for secretion and degraded matrix uptake. Traffic 4, 113–125.

Muto, A., Mizoguchi, T., Udagawa, N., Ito, S., Kawahara, I., Abiko, Y., Arai, A., Harada, S., Kobayashi, Y., Nakamichi, Y., Penninger, J.M., Noguchi, T., Takahashi, N., 2011. Lineage-committed osteoclast precursors circulate in blood and settle down into bone. J. Bone Miner. Res. 26, 2978–2990.

Nagashima, K., Sawa, S., Nitta, T., Tsutsumi, M., Okamura, T., Penninger, J.M., Nakashima, T., Takayanagi, H., 2017. Identification of subepithelial mesenchymal cells that induce IgA and diversify gut microbiota. Nat. Immunol. 18, 675–682.

Naito, A., Azuma, S., Tanaka, S., Miyazaki, T., Takaki, S., Takatsu, K., Nakao, K., Nakamura, K., Katsuki, M., Yamamoto, T., Inoue, J., 1999. Severe osteopetrosis, defective interleukin-1 signalling and lymph node organogenesis in TRAF6-deficient mice. Genes Cells 4, 353–362.

Nakamichi, Y., Mizoguchi, T., Arai, A., Kobayashi, Y., Sato, M., Penninger, J.M., Yasuda, H., Kato, S., DeLuca, H.F., Suda, T., Udagawa, N., Takahashi, N., 2012. Spleen serves as a reservoir of osteoclast precursors through vitamin D-induced IL-34 expression in osteopetrotic op/op mice. Proc. Natl. Acad. Sci. U.S.A. 109, 10006–10011.

Nakamichi, Y., Udagawa, N., Takahashi, N., 2013. IL-34 and CSF-1: similarities and differences. J. Bone Miner. Metab. 31, 486–495.

Nakamura, I., Jimi, E., Duong, L.T., Sasaki, T., Takahashi, N., Rodan, G.A., Suda, T., 1998. Tyrosine phosphorylation of p130Cas is involved in actin organization in osteoclasts. J. Biol. Chem. 273, 11144–11149.

Nakamura, I., Takahashi, N., Jimi, E., Udagawa, N., Suda, T., 2012. Regulation of osteoclast function. Mod. Rheumatol. 22, 167–177.

Nakashima, T., Hayashi, M., Fukunaga, T., Kurata, K., Oh-Hora, M., Feng, J.Q., Bonewald, L.F., Kodama, T., Wutz, A., Wagner, E.F., Penninger, J.M., Takayanagi, H., 2011. Evidence for osteocyte regulation of bone homeostasis through RANKL expression. Nat. Med. 17, 1231–1234.

Negishi-Koga, T., Gober, H.J., Sumiya, E., Komatsu, N., Okamoto, K., Sawa, S., Suematsu, A., Suda, T., Sato, K., Takai, T., Takayanagi, H., 2015. Immune complexes regulate bone metabolism through FcRgamma signalling. Nat. Commun. 6, 6637.

Nesbitt, S.A., Horton, M.A., 1997. Trafficking of matrix collagens through bone-resorbing osteoclasts. Science 276, 266–269.

Nicholson, G.C., Moseley, J.M., Sexton, P.M., Mendelsohn, F.A., Martin, T.J., 1986. Abundant calcitonin receptors in isolated rat osteoclasts. Biochemical and autoradiographic characterization. J. Clin. Investig. 78, 355–360.

Nishikawa, K., Nakashima, T., Hayashi, M., Fukunaga, T., Kato, S., Kodama, T., Takahashi, S., Calame, K., Takayanagi, H., 2010. Blimp1-mediated repression of negative regulators is required for osteoclast differentiation. Proc. Natl. Acad. Sci. U.S.A. 107, 3117–3122.

Oikawa, T., Kuroda, Y., Matsuo, K., 2013. Regulation of osteoclasts by membrane-derived lipid mediators. Cell. Mol. Life Sci. 70, 3341–3353.

Okamoto, K., Nakashima, T., Shinohara, M., Negishi-Koga, T., Komatsu, N., Terashima, A., Sawa, S., Nitta, T., Takayanagi, H., 2017. Osteoimmunology: the conceptual framework unifying the immune and skeletal systems. Physiol. Rev. 97, 1295–1349.

Okazaki, R., Toriumi, M., Fukumoto, S., Miyamoto, M., Fujita, T., Tanaka, K., Takeuchi, Y., 1999. Thiazolidinediones inhibit osteoclast-like cell formation and bone resorption *in vitro*. Endocrinology 140, 5060–5065.

Ramirez, A., Faupel, J., Goebel, I., Stiller, A., Beyer, S., Stockle, C., Hasan, C., Bode, U., Kornak, U., Kubisch, C., 2004. Identification of a novel mutation in the coding region of the grey-lethal gene OSTM1 in human malignant infantile osteopetrosis. Hum. Mutat. 23, 471–476.

Reeve, J.L., Zou, W., Liu, Y., Maltzman, J.S., Ross, F.P., Teitelbaum, S.L., 2009. SLP-76 couples Syk to the osteoclast cytoskeleton. J. Immunol. 183, 1804–1812.

Reponen, P., Sahlberg, C., Munaut, C., Thesleff, I., Tryggvason, K., 1994. High expression of 92-kD type IV collagenase (gelatinase B) in the osteoclast lineage during mouse development. J. Biol. Chem. 124, 1091–1102.

Rodan, G.A., Martin, T.J., 1981. Role of osteoblasts in hormonal control of bone resorption–a hypothesis. Calcif. Tissue Int. 33, 349–351.

Rossi, S.W., Kim, M.Y., Leibbrandt, A., Parnell, S.M., Jenkinson, W.E., Glanville, S.H., McConnell, F.M., Scott, H.S., Penninger, J.M., Jenkinson, E.J., Lane, P.J., Anderson, G., 2007. RANK signals from CD4(+)3(-) inducer cells regulate development of Aire-expressing epithelial cells in the thymic medulla. J. Exp. Med. 204, 1267–1272.

Saftig, P., Hunziker, E., Wehmeyer, O., Jones, S., Boyde, A., Rommerskirch, W., Moritz, J.D., Schu, P., von Figura, K., 1998. Impaired osteoclastic bone resorption leads to osteopetrosis in cathepsin-K-deficient mice. Proc. Natl. Acad. Sci. U.S.A. 95, 13453–13458.

Salo, J., Lehenkari, P., Mulari, M., Metsikko, K., Vaananen, H.K., 1997. Removal of osteoclast bone resorption products by transcytosis. Science 276, 270–273.

Sato, K., Suematsu, A., Nakashima, T., Takemoto-Kimura, S., Aoki, K., Morishita, Y., Asahara, H., Ohya, K., Yamaguchi, A., Takai, T., Kodama, T., Chatila, T.A., Bito, H., Takayanagi, H., 2006. Regulation of osteoclast differentiation and function by the CaMK-CREB pathway. Nat. Med. 12, 1410–1416.

Sexton, P.M., Findlay, D.M., Martin, T.J., 1999. Curr. Med. Chem. 6, 1067–1093.

Shimizu, T., Takahata, M., Kameda, Y., Endo, T., Hamano, H., Hiratsuka, S., Ota, M., Iwasaki, N., 2015. Sialic acid-binding immunoglobulin-like lectin 15 (Siglec-15) mediates periarticular bone loss, but not joint destruction, in murine antigen-induced arthritis. Bone 79, 65–70.

Shinohara, M., Koga, T., Okamoto, K., Sakaguchi, S., Arai, K., Yasuda, H., Takai, T., Kodama, T., Morio, T., Geha, R.S., Kitamura, D., Kurosaki, T., Ellmeier, W., Takayanagi, H., 2008. Tyrosine kinases Btk and Tec regulate osteoclast differentiation by linking RANK and ITAM signals. Cell 132, 794–806.

Simonet, W.S., Lacey, D.L., Dunstan, C.R., Kelley, M., Chang, M.S., Luthy, R., Nguyen, H.Q., Wooden, S., Bennett, L., Boone, T., Shimamoto, G., DeRose, M., Elliott, R., Colombero, A., Tan, H.L., Trail, G., Sullivan, J., Davy, E., Bucay, N., Renshaw-Gegg, L., Hughes, T.M., Hill, D., Pattison, W., Campbell, P., Sander, S., Van, G., Tarpley, J., Derby, P., Lee, R., Boyle, W.J., 1997. Osteoprotegerin: a novel secreted protein involved in the regulation of bone density. Cell 89, 309–319.

Sly, W.S., Whyte, M.P., Sundaram, V., Tashian, R.E., Hewett-Emmett, D., Guibaud, P., Vainsel, M., Baluarte, H.J., Gruskin, A., Al-Mosawi, M., Sakati, N., Ohlsson, A., 1985. Carbonic anhydrase II deficiency in 12 families with the autosomal recessive syndrome of osteopetrosis with renal tubular acidosis and cerebral calcification. N. Engl. J. Med. 313, 139–145.

Soriano, P., Montgomery, C., Geske, R., Bradley, A., 1991. Targeted disruption of the c-src proto-oncogene leads to osteopetrosis in mice. Cell 64, 693–702.

Stanley, E.R., Chitu, V., 2014. CSF-1 receptor signaling in myeloid cells. Cold Spring Harb. Perspect. Biol 6, a021857.

Stuible, M., Moraitis, A., Fortin, A., Saragosa, S., Kalbakji, A., Filion, M., Tremblay, G.B., 2014. Mechanism and function of monoclonal antibodies targeting siglec-15 for therapeutic inhibition of osteoclastic bone resorption. J. Biol. Chem. 289, 6498–6512.

Suda, T., Takahashi, N., Martin, T.J., 1992. Modulation of osteoclast differentiation. Endocr. Rev. 13, 66–80.

Suzuki, H., Nakamura, I., Takahashi, N., Ikuhara, T., Matsuzaki, K., Isogai, Y., Hori, M., Suda, T., 1996. Calcitonin-induced changes in the cytoskeleton are mediated by a signal pathway associated with protein kinase A in osteoclasts. Endocrinology 137, 4685–4690.

Takahashi, N., Akatsu, T., Udagawa, N., Sasaki, T., Yamaguchi, A., Moseley, J.M., Martin, T.J., Suda, T., 1988. Osteoblastic cells are involved in osteoclast formation. Endocrinology 123, 2600–2602.

Takahashi, N., Muto, A., Arai, A., Mizoguchi, T., 2010. Identification of cell cycle-arrested quiescent osteoclast precursors *in vivo*. Adv. Exp. Med. Biol. 658, 21–30.

Takahashi, N., Udagawa, N., Akatsu, T., Tanaka, H., Isogai, Y., Suda, T., 1991. Deficiency of osteoclasts in osteopetrotic mice is due to a defect in the local microenvironment provided by osteoblastic cells. Endocrinology 128, 1792–1796.

Takayanagi, H., 2007. Osteoimmunology: shared mechanisms and crosstalk between the immune and bone systems. Nat. Rev. Immunol. 7, 292–304.

Takayanagi, H., Kim, S., Koga, T., Nishina, H., Isshiki, M., Yoshida, H., Saiura, A., Isobe, M., Yokochi, T., Inoue, J., Wagner, E.F., Mak, T.W., Kodama, T., Taniguchi, T., 2002. Induction and activation of the transcription factor NFATc1 (NFAT2) integrate RANKL signaling in terminal differentiation of osteoclasts. Dev. Cell 3, 889–901.

Tanaka, S., Suzuki, H., Yamauchi, H., Nakamura, I., Nakamura, K., 2006. Signal transduction pathways of calcitonin/calcitonin receptor regulating cytoskeletal organization and bone-resorbing activity of osteoclasts. Cell. Mol. Biol. 52, 19–23.

Tanaka, S., Takahashi, N., Udagawa, N., Sasaki, T., Fukui, Y., Kurokawa, T., Suda, T., 1992. Osteoclasts express high levels of $p60^{c\text{-}src}$, preferentially on ruffled border membranes. FEBS Lett. 313, 85–89.

Tanaka, S., Takahashi, N., Udagawa, N., Tamura, T., Akatsu, T., Stanley, E.R., Kurokawa, T., Suda, T., 1993. Macrophage colony-stimulating factor is indispensable for both proliferation and differentiation of osteoclast progenitors. J. Clin. Invest. 91, 257–263.

Teitelbaum, S.L., 2007. Osteoclasts: what do they do and how do they do it? Am. J. Pathol. 170, 427–435.

Teitelbaum, S.L., 2011. The osteoclast and its unique cytoskeleton. Ann. N.Y. Acad. Sci. 1240, 14–17.

Teitelbaum, S.L., Ross, F.P., 2003. Genetic regulation of osteoclast development and function. Nat. Rev. Genet. 4, 638–649.

Teti, A., Blair, H.C., Teitelbaum, S.L., Kahn, A.J., Koziol, C., Konsek, J., Zambonin-Zallone, A., Schlesinger, P.H., 1989. Cytoplasmic pH regulation and chloride/bicarbonate exchange in avian osteoclasts. J. Clin. Invest. 83, 227–233.

Tezuka, K., Nemoto, K., Tezuka, Y., Sato, T., Ikeda, Y., Kobori, M., Kawashima, H., Eguchi, H., Hakeda, Y., Kumegawa, M., 1994a. Identification of matrix metalloproteinase 9 in rabbit osteoclasts. J. Biol. Chem. 269, 15006–15009.

Tezuka, K., Tezuka, Y., Maejima, A., Sato, T., Nemoto, K., Kamioka, H., Hakeda, Y., Kumegawa, M., 1994b. Molecular cloning of a possible cysteine proteinase predominantly expressed in osteoclasts. J. Biol. Chem. 269, 1106–1109.

Thomson, B.M., Saklatvala, J., Chambers, T.J., 1986. Osteoblasts mediate interleukin 1 stimulation of bone resorption by rat osteoclasts. J. Exp. Med. 164, 104–112.

Tondravi, M.M., McKercher, S.R., Anderson, K., Erdmann, J.M., Quiroz, M., Maki, R., Teitelbaum, S.L., 1997. Osteopetrosis in mice lacking haematopoietic transcription factor PU.1. Nature 386, 81–84.

Tsuda, E., Goto, M., Mochizuki, S., Yano, K., Kobayashi, F., Morinaga, T., Higashio, K., 1997. Isolation of a novel cytokine from human fibroblasts that specifically inhibits osteoclastogenesis. Biochem. Biophys. Res. Commun. 234, 137–142.

Udagawa, N., Takahashi, N., Akatsu, T., Tanaka, H., Sasaki, T., Nishihara, T., Koga, T., Martin, T.J., Suda, T., 1990. Origin of osteoclasts: mature monocytes and macrophages are capable of differentiating into osteoclasts under a suitable microenvironment prepared by bone marrow-derived stromal cells. Proc. Natl. Acad. Sci. U.S.A. 87, 7260–7264.

Uehara, S., Udagawa, N., Mukai, H., Ishihara, A., Maeda, K., Yamashita, T., Murakami, K., Nishita, M., Nakamura, T., Kato, S., Minami, Y., Takahashi, N., Kobayashi, Y., 2017. Protein kinase N3 promotes bone resorption by osteoclasts in response to Wnt5a-Ror2 signaling. Sci. Signal. 10, 494.

Vaananen, H.K., Laitala-Leinonen, T., 2008. Osteoclast lineage and function. Arch. Biochem. Biophys. 473, 132–138.

Wagner, E.F., Eferl, R., 2005. Fos/AP-1 proteins in bone and the immune system. Immunol. Rev. 208, 126–140.

Walsh, M.C., Lee, J., Choi, Y., 2015. Tumor necrosis factor receptor- associated factor 6 (TRAF6) regulation of development, function, and homeostasis of the immune system. Immunol. Rev. 266, 72–92.

Wan, Y., Chong, L.W., Evans, R.M., 2007. PPAR-gamma regulates osteoclastogenesis in mice. Nat. Med. 13, 1496–1503.

Wei, W., Zeve, D., Suh, J.M., Wang, X., Du, Y., Zerwekh, J.E., Dechow, P.C., Graff, J.M., Wan, Y., 2011. Biphasic and dosage-dependent regulation of osteoclastogenesis by beta-catenin. Mol. Cell Biol. 31, 4706–4719.

Weivoda, M.M., Ruan, M., Hachfeld, C.M., Pederson, L., Howe, A., Davey, R.A., Zajac, J.D., Kobayashi, Y., Williams, B.O., Westendorf, J.J., Khosla, S., Oursler, M.J., 2016. Wnt signaling inhibits osteoclast differentiation by activating canonical and noncanonical cAMP/PKA pathways. J. Bone Miner. Res. 31, 65–75.

Wiktor-Jedrzejczak, W., Bartocci, A., Ferrante Jr., A.W., Ahmed-Ansari, A., Sell, K.W., Pollard, J.W., Stanley, E.R., 1990. Total absence of colony-stimulating factor 1 in the macrophage-deficient osteopetrotic (op/op) mouse. Proc. Natl. Acad. Sci. U.S.A. 87, 4828–4832.

Wiktor-Jedrzejczak, W.W., Ahmed, A., Szczylik, C., Skelly, R.R., 1982. Hematological characterization of congenital osteopetrosis in op/op mouse. Possible mechanism for abnormal macrophage differentiation. J. Exp. Med. 156, 1516–1527.

Wong, B.R., Josien, R., Lee, S.Y., Vologodskaia, M., Steinman, R.M., Choi, Y., 1998. The TRAF family of signal transducers mediates NF-kappaB activation by the TRANCE receptor. J. Biol. Chem. 273, 28355–28359.

Wong, B.R., Rho, J., Arron, J., Robinson, E., Orlinick, J., Chao, M., Kalachikov, S., Cayani, E., Bartlett 3rd, F.S., Frankel, W.N., Lee, S.Y., Choi, Y., 1997. TRANCE is a novel ligand of the tumor necrosis factor receptor family that activates c-Jun N-terminal kinase in T cells. J. Biol. Chem. 272, 25190–25194.

Wu, J., Glimcher, L.H., Aliprantis, A.O., 2008. HCO_3^-/Cl^- anion exchanger SLC4A2 is required for proper osteoclast differentiation and function. Proc. Natl. Acad. Sci. U.S.A. 105, 16934–16939.

Xiong, J., Onal, M., Jilka, R.L., Weinstein, R.S., Manolagas, S.C., O'Brien, C.A., 2011. Matrix-embedded cells control osteoclast formation. Nat. Med. 17, 1235–1241.

Xiong, J., Cawley, K., Piemontese, M., Fujiwara, Y., Macleod, R., Goellner, J., Zhao, H., O'Brien, C., 2017. The soluble form of RANKL contributes to cancellous bone remodeling in adult mice but is dispensable for ovariectomy-induced bone loss. In: 2017 ASBMR Annul Meetinb, Abstract S14.

Xiu, Y., Xu, H., Zhao, C., Li, J., Morita, Y., Yao, Z., Xing, L., Boyce, B.F., 2014. Chloroquine reduces osteoclastogenesis in murine osteoporosis by preventing TRAF3 degradation. J. Clin. Investig. 124, 297–310.

Yagi, M., Miyamoto, T., Sawatani, Y., Iwamoto, K., Hosogane, N., Fujita, N., Morita, K., Ninomiya, K., Suzuki, T., Miyamoto, K., Suzuki, T., Miyamoto, K., Oike, Y., Takeya, M., Toyama, Y., Suda, T., 2005. DC-STAMP is essential for cell-cell fusion in osteoclasts and foreign body giant cells. J. Exp. Med. 202, 345–351.

Yamamoto, T., Kaizu, C., Kawasaki, T., Hasegawa, G., Umezu, H., Ohashi, R., Sakurada, J., Jiang, S., Shultz, L., Naito, M., 2008. Macrophage colony-stimulating factor is indispensable for repopulation and differentiation of Kupffer cells but not for splenic red pulp macrophages in osteopetrotic (op/op) mice after macrophage depletion. Cell Tissue Res. 332, 245–256.

Yamamoto, Y., Udagawa, N., Matsuura, S., Nakamichi, Y., Horiuchi, H., Hosoya, A., Nakamura, M., Ozawa, H., Takaoka, K., Penninger, J.M., Noguchi, T., Takahashi, N., 2006. Osteoblasts provide a suitable microenvironment for the action of receptor activator of nuclear factor-kappaB ligand. Endocrinology 147, 3366–3374.

Yamane, T., Kunisada, T., Tsukamoto, H., Yamazaki, H., Niwa, H., Takada, S., Hayashi, S.I., 2001. Wnt signaling regulates hemopoiesis through stromal cells. J. Immunol. 167, 765–772.

Yao, Z., Lei, W., Duan, R., Li, Y., Luo, L., Boyce, B.F., 2017. RANKL cytokine enhances TNF-induced osteoclastogenesis independently of TNF receptor associated factor (TRAF) 6 by degrading TRAF3 in osteoclast precursors. J. Biol. Chem. 292, 10169–10179.

Yao, Z., Xing, L., Boyce, B.F., 2009. NF-kappaB p100 limits TNF-induced bone resorption in mice by a TRAF3-dependent mechanism. J. Clin. Invest. 119, 3024–3034.

Yasuda, H., Shima, N., Nakagawa, N., Mochizuki, S.I., Yano, K., Fujise, N., Sato, Y., Goto, M., Yamaguchi, K., Kuriyama, M., Kanno, T., Murakami, A., Tsuda, E., Morinaga, T., Higashio, K., 1998a. Identity of osteoclastogenesis inhibitory factor (OCIF) and osteoprotegerin (OPG): a mechanism by which OPG/OCIF inhibits osteoclastogenesis *in vitro*. Endocrinology 139, 1329–1337.

Yasuda, H., Shima, N., Nakagawa, N., Yamaguchi, K., Kinosaki, M., Mochizuki, S., Tomoyasu, A., Yano, K., Goto, M., Murakami, A., Tsuda, E., Morinaga, T., Higashio, K., Udagawa, N., Takahashi, N., Suda, T., 1998b. Osteoclast differentiation factor is a ligand for osteoprotegerin/ osteoclastogenesis-inhibitory factor and is identical to TRANCE/RANKL. Proc. Natl. Acad. Sci. U.S.A. 95, 3597–3602.

Yoshida, H., Hayashi, S., Kunisada, T., Ogawa, M., Nishikawa, S., Okamura, H., Sudo, T., Shultz, L.D., Nishikawa, S., 1990. The murine mutation osteopetrosis is in the coding region of the macrophage colony stimulating factor gene. Nature 345, 442–444.

Yoshida, H., Naito, A., Inoue, J., Satoh, M., Santee-Cooper, S.M., Ware, C.F., Togawa, A., Nishikawa, S., Nishikawa, S., 2002. Different cytokines induce surface lymphotoxin-alphabeta on IL-7 receptor-alpha cells that differentially engender lymph nodes and Peyer's patches. Immunity 17, 823–833.

Yu, B., Chang, J., Liu, Y., Li, J., Kevork, K., Al-Hezaimi, K., Graves, D.T., Park, N.H., Wang, C.Y., 2014. Wnt4 signaling prevents skeletal aging and inflammation by inhibiting nuclear factor-kappaB. Nat. Med. 20, 1009–1017.

Zaidi, M., Blair, H.C., Moonga, B.S., Abe, E., Huang, C.L., 2003. Osteoclastogenesis, bone resorption, and osteoclast-based therapeutics. J. Bone Miner. Res. 18, 599–609.

Zhang, Y.H., Heulsmann, A., Tondravi, M.M., Mukherjee, A., Abu-Amer, Y., 2001. Tumor necrosis factor-alpha (TNF) stimulates RANKL-induced osteoclastogenesis via coupling of TNF type 1 receptor and RANK signaling pathways. J. Biol. Chem. 276, 563–568.

Zhao, B., Grimes, S.N., Li, S., Hu, X., Ivashkiv, L.B., 2012. TNF-induced osteoclastogenesis and inflammatory bone resorption are inhibited by transcription factor RBP-J. J. Exp. Med. 209, 319–334.

Zhao, B., Takami, M., Yamada, A., Wang, X., Koga, T., Hu, X., Tamura, T., Ozato, K., Choi, Y., Ivashkiv, L.B., Takayanagi, H., Kamijo, R., 2009. Interferon regulatory factor-8 regulates bone metabolism by suppressing osteoclastogenesis. Nat. Med. 15, 1066–1071.

Zhao, H., Ito, Y., Chappel, J., Andrews, N.W., Teitelbaum, S.L., Ross, F.P., 2008. Synaptotagmin VII regulates bone remodeling by modulating osteoclast and osteoblast secretion. Dev. Cell 14, 914–925.

Zhao, H., Laitala-Leinonen, T., Parikka, V., Vaananen, H.K., 2001. Downregulation of small GTPase Rab7 impairs osteoclast polarization and bone resorption. J. Biol. Chem. 276, 39295–39302.

Chapter 6

The osteocyte

J. Klein-Nulend[1] and L.F. Bonewald[2]

[1]*Department of Oral Cell Biology, Academic Centre for Dentistry Amsterdam (ACTA), University of Amsterdam and Vrije Universiteit Amsterdam, Amsterdam Movement Sciences, Amsterdam, The Netherlands;* [2]*Indiana Center for Musculoskeletal Health, Departments of Anatomy and Cell Biology and Orthopaedic Surgery, Indiana University, Indianapolis, IN, USA*

Chapter outline

Introduction 133
The osteocytic phenotype 134
The osteocyte network 134
Osteocyte formation and death 136
Osteocyte isolation 138
Osteocyte markers 139
Osteocytic cell lines 140
Matrix synthesis 141
The osteocyte cytoskeleton and cell–matrix adhesion 142
Hormone receptors in osteocytes 143
Osteocyte function 144
Blood–calcium/phosphate homeostasis 144
Functional adaptation, Wolff's law 145
Osteocytes as mechanosensory cells 145
Canalicular fluid flow and osteocyte mechanosensing 146
Osteocyte shape and mechanosensing 149
Response of osteocytes to fluid flow in vitro 149
Summary and conclusion 151
Acknowledgments 152
References 152
Further reading 162

Introduction

The osteocyte is the most abundant cell type in bone. There are approximately 10 times as many osteocytes as osteoblasts in adult human bone (Parfitt, 1977), and the number of osteoclasts is only a fraction of the number of osteoblasts. Our current knowledge of osteocytes lags behind what we know of the properties and functions of both osteoblasts and osteoclasts. However, the striking structural design of bone predicts an important role for osteocytes, and novel techniques have allowed this gap in knowledge to shrink rapidly in the past years.

Considering that osteocytes are located inside the bone, not on the bone surface, and spaced regularly throughout the mineralized matrix, and considering their typical morphology of stellate cells, which are connected to one another via long, slender cell processes, a parallel with the nervous system springs to one's mind. Are the osteocytes the "nerve cells" of the mineralized bone matrix, and if so, what are the stimuli that "excite" these cells? Both theoretical considerations and experimental results have strengthened the notion that osteocytes are the pivotal cells in the biomechanical regulation of bone mass and structure (Cowin et al., 1991; Mullender and Huiskes, 1994, 1995; Klein-Nulend et al., 1995b; Tatsumi et al., 2007; Klein-Nulend et al., 2013 [review]). This idea poses many questions that have to be answered. The development of osteocyte isolation techniques, the use of highly sensitive (immuno)cytochemical and in situ hybridization procedures, and the usefulness of molecular biological methods even when only small numbers of cells are available have rapidly increased our knowledge about this least understood cell type of bone, and will certainly continue to do so in the future. The use of transgenic mouse models to perform targeted deletion of genes has provided considerable information on the function and importance of this bone cell type.

Principles of Bone Biology. https://doi.org/10.1016/B978-0-12-814841-9.00006-3

The osteocytic phenotype

The osteocyte network

Mature osteocytes are stellate-shaped or dendritic cells enclosed within the lacunocanalicular network of bone. The lacunae contain the cell bodies. From these cell bodies, long, slender cytoplasmic processes, dendrites, radiate in all directions, but with the highest density perpendicular to the bone surface (Fig. 6.1). They pass through the bone matrix via small canals, the canaliculi. Processes and their canaliculi may be branched. The more mature osteocytes are connected by these cell processes to neighboring osteocytes, the most recently incorporated osteocytes to neighboring osteocytes and to the cells lining the bone surface. Some of the processes oriented to the bone surface, however, appear not to connect with the lining cells, but pass through this cell layer, thereby establishing a direct contact between the osteocyte network and the extraosseus space. This intriguing observation by Kamioka et al. (2001) suggests the existence of a signaling system between the osteocyte and the bone marrow compartment without intervention of the osteoblasts/lining cells. Osteocytes also appear to be able to retract and extend their dendritic processes, not only between cells in the bone matrix, but also into marrow spaces as shown by dynamic imaging (Veno et al., 2006). This has implications with regard to osteocytes making and breaking communication between cells.

Dramatic changes occur in the distribution of actin-binding proteins during terminal differentiation of osteoblasts to osteocytes (Kamioka et al., 2004). The typical morphology of the osteocyte was originally thought to be enforced on differentiating osteoblasts during their incorporation into the bone matrix. Osteocytes have to remain in contact with other cells and ultimately with the bone surface to ensure the access of oxygen and nutrients. Culture experiments with isolated osteocytes have shown, however, that although the cells lose their stellate shape in suspension, they reexpress this morphology as soon as they settle on a support (Van der Plas and Nijweide, 1992) (Fig. 6.2). Apparently, the typical stellate morphology and the need to establish a cellular network are intrinsic characteristics of terminal osteocyte differentiation.

In bone, gap junctions are present between the tips of the cell processes of connecting osteocytes (Doty, 1981). Within each osteon or hemiosteon (on bone surfaces), therefore, osteocytes form a network of gap junction–coupled cells. As the lacunae are connected via the canaliculi, the osteocyte network represents two network systems: an intracellular one and an extracellular one. Gap junctions are transmembrane channels connecting the cytoplasm of two adjacent cells that regulate the passage of molecules of less than 1 kDa (Goodenough et al., 1966; Bennett and Goodenough, 1978). Gap junction channels are formed by members of a family of proteins known as connexins. One of these members, connexin 43 (Cx43), appears to play an important role in bone cells, as Cx43-null mice have delayed ossification, craniofacial abnormalities, and osteoblast dysfunction (Lecanda et al., 2000). It has been proposed that gap junctions function through the propagation of intracellular signals contributing to mechanotransduction in bone, thereby regulating bone cell differentiation (Donahue, 2000). Fluid-flow-induced shear stress stimulates gap junction–mediated intercellular communication and increases Cx43 expression (Cheng et al., 2001), while oscillating fluid flow has been shown to upregulate gap junction communication by

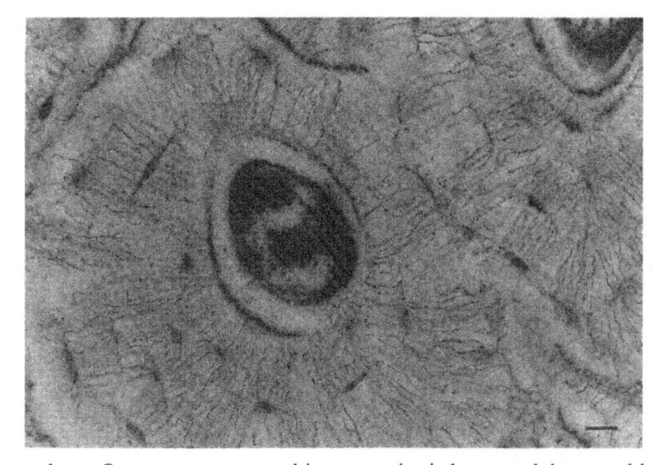

FIGURE 6.1 Osteon in mature human bone. Osteocytes are arranged in concentric circles around the central haversian channel. Note the many cell processes, radiating from the osteocyte cell bodies, in particular in the perpendicular directions. Schmorl staining. (Original magnification, ×390; bar, 25 μm.)

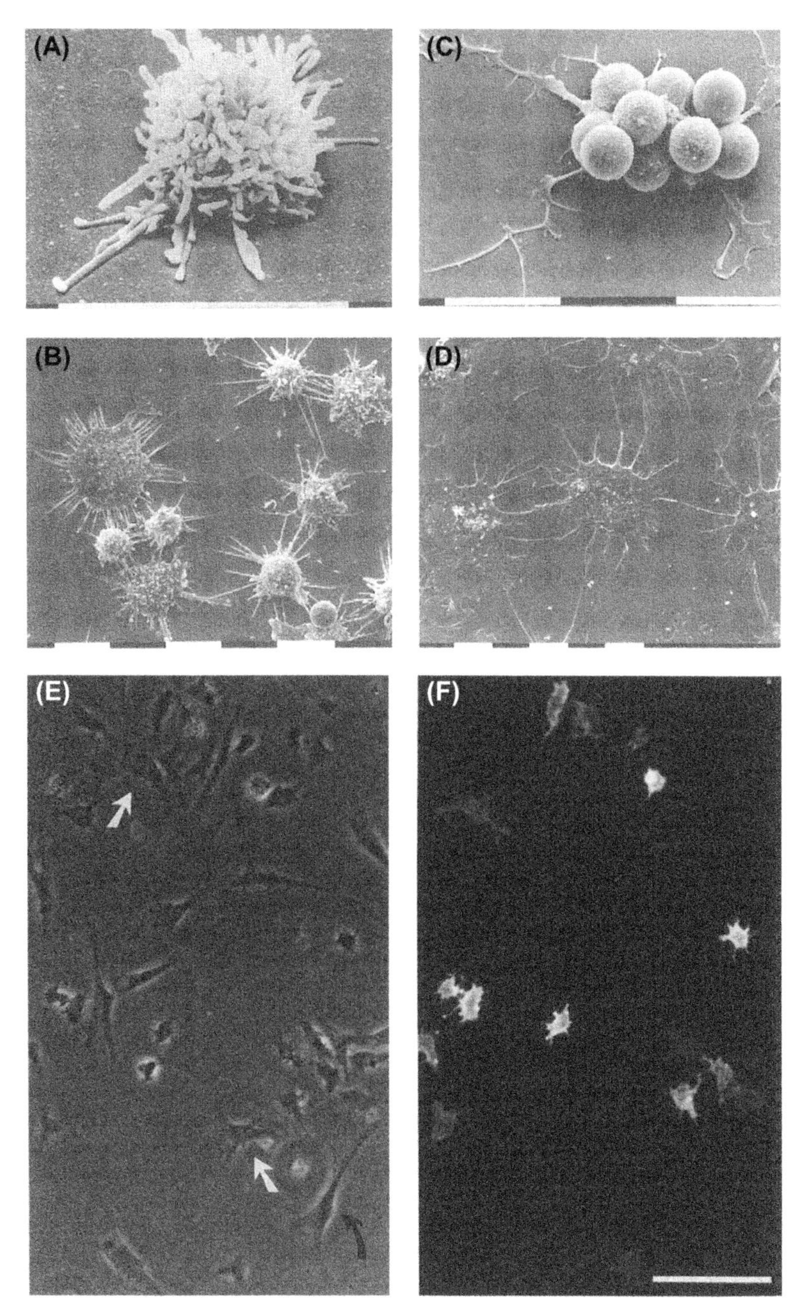

FIGURE 6.2 Isolated osteocytes in culture. Osteocytes were isolated by an immunodissection method using MAb OB7.3-coated magnetic beads. After isolation the cells were seeded on a glass support, cultured for (A) 5 min, (B) 30 min, or (C and D) 24 h and studied with a scanning electron microscope. Immediately after attachment, osteocytes form cytoplasmic extrusions in all directions (A). During subsequent culture the cell processes perpendicular on the support disappear, while the processes in the plane of the support elongate (B) and ultimately form smooth connections between neighboring cells (D). In (A), (B), and (D) the immunobeads were removed from the cells before seeding; in (C) the beads were left on the cells. (Original magnifications, (A) ×7200, (B) ×1400, (C) ×2900, and (D) ×940; bar, 10 μm.) (E and F) Cells were also isolated from periosteum-free 18-day-old chicken calvariae by collagenase digestion, seeded, and cultured for 24 h. Subsequently the osteocytes in the mixed population were specifically stained with MAb OB7.3 in combination with biotinylated horse anti-mouse IgG and streptavidin-Cy3. (E) Phase contrast. (F) Immunofluorescence. *Black arrow*, fibroblast-like cells; *white arrows*, osteoblast-like cells. (Original magnification, ×300; bar, 100 μm.)

an extracellular signal-regulated kinase (ERK) 1/2 mitogen-activated protein kinase-dependent mechanism (Alford et al., 2003) in MLO-Y4 osteocyte-like cells.

Hemichannels, unapposed halves of gap junction channels (Goodenough and Paul, 2003), have been identified in osteocytes localizing at the cell surface, independent of physical contact with adjacent cells. Primary osteocytes and MLO-

Y4 osteocyte-like cells (Kato et al., 1997) express very large amounts of Cx43 compared with other cell types such as osteoblasts, yet these cells are in contact only through the tips of their dendritic processes, raising a question regarding the function of Cx43 on the rest of the cell membrane. The opening of hemichannels results in ATP and NAD^{+} release, which in turn raises intracellular Ca^{2+} levels and wave propagation of Ca^{2+}. It has been shown that oscillating fluid flow activates hemichannels in MLO-Y4 osteocyte-like cells, but not in MC3T3-E1 osteoblast-like cells. This activation involved protein kinase C, and resulted in ATP and prostaglandin E_2 (PGE_2) release (Genetos et al., 2007). Hemichannels expressed in bone cells such as MLO-Y4 cells appear to function as essential transducers of the antiapoptotic effects of bisphosphonates (Plotkin et al., 2002) and serve as a portal for the exit of elevated intracellular PGE_2 in osteocytes induced by fluid flow shear stress (Cherian et al., 2005). Integrin $\alpha5\beta1$ interacts with Cx43 to mediate the opening of hemichannels to release prostaglandin in MLO-Y4 cells in response to mechanical stimulation independent of the integrin association with fibronectin and its interaction with the extracellular matrix (Batra et al., 2012). Therefore, gap junctions at the tips of dendrites mediate intracellular communication, while hemichannels along the dendrite and the cell body mediate extracellular communication within the osteocyte network.

In vivo studies have provided additional, sometimes conflicting, information on the functions of Cx43 gap junctions and hemichannels in osteoblasts and osteocytes. Deletion of Cx43 in osteoblasts results in animals with increased osteocyte apoptosis (Bivi et al., 2012). However, mice lacking Cx43 in osteoblasts and/or osteocytes show an increased anabolic response to loading, and decreased catabolic response to unloading (Plotkin et al., 2015). Xu and colleagues generated mice expressing a mutated Cx43 with impaired gap junctions and mice expressing a Cx43 mutant able to form functional hemichannels but unable to form gap junction channels. Mice without both functional gap junctions and functional hemichannels exhibit increased bone mass, whereas mice expressing only hemichannels were not different from wild-type littermate controls (Xu et al., 2015). This study suggests that it is the Cx43 hemichannel and not the gap junction that is responsible for the bone phenotype. It was proposed that hemichannels play a dominant role in osteocyte survival.

Osteocyte formation and death

Osteogenic cells arise from multipotential mesenchymal stem cells (see Chapter 2). These stem cells have the capacity to also differentiate into other lineages, including those of chondroblasts, fibroblasts, adipocytes, and myoblasts (Aubin et al., 1995, Chapter 2). By analogy with hemopoietic differentiation, each of these differentiation lineages is thought to originate from a different committed progenitor, which for the osteogenic lineage is called the osteoprogenitor. Osteodifferentiation progresses via a number of progenitor and precursor stages to the mature osteoblast. Osteoblasts have one of three fates: embedding in their own osteoid, differentiating into an osteocyte; quiescing into a lining cell, or undergoing apoptosis (for review see Manolagas, 2000). The mechanism by which osteoblasts differentiate into osteocytes is, however, still unknown. Imai et al. (1998) found evidence that osteocytes may stimulate osteoblast recruitment and differentiation by expressing osteoblast stimulating factor-1 (OSF-1) (Tezuka et al., 1990). The osteoblasts further differentiate into osteocytes, being surrounded by the osteoid matrix that they produce, and they then may become a new source of OSF-1 for the next round of osteoblast recruitment. The expression of OSF-1 in osteocytes may be activated by local damage to bone or local mechanical stress (Imai et al., 1998). Marotti (1996) has postulated that a newly formed osteocyte starts to produce an osteoblast inhibitory signal when its cytoplasmic processes connecting the cell with the osteoblast layer have reached their maximal length. The osteoid production of the most adjacent, most intimately connected osteoblast will be relatively more inhibited by that signal than that of its neighbors. The inactivated osteoblast then spreads over a larger bone surface area, thereby reducing its linear appositional rate of matrix production even further. A second consequence of the widening and flattening of the cell is that it may intercept more osteocytic processes carrying the inhibitory signal. This positive feedback mechanism results in the embedding of the cell in matrix produced by the neighboring osteoblasts. Ultimately, the cell will acquire the typical osteocyte morphology and the surrounding matrix will become calcified. The theory of Marotti (1996) is based entirely on morphological observations. There is no biochemical evidence on the nature or even the existence of the proposed inhibitory factor. Martin (2000) has, however, used the concept successfully in explaining mathematically the changing rates of matrix formation during bone remodeling.

Osteoid-osteocytes were described by Palumbo (1986) to be cells actively making matrix and calcifying this matrix while the cell body reduces in size in parallel with the formation of cytoplasmic processes. Bordier et al. (1976) and Nijweide et al. (1981) proposed that osteoid-osteocytes play an important role in the initiation and control of mineralization of the bone matrix. During the time in which an osteoblast has become an osteocyte, the cell has manufactured three times its own volume in matrix (Owen, 1995). Franz-Odendaal et al. (2006) propose that once a cell is surrounded by osteoid, the differentiation process has not ended, but continues.

An enzyme that is produced in high amounts by embedding osteoid-osteocytes and not by osteoblasts, casein kinase II, appears to be responsible for the phosphorylation of matrix proteins essential for mineralization (Mikuni-Takagaki et al., 1995). Phosphoproteins appear to be essential for bone mineralization as evidenced in vitro by crystal nucleation assays (Boskey, 1996) and in vivo by osteomalacia in animal models with deletion of (osteocyte-specific) genes such as dentin matrix protein 1 (DMP1) and PHEX (phosphate-regulating gene with homologies to endopeptidases on the X chromosome) (Strom et al., 1997). The roles of these proteins in mineral homeostasis will be discussed later in this chapter. An osteocyte-selective promoter, the 8-kb DMP1, driving green fluorescent protein (GFP) identifies embedding and embedded osteocytes (Kalajzic et al., 2004). With the identification of other markers selective for osteocytes, such as E11/gp38 for early osteocytes (Zhang et al., 2006) and sclerostin for late osteocytes (Poole et al., 2005), new tools have been generated for the study of osteocyte formation.

The life span of osteocytes is probably largely determined by bone turnover, when osteoclasts resorb bone and either "liberate" or destroy osteocytes. Osteocytes may have half-lives of decades if the particular bone they reside in has a slow turnover rate (Parfitt, 1977). The fate of living osteocytes that are liberated by osteoclast action is unknown as of this writing. Some of them, only half released by osteoclastic activity, may be reembedded during new bone formation that follows the resorption process (Suzuki et al., 2000). These osteocytes are then the cells that cross the cement lines between individual osteons, sometimes seen in cross sections of osteonal bone. It has been shown by Kalajzic et al. that isolated osteocytes can partially dedifferentiate into osteoblasts (Torregiani et al., 2013). Most of the osteocytes, however, will probably die by apoptosis and become phagocytosed. Phagocytosis of osteocytes by osteoclasts as part of the bone resorption process has been documented in several reports (Bronckers et al., 1996; Elmardi et al., 1990).

Apoptosis of osteocytes in their lacunae is attracting growing attention because of its expected consequence of decreased bone mechanoregulation, which may lead to osteoclastic bone resorption (Tan et al., 2007, 2008). Osteocyte apoptosis can occur with immobilization, microdamage, estrogen deprivation, elevated cytokines, glucocorticoid treatment, osteoporosis, osteoarthritis, and aging. The resulting fragility is considered to be due to loss of the ability of osteocytes to signal other bone cells for repair due to loss of the capacity to sense microdamage (Manolagas, 2000; Noble et al., 2003). Apoptotic regions around microcracks were found to be surrounded by surviving osteocytes expressing Bcl-2, whereas dying osteocytes appeared to be the target of resorbing osteoclasts (Verborgt et al., 2000, 2002). Apoptotic changes in osteocytes were shown to be associated with high bone turnover (Noble et al., 1997). However, fatigue-related microdamage in bone may cause decreased osteocyte accessibility for nutrients and oxygen, inducing osteocyte apoptosis and subsequent bone remodeling (Burger and Klein-Nulend, 1999; Verborgt et al., 2000). Lack of oxygen elevates hypoxia-inducible factor-1α, a transcription activator, by inactivation of prolyl hydroxylase leading to apoptosis and induction of the osteoclastogenic factor tumor necrosis factor α (TNFα) (Gross et al., 2001), vascular endothelial growth factor, and osteopontin, a mediator of environmental stress and a potential chemoattractant for osteoclasts (Gross et al., 2005).

In contrast to overloading, which induces microdamage, physiological mechanical loads might prevent osteocyte apoptosis in vivo. Mechanical stimulation of osteocytes in vitro, by means of a pulsating fluid flow (PFF), affects TNFα-induced apoptosis. One-hour PFF (0.70 ± 0.30 Pa, 5 Hz) inhibited (25%) TNFα-induced apoptosis in osteocytes, but not in osteoblasts or periosteal fibroblasts (Tan et al., 2006). Although the exact mechanism is not clear, loading-induced nitric oxide production by the osteocytes might be involved in the antiapoptotic effects of mechanical loading. Prostaglandin produced by osteocytes in response to fluid flow shear stress also blocks MLO-Y4 apoptosis (Kitase et al., 2006).

Also, loss of estrogen (Tomkinson et al., 1998) and chronic glucocorticoid treatment (Weinstein et al., 1998) were demonstrated to induce osteocyte apoptosis, which may, at least in part, explain the bone-deleterious effects of these conditions. Several agents, such as bisphosphonates and calcitonin (Plotkin et al., 1999), CD40 ligand (Ahuja et al., 2003), calbindin-D28k (Liu et al., 2004), and estrogen and selective estrogen receptor modulators (Kousteni et al., 2001), have been found to reduce or inhibit osteoblast and osteocyte apoptosis. Bisphosphonates inhibit apoptosis through interaction with Cx43 hemichannels and the ERK pathway (Plotkin and Bellido, 2001). Interestingly, the two antiapoptotic agents parathyroid hormone (PTH) and monocyte chemotactic protein-3 (MCP-3) have been shown to be selective for apoptosis induced by one particular agent, glucocorticoids. Unlike the agents listed earlier, both PTH (Jilka et al., 1999) and MCP-3 will inhibit only glucocorticoid-induced apoptosis, and not TNFα-induced apoptosis, of MLO-Y4 osteocyte-like cells (Kitase et al., 2006).

Osteocyte viability is crucial for the normal functioning of the skeleton and the normal function of other organs such as kidney through fibroblast growth factor 23 (FGF23) production by osteocytes, but also muscle. Osteocyte factors such as PGE_2 and Wnt3a promote myogenesis and enhance muscle function (Mo et al., 2012; Huang et al., 2017). Mice with targeted deletion of Cx43 in osteocytes have a reduced muscle phenotype (Shen et al., 2015). Conversely, secreted muscle factors prevent glucocorticoid-induced osteocyte apoptosis (Jahn et al., 2012), and β-aminoisobutyric acid produced by contracted muscle with exercise will prevent reactive oxygen–induced osteocyte apoptosis (Kitase et al., 2018). These investigators also found that this muscle metabolite will prevent bone and muscle loss with unloading.

Both mechanical loading and intracellular autophagy play important roles in bone homeostasis. Autophagy is a catabolic process that is regulated by multiple factors and is associated with skeletal diseases. It has been shown that fluid shear stress induces protective autophagy in osteocytes and that mechanically induced autophagy is associated with ATP metabolism and osteocyte survival (Zhang et al., 2018). Moreover, microRNA-199a-3p is involved in the estrogen regulatory networks that mediate MLO-Y4 osteocyte autophagy, potentially by targeting insulin-like growth factor 1 (IGF-1) and mammalian target of rapamycin (Fu et al., 2018).

In summary, osteocyte viability may play a significant role in the maintenance of bone homeostasis and integrity and other organs (Dallas et al., 2013), yet agents that block apoptosis may exacerbate conditions that require repair.

Osteocyte isolation

Analysis of osteocyte properties and functions has long been hampered by the fact that osteocytes are embedded in a mineralized matrix. Although sensitive methods are now available, such as immunocytochemistry and in situ hybridization, by which osteocytes can be studied in the tissue in some detail, osteocyte isolation and culture offer a major step forward. This approach became possible by the development of osteocyte-specific antibodies (Fig. 6.3) directed to antigenic sites on the outside of the cytoplasmic membrane (Bruder and Caplan, 1990; Nijweide and Mulder, 1986). Using an immunodissection method, Van der Plas and Nijweide (1992) subsequently succeeded in the isolation and purification of chicken osteocytes from mixed bone cell populations isolated from fetal bones by enzymatic digestion. A detailed description of the isolation procedure has been published (Semeins et al., 2012). Isolated osteocytes appeared to behave in vitro like they do in vivo in that they reacquired their stellate morphology and, when seeded sparsely, formed a network of cells coupled to one another by long, slender, often branched cell processes (Fig. 6.2). The cells retained this morphology in culture throughout the time studied (5–7 days) and even reexpressed it when passaged for a second time (Van der Plas and Nijweide, 1992).

Mikuni-Takagaki et al. (1995) isolated seven cell fractions from rat calvariae by sequential digestion. They claimed that the last fraction consisted of osteocytic cells. The cells displayed dendritic cell processes, were negative for alkaline phosphatase, had high extracellular activities of casein kinase II and ecto-5′-nucleotidase, and produced large amounts of osteocalcin. After a few days of little change in cell number, the cells of fraction VII, the osteocytic cells, proliferated, but

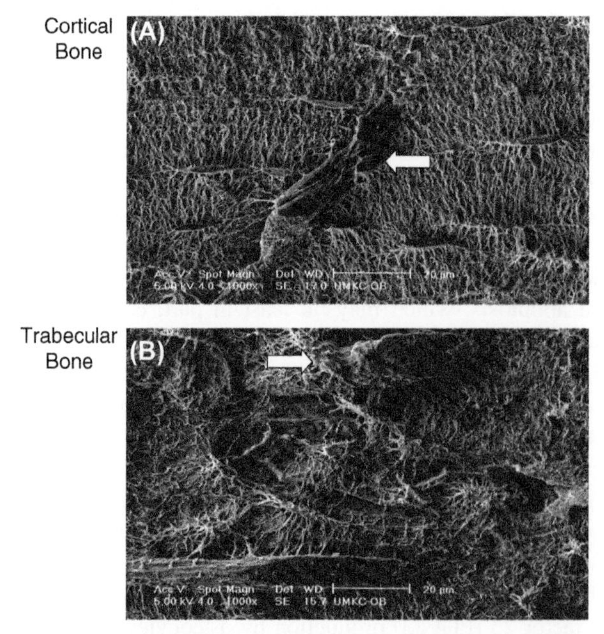

FIGURE 6.3 Images of acid-etched resin-embedded murine (A) cortical and (B) trabecular bone visualized by scanning electron microscopy showing the complexity of the osteocyte lacunocanalicular system and the intimate relationship between the lacunocanalicular system and blood vessels. In the cortical bone, note the linear alignment of the lacunae and the complexity of the canaliculi. In the trabecular bone, the lacunae are not as organized as in the cortical bone. In both sections, note the close relationship of some osteocyte lacunae with blood vessels (*arrows*).

equally fast as those of fraction III, the osteoblastic cells, in culture. With the identification of new osteocyte-selective markers such as Sost, E11/gp38, Dmp1, Phex, and Mepe, it is becoming easier to identify isolated osteocytes (see later).

A methods paper written by Stern and Bonewald (2015) provides a review of the different approaches used to isolate osteocytes from bone. Consistent approaches include collagenase digestions and calcium chelation. In this review, the authors also described isolation of osteocytes from aged bone through the use of bone particle culture. The yield of primary osteocytes decreases with the age of the animal. The isolation is easiest when the bone is hypomineralized as in young animals but becomes more difficult in hypermineralized bone as occurs in aged animals. In summary, whereas primary osteocytes can be isolated from bone, the process is still long compared with soft tissue or bone surface cells.

As of this writing, culture of human osteocytes in vitro remains a challenge. Culture of denuded human bone chips with osteocytes embedded in their native matrix has been shown to overcome this challenge, and provides a three-dimensional model that can be used to study osteocyte function and signaling (Pathak et al., 2016). Pathak and colleagues showed that these osteocytes highly express mRNA of osteocyte-specific signaling molecules (Pathak et al., 2016). The isolation of osteocytes from human trabecular bone samples acquired during surgery has been described (Prideaux et al., 2016). A protocol was used whereby the cells were digested from the bone matrix by sequential collagenase and ethylenediaminetetraacetic acid (EDTA) digestions, and the cells from later digests displayed characteristic dendritic osteocyte morphology when cultured ex vivo. These cells represent an important tool in enhancing current knowledge in human osteocyte biology.

Osteocyte markers

In bone, osteocytes are fully defined by their location within the bone matrix and their stellate morphology. One marker for isolated osteocytes is therefore their typical morphology, which they reacquire in culture (Mikuni-Takagaki et al., 1995; Van der Plas and Nijweide, 1992). Related to this stellate morphology, osteocytes have a typical cytoskeletal organization, which is important for the osteocyte's response to loading (McGarry et al., 2005a,b). It has been shown that the dendrite is the primary mechanosensing part of the osteocyte compared with the cell body (Burra et al., 2010). The prominent actin bundles in the osteocytic processes, together with the abundant presence of the actin-bundling protein fimbrin, are exemplary for osteocytes and are retained after isolation (Tanaka-Kamioka et al., 1998). In addition, osteocytes are generally found to express osteocalcin, osteonectin, and osteopontin, but show little alkaline phosphatase activity, particularly the more mature cells (Aarden et al., 1996b). As stated previously, these metabolic markers have, however, little discriminating value in mixtures of isolated cells. Franz-Odendaal et al. (2006) provide a list of molecular markers for the preosteoblast to the osteocyte.

Initially mainly morphology was used to describe osteocytes until osteocyte-specific antibodies became available. Early examples are the monoclonal antibodies MAb OB7.3 (Nijweide and Mulder, 1986) (Fig. 6.2), MAb OB37.11 (Nijweide et al., 1988), and MAb SB5 (Bruder and Caplan, 1990). All three are specific for avian osteocytes and do not cross-react with mammalian cells. The identities of two of the three antigens involved have not been reported, but that of OB7.3 has been elucidated and found to be the avian homolog of mammalian *Phex* (Westbroek et al., 2002). Using an antibody to Phex allowed purification of avian osteocytes from enzymatically isolated bone cells.

Osteoblast/osteocyte factor 45 (OF45), also known as MEPE (matrix extracellular phosphoglycoprotein), is also highly expressed in osteocytes compared with osteoblasts. Messenger RNA expression of OF45/MEPE begins at embryonic day 20 in more differentiated osteoblasts that have become encapsulated by bone matrix (Igarashi et al., 2002). *Mepe* was isolated and cloned from a TIO tumor cDNA library (Rowe et al., 2000). Cathepsin D or B can cleave MEPE, releasing the highly phosphorylated C-terminal ASARM region that is a potent inhibitor of mineralization in vitro (Bresler et al., 2004; Rowe et al., 2005). OF45/MEPE-null mice have increased bone formation, bone mass, and resistance to age-associated trabecular bone loss (Gowen et al., 2003). The authors speculate that osteocytes act directly on osteoblasts through OF45/MEPE to inhibit their bone-forming activity.

Both Toyosawa et al. (2001) and Feng et al. (2006) found Dmp1 to be highly expressed in osteocytes with very low expression in osteoblasts. DMP1 is specifically expressed along and in the canaliculi of osteocytes within the bone matrix (Feng et al., 2006). Deletion of this gene in mice results in a phenotype similar if not identical to the hyp phenotype, suggesting that Dmp1 and Phex are interactive and essential for phosphate metabolism. Potential roles for DMP1 in osteocytes may be related to the posttranslational processing and modification resulting in a highly phosphorylated protein and regulator of hydroxyapatite formation. Interestingly, Dmp1 and OF45/MEPE belong to the SIBLING (small, integrin-binding ligand, N-linked glycoprotein) family, which also includes bone sialoprotein, osteopontin, and sialophosphoprotein (Fisher and Fedarko, 2003). This family of proteins may function differently in osteocytes compared with other cell types, especially upon phosphorylation with casein kinase II, a marker of the osteoblast-to-osteocyte transition (Mikuni-Takagaki et al., 1995).

Schulze et al. (1999) and Zhang et al. (2006) have described the expression of E11/gp38 exclusively in osteocytes and not in osteoblasts in vivo. A punctate antibody reaction is observed at the interface between osteoblasts and uncalcified osteoid at the tips and along dendritic processes with less reactivity in osteocytes deeper in the bone matrix. E11/gp38 appears to be responsible for the formation of dendritic processes, as a reduction in protein expression using a short interfering RNA approach led to a decrease in dendrite extension in MLO-Y4 osteocyte-like cells in response to shear stress (Zhang et al., 2006). E11/gp38 colocalizes with ezrin, radixin, and moesin (Scholl et al., 1999), which are concentrated in cell-surface projections where they link the actin cytoskeleton to plasma membrane proteins and are involved in cell motility (Mangeat et al., 1999). CD44 is highly expressed in osteocytes compared with osteoblasts (Hughes et al., 1994), and as E11 is physically associated with CD44 in tumor vascular endothelial cells (Ohizumi et al., 2000), this association most likely occurs in osteocytes to regulate the formation of dendritic processes.

Sclerostin appears to be highly expressed in the mature and not the early osteocyte (Poole et al., 2005). Transgenic mice lacking sclerostin have increased bone mass, and the human condition sclerostosis is due to a premature termination of the SOST gene (Balemans et al., 2001). Sclerostin clearly functions as a Wnt antagonist by binding Lrp5 (Van Bezooijen et al., 2004), a receptor shown to be an important positive regulator of bone mass (Li et al., 2005). Sclerostin protein may be transported through canaliculi to the bone surface to inhibit bone-forming osteoblasts. Sclerostin is downregulated by mechanical loading (Robling et al., 2006). It has also been proposed that the anabolic effects of PTH are through inhibition of SOST expression (Bellido et al., 2005). Neutralizing antibody to sclerostin is being developed as a therapeutic to treat osteoporosis. The antibody blocks or reduces bone loss and supports bone formation, promotes fracture healing, and is a potential therapeutic for a number of conditions of low bone mass, such as osteogenesis imperfecta (Clarke, 2014). In clinical trials, despite very potent effects in increasing bone mass, the FDA rejected approval of the anti-sclerostin antibody romosozumab for osteoporosis treatment due to a higher rate of serious adverse cardiovascular events compared with alendronate (Medscape, 2017). This was a devastating event after following such a potent inducer of bone mass for over a decade. However, as of this writing, the late-phase data from this clinical study are being refiled to show the drug has a positive risk–benefit profile, while other anti-sclerostin monoclonal antibodies are being developed (MacNabb et al., 2016).

FGF23 is a major phosphate-regulating hormone responsible for autosomal dominant hypophosphatemic rickets, for phosphate wasting in tumor-induced osteomalacia, and for X-linked hypophosphatemia. FGF23 is not normally expressed at high levels in osteocytes in the healthy state but is dramatically upregulated in both DMP1- and PHEX-associated hypophosphatemic rickets (Liu et al., 2007). Clinkenbeard and White (2016) performed FGF23 deletion using Col2.3-Cre for osteoblasts and DMP1-Cre for early osteocytes and showed that both osteoblasts and osteocytes are the physiological source of FGF23. More information on the function and regulation of FGF23 is provided later.

It has been shown that receptor activator of NF-κB ligand (RANKL) is a functional marker of osteocytes. Primary osteocytes express RANKL (Kramer et al., 2010), and osteocytes express greater amounts of RANKL than osteoblasts and are better supporters of osteoclast formation (Nakashima et al., 2011; Xiong et al., 2011, 2015). Deletion of RANKL using the 10-kb Dmp1-Cre or Sost-Cre results in mice with increased bone mass. Like sclerostin, RANKL is increased with unloading (Xiong et al., 2011).

In summary, it is becoming easier to define cells as osteocytic. Even though the E11/gp38 molecule is expressed in other tissues (known as podoplanin in kidney and also known as PA2.26, T1a, etc.), it is expressed only in osteocytes, not osteoblasts, in bone (Zhang et al., 2006). While E11/gp38, Phex, and Dmp1 appear to be markers for the early osteocyte, MEPE and Sost/sclerostin are specific markers for the late osteocyte (Poole et al., 2005). Dmp1 and Phex, while expressed in low levels in osteoblasts and other tissues, are highly elevated in early osteocytes. Combining these markers with other properties such as dendricity and low or no alkaline phosphatase can be useful to define not only isolated primary cells, but also cell lines.

Osteocytic cell lines

Since the number of primary osteocytes that can be isolated from chickens each time (Van der Plas and Nijweide, 1992) is limited, several groups have tried to establish osteocytic cell lines. Basically, an osteocytic cell line is a contradiction in terms. Osteocytes are postmitotic. However, a cell line of proliferating precursor cells that would differentiate into osteocytes under specific circumstances could prove to be very valuable in the study of osteocyte properties and functions. HOB-01-C1 (Bodine et al., 1996) is a temperature-sensitive cell line, prepared from immortalized, cloned human adult bone cells. It proliferates at 34°C but stops dividing at 39°C. HOB-01-C1 cells display putative osteocytic markers, such as cellular processes, low alkaline phosphatase activity, high osteocalcin production, and the expression of CD44.

MLO-Y4 (Kato et al., 1997) is an osteocyte-like cell line that expresses high amounts of E11/gp38 (Zhang et al., 2006), CD44, osteocalcin, and osteopontin and has low alkaline phosphatase activity. Numerous investigators have used these cells to examine gap junctions, hemichannels, apoptosis, and other potential functions of osteocytes (Genetos et al., 2007; Plotkin et al., 2005; Vatsa et al., 2006, 2007; Xiao et al., 2006; Zaman et al., 2006; Zhang et al., 2006; Cherian et al., 2005; Liu et al., 2004; Alford et al., 2003; Ahuja et al., 2003; Zhao et al., 2002; Heino et al., 2002, 2004; Cheng et al., 2001; Yellowley et al., 2000) and, as of the updating of this chapter, there are over 250 references using these cells to investigate osteocyte function. Estrogen will reduce support of osteoclast formation by MLO-Y4 cells through an increase in transforming growth factor β3 (Heino et al., 2002) and induce MLO-Y4 cells to support osteoblast mesenchymal stem cell differentiation (Heino et al., 2004), supporting the hypothesis that osteocytes are orchestrators of both bone resorption and bone formation. Observations using these cells have been validated in vivo. For example, MLO-Y4 cells support osteoclast formation through RANKL expression (Zhao et al., 2002) and apoptotic bodies released from MLO-Y4 cells express RANKL (Kogianni et al., 2008). The outcomes using this cell line have been validated using transgenic mouse models generated by two different laboratories showing that RANKL is mainly functional in osteocytes (Nakashima et al., 2011; Xiong et al., 2011).

MLO-A5 cells, a postosteoblast/preosteocyte-like cell line established from the long bones of 14-day-old mice expressing the large T antigen driven by the osteocalcin promoter, differentiate into osteoid-osteocyte-like cells (Kato et al., 2001). MLO-A5 cells express all of the markers of the late osteoblast, such as high alkaline phosphatase, bone sialoprotein, PTH type 1 receptor, and osteocalcin, but begin to express markers of osteocytes such as E11/gp38 as they generate cell processes. MLO-A5 cells generate nanospherites that bud from and mineralize on their developing cellular processes. As the cellular process narrows in diameter, these mineralized structures become associated with and initiate collagen-mediated mineralization (Barragan-Adjemian et al., 2006). PTH has been shown to reduce Sost/sclerostin expression in these cells (Bellido et al., 2005). This is an excellent cell line to study collagen production, regulation, and mineralization, as they are prodigious producers of collagen (Yang et al., 2015; Lu et al., 2018).

Compared with primary osteoblasts and clonal cells, the MLO-Y4 cells show relatively high expression of osteocalcin and Cx43 with low expression of collagen type I and periostin, as well as low alkaline phosphatase activity. However, both MLO-A5 and MLO-Y4 cells have their limitations, such as the lack of sclerostin expression and low DMP1 expression by MLO-Y4 cells. This makes MLO-Y4 cells less suitable for studying signaling molecule production by mature osteocytes. Alternative cell lines that have been used to study sclerostin expression include the SaOS2 osteosarcoma cell line and the osteoblast-like UMR-106 cells. IDG-SW3 (Woo et al., 2011) and Ocy454 (Spatz et al., 2015) are osteocyte cell lines that express relatively high levels of SOST/sclerostin as well as FGF23, both key regulators of bone homeostasis, and could therefore be used to study osteocyte signaling toward other cell types.

The IDG-SW3 osteocyte cell line undergoes a temporally dependent process that replicates the osteoblast-to-osteocyte transition (Woo et al., 2011). This cell line was made from long bones of mice carrying a Dmp1 promoter driving GFP crossed with the Immortomouse. These cells can be expanded at 33°C in the presence of interferon γ (IFN-γ) and then allowed to resume their original phenotype at 37°C in the absence of IFN-γ. Another cell line has been made from the same mice, called Ocy454 (Spatz et al., 2015). Comparisons between the two cell lines suggest that sclerostin expression occurs earlier and is higher in the Ocy454 cells compared with IDG-SW3.

Cementocytes, like osteocytes, are mechanosensory cells, and when the cementocyte cell line IDG-CM6 was compared with IDG-SW3 in their response to shear stress, IDG-CM6 cells significantly increased RANKL, with no change in IDG-SW3 cells, and significantly reduced osteoprotegerin (OPG) in contrast to increased expression by IDG-SW3 cells. The higher OPG/RANKL ratio in IDG-CM6 cementocytes under fluid flow shear stress suggests a potential mechanism for the lack of cementum resorption observed during physiological use and in orthodontic tooth movement (Zhao et al., 2016). Cementocytes in vivo and in vitro express key markers known to be important in osteocyte differentiation, including *Dmp1*/DMP1, *E11/gp38*, and *Sost*/sclerostin. Loss-of-function studies in mice produced cellular cementum phenotypes resembling those in bone, showing that these factors operate in similar fashions in both tissues.

Matrix synthesis

The subcellular morphology of osteocytes and the fact that they are encased in mineralized matrix do not suggest that osteocytes partake to a large extent in matrix production. Osteocytes, especially the more mature cells, have relatively few organelles necessary for matrix production and secretion. Nevertheless, a limited secretion of specific matrix proteins may be essential for osteocyte function and survival. Several arguments are in favor of such limited matrix production. First, as the mineralization front lags behind the osteoid formation front in areas of new bone formation, osteocytes may be involved in the maturation and mineralization of the osteoid matrix by secreting specific matrix molecules. It is, however, also possible that osteocytes enable the osteoid matrix to be mineralized by phosphorylating certain matrix constituents, as was suggested by Mikuni-Takagaki et al. (1995). Mikuni-Takagaki et al. proposed that casein kinase II, produced in high amounts by embedding osteoid-osteocytes and not by osteoblasts, is responsible for phosphorylation of matrix proteins essential for mineralization. Therefore, the embedding osteoid cell and the osteocyte probably play roles in the mineralization process and potentially in phosphate metabolism (see later). Osteocytes have to inhibit mineralization of the matrix directly surrounding them to ensure the diffusion of oxygen, nutrients, and waste products through the lacunocanalicular system. Osteocalcin, which is expressed to a relatively high extent by osteocytes, may play an important role here (Aarden et al., 1996b; Ducy et al., 1996; Mikuni-Takagaki et al., 1995), as do OF45/MEPE and Sost. Osteocytes have been found positive for osteocalcin and osteonectin (Aarden et al., 1996b), molecules that are probably involved in the regulation of calcification. Osteopontin, fibronectin, and collagen type I (Aarden et al., 1996b) have also been demonstrated in and immediately around (isolated) osteocytes. These proteins may be involved in osteocyte attachment to the bone matrix (see later). Osteocytes also express Notch, which plays a critical role in mineralization (Shao et al., 2018a, 2018b). Finally, if osteocytes are the mechanosensor cells of bone (see later), the attachment of osteocytes to matrix molecules is likely to be of major importance for the transduction of stress signals into cellular signals. Production and secretion of specific matrix molecules offer a possibility for the cells to regulate their own adhesion and, thereby, sensitivity for stress signals.

In addition to collagenous and noncollagenous proteins, the bone matrix contains proteoglycans. These macromolecules consist of a core protein to which one or more glycosaminoglycan side chains are covalently bound. Early electron microscopical studies (Jande, 1971) showed that the osteocyte body, as well as its cell processes, is surrounded by a thin layer of unmineralized matrix containing collagen fibrils and proteoglycans. The proteoglycans were shown to consist of chondroitin 4-sulfate, dermatan sulfate, and keratan sulfate with immunocytochemical methods (Maeno et al., 1992; Smith et al., 1997; Takagi et al., 1997). These observations are supported by the findings of Sauren et al. (1992), who demonstrated an increased presence of proteoglycans in the pericellular matrix by staining with the cationic dye cuprolinic blue. Of special interest is the reported presence of hyaluronan in osteocyte lacunae (Noonan et al., 1996). CD44, which is highly expressed on the osteocyte membrane, is a hyaluronan-binding protein and also binds to collagen, fibronectin, and osteopontin (Nakamura and Ozawa, 1996; Yamazaki et al., 1999). The essential pericellular matrix component perlecan/HSPG2, a large monomeric heparan sulfate proteoglycan, acts as a strong but elastic tether that connects the osteocyte cell body to the bone matrix (Wijeratne et al., 2016). Solute transport in the lacunar–canalicular system plays an important role in osteocyte metabolism and cell–cell signaling. The pericellular matrix-filled lacunar–canalicular system is bone's chromatographic column, where fluid/solute transport to and from the osteocytes is regulated. A better definition of the chemical composition, deposition rate, and turnover rate of the osteocyte pericellular matrix will improve our understanding of osteocyte physiology and bone metabolism (reviewed by Wang, 2018).

The osteocyte cytoskeleton and cell–matrix adhesion

As mentioned earlier, the cell–matrix adhesion of osteocytes is of importance for the translation of biomechanical signals produced by loading of bone into chemical signals. Study of the adhesion of osteocytes to extracellular matrix molecules became feasible with the development of osteocyte isolation and culture methods (Van der Plas and Nijweide, 1992). These studies found little difference between the adhesive properties of osteocytes and osteoblasts, although the patterns of adhesion plaques (osteocytes, many small focal contacts; osteoblasts, larger adhesion plaques) were quite different (Aarden et al., 1996a). Both cell types adhered equally well to collagen type I, osteopontin, vitronectin, fibronectin, and thrombospondin. Integrin receptors are involved, as is shown by the inhibiting effects of small peptides containing an RGD sequence on the adhesion to some of these proteins. Adhesion to all aforementioned matrix molecules was blocked by an antibody reacting with the β_1-integrin subunit (Aarden et al., 1996a). The identity of the α units involved is unknown as of this writing.

Deformation of the bone matrix upon loading may cause a physical "twisting" of integrins at sites where osteocytes adhere to the matrix. Integrins are coupled to the cytoskeleton via molecules such as vinculin, talin, and α-actinin. In osteocytes, especially in the osteocytic cell processes, the actin-bundling protein fimbrin appears to play a prominent role (Tanaka-Kamioka et al., 1998). Mechanical twisting of the cell membrane via integrin-bound beads has been demonstrated to induce cytoskeletal rearrangements in cultured endothelial cells (Wang and Ingber, 1994). The integrin–cytoskeleton complex may therefore play a role as an intracellular signal transducer for stress signals (Litzenberger et al., 2010; Santos et al., 2010). Spectrin, another structural cytoskeletal protein required for the differentiation of osteoblasts to osteocytes (Bonewald, 2011), has been identified as a mechanosensitive element within the osteocyte (Wu et al., 2017). Disruption of the spectrin network promotes Ca^{2+} influx and nitric oxide (NO) secretion as a result of reduced stiffness. In addition to the integrins, the nonintegrin adhesion receptor CD44 may contribute to the attachment of osteocytes to the surrounding matrix. CD44 is present abundantly on the osteocyte surface (Hughes et al., 1994; Nakamura et al., 1995) and is also linked to the cytoskeleton.

Evidence is accumulating highlighting the crucial role of the cytoskeleton in a multitude of cellular processes. The cytoskeleton, just like our bony skeleton, provides structure and support for the cell, is actively adapted, and is highly responsive to external physical and chemical stimuli. The cytoskeleton is strongly involved in processes such as migration, differentiation, mechanosensing, and even cell death, and largely determines the material properties of the cell (i.e., stiffness).

For bone cells it was shown that the production of signaling molecules in response to an in vitro fluid shear stress (at 5 and 9 Hz) and vibration stress (5−100 Hz) correlated with the applied stress rate (Bacabac et al., 2004, 2006; Bacabac et al., 2006a,b; Mullender et al., 2006). The faster the stress was applied, the stronger the observed response of the cells. Interestingly, high-rate stimuli were found to condition bone cells to be more sensitive to high-frequency, low-amplitude loads (Bacabac et al., 2005). From the field of physics, it is known that the effects of stresses applied at different rates on an object are largely determined by the material properties of that object. This implies that bone cellular metabolic activity (e.g., the production of signaling molecules) and mechanical properties of the cell are related. Bacabac et al. (2006b) developed a novel application of two-particle microrheology, for which he devised a three-dimensional in vitro system using optical traps to quantify the forces induced by cells on attached fibronectin-coated probes (4 μm). The frequency at which the cells generate forces on the beads is related to the metabolic activity of the cell. This system can also be used to apply controlled forces on the cell using the beads, and enables the study of cells under a controlled three-dimensional morphological configuration, with possibilities for a variety of probe coatings for simulating cell−extracellular matrix attachment. Using this setup the generation of forces by different cells was probed to understand the relation between cellular metabolic activities and material properties of the cell. It was shown that at 37°C, CCL-224 fibroblasts exhibited higher force fluctuation magnitude compared with MLO-Y4 osteocytes (Bacabac et al., 2006b). The force fluctuations on the attached probes reflect intracellular movement, which might include actin (and microtubule) polymerization, as well as motor and cross-linker dynamics. Since cell migration involves these dynamic processes, the lower magnitude of force fluctuation might reflect a lower capacity of osteocytes for motility compared with fibroblasts.

Using the optical trap device, the material properties of round suspended MLO-Y4 osteocytes and flat adherent MLO-Y4 osteocytes were characterized. In addition, the cells were loaded with an NO-sensitive fluorescent dye (Vatsa et al., 2006) and the NO response of the cells to forces applied on the cell using the attached probes was studied. Osteocytes under round suspended morphology required lower force stimulation to show an NO response, even though they were an order of magnitude more elastic compared with flat adherent cells (Bacabac et al., 2006b). Apparently, elastic osteocytes require less mechanical force to respond than stiffer cells. On the other hand, flat adherent MLO-Y4 cells, primary chicken osteocytes, MC3T3-E1 osteoblasts, and primary chicken osteoblasts all showed similar elastic moduli of less than 1 kPa (Bacabac et al., 2006b), even though osteocytes are known to be more responsive to mechanical stress than osteoblasts (Klein-Nulend et al., 1995a). This indicates that differences in mechanosensitivity between cells might not be directly related to the elasticity of the cell, but might be more related to other cell-specific properties (i.e., the presence of receptors or ion channels in the membrane). Alternatively, the mechanosensitivity of a cell might be related to how cells change their material properties in relation to deformation.

Simultaneous with the increased NO release in response to mechanical stimulation, MLO-Y4 osteocytes showed increased force traction on the attached beads. In other words, the cells started to "pull harder" on the beads and generated a force up to nearly 30 pN, which interestingly is within the order of forces necessary for activating integrins. Whether there is a causal link between loading-induced NO production by the cells and force generation is under investigation at the time of writing. Since force generation and cell elasticity are (indirectly) related, these results might indicate that osteocytes adapt their elasticity in response to a mechanical stimulus. Indeed, experiments with an atomic force microscope and optical tweezers have shown that osteocytes become "stiffer" after mechanical loading (Bacabac et al., 2006b, and unpublished observations). This stiffening response was related to actual changes in material properties of the cell, suggesting that the cells actively change their cytoskeleton in response to a mechanical load.

Considering the role of the osteocytes as the professional mechanosensors of bone, and the importance of the cytoskeleton for the response of osteocytes to mechanical loading, much is to be expected from research focusing on the cytoskeletal components of the osteocyte (reviewed by Klein-Nulend et al., 2012).

Hormone receptors in osteocytes

PTH receptors have been demonstrated on rat osteocytes in situ (Fermor and Skerry, 1995) and on isolated chicken osteocytes (Van der Plas et al., 1994). Administered in vitro, PTH was reported to increase cAMP levels in isolated chicken osteocytes (Van der Plas et al., 1994; Miyauchi et al., 2000) and, administered in vivo, to increase fos protein (Takeda et al., 1999) and the mRNAs of c-fos, c-jun, and Il-6 in rat osteocytes (Liang et al., 1999). Matrix metalloproteinase 14 (MMP14) has been shown to be a novel target of PTH signaling in osteocytes that controls resorption by regulating soluble RANKL production (Delgado-Calle et al., 2018). As it is now generally accepted that osteocytes are involved in the transduction of mechanical signals into chemical signals regulating bone (re)modeling, PTH might modulate the osteocytic response to mechanical strain. Injection of PTH into rats was shown to augment the osteogenic response of bone to mechanical stimulation in vivo, whereas thyroparathyroidectomy abrogated the mechanical responsiveness of bone (Chow et al., 1998). However, such an approach cannot separate an effect at the level of osteocyte mechanosensing from one at the level of osteoprogenitor recruitment. One mechanism by which PTH may act on osteocytes was suggested by the reports of Schiller et al. (1992) and Donahue et al. (1995). These authors found that PTH increased Cx43 gene expression and gap-junctional communication in osteoblastic cells. In osteocytes, where cell-to-cell communication is so important, a similar effect might lead to more efficient communication within the osteocyte network. Osteocytes also express the receptor for the carboxy-terminal region of PTH that may play a role in osteocyte viability (Divieti et al., 2001). Activation of receptors for 1,25-dihydroxyvitamin D_3 ($1,25(OH)_2D_3$), which were also shown to be present in osteocytes by immunocytochemistry (Boivin et al., 1987) and by in situ hybridization (Davideau et al., 1996), may have similar effects.

Parfitt concluded in the 1970s that osteocytes control the rapid release of calcium (within minutes) in response to PTH (Parfitt, 1976). More recent in vivo studies have supported Parfitt's hypothesis that osteocytes are the target of PTH. There are profound skeletal effects in transgenic mice with the activation of the PTH receptor in osteocytes (O'Brien et al., 2008; Rhee et al., 2011). Mice lacking the PTH receptor in osteocytes lose less bone with lactation, and an increase in osteocyte lacunar size is absent (Qing et al., 2012). This "perilacunar remodeling" related to "osteocytic osteolysis" (Bélanger, 1969) is achieved by the expression of so-called "osteoclast-specific genes" such as carbonic anhydrases, ATPases, MMP13, cathepsin K, and TRAP (tartrate-resistant acid phosphatase) by the osteocyte. In 2017 it was shown that osteocytes reduce the pH within their lacunocanalicular network during lactation, a process necessary for the release of calcium (Jähn et al., 2017).

Another important hormone involved in bone metabolism is estrogen. Numerous studies have demonstrated that a decrease in blood estrogen levels is accompanied by a loss of bone mass. One explanation for this phenomenon is that estrogen regulates the set point for the mechanical responsiveness of bone (Frost, 1992), i.e., that lowering the ambient estrogen level increases the level of strain in bone necessary for the bone to respond with increased bone formation. If osteocytes are the main mechanosensors of bone, it is reasonable to suppose that osteocytes are the site of set-point regulation by estrogen. Estrogen receptors (ERα) were demonstrated in osteocytes with immunocytochemistry and in situ hybridization (Braidman et al., 1995; Hoyland et al., 1999) in tissue sections. In addition, Westbroek et al. (2000b) found higher levels of ERα in isolated osteocytes than in osteoblasts or osteoblast precursors. It has been shown that the anabolic response of bone to mechanical loading requires ERα (Lee et al., 2003). Studies suggest that osteocytes indeed use ERα to respond to strain, although the ERα content is regulated by estrogen (Zaman et al., 2006). In vivo studies performing targeted deletion of ERα in osteocytes using the Dmp1-Cre model have generally found a reduction in trabecular bone either in females (Kondoh et al., 2014) or in males (Windahl et al., 2013). It is not clear why these studies observed differences in the two sexes.

Receptors for PTH, $1,25(OH)_2D_3$, and estrogen, as well as the androgen receptor (Abu et al., 1997), the glucocorticoid receptor α (Abu et al., 2000; Silvestrini et al., 1999), and various prostaglandin receptors (Lean et al., 1995; Sabbieti et al., 1999), have been described in osteocytes. The prostaglandin receptors, in addition to viability, may be important for communication within the osteocyte network during mechanotransduction (see later). As outlined earlier, glucocorticoids clearly induce osteocyte apoptosis, but may have additional effects. Glucocorticoid treatment causes mature osteocytes to enlarge their lacunae and remove mineral from their microenvironment (Lane et al., 2006). Therefore, osteocytes appear to be able to modify their microenvironment in response to certain factors such as PTH/PTH-related protein and glucocorticoids.

Osteocyte function

Blood–calcium/phosphate homeostasis

The organization of osteocytes as a network of gap junction–coupled cells in each osteon represents such a unique structure that one expects it to have an important function in the metabolism and maintenance of bone. The network offers two advantages that may be exploited by the tissue:

1. a tremendous cell—bone surface contact area, about 2 orders of magnitude larger than the contact area the osteoblasts and lining cells have (Johnson, 1966),
2. an extensive intracellular and an extracellular communication system between sites within the bone and the bone surface.

The first consideration led Bélanger (1969) and others to propose the hypothesis that osteocytes are capable of local bone remodeling or osteocytic osteolysis. According to this hypothesis, osteocytes are coresponsible for blood—calcium homeostasis. Later studies (Boyde, 1980; Marotti et al., 1990) supplied alternative explanations for the observations that appeared to support the osteocytic osteolysis theory. The possibility remains, however, that osteocytes are involved in the facilitation of calcium diffusion in and out of the bone (Bonucci, 1990). Although the bulk of calcium transport in and out of the bone is apparently taken care of by osteoblasts and osteoclasts (Boyde, 1980; Marotti et al., 1990), osteocytes may have a function in the fine regulation of blood—calcium homeostasis. The major emphasis of present-day thinking is, however, on the role of the osteocyte network as a three-dimensional sensor and communication system in bone.

Osteocytes may also play a major role in phosphate homeostasis. PHEX is a metalloendoproteinase found on the plasma membrane of osteoblasts and osteocytes (Ruchon et al., 2000) whose substrate is not known. The precise function of PHEX is unclear but it clearly plays a role in phosphate homeostasis and bone mineralization. *Pex* deletion or loss of function results in X-linked hypophosphatemic rickets (The HYP Consortium, 1995). Nijweide and coworkers (Westbroek et al., 2002) were one of the first groups to propose that the osteocyte network could be considered a gland that regulates bone phosphate metabolism through expression of PHEX.

Other proteins known to regulate mineralization and mineral homeostasis are also highly expressed in osteocytes, such as DMP1, Mepe, and sclerostin. Deletion or mutation of DMP1, a gene that is highly expressed in embedding osteocytes and mature osteocytes, results in hypophosphatemic rickets (Feng et al., 2006), similar to the deletion of Phex. DMP1 is expressed along the canalicular wall, while Phex is expressed on the membrane surface of dendrites and the cell body. Other players in mineral metabolism include MEPE and FGF23, which are also highly expressed in osteocytes (Liu et al., 2006). FGF23 is highly elevated in osteocytes in DMP1-null mice, which results in hypophosphatemia in these animals. The osteocyte network might be viewed as an endocrine gland regulating mineral metabolism. Another fascinating hypothesis is that mineral metabolism is regulated by mechanical loading. DMP1 and Pex gene expression are increased in response to load (Gluhak-Heinrich et al., 2003), Sost expression is inhibited (Robling et al., 2006), while a biphasic response is observed for Mepe expression (Gluhak-Heinrich et al., 2005). Whether FGF23 expression is also regulated by loading is unknown as of this writing. Therefore proteins known to regulate mineralization and mineral homeostasis are also regulated by mechanical loading in osteocytes. These osteocyte regulators of mineral homeostasis can also be responsible for disease. For a review of the physiological and pathological functions of these molecules see Bonewald (2017).

Functional adaptation, Wolff's law

Functional adaptation is the term used to describe the ability of organisms to increase their capacity to accomplish a specific function with increased demand and to decrease this capacity with lesser demand. In the 19th century, the anatomist Julius Wolff proposed that mechanical stress is responsible for determining the architecture of bone and that bone tissue is able to adapt its mass and three-dimensional structure to the prevailing mechanical usage to obtain a higher efficiency of load bearing (Wolff, 1892). For the past century, Wolff's law has become widely accepted. Adaptation will improve an individual animal's survival chance because bone is not only hard but also heavy. Too much of it is probably as bad as too little, leading either to uneconomic energy consumption during movement (for too high a bone mass) or to an enhanced fracture risk (for too low a bone mass). This readily explains the usefulness of mechanical adaptation as an evolutionary driver, even if we do not understand how it is performed.

Osteocytes as mechanosensory cells

In principle, all cells of bone may be involved in mechanosensing, as eukaryotic cells in general are sensitive to mechanical stress (Oster, 1989). However, several features argue in favor of osteocytes as the mechanosensory cells *par excellence* of bone as discussed earlier in this chapter. From a cell biological viewpoint, therefore, bone tissue is a three-dimensional network of cells, most of which are surrounded by a very narrow sheath of unmineralized matrix, followed by a much wider layer of mineralized matrix. The sheath of unmineralized matrix is penetrated easily by macromolecules such as albumin and peroxidase (McKee et al., 1993; Tanaka and Sakano, 1985). However, others have shown that although small tracers (<6 nm) readily pass through the lacunar–canalicular porosity in the absence of mechanical loading, there appears to be an upper limit of size between 6 and 10 nm for molecular movement from bone capillaries to osteocytic lacunae in rat long bone. It was suggested that this range of pore size represents the fiber spacing that has been proposed for the annular space based on the presence of a proteoglycan fiber matrix surrounding the osteocytes (Wang et al., 2004). Therefore, there is an intracellular as well as an extracellular route for the rapid passage of ions and signal molecules. This allows for several types of cellular signaling from osteocytes lying deep within the bone tissue to surface-lining cells and vice versa (Cowin et al., 1995).

Experimental studies indicate that osteocytes are indeed sensitive to stress applied to intact bone tissue. In vivo experiments using the functionally isolated turkey ulna have shown that immediately following a 6-min period of intermittent (1 Hz) loading, the number of osteocytes expressing glucose-6-phosphate dehydrogenase activity was increased in relation to local strain magnitude (Skerry et al., 1989). The tissue strain magnitude varied between 0.05% and 0.2% (500–2000 microstrain), in line with in vivo peak strains in bone during vigorous exercise. Other models, including strained cores of adult dog cancellous bone, embryonic chicken tibiotarsi, mouse ulnae, rat caudal vertebrae, and rat tibiae, as well as experimental tooth movement in rats, have demonstrated that osteocytes in intact bone change their enzyme activity and RNA synthesis rapidly after mechanical loading (El-Haj et al., 1990; Dallas et al., 1993; Lane et al., 2006; Lean et al., 1995; Forwood et al., 1998; Terai et al., 1999). These studies show that intermittent loading produces rapid changes of metabolic activity in osteocytes and suggest that osteocytes may indeed function as mechanosensors in bone. The mechanical environment of the stress-sensitive osteocyte varies with the geometry of the osteocyte lacuna (McCreadie et al., 2004). Computer simulation studies of bone remodeling, assuming this to be a self-organizational control process, predict a role for osteocytes, rather than lining cells and osteoblasts, as stress sensors of bone (Mullender and Huiskes, 1995, 1997; Huiskes et al., 2000; Ruimerman et al., 2005). A regulating role of strain-sensitive osteocytes in basic multicellular unit (BMU) coupling has been postulated by Smit and Burger (2000). Using finite-element analysis, the subsequent activation of osteoclasts and osteoblasts during coupled bone remodeling was shown to relate to opposite strain distributions in the surrounding bone tissue. In front of the cutting cone of a forming secondary osteon, an area of *decreased* bone strain was demonstrated, whereas a layer of *increased* strain occurs around the closing cone (Smit and Burger, 2000). Osteoclasts therefore attack an area of bone where the osteocytes are underloaded, whereas osteoblasts are recruited in a bone area where the osteocytes are overloaded. Hemiosteonic remodeling of trabecular bone showed a similar strain pattern (Smit and Burger, 2000). Thus, bone remodeling regulated by load-sensitive osteocytes can explain the maintenance of osteonic and trabecular architecture as an optimal mechanical structure, as well as adaptation to alternative external loads (Huiskes et al., 2000; Smit and Burger, 2000; Van Oers et al., 2014 [review]).

Osteocytes are thought to be the major regulator of bone mechanosensation events, but little is known about how osteocytes in vivo acutely respond to tissue-level mechanical loading. A technique has been reported for the direct in vivo observation of osteocyte calcium signaling events with simultaneous whole-bone loading (Lewis et al., 2017). Osteocyte populations were found to integrate mechanical signals by altering the number of responding cells and this effect was dependent on loading frequency (Lewis et al., 2017). Ca^{2+} signaling in the osteocyte network in chick calvariae has been studied using three-dimensional time-lapse imaging (Tanaka et al., 2017). In response to flow, intracellular Ca^{2+} significantly increased in developmentally mature osteocytes in comparison with young osteocytes in the bone matrix, indicating that developmentally mature osteocytes are more responsive to mechanical stress than young osteocytes and have important functions in bone formation and remodeling (Tanaka et al., 2017).

If osteocytes are the mechanosensors of bone, how do they sense mechanical loading? This key question is, unfortunately, still open because it has not yet been established unequivocally how the loading of intact bone is transduced into a signal for the osteocytes. The application of force to bone during movement results in several potential cell stimuli. These include changes in hydrostatic pressure, direct cell strain, fluid flow, and electric fields resulting from electrokinetic effects accompanying fluid flow (Pienkowski and Pollack, 1983). Evidence has been increasing steadily for the flow of canalicular interstitial fluid as the likely stress-derived factor that informs the osteocytes about the level of bone loading (Cowin et al., 1991, 1995; Cowin, 1999; Weinbaum et al., 1994; Klein-Nulend et al., 1995b; Knothe-Tate et al., 2000; Burger and Klein-Nulend, 1999; You et al., 2000; Bakker et al., 2009). Moreover, a realistic high-resolution image-based three-dimensional model has been used to analyze the microscale fluid flow in a human osteocyte canaliculus (Kamioka et al., 2012). In this view, canaliculi are the bone porosity of interest, and the osteocytes the mechanosensor cells.

Canalicular fluid flow and osteocyte mechanosensing

In healthy, adequately adapted bone, strains as a result of physiological loads (e.g., resulting from normal locomotion) are quite small. Quantitative studies of the strain in bones of performing animals (e.g., galloping horses, fast-flying birds, even a running human volunteer) found a maximal strain not higher than 0.2%–0.3% (Rubin, 1984; Burr et al., 1996). This poses a problem in interpreting the results of in vitro studies of strained bone cells, where much higher deformations, on the order of 1%–10%, were needed to obtain a cellular response (for a review, see Burger and Veldhuijzen, 1993). In these studies, isolated bone cells were usually grown on a flexible substratum, which was then strained by stretching or bending. For instance, unidirectional cell stretching of 0.7% was required to activate PGE_2 production in primary bone cell cultures (Murray and Rushton, 1990). However, in intact bone, a 0.15% bending strain was already sufficient to activate prostaglandin-dependent adaptive bone formation in vivo (Turner et al., 1994; Forwood, 1996). If we assume that bone organ strain is somehow involved in bone cell mechanosensing, then bone tissue seems to possess a lever system whereby small matrix strains are transduced into a larger signal that is detected easily by osteocytes. The canalicular flow hypothesis proposes such a lever system. Indeed, in vitro experiments that relate the effects of fluid flow and substrate straining have shown that fluid-flow-induced shear stress induces higher release of signaling molecules (McGarry et al., 2005a,b). A numerical study showed that the deformation of cells on a two-dimensional substrate caused by fluid flow is fundamentally different from that induced by substrate straining (McGarry et al., 2005a,b). Fluid shear stress had a larger overturning effect on the bone cells, while substrate strain predominantly affected cell–substrate attachments. Whether these observations can be extrapolated toward osteocytes that are embedded in a three-dimensional matrix is a matter of debate, but they clearly indicate that substrate deformation and a flow of interstitial fluid could have differential effects on cells.

The flow of extracellular tissue fluid through the lacunocanalicular network as a result of bone tissue strains was made plausible by the theoretical study of Piekarski and Munro (1977) and has been shown experimentally by Knothe-Tate et al. (1998, 2000). This strain-derived extracellular fluid flow may help keep osteocytes healthy, particularly the deeper ones, by facilitating the exchange of nutrients and waste products between the haversian channel and the osteocyte network of an osteon (Kufahl and Saha, 1990). However, a second function of this strain-derived interstitial fluid flow could be the transmission of "mechanical information" (Fig. 6.4). The magnitude of interstitial fluid flow through the lacunocanalicular network is directly related to the amount of strain of the bone organ (Cowin et al., 1991). Because of the narrow diameter of the canaliculi, bulk bone strains of about 0.1% will produce a fluid shear stress in the canaliculi of roughly 1 Pa (Weinbaum et al., 1994), enough to produce a rapid response in endothelial cells (Frangos et al., 1985; Kamiya and Ando, 1996).

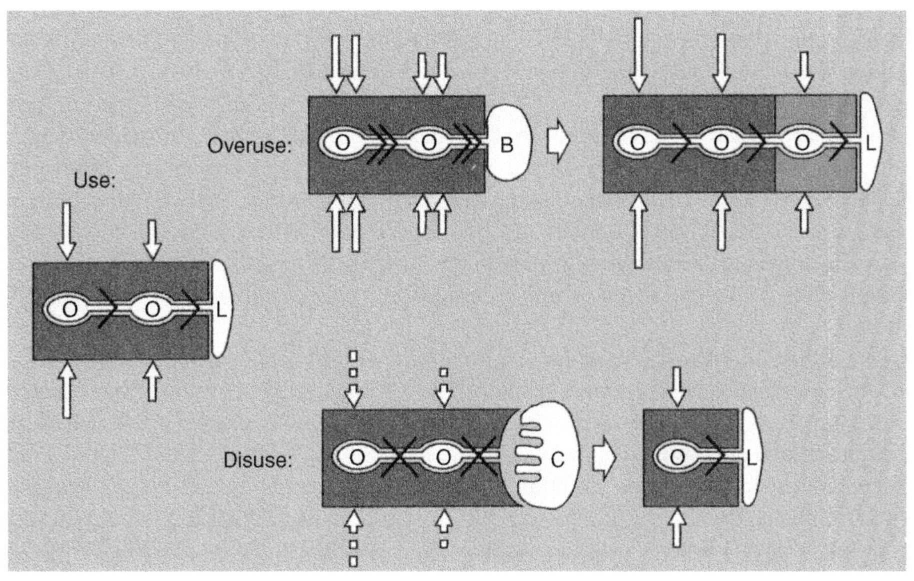

FIGURE 6.4 Schematic representation of how the osteocyte network may regulate bone modeling. In the steady state (*Use*), normal mechanical use ensures a basal level of fluid flow through the lacunocanalicular porosity, indicated by an *arrowhead* through the canaliculi. This basal flow keeps the osteocytes viable and also ensures basal osteocyte activation and signaling, thereby suppressing osteoblastic activity as well as osteoclastic attack. During (local) overuse (*Overuse*), osteocytes are overactivated by enhanced fluid flow (indicated by *double arrowheads*), leading to the release of osteoblast-recruiting signals. Subsequent osteoblastic bone formation reduces the overuse until normal mechanical use is reestablished, thereby reestablishing the steady state of basal fluid flow. During (local) disuse (*Disuse*), osteocytes are inactivated by lack of fluid flow (indicated by *crosses* through canaliculi). Inactivation leads to a release of osteoclast-recruiting signals, to a lack of osteoclast-suppressing signals, or both. Subsequent osteoclastic bone resorption reestablishes normal mechanical use (or loading) and basal fluid flow. *O*, osteocyte; *L*, lining cell; *B*, osteoblast; *C*, osteoclast; *dark gray area*, mineralized bone matrix; *light gray* area, newly formed bone matrix; *white arrows* represent direction and magnitude of loading. *Adapted from Burger, E.H., Klein-Nulend, J., 1999. Mechanotransduction in bone: role of the lacuno-canalicular network. FASEB J. 13, S101–S112.*

Experimental studies in vitro have demonstrated that osteocytes are indeed quite sensitive to the fluid shear stress of such a magnitude compared with osteoblasts and osteoprogenitor cells (Klein-Nulend et al., 1995a,b; Ajubi et al., 1996; Westbroek et al., 2000a,b; Westbroek et al., 2001). These results suggest that the combination of the cellular three-dimensional network of osteocytes and the accompanying porous network of lacunae and canaliculi acts as the mechanosensory organ of bone. Different loading-induced canalicular flow patterns around cutting cone and reversal zone during remodeling could, for instance, explain the typical alignment of haversian canals (Burger et al., 2003). Volumetric strain in the bone around a BMU cutting cone has been related to canalicular fluid flow (Smit et al., 2002), and the predicted area of low canalicular flow around the tip of the cutting cone was proposed to induce local osteocyte apoptosis. Unloading-induced osteocyte apoptosis has been shown in rat bone and was highly associated with osteoclastic bone resorption (Basso and Heersche, 2006). Osteocyte apoptosis at the tip of the cutting cone would attract osteoclasts, leading to further excavation of bone in the direction of loading. Importantly, mechanical loading by fluid shear stress has been shown to promote osteocyte survival (Bakker et al., 2004). The model by Smit et al. (2002) further predicts that at the base of the cutting cone and farther down the reversal zone, osteocytes receive enhanced fluid shear stress during loading. This could prevent osteocyte apoptosis, but may also promote the retraction and detachment of osteoclasts from the bone surface. These two mechanisms, attraction of osteoclasts to the cutting cone tip and induction of osteoclast detachment from the cutting cone base, together explain the mechanically meaningful behavior of osteoclasts during remodeling.

The flow of interstitial fluid through the bone canaliculi will have two effects: a mechanical one derived from the fluid shear stress and an electrokinetic one derived from streaming potentials (Pollack et al., 1984; Salzstein and Pollack, 1987). Either of the two, or both in combination, might activate the osteocyte. For instance, streaming potentials might modulate the movement of ions such as calcium across the cell membrane (Hung et al., 1995, 1996), whereas shear stress will pull at the macromolecular attachments between the cell and its surrounding matrix (Wang and Ingber, 1994). Both ion fluxes and cellular attachment are powerful modulators of cell behavior and therefore good conveyors of physical information (Sachs, 1989; Ingber, 1991). In vitro experiments using bone cells subjected to a flow of fluid of increasing viscosity suggest that fluid shear stress is the major activator of the bone cell response to loading (Bakker et al., 2001). However, care should be taken when extrapolating results obtained with cells seeded on a two-dimensional surface to the three-dimensional in vivo situation.

Mechanosensation and the resulting mechanotransduction of biochemical signals may be a complex event (Bonewald, 2006). Perturbation of dendrites, the cell body, and even primary cilia could be involved in combination or sequentially in this response. The osteocytes' response to different kinds of mechanical loading has predominantly been studied in cell cultures or entire bone, while knowledge of mechanosensing in osteocytes at the single-cell level is essential for understanding the complex process of bone adaptation. For instance, the mechanosensitive part of the osteocytes, the cell body or the cell processes, has not been determined yet. More information is needed on the single osteocyte's response to a localized mechanical stimulation. It has been postulated that the osteocyte processes are an order of magnitude more sensitive to mechanical loading than their cell bodies (Cabahug-Zuckermann et al., 2018). The underlying mechanisms are not clear but may be related to the infrequent $\alpha V\beta 3$ integrin sites where the osteocyte cell processes attach to canalicular walls (Cabahug-Zuckermann et al., 2018). These sites develop dramatically elevated strains during load-induced fluid flow in the lacunar canalicular system and might be primary sites for osteocyte mechanotransduction.

It has been suggested that fluid flow over dendrites in the lacunar—canalicular porosity can induce strains in the actin filament bundles of the cytoskeleton that are more than an order of magnitude larger than tissue level strains. Using the latest ultrastructural data for the cell process cytoskeleton and the tethering elements that attach the process to the canalicular wall, a realistic three-dimensional model was created for the osteocyte process. Using this model, the deformed shape of the tethering elements and the hoop strain on the central actin bundle as a result of loading-induced fluid flow were predicted. It was found that tissue-level strains of >1000 microstrain at 1 Hz resulted in a hoop strain of $>0.5\%$. The tethering elements of the osteocyte process can thus act as a strain amplifier (Han et al., 2004).

It has been published that osteocytes express a single primary cilium (Xiao et al., 2006). These authors also showed that polycystin 1 (PKD1/PC1), a known mechanosensory protein in the kidney, does play a role in normal bone structure. Primary cilia in other cells clearly function as sensors of odors, light, and movement (Singla and Reiter, 2005). It remains to be determined whether the bone defect in animals with defective PKD1 function is due to defective function of PKD1 in cilia, or whether PKD1 has a function in another location in the cell. Early data suggest that loss of cilia results in decreased sensitivity to fluid flow shear stress (Malone et al., 2006). Determining the function of PKD1 and/or cilia should lend insight into mechanotransduction in osteocytes.

The osteocyte has been viewed as either a quiescent cell acting as a "placeholder" in bone or as a mechanosensory cell as outlined in this chapter. Most observations using static two-dimensional imaging suggest that the dendritic processes are somewhat permanently anchored to the lacunar wall. Dallas et al. have shown cell body movement and the extension and retraction of dendritic processes over time using dynamic imaging of living osteocytes within their lacunae (Veno et al., 2006). Calvarial explants from transgenic mice with GFP expression targeted to osteocytes revealed that, far from being a static cell, the osteocyte is highly dynamic. Therefore, dendrites, rather than being permanent connections between osteocytes and with bone surface cells, may have the capacity to connect and disconnect and reconnect. With this new information, theories regarding shear stress within lacunae and osteocyte signaling will require modification.

Osteocyte shape and mechanosensing

The ability of osteocytes to sense and respond to mechanical stimuli depends on many factors, such as the shape of the osteocyte cell bodies, number and length of the cell processes, structure of the cytoskeleton, and presence of primary cilia (Willems et al., 2017). One intriguing factor is osteocyte lacunar shape, which has been hypothesized to affect the transduction of strain on the whole bone to the direct osteocyte microenvironment (Nicolella et al., 2006). It has been shown that considerable variations in the shape of osteocytes and their lacunae exist and that these variations may depend on anatomical location and the age of the bone (Van Hove et al., 2009; Vatsa et al., 2008a,b; Bacabac et al., 2008). Considering that a proper response of osteocytes to mechanical stimuli is highly important to maintain bone strength, the question arises whether changes in lacunar morphology could underlie the alterations in the bone adaptive response seen with aging (reviewed by Hemmatian et al., 2017).

Modifications in the morphology and orientation of osteocytes and their lacunae could result from hormonal changes, as well as from changes in mechanical loading. The alignment and shape of the osteocytes and their lacunae are related to the direction of the mechanical loading (Vatsa et al., 2008a,b; Sugawara et al., 2013). More flattened and elongated osteocytes and lacunae are observed in fibula loaded unidirectionally than in calvarial bone, which is loaded in different directions (Vatsa et al., 2008a,b). Irregularly shaped osteocytes distributed in different directions are found in the femurs of embryonic mice in the absence of mechanical loading, whereas the osteocytes in the femurs of 6-week-old mice subjected to mechanical loading are more flat and spindle shaped, and oriented parallel to the longitudinal axis of the bone (Sugawara et al., 2013). Furthermore, in neurectomized mice under little or no mechanical loading during growth, the osteocytes are round without any preferred orientation (Sugawara et al., 2013). The actin filaments in the osteocyte cytoskeleton distribute in the same direction as the mechanical loading (Vatsa et al., 2008a,b). In addition, the osteocyte morphology might vary in bone pathologies, i.e., the shape of osteocytes and their lacunae are significantly different in the tibia of individuals with osteopetrosis, osteoporosis, and osteoarthritis (Van Hove et al., 2009). Osteocyte lacunae in bone from osteoporotic persons are large and round, lacunae from osteopetrotic persons are small and discoid shaped, whereas lacunae from osteoarthritic persons are large and elongated (Van Hove et al., 2009). Furthermore, in the osteopetrotic bone, the osteocyte lacunae are less oriented to the loading direction in comparison with the orientation of osteocyte lacunae in the osteoarthritic and osteopenic bone (Van Hove et al., 2009). The variations in the shape and alignment of osteocytes and their lacunae in different bone pathologies could reflect an adaptation to the different micromechanical environments with different matrix strain associated with differences in bone mineral density.

Mechanoresponsive osteocytes regulate bone mass, which is important for, among others, dental implant success. Significant differences in osteocyte surface area and orientation seem to exist locally in maxillary bone, which may be related to the tensile strain magnitude and orientation. This might reflect local differences in the osteocyte mechanosensitivity and bone quality, suggesting differences in dental implant success based on the location in the maxilla (Wu et al., 2018).

Response of osteocytes to fluid flow in vitro

The technique of immunodissection, as discussed earlier in this chapter, made it possible to test the canalicular flow hypothesis by comparing the responsiveness of osteocytes, osteoblasts, and osteoprogenitor cells to fluid flow. The strength of the immunodissection technique is that three separate cell populations with a high (90%) degree of homogeneity are prepared, representing (1) osteocytes, with the typical "spider-like" osteocyte morphology and little matrix synthesis, (2) osteoblasts, with a high synthetic activity of bone matrix-specific proteins, and (3) (from the periosteum) osteoprogenitor cells, with a fibroblast-like morphology and very high proliferative capacity (Nijweide and Mulder, 1986). Because the cells are used within 2 days after isolation from the bone tissue, they may well represent the three differentiation steps of osteoprogenitor cell, osteoblast, and osteocyte. In contrast, mixed cell cultures derived from bone that are generally used to represent "osteoblastic" cells are likely to contain cells in various stages of differentiation. Therefore, changes in mechanosensitivity related to progressive cell differentiation cannot be studied in such cultures. The development of differentiation-stage-specific bone cell lines has facilitated the comparison of cell-type-specific responses to mechanical loading. Ponik et al. (2007) have described significant differences in the response to fluid shear between MC3T3-E1 osteoblasts and MLO-Y4 osteocytes. Stress fibers were formed and aligned within osteoblasts after 1 h of unidirectional fluid flow, but this response was not observed until greater than 5 h of oscillatory fluid flow (Ponik et al., 2007). However, due to their clonal selection, cell lines do not necessarily express the whole range of bone-specific genes characteristic of primary bone cells. Therefore, the use of primary bone cells can be preferred to the use of cell lines when comparing osteoblasts' and osteocytes' behavior.

Using immunoseparated primary cell populations, osteocytes were found to respond far more strongly to fluid flow than osteoblasts, and these stronger than osteoprogenitor cells (Klein-Nulend et al., 1995a,b; Ajubi et al., 1996; Westbroek et al., 2000a). Osteocytes also appeared to be more responsive to fluid flow than osteoblasts or osteoprogenitor cells with respect to the production of soluble signaling molecules affecting osteoblast proliferation and differentiation (Vezeridis et al., 2005). The release of these soluble factors was at least partially dependent on the activation of an NO pathway in osteocytes in response to fluid flow. PFF with a mean shear stress of 0.5 Pa (5 dyn/cm^2) with a cyclic variation of $\pm$0.2 Pa at 5 Hz stimulated the release of NO and PGE_2 and PGI_2 rapidly from osteocytes within minutes (Klein-Nulend et al., 1995a; Ajubi et al., 1996). Osteoblasts showed less response, and osteoprogenitor cells (periosteal fibroblasts) still less. Intermittent hydrostatic compression of 13,000 Pa peak pressure at 0.3 Hz (1 s compression followed by 2 s relaxation) needed more than 1 h application before prostaglandin production was increased, again more in osteocytes than in osteoblasts, suggesting that mechanical stimulation via fluid flow is more effective than hydrostatic compression (Klein-Nulend et al., 1995b). A 1-h treatment with PFF also induced a sustained release of PGE_2 from the osteocytes in the hour following PFF treatment (Klein-Nulend et al., 1995b). This sustained PGE_2 release, continuing after PFF treatment had been stopped, could be ascribed to the induction of prostaglandin G/H synthase-2 (or cyclooxygenase 2, COX-2) expression (Westbroek et al., 2000a). Again, osteocytes were much more responsive than osteoblasts and osteoprogenitor cells, as only a 15-min treatment with PFF increased COX-2 mRNA expression by threefold in osteocytes but not in the other two cell populations (Westbroek et al., 2000a,b). Upregulation of COX-2 but not COX-1 by PFF had been shown earlier in a mixed population of mouse calvarial cells (Klein-Nulend et al., 1997) and was also demonstrated in primary bone cells from elderly women (Joldersma et al., 2000), whereas the expression of COX-1 and COX-2 in osteocytes and osteoblasts in intact rat bone has been documented (Forwood et al., 1998). These in vitro experiments on immunoseparated cells suggest that as bone cells mature, they increase their capacity to produce prostaglandins in response to fluid flow (Burger and Klein-Nulend, 1999). First, their immediate production of PGE_2, PGI_2, and probably $PGF_{2\alpha}$ (Klein-Nulend et al., 1997) in response to flow increases as they develop from osteoprogenitor cell, via the osteoblastic stage, into osteocytes. Second, their capacity to increase expression of COX-2 in response to flow, and thereby to continue to produce PGE_2 even after the shear stress has stopped (Westbroek et al., 2000a), increases as they reach terminal differentiation. Since induction of COX-2 is a crucial step in the induction of bone formation by mechanical loading in vivo (Forwood, 1996), these results provide direct experimental support for the concept that osteocytes, the long-living terminal differentiation stage of osteoblasts, function as the "professional" mechanosensors in bone tissue.

PFF also rapidly induced the release of NO in osteocytes but not osteoprogenitor cells (Klein-Nulend et al., 1995a). Rapid release of NO was also found when whole rat bone rudiments were mechanically strained in organ culture (Pitsillides et al., 1995) and in human bone cells submitted to fluid flow (Sterck et al., 1998). In line with these in vitro observations, inhibition of NO production inhibited mechanically induced bone formation in animal studies (Turner et al., 1996; Fox et al., 1996). NO is a ubiquitous messenger molecule for intercellular communication, involved in many tissue reactions in which cells must collaborate and communicate with one another (Koprowski and Maeda, 1995). The intracellular upregulation of NO after mechanical stimulation has been shown in single bone cells using DAR-4M AM chromophore (Vatsa et al., 2007). It was shown that a single osteocyte can disseminate a mechanical stimulus to its surrounding osteocytes via extracellular soluble signaling factors, which reinforces the putative mechanosensory role of osteocytes and demonstrates a possible mechanism by which a single mechanically stimulated osteocyte can communicate with other cells in a bone multicellular unit, which might help us to better understand the intricacies of intercellular interactions in BMUs and thus bone remodeling (Vatsa et al., 2007). Another interesting example of the involvement of NO in tissue reactions in which cells must collaborate and communicate with one another is the adaptation of blood vessels to changes in blood flow. In blood vessels, enhanced blood flow, e.g., during exercise, leads to widening of the vessel to ensure a constant blood pressure. This response depends on the endothelial cells, which sense the increased blood flow and produce intercellular messengers such as NO and prostaglandins. In response to these messengers, the smooth muscle cells around the vessel relax to allow the vessel to increase in diameter (Kamiya and Ando, 1996). The capacity of endothelial cells to produce NO in response to fluid flow is related to a specific enzyme, endothelial NO synthase or ecNOS. Interestingly, this enzyme was found in rat bone-lining cells and osteocytes (Helfrich et al., 1997; Zaman et al., 1999) and in cultured bone cells derived from human bone (Klein-Nulend et al., 1998). Both endothelial and neuronal NOS isoforms are present in osteocytes (Caballero-Alias et al., 2004). The rapid release by mechanically stressed bone cells makes NO an interesting candidate for intercellular communication within the three-dimensional network of bone cells. Treatment with pulsatile fluid flow increased the level of ecNOS RNA transcripts in the bone cell cultures (Klein-Nulend et al., 1998), a response also described in endothelial cells (Busse and Fleming, 1998; Uematsu et al., 1995). Enhanced production of prostaglandins is also a well-described response of endothelial cells to fluid flow (Busse and Fleming, 1998; Kamiya and Ando, 1996). It seems, therefore, that endothelial cells and osteocytes possess a similar sensor system for fluid flow and that both cell types are "professional" sensors of fluid flow.

Mechanotransduction starts by the conversion of physical-loading-derived stimuli into cellular signals. Several studies suggest that the attachment complex between intracellular actin cytoskeleton and extracellular matrix macromolecules, via integrins and CD44 receptors in the cell membrane, provides the site of mechanotransduction (Wang et al., 1993; Watson, 1991; Ajubi et al., 1996, 1999; Pavalko et al., 1998). An important early response is the influx of calcium ions through mechanosensitive ion channels in the plasma membrane and the release of calcium from internal stores (Hung et al., 1995, 1996; Ajubi et al., 1999; Chen et al., 2000; You et al., 2000). The rise in intracellular calcium concentration activates many downstream signaling cascades such as protein kinase C and phospholipase A_2 and is necessary for activation of calcium/calmodulin-dependent proteins such as NOS. The activation of phospholipase A_2 results, among others, in the activation of arachidonic acid production and PGE_2 release (Ajubi et al., 1999). Other genes whose expression in osteocytes is modified by mechanical loading include c-fos, MEPE, and IGF-1 (Lean et al., 1995, 1996). MEPE has been shown to be involved in the inhibition of osteoclastogenesis by mechanically loaded osteocytes (Kulkarni et al., 2010).

Many steps in the mechanosignaling cascade are still unknown in osteocytes as well as other mechanosensory cells. One of the most important signaling pathways in response to mechanical loading in the osteocyte is the Wnt/β-catenin pathway. While important for osteoblast differentiation, proliferation, and matrix production, it was hypothesized and then shown that this pathway plays a role in transmitting signals of mechanical loading from osteocytes to cells on the bone surface (Bonewald and Johnson, 2008; Bonewald, 2011). Deletion of various components of the Wnt/β-catenin pathway has effects on bone responses to loading and unloading. For example, deletion of Lrp5 results in impaired osteogenic responses to anabolic loading. It has been shown that the Wnt family of proteins strongly modulates the anabolic response of bone to mechanical loading (Robinson et al., 2006). Mechanical loading by PFF modulates gene expression of proteins involved in Wnt signaling pathways in osteocytes (Santos et al., 2009). Whereas targeted deletion of β-catenin using the Dmp1-Cre driver results in dramatic bone loss (Kramer et al., 2010), deletion of only one allele results in mice with a normal skeleton but a completely abrogated response to anabolic loading (Javaheri et al., 2014). Mechanical loading decreases sclerostin in bone, whereas hindlimb unloading increases expression. Taking all these studies together, it appears that β-catenin signaling plays an important in bone responses to loading. More research will need to be conducted to elucidate the physiological role of Wnts in bone cell mechanotransduction. Undoubtedly, in the meantime, new players in the field of bone cell mechanotransduction will be identified.

Mechanosensitive osteocytes regulate bone mass in adults, while interleukin 6 (IL-6), such as is present during orthodontic tooth movement, also strongly regulates bone mass. Interestingly, IL-6 is produced by shear-loaded osteocytes, suggesting that IL-6 may affect bone mass by modulating osteocyte communication toward osteoblasts (Bakker et al., 2014). Multiple factors contribute to bone loss in inflammatory diseases, but circulating inflammatory factors and immobilization play a crucial role. Systemic inflammation affects human osteocyte-specific protein and cytokine expression (Pathak et al., 2016). Mechanical loading prevents bone loss in the general population, but the effects of mechanical loading in patients with inflammatory disease are less clear. Pathak et al. (2015) found that serum from patients with inflammatory disease upregulated osteocyte-to-osteoclast communication, while mechanical loading nullified this upregulation, suggesting that mechanical stimuli contribute to the prevention of osteoporosis in inflammatory disease. In addition, it has been shown that mechanical loading prevents the stimulating effect of IL-1β, of which plasma and synovial fluid levels are higher in inflammatory diseases, on osteocyte-modulated osteoclastogenesis, suggesting that mechanical loading may abolish IL-1β-induced osteoclastogenesis (Kulkarni et al., 2012).

Summary and conclusion

Tremendous progress has been made in discerning the function of osteocytes since this chapter was first written over 7–8 years ago. One only has to search the term "osteocytes" on PubMed to see the dramatic increase in references to and studies of this elusive bone cell. Markers for osteocytes have been expanded to delineate early to late stages of osteocyte differentiation and the critical role of mineralization in this process. Genomic profiling of osteocytes is a storehouse of information regarding the potential functions of these cells and their response to strain.

Molecular mechanisms and pathways involved in osteocyte mechanosensation have been identified and expanded significantly. Although some investigators hold to the concept of a single mechanoreceptor initiating a linear cascade of events, new theories suggesting simultaneous triggering events in response to load are being investigated. It remains to be determined what makes these cells more responsive to shear stress than osteoblasts and what role the cell body, dendrites, and even cilia may play in this response.

These cells make up 90%–95% of all bone cells in the adult skeleton. Collectively, any minor modulation of the entire population could have significant local and systemic effects, not only on bone, but also on other organs. Although postulated several decades ago that osteocytes could potentially play a role in calcium metabolism, it appears that this cell may have a more important role in phosphate metabolism. It remains to be determined how factors produced by osteocytes can enter the circulation, suggesting an intimate connection between the lacunocanalicular system and the blood supply.

One reason for the dramatic expansion in knowledge in the field of osteocyte biology is the generation of new tools and advancement of technology. No longer is this cell safe within its mineralized cave from "prying eyes." Transgenic technology has targeted the osteocyte with fluorescent markers and generated conditional deletions. Cell lines and osteocyte-selective and -specific antibodies and probes have been generated. Advanced instrumentation such as Raman spectroscopy and atomic force microscopy can probe and characterize the osteocyte microenvironment. Dynamic imaging has revealed osteocyte movement within their lacuna and canaliculi.

Even with these advances we may still be seeing only "the tip of the iceberg" with regard to osteocyte function. It has been proposed that the osteocyte could be the orchestrator of bone remodeling in the adult skeleton directing both osteoblast and osteoclast function. The future looks exciting and full of new discoveries with regard to the biology of the osteocyte and its role in mechanosensation.

Acknowledgments

The authors would like to acknowledge the pioneering leadership in the field of osteocyte biology of Drs. Els Burger and Peter Nijweide, original authors of this chapter. These two individuals laid the foundation upon which the present authors have founded their research. The authors wish to thank Dr. Astrid Bakker for critically reading the manuscript.

The authors would also like to acknowledge funding support from the National Institutes of Health, NIH NIA PO1AG039355 (L.F.B.) and The Netherlands Foundation for Fundamental Research on Matter (ALW/FOM/NWO project 01FB28/2) (J.K.N.) and The Netherlands Institute for Space Research (SRON grant MG-055) (J.K.N.).

References

Aarden, E.M., Nijweide, P.J., Van der Plas, A., Alblas, M.J., Mackie, E.J., Horton, M.A., Helfrich, M.H., 1996a. Adhesive properties of isolated chick osteocytes in vitro. Bone 18, 305–313.

Aarden, E.M., Wassenaar, A.M., Alblas, M.J., Nijweide, P.J., 1996b. Immunocytochemical demonstration of extracellular matrix proteins in isolated osteocytes. Histochem. Cell Biol. 106, 495–501.

Abu, E.O., Horner, A., Kusec, V., Triffitt, J.T., Compston, J.E., 1997. The localization of androgen receptors in human bone. J. Clin. Endocrinol. Metab. 82, 3493–3497.

Abu, E.O., Horner, A., Kusec, V., Triffitt, J.T., Compston, J.E., 2000. The localization of the functional glucocorticoid receptor alpha in human bone. J. Clin. Endocrinol. Metab. 85, 883–889.

Ahuja, S.S., Zhao, S., Bellido, T., Plotkin, L.I., Jimenez, F., Bonewald, L.F., 2003. CD40 ligand blocks apoptosis induced by tumor necrosis factor alpha, glucocorticoids, and etoposide in osteoblasts and the osteocyte-like cell line murine long bone osteocyte-Y4. Endocrinology 144, 1761–1769.

Ajubi, N.E., Klein-Nulend, J., Nijweide, P.J., Vrijheid-Lammers, T., Alblas, M.J., Burger, E.H., 1996. Pulsating fluid flow increases prostaglandin production by cultured chicken osteocytes—a cytoskeleton-dependent process. Biochem. Biophys. Res. Commun. 225, 62–68.

Ajubi, N.E., Klein-Nulend, J., Alblas, M.J., Burger, E.H., Nijweide, P.J., 1999. Signal transduction pathways involved in fluid flow-induced prostaglandin E_2 production by cultured osteocytes. Am. J. Physiol. 276, E171–E178.

Alford, A.I., Jacobs, C.R., Donahue, H.J., 2003. Oscillating fluid flow regulates gap junction communication in osteocytic MLO-Y4 cells by an ERK1/2 MAP kinase-dependent mechanism small star, filled. Bone 33, 64–70.

Aubin, J.E., Liu, F., Malaval, L., Gupta, A.K., 1995. Osteoblast and chondroblast differentiation. Bone 17 (Suppl. l), 77–83.

Bacabac, R.G., Smit, T.H., Mullender, M.G., Dijcks, S.J., Van Loon, J.J., Klein-Nulend, J., 2004. Nitric oxide production by bone cells is fluid shear stress rate dependent. Biochem. Biophys. Res. Commun. 315, 823–829.

Bacabac, R.G., Smit, T.H., Mullender, M.G., Van Loon, J.J., Klein-Nulend, J., 2005. Initial stress-kick is required for fluid shear stress-induced rate dependent activation of bone cells. Ann. Biomed. Eng. 33, 104–110.

Bacabac, R.G., Mizuno, D., Schmidt, C.F., MacKintosh, F.C., Smit, T.H., Van Loon, J.J.W.A., Klein-Nulend, J., 2006a. Microrheology and force traction of mechanosensitive bone cells. J. Biomech. 39 (Suppl. 1), S231–S232.

Bacabac, R.G., Smit, T.H., Van Loon, J.J., Zandieh Doulabi, B., Helder, M., Klein-Nulend, J., 2006b. Bone cell responses to high-frequency vibration stress: does the nucleus oscillate within the cytoplasm? FASEB J. 20, 858–864.

Bacabac, R.G., Mizuno, D., Schmidt, C.F., MacKintosh, F.C., Van Loon, J.J., Klein-Nulend, J., Smit, T.H., 2008. Round versus flat: bone cell morphology, elasticity, and mechanosensing. J. Biomech. 41, 1590–1598.

Bakker, A.D., Soejima, K., Klein-Nulend, J., Burger, E.H., 2001. The production of nitric oxide and prostaglandin E2 by primary bone cells is shear stress dependent. J. Biomech. 34, 671–677.

Bakker, A.D., Klein-Nulend, J., Burger, E.H., 2004. Shear stress inhibits while disuse promotes osteocyte apoptosis. Biochem. Biophys. Res. Commun. 320, 1163–1168.

Bakker, A.D., Da Silva, V.C., Krishnan, R., Bacabac, R.G., Blaauboer, M.E., Lin, Y.–C., Marcantonio, R.A., Cirelli, J.A., Klein-Nulend, J., 2009. Tumor necrosis factor alpha and interleukin-1beta modulate calcium and nitric oxide signaling in mechanically stimulated osteocytes. Arthritis Rheum. 60, 3336–3345.

Bakker, A.D., Kulkarni, R., Klein-Nulend, J., Lems, W.F., 2014. IL-6 alters osteocyte signaling toward osteoblasts but not osteoclasts. J. Dent. Res. 93, 394–399.

Balemans, W., Ebeling, M., Patel, N., Van Hul, E., Olson, P., Dioszegi, M., Lacza, C., Wuyts, W., Van Den Ende, J., Willems, P., Paes-Alves, A.F., Hill, S., Bueno, M., Ramos, F.J., Tacconi, P., Dikkers, F.G., Stratakis, C., Lindpaintner, K., Vickery, B., Foernzler, D., Van Hul, W., 2001. Increased bone density in sclerosteosis is due to the deficiency of a novel secreted protein (SOST). Hum. Mol. Genet. 10, 537–543.

Barragan-Adjemian, C., Nicolella, D.P., Dusevich, V., Dallas, M., Eick, D., Bonewald, L., 2006. Mechanism by which MLO-A5 late osteoblast/early osteocytes mineralize in culture: similarities with lamellar bone. Calcif. Tissue Int. 79, 340–353.

Basso, N., Heersche, J.N., 2006. Effects of hind limb unloading and reloading on nitric oxide synthase expression and apoptosis of osteocytes and chondrocytes. Bone 39, 807–814.

Batra, N., Burra, S., Siller-Jackson, A.J., Gu, S., Xia, X., Weber, G.F., DeSimone, D., Bonewald, L.F., Lafer, E.M., Sprague, E., Schwartz, M.A., Jiang, J.X., 2012. Mechanical stress-activated integrin $\alpha5\beta1$ induces opening of connexin 43 hemichannels. Proc. Natl. Acad. Sci. U.S.A. 109, 3359–3364.

Bélanger, L.F., 1969. Osteocytic osteolysis. Calcif. Tissue Res. 4, 1–12.

Bellido, T., Ali, A.A., Gubrij, I., Plotkin, L.I., Fu, Q., O'Brien, C.A., Manolagas, S.C., Jilka, R.L., 2005. Chronic elevation of parathyroid hormone in mice reduces expression of sclerostin by osteocytes: a novel mechanism for hormonal control of osteoblastogenesis. Endocrinology 146, 4577–4583.

Bennett, M.V., Goodenough, D.A., 1978. Gap junctions, electrotonic coupling, and intercellular communication. Neurosci. Res. Progr. Bull. 16, 1–486.

Bivi, N., Condon, K.W., Allen, M.R., Farlow, N., Passeri, G., Brun, L.R., Rhee, Y., Bellido, T., Plotkin, L.I., 2012. Cell autonomous requirement of connexin 43 for osteocyte survival: consequences for endocortical resorption and periosteal bone formation. J. Bone Miner. Res. 27, 374–389.

Bodine, P.V., Vernon, S.K., Komm, B.S., 1996. Establishment and hormonal regulation of a conditionally transformed preosteocytic cell line from adult human bone. Endocrinology 137, 4592–4604.

Boivin, G., Mesguich, P., Pike, J.W., Bouillon, R., Meunier, P.J., Haussler, M.R., Dubois, P.M., Morel, G., 1987. Ultrastructural immunocytochemical localization of endogenous 1,25-dihydroxyvitamin D and its receptors in osteoblasts and osteocytes from neonatal mouse and rat calvaria. Bone Miner. 3, 125–136.

Bonewald, L.F., 2006. Mechanosensation and transduction in osteocytes. BoneKey-Osteovision 3, 7–15.

Bonewald, L.F., 2011. The amazing osteocyte. J. Bone Miner. Res. 26, 229–238.

Bonewald, L.F., 2017. The role of the osteocyte in bone and non-bone disease. Endocrinol. Metab. Clin. North Am. 46, 1–18.

Bonewald, L.F., Johnson, M.L., 2008. Osteocytes, mechanosensing and Wnt signaling. Bone 42, 606–615.

Bonucci, E., 1990. The ultrastructure of the osteocyte. In: Bonucci, E., Motta, P.M. (Eds.), Ultrastructure of Skeletal Tissues. Kluwer Academic, Dordrecht, The Netherlands, pp. 223–237.

Bordier, P.J., Miravet, L., Ryckerwaert, A., Rasmussen, H., 1976. Morphological and morphometrical characteristics of the mineralization front. A vitamin D regulated sequence of bone remodeling. In: PJ, B. (Ed.), Bone Histomorphometry. Armour Montagu, Paris, France, pp. 335–354.

Boskey, A., 1996. Matrix proteins and mineralization: an overview. Connect. Tissue Res. 35, 357–363.

Boyde, A., 1980. Evidence against "osteocyte osteolysis. Metab. Bone Dis. Relat. Res. 2 (Suppl. l), 239–255.

Braidman, J.P., Davenport, L.K., Carter, D.H., Selby, P.L., Mawer, E.B., Freemont, A.J., 1995. Preliminary *in situ* identification of estrogen target cells in bone. J. Bone Miner. Res. 10, 74–80.

Bresler, D., Bruder, J., Mohnike, K., Fraser, W.D., Rowe, P.S., 2004. Serum MEPE-ASARM-peptides are elevated in X-linked rickets (HYP): implications for phosphaturia and rickets. J. Endocrinol. 183, R1–R9.

Bronckers, A.L., Goei, W., Luo, G., Karsenty, G., D'Souza, R.N., Lyaruu, D.M., Burger, E.H., 1996. DNA fragmentation during bone formation in neonatal rodents assessed by transferase-mediated end labeling. J. Bone Miner. Res. 11, 1281–1291.

Bruder, S.P., Caplan, A.I., 1990. Terminal differentiation of osteogenic cells in the embryonic chick tibia is revealed by a monoclonal antibody against osteocytes. Bone 11, 189–198.

Burger, E.H., Klein-Nulend, J., 1999. Mechanotransduction in bone: role of the lacuno-canalicular network. FASEB J. 13, S101–S112.

Burger, E.H., Veldhuijzen, J.P., 1993. Influence of mechanical factors on bone formation, resorption, and growth in vitro. In: Hall, B.K. (Ed.), Bone, vol. 7. CRC Press, Boca Raton, FL, pp. 37–56.

Burger, E.H., Klein-Nulend, J., Smit, T.H., 2003. Strain-derived canalicular fluid flow regulates osteoclast activity in a remodelling osteon—a proposal. J. Biomech. 36, 1453—1459.

Burr, D.B., Milgran, C., Fyhrie, D., Forwood, M.R., Nyska, M., Finestone, A., Hoshaw, S., Saiag, E., Simkin, A., 1996. In vivo measurement of human tibial strains during vigorous activity. Bone 18, 405—410.

Burra, S., Nicolella, D.P., Francis, W.L., Freitas, C.J., Mueschke, N.J., Poole, K., Jiang, J.X., 2010. Dendritic processes of osteocytes are mechanotransducers that induce the opening of hemichannels. Proc. Natl. Acad. Sci. U.S.A. 107, 13648—13653.

Busse, R., Fleming, I., 1998. Pulsatile stretch and shear stress: physical stimuli determining the production of endothelium derived relaxing factors. J. Vasc. Res. 35, 73—84.

Cabahug-Zuckermann, P., Stout Jr., R.F., Majeska, R.J., Thi, M.M., Spray, D.C., Weinbaum, S., Schaffler, M.B., 2018. Potential role for a specialized β3 integrin-based structure on osteocyte processes in bone mechanosensation. J. Orthop. Res. 36, 642—652.

Caballero-Alias, A.M., Loveridge, N., Lyon, L.E., 2004. NOS isoforms in adult human osteocytes: multiple pathways of NO regulation? Calcif. Tissue Int. 75, 78—84.

Chen, N.X., Ryder, K.D., Pavalko, F.M., Turner, C.H., Burr, D.B., Qiu, J., Duncan, R.L., 2000. Ca(2+) regulates fluid shear-induced cytoskeletal reorganization and gene expression in osteoblasts. Am. J. Physiol. 278, C989—C997.

Cheng, B., Zhao, S., Luo, J., Sprague, E., Bonewald, L.F., Jiang, J.X., 2001. Expression of functional gap junctions and regulation by fluid flow in osteocyte-like MLO-Y4 cells. J. Bone Miner. Res. 16, 249—259.

Cherian, P.P., Siller-Jackson, A.J., Gu, S., Wang, X., Bonewald, L.F., Sprague, E., Jiang, J.X., 2005. Mechanical strain opens connexin 43 hemichannels in osteocytes: a novel mechanism for the release of prostaglandin. Mol. Biol. Cell 16, 3100—3106.

Chow, J.W., Fox, S., Jagger, C.J., Chambers, T.J., 1998. Role for parathyroid hormone in mechanical responsiveness of rat bone. Am. J. Physiol. 274, E146—E154.

Clarke, B.L., 2014. Anti-sclerostin antibodies: utility in treatment of osteoporosis. Maturitas 78, 199—204.

Clinkenbeard, E.L., White, K.E., 2016. Systemic control of bone homeostasis by FGF23 signaling. Curr. Mol. Biol. Rep. 2, 62—71.

Cowin, S.C., 1999. Bone poroelasticity. J. Biomech. 32, 217—238.

Cowin, S.C., Moss-Salentijn, L., Moss, M.L., 1991. Candidates for the mechanosensory system in bone. J. Biomed. Eng. 113, 191—197.

Cowin, S.C., Weinbaum, S., Zeng, Y., 1995. A case for bone canaliculi as the anatomical site of strain generated potentials. J. Biomech. 28, 1281—1297.

Dallas, S.L., Zaman, G., Pead, M.J., Lanyon, L.E., 1993. Early strain-related changes in cultured embryonic chick tibiotarsi parallel those associated with adaptive modeling in vivo. J. Bone Miner. Res. 8, 251—259.

Dallas, S.L., Prideaux, M., Bonewald, L.F., 2013. The osteocyte: an endocrine cell … and more. Endocr. Rev. 34 (5), 658—690.

Davideau, J.L., Papagerakis, P., Hotton, D., Lezot, F., Berdal, A., 1996. In situ investigation of vitamin D receptor, alkaline phosphatase, and osteocalcin gene expression in oro-facial mineralized tissues. Endocrinology 137, 3577—3585.

Delgado-Calle, J., Hancock, B., Likine, E.F., Sato, A.Y., McAndrews, K., Sanudo, C., Bruzzaniti, A., Riancho, J.A., Tonra, J.R., Bellido, T., 2018. MMP14 is a novel target of PTH signaling in osteocytes that controls resorption by regulating soluble RANKL production. FASEB J. https://doi.org/10.1096/fj.201700919RRR [Epub ahead of print].

Divieti, P., Inomata, N., Chapin, K., Singh, R., Juppner, H., Bringhurst, F.R., 2001. Receptors for the carboxyl-terminal region of pth(1-84) are highly expressed in osteocytic cells. Endocrinology 142, 916—925.

Donahue, H.J., 2000. Gap junctions and biophysical regulation of bone cell differentiation. Bone 26, 417—422.

Donahue, H.J., McLeod, K.J., Rubin, C.T., Andersen, J., Grine, E.A., Hertzberg, E.L., Brink, P.R., 1995. Cell-to-cell communication in osteoblastic networks: cell line-dependent hormonal regulation of gap junction function. J. Bone Miner. Res. 10, 881—889.

Doty, S.B., 1981. Morphological evidence of gap junctions between bone cells. Calcif. Tissue Int. 33, 509—512.

Ducy, P., Desbois, C., Boyce, B., Pinero, G., Story, B., Dunstan, C., Smith, E., Bonadio, J., Goldstein, S., Gundberg, C., Bradley, A., Karsenty, G., 1996. Increased bone formation in osteocalcin deficient mice. Nature 382, 448—452.

El-Haj, A.J., Minter, S.L., Rawlinson, S.C., Suswillo, R., Lanyon, L.E., 1990. Cellular responses to mechanical loading in vitro. J. Bone Miner. Res. 5, 923—932.

Elmardi, A.S., Katchburian, M.V., Katchburian, E., 1990. Electron microscopy of developing calvaria reveals images that suggest that osteoclasts engulf and destroy osteocytes during bone resorption. Calcif. Tissue Int. 46, 239—245.

Feng, J.Q., Ward, L.M., Liu, S., Lu, Y., Xie, Y., Yuan, B., Yu, X., Rauch, F., Davis, S.I., Zhang, S., Rios, H., Drezner, M.K., Quarles, L.D., Bonewald, L.F., White, K.E., 2006. Loss of DMP1 causes rickets and osteomalacia and identifies a role for osteocytes in mineral metabolism. Nat. Genet. 38, 1310—1315.

Fermor, B., Skerry, T.M., 1995. PTH/PTHrP receptor expression on osteoblasts and osteocytes but not resorbing bone surfaces in growing rats. J. Bone Miner. Res. 10, 1935—1943.

Fisher, L.W., Fedarko, N.S., 2003. Six genes expressed in bones and teeth encode the current members of the SIBLING family of proteins. Connect. Tissue Res. 44 (Suppl. 1), 33—40.

Forwood, M.R., 1996. Inducible cyclooxygenase (COX-2) mediates the induction of bone formation by mechanical loading in vivo. J. Bone Miner. Res. 11, 1688—1693.

Forwood, M.R., Kelly, W.L., Worth, N.F., 1998. Localization of prostaglandin endoperoxidase H synthase (PGHS)-1 and PGHS-2 in bone following mechanical loading in vivo. Anat. Rec. 252, 580—586.

Fox, S.W., Chambers, T.J., Chow, J.W., 1996. Nitric oxide is an early mediator of the increase in bone formation by mechanical stimulation. Am. J. Physiol. 270, E955—E960.

Frangos, J.A., Eskin, S.G., McIntire, L.V., Ives, C.L., 1985. Flow effects on prostacyclin production by cultured human endothelial cells. Science 227, 1477–1479.

Franz-Odendaal, T.A., Hall, B.K., Witten, P.E., 2006. Buried alive: how osteoblasts become osteocytes. Dev. Dynam. 235, 176–190.

Frost, H.J., 1992. The role of changes in mechanical usage set points in the pathogenesis of osteoporosis. J. Bone Miner. Res. 7, 253–261.

Fu, J., Hao, L., Tian, Y., Liu, Y., Gu, Y., Wu, J., 2018. miR-199a-3p is involved in estrogen-mediated autophagy through the IGF-1/mTOR pathway in osteocyte-like MLO-Y4 cells. J. Cell. Physiol. 233, 2292–2303.

Genetos, D.C., Kephart, C.J., Zhang, Y., Yellowley, C.E., Donahue, H.J., 2007. Oscillating fluid flow activation of gap junction hemichannels induces atp release from MLO-Y4 osteocytes. J. Cell. Physiol. 212, 207–214.

Gluhak-Heinrich, J., Ye, L., Bonewald, L.F., Feng, J.Q., MacDougall, M., Harris, S.E., Pavlin, D., 2003. Mechanical loading stimulates dentin matrix protein 1 (DMP1) expression in osteocytes in vivo. J. Bone Miner. Res. 18, 807–817.

Gluhak-Heinrich, J., Yang, W., Bonewald, L.F., Robling, A.G., Turner, C.H., Harris, S.E., 2005. Mechanically induced DMP1 and MEPE expression in osteocytes: correlation to mechanical strain, osteogenic response and gene expression threshold. J. Bone Miner. Res. 20 (Suppl. 1), S73.

Goodenough, D.A., Goliger, J.A., Paul, D.L., 1996. Connexins, connexons, and intercellular communication. Annu. Rev. Biochem. 65, 475–502.

Goodenough, D.A., Paul, D.L., 2003. Beyond the gap: functions of unpaired connexon channels. Nat. Rev. Mol. Cell Biol. 4, 285–294.

Gowen, L.C., Petersen, D.N., Mansolf, A.L., Qi, H., Stock, J.L., Tkalcevic, G.T., Simmons, H.A., Crawford, D.T., Chidsey-Frink, K.L., Ke, H.Z., McNeish, J.D., Brown, T.A., 2003. Targeted disruption of the osteoblast/osteocyte factor 45 gene (OF45) results in increased bone formation and bone mass. J. Biol. Chem. 278, 1998–2007.

Gross, T.S., Akeno, N., Clemens, T.L., Komarova, S., Srinivasan, S., Weimer, D.A., Mayorov, S., 2001. Selected Contribution: osteocytes upregulate HIF-1alpha in response to acute disuse and oxygen deprivation. J. Appl. Physiol. 90, 2514–2519.

Gross, T.S., King, K.A., Rabaia, N.A., Pathare, P., Srinivasan, S., 2005. Upregulation of osteopontin by osteocytes deprived of mechanical loading or oxygen. J. Bone Miner. Res. 20, 250–256.

Han, Y., Cowin, S.C., Schaffler, M.B., Weinbaum, S., 2004. Mechanotransduction and strain amplification in osteocyte cell processes. Proc. Natl. Acad. Sci. U.S.A. 101, 16689–16694.

Heino, T.J., Hentunen, T.A., Vaananen, H.K., 2002. Osteocytes inhibit osteoclastic bone resorption through transforming growth factor-beta: enhancement by estrogen. J. Cell. Biochem. 85, 185–197.

Heino, T.J., Hentunen, T.A., Vaananen, H.K., 2004. Conditioned medium from osteocytes stimulates the proliferation of bone marrow mesenchymal stem cells and their differentiation into osteoblasts. Exp. Cell Res. 294, 458–468.

Helfrich, M.H., Evans, D.E., Grabowski, P.S., Pollock, J.S., Ohshima, H., Ralston, S.H., 1997. Expression of nitric oxide synthase isoforms in bone and bone cell cultures. J. Bone Miner. Res. 12, 1108–1115.

Hemmatian, H., Bakker, A.D., Klein-Nulend, J., van Lenthe, G.H., 2017. Aging, osteocytes, and mechanotransduction. Curr. Osteoporos. Rep. 15, 401–411.

Hoyland, J.A., Baris, C., Wood, L., Baird, P., Selby, P.L., Freemont, A.J., Braidman, I.P., 1999. Effect of ovarian steroid deficiency on oestrogen receptor alpha expression in bone. J. Pathol. 188, 294–303.

Huang, J., Romero-Suarez, S., Lara, N., Mo, C., Kaja, S., Brotto, L., Dallas, S.L., Johnson, M.L., Jähn, K., Bonewald, L.F., Brotto, M., 2017. Crosstalk between MLO-Y4 osteocytes and C2C12 muscle cells is mediated by the Wnt/β-catenin pathway. J. Bone Miner. Res. 1, 86–100.

Hughes, D.E., Salter, D.M., Simpson, R., 1994. CD44 expression in human bone: a novel marker of osteocytic differentiation. J. Bone Miner. Res. 9, 39–44.

Huiskes, R., Ruimerman, R., van Lenthe, G.H., Janssen, J.D., 2000. Effects of mechanical forces on maintenance and adaptation of form in trabecular bone. Nature 405, 704–706.

Hung, C.T., Pollack, S.R., Reilly, T.M., Brighton, C.T., 1995. Realtime calcium response of cultured bone cells to fluid flow. Clin. Orthop. Relat. Res. 313, 256–269.

Hung, C.T., Allen, F.D., Pollack, S.R., Brighton, C.T., 1996. Intracellular calcium stores and extracellular calcium are required in the real-time calcium response of bone cells experiencing fluid flow. J. Biomech. 29, 1411–1417.

Igarashi, M., Kamiya, N., Ito, K., Takagi, M., 2002. In situ localization and in vitro expression of osteoblast/osteocyte factor 45 mRNA during bone cell differentiation. Histochem. J. 34, 255–263.

Imai, S., Kaksonen, M., Raulo, E., Kinnunen, T., Fages, C., Meng, X., Lakso, M., Rauvala, H., 1998. Osteoblast recruitment and bone formation enhanced by cell matrix-associated heparin-binding growth-associated molecule (HB-GAM). J. Cell Biol. 143, 1113–1128.

Ingber, D.E., 1991. Intergrins as mechanochemical transducers. Curr. Opin. Cell Biol. 3, 841–848.

Jähn, K., Lara-Castillo, N., Brotto, L., Mo, C.L., Johnson, M.L., Brotto, M., Bonewald, L.F., 2012. Skeletal muscle secreted factors prevent glucocorticoid-induced osteocyte apoptosis through activation of beta-catenin. Eur. Cells Mater. 24, 197–210.

Jähn, K., Kelkar, S., Zhao, H., Xie, Y., Tiede-Lewis, L.M., Dusevich, V., Dallas, S.L., Bonewald, L.F., 2017. Osteocytes acidify their microenvironment in response to PTHrP in vitro and in lactating mice in vivo. J. Bone Miner. Res. PMID: 28470757.

Jande, S.S., 1971. Fine structural study of osteocytes and their surrounding bone matrix with respect to their age in young chicks. J. Ultrastruct. Res. 37, 279–300.

Javaheri, B., Stern, A.R., Lara, N., Dallas, M., Zhao, H., Bonewald, L.F., Johnson, M.L., 2014. Deletion of a single β-catenin allele in osteocytes abolishes the bone anabolic response to loading. J. Bone Miner. Res. 29, 705–715.

Jilka, R.L., Weinstein, R.S., Bellido, T., Roberson, P., Parfitt, A.M., Manolagas, S.C., 1999. Increased bone formation by prevention of osteoblast apoptosis with parathyroid hormone. J. Clin. Investig. 104, 439–446.

Johnson, L.C., 1966. The kinetics of skeletal remodeling in structural organization of the skeleton. Birth Defects 11, 66–142.

Joldersma, M., Burger, E.H., Semeins, C.M., Klein-Nulend, J., 2000. Mechanical stress induces COX-2 mRNA expression in bone cells from elderly women. J. Biomech. 33, 53–61.

Kalajzic, I., Braut, A., Guo, D., Jiang, X., Kronenberg, M.S., Mina, M., Harris, M.A., Harris, S.E., Rowe, D.W., 2004. Dentin matrix protein 1 expression during osteoblastic differentiation, generation of an osteocyte GFP-transgene. Bone 35, 74–82.

Kamioka, H., Honjo, T., Takano-Yamamoto, T., 2001. A three-dimensional distribution of osteocyte processes revealed by the combination of confocal laser scanning microscopy and differential interference contrast microscopy. Bone 28, 145–149.

Kamioka, H., Sugawara, Y., Honjo, T., Yamashiro, T., Takano-Yamamoto, T., 2004. Terminal differentiation of osteoblasts to osteocytes is accompanied by dramatic changes in the distribution of actin-binding proteins. J. Bone Miner. Res. 19, 471–478.

Kamioka, H., Kameo, Y., Imai, Y., Bakker, A.D., Bacabac, R.G., Yamada, N., Takaoka, A., Yamashiro, T., Adachi, T., Klein-Nulend, J., 2012. Microscale fluid flow analysis in a human osteocyte canaliculus using a realistic high-resolution image-based three-dimensional model. Integr. Biol. 4, 1198–1206.

Kamiya, A., Ando, J., 1996. Response of vascular endothelial cells to fluid shear stress: Mechanism. In: Hayashi, K., Kamiyn, A., Ono, K. (Eds.), Biomechanics: Functional Adaptation and Remodeling. Springer, Tokyo, pp. 29–56.

Kato, Y., Windle, J.J., Koop, B.A., Mundy, G.R., Bonewald, L.F., 1997. Establishment of an osteocyte-like cell line, MLO-Y4. J. Bone Miner. Res. 12, 2014–2023.

Kato, Y., Boskey, A., Spevak, L., Dallas, M., Hori, M., Bonewald, L.F., 2001. Establishment of an osteoid preosteocyte-like cell MLO-A5 that spontaneously mineralizes in culture. J. Bone Miner. Res. 16, 1622–1633.

Kitase, Y., Jiang, J.X., Bonewald, L.F., 2006. The anti-apoptotic effects of mechanical strain on osteocytes are mediated by PGE_2 and monocyte chemotactic protein, (MCP-3); selective protection by MCP3 against glucocorticoid (GC) and not TNF-a induced apoptosis. J. Bone Miner. Res. 21 (Suppl. 1), S48.

Kitase, Y., Vallejo, J.A., Gutheil, W., Vemula, H., Jahn, K., Yi, J., Zhou, J., Brotto, M., Bonewald, L.F., 2018. Beta-aminoisobutyric acid, l-BAIBA, is a muscle-derived osteocyte survival factor. Cell Rep. 22, 1531–1544.

Klein-Nulend, J., Semeins, C.M., Ajubi, N.E., Nijweide, P.J., Burger, E.H., 1995a. Pulsating fluid flow increases nitric oxide (NO) synthesis by osteocytes but not periosteal fibroblasts-correlation with prostaglandin upregulation. Biochem. Biophys. Res. Commun. 217, 640–648.

Klein-Nulend, J., Van der Plas, A., Semeins, C.M., Ajubi, N.E., Frangos, J.A., Nijweide, P.J., Burger, E.H., 1995b. Sensitivity of osteocytes to biomechanical stress in vitro. FASEB J. 9, 441–445.

Klein-Nulend, J., Burger, E.H., Semeins, C.M., Raisz, L.G., Pilbeam, C.C., 1997. Pulsating fluid flow stimulates prostaglandin release and inducible prostaglandin G/H synthase mRNA expression in primary mouse bone cells. J. Bone Miner. Res. 12, 45–51.

Klein-Nulend, J., Helfrich, M.H., Sterck, J.G.H., MacPherson, H., Joldersma, M., Ralston, S.H., Semeins, C.M., Burger, E.H., 1998. Nitric oxide response to shear stress by human bone cell cultures is endothelial nitric oxide synthase dependent. Biochem. Biophys. Res. Commun. 250, 108–114.

Klein-Nulend, J., Bacabac, R.G., Bakker, A.D., 2012. Mechanical loading and how it affects bone cells: the role of the osteocyte cytoskeleton in maintaining our skeleton. Eur. Cells Mater. 24, 278–291.

Klein-Nulend, J., Bakker, A.D., Bacabac, R.G., Vatsa, A., Weinbaum, S., 2013. Mechanosensation and transduction in osteocytes. Bone 54, 182–190.

Knothe-Tate, M.L., Niederer, P., Knothe, U., 1998. In vivo tracer transport throught the lacunocanalicular system of rat bone in an environment devoid of mechanical loading. Bone 22, 107–117.

Knothe-Tate, M.L., Steck, R., Forwood, M.R., Niederer, P., 2000. In vivo demonstration of load-induced fluid flow in the rat tibia and its potential implications for processes associated with functional adaptation. J. Exp. Biol. 203, 2737–2745.

Kogianni, G., Mann, V., Noble, B.S., 2008. Apoptotic bodies convey activity capable of initiating osteoclastogenesis and localised bone destruction. J. Bone Miner. Res. 23, 915–927.

Kondoh, S., Inoue, K., Igarashi, K., et al., 2014. Estrogen receptor α in osteocytes regulates trabecular bone formation in female mice. Bone 60, 68–77.

Koprowski, H., Maeda, H., 1995. The Role of Nitric Oxide in Physiology and Pathophysiology. Springer-Verlag, Berlin, Germany.

Kousteni, S., Bellido, T., Plotkin, L.I., O'Brien, C.A., Bodenner, D.L., Han, L., Han, K., DiGregorio, G.B., Katzenellenbogen, J.A., Katzenellenbogen, B.S., Roberson, P.K., Weinstein, R.S., Jilka, R.L., Manolagas, S.C., 2001. Nongenotropic, sex-nonspecific signaling through the estrogen or androgen receptors: dissociation from transcriptional activity. Cell 104, 719–730.

Kramer, I., Halleux, C., Keller, H., Pegurri, M., Gooi, J.H., Weber, P.B., Feng, J.Q., Bonewald, L.F., Kneissel, M., 2010. Osteocyte Wnt/beta-catenin signaling is required for normal bone homeostasis. Mol. Cell Biol. 30, 3071–3085.

Kufahl, R.H., Saha, S., 1990. A theoretical model for stress-generated flow in the canaliculi-lacunae network in bone tissue. J. Biomech. 23, 171–180.

Kulkarni, R., Bakker, A.D., Everts, V., Klein-Nulend, J., 2010. Inhibition of osteoclastogenesis by mechanically loaded osteocytes: involvement of MEPE. Calcif. Tissue Int. 87, 461–468.

Kulkarni, R., Bakker, A.D., Everts, V., Klein-Nulend, J., 2012. Mechanical loading prevents the stimulating effect of IL-1β on osteocyte-modulated osteoclastogenesis. Biochem. Biophys. Res. Commun. 420, 11–16.

Lane, N.E., Yao, W., Balooch, M., Nalla, R.K., Balooch, G., Habelitz, S., Kinney, J.H., Bonewald, L.F., 2006. Glucocorticoid-treated mice have localized changes in trabecular bone material properties and osteocyte lacunar size that are not observed in placebo-treated or estrogen-deficient mice. J. Bone Miner. Res. 21, 466–476.

Lean, J.M., Jagger, C.J., Chambers, T.J., Chow, J.W., 1995. Increased insulin-like growth factor I mRNA expression in rat osteocytes in response to mechanical stimulation. Am. J. Physiol. 268, E318–E327.

Lean, J.M., Mackay, A.G., Chow, J.W., Chambers, T.J., 1996. Osteocytic expression of mRNA for c-fos and IGF-I: an immediate early gene response to an osteogenic stimulus. Am. J. Physiol. 270, E937–E945.

Lecanda, F., Warlow, P.M., Sheikh, S., Furlan, F., Steinberg, T.H., Civitelli, R., 2000. Connexin43 deficiency causes delayed ossification, craniofacial abnormalities, and osteoblast dysfunction. J. Cell Biol. 151, 931–944.

Lee, K., Jessop, H., Suswillo, R., Zaman, G., Lanyon, L.E., 2003. Endocrinology: bone adaptation requires oestrogen receptor-alpha. Nature 424, 389.

Lewis, K.J., Frikha-Benayed, D., Louie, J., Stephen, S., Spray, D.C., Thi, M.M., Seref-Ferlengez, Z., Majeska, R.J., Weinbaum, S., Schaffler, M.B., 2017. Osteocyte calcium signals encode strain magnitude and loading frequency in vivo. Proc. Natl. Acad. Sci. U.S.A. 114, 11775–11780.

Li, X., Zhang, Y., Kang, H., Liu, W., Liu, P., Zhang, J., Harris, S.E., Wu, D., 2005. Sclerostin binds to LRP5/6 and antagonizes canonical Wnt signaling. J. Biol. Chem. 280, 19883–19887.

Liang, J.D., Hock, J.M., Sandusky, G.E., Santerre, R.F., Onyia, J.E., 1999. Immunohistochemical localization of selected early response genes expressed in trabecular bone of young rats given hPTH 1–34. Calcif. Tissue Int. 65, 369–373.

Litzenberger, J.B., Kim, J.-B., Tummala, P., Jacobs, C.R., 2010. Beta1 integrins mediate mechanosensitive signaling pathways in osteocytes. Calcif. Tissue Int. 86, 325–332.

Liu, Y., Porta, A., Peng, X., Gengaro, K., Cunningham, E.B., Li, H., Dominguez, L.A., Bellido, T., Christakos, S., 2004. Prevention of glucocorticoid-induced apoptosis in osteocytes and osteoblasts by calbindin-D28k. J. Bone Miner. Res. 19, 479–490.

Liu, S., Zhou, J., Tang, W., Jiang, X., Rowe, D.W., Quarles, L.D., 2006. Pathogenic role of Fgf23 in Hyp mice. Am. J. Physiol. Endocrinol. Metab. 291, E38–E49.

Liu, S., Tang, W., Zhou, J., Vierthaler, L., Quarles, L.D., 2007. Distinct roles for intrinsic osteocyte abnormalities and systemic factors in regulation of FGF23 and bone mineralization in hyp mice. Am. J. Physiol. Endocrinol. Metab. 293, E1636–E1644.

Lu, Y., Kamel-El Sayed, S.A., Grillo, M.A., Veno, P.A., Dusevich, V., Tiede-Lewis, L.M., Phillips, C.L., Bonewald, L.F., Dallas, S.L., February 20, 2018. Live imaging of type I collagen assembly dynamics in cells stably expressing GFP-collagen constructs. J. Bone Miner. Res. https://doi.org/10.1002/jbmr.3409 [Epub ahead of print].

MacNabb, C., Patton, D., Hayes, J.S., 2016. Sclerostin antibody therapy for the treatment of osteoporosis: clinical prospects and challenges. J. Osteoporos. 2016, 6217286.

Maeno, M., Taguchi, M., Kosuge, K., Otsuka, K., Takagi, M., 1992. Nature and distribution of mineral-binding, keratan sulfate-containing glycoconjugates in rat and rabbit bone. J. Histochem. Cytochem. 40, 1779–1788.

Malone, A.M.D., Anderson, C.T., Temiyasathit, S., Tang, J., Tummala, P., Sterns, T., Jacobs, C.R., 2006. Primary cilia: mechanosensory organelles in bone cells. J. Bone Miner. Res. 21 (Suppl. 1), S39.

Mangeat, P., Roy, C., Martin, M., 1999. ERM proteins in cell adhesion and membrane dynamics. Trends Cell Biol. 9, 187–192.

Manolagas, S.C., 2000. Birth and death of bone cells: basic regulatory mechanisms and implications for the pathogenesis and treatment of osteoporosis. Endocr. Rev. 21, 115–137.

Marotti, G., 1996. The structure of bone tissues and the cellular control of their deposition. Ital. J. Anat. Embryol. 101, 25–79.

Marotti, G., Cane, V., Palazzini, S., Palumbo, C., 1990. Structure-function relationships in the osteocyte. Ital. J. Miner. Electrolyte Metab. 4, 93–106.

Martin, R.B., 2000. Does osteocyte formation cause the nonlinear refilling of osteons? Bone 26, 71–78.

McCreadie, B.R., Hollister, S.J., Schaffler, M.B., Goldstein, S.A., 2004. Osteocyte lacuna size and shape in women with and without osteoporotic fracture. J. Biomech. 37, 563–572.

McGarry, J.G., Klein-Nulend, J., Mullender, M.G., Prendergast, P.J., 2005a. A comparison of strain and fluid shear stress in stimulating bone cell responses–a computational and experimental study. FASEB J. 19, 482–484.

McGarry, J.G., Klein-Nulend, J., Prendergast, P.J., 2005b. The effect of cytoskeletal disruption on pulsatile fluid flow-induced nitric oxide and prostaglandin E2 release in osteocytes and osteoblasts. Biochem. Biophys. Res. Commun. 330, 341–348.

McKee, M.D., Farach-Carson, M.C., Butler, W.T., Hauschka, P.V., Nanci, A., 1993. Ultrastructural immunolocalization of noncollagenous (osteopontin and osteocalcin) and plasma (albumin and H_2S-glycoprotein) proteins in rat bone. J. Bone Miner. Res. 8, 485–496.

Medscape FDA Rejects Romosozumab for Osteoporosis, July 17, 2017. Available at: www.medscape.com/viewarticle/882966.

Mikuni-Takagaki, Y., Kakai, Y., Satoyoshi, M., Kawano, E., Suzuki, Y., Kawase, T., Saito, S., 1995. Matrix mineralization and the differentiation of osteocyte-like cells in culture. J. Bone Miner. Res. 10, 231–242.

Miyauchi, A., Notoya, K., Mikuni-Takagaki, Y., Takagi, Y., Goto, M., Miki, Y., Takano-Yamamoto, T., Jinnai, K., Takahashi, K., Kumegawa, M., Chihara, K., Fuijita, T., 2000. Parathyroid hormone-activated volume-sensitive calcium influx pathways in mechanically loaded osteocytes. J. Biol. Chem. 275, 3335–3342.

Mo, C., Romero-Suarez, S., Bonewald, L.F., Johnson, M., Brotto, M., 2012. Prostaglandin E2: from clinical applications to its potential role in bone-muscle crosstalk and myogenic differentiation. Recent Pat. Biotechnol. 6, 223–229.

Mullender, M.G., Huiskes, R., Weinans, H., 1994. A physiological approach to the stimulation of bone remodelling as a self-organisational control process. J. Biomech. 27, 1389–1394.

Mullender, M.G., Huiskes, R., 1995. Proposal for the regulatory mechanism of Wolff's law. J. Orthop. Res. 13, 503–512.

Mullender, M.G., Huiskes, R., 1997. Osteocytes and bone lining cells: which are the best candidates for mechano-sensors in cancellous bone? Bone 20, 527–532.

Mullender, M.G., Dijcks, S.J., Bacabac, R.G., Semeins, C.M., Van Loon, J.J., Klein-Nulend, J., 2006. Release of nitric oxide, but not prostaglandin E_2, by bone cells depends on fluid flow frequency. J. Orthop. Res. 24, 1170–1177.

Murray, D.W., Rushton, N., 1990. The effect of strain on bone cell prostaglandin E2 release: a new experimental method. Calcif. Tissue Int. 47, 35–39.

Nakamura, H., Kenmotsu, S., Sakai, H., Ozawa, H., 1995. Localization of CD44, the hyaluronate receptor, on the plasma membrane of osteocytes and osteoclasts in rat tibiae. Cell Tissue Res. 280, 225–233.

Nakamura, H., Ozawa, H., 1996. Immunolocalization of CD44 and the ERM family in bone cells of mouse tibiae. J. Bone Miner. Res. 11, 1715–1722.

Nakashima, T., Hayashi, M., Fukunaga, T., Kurata, K., Oh-Hora, M., Feng, J.Q., Bonewald, L.F., Kodama, T., Wutz, A., Wagner, E.F., Penninger, J.M., Takayanagi, H., 2011. Evidence for osteocyte regulation of bone homeostasis through RANKL expression. Nat. Med. 17, 1231–1234.

Nicolella, D.P., Moravits, D.E., Galea, A.M., Bonewald, L.F., Lankford, J.L., 2006. Osteocyte lacunae tissue strain in cortical bone. J. Biomech. 39, 1735–1743.

Nijweide, P.J., Mulder, R.J., 1986. Identification of osteocytes in osteoblast-like cultures using a monoclonal antibody specifically directed against osteocytes. Histochemistry 84, 343–350.

Nijweide, P.J., van der Plas, A., Scherft, J.P., 1981. Biochemical and histological studies on various bone cell preparations. Calcif. Tissue Int. 33, 529–540.

Nijweide, P.J., Van der Plas, A., Olthof, A.A., 1988. Osteoblastic differentiation. In: Evered, D., Harnett, S. (Eds.), Cell and Molecular Biology of Vertebrate Hard Tissues, Ciba Foundation Symposium, vol. 136. Wiley, Chichester, UK, pp. 61–77.

Noble, B.S., Stevens, H., Loveridge, N., Reeve, J., 1997. Identification of apoptotic changes in osteocytes in normal and pathological human bone. Bone 20, 273–282.

Noble, B.S., Peet, N., Stevens, H.Y., Brabbs, A., Mosley, J.R., Reilly, G.C., Reeve, J., Skerry, T.M., Lanyon, L.E., 2003. Mechanical loading: biphasic osteocyte survival and targeting of osteoclasts for bone destruction in rat cortical bone. Am. J. Physiol. Cell Physiol. 284, C934–C943.

Noonan, K.J., Stevens, J.W., Tammi, R., Tammi, M., Hernandez, J.A., Midura, R.J., 1996. Spatial distribution of CD44 and hyaluronan in the proximal tibia of the growing rat. J. Orthop. Res. 14, 573–581.

O'Brien, C.A., Plotkin, L.I., Galli, C., Goellner, J.J., Gortazar, A.R., Allen, M.R., Robling, A.G., Bouxsein, M., Schipani, E., Turner, C.H., Jilka, R.L., Weinstein, R.S., Manolagas, S.C., Bellido, T., 2008. Control of bone mass and remodeling by PTH receptor signaling in osteocytes. PLoS One 3, e2942.

Ohizumi, I., Harada, N., Taniguchi, K., Tsutsumi, Y., Nakagawa, S., Kaiho, S., Mayumi, T., 2000. Association of CD44 with OTS-8 in tumor vascular endothelial cells. Biochim. Biophys. Acta 1497, 197–203.

Oster, G., 1989. Cell motility and tissue morphogenesis. In: Stein, W.D., Bronner, F. (Eds.), Cell Shape: Determinants, Regulation and Regulatory Role. Academic Press, San Diego, CA, pp. 33–61.

Owen, M., 1995. Cell population kinetics of an osteogenic tissue. I. (1963). Clin. Orthop. Relat. Res. 313, 3–7.

Palumbo, C., 1986. A three-dimensional ultrastructural study of osteoid-osteocytes in the tibia of chick embryos. Cell Tissue Res. 246, 125–131.

Parfitt, A.M., 1976. The actions of parathyroid hormone on bone: relation to bone remodeling and turnover, calcium homeostasis, and metabolic bone diseases. II. PTH and bone cells: bone turnover and plasma calcium regulation. Metabolism 25, 909–955.

Parfitt, A.M., 1977. The cellular basis of bone turnover and bone loss. Clin. Orthop. Relat. Res. 127, 236–247.

Pathak, J.L., Bravenboer, N., Luyten, F.P., Verschueren, P., Lems, W.F., Klein-Nulend, J., Bakker, A.D., 2015. Mechanical loading reduces inflammation-induced human osteocyte-to-osteoclast communication. Calcif. Tisse Int. 97, 169–178.

Pathak, J.L., Bakker, A.D., Luyten, F.P., Verschueren, P., Lems, W.F., Klein-Nulend, J., Bravenboer, N., 2016. Systemic inflammation affects human osteocyte-specific protein and cytokine expression. Calcif. Tissue Int. 98, 596–608.

Pavalko, F.M., Chen, N.X., Turner, C.H., Burr, D.B., Atkinson, S., Hsieh, Y.F., Qiu, J., Duncan, R.L., 1998. Fluid shear-induced mechanical signaling in MC3T3-E1 osteoblasts requires cytoskeletonintegrin interactions. Am. J. Physiol. 275, C1591–C1601.

Piekarski, K., Munro, M., 1977. Transport mechanism operating between blood supply and osteocytes in long bones. Nature 269, 80–82.

Pienkowski, D., Pollack, S.R., 1983. The origin of stress-generated potentials in fluid-saturated bone. J. Orthop. Res. 1, 30–41.

Pitsillides, A.A., Rawlinson, S.C., Suswillo, R.F., Bourrin, S., Zaman, G., Lanyon, L.E., 1995. Mechanical strain-induced NO production by bone cells: a possible role in adaptive bone (re)modeling? FASEB J. 9, 1614–1622.

Plotkin, L.I., Bellido, T., 2001. Bisphosphonate-induced, hemichannel-mediated, anti-apoptosis through the Src/ERK pathway: a gap junction-independent action of connexin43. Cell Commun. Adhes. 8, 377–382.

Plotkin, L.I., Weinstein, R.S., Parfitt, A.M., Roberson, P.K., Manolagas, S.C., Bellido, T., 1999. Prevention of osteocyte and osteoblast apoptosis by bisphosphonates and calcitonin. J. Clin. Investig. 104, 1363–1374.

Plotkin, L.I., Manolagas, S.C., Bellido, T., 2002. Transduction of cell survival signals by connexin-43 hemichannels. J. Biol. Chem. 277, 8648–8657.

Plotkin, L.I., Mathov, I., Aguirre, J.I., Parfitt, A.M., Manolagas, S.C., Bellido, T., 2005. Mechanical stimulation prevents osteocyte apoptosis: requirement of integrins, Src kinases and ERKs. Am. J. Physiol. Cell Physiol. 289, C633–C643.

Plotkin, L.I., Speacht, T.L., Donahue, H.J., 2015. Cx43 and mechanotransduction in bone. Curr. Osteoporos. Rep. 13, 67–72.

Pollack, S.R., Salzstein, R., Pienkowski, D., 1984. The electric double layer in bone and its influence on stress generated potentials. Calcif. Tissue Int. 36, S77–S81.

Ponik, S.M., Triplett, J.W., Pavalko, F.M., 2007. Osteoblasts and osteocytes respond differently to oscillatory and unidirectional fluid flow profiles. J. Cell. Biochem. 100, 794–807.

Poole, K.E., van Bezooijen, R.L., Loveridge, N., Hamersma, H., Papapoulos, S.E., Lowik, C.W., Reeve, J., 2005. Sclerostin is a delayed secreted product of osteocytes that inhibits bone formation. FASEB J. 19, 1842–1844.

Prideaux, M., Schutz, C., Wijenayaka, A.R., Findlay, D.M., Campbell, D.G., Solomon, L.B., Atkins, G.J., 2016. Isolation of osteocytes from human trabecular bone. Bone 88, 64–72.

Qing, H., Ardeshirpour, L., Pajevic, P.D., Dusevich, V., Jähn, K., Kato, S., Wysolmerski, J., Bonewald, L.G., 2012. Demonstration of osteocytic perilacunar/canalicular remodeling in mice during lactation. J. Bone Miner. Res. 27, 1018–1029.

Rhee, Y., Allen, M.R., Condon, K., Lezcano, V., Ronda, A.C., Galli, C., Olivos, N., Passeri, G., O'Brien, C.A., Bivi, N., Plotkin, L.I., Bellido, T., 2011. PTH receptor signaling in osteocytes governs periosteal bone formation and intra-cortical remodeling. J. Bone Miner. Res. 26, 1035–1046.

Robinson, J.A., Chatterjee-Kishore, M., Yaworsky, P.J., Cullen, D.M., Zhao, W., Li, C., Kharode, Y., Sauter, L., Babij, P., Brown, E.L., Hill, A.A., Akhter, M.P., Johnson, M.L., Recker, R.R., Komm, B.S., Bex, F.J., 2006. Wnt/beta-catenin signaling is a normal physiological response to mechanical loading in bone. J. Biol. Chem. 281, 31720–31728.

Robling, A.G., Bellido, T., Turner, C.H., 2006. Mechanical stimulation in vivo reduces osteocyte expression of sclerostin. J. Musculoskelet. Neuronal Interact. 6, 354.

Rowe, P.S., de Zoysa, P.A., Dong, R., Wang, H.R., White, K.E., Econs, M.J., Oudet, C.L., 2000. MEPE, a new gene expressed in bone marrow and tumors causing osteomalacia. Genomics 67, 54–68.

Rowe, P.S., Garrett, I.R., Schwarz, P.M., Carnes, D.L., Lafer, E.M., Mundy, G.R., Gutierrez, G.E., 2005. Surface plasmon resonance (SPR) confirms that MEPE binds to PHEX via the MEPE-ASARM motif: a model for impaired mineralization in X-linked rickets (HYP). Bone 36, 33–46.

Rubin, C.T., 1984. Skeletal strain and the functional significance of bone architecture. Calcif. Tissue Int. 36, S11–S18.

Ruchon, A.F., Tenenhouse, H.S., Marcinkiewicz, M., Siegfried, G., Aubin, J.E., DesGroseillers, L., Crine, P., Boileau, G., 2000. Developmental expression and tissue distribution of Phex protein: effect of the Hyp mutation and relationship to bone markers. J. Bone Miner. Res. 15, 1440–1450.

Ruimerman, R., Hilbers, P., van Rietbergen, B., Huiskes, R.A., 2005. Theoretical framework for strain-related trabecular bone maintenance and adaptation. J. Biomech. 38, 931–941.

Sabbieti, M.G., Marchetti, L., Abreu, C., Montero, A., Hand, A.R., Raisz, L.G., Hurley, M.M., 1999. Prostaglandins regulate the expression of fibroblast growth factor-2 in bone. Endocrinology 140, 434–444.

Sachs, F., 1989. Ion channels as mechanical transducers. In: Stein, W.D., Bronner, F. (Eds.), Cell Shape: Determinants, Regulation and Regulatory Role. Academic Press, San Diego, CA, pp. 63–94.

Salzstein, R.A., Pollack, S.R., 1987. Electromechanical potentials in cortical bone. II. Experimental analysis. J. Biomech. 20, 271–280.

Santos, A., Bakker, A.D., Zandieh-Doulabi, B., Semeins, C.M., Klein-Nulend, J., 2009. Pulsating fluid flow modulates gene expression of proteins involved in Wnt signaling pathways in osteocytes. J. Orthop. Res. 27, 1280–1287.

Santos, A., Bakker, A.D., Zandieh-Doulabi, B., de Blieck-Hogervorst, J.M.A., Klein-Nulend, J., 2010. Early activation of the ß-catenin pathway in osteocytes is mediated by nitric oxide, phosphatidyl inositol-3 kinase/Akt, and focal adhesion kinase. Biochem. Biophys. Res. Commun. 391, 364–369.

Sauren, Y.M., Mieremet, R.H., Groot, C.G., Scherft, J.P., 1992. An electron microscopic study on the presence of proteoglycans in the mineralized matrix of rat and human compact lamellar bone. Anat. Rec. 232, 36–44.

Schiller, P.C., Mehta, P.P., Roos, B.A., Howard, G.A., 1992. Hormonal regulation of intercellular communication: parathyroid hormone increases connexin 43 gene expression and gap-junctional communication in osteoblastic cells. Mol. Endocrinol. 6, 1433–1440.

Scholl, F.G., Gamallo, C., Vilar, S., Quintanilla, M., 1999. Identification of PA2.26 antigen as a novel cell-surface mucin-type glycoprotein that induces plasma membrane extensions and increased motility in keratinocytes. J. Cell Sci. 112 (Pt 24), 4601–4613.

Schulze, E., Witt, M., Kasper, M., Löwik, C.W., Funk, R.H., 1999. Immunohistochemical investigations on the differentiation marker protein E11 in rat calvaria, calvaria cell culture and the osteoblastic cell line ROS 17/2.8. Histochem. Cell Biol. 111, 61–69.

Semeins, C.M., Bakker, A.D., Klein-Nulend, J., 2012. Isolation of primary avian osteocytes. Methods Mol. Biol. 816, 43–53.

Shao, J., Zhou, Y., Lin, J., Nguyen, T.D., Huang, R., Gu, Y., Friis, T., Crawford, R., Xiao, Y., 2018a. Notch expression by osteocytes plays a critical role in mineralisation. J. Mol. Med. 96, 333–347.

Shao, J., Zhou, Y., Xiao, Y., 2018b. The regulatory roles of Notch in osteocyte differentiation via the crosstalk with canonical Wnt pathways during the transition of osteoblasts to osteocytes. Bone 108, 165–178.

Shen, H., Grimston, S., Civitelli, R., Thomopoulos, S., 2015. Deletion of connexin43 in osteoblasts/osteocytes leads to impaired muscle formation in mice. J. Bone Miner. Res. 30, 596–605.

Silvestrini, G., Mocetti, P., Ballanti, P., Di Grezia, R., Bonucci, E., 1999. Cytochemical demonstration of the glucocorticoid receptor in skeletal cells of the rat. Endocr. Res. 25, 117–128.

Singla, V., Reiter, J.F., 2005. The primary cilium as the cell's antenna: signaling at a sensory organelle. Science 313, 629–633.

Skerry, T.M., Bitensky, L., Chayen, J., Lanyon, L.E., 1989. Early strain-related changes in enzyme activity in osteocytes following bone loading *in vivo*. J. Bone Miner. Res. 4, 783–788.

Smit, T.H., Burger, E.H., 2000. Is BMU-coupling a strain-regulated phenomenon? A finite element analysis. J. Bone Miner. Res. 15, 301–307.

Smit, T.H., Burger, E.H., Huyghe, J.M., 2002. A case for strain-induced fluid flow as a regulator of BMU-coupling and osteonal alignment. J. Bone Miner. Res. 17, 2021–2029.

Smith, A.J., Singhrao, S.K., Newman, G.R., Waddington, R.J., Embery, G., 1997. A biochemical and immunoelectron microcopical analysis of chondroitin sulfate-rich proteoglycans in human alveolar bone. Histochem. J. 29, 1–9.

Spatz, J.M., Wein, M.N., Gooi, J.H., Qu, Y., Garr, J.L., Liu, S., Barry, K.J., Uda, Y., Lai, F., Dedic, C., Balcells-Camps, M., Kronenberg, H.M., Babij, P., Pajevic, P.D., 2015. The Wnt inhibitor sclerostin is up-regulated by mechanical unloading in osteocytes in vitro. J. Biol. Chem. 290, 16744–16758.

Sterck, J.G., Klein-Nulend, J., Lips, P., Burger, E.H., 1998. Response of normal and osteoporotic human bone cells to mechanical stress in vitro. Am. J. Physiol. 274, E1113–E1120.

Stern, A.R., Bonewald, L.F., 2015. Isolation of osteocytes from mature and aged murine bone. Methods Mol. Biol. 1226, 3–10.

Strom, T.M., Francis, F., Lorenz, B., Boddrich, A., Econs, M.J., Lehrach, H., Meitinger, T., 1997. Pex gene deletions in Gy and Hyp mice provide mouse models for X-linked hypophosphatemia. Hum. Mol. Genet. 6, 165–171.

Sugawara, Y., Kamioka, H., Ishihara, Y., Fujisawa, N., Kawanabe, N., Yamashiro, T., 2013. The early mouse 3D osteocyte network in the presence and absence of mechanical loading. Bone 52, 189–196.

Suzuki, R., Domon, T., Wakita, M., 2000. Some osteocytes released from thek lacunae are embedded again in the bone and not engulfed by osteoclasts during bone remodeling. Anat. Embryol. 202, 119–128.

Takagi, M., Ono, Y., Maeno, M., Miyashita, K., Omiya, K., 1997. Immunohistochemical and biochemical characterization of sulphated proteoglycans in embryonic chick bone. J. Nihon Univ. Sch. Dent. 39, 156–163.

Takeda, N., Tsuboyama, T., Kasai, R., Takahashi, K., Shimizu, M., Nakamura, T., Higuchi, K., Hosokawa, M., 1999. Expression of the c-fos gene induced by parathyroid hormone in the bones of SAMP6 mice, a murine model for senile osteoporosis. Mech. Ageing Dev. 108, 87–97.

Tan, S.D., Kuijpers-Jagtman, A.M., Semeins, C.M., Bronckers, A.L., Maltha, J.C., Von den Hoff, J.W., Everts, V., Klein-Nulend, J., 2006. Fluid shear stress inhibits TNFalpha-induced osteocyte apoptosis. J. Dent. Res. 85, 905–909.

Tan, S.D., de Vries, T.J., Kuijpers-Jagtman, A.M., Semeins, C.M., Everts, V., Klein-Nulend, J., 2007. Osteocytes subjected to fluid flow inhibit osteoclast formation and bone resorption. Bone 41, 745–751.

Tan, S.D., Bakker, A.D., Semeins, C.M., Kuijpers-Jagtman, A.M., Klein-Nulend, J., 2008. Inhibition of osteocyte apoptosi by fluid flow is mediated by nitric oxide. Biochem. Biophys. Res. Commun. 369, 1150–1154.

Tanaka, T., Sakano, A., 1985. Differences in permeability of microperoxidase and horseradish peroxidase into alveolar bone of developing rats. J. Dent. Res. 64, 870–876.

Tanaka, T., Hoshijima, M., Sunaga, J., Nishida, T., Hashimoto, M., Odagaki, N., Osumi, R., Adachi, T., Kamioka, H., October 12, 2017. Analysis of Ca2+ response of osteocyte network by three-dimensional time-lapse imaging in living bone. J. Bone Miner. Metab. https://doi.org/10.1007/s00774-017-0868-x [Epub ahead of print].

Tanaka-Kamioka, K., Kamioka, H., Ris, H., Lim, S.S., 1998. Osteocyte shape is dependent on actin filaments and osteocyte processes are unique actin-rich projections. J. Bone Miner. Res. 13, 1555–1568.

Tatsumi, S., Ishii, K., Amizuka, N., Li, M., Kobayashi, T., Kohno, K., Ito, M., Takeshita, S., Ikeda, K., 2007. Targeted ablation of osteocytes induces osteoporosis with defective mechanotransduction. Cell Metabol. 5, 464–475.

Terai, K., Takano-Yamamoto, T., Ohba, Y., Hiura, K., Sugimoto, M., Sato, M., Kawahata, H., Inaguma, N., Kitamura, Y., Nomura, S., 1999. Role of osteopontin in bone remodeling caused by mechanical stress. J. Bone Miner. Res. 14, 839–849.

Tezuka, K., Takeshita, S., Hakeda, Y., Kumegawa, M., Kikuno, R., Hashimoto-Gotoh, T., 1990. Isolation of mouse and human cDNA clones encoding a protein expressed specifically in osteoblasts and brain tissue. Biochem. Biophys. Res. Commun. 173, 246–251.

The HYP Consortium, 1995. A gene (PEX) with homologies to endopeptidases is mutated in patients with X-linked hypophosphatemic rickets. Nat. Genet. 11, 130–136.

Tomkinson, A., Gevers, E.F., Wit, J.M., Reeve, J., Noble, B.S., 1998. The role of estrogen in the control of rat osteocyte apoptosis. J. Bone Miner. Res. 13, 1243–1250.

Torreggiani, E., Matthews, B.G., Pejda, S., Matic, I., Horowitz, M.C., Grcevic, D., Kalajzic, I., 2013. Preosteocytes/osteocytes have the potential to dedifferentiate becoming a source of osteoblasts. PLoS One 8, e75204.

Toyosawa, S., Shintani, S., Fujiwara, T., Ooshima, T., Sato, A., Ijuhin, N., Komori, T., 2001. Dentin matrix protein 1 is predominantly expressed in chicken and rat osteocytes but not in osteoblasts. J. Bone Miner. Res. 16, 2017–2026.

Turner, C.H., Forwood, M.R., Otter, M.W., 1994. Mechanotransduction in bone: do bone cells act as sensors of fluid flow? FASEB J. 8, 875–878.

Turner, C.H., Takano, Y., Owan, I., Murrell, G.A., 1996. Nitric oxide inhibitor L-NAME suppresses mechanically induced bone formation in rats. Am. J. Physiol. 270, E639–E643.

Uematsu, M., Ohara, Y., Navas, J.P., Nishida, K., Murphy, T.J., Alexander, R.W., Nerem, R.M., Harrison, D.G., 1995. Regulation of endothelial nitric oxide synthase mRNA expression by shear stress. Am. J. Physiol. 269, C1371–C1378.

Van Bezooijen, R.L., Roelen, B.A., Visser, A., van der Wee-Pals, L., de Wilt, E., Karperien, M., Hamersma, H., Papapoulos, S.E., ten Dijke, P., Lowik, C.W., 2004. Sclerostin is an osteocyte-expressed negative regulator of bone formation, but not a classical BMP antagonist. J. Exp. Med. 199, 805–814.

Van der Plas, A., Nijweide, P.J., 1992. Isolation and purification of osteocytes. J. Bone Miner. Res. 7, 389–396.

Van der Plas, A., Aarden, E.M., Feyen, J.H., de Boer, A.H., Wiltink, A., Alblas, M.J., de Ley, L., Nijweide, P.J., 1994. Characteristics and properties of osteocytes in culture. J. Bone Miner. Res. 9, 1697–1704.

Van Hove, R.P., Nolte, P.A., Vatsa, A., Semeins, C.M., Salmon, P.L., Smit, T.H., Klein-Nulend, J., 2009. Osteocyte morphology in human tibiae of different bone pathologies with different bone mineral density – is there a role for mechanosensing? Bone 45, 321–329.

Van Oers, R.F.M., Klein-Nulend, J., Bacabac, R.G., 2014. The osteocyte as an orchestrator of bone remodeling: an engineers perspective. Clin. Rev. Bone Miner. Metabol. 12, 2–13.

Vatsa, A., Mizuno, D., Smit, T.H., Schmidt, C.F., MacKintosh, F.C., Klein-Nulend, J., 2006. Bio imaging of intracellular NO production in single bone cells after mechanical stimulation. J. Bone Miner. Res. 21, 1722–1728.

Vatsa, A., Smit, T.H., Klein-Nulend, J., 2007. Extracellular NO signalling from a mechanically stimulated osteocyte. J. Biomech. 40 (Suppl. 1), S89–S95.

Vatsa, A., Breuls, R.G., Semeins, C.M., Salmon, P.L., Smit, T.H., Klein-Nulend, J., 2008a. Osteocyte morphology in fibula and calvaria – is there a role for mechanosensing? Bone 43, 452–458.

Vatsa, A., Semeins, C.M., Smit, T.H., Klein-Nulend, J., 2008b. Paxillin localization in osteocytes — is it determined by the direction of loading? Biochem. Biophys. Res. Commun. 377, 1019—1024.

Veno, P., Nicolella, D.P., Sivakumar, P., Kalajzic, I., Rowe, D., Harris, S.E., Bonewald, L., Dallas, S.L., 2006. Live imaging of osteocytes within their lacunae reveals cell body and dendrite motions. J. Bone Miner. Res. 21 (Suppl. 1), S38.

Verborgt, O., Gibson, G.J., Schaffler, M.B., 2000. Loss of osteocyte integrity in association with microdamage and bone remodeling after fatigue in vivo. J. Bone Miner. Res. 15, 60—67.

Verborgt, O., Tatton, N.A., Majeska, R.J., Schaffler, M.B., 2002. Spatial distribution of Bax and Bcl-2 in osteocytes after bone fatigue: complementary roles in bone remodeling regulation? J. Bone Miner. Res. 17, 907—914.

Vezeridis, P.S., Semeins, C.M., Chen, Q., Klein-Nulend, J., 2005. Osteocytes subjected to pulsating fluid flow regulate osteoblast proliferation and differentiation. Biochem. Biophys. Res. Commun. 348, 1082—1088.

Wang, L., 2018. Solute transport in the bone lacunar-canalicular system (LCS). Curr. Osteoporos. Rep. 16, 32—41.

Wang, N., Ingber, D.E., 1994. Control of cytoskeletal mechanisms by extracellular matrix, cell shape and mechanical tension. Biophys. J. 66, 2181—2189.

Wang, N., Butler, J.P., Ingber, D.E., 1993. Mechanotransduction across the cell surface and through the cytoskeleton. Science 260, 1124—1127.

Wang, L., Ciani, C., Doty, S.B., Fritton, S.P., 2004. Delineating bone's interstitial fluid pathway in vivo. Bone 34, 499—509.

Watson, P.A., 1991. Function follows form: generation of intracellular signals by cell deformation. FASEB J. 5, 2013—2019.

Weinbaum, S., Cowin, S.C., Zeng, Y., 1994. A model for the excitation of osteocytes by mechanical loading-induced bone fluid shear stresses. J. Biomech. 27, 339—360.

Weinstein, R.S., Jilka, R.L., Parfitt, A.M., Manolagas, S.C., 1998. Inhibition of osteoblastogenesis and promotion of apoptosis of osteoblasts and osteocytes by glucocorticoids: potential mechanisms of their deleterious effects on bone. J. Clin. Investig. 102, 274—282.

Westbroek, I., Ajubi, N.E., Alblas, M.J., Semeins, C.M., Klein-Nulend, J., Burger, E.H., Nijweide, P.J., 2000a. Differential stimulation of prostaglandin G/H synthase-2 in osteocytes and other osteogenic cells by pulsating fluid flow. Biochem. Biophys. Res. Commun. 268, 414—419.

Westbroek, I., Alblas, M.J., Van der Plas, A., Nijweide, P.J., 2000b. Estrogen receptor α is preferentially expressed in osteocytes. J. Bone Miner. Res. 15 (Suppl. 1), S494.

Westbroek, I., Van der Plas, A., De Rooij, K.E., Klein-Nulend, J., Nijweide, P.J., 2001. Expression of serotonin receptors in bone. J. Biol. Chem. 276, 28961—28968.

Westbroek, I., De Rooij, K.E., Nijweide, P.J., 2002. Osteocyte-specific monoclonal antibody MAb OB7.3 is directed against Phex protein. J. Bone Miner. Res. 17, 845—853.

Wijeratne, S.S., Martinez, J.R., Grindel, B.J., Frey, E.W., Li, J., Wang, L., Farach-Carson, M.C., Kiang, C.H., 2016. Single molecule force measurements of perlecan/HSPG2: a key component of the osteocyte pericellular matrix. Matrix Biol. 50, 27—38.

Willems, H.M.E., van den Heuvel, E.G.H.M., Schoemaker, R.J.W., Klein-Nulend, J., Bakker, A.D., 2017. Diet and exercise: a match made in bone. Curr. Osteoporos. Rep. 15, 555—563.

Windahl, S.H., Borjesson, A.E., Farman, H.H., Engdahl, C., Movérare-Skrtic, S., Sjögren, K., Lagerquist, M.K., Kindblom, J.M., Koskela, A., Tuukkanen, J., Divieti Pajevic, P., Feng, J.Q., Dahlman-Wright, K., Antonson, P., Gustafsson, J.A., Ohlsson, C., 2013. Estrogen receptor-α in osteocytes is important for trabecular bone formation in male mice. Proc. Natl. Acad. Sci. U.S.A. 110, 2294—2299.

Wolff, J.D., 1892. Das Gesetz der Transformation der Knochen. A. Hirschwald, Berlin.

Woo, S.M., Rosser, J., Dusevich, V., Kalajzic, I., Bonewald, L.F., 2011. Cell line IDG-SW3 replicates osteoblast-to-late-osteocyte differentiation in vitro and accelerates bone formation in vivo. J. Bone Miner. Res. 26, 2634—2646.

Wu, X.T., Sun, L.W., Yang, X., Ding, D., Han, D., Fan, Y.B., 2017. The potential role of spectrin network in the mechanotransduction of MLO-Y4 osteocytes. Sci. Rep. 7, 40940.

Wu, V., Van Oers, R.F.M., Schulten, E.A.J.M., Helder, M.N., Bacabac, R.G., Klein-Nulend, J., 2018. Osteocyte morphology and orientation in relation to strain in the jaw bone. Int. J. Oral Sci. 10 (2), 1—8.

Xiao, Z., Zhang, S., Mahlios, J., Zhou, G., Magenheimer, B.S., Guo, D., Dallas, S.L., Maser, R., Calvet, J.P., Bonewald, L., Quarles, L.D., 2006. Cilia-like structures and polycystin-1 in osteoblasts/osteocytes and associated abnormalities in skeletogenesis and Runx2 expression. J. Biol. Chem. 281, 30884—30895.

Xiong, J., Onal, M., Jilka, R.L., Weinstein, R.S., Manolagas, S.C., O'Brien, C.A., 2011. Matrix-embedded cells control osteoclast formation. Nat. Med. 17, 1235—1241.

Xiong, J., Piemontese, M., Onal, M., Campbell, J., Goellner, J.J., Dusevich, V., Bonewald, L.F., Manolagas, S.C., O'Brien, C.A., 2015. Osteocytes, not osteoblasts or lining cells, are the main source of the RANKL required for osteoclast formation in remodeling bone. PLoS One 10, e013818.

Xu, H., Gu, S., Riquelme, M.A., Burra, S., Callaway, D., Cheng, H., Guda, T., Schmitz, J., Fajardo, R.J., Werner, S.L., Zhao, H., Shang, P., Johnson, M.L., Bonewald, L.F., Jiang, J.X., 2015. Connexin 43 channels are essential for normal bone structure and osteocyte viability. J. Bone Miner. Res. 30, 550—562.

Yamazaki, M., Nakajima, F., Ogasawara, A., Moriya, H., Majeska, R.J., Einhorn, T.A., 1999. Spatial and temporal distribution of CD44 and osteopontin in fracture callus. J. Bone Joint Surg. Br. 81, 508—515.

Yang, D., Turner, A.G., Wijenayaka, A.R., Anderson, P.H., Morris, H.A., Atkins, G.J., 2015. 1,25-Dihydroxyvitamin D3 and extracellular calcium promote mineral deposition via NPP1 activity in a mature osteoblast cell line MLO-A5. Mol. Cell. Endocrinol. 412, 140—147.

Yellowley, C.E., Li, Z., Zhou, Z., Jacobs, C.R., Donahue, H.J., 2000. Functional gap junctions between osteocytic and osteoblastic cells. J. Bone Miner. Res. 15, 209—217.

You, J., Yellowley, C.E., Donahue, H.J., Zhang, Y., Chen, Q., Jacobs, C.R., 2000. Substrate deformation levels associated with routine physical activity are less stimulatory to bone cells relative to loading induced oscillating fluid flow. J. Biomech. Eng. 122, 387–393.

Zaman, G., Pitsillides, A.A., Rawlinson, S.C., Suswillo, R.F., Mosley, J.R., Cheng, M.Z., Platts, L.A., Hukkanen, M., Polak, J.M., Lanyon, L.E., 1999. Mechanical strain stimulates nitric oxide production by rapid activation of endothelial nitric oxide synthase in osteocytes. J. Bone Miner. Res. 14, 1123–1131.

Zaman, G., Jessop, H.L., Muzylak, M., De Souza, R.L., Pitsillides, A.A., Price, J.S., Lanyon, L.L., 2006. Osteocytes use estrogen receptor alpha to respond to strain but their ERalpha content is regulated by estrogen. J. Bone Miner. Res. 21, 1297–1306.

Zhang, K., Barragan-Adjemian, C., Ye, L., Kotha, S., Dallas, M., Lu, Y., Zhao, S., Harris, M., Harris, S.E., Feng, J.Q., Bonewald, L.F., 2006. E11/gp38 selective expression in osteocytes: regulation by mechanical strain and role in dendrite elongation. Mol. Cell Biol. 26, 4539–4552.

Zhang, B., Hou, R., Zou, Z., Luo, T., Zhang, Y., Wang, L., Wang, B., 2018. Mechanically induced autophagy is associated with ATP metabolism and cellular viability in osteocytes in vitro. Redox Biol. 14, 492–498.

Zhao, S., Zhang, Y.K., Harris, S., Ahuja, S.S., Bonewald, L.F., 2002. MLO-Y4 osteocyte-like cells support osteoclast formation and activation. J. Bone Miner. Res. 17, 2068–2079.

Zhao, N., Nociti Jr., F.H., Duan, P., Prideaux, M., Zhao, H., Foster, B.L., Somerman, M.J., Bonewald, L.F., 2016. Isolation and functional analysis of an immortalized murine cementocyte cell line, IDG-CM6. J. Bone Miner. Res. 31, 430–442.

Further reading

Aubin, J.E., Turksen, K., 1996. Monoclonal antibodies as tools for studying the osteoblast lineage. Microsc. Res. Tech. 33, 128–140.

Divieti, P., Geller, A.I., Suliman, G., Juppner, H., Bringhurst, F.R., 2005. Receptors specific for the carboxyl-terminal region of parathyroid hormone on bone-derived cells: determinants of ligand binding and bioactivity. Endocrinology 146, 1863–1870.

Ikegame, M., Ishibashi, O., Yoshizawa, T., Shimomura, J., Komori, T., Ozawa, H., Kawashima, H., 2001. Tensile stress induces bone morphogenetic protein 4 in preosteoblastic and fibroblastic cells, which later differentiate into osteoblasts leading to osteogenesis in the mouse calvariae in organ culture. J. Bone Miner. Res. 16, 24–32.

Joldersma, M., Klein-Nulend, J., Oleksik, A.M., Heyligers, I.C., Burger, E.H., 2001. Estrogen enhances mechanical stress-induced prostaglandin production by bone cells from elderly women. Am. J. Physiol. 280, E436–E442.

Kaspar, D., Seidl, W., Neidlinger-Wilke, C., Ignatius, A., Claes, L., 2000. Dynamic cell stretching increases human osteoblast proliferation and CICP synthesis but decreases osteocalcin synthesis and alkaline phosphatase activity. J. Biomech. 33, 45–51.

Kawata, A., Mikuni-Takagaki, Y., 1998. Mechanotransduction in stretched osteocytes, temporal expression of immediate early and other genes. Biochem. Biophys. Res. Commun. 246, 404–408.

Mikuni-Takagaki, Y., Suzuki, Y., Kawase, T., Saito, S., 1996. Distinct responses of different populations of bone cells to mechanical stress. Endocrinology 137, 2028–2035.

Neidlinger-Wilke, C., Stall, I., Claes, L., Brand, R., Hoellen, I., Rubenacker, S., Arand, M., Kinzl, L., 1995. Human osteoblasts from younger normal and osteoporotic donors show differences in proliferation and TGF-beta release in response to cyclic strain. J. Biomech. 28, 1411–1418.

Owan, I., Burr, D.B., Turner, C.H., Qui, J., Tu, Y., Onyia, J.E., Duncan, R.L., 1997. Mechanotransduction in bone: osteoblasts are more responsive to fluid forces than mechanical strain. Am. J. Physiol. 273, C810–C815.

Petersen, D.N., Tkalcevic, G.T., Mansolf, A.L., Rivera-Gonzalez, R., Brown, T.A., 2000. Identification of osteoblast/osteocyte factor 45 (OF45), a bone-specific cDNA encoding an RGD-containing protein that is highly expressed in osteoblasts and osteocytes. J. Biol. Chem. 275, 36172–36180.

Raulo, E., Chernousov, M.A., Carey, D., Nolo, R., Rauvala, H., 1994. Isolation of a neuronal cell surface receptor of heparin-binding growth-associated molecule (HB-GAM): identification as N-syndecan (syndecan-3). J. Biol. Chem. 269, 12999–13004.

Rauvala, H., 1989. An 18-kD heparin-binding protein of developing brain that is distinct from fibroblastic growth factors. EMBO J. 8, 2933–2941.

Wetterwald, A., Hoffstetter, W., Cecchini, M.G., Lanske, B., Wagner, C., Fleisch, H., Atkinson, M., 1996. Characterization and cloning of the E11 antigen, a marker expressed by rat osteoblasts and osteocytes. Bone 18, 125–132.

Chapter 7

Transcriptional control of osteoblast differentiation and function

Gérard Karsenty
Département of Genetics and Development, Columbia University Medical Center, New York, NY, United States

Chapter outline

Runx2, a master control gene of osteoblast differentiation in bony vertebrates 163
Runx2 functions during skeletogenesis beyond osteoblast differentiation 165
Regulation of Runx2 accumulation and function 165
Osterix, a Runx2-dependent osteoblast-specific transcription factor required for bone formation 167
ATF4, a transcriptional regulator of osteoblast functions and a mediator of the neural regulation of bone mass 168
Additional transcriptional regulators of osteoblast differentiation and function 169
Transcription factors acting downstream of Wnt signaling in osteoblasts: what do they actually do in differentiated osteoblasts? 170
Regulation of osteoblast differentiation by means other than transcription factors 171
References 172
Further reading 175

As is the case for every cell differentiation process, differentiation of a mesenchymal pluripotent cell into any cell type is governed in large part by transcription factors that trigger the entire program of cell differentiation. Our knowledge about the transcriptional control of osteoblast differentiation made its main strides at the end of the 20th century and has been significantly refined since then, with the emergence of novel mechanisms regulating gene expression in addition to transcription factors. Briefly and ideally, a transcription factor that is a differentiation factor for a given cell type should (1) be expressed in progenitors of this cell type, (2) regulate the expression of all cell-specific genes in this cell type, (3) induce expression of the aforementioned genes when ectopically expressed in other cell types (sufficiency criterion), and (4) be necessary for the differentiation of this cell type in vivo, in mice, and at best in humans. As presented in this chapter, the transcription factor currently viewed as the master gene of osteoblast differentiation is one of the very few differentiation factors to fulfill all these criteria.

Runx2, a master control gene of osteoblast differentiation in bony vertebrates

The power of a combined effort between molecular biologists and human geneticists in identifying key genes regulating cell differentiation, which has been so beneficial for our understanding of skeletal biology, is best illustrated by the realization at the end of the 1990s that Runt-related transcription factor 2 (Runx2) is a master gene of osteoblast differentiation. Runx2, previously termed Pebp2a1, Aml3, or Cbfa1, was originally cloned in 1993 as one of three mammalian homologs of the *Drosophila* transcription factor Runt (Kagoshima et al., 1993; Ogawa et al., 1993). Based on the premise that it might be expressed in thymus and T cell lines, but not in B cell lines, Runx2 was thought to be involved in T cell differentiation and was deleted to study T cell differentiation (Ogawa et al., 1993; Satake et al., 1995). However, 4 years after the cloning of Runx2, its crucial role as a transcriptional determinant of osteoblast differentiation was demonstrated by several investigators working independent of one another and using different yet complementary experimental approaches (Ducy et al., 1997; Komori et al., 1997; Lee et al., 1997; Mundlos et al., 1997; Otto et al., 1997).

Principles of Bone Biology. https://doi.org/10.1016/B978-0-12-814841-9.00007-5

One approach, purely molecular, was aimed at the identification of osteoblast-specific transcription factors through the systematic analysis of the promoter of what was then the only osteoblast-specific gene, *Osteocalcin.* The analysis of a proximal promoter fragment of one of the two mouse *Osteocalcin* genes led to the identification of the only two known osteoblast-specific *cis*-acting elements, termed OSE1 and OSE2 (Ducy and Karsenty, 1995). Remarkably, as of this writing, those two *cis*-acting elements remain the only known strictly osteoblast-specific *cis*-acting elements. Sequence inspection of OSE2 revealed homology for the DNA-binding site of Runt family transcription factors, and subsequent analysis demonstrated that the factor binding to OSE2 is related immunologically to transcription factors of the Runt family (Geoffroy et al., 1995; Merriman et al., 1995). Eventually, screening of a mouse osteoblast cDNA library revealed that only one of the three mammalian *Runx* genes, namely *Runx2*, is expressed predominantly in cells of the osteoblast lineage (Ducy et al., 1997). Indeed, in situ hybridization further revealed that during mouse development, *Runx2* expression is first detected in the lateral plate mesoderm at 10.5 days postcoitus (dpc), and later is confined in cells of the mesenchymal condensations. Until 12.5 dpc these cells, which prefigure the future skeleton, represent common precursors of osteoblasts and chondrocytes. At 14.5 dpc osteoblasts first appear and maintain the expression of *Runx2*, whereas in chondrocytes *Runx2* expression decreases significantly and becomes restricted to prehypertrophic and hypertrophic chondrocytes. After birth, *Runx2* expression is strictly restricted to osteoblasts and cells of the perichondrium. This spatial and temporal expression pattern suggested Runx2 might play a critical role as a regulator of osteoblast differentiation (Ducy et al., 1997).

The demonstration that Runx2 was indeed an osteoblast differentiation factor came from several synergistic lines of molecular and genetic evidence. First, in addition to the *Osteocalcin* promoter, functional OSE2-like elements were identified in the promoter regions of most other genes that are expressed at relatively high levels in osteoblasts, such as *α1(II)-collagen*, *Osteopontin*, and *Bone sialoprotein*, and eventually many more (Ducy et al., 1997). Second, and more decisively, forced expression of *Runx2* in nonosteoblastic cell lines or primary skin fibroblasts induced osteoblast-specific gene expression in these cells, demonstrating that Runx2 acts as a transcriptional activator of osteoblast differentiation in vitro (Ducy et al., 1997). This was subsequently verified in vivo, where it was shown that the constitutive expression of Runx2 at low levels in nonhypertrophic chondrocytes triggered the entire cascade of endochondral bone formation, which normally does not occur in the cartilaginous ribs. Remarkably, however, ectopic expression of Runx in chondrocytes cannot cause transdifferentiation of chondrocytes into osteoblasts (Takeda et al., 2001). This suggests that if trans-differentiation of chondrocytes into osteoblasts can exist it remains a rather rare event. Third, the ultimate demonstration that Runx2 is an indispensable transcriptional activator of osteoblast differentiation came from genetic studies in mice and humans. At the same time, two groups deleted the *Runx2* gene from the mouse genome, both expecting an immunological phenotype based on the assumption that *Runx2* was involved in T cell differentiation (Komori et al., 1997; Otto et al., 1997). Instead, all *Runx2*-deficient mice had no skull, because intramembranous bone formation did not occur. In the rest of their skeleton, there are no osteoblasts in *Runx2*-deficient mice, an observation confirmed by the lack of expression of osteoblast marker genes.

The critical importance of Runx2 for osteoblast differentiation was further emphasized by the finding that mice lacking only one allele of *Runx2* display hypoplastic clavicles and delayed closure of the fontanelles, i.e., defects of intra-membranous ossification (Otto et al., 1997). This phenotype is identical to what is seen in a human disease termed cleidocranial dysplasia (CCD), and subsequent genetic analysis of CCD patients revealed disease-causing heterozygous mutations of the *RUNX2* gene, thereby demonstrating the relevance of Runx2 for osteoblast differentiation also in humans (Lee et al., 1997; Mundlos et al., 1997). Taken together, this overwhelming molecular and genetic evidence has led to the generally accepted view that Runx2 is a master control gene of osteoblast differentiation, providing a molecular switch inducing osteoblast-specific gene expression in mesenchymal progenitor cells (Lian and Stein, 2003). Remarkably, many, although not all, of the subsequent advances that have been made in the field are centered around the biology of Runx2.

In addition to its prominent role in osteoblast differentiation and skeletogenesis, Runx2 is also involved in the regulation of bone formation beyond development. This has been demonstrated in several ways. First, transgenic mice expressing a dominant-negative variant of Runx2 specifically in fully differentiated osteoblasts are viable, but develop severe osteopenia caused by a decreased rate of bone formation, in the face of normal osteoblast numbers (Ducy et al., 1999). This phenotype is readily explained by the finding that several Runx2 target genes encoding bone extracellular matrix proteins are expressed at much lower levels. Second, mice lacking *Stat1*, a transcription factor attenuating the nuclear translocation of Runx2, as discussed later, display a high-bone-mass phenotype that is explained not only by increased osteoblast differentiation, but also by increased bone matrix deposition (Kim et al., 2003). Third, a similar, but even more severe, phenotype is observed in mice lacking the nuclear adapter protein Shn3. Because Shn3, as discussed later, is involved in the ubiquitination and proteasomal degradation of Runx2, the increased bone formation of the Shn3-deficient mice is readily explained by increased Runx2 levels in osteoblasts that in turn lead to enhanced bone matrix deposition (Jones et al., 2006). Given these results, it came as a surprise that another transgenic mouse model, overexpressing intact *RU/u2* under the control of an osteoblast-specific 0.1 *(I)-collagen* promoter fragment, did not display the expected high-bone-mass phenotype, but a severe osteopenia accompanied by an increased fracture risk (Liu et al., 2001). Although these mice had increased numbers of osteoblasts, their bone formation rate was strikingly reduced, which likely illustrates the fact that the dosage of Runx2 needs to be tightly regulated to orchestrate proper bone formation in vivo.

Runx2 functions during skeletogenesis beyond osteoblast differentiation

Although most of these results demonstrating a key role for Runx2 in osteoblasts were already discussed in the last edition of this book, there is accumulating novel evidence that the role of Runx2 in skeletogenesis is much more complex than previously anticipated. The starting point for these findings was the observation that Runx2-deficient mice also display defects of chondrocyte hypertrophy in some skeletal elements (Inada et al., 1999; Kim et al., 1999). Moreover, because *Runx2* is transiently expressed in prehypertrophic chondrocytes of mouse embryos, there was a possibility of a function of Runx2 in chondrocyte differentiation. One way to address this possibility was the generation of a transgenic mouse model expressing *Runx2* in nonhypertrophic chondrocytes, using an *α1(II)-collagen* promoter/enhancer construct (Takeda et al., 2001).

In line with the suspected role of Runx2 as a positive regulator of chondrocyte hypertrophy, these transgenic mice displayed accelerated chondrocyte maturation in the growth plates, but also evidence of ectopic cartilage formation in the rib cage or in the trachea, among other locations. Moreover, the presence of this transgene in a Runx2-deficient genetic background prevented the absence of skeletal mineralization that is normally associated with *Runx2* deficiency. However, the skeleton of these mice contained only hypertrophic cartilage, and no bone matrix, thereby demonstrating that Runx2 induces chondrocyte hypertrophy, but not a transdifferentiation into osteoblasts (Takeda et al., 2001).

Because *Runx2* expression in prehypertrophic chondrocytes is transient, the main function of Runx2 here is probably to establish the growth plate. However, through another site of expression, namely in perichondrial cells, Runx2 has an additional function in the regulation of chondrogenesis. In these cells Runx2 positively regulates the expression of Fgfl8, a diffusible molecule that inhibits chondrocyte maturation and osteoblast differentiation (Hinoi et al., 2006; Liu et al., 2002; Ohbayashi et al., 2002). Taken together, these results establish that Runx2 is more than the master gene of osteoblast differentiation; it is, in fact, along with Sox9, the major transcriptional regulator of cell differentiation during skeletogenesis, acting positively and negatively on osteoblast and chondrocyte differentiation.

Regulation of Runx2 accumulation and function

In essence, the accumulation and demonstration that Runx2 exhibits, in vivo in mice and humans, all the characteristics of an osteoblast differentiation factor raised the usual questions: (1) What is upstream of Runx2? (2) How is Runx2 function regulated? (3) What is downstream of Runx2?

The search for mechanisms regulating *Runx2* expression is still ongoing. However, progress in this area of research came when a peculiarity of osteoblast differentiation was confronted in other progress in bone biology. This peculiarity of osteoblast biology is that the synthesis of type I collagen, the overwhelmingly most abundant protein of the bone extracellular matrix, precedes the expression of Runx2 (Wei et al., 2015). The fact that osteoblasts regulate glucose metabolism through the hormone osteocalcin prompted, of course, the study of the regulation of glucose metabolism in osteoblasts (Lee et al., 2007). This 2007 study showed that glucose is the main nutrient of osteoblasts and that, in vivo, glucose is transported in osteoblasts through Glut1, whose expression in cells of the osteoblasts lineage precedes that of Runx2. Glucose uptake favors osteoblast differentiation by preventing the proteasomal degradation of Runx2. Accordingly, Runx2 cannot induce osteoblast differentiation when glucose uptake is compromised, and raising blood glucose levels is sufficient to initiate bone formation in Runx2-deficient embryos. That Runx2 favors Glut1 expression determines the onset of osteoblast differentiation during development (Wei et al., 2015).

In addition to this strong metabolic regulation of osteoblast differentiation, other transcription factors have been shown to act upstream of Runx2. There are several lines of evidence showing that, for instance, certain homeodomain-containing transcription factors are involved in the regulation of *Runx2* expression. One of these proteins is Msx2, whose role in skeletal development was demonstrated through the identification of gain- and loss-of-function mutations in human patients suffering from Boston-type craniosynostosis or enlarged parietal foramina, respectively (Jabs et al., 1993; Wilke et al., 2000). Likewise, *Msx2*-deficient mice display defective ossification of the skull and of bones developing by endochondral ossification (Satokata et al., 2000). Moreover, because the expression of *Osteocalcin* and *Runx2* is strongly reduced in *Msx2*-deficient mice, it appears that Msx2 acts upstream of Runx2 in a transcriptional cascade regulating osteoblast differentiation. A similar observation has been described for mice lacking the homeodomain-containing transcription factor *Bpx* (Tribioli and Lufkin, 1999). These mice die at birth owing to a severe dysplasia of the axial skeleton, whereas the appendicular skeleton is virtually unaffected. *Runx2* expression in *Bpx*-deficient mice is strongly reduced in osteo-chondrogenic precursor cells of the prospective vertebral column, thereby indicating that *Bpx* is required for *Runx2* expression specifically in these skeletal elements.

There are also negative regulators of *Runx2* expression. One of them is another homeodomain-containing transcription factor, Hoxa2. *Hoxa2*-deficient mice display ectopic bone formation in the second branchial arch, which is readily explained by an induction of *Runx2* expression exactly in this region (Kanzler et al., 1998). Consistent with these observations, transgenic mice expressing *Hoxa2* in craniofacial bones under the control of an *Msx2* promoter fragment lack several bones in the craniofacial area. Beyond development there is at least one factor required to limit *Runx2* expression in osteoblast precursor cells, namely, the high-mobility group–containing transcription factor Sox8. Sox8-deficient mice display an osteopenia that is caused by accelerated osteoblast differentiation accompanied by enhanced expression of *Runx2* (Schmidt et al., 2005). Likewise, transgenic mice expressing *Sox8* under the control of an osteoblast-specific *α1(I)-collagen* promoter fragment virtually lack differentiated osteoblasts because of a decreased expression of *Runx2*. In addition to the existence of transcriptional regulators of *Runx2* expression, there are also factors interacting with the Runx2 protein, thereby activating or repressing its activity. One identified positive regulator of Runx2 action is the nuclear matrix protein Satb2. The importance of this protein in skeletogenesis was first discovered in human patients with cleft palate that carry a heterozygous chromosomal translocation inactivating the *SATB2* gene (Fitzpatrick et al., 2003). The generation of a *Satb2*-deficient mouse model confirmed the importance of this gene in craniofacial development, skeletal patterning, and osteoblast differentiation (Dobreva et al., 2006). The last function was in part attributed to an increased expression of *Hoxa2*, a negative regulator of bone formation discussed earlier, whose expression is repressed by the binding of Satb2 to an enhancer element of the *Hoxa2* gene.

In addition to this type of action, there is also a Hoxa2-independent influence of Satb2 on the transcription of *Bone sialoprotein* and *Osteocalcin*. Whereas in the case of *Bone sialoprotein*, Satb2 directly binds to an osteoblast-specific element in the promoter of this gene, the activation of *Osteocalcin* expression by Satb2 requires a physical interaction with Runx2. This was demonstrated by cotransfection assays using a *Luciferase* reporter gene under the control of an osteoblast-specific *Osteocalcin* promoter fragment, but also by coimmunoprecipitation experiments. Moreover; the synergistic action of Satb2 and Runx2 in osteoblasts was genetically confirmed through the generation and analysis of compound heterozygous mice lacking one allele of each gene (Dobreva et al., 2006). Taken together, these results identified Satb2 as an important regulator of osteoblast differentiation in mice and humans. Moreover, the finding that Satb2 also interacts with Activating transcription factor 4 (ATF4), another transcription factor involved in osteoblast differentiation and function that will be discussed later, illustrates that the transcriptional network regulating bone formation is much more complex than previously anticipated.

This is further highlighted by the discovery of several other proteins that physically interact with Runx2, thereby attenuating its activity. One of these proteins is Stat1, a transcription factor regulated by extracellular signaling molecules, such as interferons. Stat1-deficient mice, as already mentioned, are viable, but develop a high-bone-mass phenotype explained by enhanced bone formation (Kim et al., 2003). The increase in osteoblast differentiation and function in these mice is molecularly explained by the lack of a Stat1-mediated inhibition of the transcriptional activity of Runx2. Interestingly, the physical interaction of both proteins is independent of Stat1 activation by phosphorylation, and it inhibits the translocation of Runx2 into the nucleus. Thus, overexpression of Stat1 in osteoblasts leads to a cytosolic retention of Runx2, and the nuclear translocation of Runx2 is much more prominent in Stat1-deficient osteoblasts (Kim et al., 2003).

Another protein interacting with Runx2, thereby decreasing its availability in the nucleus, is Shn3. Shn3 is a zinc finger adapter protein originally thought to be involved in VDJ recombination of immunoglobulin genes (Wu et al., 1993). Unexpectedly, however, the generation of an Shn3-deficient mouse model revealed that it plays a major function in bone formation. In fact, the *Shn3*-deficient mice display a severe adult-onset osteosclerotic phenotype owing to a cell-autonomous increase in bone matrix deposition (Jones et al., 2006). Interestingly, although several Runx2 target genes are expressed at higher rates in Shn3-deficient osteoblasts, *Runx2* expression itself is not affected by the absence of Shn3. Importantly, however, the Runx2 protein level is strikingly increased in *Shn3*-deficient osteoblasts. This finding is molecularly explained by the function of Shn3 as an adapter molecule linking Runx2 to the E3 ubiquitin ligase WWP1. The Shn3-mediated recruitment of WWP1 in turn leads to an enhanced proteasomal degradation of Runx2, which is best highlighted by the finding that RNA interference–mediated downregulation of WWP1 in osteoblasts leads to increased Runx2 protein levels and enhanced extracellular matrix mineralization, thus virtually mimicking the defects observed in the absence of Shn3 (Jones et al., 2006). Taken together, these data identify Shn3 as a key negative regulator of Runx2 actions in vivo. Accordingly, and as anticipated, *Glut1* expression and glucose uptake are significantly increased in *shn3*$^{-/-}$ osteoblasts, as is *Runx2* expression (Wei et al., 2015). Moreover, given the postnatal onset of the bone phenotype of the *Shn3*-deficient mice, it has been speculated that compounds blocking the interaction of Runx2, Shn3, and WWP1 may serve as specific therapeutic agents for the treatment of bone loss diseases, such as osteoporosis. This does not exclude the possibility that *Shn3* may regulate osteoblast differentiation through additional mechanisms (Shim et al., 2013).

While Stat1 and Shn3 exemplify the importance of a negative regulation of Runx2 in postnatal bone remodeling, there is also a need for a negative regulation of Runx2 activity before bone development starts and when *Runx2* is already expressed. This is highlighted by the finding that *Runx2* expression in the lateral plate mesoderm is already detectable as early as embryonic day (E) 10 of mouse development, whereas the expression of molecular markers of differentiated osteoblasts cannot be detected before E13.5 at the earliest (Ducy, 2000). One molecular explanation for this delay between *Runx2* expression and osteoblast differentiation came from the functional analysis of Twist proteins that are transiently coexpressed with *Runx2* early during development and inhibit osteoblast differentiation by interacting with Runx2. In essence, the initiation of osteoblast differentiation occurs only when *Twist* gene expression fades away.

In brief, the fact that Runx2 expression precedes osteoblast differentiation by more than 4 days led to the assumption that Runx2 function must be regulated negatively in Runx2-expressing cells. The identification of such a factor relied on human genetic evidence. Haploinsufficiency at the *RUNX2* locus causes a lack of bone in the skull, whereas haploinsufficiency at the *TWIST1* locus causes essentially too much bone in the skull, a condition called craniosynostosis (El Ghouzzi et al., 1997; Howard et al., 1997). Because the same phenotypes are observed in the corresponding mouse models lacking one allele of either gene, it was possible to demonstrate a genetic interaction of *Runx2* and *Twist1* through the generation of compound heterozygous mice. In fact, these mice did not display any detectable defects of skull development and suture fusion (Bialek et al., 2004). In contrast, the defects of clavicle development caused by haploinsufficiency of *Runx2* were not rescued by heterozygosity of *Twist1*, but by the deletion of *Twist2*. Thus, both Twist proteins have similar functions but in different skeletal elements. This is totally consistent with their expression pattern in mouse embryos (Bialek et al., 2004).

Both Twist proteins are presumably basic helix–loop–helix (bHLH) transcription factors, yet this function of Twist is not determined by the bHLH domain, but rather by the C-terminal 20 amino acids, the so-called Twist box, whose sequence is fully conserved in both Twist proteins, in mice and humans. Through the Twist box, Twist proteins interact with the Runx2 DNA-binding domain and prevent its DNA binding. The importance of this sequence motif for Twist function in vivo was confirmed by the existence of a Twist-box mutation within the human *TWIST1* gene that causes a severe form of craniosynostosis (Gripp et al., 2000). Moreover, an ethylnitrosourea-mutagenesis approach in mice led to the identification of an amino acid substitution within the Twist box of Twist1 that causes premature osteoblast differentiation in vivo (Bialek et al., 2004).

Unexpectedly, this mouse model, termed Charlie Chaplin, also displayed decreased chondrocyte maturation, thereby suggesting an additional physiological role of Twist1, independent of its antiosteogenic function (Hinoi et al., 2006). Because *Twist1* is not expressed in chondrocytes, but in mesenchymal cells of the perichondrium, it appears that it is required to inhibit the induction of *Fgf18* expression in these cells by the action of Runx2, which was described earlier. Indeed, whereas transgenic mice overexpressing *Twist1* under the control of an osteoblast-specific 0.1 *(I)-collagen* promoter fragment displayed enhanced chondrocyte maturation, the decreased chondrocyte maturation in the *CC/CC* mice was normalized by haploinsufficiency of *Runx2* (Hinoi et al., 2006). Taken together, these data demonstrate that Twist1, through inhibition of Runx2 DNA binding, not only limits osteoblast differentiation and bone formation, but also enhances chondrocyte maturation during skeletal development. Other negative regulators of Runx2 function in vivo have been described based on their abilities to influence the CCD phenotype of *Runx2*$^{+/-}$ mice. One of them, another zinc finger–containing protein termed zinc finger protein 521 (2fp521), antagonizes Runx2 function in a histone deacetylase 3 (HDAC3)-dependent manner.

Osterix, a Runx2-dependent osteoblast-specific transcription factor required for bone formation

In addition to Runx2, there is at least one more transcription factor, termed Osterix (Osx), whose activity is absolutely required in mice for osteoblast differentiation. Osx is a zinc finger–containing transcription factor that is specifically expressed in osteoblasts of all skeletal elements. Inactivation of Osx in mice results in perinatal lethality owing to a complete absence of bone formation (Nakashima et al., 2002). Unlike the *Runx2*-deficient mice whose skeleton is completely unmineralized, the *Osx*-deficient mice lacked a mineralized matrix only in bones formed by intramembranous ossification. In contrast, the bones formed by endochondral ossification contained mineralized matrix, but this resembled calcified cartilage, not mineralized bone matrix. This finding suggested that Osx, unlike Runx2, is not required for chondrocyte hypertrophy, thereby demonstrating that it specifically induces osteoblast differentiation and bone formation in vivo. The comparative expression analysis by in situ hybridization further revealed that *Osx* is not expressed in *Runx2*-deficient embryos, and that *Runx2* is normally expressed in *Osx*-deficient embryos (Nakashima et al., 2002). These results demonstrated that Osx acts downstream of Runx2 in a transcriptional cascade of osteoblast differentiation, and its expression is apparently directly regulated by the binding of Runx2 to a responsive element in the promoter of the Osx gene (Nishio et al., 2006).

In contrast to the steadily increasing knowledge about the function of Runx2 and its regulation by other molecules, the molecular mechanisms underlying the action of Osx in osteoblasts are less well understood. Moreover, unlike for *RUNX2*, no human mutations of the *OSX* gene have yet been identified that would be associated with decreased bone formation. Nevertheless, one 2005 publication indicates that Osx contributes to the negative effects of nuclear factor of activated T cells (NFAT) inhibitors on bone mass (Koga et al., 2005). NFAT inhibitors, such as *FKS06* or cyclosporin A, are commonly used as immune suppressants, e.g., after organ transplantation. However, this treatment is often accompanied by the development of osteopenia in the patient (Rodino and Shane, 1998). Likewise, treatment of mice with *FKS06* leads to decreased bone mass owing to impaired bone formation, and the same phenotype was observed in mice lacking the transcription factor Nfatc1. The deduced role of Nfatc1 as a physiological activator of osteoblast differentiation and function can be molecularly explained by an interaction with *Osx*. In fact, both proteins synergistically activate transcription of a *Luciferase* reporter gene driven by an osteoblast-specific *α1(I)-collagen* promoter fragment, which is based on the formation of a DNA-binding complex of Nfatc1 and Osx (Koga et al., 2005). It has been shown that Osx might act as a cofactor for Dlx5 for osteoblast specification (Hojo et al., 2016).

ATF4, a transcriptional regulator of osteoblast functions and a mediator of the neural regulation of bone mass

The role of ATF4 in skeletal biology also arose from a combination of molecular biology and human and mouse genetic data. *RSK2*, which encodes a kinase, is the gene mutated in Coffin–Lowry syndrome, an X-linked mental retardation condition associated with skeletal abnormalities (Trivier et al., 1996). Like Rsk2-deficient mice, $Atf4^{-/-}$ mice display a lower bone mass owing to impaired bone formation (Yang et al., 2004). In vitro kinase assays demonstrated that ATF4 is strongly phosphorylated by Rsk2, and that this phosphorylation is undetectable in osteoblasts derived from *Rsk2*-deficient mice. The subsequent analysis of an *Atf4*-deficient mouse model revealed that this transcription factor plays a crucial role in bone formation. In fact, *Atf4*-deficient mice display a delayed skeletal development and thereafter develop a severe low-bone-mass phenotype caused by a decrease in bone formation (Yang et al., 2004).

Molecularly, ATF4 was identified as the factor binding to the osteoblast-specific element OSE1 in the *Osteocalcin* promoter, thereby directly activating the transcription of this gene. Moreover, ATF4 is required for proper synthesis of type I collagen, although this function is not mediated by a transcriptional regulation of *type I collagen* expression. In fact, because type I collagen synthesis in the absence of nonessential amino acids is specifically reduced in primary osteoblast cultures lacking ATF4, it appears that ATF4 is required for efficient amino acid import in osteoblasts, as has been described for other cell types (Harding et al., 2003). Because a reduced type I collagen synthesis was subsequently also observed in mice lacking Rsk2, these data provided evidence that the diminished ATF4 phosphorylation in the absence of Rsk2 may contribute to the skeletal defects associated with Coffin–Lowry syndrome (Yang et al., 2004).

In addition to its role in bone formation, ATF4, through its expression in osteoblasts, regulates bone resorption (Elefteriou et al., 2005). This function is molecularly explained by the fact that ATF4 binds to the promoter of the *Rankl* (receptor activator of NF-κB ligand) gene to promote osteoclast differentiation (Teitelbaum and Ross, 2003). As a result, ATF4-deficient mice have a decreased number of osteoclasts because of their reduced *Rankl* expression. Most importantly, this function of ATF4 is involved in the control of bone resorption by the sympathetic nervous system. In fact, treatment of normal osteoblasts with isoproterenol, a surrogate of sympathetic signaling, enhanced osteoclastogenesis of cocultured bone marrow macrophages through an induction of osteoblastic *Rankl* expression (Elefteriou et al., 2005). As expected, this effect was blunted when the osteoblasts were derived from mice lacking the β2-adrenergic receptor Adrb2. However, the effect of isoproterenol was also blunted by an inhibitor of protein kinase A, or by using osteoblasts derived from ATF4-deficient mice (Elefteriou et al., 2005). Taken together, these results demonstrated that ATF4 is an important mediator of extracellular signals, such as β-adrenergic stimulation, in osteoblasts.

Thus, it is not surprising that the function of ATF4 is mostly regulated posttranslationally. For example, as already mentioned, ATF4 also interacts with other proteins, such as the nuclear matrix protein Satb2 (Dobreva et al., 2006). Also as described earlier, the proximal *Osteocalcin* promoter contains two osteoblast-specific elements, termed OSE1 and OSE2, that serve as binding sites for ATF4 and Runx2, respectively (Ducy and Karsenty, 1995; Ducy et al., 1997; Schinke and Karsenty, 1999; Yang et al., 2004). Because of the proximity of both elements, there is indeed a physical interaction of the two proteins, which is stabilized by Satb2, which acts as a scaffold enhancing the synergistic activity of Runx2 and ATF4, which is required for optimum *Osteocalcin* expression (Xiao et al., 2005; Dobreva et al., 2006).

Other aspects of ATF4 biology are also regulated posttranslationally. In fact, even the osteoblast specificity of ATF4 function is not determined by osteoblast-specific *ATF4* expression, but by a selective accumulation of the ATF4 protein in osteoblasts, which itself is explained by the lack of proteasomal degradation (Yang and Karsenty, 2004). This is best demonstrated by the finding that the treatment of nonosteoblastic cell types with the proteasome inhibitor MG 115 leads to accumulation of the ATF4 protein, thereby resulting in ectopic *Osteocalcin* expression. Taken together, these data provided the first evidence for the achievement of a cell-specific function of a transcriptional activator by a posttranslational mechanism. They are therefore of general importance for our understanding of the transcriptional networks controlling cellular differentiation and function.

Remarkably, ATF4 biology further illustrates how the molecular understanding of a disease-causing gene can translate into therapeutic interventions. Indeed, an increased Rsk2-dependent phosphorylation of ATF4 may also be involved in the development of the skeletal abnormalities in human patients suffering from neurofibromatosis (Ruggieri et al., 1999; Stevenson et al., 1999). This disease, which is primarily known for tumor development within the nervous system, is caused by inactivating mutations of the *NF1* gene, which encodes a Ras-GTPase-activating protein (Klose et al., 1998). The generation of a mouse model lacking *Nf1* specifically in osteoblasts ($Nf1_{ob}^{-/-}$) led to the demonstration that this gene plays a major physiological role in bone remodeling. In fact, the $Nf1_{ob}^{-/-}$ mice displayed a high-bone-mass phenotype that is caused by increased bone turnover and is accompanied by enrichment of unmineralized osteoid (Elefteriou et al., 2006). The analysis of this phenotype revealed an increased production of type I collagen in the absence of *Nf1*, which is molecularly explained by an Rsk2-dependent activation of ATF4. Likewise, transgenic mice overexpressing *ATF4* in osteoblasts display a phenotype similar to that of the $Nf1_{ob}^{-/-}$ mice, and the increased type I collagen production and osteoid thickness in the latter are significantly reduced by haploinsufficiency of ATF4 (Elefteriou et al., 2006).

These molecular findings may also have therapeutic implications. Given the previously discussed function of ATF4 in amino acid import, it appeared reasonable to analyze whether the skeletal defects of the $Nf1_{ob}^{-/-}$ mice can be affected by dietary manipulation. Indeed, the increased bone formation and osteoid thickness of $Nf1_{ob}^{-/-}$ mice can be normalized by a low-protein diet, and the same was the case in the transgenic mice overexpressing *ATF4* in osteoblasts (Elefteriou et al., 2006). Likewise, the defects of osteoblast differentiation and bone formation observed in both the *ATF4*- and the Rsk2-deficient mice were corrected by feeding a high-protein diet. Taken together, these data not only emphasize the importance of ATF4 in osteoblast biology, but also demonstrate how knowledge about its specific functions in osteoblasts can be useful for the treatment of skeletal diseases.

Additional transcriptional regulators of osteoblast differentiation and function

Activator protein 1 (AP-1) is a heterodimeric transcription factor composed of members of the Jun and Fos families of basic leucine zipper proteins (Karin et al., 1997). These include the Jun proteins c-Jun, JunB, and JunD, as well as the Fos proteins c-Fos, Fra1, Fra2, and Fosb, respectively. Although AP-1 transcription factors have been demonstrated to fulfill various functions in different cell types, it is striking that some of the family members play specific roles in bone remodeling, as demonstrated by several loss- or gain-of-function studies in mice (Wagner and Eferl, 2005). For instance, the deletion of *c-Fos* from the mouse genome results in severe osteopetrosis owing to an arrest of osteoclast differentiation, whereas the transgenic overexpression of *c-Fos* results in osteosarcoma development (Grigoriadis et al., 1993, 1994). Moreover, transgenic mice overexpressing either *Fml* or *i1fosB*, a splice variant of *FosB*, display a severe osteosclerotic phenotype caused by increased osteoblast differentiation and function (Jochum et al., 2000; Sabatakos et al., 2000). Likewise, mice lacking Fra1 in extraplacental tissues display an osteopenia associated with reduced bone formation, indicating a physiological role of Fra1 in osteoblasts (Eferl et al., 2004). When the same approach was used to inactivate JunB in extraplacental tissues, thereby circumventing the embryonic lethality caused by a complete genomic deletion of *JunB*, the resulting mice developed a state of low bone turnover, owing to cell-autonomous defects in osteoblasts, but also in osteoclast differentiation (Kenner et al., 2004).

Taken together, these data provide evidence for a crucial role of AP-1 transcription factors in the regulation of bone formation, although their connection to the other transcriptional regulators described earlier still needs to be further investigated. For instance, it is known from other cell types that Jun proteins can also interact with ATF family members, thus raising the possibility that heterodimerization with ATF4 may be one mechanism by which these proteins can regulate osteoblast-specific gene expression (Chinenov and Kerppola, 2001). Interestingly, it has been demonstrated that the osteosarcoma development of *c-Fos* transgenic mice is dramatically decreased in an Rsk2-deficient genetic background (David et al., 2005). This observation is molecularly explained by the lack of c-Fos phosphorylation by Rsk2, thereby leading to increased proteasomal degradation. Thus, Rsk2 apparently not only is involved in the physiological regulation of bone formation via phosphorylation of ATF4, but may also have an influence on the development of osteosarcomas via phosphorylation of c-Fos.

Another potential mechanism by which AP-1 family members might be involved in the regulation of bone formation came- from the analysis of mouse models with impaired circadian regulation. These mice, which lack components of the molecular clock, namely the *Per* or *Cry* genes, display a high-bone-mass phenotype caused by increased bone formation (Fu et al., 2005). Moreover, they respond to intracerebroventricular infusion of leptin with a further increase in bone mass, suggesting that the components of the molecular clock are involved in the regulation of bone formation via the sympathetic nervous system. Interestingly, virtually all genes encoding members of the AP-1 transcription factor family were expressed at higher levels in osteoblasts derived from mice lacking either the *Per* genes or the β2-adrenergic receptor Adrb2 (Fu et al., 2005). This increase was especially pronounced in the case of the *c-Fos* gene, whose expression can also be induced by the addition of isoproterenol in wild-type osteoblasts. In turn, c-Fos leads to a direct activation of *c-Myc* transcription, thereby indirectly increasing the intracellular levels of cyclin D1 and promoting osteoblast proliferation. Taken together, these data demonstrated that the expression of AP-1 components is activated via sympathetic signaling, and that this induction is counteracted by the activity of clock gene products.

Finally, another transcription factor that, like AP-1, is not cell specific but plays a great role in osteoblast biology, is the cAMP-responsive element binding protein (CREB), a transcription factor mediating changes in gene expression caused by signaling through various G-protein-coupled receptors. The demonstration that gut-derived serotonin is an inhibitor of bone formation in mice, rats, and humans, the regulation of the synthesis of which is under the control of Lrp5 signaling in the gut, raised the question of how serotonin signals in osteoblasts (Yadav et al., 2008; Frost et al., 2011). Expression analysis and cell-specific gene deletion experiments in the mouse showed that Htr1B is the receptor of serotonin, mediating its function in osteoblasts, namely, an inhibition of proliferation, and that the transcription factor mediating this effect is CREB (Yadav et al., 2008).

Transcription factors acting downstream of Wnt signaling in osteoblasts: what do they actually do in differentiated osteoblasts?

The discovery of the *LRP5* gene as a major determinant of bone mass in humans because of its homology to a Wnt coreceptor in *Drosophila* has generated an enormous interest and hope in what the canonical Wnt signaling could do in cells of the osteoblast lineage and especially in differentiated osteoblasts. An obligatory implication of this belief, an implication that cannot be ignored or dismissed, is that ablating canonical Wnt signaling in differentiated osteoblasts should lead to an osteopetrotic phenotype. Unfortunately, when taken at face value as they should, the experiments performed to test this hypothesis not only did not provide any evidence that it is the case, but in fact clearly showed that it is not the case.

Indeed, the osteoblast-specific inactivation of β-catenin, the molecular node of the canonical Wnt signaling pathway, resulted in the expected low-bone-mass phenotype; however, this was caused by an isolated and massive increase in bone resorption (Glass et al., 2005). The molecular explanation for the regulation of osteoclast differentiation by β-catenin expression in osteoblasts came from the expression analysis of *Osteoprotegerin* (*Opg*), a well-known inhibitor of bone resorption, blocking the activity of Rankl (Simonet et al., 1997; Yasuda et al., 1998). Surprisingly, although *Opg* expression was markedly increased in osteoblasts from mice expressing the stabilized form of β-catenin, its expression was decreased in mice harboring an osteoblast-specific deletion of β-catenin (Glass et al., 2005). That similar results were observed when β-catenin was deleted from osteocytes established that it is a general rule of bone biology. Of note, these results need to be considered together with the experimental evidence gathered by two different laboratories, using for that purpose mutant mouse strains also generated in two different laboratories, that mice lacking Lrp5 signaling in osteoblasts only do not have any bone phenotype to speak of (Yadav et al., 2008; Kode et al., 2014).

One of the transcription factors activated by β-catenin in osteoblasts has already been identified as Tcfl, whose importance for the regulation of gene expression in osteoblasts is highlighted by several lines of evidence. First, in situ hybridization revealed that *Tcfl* is expressed in osteoblasts during bone development and after birth. Second, *Tcfl*-deficient mice display a low-bone-mass phenotype caused by an increase in bone resorption. Third, the compound heterozygosity of *Tcfl* and *β-catenin* in osteoblasts also results in low bone mass, which is not observed when one allele of each gene is inactivated alone. Fourth, *Opg* expression is decreased in osteoblasts of *Tcf1*-deficient mice, and the molecular analysis of the *Opg* promoter revealed the existence of a Tcf1-binding site, whose functional activity was subsequently proven by chromatin immunoprecipitation and DNA cotransfection assays (Glass et al., 2005). The implication of these results cannot be underestimated, without taking great and yet predictable risks, when proposing to harness the canonical Wnt signaling pathway for therapeutic purposes in the context of osteoporosis.

Regulation of osteoblast differentiation by means other than transcription factors

Chromatin structure, which influences by posttranslational modifications of histone proteins around which the DNA is wrapped, is a major determinant of gene expression (Allis et al., 2007). Histone acetylation promotes gene transcription by relaxing the chromatin structure, whereas deacetylation of histones by HDACs induces chromatin condensation and transcriptional repression (Berger, 2002; Verdin et al., 2003; Allis et al., 2007). Class II HDACs contain a poorly active catalytic domain and a long N-terminal extension to which transcription factors can bind. The existence of this domain has suggested that class II HDACs can link extracellular cues to the genome of a given cell (Verdin et al., 2003; Haberland et al., 2009). Several class I HDACs (Hdac1 and Hdac3) are expressed in osteoblasts, and conditional deletion of Hdac3 within osteo-chondroprogenitor cells decreases osteoblast number, and, as mentioned earlier, one mechanism whereby Zpf521 inhibits Runx2 functions during osteoblast differentiation is by interacting with HDAC3. On the other hand, one class II HDAC, HDAC4, acts as a central integrator of two extracellular cues acting on osteoblasts. One is parathyroid hormone, which targets HDAC4 for degradation and thereby releases MeF2c that can now transactivate *Rankl*, and thus favors osteoclast differentiation. The other is sympathetic tone, which instead favors HDAC4 accumulation in the nucleus, its association with ATF4, and again Rankl expression and osteoclast differentiation (Obri et al., 2014).

Another mode of regulation of osteoblast differentiation that has received much attention since the last edition of this book is the one fulfilled by small noncoding RNAs. MicroRNAs (miRNAs) are small noncoding RNAs that down-regulate expression of their target genes by either mRNA degradation or translational inhibition (Valencia-Sanchez et al., 2006; Bartel, 2009; Djuranovic et al., 2011; Huntzinger and Izaurralde, 2011). Although most miRNAs are broadly expressed, some have a more restricted pattern of expression and influence cell differentiation (Poy et al., 2004). The importance of this mode of gene regulation during skeletogenesis is inferred from the observation that inactivation of

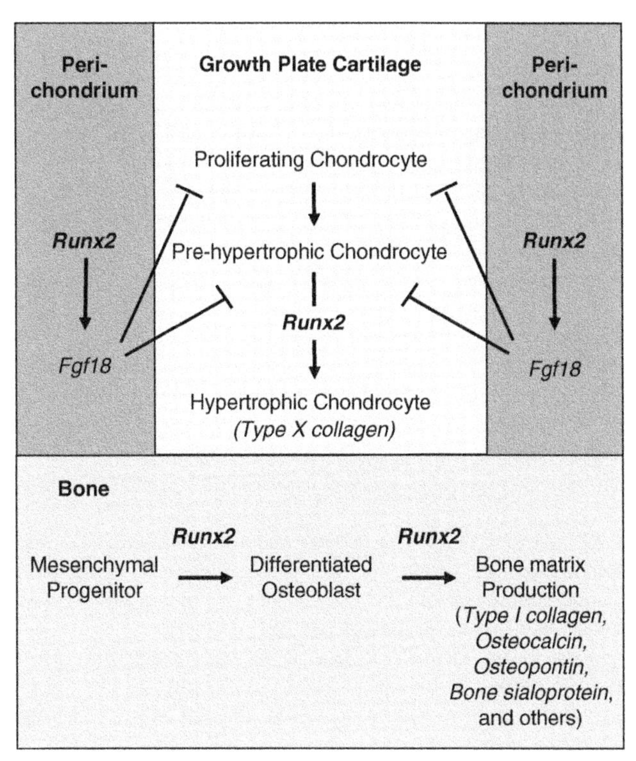

FIGURE 7.1 *Schematic representation of the functions of Runx2 during chondrogenesis and osteoblast differentiation.*

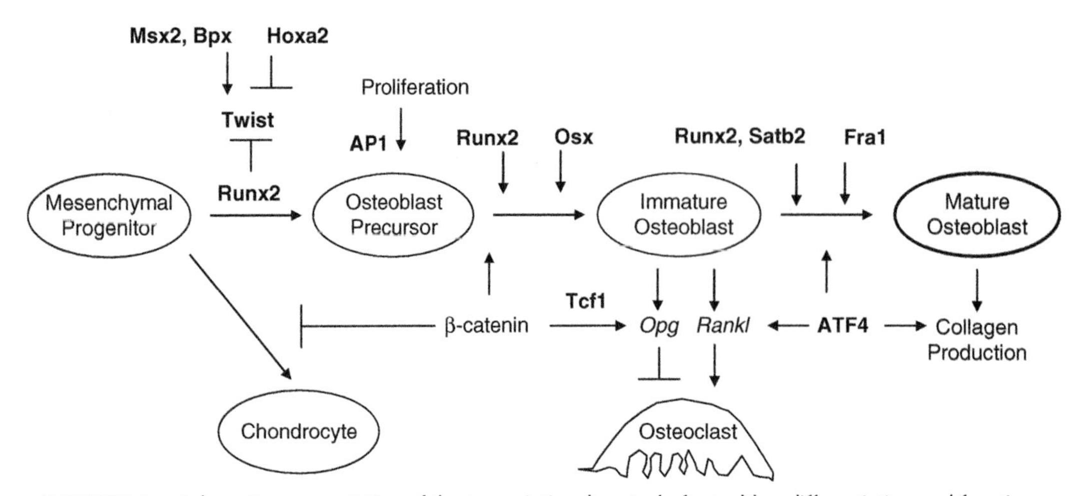

FIGURE 7.2 *Schematic representation of the transcriptional control of osteoblast differentiation and functions.*

DICER, a protein necessary for processing of miRNAs, affects osteoblast differentiation (Gaur et al., 2010). Accordingly, it has been suggested that miRNAs may be involved in osteoblast proliferation and/or differentiation (Hassan et al., 2010). However, these observations were derived from gain-of-function experiments performed in cell culture. Few miRNAs had been shown, through cell-specific loss-of-function experiments performed in the mouse, to regulate osteoblast proliferation and/or differentiation in vivo. A first example of these was miRNA-2561, whose silencing in the mouse reduces Runx2 accumulation without, however, causing a characteristic CCD phenotype (Li et al., 2009). Two other miRNAs, miR-34b and miR-34c, were shown through cell-specific gene deletion to inhibit osteoblast proliferation by suppressing cyclin D1, CDK4, and CDK6 accumulation (Wei et al., 2012). They also hampered osteoblast differentiation by inhibiting the function of SATB2, a protein interacting with Runx2 and ATF4 (Wei et al., 2012). Another miRNA, miR-124, inhibits bone formation by targeting another transcription factor implicated in osteoblast differentiation, ATF4, whereas miR-188 was shown to favor bone formation in an HDAC9-dependent manner (Li et al., 2015) (Fig. 7.1 and 7.2).

References

Allis, C.D., Berger, S.L., Cote, J., Sent, S., Jenuwein, T., Kouzarides, T., Pillus, L., Reinber, D., Shi, Y., Shiekhaltar, R., Shilatifard, A., Workman, J., Zhang, Y., 2007. New nomenclature for chromatin-modifying enzymes. Cell 131 (4), 633–636.

Bartel, D.P., 2009. MicroRNAs:target recognition and regulatory functions. Cell 136 (2), 215–233.

Berger, S.L., 2002. Histone modification in transcriptional regulation. Curr. Opin. Genet. Dev. 12 (2), 142–148.

Bialek, P., Kern, B., Yang, X., Schrock, M., Sosic, D., Hong, N., Wu, H., Yu, K., Ornitz, D.M., Olson, E.N., Justice, M.J., Karsenty, G., 2004. A twist code determines the onset of osteoblast differentiation. Dev. Cell 6, 423–435.

Chinenov, Y., Kerppola, T., 2001. Close encounters of many kinds: fos-Jun interactions that mediate transcription regulatory specificity. Oncogene 20, 2438–2452.

David, J.P., Mehic, D., Bakiri, L., Schilling, A.E., Mandic, V., Priemel, M., Idarraga, M.H., Reschke, M.O., Hoffmann, O., Amling, M., Wagner, E.E., 2005. Essential role of"RSK2 in c-Fos-dependent osteosarcoma development. J. Clin. Invest. 115, 664–672.

Djuranovic, S., Nahvi, A., Green, R., 2011. A parsimonious model of gene regulation by miRNAs. Science 331 (6017), 550–553.

Dobreva, G., Chahrour, M., Dautzenberg, M., Chirivella, L., Kanzler, B., Farinas, 1., Karsenty, G., Grosschedl, R., 2006. SATB2 is a multifunctional determinant of craniofacial patterning and osteoblast differentiation. Cell 125, 971–986.

Ducy, P., Karsenty, G., 1995. Two distinct osteoblast-specific cis-acting elements control expression of a mouse osteocalcin gene. Mol. Cell. Biol. 15, 1858–1869.

Ducy, P., Zhang, R., Geoffroy, v., Ridall, A.L., Karsenty, G., 1997. Osf2/Cbfal: a transcriptional activator of osteoblast differentiation. Cell 89, 747–754.

Ducy, P., Starbuck, M., Priemel, M., Shen, J., Pinero, G., Geoffroy, V., Amling, M., Karsenty, G., 1999. A Cbfal-dependent genetic pathway controls bone formation beyond embryonic development. Genes Dev. 13, 1025–1036.

Ducy, P., 2000. Cbfal: a molecular switch in osteoblast biology. Dev. Dynam. 219, 461–471.

Eferl, R., Hoebertz, A., Schilling, A.E., Rath, M., Karreth, E., Kenner, L., Amling, M., Wagner, E.E., 2004. The Fos-related antigen Fra-l is an activator of bone matrix formation. EMBO J. 23, 2789–2799.

Elefteriou, E., Ahn, J.D., Takeda, S., Starbuck, M., Yang, X., Liu, X., Kondo, H., Richards, W.G., Bannon, T.W., Noda, M., Clement, K., Vaisse, c., Karsenty, G., 2005. Leptin regulation of bone resorption by the sympathetic nervous system and CART. Nature 434, 514–520.

Elefteriou, E., Benson, M.D., Sowa, H., Starbuck, M., Liu, X., Ron, D., Parada, L.F., Karsenty, G., 2006. ATF4 mediation of NF1 functions in osteoblast reveals a nutritional basis for congenital skeletal dysplasiae. Cell Metabol. 4, 441–451.

Frost, M., Andersen, T., Gossiel, F., hansen, S., Bollerslev, J., Van Hul, W., Eastell, R., Kassem, M., Brixen, K., 2011. Levels of serotonin, sclerostin, bone turnover markers as well as bone density and microachitecture in patients with high bone mass phenoptype due to a mutation in Lrp5. J. Bone Miner. Res. 8, 1721–1728.

El Ghouzzi, v., Le Merrer, M., Perrin-Schmitt, F., Lt0eunie, E., Benit, P., Renier, D., Bourgeois, P., Bolcato-Bellemin, A.L., Munnich, A., Bonaventure, J., 1997. Mutations of the TWIST gene in the SaethreChotzen syndrome. Nat. Gellet. 15, 42–46.

Fitzpatrick, D.R., Carr, L.M., McLaren, L., Leek, 1. P., Wightman, P., Williamson, K., Gautier, P., McGill, N., Hayward, c., Firth, H., Markham, A.E., Fantes, 1. A., Bonthron, D.T., 2003. Identification of SATB2 as the cleft palate gene on 2q32-q33. Hum. Mol. Genet. 12, 2491–2501.

Fu, L., Patel, M.S., Bradley, A., Wagner, E.E., Karsenty, G., 2005. The molecular clock mediates leptin-regulated bone formation. Cell 122, 803–815.

Gaur, T., Hussain, S., Mudhasani, R., Parulkar, I., Colby, J.L., Frederick, D., Kream, B.E., Van Wijnen, A.J., Stein, Jl, Stein, G.S., Jones, S.N., Lian, J.B., 2010. Diver inactivation in osteoprogenitor cells compromises fetal survival and bone formation while excision in differentiated osteoblasts increases bone mass in the adult mouse. Dev. Biol. 340 (1), 1–21.

Geoffroy, v., Ducy, P., Karsenty, G., 1995. A PEBP2/AML-Related factor increases osteocalcin promoter activity through its binding to an osteoblast-specific cis-acting element. J. Biol. Chem. 270, 30973–30979.

Glass, D.A., II, Bialek, P., Ahn, J.D., Starbuck, M., Patel, M.S., Clevers, H., Taketo, M.M., Long, F., McMahon, A.P., Lang, R.A., Karsenty, G., 2005. Canonical Wnt signaling in differentiated osteoblasts controls osteoclast differentiation. Dev. Cell 751–764.

Grigoriadis, A.E., Schellander, K., Wang, Z.Q., Wagner, E.F., 1993. Osteoblasts are target cells for transformation in c-fos transgenic mice. J. Cell Biol. 122, 685–701.

Grigoriadis, A.E., Wang, Z.Q., Cecchini, M.G., Hofstetter, w., Felix, R., Fleisch, H.A., Wagner, E.E., 1994. c-Fos: a key regulator of osteoclast-macrophage lineage determination and bone remodeling. Science 266, 443–448.

Gripp, K. w., Zackai, E.H., Stolle, C.A., 2000. Mutations in the human TWIST gene. Hum. Mutat. 15, 479.

Haberland, M., Mokalled, M.H., Montgomery, R.L., Olson, E.N., 2009. Epigenetic control of skull morphogenesis by histone deacetylase 8Genes. Dev 23 (14), 1625–1630.

Harding, H.P., Zhang, y, Zeng, H., Novoa, J., Lu, P.D., Calfon, M., Sadri, N., Yun, C., Popko, B., Paules, R., Stojdl, D.F., Bell, J.C., Hettmann, T., Leiden, J.M., Ron, D., 2003. An integrated stress response regulates amino acid metabolism and resistance to oxidative stress. Mol. Cell 11, 619–633.

Hassan, M.Q., Gordon, J.A., Beloti, M.M., Croce, C.M., van Wijnen, A.J., Stein, J.L., Stein, G.S., Lian, J.B., 2010. A network connecting Runx2, SATB2 and the miR-23a-27a-24-2 cluster regulates the osteoblast differentiation program. Proc. Natl. Acad. Sci. U.S.A. 107 (46), 19879–19884.

Hinoi, E., Bialek, P., Chen, Y.T., Rached, M.T., Groner, Y., Behringer, R.R., Ornitz, D.M., Karsenty, G., 2006. Runx2 inhibits chondrocyte proliferation and hypertrophy through its expression in the perichondrium. Genes Dev. 20, 2937–2942.

Hojo, H., Ohba, S., He, X., Lai, L.P., McMahon, A.P., 2016. Sp7/Osterix is restricted to bone-forming vertebrates where it acts as a Dlx Co-factor in osteoblast specification. Dev. Cell 37 (3), 238–253.

Howard, T.D., Paznekas, W.A., Green, E.D., Chiang, L. c., Ma, N., Ortiz de Luna, R.I., Garcia Delgado, C., Gonzalez-Ramos, M., Kline, A.D., Jabs, E.W., 1997. Mutations in TWIST, a basic helix-loop-helix transcription factor, in Saethre-Chotzen syndrome. Nat. Genet. 15, 36–41.

Huntzinger E, Izaurralde E. Gene silencing by microRNAs: contributions of translational repression and mRNA decay. Nat. Rev. Genet. 12(2): 99–110

Inada, M., Yasui, T., Nomura, S., Miyake, S., Deguchi, K., Himeno, M., Sato, M., Yamagiwa, H., Kimura, T., Yasui, N., Ochi, T., Endo, N., Kitamura, Y., Kishimoto, T., Komori, T., 1999. Maturational disturbance of chondrocytes in Cbfal-deficient mice. Dev. Dynam. 214, 279–290.

Jabs, E. w., Muller, U., Li, X., Ma, L., Luo, w., Haworth, I.S., Klisak, I., Sparkes, R., Warman, M.L., Mulliken, J.B., 1993. A mutation in the home-odomain of the human MSX2 gene in a family affected with autosomal dominant craniosynostosis. Cell 75, 443–450.

Jochum, W., David, J.P., Elliott, C., Wutz, A., Plenk, H.J., Matsuo, K., Wagner, E.F., 2000. Increased bone formation and osteosclerosis in mice overexpressing the transcription factor Fra-1. Nat. Med. 6, 980–984.

Jones, D. c., Wein, M.N., Oukka, M., Hofstaetter, J.G., Glimcher, M.J., Glimcher, L.H., 2006. Regulation of adult bone mass by the zinc finger adapter protein Schnurri-3. Science 312, 1223–1227.

Kagoshima, H., Shigesada, K., Satake, M., Ito, Y., Miyoshi, H., Ohki, M., Pepling, M., Gergen, P., 1993. The Runt domain identifies a new family of heteromeric transcriptional regulators. Trends Genet. 9, 338–341.

Kanzler, B., Kuschert, S.J., Liu, Y.-H., Mallo, M., 1998. Hoxa-2 restricts the chondrogenic domain and inhibits bone formation during development of the branchial area. Development 125, 2587–2597.

Karin, M., Liu, Z., Zandi, E., 1997. AP-1 function and regulation. Curr. Opin. Cell. Bioi. 9, 240–246.

Kenner, L., Hoebertz, A., Beil, T., Keon, N., Karreth, F., Eferl, R., Scheuch, H., Szremska, A., Amling, M., Schorpp-Kistner, M., Angel, P., Wagner, E.E., 2004. Mice lacking JunB are osteopenic due to cell-autonomous osteoblast and osteoclast defects. J. Cell Biol. 164, 613–623.

Kim, I.S., Otto, F., Abel, B., Mundlos, S., 1999. Regulation of chondrocyte differentiation by Cbfal. Mech. Dev. 80, 159–170.

Kim, S., Koga, T., Isobe, M., Kern, B.E., Yokochi, T., Chin, Y.E., Karsenty, G., Taniguchi, T., Takayanagi, H., 2003. Stat! functions as a cytoplasmic attenuator of Runx2 in the transcriptional program of osteoblast differentiation. Genes Dev. 17, 1979–1991.

Klose, A., Ahmadian, M.R., Schuelke, M., Scheffzek, K., Hoffmeyer, S., Gewies, A., Schmitz, E., Kaufmann, D., Peters, H., Wittinghofer, A., Nurnberg, P., 1998. Selective disactivation of neurofibromin GAP activity in neurofibromatosis type 1. Hum. Mol. Genet. 7, 1261–1268.

Kode, A., Obri, A., Paone, R., Kousteni, S., Ducy P Karsenty, G., 2014. Lrp5 regulation of bone mass and serotonin synthesis in the gut. Nat. Med. 20, 1228–1229.

Koga, T., Matsui, Y., Asagiri, M., Kodama, T., de Crombrugghe, B., Nakashima, K., Takayanagi, H., 2005. NFAT and Osterix cooperatively regulate bone formation. Nat. Med. 11, 880–885.

Komori, T., Yagi, H., Nomura, S., Yamaguchi, A., Sasaki, K., Deguchi, K., Shimizu, Y., Bronson, R.T., Gao, Y.H., Inada, M., Sato, M., Okamoto, R., Kitamura, Y., Yoshiki, S., Kishimoto, T., 1997. Targeted disruption of Cbfal results in a complete lack of bone formation owing to maturational arrest of osteoblasts. Cell 89, 755–764.

Lee, B., Thirunavukkarasu, K., Zhou, L., Pastore, L., Baldini, A., Hecht, J., Geoffroy, V., Ducy, P., Karsenty, G., 1997. Missense mutations abolishing DNA binding of the osteoblast-specific transcription factor *OSF2/CBFAl* in cleidocranial dysplasia. Nat. Genet. 16, 307–310.

Lee, N.K., Sowa, H., Hinoi, E., Ferron, M., Ahn, J.D., Confavreux, C., Dacquin, R., Mee, P.J., McKee, M., Jung, D.Y., Zhang, Z., Kim, J.K., Mauvais-Jarvis, F., Ducy, P., Karsenty, G., 2007. Endocrine regulation of energy metabolism by the skeleton. Cell 130, 456–469.

Lian, J.B., Stein, G.S., 2003. Runx2/Cbfal: a multifunctional regulator of bone formation. Curr. Pharm. Des. 9, 2677–2685.

Li, H., Xie, H., Liu, W., Hu, R., Huang, B., Tan, Y.F., Xu, K., Sheng, Z.F., Zhou, H.D., Wu, X.P., Luo, X.H., 2009. A novel micorRNA targeting HDAC5 regulates osteoblast differentiation in mice and contributes to primary osteoporosis in humans. J. Clin. Invest. 119 (120), 3666–3677.

Li, C.J., Cheng, P., Liang, M.K., Chen, Y.S., Lu, Q., Wang, J.Y., Xia, Z.Y., Zhou, H.D., Cao, X., Xie, H., Liao, E.Y., Luo, X.H., April 2015. MicroRNA-188 regulates age-related switch between osteoblast and adipocyte differentiation. J. Clin. Investig. 125 (4), 1509–1522.

Liu, w., Toyosawa, S., Furuichi, T., Kanatani, N., YosQida, c., Liu, Y., Himeno, M., Narai, S., Yamaguchi, A., Komori, T., 2001. Overexpression of Cbfal in osteoblasts inhibits osteoblast maturation and causes osteopenia with multiple fractures. J. Cell Biol. 155, 157–166.

Liu, Z., Xu, J., Colvin, J.S., Omitz, D.M., 2002. Coordination of chondrogenesis and osteogenesis by fibroblast growth factor 18. Genes Dev. 16, 859–869.

Merriman, H.L., vanWijnen, A.J., Hiebert, S., Bidwell, J.P., Fey, E., Lian, J., Sein, J., Stein, G.S., 1995. The tissue-specific nuclear matrix protein, NMP-2, is a member of the *MAL/CBFIPEBP21Runt* domain transcription factor family: interactions with the osteocalcin gene promoter. Biochemistry 34, 13125–13132.

Mundlos, S., Otto, E., Mundlqs, c., Mulliken, J.B., Aylsworth, A.S., Albright, S., Lindhout, D., Cole, W.G., Henn, w., Knoll, J.H., Owen, M.J., Mertelsmann, R., Zabel, B.D., Olsen, B.R., 1997. Mutations involving the transcription factor CBFAI cause cleidocranial dysplasia. Cell 89, 773–779.

Nakashima, K., Zhou, X., Kunkel, G., Zhang, Z., Deng, J.M., Behringer, R.R., de Crombrugghe, B., 2002. The novel zinc finger-containing transcription factor osterix is required for osteoblast differentiation and bone formation. Cell 108, 17–29.

Nishio, Y., Dong, Y., Paris, M., O'Keefe, R.J., Schwarz, E.M., Drissi, H., 2006. Runx2-mediated regulation of the zinc finger OsterixlSp7 gene. Gene 372, 62–70.

Obri A., Mkinistoglu MP, Zhang H, Karsenty G., HDAC4 integrates PTH and sympathetic signaling in osteoblasts. J. Cell Biol. 205 (6): 771-780

Ogawa, E., Maruyama, M., Kagoshima, H., Inuzuka, M., Lu, J., Satake, M., Shigesada, K., Ito, Y., 1993. PEBP2IPEA2 represents a family of transcription factors homologous to the products of the Drosophila runt gene and the human AMLl gene. Proc. Natl. Acad. Sci. U.S.A. 90, 6859–6863.

Ohbayashi, N., Shibayama, M., Kurotaki, Y., Imanishi, M., Fujimori, T., Itoh, N., Takada, S., 2002. FGF18 is required for normal cell proliferation and differentiation during osteogenesis and chondrogenesis. Genes Dev. 16, 870–879.

Otto, E., Thronelkl, A.P., Crompton, T., Denzel, A., Gilmour, K. c., Rosewell, 1. R., Stamp, G.W.H., Beddington, R.S.P., Nundlos, S., Olsen, B.R., Selby, P.B., Owen, M.J., 1997. Cbfal, a candidate gene for cleidocranial dysplasia syndrome, is essential for osteoblast differentiation and bone development. Cell 89, 765–771.

Poy, M.N., Eliasson, L., Krutzfeldt, J., Kuwajima, S., Ma, X., Macdonald, P.E., Pfeffer, S., Tuschl, T., Rajewsky, N., Rorsman, P., Stoffel, M., 2004. A pancreatic islet-specific microRNA regulates insulin secretion. Nature 432 (7014), 226–230.

Rodino, M.A., Shane, E., 1998. Osteoporosis after organ transplantation. Am. J. Med. 104, 459–469.

Ruggieri, M., Pavone, v., De Luca, D., Franzo, A., Tine, A., Pavone, L., 1999. Congenital bone malformations in patients with neurofibromatosis type I (Nfl). J. Pediatr. Orthop. 19, 301–305.

Sabatakos, G., Sims, N.A., Chen, J., Aoki, K., Kelz, M.B., Amling, M., Bouali, Y., Mukhopadhyay, K., Ford, K., Nestler, E.J., Baron, R., 2000. Overexpression of 6FosB transcription factor(s) increases bone formation and inhibits adipogenesis. Nat. Med. 6, 985–990.

Satake, M., Nomura, S., Yamaguchi-Iwai, Y., Yousuke, T., Hashimoto, Y., Niki, M., Kitamura, Y., Ito, Y., 1995. Expression of the Runt Domain encoding PEBP2 alpha genes in T cells during thymic development. Mol. Cell. Biol. 15, 1662–1670.

Satokata, 1., Ma, L., Ohshima, H., Bei, M., Woo, I., Nishizawa, K., Maeda, T., Takano, Y., Uchiyama, M., Heaney, S., Peteres, H., Tang, Z., Maxson, R., Maas, R., 2000. Msx2 deficiency in mice causes pleotropic defects in bone growth and ectodermal organ formation. Nat. Genet. 24, 391–395.

Schmidt, K., Schinke, T., Haberland, M., Priemel, M., Schilling, A.E., Mueldner, C., Rueger, J.M., Sock, E., Wegner, M., Amling, M., 2005. The high mobility group transcription factor Sox8 is a negative regulator of osteoblast differentiation. J. Cell Biol. 168, 899–910.

Schinke, T., Karsenty, G., 1999. Characterization of Osfl, an osteoblast-specific transcription factor binding to a critical cis-acting element in the mouse osteocalcin promoter. J. Biol. Chem. 274, 30182–30189.

Shim, Jae-Hyuck, Greenblatt, Matthew B., Zou, Weiguo, Huang, Zhiwei, Wein, Marc N., Brady, Nicholas, Hu, Dorothy, et al., 2013. Schnurri-3 regulates ERK downstream of WNT signaling in osteoblasts. The Journal of clinical investigation 123 (9), 4010–4022.

Simonet, W.S., Lacey, D.L., Dunstan, C.R., Kelley, M., Chang, M.S., Luthy, R., Nguyen, H.Q., Wooden, S., Bennett, L., Boone, T., Shimamoto, G., DeRose, M., Elliott, R., Colombero, A., Tan, H.L., Trail, G., Sullivan, J., Davy, E., Bucay, N., Renshaw-Gegg, L., Hughes, T.M., Hill, D., Pattison, w., Campbell, P., Sander, S., Van, G., Tarpley, 1., Derby, P., Lee, R., Boyle, W.J., 1997. Osteoprotegerin: a novel secreted protein involved in the regulation of bone density. Cell 89, 309–319.

Stevenson, D.A., Birch, P.H., Friedman, J.M., Viskochil, D.H., Balestrazzi, P., Boni, S., Buske, A., Korf, B.R., Niimura, M., Pivnick, E.K., Schorry, E.K., Short, M.P., Tenconi, R., Tonsgard, J.H., Carey, J.C., 1999. Descriptive analysis of tibial pseudarthrosis in patients with neurofibromatosis 1. Am. J. Med. Genet. 84, 413—419.

Takeda, S., Bonnamy, J.P., Owen, M.J., Ducy, P., Karsenty, G., 2001. Continuous expression of Cbfal in nonhypertrophic chondrocytes uncovers its ability to induce hypertrophic chondrocyte differentiation and partially rescues Cbfal-deficient mice. Genes Dev. 15, 467—481.

Teitelbaum, S.L., Ross, E.P., 2003. Genetic regulation of osteoclast development and function. Nat. Rev. Genet. 4, 638—649.

Tribioli, C., Lufkin, T., 1999. The murine *Bapxl* homeobox gene plays a critical role in embryonic development of the axial skeleton and spleen. Development 126, 5699—5711.

Trivier, E., De Cesare, D., Jacquot, S., Pannetier, S., Zackai, E., Young, L., Mandel, J.L., Sassone-Corsi, P., Hanauer, A., 1996. Mutations in the kinase Rsk-2 associated with Coffin-Lowry syndrome. Nature 384, 567—570.

Valencia-Sanchez, M.A., Liu, J., Hannon, G.J., Parker, R., 2006. Control of translation and mRNA degradation by miRNAs and siRNAs. Genes Dev 20 (5), 515—524.

Verdin, E., Ott, M., 2013. Acetylphosphate: a novel link between lysine acetylation and intermediary metabolism in bacteria. Mol Cell 51 (2), 132—134.

Wagner, E.F., Eferi, R., 2005. Fos! AP-I proteins in bone and the immune system. Immunol. Rev. 208, 126—140.

Wei, J., Shi, Y., Zheng, L., Zhou, B., Inose, H., Wang, J., Guo, X.E., Grosschedl, R., Karsenty, G., May 14, 2012. miR-34s inhibit osteoblast proliferation and differentiation in the mouse by targeting SATB2. J. Cell Biol. 197 (4), 509—521.

Wei, J., Shimazu, J., Makinistoglu, M., Maurizi, A., Kajimura, D., Zong, H., Takarada, T., Iezaki, T., Pessin, J.E., Hinoi, E., Karsenty, G., 2015. Glucose uptake and Runx2 synergize to orchestrate osteoblast differentiation and bone formation. Cell 161, 1576—1591.

Wilke, A.O.M., Tang, Z., Eianko, N., Walsh, S., Twigg, S.R.E., Hurst, J.A., Wall, S.A., Chrzanowska, K.H., Maxson, R.E.J., 2000. Functional haploinsufficiency of the human homeobox gene MSX2 causes defects in skull ossification. Nat. Genet. 24, 387—390.

Wu, L.C., Mak, C.H., Dear, N., Boehm, T., Foroni, L., Rabbitts, T.H., 1993. Molecular cloning of a zinc finger protein which binds to the heptamer of the signal sequence for V(D)J recombination. Nucleic Acids Res. 21, 5067—5073.

Xiao, G., Jiang, D., Ge, C., Zhao, Z., Lai, Y., Boules, H., Phimphilai, M., Yang, X., Karsenty, G., Franceschi, R.T., 2005. Cooperative interactions between activating transcription factor 4 and Runx2! Cbfal stimulate osteoblast-specific osteocalcin gene expression. J. Biol. Chem. 280, 30689—30696.

Yadav, V.K., Ryu, J.H., Suda, N., Tanaka, K., Gingrich, J., Schutz, G., Glorieux, F.H., Insogna, K., Mann, J.J., Hen, R., Ducy, P., Karsenty, G., 2008. Lrp5 control bone mass by inhibiting serotonin synthesis in the duodenum. Cell 135, 825—837.

Yang, X., Karsenty, G., 2004. ATF4, the osteoblast accumulation of w~ich is determined post-translationally, can induce osteoblastspecific gene expression in non-osteoblastic cells. J. Biol. Chem. 279, 47109—47114.

Yang, X., Matsuda, K., Bialek, P., Jacquot, S., Masuoka, H.C., Schinke, T., Li, L., Brancorsini, S., Sassone-Corsi, P., Townes, T.M., Hanauer, A., Karsenty, G., 2004. ATF4 is a substrate of RSK2 and an essential regulator of osteoblast biology; implication for Coffin-Lowry syndrome. Cell 117, 387—398.

Yasuda, H., Shima, N., Nakagawa, N., Mochizuki, S.I., Yano, K., Fujise, N., Sato, Y., Goto, M., Yamaguchi, K., Kuriyama, M., Kanno, T., Murakami, A., Tsuda, E., Morinaga, T., Higashio, K., 1998. Identity of osteociastogenesis inhibitory factor (OCIF) and osteoprotegerin (OPG): a mechanism by which OPG/OCIF inhibits osteoclastogenesis in vitro. Endocrinology 139, 1329—1337.

Further reading

Boyden, L.M., Mao, J., Belsky, J., Mitzner, L., Farhi, A., Mitnick, M.A., Wu, D., Insogna, K., Lifton, R.P., 2002. High bone density due to a mutation in LDL-receptor-related protein 5. N. Engl. J. Med. 346, 1513—1521.

Day, T.E., Guo, X., Garrett-Beal, L., Yang, Y., 2005. Wntlbetacatenin signaling in mesenchymal progenitors controls osteoblast and chondrocyte differentiation during vertebrate skeletogenesis. Dev. Cell 8, 739—750.

Gong, Y., Slee, R.B., Fukai, N., Rawadi, G., Roman-Roman, S., Reginato, A.M., Wang, H., Cundy, T., Glorieux, E.H., Lev, D., Zacharin, M., Oexle, K., Marcelino, J., Suwairi, W., Heeger, S., Sabatakos, G., Apte, S., Adkins, W.N., Allgrove, J., Arslan-Kirchner, M., Batch, J.A., Beighton, P., Black, G.c., Boles, R.G., Boon, L.M., BOlTone, c., Brunner, H.G., Carle, G.E., Dallapiccola, B., De Paepe, A., Floege, B., Halfhide, M.L., Hall, B., Hennekam, R.C., Hirose, T., Jans, A., Juppner, H., Kim, C.A., Keppler-Noreuil, K., Kohlschuetter, A., LaCombe, D., Lambert, M., Lemyre, E., Letteboer, T., Peltonen, L., Ramesar, R.S., Romanengo, M., Somer, H., Steichen-Gersdorf, E., Steinmann, B., Sullivan, B., Superti-Furga, A., Swoboda, W., van den Boogaard, M.J., Van Hul, w., Vikkula, M., Votruba, M., Zabel, B., Garcia, T., Baron, R., Olsen, B.R., Warman, M.L., 2001. LDL receptor-related protein 5 (LRP5) affects bone accrual and eye development. Cell 107, 513—523.

Hecht, J., Seitz, V., Urban, M., Wagner, E., Robinson, P.N., Stiege, A., Dieterich, c., Kornak, U., Wilkening, U., Brieske, N., Zwingman, c., Kidess, A., Stricker, S., Mundlos, S., 2007. Detection of novel skeletogenesis target,genes by comprehensive analysis of a Runx 2(-/-) mouse model. Gene Expr. Patterns 7, 102—112.

Hill, T.P., Spater, D., Taketo, M.M., Birchmeier, W., Hartmann, C., 2005. Canonical Wntlbeta-catenin signaling prevents osteoblasts from differentiating into chondrocytes. Dev. Cell 727—738.

Huelsken, J., Birchmeier, W., 2001. New aspects of Wnt signaling pathways in higher vertebrates. Curr. Opin. Genet. Dev. 11, 547—553.

Kato, M., Patel, M.S., Levasseur, R., Lobov, I., Chang, B.H., Glass II, D.A., Hartmann, C., Li, L., Hwang, T.H., Brayton, C.F., Lang, R.A., Karsenty, G., Chan, L., 2002. Cbfa1-independent decrease in osteoblast proliferation, osteopenia, and persistent embryonic eye vascularization in mice deficient in Lrp5, a Wnt coreceptor. J. Cell Biol. 157, 303—314.

Little, R.D., Carulli, J.P., Del Mastro, R.G., Dupuis, J., Osborne, M., Folz, C., Manning, S.P., Swain, P.M., Zhao, S.C., Eustace, B., Lappe, M.M., Spitzer, L., Zweiei, S., Braunschweiger, K., Benchekroun, Y., Hu, X., Adair, R., Chee, L., FitzGerald, M.G., Tulig, C., Caruso, A., Tzellas, N., Bawa, A., Franklin, B., McGuire, S., Nogues, X., Gong, G., Allen, K.M., Anisowicz, A., Morales, A.J., Lomedico, P.T., Recker, S.M., Van Eerdewegh, P., Recker, R.R., Johnson, M.L., 2002. A mutation in the LDL receptor-related protein 5 gene results in the autosomal dominant high-bone-mass trait. Am. J. Hum. Genet. 70, 11−19.

Mao, J., Wang, J., Liu, B., Pan, W., Farr III, G.H., Flynn, C., Yuan, H., Takada, S., Kimelman, D., Li, L., Wu, D., 2001. Low-density lipoprotein receptor-related protein-5 binds to Axin and regulates the canonical Wnt signaling pathway. Mol. Cell 7, 801−809.

Vaes, B.L., Dlley, P., Sijbers, A.M., Hendriks, J.M., van Someren, E.P., de Jong, N.G., van den Heuvel, E.R., Olijve, W., van Zoelen, E.J., Dechering, K.J., 2006. Microarray analysis on Runx2-deficient mouse embryos reveals novel Runx2 functions and target genes during intrame! llbranous and endochondral bone formation. Bone 39, 724−738.

Zheng, Q., Zhou, G., Morello, R., Chen, Y., Garcia-Rojas, X., Lee, B., 2003. Type X collagen gene regulation by Runx2 contributes directly to its hypertrophic chondrocyte-specific expression in vivo. J. Cell Biol. 162, 833−842.

Chapter 8

Wnt signaling and bone cell activity

Bart O. Williams[1] and Mark L. Johnson[2]
[1]Program for Skeletal Disease and Center for Cancer and Cell Biology, Van Andel Research Institute, Grand Rapids, MI, United States;
[2]Department of Oral and Craniofacial Sciences, UMKC School of Dentistry, Kansas City, MO, United States

Chapter outline

Introduction 177
Wnt genes and proteins 178
Components of the Wnt/β-catenin signaling pathway 179
Lrp5, Lrp6, and frizzled 180
Disheveled, glycogen synthase kinase-3β, Axin, and β-catenin 181
Transcriptional regulation by β-catenin 183
Wnt signaling and bone cell function 183
Osteoblast differentiation and function 183
Osteoclast function 185
Osteocyte function 186
Interactions between Wnt/β-catenin signaling and other pathways important in bone mass regulation 186
The Wnt signaling pathway as a target for anabolic therapy in bone 188
References 189
Further Reading 199

Introduction

The first two decades of the 21st century have witnessed a literal explosion in our understanding of fundamental biological processes, catalyzed in large part by the human genome project and the vast repertoire of technologies that are now available to identify genes and study gene function. The field of bone biology has certainly benefited from these advances as illustrated by many of the chapters in this book. Not surprisingly perhaps, the genetic dissection of single-gene human disorders that present with skeletal abnormalities has led to advances in our understanding of bone biology, oftentimes in directions that could not have been anticipated by the current state of knowledge.

The Wnt/β-catenin pathway and its sister pathways (planar cell polarity, or PCP, and Wnt/Ca^{2+}) are one of five key signaling pathway families critical for normal development. They control a spectrum of cellular functions from differentiation to proliferation and growth and are involved in many aspects of patterning during development. As such it is not surprising that we now recognize that a wide spectrum of diseases have been attributed to mutations in members of the pathway or aberrant regulation of one of the Wnt pathways (Goggolidou and Wilson, 2016; Libro et al., 2016; Wang et al., 2016; Abou Ziki and Mani, 2017; Butler and Wallingford, 2017; Lodish, 2017; Nusse and Clevers, 2017). The field of bone biology was introduced to the Wnt/β-catenin signaling pathway in 2001–2 as a result of the discovery of mutations in the low-density lipoprotein receptor (LDLR)-related protein 5 (*LRP5*) gene that give rise to conditions of decreased (Gong et al., 2001) or increased bone mass (Boyden et al., 2002; Little et al., 2002). This was followed shortly after by the identification of *SOST* gene mutations as causal for sclerosteosis (Balemans et al., 2001; Brunkow et al., 2001) and, in the next year, for Van Buchem disease (Balemans et al., 2002; Staehling-Hampton et al., 2002). However, the recognition that sclerostin, the product of the *SOST* gene, was a key negative regulator of Wnt/β-catenin signaling was not made until 2005 (Li et al., 2005). Targeting sclerostin with humanized anti-sclerostin antibodies has become the focus of major clinical trials for the treatment of osteoporosis and has several other potential applications (see later). Since those initial descriptions of altered bone mass, several key components of the Wnt signaling pathway (see Fig. 8.1) have been shown to underlie a large number of monogenic bone diseases (see for review Maupin et al., 2013; Malhotra and Yang, 2014; Lara-Castillo and Johnson, 2015; Wang et al., 2016; Reppe et al., 2017). In many cases normal polymorphisms in these genes are associated

Principles of Bone Biology. https://doi.org/10.1016/B978-0-12-814841-9.00008-7

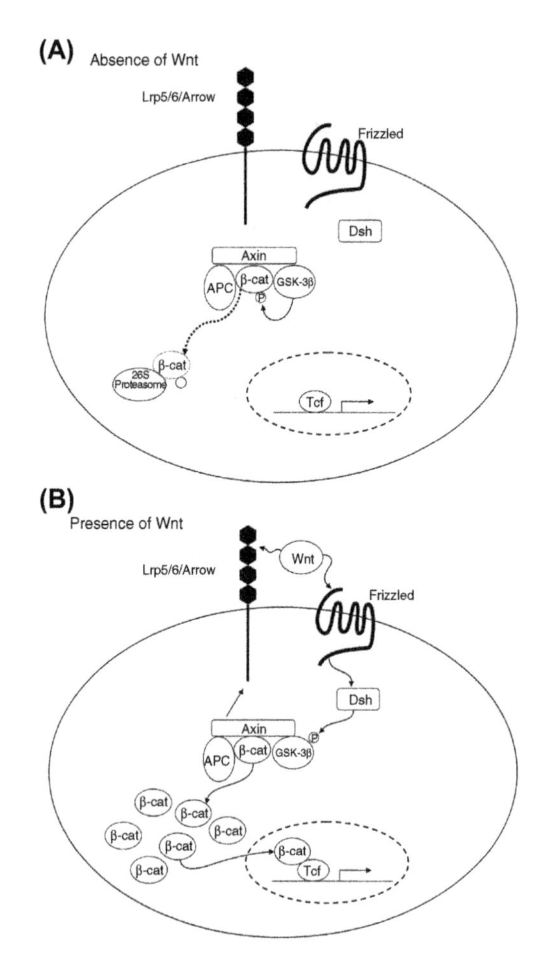

FIGURE 8.1 The Wnt/β-catenin signaling pathway. The Wnt/β-catenin signaling pathway is controlled by two cell surface coreceptors, Frizzled and one of the low-density lipoprotein receptor–related proteins (Lrp5 or Lrp6) in vertebrates or Arrow in *Drosophila*, that cooperatively bind the Wnt protein ligand. (A) In the absence of Wnt, the degradation complex consisting of the scaffolding protein Axin, adenomatous polyposis coli (*APC*), and glycogen synthase kinase-3β (*GSK-3β*), along with several other proteins (not shown), coordinates the phosphorylation (*P*) of β-catenin (*β-cat*), which is then ubiquitinated and degraded by the 26S proteasome complex. This serves to keep the intracellular levels of free β-catenin at a low level. (B) When Wnt ligand is bound by the Lrp5/6/Arrow and Frizzled coreceptors the protein Dishevelled (*Dsh*) is activated, which results in the phosphorylation of GSK-3β, thereby inhibiting its activity. The degradation complex is also induced to bind to the cytoplasmic tail of Lrp5/6/Arrow and the combination of events leads to the intracellular accumulation of high levels of free β-catenin. β-Catenin can then translocate into the nucleus, and upon binding to the T cell factor/lymphoid enhancer-binding factor (*Tcf*/LEF) family of transcription factors, regulate the expression of a number of target genes.

with or contribute to some degree of variance in complex bone traits such as bone density (see for review Urano and Inoue, 2014; Lara-Castillo and Johnson, 2015; Rocha-Braz and Ferraz-de-Souza, 2016; Reppe et al., 2017).

Prior to these human genetic studies the role of Wnt/β-catenin signaling in bone was limited mainly to skeletal development, but because of the explosion of research that occurred as a result of those seminal publications we now appreciate the role of this pathway in bone cell differentiation, proliferation, and apoptosis; bone mass regulation; and the ability of bone to respond to changes in mechanical load. While Wnts can signal through multiple pathways, in this chapter the focus will be on what is known about the Wnt/β-catenin signaling pathway and its regulation as pertains to bone.

Wnt genes and proteins

As of this writing, there are 19 known *Wnt* genes in humans. Wnt proteins are secreted, highly posttranslationally modified proteins that play key roles in development and homeostasis through their involvement in cell differentiation, proliferation, and apoptosis. Nusse and Varmus first described Wnt as the mouse protooncogene integration site/locus (*int-1*) of the murine mammary tumor virus that resulted in breast tumors (Nusse and Varmus, 1982). Because of the challenges in purifying Wnt in a biologically active form, Nusse turned to a genetic system to help characterize the downstream components of the pathway when it was found that the *Drosophila* segment polarity gene *Wingless* (*Wg*) was homologous to the *int-1* gene (Rijsewijk et al., 1987). In 1991, several of the founding researchers in this field who came from the fields of *Drosophila* genetics, mammary viral oncogenesis, and early embryonic development in *Xenopus* came together to standardize the nomenclature for the growing number of Int-1- and Wg-related genes that were being identified and coined the term "Wnt" as a combination of Wingless and Int-1 (Nusse et al., 1991).

The Wnt proteins are difficult to study because of posttranslational modifications (mainly palmitoylation) that make them extremely hydrophobic and difficult to purify in a biologically active form. In fact, it took 21 years after the initial discovery of Wnt to work out the conditions to purify recombinant Wnt protein in a manner that retained its signaling activity (Willert et al., 2003). Thus, historically, much of what we know about Wnt proteins was based on their DNA sequence analysis and genetic screens in *Drosophila* and, later, *Caenorhabditis elegans* (Sawa and Korswagen, 2013). The various Wnt genes share general homology in the range of 35%, although within subgroups the homology can be as high as 83%. The molecular weight of the various Wnt proteins ranges from 39 to 46 kDa. All Wnts contain 23 or 24 conserved cysteine residues that are spaced similarly between proteins, suggesting an important conservation of function, possibly proper folding of the protein, which is required for Wnt activity.

Nusse and colleagues successfully purified Wnt3a from mouse L cells (Willert et al., 2003). Subsequently Schulte et al. (Schulte et al., 2005) used a similar strategy to purify Wnt5a. Both groups originally thought that the addition of a palmitate to the first, most amino-terminally conserved of the cysteine residues was absolutely required for Wnt3a and Wnt5a activity. In fact, the identification of this cysteine residue as the site of palmitoylation turned out to be incorrect, as subsequent studies showed that, in fact, a serine residue was the target of palmitoylation in Wnt proteins (Takada et al., 2006; Janda et al., 2012). The implications for this are discussed in the following.

In *Drosophila* the gene *porcupine* and in *C. elegans* the gene *mom-1* encode acyltransferase enzymes that are the proteins responsible for the palmitoylation of Wnts (Kadowaki et al., 1996; Nusse, 2003). Interestingly, these proteins are membrane bound and found in the endoplasmic reticulum, indicating that the palmitoylation occurs intracellularly. While palmitoylation of secreted proteins is somewhat unusual (Dunphy and Linder, 1998), the hedgehog family of proteins is also palmitoylated and secreted in this manner (Nusse, 2003). Subsequent work revealed two functional aspects related to palmitoylation of this serine. First, it is essential for transport of Wnt ligands from the endoplasmic reticulum to the cell surface for secretion. Upon palmitoylation, a multipass transmembrane protein, Wntless, binds to the palmitoylated Wnt and acts as a chaperone protein to facilitate its transport through the Golgi to the cell surface (Banziger et al., 2006; Bartscherer et al., 2006). Loss of either Porcupine or Wntless renders a cell functionally null for the ability to secrete any Wnt ligands.

The second functional role of the palmitoylated serine residue is its requirement for facilitating the binding of Wnt to Frizzled on the cell surface. This role was identified as a result of the first successful crystal structure of the Wnt–Frizzled interaction (Janda et al., 2012). This structure revealed that the palmitoleic group attached to the serine fits into a deep, hydrophobic group in the cysteine-rich domain (CRD) of Frizzleds.

Components of the Wnt/β-catenin signaling pathway

A detailed description of the various components of the Wnt/β-catenin signaling pathway can be found at the Wnt homepage website maintained by Dr. Roel Nusse (http://web.stanford.edu/group/nusselab/cgi-bin/wnt/). The major players in the Wnt/β-catenin signaling pathway in bone are (1) the cell surface coreceptors Lrp5/6 and Frizzled; (2) the intracellular proteins Dishevelled (Dsh), glycogen synthase kinase-3β (GSK-3β), Axin, and β-catenin; and (3) the T cell factor/lymphoid enhancer-binding factor (TCF/LEF) nuclear transcription factors that bind β-catenin and regulate gene expression. In addition, there are a number of other extracellular and intracellular proteins involved in regulation of this pathway that serve to modify the functionality of one of these key proteins. Particularly important are a number of proteins that bind Lrp5/6 to modulate its binding of Wnt ligand.

The details of the Wnt/β-catenin pathway are illustrated in Fig. 8.1. Ultimately the Wnt/β-catenin signaling pathway controls the intracellular levels of free β-catenin. In the absence of Wnt ligand the intracellular levels of β-catenin are extremely low due to the activity of a degradation complex comprising Axin, GSK-3β, the adenomatous polyposis coli (APC) protein, and several other proteins. This complex, specifically GSK-3β or -3α, is responsible for the phosphorylation of β-catenin, which leads to its ubiquitination and degradation by the 26S proteasome complex (Fig. 8.1A). Wnt binding to the Lrp5/6-Frizzled coreceptors results in Dsh activation and Axin binding to the cytoplasmic tail of Lrp5/6. Dsh activation leads to GSK-3β phosphorylation and inhibition of its activity. β-Catenin is no longer phosphorylated and this, coupled with the binding of Axin to the cytoplasmic tail of Lrp5/6, releases β-catenin. β-Catenin accumulates in the cytoplasm and then translocates into the nucleus where it binds to the TCF/LEF family of transcription proteins and regulates the expression of specific target genes (Fig. 8.1B).

Lrp5, Lrp6, and frizzled

Lrp5 and Lrp6 (Arrow in *Drosophila*) are highly homologous proteins that until their role in bone biology was described were considered orphan members of the LDLR family, which includes at least 13 members (Strickland et al., 2002). One of the hallmark structural characteristics of this family is the presence of YWTD (tyrosine—tryptophan—threonine—aspartic acid) repeat motifs framed by epidermal growth factor repeats in the extracellular domain of the proteins. These repeat clusters form β-propeller structures that are important for ligand binding. Based on the amino acid sequence of Lrp5 and Lrp6 derived from cDNA cloning (Brown et al., 1998; Dong et al., 1998; Hey et al., 1998; Kim et al., 1998), Lrp5/6 each contain four of these repeat clusters. However, the YWTD motif is degenerate and one of the six repeat sequences comprising each motif in Lrp5 is substituted by LFAN (leucine—phenylalanine—alanine—asparagine) or FFTN sequences (Johnson and Summerfield, 2005). Many of these LDLR family members have a consensus sequence of NPxY (asparagine—proline—x—tyrosine) in their cytoplasmic tail that is required for endocytosis of the receptor (Chen et al., 1990). Lrp5 and Lrp6 lack an NPxY sequence. Also the orientation of their extracellular domain is reversed relative to other members of the LDLR family in that LDL-A type repeats (complement-like repeats) are adjacent to the transmembrane spanning region rather than at the amino-terminal end (or clustered throughout the extracellular domain). There are also only three of these LDL-A type repeats in Lrp5 and Lrp6.

Originally human *LRP5* was proposed as a candidate gene for susceptibility to type 1 diabetes in the insulin-dependent diabetes mellitus 4 locus that was mapped to chromosome 11q13 (Figueroa et al., 2000; Twells et al., 2001). However, subsequent genetic association studies failed to confirm this functional role (Twells et al., 2003). At the same time, Arrow (Wehrli et al., 2000) and Lrp6 (Tamai et al., 2000) were shown to mediate Wnt/β-catenin signaling and *Lrp6*$^{-/-}$ mutant mouse embryos displayed skeletal abnormalities similar to those found in mice carrying mutations in various Wnt genes (Pinson et al., 2000). Subsequently mutations in human *LRP5* were shown to cause osteoporosis pseudoglioma syndrome (OPPG) (Gong et al., 2001), which is characterized by low bone mass and progressive blindness, and the high bone mass (HBM) mutation G171V was reported in two separate kindreds (Boyden et al., 2002; Little et al., 2002). A large number of additional mutations that give rise to conditions of increased or decreased bone mass have been reported in the literature (Balemans and Van Hul, 2007). Also, allelic variants of *LRP5* have been shown to contribute to variation in bone mass in several studies, although the relative degree varies and sex differences are noted in many of the studies (Karasik et al., 2016).

The role of LRP6 in human bone density variation has been studied to a limited degree. One report indicates that the *LRP6* Ile1062Val allele contributes to a higher risk for fragility and vertebral fracture in men and that this allelic variant in combination with the *LRP5* A1330V allele accounted for 10% of the fractures in males (van Meurs et al., 2006). It has also been shown that *Lrp6*$^{+/-}$:*Lrp5*$^{-/-}$ mice have lower adult bone mineral density compared with *Lrp6*$^{+/+}$:*Lrp5*$^{-/-}$ mice, suggesting a role for Lrp6 in bone mass accrual (Holmen et al., 2004). A mutation in *LRP6* (*LRP6*^{R611C}) was linked to early-onset coronary artery disease in a family, and the five mutation carriers that were studied also had low bone density, confirming a role for LRP6 in bone mass accrual as suggested by the studies in mice (Mani et al., 2007). Also, a hypomorphic *Lrp6* mutation in the mouse (called *ringelschwanz*) has been identified, and studies in that model confirm a role for Lrp6 in somitogenesis and osteogenesis (Kokubu et al., 2004).

Given that both appear to be coexpressed in most cells/tissues that have been examined, it was of interest to determine whether Lrp5 and Lrp6 served totally redundant roles in skeletal development and homeostasis. Using the Dermo1-Cre strain to drive conditional gene deletion in mesenchymal progenitors, it was demonstrated that Lrp5 and Lrp6 played redundant roles at this stage (Joeng et al., 2011). Interestingly, deletion of the genes in mature osteoblasts (using OCN-Cre) revealed that Lrp6 and Lrp5 played overlapping, but not redundant, roles at that stage (Riddle et al., 2013). The role of Lrp5 and the Wnt/β-catenin signaling pathway in the response of bone to mechanical loading has been firmly established. Based upon analysis of the phenotype of affected members of the HBM trait kindred we had speculated that the *LRP5* G171V mutation gene might somehow have altered the sensitivity of the skeleton to mechanical loading (Johnson et al., 2002). Several lines of evidence now support a central role for Lrp5 and the Wnt/β-catenin signaling pathway in the response of bone/bone cells to mechanical loading. Sawakami et al. (Sawakami et al., 2006) have shown that the bone formation response after loading is reduced by 88% in male and 99% in female Lrp5$^{-/-}$ mice compared with wild-type loaded mice. Mice carrying the LRP5 cDNA containing the G171V mutation have increased sensitivity to mechanical loading (Cullen et al., 2004). Osteocyte β-catenin haploinsufficiency protects against bone loss in male mice subjected to hindlimb unloading (Maurel et al., 2016). Changes in the expression of Wnt/β-catenin signaling pathway target genes have been shown to occur both in vivo in bone (Robinson et al., 2006) and in vitro in bone cells (Lau et al., 2006; Robinson et al., 2006) in response to mechanical loading. Given that Lrp5 is required for the response of bone to mechanical loading, the question that next needs to be answered is, where does Lrp5/Wnt/β-catenin signaling fit into the cellular cascade events that occur in response to mechanical loading? A partial answer to this question will be discussed further later in this chapter. Another unanswered question is, does Lrp6 play a role in the response to loading? At this writing we have no definitive answer, but given the results of the ulna loading studies performed with the Lrp5$^{-/-}$ mice and that these mice have normal alleles of Lrp6, it would seem that Lrp6 plays little or no role in mechanoresponsiveness in bone.

As of this writing there are 10 known members of the Frizzled family of proteins in humans and a homologous protein called Smoothened that functions in Hedgehog signaling. The Frizzled proteins contain a CRD at the amino terminus and seven transmembrane-spanning domains. The CRD is responsible for binding of Wnt ligands (Bhanot et al., 1996). The Frizzled proteins are generally thought to be coupled to trimeric G proteins (Sheldahl et al., 1999; Katanaev et al., 2005). It also seems evident that different Frizzleds regulate different intracellular signaling cascades. However, the specific combinations are not completely understood. Specific Frizzleds functioning in combination with Lrp5 or Lrp6 in the binding of specific Wnt proteins lead to activation of β-catenin signaling; however, some Frizzleds acting alone appear to be involved in the activation of at least three other Wnt signaling pathways. These other Wnt pathways are the PCP (Mlodzik, 2002), the Wnt/Ca^{2+} involving protein kinase C (Kuhl et al., 2000), and a protein kinase A (PKA) pathway involved in muscle myogenesis (Chen et al., 2005). Dsh also appears to be involved in the first two of these Frizzled-regulated pathways. It is still not clear which of the 10 different Frizzleds are involved in the activation of each of the four Wnt pathways. However, it is known that some of the Wnt proteins preferentially activate the β-catenin signaling pathway (e.g., Wnt1 and Wnt3A), whereas other Wnts activate only one of the other pathways (e.g., Wnt5A activates the Wnt/Ca^{2+} signaling pathway) (Miller, 2001).

Recent work has begun to link specific Frizzled genes to skeletal development and homeostasis. Mouse strains carrying global deletions in most of the 10 Frizzled genes have been created (Wang et al., 2016). However, only two have been evaluated in detail for skeletal phenotypes. *Fzd8*$^{-/-}$ mice have osteopenia with normal bone formation and increased osteoclast activity (Albers et al., 2013), while *Fzd9*$^{-/-}$ mice have low bone mass associated with reduced chemokine expression but normal Wnt/β-catenin signaling (Albers et al., 2011; Heilmann et al., 2013). Based on work in human patients, Fzd2 probably plays a central role in early limb development. Specifically, deleterious heterozygous mutations in Fzd2 were associated with autosomal dominant omodysplasia or variations of Robinow syndrome, syndromes in which skeletal dysplasia is characterized by facial dysmorphism and shortening of the limbs (Saal et al., 2015; Turkmen et al., 2017). It is not clear whether the alteration of Fzd2 produces a protein that acts in a dominant negative fashion or if phenotypes are a result of haploinsufficiency for Fzd2. One report showed that an Fzd2 allele associated with autosomal dominant omodysplasia produced a protein truncated shortly after the seventh transmembrane domain of Fzd2 that was unable to activate β-catenin signaling. However, this work did not distinguish whether the underlying mechanistic effect was due to haploinsufficiency or a dominant negative effect of the truncated protein (Saal et al., 2015). Identification of the genetic alterations underlying the development of Robinow syndrome has revealed that mutations causing similar human phenotypes can be found in the following genes: WNT5A, FZD2, ROR2, DVL1, DVL3, and RAC3 (White et al., 2018). One interpretation of this is that a pathway initiated by WNT5A binding to the FZD2 and ROR2 receptors activates DVL proteins, resulting in the activation of RAC3 and its downstream targets.

The discovery that Lrp5/6/Arrow was also required for the activation of the Wnt/β-catenin signaling pathway suggested a model in which Lrp5/6/Arrow and Frizzled function as coreceptors (Tamai et al., 2000; Wehrli et al., 2000). The role of Wnt was proposed to perhaps facilitate the association/interaction of the coreceptors and this was the key event leading to pathway activation. Several lines of evidence obtained from the construction of various fusion proteins suggest that this is most likely the case (Holmen et al. 2002, 2005; Tolwinski et al., 2003). Wnt binding and the formation of this ternary complex leads to a Frizzled-dependent activation of Dsh. However, the exact molecular mechanism mediating Dsh activation is not clear. Dsh phosphorylation appears to be one common feature in all of the signaling pathways through which it is known to act. The kinases casein kinase 1 (CK1) (Peters et al., 1999) and CK2 (Willert et al., 1997) and PAR-1 (Sun et al., 2001) are the leading candidates for carrying out the phosphorylation of Dsh, which can occur through both Lrp5/6-dependent (leading to the β-catenin signaling pathway) and independent mechanisms (Wnt binding to Frizzled in the absence of Lrp5/6) (Gonzalez-Sancho et al., 2004). Three different models of how Frizzled and Dsh may interact have been proposed (He et al., 2004), but it is not clear which one functions in conjunction with Wnt and Lrp5/6/Arrow.

Dishevelled, glycogen synthase kinase-3β, Axin, and β-catenin

Dsh serves an important intermediary function between Frizzled and GSK-3β in the Wnt/β-catenin signaling pathway. Dsh also serves as an intermediary in the Wnt/Ca^{2+} signaling pathway (Kuhl et al., 2000) and the PCP pathway (Mlodzik, 2002). The PCP pathway is responsible for the proper orientation of wing hairs and thoracic bristles in *Drosophila*. Mutants with hairs/bristles that were improperly oriented led to the identification of the *Frizzled* and *Dishevelled* genes (see for review Adler, 2002). Evidence suggests that Frizzled signals through the $G\alpha_o$ subunit in both the Wnt/β-catenin pathway and the PCP pathway (Katanaev et al., 2005). While it is not fully understood how Dsh becomes activated upon Wnt binding and how it can discriminate between which pathway(s) to activate, a model has been proposed that suggests that differential intracellular localization of Dsh may be involved in determining some aspects of its function. Dsh associated with cytoplasmic vesicles has been proposed to be linked to signaling through the Wnt/β-catenin pathway, while Dsh associated with actin and the plasma membrane is proposed to signal via the PCP pathway (Capelluto et al., 2002; Povelones and Nusse, 2002). It may also be that Frizzled localization and the specific Frizzled present in the cell, along with the distribution of downstream components and the specific Wnt ligand participating in the initial binding events, are critical as well (Bejsovec, 2005; Katoh, 2005).

The mechanism responsible for the activation of Dsh is not fully understood nor is the mechanism whereby Dsh acts known. There are three domains in Dsh that appear to be critical for its function. There is a DIX domain, which is also found in Axin; a PDZ domain; a conserved stretch of basic amino acids; and a DEP domain that is found in other vertebrate proteins that interact with trimeric G proteins (Cadigan and Nusse, 1997; He et al., 2004). Activation of Dsh occurs through phosphorylation and this appears to occur in the other Wnt signaling pathways that involve Dsh. The kinases CK1 (Peters et al., 1999) and CK2 (Willert et al., 1997) and PAR-1 (Sun et al., 2001) are the leading candidates to carry out the phosphorylation of Dsh. Gonzalez-Sancho et al. (Gonzalez-Sancho et al., 2004) have suggested that Frizzled−Dsh−Axin and Lrp5/6/Arrow−Axin associations occur in parallel and both are required for stabilization of β-catenin. However, much remains to be understood about the role of Dsh in Wnt/β-catenin signaling and with regard to its role in the other Wnt signaling pathways.

GSK-3 has two isoforms, α and β. While most of the current evidence implicates GSK-3β in the control of intracellular levels of β-catenin, GSK-3α may also play a role (McManus et al., 2005; Asuni et al., 2006). GSK-3β was first identified as playing a key role in the regulation of glycogen synthesis by phosphorylating (and thereby inhibiting) glycogen synthase (see for review Hermida et al., 2017). GSK-3α and -3β are themselves both inhibited by phosphorylation at an amino-terminal serine residue, serine 21 (α) (Sutherland and Cohen, 1994) or serine-9 (β) (Sutherland et al., 1993; Stambolic and Woodgett, 1994). This phosphorylation is known to be catalyzed by a number of kinases, including mitogen-activated protein (MAP) kinase (Sutherland and Cohen, 1994), protein kinase B (PKB)/Akt (Haq et al., 2000), protein kinase C (Cook et al., 1996), and PKA (Fang et al., 2000). However, this inhibitory phosphorylation may not be linked to GSK-3-mediated regulation of β-catenin signaling (Papadopoulou et al., 2004; McManus et al., 2005). To explain this paradox, Forde and Dale (Forde and Dale, 2007) have suggested that GSK-3β phosphorylation is a redundant form of regulation with respect to β-catenin signaling. Integrin-linked kinase also regulates GSK-3β, but apparently not through phosphorylation at serine 9 (Delcommenne et al., 1998). While there are many other regulators of GSK-3β (Hardt and Sadoshima, 2002), the aforementioned kinases all have described roles in bone and, as will be discussed later, this intersection with the Wnt/β-catenin signaling pathway suggests possible cross talk between pathways.

The kinase activity of GSK-3β appears to be enhanced by a priming phosphorylation at nearby sites. In proteins such as Lrp5/6, where multiple GSK-3β phosphorylation sites are present in the cytoplasmic tail, GSK-3β can self-prime. The initial priming phosphorylation of β-catenin needed for the subsequent GSK-3β phosphorylation is carried out by CK1 (Liu et al., 2002). GSK-3β has also been shown to phosphorylate two sites within Axin (T609 and S614), which is prerequisite for the binding of β-catenin (Jho et al., 1999). It has been proposed that Wnt signaling leads to a dephosphorylation of these sites and this is part of the mechanism leading to the release of β-catenin from the degradation complex. In addition, protein phosphatase 2A (PP2A), which binds to Axin, APC, and Dsh (Hsu et al., 1999; Seeling et al., 1999; Yamamoto et al., 2001), plays an important counterbalancing role opposing GSK-3β. When CK1 phosphorylates its substrates within the degradation complex, PP2A dissociates, which favors further net phosphorylation and the release of β-catenin (Gao et al., 2002). Phosphorylated β-catenin is then ubiquitinated and degraded by the 26S proteasome complex (Aberle et al., 1997).

Axin acts as a scaffolding protein that directly interacts with/binds Dsh, GSK-3β, APC, PP2A, and β-catenin (Behrens et al., 1998; Hart et al., 1998; Ikeda et al., 1998; Itoh et al., 1998; Nakamura et al., 1998; Sakanaka et al., 1998; Yamamoto et al., 1998; Fagotto et al., 1999; Farr et al., 2000). Axin was first identified as the product of the mouse fused locus (Zeng et al., 1997). In vertebrates there are two forms of Axin: Axin 1, which is constitutively expressed and is the main form of the degradation complex, and Axin 2 (also known as Conductin). Axin 2 expression is induced by Wnt signaling and it functions as a negative feedback inhibitor of β-catenin signaling (Jho et al., 2002). Axin has several functional domains required for its interactions with the various other proteins of the degradation complex. It appears that amino acids 581−616 are responsible for binding β-catenin, while the RGS domain interacts with APC, and GSK-3β appears to bind to amino acids 444−543 (Nakamura et al., 1998), although GSK-3β was observed only when β-catenin was present. The Armadillo repeats in β-catenin mediate the interaction with Axin, while a Dix domain similar to the one found in Dsh probably mediates the interactions between Axin and Dsh. Mutations in the human AXIN2 gene have been linked to familial tooth agenesis and are predisposing for colon cancer (Lammi et al., 2004).

The collapse of the degradation complex and the release of β-catenin is a complex cascade of poorly understood events. Wnt binding to the coreceptors induces Axin to bind to the cytoplasmic tail of Lrp5/6/Arrow (after the coreceptor is phosphorylated and FRAT-1 binds). When Wnt is not present, β-catenin is bound to APC and APC to Axin as a consequence of specific phosphorylations mediated by GSK-3β (and other kinases). In the presence of Wnt and the activation of Dsh, these phosphorylation events are inhibited, and binding of Axin to the cytoplasmic tail of Lrp5/6/Arrow leads to the collapse of the complex and release of β-catenin into the cytoplasm.

β-Catenin is highly homologous to the *Drosophila* segment polarity gene *armadillo* (McCrea et al., 1991). The human β-catenin gene (*CTNNB*) is located on chromosome 3p21 and encodes a protein of 781 amino acids (~85 kDa). Two pools of β-catenin exist within the cell: one associated with E-cadherin and the other in the cytoplasm (Nelson and Nusse, 2004). β-Catenin contains a cassette of 12 Armadillo repeats in the middle of the protein that are the interaction sites with E-cadherin, APC, and the nuclear transcription factors (Cadigan and Nusse, 1997; Willert and Nusse, 1998). A number of mutations in β-catenin have been identified that give rise to human cancers. The amino-terminal end of the protein (mainly between amino acids 29–49) contains key residues whose phosphorylation by GSK-3β plays a critical role in the subsequent degradation of the protein, and when one or more of these residues are mutated the protein is more stable and the increased signaling results in tumor formation/growth (Yost et al., 1996; Polakis, 2000). Exactly how β-catenin is translocated from the cytoplasm into the nucleus is not fully understood, although there is evidence that this may require participation with estrogen receptor α (ERα) (Armstrong et al., 2007).

Transcriptional regulation by β-catenin

Once inside the nucleus, β-catenin binds to the TCF/LEF family of transcription factors and regulates the expression of a large (and ever-increasing) list of target genes. It has been shown that β-catenin can also interact with the FOXO family of transcriptional regulators and that this association is particularly important during oxidative stress (Essers et al., 2005), which may play a role in age-associated bone loss (Almeida et al., 2006). A consensus TCF/LEF core sequence has been identified that is required for binding of the TCF/LEF proteins (CCTTTGATC) (Korinek et al., 1997). The TCF/LEF proteins have DNA-binding ability but require the transactivating domain of β-catenin (at the C-terminal end) to regulate transcription.

A large number of proteins (see http://www.stanford.edu/~rnusse/ for a detailed listing) participate in the transcriptional regulation mediated by β-catenin. As of this writing, the model for how β-catenin regulates target gene transcription involves the formation of a larger complex of proteins that induces a change in chromatin structure (Barker et al., 2001). In *Drosophila*, when Wnt is not present, TCF acts as a repressor of Wnt/Wg target genes by forming a complex with Groucho (Cavallo et al., 1998), whose repressing ability is regulated by the histone deacetylase enzyme, Rpd3 (Chen et al., 1999). Studies in *Xenopus* have shown that β-catenin interacts with the acetyltransferases p300 and CBP to activate the *siamois* gene promoter (Hecht et al., 2000). Other proteins that control β-catenin signaling include the antagonist Chibby, which binds to the C-terminal end (Takemura et al., 2003), and the protein ICAT (Tago et al., 2000), which negatively regulates the interaction between β-catenin and TCF-4. TCF/LEF can also be phosphorylated by the MAP kinase–related protein NLK/Nemo (Ishitani et al., 2003), which reduces the affinity for β-catenin. Considerably less is known about the interaction of β-catenin with the FOXO family of transcription factors, but presumably it also involves cooperative and repressive interactions with a number of proteins. Also, it is not understood how choices are made as to which interactions are favored under any given set of cellular circumstances.

There are a large number of known Wnt/β-catenin target genes. Interestingly, many of these targets are components of the Wnt signaling pathway (see http://www.stanford.edu/~rnusse/ for a more complete listing), which creates a complicated set of feedback loops that can potentially serve to further amplify or inhibit signaling through the pathway. Some of these target genes will be discussed further in the following sections as they play important roles in bone.

Wnt signaling and bone cell function

Osteoblast differentiation and function

Many aspects of osteoblast differentiation have been covered in other chapters in this book (see Chapters 4 and 7). Wnt/β-catenin signaling plays important roles at several levels. A search of the PubMed database using the terms "Wnt and osteoblast and differentiation" yielded over 1000 published articles since 2002. Obviously a comprehensive, detailed review of what we now understand regarding the role of Wnt/β-catenin signaling in osteoblast differentiation is not possible, but several excellent reviews have been published that cover key aspects (Hartmann, 2006; Baron and Kneissel, 2013; Zhong et al., 2014; Lerner and Ohlsson, 2015; Ahmadzadeh et al., 2016; Kobayashi et al., 2016).

Skeletal stem cells (Robey, 2017) are capable of differentiating into adipocytes, chondrocytes, and osteoblastic lineage cells. In part, the commitment to these different cell lineages is controlled by Wnt ligands that activate either the Wnt/β-catenin or the noncanonical Wnt signaling pathway. MacDougal and colleagues (Ross et al., 2000; Bennett et al., 2005) demonstrated that Wnt10b stimulates osteoblastogenesis and inhibits adipogenesis. Subsequently, Zhou et al. (Zhou et al., 2008) suggested a model involving glucocorticoid-induced *Wnt* gene expression in osteoblasts, and the expressed Wnt proteins then induce Runx2 expression and inhibit peroxisome proliferator-activated receptor γ expression in progenitor cells, thereby leading to increased osteoblastogenesis versus adipogenesis. Control of chondrogenesis is complex, with both canonical Wnt/β-catenin and noncanonical signaling having been demonstrated in the regulation of chondrogenesis at various levels in the differentiation pathway. Wnt1 and Wnt7a have been shown to play inhibitory roles in the early prechondrogenic mesenchyme. Hartmann and Tabin (Hartmann and Tabin, 2000) demonstrated specific patterns of expression of Wnt5a, Wnt5b, and Wnt4 in the chicken limb chondrogenic regions, with Wnt5a and Wnt4 having opposing effects. Wnt5a inhibits chondrogenesis and bone collar formation, while Wnt4 accelerates these processes. Church et al. (Church et al., 2002) have shown that Wnt5a is expressed in the joints and perichondrium, with Wnt5b and Wnt11 localized to the region of prehypertrophic chondrocytes. Akiyama et al. (Akiyama et al., 2004) have shown that Sox9, a transcription factor required for chondrogenesis, directly binds to β-catenin and thereby competes for Tcf/Lef-mediated transcription by β-catenin. Reciprocally, Wnt3a has been shown to inhibit chondrocyte differentiation by upregulating the expression of *c-Jun* leading to activation of activator protein 1, which suppresses Sox-9 (Hwang et al., 2005). Hill et al. (Hill et al., 2005) proposed a model in which β-catenin is necessary, but not sufficient, for suppressing chondrogenic commitment of progenitor cells. Thus, the role of Wnt signaling in the fate decisions for differentiation between adipogenesis, chondrogenesis, and osteogenesis within the skeletal stem population is complex and fine-tuned by various Wnt ligands and is tightly controlled both spatially and temporally.

The commitment toward osteoblastogenesis is driven by the Runt domain–containing transcription factor, Runx2 (Ducy et al., 1997; Komori et al., 1997; Lee et al., 1997; Otto et al., 1997) (AML3/PEB2α/CBFA1), and the transcription factor Osterix (Osx) (Nakashima et al., 2002), which acts downstream of Runx2. The original descriptions of human LRP5 mutations in low- and high-bone-mass traits suggested an important role for Wnt signaling in osteoblast proliferation and differentiation (Gong et al., 2001; Boyden et al., 2002; Little et al., 2002). Subsequent studies with a transgenic mouse expressing the HBM $LRP5^{G171V}$ mutation provided further support for Wnt/β-catenin signaling in the regulation of osteoblast differentiation and function (Babij et al., 2003). There is a complicated relationship between activation of Wnt/β-catenin signaling and Runx2. Gaur et al. (Gaur et al., 2005) demonstrated that β-catenin induced Runx2 expression and Haxaire et al. have shown that Runx2 is a negative regulator of Wnt/β-catenin signaling (Haxaire et al., 2016). Runx2 and Osx activate the expression of SOST (Perez-Campo et al., 2016), a negative regulator of the pathway, which may partially explain the decrease in Wnt/β-catenin signaling.

Studies have also shown an important role for microRNAs (miRNAs) in the regulation of osteoblastogenesis, in part through their regulation of Wnt/β-catenin signaling. DICER, which is a key component of miRNA processing, has been shown to be significantly upregulated by Runx2 (Zheng et al., 2017). The miRNAs miR-335-5p and miR-17-92 decrease Dickkopf-1 (Dkk-1) levels (Zhang et al., 2017; Zheng et al., 2017). miR-142-3p inhibits APC, leading to increased β-catenin nuclear translocation and signaling (Hu et al., 2013). During osteoblast differentiation miR-218 is induced, which leads to activation of Wnt/β-catenin signaling by downregulating the expression of SOST, DKK-2, and sFRP2, but in a feed-forward loop the activation of the pathway subsequently further increases expression of miR-218 (Hassan et al., 2012). Similarly, miR-29 expression has also been shown to be induced by activation of Wnt/β-catenin signaling, and miR-29a and -29c repress the expression of several pathway inhibitors in a positive feedback loop (Kapinas et al. 2009, 2010).

In addition to its role in osteoblast differentiation, Wnt/β-catenin signaling also plays important roles in the proliferation, apoptosis, metabolism, and mineralization of the skeleton driven by osteoblasts. The role of Wnt signaling in regulating these functions is complex and includes both direct effects on the osteoblast itself and indirect effects through the expression of other osteoblast-derived factors that affect other bone cells and involves both canonical and noncanonical Wnt signaling pathways. For example, Wnt3a has been shown to induce matrix extracellular phosphoglycoprotein (Mepe) by stimulating expression of the *Mepe* gene. Wnt3a has also been shown to stimulate bone morphogenetic protein 2 (BMP-2) gene expression (Cho et al., 2012; Zhang et al., 2013), which also regulates *Mepe* gene expression (Cho et al., 2012). Studies have demonstrated differentiation-stage-specific effects of Wnt signaling in osteoblasts (Eijken et al., 2008; Bao et al., 2017).

Kato et al. (Kato et al., 2002) demonstrated in the Lrp5$^{-/-}$ mouse that the low-bone-mass phenotype was due to decreased osteoblast proliferation and reduced matrix deposition. Gong et al. (Gong et al., 2001) demonstrated that Wnt3a, but not Wnt5a, increased alkaline phosphatase activity in C3H10T1/2 and ST2 cell lines. Based on their data, they suggested that LRP5 and the Wnt/β-catenin signaling pathway play important roles in osteoblast differentiation and proliferation. Boyden et al. (Boyden et al., 2002) observed elevated serum osteocalcin levels, a marker of bone formation, but normal collagen type I urinary N-telopeptide levels, a marker of bone resorption. This suggested that the *LRP5*^{G171V} mutation leads to exuberant osteoblast activity resulting in increased bone density. Babij et al. (Babij et al., 2003) developed a transgenic mouse (over)expressing the human *LRP5*^{G171V} cDNA and also observed normal osteoclast activity with increased mineralizing surface and enhanced alkaline phosphatase staining in osteoblasts. However, Glass et al. (Glass et al., 2005), using mouse models with targeted deletion or overexpression of β-catenin in osteoblasts, identified another important role in the regulation of osteoclast activity through the production of osteoprotegerin (OPG). Why mutations in the Wnt receptor lead to increased osteoblast activity and alterations in β-catenin levels resulting in altered osteoclast activity is not entirely clear.

Another area of osteoblast function that has garnered attention is the role of Wnt signaling in the regulation of cellular energy metabolism in osteoblasts (Frey et al., 2017; Yao et al., 2017). The Wnt effects appear to be mediated by β-catenin in the case of fatty acid catabolism (Frey et al., 2017), mTORC2 in the case of aerobic glycolysis (Esen et al., 2013), and mTORC1 involving glutamine metabolism (Karner et al., 2015). Interestingly, both of the mTORC1/2-mediated effects appear to involve LRP5/Lrp6/Frizzled binding of a Wnt ligand (Karner and Long, 2017).

Studies have clearly shown an important role for the Wnt/β-catenin signaling pathway in the prevention of both osteoblast and osteocyte apoptosis. Kato et al. (Kato et al., 2002) demonstrated in the *Lrp5*$^{-/-}$ mouse that the hyaloid vessels failed to undergo apoptosis, resulting in the persistence of this vasculature network into the adult, which could explain the blindness associated with human OPPG patients carrying LRP5-null mutations. Babij et al. (Babij et al., 2003), using the HBM transgenic mouse, showed decreased osteoblast and osteocyte apoptosis attributable to the LRP5^{G171V} mutation. Secreted Frizzled-related protein-1 (sFRP-1) is a Wnt/β-catenin signaling pathway inhibitor (Bodine et al., 2004), and blocking sFRP with a small-molecule inhibitor caused a significant reduction in osteoblast apoptosis in cell culture (Bodine et al., 2009).

Osteoclast function

Osteoclasts are derived from hematopoietic (myeloid lineage) progenitors. Macrophage colony-stimulating factor (M-CSF) (Yoshida et al., 1990) and receptor activator of NF-κB ligand (RANKL) (Kong et al., 1999) are critical for osteoclast replication and differentiation, respectively. The early studies of osteoblast-targeted deletion or constitutive activation of β-catenin in mice surprisingly demonstrated a critical role for the pathway in the regulation of osteoclastogenesis, which was attributed to altered expressed of the RANKL ligand decoy receptor OPG (Glass et al., 2005; Holmen et al., 2005). Studies in MC3T3-E1 cells also demonstrated that activation of Wnt/β-catenin signaling downregulates the expression of RANKL (Spencer et al., 2006). Thus, by regulating the OPG/RANKL ratio, the Wnt/β-catenin signaling pathway in osteoblasts can control osteoclastogenesis. It is now recognized that the osteocyte is an important source of RANKL, and its role in osteoclastogenesis will be discussed further in a subsequent section.

Wnt signaling within osteoclasts is also an important role. The Wnt pathway inhibitor sFRP-1 has been shown to bind to RANKL and inhibit osteoclast formation (Hausler et al., 2001). Wei et al. (Wei et al., 2011) demonstrated that β-catenin-mediated Wnt signaling was induced during the M-CSF quiescence-to-replication/proliferation phase, but downregulated during the RANKL proliferation-to-differentiation phase of osteoclastogenesis. Mechanistically, Wnt/β-catenin signaling promoted proliferation of osteoclast precursors by increasing expression of GATA2 and Evi1 and blocked differentiation by inhibiting c-Jun phosphorylation. Extending this work further, Weivoda et al (Weivoda et al., 2016) have demonstrated that Lrp5/6/Frizzled binding of Wnt3a affects osteoclastogenesis through β-catenin and the noncanonical cAMP/PKA pathway. The latter pathway functions to block early osteoclast progenitors from differentiating by phosphorylation of nuclear factor of activated T cells 1.

Human genome-wide association studies have revealed that Wnt16 contributes to variation in bone density (Medina-Gomez et al., 2012; Zheng et al., 2012). Wnt16 through Wnt/β-catenin signaling has been shown to be involved in intramembranous ossification and to suppress osteoblast differentiation and mineralization (Jiang et al., 2014). Osteoblast-derived Wnt16 has been shown to stimulate OPG production by osteoblasts, which indirectly affects osteoclastogenesis, and to directly inhibit RANKL-mediated action in preosteoclasts through a c-Jun-mediated mechanism, rather than by canonical Wnt/β-catenin signaling (Weivoda et al., 2016). Additional Wnt ligands that have been implicated in osteoclast differentiation include Wnt4 and Wnt5a (Kobayashi et al. 2015a, 2015b).

Osteocyte function

In recent years, the role of the osteocyte in orchestrating the activity of osteoblasts and osteoclasts has been clearly established. Several excellent reviews have been published that detail the myriad of mechanisms through which osteocytes exert their regulation of bone formation and resorption (Bonewald and Johnson, 2008; Baron and Kneissel, 2013; Burgers and Williams, 2013; O'Brien et al., 2013; Schaffler et al., 2014; Johnson and Recker, 2017) (see also Chapter 6).

Based upon the skeletal phenotype of the HBM kindred, Johnson et al. (Johnson et al., 2002) first proposed that one of the main roles of the Wnt/β-catenin signaling pathway in bone was as a key component of the response to mechanical loading. In support of this hypothesis it was subsequently demonstrated that the $LPR5^{G171V}$ mutation resulted in increased bone formation (Bex et al., 2003; Cullen et al., 2004) and that loss of Lrp5 resulted in a lack of bone formation in response to load (Sawakami et al., 2006). Activation of Wnt/β-catenin signaling occurs following in vivo loading (Kamel et al., 2006; Lau et al., 2006; Robinson et al., 2006; Sawakami et al., 2006; Javaheri et al., 2014; Lara-Castillo et al., 2015), and targeted deletion of osteocyte β-catenin results in severe bone loss and compromised load responsiveness (Kramer et al., 2010; Javaheri et al., 2014; Kang et al., 2016). While the exact mechanism by which osteocytes (bone cells) perceive mechanical load is unresolved, a number of downstream events are now known to occur that are central to the biochemical response by the cell. Bonewald and Johnson (Bonewald and Johnson, 2008) suggested a model describing how the Wnt/β-catenin signaling pathway activates in response to mechanical load. In support of this model, Kamel et al. (Kamel et al., 2010) demonstrated that prostaglandin E_2 (PGE_2) was able to stimulate β-catenin nuclear translocation through a phosphatidylinositol 3-kinase (PI3K)/Akt-mediated mechanism. Lara-Castillo et al. (Lara-Castillo et al., 2015) subsequently demonstrated in vivo that loading rapidly activated β-catenin signaling in osteocytes and, in agreement with prior work by Sawakami et al. (Sawakami et al., 2006), found a concordant reduction in sclerostin and Dkk-1 levels in osteocytes. Pretreatment with Carpofen, a cyclooxygenase 2 inhibitor, blocked activation of Wnt/β-catenin signaling (Lara-Castillo et al., 2015), in agreement with their in vitro studies (Kamel et al., 2010).

Several other functions of the osteocyte involve aspects of Wnt/β-catenin pathway regulation, either directly or through the expression of its downstream targets. It is now recognized that the osteocyte is a key regulator of osteoblast and osteoclast function via its production of sclerostin (Poole et al., 2005; van Bezooijen et al., 2005) and RANKL (Nakashima et al., 2011; Xiong et al., 2011) and that sclerostin plays a role in osteocyte RANKL production (Wijenayaka et al., 2011). Glucocorticoids are known to induce osteocyte autophagy or apoptosis depending upon the dose (Jia et al., 2011). Protection against glucocorticoid-induced apoptosis is mitigated, in part, by the induction of PGE_2 upon mechanical loading and subsequent activation of Wnt/β-catenin signaling (Kitase et al., 2010). Parathyroid hormone (PTH) has been implicated in the regulation of sclerostin expression (Bellido et al., 2005; Keller and Kneissel, 2005) through regulation of the *Sost* gene enhancer involving MEF2 family transcription factors (Leupin et al., 2007). However, the bone anabolic effects of PTH cannot be fully explained by its regulation of sclerostin and/or interaction with Wnt/β-catenin signaling (Powell et al., 2011; Kedlaya et al., 2016; Delgado-Calle et al., 2017).

The fact that mutations in another member of the LDLR-related family, LRP4, were linked to phenotypes similar to those seen in human patients carrying loss-of-function mutations in the sclerostin gene was the catalyst for studies that eventually showed that LRP4 serves as a receptor for sclerostin, facilitating its presentation to Lrp5 or Lrp6 at the membrane and allowing it to inhibit Wnt/β-catenin signaling (Leupin et al., 2011). The importance of this interaction and its inhibition of both Lrp5 and Lrp6 signaling have been demonstrated by genetically engineered mouse models and inhibitory antibody studies (Chang et al., 2014).

Interactions between Wnt/β-catenin signaling and other pathways important in bone mass regulation

The Wnt/β-catenin signaling pathway is not the only pathway that is known to be important in the regulation of bone cell activities, and considerable effort is now being focused on understanding how all of these various pathways interact to regulate bone mass. As already discussed, GSK-3β is the critical intracellular enzyme controlling β-catenin levels. Therefore any pathway that can regulate GSK-3β activity has the potential to interact with the Wnt/β-catenin signaling pathway independent of the Lrp5/6/Frizzled coreceptors and any modulators that function at that level.

One set of well-known pathways that have the potential to cross talk with the Wnt/β-catenin signaling pathway comprises those pathways that signal through Akt/PKB. Akt/PKB is activated by PI3K, whose activity is opposed by the phosphatase and tensin homolog deleted on chromosome 10 (PTEN) tumor suppressor (Datta et al., 1999; Kandel and Hay, 1999). Akt/PKB has been shown to be stimulated by Wnt and, in association with Dsh, to phosphorylate GSK-3β (Fukumoto et al., 2001). This raises the potential for several growth factors to potentially cross talk with the Wnt/β-catenin signaling pathway. It has been shown that the growth of colon cancer cells induced by PGE_2 is partially the result of activation of the Wnt/β-catenin pathway through a dual mechanism involving Akt-mediated phosphorylation of GSK-3β and the $G\alpha_s$ subunit trimeric G proteins associated with the PGE_2 EP2 receptor binding to Axin and thereby promoting dissociation of the degradation complex (Castellone et al., 2005). PGE_2 release/production in bone cells is a well-known early response to mechanical loading. Fluid-flow shear stress has been shown to increase the phosphorylation of both Akt and GSK-3β in osteoblasts (Norvell et al., 2004). Fluid-flow shear stress results in an initial activation of the β-catenin signaling that is Lrp5 independent and probably mediated through cross talk with PGE_2 signaling (Kamel et al. 2006, 2010). As of this writing, we are testing a model in which this initial activation results in a feedback amplification loop that functions at the level of Lrp5 and leads to new bone formation. When Lrp5 is not present, the amplification loop does not occur and no new bone formation can occur, as was observed in the loading studies on the $Lrp5^{-/-}$ mouse (Sawakami et al., 2006). It has also been shown that integrin-linked kinase can regulate Akt/PKB and GSK-3β (Delcommenne et al., 1998), which potentially connects the important role of integrins in the response of bone cells to mechanical loading (Pavalko et al., 1998; Wozniak et al., 2000) with the Wnt/β-catenin signaling pathway through a cross talk mechanism.

In addition, signaling pathways that regulate the expression of activators or inhibitors (discussed previously) of the Wnt/β-catenin pathway thereby interact at this level. For example, results of studies with PTH suggest that PTH acts in part through a complementary pathway and not entirely through the Lrp5 Wnt/β-catenin signaling pathway. Continuous PTH treatment of rats in vivo or UMR 106 cells in culture results in a downregulation of *Lrp5* and *Dkk1* and an upregulation of *Lrp6* and *Fzd1* (Kulkarni et al., 2005) and, based on these studies, it was suggested that the effects of PTH are in part mediated by a cAMP/PKA pathway. In studies with the sFRP-1-knockout mouse, it was concluded that PTH and Wnt signaling may share some common components, but PTH action appears to extend beyond the Wnt pathway (Bodine et al., 2004). However, studies with the $Lrp5^{-/-}$ mouse have shown that Lrp5 is not required for the anabolic effects of PTH on bone (Iwaniec et al., 2004; Sawakami et al., 2006). Evidence has been presented that PTH infusion in mice decreases *Sost* mRNA expression and sclerostin levels in osteocytes (Bellido et al., 2005). This finding implies a negative feedback control mechanism in which sclerostin production by osteocytes opposes the actions of Wnts and/or BMPs on osteoblast precursors, and PTH, by decreasing sclerostin levels, indirectly therefore stimulates osteoblast differentiation and bone formation. Thus the picture of PTH that emerges is complex.

Likewise, several groups have examined potential cross talk between the BMP signaling pathway and Wnt/β-catenin signaling (Rawadi et al., 2003; He et al., 2004; van den Brink, 2004; Nakashima et al., 2005; Tian et al., 2005; Gaur et al., 2006; Liu et al., 2006). One model that has been proposed (Liu et al., 2006) for cross talk between these two pathways in bone marrow stromal cells involves a Dsh–Smad1 interaction in the unstimulated state that becomes disrupted when Wnt is present. When both Wnt and BMP-2 are present the phosphorylation of Smad1 stabilizes the interaction and thereby inhibits Wnt/β-catenin signaling (Liu et al., 2006). In intestinal crypt stem cell self-renewal, BMP inhibitory effects have been proposed to be mediated through the suppression of Wnt/β-catenin signaling through a mechanism involving PTEN, PI3K, and Akt (He et al., 2004). In bone, it has been proposed that BMP-2 increases the expression of Wnt proteins that then activate the Wnt/β-catenin signaling pathway in an autocrine/paracrine feedback loop (Rawadi et al., 2003). It has also been proposed that Wnts induce the expression of BMPs and this induction is necessary for expression of alkaline phosphatase (Winkler et al., 2005). This model also implicated sclerostin as imposing a level of control preferentially at the level of BMP signaling versus Wnt signaling. However, there is now clear evidence for sclerostin being a negative regulator of the Wnt/β-catenin signaling through binding to Lrp5 (Ellies et al., 2006; Semenov and He, 2006). Thus, while it is clear that these two pathways are somehow intertwined, the exact nature of their interaction(s) is complex (Marcellini et al., 2012).

As mentioned previously, there is now compelling evidence for the involvement of Lrp5 and the Wnt/β-catenin signaling pathway in the formation of new bone induced by mechanical loading of bone. Evidence also indicates that the $LRP5^{G171V}$ mutation may also protect against bone loss related to disuse, but that bone loss due to estrogen withdrawal is not affected by this mutation (Bex et al., 2003). However, it has been shown that ERα and β-catenin can/do form a functional interaction and potentially regulate gene expression (Kousmenko et al., 2004). It has also been suggested that in bone there is a convergence of the ER, kinases, BMP, and Wnt signaling pathways that regulates the differentiation of osteoblasts (Kousteni et al., 2007).

What other pathways might also interact with the Wnt/β-catenin signaling pathway? An emerging concept in the Wnt field is that there are numerous β-catenin-independent pathways that are altered upon inhibition of GSK-3. The term "Wnt-dependent stabilization of proteins," or Wnt-STOP, has been proposed for this aspect of Wnt signaling (Acebron et al., 2014; Huang et al., 2015; Koch et al., 2015; Acebron and Niehrs, 2016). One such example of this type of regulation is the GSK-3-dependent (but β-catenin-independent) regulation of the mammalian target of rapamycin (mTOR) pathway (Inoki et al., 2006; Karner and Long, 2017). This example may be particularly relevant for understanding the role of Wnt signaling in bone biology (Esen et al., 2013; Chen et al., 2014; Karner et al., 2015; Sun et al., 2016). It is likely that more Wnt-dependent pathways will be identified and that only by understanding the complex interplay of these various pathways will the control of the differentiation, proliferation, and function of bone cells be fully appreciated.

The Wnt signaling pathway as a target for anabolic therapy in bone

The LRP5^{G171V} mutation results in an HBM phenotype and a skeleton that is resistant to fracture (Johnson et al., 1997). The affected members of this kindred seem to have no other health consequences related to the presence of this mutation, although in a second kindred with the LRP5^{G171V} mutation affected members all had torus lesions in the oral cavity to varying degrees (Boyden et al., 2002). It has also been reported that some mutations in *LRP5* (Gong et al., 2001; Whyte et al., 2004) may have associated pathology. The finding that Lrp5 and Wnt/β-catenin signaling are critical components of bone responsiveness to mechanical loading (Bonewald and Johnson, 2008; Javaheri et al., 2014; Lara-Castillo et al., 2015) suggests a potentially novel pharmaceutical approach. Perhaps one strategy would be to use a pharmaceutical agent that can alter regulation of the pathway (mimicking the effect of the LRP^{G171V} mutation) in combination with an exercise regimen to produce an optimal bone anabolic effect. It might even be possible to use such a combination therapy approach with a drug dose that by itself has little consequence on bone mass and thus perhaps avoid undesirable side effects.

Given the major role of Wnt/β-catenin signaling in skeletal development and adult bone mass regulation, it is not surprising that soon after the description of the LRP5 mutations that regulate bone mass, considerable attention was focused on using key components of the pathway as targets for new drug development (Janssens et al., 2006; Baron and Rawadi, 2007; Baron and Kneissel, 2013). Caution was raised in these early discussions that misregulation of the pathway could lead to undesirable side effects such as cancer (Polakis, 2000; Moon et al., 2002; Prunier et al., 2004; Nusse, 2005; Johnson and Rajamannan, 2006). The discovery that sclerostin, the product of the *SOST* gene and a key negative regulator of the Wnt/β-catenin signaling pathway, was expressed mainly by osteocytes provided a potential solution to this paradox (Bellido et al., 2005; Li et al., 2005; Poole et al., 2005; van Bezooijen et al., 2005).

Two major pharmaceutical clinical trials using humanized monoclonal antibodies to sclerostin (romosozumab and blosozumab) have been reported in the literature (McClung, 2017). Clinical trials with romosozumab (Cosman et al., 2016; Ishibashi et al., 2017; Langdahl et al., 2017; Saag et al., 2017) consistently showed significant gains in areal bone mineral density at the lumbar spine, total hip, and femoral neck and reduced fracture risk. Studies in rats suggested that there was no significant carcinogenic risk in humans (Chouinard et al., 2016), which had been a concern identified as a result of rat studies on the use PTH(1−34) (Vahle et al., 2002). However, in a 2017 study (Saag et al., 2017) there was a slight, but significant, increase in adverse cardiovascular events that was not observed in previous studies. The basis for this cardiovascular issue is not clear. However, it is known that Wnt/β-catenin signaling plays a role in vascular calcification (Johnson and Rajamannan, 2006; Towler et al., 2006; Bostrom et al., 2011), and the question that needs to be more fully explored is whether these cardiovascular adverse events occurred in individuals with underlying vascular disease that was not excluded in the enrollment of subjects for these studies. Interestingly, the other reports did not report any observable differences in adverse events between treatment groups. Phase II clinical trials with blosozumab have also yielded similar positive gains in bone density (McColm et al., 2014; Recker et al., 2015; Recknor et al., 2015), with no significant differences in adverse events being detected in these initial studies between groups. Blosozumab has been removed from consideration for FDA approval as of this writing, and the impact of the adverse cardiovascular risk observed in the romosozumab phase III trials on FDA approval is unclear.

Are there other possibilities for exploiting this pathway pharmaceutically to treat diseases of low bone mass (Baron and Kneissel, 2013)? The majority of pharmaceutical agents on the market target either receptors or specific enzymes. In the Wnt/β-catenin pathway the obvious targets in this regard are the coreceptors Lrp5 and Frizzled and GSK-3β, which is the key enzyme controlling intracellular β-catenin levels.

Regulation of GSK-3β activity has already been reported to produce a bone anabolic effect. LiCl treatment, which inhibits GSK-3β, restores bone mass in the Lrp5$^{-/-}$ mouse (Clement-Lacroix et al., 2005), and in humans the use of lithium appears to have an associated decreased fracture risk and a trend for decreased risk of osteoporotic fractures of the spine and wrist (Colles fracture) (Vestergaard et al., 2005). GSK-3β inhibitors have also been shown to produce a bone anabolic effect in vivo and to increase expression of bone formation markers and induce osteoblast cell differentiation in vitro (Kulkarni et al., 2006; Robinson et al., 2006). In one of these studies, changes in gene expression profile that were obtained from treatment with these agents were virtually identical to changes observed in HBMtg mouse bones, in bone cells after mechanical loading, and in primary bone cells from affected members of HBM (Robinson et al., 2006). What these studies suggest is that, as we learn more about how the Wnt/β-catenin signaling pathway functions in bone, it will be possible to design better and more specific pharmaceutical agents and approaches.

Given the cross talk between other signaling pathways and the Wnt/β-catenin signaling pathway it may be that as we understand more about the nature of these complex interactions and regulation of the pathway, these studies will reveal targets/approaches for treatment (Johnson and Recker, 2017). PTH has been proposed to decrease levels of sclerostin produced by osteocytes and thus release the inhibition on osteoblast differentiation (Bellido et al., 2005). Could it be possible to regulate *Sost* gene expression through other means and thereby affect a bone anabolic effect? BMPs appear to induce the expression of Wnts (and perhaps vice versa), and so understanding what regulates Wnt gene expression in bone may provide a key to new pharmaceutical approaches. The result of β-catenin nuclear translocation is a change in expression of a number of genes, many of which function in a feedback loop as inhibitors or activators of the pathway. As we understand more about the nature of these genes/proteins and their regulation, it may be possible to design drugs that can alter the activity of these proteins and thereby tip the balance between bone formation and bone resorption in favor of formation and either increase bone mass or prevent bone loss. DKK-1 is another inhibitor of the pathway, and early animal studies have shown promise for this approach, which is similar to the anti-sclerostin antibodies therapies discussed earlier (Glantschnig et al., 2010; Li et al., 2011; Florio et al., 2016). DKK-1 antibodies have also gained considerable interest as a therapy to treat and manage myeloma bone disease (Terpos et al., 2017).

Interestingly, the importance of the Wnt pathway in bone homeostasis has also been emphasized by work in human clinical trials for cancer treatment. Several therapeutic strategies for blocking Wnt signaling to treat cancer are being tested in phase I trials (Rey and Ellies, 2010; Krishnamurthy and Kurzrock, 2018). In some instances, these trials had to be at least temporarily halted because patients displayed evidence of fracture or significant bone loss (Jimeno et al., 2017). This is consistent with several mouse models in which mice that have been genetically engineered to lack the ability to secrete Wnt ligands from osteoblasts have extremely low bone mass (Zhong et al., 2012; Wan et al., 2013; Tan et al., 2014). Strategies designed to mitigate these effects are undoubtedly being evaluated in animal models, and a greater knowledge of the specific Wnt receptors that are necessary for adult bone homeostasis may facilitate more targeted treatments.

While many of these ideas about creating new therapies targeting this pathway in bone may be considered fanciful speculation at this point, if we have learned anything since the discovery in the early years of the 21st century of the mutations in *LRP5* that give rise to low and high bone mass, it may simply be that the discoveries of the next five years may be even more enlightening in terms of our understanding of the role of Wnt/β-catenin signaling in bone.

References

Aberle, H., Bauer, A., Stappert, J., Kispert, A., Kemler, R., 1997. β-Catenin is a target for the ubiquitin-proteasome pathway. EMBO J. 16, 3797–3804.

Abou Ziki, M.D., Mani, A., 2017. Wnt signaling, a novel pathway regulating blood pressure? State of the art review. Atherosclerosis 262, 171–178.

Acebron, S.P., Niehrs, C., 2016. Beta-catenin-independent roles of wnt/LRP6 signaling. Trends Cell Biol. 26 (12), 956–967.

Acebron, S.P., Karaulanov, E., Berger, B.S., Huang, Y.L., Niehrs, C., 2014. Mitotic wnt signaling promotes protein stabilization and regulates cell size. Mol. Cell. 54 (4), 663–674.

Adler, P.N., 2002. Planar signaling and morphogenesis in *Drosophila*. Dev. Cell. 2, 525–535.

Ahmadzadeh, A., Norozi, F., Shahrabi, S., Shahjahani, M., Saki, N., 2016. Wnt/beta-catenin signaling in bone marrow niche. Cell Tissue Res. 363 (2), 321–335.

Akiyama, H., Lyons, J.P., Mori-Akiyama, Y., Yang, X., Zhang, R., Zhang, Z., Deng, J.M., Taketo, M.M., Nakamura, T., Behringer, R.R., McCrea, P.D., de Crombrugghe, B., 2004. Interactions between Sox9 and beta-catenin control chondrocyte differentiation. Genes Dev. 18 (9), 1072–1087.

Albers, J., Schulze, J., Beil, F.T., Gebauer, M., Baranowsky, A., Keller, J., Marshall, R.P., Wintges, K., Friedrich, F.W., Priemel, M., Schilling, A.F., Rueger, J.M., Cornils, K., Fehse, B., Streichert, T., Sauter, G., Jakob, F., Insogna, K.L., Pober, B., Knobeloch, K.P., Francke, U., Amling, M., Schinke, T., 2011. Control of bone formation by the serpentine receptor Frizzled-9. J. Cell Biol. 192 (6), 1057–1072.

Albers, J., Keller, J., Baranowsky, A., Beil, F.T., Catala-Lehnen, P., Schulze, J., Amling, M., Schinke, T., 2013. Canonical Wnt signaling inhibits osteoclastogenesis independent of osteoprotegerin. J. Cell Biol. 200 (4), 537–549.

Almeida, M., Han, L., Lowe, V., Warren, A., Kousteni, S., O'Brien, C.A., Manolagas, S., 2006. Reactive oxygen species antagonize the skeletal effects of wnt/β-catenin in vitro and aging mice by diverting β-catenin from TCF- to FOXO-mediated transcription. J. Bone Miner. Res. 21 (Suppl. 1), S26 (abst 1092).

Armstrong, V.J., Muzylak, M., Sunters, A., Zaman, G., Saxon, L.K., Price, J.S., Lanyon, L.E., 2007. Wnt/beta-catenin signaling is a component of osteoblastic bone cell early responses to load-bearing and requires estrogen receptor alpha. J. Biol. Chem. 282 (28), 20715–20727.

Asuni, A.A., Hooper, C., Reynolds, C.H., Lovestone, S., Anderton, B.H., Killick, R., 2006. GSK3alpha exhibits beta-catenin and tau directed kinase activities that are modulated by wnt. Eur. J. Neurosci. 24, 3387–3392.

Babij, P., Zhao, W., Small, C., Kharode, Y., Yaworsky, P., Bouxsein, M., Reddy, P., Bodine, P., Robinson, J., Bhat, B., Marzolf, J., Moran, R., Bex, F., 2003. High bone mass in mice expressing a mutant LRP5 gene. J. Bone Miner. Res. 18, 960–974.

Balemans, W., Van Hul, W., 2007. The genetics of low-density lipoprotein receptor-related protein 5 in bone: a story of extremes. Endocrinology 148 (6), 2622–2629.

Balemans, W., Ebeling, M., Patel, N., Van Hul, E., Olson, P., Dioszegi, M., Lacza, C., Wuyts, W., Van Den Ende, J., Willems, P., Paes-Alves, A.F., Hill, S., Bueno, M., Ramos, F.J., Tacconi, P., Dikkers, F.G., Stratakis, C., Lindpaintner, K., Vickery, B., Foernzler, D., Van Hul, W., 2001. Increased bone density in sclerosteosis is due to the deficiency of a novel secreted protein (SOST). Hum. Mol. Genet. 10 (5), 537–543.

Balemans, W., Patel, N., Ebeling, M., Van Hul, E., Wuyts, W., Lacza, C., Dioszegi, M., Dikkers, F.G., Hildering, P., Willems, P.J., Verheij, J.B., Lindpaintner, K., Vickery, B., Foernzler, D., Van Hul, W., 2002. Identification of a 52 kb deletion downstream of the SOST gene in patients with van Buchem disease. J. Med. Genet. 39 (2), 91–97.

Banziger, C., Soldini, D., Schutt, C., Zipperlen, P., Hausmann, G., Basler, K., 2006. Wntless, a conserved membrane protein dedicated to the secretion of wnt proteins from signaling cells. Cell 125, 509–522.

Bao, Q., Chen, S., Qin, H., Feng, J., Liu, H., Liu, D., Li, A., Shen, Y., Zhong, X., Li, J., Zong, Z., 2017. Constitutive beta-catenin activation in osteoblasts impairs terminal osteoblast differentiation and bone quality. Exp. Cell Res. 350 (1), 123–131.

Barker, N., Hurlstone, A., Musisi, H., Miles, A., Bienz, M., Clevers, H., 2001. The chromatin remodelling factor Brg-1 interacts with β-catenin to promote target gene activation. EMBO J. 20, 4935–4943.

Baron, R., Kneissel, M., 2013. WNT signaling in bone homeostasis and disease: from human mutations to treatments. Nat. Med. 19 (2), 179–192.

Baron, R., Rawadi, G., 2007. Targeting the wnt/β-catenin pathway to regulate bone formation in the adult skeleton. Endocrinology 148 (6), 2635–2643.

Bartscherer, K., Pelte, N., Ingelfinger, D., Boutros, M., 2006. Secretion of wnt ligands requires Evi, a conserved transmembrane protein. Cell 125, 523–533.

Behrens, J., Jerchow, B.A., Wurtele, M., Grimm, J., Asbrand, C., Wirtz, R., Kuhl, M., Wedlich, D., Birchmeier, W., 1998. Functional interaction of an axin homolog, conductin, with β-catenin, APC and GSK3β. Science 280, 596–599.

Bejsovec, A., 2005. Wnt pathway activation: new relations and locations. Cell 120, 11–14.

Bellido, T., Ali, A.A., Gubrij, I., Plotkin, L.I., Fu, Q., O'Brien, C.A., Manolagas, S.C., Jilka, R.L., 2005. Chronic elevation of parathyroid hormone in mice reduces expression of sclerostin by osteocytes: a novel mechanism for hormonal control of osteoblastogenesis. Endocrinology 146 (11), 4577–4583.

Bennett, C.N., Longo, K.A., Wright, W.S., Suva, L.J., Lane, T.F., Hankenson, K.D., MacDougald, O.A., 2005. Regulation of osteoblastogenesis and bone mass by Wnt10b. Proc. Natl. Acad. Sci. U. S. A. 102 (9), 3324–3329.

Bex, F., Green, p., Marsolf, J., Babij, P., Yaworsky, P., Kharode, Y., 2003. The human LRP5 G171V mutation in mice alters the skeletal response to limb unloading but not to ovariectomy. J. Bone Miner. Res. 18 (Suppl. 2), S60.

Bhanot, P., Brink, M., Harryman Samos, C., Hsieh, J.C., Wang, Y.S., Macke, J.P., Andrew, D., Nathans, J., Nusse, R., 1996. A new member of the frizzled family from *Drosophila* functions as a wingless receptor. Nature 382 (6588), 225–230.

Bodine, P.V.N., Kharode, Y.P., Seestaller-Wehr, L., Green, P., Milligan, C., Bex, F.J., 2004a. The bone anabolic effects of parathyroid hormone (PTH) are Blunted by deletion of the wnt antagonist secreted frizzled-related protein (sFRP)-1. J. Bone Miner. Res. 19 (Suppl. 1), S17 (abstract 1063).

Bodine, P.V.N., Zhao, W., Kharode, Y.P., Bex, F.J., Lambert, A.-J., Goad, M.B., Gaur, T., Stein, G.S., Lian, J.B., Komm, B.S., 2004b. The wnt antagonist secreted frizzled-related protein-1 is a negative regulator of trabecular bone formation in adult mice. Mol. Endocrinol. 18, 1222–1237.

Bodine, P.V., Stauffer, B., Ponce-de-Leon, H., Bhat, R.A., Mangine, A., Seestaller-Wehr, L.M., Moran, R.A., Billiard, J., Fukayama, S., Komm, B.S., Pitts, K., Krishnamurthy, G., Gopalsamy, A., Shi, M., Kern, J.C., Commons, T.J., Woodworth, R.P., Wilson, M.A., Welmaker, G.S., Trybulski, E.J., Moore, W.J., 2009. A small molecule inhibitor of the Wnt antagonist secreted frizzled-related protein-1 stimulates bone formation. Bone 44 (6), 1063–1068.

Bonewald, L.F., Johnson, M.L., 2008. Osteocytes, mechanosensing and Wnt signaling. Bone 42 (4), 606–615.

Bostrom, K.I., Rajamannan, N.M., Towler, D.A., 2011. The regulation of valvular and vascular sclerosis by osteogenic morphogens. Circ. Res. 109 (5), 564–577.

Boyden, L.M., Mao, J., Belsky, J., Mitzner, L., Farhi, A., Mitnick, M.A., Wu, D., Insogna, K., Lifton, R.P., 2002. High bone density due to a mutation in LDL-receptor-related protein 5. N. Engl. J. Med. 346, 1513–1521.

Brown, S.D., Twells, R.C., Hey, P.J., Cox, R.D., Levy, E.R., Soderman, A.R., Metzker, M.L., Caskey, C.T., Todd, J.A., Hess, J.F., 1998. Isolation and characterization of LRP6, a novel member of the low density lipoprotein receptor gene family. Biochem. Biophys. Res. Commun. 248, 879–888.

Brunkow, M.E., Gardner, J.C., Van-Ness, J., Paeper, B.W., Kovacevich, B.R., Proll, S., Skonier, J.E., Zhao, L., Sabo, P.J., Fu, Y., Alisch, R.S., Gillett, L., Colbert, T., Tacconi, P., Galas, D., Hamersma, H., Beighton, P., Mulligan, J., 2001. Bone dysplasia sclerosteosis results from loss of the SOST gene product, a novel cystine knot-containing protein. Am. J. Hum. Genet. 68, 577–589.

Burgers, T.A., Williams, B.O., 2013. Regulation of Wnt/beta-catenin signaling within and from osteocytes. Bone 54 (2), 244–249.

Butler, M.T., Wallingford, J.B., 2017. Planar cell polarity in development and disease. Nat. Rev. Mol. Cell Biol. 18 (6), 375–388.

Cadigan, K.M., Nusse, R., 1997. Wnt signaling: a common theme in animal development. Genes Dev. 11, 3286–3305.

Capelluto, D.G.S., Kutateladze, T.G., Habas, R., Finklestein, C.V., He, X., Overduin, M., 2002. The DIX domain targets dishevelled to actin stress fibres and vesicular membranes. Nature 419, 726–729.

Castellone, M.D., Teramoto, H., Williams, B.O., Druey, K.M., Gutkind, J.S., 2005. Prostaglandin E_2 promotes colon cancer cell growth through a G_s-axin-β-catenin signaling Axis. Science 310, 1504–1510.

Cavallo, R.A., Cox, R.T., Moline, M.M., Roose, J., Polevoy, G.A., Clevers, H., Peifer, M., Bejsovec, A., 1998. *Drosophila* Tcf and Groucho interact to repress wingless signalling activity. Nature 395, 604–608.

Chang, M.K., Kramer, I., Huber, T., Kinzel, B., Guth-Gundel, S., Leupin, O., Kneissel, M., 2014. Disruption of Lrp4 function by genetic deletion or pharmacological blockade increases bone mass and serum sclerostin levels. Proc. Natl. Acad. Sci. U. S. A. 111 (48), E5187–E5195.

Chen, W.J., Goldstein, J.L., Brown, M.S., 1990. NPXY, a sequence often found in cytoplasmic tails, is required for coated pit-mediated internalization of the low density lipoprotein receptor. J. Biol. Chem. 265, 3116–3123.

Chen, G., Fernandez, J., Mische, S., Courey, A.J., 1999. A functional interaction between the histone deacetylase Rpd3 and the corepressor Groucho in Drosophila development. Genes Dev. 13, 2218–2230.

Chen, A.E., Ginty, D.B., Fan, C.-M., 2005. Protein kinase a signalling via CREB controls myogenesis induced by wnt proteins. Nature 433, 317–322.

Chen, J., Tu, X., Esen, E., Joeng, K.S., Lin, C., Arbeit, J.M., Ruegg, M.A., Hall, M.N., Ma, L., Long, F., 2014. WNT7B promotes bone formation in part through mTORC1. PLoS Genet. 10 (1), e1004145.

Cho, Y.D., Kim, W.J., Yoon, W.J., Woo, K.M., Baek, J.H., Lee, G., Kim, G.S., Ryoo, H.M., 2012. Wnt3a stimulates Mepe, matrix extracellular phosphoglycoprotein, expression directly by the activation of the canonical Wnt signaling pathway and indirectly through the stimulation of autocrine Bmp-2 expression. J. Cell. Physiol. 227 (6), 2287–2296.

Chouinard, L., Felx, M., Mellal, N., Varela, A., Mann, P., Jolette, J., Samadfam, R., Smith, S.Y., Locher, K., Buntich, S., Ominsky, M.S., Pyrah, I., Boyce, R.W., 2016. Carcinogenicity risk assessment of romosozumab: a review of scientific weight-of-evidence and findings in a rat lifetime pharmacology study. Regul. Toxicol. Pharmacol. 81, 212–222.

Church, V., Nohno, T., Linker, C., Marcelle, C., Francis-West, P., 2002. Wnt regulation of chondrocyte differentiation. J. Cell Sci. 115 (Pt 24), 4809–4818.

Clement-Lacroix, P., Ai, M., Morvan, F., Roman-Roman, S., Vayssiere, B., Belleville, C., Estrera, K., Warman, M.L., Baron, R., Rawadi, G., 2005. Lrp5-independent activation of wnt signaling by lithium chloride increases bone formation and bone mass in mice. Proc. Natl. Acad. Sci. U.S.A. 102, 17406–17411.

Cook, D., Fry, M.J., Hughes, K., Sumathipala, R., Woodgett, J.R., Dale, T.C., 1996. Wingless inactivates glycogen synthase kinase-3 via an intracellular signalling pathway which involves a protein kinase C. EMBO J. 15, 4526–4536.

Cosman, F., Crittenden, D.B., Adachi, J.D., Binkley, N., Czerwinski, E., Ferrari, S., Hofbauer, L.C., Lau, E., Lewiecki, E.M., Miyauchi, A., Zerbini, C.A., Milmont, C.E., Chen, L., Maddox, J., Meisner, P.D., Libanati, C., Grauer, A., 2016a. Romosozumab treatment in postmenopausal women with osteoporosis. N. Engl. J. Med. 375 (16), 1532–1543.

Cosman, F., Gilchrist, N., McClung, M., Foldes, J., de Villiers, T., Santora, A., Leung, A., Samanta, S., Heyden, N., McGinnis 2nd, J.P., Rosenberg, E., Denker, A.E., 2016b. A phase 2 study of MK-5442, a calcium-sensing receptor antagonist, in postmenopausal women with osteoporosis after long-term use of oral bisphosphonates. Osteoporos. Int. 27 (1), 377–386.

Cullen, D.M., Akhter, M.P., Johnson, M.L., Morgan, S., Recker, R.R., 2004. Ulna loading response altered by the HBM mutation. J. Bone Miner. Res. 19 (Suppl. 1), S396 (abstract M217).

Datta, S.R., Brunet, A., Greenberg, M.E., 1999. Cellular survival: a play in three Akts. Genes Dev. 13, 2905–2927.

Delcommenne, M., Tan, C., Gray, V., Rue, L., Woodgett, J., Dedhar, S., 1998. Phosphoinositide-3-OH kinase-dependent regulation of glycogen synthase kinase 3 and protein kinase B/AKT by the integrin-linked kinase. Proc. Natl. Acad. Sci. U.S.A. 95, 11211–11216.

Delgado-Calle, J., Tu, X., Pacheco-Costa, R., McAndrews, K., Edwards, R., Pellegrini, G.G., Kuhlenschmidt, K., Olivos, N., Robling, A., Peacock, M., Plotkin, L.I., Bellido, T., 2017. Control of bone anabolism in response to mechanical loading and PTH by distinct mechanisms downstream of the PTH receptor. J. Bone Miner. Res. 32 (3), 522–535.

Dong, Y., Lathrop, W., Weaver, D., Qiu, Q., Cini, J., Bertolini, D., Chen, D., 1998. Molecular cloning and characterization of LR3, a novel LDL receptor family protein with mitogenic activity. Biochem. Biophys. Res. Commun. 251, 784–790.

Ducy, P., Zhang, R., Geoffroy, V., Ridall, A.L., Karsenty, G., 1997. Osf2/Cbfa1: a transcriptional activator of osteoblast differentiation. Cell 89 (5), 747–754.

Dunphy, J.T., Linder, M.E., 1998. Signalling functions of protein palmitoylation. Biochim. Biophys. Acta 1436, 245–261.

Eijken, M., Meijer, I.M., Westbroek, I., Koedam, M., Chiba, H., Uitterlinden, A.G., Pols, H.A., van Leeuwen, J.P., 2008. Wnt signaling acts and is regulated in a human osteoblast differentiation dependent manner. J. Cell. Biochem. 104 (2), 568–579.

Ellies, D.L., Viviano, B., McCarthy, J., Rey, J.-P., Itasaki, N., Saunders, S., Krumlauf, R., 2006. Bone density ligand, sclerostin, directly interacts with LRP5 but not $LRP5^{G171V}$ to modulate wnt activity. J. Bone Miner. Res. 21, 1738–1749.

Esen, E., Chen, J., Karner, C.M., Okunade, A.L., Patterson, B.W., Long, F., 2013. WNT-LRP5 signaling induces Warburg effect through mTORC2 activation during osteoblast differentiation. Cell Metabol. 17 (5), 745–755.

Essers, M.A.G., de Vries-Smits, L.M.M., Barker, N., Polderman, P.E., Burgering, B.M.T., Korswagen, H.C., 2005. Functional interactional between β-catenin and FOXO in oxidative stress signaling. Science 308, 1181–1184.

Fagotto, F., Jho, E., Zeng, L., Kurth, T., Joos, T., Kaufmann, C., Costantini, F., 1999. Domains of axin involved in protein-protein interactions, wnt pathway inhibiiton, and intracellular localization. J. Cell Biol. 145, 741–756.

Fang, X., Yu, S.X., Lu, Y., Bast, R.C., Woodgett, J.R., Mills, G.B., 2000. Phosphorylation and inactivation of glycogen synthase kinase 3 by protein kinase A. Proc. Natl. Acad. Sci. U.S.A. 97, 11960–11965.

Farr, G.H., Ferkey, D.M., Yost, C., Pierce, S.B., Weaver, C., Kimelman, D., 2000. Interaction among GSK-3, GBP, axin, and APC in *Xenopus* Axis specification. J. Cell Biol. 148, 691–701.

Figueroa, D.J., Hess, J.F., Ky, B., Brown, S.D., Sandig, V., Hermanowski-Vosatka, A., Twells, R.C., Todd, J.A., Austin, C.P., 2000. Expression of the type I diabetes-associated gene LRP5 in macrophages, vitamin A system cells, and the islets of langerhans suggests multiple potential roles in diabetes. J. Histochem. Cytochem. 48, 1357–1368.

Florio, M., Gunasekaran, K., Stolina, M., Li, X., Liu, L., Tipton, B., Salimi-Moosavi, H., Asuncion, F.J., Li, C., Sun, B., Tan, H.L., Zhang, L., Han, C.Y., Case, R., Duguay, A.N., Grisanti, M., Stevens, J., Pretorius, J.K., Pacheco, E., Jones, H., Chen, Q., Soriano, B.D., Wen, J., Heron, B., Jacobsen, F.W., Brisan, E., Richards, W.G., Ke, H.Z., Ominsky, M.S., 2016. A bispecific antibody targeting sclerostin and DKK-1 promotes bone mass accrual and fracture repair. Nat. Commun. 7, 11505.

Forde, J.E., Dale, T.C., 2007. Glycogen synthase kinase 3: a key regulator of cellular fate. Cell. Mol. Life Sci. 64 (15), 1930–1944.

Frey, J.L., Kim, S.P., Li, Z., Wolfgang, M.J., Riddle, R.C., 2018. beta-catenin directs long-chain fatty acid catabolism in the osteoblasts of male mice. Endocrinology 159 (1), 272–284.

Fukumoto, S., Hsieh, C.-M., Maemura, K., Layne, M.D., Yet, S.-F., Lee, K.-H., Matsui, T., Rosenzweig, A., Taylor, W.G., Rubin, J.S., Perrella, M.A., Lee, M.-E., 2001. Akt participation in the wnt signaling pathway through dishevelled. J. Biol. Chem. 276, 17479–17483.

Gao, Z.-H., Seeling, J.M., Hill, V., Yochum, A., Virshup, D.M., 2002. Casein kinase I phosphorylates and destabilizes the β-catenin degradation complex. Proc. Natl. Acad. Sci. U.S.A. 99, 1182–1187.

Gaur, T., Lengner, C.J., Hovhannisyan, H., Bhat, R.A., Bodine, P.V.N., Komm, B.S., Javed, A., van Wijnen, A.J., Stein, J.L., Stein, G.S., Lian, J.B., 2005. Canonical wnt signaling promotes osteogensis by directly stimulating *Runx2* gene expression. J. Biol. Chem. 280, 33132–33140.

Gaur, T., Rich, L., Lengner, C.J., Hussain, S., Trevant, B., Ayers, D., Stein, J.L., Bodine, P.V.N., Komm, B.S., Stein, G.S., Lian, J.B., 2006. Secreted frizzled related protein 1 regulates wnt signaling for BMP2 induced chondrocyte differentiation. J. Cell. Physiol. 208, 87–96.

Glantschnig, H., Hampton, R.A., Lu, P., Zhao, J.Z., Vitelli, S., Huang, L., Haytko, P., Cusick, T., Ireland, C., Jarantow, S.W., Ernst, R., Wei, N., Nantermet, P., Scott, K.R., Fisher, J.E., Talamo, F., Orsatti, L., Reszka, A.A., Sandhu, P., Kimmel, D., Flores, O., Strohl, W., An, Z., Wang, F., 2010. Generation and selection of novel fully human monoclonal antibodies that neutralize Dickkopf-1 (DKK1) inhibitory function in vitro and increase bone mass in vivo. J. Biol. Chem. 285 (51), 40135–40147.

Glass, D.A., Bialek, P., Ahn, J.D., Starbuck, M., Patel, M.S., Clevers, H., Taketo, M.M., Long, F., McMahon, A.P., Lang, R.A., Karsenty, G., 2005. Canonical wnt signaling in differentiated osteoblasts controls osteoclast differentiation. Dev. Cell 8, 751–764.

Goggolidou, P., Wilson, P.D., 2016. Novel biomarkers in kidney disease: roles for cilia, Wnt signalling and ATMIN in polycystic kidney disease. Biochem. Soc. Trans. 44 (6), 1745–1751.

Gong, Y., Slee, R.B., Fukai, N., Rawadi, G., Roman-Roman, S., Reginato, A.M., Wang, H., Cundy, T., Glorieux, F.H., Lev, D., Zacharin, M., Oexle, K., Marcelino, J., Suwairi, W., Heeger, S., Sabatakos, G., Apte, S., Adkins, W.N., Allgrove, J., Arsian-Kirchner, M., Batch, J.A., Beighton, P., Black, G.C.M., Boles, R.G., Boon, L.M., Borrone, C., Brunner, H.G., Carle, G.F., Dallapiccola, B., De Paepa, A., Floege, B., Halfide, M.L., Hall, B., Hennekam, R.C., Hirose, T., Jans, A., Juppner, H., Kim, C.A., Keppler-Noreuil, K., Kohlschuetter, A., LaCombe, D., Lambert, M., Lemyre, E., Letteboer, T., Peltonen, L., Ramesar, R.S., Romanengo, M., Somer, H., Steichen-Gersdorf, E., Steinmann, B., Sullivan, B., Superta-Furga, A., Swoboda, W., van den Boogaard, M.-J., Van Hul, W., Vikkula, M., Votruba, M., Zabel, B., Garcia, T., Baron, R., Olsen, B.R., Warman, M.L., 2001. LDL receptor-related protein 5 (LRP5) affects bone accrual and eye development. Cell 107, 513–523.

Gonzalez-Sancho, J.M., Brennan, K.R., Castelo-Soccio, L.A., Brown, A.M., 2004. Wnt proteins induce dishevelled phosporylation via an LRP5/6 independent mechanism, irrespective of their ability to stabilize β-catenin. Mol. Cell Biol. 24, 4757–4768.

Haq, S., Choukroun, G., Kang, Z.B., Ranu, H., Matsui, T., Rosenzweig, A., Molkentin, J.D., Alessandrini, A., Woodgett, J., Hajjar, R., Michael, A., Force, T., 2000. Glycogen synthase kinase-3β is a negative rgulator of cardiomyocyte hypertrophy. J. Cell Biol. 151, 117–130.

Hardt, S.E., Sadoshima, J., 2002. Glycogen synthase kinase-3b: a novel regulator of cardiac hypertrophy and development. Circ. Res. 90, 1055–1063.

Hart, M.J., de los Santos, R., Albert, I.N., Rubinfeld, B., Polakis, P., 1998. Downregulation of β-catenin by HUman axin and its association with the APC tumor suppressor, β-catenin. Curr. Biol. 8, 573–581.

Hartmann, C., 2006. A Wnt canon orchestrating osteoblastogenesis. Trends Cell Biol. 16 (3), 151–158.

Hartmann, C., Tabin, C.J., 2000. Dual roles of Wnt signaling during chondrogenesis in the chicken limb. Development 127 (14), 3141–3159.

Hassan, M.Q., Maeda, Y., Taipaleenmaki, H., Zhang, W., Jafferji, M., Gordon, J.A., Li, Z., Croce, C.M., van Wijnen, A.J., Stein, J.L., Stein, G.S., Lian, J.B., 2012. miR-218 directs a Wnt signaling circuit to promote differentiation of osteoblasts and osteomimicry of metastatic cancer cells. J. Biol. Chem. 287 (50), 42084–42092.

Hausler, K.D., Horwood, N.J., Uren, A., Ellis, J., Lengel, C., Martin, T.J., Rubin, J.S., Gillespie, M.T., 2001. Secreted frizzled-related protein (sFRP-1) binds to RANKL to inhibit osteoclast formation. J. Bone Miner. Res. 16, S153.

Haxaire, C., Hay, E., Geoffroy, V., 2016. Runx2 controls bone resorption through the down-regulation of the wnt pathway in osteoblasts. Am. J. Pathol. 186 (6), 1598–1609.

He, X., Semenov, M., Tamai, K., Zeng, X., 2004a. LDL receptor-related proteins 5 and 6 in wnt/β-catenin signaling: arrows points the way. Development 131 (8), 1663–1677.

He, X.C., Zhang, J., Tong, W.-G., Tawfik, O., Ross, J., Scoville, D.H., Tian, Q., Zeng, X., He, X., Wiedemann, L.M., Mishinia, Y., Li, L., 2004b. BMP signaling inhibits intestinal stem cell self-renewal through suppression of wnt-β-catenin signaling. Nat. Genet. 36, 1117–1121.

Hecht, A., Vleminckx, K., Stemmler, M.P., van Roy, F., Kemler, R., 2000. The p300/CBP acetyltransferases function as transcriptional coactivators of β-catenin in vertebrates. EMBO J. 19, 1839–1850.

Heilmann, A., Schinke, T., Bindl, R., Wehner, T., Rapp, A., Haffner-Luntzer, M., Nemitz, C., Liedert, A., Amling, M., Ignatius, A., 2013. The Wnt serpentine receptor Frizzled-9 regulates new bone formation in fracture healing. PLoS One 8 (12), e84232.

Hermida, M.A., Dinesh Kumar, J., Leslie, N.R., 2017. GSK3 and its interactions with the PI3K/AKT/mTOR signalling network. Adv. Biol. Regul. 65, 5–15.

Hey, P.J., Twells, R.C., Phillips, M.S., Yusuke, N., Brown, S.D., Kawaguchi, Y., Cox, R., Guochun, X., Dugan, V., Hammond, H., Metzker, M.L., Todd, J.A., Hess, J.F., 1998. Cloning of a novel member of the low-density lipoprotein receptor family. Gene 216, 103–111.

Hill, T.P., Spater, D., Taketo, M.M., Birchmeier, W., Hartmann, C., 2005. Canonical wnt/β-catenin signaling prevents osteoblasts from differentiating into chondrocytes. Dev. Cell 8, 727–738.

Holmen, S.L., Salic, A., Zylstra, C.R., Kirschner, M.W., Williams, B.O., 2002. A novel set of wnt-frizzled fusion proteins identifies receptor components that activate β-catenin-dependent signaling. J. Biol. Chem. 277, 34727–34735.

Holmen, S.L., Giambernardi, T.A., Zylstra, C.R., Buckner-Berghuis, B.D., Resau, J.H., Hess, J.F., Glatt, V., Bouxsein, M.L., Ai, M., Warman, M.L., Williams, B.O., 2004. Decreased BMD and limb deformities in mice carrying mutations in both Lrp5 and Lrp6. J. Bone Miner. Res. 19, 2033–2040.

Holmen, S.L., Robertson, S.A., Zylstra, C.R., Williams, B.O., 2005a. Wnt-independent activation of b-catenin mediated by a dkk-fz5 fusion protein. Biochem. Biophys. Res. Commun. 328, 533–539.

Holmen, S.L., Zylstra, C.R., Mukherjee, A., Sigler, R.E., Faugere, M.C., Bouxsein, M.L., Deng, L., Clemens, T.L., Williams, B.O., 2005b. Essential role of beta-catenin in postnatal bone acquisition. J. Biol. Chem. 280 (22), 21162–21168.

Hsu, W., Zeng, L., Costantini, F., 1999. Identification of a domain of axin that binds to the serine/threonine protein phosphatase 2A and a self-binding domain. J. Biol. Chem. 274, 3439–3445.

Hu, W., Ye, Y., Zhang, W., Wang, J., Chen, A., Guo, F., 2013. miR1423p promotes osteoblast differentiation by modulating Wnt signaling. Mol. Med. Rep. 7 (2), 689–693.

Huang, Y.L., Anvarian, Z., Doderlein, G., Acebron, S.P., Niehrs, C., 2015. Maternal Wnt/STOP signaling promotes cell division during early Xenopus embryogenesis. Proc. Natl. Acad. Sci. U. S. A. 112 (18), 5732–5737.

Hwang, S.G., Yu, S.S., Lee, S.W., Chun, J.S., 2005. Wnt-3a regulates chondrocyte differentiation via c-Jun/AP-1 pathway. FEBS Lett. 579 (21), 4837–4842.

Ikeda, S., Kishida, S., Yamamoto, H., Murai, H., Koyama, S., Kikuchi, A., 1998. Axin, a negative regulator of the wnt signaling pathway, forms a complex with GSK-3β and β-catenin and promotes GSK-3β-dependent phosphorylation of β-catenin. EMBO J. 17 (5), 1371–1384.

Inoki, K., Ouyang, H., Zhu, T., Lindvall, C., Wang, Y., Zhang, X., Yang, Q., Bennett, C., Harada, Y., Stankunas, K., Wang, C.Y., He, X., MacDougald, O.A., You, M., Williams, B.O., Guan, K.L., 2006. TSC2 integrates Wnt and energy signals via a coordinated phosphorylation by AMPK and GSK3 to regulate cell growth. Cell 126 (5), 955–968.

Ishibashi, H., Crittenden, D.B., Miyauchi, A., Libanati, C., Maddox, J., Fan, M., Chen, L., Grauer, A., 2017. Romosozumab increases bone mineral density in postmenopausal Japanese women with osteoporosis: a phase 2 study. Bone 103, 209–215.

Ishitani, T., Ninomiya-Tsuji, J., Matsumoto, K., 2003. Regulation of lymphoid enhancer factor/T-cell factor by mitogen-activated protein kinase-related nemo-like kinase dependent phosphorylation in wnt/β-catenin signaling. Mol. Cell Biol. 23, 1379–1389.

Itoh, K., Krupnick, V.E., Sokol, S.Y., 1998. Axis determination in Xenopus involves biochemical interactions of axin, glycogen synthase kinase 3 and β-catenin. Curr. Biol. 8, 591–598.

Iwaniec, U.T., Liu, G., Arzaga, R.R., Donovan, L.M., Brommage, R., Wronski, T.J., 2004. Lrp5 is not essential for the stimulatory effect of PTH on bone formation in mice. J. Bone Miner. Res. 19 (Suppl. 1), S18 (abstract 1064).

Janda, C.Y., Waghray, D., Levin, A.M., Thomas, C., Garcia, K.C., 2012. Structural basis of wnt recognition by frizzled. Science 337 (6090), 59–64.

Janssens, N., Janicot, M., Perera, T., 2006. The wnt-dependent signaling pathways as targets in oncology drug discovery. Investig. New Drugs 24 (4), 263–280.

Javaheri, B., Stern, A.R., Lara, N., Dallas, M., Zhao, H., Liu, Y., Bonewald, L.F., Johnson, M.L., 2014. Deletion of a single beta-catenin allele in osteocytes abolishes the bone anabolic response to loading. J. Bone Miner. Res. 29 (3), 705–715.

Jho, E., Lomvardas, S., Costanti, F., 1999. A GSK3beta phosphorylation site in axin modulates interaction with beta-catenin and tcf-mediated gene expression. Biochem. Biophys. Res. Commun. 266, 28–35.

Jho, E.-h., Zhang, T., Domon, C., Joo, C.-K., Freund, J.-N., Costantini, F., 2002. Wnt/β-Catenin/Tcf signaling induces the transcription of Axin2, a negative regulator of the signaling pathway. Mol. Cell Biol. 22, 1172–1183.

Jia, J., Yao, W., Guan, M., Dai, W., Shahnazari, M., Kar, R., Bonewald, L., Jiang, J.X., Lane, N.E., 2011. Glucocorticoid dose determines osteocyte cell fate. FASEB J. 25 (10), 3366–3376.

Jiang, Z., Von den Hoff, J.W., Torensma, R., Meng, L., Bian, Z., 2014. Wnt16 is involved in intramembranous ossification and suppresses osteoblast differentiation through the Wnt/beta-catenin pathway. J. Cell. Physiol. 229 (3), 384–392.

Jimeno, A., Gordon, M., Chugh, R., Messersmith, W., Mendelson, D., Dupont, J., Stagg, R., Kapoun, A.M., Xu, L., Uttamsingh, S., Brachmann, R.K., Smith, D.C., 2017. A first-in-human phase I study of the anticancer stem cell agent ipafricept (OMP-54F28), a decoy receptor for wnt ligands, in patients with advanced solid tumors. Clin. Cancer Res. 23 (24), 7490–7497.

Joeng, K.S., Schumacher, C.A., Zylstra-Diegel, C.R., Long, F., Williams, B.O., 2011. Lrp5 and Lrp6 redundantly control skeletal development in the mouse embryo. Dev. Biol. 359 (2), 222–229.

Johnson, M.L., Rajamannan, N.M., 2006. Diseases of wnt signaling. Rev. Endocr. Metab. Disord. 7, 41–49.

Johnson, M.L., Recker, R.R., 2017. Exploiting the WNT signaling pathway for clinical purposes. Curr. Osteoporos. Rep. 15 (3), 153–161.

Johnson, M.L., Summerfield, D.T., 2005. Parameters of LRP5 from a structural and molecular perspective. Crit. Rev. Eukaryot. Gene Expr. 15, 229–242.

Johnson, M.L., Gong, G., Kimberling, W.J., Recker, S.M., Kimmel, D.K., Recker, R.R., 1997. Linkage of a gene causing high bone mass to human chromosome 11 (11q12-13). Am. J. Hum. Genet. 60, 1326–1332.

Johnson, M.L., Picconi, J.L., Recker, R.R., 2002. The gene for high bone mass. Endocrinololgist 12, 445–453.

Kadowaki, T., Wilder, E., Klingensmith, J., Zachary, K., Perrimon, N., 1996. The segment polarity gene porcupine encodes a putative multitransmembrane protein involved in wingless processing. Genes Dev. 10 (24), 3116–3128.

Kamel, M.A., Holladay, B.R., Johnson, M.L., 2006. Potential interaction of prostaglandin and wnt signaling pathways mediating bone cell responses to fluid flow. J. Bone Miner. Res. 21 (Suppl. 1), S92 (abs F166).

Kamel, M.A., Picconi, J.L., Lara-Castillo, N., Johnson, M.L., 2010. Activation of beta-catenin signaling in MLO-Y4 osteocytic cells versus 2T3 osteoblastic cells by fluid flow shear stress and PGE2: implications for the study of mechanosensation in bone. Bone 47 (5), 872–881.

Kandel, E.S., Hay, N., 1999. The regulation and activities of the multifuncitonal serine/threonin kinase Akt/PKB. Exp. Cell Res. 253, 210–229.

Kang, K.S., Hong, J.M., Robling, A.G., 2016. Postnatal beta-catenin deletion from Dmp1-expressing osteocytes/osteoblasts reduces structural adaptation to loading, but not periosteal load-induced bone formation. Bone 88, 138–145.

Kapinas, K., Kessler, C.B., Delany, A.M., 2009. miR-29 suppression of osteonectin in osteoblasts: regulation during differentiation and by canonical Wnt signaling. J. Cell. Biochem. 108 (1), 216–224.

Kapinas, K., Kessler, C., Ricks, T., Gronowicz, G., Delany, A.M., 2010. miR-29 modulates Wnt signaling in human osteoblasts through a positive feedback loop. J. Biol. Chem. 285 (33), 25221–25231.

Karasik, D., Rivadeneira, F., Johnson, M.L., 2016. The genetics of bone mass and susceptibility to bone diseases. Nat. Rev. Rheumatol. 12 (6), 323–334.

Karner, C.M., Long, F., 2017. Wnt signaling and cellular metabolism in osteoblasts, 74 (9), 1649–1657.

Karner, C.M., Esen, E., Okunade, A.L., Patterson, B.W., Long, F., 2015. Increased glutamine catabolism mediates bone anabolism in response to WNT signaling. J. Clin. Investig. 125 (2), 551–562.

Katanaev, V.L., Ponzielli, R., Semeriva, M., Tomlinson, A., 2005. Trimeric G protein-dependent frizzled signaling in *Drosophila*. Cell 120, 111–122.

Kato, M., Patel, M.S., Levasseur, R., Lobov, I., Chang, B.H.-J., Glass, D.A., Hartmann, C., Li, L., Hwang, T.H., Brayton, C.F., Lang, R.A., Karsenty, G., Chan, L., 2002. Cbfa 1-independent decrease in osteoblast proliferation, osteopenia, and persistent embryonic eye vascularization in mice deficient in Lrp5, a wnt coreceptor. J. Cell Biol. 157, 303–314.

Katoh, M., 2005. Wnt/PCP signaling pathway and human cancer. Oncol. Rep. 14, 1583–1588.

Kedlaya, R., Kang, K.S., Hong, J.M., Bettagere, V., Lim, K.E., Horan, D., Divieti-Pajevic, P., Robling, A.G., 2016. Adult-onset deletion of beta-catenin in (10kb)dmp1-expressing cells prevents intermittent PTH-induced bone gain. Endocrinology 157 (8), 3047–3057.

Keller, H., Kneissel, M., 2005. SOST is a target gene for PTH in bone. Bone 37 (2), 148–158.

Kim, D.H., Inagaki, Y., Suzuki, T., Ioka, R.X., Yoshioka, S.Z., Magoori, K., Kang, M.J., Cho, Y., Nakano, A.Z., Liu, Q., Fujino, T., Suzuki, H., Sasano, H., Yamamoto, T.T., 1998. A new low density lipoprotein receptor related protein, LRP5, is expressed in hepatocytes and adrenal cortex, and recognizes apolipoprotein E. Eur. J. Biochem. 124, 1072–1076.

Kitase, Y., Barragan, L., Qing, H., Kondoh, S., Jiang, J.X., Johnson, M.L., Bonewald, L.F., 2010. Mechanical induction of PGE2 in osteocytes blocks glucocorticoid-induced apoptosis through both the beta-catenin and PKA pathways. J. Bone Miner. Res. 25 (12), 2657–2668.

Kobayashi, Y., Thirukonda, G.J., Nakamura, Y., Koide, M., Yamashita, T., Uehara, S., Kato, H., Udagawa, N., Takahashi, N., 2015a. Wnt16 regulates osteoclast differentiation in conjunction with Wnt5a. Biochem. Biophys. Res. Commun. 463 (4), 1278–1283.

Kobayashi, Y., Uehara, S., Koide, M., Takahashi, N., 2015b. The regulation of osteoclast differentiation by Wnt signals. Bonekey Rep. 4, 713.

Kobayashi, Y., Uehara, S., Udagawa, N., Takahashi, N., 2016. Regulation of bone metabolism by Wnt signals. J. Biochem. 159 (4), 387–392.

Koch, S., Acebron, S.P., Herbst, J., Hatiboglu, G., Niehrs, C., 2015. Post-transcriptional wnt signaling governs epididymal sperm maturation. Cell 163 (5), 1225–1236.

Kokubu, C., Heinzmann, U., Kokubu, T., Sakai, N., Kubota, N., Kawai, M., Wahl, M.B., Galceran, J., Grosschedt, R., Ozono, K., Imai, K., 2004. Skeletal defects in *ringelshwanz* mutant mice reveal that Lrp6 is required for proper somitogenesis and osteogenesis. Development 131, 5469–5480.

Komori, T., Yagi, H., Nomura, S., Yamaguchi, A., Sasaki, K., Deguchi, K., Shimizu, Y., Bronson, R.T., Gao, Y.H., Inada, M., Sato, M., Okamoto, R., Kitamura, Y., Yoshiki, S., Kishimoto, T., 1997. Targeted disruption of Cbfa1 results in a complete lack of bone formation owing to maturational arrest of osteoblasts. Cell 89 (5), 755–764.

Kong, Y.Y., Yoshida, H., Sarosi, I., Tan, H.L., Timms, E., Capparelli, C., Morony, S., Oliveira-dos-Santos, A.J., Van, G., Itie, A., Khoo, W., Wakeham, A., Dunstan, C.R., Lacey, D.L., Mak, T.W., Boyle, W.J., Penninger, J.M., 1999. OPGL is a key regulator of osteoclastogenesis, lymphocyte development and lymph-node organogenesis. Nature 397 (6717), 315–323.

Korinek, V., Barker, N., Morin, P.J., van Wichen, D., de Weger, R., Kinzler, K.W., Vogelstein, B., Clevers, H., 1997. Constitutive transcriptional activation by a β-catenin-tcf complex in $APC^{-/-}$ colon carcinoma. Science 275, 1784–1787.

Kousmenko, A.P., Takeyama, K., Ito, S., Furutani, T., Sawatsubashi, S., Maki, A., Suzuki, E., Kawasaki, Y., Akiyama, T., Tabata, T., Kato, S., 2004. Wnt/β-catenin and estrogen signaling converge *in vivo*. J. Biol. Chem. 279, 40255–40258.

Kousteni, S., Almeida, M., Han, L., Bellido, T., Jilka, R.L., Manolagas, S., 2007. Induction of osteoblast differentiation by selective activation of kinase-mediated actions of the estrogen receptor. Mol. Cell Biol. 27, 1516–1530.

Kramer, I., Halleux, C., Keller, H., Pegurri, M., Gooi, J.H., Weber, P.B., Feng, J.Q., Bonewald, L.F., Kneissel, M., 2010. Osteocyte Wnt/beta-catenin signaling is required for normal bone homeostasis. Mol. Cell Biol. 30 (12), 3071–3085.

Krishnamurthy, N., Kurzrock, R., 2018. Targeting the Wnt/beta-catenin pathway in cancer: update on effectors and inhibitors. Cancer Treat Rev. 62, 50–60.

Kuhl, M., Sheldahl, L.C., Park, M., Miller, J.R., Moon, R.T., 2000. The wnt/Ca+2 pathway: a new vertebrate wnt signaling pathway takes shape. Trends Genet. 16, 279–283.

Kulkarni, N.H., Halladay, D.L., Miles, R.R., Gilbert, L.M., Frolik, C.A., Galvin, R.J.S., Martin, T.J., Gillespie, M.T., Onyia, J.E., 2005. Effects of parathyroid hormone on wnt signaling pathway in bone. J. Cell. Biochem. 95, 1178–1190.

Kulkarni, N.H., Onyia, J.E., Zeng, Q.Q., Tian, X., Liu, M., Halladay, D.L., Frolik, C.A., Engler, T., Wei, T., Kriauciunas, A., Martin, T.J., Sato, M., Bryant, H.U., Ma, Y.L., 2006. Orally Bioavailable GSK-3α/β dual inhibitor increases markers of cellular differentiation *in vitro* and bone mass *in vivo*. J. Bone Miner. Res. 21, 910–920.

Lammi, L., Arte, S., Somer, M., Jarvinen, H., Lahermo, P., Thesleff, I., Pirinen, S., Nieminen, P., 2004. Mutations in AXIN2 cause familil tooth agenesis and predispose to colorectal cancer. Am. J. Hum. Genet. 74, 1043–1050.

Langdahl, B.L., Libanati, C., Crittenden, D.B., Bolognese, M.A., Brown, J.P., Daizadeh, N.S., Dokoupilova, E., Engelke, K., Finkelstein, J.S., Genant, H.K., Goemaere, S., Hyldstrup, L., Jodar-Gimeno, E., Keaveny, T.M., Kendler, D., Lakatos, P., Maddox, J., Malouf, J., Massari, F.E., Molina, J.F., Ulla, M.R., Grauer, A., 2017. Romosozumab (sclerostin monoclonal antibody) versus teriparatide in postmenopausal women with osteoporosis transitioning from oral bisphosphonate therapy: a randomised, open-label, phase 3 trial. Lancet 390 (10102), 1585–1594.

Lara-Castillo, N., Johnson, M.L., 2015. LRP receptor family member associated bone disease. Rev. Endocr. Metab. Disord. 16 (2), 141–148.

Lara-Castillo, N., Kim-Weroha, N.A., Kamel, M.A., Javaheri, B., Ellies, D.L., Krumlauf, R.E., Thiagarajan, G., Johnson, M.L., 2015. In vivo mechanical loading rapidly activates beta-catenin signaling in osteocytes through a prostaglandin mediated mechanism. Bone 76, 58–66.

Lau, K.-H.W., Kapur, S., Kesavan, C., Baylink, D.J., 2006. Up-regulation of the wnt, estrogen receptor, insulin-like growth factor-I, and bone morphogenetic protein pathways in C57BL/6J osteoblasts as opposed to C3H/HeJ osteoblasts in Part Contributes to the differential anabolic response to fluid shear. J. Biol. Chem. 281, 9576–9588.

Lee, B., Thirunavukkarasu, K., Zhou, L., Pastore, L., Baldini, A., Hecht, J., Geoffroy, V., Ducy, P., Karsenty, G., 1997. Missense mutations abolishing DNA binding of the osteoblast-specific transcription factor OSF2/CBFA1 in cleidocranial dysplasia. Nat. Genet. 16 (3), 307–310.

Lerner, U.H., Ohlsson, C., 2015. The WNT system: background and its role in bone. J. Intern. Med. 277 (6), 630–649.

Leupin, O., Kramer, I., Collette, N.M., Loots, G.G., Natt, F., Kneissel, M., Keller, H., 2007. Control of the SOST bone enhancer by PTH using MEF2 transcription factors. J. Bone Miner. Res. 22 (12), 1957–1967.

Leupin, O., Piters, E., Halleux, C., Hu, S., Kramer, I., Morvan, F., Bouwmeester, T., Schirle, M., Bueno-Lozano, M., Fuentes, F.J., Itin, P.H., Boudin, E., de Freitas, F., Jennes, K., Brannetti, B., Charara, N., Ebersbach, H., Geisse, S., Lu, C.X., Bauer, A., Van Hul, W., Kneissel, M., 2011. Bone overgrowth-associated mutations in the LRP4 gene impair sclerostin facilitator function. J. Biol. Chem. 286 (22), 19489–19500.

Li, X., Liu, P., Liu, W., Maye, P., Zhang, J., Zhang, Y., Hurley, M., Guo, C., Boskey, A., Sun, L., Harris, S.E., Rowe, D.W., Ke, H.Z., W. D, 2005a. Dkk2 has a role in terminal osteoblast differentiation and mineralized matrix formation. Nat. Genet. 37, 945–952.

Li, X., Zhang, Y., Kang, H., Liu, W., Liu, P., Zhang, J., Harris, S.E., Wu, D., 2005b. Sclerostin binds to LRP5/6 and antagonizes canonical wnt signaling. J. Biol. Chem. 280, 19883–19887.

Li, X., Grisanti, M., Fan, W., Asuncion, F.J., Tan, H.L., Dwyer, D., Han, C.Y., Yu, L., Lee, J., Lee, E., Barrero, M., Kurimoto, P., Niu, Q.T., Geng, Z., Winters, A., Horan, T., Steavenson, S., Jacobsen, F., Chen, Q., Haldankar, R., Lavallee, J., Tipton, B., Daris, M., Sheng, J., Lu, H.S., Daris, K., Deshpande, R., Valente, E.G., Salimi-Moosavi, H., Kostenuik, P.J., Li, J., Liu, M., Li, C., Lacey, D.L., Simonet, W.S., Ke, H.Z., Babij, P., Stolina, M., Ominsky, M.S., Richards, W.G., 2011. Dickkopf-1 regulates bone formation in young growing rodents and upon traumatic injury. J. Bone Miner. Res. 26 (11), 2610–2621.

Libro, R., Bramanti, P., Mazzon, E., 2016. The role of the Wnt canonical signaling in neurodegenerative diseases. Life Sci. 158, 78–88.

Little, R.D., Carulli, J.P., Del Mastro, R.G., Dupuis, J., Osborne, M., Folz, C., Manning, S.P., Swain, P.M., Zhao, S.C., Eustace, B., Lappe, M.M., Spitzer, L., Zweier, S., Braunschweiger, K., Benchekroun, Y., Hu, X., Adair, R., Chee, L., FitzGerald, M.G., Tulig, C., Caruso, A., Tzellas, N., Bawa, A., Franklin, B., McGuire, S., Nogues, X., Gong, G., Allen, K.M., Anisowicz, A., Morales, A.J., Lomedico, P.T., Recker, S.M., Van Eerdewegh, P., Recker, R.R., Johnson, M.L., 2002. A mutation in the LDL receptor-related protein 5 gene results in the autosomal dominant high-bone-mass trait. Am. J. Hum. Genet. 70, 11–19.

Liu, C., Li, Y., Semenov, M., Han, C., Baeg, G.-H., Tan, Y., Zhang, Z., Lin, X., He, X., 2002. Control of β-catenin phosphorylation/degradation by a dual-kinase mechanism. Cell 108, 837–847.

Liu, Z., Tang, Y., Xu Cao, T.Q., Clemens, T.L., 2006. A dishevelled-1/smad-1 interaction couples WNT and bone morphogenetic protein signaling pathways in uncommitted bone marrow stromal cells. J. Biol. Chem. 281, 17156–17163.

Lodish, M., 2017. Genetics of adrenocortical development and tumors. Endocrinol Metab. Clin. N. Am. 46 (2), 419–433.

Malhotra, D., Yang, Y., 2014. Wnts' fashion statement: from body stature to dysplasia. Bonekey Rep. 3, 541.

Mani, A., Radhakrishman, J., Wang, H., Mani, A., Mani, M.-A., Nelson-Williams, C., Carew, K.S., Mane, S., Najmabadi, H., Wu, D., Lifton, R.P., 2007. LRP6 mutation in a family with early coronary disease and metabolic risk factors. Science 315, 1278–1582.

Marcellini, S., Henriquez, J.P., Bertin, A., 2012. Control of osteogenesis by the canonical Wnt and BMP pathways in vivo: cooperation and antagonism between the canonical Wnt and BMP pathways as cells differentiate from osteochondroprogenitors to osteoblasts and osteocytes. Bioessays 34 (11), 953–962.

Maupin, K.A., Droscha, C.J., Williams, B.O., 2013. A comprehensive overview of skeletal phenotypes associated with alterations in wnt/beta-catenin signaling in humans and mice. Bone Res. 1 (1), 27–71.

Maurel, D.B., Duan, P., Farr, J., Cheng, A.L., Johnson, M.L., Bonewald, L.F., 2016. Beta-catenin haplo insufficient male mice do not lose bone in response to hindlimb unloading. PLoS One 11 (7), e0158381.

McClung, M.R., 2017. Sclerostin antibodies in osteoporosis: latest evidence and therapeutic potential. Ther Adv Musculoskelet Dis 9 (10), 263–270.

McColm, J., Hu, L., Womack, T., Tang, C.C., Chiang, A.Y., 2014. Single- and multiple-dose randomized studies of blosozumab, a monoclonal antibody against sclerostin, in healthy postmenopausal women. J. Bone Miner. Res. 29 (4), 935–943.

McCrea, P.D., Turck, C.W., Gumbiner, B., 1991. A homolog of the armadillo protein in Drosophila (plakoglobin) associated with E-cadherin. Science 254, 1359–1361.

McManus, E.J., Sakamoto, K., Armit, L.J., Ronaldson, L., Shpiro, N., Marquez, R., Alessi, D.R., 2005. Role that phosphorylation of GSK3 plays in insulin and wnt signaling defined by knockin analysis. EMBO J. 24, 1571–1583.

Medina-Gomez, C., Kemp, J.P., Estrada, K., Eriksson, J., Liu, J., Reppe, S., Evans, D.M., Heppe, D.H., Vandenput, L., Herrera, L., Ring, S.M., Kruithof, C.J., Timpson, N.J., Zillikens, M.C., Olstad, O.K., Zheng, H.F., Richards, J.B., St Pourcain, B., Hofman, A., Jaddoe, V.W., Smith, G.D., Lorentzon, M., Gautvik, K.M., Uitterlinden, A.G., Brommage, R., Ohlsson, C., Tobias, J.H., Rivadeneira, F., 2012. Meta-analysis of genome-wide scans for total body BMD in children and adults reveals allelic heterogeneity and age-specific effects at the WNT16 locus. PLoS Genet. 8 (7), e1002718.

Miller, J.R., 2001. The Wnts. Genome Biol. 3 reviews 3001.3001-3001.3015.

Mlodzik, M., 2002. Planar cell polarization: do the same mechanisms regulate Drosophila tissue polarity and vertebrate gastrulation? Trends Genet. 18, 564–571.

Moon, R.T., Bowerman, B., Boutros, M., Perrimon, N., 2002. The promise and perils of wnt signaling through β-catenin. Science 296, 1644–1646.

Nakamura, T., Hamada, F., Ihidate, T., Anai, K., Kawahara, K., Toyoshima, K., Akiyama, T., 1998. Axin, an inhibitor of the wnt signalling pathway, interacts with β-catenin, GSK-3β and APC and reduces the β-catenin level. Genes Cells 3, 395–403.

Nakashima, K., Zhou, X., Kunkel, G., Zhang, Z., Deng, J.M., Behringer, R.R., de Crombrugghe, B., 2002. The novel zinc finger-containing transcription factor osterix is required for osteoblast differentiation and bone formation. Cell 108 (1), 17–29.

Nakashima, A., Katagiri, T., Tamura, M., 2005. Cross-talk between wnt and bone morphogenetic protein 2 (BMP-2) signaling in differentiation pathway of C2C12 myoblasts. J. Biol. Chem. 280, 37660–37668.

Nakashima, T., Hayashi, M., Fukunaga, T., Kurata, K., Oh-Hora, M., Feng, J.Q., Bonewald, L.F., Kodama, T., Wutz, A., Wagner, E.F., Penninger, J.M., Takayanagi, H., 2011. Evidence for osteocyte regulation of bone homeostasis through RANKL expression. Nat. Med. 17 (10), 1231–1234.

Nelson, W.J., Nusse, R., 2004. Convergence of wnt, β-catenin, and cadherin pathways. Science 303, 1483–1487.

Norvell, S.M., Alvarez, M., Bidwell, J.P., Pavalko, F.M., 2004. Fluid shear stress induces β-catenin signaling in osteoblasts. Calcif. Tissue Int. 75, 396–404.

Nusse, R., 2003. Wnts and hedgehogs: lipid-modified proteins and similarities in signaling mechanisms at the cell surface. Development 130, 5297–5305.

Nusse, R., 2005. Wnt signaling in disease and development. Cell Res. 15, 28–32.

Nusse, R., Clevers, H., 2017. Wnt/beta-Catenin signaling, disease, and emerging therapeutic modalities. Cell 169 (6), 985–999.

Nusse, R., Varmus, H.E., 1982. Many tumors induced by the mouse mammary tumor virus contain a provirus integrated in the same region of the host genome. Cell 31, 99–109.

Nusse, R., Brown, A., Papkoff, J., Scambler, P., Shackleford, G., McMahon, A., Moon, R., Varmus, H., 1991. A new nomenclature for int-1 and related genes: the Wnt gene family. Cell 64 (2), 231.

O'Brien, C.A., Nakashima, T., Takayanagi, H., 2013. Osteocyte control of osteoclastogenesis. Bone 54 (2), 258–263.

Otto, F., Thornell, A.P., Crompton, T., Denzel, A., Gilmour, K.C., Rosewell, I.R., Stamp, G.W., Beddington, R.S., Mundlos, S., Olsen, B.R., Selby, P.B., Owen, M.J., 1997. Cbfa1, a candidate gene for cleidocranial dysplasia syndrome, is essential for osteoblast differentiation and bone development. Cell 89 (5), 765–771.

Papadopoulou, D., Bianchi, M.W., Bourouis, M., 2004. Functional studies of shaggy/glycogen synthase kinase 3 phosphorylation sites in *Drosophila melanogaster*. Mol. Cell Biol. 24 (11), 4909–4919.

Pavalko, F.M., Chen, N.X., Turner, C.H., Burr, D.B., Atkinson, S., Hsieh, Y., Qui, J., Duncan, R.L., 1998. Fluid shear-induced mechanical signaling in MC3T3-E1 osteoblasts requires cytoskeleton-integrin interactions. Am. J. Physiol. 275 (C), C1591–C1601.

Perez-Campo, F.M., Santurtun, A., Garcia-Ibarbia, C., Pascual, M.A., Valero, C., Garces, C., Sanudo, C., Zarrabeitia, M.T., Riancho, J.A., 2016. Osterix and RUNX2 are transcriptional regulators of sclerostin in human bone. Calcif. Tissue Int. 99 (3), 302–309.

Peters, J.M., McKay, R.M., McKay, J.P., Graff, J.M., 1999. Casein kinase I transduces wnt signals. Nature 401, 345–350.

Pinson, K.I., Brennan, J., Monkley, S., Avery, B.J., Skarnes, W.C., 2000. An LDL-receptor-related protein mediates wnt signalling in mice. Nature 407, 535–538.

Polakis, P., 2000. Wnt signaling and cancer. Genes Dev. 14, 1837–1851.

Poole, K.E.S., van Bezooijen, R.L., Loveridge, N., Hamersma, H., Papapoulos, S.E., Lowik, C.W., Reeve, J., 2005. Sclerostin is a delayed secreted product of osteocytes that inhibits bone formation. FASEB J. 19, 1842–1844.

Povelones, M., Nusse, R., 2002. Wnt signalling sees spots. Nat. Cell Biol. 4, E249–E250.

Powell Jr., W.F., Barry, K.J., Tulum, I., Kobayashi, T., Harris, S.E., Bringhurst, F.R., Pajevic, P.D., 2011. Targeted ablation of the PTH/PTHrP receptor in osteocytes impairs bone structure and homeostatic calcemic responses. J. Endocrinol. 209 (1), 21–32.

Prunier, C., Hocevar, B.A., Howe, P.H., 2004. Wnt signaling: physiology and pathology. Growth Factors 22, 141–150.

Rawadi, G., Vayssiere, B., Dunn, F., Baron, R., Roman-Roman, S., 2003. BMP-2 controls alkaline phosphatase expression and osteoblast mineralization by a wnt autocrine loop. J. Bone Miner. Res. 18, 1842–1953.

Recker, R.R., Benson, C.T., Matsumoto, T., Bolognese, M.A., Robins, D.A., Alam, J., Chiang, A.Y., Hu, L., Krege, J.H., Sowa, H., Mitlak, B.H., Myers, S.L., 2015. A randomized, double-blind phase 2 clinical trial of blosozumab, a sclerostin antibody, in postmenopausal women with low bone mineral density. J. Bone Miner. Res. 30 (2), 216–224.

Recknor, C.P., Recker, R.R., Benson, C.T., Robins, D.A., Chiang, A.Y., Alam, J., Hu, L., Matsumoto, T., Sowa, H., Sloan, J.H., Konrad, R.J., Mitlak, B.H., Sipos, A.A., 2015. The effect of discontinuing treatment with blosozumab: follow-up results of a phase 2 randomized clinical trial in postmenopausal women with low bone mineral density. J. Bone Miner. Res. 30 (9), 1717–1725.

Reppe, S., Datta, H.K., Gautvik, K.M., 2017. Omics analysis of human bone to identify genes and molecular networks regulating skeletal remodeling in health and disease. Bone 101, 88–95.

Rey, J.P., Ellies, D.L., 2010. Wnt modulators in the biotech pipeline. Dev. Dynam. 239 (1), 102–114.

Riddle, R.C., Diegel, C.R., Leslie, J.M., Van Koevering, K.K., Faugere, M.C., Clemens, T.L., Williams, B.O., 2013. Lrp5 and Lrp6 exert overlapping functions in osteoblasts during postnatal bone acquisition. PLoS One 8 (5), e63323.

Rijsewijk, F., Schuermann, M., Wagenaar, E., Parren, P., Weigel, D., Nusse, R., 1987. The drosphila homology of the mouse mammary oncogen int-1 is identical to the segment polarity gene wingless. Cell 50, 649–657.

Robey, P., 2017. "Mesenchymal stem cells": fact or fiction, and implications in their therapeutic use. F1000Res (6).

Robinson, J.A., Chatterjee-Kishore, M., Yaworsky, P., Cullen, D.M., Zhao, W., Li, C., Kharode, Y.P., Sauter, L., Babij, P., Brown, E.L., Hill, A.A., Akhter, M.P., Johnson, M.L., Recker, R.R., Komm, B.S., Bex, F.J., 2006. Wnt/β-Catenin signaling is a normal physiological response to mechanical loading in bone. J. Biol. Chem. 281, 31720–31728.

Rocha-Braz, M.G., Ferraz-de-Souza, B., 2016. Genetics of osteoporosis: searching for candidate genes for bone fragility. Arch Endocrinol Metab 60 (4), 391–401.

Ross, S.E., Hemati, N., Longo, K.A., Bennett, C.N., Lucas, P.C., Erickson, R.L., MacDougald, O.A., 2000. Inhibition of adipogenesis by Wnt signaling. Science 289 (5481), 950–953.

Saag, K.G., Petersen, J., Brandi, M.L., Karaplis, A.C., Lorentzon, M., Thomas, T., Maddox, J., Fan, M., Meisner, P.D., Grauer, A., 2017. Romosozumab or alendronate for fracture prevention in women with osteoporosis. N. Engl. J. Med. 377 (15), 1417–1427.

Saal, H.M., Prows, C.A., Guerreiro, I., Donlin, M., Knudson, L., Sund, K.L., Chang, C.F., Brugmann, S.A., Stottmann, R.W., 2015. A mutation in FRIZZLED2 impairs Wnt signaling and causes autosomal dominant omodysplasia. Hum. Mol. Genet. 24 (12), 3399–3409.

Sakanaka, C., Weiss, J.B., Williams, L.T., 1998. Bridging of β-catenin and glycogen syhase kinase-3b by axin and inhibiiton of β-Catenin-mediated transcription. Proc. Natl. Acad. Sci. U.S.A. 95, 3020–3030.

Sawa, H., Korswagen, H.C., 2013. Wnt signaling in *C. elegans*. Worm 1–30.

Sawakami, K., Robling, A.G., Ai, M., Pitner, N.D., Liu, D., Warden, S.J., Li, J., Maye, P., Rowe, D.W., Duncan, R.L., Warman, M.L., Turner, C.H., 2006. The wnt Co-receptor Lrp5 is essential for skeletal mechanotransduction, but not for the anabolic bone response to parathyroid hormone treatment. J. Biol. Chem. 281, 23698–23711.

Schaffler, M.B., Cheung, W.Y., Majeska, R., Kennedy, O., 2014. Osteocytes: master orchestrators of bone. Calcif. Tissue Int. 94 (1), 5–24.

Schulte, G., Bryja, V., Rawal, N., Castelo-Branco, G., Sousa, K.M., Arenas, E., 2005. Purified wnt-5a increases differentiation of midbrain dopaminergic cells and dishevelled phosphorylation. J. Neurochem. 92, 1550–1553.

Seeling, J.M., Miller, J.R., Gil, R., Moon, R.T., White, R., Virshup, D.M., 1999. Regulation of β-catenin signaling by the B56 subunit of protein phosphatase 2A. Science 283, 2089–2091.

Semenov, M., He, X., 2006. LRP5 mutations linked to high bone mass diseases cause reduced LRP5 binding and inhibition by SOST. J. Biol. Chem. 281, 38276–38284.

Sheldahl, L.C., Park, M., Malbon, C.C., Moon, R.T., 1999. Protein kinase C is differentially stimulated by wnt and frizzled homologs in a G-protein-dependent manner. Curr. Biol. 9, 695–698.

Spencer, G.J., Utting, J.S., Etheridge, S.L., Arnett, T.R., Genever, P.G., 2006. Wnt signalling in osteoblasts regulates expression of the receptor activator of NFkB ligand and inhibits osteoclastogenesis in vitro. J. Cell Sci. 119, 1283–1296.

Staehling-Hampton, K., Proll, S., Paeper, B.W., Zhao, L., Charmley, P., Brown, A., Gardner, J.C., Galas, D., Schatzman, R.C., Beighton, P., Papapoulos, S., Hamersma, H., Brunkow, M.E., 2002. A 52-kb deletion in the SOST-MEOX1 intergenic region on 17q12-q21 is associated with van Buchem disease in the Dutch population. Am. J. Med. Genet. 110 (2), 144–152.

Stambolic, V., Woodgett, J.R., 1994. Mitogen inactivation of glycogen synthase kinase-3β in intact cells via serine 9 phosphorylation. Biochem. J. 303, 701–704.

Strickland, D.K., Gonias, S.L., Argraves, W.S., 2002. Diverse roles for the LDL receptor family. Trends Endocrinol. Metabol. 13 (2), 66–74.

Sun, T.Q., Lu, B., Feng, J.J., Reinhard, C., Jan, Y.N., Fantl, W.J., Williams, L.T., 2001. PAR-1 is a dishevelled-associated kinase and a positive regulator of wnt signaling. Nat. Cell Biol. 3, 628–636.

Sun, W., Shi, Y., Lee, W.C., Lee, S.Y., Long, F., 2016a. Rictor is required for optimal bone accrual in response to anti-sclerostin therapy in the mouse. Bone 85, 1–8.

Sun, Y., Zhu, D., Chen, F., Qian, M., Wei, H., Chen, W., Xu, J., 2016b. SFRP2 augments WNT16B signaling to promote therapeutic resistance in the damaged tumor microenvironment. Oncogene 35 (33), 4321–4334.

Sutherland, C., Cohen, P., 1994. The α-isoform of glycogen synthase kinase-3 from rabbit skeletal muscle is inactivated by p70 S6 kinase or MAP kinase-activated protein kinase-1 in vitro. FEBS (Fed. Eur. Biochem. Soc.) Lett. 338, 37–42.

Sutherland, C., Leighton, I.A., Cohen, P., 1993. Inactivation of glycogen synthase kinase-3β by phosphorylation: new kinase connections in insulin and growth-factor signaling. Biochem. J. 296, 15–19.

Tago, K., Nakamura, T., Nishita, M., Hyodo, J., Nagai, S., Murata, Y., Adachi, S., Ohwada, S., Morishita, Y., Shibuya, H., Akiyama, T., 2000. Inhibition of wnt signaling by ICAT, a novel β-catenin-interacting protein. Genes Dev. 14, 1741–1749.

Takada, R., Satomi, Y., Kurata, T., Ueno, N., Norioka, S., Kondoh, H., Takao, T., Takada, S., 2006. Monounsaturated fatty acid modification of Wnt protein: its role in Wnt secretion. Dev. Cell 11 (6), 791–801.

Takemura, K., Yamaguchi, S., Lee, Y., Zhang, Y., Carthew, R.W., Moon, R.T., 2003. Chibby, a nuclear β-catenin-associated antagonist of the wnt/wingless pathway. Nature 422, 905–909.

Tamai, K., Semenov, M., Kato, Y., Spokony, R., Liu, C., Katsuyama, Y., Hess, F., Saint-Jeannet, J.P., He, X., 2000. LDL-Receptor-Related proteins in wnt signal transduction. Nature 407, 530–535.

Tan, S.H., Senarath-Yapa, K., Chung, M.T., Longaker, M.T., Wu, J.Y., Nusse, R., 2014. Wnts produced by Osterix-expressing osteolineage cells regulate their proliferation and differentiation. Proc. Natl. Acad. Sci. U. S. A. 111 (49), E5262–E5271.

Terpos, E., Christoulas, D., Gavriatopoulou, M., 2018. Biology and treatment of myeloma related bone disease. Metabolism 80, 80–90.

Tian, Q., He, X.C., Li, L., 2005. Bridging the BMP and wnt pathways by PI3 kinase/Akt and 14-3-3zeta. Cell Cycle 4, 215–216.

Tolwinski, N.S., Wehrli, M., Rives, A., Erdeniz, N., DiNardo, S., Wieschaus, E., 2003. Wg/Wnt signal can Be transmitted through arrow/LRP5,6 and axin independently of Zw3/Gsk3beta activity. Dev. Cell 4 (3), 407–418.

Towler, D.A., Shao, J.S., Cheng, S.L., Pingsterhaus, J.M., Loewy, A.P., 2006. Osteogenic regulation of vascular calcification. Ann. N. Y. Acad. Sci. 1068, 327–333.

Turkmen, S., Spielmann, M., Gunes, N., Knaus, A., Flottmann, R., Mundlos, S., Tuysuz, B., 2017. A Novel de novo FZD2 Mutation in a Patient with Autosomal Dominant Omodysplasia. Mol Syndromol 8 (6), 318–324.

Twells, R.C., Metzker, M.L., Brown, S.D., Cox, R., Garey, C., Hammond, H., Hey, P.J., Levy, E., Nakagawa, Y., Philips, M.S., Todd, J.A., Hess, J.F., 2001. The sequence and gene characterization of a 400-kb candidate region for IDDM4 on chromosome 11q13. Genomics 72, 231–242.

Twells, R.C., Mein, C.A., Payne, F., Veijola, R., Gilbey, M., Bright, M., Timms, A., Nakagawa, Y., Snook, H., Nutland, S., Rance, H.E., Carr, P., Dudridge, F., Cordell, H.J., Cooper, J., Tuomilehto-Wolf, E., Tuomilehto, J., Phillips, M., Metzker, M., Hess, J.F., Todd, J.A., 2003. Linkage and association mapping of the LRP5 locus on chromosome 11q13 in type I diabetes. Hum. Genet. 113, 99–105.

Urano, T., Inoue, S., 2014. Genetics of osteoporosis. Biochem. Biophys. Res. Commun. 452 (2), 287–293.

Vahle, J.L., Sato, M., Long, G.G., Young, J.K., Francis, P.C., Engelhardt, J.A., Westmore, M.S., Linda, Y., Nold, J.B., 2002. Skeletal changes in rats given daily subcutaneous injections of recombinant human parathyroid hormone (1-34) for 2 years and relevance to human safety. Toxicol. Pathol. 30 (3), 312–321.

van Bezooijen, R.L., ten Dijke, P., Papapoulos, S.E., Lowik, C.W., 2005. SOST/sclerostin, an osteocyte-derived negative modulator of bone formation. Cytokine Growth Factor Rev. 16, 319–327.

van den Brink, G.R., 2004. Linking pathways in colorectal cancer. Nat. Genet. 36, 1038–1039.

van Meurs, J.B., Rivadeneira, F., Jhamai, M., Hugens, W., Hofman, a., van Leeuwen, J.P., Pols, H.A., Uitterlinden, A.G., 2006. Common genetic variation of the low-density lipoprotein receptor-related protein 5 and 6 genes determine fracture risk in elderly white men. J. Bone Miner. Res. 21, 141–150.

Vestergaard, P., Rejnmark, L., Mosekilde, L., 2005. Reduced relative risk of fractures among users of lithium. Calcif. Tissue Int. 77, 1–8.

Wan, Y., Lu, C., Cao, J., Zhou, R., Yao, Y., Yu, J., Zhang, L., Zhao, H., Li, H., Zhao, J., Zhu, X., He, L., Liu, Y., Yao, Z., Yang, X., Guo, X., 2013. Osteoblastic Wnts differentially regulate bone remodeling and the maintenance of bone marrow mesenchymal stem cells. Bone 55 (1), 258–267.

Wang, Y., Chang, H., Rattner, A., Nathans, J., 2016. Frizzled receptors in development and disease. Curr. Top. Dev. Biol. 117, 113–139.

Wehrli, M., Dougan, S.T., Caldwell, K., O'Keefe, L., Schwartz, S., Vaizel-Ohayon, D., Schejter, E., Tomlinson, A., DiNardo, S., 2000. Arrow encodes an LDL-receptor-related protein essential for wingless signaling. Nature 407, 527–530.

Wei, W., Zeve, D., Suh, J.M., Wang, X., Du, Y., Zerwekh, J.E., Dechow, P.C., Graff, J.M., Wan, Y., 2011. Biphasic and dosage-dependent regulation of osteoclastogenesis by beta-catenin. Mol. Cell Biol. 31 (23), 4706–4719.

Weivoda, M.M., Ruan, M., Hachfeld, C.M., Pederson, L., Howe, A., Davey, R.A., Zajac, J.D., Kobayashi, Y., Williams, B.O., Westendorf, J.J., Khosla, S., Oursler, M.J., 2016. Wnt signaling inhibits osteoclast differentiation by activating canonical and noncanonical cAMP/PKA pathways. J. Bone Miner. Res. 31 (1), 65–75.

White, J.J., Mazzeu, J.F., Coban-Akdemir, Z., Bayram, Y., Bahrambeigi, V., Hoischen, A., van Bon, B.W.M., Gezdirici, A., Gulec, E.Y., Ramond, F., Touraine, R., Thevenon, J., Shinawi, M., Beaver, E., Heeley, J., Hoover-Fong, J., Durmaz, C.D., Karabulut, H.G., Marzioglu-Ozdemir, E., Cayir, A., Duz, M.B., Seven, M., Price, S., Ferreira, B.M., Vianna-Morgante, A.M., Ellard, S., Parrish, A., Stals, K., Flores-Daboub, J., Jhangiani, S.N., Gibbs, R.A., Brunner, H.G., Sutton, V.R., Lupski, J.R., Carvalho, C.M.B., 2018. WNT signaling perturbations underlie the genetic heterogeneity of Robinow syndrome. Am. J. Hum. Genet. 102 (1), 27–43.

Whyte, M., Reinus, W., Mumm, S., 2004. High-bone-mass disease and LRP5. N. Engl. J. Med. 350 (20), 2096–2098.

Wijenayaka, A.R., Kogawa, M., Lim, H.P., Bonewald, L.F., Findlay, D.M., Atkins, G.J., 2011. Sclerostin stimulates osteocyte support of osteoclast activity by a RANKL-dependent pathway. PLoS One 6 (10), e25900.

Willert, K., Nusse, R., 1998. β-Catenin: a key mediator of wnt signaling. Development 8, 95–102.

Willert, K., Brink, M., Wodarz, a., Varmus, H., Nusse, R., 1997. Casein kinase 2 associates with and phosphorylates dishevelled. EMBO J. 16, 3089–3096.

Willert, K., Brown, J.D., Danenberg, E., Duncan, A.W., WeissmanI, L., Reya, T., Yates, J.R., Nusse, R., 2003. Wnt proteins are lipid-modified and can act as stem cell growth factors. Nature 423 (6938), 448–452.

Winkler, D.G., Sutherland, M.S.K., Ojala, E., Turcott, E., Geoghegan, J.C., Shpektor, D., Skonier, J.E., Yu, C., Latham, J.A., 2005. Sclerostin inhibition of wnt-3a-induced C3H10t1/2 cell differentiation is indirect and mediated by bone morphogenetic proteins. J. Biol. Chem. 280, 2498–2502.

Wozniak, M., Fausto, A., Carron, C.P., Meyer, D.M., Hruska, K.A., 2000. Mechanically strained cells of the osteoblast lineage organize their extracellular matrix through unique sites of avß3-integrin expression. J. Bone Miner. Res. 15, 1731–1745.

Xiong, J., Onal, M., Jilka, R.L., Weinstein, R.S., Manolagas, S.C., O'Brien, C.A., 2011. Matrix-embedded cells control osteoclast formation. Nat. Med. 17 (10), 1235–1241.

Yamamoto, H., Kishida, S., Uochi, T., Ikeda, S., Koyama, S., Asashima, M., Kikuchi, A., 1998. Axil, a member of the axin family, interacts with both glycogen synthase kinase 3β and β-catenin and inhibits Axis formation of *Xenopus* embryos. Mol. Cell Biol. 18, 2867–2875.

Yamamoto, H., Hinoi, T., Michiue, T., Fukui, A., Usui, H., Janssens, V., Van Hoof, C., Goris, J., Asashima, M., Kikuchi, A., 2001. Inhibition of the wnt signaling pathway by the PR61 subunit of protein phosphatase 2A. J. Biol. Chem. 276, 26875–26882.

Yao, Q., Yu, C., Zhang, X., Zhang, K., Guo, J., Song, L., 2017. Wnt/beta-catenin signaling in osteoblasts regulates global energy metabolism. Bone 97, 175–183.

Yoshida, H., Hayashi, S., Kunisada, T., Ogawa, M., Nishikawa, S., Okamura, H., Sudo, T., Shultz, L.D., Nishikawa, S., 1990. The murine mutation osteopetrosis is in the coding region of the macrophage colony stimulating factor gene. Nature 345 (6274), 442–444.

Yost, C., Torres, M., Miller, J.R., Huang, E., Kimelman, D., Moon, R.T., 1996. The axis-inducing activity, stability, and subcellular distribution of b-catenin is regulated in *Xenopus* embryos by glycogen synthase kinase 3. Genes Dev. 10, 1443–1454.

Zeng, L., Fagotto, F., Zhang, t., Hsu, W., Vasicek, T.J., Perry, W.L., Gumbiner, B.M., Constantini, F., 1997. The mouse fused locus encodes axin, an inhibitor of the wnt signaling pathway that regulates embryonic Axis formation. Cell 90, 181–192.

Zhang, R., Oyajobi, B.O., Harris, S.E., Chen, D., Tsao, C., Deng, H.W., Zhao, M., 2013. Wnt/beta-catenin signaling activates bone morphogenetic protein 2 expression in osteoblasts. Bone 52 (1), 145–156.

Zhang, L., Tang, Y., Zhu, X., Tu, T., Sui, L., Han, Q., Yu, L., Meng, S., Zheng, L., Valverde, P., Tang, J., Murray, D., Zhou, X., Drissi, H., Dard, M.M., Tu, Q., Chen, J., 2017. Overexpression of MiR-335-5p promotes bone formation and regeneration in mice. J. Bone Miner. Res. 32 (12), 2466–2475.

Zheng, H.F., Tobias, J.H., Duncan, E., Evans, D.M., Eriksson, J., Paternoster, L., Yerges-Armstrong, L.M., Lehtimaki, T., Bergstrom, U., Kahonen, M., Leo, P.J., Raitakari, O., Laaksonen, M., Nicholson, G.C., Viikari, J., Ladouceur, M., Lyytikainen, L.P., Medina-Gomez, C., Rivadeneira, F., Prince, R.L., Sievanen, H., Leslie, W.D., Mellstrom, D., Eisman, J.A., Moverare-Skrtic, S., Goltzman, D., Hanley, D.A., Jones, G., St Pourcain, B., Xiao, Y., Timpson, N.J., Smith, G.D., Reid, I.R., Ring, S.M., Sambrook, P.N., Karlsson, M., Dennison, E.M., Kemp, J.P., Danoy, P., Sayers, A., Wilson, S.G., Nethander, M., McCloskey, E., Vandenput, L., Eastell, R., Liu, J., Spector, T., Mitchell, B.D., Streeten, E.A., Brommage, R., Pettersson-Kymmer, U., Brown, M.A., Ohlsson, C., Richards, J.B., Lorentzon, M., 2012. WNT16 influences bone mineral density, cortical bone thickness, bone strength, and osteoporotic fracture risk. PLoS Genet. 8 (7), e1002745.

Zheng, L., Tu, Q., Meng, S., Zhang, L., Yu, L., Song, J., Hu, Y., Sui, L., Zhang, J., Dard, M., Cheng, J., Murray, D., Tang, Y., Lian, J.B., Stein, G.S., Chen, J., 2017. Runx2/DICER/miRNA pathway in regulating osteogenesis. J. Cell. Physiol. 232 (1), 182–191.

Zhong, Z., Zylstra-Diegel, C.R., Schumacher, C.A., Baker, J.J., Carpenter, A.C., Rao, S., Yao, W., Guan, M., Helms, J.A., Lane, N.E., Lang, R.A., Williams, B.O., 2012. Wntless functions in mature osteoblasts to regulate bone mass. Proc. Natl. Acad. Sci. U. S. A. 109 (33), E2197–E2204.

Zhong, Z., Ethen, N.J., Williams, B.O., 2014. WNT signaling in bone development and homeostasis. Wiley Interdiscip. Rev. Dev. Biol. 3 (6), 489–500.

Zhou, H., Mak, W., Zheng, Y., Dunstan, C.R., Seibel, M.J., 2008. Osteoblasts directly control lineage commitment of mesenchymal progenitor cells through Wnt signaling. J. Biol. Chem. 283 (4), 1936–1945.

Further Reading

Rudnicki, J.A., Brown, A.M., 1997. Inhibition of chondrogenesis by Wnt gene expression in vivo and in vitro. Dev. Biol. 185 (1), 104–118.

Aghajanova, L., Velarde, M.C., Giudice, L.C., 2009. The progesterone receptor coactivator Hic-5 is involved in the pathophysiology of endometriosis. Endocrinology 150 (8), 3863–3870.

Bao, G.Y., Lu, K.Y., Cui, S.F., Xu, L., 2015a. DKK1 eukaryotic expression plasmid and expression product identification. Genet. Mol. Res. 14 (2), 6312–6318.

Bao, M.W., Cai, Z., Zhang, X.J., Li, L., Liu, X., Wan, N., Hu, G., Wan, F., Zhang, R., Zhu, X., Xia, H., Li, H., 2015b. Dickkopf-3 protects against cardiac dysfunction and ventricular remodelling following myocardial infarction. Basic Res. Cardiol. 110 (3), 25.

Bell, K.L., Garrahan, N., Kneissel, M., Loveridge, N., Grau, E., Stanton, M., Reeve, J., 1996. Cortical and cancellous bone in the human femoral neck: evaluation of an interactive image analysis system. Bone 19 (5), 541–548.

Betts, A.M., Clark, T.H., Yang, J., Treadway, J.L., Li, M., Giovanelli, M.A., Abdiche, Y., Stone, D.M., Paralkar, V.M., 2010. The application of target information and preclinical pharmacokinetic/pharmacodynamic modeling in predicting clinical doses of a Dickkopf-1 antibody for osteoporosis. J. Pharmacol. Exp. Ther. 333 (1), 2–13.

Binnerts, M.E., Tomasevic, N., Bright, J.M., Leung, J., Ahn, V.E., Kim, K.A., Zhan, X., Liu, S., Yonkovich, S., Williams, J., Zhou, M., Gros, D., Dixon, M., Korver, W., Weis, W.I., Abo, A., 2009. The first propeller domain of LRP6 regulates sensitivity to DKK1. Mol. Biol. Cell 20 (15), 3552–3560.

Bjorklund, P., Svedlund, J., Olsson, A.K., Akerstrom, G., Westin, G., 2009. The internally truncated LRP5 receptor presents a therapeutic target in breast cancer. PLoS One 4 (1), e4243.

Bourhis, E., Tam, C., Franke, Y., Bazan, J.F., Ernst, J., Hwang, J., Costa, M., Cochran, A.G., Hannoush, R.N., 2010. Reconstitution of a frizzled8.Wnt3a.LRP6 signaling complex reveals multiple Wnt and Dkk1 binding sites on LRP6. J. Biol. Chem. 285 (12), 9172–9179.

Bourhis, E., Wang, W., Tam, C., Hwang, J., Zhang, Y., Spittler, D., Huang, O.W., Gong, Y., Estevez, A., Zilberleyb, I., Rouge, L., Chiu, C., Wu, Y., Costa, M., Hannoush, R.N., Franke, Y., Cochran, A.G., 2011. Wnt antagonists bind through a short peptide to the first beta-propeller domain of LRP5/6. Structure 19 (10), 1433–1442.

Briot, K., Rouanet, S., Schaeverbeke, T., Etchepare, F., Gaudin, P., Perdriger, A., Vray, M., Steinberg, G., Roux, C., 2015. The effect of tocilizumab on bone mineral density, serum levels of Dickkopf-1 and bone remodeling markers in patients with rheumatoid arthritis. Joint Bone Spine 82 (2), 109–115.

Bu, G., Lu, W., Liu, C.C., Selander, K., Yoneda, T., Hall, C., Keller, E.T., Li, Y., 2008. Breast cancer-derived Dickkopf1 inhibits osteoblast differentiation and osteoprotegerin expression: implication for breast cancer osteolytic bone metastases. Int. J. Cancer 123 (5), 1034–1042.

Burton, D.W., Foster, M., Johnson, K.A., Hiramoto, M., Deftos, L.J., Terkeltaub, R., 2005. Chondrocyte calcium-sensing receptor expression is up-regulated in early Guinea pig knee osteoarthritis and modulates PTHrP, MMP-13, and TIMP-3 expression. Osteoarthr. Cartil. 13 (5), 395–404.

Campos-Obando, N., Castano-Betancourt, M.C., Oei, L., Franco, O.H., Stricker, B.H., Brusselle, G.G., Lahousse, L., Hofman, A., Tiemeier, H., Rivadeneira, F., Uitterlinden, A.G., Zillikens, M.C., 2014. Bone mineral density and chronic lung disease mortality: the rotterdam study. J. Clin. Endocrinol. Metab. 99 (5), 1834–1842.

Caraci, F., Busceti, C., Biagioni, F., Aronica, E., Mastroiacovo, F., Cappuccio, I., Battaglia, G., Bruno, V., Caricasole, A., Copani, A., Nicoletti, F., 2008. The Wnt antagonist, Dickkopf-1, as a target for the treatment of neurodegenerative disorders. Neurochem. Res. 33 (12), 2401–2406.

Clohisy, J.C., Connolly, T.J., Bergman, K.D., Quinn, C.O., Partridge, N.C., 1994. Prostanoid-induced expression of matrix metalloproteinase-1 messenger ribonucleic acid in rat osteosarcoma cells. Endocrinology 135 (4), 1447–1454.

Cook, T.F., Burke, J.S., Bergman, K.D., Quinn, C.O., Jeffrey, J.J., Partridge, N.C., 1994. Cloning and regulation of rat tissue inhibitor of metalloproteinases-2 in osteoblastic cells. Arch. Biochem. Biophys. 311 (2), 313–320.

Costa, A.G., Bilezikian, J.P., Lewiecki, E.M., 2014. Update on romosozumab : a humanized monoclonal antibody to sclerostin. Expert Opin. Biol. Ther. 14 (5), 697–707.

Culley, K.L., Dragomir, C.L., Chang, J., Wondimu, E.B., Coico, J., Plumb, D.A., Otero, M., Goldring, M.B., 2015. Mouse models of osteoarthritis: surgical model of posttraumatic osteoarthritis induced by destabilization of the medial meniscus. Methods Mol. Biol. 1226, 143–173.

Daoussis, D., Liossis, S.N., Solomou, E.E., Tsanaktsi, A., Bounia, K., Karampetsou, M., Yiannopoulos, G., Andonopoulos, A.P., 2010. Evidence that Dkk-1 is dysfunctional in ankylosing spondylitis. Arthritis Rheum. 62 (1), 150–158.

Darlavoix, T., Seelentag, W., Yan, P., Bachmann, A., Bosman, F.T., 2009. Altered expression of CD44 and DKK1 in the progression of Barrett's esophagus to esophageal adenocarcinoma. Virchows Arch. 454 (6), 629–637.

de Andres, M.C., Imagawa, K., Hashimoto, K., Gonzalez, A., Goldring, M.B., Roach, H.I., Oreffo, R.O., 2011. Suppressors of cytokine signalling (SOCS) are reduced in osteoarthritis. Biochem. Biophys. Res. Commun. 407 (1), 54–59.

Deal, C., 2009. Future therapeutic targets in osteoporosis. Curr. Opin. Rheumatol. 21 (4), 380–385.

Favero, M., Ramonda, R., Goldring, M.B., Goldring, S.R., Punzi, L., 2015. Early knee osteoarthritis. RMD Open 1 (Suppl. 1), e000062.

Fillmore, R.A., Mitra, A., Xi, Y., Ju, J., Scammell, J., Shevde, L.A., Samant, R.S., 2009. Nmi (N-Myc interactor) inhibits Wnt/beta-catenin signaling and retards tumor growth. Int. J. Cancer 125 (3), 556–564.

Fleury, D., Gillard, C., Lebhar, H., Vayssiere, B., Touitou, R., Rawadi, G., Mollat, P., 2008. Expression, purification and characterization of murine Dkk1 protein. Protein Expr. Purif. 60 (1), 74–81.

Galasso, O., Panza, S., Santoro, M., Goldring, M.B., Aquila, S., Gasparini, G., 2015. Pten elevation, autophagy and metabolic reprogramming may Be induced in human chondrocytes during steroids or nutrient depletion and osteoarthritis. J. Biol. Regul. Homeost. Agents 29 (4 Suppl. l), 1–14.

Gasser, J.A., Kneissel, M., Thomsen, J.S., Mosekilde, L., 2000. PTH and interactions with bisphosphonates. J. Musculoskelet. Neuronal Interact. 1 (1), 53–56.

Gavriatopoulou, M., Dimopoulos, M.A., Christoulas, D., Migkou, M., Iakovaki, M., Gkotzamanidou, M., Terpos, E., 2009. Dickkopf-1: a suitable target for the management of myeloma bone disease. Expert Opin. Ther. Targets 13 (7), 839–848.

Goldhahn, J., Feron, J.M., Kanis, J., Papapoulos, S., Reginster, J.Y., Rizzoli, R., Dere, W., Mitlak, B., Tsouderos, Y., Boonen, S., 2012. Implications for fracture healing of current and new osteoporosis treatments: an ESCEO consensus paper. Calcif. Tissue Int. 90 (5), 343–353.

Goldring, M.B., 2009a. The link between structural damage and pain in a genetic model of osteoarthritis and intervertebral disc degeneration: a joint misadventure. Arthritis Rheum. 60 (9), 2550–2552.

Goldring, S.R., 2009b. Needs and opportunities in the assessment and treatment of osteoarthritis of the knee and hip: the view of the rheumatologist. J. Bone Joint Surg. Am. 91 (Suppl. 1), 4–6.

Goldring, S.R., 2009c. Periarticular bone changes in rheumatoid arthritis: pathophysiological implications and clinical utility. Ann. Rheum. Dis. 68 (3), 297–299.

Goldring, M.B., 2012a. Articular cartilage degradation in osteoarthritis. HSS J. 8 (1), 7–9.

Goldring, M.B., 2012b. Chondrogenesis, chondrocyte differentiation, and articular cartilage metabolism in health and osteoarthritis. Ther. Adv. Musculoskelet. Dis. 4 (4), 269–285.

Goldring, M.B., 2012c. Do mouse models reflect the diversity of osteoarthritis in humans? Arthritis Rheum. 64 (10), 3072–3075.

Goldring, S.R., 2012d. Alterations in periarticular bone and cross talk between subchondral bone and articular cartilage in osteoarthritis. Ther. Adv. Musculoskelet Dis. 4 (4), 249–258.

Goldring, M.B., Berenbaum, F., 2015. Emerging targets in osteoarthritis therapy. Curr. Opin. Pharmacol. 22, 51–63.

Goldring, S.R., Goldring, M.B., 2006. Clinical aspects, pathology and pathophysiology of osteoarthritis. J. Musculoskelet. Neuronal Interact. 6 (4), 376–378.

Goldring, S.R., Goldring, M.B., 2010. Bone and cartilage in osteoarthritis: is what's best for one good or bad for the other? Arthritis Res. Ther. 12 (5), 143.

Goldring, M.B., Otero, M., 2011. Inflammation in osteoarthritis. Curr. Opin. Rheumatol. 23 (5), 471–478.

Goldring, S.R., Scanzello, C.R., 2012. Plasma proteins take their toll on the joint in osteoarthritis. Arthritis Res. Ther. 14 (2), 111.

Goldring, S., Wright, T., 2012. Frontiers in osteoarthritis: executive summary of the scientific meeting: executive summary of the scientific meeting. HSS J. 8 (1), 2–3.

Goldring, M.B., Otero, M., Plumb, D.A., Dragomir, C., Favero, M., El Hachem, K., Hashimoto, K., Roach, H.I., Olivotto, E., Borzi, R.M., Marcu, K.B., 2011. Roles of inflammatory and anabolic cytokines in cartilage metabolism: signals and multiple effectors converge upon MMP-13 regulation in osteoarthritis. Eur. Cells Mater. 21, 202–220.

Goldring, S., Lane, N., Sandell, L., 2012. Foreword: osteoarthritis. Bone 51 (2), 189.

Gong, Y., Bourhis, E., Chiu, C., Stawicki, S., DeAlmeida, V.I., Liu, B.Y., Phamluong, K., Cao, T.C., Carano, R.A., Ernst, J.A., Solloway, M., Rubinfeld, B., Hannoush, R.N., Wu, Y., Polakis, P., Costa, M., 2010. Wnt isoform-specific interactions with coreceptor specify inhibition or potentiation of signaling by LRP6 antibodies. PLoS One 5 (9), e12682.

Graeff, C., Campbell, G.M., Pena, J., Borggrefe, J., Padhi, D., Kaufman, A., Chang, S., Libanati, C., Gluer, C.C., 2015. Administration of romosozumab improves vertebral trabecular and cortical bone as assessed with quantitative computed tomography and finite element analysis. Bone 81, 364–369.

Granchi, D., Baglio, S.R., Amato, I., Giunti, A., Baldini, N., 2008. Paracrine inhibition of osteoblast differentiation induced by neuroblastoma cells. Int. J. Cancer 123 (7), 1526–1535.

Gregson, C.L., Wheeler, L., Hardcastle, S.A., Appleton, L.H., Addison, K.A., Brugmans, M., Clark, G.R., Ward, K.A., Paggiosi, M., Stone, M., Thomas, J., Agarwal, R., Poole, K.E., McCloskey, E., Fraser, W.D., Williams, E., Bullock, A.N., Davey Smith, G., Brown, M.A., Tobias, J.H., Duncan, E.L., 2016. Mutations in known monogenic high bone mass loci only explain a small proportion of high bone mass cases. J. Bone Miner. Res. 31 (3), 640–649.

Halleux, C., Kramer, I., Allard, C., Kneissel, M., 2012. Isolation of mouse osteocytes using cell fractionation for gene expression analysis. Methods Mol. Biol. 816, 55–66.

Hansson, M., Olesen, D.R., Peterslund, J.M., Engberg, N., Kahn, M., Winzi, M., Klein, T., Maddox-Hyttel, P., Serup, P., 2009. A late requirement for Wnt and FGF signaling during activin-induced formation of foregut endoderm from mouse embryonic stem cells. Dev. Biol. 330 (2), 286–304.

Hay, E., Nouraud, A., Marie, P.J., 2009. N-cadherin negatively regulates osteoblast proliferation and survival by antagonizing Wnt, ERK and PI3K/Akt signalling. PLoS One 4 (12), e8284.

Hernandez, L., Park, K.H., Cai, S.Q., Qin, L., Partridge, N., Sesti, F., 2007. The antiproliferative role of ERG K+ channels in rat osteoblastic cells. Cell Biochem. Biophys. 47 (2), 199–208.

Hey, F., Giblett, S., Forrest, S., Herbert, C., Pritchard, C., 2016. Phosphorylations of serines 21/9 in glycogen synthase kinase 3alpha/beta are not required for cell lineage commitment or WNT signaling in the normal mouse intestine. PLoS One 11 (6), e0156877.

Hoeppner, L.H., Secreto, F.J., Westendorf, J.J., 2009. Wnt signaling as a therapeutic target for bone diseases. Expert Opin. Ther. Targets 13 (4), 485–496.

Imagawa, K., de Andres, M.C., Hashimoto, K., Pitt, D., Itoi, E., Goldring, M.B., Roach, H.I., Oreffo, R.O., 2011. The epigenetic effect of glucosamine and a nuclear factor-kappa B (NF-kB) inhibitor on primary human chondrocytes–implications for osteoarthritis. Biochem. Biophys. Res. Commun. 405 (3), 362–367.

Jin, H., Wang, B., Li, J., Xie, W., Mao, Q., Li, S., Dong, F., Sun, Y., Ke, H.Z., Babij, P., Tong, P., Chen, D., 2015. Anti-DKK1 antibody promotes bone fracture healing through activation of beta-catenin signaling. Bone 71, 63–75.

John, M.R., Widler, L., Gamse, R., Buhl, T., Seuwen, K., Breitenstein, W., Bruin, G.J., Belleli, R., Klickstein, L.B., Kneissel, M., 2011. ATF936, a novel oral calcilytic, increases bone mineral density in rats and transiently releases parathyroid hormone in humans. Bone 49 (2), 233–241.

Johnson, K., Svensson, C.I., Etten, D.V., Ghosh, S.S., Murphy, A.N., Powell, H.C., Terkeltaub, R., 2004. Mediation of spontaneous knee osteoarthritis by progressive chondrocyte ATP depletion in Hartley Guinea pigs. Arthritis Rheum. 50 (4), 1216–1225.

Kneissel, H., 1976. Treatment of lumbar disk herniation with depot cortisone intrathecally and peridurally (author's transl). Med. Klin. 71 (37), 1506–1507.

Kneissel, H., Horcajada, J., 1970. The moment for angiography after subarachnoid hemorrhage. Wien Med. Wochenschr. 120 (47), 863–865.

Kneissel, H., Hofer, R., Horcajada, J., 1972. [Scintigraphic demonstration of the subarachnoid space for the evaluation of disorders in liquor circulation]. Wien Klin. Wochenschr. 84 (27), 457–458.

Kneissel, S., Queitsch, I., Petersen, G., Behrsing, O., Micheel, B., Dubel, S., 1999. Epitope structures recognised by antibodies against the major coat protein (g8p) of filamentous bacteriophage fd (Inoviridae). J. Mol. Biol. 288 (1), 21–28.

Kneissel, M., Boyde, A., Gasser, J.A., 2001. Bone tissue and its mineralization in aged estrogen-depleted rats after long-term intermittent treatment with parathyroid hormone (PTH) analog SDZ PTS 893 or human PTH(1-34). Bone 28 (3), 237–250.

Koh, J.M., Jung, M.H., Hong, J.S., Park, H.J., Chang, J.S., Shin, H.D., Kim, S.Y., Kim, G.S., 2004. Association between bone mineral density and LDL receptor-related protein 5 gene polymorphisms in young Korean men. J. Korean Med. Sci. 19, 407–412.

Komatsu, D.E., Mary, M.N., Schroeder, R.J., Robling, A.G., Turner, C.H., Warden, S.J., 2010. Modulation of Wnt signaling influences fracture repair. J. Orthop. Res. 28 (7), 928–936.

Kuang, H.B., Miao, C.L., Guo, W.X., Peng, S., Cao, Y.J., Duan, E.K., 2009. Dickkopf-1 enhances migration of HEK293 cell by beta-catenin/E-cadherin degradation. Front. Biosci. 14, 2212–2220.

Larsen, B.M., Hrycaj, S.M., Newman, M., Li, Y., Wellik, D.M., 2015. Mesenchymal Hox6 function is required for mouse pancreatic endocrine cell differentiation. Development 142 (22), 3859–3868.

Lee, M., Partridge, N.C., 2010. Parathyroid hormone activation of matrix metalloproteinase-13 transcription requires the histone acetyltransferase activity of p300 and PCAF and p300-dependent acetylation of PCAF. J. Biol. Chem. 285 (49), 38014–38022.

Leong, D.J., Choudhury, M., Hanstein, R., Hirsh, D.M., Kim, S.J., Majeska, R.J., Schaffler, M.B., Hardin, J.A., Spray, D.C., Goldring, M.B., Cobelli, N.J., Sun, H.B., 2014. Green tea polyphenol treatment is chondroprotective, anti-inflammatory and palliative in a mouse post-traumatic osteoarthritis model. Arthritis Res. Ther. 16 (6), 508.

Li, X., Liu, H., Qin, L., Tamasi, J., Bergenstock, M., Shapses, S., Feyen, J.H., Notterman, D.A., Partridge, N.C., 2007. Determination of dual effects of parathyroid hormone on skeletal gene expression in vivo by microarray and network analysis. J. Biol. Chem. 282 (45), 33086–33097.

Li, X., Qin, L., Partridge, N.C., 2008. In vivo parathyroid hormone treatments and RNA isolation and analysis. Methods Mol. Biol. 455, 79–87.

Li, M., Wu, X.H., Yin, G., Xie, Q.B., 2015. Correlation of RANKL/OPG, dickkopf-1 and bone marrow edema in rheumatoid arthritis with the complaint of knee pain. Sichuan Da Xue Xue Bao Yi Xue Ban 46 (2), 276–279.

Liu, J., Ho, S.C., Su, Y.X., Chen, W.Q., Zhang, C.X., Chen, Y.M., 2009. Effect of long-term intervention of soy isoflavones on bone mineral density in women: a meta-analysis of randomized controlled trials. Bone 44 (5), 948–953.

Liu, Y., Kodithuwakku, S.P., Ng, P.Y., Chai, J., Ng, E.H., Yeung, W.S., Ho, P.C., Lee, K.F., 2010. Excessive ovarian stimulation up-regulates the Wnt-signaling molecule DKK1 in human endometrium and may affect implantation: an in vitro co-culture study. Hum. Reprod. 25 (2), 479–490.

Liu, Z., Kennedy, O.D., Cardoso, L., Basta-Pljakic, J., Partridge, N.C., Schaffler, M.B., Rosen, C.J., Yakar, S., 2016. DMP-1-mediated Ghr gene recombination compromises skeletal development and impairs skeletal response to intermittent PTH. FASEB J. 30 (2), 635–652.

Loeser, R.F., Goldring, S.R., Scanzello, C.R., Goldring, M.B., 2012. Osteoarthritis: a disease of the joint as an organ. Arthritis Rheum. 64 (6), 1697–1707.

Martinez, G., Wijesinghe, M., Turner, K., Abud, H.E., Taketo, M.M., Noda, T., Robinson, M.L., de Iongh, R.U., 2009. Conditional mutations of beta-catenin and APC reveal roles for canonical Wnt signaling in lens differentiation. Investig. Ophthalmol. Vis. Sci. 50 (10), 4794–4806.

Martyn-St James, M., Carroll, S., 2009. A meta-analysis of impact exercise on postmenopausal bone loss: the case for mixed loading exercise programmes. Br. J. Sports Med. 43 (12), 898–908.

Mauriello Jr., J.A., Wasserman, B., Kraut, R., 1993. Use of Vicryl (polyglactin-910) mesh implant for repair of orbital floor fracture causing diplopia: a study of 28 patients over 5 years. Ophthalmic Plast. Reconstr. Surg. 9 (3), 191–195.

Meng, S., Zhou, F.L., Zhang, W.G., Cao, X.M., Wang, B.Y., Wang, Y., Bai, G.G., 2012. The research on the expression and localization of multiple myeloma associated antigen MMSA-1. Xi Bao Yu Fen Zi Mian Yi Xue Za Zhi 28 (1), 63–66.

Meyer, T., Kneissel, M., Mariani, J., Fournier, B., 2000. In vitro and in vivo evidence for orphan nuclear receptor RORalpha function in bone metabolism. Proc. Natl. Acad. Sci. U. S. A. 97 (16), 9197–9202.

Minisola, S., 2014. Romosozumab: from basic to clinical aspects. Expert Opin. Biol. Ther. 14 (9), 1225–1228.

Mitsuyama, H., Healey, R.M., Terkeltaub, R.A., Coutts, R.D., Amiel, D., 2007. Calcification of human articular knee cartilage is primarily an effect of aging rather than osteoarthritis. Osteoarthr. Cartil. 15 (5), 559–565.

Moors, M., Bose, R., Johansson-Haque, K., Edoff, K., Okret, S., Ceccatelli, S., 2012. Dickkopf 1 mediates glucocorticoid-induced changes in human neural progenitor cell proliferation and differentiation. Toxicol. Sci. 125 (2), 488–495.

Muka, T., de Jonge, E.A., de Jong, J.C., Uitterlinden, A.G., Hofman, A., Dehghan, A., Zillikens, M.C., Franco, O.H., Rivadeneira, F., 2016. The influence of serum uric acid on bone mineral density, hip geometry, and fracture risk: the rotterdam study. J. Clin. Endocrinol. Metab. 101 (3), 1113–1122.

Neogi, T., Booth, S.L., Zhang, Y.Q., Jacques, P.F., Terkeltaub, R., Aliabadi, P., Felson, D.T., 2006. Low vitamin K status is associated with osteoarthritis in the hand and knee. Arthritis Rheum. 54 (4), 1255–1261.

Ng, K.W., Martin, T.J., 2014. New therapeutics for osteoporosis. Curr. Opin. Pharmacol. 16, 58–63.

Nusse, R., Varmus, H.E., 1992. Wnt genes. Cell 69 (7), 1073–1087.

Olivotto, E., Otero, M., Marcu, K.B., Goldring, M.B., 2015. Pathophysiology of osteoarthritis: canonical NF-kappaB/IKKbeta-dependent and kinase-independent effects of IKKalpha in cartilage degradation and chondrocyte differentiation. RMD Open 1 (Suppl. 1), e000061.

Ootani, A., Li, X., Sangiorgi, E., Ho, Q.T., Ueno, H., Toda, S., Sugihara, H., Fujimoto, K., Weissman, I.L., Capecchi, M.R., Kuo, C.J., 2009. Sustained in vitro intestinal epithelial culture within a Wnt-dependent stem cell niche. Nat. Med. 15 (6), 701–706.

Otero, M., Goldring, M.B., 2007. Cells of the synovium in rheumatoid arthritis. Chondrocytes. Arthritis Res. Ther. 9 (5), 220.

Partridge, N.C., Alcorn, D., Michelangeli, V.P., Kemp, B.E., Ryan, G.B., Martin, T.J., 1981. Functional properties of hormonally responsive cultured normal and malignant rat osteoblastic cells. Endocrinology 108 (1), 213–219.

Partridge, N.C., Hillyard, C.J., Nolan, R.D., Martin, T.J., 1985. Regulation of prostaglandin production by osteoblast-rich calvarial cells. Prostaglandins 30 (3), 527–539.

Pettit, A.R., Walsh, N.C., Manning, C., Goldring, S.R., Gravallese, E.M., 2006. RANKL protein is expressed at the pannus-bone interface at sites of articular bone erosion in rheumatoid arthritis. Rheumatology 45 (9), 1068–1076.

Power, J., Poole, K.E., van Bezooijen, R., Doube, M., Caballero-Alias, A.M., Lowik, C., Papapoulos, S., Reeve, J., Loveridge, N., 2010. Sclerostin and the regulation of bone formation: effects in hip osteoarthritis and femoral neck fracture. J. Bone Miner. Res. 25 (8), 1867–1876.

Purro, S.A., Dickins, E.M., Salinas, P.C., 2012. The secreted Wnt antagonist Dickkopf-1 is required for amyloid beta-mediated synaptic loss. J. Neurosci. 32 (10), 3492–3498.

Qiang, Y.W., Hu, B., Chen, Y., Zhong, Y., Shi, B., Barlogie, B., Shaughnessy Jr., J.D., 2009. Bortezomib induces osteoblast differentiation via Wnt-independent activation of beta-catenin/TCF signaling. Blood 113 (18), 4319–4330.

Qin, L., Tamasi, J., Raggatt, L., Li, X., Feyen, J.H., Lee, D.C., Dicicco-Bloom, E., Partridge, N.C., 2005. Amphiregulin is a novel growth factor involved in normal bone development and in the cellular response to parathyroid hormone stimulation. J. Biol. Chem. 280 (5), 3974–3981.

Raggatt, L.J., Partridge, N.C., 2010. Cellular and molecular mechanisms of bone remodeling. J. Biol. Chem. 285 (33), 25103–25108.

Recchia, I., Rucci, N., Funari, A., Migliaccio, S., Taranta, A., Longo, M., Kneissel, M., Susa, M., Fabbro, D., Teti, A., 2004. Reduction of c-Src activity by substituted 5,7-diphenyl-pyrrolo[2,3-d]-pyrimidines induces osteoclast apoptosis in vivo and in vitro. Involvement of ERK1/2 pathway. Bone 34 (1), 65–79.

Rivadeneira, F., Styrkarsdottir, U., Estrada, K., Halldorsson, B.V., Hsu, Y.H., Richards, J.B., Zillikens, M.C., Kavvoura, F.K., Amin, N., Aulchenko, Y.S., Cupples, L.A., Deloukas, P., Demissie, S., Grundberg, E., Hofman, A., Kong, A., Karasik, D., van Meurs, J.B., Oostra, B., Pastinen, T., Pols, H.A., Sigurdsson, G., Soranzo, N., Thorleifsson, G., Thorsteinsdottir, U., Williams, F.M., Wilson, S.G., Zhou, Y., Ralston, S.H., van Duijn, C.M., Spector, T., Kiel, D.P., Stefansson, K., Ioannidis, J.P., Uitterlinden, A.G., 2009. Twenty bone-mineral-density loci identified by large-scale meta-analysis of genome-wide association studies. Nat. Genet. 41 (11), 1199–1206.

Roschger, P., Grabner, B.M., Rinnerthaler, S., Tesch, W., Kneissel, M., Berzlanovich, A., Klaushofer, K., Fratzl, P., 2001. Structural development of the mineralized tissue in the human L4 vertebral body. J. Struct. Biol. 136 (2), 126–136.
Rosi, M.C., Luccarini, I., Grossi, C., Fiorentini, A., Spillantini, M.G., Prisco, A., Scali, C., Gianfriddo, M., Caricasole, A., Terstappen, G.C., Casamenti, F., 2010. Increased Dickkopf-1 expression in transgenic mouse models of neurodegenerative disease. J. Neurochem. 112 (6), 1539–1551.
Rundle, C.H., Wang, H., Yu, H., Chadwick, R.B., Davis, E.I., Wergedal, J.E., Lau, K.H., Mohan, S., Ryaby, J.T., Baylink, D.J., 2006. Microarray analysis of gene expression during the inflammation and endochondral bone formation stages of rat femur fracture repair. Bone 38 (4), 521–529.
Sahithi, K., Swetha, M., Prabaharan, M., Moorthi, A., Saranya, N., Ramasamy, K., Srinivasan, N., Partridge, N.C., Selvamurugan, N., 2010. Synthesis and characterization of nanoscale-hydroxyapatite-copper for antimicrobial activity towards bone tissue engineering applications. J. Biomed. Nanotechnol. 6 (4), 333–339.
Sato, N., Yamabuki, T., Takano, A., Koinuma, J., Aragaki, M., Masuda, K., Ishikawa, N., Kohno, N., Ito, H., Miyamoto, M., Nakayama, H., Miyagi, Y., Tsuchiya, E., Kondo, S., Nakamura, Y., Daigo, Y., 2010. Wnt inhibitor Dickkopf-1 as a target for passive cancer immunotherapy. Cancer Res. 70 (13), 5326–5336.
Scanzello, C.R., Goldring, S.R., 2012. The role of synovitis in osteoarthritis pathogenesis. Bone 51 (2), 249–257.
Schneider, S.F., 1990. Psychology at a crossroads. Am. Psychol. 45 (4), 521–529.
Schrader, R., Kneissel, G.D., Sievert, H., Bussmann, W.D., Kaltenbach, M., 1992. [Clinical features and therapy of persistent ductus arteriosus in adults]. Dtsch. Med. Wochenschr. 117 (47), 1805–1809.
Selvamurugan, N., Shimizu, E., Lee, M., Liu, T., Li, H., Partridge, N.C., 2009. Identification and characterization of Runx2 phosphorylation sites involved in matrix metalloproteinase-13 promoter activation. FEBS Lett. 583 (7), 1141–1146.
Shalhoub, V., Conlon, D., Tassinari, M., Quinn, C., Partridge, N., Stein, G.S., Lian, J.B., 1992. Glucocorticoids promote development of the osteoblast phenotype by selectively modulating expression of cell growth and differentiation associated genes. J. Cell. Biochem. 50 (4), 425–440.
Sottile, V., Seuwen, K., Kneissel, M., 2004. Enhanced marrow adipogenesis and bone resorption in estrogen-deprived rats treated with the PPARgamma agonist BRL49653 (rosiglitazone). Calcif. Tissue Int. 75 (4), 329–337.
Streeten, E.A., Morton, H., McBride, D.J., 2003. Osteoporosis pseudoglioma syndrome: 3 siblings with a novel LRP5 mutation. J. Bone Miner. Res. 18 (Suppl. 2), S35.
Tella, S.H., Gallagher, J.C., 2014. Biological agents in management of osteoporosis. Eur. J. Clin. Pharmacol. 70 (11), 1291–1301.
Terkeltaub, R., Johnson, K., Murphy, A., Ghosh, S., 2002. Invited review: the mitochondrion in osteoarthritis. Mitochondrion 1 (4), 301–319.
Tripathi, A., Saravanan, S., Pattnaik, S., Moorthi, A., Partridge, N.C., Selvamurugan, N., 2012. Bio-composite scaffolds containing chitosan/nano-hydroxyapatite/nano-copper-zinc for bone tissue engineering. Int. J. Biol. Macromol. 50 (1), 294–299.
Tutak, W., Park, K.H., Vasilov, A., Starovoytov, V., Fanchini, G., Cai, S.Q., Partridge, N.C., Sesti, F., Chhowalla, M., 2009. Toxicity induced enhanced extracellular matrix production in osteoblastic cells cultured on single-walled carbon nanotube networks. Nanotechnology 20 (25), 255101.
Uderhardt, S., Diarra, D., Katzenbeisser, J., David, J.P., Zwerina, J., Richards, W., Kronke, G., Schett, G., 2010. Blockade of Dickkopf (DKK)-1 induces fusion of sacroiliac joints. Ann. Rheum. Dis. 69 (3), 592–597.
Ulm, C.W., Kneissel, M., Hahn, M., Solar, P., Matejka, M., Donath, K., 1997. Characteristics of the cancellous bone of edentulous mandibles. Clin. Oral Implant. Res. 8 (2), 125–130.
Vajda, E.G., Kneissel, M., Muggenburg, B., Miller, S.C., 1999. Increased intracortical bone remodeling during lactation in beagle dogs. Biol. Reprod. 61 (6), 1439–1444.
Wang, Q., Rozelle, A.L., Lepus, C.M., Scanzello, C.R., Song, J.J., Larsen, D.M., Crish, J.F., Bebek, G., Ritter, S.Y., Lindstrom, T.M., Hwang, I., Wong, H.H., Punzi, L., Encarnacion, A., Shamloo, M., Goodman, S.B., Wyss-Coray, T., Goldring, S.R., Banda, N.K., Thurman, J.M., Gobezie, R., Crow, M.K., Holers, V.M., Lee, D.M., Robinson, W.H., 2011. Identification of a central role for complement in osteoarthritis. Nat. Med. 17 (12), 1674–1679.
Weng, L.H., Wang, C.J., Ko, J.Y., Sun, Y.C., Su, Y.S., Wang, F.S., 2009. Inflammation induction of Dickkopf-1 mediates chondrocyte apoptosis in osteoarthritic joint. Osteoarthr. Cartil. 17 (7), 933–943.
Williams, B.O., 2014. Insights into the mechanisms of sclerostin action in regulating bone mass accrual. J. Bone Miner. Res. 29 (1), 24–28.
Wixted, J.J., Fanning, P., Rothkopf, I., Stein, G., Lian, J., 2010. Arachidonic acid, eicosanoids, and fracture repair. J. Orthop. Trauma 24 (9), 539–542.
Wright, T., Goldring, S., 2012. Reaching consensus and highlighting future directions for research: the osteoarthritis summit breakout sessions. HSS J. 8 (1), 80–83.
Wu, C.W., Terkeltaub, R., Kalunian, K.C., 2005. Calcium-containing crystals and osteoarthritis: implications for the clinician. Curr. Rheumatol. Rep. 7 (3), 213–219.
Xu, N., Zhou, W.J., Wang, Y., Huang, S.H., Li, X., Chen, Z.Y., 2015. Hippocampal Wnt3a is necessary and sufficient for contextual fear memory acquisition and consolidation. Cerebr. Cortex 25 (11), 4062–4075.
Yi, Q., 2009. Novel immunotherapies. Cancer J. 15 (6), 502–510.
Yoshikawa, Y., Fujimori, T., McMahon, A.P., Takada, S., 1997. Evidence that absence of wnt-3a signaling promotes neuralization instead of paraxial mesoderm development in the mouse. Dev. Biol. 183, 234–242.
Zechner, W., Kneissel, M., Kim, S., Ulm, C., Watzek, G., Plenk Jr., H., 2004. Histomorphometrical and clinical comparison of submerged and nonsubmerged implants subjected to experimental peri-implantitis in dogs. Clin. Oral Implant. Res. 15 (1), 23–33.
Zhang, S.N., Sun, A.J., Ge, J.B., Yao, K., Huang, Z.Y., Wang, K.Q., Zou, Y.Z., 2009. Intracoronary autologous bone marrow stem cells transfer for patients with acute myocardial infarction: a meta-analysis of randomised controlled trials. Int. J. Cardiol. 136 (2), 178–185.
Zhao, J., Kim, K.A., Abo, A., 2009. Tipping the balance: modulating the Wnt pathway for tissue repair. Trends Biotechnol. 27 (3), 131–136.

Zhu, Y., Sun, Z., Han, Q., Liao, L., Wang, J., Bian, C., Li, J., Yan, X., Liu, Y., Shao, C., Zhao, R.C., 2009. Human mesenchymal stem cells inhibit cancer cell proliferation by secreting DKK-1. Leukemia 23 (5), 925–933.

Zillikens, M.C., Uitterlinden, A.G., van Leeuwen, J.P., Berends, A.L., Henneman, P., van Dijk, K.W., Oostra, B.A., van Duijn, C.M., Pols, H.A., Rivadeneira, F., 2010. The role of body mass index, insulin, and adiponectin in the relation between fat distribution and bone mineral density. Calcif. Tissue Int. 86 (2), 116–125.

Zirwes, R.F., Eilbracht, J., Kneissel, S., Schmidt-Zachmann, M.S., 2000. A novel helicase-type protein in the nucleolus: protein NOH61. Mol. Biol. Cell 11 (4), 1153–1167.

Chapter 9

Vascular and nerve interactions

Ryan E. Tomlinson[1], Thomas L. Clemens[2,3] and Christa Maes[4]

[1]*Department of Orthopaedic Surgery, Thomas Jefferson University, Philadelphia, PA, United States;* [2]*Department of Orthopaedic Surgery, Johns Hopkins University School of Medicine, Baltimore, MD, United States;* [3]*Baltimore Veterans Administration Medical Center, Baltimore, MD, United States;* [4]*Laboratory of Skeletal Cell Biology and Physiology (SCEBP), Skeletal Biology and Engineering Research Center (SBE), KU Leuven, Leuven, Belgium*

Chapter outline

Introduction **205**
The vasculature of bone **206**
Vascularization of developing bone 206
Vascularization of the mature skeleton 208
Bone cells' control of skeletal vascularization and oxygenation 208
Endothelial and angiocrine signaling in bone 210
The nerve system of bone **210**
Innervation of developing bone 210
Innervation of the mature skeleton 211
Somatic nervous system 211
Autonomic nervous system 212
Conclusion **213**
References **214**

Introduction

Bone is a highly vascularized and innervated organ. During development, all the long bones of the skeleton form via an avascular and noninnervated cartilage template. The conversion of the cartilaginous model into bone is dependent on its primary invasion by blood vessels, a process that is associated with the development of a bone marrow cavity and its infiltration by various cell types populating the bone, including bone-forming osteolineage cells, bone-resorbing osteoclasts, and hematopoietic cells. The presence of an adequate vascular supply is absolutely required for further bone formation and remodeling, to supply the necessary oxygen, nutrients, and hormones and to remove waste products (Maes, 2013). Furthermore, the close physical relationship and molecular crosstalk between the endothelial cells of the blood vessels and the osteoprogenitors residing in the perivascular space are thought to contribute to angiogenic–osteogenic coupling (Maes, 2013; Dirckx et al., 2013). This term refers to the close spatial–temporal association of angiogenesis and osteogenesis, as observed during fetal skeletal development, during the juvenile growth spurt, in remodeling bone throughout adulthood, and in fracture healing (Schipani et al., 2009; Maes, 2013; Wang et al., 2007). For some time, compromised vascularity and reduced blood flow in bone have been associated with age-related declines in bone mass and osteoporosis as well as poor bone repair, emphasizing the potential therapeutic value of understanding the vasculature of bone and how to modulate it (Maes, 2013; Ramasamy et al., 2016; Tomlinson et al., 2013; Tomlinson and Silva, 2013; Mekraldi et al., 2003; Burkhardt et al., 1987; Saran et al., 2014; Carano and Filvaroff, 2003; Street et al., 2002). Interestingly, nerves accompany the blood vessels that supply bone tissue; in fact, innervation and vascularization of developing bone are tightly coordinated and possibly interdependent processes. The presence of nerves in bone has been appreciated since work of the early microscopists, who utilized routine histological methods such as toluidine blue or silver staining to reveal nerve axons in calcified tissue (Miller and Kasahara, 1963; Milgram and Robinson, 1965; Linder, 1978; Thurston, 1982). These studies helped validate the numerous clinical observations regarding skeletal pain emanating from bone tissue itself, particularly following surgery necessitated by neoplastic bone disease. Immunohistochemical studies, beginning in the mid-1980s (Hohmann et al., 1986), began to dissect the various types and subtypes of neuronal axons in bone. This work gradually revealed that the skeleton is richly innervated by sensory, sympathetic, and parasympathetic axons of the

Principles of Bone Biology. https://doi.org/10.1016/B978-0-12-814841-9.00009-9

peripheral nervous system, with axons found in the periosteum and marrow space as well as calcified tissue (Bjurholm et al., 1988; Wojtys et al., 1990; Hill and Elde, 1991; Hukkanen et al., 1992; Mach et al., 2002; Castañeda-Corral et al., 2011). Furthermore, each of these nerve types appears to serve unique functions within the bone microenvironment, many of which are under active investigation. In this chapter, we will review the current understanding of the origin, location, and function of the blood vessels and different nerve types in bone.

The vasculature of bone

Vascularization of developing bone

The development of the long bones of the skeleton by endochondral ossification is characterized by the initial formation of a cartilage template that prefigures the future bone. Cartilage is inherently avascular, in line with the typical production of angiogenic inhibitors by chondrocytes during their immature and differentiating states. When chondrocytes reach their final maturation stage and become hypertrophic chondrocytes, the balance shifts toward the predominant expression of angiogenic stimulators. At this stage, hypertrophic chondrocytes begin expressing vascular endothelial growth factor (VEGF). As a consequence, hypertrophic cartilage becomes invaded by blood vessels, which triggers the formation of the primary ossification center and the nascent bone marrow cavity. Osteoprogenitors coinvade the middiaphyseal bone region along with blood vessels originating from the perichondrium, and instigate bone formation inside the shaft (Maes et al., 2010a). The excavation of the original cartilaginous template is probably steered by both angiogenesis-related tissue remodeling and osteoclastic resorption of the calcified cartilage matrix. While the fate of the hypertrophic chondrocytes proper was classically considered to be apoptotic cell death, studies using genetic lineage tracing strategies in mice have revived the long-standing hypothesis of transdifferentiation of terminally differentiated growth plate chondrocytes into osteoblasts (Zhou et al., 2014; Yang et al., 2014; Park et al., 2015). Whether the incoming vasculature may have an instructive role in defining the fate of the hypertrophic chondrocyte, which seems highly plausible, needs to be defined.

The concept that hypertrophic chondrocytes direct the cartilage-to-bone transition through VEGF release during endochondral bone growth was established through the use of a variety of mouse models. First, administration of a soluble VEGF receptor chimeric protein (sFlt-1) that acts to inhibit VEGF was shown to impair vascular invasion of the growth plate in juvenile mice, which was associated with reduced longitudinal bone growth and trabecular bone formation (Gerber et al., 1999). Concomitantly, the mice showed a pronounced extension of the layer containing hypertrophic chondrocytes in their growth plates (Gerber et al., 1999), a feature that had also been transiently observed in knockout mice lacking the matrix metalloproteinase 9 (MMP-9) (Vu et al., 1998). In joining these findings, the theory was proposed that capillary invasion of the growth cartilage brings osteoclasts—cells of hematopoietic origin sharing a precursor with macrophages—to the chondro-osseous junction. MMP-9 production and matrix resorption by osteoclasts or postulated related cells called "chondroclasts" would progressively release matrix-bound VEGF from the degrading cartilage matrix, creating a positive-feedback system steering capillary growth, hypertrophic cartilage matrix decay and resorption, and trabecular bone formation (Karsenty and Wagner, 2002). Although the fine details of the cellular sources and situated actions of MMP-9 are as yet to be resolved, the important role of VEGF as coordinator of the progressive turnover of cartilage into bone and driver of longitudinal bone growth has been further substantiated and refined through the use of conditional knockout mice and knock-in models (reviewed in greater detail in Maes, 2013, 2017; Dirckx et al., 2013; Dirckx and Maes, 2016). These studies furthermore annotated specific functions to the major VEGF splice isoforms (VEGF120, VEGF164, and VEGF188 in the mouse), primarily based on their differential solubility versus matrix-binding characteristics. The shorter, soluble VEGF isoforms (VEGF120 and to some extent VEGF164) can diffuse and attract distant blood vessels toward the source of VEGF secretion, thus stimulating angiogenic growth of the capillary bed. The longer isoforms (VEGF164 and especially VEGF188) have strong affinity to bind heparin-containing matrix components, become sequestered in the cartilage matrix, and are released upon proteolytic degradation of the matrix, thereby serving a coordinating role in linking cartilage-to-bone turnover to the progressive vascularization of the growing bone (Maes, 2017; Maes et al., 2002, 2004; Zelzer et al., 2002) (Fig. 9.1).

VEGF plays essential roles in the initial osteo-angiogenic invasion of fetal bones during primary ossification center (POC) formation, as well as in triggering the vascularization processes that mediate the formation of the secondary ossification centers in the epiphyses later in skeletal development (Maes et al., 2004, 2012a; Duan et al., 2015). Mice engineered to express only the VEGF120 isoform or only the VEGF188 isoform showed a delay in the initial vascular invasion and development of the long bones, indicating the importance of the VEGF164 isoform or a combination of soluble and matrix-bound VEGF (Maes et al., 2002, 2004; Zelzer et al., 2002). The levels of VEGF need to be tightly

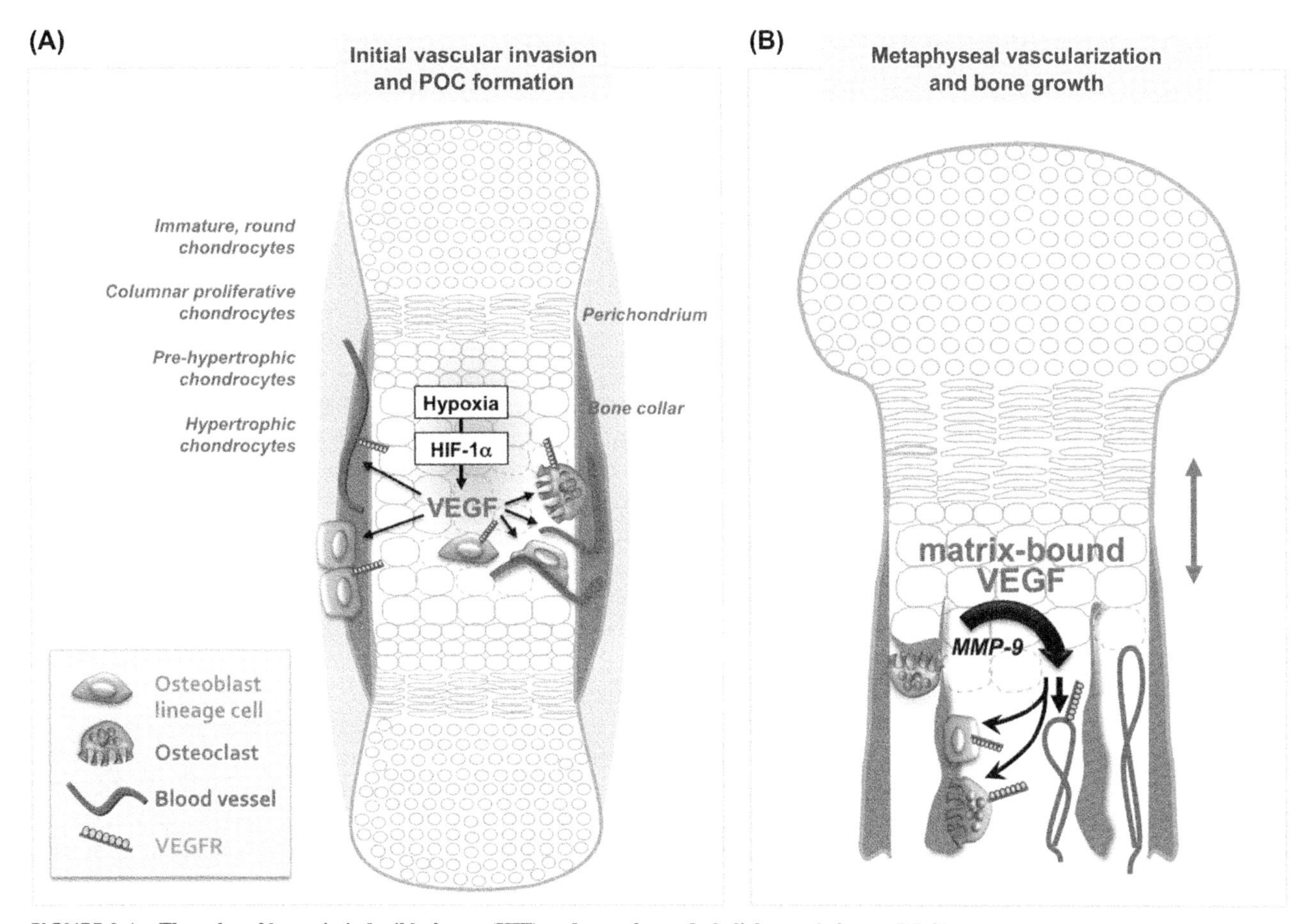

FIGURE 9.1 The roles of hypoxia-inducible factor (HIF) and vascular endothelial growth factor (VEGF) in the neovascularization of cartilage and its conversion to bone. (A) VEGF controls the initial osteo-angiogenic invasion of the endochondral bone template. During long bone development, hypertrophic chondrocytes in the middle diaphyseal region of the avascular cartilage template become hypoxic and express high levels of VEGF. Both HIF-1α and VEGF are required for the timely invasion of the template by blood vessels from the perichondrium. Along with the endothelium, osteoprogenitors move into the tissue, start to deposit bone, and establish the primary ossification center (POC). Osteoclasts, cells of hematopoietic origin, also appear, coinciding with vascular accumulation in the perichondrium, and coinvade the cartilage. All the cell types involved express VEGF receptors (*VEGFR*) and can respond directly to VEGF signaling by enhanced migration, recruitment, proliferation, and/or differentiation. **(B) VEGF actions at the chondro-osseous junction and the metaphysis of growing long bones.** The matrix-binding isoforms of VEGF, VEGF164 and VEGF188, are stored in the cartilage matrix after their secretion by hypertrophic chondrocytes. Upon cartilage resorption, mediated by osteoclasts and matrix metalloproteinase-9 (*MMP-9*), the released VEGF attracts blood vessels toward the growth plate and stimulates angiogenesis. Indirectly (via the vascular growth) and directly (via VEGFR signaling), VEGF stimulates bone formation by osteoblasts and cartilage resorption and bone remodeling by osteoclasts, thereby coordinating the conversion of cartilage into bone at the chondro-osseous junctions and stimulating the growth of the long bones. *Figure reproduced from Maes, C., 2017. Signaling pathways effecting crosstalk between cartilage and adjacent tissues: seminars in cell and developmental biology: the biology and pathology of cartilage. Semin. Cell Dev. Biol. 62, 16–33.*

controlled though, as either partial loss or local increase of VEGF is detrimental to bone development (Maes et al., 2010b; Haigh et al., 2000). In fact, Col2-Cre-driven conditional overexpression of VEGF164 in the fetal skeleton led to premature osteo-angiogenic invasion of developing bones, hypervascularization of the primary ossification center, aberrant bone deposition, and severely misshapen limbs (Maes et al., 2010b). The physiological induction of high levels of VEGF expression in hypertrophic chondrocytes appears to be fundamentally coupled to the differentiation progression of the cells and involves several transcriptional regulators, including Runx2, Osterix (Osx), and hypoxia-inducible factors (HIFs), particularly HIF-1α, the prime driver of the cellular responses to low-oxygen conditions (Chen et al., 2012; Zelzer et al., 2001; Tang et al., 2012; Cramer et al., 2004; Lin et al., 2004). HIF-1α signaling and other hypoxia-driven control mechanisms also contribute to the fine regulation of the more moderate VEGF levels in the hypoxic center of the developing growth plate and the preinvasion cartilaginous template; these signals are important for stimulating perichondrial angiogenesis as well as ensuring survival of the hypoxic chondrocytes (Maes et al., 2004, 2012a; Duan et al., 2015; Schipani et al., 2001; Provot et al., 2007; Amarilio et al., 2007; Eshkar-Oren et al., 2009).

Vascularization of the mature skeleton

Also beyond development, the skeletal vascular system remains an essential player in the regulation of bone formation and remodeling, as characterized most comprehensively in the long bones. These bones are supplied by a dense network of blood vessels comprising epiphyseal, metaphyseal, and periosteal blood vessels. The nutrient arteries provide the main entry of the blood into the bone shaft. Within the shaft, the vessels are organized in a highly branched and organized network of sinusoidal capillaries, arteries, and arterioles in the bone marrow environment, with endosteal vessels near the cortical bone and metaphyseal capillaries arranged in longitudinal orientation among the trabecular bone spicules toward the chondro-osseous junction of the growth plate (Brookes and Revell, 1998; Roche et al., 2012). Recent work has termed the endosteal and metaphyseal capillaries "H-type" blood vessels, because of their strong reactivity to antibodies recognizing the endothelial cell markers CD31/PECAM-1 and endomucin (EMCN) (CD31^{high};EMCN^{high}). The sinusoidal blood vessel network in the diaphyseal bone shaft is by these molecular criteria designated as the "L-type" endothelium in bone (CD31^{low};EMCN^{low}) (Kusumbe et al., 2014). Processes of adult bone remodeling and bone mass homeostasis, fracture repair, and bone tissue regeneration, as well as proper hematopoiesis, all rely on optimal bone and bone marrow vascularization. Yet, different types of skeletal blood vessels may provide differential support and contributions to these various processes. For instance, particularly H-type capillaries in growing and mature bones are heavily associated with perivascular osteoprogenitor cells, including Runx2-, Osx-, and platelet-derived growth factor (PDGF) receptor β-expressing cells, which are thought to be important for bone formation. The vascular system may thus help deliver these osteoprogenitor populations to bone formation sites, in addition to providing the necessary oxygen, nutrients, and growth factors for their controlled differentiation and activity (Dirckx et al., 2013; Maes et al., 2010a). Loss of these populations is observed with aging in mice and may be related to the well-documented age-related and osteoporosis-linked declines in both vascular density and bone mass (Maes, 2013; Ramasamy et al., 2016; Mekraldi et al., 2003; Burkhardt et al., 1987; Kusumbe et al., 2014). On the other hand, bone marrow arterioles and sinusoidal L-type vessels are surrounded by specific subsets of hematopoietic stem cells (HSCs) and provide critical components of the niche microenvironments that control their quiescence and differentiation balances (Ramasamy et al., 2016; Kusumbe et al., 2016; Itkin et al., 2016). Impaired revascularization has been long known to be detrimental to bone repair, but the specifics of the angiogenic processes and blood vessel types in fracture healing are only starting to be determined (Wang et al., 2017). Clearly, angiogenesis in bone and its coupling with osteogenesis provide promising therapeutic angles for low bone mass–associated diseases and for bone repair and regeneration, highlighting the importance of understanding the underlying crosstalk and regulatory mechanisms operating in osteolineage cells and in endothelial cells, as outlined next in this section (Fig. 9.2).

Bone cells' control of skeletal vascularization and oxygenation

Signaling from hypertrophic chondrocytes, osteoblast lineage cells, and osteoclasts has been found instructive toward the development, growth, and maintenance of the skeletal vascular system (Maes, 2013; Maes et al., 2012b). As the prime angiogenic factor, VEGF produced by these cell types plays a major role in adult bone. Moreover, the hypoxia-induced and HIF-mediated signaling pathways represent major regulators of the tight coupling between angiogenesis and osteogenesis in postnatal bone formation and bone remodeling, at least in part by functioning as key upstream inducers of VEGF expression (Maes et al., 2012b; Schipani et al., 2009). Some of the mouse models that exposed these roles include Cre-loxP-mediated conditional knockouts employing Osx-Cre, type I collagen-Cre, and osteocalcin (OC)-Cre driver strains to target osteoprogenitors, maturing osteoblasts, and differentiated osteoblasts, respectively. The first evidence for the importance of the HIF pathway components in the adult murine skeleton was provided by genetic targeting of HIF-1α and the upstream negative regulator of HIF activity, Von Hippel–Lindau (VHL), in OC-expressing osteoblasts (Wang et al., 2007). Mice lacking HIF-1α in osteoblasts exhibited narrower bones and reduced trabecular bone volume, associated with reduced vascular density and decreased bone formation (Wang et al., 2007; Shomento et al., 2010). Conversely, VHL inactivation, which is associated with constitutive HIF stabilization and boosted hypoxia signaling pathway activity, was associated with increased trabecular bone volume and increased VEGF levels and vascular density in bone (Wang et al., 2007). This study implicated osteoblastic HIF signaling in the regulation of angiogenesis in the bone microenvironment in response to hypoxia and postulated the principle of angiogenic–osteogenic coupling by hypoxia-sensing mechanisms, involving the indirect stimulation of osteogenesis as a consequence of skeletal vascular expansion (Wang et al., 2007). The critical oxygen-sensing prolyl hydroxylase enzymes (PHDs), which regulate HIF degradation in normoxia and control hypoxic HIF activity, have correspondingly been found to play important roles in the control of skeletal vascularization, bone formation, and bone remodeling (Wu et al., 2015). Combined inactivation of PHD1, PHD2, and PHD3 in osteoprogenitors resulted in extreme HIF signaling and excessive bone accumulation associated with overstimulation of angiogenic–osteogenic coupling (Wu et al., 2015).

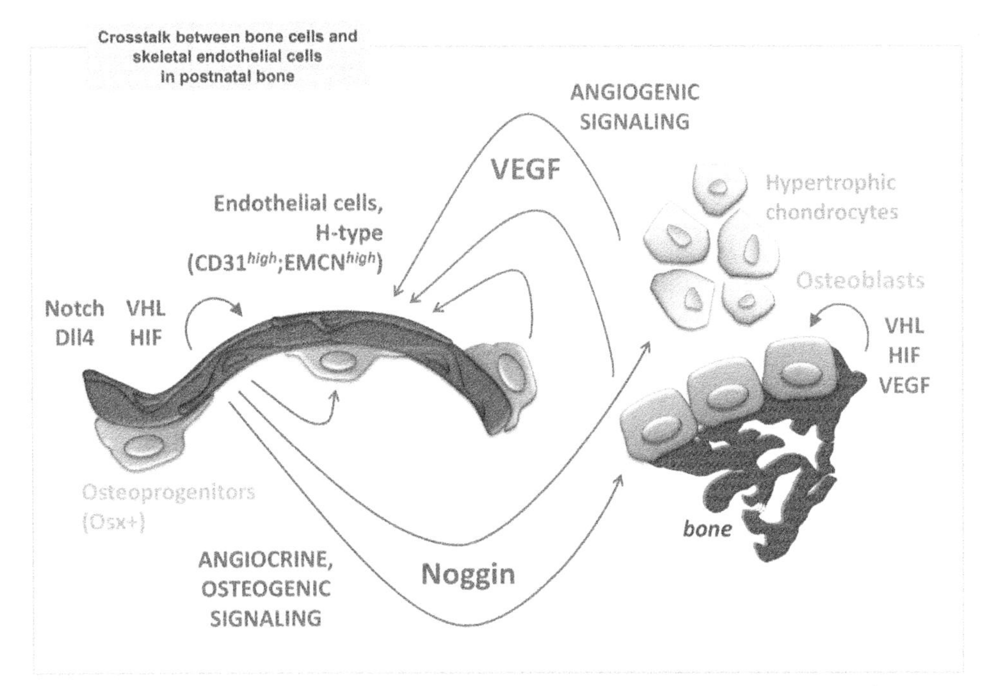

FIGURE 9.2 Bidirectional crosstalk between chondro-osteoblast lineage cells and skeletal endothelial cells, mediating angiogenic–osteogenic coupling in postnatal bone. Schematic view of the dynamic and reciprocal interplay between the different cell types in the bone environment that couples angiogenesis and osteogenesis. (Right and upper part) In the osteogenic compartment of the long bones, the signaling pathway governed by Von Hippel–Lindau (VHL), hypoxia-inducible factor (HIF), and vascular endothelial growth factor (VEGF) has been identified as a key driver of angiogenic–osteogenic coupling, by exerting both (1) cell-autonomous roles in chondrocytes, osteoprogenitors, and osteoblasts and (2) VEGF-mediated paracrine effects on the blood vessels in the bone environment, stimulating angiogenesis. (Left and lower part) In the vasculature of the bone and bone marrow environment, a specialized subtype of endothelial cells (ECs) expressing high levels of the endothelial markers CD31 and endomucin (EMCN) constitute the "type H" vessels in the metaphysis and endosteum. These ECs mediate the growth of the blood vessels in bone, through a tissue-selective mechanism of angiogenesis involving positive regulation by the VHL/HIF and Notch/Dll4 signaling pathways. Coupling of angiogenesis back to osteogenesis is mediated by osteoregulatory signals produced by ECs (i.e., angiocrine, osteogenic signals), such as Noggin. Osx, Osterix. *Figure adapted from Maes, C., Clemens, T.L., 2014. Angiogenic-osteogenic coupling: the endothelial perspective. Bonekey Rep. 3, 578.*

Work has also revealed additional roles of the hypoxia signaling pathways in the complex bone environment, as shown by using the Osx-Cre mouse to inactivate VHL in osteoprogenitors, either constitutively from fetal life onward or induced only later in the postnatal skeleton (by applying doxycycline/reverse tetracycline-controlled transactivator–mediated suppression of the activity of the Osx-Cre driver), and through other related mouse models. These studies revealed that the combined actions of the HIFs (both HIF-1 and HIF-2) increase angiogenesis and osteogenesis, alter hematopoiesis (mainly erythropoiesis), and enhance glycolytic cell metabolism in osteolineage cells (Shomento et al., 2010; Dirckx et al., 2018; Rankin et al., 2012; Regan et al., 2014). The last aspect appeared to be also associated with effects beyond bone on global energy metabolism, as hyperactivated hypoxia signaling through VHL deletion caused excessive glucose uptake and glycolysis in osteoblasts and an overall increased skeletal glucose consumption, along with impaired systemic control of glucose homeostasis (Dirckx et al., 2018). Altogether, the combined functions of the hypoxia pathway in bone are mediated by a large number of proteins encoded by direct HIF target genes (Semenza, 2012), with key roles ascribed to VEGF, erythropoietin, and genes involved in anaerobic cell metabolism, such as glucose transporters (e.g., Glut1) and key glycolytic enzymes including pyruvate dehydrogenase kinase 1, phosphoglycerate kinase 1, and lactate dehydrogenase A. As such, a growing number of effector proteins are being implicated in the responses to low oxygen in osteoblasts and other bone cell types, ensuring both cellular adaptation as well as remediation of the low oxygenation state within the tissue.

For VEGF specifically, the advantageous effects on osteogenesis are mediated partly indirectly by its paracrine proangiogenic actions—thereby stimulating the delivery of oxygen, nutrients, hormones, and angiocrine signaling factors (also see later) to the bone-forming osteoblasts—and partly directly. Indeed, most bone cell types express VEGF receptors, and ample in vitro studies have evidenced that VEGF signaling in osteolineage cells can induce their migration, differentiation, and bone-forming activity (reviewed in Maes, 2013). In vivo, a study overexpressing VEGF in the bone environment of adult mice suggested that cell-autonomous effects mediated by β-catenin signaling in osteoblast lineage cells contributed to the phenotype, which was overall characterized by high bone mass, expansion of the osteoprogenitor pool, bone marrow fibrosis, and hematopoietic defects (Maes et al., 2010b). Conversely, conditional inactivation of *VEGF* in osteoprogenitors (driven by the Osx-Cre mouse strain) was shown to result in increased bone marrow adipogenesis at the expense of osteogenesis, and implicated intracrine-acting VEGF in cell fate decisions of stromal progenitors in the bone environment (Liu et al., 2012). Proper angiogenic−osteogenic coupling thus requires tight regulation and dynamic fine-tuning of the hypoxia signaling events and VEGF levels within the bone environment.

These hypoxia-related mechanisms have also been implicated in widespread bone pathologies, and pharmacological PHD inhibitors could be of great therapeutic value, for instance, in age-related osteoporosis and compromised fracture healing. Genetic inactivation of VHL or PHDs in mice or treatment with PHD-inhibiting compounds was effective in protecting mice against age- or ovariectomy-induced bone loss (Kusumbe et al., 2014; Wu et al., 2015; Zhao et al., 2012; Weng et al., 2014; Liu et al., 2014). Furthermore, activation of HIFs can be a successful strategy to improve fracture repair and bone regeneration, as supported by experimental animal models (Wan et al., 2008; Shen et al., 2009).

Endothelial and angiocrine signaling in bone

The aforementioned studies amply showed that osteolineage cells are equipped with the molecular tools to sense oxygen tension and direct adjustments in vascularization by signaling to endothelial cells. In addition, the regulation of skeletal vascularization and angiogenic−osteogenic coupling also involves endothelial cell-intrinsic mechanisms and angiocrine signals secreted by the endothelium, including the HIF signaling pathway, Notch, and bone morphogenetic protein (BMP) family signaling. Indeed, the oxygen-regulated program mediated by HIFs is also key in the endothelial cells of the skeletal blood vessels, in line with the hypoxic status of the postnatal bone and bone marrow environment. This has been shown by endothelium-specific, inducible inactivation of VHL in mice (using the vascular endothelial−cadherin [or Cdh5] promotor to drive the expression of CreERt2), which led to a striking increase in H-type blood vessels in the bones of juvenile mice (Kusumbe et al., 2014). Concomitantly, the pool of $Runx2^{+}$ or Osx^{+} cells associated with these vessels was expanded and the bone volume increased. The converse phenotype was seen upon endothelial cell-targeted inactivation of HIF-1α, as this mutation led to severe vascular defects and reduced numbers of osteoprogenitors (Kusumbe et al., 2014). It is thought that endothelial cells of H-type blood vessels secrete pro-osteogenic factors such as fibroblast growth factors, PDGFs, and BMPs, and as such contribute to the coupling of angiogenesis and osteogenesis (Ramasamy et al., 2016; Kusumbe et al., 2014). One signaling system that was found to be very important in the local production of angiocrine signals by vascular endothelial cells in the bone microenvironment is the Notch pathway (Ramasamy et al., 2014). Loss- and gain-of-function mouse models established that Notch/Dll4 signaling in H-type capillary endothelial cells regulated angiogenesis but also osteogenesis, by influencing the endothelial production and secretion of the BMP antagonist Noggin, which surprisingly was found to positively influence osteolineage cells and bone formation in vivo (Ramasamy et al., 2014).

These studies and further work are increasingly revealing that the endothelial cells of the skeletal blood vessels exert important signaling functions contributing to the control of bone development, growth, homeostasis, and regeneration, as well as to hematopoiesis. Indeed, within the skeletal system, blood vessels also constitute critical components of the niche microenvironments for HSCs in the bone marrow; the specific localization, nature, and molecular determinants of the vascular niches for HSCs are topics of intensive ongoing investigations at the interface between bone biology and hematology, with broad therapeutic application (Kusumbe et al., 2016; Itkin et al., 2016).

The nerve system of bone

Innervation of developing bone

The development of the nervous system is a complex process of axonal growth, synapse formation, and cellular signaling, during which the spinal cord forms from the neural tube during neurulation (Ciani and Salinas, 2005). Sensory axons enter the spinal cord through the dorsal roots and then ascend through the emerging white matter to the brain, whereas sympathetic neurons migrate ventrally to coalesce in prevertebral sympathetic ganglia (Glebova and Ginty, 2005). Both sensory and sympathetic axons are guided to their appropriate peripheral target tissues, including bone, by neurotrophic factors (Pezet and McMahon, 2006). In fact, developing tissues dictate the amount and type of innervation they require by secreting the specific neurotrophin that promotes neuronal survival of its corresponding axon type (Reichardt, 2006). At least one such neurotrophin is known to be active in developing long bone: nerve growth factor (NGF). In a 2016 study, this prototypical neurotrophic factor was found to be expressed by osteoprogenitor cells during both primary and secondary ossification (Tomlinson et al., 2016). Moreover, sensory neurons located in the dorsal root ganglion project axons

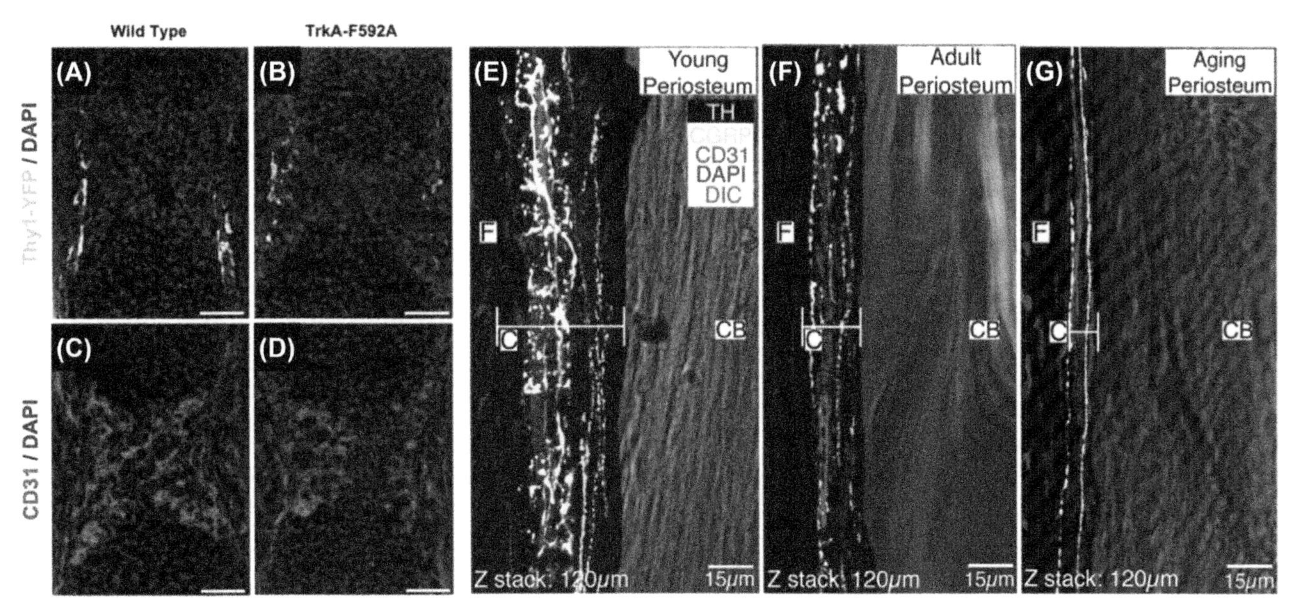

FIGURE 9.3 Skeletal nerves in the periosteum from beginning to end. (A) Thy1-YFP reporter mice were used to visualize nerve axons in the perichondrial region at embryonic day 15.5, just after primary ossification centers have formed. (B) Inhibition of nerve growth factor (NGF)−tyrosine kinase receptor 1 (TrkA) signaling in TrkA-F592A mice diminished nerve axons in the perichondrial region. (C, D) Similarly, inhibition of NGF−TrkA signaling impaired vascular invasion of the ossification center, as visualized by staining against CD31. Scale bars, 100 μm. (E) The periosteum of a young mouse is teeming with $CD31^+$ blood vessels (*red*), calcitonin gene-related peptide $(CGRP)^+$ sensory nerve axons (*green*), and tyrosine hydroxylase (TH)+ sympathetic nerve axons (*yellow*). (F, G) As the mouse ages to adulthood and beyond, the cambium layer shrinks but sensory and sympathetic nerve fibers along with blood vessels remain intact in the periosteum. Cambium (*C*) and fibrous (*F*) layers of the periosteum and cortical bone (*CB*) are labeled. *(C, D) Adapted from Tomlinson R.E., Li, Z., Zhang, Q., Goh B.C., Li, Z., Thorek D.L.J., Rajbhandari, L., Brushart T.M., Minichiello, L., Zhou, F., Venkatesan, A., Clemens T.L., 2016. NGF-TrkA signaling by sensory nerves coordinates the vascularization and ossification of developing endochondral bone. Cell Rep. 16 (10), 2723−2735. (F, G) Adapted from Chartier, S.R., Mitchell, S.A.T., Majuta, L.A., Mantyh, P.W., 2018. The changing sensory and sympathetic innervation of the young, adult and aging mouse femur. Neuroscience 387, 178−190.*

expressing neurotrophic tyrosine kinase receptor 1 (TrkA), the high-affinity receptor for NGF, which reach perichondrial bone surfaces adjacent to primary ossification centers at a time coincident with the initiation of NGF expression in osteochondral progenitor cells. Inactivation of TrkA signaling in mice impaired skeletal innervation, delayed vascular invasion, and decreased the number of osteoprogenitor cells, resulting in decreased femoral bone length and volume. These results were consistent with previous studies on the skeletal action of semaphorin 3A (Sema3A), a well-characterized inhibitor of neuronal outgrowth. Mice lacking Sema3a were observed to have gross defects in nerve, heart, and skeletal patterning (Behar et al., 1996). Subsequent work found that selective loss of Sema3a in neurons, but not osteoblasts, decreased skeletal sensory innervation and postnatal bone mass (Fukuda et al., 2013). To date, no studies have conclusively examined the timing and location of the arrival of sympathetic or parasympathetic nerves in bone. Given that sympathetic innervation of the mesenteric arteries in mouse takes place postnatally, it is unlikely that autonomic nerve fibers play a major role in primary or secondary ossification (Brunet et al., 2014) (Fig. 9.3).

Innervation of the mature skeleton

Somatic nervous system

Sensory nerves of the somatic nervous system are distributed throughout the body, including the skin, bones, joints, tendons, and muscles, and transduce signals involved in spatial orientation (proprioception), pain and noxious stimuli (nociception), temperature changes, and the perception of nonpainful tactile stimuli (Julius and Basbaum, 2001; McKemy et al., 2002; Proske and Gandevia, 2012; Jones and Smith, 2014). To accomplish these functions, nerve endings generally express a wide variety of receptors and channels that will initiate the appropriate signaling cascades in response to perturbations (Pongratz and Straub, 2013). These include Toll-like receptors, cytokine and prostaglandin receptors, transient receptor potential ion channels, and growth factor receptors, among many others (Pongratz and Straub, 2013). Surprisingly, the vast majority (>80%) of nerves in mature bone are thinly myelinated or unmyelinated sensory nerves that express

TrkA, a consequence of the expression of NGF during development (Castañeda-Corral et al., 2011; Tomlinson et al., 2016; Jimenez-Andrade et al., 2010). Furthermore, the density of sensory nerves in the adult mouse femur is highest in the periosteum and marrow space, relatively low in the mineralized bone, and essentially zero in healthy cartilage (Mach et al., 2002; Castañeda-Corral et al., 2011).

In general, the most familiar and best understood function of sensory nerves is pain sensation, which is profoundly apparent following traumatic bone fracture or in patients with bone malignancies. The central mediator of osseous pain sensation is NGF, which transmits nociceptive signals either by directly activating the abundant TrkA+ sensory nerves in bone or through indirect mechanisms that enhance the response of other nociceptive pathways (Mantyh, 2014). This upstream position of the NGF−TrkA signaling pathway in pain sensation has made it an attractive target for developing pharmaceuticals for treating pain associated with cancer and osteoarthritis as well as an active area of research. Several studies have documented upregulated expression of NGF in progenitor cells, osteoblasts, and chondrocytes during fracture repair in rodents (Asaumi et al., 2000; Grills and Schuijers, 1998). Correspondingly, blockade of NGF function has been observed to significantly decrease behavior associated with pain following fracture in mice (Jimenez-Andrade et al., 2007; Koewler et al., 2007), and additional studies demonstrated that this analgesia can be provided without affecting standard fracture healing outcomes (Rapp et al., 2015). However, others have shown that topical application of NGF to rib fractures in rats and the distraction callus in rabbit mandibles significantly decreases healing time and improves subsequent repair strength (Grills et al., 1997; Wang et al., 2006). This effect is thought to be mediated through NGF-induced release of neuropeptides such as calcitonin gene-related peptide (CGRP) and substance P (Malcangio et al., 1997; Quarcoo et al., 2004), which are both known to positively influence bone formation (Baldock et al., 2007; Goto et al., 2001, 2007; Li et al., 2010; Long et al., 2010; Lundberg et al., 2007; Sample et al., 2011). As a result, more work must be done to fully understand the role of NGF−TrkA signaling in bone fracture repair. Nonetheless, widespread therapeutic use of the highly effective anti-NGF antibodies for pain relief may be imminent (Lane et al., 2010; Jayabalan and Schnitzer, 2017; Birbara et al., 2018). Although phase III clinical trials were halted in 2010 due to an increased incidence of adverse events (Hochberg, 2015), the humanized monoclonal anti-NGF antibody tanezumab (Pfizer and Lilly) received FDA fast-track approval in 2017 as the first in a new class of nonopioid pain relievers.

In addition to pain sensation, sensory nerves in bone have also been found to play an important role in responding to mechanical loads placed on the skeleton. This line of inquiry was initiated based on the observation that primary afferent sensory nerves are present in a dense meshlike network on the periosteal and endosteal surfaces (Martin et al., 2007), which are the preferential sites for the perception of skeletal deformation. Mature, OC-expressing osteoblasts on the periosteal and endosteal surfaces of bone were observed to robustly express NGF in response to mechanical loading in mice (Tomlinson et al., 2017). Furthermore, the researchers found that inactivation of NGF−TrkA signaling significantly attenuated bone formation in response to load, via a decrease in osteocytic Wnt/β-catenin signaling. Since sensory nerves remain essentially intact in the periosteum throughout life, even as bone undergoes marked changes associated with aging (Chartier et al., 2018), targeting NGF−TrkA signaling for bone accrual may be a viable strategy in cases in which bone cell−targeted therapies may fail. Importantly, more research is necessary to determine if anti-NGF therapeutic agents for pain will substantially affect strain-adaptive bone remodeling in humans.

Autonomic nervous system

The autonomic nervous system (ANS), part of the peripheral nervous system, is divided into two antagonistic branches: the sympathetic nervous system and the parasympathetic nervous system. Whereas sympathetic nerves primarily signal through the release of norepinephrine to activate α- and β-adrenergic receptors, parasympathetic nerves release acetylcholine that activates muscarinic acetylcholine receptors (mAChRs) and nicotinic acetylcholine receptors (nAChRs). In general, these two nerve types serve to coordinate bodily functions not consciously directed, such as breathing or the regulation of blood pressure. Both nerve types are present in bone tissue, and have been primarily visualized in close contact with the main arteries that provide the blood supply to long bones (Bjurholm et al., 1988; Hill and Elde, 1991; Tabarowski et al., 1996). In particular, nerves expressing tyrosine hydroxylase (TH) (the rate-limiting enzyme in the synthesis of catecholamines) are routinely observed wrapping around blood vessels in bone with a spiral-type morphology (Mach et al., 2002; Castañeda-Corral et al., 2011). Similarly, parasympathetic nerves with immunoreactivity against vesicular acetylcholine transporter and choline acetyltransferase can be observed readily, particularly in the marrow space (Artico et al., 2002; Bajayo et al., 2012). However, the specific locations, densities, and distributions throughout the bone and marrow space are not well described.

Generally, it is held that the ANS exerts its effect on the skeleton via close contact with bone cells. However, given the limited direct interaction of ANS nerve fibers with skeletal cells (Dénes et al., 2005), others have proposed that a diffusion mechanism may have a role (Elefteriou, 2018). Nonetheless, osteoblasts and osteoclasts have both been observed to express a wide variety of adrenergic receptors, most notably the β2-adrenergic receptor (Kajimura et al., 2011), which can respond to the norepinephrine released by sympathetic nerve terminals (Togari, 2002). Work has shown that neither osteoblasts nor osteoclasts express detectable levels of mAChRs, but nAChRs are readily observable, particularly the α2 and β2 subunits, which may respond to the release of acetylcholine from parasympathetic nerve terminals (Bajayo et al., 2012).

Bone remodeling activity is at least partially restrained by signals emanating from the ANS via regulation of bone resorption (Elefteriou et al., 2005). Several studies have shown that sympathetic nervous outflow stimulates bone resorption that, in turn, negatively affects bone formation (Elefteriou et al., 2005; Katayama et al., 2006), whereas parasympathetic signaling appears to favor bone accrual by inhibiting bone resorption (Bajayo et al., 2012). This paradigm is consistent with the circadian rhythm of the ANS. Generally, sympathetic nervous activity is dominant during the day, during peak bone resorption; conversely, parasympathetic nervous activity is dominant during the night, during peak bone formation (Dudek and Meng, 2014; Komoto et al., 2012; Shao et al., 2003). Furthermore, mice lacking β2-adrenergic receptor in the osteoblast lineage have significantly increased bone formation coupled with decreased bone resorption, leading to an eventual increase in total bone mass in adulthood (Kajimura et al., 2011). However, pharmacological studies do not recapitulate this encouraging result: treatment with either a β-adrenergic receptor agonist (salbutamol) or an antagonist (isoprenaline) was associated with bone loss in mice, mainly through increased bone resorption (Bonnet et al., 2007; Kondo and Togari, 2011). Nonetheless, the known safety profile of "β blockers," which block activation of β-adrenergic receptors and have achieved widespread use for the treatment of cardiovascular disease, has resulted in significant interest in modulating ANS signaling to increase bone mass in humans. Encouragingly, a recent meta-analysis of seven cohort and nine case—control studies found that the use of β blockers was associated with a 15% decrease in overall fracture risk, with β1-specific blockers most strongly associated with the risk reduction (Toulis et al., 2014). Nonetheless, a 2015 randomized clinical trial was unable to establish an effect of β2-adrenergic agonists or antagonists on human bone turnover (Veldhuis-Vlug et al., 2015). Furthermore, preclinical studies have concluded that the contribution of sympathetic signaling to load-induced bone formation in bone is minimal (de Souza et al., 2005); no studies have investigated the role of parasympathetic signaling in this process. In total, much work remains to determine the direct and specific effects of the ANS on bone metabolism.

Conclusion

The pre-eminent role of vascularization in bone was originally reported in the 1960s (Trueta and Buhr, 1963; Trueta and Caladias, 1964) and is now well established. Moreover, much has been learned about the regulation of vascular growth in bone and the roles of VEGF and the hypoxia signaling pathways, both in physiology and in bone defect repair. Altogether, studies performed in this field are progressively uncovering the molecular underpinnings of osteo-angiogenic coupling, which obviously involves a vigorous bidirectional crosstalk between endothelial cells and osteolineage cells that is only beginning to be uncovered. Especially little is known, as of this writing, about the nature of angiocrine signaling factors emanating from the endothelium and their roles in developing, growing, remodeling, and healing bones. Many other aspects of the organization, integrity, and functioning of the vasculature in the skeletal system remain to be investigated in further depth, including the molecular specifics of the specialized vascular environments for skeletal stem/progenitor cell harboring and bone formation, and of the niches controlling the maintenance of HSCs and the support of hematopoiesis in the bone marrow environment. Increasing insights into these areas will direct future studies exploring how the bone vasculature can ultimately be modulated to achieve anabolic treatments of osteoporosis and other bone disorders, to support compromised fracture repair and tissue engineering therapies, and to improve treatments of hematological pathologies and stem cell transplantation protocols.

Similarly, the skeleton is densely innervated by a wide variety of somatic and autonomic nerve axons that carry out their unique functions in a complex microenvironment. Despite tremendous advances, much remains to be uncovered regarding the nerves in bone. These lingering questions include the timeline of sympathetic innervation in bone, the role of NGF—TrkA signaling in bone repair, the identification of nerve-derived osteogenic cues, the specific action of the ANS on bone remodeling, and the alterations in nerve signaling that occur with aging.

Finally, the mechanisms by which bone vascular and nerve systems interact during development, homeostatic bone remodeling, and fracture repair still need to be uncovered. Answers to these fundamental questions will inevitably lead to novel approaches for improving overall skeletal health.

References

Amarilio, R., Viukov, S.V., Sharir, A., Eshkar-Oren, I., Johnson, R.S., Zelzer, E., 2007. HIF-1alpha regulation of Sox9 is necessary to maintain differentiation of hypoxic prechondrogenic cells during early skeletogenesis. Development 134 (21), 3917–3928.

Artico, M., Bosco, S., Cavallotti, C., Agostinelli, E., Giuliani-Piccari, G., Sciorio, S., Cocco, L., Vitale, M., 2002. Noradrenergic and cholinergic innervation of the bone marrow. Int. J. Mol. Med. 10 (1), 77–80.

Asaumi, K., Nakanishi, T., Asahara, H., Inoue, H., Takigawa, M., 2000. Expression of neurotrophins and their receptors (TRK) during fracture healing. Bone 26 (6), 625–633.

Bajayo, A., Bar, A., Denes, A., Bachar, M., Kram, V., Attar-Namdar, M., Zallone, A., Kovács, K.J., Yirmiya, R., Bab, I., 2012. Skeletal parasympathetic innervation communicates central IL-1 signals regulating bone mass accrual. Proc. Natl. Acad. Sci. U.S.A. 109 (38), 15455–15460.

Baldock, P.A., Allison, S.J., Lundberg, P., Lee, N.J., Slack, K., Lin, E.J., Enriquez, R.F., McDonald, M.M., Zhang, L., During, M.J., Little, D.G., Eisman, J.A., Gardiner, E.M., Yulyaningsih, E., Lin, S., Sainsbury, A., Herzog, H., 2007. Novel role of Y1 receptors in the coordinated regulation of bone and energy homeostasis. J. Biol. Chem. 282 (26), 19092–19102.

Behar, O., Golden, J.A., Mashimo, H., Schoen, F.J., Fishman, M.C., 1996. Semaphorin III is needed for normal patterning and growth of nerves, bones and heart. Nature 383 (6600), 525–528.

Birbara, C., Dabezies Jr., E.J., Burr, A.M., Fountaine, R.J., Smith, M.D., Brown, M.T., West, C.R., Arends, R.H., Verburg, K.M., 2018. Safety and efficacy of subcutaneous tanezumab in patients with knee or hip osteoarthritis. J. Pain Res. 11, 151–164.

Bjurholm, A., Kreicbergs, A., Brodin, E., Schultzberg, M., 1988. Substance P- and CGRP-immunoreactive nerves in bone. Peptides 9 (1), 165–171.

Bonnet, N., Benhamou, C.L., Beaupied, H., Laroche, N., Vico, L., Dolleans, E., Courteix, D., 2007. Doping dose of salbutamol and exercise: deleterious effect on cancellous and cortical bones in adult rats. J. Appl. Physiol. 102 (4), 1502–1509.

Brookes, M., Revell, W.J., 1998. Blood Supply of Bone: Scientific Aspects. Springer, London; New York xx, 359 p.

Brunet, I., Gordon, E., Han, J., Cristofaro, B., Broqueres-You, D., Liu, C., Bouvrée, K., Zhang, J., del Toro, R., Mathivet, T., Larrivée, B., Jagu, J., Pibouin-Fragner, L., Pardanaud, L., Machado, M.J.C., Kennedy, T.E., Zhuang, Z., Simons, M., Levy, B.I., Tessier-Lavigne, M., Grenz, A., Eltzschig, H., Eichmann, A., 2014. Netrin-1 controls sympathetic arterial innervation. J. Clin. Invest. 124 (7), 3230–3240.

Burkhardt, R., Kettner, G., Bohm, W., Schmidmeier, M., Schlag, R., Frisch, B., Mallmann, B., Eisenmenger, W., Gilg, T., 1987. Changes in trabecular bone, hematopoiesis and bone marrow vessels in aplastic anemia, primary osteoporosis, and old age: a comparative histomorphometric study. Bone 8 (3), 157–164.

Carano, R.A., Filvaroff, E.H., 2003. Angiogenesis and bone repair. Drug Discov. Today 8 (21), 980–989.

Castañeda-Corral, G., Jimenez-Andrade, J.M., Bloom, A.P., Taylor, R.N., Mantyh, W.G., Kaczmarska, M.J., Ghilardi, J.R., Mantyh, P.W., 2011. The majority of myelinated and unmyelinated sensory nerve fibers that innervate bone express the tropomyosin receptor kinase a. Neuroscience 178, 196–207.

Chartier, S.R., Mitchell, S.A.T., Majuta, L.A., Mantyh, P.W., 2018. The changing sensory and sympathetic innervation of the young, adult and aging mouse femur. Neuroscience 387, 178–190.

Chen, D., Tian, W., Li, Y., Tang, W., Zhang, C., 2012. Osteoblast-specific transcription factor osterix (Osx) and HIF-1alpha cooperatively regulate gene expression of vascular endothelial growth factor (VEGF). Biochem. Biophys. Res. Commun. 424 (1), 176–181.

Ciani, L., Salinas, P.C., 2005. Wnts in the vertebrate nervous system: from patterning to neuronal connectivity. Nat. Rev. Neurosci. 6 (5), 351–362.

Cramer, T., Schipani, E., Johnson, R.S., Swoboda, B., Pfander, D., 2004. Expression of VEGF isoforms by epiphyseal chondrocytes during low-oxygen tension is HIF-1 alpha dependent. Osteoarthritis Cartilage 12 (6), 433–439.

de Souza, R.L., Pitsillides, A.A., Lanyon, L.E., Skerry, T.M., Chenu, C., 2005. Sympathetic nervous system does not mediate the load-induced cortical new bone formation. J. Bone Miner. Res. 20 (12), 2159–2168.

Dénes, A., Boldogkoi, Z., Uhereczky, G., Hornyák, A., Rusvai, M., Palkovits, M., Kovács, K.J., 2005. Central autonomic control of the bone marrow: multisynaptic tract tracing by recombinant pseudorabies virus. Neuroscience 134 (3), 947–963.

Dirckx, N., Maes, C., 2016. Hypoxia-driven pathways in endochondral bone development. In: Grässel, S., Aszodi, A. (Eds.), Cartilage, Physiology and Development, vol. 1. Springer International Publishing, Switzerland, pp. 143–168 (Chapter 6).

Dirckx, N., Van Hul, M., Maes, C., 2013. Osteoblast recruitment to sites of bone formation in skeletal development, homeostasis, and regeneration. Birth Defects Res. C Embryo Today 99 (3), 170–191.

Dirckx, N., Tower, R.J., Mercken, E.M., Vangoitsenhoven, R., Moreau-Triby, C., Breugelmans, T., Nefyodova, E., Cardoen, R., Mathieu, C., Van der Schueren, B., Confavreux, C.B., Clemens, T.L., Maes, C., 2018. VHL deletion in osteoblasts boosts cellular glycolysis and improves global glucose metabolism. J. Clin. Invest. 128 (3), 1087–1105.

Duan, X., Murata, Y., Liu, Y., Nicolae, C., Olsen, B.R., Berendsen, A.D., 2015. Vegfa regulates perichondrial vascularity and osteoblast differentiation in bone development. Development 142 (11), 1984–1991.

Dudek, M., Meng, Q.-J., 2014. Running on time: the role of circadian clocks in the musculoskeletal system. Biochem. J. 463 (1), 1–8.

Elefteriou, F., Ahn, J.D., Takeda, S., Starbuck, M., Yang, X., Liu, X., Kondo, H., Richards, W.G., Bannon, T.W., Noda, M., Clement, K., Vaisse, C., Karsenty, G., 2005. Leptin regulation of bone resorption by the sympathetic nervous system and cart. Nature 434 (7032), 514–520.

Elefteriou, F., 2018. Impact of the autonomic nervous system on the skeleton. Physiol. Rev. 98 (3), 1083–1112.

Eshkar-Oren, I., Viukov, S.V., Salameh, S., Krief, S., Oh, C.D., Akiyama, H., Gerber, H.P., Ferrara, N., Zelzer, E., 2009. The forming limb skeleton serves as a signaling center for limb vasculature patterning via regulation of VEGF. Development 136 (8), 1263–1272.

Fukuda, T., Takeda, S., Xu, R., Ochi, H., Sunamura, S., Sato, T., Shibata, S., Yoshida, Y., Gu, Z., Kimura, A., Ma, C., Xu, C., Bando, W., Fujita, K., Shinomiya, K., Hirai, T., Asou, Y., Enomoto, M., Okano, H., Okawa, A., Itoh, H., 2013. Sema3a regulates bone-mass accrual through sensory innervations. Nature 497 (7450), 490–493.

Gerber, H.P., Vu, T.H., Ryan, A.M., Kowalski, J., Werb, Z., Ferrara, N., 1999. VEGF couples hypertrophic cartilage remodeling, ossification and angiogenesis during endochondral bone formation. Nat. Med. 5 (6), 623–628.

Glebova, N.O., Ginty, D.D., 2005. Growth and survival signals controlling sympathetic nervous system development. Annu. Rev. Neurosci. 28, 191–222.

Goto, T., Kido, M.A., Yamaza, T., Tanaka, T., 2001. Substance p and substance p receptors in bone and gingival tissues. Med. Electron. Microsc. 34 (2), 77–85.

Goto, T., Nakao, K., Gunjigake, K.K., Kido, M.A., Kobayashi, S., Tanaka, T., 2007. Substance P stimulates late-stage rat osteoblastic bone formation through neurokinin-1 receptors. Neuropeptides 41 (1), 25–31.

Grills, B.L., Schuijers, J.A., 1998. Immunohistochemical localization of nerve growth factor in fractured and unfractured rat bone. Acta Orthop. Scand. 69 (4), 415–419.

Grills, B.L., Schuijers, J.A., Ward, A.R., 1997. Topical application of nerve growth factor improves fracture healing in rats. J. Orthop. Res. 15 (2), 235–242.

Haigh, J.J., Gerber, H.P., Ferrara, N., Wagner, E.F., 2000. Conditional inactivation of VEGF-A in areas of collagen2a1 expression results in embryonic lethality in the heterozygous state. Development 127 (7), 1445–1453.

Hill, E.L., Elde, R., 1991. Distribution of c CGRP-, VIP-, D beta H-, SP-, and NPY-immunoreactive nerves in the periosteum of the rat. Cell Tissue Res. 264 (3), 469–480.

Hochberg, M.C., 2015. Serious joint-related adverse events in randomized controlled trials of anti-nerve growth factor monoclonal antibodies. Osteoarthritis Cartilage 23 (Suppl. 1), S21.

Hohmann, E.L., Elde, R.P., Rysavy, J.A., Einzig, S., Gebhard, R.L., 1986. Innervation of periosteum and bone by sympathetic vasoactive intestinal peptide-containing nerve fibers. Science 232 (4752), 868–871.

Hukkanen, M., Konttinen, Y.T., Rees, R.G., Gibson, S.J., Santavirta, S., Polak, J.M., 1992. Innervation of bone from healthy and arthritic rats by substance P and calcitonin gene related peptide containing sensory fibers. J. Rheumatol. 19 (8), 1252–1259.

Itkin, T., Gur-Cohen, S., Spencer, J.A., Schajnovitz, A., Ramasamy, S.K., Kusumbe, A.P., Ledergor, G., Jung, Y., Milo, I., Poulos, M.G., Kalinkovich, A., Ludin, A., Kollet, O., Shakhar, G., Butler, J.M., Rafii, S., Adams, R.H., Scadden, D.T., Lin, C.P., Lapidot, T., 2016. Distinct bone marrow blood vessels differentially regulate haematopoiesis. Nature 532 (7599), 323–328.

Jayabalan, P., Schnitzer, T.J., 2017. Tanezumab in the treatment of chronic musculoskeletal conditions. Expert Opin. Biol. Ther. 17 (2), 245–254.

Jimenez-Andrade, J.M., Martin, C.D., Koewler, N.J., Freeman, K.T., Sullivan, L.J., Halvorson, K.G., Barthold, C.M., Peters, C.M., Buus, R.J., Ghilardi, J.R., Lewis, J.L., Kuskowski, M.A., Mantyh, P.W., 2007. Nerve growth factor sequestering therapy attenuates non-malignant skeletal pain following fracture. Pain 133 (1–3), 183–196.

Jimenez-Andrade, J.M., Mantyh, W.G., Bloom, A.P., Xu, H., Ferng, A.S., Dussor, G., Vanderah, T.W., Mantyh, P.W., 2010. A phenotypically restricted set of primary afferent nerve fibers innervate the bone versus skin: therapeutic opportunity for treating skeletal pain. Bone 46 (2), 306–313.

Jones, L.A., Smith, A.M., 2014. Tactile sensory system: encoding from the periphery to the cortex. Wiley Interdiscip. Rev. Syst. Biol. Med. 6 (3), 279–287.

Julius, D., Basbaum, A.I., 2001. Molecular mechanisms of nociception. Nature 413 (6852), 203–210.

Kajimura, D., Hinoi, E., Ferron, M., Kode, A., Riley, K.J., Zhou, B., Guo, X.E., Karsenty, G., 2011. Genetic determination of the cellular basis of the sympathetic regulation of bone mass accrual. J. Exp. Med. 208 (4), 841–851.

Karsenty, G., Wagner, E.F., 2002. Reaching a genetic and molecular understanding of skeletal development. Dev. Cell 2 (4), 389–406.

Katayama, Y., Battista, M., Kao, W.-M., Hidalgo, A., Peired, A.J., Thomas, S.A., Frenette, P.S., 2006. Signals from the sympathetic nervous system regulate hematopoietic stem cell egress from bone marrow. Cell 124 (2), 407–421.

Koewler, N.J., Freeman, K.T., Buus, R.J., Herrera, M.B., Jimenez-Andrade, J.M., Ghilardi, J.R., Peters, C.M., Sullivan, L.J., Kuskowski, M.A., Lewis, J.L., Mantyh, P.W., 2007. Effects of a monoclonal antibody raised against nerve growth factor on skeletal pain and bone healing after fracture of the C57BL/6J mouse femur. J. Bone Miner. Res. 22 (11), 1732–1742.

Komoto, S., Kondo, H., Fukuta, O., Togari, A., 2012. Comparison of β-adrenergic and glucocorticoid signaling on clock gene and osteoblast-related gene expressions in human osteoblast. Chronobiol. Int. 29 (1), 66–74.

Kondo, H., Togari, A., 2011. Continuous treatment with a low-dose β-agonist reduces bone mass by increasing bone resorption without suppressing bone formation. Calcif. Tissue Int. 88 (1), 23–32.

Kusumbe, A.P., Ramasamy, S.K., Adams, R.H., 2014. Coupling of angiogenesis and osteogenesis by a specific vessel subtype in bone. Nature 507 (7492), 323–328.

Kusumbe, A.P., Ramasamy, S.K., Itkin, T., Mae, M.A., Langen, U.H., Betsholtz, C., Lapidot, T., Adams, R.H., 2016. Age-dependent modulation of vascular niches for haematopoietic stem cells. Nature 532 (7599), 380–384.

Lane, N.E., Schnitzer, T.J., Birbara, C.A., Mokhtarani, M., Shelton, D.L., Smith, M.D., Brown, M.T., 2010. Tanezumab for the treatment of pain from osteoarthritis of the knee. N. Engl. J. Med. 363 (16), 1521–1531.

Li, J., Ahmed, M., Bergstrom, J., Ackermann, P., Stark, A., Kreicbergs, A., 2010. Occurrence of substance P in bone repair under different load comparison of straight and angulated fracture in rat tibia. J. Orthop. Res. 28 (12), 1643–1650.

Lin, C., McGough, R., Aswad, B., Block, J.A., Terek, R., 2004. Hypoxia induces HIF-1alpha and VEGF expression in chondrosarcoma cells and chondrocytes. J. Orthop. Res. 22 (6), 1175–1181.

Linder, J.E., 1978. A simple and reliable method for the silver impregnation of nerves in paraffin sections of soft and mineralized tissues. J. Anat. 127 (Pt 3), 543–551.

Liu, Y., Berendsen, A.D., Jia, S., Lotinun, S., Baron, R., Ferrara, N., Olsen, B.R., 2012. Intracellular VEGF regulates the balance between osteoblast and adipocyte differentiation. J. Clin. Invest. 122 (9), 3101–3113.

Liu, X., Tu, Y., Zhang, L., Qi, J., Ma, T., Deng, L., 2014. Prolyl hydroxylase inhibitors protect from the bone loss in ovariectomy rats by increasing bone vascularity. Cell Biochem Biophys 69 (1), 141–149.

Long, H., Ahmed, M., Ackermann, P., Stark, A., Li, J., 2010. Neuropeptide Y innervation during fracture healing and remodeling. A study of angulated tibial fractures in the rat. Acta Orthop. 81 (5), 639–646.

Lundberg, P., Allison, S.J., Lee, N.J., Baldock, P.A., Brouard, N., Rost, S., Enriquez, R.F., Sainsbury, A., Lamghari, M., Simmons, P., Eisman, J.A., Gardiner, E.M., Herzog, H., 2007. Greater bone formation of Y2 knockout mice is associated with increased osteoprogenitor numbers and altered Y1 receptor expression. J. Biol. Chem. 282 (26), 19082–19091.

Mach, D.B., Rogers, S.D., Sabino, M.C., Luger, N.M., Schwei, M.J., Pomonis, J.D., Keyser, C.P., Clohisy, D.R., Adams, D.J., O'Leary, P., Mantyh, P.W., 2002. Origins of skeletal pain: sensory and sympathetic innervation of the mouse femur. Neuroscience 113 (1), 155–166.

Maes, C., Clemens, T.L., 2014. Angiogenic-osteogenic coupling: the endothelial perspective. Bonekey Rep. 3, 578.

Maes, C., Carmeliet, P., Moermans, K., Stockmans, I., Smets, N., Collen, D., Bouillon, R., Carmeliet, G., 2002. Impaired angiogenesis and endochondral bone formation in mice lacking the vascular endothelial growth factor isoforms VEGF164 and VEGF188. Mech. Dev. 111, 61–73.

Maes, C., Stockmans, I., Moermans, K., Van Looveren, R., Smets, N., Carmeliet, P., Bouillon, R., Carmeliet, G., 2004. Soluble VEGF isoforms are essential for establishing epiphyseal vascularization and regulating chondrocyte development and survival. J. Clin. Invest. 113 (2), 188–199.

Maes, C., Kobayashi, T., Selig, M.K., Torrekens, S., Roth, S.I., Mackem, S., Carmeliet, G., Kronenberg, H.M., 2010a. Osteoblast precursors, but not mature osteoblasts, move into developing and fractured bones along with invading blood vessels. Dev. Cell 19 (2), 329–344.

Maes, C., Goossens, S., Bartunkova, S., Drogat, B., Coenegrachts, L., Stockmans, I., Moermans, K., Nyabi, O., Haigh, K., Naessens, M., Haenebalcke, L., Tuckermann, J.P., Tjwa, M., Carmeliet, P., Mandic, V., David, J.P., Behrens, A., Nagy, A., Carmeliet, G., Haigh, J.J., 2010b. Increased skeletal VEGF enhances beta-catenin activity and results in excessively ossified bones. EMBO J. 29 (2), 424–441.

Maes, C., Araldi, E., Haigh, K., Khatri, R., Van Looveren, R., Giaccia, A.J., Haigh, J.J., Carmeliet, G., Schipani, E., 2012a. VEGF-independent cell-autonomous functions of HIF-1alpha regulating oxygen consumption in fetal cartilage are critical for chondrocyte survival. J. Bone Miner. Res. 27 (3), 596–609.

Maes, C., Carmeliet, G., Schipani, E., 2012b. Hypoxia-driven pathways in bone development, regeneration and disease. Nat. Rev. Rheumatol. 8 (6), 358–366.

Maes, C., 2013. Role and regulation of vascularization processes in endochondral bones. Calcif. Tissue Int. 92 (4), 307–323.

Maes, C., 2017. Signaling pathways effecting crosstalk between cartilage and adjacent tissues: seminars in cell and developmental biology: the biology and pathology of cartilage. Semin. Cell Dev. Biol. 62, 16–33.

Malcangio, M., Garrett, N.E., Tomlinson, D.R., 1997. Nerve growth factor treatment increases stimulus-evoked release of sensory neuropeptides in the rat spinal cord. Eur. J. Neurosci. 9 (5), 1101–1104.

Mantyh, P.W., 2014. The neurobiology of skeletal pain. Eur. J. Neurosci. 39 (3), 508–519.

Martin, C.D., Jimenez-Andrade, J.M., Ghilardi, J.R., Mantyh, P.W., 2007. Organization of a unique net-like meshwork of CGRP+ sensory fibers in the mouse periosteum: implications for the generation and maintenance of bone fracture pain. Neurosci. Lett. 427 (3), 148–152.

McKemy, D.D., Neuhausser, W.M., Julius, D., 2002. Identification of a cold receptor reveals a general role for TRP channels in thermosensation. Nature 416 (6876), 52–58.

Mekraldi, S., Lafage-Proust, M.H., Bloomfield, S., Alexandre, C., Vico, L., 2003. Changes in vasoactive factors associated with altered vessel morphology in the tibial metaphysis during ovariectomy-induced bone loss in rats. Bone 32 (6), 630–641.

Milgram, J.W., Robinson, R.A., 1965. An electron microscopic demonstration of unmyelinated nerves in the haversian canals of the adult dog. Bull. Johns Hopkins Hosp. 117, 163–173.

Miller, M.R., Kasahara, M., 1963. Observations on the innervation of human long bones. Anat. Rec. (145), 13–23.

Park, J., Gebhardt, M., Golovchenko, S., Branguli, F.P., Hattori, T., Hartmann, C., Zhou, X., deCrombrugghe, B., Stock, M., Schneider, H., von der Mark, K., 2015. Dual pathways to endochondral osteoblasts: a novel chondrocyte-derived osteoprogenitor cell identified in hypertrophic cartilage. Biol. Open 4 (5), 608–621.

Pezet, S., McMahon, S.B., 2006. Neurotrophins: mediators and modulators of pain. Annu. Rev. Neurosci. 29, 507–538.

Pongratz, G., Straub, R.H., 2013. Role of peripheral nerve fibres in acute and chronic inflammation in arthritis. Nat. Rev. Rheumatol. 9 (2), 117–126.

Proske, U., Gandevia, S.C., 2012. The proprioceptive senses: their roles in signaling body shape, body position and movement, and muscle force. Physiol. Rev. 92 (4), 1651–1697.

Provot, S., Zinyk, D., Gunes, Y., Khatri, R., Le, Q., Kronenberg, H., Johnson, R., Longaker, M., Giaccia, A., Schipani, E., 2007. HIF-1alpha regulates differentiation of limb bud mesenchyme and joint development. J. Cell Biol. 177, 451–464.

Quarcoo, D., Schulte-Herbrüggen, O., Lommatzsch, M., Schierhorn, K., Hoyle, G.W., Renz, H., Braun, A., 2004. Nerve growth factor induces increased airway inflammation via a neuropeptide-dependent mechanism in a transgenic animal model of allergic airway inflammation. Clin. Exp. Allergy 34 (7), 1146–1151.

Ramasamy, S.K., Kusumbe, A.P., Wang, L., Adams, R.H., 2014. Endothelial notch activity promotes angiogenesis and osteogenesis in bone. Nature 507 (7492), 376–380.

Ramasamy, S.K., Kusumbe, A.P., Itkin, T., Gur-Cohen, S., Lapidot, T., Adams, R.H., 2016. Regulation of hematopoiesis and osteogenesis by blood vessel-derived signals. Annu. Rev. Cell Dev. Biol. 32, 649–675.

Rankin, E.B., Wu, C., Khatri, R., Wilson, T.L., Andersen, R., Araldi, E., Rankin, A.L., Yuan, J., Kuo, C.J., Schipani, E., Giaccia, A.J., 2012. The HIF signaling pathway in osteoblasts directly modulates erythropoiesis through the production of EPO. Cell 149 (1), 63–74.

Rapp, A.E., Kroner, J., Baur, S., Schmid, F., Walmsley, A., Mottl, H., Ignatius, A., 2015. Analgesia via blockade of NGF/TrkA signaling does not influence fracture healing in mice. J. Orthop. Res. 33 (8), 1235–1241.

Regan, J.N., Lim, J., Shi, Y., Joeng, K.S., Arbeit, J.M., Shohet, R.V., Long, F., 2014. Up-regulation of glycolytic metabolism is required for HIF-1alpha-driven bone formation. Proc. Natl. Acad. Sci. U.S.A. 111 (23), 8673–8678.

Reichardt, L.F., 2006. Neurotrophin-regulated signalling pathways. Philos. Trans. R. Soc. Lond. B Biol. Sci. 361 (1473), 1545–1564.

Roche, B., David, V., Vanden-Bossche, A., Peyrin, F., Malaval, L., Vico, L., Lafage-Proust, M.H., 2012. Structure and quantification of microvascularisation within mouse long bones: what and how should we measure? Bone 50 (1), 390–399.

Sample, S.J., Hao, Z., Wilson, A.P., Muir, P., 2011. Role of calcitonin gene-related peptide in bone repair after cyclic fatigue loading. PLoS One 6 (6) e20386.

Saran, U., Gemini Piperni, S., Chatterjee, S., 2014. Role of angiogenesis in bone repair. Arch. Biochem. Biophys. 561, 109–117.

Schipani, E., Ryan, H.E., Didrickson, S., Kobayashi, T., Knight, M., Johnson, R.S., 2001. Hypoxia in cartilage: HIF-1alpha is essential for chondrocyte growth arrest and survival. Genes Dev. 15 (21), 2865–2876.

Schipani, E., Maes, C., Carmeliet, G., Semenza, G.L., 2009. Regulation of osteogenesis-angiogenesis coupling by HIFs and VEGF. J. Bone Miner. Res. 24 (8), 1347–1353.

Semenza, G.L., 2012. Hypoxia-inducible factors in physiology and medicine. Cell 148 (3), 399–408.

Shao, P., Ohtsuka-Isoya, M., Shinoda, H., 2003. Circadian rhythms in serum bone markers and their relation to the effect of etidronate in rats. Chronobiol. Int. 20 (2), 325–336.

Shen, X., Wan, C., Ramaswamy, G., Mavalli, M., Wang, Y., Duvall, C.L., Deng, L.F., Guldberg, R.E., Eberhart, A., Clemens, T.L., Gilbert, S.R., 2009. Prolyl hydroxylase inhibitors increase neoangiogenesis and callus formation following femur fracture in mice. J. Orthop. Res. 27 (10), 1298–1305.

Shomento, S.H., Wan, C., Cao, X., Faugere, M.C., Bouxsein, M.L., Clemens, T.L., Riddle, R.C., 2010. Hypoxia-inducible factors 1alpha and 2alpha exert both distinct and overlapping functions in long bone development. J. Cell. Biochem. 109 (1), 196–204.

Street, J., Bao, M., deGuzman, L., Bunting, S., Peale Jr., F.V., Ferrara, N., Steinmetz, H., Hoeffel, J., Cleland, J.L., Daugherty, A., van Bruggen, N., Redmond, H.P., Carano, R.A., Filvaroff, E.H., 2002. Vascular endothelial growth factor stimulates bone repair by promoting angiogenesis and bone turnover. Proc. Natl. Acad. Sci. U.S.A. 99 (15), 9656–9661.

Tabarowski, Z., Gibson-Berry, K., Felten, S.Y., 1996. Noradrenergic and peptidergic innervation of the mouse femur bone marrow. Acta Histochem. 98 (4), 453–457.

Tang, W., Yang, F., Li, Y., de Crombrugghe, B., Jiao, H., Xiao, G., Zhang, C., 2012. Transcriptional regulation of vascular endothelial growth factor (VEGF) by osteoblast-specific transcription factor osterix (Osx) in osteoblasts. J. Biol. Chem. 287 (3), 1671–1678.

Thurston, T.J., 1982. Distribution of nerves in long bones as shown by silver impregnation. J. Anat. 134 (Pt 4), 719–728.

Togari, A., 2002. Adrenergic regulation of bone metabolism: possible involvement of sympathetic innervation of osteoblastic and osteoclastic cells. Microsc. Res. Tech. 58 (2), 77–84.

Tomlinson, R.E., Silva, M.J., 2013. Skeletal blood flow in bone repair and maintenance. Bone Research 1 (4), 311–322.

Tomlinson, R.E., McKenzie, J.A., Schmieder, A.H., Wohl, G.R., Lanza, G.M., Silva, M.J., 2013. Angiogenesis is required for stress fracture healing in rats. Bone 52 (1), 212–219.

Tomlinson, R.E., Li, Z., Li, Z., Minichiello, L., Riddle, R.C., Venkatesan, A., Clemens, T.L., 2017. NGF-TrkA signaling in sensory nerves is required for skeletal adaptation to mechanical loads in mice. Proc. Natl. Acad. Sci. U.S.A. 114 (18), E3641.

Tomlinson, R.E., Li, Z., Zhang, Q., Goh, B.C., Li, Z., Thorek, D.L.J., Rajbhandari, L., Brushart, T.M., Minichiello, L., Zhou, F., Venkatesan, A., Clemens, T.L., 2016. NGF-TrkA signaling by sensory nerves coordinates the vascularization and ossification of developing endochondral bone. Cell Rep. 16 (10), 2723–2735.

Toulis, K.A., Hemming, K., Stergianos, S., Nirantharakumar, K., Bilezikian, J.P., 2014. β-Adrenergic receptor antagonists and fracture risk: a meta-analysis of selectivity, gender, and site-specific effects. Osteoporos. Int. 25 (1), 121–129.

Trueta, J., Buhr, A.J., 1963. The vascular contribution to osteogenesis. V. The vasculature supplying the epiphysial cartilage in rachitic rats. J. Bone Joint Surg. Br. 45, 572–581.

Trueta, J., Caladias, A.X., 1964. A study of the blood supply of the long bones. Surg. Gynecol. Obstet. 118, 485–498.

Veldhuis-Vlug, A.G., Tanck, M.W., Limonard, E.J., Endert, E., Heijboer, A.C., Lips, P., Fliers, E., Bisschop, P.H., 2015. The effects of beta-2 adrenergic agonist and antagonist on human bone metabolism: a randomized controlled trial. Bone 71, 196–200.

Vu, T.H., Shipley, J.M., Bergers, G., Berger, J.E., Helms, J.A., Hanahan, D., Shapiro, S.D., Senior, R.M., Werb, Z., 1998. MMP-9/gelatinase b is a key regulator of growth plate angiogenesis and apoptosis of hypertrophic chondrocytes. Cell 93 (3), 411–422.

Wan, C., Gilbert, S.R., Wang, Y., Cao, X., Shen, X., Ramaswamy, G., Jacobsen, K.A., Alaql, Z.S., Eberhardt, A.W., Gerstenfeld, L.C., Einhorn, T.A., Deng, L., Clemens, T.L., 2008. Activation of the hypoxia-inducible factor-1alpha pathway accelerates bone regeneration. Proc. Natl. Acad. Sci. U.S.A. 105 (2), 686–691.

Wang, L., Zhou, S., Liu, B., Lei, D., Zhao, Y., Lu, C., Tan, A., 2006. Locally applied nerve growth factor enhances bone consolidation in a rabbit model of mandibular distraction osteogenesis. J. Orthop. Res. 24 (12), 2238–2245.

Wang, Y., Wan, C., Deng, L., Liu, X., Cao, X., Gilbert, S.R., Bouxsein, M.L., Faugere, M.C., Guldberg, R.E., Gerstenfeld, L.C., Haase, V.H., Johnson, R.S., Schipani, E., Clemens, T.L., 2007. The hypoxia-inducible factor alpha pathway couples angiogenesis to osteogenesis during skeletal development. J. Clin. Invest. 117 (6), 1616–1626.

Wang, J., Gao, Y., Cheng, P., Li, D., Jiang, H., Ji, C., Zhang, S., Shen, C., Li, J., Song, Y., Cao, T., Wang, C., Yang, L., Pei, G., 2017. CD31hiEmcnhi vessels support new trabecular bone formation at the frontier growth area in the bone defect repair process. Sci. Rep. 7 (1), 4990.

Weng, T., Xie, Y., Huang, J., Luo, F., Yi, L., He, Q., Chen, D., Chen, L., 2014. Inactivation of VHL in osteochondral progenitor cells causes high bone mass phenotype and protects against age-related bone loss in adult mice. J. Bone Miner. Res. 29 (4), 820–829.

Wojtys, E.M., Beaman, D.N., Glover, R.A., Janda, D., 1990. Innervation of the human knee joint by substance-P fibers. Arthroscopy 6 (4), 254–263.

Wu, C., Rankin, E.B., Castellini, L., Alcudia, J.F., LaGory, E.L., Andersen, R., Rhodes, S.D., Wilson, T.L., Mohammad, K.S., Castillo, A.B., Guise, T.A., Schipani, E., Giaccia, A.J., 2015. Oxygen-sensing PHDs regulate bone homeostasis through the modulation of osteoprotegerin. Genes Dev. 29 (8), 817–831.

Yang, L., Tsang, K.Y., Tang, H.C., Chan, D., Cheah, K.S.E., 2014. Hypertrophic chondrocytes can become osteoblasts and osteocytes in endochondral bone formation. Proc. Natl. Acad. Sci. U.S.A. 111 (33), 12097–12102.

Zelzer, E., Glotzer, D.J., Hartmann, C., Thomas, D., Fukai, N., Soker, S., Olsen, B.R., 2001. Tissue specific regulation of VEGF expression during bone development requires Cbfa1/Runx2. Mech. Dev. 106 (1–2), 97–106.

Zelzer, E., McLean, W., Ng, Y.S., Fukai, N., Reginato, A.M., Lovejoy, S., D'Amore, P.A., Olsen, B.R., 2002. Skeletal defects in VEGF (120/120) mice reveal multiple roles for VEGF in skeletogenesis. Development 129, 1893–1904.

Zhao, Q., Shen, X., Zhang, W., Zhu, G., Qi, J., Deng, L., 2012. Mice with increased angiogenesis and osteogenesis due to conditional activation of HIF pathway in osteoblasts are protected from ovariectomy induced bone loss. Bone 50 (3), 763–770.

Zhou, X., von der Mark, K., Henry, S., Norton, W., Adams, H., de Crombrugghe, B., 2014. Chondrocytes transdifferentiate into osteoblasts in endochondral bone during development, postnatal growth and fracture healing in mice. PLoS Genet. 10 (12), e1004820.

Chapter 10

Coupling of bone formation and resorption

Natalie A. Sims and T. John Martin

St. Vincent's Institute of Medical Research, Melbourne, Australia; Department of Medicine at St. Vincent's Hospital, The University of Melbourne, Melbourne, Australia

Chapter outline

Introduction:—bone modeling and bone remodeling 219
Development of the concept of coupling 220
Coupled remodeling is asynchronous throughout the skeleton 221
Coupling is unidirectional and sequential: bone formation following bone resorption 221
Coupling as a multicellular process 221
Coupling occurs locally within a basic multicellular unit 222
Coupling and balance: what is the difference? 222
The resorption phase of remodeling and its cessation in the basic multicellular unit 223
Coupling mechanisms originate from several cellular sources 223
Matrix-derived resorption products as coupling factors 227
Coupling factors synthesized and secreted by osteoclasts 228
Membrane-bound coupling factors synthesized by osteoclasts 230
Perspective on candidate osteoclast-derived coupling factors identified to date 231
How other cells contribute to coupling 231
Osteoblast lineage cells—sensing the surface and signaling to one another 232
Macrophages, immune cells, and endothelial cells 233
The reversal phase as a coupling mechanism 234
Conclusion 236
References 236

Introduction:—bone modeling and bone remodeling

Skeletal structure and function are regulated by hormones, cytokines, and the central and sympathetic nervous systems in response to a range of factors, including changes in mechanical loading and stimuli emanating from the immune and reproductive systems. Not only does the skeleton provide mechanical strength to the body, protection of internal organs, and sites of muscle attachment for locomotion, but bone is also a very significant endocrine organ with major influences on calcium, phosphate, and glucose metabolism.

Skeletal structure is determined by two processes: modeling and remodeling. The key difference between these two processes is the relationship between bone-forming osteoblasts and bone-resorbing osteoclasts. In modeling, actions of osteoblasts and osteoclasts occur in separation (i.e., on different surfaces), while in remodeling, the actions of osteoblasts and osteoclasts occur in sequence on the same bone surface.

Modeling modifies the shape of the skeleton. This includes both the construction of bone, which takes place from the beginning of skeletogenesis during fetal life until the end of the second decade, when the longitudinal growth of the skeleton is completed (Frost, 1964), and changes in bone shape associated with aging (Parfitt et al., 1983; Robling and Turner, 2009) or in response to mechanical load. In modeling bone formation occurs independently (i.e., without prior bone resorption), as does bone resorption (without subsequent bone formation). In this way bone is formed at sites of greatest mechanical load and removed where it is not required. For example, bone formed and deposited on the outer surface of the bone widens a growing bone; at the same time, bone is resorbed at a different location to enlarge the medullary cavity containing the bone marrow.

Principles of Bone Biology. https://doi.org/10.1016/B978-0-12-814841-9.00010-5

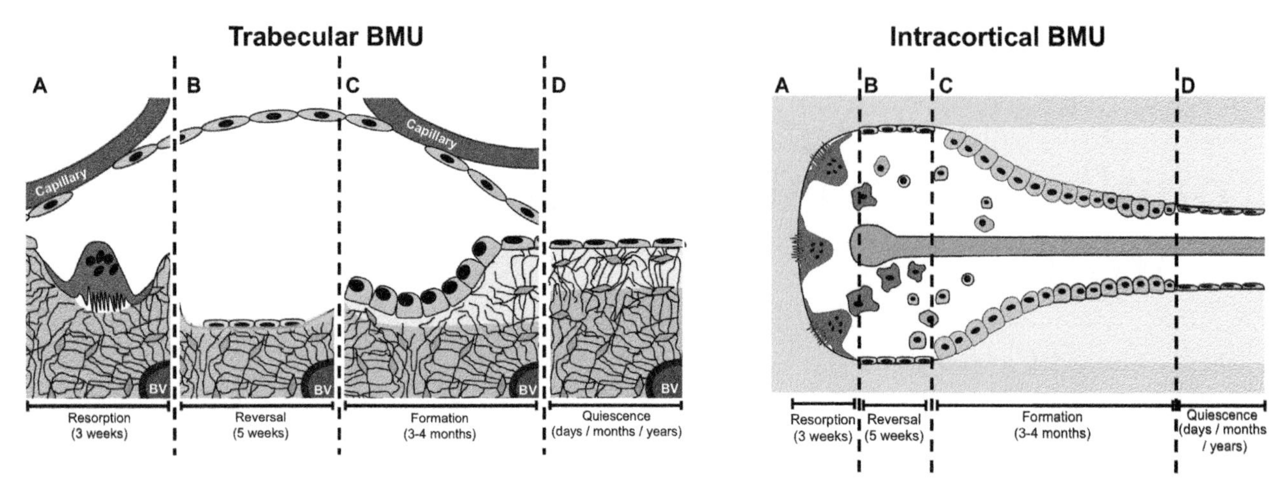

FIGURE 10.1 The basic structure of trabecular (left) and intracortical (right) basic multicellular units (*BMUs*), showing how the cell populations present on the surface change over time. (A) Initiation of remodeling, with lifting of the canopy, and osteoclast-mediated resorption on the bone surface. Osteoclast progenitors may be supplied by the capillaries adjacent to the canopy. (B) After bone resorption, the surface is covered by osteoblast lineage reversal cells. (C) Following the reversal period, osteoblasts fill the pit resorbed by osteoclasts by generating new osteoid, which is gradually mineralized over time; again, osteoblasts may be supplied by the nearby vasculature. (D) The process is complete, the pit filled, the surface is covered with bone-lining cells, and the canopy can no longer be detected. Mineralization of the osteoid continues until it reaches maximal levels, or until the next remodeling cycle occurs. *BV*, blood vessel.

Remodeling was initially described as two linked processes: (1) a destructive process, wrought by bone-resorbing osteoclasts, followed by (2) a productive process ascribed to bone-forming osteoblasts; this was based on Harold Frost's examination of multiple sections through trabecular bone from normal adult skeletons (Hattner et al., 1965). The sequence of resorption followed by formation was likened to the healing of a soft-tissue wound, and analogous progressions were noted to take place in the formation of several epithelia (e.g., Ford and Young, 1963). The tissue wound analogy is apt, since the principal functions of bone remodeling are to provide a means of adapting the skeleton to changes in loading and to remove and replace damaged or old bone. Remodeling occurs asynchronously throughout the skeleton at many anatomically distinct sites termed basic multicellular units (BMUs) (Frost, 1964; Hattner et al., 1965). In BMUs, tiny packets of bone are removed by osteoclasts and subsequently replaced by new bone matrix (osteoid) produced by osteoblasts; that matrix is mineralized to form new bone substance. In trabecular bone the BMU is located on the bone surface. In Haversian cortical bone the BMU comprises cutting zones led by osteoclasts that proceed through bone, followed by differentiating osteoblasts (Fig. 10.1) (Parfitt, 1982; Eriksen, 1986; Eriksen et al., 1993; Hauge et al., 2001). The roles of remodeling and modeling will be discussed in detail elsewhere in this book (Seeman, Chapter 11).

Development of the concept of coupling

When Frost introduced the BMU, he also introduced the concept of "coupling" as the process by which resorption of a certain amount of bone is followed by formation of the amount sufficient to replace it (Hattner et al., 1965). This was a remarkable insight given that little was known of the cells of bone at that time; their origins were not at all understood, and it was more than another decade before they could be studied directly. The events within the BMU progress from quiescence to activation, resorption, and formation, and finally return to quiescence when filling of the resorbed space is complete (Fig. 10.1).

The bone resorption activity in a BMU in adult human bone takes approximately 3 weeks (Eriksen et al., 1984b), the formation response 3–4 months (Eriksen et al., 1984a), and between the two activities there is a "reversal phase" (Tran Van et al., 1982), which takes several weeks (Eriksen et al., 1984b). The process occurs with sufficient frequency, and at sufficient sites, to replace 5%–10% of the skeleton per year, with the entire adult human skeleton replaced in around 10 years (Parfitt, 1980). In rodents the duration of this sequence is compressed, but a time delay between resorption and formation still exists: in rat alveolar bone the reversal phase lasts for approximately 3.5 days (Vignery and Baron, 1980). These numbers vary also with site, skeletal health, and treatment (Jensen et al., 2014), and in some conditions, including osteoporosis, there is an increased duration, or even arrest, of the reversal phase (Parfitt, 1982; Andersen et al., 2013).

Coupled remodeling is asynchronous throughout the skeleton

When we consider the time course over which remodeling occurs it must be noted that BMUs throughout the skeleton are active at different stages of the remodeling cycle. Indeed, it is an essential feature of bone remodeling that it does not occur uniformly throughout the skeleton, but takes place where it is needed to repair damaged or replace old bone. The events are initiated asynchronously, at sites geographically and chronologically separated from one another. Therefore, at any one time, bone is being resorbed in BMUs at some locations, while at others, a BMU is in its formation or reversal phase. Systemically administered drugs that influence bone remodeling will therefore act upon BMUs at all sites and in differing phases of their remodeling cycle. This presents a special challenge to pharmacological interventions aimed at preventing and reversing the structural decay associated with menopause and advancing age. It also needs to be borne in mind when considering ways in which circulating hormones might influence bone remodeling.

The development of a BMU in bone begins with resorption by osteoclasts, leaving scalloped contours in Howship's lacunae, followed by generation of the cement line, upon which bone formation by osteoblasts would take place (Fig. 10.1). Scalloped cement lines are used in histology to indicate that formation has taken place on previously resorbed surfaces, whereas smooth cement lines signify bone being formed on surfaces without previous resorption, i.e., modeling; while these are observed in larger mammals, they are rarely observed in murine bone, because the pits formed by mouse osteoclasts are more shallow. In adult trabecular bone, about 96% of bone formation takes place on resorbed surfaces (i.e., remodeling surfaces), the remainder on intact surfaces (modeling), with some of the latter reflecting overfilling of the resorbed lacunae (Hattner et al., 1965). The remodeling sequence proposed by Frost, of activation, resorption, and formation (ARF) (Frost, 1964), was extended by Baron, who proposed the existence of a reversal phase between resorption and formation, a stage characterized by the presence of mononuclear cells and the formation of a cement line separating new from old bone (Baron, 1977). Although the reversal phase, lasting as long as 5 weeks in human bone remodeling, has been poorly understood, it is now being investigated in some detail and will be discussed later. Frost's view of bone remodeling was readily adopted by others (Frost, 1964; Harris and Heaney, 1969; Parfitt, 1982).

Coupling is unidirectional and sequential: bone formation following bone resorption

The term "coupling" is illustrative, because it calls to mind the coupling of a train carriage to a locomotive: the osteoclast is analogous to the engine commencing the process of remodeling, with osteoblasts being the carriage coupled to it and following behind.

The first essential step in the remodeling cycle was proposed to be the generation of active osteoclasts. This suggestion was made even before it was recognized that osteoclasts develop from hemopoietic precursors. Osteoclasts are probably derived from early and late precursors available in marrow adjacent to activation sites (reviewed by Takahashi et al., Chapter 5); they may be drawn to the site by chemoattractants derived from osteocytes or from the bone matrix itself (Sims and Gooi, 2008). They may also be recruited from blood available at the bone interface through a sinus structure of bone remodeling compartments (Hauge et al., 2001; Kristensen et al., 2013). At each BMU, the resorption of a volume of bone is followed by new bone formation sufficient to fill the space. The BMU resorbs and replaces old bone at the same location so there is no change in bone size or shape. If the volumes of bone removed and replaced are equal, there will be no permanent loss of bone or compromise in bone mass.

Coupling does not include the major mechanism of intercellular communication in bone that was proposed some years later, of osteoblastic lineage cells controlling osteoclast formation (Chambers, 1980; Rodan and Martin, 1981) by their production of receptor activator of NF-κB ligand (RANKL) and osteoprotegerin (OPG) (Lacey et al., 1998; Yasuda et al., 1998; Suda et al., 1999). The Frost coupling concept preceded this by decades. The control of osteoclast formation and function by osteoblasts (or cells in that lineage) is not coupling, although this mistake is commonly made in the literature. Coupling refers strictly to *the process by which products of bone resorption and of osteoclasts and other cells within the BMU influence recruitment and differentiation of the osteoblast lineage to mature osteoblasts that synthesize bone and mineralize it.* Frost provided a great service in identifying the coupling mechanism as one that was likely to be important physiologically and important in devising therapeutic approaches; this simple concept has come to be recognized as a complex process with many participants, as will be discussed.

Coupling as a multicellular process

The balance between bone formation and resorption requires participation of a number of cell types beyond differentiated osteoclasts and osteoblasts. Coupling is therefore not simply an event taking place between osteoclasts and osteoblasts, but an extensive process of information transfer involving diverse cells and interactions aimed ultimately at regulating osteoblast differentiation/bone formation (Sims and Martin, 2015). Since the several stages of the process are prolonged in duration, any cellular communication mechanisms have to provide for the separation of these events in time.

Coupling occurs locally within a basic multicellular unit

In discussing the mechanisms of coupling, including osteoclast formation, activity, and maintenance, we are considering only local factors. They seem better equipped than circulating hormones to activate asynchronous events throughout the skeleton.

Although very little was known about osteoclasts at the time of Frost's proposal, the key initial event was agreed to be the initiation of resorption by osteoclasts. When the steps in osteoclast formation and activation were eventually revealed as being derived from precursors in the bone marrow, the question of how osteoclasts commenced BMU formation at a site of need was asked.

Much interest has developed in the data showing the development of a canopy extending over the active BMU during its initiation. This was proposed first by Rasmussen and Bordier (Rasmussen and Bordier, 1974) and suggests a mechanism to isolate the BMU and allow local communication between differing cell types. Direct evidence came from Hauge et al. (Hauge et al., 2001), who suggested that lining cells lift from the underlying bone at the start of remodeling activity and form a canopy over the site to be resorbed, whether on the surface in trabecular bone or within the substance of cortical bone undergoing haversian remodeling. The capillary blood supply closely associated with the canopy (Kristensen et al., 2013) was proposed as a source of hematopoietic precursors for osteoclast formation (Fig. 10.1). Osteoclast formation can take place rapidly in vivo; this may be made possible by partially differentiated niches of cells available for the BMU. Such partially differentiated osteoclast precursors, designated as "quiescent osteoclast precursors," have been characterized in mouse and rat (Mizoguchi et al., 2009), having arrived there in the circulation (Muto et al., 2011), but have not yet been identified in human subjects. They are reported in Takahashi et al. (Chapter 5).

Capillaries associated with the canopy also provide a mechanism for ingress of other cells, including mesenchymal precursors (Eghbali-Fatourechi et al., 2007) and immune and endothelial cells. Consistent with such supply from blood was the demonstration that red blood cells could be seen within the active cortical BMU (Andersen et al., 2009). The canopy has also been demonstrated in rabbit bone (Jensen et al., 2014), which undergoes Haversian remodeling in the cortex, although the canopy has not been convincingly shown yet in intracortical "cutting cones." Although a similar canopy has been predicted in the mouse, it has been observed only above bone-forming surfaces (Narimatsu et al., 2010). Tissue-specific macrophages ("osteomacs") have been found to form a canopy over bone-forming sites in the mouse (Chang et al., 2008), but the relationship between the two canopies remains unclear.

Osteocytes are recognized as important in mechanosensing and regulation of bone remodeling (Bonewald and Johnson, 2008). They are the most abundant cell of bone and are situated to communicate both throughout the skeleton and with bone-lining cells on the surface. Recognition of the need to remove or repair a region of bone could be achieved by osteocytes. Indeed, apoptosis of osteocytes as a result of microdamage coincides with bone resorption by osteoclasts (Verborgt et al., 2000; Noble et al., 2003; Heino et al., 2009; Kennedy et al., 2012), implicating this as a mechanism to target the initiation of remodeling for bone repair. A signal emitted by osteocytes to lining cells could lead them to lift and begin to form the canopy overlying and containing the BMU (Eghbali-Fatourechi et al., 2007; Eriksen et al., 2007). The nature of the signal to attract osteoclast precursors to the bone surface is unknown, but a number of chemoattractants have been suggested (Sims and Gooi, 2008). RANKL and macrophage colony-stimulating factor (M-CSF) production by local mesenchymal-derived cells may also play a role in the initial formation of the BMU. Since the canopy mesenchymal cells themselves have been suggested to be a source of osteoblast precursors (Delaisse, 2014), they may, in addition to lining cells, mesenchymal precursors within the BMU, and osteocytes, also be sources of RANKL and M-CSF as the necessary stimuli for osteoclast generation at the site.

Coupling and balance: what is the difference?

The amount and quality of bone is retained in the skeleton during balanced bone remodeling because the processes within BMUs are matched: the amount of bone synthesized replaces the amount removed. The two processes are "balanced" through mechanisms of intercellular communication (see later). This is the key to integrity of the remodeling process and the success of the BMU system in ensuring maintenance of normal bone during young adult life, when bone mass is maintained at a constant level. It is what led Frost to propose the coupling of bone formation and resorption processes within the remodeling BMU. At other times in life, the balance is negative (bone mass is lost) or positive (bone mass is gained), but this does not mean there a loss of coupling (although the term "uncoupling" is frequently used): formation still follows resorption, but there is an imbalance in the activities between the bone cells. This caution is especially pertinent in studies claiming (for example) that pharmacological agents or genetic alterations uncouple bone formation from resorption on the basis of serum biochemical markers; these markers represent the sum of all formation and resorption activities in the skeleton, including both modeling and remodeling.

Some forms of uncoupling or dissociation may occur at the BMU level, for example, when osteoclasts are nonfunctional or not present (and therefore are unable to initiate remodeling and provide signals to osteoblasts) (Henriksen et al., 2011) or in instances of reversal phase "arrest," when the canopy over the BMU is disrupted as in myeloma bone disease (Andersen et al., 2010) and some cases of osteoporosis (Andersen et al., 2014). This will be discussed later.

The resorption phase of remodeling and its cessation in the basic multicellular unit

Once a BMU site is activated, resorption must be limited in some way. An important unanswered question about osteoclast behavior and the control of resorption in the remodeling cycle is: how does each osteoclast know when to stop resorbing? The process is likely to finish with osteoclast death, which has been studied in vitro to some extent, but its regulation in vivo remains obscure. There are several possible mechanisms. Osteoclasts phagocytose osteocytes, which might provide a mechanism to remove the signal for resorption (Elmardi et al., 1990). The very high (millimolar) concentrations of calcium generated during mineral dissolution, capable of inhibiting osteoclasts (Zaidi et al., 1989), might come into play particularly in the localized and focused conditions of the BMU. Transforming growth factor β (TGFβ) is available either when it is resorbed from the matrix and activated by acid pH or when it is secreted by the osteoblast lineage and can promote osteoclast apoptosis (Hughes and Boyce, 1997). A direct effect of estrogen to enhance osteoclast apoptosis has been identified in mouse genetic experiments (Nakamura et al., 2007). Some insights into the control of apoptosis arise from genetic and pharmacological studies showing that inhibition of acidification of the resorption space by blockade of either chloride channel-7 (ClC-7) or the V-type H^+ ATPase of the osteoclast results in prolonged osteoclast survival (Henriksen et al., 2004; Karsdal et al. 2005, 2007). This might suggest a role for acidification in determining osteoclast life span, perhaps even through TGFβ activation. In determining how resorption ends in the BMU, there is much to be learned of mechanisms regulating osteoclast apoptosis. There might be several events limiting resorption to only a certain amount of bone at each BMU, including the possibility that accumulation of osteoblast progenitors provides a signal, but no dominant governing control has yet been identified.

Coupling mechanisms originate from several cellular sources

The development of the coupling concept led to an extensive search for possible "coupling factors"— messengers from the active osteoclast controlling the differentiation and activity of subsequent osteoblasts on the same surface. Many candidate factors have now been identified and are listed in Table 10.1. We propose, however, that rather than a single coupling factor, or even a team of coupling factors, there are a range of "coupling mechanisms" promoting and limiting bone formation in the BMU (Fig. 10.2). Some coupling mechanisms derive from osteoclasts and their activities: (1) matrix-derived signals released by osteoclasts during bone resorption and (2) factors synthesized by the mature osteoclast (which may include secreted factors, membrane-bound factors, or factors released in cell-derived vesicles). But there are also coupling mechanisms emanating from other cell types within the BMU: (3) topographical changes effected by the osteoclast on the bone surface and sensed by osteoblasts; (4) secreted and membrane-bound products of osteoblast lineage cells, including osteoblast precursors, lining cells, and osteocytes; (5) secreted products of immune and endothelial cells; and (6) neurogenic stimuli arising from the extensive innervation circuitry in bone. The last will not be considered in this chapter since they are speculative and have not been subject to direct experiment. The others will each be discussed in turn.

It is also important to note the many different contexts in which osteoclasts communicate to osteoblasts: not every signal from an osteoclast that changes osteoblast differentiation or function should be considered a coupling factor. There are three pertinent examples of this. First, during endochondral ossification, the process by which the cartilage surrounding hypertrophic chondrocytes is resorbed by osteoclast-like cells and replaced by osteoblasts is analogous to remodeling of bone; it is sequential, and both resorption of matrix and formation of new bone occur on the same surface (Baron and Sims, 2000). However, the control mechanisms existing in that anatomical location are likely to be different, because of contributions from the hypertrophic chondrocytes and adjacent invading blood vessels (Poulton et al., 2012). Second, osteoclasts appear to produce "osteotransmitter" signals transmitted through the cortical osteocyte network (either by lacunocanalicular transport or by a signal relay) to influence the function of osteoblasts on the periosteal surface (Johnson et al., 2015). Third, in murine calvariae, in which there is no remodeling and hence no coupling, osteoblasts and

TABLE 10.1 A summary of osteoclast-derived coupling factors and their other sources and influences near or in the basic multicellular unit. Influences listed are from *in vitro* studies, unless otherwise indicated. RANKL, receptor activator of NF-kappa B ligand.

Factor	Production by osteoclasts	Other potentially relevant sources	Influences on osteoblast differentiation and bone formation	Other potential influences in remodeling
Transforming growth factor β	Released from matrix (Oreffo et al., 1989)	Osteoblasts (Robey et al., 1987) T lymphocytes (Chen et al., 1998) Macrophages (Fadok et al., 1998)	Stimulates progenitor expansion (Robey et al., 1987; Hock et al., 1990) Stimulates progenitor migration and differentiation (Tang et al., 2009) Stimulates bone formation in organ culture (Hock et al., 1990)	Stimulates osteoclastogenesis by direct action on osteoclast precursors (Sells Galvin et al., 1999) Stimulates sclerostin expression (Loots et al., 2012)
Bone morphogenetic protein (BMP) 2	Released from matrix (Oreffo et al., 1989) Secreted (Garimella et al., 2008)	Osteoblasts (Robubi et al., 2014) Macrophages (Champagne et al., 2002)	Stimulates progenitor expansion and migration (Fiedler et al., 2002) Stimulates osteoblast differentiation (Rickard et al., 1994)	Stimulates osteoclast activity (Hanamura et al., 1980; Kanatani et al., 1995)
Insulin-like growth factors	Released from matrix (Centrella and Canalis, 1985a)	Osteoblasts (Canalis and Gabbitas, 1994) Macrophages (Fournier et al., 1995)	Stimulates progenitor expansion (Xian et al., 2012)	Stimulates osteoclastogenesis (Wang et al., 2006)
Platelet-derived growth factor BB	Released from matrix (Centrella and Canalis, 1985a) Secreted (Kreja et al., 2010)	Osteoblasts (Zhang et al., 1991) Endothelial cells (Daniel and Fen, 1988) Osteoclasts (Lees et al., 2001)	Promotes progenitor replication (Hock and Canalis, 1994) Promotes progenitor migration (Sanchez-Fernandez et al., 2008; Kreja et al., 2010) Stimulates bone formation in vivo (Mitlak et al., 1996) Inhibits osteoblast differentiation (Hock and Canalis, 1994; Kubota et al., 2002)	Stimulates osteoclast precursor recruitment (Hock and Canalis, 1994)
Cardiotrophin-1	Secreted (Walker et al., 2008)		Stimulates bone formation in vivo (Walker et al., 2008) Stimulates osteoblast commitment (Walker et al., 2008) Suppresses sclerostin expression (Walker et al., 2008) Bone formation is low in null mice (Walker et al., 2008)	Stimulates osteoclastogenesis (Richards et al., 2000) Bone resorption is low in null mice (Walker et al., 2008)
BMP-6	Secreted (Garimella et al., 2008)	Mesenchymal and hematopoietic stem cells (Martinovic et al., 2004; Friedman et al., 2006)	Stimulates osteoblast differentiation (Friedman et al., 2006)	Stimulates osteoclastogenesis from human marrow cells (Wutzl et al., 2006)

Wnt10b	Secreted (Pederson et al., 2008)	T cells (Terauchi et al., 2009)	Stimulates osteoblast differentiation in vivo (Bennett et al., 2007)	Stimulates osteoclast activity in vivo (Bennett et al., 2007)
Sphingosine-1-phosphate (S1P)	S1P production catalyzed by secreted sphingosine-1-kinase (Pederson et al., 2008)	Vasculature (Scariano et al., 2008) Red blood cells (Pappu et al., 2007)	Induces osteoblast precursor recruitment (Pederson et al., 2008) Promotes osteoblast survival (Ryu et al., 2006; Pederson et al., 2008) Promotes osteoblast migration (Ryu et al., 2006)	Stimulates osteoclast recruitment (Scariano et al., 2008). Stimulates RANKL expression by osteoblasts and T cells (Ryu et al., 2006) Stimulates osteoclast precursor chemotaxis (Ishii et al., 2010) Intracellular S1P inhibits osteoclast differentiation (Ryu et al., 2006)
collagen triple-helix repeat—containing 1	Secreted (Takeshita et al., 2013)	Mesenchymal cells, osteoblasts (Kimura et al., 2008), and osteocytes (Y-R Jin Bone, 2017, p153)	Stimulates osteoblast differentiation (Kimura et al., 2008) Stimulates bone formation in vivo (Kimura et al., 2008; Takeshita et al., 2013) Osteoclast-specific null mice show reduced bone formation (Takeshita et al., 2013) Osteoblast-specific overexpression causes increased bone formation (Kimura et al., 2008)	Inhibits osteoclast formation and activity (Y-R Jin Bone, 2017, p153)
Complement 3a	Secreted (Matsuoka et al., 2014)	Circulating (50 ng/mL in human serum) (Wlazlo et al., 2013)	Stimulates alkaline phosphatase activity and mineralization in calvarial osteoblasts (Matsuoka et al., 2014).	Increases cytokine output from macrophages and T cells (Carroll, 2004) Osteoclast recruitment (Sato et al., 1993)
Oncostatin M	Secreted (Fernandes et al., 2013)	Macrophages (Zarling et al., 1986) Osteoblasts (Walker et al., 2010) Osteocytes (Walker et al., 2010) T lymphocytes (Clegg et al., 1996)	Promotes osteoblast commitment (Walker et al., 2010) Stimulates bone formation in vivo (Walker et al., 2010)	Synergizes with BMP-2 (Fernandes et al., 2013) Stimulates osteoclast formation (Tamura et al., 1993)
Semaphorin 4D	Membrane bound (Negishi-Koga et al., 2011)	T lymphocytes (including soluble forms) (Wang et al., 2001)	Inhibits bone formation (Negishi-Koga et al., 2011) Gene deletion leads to enhanced bone formation (Negishi-Koga et al., 2011)	Stimulates osteoclastogenesis (Dacquin et al., 2011).
EphrinB2	Membrane bound (Zhao et al., 2006)	Osteoblasts (Zhao et al., 2006) Osteocytes (Allan et al., 2008)	Promotes osteoblast differentiation through EphB4 (Zhao et al., 2006) Suppresses osteoblast apoptosis (Tonna et al., 2014) Promotes late-stage osteoblast differentiation in vivo (Takyar et al., 2013)	Inhibits osteoclast differentiation (Zhao et al., 2006) Inhibits RANKL production by osteoblasts (Tonna et al., 2014)

Continued

TABLE 10.1 A summary of osteoclast-derived coupling factors and their other sources and influences near or in the basic multicellular unit. Influences listed are from *in vitro* studies, unless otherwise indicated. RANKL, receptor activator of NF-kappa B ligand.—cont'd

Factor	Production by osteoclasts	Other potentially relevant sources	Influences on osteoblast differentiation and bone formation	Other potential influences in remodeling
Cxcl16	Secreted (Ota et al., 2013)	Vascular smooth muscle cells (Wagsater et al., 2004) Macrophages (Barlic et al., 2009)	May stimulate osteoblast precursor migration (Ota et al., 2013)	
Leukemia inhibitory factor	Secreted (Ota et al., 2013)	Mesenchymal stem cells (Sims and Johnson, 2012)	Stimulates bone formation in vivo (Cornish et al., 1993) Stimulates osteoblast precursor expansion (Cornish et al., 1997) Promotes osteoblast differentiation (Poulton et al., 2012) Suppresses sclerostin expression (Walker et al., 2010) Gene deletion leads to low bone formation in remodeling (Poulton et al., 2012)	Stimulates osteoclastogenesis (Reid et al., 1990) Inhibits marrow adipogenesis (Poulton et al., 2012)
miR-214-3p	Exocytosed (Li et al., 2016)	Monocytes	Circulating miR-214-3p associated with less bone formation in vivo (Li et al., 2016) Suppresses osteoblast differentiation (Li et al., 2016)	
RANKL reverse signal	Membrane bound (Furuya et al., 2018)	Osteoblast precursors, osteocytes (Kartsogiannis et al., 1999; Lacey et al., 1998; Matsuzaki et al., 1998)	Promotes bone formation (Ikebuchi et al., 2018)	

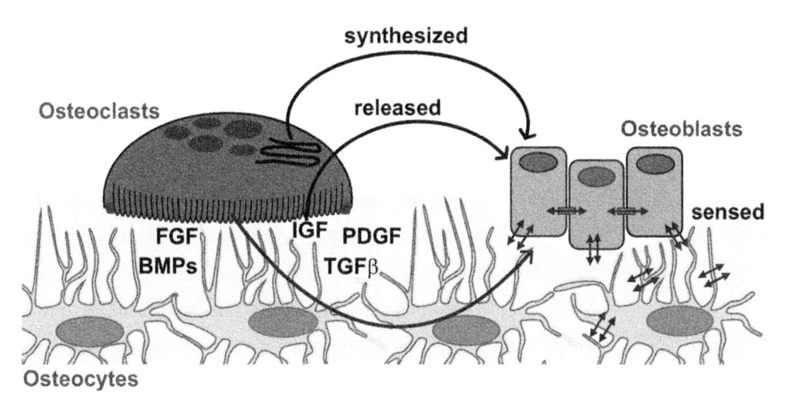

FIGURE 10.2 How do osteoblasts know how much bone to make? Signals, including proteins, are synthesized and secreted by osteoclasts and act directly on cells of the osteoblast lineage (including precursors and osteocytes). Osteoclasts also release matrix-bound growth factors such as insulin-like growth factor (*IGF*), bone morphogenetic proteins (*BMPs*), transforming growth factor β (*TGFβ*), platelet-derived growth factor (*PDGF*), and fibroblast growth factor (*FGF*) during the process of resorption; these also modify the function of osteoblast lineage cells. Osteoblasts also themselves sense the size of the pit left by the osteoclast, and signal to one another either via gap junctions or through secreted and membrane-bound proteins. Osteocytes also sense the changing mechanical strain both during the process of resorption and during bone formation, which changes the nature of the signals they release to influence bone-forming activity.

nonresorbing osteoclasts come into direct contact with each other (Furuya et al., 2018). This implies that osteoblasts may therefore inhibit osteoclast function by a direct contact mechanism in this context; this example of direct contact between the cells may also mediate communication from the osteoclast to the osteoblast, but mechanisms have not yet been identified. The range of coupling factors already identified may have also functions extending beyond coupling to these other examples of communication between these two lineages.

Matrix-derived resorption products as coupling factors

The first suggestion of a molecular mechanism for coupling came from Howard et al. (Howard et al., 1981). The bone matrix had long been known to contain a store of latent growth factors, including TGFβ, bone morphogenetic protein 2 (BMP-2), platelet-derived growth factor (PDGF), and the insulin-like growth factors (IGFs) (Hanamura et al., 1980; Centrella and Canalis, 1985a,b; Oreffo et al., 1989; Hock et al., 1990; Pfeilschifter et al., 1995). All are deposited by osteoblasts during matrix production and then released from the bone surface and activated by acidification during osteoclastic resorption, at which point they become available as agents influencing cells within the BMU, including osteoblasts and their precursors. The amount of bone resorbed by the osteoclast would determine the concentration of factors released, thereby modifying bone formation in a manner proportional to resorption.

Availability of these matrix-derived growth factors does not depend exclusively on resorption. Each is also synthesized by the osteoblasts themselves in latent forms activated by plasmin generated by plasminogen activators (Campbell et al., 1992; Yee et al., 1993) or matrix metalloproteinases (Dallas et al., 2002). In the case of PDGF-BB, its local involvement might be predominantly from a cell source, since it is produced by both osteoblasts and osteoclasts (Canalis and Ornitz, 2000), and its secretion by osteoclast precursors induces vessel formation and bone formation during both modeling and remodeling (Xie et al., 2014).

The hypothesis that coupling could be exerted by resorption-derived growth factors brought the issue before the field. It needed to take into account, though, that growth factor concentrations in matrix are high but variable in locations throughout the skeleton. The quantity and identity of growth factors contained within the matrix are determined by their production and release by osteoblasts and their incorporation into the matrix during bone formation in the previous remodeling cycle(s). Since precise replacement of bone at each site is a key requirement in coupling within the BMU, the active growth factor activities would need to be made available precisely in the quantities needed and in a spatially and temporally controlled manner. That could be difficult to achieve. Furthermore, if resorbed matrix is the main source of the growth factors, their primary availability would be at the earliest stage of the remodeling cycle (during resorption), which is temporally separate from the commencement of bone formation in the BMU (see Fig. 10.1). It seems unlikely that these growth factors would remain within the bone microenvironment for some weeks during the reversal phase until they could influence mature osteoblasts upon their arrival at the bone surface.

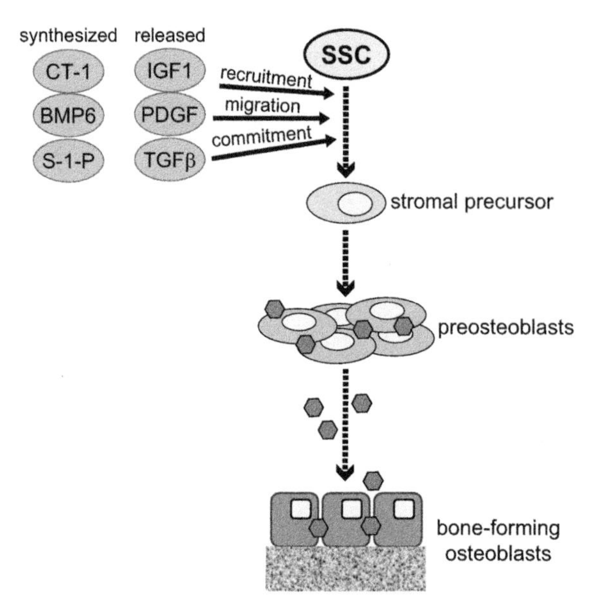

FIGURE 10.3 How do growth factors released by osteoclasts influence bone formation? Factors are synthesized by osteoclasts (such as cardiotrophin-1 [*CT-1*], bone morphogenetic protein 6 [*BMP6*], or sphingosine-1-phosphate [*S-1-P*]; see Table 10.1 for other examples) or released by osteoclast-mediated bone resorption of the matrix (such as insulin-like growth factor 1 [*IGF1*], platelet-derived growth factor [*PDGF*], and transforming growth factor β [*TGFβ*]; see Table 10.1 for other examples). These are likely to act on stromal stem cells (*SSCs*) to stimulate progenitor recruitment, to promote their migration to the bone surface, and to stimulate their early commitment to differentiate into mature, bone-forming osteoblasts. Preosteoblasts and mature osteoblasts throughout their differentiation communicate to each other through the release of factors such as TGFβ and parathyroid hormone–related protein and through the action of membrane-bound molecules such as EphrinB2 (represented by hexagons).

This conceptual difficulty with the growth factor/coupling concept has been helped by the outcome of mouse genetic experiments showing that active TGFβ release during bone resorption acts as a chemoattractant to induce migration of osteoblast precursors to prior sites of resorption (Tang et al., 2009) (Fig. 10.3). This provides osteoblast precursor cells within the BMU where they are available for stimulation by other signals promoting their differentiation and matrix production. The ability of osteoblast- and matrix-derived IGF-1 to promote osteoblast differentiation by favoring recruitment of mesenchymal stem cells (MSCs) (Xian et al., 2012) also links early release of growth factors to later differentiation of osteoblasts. These proposals for the mechanisms of growth factor action make it easier to understand how growth factors released by resorption, or from cell sources, could enhance subsequent bone formation in the BMU. Rather, their main influences would be to stimulate osteoblast progenitors, including their recruitment (Xian et al., 2012), migration (Fiedler et al., 2002; Sanchez-Fernandez et al., 2008; Tang et al., 2009; Kreja et al., 2010), and differentiation (Mitlak et al., 1996). The necessary precise quantitation could be achieved during the next stage through influences on the number and progression of stem cells after they enter the remodeling site.

Coupling factors synthesized and secreted by osteoclasts

Osteoclasts themselves might also be the source of coupling activity. This first arose during consideration of the anabolic action of intermittently injected parathyroid hormone (PTH), which was known to be achieved through activation of new and existing BMUs (Martin and Sims, 2005). This raised the question, what mechanisms determine replacement of the precise amount of bone resorbed in a BMU to preserve bone balance (Fig. 10.2)? Attention began to focus on identifying contributing osteoclast-derived factors, acting either independently or cooperatively with matrix-derived factors.

There were clues from human and mouse genetics. The failure of bone resorption results in the clinical syndrome of osteopetrosis: this may occur due to osteoclast deficiency or impaired activity. In forms of osteopetrosis with failed osteoclast formation—"osteoclast-poor" osteopetrosis—such as in the rare example of individual human subjects with lost RANKL/RANK signaling (Sobacchi et al., 2007), the bone is devoid of osteoclasts. This can be mimicked in a murine model by genetic ablation of *c-fos*, a transcription factor essential for osteoclast formation, and such mice have osteopetrosis noted by their failure to generate osteoclasts, and theirgreatly reduced bone formation (Grigoriadis et al., 1994).

The other form of osteopetrosis is "osteoclast-rich." Individuals with osteopetrosis due to defective osteoclast function, caused by mutations in either ClC-7 or the corresponding osteoclastic V-ATPase subunit A3 (also called TCIRG1), had high numbers of inactive osteoclasts (Henriksen et al., 2004; Karsdal et al., 2005). Unlike osteoclast-poor osteopetrosis, bone formation was normal or even increased, rather than diminished, as might be expected because of the greatly impaired resorption (Alatalo et al., 2004; Del Fattore et al., 2006). The level of bone formation has been linked directly to the presence of increased numbers of nonresorbing osteoclasts by direct correlation between the number of osteoclasts and the number of bone-forming osteoblasts (Del Fattore et al., 2006). Furthermore, when osteoclasts generated from the peripheral blood of a patient with osteopetrosis due to a mutation in the V-ATPase subunit A3 were cultured in vitro, conditioned medium from those osteoclasts was able to stimulate mineralized nodule formation in vitro despite their inability to resorb bone (Henriksen et al., 2012). This correlated well with earlier work in osteoclasts cultured on plastic (and therefore not resorbing bone), which were found to secrete factors promoting nodule formation (Karsdal et al., 2008; Pederson et al., 2008).

The findings from human genetics were supported by several studies in mice, whether deficient in *c-src* (Marzia et al., 2000), cathepsin K (Pennypacker et al., 2009), or *Pyk2* (Gil-Henn et al., 2007); in each case bone resorption is inhibited without inhibition of formation. In each of these knockout mice, while resorption is greatly reduced, osteoclast numbers are not. This is strikingly different from the case of *c-fos*-deficient mice that lack osteoclasts and have impaired bone formation (Grigoriadis et al., 1994). In support of the idea that the mutant osteoclasts might have impaired resorptive capacity but retain their ability to promote bone formation, transplantation of hematopoietic precursor cells deficient in the osteoclastic V-ATPase subunit a3 required for resorptive function into normal adult mice led to a significant reduction in resorption with increased osteoclast numbers but no reduction in bone formation (Lee et al., 2006). This was observed for up to 18 weeks following induction of osteopetrosis and was associated with improved bone volume and strength (Henriksen et al., 2011). A further instructive study (Thudium et al., 2014) compared osteopetrosis induced by transplanting irradiated normal mice with osteoclast precursors from oc/oc mice (osteoclast-rich osteopetrosis) or with RANK-deficient (osteoclast-poor) cells. The increase in bone volume was larger with the oc/oc cell transplantation, despite a similar reduction in bone resorption, indicating that the nonfunctional osteoclasts retained their ability to support bone formation also in vivo.

Other genetic evidence in mice indicated an intrinsic anabolic activity emanating from osteoclasts. An example is an early in vivo study showing that a signal from the active osteoclast is required for bone formation in mice lacking the osteoclast inhibitor OPG (Nakamura et al., 2003). These mice exhibit very high levels of osteoclastogenesis and resorption, resulting in severe osteopenia. They also show a high level of bone formation. Risedronate treatment reduced osteoclast activity, while osteoclast numbers remained high. The reduction in osteoclast activity was associated with a low level of bone formation, even in the presence of a BMP-2 implant, indicating that active osteoclasts produce bone formation-simulating activity. In a later study these authors (Koide et al., 2017) found that conditioned medium from osteoclasts of OPG-deficient mice included leukemia inhibitory factor as a potential osteoclast-derived coupling factor, as had been suggested by others (Pederson et al., 2008; Poulton et al., 2012). It should be emphasized, though, that it is difficult to prove whether it is a coupling factor in the context of bone remodeling or simply a factor produced by osteoclasts that can also stimulate bone formation.

Based on experiments in mice with inactivating mutations of each of the two alternative signaling pathways of gp130 it was concluded that resorption alone was insufficient to promote coupled bone formation, but that active osteoclasts are the likely source, and that the coupling pathway is interleukin-6 dependent (Sims et al., 2004; Martin and Sims, 2005). Another proposed pathway of gp130 involvement was through the gp130-signaling cytokine cardiotrophin-1 (CT-1). In mice with global deletion of CT-1, although osteoclast numbers are high, their activity is low, and so too is the activity of their osteoblasts, indicating a lack of coupling factor production (Walker et al., 2008). CT-1 was detected in resorbing osteoclasts by immunohistochemistry (IHC) and shown to stimulate osteoblast differentiation in vitro and bone formation in vivo (Walker et al., 2008).

When osteoclast conditioned medium stimulated MSC migration and osteoblastic differentiation, Pedersen et al. (Pederson et al., 2008) undertook a microarray study identifying sphingosine-1-phosphate (S1P), Wnt10b, and BMP-6 as osteoclast products, suggesting they might be coupling factors. Each of these skeletal anabolic agents was known to be secreted by osteoclasts to a greater extent than by macrophages (Baron and Rawadi, 2007; Vukicevic and Grgurevic, 2009). Whether important as an osteoclast product or not, Wnt10b has been invoked as a T cell product mediating the anabolic effect of PTH (Terauchi et al., 2009; Bedi et al., 2012). The roles of S1P in bone are also likely to be quite complex. It can have inhibitory or stimulatory effects on osteoblasts depending on the stage of cell differentiation and on the source of precursors, such as human MSCs, immortalized MSCs, and mouse calvarial osteoblasts (Ryu et al., 2006; Pederson et al., 2008; Quint et al., 2013). S1P is also expressed by cells in the vasculature and acts on its receptor, expressed in osteoclast precursors, to stimulate osteoclastic recruitment in vitro (Ishii et al., 2009). Furthermore, in vivo and in vitro studies indicated that S1P can limit bone resorption by regulating the chemotaxis and migration of osteoclast precursors, essentially resulting in increased recirculation from bone to blood (Ishii et al., 2010). In that same work, knockout of the S1P receptor S1PR1 yielded mice with excessive bone loss and enhanced osteoclast attachment to bone surfaces, and treatment with FTY720, a drug agonist of four of the five S1P receptors, including S1PR1, was effective in preventing bone loss in ovariectomized mice. Data more likely suggesting a role for osteoclast-derived S1P in the coupling mechanism came from a study in which *cathepsin K* was rendered null in osteoclasts, resulting in impaired resorption, while osteoclast numbers and bone formation were maintained (Lotinun et al., 2013). Ex vivo cultures showed that the mutated osteoclasts had a greater capability to promote aspects of osteoblast differentiation in coculture, an effect inhibited by an S1P receptor antagonist. Suggestive though this is of a role for S1P in the coupling process in the BMU, it needs to be explored further and put into the context of other actions of S1P, which has been invoked as a signaling mechanism in the actions of a number of cytokines, growth factors, and hormones (reviewed in Alvarez et al., 2007).

Other candidate osteoclast-secreted coupling factors have emerged from in vitro and ex vivo studies (listed in Table 10.1). These include afamin, a member of the albumin/vitamin D-binding protein family (Lichenstein et al., 1994); it is produced by osteoclasts and caused the recruitment of a mouse preosteoblastic cell line in vitro, in a manner that was lost by in vitro knockdown of afamin production (Kim et al., 2012). Another is PDGF-BB, produced by nonresorbing osteoclasts, which induces migration of bone marrow–derived human MSCs (Kreja et al., 2010) and mouse preosteoblasts (Sanchez-Fernandez et al., 2008), although an earlier study indicated that PDGF-BB inhibited osteoblastogenesis (Kubota et al., 2002). Perhaps more convincing evidence for a role for PDGF-BB in remodeling is from the mouse genetic studies that show its promotion of angiogenesis when mobilized by resorption and acting early in remodeling (Xie et al., 2014). Two candidate coupling activities, collagen triple-helix repeat–containing 1 and complement factor 3A (C3A), were put forward by one group (Takeshita et al., 2013; Matsuoka et al., 2014), but in each case they were produced by cells other than osteoclasts, and the high circulating level of C3A tended to exclude it from a role as a local regulator of coupling within the BMU. A more recent suggestion has been that microRNAs are released as coupling factors contained within exosomes that reach the circulation (Li et al., 2016). While it is appealing that osteoclast-derived membranous vesicles could provide a way of regulating osteoblast function, it is difficult to see how this could be controlled, either within the confines of the BMU or at the level of the whole individual.

Membrane-bound coupling factors synthesized by osteoclasts

Some of the osteoclast-derived coupling factors proposed are membrane-bound molecules, expressed on the cell surface of osteoclasts. Such molecules therefore require direct cell–cell contact with osteoblast lineage cells to produce effects that might favor osteoblast differentiation. These membrane-bound proteins have been discovered through a combination of genetic and pharmacological approaches. Factors proposed to act in this cell-contact-dependent manner include EphrinB2 (Zhao et al., 2006) and semaphorin D (Negishi-Koga et al., 2011). While plausible in vitro, and observed in the highly active murine calvarial suture that lacks BMUs (Furuya et al., 2018), such cell-contact-dependent mechanisms are problematic in the BMU because osteoclasts and osteoblasts are present on the bone surface at different times during remodeling, and therefore are rarely, if ever, in contact. If such mechanisms occur, they may exist between osteoclasts and osteoblast precursors very early in remodeling, or between osteoclasts and osteoblast lineage cells in the remodeling canopy. The latter possibility will be discussed in light of anatomic structural studies of the reversal phase, later.

The Eph family constitutes the largest family of receptor tyrosine kinases, notable because both receptors (Eph proteins) and ligands (Ephrin proteins) are membrane bound and both forward (through receptor) and reverse (through ligand) signaling takes place upon their interaction. EphrinB2 is expressed at all stages of osteoblast differentiation, as well as in osteoclasts and their precursors (Zhao et al., 2006; Allan et al., 2008). The first study of ephrin/Eph interaction in bone seemed to present an exciting prospect of EphrinB2 from the osteoclast acting as a coupling factor through a cell-contact-dependent mechanism with osteoblasts (Zhao et al., 2006). In vitro data indicated that contact between osteoclasts and osteoblasts initiated signaling in both cells such that EphrinB2 signaling in the osteoclast lineage limited their differentiation, while simultaneous EphB4 signaling in the osteoblast stimulated bone formation. However, osteoclast lineage–specific deletion of EphrinB2 presented no detectable in vivo bone phenotype (Zhao et al., 2006), nor did osteoclast precursors from an osteoclast-specific knockout of EphrinB2 show any alteration in osteoclast differentiation (Tonna et al., 2014). Further, EphrinB2/EphB4 interaction within the osteoblast lineage in vivo played a crucial role in maintaining osteoblast differentiation and attaining normal bone strength (Takyar et al., 2013). Specific blockade of EphrinB2/EphB4 signaling both in vitro and in vivo impaired osteoblast function and the anabolic response to PTH in vivo (Tonna et al., 2014). The latter was dependent on the availability of EphrinB2 for reverse signaling. Indeed, the earlier published evidence (Zhao et al., 2006) for a mild anabolic effect of transgenic EphB4 overexpression in vivo, and treatment of osteoblasts with clustered EphB4-Fc in vitro, might be explained by such an action through EphrinB2 signaling within osteoblasts. Given the extensive contact among osteoblasts that is required for bone formation (Ecarot-Charrier et al., 1983;

Abe et al., 1993; Gerber and ap Gwynn, 2001), it seems logical that such membrane-bound proteins are more likely to regulate bone formation through communication between osteoblast lineage cells than in osteoclast–osteoblast coupling activity that links two cell types without physical contact.

Semaphorins include both secreted and membrane-associated molecules that use plexins and neuropilins as their primary receptors. Plexins are the usual receptors for membrane–associated semaphorins, and neuropilins are obligate coreceptors in the case of most soluble class III receptors (Winberg et al., 1998; Takahashi et al., 1999; Tamagnone et al., 1999). Transcriptional arrays carried out on osteoclasts revealed substantial expression of semaphorin 4D (Sema4D), with none detectable in osteoblasts (Negishi-Koga et al., 2011). Targeted genetic ablation of Sema4D in osteoclasts resulted in increased trabecular bone mass, due to increased number and activity of osteoblasts, whereas osteoclast numbers were normal. Marrow transfer to wild-type mice from Sema4D-null mice resulted in increased bone formation and trabecular bone mass, and treatment of osteoblasts in vitro with recombinant soluble Sema4D-Fc decreased formation of mineralized nodules. The data point to Sema4D as an osteoclast-derived inhibitor of osteoblast differentiation and bone formation; these properties would equip it to be a fine-tuning mediator of remodeling in the BMU, acting as an inhibitor of the process. Just as OPG counters RANKL action by providing a powerful negative influence on osteoclast formation and function, it comes as no surprise that there should exist inhibitors of coupling; this raises the possibility that more such activities might exist.

A further membrane-bound coupling activity that has gained support is "outside-in" or "reverse" signaling within osteoblasts by RANK, the receptor for RANKL that signals action to generate osteoclasts (Suda et al., 1999). In addition to its actions to inhibit osteoclastogenesis, a RANKL-binding agent was found to also increase bone formation in vivo and promote osteoblast differentiation in vitro (Furuya et al., 2013). This was confirmed in an animal model of inflammatory arthritis (Kato et al., 2015), and was discussed in an editorial published in the same issue (Sims and Romas, 2015). These pharmacological studies have been supported by identification of an endogenous and potentially physiological action of RANK. RANK secreted in vesicles from maturing osteoclasts was found to increase bone formation by promoting RANKL reverse signaling to activate *runx2* (Ikebuchi et al., 2018). This is an interesting possibility that would require controlled delivery from osteoclasts early in the life of a BMU to appropriate targets in the osteoblast lineage, just as is the case with TGFβ (Tang et al., 2009) and IGF-1 (Xian et al., 2012) (*vide supra*, discussion under "Matrix-derived resorption products as coupling factors").

Perspective on candidate osteoclast-derived coupling factors identified to date

Research activity focused on the search for local osteoclast-derived coupling factors (whether by active secretion or resorption of bone matrix) has yielded many candidate molecules, summarized in Table 10.1, indicating the nature of the experiment used and the cells capable of producing those molecules. Since many of these have been identified only since 2009, most require validation by independent groups of researchers working in multiple systems. This is particularly true of those factors identified in conditioned media. These can point to factors that may be important, but they each require validation of their specificity and their in vivo relevance.

A striking feature of these published data is that none of these candidate osteoclast-derived coupling factors can be said confidently to be exclusively derived from the osteoclast. Many are also products of cells in close proximity with, and even within, the BMU, such as macrophages and other immune cells, and many circulate in concentrations high enough to exclude a primary function as local regulators at multiple sites throughout the skeleton. The frequency of contradictory reports, variability of experimental systems that have been used, and limited nature of in vitro studies have made some of these reports difficult to interpret. Observations arising out of activities discovered in conditioned medium of osteoclast cultures need to be interpreted in light of substantial macrophage contamination in such cultures that is variable between donors and laboratories, and cannot be avoided with existing methods. Furthermore, none of the in vitro studies have set out to determine whether osteoclast products influence different stages of osteoblast differentiation, which seems a likely mechanism by which they could act, given the temporal delay between bone resorption and formation at the BMU. It would seem likely that the coupling process within the BMU would require actions at different stages, as osteoblasts progress through differentiation during remodeling (Fig. 10.3).

How other cells contribute to coupling

In addition to factors released by osteoclasts through resorption, secretion, or membrane budding, other cell types may mediate the signals required to ensure that bone formed during remodeling in the BMU matches the prior level of bone resorption to maintain bone mass. Some possibilities include signaling by osteoblast lineage cells themselves, macrophages and immune cells, and the cell population resident during the reversal phase.

Osteoblast lineage cells—sensing the surface and signaling to one another

Osteoblast lineage precursors begin their differentiation in the BMU, during which they are in contact with (or at least in close proximity to) other precursors and mature osteoblasts on the bone surface. Bone-forming osteoblasts on the bone surface are also in contact, not only with one another, but also with the highly differentiated osteocytes within the bone matrix. This lends itself to communication within the lineage as part of the overall coupling process.

An example of such intralineage information transfer is the production of latent TGFβ, produced by osteoblasts, activated by proteases, and supplementing the resorption-derived TGFβ that contributes to the initial stages of coupling in the BMU (Yee et al., 1993; Pfeilschifter et al., 1995). We have drawn attention to the role of EphrinB2/EphB4 interaction between osteoblast lineage cells as an important checkpoint that must be passed for the lineage to reach late stages of osteoblast differentiation (Takyar et al., 2013; Tonna et al., 2014). EphrinB2 production remains unchanged throughout osteoblast differentiation, but responds with rapidly increased production when exposed to PTH-related protein (PTHrP) (Allan et al., 2008). Through such interactions within the osteoblast lineage, osteoblast precursors are able to respond to the cues initiated by osteoclast-derived factors to enter the BMU and differentiate to osteoblasts capable of bone matrix production (Fig. 10.3).

The osteocyte, the most abundant cell in bone by far, forms a highly complex cellular communication network through the bone matrix, with a total of ~3.7 trillion connections throughout the adult skeleton (Buenzli and Sims, 2015). This network has crucial regulatory functions by its production of sclerostin, which limits bone formation though its powerful inhibition of Wnt-stimulated osteoblast differentiation (Bonewald and Johnson, 2008). This action is likely to be important in limiting the amount of bone laid down in modeling, but its production in response to loading (Galea et al., 2011), or to local factors such as prostaglandin E_2 (Genetos et al., 2011) or PTHrP (Ansari et al., 2018), is capable of regulating the bone formation within BMUs. Just as osteoblast precursors are available in the BMU at the time when osteoclasts are present, resorbing bone and releasing or producing coupling factors, so too are osteocytes available to respond to coupling factors. Indeed, a number of osteoclast-derived coupling factors promote bone formation and suppress sclerostin production by osteocytes: these include CT-1 and Leukemia Inhibitory Factor (LIF) (Walker et al., 2008, 2010; Poulton et al., 2012).

Another mechanism to which osteoblasts can respond that would allow them to produce sufficient bone matrix to refill the cavity left by osteoclasts, is their ability to "sense" changes on the bone surface. It was shown in vitro that if rat calvarial cells were provided to bone slices with crevices made by osteoclasts or mechanically excavated grooves, the cells made bone in those defects, filling them exactly to a flat surface (Gray et al., 1996). Osteoblast precursors have been shown to respond to changes in surface topography, whether the change is much larger than the cell itself, as in the aforementioned study, or very much smaller than the cell (Dalby et al., 2006). In response to altered nanotopography, osteoblast lineage cells adhere to the surface and one another and proceed to differentiate. In this way, osteoclasts control osteoblast activity from a distance by establishing the size and shape of the resorptive pit to be filled. It is also possible that the changes brought about on the bone surface by cells in the reversal phase change the topography in a way that determines osteoblast function. Both the proposed growth factor actions and the work of Gray et al. imply that once the formation process is established, the participating cells themselves are able to sense the spatial limits and fill the space through chemical communication, possibly involving gap junctions or cell-contact-dependent communication processes both between bone-forming osteoblasts and between osteoblasts and osteocytes (Tonna and Sims, 2014). Since the in vitro study of Gray used bone that did not contain osteocytes, while they may contribute, osteocytes are not necessary for osteoblasts to respond to topographic clues, at least in vitro.

Osteocytes are the major controlling cells of bone (Bonewald, 2011; Schaffler et al., 2014), producing sclerostin as a powerful inhibitor of bone formation, whether in modeling or remodeling (van Bezooijen et al., 2004). Through their fluid-filled network of communicating channels they sense and respond to mechanical strain; this system might provide an additional coupling mechanism. Osteocytes would sense the increased strain resulting from weakening of the bone as resorption progresses (McNamara et al., 2006), and respond by producing a signal to halt resorption. They would also detect when the strain is relieved as the resorbed pit is refilled by osteoblasts. Such a strain-based model for coupling was proposed some years ago (Rodan, 1991), and as our understanding of osteocyte signaling increases, possible mediators are coming to light. As the mechanical properties of the excavated bone improve, the osteocytes may signal to osteoblasts that enough bone has been made in the BMU, a task that could be filled by sclerostin or by other osteocyte-derived regulators of bone formation such as oncostatin M (OSM) (Walker et al., 2010) or PTHrP (Ansari et al., 2018). Clearly many steps are required to achieve precision in the coupling process: precursor replication, differentiation, limiting the cell mass to exactly what is required, correct shaping, and level of mineralization.

Macrophages, immune cells, and endothelial cells

As discussed earlier, a key problem of studying osteoclastic effects on osteoblasts is the technical difficulty of obtaining sufficiently purified osteoclasts. Furthermore, the extensive overlap in gene expression between osteoclasts and macrophages suggests that factors produced by osteoclasts can be also produced by macrophages in the vicinity of the BMU.

OSM, a cytokine that signals through gp130, stimulates bone formation in mouse calvariae in vivo and in bone remodeling, as indicated by impaired bone formation in OSM-null mice, and promotes osteoblast commitment by calvarial osteoblasts in vitro (Walker et al., 2010). Although not expressed in osteoclasts, OSM was originally identified in macrophages (Zarling et al., 1986) and was present in media from enriched cultures of human osteoclasts. It was found to be the factor responsible for strongly inducing alkaline phosphatase activity and mineralization by human adipose tissue-derived MSCs (pluripotent osteoblast precursors) (Fernandes et al., 2013), where macrophage populations formed from the same source as the osteoclasts were found to be even more efficient at driving MSC maturation, probably due to higher OSM production. A cautionary finding from some of this work was that differing responses can occur with different types of macrophage activation (Guihard et al., 2012; Nicolaidou et al., 2012). These findings serve to make a broader point: factors identified as osteoclast-derived coupling factors are produced not only by osteoclasts but also by macrophages (to which they have a close ontogenic relationship).

Like the contaminating macrophages in osteoclast cultures, macrophages are present in primary osteoblast cultures from calvariae and promote osteoblast differentiation (Chang et al., 2008). This finding led to the identification of bone-resident tissue macrophages (osteal macrophages defined by F4/80 antigen expression) mingled in large numbers with osteoblasts at the endosteal and periosteal bone surfaces and forming a canopy-like layer above osteoblasts and bone-lining cells at endosteal surfaces (Chang et al., 2008). Despite their emerging importance, defining the resident osteal macrophage population remains problematic since they are primarily defined by location, and as of this writing, cell surface markers do not exist that allow the resident cells to be specifically purified. Such regulatory macrophages would be expected to be sensitive to cues from the immune system. In addition, either a change in their activation state or replacement by recruited inflammatory macrophages and availability within the BMU canopy, by modifying their secretion of factors such as OSM, might alter local bone metabolism, perhaps profoundly (Fig. 10.4).

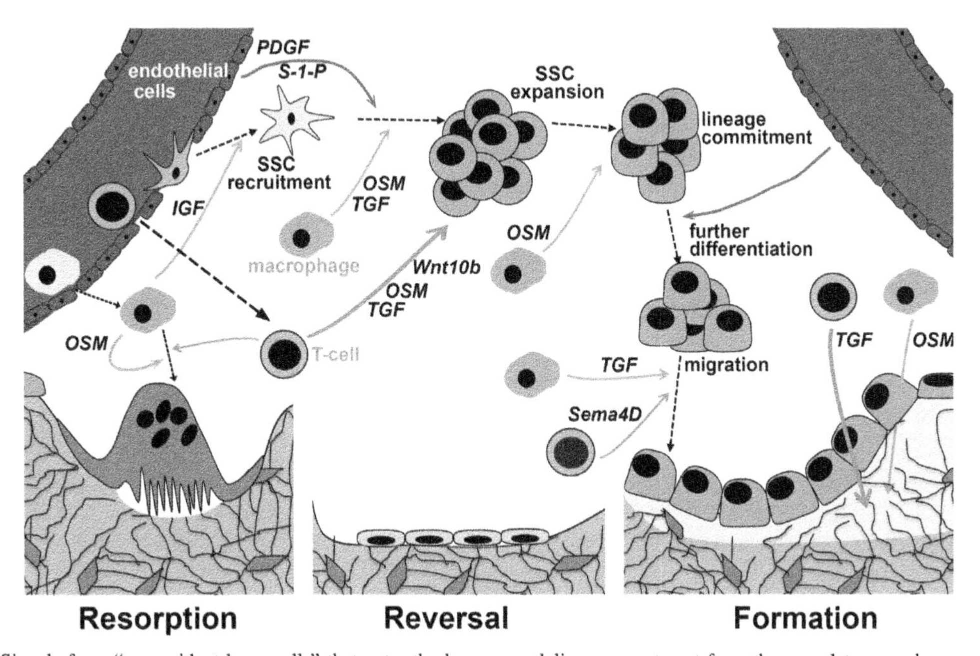

FIGURE 10.4 Signals from "nonresident bone cells" that enter the bone remodeling compartment from the vasculature, such as endothelial cells, T lymphocytes, and macrophages, also contribute to the matching of bone formation to bone resorption during the resorption, reversal, and formation phases of remodeling. Some examples of potential factors from each cell type are illustrated, such as platelet-derived growth factor (*PDGF*), sphingosine-1-phosphate (*S-1-P*), oncostatin M (*OSM*), Wnt10b, and transforming growth factor (*TGF*) are shown. See Table 10.1 for more examples. Factors act on each stage of osteoblast differentiation, including on osteocytes. *IGF*, insulin-like growth factor; *Sema4D*, semaphorin 4D; *SSC*, stromal stem cell.

Immune cells are also able to gain access to remodeling sites through the blood supply at the canopy (Fig. 10.4). Just as OSM may promote bone formation by virtue of its production by macrophages (Zarling et al., 1986), T lymphocytes (Clegg et al., 1996), osteoblasts, and osteocytes (Walker et al., 2010), the same is true of many of the other coupling factors (summarized in Table 10.1). T cells have been invoked as playing a part in normal remodeling by mediating the anabolic effect of intermittent PTH injection in the mouse (Terauchi et al., 2009; Bedi et al., 2012). In that work Wnt10b was reported as the T cell product acting upon the osteoblast lineage to achieve the PTH anabolic effect. The significance of this work might go further, though. Since the PTH anabolic effect is considered to be predominantly through promotion of remodeling, can these findings imply that T cells contribute to normal remodeling events within the BMU? It is relevant to note that Wnt10b was also one of the osteoclast conditioned medium–derived coupling factors identified by Pederson et al. (2008).

The data summarized in Table 10.1 illustrate the redundancy among cell types of secretory products that might promote osteoblast differentiation. They also illustrate the difficulty in identifying any exclusive osteoclast-derived activity. None of these considerations reduces the potential importance of these factors in bone biology. Rather, they illustrate multiple roles for these factors, the likely complexity of the coupling process, the possible involvement of multiple cell types, and the likely cross-regulation of each of these regulatory pathways.

The reversal phase as a coupling mechanism

Publications have suggested a regulatory role for the reversal phase in coupling of osteoblast to osteoclast activity. The existence of a reversal phase between the resorptive and the formative phases of remodeling was proposed originally by Baron, (1977) on the basis of studies in rat bone. Twenty years after Frost's remodeling proposals it was found that toward the end of resorption, mononuclear cells gather at the bottom of resorption pits, where they prepare the pits for the engagement of osteoblasts in bone formation (Villanueva et al., 1986). In this reversal phase of remodeling, macrophages had been long considered responsible for the postresorption digestion of collagen fragments in the BMU. The identification that cells of the osteoblast lineage were able to engulf collagen fragments (Takahashi et al., 1986) was taken further by Everts et al. (Everts et al., 2002), who identified osteoblast lineage bone-lining cells cytologically at sites of resorption, both in calvariae and in long bones, and showed that these cells actually engulf collagen fragments remaining on the bone surface after osteoclasts have resorbed and left it (Everts et al., 2002). This activity appeared to be mediated by membrane matrix metalloproteinases. Thus, although these osteoblast lineage cells are not equipped to lay down an adequate matrix, they nevertheless are able to lay down a thin layer of collagen along Howship's lacuna, closely associated with a cement line. This reversal line (cement line) contains a large abundance of osteopontin (Chen et al., 1994), which is produced by both osteoclasts and osteoblasts. Osteopontin is an RGD-containing extracellular matrix protein that interacts with integrin receptors $\alpha V\beta 3$ in osteoclasts and primarily $\alpha V\beta 5$ in osteoblasts, mediating cell attachment to the bone surface and signaling within the cell. The presence of osteopontin on the reversal line raises the possibility that it may be one of the signals for cessation of osteoclast activity or initiation of osteoblastic bone formation or possibly both.

These findings of Everts et al. were of interest in themselves, in drawing attention to a function of osteoblast lineage cells in remodeling during the period between bone resorption and formation. Further than that, though, the question of whether these lining cells might become "activated" to become matrix-producing osteoblasts after the reversal phase became an active question.

When in situ hybridization (ISH) and IHC were used in human bone biopsy samples, the reversal cells on the surface were confirmed to be of the osteoblastic lineage by virtue of their expression of osteoblast marker genes (Andersen et al., 2013). Furthermore, those next to bone-forming osteoblasts were more mature than those next to osteoclasts (Andersen et al., 2013). This was in contrast to earlier findings in human and other bones, indicating that cells near osteoclasts are tartrate-resistant acid phosphatase (TRAP)-positive mononuclear osteoclasts (Tran Van et al., 1982; Eriksen et al., 1984b; Eriksen, 1986; Bianco et al., 1988). To resolve this, sensitive ISH and IHC methods were used and showed functional and phenotypic changes in the mesenchymal reversal cells as they progressed from near the osteoclasts (TRAP positive by IHC) to near the osteoblasts (TRAP negative) (Abdelgawad et al., 2016). TRAP immunoreactivity was shown to be occurring in cells with no detectable TRAP mRNA, leading to the conclusion that TRAP positivity in the early reversal cells resulted from ingestion from nearby osteoclasts. Runx2 immunoreactivity was uniform throughout the reversal cells, reflecting their osteoblastic lineage.

Removal of the collagen remnants in Howship's lacunae is another potential coupling mechanism, since it is obligatory for bone formation to occur. It is carried out by cells close to osteoclasts, early reversal cells with enhanced collagenolytic activity mediated through production of matrix metalloproteinase-13 (MMP-13) (Abdelgawad et al., 2016) and the

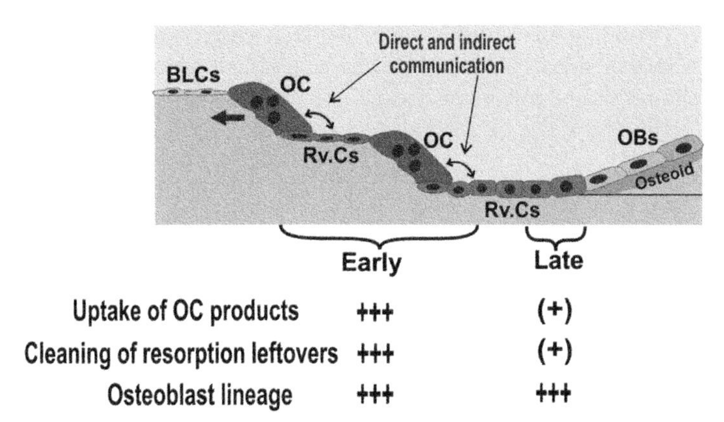

FIGURE 10.5 A new model of the reversal phase of remodeling. During quiescence, bone-lining cells (*BLC*) cover the inactive bone surface. Osteoclasts (*OC*) resorb the bone matrix and are intercalated with osteoblast lineage reversal cells (*Rv.Cs*) in the early reversal phase. In this phase osteoclast products released from the bone matrix or synthesized and secreted by osteoclasts are taken up by reversal cells, and may influence osteoblast progenitors. In the late reversal phase, reversal cells differentiate, on the bone surface, to become bone-forming osteoblasts (*OBs*) capable of synthesizing osteoid. Figure provided by Thomas L Andersen.

endocytic collagen receptor uPARAP/Endo180 (Madsen et al., 2011). With progression to late reversal cells there is much less expression of MMP-13 and uPARAP, constant Runx2, and no TRAP (Abdelgawad et al., 2016).

This evidence indicates a phenotypic change during the reversal phase in the cells on the bone surface. The cells progress from early to late reversal cells, with the latter differentiating directly into osteoid-synthesizing osteoblasts (Delaisse, 2014) (Fig. 10.5). The findings point to potential connections between osteoclasts and early osteoblast lineage cells that could transmit contact-independent or contact-dependent signals. Such contact could influence the function and progress of those osteoblast lineage cells. As an area of cell biology to study it is a fascinating one, if difficult to approach through any other than in vivo experimentation. In contrast to these potential communication points, mature osteoclasts and mature bone-forming osteoblasts are probably almost never in contact in ways that can transfer information.

Evidence consistent with these mechanisms was also obtained from remodeling in cortical bone, which is carried out with different cellular organization. The resorbing osteoclasts generate a cutting cone directed parallel to the diaphysis of long bones, with this resorption followed by the reversal zone and then osteoblasts synthesizing matrix (Eriksen, 1986; Jaworski, 1992) (Fig. 10.1).

Additional work from the same group examined the continuum of the BMU in cutting cones by IHC and ISH to identify osteoblast lineage cells and osteoclasts. In doing so, they described a reversal phase in which osteoblast progenitors differentiate until a critical mass of mature osteoblasts is reached and matrix formation follows (Abdelgawad et al., 2016; Lassen et al., 2017). They also noted the presence of sparsely distributed osteoclasts in the reversal phase in decreasing numbers as time after resorption increased. These osteoclasts are close enough to osteoblast lineage cells to be capable of transmitting signals, and seem to be responsible for a slow "secondary" bone resorption period needed to widen the osteon, as opposed to the longitudinal resorption in the cutting zone (Jaworski and Lok, 1972; Parfitt, 1983). The dynamics of these cells in the reversal phase that are modeled graphically in Fig. 10.5 confer an important role on the reversal phase in the coupling process of bone remodeling. First, it draws attention to the need for the supply of osteoblast precursors at that stage, if they are to progress to form bone. Where do they come from and how? The capillary blood supply identified near the canopy (Kristensen et al., 2013; Delaisse, 2014) is a potential source, as are the very early mesenchymal cells (pericytes) adherent to those nearby capillaries (Bianco et al., 2011). Second, while they are differentiating, these osteoblast lineage cells have osteoclasts distributed among them and are capable of signaling to them with secreted or even membrane-bound coupling activities (Lassen et al., 2017). Where do these "interstitial" osteoclasts come from? Are they newly generated or are they osteoclasts continuing to function from the initial resorption phase? Third, the onset of bone formation, with synthesis of ample osteoid, is an effective signal that dampens osteoclastic resorption, since osteoclasts are not known to attach to osteoid. Fourth, prolongation of the reversal phase for any reason, such as inadequate supply of progenitors with aging, can prolong the process and increase the osteon width, even contributing to porosity (Andreasen et al., 2017). On the other hand, the sooner bone formation starts, the faster secondary/radial bone resorption will stop.

Thus, in considering coupling of bone formation to resorption, the reversal phase is engaged in a remarkable abundance of biological interactions. In attempts to characterize coupling, intercellular communication mechanisms during the reversal phase might be a more rewarding focus of attention rather than a narrow search for signals from mature osteoclasts to osteoblasts. The many participating signals are likely to have targets and outcomes that are heavily dependent upon the stages of differentiation of cells within the osteoblast lineage.

Conclusion

The coupling of bone formation to bone resorption is the dominant biological feature of bone remodeling. It is useful to consider the relative contributions to its control of circulating hormones and of locally generated factors. The latter might come from osteoclasts, osteoblasts, osteocytes, the matrix, or other cells in the environment, such as immune cells, macrophages, and endothelial cells.

Bone remodeling takes place simultaneously and asynchronously at very many sites throughout the skeleton, with those sites being selected because damaged or old bone needs to be removed and replaced. How those sites are selected we do not know. Once a BMU is initiated, though, it needs to have its activities regulated—its osteoclast formation, activity, and death; its osteoblast differentiation from precursors as primitive as tissue stem cells—and cessation of that process is needed when the filling of the BMU is complete.

All of these are processes and mechanisms that lend themselves readily to local regulation. This makes the process of coupling such an important one to understand. It is less easy to imagine how the progression of the remodeling process could be regulated satisfactorily by circulating hormones. This is especially so when thinking of the initiation of BMUs at specific sites, control of osteoclast activity at those sites, control of the supply of mesenchymal lineage precursors, and then their differentiation.

For these reasons we ask, how could a circulating hormone be a useful contributor to the bone formation process in remodeling? The question arises with PTH, for example, where the available published information tells us that there is a circadian rhythm of circulating PTH levels, but the total change throughout 24 h is on the order of 20% (Jubiz et al., 1972; el-Hajj Fuleihan et al., 1997; Fraser et al., 1998; White et al., 2007; Redmond et al., 2016). Such constancy of circulating PTH seems to equip it poorly for the initiation and/or control of BMUs generated stochastically throughout the skeleton. Locally generated PTHrP, though, could be the way that action through the PTHR1 contributes to remodeling (see Chapter 25).

There might be quite a different view in the case of an inhibitory hormone, such as estrogen. If the function of estrogen is to be available always to maintain a constitutive brake on osteoclast formation and/or activity, complemented by a more effective local inhibitor, e.g., OPG, this could be achieved as a circulating hormone. Lowering of estrogen would be expected to result in escape of control and increased remodeling by initiation of new BMUs, which is indeed what happens. If this line of reasoning is correct, perhaps by analogy we should think of the possibility of there being circulating activity that acts as a constitutive brake on osteoblast differentiation and thence bone formation. Its relationship with the local inhibitor of formation, sclerostin, would be analogous to that between estrogen and OPG.

The process of bone remodeling is a complex one with many participating cells and effectors and many identifiable stages that we are only beginning to understand. What is becoming evident, though, is that it is much easier to inhibit remodeling than it is to stimulate it. That is why therapeutic approaches to inhibit resorption are simple compared with approaches to stimulate formation.

References

Abdelgawad, M.E., Delaisse, J.M., Hinge, M., Jensen, P.R., Alnaimi, R.W., Rolighed, L., Engelholm, L.H., Marcussen, N., Andersen, T.L., 2016. Early reversal cells in adult human bone remodeling: osteoblastic nature, catabolic functions and interactions with osteoclasts. Histochem. Cell Biol. 145 (6), 603–615.

Abe, Y., Akamine, A., Aida, Y., Maeda, K., 1993. Differentiation and mineralization in osteogenic precursor cells derived from fetal rat mandibular bone. Calcif. Tissue Int. 52 (5), 365–371.

Alatalo, S.L., Ivaska, K.K., Waguespack, S.G., Econs, M.J., Vaananen, H.K., Halleen, J.M., 2004. Osteoclast-derived serum tartrate-resistant acid phosphatase 5b in Albers-Schonberg disease (type II autosomal dominant osteopetrosis). Clin. Chem. 50 (5), 883–890.

Allan, E.H., Hausler, K.D., Wei, T., Gooi, J.H., Quinn, J.M., Crimeen-Irwin, B., Pompolo, S., Sims, N.A., Gillespie, M.T., Onyia, J.E., Martin, T.J., 2008. EphrinB2 regulation by PTH and PTHrP revealed by molecular profiling in differentiating osteoblasts. J. Bone Miner. Res. 23 (8), 1170–1181.

Alvarez, S.E., Milstien, S., Spiegel, S., 2007. Autocrine and paracrine roles of sphingosine-1-phosphate. Trends Endocrinol. Metabol. 18 (8), 300–307.

Andersen, T.L., Abdelgawad, M.E., Kristensen, H.B., Hauge, E.M., Rolighed, L., Bollerslev, J., Kjaersgaard-Andersen, P., Delaisse, J.M., 2013. Understanding coupling between bone resorption and formation: are reversal cells the missing link? Am. J. Pathol. 183 (1), 235–246.

Andersen, T.L., Hauge, E.M., Rolighed, L., Bollerslev, J., Kjaersgaard-Andersen, P., Delaisse, J.M., 2014. Correlation between absence of bone remodeling compartment canopies, reversal phase arrest, and deficient bone formation in post-menopausal osteoporosis. Am. J. Pathol. 184 (4), 1142–1151.

Andersen, T.L., Soe, K., Sondergaard, T.E., Plesner, T., Delaisse, J.M., 2010. Myeloma cell-induced disruption of bone remodelling compartments leads to osteolytic lesions and generation of osteoclast-myeloma hybrid cells. Br. J. Haematol. 148 (4), 551–561.

Andersen, T.L., Sondergaard, T.E., Skorzynska, K.E., Dagnaes-Hansen, F., Plesner, T.L., Hauge, E.M., Plesner, T., Delaisse, J.M., 2009. A physical mechanism for coupling bone resorption and formation in adult human bone. Am. J. Pathol. 174 (1), 239–247.

Andreasen, C.M., Delaisse, J.M., Cj van der Eerden, B., van Leeuwen, J.P., Ding, M., Andersen, T.L., 2018. Understanding age-induced cortical porosity in women: the accumulation and coalescence of eroded cavities upon existing intracortical canals is the main contributor. J. Bone Miner. Res. 33 (4), 606–620.

Ansari, N., Ho, P.W., Crimeen-Irwin, B., Poulton, I.J., Brunt, A.R., Forwood, M.R., Divieti Pajevic, P., Gooi, J.H., Martin, T.J., Sims, N.A., 2018. Autocrine and paracrine regulation of the murine skeleton by osteocyte-derived parathyroid hormone-related protein. J. Bone Miner. Res. 33 (1), 137–153.

Barlic, J., Zhu, W., Murphy, P.M., 2009. Atherogenic lipids induce high-density lipoprotein uptake and cholesterol efflux in human macrophages by up-regulating transmembrane chemokine CXCL16 without engaging CXCL16-dependent cell adhesion. J. Immunol. 182 (12), 7928–7936.

Baron, R., 1977. Importance of the intermediate phase between resorption and formation in the measurement and understanding of the bone remodelling sequence. In: Meunier, P. (Ed.), I. Bone Remodelling, 2nd Int Workshop. Lab Armour Montague, Paris, pp. 179–183.

Baron, R., Rawadi, G., 2007. Targeting the Wnt/beta-catenin pathway to regulate bone formation in the adult skeleton. Endocrinology 148 (6), 2635–2643.

Baron, R., Sims, N.A., 2000. Bone Cells and Their Function. Skeletal Growth Factors. E. Canalis. Lippincott Williams and Wilkins, Philadelphia, pp. 1–16.

Bedi, B., Li, J.Y., Tawfeek, H., Baek, K.H., Adams, J., Vangara, S.S., Chang, M.K., Kneissel, M., Weitzmann, M.N., Pacifici, R., 2012. Silencing of parathyroid hormone (PTH) receptor 1 in T cells blunts the bone anabolic activity of PTH. Proc. Natl. Acad. Sci. U.S.A. 109 (12), E725–E733.

Bennett, C.N., Ouyang, H., Ma, Y.L., Zeng, Q., Gerin, I., Sousa, K.M., Lane, T.F., Krishnan, V., Hankenson, K.D., MacDougald, O.A., 2007. Wnt10b increases postnatal bone formation by enhancing osteoblast differentiation. J. Bone Miner. Res. 22 (12), 1924–1932.

Bianco, P., Ballanti, P., Bonucci, E., 1988. Tartrate-resistant acid phosphatase activity in rat osteoblasts and osteocytes. Calcif. Tissue Int. 43 (3), 167–171.

Bianco, P., Sacchetti, B., Riminucci, M., 2011. Osteoprogenitors and the hematopoietic microenvironment. Best Pract. Res. Clin. Haematol. 24 (1), 37–47.

Bonewald, L.F., 2011. The amazing osteocyte. J. Bone Miner. Res. 26 (2), 229–238.

Bonewald, L.F., Johnson, M.L., 2008. Osteocytes, mechanosensing and Wnt signaling. Bone 42 (4), 606–615.

Buenzli, P.R., Sims, N.A., 2015. Quantifying the osteocyte network in the human skeleton. Bone 75, 144–150.

Campbell, P.G., Novak, J.F., Yanosick, T.B., McMaster, J.H., 1992. Involvement of the plasmin system in dissociation of the insulin-like growth factor-binding protein complex. Endocrinology 130 (3), 1401–1412.

Canalis, E., Gabbitas, B., 1994. Bone morphogenetic protein 2 increases insulin-like growth factor I and II transcripts and polypeptide levels in bone cell cultures. J. Bone Miner. Res. 9 (12), 1999–2005.

Canalis, E., Ornitz, D.M., 2000. Biology of Platelet-Derived Growth Factor. Skeletal Growth Factors. E. Canalis. Lippincott Williams and Wilkins, Philadelphia, USA, pp. 153–166.

Carroll, M.C., 2004. The complement system in regulation of adaptive immunity. Nat. Immunol. 5 (10), 981–986.

Centrella, M., Canalis, E., 1985a. Local regulators of skeletal growth: a perspective. Endocr. Rev. 6 (4), 544–551.

Centrella, M., Canalis, E., 1985b. Transforming and nontransforming growth factors are present in medium conditioned by fetal rat calvariae. Proc. Natl. Acad. Sci. U.S.A. 82 (21), 7335–7339.

Chambers, T.J., 1980. The cellular basis of bone resorption. Clin. Orthop. Relat. Res. 151, 283–293.

Champagne, C.M., Takebe, J., Offenbacher, S., Cooper, L.F., 2002. Macrophage cell lines produce osteoinductive signals that include bone morphogenetic protein-2. Bone 30 (1), 26–31.

Chang, M.K., Raggatt, L.J., Alexander, K.A., Kuliwaba, J.S., Fazzalari, N.L., Schroder, K., Maylin, E.R., Ripoll, V.M., Hume, D.A., Pettit, A.R., 2008. Osteal tissue macrophages are intercalated throughout human and mouse bone lining tissues and regulate osteoblast function in vitro and in vivo. J. Immunol. 181 (2), 1232–1244.

Chen, J., McKee, M.D., Nanci, A., Sodek, J., 1994. Bone sialoprotein mRNA expression and ultrastructural localization in fetal porcine calvarial bone: comparisons with osteopontin. Histochem. J. 26 (1), 67–78.

Chen, W., Jin, W., Wahl, S.M., 1998. Engagement of cytotoxic T lymphocyte-associated antigen 4 (CTLA-4) induces transforming growth factor beta (TGF-beta) production by murine CD4(+) T cells. J. Exp. Med. 188 (10), 1849–1857.

Clegg, C.H., Rulffes, J.T., Wallace, P.M., Haugen, H.S., 1996. Regulation of an extrathymic T-cell development pathway by oncostatin M. Nature 384 (6606), 261–263.

Cornish, J., Callon, K., King, A., Edgar, S., Reid, I.R., 1993. The effect of leukemia inhibitory factor on bone in vivo. Endocrinology 132 (3), 1359–1366.

Cornish, J., Callon, K.E., Edgar, S.G., Reid, I.R., 1997. Leukemia inhibitory factor is mitogenic to osteoblasts. Bone 21 (3), 243–247.

Dacquin, R., Domenget, C., Kumanogoh, A., Kikutani, H., Jurdic, P., Machuca-Gayet, I., 2011. Control of bone resorption by semaphorin 4D is dependent on ovarian function. PLoS One 6 (10), e26627.

Dalby, M.J., McCloy, D., Robertson, M., Wilkinson, C.D., Oreffo, R.O., 2006. Osteoprogenitor response to defined topographies with nanoscale depths. Biomaterials 27 (8), 1306–1315.

Dallas, S.L., Rosser, J.L., Mundy, G.R., Bonewald, L.F., 2002. Proteolysis of latent transforming growth factor-beta (TGF-beta)-binding protein-1 by osteoclasts. A cellular mechanism for release of TGF-beta from bone matrix. J. Biol. Chem. 277 (24), 21352–21360.

Daniel, T.O., Fen, Z., 1988. Distinct pathways mediate transcriptional regulation of platelet-derived growth factor B/c-sis expression. J. Biol. Chem. 263 (36), 19815–19820.

Del Fattore, A., Peruzzi, B., Rucci, N., Recchia, I., Cappariello, A., Longo, M., Fortunati, D., Ballanti, P., Iacobini, M., Luciani, M., Devito, R., Pinto, R., Caniglia, M., Lanino, E., Messina, C., Cesaro, S., Letizia, C., Bianchini, G., Fryssira, H., Grabowski, P., Shaw, N., Bishop, N., Hughes, D., Kapur, R.P., Datta, H.K., Taranta, A., Fornari, R., Migliaccio, S., Teti, A., 2006. Clinical, genetic, and cellular analysis of 49 osteopetrotic patients: implications for diagnosis and treatment. J. Med. Genet. 43 (4), 315–325.

Delaisse, J.M., 2014. The reversal phase of the bone-remodeling cycle: cellular prerequisites for coupling resorption and formation. Bonekey Rep. 3, 561.

Ecarot-Charrier, B., Glorieux, F.H., van der Rest, M., Pereira, G., 1983. Osteoblasts isolated from mouse calvaria initiate matrix mineralization in culture. J. Cell Biol. 96 (3), 639–643.

Eghbali-Fatourechi, G.Z., Modder, U.I., Charatcharoenwitthaya, N., Sanyal, A., Undale, A.H., Clowes, J.A., Tarara, J.E., Khosla, S., 2007. Characterization of circulating osteoblast lineage cells in humans. Bone 40 (5), 1370–1377.

el-Hajj Fuleihan, G., Klerman, E.B., Brown, E.N., Choe, Y., Brown, E.M., Czeisler, C.A., 1997. The parathyroid hormone circadian rhythm is truly endogenous–a general clinical research center study. J. Clin. Endocrinol. Metab. 82 (1), 281–286.

Elmardi, A.S., Katchburian, M.V., Katchburian, E., 1990. Electron microscopy of developing calvaria reveals images that suggest that osteoclasts engulf and destroy osteocytes during bone resorption. Calcif. Tissue Int. 46 (4), 239–245.

Eriksen, E.F., 1986. Normal and pathological remodeling of human trabecular bone: three dimensional reconstruction of the remodeling sequence in normals and in metabolic bone disease. Endocr. Rev. 7 (4), 379–408.

Eriksen, E.F., Eghbali-Fatourechi, G.Z., Khosla, S., 2007. Remodeling and vascular spaces in bone. J. Bone Miner. Res. 22 (1), 1–6.

Eriksen, E.F., Gundersen, H.J., Melsen, F., Mosekilde, L., 1984a. Reconstruction of the formative site in iliac trabecular bone in 20 normal individuals employing a kinetic model for matrix and mineral apposition. Metab. Bone Dis. Relat. Res. 5 (5), 243–252.

Eriksen, E.F., Melsen, F., Mosekilde, L., 1984b. Reconstruction of the resorptive site in iliac trabecular bone: a kinetic model for bone resorption in 20 normal individuals. Metab. Bone Dis. Relat. Res. 5 (5), 235–242.

Eriksen, E.F., Vesterby, A., Kassem, M., Melsen, F., Mosekilde, L. (Eds.), 1993. Bone Remodeling and Bone Structure. Handbook of Experimental Pharmacology. Springer Verlag, Berlin.

Everts, V., Delaisse, J.M., Korper, W., Jansen, D.C., Tigchelaar-Gutter, W., Saftig, P., Beertsen, W., 2002. The bone lining cell: its role in cleaning Howship's lacunae and initiating bone formation. J. Bone Miner. Res. 17 (1), 77–90.

Fadok, V.A., Bratton, D.L., Konowal, A., Freed, P.W., Westcott, J.Y., Henson, P.M., 1998. Macrophages that have ingested apoptotic cells in vitro inhibit proinflammatory cytokine production through autocrine/paracrine mechanisms involving TGF-beta, PGE2, and PAF. J. Clin. Investig. 101 (4), 890–898.

Fernandes, T.J., Hodge, J.M., Singh, P.P., Eeles, D.G., Collier, F.M., Holten, I., Ebeling, P.R., Nicholson, G.C., Quinn, J.M., 2013. Cord blood-derived macrophage-lineage cells rapidly stimulate osteoblastic maturation in mesenchymal stem cells in a glycoprotein-130 dependent manner. PLoS One 8 (9), e73266.

Fiedler, J., Roderer, G., Gunther, K.P., Brenner, R.E., 2002. BMP-2, BMP-4, and PDGF-bb stimulate chemotactic migration of primary human mesenchymal progenitor cells. J. Cell. Biochem. 87 (3), 305–312.

Ford, J.K., Young, R.W., 1963. Cell proliferation and displacement in the adrenal cortex of young rats injected with tritiated thymidine. Anat. Rec. 146, 125–137.

Fournier, T., Riches, D.W., Winston, B.W., Rose, D.M., Young, S.K., Noble, P.W., Lake, F.R., Henson, P.M., 1995. Divergence in macrophage insulin-like growth factor-I (IGF-I) synthesis induced by TNF-alpha and prostaglandin E2. J. Immunol. 155 (4), 2123–2133.

Fraser, W.D., Logue, F.C., Christie, J.P., Gallacher, S.J., Cameron, D., O'Reilly, D.S., Beastall, G.H., Boyle, I.T., 1998. Alteration of the circadian rhythm of intact parathyroid hormone and serum phosphate in women with established postmenopausal osteoporosis. Osteoporos. Int. 8 (2), 121–126.

Friedman, M.S., Long, M.W., Hankenson, K.D., 2006. Osteogenic differentiation of human mesenchymal stem cells is regulated by bone morphogenetic protein-6. J. Cell. Biochem. 98 (3), 538–554.

Frost, H.M., 1964. Dynamics of bone remodeling. Bone Biodynamics 315–333.

Furuya, M., Kikuta, J., Fujimori, S., Seno, S., Maeda, H., Shirazaki, M., Uenaka, M., Mizuno, H., Iwamoto, Y., Morimoto, A., Hashimoto, K., Ito, T., Isogai, Y., Kashii, M., Kaito, T., Ohba, S., Chung, U.I., Lichtler, A.C., Kikuchi, K., Matsuda, H., Yoshikawa, H., Ishii, M., 2018. Direct cell-cell contact between mature osteoblasts and osteoclasts dynamically controls their functions in vivo. Nat. Commun. 9 (1), 300.

Furuya, Y., Inagaki, A., Khan, M., Mori, K., Penninger, J.M., Nakamura, M., Udagawa, N., Aoki, K., Ohya, K., Uchida, K., Yasuda, H., 2013. Stimulation of bone formation in cortical bone of mice treated with a receptor activator of nuclear factor-kappaB ligand (RANKL)-binding peptide that possesses osteoclastogenesis inhibitory activity. J. Biol. Chem. 288 (8), 5562–5571.

Galea, G.L., Sunters, A., Meakin, L.B., Zaman, G., Sugiyama, T., Lanyon, L.E., Price, J.S., 2011. Sost down-regulation by mechanical strain in human osteoblastic cells involves PGE2 signaling via EP4. FEBS Lett. 585 (15), 2450–2454.

Garimella, R., Tague, S.E., Zhang, J., Belibi, F., Nahar, N., Sun, B.H., Insogna, K., Wang, J., Anderson, H.C., 2008. Expression and synthesis of bone morphogenetic proteins by osteoclasts: a possible path to anabolic bone remodeling. J. Histochem. Cytochem. 56 (6), 569–577.

Genetos, D.C., Yellowley, C.E., Loots, G.G., 2011. Prostaglandin E2 signals through PTGER2 to regulate sclerostin expression. PLoS One 6 (3), e17772.

Gerber, I., ap Gwynn, I., 2001. Influence of cell isolation, cell culture density, and cell nutrition on differentiation of rat calvarial osteoblast-like cells in vitro. Eur. Cells Mater. 2, 10–20.

Gil-Henn, H., Destaing, O., Sims, N.A., Aoki, K., Alles, N., Neff, L., Sanjay, A., Bruzzaniti, A., De Camilli, P., Baron, R., Schlessinger, J., 2007. Defective microtubule-dependent podosome organization in osteoclasts leads to increased bone density in Pyk2(-/-) mice. J. Cell Biol. 178 (6), 1053–1064.

Gray, C., Boyde, A., Jones, S.J., 1996. Topographically induced bone formation in vitro: implications for bone implants and bone grafts. Bone 18 (2), 115–123.

Grigoriadis, A.E., Wang, Z.Q., Cecchini, M.G., Hofstetter, W., Felix, R., Fleisch, H.A., Wagner, E.F., 1994. c-Fos: a key regulator of osteoclast-macrophage lineage determination and bone remodeling. Science 266 (5184), 443–448.

Guihard, P., Danger, Y., Brounais, B., David, E., Brion, R., Delecrin, J., Richards, C.D., Chevalier, S., Redini, F., Heymann, D., Gascan, H., Blanchard, F., 2012. Induction of osteogenesis in mesenchymal stem cells by activated monocytes/macrophages depends on oncostatin M signaling. Stem Cell. 30 (4), 762–772.

Hanamura, H., Higuchi, Y., Nakagawa, M., Iwata, H., Nogami, H., Urist, M.R., 1980. Solubilized bone morphogenetic protein (BMP) from mouse osteosarcoma and rat demineralized bone matrix. Clin. Orthop. Relat. Res. 148, 281–290.

Harris, W.H., Heaney, R.P., 1969. Skeletal renewal and metabolic bone disease. N. Engl. J. Med. 280 (6), 303–311 concl.

Hattner, R., Epker, B.N., Frost, H.M., 1965. Suggested sequential mode of control of changes in cell behaviour in adult bone remodelling. Nature 206 (983), 489–490.

Hauge, E.M., Qvesel, D., Eriksen, E.F., Mosekilde, L., Melsen, F., 2001. Cancellous bone remodeling occurs in specialized compartments lined by cells expressing osteoblastic markers. J. Bone Miner. Res. 16 (9), 1575–1582.

Heino, T.J., Kurata, K., Higaki, H., Vaananen, H.K., 2009. Evidence for the role of osteocytes in the initiation of targeted remodeling. Technol. Health Care 17 (1), 49–56.

Henriksen, K., Andreassen, K.V., Thudium, C.S., Gudmann, K.N., Moscatelli, I., Cruger-Hansen, C.E., Schulz, A.S., Dziegiel, M.H., Richter, J., Karsdal, M.A., Neutzsky-Wulff, A.V., 2012. A specific subtype of osteoclasts secretes factors inducing nodule formation by osteoblasts. Bone 51 (3), 353–361.

Henriksen, K., Flores, C., Thomsen, J.S., Bruel, A.M., Thudium, C.S., Neutzsky-Wulff, A.V., Langenbach, G.E., Sims, N., Askmyr, M., Martin, T.J., Everts, V., Karsdal, M.A., Richter, J., 2011. Dissociation of bone resorption and bone formation in adult mice with a non-functional V-ATPase in osteoclasts leads to increased bone strength. PLoS One 6 (11), e27482.

Henriksen, K., Gram, J., Schaller, S., Dahl, B.H., Dziegiel, M.H., Bollerslev, J., Karsdal, M.A., 2004. Characterization of osteoclasts from patients harboring a G215R mutation in ClC-7 causing autosomal dominant osteopetrosis type II. Am. J. Pathol. 164 (5), 1537–1545.

Hock, J.M., Canalis, E., 1994. Platelet-derived growth factor enhances bone cell replication, but not differentiated function of osteoblasts. Endocrinology 134 (3), 1423–1428.

Hock, J.M., Canalis, E., Centrella, M., 1990. Transforming growth factor-beta stimulates bone matrix apposition and bone cell replication in cultured fetal rat calvariae. Endocrinology 126 (1), 421–426.

Howard, G.A., Bottemiller, B.L., Turner, R.T., Rader, J.I., Baylink, D.J., 1981. Parathyroid hormone stimulates bone formation and resorption in organ culture: evidence for a coupling mechanism. Proc. Natl. Acad. Sci. U.S.A. 78 (5), 3204–3208.

Hughes, D.E., Boyce, B.F., 1997. Apoptosis in bone physiology and disease. Mol. Pathol. 50 (3), 132–137.

Ikebuchi, Y., Aoki, S., Honma, M., Hayashi, M., Sugamori, Y., Khan, M., Kariya, Y., Kato, G., Tabata, Y., Penninger, J.M., Udagawa, N., Aoki, K., Suzuki, H., 2018. Coupling of bone resorption and formation by RANKL reverse signalling. Nature 561 (7722), 195–200.

Ishii, M., Egen, J.G., Klauschen, F., Meier-Schellersheim, M., Saeki, Y., Vacher, J., Proia, R.L., Germain, R.N., 2009. Sphingosine-1-phosphate mobilizes osteoclast precursors and regulates bone homeostasis. Nature 458 (7237), 524–528.

Ishii, M., Kikuta, J., Shimazu, Y., Meier-Schellersheim, M., Germain, R.N., 2010. Chemorepulsion by blood S1P regulates osteoclast precursor mobilization and bone remodeling in vivo. J. Exp. Med. 207 (13), 2793–2798.

Jaworski, Z.F., 1992. Haversian Systems and Haversian Bone. CRC, London.

Jaworski, Z.F., Lok, E., 1972. The rate of osteoclastic bone erosion in Haversian remodeling sites of adult dog's rib. Calcif. Tissue Res. 10 (2), 103–112.

Jensen, P.R., Andersen, T.L., Pennypacker, B.L., Duong, L.T., Engelholm, L.H., Delaisse, J.M., 2014. A supra-cellular model for coupling of bone resorption to formation during remodeling: lessons from two bone resorption inhibitors affecting bone formation differently. Biochem. Biophys. Res. Commun. 443 (2), 694–699.

Jin YR, Stohn JP, Wang Q, Nagano K, Baron R, Bouxsein ML, Rosen CJ, Adarichev VA, Lindner V. Inhibition of osteoclast differentiation and collagen antibody-induced arthritis by CTHRC1, Bone 97 (4), 153–167.

Johnson, R.W., McGregor, N.E., Brennan, H.J., Crimeen-Irwin, B., Poulton, I.J., Martin, T.J., Sims, N.A., 2015. Glycoprotein130 (Gp130)/interleukin-6 (IL-6) signalling in osteoclasts promotes bone formation in periosteal and trabecular bone. Bone 81, 343–351.

Jubiz, W., Canterbury, J.M., Reiss, E., Tyler, F.H., 1972. Circadian rhythm in serum parathyroid hormone concentration in human subjects: correlation with serum calcium, phosphate, albumin, and growth hormone levels. J. Clin. Investig. 51 (8), 2040–2046.

Kanatani, M., Sugimoto, T., Kaji, H., Kobayashi, T., Nishiyama, K., Fukase, M., Kumegawa, M., Chihara, K., 1995. Stimulatory effect of bone morphogenetic protein-2 on osteoclast-like cell formation and bone-resorbing activity. J. Bone Miner. Res. 10 (11), 1681–1690.

Karsdal, M.A., Henriksen, K., Sorensen, M.G., Gram, J., Schaller, S., Dziegiel, M.H., Heegaard, A.M., Christophersen, P., Martin, T.J., Christiansen, C., Bollerslev, J., 2005. Acidification of the osteoclastic resorption compartment provides insight into the coupling of bone formation to bone resorption. Am. J. Pathol. 166 (2), 467–476.

Karsdal, M.A., Martin, T.J., Bollerslev, J., Christiansen, C., Henriksen, K., 2007. Are nonresorbing osteoclasts sources of bone anabolic activity? J. Bone Miner. Res. 22 (4), 487–494.

Karsdal, M.A., Neutzsky-Wulff, A.V., Dziegiel, M.H., Christiansen, C., Henriksen, K., 2008. Osteoclasts secrete non-bone derived signals that induce bone formation. Biochem. Biophys. Res. Commun. 366 (2), 483–488.

Kartsogiannis, V., Zhou, H., Horwood, N.J., Thomas, R.J., Hards, D.K., Quinn, J.M., Niforas, P., Ng, K.W., Martin, T.J., Gillespie, M.T., 1999. Localization of RANKL (receptor activator of NF kappa B ligand) mRNA and protein in skeletal and extraskeletal tissues. Bone 25, 525–534.

Kato, G., Shimizu, Y., Arai, Y., Suzuki, N., Sugamori, Y., Maeda, M., Takahashi, M., Tamura, Y., Wakabayashi, N., Murali, R., Ono, T., Ohya, K., Mise-Omata, S., Aoki, K., 2015. The inhibitory effects of a RANKL-binding peptide on articular and periarticular bone loss in a murine model of collagen-induced arthritis: a bone histomorphometric study. Arthritis Res. Ther. 17, 251.

Kennedy, O.D., Herman, B.C., Laudier, D.M., Majeska, R.J., Sun, H.B., Schaffler, M.B., 2012. Activation of resorption in fatigue-loaded bone involves both apoptosis and active pro-osteoclastogenic signaling by distinct osteocyte populations. Bone 50 (5), 1115–1122.

Kim, B.J., Lee, Y.S., Lee, S.Y., Park, S.Y., Dieplinger, H., Ryu, S.H., Yea, K., Choi, S., Lee, S.H., Koh, J.M., Kim, G.S., 2012. Afamin secreted from nonresorbing osteoclasts acts as a chemokine for preosteoblasts via the Akt-signaling pathway. Bone 51 (3), 431–440.

Kimura, H., Kwan, K.M., Zhang, Z., Deng, J.M., Darnay, B.G., Behringer, R.R., Nakamura, T., de Crombrugghe, B., Akiyama, H., 2008. Cthrc1 is a positive regulator of osteoblastic bone formation. PLoS One 3 (9), e3174.

Koide, M., Kobayashi, Y., Yamashita, T., Uehara, S., Nakamura, M., Hiraoka, B.Y., Ozaki, Y., Iimura, T., Yasuda, H., Takahashi, N., Udagawa, N., 2017. Bone formation is coupled to resorption via suppression of sclerostin expression by osteoclasts. J. Bone Miner. Res. 32 (10), 2074–2086.

Kreja, L., Brenner, R.E., Tautzenberger, A., Liedert, A., Friemert, B., Ehrnthaller, C., Huber-Lang, M., Ignatius, A., 2010. Non-resorbing osteoclasts induce migration and osteogenic differentiation of mesenchymal stem cells. J. Cell. Biochem. 109 (2), 347–355.

Kristensen, H.B., Andersen, T.L., Marcussen, N., Rolighed, L., Delaisse, J.M., 2013. Increased presence of capillaries next to remodeling sites in adult human cancellous bone. J. Bone Miner. Res. 28 (3), 574–585.

Kubota, K., Sakikawa, C., Katsumata, M., Nakamura, T., Wakabayashi, K., 2002. Platelet-derived growth factor BB secreted from osteoclasts acts as an osteoblastogenesis inhibitory factor. J. Bone Miner. Res. 17 (2), 257–265.

Lacey, D.L., Timms, E., Tan, H.L., Kelley, M.J., Dunstan, C.R., Burgess, T., Elliott, R., Colombero, A., Elliott, G., Scully, S., Hsu, H., Sullivan, J., Hawkins, N., Davy, E., Capparelli, C., Eli, A., Qian, Y.X., Kaufman, S., Sarosi, I., Shalhoub, V., Senaldi, G., Guo, J., Delaney, J., Boyle, W.J., 1998. Osteoprotegerin ligand is a cytokine that regulates osteoclast differentiation and activation. Cell 93 (2), 165–176.

Lassen, N.E., Andersen, T.L., Ploen, G.G., Soe, K., Hauge, E.M., Harving, S., Eschen, G.E.T., Delaisse, J.M., 2017. Coupling of bone resorption and formation in real time: new knowledge gained from human haversian BMUs. J. Bone Miner. Res. 32 (7), 1395–1405.

Lee, S.H., Rho, J., Jeong, D., Sul, J.Y., Kim, T., Kim, N., Kang, J.S., Miyamoto, T., Suda, T., Lee, S.K., Pignolo, R.J., Koczon-Jaremko, B., Lorenzo, J., Choi, Y., 2006. v-ATPase V0 subunit d2-deficient mice exhibit impaired osteoclast fusion and increased bone formation. Nat. Med. 12 (12), 1403–1409.

Lees, R.L., Sabharwal, V.K., Heersche, J.N., 2001. Resorptive state and cell size influence intracellular pH regulation in rabbit osteoclasts cultured on collagen-hydroxyapatite films. Bone 28 (2), 187–194.

Li, D., Liu, J., Guo, B., Liang, C., Dang, L., Lu, C., He, X., Cheung, H.Y., Xu, L., Lu, C., He, B., Liu, B., Shaikh, A.B., Li, F., Wang, L., Yang, Z., Au, D.W., Peng, S., Zhang, Z., Zhang, B.T., Pan, X., Qian, A., Shang, P., Xiao, L., Jiang, B., Wong, C.K., Xu, J., Bian, Z., Liang, Z., Guo, D.A., Zhu, H., Tan, W., Lu, A., Zhang, G., 2016. Osteoclast-derived exosomal miR-214-3p inhibits osteoblastic bone formation. Nat. Commun. 7, 10872.

Lichenstein, H.S., Lyons, D.E., Wurfel, M.M., Johnson, D.A., McGinley, M.D., Leidli, J.C., Trollinger, D.B., Mayer, J.P., Wright, S.D., Zukowski, M.M., 1994. Afamin is a new member of the albumin, alpha-fetoprotein, and vitamin D-binding protein gene family. J. Biol. Chem. 269 (27), 18149–18154.

Loots, G.G., Keller, H., Leupin, O., Murugesh, D., Collette, N.M., Genetos, D.C., 2012. TGF-beta regulates sclerostin expression via the ECR5 enhancer. Bone 50 (3), 663–669.

Lotinun, S., Kiviranta, R., Matsubara, T., Alzate, J.A., Neff, L., Luth, A., Koskivirta, I., Kleuser, B., Vacher, J., Vuorio, E., Horne, W.C., Baron, R., 2013. Osteoclast-specific cathepsin K deletion stimulates S1P-dependent bone formation. J. Clin. Investig. 123 (2), 666–681.

Madsen, D.H., Ingvarsen, S., Jurgensen, H.J., Melander, M.C., Kjoller, L., Moyer, A., Honore, C., Madsen, C.A., Garred, P., Burgdorf, S., Bugge, T.H., Behrendt, N., Engelholm, L.H., 2011. The non-phagocytic route of collagen uptake: a distinct degradation pathway. J. Biol. Chem. 286 (30), 26996–27010.

Martin, T.J., Sims, N.A., 2005. Osteoclast-derived activity in the coupling of bone formation to resorption. Trends Mol. Med. 11 (2), 76–81.

Martinovic, S., Mazic, S., Kisic, V., Basic, N., Jakic-Razumovic, J., Borovecki, F., Batinic, D., Simic, P., Grgurevic, L., Labar, B., Vukicevic, S., 2004. Expression of bone morphogenetic proteins in stromal cells from human bone marrow long-term culture. J. Histochem. Cytochem. 52 (9), 1159–1167.

Marzia, M., Sims, N.A., Voit, S., Migliaccio, S., Taranta, A., Bernardini, S., Faraggiana, T., Yoneda, T., Mundy, G.R., Boyce, B.F., Baron, R., Teti, A., 2000. Decreased c-Src expression enhances osteoblast differentiation and bone formation. J. Cell Biol. 151 (2), 311–320.

Matsuoka, K., Park, K.A., Ito, M., Ikeda, K., Takeshita, S., 2014. Osteoclast-derived complement component 3a stimulates osteoblast differentiation. J. Bone Miner. Res. 29 (7), 1522–1530.

Matsuzaki, K., Udagawa, N., Takahashi, N., Yamaguchi, K., Yasuda, H., Shima, N., Morinaga, T., Toyama, Y., Yabe, Y., Higashio, K., Suda, T., 1998. Osteoclast differentiation factor (ODF) induces osteoclast-like cell formation in human peripheral blood mononuclear cell cultures. Biochem Biophys Res Commun 246, 199–204.

McNamara, L.M., Van der Linden, J.C., Weinans, H., Prendergast, P.J., 2006. Stress-concentrating effect of resorption lacunae in trabecular bone. J. Biomech. 39 (4), 734–741.

Mitlak, B.H., Finkelman, R.D., Hill, E.L., Li, J., Martin, B., Smith, T., D'Andrea, M., Antoniades, H.N., Lynch, S.E., 1996. The effect of systemically administered PDGF-BB on the rodent skeleton. J. Bone Miner. Res. 11 (2), 238–247.

Mizoguchi, T., Muto, A., Udagawa, N., Arai, A., Yamashita, T., Hosoya, A., Ninomiya, T., Nakamura, H., Yamamoto, Y., Kinugawa, S., Nakamura, M., Nakamichi, Y., Kobayashi, Y., Nagasawa, S., Oda, K., Tanaka, H., Tagaya, M., Penninger, J.M., Ito, M., Takahashi, N., 2009. Identification of cell cycle-arrested quiescent osteoclast precursors in vivo. J. Cell Biol. 184 (4), 541–554.

Muto, A., Mizoguchi, T., Udagawa, N., Ito, S., Kawahara, I., Abiko, Y., Arai, A., Harada, S., Kobayashi, Y., Nakamichi, Y., Penninger, J.M., Noguchi, T., Takahashi, N., 2011. Lineage-committed osteoclast precursors circulate in blood and settle down into bone. J. Bone Miner. Res. 26 (12), 2978–2990.

Nakamura, M., Udagawa, N., Matsuura, S., Mogi, M., Nakamura, H., Horiuchi, H., Saito, N., Hiraoka, B.Y., Kobayashi, Y., Takaoka, K., Ozawa, H., Miyazawa, H., Takahashi, N., 2003. Osteoprotegerin regulates bone formation through a coupling mechanism with bone resorption. Endocrinology 144 (12), 5441–5449.

Nakamura, T., Imai, Y., Matsumoto, T., Sato, S., Takeuchi, K., Igarashi, K., Harada, Y., Azuma, Y., Krust, A., Yamamoto, Y., Nishina, H., Takeda, S., Takayanagi, H., Metzger, D., Kanno, J., Takaoka, K., Martin, T.J., Chambon, P., Kato, S., 2007. Estrogen prevents bone loss via estrogen receptor alpha and induction of Fas ligand in osteoclasts. Cell 130 (5), 811–823.

Narimatsu, K., Li, M., de Freitas, P.H., Sultana, S., Ubaidus, S., Kojima, T., Zhucheng, L., Ying, G., Suzuki, R., Yamamoto, T., Oda, K., Amizuka, N., 2010. Ultrastructural observation on cells meeting the histological criteria for preosteoblasts–a study in the mouse tibial metaphysis. J. Electron. Microsc. 59 (5), 427–436.

Negishi-Koga, T., Shinohara, M., Komatsu, N., Bito, H., Kodama, T., Friedel, R.H., Takayanagi, H., 2011. Suppression of bone formation by osteoclastic expression of semaphorin 4D. Nat. Med. 17 (11), 1473–1480.

Nicolaidou, V., Wong, M.M., Redpath, A.N., Ersek, A., Baban, D.F., Williams, L.M., Cope, A.P., Horwood, N.J., 2012. Monocytes induce STAT3 activation in human mesenchymal stem cells to promote osteoblast formation. PLoS One 7 (7), e39871.

Noble, B.S., Peet, N., Stevens, H.Y., Brabbs, A., Mosley, J.R., Reilly, G.C., Reeve, J., Skerry, T.M., Lanyon, L.E., 2003. Mechanical loading: biphasic osteocyte survival and targeting of osteoclasts for bone destruction in rat cortical bone. Am. J. Physiol. Cell Physiol. 284 (4), C934–C943.

Oreffo, R.O., Mundy, G.R., Seyedin, S.M., Bonewald, L.F., 1989. Activation of the bone-derived latent TGF beta complex by isolated osteoclasts. Biochem. Biophys. Res. Commun. 158 (3), 817–823.

Ota, K., Quint, P., Weivoda, M.M., Ruan, M., Pederson, L., Westendorf, J.J., Khosla, S., Oursler, M.J., 2013. Transforming growth factor beta 1 induces CXCL16 and leukemia inhibitory factor expression in osteoclasts to modulate migration of osteoblast progenitors. Bone 57 (1), 68–75.

Pappu, R., Schwab, S.R., Cornelissen, I., Pereira, J.P., Regard, J.B., Xu, Y., Camerer, E., Zheng, Y.W., Huang, Y., Cyster, J.G., Coughlin, S.R., 2007. Promotion of lymphocyte egress into blood and lymph by distinct sources of sphingosine-1-phosphate. Science 316 (5822), 295–298.

Parfitt, A., 1980. Morphological basis of bone mineral measurements: transient and steady state effects of treatment in osteoporosis. Mineral and Elecrolyte Metabolism 4, 273–287.

Parfitt, A., 1983. Bone histomorphometry: techniques and interpretations. In: Recker, R.R. (Ed.), Histomorphometry. CRC Press, Baton Rouge, USA, pp. 142–221.

Parfitt, A.M., 1982. The coupling of bone formation to bone resorption: a critical analysis of the concept and of its relevance to the pathogenesis of osteoporosis. Metab. Bone Dis. Relat. Res. 4 (1), 1–6.

Parfitt, A.M., Mathews, C.H., Villanueva, A.R., Kleerekoper, M., Frame, B., Rao, D.S., 1983. Relationships between surface, volume, and thickness of iliac trabecular bone in aging and in osteoporosis. Implications for the microanatomic and cellular mechanisms of bone loss. J. Clin. Investig. 72 (4), 1396–1409.

Pederson, L., Ruan, M., Westendorf, J.J., Khosla, S., Oursler, M.J., 2008. Regulation of bone formation by osteoclasts involves Wnt/BMP signaling and the chemokine sphingosine-1-phosphate. Proc. Natl. Acad. Sci. U.S.A. 105 (52), 20764–20769.

Pennypacker, B., Shea, M., Liu, Q., Masarachia, P., Saftig, P., Rodan, S., Rodan, G., Kimmel, D., 2009. Bone density, strength, and formation in adult cathepsin K (-/-) mice. Bone 44 (2), 199–207.

Pfeilschifter, J., Laukhuf, F., Muller-Beckmann, B., Blum, W.F., Pfister, T., Ziegler, R., 1995. Parathyroid hormone increases the concentration of insulin-like growth factor-I and transforming growth factor beta 1 in rat bone. J. Clin. Investig. 96 (2), 767–774.

Poulton, I.J., McGregor, N.E., Pompolo, S., Walker, E.C., Sims, N.A., 2012. Contrasting roles of leukemia inhibitory factor in murine bone development and remodeling involve region-specific changes in vascularization. J. Bone Miner. Res. 27 (3), 586–595.

Quint, P., Ruan, M., Pederson, L., Kassem, M., Westendorf, J.J., Khosla, S., Oursler, M.J., 2013. Sphingosine 1-phosphate (S1P) receptors 1 and 2 coordinately induce mesenchymal cell migration through S1P activation of complementary kinase pathways. J. Biol. Chem. 288 (8), 5398–5406.

Rasmussen, H., Bordier, P., 1974. The Physiological Basis of Metabolic Bone Disease. Williams and Wilkins, Waverley Press, Baltimore.

Redmond, J., Fulford, A.J., Jarjou, L., Zhou, B., Prentice, A., Schoenmakers, I., 2016. Diurnal rhythms of bone turnover markers in three ethnic groups. J. Clin. Endocrinol. Metab. 101 (8), 3222–3230.

Reid, L.R., Lowe, C., Cornish, J., Skinner, S.J., Hilton, D.J., Willson, T.A., Gearing, D.P., Martin, T.J., 1990. Leukemia inhibitory factor: a novel bone-active cytokine. Endocrinology 126 (3), 1416–1420.

Richards, C.D., Langdon, C., Deschamps, P., Pennica, D., Shaughnessy, S.G., 2000. Stimulation of osteoclast differentiation in vitro by mouse oncostatin M, leukaemia inhibitory factor, cardiotrophin-1 and interleukin 6: synergy with dexamethasone. Cytokine 12 (6), 613–621.

Rickard, D.J., Sullivan, T.A., Shenker, B.J., Leboy, P.S., Kazhdan, I., 1994. Induction of rapid osteoblast differentiation in rat bone marrow stromal cell cultures by dexamethasone and BMP-2. Dev. Biol. 161 (1), 218–228.

Robey, P.G., Young, M.F., Flanders, K.C., Roche, N.S., Kondaiah, P., Reddi, A.H., Termine, J.D., Sporn, M.B., Roberts, A.B., 1987. Osteoblasts synthesize and respond to transforming growth factor-type beta (TGF-beta) in vitro. J. Cell Biol. 105 (1), 457–463.

Robling, A.G., Turner, C.H., 2009. Mechanical signaling for bone modeling and remodeling. Crit. Rev. Eukaryot. Gene Expr. 19 (4), 319–338.

Robubi, A., Berger, C., Schmid, M., Huber, K.R., Engel, A., Krugluger, W., 2014. Gene expression profiles induced by growth factors in in vitro cultured osteoblasts. Bone Joint Res. 3 (7), 236–240.

Rodan, G.A., 1991. Mechanical loading, estrogen deficiency, and the coupling of bone formation to bone resorption. J. Bone Miner. Res. 6 (6), 527–530.

Rodan, G.A., Martin, T.J., 1981. Role of osteoblasts in hormonal control of bone resorption–a hypothesis. Calcif. Tissue Int. 33 (4), 349–351.

Ryu, J., Kim, H.J., Chang, E.J., Huang, H., Banno, Y., Kim, H.H., 2006. Sphingosine 1-phosphate as a regulator of osteoclast differentiation and osteoclast-osteoblast coupling. EMBO J. 25 (24), 5840–5851.

Sanchez-Fernandez, M.A., Gallois, A., Riedl, T., Jurdic, P., Hoflack, B., 2008. Osteoclasts control osteoblast chemotaxis via PDGF-BB/PDGF receptor beta signaling. PLoS One 3 (10), e3537.

Sato, T., Abe, E., Jin, C.H., Hong, M.H., Katagiri, T., Kinoshita, T., Amizuka, N., Ozawa, H., Suda, T., 1993. The biological roles of the third component of complement in osteoclast formation. Endocrinology 133 (1), 397–404.

Scariano, J.K., Emery-Cohen, A.J., Pickett, G.G., Morgan, M., Simons, P.C., Alba, F., 2008. Estrogen receptors alpha (ESR1) and beta (ESR2) are expressed in circulating human lymphocytes. J. Recept. Signal Transduct. Res. 28 (3), 285–293.

Schaffler, M.B., Cheung, W.Y., Majeska, R., Kennedy, O., 2014. Osteocytes: master orchestrators of bone. Calcif. Tissue Int. 94 (1), 5–24.

Sells Galvin, R.J., Gatlin, C.L., Horn, J.W., Fuson, T.R., 1999. TGF-beta enhances osteoclast differentiation in hematopoietic cell cultures stimulated with RANKL and M-CSF. Biochem. Biophys. Res. Commun. 265 (1), 233–239.

Sims, N.A., Gooi, J.H., 2008. Bone remodeling: multiple cellular interactions required for coupling of bone formation and resorption. Semin. Cell Dev. Biol. 19 (5), 444–451.

Sims, N.A., Jenkins, B.J., Quinn, J.M., Nakamura, A., Glatt, M., Gillespie, M.T., Ernst, M., Martin, T.J., 2004. Glycoprotein 130 regulates bone turnover and bone size by distinct downstream signaling pathways. J. Clin. Investig. 113 (3), 379–389.

Sims, N.A., Johnson, R.W., 2012. Leukemia inhibitory factor: a paracrine mediator of bone metabolism. Growth Factors. 30 (2), 76–87.

Sims, N.A., Martin, T.J., 2015. Coupling signals between the osteoclast and osteoblast: how are messages transmitted between these temporary visitors to the bone surface? Front. Endocrinol. 6, 41.

Sims, N.A., Romas, E., 2015. Is RANKL inhibition both anti-resorptive and anabolic in rheumatoid arthritis? Arthritis Res. Ther. 17, 328.

Sobacchi, C., Frattini, A., Guerrini, M.M., Abinun, M., Pangrazio, A., Susani, L., Bredius, R., Mancini, G., Cant, A., Bishop, N., Grabowski, P., Del Fattore, A., Messina, C., Errigo, G., Coxon, F.P., Scott, D.I., Teti, A., Rogers, M.J., Vezzoni, P., Villa, A., Helfrich, M.H., 2007. Osteoclast-poor human osteopetrosis due to mutations in the gene encoding RANKL. Nat. Genet. 39, 960.

Suda, T., Takahashi, N., Udagawa, N., Jimi, E., Gillespie, M.T., Martin, T.J., 1999. Modulation of osteoclast differentiation and function by the new members of the tumor necrosis factor receptor and ligand families. Endocr. Rev. 20 (3), 345–357.

Takahashi, T., Fournier, A., Nakamura, F., Wang, L.H., Murakami, Y., Kalb, R.G., Fujisawa, H., Strittmatter, S.M., 1999. Plexin-neuropilin-1 complexes form functional semaphorin-3A receptors. Cell 99 (1), 59–69.

Takahashi, T., Kurihara, N., Takahashi, K., Kumegawa, M., 1986. An ultrastructural study of phagocytosis in bone by osteoblastic cells from fetal mouse calvaria in vitro. Arch. Oral Biol. 31 (10), 703–706.

Takeshita, S., Fumoto, T., Matsuoka, K., Park, K.A., Aburatani, H., Kato, S., Ito, M., Ikeda, K., 2013. Osteoclast-secreted CTHRC1 in the coupling of bone resorption to formation. J. Clin. Investig. 123 (9), 3914–3924.

Takyar, F.M., Tonna, S., Ho, P.W., Crimeen-Irwin, B., Baker, E.K., Martin, T.J., Sims, N.A., 2013. EphrinB2/EphB4 inhibition in the osteoblast lineage modifies the anabolic response to parathyroid hormone. J. Bone Miner. Res. 28 (4), 912–925.

Tamagnone, L., Artigiani, S., Chen, H., He, Z., Ming, G.I., Song, H., Chedotal, A., Winberg, M.L., Goodman, C.S., Poo, M., Tessier-Lavigne, M., Comoglio, P.M., 1999. Plexins are a large family of receptors for transmembrane, secreted, and GPI-anchored semaphorins in vertebrates. Cell 99 (1), 71–80.

Tamura, T., Udagawa, N., Takahashi, N., Miyaura, C., Tanaka, S., Yamada, Y., Koishihara, Y., Ohsugi, Y., Kumaki, K., Taga, T., Kishimoto, T., Suda, T., 1993. Soluble interleukin-6 receptor triggers osteoclast formation by interleukin 6. Proc. Natl. Acad. Sci. U. S. A. 90 (24), 11924–11928.

Tang, Y., Wu, X., Lei, W., Pang, L., Wan, C., Shi, Z., Zhao, L., Nagy, T.R., Peng, X., Hu, J., Feng, X., Van Hul, W., Wan, M., Cao, X., 2009. TGF-beta1-induced migration of bone mesenchymal stem cells couples bone resorption with formation. Nat. Med. 15 (7), 757–765.

Terauchi, M., Li, J.Y., Bedi, B., Baek, K.H., Tawfeek, H., Galley, S., Gilbert, L., Nanes, M.S., Zayzafoon, M., Guldberg, R., Lamar, D.L., Singer, M.A., Lane, T.F., Kronenberg, H.M., Weitzmann, M.N., Pacifici, R., 2009. T lymphocytes amplify the anabolic activity of parathyroid hormone through Wnt10b signaling. Cell Metabol. 10 (3), 229–240.

Thudium, C.S., Moscatelli, I., Flores, C., Thomsen, J.S., Bruel, A., Gudmann, N.S., Hauge, E.M., Karsdal, M.A., Richter, J., Henriksen, K., 2014. A comparison of osteoclast-rich and osteoclast-poor osteopetrosis in adult mice sheds light on the role of the osteoclast in coupling bone resorption and bone formation. Calcif. Tissue Int. 95 (1), 83–93.

Tonna, S., Sims, N.A., 2014. Talking among ourselves: paracrine control of bone formation within the osteoblast lineage. Calcif. Tissue Int. 94 (1), 35–45.

Tonna, S., Takyar, F.M., Vrahnas, C., Crimeen-Irwin, B., Ho, P.W., Poulton, I.J., Brennan, H.J., McGregor, N.E., Allan, E.H., Nguyen, H., Forwood, M.R., Tatarczuch, L., Mackie, E.J., Martin, T.J., Sims, N.A., 2014. EphrinB2 signaling in osteoblasts promotes bone mineralization by preventing apoptosis. FASEB J. 28 (10), 4482–4496.

Tran Van, P.T., Vignery, A., Baron, R., 1982. Cellular kinetics of the bone remodeling sequence in the rat. Anat. Rec. 202 (4), 445–451.

van Bezooijen, R.L., Roelen, B.A., Visser, A., van der Wee-Pals, L., de Wilt, E., Karperien, M., Hamersma, H., Papapoulos, S.E., ten Dijke, P., Lowik, C.W., 2004. Sclerostin is an osteocyte-expressed negative regulator of bone formation, but not a classical BMP antagonist. J. Exp. Med. 199 (6), 805–814.

Verborgt, O., Gibson, G.J., Schaffler, M.B., 2000. Loss of osteocyte integrity in association with microdamage and bone remodeling after fatigue in vivo. J. Bone Miner. Res. 15 (1), 60–67.

Vignery, A., Baron, R., 1980. Dynamic histomorphometry of alveolar bone remodeling in the adult rat. Anat. Rec. 196 (2), 191–200.

Villanueva, A.R., Sypitkowski, C., Parfitt, A.M., 1986. A new method for identification of cement lines in undecalcified, plastic embedded sections of bone. Stain Technol. 61 (2), 83–88.

Vukicevic, S., Grgurevic, L., 2009. BMP-6 and mesenchymal stem cell differentiation. Cytokine Growth Factor Rev. 20 (5–6), 441–448.

Wagsater, D., Olofsson, P.S., Norgren, L., Stenberg, B., Sirsjo, A., 2004. The chemokine and scavenger receptor CXCL16/SR-PSOX is expressed in human vascular smooth muscle cells and is induced by interferon gamma. Biochem. Biophys. Res. Commun. 325 (4), 1187–1193.

Walker, E.C., McGregor, N.E., Poulton, I.J., Pompolo, S., Allan, E.H., Quinn, J.M., Gillespie, M.T., Martin, T.J., Sims, N.A., 2008. Cardiotrophin-1 is an osteoclast-derived stimulus of bone formation required for normal bone remodeling. J. Bone Miner. Res. 23 (12), 2025–2032.

Walker, E.C., McGregor, N.E., Poulton, I.J., Solano, M., Pompolo, S., Fernandes, T.J., Constable, M.J., Nicholson, G.C., Zhang, J.G., Nicola, N.A., Gillespie, M.T., Martin, T.J., Sims, N.A., 2010. Oncostatin M promotes bone formation independently of resorption when signaling through leukemia inhibitory factor receptor in mice. J. Clin. Investig. 120 (2), 582–592.

Wang, X., Kumanogoh, A., Watanabe, C., Shi, W., Yoshida, K., Kikutani, H., 2001. Functional soluble CD100/Sema4D released from activated lymphocytes: possible role in normal and pathologic immune responses. Blood 97 (11), 3498–3504.

Wang, Y., Nishida, S., Elalieh, H.Z., Long, R.K., Halloran, B.P., Bikle, D.D., 2006. Role of IGF-I signaling in regulating osteoclastogenesis. J. Bone Miner. Res. 21 (9), 1350–1358.

White, H.D., Ahmad, A.M., Durham, B.H., Peter, R., Prabhakar, V.K., Corlett, P., Vora, J.P., Fraser, W.D., 2007. PTH circadian rhythm and PTH target-organ sensitivity is altered in patients with adult growth hormone deficiency with low BMD. J. Bone Miner. Res. 22 (11), 1798–1807.

Winberg, M.L., Noordermeer, J.N., Tamagnone, L., Comoglio, P.M., Spriggs, M.K., Tessier-Lavigne, M., Goodman, C.S., 1998. Plexin A is a neuronal semaphorin receptor that controls axon guidance. Cell 95 (7), 903–916.

Wlazlo, N., van Greevenbroek, M.M., Ferreira, I., Jansen, E.H., Feskens, E.J., van der Kallen, C.J., Schalkwijk, C.G., Bravenboer, B., Stehouwer, C.D., 2013. Activated complement factor 3 is associated with liver fat and liver enzymes: the CODAM study. Eur. J. Clin. Investig. 43 (7), 679–688.

Wutzl, A., Brozek, W., Lernbass, I., Rauner, M., Hofbauer, G., Schopper, C., Watzinger, F., Peterlik, M., Pietschmann, P., 2006. Bone morphogenetic proteins 5 and 6 stimulate osteoclast generation. J. Biomed. Mater. Res. A 77 (1), 75–83.

Xian, L., Wu, X., Pang, L., Lou, M., Rosen, C.J., Qiu, T., Crane, J., Frassica, F., Zhang, L., Rodriguez, J.P., Xiaofeng, J., Shoshana, Y., Shouhong, X., Argiris, E., Mei, W., Cao, X., 2012. Matrix IGF-1 maintains bone mass by activation of mTOR in mesenchymal stem cells. Nat. Med. 18 (7), 1095–1101.

Xie, H., Cui, Z., Wang, L., Xia, Z., Hu, Y., Xian, L., Li, C., Xie, L., Crane, J., Wan, M., Zhen, G., Bian, Q., Yu, B., Chang, W., Qiu, T., Pickarski, M., Duong, L.T., Windle, J.J., Luo, X., Liao, E., Cao, X., 2014. PDGF-BB secreted by preosteoclasts induces angiogenesis during coupling with osteogenesis. Nat. Med. 20 (11), 1270–1278.

Yasuda, H., Shima, N., Nakagawa, N., Yamaguchi, K., Kinosaki, M., Mochizuki, S., Tomoyasu, A., Yano, K., Goto, M., Murakami, A., Tsuda, E., Morinaga, T., Higashio, K., Udagawa, N., Takahashi, N., Suda, T., 1998. Osteoclast differentiation factor is a ligand for osteoprotegerin/osteoclastogenesis-inhibitory factor and is identical to TRANCE/RANKL. Proc. Natl. Acad. Sci. U.S.A. 95 (7), 3597–3602.

Yee, J.A., Yan, L., Dominguez, J.C., Allan, E.H., Martin, T.J., 1993. Plasminogen-dependent activation of latent transforming growth factor beta (TGF beta) by growing cultures of osteoblast-like cells. J. Cell. Physiol. 157 (3), 528–534.

Zaidi, M., Datta, H.K., Patchell, A., Moonga, B., MacIntyre, I., 1989. Calcium-activated' intracellular calcium elevation: a novel mechanism of osteoclast regulation. Biochem. Biophys. Res. Commun. 163 (3), 1461–1465.

Zarling, J.M., Shoyab, M., Marquardt, H., Hanson, M.B., Lioubin, M.N., Todaro, G.J., 1986. Oncostatin M: a growth regulator produced by differentiated histiocytic lymphoma cells. Proc. Natl. Acad. Sci. U.S.A. 83 (24), 9739–9743.

Zhang, L., Leeman, E., Carnes, D.C., Graves, D.T., 1991. Human osteoblasts synthesize and respond to platelet-derived growth factor. Am. J. Physiol. 261 (2 Pt 1), C348–C354.

Zhao, C., Irie, N., Takada, Y., Shimoda, K., Miyamoto, T., Nishiwaki, T., Suda, T., Matsuo, K., 2006. Bidirectional ephrinB2-EphB4 signaling controls bone homeostasis. Cell Metabol. 4 (2), 111–121.

Chapter 11

Modeling and remodeling: the cellular machinery responsible for bone's material and structural strength during growth, aging, and drug therapy

Ego Seeman

Department of Endocrinology and Medicine, Austin Health, University of Melbourne, Melbourne, VIC, Australia; Mary MacKillop Institute for Health Research, Australian Catholic University, Melbourne, VIC, Australia

Chapter Outline

Summary 245
Bone modeling and remodeling during growth and the attainment of bone's peak material and structural strength 246
Definition of bone modeling and remodeling 246
Bone's material and structural strength 246
Trait variances in adulthood originate before puberty 251
Sex and racial differences in bone structure 252
Bone remodeling by the basic multicellular unit 254
Osteocyte death in signaling bone remodeling 254
The bone remodeling compartment 254
The multidirectional steps of the remodeling cycle 256
Bone remodeling and microstructure during young adulthood, menopause, and advanced age 256
Young adulthood: reversible bone loss and microstructural deterioration 256
Menopause: reversible and irreversible bone loss and microstructural deterioration 258
Advanced age: the predominance of cortical bone loss 259
The net effects of reduced periosteal apposition and endosteal bone loss 261
Sexual dimorphism in trabecular and cortical bone loss 263
The heterogeneous material and structural basis of bone fragility in patients with fractures 263
Bone modeling, remodeling, and drug therapy 263
Antiresorptive therapy reduces the reversible but not the irreversible deficit in mineralized matrix volume 263
Anabolic therapy: restoring the irreversible deficit in mineralized matrix volume and microstructural deterioration by remodeling- and modeling-based bone formation 266
Combined antiresorptive and anabolic therapy 267
Conclusion 268
References 269

Summary

During growth, the cellular machinery of bone modeling and remodeling assembles bone's size, shape, and microstructure. Bone matrix is synthesized as a composite of organic and inorganic material configured with varying volumes of extracellular fluid—containing void to form its cortical and trabecular compartments. Endocortical resorptive modeling excavates a medullary cavity void volume, which shifts the cortex radially, increasing resistance to bending exponentially while minimizing cortical thickening and so avoiding bulk; larger bone cross sections are assembled with a relatively thinner cortex.

During young adulthood, balanced remodeling renews the mineralized bone matrix volume without permanently compromising bone microstructure. As modeling and remodeling are surface dependent, the large surface area/matrix volume configuration of interconnected thin trabecular plates facilitates accessibility of the matrix to being renewed; this is

Principles of Bone Biology. https://doi.org/10.1016/B978-0-12-814841-9.00011-7

an advantage provided that remodeling remains balanced. The lower surface area/matrix volume of compact cortical bone makes it less accessible to being renewed and so liable to damage accumulation.

Around midlife, remodeling becomes unbalanced in both sexes and rapid in postmenopausal women. Unbalanced remodeling transactions deposit less bone than was resorbed, producing permanent cortical thinning, porosity, trabecular thinning, perforation, and loss of connectivity, which compromise bone strength disproportionate to the bone loss producing this deterioration.

Antiresorptive agents do not reverse microstructural deterioration or bone fragility present at the time of initiating treatment. Most antiresorptives only slow remodeling, they do not abolish it, so that total bone matrix volume and microstructure continue to deteriorate, albeit more slowly than without treatment. The slowly declining bone matrix volume is less often remodeled and so becomes more fully mineralized and less ductile, predisposing to damage accumulation. Teriparatide, the only available anabolic agent, produces predominantly remodeling-based bone formation and appears to reduce vertebral and nonvertebral fracture risk more efficaciously than antiresorptive therapy. Encouraging observations have been reported using modeling-based anabolic therapy and combined antiresorptive and anabolic therapy, but reducing the population burden of fractures remains a challenging unmet need.

Bone modeling and remodeling during growth and the attainment of bone's peak material and structural strength

Definition of bone modeling and remodeling

Bone *modeling* or construction is usually thought of as bone formation. Bone modeling is both formative and resorptive. Formative modeling is carried out by osteoblasts, cells that synthesize and deposit a volume of osteoid upon a quiescent bone surface that has not undergone prior bone resorption (Parfitt, 1996). Resorptive modeling is carried out by osteoclasts, cells that resorb a volume of bone without a subsequent phase of bone formation.

Formative and resorptive modeling occur mainly during growth and confer the paradoxical properties of strength for loading, yet lightness for mobility, and resistance to deformation, yet flexibility to allow energy absorption. These properties are conferred by the deposition of bone at locations where it is needed and by removal of bone from locations where it is not needed, thereby changing bone's external and internal size and shape.

Bone *remodeling* or reconstruction maintains the composition and properties of the mineralized bone matrix during young adulthood. Remodeling is carried out by teams of osteoclasts and osteoblasts, basic multicellular units (BMUs), which respectively resorb and replace a volume of damaged or older bone with the same volume of newly synthesized osteoid at the same location (see chapters 4 and 10 for further detail).

The osteoid undergoes rapid primary mineralization within days of deposition to become "bone" and then slower secondary mineralization, a physicochemical process of enlargement of crystals of calcium hydroxyapatite deposited during primary mineralization. The crystals enlarge and displace water without change in the dimensions of the collagen fibrils (Akkus, 2003).

Bone's material and structural strength

Bone is a specialized connective tissue. Type I collagen is tough; it is distensible in tension but lacks resistance to bending, i.e., stiffness. Stiffening is achieved by mineralizing the collagen with platelets of calcium hydroxyapatite. The mineral confers material stiffness but sacrifices the ability to absorb and store energy by deforming. For a given increase in the percentage of mineral ash, stiffness increases fivefold, but the amount of work needed to produce a fracture decreases 14-fold, almost three times more (Currey, 2002) (Fig. 11.1, top).

Nature selects a mineralization density that is suited to the function a bone usually performs. Ossicles of the ear are over 80% mineral, a feature conferring stiffness that is selected for so that they can vibrate like tuning forks and transmit sound with high fidelity. The ability to store energy by deforming is sacrificed, but it is unnecessary (Fig. 11.1, bottom). Cracking is unlikely because ossicles are housed safely in the skull. Deer antlers are less densely mineralized to facilitate deformation, so they can absorb energy like springs during head butting in mating season. Greater energy-absorbing ability of antlers, "toughness" or resilience, is selected for over stiffness, but stiffness is unnecessary because antlers are not load bearing (Currey, 1969).

Although mineral stiffens bone, it is also the most brittle component of bone; an increase in mineral density decreases toughness more than it increases stiffness. So to defend against loss of toughness, the platelets of mineral are protected from being excessively loaded by noncollagenous proteins, glue-like material that releases energy imposed during loading

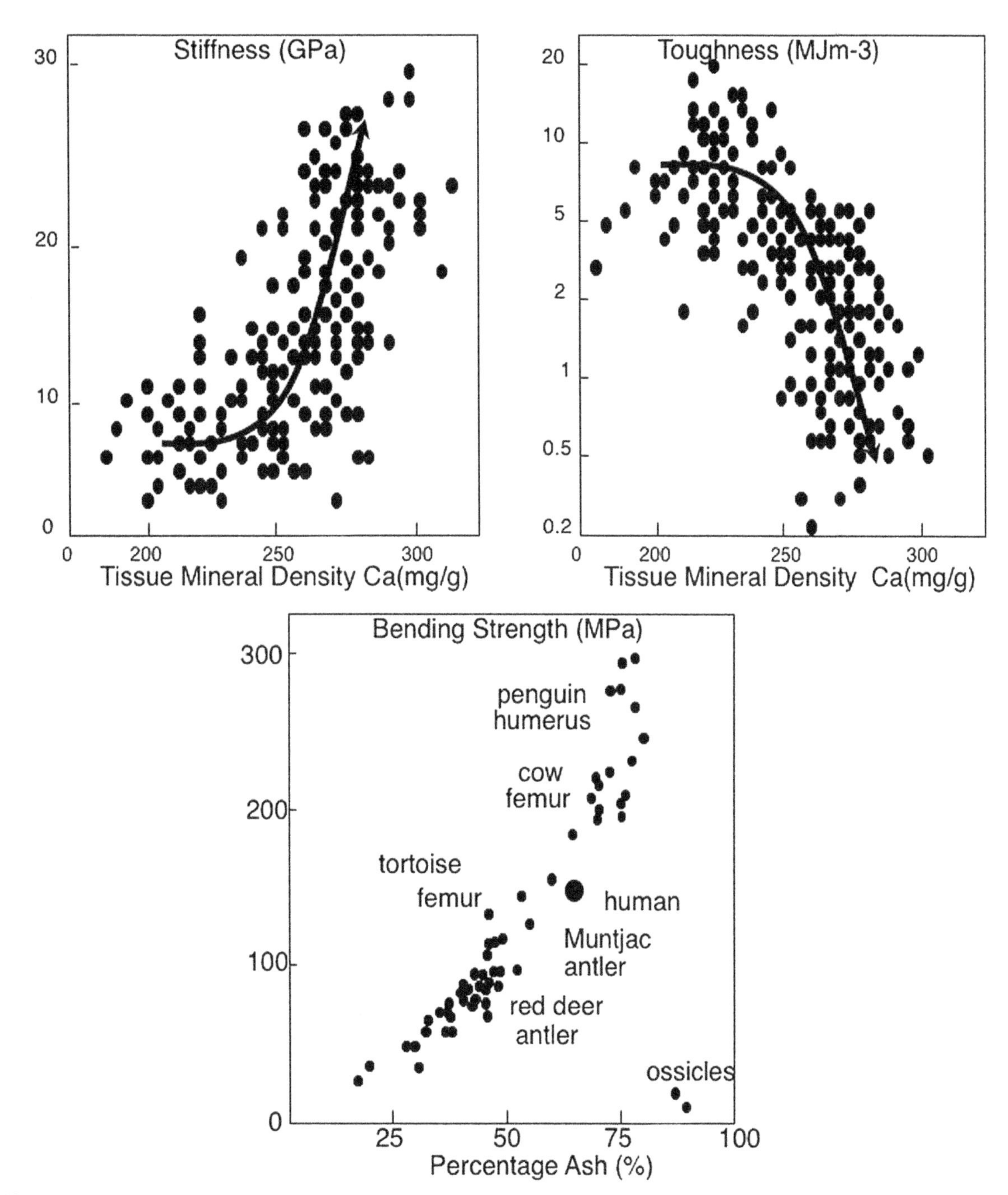

FIGURE 11.1 (Top) As tissue mineral content increases, stiffness increases, but toughness decreases disproportionately. (Bottom) Ossicles of the ear are 90% mineral. They have little resistance to bending. Antlers are about 40% mineral, allowing them to deform without cracking. *(Top) Adapted from Currey, J.D., 2002. Bones. Structure and Mechanics. Princeton UP, New Jersey, pp. 1–380. (Bottom) Adapted from Curry, J.D., 1969. Mechanical consequences of variation in the mineral content of bone. J. Biomechan. 2, 1–11.*

by breaking intrahelical "sacrificial" bonds, which allows uncoiling of the noncollagenous proteins. This provides "hidden" length. Stresses at the tissue, fiber, and mineral levels decrease in proportions of 12:5:2 (Fantner et al., 2005, Gupt et al., 2006) (Fig. 11.2).

Advancing age compromises this mechanism as advanced glycation end products accumulate, reducing the ability of noncollagenous proteins to protect the mineral platelets (Ural and Vashishth, 2014). This predisposes to the development of diffuse damage, which differs from microcracking. Diffuse damage occurs within the osteon. Microcracks occur in the interosteonal (interstitial) matrix. Diffuse damage is reversible and its repair does not involve bone remodeling.

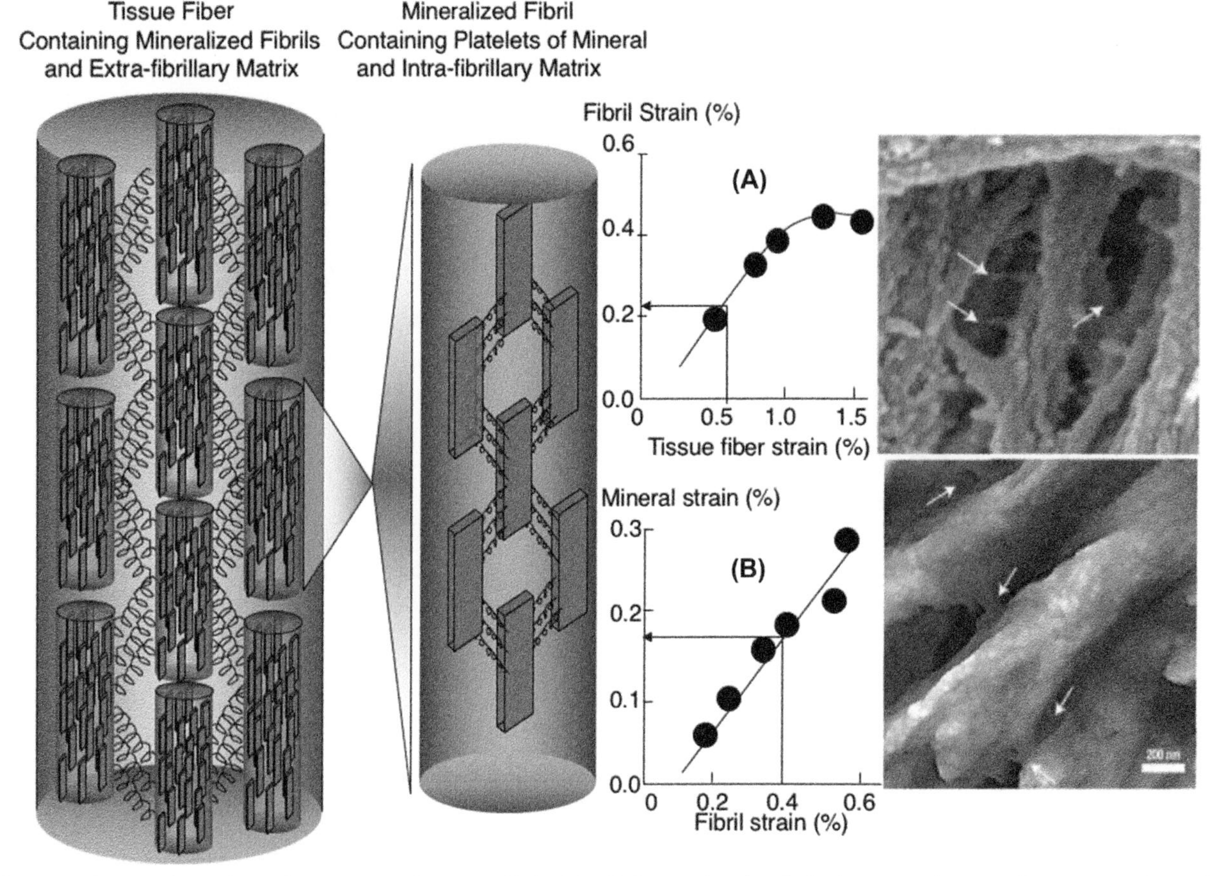

FIGURE 11.2 A collagen tissue fiber contains mineralized fibrils. The fibrils contain mineral platelets bound by noncollagenous proteins, helical structures that can lengthen via the breakage of sacrificial intrahelical bonds to provide "hidden length" and avoid overloading the mineral platelets. (Graph A) Tissue fiber strain distributed to fibrils. (Graph B) Fibrils absorb strain, minimizing mineral strain. *Adapted from Gupta et al. (2005). Images of non-collagenous glue like proteins. From Fantner, G., Hassen kam, T., Kindt, J.H., Weaver, J.C., Birkedal, H., Pechenik, L., Cutroni, J.A., Cidade, G.C., Stucky, G.D., Morse, D.E., Hansma, P.K., 2005. Nat. Mater. 4, 612–616.*

During growth, the mineralized bone matrix volume is fashioned into three-dimensional structures. Although there is variability in the material composition of bone, this composition is similar among land-dwelling mammals (Keaveney et al., 1998). Most of the diversity in bone strength is the result of structural diversity, which is obvious at the macroscopic level from bone to bone and from species to species, but how this diversity in size and shape is achieved is neither obvious nor intuitive.

If bone had only to be strong, this could be achieved by bulk alone—more mass. But mass takes time to grow, is costly to maintain, and limits mobility. Bone also must serve as a lever to facilitate mobility and so it must be light. Longer tubular bones need more mass than shorter bones to construct their length, but in an individual, the diversity in total external and internal cross-sectional areas (CSAs) and varying shapes of a cross section, from cross section to adjacent cross section along the length of a bone is achieved using *similar* amounts of mineralized matrix volume.

In an individual, differences in the total cross section size, shape, and internal architecture of adjacent cross sections of a long bone are achieved by varying the volume of void, not the volume of mineralized bone matrix. For example, at the radius, the large total CSA of the distal metaphysis is assembled using more void volume forming a large medullary canal, not more mineralized matrix volume (Fig. 11.3, left). Most of the mineralized matrix is fashioned as trabecular bone, a porous sponge-like configuration of intersecting thin plates of mineralized matrix. Only a small amount of mineralized matrix is assembled as cortical bone. The cortex is thin and porous with a low matrix mineral density (Ghasem-Zadeh et al., 2017).

More proximally, the *same* mineralized matrix volume is contained within the smaller total CSA of diaphyseal shaft achieved, by assembling it with a smaller medullary void volume, not less material. The constant amount of mineralized bone matrix is fashioned here as a thicker compact cortex with a low porosity and high matrix mineral density. There is little, if any, trabecular bone. This design serves the need for the lever function of a long bone or a cantilever function of a

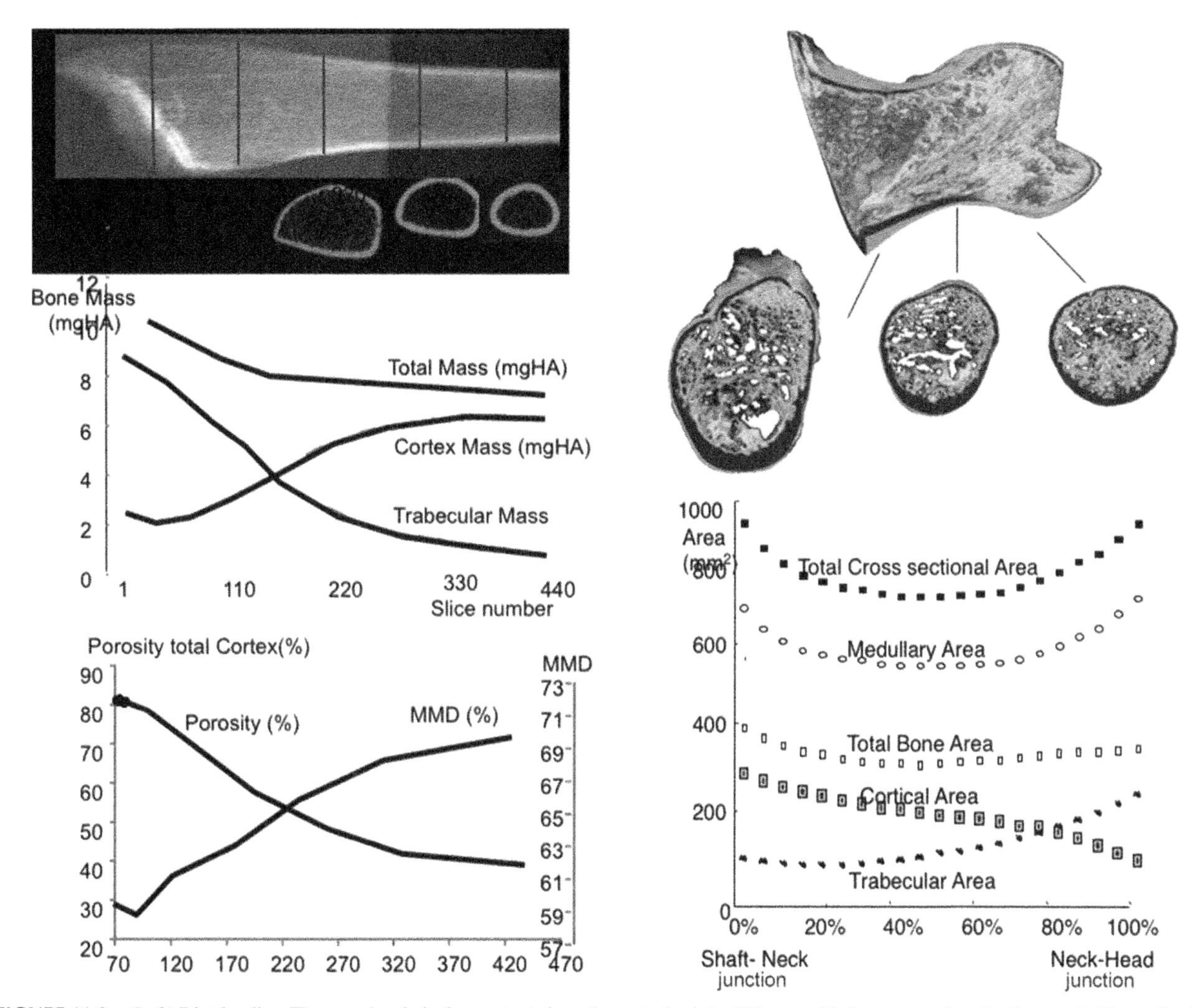

FIGURE 11.3 (Left) Distal radius. The mass is relatively constant along the metaphysis but it is assembled more as trabecular than cortical bone distally, with cortices of higher porosity and lower matrix mineral density. Proximally, the similar amount of material is mainly cortical with low porosity and higher matrix mineral density. (Right) The femoral neck differs in size but each cross section is assembled with a similar amount of material, forming more cortical bone adjacent to the femoral shaft and more trabecular bone nearer the femoral head. *MMD*, Matrix mineral density. *Adapted from Zebaze, R.M., Jones, A., Knackstedt, M., Maalouf, G., Seeman, E., 2007. Construction of the femoral neck during growth determines its strength in old age. J. Bone Miner. Res. 22 (7), 1055–1061.*

shorter bone like the femoral neck. Similar observations have been reported in studies of the femoral neck (Fig. 11.3, right panel).

Thus, the total CSA of a tubular bone and its mineralized matrix mass are inversely associated; larger cross sections are assembled with less material relative to their total CSA, producing a lower apparent volumetric bone mineral density (vBMD); there is less bone within the periosteal envelope of a bigger bone, avoiding bulk. Smaller cross sections are assembled with more material relative to their size, producing a higher apparent vBMD; there is more bone within the periosteal envelope of a smaller bone, minimizing the liability to fracture of slenderness.

Bulk is avoided in larger cross sections by greater modeling-based endocortical resorption during growth. A larger medullary cavity is excavated so that thickening of the cortex by periosteal apposition is offset, thereby minimizing mass, yet radial displacement by this radial modeling drift achieves the same cortical bone area and compressive strength because the thinner "ribbon" of cortex is distributed around a larger perimeter. The radial drift of the cortex increases resistance to bending even though the cortex is thinner relative to the total CSA (Ruff and Hayes, 1988).

Thus, long bones are not drinking straws; they do not have a single cross-sectional diameter, a single cortical thickness, or medullary cavity diameter. Group means are used to express these dimensions but they obscure the diversity in structure and matrix mineral distribution so critical to determining diversity in bone strength. Diameters of a bone cross section differ at each degree around the periosteal perimeter, creating differences in the external shape. Differences in the medullary diameters at corresponding points around the endocortical perimeter determine the shape of the medullary cavity, and the proximity of the periosteal and endocortical envelopes determines the cortical thicknesses around the perimeter of the cross section and the distances at which the cortical mass at each point around a cross section is placed from the neutral axis (Zebaze et al. 2005, 2007).

This diversity in bone size, shape, and mass distribution is achieved by differing degrees of focal formative and resorptive modeling at each point around the periosteal and endocortical perimeters. Intracortical remodeling assembles osteons, which are bone structural units with a central canal and concentric lamellae of differently oriented mineralized collagen fibers. A cement line delineates each osteon from the interstitial (interosteonal) bone matrix and other osteons.

An example of this regional specificity of remodeling is seen in studies of the femoral neck, the location of fractures associated with high morbidity and mortality. Remodeling upon the periosteal, intracortical, and endocortical surfaces varies according to the location chosen (Fig. 11.4, left). This heterogeneity produces the varying shape of the femoral neck cross sections. Greater periosteal apposition superiorly and inferiorly relative to mediolaterally produces the elliptical shape. Differences in periosteal apposition and endocortical resorption produce a thicker cortex inferiorly and a thinner cortex superiorly (Zebaze et al., 2007). The loss of bone during aging is also heterogeneous and varies depending on the location measured. It is greatest at the superior segment where the cortex is thinnest.

This principle of optimizing strength and minimizing mass is illustrated in a prospective study of the growth of a tibial cross section (Wang et al., 2005a, 2005b). In prepubertal girls, tibial cross-sectional shape was already elliptical at 10 years of age. During 2 years, focal periosteal apposition increased the ellipticity by adding twice the amount of bone anteroposteriorly than mediolaterally. Consequently, estimates of bending strength increased more in the anteroposterior (I_{max})

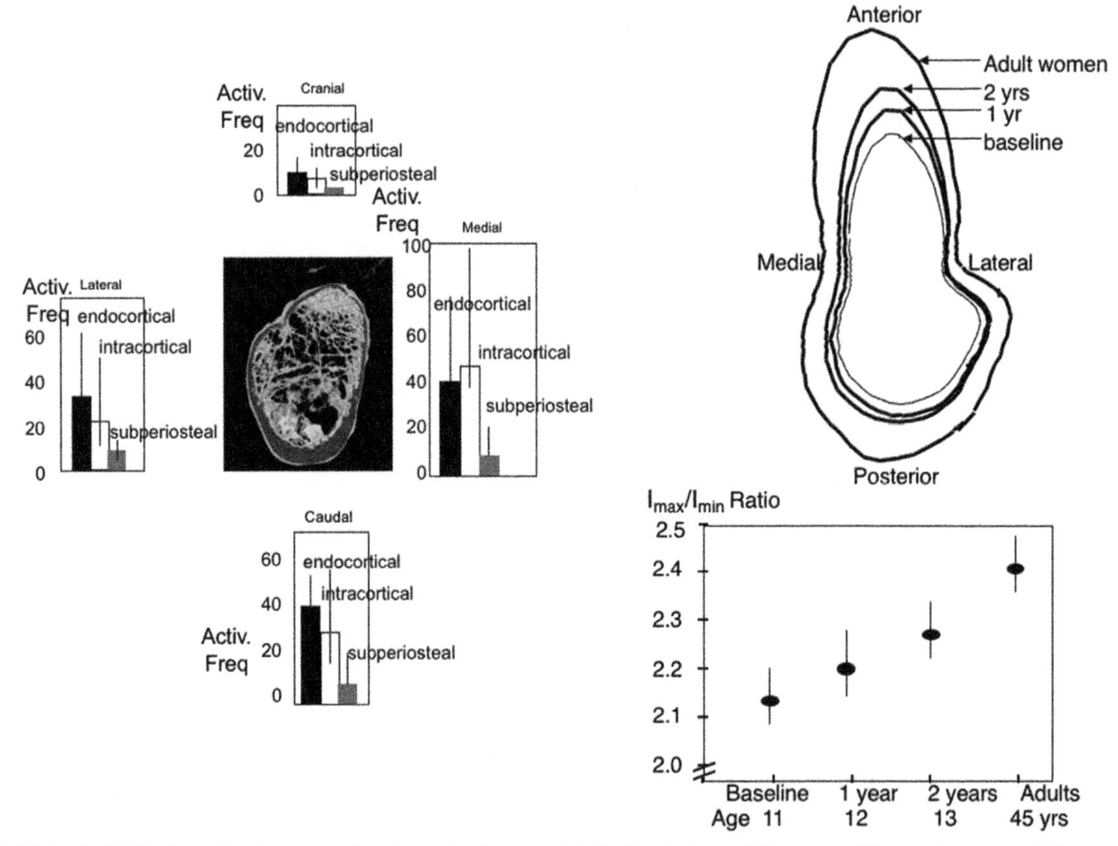

FIGURE 11.4 (Left) The femoral neck cross section. Activation frequency (*Activ. Freq*) of remodeling events differs on the periosteal, intracortical, and endocortical surface depending on the location studied. (Right) Bone mass distribution around the center of the tibial cross section. More bone is deposited anteriorly and posteriorly than medially and laterally during 2 years of growth, increasing the ellipticity of the cross section. Bending resistance increases more along the anteroposterior axis (I_{max}) than the mediolateral axis (I_{min}) as reflected in the increasing ratio. *Adapted from Wang, Q., Alen, M., Nicholson, P., Lyytikainen, A., Suurubuenu, M., Helkala, E., Suominen, H., Cheng, S., 2005a. Growth patterns at distal radius and tibial shaft in pubertal girls: a 2-year longitudinal study. J. Bone Miner. Res. 20 (6), 954–961, Wang, X.F., Duan, Y., Beck, T., Seeman, E.R., 2005b. Varying contributions of growth and ageing to racial and sex differences in femoral neck structure and strength in old age. Bone 36 (6), 978–986.*

than the mediolateral direction (I_{min}) (Fig. 11.4, right). Marrow area changed little, so more mass was distributed as a thicker cortex anteroposteriorly due to periosteal apposition without concurrent endocortical resorption. Resistance to bending increased by 44% along the principal axis (I_{max}) with a 22% increase in mass. If cortical thickness increased by the same amount of periosteal apposition at each point around the tibial perimeter, the amount of bone producing the same increase in bending resistance would be 205 mg, four-fold more than observed.

While it is intuitive that a bone with a larger CSA must be constructed with more periosteal bone than a smaller cross section, the contrary is observed. During 2 years, the absolute amount of bone deposited on the periosteal surface of the tibial cross section was similar in children with baseline tibial total CSA in the upper, middle, and lower tertile at 10 years of age. Thus, larger cross sections were assembled with less mass relative to their starting cross-sectional size, avoiding bulk, and smaller cross sections were assembled with more mass relative to their starting total CSA, offsetting the fragility associated with slenderness.

Deposition of similar amounts of bone on the periosteal surface of larger and smaller cross sections (and so less in relative terms on the former and more on the latter) was possible because the differences in bone size were established early, probably in utero (see later). Consequently, the deposition of an amount of bone on the periosteal surface of a larger cross section confers more bending resistance than the deposition of the same amount of bone on the periosteal surface of a smaller cross section, because resistance to bending is proportional to the fourth power of the distance from the neutral axis (Ruff and Hayes, 1988).

The increase in strength of a bone achieved by modifying its size and shape rather than increasing its mass is convincingly documented in racket sports. During growth, greater loading of the playing arm achieves greater bone strength by modifying its external size, its shape, and its internal architecture. Focal periosteal apposition and endocortical resorption at some locations, but endocortical bone formation at others, change the distribution of bone in space *without* a net change in its mass to accommodate loading patterns; apparent vBMD does not change; bending strength increases without increasing bulk (Haapsalo et al., 2000; Bass et al., 2002).

Trait variances in adulthood originate before puberty

Although adults have larger skeletons than children, differences in bone size and mass in adult life probably begin early in life. In a 3-year prospective study of growth in 40 boys and girls, Loro et al. reported that the variance at Tanner stage 2 (prepuberty) in vertebral CSA and volumetric trabecular BMD, femoral shaft CSA, and cortical area was no less than at Tanner stage 5 (maturity); 60%−90% of the variance at maturity was accounted for by the variance present before puberty. Thus, the magnitude of trait variances (dispersion around the age-specific mean) is largely established before puberty (Loro et al., 2000).

The ranking of individual values at Tanner stage 2 was unchanged during 3 years in girls. These traits tracked, so that an individual with a large vertebral or femoral shaft cross section, or higher vertebral vBMD or femoral cortical area, before puberty retained this position at maturity. The regression lines for each of the quartiles did not cross during 3 years. Similar observations were made in boys.

Similar observations have been reported using peripheral computed tomography of the tibia in 258 girls. The magnitude of variance at 10−13 years of age did not differ from that 2 years later, and did not differ from that of their premenopausal mothers. Likewise, in a study monitoring 744 women and men during 25 years, about 90% of the variance in cortical thickness in adulthood was accounted for by variance at completion of growth 25 years earlier (Garn et al., 1992). Similarly, studies from Tromso suggest that distal and ultradistal radial size and mass tracked during 6.5 years of follow-up of 5366 women and men ages 45−84 years (Emaus et al. 2005, 2006).

Finding that the magnitude of the trait variances at maturity is no different from the magnitude of their variances before puberty suggests that growth in larger and smaller bones occurs the same rate. (If larger bones deposit more bone during growth than smaller bones, variance will increase.) In addition, the constant variance and tracking also suggest that environmental factors are likely to contribute little to total variance of a trait in the population.

In infants and children at ages between 1 and 10 years, variances in diaphyseal diameter and muscle diameter were established at 1−2 years of age (Maresh, 1961). In a cross-sectional study of 146 stillborn fetuses of 20−41 weeks gestation, the percentage of a femur, tibia, and humerus diaphyseal cross section that was cortical area was about 80%−90% at 20 weeks gestation and remained so across the 20 weeks of intrauterine life, suggesting that as bone size increased during advancing intrauterine life, the proportion of bone within the cross section was established prior to 20 weeks gestation (Rodriguez et al., 1992).

The inference from the early establishment and constancy of trait variances is that genetic rather than environmental factors account for this variance. Studies in family members, twins, and birth cohorts followed for many decades and studies of fetal limb buds grown in vitro support this view (Murray and Huxley, 1925; Pocock et al., 1987; Seeman et al., 1996).

This does not mean that traits *in an individual* are immutably fixed. This flawed notion confuses the proportion of the *population* variance in a trait attributable to genetic or environmental differences in that population with the effect of an environmental factor or disease on a trait in an individual. Muscle paralysis in utero, exercise during growth, or effects of disease in adulthood all have profound effects on bone structure in individuals (Bass et al., 2002; Pitsillides, 2006). Lifestyle changes influence the population mean of a trait as documented many times by secular increases in height, a highly heritable trait (Bakwin, 1964; Meredith, 1978; Cameron et al., 1982; Tanner et al., 1982; Malina and Brown, 1987). However, under stable conditions, lifestyle *differences* within a population make only a small contribution to trait variances compared with genetic *differences* in that population.

Sex and racial differences in bone structure

For the vertebrae, increasing bone size by periosteal apposition builds a wider vertebral body in males than in females and in some races than in others (Seeman, 1998). Trabecular number per unit area is constant during growth (Fig. 11.5, bottom left). Therefore, individuals with a low trabecular number in young adulthood are likely to have lower trabecular numbers in childhood (Parfitt et al., 2000). The age-related increase in trabecular density is the result of increased thickness of existing trabeculae (Fig. 11.5, top left). Before puberty there is no difference in trabecular density in boys and girls of either Caucasian or African American origin (Gilsanz et al 1988, 1991) (Fig. 11.5, right). This suggests that both vertebral body size and the mass within its periosteal envelope increase in proportion until Tanner stage 3.

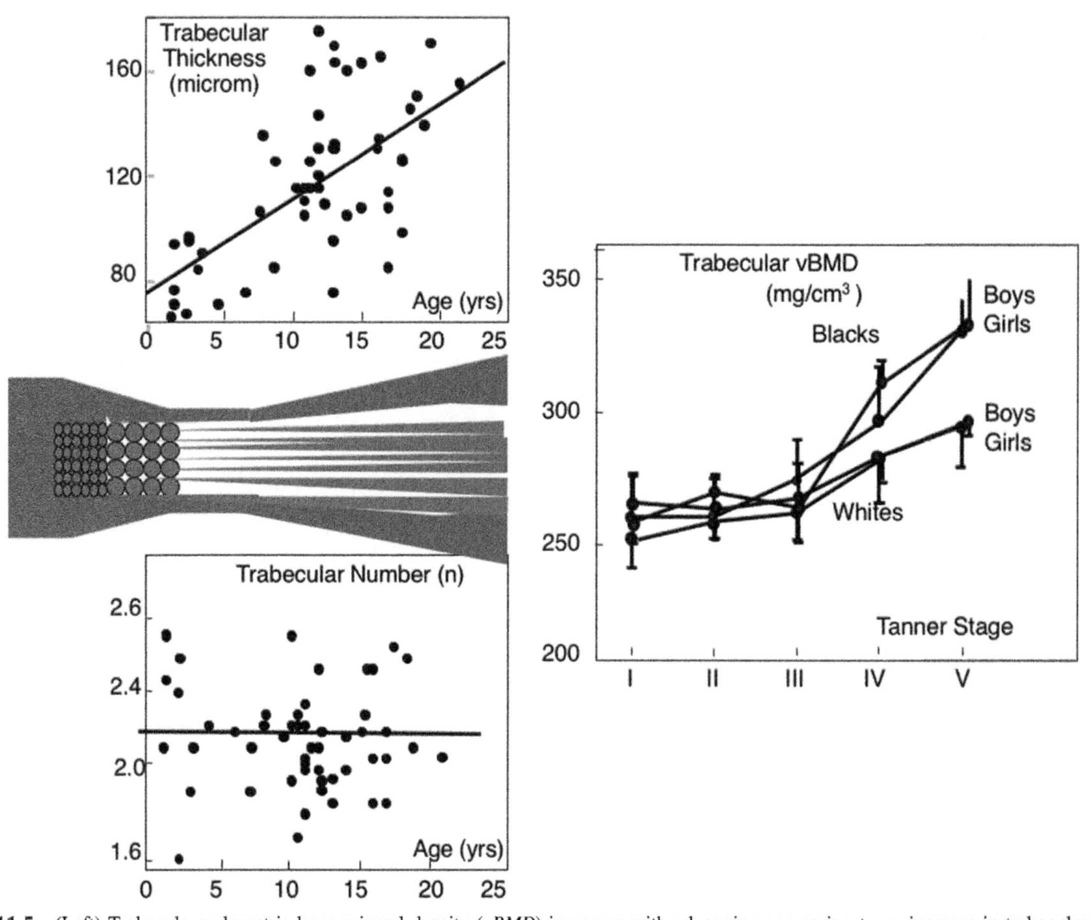

FIGURE 11.5 (Left) Trabecular volumetric bone mineral density (*vBMD*) increases with advancing age owing to an increase in trabecular thickness (top), not number (bottom). (Right) Before puberty, trabecular vBMD is no different by sex or race and increases at Tanner stage 3 similarly by sex within a race, but more greatly in blacks than in whites. *(Left) Adapted from Parfitt, A.M., Travers, R., Rauch, F., Glorieux, F.H., 2000. Structural and cellular changes during bone growth in healthy children. Bone 27, 487–494. (Right) Adapted from Gilsanz, V., Roe, T.F., Stefano, M., Costen, G., Goodman, W.G., 1991. Changes in vertebral bone density in black girls and white girls during childhood and puberty. N. Engl. J. Med. 325, 1597–1600.*

At puberty, trabecular density increases by race and sex, but there is no sex difference in trabecular density within a race. This increase is probably the result of the cessation of external growth in bone size but continued bone formation upon trabecular and endocortical surfaces, resulting in more bone within the periosteal surface of the bone—higher apparent vBMD. Thus, growth does not build a "denser" vertebral body in males than females, it builds a bigger vertebral body in males. Strength of the vertebral body is greater in young males than females because of size differences. Within a sex, African Americans have a higher trabecular density than whites due to a greater increase in trabecular thickness (Han et al., 1996). The mechanisms responsible for the racial dimorphism in trabecular density but lack of sexual dimorphism within a race are not known. The greater trabecular thickness in African Americans partly accounts for the lower remodeling rate in adulthood because there is less surface available upon which remodeling can occur (Han et al., 1996).

Sex differences in appendicular growth are largely the result of differences in timing of puberty (Fig. 11.6, left). Before puberty, there are already sex differences in diaphyseal diameter (Iuliano-Burns et al., 2008). As long bones increase in length by endochondral apposition, periosteal apposition widens the lengthening long bone. Concurrent endocortical resorption excavates the medullary cavity. As periosteal apposition is greater than endocortical resorption, the cortex thickens. In females, earlier completion of longitudinal growth with epiphyseal fusion and earlier inhibition of periosteal apposition produces a smaller bone.

Bone length continues to increase in males and periosteal apposition increases cortical thickness. However, cortical thickness is similar in males and females because endocortical apposition in females contributes to final cortical thickness (Garn, 1970; Bass et al., 1999). Cortical thickness is similar by race and sex (Fig. 11.6, right). What differs is the position of the cortex in relationship to the long axis of the long bone (Wang et al., 2005a, 2005b; Duan et al., 2005). It is not clear

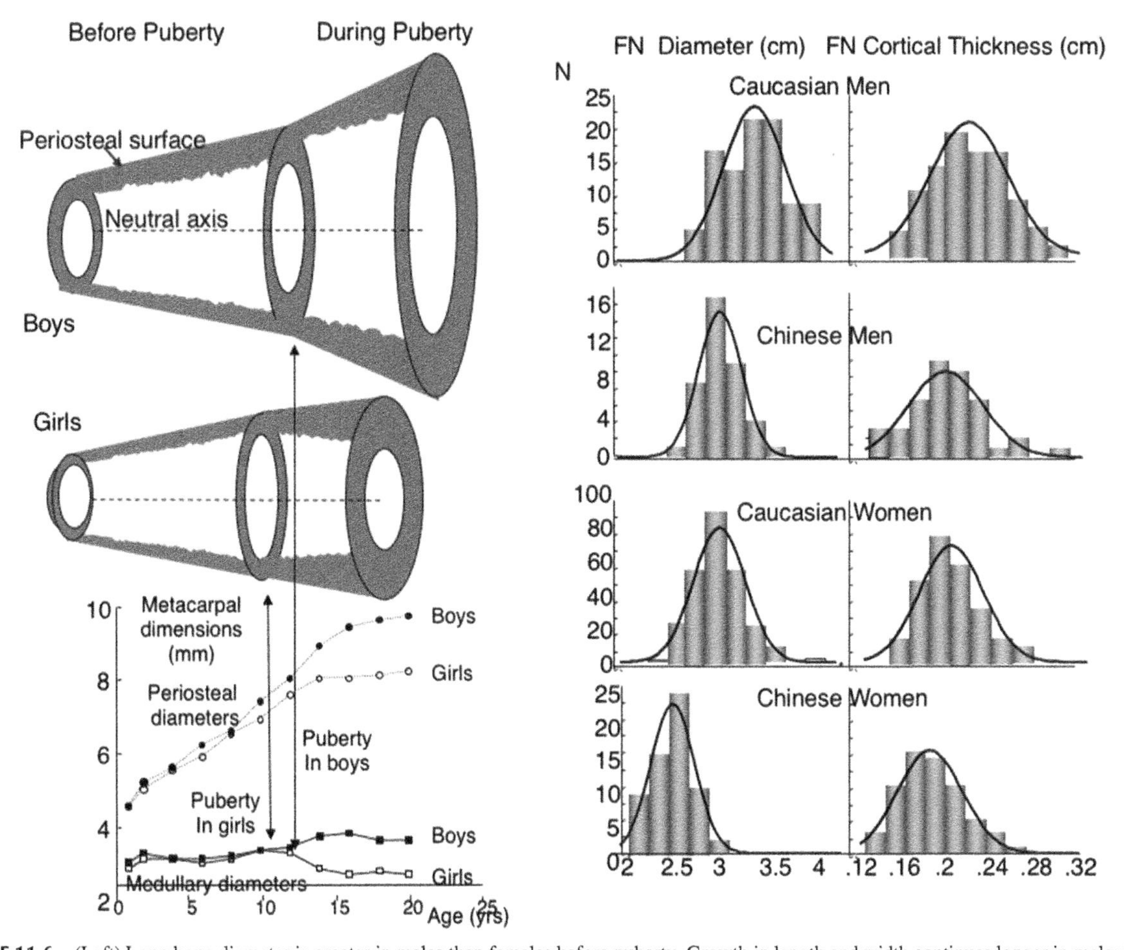

FIGURE 11.6 (Left) Long bone diameter is greater in males than females before puberty. Growth in length and width continues longer in males because puberty occurs later. Endocortical apposition in females contributes to cortical thickness so that final cortical thickness is similar in males and females but displaced further radially in males. (Right) Distribution of femoral neck (*FN*) diameter differs by sex and race but FN cortical thickness is similar by sex and race. *(Left) Adapted from Garn, S., 1970. The Earlier Gain and Later Loss of Cortical Bone. Nutritional Perspectives. Charles C. Thomas, Springfield, IL, pp. 3–120. (Right) E. Seeman with permission.*

whether the wider diaphysis in males than in females is the result of more rapid periosteal apposition in males, as commonly believed, or more protracted longitudinal growth at the same rate, as males enter puberty 1–2 years after females (Garn, 1970).

Bone remodeling by the basic multicellular unit

Osteocyte death in signaling bone remodeling

Bone, like roads, buildings, and bridges, develops fatigue damage during repeated loading, but only bone has a mechanism enabling it to detect the location and magnitude of the damage, remove it, replace it with new bone, and restore the bone's material composition while maintaining its micro- and macroarchitecture (Parfitt 1996, 2002).

Bone resorption is not necessarily bad for bone. On the contrary, the resorptive phase of remodeling removes old or damaged mineralized matrix. The formation phase of the remodeling cycle renews the material composition of bone and restores bone's microstructure provided that the volume of bone formed and deposited is the same as the volume of damaged bone removed. This process depends on the normal production, work, and life span of osteoclasts and osteoblasts. While these are the two executive cells of the BMU, the BMU consists of a range of cells that participate in the renewal of bone matrix composition, as discussed elsewhere (see chapters 4 and 10 for further detail).

It is likely that the osteocyte plays a pivotal role in bone modeling and remodeling. Osteocytes are the most numerous, longest-lived cells of bone. There are about 10,000 cells per cubic millimeter of mineralized bone matrix volume and each cell possesses about 50 neuron-like processes that connect osteocytes with one another and with flattened lining cells on the endosteal surface (Marotti et al., 1990) (Fig. 11.7A, see legend also).

The dense lacelike network of osteocytes with their processes ensures that no part of bone is more than several micrometers from a lacuna containing its osteocyte, suggesting that these cells are part of the machinery guarding the integrity of the composition and structure of bone (Parfitt, 2002). Microcracks sever osteocyte processes in their canaliculi, producing osteocyte apoptosis (Hazenberg et al., 2006) (Fig. 11.7B and C). Prevention of osteocyte death may be an attractive therapeutic target if they are the result of damage, if they become a form of damage themselves when they become apoptotic, or if they produce damage (O'Brien et al., 2004; Keller and Kneissel, 2005; Manolagas, 2006).

Apoptotic osteocytes, for example, may be a form of damage themselves, perhaps reducing the energy-absorbing/dissipating capacity of bone when lacunae mineralize. Estrogen deficiency and corticosteroid therapy result in their apoptosis (Manolagas, 2006). Osteocyte apoptosis may damage surrounding mineralized matrix, producing bone fragility (independent of bone loss). Corticosteroid-treated mice have large osteocyte lacunae surrounded by matrix with a 40% reduction in mineral and reduced elastic modulus (Lane et al., 2006). Genetic ablation of osteocytes produces bone fragility (Tatsumi et al., 2007). Whether the increased rate of remodeling in midlife in women is partly the result of osteocyte death is not known.

The number of dead osteocytes provides the topographical information needed to identify the location and size of damage (Verborgt et al., 2000; Taylor, 1997; Schaffler and Majeska, 2005) (Fig. 11.11C). Osteocyte apoptosis is likely to be one of the first events signaling the need for remodeling. It precedes osteoclastogenesis (Clark et al., 2005). In vivo, osteocyte apoptosis occurs within 3 days of immobilization and is followed within 2 weeks by osteoclastogenesis (Aguirre et al., 2006). In vitro, death of the osteocyte-like MLO-Y4 cells, induced by scratching, results in the formation of TRACP-positive (osteoclast-like) cells along the scratching path (Kurata et al., 2006).

Thus, just as the spider knows the location and size of its wriggling prey by signals sent along its vibrating web, the need for reparative remodeling is likely to be signaled by osteocyte death. This takes place via their processes connected to viable osteocytes and to flattened osteoblasts lining the three intracortical, endocortical, and trabecular components of the inner (endosteal) surface of bone upon which remodeling is initiated.

The bone remodeling compartment

Bone remodeling is initiated at points upon on the endocortical, trabecular, and intracortical components of the endosteal envelope. The endocortical and trabecular surfaces are adjacent to marrow. The intracortical surface is formed by the surface of Haversian and Volkmann's canals. While remodeling is initiated at points upon these surfaces, damage occurs deep to them, within the matrix of osteons or the interstitial (interosteonal) bone in the case of cortical bone or within hemiosteons in the case of trabecular bone. Information concerning the location and size of damage must reach these surfaces, and cells involved in remodeling must reach the site of damage beneath the endosteal surface. This anatomical

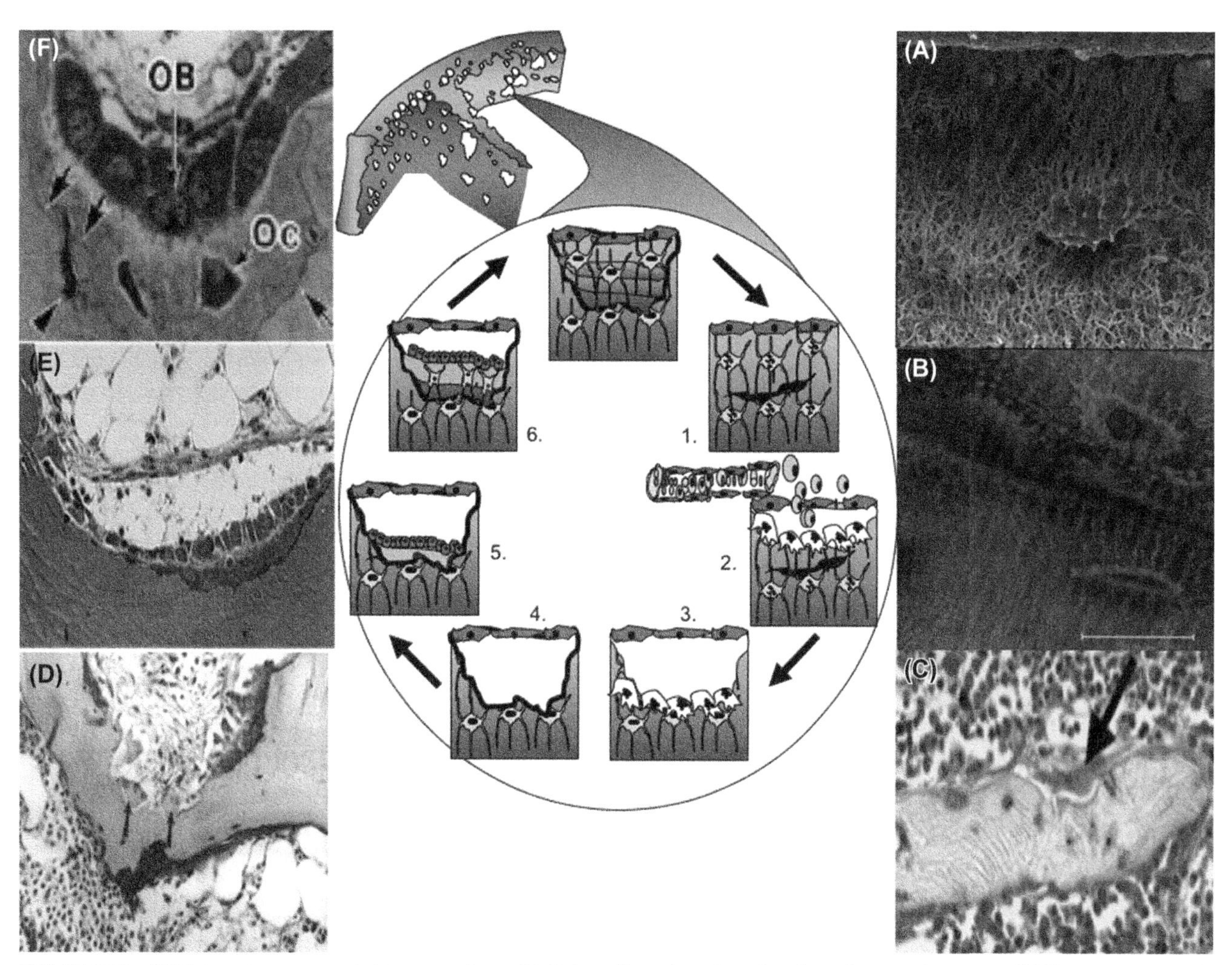

FIGURE 11.7 (A) Osteocytes are connected to one another and to lining cells on the endosteal surface adjacent to the marrow. (B) Damage to osteocytic processes by a microcrack produces osteocyte apoptosis. (C) The distribution of apoptotic osteocytes provides the topographical information needed to target osteoclasts (identified by arrow) to the damage. (D) Osteoclasts (arrows) resorb bone. (E) Osteoblasts deposit osteoid. (F) Some osteoblasts are entombed in the osteoid (arrows) they deposit. Central cartoons depict the remodeling events with (1) damage causing osteocytes to apoptose; (2) formation of a bone remodeling compartment with a vascular supply and osteoclasts resorbing bone; (3 and 4) osteoclasts resorbing damage; (5) a cement line being formed; and (6) bone formation following refilling of the cavity with entombment of osteoblasts that become osteocytes. *OB*, osteoblast; *Oc*, osteocyte. *(B) Courtesy J. Hazenberg. (C) Courtesy M. Schaffler. (F) From Suzuki, R., Domon, T., Wakita, M., 2000. Some osteocytes released from their lacunae are embedded again in the bone and not engulfed by osteoclasts during remodelling. Anat. Embrol. 202, 119–128.*

arrangement makes the flattened lining cells likely conduits transmitting the health status of the bone matrix to the bone marrow environment, which in turn is a source of the cells of the BMU, but not the only source.

Apoptotic osteocytes signal the location and size of the damage burden to the flattened lining cells of the endosteal surface, leading to the formation of a bone remodeling compartment (BRC), which confines and targets remodeling to the damage that is removed by osteoclasts (Hauge et al., 2001) (Fig. 11.7D). The regulatory steps between osteocyte apoptotic death and creation of the BRC are not known. Bone-lining cells express collagenase mRNA (Fuller and Chambers, 1995). An early event creating the BRC may be collagenase digestion of unmineralized osteoid to expose mineralized bone, a requirement for osteoclastic bone resorption.

The flattened bone-lining cells express markers of the osteoblast lineage, particularly lining cells forming the BRC canopy (Hauge et al., 2001; Parfitt, 2001). These canopy cells also express markers for growth factors and regulators of osteoclastogenesis, such as receptor activator of NF-κB ligand (RANKL), suggesting that the canopy has a central role in the differentiation of precursor cells of marrow stromal origin, monocyte–macrophage origin, and vascular origin toward their respective osteoblast, osteoclast, or vascular phenotype.

The multidirectional steps of the remodeling cycle

While the two classical events of remodeling—resorption of a volume of bone by osteoclasts and formation of a similar volume of bone by osteoblasts—occur sequentially (Hattner et al., 1965), the cellular and molecular regulatory events leading to these two fully differentiated functions may not be sequential. Some may be contemporaneous and multidirectional; osteoblastogenesis and its regulators determine osteoclastogenesis, and so the volume of bone to be resorbed, while osteoclastogenesis and the products of the resorbed matrix regulate osteoblastogenesis. Both pathways may be regulated to some extent by osteocytes and their products (e.g., sclerostin). How this cellular and molecular traffic is orchestrated from beginning to end remains unclear (see chapters 4 and 10 for further detail).

Signaling from apoptotic osteocytes to cells in the canopy expressing the osteoblast phenotype may influence further differentiation toward osteoblast precursors expressing RANKL and fully differentiated osteoblasts producing osteoid. Even at this stage, regulation of osteoclastogenesis and osteoblastogenesis occurs simultaneously through osteoblast precursors. In the MLO-Y4 cell line, damaged osteocyte-like cells have been reported to secrete macrophage colony-stimulating factor and RANKL (Kurata et al., 2006). Whether this occurs in human subjects in vivo is not known, but it raises the possibility that osteocytes participate in the differentiation of monocyte–macrophage precursor cells toward the osteoclast lineage. Both osteoblast and osteoclast precursors circulate and so may arrive at the BRC via the circulation and via capillaries penetrating the canopy (Eghbai-Fatourechi et al 2005, 2007; Fujikawa et al., 1996).

The contribution of precursors from the canopy or the marrow via sinusoids or capillaries is not well defined. Angiogenesis is essential to bone remodeling. Osteoprogenitor cells are associated with vascular structures in the marrow and several studies suggest there may be common progenitors giving rise to cells forming the blood vessels and the perivascular cells that can differentiate toward cells of multiple lineages (Doherty et al., 1998; Howson et al., 2005; Sacchetti et al., 2007; Matsumoto et al., 2006; Kholsa, 2007; Otsura et al., 2007; Khosla et al., 2008).

Little is known about the factors determining the volume of bone resorbed or how resorption stops after the damaged region has been resorbed. Osteoclasts phagocytose osteocytes and this may be one way the signal for resorption is removed (Elmardi et al., 1990). Products from the osteoclasts, independent of their resorption activity, and products from the resorbed matrix partly regulate osteoblastogenesis and bone formation (Suda et al., 1999; Martin and Sims, 2005; Lorenzo, 2000). Sclerostin, a negative regulator of bone formation, is an osteocyte product that inhibits bone formation and its inhibition is permissive for bone formation.

After the reversal phase, osteoblasts deposit osteoid, partly or completely filling the cavity (establishing the size of any negative BMU balance) (Fig. 11.7E). Osteoblasts deposit type I collagen, which is configured as the lamellae that undergo primary and secondary mineralization. How the osteoblasts change polarity to produce the differently oriented collagen fibers from lamella to lamella is not known. Most osteoblasts die, others become lining cells, while others are entombed in the osteoid they form, leaving reconstruction and "rewiring" of the osteocytic canalicular communicating system for later mechanotransduction, damage detection, and repair (Fig. 11.7F) (Han et al., 2004).

Perhaps the most fundamental and challenging question remains unanswered. Why is remodeling initiated at a given location and at a given time? While it is commonly stated that remodeling is initiated by "damaged" or "old" bone, the definitions of "damaged" and "old" remain enigmatic. Damage at the nano- or microstructural level has not been categorized in morphological terms, so the causes of damage, the biomechanical effects, and the biochemical and structural means of detecting, signaling, and repairing different types of damage remain uncertain (Akkus et al., 2004; Burr et al., 1998; Danova et al., 2003; Diab et al., 2006; Diab T and Vashisha D 2005; Garnero et al., 2006; Landis 2002; Ruppel et al., 2006; Silva et al., 2006; Taylor et al., 1997).

Bone remodeling and microstructure during young adulthood, menopause, and advanced age

Young adulthood: reversible bone loss and microstructural deterioration

Remodeling is balanced during young adulthood; the same volumes of bone are removed and eventually replaced at specific locations upon the endocortical, intracortical, and trabecular components of bone's endosteal envelope (Hattner et al., 1965). There is no net focal gain or loss of bone so that the external and internal dimensions of the mineralized bone matrix volume remain unchanged. Remodeling also occurs upon the outer (periosteal) surface but to a minimal extent (Orwoll 2003; Blizoites et al., 2006).

The resorption of a volume of bone, its replacement with an identical volume of osteoid, and primary and secondary mineralization of osteoid are not instantaneous events. The resorptive phase of remodeling takes ~3 weeks, the reversal phase takes ~1 week, the formation phase takes ~3 months, and completion of mineralization takes many months, if not years (Baron et al., 1984; Hattner et al., 1965; Parfitt, 1996).

Remodeling occurs simultaneously at different locations but it is asynchronous. BMUs at different locations are at different phases of their remodeling cycle. While remodeling is balanced, the deficit in matrix and its mineral content is *focally* transient and fully reversible. However, as concurrent events are asynchronous, there is a *globally* ever-present deficit in matrix and its mineral content formed by the BMUs at different stages of their remodeling cycle. The more rapid the remodeling, the greater the number of BMUs, and so the greater size of the reversible deficit in matrix and its mineral content.

In morphological terms, the ever-present reversible global deficit in matrix and its mineral content consists of BMUs in their resorption phase, BMUs with cavities in their reversal phase, cavities containing osteoid that has undergone primary mineralization, and other cavities containing matrix at various stages of secondary mineralization (Parfitt, 1996). More BMUs are at varying stages of their formation or mineralization phase than in their resorption or reversal phase because the matrix formation phase is longer than the resorption phase and the matrix mineralization phase is very much longer than the matrix deposition phase. As secondary mineralization is so much slower than the matrix deposition, the ever-present global deficit in mineralized bone matrix volume is largely the result of a deficit in the mineral content of the matrix, not the matrix. New osteons in cortical bone or hemiosteons in trabecular bone are fully reconstructed by the cellular activity of osteoblastic bone formation many months before they become fully mineralized.

In premenopausal women, the ever-present global deficit is ~10% of the total mineralized bone matrix volume. After menopause, when remodeling rate increases, the deficit is ~20% of the total mineralized bone matrix volume. Of the 20% of the skeleton being remodeled—"turned over"—annually, not all is reversible because of the emergence of the remodeling imbalance and the appearance of the irreversible deficit in bone matrix and its mineral content.

The recognition of the existence of the reversible and irreversible deficits in matrix volume and its mineral content and the differing time course of completion of matrix resorption, matrix deposition, and the much slower secondary mineralization of the matrix is important because each event contributes differently to aspects of the pathogenesis of menopause and age-related loss of bone matrix, its mineral content, and so bone's material and structural strength and whole-bone strength.

This "deconstruction" into the reversible and irreversible deficits of bone matrix and mineral content is also relevant to understanding the morphological changes produced by antiresorptive and anabolic therapy, particularly when these two classes of drugs are combined. These treatments have opposite effects on matrix mineral density and so have effects on material and microstructural strength not captured by the BMD measurement, and indeed obscured by the measurement, which is used as a surrogate of "bone strength." Several examples of the effects of the reversible and irreversible deficits in matrix and mineral follow here and in the section concerning modeling and remodeling during drug therapy.

The reversible deficit in matrix and its mineral content produces no permanent microstructural deterioration before menopause. If remodeling is rapid, the presence of many excavated cavities may increase bone fragility independent of the temporary small deficit in matrix and its mineral content produced by the delay and slowness of the formation phase. These cavities form stress "risers" or stress "concentrators," which are likely to contribute to the pathogenesis of fractures (Hernandez, 2006).

Before menopause, remodeling is balanced and the deficit in matrix and mineral at a given location is transitory, yet several studies suggest that bone loss occurs before menopause. For example, Riggs et al. reported that women lose 37% and men 42% of the total trabecular bone before age 50 years and 6% and 15% of lifetime cortical bone is lost (Riggs et al 1986, 2007; Gilsanz et al., 1987). However, a prospective study of female twins remaining premenopausal or entering peri- or postmenopause did not support the occurrence of bone loss or microstructural deterioration before menopause (Bjørnerem et al., 2018).

The loss of bone (measured as a decrease in BMD) may be an artifact produced by an increase in medullary cavity fat, which attenuates photons less than water or cells, producing a seeming decline in BMD giving the impression of bone "loss" (Bolotin and Sievänen, 2001). Alternatively, a remodeling imbalance may exist before menopause but is below the detection limits of current methods. If bone loss does occur before menopause, the structural and biomechanical consequences are likely to be less than those of bone loss later in life because the remodeling rate is slow, and trabecular bone loss probably proceeds by reduced bone formation and thinning rather than increased bone resorption with perforation and loss of connectivity. Thinning is less deleterious to loss of strength than perforation and loss of connectivity (Van der Linden et al., 2001). Moreover, continued periosteal apposition partly offsets endocortical bone loss, shifting the cortices radially, maintaining cortical area and resistance to bending (Szulc et al., 2006).

Menopause: reversible and irreversible bone loss and microstructural deterioration

At the time of menopause, there is an early accelerated phase of bone loss and an accelerated decrease in BMD. The worsening of the reversible deficit in matrix and mineral is the net result of the concurrent appearance of many more BMUs in their resorptive phase than the BMUs generated before menopause only now entering their refilling phase. The deficit in matrix and mineral produced by perturbation of steady-state surface level remodeling is fully reversible.

If the increase in the rate of bone remodeling was the only effect of menopause, BMD would decrease to a level determined by the higher rate of remodeling at the new steady state with BMUs in varying states of incompleteness of their remodeling cycle. No further bone loss would occur. Bone fragility would increase as a function of numbers of stress concentrators. If estrogen deficiency was reversed and remodeling returned to its premenopausal rate, BMD would be fully restored to its premenopausal level and there would be no permanent microstructural deterioration. This is not the case.

The percentages of the endosteal surface participating in bone resorptive activity and in formative activity at other locations are increased because the birth rate of BMUs increases after menopause. This alone does not determine whether bone is irreversibly lost from the skeleton. Whether bone loss occurs depends on the presence of a remodeling imbalance at the level of each BMU. Menopause is accompanied by the appearance of remodeling imbalance; less bone is deposited than was resorbed during each remodeling transaction. This differs from the reversible deficit and reversible microstructural deterioration produced by the normal delay and slowness of the refilling phase of remodeling.

Remodeling imbalance with less bone deposited than formed may occur in a range of ways. The most common and demonstrated by histomorphometry is a reduction in *both* the volumes of bone resorbed and the volumes deposited by each BMU, but a greater reduction in the volume of bone formed (Lips et al., 1978; Vedi et al., 1984). The decrease in resorption by the BMU is reflected in a smaller resorption cavity and an age-related increase in interstitial thickness (Croucher et al., 1991; Ericksen et al., 1999). (If resorption depth increased, interstitial wall thickness, the distance between cement lines of adjacent hemiosteons in trabecular bone, should decrease.) The reduction in the volume of bone resorbed results in the formation of smaller osteons, as the diameter of the resorption cavity defines the outer diameter of the osteon to be reassembled, while the reduction in the volume of bone deposited in the smaller cavity produces fewer lamellae and a larger central Haversian canal (porosity).

Another mechanism producing remodeling imbalance is an increase (not decrease) in the volume of bone resorbed and a decrease in the volume of bone formed due to an increase in the life span of osteoclasts and decrease in the life span of the osteoblasts, respectively (Manolagas Bonyadi et al., 2003; Nishida et al., 1999; Stenderup et al., 2001; Oreffo et al., 1998). The observation that menopause is accompanied by reduced numbers of trabeculae not thinning is consistent with larger cavities perforating these trabeculae (Aaron et al., 1987; Bjornerem et al., 2018). The observation of reduced cavity depth is consistent with a reduction in bone resorption. The differing observations can be reconciled if the increased life span of osteoclasts produced by estrogen deficiency is transitory (Ericksen et al 1986, 1999; Manolagas, 2000; Compston et al., 1995).

The factor driving the early accelerated bone loss is the perturbation of the surface extents of resorption and formation, not remodeling imbalance. Whether the early accelerated loss of bone at menopause is partly due to a reduction in the volume of bone formed by BMUs generated just *before* menopause is uncertain, because osteoblasts of these BMUs may not yet have a reduced life span. The reduction in osteoblast life span may emerge only during menopause.

Remodeling returns to steady state 6–12 months postmenopause at a higher remodeling rate than the steady-state remodeling before menopause, but at a slower rate than in early menopause. This slowing is not the result of slowing of the rate of remodeling. On the contrary, the rate of remodeling continues to be rapid. Bone loss is driven by the size of the remodeling imbalance and the birth rate of the new BMUs, but no longer by perturbation of the surface extents of resorption and formation. Now, the great many new BMUs generated in early menopause refill incompletely, and concurrently, a similarly large number of BMUs in their resorptive phase are generated. The rates of increase in cortical porosity, decrease in trabecular density due to reduced numbers of trabeculae, and reduction in matrix mineral density accelerate in women transitioning from peri- to postmenopause, and then decelerate (Bjørnerem et al., 2018) (Fig. 11.8).

Whatever the mechanisms producing remodeling imbalance, the imbalance in the volumes of bone resorbed and deposited by each BMU is the necessary and sufficient morphological basis for bone loss, microstructural deterioration, and bone fragility. Bone can no longer accommodate its loading circumstance by adaptive modeling and remodeling because each time a remodeling event occurs in an attempt to maintain bone's material composition, there is a loss of bone and microstructural deterioration. Most importantly, the loss of strength is out of proportion to the bone loss producing this deterioration; an increase in porosity of a compact structure like cortical bone reduces resistance to bending to the seventh power of the rise in porosity. A rise in porosity of an already porous structure like cortical bone reduces its resistance to bending to the third power (Schaffler and Burr, 1988).

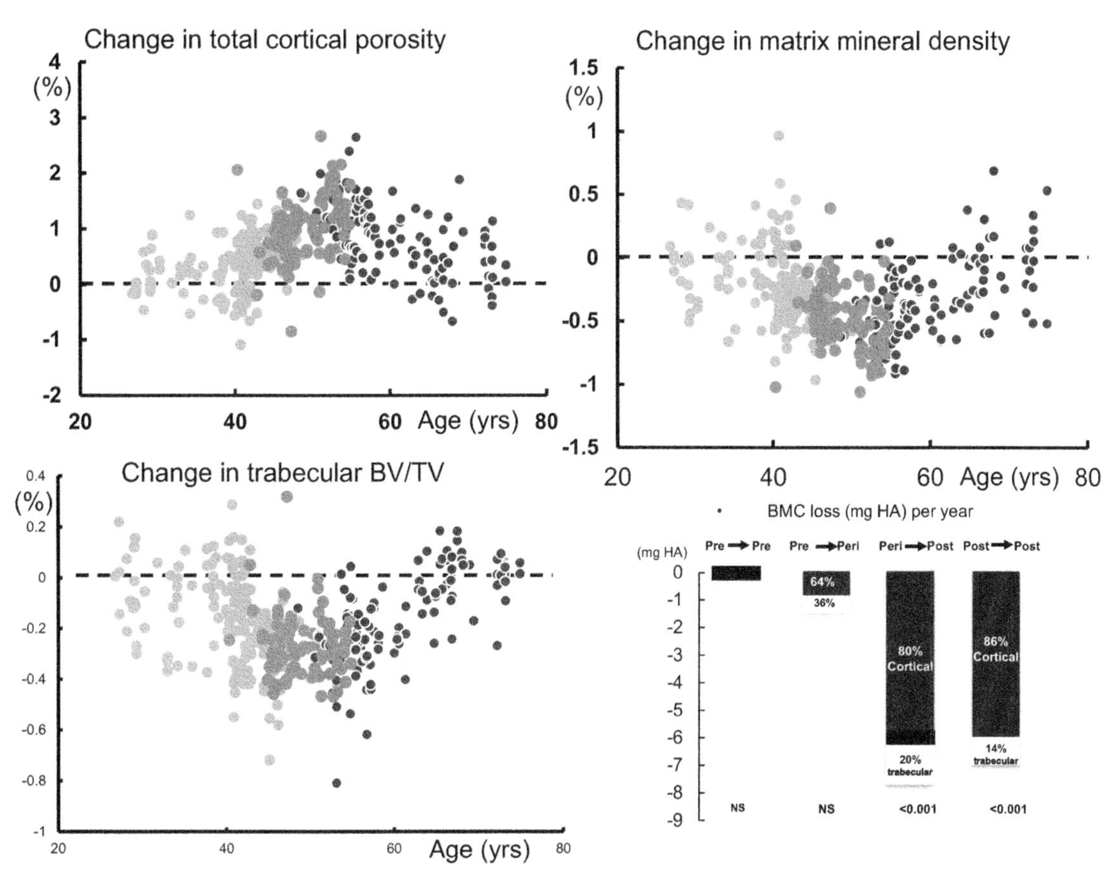

FIGURE 11.8 Annual changes in distal tibia total cortical porosity, trabecular bone volume/total volume (*BV/TV*), and matrix mineral density as a function of baseline age in women remaining premenopausal (*light orange dots* [light gray in print version]), becoming perimenopausal (*green dots* [medium gray in print version]), becoming postmenopausal (*dark orange dots* [dark gray in print version]), and remaining postmenopausal (*blue dots* [black in print version]). The bottom right shows the amount of tibial cortical and trabecular bone loss as a percentage of annual loss of total bone mineral content (*BMC*; mg hydroxyapatite [*HA*]) in women remaining premenopausal (*Pre → Pre*), becoming perimenopausal (*Pre → Peri*), moving from peri- to postmenopausal (*Peri → Post*), and remaining postmenopausal (*Post → Post*). *P* values within each group tested whether the total annual loss was different from zero.

Thus, BMD may decrease only into the so-called "low normal" range or osteopenic range, but bone fragility is present and is obscured by these modest reductions in BMD, which misleadingly suggest fracture risk is low. Fracture risk is lower than in women with BMD within the osteoporosis range (T score < -2.5 SD), but osteopenia or normal BMD in a postmenopausal woman is no assurance of absence of fracture risk; on the contrary, most fractures in the community arise among women and men with osteopenia (Siris et al., 2004).

Advanced age: the predominance of cortical bone loss

Modeling and remodeling are surface-dependent cellular events. For remodeling to occur there must be a surface upon which it is initiated. The larger the surface area, the greater the likelihood that remodeling can be initiated by events occurring deep to that surface, whether this event is matrix damage, osteocyte death, or other still undefined factors. Trabecular bone is fashioned as thin plates with a large surface area and a high surface area/matrix volume configuration. This is an advantage for matrix damage repair because damage can be easily signaled to the trabecular surface and the initiated remodeling can then easily be targeted to return to the matrix damage, remove it, and replace it with new bone. Hence, trabecular bone is more readily "turned over," a liability when remodeling becomes unbalanced. Indeed, when this occurs, rapid loss of complete trabeculae removes them and their surface so that trabecular remodeling diminishes as trabeculae disappear. In addition, loss of strength is greater with perforation of trabeculae, as found in women, rather than thinning, as occurs in men (Van der Linden, 2001) (Fig. 11.9).

As trabeculae are lost with their surfaces, bone loss from this compartment slows as less trabecular surface is available for resorption, but remodeling upon the endocortical and intracortical surfaces continues and increases as the endocortical

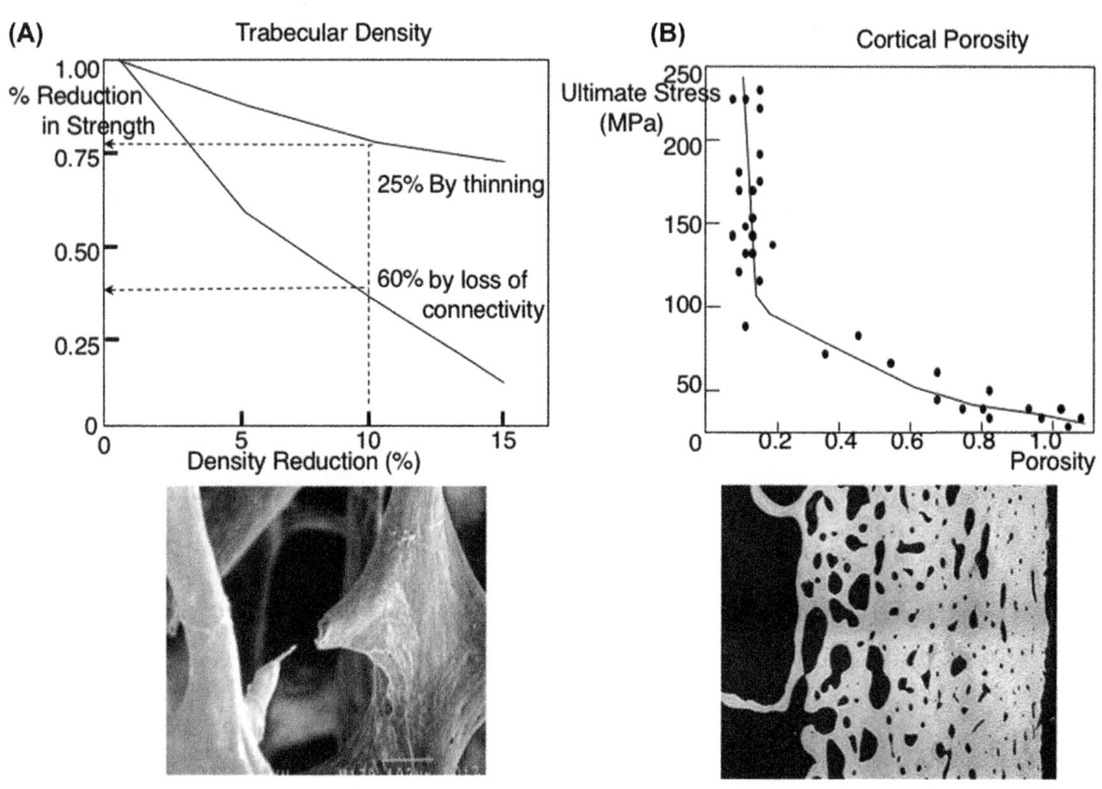

FIGURE 11.9 (A) Reduction in strength produced by a 10% deficit in trabecular density is greater when this deficit is produced by loss of trabecular connectivity than by thinning. Image shows loss of connectivity (Mosekilde et al., 1990). (B) A small increase in cortical porosity by a few percentage points is associated with a decline in ultimate stress and reduction in toughness. Scanning electron microscopic image of irregularly shaped enlarged pores in cortical bone. *(A) Adapted from Van der Linden, J.C., Homminga, J., Verhaar, J.A.N., Weinans, H., 2001. Mechanical consequences of bone loss in cancellous bone. J. Bone Miner. Res. 16, 457–465. (B) Adapted from Martin, R.B., 1984. Porosity and specific surface of bone. CRC Critical Rev. Biomed. Eng. 10, 179–221; and Yeni, Y.N., Brown, C.U., Wang, Z., Norman, T.L., 1997. The Influence of bone morphology on fracture toughness of the human femur and tibia. Bone 21, 453–459.*

surface undergoes bone resorption, so that its surface area increases like the folds of a curtain (Parfitt, 1984; Brown et al., 1987; Arlot et al., 1990; Foldes et al., 1991). The main source of bone loss is intracortical remodeling initiated upon the surface of the myriads of Haversian canals traversing the cortex (Martin, 1984; Brockstedt et al., 1993; Yeni et al., 1997). Increased porosity is mainly due to increased size of existing canals that coalesce, forming irregularly shaped pores (canals seen in cross section). As the canals enlarge, the surface area of the canals also enlarges as does the surface area/matrix volume configuration, making cortical matrix accessible to being turned over and lost by the unbalanced remodeling (Bui et al., 2013). Total bone surface area either does not change (increasing in cortical bone, decreasing in trabecular bone) or increases (in regions of cortical bone only) so that late in life, bone loss is more cortical than trabecular in origin (Zebaze et al., 2010).

Increased intracortical remodeling of cortex adjacent to the medullary canal fragments the cortex, causing it to "trabecularize"; the cortical fragments look like trabecular bone and the porosity is erroneously "seen" as part of an enlarging medullary canal. Apportioning the intracortical porosity to the medullary compartment underestimates the age-related increase in cortical porosity and the trabecularized cortical fragments are regarded as trabecular bone within the seemingly expanded medullary canal, leading to an underestimation of the decrease in trabecular density with age (Zebaze et al., 2010). Both errors underestimate fracture risk in individuals and so fail to signal the imperative to initiate treatment.

Solutions to these problems have been found by segmenting a corticotrabecular outer and inner transitional zone using non-threshold-based image analysis. This approach allocates and confines cortical porosity to this "third" compartment, rather than erroneously expanding the medullary canal void, and cortical fragments to the outer transitional zone and partly to the inner transitional zone, which also contains true trabecular fragments (Zebaze et al., 2013).

As age advances and rapid unbalanced remodeling continues due to permanent estrogen deficiency and perhaps emerging secondary hyperparathyroidism, the extent of coalescence of pores increases, so the number of intracortical pores decreases, but the total area of porosity increases, as reported in patients with hip fractures (Bell et al., 1999). Cortical porosity reduces the ability of bone to limit crack propagation so that bone cannot absorb the energy imparted by the impact of a fall. This energy is released in the worst possible way, by fracture (Martin, 1984; Yeni et al., 1997). The continued unbalanced remodeling at a similar intensity removes the same volume of bone from an ever-decreasing amount of bone, accelerating the rate of bone loss and microstructural deterioration (Zebaze et al. 2013).

While the unbalanced remodeling produces bone loss and microstructural deterioration, remodeling is a transaction. Older, more completely mineralized bone is replaced with a smaller volume of younger bone that is less completely mineralized. This results in a decrease in mean matrix mineral density and an increase in the heterogeneity in the degree of mineralization of adjacent osteons and the interstitial (interosteonal) bone between osteons (which is more fully mineralized) (Boivin and Meunier, 2002; Boivin et al., 2003). This heterogeneity may limit microcrack propagation because more energy is needed to propagate a microcrack through a heterogeneously than a homogeneously mineralized material (Ural and Vashishth, 2014). However, interstitial (interosteonal) bone is less remodeled and becomes more highly mineralized and cross-linked with advanced glycation end products like pentosidine, both of which reduce the ductility of the matrix, facilitating microcrack propagation (Bailey et al., 1999; Banse et al., 2002; Nalla et al., 2004; Qui et al., 2005; Yeni et al., 1997; Viguet-Carrin et al., 2006).

The net effects of reduced periosteal apposition and endosteal bone loss

Over 80 years ago, Fuller Albright suggested that osteoporosis was a disorder of reduced bone formation (Albright et al., 1941). Research into the pathogenesis of bone fragility has focused on the role of bone resorption. During aging, reduced bone formation plays a central role in producing remodeling imbalance and so net bone loss takes place from the three components of the endosteal surface and reduced periosteal bone formation.

As age advances bone modeling by periosteal apposition continues, but much more slowly than during growth and much more slowly than endosteal bone loss. The net effect is an imperceptibly small increase in bone size but a reduction in total mineralized bone matrix volume and microstructural deterioration, with eventual almost complete loss of trabecular bone, cortical thinning, and a reduction in the number of cortical pores, but an increase in the cortical porosity as enlarging pores coalesce, so that all that is left is an eggshell rim of cortical bone (Balena et al., 1992; Seeman et al., 2003; Szulc et al., 2006).

Periosteal apposition is believed to increase as an adaptive response to compensate for the loss of strength produced by endocortical bone loss, so there will be no *net* loss of bone, no cortical thinning, and no loss of bone strength (Ahlborg et al., 2003). While this is often claimed, there are formidable challenges in identifying the existence of periosteal apposition during adulthood, its site specificity, its magnitude, and sex differences. In cross-sectional studies, secular changes in bone size may obscure or exaggerate periosteal apposition. Secular increases in stature are variously reported in one or both sexes, in some races but not others, and in the skeleton of the upper or lower body or both (Bakwin, 1964; Meredith, 1978; Cameron et al., 1982; Tanner et al., 1982; Malina and Brown, 1987).

These secular trends can produce misleading inferences when increments or lack of increments in bone diameters are used as surrogates of periosteal apposition. For example, in cross-sectional studies, the absence of an increment in periosteal diameter across age may not mean periosteal apposition failed to occur. Earlier born individuals (the elderly in a cross-sectional sample) are likely to have been shorter and to have had more slender bones than later born individuals (young normals in a cross-sectional sample). When periosteal apposition occurs, earlier born persons (forming the older subjects in a cross-sectional sample) with more slender bones have an increase in bone diameter that comes to equal that in later born persons (who have not yet had age-related periosteal apposition), leading to the flawed inference that there was no periosteal apposition.

When comparisons are made between sexes (or races) in cross-sectional studies, if the truth is that periosteal apposition is greater in men than in women but men have a secular increase in bone size and women do not, then the secular increase in men will blunt the increment in bone width across age in men and make it appear that the age-related increase in vertebral and femoral neck diameters (and so periosteal apposition) is similar in women and men.

Longitudinal studies are also problematic because changes in periosteal apposition during aging are small (Balena et al., 1992). The precision of methods to determine bone diameter, usually bone densitometry, and problems with edge detection when BMD is changing, limit the credibility of these measurements. Nevertheless, in one prospective study of over 600 women, Szulc et al. reported that endocortical bone loss occurred in premenopausal women with concurrent periosteal apposition (Szulc et al., 2006) (Fig. 11.10). As periosteal apposition was less than endocortical resorption, the cortices thinned but there was no *net* bone loss because the thinner cortex was now distributed around a larger perimeter, conserving total bone mass. Resistance to bending increased despite bone loss and cortical thinning, because this same amount of bone was now distributed farther from the neutral axis. So bone mass alone is a poor predictor of strength,

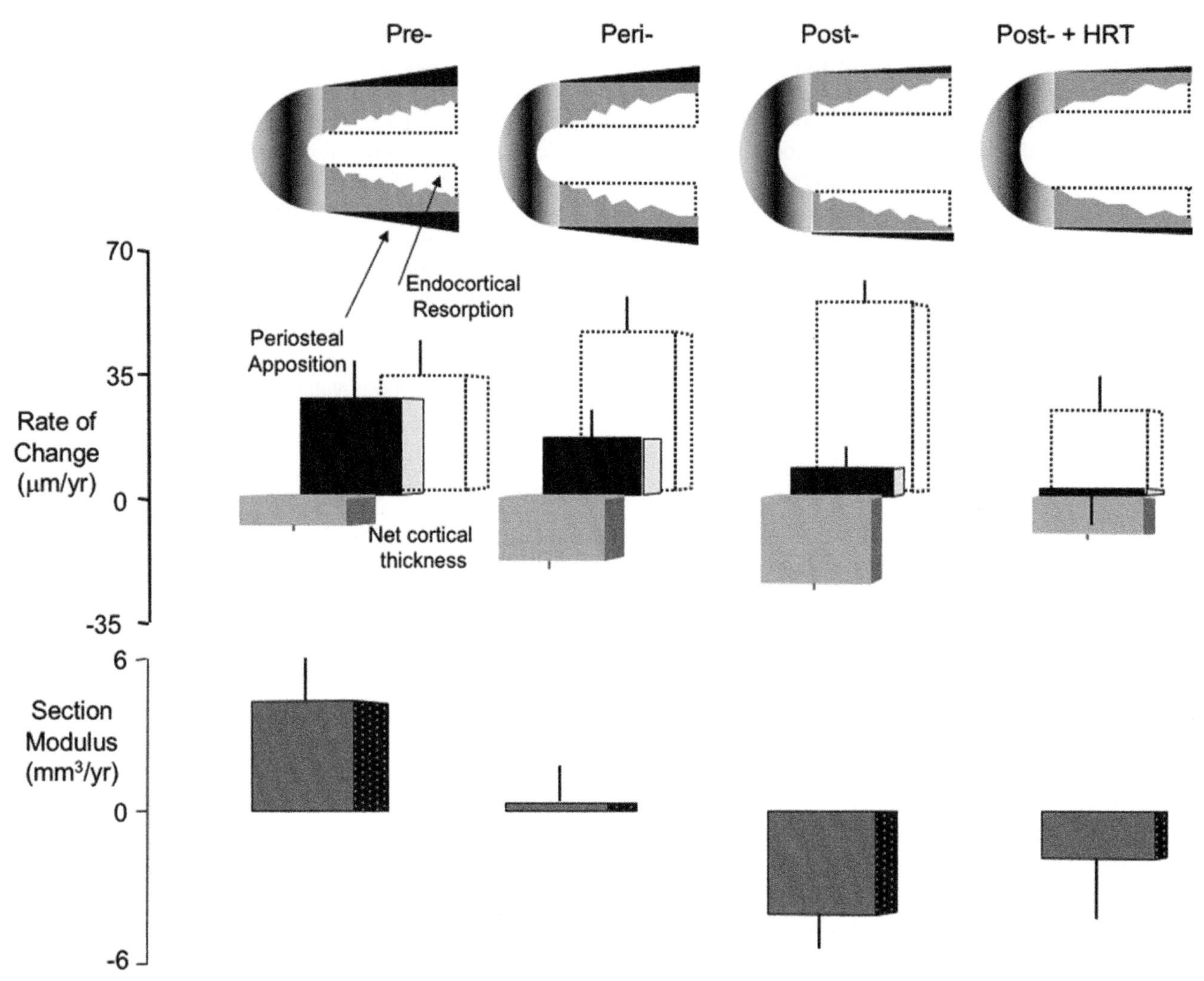

FIGURE 11.10 The amount of bone resorbed by endocortical resorption (*open bar*) increases with age. The amount deposited by periosteal apposition (*black bar*) decreases. The net effect is a decline in cortical thickness (*gray bar*). In premenopausal women, the thinner cortex is displaced radially, increasing section modulus (Z). In perimenopausal women Z does not decrease, despite cortical thinning, because periosteal apposition still produces radial displacement. In postmenopausal women, Z decreases because endocortical resorption continues, periosteal apposition declines, and little radial displacement occurs. In women treated with hormone replacement therapy (*HRT*), resorption is decreased with no effect on periosteal apposition. Z is less reduced than in untreated women. *Adapted from Szulc, P., Seeman, P., Duboeuf, F., Sornay-Rendu, E., Delmas, P.D., 2006. Bone fragility: failure of periosteal apposition to compensate for increased endocortical resorption in postmenopausal women. J. Bone Miner. Res. 21, 1856–1863.*

because resistance to bending is determined by the spatial distribution of the bone and increases as a fourth-power function of the radial distance a volume of bone is positioned from the neutral axis.

Endocortical resorption increased during the perimenopausal period, yet periosteal apposition decreased: it did not increase as expected if periosteal apposition is compensatory. The cortices thinned as periosteal apposition declined further. Nevertheless, bending strength remained unchanged, despite bone loss and cortical thinning, because periosteal apposition was still sufficient to shift the thinning cortex outward (Fig. 11.10).

Bone fragility emerged after menopause when acceleration in endocortical bone resorption and deceleration in periosteal apposition produced further net cortical thinning. As periosteal apposition was now minimal, there was little outward displacement of the thinning cortex, so cortical area now declined as did resistance to bending. Endocortical resorption was reduced but not abolished in women receiving hormone replacement therapy, while periosteal apposition was no different from that of untreated women; cortical thinning was reduced and the resistance to bending occurred, but less than in untreated women.

Periosteal envelope is not an exclusively bone-forming surface. During growth, bone resorption is critical for the in-wasting that produces the fan-shaped metaphyses (Rauch et al., 2001). Bliziotes and colleagues report that bone resorption occurs in adult nonhuman primates (Blizoites et al., 2006). Femur specimens from 16 intact adult male and female nonhuman primates showed that periosteal remodeling of the femoral neck in intact animals was slower than in cancellous bone but more rapid than at the femoral shaft. Gonadectomized females showed an increase in osteoclast number on the periosteal surface compared with intact controls. If these data are correct, adult skeletal dimensions may decrease in size as age advances.

Sexual dimorphism in trabecular and cortical bone loss

A greater proportion of women than men sustain fragility fractures. Men have a larger skeleton than women so that resistance to bending is greater in men. Bone loss in both sexes is the result of remodeling imbalance but remodeling rate does not increase in midlife in men. If there is a transitory increase in the volume of bone resorbed by each BMU in women, this does not seem to be the case in men. So, trabecular thinning occurs in men, trabecular perforation with greater loss of trabecular strength occurs in women (Aaron et al., 1987; Van der Linden et al., 2001). Net trabecular bone loss across age is reported to be only slightly greater in women than in men (Riggs et al., 2004), or is similar (Aaron et al., 1987; Meunier et al., 1990; Kalender et al., 1989; Mosekilde and Mosekilde, 1990; Seeman, 1997; Seeman et al., 2001). However, measurement error is likely to produce this observation because greater intracortical remodeling with trabecularization of the cortex in women underestimates their loss of trabecular bone. Cortical porosity increases less in men than in women because remodeling rate is lower in men and so crack propagation in cortical bone is probably better resisted in men than in women. Research is needed because cortical porosity is underestimated by imaging methods that use threshold-based image analysis (Zebaze et al., 2013).

Thus, several methodological issues leave the question of the morphological basis of sex differences in bone fragility unanswered (Seeman et al., 2004). The absolute risks for fracture in women and men of the same age and BMD are similar (Kanis et al., 2001, 2005). If this is correct, then the reason fewer men than women suffer fractures in their lifetime is likely to be that fewer men than women have material and structural properties that cause bone fragility, such as high cortical porosity and low trabecular density. Structural failure occurs less often in men because the relationship between load and bone strength is better maintained in men than in women (Riggs et al., 2006; Bouxsein et al., 2006).

The heterogeneous material and structural basis of bone fragility in patients with fractures

Patients with fractures are grouped by having "one or more minimal trauma fractures" or sustaining a fall from "no greater than the standing position." However, the pathogenesis and structural basis of the bone fragility underlying the fractures are heterogeneous. Patients with fractures may have high, normal, or low remodeling rates (Brown et al., 1984; Arlot et al., 1990; Delmas, 2000). Some have a negative BMU balance due to reduced formation, increased resorption, or both, or no negative BMU balance (Ericksen et al., 1990). Some patients with fractures have increased, while others have reduced, matrix mineral density (Ciarelli et al., 2003) (Fig 11.11). Some patients have reduced osteocyte density, others do not (Qui et al 2003, 2005). Contemporary therapeutics gives no consideration to selecting a treatment based on the underlying pathogenesis or structural abnormalities of bone fragility. Consequently, there are no data to support the notion that selecting treatment according to the causes of bone fragility will lead to reducing the number needed to treat to avert one person from having a fracture.

Bone modeling, remodeling, and drug therapy

Antiresorptive therapy reduces the reversible but not the irreversible deficit in mineralized matrix volume

The effects of antiresorptive therapy are the reciprocal of the effects of menopause in several ways. When an antiresorptive agent is administered, the early accelerated increase in BMD is largely due to the rapid reduction in the reversible deficit in matrix and its mineral content. At the time of administration, the number of BMUs in their resorptive phase decreases in proportion to the antiresorptive efficacy of the drug. Bisphosphonates reduce remodeling by 50%–60% (as estimated using remodeling markers). The net effect of fewer new BMUs excavating resorption cavities and the many more resorption cavities excavated shortly before treatment concurrently entering their refilling phase is the rapid early net increase in BMD and reduction in cortical porosity reported during the first 6–12 months of treatment (Seeman, 2010).

The rise in BMD is often mistakenly interpreted as an increase in bone "mass" or "volume" and a restoration of microstructural deterioration. This is not the case; the incomplete refilling of cavities excavated just before treatment is the

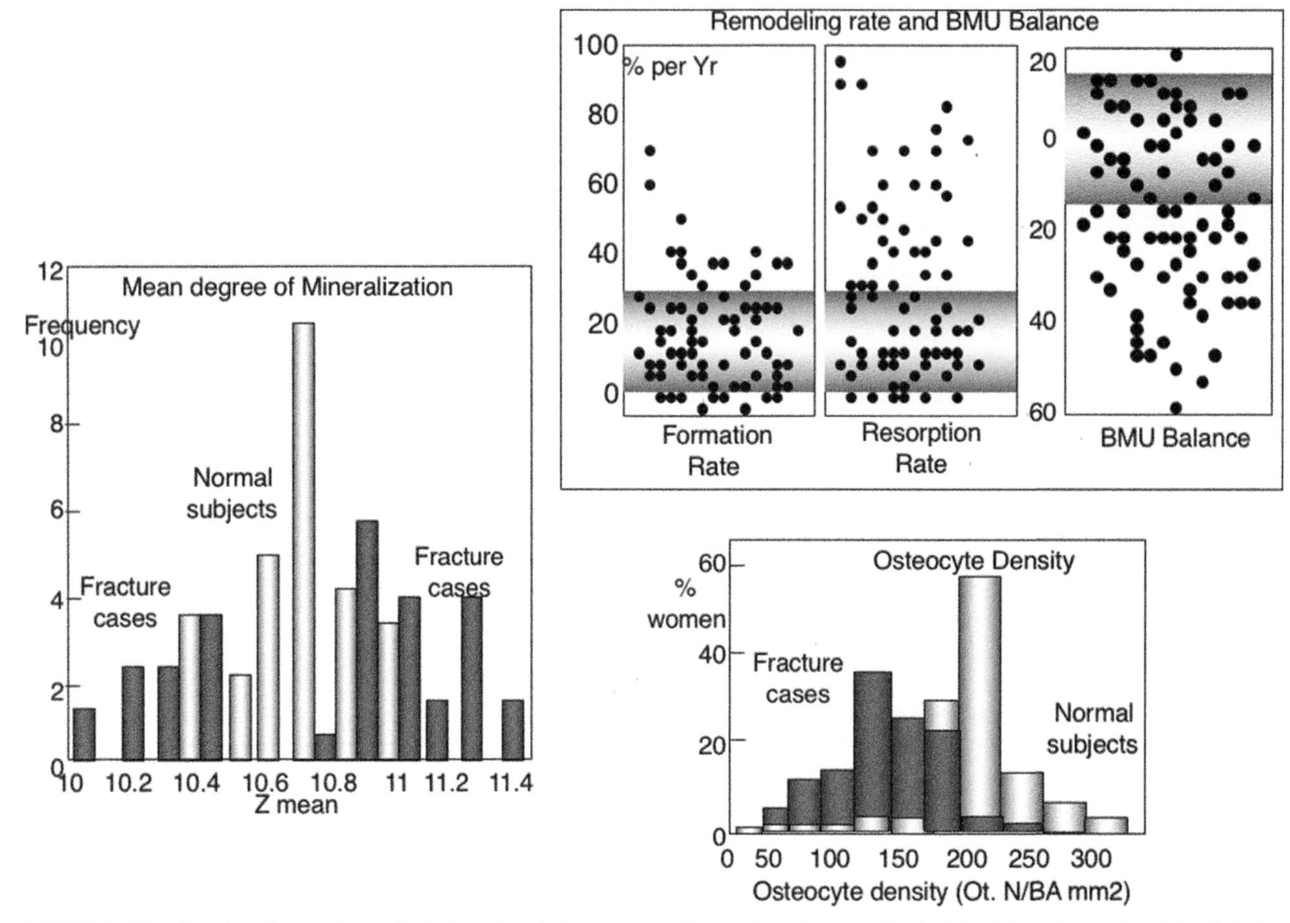

FIGURE 11.11 Bone fragility in patients with fractures has a heterogeneous pathogenesis and structural basis. Patients have tissue mineral density in the upper or lower part of the normal distribution. Some have reduced or normal osteocyte density. Formation and resorption rates may be lower normal or high; bone balance in the basic multicellular level (*BMU*) may be normal or negative. *Adapted from Ciarelli, T.E., Fyhrie, D.P., Parfitt, A.M., 2003. Effects of vertebral bone fragility and bone formation rate on the mineralization levels of cancellous bone from white females. Bone 32, 311–315. Qui, S., Rao, R.D., Saroj, I., Sudhaker, I., Palnitkar, S., Parfitt, A.M., 2003. Reduced iliac cancellous osteocyte density in patients with osteoporotic vertebral fracture. J. Bone Miner. Res. 18, 1657–1663. Eriksen et al. (1990).*

same in a treated and a placebo group. The reason BMD increases in the treated group is that fewer cavities are being excavated than are being incompletely refilled. BMD decreases in the control group because similar numbers of cavities are being excavated and incompletely refilled. In the treated group there is no change in the external or internal dimensions of bone; the periosteal perimeter does not increase, as occurs during growth or anabolic therapy; the endocortical perimeter does not decrease, as occurs during endocortical apposition using anabolic therapy; and trabeculae do not thicken.

There may be focal trabecular thickening at the remodeled location as incomplete refilling of the resorption cavity occurs due to the reversible component of the deficit. The irreversible deficit is not corrected, so that there is no focal restitution of the dimensions of a trabecular plate or cortical thickness to the level before menopause, because the remodeling imbalance is not corrected. Thickening above the premenopausal level would require overfilling of the resorption cavity. As antiresorptives may reduce the size of the resorption cavity, it is plausible that refilling or overfilling might occur (Allen et al., 2010). Even so, antiresorptives cannot reconstruct the skeleton because the remodeling rate is suppressed. Only about 5%–7% of the skeleton is remodeled annually.

The trabecular and cortical "thickening" often reported using imaging methods is an artifact produced by an increase in matrix mineral density (part of the reversible deficit in mineral), which leads to edge detection as photons are attenuated by the same deficit matrix that is now more densely mineralized (Seeman, 2010). Nevertheless, incomplete refilling of cavities may reduce stress risers and this may partly account for the rapid reduction in vertebral fracture risk achieved when remodeling is suppressed by antiresorptive agents.

As steady-state remodeling is restored at the new slower rate determined by the drug's antiresorptive efficacy, bone loss continues because the unsuppressed unbalanced remodeling continues to deteriorate the bone. Total bone matrix volume decreases despite treatment. Moreover, matrix mineral density of the declining matrix volume increases due to continued secondary mineralization, which is part of the *reversible* deficit in *mineral*. So BMD continues to increase, obscuring the decrease in total bone matrix volume, and this increase may be mistaken as an increase in bone matrix mass or volume and a reversal of microstructural deterioration (Fig. 11.12).

The only antiresorptive that virtually abolishes bone loss is denosumab, because it is widely distributed and reduces osteoclast synthesis from its precursors and reduces the life span of osteoclasts existing at the time of treatment. Denosumab profoundly suppresses serum C-terminal telopeptide of type I collagen (CTX), so there is almost complete separation of the frequency distribution curves of serum CTX in treated and untreated subjects (Zebaze et al., 2014). This is not the case when alendronate is administered. About half of the women receiving alendronate had serum CTX no different from that of untreated controls, probably because of continued cortical bone loss despite alendronate treatment. Bisphosphonates bind avidly to mineral and fail to penetrate deep peri-Haversian cortical matrix (Smith, 2003). The concentrations of bisphosphonates are lower in cortical than in trabecular bone. When an osteoclast imbibes deeper cortical matrix its resorptive activity is not inhibited. In studies of nonhuman primates, ibandronate reduces remodeling upon endocortical and trabecular, but not Haversian canal, surfaces and improves trabecular, not cortical, bone strength (Smith et al., 2003).

As reported in human subjects and nonhuman primates, upon the return of steady state at a slow remodeling rate in the second 6 months of therapy, continued intracortical remodeling with alendronate probably accounts for cortical porosity no longer being lower than in controls by 12 months (Zebaze et al., 2014). With weak antiresorptive agents like calcium supplements or selective estrogen receptor modulators, the rate of remodeling is only modestly reduced relative to pretreatment, perhaps by 20%–30%; so after steady state is restored from its initial perturbed state, bone loss and microstructural deterioration continue, but only slightly more slowly than prior therapy (Reid et al., 2006; Silverman et al., 2012).

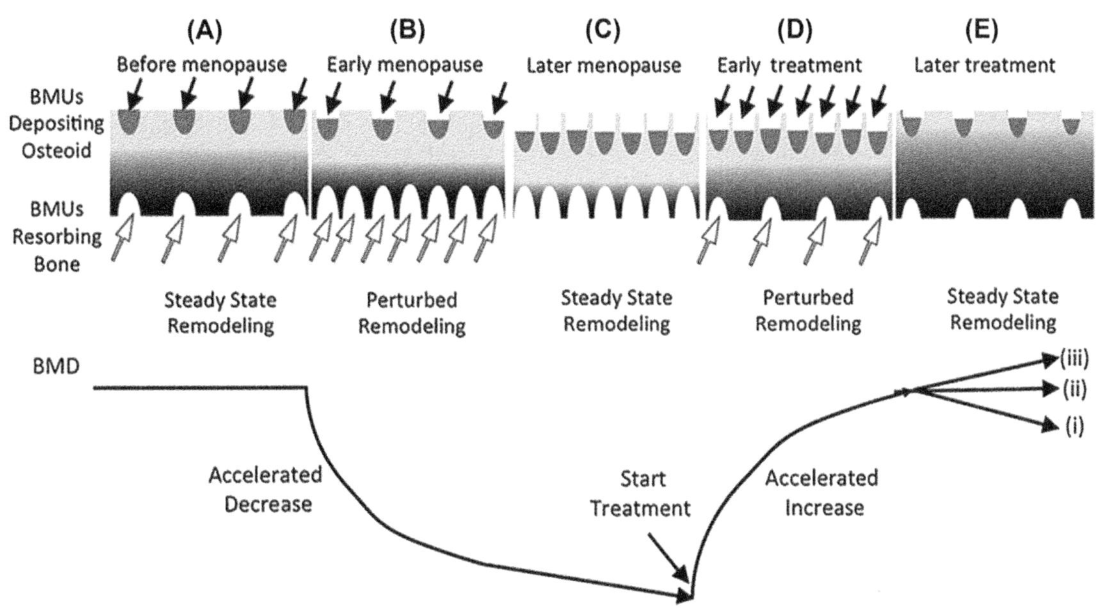

FIGURE 11.12 (A) Before menopause, remodeling is slow and in steady state. Similar numbers of sites are excavated (*white arrows*) and completely refilled (*black arrows*). No net bone loss occurs. (B) During early menopause, surface level remodeling is perturbed as more basic multicellular units (BMUs) resorb bone and each resorbs more bone (*white arrows*), while concurrently, the fewer cavities excavated before menopause now refill but do so incompletely (*black arrows*). Bone mineral density (*BMD*) decreases rapidly (bottom). (C) During later menopause, surface level remodeling returns to steady state but at a higher rate. The number of BMUs excavating bone approximately equals the number excavated in early menopause, only now refilling but doing so incompletely. Bone loss continues but more slowly. (D) During early antiresorptive therapy remodeling now becomes perturbed at the surface level but in a reciprocal fashion to early menopause. Fewer BMUs excavate smaller cavities (*white arrows*), while the many more BMUs excavating in later menopause refill incompletely (*black arrows*). BMD increases rapidly. (E) During later antiresorptive therapy, remodeling returns to steady state at the surface level at a slower rate, much like premenopause, but remodeling is unbalanced. The fewer and smaller cavities excavated during early antiresorptive treatment refill incompletely as similarly few new BMUs excavate smaller cavities. (i) If the negative BMU balance remains, BMD slowly declines from its higher level. (ii) If BMU balance is restored, there is no change in BMD. (iii) If remodeling remains unbalanced, secondary mineralization increases BMD, obscuring continued structural deterioration and the slow decrease in bone matrix volume.

With protracted remodeling suppression beyond the first 3 years, there is continued slow increase in BMD due to slow completion of secondary mineralization, which may take years (Akkus, 2004). Increases in matrix mineral density should become asymptotic with full mineralization of matrix after about 3–5 years of treatment (Dempster et al., 2018). During denosumab therapy there is continued increase in BMD reported after 5 years of treatment. This is not an anabolic effect of this antiresorptive agent. Studies of nonhuman primates suggest that age-related bone modeling upon the endocortical surface remains obscured by rapid remodeling prior to treatment and becomes detectable when remodeling is markedly suppressed by denosumab (Ominsky et al., 2015). Whether this occurs during treatment in human subjects remains uncertain.

Anabolic therapy: restoring the irreversible deficit in mineralized matrix volume and microstructural deterioration by remodeling- and modeling-based bone formation

Thus, the reversible deficit in matrix produced by the slowness of its deposition (~3 months) and the slowness of matrix mineralization (taking ~2 years), both respond to antiresorptive therapy. Slowing remodeling allows completion of the deposition of matrix and its complete mineralization. The irreversible deficit is not responsive to antiresorptive therapy. Reconstruction of the skeleton, "curing" bone fragility, requires anabolic therapy. Anabolic agents produce modeling- or remodeling-based bone formation.

Modeling-based bone formation occurs upon quiescent bone surfaces. Remodeling-based bone formation occurs within BMUs present at the time of treatment or by the initiation of new BMUs. Of necessity, the latter is preceded by the initial resorptive phase of remodeling removing mineralized bone by excavating a cavity within cortical bone or upon trabecular surfaces. Refilling of the cavity may correct the remodeling imbalance. If achieved, this will allow continued remodeling without causing microstructural deterioration. This will not reconstruct the skeleton.

Overfilling of existing and newly created cavities is needed to begin reconstruction of the skeleton, but this will result in focal reconstruction only. This approach is limited in scope because, at best, only 10%–20% of the skeleton is remodeled annually. Thus, reconstruction of an already deteriorated microstructure is a formidable challenge and probably requires sustained modeling-based bone formation.

Modeling upon the periosteal surface increases bone's total cross-sectional area, modeling upon the intracortical surface reduces intracortical porosity, modeling upon the endocortical surface thickens the cortex and increases its CSA. Modeling upon trabecular surfaces thickens them and may improve their connectivity, provided some connectivity is preserved at the time of treatment. Modeling upon surfaces of trabeculae abutting the cortex may "corticalize" them, thickening the cortex in a way that is the opposite of cortical trabecularization and cortical thinning from "within" the cortex during aging. Corticalization of trabeculae is similar to the formation of the metaphyseal cortex, which is the result of the coalescence of trabeculae emerging from the periphery of the growth plate (Cadet et al., 2003).

If successful, modeling produces an absolute increase in mineralized bone matrix volume, an increase in periosteal perimeter, a decrease in endocortical perimeter and medullary area, cortical thickening, a reduction in cortical porosity, and an increase in trabecular density. This differs from the increase in mineralized matrix density of a slowly diminishing total bone matrix volume produced by remodeling suppression using antiresorptive agents: both treatments increase BMD, but the morphological basis of the increase in BMD is of course very different and is likely to have different effects on bone strength.

About 80% of the total osteoid formed by parathyroid hormone (PTH) (1–34)- or PTH(1–84)-mediated bone formation is remodeling based. This is likely to be the same using abaloparatide, a peptide acting on the PTHR1 receptor that shares some of its amino acid sequence with PTH(1–34) and PTH-related protein (PTHrP) (Miller et al., 2016; Martin and Seeman, 2017). The proportion of the remodeling-based bone formation derived from anabolic agents acting on existing BMUs and newly generated BMUs resulting from PTH(1–34) treatment is uncertain, but the more rapid the remodeling, the greater the surface extent of remodeling, and the larger the number of existing BMUs at various stages of remodeling available to PTH.

During the resorptive phase of a remodeling cycle, PTH is likely to promote osteocyte and osteoblast precursor production of RANKL, osteoclastogenesis, bone resorption, and an increase in cortical porosity, until production of local factors from osteoclasts and the matrix they resorb influences bone formation. PTH acting on BMUs in their reversal phase may promote differentiation of osteoblast lineage cells into mature osteoid-producing forms. PTH acting on BMUs in their formation phase is likely to increase matrix production by inhibiting osteoblast apoptosis. Of necessity, BMU-based bone formation must follow resorption; the earlier increase in resorption markers is not a signal of the closure of a mythological "anabolic window." Excavated cavities upon endocortical or trabecular surfaces may refill or overfill, the latter thickening cortices and trabeculae focally.

Periosteal apposition is modest during adulthood and is advantageous because of the disproportionate increase in bending strength achieved by a small increment in bone diameter. However, there is no evidence that intermittent PTH produces measurable changes in periosteal circumference, even though there is evidence of increased apposition using quadruple labeling procedures (Lindsay et al., 2006).

Modeling-based bone formation upon intracortical canal surfaces may reduce canal diameter focally. However, many studies suggest that cortical porosity increases during early intermittent PTH therapy. Even though this increase in porosity is likely to be transient, bone fragility may increase transiently. Whether this occurs is not known. Trabecular thickening is likely to increase connectivity, provided some connectivity is present, but thickening may remain undetected by noninvasive imaging methods (that depend on photon attenuation by mineral) because replacement of older mineralized matrix with newly synthesized osteoid transmits rather than attenuates photons.

Abaloparatide, a peptide with partial amino acid homology to PTH(1−34) and PTHrP, has been reported to reduce the incidence of vertebral and nonvertebral fractures (Miller et al., 2016). There is no evidence quantifying the proportions of any anabolic effect due to existing BMUs at the time of treatment, newly generated BMUs in response to treatment, or modeling-based bone formation. A claim is made that the anabolic effect occurs with relatively less resorptive effect than observed using PTH(1−34). This is partly based on finding less of an increase in serum CTX relative to the increase in procollagen type 1 N-terminal propeptide (P1NP) in clinical trials in human subjects. The higher ratio of P1NP to CTX is inferred to be a surrogate of greater net bone formation relative to resorption, which in turn is responsible for the 1%−2% greater gain in BMD with abaloparatide than with PTH(1−34). These inferences are problematic for several reasons that are discussed elsewhere (Seeman, 2016; Martin and Seeman, 2017).

About 80% of the intracortical, endocortical, and trabecular components of the endosteal surface are quiescent and provide a vast surface area upon which osteoid can be deposited. Sclerostin antibodies (Scl-Abs) like romosozumab block the action of sclerostin, an inhibitor of bone formation. The anabolic effect is largely modeling based and is accompanied by a transient reduction in circulating serum CTX, suggesting that there is also a reduction in bone resorption. Ominsky et al. (2017) quantified the bone modeling and remodeling activity on bone surfaces in OVX rats administered vehicle or Scl-Ab (25 mg/kg) twice a week for 5 weeks, and in adolescent cynomolgus monkeys administered vehicle or Scl-Ab (30 mg/kg) subcutaneously every 2 weeks for 10 weeks. In OVX rats, Scl-Ab increased modeling eightfold from 7% to 63% of the bone surface. In cynomolgus monkeys, Scl-Ab increased modeling-based formation on trabeculae from 0.6% to 34% and on the endocortical surface from 7% to 77%. Scl-Ab did not increase remodeling-based bone formation despite decreased resorption surface in both species.

These observations have been confirmed by unpublished data reported by Chavassieux et al. (JBMR, 2019). The effects of 210 mg romosozumab on iliac crest histomorphometry at 2 and 12 months in postmenopausal women with osteoporosis were reported. For histomorphometry, 29 women had quadruple labeling at 2 m and 70 double labeling at 12 m. For microcomputed tomography (microCT), 28 women were evaluable at 2 m, and 71 at 12 m. At 2 months, bone formation rate increased in cancellous bone by 328% and by 233% upon endocortical surfaces. At 2 and 12 months, eroded surface decreased in cancellous and endocortical bone. At 12 months, wall thickness, bone mass, and trabecular and cortical thickness increased with romosozumab. At 2 months, microCT showed a decrease in trabecular separation; at 12 months, trabecular BMD, matrix density, and bone volume increased. There is robust evidence of vertebral and nonvertebral fracture risk reduction that appears to be more effective than that obtained with antiresorptive therapy or PTH(1−34) (Saag et al., 2017; Kendler et al., 2018). Details of the anti-fracture efficacy are presented elsewhere (Reid, 2015; Ramchand and Seeman, 2018).

Combined antiresorptive and anabolic therapy

Combining antiresorptive and anabolic therapy makes sense. The justification for any additional cost and exposure to the side effects of two drugs is evidence of greater fracture risk reduction than achieved by either drug alone. A prospective randomized blinded placebo-controlled trial is a formidable challenge to execute and has never been done. A lower level of evidence comes from animal experiments demonstrating that combined therapy increases the breaking strength of bone ex vivo more greatly than either drug alone. Only one study comparing PTH(1−34)/alendronate and PTH(1−34)/osteoprotegerin (OPG) versus PTH(1−34) alone has been reported. While trabecular bone volume increased more greatly than PTH(1−34) alone, bone strength assessed ex vivo in this study of rodents was not greater than that produced by PTH(1−34) alone (Samadfam et al., 2007).

Short of evidence of improved bone strength is morphological data showing that combined therapy produces a greater net increase in the volumes of bone deposited upon the periosteal surface and the three (intracortical, endocortical, and trabecular) components of the endosteal surface than therapy with either drug alone. Combined therapy has not been reported to produce a greater periosteal perimeter, a smaller endocortical perimeter, increased cortical thickness and area, lower cortical porosity, or thicker, more connected, or greater numbers of trabeculae than therapy with either drug alone.

Most comparator studies have been done using changes in BMD, bone remodeling markers, or bone microstructure as the outcome. This approach is fraught with challenges because a change in BMD may be the result of a change in bone matrix volume, its mineral content, or both, and often, these traits change in the opposite direction, even with single therapy. For example, antiresorptives slow unbalanced remodeling; matrix volume continues to decrease and microstructure continues to deteriorate, but matrix mineral density increases, producing a net increase in BMD of the deteriorating structure. PTH(1–34) increases bone matrix volume but decreases its matrix mineral density as remodeling-based bone formation replaces older more mineralized bone with younger less mineralized bone. BMD has been reported variously to increase, decrease, or remain unchanged. Thus, the behavior of these traits even during single-drug therapy makes inferences about the effects on bone strength challenging, but more so when these drugs are combined.

Indeed, combined therapy has been reported to blunt the BMD response to PTH in some, but not all, studies. Blunting is held to be due to antiresorptive treatment suppressing remodeling so that remodeling-based bone formation by PTH(1–34) is prevented. Scrutiny of the data does not support this notion. If blunting of the BMD response was due to fewer BMUs, then blunting should be *more* severe with coadministration of PTH(1–34) with zoledronate, denosumab, or OPG than with alendronate. The opposite is reported (Tsai et al., 2015), and many studies report additive effects (Seeman and Martin, 2015). Blunting is not *greater* with denosumab/PTH(1–34) than with alendronate/PTH, even though denosumab suppresses remodeling more greatly than alendronate. Additive effects on BMD are reported with PTH/denosumab relative to PTH(1–34) alone (Tsai et al., 2015; Leder et al., 2014) and comparing PTH/OPG versus PTH (Kostenuik et al., 2001).

The difficulties using bone densitometry are only partly overcome by independently measuring changes in matrix mineral density and microstructure, because image acquisition and analysis using high-resolution peripheral computed tomography also depends on photon attenuation by a region's mineral content. Challenges in the interpretation of the effects of combined therapy are illustrated in the study by (Tsai et al., 2015; Leder et al., 2014).

These investigators report that combined therapy increased cortical vBMD. However, PTH reduced cortical vBMD, while denosumab had no detectable effect, leaving unexplained the increased cortical vBMD using combined therapy. Combined therapy increased cortical matrix mineral density, yet PTH decreased it, and denosumab had no detectable effect, again leaving unexplained the increased cortical matrix density using combined therapy. Combined therapy had no effect on porosity, yet PTH increased it, while denosumab had no detectable effect, leaving the combined effect unexplained. Finally, combined therapy increased cortical thickness, PTH had no detectable effect, but denosumab increased it; the opposite of what was expected.

The findings reasonably to be expected are as follows: for cortical vBMD combined therapy should have no net effect, because PTH increases porosity and reduces matrix mineral density; denosumab does the reverse. Combined therapy therefore should also have no net effect on cortical porosity or matrix mineral density, but should increase cortical thickness. Failure to detect an increase in cortical thickness is probably the result of replacement of mineralized bone with younger bone. Finding an increase in cortical thickness with denosumab is likely to be due to an edge detection error, as antiresorptives increase matrix mineral density, they do not thicken cortices. Most pores are <100 μm. At a resolution of 130 μm, voxels containing a pore or part of a pore also contain matrix and so attenuate photons above the threshold designated as "porosity."

Conclusion

Deterioration of the cellular machinery of modeling and remodeling compromises the many qualities of bone conferred by its material composition and structural design. For the past 60 years, the bone strength conferred by these qualities and the bone fragility resulting from their deterioration have been inferred using BMD. This was a good beginning because it provided a quantitative measure of fracture risk, but BMD is insensitive; most fractures occur in individuals with modest deficits in BMD. These individuals have fragile bones caused by microstructural deterioration. BMD does not capture this fragility because the loss of strength produced by microstructural deterioration is disproportionate to the bone loss producing it and the modest reductions in BMD. There is progress in image acquisition and quantification of microstructure, but whether measurement of microstructural deterioration improves detection of individuals at imminent risk for a first fracture and whether treatment allocation based on the severity of microstructural deterioration reduces the number of persons that need to be treated to avert one event remain unknown. Answers to these questions are needed because longevity and the burden of fractures are increasing. Antiresorptive agents neither reverse existing microstructural deterioration nor abolish its progression. These drugs only slow continued microstructural deterioration and reduce vertebral and hip fracture risk by ~50%. For nonvertebral fractures, 80% of all fractures, the risk reduction is only 20%–30%. Teriparatide, the only anabolic agent widely available for clinical use, produces predominantly remodeling-based bone formation and transitory cortical porosity but may have better vertebral and nonvertebral anti-fracture efficacy than antiresorptives. The recently marketed abaloparatide reduces vertebral and nonvertebral fractures and is also likely to produce remodeling-based bone formation. Neither drug reduces the risk of hip fracture, the most devastating fracture. Modeling-based bone formation using romosozumab, an Scl-Ab, has antivertebral and nonvertebral fracture efficacy within 12 months. Whether anabolic agents in combination with antiresorptive agents offer better anti-fracture efficacy than either drug alone is not known. The challenge of reducing the burden of fractures remains an unmet need.

References

Aaron, J.E., Makins, N.B., Sagreiy, K., 1987. The microanatomy of trabecular bone loss in normal aging men and women. Clin. Orthop. Relat. Res. 215, 260–271.

Aguirre, J.I., Plotkin, L.I., Stewart, S.A., Weinstein, R.S., Parfitt, A.M., Manolagas, S.C., Bellido, T., 2006. Osteocyte apoptosis is induced by weightlessness in mice and precedes osteoclast recruitment and bone loss. J. Bone Miner. Res. 21, 605–615.

Ahlborg, H.G., Johnell, O., Turner, C.H., Rannevik, G., Karlsson, M.K., 2003. Bone loss and bone size after the menopause. N. Engl. J. Med. 349, 327–334.

Akkus, O., Polyakova-Akkus, A., Adar, F., Schaffler, M.B., 2003. Aging of microstructural compartments in human compact bone. J. Bone Miner. Res. 18, 1012–1019.

Akkus, O., Adar, F., Schaffler, M.B., 2004. Age-related changes in physicochemical properties of mineral crystals are related to impaired mechanical function of cortical bone. Bone 34, 443–453.

Albright, F., Smith, P.H., Richardson, A.M., 1941. Postmenopausal osteoporosis. J. Am. Med. Assoc. 116, 2465–2474.

Allen, M.R., Erickson, A.M., Wang, X., Burr, D.B., Martin, R.B., Hazelwood, S.J., 2010. Morphological assessment of basicmulticellular unit resorption parameters in dogs shows additional mechanisms of bisphosphonate effects on bone. Calcif. Tissue Int. 86 (1), 67–71.

Arlot, M.E., Delmas, P.D., Chappard, D., Meunier, P.J., 1990. Trabecular and endocortical bone remodelling in postmenopausal osteoporosis: comparison with normal postmenopausal women. Osteoporos. Int. 1, 41–49.

Bailey, A.J., Sims, T.J., Ebbesen, E.N., Mansell, J.P., Thomsen, J.S., Mosekilde, L., 1999. Age-related changes in the biochemical properties of human cancellous bone collagen: relationship to bone strength. Calcif. Tissue Int. 65, 203–210.

Bakwin, H., 1964. Secular increase in height: is the end in sight? Lancet 2, 1195–1196.

Balena, R., Shih, M.-S., Parfitt, 1992. Bone resorption and formation on the periosteal envelope of the ilium: a histomorphometric study in healthy women. J. Bone Miner. Res. 7, 1475–1482.

Banse, X., Sims, T.J., Bailey, A.J., 2002. Mechanical properties of adult vertebral cancellous bone: correlation with collagen intermolecular cross-links. J. Bone Miner. Res. 17, 1621–1628.

Baron, R., Tross, R., Vignery, A., 1984. Evidence of sequential remodeling in rat trabecular bone: morphology, dynamic histomorphometry, and changes during skeletal maturation. Anat. Rec. 208 (1), 137–145.

Bass, S., Delmas, P.D., Pearce, G., Hendrich, E., Tabensky, A., Seeman, E., 1999. The differing tempo of growth in bone size, mass and density in girls is region-specific. J. Clin. Investig. 104, 795–804.

Bass, S.L., Saxon, L., Daly, R., Turner, C.H., Robling, A.G., Seeman, E., 2002. The effect of mechanical loading on the size and shape of bone in pre-, peri- and post-pubertal girls: a study in tennis players. J. Bone Miner. Res. 17 (12), 2274–2280.

Bell, K.L., Loveridge, N., Power, J., Garrahan, N., Meggitt, B.F., Reeve, J., 1999. Regional differences in cortical porosity in the fractured femoral neck. Bone 24, 57–64.

Bjørnerem, Å., Wang, X., Bui, M., Ghasem-Zadeh, A., Hopper, J.L., Zebaze, R., Seeman, E., 2018. Menopause-Related Appendicular Bone Loss is Mainly Cortical and Results in Increased Cortical Porosity. J. Bone Miner. Res. 33 (4), 598–605.

Blizoites, M., Sibonga, J.D., Turner, R.T., Orwoll, E., 2006. Periosteal remodeling at the femoral neck in nonhuman primates. J. Bone Miner. Res. 21, 1060–1067.

Boivin, G., Lips, P., Ott, S.M., Harper, K.D., Sarkar, S., Pinette, K.V., Meunier, P.J., 2003. Contribution of raloxifene and calcium and vitamin D supplementation to the increase of the degree of mineralization of bone in postmenopausal women. J. Clin. Endocrinol. Metab. 88, 4199–4205.

Boivin, G., Meunier, P.J., 2002. Changes in bone remodeling rate influence the degree of mineralization of bone. Connect. Tissue Res. 43, 535–537.

Bolotin, H.H., Sievänen, H., 2001. Inaccuracies inherent in dual-energy x-ray absorptiometry in vivo bone mineral density can seriously mislead diagnostic/prognostic interpretations of patient –specific bone fragility. J. Bone Miner. Res. 16, 799–805.

Bouxsein, M.L., Melton 3rd, L.J., Riggs, B.L., Muller, J., Atkinson, E.J., Oberg, A.L., Robb, R.A., Camp, J.J., Rouleau, P.A., McCollough, C.H., Khosla, S., 2006. Age- and sex-specific differences in the factor of risk for vertebral fracture: a population-based study using QCT. J. Bone Miner. Res. 21 (9), 1475–1482.

Brockstedt, H., Kassem, M., Eriksen, E.F., Mosekilde, L., Melsen, F., 1993. Age- and sex-related changes in iliac cortical bone mass and remodeling. Bone 14 (4), 681–691.

Brown, J.P., Delmas, P.D., Arlot, M., Meunier, P.J., 1987. Active bone turnover of the cortico-endosteal envelope in postmenopausal osteoporosis. J. Clin. Endocrinol. Metab. 64, 954–959.

Brown, J.P., Delmas, P.D., Malaval, L., Edouard, C., Chapuy, M.C., Meunier, P.J., 1984. Serum bone gla-protein: a specific marker for bone formation in postmenopausal osteoporosis. Lancet i, 1091–1093.

Bui, M., Bjornerem, A., Ghasem-Zadeh, A., Dite, G.S., Hopper, J.L., Seeman, E., 2013. Architecture of cortical bone determines in part its remodelling and structural decay. Bone 55 (2), 353–358.

Burr, D.B., Turner, C.H., Naick, P., Forwood, M.R., Ambrosius, W., Hasan, S., Pidaparti, R., 1998. Does microdamage accumulation affect the mechanical properties of bone? J. Biomech. 31, 337–345.

Cadet, E.R., Gafni, R.I., McCarthy, E.F., McCray, D.R., Bacher, J.D., Barnes, K.M., et al., 2003. Mechanisms responsible for longitudinal growth of the cortex: coalescence of trabecular bone into cortical bone. J. Bone Joint Surg. Am. 85-A (9), 1739–1748.

Cameron, N., Tanner, J.M., Whitehouse, R.H., 1982. A longitudinal analysis of the growth of limb segments in adolescence. Ann. Hum. Biol. 9, 211–220.

Chavassieux, P., Chapurlat, R., Portero-Muzy, N., Roux, J.P., Garcia, P., Brown, J.P., Libanati, C., Boyce, R.W., Wang, A., Grauer, A., 2019. Bone-forming and antiresorptive effects of romosozumab in postmenopausal women with osteoporosis: bone histomorphometry and microcomputed tomography analysis after 2 and 12 months of treatment. J. Bone Miner. Res. doi: 10.1002/jbmr.3735. [Epub ahead of print]

Ciarelli, T.E., Fyhrie, D.P., Parfitt, A.M., 2003. Effects of vertebral bone fragility and bone formation rate on the mineralization levels of cancellous bone from white females. Bone 32, 311–315.

Clark, W.D., Smith, E.L., Linn, K.A., Paul-Murphy, J.R., Muir, P., Cook, M.E., 2005. Osteocyte apoptosis and osteoclast presence in chicken radii 0–4 days following osteotomy. Calcif. Tissue Int. 77, 327–336.

Compston, J.E., Yamaguchi, K., Croucher, P.I., Garrahan, N.J., Lindsay, P.E., Shaw, R.W., 1995. The effects of gonadotrophin releasing hormone agonists on iliac crest cancellous bone structure in women with endometriosis. Bone 16, 261–267.

Croucher, P.I., Garrahan, N.J., Mellish, R.W.E., Compston, J.E., 1991. Age-related changes in resorption cavity characteristics in human trabecular bone. Osteoporos. Int. 1, 257–261.

Currey, J.D., 2002. Bones. Structure and Mechanics. Princeton UP, New Jersey, pp. 1–380.

Curry, J.D., 1969. Mechanical consequences of variation in the mineral content of bone. J. Biomech. 2, 1–11.

Danova, N.A., Colopy, S.A., Radtke, C.L., Kalscheur, V.L., Markel, M.D., Vanderby Jr., R., McCabe, R.P., Escarcega, A.J., Muir, P., 2003. Degradation of bone structural properties by accumulation and coalescence of microcracks. Bone 33, 197–205.

Delmas, P.D., 2000. The use of biochemical markers in the evaluation of fracture risk and treatment response. Osteoporos. Int. 11 (Suppl. 1), S5–S6.

Dempster, D.W., Brown, J.P., Fahrleitner-Pammer, A., Kendler, D., Rizzo, S., Valter, I., Wagman, R.B., Yin, X., Yue, S.V., Boivin, G., 2018. Effects of long- term denosumab on bone histomorphometry and mineralization in women with postmenopausal osteoporosis. J. Clin. Endocrinol. Metab. 103, 2498–2509.

Diab, T., Condon, K.W., Burr, D.B., Vashishth, D., 2006. Age-related change in the damage morphology of human cortical bone and its role in bone fragility. Bone 38, 427–431.

Diab, T., Vashisha, D., 2005. Effects of damage morphology on cortical bone fragility. Bone 37, 96–102.

Doherty, M.,J., Ashton, B.A., Walsh, S., Beresford, J.N., Grant, M.E., Canfield, A.E., 1998. Vascular pericytes express osteogenic potential in vitro and in vivo. J. Bone Miner. Res. 13, 828–838.

Duan, Y., Wang, X.F., Evans, A., Seeman, E., 2005. Structural and biomechanical basis of racial and sex differences in vertebral fragility in Chinese and Caucasians. Bone 36, 987–998.

Eghbali-Fatourechi, G.Z., Lamsam, J., Fraser, D., Nagel, D.A., Riggs, B.L., Khosla, S., 2005. Circulating osteoblast lineage cells in humans. N. Engl. J. Med. 352, 1959–1966.

Eghbali-Fatourechi, G.Z., Moedder, U.I., Charatcharoenwitthaya, N., Sanyal, A., Undale, A.H., Clowes, J.A., Tarara, J.E., Khosla, S., 2007. Characterization of circulating osteoblast lineage cells in humans. Bone 40, 1370–1377.

Elmardi, A.S., Katchburian, M.V., Katchburian, E., 1990. Electron microscopy of developing calvaria reveal images that suggest that osteoclasts engulf and destroy osteocytes during bone resorption. Calcif. Tissue Int. 46, 239–245.

Emaus, N., Berntsen, G.K., Joakimsen, R., Fonnebo, V., 2005. Longitudinal changes in forearm bone mineral density in women and men aged 25–44 years: the Tromso Study, a population-based study. Am. J. Epidemiol. 162, 633–643.

Emaus, N., Berntsen, G.K., Joakimsen, R., Fonnebo, V., 2006. Longitudinal changes in forearm bone mineral density in women and men aged 45–84 years: the Tromso Study, a population-based study. Am. J. Epidemiol. 163 (5), 441–449.

Ericksen, E.F., 1986. Normal and pathological remodeling of human trabecular bone: three dimensional reconstruction of the remodelling sequence in normals and in metabolic disease. Endocr. Rev. 4, 379–408.

Eriksen, E.F., Hodgson, S.F., Eastell, R., Cedel, S.L., O'Fallon, W.M., Riggs, B.L., 1990. Cancellous bone remodeling in type I (postmenopausal) osteoporosis: quantitative assessment of rates of formation, resorption, and bone loss at tissue and cellular levels. J. Bone Miner. Res. 5, 311–319.

Eriksen, E.F., Langdahl, B., Vesterby, A., Rungby, J., Kassem, M., 1999. Hormone replacement therapy prevents osteoclastic hyperactivity: a histomorphometric study in early postmenopausal women. J. Bone Miner. Res. 14, 1217–1221.

Fantner, G., Hassen kam, T., Kindt, J.H., Weaver, J.C., Birkedal, H., Pechenik, L., Cutroni, J.A., Cidade, G.C., Stucky, G.D., Morse, D.E., Hansma, P.K., 2005. Nat. Mater. 4, 612–616.

Foldes, J., Parfitt, A.M., Shih, M.-S., Rao, D.S., Kleerekoper, M., 1991. Structural and geometric changes in iliac bone: relationship to normal aging and osteoporosis. J. Bone Miner. Res. 6, 759–766.

Fujikawa, Y., Quinn, J.M., Sabokbar, A., McGee, J.O., Athanasou, N.A., 1996. The human osteoclast precursor circulates in the monocyte fraction. Endocrinology 137, 4058–4060.

Fuller, K., Chambers, T.J., 1995. Localisation of mRNA for collagenase in osteocytic, bone surface and chondrocytic cells but not osteoclasts. J. Cell Sci. 106, 2221–2230.

Garn, S., 1970. The Earlier Gain and Later Loss of Cortical Bone. Nutritional Perspectives. Charles C. Thomas, Springfield, IL, pp. 3–120.

Garn, S.M., Sullivan, T.V., Decker, S.A., Larkin, F.A., Hawthorne, V.M., 1992. Continuing bone expansion and increasing bone loss over a two-decade period in men and women from a total community sample. Am. J. Hum. Biol. 4 (1), 57–67.

Garnero, P., Borel, O., Gineyts, E., Duboeuf, F., Solberg, H., Bouxsein, M.L., Christiansen, C., Delmas, P.D., 2006. Extracellular post-translational modifications of collagen are major determinants of biomechanical properties of fetal bovine cortical bone. Bone 38, 300–309.

Ghasem-Zadeh, A., Burghardt, A., Wang, X.F., Iuliano, S., Bonaretti, S., Bui, M., et al., 2017. Quantifying sex, race, and age specific differences in bone microstructure requires measurement of anatomically equivalent regions. Bone 101, 206–213.

Gilsanz, V., Gibbens, D.T., Carlson, M., Boechat, I., Cann, C.E., Schulz, E.S., 1987. Peak trabecular bone density: a comparison of adolescent and adult. Calcif. Tissue Int. 43, 260–262.

Gilsanz, V., Gibbens, D.T., Roe, T.F., Carlson, M., Senac, M.O., 1988. Vertebral bone density in children: effect of puberty. Radiology 166, 847–850.

Gilsanz, V., Roe, T.F., Stefano, M., Costen, G., Goodman, W.G., 1991. Changes in vertebral bone density in black girls and white girls during childhood and puberty. N. Engl. J. Med. 325, 1597–1600.

Gupta, H.S., Seto, J., Wagermier, W., Zaslansky, P., Boesecke, P., Fratzl, P., 2006. Cooperative deformation of mineral and collagen in bone at the nanoscale. Proc. Natl. Acad. Sci. U. S. A. 103 (47), 17741–17746.

Gupta, S.H., Wagermaier, W., Zickler, G.A., Raz-Ben Aroush, D., Funari, S.S., Roschger, P., Wagner, H.D., Fratzl, P., 2005. Nanoscale Deformation Mechanisms in Bone. Nano Lett. 5 (10), 2108–2111.

Haapasalo, H., Kontulainen, S., Sievanen, H., Kannus, P., Jarvinen, M., Vuori, I., 2000. Exercise-induced bone gain is due to enlargement in bone size without a change in volumetric bone density: a peripheral quantitative computed tomography study of the upper arms of male tennis players. Bone 27 (3), 351–357.

Han, Y., Cowin, S.C., Schaffler, M.B., Weinbaum, S., 2004. Mechanotransduction and strain amplification in osteocyte cell processes. Proc. Natl. Acad. Sci. U. S. A. 101 (47), 16689–16694.

Han, Z.-H., Palnitkar, S., Rao, D.S., Nelson, D., Parfitt, A.M., 1996. Effect of ethnicity and age or menopause on the structure and geometry of iliac bone. J. Bone Miner. Res. 11, 1967–1975.

Hattner, R., Epker, B.N., Frost, H.M., 1965. Suggested sequential mode of control of changes in cell behaviour in adult bone remodelling. Nature 963, 489–490.

Hauge, E.M., Qvesel, D., Eriksen, E.F., Mosekilde, I., Melsen, F., 2001. Cancellous bone remodelling occurs in specialized compartments lined by cells expressing osteoblastic markers. J. Bone Miner. Res. 16, 1575–1582.

Hazenberg, J.G., Freeley, M., Foran, E., Lee, T.C., Taylor, D., 2006. Microdamage: a cell transducing mechanism based on ruptured osteocyte processes. J. Biomech. 39, 2096–2103.

Hernandez, C.J., Gupt, A., Keaveny, T.M., 2006. A biomechanical analysis of the effects of resorption cavities on cancellous bone strength. J. Bone Miner. Res. 21, 1248–1255.

Howson, K.M., Aplin, A.C., Gelati, M., Alessandri, E.A., Nicosia, R.F., 2005. The postnatal rat aorta contains pericyte progenitor cells that form spheroidal colonies in suspension culture. Am. J. Cell Physiol. 289, 1396–1407.

Iuliano-Burns, S., Hopper, J., Seeman, E., 2008. Sexual Dimorphism in Bone Structure Is Present before Puberty: A Male:female Co-twin Study (Submitted for publication).

Kalender, W.A., Felsenberg, D., Louis, O., Lopez, O., Lopez, P., Klotz, E., Osteaux, M., Fraga, J., 1989. Reference values for trabecular and cortical vertebral bone density in single and dual-energy quantitative computed tomography. Eur. J. Radiol. 9, 75–80.

Kanis, J.A., Borgstrom, F., Zethraeus, Z., Johmell, O., Oden, A., Jonsson, B., 2005. Intervention thresholds for osteoporosis in men and women. Bone 36, 22–32.

Kanis, J.A., Johnell, O., Oden, A., Dawson, A., De laet, C., Jonsson, B., 2001. Ten year probabilities of osteoporotic fractures according to BMD and diagnostic thresholds. Osteoporos. Int. 12, 989–995.

Keaveney, T.M., 1998. Cancellous bone. In: Black, J., Hastings, G. (Eds.), Handbook of Biomaterials Properties. Chapman and Hall, London.

Keller, H., Kneissel, M., 2005. SOST is a target gene for PTH in bone. Bone 37, 148–158.

Kendler, D.L., Marin, F., Zerbini, C.A.F., Russo, L.A., Greenspan, S.L., Zikan, V., Bagur, A., Malouf-Sierra, J., Lakatos, P., Fahreieitner-Pammer, A., et al., 2018. Effects of teriparatide and risedronate on new fractures in post- menopausal women with severe osteoporosis (VERO): a multicentre, double- blind, double- dummy, randomised controlled trial. Lancet 391, 230–240.

Kholsa, S., Westendorf, J.J., Oursler, M.J., 2008. Building bone to reverse osteoporosis and repair fractures. J. Clin. Investig. 118 (2), 421–428.

Khosla, S., 2008. Building bone to rcverse osteoporosis and repair fractures. J. Clin. Investig. 118 (2), 421–428.

Kostenuik, P.J., Capparelli, C., Morony, S., Adamu, S., Shimamoto, G., Shen, V., Lacey, D.L., Dunstan, C.R., 2001. OPG and PTH(1-34) have additive effects on bone density and mechanical strength in osteopenic ovariectomized rats. Endocrinology 142, 4295–4304.

Kurata, K., Heino, T.J., Higaki, H., Väänänen, H.K., 2006. Bone marrow cell differentiation induced by mechanically damaged osteocytes in 3D gel-embedded culture. J. Bone Miner. Res. 21, 616–625.

Landis, W.J., 2002. The strength of a calcified tissue depends in part on the molecular structure and organization of its constituent mineral crystals in their organic matrix. Bone 30, 492–497.

Lane, N.E., Yao, W., Balooch, M., Nalla, R.K., Balooch, G., Habelitz, S., Kinney, J.H., Bonewald, L.F., 2006. Glucocorticoid-treated mice have localized changes in trabecular bone material properties and osteocyte lacunar size that are not observed in placebo-treated or estrogen-deficient mice. J. Bone Miner. Res. 21, 466–476.

Leder, B.Z., Tsai, J.N., Uihlein, A.V., Burnett-Bowie, S.A., Zhu, Y., Foley, K., Lee, H., Neer, R.M., 2014. Two years of denosumab and teriparatide administration in postmenopausal women with osteoporosis (The DATA Extension Study): a randomized controlled trial. J. Clin. Endocrinol Metab. 99, 1694–1700.

Lindsay, R., Cosman, F., Zhou, H., Bostrom, M.P., Shen, V.W., Cruz, J.D., Nieves, J.W., Dempster, D.W., 2006. A novel tetracycline labeling schedule for longitudinal evaluation of the short- term effects of anabolic therapy with a single iliac crest bone biopsy: early actions of teriparatide. J. Bone Miner. Res. 21, 366–373.

Lips, P., Courpron, P., Meunier, P.J., 1978. Mean wall thickness of trabecular bone packets in the human iliac crest: changes with age. Calcif. Tissue Res. 10, 13–17.

Lorenzo, J., 2000. Interactions between immune and bone cells: new insights with many remaining questions. J. Clin. Investig. 106, 749–752.

Loro, M.L., Sayre, J., Roe, T.F., Goran, M.I., Kaufman, F.R., Gilsanz, V., 2000. Early identification of children predisposed to low peak bone mass and osteoporosis later in life. J. Clin. Endocrinol. Metab. 85 (10), 3908–3918.

Malina, R.M., Brown, K.H., 1987. Relative lower extremity length in Mexican American and in American black and white youth. Am. J. Phys. Anthropol. 72, 89–94.

Manolagas, S.C., 2000. Birth and death of bone cells: basic regulatory mechanisms and implications for the pathogenesis and treatment of osteoporosis. Endocr. Rev. 21, 115–137.

Manolagas, S.C., 2006. Choreography from the tomb: an emerging role of dying osteocytes in the purposeful, and perhaps not so purposeful, targeting of bone remodeling. BoneKEy Osteovision 3 (1), 5–14.

Maresh, M.M., 1961. Bone, muscle and fat measurements. Longitudinal measurements of the bone, muscle and fat widths from roentgenograms of the extremities during the first six years of life. Pediatrics 28, 971–984.

Marotti, G., Cane, V., Palazzini, S., Palumbo, C., 1990. Structure-function relationships in the osteocyte. Ital. J. Miner. Electrolyte Metab. 4, 93–106.

Martin, R.B., 1984. Porosity and specific surface of bone. CRC Crit. Rev. Biomed. Eng. 10, 179–221.

Martin, T.J., Seeman, E., 2017. Abaloparatide is an anabolic, but does it spare resorption? J. Bone Miner. Res. 32, 11–16.

Martin, T.J., Sims, N.A., 2005. Osteoclast-derived activity in the coupling of bone formation to resorption. Trends Mol. Med. 11, 76–81.

Matsumoto, T., Kawamoto, A., Kuroda, R., Ishikawa, M., Mifune, Y., Iwasaki, H., Miwa, M., Horii, M., Hayashi, S., Oyamada, A., Nishimura, H., Murasawa, S., Doita, M., Kurosaka, M., Asahara, T., 2006. Therapeutic potential of vasculogenesis and osteogenesis promoted by peripheral blood CD34 positive cells for functional bone healing. Am. J. Pathol. 169, 1440–1457.

Meredith, H.V., 1978. Secular change in sitting height and lower limb height of children, youths, and young adults of Afro-black, European, and Japanese ancestry. Growth 42, 37–41.

Meunier, P.J., Sellami, S., Briancon, D., Edouard, C., 1990. Histological heterogeneity of apparently idiopathic osteoporosis. In: Deluca, H.F., Frost, H.M., Jee, W.S.S., Johnston, C.C., Parfitt, A.M.U.P.P. (Eds.), *Osteoporosis*. Recent Advances in Pathogenesis and Treatment, pp. 293–301. Baltimore.

Miller, P.D., Hattersley, G., Riis, B.J., Williams, G.C., Lau, E., Russo, L.A., Alexandersen, P., Zerbini, C.A.F., Hu, Ming-yi, Harris, A.G., Fitzpatrick, L.A., Cosman, F., Christiansen, C., 2016. Effect of abaloparatide versus placebo on new vertebral fractures in postmenopausal women with osteoporosis: a randomized clinical trial. JAMA 316, 722–733.

Mödder, U.I., Khosla, S., 2008. Skeletal stem cell/osteoprogenitor cells: current concepts, alternate hypotheses and relationsjip to the bone remodelling compartment. J. Cell. Biochem. 103 (2), 393–400.

Mosekilde, L., Mosekilde, L., 1990. Sex differences in age-related changes in vertebral body size, density and biochemical competence in normal individuals. Bone 11, 67–73.

Murray, P.D.F., Huxley, J.S., 1925. Self-differentiation in the grafted limb bud of the chick. J. Anat. 59, 379–384.

Nalla, R.K., Kruzic, J.J., Kinney, J.H., Ritchie, R.O., 2004. Effect of aging on the toughness of human cortical bone: evaluation by R-curves. Bone 35, 1240–1246.

Nishida, S., Endo, N., Yamagiwa, H., Tanizawa, T., Takahashi, H.E., 1999. Number of osteoprogenitor cells in human bone marrow markedly decreases after skeletal maturation. J. Bone Miner. Metab. 17, 171–177.

O'Brien, C.A., Jia, D., Plotkin, L.I., Bellido, T., Powers, C.C., Steward, S.Q., Manolagas, S.C., Weinstein, R.S., 2004. Glucocorticoids act directly on osteoblasts and osteocytes to induce their apoptosis and reduce bone formation and strength. Endocrinology 145, 1925–1941.

Ominsky, M.S., Boyce, R.W., Li, X., Ke, H.Z., 2017. Effects of sclerostin antibodies in animal models of osteoporosis. Bone 96, 63–75.

Ominsky, M.S., Libanati, C., Niu, Qing-Tian, Boyce, R.W., Kosteniol, P.J., Baron, R., Dempster, D.W., 2015. Sustained modeling-based bone formation during adulthood in cynomolgus monkeys may contribute to continuous BMD gains with denosumab. J. Bone Miner. Res. 30, 1280–1289.

Oreffo, R.O., Bord, S., Triffitt, J.T., 1998. Skeletal progenitor cells and ageing human populations. Clin. Sci. 94, 549–555.

Orwoll, E.S., 2003. Toward an expanded understanding of the role of the periosteum in skeletal health. J. Bone Miner. Res. 18, 949–954.

Otsura, S., Tamai, K., Yamazaki, T., Yoshjkawa, H., Kaneda, Y., 2007. Bone marrow-derived osteoblast progenitor cells in circulating blood contribute to ectopic bone formation in mice. Biochem. Biophys. Res. Commun. 354, 453–458.

Parfitt, A.M., 1984. Age-related structural changes in trabecular and cortical bone: cellular mechanisms and biomechanical consequences. Calcif. Tissue Int. 36, S123–S128.

Parfitt, A.M., 1996. Skeletal heterogeneity and the purposes of bone remodelling: implications for the understanding of osteoporosis. In: Marcus, R., Feldman, D., Kelsey, J. (Eds.), Osteoporosis. Academic, San Diego, CA, pp. 315–339.

Parfitt, A.A., 2001. The bone remodelling compartment: a circulatory function of bone lining cells. J. Bone Miner. Res. 16 (9), 1583–1585.

Parfitt, A.M., 2002. Targeted and non-targeted bone remodeling: relationship to basic multicellular unit origination and progression. Bone 30, 5–7.
Parfitt, A.M., Travers, R., Rauch, F., Glorieux, F.H., 2000. Structural and cellular changes during bone growth in healthy children. Bone 27, 487–494.
Pitsillides, A.A., 2006. Early effects of embryonic movement: 'a shot out of the dark'. J. Anat. 206, 417–431.
Pocock, N.A., Eisman, J.A., Hopper, J.L., Yeates, M.G., Sambrook, P.N., Eberl, S., 1987. Genetic determinants of bone mass in adults. A twin study. J. Clin. Investig. 80 (3), 706–710.
Qiu, S., Rao, D.S., Fyhrie, D.P., Palnitkar, S., Parfitt, A.M., 2005. The morphological association between microcracks and osteocyte lacunae in human cortical bone. Bone 37, 10–15.
Qui, S., Rao, R.D., Saroj, I., Sudhaker, 1., Palnitkar, S., Parfitt, A.M., 2003. Reduced iliac cancellous osteocyte density in patients with osteoporotic vertebral fracture. J. Bone Miner. Res. 18, 1657–1663.
Ramchand, S.K., Seeman, E., 2018. Advances and unmet needs in the therapeutics of bone fragily. Front Endocrinol. https://doi.org/10.3389/fendo.2018.00505.
Rauch, F., Neu, C., Manz, F., Schoenau, E., 2001. The development of metaphyseal cortex – implications for distal radius fractures during growth. J. Bone Miner. Res. 16, 1547–1555.
Reid, I.R., 2015. Short-term and long-term effects of osteoporosis therapies. Nat. Rev. Endocrinol. 11, 418–428.
Reid, I.R., Mason, B., Horne, A., Ames, R., Reid, H.E., Bava, U., Bolland, M.J., Gamble, G.D., 2006. Randomized controlled trial of calcium in healthy older women. Am. J. Med. 119, 777–785.
Riggs, B.L., Melton, L.J., Robb, R., Camp, J.J., Atkinson, E.J., McDaniel, L., Amin, S., Rouleau, P.A., Khosla, S., 2007. A population-based assessment of rates of bone loss at multiple skeletal sites: evidence for substantial trabecular bone loss in young women and men. J. Bone Miner. Res. 23, 205–214.
Riggs, B.L., Melton 3rd, L.J., Robb, R.A., Camp, J.J., Atkinson, E.J., Oberg, A.L., Rouleau, P.A., McCollough, C.H., Khosla, S., Bouxsein, M.L., 2006. Population-based analysis of the relationship of whole bone strength indices and fall-related loads to age- and sex-specific patterns of hip and wrist fractures. J. Bone Miner. Res. 21 (2), 315–323.
Riggs, B.L., Melton III., L.J., Robb, R.A., Camp, J.J., Atkinson, E.J., Peterson, J.M., Rouleau, P.A., McCollough, C.H., Bouxsein, M.L., Khosla, S., 2004. A population-based study of age and sex differences in bone volumetric density, size, geometry and structure at different skeletal sites. J. Bone Miner. Res. 19, 1945–1954.
Riggs, B.L., Wahner, H.W., Melton, L.J., Richelson, L.S., Judd, H.L., Offord, K.P., 1986. Rates of bone loss in the appendicular and axial skeletons of women: evidence of substantial vertebral bone loss before menopause. J. Clin. Investig. 77, 1487–1491.
Rodriguez, I., Palacios, J., Rodriguez, S., 1992. Transverse bone growth and cortical bone mass in the human prenatal period. Biol. Neonate 62–69.
Ruff, C.B., Hayes, W.C., 1988. Sex differences in age-related remodeling of the femur and tibia. J. Orthop. Res. 6, 886–896.
Ruppel, M.E., Burr, D.B., Miller, L.M., 2006. Chemical makeup of micro-damaged bone differs from undamaged bone. Bone 39, 318–324.
Saag, K.G., Petersen, J., Brandi, M.L., Karaplis, A.C., Lorentzon, M., Thomas, T., Maddox, J., Fan, M., Meisner, P.D., Grauer, A., 2017. Romosozumab or alendronate for fracture prevention in women with osteoporosis. N. Engl. J. Med. 377, 1417–1427.
Sacchetti, B., Funari, A., Michienzi, S., Di Cesare, S., Piersanti, S., Saggio, I., Tagliafico, E., Ferrari, S., Robey, P.G., Riminucci, M., Bianco, P., 2007. Marrow sinusoids can organise a hematopoietic microenvironment. Cell 131, 324–336.
Samadfam, R., Xia, Q., Goltzman, D., 2007. Co-treatment of PTH with osteoprotegerin or alendronate increases its anabolic effect on the skeleton of oophorectomized mice. J. Bone Miner. Res. 22, 55–63.
Schaffler, M.B., Burr, D.B., 1988. Stiffness of compact bone: effects of porosity and density. J. Biomech. 21, 13–16.
Schaffler, M.B., Majeska, R.J., 2005. Role of the osteocyte in mechanotransduction and skeletal fragility. Abst 20, p. 12. In: Proceedings of Meeting. Bone Quality: What Is It and Can We Measure It? Besthesda, Maryland May 2–3.
Seeman, E., 1997. From density to structure: growing up and growing old on the surfaces of bone. J. Bone Miner. Res. 12, 1–13.
Seeman, E., 1998. Growth in bone mass and size–are racial and gender differences in bone mineral density more apparent than real? J. Clin. Endocrinol. Metab. 83 (5), 1414–1419.
Seeman, E., 2003. Periosteal bone formation – a neglected determinant of bone strength. N. Engl. J. Med. 349, 320–323.
Seeman, E., 2010. Bone Morphology in Response to Alendronate as Seen by High-Resolution Computed Tomography: Through a Glass Darkly. J. Bone Miner. Res. 25 (12), 2277–2281.
Seeman, E., Bianchi, G., Adami, S., Kanis, J., Khosla, S., Orwoll, E., 2004. Osteoporosis in men-consensus is premature. Calcif. Tissue Int. 75, 120–122.
Seeman, E., Duan, Y., Fong, C., Edmonds, J., 2001. Fracture site-specific deficits in bone size and volumetric density in men with spine or hip fractures. J. Bone Miner. Res. 16 (1), 120–127.
Seeman, E., Hopper, J.L., Young, N.R., Formica, C., Goss, P., Tsalamandris, C., 1996. Do genetic factors explain associations between muscle strength, lean mass, and bone density? A twin study. Am. J. Physiol. 270 (2 Pt 1), E320–E327.
Seeman, E., Martin, T.J., 2015. Combined antiresorptive and anabolic therapy: a missed opportunity. J. Bone Miner. Res. 30, 753–764.
Silva, M.J., Brodt, M.D., Wopenka, B., Thomopoulos, S., Williams, D., Wassen, M.H., Ko, M., Kusano, N., Bank, R.A., 2006. Decreased collagen organization and content are associated with reduced strength of demineralized and intact bone in the SAMP6 mouse. J. Bone Miner. Res. 21, 78–88.
Silverman, S.L., Chines, A.A., Kendler, D.L., Kung, A.W., Teglbjærg, C.S., Felsenberg, D., Mairon, N., Constantine, G.D., Adachi, J.D., 2012. Sustained efficacy and safety of bazedoxifene in preventing fractures in postmenopausal women with osteoporosis: results of a 5-year, randomized, placebo-controlled study. Osteoporos Int. 23, 351–363.

Siris, E.S., Chen, Y.T., Abbott, T.A., Barrett-Connor, E., Miller, P.D., Wehren, L.E., Berger, M.L., 2004. Bone mineral density thresholds for pharmacological intervention to prevent fractures. Arch. Intern. Med. 164, 1108–1112.

Smith, S.Y., Recker, R.R., Hannan, M., Müller, R., Bauss, F., 2003. Intermittent intravenous administration of the bisphosphonate ibandronate prevents bone loss and maintains bone strength and quality in ovariectomized cynomolgus monkeys. Bone 32, 45–55.

Stenderup, K., Justesen, J., Eriksen, E.F., Rattan, S.I., Kassem, M., 2001. Number and proliferative capacity of osteogenic stem cells are maintained during aging and in patients with osteoporosis. J. Bone Miner. Res. 16, 1120–1129.

Suda, T., Takahashi, N., Udagawa, N., Jimi, E., Gillespie, M.T., Martin, T.J., 1999. Modulation of osteoclast differentiation and function by the new members of the tumor necrosis factor receptor and ligand families. Endocr. Rev. 20 (3), 345–357.

Suzuki, R., Domon, T., Wakita, M., 2000. Some osteocytes released from their lacunae are embedded again in the bone and not engulfed by osteoclasts during remodelling. Anat. Embrol. 202, 119–128.

Szulc, P., Seeman, P., Duboeuf, F., Sornay-Rendu, E., Delmas, P.D., 2006. Bone fragility: failure of periosteal apposition to compensate for increased endocortical resorption in postmenopausal women. J. Bone Miner. Res. 21, 1856–1863.

Tanner, J.M., Hayashi, T., Preece, M.A., Cameron, N., 1982. Increase in length of leg relative to trunk in Japanese children and adults from 1957 to 1977: comparison with British and with Japanese Americans. Ann. Hum. Biol. 9, 411–423.

Tatsumi, S., Ishii, K., Amizuka, N., Li, M., Kobayashi, T., Kohno, K., Ito, M., Takeshita, S., Ikeda, K., 2007. Targeted ablation of osteocytes induces osteoporosis with defective mechanotransduction. Cell Metabol. 5, 464–475.

Taylor, D., 1997. Bone maintenance and remodeling: a control system based on fatigue damage. J. Orthop. Res. 15, 601–606.

Tsai, J.N., Uihlein, A.V., Burnett-Bowie, S.A., Neer, R.M., Zhu, Y., Derrico, N., Lee, H., Bouxsein, M.L., Leder, B.Z., 2015. Comparative effects of teriparatide, denosumab, and combination therapy on peripheral compartmental bone density, microarchitecture, and estimated strength: the DATA-HRpQCT study. J. Bone Miner. Res. 30, 39–45.

Ural, A., Vashishth, D., 2014. Hierarchical perspective of bone toughness – from molecules to fracture. Int. Mater. Rev. 59 (5), 245–263.

Van der Linden, J.C., Homminga, J., Verhaar, J.A.N., Weinans, H., 2001. Mechanical consequences of bone loss in cancellous bone. J. Bone Miner. Res. 16, 457–465.

Vedi, S., Compston, J.E., Webb, A., Tighe, J.R., 1984. Histomorphometric analysis of dynamic parameters of trabecular bone formation in the iliac crest of normal British subjects. Metab. Bone Dis. Relat. Res. 5, 69–74.

Verborgt, O., Gibson, G.J., Schaffler, M.B., 2000. Loss of osteocyte integrity in association with microdamage and bone remodeling after fatigue damage in vivo. J. Bone Miner. Res. 15, 60–67.

Viguet-Carrin, S., Garnero, S.P., Delmas, P.D.D., 2006. The role of collagen in bone strength. Osteoporos. Int. 17, 319–336.

Wang, Q., Alen, M., Nicholson, P., Lyytikainen, A., Suurubuenu, M., Helkala, E., Suominen, H., Cheng, S., 2005a. Growth patterns at distal radius and tibial shaft in pubertal girls: a 2-year longitudinal study. J. Bone Miner. Res. 20 (6), 954–961.

Wang, X.F., Duan, Y., Beck, T., Seeman, E.R., 2005b. Varying contributions of growth and ageing to racial and sex differences in femoral neck structure and strength in old age. Bone 36 (6), 978–986.

Yeni, Y.N., Brown, C.U., Wang, Z., Norman, T.L., 1997. The Influence of bone morphology on fracture toughness of the human femur and tibia. Bone 21, 453–459.

Zebaze, R., Ghasem-Zadeh, A., Bohte, A., Iuliano-Burns, S., Mirams, M., Price, R.I., et al., 2010. Intracortical remodelling and porosity in the distal radius and post-mortem femurs of women: a cross-sectional study. Lancet 375 (9727), 1729–1736.

Zebaze, R., Ghasem-Zadeh, A., Mbala, A., Seeman, E., 2013. A new method of segmentation of compact-appearing, transitional and trabecular compartments and quantification of cortical porosity from high resolution peripheral quantitative computed tomographic images. Bone 54 (1), 8–20.

Zebaze, R.M., Jones, A., Knackstedt, M., Maalouf, G., Seeman, E., 2007. Construction of the femoral neck during growth determines its strength in old age. J. Bone Miner. Res. 22 (7), 1055–1061.

Zebaze, R.M., Jones, A., Welsh, F., Knackstedt, M., Seeman, E., 2005. Femoral neck shape and the spatial distribution of its mineral mass varies with its size: clinical and biomechanical implications. Bone 37 (2), 243–252.

Zebaze, R.M., Libanati, C., Austin, M., Ghasem-Zadeh, A., Hanley, D.A., Zanchetta, J.R., Thomas, T., Boutroy, S., Bogado, C.E., Bilezikian, J.P., Seeman, E., 2014. Differing Effects of Denosumab and Alendronate on Cortical and Trabecular Bone. Bone 59, 173–179.

Chapter 12

Aging and bone

Maria Almeida[1,2] and Stavros Manolagas[1,2]
[1]Division of Endocrinology and Metabolism, Center for Osteoporosis and Metabolic Bone Diseases, University of Arkansas for Medical Sciences, Little Rock, AR, United States; [2]The Central Arkansas Veterans Healthcare System, Little Rock, AR, United States

Chapter outline

Characteristics of the aged skeleton **275**
Human 275
Rodents 276
Bone cell aging **276**
Osteoblast progenitors 277
Osteocytes 277
Molecular mechanisms of aging **278**
Mitochondrial dysfunction 278
Cellular senescence 280
Loss of autophagy 281
Contribution of bone extrinsic mechanisms to skeletal aging **282**
Loss of sex steroids 282
Lipid peroxidation and declining innate immunity 283
Decreased physical activity 284
Future directions **284**
References **285**

Characteristics of the aged skeleton

Human

The proportion of elderly humans among the global population is now higher than at any time in history. Old age is a major risk factor for several chronic diseases, including osteoporosis, and increases exponentially the risk of fractures (Almeida et al., 2017; Niccoli and Partridge, 2012). Soon after the attainment of peak bone mass the balance between bone formation and bone resorption begins to progressively tilt in favor of the latter, in both women and men (Looker et al., 1998). This process is slowed by the presence of sex steroids and accelerates following menopause. The rate of bone loss due to the menopause is followed, within 5−10 years, by a slower phase of bone loss (Black and Rosen, 2016; Seeman, 2013). This later phase occurs also in men and causes structural deterioration of cortical bone, a significant portion of which is due to increased intracortical porosity (Bala et al., 2015). Postmenopausal women have greater cortical porosity than men over the age of 50 (Nirody et al., 2015; Shanbhogue et al., 2016). It is unknown, however, whether this difference is due to decreased estrogen levels. Importantly, about 80% of fractures occur in the appendicular skeleton, at regions containing large amounts of cortical bone.

A histological hallmark of aged human bone is decreased wall width, an index of the reduced amount of work performed by teams of osteoblasts (Lips et al., 1978; Parfitt, 1990). This is due primarily to an insufficient number of osteoblasts relative to the need for the replacement of bone created by increased osteoclastic resorption (Manolagas, 2000). Another common histologic feature of aged human bone is decreased osteocyte density (Manolagas and Parfitt, 2010; Qiu et al., 2002b). The decrease in osteocyte lacunar density in cortical bone with age correlates with microcrack accumulation (Vashishth et al., 2000). In contrast, a dense osteocyte network is associated with better bone material quality (Kerschnitzki et al., 2013). Mineralization of osteocyte lacunae, a process known as micropetrosis, may contribute to the decrease in osteocytes with age (Busse et al., 2010; Frost, 1960). Micropetrosis may be one potential outcome of osteocyte death; however, the conditions that cause it are unclear. Be that as it may, it is now abundantly clear that the integrity or lack thereof of the osteocyte network plays a major role in skeletal health and disease.

Principles of Bone Biology. https://doi.org/10.1016/B978-0-12-814841-9.00012-9

Rodents

Like humans, both female and male mice lose bone mass and strength with age. Aging female mice do not experience menopause, but become acyclic while retaining functional levels of estrogens (Almeida et al., 2017; Ucer et al., 2017). Androgen levels in aged male mice are maintained at a 20-fold higher level than in females (Nilsson et al., 2015). Yet, both female and male mice exhibit all of the major features of skeletal aging, including the decline of cancellous and cortical bone mass and the development of cortical porosity by 18 months of age (Almeida et al., 2007b; Ferguson et al., 2003; Glatt et al., 2007; Halloran et al., 2002; Jilka et al., 2010; Ucer et al., 2017). In C57BL/6J (B6) mice the accrual of bone mass occurs up to 6–7 months of age, and soon after bone mineral density starts to slowly decline at a constant rate until the end of life. Micro-computed tomography (microCT) analysis performed at about 12 months (equivalent to 40 years in humans; Dutta and Sengupta, 2016) shows that the marrow begins to expand and additional bone is slowly added to the periosteum, but the former exceeds the latter, leading to a thinner and more fragile cortex. The number of osteoclasts also decreases in cancellous bone with advancing age.

The structural and cellular features of aged bone in mice have been well characterized. Old age causes a decline in osteoblast number, bone formation rate, and wall width, which is seen in cancellous and endocortical bone surfaces of both female and male mice (Almeida et al., 2007b; Tiede-Lewis et al., 2017; Ucer et al., 2017). Osteoclast number, however, also declines with age in the cancellous compartment (Almeida et al., 2007b). Thus, an insufficient number of osteoblasts must be the key mechanism for the unbalanced remodeling and the age-dependent loss of cancellous bone.

It has been argued earlier that the endosteal surface of adult mice undergoes osteoclastic modeling but not basic multicellular unit (BMU)-based remodeling. Several lines of evidence refute this idea. Specifically, the endosteal surface of both 7- and 21-month-old B6 mice exhibits scalloped cement lines, which are associated with fluorochrome labeling, indicating that BMU-based remodeling of the endosteal surface persists with advancing age. Moreover, administration of osteoprotegerin to adult mice ablates both endosteal bone resorption and formation. The same is true in adult rats, as shown by histologic analysis of 12-month-old female Fischer-344 rats (Erben, 1996) and a strong inhibitory effect of risedronate on endosteal fluorochrome labeling in 15-month-old ovariectomized Sprague–Dawley rats (Baumann, 1995). The loss of bone at the endosteal surface is caused by inadequate filling of resorption cavities due to an increase in osteoclasts and decrease in osteoblasts (Li et al., 2015; Ucer et al., 2017). As is the case in humans (Han et al., 1997; Power et al., 2003), cortical thinning with aging in mice is due to loss of bone from the endosteal surface that exceeds the amount of bone added to the periosteal surface.

Another feature of skeletal aging in both female and male mice is an increase in cortical porosity (Ferguson et al., 2003; Jilka et al., 2014; Tiede-Lewis et al., 2017). Like humans, aged female mice exhibit higher cortical porosity than aged males (Piemontese et al., 2017). In female B6 mice, the number and size of pores are much higher in the femoral cortex of 21-month-old mice than in 7-month-old young adults. Large pores are especially prominent in the metaphyseal cortex. Unbalanced remodeling is most evident near the endosteal surface and leads to the penetration of the endosteal boundary and the trabecularization of the endocortical part of the cortex, similar to the situation in humans. In young adult mice, cortical capillaries formed during development persist in the cortex, and osteons arise only occasionally from these capillaries (Piemontese et al., 2017; Schneider et al., 2009). With advancing age, however, osteons exhibit several histologic hallmarks of remodeling activity, including osteoclasts and fluorochrome-labeled bone matrix adjacent to scalloped cement lines. Histologic and microCT imaging studies in aged mice suggest that new osteons may arise from preexisting cortical capillaries, from redirection of endosteal BMUs from the endosteal surface into the cortex in tandem with vascular invasion from the marrow, or from both mechanisms. However, murine osteons do not exhibit the numerous concentric lamellae that characterize the much larger osteons of human bone. Instead, they resemble rabbit osteons (Pazzaglia et al., 2015). The highly organized networks seen in humans are not present in murine osteons, perhaps due to the lack of a preexisting haversian system. Nevertheless, this is probably a consequence of body size rather than phylogenetic determinants.

Interestingly, genetic background greatly influences the gain or loss of murine cortical bone mass during growth, in old age, or following sex steroid deficiency (Li et al., 2005; Piemontese et al., 2017; Price et al., 2005). In contrast, genetic background does not influence trabecular bone loss (Piemontese et al., 2017). This and the evidence that an increase in bone resorption accounts for the loss of cortical, but not cancellous, bone with age, indicates that distinct molecular mechanisms underlie the age-dependent dysregulation of endosteal and trabecular remodeling in mice.

Bone cell aging

Cellular aging is frequently described as a decline in function due to the accumulation of damage to lipids, proteins, and DNA (Droge, 2002). Damage of long-lived cells, together with the cellular response to damage, is thought to be involved in the functional deterioration of many tissues with advancing age (55). Mesenchymal progenitors and osteocytes are long-lived cells and, most likely, are more prone to suffer the damaging effects of aging. The declining osteoblast number in the aging skeleton has been attributed to a decrease in the number of mesenchymal stem cells (MSCs), defective proliferation/differentiation of progenitor cells, or diversion of these progenitors toward the adipocyte lineage (Almeida and O'Brien, 2013; Kassem and Marie, 2011).

The loss of functional adult stem/progenitor cells (Rossi et al., 2008; Sharpless and DePinho, 2007) might contribute to the defective regenerative capacity of different tissues with age. Indeed, old mice exhibit a reduction in the number, proliferative capacity, or differentiation potential of distinct stem cells such as germ-line (Ryu et al., 2006; Zhang et al., 2006), neuronal (Kuhn et al., 1996; Molofsky et al., 2006), hematopoietic (Morrison et al., 1996), and muscle progenitor cells (Conboy et al., 2003; Shefer et al., 2006). Genetic mutations, epigenetic changes, and the extrinsic environmental milieu alter stem cell function with aging (Goodell and Rando, 2015).

Osteocytes orchestrate bone resorption and formation via the production of receptor activator of nuclear factor κB (NF-κB) ligand (RANKL), sclerostin (Sost), and other factors. Osteocytes formed during growth are still present in the cortical bone of skeletally mature and of aged mice (Piemontese et al., 2017). Findings indicate that aging of osteocytes can affect their expression profile and, consequently, alter both bone resorption and bone formation (Farr et al., 2016; Piemontese et al., 2017).

Osteoblast progenitors

MSCs are generally defined by their ability to self-replicate and differentiate into chondrocytes, osteoblasts, and adipocytes (Caplan, 1991; Friedenstein et al., 1974). MSCs are found in the bone marrow and in the perivascular niche in multiple human organs (Crisan et al., 2008). Mesenchymal cells in the bone marrow of long bones in mice express smooth muscle α-actin, myxovirus resistance-1, leptin receptor, and/or collagen II, and give rise to osteoblasts on the bone surfaces (Grcevic et al., 2012; Ono et al., 2014; Park et al., 2012; Zhou et al., 2014). Of note, cells expressing leptin receptor are the source of the majority of osteoblasts and osteocytes in adult mice. However, the effects of aging in these or other adult skeletal MSC populations remain unknown.

Because of the very limited number of MSCs, and the laborious procedures required to isolate pure populations, cultures of bone marrow stromal cells from humans and rodents are commonly used as surrogates for progenitor cells. These cultures have been used to elucidate changes in MSCs with aging in both humans and rodents (Coipeau et al., 2009; Kasper et al., 2009; Sethe et al., 2006; Stenderup et al., 2003). Changes in the behavior of bone marrow—derived MSCs with aging, such as loss of potential to proliferate and differentiate, loss of capacity to form bone in vivo, and increased senescence, have been also reported (Coipeau et al., 2009; Kasper et al., 2009; Sethe et al., 2006; Stenderup et al., 2003). Multipotent cells derived from adipose tissue similarly show an age-dependent loss of self-renewal capacity as well as an increased propensity for adipogenesis (Huang et al., 2010). Furthermore, cultures of MSCs from patients with Hutchinson—Gilford progeria syndrome, a disease characterized by accelerated aging, exhibit defective ability to differentiate (Scaffidi and Misteli, 2008). Similar findings have been reported in murine models of progeria (Chen et al., 2013; Diderich et al., 2012).

The transcription factor Osterix1 (Osx1) is required for osteoblast differentiation. Bone marrow mesenchymal progenitor cells expressing Osx1 give rise to all osteoblast and adipocytes in bone (Horowitz et al., 2017; Xiong et al., 2011), suggesting that expression of Osx1 marks a bipotent progenitor. In a mouse model in which Osx1-expressing cells are labeled with a fluorescent protein, the number of these cells in the bone marrow greatly decreases between adulthood and old age (Kim et al., 2017). Moreover, the cells from old mice exhibit decreased proliferation and several markers of cellular senescence, as determined in freshly isolated cells. Notably, the decline in the number of Osx1-expressing cells with age is associated with a decrease in type H capillaries at the distal end of the arterial network, which are critical for the maintenance of perivascular osteoprogenitors (Kusumbe et al., 2014). Aging leads to an increase in the expression of peroxisome proliferator-activated receptor γ—the master transcription factor for adipogenesis—in Osx1-expressing progenitor cells (Kim et al., 2017), in line with their propensity to differentiate toward adipocytes. Indeed, an increase in marrow adipocytes is a well-established consequence of skeletal aging in humans and mice (Horowitz et al., 2017). It remains unclear, however, whether any of the changes noted in osteoblast progenitors contribute to the decrease in bone formation with age and whether they are functionally related to skeletal involution.

Osteocytes

Depending on their anatomical location, osteocyte life span in humans can range from a few months to several decades. Osteocytes are as old as the bone matrix in which they are embedded and this depends on the rate of remodeling at that site.

Within cancellous bone and remodeling cortical bone, a subset of osteoblasts becomes embedded in bone matrix with each remodeling cycle, thereby generating a fresh population of osteocytes within the new packet of bone. Osteocytes near the periosteum are derived from the osteoblasts that continuously expand the periosteum via modeling. As mentioned earlier, osteocyte survival and number decline with age in rodents and humans (Jilka et al., 2013; Noble et al., 1997; Qiu et al., 2002a). In aged mice, there is a concomitant and dramatic decrease in dendrite number per osteocyte. This effect is more severe in females compared with males (Tiede-Lewis et al., 2017). A loss of dendrites with age is also seen in neuronal cells and is associated with impaired nervous system function in a variety of disorders such as Parkinson's and Alzheimer's disease (Koleske, 2013; McNeill et al., 1988). Dendrite loss in neuronal disorders has been linked to defective autophagy (Coleman, 2013; Tang et al., 2014). Interestingly, young mice with defective autophagy in osteocytes show an aging bone phenotype, including reduced trabecular and cortical bone volume, increased cortical porosity, and loss of dendrites (Piemontese et al., 2016). Nevertheless, it remains unknown whether loss of dendrites is causally related to the low bone mass in this model or in wild-type aged mice.

Because of the difficulty in isolating pure populations of osteocytes, there has been little quantitative information on molecular damage in osteocytes from aged mice. Based on the evidence that osteocytes control bone resorption via RANKL and bone formation via sclerostin, reduced osteocyte number might lead to altered production of these factors, leading to changes in bone remodeling. The levels of sclerostin decrease in cortical bone with age (Piemontese et al., 2017). However, the low sclerostin levels should promote Wnt signaling and thereby osteoblast formation; yet osteoblast number is reduced with age. In contrast to sclerostin, RANKL expression levels in cortical bone increase with age, consistent with the increase in endocortical bone resorption. The increase of RANKL along with an increase in other cytokines, chemokines, and metalloproteinases indicates that aging alters the synthetic capacity of osteocytes.

Although the cellular and molecular mechanisms responsible for the increase in cortical porosity with age remain unclear, there is evidence to suggest that osteocytes are the mediators of this effect. Increased osteocyte apoptosis, perhaps in response to age-dependent bone microdamage, is thought to contribute to the development of cortical porosity (Bentolila et al., 1998; Cardoso et al., 2009; Jilka and O'Brien, 2016). This notion is based on evidence that apoptotic osteocytes release factors that stimulate RANKL production by neighboring viable osteocytes. The age-dependent increases in RANKL and cortical porosity are exacerbated in female mice with apoptosis-resistant osteocytes (Jilka et al., 2014), probably due to accumulation of damaged osteocytes that cannot complete the apoptotic death program. Senescent osteocytes might be also involved in the activation of de novo intracortical remodeling via the senescence-associated secretory phenotype (SASP) (see section on "Cellular senescence"). As of this writing, there is no explanation for the lower porosity in male compared with female mice, in the face of a seemingly equivalent increase in osteocyte senescence in both sexes. Similarly, the highly variable magnitude and preferential development of cortical porosity near the metaphyses remain unexplained. In addition, it remains unclear why these mechanisms have different effects on the rate of remodeling in cancellous versus cortical bone.

Molecular mechanisms of aging

The time-dependent accumulation of cellular damage is widely considered the general cause of aging (Gems and Partridge, 2013; Lithgow and Kirkwood, 1996; Vijg and Campisi, 2008). Nevertheless, the sources of aging-caused damage, the compensatory responses that attempt to reestablish homeostasis, the interconnection between the different types of damage and compensatory responses, and the optimal means to intervene exogenously to delay aging remain unknown. Nonetheless, several cellular and molecular hallmarks of aging have been identified (Kennedy et al., 2014; Lopez-Otin et al., 2013). A hallmark should ideally fulfill the following criteria: it should manifest during normal aging, its experimental aggravation should accelerate aging, and its experimental amelioration should delay aging. Below, we review evidence that aging hallmarks such as mitochondrial dysfunction, cell senescence, and loss of proteostasis contribute to skeletal involution with old age.

Mitochondrial dysfunction

The mitochondrial free radical theory of aging proposes that as cells and organisms age, the efficacy of the respiratory chain diminishes, thus reducing ATP generation and increasing electron leakage and production of reactive oxygen species (ROS) (Green et al., 2011; Harman, 1956). The respiration-produced free radicals cause damage to macromolecules, resulting in cellular and tissue loss of function observed during aging (Balaban et al., 2005; Droge, 2002). The majority of cellular ROS are generated by the mitochondria during oxidative phosphorylation. H_2O_2 is the most stable and abundant form of ROS and is produced from the conversion of superoxide by superoxide dismutase enzymes (Balaban et al., 2005;

Chance et al., 1979; D'Autreaux and Toledano, 2007). ROS are also produced in other cellular compartments by NADPH oxidases, lipoxygenases, and other enzymes, in response to growth factors and cytokines (Janssen-Heininger et al., 2008). To prevent excessive ROS production, cells scavenge ROS by multiple mechanisms, including superoxide dismutases and catalase as well as thiol-containing oligopeptides with redox-active sulfhydryl moieties. The most abundant of the last are glutathione and thioredoxin (Dickinson and Forman, 2002).

Several lines of evidence have called into question the idea that an increase in ROS limits life span at the organismal level. Specifically, mouse models of increased mitochondrial ROS and oxidative damage do not exhibit accelerated aging (Van Remmen et al., 2003; Zhang et al., 2009). In addition, increased ROS may prolong life span in lower organisms (Doonan et al., 2008; Mesquita et al., 2010). In line with these observations, ROS can trigger proliferation and survival in response to physiological signals and stress conditions (Sena and Chandel, 2012). Nevertheless, many other studies indicate that oxidative stress is responsible for age-related pathologies, including diabetes, cardiovascular disease, cancer, and neurodegeneration (Dai et al., 2014; Lee et al., 2010; Schriner et al., 2005; Wanagat et al., 2010). A reevaluation of these apparent contradictory lines of evidence posits that the primary effect of ROS is the activation of compensatory homeostatic responses. Above certain threshold levels, ROS betray their original homeostatic purpose and exacerbate, rather than ameliorate, the age-associated damage (Hekimi et al., 2011; Salmon et al., 2010). This new conceptual framework accommodates seemingly conflicting evidence regarding the positive, negative, or neutral effects of ROS on aging.

It has been well established that ROS play a role in osteoclast differentiation and bone resorption (Garrett et al., 1990). RANKL and macrophage colony-stimulating factor—the two critical cytokines for osteoclast generation—promote mitochondrial biogenesis and the accumulation of H_2O_2 in osteoclasts, and H_2O_2 stimulates osteoclastogenesis (Bartell et al., 2014; Garrett et al., 1990; Ha et al., 2004; Ishii et al., 2009; Lee et al., 2005). The increase in H_2O_2 by proosteoclastogenic cytokines is due to an Akt-mediated inhibition of FoxO transcription factors, which causes a decrease in the expression of antioxidant enzymes (Bartell et al., 2014; Liu et al., 2005; Tan et al., 2015). Importantly, lowering H_2O_2 generation in osteoclasts, by overexpressing human catalase targeted to the mitochondria, reduces osteoclast numbers and increases bone mass (Bartell et al., 2014).

As in other tissues, skeletal aging is associated with an increase in ROS levels in bone cells (Almeida et al., 2007b; Jilka et al., 2010; Lean et al., 2003, 2005; Manolagas, 2010). The loss of bone mass caused by sex steroid deficiency in mice is also associated with an increase in ROS (Almeida et al., 2007b; Lean et al., 2003; Manolagas, 2010; Yamasaki et al., 2009). These findings raised the possibility that elevated ROS may be a common mechanism of the adverse effects of sex steroid deficiency and old age on bone; and that sex steroid deficiency may accelerate the effects of aging. Studies involving administration of antioxidants or decreasing mitochondrial levels of H_2O_2 (by expressing a mitochondria-targeted catalase transgene) have revealed that mitochondrial ROS are required for the increase in osteoclast number and the loss of bone caused by ovariectomy or orchidectomy. Surprisingly, however, mitochondrial ROS in osteoclasts do not contribute to the adverse effects of aging on either cortical or cancellous bone. Attenuation of H_2O_2 generation in cells of the mesenchymal lineage, on the other hand, ameliorates the age-dependent decline in mineralizing surfaces and prevents the loss of cortical bone. Together, these studies have indicated a major role of ROS in the bone loss associated with aging and estrogen deficiency. Yet, the evidence that distinct cell types are implicated in the effects of ROS within each condition suggests that the cellular and molecular mechanisms responsible for the loss of bone mass with aging and estrogen deficiency are distinct.

Oxidative stress decreases the life span of an osteoblast, as evidenced by the observation that administration of antioxidants abrogates osteoblast apoptosis in ovariectomized or aged mice (Almeida et al., 2007b; Jilka et al., 2010). ROS also inhibit the Wnt/β-catenin signaling pathway, which is indispensable for osteoblastogenesis during development and adulthood (Baron and Kneissel, 2013). Wnt proteins bind to the Frizzled/low-density lipoprotein (LDL) receptor–related protein 5 (LRP5) or LRP6 receptor complex, thereby preventing the degradation of the transcriptional coactivator β-catenin (Clevers and Nusse, 2012). In the nucleus β-catenin associates with the T cell factor (TCF) lymphoid-enhancer binding factor family of transcription factors and regulates the expression of Wnt target genes. In the setting of oxidative stress or nutrient depletion, FoxOs divert the limited pool of active β-catenin from TCF- to FoxO-mediated transcription in diverse cell types, including osteoblasts, colon cancer cells, and hepatocytes (Almeida et al., 2007a; Hoogeboom et al., 2008; Liu et al., 2011). In murine models of targeted deletion of FoxOs in osteoblast progenitors FoxOs restrain the pro-proliferative effects of Wnt signaling and attenuate bone formation (Iyer et al., 2013). Likewise, FoxOs in enteroendocrine progenitors or in neuronal progenitors suppress β-catenin/TCF-mediated transcription and proliferation (Paik et al., 2009; Talchai et al., 2012). Thus, diversion of β-catenin from TCF- to FoxO-mediated transcription in response to stressful conditions, such as increased ROS and growth factor deficiency, may represent a pathogenetic mechanism for osteoporosis. Support for this idea is provided by the evidence that mice that lack FoxOs in the osteoblast lineage maintain high bone mass throughout life (Iyer et al., 2013).

Cellular senescence

Cellular senescence is a process in which cells stop dividing and initiate a gene expression pattern known as SASP (Campisi, 2013; Kuilman et al., 2010; Lopez-Otin et al., 2013; Newgard and Sharpless, 2013). An increase in senescent cells occurs in most tissues with age; however, its impact on age-related diseases and longevity has remained unknown until recently. Senescent cells can be identified by several features, including cell cycle arrest, DNA damage foci that contain phosphorylated histone H2AX, activated ataxia–telangiectasia mutated (ATM) kinase ATM(Ser^{1981}), p53 binding protein, activated p53 (phospho-Ser^{15}), elevated levels of the cell cycle inhibitor $p21^{CIP1}$ or $p16^{INK4a}$, and high senescence-associated β-galactosidase activity. Although none of these markers is, in and of itself, specific or universal for all senescent types, there is a clear consensus that senescent cells express most of them (Campisi, 2013; Childs et al., 2015; Munoz-Espin and Serrano, 2014). Another characteristic of senescent cells is a resistance to apoptosis. Notably, postmitotic cells also exhibit key characteristics of senescence. For example, neurons in various parts of human and mouse brains are known to accumulate high amounts of DNA damage (Sedelnikova et al., 2004) along with other senescence-associated markers, including heterochromatinization, accumulation of GATA4 and activation of NF-κB, synthesis of proinflammatory interleukins, and senescence-associated β-galactosidase activity (Jurk et al., 2012). Senescence-like features are similarly seen in adipocytes of mice on a high-fat diet (Minamino et al., 2009).

Selective elimination of senescent cells expressing p16 in mice, through an INK-ATTAC or p16-3MR transgene, has revealed that senescent cells reduce organismal life and health span (reviewed in Childs et al., 2017). For example, genetic depletion of senescent cells rejuvenates the hematopoietic stem cells (HSCs), counters frailty and the loss of renal function in aged mice, diminishes cancer relapse, and prevents the development of atherosclerosis and osteoarthritis (Baar et al., 2017; Chang et al., 2016; Childs et al., 2016; Demaria et al., 2017; Jeon et al., 2017). The evidence that genetic ablation of senescent cells counters the effects of aging has prompted a rush to the discovery of small molecules that can selectively kill senescent cells (Baker et al., 2011; Zhu et al., 2015). These drugs, called senolytics, have the potential to be novel antiaging agents. Ablation of senescent cells using senolytics delays age-related pathologies in naturally aged mice, similar to the elimination of senescence with genetic means (Baar et al., 2017; Baker et al., 2011, 2016; Jeon et al., 2017; Ogrodnik et al., 2017; Schafer et al., 2017).

In vitro studies using serially passaged bone marrow–derived stromal cells have suggested that cells from aged humans become senescent at earlier passages compared with cells from young individuals (Stenderup et al., 2003; Zhou et al., 2008). However, because serial passaging by itself causes replicative senescence (Hayflick and Moorhead, 1961), it has remained unclear whether osteoblast progenitors become senescent with old age in vivo. As we discussed earlier, studies using freshly isolated osteoprogenitor cells have elucidated that the age-related decline in number of these cells is associated with several markers of senescence, including DNA damage foci, G1 cell cycle arrest, activated p53, and elevated levels of $p21^{CIP1}$. Markers of senescence, including elevated p16 and expression of the SASP, were also present in osteocyte-enriched bone fractions from old mice. Elimination of p16-expressing cells in 20-month-old mice, using the INK-ATTAC transgene or senolytics, increases bone mass (Farr et al., 2017). This effect is associated with a decrease in osteoclasts and an increase in osteoblasts at the endocortical surfaces.

One of the causes of senescence is DNA damage. Of all DNA lesions, DNA double-strand breaks are the most harmful. DNA double-strand breaks can be induced by radiation, radiomimetic chemicals, or ROS, but also during DNA replication when a polymerase encounters a single-strand lesion at a replication fork (Jackson, 2002). DNA double-strand breaks pose problems for cells because their immediate and efficient repair by ligation is often constrained by their physical separation and/or the need to process damaged DNA termini (Mine-Hattab and Rothstein, 2012; Soutoglou and Misteli, 2008). In the absence of repair, damaged cells can be eliminated by apoptosis. Alternatively, mitotically active cells respond to DNA double-strand breaks by becoming senescent. DNA damage triggers a repair response known as the DNA damage response (DDR). This response is characterized by the activation of ATM and the recruitment of RAD-3-related kinase (or ATR) (Rouse and Jackson, 2002) to the site of damage, leading to phosphorylation of Ser^{139} of histone H2AX molecules adjacent to the site of DNA damage. The phosphorylation of histone H2AX activates the transducer kinases Chk1 and Chk2, which converge on p53/p21 and p16 (Smogorzewska and de, 2002). In addition, the DDR stimulates the SASP by upregulating GATA4 and NF-κB (Kang et al., 2015).

Markers of DNA damage increase with age in both osteoprogenitors and osteocytes and are associated with cell senescence (Kim et al., 2017; Piemontese et al., 2017). While it remains unclear whether DNA damage is indeed the cause of skeletal cell senescence, DNA damage induced by irradiation causes changes in osteoprogenitors that are similar to the ones seen with aging (Kim et al., 2017). In addition, DNA damage due to focal irradiation in long bones causes senescence of osteoblast lineage cells, decreases bone formation, and leads to bone loss in mice (Chandra et al., 2017). Additional support for the deleterious effects of DNA damage on the skeleton is provided from the phenotype of murine models of progeroid syndromes due to defective DNA damage repair. For example, a mouse model of trichothiodystrophy (Diderich et al., 2012; Nicolaije et al., 2012; Wijnhoven et al., 2005) exhibits features of accelerated aging, including loss of trabecular and cortical bone mass and a reduction in bone strength, from 9 months onward. Similar observations have been made in a murine model of Werner syndrome (Brennan et al., 2014). Bone marrow–derived osteoblastic cell cultures from these models exhibit increased senescence (Chen et al., 2013; Saeed et al., 2011). Nevertheless, caution should be exercised in interpreting results from progeroid mice, as several of these models exhibit impaired growth, poor health, and short life spans (Chen et al., 2013; Niedernhofer et al., 2006; Saeed et al., 2011). Furthermore, not all the cellular features of natural aging are replicated in the progeroid models. Specifically, the loss of trabecular bone mass in several progeroid models is associated with an increase in osteoclast number, which is not seen with natural aging (Almeida et al., 2007b; Brennan et al., 2014; Chen et al., 2013; Saeed et al., 2011).

Together these lines of evidence implicate cellular senescence in the loss of bone mass with age. Future work should elucidate the cell targets and the molecular mechanisms mediating the effects of senescence in skeletal aging. These studies should also clarify the extent to which elimination of cell senescence ameliorates the loss of bone mass with age.

Loss of autophagy

Proteostasis, the process by which cells control the abundance and folding of the proteome, is maintained by proteolytic systems, such as the ubiquitin–proteasome system and the autophagy–lysosomal system, both of which decrease in old age (Cuervo, 2008; Morimoto and Cuervo, 2014). These changes occur in age-related diseases characterized by the accumulation of intracellular or extracellular protein aggregates, including neurodegeneration. These observations have led to the idea that a decline in proteostasis is a common mechanism of aging.

Autophagy is a quality control system that removes defective organelles or protein aggregates to maintain the health and viability of the cell (Yang and Klionsky, 2010). However, under stress conditions, for example, hypoxia or nutrient deprivation, autophagy increases to break down cellular components that can be reused as an energy source. During macroautophagy large components of the cytoplasm, including protein aggregates and mitochondria, are surrounded by a double-membrane structure to form a vacuole known as an autophagosome. This structure then fuses with the lysosome to allow the degradation of its components. Autophagosome formation is controlled by a ubiquitin-like protein known as LC3 (Geng and Klionsky, 2008), which, in turn, is activated by the E1-like enzyme Atg7 and is eventually conjugated to phosphatidylethanolamine in the double-membrane structure. This process is dependent on Atg7, and genetic inactivation of Atg7 effectively abrogates autophagy (Komatsu et al., 2007). Other forms of autophagy include chaperone-mediated autophagy and microautophagy (Levine and Kroemer, 2008). Chaperone-mediated autophagy targets proteins containing a specific five-amino-acid motif directly to the lysosome for degradation. Microautophagy is characterized by invagination of the lysosomal membrane to engulf small portions of the cytoplasm. Because macroautophagy is the most prevalent form of autophagy in many cell types, hereafter, we will use the general term autophagy to refer to this particular process.

Many organisms show signs of decreased autophagic capacity with aging in tandem with the accumulation of damaged cellular components, suggesting that a decline in autophagy may contribute to organismal aging (Cuervo, 2008; Hansen et al., 2018; Lopez-Otin et al., 2013). Specifically, expression of autophagy-related genes declines in muscle tissue from aged humans and in the hypothalamus, muscle, liver, and osteoarthritic bone chondrocytes from aged rodents (Carames et al., 2015; Carnio et al., 2014; Cuervo and Dice, 2000). Furthermore, suppression of autophagy in neurons or myocytes mimics the effects of aging on the nervous or muscle tissue, respectively (Carnio et al., 2014; Komatsu et al., 2007).

Deletion of ATG7 in osteoblasts and osteocytes decreases cancellous and cortical bone mass (Onal et al., 2013). These changes are associated with lower osteoclast and osteoblast numbers in cancellous bone. In addition, ROS levels are increased in the bones of mice lacking autophagy in osteocytes. This is probably due to the accumulation of damaged mitochondria, which in turn produce more ROS. Nevertheless, autophagy exerts positive effects on the skeleton by mechanisms other than suppression of H_2O_2 levels in the mitochondria of osteoblasts and osteocytes (Piemontese et al., 2016). Strikingly, deletion of Atg7 in the entire osteoblast lineage using an Osx1-Cre transgene causes fractures and decreases bone mass. The low-bone-mass phenotype is more pronounced than the one of mice in which ATG7 is deleted from mature osteoblasts and osteocytes (Piemontese et al., 2016). Lack of autophagy in early progenitors of osteoblasts alters osteocyte cell body morphology and reduces the extent of osteocyte cellular projections, suggesting that autophagy contributes to the transition of osteoblasts into osteocytes.

A critical role for autophagy in bone homeostasis is further supported by evidence that mice with deletion of ATG5 or the focal adhesion kinase family–interacting protein of 200 kDa (FIP200) in cells of the osteoblast lineage have compromised autophagy and low bone mass (Liu et al., 2013; Nollet et al., 2014).

Autophagy is important for the long-term health of stem cells as well as fully differentiated long-lived cell types. For example, autophagy is required for the maintenance of the HSC compartment in adult mice (Mortensen et al., 2011; Warr et al., 2013). The FoxO transcription factors are critical for the expression of autophagy genes and the induction of autophagy in response to stress in HSCs and other cell types (Warr et al., 2013; Zhao et al., 2008). It is possible that autophagy plays a similar role in the maintenance of MSCs in bone and this may underlie changes in MSC behavior with age.

In conclusion, suppression of autophagy in the osteoblast lineage is sufficient to replicate in young adult mice the skeletal changes observed in aged wild-type mice. Future studies will be required to identify the molecular mechanisms by which autophagy in cells of the osteoblast lineage controls bone remodeling and bone mass and whether autophagy does indeed decline with age in osteocytes or other cell types of the lineage.

Contribution of bone extrinsic mechanisms to skeletal aging

Loss of sex steroids

A decrease in estrogen levels at menopause, or both estrogens and androgens in elderly men, causes an imbalance between bone resorption and formation, leading to loss of bone mass and strength and increased risk of osteoporotic fractures (Almeida et al., 2017; Khosla, 2010; Manolagas, 2000; Manolagas et al., 2013; Vanderschueren et al., 2014). Within 5−10 years after menopause, the slope of the decline of bone mass in women slows and becomes indistinguishable from the slope of bone loss seen in elderly males. This slower phase of bone loss in both sexes later in life affects primarily the cortical compartment and is characterized by cortical thinning as well as increased cortical porosity (Nicks et al., 2012; Zebaze et al., 2010). Consequently, after the age of 65 most fractures are nonvertebral and predominantly occur at cortical sites (Zebaze et al., 2010).

Estrogens and androgens slow the rate of bone remodeling, and estrogen or androgen deficiency accelerates it. In addition, cell and biochemical studies have revealed that the antiremodeling effects of these two hormones result from their ability to restrain the birth rate of osteoclasts and shorten their life span (Almeida et al., 2017; Hughes et al., 1996; Manolagas, 2000, 2010, 2013). Conditional deletion models of the estrogen receptor α or the androgen receptor have provided critical new insights into the cellular targets of sex steroid action in vivo (Almeida et al., 2013; Chiang et al., 2009; Maatta et al., 2013; Martin-Millan et al., 2010; Nakamura et al., 2007; Notini et al., 2007; Sinnesael et al., 2012; Ucer et al., 2015). This body of work, reviewed elsewhere (Almeida et al., 2017), has produced two seminal but unexpected findings. First, the effects of estrogens and androgens on the cancellous bone compartment are mediated via different cell types. In females, estrogens protect against the loss of cancellous bone via direct actions on cells of the osteoclast lineage. In males, on the other hand, androgens protect against the loss of cancellous bone via actions on mature osteoblasts or osteocytes. Second, in both females and males, estrogens alone protect against the resorption of cortical bone, via actions on osteoprogenitors. In males, the estrogens responsible for this effect derive from the aromatization of androgens. In support of this conclusion, administration of estrogens to orchidectomized B6 mice prevents the loss of cortical thickness, while administration of dihydrotestosterone has no effect (Ucer et al., 2017). Notably, in men, estrogens account for ~70% and testosterone for at most ~30% of the protective effect of sex steroids on bone resorption (Khosla, 2015), remarkably consistent with the fact that the skeleton is ~80% cortical and ~20% cancellous.

Because of the abrupt decline in ovarian function at menopause in women and a slower decline of both androgen and estrogen levels in men with advancing age, the two conditions inexorably overlap, making it impossible to dissect their independent contributions to the cumulative anatomic deficit. Experiments using B6 mice ovariectomized at 4 or 18 months of age have elucidated that at least up to the age of 19.5 months mice remain functionally estrogen sufficient (Ucer et al., 2017). Six weeks following the loss of ovarian function at either age, we observed the expected decrease in uterine weight and increase in body weight in both young and old mice. Nonetheless, femoral cortical thickness and vertebral cancellous bone volume declined and cortical porosity increased in the estrogen-sufficient mice between 5.5 and 19.5 months of age. These findings clearly show that in mice the adverse effects of aging are independent of sex steroid status.

The canonical inhibitor of NF-κB kinase β/NF-κB pathway plays a seminal role in the SASP (Salminen et al., 2012). As discussed earlier under "Cellular senescence," the transcription factors GATA4 and NF-κB are major activators of the SASP (Kang et al., 2015). GATA4 protein levels, NF-κB activity, and the expression of common SASP genes are increased in both bone marrow stromal cells and osteocytes from old B6 mice (Farr et al., 2016; Kim et al., 2017; Piemontese et al., 2017). In line with these findings, the number of osteoclasts formed in cocultures of macrophages with bone marrow–derived stromal cells is higher when stromal cells originate from old mice compared with cells from young mice (Cao et al., 2005; Kim et al., 2017). Estrogens suppress NF-κB activation via direct interactions of the estrogen receptor α with NF-κB (De Bosscher et al., 2006; Stice and Knowlton, 2008). Moreover, some of the same cytokines of the SASP have been implicated in the loss of cortical bone caused by estrogen deficiency (Almeida et al., 2017). It is, therefore, quite likely that in the aged skeleton, senescence of mesenchymal lineage cells and the SASP, driven by NF-κB activation, are the predominant mechanism of the increased osteoclast number and resorption in the endosteal surface. Because NF-κB is also upregulated by estrogen deficiency, the two pathologies may act in concert. In other words, NF-κB may be a key molecular nexus where the two pathologies intersect, and as a result estrogen deficiency accelerates the adverse effects of aging on cortical bone in both sexes.

Lipid peroxidation and declining innate immunity

ROS-induced oxidative damage causes lipid peroxidation and thereby produces highly reactive degradation products, including oxidized phosphatidylcholine, malondialdehyde (MDA), and 4-hydroxynonenal (Witztum and Lichtman, 2014). These neoepitopes interact with amino groups of proteins and other lipids to create oxidation-specific epitopes (OSEs). The same neoepitopes occur on apoptotic cell bodies. OSEs are part of a larger group of proinflammatory molecules, collectively known as damage-associated molecular patterns (DAMPs), that are produced in response to cellular stressors, including oxidative stress. DAMPs share structural homology with another group of moieties present on microbes and collectively known as pathogen-associated molecular patterns (PAMPs). Because of their shared molecular signatures, PAMPs and DAMPs bind to evolutionarily conserved pattern recognition receptors (PRRs) of the innate immune system.

Natural antibodies (nAbs) of the innate immune system are the first line of defense against microbial pathogens. nAbs are produced by B-1 lymphocytes and are predominantly of the IgM class. The antigen binding sites of nAbs are generated by rearrangement of germ-line-encoded variable-region genes in the complete absence of foreign antigen exposure, hence the term "natural" as opposed to antibodies produced by the "adaptive" immune system. As opposed to the adaptive immune system, nAbs have only a limited repertoire of binding specificities (Ehrenstein and Notley, 2010; Vas et al., 2013). nAbs are soluble PRRs that recognize OSEs and block their adverse effects.

When OSEs are not neutralized and/or their production is excessive, relative to the capacity of the innate immune system, the physiologic response is overwhelmed and becomes instead a disease-causing mechanism (Witztum and Lichtman, 2014). A series of clinical studies has provided compelling evidence that anti-OSE-specific IgM antibodies made by B-1 cells play a role in the development of disease in humans, as they do in mice (Binder et al., 2016; Griffin et al., 2011). Indeed, OSE-specific natural IgM antibodies protect against cardiovascular disease (Tsiantoulas et al., 2014), and titers of IgM antibody against oxidized LDL (OxLDL) and MDA–LDL are inversely correlated with cardiovascular disease (Karvonen et al., 2003; Tsimikas et al., 2007). In particular, anti-PC (phosphocholine) IgM levels are inversely correlated with cardiovascular disease risk in patients with systemic lupus erythematous (Anania et al., 2010; Gronwall et al., 2012), stroke (Fiskesund et al., 2010), and heart attacks (de Faire et al., 2010; Fiskesund et al., 2012; Gronlund et al., 2009; Imhof et al., 2015). Notably, in humans, B-1 cells decline with age (Griffin et al., 2011), raising the possibility that declining levels of nAbs against OSEs contribute to the pathogenesis of age-related diseases.

Hyperlipidemia leads to atherogenesis by promoting the formation of oxidized forms of LDL and phospholipids. These oxidation-modified moieties, in turn, induce potent inflammatory responses in the subendothelial matrix of the arteries, thereby triggering the pathogenetic changes that are ultimately responsible for the generation of the atherosclerotic lesions (Steinberg and Witztum, 2010). Consistent with this mechanism, genetic animal models of hyperlipidemia exhibit disruption of normal lipoprotein regulation and metabolism. Indeed, B6 mice maintained on an atherogenic high-fat diet, or genetically modified B6 mice lacking the LDL receptor or its ligand (ApoE), develop hyperlipidemia and atherosclerosis as a result of reduced clearance of non–high-density lipoproteins (Tintut and Demer, 2014).

Osteoporosis and atherosclerosis are epidemiologically linked (Makovey et al., 2009; Tintut and Demer, 2014). Moreover, bone mineral density and LDL-cholesterol levels are inversely correlated in mice as well as rats (Liu et al., 2016; Tintut and Demer, 2014). Diet-induced hyperlipidemia adversely affects bone growth and bone mineral density in these models. In line with the evidence for an adverse effect of lipids on bone, lipid accumulation has been demonstrated by histochemical staining in the subendothelial spaces of haversian canals from patients with osteoporosis (Tintut et al., 2004). Based on this evidence, it has been hypothesized that, similar to the situation in the subendothelial space of the vasculature, lipids accumulating in the subendothelial spaces of the haversian canals undergo nonenzymatic modifications, such as oxidation by ROS, rendering them capable of inducing inflammatory responses. Oxidized lipids may then induce bone loss by inhibiting the differentiation of bone-forming osteoblasts and promoting the differentiation of bone-resorbing osteoclasts (Almeida et al., 2009; Liu et al., 2016).

Witztum and colleagues have generated transgenic mice that overexpress a single-chain variable fragment of the IgM nAb E06 (Que et al., 2018). The fragment is secreted from liver and macrophages and achieves sufficient plasma levels to inhibit macrophage uptake of OxLDL. High-fat diet–fed LDL receptor–null mice expressing E06 develop less atherosclerosis compared with controls. Importantly, high-fat diet–induced bone loss is also attenuated in mice expressing the E06 transgene due to an increase in osteoblast number and bone formation (Ambrogini et al., 2018). More strikingly, E06-scFv increases bone mass in mice fed a normal diet. Moreover, the levels of anti-PC IgM decrease in aged mice. These results indicate that OSEs chronically occurring with or without a high-fat diet in mice exert a restraining effect on bone formation. Moreover, the age-related bone loss might be due in part to diminished innate immune system defense against OSEs. Taken together with the evidence that the same nAbs prevent atherosclerosis, these discoveries demonstrate that anti-OSE nAbs may well represent a novel therapeutic approach against atherosclerosis and osteoporosis simultaneously.

Decreased physical activity

Mechanical strains are critical signals for bone mass accrual and strength (Frost, 2003). Mechanical loading tilts the balance between bone formation and resorption in favor of the former, by stimulating bone formation and suppressing bone resorption (Robling et al., 2006; Rubin and Lanyon, 1985; Zhao et al., 2013). For example, exposure to frequent high-impact loading increased bone mass in the dominant arm of tennis players (Ireland et al., 2013; Jones et al., 1977). Conversely, unloading due to lack of ambulation or weightlessness during space flights causes a marked reduction in bone mass, due to decreased bone formation and increased bone resorption (Lang et al., 2004; LeBlanc et al., 1990). A decrease in physical activity with aging, most likely, contributes to skeletal fragility. Exercise that is beneficial to bone in premenopausal women is typically not effective in elderly women (Korpelainen et al., 2006; Vainionpaa et al., 2007). Likewise, in animals the stimulatory effects of loading on bone are greater in growing or young adult animals compared with older ones (Holguin et al., 2014; Meakin et al., 2014; Razi et al., 2015; Rubin et al., 1992; Turner et al., 1995). These findings indicate that with aging, there is reduced responsiveness of the skeleton to mechanical loading (Srinivasan et al., 2012).

Osteocytes are thought of as critical mediators of the effects of mechanical loading by virtue of their location and connectivity with one another and with cells at the bone surface. Because osteocyte dendrites are critical for mechanotransduction (Burra et al., 2010), reduced dendrite connectivity may contribute to the impaired bone response to mechanical loading in aged animals. In addition, altered cell proliferation and decreased Wnt signaling have been implicated in the age-related decline in bone mechanoresponsiveness. The importance of Wnt signaling in loading-induced bone formation is well documented. Mechanical loading in young adult animals downregulates Sost/sclerostin at sites of subsequent bone formation and stimulates canonical Wnt signaling (Lara-Castillo et al., 2015; Moustafa et al., 2012; Robinson et al., 2006; Robling et al., 2008). Moreover, the osteogenic response to loading in mice is disrupted by overexpression of Sost, deletion of Lrp5, or heterozygous deletion of β-catenin in bone cells (Javaheri et al., 2014; Sawakami et al., 2006; Tu et al., 2012; Zhao et al., 2013). Activation of Wnt signaling by mechanical loading is impaired in bone of aged mice (Holguin et al., 2016). Meakin et al. reported that osteoblasts from old mice are less proliferative after in vitro stretching, and that tibial compression in old mice produces a smaller increase in periosteal osteoblast number than in young mice (Meakin et al., 2014). The age-related decline in osteoprogenitor number might contribute to the deficient response of the skeleton to mechanical loading (Kim et al., 2017).

Future directions

Research in animal models has revealed that the rate of physiological aging can be ameliorated by a variety of behavioral, genetic, and pharmacological means. There is great enthusiasm that this can also be accomplished in humans. Most importantly, decreased rate of aging in animal models is often accompanied by a delay (and decreased severity) of a number of age-associated diseases. This evidence strongly suggests that therapeutic interventions that can either delay the onset or decrease the rate of aging could have a beneficial effect on the health of the elderly. As a result, there is a race to develop drugs that may alleviate several degenerative diseases simultaneously. Current antiosteoporotic treatments have limited impact on public health, despite their effectiveness at the individual level. It is, therefore, imperative to elucidate mechanisms of skeletal aging so that osteoporosis is added to the list of degenerative diseases that are amenable to treatment with antiaging drugs.

References

Almeida, M., Ambrogini, E., Han, L., Manolagas, S.C., Jilka, R.L., 2009. Increased lipid oxidation causes oxidative stress, increased PPAR{gamma} expression and diminished pro-osteogenic Wnt signaling in the skeleton. J. Biol. Chem. 284, 27438–27448.

Almeida, M., Han, L., Martin-Millan, M., O'Brien, C.A., Manolagas, S.C., 2007a. Oxidative stress antagonizes Wnt signaling in osteoblast precursors by diverting beta-catenin from T cell factor- to forkhead box O-mediated transcription. J. Biol. Chem. 282, 27298–27305.

Almeida, M., Han, L., Martin-Millan, M., Plotkin, L.I., Stewart, S.A., Roberson, P.K., Kousteni, S., O'Brien, C.A., Bellido, T., Parfitt, A.M., et al., 2007b. Skeletal involution by age-associated oxidative stress and its acceleration by loss of sex steroids. J. Biol. Chem. 282, 27285–27297.

Almeida, M., Iyer, S., Martin-Millan, M., Bartell, S.M., Han, L., Ambrogini, E., Onal, M., Xiong, J., Weinstein, R.S., Jilka, R.L., et al., 2013. Estrogen receptor-alpha signaling in osteoblast progenitors stimulates cortical bone accrual. J. Clin. Investig. 123, 394–404.

Almeida, M., Laurent, M.R., Dubois, V., Claessens, F., O'Brien, C.A., Bouillon, R., Vanderschueren, D., Manolagas, S.C., 2017. Estrogens and androgens in skeletal physiology and pathophysiology. Physiol. Rev. 97, 135–187.

Almeida, M., O'Brien, C.A., 2013. Basic biology of skeletal aging: role of stress response pathways. J. Gerontol. A Biol. Sci. Med. Sci. 68, 1197–1208.

Ambrogini, E., Que, X., Wang, S., Yamaguchi, F., Weinstein, R.S., Tsimikas, S., Manolagas, S.C., Witztum, J.L., Jilka, R.L., 2018. Oxidation-specific epitopes restrain bone formation. Nat. Commun. 9, 2193.

Anania, C., Gustafsson, T., Hua, X., Su, J., Vikstrom, M., de Faire, U., Heimburger, M., Jogestrand, T., Frostegard, J., 2010. Increased prevalence of vulnerable atherosclerotic plaques and low levels of natural IgM antibodies against phosphorylcholine in patients with systemic lupus erythematosus. Arthritis Res. Ther. 12, R214.

Baar, M.P., Brandt, R.M., Putavet, D.A., Klein, J.D., Derks, K.W., Bourgeois, B.R., Stryeck, S., Rijksen, Y., van Willigenburg, H., Feijtel, D.A., et al., 2017. Targeted apoptosis of senescent cells restores tissue homeostasis in response to chemotoxicity and aging. Cell 169, 132–147 e116.

Baker, D.J., Childs, B.G., Durik, M., Wijers, M.E., Sieben, C.J., Zhong, J., Saltness, R.A., Jeganathan, K.B., Verzosa, G.C., Pezeshki, A., et al., 2016. Naturally occurring p16(Ink4a)-positive cells shorten healthy lifespan. Nature 530, 184–189.

Baker, D.J., Wijshake, T., Tchkonia, T., LeBrasseur, N.K., Childs, B.G., van de Sluis, B., Kirkland, J.L., Van Deursen, J.M., 2011. Clearance of p16Ink4a-positive senescent cells delays ageing-associated disorders. Nature 479, 232–236.

Bala, Y., Zebaze, R., Seeman, E., 2015. Role of cortical bone in bone fragility. Curr. Opin. Rheumatol. 27, 406–413.

Balaban, R.S., Nemoto, S., Finkel, T., 2005. Mitochondria, oxidants, and aging. Cell 120, 483–495.

Baron, R., Kneissel, M., 2013. WNT signaling in bone homeostasis and disease: from human mutations to treatments. Nat. Med. 19, 179–192.

Bartell, S.M., Kim, H.N., Ambrogini, E., Han, L., Iyer, S., Serra, U.S., Rabinovitch, P., Jilka, R.L., Weinstein, R.S., Zhao, H., et al., 2014. FoxO proteins restrain osteoclastogenesis and bone resorption by attenuating H_2O_2 accumulation. Nat. Commun. 5, 3773.

Baumann, B.D., Wronski, T.J., 1995. Response of cortical bone to antiresorptive agents and parathyroid hormone in aged ovariectomized rats. Bone 16, 247–253.

Bentolila, V., Boyce, T.M., Fyhrie, D.P., Drumb, R., Skerry, T.M., Schaffler, M.B., 1998. Intracortical remodeling in adult rat long bones after fatigue loading. Bone 23, 275–281.

Binder, C.J., Papac-Milicevic, N., Witztum, J.L., 2016. Innate sensing of oxidation-specific epitopes in health and disease. Nat. Rev. Immunol. 16, 485–497.

Black, D.M., Rosen, C.J., 2016. Clinical practice. Postmenopausal osteoporosis. N. Engl. J. Med. 374, 254–262.

Brennan, T.A., Egan, K.P., Lindborg, C.M., Chen, Q., Sweetwyne, M.T., Hankenson, K.D., Xie, S.X., Johnson, F.B., Pignolo, R.J., 2014. Mouse models of telomere dysfunction phenocopy skeletal changes found in human age-related osteoporosis. Dis. Model. Mech. 7, 583–592.

Burra, S., Nicolella, D.P., Francis, W.L., Freitas, C.J., Mueschke, N.J., Poole, K., Jiang, J.X., 2010. Dendritic processes of osteocytes are mechanotransducers that induce the opening of hemichannels. Proc. Natl. Acad. Sci. U. S. A. 107, 13648–13653.

Busse, B., Djonic, D., Milovanovic, P., Hahn, M., Puschel, K., Ritchie, R.O., Djuric, M., Amling, M., 2010. Decrease in the osteocyte lacunar density accompanied by hypermineralized lacunar occlusion reveals failure and delay of remodeling in aged human bone. Aging Cell 9, 1065–1075.

Campisi, J., 2013. Aging, cellular senescence, and cancer. Annu. Rev. Physiol. 75, 685–705.

Cao, J.J., Wronski, T.J., Iwaniec, U., Phleger, L., Kurimoto, P., Boudignon, B., Halloran, B.P., 2005. Aging increases stromal/osteoblastic cell-induced osteoclastogenesis and alters the osteoclast precursor pool in the mouse. J. Bone Miner. Res. 20, 1659–1668.

Caplan, A.I., 1991. Mesenchymal stem cells. J. Orthop. Res. 9, 641–650.

Carames, B., Olmer, M., Kiosses, W.B., Lotz, M.K., 2015. The relationship of autophagy defects to cartilage damage during joint aging in a mouse model. Arthritis Rheum. 67, 1568–1576.

Cardoso, L., Herman, B.C., Verborgt, O., Laudier, D., Majeska, R.J., Schaffler, M.B., 2009. Osteocyte apoptosis controls activation of intracortical resorption in response to bone fatigue. J. Bone Miner. Res. 24, 597–605.

Carnio, S., LoVerso, F., Baraibar, M.A., Longa, E., Khan, M.M., Maffei, M., Reischl, M., Canepari, M., Loefler, S., Kern, H., et al., 2014. Autophagy impairment in muscle induces neuromuscular junction degeneration and precocious aging. Cell Rep. 8, 1509–1521.

Chance, B., Sies, H., Boveris, A., 1979. Hydroperoxide metabolism in mammalian organs. Physiol. Rev. 59, 527–605.

Chandra, A., Lin, T., Young, T., Tong, W., Ma, X., Tseng, W.J., Kramer, I., Kneissel, M., Levine, M.A., Zhang, Y., et al., 2017. Suppression of sclerostin alleviates radiation-induced bone loss by protecting bone-forming cells and their progenitors through distinct mechanisms. J. Bone Miner. Res. 32, 360–372.

Chang, J., Wang, Y., Shao, L., Laberge, R.M., Demaria, M., Campisi, J., Janakiraman, K., Sharpless, N.E., Ding, S., Feng, W., et al., 2016. Clearance of senescent cells by ABT263 rejuvenates aged hematopoietic stem cells in mice. Nat. Med. 22, 78–83.

Chen, Q., Liu, K., Robinson, A.R., Clauson, C.L., Blair, H.C., Robbins, P.D., Niedernhofer, L.J., Ouyang, H., 2013. DNA damage drives accelerated bone aging via an NF-kappaB-dependent mechanism. J. Bone Miner. Res. 28, 1214–1228.

Chiang, C., Chiu, M., Moore, A.J., Anderson, P.H., Ghasem-Zadeh, A., McManus, J.F., Ma, C., Seeman, E., Clemens, T.L., Morris, H.A., et al., 2009. Mineralization and bone resorption are regulated by the androgen receptor in male mice. J. Bone Miner. Res. 24, 621–631.

Childs, B.G., Baker, D.J., Wijshake, T., Conover, C.A., Campisi, J., van Deursen, J.M., 2016. Senescent intimal foam cells are deleterious at all stages of atherosclerosis. Science 354, 472–477.

Childs, B.G., Durik, M., Baker, D.J., Van Deursen, J.M., 2015. Cellular senescence in aging and age-related disease: from mechanisms to therapy. Nat. Med. 21, 1424–1435.

Childs, B.G., Gluscevic, M., Baker, D.J., Laberge, R.M., Marquess, D., Dananberg, J., van Deursen, J.M., 2017. Senescent cells: an emerging target for diseases of ageing. Nat. Rev. Drug Discov. 16, 718–735.

Clevers, H., Nusse, R., 2012. Wnt/beta-Catenin signaling and disease. Cell 149, 1192–1205.

Coipeau, P., Rosset, P., Langonne, A., Gaillard, J., Delorme, B., Rico, A., Domenech, J., Charbord, P., Sensebe, L., 2009. Impaired differentiation potential of human trabecular bone mesenchymal stromal cells from elderly patients. Cytotherapy 11, 584–594.

Coleman, M.P., 2013. The challenges of axon survival: introduction to the special issue on axonal degeneration. Exp. Neurol. 246, 1–5.

Conboy, I.M., Conboy, M.J., Smythe, G.M., Rando, T.A., 2003. Notch-mediated restoration of regenerative potential to aged muscle. Science 302, 1575–1577.

Crisan, M., Yap, S., Casteilla, L., Chen, C.W., Corselli, M., Park, T.S., Andriolo, G., Sun, B., Zheng, B., Zhang, L., et al., 2008. A perivascular origin for mesenchymal stem cells in multiple human organs. Cell Stem Cell 3, 301–313.

Cuervo, A.M., 2008. Autophagy and aging: keeping that old broom working. Trends Genet. 24, 604–612.

Cuervo, A.M., Dice, J.F., 2000. Age-related decline in chaperone-mediated autophagy. J. Biol. Chem. 275, 31505–31513.

D'Autreaux, B., Toledano, M.B., 2007. ROS as signalling molecules: mechanisms that generate specificity in ROS homeostasis. Nat. Rev. Mol. Cell Biol. 8, 813–824.

Dai, D.F., Chiao, Y.A., Marcinek, D.J., Szeto, H.H., Rabinovitch, P.S., 2014. Mitochondrial oxidative stress in aging and healthspan. Longev. Health. 3, 6.

De Bosscher, K., Vanden Berghe, W., Haegeman, G., 2006. Cross-talk between nuclear receptors and nuclear factor kappaB. Oncogene 25, 6868–6886.

de Faire, U., Su, J., Hua, X., Frostegard, A., Halldin, M., Hellenius, M.L., Wikstrom, M., Dahlbom, I., Gronlund, H., Frostegard, J., 2010. Low levels of IgM antibodies to phosphorylcholine predict cardiovascular disease in 60-year old men: effects on uptake of oxidized LDL in macrophages as a potential mechanism. J. Autoimmun. 34, 73–79.

Demaria, M., O'Leary, M.N., Chang, J., Shao, L., Liu, S., Alimirah, F., Koenig, K., Le, C., Mitin, N., Deal, A.M., et al., 2017. Cellular senescence promotes adverse effects of chemotherapy and cancer relapse. Cancer Discov. 7, 165–176.

Dickinson, D.A., Forman, H.J., 2002. Glutathione in defense and signaling: lessons from a small thiol. Ann. N. Y. Acad. Sci. 973, 488–504.

Diderich, K.E., Nicolaije, C., Priemel, M., Waarsing, J.H., Day, J.S., Brandt, R.M., Schilling, A.F., Botter, S.M., Weinans, H., van der Horst, G.T., et al., 2012. Bone fragility and decline in stem cells in prematurely aging DNA repair deficient trichothiodystrophy mice. Age 34, 845–861.

Doonan, R., McElwee, J.J., Matthijssens, F., Walker, G.A., Houthoofd, K., Back, P., Matscheski, A., Vanfleteren, J.R., Gems, D., 2008. Against the oxidative damage theory of aging: superoxide dismutases protect against oxidative stress but have little or no effect on life span in *Caenorhabditis elegans*. Genes Dev. 22, 3236–3241.

Droge, W., 2002. Free radicals in the physiological control of cell function. Physiol. Rev. 82, 47–95.

Dutta, S., Sengupta, P., 2016. Men and mice: relating their ages. Life Sci. 152, 244–248.

Ehrenstein, M.R., Notley, C.A., 2010. The importance of natural IgM: scavenger, protector and regulator. Nat. Rev. Immunol. 10, 778–786.

Erben, R.G., 1996. Trabecular and endocortical bone surfaces in the rat: modeling or remodeling? Anat. Rec. 246, 39–46.

Farr, J.N., Fraser, D.G., Wang, H., Jaehn, K., Ogrodnik, M.B., Weivoda, M.M., Drake, M.T., Tchkonia, T., LeBrasseur, N.K., Kirkland, J.L., et al., 2016. Identification of senescent cells in the bone microenvironment. J. Bone Miner. Res. 31, 1920–1929.

Farr, J.N., Xu, M., Weivoda, M.M., Monroe, D.G., Fraser, D.G., Onken, J.L., Negley, B.A., Sfeir, J.G., Ogrodnik, M.B., Hachfeld, C.M., et al., 2017. Targeting cellular senescence prevents age-related bone loss in mice. Nat. Med. 23, 1072–1079.

Ferguson, V.L., Ayers, R.A., Bateman, T.A., Simske, S.J., 2003. Bone development and age-related bone loss in male C57BL/6J mice. Bone 33, 387–398.

Fiskesund, R., Stegmayr, B., Hallmans, G., Vikstrom, M., Weinehall, L., de Faire, U., Frostegard, J., 2010. Low levels of antibodies against phosphorylcholine predict development of stroke in a population-based study from northern Sweden. Stroke 41, 607–612.

Fiskesund, R., Su, J., Bulatovic, I., Vikstrom, M., de Faire, U., Frostegard, J., 2012. IgM phosphorylcholine antibodies inhibit cell death and constitute a strong protection marker for atherosclerosis development, particularly in combination with other auto-antibodies against modified LDL. Results Immunol. 2, 13–18.

Friedenstein, A.J., Chailakhjan, R.K., Latsinik, N.V., Panasyuk, A.F., Keiliss-Borok, I.V., 1974. Stromal cells responsible for transferring the microenvironment of the hemopoietic tissues. Cloning in Vitro and Retransplantation in Vivo. Transplantation 17, 331–340.

Frost, H.M., 1960. Micropetrosis. J. Bone Joint. Surg. Am. 42A, 144–150.

Frost, H.M., 2003. Bone's mechanostat: a 2003 update. Anat. Rec. A Discover. Mol. Cel. Evolut. Biol. 275, 1081–1101.

Garrett, I.R., Boyce, B.F., Oreffo, R.O., Bonewald, L., Poser, J., Mundy, G.R., 1990. Oxygen-derived free radicals stimulate osteoclastic bone resorption in rodent bone in vitro and in vivo. J. Clin. Investig. 85, 632–639.

Gems, D., Partridge, L., 2013. Genetics of longevity in model organisms: debates and paradigm shifts. Annu. Rev. Physiol. 75, 621–644.

Geng, J., Klionsky, D.J., 2008. The Atg8 and Atg12 ubiquitin-like conjugation systems in macroautophagy. 'Protein modifications: beyond the usual suspects' review series. EMBO Rep. 9, 859–864.

Glatt, V., Canalis, E., Stadmeyer, L., Bouxsein, M.L., 2007. Age-related changes in trabecular architecture differ in female and male C57BL/6J mice. J. Bone Miner. Res. 22, 1197–1207.

Goodell, M.A., Rando, T.A., 2015. Stem cells and healthy aging. Science 350, 1199–1204.

Grcevic, D., Pejda, S., Matthews, B.G., Repic, D., Wang, L., Li, H., Kronenberg, M.S., Jiang, X., Maye, P., Adams, D.J., et al., 2012. In vivo fate mapping identifies mesenchymal progenitor cells. Stem Cell. 30, 187–196.

Green, D.R., Galluzzi, L., Kroemer, G., 2011. Mitochondria and the autophagy-inflammation-cell death axis in organismal aging. Science 333, 1109–1112.

Griffin, D.O., Holodick, N.E., Rothstein, T.L., 2011. Human B1 cells in umbilical cord and adult peripheral blood express the novel phenotype $CD20^{+}$ $CD27^{+}$ $CD43^{+}$ $CD70^{-}$. J. Exp. Med. 208, 67–80.

Gronlund, H., Hallmans, G., Jansson, J.H., Boman, K., Wikstrom, M., de Faire, U., Frostegard, J., 2009. Low levels of IgM antibodies against phosphorylcholine predict development of acute myocardial infarction in a population-based cohort from northern Sweden. Eur. J. Cardiovasc. Prev. Rehabil. 16, 382–386.

Gronwall, C., Akhter, E., Oh, C., Burlingame, R.W., Petri, M., Silverman, G.J., 2012. IgM autoantibodies to distinct apoptosis-associated antigens correlate with protection from cardiovascular events and renal disease in patients with SLE. Clin. Immunol. 142, 390–398.

Ha, H., Kwak, H.B., Lee, S.W., Jin, H.M., Kim, H.M., Kim, H.H., Lee, Z.H., 2004. Reactive oxygen species mediate RANK signaling in osteoclasts. Exp. Cell Res. 301, 119–127.

Halloran, B.P., Ferguson, V.L., Simske, S.J., Burghardt, A., Venton, L.L., Majumdar, S., 2002. Changes in bone structure and mass with advancing age in the male C57BL/6J mouse. J. Bone Miner. Res. 17, 1044–1050.

Han, Z.H., Palnitkar, S., Rao, D.S., Nelson, D., Parfitt, A.M., 1997. Effects of ethnicity and age or menopause on the remodeling and turnover of iliac bone: implications for mechanisms of bone loss. J. Bone Miner. Res. 12, 498–508.

Hansen, M., Rubinsztein, D.C., Walker, D.W., 2018. Autophagy as a promoter of longevity: insights from model organisms. Nat. Rev. Mol. Cell Biol. 19, 579–593.

Harman, D., 1956. Aging: a theory based on free radical and radiation chemistry. J. Gerontol. 11, 298–300.

Hayflick, L., Moorhead, P.S., 1961. The serial cultivation of human diploid cell strains. Exp. Cell Res. 25, 585–621.

Hekimi, S., Lapointe, J., Wen, Y., 2011. Taking a "good" look at free radicals in the aging process. Trends Cell Biol. 21, 569–576.

Holguin, N., Brodt, M.D., Sanchez, M.E., Silva, M.J., 2014. Aging diminishes lamellar and woven bone formation induced by tibial compression in adult C57BL/6. Bone 65, 83–91.

Holguin, N., Brodt, M.D., Silva, M.J., 2016. Activation of Wnt signaling by mechanical loading is impaired in the bone of old mice. J. Bone Miner. Res. 31, 2215–2226.

Hoogeboom, D., Essers, M.A., Polderman, P.E., Voets, E., Smits, L.M., Burgering, B.M., 2008. Interaction of FOXO with beta-catenin inhibits beta-catenin/T cell factor activity. J. Biol. Chem. 283, 9224–9230.

Horowitz, M.C., Berry, R., Holtrup, B., Sebo, Z., Nelson, T., Fretz, J.A., Lindskog, D., Kaplan, J.L., Ables, G., Rodeheffer, M.S., et al., 2017. Bone marrow adipocytes. Adipocyte 6, 193–204.

Huang, S.C., Wu, T.C., Yu, H.C., Chen, M.R., Liu, C.M., Chiang, W.S., Lin, K.M., 2010. Mechanical strain modulates age-related changes in the proliferation and differentiation of mouse adipose-derived stromal cells. BMC Cell Biol. 11, 18.

Hughes, D.E., Dai, A., Tiffee, J.C., Li, H.H., Mundy, G.R., Boyce, B.F., 1996. Estrogen promotes apoptosis of murine osteoclasts mediated by TGF-b. Nat. Med. 2, 1132–1136.

Imhof, A., Koenig, W., Jaensch, A., Mons, U., Brenner, H., Rothenbacher, D., 2015. Long-term prognostic value of IgM antibodies against phosphorylcholine for adverse cardiovascular events in patients with stable coronary heart disease. Atherosclerosis 243, 414–420.

Ireland, A., Maden-Wilkinson, T., McPhee, J., Cooke, K., Narici, M., Degens, H., Rittweger, J., 2013. Upper limb muscle-bone asymmetries and bone adaptation in elite youth tennis players. Med. Sci. Sports Exerc. 45, 1749–1758.

Ishii, K.A., Fumoto, T., Iwai, K., Takeshita, S., Ito, M., Shimohata, N., Aburatani, H., Taketani, S., Lelliott, C.J., Vidal-Puig, A., et al., 2009. Coordination of PGC-1beta and iron uptake in mitochondrial biogenesis and osteoclast activation. Nat. Med. 15, 259–266.

Iyer, S., Ambrogini, E., Bartell, S.M., Han, L., Roberson, P.K., de Cabo, R., Jilka, R.L., Weinstein, R.S., O'Brien, C.A., Manolagas, S.C., et al., 2013. FOXOs attenuate bone formation by suppressing Wnt signaling. J. Clin. Investig. 123, 3409–3419.

Jackson, S.P., 2002. Sensing and repairing DNA double-strand breaks. Carcinogenesis 23, 687–696.

Janssen-Heininger, Y.M., Mossman, B.T., Heintz, N.H., Forman, H.J., Kalyanaraman, B., Finkel, T., Stamler, J.S., Rhee, S.G., van der Vliet, A., 2008. Redox-based regulation of signal transduction: principles, pitfalls, and promises. Free Radic. Biol. Med. 45, 1–17.

Javaheri, B., Stern, A.R., Lara, N., Dallas, M., Zhao, H., Liu, Y., Bonewald, L.F., Johnson, M.L., 2014. Deletion of a single beta-catenin allele in osteocytes abolishes the bone anabolic response to loading. J. Bone Miner. Res. 29, 705–715.

Jeon, O.H., Kim, C., Laberge, R.M., Demaria, M., Rathod, S., Vasserot, A.P., Chung, J.W., Kim, D.H., Poon, Y., David, N., et al., 2017. Local clearance of senescent cells attenuates the development of post-traumatic osteoarthritis and creates a pro-regenerative environment. Nat. Med. 23, 775–781.

Jilka, R.L., Almeida, M., Ambrogini, E., Han, L., Roberson, P.K., Weinstein, R.S., Manolagas, S.C., 2010. Decreased oxidative stress and greater bone anabolism in the aged, when compared to the young, murine skeleton with parathyroid hormone administration. Aging Cell 9, 851–867.

Jilka, R.L., Noble, B., Weinstein, R.S., 2013. Osteocyte apoptosis. Bone 54, 264–271.

Jilka, R.L., O'Brien, C.A., 2016. The role of osteocytes in age-related bone loss. Curr. Osteoporos. Rep. 14, 16–25.

Jilka, R.L., O'Brien, C.A., Roberson, P.K., Bonewald, L.F., Weinstein, R.S., Manolagas, S.C., 2014. Dysapoptosis of osteoblasts and osteocytes increases cancellous bone formation but exaggerates cortical porosity with age. J. Bone Miner. Res. 29, 103–117.

Jones, H.H., Priest, J.D., Hayes, W.C., Tichenor, C.C., Nagel, D.A., 1977. Humeral hypertrophy in response to exercise. J. Bone Joint Surg. Am. 59, 204–208.

Jurk, D., Wang, C., Miwa, S., Maddick, M., Korolchuk, V., Tsolou, A., Gonos, E.S., Thrasivoulou, C., Saffrey, M.J., Cameron, K., et al., 2012. Postmitotic neurons develop a p21-dependent senescence-like phenotype driven by a DNA damage response. Aging Cell 11, 996–1004.

Kang, C., Xu, Q., Martin, T.D., Li, M.Z., Demaria, M., Aron, L., Lu, T., Yankner, B.A., Campisi, J., Elledge, S.J., 2015. The DNA damage response induces inflammation and senescence by inhibiting autophagy of GATA4. Science 349 aaa5612.

Karvonen, J., Paivansalo, M., Kesaniemi, Y.A., Horkko, S., 2003. Immunoglobulin M type of autoantibodies to oxidized low-density lipoprotein has an inverse relation to carotid artery atherosclerosis. Circulation 108, 2107–2112.

Kasper, G., Mao, L., Geissler, S., Draycheva, A., Trippens, J., Kuhnisch, J., Tschirschmann, M., Kaspar, K., Perka, C., Duda, G.N., et al., 2009. Insights into mesenchymal stem cell aging: involvement of antioxidant defense and actin cytoskeleton. Stem Cell. 27, 1288–1297.

Kassem, M., Marie, P.J., 2011. Senescence-associated intrinsic mechanisms of osteoblast dysfunctions. Aging Cell 10, 191–197.

Kennedy, B.K., Berger, S.L., Brunet, A., Campisi, J., Cuervo, A.M., Epel, E.S., Franceschi, C., Lithgow, G.J., Morimoto, R.I., Pessin, J.E., et al., 2014. Geroscience: linking aging to chronic disease. Cell 159, 709–713.

Kerschnitzki, M., Kollmannsberger, P., Burghammer, M., Duda, G.N., Weinkamer, R., Wagermaier, W., Fratzl, P., 2013. Architecture of the osteocyte network correlates with bone material quality. J. Bone Miner. Res. 28, 1837–1845.

Khosla, S., 2010. Update on estrogens and the skeleton. J. Clin. Endocrinol. Metab. 95, 3569–3577.

Khosla, S., 2015. New insights into androgen and estrogen receptor regulation of the male skeleton. J. Bone Miner. Res. 30, 1134–1137.

Kim, H.N., Chang, J., Shao, L., Han, L., Iyer, S., Manolagas, S.C., O'Brien, C.A., Jilka, R.L., Zhou, D., Almeida, M., 2017. DNA damage and senescence in osteoprogenitors expressing Osx1 may cause their decrease with age. Aging Cell.

Koleske, A.J., 2013. Molecular mechanisms of dendrite stability. Nat. Rev. Neurosci. 14, 536–550.

Komatsu, M., Wang, Q.J., Holstein, G.R., Friedrich Jr., V.L., Iwata, J., Kominami, E., Chait, B.T., Tanaka, K., Yue, Z., 2007. Essential role for autophagy protein Atg7 in the maintenance of axonal homeostasis and the prevention of axonal degeneration. Proc. Natl. Acad. Sci. U. S. A. 104, 14489–14494.

Korpelainen, R., Keinanen-Kiukaanniemi, S., Heikkinen, J., Vaananen, K., Korpelainen, J., 2006. Effect of impact exercise on bone mineral density in elderly women with low BMD: a population-based randomized controlled 30-month intervention. Osteoporos. Int. 17, 109–118.

Kuhn, H.G., Dickinson-Anson, H., Gage, F.H., 1996. Neurogenesis in the dentate gyrus of the adult rat: age-related decrease of neuronal progenitor proliferation. J. Neurosci. 16, 2027–2033.

Kuilman, T., Michaloglou, C., Mooi, W.J., Peeper, D.S., 2010. The essence of senescence. Genes Dev. 24, 2463–2479.

Kusumbe, A.P., Ramasamy, S.K., Adams, R.H., 2014. Coupling of angiogenesis and osteogenesis by a specific vessel subtype in bone. Nature 507, 323–328.

Lang, T., LeBlanc, A., Evans, H., Lu, Y., Genant, H., Yu, A., 2004. Cortical and trabecular bone mineral loss from the spine and hip in long-duration spaceflight. J. Bone Miner. Res. 19, 1006–1012.

Lara-Castillo, N., Kim-Weroha, N.A., Kamel, M.A., Javaheri, B., Ellies, D.L., Krumlauf, R.E., Thiagarajan, G., Johnson, M.L., 2015. In vivo mechanical loading rapidly activates beta-catenin signaling in osteocytes through a prostaglandin mediated mechanism. Bone 76, 58–66.

Lean, J.M., Davies, J.T., Fuller, K., Jagger, C.J., Kirstein, B., Partington, G.A., Urry, Z.L., Chambers, T.J., 2003. A crucial role for thiol antioxidants in estrogen-deficiency bone loss. J. Clin. Investig. 112, 915–923.

Lean, J.M., Jagger, C.J., Kirstein, B., Fuller, K., Chambers, T.J., 2005. Hydrogen peroxide is essential for estrogen-deficiency bone loss and osteoclast formation. Endocrinology 146, 728–735.

LeBlanc, A.D., Schneider, V.S., Evans, H.J., Engelbretson, D.A., Krebs, J.M., 1990. Bone mineral loss and recovery after 17 weeks of bed rest. J. Bone Miner. Res. 5, 843–850.

Lee, H.Y., Choi, C.S., Birkenfeld, A.L., Alves, T.C., Jornayvaz, F.R., Jurczak, M.J., Zhang, D., Woo, D.K., Shadel, G.S., Ladiges, W., et al., 2010. Targeted expression of catalase to mitochondria prevents age-associated reductions in mitochondrial function and insulin resistance. Cell Metabol. 12, 668–674.

Lee, N.K., Choi, Y.G., Baik, J.Y., Han, S.Y., Jeong, D.W., Bae, Y.S., Kim, N., Lee, S.Y., 2005. A crucial role for reactive oxygen species in RANKL-induced osteoclast differentiation. Blood 106, 852–859.

Levine, B., Kroemer, G., 2008. Autophagy in the pathogenesis of disease. Cell 132, 27–42.

Li, C.J., Cheng, P., Liang, M.K., Chen, Y.S., Lu, Q., Wang, J.Y., Xia, Z.Y., Zhou, H.D., Cao, X., Xie, H., et al., 2015. MicroRNA-188 regulates age-related switch between osteoblast and adipocyte differentiation. J. Clin. Investig. 125, 1509–1522.

Li, C.Y., Schaffler, M.B., Wolde-Semait, H.T., Hernandez, C.J., Jepsen, K.J., 2005. Genetic background influences cortical bone response to ovariectomy. J. Bone Miner. Res. 20, 2150–2158.

Lips, P., Courpron, P., Meunier, P.J., 1978. Mean wall thickness of trabecular bone packets in the human iliac crest: changes with age. Calcif. Tissue Res. 26, 13–17.

Lithgow, G.J., Kirkwood, T.B.L., 1996. Mechanisms and evolution of aging. Science 273, 80.

Liu, F., Fang, F., Yuan, H., Yang, D., Chen, Y., Williams, L., Goldstein, S.A., Krebsbach, P.H., Guan, J.L., 2013. Suppression of autophagy by FIP200 deletion leads to osteopenia in mice through the inhibition of osteoblast terminal differentiation. J. Bone Miner. Res. 28, 2414–2430.

Liu, H., Fergusson, M.M., Wu, J.J., Rovira, I.I., Liu, J., Gavrilova, O., Lu, T., Bao, J., Han, D., Sack, M.N., et al., 2011. Wnt signaling regulates hepatic metabolism. Sci. Signal. 4, ra6.

Liu, W., Wang, S., Wei, S., Sun, L., Feng, X., 2005. Receptor activator of NF-kappaB (RANK) cytoplasmic motif, 369PFQEP373, plays a predominant role in osteoclast survival in part by activating Akt/PKB and its downstream effector AFX/FOXO4. J. Biol. Chem. 280, 43064–43072.

Liu, Y., Almeida, M., Weinstein, R.S., O'Brien, C.A., Manolagas, S.C., Jilka, R.L., 2016. Skeletal inflammation and attenuation of Wnt signaling, Wnt ligand expression, and bone formation in atherosclerotic ApoE-null mice. Am. J. Physiol. Endocrinol. Metab. 310, E762–E773.

Looker, A.C., Wahner, H.W., Dunn, W.L., Calvo, M.S., Harris, T.B., Heyse, S.P., Johnston Jr., C.C., Lindsay, R., 1998. Updated data on proximal femur bone mineral levels of US adults. Osteoporos. Int. 8, 468–489.

Lopez-Otin, C., Blasco, M.A., Partridge, L., Serrano, M., Kroemer, G., 2013. The hallmarks of aging. Cell 153, 1194–1217.

Maatta, J.A., Buki, K.G., Ivaska, K.K., Nieminen-Pihala, V., Elo, T.D., Kahkonen, T., Poutanen, M., Harkonen, P., Vaananen, K., 2013. Inactivation of the androgen receptor in bone-forming cells leads to trabecular bone loss in adult female mice. BoneKEy Rep. 2, 440.

Makovey, J., Chen, J.S., Hayward, C., Williams, F.M., Sambrook, P.N., 2009. Association between serum cholesterol and bone mineral density. Bone 44, 208–213.

Manolagas, S.C., 2000. Birth and death of bone cells: basic regulatory mechanisms and implications for the pathogenesis and treatment of osteoporosis. Endocr. Rev. 21, 115–137.

Manolagas, S.C., 2010. From estrogen-centric to aging and oxidative stress: a revised perspective of the pathogenesis of osteoporosis. Endocr. Rev. 31, 266–300.

Manolagas, S.C., 2013. Steroids and osteoporosis: the quest for mechanisms. J. Clin. Investig. 123, 1919–1921.

Manolagas, S.C., O'Brien, C.A., Almeida, M., 2013. The role of estrogen and androgen receptors in bone health and disease. Nat. Rev. Endocrinol. 9, 699–712.

Manolagas, S.C., Parfitt, A.M., 2010. What old means to bone. Trends Endocrinol. Metab. 21, 369–374.

Martin-Millan, M., Almeida, M., Ambrogini, E., Han, L., Zhao, H., Weinstein, R.S., Jilka, R.L., O'Brien, C.A., Manolagas, S.C., 2010. The estrogen receptor-alpha in osteoclasts mediates the protective effects of estrogens on cancellous but not cortical bone. Mol. Endocrinol. 24, 323–334.

McNeill, T.H., Brown, S.A., Rafols, J.A., Shoulson, I., 1988. Atrophy of medium spiny I striatal dendrites in advanced Parkinson's disease. Brain Res. 455, 148–152.

Meakin, L.B., Galea, G.L., Sugiyama, T., Lanyon, L.E., Price, J.S., 2014. Age-related impairment of bones' adaptive response to loading in mice is associated with sex-related deficiencies in osteoblasts but no change in osteocytes. J. Bone Miner. Res. 29, 1859–1871.

Mesquita, A., Weinberger, M., Silva, A., Sampaio-Marques, B., Almeida, B., Leao, C., Costa, V., Rodrigues, F., Burhans, W.C., Ludovico, P., 2010. Caloric restriction or catalase inactivation extends yeast chronological lifespan by inducing H_2O_2 and superoxide dismutase activity. Proc. Natl. Acad. Sci. U. S. A. 107, 15123–15128.

Minamino, T., Orimo, M., Shimizu, I., Kunieda, T., Yokoyama, M., Ito, T., Nojima, A., Nabetani, A., Oike, Y., Matsubara, H., et al., 2009. A crucial role for adipose tissue p53 in the regulation of insulin resistance. Nat. Med. 15, 1082–1087.

Mine-Hattab, J., Rothstein, R., 2012. Increased chromosome mobility facilitates homology search during recombination. Nat. Cell Biol. 14, 510–517.

Molofsky, A.V., Slutsky, S.G., Joseph, N.M., He, S., Pardal, R., Krishnamurthy, J., Sharpless, N.E., Morrison, S.J., 2006. Increasing p16INK4a expression decreases forebrain progenitors and neurogenesis during ageing. Nature 443, 448–452.

Morimoto, R.I., Cuervo, A.M., 2014. Proteostasis and the aging proteome in health and disease. J. Gerontol. A Biol. Sci. Med. Sci. 69 (Suppl. 1), S33–S38.

Morrison, S.J., Wandycz, A.M., Akashi, K., Globerson, A., Weissman, I.L., 1996. The aging of hematopoietic stem cells. Nat. Med. 2, 1011–1016.

Mortensen, M., Soilleux, E.J., Djordjevic, G., Tripp, R., Lutteropp, M., Sadighi-Akha, E., Stranks, A.J., Glanville, J., Knight, S., Jacobsen, S.E., et al., 2011. The autophagy protein Atg7 is essential for hematopoietic stem cell maintenance. J. Exp. Med. 208, 455–467.

Moustafa, A., Sugiyama, T., Prasad, J., Zaman, G., Gross, T.S., Lanyon, L.E., Price, J.S., 2012. Mechanical loading-related changes in osteocyte sclerostin expression in mice are more closely associated with the subsequent osteogenic response than the peak strains engendered. Osteoporos. Int. 23, 1225–1234.

Munoz-Espin, D., Serrano, M., 2014. Cellular senescence: from physiology to pathology. Nat. Rev. Mol. Cell Biol. 15, 482–496.

Nakamura, T., Imai, Y., Matsumoto, T., Sato, S., Takeuchi, K., Igarashi, K., Harada, Y., Azuma, Y., Krust, A., Yamamoto, Y., et al., 2007. Estrogen prevents bone loss via estrogen receptor alpha and induction of Fas ligand in osteoclasts. Cell 130, 811–823.

Newgard, C.B., Sharpless, N.E., 2013. Coming of age: molecular drivers of aging and therapeutic opportunities. J. Clin. Investig. 123, 946–950.

Niccoli, T., Partridge, L., 2012. Ageing as a risk factor for disease. Curr. Biol. 22, R741–R752.

Nicks, K.M., Amin, S., Atkinson, E.J., Riggs, B.L., Melton III, L.J., Khosla, S., 2012. Relationship of age to bone microstructure independent of areal bone mineral density. J. Bone Miner. Res. 27, 637–644.

Nicolaije, C., Diderich, K.E., Botter, S.M., Priemel, M., Waarsing, J.H., Day, J.S., Brandt, R.M., Schilling, A.F., Weinans, H., Van der Eerden, B.C., et al., 2012. Age-related skeletal dynamics and decrease in bone strength in DNA repair deficient male trichothiodystrophy mice. PLoS One 7, e35246.

Niedernhofer, L.J., Garinis, G.A., Raams, A., Lalai, A.S., Robinson, A.R., Appeldoorn, E., Odijk, H., Oostendorp, R., Ahmad, A., van, L.W., et al., 2006. A new progeroid syndrome reveals that genotoxic stress suppresses the somatotroph axis. Nature 444, 1038–1043.

Nilsson, M.E., Vandenput, L., Tivesten, A., Norlen, A.K., Lagerquist, M.K., Windahl, S.H., Borjesson, A.E., Farman, H.H., Poutanen, M., Benrick, A., et al., 2015. Measurement of a comprehensive sex steroid profile in rodent serum by high-sensitive gas chromatography-tandem mass spectrometry. Endocrinology 156, 2492–2502.

Nirody, J.A., Cheng, K.P., Parrish, R.M., Burghardt, A.J., Majumdar, S., Link, T.M., Kazakia, G.J., 2015. Spatial distribution of intracortical porosity varies across age and sex. Bone 75, 88–95.

Noble, B.S., Stevens, H., Loveridge, N., Reeve, J., 1997. Identification of apoptotic changes in osteocytes in normal and pathological human bone. Bone 20, 273–282.

Nollet, M., Santucci-Darmanin, S., Breuil, V., Al-Sahlanee, R., Cros, C., Topi, M., Momier, D., Samson, M., Pagnotta, S., Cailleteau, L., et al., 2014. Autophagy in osteoblasts is involved in mineralization and bone homeostasis. Autophagy 10, 1965–1977.

Notini, A.J., McManus, J.F., Moore, A., Bouxsein, M., Jimenez, M., Chiu, W.S., Glatt, V., Kream, B.E., Handelsman, D.J., Morris, H.A., et al., 2007. Osteoblast deletion of exon 3 of the androgen receptor gene results in trabecular bone loss in adult male mice. J. Bone Miner. Res. 22, 347–356.

Ogrodnik, M., Miwa, S., Tchkonia, T., Tiniakos, D., Wilson, C.L., Lahat, A., Day, C.P., Burt, A., Palmer, A., Anstee, Q.M., et al., 2017. Cellular senescence drives age-dependent hepatic steatosis. Nat. Commun. 8, 15691.

Onal, M., Piemontese, M., Xiong, J., Wang, Y., Han, L., Ye, S., Komatsu, M., Selig, M., Weinstein, R.S., Zhao, H., et al., 2013. Suppression of autophagy in osteocytes mimics skeletal aging. J. Biol. Chem. 288, 17432–17440.

Ono, N., Ono, W., Nagasawa, T., Kronenberg, H.M., 2014. A subset of chondrogenic cells provides early mesenchymal progenitors in growing bones. Nat. Cell Biol. 16, 1157–1167.

Paik, J.H., Ding, Z., Narurkar, R., Ramkissoon, S., Muller, F., Kamoun, W.S., Chae, S.S., Zheng, H., Ying, H., Mahoney, J., et al., 2009. FoxOs cooperatively regulate diverse pathways governing neural stem cell homeostasis. Cell Stem Cell 5, 540–553.

Parfitt, A.M., 1990. Bone-forming cells in clinical conditions. In: Hall, B.K. (Ed.), Bone, The Osteoblast and Osteocyte, vol. 1. Telford Press and CRC Press, Boca Raton, FL, pp. 351–429.

Park, D., Spencer, J.A., Koh, B.I., Kobayashi, T., Fujisaki, J., Clemens, T.L., Lin, C.P., Kronenberg, H.M., Scadden, D.T., 2012. Endogenous bone marrow MSCs are dynamic, fate-restricted participants in bone maintenance and regeneration. Cell Stem Cell 10, 259–272.

Pazzaglia, U.E., Sibilia, V., Congiu, T., Pagani, F., Ravanelli, M., Zarattini, G., 2015. Setup of a bone aging experimental model in the rabbit comparing changes in cortical and trabecular bone: morphological and morphometric study in the femur. J. Morphol. 276, 733–747.

Piemontese, M., Almeida, M., Robling, A.G., Kim, H.N., Xiong, J., Thostenson, J.D., Weinstein, R.S., Manolagas, S.C., O'Brien, C.A., Jilka, R.L., 2017. Old age causes de novo intracortical bone remodeling and porosity in mice. JCI Insight 2.

Piemontese, M., Onal, M., Xiong, J., Han, L., Thostenson, J.D., Almeida, M., O'Brien, C.A., 2016. Low bone mass and changes in the osteocyte network in mice lacking autophagy in the osteoblast lineage. Sci. Rep. 6, 24262.

Power, J., Loveridge, N., Lyon, A., Rushton, N., Parker, M., Reeve, J., 2003. Bone remodeling at the endocortical surface of the human femoral neck: a mechanism for regional cortical thinning in cases of hip fracture. J. Bone Miner. Res. 18, 1775–1780.

Price, C., Herman, B.C., Lufkin, T., Goldman, H.M., Jepsen, K.J., 2005. Genetic variation in bone growth patterns defines adult mouse bone fragility. J. Bone Miner. Res. 20, 1983–1991.

Qiu, S., Rao, D.S., Palnitkar, S., Parfitt, A.M., 2002a. Age and distance from the surface but not menopause reduce osteocyte density in human cancellous bone. Bone 31, 313–318.

Qiu, S., Rao, D.S., Palnitkar, S., Parfitt, A.M., 2002b. Relationships between osteocyte density and bone formation rate in human cancellous bone. Bone 31, 709–711.

Que, X., Hung, M.Y., Yeang, C., Gonen, A., Prohaska, T.A., Sun, X., Diehl, C., Maatta, A., Gaddis, D.E., Bowden, K., et al., 2018. Oxidized phospholipids are proinflammatory and proatherogenic in hypercholesterolaemic mice. Nature 558, 301–306.

Razi, H., Birkhold, A.I., Weinkamer, R., Duda, G.N., Willie, B.M., Checa, S., 2015. Aging leads to a dysregulation in mechanically driven bone formation and resorption. J. Bone Miner. Res. 30, 1864–1873.

Robinson, J.A., Chatterjee-Kishore, M., Yaworsky, P.J., Cullen, D.M., Zhao, W., Li, C., Kharode, Y., Sauter, L., Babij, P., Brown, E.L., et al., 2006. Wnt/beta-catenin signaling is a normal physiological response to mechanical loading in bone. J. Biol. Chem. 281, 31720–31728.

Robling, A.G., Castillo, A.B., Turner, C.H., 2006. Biomechanical and molecular regulation of bone remodeling. Annu. Rev. Biomed. Eng. 8, 455–498.

Robling, A.G., Niziolek, P.J., Baldridge, L.A., Condon, K.W., Allen, M.R., Alam, I., Mantila, S.M., Gluhak-Heinrich, J., Bellido, T.M., Harris, S.E., et al., 2008. Mechanical stimulation of bone in vivo reduces osteocyte expression of Sost/sclerostin. J. Biol. Chem. 283, 5866–5875.

Rossi, D.J., Jamieson, C.H., Weissman, I.L., 2008. Stems cells and the pathways to aging and cancer. Cell 132, 681–696.

Rouse, J., Jackson, S.P., 2002. Interfaces between the detection, signaling, and repair of DNA damage. Science 297, 547–551.

Rubin, C.T., Bain, S.D., McLeod, K.J., 1992. Suppression of the osteogenic response in the aging skeleton. Calcif. Tissue Int. 50, 306–313.

Rubin, C.T., Lanyon, L.E., 1985. Regulation of bone mass by mechanical strain magnitude. Calcif. Tissue Int. 37, 411–417.

Ryu, B.Y., Orwig, K.E., Oatley, J.M., Avarbock, M.R., Brinster, R.L., 2006. Effects of aging and niche microenvironment on spermatogonial stem cell self-renewal. Stem Cell. 24, 1505–1511.

Saeed, H., Abdallah, B.M., Ditzel, N., Catala-Lehnen, P., Qiu, W., Amling, M., Kassem, M., 2011. Telomerase-deficient mice exhibit bone loss owing to defects in osteoblasts and increased osteoclastogenesis by inflammatory microenvironment. J. Bone Miner. Res. 26, 1494–1505.

Salminen, A., Kauppinen, A., Kaarniranta, K., 2012. Emerging role of NF-kappaB signaling in the induction of senescence-associated secretory phenotype (SASP). Cell. Signal. 24, 835–845.

Salmon, A.B., Richardson, A., Perez, V.I., 2010. Update on the oxidative stress theory of aging: does oxidative stress play a role in aging or healthy aging? Free Radic. Biol. Med. 48, 642–655.

Sawakami, K., Robling, A.G., Ai, M., Pitner, N.D., Liu, D., Warden, S.J., Li, J., Maye, P., Rowe, D.W., Duncan, R.L., et al., 2006. The Wnt co-receptor LRP5 is essential for skeletal mechanotransduction but not for the anabolic bone response to parathyroid hormone treatment. J. Biol. Chem. 281, 23698–23711.

Scaffidi, P., Misteli, T., 2008. Lamin A-dependent misregulation of adult stem cells associated with accelerated ageing. Nat. Cell Biol. 10, 452–459.

Schafer, M.J., White, T.A., Iijima, K., Haak, A.J., Ligresti, G., Atkinson, E.J., Oberg, A.L., Birch, J., Salmonowicz, H., Zhu, Y., et al., 2017. Cellular senescence mediates fibrotic pulmonary disease. Nat. Commun. 8, 14532.

Schneider, P., Krucker, T., Meyer, E., Ulmann-Schuler, A., Weber, B., Stampanoni, M., Muller, R., 2009. Simultaneous 3D visualization and quantification of murine bone and bone vasculature using micro-computed tomography and vascular replica. Microsc. Res. Tech. 72, 690–701.

Schriner, S.E., Linford, N.J., Martin, G.M., Treuting, P., Ogburn, C.E., Emond, M., Coskun, P.E., Ladiges, W., Wolf, N., Van, R.H., et al., 2005. Extension of murine life span by overexpression of catalase targeted to mitochondria. Science 308, 1909–1911.

Sedelnikova, O.A., Horikawa, I., Zimonjic, D.B., Popescu, N.C., Bonner, W.M., Barrett, J.C., 2004. Senescing human cells and ageing mice accumulate DNA lesions with unrepairable double-strand breaks. Nat. Cell Biol. 6, 168–170.

Seeman, E., 2013. Age- and menopause-related bone loss compromise cortical and trabecular microstructure. J. Gerontol. A Biol. Sci. Med. Sci. 68, 1218–1225.

Sena, L.A., Chandel, N.S., 2012. Physiological roles of mitochondrial reactive oxygen species. Mol. Cell. 48, 158–167.

Sethe, S., Scutt, A., Stolzing, A., 2006. Aging of mesenchymal stem cells. Ageing Res. Rev. 5, 91–116.

Shanbhogue, V.V., Brixen, K., Hansen, S., 2016. Age- and sex-related changes in bone microarchitecture and estimated strength: a three-year prospective study using HRpQCT. J. Bone Miner. Res. 31, 1541–1549.

Sharpless, N.E., DePinho, R.A., 2007. How stem cells age and why this makes us grow old. Nat. Rev. Mol. Cell Biol. 8, 703–713.

Shefer, G., Van de Mark, D.P., Richardson, J.B., Yablonka-Reuveni, Z., 2006. Satellite-cell pool size does matter: defining the myogenic potency of aging skeletal muscle. Dev. Biol. 294, 50–66.

Sinnesael, M., Claessens, F., Laurent, M., Dubois, V., Boonen, S., Deboel, L., Vanderschueren, D., 2012. Androgen receptor (AR) in osteocytes is important for the maintenance of male skeletal integrity: evidence from targeted AR disruption in mouse osteocytes. J. Bone Miner. Res. 27, 2535–2543.

Smogorzewska, A., de Lange, T., 2002. Different telomere damage signaling pathways in human and mouse cells. EMBO J. 21, 4338–4348.

Soutoglou, E., Misteli, T., 2008. On the contribution of spatial genome organization to cancerous chromosome translocations. J. Natl. Cancer Inst. Monogr. 16–19.

Srinivasan, S., Gross, T.S., Bain, S.D., 2012. Bone mechanotransduction may require augmentation in order to strengthen the senescent skeleton. Ageing Res. Rev. 11, 353–360.

Steinberg, D., Witztum, J.L., 2010. Oxidized low-density lipoprotein and atherosclerosis. Arterioscler. Thromb. Vasc. Biol. 30, 2311–2316.

Stenderup, K., Justesen, J., Clausen, C., Kassem, M., 2003. Aging is associated with decreased maximal life span and accelerated senescence of bone marrow stromal cells. Bone 33, 919–926.

Stice, J.P., Knowlton, A.A., 2008. Estrogen, NFkappaB, and the heat shock response. Mol. Med. 14, 517–527.

Talchai, C., Xuan, S., Kitamura, T., DePinho, R.A., Accili, D., 2012. Generation of functional insulin-producing cells in the gut by Foxo1 ablation. Nat. Genet. 44, 406–412. S401.

Tan, P., Guan, H., Xie, L., Mi, B., Fang, Z., Li, J., Li, F., 2015. FOXO1 inhibits osteoclastogenesis partially by antagonizing MYC. Sci. Rep. 5, 16835.

Tang, G., Gudsnuk, K., Kuo, S.H., Cotrina, M.L., Rosoklija, G., Sosunov, A., Sonders, M.S., Kanter, E., Castagna, C., Yamamoto, A., et al., 2014. Loss of mTOR-dependent macroautophagy causes autistic-like synaptic pruning deficits. Neuron 83, 1131–1143.

Tiede-Lewis, L.M., Xie, Y., Hulbert, M.A., Campos, R., Dallas, M.R., Dusevich, V., Bonewald, L.F., Dallas, S.L., 2017. Degeneration of the osteocyte network in the C57BL/6 mouse model of aging. Aging 9, 2190–2208.

Tintut, Y., Demer, L.L., 2014. Effects of bioactive lipids and lipoproteins on bone. Trends Endocrinol. Metab. 25, 53–59.

Tintut, Y., Morony, S., Demer, L.L., 2004. Hyperlipidemia promotes osteoclastic potential of bone marrow cells ex vivo. Arterioscler. Thromb. Vasc. Biol. 24, e6–10.

Tsiantoulas, D., Diehl, C.J., Witztum, J.L., Binder, C.J., 2014. B cells and humoral immunity in atherosclerosis. Circ. Res. 114, 1743–1756.

Tsimikas, S., Brilakis, E.S., Lennon, R.J., Miller, E.R., Witztum, J.L., McConnell, J.P., Kornman, K.S., Berger, P.B., 2007. Relationship of IgG and IgM autoantibodies to oxidized low density lipoprotein with coronary artery disease and cardiovascular events. J. Lipid Res. 48, 425–433.

Tu, X., Rhee, Y., Condon, K.W., Bivi, N., Allen, M.R., Dwyer, D., Stolina, M., Turner, C.H., Robling, A.G., Plotkin, L.I., et al., 2012. Sost down-regulation and local Wnt signaling are required for the osteogenic response to mechanical loading. Bone 50, 209–217.

Turner, C.H., Takano, Y., Owan, I., 1995. Aging changes mechanical loading thresholds for bone formation in rats. J. Bone Miner. Res. 10, 1544–1549.

Ucer, S., Iyer, S., Bartell, S.M., Martin-Millan, M., Han, L., Kim, H.N., Weinstein, R.S., Jilka, R.L., O'Brien, C.A., Almeida, M., et al., 2015. The effects of androgens on murine cortical bone do not require AR or ERalpha signaling in osteoblasts and osteoclasts. J. Bone Miner. Res. 30, 1138–1149.

Ucer, S., Iyer, S., Kim, H.N., Han, L., Rutlen, C., Allison, K., Thostenson, J.D., de, C.R., Jilka, R.L., O'Brien, C., et al., 2017. The effects of aging and sex steroid deficiency on the murine skeleton are independent and mechanistically distinct. J. Bone Miner. Res. 32, 560–574.

Vainionpaa, A., Korpelainen, R., Sievanen, H., Vihriala, E., Leppaluoto, J., Jamsa, T., 2007. Effect of impact exercise and its intensity on bone geometry at weight-bearing tibia and femur. Bone 40, 604–611.

Van Remmen, H., Ikeno, Y., Hamilton, M., Pahlavani, M., Wolf, N., Thorpe, S.R., Alderson, N.L., Baynes, J.W., Epstein, C.J., Huang, T.T., et al., 2003. Life-long reduction in MnSOD activity results in increased DNA damage and higher incidence of cancer but does not accelerate aging. Physiol. Genom. 16, 29–37.

Vanderschueren, D., Laurent, M.R., Claessens, F., Gielen, E., Lagerquist, M.K., Vandenput, L., Borjesson, A.E., Ohlsson, C., 2014. Sex steroid actions in male bone. Endocr. Rev. 35, 906–960.

Vas, J., Gronwall, C., Silverman, G.J., 2013. Fundamental roles of the innate-like repertoire of natural antibodies in immune homeostasis. Front. Immunol. 4, 4.

Vashishth, D., Verborgt, O., Divine, G., Schaffler, M.B., Fyhrie, D.P., 2000. Decline in osteocyte lacunar density in human cortical bone is associated with accumulation of microcracks with age. Bone 26, 375–380.

Vijg, J., Campisi, J., 2008. Puzzles, promises and a cure for ageing. Nature 454, 1065–1071.

Wanagat, J., Dai, D.F., Rabinovitch, P., 2010. Mitochondrial oxidative stress and mammalian healthspan. Mech. Ageing Dev. 131, 527–535.

Warr, M.R., Binnewies, M., Flach, J., Reynaud, D., Garg, T., Malhotra, R., Debnath, J., Passegue, E., 2013. FOXO3A directs a protective autophagy program in haematopoietic stem cells. Nature 494, 323–327.

Wijnhoven, S.W., Beems, R.B., Roodbergen, M., van den Berg, J., Lohman, P.H., Diderich, K., van der Horst, G.T., Vijg, J., Hoeijmakers, J.H., van Steeg, H., 2005. Accelerated aging pathology in ad libitum fed Xpd(TTD) mice is accompanied by features suggestive of caloric restriction. DNA Repair 4, 1314–1324.

Witztum, J.L., Lichtman, A.H., 2014. The influence of innate and adaptive immune responses on atherosclerosis. Annu. Rev. Pathol. 9, 73–102.

Xiong, J., Onal, M., Jilka, R.L., Weinstein, R.S., Manolagas, S.C., O'Brien, C.A., 2011. Matrix-embedded cells control osteoclast formation. Nat. Med. 17, 1235–1241.

Yamasaki, N., Tsuboi, H., Hirao, M., Nampei, A., Yoshikawa, H., Hashimoto, J., 2009. High oxygen tension prolongs the survival of osteoclast precursors via macrophage colony-stimulating factor. Bone 44, 71–79.

Yang, Z., Klionsky, D.J., 2010. Eaten alive: a history of macroautophagy. Nat. Cell Biol. 12, 814–822.

Zebaze, R.M., Ghasem-Zadeh, A., Bohte, A., Iuliano-Burns, S., Mirams, M., Price, R.I., Mackie, E.J., Seeman, E., 2010. Intracortical remodelling and porosity in the distal radius and post-mortem femurs of women: a cross-sectional study. Lancet 375, 1729–1736.

Zhang, X., Ebata, K.T., Robaire, B., Nagano, M.C., 2006. Aging of male germ line stem cells in mice. Biol. Reprod. 74, 119–124.

Zhang, Y., Ikeno, Y., Qi, W., Chaudhuri, A., Li, Y., Bokov, A., Thorpe, S.R., Baynes, J.W., Epstein, C., Richardson, A., et al., 2009. Mice deficient in both Mn superoxide dismutase and glutathione peroxidase-1 have increased oxidative damage and a greater incidence of pathology but no reduction in longevity. J. Gerontol. A Biol. Sci. Med. Sci. 64, 1212–1220.

Zhao, J., Brault, J.J., Schild, A., Goldberg, A.L., 2008. Coordinate activation of autophagy and the proteasome pathway by FoxO transcription factor. Autophagy 4, 378–380.

Zhao, L., Shim, J.W., Dodge, T.R., Robling, A.G., Yokota, H., 2013. Inactivation of Lrp5 in osteocytes reduces young's modulus and responsiveness to the mechanical loading. Bone 54, 35–43.

Zhou, B.O., Yue, R., Murphy, M.M., Peyer, J.G., Morrison, S.J., 2014. Leptin-receptor-expressing mesenchymal stromal cells represent the main source of bone formed by adult bone marrow. Cell Stem Cell 15, 154–168.

Zhou, S., Greenberger, J.S., Epperly, M.W., Goff, J.P., Adler, C., Leboff, M.S., Glowacki, J., 2008. Age-related intrinsic changes in human bone-marrow-derived mesenchymal stem cells and their differentiation to osteoblasts. Aging Cell 7, 335–343.

Zhu, Y., Tchkonia, T., Pirtskhalava, T., Gower, A.C., Ding, H., Giorgadze, N., Palmer, A.K., Ikeno, Y., Hubbard, G.B., Lenburg, M., et al., 2015. The Achilles' heel of senescent cells: from transcriptome to senolytic drugs. Aging Cell 14, 644–658.

Section B

Biochemistry

Chapter 13

Type I collagen structure, synthesis, and regulation

George Bou-Gharios[1], David Abraham[2] and Benoit de Crombrugghe[3]

[1]Institute of Ageing and Chronic Disease, University of Liverpool, Liverpool, United Kingdom; [2]Centre for Rheumatology and Connective Tissue Diseases, University College London, London, United Kingdom; [3]The University of Texas M.D. Anderson Cancer Center, Houston, TX, United States

Chapter outline

Introduction 295
The family of fibrillar collagens 296
Structure, biosynthesis, transport, and assembly of type I collagen 297
Structure 297
Regulation of transcription 297
Control of translation 299
Intracellular transport 300
Fibrillogenesis 301
Assembly 301
Consequences of genetic mutations on type I collagen formation 302
Collagen type I degradation and catabolism 303
Collagen type I and bone pathologies 304
Transcriptional regulation of type I collagen genes 305
Proximal promoters of type I collagen genes 305
Transcription factors binding to the pro-α1(I) proximal promoter 305
Factors binding to the pro-COL1A2 proximal promoter 307
Structure and functional organization of upstream segments of type I collagen genes 308
Upstream elements in the pro-*Col1a1* gene 308
Upstream elements of the mouse pro-*Col1a2* gene 310
Delineating the mode of action of tissue-specific elements 311
Role of the first intronic elements in regulating collagen type I 312
First intron of the pro-*COL1A1* gene 312
First intron of the pro-*COL1A2* gene 313
Posttranscriptional regulation of type I collagen 313
Critical factors involved in type I collagen gene regulation 314
Growth factors 314
Transforming growth factor β 314
Connective tissue growth factor 316
Fibroblast growth factor 317
Insulin-like growth factor 318
Cytokines 319
Tumor necrosis factor α 319
Interferon γ 319
Other cytokines 320
Arachidonic acid derivatives 321
Hormones and vitamins 322
Corticosteroids 322
Thyroid hormones 323
Parathyroid hormone 323
Vitamin D 323
References 324

Introduction

It has been over 150 years since the term "collagen" was first adopted in the English language. This ropelike structure that yields gelatin upon boiling made its early appearance in evolution in primitive animals such as jellyfish, coral, and sea anemones (Bergeon, 1967). Today, the collagen family of proteins has grown to 28 different types and is used as a versatile biomaterial for delivery of drugs as well as for cosmetic purposes.

The work of Nageotte in the early 1920s used acid solubilization to reveal the fibers that histologists had earlier described in sections of connective tissues (Nageotte, 1927); X-ray diffraction and then electron microscopy characterized those fibers that made up the collagen molecule. In addition to collagens involved in fibril formation, several other groups

Principles of Bone Biology. https://doi.org/10.1016/B978-0-12-814841-9.00013-0

of nonfibrillar collagen have been discovered. Among these, some are involved in the formation of membranes that surround tissues, such as basement and Descemet's membranes, the cuticle of worms, and the skeleton of sponges. During those 150 years our understanding of collagen has evolved with advances in techniques and technology. This led to the discovery that collagens mediate adulthood extracellular matrix (ECM) remodeling and are needed for aging to be delayed (Ewald et al., 2015). The importance of collagen production in diverse antiaging interventions implies that ECM remodeling is a generally essential signature of longevity assurance.

In this chapter, we are focusing on fibrillar collagens and, in particular, collagen type I, the most abundant extracellular protein, especially in bone where it is essential for bone strength and in soft connective tissues where it confers compliance, flexibility, and resilience. We will discuss the structure and biosynthesis, regulation, and degradation of type I collagen and associated proteins that maintain its homeostasis and recent insights into the organization of regulatory elements in type I collagen genes, many of which are based on studies in transgenic mice. Then we will address how collagen synthesis is regulated by cytokines and growth factors.

The accepted definition of collagens is "structural proteins of the extracellular matrix which contain one or more domains harboring the conformation of a collagen triple helix" (Myllyharju and Kivirikko, 2004; van der Rest and Garrone, 1991). The triple-helix motif is composed of three polypeptide chains whose amino acid sequence consists of Gly−X−Y repeats. Because of this particular peptide sequence, each chain is coiled in a left-handed helix, and the three chains assemble in a right-handed triple helix, where Gly residues are in the center of the triple helix and the lateral chains of X and Y residues are on the surface of the helix (van der Rest and Garrone, 1991). In about one-third of the cases, X is a proline and Y is a hydroxyproline; the presence of hydroxyproline is essential to stabilize the triple helix and is a unique characteristic of collagen molecules. At the time of this review, 28 different types of collagens have been described, which are grouped in subfamilies depending on their structure and/or their function. The Roman numerals denoting collagen types follow the order in which they were reported. For each collagen type, the α chains are identified with Arabic numerals (Myllyharju and Kivirikko, 2001). Although a standard nomenclature has been agreed on, the representation of the collagen names can be sometimes confusing. Throughout this chapter, we will address the unassembled collagen molecules as procollagens, the mouse genes as Col1a1 or Col1a2 (lower case), and the human genes in capital letters (COL1A1 or COL1A2).

The family of fibrillar collagens

Types I, II, III, V, and XI and the newly described types XXIV and XXVII collagens (Boot-Handford et al., 2003; Koch et al., 2003) form the group of fibrillar collagens. The characteristic feature of fibrillar collagens is that they consist of a long continuous triple helix that self-assembles into highly organized fibrils. These fibrils have a very high tensile strength and play a key role in providing a structural framework for body structures such as skeleton, skin, blood vessels, intestines, or fibrous capsules of organs. Type I collagen, which is the most abundant protein in vertebrates, is present in many organs and is a major constituent of bone, tendons, ligaments, and skin. Type III collagen is less abundant than type I collagen, but its distribution essentially parallels that of type I collagen with the exception of bones and tendons, which contain virtually no type III collagen. Moreover, type III collagen is relatively more abundant in distensible tissues, such as blood vessels, than in nondistensible tissues. Type V collagen is present in tissues that also contain type I collagen. Type II collagen is a major constituent of cartilage and is also present in the vitreous body. Like type II, type XI and type XXVII collagens are present in cartilage. However, unlike other collagens, type XXVII appears to express in epithelial cells of cochlea, lung, gonad, and stomach (Boot-Handford et al., 2003), suggesting that its function in these epithelial layers cannot depend on the copolymerization with other collagens. Collagen type XXIV displays unique structural features of invertebrate fibrillar collagens and is expressed predominantly in bone tissue (Matsuo et al., 2006).

Bone formation is a complex and tightly regulated genetic program that involves two distinct pathways at different anatomical locations (de Crombrugghe et al., 2001; Karsenty and Wagner, 2002; Olsen et al., 2000). In intramembranous ossification, mesenchymal cells condense and differentiate directly into mainly collagen type I−producing osteoblasts, whereas in endochondral bone formation, a cartilage model that is initially rich in type II and type XI collagens, which are secreted by chondrocytes, is replaced by an ostium rich in collagen type I matrix. Cartilage formation in endochondral skeletal elements is initiated by the condensation of chondrogenic mesenchymal cells followed by the overt differentiation of cells in these condensations. After undergoing a unilateral form of proliferation, these cells gradually become hypertrophic. At the same time, cells around the condensations form the perichondrial layer that gives rise to the osteoblast-forming periosteum and ultimately to cortical bone. The process of cartilage replacement by a bone matrix involves invasion by preosteoblasts in the periosteum as well as blood vessels and hematopoietic cells of the zone of hypertrophic chondrocytes. Expression of the genes for collagen type I and those for collagen types II and XI follows distinct transcriptional codes that control osteoblastogenesis and chondrogenesis (Bridgewater et al., 1998; de Crombrugghe et al., 2001; Karsenty and Wagner, 2002; Lefebvre et al., 2001; Lefebvre and de Crombrugghe, 1998). In addition to fibrillar collagens, collagen type X has been implicated in the morphogenic events of hypertrophic cartilage prior to its replacement by bone. Although knockout mice for collagen X showed no apparent phenotype (Rosati et al., 1994), significant reduction in the amount and quality of bone minerals was evident (Paschalis et al., 1996).

Collagen types XXIV and XXVII display mutually exclusive patterns of expression in the developing and adult mouse skeleton. Gene expression studies have shown that, whereas Col24a1 transcripts accumulate at ossification centers of the craniofacial, axial, and appendicular skeleton, Col27a1 activity is instead confined to the cartilaginous anlagen of skeletal elements (Boot-Handford et al., 2003; Koch et al., 2003; Pace et al., 2003). In addition, structural considerations have suggested that collagens XXIV and XXVII are likely to form distinct homotrimers (Koch et al., 2003). Together these observations have been interpreted to indicate that these newly discovered fibrillar collagens may participate in the control of important physiological processes in bone and cartilage, such as collagen fibrillogenesis and/or matrix calcification and mineralization (Boot-Handford et al., 2003; Koch et al., 2003; Pace et al., 2003).

Structure, biosynthesis, transport, and assembly of type I collagen

Structure

Fibril-forming collagens are synthesized in precursor form, procollagens. Each molecule of type I collagen is typically composed of two α1 chains and one α2 chain (α1(I)2−α2(I)) coiled around one another in a characteristic triple helix. Both the α1 chain and the α2 chain consist of a long helical domain preceded by a short N-terminal peptide and followed by a short C-terminal peptide (for reviews, see Myllyharju and Kivirikko, 2001, 2004, and van der Rest and Garrone, 1991).

The mechanism that controls the 2:1 stoichiometry of the collagen chains in type I collagen is not well understood. It is evident that a number of type I collagen molecules can be formed by three α1 chains (α1(I)3). The homotrimeric type I collagen isotype containing three pro-α1(I) collagen chains (α1(I)3) is a minor isotype, whose role is not well understood. Homotrimers are found embryonically (Jimenez et al., 1977; Rupard et al., 1988), in small amounts in skin (Uitto, 1979), in certain tumors and cultured cancer cell lines (Moro and Smith, 1977; Rupard et al., 1988), and also during wound healing (Haralson et al., 1987). Mesangial cells, which do not synthesize collagen type I in vivo, produce homotrimeric type I collagen in culture, further suggesting that homotrimers play a role in wound healing (Johnson et al., 1992). The collagen I α2-deficient mouse, otherwise known as the oim mouse (osteogenesis imperfecta [OI] model), is homozygous for a spontaneous nucleotide deletion in the Col1a2 gene, resulting in a frameshift altering the carboxy propeptide of the pro-Col1a2 chain. Although the carboxy propeptide is not present in mature type I collagen, it is responsible for association of the Col1a2 chain with the Col1a1 chains during assembly of the triple helix (Chipman et al., 1993; Deak et al., 1983; McBride et al., 1997) (see sections on collagen diseases).

Type I collagen is secreted as a propeptide, but the N telopeptide and the C telopeptide are cleaved rapidly by specific proteases, ADAMTS- 2 and bone morphogenetic protein 1 (BMP-1), respectively, so that shorter molecules assemble to form fibrils (Canty and Kadler, 2005). In fibrils, molecules of collagen are parallel to one another (Fig. 13.1); they overlap one another by multiples of 67 nm (distance D), with each molecule being 4.4 D (300 nm) long; there is a 40-nm (0.6 D) gap between the end of a molecule and the beginning of the next (see Fig. 13.1). This quarter-staggered assembly explains the banded aspect displayed by type I collagen fibrils in electron microscopy. In tissues, type I collagen fibrils can be parallel to one another and form bundles (or fibers), as in tendons, or they can be oriented randomly and form a complex network of interlaced fibrils, as in skin. In bone, hydroxyapatite crystals seem to lie in the gaps between collagen molecules.

Regulation of transcription

In humans the gene coding for the α1 chain of type I collagen is located on the long arm of chromosome 17 (17q21.3−q22; chromosome 11 in mouse), and the gene coding for the α2 chain is located on the long arm of chromosome 7 (7q21.3−q22; chromosome 6 in mouse). Both genes have very similar structures (Chu et al., 1984; D'Alessio et al., 1988), and this structure is also very similar to that of genes coding for other fibrillar collagens (Vuorio and de Crombrugghe, 1990). The difference in size between the two genes (18 kb for the Col1a1 gene and 38 kb for the Col1a2 gene) is explained by differences in the sizes of the introns.

The triple-helical domain of the α1 chain is coded by 41 exons, which code for Gly−X−Y repeats, and by two so-called joining exons. These joining exons code in part for the telopeptides and in part for Gly−X−Y repeats, which

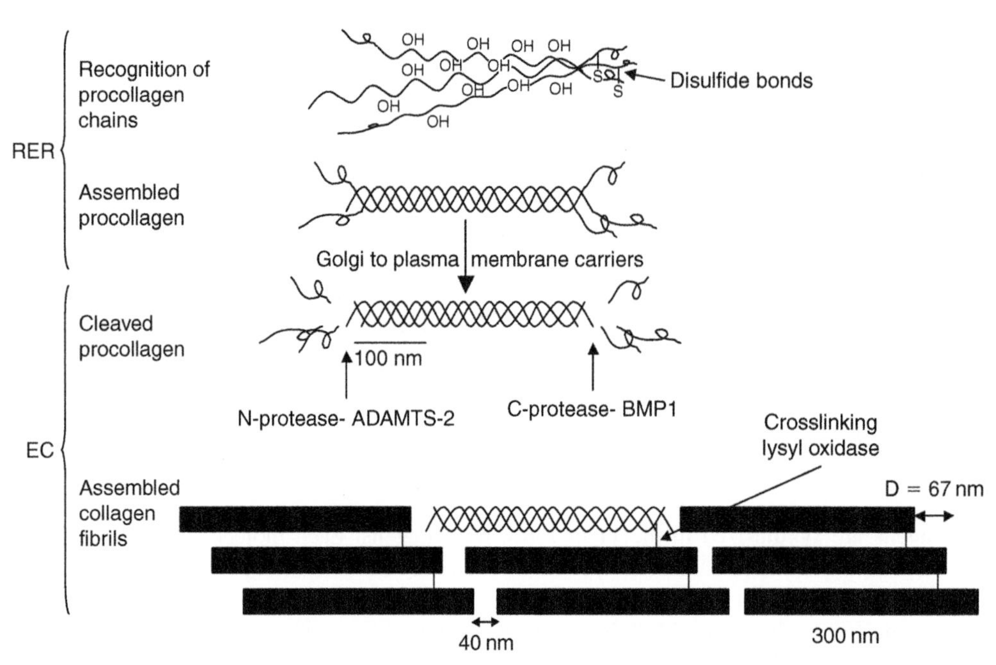

FIGURE 13.1 Schematic diagram of collagen assembly depicting the recognition of the α chains in the rough endoplasmic reticulum (*RER*) and the helix formation strengthened by the hydroxylated proline and lysine residues (*OH*). The collagen is then transported by the Golgi to plasma membrane vesicles to the extracellular space (*EC*) where the propeptides are removed by proteinases and the fibrils are assembled into collagen fibers that are cross-linked by lysyl oxidase. *ADAMTS-2*, a disintegrin and metalloproteinase with thrombospondin motifs-2; *BMP1*, bone morphogenetic protein-1.

are part of the triple-helical domain. The triple-helical domain of the α2 chain is coded by 42 exons, plus two joining exons. The corresponding exons coding for the triple-helical domain of the α1 chain and for the triple-helical domain of the α2 chain have similar lengths (Table 13.1). The only exception is that exons 34 and 35 in Col1a2, which are 54 bp long each, correspond to a single 108-bp 34/35 exon in the a1 gene. Except for the two joining exons, each exon starts exactly with a G codon and ends precisely with a Y codon, and all the exons are 54, 108 (54 3 2), 162 (54 3 3), 45, or 99 bp long (see Table 13.1). This organization suggests that exons coding for triple-helical domains could have originated from the amplification of a DNA unit containing a 54-bp exon embedded in intron sequences. Exons of 108 and 162 bp would result from a loss of intervening introns. Exons of 45 and 99 bp would result from recombinations between two 54-bp exons (Vuorio and de Crombrugghe, 1990).

For both the α1 chain and the α2 chain, the C propeptide plus the C telopeptide are coded by four exons (exons 48–51 of the Col1a1 gene, exons 49–52 of the Col1a2 gene). The first of these exons codes for the end of the triple-helical domain, the C-terminal telopeptide, and the beginning of the C-terminal propeptide. The three other exons code for the rest of the C-terminal propeptide. The C-terminal propeptide has a globular structure that is stabilized by two intrachain disulfide bonds (see Fig. 13.1). It contains three (α2 chain) and four (α1 chain) additional cysteine residues that form interchain disulfide bonds. The formation of disulfide bonds precedes the triple-helix formation and plays an essential role in the intracellular assembly of the three α chains (see sections on translational and posttranslational modifications). Prdm5 is a member of a family of transcriptional regulators that predominantly bind exonic regions of collagen genes and associate with RNA polymerase II to sustain collagen I transcription. Prdm5 targets all mouse collagen genes as well as several small leucine-rich proteoglycan (SLRP) genes. Prdm5 controls both collagen I transcription and fibrillogenesis by binding inside the Col1a1 gene body and maintaining RNA polymerase II occupancy. In vivo, Prdm5 loss results in delayed ossification involving a pronounced impairment in the assembly of fibrillar collagens (Galli et al., 2012).

The signal peptide, the N propeptide, and the N telopeptide of the 1 chain, as well as of the α2 chain, are coded by the first six exons. The N propeptide of the α1 chain contains a cysteine-rich (10-cysteine-residue) globular domain, a short triple-helical domain, and a short globular domain, which harbors the N-terminal peptidase cleavage site (see Fig. 13.1). The N-terminal propeptide of the α2 chain does not contain a cysteine-rich domain but a short globular domain. The 3′-untranslated region (UTR) of both the Col1a1 gene and the Col1a2 gene contains more than one polyadenylation site,

TABLE 13.1 Sizes of exons coding for the triple-helical domain of type I collagen[a].

Exon	Size (bp)	Exon	Size (bp)	Exon	Size (bp)
7	45	21	108	35	54
8	54	22	54	36	54
9	54	23	99	37	108
10	54	24	54	38	54
11	54	25	99	39	54
12	54	26	54	40	162
13	45	27	54	41	108
14	54	28	54	42	108
15	45	29	54	43	54
16	54	30	45	44	108
17	99	31	99	45	54
18	45	32	108	46	108
19	99	33	54	47	54
20	54	34	54	48	108

[a]*In the pro-α1(I) collagen gene, exons 33 (54 bp) and 34 (54 bp) are replaced by a single 108-bp 33/34 exon. The two joining exons (exons 6 and 49) are not considered in this table (see text for details).*

which indicates that mRNAs with different sizes will be generated. As in many other genes, the functional role of the different polyadenylation sites is still unknown.

Control of translation

After being transcribed, the pre-mRNA undergoes exon splicing, capping, and addition of a poly(A) tail, which gives rise to a mature mRNA. These mature mRNAs are then translated in polysomes, and the resulting proteins undergo extensive posttranslational modifications before being assembled into a triple helix and released in the extracellular space (for reviews, see Lamande and Bateman, 1999; Myllyharju and Kivirikko, 2001).

Signal peptides are cleaved from the chains when their N-terminal end enters the cisternae of the rough endoplasmic reticulum (ER). Both the pro-α1 chain and the pro-α2 chain undergo hydroxylation and glycosylation, and these modifications are essential for the assembly of type I collagen chains in a triple helix. About 100 proline residues in the Y position of the Gly−X−Y repeats, a few proline residues in the X position, and about 10 lysine residues in the Y position undergo hydroxylation, respectively, by a prolyl 4-hydroxylase, a prolyl 3-hydroxylase, and a lysyl hydroxylase. Hydroxylation of proline to hydroxyproline is critical to obtain a stable triple helix, and at 37°C, stable folding in a triple-helical conformation cannot be obtained before at least 90 prolyl residues have been hydroxylated. These hydroxylases have different requirements to be active, and, in particular, they can act only when prolyl or lysyl residues occupy the correct position in the amino acid sequence of the α chain and when peptides are not in a triple-helical configuration. Moreover, these enzymes require ferrous ions, molecular oxygen, α-ketoglutarate, and ascorbic acid to be active. This requirement for ascorbic acid could explain some of the consequences of scurvy on wound healing. When lysyl residues become hydroxylated, they serve as a substrate for a glycosyltransferase and for a galactosyltransferase, which add glucose and galactose, respectively, to the E−OH group. As for hydroxylases, glycosylating enzymes are active only when the collagen chains are not in a triple-helical conformation. Glycosylation interferes with the packaging of mature molecules into fibrils, and increased glycosylation tends to decrease the diameter of fibrils.

While hydroxylations and glycosylations described previously occur, after a mannose-rich oligosaccharide is added to the C propeptide of each pro-α chain, C propeptides from two α1 chains and one α2 chain associate, with the formation of intrachain and interchain disulfide bonds. After prolyl residues have been hydroxylated, and the three C propeptides have associated, a triple helix will form at the C-terminal end of the molecule and then extend toward the N-terminal end (see Fig. 13.1). This propagation of the triple-helical configuration occurs in a "zipper-like" fashion (Prockop, 1990). If prolyl residues are not hydroxylated or if interchain disulfide bonds are not formed between the C propeptides, the α chain will not fold into a triple helix. Although the functions of the C-terminal sequences, which have been associated with initiation of triple-helix formation, are thought to be well established, those of the N-terminal propeptide are poorly understood (Bornstein, 2002). The N propeptide of type I procollagen, as released physiologically by procollagen N-protease (ADAMTS-2), contains a globular domain largely encoded by exon 2 in the Col1a1 gene and a short triple helix that terminates in a non-triple-helical telopeptide sequence, which separates this helix from the major collagen helix. Bornstein and colleagues generated a mouse with a targeted deletion of exon 2 in the Col1a1 gene, thus replicating the type IIB splice form of type II procollagen in type I procollagen (Bornstein et al., 2002); surprisingly, homozygous mutant mice were essentially normal. In particular, none of the steps in collagen biogenesis thought to be dependent on the N propeptide were defective. However, there was a significant, but background-dependent, fetal mortality, which suggested a role for the type I collagen N propeptide in developmental processes.

Toman and colleagues have gone even further to demonstrate that propeptide may not be necessary for the selection and folding of procollagen. They engineered type I collagen genes that encode the N and C telopeptides with the entire triple-helical domain and showed that these sequences are sufficient for the assembly of a triple helix in *Saccharomyces cerevisiae* (Olsen et al., 2001). Other fibrillar collagens (types II, III, V, and XI) have similar structures and thus would be expected to fold into triple helices without the propeptide regions in an analogous system.

Intracellular transport

The transport of newly synthesized secretory proteins begins at their site of synthesis, the ER, a network of dynamically interconnected membrane tubules and cisternae (Fig. 13.1). All vesicles then detach from the ER through membrane fission and move to the ER–Golgi intermediate compartment. From there, carriers containing secretory cargoes are transported forward to the Golgi complex.

The newly formed triple-helical forms are then stabilized by heat shock protein 47 (Hsp47), a molecular chaperone of type I collagen molecules (Nagai et al., 2000; Tasab et al., 2000). This protein belongs to the serine protease inhibitor (serpin) superfamily containing a serpin signature sequence. Hsp47 resides in the ER, as inferred from the presence of a carboxyl-terminal RDEL sequence similar to the ER retrieval signal, KDEL. After folding, proteins enter the exit sites of the ER, where they are sorted into large pleomorphic budding vesicles that are generated through the membrane-bending properties of coat protein complex II (COPII). Disrupting the Hsp47 gene in mice resulted in embryonic lethality in mice by 11.5 days postcoitus and caused a molecular abnormality in procollagens (Nagai et al., 2000). Type I procollagen chains containing propeptides accumulated in the tissues, but the mature collagen chains normally processed were scarcely observed, suggesting that HSP47 is essential as a collagen-specific molecular chaperone for the proper processing of procollagen molecules, and the Hsp47 gene is needed for the normal development of mouse embryo (Nagai et al., 2000). However, the transfer from ER to the Golgi was unexplained since the mean fibrillar collagen, when formed into a trimer, adopts a rigid, rodlike structure of >300 nm in length and could not fit into a generic COPII vesicle of 60–90 nm diameter.

A genome-wide screen was performed in *Drosophila* tissue culture S2 cells to identify transport components (Bard et al., 2006). This screen revealed a number of previously unidentified genes required for transport and Golgi organization (TANGO genes). TANGO1 was shown to transport carriers without following the cargo into the vesicle itself. Thus, the mechanism of TANGO1-mediated cargo loading of collagen VII into COPII carriers (j.cell. 2008.12.025). A null allele of Mia3 in the mouse. Melanoma inhibitory activity member 3 (MIA3/TANGO1) is an evolutionarily conserved ER resident transmembrane protein.

Mia3 knockouts display a chondrodysplasia that causes dwarfing of the fetus, peripheral edema, and perinatal lethality. Thus far, our understanding of Mia3 (*Drosophila melanogaster* homolog of the vertebrate gene MIA3, also known as TANGO1) in escorting all collagens examined to date, including collagens I, II, III, IV, VII, and IX, but not other ECM components, such as fibronectin or aggrecan, indicates that this protein plays a unique role within the cell to facilitate the nucleation of large ER transport vesicles. However, collagen secretion indeed still occurs in Mia3-null cells. Either an alternate independent pathway of egress remains open to large ECM cargoes, or these proteins can still exit via large COPII vesicles (Wilson et al., 2011). Interestingly, studies in mice focusing on liver fibrogenesis and hepatic stellate cell (HSC) function noted that the loss of MIA3/TANGO1 resulted in retention of procollagen I in the ER and promotion of the unfolded protein response. In vivo this was manifest as a reduction in hepatic fibrosis following liver injury and a dysregulation in HSC homeostasis leading to enhanced apoptosis (Maiers et al., 2017). A critical role for correct intracellular transport of procollagen I has also been highlighted in HSCs deficient in the ER oxidase 1α enzyme (ERO1α), which mediates protein modifications crucial to the transport of secreted proteins. Silencing of ERO1α caused impaired secretion of collagen type I and intracellular accumulation severely inhibiting cell proliferation (Fujii et al., 2017). It is notable that mutations in the ubiquitous COPII component Sec23a or in the transport protein particle complex subunit TRAPPC2 (which is involved in trafficking between the ER and the Golgi complex) can selectively affect osteocytes and chondrocytes, resulting in craniolenticulosutural dysplasia (Boyadjiev et al., 2006) and spondyloepiphyseal dysplasia tarda (Gedeon et al.,1999), respectively.

After passing through the Golgi complex and reaching the *trans*-Golgi network (TGN), different cargoes are packaged in specialized membranous carriers, within which they are shipped out to the plasma membrane. There they are secreted into the extracellular space, where removal of the N and C propeptides from procollagens I, II, and III is catalyzed by specific Zn^{2+}-dependent metalloendopeptidases, procollagen N- and C-proteinase, respectively (Prockop, 1998). Procollagen C-proteinase, also known as BMP-1, is a member of the astacin family of proteases and triggers spontaneous self-assembly of collagen molecules into fibrils, giving rise to mature collagen molecules (Kadler, 2004; Kadler et al., 1990). Cleavage of the propeptide decreases the solubility of collagen molecules dramatically. Thus a major extracellular function of C propeptides is thought to prevent fibril formation, while N propeptides influence fibril shape and diameter (Hulmes, 2002).

The free propeptides are believed to be involved in feedback regulation of the synthesis of types I and III collagens by fibroblasts in culture (Wiestner et al., 1979). However, the mechanism of this inhibition remained elusive despite attempts by several groups to characterize it.

Fibrillogenesis

In the extracellular space, the molecules of mature collagen assemble spontaneously into quarter-staggered fibrils; this assembly is directed by the presence of clusters of hydrophobic and of charged amino acids on the surface of the molecules. Fibril formation has been compared with crystallization in that it follows the principle of "nucleated growth" (Prockop, 1990). Once a small number of molecules have formed a nucleus, it grows rapidly to form large fibrils. During fibrillogenesis, some lysyl and hydroxylysyl residues are deaminated by a lysine oxidase, which deaminates the «-NH_2 group, giving rise to aldehyde derivatives. These aldehydes will associate spontaneously with «-NH_2 groups from a lysyl or hydroxylysyl residue of adjacent molecules, forming interchain cross-links. These cross-links will increase the tensile strength of the fibrils considerably (see Fig. 13.1). In vitro studies have shown that procollagen molecules and their various structural domains have a remarkable capacity to control all stages of collagen assembly, from intracellular assembly to extracellular suprafibrillar assembly at a micrometer scale. The in vivo process is much more complex, but we are beginning to understand some of this in particular cell types. Kadler and coworkers have shown that GPCs are indeed present in vivo in embryonic tendon fibroblasts and that some GPCs contain 28-nm-diameter collagen fibrils. Moreover, GPCs are targeted to novel plasma membrane protrusions, which they have termed "fibripositors" (fibril depositors). What was intriguing in this study is the fact that procollagen can be converted to collagen within the confines of the cell membrane, which is consistent with the observation of collagen fibrils in some GPCs and the known intracellular activation of BMP-1 in the TGN compartment (Canty-Laird et al., 2012). In addition, fibripositors were shown to be always oriented along the tendon axis, which establishes a link between intracellular transport and the organization of the ECM (Canty et al., 2004). Interestingly, fibripositor formation is not a constitutive process in procollagen-secreting cells. It is absent in postnatal development despite active procollagen synthesis, and occurs only during a narrow window of embryonic development when tissue architecture is being established. It is not known whether this phenomenon occurs in other types of specialized collagen-secreting cells. More recently, the Kadler group has shown that fibripositors are specialized sites of fibril assembly and fibril transport. The fibripositors form an extended mechanical interface between the cytoskeleton and extracellular collagen fibrils. This interface serves to transmit cell-derived tensioning force to the tissue, which is critical in maintaining fibril alignment. The dynamic transport of new collagen fibrils occurs at the plasma membrane and this process is driven by nonmuscle myosin II, since inhibition of dynamin-dependent endocytosis with dynasore blocked fibricarrier formation and caused accumulation of fibrils in fibripositors (Kalson et al., 2013).

Assembly

The final assembly of fibrillar collagen involves the direct interaction of several molecules, which include other collagens, SLRPs, and others. These interactions shape the diameter of the fibrils (Kuc and Scott, 1997; Vogel and Trotter, 1987) and patterning of the final matrix. SLRPs are a group of secreted proteins, including decorin, biglycan, fibromodulin, lumican, and keratocan, among others, that play major roles in tissue development and assembly, especially in collagen fibrillogenesis (Iozzo, 1999). Biglycan and decorin are highly expressed in extracellular bone matrix and there is now substantial evidence to support an increasing role for biglycan and decorin in influencing bone cell differentiation and proliferative activity (Waddington et al., 2003). The ability of decorin and biglycan to interact with collagen molecules and to facilitate fibril formation has implicated these macromolecules in important roles in the provision of a collagenous framework in bone, which eventually allows for mineral deposition. Initial mineral deposition is proposed to occur within or near the gap zones along the collagen fibers, and the structural architecture of the collagen fibers along with interacting noncollagenous proteins is likely to play a key role in directing placement of the mineral crystals (Dahl and Veis, 2003).

Molecular modeling techniques have led to the proposal that decorin and biglycan adopt an open-horseshoe structure (Weber et al., 1996) where the inner cavity interacts with a single triple-helical molecule. The generation of mutated forms of decorin has demonstrated the importance of leucine-rich sequences 4— to 6 in mediating this interaction (Kresse et al., 1997). In addition, reduction in the disulfide bridges at the C- and N-terminal ends of decorin also abolished interaction with type I collagen (Ramamurthy et al., 1996) and this led to the proposal that the disulfide loop at the C terminus binds to adjacent collagen fibrils, thereby facilitating the lateral assembly and stabilization of the fibrils.

Interestingly, molecular analysis data have put forward the idea that decorin exists as a dimer in solution (Scott et al., 2003), and if this is the case in vivo, then the nature of this interaction will be important when considering the mechanistic role of decorin in fibril assembly. The glycosaminoglycan moieties of decorin and biglycan have also been deemed to play an important role in collagen fibrillogenesis, where the interaction of glycosylated forms of these SLRPs with collagen appeared to be greater than that of nonglycosylated forms (Bittner et al., 1996). Adding to the macro- and microstructural information of collagen type I, three important studies using high-resolution mapping, crystallographic techniques, and modeling have highlighted the spatial arrangement and packaging of the collagen molecules in fibrils and fibril domain architecture. These investigations have helped to evolve our understanding of cell—collagen—matrix interactions via the presentation of organized domains and provide novel solutions and information relevant to the functional interactions of decorin in assembly and matrix metalloproteinases (MMPs) in collagenolysis (Orgel et al., 2006; Sweeney et al., 2008; Varma et al., 2016). Further evidence for the role of decorin and biglycan in bone formation is provided by targeted deletion of the genes. The biglycan-knockout (Bgn 2/2) mouse (Xu et al., 1998), unlike the Dcn 2/2 mouse (Danielson et al., 1997), showed no gross skin abnormalities but rather a reduction in bone density. These mice were seen to develop an osteoporotic phenotype, failing to achieve peak bone mass, owing to decreased bone formation, with significantly shorter femurs (Ameye et al., 2002). Within these animals lower osteoblast numbers and osteoblast activity were observed. In vitro experiments demonstrated that the number and responsiveness of bone marrow stromal cells to transforming growth factor β (TGFβ), and hence osteogenic precursor cells, decreased dramatically with age, but apoptosis rates increased (Chen et al., 2002a,b). The effects were not confined only to the skeletal tissues. Within the teeth the transition of predentin to dentin appeared to be impaired and the thickness of the enamel was dramatically increased (Goldberg et al., 2002). Taken together these results would suggest that biglycan plays an important role in the formation of mineralized tissue. Furthermore, despite high sequence identity and somewhat similar patterns of localization, decorin and biglycan are not interchangeable in function and do not have the ability to rescue each other's knockout phenotypes. Notably, Bgn 2/2 and Dcn 2/2 double-knockout animals revealed that the effects of both gene deficiencies were additive in the dermis and synergistic in bone (Corsi et al., 2002). The lack of both genes caused a phenotype with severe skin fragility and osteopenia, resembling a rare progeroid variant of Ehlers—Danlos syndrome.

Consequences of genetic mutations on type I collagen formation

OI (also known as "brittle bone disease") is a genetic disease characterized by an extreme fragility of bones. Genetic studies have shown that it is due to a mutation in the coding sequence of either the pro-COL1A1 gene or the pro-COL1A2 gene, and more than 150 mutations have been identified (for review, see Byers, 2001). Most severe cases of OI result from mutations that lead to the synthesis of normal amounts of an abnormal chain, which can have three consequences (Marini et al., 2007). First, the structural abnormality can prevent the complete folding of the three chains in a triple helix, e.g., if a glycine is substituted by a bulkier amino acid that will not fit in the center of the triple helix. In this case, the incompletely folded triple-helical molecules will be degraded intracellularly, resulting in a phenomenon known as "procollagen suicide." Second, some mutations appear not to prevent folding of the three chains in a triple helix, but presumably prevent proper fibril assembly. For example, Prockop's group has shown that a mutation of the pro-Col1a1 gene that changed the cysteine at position 748 to a glycine produced a kink in the triple helix (Kadler et al., 1991). Finally, some mutations will not prevent triple-helical formation or fibrillogenesis but might modify the structural characteristics of the fibrils slightly and thus affect their mechanical properties. In all these cases, the consequences on the mechanical properties of bone are probably similar. Mild forms of OI most often result from a functionally null allele, which decreases the production of normal type I collagen. Null mutations are usually the result of the existence of a premature stop codon or of an abnormality in mRNA splicing. In these cases, the abnormal mRNA appears to be retained in the nucleus (Johnson et al., 2000). A mouse model of OI has been obtained by using a knock-in strategy that introduced a $Gly^{349} \rightarrow$ Cys mutation into the pro-Col1a1 gene (Forlino et al., 1999). This model faithfully reproduced the human disease. Another spontaneous mouse mutation, the oim/oim mouse, analogous to human type III OI, carries a spontaneous deletion of G at nucleotide 3983 in the α2 chain of collagen type I, which alters the reading frame to result in the final 48 amino acids of the COOH-terminal propeptide generating a new stop codon, with addition of an extra amino acid (Chipman et al., 1993). In these mice, collagen type I is made of α(I)3 homotrimers in place of the normal α1(I)2 α2(I)1 heterotrimers (Chipman et al., 1993; Kuznetsova et al., 2004; Miles et al., 2002), which results in marked skeletal fragility, with thinning of the long bones and reduced mechanical strength (Chipman et al., 1993; Pereira et al., 1995). An additional organ pathology has been described by Phillips and coworkers, namely a glomerulopathy characterized by abnormal renal collagen deposition (Brodeur et al., 2007; Phillips et al., 2002). Interestingly, data have also provided an insight into a fourth mechanism potentially resulting in OI. Here, a mutation at the signal peptide cleavage site of collagen type I, a domain not well characterized, caused reduced production and the intracellular retention of collagen I, leading to a severe OI phenotype (Lindert et al., 2018). This study is consistent with the notion that mechanisms influencing intracellular transport of collagen I may exert a profound impact on bone and other connective tissues in which collagen type I plays an important structural role.

Ehlers–Danlos syndrome types VIIA and VIIB are two rare dominant genetic diseases characterized mainly by an extreme joint laxity. They result from mutations in the pro-COL1A1 gene (Ehlers–Danlos syndrome type VIIA) or in the pro-COL1A2 gene (Ehlers–Danlos syndrome type VIIB) that interfere with the normal splicing of exon 6, and a little fewer than 20 mutations have been described. These mutations can affect the splice donor site of intron 7 or the splice acceptor site of intron 5; in the latter case, there is efficient recognition of a cryptic site in exon 6 (Byers et al., 1997). Thus, these mutations induce a partial or complete excision of exon 6. They do not appear to affect the secretion of the abnormal procollagen molecules, but they are responsible for the disappearance of the cleavage site of the N-terminal propeptide and thus for the presence of partially processed collagen molecules in fibrils that fail to provide normal tensile strength to tissues (Byers, 2001). Nevertheless, these mutations seem to affect the rate of cleavage of the N-terminal propeptide rather than completely prevent it, which explains why the phenotype is less severe than for patients who do not have a functional N-proteinase (Ehlers–Danlos syndrome type VIIC).

Collagen type I degradation and catabolism

The absolute requirement to critically regulate collagen type I synthesis and assembly in health is paralleled by the necessity to control the level and deposition of collagen type I once produced at the level of degradation and catabolism (for reviews see Fields 2013; Van Doren 2015; Zigrino et al., 2016). Furthermore, work over the past 5–10 years has begun to unravel the molecular pathways mediating degradation of collagen (Mi et al., 2017; Sprangers et al., 2019). These studies have shown that several key intracellular and extracellular pathways exist, and both mechanisms are important for maintaining collagen homeostasis. Different pathways for extracellular and intracellular degradation are employed by a wide range of tissue and cell types (i.e., bone, cartilage, and skin). There are at least three main types of enzymes that possess collagenolytic activity, including the MMPs (Nagase et al., 2006), cathepsins (Novinec et al., 2013), and neutrophil elastase (Fields, 2013). One of the most typical vertebrate collagenases and the most studied with respect to collagen type I cleavage is MMP-1 (Arakaki et al., 2009; Fields, 2013). MMP-1 cleaves collagen type I into one-quarter- and three-quarter-length fragments (Lauer-Fields, 2002). Once cleaved, a number of factors appear to influence further collagenolysis and sequential degradation. These include conformational dynamics and the mechanism by which the MMPs disrupt the stable collagen triple helix (Adhikari, 2012), cross-linking (Kwansa et al., 2014), cell surface binding that facilitates collagen unwinding (Adhikari et al., 2011), and the presence of membrane-bound proteases such as MT1-MMP (Yañez-Mó et al., 2008). Several other MMPs, such as MMP-2, MMP-8, MMP-9, MMP-12, MMP-13, and MMP-14, are also known to degrade collagen, but in general the precise modes of action of many have not been studied in detail. For MMP-12, the catalytic domain has been studied, and it recognizes, binds, and cleaves at specific sites in the triple helix in regions of relatively less proline, notably between Gly^{775} and Leu/Ile^{776} in the α1 and α2 collagen chains, respectively, in a manner that is characteristic of the unique collagenolytic activity of MMP collagenases (Bigg et al., 2007; Tam et al., 2004; Taddese et al., 2010).

It is clear that correct proteolytic processing and collagenolysis are critical in development and for normal tissue homeostasis, and abnormalities in these processes or their dysregulation in disease and aging can significantly contribute to many severe pathologies (reviewed in 2017 by Amar et al.,). It has also become apparent that many studies aimed at defining the molecular mechanism(s) have taken advantage of animal systems or in vitro analysis. For instance, in models of diabetes in rodents, reduced MMP (MMP-2 and MMP-9) activity was noted to be associated with increased collagen deposition and stromal growth (Santos et al., 2017), and an imbalance in MMP/tissue inhibitor of metalloproteinases (TIMP) levels (creating reduced MMP-1 and MMP-3) appears to be responsible for the elevated collagen type I production in an in vitro model of chronic liver disease using HSCs (Robert et al., 2016). Enhanced type I collagen levels can directly promote fibroblast differentiation by altering integrin expression in models of cardiac fibrosis (Hong et al., 2017). Abnormally high collagen turnover by enhanced MMP activity can be equally disruptive such as aortic aneurysms, where elevated MMP-3, MMP-9, and MMP-12 are present and associated with collagen degradation (Klaus et al., 2017). Another important aspect of dysregulated collagen catabolism is the direct impact excessive levels of collagen type I can exert on cells via surface receptors. Studies have shown type I collagen promotes differentiation, proliferation, and scar formation by cardiac and ligament fibroblasts and astrocytes via interactions with specific integrins, cadherins, and SPARC (secreted protein acidic and cysteine rich) (Hara et al., 2017; Hong et al., 2017; Rosset et al., 2017), and plays a proinflammatory role by altering the release of cytokines by monocytes (Schultz et al., 2016). These data provide an insight into the distinct and pivotal role of collagen type I in regulating critical cell functions and the deleterious impact resulting from failure of correct collagen type I degradation and homeostasis. A greater understanding of role of MMPs and the molecule mechanisms that underlie their distinct roles in collagenolysis will enable the development of selective, specific, or targeted approaches for the effective treatment of an array of human conditions.

Collagen type I and bone pathologies

Because type I collagen is the most abundant protein in bone and because mutations in the COL1A1 gene are a major cause of OI, this gene has been considered a strong candidate for susceptibility to osteoporosis (Ralston and de Crombrugghe, 2006). Indeed polymorphisms in both the promoter (Garcia-Giralt et al., 2002) and the first intron of COL1A1 (Grant et al., 1996) have been associated with changes in bone mineral density (BMD). The polymorphism of intron 1 is located in a binding site for the transcription factor Sp1 and has been associated with various osteoporosis-related symptoms such as bone density and fractures (Grant et al., 1996; Uitterlinden et al., 1998), postmenopausal bone loss (Harris et al., 2000; MacDonald et al., 2001), bone geometry (Qureshi et al., 2001), bone quality (Mann et al., 2001), and bone mineralization (Stewart et al., 2005). The osteoporosis-associated Sp1 polymorphism causes increased Sp1 binding, enhanced transcription, and abnormally high production of COLIA1 mRNA and protein (Mann et al., 2001), which result in an imbalance between the COLIA1 and the COLIA2 chains. This is thought to lead to impairment of bone strength and reduced bone mass in carriers of the Sp1 polymorphism (Stewart et al., 2005). Retrospective meta-analyses of previous studies have indicated that the Sp1 polymorphic allele is associated with reduced BMD and with vertebral fractures (Efstathiadou et al., 2001; Mann and Ralston, 2003). In a large prospective meta-analysis of more than 20,000 participants from several European countries in the GENOMOS study, homozygotes for the Sp1 polymorphism were found to be associated with lower BMD at the lumbar spine and femoral neck and a predisposition to vertebral fractures (Ralston et al., 2006). In this study, however, the BMD association was observed only for homozygotes of the Sp1 polymorphism, in contrast to previous studies in which heterozygotes also showed a reduction in BMD (Mann and Ralston, 2003). It should be noted that the association between COLIA1 alleles and vertebral fracture reported in GENOMOS and other studies was not fully accounted for by the reduced bone density, suggesting that the Sp1 allele may also be a measure of bone quality. Furthermore, the existence of an extended haplotype defined by the Sp1 polymorphism and other promoter polymorphisms has been proposed to exert stronger effects on BMD than the individual polymorphisms (Garcia-Giralt et al., 2002; Stewart et al., 2006). Evidence has been presented that suggests that the promoter polymorphism at position 21,663 interacts with the transcription factor NMP4, which plays a role in osteoblast differentiation by interacting with Smads (Garcia-Giralt et al., 2005).

Although collagen type I assembly stipulates a heterotrimer assembly, a homotrimer assembly appears to occur in fetal tissues, fibrosis, and cancer (Han et al., 2010). Genetic studies indicate that collagen (I) homotrimers may be detrimental to bone structure, as in the mouse model of OI (oim), in which only (α1)3 homotrimers are present and the bone contains smaller and more tightly packed apatite crystals, but has a lower BMD due to altered collagen organization and bone structure (Vanleene et al., 2012).

However, the homotrimer can reduce the risk of tendon/ligament rupture, possibly because of the laxity in the fibrils. Translational mechanisms may also play a role in collagen (I) homotrimer biosynthesis, as nonmuscle myosin has been reported to coordinate cotranslational translocation of collagen (I) heterotrimers (Cai et al., 2010). However, disassembly of nonmuscle myosin filaments appeared to result in collagen (I) homotrimer production in lung fibroblasts but not in scleroderma skin fibroblasts, indicating that translational control may be specific to particular cell or tissue types. Collagen (I) homotrimer is resistant to proteolytic degradation by MMPs through its resistance to local unwinding at the MMP cleavage site (Han et al., 2010) and may therefore persist in tissues during remodeling and contribute to tissue sclerosis. In cancer, invasive cancer cells may use homotrimers for building MMP-resistant invasion paths, supporting local proliferation and directed migration of the cells, whereas surrounding normal stromal collagens are cleaved.

Transcriptional regulation of type I collagen genes

Expression of the pro-Col1a1 gene and the pro-Col1a2 gene is coordinately regulated in a variety of physiological and pathological situations. In many of these instances it is likely that the control of expression of these two genes is mainly exerted at the level of transcription, suggesting that similar transcription factors control the transcription of both genes.

This section considers successively the proximal promoter elements of these genes and then the nature of cell-specific enhancers located in other areas of these genes, including intronic sequences. Information about the various DNA elements has come from transient expression experiments in tissue culture cells, in vitro transcription experiments, and experiments in transgenic mice. In vitro transcription experiments and, in large measure, transient expression experiments identify DNA elements that have the potential of activating or inhibiting promoter activity. These DNA elements can be used as probes to detect DNA-binding proteins. However, transient expression and in vitro transcription experiments do not take into account the role of the chromatin structure in the control of gene expression. Transgenic mice are clearly the most physiological system to identify tissue-specific elements; the DNAs that are tested are integrated into the mouse genome and their activities are presumably also influenced by their chromatin environment. In transgenic mice experiments, reporter genes such as green fluorescent protein, luciferase, or the *Escherichia coli* β-galactosidase offer the advantage that their activity is indicative of promoter activation and location of the expression. All three transgenes can be detected in vivo without having to kill the mouse. In addition, X-Gal histochemical stain for β-galactosidase can identify the cell types in which the transgene is active by histology.

Transient transfection experiments using various sequences upstream of the transcription start site of either the pro-α1(I) or the pro-α2(I) promoter, cloned upstream of a reporter gene, and introduced into fibroblasts have delineated positive *cis*-acting regulatory segments in these two sequences that were designated as minimal proximal promoters. Footprint experiments and gel-shift assays performed using these regulatory elements as DNA templates have also characterized sequences that interact with DNA-binding proteins to modulate the expression of the two genes.

Proximal promoters of type I collagen genes

Transcription factors binding to the pro-α1(I) proximal promoter

In the pro-α1(1) collagen gene, a close homology exists between human and mouse promoters. Brenner and colleagues demonstrated by progressively deleting the mouse pro-Col1a1 promoter that sequences downstream of 2181 bp are needed for high-level transient transfection (Brenner et al., 1989). Therefore, the sequence between 2220 and 1110 has been used as the proximal promoter (Fig. 13.2) and contains binding sites for DNA-binding factors that also bind to the proximal pro-Col1a2 promoter (Ghosh, 2002). These DNA elements include an RFX consensus binding site surrounding the transcription start site (−11 to +10) that contains three methylation sites, rather than one in the COL1A2 gene RFX-binding site. RFX1 interacts weakly with the unmethylated COL1A1 site, and binds with higher affinity to the methylated site (Sengupta et al., 2005).

RFX1 represses the unmethylated COL1A1 less efficiently than COL1A2, while RFX5 interacts with both collagen type I genes with similar binding affinity and represses both promoters equally.

The mouse pro-Col1a1 proximal promoter includes a binding site for the CCAAT-binding protein CBF between 290 and 2115 (Karsenty and de Crombrugghe, 1990). A second CCAAT box located slightly more upstream is, however, unable to bind CBF, suggesting that sequences surrounding the CCAAT motif also have a role in CBF binding. Similar results were observed in the human COL1A1 promoter (Saitta et al., 2000). DNA transfection experiments with the pro-Col1a1 promoter showed that point mutations in the CBF-binding site decreased promoter activity (Karsenty and de Crombrugghe, 1990). The CBF-binding site is flanked by two identical 12-bp repeat sequences that are binding sites for Sp1 and probably other GC-rich binding proteins (Nehls et al., 1991). In transient transfection experiments, a mutation in the binding site that prevents the binding of Sp1 surprisingly increased the activity of the promoter, and overexpression of Sp1 decreased the activity of the promoter (Nehls et al., 1991). It is possible that several transcription factors with different activating potentials bind to overlapping binding sites and compete with one another for binding to these sites; the overall activity of the promoter could then depend on the relative occupancy of the different factors on the promoter DNA. Substitution mutations in two apparently redundant sites between 2190 and 2170 and between 2160 and 2130 that abolished DNA binding resulted in an increase in transcription (Karsenty and de Crombrugghe, 1990). Formation of a

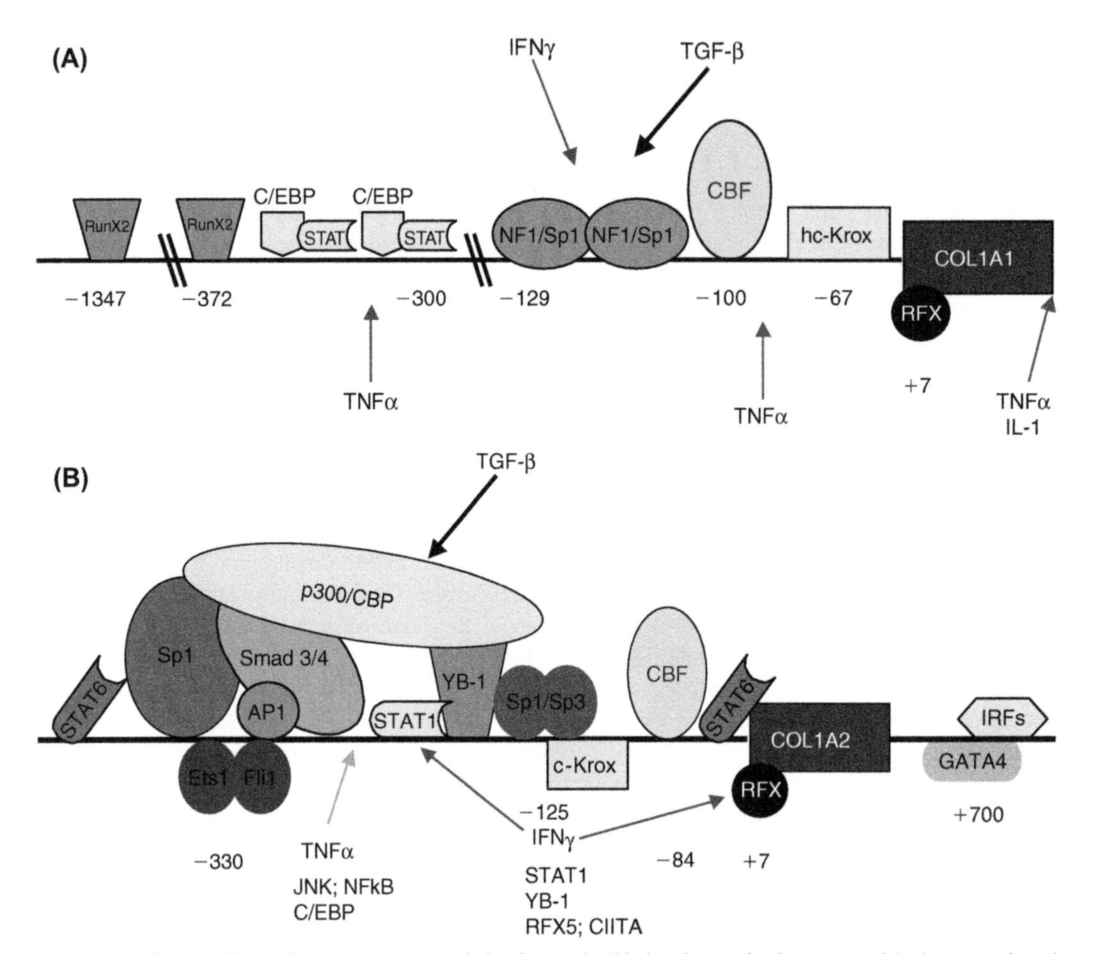

FIGURE 13.2 Schematic diagrams illustrating the known transcription factors that bind to the proximal promoter of the human and murine (A) Col1a1 and (B) Col1a2 promoters. Also shown are the cytokines that modulate the expression of these genes. The data are accumulated from several papers as cited in the text. *AP1*, activator protein 1; *CBF*, CCAAT-binding factor; *CBP*, CREB-binding protein; *C/EBP*, CCAAT/enhancer-binding protein; *CIITA*, class II, major histocompatibility complex, transactivator; *hc-Krox*, human collagen Krox; *IFNγ*, interferon-γ; *IL-1*, interleukin 1; *IRF*, interferon regulatory factor; *JNK*, Jun N-terminal kinase; *NF1*, nuclear factor 1; *NF-κB*, nuclear factor κB; *STAT*, signal transducer and activator of transcription; *TGF-β*, transforming growth factor β; *TNFα*, tumor necrosis factor α; *YB-1*, Y-box binding factor.

DNA−protein complex with these two redundant elements in the pro-Col1a1 promoter was also shown to be completed by the sequence of the pro-Col1a2 promoter between 2173 and 2143, suggesting that both type I promoters contained similar binding sites (Karsenty and de Crombrugghe, 1991). A member of the Krox family, designated c-Krox, binds to these two sites in the pro-Col1a1 promoter (Galera et al., 1994). In addition, c-Krox binds to a site located near the CCAAT box in the pro-Col1a1 promoter and to three GC-rich sequences in the pro-Col1a2 proximal promoter, located between 2277 and 2264 bp, between 2175 and 2143 bp, and near the CCAAT box, respectively (Galera et al., 1996). More recently, c-Krox has been shown to exert its action by synergistic association with SP1 (Kypriotou et al., 2007).

The Col1a1 promoter region spanning bp 29 to 256 bound purified recombinant YY1, and the corresponding binding activity in nuclear extracts was supershifted using a YY1-specific antibody. Mutation of the TATA box to TgTA enhanced YY1 complex formation. YY1 functions as a positive regulator of constitutive activity in fibroblasts. Although YY1 is not sufficient for transcriptional initiation, it is a required component of the transcription machinery in this promoter (Riquet et al., 2001).

A RunX2/Cbfa1-binding element is present in the Col1a1 promoter of mouse, rat, and human at approximately position 2372 (Kern et al., 2001). This site binds RunX2/Cbfa1 only weakly and does not act as a *cis*-acting activator of transcription when tested in DNA transfection experiments. These may interact with the upstream osteoblast-specific element (OSF2) site at 21,347 (Kern et al., 2001), as discussed under "Upstream elements in the pro-*Col1a1* gene" later.

A study exploring the impact of hypoxia on human cartilage identified the role of Sp3 transcription factor in down-regulating collagen type I by binding to a region in the human COL1A1 promoter (between −2300 and −1816) upstream of the initiation site. This motif contains two GC boxes and the data support the competition between Sp1 and Sp3 for binding, and the accumulation of Sp3 in response to cartilage hypoxia switches from Sp1 binding, a positive regulator toward transcriptional suppression via Sp3 (Duval et al., 2016).

It is likely that the transcriptional function of various transcription factors and eventually their DNA-binding properties offer opportunities for regulation by intracellular signaling pathways. These pathways are triggered by a variety of cytokines such as tumor necrosis factor α (TNFα), which exerts its inhibitory action on Col1a1 expression through nuclear factor κB (NF-κB) (Rippe et al., 1999).

Factors binding to the pro-COL1A2 proximal promoter

Several functional *cis*-acting elements have been identified in the approximately 400-bp proximal promoter of the mouse pro-Col1a2 gene (154–2350 bp) and the human minimal sequence between 152 and 2378 bp. The first transcription factor found to bind to this promoter was the ubiquitous CCAAT-binding protein CBF. This transcription factor is formed by three separate subunits, named A, B, and C, which have all been cloned and sequenced (Maity and de Crombrugghe, 1998). All three subunits are needed for CBF to bind to the sequence containing the CCAAT box located between 284 and 280 and activate transcription in both human and mouse genes (see Fig. 13.2B). In vitro data suggest that the A and C subunits first associate to form an A–C complex and that this complex then forms a heteromeric molecule with the B subunit (Sinha et al., 1995). Mutations in the CCAAT box that prevent the binding of CBF decrease the transcriptional activity of the pro-Col1a2 proximal promoter three to five times in transient transfection experiments of fibroblastic cell lines (Coustry et al., 1995). Purified CBF as well as CBF composed of its three recombinant subunits also activate the pro-Col1a2 promoter in cell-free nuclear extracts previously depleted of CBF (Coustry et al., 1995). Two of the three subunits of CBF contain transcriptional activation domains. More recently, a single substitution of T to an A in vivo in the presence of an upstream enhancer of the human sequence suggested that CBF is involved in patterning of the collagen type I expression in the dorsoventral as well as the rostrocaudal axis of mouse skin fibroblasts (Tanaka et al., 2004).

In addition to the binding site for CBF, footprinting experiments and gel-shift studies identified other binding sites in the first 350 bp of the mouse pro-Col1a2 promoter. Three GC-rich sequences, located at about 2160 bp (between 2176 and 2152 bp) and 2120 bp (between 2131 and 2114 bp) have been shown to interact with DNA-binding proteins by footprint experiments and gel-shift assays (Hasegawa et al., 1996). A deletion in the mouse promoter encompassing these three footprinted sequences completely abolished the transcriptional activity of the pro-Col1a2 proximal promoter in transient transfection experiments using fibroblastic cell lines. The corresponding regions in the human pro-COL1A2 promoter were also protected in in vivo and in vitro footprinting experiments (Ihn et al., 1996) and bound transcriptional activators. However, gel-shift assays performed using the human pro-COL1A2 promoter have suggested that a *cis*-acting element located at 2160 bp represents a repressor element (Ihn et al., 1996), implying that interactions between proteins interacting with the activator elements at 2300, 2125, and 280 bp and proteins binding to a repressor element at 2160 bp regulate this gene. More recently it was shown that CUX1, a CCAAT displacement protein, is associated with reduced expression of type I collagen both in vivo and in vitro and that enhancing the expression of CUX1 results in effective suppression of type I collagen by interfering with CBF binding (Fragiadaki et al., 2011).

Sp1 and Sp3 have been shown to bind the TCCTCC motif located between 2123 and 2128 bp; both transcription factors activate this promoter (Ihn et al., 2001) (see Fig. 13.2B). Furthermore, proteins that bind to the two proximal segments also bind to the most upstream GC-rich segment, at 2300 bp, with the exception of CBF, suggesting a redundancy among functionally active DNA segments of the pro-COL1A2 proximal promoter.

TGFβ mediates its action in the human promoter through the combination of the ubiquitous transcription factor Sp1, the Smad3/4 complex, and the coactivators p300/CREB-binding protein (CBP) in what is termed a TGFβ-responsive element (TβRE) (Zhang et al., 2000). This segment contains three GC-rich motifs between 2330 and 2255, capable of binding Sp1, CCAAT/enhancer-binding protein (C/EBP), activator protein 1 (AP1), and Smad complexes (Chen et al., 1999, 2000a,b; Kanamaru et al., 2003; Tamaki et al., 1995; Zhang et al., 2000). Interaction among these nuclear factors is synergistic and requires binding of Sp1 and Smad3/4 to the GC-rich elements and the downstream CAGAC Smad site, respectively (Ghosh et al., 2000; Poncelet and Schnaper, 2001; Zhang et al., 2000). At the turn of this century, there was a great debate over whether Smads are more important than AP1 in the proximal promoter. This was resolved some 10 years later when it was shown that TGFβ also activates the COL1A2 gene via a noncanonical (Smad-independent) signaling pathway, which requires enhancer/promoter cooperation involving exchange of c-Jun/JunB transcription factor occupancy of a critical enhancer site, resulting in the stabilization of enhancer/promoter coalescence (Ponticos et al., 2009). A report exploring the effects of hypoxia and TGFβ on the COL1A2 gene has added to the potential role of Smads, in particular Smad3, in regulating collagen expression. These studies in human mesangial cells suggest that under hypoxic conditions and in the presence of TGFβ, hypoxia-inducible factor 1α (HIF-1α) and Smad3 can form transcriptional complexes and preferentially enhance binding to one of the three hypoxia response elements within the human COL1A2 promoter at position −335 bp (Baumann et al., 2016). The authors suggest that this novel interaction provides a mechanism that accounts for the synergy observed between HIF and TGFβ in kidney fibrosis.

TNFα and interferon-γ (IFN-γ) suppress matrix production. In contrast to the single DNA element that mediates TGFβ stimulation of the COL1A2 proximal promoter, both TNFα and IFN-γ have been shown to inhibit COL1A2 transcription by interfering with the formation of the TGFβ-induced complex, and by stimulating the interaction of negative factors with responsive DNA elements located 5′ and 3′ of it. This so-called "cytokine responsive element" plays an important role in maintaining homeostasis. Involvement of AP1 and NF-κB in transducing the inhibitory action of TNFα on COL1A2 gene expression has been shown using immortalized fibroblasts from mice that lack either the AP1 activator Jun N-terminal kinase 1 (JNK1) or the NF-κB essential modulator NEMO. Specifically, the loss of JNK1 prevented the TNFα antagonism of TGFβ, but preserved TNFα inhibition of constitutive COL1A2 expression. TGFβ antagonism by TNFα involves JNK1 phosphorylation of c-jun, leading to off-DNA competition of the latter molecule for Smad3 binding to the cognate DNA site and/or for interaction with the p300/CBP coactivators (Kouba et al., 1999; Verrecchia et al., 2001, 2002).

Binding of IFN-γ to its receptors leads to tyrosine phosphorylation of janus kinase (JAK) tyrosine kinases and this in turn results in signal transducer and activator of transcription (STAT1) phosphorylation. In COL1A2, STAT1 activation results in competition with Smad3 for interaction with p300/CBP (Inagaki et al., 2003). In addition, JAK1 can also activate transcription factor Y-box binding factor (YB-1); this activation results in both inhibition of constitutive promoter activity through YB-1 binding of the 2125TC box and antagonism of TGFβ signals through YB-1 competition with Smad3 and/or p300/CBP (Ghosh et al., 2001; Higashi et al., 2003). For more details, see "Growth factors" and "Cytokines."

Est1/Fli1 have also been shown to bind the same sequence with opposite effects on collagen type I transcription. A functional Ets transcription factor was identified in COL1A2 in close proximity of Sp1 sites. Ets1 stimulated, whereas Fli1 inhibited, promoter activity. Sp1 binding was essential for inhibition of Fli1. Moreover, overexpression of Fli1 in dermal fibroblasts led to a decrease in COL1A2 mRNA and protein levels (Czuwara-Ladykowska et al., 2001). Furthermore, TGFβ treatment of dermal fibroblasts leads to dissociation of Ets1 from the CBP/p300 complexes and alters their response to TGFβ in favor of matrix degradation (Czuwara-Ladykowska et al., 2002).

Furthermore, a CpG motif at 17 bp has been shown to be preferentially methylated and bound by RFX proteins in cells acquiring a collagen I–negative state. Cell transfection experiments in conjunction with DNA-binding assays have assigned positive or negative properties to each of the proteins binding to the proximal promoter of pro-COL1A2 (Sengupta et al., 2002). The degree of methylation at the 17 CpG site, on the other hand, has been shown to modulate the binding affinity of RFX proteins and, consequently, their ability to downregulate promoter activity by recruiting associated proteins that interfere with the assembly of positively acting transcriptional complexes (Xu et al., 2003, 2004).

Structure and functional organization of upstream segments of type I collagen genes

Upstream elements in the pro-*Col1a1* gene

In complete contrast with its high-level activity in transient expression and in vitro transcription experiments, the 220-bp pro-Col1a1 proximal promoter is almost completely inactive in stable transfection experiments and in transgenic mice (Rossert et al., 1996). By increasing the length of this promoter, evidence from different groups showed that the 2.3-kb human promoter linked to different transgenes resulted in a high degree of tissue-specific expression of the reporter gene in bone, tail, and skin. However, this expression of the transgene was not identical to that of the endogenous gene (Slack et al., 1991). Indeed, in later experiments using embryos from the same mice, in situ hybridization assays showed no expression of the transgene in perichondrium and in skeletal muscle, implying that additional sequences might be needed to obtain expression of the gene in all type I collagen–producing cells (Liska et al., 1994). In newborn mice carrying a 3.6-kb rat Col1a1 promoter linked to a chloramphenicol acetyltransferase (CAT) reporter gene, high levels of transgene expression were found in extracts of bone, tooth, and tendon (Bogdanovic et al., 1994; Pavlin et al., 1992). Similarly, the mouse 3.6-kb promoter fused to β-galactosidase transgene resulted in a similar pattern, but localized to three distinct *cis*-acting sequences that directed expression in the skin, bone, and tendon (Rossert et al., 1995). The skin element was between 2220 and 2900 bp. The bone element was further delineated to 117 bp between 21,656 and 21,540 bp, and the tendon to a sequence between 22,300 and 23,600 bp (Rossert et al., 1996) (Fig. 13.3). These experiments suggested a modular arrangement of separate *cis*-acting elements that activate the pro-Col1a1 gene in different type I collagen-

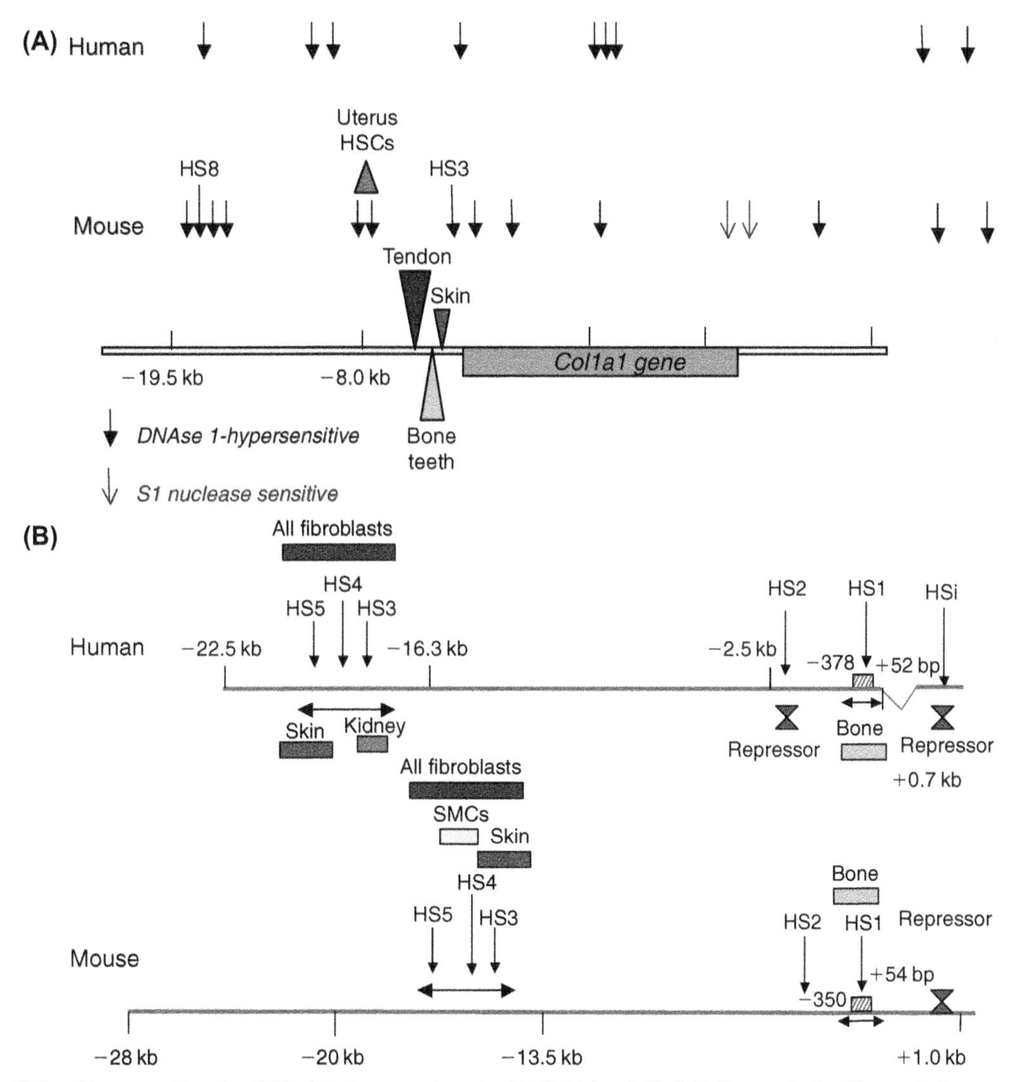

FIGURE 13.3 DNase I hypersensitive sites (*HS*) of the human and murine (A) Col1a1 and (B) Col1a2 genes depicted by *vertical arrows*, which indicate the position within the gene where the sequence may be involved in transcriptional regulation. Areas of high homology in the promoter (0.85%) are indicated by *horizontal arrows*. Also indicated are the regions in which *cis*-acting sequences have been shown to be expressed in transgenic mice, and the tissues or cells in which they are expressed, and which are called "elements" SMCs, smooth muscle cells.

producing cells. The direct consequence of such a modular arrangement is that it should be possible to selectively modulate the activation of type I collagen genes in well-defined subpopulations of type I collagen–producing cells. Perhaps the best illustration of a *cis*-acting element responsible for the activation of the mouse pro-Col1a1 gene is the osteoblast element. This 117-bp segment is very well conserved among species. In the mouse promoter, located between 21,656 and 21,540 bp, is a minimal sequence able to induce high levels of expression of the reporter gene exclusively in osteoblasts. In these mice the transgene becomes active at the same time during embryonic development when osteoblasts first appear in the different ossification centers. This so-called "osteoblast-specific element" can be divided into three subsegments that have different functions. The 29-bp A segment, which is located most 5′ (21,656–21,628 bp), is required to activate the gene in osteoblasts. A deletion of the A element or a 4-bp mutation in the TAAT sequence of this segment completely abolished the expression of the reporter gene in osteoblasts of transgenic mice. The C segment, which is located at the 3′ end of the 117-bp sequence (21,575–21,540 bp), is required to obtain consistent high-level expression of the reporter gene in transgenic mice (Rossert et al., 1996). When this C segment was deleted, the lacZ gene was expressed at very low levels and only in a small proportion of transgenic mice. The function of the intermediary segment (B segment) is still poorly understood, but it could be to prevent a promiscuous expression of the gene and, in particular, expression of the gene in the nervous system.

In addition to the 117-bp region, a consensus Cbfa1-binding site, termed oxidation-specific epitope 2 (OSE2), is present at approximately 21,347 bp of the rat, mouse, and human genes (Kern et al., 2001). RunX2/Cbfa1 can bind to this site, as demonstrated by electrophoretic mobility shift assay and supershift experiments using an anti-Cbfa1 antibody. Mutagenesis of the Col1a1 OSE2 at 21,347 bp reduces the activity of a Col1a1 promoter fragment - to threefold. Moreover, multimers of this OSE2 at 21,347 bp confer osteoblast-specific activity to a minimum Col1a1 collagen promoter fragment in DNA transfection experiments. However, this region did not confer expression in transgenic mice, suggesting that other specific factors contribute to expression of collagen type I in osteoblasts. Indeed, Osterix (Osx), a zinc finger transcription factor, was found to be essential for bone formation and to act downstream of RunX2/Cbfa1. It was shown to bind to segment C of the osteoblast element. The tendon element was found to be a combination of sequences between 23.2 and 23.6 kb termed TSE1 and TSE2 (Terraz et al., 2002). Gel-shift experiments of this region showed that scleraxis, which is a basic helix–loop–helix transcription factor that is expressed selectively in tendon fibroblasts, binds TSE2, whereas TSE1 binds NFATc (nuclear factor of activated T cells) transcription factors (Lejard et al., 2007).

In pursuit of the controlling regions of this gene, a chromatin structure analysis of the human COL1A2 gene has revealed the presence of several proximal and distal 5′ and 3′ DNase I hypersensitive sites (HSs) (Barsh et al., 1984). Breindl's team used the same technique and mapped a 55-kb sequence spanning the mouse gene, including 24 kb of 5′- and 13 kb of 3′-flanking sequences (Salimi-Tari et al., 1997). Nine HSs were found, seven of which were in the promoter sequence (HS3–HS9) up to 220 kb (see Fig. 13.3A). All of these HSs were present in collagen type I–producing cells except for HS8. When these sequences were fused to a reporter gene in transgenic animals, they revealed an additional *cis*-acting sequence in uterine smooth muscle cells between 27 and 28 kb (Krempen et al., 1999) and in liver stellate cells (Yata et al., 2003).

Upstream elements of the mouse pro-*Col1a2* gene

The activity of the mouse 350-bp pro-Col1a2 proximal promoter in transgenic mice is very low compared with that of the corresponding endogenous gene. Although this low-level activity appears to be present selectively in fibroblasts and mesenchymal cells (Niederreither et al., 1992), the precise sequences and the factors responsible for this tissue specificity have not yet been identified. An element located 13.5–17 kb upstream of the start site of transcription, and named "far-upstream enhancer," increased the levels of expression of the lacZ and luciferase reporter genes considerably when it was cloned upstream of the 350-bp mouse pro-Col1a2 proximal promoter. Moreover, this element by itself contributed to the tissue-specific expression of a reporter gene. Indeed, when it was cloned upstream of a minimal promoter that has no tissue-specific expression by itself, i.e., the first 220 bp of the pro-Col1a1 proximal promoter, it conferred a tissue-specific expression to the lacZ reporter gene in transgenic mice (Bou-Gharios et al., 1996). The enhancer was reduced in size to 1.5 kb between −15.5 and −17 kb and was shown to depend on HS4 around 250 bp to drive the expression. Interestingly, in transgenic mice harboring the lacZ reporter gene cloned downstream of a pro-Col1a2 promoter segment containing the far-upstream enhancer, fibroblastic cells expressed the lacZ reporter gene at very high levels, but only a subset of osteoblastic cells expressed this reporter gene, and odontoblasts and fully differentiated tendon fibroblasts did not (Bou-Gharios et al., 1996; Li et al., 2017), suggesting that a *cis*-acting element active in all osteoblasts is missing from this promoter/enhancer sequence. Other than this heterogeneous expression of the lacZ reporter gene in bone, the lack of expression of the lacZ gene in odontoblasts suggests that other elements exist that control expression in these cells and strongly supports a modular organization of different regulatory domains in the mouse pro-Col1a2 promoter, as described for the pro-Col1a1 promoter.

To verify the transgene staining results, both luciferase and β-galactosidase activity in tissue homogenate was measured, because the transgenic line used harbored both β-galactosidase and luciferase transgenes. Although β-galactosidase staining was no longer present in many tissues by the time transgenic mice were weaned, luciferase activity of the same tissues remained significantly higher than in controls. The endogenous mRNA level followed the overall trend of the transgene, and very little transgene mRNA was detected by 3 months of age. The skin showed the highest level of endogenous mRNA during development, which was matched by transgene expression (Ponticos et al., 2004a, b), and although mRNA level dropped by day 10 after birth, the transgene remained significantly high until 3 weeks of age, suggesting perhaps a specific role of this enhancer in the dermis.

The role of the enhancer in the regulation of the collagen type I gene was also revealed by measuring activation of the transgenes after injury to tissues, either by injection of fibrogenic cytokines such as TGFβ1 or by physical injury. Such challenge appears to stimulate a significant increase in transcription of both the endogenous collagen gene and the transgenes. Although mRNA levels of collagen have been shown to diminish in many tissues after birth and throughout adulthood (Goldberg et al., 1992), acute injury reactivates the lacZ transgene to a level that is visually apparent

(Ponticos et al., 2004a, b). This finding makes such transgenic animals an important tool and provides a model to address questions about the nature of cells that are active in fibrosis. For example, in muscular dystrophy, where fibrosis bears the hallmark of the disease (Morrison et al., 2000), this model can be used to identify those cells that are active in producing collagen type I as shown with glomerulosclerosis (Chatziantoniou et al., 1998) and the Tsk mouse model bred with these transgenic mice (Denton et al., 2001). More importantly, this will allow us to test the direct effects of various antifibrotic agents on the cells that are actively synthesizing collagen type I.

In the human COL1A2 gene, six distinct DNase I–HSs were mapped within 22 kb upstream and around the start site of transcription. Their spatial arrangement, cell type specificity, and relative availability to DNase I digestion are comparable to those of the mouse gene (Antoniv et al., 2001). The more striking finding of the chromatin survey was the identification of three strong cell-type-specific HSs (HS3–5), at a location comparable to those of the mouse far-upstream enhancer and residing within nearly identical sequences (see Fig. 13.3B). DNase I footprinting correlated areas of sequence identity with 12 distinct binding sites of nuclear proteins, the majority of which are likely not to be tissue specific. Using transgenic analysis, it was found that the proximal 2378 promoter contains elements that drive tissue-specific transcription in subsets of fibroblastic and osteoblastic cells, and this activity is significantly augmented when the proximal promoter is linked to the far-upstream enhancer. The results of the human transgenes thus indicate that the predominant function of the upstream element is to broaden and intensify tissue-specific transcription from the proximal promoter. Unlike the mouse enhancer that was active only in a subset of osteoblasts, the human sequence was active in all osteoblasts and it is most likely that the osteoblast element resides in the proximal promoter (Antoniv et al., 2001). Focusing on the mouse Col1a2 gene, a study by Marongiu et al. has identified a potentially new long-range autoregulatory mechanism controlling collagen transcription. The group investigated FOXL2, a member of the evolutionarily conserved forkhead box transcription factors, and found, using chromatin immunoprecipitation experiments followed by genomic DNA sequencing (ChIP-seq) and electrophoretic mobility shift assays, that this factor bound to a DNA element ~65 kb upstream of the collagen gene. This region was found to contain a putative forkhead transcription factor binding site (ACAAACA), thereby providing a mechanism antagonizing transcription from the promoter. These experiments were centered on studies of the mouse reproductive system and therefore whether this represents a general automodulatory process governing Col1a2 transcription or is a tissue-restricted process is unclear as of this writing (Marongiu et al., 2016).

Delineating the mode of action of tissue-specific elements

The mode of action of the different lineage-specific transcription elements and their postulated cognate-binding proteins is still unknown, but studies of HSs (Antoniv et al., 2001; Bou-Gharios et al., 1996) and in vivo footprinting experiments (Chen et al., 1997) strongly suggest that the chromatin structure of discrete areas in the regulatory regions of type I collagen genes is different in cells when these genes are being transcribed actively, compared with cells in which they are silent. These experiments suggest that, in intact cells, transcription factors such as CBF bind to the proximal promoters of type I collagen genes only in cells in which the genes are transcribed actively (Coustry et al., 2001). Although in vivo footprint experiments show a protection of the CCAAT box in different fibroblastic cell lines, such a protection does not exist in cell lines that do not produce type I collagen. Similarly, HSs corresponding to the far-upstream enhancer of the pro-Col1a2 promoter can be detected only in cells that express type I collagen. The importance of chromatin structure is also highlighted by comparison of transient and stable transfection experiments. When a chimeric construct harboring the pro-Col1a1 osteoblast-specific element cloned upstream of a minimal promoter and of the lacZ reporter gene was transfected stably in different cell lines, it was expressed in the ROS17/2 osteoblastic cell line, but not in two fibroblastic cell lines or in a cell line that does not produce type I collagen. In contrast, in transient transfection experiments, the same chimeric construction was expressed at high levels by all cell lines. These results suggest a model in which the binding of a lineage-specific transcription factor to specific enhancer segments of type I collagen genes would result in opening the chromatin around the promoter and allow ubiquitous transcription factors to bind to the proximal promoter and to activate transcription of the genes.

The discovery of the far-upstream enhancer raised the question of whether the enhancer could confer tissue specificity on its own, and if individual elements of the proximal promoter contribute differently to COL1A2 transcription in vivo. A series of mutations introduced into each of the transcription factors' binding sites within the proximal promoter, such as the three Sp1 sites, GC-rich Sp1/Sp3-binding sites, and the CBF-binding site, reduced transgene activity, providing strong evidence that these protein interactions operate in vivo as well, and in concert with the enhancer-bound complex (Tanaka et al., 2004). The unique expression pattern of transgenic mice harboring mutations in the binding site of transcription factor CBF/NFY led to loss of lacZ expression in the ventral and head regions of the dermis, as well as in the muscles of the forelimbs. The patterning of specific cell lineages implies that CBF/NFY may be essential for COL1A2 activity in those cells that do not express the transgene. These transgenic data therefore raise the intriguing possibility that CBF/NFY may be implicated in patterning COL1A2 expression in some mesenchyme cell lineages through a yet-to-be-defined mechanism. The results suggested cooperativity between the far-upstream enhancer and the proximal promoter in assembling tissue-specific protein complexes. They confirmed in vitro observations indicating that interactions among proximal promoter elements are required for optimal transcription. They also indicated that transcription factors binding to individual promoter elements are responsible for distinct properties of COL1A2 expression in vivo (Ramirez et al., 2006). More importantly, the upstream enhancer was shown to play a part in the fibrotic process of connective tissue disease, such as diffuse cutaneous systemic sclerosis (SSc), where the accumulation of JunB in the COL1A2 enhancer in fibroblast DNA resulted in altered mammalian target of rapamycin (mTOR)/protein kinase B (AKT) signaling and a failure of proteolytic degradation of collagen, which underpins the aberrant overexpression of type I collagen in this disease (Ponticos et al., 2015).

In the enhancer sequence of the mouse Col1a2 gene, the first tissue-specific element to be characterized was a vascular smooth muscle cell (vSMC) element: a 100-bp sequence at about 216.6 kb upstream of the transcription start site regulates collagen expression exclusively in vSMCs (Ponticos et al., 2004a, b). This expression was shown to be activated through the binding of the homeodomain protein Nkx2.5, and further potentiated by the presence of GATA6. In contrast, this element was repressed by the binding of the zinc finger protein δ-EF1/ZEB1. A model of regulation was proposed in which the activating transcription factor, Nkx2.5, and the repressor, δ-EF1/ZEB1, compete for an overlapping DNA-binding site. This element is important in understanding the molecular mechanisms of vessel remodeling and is a potential target for intervention in vascular diseases in which there is excessive deposition of collagen in the vessel wall.

Role of the first intronic elements in regulating collagen type I

First intron of the pro-*COL1A1* gene

Different negative or positive regulatory segments have been identified within the first intron, but most of the transcription factors binding to these regulatory segments are still unknown. A sequence of the first intron of the human COL1A2 gene located about 1600 bp downstream of the transcription start site binds AP1, and a mutation that abolished this binding diminished the expression of a reporter gene in transient transfection experiments (Liska et al., 1990) This AP1 was also found to mediate the repressive effect of ras in fibroblasts (Slack et al., 1995). Another segment of the first intron of the human gene, which extends from 820 to 1093 bp, has been shown to inhibit the activity of a reporter gene in transient transfection experiments (Liska et al., 1992). This sequence contains two binding sites for an Sp1-like transcription factor, and mutations in these two Sp1-binding sites tended to increase the activity of the reporter gene (Liska et al., 1992). An Sp1-binding site is also located at about 1240 bp, in the human gene, and a frequent G-to-T polymorphism in this Sp1-binding site (G1242T) has been linked with low BMD and increased risk of osteoporotic vertebrate fracture (Grant et al., 1996), which suggests that it may be important for normal levels of type I collagen synthesis by osteoblasts.

The phenotype of Mov 13 mice suggested that the first intron of the pro-α1(I) collagen gene could play a role in the expression of this gene. These mice, which harbor a retrovirus in the first intron of the pro-Col1a1 gene (Harbers et al., 1984), express this gene in osteoblasts and odontoblasts, but not in fibroblastic cells (Schwarz et al., 1990). Nevertheless, the presence of tissue-specific regulatory elements in the first intron of the pro-Col1a1 gene has long been controversial. Two groups have reported that, in transgenic mice harboring the proximal promoter of either the human or the rat pro-Col1a1 gene, the pattern of expression of the reporter gene was the same whether or not these mice harbored the first intron of the pro-Col1a1 collagen gene (Sokolov et al., 1993). In contrast, data obtained by in situ hybridization in transgenic mice harboring 2.3 kb of the human pro-Col1a1 proximal promoter suggested that the first intron of this gene was necessary to obtain high-level expression of the transgene in the dermis of skin (Liska et al., 1994). Only mice harboring the first intron, in addition to the 2.3-kb proximal promoter segment, expressed the human growth hormone reporter gene at high levels in skin. To clarify this issue, Bornstein's group generated knock-in mice with a targeted deletion of most of the first intron (Hormuzdi et al., 1998). Mice homozygous for the mutated allele developed normally and showed no apparent abnormalities. Nevertheless, in heterozygous mice, the mutated allele was expressed at normal levels in skin, but at lower levels in lung and muscle, and its levels of expression decreased with age in these two tissues more than in wild-type mice. Thus, the first intron does not play a role in the tissue-specific expression of the pro-Col1a1 gene, but it seems to be important for maintaining normal transcriptional levels of this gene in certain tissues (Hormuzdi et al., 1999).

First intron of the pro-*COL1A2* gene

The first intron of the mouse pro-Col1a2 gene has also been shown to contain a tissue-specific enhancer in transient transfection experiments (Rossi and de Crombrugghe, 1987). In transgenic mice, however, the presence of this tissue-specific enhancer apparently had no effect on the pattern of expression of a CAT reporter gene (Goldberg et al., 1992). However, transgenic analysis of the human gene suggested that, despite the homologies between the two genes, species-specific differences have been reported regarding the function of individual *cis*-acting elements, such as the first intron sequence (Antoniv et al., 2005). In vitro DNase I footprinting of the sequence corresponding to the open chromatin site identified a cluster of three distinct areas of nuclease protection that span from nucleotide 1647 to 1759 of the COL1A2 gene. These areas contain consensus sequences for GATA and interferon regulatory factor (IRF) transcription factors. Gel mobility shift and chromatin immunoprecipitation assays corroborated this finding by documenting binding of GATA4 and IRFs 1and 2 to the first intron sequence (Antoniv et al., 2005). Moreover, a short sequence encompassing the three footprints was found to inhibit expression of transgenic constructs containing the COL1A2 proximal promoter and far-upstream enhancer in a position-independent manner. Mutations inserted into each of the footprints restored transgenic expression to different extents. These results therefore indicated that the intronic sequence corresponds to a repressor element, the activity of which seems to be mediated by the concerted action of GATA and IRF proteins (see Fig. 13.2B). Furthermore, GATA4 was found to bind the promoter sequence of HS2 at 22.3 kb in the COL1A2 gene. Overexpression of this transcription factor inhibited the expression of the transgene as well as the endogenous gene in fibroblasts (Wang et al., 2005). The presence of a common repressor transcription factor in both HS2 and HS intron led to the suggestion that the "switch-off" of collagen in adults may occur via cooperation of the intronic sequence and HS2 to block the enhancer activity from exerting its effect through its cooperation with the proximal promoter in HS1 (Ramirez et al., 2006).

ChIP-seq with chromatin of mouse calvarial cells has been used to identify Osx/Sp7 interaction sites (Hojo et al., 2016). Osx was previously shown to be a transcription factor completely required for osteoblast differentiation and for bone formation (Nakashima *et al.* 2002). In this ChIP-seq study one major peak of Osx/Sp7 interaction in the chromatin of the Col1a1 gene was localized in the segment that was previously identified to be osteoblast specific in transgenic mice (Rossert et al., 1996; Bedalov et al., 1995). Other peaks of Osx/Sp7 interactions were also identified farther upstream of the sequence previously identified as osteoblast specific. The same ChIP-seq study also provided evidence that Osx/Sp7, which belongs to the Sp1 family of transcription factors, was itself not strongly binding to a GC-rich DNA target, unlike the other members of the SP1 family, but interacted as a coactivator with another DNA-binding protein, most likely a member of the Dlx family of transcription factors (Hojo et al., 2016). ChIP-seq is a powerful method to identify interaction sites for specific transcription factors in the genome of specific cell types. These sites can then further be characterized either by reporter assays in vivo in transgenic mice or by transfection of tissue culture cells. In addition, ChIP-seq can be used to characterize epigenomic modifications of histones or other chromatin proteins in the genome of specific cells.

In other ChIP-seq experiments interactions of the Runx2 transcription factor, which is also required for osteoblast differentiation and for bone formation, with the Col1a1 gene were examined (Wu et al., 2014). A major site of interactions was found at a promoter-proximal site of this gene. This study compared the interactions of Runx2 in the genome of a preosteoblastic cell line at three stages of osteoblast differentiation.

Posttranscriptional regulation of type I collagen

Even if the control of expression of type I collagen genes appears to be mainly exerted at the level of transcription, type I collagen production is also regulated at a posttranscriptional level. For example, TGFβ and IFN-γ modulate not only the levels of transcription of type I collagen genes, but also the stability of the corresponding mRNAs. Similarly, activation of HSCs is associated with a dramatic increase in the stability of the pro-Col1a1 mRNA. Run-on experiments have shown that the half-life of this mRNA was increased about 15-fold in activated rat HSCs compared with quiescent ones (Stefanovic et al., 1997). Several lines of evidence indicate that expression of the genes coding for the two collagens I and III is controlled predominantly posttranscriptionally at the mRNA level, as the 3′-UTR of the collagen family mRNA has been shown to be a target for mRNA-binding protein. Heterogeneous nuclear ribonucleoproteins (hnRNPs) A1, E1, and K substantially participate in the coordinative upregulation of collagen I and III synthesis, which is involved in mRNA stabilization (Lindquist et al., 2000). Indeed, all three COL1A1, COL1A2, and COL3A1 3′-UTRs contain hnRNP K-like consensus motifs and do actually bind it. COL1A1 and COL1A2 contain a classical CU-rich hnRNP E1-binding site, and it is also able to bind hnRNP E1, although not in COL3A1, which lacks a more extended CU-rich region. Furthermore, these mRNA-binding proteins have been shown to increase in fibrotic conditions and are suggested to coordinate expression of type I and III collagens by common posttranscriptional mechanisms targeted at the mRNA 3′-UTR level (Thiele et al., 2004).

The sequence of the mouse pro-Col1a1 mRNA surrounding the start codon could also play a role in regulating mRNA stability. This sequence has been shown to form a stem−loop structure, and mutations that prevent its formation decreased the stability of the mRNA dramatically (Stefanovic et al., 1999). Indeed, the 5′-UTRs of COL1A1, COL1A2, and COL3A1 mRNAs are involved in such translational control. These 5′ stem−loops, together with their cognate binding proteins, should help coordinate translation and couple the translation apparatus to the rest of the collagen biosynthetic pathway (Stefanovic and Brenner, 2003).

MicroRNAs (miRNAs) mediate the posttranscriptional regulation of gene expression by binding to partially complementary sites in the 3′-UTR of target mRNA. The formation of miRNA−mRNA duplexes leads to mRNA degradation or translational repression (Ambros, 2003) The first miRNAs that were shown to be involved in collagen regulation were members of the miR-29 family. TGF is activated and triggers the downregulation of miR-29 in cardiac fibroblasts and consequent upregulation of the expression of collagens and other ECM proteins involved in fibrosis (pnas.0805038105). This was also evident in other fibrotic disease. Moreover, miR-185 and miR-98 have been shown to regulate TGFβ and collagen type I in hypertrophic scar tissue fibroblasts. miR-185 loss of function resulted in elevated TGFβ1 and collagen expression, suggesting miR-185 might serve as a potential therapeutic target for scarring. miR-98 was shown to directly target the Col1A1 3′-UTR, resulting in reduced expression, and therefore likely to have an impact on scar formation. In addition to miRs that directly interact and inhibit collagen and TGFβ, the study of keloid fibroblasts has highlighted the role of miR-21in promoting Col1A1 production via the targeting of Smad7. Negatively regulating the expression of the inhibitory Smad, Smad 7, appears to alter the TGFβ/Smad7 signaling pathway favoring increased collagen production and keloid scarring. Similarly, downregulation of miRNA let-7a or miR-196a was shown to contribute to scarring in SSc by the intrinsic activation of TGFβ (jimmunol.1200822) and the expression of miR-21 or miR-145 and decreased mRNA levels of Smad7 or Smad3, respectively. Similarly, the cytokine decreased the expression of miR-29b and increased α1(I) collagen mRNA. Accordingly, these miRNAs may contribute to the pathogenesis of SSc via the regulation of Smad7, Smad3, and α1(I) collagen (Zhu et al., 2012).

Critical factors involved in type I collagen gene regulation

Various cytokines, hormones, vitamins, and growth factors can modify type I collagen synthesis by osteoblasts and/or fibroblasts. The effects of these molecules have been studied mainly in vitro by using either bone organ cultures or cell cultures. However, in some instances the in vivo effects of these factors on type I collagen synthesis have also been studied. A degree of complexity is due to the fact that some factors can act on type I collagen synthesis but can also act indirectly by modifying the secretion of other factors, which will themselves affect type I collagen synthesis. For example, TNFα will inhibit type I collagen production through the NF-κB pathway, but it will also induce the secretion of prostaglandin E_2 (PGE_2) and of interleukin 1 (IL-1), which themselves affect type I collagen production. Furthermore, PGE_2 will induce the production of insulin-like growth factor 1 (IGF-1), which also modifies the rate of type I collagen synthesis.

Growth factors

Transforming growth factor β

In mammals, the TGFβ family consists of three members (TGFβ1, TGFβ2, and TGFβ3), which have similar biological effects but different spatial and temporal patterns of expression. These three molecules are part of a large family, the TGFβ superfamily, which also contains proteins such as BMPs, activins/inhibins, and growth and differentiation factors, as well as Nodal and its related proteins (Chang et al., 2002; Massague and Chen, 2000). Members of the TGFβ family are multifunctional proteins that regulate cell proliferation and differentiation, ECM modification, angiogenesis, apoptosis, and immunosuppression (Feng and Derynck, 2005; Jones et al., 2006; Schier and Shen, 2000). A TGFβ protein exerts its function by binding to and bringing together on the cell surface type I and II receptors to form a ligand−receptor complex (Leask and Abraham, 2004). Five members of type II and seven members of type I receptors (Activin receptor-like kinase 1−7) have been characterized in mammals (Graham and Peng, 2006). Upon phosphorylation by the type II receptor, the type I receptor phosphorylates and activates Smads, which are intracellular signaling molecules for members of the TGFβ superfamily. Phosphorylated Smads are released from the receptors and form oligomeric complexes with a common-partner Smad, Smad4, and translocate into the nucleus where they bind to specific DNA sequences called Smad-binding elements and act as transcription factors to regulate the transcription of target genes. Smad2 and Smad3 respond to TGFβs, Activins, Nodal, and Lefty, whereas Smad1, Smad5, and Smad8 mediate BMP signaling (Miyazawa et al., 2002). Several transcription factors and transcriptional coactivators have been shown to cooperate with Smad complexes, such as Sp1, AP1, PEBP2/CBF, TFE-3, ATF-2, and CBP/p300 (reviewed in Attisano and Wrana, 2000).

TGFβs are secreted by many cell types, including monocytes/macrophages, lymphocytes, platelets, fibroblasts, osteoblasts, and osteoclasts (Leask and Abraham, 2004). Synthesis by bone cells is quantitatively important because, in vivo, the highest levels of TGFβ are found in platelets and bone (Seyedin et al., 1985). Nearly all cells, including osteoblasts and fibroblasts, have TGFβ receptors (Verrecchia and Mauviel, 2007).

The three mammalian TGFβ isoforms are synthesized as homodimeric proproteins (pro-TGFβ) that have a mass of 75 kDa. The dimeric propeptides are cleaved from the mature TGFβ 24-kDa dimer in the *trans*-Golgi by furin-type enzymes. Mature TGFβ1 then associates noncovalently with a dimer of its N-terminal propeptide (also called LAP, for latency-associated peptide). This complex is referred to as the small latent complex (Annes et al., 2003). Early in the assembly of TGFβ and LAP, disulfide linkages are formed between cysteine residues of LAP and specific cysteine residues in another protein, the latent-TGFβ-binding protein (LTBP), to form the large latent complex (Annes et al., 2003). The LTBP is removed extracellularly by either proteolytic cleavage by various proteases, such as plasmin, thrombin, plasma transglutaminase, or endoglycosylases, or by physical interactions of the LAP with other proteins, such as thrombospondin-1, which is able to transform latent TGFβ1 into active TGFβ1 in vitro. Analysis of mice harboring a targeted disruption of the thrombospondin-1 gene suggests that it probably plays an important role in vivo (Crawford et al., 1998). In particular, the phenotype of thrombospondin-1-null mice was relatively similar to that of TGFβ1-null mice, and fibroblasts isolated from the former mice had a decreased ability to activate TGFβ1. Nevertheless, thrombospondin-1 is probably not the only molecule that activates TGFβ in vivo (Abdelouahed et al., 2000).

The four LTBPs (LTBP1−4) belong to the LTBP/fibrillin family of large extracellular glycoproteins (Todorovic et al., 2005). LTBP1, LTBP3, and LTBP4 form a subgroup within the family, because they covalently interact with latent TGFβ and have an important role as regulators of TGFβ function: LTBPs facilitate the secretion of latent TGFβ, direct its localization in the ECM, and regulate the activation of the cytokine. In mice ablation of the Ltbp3 gene and attenuation of Ltbp4 expression both result in developmental defects associated with reduced TGFβ activity (Dabovic et al., 2002; Sterner-Kock et al., 2002), further demonstrating that Ltbp3 and Ltbp4 modulate extracellular TGFβ levels in a specific and nonredundant manner.

The role of TGFβ on type I collagen synthesis has been demonstrated both in vivo and in vitro. In vivo, subcutaneous injections of platelet-derived TGFβ in newborn mice increased type I collagen synthesis by dermal fibroblasts with formation of granulation tissue (Roberts et al., 1986). Injections of platelet-derived TGFβ onto the periostea of parietal bone of newborn rats stimulated bone formation, and thus accumulation of ECM (Noda and Camilliere, 1989). Transgenic mice that overexpressed mature TGFβ1 developed hepatic fibrosis and renal fibrosis (Sanderson et al., 1995). Increased expression of TGFβ2 in osteoblasts in transgenic mice resulted in an osteoporosis-like phenotype with progressive bone loss. The bone loss was associated with an increase in osteoblastic matrix deposition and osteoclastic bone resorption (Erlebacher and Derynck, 1996). Conversely, expression of a dominant-negative TGFβ receptor mutant in osteoblasts led to decreased bone remodeling and increased trabecular bone mass (Filvaroff et al., 1999). In humans, mutations in the LAP of TGFβ1 cause Camurati−Engelmann disease, a rare sclerosing bone dysplasia inherited in an autosomal-dominant matter. It is unclear whether these mutations impair the ability of the LAP to inhibit TGFβ activity or cause accelerated degradation of TGFβ (Janssens et al., 2000; Kinoshita et al., 2000).

More recently, transgenic mice expressing a kinase-deficient type II TGF receptor (T-RIIΔk) in fibroblasts demonstrated unexpected skin and lung fibrosis (Denton et al., 2003). Moreover, the fibrotic phenotype of explanted dermal fibroblasts from the T-RIIΔk-deficient transgenic mice showed that expression of the mutant receptor leads to multilevel activation of the TGFβ ligand−receptor axis and that activation depends on endogenous T-RI receptor kinase activity (Denton et al., 2005). This paradoxical increased activity of the TGFβ ligand−receptor axis in these transgenic mice occurs despite previous findings that the kinase-deficient TGFβ receptors are dominant-negative inhibitors of signaling in several experimental systems, when expressed at high levels compared with wild-type receptors. Further evidence that TGFβ and its signaling pathways significantly influence collagen gene regulation is manifested in Smad3 2/2 mice exposed to a single dose of 30−50 Gy of γ-irradiation. These mice showed significantly less epidermal acanthosis and dermal influx of mast cells, macrophages, and neutrophils and decreased expression of TGFβ compared with skin from wild-type littermates, suggesting that inhibition of Smad3 could decrease tissue damage and reduce fibrosis after exposure to ionizing radiation (Flanders et al., 2002). Conversely, C57BL/6 mice with bleomycin-induced lung fibrosis receiving an intratracheal injection of a recombinant adenovirus expressing Smad7 demonstrated suppression of type I procollagen mRNA, reduced hydroxyproline content, and no morphological fibrotic responses in the lungs, indicating that gene transfer of Smad7 prevents bleomycin-induced lung fibrosis (Nakao et al., 1999). More recently, using mice with a targeted deletion of

Smad3, Roberts et al. (2006) demonstrated that lack of Smad3 prevents the epithelial-to-mesenchymal transition of lens epithelial cells following injury, and attenuates the development of fibrotic sequelae. Together, these various experimental approaches demonstrate the direct implication of Smad3 and TGFβ and molecular changes in collagen type I in fibrotic diseases.

In vitro, TGFβ stimulates the synthesis of most of the structural components of the ECM by fibroblasts and osteoblasts, including type I collagen (for a review, see Leask and Abraham, 2004). It also decreases ECM degradation by repressing the synthesis of collagenases and stromelysins and by increasing the synthesis of TIMPs. It increases lysyl-oxidase activity, which may favor interchain cross-linking in collagen fibrils (Feres-Filho et al., 1995). Finally, it stimulates the proliferation of both fibroblasts and osteoblasts, in contrast to its inhibitory effect on the proliferation of epithelial cells. Moreover, TGFβ may have a role in controlling the lineage-specific expression of type I collagen genes during embryonic development, because there is an excellent temporal and spatial correlation between the activation of type I collagen genes and the presence of immunoreactive TGFβ in the extracellular environment (Niederreither et al., 1992). As noted earlier, Fli1 is a key transcriptional repressor of the *COL1A2* gene. Interestingly, TGFβ induces a sequential displacement of Fli1 from the COL1A2 promoter by posttranslational modifications. The molecular mechanism underlying this process has been delineated and involves the TGFβ-induced phosphorylation of Fli1 leading to the dissociation of the transcriptional repressor protein complex composed of Fli1, histone deacetylase (HDAC), and p300. Thus by alleviating the chromatin remodeling by histone deacetylation, the repressor complex that occupies the COL1A2 promoter (at −404 to −237) is disassembled, leading to transcriptional activation (Asano et al., 2013).

Another way in which TGFβ exerts its action is at a pretranslational level, by increasing mRNA levels of the pro-Col1a1 and pro-Col1a2 transcripts. This increase in type I collagen mRNA levels can be caused by an increase in the transcription rate of type I collagen genes and/or by an increase in procollagen mRNA stability, with the relative contribution of these two mechanisms depending on cell type and on culture conditions. The second effect is through Smad proteins as indicated later: The effect of TGFβ on transcription of the human pro-COL1A2 gene involves Smad complexes (reviewed by Ramirez et al., 2006). This effect is mediated through a sequence of the promoter located between 2378 and 2183 bp upstream of the start site of transcription and is called the TβRE, as demonstrated by transfection experiments (Inagaki et al., 1994). Footprinting experiments performed with this sequence revealed two distinct segments interacting with DNA-binding proteins (Inagaki et al., 1994). It contains two binding sites for Sp1, as well as a binding site for C/EBP (Greenwel et al., 1997; Inagaki et al., 1994). The DNA sequence between 2271 and 2250 contains a CAGA box that binds Smad 3/Smad4 complexes (Chen et al., 1999; Zhang et al., 2000), as well as a binding site for AP1 (Chung et al., 1996). Data have shown that Smad 3/Smad4 complexes can bind to the CAGA box and mediate TGFβ-induced stimulation of the pro-COL1A2 gene, in cooperation with Sp1 proteins (Zhang et al., 2000). Other studies have suggested a role of AP1 in mediating the effects of TGFβ (Chung et al., 1996). This Smad/AP1 interaction was later found to take place "off DNA" (Verrecchia et al., 2001). The transcriptional coactivator CBP/p300, which can bind to Smad complexes, also plays an important role in mediating the effects of TGFβ on the transcriptional activity of the pro-COL1A2 gene (Ghosh et al., 2000). Thus, TGFβ probably activates the transcription of the pro-COL1A2 gene through the binding of a multimeric complex, which includes Smad3/Smad4, Sp1, CBP/p300, and possibly AP1. More recently, using Affymetrix microarrays to detect cellular genes whose expression is regulated by TGFβ through Smad3, Chen et al. identified the gene for early growth response factor-1 (EGR-1) as a Smad3-inducible gene (Chen et al., 2006a,b). It was also found that TGFβ enhanced endogenous Egr-1 interaction with a consensus Egr-1-binding-site element and with GC-rich DNA sequences of the human COL1A2 promoter in vitro and in vivo. Furthermore, forced expression of Egr-1 by itself caused dose-dependent upregulation of COL1A2 promoter activity and further enhanced the stimulation induced by TGFβ (Chen et al., 2006a,b). Other transcriptional coactivators such as the steroid receptor coactivator-1 may also participate in TGFβ effects (Dennler et al., 2005).

In the rat pro-Col1a1 gene, a TβRE has been described at about 1.6 kb upstream of the start site of transcription (Ritzenthaler et al., 1993) and between 2174 and 284 bp in the human pro-α1 collagen gene (Jimenez et al., 1994). The latter sequence contains an Sp1-like binding site, but neither of these two sequences seems to contain a potential Smad-binding site. In addition to direct action of TGFβ, part of the profibrotic properties may be indirect, mediated by increased production of a cysteine-rich protein called connective tissue growth factor (CTGF) (Leask and Abraham, 2003).

Connective tissue growth factor

CTGF (also called CCN2), a member of the CCN family of matricellular proteins, has long been known to promote differentiation and proliferation of chondrocytes and osteoblasts (for a review, see Takigawa et al., 2003). CCN2 promotes fibroblast proliferation, matrix production, and granulation tissue formation, as well as cell adhesion and migration.

Experiments using recombinant CTGF and neutralizing antibodies targeting CTGF have suggested that CTGF mediates at least some of the effects of TGFβ on fibroblast proliferation, adhesion, and ECM production, including collagen and fibronectin (Crean et al., 2002; Weston et al., 2003). Moreover, an expression vector encoding CTGF transfected into fibroblasts can activate a reporter driven by the type I collagen promoter, suggesting that a CTGF response element exists in the promoter of type I collagen (Shi-wen et al., 2000). Perhaps the most significant recent insights into the specific physiological roles of CCN2 have come from the generation of mutant mice lacking CCN2 (Ivkovic et al., 2003). Ccn2 2/2 mice display severely malformed rib cages and die soon after birth owing to a failure to breathe. These mice exhibit impaired chondrocyte proliferation and proteoglycan production within the hypertrophic zone. Excessive chondrocytic hypertrophy and a concomitant reduction in endochondral ossification are also observed. Further support for the idea that CCN2 regulates bone formation in development comes from studies of transgenic mice that overproduce CCN2 under the control of the mouse type XI collagen promoter. These mice develop normally but show dwarfism within a few months of birth owing to a reduced bone density (Nakanishi et al., 2001). The molecular basis for this deformity has not yet been explored; however, a possible explanation is that CCN2 overexpression results in abnormally premature ossification, before proper chondrocyte maturation.

CCN2 is constitutively expressed in fibrotic and embryonic fibroblasts independent of TGFβ (Chen et al., 2006a,b; Holmes et al., 2001, 2003). Experiments using Ccn2 2/2 mouse embryonic fibroblasts (MEFs) have shown that loss of CCN2 results in an inability of TGFβ to induce expression of approximately one-third of those mRNAs induced in Ccn2 1/1 MEFs (Shi-wen et al., 2006). Consistent with the fact that CCN2 is required for only a subset of TGF responses, Ccn2 2/2 MEFs show no impairment of the generic Smad pathway, emphasizing the relative selectivity of CCN2-dependent action. In contrast to the lack of effect of loss of CCN2 expression on basal type I collagen and α-smooth muscle actin expression, the ability of TGFβ to induce these proteins is impaired in Ccn2 2/2 MEFs (Shi-wen et al., 2006).

Expression of the CTGF gene in fibroblasts is strongly induced by TGFβ but not by other growth factors, and intradermal injections of TGFβ in neonatal mice induced an overexpression of CTGF in skin fibroblasts (Frazier et al., 1996; Igarashi et al., 1993). The ability of TGFβ to induce CCN2 also requires protein kinase C and the Ras/MEK/ERK MAP kinase cascade (Chen et al., 2002a,b; Stratton et al., 2002). As in the case of other TGFβ-responsive promoters that do not require the transcription factor AP1, the induction of CCN2 by TGFβ is antagonized by hyperactive AP1 or JNK (Leask et al., 2003) because of the ability of active Jun to bind to Smads off DNA and inhibit Smads from interacting with the target DNA sequences (Verrecchia et al., 2001). The role of CTGF in scarring and fibrosis regulating collagen type I also appears to be via mediating the activity of other factors such as adenosine or by modulating critical signaling pathways such as the AKT/mTOR pathways. For instance, activation of the adenosine (2A) receptor was shown to suppress the expression of the transcriptional repressor Fli1, yet induce the secretion of CTGF. Together, the release of the inhibitory activity of Fli1 (able to suppress both collagen type I and CTGF) and the elevation in CTGF level resulted in a significant enhancement of collagen type I production (Chan et al., 2013). In addition, the targeting of the CTGF 3′-UTR by an endogenous miRNA (miR-143-3p) reduced CTGF levels and inhibited collagen expression. This observation was further refined to reveal that reducing CTGF concomitantly attenuated AKT/mTOR signaling, and suggested that restraining the AKT/mTOR pathway by targeting CTGF represents a mechanism to effectively block collagen ECM production (Mu et al., 2016). Interestingly, platelet-derived growth factor receptor β also uses the AKT/mTOR signaling pathway to promote collagen (1α2) expression in an in vitro model of tubulointerstitial fibrosis (Das et al., 2017).

Although a specific CTGF receptor has yet to be identified, CTGF appears to perform many of its functions through integrins, heparan sulfate-containing proteoglycans, and the low-density lipoprotein receptor-related protein (Gao and Brigstock, 2003; Segarini et al., 2001; Weston et al., 2003).

Fibroblast growth factor

Fibroblasts growth factors (FGFs) comprise a family of 23 genes encoding structurally related proteins divided into six subfamilies of FGFs (Ornitz and Marie, 2002). The most recent member is FGF23, which is mainly produced by osteocytes in bone and acts as a hormone primarily to inhibit phosphate reabsorption in the proximal tubules of the kidney (Liu et al., 2007; Razzaque and Lanske, 2007). Most members of the FGF family bind to four distinct FGF receptor tyrosine kinase molecules (FGFRs) and activate the receptors. Historically, FGF2 (basic FGF) was the first FGF ligand to be isolated from growth-plate chondrocytes (Sullivan and Klagsbrun, 1985). Subsequently, Fgf2 gene expression has also been observed in periosteal cells and in osteoblasts (Hurley et al., 1999; Sabbieti et al., 1999). FGF signaling affects the expression and activity of several transcription factors that are required for calvarial osteogenesis. In rat or mouse calvarial cells, FGF2 activates osteocalcin transcription. This activity is inhibited by the transcription factor MSX2 and is activated by DLX5

(Newberry et al., 1997, 1999). FGF2 can upregulate Twist expression in mouse calvarial mesenchyme (Rice et al. 2000). Twist-heterozygous mice show altered FGFR protein expression (Rice et al., 2000), suggesting that Twist acts upstream of FGF signaling pathways. In addition, Twist could be a potential transcriptional regulator that mediates the negative effect of FGF2 on type I collagen expression in calvarial cells (Fang et al., 2001). Thus, FGF/FGFR, MSX, and Twist functionally interact to control cranial suture development in a coordinated manner. FGF2 production by calvarial osteoblasts is upregulated by FGF2 itself and by parathyroid hormone (PTH), PGE_2, and TGFβ. Thus, the balance between high and low levels of endogenous FGF2 may constitute a mechanism to control proliferation and ensure normal cranial vault development (Moore et al., 2002).

Bone resorption by osteoclasts is required to maintain the shape of craniofacial bones during development. It is therefore significant that FGF2 can increase the formation of osteoclast-like cells and activate mature osteoclasts through FGFR1 (Chikazu et al., 2000). In addition, FGF2 increases the expression of metalloproteinases, collagenases 1 and 3 (Newberry et al., 1997; Tang et al., 1996; Varghese et al., 2000), TIMPs 1 and 3 (Varghese et al., 1995), and stromelysin-3, which regulates collagenase activity in calvarial cells (Delany and Canalis, 1998). These mechanisms may control bone matrix degradation and remodeling by FGFs during calvarial expansion. Targeted deletion of FGF2 causes a relatively subtle defect in osteoblastogenesis, leading to decreased bone growth and bone density. However, no defects in chondrogenesis were observed (Montero et al., 2000).

Insulin-like growth factor

IGF-1 is synthesized by many cells, including cells of the osteoblastic lineage, as well as chondroblasts and osteoclasts (Rajaram et al., 1997). IGF-1 plays an important role in regulating peak BMD and bone size (Mohan and Baylink, 2005; Mohan et al., 2003). The IGFs interact with specific cell surface receptors, designated type 1 and type 2 IGF receptors (IGF-1R and IGF-2R). Most of the actions of IGF-1 and IGF-2 are mediated by the IGF-1R, which is a transmembrane receptor with tyrosine kinase activity.

The IGFs in serum and other extracellular environments are bound to specific IGF-binding proteins (IGFBPs), which represent a family of six secreted proteins with a common domain organization. Each IGFBP has unique properties and exhibits specific functions. IGFBPs 2– to 5 are present in bone; IGFBPs 4 and 5 are expressed at the highest levels. Unlike IGFBP-4, which has an inhibiting effect on IGF actions, IGFBP-5 has a potentiating effect, both in vivo and in vitro, probably by binding directly to sites that are independent of the IGF receptor (Govoni et al., 2005).

IGF-1 and IGF-2 can stimulate osteoblast and fibroblast proliferation and increase type I collagen production by these cells. Their effect on type I collagen production by osteoblasts has been demonstrated by using both fetal rat calvariae and osteoblastic cells, and it is related to an increase in corresponding mRNA transcripts (McCarthy et al., 1989; Thiebaud et al., 1994; Woitge and Kream, 2000). IGF-1 can also promote osteogenic differentiation and Col1a2 synthesis via induction of the mRNA-binding La ribonucleoprotein 6 gene (LARP6). Here, IGF-1 treatment of primary osteoblasts caused an increase in LARP6, which bound to the Col1a2 mRNA, promoting expression. Although in this study the precise mechanism by which LARP6 binding promoted COL1A2 expression was not delineated, a previous series of experiments by another group had noted that IGF-1 rapidly increased LARP6 expression, which was able to bind to the 5′ stem–loop motif structures of both the COL1a1 and the COL1a2 mRNAs. This suggested that increased synthesis of collagen would result from alterations in mRNA stability/accessibility or the ability of LARP6 to enhance or coordinate its translation into the heterotrimeric collagen type I molecule (Blackstock et al., 2014).

In vivo, administration of IGF-1 to hypophysectomized rats increased mRNA transcripts for pro-Col1a1 and pro-Col1a2 genes in parietal bones (Schmid et al., 1989). Furthermore, in calvariae of mice in which the Igf1 gene was disrupted, there was a reduced rate of collagen synthesis. By using the Cre-LoxP model to disrupt IGF-1 in all Col1a2-expressing cells it was demonstrated that locally produced IGF-1 plays a critical role in embryonic and postnatal growth. Local IGF-1 from Col1a2-producing cells is required for bone matrix mineralization during embryonic development and postnatal growth (Govoni et al., 2007). Moreover, to determine the effects of locally expressed IGF-1 on bone remodeling, a transgene was produced in which murine IGF-1 cDNA was cloned downstream of a gene fragment comprising 3.6 kb of 5′ upstream regulatory sequence and most of the first intron of the rat Col1a1 gene. Transgenic calvariae showed an increase in osteoclast numbers per bone surface, as well as increased collagen synthesis and cell proliferation. Femur length, cortical width, and cross-sectional area were also increased in transgenic femurs, whereas femoral trabecular bone volume displayed little change. Thus, broad overexpression of IGF-1 in cells of the osteoblast lineage increased indices of bone formation and resorption (Jiang et al., 2006). Furthermore, in calvariae of mice in which the IGF-1 gene was disrupted, there was a reduced rate of collagen synthesis (Woitge and Kream, 2000).

Cytokines

Tumor necrosis factor α

TNFα is a cytokine secreted mainly by monocytes/macrophages, but osteoblasts seem to be able to produce TNFα under certain conditions (Gowen et al., 1990). After being cleaved from its propeptide, TNFα undergoes trimerization and binds to type I receptors, which transduce most of the effects of TNFα, or to type II receptors, which activates them and transmits signals to the nucleus via different transcription factors.

TNFα levels are elevated in various bone disorders, such as rheumatoid arthritis and osteoporosis (Beutler and Cerami, 1988; Pacifici, 1996). In bone tissue, TNFα inhibits osteoblast function and increases osteoclastogenesis, thus favoring net matrix destruction (Centrella et al., 1988; Chou et al., 1996; Panagakos et al., 1996). Similarly, TNFα stimulates fibroblast and osteoblast proliferation and inhibits the production of ECM components, including type I collagen and its modifying enzyme, lysyl oxidase (Kouba et al., 1999; Pischon et al., 2004; Verrecchia et al., 2000), as well as the transcription factor Sox9 (Murakami et al., 2000), and increases collagenase production and thus ECM degradation (Iraburu et al., 2000).

In vivo, inoculation of nude mice with TNFα-producing cells decreased type I collagen production in skin and liver, impaired wound healing, and decreased TGFβ1 synthesis in skin (Houglum et al., 1998). TNFα also increases PGE_2 and interleukin 1 production by osteoblasts and fibroblasts, which themselves modulate type I collagen synthesis.

In dermal fibroblasts, inhibition of collagen synthesis in fibroblasts by TNFα is associated with a decrease in mRNA levels for the pro-Col1a1 and pro-Col1a2 transcripts and in the transcription of type I collagen genes (Solis-Herruzo et al., 1988). Inagaki et al. (1995) showed that TNFα increases the binding of a protein complex that recognizes the negative *cis*-acting element located immediately next to the TβRE, and postulated that TNFα counteracts the TGFβ-elicited stimulation of collagen gene expression through overlapping nuclear signaling pathways. Kouba et al. (1999) characterized this specific TNFα-response element between nucleotides 2271 and 2235 relative to the transcription initiation site and termed it TαRE. Electrophoretic mobility supershift assays identified the NF-κB family members NF-κB1 and RelA as transcription factors binding the TαRE and mediating TNFα repression of COL1a2 promoter activity. By using a gene-knockout approach, the same group showed that, in primary fibroblasts from NEMO-knockout mice, lack of NF-κB activation prevented repression of basal COL1a2 gene expression by TNFα. Similar regulatory mechanisms take place in dermal fibroblasts transfected with dominant-negative forms of IKK-α, a critical kinase upstream of NF-κB (Verrecchia et al., 2002). More interestingly, the antagonist activities of TGFβ and TNFα in COL1a2 may be the result of steric interactions between transcription factors binding to TβRE and TαRE, respectively (Greenwel et al., 2000; Verrecchia et al., 2001).

The inhibitory effects of TNFα on the rat pro-Col1a1 proximal promoter were shown to be mediated by a sequence located between 2378 and 2345 bp through the binding of proteins of the C/EBP family, such as C/EBPδ and p20C/EBPβ (Iraburu et al., 2000). Other TNFα response elements have been identified within the pro-Col1a1 gene between 2101 and 238 bp and between 68 and 86 bp, in dermal fibroblasts and HSCs, respectively (Hernandez et al., 2000; Mori et al., 1996). The latter *cis*-acting element binds proteins of the Sp1 family, whereas the proteins binding to the former have not been identified (Hernandez et al., 2000; Mori et al., 1996). Using two lines of transgenic mice harboring a growth hormone reporter gene under the control of either at 2.3 kb, the human pro-COL1A1 proximal promoter plus the first intron, or at 440 bp, this promoter plus the first intron, Chojkier's group reported that different *cis*-acting elements mediate the inhibitory effects of TNFα, depending on the tissue (Houglum et al., 1998). In skin, the inhibitory effect of TNFα on the activity of the reporter gene was mediated through a *cis*-acting element located between 22.3 kb and 2440 bp (Buck et al., 1996). In contrast, this inhibitory effect of TNFα was mediated in liver through an element located between 2440 and 11,607 bp (Houglum et al., 1998).

Interferon γ

IFN-γ is a cytokine produced both by monocytes/macrophages and by type 1 helper T cells. IFN-γ binds to the IFN-γ receptor complex (IFNGR1 and IFNGR2), followed by the activation of receptor-associated JAK, which in turn phosphorylates and activates STAT1. Once phosphorylated, STAT1 molecules dimerize and translocate to the nucleus where they modulate target gene transcription either by direct interaction with specific sequences or through protein−protein interactions (Darnell et al., 1994; Horvai et al., 1997).

In vitro, IFN-γ decreases osteoblast and fibroblast proliferation and type I collagen synthesis by these cells. This effect seems to be caused both by a decrease in type I collagen mRNA stability (Czaja et al., 1987; Kahari et al., 1990) and by a decrease in the transcription rate of the pro-Col1a1 gene (Diaz and Jimenez, 1997; Yuan et al., 1999). Transfection studies using different segments of the human pro-COL1A1 proximal promoter have shown the existence of an IFN-γ response element between 2129 and 2107 bp that can bind transcription factors NF1 and members of the Sp1 family (Yuan et al., 1999). However, mutations in NF1 and Sp1 sites, which abrogate the binding of these factors, repress the basal COL1A1 promoter activity but are unable to abrogate the IFN-γ-mediated inhibition, suggesting that NF1 and Sp1 are not involved in this inhibitory action of IFN-γ (Yuan et al., 1999). The activation of major histocompatibility complex II by IFN-γ requires activation of a trimeric DNA-binding transcriptional complex, the RFX5 complex, containing RFXB (also called RFXANK or Tvl-1), RFXAP, and RFX5 protein. RFX5 was later shown to bind to the collagen transcription start site and repress collagen gene expression (Sengupta et al., 2002). Moreover, two studies have provided further evidence for the role of the RFX5 repressor complex in suppressing COL1A2 transcription and have noted a prominent role of epigenetics in this mode of regulation of collagen synthesis. Both studies identified components of the HDAC complex (SIRT1 and SIN3B) acting upon RFX5, altering its acetylation status. In smooth muscle cells IFN-γ repressed COL1A2 transcription by downregulating the sirtuin SIRT1, leading to acetylation, preventing nuclear expulsion and degradation and thereby modulating RFX5 repressor activity. IFN-γ also promoted the recruitment of SIN3B to the COL1A2 transcription start site, where it cooperated with G9a, a histone H3K9 methyltransferase, to effect a repressive chromatin structure. Taken together these studies strongly suggest that IFN-γ coordinates the assembly of a potent repressor complex containing RFX5, HDACs, Sirt1, Sin3B, and G9a on the COL1A2 promoter. In the human COL1A2 gene, an IFN-γ response element has been identified between 2161 and 125 bp by using transfection experiments in dermal fibroblasts (Higashi et al., 1998). Similar to the action of TNFα, Ulloa et al. (1999) reported that IFN-γ abrogates the TGFβ stimulation of TβRE-containing reporter constructs in fibrosarcoma cells by inducing the level of Smad7 via the JAK–STAT1 signaling pathway. To account for the antagonistic action of IFN-γ and TGFβ on the expression of the COL1a2 gene, however, evidence suggests another mechanism involving p300/CBP. This evidence indicates that IFN-γ-activated STAT1 sequesters the endogenous p300 and reduces its interaction with Smad3, thus inhibiting the TGFβ/Smad-induced collagen gene transcription (Ghosh et al., 2001). Moreover, Higashi et al. (2003) showed that YB-1 binds to the IFN-γ responsive element at 2165 to 2150 and mediates transcriptional repression of COL1A2. YB-1 was used successfully as a therapeutic target to downregulate collagen type I as a strategy to treat liver fibrosis in vivo. This effect was potentiated by the addition of exogenous IFN-γ (Inagaki et al., 2005).

Other cytokines

Interleukin 1

IL-1 is a cytokine secreted mainly by monocytes/macrophages, but also by other cells, including fibroblasts, osteoblasts, synoviocytes, and chondrocytes. Two forms of IL-1 have been described, IL-1α and IL-1β, which have little primary structure homology but bind to the same receptor and have similar biological activities (Stylianou and Saklatvala, 1998).

IL-1 modulates the type I collagen synthesis at different levels, and the variations may be attributable to cell type, species, and age differences, but, at this writing, the molecular mechanisms by which IL-1 modulates type I collagen gene transcription are not clear. In vitro, IL-1 has an inhibitory effect on type I collagen production by osteoblasts, which is due to an inhibition of type I collagen gene transcription (Harrison et al., 1990). Nevertheless, it can be masked when low doses of IL-1 are used, as this cytokine stimulates the production of PGE_2, which in turn can modulate type I collagen synthesis. Slack et al. (1995) have suggested that the inhibitory effects of IL-1 on the transcription of the human pro-COL1A1 collagen gene by osteoblasts could be mediated through the binding of AP1 to the first intron of the gene (Slack et al., 1995), but this has not been confirmed. In vitro, IL-1β inhibits the expression of the COL1A2 in human lung fibroblasts at the transcriptional level by a PGE_2-independent effect, as well as through the effect of endogenous fibroblast PGE_2 released under the stimulus of the cytokine (Diaz et al., 1993). More recently, in SSc fibroblasts, pre-IL-1α was shown to form a complex with IL-1β-binding proteins that are translocated into the nuclei of fibroblasts via HAX-1 (HS1-associated protein X-1) and IL-1 receptor type II to increase production of collagen and IL-6 (Kawaguchi et al., 2006). In contrast, the biological impact of IL-1β on tendon fibroblasts showed that the presence of IL-1β significantly decreased the level of collagen type I mRNA. These effects were found to be mediated by selective upregulation of the EP(4) receptor, which is a member of the G-protein-coupled receptors that transduces the PGE_2 signal via the p38 MAP kinase pathway (Thampatty et al., 2007).

Interleukin 13

IL-13 is a major inducer of fibrosis. Indeed, IL-13 induces expression of TGF 2 β1 in macrophages. The increase in TGFβ expression requires both TNFα and signaling through the IL-13 receptor α2 (IL-13Ra2) to activate an AP1 variant, which stimulates the TGFβ promoter. Prevention of IL-13Ra2 expression, IL-13Ra2 gene silencing, or blockade of IL-13Ra2 signaling leads to marked downregulation of TGFβ1 production and collagen deposition in bleomycin-induced lung fibrosis (Fichtner-Feigl et al., 2006). Thus, IL-13Ra2 signaling during prolonged inflammation could be an important therapeutic target for the prevention of TGFβ1-mediated fibrosis. Evidence has also suggested that IL-13 can mediate the increased deposition of collagen type I via TGFβ (Firszt et al., 2014) and independent of the action of TGFβ via STAT6. By showing a direct effect of IL-13 on collagen expression using specific STAT6 small interfering RNAs and miR-135b, an endogenous miRNA that targets STAT6, and methylation inhibitors, the authors were able to show a clear role for STAT6 and epigenetic mechanisms regulating IL-13-mediated collagen production (O'Reilly et al., 2016).

Interleukin 4

IL-4 is secreted by type 2 helper T cells and by mastocytes. In vitro, IL-4 increases type I collagen production by human fibroblasts by increasing both transcriptional levels of type I collagen genes and the stability of the corresponding mRNAs (Serpier et al., 1997). In bronchial fibroblasts, IL-4 positively regulates pro-COL1A1 transcription by direct promoter activation and increases the TIMP2/MMP-2 ratio, thereby supporting the profibrotic effect of this cytokine. Furthermore, a combined action of SP1, NF-κB, and STAT6 contributes to the IL-4-mediated COL1A2 gene activation. An AP2 site adjacent to a STAT6 consensus motif, TTC N(3/4) GCT, is located within 205 bases of the transcription start site and seems to support the moderate IL-4 induction of COL1A1 gene expression (Bergeron et al., 2003; Buttner et al., 2004).

Interleukin 6 IL-6 has been shown to induce collagen type I and is elevated in patients with SSc (Kawaguchi et al., 1999) and in cultured HSCs (Nieto, 2006). Two studies have indicated that IL-6 can regulate collagen type I via STAT3 by two distinct mechanisms. First, IL-6 induction of STAT3 (phospho-STAT3/S727) led to noncanonical STAT3 activation and to increased TGFβ and collagen expression (Li et al., 2015). Second, at both the transcriptional level, through the upstream elements of the COL1A2 enhancer and their interaction with the proximal promoter, and the posttranscriptional level, where IL-6 *trans* signaling and high levels of STAT3 activation led to increased COL1A2 protein expression by (Papaioannou et al., 2018).

Other interleukins

IL-10 is secreted mainly by monocytes/macrophages and inhibits type I collagen gene transcription and type I collagen production by skin fibroblasts (Reitamo et al., 1994). In addition, several other cytokines have also been reported to regulated type I collagen production, for instance, IL-17, IL-18, and the IL-12 family of cytokines, such as IL-23. Although the molecular mechanisms responsible for changes in collagen have not been fully explored, they often show alterations in the levels of transcription factors, upstream signaling pathways, and miRNA expression. For example, IL-18 appears to involved in the activation of the ERK signaling cascade leading to the induction of the collagen type I transcriptional repressor Ets1 (Kim et al., 2010). With respect to IL-17, this cytokine exerts a direct impact on the expression of TGFβ type II receptors (TGFβRIIs) via JNK-dependent signaling pathways, with the resultant upregulation and stabilization of TGFβRII leading to enhance Smad2/3 signaling and the induction of collagen gene expression (Fabre et al., 2014). The dysregulation in IL-23 signaling reported in fibrosis and contributing to increased collagen production has been attributed to an IL-23-dependent imbalance between two miRNAs (miR-4458 and miR-18a) that target type I collagen, and in particular the IL-23 downregulation of miR-18a resulting in collagen upregulation (Nakayama et al., 2017).

Oncostatin M is produced mainly by activated T cells and monocytes/macrophages and belongs to the hematopoietic cytokine family. It is mitogenic for fibroblasts and stimulates type I collagen production by fibroblasts by increasing transcriptional levels of type I collagen genes (Ihn et al., 1997). Transfection studies performed using different segments of the human pro-α2(I) collagen gene have shown that a 12-bp segment located between 2131 and 2120 bp, and that contains a TCCTCC motif, mediated the stimulatory effects of oncostatin M (Ihn et al., 1997).

Arachidonic acid derivatives

PGE_2, a product of the cyclooxygenase pathway, is synthesized by various cell types, including endothelial cells, monocytes/macrophages, osteoblasts, and fibroblasts. Its production by fibroblasts is increased by IL-1 and TNFα. PGE_2 has a biphasic effect on type I collagen synthesis by bone organ cultures and by osteoblastic cells (Ono et al., 2005). At low concentrations, it increases type I collagen synthesis, whereas at higher concentrations it decreases type I collagen synthesis. PGE_2 induces the production of IGF-1 by osteoblastic cells, and part of the stimulatory effect of low doses of PGE_2 on type I collagen production seems to be indirect, mediated by a stimulation of IGF-1 production (Raisz et al., 1993a, b). Nevertheless, part of this stimulatory effect is independent of IGF-1 production and persists after blocking the effects of IGF-1 (Raisz et al., 1993a, b). It is notable that when PGE_2 is added to fibroblasts in culture, it inhibits type I collagen synthesis and decreases the levels of the corresponding mRNAs. Most of the effects of PGE_2 are mediated through an increase in cAMP levels (Sakuma et al., 2004) and the activation of collagen synthesis by low doses of PGE_2 could be caused by such a mechanism, because cAMP analogs can also increase collagen synthesis in bone (Fall et al., 1994). In contrast, the inhibitory effect of PGE_2, which has been shown to be caused by an inhibition of transcription of type I collagen genes, is not mediated through a cAMP-dependent pathway but through a pathway involving the activation of protein kinase C (Sakuma et al., 2004). A study using an osteoblastic cell line transfected stably with various segments of the rat pro-Col1a1 promoter cloned upstream of a CAT reporter gene has shown that PGE_2 acts through an element located more than 2.3 kb upstream of the start site of transcription (Raisz et al., 1993a, b). A more recent study using fibroblasts transfected transiently with a construct containing 220 bp of the mouse pro-Col1a1 proximal promoter has shown that PGE_2 can also act through a *cis*-acting element located within this promoter segment (Riquet et al., 2000).

Hormones and vitamins

Corticosteroids

It has been known for many years that the administration of corticosteroids to patients results in osteoporosis and growth retardation. In mice, corticosteroids have also been shown to decrease collagen production in calvariae (Advani et al., 1997). In vitro, incubation of fetal rat calvariae with high doses of corticosteroids, or with lower doses but for a prolonged period of time, decreased the synthesis of type I collagen (Canalis, 1983; Dietrich et al., 1979); this inhibitory effect could also be observed with osteoblastic cell lines (Hodge and Kream, 1988). Moreover, cortisol increases interstitial collagenase transcript levels by posttranscriptional mechanisms in osteoblastic cells (Delany et al., 1995a, b). Nuclear run-off experiments performed using osteoblasts derived from fetal rat calvariae showed that glucocorticoids downregulate transcriptional levels of the pro-Col1a1 gene, as well as the stability of the corresponding mRNA (Delany et al., 1995a, b). Because corticosteroids inhibit the secretion of IGF-1, part of their inhibitory effect on type I collagen synthesis could be indirect, but calvariae from IGF-1-null mice maintain their responsiveness to glucocorticoids (Woitge and Kream, 2000).

When added to fibroblasts in culture, corticosteroids usually decrease type I collagen synthesis by acting at a pretranslational level, which is in agreement with their in vivo effect on wound healing (Cockayne et al., 1986). Stable transfection experiments using the mouse pro-Col1a2 proximal promoter fused to a CAT reporter gene and transfected into fibroblasts have shown that sequences located between 22,048 and 2981 bp and between 2506 and 2351 bp were important for the corticosteroid-mediated inhibition of transcription (Perez et al., 1992). However, the *cis*-acting element(s) responsible was not sufficient to block the effect alone (Meisler et al., 1995), and further experiments from the same group indicated that glucocorticoids coordinately regulate procollagen gene expression through TGFβ elements. Depression of procollagen gene expression by glucocorticoids through the TGFβ element is mediated by decreased TGFβ secretion, possibly involving a secondary effect on regulatory proteins encoded by noncollagenous protein genes (Cutroneo and Sterling, 2004).

BMD decreases by 2%−4.5% after just 6 months of glucocorticoid administration to healthy men, but subsequently the rate of bone loss declines (LoCascio et al., 1990). Considerable evidence indicates that glucocorticoid-induced bone loss occurs in two phases in both humans and mice: an early, rapid phase in which bone mass is lost because of excessive bone resorption and a slower, later phase in which bone is lost because of inadequate bone formation (LoCascio et al., 1990; Weinstein et al., 1998). Transgenic mice overexpressing 11β-hydroxysteroid dehydrogenase type 2 (11β-HSD2) using the osteoclast-specific murine tartrate-resistant acid phosphatase promoter (Reddy et al., 1995) exhibited decreased cancellous osteoclasts after glucocorticoid administration and were protected from the glucocorticoid-induced early, rapid loss of BMD (Jia et al., 2006), suggesting that direct effects of glucocorticoids on osteoclasts are more important than these mediators in the early, rapid loss of bone mass that follows glucocorticoid administration. Similarly, when an osteoblast-specific 2.3-kb Col1a1 promoter drives 11β-HSD2 expression in mature osteoblasts, 11β-HSD2 should metabolically inactivate endogenous glucocorticoids in the targeted cells, thereby reducing glucocorticoid signaling; collagen synthesis rates were lower in calvarial organ cultures of transgenic mice than in wild-type. Furthermore, the inhibitory effect of 300 nM hydrocortisone on collagen synthesis was blunted in transgenic calvariae. Trabecular bone parameters measured by microcomputed tomography were also reduced in L3 vertebrae, but not femurs, of 7- and 24-week-old transgenic females. This effect was not seen in male mice, suggesting that endogenous glucocorticoid signaling is required for normal vertebral trabecular bone volume and architecture in female mice (Sher et al., 2004).

Thyroid hormones

Thyroid hormones have been shown to inhibit type I collagen production by cardiac fibroblasts, and this effect was associated with a decrease in the levels of pro-Col1a1 mRNA (Chen et al., 2000a,b). Transfection studies have shown that thyroid hormones modulate transcriptional levels of the pro-Col1a1 gene through a *cis*-acting element located between 2224 and 115 bp (Chen et al., 2000a,b). In addition, thyroid hormone (T3) regulates the FGFR1 promoter in osteoblasts through a thyroid receptor-binding site at position 2279/2264 (O'Shea P et al., 2007). With respect to chondrocyte differentiation, a study has indicated that thyroid hormone receptor-associated protein 3 negatively regulates SOX9 transcriptional activity as a cofactor of a SOX9 transcriptional complex during chondrogenesis (Sono et al., 2018).

Parathyroid hormone

PTH binds to specific receptors in osteoblasts and upregulates the expression of receptor activator of NF-κB ligand, a protein essential for osteoclast development and survival. PTH signaling occurs via a PTH receptor 1/cAMP/protein kinase A/CREB cascade. Runx2 may contribute to the osteoblast specificity of PTH signaling (Fu et al., 2006) by downregulating osteoprotegerin expression (Boyle et al., 2003). In vitro, PTH inhibits type I collagen synthesis by osteoblastic cell lines as well as by bone organotypic cultures (Kream et al., 1986). This inhibitory effect is associated with a decrease in the levels of procollagen mRNAs (Kream et al., 1986). When calvariae of transgenic mice harboring a 1.7-, 2.3-, or 3.6-kb segment of the rat pro-Col1a1 proximal promoter were cultured in the presence of PTH, there was a parallel decrease in the incorporation of [^{3}H]proline and in the activity of the reporter gene, suggesting that the pro-Col1a1 promoter contains a *cis*-acting element located downstream of 21.7 kb, which mediates the inhibition of the pro-Col1a1 gene expression induced by PTH (Bogdanovic et al., 2000; Kream et al., 1993). Furthermore, the effect of PTH on the levels of expression of the reporter gene were mimicked by cAMP and potentiated by a phosphodiesterase inhibitor, suggesting that the inhibitory effects of PTH are mediated mainly by a cAMP signaling pathway (Bogdanovic et al., 2000).

Vitamin D

The classic role of the vitamin D endocrine system is to stimulate calcium absorption in the intestine, thus maintaining normocalcemia and indirectly regulating bone mineralization (van Driel et al., 2004). The actions of vitamin D are mediated through the vitamin D receptor (VDR), which acts as a ligand-activated transcription factor to regulate the expression of target genes. The VDR heterodimerizes with retinoid X receptor (RXR) and associates with the transcriptional complex on promoters of target genes. In vitro, the active metabolite of vitamin D3, 1,25$(OH)_2D_3$, has been shown to inhibit type I collagen synthesis by bone organ cultures and by osteoblastic cells, and this inhibitory effect is caused by an inhibition of the transcription of type I collagen genes (Bedalov et al., 1998; Harrison et al.,1989). Transfection studies performed with the rat pro-Col1a1 proximal promoter led to the identification of a vitamin D-responsive element between 22.3 and 21.6 kb (Pavlin et al., 1994). Nevertheless, when transgenic mice harboring a 1.7-kb segment of the rat pro-Col1a1 promoter cloned upstream of a CAT reporter gene were treated with 1,25$(OH)_2D_3$, the levels of expression of the CAT reporter gene decreased, which suggests that a vitamin D-response element is located downstream of 21.7 kb (Bedalov et al., 1998). Similarly, when calvariae from these transgenic mice were cultured in the presence of 1,25$(OH)_2D_3$, it inhibited reporter gene expression (Bedalov et al., 1998). It is of note that part of the effects of vitamin D on type I collagen could be mediated through an inhibition of the production of IGF-1, because vitamin D has been shown to inhibit IGF-1 production and increase the concentrations of inhibitory IGFBP-4 (Scharla et al., 1991). Indeed, the effects of 1,25$(OH)_2D_3$ on release of the IGFs were independent of bone resorption and support the conclusion that 1,25$(OH)_2D_3$ modulated the production and secretion of IGF-1 and IGF-2 in calvarial cells (Linkhart and Keffer, 1991). The results of these and similar studies on PTH in calvarial cells suggest that PTH, TGFβ, and 1,25$(OH)_2D_3$ differentially regulate mouse calvarial cell IGF-1 and IGF-2 production. Furthermore, IGFBP-5 has been shown to reduce the effects of 1,25$(OH)_2D_3$ by blocking cell cycle progression at G0/G1 in osteoblasts and by decreasing the expression of cyclin D1. Moreover, IGFBP-5 can interact with VDR to prevent RXR:VDR heterodimerization and IGFBP-5 may attenuate the 1,25$(OH)_2D_3$-induced expression of bone differentiation markers (Schedlich et al., 2007).

Many studies have reported an inverse relationship between vitamin D and PTH concentrations, where PTH plateaued at concentrations of 25OHD between 75 and 100 nmol/L (30–40 ng/mL). Biochemical markers of bone metabolism are increasingly used in the management of patients with osteoporosis and other metabolic bone diseases.

Among bone formation markers is procollagen type 1 amino-terminal propeptide (P1NP), which is released into the circulation during bone formation. Bone resorption markers include carboxy-terminal cross-linked telopeptide of type I collagen (CTX). CTX is a degradation product of type I collagen and is released into the circulation during bone resorption

(Vasikaran et al., 2011). The Bone Marker Standards Working Group recommends P1NP and CTX as reference standard markers of bone formation and resorption (Vasikaran et al., 2011; Stokes et al., 2011). In humans, many studies have reported deficiencies or insufficiencies in vitamin D, even in equatorial countries (Tan et al., 2013), and a confusing picture of the markers of bone turnover, P1NP and CTX, and vitamin D levels, which may be influenced by reduced PTH, has emerged.

References

Abdelouahed, M., Ludlow, A., Brunner, G., Lawler, J., 2000. Activation of platelet-transforming growth factor beta-1 in the absence of thrombospondin-1. J. Biol. Chem. 275, 17933–17936.

Adhikari, A.S., Chai, J., Dunn, A.R., 2011. Mechanical load induces a 100-fold increase in the rate of collagen proteolysis by MMP-1. J. Am. Chem. Soc. 133 (6), 1686–1689.

Adhikari, A.S., Glassey, E., Dunn, A.R., 2012. Conformational dynamics accompanying the proteolytic degradation of trimeric collagen I by collagenases. J. Am. Chem. Soc. 134 (32), 13259–13265. https://doi.org/10.1021/ja212170b.

Advani, S., LaFrancis, D., Bogdanovic, E., Taxel, P., Raisz, L.G., Kream, B.E., 1997. Dexamethasone suppresses in vivo levels of bone collagen synthesis in neonatal mice. Bone 20, 41–46.

Amar, S., Smith, L., Fields, G.B., 2017. Matrix metalloproteinase collagenolysis in health and disease. Biochim. Biophys. Acta. Mol. Cell. Res. 1864 (11 Pt A), 1940–1951. https://doi.org/10.1016/j.bbamcr.2017.04.015.

Ambros, 2003. Cell 113, 673–676.

Ameye, L., Aria, D., Jepsen, K., Oldberg, A., Xu, T., Young, M.F., 2002. Abnormal collagen fibrils in tendons of biglycan/fibromodulin-deficient mice lead to gait impairment, ectopic ossification, and osteoarthritis. FASEB J. 16, 673–680.

Annes, J.P., Munger, J.S., Rifkin, D.B., 2003. Making sense of latent TGFbeta activation. J. Cell Sci. 116, 217–224.

Antoniv, T.T., De Val, S., Wells, D., Denton, C.P., Rabe, C., de Crombrugghe, B., Ramirez, F., Bou-Gharios, G., 2001. Characterization of an evolutionarily conserved far-upstream enhancer in the human alpha 2(I) collagen (COL1A2) gene. J. Biol. Chem. 276, 21754–21764.

Antoniv, T.T., Tanaka, S., Sudan, B., De Val, S., Liu, K., Wang, L., Wells, D.J., Bou-Gharios, G., Ramirez, F., 2005. Identification of a repressor in the first intron of the human alpha 2(I) collagen gene (COL1A2). J. Biol. Chem.

Arakaki, P.A., Marques, M.R., Santos, M.C., 2009. MMP-1 polymorphism and its relationship to pathological processes. J. Biosci. 34 (2), 313–320.

Asano, Y., Trojanowska, M., 2013. Fli1 represses transcription of the human α2(I) collagen gene by recruitment of the HDAC1/p300 complex. PLoS One 8 (9), e74930.

Attisano, L., Wrana, J.L., 2000. Smads as transcriptional co-modulators. Curr. Opin. Cell Biol. 12, 235–243.

Barsh, G.S., Roush, C.L., Gelinas, R.E., 1984. DNA and chromatin structure of the human alpha 1 (I) collagen gene. J. Biol. Chem. 259, 14906–14913.

Baumann, B., Hayashida, T., Liang, X., Schnaper, H.W., 2016. Hypoxia-inducible factor-1α promotes glomerulosclerosis and regulates COL1A2 expression through interactions with Smad3. Kidney Int. 90 (4), 797–808.

Bedalov, A., Salvatori, R., Dodig, M., Kapural, B., Pavlin, D., Kream, B.E., Clark, S.H., Woody, C.O., Rowe, D.W., Lichtler, A.C., 1998. 1,25-Dihydroxyvitamin D3 inhibition of col1a1 promoter expression in calvariae from neonatal transgenic mice. Biochim. Biophys. Acta 1398, 285–293.

Bergeon, M.T., 1967. Collagen: a review. J. Okla. State Med. Assoc. 60, 330–332.

Bergeron, C., Page, N., Joubert, P., Barbeau, B., Hamid, Q., Chakir, J., 2003. Regulation of procollagen I (alpha1) by interleukin-4 in human bronchial fibroblasts: a possible role in airway remodelling in asthma. Clin. Exp. Allergy 33, 1389–1397.

Beutler, B., Cerami, A., 1988. Cachectin (tumor necrosis factor): a macrophage hormone governing cellular metabolism and inflammatory response. Endocr. Rev. 9, 57–66.

Bigg, H.F., Rowan, A.D., Barker, M.D., Cawston, T.E., 2007. Activity of matrix metalloproteinase-9 against native collagen types I and III. FEBS J. 274 (5), 1246–1255.

Bittner, K., Liszio, C., Blumberg, P., Schonherr, E., Kresse, H., 1996. Modulation of collagen gel contraction by decorin. Biochem. J. 314 (Pt 1), 159–166.

Blackstock, C.D., Higashi, Y., Sukhanov, S., Shai, S.Y., Stefanovic, B., Tabony, A.M., Yoshida, T., Delafontaine, P., 2014. Insulin-like growth factor-1 increases synthesis of collagen type I via induction of the mRNA-binding protein LARP6 expression and binding to the 5' stem-loop of COL1a1 and COL1a2 mRNA. J. Biol. Chem. 289 (11), 7264–7274.

Bogdanovic, Z., Bedalov, A., Krebsbach, P.H., Pavlin, D., Woody, C.O., Clark, S.H., Thomas, H.F., Rowe, D.W., Kream, B.E., Lichtler, A.C., 1994. Upstream regulatory elements necessary for expression of the rat COL1A1 promoter in transgenic mice. J. Bone Miner. Res. 9, 285–292.

Bogdanovic, Z., Huang, Y.F., Dodig, M., Clark, S.H., Lichtler, A.C., Kream, B.E., 2000. Parathyroid hormone inhibits collagen synthesis and the activity of rat col1a1 transgenes mainly by a cAMP-mediated pathway in mouse calvariae. J. Cell. Biochem. 77, 149–158.

Boot-Handford, R.P., Tuckwell, D.S., Plumb, D.A., Rock, C.F., Poulsom, R., 2003. A novel and highly conserved collagen (pro(alpha)1(XXVII)) with a unique expression pattern and unusual molecular characteristics establishes a new clade within the vertebrate fibrillar collagen family. J. Biol. Chem. 278, 31067–31077.

Bornstein, P., 2002. The NH(2)-terminal propeptides of fibrillar collagens: highly conserved domains with poorly understood functions. Matrix Biol. 21, 217–226.

Bornstein, P., Walsh, V., Tullis, J., Stainbrook, E., Bateman, J.F., Hormuzdi, S.G., 2002. The globular domain of the proalpha 1(I) N-propeptide is not required for secretion, processing by procollagen N-proteinase, or fibrillogenesis of type I collagen in mice. J. Biol. Chem. 277, 2605–2613.

Bou-Gharios, G., Garrett, L.A., Rossert, J., Niederreither, K., Eberspaecher, H., Smith, C., Black, C., Crombrugghe, B., 1996. A potent far-upstream enhancer in the mouse pro alpha 2(I) collagen gene regulates expression of reporter genes in transgenic mice. J. Cell Biol. 134, 1333–1344.

Boyadjiev, S.A., et al., 2006. Cranio-lenticulo-sutural dysplasia is caused by a SEC23A mutation leading to abnormal endoplasmic-reticulum-to-Golgi trafficking. Nat Genet 38, 1192–1197.

Boyle, W.J., Simonet, W.S., Lacey, D.L., 2003. Osteoclast differentiation and activation. Nature 423, 337–342.

Brenner, D.A., Rippe, R.A., Veloz, L., 1989. Analysis of the collagen alpha 1(I) promoter. Nucleic Acids Res. 17, 6055–6064.

Bridgewater, L.C., Lefebvre, V., de Crombrugghe, B., 1998. Chondrocyte-specific enhancer elements in the Col11a2 gene resemble the Col2a1 tissue-specific enhancer. J. Biol. Chem. 273, 14998–15006.

Brodeur, A.C., Wirth, D.A., Franklin, C.L., Reneker, L.W., Miner, J.H., Phillips, C.L., 2007. Type I collagen glomerulopathy: postnatal collagen deposition follows glomerular maturation. Kidney Int. 71, 985–993.

Buck, M., Houglum, K., Chojkier, M., 1996. Tumor necrosis factor-alpha inhibits collagen alpha1(I) gene expression and wound healing in a murine model of cachexia. Am. J. Pathol. 149, 195–204.

Buttner, C., Skupin, A., Rieber, E.P., 2004. Transcriptional activation of the type I collagen genes COL1A1 and COL1A2 in fibroblasts by interleukin-4: analysis of the functional collagen promoter sequences. J. Cell. Physiol. 198, 248–258.

Byers, P.H., 2001. Folding defects in fibrillar collagens. Philos. Trans. R. Soc. Lond. B Biol. Sci. 356, 151–157 discussion 157–158.

Byers, P.H., Duvic, M., Atkinson, M., Robinow, M., Smith, L.T., Krane, S.M., Greally, M.T., Ludman, M., Matalon, R., Pauker, S., Quanbeck, D., Schwarze, U., 1997. Ehlers–Danlos syndrome type VIIA and VIIB result from splice-junction mutations or genomic deletions that involve exon 6 in the COL1A1 and COL1A2 genes of type I collagen. Am. J. Med. Genet. 72, 94–105.

Cai, et al., 2010. J Mol Biol. 401, 564–578.

Canty, E.G., Kadler, K.E., 2005. Procollagen trafficking, processing and fibrillogenesis. J. Cell Sci. 118, 1341–1353.

Canty, E.G., Lu, Y., Meadows, R.S., Shaw, M.K., Holmes, D.F., Kadler, K.E., 2004. Coalignment of plasma membrane channels and protrusions (fibripositors) specifies the parallelism of tendon. J. Cell Biol. 165, 553–563.

Canty-Laird, E.G., Lu, Y., Kadler, K.E., 2012. Stepwise proteolytic activation of type I procollagen to collagen within the secretory pathway of tendon fibroblasts in situ. Biochem. J. 441 (2), 707–717. https://doi.org/10.1042/BJ20111379.

Centrella, M., McCarthy, T.L., Canalis, E., 1988. Tumor necrosis factor-alpha inhibits collagen synthesis and alkaline phosphatase activity independently of its effect on deoxyribonucleic acid synthesis in osteoblast-enriched bone cell cultures. Endocrinology 123, 1442–1448.

Chan, E.S., Liu, H., Fernandez, P., Luna, A., Perez-Aso, M., Bujor, A.M., Trojanowska, M., Cronstein, B.N., 2013. Adenosine A(2A) receptors promote collagen production by a Fli1- and CTGF-mediated mechanism. Arthritis Res. Ther. 15 (3), R58.

Chang, H., Brown, C.W., Matzuk, M.M., 2002. Genetic analysis of the mammalian transforming growth factor-beta superfamily. Endocr. Rev. 23, 787–823.

Chatziantoniou, C., Boffa, J.J., Ardaillou, R., Dussaule, J.C., 1998. Nitric oxide inhibition induces early activation of type I collagen gene in renal resistance vessels and glomeruli in transgenic mice. Role of endothelin. J. Clin. Investig. 101, 2780–2789.

Chen, Y., Blom, I.E., Sa, S., Goldschmeding, R., Abraham, D.J., Leask, A., 2002. CTGF expression in mesangial cells: involvement of SMADs, MAP kinase, and PKC. Kidney Int. 62, 1149–1159.

Chen, W.J., Lin, K.H., Lee, Y.S., 2000. Molecular characterization of myocardial fibrosis during hypothyroidism: evidence for negative regulation of the pro-alpha1(I) collagen gene expression by thyroid hormone receptor. Mol. Cell. Endocrinol. 162, 45–55.

Chen, S.J., Ning, H., Ishida, W., Sodin-Semrl, S., Takagawa, S., Mori, Y., Varga, J., 2006. The early-immediate gene EGR-1 is induced by transforming growth factor-beta and mediates stimulation of collagen gene expression. J. Biol. Chem. 281, 21183–21197.

Chen, S.S., Ruteshouser, E.C., Maity, S.N., de Crombrugghe, B., 1997. Cell-specific in vivo DNA-protein interactions at the proximal promoters of the pro alpha 1(I) and the pro alpha 2(I) collagen genes. Nucleic Acids Res. 25, 3261–3268.

Chen, X.D., Shi, S., Xu, T., Robey, P.G., Young, M.F., 2002. Age-related osteoporosis in biglycan-deficient mice is related to defects in bone marrow stromal cells. J. Bone Miner. Res. 17, 331–340.

Chen, Y., Shi-wen, X., Eastwood, M., Black, C.M., Denton, C.P., Leask, A., Abraham, D.J., 2006. Contribution of activin receptor-like kinase 5 (transforming growth factor beta receptor type I) signaling to the fibrotic phenotype of scleroderma fibroblasts. Arthritis Rheum. 54, 1309–1316.

Chen, S.J., Yuan, W., Lo, S., Trojanowska, M., Varga, J., 2000. Interaction of smad3 with a proximal smad-binding element of the human alpha2(I) procollagen gene promoter required for transcriptional activation by TGF-beta. J. Cell. Physiol. 183, 381–392.

Chen, S.J., Yuan, W., Mori, Y., Levenson, A., Trojanowska, M., Varga, J., 1999. Stimulation of type I collagen transcription in human skin fibroblasts by TGF-beta: involvement of Smad 3. J. Investig. Dermatol. 112, 49–57.

Chikazu, D., Hakeda, Y., Ogata, N., Nemoto, K., Itabashi, A., Takato, T., Kumegawa, M., Nakamura, K., Kawaguchi, H., 2000. Fibroblast growth factor (FGF)-2 directly stimulates mature osteoclast function through activation of FGF receptor 1 and p42/p44 MAP kinase. J. Biol. Chem. 275, 31444–31450.

Chipman, S.D., Sweet, H.O., McBride Jr., D.J., Davisson, M.T., Marks Jr., S.C., Shuldiner, A.R., Wenstrup, R.J., Rowe, D.W., Shapiro, J.R., 1993. Defective pro alpha 2(I) collagen synthesis in a recessive mutation in mice: a model of human osteogenesis imperfecta. Proc. Natl. Acad. Sci. U.S.A. 90, 1701–1705.

Chou, D.H., Lee, W., McCulloch, C.A., 1996. TNF-alpha inactivation of collagen receptors: implications for fibroblast function and fibrosis. J. Immunol. 156, 4354–4362.

Chu, M.L., de Wet, W., Bernard, M., Ding, J.F., Morabito, M., Myers, J., Williams, C., Ramirez, F., 1984. Human pro alpha 1(I) collagen gene structure reveals evolutionary conservation of a pattern of introns and exons. Nature 310, 337–340.

Chung, K.Y., Agarwal, A., Uitto, J., Mauviel, A., 1996. An AP-1 binding sequence is essential for regulation of the human alpha 2(I) collagen (COL1A2) promoter activity by transforming growth factor-beta. J. Biol. Chem. 271, 3272–3278.
Cockayne, D., Sterling Jr., K.M., Shull, S., Mintz, K.P., Illeyne, S., Cutroneo, K.R., 1986. Glucocorticoids decrease the synthesis of type I procollagen mRNAs. Biochemistry 25, 3202–3209.
Corsi, A., Xu, T., Chen, X.D., Boyde, A., Liang, J., Mankani, M., Sommer, B., Iozzo, R.V., Eichstetter, I., Robey, P.G., Bianco, P., Young, M.F., 2002. Phenotypic effects of biglycan deficiency are linked to collagen fibril abnormalities, are synergized by decorin deficiency, and mimic Ehlers–Danlos-like changes in bone and other connective tissues. J. Bone Miner. Res. 17, 1180–1189.
Coustry, F., Hu, Q., de Crombrugghe, B., Maity, S.N., 2001. CBF/NF-Y functions both in nucleosomal disruption and transcription activation of the chromatin-assembled topoisomerase IIalpha promoter. Transcription activation by CBF/NF-Y in chromatin is dependent on the promoter structure. J. Biol. Chem. 276, 40621–40630.
Coustry, F., Maity, S.N., de Crombrugghe, B., 1995. Studies on transcription activation by the multimeric CCAAT-binding factor CBF. J. Biol. Chem. 270, 468–475.
Crean, J.K., Finlay, D., Murphy, M., Moss, C., Godson, C., Martin, F., Brady, H.R., 2002. The role of p42/44 MAPK and protein kinase B in connective tissue growth factor induced extracellular matrix protein production, cell migration, and actin cytoskeletal rearrangement in human mesangial cells. J. Biol. Chem. 277, 44187–44194.
Cutroneo, K.R., Sterling, K.M., 2004. How do glucocorticoids compare to oligo decoys as inhibitors of collagen synthesis and potential toxicity of these therapeutics? J. Cell. Biochem. 92, 6–15.
Czaja, M.J., Weiner, F.R., Eghbali, M., Giambrone, M.A., Eghbali, M., Zern, M.A., 1987. Differential effects of gamma-interferon on collagen and fibronectin gene expression. J. Biol. Chem. 262, 13348–13351.
Czuwara-Ladykowska, J., Sementchenko, V.I., Watson, D.K., Trojanowska, M., 2002. Ets1 is an effector of the transforming growth factor beta (TGF-beta) signaling pathway and an antagonist of the profibrotic effects of TGF-beta. J. Biol. Chem. 277, 20399–20408.
Czuwara-Ladykowska, J., Shirasaki, F., Jackers, P., Watson, D.K., Trojanowska, M., 2001. Fli-1 inhibits collagen type I production in dermal fibroblasts via an Sp1-dependent pathway. J. Biol. Chem. 276, 20839–20848.
D'Alessio, M., Bernard, M., Pretorius, P.J., de Wet, W., Ramirez, F., 1988. Complete nucleotide sequence of the region encompassing the first twenty-five exons of the human pro alpha 1(I) collagen gene (COL1A1). Gene 67, 105–115.
Dabovic, B., Chen, Y., Colarossi, C., Obata, H., Zambuto, L., Perle, M.A., Rifkin, D.B., 2002. Bone abnormalities in latent TGF-[beta] binding protein (Ltbp)-3-null mice indicate a role for Ltbp-3 in modulating TGF-[beta] bioavailability. J. Cell Biol. 156, 227–232.
Dahl, T., Veis, A., 2003. Electrostatic interactions lead to the formation of asymmetric collagen-phosphophoryn aggregates. Connect. Tissue Res. 44 (Suppl. 1), 206–213.
Danielson, K.G., Baribault, H., Holmes, D.F., Graham, H., Kadler, K.E., Iozzo, R.V., 1997. Targeted disruption of decorin leads to abnormal collagen fibril morphology and skin fragility. J. Cell Biol. 136, 729–743.
Darnell Jr., J.E., Kerr, I.M., Stark, G.R., 1994. Jak-STAT pathways and transcriptional activation in response to IFNs and other extracellular signaling proteins. Science 264, 1415–1421.
Das, F., Ghosh-Choudhury, N., Venkatesan, B., Kasinath, B.S., Ghosh Choudhury, G., 2017. PDGF receptor-β uses Akt/mTORC1 signaling node to promote high glucose-induced renal proximal tubular cell collagen I (α2) expression. Am. J. Physiol. Renal. Physiol. 313 (2), F291–F307.
de Crombrugghe, B., Lefebvre, V., Nakashima, K., 2001. Regulatory mechanisms in the pathways of cartilage and bone formation. Curr. Opin. Cell Biol. 13, 721–727.
Deak, S.B., Nicholls, A., Pope, F.M., Prockop, D.J., 1983. The molecular defect in a nonlethal variant of osteogenesis imperfecta. Synthesis of pro-alpha 2(I) chains which are not incorporated into trimers of type I procollagen. J. Biol. Chem. 258, 15192–15197.
Delany, A.M., Canalis, E., 1998. Dual regulation of stromelysin-3 by fibroblast growth factor-2 in murine osteoblasts. J. Biol. Chem. 273, 16595–165600.
Delany, A.M., Gabbitas, B.Y., Canalis, E., 1995. Cortisol downregulates osteoblast alpha 1 (I) procollagen mRNA by transcriptional and post-transcriptional mechanisms. J. Cell. Biochem. 57, 488–494.
Delany, A.M., Jeffrey, J.J., Rydziel, S., Canalis, E., 1995. Cortisol increases interstitial collagenase expression in osteoblasts by post-transcriptional mechanisms. J. Biol. Chem. 270, 26607–26612.
Dennler, S., Pendaries, V., Tacheau, C., Costas, M.A., Mauviel, A., Verrecchia, F., 2005. The steroid receptor co-activator-1 (SRC-1) potentiates TGF-beta/Smad signaling: role of p300/CBP. Oncogene 24, 1936–1945.
Denton, C.P., Lindahl, G.E., Khan, K., Shiwen, X., Ong, V.H., Gaspar, N.J., Lazaridis, K., Edwards, D.R., Leask, A., Eastwood, M., Leoni, P., Renzoni, E.A., Bou Gharios, G., Abraham, D.J., Black, C.M., 2005. Activation of key profibrotic mechanisms in transgenic fibroblasts expressing kinase-deficient type II Transforming growth factor-{beta} receptor (T{beta}RII{delta}k). J. Biol. Chem. 280, 16053–16065.
Denton, C.P., Zheng, B., Evans, L.A., Shi-wen, X., Ong, V.H., Fisher, I., Lazaridis, K., Abraham, D.J., Black, C.M., de Crombrugghe, B., 2003. Fibroblast-specific expression of a kinase-deficient type II transforming growth factor beta (TGFbeta) receptor leads to paradoxical activation of TGFbeta signaling pathways with fibrosis in transgenic mice. J. Biol. Chem. 278, 25109–25119.
Denton, C.P., Zheng, B., Shiwen, X., Zhang, Z., Bou-Gharios, G., Eberspaecher, H., Black, C.M., de Crombrugghe, B., 2001. Activation of a fibroblast-specific enhancer of the proalpha2(I) collagen gene in tight-skin mice. Arthritis Rheum. 44, 712–722.
Diaz, A., Jimenez, S.A., 1997. Interferon-gamma regulates collagen and fibronectin gene expression by transcriptional and post-transcriptional mechanisms. Int. J. Biochem. Cell Biol. 29, 251–260.
Diaz, A., Munoz, E., Johnston, R., Korn, J.H., Jimenez, S.A., 1993. Regulation of human lung fibroblast alpha 1(I) procollagen gene expression by tumor necrosis factor alpha, interleukin-1 beta, and prostaglandin E2. J. Biol. Chem. 268, 10364–10371.

Duval, E., Bouyoucef, M., Leclercq, S., Baugé, C., Boumédiene, K., 2016. Hypoxia inducible factor 1 alpha down-regulates type i collagen through Sp3 transcription factor in human chondrocytes. IUBMB Life 68 (9), 756–763.

Efstathiadou, Z., Tsatsoulis, A., Ioannidis, J.P., 2001. Association of collagen Ialpha 1 Sp1 polymorphism with the risk of prevalent fractures: a meta-analysis. J. Bone Miner. Res. 16, 1586–1592.

Erlebacher, A., Derynck, R., 1996. Increased expression of TGF-beta 2 in osteoblasts results in an osteoporosis-like phenotype. J. Cell Biol. 132, 195–210.

Ewald, C.Y., Landis, J.N., Porter Abate, J., Murphy, C.T., Blackwell, T.K., 2015. Dauer-independent insulin/IGF-1-signalling implicates collagen remodelling in longevity. Nature 519 (7541), 97–101.

Fabre, T., Kared, H., Friedman, S.L., Shoukry, N.H., 2014. IL-17A enhances the expression of profibrotic genes through upregulation of the TGF-β receptor on hepatic stellate cells in a JNK-dependent manner. J. Immunol. 193 (8), 3925–3933. https://doi.org/10.4049/jimmunol.1400861.

Fang, M.A., Glackin, C.A., Sadhu, A., McDougall, S., 2001. Transcriptional regulation of alpha 2(I) collagen gene expression by fibroblast growth factor-2 in MC3T3-E1 osteoblast-like cells. J. Cell. Biochem. 80, 550–559.

Feng, X.H., Derynck, R., 2005. Specificity and versatility in tgf-beta signaling through Smads. Annu. Rev. Cell Dev. Biol. 21, 659–693.

Feres-Filho, E.J., Choi, Y.J., Han, X., Takala, T.E., Trackman, P.C., 1995. Pre- and post-translational regulation of lysyl oxidase by transforming growth factor-beta 1 in osteoblastic MC3T3-E1 cells. J. Biol. Chem. 270, 30797–30803.

Fichtner-Feigl, S., Strober, W., Kawakami, K., Puri, R.K., Kitani, A., 2006. IL-13 signaling through the IL-13alpha2 receptor is involved in induction of TGF-beta1 production and fibrosis. Nat. Med. 12, 99–106.

Fields, G.B., 2013 Mar 29. Interstitial collagen catabolism. J. Biol. Chem. 288 (13), 8785–8793. https://doi.org/10.1074/jbc.R113.451211.

Filvaroff, E., Erlebacher, A., Ye, J., Gitelman, S.E., Lotz, J., Heillman, M., Derynck, R., 1999. Inhibition of TGF-beta receptor signaling in osteoblasts leads to decreased bone remodeling and increased trabecular bone mass. Development 126, 4267–4279.

Firszt, R., Francisco, D., Church, T.D., Thomas, J.M., Ingram, J.L., Kraft, M., 2014. Interleukin-13 induces collagen type-1 expression through matrix metalloproteinase-2 and transforming growth factor-β1 in airway fibroblasts in asthma. Eur. Respir. J. 43 (2), 464–473.

Flanders, K.C., Sullivan, C.D., Fujii, M., Sowers, A., Anzano, M.A., Arabshahi, A., Major, C., Deng, C., Russo, A., Mitchell, J.B., Roberts, A.B., 2002. Mice lacking Smad3 are protected against cutaneous injury induced by ionizing radiation. Am. J. Pathol. 160, 1057–1068.

Forlino, A., Porter, F.D., Lee, E.J., Westphal, H., Marini, J.C., 1999. Use of the Cre/lox recombination system to develop a non-lethal knock-in murine model for osteogenesis imperfecta with an alpha1(I) G349C substitution. Variability in phenotype in BrtlIV mice. J. Biol. Chem. 274, 37923–37931.

Fragiadaki, M., Ikeda, T., Witherden, A., Mason, R.M., Abraham, D., Bou-Gharios, G., 2011. High doses of TGF-β potently suppress type I collagen via the transcription factor CUX1. Mol. Biol. Cell 22 (11), 1836–1844.

Frazier, K., Williams, S., Kothapalli, D., Klapper, H., Grotendorst, G.R., 1996. Stimulation of fibroblast cell growth, matrix production, and granulation tissue formation by connective tissue growth factor. J. Investig. Dermatol. 107, 404–411.

Fu, Q., Manolagas, S.C., O'Brien, C.A., 2006. Parathyroid hormone controls receptor activator of NF-kappaB ligand gene expression via a distant transcriptional enhancer. Mol. Cell Biol. 26, 6453–6468.

Fujii, M., Yoneda, A., Takei, N., Sakai-Sawada, K., Kosaka, M., Minomi, K., Yokoyama, A., Tamura, Y., 2017. Endoplasmic reticulum oxidase 1α is critical for collagen secretion from and membrane type 1-matrix metalloproteinase levels in hepatic stellate cells. J. Biol. Chem. 292 (38), 15649–15660. https://doi.org/10.1074/jbc.M117.783126.

Galera, P., Musso, M., Ducy, P., Karsenty, G., 1994. c-Krox, a transcriptional regulator of type I collagen gene expression, is preferentially expressed in skin. Proc. Natl. Acad. Sci. U.S.A. 91, 9372–9376.

Galera, P., Park, R.W., Ducy, P., Mattei, M.G., Karsenty, G., 1996. c-Krox binds to several sites in the promoter of both mouse type I collagen genes. Structure/function study and developmental expression analysis. J. Biol. Chem. 271, 21331–21339.

Galli, G.G., Honnens de Lichtenberg, K., Carrara, M., Hans, W., Wuelling, M., Mentz, B., Multhaupt, H.A., Fog, C.K., Jensen, K.T., Rappsilber, J., Vortkamp, A., Coulton, L., Fuchs, H., Gailus-Durner, V., Hrabě de Angelis,, M., Calogero, R.A., Couchman, J.R., Lund, A.H., 2012. Prdm5 regulates collagen gene transcription by association with RNA polymerase II in developing bone. PLoS Genet. 8 (5) e1002711. Epub 2012 May 10.

Gao, R., Brigstock, D.R., 2003. Low density lipoprotein receptor-related protein (LRP) is a heparin-dependent adhesion receptor for connective tissue growth factor (CTGF) in rat activated hepatic stellate cells. Hepatol. Res. 27, 214–220.

Garcia-Giralt, N., Enjuanes, A., Bustamante, M., Mellibovsky, L., Nogues, X., Carreras, R., Diez-Perez, A., Grinberg, D., Balcells, S., 2005. In vitro functional assay of alleles and haplotypes of two COL1A1-promoter SNPs. Bone 36, 902–908.

Garcia-Giralt, N., Nogues, X., Enjuanes, A., Puig, J., Mellibovsky, L., Bay-Jensen, A., Carreras, R., Balcells, S., Diez-Perez, A., Grinberg, D., 2002. Two new single-nucleotide polymorphisms in the COL1A1 upstream regulatory region and their relationship to bone mineral density. J. Bone Miner. Res. 17, 384–393.

Gedeon, A.K., et al., 1999. Nat Genet 22, 400–404.

Ghosh, A.K., 2002. Factors involved in the regulation of type I collagen gene expression: implication in fibrosis. Exp. Biol. Med. 227, 301–314.

Ghosh, A.K., Yuan, W., Mori, Y., Chen, S., Varga, J., 2001. Antagonistic regulation of type I collagen gene expression by interferon-gamma and transforming growth factor-beta. Integration at the level of p300/CBP transcriptional coactivators. J. Biol. Chem. 276, 11041–11048.

Ghosh, A.K., Yuan, W., Mori, Y., Varga, J., 2000. Smad-dependent stimulation of type I collagen gene expression in human skin fibroblasts by TGF-beta involves functional cooperation with p300/CBP transcriptional coactivators. Oncogene 19, 3546–3555.

Goldberg, H., Helaakoski, T., Garrett, L.A., Karsenty, G., Pellegrino, A., Lozano, G., Maity, S., de Crombrugghe, B., 1992. Tissue-specific expression of the mouse alpha 2(I) collagen promoter. Studies in transgenic mice and in tissue culture cells. J. Biol. Chem. 267, 19622–19630.

Goldberg, M.D., Septier, O., Rapoport, M., Young, L., Ameye, 2002. Biglycan is a repressor of amelogenin expression and enamel formation: an emerging hypothesis. J. Dent. Res. 81, 520–524.
Govoni, K.E., Baylink, D.J., Mohan, S., 2005. The multi-functional role of insulin-like growth factor binding proteins in bone. Pediatr. Nephrol. 20, 261–268.
Govoni, K.E., Wergedal, J.E., Florin, L., Angel, P., Baylink, D.J., Mohan, S., 2007. Conditional deletion of IGF-I in collagen type 1{alpha}2 (Col1{alpha}2) expressing cells results in postnatal lethality and a dramatic reduction in bone accretion. Endocrinology.
Gowen, M., Chapman, K., Littlewood, A., Hughes, D., Evans, D., Russell, G., 1990. Production of tumor necrosis factor by human osteoblasts is modulated by other cytokines, but not by osteotropic hormones. Endocrinology 126, 1250–1255.
Graham, H., Peng, C., 2006. Activin receptor-like kinases: structure, function and clinical implications. Endocr. Metab. Immune Disord. - Drug Targets 6, 45–58.
Grant, S.F., Reid, D.M., Blake, G., Herd, R., Fogelman, I., Ralston, S.H., 1996. Reduced bone density and osteoporosis associated with a polymorphic Sp1 binding site in the collagen type I alpha 1 gene. Nat. Genet. 14, 203–205.
Greenwel, P., Inagaki, Y., Hu, W., Walsh, M., Ramirez, F., 1997. Sp1 is required for the early response of alpha2(I) collagen to transforming growth factor-beta1. J. Biol. Chem. 272, 19738–19745.
Greenwel, P., Tanaka, S., Penkov, D., Zhang, W., Olive, M., Moll, J., Vinson, C., Di Liberto, M., Ramirez, F., 2000. Tumor necrosis factor alpha inhibits type I collagen synthesis through repressive CCAAT/enhancer-binding proteins. Mol. Cell Biol. 20, 912–918.
Han, et al., 2010. J Biol Chem. 285, 22276–22281.
Hara, M., Kobayakawa, K., Ohkawa, Y., Kumamaru, H., Yokota, K., Saito, T., Kijima, K., Yoshizaki, S., Harimaya, K., Nakashima, Y., Okada, S., 2017. Interaction of reactive astrocytes with type I collagen induces astrocytic scar formation through the integrin-N-cadherin pathway after spinal cord injury. Nat. Med. 23 (7), 818–828.
Haralson, M.A., Jacobson, H.R., Hoover, R.L., 1987. Collagen polymorphism in cultured rat kidney mesangial cells. Lab. Invest. 57, 513–523.
Harbers, K., Kuehn, M., Delius, H., Jaenisch, R., 1984. Insertion of retrovirus into the first intron of alpha 1(I) collagen gene to embryonic lethal mutation in mice. Proc. Natl. Acad. Sci. U.S.A. 81, 1504–1508.
Harris, S.S., Patel, M.S., Cole, D.E., Dawson-Hughes, B., 2000. Associations of the collagen type Ialpha1 Sp1 polymorphism with five-year rates of bone loss in older adults. Calcif. Tissue Int. 66, 268–271.
Harrison, J.R., Petersen, D.N., Lichtler, A.C., Mador, A.T., Rowe, D.W., Kream, B.E., 1989. 1,25-Dihydroxyvitamin D3 inhibits transcription of type I collagen genes in the rat osteosarcoma cell line ROS 17/2.8. Endocrinology 125, 327–333.
Harrison, J.R., Vargas, S.J., Petersen, D.N., Lorenzo, J.A., Kream, B.E., 1990. Interleukin-1 alpha and phorbol ester inhibit collagen synthesis in osteoblastic MC3T3-E1 cells by a transcriptional mechanism. Mol. Endocrinol. 4, 184–190.
Hasegawa, T., Zhou, X., Garrett, L.A., Ruteshouser, E.C., Maity, S.N., de Crombrugghe, B., 1996. Evidence for three major transcription activation elements in the proximal mouse proalpha2(I) collagen promoter. Nucleic Acids Res. 24, 3253–3560.
Hernandez, I., de la Torre, P., Rey-Campos, J., Garcia, I., Sanchez, J.A., Munoz, R., Rippe, R.A., Munoz-Yague, T., Solis-Herruzo, J.A., 2000. Collagen alpha1(I) gene contains an element responsive to tumor necrosis factor-alpha located in the 5' untranslated region of its first exon. DNA Cell Biol. 19, 341–352.
Higashi, K., Inagaki, Y., Suzuki, N., Mitsui, S., Mauviel, A., Kaneko, H., Nakatsuka, I., 2003. Y-box-binding protein YB-1 mediates transcriptional repression of human alpha 2(I) collagen gene expression by interferon-gamma. J. Biol. Chem. 278, 5156–5162.
Higashi, K., Kouba, D.J., Song, Y.J., Uitto, J., Mauviel, A., 1998. A proximal element within the human alpha 2(I) collagen (COL1A2) promoter, distinct from the tumor necrosis factor-alpha response element, mediates transcriptional repression by interferon-gamma. Matrix Biol. 16, 447–456.
Hojo, et al., 2016. Dev Cell. 37, 238–253.
Holmes, A., Abraham, D.J., Chen, Y., Denton, C., Shi-wen, X., Black, C.M., Leask, A., 2003. Constitutive connective tissue growth factor expression in scleroderma fibroblasts is dependent on Sp1. J. Biol. Chem. 278, 41728–41733.
Holmes, A., Abraham, D.J., Sa, S., Shiwen, X., Black, C.M., Leask, A., 2001. CTGF and SMADs, maintenance of scleroderma phenotype is independent of SMAD signaling. J. Biol. Chem. 276, 10594–10601.
Hong, J., Chu, M., Qian, L., Wang, J., Guo, Y., Xu, D., 2017. Fibrillar type I collagen enhances the differentiation and proliferation of myofibroblasts by lowering α2β1 integrin expression in cardiac fibrosis. Biomed. Res Int. 2017, 1790808. https://doi.org/10.1155/2017/1790808.
Hormuzdi, S.G., Penttinen, R., Jaenisch, R., Bornstein, P., 1998. A gene-targeting approach identifies a function for the first intron in expression of the alpha1(I) collagen gene. Mol. Cell Biol. 18, 3368–3375.
Hormuzdi, S.G., Strandjord, T.P., Madtes, D.K., Bornstein, P., 1999. Mice with a targeted intronic deletion in the Col1a1 gene respond to bleomycin-induced pulmonary fibrosis with increased expression of the mutant allele. Matrix Biol. 18, 287–294.
Horvai, A.E., Xu, L., Korzus, E., Brard, G., Kalafus, D., Mullen, T.M., Rose, D.W., Rosenfeld, M.G., Glass, C.K., 1997. Nuclear integration of JAK/STAT and Ras/AP-1 signaling by CBP and p300. Proc. Natl. Acad. Sci. U.S.A. 94, 1074–1079.
Houglum, K., Buck, M., Kim, D.J., Chojkier, M., 1998. TNF-alpha inhibits liver collagen-alpha 1(I) gene expression through a tissue-specific regulatory region. Am. J. Physiol. 274, G840–G847.
Hulmes, D.J., 2002. Building collagen molecules, fibrils, and suprafibrillar structures. J. Struct. Biol. 137, 2–10.
Hurley, M.M., Tetradis, S., Huang, Y.F., Hock, J., Kream, B.E., Raisz, L.G., Sabbieti, M.G., 1999. Parathyroid hormone regulates the expression of fibroblast growth factor-2 mRNA and fibroblast growth factor receptor mRNA in osteoblastic cells. J. Bone Miner. Res. 14, 776–783.
Igarashi, A., Okochi, H., Bradham, D.M., Grotendorst, G.R., 1993. Regulation of connective tissue growth factor gene expression in human skin fibroblasts and during wound repair. Mol. Biol. Cell 4, 637–645.

Ihn, H., Ihn, Y., Trojanowska, M., 2001. Spl phosphorylation induced by serum stimulates the human alpha2(I) collagen gene expression. J. Investig. Dermatol. 117, 301–308.

Ihn, H., LeRoy, E.C., Trojanowska, M., 1997. Oncostatin M stimulates transcription of the human alpha2(I) collagen gene via the Spl/Sp3-binding site. J. Biol. Chem. 272, 24666–24672.

Ihn, H., Ohnishi, K., Tamaki, T., LeRoy, E.C., Trojanowska, M., 1996. Transcriptional regulation of the human alpha2(I) collagen gene. Combined action of upstream stimulatory and inhibitory cis-acting elements. J. Biol. Chem. 271, 26717–26723.

Inagaki, Y., Kushida, M., Higashi, K., Itoh, J., Higashiyama, R., Hong, Y.Y., Kawada, N., Namikawa, K., Kiyama, H., Bou-Gharios, G., Watanabe, T., Okazaki, I., Ikeda, K., 2005. Cell type-specific intervention of transforming growth factor beta/Smad signaling suppresses collagen gene expression and hepatic fibrosis in mice. Gastroenterology 129, 259–268.

Inagaki, Y., Nemoto, T., Kushida, M., Sheng, Y., Higashi, K., Ikeda, K., Kawada, N., Shirasaki, F., Takehara, K., Sugiyama, K., Fujii, M., Yamauchi, H., Nakao, A., de Crombrugghe, B., Watanabe, T., Okazaki, I., 2003. Interferon alfa down-regulates collagen gene transcription and suppresses experimental hepatic fibrosis in mice. Hepatology 38, 890–899.

Inagaki, Y., Truter, S., Ramirez, F., 1994. Transforming growth factor-beta stimulates alpha 2(I) collagen gene expression through a cis-acting element that contains an Spl-binding site. J. Biol. Chem. 269, 14828–14834.

Inagaki, Y., Truter, S., Tanaka, S., Di Liberto, M., Ramirez, F., 1995. Overlapping pathways mediate the opposing actions of tumor necrosis factor-alpha and transforming growth factor-beta on alpha 2(I) collagen gene transcription. J. Biol. Chem. 270, 3353–3358.

Iozzo, R.V., 1999. The biology of the small leucine-rich proteoglycans. Functional network of interactive proteins. J. Biol. Chem. 274, 18843–18846.

Iraburu, M.J., Dominguez-Rosales, J.A., Fontana, L., Auster, A., Garcia-Trevijano, E.R., Covarrubias-Pinedo, A., Rivas-Estilla, A.M., Greenwel, P., Rojkind, M., 2000. Tumor necrosis factor alpha down-regulates expression of the alpha1(I) collagen gene in rat hepatic stellate cells through a p20C/EBPbeta- and C/EBPdelta-dependent mechanism. Hepatology 31, 1086–1093.

Ivkovic, S., Yoon, B.S., Popoff, S.N., Safadi, F.F., Libuda, D.E., Stephenson, R.C., Daluiski, A., Lyons, K.M., 2003. Connective tissue growth factor coordinates chondrogenesis and angiogenesis during skeletal development. Development 130, 2779–2791.

Janssens, K., Gershoni-Baruch, R., Guanabens, N., Migone, N., Ralston, S., Bonduelle, M., Lissens, W., Van Maldergem, L., Vanhoenacker, F., Verbruggen, L., Van Hul, W., 2000. Mutations in the gene encoding the latency-associated peptide of TGF-beta 1 cause Camurati-Engelmann disease. Nat. Genet. 26, 273–275.

Jia, D., O'Brien, C.A., Stewart, S.A., Manolagas, S.C., Weinstein, R.S., 2006. Glucocorticoids act directly on osteoclasts to increase their life span and reduce bone density. Endocrinology 147, 5592–5599.

Jiang, J., Lichtler, A.C., Gronowicz, G.A., Adams, D.J., Clark, S.H., Rosen, C.J., Kream, B.E., 2006. Transgenic mice with osteoblast-targeted insulin-like growth factor-I show increased bone remodeling. Bone 39, 494–504.

Jimenez, S.A., Bashey, R.I., Benditt, M., Yankowski, R., 1977. Identification of collagen alpha1(I) trimer in embryonic chick tendons and calvaria. Biochem. Biophys. Res. Commun. 78, 1354–1361.

Jimenez, S.A., Varga, J., Olsen, A., Li, L., Diaz, A., Herhal, J., Koch, J., 1994. Functional analysis of human alpha 1(I) procollagen gene promoter, Differential activity in collagen-producing and -nonproducing cells and response to transforming growth factor beta 1. J. Biol. Chem. 269, 12684–12691.

Johnson, R.J., Floege, J., Yoshimura, A., Iida, H., Couser, W.G., Alpers, C.E., 1992. The activated mesangial cell: a glomerular "myofibroblast"? J. Am. Soc. Nephrol. 2, S190–S197.

Johnson, C., Primorac, D., McKinstry, M., McNeil, J., Rowe, D., Lawrence, J.B., 2000. Tracking COL1A1 RNA in osteogenesis imperfecta, splice-defective transcripts initiate transport from the gene but are retained within the SC35 domain. J. Cell Biol. 150, 417–432.

Jones, R.L., Stoikos, C., Findlay, J.K., Salamonsen, L.A., 2006. TGF-beta superfamily expression and actions in the endometrium and placenta. Reproduction 132, 217–232.

Kadler, K., 2004. Matrix loading: assembly of extracellular matrix collagen fibrils during embryogenesis. Birth Defects Res. C Embryo Today 72, 1–11.

Kadler, K.E., Hulmes, D.J., Hojima, Y., Prockop, D.J., 1990. Assembly of type I collagen fibrils de novo by the specific enzymic cleavage of pC collagen. The fibrils formed at about 37 degrees C are similar in diameter, roundness, and apparent flexibility to the collagen fibrils seen in connective tissue. Ann. N. Y. Acad. Sci. 580, 214–224.

Kadler, K.E., Torre-Blanco, A., Adachi, E., Vogel, B.E., Hojima, Y., Prockop, D.J., 1991. A type I collagen with substitution of a cysteine for glycine-748 in the alpha 1(I) chain copolymerizes with normal type I collagen and can generate fractallike structures. Biochemistry 30, 5081–5088.

Kahari, V.M., Chen, Y.Q., Su, M.W., Ramirez, F., Uitto, J., 1990. Tumor necrosis factor-alpha and interferon-gamma suppress the activation of human type I collagen gene expression by transforming growth factor-beta 1. Evidence for two distinct mechanisms of inhibition at the transcriptional and posttranscriptional levels. J. Clin. Investig. 86, 1489–1495.

Kalson, N.S., Starborg, T., Lu, Y., Mironov, A., Humphries, S.M., Holmes, D.F., Kadler, K.E., 2013. Nonmuscle myosin II powered transport of newly formed collagen fibrils at the plasma membrane. Proc. Natl. Acad. Sci. U.S.A. 110 (49), E4743–E4752. https://doi.org/10.1073/pnas.1314348110.

Kanamaru, Y., Nakao, A., Tanaka, Y., Inagaki, Y., Ushio, H., Shirato, I., Horikoshi, S., Okumura, K., Ogawa, H., Tomino, Y., 2003. Involvement of p300 in TGF-beta/Smad-pathway-mediated alpha 2(I) collagen expression in mouse mesangial cells. Nephron Exp. Nephrol. 95, e36–e42.

Karsenty, G., de Crombrugghe, B., 1990. Two different negative and one positive regulatory factors interact with a short promoter segment of the alpha 1 (I) collagen gene. J. Biol. Chem. 265, 9934–9942.

Karsenty, G., de Crombrugghe, B., 1991. Conservation of binding sites for regulatory factors in the coordinately expressed alpha 1 (I) and alpha 2 (I) collagen promoters. Biochem. Biophys. Res. Commun. 177, 538–544.

Karsenty, G., Wagner, E.F., 2002. Reaching a genetic and molecular understanding of skeletal development. Dev. Cell 2, 389–406.

Kawaguchi, Y., Hara, M., Wright, T.M., 1999. Endogenous IL-1alpha from systemic sclerosis fibroblasts induces IL-6 and PDGF-A. J. Clin. Investig. 103, 1253–1260.

Kawaguchi, Y., Nishimagi, E., Tochimoto, A., Kawamoto, M., Katsumata, Y., Soejima, M., Kanno, T., Kamatani, N., Hara, M., 2006. Intracellular IL-1alpha-binding proteins contribute to biological functions of endogenous IL-1alpha in systemic sclerosis fibroblasts. Proc. Natl. Acad. Sci. U.S.A. 103, 14501–14506.

Kern, B., Shen, J., Starbuck, M., Karsenty, G., 2001. Cbfa1 contributes to the osteoblast-specific expression of type I collagen genes. J. Biol. Chem. 276, 7101–7107.

Kim, H.J., Song, S.B., Choi, J.M., Kim, K.M., Cho, B.K., Cho, D.H., Park, H.J., 2010. IL-18 downregulates collagen production in human dermal fibroblasts via the ERK pathway. J. Invest. Dermatol. 130 (3), 706–715.

Kinoshita, A., Saito, T., Tomita, H., Makita, Y., Yoshida, K., Ghadami, M., Yamada, K., Kondo, S., Ikegawa, S., Nishimura, G., Fukushima, Y., Nakagomi, T., Saito, H., Sugimoto, T., Kamegaya, M., Hisa, K., Murray, J.C., Taniguchi, N., Niikawa, N., Yoshiura, K., 2000. Domain-specific mutations in TGFB1 result in Camurati-Engelmann disease. Nat. Genet. 26, 19–20.

Klaus, V., Tanios-Schmies, F., Reeps, C., Trenner, M., Matevossian, E., Eckstein, H.H., Pelisek, J., 2017. Association of matrix metalloproteinase levels with collagen degradation in the context of abdominal aortic aneurysm. Eur. J. Vasc. Endovasc. Surg. 53 (4), 549–558.

Koch, M., Laub, F., Zhou, P., Hahn, R.A., Tanaka, S., Burgeson, R.E., Gerecke, D.R., Ramirez, F., Gordon, M.K., 2003. Collagen XXIV, a vertebrate fibrillar collagen with structural features of invertebrate collagens: selective expression in developing cornea and bone. J. Biol. Chem. 278, 43236–43244.

Kouba, D.J., Chung, K.Y., Nishiyama, T., Vindevoghel, L., Kon, A., Klement, J.F., Uitto, J., Mauviel, A., 1999. Nuclear factor-kappa B mediates TNF-alpha inhibitory effect on alpha 2(I) collagen (COL1A2) gene transcription in human dermal fibroblasts. J. Immunol. 162, 4226–4234.

Kream, B.E., LaFrancis, D., Petersen, D.N., Woody, C., Clark, S., Rowe, D.W., Lichtler, A., 1993. Parathyroid hormone represses alpha 1(I) collagen promoter activity in cultured calvariae from neonatal transgenic mice. Mol. Endocrinol. 7, 399–408.

Kream, B.E., Rowe, D., Smith, M.D., Maher, V., Majeska, R., 1986. Hormonal regulation of collagen synthesis in a clonal rat osteosarcoma cell line. Endocrinology 119, 1922–1928.

Krempen, K., Grotkopp, D., Hall, K., Bache, A., Gillan, A., Rippe, R.A., Brenner, D.A., Breindl, M., 1999. Far upstream regulatory elements enhance position-independent and uterus-specific expression of the murine alpha1(I) collagen promoter in transgenic mice. Gene Expr. 8, 151–163.

Kresse, H., Liszio, C., Schonherr, E., Fisher, L.W., 1997. Critical role of glutamate in a central leucine-rich repeat of decorin for interaction with type I collagen. J. Biol. Chem. 272, 18404–18410.

Kuc, I.M., Scott, P.G., 1997. Increased diameters of collagen fibrils precipitated in vitro in the presence of decorin from various connective tissues. Connect. Tissue Res. 36, 287–296.

Kuznetsova, N.V., Forlino, A., Cabral, W.A., Marini, J.C., Leikin, S., 2004. Structure, stability and interactions of type I collagen with GLY349-CYS substitution in alpha 1(I) chain in a murine Osteogenesis Imperfecta model. Matrix Biol. 23, 101–112.

Kwansa, A.L., De Vita, R., Freeman, J.W., 2014. Mechanical recruitment of N- and C-crosslinks in collagen type I. Matrix Biol. 34, 161–169.

Kypriotou, M., Beauchef, G., Chadjichristos, C., Widom, R., Renard, E., Jimenez, S., Korn, J., Maquart, F.X., Oddos, T., Von Stetten, O., Pujol, J.P., Galera, P., 2007. Human collagen-Krox (hc-Krox) up-regulates type I collagen expression in normal and scleroderma fibroblasts through interaction with Sp1 and Sp3 transcription factors. J. Biol. Chem.

Lamande, S.R., Bateman, J.F., 1999. Procollagen folding and assembly: the role of endoplasmic reticulum enzymes and molecular chaperones. Semin. Cell Dev. Biol. 10, 455–464.

Lauer-Fields, J.L., Juska, D., Fields, G.B., 2002. Matrix metalloproteinases and collagen catabolism. Biopolymers 66 (1), 19–32.

Leask, A., Abraham, D.J., 2003. The role of connective tissue growth factor, a multifunctional matricellular protein, in fibroblast biology. Biochem. Cell Biol. 81, 355–363.

Leask, A., Abraham, D.J., 2004. TGF-beta signaling and the fibrotic response. FASEB J. 18, 816–827.

Leask, A., Holmes, A., Black, C.M., Abraham, D.J., 2003. Connective tissue growth factor gene regulation. Requirements for its induction by transforming growth factor-beta 2 in fibroblasts. J. Biol. Chem. 278, 13008–13015.

Lefebvre, V., Behringer, R.R., de Crombrugghe, B., 2001. L-Sox5, Sox6 and Sox9 control essential steps of the chondrocyte differentiation pathway. Osteoarthritis Cartilage 9 (Suppl. A), S69–S75.

Lefebvre, V., de Crombrugghe, B., 1998. Toward understanding SOX9 function in chondrocyte differentiation. Matrix Biol. 16, 529–540.

Lejard, V., Brideau, G., Blais, F., Salingcarnboriboon, R., Wagner, G., Roehrl, M.H., Noda, M., Duprez, D., Houillier, P., Rossert, J., 2007. Scleraxis and NFATc regulate the expression of the pro-alpha1(I) collagen gene in tendon fibroblasts. J. Biol. Chem. 282, 17665–17675.

Li, I.M.H., Horwell, A.L., Chu, G., de Crombrugghe, B., Bou-Gharios, G., 2017. Characterization of mesenchymal-fibroblast cells using the Col1a2 promoter/enhancer. Methods Mol. Biol. 1627, 139–161.

Li, C., Iness, A., Yoon, J., Grider, J.R., Murthy, K.S., Kellum, J.M., Kuemmerle, J.F., 2015. Noncanonical STAT3 activation regulates excess TGF-β1 and collagen I expression in muscle of stricturing Crohn's disease. J. Immunol. 194 (7), 3422–3431.

Lindert, U., Gnoli, M., Maioli, M., Bedeschi, M.F., Sangiorgi, L., Rohrbach, M., Giunta, C., 2018. Insight into the pathology of a COL1A1 signal peptide heterozygous mutation leading to severe osteogenesis imperfecta. Calcif. Tissue Int. 102 (3), 373–379.

Lindquist, J.N., Marzluff, W.F., Stefanovic, B., 2000. Fibrogenesis. III. Posttranscriptional regulation of type I collagen. Am. J. Physiol. 279, G471–G476.

Linkhart, T.A., Keffer, M.J., 1991. Differential regulation of insulin-like growth factor-I (IGF-I) and IGF-II release from cultured neonatal mouse calvaria by parathyroid hormone, transforming growth factor-beta, and 1,25-dihydroxyvitamin D3. Endocrinology 128, 1511–1518.

Liska, D.J., Reed, M.J., Sage, E.H., Bornstein, P., 1994. Cell-specific expression of alpha 1(I) collagen-hGH minigenes in transgenic mice. J. Cell Biol. 125, 695–704.

Liska, D.J., Robinson, V.R., Bornstein, P., 1992. Elements in the first intron of the alpha 1(I) collagen gene interact with Sp1 to regulate gene expression. Gene Expr. 2, 379–389.

Liska, D.J., Slack, J.L., Bornstein, P., 1990. A highly conserved intronic sequence is involved in transcriptional regulation of the alpha 1(I) collagen gene. Cell Regul. 1, 487–498.

Liu, S., Gupta, A., Quarles, L.D., 2007. Emerging role of fibroblast growth factor 23 in a bone-kidney axis regulating systemic phosphate homeostasis and extracellular matrix mineralization. Curr. Opin. Nephrol. Hypertens. 16, 329–335.

LoCascio, V., Bonucci, E., Imbimbo, B., Ballanti, P., Adami, S., Milani, S., Tartarotti, D., DellaRocca, C., 1990. Bone loss in response to long-term glucocorticoid therapy. Bone Miner. 8, 39–51.

MacDonald, H.M., McGuigan, F.A., New, S.A., Campbell, M.K., Golden, M.H., Ralston, S.H., Reid, D.M., 2001. COL1A1 Sp1 polymorphism predicts perimenopausal and early postmenopausal spinal bone loss. J. Bone Miner. Res. 16, 1634–1641.

Maiers, J.L., Kostallari, E., Mushref, M., deAssuncao, T.M., Li, H., Jalan-Sakrikar, N., Huebert, R.C., Cao, S., Malhi, H., Shah, V.H., 2017. The unfolded protein response mediates fibrogenesis and collagen I secretion through regulating TANGO1 in mice. Hepatology 65 (3), 983–998. https://doi.org/10.1002/hep.28921.

Maity, S.N., de Crombrugghe, B., 1998. Role of the CCAAT-binding protein CBF/NF-Y in transcription. Trends Biochem. Sci. 23, 174–178.

Mann, V., Hobson, E.E., Li, B., Stewart, T.L., Grant, S.F., Robins, S.P., Aspden, R.M., Ralston, S.H., 2001. A COL1A1 Sp1 binding site polymorphism predisposes to osteoporotic fracture by affecting bone density and quality. J. Clin. Investig. 107, 899–907.

Mann, V., Ralston, S.H., 2003. Meta-analysis of COL1A1 Sp1 polymorphism in relation to bone mineral density and osteoporotic fracture. Bone 32, 711–717.

Marini, J.C., Forlino, A., Cabral, W.A., Barnes, A.M., San Antonio, J.D., Milgrom, S., Hyland, J.C., Korkko, J., Prockop, D.J., De Paepe, A., Coucke, P., Symoens, S., Glorieux, F.H., Roughley, P.J., Lund, A.M., Kuurila-Svahn, K., Hartikka, H., Cohn, D.H., Krakow, D., Mottes, M., Schwarze, U., Chen, D., Yang, K., Kuslich, C., Troendle, J., Dalgleish, R., Byers, P.H., 2007. Consortium for osteogenesis imperfecta mutations in the helical domain of type I collagen: regions rich in lethal mutations align with collagen binding sites for integrins and proteoglycans. Hum. Mutat. 28, 209–221.

Marongiu, M., Deiana, M., Marcia, L., Sbardellati, A., Asunis, I., Meloni, A., Angius, A., Cusano, R., Loi, A., Crobu, F., Fotia, G., Cucca, F., Schlessinger, D., Crisponi, L., 2016. Novel action of FOXL2 as mediator of Col1a2 gene autoregulation. Dev. Biol. 416 (1), 200–211.

Massague, J., Chen, Y.G., 2000. Controlling TGF-beta signaling. Genes Dev. 14, 627–644.

Matsuo, N., Tanaka, S., Gordon, M.K., Koch, M., Yoshioka, H., Ramirez, F., 2006. CREB-AP1 protein complexes regulate transcription of the collagen XXIV gene (Col24a1) in osteoblasts. J. Biol. Chem. 281, 5445–5452.

McBride Jr., D.J., Choe, V., Shapiro, J.R., Brodsky, B., 1997. Altered collagen structure in mouse tail tendon lacking the alpha 2(I) chain. J. Mol. Biol. 270, 275–284.

McCarthy, T.L., Centrella, M., Canalis, E., 1989. Regulatory effects of insulin-like growth factors I and II on bone collagen synthesis in rat calvarial cultures. Endocrinology 124, 301–309.

Meisler, N., Shull, S., Xie, R., Long, G.L., Absher, M., Connolly, J.P., Cutroneo, K.R., 1995. Glucocorticoids coordinately regulate type I collagen pro alpha 1 promoter activity through both the glucocorticoid and transforming growth factor beta response elements: a novel mechanism of glucocorticoid regulation of eukaryotic genes. J. Cell. Biochem. 59, 376–388.

Mi, Y., Wang, W., Zhang, C., Liu, C., Lu, J., Li, W., Zuo, R., Myatt, L., Sun, K., 2017. Autophagic degradation of collagen 1A1 by cortisol in human amnion fibroblasts. Endocrinology 158 (4), 1005–1014. https://doi.org/10.1210/en.2016-1829.

Miles, C.A., Sims, T.J., Camacho, N.P., Bailey, A.J., 2002. The role of the alpha2 chain in the stabilization of the collagen type I heterotrimer: a study of the type I homotrimer in oim mouse tissues. J. Mol. Biol. 321, 797–805.

Miyazawa, K., Shinozaki, M., Hara, T., Furuya, T., Miyazono, K., 2002. Two major Smad pathways in TGF-beta superfamily signalling. Genes Cells 7, 1191–1204.

Mohan, S., Baylink, D.J., 2005. Impaired skeletal growth in mice with haploinsufficiency of IGF-I: genetic evidence that differences in IGF-I expression could contribute to peak bone mineral density differences. J. Endocrinol. 185, 415–420.

Mohan, S., Richman, C., Guo, R., Amaar, Y., Donahue, L.R., Wergedal, J., Baylink, D.J., 2003. Insulin-like growth factor regulates peak bone mineral density in mice by both growth hormone-dependent and -independent mechanisms. Endocrinology 144, 929–936.

Montero, A., Okada, Y., Tomita, M., Ito, M., Tsurukami, H., Nakamura, T., Doetschman, T., Coffin, J.D., Hurley, M.M., 2000. Disruption of the fibroblast growth factor-2 gene results in decreased bone mass and bone formation. J. Clin. Investig. 105, 1085–1093.

Moore, R., Ferretti, P., Copp, A., Thorogood, P., 2002. Blocking endogenous FGF-2 activity prevents cranial osteogenesis. Dev. Biol. 243, 99–114.

Mori, K., Hatamochi, A., Ueki, H., Olsen, A., Jimenez, S.A., 1996. The transcription of human alpha 1(I) procollagen gene (COL1A1) is suppressed by tumour necrosis factor-alpha through proximal short promoter elements: evidence for suppression mechanisms mediated by two nuclear–factor binding sites. Biochem. J. 319 (Pt 3), 811–816.

Moro, L., Smith, B.D., 1977. Identification of collagen alpha1(I) trimer and normal type I collagen in a polyoma virus-induced mouse tumor. Arch. Biochem. Biophys. 182, 33–41.

Morrison, J., Lu, Q.L., Pastoret, C., Partridge, T., Bou-Gharios, G., 2000. T-cell-dependent fibrosis in the mdx dystrophic mouse. Lab. Invest. 80, 881–891.

Mu, S., Kang, B., Zeng, W., Sun, Y., Yang, F., 2016. MicroRNA-143-3p inhibits hyperplastic scar formation by targeting connective tissue growth factor CTGF/CCN2 via the Akt/mTOR pathway. Mol. Cell Biochem. 416 (1-2), 99–108.

Murakami, S., Lefebvre, V., de Crombrugghe, B., 2000. Potent inhibition of the master chondrogenic factor Sox9 gene by interleukin-1 and tumor necrosis factor-alpha. J. Biol. Chem. 275, 3687–3692.

Myllyharju, J., Kivirikko, K.I., 2001. Collagens and collagen-related diseases. Ann. Med. 33, 7–21.

Myllyharju, J., Kivirikko, K.I., 2004. Collagens, modifying enzymes and their mutations in humans, flies and worms. Trends Genet. 20, 33–43.

Nagai, N., Hosokawa, M., Itohara, S., Adachi, E., Matsushita, T., Hosokawa, N., Nagata, K., 2000. Embryonic lethality of molecular chaperone hsp47 knockout mice is associated with defects in collagen biosynthesis. J. Cell Biol. 150, 1499–1506.

Nagase, H., Visse, R., Murphy, G., 2006. Structure and function of matrix metalloproteinases and TIMPs. Cardiovasc. Res. 69 (3), 562–573.

Nageotte, J., 1927. Action des sels neutres sur la formation du caillot artificiel de collagene. C. R. Soc. Biol. 96, 828–830.

Nakanishi, T., Yamaai, T., Asano, M., Nawachi, K., Suzuki, M., Sugimoto, T., Takigawa, M., 2001. Overexpression of connective tissue growth factor/hypertrophic chondrocyte-specific gene product 24 decreases bone density in adult mice and induces dwarfism. Biochem. Biophys. Res. Commun. 281, 678–681.

Nakao, A., Fujii, M., Matsumura, R., Kumano, K., Saito, Y., Miyazono, K., Iwamoto, I., 1999. Transient gene transfer and expression of Smad7 prevents bleomycin-induced lung fibrosis in mice. J. Clin. Investig. 104, 5–11.

Nakayama, W., Jinnin, M., Tomizawa, Y., Nakamura, K., Kudo, H., Inoue, K., Makino, K., Honda, N., Kajihara, I., Fukushima, S., Ihn, H., 2017. Dysregulated interleukin-23 signalling contributes to the increased collagen production in scleroderma fibroblasts via balancing microRNA expression. Rheumatology (Oxford) 56 (1), 145–155.

Nehls, M.C., Rippe, R.A., Veloz, L., Brenner, D.A., 1991. Transcription factors nuclear factor I and Sp1 interact with the murine collagen alpha 1 (I) promoter. Mol. Cell Biol. 11, 4065–4073.

Newberry, E.P., Latifi, T., Towler, D.A., 1999. The RRM domain of MINT, a novel Msx2 binding protein, recognizes and regulates the rat osteocalcin promoter. Biochemistry 38, 10678–10690.

Newberry, E.P., Willis, D., Latifi, T., Boudreaux, J.M., Towler, D.A., 1997. Fibroblast growth factor receptor signaling activates the human interstitial collagenase promoter via the bipartite Ets-AP1 element. Mol. Endocrinol. 11, 1129–1144.

Niederreither, K., D'Souza, R.N., De Crombrugghe, B., 1992. Minimal DNA sequences that control the cell lineage-specific expression of the pro alpha 2(I) collagen promoter in transgenic mice. J. Cell Biol. 119, 1361–1370.

Nieto, N., 2006. Oxidative-stress and IL-6 mediate the fibrogenic effects of [corrected] Kupffer cells on stellate cells. Hepatology 44, 1487–1501.

Noda, M., Camilliere, J.J., 1989. In vivo stimulation of bone formation by transforming growth factor-beta. Endocrinology 124, 2991–2994.

Novinec, M., Lenarčič, B., 2013. Cathepsin K: a unique collagenolytic cysteine peptidase. Biol. Chem. 394 (9), 1163–1179. https://doi.org/10.1515/hsz-2013-0134.

O'Reilly, S., Ciechomska, M., Fullard, N., Przyborski, S., van Laar, J.M., 2016. IL-13 mediates collagen deposition via STAT6 and microRNA-135b: a role for epigenetics. Sci. Rep. 6, 25066.

O'Shea, P.J., Guigon, C.J., Williams, G.R., Cheng, S.Y., 2007. Regulation of fibroblast growth factor receptor-1 by thyroid hormone: identification of a thyroid hormone response element in the murine Fgfr1 promoter. Endocrinology.

Olsen, D.R., Leigh, S.D., Chang, R., McMullin, H., Ong, W., Tai, E., Chisholm, G., Birk, D.E., Berg, R.A., Hitzeman, R.A., Toman, P.D., 2001. Production of human type I collagen in yeast reveals unexpected new insights into the molecular assembly of collagen trimers. J. Biol. Chem. 276, 24038–24043.

Olsen, B.R., Reginato, A.M., Wang, W., 2000. Bone development. Annu. Rev. Cell Dev. Biol. 16, 191–220.

Ono, K., Kaneko, H., Choudhary, S., Pilbeam, C.C., Lorenzo, J.A., Akatsu, T., Kugai, N., Raisz, L.G., 2005. Biphasic effect of prostaglandin E2 on osteoclast formation in spleen cell cultures: role of the EP2 receptor. J. Bone Miner. Res. 20, 23–29.

Orgel, J.P., Irving, T.C., Miller, A., Wess, T.J., 2006. Microfibrillar structure of type I collagen in situ. Proc. Natl. Acad. Sci. U.S.A. 103 (24), 9001–9005. Epub 2006 Jun 2.

Ornitz, D.M., Marie, P.J., 2002. FGF signaling pathways in endochondral and intramembranous bone development and human genetic disease. Genes Dev. 16, 1446–1465.

Pace, J.M., Corrado, M., Missero, C., Byers, P.H., 2003. Identification, characterization and expression analysis of a new fibrillar collagen gene, COL27A1. Matrix Biol. 22, 3–14.

Pacifici, R., 1996. Estrogen, cytokines, and pathogenesis of postmenopausal osteoporosis. J. Bone Miner. Res. 11, 1043–1051.

Panagakos, F.S., Fernandez, C., Kumar, S., 1996. Ultrastructural analysis of mineralized matrix from human osteoblastic cells: effect of tumor necrosis factor-alpha. Mol. Cell. Biochem. 158, 81–89.

Papaioannou, I., Xu, S., Denton, C.P., Abraham, D.J., Ponticos, M., 2018. STAT3 controls COL1A2 enhancer activation cooperatively with JunB, regulates type I collagen synthesis posttranscriptionally, and is essential for lung myofibroblast differentiation. Mol. Biol. Cell 29 (2), 84–95.

Paschalis, E.P., Jacenko, O., Olsen, B., deCrombrugghe, B., Boskey, A.L., 1996. The role of type X collagen in endochondral ossification as deduced by Fourier transform infrared microscopy analysis. Connect. Tissue Res. 35, 371–377.

Pavlin, D., Bedalov, A., Kronenberg, M.S., Kream, B.E., Rowe, D.W., Smith, C.L., Pike, J.W., Lichtler, A.C., 1994. Analysis of regulatory regions in the COL1A1 gene responsible for 1,25-dihydroxyvitamin D3-mediated transcriptional repression in osteoblastic cells. J. Cell. Biochem. 56, 490–501.

Pavlin, D., Lichter, A., Bedalov, A., Kream, B., Harrison, J., Thomas, H., Gronowicz, G., Clark, S., Woody, C., Rowe, D., 1992. Differential utilization of regulatory domains within the a(1) collagen promoter in osseous and fibroblastic cells. J. Cell Biol. 116, 227–236.

Pereira, R.F., Hume, E.L., Halford, K.W., Prockop, D.J., 1995. Bone fragility in transgenic mice expressing a mutated gene for type I procollagen (COL1A1) parallels the age-dependent phenotype of human osteogenesis imperfecta. J. Bone Miner. Res. 10, 1837–1843.

Perez, J.R., Shull, S., Gendimenico, G.J., Capetola, R.J., Mezick, J.A., Cutroneo, K.R., 1992. Glucocorticoid and retinoid regulation of alpha-2 type I procollagen promoter activity. J. Cell. Biochem. 50, 26–34.

Phillips, C.L., Pfeiffer, B.J., Luger, A.M., Franklin, C.L., 2002. Novel collagen glomerulopathy in a homotrimeric type I collagen mouse (oim). Kidney Int. 62, 383–391.

Pischon, N., Darbois, L.M., Palamakumbura, A.H., Kessler, E., Trackman, P.C., 2004. Regulation of collagen deposition and lysyl oxidase by tumor necrosis factor-alpha in osteoblasts. J. Biol. Chem. 279, 30060–30065.

Poncelet, A.C., Schnaper, H.W., 2001. Sp1 and Smad proteins cooperate to mediate transforming growth factor-beta 1-induced alpha 2(I) collagen expression in human glomerular mesangial cells. J. Biol. Chem. 276, 6983–6992.

Ponticos, M., Abraham, D., Alexakis, C., Lu, Q.L., Black, C., Partridge, T., Bou-Gharios, G., 2004. Col1a2 enhancer regulates collagen activity during development and in adult tissue repair. Matrix Biol. 22, 619–628.

Ponticos, M., Partridge, T., Black, C.M., Abraham, D.J., Bou-Gharios, G., 2004. Regulation of collagen type I in vascular smooth muscle cells by competition between Nkx2.5 and deltaEF1/ZEB1. Mol. Cell Biol. 24, 6151.

Ponticos, M., Harvey, C., Ikeda, T., Abraham, D., Bou-Gharios, G., 2009. JunB mediates enhancer/promoter activity of COL1A2 following TGF-beta induction. Nucleic Acids Res. 37 (16), 5378–5389.

Ponticos, M., Papaioannou, I., Xu, S., Holmes, A.M., Khan, K., Denton, C.P., Bou-Gharios, G., Abraham, D.J., 2015. Failed degradation of JunB contributes to overproduction of type I collagen and development of dermal fibrosis in patients with systemic sclerosis. Arthritis & Rheumato. (Hoboken, N.J) 67 (1), 243–253.

Prockop, D.J., 1990. Mutations that alter the primary structure of type I collagen. The perils of a system for generating large structures by the principle of nucleated growth. J. Biol. Chem. 265, 15349–15352.

Qureshi, A.M., McGuigan, F.E., Seymour, D.G., Hutchison, J.D., Reid, D.M., Ralston, S.H., 2001. Association between COLIA1 Sp1 alleles and femoral neck geometry. Calcif. Tissue Int. 69, 67–72.

Raisz, L.G., Fall, P.M., Gabbitas, B.Y., McCarthy, T.L., Kream, B.E., Canalis, E., 1993. Effects of prostaglandin E2 on bone formation in cultured fetal rat calvariae: role of insulin-like growth factor-I. Endocrinology 133, 1504–1510.

Raisz, L.G., Fall, P.M., Petersen, D.N., Lichtler, A., Kream, B.E., 1993. Prostaglandin E2 inhibits alpha 1(I)procollagen gene transcription and promoter activity in the immortalized rat osteoblastic clonal cell line Py1a. Mol. Endocrinol. 7, 17–22.

Rajaram, S., Baylink, D.J., Mohan, S., 1997. Insulin-like growth factor-binding proteins in serum and other biological fluids: regulation and functions. Endocr. Rev. 18, 801–831.

Ralston, S.H., de Crombrugghe, B., 2006. Genetic regulation of bone mass and susceptibility to osteoporosis. Genes Dev. 20, 2492–2506.

Ralston, S.H., Uitterlinden, A.G., Brandi, M.L., Balcells, S., Langdahl, B.L., Lips, P., Lorenc, R., Obermayer-Pietsch, B., Scollen, S., Bustamante, M., Husted, L.B., Carey, A.H., Diez-Perez, A., Dunning, A.M., Falchetti, A., Karczmarewicz, E., Kruk, M., van Leeuwen, J.P., van Meurs, J.B., Mangion, J., McGuigan, F.E., Mellibovsky, L., del Monte, F., Pols, H.A., Reeve, J., Reid, D.M., Renner, W., Rivadeneira, F., van Schoor, N.M., Sherlock, R.E., Ioannidis, J.P., 2006. Large-scale evidence for the effect of the COLIA1 Sp1 polymorphism on osteoporosis outcomes: the GENOMOS study. PLoS Med. 3, e90.

Ramamurthy, P., Hocking, A.M., McQuillan, D.J., 1996. Recombinant decorin glycoforms. Purification and structure. J. Biol. Chem. 271, 19578–19584.

Ramirez, F., Tanaka, S., Bou-Gharios, G., 2006. Transcriptional regulation of the human alpha 2(I) collagen gene (COL1A2), an informative model system to study fibrotic diseases. Matrix Biol. 25, 365.

Razzaque, M.S., Lanske, B., 2007. The emerging role of the fibroblast growth factor-23-klotho axis in renal regulation of phosphate homeostasis. J. Endocrinol. 194, 1–10.

Reddy, S.V., Hundley, J.E., Windle, J.J., Alcantara, O., Linn, R., Leach, R.J., Boldt, D.H., Roodman, G.D., 1995. Characterization of the mouse tartrate-resistant acid phosphatase (TRAP) gene promoter. J. Bone Miner. Res. 10, 601–606.

Reitamo, S., Remitz, A., Tamai, K., Uitto, J., 1994. Interleukin-10 modulates type I collagen and matrix metalloprotease gene expression in cultured human skin fibroblasts. J. Clin. Investig. 94, 2489–2492.

Rice, D.P., Aberg, T., Chan, Y., Tang, Z., Kettunen, P.J., Pakarinen, L., Maxson, R.E., Thesleff, I., 2000. Integration of FGF and TWIST in calvarial bone and suture development. Development 127, 1845–1855.

Rippe, R.A., Schrum, L.W., Stefanovic, B., Solis-Herruzo, J.A., Brenner, D.A., 1999. NF-kappaB inhibits expression of the alpha1(I) collagen gene. DNA Cell Biol. 18, 751–761.

Riquet, F.B., Lai, W.F., Birkhead, J.R., Suen, L.F., Karsenty, G., Goldring, M.B., 2000. Suppression of type I collagen gene expression by prostaglandins in fibroblasts is mediated at the transcriptional level. Mol. Med. 6, 705–719.

Riquet, F.B., Tan, L., Choy, B.K., Osaki, M., Karsenty, G., Osborne, T.F., Auron, P.E., Goldring, M.B., 2001. YY1 is a positive regulator of transcription of the Col1a1 gene. J. Biol. Chem. 276, 38665–38672.

Ritzenthaler, J.D., Goldstein, R.H., Fine, A., Smith, B.D., 1993. Regulation of the alpha 1(I) collagen promoter via a transforming growth factor-beta activation element. J. Biol. Chem. 268, 13625–13631.

Robert, S., Gicquel, T., Bodin, A., Lagente, V., Boichot, E., 2016. Characterization of the MMP/TIMP imbalance and collagen production induced by IL-1β or TNF-α release from human hepatic stellate cells. PLoS One 11 (4), e0153118.

Roberts, A.B., Sporn, M.B., Assoian, R.K., Smith, J.M., Roche, N.S., Wakefield, L.M., Heine, U.I., Liotta, L.A., Falanga, V., Kehrl, J.H., et al., 1986. Transforming growth factor type beta: rapid induction of fibrosis and angiogenesis in vivo and stimulation of collagen formation in vitro. Proc. Natl. Acad. Sci. U.S.A. 83, 4167–4171.

Roberts, A.B., Tian, F., Byfield, S.D., Stuelten, C., Ooshima, A., Saika, S., Flanders, K.C., 2006. Smad3 is key to TGF-beta-mediated epithelial-to-mesenchymal transition, fibrosis, tumor suppression and metastasis. Cytokine Growth Factor Rev. 17, 19–27.

Rosati, R., Horan, G.S., Pinero, G.J., Garofalo, S., Keene, D.R., Horton, W.A., Vuorio, E., de Crombrugghe, B., Behringer, R.R., 1994. Normal long bone growth and development in type X collagen-null mice. Nat. Genet. 8, 129–135.

Rossert, J.A., Chen, S.S., Eberspaecher, H., Smith, C.N., de Crombrugghe, B., 1996. Identification of a minimal sequence of the mouse pro-alpha 1(I) collagen promoter that confers high-level osteoblast expression in transgenic mice and that binds a protein selectively present in osteoblasts. Proc. Natl. Acad. Sci. U.S.A. 93, 1027–1031.

Rossert, J., Eberspaecher, H., de Crombrugghe, B., 1995. Separate cis-acting DNA elements of the mouse pro-alpha 1(I) collagen promoter direct expression of reporter genes to different type I collagen-producing cells in transgenic mice. J. Cell Biol. 129, 1421–1432.

Rosset, E.M., Trombetta-eSilva, J., Hepfer, G., Yao, H., Bradshaw, A.D., 2017. SPARC and the N-propeptide of collagen I influence fibroblast proliferation and collagen assembly in the periodontal ligament. PLoS One 12 (2), e0173209.

Rossi, P., de Crombrugghe, B., 1987. Identification of a cell-specific transcriptional enhancer in the first intron of the mouse alpha 2 (type I) collagen gene. Proc. Natl. Acad. Sci. U.S.A. 84, 5590–5594.

Rupard, J.H., Dimari, S.J., Damjanov, I., Haralson, M.A., 1988. Synthesis of type I homotrimer collagen molecules by cultured human lung adenocarcinoma cells. Am. J. Pathol. 133, 316–326.

Sabbieti, M.G., Marchetti, L., Abreu, C., Montero, A., Hand, A.R., Raisz, L.G., Hurley, M.M., 1999. Prostaglandins regulate the expression of fibroblast growth factor-2 in bone. Endocrinology 140, 434–444.

Saitta, B., Gaidarova, S., Cicchillitti, L., Jimenez, S.A., 2000. CCAAT binding transcription factor binds and regulates human COL1A1 promoter activity in human dermal fibroblasts: demonstration of increased binding in systemic sclerosis fibroblasts. Arthritis Rheum. 43, 2219–2229.

Sakuma, Y., Li, Z., Pilbeam, C.C., Alander, C.B., Chikazu, D., Kawaguchi, H., Raisz, L.G., 2004. Stimulation of cAMP production and cyclooxygenase-2 by prostaglandin E(2) and selective prostaglandin receptor agonists in murine osteoblastic cells. Bone 34, 827–834.

Salimi-Tari, P., Cheung, M., Safar, C.A., Tracy, J.T., Tran, I., Harbers, K., Breindl, M., 1997. Molecular cloning and chromatin structure analysis of the murine alpha1(I) collagen gene domain. Gene 198, 61–72.

Sanderson, N., Factor, V., Nagy, P., Kopp, J., Kondaiah, P., Wakefield, L., Roberts, A.B., Sporn, M.B., Thorgeirsson, S.S., 1995. Hepatic expression of mature transforming growth factor beta 1 in transgenic mice results in multiple tissue lesions. Proc. Natl. Acad. Sci. U.S.A. 92, 2572–2576.

Santos, S.A.A.D., Porto Amorim, E.M., Ribeiro, L.M., Rinaldi, J.C., Delella, F.K., Justulin, L.A., Felisbino, S.L., 2017. Hyperglycemic condition during puberty increases collagen fibers deposition in the prostatic stroma and reduces MMP-2 activity. Biochem. Biophys. Res. Commun. 493 (4), 1581–1586.

Schedlich, L.J., Muthukaruppan, A., O'Han, M.K., Baxter, R.C., 2007. Insulin-like growth factor binding protein-5 interacts with the vitamin D receptor and modulates the vitamin D response in osteoblasts. Mol. Endocrinol. 21, 2378–2390.

Schier, A.F., Shen, M.M., 2000. Nodal signalling in vertebrate development. Nature 403, 385–389.

Schmid, C., Guler, H.P., Rowe, D., Froesch, E.R., 1989. Insulin-like growth factor I regulates type I procollagen messenger ribonucleic acid steady state levels in bone of rats. Endocrinology 125, 1575–1580.

Schultz, H.S., Guo, L., Keller, P., Fleetwood, A.J., Sun, M., Guo, W., Ma, C., Hamilton, J.A., Bjørkdahl, O., Berchtold, M.W., Panina, S., 2016. OSCAR-collagen signaling in monocytes plays a proinflammatory role and may contribute to the pathogenesis of rheumatoid arthritis. Eur. J. Immunol. 46 (4), 952–963.

Schwarz, M., Harbers, K., Kratochwil, K., 1990. Transcription of a mutant collagen I gene is a cell type and stage-specific marker for odontoblast and osteoblast differentiation. Development 108, 717–726.

Scott, P.G., Grossmann, J.G., Dodd, C.M., Sheehan, J.K., Bishop, P.N., 2003. Light and X-ray scattering show decorin to be a dimer in solution. J. Biol. Chem. 278, 18353–18359.

Segarini, P.R., Nesbitt, J.E., Li, D., Hays, L.G., Yates III, J.R., Carmichael, D.F., 2001. The low density lipoprotein receptor-related protein/alpha2-macroglobulin receptor is a receptor for connective tissue growth factor. J. Biol. Chem. 276, 40659–40667.

Sengupta, P.K., Fargo, J., Smith, B.D., 2002. The RFX family interacts at the collagen (COL1A2) start site and represses transcription. J. Biol. Chem. 277, 24926–24937.

Sengupta, P., Xu, Y., Wang, L., Widom, R., Smith, B.D., 2005. Collagen alpha1(I) gene (COL1A1) is repressed by RFX family. J. Biol. Chem. 280 (22), 21004–21014.

Serpier, H., Gillery, P., Salmon-Ehr, V., Garnotel, R., Georges, N., Kalis, B., Maquart, F.X., 1997. Antagonistic effects of interferon-gamma and interleukin-4 on fibroblast cultures. J. Investig. Dermatol. 109, 158–162.

Seyedin, S.M., Thomas, T.C., Thompson, A.Y., Rosen, D.M., Piez, K.A., 1985. Purification and characterization of two cartilage-inducing factors from bovine demineralized bone. Proc. Natl. Acad. Sci. U.S.A. 82, 2267–2271.

Sher, L.B., Woitge, H.W., Adams, D.J., Gronowicz, G.A., Krozowski, Z., Harrison, J.R., Kream, B., 2004. Transgenic expression of 11beta-hydroxysteroid dehydrogenase type 2 in osteoblasts reveals an anabolic role for endogenous glucocorticoids in bone. Endocrinology 145, 922–929.

Shi-wen, X., Pennington, D., Holmes, A., Leask, A., Bradham, D., Beauchamp, J.R., Fonseca, C., du Bois, R.M., Martin, G.R., Black, C.M., Abraham, D.J., 2000. Autocrine overexpression of CTGF maintains fibrosis: RDA analysis of fibrosis genes in systemic sclerosis. Exp. Cell Res. 259, 213–224.

Shi-wen, X., Stanton, L.A., Kennedy, L., Pala, D., Chen, Y., Howat, S.L., Renzoni, E.A., Carter, D.E., Bou-Gharios, G., Stratton, R.J., Pearson, J.D., Beier, F., Lyons, K.M., Black, C.M., Abraham, D.J., Leask, A., 2006. CCN2 is necessary for adhesive responses to transforming growth factor-beta1 in embryonic fibroblasts. J. Biol. Chem. 281, 10715–10726.

Sinha, S., Maity, S.N., Lu, J., de Crombrugghe, B., 1995. Recombinant rat CBF-C, the third subunit of CBF/NFY, allows formation of a protein-DNA complex with CBF-A and CBF-B and with yeast HAP2 and HAP3. Proc. Natl. Acad. Sci. U.S.A. 92, 1624–1628.

Slack, J.L., Liska, D.J., Bornstein, P., 1991. An upstream regulatory region mediates high-level, tissue-specific expression of the human alpha 1(I) collagen gene in transgenic mice. Mol. Cell Biol. 11, 2066–2074.

Slack, J.L., Parker, M.I., Bornstein, P., 1995. Transcriptional repression of the alpha 1(I) collagen gene by ras is mediated in part by an intronic AP1 site. J. Cell. Biochem. 58, 380–392.

Sokolov, B.P., Mays, P.K., Khillan, J.S., Prockop, D.J., 1993. Tissue- and development-specific expression in transgenic mice of a type I procollagen (COL1A1) minigene construct with 2.3 kb of the promoter region and 2 kb of the 39-flanking region. Specificity is independent of the putative regulatory sequences in the first intron. Biochemistry 32, 9242–9249.

Solis-Herruzo, J.A., Brenner, D.A., Chojkier, M., 1988. Tumor necrosis factor alpha inhibits collagen gene transcription and collagen synthesis in cultured human fibroblasts. J. Biol. Chem. 263, 5841–5845.

Sono, T., Akiyama, H., Miura, S., Deng, J.M., Shukunami, C., Hiraki, Y., Tsushima, Y., Azuma, Y., Behringer, R.R., Matsuda, S., 2018. THRAP3 interacts with and inhibits the transcriptional activity of SOX9 during chondrogenesis. J. Bone Miner. Metab. 36 (4), 410–419.

Sprangers, S., Everts, V., 2019. Molecular pathways of cell-mediated degradation of fibrillar collagen. Matrix Biol. 75-76, 190–200. https://doi.org/10.1016/j.matbio.2017.11.008.

Stefanovic, B., Brenner, D.A., 2003. 5' stem-loop of collagen alpha 1(I) mRNA inhibits translation in vitro but is required for triple helical collagen synthesis in vivo. J. Biol. Chem. 278, 927–933.

Stefanovic, B., Hellerbrand, C., Holcik, M., Briendl, M., Aliebhaber, S., Brenner, D.A., 1997. Posttranscriptional regulation of collagen alpha1(I) mRNA in hepatic stellate cells. Mol. Cell Biol. 17, 5201–5209.

Sterner-Kock, A., Thorey, I.S., Koli, K., Wempe, F., Otte, J., Bangsow, T., Kuhlmeier, K., Kirchner, T., Jin, S., Keski-Oja, J., von Melchner, H., 2002. Disruption of the gene encoding the latent transforming growth factor-beta binding protein 4 (LTBP-4) causes abnormal lung development, cardiomyopathy, and colorectal cancer. Genes Dev. 16, 2264–2273.

Stewart, T.L., Jin, H., McGuigan, F.E., Albagha, O.M., Garcia-Giralt, N., Bassiti, A., Grinberg, D., Balcells, S., Reid, D.M., Ralston, S.H., 2006. Haplotypes defined by promoter and intron 1 polymorphisms of the COLIA1 gene regulate bone mineral density in women. J. Clin. Endocrinol. Metab. 91, 3575–3583.

Stewart, T.L., Roschger, P., Misof, B.M., Mann, V., Fratzl, P., Klaushofer, K., Aspden, R., Ralston, S.H., 2005. Association of COLIA1 Sp1 alleles with defective bone nodule formation in vitro and abnormal bone mineralization in vivo. Calcif. Tissue Int. 77, 113–118.

Stokes, F.J., Ivanov, P., Bailey, L.M., Fraser, W.D., 2011. The effects of sampling procedures and storage conditions on short-term stability of blood-based biochemical markers of bone metabolism. Clin. Chem. 57 (1), 138–140.

Stratton, R., Rajkumar, V., Ponticos, M., Nichols, B., Shiwen, X., Black, C.M., Abraham, D.J., Leask, A., 2002. Prostacyclin derivatives prevent the fibrotic response to TGF-beta by inhibiting the Ras/MEK/ERK pathway. FASEB J. 16, 1949–1951.

Stylianou, E., Saklatvala, J., 1998. Interleukin-1. Int. J. Biochem. Cell Biol. 30, 1075–1079.

Sullivan, R., Klagsbrun, M., 1985. Purification of cartilage-derived growth factor by heparin affinity chromatography. J. Biol. Chem. 260, 2399–2403.

Sweeney, S.M., Orgel, J.P., Fertala, A., McAuliffe, J.D., Turner, K.R., Di Lullo, G.A., Chen, S., Antipova, O., Perumal, S., Ala-Kokko, L., Forlino, A., Cabral, W.A., Barnes, A.M., Marini, J.C., San Antonio, J.D., 2008. Candidate cell and matrix interaction domains on the collagen fibril, the predominant protein of vertebrates. J. Biol. Chem. 283 (30), 21187–21197. https://doi.org/10.1074/jbc.M709319200.

Taddese, S., Jung, M.C., Ihling, C., Heinz, A., Neubert, R.H., Schmelzer, C.E., 2010. MMP-12 catalytic domain recognizes and cleaves at multiple sites in human skin collagen type I and type III. Biochim. Biophys. Acta. 1804 (4), 731–739.

Takigawa, M., Nakanishi, T., Kubota, S., Nishida, T., 2003. Role of CTGF/HCS24/ecogenin in skeletal growth control. J. Cell. Physiol. 194, 256–266.

Tam, E.M., Moore, T.R., Butler, G.S., Overall, C.M., 2004. Characterization of the distinct collagen binding, helicase and cleavage mechanisms of matrix metalloproteinase 2 and 14 (gelatinase A and MT1-MMP): the differential roles of the MMP hemopexin c domains and the MMP-2 fibronectin type II modules in collagen triple helicase activities. J. Biol. Chem. 279 (41), 43336–43344.

Tamaki, T., Ohnishi, K., Hartl, C., LeRoy, E.C., Trojanowska, M., 1995. Characterization of a GC-rich region containing Sp1 binding site(s) as a constitutive responsive element of the alpha 2(I) collagen gene in human fibroblasts. J. Biol. Chem. 270, 4299–4304.

Tan, K.M., Saw, S., Sethi, S.K., 2013. Vitamin D and its relationship with markers of bone metabolism in healthy Asian women. J. Clin. Lab. Anal. 27 (4), 301–304.

Tanaka, S., Antoniv, T.T., Liu, K., Wang, L., Wells, D.J., Ramirez, F., Bou-Gharios, G., 2004. Cooperativity between far upstream enhancer and proximal promoter elements of the human {alpha}2(I) collagen (COL1A2) gene instructs tissue specificity in transgenic mice. J. Biol. Chem. 279, 56024–56031.

Tang, K.T., Capparelli, C., Stein, J.L., Stein, G.S., Lian, J.B., Huber, A.C., Braverman, L.E., DeVito, W.J., 1996. Acidic fibroblast growth factor inhibits osteoblast differentiation in vitro: altered expression of collagenase, cell growth-related, and mineralization-associated genes. J. Cell. Biochem. 61, 152–166.

Tasab, M., Batten, M.R., Bulleid, N.J., 2000. Hsp47: a molecular chaperone that interacts with and stabilizes correctly-folded procollagen. EMBO J. 19, 2204–2211.

Terraz, C., Brideau, G., Ronco, P., Rossert, J., 2002. A combination of cis-acting elements is required to activate the pro-alpha 1(I) collagen promoter in tendon fibroblasts of transgenic mice. J. Biol. Chem. 277, 19019–19026.

Thampatty, B.P., Li, H., Im, H.J., Wang, J.H., 2007. EP4 receptor regulates collagen type-I, MMP-1, and MMP-3 gene expression in human tendon fibroblasts in response to IL-1 beta treatment. Gene 386, 154–161.

Thiebaud, D., Guenther, H.L., Porret, A., Burckhardt, P., Fleisch, H., Hofstetter, W., 1994. Regulation of collagen type I and biglycan mRNA levels by hormones and growth factors in normal and immortalized osteoblastic cell lines. J. Bone Miner. Res. 9, 1347–1354.

Thiele, B.J., Doller, A., Kahne, T., Pregla, R., Hetzer, R., Regitz-Zagrosek, V., 2004. RNA-binding proteins heterogeneous nuclear ribonucleoprotein A1, E1, and K are involved in post-transcriptional control of collagen I and III synthesis. Circ. Res. 95, 1058–1066.

Todorovic, V., Jurukovski, V., Chen, Y., Fontana, L., Dabovic, B., Rifkin, D.B., 2005. Latent TGF-beta binding proteins. Int. J. Biochem. Cell Biol. 37, 38–41.

Uitterlinden, A.G., Burger, H., Huang, Q., Yue, F., McGuigan, F.E., Grant, S.F., Hofman, A., van Leeuwen, J.P., Pols, H.A., Ralston, S.H., 1998. Relation of alleles of the collagen type Ialpha1 gene to bone density and the risk of osteoporotic fractures in postmenopausal women. N. Engl. J. Med. 338, 1016–1021.

Uitto, J., 1979. Collagen polymorphism: isolation and partial characterization of alpha 1(I)-trimer molecules in normal human skin. Arch. Biochem. Biophys. 192, 371–379.

Ulloa, L., Doody, J., Massague, J., 1999. Inhibition of transforming growth factor-beta/SMAD signalling by the interferon-gamma/STAT pathway. Nature 397, 710–713.

van der Rest, M., Garrone, R., 1991. Collagen family of proteins. FASEB J. 5, 2814–2823.

Van Doren, S.R., 2015. Matrix metalloproteinase interactions with collagen and elastin. Matrix Biol. 44-46, 224–231. https://doi.org/10.1016/j.matbio.2015.01.005.

van Driel, M., Pols, H.A., van Leeuwen, J.P., 2004. Osteoblast differentiation and control by vitamin D and vitamin D metabolites. Curr. Pharmaceut. Des. 10, 2535–2555.

Vanleene, et al., 2012. Bone 50, 1317–1323.

Varghese, S., Ramsby, M.L., Jeffrey, J.J., Canalis, E., 1995. Basic fibroblast growth factor stimulates expression of interstitial collagenase and inhibitors of metalloproteinases in rat bone cells. Endocrinology 136, 2156–2162.

Varghese, S., Rydziel, S., Canalis, E., 2000. Basic fibroblast growth factor stimulates collagenase-3 promoter activity in osteoblasts through an activator protein-1-binding site. Endocrinology 141, 2185–2191.

Varma, S., Orgel, J.P., Schieber, J.D., 2016. Nanomechanics of type I collagen. Biophys J. 111 (1), 50–56. https://doi.org/10.1016/j.bpj.2016.05.038.

Vasikaran, S., Eastell, R., Bruyere, O., Foldes, A.J., Garnero, P., Griesmacher, A., et al., 2011. Markers of bone turnover for the prediction of fracture risk and monitoring of osteoporosis treatment: a need for international reference standards. Osteoporos. Int. 22 (2), 391–420.

Verrecchia, F., Mauviel, A., 2007. Transforming growth factor-beta and fibrosis. World J. Gastroenterol. 13, 3056–3062.

Verrecchia, F., Pessah, M., Atfi, A., Mauviel, A., 2000. Tumor necrosis factor-alpha inhibits transforming growth factor-beta/Smad signaling in human dermal fibroblasts via AP-1 activation. J. Biol. Chem. 275, 30226–30231.

Verrecchia, F., Vindevoghel, L., Lechleider, R.J., Uitto, J., Roberts, A.B., Mauviel, A., 2001. Smad3/AP-1 interactions control transcriptional responses to TGF-beta in a promoter-specific manner. Oncogene 20, 3332–3340.

Verrecchia, F., Wagner, E.F., Mauviel, A., 2002. Distinct involvement of the Jun-N-terminal kinase and NF-kappaB pathways in the repression of the human COL1A2 gene by TNF-alpha. EMBO Rep. 3, 1069–1074.

Vogel, K.G., Trotter, J.A., 1987. The effect of proteoglycans on the morphology of collagen fibrils formed in vitro. Collagen Relat. Res. 7, 105–114.

Vuorio, E., de Crombrugghe, B., 1990. The family of collagen genes. Annu. Rev. Biochem. 59, 837–872.

Waddington, R.J., Roberts, H.C., Sugars, R.V., Schonherr, E., 2003. Differential roles for small leucine-rich proteoglycans in bone formation. Eur. Cells Mater. 6, 12–21 discussion 21.

Wang, L., Tanaka, S., Ramirez, F., 2005. GATA-4 binds to an upstream element of the human alpha2(I) collagen gene (COL1A2) and inhibits transcription in fibroblasts. Matrix Biol. 24, 333–340.

Weber, I.T., Harrison, R.W., Iozzo, R.V., 1996. Model structure of decorin and implications for collagen fibrillogenesis. J. Biol. Chem. 271, 31767–31770.

Weinstein, R.S., Jilka, R.L., Parfitt, A.M., Manolagas, S.C., 1998. Inhibition of osteoblastogenesis and promotion of apoptosis of osteoblasts and osteocytes by glucocorticoids. Potential mechanisms of their deleterious effects on bone. J. Clin. Investig. 102, 274–282.

Weston, B.S., Wahab, N.A., Mason, R.M., 2003. CTGF mediates TGF-beta-induced fibronectin matrix deposition by upregulating active alpha5beta1 integrin in human mesangial cells. J. Am. Soc. Nephrol. 14, 601–610.

Wiestner, M., Krieg, T., Horlein, D., Glanville, R.W., Fietzek, P., Muller, P.K., 1979. Inhibiting effect of procollagen peptides on collagen biosynthesis in fibroblast cultures. J. Biol. Chem. 254, 7016–7023.

Wilson, et al., 2011. J Cell Biol. 194 (2), 347.

Woitge, H.W., Kream, B.E., 2000. Calvariae from fetal mice with a disrupted Igf1 gene have reduced rates of collagen synthesis but maintain responsiveness to glucocorticoids. J. Bone Miner. Res. 15, 1956–1964.

Wu, et al., 2014. Genome Biol 15, R52.

Xu, T., Bianco, P., Fisher, L.W., Longenecker, G., Smith, E., Goldstein, S., Bonadio, J., Boskey, A., Heegaard, A.M., Sommer, B., Satomura, K., Dominguez, P., Zhao, C., Kulkarni, A.B., Robey, P.G., Young, M.F., 1998. Targeted disruption of the biglycan gene leads to an osteoporosis-like phenotype in mice. Nat. Genet. 20, 78–82.

Xu, Y., Wang, L., Buttice, G., Sengupta, P.K., Smith, B.D., 2003. Interferon gamma repression of collagen (COL1A2) transcription is mediated by the RFX5 complex. J. Biol. Chem. 278, 49134–49144.

Xu, Y., Wang, L., Buttice, G., Sengupta, P.K., Smith, B.D., 2004. Major histocompatibility class II transactivator (CIITA) mediates repression of collagen (COL1A2) transcription by interferon gamma (IFN-gamma). J. Biol. Chem. 279, 41319–41332.

Yañez-Mó, M., Barreiro, O., Gonzalo, P., Batista, A., Megías, D., Genís, L., Sachs, N., Sala-Valdés, M., Alonso, M.A., Montoya, M.C., Sonnenberg, A., Arroyo, A.G., Sánchez-Madrid, F., 2008. MT1-MMP collagenolytic activity is regulated through association with tetraspanin CD151 in primary endothelial cells. Blood 112 (8), 3217–3226.

Yata, Y., Scanga, A., Gillan, A., Yang, L., Reif, S., Breindl, M., Brenner, D.A., Rippe, R.A., 2003. DNase I-hypersensitive sites enhance alpha1(I) collagen gene expression in hepatic stellate cells. Hepatology 37, 267–276.

Yuan, W., Yufit, T., Li, L., Mori, Y., Chen, S.J., Varga, J., 1999. Negative modulation of alpha1(I) procollagen gene expression in human skin fibroblasts: transcriptional inhibition by interferon-gamma. J. Cell. Physiol. 179, 97–108.

Zhang, W., Ou, J., Inagaki, Y., Greenwel, P., Ramirez, F., 2000. Synergistic cooperation between Sp1 and Smad3/Smad4 mediates transforming growth factor beta1 stimulation of alpha 2(I)-collagen (COL1A2) transcription. J. Biol. Chem. 275, 39237–39245.

Zhu, et al., 2012. J clin Immunol 32, 514–522.

Zigrino, P., Brinckmann, J., Niehoff, A., Lu, Y., Giebeler, N., Eckes, B., Kadler, K.E., Mauch, C., 2016. Fibroblast-derived MMP-14 regulates collagen homeostasis in adult skin. J. Invest. Dermatol. 136 (8), 1575–1583. https://doi.org/10.1016/j.jid.2016.03.036.

Chapter 14

Collagen cross-linking and bone pathobiology

David M. Hudson, MaryAnn Weis and David R. Eyre

Department of Orthopaedics and Sports Medicine, University of Washington, Seattle, WA, United States

Chapter outline

Introduction 339
Advances in collagen cross-link analysis 340
Mature cross-link analysis 340
Divalent cross-link analysis 340
Electrospray mass spectrometry 340
Cross-link formation 340
Bone collagen cross-linking 340
Cross-link structures 342
Divalent cross-links 342
Pyridinium cross-links 343
Pyrrole cross-links 343
Pyridinoline and pyrrolic cross-linked peptides in urine 344
Histidine-containing collagen cross-links and other maturation products 344
Glycosylations and glycations 345
Enzymatic glycosylation 345
Nonspecific glycations 345
Advanced glycation end products 346
Potential consequences 347
Cross-linking lysine-modifying enzymes 347
Lysyl hydroxylases 347
Consequences of lysyl hydroxylase gene mutations 347
Lysyl oxidases 348
Heritable disorders and mouse models 348
Heritable disorders 348
Collagen posttranslational modifications 348
CRTAP, LEPRE1, PPIB 348
TMEM38B 350
PLOD2 and FKBP10 350
SC65 and P3H3 350
MBTPS2 351
Collagen processing 351
Bone morphogenetic protein 1 351
Collagen chaperone 351
SERPINH1 351
Bone mineralization 352
IFITM5 352
SERPINF1 352
Implications for bone fragility and mineral deposition 352
Future challenges 353
References 353

Introduction

Lysyl oxidase (LOX)-mediated covalent cross-links between individual collagen molecules are essential for the strength of collagen fibrils. They appeared at the dawn of metazoan evolution (Rodriguez-Pascual and Slatter, 2016) and can be argued to have been a critical step in allowing larger animals to evolve (Boot-Handford and Tuckwell, 2003). Divergencies in the content, placement, and chemical stability of these intermolecular cross-links have clearly evolved to modulate the properties of collagen fibrils in different tissue types.

Collagen accounts for about a third of all protein in the human body (Verzár, 1964) and is the main organic component of bone (Myllyharju and Kivirikko, 2004; Eyre, 1980). Fibrillar collagen molecules consist of three polypeptide α-chains, each about 1000 residues of a single repeating Gly−Xaa−Yaa primary amino acid sequence, which folds into the defining triple-helical conformation of a collagen molecule. Variations in the extensive posttranslational modifications of the major fibril-forming collagen type I contribute to the structural and functional differences between tissues (Eyre et al., 1984a,b;

Principles of Bone Biology. https://doi.org/10.1016/B978-0-12-814841-9.00014-2

Hudson and Eyre, 2013). Indeed, cross-linking chemistry differs fundamentally between skin, tendon, and bone type I collagen, with distinct changes continuing during tissue growth and maturation (Light and Bailey, 1979).

LOXs initiate fibrillar collagen cross-linking, in the sole enzymatic step, by forming aldehyde side-chains from specific telopeptide lysine and hydroxylysine residues. From then on, further reactions are driven by the local protein environment at the cross-linking sites. All the fibril-forming collagens (types I, II, III, V, and XI) can have up to four cross-linking loci per molecule, one per telopeptide domain and two placed symmetrically at opposite ends of the triple helix. Certain chains (α2(I), α1(V), α2(V), α1(XI) and α2(XI)) lack C-telopeptide cross-linking lysines. Upon collagen fibril assembly, the triple-helical cross-linking lysines align and spontaneously react with telopeptide lysine and hydroxylysine aldehydes in adjacent molecules staggered axially by 4D periods (where D = 67nm, the axial stagger of adjacent collagen molecules packed in a fibril). The mature cross-linking chemistry in bone collagen is distinctive, with products containing both lysine and hydroxylysine aldehyde precursors (Fig. 14.1).

Advances in collagen cross-link analysis

Mature cross-link analysis

The content of pyridinoline cross-links, readily quantified by their natural fluorescence, has been used as a gold standard for assessing collagen cross-linking quality for decades (Eyre et al., 1984a). Specifically, high-performance liquid chromatography-based assays were developed to measure the quantitative distribution of hydroxylysyl pyridinoline (HP) and lysyl pyridinoline (LP) across tissues and collagen genetic types (Eyre et al., 1984b; Wu and Eyre, 1985; Fujimoto, 1977). Variances in the total amount and/or HP/LP ratio of these trifunctional residues have been one of the traditional measures of collagen quality and pathobiology (Eyre et al., 1984a,b). Although informative, the content of pyridinolines is only a fraction of the total LOX-mediated cross-links in most tissues, including bone.

Even mature bone contains more divalent cross-links and more pyrrolic cross-links than total pyridinolines. Trivalent pyrrole cross-links are measured in collagenous tissues using a colorimetric assay. Ehrlich's reagent (*p*-dimethylaminobenzaldehyde) reacts with pyrrole residues to yield a quantifiable pink color (Scott et al., 1981). Pyrrole cross-links appear to be limited mostly to fibrillar collagens of bone and load-bearing tendons (Hanson and Eyre, 1996; Horgan et al., 1990).

Divalent cross-link analysis

Divalent cross-linking amino acids were originally profiled in tissues as tritiated products after sodium borotritide reduction on the amino acid analyzer, with quantitation relative to nonreducible cross-links by ninhydrin detection (Eyre, 1987; Avery et al., 2009). Mass spectrometric approaches have begun to supplement the traditional methods of amino acid analysis and N-terminal sequencing for studying divalent collagen cross-links (Hudson et al., 2017; Eyre et al., 2004; Kalamajski et al., 2014; Terajima et al., 2014; Naffa et al., 2016).

Electrospray mass spectrometry

Early mass spectrometric methods were used to structurally characterize collagen cross-links beginning in the 1960s (Blumenfeld and Gallop, 1966; Schneider et al., 1967). However, using ion-trap mass spectrometry, collagen cross-linked peptides could be identified by their peptide fragmentation patterns (Fig. 14.2) (Eyre et al., 2004). Collagen peptides can be generated for mass spectrometric analysis using several approaches, including sodium dodecyl sulfate–polyacrylamide gel electrophoresis of whole chains and CNBr peptides and in-gel protease digestion (e.g., trypsin, endo-Asp) or bacterial collagenase digestion followed by liquid chromatography, or directly from urine after Sep-Pak enrichment (Eyre et al., 2008). By using sequence-specific antibodies for detection followed by mass spectrometric analysis, heterotypic cross-links between α-chains of different collagen types were identified (Eyre et al., 2008).

Cross-link formation

Bone collagen cross-linking

The distinctive profile of mature cross-links in bone collagen is controlled by the partial hydroxylation of lysine residues at both telopeptide and triple-helical cross-linking sites (Fig. 14.3). Telopeptide lysines at both ends are about 50%

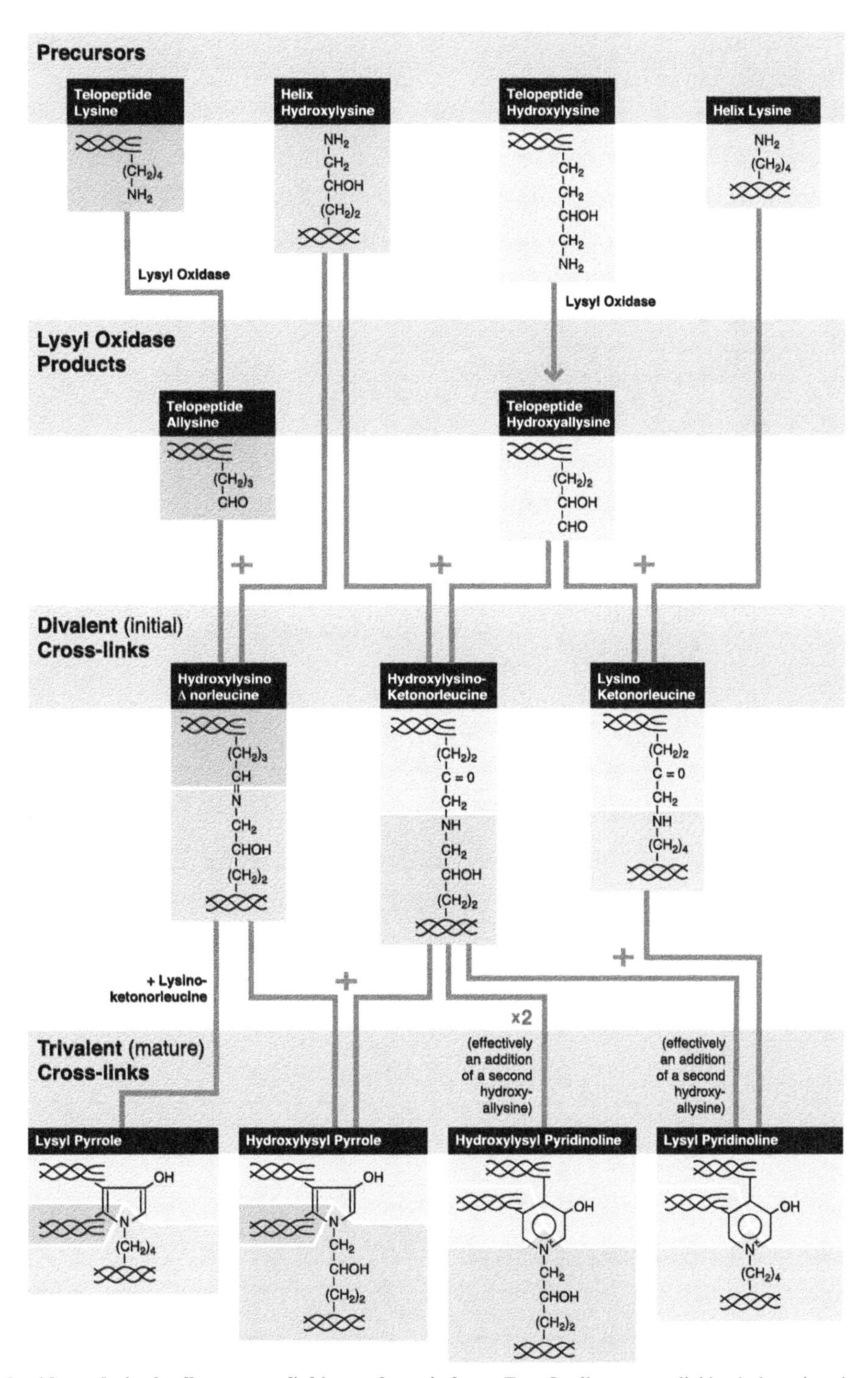

FIGURE 14.1 **Lysyl oxidase–derived collagen cross-linking pathway in bone.** Type I collagen cross-linking in bone is unique in using both lysine (allysine) and hydroxylysine (hydroxyallysine) aldehyde precursors in the telopeptide domains. In growing fibrils, the aldehydes interact with specific lysine or hydroxylysine residues at helical sites (K87 and K930) in neighboring collagen molecules. The resulting divalent cross-links are the predominant cross-links found in bone. These initial divalent cross-links can add an additional telopeptide aldehyde (from a second divalent cross-link) to form trivalent (mature) cross-links. In adult human bone, the trivalent cross-links are a mixture of about equal amounts of pyridinolines and pyrroles.

hydroxylated (human, bovine) compared with 0% in skin type I collagen and 100% in cartilage type II collagen (Eyre et al., 2008). This results in a unique pattern of both divalent and mature, trivalent cross-linked peptides from bone collagen including a mixture of mature pyrrole and pyridinoline structures (Hanson and Eyre, 1996). The formation of the trifunctional cross-links requires the precise lateral packing of collagen molecules in the native staggered structure that

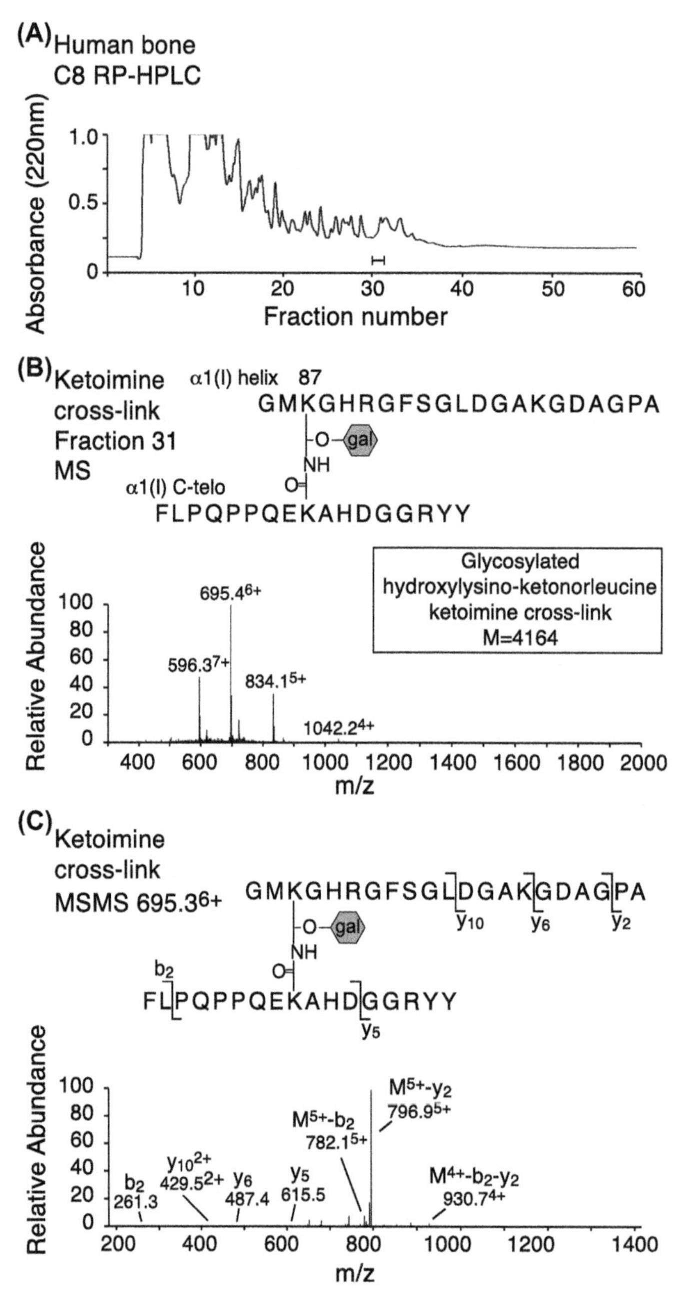

FIGURE 14.2 The identification of a divalent cross-linked peptide from bone collagen (Lys-87 to C-telopeptide) using mass spectrometry. (A) Reverse-phase high-performance liquid chromatography resolution of cross-linked peptides from human bone collagen digested with bacterial collagenase. (B) Mass spectrometry (*MS*) profile of fraction 31 reveals a glycosylated (galactosyl-) ketoimine structure from human bone (596.3^{7+}, 695.4^{6+}, 834.1^{5+}, and 1042.2^{4+}). (C) Tandem MS fragmentation spectrum of the parent ion (695.3^{6+}) confirms the mass of the glycosylated ketoimine cross-link.

provides the appropriate nearest-neighbor intermolecular relationships for reactions to occur. The well-characterized chemical pathway of cross-linking interactions in bone collagen is shown in Fig. 14.1.

Cross-link structures

Divalent cross-links

The predominant cross-links in mature bone collagen are divalent residues (Eyre et al., 1984a; Eyre, 1981, 1987). With increasing age the mean sum of divalent cross-links (dihydroxylysinonorleucine [DHLNL] and hydroxylysinonorleucine

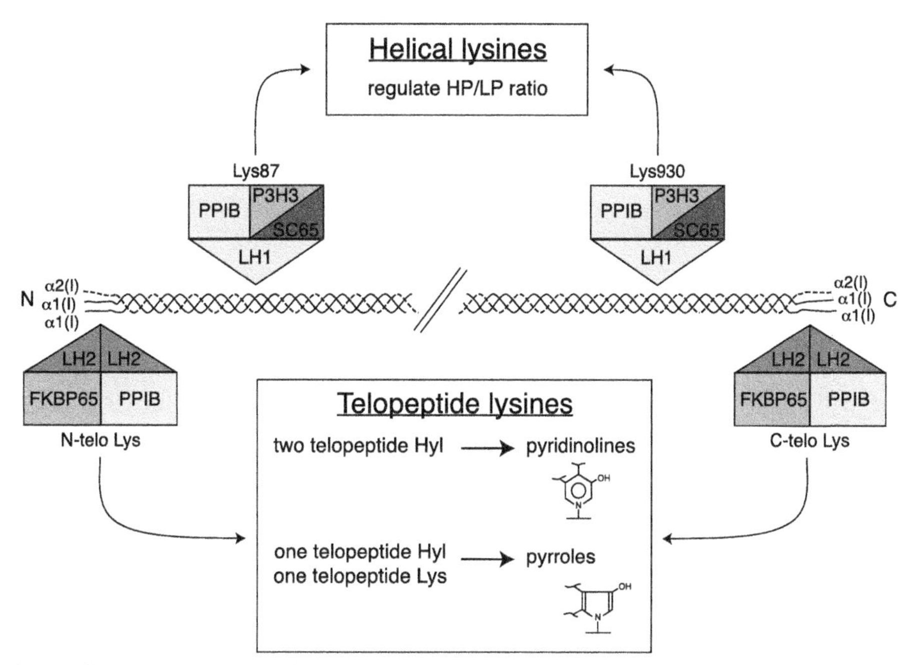

FIGURE 14.3 Regulation of collagen cross-linking in bone by the lysyl hydroxylase (*LH*) complexes. The four lysine sites of cross-linking are symmetrically located in type I collagen (two telopeptide sites and two triple-helical sites). The LH1 complex (LH1, PPIB, SC65, P3H3) catalyzes the hydroxylation of helical lysines 87 and 930. The LH2 complex (LH2 dimer, PPIB, FKBP65) catalyzes the hydroxylation of N- and C-telopeptide lysines. The pattern of lysyl hydroxylation determines the chemical nature of the cross-links, which varies between tissue types. In bone collagen, LH2 partially hydroxylates the telopeptide lysines and LH1 partially hydroxylates the two helical-site lysines of both α-chains. This pattern is a distinguishing post-translational feature of bone type I collagen compared with type I collagen of skin, tendon, and other soft tissues. *FKBP65*, 65-kDa FK506-binding protein; *HP*, hydroxylysyl pyridinoline; *Hyl*, hydroxylysine; *LP*, lysyl pyridinoline; *P3H3*, prolyl 3-hydroxylase 3; *PPIB*, cyclophilin B (peptidyl-prolyl isomerase B); *SC65*, synaptonemal complex 65.

[HLNL]) decreases from 1.9 mol/mol in young bone (13 years) to 0.8 mol/mol in adult bone (41 years) (Eyre, 1987). This decrease coincides with cross-link maturation from bifunctional to trifunctional structures. It is important to note that the physiological bone cross-links, such as dehydro-DHLNL and dehydro-HLNL, and their respective ketoamine forms are reduced to stable secondary amines, DHLNL and HLNL, upon borohydride treatment, allowing their quantitation after acid hydrolysis. Several studies have applied mass spectrometry to further investigate the divalent cross-links in normal tissue (Terajima et al., 2014) and in mouse models that alter tendon (Kalamajski et al., 2014), skin and bone cross-linking chemistry (Hudson et al., 2017).

Pyridinium cross-links

Telopeptide hydroxylysines are a prerequisite for pyridinoline formation (Bank et al., 1999), and the degree of helical cross-linking lysine hydroxylation determines the ratio of HP to LP (Eyre and Wu, 2005). HP and LP each show molecular site and α-chain selectivity in their concentrations (Hanson and Eyre, 1996). Their levels in urine relative to creatinine offer a convenient measure of bone resorption (Beardsworth et al., 1990). Even more specific to osteoclastic bone resorption are the NTx and CTx urine and serum immunoassays directed at short type I collagen telopeptide fragments that were protected by the cross-links through to excretion (Hanson et al., 1992; Singer and Eyre, 2008).

Pyrrole cross-links

Pyrroles are the product of an interaction between a telopeptide lysine aldehyde, a telopeptide hydroxylysine aldehyde, and a triple-helical-domain lysine or hydroxylysine. We believe pyrrole formation involves an interaction between an initial ketoimine and an aldimine divalent cross-link, analogous to that proposed between two divalent cross-links to form HP and LP (Eyre and Oguchi, 1980). This mechanism produces the trivalent structure and releases one triple-helical domain donor sequence. The highest pyrrole cross-link levels will occur when the telopeptide lysines are 50% hydroxylated as they roughly are in normal human bone collagen (Hanson and Eyre, 1996). There is evidence that the content of pyrrole cross-links is positively associated with bone strength (Knott and Bailey, 1998), perhaps by aiding in the assembly of the highly ordered mineral nanocrystal/protein composite of normal bone. Thus the ratio of pyrrole/pyridinoline cross-links has been shown to be associated with bone quality, with a lower pyrrole content and altered trabecular organization in osteoporotic bone compared with control bone (Banse et al., 2002).

Pyridinoline and pyrrolic cross-linked peptides in urine

The 3-hydroxy pyrrole cross-linking structures, unlike pyridinolines, are labile to acid hydrolysis and so cannot be measured in the same tissue acid hydrolysates as pyridinolines. They can, however, be recovered in cross-linked peptides from mild, proteolytic digests of bone matrix (Hanson and Eyre, 1996) and also in cross-linked peptides excreted in urine as products of osteoclastic bone resorption along with their pyridinoline counterparts.

Fig. 14.4 shows an example of liquid chromatography—mass spectrometry analysis of the urinary pool of cross-linked peptides from osteoclastic resorption that were derived from the N-telopeptide cross-linking site. The results shown reveal a mix of pyridinoline- and pyrrole-linked, otherwise identical trivalent peptides from the main site of pyrrole cross-linking (Hanson and Eyre, 1996). The resulting estimate of pyridinoline-to-pyrrole ratio can provide a potentially useful, noninvasive index of the posttranslational quality of a patient's bone.

Histidine-containing collagen cross-links and other maturation products

It is becoming increasingly evident that the diversity in biomechanical properties between different collagen-rich tissues is very much a function of evolved differences in cross-linking chemistry. Although beyond the scope of this review, collagen cross-linking variations include the apparent involvement of histidine residues in the pathway of lysine aldehyde cross-linking in skin collagen (Yamauchi et al., 1987); an arginine-bound adduct, arginoline, from divalent ketoimine cross-links in cartilages (Eyre et al., 2010); and the sulfilimine cross-links created by the enzyme peroxidasin between specific hydroxylysine and methionine residues in all forms of type IV collagen that are crucial for the biological function of basement membranes (Vanacore et al., 2009; Bhave et al., 2012).

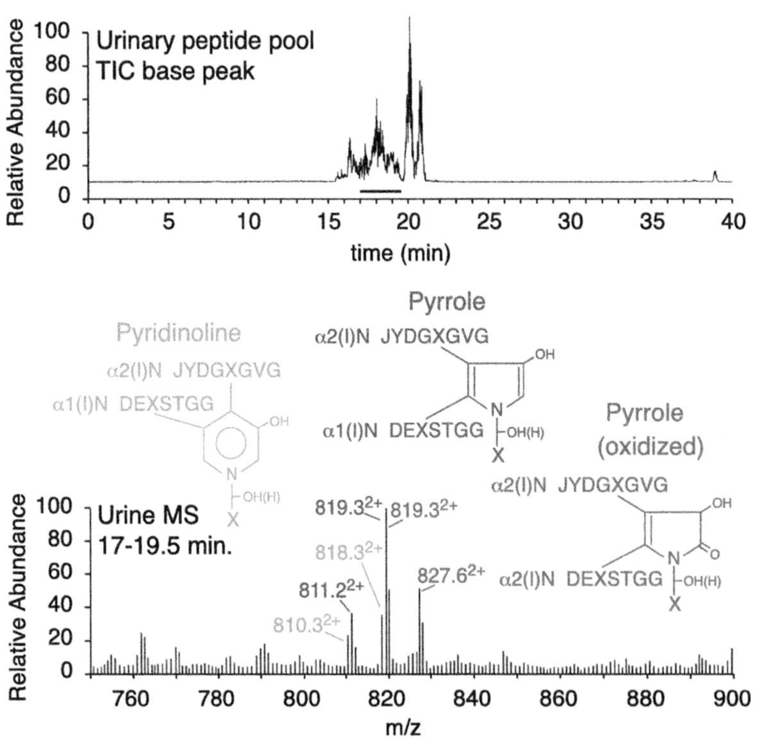

FIGURE 14.4 Liquid chromatography—mass spectrometry (LC—MS) analysis of cross-linked collagen peptides in human urine. The top shows the total ion current (*TIC*) elution profile of peptides on reverse-phase LC and the bottom the ion-trap mass spectral analysis of the *underlined* peptide component peak. The pool of peptides was initially enriched for analysis by ion-exchange and reverse-phase cartridge extraction. The structures shown were identified by their MS/MS fragmentation patterns on tandem mass spectrometry and fitting of the masses of the trivalent cross-linking residues shown. The 3-hydroxy pyrrole ring is readily oxidizable to the +16 Da oxo C2 ring form (819.3^{2+} and 827.6^{2+}), and so gives two pairs of peptides for the lysyl- and hydroxylysyl-derived variants.

In recent work, we have shown that the stable maturation product histidinohydroxylysinonorleucine (HHL), believed to form naturally in the lysine aldehyde pathway of cross-linking that dominates in skin, cornea, and other soft tissue collagens (Yamauchi et al., 1987), is an artifactual product of acid hydrolysis. It is a reaction product formed at sites of C-telopeptide allysine aldol in species with a neighboring C-telopeptide histidine residue (Eyre et al., 2019). This and other evidence is consistent with a role for nearby histidines in catalyzing cross-link formation. Histidine incorporation into mature cross-linking structures is seen only after borohydride reduction (HHMD, histidinohydroxymerodesmosine) or acid hydrolysis (HHL), not on mass spectral analysis of proteolytic digests of tissue collagen (Eyre et al., 2019).

Glycosylations and glycations

Enzymatic glycosylation

Hydroxylysine glycoside content can vary significantly between tissues and collagen types. Bone type I collagen is particularly limited. Only one site, hydroxylysine (Hyl)-87, in each α-chain appears to be fully glycosylated, and Gal-Hyl dominates over Glc-Gal-Hyl. The Gal-Hyl residue participates in cross-link formation with α1(I) C-telopeptide aldehydes (Hanson and Eyre, 1996). Additional partial glycosylation sites include Hyl-174 and Hyl-219 in both α-chains, but only at low occupancy (Terajima et al., 2014; Pokidysheva et al., 2013).

Enzymatic glycosylation is a specific and regulated intracellular modification of nascent collagen α-chains in the endoplasmic reticulum (ER) prior to triple-helix formation (Myllyharju and Kivirikko, 2004; Kivirikko and Myllyla, 1982). The predominant hydroxylysyl galactosyltransferase for bone type I collagen, glycosyltransferase 25 domain 1, is encoded by the GLT25D1 gene (Schegg et al., 2009; Sricholpech et al., 2011). Interestingly, a member of the lysyl hydroxylase (LH) family, the multifunctional LH3, is the principal galactosyl−hydroxylysyl glucosyltransferase in basement membrane collagen (Rautavuoma et al., 2004; Ruotsalainen et al., 2006) and potentially bone type I collagen biosynthesis (Sricholpech et al., 2011).

Tissue-dependent variations in cross-link glycosylation

The ratio of glycosylated to nonglycosylated cross-links can reflect maturational age, tissue type, and even disease state of a collagen sample. Such variation was demonstrated by a 2014 study of adult bovine bone collagen peptides using tandem mass spectrometry (Terajima et al., 2014). Glycosylated pyridinolines and pyrroles were more prevalent than their non-glycosylated forms, and the Gal form dominated. Divalent cross-links were mostly glycosylated too, but with about equal Gal and Glc-Gal forms. The results suggest a steric selectivity for the Gal-Hyl-87 ketoimine cross-link being incorporated into the trivalent structure.

This is consistent with a preference for nonglycosylated pyridinolines being formed during type II collagen maturation in cartilage (Eyre et al., 2008). Divalent ketoimines containing Glc-Gal-Hyl-87 are the primary cross-links in fetal cartilage, whereas nonglycosylated HP is the main pyridinoline in adult articular cartilage (Eyre et al., 2008). An attached disaccharide may therefore sterically hinder a ketoimine cross-link from being the precursor of Glc-Gal HP. A preference for nonglycosylated $>$ Gal $>$ Glc-Gal precursors is indicated.

Potential functions

Potentially, glycosyl groups on bone collagen could either aid or interfere with fibril mineralization. A role for type I collagen overglycosylation in the pathogenesis of osteogenesis imperfecta (OI) has been suggested (Bateman et al., 1984; Tenni et al., 1993; Taga et al., 2013). Altered levels of bone collagen glycosylation have been noted in several musculoskeletal disorders, including postmenopausal osteoporosis (Michalsky et al., 1993), osteosarcoma, and osteofibrous dysplasia (Lehmann et al., 1995). Although in general the functional effects of sugar residues on collagen α-chains are unclear, a broad spectrum of roles have been proposed. They include activities in matrix remodeling (Yang et al., 1993; Jürgensen et al., 2011), collagen fibrillogenesis (Notbohm et al., 1999; Bätge et al., 1997; Torre-Blanco et al., 1992), collagen cross-linking (Rautavuoma et al., 2004; Ruotsalainen et al., 2006; Wang et al., 2002; Moro et al., 2000), and bone mineralization (Sricholpech et al., 2012).

Nonspecific glycations

Among the known and proposed cross-links in collagens, perhaps the least understood but most speculative pathologically are those identified as advanced glycation end products (AGEs). Collagens are frequently cited targets for these adventitious sugar additions since they are typically very long-lived proteins with half-lives ranging for humans from about

5 years in bone (Manolagas, 2000) and 10 years in skin (Avery and Bailey, 2006) to 100 years or more in tendon (Thorpe et al., 2010; Heinemeier et al., 2013) and cartilage (Verzijl et al., 2000). Interestingly, whereas LOX-controlled collagen cross-links appear to plateau in content at a relatively young tissue age (Haut et al., 1992; Eyre et al., 1988), these nonenzymatic cross-links seem to increase steadily over time and cause host tissues to stiffen (Bank et al., 1998; Snedeker and Gautieri, 2014).

Being generally stochastic in nature, AGE cross-linking is proposed to result from randomly accumulated lysine, and perhaps arginine, side-chain glycation adducts that by their placement have potential to go on to form intra- and intermolecular cross-links (Monnier et al., 2005). There is evidence, however, that collagen glycations and AGE formation sites are not random, but will tend to occur at restricted sites (Reiser et al., 1992; Hudson et al., 2018).

Advanced glycation end products

AGEs are formed following the reaction of the carbonyl of a reducing sugar (commonly glucose) with a lysine residue (Fig. 14.5). This nonenzymatic process is the first step in the classical Maillard reaction pathway. The initial product is a relatively unstable Schiff base with the ε-amine of the lysine, which Amadori rearranges to the ketoimine fructosyllysine. This can undergo dehydration, condensation, fragmentation, and cross-linking to form various AGE products. The most prominent is glucosepane, formed by dehydration, carbonyl migration along the sugar carbon chain, and arginine side-chain addition. For this to produce interchain or intrachain protein cross-links, arginine and lysine side-chains need to be 7 Å or less apart (Dai et al., 2008). Several potential sites of glucosepane formation in collagen type I fibrils have been predicted on theoretical grounds (Gautieri et al., 2014; Collier et al., 2015, 2016), but direct analytical proof on connective tissue samples is limited at best. The potential for hydroxylysine as well as lysine residues to be glycated in collagens tends also to have been largely ignored.

Types of advanced glycation end products

From the analysis of complete protease hydrolysates, glucosepane is by far the most prevalent of the potential AGE cross-links in connective tissues (Biemel et al., 2002). Although many more glycation-based structures and potential cross-links have been predicted from experiments in vitro (Snedeker and Gautieri, 2014; Monnier et al., 2005; Sell and Monnier, 2012), their natural tissue levels and significance are less clear. For example, most AGEs arise as monovalent side-chain

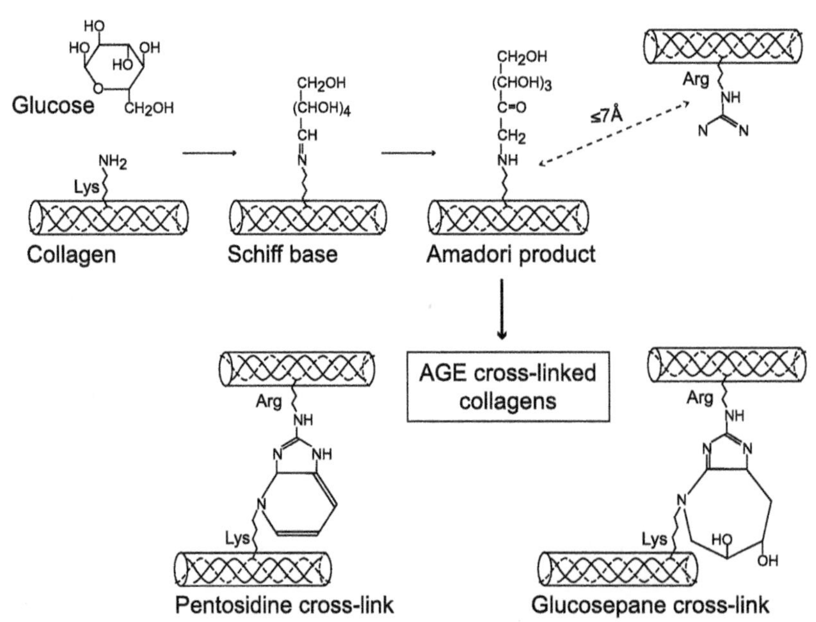

FIGURE 14.5 Glycation products as potential collagen cross-links in bone. The covalent attachment of glucose to collagens and other long-lived proteins leads to advanced glycation end products (*AGEs*) that include candidate cross-links. The Maillard reaction pathway is initiated when the carbonyl of a reducing sugar (commonly glucose) reacts with lysine residues (commonly collagen). The relatively unstable initial Schiff base Amadori rearranges to the ketoimine fructosyllysine. This product can undergo dehydration, condensation, fragmentation, and cross-linking to form various AGE products, which include glucosepane and pentosidine. Such pathological cross-links are thought to contribute to age-related soft tissue stiffening, but the significance for bone is largely unknown.

modifications, which include the arginine product, ornithine; the lysine adduct, fructosyllysine; and the lysine product, carboxymethyllysine. The last has been proposed to cause tissue damage through its ability to chelate transitional metals that can produce reactive oxygenated species (Saxena et al., 1999).

Glucosepane can in theory form intrachain, interchain, and intermolecular collagen cross-links. Based on quantitation in whole-tissue proteolytic digests, glucosepane has been proposed to be the most significant AGE cross-link in older tissues (~0.3 mol/mol of collagen (age 80 skin) (Monnier et al., 2005). Pentosidine, the fluorescent AGE, is quantitatively minor (0.019 mol/mol of collagen (age 80 bone) (Viguet-Carrin et al., 2009). However, to put this in perspective, even if all the measured tissue glucosepane was in collagen as intermolecular cross-links, the levels reached are still an order of magnitude lower than those of the main LOX-based cross-links introduced at synthesis.

Potential consequences

Diabetes mellitus seems to accelerate the age-related decline in bone quality. Increased risk of bone fractures has long been associated with both type 1 and type 2 diabetes (Vestergaard et al., 2009). Bone biopsies from patients with type 1 diabetes revealed a positive association between bone fracture incidence and pentosidine content (Farlay et al., 2016). However, despite decades of research, the pathogenesis of the skeletal effects in diabetes is still poorly understood.

Cross-linking lysine-modifying enzymes

Lysyl hydroxylases

The LH isoenzymes (LH1−3) are encoded by procollagen-lysine, 2-oxoglutarate 5-dioxygenases (*PLOD1−3*). These LH isoenzymes catalyze the hydroxylation of specific collagen lysine residues in a reaction requiring several cofactors (Fe^{2+}, 2-oxoglutarate, O_2, and ascorbate). Unlike prolyl hydroxylation, the extent of lysine hydroxylation varies significantly between tissues and collagen types.

The exact substrate sequence specificity of each isoenzyme has not been fully resolved. It is becoming increasingly clear, however, that LH1 is mainly responsible for catalyzing the hydroxylation of lysine residues within the triple-helical sequence Gly-Xaa-Lys of type I collagen, with particular preference for the cross-linking sites. LH2 is responsible for hydroxylating cross-linking lysine residues in the telopeptide sequences Xaa-Lys-Gly, Xaa-Lys-Ala, and Xaa-Lys-Ser of fibrillar collagens. *PLOD2* encodes two splice variants of LH2, a short form termed LH2a and a long form with an extra exonic 21 amino acids termed LH2b. It has long been thought that LH2b is responsible for collagen telopeptide lysine hydroxylation in bone (Yamauchi and Shiiba, 2008) and hence is the splice variant that controls expression of a hydroxylysine−aldehyde cross-linking tissue phenotype. However, findings in our laboratory are consistent with each splice variant having a unique tissue-specific function. Indeed, although LH2b is ubiquitously expressed in many tissues (Yeowell and Walker, 1999), we suspect that its predominant activity is in bone, tendon, and ligaments, and LH2a is the likely active isoform in connective tissues other than bone.

LH3 is the only multifunctional PLOD isoenzyme, which not only has LH activity, but also hydroxylysyl−galactosyltransferase and galactosylhydroxylysyl−glucosyltransferase activities (Wang et al., 2002). The LH activity of LH3 may be more important functionally for basement membrane collagens, such as type IV collagen, than for fibrillar collagens (Ruotsalainen et al., 2006).

Consequences of lysyl hydroxylase gene mutations

The kyphoscoliotic type of Ehlers−Danlos syndrome (EDS VIA) is caused by homozygous or compound heterozygous mutations in *PLOD1*. *Plod1*-null mice, which exhibit muscle hypotonia and aortic ruptures but not kyphoscoliosis or skin laxity, have a milder phenotype than the human *PLOD1* mutations (Takaluoma et al., 2007; Steinmann et al., 2002). A lack of helical lysine hydroxylation will systemically alter normal collagen cross-linking. In bone, for example, the HP/LP ratio is severely reduced due to the disruption of HP formation. Mutations in *PLOD2* have been identified in Bruck syndrome, an autosomal recessive disorder characterized by bone fragility and congenital joint contractures (Ha-Vinh et al., 2004). Mutations in *PLOD3* have been shown to cause epidermolysis bullosa simplex and other severe connective tissue defects (Salo et al., 2008). The *Plod3*-null mouse is embryonically lethal, which was attributed to an intracellular accumulation of type IV collagen lacking Glc-Gal-Hyl (Rautavuoma et al., 2004). However, if Glc-Gal-Hyl was also lost from some tissue fibrillar collagens, their cross-linking chemistry and supramolecular assembly may also have been affected.

Lysyl oxidases

LOX is responsible for the oxidative deamination of side-chain lysine and hydroxylysine primary amine substrates to produce reactive aldehydes (Siegel et al., 1970). The LOX-driven conversion of telopeptide lysine and hydroxylysine residues to allysines is the sole enzymatic step driving collagen cross-linking chemistry (Siegel and Fu, 1976). This enzyme reaction occurs outside the cell, specifically on the extracellular matrix substrates elastin and collagen (Eyre et al., 1984a). Five additional, LOX-like (LOXL) isoenzymes have been identified in eukaryotes (Grau-Bové et al., 2015; Wordinger and Clark, 2014). Despite all having a similar, highly conserved catalytic domain, two LOXL subfamilies can be classified, LOXL2/L3/L4 and LOX/L1/L5 (Grau-Bové et al., 2015). The former is proposed to influence collagen IV cross-linking in basement membranes and the latter to be involved in chordate/vertebrate-specific matrix remodeling (Grau-Bové et al., 2015).

A range of disease states, from glaucoma to thoracic aortic aneurysm and dissection, have been associated with mutations in the *LOX* and *LOXL* genes (Lee et al., 2016; Thorleifsson et al., 2007). The *Lox*$^{-/-}$ mice are perinatal lethal and exhibit cardiovascular instability with arterial and diaphragmatic ruptures (Hornstra et al., 2003; Maki, 2002). Desmosine cross-links in elastin were shown to be reduced by approximately 60% in the aorta and lungs of these knockout mice (Maki, 2002).

Heritable disorders and mouse models

Heritable disorders

OI, also known as brittle bone disease, is a generalized connective tissue disorder characterized by low bone mass and bone fragility. An array of systemic features is associated with the disease, but none are as pronounced as the skeletal phenotype. Approximately 90% of OI cases are caused by dominantly inherited mutations in the type I collagen gene *COL1A1* or *COL1A2*. The most common of these mutations substitute glycine residues with bulkier or charged residues. Generally, severity increases the more C-terminal the mutation is located, consistent with triple-helix folding from the C terminus, but with many site-related inconsistencies, suggesting a regional severity model (Marini et al., 2007).

In 2006, the first recessively inherited OI, caused by CRTAP mutation, was characterized (Morello et al., 2006). A long list of additional genes quickly followed (Table 14.1) (Kang et al., 2017; Eyre and Weis, 2013). The new insights into OI pathological mechanisms also revealed previously unappreciated genes that regulate collagen cross-linking chemistry and homeostasis. A decade later, new genes linked to rare musculoskeletal disorders, including OI, Bruck syndrome, and EDS, continued to emerge. Many of the new OI-causing genes (>10) encode proteins with essential roles in collagen posttranslational modification, cross-linking, and mineralization, and some function as collagen chaperones and transporters through the ER and Golgi. The expanding list of OI-causing genes includes X-linked, dominant, and recessive variants that affect osteoblast function without any detected effect on collagen cross-linking. Although beyond the present scope, several comprehensive reviews cover them (Cundy, 2012; Forlino and Marini, 2016; Marini et al., 2017).

Collagen posttranslational modifications

CRTAP, LEPRE1, PPIB

An association between a collagen posttranslational modification and OI was discovered from studying the cartilage-associated protein (CRTAP)-null mouse (Morello et al., 2006). This led to the identification of other OI-causing genes encoding subunits of the prolyl 3-hydroxylase enzyme complex, which is composed of CRTAP, cyclophilin B (also called peptidylprolyl isomerase B; PPIB), and prolyl 3-hydroxylase 1 (P3H1) in a 1:1:1 ratio (Vranka et al., 2004). P3H1 (*LEPRE1*) and CRTAP (*LEPREL3*) are both members of the leprecan family of genes; however, only P3H1 has enzyme activity. P3H1 is the enzyme responsible for 3-hydroxylation of specific Pro residues in the X-position of the collagen Gly-Xaa-Yaa triplet repeat, to form 3-hydroxyproline (3*S*,2*S*-L-hydroxyproline; 3Hyp) (Vranka et al., 2004; Ogle et al., 1962; Ogle et al., 1961). The other two proteins in the enzyme complex, PPIB and CRTAP, act as a peptidylprolyl isomerase and an essential "helper" protein, respectively. Disruption of the P3H1 complex results in the loss or reduction of prolyl 3-hydroxylation at two sites in type I collagen, α1(I)Pro986 and α2(I)Pro707 (Morello et al., 2006).

Gene mutations that disrupt expression of any protein in the P3H1 complex (P3H1, CRTAP, and PPIB) have been shown to cause recessive OI (Morello et al., 2006; Barnes et al., 2006, 2010; Cabral et al., 2007; van Dijk et al., 2009). A common consequence of the gene mutations includes altered bone collagen cross-linking (Eyre and Weis, 2013). Bone from the *Crtap*-null mouse and a *LEPRE1*-mutant OI patient had abnormally high HP/LP ratios and overhydroxylated

TABLE 14.1 Noncollagen genes in which mutations cause osteogenesis imperfecta variants or related bone collagen abnormalities.

Gene	Protein	Phenotype	Bone collagen abnormalities
CRTAP *LEPRE1* *PPIB*	CRTAP P3H1, prolyl hydroxylase CYPB, cyclophilin B } P3H1 Complex	Mild to severe OI, reduced mineral density	α1(I)P986 and α2(I)P707 under prolyl-3-hydroxylation, high HP/LP (CRTAP and LEPRE1), low HP/LP (PPIB)
TMEM38B	TRIC-B	Bone fragility, moderate/severe OI	Reduced helical Lys hydroxylation, increased telopeptide hydroxylation
FKPB10 *PLOD2*	FKBP65 LH2, lysyl hydroxylase 2 }	Bruck syndrome: bone fragility, joint contractures	Lack of telopeptide hydroxylysines produces skin-like cross-links
LEPREL2 *LEPREL4*	P3H3 Sc65 }	Low bone mass, skin fragility	Low HP/LP, reduced helical Lys hydroxylation, altered divalent cross-links
MBTPS2	SP2, site-2 metalloprotease	Moderate/severe X-linked OI	Reduced helical Lys-87 hydroxylation and glycosylation
BMP1	Procollagen type I C-propeptidase	High mineral density, mild to severe OI	Defective C-propeptide removal, potential cross-linking defects
SERPINH1	HSP47, heat shock protein 47	Moderate/severe OI, bone fragility	High HP/LP and abnormal arrangement of cross-linking bonds
IFITM5	Bril, osteoblast-specific small transmembrane protein	Normal to severe OI, bone fragility	Altered mineralization, no other collagen abnormalities
SERPINF1	PEDF, pigment epithelium-derived factor	Moderate/severe OI, low mineral density, osteoid seams	Failed mineralization, no other collagen abnormalities

HP, hydroxylysyl pyridinoline; *LP*, lysyl pyridinoline; *OI*, osteogenesis imperfecta.

telopeptide lysines (Eyre and Weis, 2013). In contrast, bone and tendon from the *PPIB*-null mouse and bone from a *PPIB*-mutant OI patient had abnormally low HP/LP ratios but overhydroxylated telopeptide lysines in type I collagen (Eyre and Weis, 2013; Terajima et al., 2016). How an absent P3H1 complex can bring about these differential effects on collagen cross-linking is not fully clear. If it delays triple-helix folding, then helix lysine overmodification (higher HP/LP) might be expected. In the case of PPIB, which is a vital component of several ER complexes, including the LH1 complex (Cabral et al., 2014), both the P3H1 complex and LH1 activity are knocked down. The overhydroxylation of telopeptide lysines in the absence of a P3H1 complex is harder to explain, but may involve recently recognized interactions between prolyl 3-hydroxylase and LH complexes in the ER (see "SC65 and P3H3"). The effects on cross-linking are therefore most likely due mainly to an absence of the P3H1 complex and consequences of ER distress, not the absence of 3Hyp residues at α1(I) P986 and α2(I)P707 (Cabral et al., 2014; Pyott et al., 2011).

Solid-phase binding studies do, however, support a role for 3Hyp in collagen intermolecular recognition and binding during assembly (Hudson et al., 2012). Loss of prolyl 3-hydroxylation in CRTAP, P3H1, and PPIB forms of OI could potentially disrupt the fundamental short-range order in the supramolecular assembly within collagen fibrils of bone (Hudson et al., 2012). A compromised polymeric architecture might hinder the ordered hydroxyapatite nanocrystal growth within collagen fibrils (Eyre and Weis, 2013).

TMEM38B

TMEM38B encodes the ER-membrane monovalent cation channel TRIC-B. Mutations in *TMEM38B* have been shown to cause moderately severe recessive OI (Volodarsky et al., 2013; Shaheen et al., 2012). Disruption of TRIC-B-driven intracellular calcium homeostasis was proposed to dysregulate collagen synthesis and cause OI (Cabral et al., 2016). Type I collagen exhibited altered posttranslational modifications as a consequence of altering the function and expression of vital modifying enzymes. For example, cross-linking lysine hydroxylation was reduced in the helix and increased in telopeptides (Cabral et al., 2016). Much like P3H1 complex mutations, TMEM38B defects were proposed to cause intracellular accumulation of collagen, which resulted in ER stress and the unfolded protein response.

PLOD2 and FKBP10

PLOD2 and *FKBP10* mutations have both been linked to Bruck syndrome (Bank et al., 1999; Ha-Vinh et al., 2004). *FKBP10* encodes the 65-kDa FK506-binding protein FKBP65, a collagen chaperone protein with peptidylprolyl isomerase activity (Ishikawa et al., 2008). *FKBP10* mutations were also shown to cause both Bruck syndrome with severe OI (Kelley et al., 2011; Schwarze et al., 2013; Alanay et al., 2010) and Kuskokwim syndrome with minimal skeletal defects (Barnes et al., 2013). Notably, an *FKBP10* homozygous null mutation, which yields bone collagen with only lysine aldehyde-derived cross-links, does not seem to affect cartilage or ligament collagen (Bank et al., 1999). The lack of telopeptide hydroxylysines precludes pyridinoline formation in bone, and results in skin-like cross-links. The reduction of hydroxyallysine-derived cross-links observed in bone collagen is phenocopied in patients with *PLOD2* and *FKBP10* mutations (Bank et al., 1999; Schwarze et al., 2013). Interestingly, the *Fkbp10*-null mouse exhibited severe embryonic growth defects and did not survive birth (Lietman et al., 2014). The *Plod2*-null mouse is also embryonic lethal (Hyry et al., 2009); however, a zebrafish model carrying a biallelic nonsense mutation in *plod2* displayed severe musculoskeletal abnormalities resembling Bruck syndrome (Gistelinck et al., 2016). The bone collagen defects in *PLOD2* and *FKBP10*—derived Bruck syndrome can be detected as abnormal ratios of HP/LP in patients' urine (Schwarze et al., 2013). Indeed, biochemical data revealed that FKBP65 was part of an enzyme complex with LH2, as a deficiency of FKBP65 resulted in reduced telopeptide cross-linking lysine hydroxylation (Schwarze et al., 2013; Barnes et al., 2012). Consequently, FKBP65-deficient bone produced skin-like collagen cross-links with complete absence of pyridinolines (Schwarze et al., 2013).

SC65 and P3H3

Synaptonemal complex 65 (SC65 or P3H4) is another member of the leprecan gene family and a homolog of CRTAP. Like CRTAP, SC65 has no known enzymatic activity and appears to also function as a "helper" protein. SC65 operates in an enzymatic multiprotein complex, comprising LH1, PPIB, and prolyl 3-hydroxylase 3 (P3H3) (Heard et al., 2016). Although no known human mutation has yet been described, two *Sc65*-null mice have been characterized as having low bone mass and skin fragility (Heard et al., 2016; Gruenwald et al., 2014). *P3h3*-null mouse skin was similarly characterized as having a lack of structural integrity and fragile collagen fabric (Hudson et al., 2017). Both the *Sc65*-null and the *P3h3*-null mice were found to have decreased collagen lysine hydroxylation and abnormal cross-linking (Hudson et al., 2017).

The ratio of mature HP/LP cross-links in bone of both *Sc65*-null and *P3h3*-null mice was reversed compared with wild-type, consistent with the level of lysine underhydroxylation seen in individual chains at cross-linking sites. This was the first direct evidence of an ER complex controlling collagen modification that combines both a lysyl- and a prolyl-hydroxylase activity.

Mechanistically, lysine underhydroxylation at the helical-domain cross-linking sites altered the divalent aldimine cross-link chemistry of the mutant collagen. In normal skin, fully glycosylated Hyl-87 preferentially forms an intermolecular aldimine cross-link with a C-telopeptide lysine aldehyde. However, in the *Sc65*-null and *P3h3*-null mice, the type I collagen LH1 substrate sites are underhydroxylated and subsequently underglycosylated. As a result, the C-telopeptide lysine aldehydes preferentially form intramolecular aldol cross-links (as opposed to intermolecular aldols). We predict that under normal conditions the presence of the disaccharide on Hyl-87 favors aldimine formation with a single C-telopeptide aldehyde. This glycosylated aldimine then sterically hinders intramolecular aldol formation with the second α1(I) C-telopeptide aldehyde of the molecule, which favors intermolecular aldol formation. So, the net effect of under-hydroxylated Lys-87 is fewer aldol intermolecular cross-links. A very similar underhydroxylation of triple-helical domain lysines occurs in human EDS VIA and *Plod1*-null mouse tissues (Takaluoma et al., 2007; Steinmann et al., 2002; Eyre et al., 2002), suggesting that *P3H3* and *SC65* mutations may cause as yet undefined EDS variants.

MBTPS2

The regulated intramembrane proteolysis of transmembrane proteins is an essential cellular process that releases functional protein domains into signaling pathways (Lal and Caplan, 2011). *MBPTS2* (membrane-bound transcription factor peptidase, site 2) encodes the serine protease S2P (site-2 metalloprotease), which spans the Golgi membrane (Ye et al., 2000). Missense mutations in MBTPS2 have been shown to cause moderately severe X-linked OI (Lindert et al., 2016). S2P was found to have impaired function, which ultimately resulted in reduced secretion of type I collagen in fibroblast cultures. Type I collagen also exhibited altered posttranslational modifications with decreased lysine hydroxylation and glycosylation on the K87 cross-linking lysines of both α-chains (Lindert et al., 2016). The main contributor to the bone phenotype from loss of S2P may be the impaired activation of several transcription factors fundamental to normal bone development (Lindert et al., 2016).

Collagen processing

Bone morphogenetic protein 1

Bone morphogenetic protein 1 is the type I procollagen C-propeptide protease. Homozygous missense mutations in *BMP1* have been shown to cause a recessively inherited form of OI with a severe skeletal phenotype (Martinez-Glez et al., 2012). Missense mutations at the type I collagen propeptide cleavage site result in a phenotypically similar but less severe form of OI characterized by unusually high mineral density bone (Lindahl et al., 2011, Cundy et al., 2018). Although not established, retention of the type I collagen C-propeptide presumably disrupts fibril assembly and could alter collagen cross-linking (Eyre and Weis, 2013). It is tempting to speculate that the retained propeptide could alter collagen fibril three-dimensional packing so that the ratio of bone mineral to collagen and bone density are increased.

Collagen chaperone

SERPINH1

SERPINH1 encodes heat shock protein 47 (HSP47), an ER-resident and collagen-specific chaperone protein. HSP47 functions as a quality control protein that prevents premature aggregation of native procollagen molecules during transport through the ER (Widmer et al., 2012). Mutations in *SERPINH1* cause OI with severe skeletal deformity (Christiansen et al., 2010). Loss of HSP47 chaperone function resulted in protein misfolding and intracellular retention of type I collagen with subsequent ER stress (Drogemuller et al., 2009). The importance of *SERPINH1* was highlighted by the embryonic lethality observed in *Hsp47*-null mice (Nagai et al., 2000). An OI dachshund with a loss-of-function mutation in *SERPINH1* (Drogemuller et al., 2009) revealed bone collagen cross-linking abnormalities consistent with a defective molecular assembly and posttranslational chemistry (Lindert et al., 2015). Mutations in *SERPINH1* yield brittle bones with a high HP/LP ratio, telopeptide lysine overhydroxylation, and resulting overabundance of stable pyridinoline cross-links (Eyre and Weis, 2013).

Bone mineralization

IFITM5

Mutations in *IFITM5* yield an autosomal dominant form of OI with a moderate to severe skeletal phenotype (Glorieux et al., 2000). The *IFITM5* gene encodes interferon-induced transmembrane protein 5 (IFITM5), a 14.8-kDa transmembrane protein thought to function in early bone mineralization (Moffatt et al., 2008). The *IFITM5*-null mouse had stunted growth with smaller long bones (Hanagata et al., 2011). No defects in collagen posttranslational chemistry have been found in the bone of mice expressing the type V OI-causing IFITM5 mutation (Lietman et al., 2015).

SERPINF1

OI caused by homozygous null mutations in *SERPINF1* result in bone mineralization defects without any known effect on collagen structure (Glorieux et al., 2002). *SERPINF1* encodes pigment epithelium-derived factor (PEDF), a member of the serine protease inhibitor serpin superfamily (Rauch et al., 2012). PEDF is a 50-kDa glycoprotein known as a potent inhibitor of angiogenesis (Becker et al., 2011). PEDF has been shown to bind to type I collagen from bone and is thought to help regulate bone angiogenesis and matrix remodeling (Tombran-Tink and Barnstable, 2004). Indeed, the absence of PEDF results in large quantities of unmineralized osteoid and an abnormal lamellar pattern on histology of patient bone (Homan et al., 2011). It is likely that the bone phenotype is the result of a dysregulated step in the mineral deposition machinery.

Implications for bone fragility and mineral deposition

Abnormal collagen cross-linking emerges as a recurring theme in many of these rare musculoskeletal disorders. We end with a hypothesis that the abnormal collagen cross-linking seen in bone from many variants of OI is a responsible contributor to the underlying brittleness. Clearly, collagen strength rests heavily on its cross-linking, but other properties of bone, including ductility and resistance to microdamage and crack propagation, may depend on the unique pattern of LOX-mediated cross-links that characterize normal bone collagen of higher vertebrates. The placement of cross-links and the chemical lability of a significant fraction of them (those formed from lysine aldehydes) may be required to produce a malleable framework in which highly ordered hydroxyapatite nanocrystals can grow optimally within the fibrils (Eyre and Weis, 2013).

Fig. 14.6 illustrates this concept of labile, aldimine cross-links breaking as mineral crystallites grow in the interior of bone collagen fibrils. If, as the data suggest, lysine aldehyde and hydroxylysine aldehyde cross-links are each distributed in an orderly (nonrandom) pattern in the packed collagen lattice, this could channel a parallel, ordered growth of

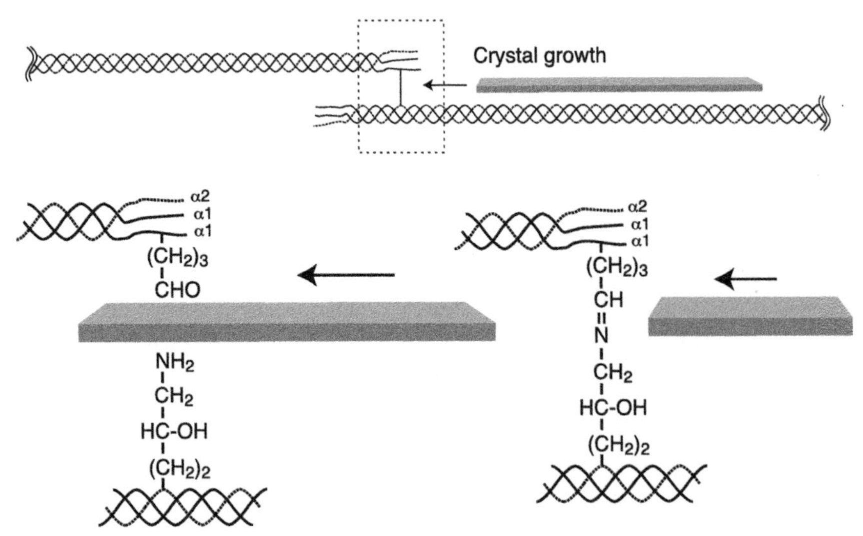

FIGURE 14.6 Illustrated concept that labile (aldimine) cross-links break as part of the orderly growth of hydroxyapatite crystal in mineralizing collagen fibrils.

hydroxyapatite crystals. In theory, for human bone, up to half the initial divalent cross-links (aldimines) would break, reverting back to their aldehyde plus amine precursors. The result is potentially increased intermolecular packing space for growing crystallites in a still stably cross-linked collagen fibril lattice (Eyre and Weis, 2013).

Future challenges

After more than half a century of study, we still do not understand at the molecular level the mechanism that drives the orderly mineralization of bone collagen. To what degree mineralization is driven by bone-specific features of the fibrils themselves, particularly their unique cross-linking chemistry, or is a process driven by extrafibrillar factors is still unclear. Understanding more broadly how differences in cross-linking chemistry between tissues and during development confer functional advantages will also help in understanding how defects in bone collagen cross-linking cause a compromised bone structure. Whether pyrrole cross-links contribute to the unique material properties of normal bone and how their loss might increase bone fragility in OI are also important questions. It should also not escape attention that the mechanisms causing bone collagen cross-linking abnormalities revealed by the study of rare genetic disorders may well be operative in more common brittle bone disorders, including osteoporosis and diabetes.

References

Alanay, Y., Avaygan, H., Camacho, N., Utine, G.E., Boduroglu, K., Aktas, D., et al., April 9, 2010. Mutations in the gene encoding the RER protein FKBP65 cause autosomal-recessive osteogenesis imperfecta. Am. J. Hum. Genet. 86 (4), 551–559.

Avery, N.C., Bailey, A.J., September 2006. The effects of the Maillard reaction on the physical properties and cell interactions of collagen. Pathol. Biol. 54 (7), 387–395.

Avery, N.C., Sims, T.J., Bailey, A.J., 2009. Quantitative determination of collagen cross-links. In: Extracellular Matrix Protocols. Methods in Molecular Biology, vol. 522. Humana Press, Totowa, NJ, pp. 103–121.

Bank, R.A., Bayliss, M.T., Lafeber, F.P., Maroudas, A., Tekoppele, J.M., February 15, 1998. Ageing and zonal variation in post-translational modification of collagen in normal human articular cartilage. The age-related increase in non-enzymatic glycation affects biomechanical properties of cartilage. Portland Press Limited Biochem. J. 330 (Pt 1), 345–351.

Bank, R.A., Robins, S.P., Wijmenga, C., Breslau-Siderius, L.J., Bardoel, A.F., van der Sluijs, H.A., et al., February 2, 1999. Defective collagen crosslinking in bone, but not in ligament or cartilage, in Bruck syndrome: indications for a bone-specific telopeptide lysyl hydroxylase on chromosome 17. National Academy of Sciences Proc. Natl. Acad. Sci. U. S. A. 96 (3), 1054–1058.

Banse, X., Devogelaer, J.P., Lafosse, A., Sims, T.J., Grynpas, M., Bailey, A.J., July 2002. Cross-link profile of bone collagen correlates with structural organization of trabeculae. Bone 31 (1), 70–76.

Barnes, A.M., Chang, W., Morello, R., Cabral, W.A., Weis, M., Eyre, D.R., et al., December 28, 2006. Deficiency of cartilage-associated protein in recessive lethal osteogenesis imperfecta. N. Engl. J. Med. 355 (26), 2757–2764.

Barnes, A.M., Carter, E.M., Cabral, W.A., Weis, M., Chang, W., Makareeva, E., et al., February 11, 2010. Lack of cyclophilin B in osteogenesis imperfecta with normal collagen folding. N. Engl. J. Med. 362 (6), 521–528.

Barnes, A.M., Cabral, W.A., Weis, M., Makareeva, E., Mertz, E.L., Leikin, S., et al., November 2012. Absence of FKBP10 in recessive type XI osteogenesis imperfecta leads to diminished collagen cross-linking and reduced collagen deposition in extracellular matrix. Wiley Subscription Services, Inc., A Wiley Company Hum. Mutat. 33 (11), 1589–1598.

Barnes, A.M., Duncan, G., Weis, M., Paton, W., Cabral, W.A., Mertz, E.L., et al., September 2013. Kuskokwim syndrome, a recessive congenital contracture disorder, extends the phenotype of FKBP10 mutations. Hum. Mutat. 34 (9), 1279–1288.

Bateman, J.F., Mascara, T., Chan, D., Cole, W.G., January 1, 1984. Abnormal type I collagen metabolism by cultured fibroblasts in lethal perinatal osteogenesis imperfecta. Portland Press Ltd Biochem J. 217 (1), 103–115.

Bätge, B., Winter, C., Notbohm, H., Acil, Y., Brinckmann, J., Müller, P.K., July 1997. Glycosylation of human bone collagen I in relation to lysylhydroxylation and fibril diameter. J. Biochem. 122 (1), 109–115.

Beardsworth, L.J., Eyre, D.R., Dickson, I.R., July 1990. Changes with age in the urinary excretion of lysyl- and hydroxylysylpyridinoline, two new markers of bone collagen turnover. John Wiley and Sons and The American Society for Bone and Mineral Research (ASBMR). J. Bone Miner. Res. 5 (7), 671–676.

Becker, J., Semler, O., Gilissen, C., Li, Y., Bolz, H.J., Giunta, C., et al., March 11, 2011. Exome sequencing identifies truncating mutations in human SERPINF1 in autosomal-recessive osteogenesis imperfecta. Am. J. Hum. Genet. 88 (3), 362–371.

Bhave, G., Cummings, C.F., Vanacore, R.M., Kumagai-Cresse, C., Ero-Tolliver, I.A., Rafi, M., et al., September 2012. Peroxidasin forms sulfilimine chemical bonds using hypohalous acids in tissue genesis. Nat. Chem. Biol. 8 (9), 784–790.

Biemel, K.M., Friedl, D.A., Lederer, M.O., July 12, 2002. Identification and quantification of major maillard cross-links in human serum albumin and lens protein. Evidence for glucosepane as the dominant compound. American Society for Biochemistry and Molecular Biology J. Biol. Chem. 277 (28), 24907–24915.

Blumenfeld, O.O., Gallop, P.M., October 1966. Amino aldehydes in tropocollagen: the nature of a probable cross-link. National Academy of Sciences Proc. Natl. Acad. Sci. U. S. A. 56 (4), 1260–1267.

Boot-Handford, R.P., Tuckwell, D.S., February 2003. Fibrillar collagen: the key to vertebrate evolution? A tale of molecular incest. Bioessays 25 (2), 142–151.

Cabral, W.A., Chang, W., Barnes, A.M., Weis, M., Scott, M.A., Leikin, S., et al., March 2007. Prolyl 3-hydroxylase 1 deficiency causes a recessive metabolic bone disorder resembling lethal/severe osteogenesis imperfecta. Nat. Genet. 39 (3), 359–365.

Cabral, W.A., Perdivara, I., Weis, M., Terajima, M., Blissett, A.R., Chang, W., et al., June 2014. Abnormal type I collagen post-translational modification and crosslinking in a cyclophilin B KO mouse model of recessive osteogenesis imperfecta. Public Library of Science. In: Cohn, D. (Ed.), PLoS Gen. 10 (6), e1004465.

Cabral, W.A., Ishikawa, M., Garten, M., Makareeva, E.N., Sargent, B.M., Weis, M., et al., July 2016. Absence of the ER cation channel TMEM38B/TRIC-B disrupts intracellular calcium homeostasis and dysregulates collagen synthesis in recessive osteogenesis imperfecta. PLoS Gen. 12 (7), e1006156. Bateman, J.F., editor.

Christiansen, H.E., Schwarze, U., Pyott, S.M., AlSwaid, A., Balwi Al, M., Alrasheed, S., et al., March 12, 2010. Homozygosity for a missense mutation in SERPINH1, which encodes the collagen chaperone protein HSP47, results in severe recessive osteogenesis imperfecta. Am. J. Hum. Genet. 86 (3), 389–398.

Collier, T.A., Nash, A., Birch, H.L., de Leeuw, N.H., October 2015. Preferential sites for intramolecular glucosepane cross-link formation in type I collagen: a thermodynamic study. Matrix Biol. 48, 78–88.

Collier, T.A., Nash, A., Birch, H.L., de Leeuw, N.H., November 2016. Intra-molecular lysine-arginine derived advanced glycation end-product cross-linking in Type I collagen: a molecular dynamics simulation study. Biophys. Chem. 218, 42–46.

Cundy, T., June 2012. Recent advances in osteogenesis imperfecta. Calcif. Tissue Int. 90 (6), 439–449.

Cundy, T., Dray, M., Delahunt, J., Hald, J.D., Langdahl, B., Li, C., et al., July 2018. Mutations that alter the carboxy-terminal-propeptide cleavage site of the chains of type I procollagen are associated with a unique osteogenesis imperfecta phenotype. J. Bone. Miner. Res. 33 (7), 1260–1271. https://doi.org/10.1002/jbmr.3424. Epub 2018 Apr 18. PubMed PMID: 29669177; PubMed Central PMCID: PMC6031457.

Dai, Z., Wang, B., Sun, G., Fan, X., Anderson, V.E., Monnier, V.M., July 2008. Identification of glucose-derived cross-linking sites in ribonuclease A. American Chemical Society J. Proteome Res. 7 (7), 2756–2768.

Drogemuller, C., Becker, D., Brunner, A., Haase, B., Kircher, P., Seeliger, F., et al., July 2009. A missense mutation in the SERPINH1 gene in Dachshunds with osteogenesis imperfecta. PLoS Gen. 5 (7), e1000579.

Eyre, D.R., Oguchi, H., January 29, 1980. The hydroxypyridinium crosslinks of skeletal collagens: their measurement, properties and a proposed pathway of formation. Biochem. Biophys. Res. Commun. 92 (2), 403–410.

Eyre, D.R., Weis, M.A., October 2013. Bone collagen: new clues to its mineralization mechanism from recessive osteogenesis imperfecta. Springer US Calcif. Tissue Int. 93 (4), 338–347.

Eyre, D.R., Wu, J.-J., 2005. Collagen cross-links. In: Collagen. Topics in Current Chemistry, vol. 247. Springer Berlin Heidelberg, Berlin, Heidelberg, pp. 207–229.

Eyre, D.R., Paz, M.A., Gallop, P.M., 1984a. Cross-linking in collagen and elastin. Annu. Rev. Biochem. 53, 717–748.

Eyre, D.R., Koob, T.J., Van Ness, K.P., March 1984. Quantitation of hydroxypyridinium crosslinks in collagen by high-performance liquid chromatography. Anal. Biochem. 137 (2), 380–388.

Eyre, D.R., Dickson, I.R., Van Ness, K., June 1, 1988. Collagen cross-linking in human bone and articular cartilage. Age-related changes in the content of mature hydroxypyridinium residues. Portland Press Ltd Biochem. J. 252 (2), 495–500.

Eyre, D., Shao, P., Weis, M.A., Steinmann, B., July 2002. The kyphoscoliotic type of Ehlers-Danlos syndrome (type VI): differential effects on the hydroxylation of lysine in collagens I and II revealed by analysis of cross-linked telopeptides from urine. Mol. Genet. Metabol. 76 (3), 211–216.

Eyre, D.R., Pietka, T., Weis, M.A., Wu, J.-J., January 23, 2004. Covalent cross-linking of the NC1 domain of collagen type IX to collagen type II in cartilage. J. Biol. Chem. 279 (4), 2568–2574.

Eyre, D.R., Weis, M.A., Wu, J.-J., May 2008. Advances in collagen cross-link analysis. Methods 45 (1), 65–74.

Eyre, D.R., Weis, M.A., Wu, J.J., May 28, 2010. Maturation of collagen Ketoimine cross-links by an alternative mechanism to pyridinoline formation in cartilage. J. Biol. Chem. 285 (22), 16675–16682.

Eyre, D.R., March 21, 1980. Collagen: molecular diversity in the body's protein scaffold. Science 207 (4437), 1315–1322.

Eyre, D.R., Weis, M., Rai, J., February 7, 2019. Analyses of lysine aldehyde cross-linking in collagen reveal that the mature cross-link histidinohydroxylysinonorleucine is an artifact. J. Biol. Chem. pii: jbc.RA118.007202. doi: 10.1074/jbc.RA118.007202. [Epub ahead of print] PubMed PMID: 30733334.

Eyre, D.R., 1981. Crosslink maturation in bone collagen. Dev. Biochem. 22, 51–55.

Eyre, D., 1987. Collagen cross-linking amino acids. Methods Enzymol. 144, 115–139.

Farlay, D., Armas, L.A.G., Gineyts, E., Akhter, M.P., Recker, R.R., Boivin, G., January 2016. Nonenzymatic glycation and degree of mineralization are higher in bone from fractured patients with type 1 diabetes mellitus. J. Bone Miner. Res. 31 (1), 190–195.

Forlino, A., Marini, J.C., April 16, 2016. Osteogenesis imperfecta. Lancet 387 (10028), 1657–1671.

Fujimoto, D., June 20, 1977. Isolation and characterization of a fluorescent material in bovine achilles tendon collagen. Biochem. Biophys. Res. Commun. 76 (4), 1124–1129.

Gautieri, A., Redaelli, A., Buehler, M.J., Vesentini, S., February 2014. Age- and diabetes-related nonenzymatic crosslinks in collagen fibrils: candidate amino acids involved in Advanced Glycation End-products. Matrix Biol. 34, 89–95.

Gistelinck, C., Witten, P.E., Huysseune, A., Symoens, S., Malfait, F., Larionova, D., et al., November 2016. Loss of type I collagen telopeptide lysyl hydroxylation causes musculoskeletal abnormalities in a zebrafish model of Bruck syndrome. J. Bone Miner. Res. 31 (11), 1930–1942.

Glorieux, F.H., Rauch, F., Plotkin, H., Ward, L., Travers, R., Roughley, P., et al., September 2000. Type V osteogenesis imperfecta: a new form of brittle bone disease. John Wiley and Sons and The American Society for Bone and Mineral Research (ASBMR). J. Bone Miner. Res. 15 (9), 1650–1658.

Glorieux, F.H., Ward, L.M., Rauch, F., Lalic, L., Roughley, P.J., Travers, R., January 2002. Osteogenesis imperfecta type VI: a form of brittle bone disease with a mineralization defect. John Wiley and Sons and The American Society for Bone and Mineral Research (ASBMR). J. Bone Miner. Res. 17 (1), 30–38.

Grau-Bové, X., Ruiz-Trillo, I., Rodriguez-Pascual, F., May 29, 2015. Origin and evolution of lysyl oxidases. Nature Publishing Group Sci. Rep. 5 (1), 651.

Gruenwald, K., Castagnola, P., Besio, R., Dimori, M., Chen, Y., Akel, N.S., et al., March 2014. Sc65 is a novel endoplasmic reticulum protein that regulates bone mass homeostasis. J. Bone Miner. Res. 29 (3), 666–675.

Hanagata, N., Li, X., Morita, H., Takemura, T., Li, J., Minowa, T., May 2011. Characterization of the osteoblast-specific transmembrane protein IFITM5 and analysis of IFITM5-deficient mice. J. Bone Miner. Metab. 29 (3), 279–290.

Hanson, D.A., Eyre, D.R., October 25, 1996. Molecular site specificity of pyridinoline and pyrrole cross-links in type I collagen of human bone. J. Biol. Chem. 271 (43), 26508–26516.

Hanson, D.A., Weis, M.A., Bollen, A.M., Maslan, S.L., Singer, F.R., Eyre, D.R., November 1992. A specific immunoassay for monitoring human bone resorption: quantitation of type I collagen cross-linked N-telopeptides in urine. J. Bone Miner. Res. 7 (11), 1251–1258.

Haut, R.C., Lancaster, R.L., DeCamp, C.E., February 1992. Mechanical properties of the canine patellar tendon: some correlations with age and the content of collagen. J. Biomech. 25 (2), 163–173.

Ha-Vinh, R., Alanay, Y., Bank, R.A., Campos-Xavier, A.B., Zankl, A., Superti-Furga, A., et al., December 1, 2004. Phenotypic and molecular characterization of Bruck syndrome (osteogenesis imperfecta with contractures of the large joints) caused by a recessive mutation in PLOD2. Am. J. Med. Genet. 131 (2), 115–120.

Heard, M.E., Besio, R., Weis, M., Rai, J., Hudson, D.M., Dimori, M., et al., April 2016. Sc65-Null mice provide evidence for a novel endoplasmic reticulum complex regulating collagen lysyl hydroxylation. Public Library of Science. In: Bateman, J.F. (Ed.), PLoS Gen. 12 (4), e1006002.

Heinemeier, K.M., Schjerling, P., Heinemeier, J., Magnusson, S.P., Kjaer, M., May 2013. Lack of tissue renewal in human adult Achilles tendon is revealed by nuclear bomb (14)C. FASEB J. 27 (5), 2074–2079.

Homan, E.P., Rauch, F., Grafe, I., Lietman, C., Doll, J.A., Dawson, B., et al., December 2011. Mutations in SERPINF1 cause osteogenesis imperfecta type VI. Wiley Subscription Services, Inc., A Wiley Company J. Bone Miner. Res. 26 (12), 2798–2803.

Horgan, D.J., King, N.L., Kurth, L.B., Kuypers, R., August 1990. Collagen crosslinks and their relationship to the thermal properties of calf tendons. Arch. Biochem. Biophys. 281 (1), 21–26.

Hornstra, I.K., Birge, S., Starcher, B., Bailey, A.J., Mecham, R.P., Shapiro, S.D., April 18, 2003. Lysyl oxidase is required for vascular and diaphragmatic development in mice. American Society for Biochemistry and Molecular Biology J. Biol. Chem. 278 (16), 14387–14393.

Hudson, D.M., Archer, M., King, K.B., Eyre, D.R., October 5, 2018. Glycation of type I collagen selectively targets the same helical domain lysine sites as lysyl oxidase-mediated cross-linking. J. Biol. Chem. 293 (40), 15620–15627.

Hudson, D.M., Eyre, D.R., 2013. Collagen prolyl 3-hydroxylation: a major role for a minor post-translational modification? Connect. Tissue Res. 54 (4–5), 245–251.

Hudson, D.M., Kim, L.S., Weis, M., Cohn, D.H., Eyre, D.R., March 27, 2012. Peptidyl 3-hydroxyproline binding properties of type I collagen suggest a function in fibril supramolecular assembly. Biochemistry (Mosc.) 51 (12), 2417–2424.

Hudson, D.M., Weis, M., Rai, J., Joeng, K.S., Dimori, M., Lee, B.H., et al., March 3, 2017. P3h3-null and Sc65-null mice phenocopy the collagen lysine under-hydroxylation and cross-linking abnormality of Ehlers-danlos syndrome type VIA. J. Biol. Chem. 292 (9), 3877–3887.

Hyry, M., Lantto, J., Myllyharju, J., November 6, 2009. Missense mutations that cause Bruck syndrome affect enzymatic activity, folding, and oligomerization of lysyl hydroxylase 2. J. Biol. Chem. 284 (45), 30917–30924.

Ishikawa, Y., Vranka, J., Wirz, J., Nagata, K., Bachinger, H.P., November 14, 2008. The rough endoplasmic reticulum-resident FK506-binding protein FKBP65 is a molecular chaperone that interacts with collagens. J. Biol. Chem. 283 (46), 31584–31590.

Jürgensen, H.J., Madsen, D.H., Ingvarsen, S., Melander, M.C., Gårdsvoll, H., Patthy, L., et al., September 16, 2011. A novel functional role of collagen glycosylation: interaction with the endocytic collagen receptor uparap/ENDO180. J. Biol. Chem. 286 (37), 32736–32748.

Kalamajski, S., Liu, C., Tillgren, V., Rubin, K., Oldberg, Å., Rai, J., et al., July 4, 2014. Increased C-telopeptide cross-linking of tendon type I collagen in fibromodulin-deficient mice. American Society for Biochemistry and Molecular Biology J. Biol. Chem. 289 (27), 18873–18879.

Kang, H., Aryal, A.C.S., Marini, J.C., March 2017. Osteogenesis imperfecta: new genes reveal novel mechanisms in bone dysplasia. Transl. Res. 181, 27–48.

Kelley, B.P., Malfait, F., Bonafe, L., Baldridge, D., Homan, E., Symoens, S., et al., March 2011. Mutations in FKBP10 cause recessive osteogenesis imperfecta and Bruck syndrome. J. Bone Miner. Res. 26 (3), 666–672.

Kivirikko, K.I., Myllyla, R., 1982. Posttranslational enzymes in the biosynthesis of collagen: intracellular enzymes. Methods Enzymol. 82 (Pt A), 245–304.

Knott, L., Bailey, A.J., March 1998. Collagen cross-links in mineralizing tissues: a review of their chemistry, function, and clinical relevance. Bone 22 (3), 181–187.

Lal, M., Caplan, M., February 2011. Regulated intramembrane proteolysis: signaling pathways and biological functions. Physiology 26 (1), 34–44.

Lee, V.S., Halabi, C.M., Hoffman, E.P., Carmichael, N., Leshchiner, I., Lian, C.G., et al., August 2, 2016. Loss of function mutation in LOX causes thoracic aortic aneurysm and dissection in humans. Proc. Natl. Acad. Sci. U. S. A. 113 (31), 8759–8764.

Lehmann, H.W., Wolf, E., Röser, K., Bodo, M., Delling, G., Müller, P.K., 1995. Composition and posttranslational modification of individual collagen chains from osteosarcomas and osteofibrous dysplasias. J. Cancer Res. Clin. Oncol. 121 (7), 413–418.

Lietman, C.D., Rajagopal, A., Homan, E.P., Munivez, E., Jiang, M.-M., Bertin, T.K., et al., September 15, 2014. Connective tissue alterations in Fkbp10-/- mice. Oxford University Press Hum. Mol. Genet. 23 (18), 4822–4831.

Lietman, C.D., Marom, R., Munivez, E., Bertin, T.K., Jiang, M.-M., Chen, Y., et al., March 2015. A transgenic mouse model of OI type V supports a neomorphic mechanism of the IFITM5 mutation. J. Bone Miner. Res. 30 (3), 489–498.

Light, N.D., Bailey, A.J., January 1, 1979. Changes in crosslinking during aging in bovine tendon collagen. FEBS Lett. 97 (1), 183–188.

Lindahl, K., Barnes, A.M., Fratzl-Zelman, N., Whyte, M.P., Hefferan, T.E., Makareeva, E., et al., June 2011. COL1 C-propeptide cleavage site mutations cause high bone mass osteogenesis imperfecta. Hum. Mutat. 32 (6), 598–609.

Lindert, U., Weis, M.A., Rai, J., Seeliger, F., Hausser, I., Leeb, T., et al., July 17, 2015. Molecular consequences of the SERPINH1/HSP47 mutation in the Dachshund natural model of osteogenesis imperfecta. J. Biol. Chem. 290 (29), 17679–17689.

Lindert, U., Cabral, W.A., Ausavarat, S., Tongkobpetch, S., Ludin, K., Barnes, A.M., et al., July 6, 2016. MBTPS2 mutations cause defective regulated intramembrane proteolysis in X-linked osteogenesis imperfecta. Nature Publishing Group Nat. Commun. 7, 11920.

Maki, J.M., October 14, 2002. Inactivation of the lysyl oxidase gene lox leads to aortic aneurysms, cardiovascular dysfunction, and perinatal death in mice. American Heart Association, Inc Circulation 106 (19), 2503–2509.

Manolagas, S.C., April 2000. Birth and death of bone cells: basic regulatory mechanisms and implications for the pathogenesis and treatment of osteoporosis. Endocr. Rev. 21 (2), 115–137.

Marini, J.C., Forlino, A., Cabral, W.A., Barnes, A.M., San Antonio, J.D., Milgrom, S., et al., March 2007. Consortium for osteogenesis imperfecta mutations in the helical domain of type I collagen: regions rich in lethal mutations align with collagen binding sites for integrins and proteoglycans. Hum. Mutat. 28 (3), 209–221.

Marini, J.C., Forlino, A., Bächinger, H.P., Bishop, N.J., Byers, P.H., Paepe, A.D., et al., August 18, 2017. Osteogenesis imperfecta. Primers Nat. Rev. Dis. 3, 17052.

Martinez-Glez, V., Valencia, M., Caparros-Martin, J.A., Aglan, M., Temtamy, S., Tenorio, J., et al., February 2012. Identification of a mutation causing deficient BMP1/mTLD proteolytic activity in autosomal recessive osteogenesis imperfecta. Hum. Mutat. 33 (2), 343–350.

Michalsky, M., Norris-Suarez, K., Bettica, P., Pecile, A., Moro, L., May 14, 1993. Rat cortical and trabecular bone collagen glycosylation are differently influenced by ovariectomy. Biochem. Biophys. Res. Commun. 192 (3), 1281–1288.

Moffatt, P., Gaumond, M.-H., Salois, P., Sellin, K., Bessette, M.-C., Godin, E., et al., September 2008. Bril: a novel bone-specific modulator of mineralization. John Wiley and Sons and The American Society for Bone and Mineral Research (ASBMR). J. Bone Miner. Res. 23 (9), 1497–1508.

Monnier, V.M., Mustata, G.T., Biemel, K.L., Reihl, O., Lederer, M.O., Zhenyu, D., et al., June 2005. Cross-linking of the extracellular matrix by the maillard reaction in aging and diabetes: an update on "a puzzle nearing resolution". Blackwell Publishing Ltd Ann. N. Y. Acad. Sci. 1043 (1), 533–544.

Morello, R., Bertin, T.K., Chen, Y., Hicks, J., Tonachini, L., Monticone, M., et al., October 20, 2006. CRTAP is required for prolyl 3- hydroxylation and mutations cause recessive osteogenesis imperfecta. Cell 127 (2), 291–304.

Moro, L., Romanello, M., Favia, A., Lamanna, M.P., Lozupone, E., February 2000. Posttranslational modifications of bone collagen type I are related to the function of rat femoral regions. Calcif. Tissue Int. 66 (2), 151–156.

Myllyharju, J., Kivirikko, K.I., January 2004. Collagens, modifying enzymes and their mutations in humans, flies and worms. Trends Genet. 20 (1), 33–43.

Naffa, R., Holmes, G., Ahn, M., Harding, D., Norris, G., December 23, 2016. Liquid chromatography-electrospray ionization mass spectrometry for the simultaneous quantitation of collagen and elastin crosslinks. J. Chromatogr. A 1478, 60–67.

Nagai, N., Hosokawa, M., Itohara, S., Adachi, E., Matsushita, T., Hosokawa, N., et al., September 18, 2000. Embryonic lethality of molecular chaperone hsp47 knockout mice is associated with defects in collagen biosynthesis. The Rockefeller University Press J. Cell Biol. 150 (6), 1499–1506.

Notbohm, H., Nokelainen, M., Myllyharju, J., Fietzek, P.P., Müller, P.K., Kivirikko, K.I., March 26, 1999. Recombinant human type II collagens with low and high levels of hydroxylysine and its glycosylated forms show marked differences in fibrillogenesis in vitro. J. Biol. Chem. 274 (13), 8988–8992.

Ogle, J.D., Arlinghaus, R.B., Logan, M.A., July 1961. Studies on peptides obtained from enzymic digests of collagen with evidence for the presence of an unidentified compound in this protein. Arch. Biochem. Biophys. 94, 85–93.

Ogle, J.D., Arlinghaus, R.B., Logan, M.A., December 1962. 3-Hydroxyproline, a new amino acid of collagen. J. Biol. Chem. 237, 3667–3673.

Pokidysheva, E., Zientek, K.D., Ishikawa, Y., Mizuno, K., Vranka, J.A., Montgomery, N.T., et al., August 23, 2013. Posttranslational modifications in type I collagen from different tissues extracted from wild type and prolyl 3-hydroxylase 1 null mice. American Society for Biochemistry and Molecular Biology J. Biol. Chem. 288 (34), 24742–24752.

Pyott, S.M., Schwarze, U., Christiansen, H.E., Pepin, M.G., Leistritz, D.F., Dineen, R., et al., April 15, 2011. Mutations in PPIB (cyclophilin B) delay type I procollagen chain association and result in perinatal lethal to moderate osteogenesis imperfecta phenotypes. Hum. Mol. Genet. 20 (8), 1595–1609.

Rauch, F., Husseini, A., Roughley, P., Glorieux, F.H., Moffatt, P., August 2012. Lack of circulating pigment epithelium-derived factor is a marker of osteogenesis imperfecta type VI. J. Clin. Endocrinol. Metab. 97 (8), E1550–E1556.

Rautavuoma, K., Takaluoma, K., Sormunen, R., Myllyharju, J., Kivirikko, K.I., Soininen, R., September 28, 2004. Premature aggregation of type IV collagen and early lethality in lysyl hydroxylase 3 null mice. National Acad Sciences Proc. Natl. Acad. Sci. U. S. A. 101 (39), 14120–14125.

Reiser, K.M., Amigable, M.A., Last, J.A., December 5, 1992. Nonenzymatic glycation of type I collagen. The effects of aging on preferential glycation sites. J. Biol. Chem. 267 (34), 24207–24216.

Rodriguez-Pascual, F., Slatter, D.A., November 23, 2016. Collagen cross-linking: insights on the evolution of metazoan extracellular matrix. Nature Publishing Group Sci. Rep. 6 (1), 1955.

Ruotsalainen, H., Sipilä, L., Vapola, M., Sormunen, R., Salo, A.M., Uitto, L., et al., February 15, 2006. Glycosylation catalyzed by lysyl hydroxylase 3 is essential for basement membranes. J. Cell Sci. 119 (Pt 4), 625–635.

Salo, A.M., Cox, H., Farndon, P., Moss, C., Grindulis, H., Risteli, M., et al., October 2008. A connective tissue disorder caused by mutations of the lysyl hydroxylase 3 gene. Am. J. Hum. Genet. 83 (4), 495–503.

Saxena, A.K., Saxena, P., Wu, X., Obrenovich, M., Weiss, M.F., Monnier, V.M., July 5, 1999. Protein aging by carboxymethylation of lysines generates sites for divalent metal and redox active copper binding: relevance to diseases of glycoxidative stress. Biochem. Biophys. Res. Commun. 260 (2), 332–338.

Schegg, B., Hülsmeier, A.J., Rutschmann, C., Maag, C., Hennet, T., February 2009. Core glycosylation of collagen is initiated by two beta(1-O)galactosyltransferases. American Society for Microbiology Mol. Cell. Biol. 29 (4), 943–952.

Schneider, A., Henson, E., Blumenfeld, O.O., Gallop, P.M., June 9, 1967. The presence of lysinal (2,6-diaminohexanal) in tropocollagen. Biochem. Biophys. Res. Commun. 27 (5), 546–551.

Schwarze, U., Cundy, T., Pyott, S.M., Christiansen, H.E., Hegde, M.R., Bank, R.A., et al., January 1, 2013. Mutations in FKBP10, which result in Bruck syndrome and recessive forms of osteogenesis imperfecta, inhibit the hydroxylation of telopeptide lysines in bone collagen. Hum. Mol. Genet. 22 (1), 1–17.

Scott, J.E., Hughes, E.W., Shuttleworth, A., August 1981. A collagen-associated Ehrlich chromogen: a pyrrolic cross-link? Portland Press Limited Biosci. Rep. 1 (8), 611–618.

Sell, D.R., Monnier, V.M., 2012. Molecular basis of arterial stiffening: role of glycation - a mini-review. Karger Publishers Gerontology 58 (3), 227–237.

Shaheen, R., Alazami, A.M., Alshammari, M.J., Faqeih, E., Alhashmi, N., Mousa, N., et al., October 2012. Study of autosomal recessive osteogenesis imperfecta in Arabia reveals a novel locus defined by TMEM38B mutation. J. Med. Genet. 49 (10), 630–635.

Siegel, R.C., Fu, J.C., September 25, 1976. Collagen cross-linking. Purification and substrate specificity of lysyl oxidase. J. Biol. Chem. 251 (18), 5779–5785.

Siegel, R.C., Pinnell, S.R., Martin, G.R., November 10, 1970. Cross-linking of collagen and elastin. Properties of lysyl oxidase. Biochemistry 9 (23), 4486–4492.

Singer, F.R., Eyre, D.R., October 2008. Using biochemical markers of bone turnover in clinical practice. Clevel. Clin. J. Med. 75 (10), 739–750.

Snedeker, J.G., Gautieri, A., July 2014. The role of collagen crosslinks in ageing and diabetes - the good, the bad, and the ugly. CIC Edizioni Internazionali Muscles Ligaments Tendons J. 4 (3), 303–308.

Sricholpech, M., Perdivara, I., Nagaoka, H., Yokoyama, M., Tomer, K.B., Yamauchi, M., March 18, 2011. Lysyl hydroxylase 3 glucosylates galactosylhydroxylysine residues in type I collagen in osteoblast culture. American Society for Biochemistry and Molecular Biology J. Biol. Chem. 286 (11), 8846–8856.

Sricholpech, M., Perdivara, I., Yokoyama, M., Nagaoka, H., Terajima, M., Tomer, K.B., et al., June 29, 2012. Lysyl hydroxylase 3-mediated glucosylation in type I collagen: molecular loci and biological significance. American Society for Biochemistry and Molecular Biology J. Biol. Chem. 287 (27), 22998–23009.

Steinmann, B., Royce, P.M., Superti-Furga, A., 2002. The Ehlers-danlos syndrome. In: Connective Tissue and its Heritable Disorders. John Wiley & Sons, Inc, Hoboken, NJ, USA, pp. 431–523.

Taga, Y., Kusubata, M., Ogawa-Goto, K., Hattori, S., May 3, 2013. Site-specific quantitative analysis of overglycosylation of collagen in osteogenesis imperfecta using hydrazide chemistry and SILAC. American Chemical Society J. Proteome Res. 12 (5), 2225–2232.

Takaluoma, K., Hyry, M., Lantto, J., Sormunen, R., Bank, R.A., Kivirikko, K.I., et al., March 2, 2007. Tissue-specific changes in the hydroxylysine content and cross-links of collagens and alterations in fibril morphology in lysyl hydroxylase 1 knock-out mice. American Society for Biochemistry and Molecular Biology J. Biol. Chem. 282 (9), 6588–6596.

Tenni, R., Valli, M., Rossi, A., Cetta, G., January 15, 1993. Possible role of overglycosylation in the type I collagen triple helical domain in the molecular pathogenesis of osteogenesis imperfecta. Wiley Subscription Services, Inc., A Wiley Company Am. J. Med. Genet. 45 (2), 252–256.

Terajima, M., Perdivara, I., Sricholpech, M., Deguchi, Y., Pleshko, N., Tomer, K.B., et al., August 15, 2014. Glycosylation and cross-linking in bone type I collagen. American Society for Biochemistry and Molecular Biology J. Biol. Chem. 289 (33), 22636–22647.

Terajima, M., Taga, Y., Chen, Y., Cabral, W.A., Hou-Fu, G., Srisawasdi, S., et al., March 2, 2016. Cyclophilin-B modulates collagen cross-linking by differentially affecting lysine hydroxylation in the helical and telopeptidyl domains of tendon type I collagen. American Society for Biochemistry and Molecular Biology J. Biol. Chem. jbc.M115.699470.

Thorleifsson, G., Magnusson, K.P., Sulem, P., Walters, G.B., Gudbjartsson, D.F., Stefansson, H., et al., September 7, 2007. Common sequence variants in the LOXL1 gene confer susceptibility to exfoliation glaucoma. Science 317 (5843), 1397–1400.

Thorpe, C.T., Streeter, I., Pinchbeck, G.L., Goodship, A.E., Clegg, P.D., Birch, H.L., May 21, 2010. Aspartic acid racemization and collagen degradation markers reveal an accumulation of damage in tendon collagen that is enhanced with aging. J. Biol. Chem. 285 (21), 15674–15681.

Tombran-Tink, J., Barnstable, C.J., April 2, 2004. Osteoblasts and osteoclasts express PEDF, VEGF-A isoforms, and VEGF receptors: possible mediators of angiogenesis and matrix remodeling in the bone. Biochem. Biophys. Res. Commun. 316 (2), 573–579.

Torre-Blanco, A., Adachi, E., Hojima, Y., Wootton, J.A., Minor, R.R., Prockop, D.J., February 5, 1992. Temperature-induced post-translational overmodification of type I procollagen. Effects of over-modification of the protein on the rate of cleavage by procollagen N-proteinase and on self-assembly of collagen into fibrils. J. Biol. Chem. 267 (4), 2650–2655.

van Dijk, F.S., Nesbitt, I.M., Zwikstra, E.H., Nikkels, P.G., Piersma, S.R., Fratantoni, S.A., et al., October 2009. PPIB mutations cause severe osteogenesis imperfecta. Am. J. Hum. Genet. 85 (4), 521–527.

Vanacore, R., Ham, A.J., Voehler, M., Sanders, C.R., Conrads, T.P., Veenstra, T.D., et al., September 4, 2009. A sulfilimine bond identified in collagen IV. Science 325 (5945), 1230–1234.

Verzár, F., 1964. Aging of the collagen fiber. Elsevier Int. Rev. Connect. Tissue Res. 2, 243–300.

Verzijl, N., DeGroot, J., Thorpe, S.R., Bank, R.A., Shaw, J.N., Lyons, T.J., et al., December 15, 2000. Effect of collagen turnover on the accumulation of advanced glycation end products. J. Biol. Chem. 275 (50), 39027–39031.

Vestergaard, P., Rejnmark, L., Mosekilde, L., January 2009. Diabetes and its complications and their relationship with risk of fractures in type 1 and 2 diabetes. Calcif. Tissue Int. 84 (1), 45–55.

Viguet-Carrin, S., Gineyts, E., Bertholon, C., Delmas, P.D., January 1, 2009. Simple and sensitive method for quantification of fluorescent enzymatic mature and senescent crosslinks of collagen in bone hydrolysate using single-column high performance liquid chromatography. J. Chromatogr. B, Anal. Technol. Biomed. Life Sci. 877 (1–2), 1–7.

Volodarsky, M., Markus, B., Cohen, I., Staretz-Chacham, O., Flusser, H., Landau, D., et al., January 2013. A deletion mutation in TMEM38B associated with autosomal recessive osteogenesis imperfecta. Hum. Mutat. 34 (4) (n/a–n/a).

Vranka, J.A., Sakai, L.Y., Bächinger, H.P., May 28, 2004. Prolyl 3-hydroxylase 1, enzyme characterization and identification of a novel family of enzymes. American Society for Biochemistry and Molecular Biology J. Biol. Chem. 279 (22), 23615–23621.

Wang, C., Risteli, M., Heikkinen, J., Hussa, A.-K., Uitto, L., Myllyla, R., May 24, 2002. Identification of amino acids important for the catalytic activity of the collagen glucosyltransferase associated with the multifunctional lysyl hydroxylase 3 (LH3). J. Biol. Chem. 277 (21), 18568–18573.

Widmer, C., Gebauer, J.M., Brunstein, E., Rosenbaum, S., Zaucke, F., Drogemuller, C., et al., August 14, 2012. Molecular basis for the action of the collagen-specific chaperone Hsp47/SERPINH1 and its structure-specific client recognition. Proc. Natl. Acad. Sci. U. S. A. 109 (33), 13243–13247.

Wordinger, R.J., Clark, A.F., October 2014. Lysyl oxidases in the trabecular meshwork. J. Glaucoma 23 (8 Suppl. 1), S55–S58.

Wu, J.J., Eyre, D.R., December 30, 1985. Studies on the distribution of hydroxypyridinium cross-links in different collagen types. Ann. N. Y. Acad. Sci. 460, 520–523.

Yamauchi, M., Shiiba, M., 2008. Lysine hydroxylation and cross-linking of collagen. In: Post-translational Modifi Cations of Proteins. Methods in Molecular Biology?, vol. 446. Humana Press, Totowa, NJ, pp. 95–108.

Yamauchi, M., London, R.E., Guenat, C., Hashimoto, F., Mechanic, G.L., August 25, 1987. Structure and formation of a stable histidine-based trifunctional cross-link in skin collagen. J. Biol. Chem. 262 (24), 11428–11434.

Yang, C.L., Rui, H., Mosler, S., Notbohm, H., Sawaryn, A., Müller, P.K., May 1, 1993. Collagen II from articular cartilage and annulus fibrosus. Structural and functional implication of tissue specific posttranslational modifications of collagen molecules. Eur. J. Biochem. 213 (3), 1297–1302.

Ye, J., Rawson, R.B., Komuro, R., Chen, X., Davé, U.P., Prywes, R., et al., December 2000. ER stress induces cleavage of membrane-bound ATF6 by the same proteases that process SREBPs. Mol. Cell. 6 (6), 1355–1364.

Yeowell, H.N., Walker, L.C., April 1999. Tissue specificity of a new splice form of the human lysyl hydroxylase 2 gene. Matrix Biol. 18 (2), 179–187.

Chapter 15

Secreted noncollagenous proteins of bone

Jeffrey P. Gorski[1] and Kurt D. Hankenson[2]

[1]*Department of Oral and Craniofacial Sciences, School of Dentistry, and Center for Excellence in Mineralized and Dental Tissues, University of Missouri–Kansas City, Kansas City, MO, United States;* [2]*Department of Orthopaedic Surgery, University of Michigan Medical School, Ann Arbor, MI, United States*

Chapter outline

Introduction **359**
Proteoglycans **360**
Aggrecan and versican (PG-100) 360
Decorin (PG-II) and biglycan (PG-I) 360
Other leucine-rich repeat sequence proteins and proteoglycans 362
Hyaluronan **363**
Glycoproteins **363**
Alkaline phosphatase 363
Sclerostin 363
Periostin 365
Osteonectin (SPARC, culture shock protein, and BM40) 365
Tetranectin 365
RGD-containing glycoproteins **365**
Thrombospondins 367
Fibronectin 367
Irisin (FRCP2, fibronectin type III domain–containing protein 5) 367
Vitronectin 368
Fibrillins 368
Bone acidic glycoprotein-75 368
Small integrin-binding ligands with N-linked glycosylation 368
Osteopontin (spp, BSP-I) 368
Bone sialoprotein (BSP-II) 369
Dentin matrix protein 1 369
Matrix extracellular phosphoglycoprotein 369
γ-Carboxyglutamic acid-containing proteins 370
Matrix Gla protein 370
Osteocalcin 371
Serum proteins **371**
Other proteins **372**
Control of gene expression **372**
Bone matrix glycoproteins and ectopic calcifications **372**
Summary **373**
References **373**

Introduction

Bone differs from soft connective tissues in that it is composed of a mineralized hydroxyapatite phase and an organic phase. Although the organic matrix of bone consists primarily of collagen(s) (as reviewed early in this volume), the existence of other secreted noncollagenous components was first proposed by Herring and coworkers in the 1960s (Herring and Ashton, 1974). Historically, using degradative techniques, a variety of carbohydrate-containing moieties were extracted and partially characterized. A major breakthrough in the isolation and characterization of noncollagenous proteins of bone came with the development of "dissociative" techniques whereby proteins could be extracted in an intact, nonaggregated form (Termine et al., 1980, 1981a,b). Although they are not as abundant as so-called structural components such as collagen, their importance in bone physiology cannot be underestimated. This is emphasized by the identification of mutations in a number of these proteins that result in abnormal bone (Gorski, 2011). This chapter discusses secreted noncollagenous proteins found in bone matrix. Since the publication of the previous edition of this work in 2008, substantial additions have been made to include recently reported functional studies, inclusion of noncollagenous proteins that were not previously considered, and new information regarding the developing area of bone–muscle/fat/pancreas/testis cross-talk signaling and the role that these proteins play in these interactions.

Principles of Bone Biology. https://doi.org/10.1016/B978-0-12-814841-9.00015-4

Proteoglycans

This class of molecules is characterized by the covalent attachment of long-chain polysaccharides (glycosaminoglycans, GAGs) to core protein molecules. GAGs are composed of repeating disaccharide units that are sulfated to varying degrees and include chondroitin sulfate, dermatan sulfate, keratan sulfate, and heparan sulfate (HS). Hyaluronic acid is also a GAG found in bone, but itself is not attached to a core protein. Different subclasses of proteoglycans are generally characterized by the structure of the core protein and by the nature of the GAG (Table 15.1). Although other types of molecules can be sulfated, proteoglycans bear greater than 95% of the sulfate groups within any organic matrix (Lamoureux et al., 2007).

Aggrecan and versican (PG-100)

There are two large chondroitin sulfate proteoglycans associated with skeletal tissue that are characterized by core proteins with globular domains at the amino and carboxy termini and by binding to hyaluronan to form large aggregates. Aggrecan was assumed to be virtually cartilage specific, but its mRNA is detected in developing bone (Wong et al., 1992) and is present in high-throughput gene expression studies of osteoblasts. It should also be noted that aggrecan is an important component of the primary spongiosa that forms during endochondral ossification. In the nanomelic chick, there is a mutation in the aggrecan core protein such that it is not expressed in cartilage (Primorac et al., 1999). However, there is an impact on bones that form via the intramembranous pathway, an unanticipated finding because these bones would not be expected to be affected by abnormal cartilage development. This suggests a role for aggrecan in bone matrix separate from its role in cartilage. Mutation of the aggregan gene results in a form of spondyloepiphyseal dysplasia in humans leading to premature osteoarthritis (Gleghorn et al., 2005), and in mice, a mutation leading to a null allele results in cartilage matrix deficiency with severe skeletogenesis defects and perinatal death (*cdm* mice) (Watanabe et al., 1994). While not specifically examined, there also appear to be defects in bones that form via intramembranous ossification. However, conditional mouse models that disrupt the aggrecan gene in an osteoblast-specific manner have not been described.

Closely related, but not identical, is a soft connective tissue–enriched proteoglycan termed versican, which is mostly localized to loose, interstitial mesenchyme in developing bone. It has been hypothesized that it captures space that will ultimately become bone. It is this proteoglycan that is destroyed as osteogenesis progresses. It is noteworthy that the core protein of versican contains epidermal growth factor (EGF)-like sequences (Wight, 2002), and release of these sequences may influence the metabolism of cells in the osteoblastic lineage. As osteogenesis progresses, versican is replaced by two members of another class of proteoglycans that contain core proteins of a different chemical nature (decorin and biglycan; Fisher et al., 1989). In the *hdf* (heart defect) mouse, the versican gene is disrupted and the protein is not expressed. The mutation is embryonic lethal, and in vitro studies have shown defective chondrogenesis by limb mesenchyme (Williams et al., 2005). A conditional versican allele has been generated and mice with limb bud mesenchymal disruption show altered digit formation and alterations in transforming growth factor β (TGFβ) localization, but an effect on bone has not been described (Choocheep et al., 2010).

Decorin (PG-II) and biglycan (PG-I)

The two small proteoglycans that are heavily enriched in bone matrix are decorin and biglycan, both of which contain chondroitin sulfate chains in bone, but bear dermatan sulfate in soft connective tissues. They are characterized by core proteins that contain a leucine-rich repeat (LRR) sequence, a property shared with proteins that are associated with morphogenesis such as *Drosophila* toll protein and chaoptin, the leucine-rich protein of serum, and adenylyl cyclase (Fisher et al., 1989). As leucine-rich proteoglycans, they are included in a family referred to as small leucine-rich proteoglycans (SLRPs). The crystal structures of decorin (McEwan et al., 2006) and biglycan (Scott et al., 2006) have been determined, and based on this analysis, it appears that the SLRPs are not as curved as ribonuclease and that they may dimerize.

Although decorin and biglycan share many properties owing to the high degree of homology of their core proteins, they are also quite distinct, as best demonstrated by their patterns of expression (Bianco et al., 1990). In cartilage, decorin is found in the interterritorial matrix away from the chondrocytes, whereas biglycan is in the intraterritorial matrix. In keeping with this pattern, during endochondral bone formation, decorin is widely distributed in a pattern that is virtually indistinguishable from that of type I collagen. It first appears in preosteoblasts, is maintained in fully mature osteoblasts, and is subsequently downregulated as cells become buried in the extracellular matrix to become osteocytes. However, biglycan exhibits a much more distinctive pattern of distribution. It is found in a pericellular location in unique areas undergoing morphological delineation. It is upregulated in osteoblasts and, interestingly, its expression is maintained in osteocytes and

TABLE 15.1 Proteoglycans in bone matrix.

Molecule	Structural characteristics	Gene ontology biological processes
Aggrecan	1×10^6 to 3×10^6 MW; 210- to 250-kDa core protein; 100–150 kDa GAG chains (CS and KS); two globular domains (G1 and G2) and a third globular domain (G3)	Skeletal system development; chondrocyte development; proteolysis; cell adhesion; central nervous system development; heart development; keratan sulfate biosynthetic process; extracellular matrix disassembly; proteoglycan biosynthetic process; extracellular matrix organization; collagen fibril organization; keratan sulfate catabolic process
Versican (PG-100)	1×10^6 MW intact protein ; 360-kDa core; 12 CS chains of 45 kDa, G1 and G3 globular domains with hyaluronan-binding sites; EGF- and CRP-like sequences	Skeletal system development; osteoblast differentiation; cell adhesion; multicellular organism development; central nervous system development; cell recognition; extracellular matrix organization; GAG metabolic process; CS metabolic, biosynthetic, and catabolic processes; dermatan sulfate biosynthetic process; posttranslational protein modification; cellular protein metabolic process
Decorin	130-kDa intact protein; 38- to 45-kDa core with 10 leucine-rich repeat sequences, 1 CS chain of 40 kDa	RNA binding; protein kinase inhibitor activity; protein binding; collagen binding; GAG binding; protein N-terminus binding; extracellular matrix binding
Biglycan	270-kDa intact protein, 38- to 45-kDa core protein with 12 leucine-rich repeat sequences, two CS chains of 40 kDa	May bind to collagen; may bind to TGFβ; pericellular environment; a genetic determinant of peak bone mass
Fibromodulin	59-kDa intact protein, 42-kDa core protein with leucine-rich repeat sequences, one N-linked KS chain	TGFβ receptor complex assembly; NOT axonogenesis; KS biosynthetic process; collagen fibril organization; KS catabolic process
Osteoglycin (mimecan)	299-aa precursor, 105-aa mature protein, leucine-rich repeat sequences	Axonogenesis; regulation of receptor activity; KS biosynthetic process; KS catabolic process; negative regulation of smooth muscle cells
Asporin	43-kDa core protein	Negative regulation of protein kinase activity; cytokine-mediated signaling pathway; bone mineralization; negative regulation of TGFβ receptor signaling pathway; biomineral tissue development; negative regulation of JAK–STAT cascade; negative regulation of tooth mineralization
Osteoadherin (osteomodulin)	85-kDa intact protein, 47-kDa core protein, 11 leucine-rich repeat sequences, RGD sequence	Cell adhesion; NOT axonogenesis; KS biosynthetic process; regulation of bone mineralization; KS catabolic process
Syndecan	32-kDa intact protein	Retinoid metabolic process; ureteric bud development; GAG biosynthetic process; GAG catabolic process; inflammatory response; response to toxic substance; response to organic substance; cell migration; GAG metabolic process; wound healing; odontogenesis; response to hydrogen peroxide; myoblast development; leukocyte migration; response to glucocorticoid; response to cAMP; response to calcium ion; striated muscle cell development; Sertoli cell development; canonical Wnt signaling; positive regulation of exosomal secretion; positive regulation of extracellular exosome assembly
Perlecan	660-kDa intact protein, 450-kDa core protein, three 70- to 100-kDa HS GAGs	Interacts with matrix components, regulates cell signaling
Glypican	Lipid-linked HSPG, 66-kDa core, two HS GAGs	Copper ion binding; fibroblast growth factor binding; laminin binding; HS proteoglycan binding
Hyaluronan	Multiple proteins associated outside of the cell, structure unknown	May work with versican-like molecule to capture space destined to become bone

CRP, C-reactive protein; *CS*, chondroitin sulfate; *EGF*, epidermal growth factor; *GAG*, glycosaminoglycan; *HS*, heparan sulfate; *HSPG*, heparan sulfate proteoglycan; *KS*, keratan sulfate; *MW*, molecular weight; *TGFβ*, transforming growth factor β.
Gene ontologies listed are derived in large part from one of two on-line sources: www.informatics.jax.org/vocab/gene_ontology or www.genecards.org/search/keyword.

is localized to lacunae and canaliculi. Osteocytes act as mechanoreceptors within the bone matrix (Bonewald, 2006), and proteoglycans, possibly biglycan or cell surface–associated molecules (such as HS proteoglycans), may play a role as force transducers.

Transgenic mice that are deficient in decorin have primarily thin skin (Danielson et al., 1997), whereas mice deficient in biglycan fail to achieve peak bone mass and develop osteopenia (Xu et al., 1998). Animals that are double deficient for decorin and biglycan exhibit a phenotype reminiscent of the progeroid form of Ehlers–Danlos syndrome, whereby collagen fibrils are highly disrupted in both the dermis and the bone (Corsi et al., 2002). Disruption of biglycan in mice also impairs fracture healing, possibly through regulation of endostatin (Myren et al., 2016).

Both decorin and biglycan have been found to bind to TGFβ and to regulate its availability and activity (Markmann et al., 2000). In view of current interest in bone muscle cross talk, it is interesting that decorin also binds specifically to myostatin, another activin superfamily member, and is able to block its inhibitory actions on myoblasts by sequestration (Miura et al., 2006). Subsequent work has shown that decorin's anti-myostatin action can be largely mimicked by an N-terminal Nos. 48–71 peptide, which is capable of blocking myostatin's binding to the activin type II receptor (El Shafey et al., 2016). Interestingly, anti-myostatin peptides derived from decorin are associated with human atrial fibrillation (Barallobre-Barreiro et al., 2016). Further work is needed to determine if decorin and decorin N-terminal peptides can modulate the receptor-mediated actions of TGFβ on osteoblastic cells in a similar manner.

A novel role for biglycan, and another SLRP, fibromodulin, has also been demonstrated in regulating osteoclastogenesis (Kram et al., 2017). Double-knockout mice for both genes show significant osteopenia because of enhanced osteoclastogenesis. Biglycan and fibromodulin bind to both tumor necrosis factor α and receptor activator of NF-κB ligand (RANKL) and thereby regulate osteoclastogenesis. In the absence of the proteins, osteoblasts are unable to sequester these proosteoclastic factors, and osteoclastogenesis is unregulated.

Decorin also binds to collagen (decorating collagen fibrils), as does biglycan. This interaction has dramatic effects on collagen fibril assembly and on the mechanical properties, including tensile strength, of the resultant matrix (Reese et al., 2013). Another activity has been demonstrated by in vitro cell attachment assays in which decorin and biglycan were both found to inhibit bone cell attachment, presumably by binding to fibronectin and inhibiting its cell–matrix binding capabilities (Grzesik and Robey, 1994).

Other leucine-rich repeat sequence proteins and proteoglycans

Interestingly, there are at least 60 proteins that have been found to contain LRRs, and many of them are also SLRPs (Matsushima et al., 2000). Another SLRP found in bone is osteoglycin, previously termed "osteoinductive factor" and later found to be a protein bound to TGFβ (Ujita et al., 1995). This molecule is similar but not identical to the proteoglycan PG-Lb, which has now been found to be epiphycan, localized primarily in epiphyseal cartilage. More recently, another LRR, asporin, has been localized to developing bone. Unlike other LRRs, it has an aspartic acid–rich amino sequence (Henry et al., 2001). Asporin-knockout mice have been generated and a defect in bone was not reported (Maccarana et al., 2017).

Other members of the SLRP family found in bone include fibromodulin, as previously mentioned, which contains keratan sulfate and binds to collagen fibrils in regions distinctly different from those of decorin (Hedbom and Heinegard, 1993); osteoadherin, also known as osteomodulin, which contains the cell attachment sequence Arg-Gly-Asp (RGD) (Sommarin et al., 1998); and lumican, which may regulate collagen fibril formation. Osteomodulin is highly upregulated during bone morphogenetic protein (BMP)-induced osteoblast differentiation, but it's impacts on bone in knockout mice are minimal.

Although SLRPs appear to be "born to bind," definitive functions are not known. The mouse deficient for both biglycan and fibromodulin exhibits initial joint laxity, followed by development of extra sesamoid bones in many tendons and development of an osteoarthritis-like condition (Ameye and Young, 2002). The lumican/fibromodulin-knockout mouse resembles a variant of Ehlers–Danlos syndrome (OMIM: 130000) and has ectopic calcification (Chakravarti et al., 2003). Other proteoglycans have been isolated from a variety of animal species by using varying techniques, for example, HAPGIII (so named for its ability to bind to hydroxyapatite) and PG-100, which has been shown subsequently to be homologous to versican, as reviewed previously (Zu et al., 2007).

Although not generally found in the extracellular matrix, HS proteoglycans found associated with, or intercalated into, cell membranes may be very influential in regulating bone cell metabolism. Receptors for several growth factors (TGFβ and fibroblast growth factor 2 [FGF2], to name two) have been found to associate with HS (either bound covalently to core proteins or as free GAGs). Deficiency in the *EXT1* or *EXT2* genes in humans, which encode Golgi glycosyltransferases that are responsible for HS synthesis, results in hereditary multiple exostoses (HME) (OMIM: 133700). In HME, patients develop osteochondromas because of dysregulated BMP signaling (Pacifici, 2017).

Intercalated HS proteoglycans (the syndecan family) have been postulated to regulate cell growth, perhaps through modulation of growth factor and receptor activity (Markmann et al., 2000). Work shows that over-expression of syndecan-2 in mouse osteoblasts results in alterations in Wnt signaling and a reduction in the osteoclast regulatory protein RANKL (Mansouri et al., 2017). Perlecan is highly expressed in cartilage and has been found to interact with matrix components and to regulate cell signaling proteins, including FGF. Knockout mice have a phenotype that resembles thanatophoric dysplasia type I (OMIM: 187600) (Arikawa-Hirasawa et al., 1999), and mutations in humans are associated with Schwartz–Jampel syndrome (OMIM: 142461). Glypicans are another class of HS proteoglycans linked to cell membranes by phosphoinositol linkages that are cleavable by phospholipase C. Consequently, their activity may be in the pericellular environment or in the extracellular matrix. Mutations in glypican 3 give rise to Simpson–Golabi–Behmel syndrome (OMIM: 300037), and the knockout mouse has delayed endochondral bone formation and impaired osteoclastic development (Viviano et al., 2005). The complete cast of HS proteoglycans present in the cellular and pericellular environment in bone is not yet complete (see Zu et al., 2007, for a review).

Hyaluronan

This unsulfated GAG is not attached to a protein core and is synthesized by a completely different pathway. Hyaluronan is synthesized in the extracellular environment by a group of three plasma membrane hyaluronic acid synthases, which use cytoplasmic UDP-glucuronic acid and UDP-*N*-acetylglucosamine as substrates (Vigetti et al., 2014). Large amounts of hyaluronan are synthesized during the early stages of bone formation and may associate with versican to form high molecular weight aggregates (Fedarko et al., 1992; Falconi and Aubin, 2007). Interestingly, the long bones of hyaluronic acid synthase 2 (Has2)-deficient mice are severely shortened, implying that hyaluronic acid synthesis is required for normal longitudinal growth of all limbs (Matsumoto et al., 2009). A series of changes in the cartilage of *Has2*-null mice, e.g., reduced number of hypertrophic chondrocytes, disorganized growth plates, and lower expression of hypertrophic differentiation markers, indicate that hyaluronic acid is also needed for normal chondrocyte maturation. In other tissues, hyaluronic acid is believed to participate in cell migration and differentiation (Fedarko et al., 1992).

Glycoproteins

Virtually all noncollagenous bone matrix proteins are modified post-translationally to contain either N- or O-linked oligosaccharides, many of which can be modified further by the addition of phosphate and/or sulfate (Table 15.2). In general, compared with their soft connective tissue counterparts, bone matrix proteins are modified more extensively and in a different pattern. In some cases, differences in posttranslational modifications result from differential splicing of heterogeneous nuclear RNA, but, in general, they result from differences in the activities of enzymes located along the intracellular pathway of secretion. The pattern of post-translational modifications may be cell type specific and consequently may be of use in distinguishing protein metabolism from one tissue type versus another.

The number of secreted glycoproteins that have been identified in bone matrix grows every year. This, in part, is because of the explosion of information from high-throughput expression techniques, including RNA sequencing. What follows next is a brief description of secreted bone proteins that most likely play structural as well as metabolic roles. Other glycoprotein species have been identified primarily as growth factors and will be covered in more detail elsewhere in this volume.

Alkaline phosphatase

Tissue nonspecific alkaline phosphatase (TNAP) is a product of the *ALPL* gene. TNAP is localized to the cell membrane of osteoblasts and plays an essential role in bone mineralization. Human mutations in *ALPL* are responsible for hypophosphatasia (OMIM: 146300, 241510, and 241500), a disease condition characterized by hypomineralized bones, abnormal bone formation and skeleton growth, and dental abnormalities. TNAP is present in abundance on matrix vesicles and thus in developing bone matrix, where it degrades extracellular pyrophosphate, a potent inhibitor of bone mineralization. Mice with deficiency in TNAP have severe bone abnormalities (Millan and Whyte, 2016).

Sclerostin

Sclerostin is a product of the *SOST* gene and is secreted by kidney, vascular cells, and bone osteocytes. Early work suggested that sclerostin may work in combination with noggin to form complexes that antagonize BMP mitogenic activity

TABLE 15.2 Characteristics of glycoproteins in bone matrix.

Molecule	Structural characteristics	Gene ontology biological processes
Alkaline phosphatase	Two identical subunits of 80 kDa, disulfide-bonded, tissue-specific posttranslational modifications	Skeletal system development; osteoblast differentiation; endochondral ossification; developmental process involved in reproduction; C-terminal protein lipidation; metabolic process; dephosphorylation; biomineral tissue development; response to lipopolysaccharide; response to vitamin D; response to antibiotic; response to glucocorticoid; cellular response to organic cyclic compound; cementum mineralization
Sclerostin	24 kDa, C-terminal cysteine knot-like domain similar to BMP antagonists	Ossification; response to mechanical stimulus; Wnt signaling; negative regulation of ossification; negative regulation of BMP signaling pathway; negative regulation of protein complex assembly; positive regulation of transcription, DNA-templated; cellular response to parathyroid hormone stimulus; negative regulation of canonical Wnt signaling pathway; negative regulation of Wnt signaling pathway involved in dorsal/ventral axis specification
Periostin	93 kDa	Skeletal system development; response to hypoxia; negative regulation of cell–matrix adhesion; regulation of systemic arterial blood pressure; cell adhesion; regulation of notch signaling pathway; response to mechanical stimulus; tissue development; response to muscle activity; positive regulation of smooth muscle cell migration; extracellular matrix organization; response to estradiol; wound healing; cellular response to fibroblast growth factor stimulus; cellular response to vitamin K; cellular response to tumor necrosis factor; cellular response to transforming growth factor β stimulus; negative regulation of substrate adhesion-dependent cell spreading; positive regulation of chemokine (C–C motif) ligand 2 secretion; neuron projection extension; bone regeneration
Osteonectin/ SPARC	35–45 kDa, intramolecular disulfide bonds; α-helical amino terminus with multiple low-affinity, Ca^{2+} binding sites; two EF-hand high-affinity, Ca^{2+} sites; ovomucoid homology; glycosylated, phosphorylated, tissue-specific modifications	Ossification; negative regulation of endothelial cell proliferation; platelet degranulation; receptor-mediated endocytosis; signal transduction; heart development; response to gravity; response to lead ion; positive regulation of endothelial cell migration; negative regulation of angiogenesis; regulation of cell morphogenesis; extracellular matrix organization; lung development; response to lipopolysaccharide; response to L-ascorbic acid; response to cytokine; wound healing; regulation of cell proliferation; response to peptide hormone; pigmentation; response to ethanol; response to cadmium ion; inner ear development; response to glucocorticoid; response to cAMP; response to calcium ion; bone development; cellular response to growth factor stimulus
Tetranectin	21-kDa protein composed of four identical subunits of 5.8 kDa, sequence homologies with asialoprotein receptor and G3 domain of aggrecan	Ossification; platelet degranulation; positive regulation of plasminogen activation; bone mineralization; cellular response to organic substance; cellular response to transforming growth factor β stimulus

BMP, bone morphogenetic protein.
Gene ontologies listed are derived in large part from one of two on-line sources: www.informatics.jax.org/vocab/gene_ontology or www.genecards.org/search/keyword.

by physically sequestering the latter (Winkler et al., 2004). However, more recently, sclerostin is most recognized as an inhibitor of Wnt signaling. Sclerostin expression has been shown to be important for skeletal function because deficient individuals homozygous for a nonsense mutation in the N-terminal coding region exhibit characteristics of sclerosteosis (severe bone overgrowth often including syndactyly; Brunkow et al., 2001). Van Buchem's disease represents a milder form of sclerosteosis caused by a regulatory mutation that reduces the expression of sclerostin and results in abnormal bone growth (Vanhoenacker et al., 2003). These sclerosteosis conditions are similar to a high-bone-mass disorder caused by an activating mutation in the Wnt co-receptor, *LRP5* (Gong et al., 2001). Sclerostin is recognized as a key osteocyte-produced feedback mechanism that serves to modulate the development and activity of osteoblasts.

The functional importance of sclerostin as an inhibitor of bone formation has been reinforced through many mouse knockout and transgenic studies, as well through clinical and translational studies (Ominsky et al., 2010; Padhi et al., 2011). For example, multiple myeloma is associated with increased fracture risk and low bone density, due to the production of paracrine factors that stimulate osteoclastic bone resorption (Roodman, 2011). Blocking osteoclast action with bisphosphonates has been shown to reduce bone loss; however, it does not stimulate replacement of bone and treated patients still experience low bone mass and increased fracture risk. It has now been shown that anti-sclerostin antibodies are able to induce new bone formation in a mouse model of multiple myeloma (McDonald et al., 2017) and, importantly, tumor-bearing mice treated with both bisphosphonate and anti-sclerostin antibodies gained more bone mass and displayed better vertebral fracture resistance than bisphosphonate-treated tumor control mice. These results suggest that antibody therapy against sclerostin may ultimately be beneficial in stimulating new bone formation. However, a number of issues remain to be resolved before therapeutic anti-sclerostin antibodies can be prescribed safely for building bone mass in multiple myeloma or osteoporosis patients (Bhattacharya et al., 2018).

Periostin

The *POSTN* gene encodes a secreted $M_r = 90$ kDa extracellular protein involved in integrin-mediated adhesion of epithelial cells in heart, bone, and teeth. Deletion of *POSTN* leads to cardiac valve disease, as well as skeletal and dental abnormalities including low bone mineral density (BMD) and reduced cortical bone strength (Rios et al., 2005). As it is highly expressed in periosteal bone cells, work by Bonnet et al., 2017) suggests that periostin plays a positive role in the formation of cortical bone. They showed that periostin is a substrate and direct target of cathepsin K both in vivo and in vitro. Importantly, when cathepsin K was inhibited either pharmacologically or through genetic ablation, expression of periostin was increased in vivo, especially in the periosteum. Interestingly, the bone formation response to mechanotransduction signals was increased in cortical bone in mice treated with cathepsin K functional inhibitors or in cathepsin K–deficient mice (Bonnet et al., 2017). In a commensurate way, the mechanotransduction response was blocked in periostin-deficient mice. These results indicate that periostin may play a positive role in regulating growth and remodeling of cortical bone during development, e.g., growth in the width of long bones. It is highly likely that the ability of periostin to bind to multiple extracellular matrix proteins (fibronectin, laminin, collagens I and V, tenascin, and lysyl oxidase) contributes to its capacity to influence the transmission of mechanotransduction signals in bone. Further work is necessary to assess whether the actions of periostin on skeletal stem cells within the periosteum also contribute to this growth (Duchamp de Lageneste et al., 2018).

Osteonectin (SPARC, culture shock protein, and BM40)

With the development of procedures to demineralize and extract bone matrix proteins without the use of degradative enzymes, osteonectin was one of the first proteins isolated in intact form. This molecule was so named owing to its ability to bind to Ca^{2+}, hydroxyapatite, and collagen and to nucleate hydroxyapatite deposition (Termine et al., 1981a,b). Osteonectin is çommonly referred to as SPARC (secreted protein acidic and rich in cysteine). It was one of the first extracellular proteins to be characterized as a "matricellular" protein by Sage and Bornstein (Sage and Bornstein, 1991). The osteonectin molecule contains several different structural features, the most notable of which is the presence of two EF-hand high-affinity calcium-binding sites. These structures are usually found in intracellular proteins, such as calmodulin, that function in calcium metabolism (Bhattacharya et al., 2004).

Although osteonectin is highly enriched in bone, it is also expressed in a variety of other connective tissues at specific points during development, maturation, or repair processes in vivo. It was identified after induction by cAMP in teratocarcinoma cells and was found to be produced at very early stages of embryogenesis. Interestingly, if osteonectin is inactivated by the use of blocking antibodies during tadpole development, there is a disruption of somite formation and subsequent malformations in the head and trunk (Purcell et al., 1993). Mice that are deficient in osteonectin present with severe cataracts (Bassuk et al., 1999) and develop severe osteopenia characterized by decreased trabecular connectivity, decreased mineral content, but increased apatite crystal size (Mansergh et al., 2007).

Tetranectin

This tetrameric protein has been identified in woven bone and in tumors undergoing mineralization (Wewer et al., 1994). This protein shares sequence homologies with globular domains of aggrecan and asialoprotein receptor. Mice deficient in tetranectin develop spinal deformities (Iba et al., 2001), but the specific functional role that tetranectin plays in osteogenesis is unknown.

RGD-containing glycoproteins

Some of the major glycoproteins in bone matrix (collagens, thrombospondin (TSP), fibronectin, vitronectin, fibrillins, osteoadherin, osteopontin (OPN), bone sialoprotein (BSP), dentin matrix protein 1 (DMP1), and proteins derived from the dentin sialophosphoprotein gene) contain the amino acid sequence RGD, which conveys the ability of the extracellular matrix protein to bind to the integrin class of cell surface receptors (Table 15.3). This binding is the basis of many cell attachment activities that have been identified by in vitro analysis; however, it should be noted that in many cases it is not clear how this in vitro activity translates into in vivo physiology.

TABLE 15.3 Glycoproteins in bone matrix, continued: RGD-containing glycoproteins.

Molecule	Structural characteristics	Gene ontology biological processes
Thrombospondins	450-kDa molecule; three identical disulfide-linked subunits of 150–180 kDa; homologies to fibrinogen, properdin, EGF, collagen, von Willebrand Factor , *Plasmodium falciparum*, and calmodulin; RGD at the C-terminal globular domain	Apoptosis-related network due to altered Notch3 in ovarian cancer; bladder cancer; focal adhesion; inflammatory response pathway; microRNA targets in ECM and membrane receptors; senescence and autophagy in cancer; syndecan-4-mediated signaling events; TGFβ receptor signaling; TGFβ signaling pathway
Fibronectin	400 kDa with two nonidentical subunits of 200 kDa, composed of type I, II, and III repeats; RGD in the 11th type III repeat 2/3 from the N terminus	MicroRNA targets in ECM and membrane receptors; nanoparticle-mediated activation of receptor signaling overview of nanoparticle effects; Rac1/Pak1/p38/MMP-2 pathway; regulation of actin cytoskeleton; senescence and autophagy in cancer; simplified interaction map between LOXL4 and oxidative stress pathway; syndecan-2-mediated signaling events; syndecan-4-mediated signaling events; TGFβ signaling in thyroid cells for epithelial–mesenchymal transition; TGFβ signaling pathway; VEGFR3 signaling in lymphatic endothelium
Irisin	24-kDa protein with fibronectin III domain, N-glycosylated, extracellular domain is membrane cleaved	Biological process; regulation of receptor activity; response to muscle activity; positive regulation of brown fat cell differentiation
Vitronectin	Cell attachment protein, terminus, homology to somatomedin B, rich in cysteines, sulfated, phosphorylated	αMβ2 integrin signaling; focal adhesion; FOXA1 transcription factor network inflammatory response pathway; integrins in angiogenesis; primary focal segmental glomerulosclerosis; senescence and autophagy in cancer
Fibrillin1	350 kDa, EGF-like domains, RGD, cysteine motifs	Integrin binding; hormone activity; ECM structural constituent; calcium ion binding; protein binding; heparin binding; ECM constituent conferring elasticity; protein complex binding; identical protein binding
Osteopontin	44–75 kDa, polyaspartyl stretches, no disulfide bonds, glycosylated, phosphorylated, RGD located 2/3 from the N terminus	Brain-derived neurotrophic factor signaling pathway; direct *p53* effectors; endochondral ossification; FGF signaling pathway; focal adhesion; integrins in angiogenesis; lung fibrosis; osteoclast signaling; osteopontin signaling; osteopontin-mediated events; TGFβ receptor signaling; Toll-like receptor signaling pathway
Bone sialoprotein	46–75 kDa, polyglutamyl stretches, no disulfide bonds, 50% carbohydrate, tyrosine-sulfated, RGD near the C terminus	Focal adhesion; interleukin-11 signaling pathway; osteoblast signaling
Matrix extracellular phosphoglycoprotein	58 kDa	Skeletal system development; biomineral tissue development; posttranslational protein modification; cellular protein metabolic process
BAG-75	75 kDa; sequence homologies to dentin matrix protein-1, osteopontin, and bone sialoprotein; 7% sialic acid, 8% phosphate, 29% acidic amino acid content	Binds large number of calcium ions/mole, may act as a cell attachment protein (RGD sequence not yet confirmed), may regulate bone resorption

ECM, extracellular matrix; *EGF*, epidermal growth factor; *FGF*, fibroblast growth factor; *FOXA1*, forkhead box A1; *LOXL4*, lysyl oxidase-like 4; *MMP-2*, matrix metalloproteinase 2; *TGFβ*, transforming growth factor β; *VEGFR3*, vascular endothelial growth factor receptor 3.
Gene ontologies listed are derived in large part from one of two on-line sources: www.informatics.jax.org/vocab/gene_ontology or www.genecards.org/search/keyword.

Thrombospondins

TSPs are a family of five multidomain, matricellular proteins. TSP-1 was first identified as the most abundant protein in platelet alpha granules, but is found in many tissues during development, including bone, where it is highly expressed (Robey et al., 1989). Subsequently, four other members have been described, including COMP (cartilage oligomeric matrix protein) as TSP-5 (reviewed in Adams, 2004). In bone, all five proteins are present, synthesized by different cell types at different stages of maturation and development (Carron et al., 1999), but TSP-1 and TSP-2 are the most highly expressed by osteoblast lineage cells and the best characterized TSP proteins in bone. TSP-1 and TSP-2 are trimeric proteins, whereas TSP-3 to 5 are pentameric. TSP-1 and 2 are able to bind to a variety of cell surface receptors, including CD36, CD47, LRP1, and integrins, but in addition can bind to other matrix proteins, matrix metalloproteinases (MMPs), and growth factors, particularly TGFβ. The bone phenotype of TSP-2-deficient mice has been extensively characterized. These mice exhibit an increase in osteoblast progenitors, but have defective mineralization. Cortical bone is thicker, but the bones are reduced in periosteal diameter. Manley et al. have shown that cells deficient in TSP-2 are unable to effectively process collagen and the mice have a brittle bone phenotype (Manley et al., 2015). Mice deficient in TSP-2 also show accelerated bone regeneration characterized by reduced chondrogenesis and accelerated intramembranous bone formation. This phenotype is linked to the well-described role of TSPs in regulating vascularization.

Fibronectin

Fibronectin is synthesized by virtually all connective tissue cells and is a major component of serum. There are a large number of different mRNA splice variants such that the number of potential forms is quite high. Consequently, bone matrix could contain fibronectin that originates from exogenous as well as endogenous sources (Pankov and Yamada, 2002). The precise form that is present in cells of the osteoblastic lineage is unknown. Fibronectin is produced during early stages of bone formation and has been found to be highly upregulated in the osteoblastic cell layer. Interestingly, bone cell attachment to fibronectin in vitro takes place in an RGD-independent fashion (Grzesik and Robey, 1994). However, this correlates well with the expression of the fibronectin receptor α4β1, which binds to a sequence other than RGD in the fibronectin molecule and is also expressed by some osteoblastic cells. Cell–matrix interactions mediated by fibronectin receptor–α4β1 binding may play a role in the maturation sequence of cells in the osteoblastic lineage.

In 2017, Lee et al. showed that mutations in fibronectin cause a subtype of spondylometaphyseal dysplasia (SMD) with "corner fractures" at metaphyses. SMDs are often characterized by growth plate abnormalities, short stature, and vertebral problems. Comparisons of the exomes of three individuals with SMD with corner fractures yielded fibronectin mutations in highly conserved residues. The functional consequences of these mutations were examined by expressing the three missense variants into a recombinant secreted N-terminal 70-kDa fragment and into full-length fibronectin. While wild-type forms were secreted into the culture medium as expected, all mutant proteins were either not secreted or secreted only at dramatically lower amounts (Lee et al., 2017). Thus, SMD with corner fractures is caused by defective fibronectin secretion and reinforces the importance of fibronectin in cartilage and bone, two tissues whose structure is altered with functional consequences.

Irisin (FRCP2, fibronectin type III domain–containing protein 5)

Irisin is encoded by a distinct gene that generates several different transcripts produced by alternative splicing. Fibronectin type III domain–containing protein 5 (FNDC5)/irisin, a novel secreted energy-regulating hormone, has been associated with bone lipid and carbohydrate metabolism. Expressed at high levels by muscle, particularly when exercised, irisin has been implicated in muscle–bone cross-talk signalling reactions. Several specific examples illustrate this point.

For one, the use of models of mechanical unloading (hindlimb unloading and sciatic neurectomy) reduce both tibial trabecular bone mineral and surrounding muscle volume. Interestingly, *irisin* mRNA in soleus muscle correlated positively with bone BMD in respective controls and in unloaded mice (Kawao et al., 2018). Since irisin also negatively correlated with RANKL mRNA levels in bone of control and hindlimb-unloaded mice, these authors suggested that irisin is directly implicated in the suppression of osteoclasts and osteoclastic bone resorption, particularly in situations where mechanical stress regulates muscle and bone structure. Using similar mechanical-unloading models in mice, Colaianni et al. showed also that irisin was able not only to prevent but also to restore bone and muscle loss (Colaianni et al., 2017). Treatment maintained muscle volume as well as fiber cross-sectional area and myosin type II composition. These data followed up on earlier pioneering work showing that irisin had positive effects on anabolic bone formation in vivo in mice (Colaianni et al., 2015). Finally, intraperitoneal injection of irisin has also been shown to increase trabecular and cortical bone volume in mice, a result mimicked by intraperitoneal FNDC5.

Vitronectin

This serum protein, first identified as S-protein because of its cell-spreading activity, is found at low levels in mineralized matrix (Grzesik and Robey, 1994). Its cell surface receptor, $\alpha V\beta 3$, is distributed broadly throughout bone tissue. In addition to cell attachment activity, it also binds to and affects the activity of plasminogen activator inhibitor (Schvartz et al., 1999). Osteoclasts utilize vitronectin as a cell adhesion substrate when bound to bone.

Fibrillins

In addition to their RGD sequences, fibrillin-1 and fibrillin-2 are glycoproteins that also contain multiple EGF-like repeats. Although generally minor constituents of skeletal tissues in terms of mass, fibrillins are major components of microfibrils, and mutations in these genes lead to Marfan's syndrome and congenital contractural arachnodactyly, which exhibit abnormalities in bone growth (reviewed in Ramirez and Dietz, 2007). Fibrillin-1 and fibrillin-2 associate with LTBP (latent TGFβ-binding protein) in microfibrils and regulate TGFβ bioactivity (Chaudhry et al., 2007). Deletion of fibrillin-1 with Prx1 Cre leads to abnormal mesenchymal stem cell differentiation with resultant bone loss (Smaldone et al., 2016). Work also demonstrates that fibrillin-2 exhibits a special capacity to regulate BMP-7 activity during limb patterning (Nistala et al., 2010). In addition, both fibrillin-1 and fibrillin-2 regulate commitment and differentiation of bone marrow skeletal stem cells (Smaldone et al., 2016). Taken together, the data show that fibrillins function as critical components of the extracellular matrix where, in combination with activin growth and differentiation factors, they regulate the differentiation of skeletal stem cells directly and hemopoietic stem cells indirectly.

Bone acidic glycoprotein-75

The identity of bone acidic glyprotein-75 (BAG-75) relies upon its detection and recognition with three different antibodies, e.g., a monoclonal antibody (HTP IV-#1), a rabbit polyclonal antibody (#704), and an anti-peptide #3−13 antibody against its N-terminal sequence, and its unique N-terminal sequence as well as accumulated internal peptide sequences totaling about 100 residues (Gorski and Shimizu, 1988; Gorski et al., 1990; Wang et al., 2009) (J.P. Gorski, personal communication). Functionally, direct comparisons with OPN and BSP revealed that BAG-75 bound more calcium ions per mole (139 atoms/molecule) than either of the other bone glycoproteins (Chen et al., 1992). When added to osteoclast cultures, BAG-75 was also able to block resorption, presumably by blocking or interfering with the adhesion of osteoclasts with mineralized bone matrix (Sato et al., 1992). Finally, an intriguing property of BAG-75 remains its capacity to identify matrix sites within developing primary bone or in osteoblastic cultures that will be subsequently mineralized (Gorski et al., 2004; Midura et al., 2004). In this sense, it remains an unusual biomarker of sites that are about to become calcified.

Small integrin-binding ligands with N-linked glycosylation

Several bone matrix proteins are characterized not only by the inclusion of RGD within their sequences, but also by the presence of relatively large amounts of sialic acid. Interestingly, they are clustered at a distinct gene locus, 4q21−q23 and appear to have arisen by gene duplication. For this reason, the family has been termed SIBLINGs (small integrin-binding ligands with N-linked glycosylation) (Fisher and Fedarko, 2003) (Table 15.3). The family includes OPN and BSP, the two best characterized proteins of the family; along with DMP1, dentin sialoprotein, and dentin phosphoprotein, which are coded for by the same gene, now termed dentin sialophosphoprotein; matrix extracellular phosphoglycoprotein (MEPE); and the more distantly related protein, enamelin. Although the SIBLINGs were initially thought to be specific for mineralized tissues, it is now apparent that many of them are expressed in metabolically active epithelial cells (Ogbureke and Fisher, 2005). Interestingly, three of the family members bind to and activate specific MMPs (BSP−MMP-2, OPN−MMP-3, and DMP1−MMP-9) (Fedarko et al., 2004).

Osteopontin (spp, BSP-I)

This sialoprotein was first identified in bone matrix extracts, but it was also identified as the primary protein induced by cellular transformation. In bone, it is produced at late stages of osteoblastic maturation corresponding to stages of matrix formation just prior to mineralization. In vitro, it mediates the attachment of many cell types, including osteoclasts. In osteoclasts, it has also been reported to induce intracellular signaling pathways as well. In addition to the RGD sequence, it also contains stretches of polyaspartic acid and it has a fairly high capacity for binding Ca^{2+}ions; however, it does not appear to nucleate hydroxyapatite formation in several different assays. The OPN-deficient mouse develops normally, but has increased mineral content, although the crystals are smaller than normal. The role of OPN in skeletal homeostasis is covered elsewhere in this volume, and there are numerous reviews available for its role in cancer and immune function (for example, Scatena et al., 2007).

Bone sialoprotein (BSP-II)

The other major sialoprotein is BSP, composed of 50% carbohydrate (12% is sialic acid) and stretches of polyglutamic acid (as opposed to polyaspartic acid in OPN). The RGD sequence is located at the carboxy terminus of the molecule, whereas it is located centrally in OPN. The sequence is also characterized by multiple tyrosine sulfation consensus sequences found throughout the molecule, in particular, in regions flanking the RGD (Fisher et al., 1990). Sulfated BSP has been isolated in a number of animal species, but the levels appear to be variable.

BSP exhibits a more limited pattern of expression than OPN. In general, its expression is tightly associated with mineralization phenomena (although there are exceptions). In the skeleton, it is found at low levels in chondrocytes, in hypertrophic cartilage, in a subset of osteoblasts at the onset of matrix mineralization, and in osteoclasts (Bianco et al., 1991). Consequently, BSP expression marks a late stage of osteoblastic differentiation and an early stage of matrix mineralization. Outside of the skeleton, BSP is found in trophoblasts in placental membranes, which in late stages of gestation fuse and form mineralized foci.

BSP may be multifunctional in osteoblastic metabolism. It is very clear that it plays a role in matrix mineralization as supported by the timing of its appearance in relationship to the appearance of mineral and its Ca^{2+}-binding properties. BSP has a high capacity for binding calcium ions. The polyglutamyl stretches were thought to be solely responsible for this property; however, studies using recombinant peptides suggest that although the polyglutamyl stretches are required, they are not the sole determinants (Stubbs et al., 1997). Unlike OPN, BSP does nucleate hydroxyapatite deposition in a variety of in vitro assays.

It is also clear from in vitro assays that BSP is capable of mediating cell attachment, most likely through interaction with the somewhat ubiquitous $\alpha V\beta 3$ (vitronectin) receptor. Bone cells attach to the intact molecule in an RGD-dependent fashion. However, when BSP is fragmented, either endogenously by cells or using commercially available enzymes, the fragment most active in cell attachment does not contain the RGD sequence (Mintz et al., 1993). Studies indicate that the sequence upstream from the RGD mediates attachment (in an RGD-independent fashion) and suggest that the integrin-binding site is more extended than had been envisioned previously. Sequences flanking the RGD site are often tyrosine sulfated. However, it is not known how sulfation influences BSP activity, as in vitro, unsulfated BSP appears to be equivalent in its activity. Once again, it is not clear if currently available in vitro assays are sufficiently sophisticated to determine what influence post-translational modifications, such as sulfation, have on the biological activity. In addition to sulfation, conformation of the RGD site may also influence the activity of the protein. Although the RGD region in fibronectin is found in a looped-out region that is stabilized by disulfide bonding, there are no disulfide bonds in BSP. However, the flanking sequences most likely influence the conformation of the region. The cyclic conformations also appear to have a higher affinity for cell surface receptors than linear sequences (van der Pluijm et al., 1996).

BSP-knockout mice have been extensively studied by the Malaval laboratory (Bouleftour et al., 2016). Initial studies of the knockout mouse clearly implicate BSP in playing an important role in mineralization and cortical or primary bone formation, but not endochondral bone formation (Bouleftour et al., 2014). More recent studies have demonstrated that BSP can also influence hematopoiesis and vascularization (Granito et al., 2015). Interestingly, in the absence of BSP, OPN is upregulated and it is speculated that some aspects of the BSP-deficient phenotype may be secondary to upregulation of OPN (Granito et al., 2015; Bouleftour et al., 2016)

Dentin matrix protein 1

Although DMP1 was originally thought to be specific to dentin, it was subsequently found to be synthesized by osteoblasts as well (D'Souza et al., 1997). However, its function in bone metabolism is not known as of this writing. The DMP1-deficient mouse has craniofacial and growth plate abnormalities, along with rickets and osteomalacia owing to increased secretion of the phosphate-regulating hormone FGF23. The overproduction of FGF23 is hypothesized to be caused by abnormal osteocyte function (Feng et al., 2006). Mutations in *DMP1* have been identified in patients with dentinogenesis imperfecta (OMIM: 600980) and in forms of autosomal recessive hypophosphatemic rickets (OMIM: 241520). *DMP1* is well recognized as a gene highly expressed by early osteocytes.

Matrix extracellular phosphoglycoprotein

MEPE is expressed in bone and bone marrow, but also at high levels in the brain and low levels in the lung, kidney, and placenta. This protein is also highly expressed by tumors that induce osteomalacia, and it has been regarded as a potential phosphate-regulating hormone (Rowe et al., 2000). It is thought that the C-terminal portion of MEPE (ASARM), together with PHEX, regulates mineralization and renal phosphate metabolism (Rowe, 2004). Conversely, animals deficient in MEPE have significantly increased bone mass owing to increased numbers of trabeculae and increased cortical thickness (Gowen et al., 2003).

γ-Carboxyglutamic acid-containing proteins

In bone matrix, there are three proteins that undergo γ-carboxylation via vitamin K-dependent enzymes: matrix γ-carboxyglutamic acid (Gla) protein (MGP) (Price et al., 1983) and osteocalcin (Gla protein) (Price et al., 1976), both of which are made by bone cells, and protein S (made primarily in the liver but also made by osteogenic cells) (Maillard et al., 1992) (Table 15.4). The presence of dicarboxylic glutamyl residues confers low-affinity calcium-binding properties to these proteins.

Matrix Gla protein

MGP is found in many connective tissues and is highly expressed in cartilage. It appears that the physiological role of MGP is to act as an inhibitor of mineral deposition. Similar to secreted proteins in other calcium-rich environments, e.g., casein, MGP possesses three highly conserved phosphorylation sequence motifs (Ser-X-Glu/Ser(P)) within its N-terminal region (Price et al., 1994). MGP-deficient mice develop calcification in extraskeletal sites such as in the aorta (Luo et al., 1997). Interestingly, the vascular calcification proceeds via transition of vascular smooth muscle cells into chondrocytes, which subsequently hypertrophy (El-Maadawy et al., 2003). In humans, mutations in *MGP* have also been associated with excessive cartilage calcification (Keutel syndrome, OMIM: 245150). And it has been reported that deficiency of MGP in mice closely mimics the phenotype of patients with Keutel syndrome (Marulanda et al., 2017). A 2017 meta-analysis of 23 different case−control studies now shows that there is a significant association between the *MGP* gene rs1800801 polymorphism and vascular calcification and atherosclerotic disease, particularly in the Caucasian subgroup (Sheng et al., 2017).

Functionally, MGP is an antagonist of BMPs and is highly expressed by vascular endothelial cells. However, another consequence of deletion of *MGP* in mice is the dysregulation of endothelial cell differentiation from mouse embryonic stem cells (Yao et al., 2016). These authors have subsequently shown that this effect may have broader downstream effects,

TABLE 15.4 Characteristics of γ-carboxyglutamic acid−containing proteins in bone matrix.

Molecule	Structural characteristics	Gene ontology biological processes
Matrix Gla protein	15 kDa, five Gla residues, one disulfide bridge, phosphoserine residues	Cartilage condensation; ossification; multicellular organism development; cell differentiation; regulation of bone mineralization; cartilage development
Osteocalcin	5 kDa, one disulfide bridge, Gla residues located in α-helical region	Skeletal system development; ossification; osteoblast differentiation; osteoblast development; signal peptide processing; ER-to-Golgi vesicle-mediated transport; cell adhesion; aging; cell aging; response to mechanical stimulus; response to gravity; response to inorganic substance; response to zinc ion; response to organic cyclic compound; response to activity; peptidylglutamic acid carboxylation; bone mineralization; regulation of bone mineralization; biomineral tissue development; response to nutrient levels; response to vitamin K; response to vitamin D; response to testosterone; response to hydroxyisoflavone; odontogenesis; response to drug; response to estrogen; regulation of bone resorption; response to ethanol; regulation of osteoclast differentiation; response to glucocorticoid; bone development; cellular response to vitamin D; cellular response to growth factor stimulus
Protein S	75 kDa	Platelet degranulation; signal peptide processing; ER-to-Golgi vesicle-mediated transport; blood coagulation; hemostasis; negative regulation of endopeptidase activity; peptidylglutamic acid carboxylation; negative regulation of blood coagulation; regulation of complement activation; fibrinolysis; leukocyte migration

ER, endoplasmic reticulum.
Gene ontologies listed are derived in large part from one of two on-line sources: www.informatics.jax.org/vocab/gene_ontology or www.genecards.org/search/keyword.

since dysregulated *MGP*-deficient endothelial cells were subsequently shown to stimulate epithelial cell precursors in the lung toward a hepatic rather than pulmonary phenotype via altered cross-talk signaling (Yao et al., 2016). Finally, MGP has been suggested to act as an inhibitor of BMPs. While not providing mechanistic details, the work of Malhotra et al. (2015) demonstrates that BMPs play a key role in the signal transduction pathway regulating medial vascular aortic calcification in MGP-deficient mice. Specifically, two separate BMP inhibitors lowered calcification by 80% and prolonged life span compared with vehicle controls (Malhotra et al., 2015).

Osteocalcin

Whereas MGP is broadly expressed, osteocalcin is somewhat bone specific, although mRNA has been found in platelets and megakaryocytes (Thiede et al., 1994). Osteocalcin-deficient mice were first reported to have increased BMD compared with normal controls (Ducy et al., 1996). In human bone, it is concentrated in osteocytes, and its release may be a signal in the bone-turnover cascade (Kasai et al., 1994). Osteocalcin measurements in serum have proved valuable as a marker of bone turnover in metabolic disease states.

Importantly, it has been demonstrated that osteocalcin also acts as a bone-derived hormone that influences energy metabolism by muscle and fat cells, and, that regulates male reproductive function (Lee et al., 2007; Karsenty and Mera, 2017). This topic is considered in detail in Chapter 86. By specifically binding to the G-coupled receptor Gprc6a (Pi and Quarles, 2012), decarboxylated osteocalcin increases insulin secretion by pancreatic cells, promotes glucose homeostasis by muscle and fat tissues, and stimulates testosterone synthesis by Leydig cells of the testis (Ferron et al., 2012; Karsenty and Oury, 2014; Oury et al., 2015). These data provide clear evidence that bone can act as an endocrine organ, regulating systemic functions via receptor-mediated signaling pathways in metabolically active tissues such as bone, muscle, pancreas, fat, and testis. It is expected that new discoveries will help further define the detailed regulatory loops and feedback networks that are likely to exist among and between these skeletal tissues.

Serum proteins

The presence of hydroxyapatite in the bone matrix accounts for the adsorption of a large number of proteins that are synthesized elsewhere and brought in contact with bone via the circulation (Delmas et al., 1984). Most of these proteins are synthesized in the liver and hematopoietic tissue and represent classes of immunoglobulins, carrier proteins, cytokines, chemokines, and growth factors. Interestingly, some of these proteins are also synthesized endogenously by cells of the osteoblastic lineage. It is not known if the origin of a particular factor (and hence proteins with potentially different post-translational modifications) affects its function.

Although serum proteins are not synthesized locally, they may have a significant impact on bone metabolism (Table 15.5). Albumin, which is synthesized by the liver, is concentrated in bone severalfold above levels found in the

TABLE 15.5 Characteristics of serum proteins found in bone matrix.

Molecule	Structural characteristics	Gene ontology biological processes
Albumin	69 kDa, nonglycosylated, one sulfhydryl, 17 disulfide bonds, high-affinity hydrophobic binding pocket	Retina homeostasis; platelet degranulation; receptor-mediated endocytosis; cellular response to starvation; bile acid and bile salt transport; hemolysis by symbiont of host erythrocytes; high-density lipoprotein particle remodeling; negative regulation of apoptotic process; negative regulation of programmed cell death; sodium-independent organic anion transport; posttranslational protein modification; cellular protein metabolic process; maintenance of mitochondrion location; cellular oxidant detoxification
α_2-HS-glycoprotein	Precursor protein of fetuin, cleaved to form A and B chains that are disulfide linked, Ala-Ala and Pro-Pro repeat sequences, N-linked oligosaccharides, cystatin-like domains	Skeletal system development; ossification; platelet degranulation; pinocytosis; acute-phase response; negative regulation of endopeptidase activity; regulation of bone mineralization; negative regulation of bone mineralization; negative regulation of phosphorylation; neutrophil degranulation; post-translational protein modification; cellular protein metabolic process; negative regulation of insulin receptor signaling pathway; regulation of inflammatory response; positive regulation of phagocytosis; negative regulation of biomineral tissue development

Gene ontologies listed are derived in large part from one of two on-line sources: www.informatics.jax.org/vocab/gene_ontology or www.genecards.org/search/keyword.

circulation. It is not known whether it plays a structural role in bone matrix formation but it does have an influence on hydroxyapatite formation. In in vitro assays, albumin inhibits hydroxyapatite growth by binding to several faces of the seed crystal (Garnett and Dieppe, 1990). In addition to this inhibitory activity, it also inhibits crystal aggregation.

Another serum protein, α_2-HS-glycoprotein, is even more highly concentrated in bone than albumin (up to 100 times more concentrated). It is known that α_2-HS-glycoprotein is the human analog of bovine fetuin (Ohnishi et al., 1993). This protein is synthesized as a precursor that contains a disulfide bond linking the amino- and carboxy-terminal regions. Subsequently, the midregion is cleaved and removed from the molecule, yielding the A and B peptides (much in the same way that insulin is processed). In rat, the midregion is not removed and the molecule consists of a single polypeptide. This protein also contains cystatin-like domains (disulfide-linked loop regions) and another member of this family has been identified in bone matrix extracts.

α_2-HS-glycoprotein has many proposed functions that may also be operative in bone cell metabolism. It is a chemoattractant for monocytic cells, and consequently, it may influence the influx of osteoclastic precursor cells into a particular area (Nakamura et al., 1999). Furthermore, it is a TGFβ type II receptor mimic and cytokine antagonist (Demetriou et al., 1996). Fetuin, the bovine homolog, has been found to be a major growth-promoting factor in serum, and results in vitro suggest that fetuin, along with TGFβ, inhibits osteogenesis (Binkert et al., 1999). Interestingly, deletion of the fetuin gene in mice leads to widespread ectopic calcification throughout the body (Jahnen-Dechent et al., 1997). Consequently, this protein may play a very important role in bone cell metabolism irrespective of whether or not it is synthesized locally.

Other proteins

In addition to the proteins already described, there are representatives of many other classes of proteins in the bone matrix, including proteolipids, enzymes and their inhibitors (including metalloproteinases and tissue inhibitors of metalloproteinase, plasminogen activator and plasminogen activator inhibitor, matrix phosphoprotein kinases, and lysosomal enzymes), morphogenetic proteins, and growth factors (Zu et al., 2007). Although their influence on bone cell metabolism is highly significant and they may cause important alterations of the major structural elements of bone matrix, they are not necessarily part of the bone matrix (with the possible exception of proteolipids). Important aspects of many of these classes of proteins are reviewed elsewhere.

Control of gene expression

Even a cursory analysis of noncollagenous protein expression reveals a complex regulatory pathway of interacting environmental, hormonal, and nuclear factors by which cells of the osteoblastic lineage secrete these factors in a temporal and distribution-sensitive manner to produce biomechanically functional mineralized bone (Stein et al., 1996; Wu et al., 2014).

Bone matrix glycoproteins and ectopic calcifications

The development of sensitive radiographic techniques, in addition to histological observations, has led to the description of ectopic calcifications in many different pathological disorders. Although dystrophic mineralization has long been noted, it was not thought that bone matrix proteins played a role in generating this type of mineralization. Dystrophic mineralization (such as in traumatic muscle injury) is brought about by cell death (perhaps in the form of apoptosis) and not by the physiological pathways mediated by collagen or matrix vesicles. However, bone matrix proteins have been identified in mineralized foci in several different pathological states. Osteonectin, OPN, and BSP have been found in mineralized foci in primary breast cancer (Bellahcene and Castronovo, 1997). BSP has also been found in other cancers, such as prostate, thyroid, and lung (Bellahcene et al., 1997; Bellahcene et al., 1998; Waltregny et al., 1998). Although it is possible that, in some cases, the area mineralizes dystrophically and bone matrix proteins are adsorbed from the circulation because of their affinity to hydroxyapatite, mRNAs for the bone matrix proteins are expressed and it appears that the proteins are actually synthesized by resident cells that have been triggered by factors that have yet to be identified.

Given the fact that bone matrix proteins are expressed by a number of different cancers, the obvious question is why. The processes by which cancer cells invade the surrounding normal tissue, gain entry into the circulation, and metastasize to other tissues are complex, but members of the SIBLING family have emerged as active players in tumorigenesis and metastasis. SIBLINGs and their proteolytic fragments, along with their partner MMPs (for BSP, OPN, and DMP1, at least), may modulate a tumor cell's adhesion via specific integrins, matrix degradation, and migration (reviewed in Bellahcene et al., 2008).

Another example of ectopic calcification is seen in atherosclerosis, again, associated with the production of bone matrix proteins (Bini et al., 1999). Unlike dystrophic calcification, vascular calcification appears to form in a regulated fashion similar to what is seen in bone formation in the skeleton. As mentioned earlier in relationship to the *MGP*-deficient mouse, it appears that vascular smooth muscle cells are emerging as the culprit. Factors that stimulate the expression of bone matrix proteins are not yet known but may include changes in cell—cell interactions, serum lipid composition, and phosphate concentrations, and apoptosis of vascular smooth muscle cells may initiate the process (Johnson et al., 2007). It may also be that there exists a population of stem cells that are normally quiescent, but then are induced to become osteogenic again, by factors that are not known. The aorta has its own vasculature, which may harbor these stem cells. Supporting this hypothesis, pericytes from the retinal vasculature have been shown to undergo bone formation in vitro and in vivo (Canfield et al., 2000).

Summary

Bone matrix proteoglycans and glycoproteins are proportionally the most abundant constituents of the noncollagenous proteins in bone matrix. Proteoglycans with protein cores composed of the LRR sequences (decorin, biglycan, fibromodulin, and osteoadherin) are the predominant form found in mineralized matrix, although hyaluronan-binding forms (in particular, versican) are present during early stages of osteogenesis. They participate in matrix organization and in regulating growth factor activity. Glycoproteins such as alkaline phosphatase, osteonectin, RGD-containing proteins (osteoadherin, TSP, fibronectin, vitronectin, OPN, and BSP), irisin, fibrillin, and tetranectin are produced at different stages of osteoblastic maturation. They exhibit a broad array of functions ranging from control of cell proliferation to cell—matrix interactions, mediation of hydroxyapatite deposition, and bone—muscle/adipocyte cross-talk signaling. Finally, sclerostin and DMP1 are preferentially expressed by osteocytic cells in bone where they regulate osteogenesis by feeding back to osteoblastic cells and (together with PHEX) regulate phosphate metabolism via FGF23 production by osteocytes, respectively. The ectopic expression of bone cell secreted noncollagenous proteins may also play a significant role in, normal systemic regulation of glucose and energy metabolism, and of male reproductive function, as well as in pathological processes such as bone metastasis and in atherosclerosis.

References

Adams, J.C., 2004. Functions of the conserved thrombospondin carboxy-terminal cassette in cell-extracellular matrix interactions and signaling. Int. J. Biochem. Cell Biol. 36 (6), 1102—1114.

Ameye, L., Young, M.F., 2002. Mice deficient in small leucine-rich proteoglycans: novel in vivo models for osteoporosis, osteoarthritis, Ehlers-Danlos syndrome, muscular dystrophy, and corneal diseases. Glycobiology 12 (9), 107R—116R.

Arikawa-Hirasawa, E., Watanabe, H., Takami, H., Hassell, J.R., Yamada, Y., 1999. Perlecan is essential for cartilage and cephalic development. Nat. Genet. 23 (3), 354—358.

Barallobre-Barreiro, J., Gupta, S.K., Zoccarato, A., Kitazume-Taneike, R., Fava, M., Yin, X., Werner, T., Hirt, M.N., Zampetaki, A., Viviano, A., Chong, M., Bern, M., Kourliouros, A., Domenech, N., Willeit, P., Shah, A.M., Jahangiri, M., Schaefer, L., Fischer, J.W., Iozzo, R.V., Viner, R., Thum, T., Heineke, J., Kichler, A., Otsu, K., Mayr, M., 2016. Glycoproteomics reveals decorin peptides with anti-myostatin activity in human atrial fibrillation. Circulation 134 (11), 817—832.

Bassuk, J.A., Birkebak, T., Rothmier, J.D., Clark, J.M., Bradshaw, A., Muchowski, P.J., Howe, C.C., Clark, J.I., Sage, E.H., 1999. Disruption of the Sparc locus in mice alters the differentiation of lenticular epithelial cells and leads to cataract formation. Exp. Eye Res. 68 (3), 321—331.

Bellahcene, A., Albert, V., Pollina, L., Basolo, F., Fisher, L.W., Castronovo, V., 1998. Ectopic expression of bone sialoprotein in human thyroid cancer. Thyroid 8 (8), 637—641.

Bellahcene, A., Castronovo, V., 1997. Expression of bone matrix proteins in human breast cancer: potential roles in microcalcification formation and in the genesis of bone metastases. Bull. Cancer 84 (1), 17—24.

Bellahcene, A., Maloujahmoum, N., Fisher, L.W., Pastorino, H., Tagliabue, E., Menard, S., Castronovo, V., 1997. Expression of bone sialoprotein in human lung cancer. Calcif. Tissue Int. 61 (3), 183—188.

Bhattacharya, S., Bunick, C.G., Chazin, W.J., 2004. Target selectivity in EF-hand calcium binding proteins. Biochim. Biophys. Acta 1742 (1—3), 69—79.

Bhattacharya, S., Pal, S., Chattopadhyay, N., 2018. Targeted inhibition of sclerostin for post-menopausal osteoporosis therapy: a critical assessment of the mechanism of action. Eur. J. Pharmacol. 826, 39—47.

Bianco, P., Fisher, L.W., Young, M.F., Termine, J.D., Robey, P.G., 1990. Expression and localization of the two small proteoglycans biglycan and decorin in developing human skeletal and non-skeletal tissues. J. Histochem. Cytochem. 38 (11), 1549—1563.

Bianco, P., Fisher, L.W., Young, M.F., Termine, J.D., Robey, P.G., 1991. Expression of bone sialoprotein (BSP) in developing human tissues. Calcif. Tissue Int. 49 (6), 421—426.

Bini, A., Mann, K.G., Kudryk, B.J., Schoen, F.J., 1999. Noncollagenous bone matrix proteins, calcification, and thrombosis in carotid artery atherosclerosis. Arterioscler. Thromb. Vasc. Biol. 19 (8), 1852—1861.

Binkert, C., Demetriou, M., Sukhu, B., Szweras, M., Tenenbaum, H.C., Dennis, J.W., 1999. Regulation of osteogenesis by fetuin. J. Biol. Chem. 274 (40), 28514–28520.

Bonewald, L.F., 2006. Mechanosensation and transduction in osteocytes. BoneKEy Osteovision 3 (10), 7–15.

Bonnet, N., Brun, J., Rousseau, J.C., Duong, L.T., Ferrari, S.L., 2017. Cathepsin K controls cortical bone formation by degrading periostin. J. Bone Miner. Res. 32 (7), 1432–1441.

Bouleftour, W., Boudiffa, M., Wade-Gueye, N.M., Bouet, G., Cardelli, M., Laroche, N., Vanden-Bossche, A., Thomas, M., Bonnelye, E., Aubin, J.E., Vico, L., Lafage-Proust, M.H., Malaval, L., 2014. Skeletal development of mice lacking bone sialoprotein (BSP)–impairment of long bone growth and progressive establishment of high trabecular bone mass. PLoS One 9 (5), e95144.

Bouleftour, W., Juignet, L., Bouet, G., Granito, R.N., Vanden-Bossche, A., Laroche, N., Aubin, J.E., Lafage-Proust, M.H., Vico, L., Malaval, L., 2016. The role of the SIBLING, Bone Sialoprotein in skeletal biology - contribution of mouse experimental genetics. Matrix Biol. 52–54, 60–77.

Brunkow, M.E., Gardner, J.C., Van Ness, J., Paeper, B.W., Kovacevich, B.R., Proll, S., Skonier, J.E., Zhao, L., Sabo, P.J., Fu, Y., Alisch, R.S., Gillett, L., Colbert, T., Tacconi, P., Galas, D., Hamersma, H., Beighton, P., Mulligan, J., 2001. Bone dysplasia sclerosteosis results from loss of the SOST gene product, a novel cystine knot-containing protein. Am. J. Hum. Genet. 68 (3), 577–589.

Canfield, A.E., Doherty, M.J., Kelly, V., Newman, B., Farrington, C., Grant, M.E., Boot-Handford, R.P., 2000. Matrix Gla protein is differentially expressed during the deposition of a calcified matrix by vascular pericytes. FEBS Lett. 487 (2), 267–271.

Carron, J.A., Bowler, W.B., Wagstaff, S.C., Gallagher, J.A., 1999. Expression of members of the thrombospondin family by human skeletal tissues and cultured cells. Biochem. Biophys. Res. Commun. 263 (2), 389–391.

Chakravarti, S., Paul, J., Roberts, L., Chervoneva, I., Oldberg, A., Birk, D.E., 2003. Ocular and scleral alterations in gene-targeted lumican-fibromodulin double-null mice. Investig. Ophthalmol. Vis. Sci. 44 (6), 2422–2432.

Chaudhry, S.S., Cain, S.A., Morgan, A., Dallas, S.L., Shuttleworth, C.A., Kielty, C.M., 2007. Fibrillin-1 regulates the bioavailability of TGFbeta1. J. Cell Biol. 176 (3), 355–367.

Chen, Y., Bal, B.S., Gorski, J.P., 1992. Calcium and collagen binding properties of osteopontin, bone sialoprotein, and bone acidic glycoprotein-75 from bone. J. Biol. Chem. 267 (34), 24871–24878.

Choocheep, K., Hatano, S., Takagi, H., Watanabe, H., Kimata, K., Kongtawelert, P., Watanabe, H., 2010. Versican facilitates chondrocyte differentiation and regulates joint morphogenesis. J. Biol. Chem. 285 (27), 21114–21125.

Colaianni, G., Cuscito, C., Mongelli, T., Pignataro, P., Buccoliero, C., Liu, P., Lu, P., Sartini, L., Di Comite, M., Mori, G., Di Benedetto, A., Brunetti, G., Yuen, T., Sun, L., Reseland, J.E., Colucci, S., New, M.I., Zaidi, M., Cinti, S., Grano, M., 2015. The myokine irisin increases cortical bone mass. Proc. Natl. Acad. Sci. U. S. A. 112 (39), 12157–12162.

Colaianni, G., Mongelli, T., Cuscito, C., Pignataro, P., Lippo, L., Spiro, G., Notarnicola, A., Severi, I., Passeri, G., Mori, G., Brunetti, G., Moretti, B., Tarantino, U., Colucci, S.C., Reseland, J.E., Vettor, R., Cinti, S., Grano, M., 2017. Irisin prevents and restores bone loss and muscle atrophy in hind-limb suspended mice. Sci. Rep. 7 (1), 2811.

Corsi, A., Xu, T., Chen, X.D., Boyde, A., Liang, J., Mankani, M., Sommer, B., Iozzo, R.V., Eichstetter, I., Robey, P.G., Bianco, P., Young, M.F., 2002. Phenotypic effects of biglycan deficiency are linked to collagen fibril abnormalities, are synergized by decorin deficiency, and mimic Ehlers-Danlos-like changes in bone and other connective tissues. J. Bone Miner. Res. 17 (7), 1180–1189.

D'Souza, R.N., Cavender, A., Sunavala, G., Alvarez, J., Ohshima, T., Kulkarni, A.B., MacDougall, M., 1997. Gene expression patterns of murine dentin matrix protein 1 (Dmp1) and dentin sialophosphoprotein (DSPP) suggest distinct developmental functions in vivo. J. Bone Miner. Res. 12 (12), 2040–2049.

Danielson, K.G., Baribault, H., Holmes, D.F., Graham, H., Kadler, K.E., Iozzo, R.V., 1997. Targeted disruption of decorin leads to abnormal collagen fibril morphology and skin fragility. J. Cell Biol. 136 (3), 729–743.

Delmas, P.D., Tracy, R.P., Riggs, B.L., Mann, K.G., 1984. Identification of the noncollagenous proteins of bovine bone by two-dimensional gel electrophoresis. Calcif. Tissue Int. 36 (3), 308–316.

Demetriou, M., Binkert, C., Sukhu, B., Tenenbaum, H.C., Dennis, J.W., 1996. Fetuin/alpha2-HS glycoprotein is a transforming growth factor-beta type II receptor mimic and cytokine antagonist. J. Biol. Chem. 271 (22), 12755–12761.

Duchamp de Lageneste, O., Julien, A., Abou-Khalil, R., Frangi, G., Carvalho, C., Cagnard, N., Cordier, C., Conway, S.J., Colnot, C., 2018. Periosteum contains skeletal stem cells with high bone regenerative potential controlled by Periostin. Nat. Commun. 9 (1), 773.

Ducy, P., Desbois, C., Boyce, B., Pinero, G., Story, B., Dunstan, C., Smith, E., Bonadio, J., Goldstein, S., Gundberg, C., Bradley, A., Karsenty, G., 1996. Increased bone formation in osteocalcin-deficient mice. Nature 382 (6590), 448–452.

El-Maadawy, S., Kaartinen, M.T., Schinke, T., Murshed, M., Karsenty, G., McKee, M.D., 2003. Cartilage formation and calcification in arteries of mice lacking matrix Gla protein. Connect. Tissue Res. 44 (Suppl. 1), 272–278.

El Shafey, N., Guesnon, M., Simon, F., Deprez, E., Cosette, J., Stockholm, D., Scherman, D., Bigey, P., Kichler, A., 2016. Inhibition of the myostatin/Smad signaling pathway by short decorin-derived peptides. Exp. Cell Res. 341 (2), 187–195.

Falconi, D., Aubin, J.E., 2007. LIF inhibits osteoblast differentiation at least in part by regulation of HAS2 and its product hyaluronan. J. Bone Miner. Res. 22 (8), 1289–1300.

Fedarko, N.S., Jain, A., Karadag, A., Fisher, L.W., 2004. Three small integrin binding ligand N-linked glycoproteins (SIBLINGs) bind and activate specific matrix metalloproteinases. FASEB J. 18 (6), 734–736.

Fedarko, N.S., Vetter, U.K., Weinstein, S., Robey, P.G., 1992. Age-related changes in hyaluronan, proteoglycan, collagen, and osteonectin synthesis by human bone cells. J. Cell. Physiol. 151 (2), 215–227.

Feng, J.Q., Ward, L.M., Liu, S., Lu, Y., Xie, Y., Yuan, B., Yu, X., Rauch, F., Davis, S.I., Zhang, S., Rios, H., Drezner, M.K., Quarles, L.D., Bonewald, L.F., White, K.E., 2006. Loss of DMP1 causes rickets and osteomalacia and identifies a role for osteocytes in mineral metabolism. Nat. Genet. 38 (11), 1310–1315.

Ferron, M., McKee, M.D., Levine, R.L., Ducy, P., Karsenty, G., 2012. Intermittent injections of osteocalcin improve glucose metabolism and prevent type 2 diabetes in mice. Bone 50 (2), 568–575.

Fisher, L.W., Fedarko, N.S., 2003. Six genes expressed in bones and teeth encode the current members of the SIBLING family of proteins. Connect. Tissue Res. 44 (Suppl. 1), 33–40.

Fisher, L.W., McBride, O.W., Termine, J.D., Young, M.F., 1990. Human bone sialoprotein. Deduced protein sequence and chromosomal localization. J. Biol. Chem. 265 (4), 2347–2351.

Fisher, L.W., Termine, J.D., Young, M.F., 1989. Deduced protein sequence of bone small proteoglycan I (biglycan) shows homology with proteoglycan II (decorin) and several nonconnective tissue proteins in a variety of species. J. Biol. Chem. 264 (8), 4571–4576.

Garnett, J., Dieppe, P., 1990. The effects of serum and human albumin on calcium hydroxyapatite crystal growth. Biochem. J. 266, 863–868.

Gleghorn, L., Ramesar, R., Beighton, P., Wallis, G., 2005. A mutation in the variable repeat region of the aggrecan gene (AGC1) causes a form of spondyloepiphyseal dysplasia associated with severe, premature osteoarthritis. Am. J. Hum. Genet. 77 (3), 484–490.

Gong, Y., Slee, R.B., Fukai, N., Rawadi, G., Roman-Roman, S., Reginato, A.M., Wang, H., Cundy, T., Glorieux, F.H., Lev, D., Zacharin, M., Oexle, K., Marcelino, J., Suwairi, W., Heeger, S., Sabatakos, G., Apte, S., Adkins, W.N., Allgrove, J., Arslan-Kirchner, M., Batch, J.A., Beighton, P., Black, G.C., Boles, R.G., Boon, L.M., Borrone, C., Brunner, H.G., Carle, G.F., Dallapiccola, B., De Paepe, A., Floege, B., Halfhide, M.L., Hall, B., Hennekam, R.C., Hirose, T., Jans, A., Juppner, H., Kim, C.A., Keppler-Noreuil, K., Kohlschuetter, A., LaCombe, D., Lambert, M., Lemyre, E., Letteboer, T., Peltonen, L., Ramesar, R.S., Romanengo, M., Somer, H., Steichen-Gersdorf, E., Steinmann, B., Sullivan, B., Superti-Furga, A., Swoboda, W., van den Boogaard, M.J., Van Hul, W., Vikkula, M., Votruba, M., Zabel, B., Garcia, T., Baron, R., Olsen, B.R., Warman, M.L., 2001. LDL receptor-related protein 5 (LRP5) affects bone accrual and eye development. Cell 107 (4), 513–523.

Gorski, J.P., 2011. Biomineralization of bone: a fresh view of the roles of non-collagenous proteins. Front. Biosci. 17, 2598–2621.

Gorski, J.P., Griffin, D., Dudley, G., Stanford, C., Thomas, R., Huang, C., Lai, E., Karr, B., Solursh, M., 1990. Bone acidic glycoprotein-75 is a major synthetic product of osteoblastic cells and localized as 75- and/or 50-kDa forms in mineralized phases of bone and growth plate and in serum. J. Biol. Chem. 265 (25), 14956–14963.

Gorski, J.P., Shimizu, K., 1988. Isolation of new phosphorylated glycoprotein from mineralized phase of bone that exhibits limited homology to adhesive protein osteopontin. J. Biol. Chem. 263 (31), 15938–15945.

Gorski, J.P., Wang, A., Lovitch, D., Law, D., Powell, K., Midura, R.J., 2004. Extracellular bone acidic glycoprotein-75 defines condensed mesenchyme regions to be mineralized and localizes with bone sialoprotein during intramembranous bone formation. J. Biol. Chem. 279 (24), 25455–25463.

Gowen, L.C., Petersen, D.N., Mansolf, A.L., Qi, H., Stock, J.L., Tkalcevic, G.T., Simmons, H.A., Crawford, D.T., Chidsey-Frink, K.L., Ke, H.Z., McNeish, J.D., Brown, T.A., 2003. Targeted disruption of the osteoblast/osteocyte factor 45 gene (OF45) results in increased bone formation and bone mass. J. Biol. Chem. 278 (3), 1998–2007.

Granito, R.N., Bouleftour, W., Sabido, O., Lescale, C., Thomas, M., Aubin, J.E., Goodhardt, M., Vico, L., Malaval, L., 2015. Absence of bone sialoprotein (BSP) alters profoundly hematopoiesis and upregulates osteopontin. J. Cell. Physiol. 230 (6), 1342–1351.

Grzesik, W.J., Robey, P.G., 1994. Bone matrix RGD glycoproteins: immunolocalization and interaction with human primary osteoblastic bone cells in vitro. J. Bone Miner. Res. 9 (4), 487–496.

Hedbom, E., Heinegard, D., 1993. Binding of fibromodulin and decorin to separate sites on fibrillar collagens. J. Biol. Chem. 268 (36), 27307–27312.

Henry, S.P., Takanosu, M., Boyd, T.C., Mayne, P.M., Eberspaecher, H., Zhou, W., de Crombrugghe, B., Hook, M., Mayne, R., 2001. Expression pattern and gene characterization of asporin. a newly discovered member of the leucine-rich repeat protein family. J. Biol. Chem. 276 (15), 12212–12221.

Herring, G.M., Ashton, B.A., 1974. The isolation of soluble proteins, glycoproteins, and proteoglycans from bone. Prep. Biochem. 4 (2), 179–200.

Iba, K., Durkin, M.E., Johnsen, L., Hunziker, E., Damgaard-Pedersen, K., Zhang, H., Engvall, E., Albrechtsen, R., Wewer, U.M., 2001. Mice with a targeted deletion of the tetranectin gene exhibit a spinal deformity. Mol. Cell Biol. 21 (22), 7817–7825.

Jahnen-Dechent, W., Schinke, T., Trindl, A., Muller-Esterl, W., Sablitzky, F., Kaiser, S., Blessing, M., 1997. Cloning and targeted deletion of the mouse fetuin gene. J. Biol. Chem. 272 (50), 31496–31503.

Johnson, W.E., Li, C., Rabinovic, A., 2007. Adjusting batch effects in microarray expression data using empirical Bayes methods. Biostatistics 8 (1), 118–127.

Karsenty, G., Mera, P., 2017. Molecular bases of the crosstalk between bone and muscle. Bone 115, 43–49.

Karsenty, G., Oury, F., 2014. Regulation of male fertility by the bone-derived hormone osteocalcin. Mol. Cell. Endocrinol. 382 (1), 521–526.

Kasai, R., Bianco, P., Robey, P.G., Kahn, A.J., 1994. Production and characterization of an antibody against the human bone GLA protein (BGP/osteocalcin) propeptide and its use in immunocytochemistry of bone cells. Bone Miner. 25 (3), 167–182.

Kawao, N., Moritake, A., Tatsumi, K., Kaji, H., 2018. Roles of irisin in the linkage from muscle to bone during mechanical unloading in mice. Calcif. Tissue Int. 103 (1), 24–34.

Kram, V., Kilts, T.M., Bhattacharyya, N., Li, L., Young, M.F., 2017. Small leucine rich proteoglycans, a novel link to osteoclastogenesis. Sci. Rep. 7 (1), 12627.

Lamoureux, F., Baud'huin, M., Duplomb, L., Heymann, D., Redini, F., 2007. Proteoglycans: key partners in bone cell biology. Bioessays 29 (8), 758–771.

Lee, C.S., Fu, H., Baratang, N., Rousseau, J., Kumra, H., Sutton, V.R., Niceta, M., Ciolfi, A., Yamamoto, G., Bertola, D., Marcelis, C.L., Lugtenberg, D., Bartuli, A., Kim, C., Hoover-Fong, J., Sobreira, N., Pauli, R., Bacino, C., Krakow, D., Parboosingh, J., Yap, P., Kariminejad, A., McDonald, M.T., Aracena, M.I., Lausch, E., Unger, S., Superti-Furga, A., Lu, J.T., Baylor-Hopkins Center for Mendelian, G., Cohn, D.H., Tartaglia, M., Lee, B.H., Reinhardt, D.P., Campeau, P.M., 2017. Mutations in fibronectin cause a subtype of spondylometaphyseal dysplasia with "corner fractures". Am. J. Hum. Genet. 101 (5), 815–823.

Lee, N.K., Sowa, H., Hinoi, E., Ferron, M., Ahn, J.D., Confavreux, C., Dacquin, R., Mee, P.J., McKee, M.D., Jung, D.Y., Zhang, Z., Kim, J.K., Mauvais-Jarvis, F., Ducy, P., Karsenty, G., 2007. Endocrine regulation of energy metabolism by the skeleton. Cell 130 (3), 456–469.

Luo, G., Ducy, P., McKee, M.D., Pinero, G.J., Loyer, E., Behringer, R.R., Karsenty, G., 1997. Spontaneous calcification of arteries and cartilage in mice lacking matrix GLA protein. Nature 386 (6620), 78–81.

Maccarana, M., Svensson, R.B., Knutsson, A., Giannopoulos, A., Pelkonen, M., Weis, M., Eyre, D., Warman, M., Kalamajski, S., 2017. Asporin-deficient mice have tougher skin and altered skin glycosaminoglycan content and structure. PLoS One 12 (8), e0184028.

Maillard, C., Berruyer, M., Serre, C.M., Dechavanne, M., Delmas, P.D., 1992. Protein-S, a vitamin K-dependent protein, is a bone matrix component synthesized and secreted by osteoblasts. Endocrinology 130 (3), 1599–1604.

Malhotra, R., Burke, M.F., Martyn, T., Shakartzi, H.R., Thayer, T.E., O'Rourke, C., Li, P., Derwall, M., Spagnolli, E., Kolodziej, S.A., Hoeft, K., Mayeur, C., Jiramongkolchai, P., Kumar, R., Buys, E.S., Yu, P.B., Bloch, K.D., Bloch, D.B., 2015. Inhibition of bone morphogenetic protein signal transduction prevents the medial vascular calcification associated with matrix Gla protein deficiency. PLoS One 10 (1), e0117098.

Manley Jr., E., Perosky, J.E., Khoury, B.M., Reddy, A.B., Kozloff, K.M., Alford, A.I., 2015. Thrombospondin-2 deficiency in growing mice alters bone collagen ultrastructure and leads to a brittle bone phenotype. J. Appl. Physiol. (1985) 119 (8), 872–881.

Mansergh, F.C., Wells, T., Elford, C., Evans, S.L., Perry, M.J., Evans, M.J., Evans, B.A., 2007. Osteopenia in Sparc (osteonectin)-deficient mice: characterization of phenotypic determinants of femoral strength and changes in gene expression. Physiol. Genom. 32 (1), 64–73.

Mansouri, R., Jouan, Y., Hay, E., Blin-Wakkach, C., Frain, M., Ostertag, A., Le Henaff, C., Marty, C., Geoffroy, V., Marie, P.J., Cohen-Solal, M., Modrowski, D., 2017. Osteoblastic heparan sulfate glycosaminoglycans control bone remodeling by regulating Wnt signaling and the crosstalk between bone surface and marrow cells. Cell Death Dis. 8 (6), e2902.

Markmann, A., Hausser, H., Schonherr, E., Kresse, H., 2000. Influence of decorin expression on transforming growth factor-beta-mediated collagen gel retraction and biglycan induction. Matrix Biol. 19 (7), 631–636.

Marulanda, J., Eimar, H., McKee, M.D., Berkvens, M., Nelea, V., Roman, H., Borras, T., Tamimi, F., Ferron, M., Murshed, M., 2017. Matrix Gla protein deficiency impairs nasal septum growth, causing midface hypoplasia. J. Biol. Chem. 292 (27), 11400–11412.

Matsumoto, K., Li, Y., Jakuba, C., Sugiyama, Y., Sayo, T., Okuno, M., Dealy, C.N., Toole, B.P., Takeda, J., Yamaguchi, Y., Kosher, R.A., 2009. Conditional inactivation of Has2 reveals a crucial role for hyaluronan in skeletal growth, patterning, chondrocyte maturation and joint formation in the developing limb. Development 136 (16), 2825–2835.

Matsushima, N., Ohyanagi, T., Tanaka, T., Kretsinger, R.H., 2000. Super-motifs and evolution of tandem leucine-rich repeats within the small proteoglycans–biglycan, decorin, lumican, fibromodulin, PRELP, keratocan, osteoadherin, epiphycan, and osteoglycin. Proteins 38 (2), 210–225.

McDonald, M.M., Reagan, M.R., Youlten, S.E., Mohanty, S.T., Seckinger, A., Terry, R.L., Pettitt, J.A., Simic, M.K., Cheng, T.L., Morse, A., Le, L.M.T., Abi-Hanna, D., Kramer, I., Falank, C., Fairfield, H., Ghobrial, I.M., Baldock, P.A., Little, D.G., Kneissel, M., Vanderkerken, K., Bassett, J.H.D., Williams, G.R., Oyajobi, B.O., Hose, D., Phan, T.G., Croucher, P.I., 2017. Inhibiting the osteocyte-specific protein sclerostin increases bone mass and fracture resistance in multiple myeloma. Blood 129 (26), 3452–3464.

McEwan, P.A., Scott, P.G., Bishop, P.N., Bella, J., 2006. Structural correlations in the family of small leucine-rich repeat proteins and proteoglycans. J. Struct. Biol. 155 (2), 294–305.

Midura, R.J., Wang, A., Lovitch, D., Law, D., Powell, K., Gorski, J.P., 2004. Bone acidic glycoprotein-75 delineates the extracellular sites of future bone sialoprotein accumulation and apatite nucleation in osteoblastic cultures. J. Biol. Chem. 279 (24), 25464–25473.

Millan, J.L., Whyte, M.P., 2016. Alkaline phosphatase and hypophosphatasia. Calcif. Tissue Int. 98 (4), 398–416.

Mintz, K.P., Grzesik, W.J., Midura, R.J., Robey, P.G., Termine, J.D., Fisher, L.W., 1993. Purification and fragmentation of nondenatured bone sialoprotein: evidence for a cryptic, RGD-resistant cell attachment domain. J. Bone Miner. Res. 8 (8), 985–995.

Miura, T., Kishioka, Y., Wakamatsu, J., Hattori, A., Hennebry, A., Berry, C.J., Sharma, M., Kambadur, R., Nishimura, T., 2006. Decorin binds myostatin and modulates its activity to muscle cells. Biochem. Biophys. Res. Commun. 340 (2), 675–680.

Myren, M., Kirby, D.J., Noonan, M.L., Maeda, A., Owens, R.T., Ricard-Blum, S., Kram, V., Kilts, T.M., Young, M.F., 2016. Biglycan potentially regulates angiogenesis during fracture repair by altering expression and function of endostatin. Matrix Biol. 52–54, 141–150.

Nakamura, I., Pilkington, M.F., Lakkakorpi, P.T., Lipfert, L., Sims, S.M., Dixon, S.J., Rodan, G.A., Duong, L.T., 1999. Role of alpha(v)beta(3) integrin in osteoclast migration and formation of the sealing zone. J. Cell Sci. 112 (Pt 22), 3985–3993.

Nistala, H., Lee-Arteaga, S., Smaldone, S., Siciliano, G., Carta, L., Ono, R.N., Sengle, G., Arteaga-Solis, E., Levasseur, R., Ducy, P., Sakai, L.Y., Karsenty, G., Ramirez, F., 2010. Fibrillin-1 and -2 differentially modulate endogenous TGF-beta and BMP bioavailability during bone formation. J. Cell Biol. 190 (6), 1107–1121.

Ogbureke, K.U., Fisher, L.W., 2005. Renal expression of SIBLING proteins and their partner matrix metalloproteinases (MMPs). Kidney Int. 68 (1), 155–166.

Ohnishi, T., Nakamura, O., Ozawa, M., Arakaki, N., Muramatsu, T., Daikuhara, Y., 1993. Molecular cloning and sequence analysis of cDNA for a 59 kD bone sialoprotein of the rat: demonstration that it is a counterpart of human alpha 2-HS glycoprotein and bovine fetuin. J. Bone Miner. Res. 8 (3), 367–377.

Ominsky, M.S., Vlasseros, F., Jolette, J., Smith, S.Y., Stouch, B., Doellgast, G., Gong, J., Gao, Y., Cao, J., Graham, K., Tipton, B., Cai, J., Deshpande, R., Zhou, L., Hale, M.D., Lightwood, D.J., Henry, A.J., Popplewell, A.G., Moore, A.R., Robinson, M.K., Lacey, D.L., Simonet, W.S., Paszty, C., 2010. Two doses of sclerostin antibody in cynomolgus monkeys increases bone formation, bone mineral density, and bone strength. J. Bone Miner. Res. 25 (5), 948–959.

Oury, F., Ferron, M., Huizhen, W., Confavreux, C., Xu, L., Lacombe, J., Srinivas, P., Chamouni, A., Lugani, F., Lejeune, H., Kumar, T.R., Plotton, I., Karsenty, G., 2015. Osteocalcin regulates murine and human fertility through a pancreas-bone-testis axis. J. Clin. Investig. 125 (5), 2180.

Pacifici, M., 2017. The pathogenic roles of heparan sulfate deficiency in hereditary multiple exostoses. Matrix Biol. 71–72, 28–39.

Padhi, D., Jang, G., Stouch, B., Fang, L., Posvar, E., 2011. Single-dose, placebo-controlled, randomized study of AMG 785, a sclerostin monoclonal antibody. J. Bone Miner. Res. 26 (1), 19–26.

Pankov, R., Yamada, K.M., 2002. Fibronectin at a glance. J. Cell Sci. 115 (Pt 20), 3861–3863.

Pi, M., Quarles, L.D., 2012. Multiligand specificity and wide tissue expression of GPRC6A reveals new endocrine networks. Endocrinology 153 (5), 2062–2069.

Price, P.A., Otsuka, A.A., Poser, J.W., Kristaponis, J., Raman, N., 1976. Characterization of a gamma-carboxyglutamic acid-containing protein from bone. Proc. Natl. Acad. Sci. U. S. A. 73 (5), 1447–1451.

Price, P.A., Rice, J.S., Williamson, M.K., 1994. Conserved phosphorylation of serines in the Ser-X-Glu/Ser(P) sequences of the vitamin K-dependent matrix Gla protein from shark, lamb, rat, cow, and human. Protein Sci. 3 (5), 822–830.

Price, P.A., Urist, M.R., Otawara, Y., 1983. Matrix Gla protein, a new gamma-carboxyglutamic acid-containing protein which is associated with the organic matrix of bone. Biochem. Biophys. Res. Commun. 117 (3), 765–771.

Primorac, D., Johnson, C.V., Lawrence, J.B., McKinstry, M.B., Stover, M.L., Schanfield, M.S., Andjelinovic, S., Tadic, T., Rowe, D.W., 1999. Premature termination codon in the aggrecan gene of nanomelia and its influence on mRNA transport and stability. Croat. Med. J. 40 (4), 528–532.

Purcell, L., Gruia-Gray, J., Scanga, S., Ringuette, M., 1993. Developmental anomalies of Xenopus embryos following microinjection of SPARC antibodies. J. Exp. Zool. 265 (2), 153–164.

Ramirez, F., Dietz, H.C., 2007. Marfan syndrome: from molecular pathogenesis to clinical treatment. Curr. Opin. Genet. Dev. 17 (3), 252–258.

Reese, S.P., Underwood, C.J., Weiss, J.A., 2013. Effects of decorin proteoglycan on fibrillogenesis, ultrastructure, and mechanics of type I collagen gels. Matrix Biol. 32 (7–8), 414–423.

Rios, H., Koushik, S.V., Wang, H., Wang, J., Zhou, H.M., Lindsley, A., Rogers, R., Chen, Z., Maeda, M., Kruzynska-Frejtag, A., Feng, J.Q., Conway, S.J., 2005. Periostin null mice exhibit dwarfism, incisor enamel defects, and an early-onset periodontal disease-like phenotype. Mol. Cell Biol. 25 (24), 11131–11144.

Robey, P.G., Young, M.F., Fisher, L.W., McClain, T.D., 1989. Thrombospondin is an osteoblast-derived component of mineralized extracellular matrix. J. Cell Biol. 108 (2), 719–727.

Roodman, G.D., 2011. Osteoblast function in myeloma. Bone 48 (1), 135–140.

Rowe, P.S., 2004. The wrickkened pathways of FGF23, MEPE and PHEX. Crit. Rev. Oral Biol. Med. 15 (5), 264–281.

Rowe, P.S., de Zoysa, P.A., Dong, R., Wang, H.R., White, K.E., Econs, M.J., Oudet, C.L., 2000. MEPE, a new gene expressed in bone marrow and tumors causing osteomalacia. Genomics 67 (1), 54–68.

Sage, E.H., Bornstein, P., 1991. Extracellular proteins that modulate cell-matrix interactions. SPARC, tenascin, and thrombospondin. J. Biol. Chem. 266 (23), 14831–14834.

Sato, M., Grasser, W., Harm, S., Fullenkamp, C., Gorski, J.P., 1992. Bone acidic glycoprotein 75 inhibits resorption activity of isolated rat and chicken osteoclasts. FASEB J. 6 (11), 2966–2976.

Scatena, M., Liaw, L., Giachelli, C.M., 2007. Osteopontin: a multifunctional molecule regulating chronic inflammation and vascular disease. Arterioscler. Thromb. Vasc. Biol. 27 (11), 2302–2309.

Schvartz, I., Seger, D., Shaltiel, S., 1999. Vitronectin. Int. J. Biochem. Cell Biol. 31 (5), 539–544.

Scott, P.G., Dodd, C.M., Bergmann, E.M., Sheehan, J.K., Bishop, P.N., 2006. Crystal structure of the biglycan dimer and evidence that dimerization is essential for folding and stability of class I small leucine-rich repeat proteoglycans. J. Biol. Chem. 281 (19), 13324–13332.

Sheng, K., Zhang, P., Lin, W., Cheng, J., Li, J., Chen, J., 2017. Association of Matrix Gla protein gene (rs1800801, rs1800802, rs4236) polymorphism with vascular calcification and atherosclerotic disease: a meta-analysis. Sci. Rep. 7 (1), 8713.

Smaldone, S., Clayton, N.P., del Solar, M., Pascual, G., Cheng, S.H., Wentworth, B.M., Schaffler, M.B., Ramirez, F., 2016. Fibrillin-1 regulates skeletal stem cell differentiation by modulating TGFbeta activity within the marrow Niche. J. Bone Miner. Res. 31 (1), 86–97.

Sommarin, Y., Wendel, M., Shen, Z., Hellman, U., Heinegard, D., 1998. Osteoadherin, a cell-binding keratan sulfate proteoglycan in bone, belongs to the family of leucine-rich repeat proteins of the extracellular matrix. J. Biol. Chem. 273 (27), 16723–16729.

Stein, G.S., Lian, J.B., Stein, J.L., Van Wijnen, A.J., Montecino, M., 1996. Transcriptional control of osteoblast growth and differentiation. Physiol. Rev. 76 (2), 593–629.

Stubbs 3rd, J.T., Mintz, K.P., Eanes, E.D., Torchia, D.A., Fisher, L.W., 1997. Characterization of native and recombinant bone sialoprotein: delineation of the mineral-binding and cell adhesion domains and structural analysis of the RGD domain. J. Bone Miner. Res. 12 (8), 1210–1222.

Termine, J.D., Belcourt, A.B., Christner, P.J., Conn, K.M., Nylen, M.U., 1980. Properties of dissociatively extracted fetal tooth matrix proteins. I. Principal molecular species in developing bovine enamel. J. Biol. Chem. 255 (20), 9760–9768.

Termine, J.D., Belcourt, A.B., Conn, K.M., Kleinman, H.K., 1981a. Mineral and collagen-binding proteins of fetal calf bone. J. Biol. Chem. 256 (20), 10403–10408.

Termine, J.D., Kleinman, H.K., Whitson, S.W., Conn, K.M., McGarvey, M.L., Martin, G.R., 1981b. Osteonectin, a bone-specific protein linking mineral to collagen. Cell 26 (1 Pt 1), 99–105.

Thiede, M.A., Smock, S.L., Petersen, D.N., Grasser, W.A., Thompson, D.D., Nishimoto, S.K., 1994. Presence of messenger ribonucleic acid encoding osteocalcin, a marker of bone turnover, in bone marrow megakaryocytes and peripheral blood platelets. Endocrinology 135 (3), 929–937.

Ujita, M., Shinomura, T., Kimata, K., 1995. Molecular cloning of the mouse osteoglycin-encoding gene. Gene 158 (2), 237–240.

van der Pluijm, G., Vloedgraven, H.J., Ivanov, B., Robey, F.A., Grzesik, W.J., Robey, P.G., Papapoulos, S.E., Lowik, C.W., 1996. Bone sialoprotein peptides are potent inhibitors of breast cancer cell adhesion to bone. Cancer Res. 56 (8), 1948–1955.

Vanhoenacker, F.M., Balemans, W., Tan, G.J., Dikkers, F.G., De Schepper, A.M., Mathysen, D.G., Bernaerts, A., Hul, W.V., 2003. Van Buchem disease: lifetime evolution of radioclinical features. Skeletal Radiol. 32 (12), 708–718.

Vigetti, D., Karousou, E., Viola, M., Deleonibus, S., De Luca, G., Passi, A., 2014. Hyaluronan: biosynthesis and signaling. Biochim. Biophys. Acta 1840 (8), 2452–2459.

Viviano, B.L., Silverstein, L., Pflederer, C., Paine-Saunders, S., Mills, K., Saunders, S., 2005. Altered hematopoiesis in glypican-3-deficient mice results in decreased osteoclast differentiation and a delay in endochondral ossification. Dev. Biol. 282 (1), 152–162.

Waltregny, D., Bellahcene, A., Van Riet, I., Fisher, L.W., Young, M., Fernandez, P., Dewe, W., de Leval, J., Castronovo, V., 1998. Prognostic value of bone sialoprotein expression in clinically localized human prostate cancer. J. Natl. Cancer Inst. 90 (13), 1000–1008.

Wang, C., Wang, Y., Huffman, N.T., Cui, C., Yao, X., Midura, S., Midura, R.J., Gorski, J.P., 2009. Confocal laser Raman microspectroscopy of biomineralization foci in UMR 106 osteoblastic cultures reveals temporally synchronized protein changes preceding and accompanying mineral crystal deposition. J. Biol. Chem. 284 (11), 7100–7113.

Watanabe, H., Kimata, K., Line, S., Strong, D., Gao, L.Y., Kozak, C.A., Yamada, Y., 1994. Mouse cartilage matrix deficiency (cmd) caused by a 7 bp deletion in the aggrecan gene. Nat. Genet. 7 (2), 154–157.

Wewer, U.M., Ibaraki, K., Schjorring, P., Durkin, M.E., Young, M.F., Albrechtsen, R., 1994. A potential role for tetranectin in mineralization during osteogenesis. J. Cell Biol. 127 (6 Pt 1), 1767–1775.

Wight, T.N., 2002. Versican: a versatile extracellular matrix proteoglycan in cell biology. Curr. Opin. Cell Biol. 14 (5), 617–623.

Williams Jr., D.R., Presar, A.R., Richmond, A.T., Mjaatvedt, C.H., Hoffman, S., Capehart, A.A., 2005. Limb chondrogenesis is compromised in the versican deficient hdf mouse. Biochem. Biophys. Res. Commun. 334 (3), 960–966.

Winkler, D.G., Yu, C., Geoghegan, J.C., Ojala, E.W., Skonier, J.E., Shpektor, D., Sutherland, M.K., Latham, J.A., 2004. Noggin and sclerostin bone morphogenetic protein antagonists form a mutually inhibitory complex. J. Biol. Chem. 279 (35), 36293–36298.

Wong, M., Lawton, T., Goetinck, P.F., Kuhn, J.L., Goldstein, S.A., Bonadio, J., 1992. Aggrecan core protein is expressed in membranous bone of the chick embryo. Molecular and biomechanical studies of normal and nanomelia embryos. J. Biol. Chem. 267 (8), 5592–5598.

Wu, H., Whitfield, T.W., Gordon, J.A., Dobson, J.R., Tai, P.W., van Wijnen, A.J., Stein, J.L., Stein, G.S., Lian, J.B., 2014. Genomic occupancy of Runx2 with global expression profiling identifies a novel dimension to control of osteoblastogenesis. Genome Biol. 15 (3), R52.

Xu, T., Bianco, P., Fisher, L.W., Longenecker, G., Smith, E., Goldstein, S., Bonadio, J., Boskey, A., Heegaard, A.M., Sommer, B., Satomura, K., Dominguez, P., Zhao, C., Kulkarni, A.B., Robey, P.G., Young, M.F., 1998. Targeted disruption of the biglycan gene leads to an osteoporosis-like phenotype in mice. Nat. Genet. 20 (1), 78–82.

Yao, J., Guihard, P.J., Blazquez-Medela, A.M., Guo, Y., Liu, T., Bostrom, K.I., Yao, Y., 2016. Matrix Gla protein regulates differentiation of endothelial cells derived from mouse embryonic stem cells. Angiogenesis 19 (1), 1–7.

Zu, W., Robey, P.G., Boskey, A.L., 2007. The biochemistry of bone. In: Marcus, R., Feldman, D., Nelson, D.A., Rosen, C.J. (Eds.), Osteoporosis. Elsevier Science and Technology, Burlington, MA, pp. 191–240.

Chapter 16

Bone proteinases

Teruyo Nakatani and Nicola C. Partridge
Department of Basic Science and Craniofacial Biology, New York University College of Dentistry, New York, NY, United States

Chapter outline

Introduction	379	Urokinase-type plasminogen activator	388
Metalloproteinases	379	Tissue-type plasminogen activator	388
Stromelysin	381	Plasminogen activators in bone	388
Type IV collagenases (gelatinases)	381	Cysteine proteinases	389
Membrane-type matrix metalloproteinases	382	Aspartic proteinases	390
Collagenases	383	Conclusions	390
Collagenase-3/MMP-13	384	References	390
Plasminogen activators	387		

Introduction

This chapter surveys our knowledge of the proteinases expressed in bone. Although for a long time the osteoclast had been considered the main producer of proteinases in bone, it became increasingly clear that the osteoblast lineage plays a significant role in the production of many of these proteinases. For example, it is true that the osteoclast secretes abundant lysosomal cysteine proteinases, especially cathepsin K (Vaes, 1988; Xia et al., 1999; Sahara et al., 2003), and produces some of the neutral proteinases, e.g., matrix metalloproteinase-9 (MMP-9; Wucherpfennig et al., 1994; Delaissé et al., 2000). However, osteoblasts and osteocytes, like their related cells, fibroblasts, are able to secrete a host of proteinases, including neutral proteinases such as serine proteinases, plasminogen activators (PAs), and metalloproteinases such as MMP-13, as well as lysosomal proteinases, e.g., cathepsins. Thus, osteoblasts and osteocytes, like fibroblasts, have the capacity not only to synthesize a range of matrix proteins, including type I collagen, but also to remodel their own extracellular matrix via the secretion of a range of proteinases.

Proteinases can be classified into four groups: metalloproteinases, e.g., MMP-13 (also known as collagenase-3); serine proteinases, e.g., PA; cysteine proteinases, e.g., cathepsin K; and aspartic proteinases, e.g., cathepsin D. This subdivision is based on the structure and the catalytic mechanism of the active site involving particular amino acid residues and/or zinc. In the following review of the proteinases synthesized in bone, we deal with each group according to this subdivision in the order just given. For some, much more is known than for others and they have warranted their own section.

Metalloproteinases

MMPs are a family of zinc- and calcium-dependent neutral endoproteinases. MMPs are responsible for remodeling the extracellular matrix (ECM), which is necessary for physiological events such as angiogenesis, wound healing, and bone development (Page-McCaw et al., 2007). Abnormal expression and activation of MMPs lead to the development of diseases such as cirrhosis, cancer, and arthritis (Gong et al., 2014). To maintain homeostasis, MMPs are tightly regulated at the levels of transcription, posttranslational modification, production of the enzymes as inactive zymogens requiring activation, coexpression of tissue inhibitors of metalloproteinases (TIMPs), and receptors to regulate their extracellular abundance (Varghese, 2006; Cerda-Costà and Gomis-Rüth, 2014). At this writing, the MMP family comprises ~24

Principles of Bone Biology. https://doi.org/10.1016/B978-0-12-814841-9.00016-6

structurally and functionally related members in mammals (23 in humans, 22 in mice). According to their structural and functional characteristics, human MMPs can be classified into at least six different subfamilies of closely related members: collagenases (MMP-1, -8, and -13), type IV collagenases (gelatinases, MMP-2, and MMP-9), stromelysins (MMP-3, -10, and -11), matrilysins (MMP-7 and -26), membrane-type MMPs (MT-MMPs; MMP-14, -15, -16, -17, -24, and -25), and other MMPs (MMP-12, -19, -20, -21, -23, -27, and -28) (Matrisian, 1992; Vu and Werb, 2000; Visse and Nagase, 2003). All MMPs are active at neutral pH, require Ca^{2+} for activity, and contain Zn^{2+} in their active site. Mammalian MMPs share a conserved domain structure that consists of a catalytic domain and an autoinhibitory prodomain. The prodomain contains a conserved Cys residue that coordinates the active-site zinc to inhibit catalysis. The catalytic domain of MMPs contains the conserved sequence HEXGH, which is believed to be the zinc-binding site. Metalloproteinases are secreted or inserted into the cell membrane in a latent form caused by the presence of a conserved cysteine residue in the prosegment, which completes the tetrad of zinc bound to three other residues in the active site. Cleavage of this propiece by other proteolytic enzymes (e.g., trypsin, plasmin, cathepsins, or other unknown activators) causes a loss of ~10 kDa of the propiece; this disrupts the cysteine's association with the zinc and results in a conformational change in the enzyme yielding activation. Metalloproteinases all have homology to human fibroblast collagenase (collagenase-1, MMP-1) (Varghese, 2006). It should be noted that MMPs, named "gelatinases" when originally characterized, clearly function as collagenases in vitro, and could potentially function as collagenases in vivo. MMP-2, for example, when free of TIMPs, cleaves native collagens to yield the typical 3/4—1/4 fragments (Ames and Quigley, 1995; Seandel et al., 2001). There are four members of the TIMP family, TIMP-1, -2, -3, and -4, each inhibiting the activities of MMPs with varying efficiency (Gong et al., 2014; Brew et al., 2010).

MMPs play an important role in tissue remodeling associated with various physiological processes; however, abnormal expression and activation of MMPs is implicated in the pathogenesis and pathological progression of multiple diseases, including cirrhosis, arthritis, and cancer (Gong et al., 2014; Fingleton, 2007). The MMPs are promising drug targets in diverse pathologies (Nam et al., 2016). Although many synthetic inhibitors of MMPs (MMPIs) were designed and tested in animal models and in human clinical trials, all of these trials failed (to date, the only approved MMPI is Periostat, for the treatment of chronic periodontal disease). Broad-spectrum MMPIs have failed in clinical trials due to their very strong targeting of the catalytic zinc ion but low specificity (Zucker et al., 2000). In 2017, a novel strategy was used to develop JNJ0966, which binds zymogen and prevents generation of active MMP-9. The agent did not inhibit the production of the mature, active forms of MMP-1, MMP-2, MMP-3, and MMP-14. There is a hope that there will be the development of a next generation of MMP drugs: specific and without off-target effects (Scannevin et al., 2017).

Activation of MMPs can occur via the PA/plasmin pathway. PAs convert plasminogen to plasmin, which subsequently can activate prostromelysin to stromelysin and procollagenase to collagenase. The activated MMPs can then degrade collagens and other ECM proteins. MT-MMP is necessary for MMP-2 activation in fibroblasts (Ruangpanit et al., 2001), and MT-1 MMP (MMP-14)—mediated MMP-2 activation is important for invasion and metastasis of tumors (Mitra et al., 2006).

Apart from the regulation of secretion, activation, and/or inhibition, MMPs are substantially regulated at the transcriptional level (Matrisian, 1992; Crawford and Matrisian, 1996). Several *MMPs* contain specific regulatory elements in their promoter sequences. Human and rat stromelysin-1 and -2 contain activator protein-1 (AP-1)- and polyoma enhancer activator-3 (PEA-3)-binding sites that may be important for basal levels and inducibility. AP-1 and PEA-3 consensus sequences have also been found in human, rabbit, and rat collagenase genes (Brinckerhoff, 1992; Selvamurugan et al., 1998; Tardif et al., 2004). The transcription factors Fos and Jun form heterodimers and act through the AP-1 sequence (Lee et al., 1987; Chiu et al., 1988), whereas c-*ets* family members bind at the PEA-3 sequence (Wasylyk et al., 1993). The urokinase PA gene also contains AP-1- and PEA-3-binding sites and, as a result, agents acting through these sites could lead to coordinate expression of many of these genes (Matrisian, 1992; Hsieh et al., 2007). Moreover, TIMP-1 is controlled by AP-1 transcription factor in brain, and there is a role for AP-1 in regulation of the neuronal *Mmp9* gene (Kaczmarek et al., 2002). Glucocorticoids and retinoids can suppress metalloproteinase synthesis at the transcriptional level (Brinckerhoff, 1992) by forming a complex with AP-1 transcription factors and inhibiting their action (Schroen et al., 1996).

A second transcription factor-binding site was identified in the *Mmp13* (collagenase-3) promoter as well as in bone-specific genes such as osteocalcin (Shah et al., 2004; Selvamurugan et al., 2006). This site is referred to as the runt domain (RD)-binding site or polyomavirus enhancer-binding protein-2A/osteoblast-specific element-2/nuclear matrix protein-2-binding site (Geoffroy et al., 1995; Merriman et al., 1995). Members of the core-binding factor (CBF) protein family (renamed Runx by the Human Genome Organization), such as the osteoblastic transcription factor Runx2 (Cbfa1), bind to these RD sites (Kagoshima et al., 1993). Runx proteins are capable of binding to DNA as monomers, but can also heterodimerize with CBF subunit B, a ubiquitously expressed nuclear factor (Kanno et al., 1998). Runx2 is essential for the maturation of osteoblasts, and targeted disruption of the *Runx2* gene in mice produces skeletal defects that are essentially identical to those found in human cleidocranial dysplasia (Banerjee et al., 1997; Ducy et al., 1997; Mundlos et al., 1997; Otto et al., 1997).

Stromelysin

Stromelysin-1 (MMP-3) degrades fibronectin, gelatin, proteoglycans, denatured type I collagen, laminin, and other ECM components (Chin et al., 1985). Mesenchymal cells, such as chondrocytes and fibroblasts, are commonly found to secrete stromelysin-1 (Matrisian, 1992). Transin, the rat homolog of human stromelysin, was originally discovered in fibroblasts transformed with the polyomavirus (Matrisian et al., 1985). One importance of stromelysin comes from its implication in the direct activation of procollagenases, including MMP-1, -8, -9, and -13 (Murphy et al., 1987; Knauper et al., 1993), and the enzyme is thought to play a role, together with collagenases, in the destruction of connective tissues during disease states (Brinckerhoff, 1992; Posthumus et al., 2000). It has also been identified to have a crucial role in MMP-mediated cartilage damage in osteoarthritis, as *Mmp3*-knockout mice were shown to have reduced MMP-mediated cartilage breakdown after induction of osteoarthritis (Blom et al., 2007).

Stromelysin-1 is regulated by growth factors, oncogenes, cytokines, and tumor promoters. Epidermal growth factor (EGF) has been shown to increase stromelysin transcription through the induction of Fos and Jun, which interact at the AP-1 site in the promoter (McDonnell et al., 1990). Platelet-derived growth factor is also important in the induction of stromelysin (Kerr et al., 1988a). The protein kinase C (PKC) activator, phorbol myristate acetate (PMA), is a notable stimulator of stromelysin transcription (Brinckerhoff, 1992; Prontera et al., 1996). Transforming growth factor β (TGFβ), however, causes an inhibition of transin (rat stromelysin) expression (Matrisian et al., 1986; Kerr et al., 1988b) through a TGFβ inhibitory element (Kerr et al., 1990). Other studies have shown that bone morphogenetic protein (BMP-4) represses *Mmp3* and *Mmp13* gene expression, but does not induce adipocyte differentiation in C3H10T1/2 cells (Otto et al., 2007). Interleukin-4 (IL-4) and IL-13 were shown to inhibit MMP-3 synthesis in human conjunctival fibroblasts (Fukuda et al., 2006; Stewart et al., 2007).

In bone, stromelysin-1 has been shown to be produced by normal human osteoblasts (Meikle et al., 1992) after stimulation with parathyroid hormone (PTH) or monocyte-conditioned medium (cytokine-rich). Similarly, Rifas et al. (1994) have shown that two human osteosarcoma cell lines (MG-63 and U2OS) secrete stromelysin and this may be increased by treatment with PMA, IL-1β, and tumor necrosis factor-α (TNFα), but these authors were not able to find the enzyme in medium conditioned by cultured normal human osteoblasts. Mouse osteoblasts and osteoblastic cell lines also produce stromelysin-1 and demonstrate enhanced expression with $1,25(OH)_2D_3$, IL-1, or IL-6 treatment (Thomson et al., 1989; Breckon et al., 1999; Kusano et al., 1998; Le Maitre et al., 2005). There have also been reports that this stromelysin is expressed by osteoclasts (Witty et al., 1992). Despite the many papers on stromelysin-1 it is not clear what its physiological substrate is, nor is its role in physiological and pathophysiological skeletal resorption established.

Type IV collagenases (gelatinases)

Type IV collagenases or gelatinases are neutral metalloproteinases requiring Ca^{2+} for activity and are involved in the proteolysis and disruption of basement membranes by degradation of type IV, type V, and denatured collagens. There are two types of gelatinases, 72-kDa gelatinase (gelatinase A), or MMP-2 (Collier et al., 1988), and 92-kDa gelatinase (gelatinase B), or MMP-9 (Wilhelm et al., 1989). There are very distinct differences between the two gelatinases. The 72-kDa gelatinase has been found complexed to TIMP-2 (Stetler-Stevenson et al., 1989), whereas the 92-kDa gelatinase has been found complexed to TIMP-1 (Wilhelm et al., 1989). Regulation of the two gelatinases is also very distinct. Analysis of the genomic structure and promoter of the 72-kDa gelatinase has revealed that this gene does not have an AP-1 site or a TATA box in the 5′ promoter region as all the other *MMPs* have been shown to have (Huhtala et al., 1990). This enzyme is also not regulated by PMA and, in many cases, seems to be expressed constitutively rather than in a regulated fashion. In contrast, the 92-kDa gelatinase has a promoter very similar to that of the other *MMPs* and is regulated similarly (Huhtala et al., 1991). Nevertheless, expression and activity of both types of gelatinase are markedly stimulated by IL-1 (Kusano et al., 1998).

In bone, as is to be expected, MMP-2 is expressed constitutively by many osteoblastic preparations (Overall et al., 1989; Rifas et al., 1989, 1994; Meikle et al., 1992) and is unchanged by treatment with any of the agents tested. The zymogen form of MMP-2 is also resistant to activation by serine proteases, but MT1-MMP (MMP-14) can initiate the activation of MMP-2 by cleavage of the Asn66−Leu peptide bond (Sato et al., 1994). In 2001, Martignetti et al. (2001) described a form of multicentric osteolysis (Winchester/Torg syndrome) with striking tarsal and carpal bone resorption, accompanied by arthropathy, osteoporosis, subcutaneous nodules, and a distinctive facies in large, consanguineous Saudi

Arabian families. They localized the gene to 16q12–q21 and demonstrated two homoallelic, family-specific, mutations in the region that encodes *MMP2*. Nonsense and missense mutations have been documented that are consistent with decreased levels of MMP-2 (Al-Aqeel, 2005). *Mmp2*-null mice were first reported by Itoh et al. (1997) to have no phenotype except for some shortening of limb bones. Subsequent work by this group documented osteoporosis in older mice as well as altered remodeling of the canalicular system with osteocyte apoptosis, fewer canaliculi, and decreased canalicular connectivity (Inoue et al., 2006). They also described striking defects in formation/maintenance of osteocyte networks and connectivity in collagenase-resistant (r/r) mice; the r/r mice had previously been shown to have osteocyte and osteoblast apoptosis and prominent emptying of osteocyte lacunae (Zhao et al., 2000). We emphasize, however, that although the *Mmp2*-null mice had osteoporosis, the characteristic nodulosis and severe focal osteolysis of the human NAO (nodulosis, arthropathy and osteolysis) syndrome with mutations in *MMP2* were not found in the $Mmp2^{-/-}$ mice. Later, Mosig et al. (2007) reported that *Mmp2*- null mice obtained from Itoh et al. (1997), and described above, display progressive loss of bone mineral density, articular cartilage destruction, and abnormal long bone and craniofacial development. These mice had 50% fewer osteoblasts and osteoclasts compared with control littermates at 4 days, while there was almost no difference after 4 weeks of age. In addition, inhibition of MMP-2 via small interfering RNA in human SaOS2 and murine MC3T3 osteoblast cell lines caused a decrease in cell proliferation rates. These findings imply that MMP-2 is critical for normal skeletal and craniofacial development, as well as bone cell growth and proliferation. Mosig et al. (2007) did not comment on focal osteolysis of the NAO human syndrome in the *Mmp2*-null mice they studied, nor did they examine the canalicular networks using approaches similar to those of Inoue et al. (2006).

$Mmp7^{-/-}$ mice were reported to have several abnormalities, such as decreased intestinal tumorigenesis, but no obvious skeletal defects (Wilson et al., 1997). Later work from this group demonstrated in a prostate cancer model that MMP-7 produced by osteoclasts at the tumor–bone interface has the capacity to process cell-bound receptor activator of NF-κB ligand (RANKL) to a soluble form that further promotes osteoclast activation (Lynch et al., 2005). In *Mmp7*-deficient mice, there was reduced RANKL processing and reduced tumor-induced osteolysis. It appears, however, that $Mmp7^{-/-}$ mice have no physiological abnormality in physiological skeletal remodeling (i.e., no abnormality in the absence of bone metastasis).

The 92-kDa gelatinase (MMP-9) is secreted by three osteosarcoma cell lines (TE-85, U2OS, and MG-63) (Rifas et al., 1994) and, in some of the cell lines, can be stimulated by PMA, IL-1β, and TNFα, analogous to these authors' observations regarding stromelysin. Similarly, they were unable to identify secreted MMP-9 in the media of normal human osteoblasts or the human osteosarcoma cell line SaOS-2, which has been shown to have retained many characteristics of highly differentiated osteoblasts. Likewise, Meikle et al. (1992) found very little immunohistochemical staining for MMP-9 in normal human osteoblasts. In fact, this enzyme has been found to be highly expressed by rabbit and human osteoclasts (Tezuka et al., 1994a; Wucherpfennig et al., 1994; Vu et al., 1998). Indeed, a lack of expression of MMP-9 in mature osteoclasts of c-*fos*-null mice may be one of the reasons the animals exhibit an osteopetrotic phenotype (Grigoriadis et al., 1994). Furthermore, studies of mice with a targeted inactivation of the gene indicate that MMP-9 plays a role in regulating endochondral bone formation, particularly of the primary spongiosa, possibly by mediating capillary invasion. Mice containing a null mutation in the *Mmp9* gene exhibit delays in vascularization, ossification, and apoptosis of the hypertrophic chondrocytes at the skeletal growth plates (Vu et al., 1998). These defects result in an accumulation of hypertrophic cartilage in the growth plate and lengthening of the growth plate. The defects are reversible, and by several months of age the affected mice have an axial skeleton of normal appearance. It was postulated that MMP-9 is somehow involved in releasing angiogenic factors such as vascular endothelial growth factor, which is normally sequestered in the ECM (Gerber et al., 1999). Extracellular galectin-3 could be an endogenous substrate of MMP-9 that acts downstream to regulate hypertrophic chondrocyte death and osteoclast recruitment during endochondral bone formation. Thus, the disruption of growth plate homeostasis in *Mmp9*-null mice links galectin-3 and MMP-9 in the regulation of the clearance of late chondrocytes through regulation of their terminal differentiation (Ortega et al., 2005).

Membrane-type matrix metalloproteinases

While most MMPs are secreted, a subtype called MT-MMPs are inserted into the cell membrane (Sato et al., 1997; Pei, 1999). These proteases contain a single transmembrane domain and an extracellular catalytic domain. Characteristically, MT-MMPs have the potential to be activated intracellularly by furin or furin-like proteases through recognition of a unique amino acid sequence: Arg–Arg–Lys–Arg111 (Sato et al., 1996). To date, six MT-MMPs have been described, four transmembrane proteins (MMP-14, -15, -16, and -24) and two glycosylphosphatidylinositol-anchored ones (MMP-17 and -25). MT1-MMP (MMP-14), MT2-MMP (MMP-15), and MT3-MMP (MMP-16) have been shown to have a wide range of activities against ECM proteins (Pei and Weiss, 1996; Velasco et al., 2000). MT1-MMP is involved in endothelial cell migration and invasion (Galvez et al., 2000; Collen et al., 2003), and MT2-MMP and MT3-MMP are also involved in cell migration and invasion, depending on the cell type (Hotary et al., 2000; Shofuda et al., 2001). In a collagen-invasion model, MT1-MMP appears to be the critical MMP (Sabeh et al., 2004).

MT4-MMP (MMP-17) has the smallest degree of sequence identity to the other family members and has TNFα convertase activity, but does not activate pro-MMP-2 (Puente et al., 1996; English et al., 2000). Conversely, MT5-MMP (MMP-24) and MT6-MMP (MMP-25) may facilitate tumor progression through their ability to activate pro-MMP-2 at the membrane of cells from tumor tissue (Llano et al., 1999; Velasco et al., 2000). As mentioned earlier, MT1-MMP (MMP-14) serves as a membrane receptor or activator of MMP-2 and possibly other secreted MMPs (Sato et al., 1994). Further, studies indicate that MT1-MMP may also function as a fibrinolytic enzyme in the absence of plasmin and facilitate the angiogenesis of endothelial cells (Hiraoka et al., 1998). MT1-MMP is highly expressed in embryonic skeletal and peri-skeletal tissues and has been identified in osteoblasts by in situ hybridization and immunohistochemistry (Apte et al., 1997; Kinoh et al., 1996). Targeted inactivation of the *Mmp14* gene in mice produces several skeletal defects that result in osteopenia, craniofacial dysmorphisms, arthritis, and dwarfism (Holmbeck et al., 1999; Zhou et al., 2000). Several of the notable defects in bone formation include delayed ossification of the membranous calvarial bones, persistence of the parietal cartilage vestige, incomplete closure of the sutures, and marked delay in the postnatal development of the epiphyseal ossification centers characterized by impaired vascular invasion. Histological observation suggested that the progressive osteopenia noted in these animals may be attributed to excessive osteoclastic resorption and diminished bone formation. This finding was supported by evidence that osteoprogenitor cells isolated from the bone marrow of these mutant mice demonstrate defective osteogenic activity. A similar human disease of "vanishing bone" was observed in two sisters with mutations in MT1-MMP preventing its membrane localization (Evans et al., 2012). This raises the question of a protective effect in bone exerted by MT1-MMP and/or activation of MMP-2. In contrast, a 2018 paper (Delgado-Calle et al., 2018) has shown that MT1-MMP is stimulated by constitutively active PTH receptor 1 signaling in bone and appears to be partly responsible for the high bone turnover phenotype of these mice. The authors showed that MT1-MMP is, in some way, associated with increases in soluble RANKL production. MT1-MMP is also associated with osteoclast-mediated bone resorption in rheumatoid arthritis (Pap et al., 2000).

Collagenases

Collagenases generally cleave fibrillar native collagens I–III at a single helical site at neutral pH (Matrisian, 1992). The resultant cleavage products denature spontaneously at 37°C and become substrates for many enzymes, particularly gelatinases. The collagenase subfamily of human MMPs consists of three distinct members: fibroblast collagenase-1 (MMP-1), neutrophil collagenase-2 (MMP-8), and collagenase-3 (MMP-13) (Goldberg et al., 1986; Freije et al., 1994). An additional collagenase, called collagenase-4 (initially called MMP-18), was identified in *Xenopus laevis* (Stolow et al., 1996), and a human homolog of this enzyme was identified and was given other names, such as MMP RASI-1, and has now been designated MMP-19. As of this writing, only one rat/mouse interstitial collagenase has been studied thoroughly and shown to be expressed by a range of cells, including osteoblasts and osteocytes. This collagenase has a high degree of homology (86%) to human collagenase-3 and is aptly given the same name (Quinn et al., 1990), and both are now called MMP-13. MMP-13 is secreted by osteoblasts and osteocytes, hypertrophic chondrocytes, smooth muscle cells, and fibroblasts, in proenzyme form at 58 kDa, and is subsequently cleaved to its active form of 48 kDa (Roswit et al., 1983). Two murine orthologs of human collagenase-1 (MMP-1), called murine collagenase-like A (Mcol-A) and murine collagenase-like B (Mcol-B), were first identified by nucleotide sequence similarity to human MMP-1, but only Mcol-A was able to degrade native type I and II collagens, casein, and gelatins (Balbin et al., 2001). In this report, the expression of Mcol-A was limited to early embryos. It should be noted here that studies of mouse and rat tissues that report expression of MMP-1 by immunohistochemistry or in situ hybridization are probably not detecting MMP-1. We also made the mistake of calling rat collagenase-3 (i.e., MMP-13) MMP-1 when the nomenclature of the MMPs was being thrashed out (Omura et al., 1994). A murine ortholog of collagenase-2 (MMP-8) has been identified by two groups (Lawson et al., 1998; Balbin et al., 1998). A role for Mcol-A, Mcol-B, or murine collagenase-2 (MMP-8) in bone cell function has not been demonstrated, although human MMP-8 is expressed in chondrocytes and other skeletal cells. *Mmp8*$^{-/-}$ mice (Balbin et al., 2003) have no skeletal abnormalities during development; skeletal changes in adults have not yet been reported.

As noted previously, it has also been shown that other MMPs (MMP-2 [gelatinase A (GelA or 72-kDa gelatinase)] and MMP-14 [MT1-MMP]) can function as collagenases in vitro (Aimes and Quigley, 1995; Ohuchi et al., 1997). These MMPs (-1, -2, -8, -13, and -14) all cleave each of the triple-helical interstitial collagens at the same locus and therefore must also be considered collagenases.

Collagenase-3/MMP-13

In developing rat calvariae, we have found ample amounts of MMP-13 by immunohistochemistry 14 days after birth (Davis et al., 1998). These are always in select areas, mostly associated with sites of active modeling. At the cellular level, staining is associated with osteocytes and bone-lining cells that have the appearance of osteoblasts. Originally, there was controversy regarding the cellular origin of bone collagenase. The osteoclast was reported to show immunohistochemical staining for collagenase (Delaissé et al., 1993), but it was not determined whether this was a gene product of the osteoclast or was, perhaps, produced by osteoblasts/osteocytes and bound by the osteoclast through a receptor (see later). However, in situ hybridization of 17- to 19-gestational-day rat fetal long bones showed *Mmp13* expression only in chondrocytes, bone surface mononuclear cells, and osteocytes adjacent to osteoclasts; there was no evidence of expression in osteoclasts (Fuller and Chambers, 1995). Similarly, Mattot et al. (1995) showed expression of mouse *Mmp13* in hypertrophic chondrocytes and in cells of forming bone from humeri of mice at the 18th gestational day. In human fetal cartilage and calvaria, *MMP13* transcripts were detected in hypertrophic chondrocytes, osteoblasts, and periosteal cells by in situ hybridization, whereas no expression of *MMP13* was detected in osteoclasts (Johansson et al., 1997). In addition, it has been known for some time that bone explants from osteopetrotic mice (lacking active osteoclasts) continue to produce abundant collagenolytic activity, either unstimulated or stimulated by bone-resorbing hormones (Jilka and Cohn, 1983; Heath et al., 1990). These studies demonstrate that osteoblasts/osteocytes and hypertrophic chondrocytes are the sources of collagenases in skeletal tissue, whereas the osteoclast does not appear to express these genes. It should also be noted that the expression of *Mmp13* assayed by in situ hybridization was strikingly reduced (Lanske et al., 1996) in the distal growth plate and midshafts of bones from $Pthr1^{-/-}$ mouse embryos (Lanske et al., 1998).

The remodeling of the fracture callus mimics the developmental process of endochondral bone formation. Excess tissue accumulates as callus prior to endochondral ossification followed by osteoclast repopulation. In collaboration with Dr. Mark Bolander, we demonstrated profuse concentrations of metalloproteinases in the fracture callus of adult rat long bones (Partridge et al., 1993). The predominant cells observed to stain for MMP-13 are hypertrophic chondrocytes during the phase of endochondral ossification, marrow stromal cells (putative osteoblasts) when the primary spongiosa is remodeled, and osteoblasts/osteocytes at a time when newly formed woven bone is being remodeled to lamellar bone. This indicates that the adult long bone has the ability to produce profuse levels of MMP-13, but only when challenged, e.g., by a wound-healing situation or an osteotropic hormone. As well, it was shown that *Mmp13*-null mice have delayed bone fracture healing, characterized by a retarded cartilage response in the fracture callus (Kosaki et al., 2007). The consistent observation here is a role for this enzyme when a collagenous matrix must undergo substantial, rapid remodeling.

Liu et al. (1995) have demonstrated that targeted mutation around the collagenase cleavage site in both alleles of the endogenous mouse type I collagen gene Colla1, which results in resistance to collagenase cleavage, leads to dermal fibrosis and uterine collagenous nodules. These animals are able to develop normally to adulthood, and some of the major abnormalities become apparent only with increasing age. Studies of these mice revealed that homozygous mutant (r/r) mice have diminished PTH-induced bone resorption, diminished PTH-induced calcemic responses, and thicker bones (Zhao et al., 1999). These observations imply that collagenase activity is necessary not only in older animals for rapid collagen turnover, but also for PTH-stimulated bone resorption. In the r/r mice, as early as 2 weeks of age, empty osteocyte lacunae were evident in the calvariae and long bones, with the number of empty lacunae increasing with age, and an increase in apoptosis was observed in osteocytes, as well as periosteal cells (Zhao et al., 2000). Thus, normal osteocytes (and osteoblasts) and osteoclasts might bind to cryptic epitopes that are revealed by the collagenase cleavage of type I collagen by liganding the $\alpha V\beta 3$ integrin and, if such signals are not induced (as postulated for the osteoclastic defect in r/r mice), they would undergo apoptosis and their lacunae would empty. Young r/r mice are also observed to develop thickening of the calvariae through the deposition of new bone predominantly at the inner periosteal surface; an increased deposition of endosteal trabecular bone was found in long bones in older r/r mice. Thus, the failure of collagenase to cleave type I collagen in r/r mice was associated with increased osteoblast and osteocyte apoptosis and, paradoxically, increased bone deposition as well.

To elucidate the functional roles of MMP-13 during skeletal development in vivo, Inada et al. (2004) generated *Mmp13*-null mice. These mice were found to have lengthened growth plates, due to an increase in the hypertrophic chondrocyte zone, as well as delayed ossification at the primary centers. Abnormalities of growth plates were apparent in the early stages of embryonic development and persisted throughout adulthood. This abnormality is most likely due to a decrease in degradation of ECM cartilage, as was shown by the significant increase in the area of type X collagen deposition, although an increase in the synthesis of type X collagen also remains a possible cause. These observations suggest that MMP-13 plays a critical role in collagen degradation during growth plate development and endochondral ossification. Similar to these findings, Stickens et al. (2004) found that deletion of MMP-13 caused abnormal endochondral bone development as a result of impaired ECM remodeling. These *Mmp13*-deficient mice were viable, were fertile, and had a normal life span, with no gross phenotypic abnormalities. However, an increase in the hypertrophic chondrocyte zone of the skeletal growth plate was observed, as a result of the delayed exit of chondrocytes from the growth plate. In addition, unlike the late phenotype of the collagenase-resistant mice, these mice showed an early increase in trabecular bone that persisted for months. This was due to the absence of collagenase expression in osteoblasts, not chondrocytes, as was shown by tissue-specific knockouts. The crucial role of MMP-13 in bone formation and remodeling is further demonstrated by a missense mutation, F56S, in the proregion domain of MMP-13 in a form of chondrodysplasia in humans. This mutation, the substitution of an evolutionarily conserved phenylalanine residue for a serine, causes the Missouri type of spondyloepimetaphyseal dysplasia, an autosomal dominant disorder characterized by defective growth and modeling of vertebrae and long bones (Kennedy et al., 2005). This is thought to be due to intracellular autoactivation and degradation of the mutant enzyme.

Related to work in whole animals, we have shown, together with Drs. Jane Lian and Gary Stein, that *Mmp13* is expressed late in differentiation in in vitro mineralizing rat osteoblast cultures (Shalhoub et al., 1992; Winchester et al., 1999, 2000). The appearance of the enzyme in late differentiated osteoblasts may correlate with a period of remodeling of the collagenous ECM. These observations regarding the differentiation of rat osteoblasts may explain the very low levels of MMP-1 observed in cultures of normal human osteoblasts (Rifas et al., 1989), where mRNAs and proteins were isolated from cells at confluence, but apparently not from mineralized cultures. Alternatively, the cultures may predominantly express MMP-13 (rather than MMP-1), which has been shown to be expressed by osteoblasts, by chondrocytes, and in synovial tissue, particularly in pathological conditions such as osteoarthritis (Johansson et al., 1997; Mitchel et al., 1996; Reboul et al., 1996; Wernicke et al., 1996). At the time that Rifas and colleagues conducted the work on human osteoblasts, human MMP-13 had not been identified.

Canalis's group has conducted considerable research on the hormonal regulation of MMP-13 in rat calvarial osteoblasts, including demonstrating stimulation by retinoic acid (Varghese et al., 1994). They have also demonstrated that triiodothyronine, platelet-derived growth factor, and basic fibroblast growth factor (bFGF) all stimulate *Mmp13* transcription (Pereira et al., 1999; Rydziel et al., 2000; Varghese et al., 2000). Interestingly, they have also shown that insulin-like growth factors (IGFs) inhibit both basal and retinoic-stimulated collagenase expression (Canalis et al., 1995) by these cells. We have shown that TGFβ1 stimulates *Mmp13* expression in the rat osteosarcoma cell line UMR 106-01 (Selvamurugan et al., 2004).

We have conducted many studies with the clonal rat osteosarcoma line UMR 106-01, which has been described as osteoblastic in phenotype (Partridge et al., 1980, 1983). This cell line responds to all of the bone-resorbing hormones by synthesizing MMP-13 (Partridge et al., 1987; Civitelli et al., 1989). In contrast to the physiological regulation of collagenase in fibroblasts (Woessner, 1991), synoviocytes (Brinckerhoff and Harris, 1981), and uterine smooth muscle cells (Wilcox et al., 1994), the control of expression of MMP-13 in bone and osteoblastic cells appears to have some distinct differences. First, it is stimulated by all the bone-resorbing hormones (Partridge et al., 1987; Delaissé et al., 1988), which act through different pathways, including protein kinase A (PKA; PTH and prostaglandins [PGs]), PKC (PTH and PGs), tyrosine phosphorylation (EGF), and direct nuclear action ($1,25(OH)_2D_3$, retinoic acid). Second, glucocorticoids do not inhibit stimulation by PTH (Delaissé et al., 1988; T.J. Connolly, N.C. Partridge, and C.O. Quinn, unpublished observations), whereas retinoic acid stimulates MMP-13 expression rather than inhibiting it (Delaissé et al., 1988; Connolly et al., 1994; Varghese et al., 1994). Last, in rat osteosarcoma cells, PMA is unable to elicit a pronounced stimulatory effect on *Mmp13* gene expression.

Among the bone-resorbing agents tested, PTH is the most effective in stimulating MMP-13 production by UMR cells. A single 10^{-7} M PTH dose significantly stimulates transient MMP-13 secretion, with maximal extracellular enzyme concentrations achieved between 12 and 24 h (Partridge et al., 1987; Civitelli et al., 1989). This level is maintained at 48 h, decreases to 20% of the maximum by 72 h, and is ultimately undetectable by 96 h. We hypothesized that MMP-13 was removed from the medium through a cell-mediated binding process because the enzyme is stable in conditioned medium, and experiments showed that this disappearance was not due to extracellular enzymatic degradation. Binding studies conducted with ^{125}I-MMP-13 revealed a specific receptor with high affinity ($K_d = 5$ nM) for rat MMP-13 (Omura et al., 1994). Further studies showed that binding of MMP-13 in this fashion is responsible for its rapid internalization and degradation. The processing of MMP-13 in this system requires receptor-mediated endocytosis and involves sequential processing by endosomes and lysosomes (Walling et al., 1998). In addition to UMR cells, we identified a very similar MMP-13 receptor on normal, differentiated rat osteoblasts, rat and mouse embryonic fibroblasts, and human and rabbit chondrocytes (Walling et al., 1998; Barmina et al., 1999; Raggatt et al., 2006). These results indicate that the function of these receptors is to limit the extracellular abundance of MMP-13 and, consequently, breakdown of the ECM.

Further investigation of the MMP-13 receptor system led us to conclude that MMP-13 binding and internalization required a two-step mechanism involving both a specific MMP-13 receptor and a member of the low-density lipoprotein (LDL) receptor—related superfamily. Ligand blot analyses demonstrated that ^{125}I-labeled MMP-13 specifically bound two proteins (approximately 170 and 600 kDa) in UMR 106-01 cells (Barmina et al., 1999). Of these two binding proteins, the 170-kDa protein appeared to be a high-affinity primary binding site, and the 600-kDa protein appeared to be the LDL receptor—related protein-1 (LRP-1) responsible for mediating internalization. The LDL receptor superfamily represents a diverse group of receptors, including the LDL receptor; the VLDL receptor; LRP-1; megalin (LRP-2); and LRP-4, -5, and -6 (Herz and Bock, 2002; Pohlkamp et al., 2017). All of these plasma membrane receptors have a single membrane-spanning domain and several stereotyped repeats, ligand-binding type and EGF-precursor-like. Most receptors in this family participate in receptor-mediated endocytosis, whereby the receptor—ligand complex is directed (via an NPXY signal in the receptor) to clathrin-coated pits and then internalized. Ligands of these receptors include LDL, VLDL, urokinase-type PA (uPA)— or tissue-type PA (tPA)—PA inhibitor-1 (PAI-1) complexes, tPA, lactoferrin, activated α_2-macroglobulin/proteinase complexes, apolipoprotein E—enriched β-VLDL, lipoprotein lipase, *Pseudomonas* exotoxin A, Wnts, Wnt inhibitors, BMP-4, vitamin D—binding protein, and vitellogenin (Herz and Bock, 2002; Pohlkamp et al., 2017; Yang and Williams, 2017).

The striking stimulation of MMP-13 secretion by bone-resorbing agents in UMR cells was shown to be paralleled by an even more striking induction of *Mmp13* mRNA. Northern blots showed an ~180-fold induction of *Mmp13* mRNA 4 h after PTH treatment (Scott et al., 1992), with an initial lag period between 0.5 and 2 h before *Mmp13* mRNA levels rose above basal. Nuclear run-on studies showed a comparable increase in transcription of the gene 2 h after treatment with PTH. The PTH-induced increase in *Mmp13* transcription was completely inhibited by cycloheximide, whereas the transcriptional rate of β-actin was unaffected by inclusion of the protein synthesis inhibitor (Scott et al., 1992). These results demonstrate that the PTH-mediated stimulation of MMP-13 involves transcription and requires de novo synthesis of a protein factor(s), i.e., it is a secondary response gene.

PTH treatment was found to increase the transcription of *Mmp13* in rat osteoblastic osteosarcoma cells primarily by stimulation of the cAMP signal transduction pathway (Scott et al., 1992). Second-messenger analogs were used to test which signal transduction pathway is of primary importance in the PTH-mediated transcriptional induction of the *Mmp13* gene. The cAMP analog 8BrcAMP was capable of inducing *Mmp13* transcription to levels close to those of PTH. In contrast, neither the PKC activator, PMA, nor the calcium ionophore, ionomycin, when used alone, resulted in any increase in *Mmp13* gene transcription similar to that elicited by PTH after 2 h of treatment. Furthermore, this effect requires protein synthesis and a 1- to 1.5-h lag period, suggesting that the transcriptional activation of the *Mmp13* gene may be the result of interactions with immediate early gene products. PTH treatment was also found to transiently increase the mRNA expression of the AP-1 protein subunits c-Fos and c-Jun (Clohisy et al., 1992). Both mRNA species were maximally induced within 30 min, well before the maximal transcription rate at 90 min for *Mmp13*. Later, it was determined that PTH is responsible for phosphorylation of the cAMP response element—binding (CREB) protein at serine 133 (Tyson et al., 1999). Once phosphorylated, the CREB protein binds a cAMP response element in the c-Fos promoter and activates transcription (Pearman et al., 1996).

The *Mmp13* gene has 10 exons (Rajakumar et al., 1993), encoding an mRNA of ~2.9 kb, which in turn encodes the proenzyme, with a predicted core protein molecular weight of 52 kDa (Quinn et al., 1990). A series of deletion and point mutants of the promoter region was generated to identify the PTH-responsive region and subsequently the primary response factors, which convey the hormonal signal and bind to this region(s) of the *Mmp13* gene. The minimum PTH regulatory region was found to be within 148 bp upstream of the transcriptional start site (Selvamurugan et al., 1998). This region contains several consensus transcription factor recognition sequences, including SBE (Smad binding element), C/EBP (CCAAT enhancer-binding protein site), the RD binding sequence, p53, PEA-3, and AP-1 and -2. The AP-1 site is a major target for the Fos and Jun families of oncogenic transcription factors (Chiu et al., 1988; Lee et al., 1987; Angel and Karin, 1991). The RD site is a target for CBF proteins, specifically Runx2. Mice containing a targeted disruption of the Runx2 gene die at birth and lack both skeletal ossification and mature osteoblasts (Ducy et al., 1997; Komori et al., 1997; Otto et al., 1997). These mutant mice also do not express *Mmp13* during fetal development, indicating that *Mmp13* is one of the target genes regulated by Runx2 (Jimenez et al., 1999).

Additional experiments on the *Mmp13* promoter determined that both native AP-1 and RD sites and their corresponding binding proteins, AP-1 and Runx2-related proteins, are involved in PTH regulation of *Mmp13* transcription. Using gel-shift analysis, we further showed enhanced binding of c-Fos and c-Jun proteins at the AP-1 site upon treatment with PTH (Selvamurugan et al., 1998), although there was no significant change in the level of Runx2 binding to the RD site. We determined that PTH induces PKA-mediated posttranslational modification of Runx2 and leads to enhanced *Mmp13* promoter activity in UMR cells (Selvamurugan et al., 2000b). The binding of members of the AP-1 and Runx families to their corresponding binding sites in the *Mmp13* promoter also appears to regulate *Mmp13* gene expression during osteoblast differentiation (Winchester et al., 2000). As discussed earlier, MMP-13 expression is regulated by a variety of growth factors, hormones, and cytokines, but the effects of these compounds appear to be cell-type specific. Data obtained in breast cancer and other cell lines suggest that the differential expression of and regulation of MMP-13 in osteoblastic compared with nonosteoblastic cells may depend on the expression of AP-1 factors and posttranslational modifications of Runx2 (Selvamurugan and Partridge, 2000; Selvamurugan et al., 2000a). The close proximity of the AP-1 and RD sites and their cooperative involvement in the activation of the *Mmp13* promoter suggest that the proteins binding to these sites physically interact; Runx2 was found to directly bind c-Fos and c-Jun in both in vitro and in vivo experiments (D'Alonzo et al., 2002). To determine the importance of these regulatory sites in the expression of *Mmp13* in vivo, transgenic mice containing the *Escherichia coli* lacZ reporter fused to either the wild-type *Mmp13* promoter or that with mutated AP-1 and RD sites were generated (Selvamurugan et al., 2006). The wild-type transgenic lines expressed higher levels of bacterial β-galactosidase in bone, teeth, and skin compared with the mutant and transgenic lines. Thus, the AP-1 and RD sites of the promoter most likely regulate and are necessary for gene expression in vivo in bone, as well as teeth and skin.

We have shown that the RD and AP-1 sites in the proximal promoter region are essential for PTH stimulation of *Mmp13* promoter activity (Selvamurugan et al., 1998; Winchester et al., 2000; D'Alonzo et al., 2002). At the *Mmp13* promoter in UMR 106-01 cells, Runx2 binds histone deacetylase 4 (HDAC4) at the RD binding site, resulting in a repression of transcription under basal conditions (Shimizu et al., 2010). After PTH treatment, PKA-dependent phosphorylated HDAC4 dissociates from Runx2, which is then free to recruit histone acetyltransferases, especially p300 and p300/CBP-associated factor, to activate transcription (Boumah et al., 2009, Lee and Partridge, 2010). We also demonstrated that HDAC4 associates with myocyte enhancer factor 2C (MEF2C) and MEF2C participates in PTH-stimulated *Mmp13* gene expression by increased binding to c-Fos at the AP-1 site in the *Mmp13* promoter. As with Runx2, PTH causes release of HDAC4 from MEF2C and p300 is recruited to bind MEF2C (Nakatani and Partridge, 2017).

In vivo, we have found that global *Hdac4*$^{-/-}$ mice have increased MMP-13 mRNA and protein expression in hypertrophic chondrocytes and trabecular osteoblasts (Shimizu et al., 2010). Global *Hdac4*$^{-/-}$ mice are runted in size and do not survive to weaning. This phenotype is primarily due to the acceleration of onset of chondrocyte hypertrophy and, as a consequence, inappropriate endochondral mineralization. MMP-13 is thought to be involved in endochondral ossification and bone remodeling. To identify whether the phenotype of *Hdac4*$^{-/-}$ mice was due to upregulation of MMP-13, we generated *Hdac4/Mmp13* double-knockout mice and determined the ability of deletion of MMP-13 to rescue the *Hdac4*$^{-/-}$ mouse phenotype (Nakatani et al., 2016). *Mmp13*$^{-/-}$ mice have normal body size. The double-knockout mice were significantly heavier and larger than *Hdac4*$^{-/-}$ mice, survived longer, and recovered the thickness of their growth plate zones. Micro-computed tomographic analysis revealed that *Hdac4*$^{-/-}$ mice had significantly decreased cortical bone area compared with the wild-type mice. In addition, bone porosity was significantly decreased. The double-knockout mice recovered these cortical parameters. Likewise, their trabecular bone recovered toward normal for this age. Taken together, our findings indicate that the phenotype seen in the *Hdac4*$^{-/-}$ mice is partially derived from elevation of MMP-13 and may be due to a bone remodeling disorder caused by overexpression of this enzyme.

Plasminogen activators

The PA/plasmin pathway is involved in several processes, including tissue inflammation, fibrinolysis, ovulation, tumor invasion, malignant transformation, tissue remodeling, and cell migration. The PA/plasmin pathway is also thought to be involved in bone remodeling by osteoblasts and osteoclasts. The pathway results in the formation of plasmin, another neutral serine proteinase, which degrades fibrin and the ECM proteins fibronectin, laminin, and proteoglycans. In addition, plasmin can convert MMPs, procollagenase, and prostromelysin to their active forms (Eeckhout and Vaes, 1977). Plasminogen has been localized to the cell surface of the human osteosarcoma line MG63, where its activity was enhanced by endogenous cell-bound uPA (Campbell et al., 1994).

The PA/plasmin pathway is regulated by members of the serpin family in addition to various hormones and cytokines. The primary function of this family of inhibitors is to neutralize serine proteinases by specific binding to the target enzyme. Serpins are involved in the regulation of several processes, including fibrinolysis, cell migration, tumor suppression, blood coagulation, and ECM remodeling (Potempa et al., 1994). Members of this pathway involved in the regulation of the PA/plasmin pathway are PAI-1 and PAI-2, which regulate uPA and tPA; protease nexin-1, which regulates thrombin, plasmin, and uPA; and α_2-antiplasmin, which regulates plasmin. Active PAI-1 combines with uPA and tPA, forming an equimolar complex, exerting its inhibition through interactions with the active-site serine. PAI-1 has been detected in media of cultured human fibrosarcoma cells (Andreasen et al., 1986) and primary cultures of rat hepatocytes and hepatoma cells.

PAI-1 was also detected in conditioned medium of rat osteoblast-like cells and rat osteosarcoma cells (Allan et al., 1990). The expression of uPA, tPA, type I receptor for uPA, PAI-1, PAI-2, and the broad-spectrum serine proteinase inhibitor protease nexin-1 is induced by PTH treatment in primary mouse osteoblasts. The regulation of these various enzymes within bone tissue may determine the sites where bone resorption will be initiated (Tumber et al., 2003).

Urokinase-type plasminogen activator

The uPA is secreted as a precursor form of ~55 kDa (Nielsen et al., 1988; Wun et al., 1982). It is activated by cleavage into a 30-kDa heavy chain and a 24-kDa light chain, joined by a disulfide bond, with the active site residing in the 30-kDa fragment. Urokinase has a Kringle domain, a serine proteinase-like active site, and a growth factor domain (GFD). The noncatalytic NH_2-terminal fragment contains the GFD and the Kringle domain and is referred to as the amino-terminal fragment (ATF). The uPA and PAI-1 are involved in regulation of the first steps of angiogenesis (Pepper, 1997). Rabbani et al. (1990) demonstrated that ATF stimulated proliferation and was involved in mitogenic activity in primary rat osteoblasts and the human osteosarcoma cell line SaOS-2. The GFD of the ATF is necessary for the binding of uPA to its specific receptor.

Tissue-type plasminogen activator

The tPA is secreted as a single-chain glycosylated 72-kDa polypeptide. This enzyme has been found in human plasma and various tissue extracts, as well as in normal and malignant cells. The cleavage of tPA forms a 39-kDa heavy chain and a 33-kDa light chain linked by a disulfide bond. The heavy chain has no proteinase activity, but contains two Kringle domains that assist in binding fibrin to plasminogen (Banyai et al., 1983; Pennica et al., 1983). Furthermore, the heavy chain contains a finger domain involved in fibrin binding (van Zonneveld et al., 1986) and a GFD with homology to human and murine EGF. tPA has been shown to be expressed in osteoblastic cells after nicotine or PTH treatment (Katano et al., 2006; Tumber et al., 2003).

Plasminogen activators in bone

PA activity is increased in normal and malignant osteoblasts as well as calvariae by many agents, including PTH, $1,25(OH)_2D_3$, PGE_2, IL-1α, fibroblast growth factor, and EGF (Hamilton et al., 1984, 1985; Thomson et al., 1989; Pfeilschifter et al., 1990; Cheng et al., 1991; Leloup et al., 1991; De Bart et al., 1995). It should be noted that other work suggests that PAs are not necessary for PTH- and $1,25(OH)_2D_3$-induced bone resorption (Leloup et al., 1994). Expression of tPA, uPA, PAI-1 and -2, protease nexin, and urokinase receptor isoform 1 was detected in microdissected mouse osteoclasts (Yang et al., 1997). Deletion of tPA, uPA, PAI-1, and plasminogen genes in mice can lead to fibrin deposition, some growth retardation, and inhibition of the osteoclast's ability to remove noncollagenous proteins in vitro (Carmeliet et al., 1993, 1994; Bugge et al., 1995; Daci et al., 1999). Moreover, lack of both PAs leads to elongation of neonatal bones and increased bone mass. Osteoblast differentiation and formation of a mineralized bone matrix are enhanced in osteoblast cultures derived from $tPA^{-/-}/uPA^{-/-}$ neonatal mice (Daci et al., 2003). In a more recent paper (Kawao et al., 2014) the authors showed that bone repair after a femoral bone defect was significantly delayed in global $tPA^{-/-}$ mice but not in global $uPA^{-/-}$ mice and concluded that this was due to reduced proliferation of osteoblasts. In another study of global $uPA^{-/-}$ mice (Popa et al., 2014), fracture healing was slightly delayed with increased cartilage area and decreased tartrate-resistant acid phosphatase staining, suggesting a role for osteoclastic uPA expression.

There are conflicting data as to whether the increase in osteoblastic PA activity in vitro is due to an increase in the total amount of one or both of the PAs or is due to a decline in the amount of PAI-1. All possible results have been observed, depending on which osteoblastic cell culture system is used or the method of identification of the enzymes. The latter have been difficult to assay categorically because there have not been abundant amounts of specific antibodies available for each of the rat PAs. Similarly, different groups have found the predominant osteoblastic PA to be uPA, whereas others have obtained results indicating it to be tPA.

A range of agents have also been found to inhibit the amount of osteoblastic PA activity. These include glucocorticoids, TGFβ, bFGF, leukemia inhibitory factor, and IGF-1 (Allan et al., 1990, 1991; Cheng et al., 1991; Forbes et al., 2003; Lalou et al., 1994; Pfeilschifter et al., 1990). Where it has been examined, in many of these cases the decline is due to a substantial increase in PAI-1 mRNA and protein. Nevertheless, some of these agents also markedly enhance the abundance of mRNA for the PAs (Allan et al., 1991), although the net effect is a decline in PA activity.

Cysteine proteinases

The major organic constituent of the ECM of bone is fibrillar type I collagen, which is deposited in intimate association with an inorganic calcium/phosphate mineral phase. The presence of the mineral phase protects the collagen not only from thermal denaturation but also from attack by proteolytic enzymes (Glimcher, 1998). The mature osteoclast, the bone-resorbing cell, has the capacity to degrade bone collagen through the production of a unique acid environment adjacent to the ruffled border through the concerted action of a vacuolar proton pump ([V]-type H^+-ATPase) (Chakraborty et al., 1994; Bartkowicz et al., 1995; Teitelbaum, 2000) and a chloride channel of the Cl-7 type (Kornak et al., 2001). Loss-of-function mutations in the genes that encode either this proton pump (Li et al., 1999; Frattini et al., 2000; Kornak et al., 2000) or the chloride channel (Kornak et al., 2001) lead to osteopetrosis. At the low pH in this extracellular space adjacent to the ruffled border, it is possible to leach the mineral phase from the collagen and permit proteinases that act at acid pH to cleave the collagen (Blair et al., 1993). Candidate acid-acting proteinases are cysteine proteinases such as cathepsin K. Cathepsin K is highly expressed in osteoclasts (Drake et al., 1996; Bossard et al., 1996). Cysteine proteinases contain an essential cysteine residue at their active site that is involved in forming a covalent intermediate complex with their substrates (Bond and Butler, 1987). The enzymes are either cytosolic or lysosomal. The latter have an acidic pH optimum and make up the majority of the cathepsins. These enzymes are regulated by a variety of protein inhibitors, including the cystatin superfamily (Turk and Bode, 1991) and α_2-macroglobulin (Barrett, 1986). Their extracellular abundance must consequently be regulated by cell surface receptors for α_2-macroglobulin as well as the lysosomal enzyme targeting mannose-6-phosphate/IGF-2 receptors.

The involvement of lysosomal cysteine proteinases in bone resorption has been indicated by many studies showing that inhibition of these enzymes prevents bone resorption in vitro as well as lowering serum calcium in vivo (Delaissé et al., 1984; Montenez et al., 1994). In particular, cathepsin K (Tezuka et al., 1994b) was found to have substantial effects on bone. Mice containing a targeted disruption of cathepsin K were developed and found to exhibit an osteopetrotic phenotype characterized by excessive trabeculation of the bone marrow space (Saftig et al., 1998, 2000). Cathepsin K–deficient osteoclasts are capable of demineralizing the ECM, but are unable to fully remove the demineralized bone (Gowen et al., 1999). In addition, cathepsin K mutations have been linked to pycnodysostosis, a hereditary bone disorder characterized by osteosclerosis, short stature, and defective osteoclast function (Gelb et al., 1996). Further studies show that the expression of MMP-9, TRACP (tartrate-specific acid phosphatase) for osteoclastic enzymes and osteoblastic proteases (MMP-13, MMP-14), and RANKL is increased in cathepsin K–deficient mice (Kiviranta et al., 2005). Moreover, cathepsin K–deficient osteoclasts compensate for the lack of this enzyme by using MMPs in the resorption of bone matrix (Everts et al., 2006). It has been shown that cathepsin K is responsible for the activation of pro-MMP-9 in acidic environments such as are seen in tumors and during bone resorption (Christensen et al., 2015). Importantly, cathepsin K not only degrades type I collagen, it also degrades periostin, bradykinin, TGFβ1, and IGF-1, among other proteins (Bonnet et al., 2017; Godat et al., 2008; Fuller et al., 2004; Panwar et al., 2016). Cathepsin K inhibition increases periostin levels and bone formation in vivo (Bonnet et al., 2017). Its inhibition also leads to increases in serum TGFβ and evident fibrosis in a number of tissues (Runger et al., 2012).

Cathepsin inhibitors may be therapeutically beneficial in the treatment of osteoporosis and rheumatoid arthritis to stimulate cortical bone formation and inhibit bone resorption (Xiang et al., 2007). However, cathepsin K deficiency reduces atherosclerotic plaque and induces plaque fibrosis (Lutgens et al., 2006a). Use of a cathepsin K inhibitor as a possible therapeutic target for atherosclerosis should have been evaluated with care because cathepsin K inhibition probably leads to a profibrotic, but also to a more lipogenic, plaque phenotype (Lutgens et al., 2006b). In fact, this may have been the reason the cathepsin K active-site inhibitor odanacatinib failed in osteoporosis treatment clinical trials due to strokes and other cardiovascular adverse events (Mullard, 2016). However, other non-active-site inhibitors, ectosteric inhibitors, may be promising alternatives (Panwar et al., 2016) that do not have the off-target effects of odanacatinib, since they appear to inhibit only the cathepsin K–mediated degradation of type I collagen.

Immunohistochemistry revealed that the majority of the cysteine proteinases (cathepsins B, K, and L) and the aminopeptidases (cathepsins C and H) are products of osteoclasts (Ohsawa et al., 1993; Yamaza et al., 1998; Littlewood-Evans et al., 1997), although immunoreactive staining for cathepsins B, C, and H was also seen in osteoblasts and osteocytes. It is notable that the most potent collagenolytic cathepsin at acid pH, cathepsin L, was strongly expressed in osteoclasts and very weakly in osteoblasts. Mathieu et al. (1994), however, have detected both cathepsins B and L as proteins secreted by their immortalized osteogenic stromal cell line MN7. Everts et al. (2006) have shown that cathepsin L is involved in modulating MMP-mediated resorption by calvarial osteoclasts.

There has been work demonstrating that osteocytes produce cathepsin K (Qing et al.,2012), and this may have an important physiological role in osteocytic osteolysis, particularly in the lactating mouse, when serum PTH-related protein levels are high and demand for calcium for milk production is very high. Presumably, osteocytic cathepsin K also degrades growth factors, or perhaps activates them, and thus, could have an added role in bone turnover. The knowledge that the osteocyte expresses cathepsin K is also a cautionary note in the use of the cathepsin K—Cre mouse, since this is unlikely to be specific for osteoclasts.

Oursler et al. (1993) have also demonstrated that normal human osteoblast-like cells produce cathepsin B and that dexamethasone can increase expression and secretion of this lysosomal enzyme by these cells. Interestingly, they also showed that dexamethasone treatment causes activation of TGFβ and, by the use of lysosomal proteinase inhibitors, ascribed a role for cathepsins B and D in the activation of this growth factor.

Aspartic proteinases

These lysosomal proteinases contain an aspartic acid residue at their active site and act at acid pH. Very little investigation has been conducted on these enzymes in bone cells, except for the observations that cathepsin D, a member of this family, can be found by immunohistochemical staining in osteoblasts and osteocytes (Ohsawa et al., 1993), and expression of this enzyme is increased markedly by dexamethasone treatment of human osteoblasts in culture (Oursler et al., 1993). Cathepsin D is secreted into the resorbing area of human odontoclasts in order to participate in degradation of mineralized tooth matrix (Gotz et al., 2000).

Conclusions

The osteoblast lineage has the ability to produce proteinases of all four classes, but far more is known about their production of collagenase and PAs, at least in vitro. We still do not know the absolute role of any of these osteoblastic enzymes in vivo. Further work with global and conditional knockouts of the respective enzymes is likely to be the only way we will determine their required functions. These roles may not be restricted to assisting in the resorption process but may include functions to regulate bone development. In addition, the osteoclast produces MMP-9 and cathepsin K, which appear to have similar roles in the two diverse processes.

References

Aimes, R.T., Quigley, J.P., 1995. Matrix metalloproteinase-2 is an interstitial collagenase: inhibitor-free enzyme catalyzes the cleavage of collagen fibrils and soluble native type I collagen generating the specific 3/4- and 1/4-length cleavage fragments. J. Biol. Chem. 270, 5872—5876.

Al-Aqeel, A.I., 2005. Al-Aqeel Sewairi syndrome, a new autosomal recessive disorder with multicentric osteolysis, nodulosis and arthropathy. The first genetic defect of matrix metalloproteinase-2 gene. Saudi Med. J. 26, 24—30.

Allan, E.H., Hilton, D.J., Brown, M.A., Evely, R.S., Yumita, S., Medcalf, D., Gough, N.M., Ng, K.W., Nicola, N.A., Martin, T.J., 1990. Osteoblasts display receptors for and responses to leukemia-inhibitory factor. J. Cell. Physiol. 145, 110—119.

Allan, E.H., Zeheb, R., Gelehrter, T.D., Heaton, J.H., Fukumoto, S., Yee, J.A., Martin, T.J., 1991. Transforming growth factor beta inhibits plasminogen activator (PA) activity and stimulates production of urokinase-type PA, PA inhibitor-1 mRNA, and protein in rat osteoblast-like cells. J. Cell. Physiol. 149, 34—43.

Andreasen, P.A., Nielsen, L.S., Kristensen, P., Grondahl-Hansen, J., Skriver, L., Dano, K., 1986. Plasminogen activator inhibitor from human fibrosarcoma cells binds urokinase-type plasminogen activator, but not its proenzyme. J. Biol. Chem. 261, 7644—7651.

Angel, P., Karin, M., 1991. The role of *jun*, *fos* and the AP-1 complex in cell-proliferation and transformation. Biochim. Biophys. Acta 1072, 129—157.

Apte, S.S., Fukai, N., Beier, D.R., Olsen, B.R., 1997. The matrix metalloproteinase-14 (MMP-14) gene is structurally distinct from other MMP genes and is co-expressed with the TIMP-2 gene during mouse embryogenesis. J. Biol. Chem. 272, 25511—25517.

Balbín, M., Fueyo, A., Knauper, V., Lopez, J.M., Alvarez, J., Sanchez, L.M., Quesada, V., Bordallo, J., Murphy, G., Lopez-Otín, C., 2001. Identification and enzymatic characterization of two diverging murine counterparts of human interstitial collagenase (MMP-1) expressed at sites of embryo implantation. J. Biol. Chem. 276, 10253—10262.

Balbín, M., Fueyo, A., Knauper, V., Pendas, A.M., Lopez, J.M., Jimenez, M.G., Murphy, G., Lopez-Otín, C., 1998. Collagenase 2 (MMP-8) expression in murine tissue-remodeling processes: analysis of its potential role in postpartum involution of the uterus. J. Biol. Chem. 273, 23959—23968.

Balbin, M., Fueyo, A., Tester, A.M., Pendas, A.M., Pitiot, A.S., Astudillo, A., Overall, C.M., Shapiro, S.D., López-Otín, C., 2003. Loss of collagenase-2 confers increased skin tumor susceptibility to male mice. Nat. Genet. 35, 252—257.

Banerjee, C., McCabe, L.R., Choi, J.Y., Hiebert, S.W., Stein, J.L., Stein, G.S., Lian, J.B., 1997. An AML-1 consensus sequence binds an osteoblast-specific complex and transcriptionally activates the osteocalcin gene. J. Cell. Biochem. 66, 1—8.

Banyai, L., Varadi, A., Patthy, L., 1983. Common evolutionary origin of the fibrin-binding structures of fibronectin and tissue-type plasminogen activator. FEBS Lett. 163, 37—41.

Barmina, O.Y., Walling, H.W., Fiacco, G.J., Freije, J.M., López-Otín, C., Jeffrey, J.J., Partridge, N.C., 1999. Collagenase-3 binds to a specific receptor and requires the low density lipoprotein receptor-related protein for internalization. J. Biol. Chem. 274, 30087—30093.

Barrett, A.J., 1986. Physiological inhibitors of the human lysosomal cysteine proteinases. In: Ogawa, H., Lazarus, G.S., Hopsu-Havu, V.K. (Eds.), The Biological Role of Proteinases and Their Inhibitors in Skin. Elsevier, New York, pp. 13–26.

Bartkiewicz, M., Hernando, N., Reddy, S.V., Roodman, G.D., Baron, R., 1995. Characterization of the osteoclast vacuolar H(+)-ATPase B-subunit. Gene 160, 157–164.

Blair, H.D., Teitelbaum, S.L., Grosso, L.E., Lacey, D.L., Tan, H.-L., McCourt, D.W., Jeffrey, J.J., 1993. Extracellular-matrix degradation at acid pH: avian osteoclast acid collagenase isolation and characterization. Biochem. J. 29, 873–874.

Blom, A.B., van Lent, P.L., Libregts, S., Holthuysen, A.E., van der Kraan, P.M., van Rooijen, N., van den Berg, W.B., 2007. Crucial role of macrophages in matrix metalloproteinase-mediated cartilage destruction during experimental osteoarthritis: involvement of matrix metalloproteinase 3. Arthritis Rheum. 56, 147–157.

Bond, J.S., Butler, P.E., 1987. Intracellular proteases. Annu. Rev. Biochem. 56, 333–364.

Bonnet, N., Brun, J., Rousseau, J.-C., Duong, L.T., Ferrari, S.L., 2017. Cathepsin K controls cortical bone formation by degrading periostin. J. Bone Miner. Res. 32, 1432–1441.

Bossard, M.J., Tomaszek, T.A., Thompson, S.K., Amegadzie, B.Y., Hanning, C.R., Jones, C., Kurdyla, J.T., McNulty, D.E., Drake, F.H., Gowen, M., Levey, M.A., 1996. Proteolytic activity of human osteoclast cathepsin K: expression, purification, activation, and substrate identification. J. Biol. Chem. 271, 12517–12524.

Boumah, C.E., Lee, M., Selvamurugan, N., Shimizu, E., Partridge, N.C., 2009. Runx2 recruits p300 to mediate parathyroid hormone's effects on histone acetylation and transcriptional activation of the matrix metalloproteinase-13 gene. Mol. Endocrinol. 23, 1255–1263.

Breckon, J.J., Papaioannou, S., Kon, L.W., Tumber, A., Hembry, R.M., Murphy, G., Reynolds, J.J., Meikle, M.C., 1999. Stromelysin (MMP-3) synthesis is up-regulated in estrogen-deficient mouse osteoblasts in vivo and in vitro. J. Bone Miner. Res. 14, 1880–1890.

Brew, K., Nagase, H., 2010. The tissue inhibitors of metalloproteinases (TIMPs): an ancient family with structural and functional diversity. Biochim. Biophys. Acta 1803, 55–71.

Brinckerhoff, C.E., 1992. Regulation of metalloproteinase gene expression: implications for osteoarthritis. Crit. Rev. Eukaryot. Gene Expr. 2, 145–164.

Brinckerhoff, C.E., Harris Jr., E.D., 1981. Modulation by retinoic acid and corticosteroids of collagenase production by rabbit synovial fibroblasts treated with phorbol myristate acetate or poly(ethylene glycol). Biochem. Biophys. Acta 677, 424–432.

Bugge, T.H., Flick, M.T., Daugherty, C.C., Degen, J.L., 1995. Plasminogen deficiency causes severe thrombosis but is compatible with development and reproduction. Genes Dev. 9, 794–807.

Campbell, P.G., Wines, K., Yanosick, T.B., Novak, J.F., 1994. Binding and activation of plasminogen on the surface of osteosarcoma cells. J. Cell. Physiol. 159, 1–10.

Canalis, E., Rydziel, S., Delany, A.M., Varghese, S., Jeffrey, J.J., 1995. Insulin-like growth factors inhibit interstitial collagenase synthesis in bone cultures. Endocrinology 136, 1348–1354.

Carmeliet, P., Kieckens, L., Schoonjans, L., Ream, B., van Nuffelen, A., Prendergast, G., Cole, M., Bronson, R., Collen, D., Mulligan, R.C., 1993. Plasminogen activator inhibitor-1 gene-deficient mice. I. Generation by homologous recombination and characterization. J. Clin. Investig. 92, 2746–2755.

Carmeliet, P., Schoonjans, L., Kieckens, L., Ream, B., Degen, J., Bronson, R., De Vos, R., van den Oord, J.J., Collen, D., Mulligan, R.C., 1994. Physiological consequences of loss of plasminogen activator gene function in mice. Nature 368, 419–424.

Cerda-Costà, N., Gomis-Rüth, X., 2014. Architecture and function of metallopeptidase catalytic domains. Protein Sci. 23, 123–144.

Chakraborty, M., Chatterjee, D., Gorelick, F.S., Baron, R., 1994. Cell cycle-dependent and kinase-specific regulation of the apical Na/H exchanger and the Na, K-ATPase in the kidney cell line LLC-PK_1 by calcitonin. Proc. Natl. Acad. Sci. U.S.A. 91, 2115–2119.

Cheng, S.-L., Shen, V., Peck, W.A., 1991. Regulation of plasminogen activator and plasminogen activator inhibitor production by growth factors and cytokines in rat calvarial cells. Calcif. Tissue Int. 49, 321–327.

Chin, J.R., Murphy, G., Werb, Z., 1985. Stromelysin, a connective tissue-degrading metalloendopeptidase secreted by stimulated rabbit synovial fibroblasts in parallel with collagenase. J. Biol. Chem. 260, 12367–12376.

Chiu, R., Boyle, W.J., Meek, J., Smeal, T., Hunter, T., Karin, M., 1988. The c-fos protein interacts with c-Jun/AP-1 to stimulate transcription of AP-1 responsive genes. Cell 54, 541–542.

Christensen, J., Shastri, V.P., 2015. Matrix-metalloproteinase-9 is cleaved and activated by cathepsin K. BMC Res. Notes 8, 322.

Civitelli, R., Hruska, K.A., Jeffrey, J.J., Kahn, A.J., Avioli, L.V., Partridge, N.C., 1989. Second messenger signaling in the regulation of collagenase production by osteogenic sarcoma cells. Endocrinology 124, 2928–2934.

Clohisy, J.C., Scott, D.K., Brakenhoff, K.D., Quinn, C.O., Partridge, N.C., 1992. Parathyroid hormone induces *c-fos* and *c-jun* messenger RNA in rat osteoblastic cells. Mol. Endocrinol. 6, 1834–1842.

Collen, A., Hanemaaijer, R., Lupu, F., Quax, P.H., van Lent, N., Grimbergen, J., Peters, E., Koolwijk, P., van Hinsbergh, V.W.M., 2003. Membrane-type matrix metalloproteinase-mediated angiogenesis in a fibrin-collagen matrix. Blood 101, 1810–1817.

Collier, I.E., Wilhelm, S.M., Eisen, A.Z., Marmer, B.L., Grant, G.A., Seltzer, J.L., Kronberger, A., He, C., Bauer, E.A., Goldberg, G.I., 1988. H-ras oncogene-transformed human bronchial epithelial cells (TBE-1) secrete a single metalloprotease capable of degrading basement membrane collagen. J. Biol. Chem. 263, 6579–6587.

Connolly, T.J., Clohisy, J.C., Bergman, K.D., Partridge, N.C., Quinn, C.O., 1994. Retinoic acid stimulates interstitial collagenase mRNA in osteosarcoma cells. Endocrinology 135, 2542–2548.

Crawford, H.C., Matrisian, L.M., 1996. Mechanisms controlling the transcription of matrix metalloproteinase genes in normal and neoplastic cells. Enzyme Protein 49, 20–37.

Daci, E., Udagawa, N., Martin, T.J., Bouillon, R., Carmeliet, G., 1999. The role of the plasminogen system in bone resorption in vitro. J. Bone Miner. Res. 14, 946–952.

Daci, E., Everts, V., Torrekens, S., van Herck, E., Tigchelaar-Gutterr, W., Bouillon, R., Carmeliet, G., 2003. Increased bone formation in mice lacking plasminogen activators. J. Bone Miner. Res. 18, 1167–1176.

D'Alonzo, R.C., Selvamurugan, N., Karsenty, G., Partridge, N.C., 2002. Physical interaction of the activator protein-1 factors c-Fos and c-Jun with Cbfa1 for collagenase promoter activation. J. Biol. Chem. 277, 816–822.

Davis, B.A., Sipe, B., Gershan, L.A., Fiacco, G.J., Lorenz, T.C., Jeffrey, J.J., Partridge, N.C., 1998. Collagenase and tissue plasminogen activator production in developing rat calvariae: normal progression despite fetal exposure to microgravity. Calcif. Tissue Int. 63, 416–422.

De Bart, A.C.W., Quax, P.H.A., Lowik, C.W.G.M., Verheijen, J.H., 1995. Regulation of plasminogen activation, matrix metalloproteinases and urokinase-type plasminogen activator-mediated extracellular matrix degradation in human osteosarcoma cell line MG63 by interleukin-1 alpha. J. Bone Miner. Res. 10, 1374–1384.

Delaissé, J.-M., Eeckhout, Y., Vaes, G., 1984. *In vivo* and *in vitro* evidence for the involvement of cysteine proteinases in bone resorption. Biochem. Biophys. Res. Commun. 125, 441–447.

Delaissé, J.-M., Eeckhout, Y., Vaes, G., 1988. Bone-resorbing agents affect the production and distribution of procollagenase as well as the activity of collagenase in bone tissue. Endocrinology 123, 264–276.

Delaissé, J.-M., Eeckhout, Y., Neff, L., Francois-Gillet, C.H., Henriet, P., Su, Y., Vaes, G., Baron, R., 1993. (Pro) collagenase (matrix metalloproteinase-1) is present in rodent osteoclasts and in the underlying bone-resorbing compartment. J. Cell Sci. 106, 1071–1082.

Delaissé, J.-M., Engsig, M.T., Everts, V., del Carmen Ovejero, M., Ferreras, M., Lund, L., Vu, T.H., Werb, Z., Winding, B., Lochter, A., Karsdal, M.A., Kirkegaard, T., Lenhard, T., Heegaard, A.M., Neff, L., Baron, R., Foged, N.T., 2000. Proteinases in bone resorption: obvious and less obvious roles. Clin. Chim. Acta 291, 223–234.

Delgado-Calle, J., Hancock, B., Likine, E.F., Sato, A.Y., McAndrews, K., Sanudo, C., Bruzzaniti, A., Riancho, J.A., Tonra, J.R., Bellido, T., 2018. MMP14 is a novel target of PTH signaling in osteocytes that controls resorption by regulating soluble RANKL production. FASEB J. 32, 2878–2890.

Drake, F.H., Dodds, R.A., James, I.E., Connor, J.R., Debouck, S., Richardson, E., Lee-Rykaczewski, L., Coleman, D., Rieman, R., Barthlow, G., Gowen, M., 1996. Cathepsin K, but not cathepsin B, L, or S, is abundantly expressed in human osteoclasts. J. Biol. Chem. 271, 12511–12516.

Ducy, P., Zhang, R., Geoffroy, V., Ridall, A.L., Karsenty, G., 1997. Osf2/Cbfa1: a transcriptional activator of osteoblast differentiation. Cell 89, 747–754.

Eeckhout, Y., Vaes, G., 1977. Further studies on the activation of procollagenase, the latent precursor of bone collagenase: effects of lysosomal cathepsin B, plasmin and kallikrein and spontaneous activation. Biochem. J. 166, 21–31.

English, W.R., Puente, X.S., Freije, J.M., Knauper, V., Amour, A., Merryweather, A., López-Otín, C., Murphy, G., 2000. Membrane type 4 matrix metalloproteinase (MMP17) has tumor necrosis factor-alpha convertase activity but does not activate pro-MMP2. J. Biol. Chem. 275, 14046–14055.

Evans, B.R., Mosig, R.A., Lobl, M., Martignetti, C.R., Camacho, C., Grum-Tokars, V., Glucksman, M.J., Martignetti, J.A., 2012. Mutation of membrane type-1 metalloproteinase, MT1-MMP, causes the multicentric osteolysis and arthritis disease Winchester syndrome. Am. J. Hum. Genet. 91, 572–576.

Everts, V., Korper, W., Hoeben, K.A., Jansen, I.D., Bromme, D., Cleutiens, K.B., Heeneman, S., Peters, C., Reinheckel, T., Saftig, P., Beertsen, W., 2006. Osteoclastic bone degradation and the role of different cysteine proteinases and matrix metalloproteinases: differences between calvaria and long bone. J. Bone Miner. Res. 21, 1399–1408.

Fingleton, B., 2007. Matrix metalloproteinases as valid clinical target. Curr. Pharmaceut. Des. 13, 333–346.

Forbes, K., Webb, M.A., Sehgal, I., 2003. Growth factor regulation of secreted matrix metalloproteinase and plasminogen activators in prostate cancer cells, normal prostate fibroblasts and normal osteoblasts. Prostate Cancer Prostatic Dis. 6, 148–153.

Frattini, A., Orchard, P.J., Sobacchi, C., Giliani, S., Abinun, M., Matsson, J.P., Kieling, D.J., Andersson, A.K., Wallbrandt, P., Zecca, L., Notarangelo, L.D., Vezzoni, P., Villa, A., 2000. Defects in TCIRG1 subunit of the vacuolar proton pump are responsible for a subset of human autosomal recessive osteopetrosis. Nat. Genet. 25, 343–346.

Freije, J.M.P., Diez-Itza, I., Balbín, M., Sanchez, L.M., Blasco, R., Tolivia, J., López-Otín, C., 1994. Molecular cloning and expression of collagenase-3, a novel human matrix metalloproteinase produced by breast carcinomas. J. Biol. Chem. 269, 16766–16773.

Fuller, K., Chambers, T.J., 1995. Localisation of mRNA for collagenase in osteocytic, bone surface, and chondrocytic cells but not osteoclasts. J. Cell Sci. 108, 2221–2230.

Fuller, K., Lawrence, K.M., Ross, J.L., Grabowska, U.B., Shiroo, M., Samuelsson, B., Chambers, T.J., 2008. Cathepsin K inhibitors prevent matrix-derived growth factor degradation by human osteoclasts. Bone 42, 200–211.

Fukuda, K., Fujitsu, Y., Kumagai, N., Nishida, T., 2006. Inhibition of matrix metalloproteinase-3 synthesis in human conjunctival fibroblasts by interleukin-4 or interleukin-13. Invest Ophthalmol Vis Sci 47, 2857–2864.

Galvez, B.G., Matias-Roman, S., Alber, J.P., Sanchez-Madrid, F., Arroyo, A.G., 2001. Membrane type 1-matrix metalloproteinase is activated during migration of human endothelial cells and modulates endothelial motility and matrix remodeling. J. Biol. Chem. 276, 37491–37500.

Gelb, B.D., Shi, G.P., Chapman, H.A., Desnick, R.J., 1996. Pycnodysostosis, a lysosomal disease caused by cathepsin K deficiency. Science 273, 1236–1238.

Geoffroy, V., Ducy, P., Karsenty, G., 1995. A PEBP2a/AML-1-related factor increases osteocalcin promoter activity through its binding to an osteoblast-specific cis-acting element. J. Biol. Chem. 270, 30973–30979.

Gerber, H.P., Vu, T.H., Ryan, A.M., Kowalski, J., Werb, Z., Ferrara, N., 2000. VEGF couples hypertrophic cartilage remodeling, ossification and angiogenesis during endochondral bone formation. Nat. Med. 5, 623–628.

Glimcher, M.J., 1998. The nature of the mineral component of bone: biological and clinical implications. In: Avioli, L.V., Krane, S.M. (Eds.), Metabolic Bone Disease and Clinically Related Disorders. Academic Press, San Diego, pp. 23–50.

Godat, E., Lecaille, F., Desmazes, C., Duchene, S., Weidauer, E., Saftig, P., Brömme, D., Vandier, C., Lalmanach, G., 2004. Cathepsin K: a cysteine protease with unique kinin-degrading properties. Biochem. J. 383, 501–506.

Goldberg, G.I., Wilhelm, S.M., Kronberger, A., Bauer, E.A., Grant, G.A., Eisen, A.Z., 1986. Human fibroblast collagenase: complete primary structure and homology to an oncogene transformation-induced rat protein. J. Biol. Chem. 261, 6600–6605.

Gotz, W., Quondamatteo, F., Ragotzki, S., Affeldt, J., Jager, A., 2000. Localization of cathepsin D in human odontoclasts. A light and electron microscopical immunocytochemical study. Connect. Tissue Res. 41, 185–194.

Gong, Y., Chippa-Vankata, U.D., Oh, W.K., 2014. Role of matrix metalloproteinases and their natural inhibitor in prostate cancer progression. Cancers 6, 1298–1327.

Grigoriadis, A.E., Wang, Z.-Q., Cecchini, M.G., Hofstetter, W., Felix, R., Fleisch, H.A., Wagner, E.F., 1994. c-Fos: a key regulator of osteoclast-macrophage lineage determination and bone remodeling. Science 266, 443–448.

Gowen, M., Lazner, R., Dodds, R., Kapadia, R., Field, J., Tavaria, M., 1999. Cathepsin K knockout mice develop osteopetrosis due to a deficit in matrix degradation but not demineralization. J. Bone Miner. Res. 14, 1654–1663.

Hamilton, J.A., Lingelbach, S.R., Partridge, N.C., Martin, T.J., 1984. Stimulation of plasminogen activator in osteoblast-like cells by bone resorbing-hormones. Biochem. Biophys. Res. Commun. 122, 230–236.

Hamilton, J.A., Lingelbach, S., Partridge, N.C., Martin, T.J., 1985. Regulation of plasminogen activator production by bone-resorbing hormones in normal and malignant osteoblasts. Endocrinology 116, 2186–2191.

Heath, J.K., Reynolds, J.J., Meikle, M.C., 1990. Osteopetrotic (grey-lethal) bone produces collagenase and TIMP in organ culture: regulation by vitamin A. Biochem. Biophys. Res. Commun. 168, 1171–1176.

Herz, J., Bock, H.H., 2002. Lipoprotein receptors in the nervous system. Annu. Rev. Biochem. 71, 405–434.

Hiraoka, N., Allen, E., Apel, I.J., Gyetko, M.R., Weiss, S.J., 1998. Matrix metalloproteinases regulate neovascularization by acting as pericellular fibrinolysins. Cell 95, 365–377.

Holmbeck, K., Bianco, P., Caterina, J., Yamada, S., Kromer, M., Kuznetsov, S.A., Mankani, M., Robey, P.G., Poole, A.R., Pidoux, I., Ward, J.M., Birkedal-Hansen, H., 1999. MT1-MMP-deficient mice develop dwarfism, osteopenia, arthritis, and connective tissue disease due to inadequate collagen turnover. Cell 99, 81–92.

Hotary, K., Allen, E., Punturieri, A., Yana, I., Weiss, S.J., 2000. Regulation of cell invasion and morphogenesis in a three-dimensional type I collagen matrix by membrane-type matrix metalloproteinases 1, 2, and 3. J. Cell Biol. 149, 1309–1323.

Hsieh, Y.S., Chu, S.C., Yang, S.F., Chen, P.N., Liu, Y.C., Lu, K.H., 2007. Silibinin suppresses human osteosarcoma MG-63 cell invasion by inhibiting the ERK-dependent c-Jun/AP-1 induction of MMP-2. Carcinogenesis 28, 977–987.

Huhtala, P., Chow, L.T., Tryggvason, K., 1990. Structure of the human type IV collagenase gene. J. Biol. Chem. 265, 11077–11082.

Huhtala, P., Tuuttila, A., Chow, L.T., Lohi, J., Keski-Oja, J., Tryggvason, K., 1991. Complete structure of the human gene for 92-kDa type IV collagenase. Divergent regulation of expression for the 92- and 72-kiloDalton enzyme genes in HT-1080 cells. J. Biol. Chem. 266, 16485–16490.

Inada, M., Wang, Y., Byrne, M.H., Rahman, M.U., Miyaura, C., Lopez-Otin, C., Krane, S.M., 2004. Critical roles for collagenase-3 (Mmp-13) in development of growth plate cartilage and in endochondral ossification. Proc. Natl. Acad. Sci. U.S.A. 101, 17192–17197.

Inoue, K., Mikuni-Takagaki, Y., Oikawa, K., Itoh, T., Inada, M., Noguchi, T., Park, J.S., Onodera, T., Krane, S.M., Noda, M., Itohara, S., 2006. A crucial role for matrix metalloproteinase 2 in osteocytic canalicular formation and bone metabolism. J. Biol. Chem. 281, 33814–33824.

Itoh, T., Ikeda, T., Gomi, H., Nakao, S., Suzuki, T., Itohara, S., 1997. Unaltered secretion of beta-amyloid precursor protein in gelatinase A (matrix metalloproteinase 2)-deficient mice. J. Biol. Chem. 272, 22389–22392.

Jilka, R.L., Cohn, D.V., 1983. A collagenolytic response to parathormone, 1,25-dihydroxycholecalciferol D_3, and prostaglandin E_2 in bone of osteopetrotic (mi/mi) mice. Endocrinology 112, 945–950.

Jimenez, M.J.G., Balbín, M., Lopez, J.M., Alvarez, J., Komori, T., López-Otín, C., 1999. Collagenase 3 is a target of Cbfa1, a transcription factor of the runt gene family involved in bone formation. Mol. Cell Biol. 19, 4431–4442.

Johansson, N., Saarialho-Kere, U., Airola, K., Herva, R., Nissinen, L., Westermarck, J., Vuorio, E., Heino, J., Kahari, V.M., 1997. Collagenase-3 (MMP-13) is expressed by hypertrophic chondrocytes, periosteal cells, and osteoblasts during human fetal bone development. Dev. Dynam. 208, 387–397.

Kaczmarek, L., Lapinska-Dzwonek, J., Szymczak, S., 2002. Matrix Metalloproteinases in the adult brain physiology: a link between c-Fos, AP-1 and remodeling of neuronal connections? EMBO J. 21, 6643–6648.

Kagoshima, H., Satake, M., Miyoshi, H., Ohki, M., Pepling, M., Gergen, J.P., Shigesada, K., Ito, Y., 1993. The runt domain identifies a new family of heteromeric transcriptional regulators. Trends Genet. 9, 338–341.

Kanno, T., Kanno, Y., Chen, L.F., Ogawa, E., Kim, W.Y., Ito, Y., 1998. Intrinsic transcriptional activation-inhibition domains of the polyomavirus enhancer binding protein 2/core binding factor alpha subunit revealed in the presence of the beta subunit. Mol. Cell Biol. 18, 2444–2454.

Katono, T., Kawato, T., Tanabe, N., Suzuki, N., Yamanaka, K., Oka, H., Motohashi, M., Maeno, M., 2006. Nicotine treatment induces expression of matrix metalloproteinases in human osteoblastic Saos-2 cells. Acta Biochim. Biophys. Sin. 38, 874–882.

Kawao, N., Tamura, Y., Okumoto, K., Yano, M., Okada, K., Matsuo, O., Kaji, H., 2014. Tissue-type plasminogen activator deficiency delays bone repair: roles of osteoblastic proliferation and vascular endothelial growth factor. Am. J. Physiol. Endocrinol. Metab. 307, E278–E288.

Kennedy, A.M., Inada, M., Krane, S.M., Christie, P.T., Harding, B., Lopez-Otin, C., Sanchez, L.M., Pannett, A.A., Dearlove, A., Hartley, C., Byrne, M.H., Reed, A.A., Nesbit, M.A., Whyte, M.P., Thakker, R.V., 2005. MMP13 mutation causes spondyloepimetaphyseal dysplasia, Missouri type (SEMD(MO). J. Clin. Investig. 115, 2832–2842.

Kerr, L.D., Holt, J.T., Matrisian, L.M., 1988a. Growth factors regulate transin gene expression by c-fos-dependent and c-fos-independent pathways. Science 242, 1424–1427.

Kerr, L.D., Olashaw, N.E., Matrisian, L.M., 1988b. Transforming growth factor $\beta 1$ and and cAMP inhibit transcription of the epidermal growth factor and oncogene-induced transin RNA. J. Biol. Chem. 263, 16999–17005.

Kerr, L.D., Miller, D.B., Matrisian, L.M., 1990. TGF-$\beta 1$ inhibition of transin/stromelysin gene expression is mediated through a fos binding sequence. Cell 61, 267–278.

Kinoh, H., Sato, H., Tsunezuka, Y., Takino, T., Kawashima, A., Okada, Y., Seiki, M., 1996. MT-MMP, the cell surface activator of proMMP-2 (pro-gelatinase A), is expressed with its substrate in mouse tissue during embryogenesis. J. Cell Sci. 109, 953–959.

Kiviranta, R., Morko, J., Alartalo, S.L., NicAmhlaoibh, R., Risteli, J., Laitala-Leinonen, T., Vuorio, E., 2005. Impared bone resorption in cathepsin K-deficient mice is partially compensated for by enhanced osteoclastogenesis and increased expression of other proteases via an increased RANKL/OPG ratio. Bone 36, 159–172.

Knauper, V., Wilhelm, S.M., Seperack, P.K., DeClerck, Y.A., Osthues, A., Tschesche, H., 1993. Direct activation of human neutrophil procollagenase by recombinant stromelysin. Biochem. J. 295, 581–586.

Komori, T., Yagi, H., Nomura, S., Yamaguchi, A., Sasaki, K., Deguchi, K., Shimizu, Y., Bronson, R.T., Gao, Y.-H., Inada, M., Sato, M., Okamoto, R., Kitamura, Y., Yoshiki, S., Kishimoto, T., 1997. Targeted disruption of *Cbfa1* results in a complete lack of bone formation owing to maturational arrest of osteoblasts. Cell 89, 755–764.

Kornak, U., Schulz, A., Friedrich, W., Uhlhaas, S., Kremens, B., Voit, T., Hasan, C., Bole, U., Jentsch, T.J., Kubisch, C., 2000. Mutations in the a3 subunit of the vacuolar H^+-ATPase cause infantile malignant osteopetrosis. Hum. Mol. Genet. 9, 2059–2063.

Kornak, U., Kasper, D., Bösl, M.R., Kaiser, E., Schweizer, M., Schulz, A., Friedrich, W., Delling, G., Jentsch, T.J., 2001. Loss of the C1C-7 chloride channel leads to osteopetrosis in mice and man. Cell 104, 205–215.

Kosaki, N., Takaishi, H., Kamekura, S., Kimura, T., Okada, Y., Minqi, L., Amizuka, N., Chung, U.I., Nakamura, K., Kawaguchi, H., Toyama, Y., D'Armiento, J., 2007. Impaired bone fracture healing in matrix metalloproteinase-13 deficient mice. Biochem. Biophys. Res. Commun. 354, 846–851.

Kusano, K., Miyaura, C., Inada, M., Tamura, T., Ito, A., Nagase, H., Kamoi, K., Suda, T., 1998. Regulation of matrix metalloproteinases (MMP-2, -3, -9, and -13) by interleukin-1 and interleukin-6 in mouse calvaria: association of MMP induction with bone resorption. Endocrinology 139, 1338–1345.

Lalou, C., Silve, C., Rosato, R., Segovia, B., Binoux, M., 1994. Interactions between insulin-like growth factor-I (IGF-I) and the system of plasminogen activators and their inhibitors in the control of IGF-binding protein-3 production and proteolysis in human osteosarcoma cells. Endocrinology 135, 2318–2326.

Lanske, B., Divieti, P., Kovacs, C.S., Pirro, A., Landis, W.J., Krane, S.M., Bringhurst, F.R., Kronenberg, H.M., 1998. The parathyroid hormone (PTH)/PTH-related peptide receptor mediates actions of both ligands in murine bone. Endocrinology 139, 5194–5204.

Lanske, B., Karaplis, C.A.C., Lee, K., Luz, A., Vortkamp, A., Pirro, A., Karperien, M., Defize, L.H.K., Ho, C., Mulligan, R.C., Abou-Samra, A.B., Jüppner, H., Segre, G.V., Kronenberg, H.M., 1996. PTH/PTHrP receptor in early development and indian hedgehog-regulated bone growth. Science 273, 663–666.

Lawson, N.D., Khanna-Gupta, A., Berliner, N., 1998. Isolation and characterization of the cDNA for mouse neutrophil collagenase: demonstration of shared negative regulatory pathway neutrophil secondary granule protein gene expression. Blood 91, 2517–2524.

Lee, W., Mitchell, P., Tjian, R., 1987. Purified transcription factor AP-1 interacts with TPA-inducible enhancer elements. Cell 49, 741–752.

Lee, M., Partridge, N.C., 2010. Parathyroid hormone activation of matrix metalloproteinase-13 transcription requires the histone acetyltransferase activity of p300 and PCAF and p300-dependent acetylation of PCAF. J. Biol. Chem. 285, 38014–38022.

Leloup, G., Delaissé, J.-M., Vaes, G., 1994. Relationship of the plasminogen activator/plasmin cascade to osteoclast invasion and mineral resorption in explanted fetal metatarsal bones. J. Bone Miner. Res. 9, 891–902.

Leloup, G., Peeters-Joris, C., Delaissé, J.-M., Opdenakker, G., Vaes, G., 1991. Tissue and urokinase plasminogen activators in bone tissue and their regulation by parathyroid hormone. J. Bone Miner. Res. 6, 1081–1090.

Le Maitre, C.L., Freemont, A.J., Hoyland, J.A., 2005. The role of interleukin-1 in the pathogenesis of human intervertebral disc degradation. Arthritis Res. Ther. 7, R732–R745.

Li, Y.P., Chen, W., Liang, Y., Li, E., Stashenko, P., 1999. Atp6i-deficient mice exhibit severe osteopetrosis due to loss of osteoclast-mediated extracellular acidification. Nat. Genet. 23, 447–451.

Littlewood-Evans, A., Kokubo, T., Ishibashi, O., Inaoka, T., Wlodarski, B., Gallagher, J.A., Bilbe, G., 1997. Localization of cathepsin K in human osteoclasts by in situ hybridization and immunohistochemistry. Bone 20, 81–86.

Liu, X., Wu, H., Byrne, M., Jeffrey, J., Krane, S., Jaenisch, R., 1995. A targeted mutation at the known collagenase cleavage site in mouse type I collagen impairs tissue remodeling. J. Cell Biol. 130, 227–237.

Llano, E., Pendas, A.M., Freije, J.P., Nakano, A., Knauper, V., Murphy, G., López-Otín, C., 1999. Identification and characterization of human MT5-MMP, a new membrane-bound activator of progelatinase a overexpressed in brain tumors. Cancer Res. 59, 2570–2576.

Lutgens, E., Lutgens, S.P., Faber, B.C., Heeneman, S., Gijbels, M.M., de Winther, M.P., Frederik, P., van der Made, I., Daugherty, A., Sijbergs, A.M., Fisher, A., Long, C.J., Saftig, P., Black, D., Daemen, M.J., Cleutjens, K.B., 2006a. Disruption of cathpsin K gene reduces atherosclerosis progression and induces plaque fibrosis but accelerates macrophage foam cell formation. Circulation 113, 98–107.

Lutgens, S.P., Kisters, N., Lutgens, E., van Haaften, R.I., Evelo, C.T., de Winther, M.P., Saftig, P., Daemen, M.J., Heenaman, S., Cleutjens, K.B., 2006b. Gene profiling of cathepsin K deficiency in atherogenesis: profibrotic but lipogenic. J. Pathol. 210, 334–343.

Lynch, C.C., Hikosaka, A., Acuff, H.B., Martin, M.D., Kawai, N., Singh, R.K., Vargo-Gogola, T.C., Begtrup, J.L., Peterson, T.E., Fingleton, B., Shirai, T., Matrisian, L.M., Futakuchi, M., 2005. MMP-7 promotes prostate cancer-induced osteolysis via the solubilization of RANKL. Cancer Cell 5, 485–496.

Martignetti, J.A., Aqeel, A.A., Sewairi, W.A., Boumah, C.E., Kambouris, M., Mayouf, S.A., Sheth, K.V., Eid, W.A., Dowling, O., Harris, J., Glucksman, M.J., Bahabri, S., Meyer, B.F., Desnick, R.J., 2001. Mutation of the matrix metalloproteinase 2 gene (MMP2) causes a multicentric osteolysis and arthritis syndrome. Nat. Genet. 28, 261–265.

Mathieu, E., Meheus, L., Raymackers, J., Merregaert, J., 1994. Characterization of the osteogenic stromal cell line MN7: identification of secreted MN7 proteins using two-dimensional polyacrylamide gel electrophoresis, Western blotting, and microsequencing. J. Bone Miner. Res. 9, 903–913.

Matrisian, L.M., 1992. The matrix-degrading metalloproteinases. Bioessays 14, 455–463.

Matrisian, L.M., Glaichenhaus, N., Gesnel, M.C., Breathnach, R., 1985. Epidermal growth factor and oncogenes induce transcription of the same cellular mRNA in rat fibroblasts. EMBO J. 4, 1435–1440.

Matrisian, L.M., Leroy, P., Ruhlmann, C., Gesnel, M.C., Breathnach, R., 1986. Isolation of the oncogene and epidermal growth factor-induced transin gene: complex control in rat fibroblasts. Mol. Cell Biol. 6, 1679–1686.

Mattot, V., Raes, M.B., Henriet, P., Eeckhout, Y., Stehelin, D., Vandenbunder, B., Desbiens, X., 1995. Expression of *interstitial collagenase* is restricted to skeletal tissue during mouse embryogenesis. J. Cell Sci. 108, 529–535.

McDonnell, S.E., Kerr, L.D., Matrisian, L.M., 1990. Epidermal growth factor stimulation of stromelysin mRNA in rat fibroblasts requires induction of proto-oncogenes c-fos and c-jun and activation of protein kinase C. J. Biol. Chem. 10, 4284–4293.

Meikle, M.C., Bond, S., Hembry, R.M., Compston, J., Croucher, P.I., Reynolds, J.J., 1992. Human osteoblasts in culture synthesize collagenase and other matrix metalloproteinases in response to osteotrophic hormones and cytokines. J. Cell Sci. 103, 1093–1099.

Merriman, H.L., van Wijnen, A.J., Hiebert, S., Bidwell, J.P., Fey, E., Lian, J., Stein, J., Stein, G.S., 1995. The tissue-specific nuclear matrix protein, NMP-2, is a member of the AML/CBF/PEBP2/*runt domain* transcription factor family: interactions with the osteocalcin gene promoter. Biochemistry 34, 13125–13132.

Mitra, A., Charkrabarti, J., Banerji, A., Chatterjee, A., 2006. Cell membrane-associated MT1-MMP dependent activation of MMP-2 in SiHa (human cervical cancer) cells. J. Environ. Pathol. Toxicol. Oncol. 25, 655–666.

Mitchell, P.G., Magna, H.A., Reeves, L.M., Lopresti-Morrow, L.L., Yocum, S.A., Rosner, P.J., Geoghegan, K.F., Hambor, J.E., 1996. Cloning, expression, and type II collagenolytic activity of matrix metalloproteinase-13 from human osteoarthritic cartilage. J. Clin. Investig. 97, 761–768.

Montenez, J.P., Delaissé, J.-M., Tulkens, P.M., Kishore, B.K., 1994. Increased activities of cathepsin B and other lysosomal hydrolases in fibroblasts and bone tissue cultured in the presence of cysteine proteinases inhibitors. Life Sci. 55, 1199–1209.

Mosig, R.A., Dowling, O., Difeo, A., Ramirez, M.C., Parker, I.C., Abe, E., Diouri, J., Ageel, A.A., Wylie, J.D., Oblander, S.A., Madri, J., Bianco, P., Apte, S.S., Zaidi, M., Doty, S.B., Majeska, R.J., Schaffler, M.B., Martignetti, J.A., 2007. Loss of MMP-2 disrupts skeletal and craniofacial development, and results in decreased bone mineralization, joint erosion, and defects in osteoblast and osteoclast growth. Hum. Mol. Genet. 16, 1113–1123.

Mudgett, J.S., Hutchinson, N.I., Chartrain, N.A., Forsyth, A.J., McDonnell, J., Singer, I.I., Bayne, E.K., Flanagan, J., Kawka, D., Shen, C.F., Stevens, K., Chen, H., Trumbauer, M., Visco, D.M., 1998. Susceptibility of stromelysin 1-deficient mice to collagen-induced arthritis and cartilage destruction. Arthritis Rheum. 41, 110–121.

Mullard, A., 2016. Merck &Co. drops osteoporosis drug odanacatib. Nat. Rev. Drug Discov. 15, 669.

Mundlos, S., Otto, F., Mundlos, C., Mulliken, J.B., Aylsworth, A.S., Albright, S., Lindhout, D., Cole, W.G., Henn, W., Knoll, J.H., Owen, M.J., Mertelsmann, R., Zabel, B.U., Olsen, B.R., 1997. Mutations involving the transcription factor CBFA1 cause cleidocranial dysplasia. Cell 89, 773–779.

Murphy, G., Cockett, M.I., Stephens, P.E., Smith, B.J., Docherty, A.J.P., 1987. Stromelysin is an activator of procollagenase. Biochem. J. 248, 265–268.

Nakatani, T., Chen, T., Partridge, N.C., 2016. MMP-13 is one of the critical mediators of the effect of HDAC4 deletion on the skeleton. Bone 90, 142–151.

Nakatani, T., Partridge, N.C., 2017. MEF2C interacts with c-FOS in PTH-stimulated *Mmp13* gene expression in osteoblastic cells. Endocrinology 158, 3778–3791.

Nam, D.H., Rodriguez, C., Remade, A.G., Strongin, A.Y., Ge, X., 2016. Active-site MMP-selective antibody inhibitors discovered from convex paratope synthetic libraries. Proc. Natl. Acad. Sci. U.S.A. 27, 14970–14975.

Nielsen, L.S., Kellerman, G.M., Behrendt, N., Picone, R., Dano, K., Blasi, F., 1988. A 55,000–60,000 Mr receptor protein for urokinase-type plasminogen activator. J. Biol. Chem. 263, 2358–2363.

Ohsawa, Y., Nitatori, T., Higuchi, S., Kominami, E., Uchiyama, Y., 1993. Lysosomal cysteine and aspartic proteinases, acid phosphatase, and endogenous cysteine proteinase inhibitor, cystatin-β, in rat osteoclasts. J. Histochem. Cytochem. 41, 1075–1083.

Ohuchi, E., Imai, K., Fuji, Y., Sato, H., Seiki, M., Okada, Y., 1997. Membrane type 1 matrix metalloproteinase digests interstitial collagens and other extracellular matrix macromolecules. J. Biol. Chem. 272, 2446–2451.

Omura, T.H., Noguchi, A., Johanns, C.A., Jeffrey, J.J., Partridge, N.C., 1994. Identification of a specific receptor for interstitial collagenase on osteoblastic cells. J. Biol. Chem. 269, 24994–24998.

Ortega, N., Behonick, D.J., Colnot, C., Cooper, D.N., Werb, Z., 2005. Galectin-3 is a downstream regulator of matrix metalloproteinase-9 function during endochondral bone formation. Mol. Biol. Cell 16, 3028–3039.

Otto, F., Thornell, A.P., Crompton, T., Denzel, A., Gilmour, K.C., Rosewell, I.R., Stamp, G.W.H., Beddington, R.S.P., Mundlos, S., Olsen, B.R., Selby, P.B., Owen, M.J., 1997. *Cbfa1*, a candidate gene for cleidocranial dysplasia syndrome, is essential for osteoblast differentiation and bone development. Cell 89, 765–771.

Otto, T.C., Bowers, R.R., Lane, M.D., 2007. BMP-4 treatment of C3H10T1/2 stem cells blocks expression of MMP-3 and MMP-13. Biochem. Biophys. Res. Commun. 353, 1097–1104.

Oursler, M.J., Riggs, B.L., Spelsberg, T.C., 1993. Glucocorticoid-induced activation of latent transforming growth factor-β by normal human osteoblast-like cells. Endocrinology 133, 2187–2196.

Overall, C.M., Wrana, J.F., Sodek, J., 1989. Transforming growth factor-β regulation of collagenase, 72 kDa- progelatinase, TIMP and PAI-1 expression in rat bone cell populations and human fibroblasts. Connect. Tissue Res. 20, 289–294.

Page-McCaw, A., Ewald, A.J., Werb, Z., 2007. Matrix metalloproteinases and the regulation of tissue remodeling. Nat. Rev. Mol. Cell Biol. 8, 221–233.

Panwar, P., Soe, K., Guido, R.V., Bueno, R.V., Delaisse, J.M., Bromme, D., 2016. A novel approach to inhibit bone resorption: exosite inhibitors against cathepsin K. Br. J. Pharmacol. 173, 396–410.

Pap, T., Shigeyama, Y., Kuchen, S., Fernihough, J.K., Simmen, B., Gay, R.E., Billingham, M., Gay, S., 2000. Differential expression pattern of membrane-type matrix metalloproteinases in rheumatoid arthritis. Arthritis Rheum. 43, 1226–1232.

Partridge, N.C., Frampton, R.J., Eisman, J.A., Michelangeli, V.P., Elms, E., Bradley, T.R., Martin, T.J., 1980. Receptors for 1,25$(OH)_2$-vitamin D_3 enriched in cloned osteoblast-like rat osteogenic sarcoma cells. FEBS Lett. 115, 139–142.

Partridge, N.C., Alcorn, D., Michelangeli, V.P., Ryan, G., Martin, T.J., 1983. Morphological and biochemical characterization of four clonal osteogenic sarcoma cell lines of rat origin. Cancer Res. 43, 4308–4314.

Partridge, N.C., Jeffrey, J.J., Ehlich, L.S., Teitelbaum, S.L., Fliszar, C., Welgus, H.G., Kahn, A.J., 1987. Hormonal regulation of the production of collagenase and a collagenase inhibitor activity by rat osteogenic sarcoma cells. Endocrinology 120, 1956–1962.

Partridge, N.C., Scott, D.K., Gershan, L.A., Omura, T.H., Burke, J.S., Jeffrey, J.J., Bolander, M.E., Quinn, C.O., 1993. Collagenase production by normal and malignant osteoblastic cells. In: Novak, J.F., McMaster, J.H. (Eds.), Frontiers of Osteosarcoma Research. Hogrefe and Huber Publishers, Seattle, pp. 269–276.

Pearman, A.T., Chou, W.Y., Bergman, K.D., Pulumati, M.R., Partridge, N.C., 1996. Parathyroid hormone induces c-fos promoter activity in osteoblastic cells through phosphorylated cAMP response element (CRE)-binding protein binding to the major CRE. J. Biol. Chem. 271, 25715–25721.

Pei, D., 1999. Identification and characterization of the fifth membrane-type matrix metalloproteinase MT5-MMP. J. Biol. Chem. 274, 8925–8932.

Pei, D., Weiss, S.J., 1996. Transmembrane-deletion mutants of the membrane-type matrix metalloproteinase-1 process progelatinase A and express intrinsic matrix-degrading activity. J. Biol. Chem. 271, 9135–9140.

Pennica, D., Holmes, W.E., Kohr, W.T., Harkins, R.N., Vehar, G.A., Ward, C.A., Bennett, W.F., Yelverton, E., Seeburg, P.H., Heyneker, H.L., Goeddel, D.V., Collen, D., 1983. Cloning and expression of human tissue-type plasminogen activator cDNA in *E. coli*. Nature 301, 214–221.

Pepper, M.S., 1997. Manipulating angiogenesis: from basic science to the bedside. Arterioscler. Thromb. Vasc. Biol. 17, 605–619.

Pereira, R.C., Jorgetti, V., Canalis, E., 1999. Triiodothyronine induces collagenase-3 and gelatinase B expression in murine osteoblasts. Am J Physiol 277, E496–504.

Pfeilschifter, J., Erdmann, J., Schmidt, W., Naumann, A., Minne, H.W., Ziegler, R., 1990. Differential regulation of plasminogen activator and plasminogen activator inhibitor by osteotropic factors in primary cultures of mature osteoblasts and osteoblast precursors. Endocrinology 126, 703–711.

Pohlkamp, T., Wasser, C.R., Herz, J., 2017. Functional roles of the interaction of APP and lipoprotein receptors. Front. Mol. Neurosci. 10, 54. https://doi.org/10.3389/fnmol.2017.00054.

Popa, N.L., Wergedal, J.E., Lau, K.H., Mohan, S., Rundle, C.H., 2014. Urokinase plasminogen activator gene deficiency inhibits fracture cartilage remodeling. J. Bone Miner. Metab. 32, 124–135.

Posthumus, M.D., Limburg, P.C., Westra, J., van Leeuwen, M.A., van Rijswijk, M.H., 2000. Serum matrix metalloproteinase 3 in early rheumatoid arthritis is correlated with disease activity and radiological progression. J. Rheumatol. 27, 2761–2768.

Potempa, J., Korzos, E., Travis, J., 1994. The serpin superfamily of proteinase inhibitors: structure, function and regulation. J. Biol. Chem. 269, 15957–15960.

Prontera, C., Crescenzi, G., Rotilio, D., 1996. Inhibition by interleukin-4 of stromelysin expression in human skin fibroblasts: role of PKC. Exp. Cell Res. 224, 183–188.

Puente, X.S., Pendás, A.M., Llano, E., Velasco, G., López-Otín, C., 1996. Molecular cloning of a novel membrane-type matrix metalloproteinase from a human breast carcinoma. Cancer Res. 56, 944–949.

Qing, H., Ardeshirpour, L., Pajevic, P.D., Dusevich, V., Jahn, K., Kato, S., Wysolmerski, J., Bonewald, L.F., 2012. Demonstration of osteocytic perilacunar/canalicular remodeling in mice during lactation. J. Bone Miner. Res. 27, 1018–1029.

Quinn, C.O., Scott, D.K., Brinckerhoff, C.E., Matrisian, L.M., Jeffrey, J.J., Partridge, N.C., 1990. Rat collagenase: cloning, amino acid sequence comparison, and parathyroid hormone regulation in osteoblastic cells. J. Biol. Chem. 265, 22342–22347.

Rabbani, S.A., Desjardins, J., Bell, A.W., Banville, D., Mazar, A., Henkin, J., Goltzman, D., 1990. An amino terminal fragment of urokinase isolated from a prostate cancer cell line (PC-3) is mitogenic for osteoblast-like cells. Biochem. Biophys. Res. Commun. 173, 1058–1064.

Raggatt, L.J., Jefcoat Jr., S.C., Choudhury, I., Williams, S., Tiku, M., Partridge, N.C., 2006. Matrix metalloproteinase-13 influences ERK signalling in articular rabbit chondrocytes. Osteoarthritis Cartilage 14, 680–689.

Rajakumar, R.A., Partridge, N.C., Quinn, C.O., 1993. Transcriptional induction of collagenase in rat osteosarcoma cells is mediated by sequence 5 prime of the gene. J. Bone Miner. Res. 8 (Suppl. 1), S294.

Reboul, P., Pelletier, J.P., Tardif, G., Cloutier, J.M., Martel-Pelletier, J., 1996. The new collagenase, collagenase-3, is expressed and synthesized by human chondrocytes but not by synoviocytes: a role in osteoarthritis. J. Clin. Investig. 97, 2011–2019.

Rifas, L., Halstead, L.R., Peck, W.A., Avioli, L.V., Welgus, H.G., 1989. Human osteoblasts *in vitro* secrete tissue inhibitor of metalloproteinases and gelatinase but not interstitial collagenase as major cellular products. J. Clin. Investig. 84, 686–694.

Rifas, L., Fausto, A., Scott, M.J., Avioli, L.V., Welgus, H.G., 1994. Expression of metalloproteinases and tissue inhibitors of metallopro-teinases in human osteoblast-like cells: differentiation is associated with repression of metalloproteinase biosynthesis. Endocrinology 134, 213–221.

Roswit, W.T., Halme, J., Jeffrey, J.J., 1983. Purification and properties of rat uterine procollagenase. Arch. Biochem. Biophys. 225, 285–295.
Ruangpanit, N., Chan, D., Holmbeck, K., Birkedal-Hansen, H., Polarek, J., Yang, C., Bateman, J.F., Thompson, E.W., 2001. Gelatinase A (MMP-2) activation by skin fibroblasts: dependence on MT-1MMP expression and fibrillar collagen form. Matrix Biol. 20, 193–203.
Runger, T.M., Adami, S., Benhamou, C.L., Czerwinski, E., Farrerons, J., Kendler, D.L., Mindeholm, L., Realdi, G., Roux, C., Smith, V., 2012. Morphea-like skin reactions in patients treated with the cathepsin K inhibitor balicatib. J. Am. Acad. Dermatol. 66, e89–96.
Rydziel, S., Durant, D., Canalis, E., 2000. Platelet-derived growth factor induces collagenase 3 transcription in osteoblasts through the activator protein 1 complex. J. Cell. Physiol. 184, 326–333.
Sabeh, F., Ota, I., Holmbeck, K., Birkedal-Hansen, H., Soloway, P., Balbin, M., Lopez-Otin, C., Shapiro, S., Inada, M., Krane, S., Allen, E., Chung, D., Weiss, S.J., 2004. Tumor cell traffic through the extracellular matrix is controlled by the membrane-anchored collagenase MT1-MMP. J. Cell Biol. 167, 769–781.
Saftig, P., Hunziker, E., Wehmeyer, O., Jones, S., Boyde, A., Rommerskirch, W., Moritz, J.D., Schu, P., von Figura, K., 1998. Impaired osteoclastic bone resorption leads to osteopetrosis in cathepsin-K-deficient mice. Proc. Nat. Acad. Sci. USA 95, 13453–13458.
Saftig, P., Hunziker, E., Everts, V., Jones, S., Boyde, A., Wehmeyer, O., Suter, A., Figura, K., 2000. Functions of cathepsin K in bone resorption: lessons from cathepsin K deficient mice. Adv. Exp. Biol. Med. 477, 293–303.
Sahara, T., Itoh, K., Debari, K., Sasaki, T., 2003. Specific biological functions of vacuolar-type H(+)-ATPase and lysosomal cystein proteinase, cathepsin K, in osteoclasts. Anat Rec A Discov Mol. Cell Evol. Biol. 270, 152–161.
Sato, H., Takino, T., Okada, Y., Cao, J., Shinagawa, A., Yamamoto, E., Seiki, M., 1994. A matrix metalloproteinase expressed on the surface of invasive tumour cells. Nature 370, 61–65.
Sato, H., Kinoshita, T., Takino, T., Nakayama, K., Seiki, M., 1996. Activation of a recombinant membrane type 1-matrix metalloproteinase (MT1-MMP) by furin and its interaction with tissue inhibitor of metalloproteinases (TIMP)-2. FEBS Lett. 393, 101–104.
Sato, H., Okada, Y., Seiki, M., 1997. Membrane-type matrix metalloproteinases (MT-MMPs) in cell invasion. Thromb. Haemostasis 78, 497–500.
Scannevin, R.H., Alexander, R., Mezzasalma, T., Burke, S.L., Singer, M., Huo, C., Zhang, Y.-M., Maguire, D., Spurlino, J., Deckman, I., Carroll, K.I., Lewandowski, F., Devine, E., Dzordzorme, K., Tounge, B., Milligan, C., Bayoumy, S., Williams, R., Schalk-Hihi, C., Leonard, K., Jackson, P., Todd, M., Kuo, L.C., Rhodes, K.J., 2017. Discovery of highly selective chemical inhibitor of matrix metalloproteinase-9 (MMP-9) that allosterically inhibits zymogen activation. J. Biol. Chem. 292, 17963–17974.
Schroen, D.J., Brinckerhoff, C.E., 1996. Nuclear hormone receptors inhibit matrix metalloproteinase (MMP) gene expression through diverse mechanisms. Gene Expr. 6, 197–207.
Scott, D.K., Brakenhoff, K.D., Clohisy, J.C., Quinn, C.O., Partridge, N.C., 1992. Parathyroid hormone induces transcription of collagenase in rat osteoblastic cells by a mechanism using cyclic adenosine 3′,5′-monophosphate and requiring protein synthesis. Mol. Endocrinol. 6, 2153–2159.
Seandel, M., Noack-Kunnmann, K., Zhu, D., Aimes, R.T., Quigley, J.P., 2001. Growth factor-induced angiogenesis in vivo requires specific cleavage of fibrillar type I collagen. Blood 97, 2323–2332.
Selvamurugan, N., Chou, W.Y., Pearman, A.T., Pulumati, M.R., Partridge, N.C., 1998. Parathyroid hormone regulates the rat collagenase-3 promoter in osteoblastic cells through the cooperative interaction of the activator protein-1 site and the runt domain binding sequence. J. Biol. Chem. 273, 10647–10657.
Selvamurugan, N., Brown, R.J., Partridge, N.C., 2000a. Constitutive expression and regulation of collagenase-3 in human breast cancer cells. J. Cell. Biochem. 79, 182–190.
Selvamurugan, N., Pulumati, M.R., Tyson, D.R., Partridge, N.C., 2000b. Parathyroid hormone regulation of the rat collagenase-3 promoter by protein kinase A-dependent transactivation of core binding factor A1. J. Biol. Chem. 275, 5037–5042.
Selvamurugan, N., Kwok, S., Alliston, T., Reiss, M., Partridge, N.C., 2004. Transforming growth factor-beta 1 regulation of collagenase-3 expression in osteoblastic cells by cross-talk between the Smad and MAPK signaling pathways and their components, Smad2 and Runx2. J. Biol. Chem. 279, 19327–19334.
Selvamurugan, N., Jefcoat, S.C., Kwok, S., Kowalewski, R., Tamasi, J.A., Partridge, N.C., 2006. Overexpression of Runx2 directed by the matrix metalloproteinase-13 promoter containing the AP-1 and Runx2/RD/Cbfa sites alters bone remodeling in vivo. J. Cell. Biochem. 99, 545–557.
Selvamurugan, N., Partridge, N.C., 2000. Constitutive expression and regulation of collagenase-3 in human breast cancer cells. Mol Cell Biol Res Commun 3, 218–223.
Shah, R., Alvarez, M., Joes, D.R., Torrungruang, K., Wat, A.J., Selvamurugan, N., Partridge, N.C., Quinn, C.O., Pavalko, F.M., Rhodes, S.J., Bidwell, J.P., 2004. Nmp4/CIZ regulation of matrix metalloproteinase 13 (MMP-13) response to parathyroid hormone in osteoblasts. Am. J. Physiol. Endocrinol. Metab. 287, E289–E296.
Shalhoub, V., Conlon, D., Tassinari, M., Quinn, C., Partridge, N., Stein, G.S., Lian, J.B., 1992. Glucocorticoids promote development of the osteoblast phenotype by selectively modulating expression of cell growth and differentiation associated genes. J. Cell. Biochem. 50, 425–440.
Shimizu, E., Selvamurugan, N., Westendorf, J.J., Olson, E.N., Partridge, N.,C., 2010. HDAC4 represses matrix metalloproteinase-13 transcription in osteoblastic cells, and parathyroid hormone controls this repression. J. Biol. Chem. 285, 9616–9626.
Shofuda, K.I., Hasenstab, D., Kenagy, R., Shofuda, T., Li, Z.Y., Lieber, A., Clowes, A.W., 2001. Membrane-type matrix metalloproteinase-1 and -3 activity in primate smooth muscle cells. FASEB J. 15, 2010–2012.
Stetler-Stevenson, W.G., Krutzsch, H.C., Liotta, L.A., 1989. Tissue inhibitor of metalloproteinase (TIMP-2): a new member of the metalloproteinase family. J. Biol. Chem. 264, 17374–17378.
Stewart, D., Javadi, M., Chambers, M., Gunsolly, C., Gorski, G., Borghaei, R.C., 2007. Interleukin-4 inhibition of interleukin-1-induced expression of matrix metalloproteinase-3 (MMP-3) is independent of lipoxygenase and PPARgamma activation in human gingival fibroblasts. BMC Mol. Biol. 8 (12), 1–13.

Stickens, D., Behonick, D.J., Ortega, N., Heyer, B., Hartenstein, B., Yu, Y., Fosang, A.J., Schorpp-Kistner, M., Angel, P., Werb, Z., 2004. Altered endochondral bone development in matrix metalloproteinase 13-deficient mice. Development 131, 5883–5895.

Stolow, M.A., Bauzon, D.D., Li, J., Sedgwick, T., Liang, V.C., Sang, Q.A., Shi, Y.B., 1996. Identification and characterization of a novel collagenase in *Xenopus laevis:* possible roles during frog development. Mol. Biol. Cell 7, 1471–1483.

Tardif, G., Reboul, P., Pelletier, J.P., Martel-Pelletier, J., 2004. Ten years in the life of an enzyme: the story of the human MMP-13 (collagenase-3). Mod. Rheumatol. 14, 197–204.

Teitelbaum, S.L., 2000. Bone resorption by osteoclasts. Science 289, 1504–1508.

Tezuka, K.-I., Nemoto, K., Tezuka, Y., Sato, T., Ikeda, Y., Kobori, M., Kawashima, H., Eguchi, H., Hakeda, Y., Kumegawa, M., 1994a. Identification of matrix metalloproteinase 9 in rabbit osteoclasts. J. Biol. Chem. 269, 15006–15009.

Tezuka, K.-I., Tezuka, Y., Maejima, A., Sato, T., Nemoto, K., Kamioka, H., Hakeda, Y., Kumegawa, M., 1994b. Molecular cloning of a possible cysteine proteinase predominantly expressed in osteoclasts. J. Biol. Chem. 269, 1106–1109.

Thomson, B.M., Atkinson, S.J., McGarrity, A.M., Hembry, R.M., Reynolds, J.J., Meikle, M.C., 1989. Type I collagen degradation by mouse calvarial osteoblasts stimulated with 1,25-dihydroxyvitamin D-3: evidence for a plasminogen-plasmin-metalloproteinase activation cascade. Biochim. Biophys. Acta 1014, 125–132.

Tumber, A., Papaioannou, S., Breckon, J., Meikle, M.C., Reynolds, J.J., Hill, P.A., 2003. The effects of serine protease inhibitors on bone resorption in vitro. J. Endocrinol. 178, 437–447.

Turk, V., Bode, W., 1991. The cystatins: protein inhibitors of cysteine proteinases. FEBS Lett. 285, 213–219.

Tyson, D.R., Swarthout, J.T., Partridge, N.C., 1999. Increased osteoblastic c-fos expression by parathyroid hormone requires protein kinase A phosphorylation of the cyclic adenosine 3′,5′-monophosphate response element-binding protein at serine 133. Endocrinology 140, 1255–1261.

Vaes, G., 1988. Cellular biology and biochemical mechanism of bone resorption. A review of recent developments on the formation, activation and mode of action of osteoclasts. Clin. Orthop. Relat. Res. 231, 239–271.

van Zonneveld, A.-J., Veerman, H., Pannekoek, H., 1986. On the interaction of the finger and the kringle-2 domain of tissue-type plasminogen activator with fibrin. J. Biol. Chem. 261, 14214–14218.

Varghese, S., Rydziel, S., Jeffrey, J.J., Canalis, E., 1994. Regulation of interstitial collagenase expression and collagen degradation by retinoic acid in bone cells. Endocrinology 134, 2438–2444.

Varghese, S., Rydziel, S., Canalis, E., 2000. Basic fibroblast growth factor stimulates collagenase-3 promoter activity in osteoblasts through an activator protein-1-binding site. Endocrinology 141, 2185–2191.

Varghese, S., 2006. Matrix metalloproteinases and their inhibitors in bone: an overview of regulation and functions. Front. Biosci. 11, 2949–2966.

Velasco, G., Cal, S., Merlos-Suárez, A., Ferrando, A.A., Alvarez, S., Nakano, A., Arribas, J., López-Otín, C., 2000. Human MT6-matrix metalloproteinase: identification, progelatinase A activation, and expression in brain tumors. Cancer Res. 60, 877–882.

Visse, R., Nagase, H., 2003. Matrix metalloproteinases and tissue inhibitors of metalloproteinases. Circ. Res. 92, 827–839.

Vu, T.H., Shipley, J.M., Bergers, G., Berger, J.E., Helms, J.A., Hanahan, D., Shapiro, S.D., Senior, R.M., Werb, Z., 1998. MMP-9/gelatinase B is a key regulator of growth plate angiogenesis and apoptosis of hypertrophic chondrocytes. Cell 93, 411–422.

Vu, T.H., Werb, Z., 2000. Matrix metalloproteinases: effectors of development and normal physiology. Genes Dev. 14, 2123–2133.

Walling, H.W., Chan, P.T., Omura, T.H., Barmina, O.Y., Fiacco, G.J., Jeffrey, J.J., Partridge, N.C., 1998. Regulation of the collagenase-3 receptor and its role in intracellular ligand processing in rat osteoblastic cells. J. Cell. Physiol. 177, 563–574.

Wasylyk, B., Hahn, S.L., Giovane, A., 1993. The Ets family of transcription factors. Eur. J. Biochem. 211, 7–18.

Wernicke, D., Seyfert, C., Hinzmann, B., Gromnica-Ihle, E., 1996. Cloning of collagenase 3 from the synovial membrane and its expression in rheumatoid arthritis and osteoarthritis. J. Rheumatol. 23, 590–595.

Wilcox, B.D., Dumin, J.A., Jeffrey, J.J., 1994. Serotonin regulation of interleukin-1 messenger RNA in rat uterine smooth muscle cells. J. Biol. Chem. 269, 29658–29664.

Wilson, C.L., Heppner, K.J., Labosky, P.A., Hogan, B.L.M., Matrisian, L.M., 1997. Intestinal tumorigenesis is suppressed in mice lacking the metalloproteinase, matrilysin. Proc. Natl. Acad. Sci. U.S.A. 94, 1402–1407.

Wilhelm, S.M., Collier, I.E., Marmer, B.L., Eisen, A.Z., Grant, G.A., Goldberg, G.I., 1989. SV40-transformed human lung fibroblasts secrete a 92-kDa type IV collagenase which is identical to that secreted by normal human macrophages. J. Biol. Chem. 264, 17213–17221.

Winchester, S.K., Bloch, S.R., Fiacco, G.J., Partridge, N.C., 1999. Regulation of expression of collagenase-3 in normal, differentiating rat osteoblasts. J. Cell. Physiol. 181, 479–488.

Winchester, S.K., Selvamurugan, N., D'Alonzo, R.C., Partridge, N.C., 2000. Developmental regulation of collagenase-3 mRNA in normal, differentiating osteoblasts through the activator protein-1 and the runt domain binding sites. J. Biol. Chem. 275, 23310–23318.

Witty, J.P., Matrisian, L., Foster, S., Stern, P.H., 1992. Stromelysin in PTH-stimulated bones *in vitro*. J. Bone Miner. Res. 7 (Suppl. 1), S103.

Woessner Jr., J.F., 1991. Matrix metalloproteinases and their inhibitors in connective tissue remodeling. FASEB J 5, 2145–2154.

Wucherpfennig, A.L., Li, Y.-P., Stetler-Stevenson, W.G., Rosenberg, A.E., Stashenko, P., 1994. Expression of 92 kD type IV collagenase/gelatinase B in human osteoclasts. J. Bone Miner. Res. 9, 549–556.

Wun, T.-C., Ossowski, L., Reich, E., 1982. A proenzyme form of human urokinase. J. Biol. Chem. 257, 7262–7268.

Xia, L., Kilb, J., Wex, H., Li, Z., Lipyansky, A., Breuil, V., Stein, L., Palmer, J.T., Dempster, D.W., Bromme, D., 1999. Localization of rat cathepsin K in osteoclasts and resorption pits: inhibition of bone resorption and cathepsin K-activity by peptidyl vinyl sulfones. Biol. Chem. 380, 679–687.

Xiang, A., Kanematsu, M., Kumar, S., Yamashita, D., Kaise, T., Kikkawa, H., Asano, S., Kinoshita, M., 2007. Changes in micro-CT 3D bone parameters reflect effects of a potent cathepsin K inhibitor (SB-553484) on bone resorption and cortical bone formation in ovariectomized mice. Bone 40, 1231–1237.

Yamaza, T., Goto, T., Kamiya, T., Kobayashi, Y., Sakai, H., Tanaka, T., 1998. Study of immunoelectron microscopic localization of cathepsin K in osteoclasts and other bone cells in the mouse femur. Bone 23, 499–509.

Yang, J.N., Allan, E.H., Anderson, G.I., Martin, T.J., Minkin, C., 1997. Plasminogen activator system in osteoclasts. J. Bone Miner. Res. 12, 761–768.

Yang, T., Williams, B.O., 2017. Low-density lipoprotein receptor-related proteins in skeletal development and disease. Physiol. Rev. 97, 1211–1228.

Zhao, W., Byrne, M.H., Boyce, B.F., Krane, S.M., 1999. Bone resorption induced by parathyroid hormone is strikingly diminished in collagenase-resistant mice. J. Clin. Investig. 103, 517–524.

Zhao, W., Byrne, M.H., Wang, Y., Krane, S.M., 2000. Inability of collagenase to cleave type I collagen *in vivo* is associated with osteocyte and osteoblast apoptosis and excessive bone deposition. J. Clin. Investig. 106, 841–849.

Zhou, Z., Apte, S.S., Soininen, R., Cao, R., Baaklini, G.Y., Rauser, R.W., Wang, J., Cao, Y., Tryggvason, K., 2000. Impaired endochondral ossification and angiogenesis in mice deficient in membrane-type matrix metalloproteinase I. Proc. Natl. Acad. Sci. U.S.A. 97, 4052–4057.

Zucker, S., Cao, J., Chen, W.-T., 2000. Critical appraisal of the use of matrix metalloproteinase inhibitors in cancer treatment. Oncogene 19, 6642–6650.

Chapter 17

Integrins and other cell surface attachment molecules of bone cells

Pierre J. Marie[1] and Anna Teti[2]

[1]*UMR-1132 Inserm (Institut national de la Santé et de la Recherche Médicale) and University Paris Diderot, Sorbonne Paris Cité, Paris, France;*
[2]*Department of Biotechnological and Applied Clinical Sciences, University of L'Aquila, L'Aquila, Italy*

Chapter outline

Introduction **401**
Role of integrins in bone cells **401**
Osteoblasts and osteocytes 401
Osteoclasts 404
Chondrocytes 406
Role of cadherins in bone cells **406**
Osteoblasts and osteocytes 406
Osteoclasts 408
Chondrocytes 409
Roles of other attachment molecules in bone cells **410**
Syndecans 410
Glypicans and perlecan 410
CD44 411
Immunoglobulin superfamily members 411
Osteoactivin 412
Chondroadherin 412
Conclusion **413**
Acknowledgments **413**
References **413**

Introduction

Cell adhesion molecules play important roles in bone cell functions and fate. Integrins are a family of cell surface adhesion transmembrane molecules composed of α chains and β chains that assemble noncovalently as heterodimers (Campbell and Humphries, 2011), allowing the attachment of osteoclasts and osteoblasts to the extracellular matrix (ECM). Binding of integrins to ECM proteins is essential for the function of bone cells (Bennett et al., 2001; Horton, 2001), and contributes to activating a number of intracellular signals that govern bone cell fate and activity (Marie et al., 2014a). Cadherins are other attachment molecules that control the functions of bone cells through cell–cell adhesion, interactions with other cell surface molecules, and modulation of intracellular pathways (Marie et al., 2014b). During recent years, progress in cell biology and mouse genetics has led to significant advances in our understanding of the roles of integrins, cadherins, and other cell attachment molecules in the control of bone cell activity in vitro and in vivo, and the mechanisms involved in these functions are now better understood. This chapter updates our knowledge of the role of these molecules in bone cell functions and discusses potential therapeutic strategies that emerged from these findings.

Role of integrins in bone cells

Osteoblasts and osteocytes

Osteoblasts and osteocytes express several integrins, although the pattern of expression varies with the stage of cell differentiation (Clover et al., 1992; Hughes et al., 1993; Hultenby et al., 1993; Grzesik and Robey, 1994). Integrin binds to the Arg–Gly–Asp (RGD) sequence present in bone matrix proteins such as fibronectin, type I collagen, bone sialoprotein, and osteopontin (Gronthos et al., 2001; Grzesik, 1997; Puleo and Bizios, 1991; Schaffner and Dard, 2003). Integrins control cell adherence to the ECM through the assembly of intracellular proteins linked to the cytoskeleton. In addition to allowing

Principles of Bone Biology. https://doi.org/10.1016/B978-0-12-814841-9.00017-8

cell−matrix adherence, integrins play a key role in osteoblast differentiation (Moursi et al., 1997; Jikko et al., 1999; Mizuno et al., 2000), an effect that is mediated by intracellular signals generated by ECM−osteoblast interactions. Integrin−ECM interaction leads to the phosphorylation of focal adhesion kinase (FAK) and subsequent activation of mitogen-activated protein kinase (MAPK) extracellular signal-regulated kinase 1/2 (ERK1/2), phosphatidylinositol 3-kinase (PI3K), or GTPases of the Rho family (Lai et al., 2001; Salasznyk et al., 2007; Khatiwala et al., 2009). These signals converge to promote specific gene expression and osteoblast differentiation. In mesenchymal skeletal cells (MSCs), ERK1/2 phosphorylation induced by ECM−integrin binding leads to RUNX2 activation and osteoblast differentiation (Ge et al., 2012). FAK also activates Wnt/β-catenin signaling (Sun et al., 2016), making the activation of FAK an essential step for osteogenic differentiation (Tamura et al., 2001). In vivo, deletion of FAK in type I collagen-expressing osteoblasts reduced reparative bone formation in mice (Rajshankar et al., 2017). However, the lack of FAK in osteoblasts may be in part compensated for by proline-rich tyrosine kinase 2 (PYK2), a tyrosine kinase highly homologous to FAK, in the focal adhesion sites (Kim et al., 2007). One important regulator of integrin-mediated signaling is integrin-linked kinase (ILK), a linker between integrins and the cytoskeleton. In osteoblasts, ILK-dependent phosphorylation of α-nascent polypeptide−associated complex, a c-JUN transcriptional coactivator, potentiates c-JUN-dependent transcription and osteoblast maturation (Meury et al., 2010). ILK also interacts with Wnt signaling by phosphorylating glycogen synthase kinase-3β, leading to β-catenin− lymphoid enhancer factor transcriptional activity and expression of Wnt target genes (Dejaeger et al., 2017). Data indicate that conditional inactivation of ILK in osteoprogenitors impaired bone formation associated with reduced bone morphogenetic protein (BMP)/Smad and Wnt/β-catenin signaling, showing the important role of this linker between integrins and the cytoskeleton in bone formation (Dejaeger et al., 2017). In addition to control of cell differentiation, osteoblast attachment to the ECM is essential for cell survival (Grigoriou et al., 2005; Triplett and Pavalko, 2006). Disruption of RGD−integrin binding leads to altered osteoblast adhesion in vitro (Gronthos et al., 2001), and conversely integrin−ECM interaction suppresses cell apoptosis through PI3K activation (Frisch and Ruoslahti, 1997). Thus, integrins play an important role in the control of bone formation through activation of signaling mechanisms regulating both osteoblast differentiation and survival (Fig. 17.1).

The specific role of integrins in osteoblast function and fate is now better understood. The β1 integrin is the main adhesion receptor required for adhesion to fibronectin and type I collagen in vitro. In vivo, impairment of β1 integrin in mature osteoblasts led to decreased osteoblast activity, bone formation, and bone mass in growing mice (Zimmerman et al., 2000). Disruption of β1 integrin signaling by overexpression of its cytoplasmic tail resulted in skeletal defects, showing the important role of β1 integrin in osteogenesis (Globus et al., 2005). In support of the role of β1 integrin in bone formation, ablation of the specific β1 integrin regulator ICAP-1 in osteoblasts resulted in defective osteoblast proliferation, differentiation, and function; decreased type I collagen deposition; and delayed bone formation in mice (Bouvard et al., 2007; Brunner et al., 2011). Data showed that conditional β1 integrin deletion in early osteogenic mesenchymal cells or in preosteoblasts caused abnormal skeletal ossification or affected intramembranous and

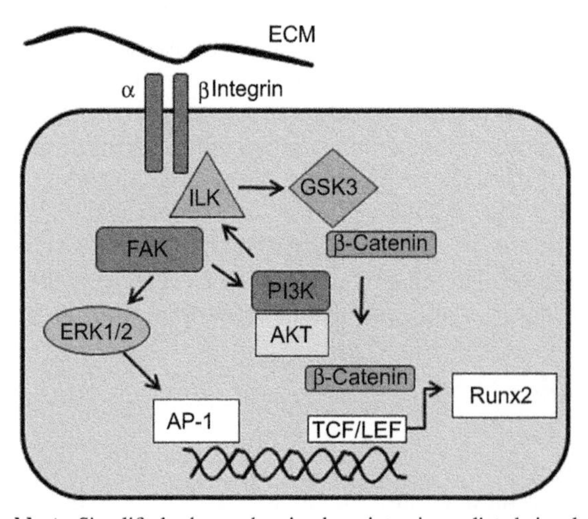

FIGURE 17.1 Role of integrins in osteoblasts. Simplified scheme showing how integrin-mediated signals regulate osteoblast function. Extracellular matrix (*ECM*)−integrin interaction via RGD interacts with integrin-linked kinase (ILK) and activates focal adhesion kinase (FAK), extracellular signal-regulated kinase 1/2 (ERK1/2), and Wnt/β-catenin signaling, leading to Runx2 expression and osteogenic differentiation, and with phosphatidylinositol 3-kinase/protein kinase B (PI3K/AKT) to trigger osteoblast survival (see text for more details). *AP-1*, activator protein 1; *GSK3*, glycogen synthase kinase-3; *TCF/LEF*, T cell factor/lymphoid enhancer factor.

endochondral bone formation in young mice. In contrast, osteocalcin-specific β1 integrin deletion had only minor effects on the skeletal phenotype, indicating that the β1 integrin is essential for early stages of osteogenesis in vivo (Shekaran et al., 2014). In addition to β1, other integrins were shown to control osteoblast function and fate. Early in vitro studies showed that the αVβ3 integrin is involved in dexamethasone- and BMP-2-induced osteoblast differentiation (Cheng et al., 2000, 2001). More recent data indicate that the αVβ3 integrin mediates BMP-2-induced osteoblast differentiation in vitro through activation of the ILK/ERK1/2 signaling pathway (Su et al., 2010). Another integrin, αVβ5, was found to mediate osteoblast attachment to vitronectin in vitro (Lai et al., 2000). The αVβ1 integrin was reported to promote osteoblast differentiation and to inhibit adipocyte differentiation in MSCs through its interaction with RGD present in osteopontin (Chen et al., 2014). Another integrin, α2β1, a major receptor for collagen type 1, is involved in osteoblast differentiation (Takeuchi et al., 1997) by activating ERK signaling and RUNX2 expression in vitro (Xiao et al., 1998). In vivo, α2 integrin deficiency resulted in altered bone properties in mice (Stange et al., 2013). This integrin was found to play a key role in MSC osteogenic differentiation and survival through activation of Rho-associated protein kinase (ROCK) and FAK signaling (Popov et al., 2011; Sens et al., 2017). The α4β1 integrin, which binds to fibronectin, was also found to be involved in MSC homing and osteoblast differentiation in vivo (Kumar and Ponnazhagan, 2007). Both MSCs and osteoblasts express α5β1 integrin, which interacts with fibronectin and is involved in the differentiation of osteoblast precursor cells (Hamidouche et al., 2009). The downregulation of α5 integrin subunit blunts osteoblast differentiation, while its overexpression promotes osteoblast differentiation in MSCs, an effect that is mediated by FAK and ERK1/2 signaling leading to RUNX2 activation (Hamidouche et al., 2009). α5β1 integrin activation also promotes insulin-like growth factor 2 (IGF-2) expression and signaling in MSCs, which contributes to osteogenic differentiation (Hamidouche et al., 2010). In addition, α5β1 integrin activation cross talks with Wnt signaling. α5β1 integrin priming by a monoclonal antibody, or a high-affinity peptide ligand, leads to increased osteogenic differentiation (Hamidouche et al., 2009) in part through activation of PI3K/protein kinase B (AKT) and Wnt/β-catenin signaling, indicating that Wnt/β-catenin signaling is involved in osteoblast differentiation mediated by α5β1 integrin (Saidak et al., 2015). In pluripotent cells, other integrins may be involved in osteogenic differentiation. In vitro data indicate that fibrinogen binds to the α9β1 integrin in human embryonic stem cells to induce pluripotent stem cell osteogenic differentiation mediated by SMAD1/5/8 signaling and *Runx2* expression (Kidwai et al., 2016).

There is ample evidence that integrins are critically involved in the response of osteoblasts and osteocytes to mechanical loading via cytoskeleton–integrin interactions at focal adhesion sites (Katsumi et al., 2004; Bonewald and Johnson, 2008; Turner et al., 2009) and induction of intracellular signaling mechanisms, including FAK, ERK, ROCK, PI3K/AKT, and Wnt signaling (Pommerenke et al., 2002; Wu et al., 2013; Du et al., 2016; Plotkin et al., 2005; Wang et al., 2007; Lee et al., 2010; Chen and Jacobs, 2013; Uda et al., 2017). Several integrins may be involved in the mechanisms mediating mechanotransduction, depending on the context. As expected from the aforementioned important role of β1 integrin in osteoblast differentiation, β1 integrin expressed by osteogenic cells was found to play a major role in mechanotransduction (Iwaniec et al., 2005; Phillips et al., 2008; Litzenberger et al., 2010). In osteocytes, in addition to the β1 integrin, the αVβ3 integrin is required for mechanotransduction (Thi et al., 2013), which involves calcium influx (Miyauchi et al., 2006), IGF-1 expression (Dai et al., 2014), and inhibition of the Wnt inhibitor sclerostin by periostin (Bonnet et al., 2012). In vivo, skeletal unloading in mice decreased αVβ3 integrin expression in osteoblasts, resulting in altered activation of IGF-1 signaling (Bikle, 2008). Skeletal unloading also reduced α5β1 integrin expression in osteoblasts and osteocytes, resulting in decreased PI3K signaling and altered bone formation (Dufour et al., 2007). Conversely, mechanical strain activates α5β1 integrin (Yan et al., 2012) and PI3K/AKT signaling in osteoblasts (Watabe et al., 2011). The activation of α5β1 integrin by mechanical stress in osteoblasts also causes opening of connexin 43 hemichannels and the subsequent release of anabolic factors (Batra et al., 2012). Other integrins were found to be involved in mechanotransduction. In vitro, the α2 integrin subunit is upregulated during induction on stiffer matrices and mediates the osteogenic differentiation of MSCs (Shih et al., 2011). However, mechanosensitivity appears to depend on the ligation between α2 or α5 subunit and ECM proteins to induce FAK activation (Seong et al., 2013), suggesting that specific integrin subunits may mediate mechanosensitivity in osteoblasts. Thus, several mechanisms generated by integrin–ECM interactions in osteoblasts and osteocytes are involved in the bone cell response to mechanotransduction.

The essential role of integrins in both bone formation and mechanotransduction suggests that targeting specific integrins may be an efficient way to promote osteogenic differentiation and bone regeneration. Several strategies were used to target integrins. One example is the ectopic expression of the α4 integrin in MSCs (Kumar and Ponnazhagan, 2007). While the injection of modified MSCs into mice led to increased cell homing in bone and their subsequent differentiation into osteoblasts (Guan et al., 2012), the systemic injection of a peptidomimetic ligand (LLP2A) that binds the α4β1 integrin, conjugated to a bisphosphonate to allow bone binding, led to increased osteoblast differentiation and bone formation in mice with established osteopenia (Guan et al., 2012; Yao et al., 2013). Another strategy is to target the α5β1 integrin (Marie, 2013). Lentiviral-mediated expression of the α5 integrin subunit in human MSCs promoted osteogenic differentiation, bone formation (Hamidouche et al., 2009), and bone repair in critical-size bone defects in mice (Srouji et al., 2012). Moreover, α5 integrin activation by a synthetic cyclic peptide, which activates FAK and ERK1/2–MAPK signaling in osteoprogenitor cells, led to increased osteoblast differentiation and to reduced cell apoptosis in vitro (Fromigué et al., 2012). In vivo, the local injection of an α5 integrin agonist peptide into adult mice increased parietal bone formation, indicating that pharmacological activation of α5 integrin in osteoprogenitor cells is effective in promoting de novo bone formation (Fromigué et al., 2012). An alternative strategy to target integrins may be the use of recombinant NELL-1, a secreted osteoinductive protein that binds to β1 integrin. In vitro, this strategy leads to activation of Wnt/β-catenin signaling and increased osteoblast differentiation. In vivo, delivery of NELL-1 to ovariectomized mice or osteopenic sheep improved bone mineral density (James et al., 2015). These studies showed that targeting integrins may be a promising approach for promoting bone formation and repair in skeletal disorders.

Osteoclasts

Osteoclasts are the multinucleated cells that resorb the bone matrix to promote skeletal modeling and renewal (Soysa and Alles, 2016; Cappariello et al., 2014). They rely on a tight and dynamic mechanism of adhesion to the mineralized matrix, which ensures cell changes indispensable for bone resorption. These are (1) the polarization of the osteoclast and (2) the sealing of the extracellular space between the cell and the bone matrix to be removed, called resorption lacuna (Soysa and Alles, 2016; Cappariello et al., 2014) (Fig. 17.2). The principal integrin involved in the adhesion of osteoclasts to the substrate is the αVβ3 receptor. While αV-deleted mice are embryonically lethal, β3-deleted mice showed a phenotype mimicking Glanzmann thrombasthenia (Morgan et al., 2010) and osteopetrosis (Zou and Teitelbaum, 2015; McHugh et al., 2000). Osteopetrosis is caused by osteoclast dysfunction, and these observations revealed the relevance of the αVβ3 integrin in osteoclast biology. Specifically, β3-null osteoclasts form normally but are prevented from resorbing bone. Their cytoskeletal array is disrupted, adhesion to substrate is reduced, and cell spreading is impeded (McHugh et al., 2000). These results suggest that the αVβ3 integrin is not involved in the molecular mechanisms of osteoclastogenesis, but it rather affects the morphological and molecular changes that make osteoclasts capable of resorbing bone. αVβ3 heterodimers are clustered in specific adhesion areas, called podosomes (Marchisio et al., 1984), that are typical of osteoclasts and other highly motile cells, such as monocytes, macrophages (Calle et al., 2006), and invasive cancer cells, in which they are called invadopodia (Eddy et al., 2017). Compared with classical focal adhesions, podosomes and invadopodia are much more dynamic, with a turnover of a few minutes rather than the hours observed in focal adhesions (Georgess et al., 2014). A typical feature of osteoclast podosomes is their clustering in a continuous peripheral annulus, called actin ring

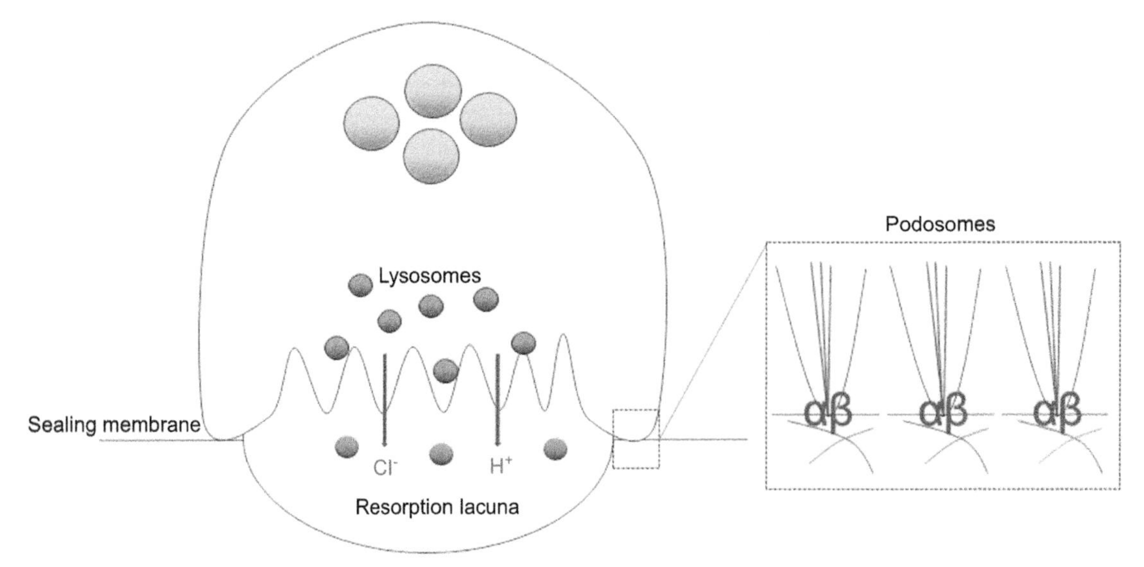

FIGURE 17.2 Role of integrins in osteoclasts. Cartoon showing a polarized osteoclast sitting on the resorption lacuna. Release of protons and chlorides acidifies the lacuna microenvironment, solubilizing the hydroxyapatite and allowing the degradation of the organic matrix by lysosomal enzymes therein secreted. Adhesion to the mineralized matrix occurs through podosomes located in the sealing membrane, which establish contacts with the extracellular molecules through integrin receptors (see text for more details).

(Cappariello et al., 2014), representing the sealing zone of the apical membrane. Here, the number of podosomes and the extension of the acting ring are proportional to the activity of the osteoclasts, reaching the maximum development during bone resorption, when multiple actin rings/osteoclast can be observed (Cappariello et al., 2014). For their high dynamism, podosomes and their associated αVβ3 integrins change rapidly in terms of number and distribution, contributing not only to the sealing of the resorption lacuna and the subsequent organization of the inner irregular domain of the apical membrane lodging the molecular mechanisms of bone resorption, called the ruffled border, but also to the osteoclast motility on the mineralized matrix during and between the bone resorption cycles (Cappariello et al., 2014; Rucci and Teti, 2016). In osteoclast podosomes, the αVβ3 integrin is largely represented and concentrated in a circular podosomal area of membrane adhesion to substrate, whose cytosolic center lodges a core of microfilaments; actin-binding proteins, including α-actinin; and actin-branching proteins, including cortactin, the ARP2/3 complex, and WASP and WASP-interacting protein (Georgess et al., 2014; Rucci and Teti, 2016). The αVβ3 integrin circular "rosette" that surrounds this microfilament core is associated with several intracellular adhesion proteins, such as vinculin, talin, paxillin and tensin, that link the integrins to the microfilaments (Saltel et al., 2008; Correia et al., 1999; Evans and Matsudaira, 2006; Marchisio et al., 1988; Akisaka et al., 2001). Importantly, this complex structure includes microfilament-severing proteins (gelsolin and cofilin) (Ory et al., 2008), as well as the small GTP-binding protein RHO (Teitelbaum, 2011) and the guanine exchange factors DOCK5 and VAV3, which provide rapid podosome remodeling indispensable for their dynamism (Purev et al., 2009). This is also ensured by the localization in podosomes of the large GTP-binding protein dynamin, which regulates E3 ubiquitin ligase c-CBL-dependent degradation or receptors and signaling factors (Bruzzaniti et al., 2005). GTPase activity of dynamin is enhanced by GRB2, which also increases WASP-mediated ARP2/3-dependent actin nucleation (Spinardi and Marchisio, 2006). Along with the microfilament-severing function of gelsolin and cofilin, this molecular cascade ensures rapid microfilament turnover and podosome disassembly and reassembly (Luxenburg et al., 2012). Moreover, ubiquitin ligases, including c-CBL and CBL-b (Horne et al., 2005), phosphoinositide kinases such as PI3K (Chellaiah et al., 2001), nonreceptor tyrosine kinases such as c-SRC and PYK2 (Duong et al., 1998; Destaing et al., 2008), ABL and FAK (Ray et al., 2012), and phosphatases, including PTPα, PTPε, and SHP2 (Granot-Attas and Elson, 2008), associate with the αVβ3 integrin complex to further regulate osteoclast adhesion and podosome dynamics.

Genetic studies have been instrumental in demonstrating the relevance of the αVβ3 receptor signaling in osteoclast activity. The analysis of *c-Src* deletion in mice showed that osteoclasts were not able to polarize and became immotile (Lowe et al., 1993). They exhibited a severe osteopetrotic phenotype, which was observed, in milder form, also in *Pyk2*- and *c-Cbl*-null mice (Gil-Henn et al., 2007; Tanaka et al., 1996). c-SRC coprecipitates with the αVβ3 integrin and through its c-CBL tyrosine phosphorylation activity, promotes substrate ubiquitination and prolonged cell survival (Horne et al., 2005). Furthermore, c-SRC contributes to the metabolism of PYK2 and p130CAS, involved in the organization of the sealing membrane (Nakamura et al., 1998). Therefore, given that c-SRC is upstream of the tyrosine phosphorylation cascade involving PYK2, c-CBL and p130CAS, it is likely that these intracellular signaling molecules act synergistically to determine correct cytoskeletal arrangement, podosome formation, and osteoclast polarization and survival.

On the extracellular side, the osteoclast αVβ3 integrin binds the RGD sequence of a number of matrix proteins that contribute to osteoclast adhesion and outside-in signaling (Rucci and Teti, 2016). Many proteins are recognized by the αVβ3 integrin, including osteopontin, bone sialoprotein 2, vitronectin, fibronectin, and von Willebrand factor (Zou and Teitelbaum, 2010). αVβ3 integrin recognizes also collagen type I but only upon molecular remodeling, the collagen type I RGD sequence being lodged in a cryptic domain. In osteoclasts, osteopontin, bone sialoprotein II, synthetic RGD peptides, echistatin (an RGD-disintegrin peptide extracted from the poison of viper venom), and the αvβ3 integrin activating LM609 monoclonal antibody mobilize intracellular calcium through the αVβ3 receptor, albeit with some species specificity. In fact, in rat osteoclasts, αVβ3 integrin activation triggers intracellular calcium transients and cell retraction (Shankar et al., 1993), while in chicken osteoclasts it induces a calmodulin/Ca^{2+}-ATPase-dependent calcium efflux and intracellular reduction (Miyauchi et al., 1991). The molecular mechanisms underlying these opposite events have not been investigated, leaving this question still open. Finally, in mouse osteoclasts, osteopontin induces RGD-dependent nuclear translocation of nuclear factor of activated T cells 1 (NFATc1), a transcription factor essential for osteoclastogenesis (Tanabe et al., 2011), whose activation is triggered by calcium oscillations (Negishi-Koga and Takayanagi, 2009). The αVβ3 integrin is relatively infrequent in the organism, and osteoclasts are among the cells mostly enriched in these receptors. The αVβ3 integrin has been considered a pharmacological target to combat cancer-induced osteolytic lesions, osteoporosis, and rheumatoid arthritis (Desgrosellier and Cheresh, 2010; Schneider et al., 2011; Tian et al., 2015). αVβ3 integrin targeting molecules include cyclic peptides carrying the αVβ3 integrin binding site (Auzzas et al., 2010), neutralizing antibodies (Hsu et al., 2007) and nonpeptide antagonists (Hsu et al., 2007; Sheldrake and Patterson, 2014). Several of them have been tested in preclinical studies, and one has been tested in clinical trials to treat cancer (Danhier et al., 2012).

Other integrin members are also expressed by osteoclasts. αVβ1, αMβ2, and αVβ5 receptors are expressed by osteoclast mononuclear precursors and switch to αVβ3 during preosteoclast maturation in polykarya (Inoue et al., 1998). The α2β1 integrin is engaged in the binding to collagen type I. It does not recognize vitronectin, fibronectin, or fibrinogen and promotes migration and fusion of precursors into mature cells (Townsend et al., 1999). However, these integrins have not been investigated any further and their relevance in osteoclast biology is probably underestimated.

Chondrocytes

Chondrocytes express a panel of at least seven integrin heterodimers recognizing molecules of the basal lamina and the territorial cartilage matrix. α5β1, αNβ3, and αNβ5 are fibronectin receptors expressed by chondrocytes, which, along with the α6β1 receptor recognizing laminin, and the α1β1, α2β1 and α10β1 integrins binding collagen, ensure tight chondrocyte interaction with the substrate (Loeser, 2002; Aszodi et al., 2003). The dominant integrin subunit in chondrocytes is the β1 chain (Woods et al., 2007a). Consequently, conditional chondrocyte knockout (KO) of *β1 integrin* resulted in a severe cartilage phenotype (Woods et al., 2007a). Typical chondrodysplasia malformations were noticed in these mice, with shorter long bones and reduced growth plate hypertrophic zone. No vascularization of the primary ossification centers was observed, with vessel present only in the periosteum. The growth plate appeared broadened, the chondrocyte columnar array was disrupted, and the hypertrophic zone mineralization was reduced and patchy (Woods et al., 2007a). Given the broad presence of the β1 chain in at least five chondrocyte integrin receptors, such a severe cartilage phenotype is not surprising. Furthermore, a similar outcome was observed in conditional *Il1*-null chondrocytes in mice (Grashoff et al., 2003; Terpstra et al., 2003).

The expression of chondrocyte β1 integrins is regulated by mechanical forces. Tension forces stimulate their expression with a mechanism that appears to be mediated by FAK, and this effect leads to inhibition of chondrogenesis, especially through α2β1 and α5β1 receptors (Takahashi et al., 2003; Onodera et al., 2005). This result is rather surprising and while it is confirmed by the observation that plating mesenchymal stem cells on integrin-activating RGD peptides results in their reduced commitment to chondrogenesis (Connelly et al., 2007), it is contradicted by the discovery that inhibition of β1 integrin activity reduces chondrogenesis and cartilage production (Shakibaei, 1998). These results suggest that integrins have dual roles in cartilage development, inhibiting or inducing chondrogenesis in unlike contexts and in response to different environmental regulations. Functionally, Wang and Kirsch (Wang and Kirsch, 2006) showed that β1 and β5 integrins promote cell survival, while their deletion or blockage by neutralizing antibodies reduces hypertrophic differentiation (Hirsch et al., 1997). Proliferative and hypertrophic chondrocytes of the growth plate express the highest levels of α5β1 receptors, and neutralization of this integrin results in impairment of chondrocyte proliferation (Enomoto-Iwamoto et al., 1997). The α5β1 integrin is also important for the development of the appendicular skeleton and joints. For instance, premature formation of prehypertrophic chondrocyte and joint fusion were observed when *α5β1 integrin* was misexpressed (Garciadiego-Cázares et al., 2004). Finally, deletion of *α10β1 integrin* caused dysfunctions of the growth plate, inducing growth retardation (Bengtsson et al., 2005), with increased apoptosis and altered chondrocyte morphology. Instead deletion of the α1 chain caused osteoarthritis but not growth defects (Zemmyo et al., 2003).

In addition to integrins, the adhesion of chondrocytes to the cartilage matrix is ensured by other types of molecules. The discoidin domain receptor, *Ddr2*, is highly expressed by articular chondrocytes in osteoarthritis, and its conditional deficiency in growth plate chondrocytes was associated with decreased chondrocyte proliferation and subsequent dwarfism (Labrador et al., 2001). Annexin V instead mediates adhesion to the N telopeptide of collagen II in articular cartilage and growth plate, where it appears to regulate apoptosis and matrix mineralization (Wang and Kirsch, 2006; Jennings et al., 2001; Kirsch, 2005). Finally, CD44, an important membrane glycoprotein receptor recognizing collagens and hyaluronic acid, increases during chondrogenesis and contributes to the organization of the territorial matrix and the regulation of chondrocyte survival (Nicoll et al., 2002; Knudson, 2003).

Role of cadherins in bone cells

Osteoblasts and osteocytes

Cadherins are transmembrane glycoproteins that mediate calcium-dependent cell—cell adhesion through homophilic interactions of their extracellular domains (Gumbiner, 2005). The intracellular domain of cadherins interacts with cytoskeletal proteins at adherens junction sites and with signaling molecules such as vinculin, α- and β-catenin, and other molecules involved in cellular signaling processes (Nelson and Nusse, 2004). Osteoblasts express Cadh1 (E-cadherin), Cadh2 (N-cadherin), Cadh3 (P-cadherin), and Cadh11 ("osteoblast cadherin") (Cheng et al., 1998; Goomer et al., 1998;

Ferrari et al., 2000; Lemonnier et al., 2000; Kawaguchi et al., 2001a), although the expression of these cadherins changes with the stage of osteoblast differentiation. Cadh2 expression initially increases during osteogenic differentiation in vitro but declines thereafter (Ferrari et al., 2000; Greenbaum et al., 2012), whereas Cadh11 is upregulated throughout the osteoblast differentiation program (Kawaguchi et al., 2001b; Shin et al., 2000). In vivo, *Cadh2* is expressed in lining cells but not in osteocytes (Shin et al., 2000). In vitro and in vivo studies suggest that cadherins may control precursor cell lineage determination (Marie et al., 2014a). Both *Cadh2* and *Cadh11* are downregulated with commitment of mesenchymal progenitors to adipogenesis in vitro (Kawaguchi et al., 2001b; Shin et al., 2000). Adult *Cadh11*-KO mice showed increased number of adipogenic precursors and adipogenic differentiation in vitro, suggesting that Cadh11 blocks adipogenesis (Di Benedetto et al., 2010). However, deletion of *Cadh11* does not result in a major skeletal defect in adult mice (Di Benedetto et al., 2010) and causes only modest osteopenia in young mice (Castro et al., 2004), presumably due to compensation by Cadh2. Indeed, ablation of one *Cadh2* allele in *Cadh11*-null mice resulted in severe osteopenia associated with reduced bone strength (Di Benedetto et al., 2010). Consistent with a role for Cadh2 in osteogenesis, overexpression of a dominant-negative *Cadh2* mutant in mature osteoblasts using the osteocalcin promoter resulted in reduced osteoblast differentiation, bone formation, and bone mass (Castro et al., 2004). However, *Cadh2* haploinsufficiency, or conditional deletion of *Cadh2*, in osteoblasts led to increased differentiation of bone marrow stromal cells in vitro, suggesting that Cadh2 may inhibit terminal osteoblast differentiation. Thus, although Cadh2 and Cadh11 contribute to early stages of osteoblast differentiation, Cadh2, but not Cadh11, downregulation is required for terminal differentiation.

How do cadherins control osteoblast differentiation and fate? As they behave as adhesive molecules, cadherins can control osteoblast precursor cell fate in part through cell—cell adhesion (Haÿ et al., 2000). Blocking Cadh2-mediated cell—cell adhesion using specific peptides or antibody reduces osteoblast differentiation in vitro (Ferrari et al., 2000; Haÿ et al., 2000). However, overexpression of *Cadh2* in osteoblasts driven by the Col1A1 promoter reduced osteoblast activity and peak bone mass, indicating that Cadh2 controls bone formation through other mechanisms than cell—cell adhesion (Haÿ et al., 2009a). Cadherins are known modulators of intracellular signaling related to Wnt signaling (Heuberger and Birchmeier, 2010). Cadherins can bind to β-catenin, which leads to increased stabilization of the adhesion structure. Conversely, the release of β-catenin from cadherins destabilizes the adhesion complex and allows cell mobility (Gottardi and Gumbiner, 2001). In addition, cell—cell adhesion mediated by cadherins promotes β-catenin phosphorylation and inactivation of Wnt/β-catenin signaling (Lilien and Balsamo, 2005). Furthermore, cadherins can sequestrate β-catenin at the cell membrane, resulting in decreased β-catenin pool availability for nuclear translocation (Gottardi and Gumbiner, 2001). In osteoblasts, several interactions between cadherins and Wnt signaling molecules were shown to control osteogenic differentiation (Lilien and Balsamo, 2005). The decreased β-catenin abundance at cell—cell contacts induced by deletion of *Cadh2* or *Cadh11* led to decreased cell—cell adhesion with a negative effect on osteoblastogenesis (Di Benedetto et al., 2010). In addition to this mechanism, Cadh2 can bind to low-density lipoprotein receptor—related protein (LRP) 5 or LRP6 via Axin, leading to LRP5/6 sequestration at the cell membrane, reduced LRP5/6 availability, and decreased Wnt/β-catenin signaling (Haÿ et al., 2009a). Cadh2/Axin/LRP5/6 interaction in osteoblasts also reduces ERK1/2 and PI3K/AKT signaling, resulting in decreased osteoblast proliferation, differentiation, and survival (Marie et al., 2014a). Furthermore, Cadh2/LRP5/6 interaction in osteogenic cells causes decreased endogenous Wnt3a expression, which contributes to reduced Wnt signaling and osteoblastogenesis (Marie et al., 2014a). The resulting effect of the negative interaction of Cadh2 overexpression on Wnt signaling in osteogenic cells is to decrease bone formation and delay peak bone mass in young mice (Haÿ et al., 2009a). Consistent with the finding that Cadh2 negatively regulates Wnt/β-catenin signaling in osteoblasts (Marie et al., 2014a; Haÿ et al., 2009a), Cadh2 was found to restrain the bone anabolic action of intermittent parathyroid hormone (iPTH). In vitro, the ablation of *Cadh2* in osteogenic cells results in increased LRP6/PTHR1 interaction and enhanced iPTH-induced protein kinase-dependent Wnt/β-catenin signaling. In mice, conditional *Cadh2* deletion in osteoprogenitor cells led to a greater than normal osteoblast activity and bone mass in response to iPTH, indicating that Cadh2—LRP6 interaction restrains PTH-induced β-catenin signaling (Haÿ et al., 2009b). Consistently, overexpression of *Cadh2* blunted the suppressive effect of PTH on sclerostin/SOST expression in vitro and in vivo, further indicating that Cadh2 expression influences the anabolic effect of iPTH in mice (Revollo et al., 2015). However, the influence of Cadh2 on Wnt signaling and osteogenic cell commitment and differentiation varies with aging. While Cadh2 overexpression in osteoblasts decreased osteoblast differentiation and increased bone marrow adipocyte differentiation in young mice, this phenotype was fully reversed with aging (Yang et al., 2016), which is consistent with the downregulation of Cadh2 during osteoblast maturation. This phenotype was linked to reversal with age of endogenous Wnt5a and Wnt10b signals that are key factors controlling osteogenic cell lineage commitment (Yang et al., 2016). In addition to aging, the negative effect of Cadh2 on Wnt/β-catenin signaling depends on the osteogenic cell differentiation stage. Conditional deletion of *Cadh2* in osteoprogenitors at embryonic and perinatal age was detrimental to bone accrual, whereas loss of *Cadh2* in osteolineage cells in adult mice favored bone formation (Haÿ et al., 2014), indicating that Wnt/β-catenin

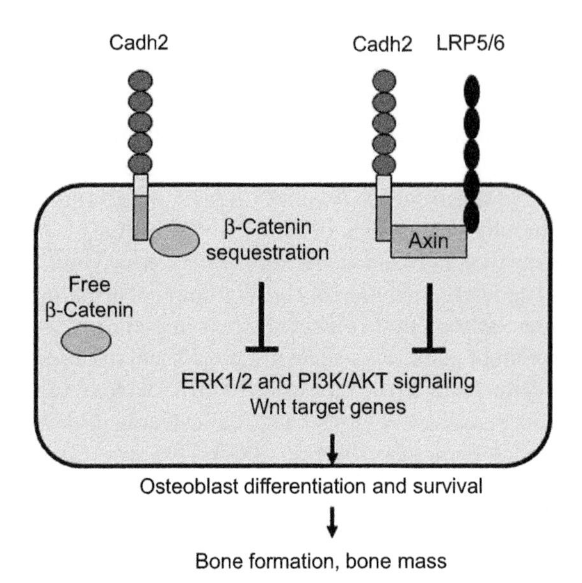

FIGURE 17.3 Role of cadherins in osteoblasts. Cadherin 2 (*Cadh2*) can interact with β-catenin at the cell surface, resulting in β-catenin sequestration, decreased free β-catenin, and reduced Wnt signaling. Cadh2 can also interact with the Wnt coreceptors low-density lipoprotein receptor–related protein 5 (*LRP5*) and *LRP6*, resulting in reduction of Wnt/β-catenin signaling and other signaling pathways that control osteoblast function and survival, bone formation, and bone mass (see text for more details). *ERK1/2*, extracellular signal-regulated kinase 1/2; *PI3K/AKT*, phosphatidylinositol 3-kinase/protein kinase B.

signaling, bone formation, and bone mass are directly influenced by the level of expression of Cadh2 at all stages of the osteoblast lineage. Cadherins also modulate Wnt/β-catenin signaling induced in response to mechanotransduction in bone cells. Mechanical stimulation induced by oscillatory fluid flow in osteoblastic cells leads to reduced Cadh2/β-catenin binding at the cell membrane, resulting in β-catenin nuclear translocation and osteogenic differentiation (Fontana et al., 2017). In addition to the impact on Wnt signaling, Cadh2 interacts with PI3K signaling in osteoblastic cells. Cadh2-mediated adherens junctions activate the PI3K signaling cascade in osteoblasts, which contributes to osteoblast differentiation and osteogenesis in the perichondrium (Arnsdorf et al., 2009). Conversely, *Cadh2* overexpression in osteoblasts reduced Wnt-dependent PI3K/AKT signaling, resulting in decreased osteoblast survival in mice (Marie et al., 2014a; Haÿ et al., 2009a). Thus, cadherins control bone formation through multiple mechanisms that are directly or indirectly linked to Wnt/β-catenin signaling (Fig. 17.3).

Based on the finding that Cadh2 interacts with LRP5/6 to reduce β-catenin/Wnt signaling and osteoblast function (Haÿ et al., 2009a), therapeutic approaches targeting Cadh2/LRP5/6 interaction were developed for promoting bone formation (Guntur et al., 2012). In vitro, deletion of the Cadh2 domain interacting with Axin and LRP5/6 leads to promotion of Wnt/β-catenin signaling and osteoblast differentiation (Fiorino and Harrison, 2016). Moreover, disruption of the Cadh2/LRP5/6 interaction using a competitor peptide that binds to the Cadh2/Axin-interacting domain of LRP5/6 results in enhanced Wnt/β-catenin signaling and osteoblast function in vitro, and increased calvaria bone formation in mice (Marie and Haÿ, 2013). In senescent osteopenic mice, blocking the Cadh2/LRP5/6 interaction led to increased endogenous Wnt5a and Wnt10b expression, osteogenic differentiation, bone formation, and bone mass (Haÿ et al., 2012), suggesting a therapeutic approach targeting the Cadh2/LRP5/6 interaction for promoting bone formation in the aging skeleton.

Osteoclasts

Osteoclasts have been investigated for cadherin expression, especially to understand the mechanisms underlying the fusion process of mononuclear precursors into polykarya (Mbalaviele et al., 1995). Osteoclasts express Cadh1, while they are negative for Cadh2 and Cadh3. Interestingly, Cadh1 expression peaks at the time of preosteoclast fusion, a process that is largely inhibited by Cadh1-neutralizing antibodies. Cadh1 inhibition blocks migration of osteoclast precursors, which is essential for mononuclear cell clustering and fusion of their plasma membranes (Fiorino and Harrison, 2016). In contrast, neutralization of Cadh1 fails to affect proliferation of precursors or adhesion to substrate, confirming a specific role of this cell adhesion molecule in the maturation of polykarya (Mbalaviele et al., 1995). In osteoclast precursors, Cadh1 is localized in areas of membrane protrusions (Fiorino and Harrison, 2016) that are distributed throughout the entire cell surface, but its expression dramatically declines in mature multinucleated osteoclasts. Cadh1 engagement in osteoclast precursors activates gene expression, and several osteoclast genes are under the control of the Cadh1 pathway. Blocking Cadh1 function, before the preosteoclasts fuse into polykarya, retards the expression of the osteoclast transcription factor NFATc1, the fusion protein dendritic cell-specific transmembrane protein, and the osteoclast enzymes cathepsin K and tartrate-resistant acid phosphatase. Since these genes are essential for osteoclast multinucleation and maturation, these observations highlight the important role of Cadh1 in the late stage of osteoclastogenesis. Conversely, overexpression of *Cadh1* in osteoclast precursors brought forward the activation of NFATc1, which translocates to the nucleus earlier than in control cells (Fiorino and Harrison, 2016). Given that the studies on osteoclast cadherins are scanty, we have no specific clues on the signals induced by these molecules during osteoclastogenesis. However, Cadh1 is implicated in the maturation of a number of other macrophage subtypes. For instance, it is involved in the differentiation and motility of Langerhans cells and regulates the activity of dendritic cells via cooperation of the Wnt/β-catenin signaling (Van den Bossche et al., 2012). Because of the tight interaction of Cadh1 with β-catenin and the role of Wnt/β-catenin signaling in osteoclast formation (Kobayashi et al., 2015), it is likely that these two pathways cooperate for the regulation of osteoclastogenesis as well. Another cadherin isoform identified in osteoclasts is Cadh6/2 (Mbalaviele et al., 1998). Dominant negative or antisense *Cadh6/2* prevents osteoclast precursors from interacting with ST2 cells, known to support osteoclastogenesis, suggesting this cell adhesion molecule is involved in heterotypic interaction between these two cell types, leading to full maturation of the osteoclasts (Mbalaviele et al., 1998). Unfortunately, there are no other data supporting this observation; therefore the underlying molecular mechanisms remain elusive.

Chondrocytes

The most prominent cadherin in chondrocytes is Cadh2. It is strongly expressed during mesenchymal condensation, which allows the formation of the primordial tissue from which limb cartilage originates (Oberlender and Tuan, 1994). This process can be impaired using Cadh2 blocking antibody, which prevents mesenchymal condensation and subsequent chondrogenesis. Cadh2 was upregulated in an in vitro model of chondrocyte differentiation. It was detected in prechondrogenic cells and transiently increased during cell aggregation, disappearing in hypertrophic chondrocytes (Tavella et al., 1994). Cadh2 appears to be essential for cellular condensation (Woodward and Tuan, 1999). Using high-density micromass cultures, Woodward and Tuan (Woodward and Tuan, 1999) induced chick limb mesenchymal cell condensation by ion cross-links promoted by poly-L-lysine. This condition unveiled a time-dependent increase in Cadh2, while its neutralization by a specific antibody inhibited the effect of poly-L-lysine on chondrogenesis. Forced overexpression of *Cadh2* induced precartilage cellular condensation and enhanced chondrogenesis in vitro, with a mechanism that required both extracellular and intracellular domains of Cadh2 (Delise and Tuan, 2002). Cadh2 also appears to be a target for the antichondrogenic effect of retinoic acid. Retinoic acid inhibits the progression of condensed precartilage tissue to cartilage nodules. This progression occurs upon suppression of Cadh2 expression, which is prevented by retinoic acid along with the downregulation of the associated α- and β-catenins. This effect of retinoic acid is blocked by cytochalasin D, a molecule that disrupts the microfilaments implicated in the Cadh2 adhesion function (Cho et al., 2003).

Another important regulator of chondrogenesis that requires Cadh2 is transforming growth factor β (TGFβ). TGFβ induces chondrogenesis through activation of MAPKs, especially p38 and ERK1. These signaling kinases transiently upregulate Cadh2 to allow cellular condensation, followed by its downregulation to induce progression toward chondrogenesis (Tuli et al., 2003). Cellular condensation is also regulated by the small GTP-binding protein RAC1, which triggers upregulation of Cadh2 as well (Woods et al., 2007b). TGFβ also induces the expression of Wnt7. Chondrogenesis is regulated by the Wnt pathway and Wnt7a is a lead chondrogenic signaling t that acts in concert with Cadh2. Using the chick limb mesenchymal cell micromass cultures, Tufan et al. (Tufan et al., 2002) investigated the role of *Chfz-1* and *Chfz-7*, which are members of the Wnt pathway encoded by the Frizzled genes. While CHFZ-1 surrounded the nascent cartilage rudiment, CHFZ-7 was expressed in the area of cell condensation, with progressive downregulation toward the peripheral area. This pattern of expression was similar to the expression of Cadh2, and misexpression of *Chfz-7* impaired chondrogenesis at the early stage of formation of the precartilage aggregates. Notably, *Cadh2* expression was downregulated by *Chfz-7* misexpression, suggesting a functional link between Cadh2 and the Wnt pathway during mesenchymal condensation and subsequent chondrogenesis (Tufan et al., 2002).

To induce chondrogenesis, Cadh2 must be cleaved by ADAM10. Cadh2 cleavage mutants failed to induce cartilage formation, impeding the organization of cartilage aggregates as well as the synthesis of proteoglycans. Furthermore, overexpression of these mutants downregulated type II collagen, aggrecan, and type X collagen (Nakazora et al., 2010), confirming a pivotal role for enzymatic cleavage in the activation of cartilage matrix production. Another factor involved in chondrogenesis that affects Cadh2 is the C-type natriuretic peptide. Treatment of micromass cultures of chick limb mesenchymal cells with this factor upregulated *Cadh2* along with collagen type X and glycosaminoglycans, ending up with enhanced chondrogenesis (Alan and Tufan, 2008). Despite these striking observations, *Cadh2*-null mice have no skeletal alterations. Given the redundant functions of cadherins, it was hypothesized that, in this mouse model, *Cadh2* deficiency is compensated for by the activity of Cadh11 in cartilage development. *Cadh11* is strongly upregulated in response to chondrogenic conditions and is expressed in normal and osteoarthritic articular cartilage (Stokes et al., 2002). In the growth plate, *Cadh11* is expressed in the late hypertrophic zone. Knockdown of *Cadh11* inhibited the formation of calcified nodules in a growth plate–derived chondrocyte cell line (Matsusaki et al., 2006). The function of Cadh11 has been investigated during the differentiation of mesenchymal stem cells. Kii et al. (Kii et al., 2004) observed that teratomas derived from embryonic stem cells transfected with *Cadh11* formed preferentially bone and cartilage. Using tridimensional hanging drop cultures, it was noticed that *Cadh11*-transfected cells formed sheetlike aggregates, as opposed to *Cadh2*-overexpressing cells, which formed spherical structures, suggesting independent and nonoverlapping functions of Cadh11 and Cadh2 in this context (Kii et al., 2004).

Roles of other attachment molecules in bone cells

Syndecans

Cell surface proteoglycans are composed of a membrane-associated core protein to which glycosaminoglycan (heparan sulfate [HS] or chondroitin sulfate) chains are covalently attached. HS chains can bind several proteins, including growth factors, signaling proteins, membrane receptors, and ECM proteins (Bishop et al., 2007; Bernfield et al., 1999), thus allowing regulation of the availability and function of signaling proteins (Mitsou et al., 2017; Billings and Pacifici, 2015). Notably, binding of extracellular ligands on HS chains increases the probability for ligands to interact with their high-affinity receptors (Billings and Pacifici, 2015). In the growth plate, HS proteoglycans (HSPGs) bind members of the hedgehog, BMP, fibroblast growth factor (FGF), and Wnt protein families (Huegel et al., 2013). Syndecans are an HSPG family composed of four members (syndecan-1 to -4) (Dews and Mackenzie, 2007). In bone, syndecan-1 is expressed transiently during mesenchymal condensation, syndecan-2 is expressed by mesenchymal cells and persists during bone development and osteoblast differentiation, syndecan-3 is mainly expressed in cartilage, whereas syndecan-4 expression is more ubiquitous (David et al., 1993). In osteoblasts, syndecan-2 is strongly associated with FGF/FGFRs (Molténi et al., 1999a; Song et al., 2007) and granulocyte/macrophage colony-stimulating factor (Modrowski et al., 2000), and contributes to the activity of these ligands. Syndecan-2 expression is regulated by Runx2, Wnts, FGF, and TGFβ in osteoblastic cells (Teplyuk et al., 2009; Worapamorn et al., 2002; Dieudonné et al., 2010). BMP-2 increases the synthesis of syndecans (Gutierrez et al., 2006) and interacts with HS complexes to regulate osteoblast differentiation in mesenchymal stromal cells (Manton et al., 2007). In osteogenic cells, syndecan-2 acts as a coreceptor for FGFRs, which is essential for the response to FGF2 (Molténi et al., 1999b; Steinfeld et al., 1996). FGF binding to HS chains results in growth factor dimerization and formation of a tertiary complex with FGFR (Matsuo and Kimura-Yoshida, 2013). Syndecans also interact with Wnt molecules via HS chains to modulate Wnt signaling (Baeg et al., 2001). This leads to the modulation of Wnt molecule concentration at the cell surface, which stabilizes the signaling activity (Fuerer et al., 2010). In bone, syndecan-2 controls the extracellular availability of Wnt effectors and modulates intracellular signals linked to Wnt signaling (Mansouri et al., 2015). Transgenic mice overexpressing *syndecan-2* in osteoblasts showed decreased osteogenesis associated with increased mesenchymal osteoprogenitor cell apoptosis. This phenotype results from inhibition of Wnt/β-catenin signaling and decreased production of Wnt ligands, supporting a role of syndecan-2/Wnt signaling interaction in the control of osteoblastogenesis in vivo (Mansouri et al., 2017). In addition to interacting with signaling factors that regulate osteoblasts, syndecans interact with fibronectin to facilitate cell adhesion through fibronectin–transglutaminase complexes induced by syndecan-4-dependent activation of protein kinase Cα (Wang et al., 2010). This complex supports osteoblast adhesion and rescues from cell death by anoikis in a syndecan- and β1 integrin-dependent manner (Wang et al., 2011). Thus, syndecans control osteoblast adhesion and response to exogenous factors by interacting with both ECM and signaling factors, resulting in the modulation of intracellular factors controlling cell fate.

Glypicans and perlecan

Glypicans are other cell surface proteoglycans expressed in bone. The glycosaminoglycan-bearing perlecan domain I interacts with ligands such as BMP, FGF, hedgehog, and Wnt proteins (Dwivedi et al., 2013) and thereby supports early chondrogenesis (Farach-Carson et al., 2008). In vitro, glypican-3 is involved in osteogenic commitment, as reduced *glypican-3* expression leads to decreased *Runx2* expression and osteoblast differentiation in murine osteoblastic cells

(Haupt et al., 2009). In vivo, *glypican*-KO mice showed decreased trabecular bone mass and delayed endochondral ossification due to reduced osteoclastogenesis, indicating that this HSPG plays a role in bone growth (Viviano et al., 2005). Another HSPG, perlecan, interacts with the ECM, growth factors, and receptors and influences cellular signaling (Whitelock et al., 2008). In bone, perlecan is expressed in cartilage and strongly potentiates chondrogenic differentiation in vitro (French et al., 1999). Perlecan promotes chondrocyte attachment to the matrix (SundarRaj et al., 1995) and is involved in chondrogenic differentiation in vitro (Gomes Jr et al., 2006; French et al., 2002). In vivo, *perlecan*-KO mice showed defective endochondral ossification due to decreased proliferation of chondrocytes and reduced prehypertrophic zone (Arikawa-Hirasawa et al., 1999). Consistent with a role of perlecan in growth plate development, reduced perlecan secretion resulted in achondroplasia in mice (Rodgers et al., 2007). Mechanistically, perlecan binds to FGF2 by its HS chains and enhances FGF2 binding to FGFR-1 and FGFR-3 receptors in the growth plate (Smith et al., 2007). Perlecan may also be involved in the control of osteoblastogenesis and bone formation (Lowe et al., 2014). In the bone marrow, cell-derived ECM that contains perlecan, among other matrix molecules, preserves the ability of mesenchymal stromal cells to differentiate into osteoblasts or adipocytes (Chen et al., 2007). In addition, exogenous perlecan suppresses adipogenic differentiation and promotes osteogenic differentiation of mesenchymal stem cells in vitro (Nakamura et al., 2014). This may be due in part to enhanced interaction with BMP-2, leading to increased BMP-2 bioactivity (Decarlo et al., 2012). At a later stage of osteoblast differentiation, perlecan is localized in the pericellular space of osteocytes in the lacunocanalicular system in cortical bone (Thompson et al., 2011), where it regulates solute transport and mechanosensing within the osteocyte lacunar–canalicular system (Wang et al., 2014). These studies support the notion that, in addition to being involved in cell attachment to the ECM, HSPGs control bone cell functions by interacting with signaling proteins involved in the response to extracellular signals.

CD44

CD44 is a cellular surface adhesion molecule involved in various processes. CD44 binds to hyaluronan (HA), osteopontin, fibronectin, and collagen type I and thereby may regulate bone cell function (Goodison et al., 1999). As noted earlier, CD44 is involved in chondrocyte survival (Nicoll et al., 2002; Knudson, 2003). Osteoclasts express CD44 (Nakamura et al., 1995; Suzuki et al., 2002), and the interaction of CD44 with HA or osteopontin induces intracellular signals in preosteoclasts, leading to osteoclast formation (Spessotto et al., 2002; Chellaiah et al., 2003; Chellaiah and Hruska, 2003). In vitro, osteopontin signals through calcium and NFATc in osteoclasts (Tanabe et al., 2011). Consistently, *CD44* deficiency led to inhibition of osteoclast activity and function by downregulating NF-κB/NFATc1-mediated signaling (Li et al., 2015). Moreover, receptor activator of NF-κB ligand (RANKL) induces *CD44* expression, and CD44 promotes the activation of RANKL–RANK–NF-κB-mediated signaling during osteoclastogenesis (Li et al., 2015). In vivo, *CD44*-KO mice showed normal trabecular bone volume but increased cortical thickness, suggesting a site-specific effect of *CD44* deficiency (Cao et al., 2005). Accordingly, the reduced osteoclastogenesis and osteoclast function induced by *CD44* deficiency counteracts the cortical, but not trabecular, bone loss induced by hindlimb unloading in mice (Li et al., 2015). In contrast to osteoclasts, a role of CD44 in osteoblast function is not firmly established. In vitro, galectin-9 binding to CD44 was reported to induce the formation of a CD44/BMP receptor complex, leading to Smad1/5/8 phosphorylation and osteoblast differentiation (Tanikawa et al., 2010). However, *CD44* deficiency inhibited osteoclast but not osteoblast function in hindlimb-unloading-induced bone loss in mice (Li et al., 2015), suggesting a role of CD44 in bone resorption rather than in bone formation.

Immunoglobulin superfamily members

Neural cell adhesion molecule**,** a member of the immunoglobulin superfamily, is a cell surface molecule expressed transiently during osteoblast lineage (Haÿ et al., 2000; Tanikawa et al., 2010), and its expression is associated with the osteogenic phenotype (Rundus et al., 1998). Activated leukocyte cell adhesion molecule (ALCAM or CD166) is another immunoglobulin member expressed by osteoblasts. *CD166*-deficient mice show increased osteoblast differentiation and bone formation with no change in bone resorption, suggesting that CD166 regulates bone formation (Hooker et al., 2015). Osteoblasts also express intercellular adhesion molecule (ICAM-1) and vascular cell adhesion molecule 1 (VCAM-1), and the cross talk of these molecules induces interleukin-6 (IL-6) secretion by osteoblasts, suggesting that these adhesion molecules transduce activation signals that induce the production of bone-resorbing cytokines (Tanaka et al., 1995). In addition, ICAM-1 and VCAM-1 mediate cell–cell adhesion between osteoclastic precursors and bone marrow stromal cells or osteoblasts, which controls osteoclastogenesis. Neutralization of VCAM-1 in bone marrow stromal cells inhibits the formation of osteoclasts in vitro, indicating that VCAM-1 expression by stromal cells is required for osteoclastogenesis

(Feuerbach and Feyen, 1997). A fraction of osteoblasts that highly express ICAM-1 strongly adhere to osteoclast precursors, resulting in multinuclear osteoclast-like cell formation, indicating that a subpopulation of ICAM-1-expressing osteoblasts controls osteoclastogenesis (Tanaka et al., 2000). Consistent with a role of ICAM-1 in osteoclastogenesis, ICAM-1-mediated cell-to-cell adhesion of osteoblasts and osteoclast precursors was involved in RANKL-dependent osteoclast maturation stimulated by 1,25-dihydroxyvitamin D, PTH, and IL-1α (Okada et al., 2002).

Osteoactivin

Osteoactivin is a glycoprotein expressed by osteoblasts and its expression upregulates osteoblast differentiation in vitro (Abdelmagid et al., 2008). Mice with a loss-of-function mutation in *Gpnmb*, which encodes osteoactivin, showed decreased trabecular bone mass due to reduced osteoblast differentiation, confirming the positive role of osteoactivin in bone formation (Abdelmagid et al., 2014). Osteoactivin is also expressed in osteoclasts and acts as a negative regulator of osteoclast differentiation and survival, but not function (Abdelmagid et al., 2015). Consistent with these findings, transgenic mice overexpressing *osteoactivin* under the cytomegalovirus promoter showed increased trabecular bone mass and bone formation, and decreased bone resorption (Frara et al., 2016). Mechanistically, osteoactivin can bind to CD44 in osteoclasts, leading to inhibition of ERK phosphorylation and RANKL-induced osteolysis (Sondag et al., 2016). In murine osteoblastic cells, recombinant osteoactivin stimulates cell adhesion and spreading through its binding to $\alpha V\beta 1$ integrin and HSPGs at the cell surface. This interaction results in FAK and ERK activation and osteoblast differentiation, suggesting a mechanism by which osteoactivin may control osteoblast differentiation (Moussa et al., 2014).

Chondroadherin

Chondroadherin belongs to the family of leucine-rich repeat proteins and was identified in the cartilage matrix, where it promotes attachment of chondrocytes (Larsson et al., 1991). Chondroadherin localizes near the cell surface and is highly expressed in the proliferative and hypertrophic zones of the growth plate. It is a small molecule of 38 kDa molecular weight, containing 359 amino acids. Relevant domains of chondroadherin are a putative signal peptide, a cysteine-rich region at the N-terminal tail, 11 leucine-rich repeats, and a double cysteine loop in the C-terminal tail. Among the members of the leucine-rich repeat protein family, chondroadherin exhibits the unique feature of no posttranslational glycosylation (Neam et al., 1994). The very C terminus of the protein includes a heparin-binding consensus sequence that allows its interaction with heparin (Haglund et al., 2013). This heparin-binding domain recognizes cell surface syndecans and triggers ERK1/2 phosphorylation (Haglund et al., 2013). Chondroadherin is also recognized by the $\alpha 2\beta 1$ integrin expressed by chondrocytes (Camper et al., 1997). The $\alpha 2\beta 1$ integrin binding site was identified in the region carrying the amino acid residues 306−318 at the C terminus of the protein. By affinity purification procedure, $\alpha 2\beta 1$ integrin was confirmed to bind the chondroadherin CQLRGLRRWLEAK^{318} peptide. A longer chondroadherin peptide, spanning amino acid residues 306−326 (CQLRGLRRWEKLAASRPDATC^{326}) was made cyclic and stable through a disulfide bond occurring between the two terminal cysteines and was largely used to investigate the functional role of chondroadherin in vitro and in vivo. The peptide was confirmed to induce cell adhesion and spreading in an $\alpha 2\beta 1$ integrin-dependent manner, activating ERK1/2 phosphorylation (Haglund et al., 2011).

Chondroadherin is expressed also by osteoblasts. It was found to be 50% less expressed in bone biopsies of relatively young female osteoporotic patients (ages between 50 and 65 years), and in ovariectomized mice, a model of estrogen deficiency-induced osteoporosis (Capulli et al., 2014). The cyclic CQLRGLRRWEKLAASRPDATC^{326} peptide was inactive on osteoblasts, but strongly impaired osteoclastogenesis at the late stage of the process. Specifically, the major effect was exerted by the cyclic peptide on migration of osteoclast precursors, which is mandatory for cell clustering and fusion into mature polykarya. The underlying molecular mechanism involved the decreased expression of migfilin and vasodilator-stimulated phosphoprotein (VASP) (Capulli et al., 2014). Migfilin is associated with adhesion sites and binds filamins, VASP, kindling-2 and the transcription factor CSX/NKX2-5, recruiting acting cytoskeleton and promoting cell adhesion, shape modulation, motility and gene expression (Tu et al., 2003). *Migfilin* inactivation in mice induced a severe osteopenic phenotype (Xiao et al., 2012). However, this effect was mostly due to reduced osteoblast differentiation with a parallel increase in the proosteoclastogenic cytokine RANKL, which exacerbated osteoclast differentiation. VASP is known to induce monomeric actin recruitment to the barbed end of microfilaments, preventing capping and regulating filament bundling (Krause et al., 2003). It is implicated in cell motility, adhesion, and sensory capacity, localizing in the tips of filopodia and in adhesion structures (Tokuo and Ikebe, 2004). In osteoclasts, VASP is associated with the $\alpha V\beta 3$ integrin and is activated by NO, which promotes osteoclast motility. Consistently, knockdown of *Vasp* reduced osteoclast migration on substrate (Yaroslavskiy et al., 2005). Interestingly, cyclic chondroadherin downregulated NO synthase (*Nos2*) in osteoclasts, and treatment with an α2β1 integrin blocking antibody increased osteoclast *Nos2* expression, suggesting an inhibitory role of the α2β1 integrin on *Nos2* triggered by chondroadherin (Capulli et al., 2014). Overall, the results obtained using the cyclic α2β1 integrin-binding domain of chondroadherin suggest that this ECM component affects osteoclastogenesis via transcriptional downregulation of *Nos2* and decreased NO, which regulates migfilin and *Vasp* expression and is required for preosteoclast migration, clustering, and fusion into multinucleated osteoclasts.

Cyclic chondroadherin peptide showed also the ability to impair osteoclastogenesis in vivo. Mice injected with cyclic chondroadherin showed enhanced bone mass and reduced osteoclast variables, while osteoblasts and bone formation were not affected (Capulli et al., 2014). This effect was evident both in normal mice and in mice subjected to ovariectomy. In this circumstance, cyclic chondroadherin blocked the increase in serum level of bone resorption biomarkers and prevented bone loss induced by estrogen deficiency, with an improvement of bone quality. These effects were observed in young and old mice and were mimicked by treatment with the NOS2 activity inhibitor (L-N^6-(1-iminoethyl)lysine dihydrochloride). Furthermore, cyclic chondroadherin was effective not only in preventative treatment started at the time of ovariectomy, but also in a curative setting, with the treatment started 5 weeks after ovariectomy, a time at which the osteoporotic phenotype was overt (Capulli et al., 2014). Given the positive response of the bone to treatment with cyclic chondroadherin, the peptide was also tested against bone metastasis-induced osteolysis, which is caused by exacerbated osteoclast activity (Rucci et al., 2015). The results of this study complemented the observations on ovariectomized mice, showing a beneficial effect of the peptide in mice injected intracardiacally or orthotopically with osteotropic breast cancer cells. The peptide reduced the in vitro motility of tumor cells and in vivo tumor growth. It also inhibited the process of tumor-induced osteoclast formation, with consequent reduction of bone resorption, development of osteolytic lesions, and cachexia. Interestingly, cyclic chondroadherin synergistically enhanced the antitumoral effect of the chemotherapeutic doxorubicin, which achieved maximal efficacy at half of the effective dose when administered alone (Rucci et al., 2015). The cyclic chondroadherin−doxorubicin treatment affected tumor cells also in vitro, confirming a synergistic impairment of cell motility at lower doses. Taken together, these results demonstrated that chondroadherin is an important regulator of cell migration, exerting this effect by binding to the α2β1 integrin and impairing the late stage of osteoclast formation and the development of metastatic osteolysis.

Conclusion

Multiple in vitro and in vivo studies have shown that integrins, cadherins, and several other adhesion molecules control bone cell function and fate during chondrogenesis, osteoblastogenesis, and osteoclastogenesis. Genetic studies in mice confirmed the importance of some of these adhesion molecules in the control of bone resorption and formation in vivo. These effects are mediated by complex interactions of these adhesion molecules with bone matrix proteins or cell surface molecules, leading to the modulation of intracellular signaling pathways controlling bone cell differentiation, function, and survival. Studies have revealed that the signaling pathways mediated by integrins, cadherins, and other cell adhesion molecules can cross talk with Wnt/β-catenin signaling to regulate osteogenic differentiation and mechanotransduction, and with other signaling mechanisms to control osteoclastogenesis. These advances led to a more comprehensive view of the role of these adhesion molecules in the signaling mechanisms controlling bone cell recruitment and function. Future studies will have to confirm that targeting specific adhesion molecules and their downstream signals may have potential therapeutic implications in reducing bone resorption or promoting bone formation in skeletal disorders.

Acknowledgments

The authors thank all collaborators who contributed to the work reviewed in this chapter. This work was supported by the Institut National de la Recherche Médicale (Inserm), the Agence Nationale de la Recherche, and the European Commission FP6 and FP7 programs (P.J.M.), and by the Telethon, the Italian Association of Cancer Research, the PRIN-MIUR, and the European Commission FP6, FP7, and H2020 programs (A.T.).

References

Abdelmagid, S.M., Barbe, M.F., Rico, M.C., Salihoglu, S., Arango-Hisijara, I., Selim, A.H., et al., 2008. Osteoactivin, an anabolic factor that regulates osteoblast differentiation and function. Exp. Cell Res. 314, 2334−2351.

Abdelmagid, S.M., Belcher, J.Y., Moussa, F.M., Lababidi, S.L., Sondag, G.R., Novak, K.M., et al., 2014. Mutation in osteoactivin decreases bone formation in vivo and osteoblast differentiation in vitro. Am. J. Pathol. 184, 697−713.

Abdelmagid, S.M., Sondag, G.R., Moussa, F.M., Belcher, J.Y., Yu, B., Stinnett, H., et al., 2015. Mutation in osteoactivin promotes receptor activator of NFκB ligand (RANKL)-mediated osteoclast differentiation and survival but inhibits osteoclast function. J. Biol. Chem. 290, 20128−20146.

Akisaka, T., Yoshida, H., Inoue, S., Shimizu, 2001. Organization of cytoskeletal F-actin, G-actin, and gelsolin in the adhesion structures in cultured osteoclast. J. Bone Miner. Res. 16, 1248–1255.

Alan, T., Tufan, A.C., 2008. C-type natriuretic peptide regulation of limb mesenchymal chondrogenesis is accompanied by altered N-cadherin and collagen type X-related functions. J. Cell. Biochem. 105, 227–235.

Arikawa-Hirasawa, E., Watanabe, H., Takami, H., Hassell, J.R., Yamada, Y., 1999. Perlecan is essential for cartilage and cephalic development. Nat. Genet. 23, 354–358.

Arnsdorf, E.J., Tummala, P., Jacobs, C.R., 2009. Non-canonical Wnt signaling and N-cadherin related beta-catenin signaling play a role in mechanically induced osteogenic cell fate. PLoS One 4, e5388.

Aszodi, A., Hunziker, E.B., Brakebusch, C., Fässler, R., 2003. Beta1 integrins regulate chondrocyte rotation, G1 progression, and cytokinesis. Genes Dev. 17, 2465–2479.

Auzzas, L., Zanardi, F., Battistini, L., Burreddu, P., Carta, P., Rassu, G., et al., 2010. Targeting alphavbeta3 integrin: design and applications of mono- and multifunctional RGD-based peptides and semipeptides. Curr. Med. Chem. 17, 1255–1299.

Baeg, G.H., Lin, X., Khare, N., Baumgartner, S., Perrimon, N., 2001. Heparan sulfate proteoglycans are critical for the organization of the extracellular distribution of Wingless. Development 128, 87–94.

Batra, N., Burra, S., Siller-Jackson, A.J., Gu, S., Xia, X., Weber, G.F., et al., 2012. Mechanical stress-activated integrin $\alpha5\beta1$ induces opening of connexin 43 hemichannels. Proc. Natl. Acad. Sci. U. S. A. 109, 3359–3364.

Bengtsson, T., Aszodi, A., Nicolae, C., Hunziker, E.B., Lundgren-Akerlund, E., Fässler, R., 2005. Loss of alpha10beta1 integrin expression leads to moderate dysfunction of growth plate chondrocytes. J. Cell Sci. 118, 929–936.

Bennett, J.H., Moffatt, S., Horton, M., 2001. Cell adhesion molecules in human osteoblasts: structure and function. Histol. Histopathol. 16, 603–611.

Bernfield, M., Götte, M., Park, P.W., Reizes, O., Fitzgerald, M.L., Lincecum, J., et al., 1999. Functions of cell surface heparan sulfate proteoglycans. Annu. Rev. Biochem. 68, 729–777.

Bikle, D.D., 2008. Integrins, insulin like growth factors, and the skeletal response to load. Osteoporos. Int. 19, 1237–1246.

Billings, P.C., Pacifici, M., 2015. Interactions of signaling proteins, growth factors and other proteins with heparan sulfate: mechanisms and mysteries. Connect. Tissue Res. 56, 272–280.

Bishop, J.R., Schuksz, M., Esko, J.D., 2007. Heparan sulphate proteoglycans fine-tune mammalian physiology. Nature 446, 1030–1037.

Bonewald, L.F., Johnson, M.L., 2008. Osteocytes, mechanosensing and Wnt signaling. Bone 42, 606–615.

Bonnet, N., Conway, S.J., Ferrari, S.L., 2012. Regulation of beta catenin signaling and parathyroid hormone anabolic effects in bone by the matricellular protein periostin. Proc. Natl. Acad. Sci. U. S. A. 109, 15048–15053.

Bouvard, D., Aszodi, A., Kostka, G., Block, M.R., Albigès-Rizo, C., Fässler, R., 2007. Defective osteoblast function in ICAP-1-deficient mice. Development 134, 2615–2625.

Brunner, M., Millon-Frémillon, A., Chevalier, G., Nakchbandi, I.A., Mosher, D., Block, M.R., et al., 2011. Osteoblast mineralization requires beta1 integrin/ICAP-1-dependent fibronectin deposition. J. Cell Biol. 194, 307–322.

Bruzzaniti, A., Neff, L., Sanjay, A., Horne, W.C., De Camilli, P., Baron, R., 2005. Dynamin forms a Src kinase-sensitive complex with Cbl and regulates podosomes and osteoclast activity. Mol. Biol. Cell 16, 3301–3313.

Calle, Y., Burns, S., Thrasher, A.J., Jones, G.E., 2006. The leukocyte podosome. Eur. J. Cell Biol. 85, 151–157.

Campbell, I.D., Humphries, M.J., 2011. Integrin structure, activation, and interactions. Cold Spring Harb Perspect Biol 3 (3).

Camper, L., Heinegard, D., Lundgren-Åkerlund, E., 1997. Integrin alpha2beta1 is a receptor for the cartilage matrix protein chondroadherin. J. Cell Biol. 138, 1159–1167.

Cao, J.J., Singleton, P.A., Majumdar, S., Boudignon, B., Burghardt, A., Kurimoto, P., et al., 2005. Hyaluronan increases RANKL expression in bone marrow stromal cells through CD44. J. Bone Miner. Res. 20, 30–40.

Cappariello, A., Maurizi, A., Veeriah, V., Teti, A., 2014. The great beauty of the osteoclast. Arch. Biochem. Biophys. 558, 70–78.

Capulli, M., Olstad, O.K., Önnerfjord, P., Tillgren, V., Muraca, M., Gautvik, K.M., et al., 2014. The C-terminal domain of chondroadherin: a new regulator of osteoclast motility counteracting bone loss. J. Bone Miner. Res. 29, 1833–1846.

Castro, C.H., Shin, C.S., Stains, J.P., Cheng, S.L., Sheikh, S., Mbalaviele, G., et al., 2004. Targeted expression of a dominant-negative N-cadherin in vivo delays peak bone mass and increases adipogenesis. J. Cell Sci. 117, 2853–2864.

Chellaiah, M.A., Hruska, K.A., 2003. The integrin alpha(v)beta(3) and CD44 regulate the actions of osteopontin on osteoclast motility. Calcif. Tissue Int. 72, 197–205.

Chellaiah, M.A., Biswas, R.S., Yuen, D., Alvarez, U.M., Hruska, K.A., 2001. Phosphatidylinositol 3,4,5-trisphosphate directs association of Src homology 2-containing signaling proteins with gelsolin. J. Biol. Chem. 276, 47434–47444.

Chellaiah, M.A., Kizer, N., Biswas, R., Alvarez, U., Strauss-Schoenberger, J., Rifas, L., et al., 2003. Osteopontin deficiency produces osteoclast dysfunction due to reduced CD44 surface expression. Mol. Biol. Cell 14, 173–189.

Chen, J.C., Jacobs, C.R., 2013. Mechanically induced osteogenic lineage commitment of stem cells. Stem Cell Res. Ther. 4, 107.

Chen, X.D., Dusevich, V., Feng, J.Q., Manolagas, S.C., Jilka, R.L., 2007. Extracellular matrix made by bone marrow cells facilitates expansion of marrow-derived mesenchymal progenitor cells and prevents their differentiation into osteoblasts. J. Bone Miner. Res. 22, 1943–1956.

Chen, Q., Shou, P., Zhang, L., Xu, C., Zheng, C., Han, Y., Li, W., et al., 2014. An osteopontin-integrin interaction plays a critical role in directing adipogenesis and osteogenesis by mesenchymal stem cells. Stem Cell. 32, 327–337.

Cheng, S.L., Lecanda, F., Davidson, M.K., Warlow, P.M., Zhang, S.F., Zhang, L., et al., 1998. Human osteoblasts express a repertoire of cadherins, which are critical for BMP-2-induced osteogenic differentiation. J. Bone Miner. Res. 13, 633–644.

Cheng, S.L., Lai, C.F., Fausto, A., Chellaiah, M., Feng, X., McHugh, K.P., et al., 2000. Regulation of alphaVbeta3 and alphaVbeta5 integrins by dexamethasone in normal human osteoblastic cells. J. Cell. Biochem. 77, 265–276.

Cheng, S.L., Lai, C.F., Blystone, S.D., Avioli, L.V., 2001. Bone mineralization and osteoblast differentiation are negatively modulated by integrin alpha(v)beta3. J. Bone Miner. Res. 16, 277–288.

Cho, S.H., Oh, C.D., Kim, S.J., Kim, I.C., Chun, J.S., 2003. Retinoic acid inhibits chondrogenesis of mesenchymal cells by sustaining expression of N-cadherin and its associated proteins. Cell Biochem 89, 837–847.

Clover, J., Dodds, R.A., Gowen, M., 1992. Integrin subunit expression by human osteoblasts and osteoclasts in situ and in culture. J. Cell Sci. 103, 267–271.

Connelly, J.T., García, A.J., Levenston, M.E., 2007. Inhibition of in vitro chondrogenesis in RGD-modified three-dimensional alginate gels. Biomaterials 28, 1071–1083.

Correia, I., Chu, D., Chou, Y.H., Goldman, R.D., Matsudaira, P., 1999. Integrating the actin and vimentin cytoskeletons. Adhesion dependent formation of fimbrin–vimentin complexes in macrophages. J. Cell Biol. 146, 831–842.

Dai, Z., Guo, F., Wu, F., Xu, H., Yang, C., Li, J., et al., 2014. Integrin αvβ3 mediates the synergetic regulation of core-binding factor α1 transcriptional activity by gravity and insulin-like growth factor-1 through phosphoinositide 3-kinase signaling. Bone 69, 126–132.

Danhier, F., Le Breton, A., Préat, V., 2012. RGD-based strategies to target alpha(v) beta(3) integrin in cancer therapy and diagnosis. Mol. Pharm. 9, 2961–2973.

David, G., Bai, X.M., Van der Schueren, B., Marynen, P., Cassiman, J.J., Van den Berghe, H., 1993. Spatial and temporal changes in the expression of fibroglycan (syndecan-2) during mouse embryonic development. Development 119, 841–854.

Decarlo, A.A., Belousova, M., Ellis, A.L., Petersen, D., Grenett, H., Hardigan, P., et al., 2012. Perlecan domain 1 recombinant proteoglycan augments BMP-2 activity and osteogenesis. BMC Biotechnol. 12, 60.

Dejaeger, M., Böhm, A.M., Dirckx, N., Devriese, J., Nefyodova, E., Cardoen, R., et al., 2017. Integrin-linked kinase regulates bone formation by controlling cytoskeletal organization and modulating BMP and Wnt signaling in osteoprogenitors. J. Bone Miner. Res. 32, 2087–2102.

Delise, A.M., Tuan, R.S., 2002. Analysis of N-cadherin function in limb mesenchymal chondrogenesis in vitro. Dev. Dynam. 225, 195–204.

Desgrosellier, J.S., Cheresh, D.A., 2010. Integrins in cancer: biological implications and therapeutic opportunities. Nat. Rev. Canc. 10, 9–22.

Destaing, O., Sanjay, A., Itzstein, C., Horne, W.C., Toomre, D., De Camilli, P., et al., 2008. The tyrosine kinase activity of c-Src regulates actin dynamics and organization of podosomes in osteoclasts. Mol. Biol. Cell 19, 394–404.

Dews, I.C., Mackenzie, K.R., 2007. Transmembrane domains of the syndecan family of growth factor coreceptors display a hierarchy of homotypic and heterotypic interactions. Proc. Natl. Acad. Sci. U. S. A. 104, 20782–20787.

Di Benedetto, A., Watkins, M., Grimston, S., Salazar, V., Donsante, C., Mbalaviele, G., et al., 2010. N-cadherin and cadherin 11 modulate postnatal bone growth and osteoblast differentiation by distinct mechanisms. J. Cell Sci. 123, 2640–2648.

Dieudonné, F.X., Marion, A., Haÿ, E., Marie, P.J., Modrowski, D., 2010. High Wnt signaling represses the proapoptotic proteoglycan syndecan-2 in osteosarcoma cells. Cancer Res. 70, 5399–5408.

Du, J., Zu, Y., Li, J., Du, S., Xu, Y., Zhang, L., et al., 2016. Extracellular matrix stiffness dictates Wnt expression through integrin pathway. Sci. Rep. 6, 20395.

Dufour, C., Holy, X., Marie, P.J., 2007. Skeletal unloading induces osteoblast apoptosis and targets alpha5beta1-PI3K-Bcl-2 signaling in rat bone. Exp. Cell Res. 313, 394–403.

Duong, L.T., Lakkakorpi, P.T., Nakamura, I., Machwate, M., Nagy, R.M., Rodan, G.A., 1998. PYK2 in osteoclasts is an adhesion kinase, localized in the sealing zone, activated by ligation of alpha(v)beta3 integrin, and phosphorylated by src kinase. J. Clin. Investig. 102, 881–892.

Dwivedi, P.P., Lam, N., Powell, B.C., 2013. Boning up on glypicans-Opportunities for new insights into bone biology. Cell Biochem. Funct. 31, 91–114.

Eddy, R.J., Weidmann, M.D., Sharma, V.P., Condeelis, J.S., 2017. Tumor cell invadopodia: invasive protrusions that orchestrate metastasis. Trends Cell Biol. 27, 595–607.

Enomoto-Iwamoto, M., Iwamoto, M., Nakashima, K., Mukudai, Y., Boettiger, D., Pacifici, M., et al., 1997. Involvement of alpha5beta1 integrin in matrix interactions and proliferation of chondrocytes. J. Bone Miner. Res. 12, 1124–1132.

Evans, J.G., Matsudaira, P., 2006. Structure and dynamics of macrophage podosomes. Eur. J. Cell Biol. 85, 145–149.

Farach-Carson, M.C., Brown, A.J., Lynam, M., Safran, J.B., Carson, D.D., 2008. A novel peptide sequence in perlecan domain IV supports cell adhesion, spreading and FAK activation. Matrix Biol. 27, 150–160.

Ferrari, S.L., Traianedes, K., Thorne, M., Lafage-Proust, M.H., Genever, P., Cecchini, M.G., et al., 2000. A role for N-cadherin in the development of the differentiated osteoblastic phenotype. J. Bone Miner. Res. 15, 198–208.

Feuerbach, D., Feyen, J.H., 1997. Expression of the cell-adhesion molecule VCAM-1 by stromal cells is necessary for osteoclastogenesis. FEBS Lett. 402, 21–24.

Fiorino, C., Harrison, R.E., 2016. E-cadherin is important for cell differentiation during osteoclastogenesis. Bone 86, 106–118.

Fontana, F., Hickman-Brecks, C.L., Salazar, V.S., Revollo, L., Abou-Ezzi, G., Grimston, S.K., et al., 2017. N-cadherin regulation of bone growth and homeostasis is osteolineage stage-specific. J. Bone Miner. Res. 32, 1332–1342.

Frara, N., Abdelmagid, S.M., Sondag, G.R., Moussa, F.M., Yingling, V.R., Owen, T.A., et al., 2016. Transgenic expression of osteoactivin/gpnmb enhances bone formation in vivo and osteoprogenitor differentiation ex vivo. J. Cell. Physiol. 231, 72–83.

French, M.M., Smith, S.E., Akanbi, K., Sanford, T., Hecht, J., Farach-Carson, M.C., et al., 1999. Expression of the heparan sulfate proteoglycan, perlecan, during mouse embryogenesis and perlecan chondrogenic activity in vitro. J. Cell Biol. 145, 1103–1115.

French, M.M., Gomes Jr., R.R., Timpl, R., Höök, M., Czymmek, K., Farach-Carson, M.C., et al., 2002. Chondrogenic activity of the heparan sulfate proteoglycan perlecan maps to the N-terminal domain I. J. Bone Miner. Res. 17, 48–55.

Frisch, S.M., Ruoslahti, E., 1997. Integrins and anoikis. Curr. Opin. Cell Biol. 9, 701–706.

Fromigué, O., Brun, J., Marty, C., Da Nascimento, S., Sonnet, P., Marie, P.J., 2012. Peptide-based activation of alpha5 integrin for promoting osteogenesis. J. Cell. Biochem. 113, 3029–3038.

Fuerer, C., Habib, S.J., Nusse, R., 2010. A study on the interactions between heparan sulfate proteoglycans and Wnt proteins. Dev. Dynam. 239, 184–190.

Garciadiego-Cázares, D., Rosales, C., Katoh, M., Chimal-Monroy, J., 2004. Coordination of chondrocyte differentiation and joint formation by alpha5beta1 integrin in the developing appendicular skeleton. Development 131, 4735–4742.

Ge, C., Yang, Q., Zhao, G., Yu, H., Kirkwood, K.L., Franceschi, R.T., 2012. Interactions between extracellular signal-regulated kinase 1/2 and P38 MAP kinase pathways in the control of RUNX2 phosphorylation and transcriptional activity. J. Bone Miner. Res. 27, 538–551.

Georgess, D., Machuca-Gayet, I., Blangy, A., Jurdic, P., 2014. Podosome organization drives osteoclast-mediated bone resorption. Cell Adhes. Migrat. 8, 191–204.

Gil-Henn, H., Destaing, O., Sims, N.A., Aoki, K., Alles, N., Neff, L., et al., 2007. Defective microtubule-dependent podosome organization in osteoclasts leads to increased bone density in Pyk2(-/-) mice. J. Cell Biol. 178, 1053–1064.

Globus, R.K., Amblard, D., Nishimura, Y., Iwaniec, U.T., Kim, J.B., Almeida, E.A., et al., 2005. Skeletal phenotype of growing transgenic mice that express a function-perturbing form of beta1 integrin in osteoblasts. Calcif. Tissue Int. 76, 39–49.

Gomes Jr., R.R., Joshi, S.S., Farach-Carson, M.C., Carson, D.D., 2006. Ribozyme-mediated perlecan knockdown impairs chondrogenic differentiation of C3H10T1/2 fibroblasts. Differentiation 74, 53–63.

Goodison, S., Urquidi, V., Tarin, D., 1999. CD44 cell adhesion molecules. Mol. Pathol. 52, 189–196.

Goomer, R.S., Maris, T., Amiel, D., 1998. Age-related changes in the expression of cadherin-11, the mesenchyme specific calcium-dependent cell adhesion molecule. Calcif. Tissue Int. 62, 532–537.

Gottardi, C.J., Gumbiner, B.M., 2001. Adhesion signaling: how beta-catenin interacts with its partners. Curr. Biol. 11, R792–R794.

Granot-Attas, S., Elson, A., 2008. Protein tyrosine phosphatases in osteoclast differentiation, adhesion, and bone resorption. Eur. J. Cell Biol. 87, 479–490.

Grashoff, C., Aszódi, A., Sakai, T., Hunziker, E.B., Fässler, R., 2003. Integrin-linked kinase regulates chondrocyte shape and proliferation. EMBO Rep. 4, 432–438.

Greenbaum, A.M., Revollo, L.D., Woloszynek, J.R., Civitelli, R., Link, D.C., 2012. N-cadherin in osteolineage cells is not required for maintenance of hematopoietic stem cells. Blood 120, 295–302.

Grigoriou, V., Shapiro, I.M., Cavalcanti-Adam, E.A., Composto, R.J., Ducheyne, P., Adams, C.S., 2005. Apoptosis and survival of osteoblast-like cells are regulated by surface attachment. J. Biol. Chem. 280, 1733–1739.

Gronthos, S., Simmons, P.J., Graves, S.E., Robey, P.G., 2001. Integrin-mediated interactions between human bone marrow stromal precursor cells and the extracellular matrix. Bone 28, 174–178.

Grzesik, W.J., Robey, P.G., 1994. Bone matrix RGD glycoproteins: immunolocalization and interaction with human primary osteoblastic bone cells in vitro. J. Bone Miner. Res. 9, 487–496.

Grzesik, W.J., 1997. Integrins and bone–cell adhesion and beyond. Arch. Immunol. Ther. Exp. 45, 271–275.

Guan, M., Yao, W., Liu, R., Lam, K.S., Nolta, J., Jia, J., et al., 2012. Directing mesenchymal stem cells to bone to augment bone formation and increase bone mass. Nat. Med. 18, 456–462.

Gumbiner, B.M., 2005. Regulation of cadherin-mediated adhesion in morphogenesis. Nat. Rev. Mol. Cell Biol. 6, 622–634.

Guntur, A.R., Rosen, C.J., Naski, M.C., 2012. N-cadherin adherens junctions mediate osteogenesis through PI3K signaling. Bone 50, 54–62.

Gutierrez, J., Osses, N., Brandan, E., 2006. Changes in secreted and cell associated proteoglycan synthesis during conversion of myoblasts to osteoblasts in response to bone morphogenetic protein-2: role of decorin in cell response to BMP-2. J. Cell. Physiol. 206, 58–67.

Haglund, L., Tillgren, V., Addis, L., Wenglén, C., Recklies, A., Heinegård, D., 2011. Identification and characterization of the integrin $\alpha2\beta1$ binding motif in chondroadherin mediating cell attachment. J. Biol. Chem. 286, 3925–3934.

Haglund, L., Tillgren, V., Önnerfjord, P., Heinegård, D., 2013. The C-terminal peptide of chondroadherin modulate cellular activity by selectively binding to heparan sulfate chains. J. Biol. Chem. 288, 995–1008.

Hamidouche, Z., Fromigué, O., Ringe, J., Häupl, T., Vaudin, P., Pagès, J.C., et al., 2009. Priming integrin alpha5 promotes human mesenchymal stromal cell osteoblast differentiation and osteogenesis. Proc. Natl. Acad. Sci. U. S. A. 106, 18587–18591.

Hamidouche, Z., Fromigué, O., Ringe, J., Häupl, T., Marie, P.J., 2010. Crosstalks between integrin alpha 5 and IGF2/IGFBP2 signalling trigger human bone marrow-derived mesenchymal stromal osteogenic differentiation. BMC Cell Biol. 11, 44.

Haupt, L.M., Murali, S., Mun, F.K., Teplyuk, N., Mei, L.F., Stein, G.S., et al., 2009. The heparan sulfate proteoglycan (HSPG) glypican-3 mediates commitment of MC3T3-E1 cells toward osteogenesis. J. Cell. Physiol. 220, 780–791.

Haÿ, E., Lemonnier, J., Modrowski, D., Lomri, A., Lasmoles, F., Marie, P.J., 2000. N- and E-cadherin mediate early human calvaria osteoblast differentiation promoted by bone morphogenetic protein-2. J. Cell. Physiol. 183, 117–128.

Haÿ, E., Laplantine, E., Geoffroy, V., Frain, M., Kohler, T., Muller, R., et al., 2009a. N-cadherin interacts with axin and LRP5 to negatively regulate Wnt/beta-catenin signaling, osteoblast function, and bone formation. Mol. Cell Biol. 29, 953–964.

Haÿ, E., Nouraud, A., Marie, P.J., 2009b. N-cadherin negatively regulates osteoblast proliferation and survival by antagonizing Wnt, ERK and PI3K/Akt signalling. PLoS One 4, e8284.

Haÿ, E., Buczkowski, T., Marty, C., Da Nascimento, S., Sonnet, P., Marie, P.J., 2012. Peptide-based mediated disruption of N-cadherin-LRP5/6 interaction promotes Wnt signaling and bone formation. J. Bone Miner. Res. 27, 1852–1863.

Haÿ, E., Dieudonné, F.X., Saidak, Z., Marty, C., Brun, J., Da Nascimento, S., et al., 2014. N-cadherin/Wnt interaction controls bone marrow mesenchymal cell fate and bone mass during aging. J. Cell. Physiol. 229, 1765–1767.

Heuberger, J., Birchmeier, W., 2010. Interplay of cadherin-mediated cell adhesion and canonical Wnt signaling. Cold Spring Harb. Perspect. Biol. 2, a002915.

Hirsch, M.S., Lunsford, L.E., Trinkaus-Randall, V., Svoboda, K.K., 1997. Chondrocyte survival and differentiation in situ are integrin mediated. Dev. Dynam. 210, 249–263.

Hooker, R.A., Chitteti, B.R., Egan, P.H., Cheng, Y.H., Himes, E.R., Meijome, T., et al., 2015. Activated leukocyte cell adhesion molecule (ALCAM or CD166) modulates bone phenotype and hematopoiesis. J. Musculoskelet. Neuronal Interact. 15, 83–94.

Horne, W.C., Sanjay, A., Bruzzaniti, A., Baron, R., 2005. The role(s) of Src kinase and Cbl proteins in the regulation of osteoclast differentiation and function. Immunol. Rev. 208, 106–125.

Horton, M.A., 2001. Integrin antagonists as inhibitors of bone resorption: implications for treatment. Proc. Nutr. Soc. 60, 275–281.

Hsu, A.R., Veeravagu, A., Cai, W., Hou, L.C., Tse, V., Chen, X., 2007. Integrin alpha v beta 3 antagonists for anti-angiogenic cancer treatment. Recent Pat. Anti-Cancer Drug Discov. 2, 143–158.

Huegel, J., Sgariglia, F., Enomoto-Iwamoto, M., Koyama, E., Dormans, J.P., Pacifici, M., 2013. Heparan sulfate in skeletal development, growth, and pathology: the case of hereditary multiple exostoses. Dev. Dynam. 242, 1021–1032.

Hughes, D.E., Salter, D.M., Dedhar, S., Simpson, R., 1993. Integrin expression in human bone. J. Bone Miner. Res. 8, 527–533.

Hultenby, K., Reinholt, F.P., Heinegård, D., 1993. Distribution of integrin subunits on rat metaphyseal osteoclasts and osteoblasts. Eur. J. Cell Biol. 62, 86–93.

Inoue, M., Namba, N., Chappel, J., Teitelbaum, S.L., Ross, F.P., 1998. Granulocyte macrophage-colony stimulating factor reciprocally regulates alphav-associated integrins on murine osteoclast precursors. Mol. Endocrinol. 12, 1955–1962.

Iwaniec, U.T., Wronski, T.J., Amblard, D., Nishimura, Y., van der Meulen, M.C., Wade, C.E., et al., 2005. Effects of disrupted beta1-integrin function on the skeletal response to short-term hindlimb unloading in mice. J. Appl. Physiol. 98, 690–696.

James, A.W., Shen, J., Zhang, X., Asatrian, G., Goyal, R., Kwak, J.H., et al., 2015. NELL-1 in the treatment of osteoporotic bone loss. Nat. Commun. 6, 7362.

Jennings, L., Wu, L., King, K.B., Hämmerle, H., Cs-Szabo, G., Mollenhauer, J., 2001. The effects of collagen fragments on the extracellular matrix metabolism of bovine and human chondrocytes. Connect. Tissue Res. 42, 71–86.

Jikko, A., Harris, S.E., Chen, D., Mendrick, D.L., Damsky, C.H., 1999. Collagen integrin receptors regulate early osteoblast differentiation induced by BMP-2. J. Bone Miner. Res. 14, 1075–1083.

Katsumi, A., Orr, A.W., Tzima, E., Schwartz, M.A., 2004. Integrins in mechanotransduction. J. Biol. Chem. 279, 12001–12004.

Kawaguchi, J., Kii, I., Sugiyama, Y., Takeshita, S., Kudo, A., 2001a. The transition of cadherin expression in osteoblast differentiation from mesenchymal cells: consistent expression of cadherin-11 in osteoblast lineage. J. Bone Miner. Res. 16, 260–269.

Kawaguchi, J., Kii, I., Sugiyama, Y., Takeshita, S., Kudo, A., 2001b. The transition of cadherin expression in osteoblast differentiation from mesenchymal cells: consistent expression of cadherin-11 in osteoblast lineage. J. Bone Miner. Res. 16, 260, 26.

Khatiwala, C.B., Kim, P.D., Peyton, S.R., Putnam, A.J., 2009. ECM compliance regulates osteogenesis by influencing MAPK signaling downstream of RhoA and ROCK. J. Bone Miner. Res. 24, 886–898.

Kidwai, F., Edwards, J., Zou, L., Kaufman, D.S., 2016. Fibrinogen induces Runx2 activity and osteogenic development from human pluripotent stem cells. Stem Cell. 34, 2079–2089.

Kii, I., Amizuka, N., Shimomura, J., Saga, Y., Kudo, A., 2004. Cell-cell interaction mediated by cadherin-11 directly regulates the differentiation of mesenchymal cells into the cells of the osteo-lineage and the chondro-lineage. J. Bone Miner. Res. 19, 1840–1849.

Kim, J.B., Leucht, P., Luppen, C.A., Park, Y.J., Beggs, H.E., Damsky, C.H., et al., 2007. Reconciling the roles of FAK in osteoblast differentiation, osteoclast remodeling, and bone regeneration. Bone 41, 39–51.

Kirsch, T., 2005. Annexins – their role in cartilage mineralization. Front. Biosci. 10, 576–581.

Knudson, C.B., 2003. Hyaluronan and CD44: strategic players for cell-matrix interactions during chondrogenesis and matrix assembly. Birth Defects Res C Embryo Today 69, 174–196.

Kobayashi, Y., Uehara, S., Koide, M., Takahashi, N., 2015. The regulation of osteoclast differentiation by Wnt signals. Bonekey Rep. 4, 713.

Krause, M., Dent, E.W., Bear, J.E., Loureiro, J.J., Gertler, F.B., 2003. Ena/VASP proteins: regulators of the actin cytoskeleton and cell migration. Annu. Rev. Cell Dev. Biol. 19, 541–564.

Kumar, S., Ponnazhagan, S., 2007. Bone homing of mesenchymal stem cells by ectopic alpha 4 integrin expression. FASEB J. 21, 3917–3927.

Labrador, J.P., Azcoitia, V., Tuckermann, J., Lin, C., Olaso, E., Mañes, S., et al., 2001. The collagen receptor DDR2 regulates proliferation and its elimination leads to dwarfism. EMBO Rep. 2, 446–452.

Lai, C.F., Feng, X., Nishimura, R., Teitelbaum, S.L., Avioli, L.V., Ross, F.P., et al., 2000. Transforming growth factor-beta up-regulates the beta 5 integrin subunit expression via Sp1 and Smad signaling. J. Biol. Chem. 275, 36400–36406.

Lai, C.F., Chaudhary, L., Fausto, A., Halstead, L.R., Ory, D.S., Avioli, L.V., et al., 2001. Erk is essential for growth, differentiation, integrin expression, and cell function in human osteoblastic cells. J. Biol. Chem. 276, 14443–14450.

Larsson, T., Sommarin, Y., Paulsson, M., Antonsson, P., Hedbom, E., Wendel, M., et al., 1991. Cartilage matrix proteins. A basic 36-kDa protein with a restricted distribution to cartilage and bone. J. Biol. Chem. 266, 20428–20433.

Lee, Y.S., Chuong, C.M., 1992. Adhesion molecules in skeletogenesis: I. Transient expression of neural cell adhesion molecules (NCAM) in osteoblasts during endochondral and intramembranous ossification. J. Bone Miner. Res. 7, 1435–1446.

Lee, D.Y., Li, Y.S., Chang, S.F., Zhou, J., Ho, H.M., Chiu, J.J., et al., 2010. Oscillatory flow-induced proliferation of osteoblast-like cells is mediated by alphavbeta3 and beta1 integrins through synergistic interactions of focal adhesion kinase and Shc with phosphatidylinositol 3-kinase and the Akt/mTOR/p70S6K pathway. J. Biol. Chem. 285, 30–42.

Lemonnier, J., Delannoy, P., Hott, M., Lomri, A., Modrowski, D., Marie, P.J., 2000. The Ser252Trp fibroblast growth factor receptor-2 (FGFR-2) mutation induces PKC-independent downregulation of FGFR-2 associated with premature calvaria osteoblast differentiation. Exp. Cell Res. 256, 158–167.

Li, Y., Zhong, G., Sun, W., Zhao, C., Zhang, P., Song, J., et al., 2015. CD44 deficiency inhibits unloading-induced cortical bone loss through down-regulation of osteoclast activity. Sci. Rep. 5, 16124.

Lilien, J., Balsamo, J., 2005. The regulation of cadherin-mediated adhesion by tyrosine phosphorylation/dephosphorylation of beta-catenin. Curr. Opin. Cell Biol. 17, 459–465.

Litzenberger, J.B., Kim, J.B., Tummala, P., Jacobs, C.R., 2010. Beta1 integrins mediate mechanosensitive signaling pathways in osteocytes. Calcif. Tissue Int. 86, 325–332.

Loeser, R.F., 2002. Integrins and cell signaling in chondrocytes. Biorheology 39, 119–124.

Lowe, C., Yoneda, T., Boyce, B.F., Chen, H., Mundy, G.R., Soriano, P., 1993. Osteopetrosis in Src-deficient mice is due to an autonomous defect of osteoclasts. Proc. Natl. Acad. Sci. U. S. A. 90, 4485–4489.

Lowe, D.A., Lepori-Bui, N., Fomin, P.V., Sloofman, L.G., Zhou, X., Farach-Carson, M.C., et al., 2014. Deficiency in perlecan/HSPG2 during bone development enhances osteogenesis and decreases quality of adult bone in mice. Calcif. Tissue Int. 95, 29–38.

Luxenburg, C., Winograd-Katz, S., Addadi, L., Geiger, B., 2012. Involvement of actin polymerization in podosome dynamics. J. Cell Sci. 125, 1666–1672.

Mansouri, R., Haÿ, E., Marie, P.J., Modrowski, D., 2015. Role of syndecan-2 in osteoblast biology and pathology. Bonekey Rep. 4, 666.

Mansouri, R., Jouan, Y., Haÿ, E., Blin-Wakkach, C., Frain, M., Ostertag, A., et al., 2017. Osteoblastic heparan sulfate glycosaminoglycans control bone remodeling by regulating Wnt signaling and the crosstalk between bone surface and marrow cells. Cell Death Dis. 8, e2902.

Manton, K.J., Leong, D.F., Cool, S.M., Nurcombe, V., 2007. Disruption of heparan and chondroitin sulfate signaling enhances mesenchymal stem cell-derived osteogenic differentiation via bone morphogenetic protein signaling pathways. Stem Cell. 25 (11), 2845–2854.

Marchisio, P.C., Cirillo, D., Naldini, L., Primavera, M.V., Teti, A., Zambonin-Zallone, A., 1984. Cell-substratum interaction of cultured avian osteoclasts is mediated by specific adhesion structures. J. Cell Biol. 99, 1696–1705.

Marchisio, P.C., Bergui, L., Corbascio, G.C., Cremona, O., D'Urso, N., Schena, M., et al., 1988. Vinculin, talin, and integrins are localized at specific adhesion sites of malignant B lymphocytes. Blood 66, 830–838.

Marie, P.J., Haÿ, E., 2013. Cadherins and Wnt signalling: a functional link controlling bone formation. Bonekey Rep. 2, 330.

Marie, P.J., Haÿ, E., Saidak, Z., 2014a. Integrin and cadherin signaling in bone : role and potential therapeutic targets. Trends Endocrinol. Metabol. 25, 567–575.

Marie, P.J., Haÿ, E., Modrowski, D., Revollo, L., Mbalaviele, G., Civitelli, R., 2014b. Cadherin-mediated cell-cell adhesion and signaling in the skeleton. Calcif. Tissue Int. 94, 46–54.

Marie, P.J., 2013. Targeting integrins to promote bone formation and repair. Nat. Rev. Endocrinol. 9, 288–295.

Matsuo, I., Kimura-Yoshida, C., 2013. Extracellular modulation of Fibroblast Growth Factor signaling through heparan sulfate proteoglycans in mammalian development. Curr. Opin. Genet. Dev. 23, 399–407.

Matsusaki, T., Aoyama, T., Nishijo, K., Okamoto, T., Nakayama, T., Nakamura, T., et al., 2006. Expression of the cadherin-11 gene is a discriminative factor between articular and growth plate chondrocytes. Osteoarthritis Cartilage 14, 353–366.

Mbalaviele, G., Chen, H., Boyce, B.F., Mundy, G.R., Yoneda, T., 1995. The role of cadherin in the generation of multinucleated osteoclasts from mononuclear precursors in murine marrow. J. Clin. Investig. 95, 2757–2765.

Mbalaviele, G., Nishimura, R., Myoi, A., Niewolna, M., Reddy, S.V., Chen, D., et al., 1998. Cadherin-6 mediates the heterotypic interactions between the hemopoietic osteoclast cell lineage and stromal cells in a murine model of osteoclast differentiation. J. Cell Biol. 141, 1467–1476.

McHugh, K.P., Hodivala-Dilke, K., Zheng, M.H., Namba, N., Lam, J., Novack, D., et al., 2000. Mice lacking beta3 integrins are osteosclerotic because of dysfunctional osteoclasts. J. Clin. Investig. 105, 433–440.

Meury, T., Akhouayri, O., Jafarov, T., Mandic, V., St-Arnaud, R., 2010. Nuclear alpha NAC influences bone matrix mineralization and osteoblast maturation in vivo. Mol. Cell Biol. 30, 43–53.

Mitsou, I., Multhaupt, H.A.B., Couchman, J.R., 2017. Proteoglycans, ion channels and cell-matrix adhesion. Biochem. J. 474, 1965–1979.

Miyauchi, A., Alvarez, J., Greenfield, E.M., Teti, A., Grano, M., Colucci, S., et al., 1991. Recognition of osteopontin and related peptides by an alpha v beta 3 integrin stimulates immediate cell signals in osteoclasts. J. Biol. Chem. 266, 20369–20374.

Miyauchi, A., Gotoh, M., Kamioka, H., Notoya, K., Sekiya, H., Takagi, Y., et al., 2006. AlphaVbeta3 integrin ligands enhance volume-sensitive calcium influx in mechanically stretched osteocytes. J. Bone Miner. Metab. 24, 498–504.

Mizuno, M., Fujisawa, R., Kuboki, Y., 2000. Type I collagen-induced osteoblastic differentiation of bone-marrow cells mediated by collagen-alpha2beta1 integrin interaction. J. Cell. Physiol. 184, 207–213.

Modrowski, D., Baslé, M., Lomri, A., Marie, P.J., 2000. Syndecan-2 is involved in the mitogenic activity and signaling of granulocyte-macrophage colony-stimulating factor in osteoblasts. J. Biol. Chem. 275, 9178–9185.

Molténi, A., Modrowski, D., Hott, M., Marie, P.J., 1999a. Differential expression of fibroblast growth factor receptor-1, -2, and -3 and syndecan-1, -2, and -4 in neonatal rat mandibular condyle and calvaria during osteogenic differentiation in vitro. Bone 24, 337–347.

Molténi, A., Modrowski, D., Hott, M., Marie, P.J., 1999b. Alterations of matrix- and cell-associated proteoglycans inhibit osteogenesis and growth response to fibroblast growth factor-2 in cultured rat mandibular condyle and calvaria. Cell Tissue Res. 295, 523–536.

Morgan, E.A., Schneider, J.G., Baroni, T.E., Uluçkan, O., Heller, E., Hurchla, M.A., et al., 2010. Dissection of platelet and myeloid cell defects by conditional targeting of the beta3-integrin subunit. FASEB J. 24, 1117–1127.

Moursi, A.M., Globus, R.K., Damsky, C.H., 1997. Interactions between integrin receptors and fibronectin are required for calvarial osteoblast differentiation in vitro. J. Cell Sci. 110, 2187–2196.

Moussa, F.M., Hisijara, I.A., Sondag, G.R., Scott, E.M., Frara, N., Abdelmagid, S.M., et al., 2014. Osteoactivin promotes osteoblast adhesion through HSPG and αvβ1 integrin. J. Cell. Biochem. 115, 1243–1253.

Nakamura, H., Kenmotsu, S., Sakai, H., Ozawa, H., 1995. Localization of CD44, the hyaluronate receptor, on the plasma membrane of osteocytes and osteoclasts in rat tibiae. Cell Tissue Res. 280, 225–233.

Nakamura, I., Jimi, E., Duong, L.T., Sasaki, T., Takahashi, N., Rodan, G.A., et al., 1998. Tyrosine phosphorylation of p130Cas is involved in actin organization in osteoclasts. J. Biol. Chem. 273, 11144–11149.

Nakamura, R., Nakamura, F., Fukunaga, S., 2014. Contrasting effect of perlecan on adipogenic and osteogenic differentiation of mesenchymal stem cells in vitro. Anim. Sci. J. 85, 262–270.

Nakazora, S., Matsumine, A., Iino, T., Hasegawa, M., Kinoshita, A., Uemura, K., et al., 2010. The cleavage of N-cadherin is essential for chondrocyte differentiation. Biochem. Biophys. Res. Commun. 400, 493–499.

Neam, P.J., Sommarin, Y., Boynton, R.E., Heinegård, D., 1994. The structure of a 38-kDa leucine-rich protein (chondroadherin) isolated from bovine cartilage. J. Biol. Chem. 269, 21547–21554.

Negishi-Koga, T., Takayanagi, H., 2009. Ca2+-NFATc1 signaling is an essential axis of osteoclast differentiation. Immunol. Rev. 231, 241–256.

Nelson, W.J., Nusse, R., 2004. Convergence of Wnt, beta-catenin, and cadherin pathways. Science 303, 1483–1487.

Nicoll, S.B., Barak, O., Csóka, A.B., Bhatnagar, R.S., Stern, R., 2002. Hyaluronidases and CD44 undergo differential modulation during chondrogenesis. Biochem. Biophys. Res. Commun. 292, 819–825.

Oberlender, S.A., Tuan, R.S., 1994. Expression and functional involvement of N-cadherin in embryonic limb chondrogenesis. Development 120, 177–187.

Okada, Y., Morimoto, I., Ura, K., Watanabe, K., Eto, S., Kumegawa, M., et al., 2002. Cell-to-Cell adhesion via intercellular adhesion molecule-1 and leukocyte function-associated antigen-1 pathway is involved in 1alpha,25(OH)2D3, PTH and IL-1alpha-induced osteoclast differentiation and bone resorption. Endocr. J. 49, 483–495.

Onodera, K., Takahashi, I., Sasano, Y., Bae, J.W., Mitani, H., Kagayama, M., et al., 2005. Stepwise mechanical stretching inhibits chondrogenesis through cell-matrix adhesion mediated by integrins in embryonic rat limb-bud mesenchymal cells. Eur. J. Cell Biol. 84, 45–58.

Ory, S., Brazier, H., Pawlak, G., Blangy, A., 2008. Rho-GTPases I osteoclasts: orchestrators of podosome arrangement. Eur. J. Cell Biol. 87, 469–477.

Phillips, J.A., Almeida, E.A., Hill, E.L., Aguirre, J.I., Rivera, M.F., Nachbandi, I., et al., 2008. Role for beta1 integrins in cortical osteocytes during acute musculoskeletal disuse. Matrix Biol. 27, 609–618.

Plotkin, L.I., Mathov, I., Aguirre, J.I., Parfitt, A.M., Manolagas, S.C., Bellido, T., 2005. Mechanical stimulation prevents osteocyte apoptosis: requirement of integrins, Src kinases, and ERKs. Am. J. Physiol. Cell Physiol. 289, C633–C643.

Pommerenke, H., Schmidt, C., Dürr, F., Nebe, B., Lüthen, F., Muller, P., et al., 2002. The mode of mechanical integrin stressing controls intracellular signaling in osteoblasts. J. Bone Miner. Res. 17, 603–611.

Popov, C., Radic, T., Haasters, F., Prall, W.C., Aszodi, A., Gullberg, D., Schieker, M., et al., 2011. Integrins alpha2beta1 and alpha11beta1 regulate the survival of mesenchymal stem cells on collagen I. Cell Death Dis. 2, e186.

Puleo, D.A., Bizios, R., 1991. RGDS tetrapeptide binds to osteoblasts and inhibits fibronectin-mediated adhesion. Bone 12, 271–276.

Purev, E., Neff, L., Horne, W.C., Baron, R., 2009. c-Cbl and Cbl-b act redundantly to protect osteoclasts from apoptosis and to displace HDAC6 from beta-tubulin, stabilizing microtubules and podosomes. Mol. Biol. Cell 20, 4021–4030.

Rajshankar, D., Wang, Y., McCulloch, C.A., 2017. Osteogenesis requires FAK-dependent collagen synthesis by fibroblasts and osteoblasts. FASEB J. 31, 937–953.

Ray, B.J., Thomas, K., Huang, C.S., Gutknecht, M.F., Botchwey, E.A., Bouton, A.H., 2012. Regulation of osteoclast structure and function by FAK family kinases. J. Leukoc. Biol. 92, 1021–1028.

Revollo, L., Kading, J., Jeong, S.Y., Li, J., Salazar, V., Mbalaviele, G., et al., 2015. N-cadherin restrains PTH activation of Lrp6/β-catenin signaling and osteoanabolic action. J. Bone Miner. Res. 30, 274–285.

Rodgers, K.D., Sasaki, T., Aszodi, A., Jacenko, O., 2007. Reduced perlecan in mice results in chondrodysplasia resembling Schwartz-Jampel syndrome. Hum. Mol. Genet. 16, 515–528.

Rucci, N., Teti, A., 2016. The "love-hate" relationship between osteoclasts and bone matrix. Matrix Biol. 52–54, 176–190.

Rucci, N., Capulli, M., Olstad, O.K., Önnerfjord, P., Tillgren, V., Gautvik, K.M., Heinegård, D., Teti, A., 2015. The α2β1 binding domain of chondroadherin inhibits breast cancer-induced bone metastases and impairs primary tumour growth: a preclinical study. Cancer Lett. 358, 67–75.

Rundus, V.R., Marshall, G.B., Parker, S.B., Bales, E.S., Hertzberg, E.L., Minkoff, R., 1998. Association of cell and substrate adhesion molecules with connexin43 during intramembranous bone formation. Histochem. J. 30, 879–896.

Saidak, Z., Le Henaff, C., Azzi, S., Marty, C., Da Nascimento, S., Sonnet, P., et al., 2015. Wnt/β-catenin signaling mediates osteoblast differentiation triggered by peptide-induced α5β1 integrin priming in mesenchymal skeletal cells. J. Biol. Chem. 290, 6903–6912.

Salasznyk, R.M., Klees, R.F., Boskey, A., Plopper, G.E., 2007. Activation of FAK is necessary for the osteogenic differentiation of human mesenchymal stem cells on laminin-5. J. Cell. Biochem. 100, 499–514.

Saltel, F., Chabadel, A., Bonnelye, E., Jurdic, P., 2008. Actin cytoskeletal organisation in osteoclasts: a model to decipher transmigration and matrix degradation. Eur. J. Cell Biol. 87, 459–468.

Schaffner, P., Dard, M.M., 2003. Structure and function of RGD peptides involved in bone biology. Cell. Mol. Life Sci. 60, 119–132.

Schneider, J.G., Amend, S.R., Weilbaecher, K.N., 2011. Integrins and bone metastasis: integrating tumor cell and stromal cell interactions. Bone 48, 54–65.

Sens, C., Huck, K., Pettera, S., Uebel, S., Wabnitz, G., Moser, M., et al., 2017. Fibronectins containing extradomain A or B enhance osteoblast differentiation via distinct integrins. J. Biol. Chem. 292, 7745–7760.

Seong, J., Tajik, A., Sun, J., Guan, J.L., Humphries, M.J., Craig, S.E., et al., 2013. Distinct biophysical mechanisms of focal adhesion kinase mechanoactivation by different extracellular matrix proteins. Proc. Natl. Acad. Sci. U. S. A. 110, 19372–19377.

Shakibaei, M., 1998. Inhibition of chondrogenesis by integrin antibody in vitro. Exp. Cell Res. 240, 95–106.

Shankar, G., Davison, I., Helfrich, M.H., Mason, W.T., Horton, M.A., 1993. Integrin receptor-mediated mobilisation of intranuclear calcium in rat osteoclasts. J. Cell Sci. 105, 61–68.

Shekaran, A., Shoemaker, J.T., Kavanaugh, T.E., Lin, A.S., LaPlaca, M.C., Fan, Y., et al., 2014. The effect of conditional inactivation of beta 1 integrins using twist 2 Cre, Osterix Cre and osteocalcin Cre lines on skeletal phenotype. Bone 68, 131–141.

Sheldrake, H.M., Patterson, L.H., 2014. Strategies to inhibit tumor associated integrin receptors: rationale for dual and multiantagonists. J. Med. Chem. 57, 6301–6315.

Shih, Y.R., Tseng, K.F., Lai, H.Y., Lin, C.H., Lee, O.K., 2011. Matrix stiffness regulation of integrin-mediated mechanotransduction during osteogenic differentiation of human mesenchymal stem cells. J. Bone Miner. Res. 26, 730–738.

Shin, C.S., Lecanda, F., Sheikh, S., Weitzmann, L., Cheng, S.L., Civitelli, R., 2000. Relative abundance of different cadherins defines differentiation of mesenchymal precursors into osteogenic, myogenic, or adipogenic pathways. J. Cell. Biochem. 78, 566–577.

Smith, S.M., West, L.A., Govindraj, P., Zhang, X., Ornitz, D.M., Hassell, J.R., 2007. Heparan and chondroitin sulfate on growth plate perlecan mediate binding and delivery of FGF-2 to FGF receptors. Matrix Biol. 26, 175–184.

Sondag, G.R., Mbimba, T.S., Moussa, F.M., Novak, K., Yu, B., Jaber, F.A., et al., 2016. Osteoactivin inhibition of osteoclastogenesis is mediated through CD44-ERK signaling. Exp. Mol. Med. 48, e257.

Song, S.J., Cool, S.M., Nurcombe, V., 2007. Regulated expression of syndecan-4 in rat calvaria osteoblasts induced by fibroblast growth factor-2. J. Cell. Biochem. 100, 402–411.

Soysa, N.S., Alles, N., 2016. Osteoclast function and bone-resorbing activity: an overview. Biochem. Biophys. Res. Commun. 476, 115–120.

Spessotto, P., Rossi, F.M., Degan, M., Di Francia, R., Perris, R., Colombatti, A., et al., 2002. Hyaluronan-CD44 interaction hampers migration of osteoclast-like cells by down-regulating MMP-9. J. Cell Biol. 158, 1133–1144.

Spinardi, L., Marchisio, P.C., 2006. Podosomes as smart regulators of cellular adhesion. Eur. J. Cell Biol. 85, 191–194.

Srouji, S., Ben-David, D., Fromigué, O., Vaudin, P., Kuhn, G., Müller, R., et al., 2012. Lentiviral-mediated integrin α5 expression in human adult mesenchymal stromal cells promotes bone repair in mouse cranial and long-bone defects. Hum. Gene Ther. 23, 167–172.

Stange, R., Kronenberg, D., Timmen, M., Everding, J., Hidding, H., Eckes, B., et al., 2013. Age-related bone deterioration is diminished by disrupted collagen sensing in integrin α2β1 deficient mice. Bone 56, 48–54.

Steinfeld, R., Van Den Berghe, H., David, G., 1996. Stimulation of fibroblast growth factor receptor-1 occupancy and signaling by cell surface-associated syndecans and glypican. J. Cell Biol. 133, 405–416.

Stokes, D.G., Liu, G., Coimbra, I.B., Piera-Velazquez, S., Crowl, R.M., Jiménez, S.A., 2002. Assessment of the gene expression profile of differentiated and dedifferentiated human fetal chondrocytes by microarray analysis. Arthritis Rheum. 46, 404–419.

Su, J.L., Chiou, J., Tang, C.H., Zhao, M., Tsai, C.H., Chen, P.S., et al., 2010. CYR61 regulates BMP-2-dependent osteoblast differentiation through the {alpha}v{beta}3 integrin/integrin-linked kinase/ERK pathway. J. Biol. Chem. 285, 31325–31336.

Sun, C., Yuan, H., Wang, L., Wei, X., Williams, L., Krebsbach, P.H., et al., 2016. FAK promotes osteoblast progenitor cell proliferation and differentiation by enhancing Wnt signaling. J. Bone Miner. Res. 31, 2227–2238.

SundarRaj, N., Fite, D., Ledbetter, S., Chakravarti, S., Hassell, J.R., 1995. Perlecan is a component of cartilage matrix and promotes chondrocyte attachment. J. Cell Sci. 108, 2663–2672.

Suzuki, K., Zhu, B., Rittling, S.R., Denhardt, D.T., Goldberg, H.A., McCulloch, C.A., et al., 2002. Colocalization of intracellular osteopontin with CD44 is associated with migration, cell fusion, and resorption in osteoclasts. J. Bone Miner. Res. 17, 1486–1497.

Takahashi, I., Onodera, K., Sasano, Y., Mizoguchi, I., Bae, J.W., Mitani, H., et al., 2003. Effect of stretching on gene expression of beta1 integrin and focal adhesion kinase and on chondrogenesis through cell-extracellular matrix interactions. Eur. J. Cell Biol. 82, 182–192.

Takeuchi, Y., Suzawa, M., Kikuchi, T., Nishida, E., Fujita, T., Matsumoto, T., 1997. Differentiation and transforming growth factor-beta receptor down-regulation by collagen-alpha2beta1 integrin interaction is mediated by focal adhesion kinase and its downstream signals in murine osteoblastic cells. J. Biol. Chem. 14 (272), 29309–29316.

Tamura, Y., Takeuchi, Y., Suzawa, M., Fukumoto, S., Kato, M., Miyazono, K., et al., 2001. Focal adhesion kinase activity is required for bone morphogenetic protein-Smad1 signaling and osteoblastic differentiation in murine MC3T3-E1 cells. J. Bone Miner. Res. 16, 1772–1779.

Tanabe, N., Wheal, B.D., Kwon, J., Chen, H.H., Shugg, R.P., Sims, S.M., et al., 2011. Osteopontin signals through calcium and nuclear factor of activated T cells (NFAT) in osteoclasts: a novel RGD-dependent pathway promoting cell survival. J. Biol. Chem. 286, 39871–39881.

Tanaka, Y., Morimoto, I., Nakano, Y., Okada, Y., Hirota, S., Nomura, S., et al., 1995. Osteoblasts are regulated by the cellular adhesion through ICAM-1 and VCAM-1. J. Bone Miner. Res. 10, 1462–1469.

Tanaka, S., Amling, M., Neff, L., Peyman, A., Uhlmann, E., Levy, J.B., et al., 1996. c-Cbl is downstream of c-Src in a signalling pathway necessary for bone resorption. Nature 383, 528–531.

Tanaka, Y., Maruo, A., Fujii, K., Nomi, M., Nakamura, T., Eto, S., et al., 2000. Intercellular adhesion molecule 1 discriminates functionally different populations of human osteoblasts: characteristic involvement of cell cycle regulators. J. Bone Miner. Res. 15, 1912–1923.

Tanikawa, R., Tanikawa, T., Hirashima, M., Yamauchi, A., Tanaka, Y., 2010. Galectin-9 induces osteoblast differentiation through the CD44/Smad signaling pathway. Biochem. Biophys. Res. Commun. 394, 317–322.

Tavella, S., Raffo, P., Tacchetti, C., Cancedda, R., Castagnola, P., 1994. N-CAM and N-cadherin expression during in vitro chondrogenesis. Exp. Cell Res. 215, 354–362.

Teitelbaum, S.L., 2011. The osteoclast and its unique cytoskeleton. Ann NY Acad Sci 1240, 14–17.

Teplyuk, N.M., Haupt, L.M., Ling, L., Dombrowski, C., Mun, F.K., Nathan, S.S., et al., 2009. The osteogenic transcription factor Runx2 regulates components of the fibroblast growth factor/proteoglycan signaling axis in osteoblasts. J. Cell. Biochem. 107, 144–154.

Terpstra, L., Prud'homme, J., Arabian, A., Takeda, S., Karsenty, G., Dedhar, S., et al., 2003. Reduced chondrocyte proliferation and chondrodysplasia in mice lacking the integrin-linked kinase in chondrocytes. J. Cell Biol. 162, 139–148.

Thi, M.M., Suadicani, S.O., Schaffler, M.B., Weinbaum, S., Spray, D.C., 2013. Mechanosensory responses of osteocytes to physiological forces occur along processes and not cell body and require $\alpha V\beta 3$ integrin. Proc. Natl. Acad. Sci. U. S. A. 110, 21012–21017.

Thompson, W.R., Modla, S., Grindel, B.J., Czymmek, K.J., Kirn-Safran, C.B., Wang, L., et al., 2011. Perlecan/Hspg2 deficiency alters the pericellular space of the lacunocanalicular system surrounding osteocytic processes in cortical bone. J. Bone Miner. Res. 26, 618–629.

Tian, J., Zhang, F.J., Lei, G.H., 2015. Role of integrins and their ligands in osteoarthritic cartilage. Rheumatol. Int. 35, 787–798.

Tokuo, H., Ikebe, M., 2004. Myosin X transports Mena/VASP to the tip of filopodia. Biochem. Biophys. Res. Commun. 319, 214–220.

Townsend, P.A., Villanova, I., Teti, A., Horton, M.A., 1999. Beta1 integrin antisense oligodeoxy-nucleotides: utility in controlling osteoclast function. Eur. J. Cell Biol. 78, 485–496.

Triplett, J.W., Pavalko, F.M., 2006. Disruption of alpha-actinin-integrin interactions at focal adhesions renders osteoblasts susceptible to apoptosis. Am. J. Physiol. Cell Physiol. 291, C909–C921.

Tu, S., Wu, S., Shi, X., Chen, K., Wu, C., 2003. Migfilin and Mig-2 link focal adhesions to filamin and the actin cytoskeleton and function in cell shape modulation. Cell 113, 37–47.

Tufan, A.C., Daumer, K.M., Tuan, R.S., 2002. Frizzled-7 and limb mesenchymal chondrogenesis: effect of misexpression and involvement of N-cadherin. Dev. Dynam. 223, 241–253.

Tuli, R., Tuli, S., Nandi, S., Huang, X., Manner, P.A., Hozack, W.J., et al., 2003. Transforming growth factor-beta-mediated chondrogenesis of human mesenchymal progenitor cells involves N-cadherin and mitogen-activated protein kinase and Wnt signaling cross-talk. J. Biol. Chem. 278, 41227–41236.

Turner, C.H., Warden, S.J., Bellido, T., Plotkin, L.I., Kumar, N., Jasiuk, I., et al., 2009. Mechanobiology of the skeleton. Sci. Signal. 2, pt.3.

Uda, Y., Azab, E., Sun, N., Shi, C., Pajevic, P.D., 2017. Osteocyte mechanobiology. Curr. Osteoporos. Rep. 15, 318–332.

Van den Bossche, J., Malissen, B., Mantovani, A., De Baetselier, P., Van Ginderachter, J.A., 2012. Regulation and function of the E-cadherin/catenin complex in cells of the monocyte-macrophage lineage and DCs. Blood 119, 1623–1633.

Viviano, B.L., Silverstein, L., Pflederer, C., Paine-Saunders, S., Mills, K., Saunders, S., 2005. Altered hematopoiesis in glypican-3-deficient mice results in decreased osteoclast differentiation and a delay in endochondral ossification. Dev. Biol. 282, 152–162.

Wang, W., Kirsch, T., 2006. Annexin V/beta5 integrin interactions regulate apoptosis of growth plate chondrocytes. J. Biol. Chem. 281, 30848–30856.

Wang, Y., McNamara, L.M., Schaffler, M.B., Weinbaum, S., 2007. A model for the role of integrins in flow induced mechanotransduction in osteocytes. Proc. Natl. Acad. Sci. U. S. A. 104, 15941–15946.

Wang, Z., Collighan, R.J., Gross, S.R., Danen, E.H., Orend, G., Telci, D., et al., 2010. RGD-independent cell adhesion via a tissue transglutaminase-fibronectin matrix promotes fibronectin fibril deposition and requires syndecan-4/2 alpha5beta1 integrin co-signaling. J. Biol. Chem. 285, 40212–40229.

Wang, Z., Telci, D., Griffin, M., 2011. Importance of syndecan-4 and syndecan -2 in osteoblast cell adhesion and survival mediated by a tissue transglutaminase-fibronectin complex. Exp. Cell Res. 317, 367–381.

Wang, B., Lai, X., Price, C., Thompson, W.R., Li, W., Quabili, T.R., et al., 2014. Perlecan-containing pericellular matrix regulates solute transport and mechanosensing within the osteocyte lacunar-canalicular system. J. Bone Miner. Res. 29, 878–891.

Watabe, H., Furuhama, T., Tani-Ishii, N., Mikuni-Takagaki, Y., 2011. Mechanotransduction activates $\alpha_5\beta_1$ integrin and PI3K/Akt signaling pathways in mandibular osteoblasts. Exp. Cell Res. 317, 2642–2649.

Whitelock, J.M., Melrose, J., Iozzo, R.V., 2008. Diverse cell signaling events modulated by perlecan. Biochemistry 47, 11174–11183.

Woods, A., Wang, G., Beier, F., 2007a. Regulation of chondrocyte differentiation by the actin cytoskeleton and adhesive interactions. J. Cell. Physiol. 213, 1–8.

Woods, A., Wang, G., Dupuis, H., Shao, Z., Beier, F., 2007b. Rac1 signaling stimulates N-cadherin expression, mesenchymal condensation, and chondrogenesis. J. Biol. Chem. 282, 23500–23508.

Woodward, W.A., Tuan, R.S., 1999. N-Cadherin expression and signaling in limb mesenchymal chondrogenesis: stimulation by poly-L-lysine. Dev. Genet. 24, 178–187.

Worapamorn, W., Tam, S.P., Li, H., Haase, H.R., Bartold, P.M., 2002. Cytokine regulation of syndecan-1 and -2 gene expression in human periodontal fibroblasts and osteoblasts. J. Periodontal. Res. 37, 273–278.

Wu, D., Schaffler, M.B., Weinbaum, S., Spray, D.C., 2013. Matrix-dependent adhesion mediates network responses to physiological stimulation of the osteocyte cell process. Proc. Natl. Acad. Sci. U. S. A. 110, 12096–12101.

Xiao, G., Wang, D., Benson, M.D., Karsenty, G., Franceschi, R.T., 1998. Role of the alpha2-integrin in osteoblast-specific gene expression and activation of the Osf2 transcription factor. J. Biol. Chem. 273, 32988–32994.

Xiao, G., Cheng, H., Cao, H., Chen, K., Tu, Y., Yu, S., et al., 2012. Critical role of filamin-binding LIM protein 1 (FBLP-1)/migfilin in regulation of bone remodeling. J. Biol. Chem. 287, 21450–21460.

Yan, Y.X., Gong, Y.W., Guo, Y., Lv, Q., Guo, C., Zhuang, Y., et al., 2012. Mechanical strain regulates osteoblast proliferation through integrin-mediated ERK activation. PLoS One 7, e35709.

Yang, H., Dong, J., Xiong, W., Fang, Z., Guan, H., Li, F., 2016. N-cadherin restrains PTH repressive effects on sclerostin/SOST by regulating LRP6-PTH1R interaction. Ann. N. Y. Acad. Sci. 1385, 41–52.

Yao, W., Guan, M., Jia, J., Dai, W., Lay, Y.A., Amugongo, S., et al., 2013. Reversing bone loss by directing mesenchymal stem cells to bone. Stem Cell. 31, 2003–2014.

Yaroslavskiy, B.B., Zhang, Y., Kalla, S.E., García Palacios, V., Sharrow, A.C., Li, Y., et al., 2005. NO-dependent osteoclast motility: reliance on cGMP-dependent protein kinase I and VASP. J. Cell Sci. 118, 5479–5487.

Zemmyo, M., Meharra, E.J., Kühn, K., Creighton-Achermann, L., Lotz, M., 2003. Accelerated, aging-dependent development of osteoarthritis in alpha1 integrin-deficient mice. Arthritis Rheum. 48, 2873–2880.

Zimmerman, D., Jin, F., Leboy, P., Hardy, S., Damsky, C., 2000. Impaired bone formation in transgenic mice resulting from altered integrin function in osteoblasts. Dev. Biol. 220, 2–15.

Zou, W., Teitelbaum, S.L., 2010. Integrins, growth factors, and the osteoclast cytoskeleton. Ann. N. Y. Acad. Sci. 1192, 27–31.

Zou, W., Teitelbaum, S.L., 2015. Absence of Dap12 and the αvβ3 integrin causes severe osteopetrosis. J. Cell Biol. 208, 125–136.

Chapter 18

Intercellular junctions and cell–cell communication in the skeletal system

Joseph P. Stains[1], Francesca Fontana[2] and Roberto Civitelli[2]
[1]*Department of Orthopaedics, University of Maryland School of Medicine, Baltimore, MD, United States;* [2]*Washington University in St. Louis, Department of Medicine, Division of Bone and Mineral Diseases, St. Louis, MO, United States*

Chapter outline

Introduction **423**
Adherens junctions and the cadherin superfamily of cell adhesion molecules **423**
Cadherins in commitment and differentiation of chondro-osteogenic cells 425
Cadherins in skeletal development, growth, and maintenance 426
N-cadherin modulation of Wnt/β-catenin signaling in osteoblastogenesis and osteoanabolic responses 426
Connexins and gap-junctional intercellular communication **428**
Connexin diseases affecting the skeleton 429
Connexins in the skeleton across the life span 431
Function of connexin43 in bone cells 432
Mechanisms of connexin43 control of bone cell function 434
Conclusions **436**
Acknowledgments **436**
References **436**

Introduction

The organization of cells in tissues and organs is controlled by molecular programs that afford cells the ability to recognize other cells and the extracellular matrix and to communicate with their neighbors. Adhesive interactions are essential not only in embryonic development, but also in the differentiation and maintenance of tissue architecture and cell polarity, the immune response and the inflammatory process, cell division and death, and tumor progression and metastases. Cell–cell and cell–matrix adhesion is mediated by four major groups of molecules: cadherins, immunoglobulin-like molecules, integrins, and selectins. Cadherins are an integral part of *adherens junctions*, which anchor cells to one another by linking their cytoskeletons. Thus, along with tight junctions and desmosomes, adherens junctions constitute the so-called anchoring junctions (Halbleib and Nelson, 2006). A different type of intercellular junction is the *gap junction*, which does not provide cell anchorage but allows direct communication via specialized intercellular channels formed by connexins (Goodenough et al., 1996).

In the adult skeleton, bone remodeling occurs via repeated sequences of bone resorptive and formative cycles, which require continuous recruitment and differentiation of bone marrow precursors. The cooperative nature of bone remodeling requires efficient means of intercellular recognition, adhesion, and communication that allow cells to sort and migrate, synchronize their activity, equalize hormonal responses, and diffuse locally generated signals. This chapter reviews current knowledge about the role of direct cell–cell interactions in the development and remodeling of the skeletal tissue, focusing on cell–cell adhesion via cadherins and cell–cell communication via gap junctions.

Adherens junctions and the cadherin superfamily of cell adhesion molecules

Cadherins are transmembrane glycoproteins that mediate calcium-dependent cell–cell adhesion through homophilic interactions (Harris and Tepass, 2010; Pokutta et al., 1994). In mammals, there are over 80 members of this large

Principles of Bone Biology. https://doi.org/10.1016/B978-0-12-814841-9.00018-X

superfamily, which also includes protocadherins, desmocollins, and desmogleins (Harris and Tepass, 2010; Cavallaro and Dejana, 2011). Classic cadherins are composed of five repeats of a calcium-binding extracellular domain (EC), a single transmembrane domain, and a cytoplasmic C-terminal region containing two catenin-binding domains (Ishiyama et al., 2010; Obata et al., 1998) (Fig. 18.1). Classical cadherins are further categorized as type I cadherins, including E-, M-, and N-cadherin and cadherin-4 (the human ortholog of mouse R-cadherin), and type II cadherins, comprising cadherin-5 through -12 and VE-cadherin. Typical of type I, but not type II, cadherins is a HAV (His-Ala-Val) motif in the EC1 domain (Blaschuk et al., 1990; Patel et al., 2006), which is important for their adhesive function. At the interface between the five EC domains, four Ca^{2+}-binding pockets allow classical cadherins to assume a rodlike conformation in the presence of Ca^{2+}, thus forming *cis*-dimers (through EC1–2) with adjacent cadherins and *trans*-dimers with cadherins on opposing cells (Pertz et al., 1999). *Trans*-dimerization is necessary for cell–cell adhesion, and it occurs by "strand exchange" or "β-strand swapping" between the EC1 of opposing cadherins (Vendome et al., 2011; Harrison et al., 2005). This interaction requires calcium and docking of a highly conserved Trp residue into the complementary hydrophobic pocket of the opposing cadherin (Vendome et al., 2011; Harrison et al., 2005): for type I cadherins Trp2 mediates the swap, while in type II cadherins both Trp2 and Trp4 are swapped (Patel et al., 2006; Brasch et al., 2012). Accordingly, type I and type II cadherins do not usually engage in heterotypic binding (Patel et al., 2006), while heterotypic dimerization can occur between members of the same cadherin type, for example, between N- and E-cadherin (Katsamba et al., 2009). While *trans*-dimerization is responsible for the relatively weak binding between individual cadherins on opposing cells (Harris and Tepass, 2010), *cis*-interactions account for the assembly of multiple cadherin molecules at sites of cell–cell contact (Brasch et al., 2012), thus creating cadherin clusters that are further stabilized by linkage to the cytoskeleton (Harris and Tepass, 2010; Troyanovsky, 1999).

Cadherins are linked to the cytoskeleton via binding to catenins, specifically, α-, β-, and γ-catenin (the last also known as plakoglobin) and $p120^{ctn}$, which links cadherins to actin filaments at their intracellular cytoplasmic tail. In the classic view, the cadherin intracellular domain binds to β-catenin and plakoglobin, which in turn bind to α-catenin, either directly or through other proteins, including actinin, vinculin, or zonula occludens-1 (ZO-1) (Yamada and Geiger, 1997). Importantly, α-catenin binding either to the cadherin/β-catenin complex or to actin filaments each excludes the other (Drees et al., 2005), implying that the connection between the adhesion structure and the cytoskeleton is dynamic rather than static (Yamada et al., 2005). Studies also suggest that a minimal cadherin–catenin complex, in which both α- and β-catenin can participate, can directly bind filamentous actin upon mechanical strain (Buckley et al., 2014), and that tissue-specific

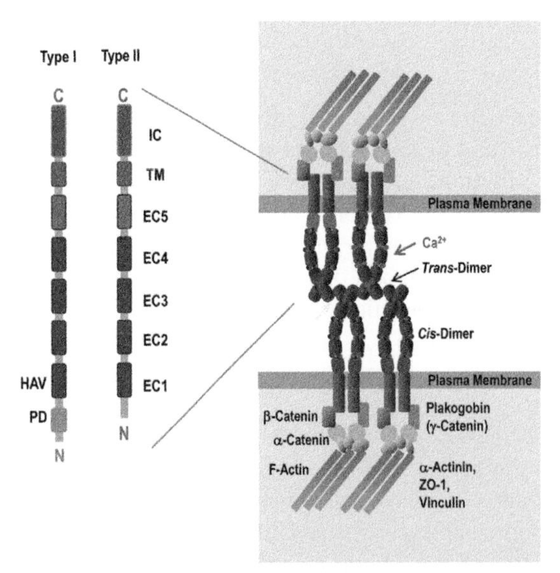

FIGURE 18.1 Cadherins and the adherens junction. (Right) Cadherins are shown with their cytoplasmic (IC), transmembrane (*TM*), and five extracellular domains (*EC1–5*). Small circles between the EC domains symbolize calcium ions. Two complete cadherin *cis*-dimers are shown side by side. These engage in binding with similar complexes from an apposing cell, thus forming *trans*-dimers. The alignment of EC1 domains forms the so-called "zipper" structure of the adhesion complex. Through their CT domain, cadherins bind to β-catenin and plakoglobin, which in turn tether the complex to the actin cytoskeleton via α-catenin and other interacting proteins, including α-actinin, zonula occludens-1 (*ZO-1*), and vinculin. Binding to $p120^{ctn}$ also contributes to stabilizing the cadherin-based adhesion structure. (Left) Enlarged schematic structure of a type I (left) and a type II (right) cadherin. Type I cadherins, such as N-cadherin, have a consensus HAV domain in EC1 important for cell–cell adhesion, and a cadherin-like prodomain (*PD*) at the extreme N terminus. Type II cadherins, such as cadherin-11, engage in cell–cell adhesion via different domains.

catenin isoforms such as αT-catenin may constitutively bind actin (Wickline et al., 2016). Other proteins, such as Lima-1 (Abe and Takeichi, 2008) and ZO-1, can mediate binding of cadherin–catenin complexes to the cytoskeleton (Itoh et al., 1997). The adhesion properties of cadherins are also dependent on the binding of p120^{ctn} to other proteins, such as Rho-family GTPases (Grosheva et al., 2001), which affect membrane dynamics, signaling, and stability of the cadherin complex (Wildenberg et al., 2006). The assembly of cadherins and their associated cytoskeletal elements forms the junctional structures known as adherens junctions (Gumbiner, 2005) (Fig. 18.1).

In addition to their role in cell adhesion, catenins are also components of a signaling system, a function that may evolutionarily predate the appearance of cadherins, as catenin complexes determine cell polarity in *Dictyostelium discoideum* (Dickinson et al., 2011). As detailed later, β-catenin, plakoglobin, and p120^{ctn} modulate the activity of nuclear transcription factors primarily involved in canonical Wnt signaling, a key regulator of osteolineage cell differentiation, function, and bone formation (Krishnan et al., 2006).

Cadherins in commitment and differentiation of chondro-osteogenic cells

Skeletal cells and their precursors express class I and class II cadherins, though their expression levels and functions vary according to developmental stage, lineage, and phase of cell differentiation (Marie et al., 2014; Mbalaviele et al., 2006). During mesenchyme condensation, N-cadherin (gene: *Cdh2*) and cadherin-11 (gene: *Cdh11*) are both expressed in the developing limb buds; and N-cadherin expression increases during chondrogenic differentiation, promoting cell–cell contacts and mesenchymal condensation. N-cadherin then decreases to allow chondrogenic differentiation to proceed (DeLise and Tuan, 2002). These changes in N-cadherin expression during early chondrogenesis are probably modulated by Sox9 (Panda et al., 2001) and bone morphogenetic protein-2 (BMP-2) signaling (Haas and Tuan, 1999). In vitro, inhibition of N-cadherin adhesive functions disrupts mesenchymal condensation in micromass culture and chick limb development, and a dominant-negative N-cadherin mutant is sufficient to impair BMP-2-dependent chondrogenesis (Haas and Tuan, 1999). *Cdh2*-null limb bud cultures, however, do not show major anomalies in chondrogenesis and maintain membrane-bound catenin complexes, suggesting that other cadherins—such as cadherin-11—could be compensating for N-cadherin adhesive function (Luo et al., 2005). Regulation of Wnt signaling by N-cadherin is important in chondrogenesis and in osteogenesis, as detailed later (Modarresi et al., 2005). Indeed, N-cadherin-functionalized hydrogels promote chondrogenesis from human mesenchymal precursors by favoring micromass formation and reducing Wnt signaling via decreased nuclear β-catenin translocation, transcriptional regulation of Wnt receptors, and activating kinases (Li et al., 2017).

Early osteogenic precursors express a broader range of cadherins, including P-cadherin, K-cadherin, E-cadherin, N-cadherin, and cadherin-11 (Cheng et al., 1998; Turel and Rao, 1998; Kawaguchi et al., 2001a; Shin et al., 2005; Hay et al., 2000). Upon osteogenic differentiation, R-cadherin is rapidly downregulated (Cheng et al., 1998), while cadherin-11 is upregulated and maintained through full differentiation (Cheng et al., 1998; Kawaguchi et al., 2001a; Shin et al., 2000). In fact, *Cdh11* was initially cloned in mouse osteoblasts and osteogenic cell lines (Okazaki et al., 1994) and provisionally named OB-cadherin, but it is present in multiple adult and embryonic tissues of mesenchymal origin (Alimperti and Andreadis, 2015). Supporting cadherin-11 proosteogenic action, teratomas originating from cells overexpressing *Cdh11* form preferentially bone and cartilage tissue (Kawaguchi et al., 2001a). As they differentiate into adipocytes, undifferentiated precursors downregulate both *Cdh2* and *Cdh11*, which are absent in adipocytes (Kawaguchi et al., 2001a; Shin et al., 2005), suggesting that coexpression of *Cdh2* and *Cdh11* may be permissive of commitment to osteogenic differentiation while hindering adipogenesis. After the initial increase, N-cadherin is downregulated upon in vitro osteoblast differentiation (Greenbaum et al., 2012), and it is undetectable in osteocytic cells (Kawaguchi et al., 2001a). Accordingly, in histologic sections N-cadherin is present on cells in the bone surface in both humans and rodents (Cheng et al., 1998; Hay et al., 2000; Shin et al., 2000; Castro et al., 2004), but it is absent in osteocytes (Greenbaum et al., 2012). Indeed, for the osteogenic program to proceed to terminal differentiation *Cdh2* must be downregulated (Greenbaum et al., 2012). These observations highlight partially overlapping but distinct functions of *Cdh2* and *Cdh11* at different stages of osteogenesis. Similarly, dental bud stem cells express N-cadherin and cadherin-11, whose expression is modulated during odontogenic differentiation, and they also maintain a low but stable expression of E-cadherin and P-cadherin (Di Benedetto et al., 2015). Cadherin expression in the osteogenic lineage is regulated by local and hormonal factors. BMP-2 and parathyroid hormone (PTH) upregulate both *Cdh2* (Cheng et al., 1998; Hay et al., 2000; Ferrari et al., 2000) and *Cdh11* (Cheng et al., 1998; Yao et al., 2014). *Cdh2* expression is also stimulated by fibroblast growth factor 2 (FGF2) receptor activation via protein kinase Cα (PKCα) in osteoblasts (Delannoy et al., 2001; Debiais et al., 2001), as it occurs in Apert syndrome, where FGFR2 is constitutively activated (Lemonnier et al., 2001). Conversely, stimuli that inhibit osteogenesis, such as dexamethasone, inhibit both *Cdh2* and *Cdh11* (Lecanda et al., 2000a), while proinflammatory cytokines downregulate *Cdh2* only (Tsutsumimoto et al., 1999; Tsuboi et al., 1999).

The function of N-cadherin in osteoblasts also extends to mediating interactions with tumor cells. Indeed, N-cadherin is the hallmark of epithelial-to-mesenchymal transition, which favors tumor growth and metastasis (Tran et al., 1999). Accordingly, N-cadherin has been proposed as a key factor for the establishment of the "preosteolytic" micrometastasis; and inhibition of N-cadherin in tumor cells reduces metastasis in xenograft models (Wang et al., 2015). Preliminary data from our group suggest that the action of N-cadherin in bone cells is actually inhibitory for cancer growth, and it occurs via paracrine and systemic effects (Fontana et al., 2017a). It seems likely that this "niche" function of N-cadherin might also be used for therapeutic targeting.

Cadherins in skeletal development, growth, and maintenance

Mouse models have helped establish the relevance of cadherins in bone formation and homeostasis. Germ-line deletion of *Cdh2* is lethal by day 10 of gestation, with mutant embryos displaying small and irregularly shaped somites and cardiac dysfunction (Radice et al., 1997). Heterozygous *Cdh2*$^{-/+}$ mice have normal skeletal morphology, but upon ovariectomy they undergo accentuated bone loss with reduced bone formation (Lai et al., 2006). Despite strong in vitro evidence of a proosteogenic action, *Cdh11*-null mice are morphologically normal, except for some widening of calvarial sutures and metaphyseal osteopenia (Kawaguchi et al., 2001b), suggesting some degree of redundancy among cadherins in osteogenesis. Indeed, heterozygous loss of *Cdh2* in a *Cdh11*-null background results in severely reduced trabecular bone, smaller and shorter diaphyses, severe osteopenia, and increased bone fragility (Di Benedetto et al., 2010). Tissue-specific overexpression and deletion models have provided more detailed evidence of the role of *Cdh2* in bone. Conditional ablation of *Cdh2* in osteogenic precursors and committed osteoblasts, using *Osx-Cre*, *Prx-Cre*, and *Col1a1-Cre* consistently results in smaller bones and reduced trabecular bone mineral density (Greenbaum et al., 2012; Di Benedetto et al., 2010; Fontana et al., 2017b), suggesting that N-cadherin has an overall positive role in the osteolineage. However, *Cdh2* overexpression in differentiation osteoblasts using *Bglap-Cre* also leads to low bone mass and delayed bone mass acquisition (Hay et al., 2009a). Work has contributed to resolving this apparent paradox by demonstrating a dual role of N-cadherin in the osteolineage: while N-cadherin supports osteogenic precursor number early in osteogenic differentiation, it also restrains progression through the osteogenic program, a function that can be linked to Wnt signaling inhibition, as discussed later. Further supporting this model, delaying *Cdh2* ablation in osteogenic cells until the postnatal stage using the *tet*-suppressible *Osx-Cre* model prevents the low-bone-mass phenotype, actually leading to expanded metaphyseal trabecularization (Fontana et al., 2017b). Indeed, transgenic mice overexpressing *Cdh2* in osteoblasts exhibit increased osteoprogenitor numbers and bone formation with aging (Hay et al., 2014), despite low bone mass in juvenile animals (Hay et al., 2009a). This observation shows that interference with N-cadherin in the adult skeleton could be targeted to enhance bone formation without potential effects on growth.

In vitro studies corroborate the model of a dual action of N-cadherin in osteogenic differentiation. Interference with N-cadherin-mediated cell–cell adhesion using a HAV inhibitory peptide (specific for type I cadherins, thus not interfering with cadherin-11 adhesion), neutralizing antibodies, antisense oligonucleotides, or transfection of a *Cdh2* mutant with dominant-negative action reduces osteoblast differentiation in vitro, probably via reduced cell–cell adhesion. Cadherin adhesive function might be important for keeping osteoprogenitors in a niche environment and preventing commitment to a lineage (Hay et al., 2000; Ferrari et al., 2000; Cheng et al., 2000). On the other hand, bone marrow stromal cells and calvaria cells isolated from *Cdh2*-haploinsufficient mice or from conditionally osteoblast/osteocyte *Cdh2*-deleted mice differentiate faster in vitro (Greenbaum et al., 2012; Fang et al., 2006); however, the number of osteoprogenitors is decreased in *Cdh2*-haploinsufficient or osteoblast *Cdh2*-deficient mice (Di Benedetto et al., 2010; Fang et al., 2006), consistent with a dual function of N-cadherin in supporting osteoprogenitor number but restraining full osteogenic differentiation (Fig. 18.2). More recent data have confirmed that the number of bone marrow Sca1^{+}PDGFα^{+}CD45^{-}Ter113^{-} cells, representing mesenchymal stem and progenitor cells, is very low in conditionally *Cdh2*-ablated mice. However, delaying *Osx-Cre*-driven *Cdh2* recombination until after weaning, when only committed osteoblasts are targeted (Mizoguchi et al., 2014), prevents such a decrease and associated stunted growth (Fontana et al., 2017b).

N-cadherin modulation of Wnt/β-catenin signaling in osteoblastogenesis and osteoanabolic responses

In addition to their function in the adherens junction, cadherins interfere with intracellular signaling pathways in several cell types (Knudsen and Soler, 2000; Wheelock and Johnson, 2003). Indeed, cadherins may retain signaling functions even after modifications that abrogate their adhesive properties by cleavage, shedding, or mutations (Cavallaro and Dejana, 2011). It is rather intuitive that cadherins may affect the Wnt pathway simply because cadherins bind to β-catenin, a key

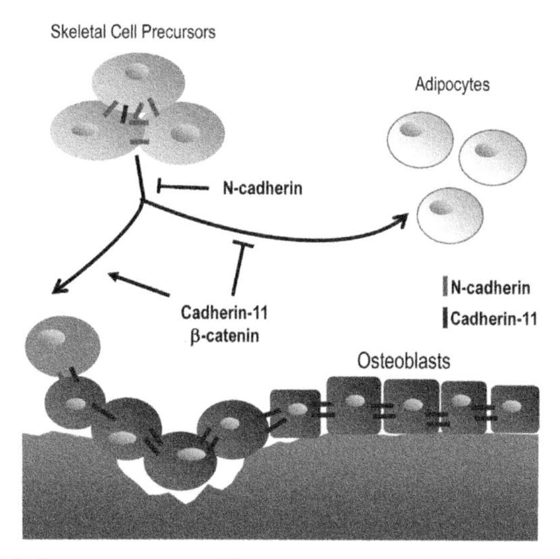

FIGURE 18.2 Cadherins in the osteogenic lineage. In their undifferentiated stage, skeletal cell precursors express a relatively wide repertoire of cadherins, including N-cadherin, cadherin-11, and R-cadherin, with the first being the most abundant. Precursor cells are held together, presumably in a niche, to which N-cadherin is an important contributor. Differentiation of precursors requires downregulation of N-cadherin, which acts as an inhibitor of lineage commitment. By contrast, upregulation of cadherin-11 favors osteogenesis, while adipogenesis is associated with downregulation of cadherin-11. Loss of N-cadherin in the chondro-osteogenic lineage leads to loss of uncommitted precursors. The switch in cadherin expression may allow more β-catenin to become available for transcriptional regulation, thus contributing to osteogenesis and anti-adipogenesis.

component of the canonical Wnt signaling pathway (Daugherty and Gottardi, 2007; Heuberger and Birchmeier, 2010). The Wnt/β-catenin pathway is activated by binding of Wnt ligands to low-density lipoprotein receptor−related protein (LRP) 5/6 and Frizzled coreceptors, followed by recruitment of a complex of Axin/Frat1/adenomatous polyposis coli/glycogen synthase kinase-3β (GSK-3β). This is turn leads to inhibition of GSK-3β, decreased β-catenin phosphorylation at specific sites near its N terminus that prevents β-catenin proteasomal degradation, cytoplasmic accumulation of β-catenin, and subsequent translocation to the nucleus where it binds T cell factor/lymphoid enhancer factor transcription factors to activate gene transcription (Logan and Nusse, 2004). Association of β-catenin with cadherins contributes to stabilizing the adhesion structure (Lilien and Balsamo, 2005), and cell−cell adhesion via cadherins promotes β-catenin phosphorylation and inactivation (Maher et al., 2009). Conversely, release of β-catenin from cadherins destabilizes the adhesion complex, thus facilitating cell movement (Nelson and Nusse, 2004). Thus, an inverse relationship exists between cell adhesion and Wnt/β-catenin signaling. More recent data demonstrate that the interplay between cadherins and signaling is more complex than predicted by a simple interaction with β-catenin at the adherens junction. Certainly, increased cadherin abundance on the cell surface results in decreased β-catenin nuclear translocation and transcriptional activity (Nelson and Nusse, 2004; Gottardi and Gumbiner, 2001; Rhee et al., 2007). As already noted, overexpression of a dominant-negative *Cdh2* mutant in differentiated osteoblasts sequesters β-catenin at the cell surface and decreases bone formation (Castro et al., 2004), as does *Cdh2* overexpression in differentiated osteoblasts (Hay et al., 2012). By contrast, deletion of either *Cdh2* or *Cdh11* or both reduces β-catenin abundance at cell−cell contacts and cell−cell adhesion with a negative effect on osteoblastogenesis and bone growth (Di Benedetto et al., 2010). N-cadherin/β-catenin association can also be modulated by mechanical stimuli, as oscillatory fluid flow releases β-catenin from adherens junctions, resulting in nuclear translocation and transcriptional activation, in turn contributing to osteogenic differentiation (Arnsdorf et al., 2009).

Skeletal biology studies have disclosed an alternative mechanism of N-cadherin interaction with Wnt signaling components via interaction with Lrp5 or Lrp6 (Hay et al., 2009a). Specifically, N-cadherin forms a complex with Lrp5/6 via Axin, thus preventing Lrp5/6 activation of β-catenin, which is instead targeted for proteasomal degradation. This results in decreased Wnt signaling and reduced osteogenic differentiation, function, and bone formation (Marie et al., 2014; Hay et al., 2009a). N-cadherin−Lrp5/6 interaction controls osteoblast differentiation and survival also by reducing endogenous Wnt3a expression and by inhibiting Wnt-dependent phosphatidylinositol 3-kinase (PI3K)/protein kinase B (Akt) signaling (Hay et al., 2009b). Indeed, N-cadherin-mediated adhesion has been linked to activation of PI3K signaling in osteogenic cells (Guntur et al., 2012). Considering the negative action of N-cadherin on Wnt/β-catenin signaling, interference with N-cadherin/Axin/Lrp5/6 binding could potentially lead to osteogenesis by facilitating canonical Wnt signaling. Indeed, deletion of the N-cadherin domain interacting with Axin and Lrp5/6 promotes Wnt/β-catenin signaling and osteoblast differentiation without affecting cell−cell adhesion (Hay et al., 2012). And peptides that bind to the

N-cadherin/Axin-interacting domain of Lrp5 disrupt N-cadherin–Lrp5/6 interaction, enhance Wnt/β-catenin signaling, stimulate osteoblast function in vitro, and promote calvaria bone formation in vivo (Hay et al., 2012).

N-cadherin restraining action on Wnt signaling and its potential for therapeutic targeting to stimulate bone anabolism has been further refined. Our group has reported that the bone anabolic action of PTH is enhanced in mice lacking *Cdh2* in Osterix (Osx)-targeted cells, an effect linked to "noncanonical" β-catenin activation by PTH via the protein kinase A (PKA) pathway and increased Lrp6 association with the PTH/PTH-related protein receptor (Revollo et al., 2015). Enhanced PTH anabolism was reproduced by another group using *DMP1-Cre* to ablate *Cdh2* (Yang et al., 2016a), but was not noted in a study in which *Col1a1-Cre* was used (Bromberg et al., 2012). The reasons for such discrepancies remain unclear, but in vitro work supports a restraining effect of N-cadherin on PTH-induced activation of Wnt signaling (Revollo et al., 2015). More recently, an enhanced osteoanabolic effect of an antibody against Dkk1 was observed in *Col1a1-Cre*-driven *Cdh2*-ablated mice (Fontana et al., 2017b). In the same study, the presence of a Dkk1- and sclerostin-insensitive *Lrp5* mutant in conditional *Cdh2*-ablated mice prevented development of osteopenia but not delayed growth, further supporting the notion that regulation of bone mass accrual, but not longitudinal bone growth, by N-cadherin in young mice is dependent on Wnt signaling (Fontana et al., 2017b). If these data can be corroborated and translated to humans, they would further expand the N-cadherin biologic role in bone-forming cells and further prove that cadherins represent legitimate targets for pharmacological intervention (Fig. 18.3).

Connexins and gap-junctional intercellular communication

The need to maintain tissue integrity during skeletal remodeling demands the temporal and spatial control of bone formation and resorption. Direct intercellular communication through gap junctions allows osteoblasts, osteoprogenitors, and osteocytes to share and amplify signals that permit coordinated function. Gap junctions are intercellular channels that link the cytoplasm of adjacent cells and permit the exchange of second messengers, small molecules, ions, and metabolites that are less than about 1 kDa in molecular mass. In addition, gap junctions can also allow sharing of microRNAs among communicating cells (Valiunas et al., 2005; Plotkin et al., 2017; Zong et al., 2016; Liu et al., 2015). The monomeric subunits of a gap junction are connexins, an evolutionarily conserved family of proteins. There are 21 human connexin genes of which at least four are expressed in bone cells. Osteoblast lineage cells express connexin43 (gene: *Gja1*), connexin45 (*Gjc1*), connexin37 (*Gja4*), and connexin40 (*Gja5*). Of these, connexin43 is the most abundant. On the other hand, osteoclast lineage cells express both connexin43 and connexin37. Mutations or deletions of any these connexins in human or mouse skeletal tissue result in diseases with skeletal manifestations (see later). Pannexins (*Panx1*, *Panx2*, and *Panx3*) are a family of structurally closely related proteins that form a direct channel between the cytoplasm and the extracellular fluid, rather than between two cells. Despite their structural similarities and common sensitivity to pharmacological inhibitors, the mechanisms of action, regulation, and biological role of pannexins in bone are distinct from those of connexins (Plotkin; Stains, 2015; Plotkin et al., 2016; Ishikawa et al., 2016; Ishikawa and Yamada, 2017), thus they will not be reviewed here.

Connexin monomers have a common structure, with four transmembrane-spanning domains, two extracellular loops, an intracellular loop, and both the N and the C termini on the cytoplasmic side of the plasma membrane (Fig. 18.4). Hexamers of connexins form connexons, also called "hemichannels," which dock with connexons on adjacent cells to form an aqueous gap junction channel linking the cytoplasmic compartment of coupled cells. Under some circumstances, undocked hemichannels can also function as a plasma membrane channel that permits exchange of small molecules, such as ATP, between the cytoplasm and the extracellular fluid, acting similar to pannexins. Notably, gap junction channels aggregate into large plaques composed of up to thousands of gap junctions at cell–cell junctions. Thus, the capacity for direct intercellular exchange of small molecules can be quite large.

The size and charge permeability of gap junctions are determined by the connexin composition of a gap junction channel. For example, connexin43 permits the passage of molecules up to 1.2 kDa with a preference for negatively charged molecules (Kanaporis et al., 2011; Harris, 2009). By contrast, connexin45 permits the passage of molecules smaller than 0.3 kDa with a preference for positively charged molecules (Kanaporis et al., 2011). Connexin40 forms gap junctions of size selectivity similar to that of connexin45, while connexin37 is even more restrictive in its size exclusion limit (Kanaporis et al., 2011; Weber et al., 2004). Connexons can be composed of multiple connexin monomeric units (heteromeric gap junctions), and connexins of one type can dock with connexins of a different type, forming a heterotypic gap junction (Fig. 18.4). This variety of heterotypic and heteromeric channels can confer unique biophysical properties to the resulting channel. As most cells express multiple connexins, pairing of different heteromeric channels can tune the resultant gap junction through a wide range of molecular permeabilities. Thus, the composition of a gap junction channel directly regulates its permeability. Similarly, connexins can bind a diverse interactome of

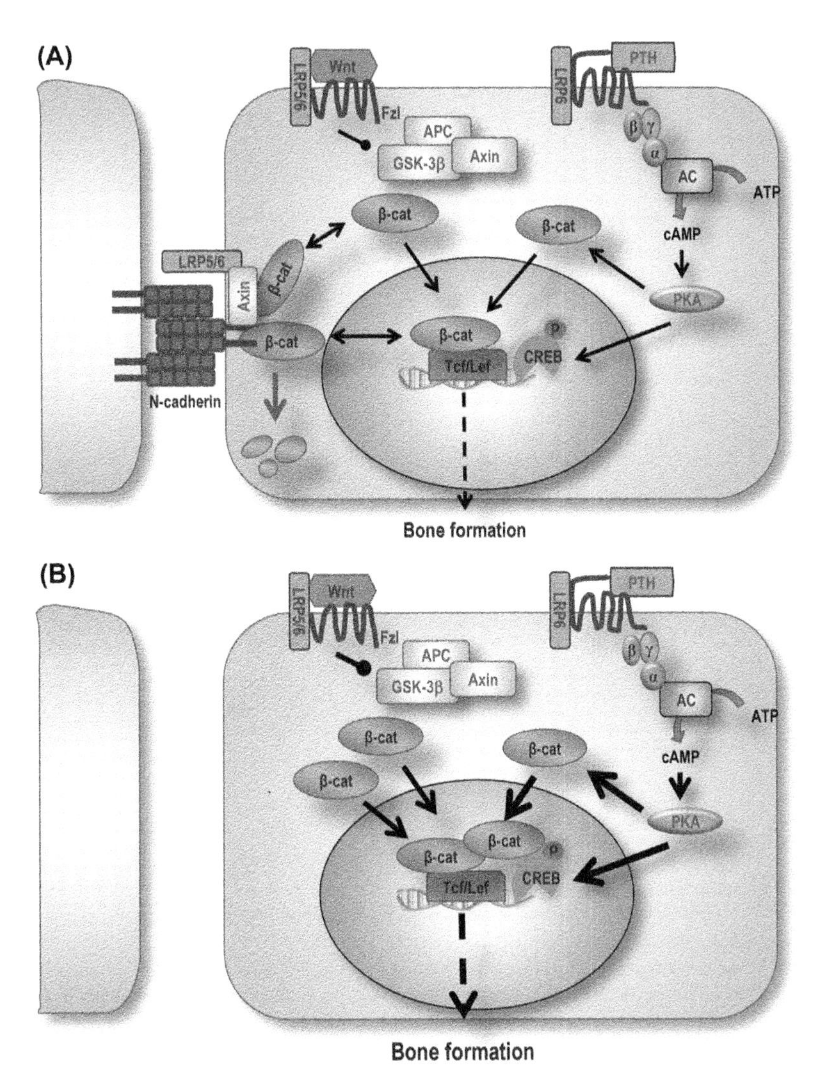

FIGURE 18.3 Cadherins and the Wnt signaling system. (A) Upon Wnt agonist activation of the LRP5/6–Fzl coreceptor complex, GSK-3β is inhibited by binding in an APC/Axin/GSK-3β complex, thus allowing β-catenin nuclear translocation and transcriptional activation. Stabilization of β-catenin can also occur by direct, "noncanonical" phosphorylation at non-GSK-3β sites by PKA, as occurs with PTH binding to PTHR1. PTHR1 can also bind LRP6, presumably also activating the canonical Wnt pathway. Binding to the adherens junction complex keeps β-catenin away from transcriptionally active pools, thus reducing signaling. Another β-catenin pool is kept on the cell surface by binding to N-cadherin via an Axin bridge, which facilitates proteasomal degradation. (B) Absence of N-cadherin releases the system from an inhibitory control mechanism, resulting in increased abundance of active β-catenin and enhanced transcriptional activity. Lack of N-cadherin also releases Lrp5/6 from inactive pools, making more of these coreceptors available for binding to Fzl or PTHR1, and in turn leading to increased β-catenin signaling. In osteogenic cells, this results in proliferation and increased bone-forming activity, in addition to inhibition of osteoclastogenesis. *AC*, adenylyl cyclase; *APC*, adenomatous polyposis coli; *β-cat*, β-catenin; *CREB*, cyclic-AMP response element binding protein; *Fzl*, frizzled; *GSK-3β*, glycogen synthase kinase-3β; *LRP5/6*, low-density lipoprotein receptor–related protein 5/6; *PKA*, protein kinase A; *PTH*, parathyroid hormone; *PTHR1*, PTH/parathyroid hormone related peptide receptor-1; *Tcf/Lef*, T-cell factor/lymphoid enhancer factor.

structural and signaling proteins, which can be connexin specific and can influence downstream signaling (Herve et al., 2012; Moorer and Stains, 2017; Leithe et al., 2018). Thus, the size, number, and composition of a gap junction channel or plaque can lend great plasticity to the ability of the cells to exchange small molecules, thus affecting their function.

Connexin diseases affecting the skeleton

Bone cells are extensively coupled by gap junctions (Yellowley et al., 2000; Doty, 1981; Civitelli et al., 1993; Wang et al., 2016). This permits rapid and efficient exchange of signals via direct intercellular communication among all cells of the bone network, thus providing a key biological mechanism for regulation of bone as a tissue (Fujita et al., 2014;

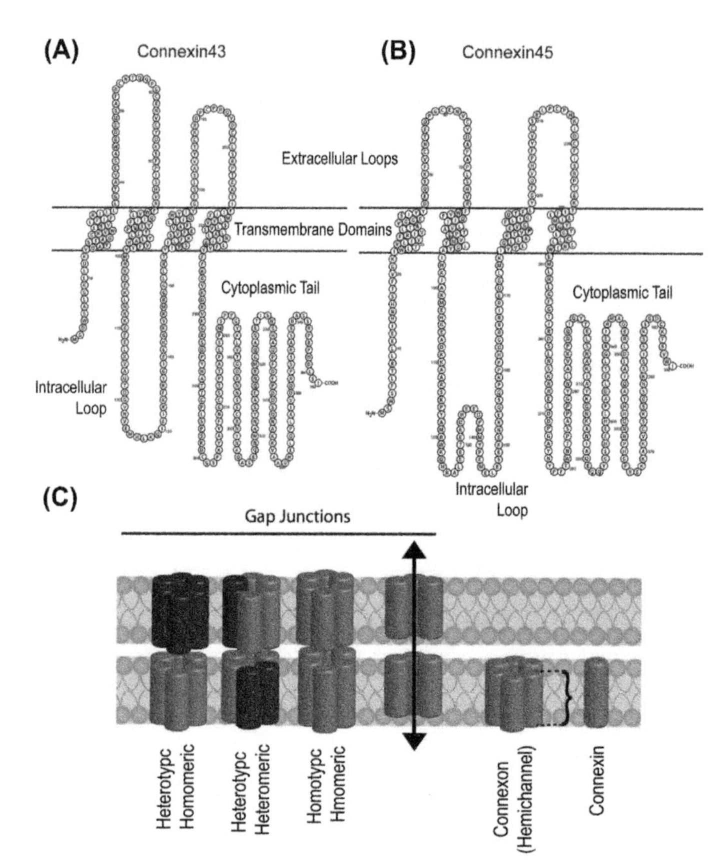

FIGURE 18.4 Connexins and gap junctions. Structure of (A) connexin43 and (B) connexin45, showing conserved topology including four transmembrane spanning domains, two extracellular loops, an intracellular loop, and both the N and C-termini on the cytoplasmic side of the plasma membrane. Images generated using Protter version 1.0 software. (C) Gap junction hemichannels can be formed by hexameric arrays of the same connexin (homomeric) or by mixed connexin hexamers (heteromeric). Likewise, hemichannels in the plasma membrane of one cell can pair with a hemichannel composed of the same (homotypic) or different connexin composition (heterotypic) in an adjacent cell. These combinations can alter the biophysical properties of the resultant gap junction channel.

Nishikawa et al., 2015; Vazquez et al., 2014; Suswillo et al., 2017). The consequences of disrupting this intercellular communication network can be profound. In humans, numerous distinct point mutations in *GJA1* (>62 reported) can lead to the pleiotropic, predominantly autosomal dominant disorder oculodentodigital dysplasia (ODDD) (Paznekas et al., 2009). Skeletal manifestations are prevalent in this disorder and include craniofacial abnormalities (pointed nose, microphthalmia), dental abnormalities (enamel hypoplasia), missing phalanges, syndactyly or camptodactyly, and broad, tubular long bones (Paznekas et al., 2009). These mutations in most cases are dominant negative for connexin43 function, resulting in abnormal gap junction channel function and impaired intercellular communication (Gong et al., 2007; McLachlan et al., 2008). Mice carrying a *Gja1* G60S missense mutation phenocopy many aspects of human ODDD, including cardiac dysfunction, craniofacial abnormalities, syndactyly, tooth enamel hypoplasia, and an alteration in the geometry of the long bones with a broader cross-sectional area at the mid-diaphysis. In addition, these mice display decreased trabecular bone volume fraction (Flenniken et al., 2005; Zappitelli et al., 2013), a feature not reported in ODDD patients, who instead have an osteosclerotic skull (Kjaer et al., 2004). Such discrepancy may reflect species differences or an effect of aging. The changes in skeletal geometry and bone mass are autonomous to the osteoblast lineage, as tissue-specific expression of a *Gja1* G138R mutant, causative of ODDD, in mice reproduces these skeletal features, except, again, the thickened skull (Dobrowolski et al., 2008; Watkins et al., 2011).

A missense mutation in the cytoplasmic tail of the human connexin43 gene (*GJA1* R239Q) has been found in patients with recessive craniometaphyseal dysplasia (CMD), a condition sharing several skeletal manifestations with ODDD, including cranial hyperostosis and widening of long bones with metaphyseal flaring (Hu et al., 2013), but lacking the syndactyly and ocular and dental abnormalities observed in ODDD. Intriguingly, the less rare dominant form of CMD is caused by mutation of a different gene, *Ank*, which encodes the progressive ankylosis protein (ANK), an inorganic pyrophosphate transporter, raising the intriguing possibility that cell−cell diffusion or secretion of pyrophosphate via hemichannels may contribute to the mechanism of connexin43 regulation of bone cell function. In any case, because of their phenotypic and genetic similarities, ODDD and recessive CMD probably represent two manifestations of the spectrum of connexin43 disease, whose main features are a cortical modeling defect leading to metaphyseal and diaphyseal widening and craniofacial malformations. It is tempting to speculate that the more subtle phenotypic differences between ODDD and recessive CMD may be linked to selective interference by the different mutations with either gap junction channel or signaling, the two main functions of connexins.

A loss-of-function mutation (R75H) in the connexin45 gene (*GJC1*) has been linked to cardiac atrial conduction defects and severe arrhythmias, but also to skeletal manifestations, including clinodactyly and camptodactyly and dental and craniofacial abnormalities (Seki et al., 2017). Thus, despite the different biophysical properties of the gap junctions they form (Kanaporis et al., 2011), there is clear overlap in skeletal features between ODDD-causing *GJA1* mutations and the *GJC1* R75H mutation. These similarities could be related to the existence of heteromeric/heterotypic channels formed by connexin43 and connexin45 in bone cells. Hence, loss of one of the connexins would be sufficient to alter the biophysical properties of such heteromeric/heterotypic gap junctions. Ongoing studies should clarify the relative roles of the two connexins in the skeleton.

Connexins in the skeleton across the life span

As anticipated by human genetics, deletion of *Gja1* results in substantial skeletal abnormalities in mice. Germ-line *Gja1* ablation ($Gja1^{-/-}$) is postnatally lethal, but newborn mice have delayed intramembranous and endochondral ossification; brittle, misshapen ribs; and skull abnormalities that include a hypomineralized cranial vault and open parietal foramen (Lecanda et al., 2000b). These defects are consequent to a cell-autonomous defect in osteoblast differentiation with significant delays in the expression of most osteoblast genes (Lecanda et al., 2000b). Likewise, conditional deletion of *Gja1* in the osteoblast lineage leads to changes in cortical bone geometry, including an expansion of the marrow cavity with a corresponding increase in periosteal and endocortical perimeters, resulting in decreased cortical bone cross-sectional thickness (Stains et al., 2014; Grimston et al., 2013). This skeletal phenotype is more severe the earlier in the osteoblast lineage that *Gja1* is deleted in vivo (Stains et al., 2014). Hence, *Gja1* deletion in chondro-osteoblast progenitors using *Dermo1/Twist2-Cre* results in the most severe cortical bone phenotype, with a 40% greater cross-sectional area at the femoral mid-diaphysis (Watkins et al., 2011). Early lineage deletion of *Gja1* is also accompanied by increased number of endocortical osteoclasts, which are responsible for the expanded marrow cavity, but also for cortical porosity and cortical thinning. Enhanced periosteal bone formation only partially compensates for cortical thinning. Similar alterations in cortical bone are present when *Osx-Cre* is used to delete *Gja1* (Hashida et al., 2014). In contrast, deletion of *Gja1* in late osteoblasts and osteocytes using *Bglap-Cre* or *Dmp1-Cre* results a milder skeletal phenotype with only a 20%–25% increase in middiaphyseal cross-sectional area and little or no change in cortical thickness or cortical porosity (Bivi et al., 2012a,b; Zhang et al., 2011). Compromised bone strength and material properties are also common features of *Gja1* ablation models, demonstrating that connexin43 expression throughout the lineage is crucial for maintaining bone quality (Flenniken et al., 2005; Watkins et al., 2011; Bivi et al., 2012b). Accordingly, loss-of-function *Gja1* mutations phenocopy quite well the skeletal abnormalities of connexin43 diseases, with the exception of the craniofacial changes. Aging is associated with loss of osteocyte network connectivity, as the numbers of viable osteocytes and osteocyte processes and connexin43 expression decrease (Davis et al., 2017; Tiede-Lewis et al., 2017). Not surprisingly, the changes in bone architecture during aging closely resemble those seen in *Gja1* deletion or mutation (Zappitelli et al., 2013; Watkins et al., 2011; Davis et al., 2017).

Far less is known about the role of other connexins in skeletal biology. Genetic association studies have evidenced polymorphisms in *Gja4* segregating with bone mineral density in both humans (Yamada et al., 2007) and mice (Xiong et al., 2009). In mice, germ-line *Gja4* deletion results in increased trabecular bone volume, though connexin37 seems to act directly on the osteoclast rather than the osteoblast lineage, as lack of connexin37 decreases osteoclast number in vivo and impairs osteoclast fusion in vitro (Pacheco-Costa et al., 2014). A subtle and paradoxical defect in cortical bone is also observed in these *Gja4*-knockout mice, which exhibit higher cortical bone strength despite decreased cortical bone thickness. This may result from increased mineralization in a low-turnover state, or from changes in the bone extracellular matrix (Pacheco-Costa et al., 2017). As noted earlier, in humans a loss-of-function mutation of *GJC1* results in skeletal phenotypic features resembling, in part, those of *GJA1*-linked mutations (Seki et al., 2017), suggesting either overlapping function or a dominant-negative action of connexin45 on connexin43, presumably in heteromeric or heterotypic channels. However, a preliminary finding on a mouse model of osteolineage-specific *Gjc1* ablation reveals a surprising high trabecular bone mass but no modeling defects, which would suggest distinct functions of the two connexins in bone and support the notion that the overlapping skeletal features of the connexin45 mutant may be related to a dominant-negative action on connexin43 (Watkins et al., 2014). While a full report has yet to be published, these results support a role of connexin45 in bone homeostasis. Finally, connexin40 is expressed early in the developing skeleton, where it participates in endochondral bone development; germ-line deletion or haploinsufficiency of *Gja5* results in bony fusions in the forelimbs, elongated metacarpals and phalanges, and malformations of the sternum (Pizard et al., 2005). However, the role of connexin40 in postnatal bone is less clear.

Function of connexin43 in bone cells

Gap junctions propagate signals through the osteoblast and osteocyte network to disseminate, coordinate, and equilibrate their function throughout the skeletal tissue. Indeed, a primary function of gap junctions is to help groups of cells integrate signals from various stimuli to enhance the signal-to-noise ratio (Ellison et al., 2016). The biologic effects of gap junction–communicated signals across the osteogenic lineage are multiple, including coordinating osteogenic differentiation and function, modulating osteoclastogenesis, maintaining osteocyte viability, and processing, organizing, and orienting collagen in the bone extracellular matrix (Fig. 18.5). Indeed, even small changes in this cell network connectivity can have a profound impact on the tissue level response to regulatory cues. A computer simulation of the osteocyte and bone-lining cell network predicted that a 5% loss of gap junction–coupled osteocytes in the network may reduce maximum signaling to the bone surface by 25% (Jahani et al., 2012). Connexin43 is functionally involved in the entire osteogenic program; lack of connexin43 disrupts the expression of most osteoblast genes and reduces mineralization capacity (Lecanda et al., 1998, 2000; Li et al., 1999; Gramsch et al., 2001; Schiller et al., 2001). These actions of connexin43 have been established by multiple models of *Gja1* ablation, or by induction of an ODDD-causing mutation (G138R) in mesenchymal progenitor cells. Mice carrying such mutations have defective osteogenic differentiation (McLachlan et al., 2008; Mikami et al., 2015; Talbot et al., 2018; Wagner et al., 2017; Damaraju et al., 2015; Esseltine et al., 2017), associated with a reciprocal increase in cell proliferation and expansion of the osteoblast progenitor pool, resulting in the accumulation of immature osteoblasts (Watkins et al., 2011; Moorer et al., 2017; Buo et al., 2017). Expansion of immature osteoblasts and consequent production of an abnormal matrix may contribute to increased periosteal bone formation, but also to poor bone quality and material properties despite enhanced osteoblast activity (Watkins et al., 2011; Bivi et al., 2012a,b; Zhang et al., 2011; Moorer et al., 2017). Indeed, the organic matrix of conditional *Gja1*-ablated mice exhibits a highly disorganized fibrillar collagen network, associated with altered expression of collagen-modifying genes, such as lysyl oxidase (*Lox*) and SerpinH1 (*Hsp47*) (Watkins et al., 2011; Bivi et al., 2012b; Moorer et al., 2017; Pacheco-Costa et al., 2016). Among the broad range of osteogenic cell activities regulated by

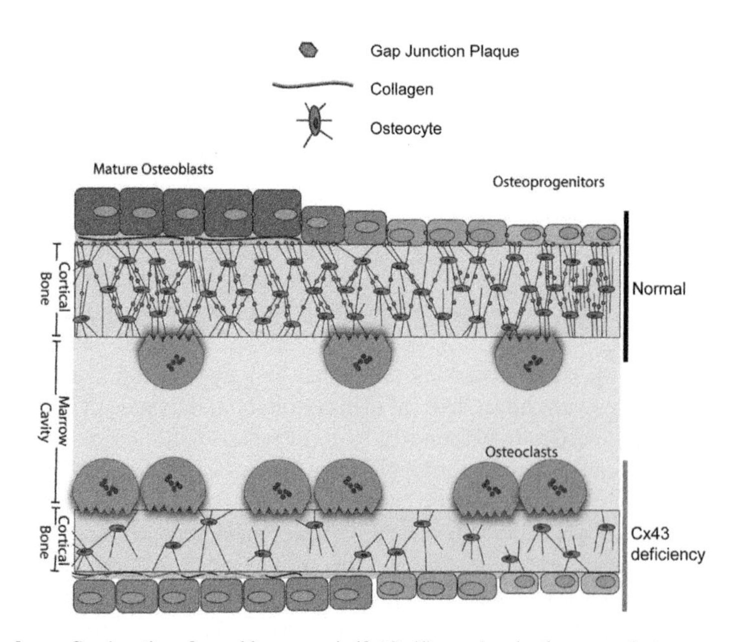

FIGURE 18.5 Connexin43 in bone. Gap junctions formed by connexin43 (*Cx43*) are abundantly present between osteoblasts and their progenitors on the bone surface, between osteoblasts and osteocytes, and at the intersection of osteocytic processes. The presence of connexin43 gap junctions within the osteoblast–osteocyte network is crucial for regulation of osteoblast differentiation, processing and organization of collagen in the bone extracellular matrix, control of osteoclastogenesis, and maintenance of osteocyte viability (top of cartoon). In the absence of connexin43 (bottom of cartoon), osteoblast differentiation is delayed with an expansion of earlier precursors, resulting in production of a disorganized and less mature collagen matrix. Osteocyte viability is reduced, leading to reduced osteocyte number. Endocortical osteoclast number is increased, resulting in expansion of the marrow cavity and thinning of cortical bone.

connexin43 is paracrine regulation of osteoclastogenesis, an action that is primarily responsible for connexin43 modulation of cortical modeling. In vivo and in vitro evidence demonstrates that connexin43 upregulates osteoprotegerin (OPG) (*Tnfrsf11b*) and downregulates receptor activator of NF-κB ligand (RANKL) (*Tnfsf11*) gene expression, resulting in higher RANKL relative to OPG and enhanced osteoclastogenesis (Zappitelli et al., 2013; Watkins et al., 2011; Zhang et al., 2011; Davis et al., 2017; Moorer et al., 2017; Lloyd et al., 2013). The reason why enhanced bone resorption in vivo occurs primarily on the endocortical surfaces, despite broad expression of connexin43 in the trabecular compartment (Watkins et al., 2011), remains to be determined. Increased osteocytic osteolysis (Lloyd et al., 2014a) and osteocyte apoptosis following *Gja1* ablation (Bivi et al., 2012a; Davis et al., 2017; Lloyd et al., 2013; Xu et al., 2015) may contribute to increased osteoclast recruitment and bone resorption, though these effects have not been universally observed. Furthermore, in adult mice, cortical thinning and low bone mass can be corrected by administration of bisphosphonates (Watkins et al., 2012), corroborating the notion that enhanced endocortical bone resorption is indeed the main mechanism causing the cortical abnormalities in conditional connexin43-deficient mice.

Cortical widening, thinning, and porosity present in *Gja1*-deficient bones recapitulate the changes in cortical bone architecture and geometry that develop in disuse osteoporosis and aging (Grimston et al., 2011; Peres-Ueno et al., 2017; Lloyd et al., 2014b). While this phenotype would suggest that connexin43 is involved in elaboration of bone anabolic signals in response to mechanical load, in vivo studies reveal a more complex mechanism. Skeletal mechanical loading by axial tibial compression or cantilever bending consistently shows that lack of connexin43 at different stages of the osteoblast lineage enhances periosteal bone apposition (Zhang et al., 2011; Bivi et al., 2013; Grimston et al., 2012) and accentuates the load-induced decrease of endocortical bone formation (Grimston et al., 2008, 2012). These results can be interpreted as suggesting that absence of connexin43 alters mechanosensing, so that bone-forming cells perceive normal mechanical loading as a disuse scenario, resulting in accentuated endocortical bone resorption, expansion of the marrow cavity, and cortical thinning. This view is further supported by attenuation of the bone-catabolic effects of experimentally induced skeletal unloading by either muscle paralysis or hindlimb suspension in conditional *Gja1*-ablated mice (Lloyd et al., 2012, 2013; Grimston et al., 2011), a finding consistent with the notion that these mice are insensitive to unloading, presumably because endocortical bone resorption is already hyperactivated as an adaptation to abnormal mechanosensing under normal loading conditions in connexin43 deficiency. Further support for this hypothesis is provided by the observation that the cortical modeling abnormalities develop postnatally, after the animal has been able to begin autonomous life (Grimston et al., 2013). Mechanistically, it has been proposed that the accentuated cortical expansion and thinning may be linked to dysregulated sclerostin expression and aberrant Wnt/β-catenin signaling, consistently observed in osteolineage-specific *Gja1*-knockout mice (Watkins et al., 2011; Bivi et al., 2012a; Lloyd et al., 2012, 2013). In support of a role of Wnt/β-catenin signaling in connexin43 action, inhibition of GSK-3β and consequent β-catenin activation improves fracture healing (Loiselle et al., 2013). However, in a genetic interaction study, double *Gja1/Sost* heterozygous mice did not show the cortical phenotype seen in *Gja1*-deficient mice, arguing against a strong interaction between *Sost* and *Gja1* in driving cortical modeling, at least under normal mechanical stimulation (Grimston et al., 2017). Other potential mechanisms may emerge from in vitro studies on mechanically activated osteocytes or osteoblasts, which suggest that multiple paracrine mediators might be involved in connexin43-mediated mechanotransduction (Vazquez et al., 2014; Suswillo et al., 2017; Romanello & D'Andrea, 2001; Genetos et al., 2007; Cherian et al., 2005).

Expression of connexin43 is also required for the bone anabolic action of PTH (Pacheco-Costa et al., 2016; Chung et al., 2006). Intriguingly, *Gja1* gene deletion driven by the 2.3-kb *Col1a1* promoter fully abrogates the anabolic action of PTH, while deletion of *Gja1* using the *Dmp1* promoter does not (Pacheco-Costa et al., 2016; Chung et al., 2006), suggesting that the osteoanabolic effect of PTH is dependent on connexin43 at earlier stages of osteogenesis, perhaps during commitment of undifferentiated precursors. This premise would be consistent with connexin43 key action at the Runx2/Osx transition, as discussed later. Similarly, the C-terminal domain of connexin43 is important for modulating the PTH response of bone cells, as such domain binds to and sequesters β-arrestin to promote PKA-dependent signaling (Bivi et al., 2011). Furthermore, the *Gja1* K258Stop mouse—a model of connexin43 C terminus truncation—failed to increase endocortical bone formation or bone strength in response to anabolic PTH administration (Pacheco-Costa et al., 2015).

Although much less studied, connexins are present and biologically relevant in the osteoclast lineage. Pharmacologic disruption of connexin43 in cell culture reduces fusion of osteoclast progenitors and bone resorption activity (Schilling et al., 2008; Glenske et al., 2014; Kylmaoja et al., 2013; Hobolt-Pedersen et al., 2014; Ransjo et al., 2003; Ilvesaro et al., 2001). However, as of this writing, connexin43 direct action in osteoclasts or their precursors has not been tested in vivo. On the other hand, connexin37 (*Gja4*) does have a function in osteoclasts, as germ-line *Gja4*-null mice have a high bone mass phenotype driven almost exclusively by a defect in osteoclastogenesis (Pacheco-Costa et al., 2014). Accordingly, all osteoclast lineage markers, including RANK, TRAP, cathepsin K, NFATc1, and DC-STAMP, are reduced in connexin37-deficient osteoclasts, as are osteoclast fusion and activity (Pacheco-Costa et al., 2014).

Mechanisms of connexin43 control of bone cell function

Connexin43 expression increases during osteoblast differentiation (Schiller et al., 2001; Donahue et al., 2000), and modulates the activity of the early osteoblast transcription factors Runx2 (Talbot et al., 2018; Lima et al., 2009; Niger et al., 2012; Li et al., 2015; Yang et al., 2016b) and Osx, indirectly (Niger et al., 2011; Stains et al., 2003), resulting in downregulation of downstream genes (Watkins et al., 2011). Functional interaction between connexin43 and Runx2 has been corroborated by in vivo studies, demonstrating that mice heterozygous for both *Gja1* and *Runx2* null mutations ($Gja^{-/+}$;$Runx2^{-/+}$) exhibit both an exacerbated cleidocranial dysplasia-like phenotype than Runx2-/+ mice and a skeletal phenotype that closely resembles that of osteoblast-lineage-specific *Gja1*-knockout mice, including increased middiaphyseal cross-sectional area with periosteal and endocortical expansion, increased endocortical osteoclast number, cortical thinning and porosity, defective osteoblastogenesis, and increased progenitor proliferation (Buo et al., 2017). Hence, connexin43-dependent signals that converge on Runx2 activation are a key mechanism by which gap junctions regulate skeletal homeostasis. Exemplifying such notion, connexin43-dependent parallel activation of the extracellular signal-regulated kinase 1/2 (ERK1/2) and PKCδ pathways converge on regulating Runx2 transcriptional activity and drive osteoblast lineage progression, an effect partly dependent on intercellular diffusion of inositol polyphosphates (Lima et al., 2009; Niger et al., 2012, 2013). Gap junctions also permit intercellular exchange of other second messengers, such as cAMP, Ca^{2+} ions, and ATP, as well as microRNAs (Davis et al., 2017; Romanello & D'Andrea, 2001; Genetos et al., 2007; Gupta et al., 2016; Ishihara et al., 2013; Guo et al., 2006; Jorgensen et al., 2003). These gap junction–communicated second messengers modulate numerous signaling pathways, including ERK1/2, Src, PI3K, 14-3-3θ, PKA, PKCδ, and β-catenin, in osteoblasts and osteocytes (Moorer et al., 2017; Loiselle et al., 2013; Bivi et al., 2011; Gupta et al., 2016; Batra et al., 2012, 2014; Plotkin et al., 2002; Tu et al., 2016). These pathways in turn affect diverse functions, from cell differentiation and survival to fracture healing and expression of paracrine regulators (Watkins et al., 2011; Lecanda et al., 2000b; Bivi et al., 2012a; Lecanda et al., 1998; Loiselle et al., 2013; Cherian et al., 2005; Chung et al., 2006; Stains et al., 2003; Gupta et al., 2016; Plotkin et al., 2002; Plotkin et al., 2008; Stains and Civitelli, 2005; Saunders & et al., 2003; Li et al., 2013; York et al., 2016). Notably, there is dynamic reciprocity, in that stimuli that signal through gap junctions also regulate the expression of gap junction proteins. For example, PTH, prostaglandin E_2 (PGE_2), mechanical load, vitamin D_3, and neuropeptides all increase connexin43 protein abundance (Donahue et al., 1995; Civitelli et al., 1998; Cheng et al., 2001; Cherian et al., 2003; Ziambaras et al., 1998; Schiller et al., 1992; Shen et al., 1986; Saunders et al., 2001; Schirrmacher and Bingmann, 1998; Ma et al., 2013). Conversely, glucocorticoids and 17β-estradiol, as well as aging, decrease connexin43 expression (Davis et al., 2017; Schirrmacher and Bingmann, 1998; Joiner et al., 2014; Roforth et al., 2015; Massas et al., 1998; Gao et al., 2016; Shen et al., 2016). In addition to small molecules, gap junctions formed by connexin43 can transmit microRNAs from cell to cell. In osteocytes, lack of connexin43 decreases the abundance of miR-21, a prosurvival signal that suppresses PTEN-Akt activity (Davis et al., 2017). The resultant increase in osteocyte apoptosis leads to enhanced osteoclastogenesis via increased RANKL/OPG ratio; extracellular release of the high-mobility-group box 1 protein, a pro-osteoclastogenic factor; and consequent bone loss (Davis et al., 2017; Zhou et al., 2008; Yang et al., 2008).

Connexin43 function is not limited to forming transcellular channels; its biologic action is in part mediated by direct interaction with components of cell signaling systems. More than 40 proteins can interact directly or indirectly with the C-terminal tail domain of the connexin43 monomer (Herve et al., 2012; Leithe et al., 2018; Laird, 2006). In bone, PKCδ, ERK1/2, β-catenin, β-arrestin, α5β1 integrin, and Src are part of the characterized connexin43 interactome, where they contribute to regulation of osteoblast signaling processes (Moorer et al., 2017; Bivi et al., 2011, 2013; Batra et al., 2012; Plotkin et al., 2002; Hebert and Stains, 2013; Niger et al., 2010). The assembly of signaling molecules at the gap junction plaque permits efficient modulation of the downstream signaling response to intracellularly communicated signals (Moorer et al., 2017; Batra et al., 2012; Hebert and Stains, 2013). In vivo experiments have confirmed the functional relevance of the C-terminal domain of connexin43 in bone. Genetic deletion of the last 124 amino acids at the connexin43 C-terminal domain (*Gja1* K258Stop) in male mice recapitulates key aspects of *Gja1* gene ablation in bone, including a reciprocal increase in osteoblast proliferation and decrease in osteoblast differentiation, exuberant periosteal osteoblast activity with poorly organized collagenous matrix, increased endocortical osteoclast number, decreased cortical thickness, increased cortical porosity, and widened marrow area (Moorer et al., 2017). Importantly, this C-terminal truncation is associated with a failure to recruit signaling molecules, such as PKCδ, ERK1/2, and β-catenin, to the mutant connexin43, consistent with the notion that the connexin43 C-terminal domain serves as a docking platform for efficient signal transduction (Moorer et al., 2017; Hebert and Stains, 2013). Interestingly, a distinct skeletal phenotype is observed in *Gja1* K258Stop female mice, which exhibit reduced trabecular bone mass (Moorer et al., 2017; Pacheco-Costa et al., 2015; Hammond et al., 2016), although subtle changes in cortical geometry, bone strength, mineral crystallinity, and collagen spacing are also present

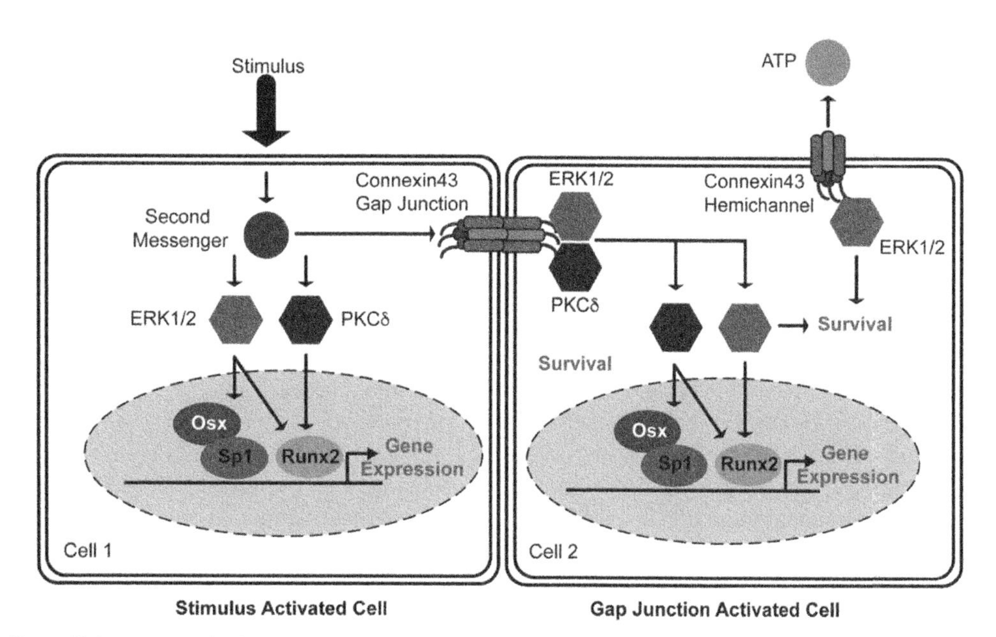

FIGURE 18.6 Intercellular communication and signaling among bone cells. Connexin43-containing gap junctions permit cells that may not be activated by a stimulus to respond as a result of second-messenger diffusion through gap junction channels. There, locally recruited signaling molecules, such as extracellular signal-regulated kinase 1/2 (*ERK1/2*) and protein kinase Cδ (*PKCδ*), are activated by the communicated second messenger resulting in activation of specific signaling cascades, which in turn regulate osteoblast gene expression by modulating the transcriptional activity of specific transcription factors, notably, Runx2, Osterix (*Osx*), and Sp-family members, and by promoting cell survival. Under certain conditions (mechanical strain), connexin43 hemichannels may open and activate signaling independent of gap-junctional communication.

(Moorer et al., 2017; Hammond et al., 2016). These data strongly support the notion that the signal modulation function of connexin43 is necessary for skeletal modeling and homeostasis, independent of or in addition to, gap junction channel function (Fig. 18.6). It should be noted that this biologic model may not apply to all systems and organs; for example, the cardiovascular malformations present in *Gja1*-null mice ($Gja1^{-/-}$) are not present in a C-terminus *Gja1*-truncated mutant, although lethal arrhythmias develop (Lubkemeier et al., 2013).

In cultured osteocytes, unpaired connexin43 hemichannels, which form a direct channel between the cytoplasm and the extracellular compartment of a cell, open in response to fluid shear stress and result in PGE_2 and ATP release into the extracellular fluid (Romanello & D'Andrea, 2001; Genetos et al., 2007; Cherian et al., 2005; Plotkin and Bellido, 2001; Riquelme et al., 2015). This process requires an interaction between connexin43 and α5β1 integrin, triggered by mechanical forces and activation of the Akt pathway (Batra et al., 2012). The subsequent actions of extracellular ATP and PGE_2 then regulate osteocyte responses to mechanical load (Blackwell et al., 2010; Kitase et al., 2010; Lenertz et al., 2015; Orriss et al., 2010; Agrawal and Gartland, 2015). However, there is some controversy as to whether the release of extracellular ATP from mechanically activated bone cells is indeed mediated by connexin43 hemichannels or instead by pannexins, a closely related family of proteins. In vitro, both gap junctions and pannexin channels can conduct ATP and are sensitive to the same pharmacologic inhibitors (Thi et al., 2012; Cheung et al., 2016). In any case, connexin43 hemichannels, but not gap junctions, seem to be required for an antiapoptotic effect of bisphosphonates and protection from oxidative stress in osteocytes (Plotkin et al., 2002, 2005, 2008; Kar et al., 2013). Transgenic mice overexpressing either a *Gja1* mutant that exhibits dominant-negative interference with endogenous gap-junctional communication but not hemichannel activity (R76W) or a *Gja1* mutant that exhibits dominant-negative interference with both functions (Δ130−136) have been developed to test the role of connexin43 hemichannels in vivo (Xu et al., 2015). While Δ130−136-expressing mice mimicked the skeletal phenotype observed in models of connexin43 deficiency, cortical modeling defects were less obvious in connexin43 R76W transgenic mice, suggesting that hemichannels, rather than gap-junctional communication, mediate connexin43 action on cortical modeling (Xu et al., 2015). However, these data are in sharp contrast with the evidence of a full skeletal phenotype, with cortical expansion, thinning, and increased endocortical bone resorption, in mice in which one *Gja1* allele was replaced by the G138R *Gja1* ODDD mutant (Watkins et al., 2011). Since this mutant abolishes gap junction function but allows extracellular ATP release (Dobrowolski et al., 2008), the results support the conclusion that gap-junctional communication is a key driver of the skeletal phenotype. Furthermore, ATP can be released by the opening of pannexin channels, but mice with a germ-line null mutation of *Pnx3*, which is highly expressed in bone cells, show no cortical phenotype (Caskenette et al., 2016). Overall, data from these gene replacement or ablation models argue against a role of ATP release via either connexin43 hemichannels or pannexin channels in the genesis of the cortical modeling defect. Interference with multiple connexins or connexin-binding partners by overexpressed connexin43 mutants may complicate interpretation of the contrasting data emerging from the Δ130–136 transgenic mice.

Conclusions

The unraveling of many of the molecular mechanisms by which cadherins and connexins control skeletal biology has established the importance of direct cell–cell interactions for the achievement of adequate bone mass and maintenance of bone quality. While the necessity of an interconnected cellular communication network in bone is established, much remains to be discovered about the context-specific signals and the modulatory action of cadherins and connexins in elaborating hormonal, paracrine, and physical factors that drive bone modeling and homeostasis. Key questions about how cadherins in osteogenic cells maintain the skeletal stem cell niche, in normal homeostasis and in tumor metastases, and how N-cadherin and cadherin-11 can be leveraged as potential therapeutic targets remain to be addressed. Likewise, the mechanisms underlying mechanotransduction by connexin43, and its exquisite action on cortical modeling, via differential regulation of endocortical and periosteal bone modeling; the role of other connexins in skeletal homeostasis; and the sexual dysmorphism of some connexin actions remain exciting areas of investigation. Defining the signals communicated by gap junction networks adds a new dimension to our understanding of bone modeling and remodeling at the tissue level, opening new avenues for therapeutic intervention to affect bone quality.

Acknowledgments

Part of the work described in this chapter was supported by the National Institute for Musculoskeletal and Skin Diseases, National Institutes of Health (grants AR041255 and AR055913 to R.C., AR063631 and AR071614 to J.P.S.). R.C. receives grant support from Mereo BioPharma and Amgen, and owns stock in Amgen, Eli-Lilly, and Merck & Co. The other authors have no conflicts to disclose.

References

Abe, K., Takeichi, M., 2008. EPLIN mediates linkage of the cadherin catenin complex to F-actin and stabilizes the circumferential actin belt. Proc. Natl. Acad. Sci. U. S. A. 105 (1), 13–19.

Agrawal, A., Gartland, A., 2015. P2X7 receptors: role in bone cell formation and function. J. Mol. Endocrinol. 54 (2), R75–R88.

Alimperti, S., Andreadis, S.T., 2015. CDH2 and CDH11 act as regulators of stem cell fate decisions. Stem Cell Res. 14 (3), 270–282.

Arnsdorf, E.J., Tummala, P., Jacobs, C.R., 2009. Non-canonical Wnt signaling and N-cadherin related beta-catenin signaling play a role in mechanically induced osteogenic cell fate. PLoS One 4 (4), e5388.

Batra, N., et al., 2012. Mechanical stress-activated integrin alpha5beta1 induces opening of connexin43 hemichannels. Proc. Natl. Acad. Sci. U. S. A. 109 (9), 3359–3364.

Batra, N., et al., 2014. 14-3-3theta facilitates plasma membrane delivery and function of mechanosensitive connexin43 hemichannels. J. Cell Sci. 127 (Pt 1), 137–146.

Bivi, N., et al., 2011. Connexin43 interacts with betaarrestin: a pre-requisite for osteoblast survival induced by parathyroid hormone. J. Cell. Biochem. 112 (10), 2920–2930.

Bivi, N., et al., 2012a. Cell autonomous requirement of connexin43 for osteocyte survival: consequences for endocortical resorption and periosteal bone formation. J. Bone Miner. Res. 27 (2), 374–389.

Bivi, N., et al., 2012b. Deletion of Cx43 from osteocytes results in defective bone material properties but does not decrease extrinsic strength in cortical bone. Calcif. Tissue Int. 91 (3), 215–224.

Bivi, N., et al., 2013. Absence of Cx43 selectively from osteocytes enhances responsiveness to mechanical force in mice. J. Orthop. Res. 31 (7), 1075–1081.

Blackwell, K.A., Raisz, L.G., Pilbeam, C.C., 2010. Prostaglandins in bone: bad cop, good cop? Trends Endocrinol. Metabol. 21 (5), 294–301.

Blaschuk, O.W., et al., 1990. Identification of a cadherin cell adhesion recognition sequence. Dev. Biol. 139 (1), 227–229.

Brasch, J., et al., 2012. Thinking outside the cell: how cadherins drive adhesion. Trends Cell Biol. 22 (6), 299–310.

Bromberg, O., et al., 2012. Osteoblastic N-cadherin is not required for microenvironmental support and regulation of hematopoietic stem and progenitor cells. Blood 120 (2), 303–313.

Buckley, C.D., et al., 2014. Cell adhesion. The minimal cadherin-catenin complex binds to actin filaments under force. Science 346 (6209), 1254211.

Buo, A.M., et al., 2017. Connexin43 and Runx2 interact to affect cortical bone geometry, skeletal development, and osteoblast and osteoclast function. J. Bone Miner. Res. 32 (8), 1727–1738.

Caskenette, D., et al., 2016. Global deletion of Panx3 produces multiple phenotypic effects in mouse humeri and femora. J. Anat. 228 (5), 746–756.

Castro, C.H., et al., 2004. Targeted expression of a dominant-negative N-cadherin in vivo delays peak bone mass and increases adipogenesis. J. Cell Sci. 117 (Pt 13), 2853–2864.

Cavallaro, U., Dejana, E., 2011. Adhesion molecule signalling: not always a sticky business. Nat. Rev. Mol. Cell Biol. 12 (3), 189–197.

Cheng, S.L., et al., 1998. Human osteoblasts express a repertoire of cadherins, which are critical for BMP-2-induced osteogenic differentiation. J. Bone Miner. Res. 13 (4), 633–644.

Cheng, S.L., et al., 2000. A dominant negative cadherin inhibits osteoblast differentiation. J. Bone Miner. Res. 15 (12), 2362–2370.

Cheng, B., et al., 2001. PGE(2) is essential for gap junction-mediated intercellular communication between osteocyte-like MLO-Y4 cells in response to mechanical strain. Endocrinology 142 (8), 3464–3473.

Cherian, P.P., et al., 2003. Effects of mechanical strain on the function of Gap junctions in osteocytes are mediated through the prostaglandin EP2 receptor. J. Biol. Chem. 278 (44), 43146–43156.

Cherian, P.P., et al., 2005. Mechanical strain opens connexin43 hemichannels in osteocytes: a novel mechanism for the release of prostaglandin. Mol. Biol. Cell 16 (7), 3100–3106.

Cheung, W.Y., et al., 2016. Pannexin-1 and P2X7-receptor are required for apoptotic osteocytes in fatigued bone to trigger RANKL production in neighboring bystander osteocytes. J. Bone Miner. Res. 31 (4), 890–899.

Chung, D.J., et al., 2006. Low peak bone mass and attenuated anabolic response to parathyroid hormone in mice with an osteoblast-specific deletion of connexin43. J. Cell Sci. 119 (Pt 20), 4187–4198.

Civitelli, R., et al., 1993. Connexin43 mediates direct intercellular communication in human osteoblastic cell networks. J. Clin. Investig. 91 (5), 1888–1896.

Civitelli, R., et al., 1998. Regulation of connexin43 expression and function by prostaglandin E2 (PGE2) and parathyroid hormone (PTH) in osteoblastic cells. J. Cell. Biochem. 68 (1), 8–21.

Damaraju, S., et al., 2015. The role of gap junctions and mechanical loading on mineral formation in a collagen-I scaffold seeded with osteoprogenitor cells. Tissue Eng. 21 (9–10), 1720–1732.

Daugherty, R.L., Gottardi, C.J., 2007. Phospho-regulation of Beta-catenin adhesion and signaling functions. Physiology 22, 303–309.

Davis, H.M., et al., 2017. Disruption of the Cx43/miR21 pathway leads to osteocyte apoptosis and increased osteoclastogenesis with aging. Aging Cell 16 (3), 551–563.

Debiais, F., et al., 2001. Fibroblast growth factor-2 (FGF-2) increases N-cadherin expression through protein kinase C and Src-kinase pathways in human calvaria osteoblasts. J. Cell. Biochem. 81 (1), 68–81.

Delannoy, P., et al., 2001. Protein kinase C-dependent upregulation of N-cadherin expression by phorbol ester in human calvaria osteoblasts. Exp. Cell Res. 269 (1), 154–161.

DeLise, A.M., Tuan, R.S., 2002. Alterations in the spatiotemporal expression pattern and function of N-cadherin inhibit cellular condensation and chondrogenesis of limb mesenchymal cells in vitro. J. Cell. Biochem. 87 (3), 342–359.

Di Benedetto, A., et al., 2010. N-cadherin and cadherin 11 modulate postnatal bone growth and osteoblast differentiation by distinct mechanisms. J. Cell Sci. 123 (Pt 15), 2640–2648.

Di Benedetto, A., et al., 2015. Osteogenic differentiation of mesenchymal stem cells from dental bud: role of integrins and cadherins. Stem Cell Res. 15 (3), 618–628.

Dickinson, D.J., Nelson, W.J., Weis, W.I., 2011. A polarized epithelium organized by beta- and alpha-catenin predates cadherin and metazoan origins. Science 331 (6022), 1336–1339.

Dobrowolski, R., et al., 2008. The conditional connexin43G138R mouse mutant represents a new model of hereditary oculodentodigital dysplasia in humans. Hum. Mol. Genet. 17 (4), 539–554.

Donahue, H.J., et al., 1995. Cell-to-cell communication in osteoblastic networks: cell line-dependent hormonal regulation of gap junction function. J. Bone Miner. Res. 10 (6), 881–889.

Donahue, H.J., et al., 2000. Differentiation of human fetal osteoblastic cells and gap junctional intercellular communication. Am. J. Physiol. Cell Physiol. 278 (2), C315–C322.

Doty, S.B., 1981. Morphological evidence of gap junctions between bone cells. Calcif. Tissue Int. 33 (5), 509–512.

Drees, F., et al., 2005. Alpha-catenin is a molecular switch that binds E-cadherin-beta-catenin and regulates actin-filament assembly. Cell 123 (5), 903–915.

Ellison, D., et al., 2016. Cell-cell communication enhances the capacity of cell ensembles to sense shallow gradients during morphogenesis. Proc. Natl. Acad. Sci. U. S. A. 113 (6), E679–E688.

Esseltine, J.L., et al., 2017. Connexin43 mutant patient-derived induced pluripotent stem cells exhibit altered differentiation potential. J. Bone Miner. Res. 32 (6), 1368–1385.

Fang, L.C., et al., 2006. Accentuated ovariectomy induced bone loss and altered osteogenesis in hyeterozygous N-cadherin null mice. J. Bone Miner. Res. 21 (12), 1897–1906.

Ferrari, S.L., et al., 2000. A role for N-cadherin in the development of the differentiated osteoblastic phenotype. J. Bone Miner. Res. 15 (2), 198–208.

Flenniken, A.M., et al., 2005. A Gja1 missense mutation in a mouse model of oculodentodigital dysplasia. Development 132 (19), 4375–4386.

Fontana, F., et al., 2017a. N-cadherin in extra-skeletal osterix (Osx) positive cells modulates tumor growth independently of cell-cell adhesion. J. Bone Miner. Res. 32 (S1).

Fontana, F., et al., 2017b. N-cadherin regulation of bone growth and homeostasis is osteolineage stage-specific. J. Bone Miner. Res. 32 (6), 1332–1342.

Fujita, K., et al., 2014. Mutual enhancement of differentiation of osteoblasts and osteocytes occurs through direct cell-cell contact. J. Cell. Biochem. 115 (11), 2039–2044.

Gao, J., et al., 2016. Glucocorticoid impairs cell-cell communication by autophagy-mediated degradation of connexin43 in osteocytes. Oncotarget 7 (19), 26966–26978.

Genetos, D.C., et al., 2007. Oscillating fluid flow activation of gap junction hemichannels induces ATP release from MLO-Y4 osteocytes. J. Cell. Physiol. 212 (1), 207–214.

Glenske, K., et al., 2014. Bioactivity of xerogels as modulators of osteoclastogenesis mediated by connexin43. Biomaterials 35 (5), 1487–1495.

Gong, X.Q., et al., 2007. Differential potency of dominant negative connexin43 mutants in oculodentodigital dysplasia. J. Biol. Chem. 282 (26), 19190–19202.

Goodenough, D.A., Goliger, J.A., Paul, D.L., 1996. Connexins, connexons, and intercellular communication. Annu. Rev. Biochem. 65, 475–502.

Gottardi, C.J., Gumbiner, B.M., 2001. Adhesion signaling: how beta-catenin interacts with its partners. Curr. Biol. 11 (19), R792–R794.

Gramsch, B., et al., 2001. Enhancement of connexin43 expression increases proliferation and differentiation of an osteoblast-like cell line. Exp. Cell Res. 264 (2), 397–407.

Greenbaum, A.M., et al., 2012. N-cadherin in osteolineage cells is not required for maintenance of hematopoietic stem cells. Blood 120 (2), 295–302.

Grimston, S.K., et al., 2008. Attenuated response to in vivo mechanical loading in mice with conditional osteoblast ablation of the connexin43 gene (Gja1). J. Bone Miner. Res. 23 (6), 879–886.

Grimston, S.K., et al., 2011. Connexin43 deficiency reduces the sensitivity of cortical bone to the effects of muscle paralysis. J. Bone Miner. Res. 26 (9), 2151–2160.

Grimston, S.K., et al., 2012. Enhanced periosteal and endocortical responses to axial tibial compression loading in conditional connexin43 deficient mice. PLoS One 7 (9), e44222.

Grimston, S.K., et al., 2013. Connexin43 modulates post-natal cortical bone modeling and mechano-responsiveness. BoneKEy Rep. 2, 446.

Grimston, S.K., et al., 2017. Heterozygous deletion of both sclerostin (Sost) and connexin43 (Gja1) genes in mice is not sufficient to impair cortical bone modeling. PLoS One 12 (11), e0187980.

Grosheva, I., et al., 2001. p120 catenin affects cell motility via modulation of activity of Rho-family GTPases: a link between cell-cell contact formation and regulation of cell locomotion. J. Cell Sci. 114 (Pt 4), 695–707.

Gumbiner, B.M., 2005. Regulation of cadherin-mediated adhesion in morphogenesis. Nat. Rev. Mol. Cell Biol. 6 (8), 622–634.

Guntur, A.R., Rosen, C.J., Naski, M.C., 2012. N-cadherin adherens junctions mediate osteogenesis through PI3K signaling. Bone 50 (1), 54–62.

Guo, X.E., et al., 2006. Intracellular calcium waves in bone cell networks under single cell nanoindentation. Mol. Cell. BioMech. 3 (3), 95–107.

Gupta, A., et al., 2016. Communication of cAMP by connexin43 gap junctions regulates osteoblast signaling and gene expression. Cell. Signal. 28 (8), 1048–1057.

Haas, A.R., Tuan, R.S., 1999. Chondrogenic differentiation of murine C3H10T1/2 multipotential mesenchymal cells: II. Stimulation by bone morphogenetic protein-2 requires modulation of N-cadherin expression and function. Differentiation 64 (2), 77–89.

Halbleib, J.M., Nelson, W.J., 2006. Cadherins in development: cell adhesion, sorting, and tissue morphogenesis. Genes Dev. 20 (23), 3199–3214.

Hammond, M.A., et al., 2016. Removing or truncating connexin43 in murine osteocytes alters cortical geometry, nanoscale morphology, and tissue mechanics in the tibia. Bone 88, 85–91.

Harris, T.J., Tepass, U., 2010. Adherens junctions: from molecules to morphogenesis. Nat. Rev. Mol. Cell Biol. 11 (7), 502–514.

Harris, A.L.L.,D., 2009. Permeability of connexin channels. In: Harris, A.L.L. (Ed.), Connexins: A Guide. Humana Press, pp. 165–206.

Harrison, O.J., et al., 2005. The mechanism of cell adhesion by classical cadherins: the role of domain 1. J. Cell Sci. 118 (Pt 4), 711–721.

Hashida, Y., et al., 2014. Communication-dependent mineralization of osteoblasts via gap junctions. Bone 61, 19–26.

Hay, E., et al., 2000. N- and E-cadherin mediate early human calvaria osteoblast differentiation promoted by bone morphogenetic protein-2. J. Cell. Physiol. 183 (1), 117–128.

Hay, E., et al., 2009a. N-cadherin interacts with axin and LRP5 to negatively regulate Wnt/beta-catenin signaling, osteoblast function, and bone formation. Mol. Cell Biol. 29 (4), 953–964.

Hay, E., et al., 2012. Peptide-based mediated disruption of N-cadherin-LRP5/6 interaction promotes Wnt signaling and bone formation. J. Bone Miner. Res. 27 (9), 1852–1863.

Hay, E., et al., 2014. N-cadherin/wnt interaction controls bone marrow mesenchymal cell fate and bone mass during aging. J. Cell. Physiol. 229 (11), 1765–1775.

Hay, E., Nouraud, A., Marie, P.J., 2009b. N-cadherin negatively regulates osteoblast proliferation and survival by antagonizing Wnt, ERK and PI3K/Akt signalling. PLoS One 4 (12), e8284.

Hebert, C., Stains, J.P., 2013. An intact connexin43 is required to enhance signaling and gene expression in osteoblast-like cells. J. Cell. Biochem. 114 (11), 2542–2550.

Herve, J.C., et al., 2012. Gap junctional channels are parts of multiprotein complexes. Biochim. Biophys. Acta 1818 (8), 1844–1865.

Heuberger, J., Birchmeier, W., 2010. Interplay of cadherin-mediated cell adhesion and canonical Wnt signaling. Cold Spring Harb Perspect Biol 2 (2), a002915.

Hobolt-Pedersen, A.S., Delaisse, J.M., Soe, K., 2014. Osteoclast fusion is based on heterogeneity between fusion partners. Calcif. Tissue Int. 95 (1), 73–82.

Hu, Y., et al., 2013. A novel autosomal recessive GJA1 missense mutation linked to Craniometaphyseal dysplasia. PLoS One 8 (8), e73576.

Ilvesaro, J., Tavi, P., Tuukkanen, J., 2001. Connexin-mimetic peptide Gap 27 decreases osteoclastic activity. BMC Muscoskelet. Disord. 2, 10.

Ishihara, Y., et al., 2013. Ex vivo real-time observation of Ca(2+) signaling in living bone in response to shear stress applied on the bone surface. Bone 53 (1), 204–215.

Ishikawa, M., et al., 2016. Pannexin 3 and connexin43 modulate skeletal development through their distinct functions and expression patterns. J. Cell Sci. 129 (5), 1018–1030.

Ishikawa, M., Yamada, Y., 2017. The role of pannexin 3 in bone biology. J. Dent. Res. 96 (4), 372–379.

Ishiyama, N., et al., 2010. Dynamic and static interactions between p120 catenin and E-cadherin regulate the stability of cell-cell adhesion. Cell 141 (1), 117–128.

Itoh, M., et al., 1997. Involvement of ZO-1 in cadherin-based cell adhesion through its direct binding to alpha catenin and actin filaments. J. Cell Biol. 138 (1), 181–192.

Jahani, M., et al., 2012. The effect of osteocyte apoptosis on signalling in the osteocyte and bone lining cell network: a computer simulation. J. Biomech. 45 (16), 2876–2883.

Joiner, D.M., et al., 2014. Aged male rats regenerate cortical bone with reduced osteocyte density and reduced secretion of nitric oxide after mechanical stimulation. Calcif. Tissue Int. 94 (5), 484–494.

Jorgensen, N.R., et al., 2003. Activation of L-type calcium channels is required for gap junction-mediated intercellular calcium signaling in osteoblastic cells. J. Biol. Chem. 278 (6), 4082–4086.

Kanaporis, G., Brink, P.R., Valiunas, V., 2011. Gap junction permeability: selectivity for anionic and cationic probes. Am. J. Physiol. Cell Physiol. 300 (3), C600–C609.

Kar, R., et al., 2013. Connexin43 channels protect osteocytes against oxidative stress-induced cell death. J. Bone Miner. Res. 28 (7), 1611–1621.

Katsamba, P., et al., 2009. Linking molecular affinity and cellular specificity in cadherin-mediated adhesion. Proc. Natl. Acad. Sci. U. S. A. 106 (28), 11594–11599.

Kawaguchi, J., et al., 2001a. The transition of cadherin expression in osteoblast differentiation from mesenchymal cells: consistent expression of cadherin-11 in osteoblast lineage. J. Bone Miner. Res. 16 (2), 260–269.

Kawaguchi, J., et al., 2001b. Targeted disruption of cadherin-11 leads to a reduction in bone density in calvaria and long bone metaphyses. J. Bone Miner. Res. 16 (7), 1265–1271.

Kitase, Y., et al., 2010. Mechanical induction of PGE2 in osteocytes blocks glucocorticoid-induced apoptosis through both the beta-catenin and PKA pathways. J. Bone Miner. Res. 25 (12), 2657–2668.

Kjaer, K.W., et al., 2004. Novel Connexin43 (GJA1) mutation causes oculo-dento-digital dysplasia with curly hair. Am. J. Med. Genet. 127A (2), 152–157.

Knudsen, K.A., Soler, A.P., 2000. Cadherin-mediated cell-cell interactions. Methods Mol. Biol. 137, 409–440.

Krishnan, V., Bryant, H.U., Macdougald, O.A., 2006. Regulation of bone mass by Wnt signaling. J. Clin. Investig. 116 (5), 1202–1209.

Kylmaoja, E., et al., 2013. Osteoclastogenesis is influenced by modulation of gap junctional communication with antiarrhythmic peptides. Calcif. Tissue Int. 92 (3), 270–281.

Lai, C.F., et al., 2006. Accentuated ovariectomy-induced bone loss and altered osteogenesis in heterozygous N-cadherin null mice. J. Bone Miner. Res. 21 (12), 1897–1906.

Laird, D.W., 2006. Life cycle of connexins in health and disease. Biochem. J. 394 (Pt 3), 527–543.

Lecanda, F., et al., 1998. Gap junctional communication modulates gene expression in osteoblastic cells. Mol. Biol. Cell 9 (8), 2249–2258.

Lecanda, F., et al., 2000a. Connexin43 deficiency causes delayed ossification, craniofacial abnormalities, and osteoblast dysfunction. J. Cell Biol. 151 (4), 931–944.

Lecanda, F., et al., 2000b. Differential regulation of cadherins by dexamethasone in human osteoblastic cells. J. Cell. Biochem. 77 (3), 499–506.

Leithe, E., Mesnil, M., Aasen, T., 2018. The connexin43 C-terminus: a tail of many tales. Biochim. Biophys. Acta 1860 (1), 48–64.

Lemonnier, J., et al., 2001. Role of N-cadherin and protein kinase C in osteoblast gene activation induced by the S252W fibroblast growth factor receptor 2 mutation in Apert craniosynostosis. J. Bone Miner. Res. 16 (5), 832–845.

Lenertz, L.Y., et al., 2015. Control of bone development by P2X and P2Y receptors expressed in mesenchymal and hematopoietic cells. Gene 570 (1), 1–7.

Li, Z., et al., 1999. Inhibiting gap junctional intercellular communication alters expression of differentiation markers in osteoblastic cells. Bone 25 (6), 661–666.

Li, X., et al., 2013. Connexin43 is a potential regulator in fluid shear stress-induced signal transduction in osteocytes. J. Orthop. Res. 31 (12), 1959–1965.

Li, S., et al., 2015. Connexin43 and ERK regulate tension-induced signal transduction in human periodontal ligament fibroblasts. J. Orthop. Res. 33 (7), 1008–1014.

Li, R., et al., 2017. Self-assembled N-cadherin mimetic peptide hydrogels promote the chondrogenesis of mesenchymal stem cells through inhibition of canonical Wnt/beta-catenin signaling. Biomaterials 145, 33–43.

Lilien, J., Balsamo, J., 2005. The regulation of cadherin-mediated adhesion by tyrosine phosphorylation/dephosphorylation of beta-catenin. Curr. Opin. Cell Biol. 17 (5), 459–465.

Lima, F., et al., 2009. Connexin43 potentiates osteoblast responsiveness to fibroblast growth factor 2 via a protein kinase C-delta/Runx2-dependent mechanism. Mol. Biol. Cell 20 (11), 2697–2708.

Liu, S., et al., 2015. Connexin43 mediated delivery of ADAMTS5 targeting siRNAs from mesenchymal stem cells to synovial fibroblasts. PLoS One 10 (6), e0129999.

Lloyd, S.A., et al., 2012. Connexin43 deficiency attenuates loss of trabecular bone and prevents suppression of cortical bone formation during unloading. J. Bone Miner. Res. 27 (11), 2359–2372.

Lloyd, S.A., et al., 2013. Connexin43 deficiency desensitizes bone to the effects of mechanical unloading through modulation of both arms of bone remodeling. Bone 57 (1), 76–83.

Lloyd, S.A., et al., 2014a. Evidence for the role of connexin43-mediated intercellular communication in the process of intracortical bone resorption via osteocytic osteolysis. BMC Muscoskelet. Disord. 15, 122.

Lloyd, S.A., et al., 2014b. Interdependence of muscle atrophy and bone loss induced by mechanical unloading. J. Bone Miner. Res. 29 (5), 1118–1130.

Logan, C.Y., Nusse, R., 2004. The Wnt signaling pathway in development and disease. Annu. Rev. Cell Dev. Biol. 20, 781–810.

Loiselle, A.E., et al., 2013. Inhibition of GSK-3beta rescues the impairments in bone formation and mechanical properties associated with fracture healing in osteoblast selective connexin43 deficient mice. PLoS One 8 (11), e81399.

Lubkemeier, I., et al., 2013. Deletion of the last five C-terminal amino acid residues of connexin43 leads to lethal ventricular arrhythmias in mice without affecting coupling via gap junction channels. Basic Res. Cardiol. 108 (3), 348.

Luo, Y., Kostetskii, I., Radice, G.L., 2005. N-cadherin is not essential for limb mesenchymal chondrogenesis. Dev. Dynam. 232 (2), 336–344.

Ma, W., et al., 2013. Neuropeptides stimulate human osteoblast activity and promote gap junctional intercellular communication. Neuropeptides 47 (3), 179–186.

Maher, M.T., et al., 2009. Activity of the beta-catenin phosphodestruction complex at cell-cell contacts is enhanced by cadherin-based adhesion. J. Cell Biol. 186 (2), 219–228.

Marie, P.J., et al., 2014. Cadherin-mediated cell-cell adhesion and signaling in the skeleton. Calcif. Tissue Int. 94 (1), 46–54.

Massas, R., Korenstein, R., Benayahu, D., 1998. Estrogen modulation of osteoblastic cell-to-cell communication. J. Cell. Biochem. 69 (3), 282–290.

Mbalaviele, G., Shin, C.S., Civitelli, R., 2006. Cell-cell adhesion and signaling through cadherins: connecting bone cells in their microenvironment. J. Bone Miner. Res. 21 (12), 1821–1827.

McLachlan, E., et al., 2008. ODDD-linked Cx43 mutants reduce endogenous Cx43 expression and function in osteoblasts and inhibit late stage differentiation. J. Bone Miner. Res. 23 (6), 928–938.

Mikami, Y., et al., 2015. Osteogenic gene transcription is regulated via gap junction-mediated cell-cell communication. Stem Cell. Dev. 24 (2), 214–227.

Mizoguchi, T., et al., 2014. Osterix marks distinct waves of primitive and definitive stromal progenitors during bone marrow development. Dev. Cell 29 (3), 340–349.

Modarresi, R., et al., 2005. N-cadherin mediated distribution of beta-catenin alters MAP kinase and BMP-2 signaling on chondrogenesis-related gene expression. J. Cell. Biochem. 95 (1), 53–63.

Moorer, M.C., et al., 2017. Defective signaling, osteoblastogenesis and bone remodeling in a mouse model of connexin43 C-terminal truncation. J. Cell Sci. 130 (3), 531–540.

Moorer, M.C., Stains, J.P., 2017. Connexin43 and the intercellular signaling network regulating skeletal remodeling. Curr. Osteoporos. Rep. 15 (1), 24–31.

Nelson, W.J., Nusse, R., 2004. Convergence of Wnt, beta-catenin, and cadherin pathways. Science 303 (5663), 1483–1487.

Niger, C., et al., 2011. The transcriptional activity of osterix requires the recruitment of Sp1 to the osteocalcin proximal promoter. Bone 49 (4), 683–692.

Niger, C., et al., 2012. ERK acts in parallel to PKCdelta to mediate the connexin43-dependent potentiation of Runx2 activity by FGF2 in MC3T3 osteoblasts. Am. J. Physiol. Cell Physiol. 302 (7), C1035–C1044.

Niger, C., et al., 2013. The regulation of runt-related transcription factor 2 by fibroblast growth factor-2 and connexin43 requires the inositol polyphosphate/protein kinase Cdelta cascade. J. Bone Miner. Res. 28 (6), 1468–1477.

Niger, C., Hebert, C., Stains, J.P., 2010. Interaction of connexin43 and protein kinase C-delta during FGF2 signaling. BMC Biochem. 11, 14.

Nishikawa, Y., et al., 2015. Osteocytes up-regulate the terminal differentiation of pre-osteoblasts via gap junctions. Biochem. Biophys. Res. Commun. 456 (1), 1–6.

Obata, S., et al., 1998. A common protocadherin tail: multiple protocadherins share the same sequence in their cytoplasmic domains and are expressed in different regions of brain. Cell Adhes. Commun. 6 (4), 323–333.

Okazaki, M., et al., 1994. Molecular cloning and characterization of OB-cadherin, a new member of cadherin family expressed in osteoblasts. J. Biol. Chem. 269 (16), 12092–12098.

Orriss, I.R., Burnstock, G., Arnett, T.R., 2010. Purinergic signalling and bone remodelling. Curr. Opin. Pharmacol. 10 (3), 322–330.

Pacheco-Costa, R., et al., 2014. High bone mass in mice lacking Cx37 because of defective osteoclast differentiation. J. Biol. Chem. 289 (12), 8508–8520.

Pacheco-Costa, R., et al., 2015. Defective cancellous bone structure and abnormal response to PTH in cortical bone of mice lacking Cx43 cytoplasmic C-terminus domain. Bone 81, 632–643.

Pacheco-Costa, R., et al., 2016. Osteocytic connexin43 is not required for the increase in bone mass induced by intermittent PTH administration in male mice. J. Musculoskelet. Neuronal Interact. 16 (1), 45–57.

Pacheco-Costa, R., et al., 2017. Connexin37 deficiency alters organic bone matrix, cortical bone geometry, and increases Wnt/beta-catenin signaling. Bone 97, 105–113.

Panda, D.K., et al., 2001. The transcription factor SOX9 regulates cell cycle and differentiation genes in chondrocytic CFK2 cells. J. Biol. Chem. 276 (44), 41229–41236.

Patel, S.D., et al., 2006. Type II cadherin ectodomain structures: implications for classical cadherin specificity. Cell 124 (6), 1255–1268.

Paznekas, W.A., et al., 2009. GJA1 mutations, variants, and connexin43 dysfunction as it relates to the oculodentodigital dysplasia phenotype. Hum. Mutat. 30 (5), 724–733.

Peres-Ueno, M.J., et al., 2017. Model of hindlimb unloading in adult female rats: characterizing bone physicochemical, microstructural, and biomechanical properties. PLoS One 12 (12), e0189121.

Pertz, O., et al., 1999. A new crystal structure, Ca2+ dependence and mutational analysis reveal molecular details of E-cadherin homoassociation. EMBO J. 18 (7), 1738–1747.

Pizard, A., et al., 2005. Connexin40, a target of transcription factor Tbx5, patterns wrist, digits, and sternum. Mol. Cell Biol. 25 (12), 5073–5083.

Plotkin, L.I., Bellido, T., 2001. Bisphosphonate-induced, hemichannel-mediated, anti-apoptosis through the Src/ERK pathway: a gap junction-independent action of connexin43. Cell Commun. Adhes. 8 (4–6), 377–382.

Plotkin, L.I., et al., 2005. Bisphosphonates and estrogens inhibit osteocyte apoptosis via distinct molecular mechanisms downstream of extracellular signal-regulated kinase activation. J. Biol. Chem. 280 (8), 7317–7325.

Plotkin, L.I., et al., 2008. Connexin43 is required for the anti-apoptotic effect of bisphosphonates on osteocytes and osteoblasts in vivo. J. Bone Miner. Res. 23 (11), 1712–1721.

Plotkin, L.I., Stains, J.P., 2015. Connexins and pannexins in the skeleton: gap junctions, hemichannels and more. Cell. Mol. Life Sci. 72 (15), 2853–2867.

Plotkin, L.I., Manolagas, S.C., Bellido, T., 2002. Transduction of cell survival signals by connexin-43 hemichannels. J. Biol. Chem. 277 (10), 8648–8657.

Plotkin, L.I., Laird, D.W., Amedee, J., 2016. Role of connexins and pannexins during ontogeny, regeneration, and pathologies of bone. BMC Cell Biol. 17 (Suppl. 1), 19.

Plotkin, L.I., Pacheco-Costa, R., Davis, H.M., 2017. microRNAs and connexins in bone: interaction and mechanisms of delivery. Curr Mol Biol Rep 3 (2), 63–70.

Pokutta, S., et al., 1994. Conformational changes of the recombinant extracellular domain of E-cadherin upon calcium binding. Eur. J. Biochem. 223 (3), 1019–1026.

Radice, G.L., et al., 1997. Developmental defects in mouse embryos lacking N-cadherin. Dev. Biol. 181 (1), 64–78.

Ransjo, M., Sahli, J., Lie, A., 2003. Expression of connexin43 mRNA in microisolated murine osteoclasts and regulation of bone resorption in vitro by gap junction inhibitors. Biochem. Biophys. Res. Commun. 303 (4), 1179–1185.

Revollo, L., et al., 2015. N-cadherin restrains PTH activation of Lrp6/beta-catenin signaling and osteoanabolic action. J. Bone Miner. Res. 30 (2), 274–285.

Rhee, J., et al., 2007. Cables links Robo-bound Abl kinase to N-cadherin-bound beta-catenin to mediate Slit-induced modulation of adhesion and transcription. Nat. Cell Biol. 9 (8), 883–892.

Riquelme, M.A., et al., 2015. Mitogen-activated protein kinase (MAPK) activated by prostaglandin E2 phosphorylates connexin43 and closes osteocytic hemichannels in response to continuous flow shear stress. J. Biol. Chem. 290 (47), 28321–28328.

Roforth, M.M., et al., 2015. Global transcriptional profiling using RNA sequencing and DNA methylation patterns in highly enriched mesenchymal cells from young versus elderly women. Bone 76, 49–57.

Romanello, M., D'Andrea, P., 2001. Dual mechanism of intercellular communication in HOBIT osteoblastic cells: a role for gap-junctional hemichannels. J. Bone Miner. Res. 16 (8), 1465–1476.

Saunders, M.M., et al., 2001. Gap junctions and fluid flow response in MC3T3-E1 cells. Am. J. Physiol. Cell Physiol. 281 (6), C1917–C1925.

Saunders, M.M., et al., 2003. Fluid flow-induced prostaglandin E2 response of osteoblastic ROS 17/2.8 cells is gap junction-mediated and independent of cytosolic calcium. Bone 32 (4), 350–356.

Schiller, P.C., et al., 1992. Hormonal regulation of intercellular communication: parathyroid hormone increases connexin43 gene expression and gap-junctional communication in osteoblastic cells. Mol. Endocrinol. 6 (9), 1433–1440.

Schiller, P.C., et al., 2001. Gap-junctional communication is required for the maturation process of osteoblastic cells in culture. Bone 28 (4), 362–369.

Schilling, A.F., et al., 2008. Gap junctional communication in human osteoclasts in vitro and in vivo. J. Cell Mol. Med. 12 (6A), 2497–2504.

Schirrmacher, K., Bingmann, D., 1998. Effects of vitamin D3, 17beta-estradiol, vasoactive intestinal peptide, and glutamate on electric coupling between rat osteoblast-like cells in vitro. Bone 23 (6), 521–526.

Seki, A., et al., 2017. Progressive atrial conduction defects associated with bone malformation caused by a connexin45 mutation. J. Am. Coll. Cardiol. 70 (3), 358–370.

Shen, V., et al., 1986. Prostaglandins change cell shape and increase intercellular gap junctions in osteoblasts cultured from rat fetal calvaria. J. Bone Miner. Res. 1 (3), 243–249.

Shen, C., et al., 2016. Glucocorticoid suppresses connexin43 expression by inhibiting the Akt/mTOR signaling pathway in osteoblasts. Calcif. Tissue Int. 99 (1), 88–97.

Shin, C.S., et al., 2000. Relative abundance of different cadherins defines differentiation of mesenchymal precursors into osteogenic, myogenic, or adipogenic pathways. J. Cell. Biochem. 78 (4), 566–577.

Shin, C.S., et al., 2005. Dominant negative N-cadherin inhibits osteoclast differentiation by interfering with beta-catenin regulation of RANKL, independent of cell-cell adhesion. J. Bone Miner. Res. 20 (12), 2200–2212.

Stains, J.P., Civitelli, R., 2005. Gap junctions regulate extracellular signal-regulated kinase signaling to affect gene transcription. Mol. Biol. Cell 16 (1), 64–72.

Stains, J.P., et al., 2003. Gap junctional communication modulates gene transcription by altering the recruitment of Sp1 and Sp3 to connexin-response elements in osteoblast promoters. J. Biol. Chem. 278 (27), 24377–24387.

Stains, J.P., et al., 2014. Molecular mechanisms of osteoblast/osteocyte regulation by connexin43. Calcif. Tissue Int. 94 (1), 55–67.

Suswillo, R.F., et al., 2017. Strain uses gap junctions to reverse stimulation of osteoblast proliferation by osteocytes. Cell Biochem. Funct. 35 (1), 56–65.

Talbot, J., et al., 2018. Connexin43 intercellular communication drives the early differentiation of human bone marrow stromal cells into osteoblasts. J. Cell. Physiol. 233 (2), 946–957.

Thi, M.M., et al., 2012. Connexin43 and pannexin1 channels in osteoblasts: who is the "hemichannel"? J. Membr. Biol. 245 (7), 401–409.
Tiede-Lewis, L.M., et al., 2017. Degeneration of the osteocyte network in the C57BL/6 mouse model of aging. Aging 9 (10), 2190–2208.
Tran, N.L., et al., 1999. N-Cadherin expression in human prostate carcinoma cell lines. An epithelial-mesenchymal transformation mediating adhesion withStromal cells. Am. J. Pathol. 155 (3), 787–798.
Troyanovsky, S.M., 1999. Mechanism of cell-cell adhesion complex assembly. Curr. Opin. Cell Biol. 11 (5), 561–566.
Tsuboi, M., et al., 1999. Tumor necrosis factor-alpha and interleukin-1beta increase the Fas-mediated apoptosis of human osteoblasts. J. Lab. Clin. Med. 134 (3), 222–231.
Tsutsumimoto, T., et al., 1999. TNF-alpha and IL-1beta suppress N-cadherin expression in MC3T3-E1 cells. J. Bone Miner. Res. 14 (10), 1751–1760.
Tu, B., et al., 2016. Inhibition of connexin43 prevents trauma-induced heterotopic ossification. Sci. Rep. 6, 37184.
Turel, K.R., Rao, S.G., 1998. Expression of the cell adhesion molecule E-cadherin by the human bone marrow stromal cells and its probable role in CD34(+) stem cell adhesion. Cell Biol. Int. 22 (9–10), 641–648.
Valiunas, V., et al., 2005. Connexin-specific cell-to-cell transfer of short interfering RNA by gap junctions. J. Physiol. 568 (Pt 2), 459–468.
Vazquez, M., et al., 2014. A new method to investigate how mechanical loading of osteocytes controls osteoblasts. Front. Endocrinol. 5, 208.
Vendome, J., et al., 2011. Molecular design principles underlying beta-strand swapping in the adhesive dimerization of cadherins. Nat. Struct. Mol. Biol. 18 (6), 693–700.
Wagner, A.S., et al., 2017. Osteogenic differentiation capacity of human mesenchymal stromal cells in response to extracellular calcium with special regard to connexin43. Ann. Anat. 209, 18–24.
Wang, H., et al., 2015. The osteogenic niche promotes early-stage bone colonization of disseminated breast cancer cells. Cancer Cell 27 (2), 193–210.
Wang, Z., et al., 2016. Alternation in the gap-junctional intercellular communication capacity during the maturation of osteocytes in the embryonic chick calvaria. Bone 91, 20–29.
Watkins, M., et al., 2011. Osteoblast connexin43 modulates skeletal architecture by regulating both arms of bone remodeling. Mol. Biol. Cell 22 (8), 1240–1251.
Watkins, M.P., et al., 2012. Bisphosphonates improve trabecular bone mass and normalize cortical thickness in ovariectomized, osteoblast connexin43 deficient mice. Bone 51 (4), 787–794.
Watkins, M.G., Wang, S., Zhang, B., Civitelli, R., 2014. Connexin45 is involved in cancellous but not cortical bone homeostasis. J. Bone Miner. Res. 29 (Suppl. 1).
Weber, P.A., et al., 2004. The permeability of gap junction channels to probes of different size is dependent on connexin composition and permeant-pore affinities. Biophys. J. 87 (2), 958–973.
Wheelock, M.J., Johnson, K.R., 2003. Cadherin-mediated cellular signaling. Curr. Opin. Cell Biol. 15 (5), 509–514.
Wickline, E.D., et al., 2016. alphaT-catenin is a constitutive actin-binding alpha-catenin that directly couples the Cadherin.Catenin complex to actin filaments. J. Biol. Chem. 291 (30), 15687–15699.
Wildenberg, G.A., et al., 2006. p120-catenin and p190RhoGAP regulate cell-cell adhesion by coordinating antagonism between Rac and Rho. Cell 127 (5), 1027–1039.
Xiong, Q., et al., 2009. Quantitative trait loci, genes, and polymorphisms that regulate bone mineral density in mouse. Genomics 93 (5), 401–414.
Xu, H., et al., 2015. Connexin43 channels are essential for normal bone structure and osteocyte viability. J. Bone Miner. Res. 30 (3), 436–448.
Yamada, S., et al., 2005. Deconstructing the cadherin-catenin-actin complex. Cell 123 (5), 889–901.
Yamada, K.M., Geiger, B., 1997. Molecular interactions in cell adhesion complexes. Curr. Opin. Cell Biol. 9 (1), 76–85.
Yamada, Y., Ando, F., Shimokata, H., 2007. Association of candidate gene polymorphisms with bone mineral density in community-dwelling Japanese women and men. Int. J. Mol. Med. 19 (5), 791–801.
Yang, J., et al., 2008. HMGB1 is a bone-active cytokine. J. Cell. Physiol. 214 (3), 730–739.
Yang, H., et al., 2016a. Connexin43 affects osteogenic differentiation of the posterior longitudinal ligament cells via regulation of ERK activity by stabilizing Runx2 in ossification. Cell. Physiol. Biochem. 38 (1), 237–247.
Yang, H., et al., 2016b. N-cadherin restrains PTH repressive effects on sclerostin/SOST by regulating LRP6-PTH1R interaction. Ann. N. Y. Acad. Sci. 1385 (1), 41–52.
Yao, H., et al., 2014. Parathyroid hormone enhances hematopoietic expansion via upregulation of cadherin-11 in bone marrow mesenchymal stromal cells. Stem Cell. 32 (8), 2245–2255.
Yellowley, C.E., et al., 2000. Functional gap junctions between osteocytic and osteoblastic cells. J. Bone Miner. Res. 15 (2), 209–217.
York, S.L., Sethu, P., Saunders, M.M., 2016. Impact of gap junctional intercellular communication on MLO-Y4 sclerostin and soluble factor expression. Ann. Biomed. Eng. 44 (4), 1170–1180.
Zappitelli, T., et al., 2013. The G60S connexin43 mutation activates the osteoblast lineage and results in a resorption-stimulating bone matrix and abrogation of old-age-related bone loss. J. Bone Miner. Res. 28 (11), 2400–2413.
Zhang, Y., et al., 2011. Enhanced osteoclastic resorption and responsiveness to mechanical load in gap junction deficient bone. PLoS One 6 (8), e23516.
Zhou, Z., et al., 2008. HMGB1 regulates RANKL-induced osteoclastogenesis in a manner dependent on RAGE. J. Bone Miner. Res. 23 (7), 1084–1096.
Ziambaras, K., et al., 1998. Cyclic stretch enhances gap junctional communication between osteoblastic cells. J. Bone Miner. Res. 13 (2), 218–228.
Zong, L., et al., 2016. Gap junction mediated miRNA intercellular transfer and gene regulation: a novel mechanism for intercellular genetic communication. Sci. Rep. 6, 19884.

Section C

Bone remodeling and mineral homeostasis

Chapter 19

Histomorphometric analysis of bone remodeling

Carolina A. Moreira[1] and David W. Dempster[2,3]

[1]*Bone Unit of Endocrine Division of Federal University of Parana, Laboratory PRO, Section of Bone Histomorphometry, Pro Renal Foundation, Curitiba, Parana, Brazil;* [2]*Regional Bone Center, Helen Hayes Hospital, West Haverstraw, NY, United States;* [3]*Department of Pathology and Cell Biology, College of Physicians and Surgeons, Columbia University, New York, NY, United States*

Chapter outline

Introduction **445**
Tetracycline labeling and the surgical procedure **445**
Sample preparation and analysis 447
Routine histomorphometric variables 447
Static parameters 448
Dynamic parameters 449
Normal bone 449
Hyperparathyroidism 449
Osteomalacia 451
Renal osteodystrophy 451
Osteoporosis 452
Clinical indications for bone biopsy 453
Histomorphometric studies of the effects of osteoporosis drugs **454**
Anticatabolic agents 454
Calcitonin 454
Hormone therapy 454
Selective estrogen receptor modulators 454
Bisphosphonates 455
Denosumab 455
Osteoanabolic therapies 456
PTH(1−34) and PTH(1−84) 456
Abaloparatide 457
Romosozumab 460
Comparative studies of anabolic and anticatabolic drugs 460
SHOTZ 460
AVA study: differential effects of teriparatide and denosumab on intact parathyroid hormone and bone formation indices 461
Conclusion **461**
References **461**

Introduction

Bone histomorphometry is a gold standard technique for evaluation of bone remodeling. It has provided great information about the metabolic bone diseases, including the changes on bone structure and remodeling following treatment.

Tetracycline labeling and the surgical procedure

Prelabeling the patient with tetracycline prior to biopsy allows the histomorphometrist to quantify precisely the rate of bone formation at the time of the biopsy (Frost, 1983). About 3 weeks prior to the biopsy, the patient is given a 3-day course of tetracycline. This is followed by 12 drug-free days and then another 3-day course of tetracycline. This is termed a 3:12:3 sequence. The biopsy should not be performed until at least 5 days after the last tetracycline dose to prevent the last label from leaching out during the processing of the biopsy. This is often denoted as a 3:12:3:5 sequence. The tetracycline binds irreversibly to recently formed hydroxyapatite crystals at sites undergoing new deposition. When the histomorphometrist cuts and visualizes thin sections of the biopsy in a microscope equipped with ultraviolet illumination, the tetracycline fluoresces to label the sites of new bone formation (Fig. 19.1). The labels can be either double labels, if bone formation at that site is continuous throughout the labeling sequence, or single labels, if formation starts after the first or stops before the second label is administered. Demeclocycline, tetracycline, and oxytetracycline can all be used as fluorochrome labels. In

Principles of Bone Biology. https://doi.org/10.1016/B978-0-12-814841-9.00019-1

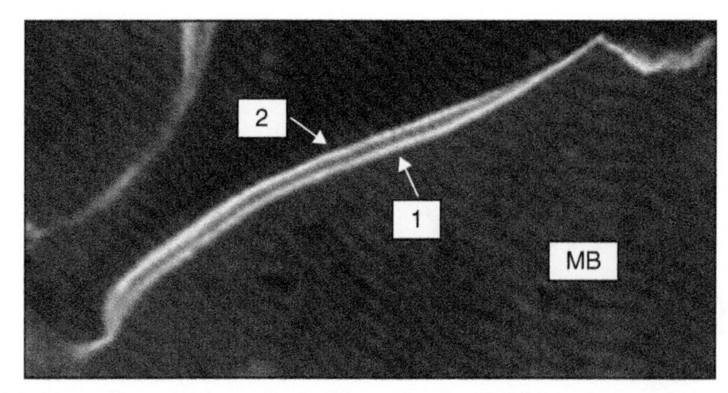

FIGURE 19.1 Double tetracycline label in an iliac crest bone biopsy. The patient was labeled with demeclocycline in a 3:12:3 sequence. *1*, label 1; *2*, label 2; *MB*, mineralized bone.

our laboratory, our preference is for demeclocycline, 600 mg/day (4 × 150-mg tablets), taken on an empty stomach. Dairy products and antacids containing aluminum, calcium, or magnesium should be avoided because they impair absorption. Tetracyclines can cause nausea, vomiting, and diarrhea in some patients, and all patients should be cautioned to avoid excessive exposure to sunlight and UV light because tetracyclines can cause skin phototoxicity. Tetracyclines should not be given to children less than 8 years of age or to pregnant women because, just as it is incorporated into bones, it is incorporated into growing teeth and discolors them.

Although the original site for bone biopsy was the rib, it is now performed exclusively at the anterior iliac crest, which is easily accessible, and the biopsy can be performed with minimal complications. This site also allows one to sample both cancellous and cortical bone in a single biopsy (Figs. 19.2 and 19.3). The structure and cellular activity at this site have been well characterized in a number of laboratories and have been shown to correlate with other clinically relevant skeletal sites, such as the spine and the hip (Bordier et al., 1964; Parfitt, 1983a; Rao, 1983; Dempster, 1988).

The biopsy generally is performed with a standard trephine with an internal diameter of at least 8 mm to obtain sufficient tissue and to minimize damage to the sample (Bordier et al., 1964; Rao, 1983). Immediately before the procedure, the patient should be sedated, usually with intravenous meperidine hydrochloride (Demerol) and diazepam (Valium). The skin, subcutaneous tissue, muscle, and, in particular, the periosteum covering both the lateral and the medial aspects of the

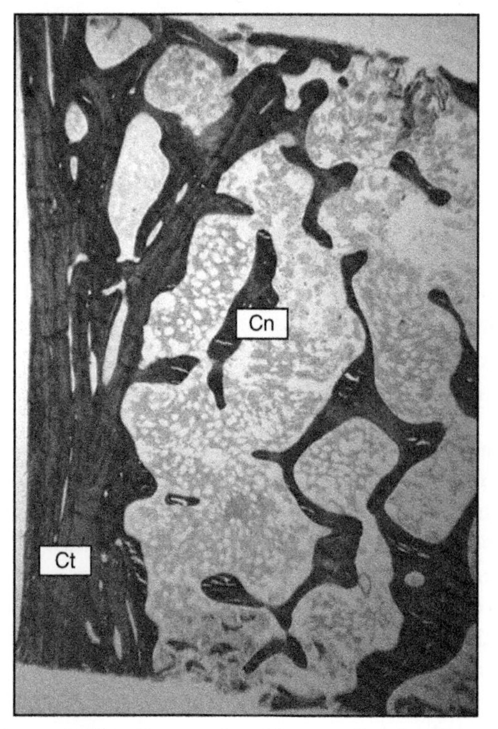

FIGURE 19.2 Low-power photomicrograph of an iliac crest bone biopsy section showing cancellous (*Cn*) and cortical (*Ct*) bone.

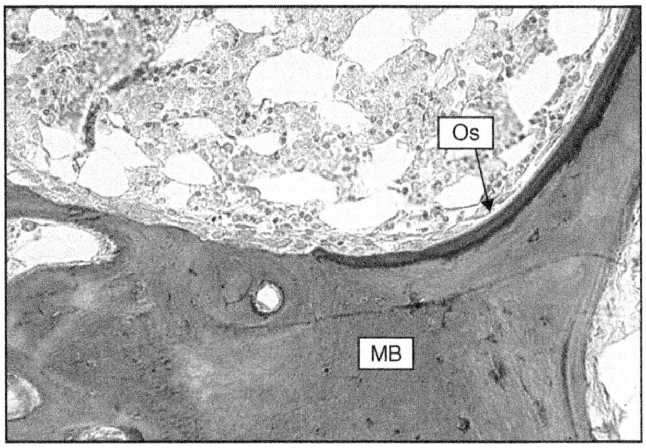

FIGURE 19.3 Photomicrograph of an iliac crest bone biopsy section showing an osteoid seam (*Os*) on the surface of mineralized bone (*MB*).

ilium, should be thoroughly anesthetized with local anesthetic. Access to the iliac crest is achieved through a 2- to 3-cm skin incision made at a point 2 cm posterior and 2 cm inferior to the anterior superior iliac spine. It is important to locate this site carefully because there is considerable variation in bone structure around this location. To avoid damage to the biopsy, which could render it uninterpretable, the trephine should be rotated back and forth with gentle but firm pressure so that it cuts rather than pushes through the ilium. The patient should refrain from excessive activity for 24 h after the procedure and a mild analgesic may be required. Significant complications from transiliac bone biopsy are rare. In an international multicenter study involving 9131 transiliac biopsies, complications were recorded in 64 patients (0.7%) (Rao, 1983). The most common complications were hematoma and pain at the biopsy site that persisted for more than 7 days; rarer complications included wound infection, fracture through the iliac crest, and osteomyelitis.

Sample preparation and analysis

The biopsy should be fixed in 70% ethanol because more aqueous fixatives may leach the tetracycline from the bone. After a fixation period of 4–7 days, the biopsy is dehydrated in ethanol, cleared in toluene, and embedded in methyl methacrylate without decalcification. The polymerized methyl methacrylate allows good-quality, thin (5–10 μm) sections to be cut on a heavy-duty microtome. The sections are then stained with a variety of dyes to allow good discrimination between mineralized and unmineralized bone matrix, which is termed "osteoid" (Figs. 19.4 and 19.5), and clear visualization of the cellular components of bone and marrow (see Fig. 19.7). Unstained sections are also mounted to allow observation of the tetracycline labels by fluorescence microscopy (see Fig. 19.1) (Baron et al., 1983a,b; Weinstein, 2002). The sections are subjected to morphometric analysis, according to standard stereological principles, using either simple "point-counting" techniques or computer-aided image analysis (Parfitt, 1983b; Malluche and Faugere, 1987; Compston, 1997).

Routine histomorphometric variables

A large number of histomorphometric variables can be measured or derived. Because the morphometric analysis is extremely time consuming, the number of variables evaluated depends on whether the biopsy specimen is being analyzed for diagnostic or research purposes. Listed next are eight indices of trabecular bone that are of particular clinical relevance. For a detailed account of more theoretical aspects of bone biopsy analysis, see Parfitt (1983a) and Frost (1983).

More recently, increased attention has been given to cortical bone, as it comprises 80% of the human skeleton and its failure contributes significantly to the nonvertebral fracture burden.

It is conventional to divide histomorphometric parameters into two categories. Static variables yield information on the amount of bone present and the proportion of bone surface engaged in the different phases of the remodeling cycle. Dynamic variables provide information on the rate of cell-mediated processes involved in remodeling. This category can be evaluated only in tetracycline-labeled biopsies. By measuring the extent of tetracycline-labeled surface and the distance between double tetracycline labels, the bone formation rate can be computed directly in a single biopsy. Conversely, the resorption rate can be calculated only indirectly, using certain indices of bone formation, in a single biopsy specimen or from two sequential biopsy specimens (Frost, 1983; Eriksen, 1986).

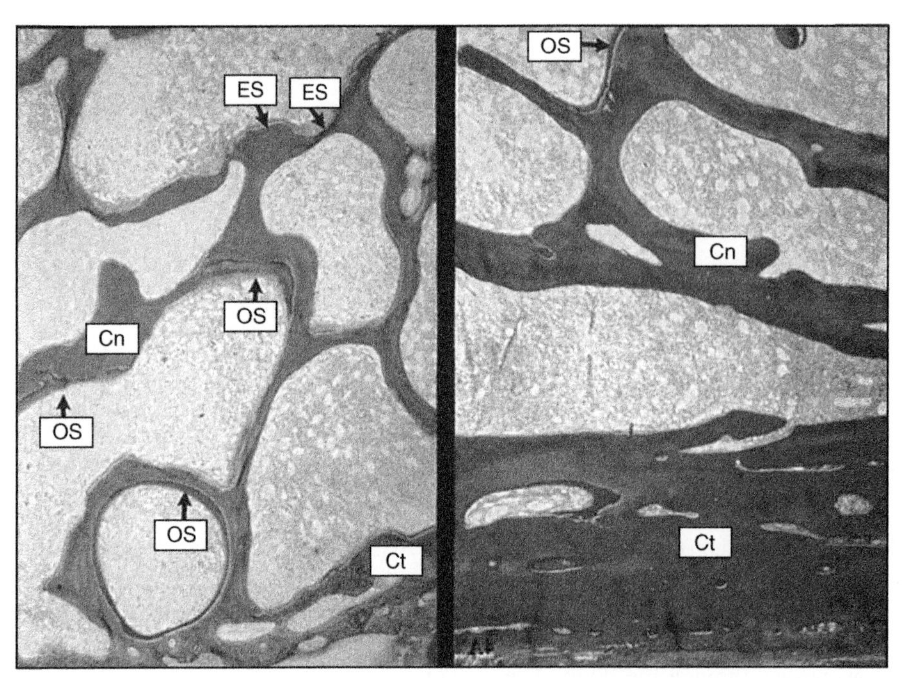

FIGURE 19.4 Photomicrographs of iliac bone biopsy sections from a subject with primary hyperparathyroidism (PHPT) (left), compared with a control subject (right). Note preservation of cancellous bone (*Cn*) and loss of cortical bone (*Ct*) in the subject with PHPT. Also note the marked extension of eroded surface (*ES*) and osteoid surface (*OS*) in PHPT.

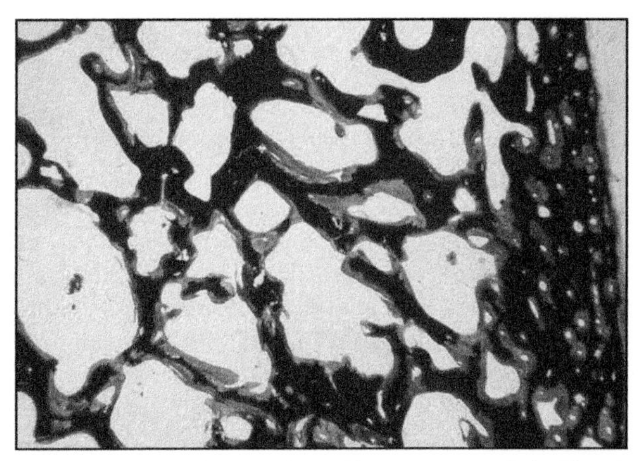

FIGURE 19.5 Photomicrograph of a bone biopsy section from a patient with severe osteomalacia. Note the dramatic extension of osteoid surface (stained *light gray*) and the increase in osteoid thickness.

Static parameters

A list of five commonly used static variables is given here. The terms and abbreviations for all histomorphometric variables have been standardized by the Histomorphometry Nomenclature Committee of the American Society for Bone and Mineral Research, whose recommendations have been widely adopted (Parfitt et al., 1987; Dempster et al., 2013).

Cancellous bone volume (Cn-BV/TV, %) is the fraction of a given volume of whole cancellous bone tissue (i.e., bone + bone marrow) that consists of mineralized and nonmineralized bone.

Osteoid volume (OV/BV, %) is the fraction of a given volume of bone tissue (mineralized bone + osteoid) that is osteoid (i.e., unmineralized matrix).

Osteoid surface (OS/BS, %) is the fraction of the entire trabecular surface that is covered by osteoid seams.

Osteoid thickness (O.Th, μm) is the average thickness of osteoid seams.
Eroded surface (ES/BS, %) is the fraction of the entire trabecular surface that is occupied by resorption bays (Howship lacunae), including both those with and those without osteoclasts.

The main cortical bone parameters are as follows.

Cortical area (Ct.Ar) is the total area of cortical bone, including inner and outer cortices.
Cortical width (Ct.Wi) is the average width of both inner and outer cortices. In growing individuals, however, in whom there are differences in bone cell activity at the internal and external cortices, they are recorded separately.
Cortical porosity number (Ct.Po.N) is the total number of pores of both inner and outer cortices.
Cortical porosity area (Ct.Po.Ar) is the total area of pores.

Dynamic parameters

Following is a list of commonly used dynamic parameters.

Mineral apposition rate (MAR, μm/day): this is calculated by dividing the average distance between the first and the second tetracycline label by the time interval (e.g., 15 days) separating them. It is a measure of linear rate of production of calcified bone matrix by the osteoblasts.
Mineralizing surface (MS/BS, %): this is the fraction of trabecular surface bearing double tetracycline label plus one-half of the singly labeled surface. It is a measure of the proportion of bone surface on which new mineralized bone was being deposited at the time of tetracycline labeling.
Bone formation rate (BFR/BS, $\mu m^3/\mu m^2$ per day): this is the volume of mineralized bone made per unit surface of trabecular bone per year. It is calculated by multiplying the mineralizing surface by the mineral apposition rate.
Many of these static and dynamic parameters described for the cancellous envelope can also be measured in the other three bone envelopes: endocortical, intracortical, and periosteal (Dempster et al., 2012, 2016b).

In the following section, we will briefly review the remodeling process in normal bone and the changes that occur in a number of common disease states, as assessed by bone histomorphometry (Table 19.1).

Normal bone

Bone undergoes a continuous process of renewal, with approximately 25% of trabecular bone and 3% of cortical bone being replaced annually. This remodeling process is referred to as a quantum phenomenon because it occurs in discrete units or "packets." Osteoclasts resorb the old bone and osteoblasts replace it. The group of cells that work cooperatively to create one new packet of bone is called a bone remodeling unit. In normal trabecular bone, approximately 900 bone remodeling units are initiated each day. In cortical bone, about 180 remodeling units are initiated per day (Frost, 1973; Parfitt, 1983a, 1988; Dempster, 2002).

Hyperparathyroidism

Increased circulating parathyroid hormone (PTH) levels increase the activation frequency of bone remodeling units, resulting in increased osteoclast and osteoblast numbers. As a result, histomorphometric analysis of a biopsy from a patient with either primary or secondary hyperparathyroidism reveals increases in eroded surface, osteoid surface, and mineralizing surface (see Fig. 19.4) (Melsen et al., 1983; Malluche and Faugere, 1987; Parisien et al., 1990; Silverberg et al., 1990). Mineralizing surface is increased, but mineral apposition rate is slightly reduced. However, the increased mineralizing surface overcompensates for the decrease in mineral apposition rate, so that the bone formation rate, the product of these two variables, is increased. Bone turnover is higher in hyperparathyroid patients with vitamin D insufficiency (Silverberg et al., 1990). Cancellous bone volume and trabecular connectivity are preserved in primary hyperparathyroidism (Parisien et al., 1992). The elevated bone turnover is often accompanied by increased deposition of immature (woven) bone and marrow fibrosis, in particular, in cases of severe secondary hyperparathyroidism. In cortical bone, there is an increase in cortical porosity and a decrease in cortical width (Fig. 19.4).

Because the biopsy reflects the long-term effects of excessive remodeling activity (e.g., increased eroded surface) it can be a sensitive indicator of parathyroid gland hyperactivity, especially when this is mild or intermittent. However, examination of the biopsy alone does not allow one to distinguish between primary and secondary hyperparathyroidism.

TABLE 19.1 Bone biopsy variables in a variety of disease states.[a]

Disease state	Cancellous bone volume	Osteoid volume	Osteoid surface	Osteoid thickness	Eroded surface	Mineral apposition rate	Mineralizing surface	Bone formation rate
Hyperparathyroidism[b]	N or ↑	↑	↑	N	↑	↓	↑	↑
Osteomalacia[c]	N	↑	↑	↑	↑	↓	↓	↓
Renal osteodystrophy/dialysis[d]	↓ or ↑	↓ or ↑	↓ or ↑	↓ or ↑	↑	↓ or ↑	↓ or ↑	↓ or ↑
Postmenopausal or senile osteoporosis[e]	↓ or N	N or ↑	N or ↑	N or ↓	N or ↑	N or ↓	N, ↑, or ↓	N, ↑, or ↓
Cushing syndrome and corticosteroid-induced osteoporosis[f]	↓ or N	N	↑	↓	↑	↓	↓	↓
Paget disease[g]	↑	↑	↑	↓	↑	↑	↑	↑
Thyrotoxicosis[h]	↓	↑	↑	↓	↑	↑	↑	↑
Hypothyroidism[h]	N	↓	N	↓	N	↓	↓	↓
Medullary thyroid carcinoma[h]	N	↑	↑	N	↑	↓	↑	N
Multiple myeloma[i]	N, ↑, or ↓	↑	↑	↓	↑	↓	↑	↑
Osteogenesis imperfect tarda[j]	↓	N	↑	↓	↑ or N	↓	N	↓

[a] *N, normal; ↑, increased; ↓, decreased.*
[b] *Melsen et al. (1983), Malluche and Faugere (1987), Parisien et al. (1990), 1992, Silverberg et al. (1990).*
[c] *Teitelbaum (1980), Jaworski (1983), Malluche and Faugere (1987), Siris et al. (1987).*
[d] *Malluche et al. (1976), Boyce et al. (1982), Hodsman et al. (1982), Charhon et al. (1985), Dunstan et al. (1985), Parisien et al. (1988), Salusky et al. (1988), Felsenfeld et al. (1991), Sherrard et al. (1993), Coburn and Salusky (2001), Slatopolsky and Delmez (2002).*
[e] *Meunier et al. (1981), Parfitt et al. (1982), Whyte et al. (1982), Civitelli et al. (1988), Meunier (1988), Garcia Carasco et al., 1989, Arlot et al. (1990), Eriksen et al. (1990), Kimmel et al. (1990), Steiniche et al. (1994), Dempster (2000).*
[f] *Bressot et al. (1979), Dempster (1989).*
[g] *Meunier et al. (1980).*
[h] *Melsen et al. (1983).*
[i] *Valentin-Opran et al. (1982).*
[j] *Baron et al. (1983), Ste-Marie et al. (1984).*

Osteomalacia

The hallmark of osteomalacia, regardless of the underlying pathogenetic mechanism, is inhibition of bone mineralization. Although mineralization is inhibited, the osteoblasts continue to synthesize and secrete organic matrix, leading to an accumulation of osteoid (see Fig. 19.5). Although the cancellous bone volume is normal in osteomalacia, the amount of mineralized bone is actually reduced.

Careful analysis of the dynamic parameters is called for in suspected cases of osteomalacia. At some formation sites, mineral is still deposited, but at a reduced rate, resulting in low values for mineral apposition rate. At other sites, mineralization may be completely inhibited, resulting in reduced mineralizing surface. The decrease in both these variables markedly reduces bone formation rate. The accumulation of osteoid is reflected in increased osteoid thickness, osteoid surface, and osteoid volume. If PTH secretion is elevated, the activation frequency of bone remodeling units is enhanced and the biopsy may show an increase in eroded surface. However, as osteoid surface increases, eroded surface often declines, because osteoid is resistant to osteoclastic resorption. An elevated activation frequency, when accompanied by mineralization failure, will enhance the rate at which osteoid is deposited (Teitelbaum, 1980; Jaworski, 1983; Malluche and Faugere, 1987; Siris et al., 1987).

Renal osteodystrophy

Chronic renal failure is usually accompanied by phosphate retention and hyperphosphatemia, which leads to a reciprocal decrease in serum ionized calcium concentration and secondary hyperparathyroidism. Furthermore, as functional renal mass decreases, the plasma 1,25-dihydroxyvitamin D level falls, leading to impaired intestinal calcium absorption, which exacerbates hypocalcemia and ultimately may impair bone mineralization. As a result of these marked disturbances in metabolism, it is perhaps not surprising that the bone biopsy findings in renal osteodystrophy are heterogeneous (Malluche et al., 1976, 2011; Boyce et al., 1982; Hodsman et al., 1982; Charhon et al., 1985; Dunstan et al., 1985; Parisien et al., 1988; Salusky et al., 1988; Moriniere et al., 1989; Felsenfeld et al., 1991; Hercz et al., 1993; Sherrard et al., 1993; Coburn and Salusky, 2001; Slatopolsky and Delmez, 2002). Indeed, in allowing a better understanding of the skeletal status in patients with chronic renal failure, the bone biopsy continues to play an important role in the management of this disease (Ketteler et al., 2017). Thus, renal osteodystrophy has been subdivided into two broad types, primarily on the basis of histomorphometric features. One type is characterized by normal or high bone turnover and a second is characterized by low bone turnover.

The most frequently observed biopsy changes in patients with end-stage renal disease are the result of chronic excess PTH secretion on the skeleton. These are classified as normal/high turnover, and include osteitis fibrosa, mild hyperparathyroidism, and mixed bone disease (Fig. 19.6). These features are characterized histomorphometrically by increased eroded surface, osteoid surface, and mineralizing surface. In osteitis fibrosa, however, woven osteoid is often present and there are variable amounts of peritrabecular marrow fibrosis in contrast to the minimal or absent fibrosis observed in mild

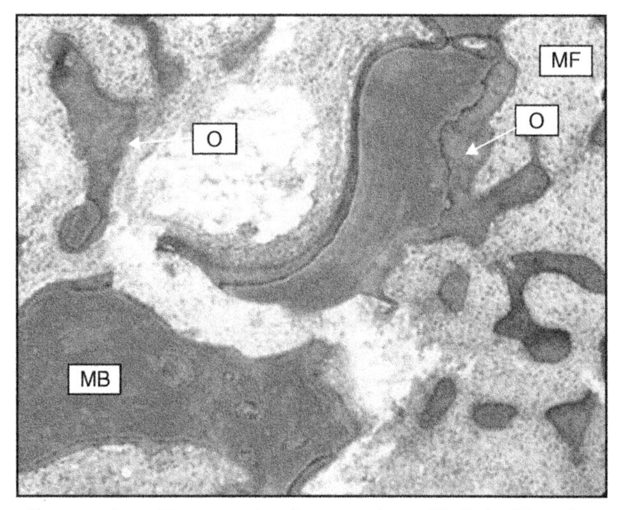

FIGURE 19.6 Photomicrograph of an iliac crest bone biopsy section from a patient with "mixed" renal osteodystrophy. Note deposition of woven osteoid (*O*) and marrow fibrosis (*MF*). *MB*, mineralized bone.

hyperparathyroidism. On the other end of the spectrum, low bone turnover is frequently observed in patients undergoing dialysis, albeit less often than high-turnover disease. The low-turnover states are classified as osteomalacia and aplastic or adynamic bone disease. Patients with osteomalacia have evidence of reduced values for dynamic variables accompanied by the accumulation of excess osteoid, whereas patients with aplastic or adynamic disease have a reduced tetracycline-based bone formation rate, but normal or reduced osteoid volume. In the 1970s and 1980s, most symptomatic patients with osteomalacia or adynamic bone disease showed evidence of aluminum accumulation, with more than 25% of surfaces displaying aluminum stain. They were considered to have aluminum-related bone disease (Boyce et al., 1982; Hodsman et al., 1982; Dunstan et al., 1985; Parisien et al., 1988). The aluminum was primarily derived from aluminum-containing phosphate binders used to control hyperphosphatemia and dialysis solutions that were contaminated with aluminum. Like tetracycline, aluminum accumulates at sites of new bone formation, where it may directly inhibit mineralization, which is manifested in an increase in osteoid thickness and osteoid surface. However, aluminum also is toxic to osteoblasts and may impair their ability both to synthesize and to mineralize bone matrix, resulting in a decrease in mineral apposition rate and the mineralizing surface. However, if matrix production is also reduced, osteoid thickness will not be elevated. With appreciation of the sources of aluminum contamination and increased use of calcium-containing phosphate binders, aluminum-related bone disease has become much less common in recent years.

Another form of low-turnover bone disease has been described that is not accompanied by significant aluminum accumulation. This is called idiopathic aplastic or adynamic bone disease. Its pathogenesis is unclear but may be related to various therapeutic maneuvers designed to prevent or reverse hyperparathyroidism in patients undergoing dialysis, including the use of dialysates with higher calcium concentrations (3.0–3.5 mEq/L), large doses of calcium-containing phosphate binders, and calcitriol therapy. As a rule, these patients have few symptoms of bone disease, and this "disease" ultimately may prove to be a histological rather than a clinically relevant form of bone disorder. However, it is unknown whether patients with adynamic bone disease are at increased risk of the development of skeletal problems in the future.

A 2011 large study by Malluche et al. (2011) provided histomorphometric data on 635 adult patients with chronic kidney disease stage 5 on dialysis. The authors employed the so-called TMV classification (turnover [T], mineralization [M], and volume [V]), which they had previously proposed (Malluche and Monier-Faugere, 2006). They reported that a mineralization defect was observed in only 3% of the subjects. This is probably due to the lower use of aluminum-containing phosphate binders nowadays, compared with the aforementioned earlier studies. The authors also noted distinct racial differences. For example, 62% of white subjects exhibited low turnover, whereas high turnover was observed in 68% of black subjects.

Osteoporosis

The classic feature of bone biopsies in osteoporosis is the reduction in cancellous bone volume. Approximately 80% of patients with vertebral crush fractures have values that are lower than normal. In postmenopausal osteoporosis the reduction in cancellous bone volume is primarily caused by the loss of entire trabeculae and, to a lesser degree, by the thinning of those that remain (Fig. 19.7) (Meunier et al., 1981; Parfitt et al., 1982; Whyte et al., 1982; Meunier, 1988; Dempster, 2000).

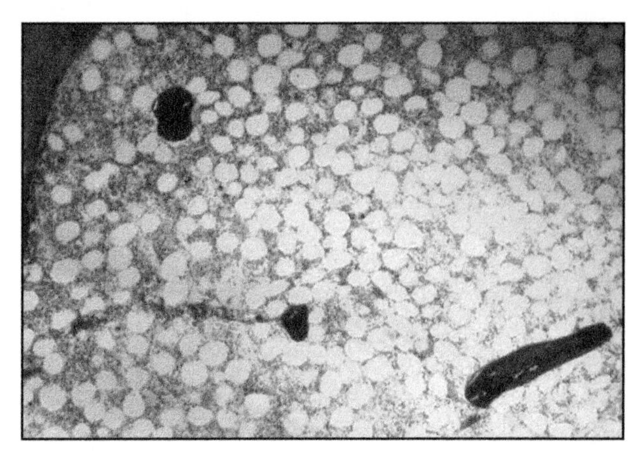

FIGURE 19.7 Photomicrograph of an iliac crest bone biopsy section from a patient with severe osteoporosis. Note marked reduction in cancellous bone volume in isolated trabecular profiles (*arrowheads*), which are cross sections of thin, rodlike structures in three dimensions.

With respect to the changes in the other static and dynamic variables in osteoporosis, there has been debate over whether patients can be stratified into high-, normal-, or low-turnover groups. Even if they can, the pathogenetic and clinical significance of this so-called histological heterogeneity in patients with osteoporosis is unclear. In a study of 50 postmenopausal women with untreated osteoporosis, two subsets of patients were identified: one with normal turnover and one with high turnover, with the high turnover representing 30% of the cases (Arlot et al., 1990). However, this conclusion was based on the finding of a bimodal distribution in the static parameter osteoid surface. The tetracycline-based bone formation rate, a dynamic measure of turnover rate, displayed a normal distribution. Based on the interval between the 10th percentile and the 90th percentile for calculated bone resorption rate in a group of normal postmenopausal women, another study classified 30% of women with untreated postmenopausal osteoporosis as having high turnover, whereas 64% and 6% had normal and low turnover, respectively (Eriksen et al., 1990). When bone formation rate was used as the discriminant variable, 19% were classified as having high turnover, 72% as having normal turnover, and 9% as having low turnover. On the other hand, in two studies of postmenopausal women with osteoporosis and their normal counterparts, the same wide variation in turnover indices was found in both groups, leading the investigators to conclude that there were no important subsets of patients with osteoporosis (Garcia Carasco et al., 1989; Kimmel et al., 1990). These studies, however, confirmed earlier observations that some patients with osteoporosis show profoundly depressed formation with little or no tetracycline uptake (Whyte et al., 1982).

From a clinical viewpoint, the desire to classify patients with osteoporosis according to their turnover status stems from the notion that the turnover rate may influence the response to particular therapeutic agents. For example, patients with high turnover rates may respond better to anticatabolic treatments. There was early evidence that this was the case for calcitonin (Civitelli et al., 1988). However, in clinical practice, the biopsy is an impractical way to determine turnover status. It was once believed that biochemical markers of bone resorption and formation would be useful in this regard, but this has yet to materialize.

Note that, in most cases, bone biopsy is performed when the disease is severe, with multiple fractures having already occurred. It is probable that, in many cases, the disturbances in bone metabolism that led to the reduction in bone mass and strength took place several years before the time of the biopsy and are no longer evident (Steiniche et al., 1994). Another confounding factor is that most patients who undergo biopsy for osteoporosis have already received treatment with one or more pharmaceutical agents.

Clinical indications for bone biopsy

In general, a bone biopsy is helpful only in metabolic bone diseases. Only rarely is a biopsy indicated in patients with localized skeletal disease such as Paget disease of bone, primary bone tumors, or bone metastases involving the iliac crest. The biopsy usually does not provide significantly greater insight into the disease process in postmenopausal women with osteoporosis. However, bone biopsy can be useful in patients who are less frequently affected by osteoporosis, such as young men and premenopausal women. Patients with osteopenia or women with postmenopausal osteoporosis should not have biopsies simply to measure cancellous bone volume to confirm the diagnosis of osteoporosis. The intraindividual and interindividual variability in cancellous bone volume is too great, and there is too much overlap between cancellous bone volume in patients with clinical osteoporosis and normal subjects to make this useful. However, bone biopsy is useful to exclude subclinical osteomalacia. In one study, 5% of patients with vertebral fractures displayed definitive evidence of osteomalacia on biopsy despite normal biochemical and radiological findings (Meunier, 1981). Moreover, the biopsy can be useful in identifying more precisely the probable cause of bone loss in individual patients with osteoporosis. For example, if the biopsy reveals or confirms a high bone turnover rate, it would be important to rule out endocrine disorders, such as hyperthyroidism and hyperparathyroidism. Finally, the biopsy is the best available way to evaluate the effect of various therapeutic maneuvers on bone cell function (e.g., Holland et al., 1994a,b; Marcus et al., 2000). This is discussed in detail in the context of therapies for osteoporosis in the following section.

As noted earlier, the bone biopsy can be extremely useful in patients with renal osteodystrophy (Ketteler et al., 2017), although the large number of patients with renal disease precludes its use in every case. In general, if a symptomatic patient has biochemical evidence of secondary hyperparathyroidism (hyperphosphatemia, hypocalcemia, and markedly elevated intact PTH levels), biopsy is not necessary because one can predict with reasonable certainty that it would reveal osteitis fibrosa. However, patients with renal disease who have bone pain and fractures without the biochemical profile of secondary hyperparathyroidism should undergo biopsy to determine whether they have osteomalacia or idiopathic, aplastic bone disease. Although aluminum accumulation is much less common nowadays, the biopsy will also permit the physician to determine whether it is a significant contributory factor. Although the biopsy can be useful in the clinical management of certain patients with bone disease, its principal use today is as a research tool.

Histomorphometric studies of the effects of osteoporosis drugs

In this section we will review what histomorphometry has revealed about the effects of drugs used to treat osteoporosis. The drugs will be covered under the headings of their two principal mechanisms of action: anticatabolic, also known as antiresorptive, and anabolic (Riggs and Parfitt, 2005).

Anticatabolic agents

Calcitonin

Intranasal calcitonin is approved to reduce the risk of vertebral fractures, but its efficacy in nonvertebral fractures has not been established (Silverman, 2003). There have been several histomorphometric studies of the effects of calcitonin in subjects with osteoporosis or with rheumatoid arthritis (Gruber et al., 1984, 2000; Marie and Caulin, 1986; Alexandre et al., 1988; Palmieri et al., 1989; Kroger et al., 1992; Pepene et al., 2004; Chesnut et al., 2005). In cancellous bone, calcitonin treatment reduced eroded surface (Kroger et al., 1982; Gruber et al., 1984) and active resorption surface (Alexandre et al., 1988) and mean resorption rate (Chesnut et al., 2005), with no observed decrease in osteoclasts (Marie and Caulin, 1986; Palmieri et al., 1989; Gruber et al., 2000). Most studies failed to reveal any differences in bone formation parameters, e.g., osteoblast number and perimeter, osteoid perimeter and thickness, mineralized perimeter, or mineral apposition rate (Gruber et al., 1984; Marie and Caulin, 1986; Alexandre et al., 1988; Chesnut et al., 2005). However, one report (Gruber et al., 2000) suggested that bone formation was not reduced to the same extent as resorption. Cancellous bone volume was shown to be unchanged (Alexandre et al., 1988; Gruber et al., 2000; Chesnut et al., 2005) or increased (Gruber et al., 1984; Alexandre et al., 1988; Palmieri et al., 1989; Marie and Caulin, 1986; Kroger et al., 1992).

Hormone therapy

Bone histomorphometry has been used by several investigators to assess the effects of hormone therapy (HT) on both cancellous and, in some studies, cortical bone of the ilium (Steiniche et al., 1989; Lufkin et al., 1992; Holland et al., 1994a,b; Eriksen et al., 1999; Vedi and Compston, 1996, Vedi et al., 2003). One of the most interesting studies was by Eriksen et al. (1999), who showed that 2 years of HT decreased resorption parameters, reducing bone formation at the basic multicellular unit level. Cancellous wall thickness was similar in treated and placebo groups, but there was a significant reduction in resorption rate in the HT group. This was in contrast to the placebo group, which showed a significant increase in erosion depth and a modest increase in resorption rate. The reduction in the size of the resorption cavity with HT was confirmed in a later study (Vedi and Compston, 1996), although that study also demonstrated a compensatory decrease in the wall width of trabecular bone packets. These findings were not replicated in a study (Steiniche et al., 1989) in which HT was given for only 1 year.

Estrogen treatment has been shown to stimulate bone formation in animal models (Chow et al., 1992a, 1992b; Edwards et al., 1992), but this remains controversial in humans (Steiniche et al., 1989; Lufkin et al., 1992; Holland et al., 1994a,b; Vedi and Compston, 1996; Wahab et al., 1997; Eriksen et al., 1999; Patel et al., 1999; Vedi et al., 1999, 2003). Standard doses of HT reduce osteoid and mineralizing surfaces and bone formation rate, with no change or a decrease in wall width (Steiniche et al., 1989; Lufkin et al., 1992; Holland et al., 1994a,b; Vedi and Compston, 1996; Eriksen et al., 1999; Patel et al., 1999; Vedi et al., 2003). On the other hand, long-term, high-dose HT was reported to increase cancellous wall width and to decrease eroded cavity area. Similarly, 6 years of subcutaneous HT increased cancellous bone volume with an increment in trabecular thickness and number as well as wall width (Khastgir et al., 2001a,b). Such anabolic actions of HT have also been reported in Turner syndrome treated with HT (Khastgir et al., 2003). The improvements in bone structure demonstrated by two-dimensional histomorphometric analysis are supported by micro-computed tomography findings of a higher ratio of plate- to rodlike structures (Jiang et al., 2005). In addition to these changes in histomorphometric variables, HT has also been shown to increase the degree of collagen cross-linking and bone mineralization, consistent with its primary action to lower bone turnover (Walters and Eyre, 1980; Holland et al., 1994a,b; Rey et al., 1995; Yamauchi, 1996; Khastgir et al., 2001a,b; Boivin and Meunier, 2002; Burr et al., 2003; Paschalis et al., 2003; Boivin et al., 2005).

Selective estrogen receptor modulators

Selective estrogen receptor modulators (SERMs) bind to the estrogen receptor and exhibit agonist actions in some tissues, such as bone, and antagonist actions in others, such as breast (Lindsay et al., 1997a,b). Bone histomorphometry studies are primarily limited to raloxifene. Two years of raloxifene treatment in the MORE trial (Ettinger et al., 1999) decreased the bone formation rate, without changes in eroded surface or osteoclast number at the 60-mg dose, whereas the dose of 120 mg also decreased the bone formation rate and showed a trend toward a decrease in eroded surface and osteoclast number (Ott et al., 2000). Cancellous bone volume, trabecular thickness, and cortical width were unchanged compared with baseline and the placebo group (Ott et al., 2000). A significant decrease in activation frequency was observed with a higher dose (150 mg) of raloxifene (Weinstein et al., 2003). In that study, raloxifene was shown to have effects similar to those of HT. However, a 6-month treatment with 60 mg of raloxifene did not suppress activation frequency and bone formation rate to the same extent as HT (Prestwood et al., 2000). Reductions in activation frequency, bone formation rate, and resorption cavity area have also been demonstrated for another SERM, tamoxifen (Wright et al., 1994). In contrast to HT, raloxifene had little effect on mineralization density as assessed by quantitative microradiography of the biopsy sections (Boivin et al., 2003).

Bisphosphonates

The bisphosphonates have been the mainstay of osteoporosis therapy and will continue to be so for some time to come (Fleisch, 1998). The effects of alendronate, the first bisphosphonate to be approved in the United States, have been investigated in patients with postmenopausal osteoporosis (Bone et al., 1997; Chavassieux et al., 1997; Arlot et al., 2005), as well as in patients with glucocorticoid-induced osteoporosis (Chavassieux et al., 2000). Alendronate reduced osteoid surface and thickness, mineralizing surface, bone formation rate, and activation frequency. The mineral apposition rate was unchanged (Bone et al., 1997; Chavassieux et al., 1997; Arlot et al., 2005). Although the primary target of bisphosphonates is the osteoclast, alendronate, like other anticatabolic agents, had little, if any effect on histomorphometric variables of bone resorption, including eroded surface and volume, osteoclast number, and erosion depth. This is inconsistent with the marked reductions seen in biochemical markers of bone resorption. The discrepancy is most likely explained by the fact that histomorphometric indices of bone resorption are static parameters, in contrast to bone formation indices, which are dynamic. In one study (Chavassieux et al., 1997), wall thickness of trabecular bone packets was increased after 2 years of treatment, but this effect was not observed after 3 years. Histomorphometric studies failed to show an improvement in cancellous bone microarchitecture compared with placebo-treated subjects, but such an effect has been reported for three-dimensional structural parameters obtained by micro-computed tomography (Recker et al., 2005), with the assumption that alendronate prevented the loss of structural integrity experienced by the placebo-treated patients. Consistent with its primary action to reduce the activation frequency, alendronate increased the degree of mineralization of the matrix (Meunier and Boivin, 1997; Boivin et al., 2000; Hernandez et al., 2001; Roschger et al., 1997, 2001).

There have also been extensive studies of the effects of risedronate on the bone biopsy. Here, a paired biopsy design was employed, with biopsies being obtained before and after treatment in the same subjects (Eriksen et al., 2002; Dufresne et al., 2003; Borah et al., 2004, 2005, 2006; Seeman and Delmas, 2006; Zoehrer et al., 2006). Like alendronate, 3 years of risedronate treatment decreased mineralizing surface, bone formation rate, and activation frequency (Eriksen et al., 2002). Again, no significant change was noted in eroded surface and depth, but there was a significant decrease in resorption rate after risedronate treatment, with a significant increase in erosion depth in placebo-treated subjects. Also similar to alendronate's effects, risedronate preserved cancellous microarchitecture, as assessed by micro-computed tomography (Dufresne et al., 2003; Borah et al., 2004, 2005). No significant changes were seen in three-dimensional structural variables compared with baseline in risedronate-treated women, whereas trabecular microstructure deteriorated significantly in a subset of placebo-treated women who exhibited higher bone turnover at baseline (Borah et al., 2004). Furthermore, the degree of structural deterioration was positively correlated with the bone turnover, confirming that high turnover has a deleterious effect on bone structure. Similar results were reported for early postmenopausal women who were treated for just 1 year with risedronate (Dufresne et al., 2003). The reduction in bone turnover was associated with an increase in bone mineralization density, but there was no evidence of an abnormally high degree of mineralization, even when treatment was extended to 5 years (Borah et al., 2006; Durchschlag et al., 2006; Seeman and Delmas, 2006; Zoehrer et al., 2006).

There are a number of other studies on the effects of different bisphosphonates, such as zoledronate and ibandronate, on iliac bone (Recker et al., 2004, 2008). In general, the data obtained in patients with osteoporosis treated with these bisphosphonates are similar to those obtained with alendronate and risedronate (Recker et al., 2004, 2008). It should also be noted that there is evidence of dramatic improvements in bone structure and turnover in children with osteogenesis imperfecta treated with bisphosphonates (Munns et al., 2005).

Denosumab

Bone histomorphometry substudies in the FREEDOM trial have demonstrated the potent antiresorptive mechanism of action of denosumab. Denosumab is a human monoclonal antibody against receptor activator of NF-κB ligand, which reversibly inhibits osteoclast-mediated bone resorption. Its effects on bone histomorphometry were published for the first time by Reid et al. (2010) after 24 and/or 36 months of treatment with denosumab in postmenopausal women with osteoporosis. The results demonstrated a significant reduction in eroded surface by more than 80%, whereas osteoclasts were absent from more than 50% of biopsies in the denosumab group. Double labeling in trabecular bone was observed in 19% of those treated with denosumab, while it was present in 94% of the placebo group. Bone-formation rate was reduced by 97% in the denosumab group (Reid, 2010). However, even with this important reduction at bone remodeling, the qualitative histologic evaluation of biopsies was unremarkable, showing normally mineralized lamellar bone. Furthermore, bone biopsies after denosumab treatment were compared with biopsies of subjects after treatment with alendronate in the STAND trial. In this study, dynamic indices of bone turnover tended to be lower in the denosumab group than in the alendronate group. The presence of double labeling in trabecular bone was seen in 20% of the denosumab biopsies and in 90% of the alendronate samples, suggesting that the antiresorptive action of denosumab is more potent than that of the bisphosphonate.

Data from the FREEDOM extension trial after 5 or 10 years of denosumab treatment demonstrated that long-term treatment with denosumab maintained normal bone histology and structure (Brown et al., 2014; Bone et al., 2018), Furthermore, denosumab continued to induce a low turnover state, consistent with its mechanism of action.

Osteoanabolic therapies

Bone histomorphometry has confirmed that the mechanism of action of anabolic agents is fundamentally different from that of anticatabolic drugs (Riggs and Parfitt, 2005). Rather than reducing the activation frequency of bone remodeling, anabolic agents elevate it with a positive bone balance. In each bone remodeling unit, more bone is formed than was resorbed. Bone formation is increased prior to the increase in bone resorption. Consequently, anabolic agents are able to improve, rather than simply preserve, cancellous and cortical bone microarchitecture.

PTH(1−34) and PTH(1−84)

The first bone biopsy studies of the effects of PTH(1−34) (teriparatide) were conducted in postmenopausal women with osteoporosis who were treated concurrently with PTH(1−34) and HT for 6 or 12 months (Reeve et al., 1980, 1991; Bradbeer et al., 1992). Hodsman et al. (1993, 2000) also used bone biopsy to study the effects of a cyclical regimen of 28 days of PTH(1−34) every 3 months, with or without sequential calcitonin, for 2 years. Dempster et al. (2001) and Misof et al. (2003) performed paired biopsies in men with osteoporosis treated with PTH(1−34) for 18 months, as well as in postmenopausal women treated with a combination of PTH(1−34) and HT for 3 years. Biopsy studies of the effects of monotherapy with PTH(1−34) were completed as part of a multinational fracture trial (Neer et al., 2001; Jiang et al., 2003; Dobnig et al., 2005; Paschalis et al., 2005; Ma et al., 2006). Arlot et al. (2005) compared the effects of PTH(1−34) with those of alendronate in postmenopausal women with osteoporosis. The effects of the two agents on activation frequency and bone formation rate were diametrically opposed. Compared with appropriate reference ranges (Arlot et al., 1990; Chavassieux et al., 1997), the bone formation rate was 10% of normal in the alendronate-treated group and 150% higher than normal in the PTH(1−34)-treated group. The higher activation frequency in the PTH(1−34) group led to an increase in cortical porosity, which may explain observations of transient reductions in bone mineral density following treatment with PTH(1−34) (Neer et al., 2001; Finkelstein et al., 2003; Ettinger et al., 2004). A novel labeling regimen, quadruple tetracycline, has been perform in a longitudinal study in order to evaluate of the early effects of teriparatide treatment on bone formation. Within 4 weeks of treatment, PTH(1−34) increased mineralized perimeter, mineral apposition rate, and bone formation rate (Hodsman et al., 1993, 2000; Lindsay et al., 2006). The study of Lindsay et al. (2006) suggested that PTH(1−34) stimulates osteoblastic activity in preexisting remodeling units. This could be accomplished by a variety of mechanisms, including an increase in the work rate of preexisting osteoblasts, enhanced recruitment of new osteoblasts, or a prolongation of osteoblast life span (Jilka et al., 1999). Regardless of the mechanism, one noteworthy feature of PTH(1−34) is its ability to extend formation to quiescent surfaces surrounding the original remodeling unit (Fig. 19.8) (Lindsay et al., 2006). Bone-remodeling indices were increased after 1 month (Holland et al., 1994b), 2 months (Chow et al., 1992b), and 6 months (Reeve et al., 1980; Arlot et al., 2005) of treatment, and they returned toward baseline between 12 (Reeve et al., 1991) and 36 months of continuous treatment (Hodsman et al., 2000; Dempster et al., 2001). This temporal sequence of remodeling activation and deactivation, derived from biopsy studies, is confirmed by parallel changes in bone markers (Lindsay et al., 1997a,b; Kurland et al., 2000; Cosman et al., 2001; Arlot et al., 2005; McClung et al., 2005).

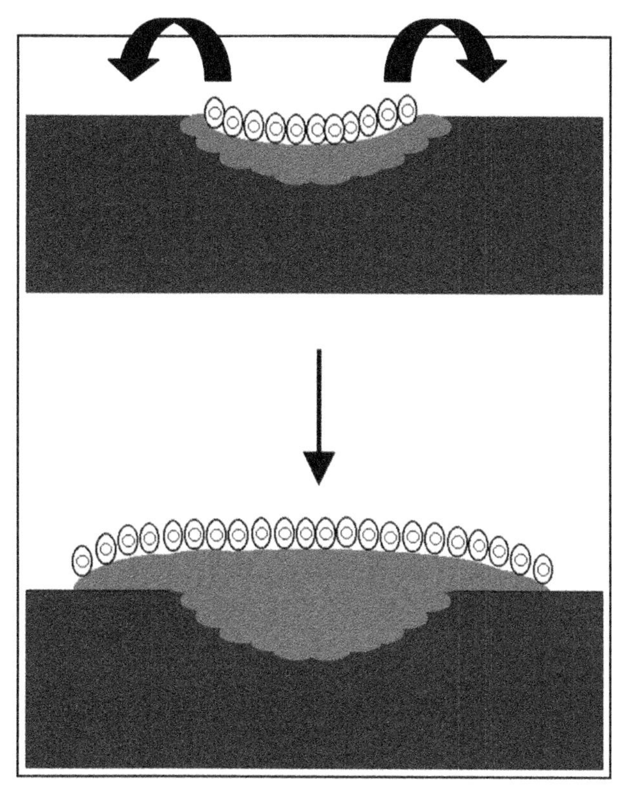

FIGURE 19.8 Proposed mechanism whereby PTH(1–34) could extend bone formation beyond the limits of the remodeling unit to annex the surrounding bone surface. *Reproduced with permission from Lindsay, R., Cosman, F., Zhou, H., Bostrom, M.P., Shen, V.W., Cruz, J.D., Nieves, J.W., Dempster, D.W., 2006. A novel tetracycline labeling schedule for longitudinal evaluation of the short-term effects of anabolic therapy with a single iliac crest bone biopsy: early actions of teriparatide. J. Bone Miner. Res. 21 (3), 366–373.*

The striking stimulation of bone formation by PTH(1–34) provides a mechanism for the reported increases in wall thickness of bone packets on cancellous and endocortical surfaces (Bradbeer et al., 1992; Hodsman et al., 2000; Dempster et al., 2001; Ma et al., 2006), which in turn leads to improvements in cancellous bone mass, trabecular connectivity, and cortical thickness (Dempster et al., 2001; Jiang et al., 2003) (Figs. 19.9 and 19.10). These improvements in cancellous bone structure were correlated with the early increases in bone formation markers (Dobnig et al., 2005). The deposition of new bone brought about an increase in the proportion of bone matrix with lower mineralization, mineral crystallinity, and collagen cross-link ratio (Misof et al., 2003; Paschalis et al., 2005).

The first study of the effects of PTH(1–34) raised the specter that the improvement in cancellous bone mass and structure may have been gained at the expense of cortical bone (Reeve et al., 1980). This was not confirmed in animal models in which cortical thickness and diameter were improved by PTH(1–34) treatment (Hirano et al., 1999, 2000; Jerome et al., 1999; Burr et al., 2001; Mashiba et al., 2001). Histomorphometric and micro-computed tomographic studies in humans revealed an increase in cortical thickness at the iliac crest, which was accompanied by stimulation of bone formation on the endosteal surface (see Figs. 19.9 and 19.10) (Dempster et al., 2001; Jiang et al., 2003; Lindsay et al., 2006). However, whether PTH(1–34) is able to stimulate periosteal bone formation in humans as it does in animals is not yet clear. Noninvasive techniques have yielded conflicting data on the effects of PTH(1–34) on bone diameter in humans (Zanchetta et al., 2003; Uusi-Rasi et al., 2005). However, biopsy studies suggest that PTH(1–34) can enhance periosteal bone formation (Ma et al., 2006; Lindsay et al., 2007). Although there are few data as of this writing, the effects of PTH(1–84) on the human ilium appear to be broadly similar to those of PTH(1–34) (Fox et al., 2005).

Abaloparatide

Abaloparatide (ABL) is a peptide designed with amino acid sequence identical to that of PTH-related peptide in the first 20 amino acids, with strategic insertions of different amino acids between residues 22 and 34 (Hattersley et al., 2016). The resulting peptide is a selective activator of the PTH type 1 receptor signaling pathway with the ability to produce anabolic

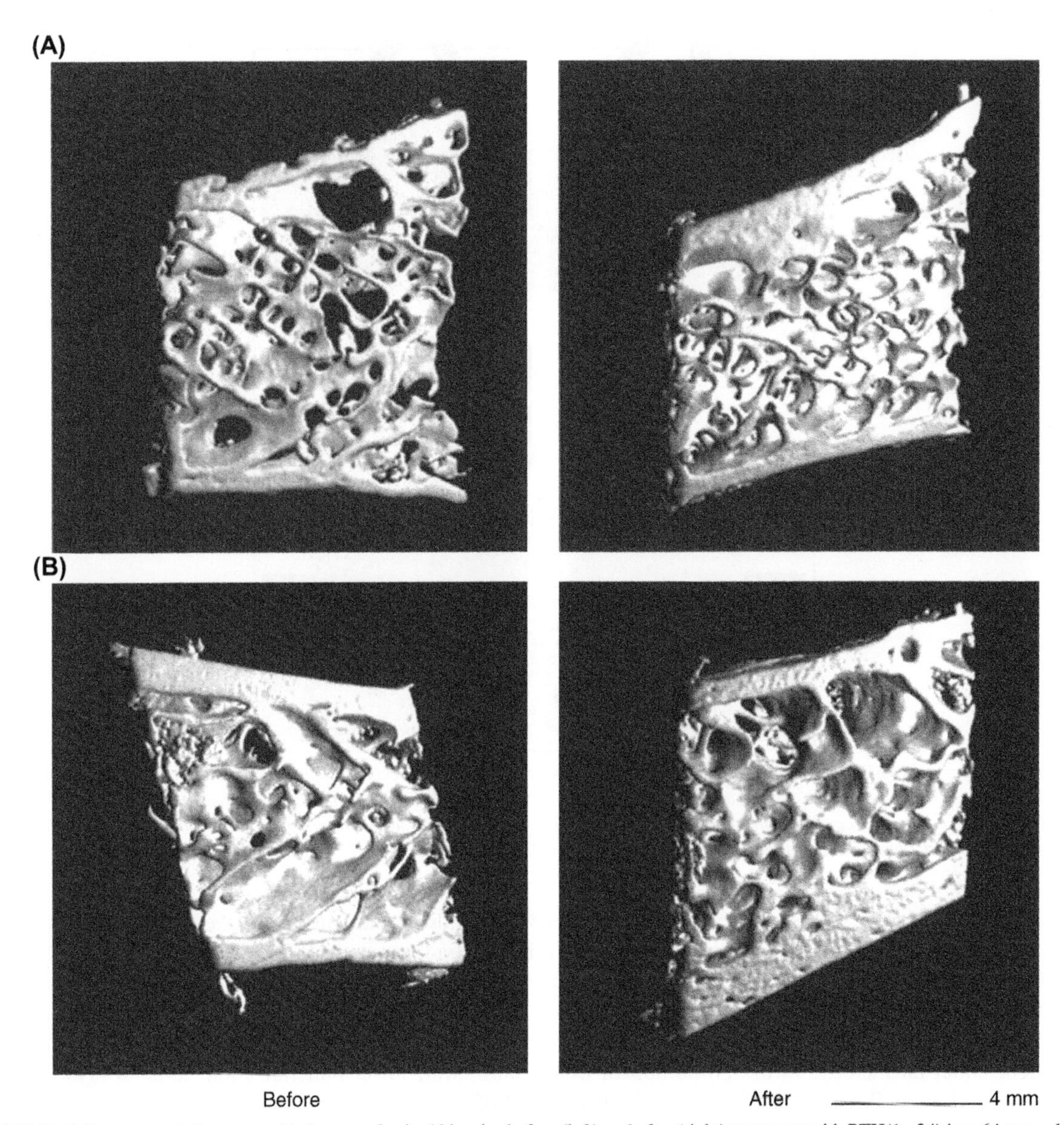

FIGURE 19.9 Micro-computed tomographic images of paired biopsies before (left) and after (right) treatment with PTH(1−34) in a 64-year-old woman (A) and a 47-year-old man (B). Note improvement in cancellous and cortical bone structure after treatment. *Reproduced with permission from Dempster, D.W., Cosman, F., Kurland, E.S., Zhou, H., Nieves, J., Woelfert, L., Shane, E., Plavetic, K., Muller, R., Bilezikian, J., Lindsay, R., 2001. Effects of daily treatment with parathyroid hormone on bone microarchitecture and turnover in patients with osteoporosis: a paired biopsy study. J. Bone Miner. Res. 16 (10), 1846−1853.*

effects with modest stimulation of bone resorption compared with PTH(1−34). In fact, ABL binds to the PTH type 1 receptor with different affinity for the R0 confirmation in comparison to PTH(1−34), resulting in less stimulation of bone resorption and formation. The Abaloparatide-SC Comparator Trial in Vertebral Endpoints (ACTIVE) was a comparative phase III, randomized, double-blind, placebo-controlled, multicenter, international study (Miller et al., 2016). Postmenopausal women were randomized to receive, blinded, daily subcutaneous injections of placebo or ABL 80 μg or open-label teriparatide 20 μg for 18 months.

Iliac bone biopsies were obtained in a subset of patients treated with placebo (n = 35), ABL(n = 36), or teriparatide (n = 34) for between 12 and 18 months.

Histological analysis revealed normal lamellar bone with normal microstructure and bone cell morphology in all three treatment groups (Moreira et al., 2016). There was no evidence of mineralization abnormality, excess woven bone or osteoid, marrow fibrosis, or marrow abnormalities. Histomorphometric analysis was performed on 78 (74.3%) of the 105 specimens. The remaining specimens were not evaluable due to significant damage or fragmentation that occurred in the

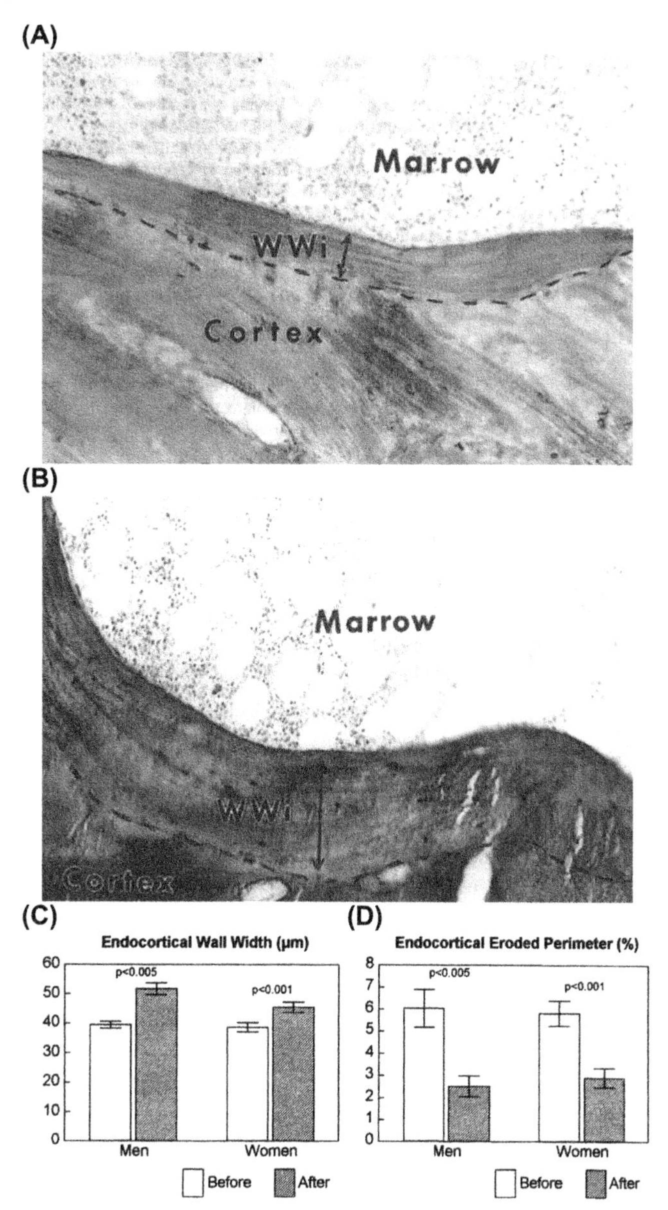

FIGURE 19.10 Bone packets on the endocortical surface of a 52-year-old man before (A) and after (B) 18 months of treatment with PTH(1−34). Note that after treatment the packet is almost twice as wide as the one before treatment. (C and D) Endocortical wall width and eroded perimeter before and after treatment of men and women with teriparatide. Note the increase in wall width and reduction in eroded perimeter after treatment. *WWi*, wall width. *Reproduced with permission from Dempster, D.W., Cosman, F., Kurland, E.S., Zhou, H., Nieves, J., Woelfert, L., Shane, E., Plavetic, K., Muller, R., Bilezikian, J., Lindsay, R., 2001. Effects of daily treatment with parathyroid hormone on bone microarchitecture and turnover in patients with osteoporosis: a paired biopsy study. J. Bone Miner. Res. 16 (10), 1846–1853.*

process of obtaining the biopsy. All specimens displayed tetracycline labels; however, double labels were not observed in three, five, and four specimens in the placebo, ABL, and teriparatide groups, respectively. There were only a few significant differences among the three groups. MAR was significantly higher in the teriparatide-treated group than in the placebo-treated group. The eroded surface was lower in the ABL-treated group than in the placebo group, suggesting a lower resorption rate. This was consistent with the biochemical marker data wherein the increase in carboxy-terminal collagen crosslinks (s-CTX) was less pronounced with ABL than with teriparatide, as well as with the lower incidence of hypercalcemia with ABL compared with teriparatide. Cortical porosity was significantly higher in both treated groups compared to placebo.

Surprisingly, activation frequency and bone formation rate were not different among the three groups, despite the fact that bone mineral density increased in the two treated groups relative to baseline and placebo in the bone biopsy cohort.

One explanation for this result was the timing of the bone biopsies that were obtained after at least 12 months of treatment. Previous histomorphometric studies with similar duration of teriparatide treatment have also not shown increases in bone turnover parameters. In fact, the largest increases in cancellous bone formation rate with teriparatide treatment have been seen with biopsies taken earlier in the course of treatment (Arlot et al., 2005; Lindsay et al., 2006). Furthermore, skeletal sites that exhibit a robust response to anabolic agents, namely the endocortical, intracortical, and periosteal envelopes (Lindsay et al., 2006; Dempster et al., 2012, 2016a), were not analyzed in this study.

Romosozumab

Romosozumab is a sclerostin antibody that has an anabolic effect (Chavassieux et al., 2017). The phase III FRAME study enrolled 7180 postmenopausal women at high risk of fracture and compared the effects of romosozumab with placebo at 12 months and after transition to denosumab at 24 months (Cosman et al., 2017). Vertebral fractures were significantly reduced in the romosozumab compared with the placebo arm at 12 months. At this writing, bone histomorphometry data following romosozumab treatment have not yet been published; however, some results have been presented (Chavassieux et al., 2017). In quadruple-labeled biopsies taken at 2 months, the bone formation rate was dramatically increased compared with pretreatment levels. However, in biopsies taken at 12 months with conventional double labeling, bone formation rate was reduced compared with placebo treatment. If the quadruple-labeling protocol had not been employed, it would have been difficult to explain the large increments in bone mineral density, cancellous bone volume, trabecular connectivity, and cortical thickness achieved by romosozumab treatment.

Comparative studies of anabolic and anticatabolic drugs

SHOTZ

The Skeletal Histomorphometry in Patients on Teriparatide or Zoledronic Acid Therapy study (SHOTZ) assessed the progressive effects of 2 years of treatment with teriparatide 20 μg/day and zoledronic acid 5 mg/year on bone remodeling in the cancellous, endocortical, intracortical, and periosteal envelopes in postmenopausal women with osteoporosis (Dempster et al., 2012, 2016b). Bone biopsies from both groups were performed at 6 and 24 months. A marked difference in the mechanism of action of these two drugs was observed, whereby teriparatide demonstrated higher bone formation indices than zoledronic acid in all four bone envelopes at both early and later time points. In the intracortical envelope, higher cortical porosity rate was observed with teriparatide compared with zoledronic acid, which is attributed to a higher intracortical remodeling rate with the pro-remodeling anabolic agent. However, the 24-month biopsies showed that haversian systems that were open at 6 months were completely filled with new bone at 24 months. In addition, the increase at BFR/BS with teriparatide seen on periosteal and, to a greater extent, on endocortical envelopes at 24 months provides a mechanistic basis for the reported increase in cortical thickness with this drug (Dempster et al., 2001; Jiang et al., 2003).

This study was able to show longitudinal changes within treatment groups with paired biopsies. In the teriparatide group, for example, dynamic indices declined from month 6 to month 24 only in the endocortical envelope. However, the median values for MS/BS and BFR/BS in the endocortical and intracortical envelopes were greater or similar to the values observed in the cancellous envelopes at both time points. Indeed, these results suggest that the anabolic effect of teriparatide is highest in the endocortical and intracortical envelopes and that the anabolic actions continue through 24 months of therapy.

In contrast to the teriparatide group, the low bone formation rate seen at month 6 in the zoledronic acid group persisted through 24 months. In fact, no difference in bone histomorphometry parameters were seen within individual bone envelopes in the zoledronic acid group over time. The only exception was noted in the intracortical envelope, where the median values for MS/BS and BFR/BS were higher compared with the cancellous envelope. This finding suggests that zoledronic acid may have limited access to the intracortical envelope, unlike teriparatide and denosumab as described earlier.

More recently, the SHOTZ study confirmed earlier studies (Lindsay et al., 2006; Ma et al., 2006) showing that teriparatide increases modeling-based formation as well as remodeling-based formation (Dempster et al., 2018). This effect was most prominent during the early course of treatment. In contrast, very little modeling-based formation was seen with zoledronic acid treatment.

AVA study: differential effects of teriparatide and denosumab on intact parathyroid hormone and bone formation indices

This open-label study was designed to evaluate whether denosumab-induced increases in endogenous PTH are able to exert some anabolic effect at the tissue level and on bone-turnover markers. The study included 69 postmenopausal women with osteoporosis and employed quadruple tetracycline labeling to enable longitudinal assessment of bone formation parameters in a single biopsy sample, which was taken at 3 months (Dempster, 2016a). Patients were randomized to teriparatide 20 μg daily for 6 months or denosumab 60 mg in a single infusion. Histomorphometric indices in all four envelopes were analyzed, as well as bone-turnover markers and endogenous intact PTH at baseline and 1, 3, and 6 months.

The results demonstrated that, while denosumab treatment resulted in a significant and prolonged increase in endogenous PTH levels, bone formation indices were decreased. By contrast, teriparatide decreased endogenous PTH but stimulated bone formation indices. These effects were reflected in the changes in bone-turnover markers, which increased with teriparatide and decreased with denosumab treatment. As in the SHOTZ study, stimulation of modeling-based formation accounted for a significant proportion of the total amount of new bone formation induced by teriparatide treatment (Dempster et al., 2017a). On the other hand, the small amount of modeling-based formation seen at baseline in the denosumab group was not affected by treatment, with the exception of a small but significant increase in the cancellous envelope.

Conclusion

Iliac crest bone biopsy and histomorphometry remains a powerful research tool, providing unique information on bone structure and remodeling activity at the tissue level. It is still the single most useful tool for understanding the mechanism of action (MOA) of bone active agents. Although it is rarely used as a diagnostic tool nowadays, a key exception is in the evaluation of renal osteodystrophy.

References

Alexandre, C., Chappard, D., Caulin, F., Bertrand, A., Palle, S., Riffat, G., 1988. Effects of a one-year administration of phosphate and intermittent calcitonin on bone-forming and bone-resorbing cells in involutional osteoporosis: a histomorphometric study. Calcif. Tissue Int. 42 (6), 345–350.

Arlot, M.E., Delmas, P.D., Chappard, D., Meunier, P.J., 1990. Trabecular and endocortical bone remodeling in postmenopausal osteoporosis: comparison with normal postmenopausal women. Osteoporos. Int. 1, 41.

Arlot, M., Meunier, P.J., Boivin, G., Haddock, L., Tamayo, J., Correa-Rotter, R., Jasqui, S., Donley, D.W., Dalsky, G.P., Martin, J.S., Eriksen, E.F., 2005. Differential effects of teriparatide and alendronate on bone remodeling in postmenopausal women assessed by histomorphometric parameters. J. Bone Miner. Res. 20 (7), 1244–1253.

Baron, R., Gertner, J.M., Lang, R., Vignery, A., 1983a. Increased bone turnover with decreased bone formation by osteoblasts in children with osteogenesis imperfect tarda. Pediatr. Res. 17, 204.

Baron, R., Vignery, A., Neff, L., et al., 1983b. Processing of undecalcified bone specimens for bone histomorphometry. In: Recker, R.R. (Ed.), Bone Histomorphometry: Techniques and Interpretation. CRC Press, Boca Raton, FL, p. 13.

Boivin, G., Meunier, P.J., 2002. The degree of mineralization of bone tissue measured by computerized quantitative contact microradiography. Calcif. Tissue Int. 70, 503–511.

Boivin, G.Y., Chavassieux, P.M., Santora, A.C., Yates, J., Meunier, P.J., 2000. Alendronate increases bone strength by increasing the mean degree of mineralization of bone tissue in osteoporotic women. Bone 27 (5), 687–694.

Boivin, G., Lips, P., Ott, S.M., Harper, K.D., Sarkar, S., Pinette, K.V., Meunier, P.J., 2003. Contribution of raloxifene and calcium and vitamin D3 supplementation to the increase of the degree of mineralization of bone in postmenopausal women. J. Clin. Endocrinol. Metab. 88 (9), 4199–4205.

Boivin, G., Vedi, S., Purdie, D.W., Compston, J.E., Meunier, P.J., 2005. Influence of estrogen therapy at conventional and high doses on the degree of mineralization of iliac bone tissue: a quantitative microradiographic analysis in postmenopausal women. Bone 36 (3), 562–567.

Bone, H.G., Downs Jr., R.W., Tucci, J.R., Harris, S.T., Weinstein, R.S., Licata, A.A., McClung, M.R., Kimmel, D.B., Gertz, B.J., Hale, E., Polvino, W.J., 1997. Dose-response relationships for alendronate treatment in osteoporotic elderly women. Alendronate Elderly Osteoporosis Study Centers. J. Clin. Endocrinol. Metab. 82 (1), 265–274.

Bone, H.G., Wagman, R.B., Brandi, M.L., Brown, J.P., Chapurlat, R., Cummings, S.R., Czerwiński, E., Fahrleitner-Pammer, A., Kendler, D.L., Lippuner, K., Reginster, J.Y., Roux, C., Malouf, J., Bradley, M.N., Daizadeh, N.S., Wang, A., Dakin, P., Pannacciulli, N., Dempster, D.W., Papapoulos, S., July 2017. 10 Years of denosumab treatment in postmenopausal women with osteoporosis:results from the phase 3 randomised FREEDOM trial and open-label extension. Lancet Diabetes Endocrinol. 5 (7), 513–523.

Borah, B., Dufresne, T.E., Chmielewski, P.A., Johnson, T.D., Chines, A., Manhart, M.D., 2004. Risedronate preserves bone architecture in postmenopausal women with osteoporosis as measured by three-dimensional microcomputed tomography. Bone 34 (4), 736–746.

Borah, B., Ritman, E.L., Dufresne, T.E., Jorgensen, S.M., Liu, S., Sacha, J., Phipps, R.J., Turner, R.T., 2005. The effect of risedronate on bone mineralization as measured by micro-computed tomography with synchrotron radiation: correlation to histomorphometric indices of turnover. Bone 37 (1), 1–9.

Borah, B., Dufresne, T.E., Ritman, E.L., Jorgensen, S.M., Liu, S., Chmielewski, P.A., Phipps, R.J., Zhou, X., Sibonga, J.D., Turner, R.T., 2006. Long-term risedronate treatment normalizes mineralization and continues to preserve trabecular architecture: sequential triple biopsy studies with micro-computed tomography. Bone 39, 345–352.

Bordier, P., Matrajt, H., Miravet, B., Hioco, D., 1964. Mesure histologique de la masse et de la résorption des través osseuse. Pathol. Biol. 12, 1238.

Boyce, B.F., Fell, G.S., Elder, H.Y., et al., 1982. Hypercalcemic osteomalacia due to aluminum toxicity. Lancet 2, 1009.

Bradbeer, J.N., Arlot, M.E., Meunier, P.J., Reeve, J., 1992. Treatment of osteoporosis with parathyroid peptide (hPTH 1–34) and oestrogen: increase in volumetric density of iliac cancellous bone may depend on reduced trabecular spacing as well as increased thickness of packets of newly formed bone. Clin. Endocrinol. 37 (3), 282–289.

Bressot, C., Meunier, P.J., Chapuy, M.C., et al., 1979. Histomorphometric profile, pathophysiology and reversibility of corticosteroid-induced osteoporosis. Metab. Bone Dis. Relat. Res. 1, 1303.

Brown [1], J.P., Reid, I.R., Wagman, R.B., Kendler, D., Miller, P.D., Jensen, J.E., Bolognese, M.A., Daizadeh, N., Valter, I., Zerbini, C.A., September 2014. Dempster DW Effects of up to 5 years of denosumab treatment on bone histology and histomorphometry: the FREEDOM study extension. J. Bone Miner. Res. 29 (9), 2051–2056.

Burr, D., Hirano, T., Turner, C., Hotchkiss, C., Brommage, R., Hock, J., 2001. Intermittently administered human parathyroid hormone (1–34) treatment increases intracortical bone turnover and porosity without reducing bone strength in the humerus of ovariectomized cynomolgus monkeys. J. Bone Miner. Res. 16, 157–165.

Burr, D.B., Miller, L., Grynpas, M., Li, J., Boyde, A., Mashia, T., Hirano, T., Johnston, C.C., 2003. Tissue mineralization is increased following 1-year treatment with high doses of bisphosphonates in dogs. Bone 33 (6), 960–969.

Charhon, S.A., Berland, Y.F., Olmer, M.J., et al., 1985. Effects of parathyroidectomy on bone formation and mineralization in hemodialized patients. Kidney Int. 27, 426.

Chavassieux, P.M., Arlot, M.E., Reda, C., Wei, L., Yates, A.J., Meunier, P.J., 1997. Histomorphometric assessment of the long-term effects of alendronate on bone quality and remodeling in patients with osteoporosis. J. Clin. Invest. 100 (6), 1475–1480.

Chavassieux, P.M., Arlot, M.E., Roux, J.P., Portero, N., Daifotis, A., Yates, A.J., Hamdy, N.A., Malice, M.P., Freedholm, D., Meunier, P.J., 2000. Effects of alendronate on bone quality and remodeling in glucocorticoid-induced osteoporosis: a histomorphometric analysis of transiliac biopsies. J. Bone Miner. Res. 15 (4), 754–762.

Chavassieux, P., Chapurlat, R., Portero-Muzy, N., Brown, J.P., Horlait, S., Libinati, C., Boyce, R., Wang, A., Grauer, A., September 2017. Effects of romozosumab in postmenopausal women with osteoporosis after 2 and 12 months: Bone histomorphometry substudy. In: ASBMR abstract no. 1072, ASBMR Meeting, Denver, CO.

Chesnut III, C.H., Majumdar, S., Newitt, D.C., Shields, A., Van Pelt, J., Laschansky, E., Azria, M., Kriegman, A., Olson, M., Eriksen, E.F., Mindeholm, L., 2005. Effects of salmon calcitonin on trabecular microarchitecture as determined by magnetic resonance imaging: results from the QUEST study. J. Bone Miner. Res. 20 (9), 1548–1561.

Chow, J., Tobias, J.H., Colston, K.W., Chambers, T.J., 1992a. Estrogen maintains trabecular bone volume in rats not only by suppression of resorption but also by stimulation of bone formation. J. Clin. Invest. 89, 74–78.

Chow, J.W.M., Lean, J.M., Chambers, T.J., 1992b. 17β-Estradiol stimulates cancellous bone formation in female rats. Endocrinology 130, 3025–3032.

Civitelli, R., Gonnelli, S., Zacchei, F., et al., 1988. Bone turnover in postmenopausal osteoporosis. Effect of calcitonin treatment. J. Clin. Invest. 82, 1268.

Coburn, J.W., Salusky, I.B., 2001. Renal bone diseases: clinical features, diagnosis, and management. In: Bilezikian, J.P., Marcus, R., Levine, M.A. (Eds.), The Parathyroids. Basic and Clinical Concepts, second ed. Academic Press, New York, p. 635.

Compston, J., 1997. Bone histomorphometry. In: Arnett, T.R., Henderson, B. (Eds.), Methods in Bone Biology. Chapman and Hall, London, p. 177.

Cosman, F., Nieves, J., Woelfert, L., Formica, C., Gordon, S., Shen, V., Lindsay, R., 2001. Parathyroid hormone added to established hormone therapy: effects on vertebral fracture and maintenance of bone mass after parathyroid hormone withdrawal. J. Bone Miner. Res. 16, 925–931.

Cosman, F., Crittenden, D.B., Adachi, J.D., Binkley, N., Czerwinski, E., Ferrari, S., Hofbauer, L.C., Lau, E., Lewiecki, E.M., Miyauchi, A., Zerbini, C.A., Milmont, C.E., Chen, L., Maddox, J., Meisner, P.D., Libanati, C., Grauer, A., October 20, 2016. Romosozumab treatment in postmenopausal women with osteoporosis. N. Engl. J. Med. 375 (16), 1532–1543.

Dempster, D.W., 1988. The relationship between the iliac crest bone biopsy and other skeletal sites. In: Kleerekoper, M., Krane, S. (Eds.), Clinical Disorders of Bone and Mineral Metabolism. Mary Ann Liebert, Inc., New York, p. 247.

Dempster, D.W., 1989. Bone histomorphometry in glucocorticoid-induced osteoporosis. J. Bone Miner. Res. 4, 137.

Dempster, D.W., 2000. The contribution of trabecular architecture to cancellous bone quality. J. Bone Miner. Res. 15, 20.

Dempster, D.W., 2002. Bone remodeling. In: Coe, F.L., Favus, M.J. (Eds.), Disorders of Bone and Mineral Metabolism, second ed. Lippincott Williams & Wilkins, New York.

Dempster, D.W., Cosman, F., Kurland, E.S., Zhou, H., Nieves, J., Woelfert, L., Shane, E., Plavetic, K., Muller, R., Bilezikian, J., Lindsay, R., 2001. Effects of daily treatment with parathyroid hormone on bone microarchitecture and turnover in patients with osteoporosis: a paired biopsy study. J. Bone Miner. Res. 16 (10), 1846–1853.

Dempster, D.W., Zhou, H., Recker, R.R., Brown, J.P., Bolognese, M.A., Recknor, C.P., Kendler, D.L., Lewiecki, E.M., Hanley, D.A., Rao, D.S., Miller, P.D., Woodson 3rd, G.C., Lindsay, R., Binkley, N., Wan, X., Ruff, V.A., Janos, B., Taylor, K.A., August 2012. Skeletal histomorphometry in subjects on teriparatide or zoledronic acid therapy (SHOTZ) study: a randomized controlled trial. J. Clin. Endocrinol. Metab. 97 (8), 2799–2808.

Dempster, D.W., Compston, J.E., Drezner, M.K., Glorieux, F.H., Kanis, J.A., Malluche, H., Meunier, P.J., Ott, S.M., Recker, R.R., Parfitt, A.M., January 2013. Standardized nomenclature, symbols, and units for bone histomorphometry: a 2012 update of the report of the ASBMR Histomorphometry Nomenclature Committee. J. Bone Miner. Res. 28 (1), 2–17.

Dempster, D.W., Zhou, H., Recker, R.R., Brown, J.P., Recknor, C.P., Lewiecki, E.M., Miller, P.D., Rao, S.D., Kendler, D.L., Lindsay, R., Krege, J.H., Alam, J., Taylor, K.A., Janos, B., Ruff, V.A., 2016a. Differential effects of teriparatide and denosumab on intact PTH and bone formation indices: AVA osteoporosis study. J. Clin. Endocrinol. Metab. 101 (4), 1353–1363.

Dempster, D.W., Zhou, H., Recker, R.R., Brown, J.P., Bolognese, M.A., Recknor, C.P., Kendler, D.L., Lewiecki, E.M., Hanley, D.A., Rao, S.D., Miller, P.D., Woodson 3rd, G.C., Lindsay, R., Binkley, N., Alam, J., Ruff, V.A., Gallagher, E.R., Taylor, K.A., 2016b. A longitudinal study of skeletal histomorphometry at 6 and 24 Months across four bone envelopes in postmenopausal women with osteoporosis receiving teriparatide or zoledronic acid in the SHOTZ trial. J. Bone Miner. Res. 31 (7), 1429–1439.

Dempster, D.W., Zhou, H., Ruff, V.A., Melby, T.E., Alam, J., Taylor, K.A., 2017. Longitudinal effects of teriparatide or zoledronic acid on bone modeling- and remodeling-based formation in the SHOTZ study. J. Bone Miner. Res. https://doi.org/10.1002/jbmr.3350 [Epub ahead of print] PubMed PMID: 29194749.

Dempster, D.W., Zhou, H., Recker, R.R., Brown, J.P., Recknor, C.P., Lewiecki, E.M., Miller, P.D., Rao, S.D., Kendler, D.L., Lindsay, R., Krege, J.H., Alam, J., Taylor, K.A., Melby, T.E., Ruff, V.A., February 2018. Remodeling- and modeling-based bone formation with teriparatide versus denosumab:A longitudinal analysis from baseline to 3 months in the AVA study. J. Bone Miner. Res. 33 (2), 298–306.

Dobnig, H., Sipos, A., Jiang, Y., Fahrleitner-Pammer, A., Ste-Marie, L.G., Gallagher, J.C., Pavo, I., Wang, J., Eriksen, E.F., 2005. Early changes in biochemical markers of bone formation correlate with improvements in bone structure during teriparatide therapy. J. Clin. Endocrinol. Metab. 90 (7), 3970–3977.

Dufresne, T.E., Chmielewski, P.A., Manhart, M.D., Johnson, T.D., Borah, B., 2003. Risedronate preserves bone architecture in early postmenopausal women in 1 year as measured by three-dimensional microcomputed tomography. Calcif. Tissue Int. 73 (5), 423–432.

Dunstan, C.R., Hills, E., Norman, A., et al., 1985. The pathogenesis of renal osteodystrophy: role of vitamin D, aluminum, parathyroid hormone, calcium and phosphorus. Q. J. Med. 55, 127.

Durchschlag, E., Paschalis, E.P., Zoehrer, R., Roschger, P., Fratzl, P., Recker, R., Phipps, R., Klaushofer, K., 2006. Bone material properties in trabecular bone from human iliac crest biopsies after 3- and 5-year treatment with risedronate. J. Bone Miner. Res. 21, 1581–1590.

Edwards, M.W., Bain, S.D., Bailey, M.C., Lantry, M.M., Howard, G.A., 1992. 17β-Estradiol stimulation of endosteal bone formation in the ovarietomised mouse: an animal model for the evaluation of bone -targeted estrogens. Bone 13, 29–34.

Eriksen, E.F., 1986. Normal and pathological remodeling of human trabecular bone: three dimensional reconstruction of the remodeling sequence in normals and metabolic bone disease. Endocr. Rev. 7, 379.

Eriksen, E.F., Hodgson, S.F., Eastell, R., et al., 1990. Cancellous bone remodeling in type I (postmenopausal) osteoporosis: quantitative assessment of rates of formation, resorption, and bone loss at tissue and cellular levels. J. Bone Miner. Res. 5, 311.

Eriksen, E.F., Langdahl, B., Vesterby, A., Rungby, J., Kassem, M., 1999. Hormone replacement therapy prevents osteoclastic hyperactivity: a histomorphometric study in early postmenopausal women. J. Bone Miner. Res. 14 (7), 1217–1221.

Eriksen, E.F., Melsen, F., Sod, E., Barton, I., Chines, A., 2002. Effects of long-term risedronate on bone quality and bone turnover in women with postmenopausal osteoporosis. Bone 31 (5), 620–625.

Ettinger, B., Black, D.M., Mitlak, B.H., Knickerbocker, R.K., Nickelsen, T., Genant, H.K., Christiansen, C., Delmas, P.D., Zanchetta, J.R., Stakkestad, J., Gluer, C.C., Krueger, K., Cohen, F.J., Eckert, S., Ensrud, K.E., Avioli, L.V., Lips, P., Cummings, S.R., 1999. Reduction of vertebral fracture risk in postmenopausal women with osteoporosis treated with raloxifene. J. Am. Med. Assoc. 282, 637–645.

Ettinger, B., San Martin, J., Crans, G., Pavo, I., 2004. Differential effects of teriparatide on BMD after treatment with raloxifene or alendronate. J. Bone Miner. Res. 19 (5), 745–751.

Felsenfeld, A.J., Rodriguez, M., Dunlay, R., Llach, F., 1991. A comparison of parathyroid gland function in hemodialysis patients with different forms of renal osteodystrophy. Nephrol. Dial. Transplant. 6, 244.

Finkelstein, J.S., Hayes, A., Hunzelman, J.L., Wyland, J.J., Lee, H., Neer, R.M., 2003. The effects of parathyroid hormone, alendronate, or both in men with osteoporosis. N. Engl. J. Med. 349 (13), 1216–1226.

Fleisch, H., 1998. Bisphosphonates: mechanism of action. Endocr. Rev. 19, 80–100.

Fox, J., Miller, M.A., Recker, R.R., Bare, S.P., Smith, S.Y., Moreau, I., 2005. Treatment of postmenopausal osteoporotic women with parathyroid hormone 1-84 for 18 months increases cancellous bone formation and improves cancellous architecture: a study of iliac crest biopsies using histomorphometry and microcomputed tomography. J. Musculoskelet. Neuronal Interact. 5, 356–357.

Frost, H.M., 1973. Bone Remodeling and Its Relationship to Metabolic Bone Diseases. Charles C Thomas, Springfield, IL.

Frost, H.M., 1983a. Bone histomorphometry: analysis of trabecular bone dynamics. In: Recker, R.R. (Ed.), Bone Histomorphometry: Techniques and Interpretation. CRC Press, Boca Raton, FL, p. 109.

Frost, H.M., 1983b. Bone histomorphometry: choice of marking agent and labeling schedule. In: Recker, R.R. (Ed.), Bone Histomorphometry: Techniques and Interpretation. CRC Press, Boca Raton, FL, p. 37.

Garcia Carasco, M., de Vernejoul, M.C., Sterkers, Y., et al., 1989. Decreased bone formation in osteoporotic patients compared with age-matched controls. Calcif. Tissue Int. 44, 173.

Gruber, H.E., Ivey, J.L., Baylink, D.J., Matthews, M., Nelp, W.B., Sisom, K., Chesnut III, C.H., 1984. Long-term calcitonin therapy in postmenopausal osteoporosis. Metabolism 33 (4), 295–303.

Gruber, H.E., Grigsby, J., Chesnut III, C.H., 2000. Osteoblast numbers after calcitonin therapy: a retrospective study of paired biopsies obtained during long-term calcitonin therapy in postmenopausal osteoporosis. Calcif. Tissue Int. 66 (1), 29–34.

Hattersley, G., Dean, T., Corbin, B.A., Bahar, H., Gardella, T.J., January 2016. Binding selectivity of abaloparatide for PTH-type-1-receptor conformations and effects on downstream signaling. Endocrinology 157 (1), 141–149.

Hercz, F., Pei, Y., Greenwood, C., et al., 1993. Low turnover osteodystrophy without aluminum; the role of "suppressed" parathyroid function. Kidney Int. 44, 860.

Hernandez, C.J., Beaupre, G.S., Marcus, R., Carter, D.R., 2001. A theoretical analysis of the contributions of remodeling space, mineralization, and bone balance to changes in bone mineral density during alendronate treatment. Bone 29 (6), 511–516.

Hirano, T., Burr, D.B., Turner, C.H., Sato, M., Cain, R.L., Hock, J.M., 1999. Anabolic effects of human biosynthetic parathyroid hormone fragment (1-34), LY333334, on remodeling and mechanical properties of cortical bone in rabbits. J. Bone Miner. Res. 14, 536–545.

Hirano, T., Burr, D.B., Cain, R.L., Hock, J.M., 2000. Changes in geometry and cortical porosity in adult, ovary-intact rabbits after 5 months treatment with LY333334 (hPTH 1–34). Calcif. Tissue Int. 66, 456–460.

Hodsman, A.B., Sherrard, D.J., Alfrey, A.C., et al., 1982. Bone aluminum and histomorphometric features of renal osteodystrophy. J. Clin. Endocrinol. Metab. 54, 539.

Hodsman, A.B., Fraher, L.J., Ostbye, T., Adachi, J.D., Steer, B.M., 1993. An evaluation of several biochemical markers for bone formation and resorption in a protocol utilizing cyclical parathyroid hormone and calcitonin therapy for osteoporosis. J. Clin. Invest. 91 (3), 1138–1148.

Hodsman, A.B., Kisiel, M., Adachi, J.D., Fraher, L.J., Watson, P.H., 2000. Histomorphometric evidence for increased bone turnover without change in cortical thickness or porosity after 2 years of cyclical hPTH(1–34) therapy in women with severe osteoporosis. Bone 27 (2), 311–318.

Holland, E.F.N., Studd, J.W.W., Mansell, J.P., Leather, A.T., Bailey, A.J., 1994a. Changes in collage composition and cross-links in bone and skin of osteoporosis postmenopausal women treated with percutaneous estradiol implants. Obstet. Gynecol. 83, 180–183.

Holland, E.F.N., Chow, J.W.M., Studd, J.W.W., Leather, A.T., Chambers, T.J., 1994b. Histomorphometric changes in the skeleton of postmenopausal women with low bone mineral density treated with percutaneous estradiol implants. Obstet. Gynecol. 83 (3), 387–391.

Jaworski, Z.F.G., 1983. Histomorphometric characteristics of metabolic bone disease. In: Recker, R.R. (Ed.), Bone Histomorphometry: Techniques and Interpretation. CRC Press, Boca Raton, FL, p. 241.

Jerome, C.P., Johnson, C.S., Vafai, H.T., Kaplan, K.C., Bailey, J., Capwell, B., Fraser, F., Hansen, L., Ramsay, H., Shadoan, M., Lees, C.J., Thomsen, J.S., Mosekilde, L., 1999. Effect of treatment for 6 months with human parathyroid hormone (1-34) peptide in ovariectomized cynomolgus monkeys (*Macaca fascicularis*). Bone 25, 301–309.

Jiang, Y., Zhao, J.J., Mitlak, B.H., Wang, O., Genant, H.K., Eriksen, E.F., 2003. Recombinant human parathyroid hormone (1-34) [teriparatide] improves both cortical and cancellous bone structure. J. Bone Miner. Res. 18 (11), 1932–1941.

Jiang, Y., Zhao, J., Liao, E.Y., Dai, R.C., Wu, X.P., Genant, H.K., 2005. Application of micro-CT assessment of 3-D bone microstructure in preclinical and clinical studies. J. Bone Miner. Metab. 23 (Suppl. l), 122–131.

Jilka, R.L., Weinstein, R.S., Bellido, T., Roberson, P., Parfitt, A.M., Manolagas, S.C., 1999. Increased bone formation by prevention of osteoblast apoptosis with parathyroid hormone. J. Clin. Invest. 104 (4), 439–446.

Ketteler, M., Block, G.A., Evenepoel, P., Fukagawa, M., Herzog, C.A., McCann, L., Moe, S.M., Shroff, R., Tonelli, M.A., Toussaint, N.D., Vervloet, M.G., Leonard, M.B., February 20, 2018. Diagnosis, evaluation, prevention, and treatment of chronic kidney disease-mineral and bone disorder: synopsis of the kidney disease: improving global outcomes 2017 clinical practice guideline update. Ann. Intern. Med. https://doi.org/10.7326/M17-2640 [Epub ahead of print] PubMed PMID: 29459980.

Khastgir, G., Studd, J., Holland, N., Alaghband-Zadeh, J., Fox, S., Chow, J., 2001a. Anabolic effect of estrogen replacement on bone in postmenopausal women with osteoporosis: histomorphometric evidence in a longitudinal study. J. Clin. Endocrinol. Metab. 86 (1), 289–295.

Khastgir, G., Studd, J., Holland, N., Alaghband-Zadeh, J., Sims, T.J., Bailey, A.J., 2001b. Anabolic effect of long-term estrogen replacement on bone collagen in elderly postmenopausal women with osteoporosis. Osteoporos. Int. 12 (6), 465–470.

Khastgir, G., Studd, J.W., Fox, S.W., Jones, J., Alaghband-Zadeh, J., Chow, J.W., 2003. A longitudinal study of the effect of subcutaneous estrogen replacement on bone in young women with Turner's syndrome. J. Bone Miner. Res. 18 (5), 925–932.

Kimmel, D.B., Recker, R.R., Gallagher, J.C., et al., 1990. A comparison of iliac bone histomorphometric data in post-menopausal osteoporotic and normal subjects. Bone Miner. 11, 217.

Kroger, H., Arnala, I., Alhava, E.M., 1992. Effect of calcitonin on bone histomorphometry and bone metabolism in rheumatoid arthritis. Calcif. Tissue Int. 50 (1), 11–13.

Kurland, E.S., Cosman, F., McMahon, D.J., Rosen, C.J., Lindsay, R., Bilezikian, J.P., 2000. Parathyroid hormone as a therapy for idiopathic osteoporosis in men: effects on bone mineral density and bone markers. J. Clin. Endocrinol. Metab. 85, 3069–3076.

Lindsay, R., Dempster, D.W., Jordan, C.V. (Eds.), 1997a. Estrogens and Anti-estrogens: Basic and Clinical Aspects. Lipincott-Raven Publishers, Philadelphia.

Lindsay, R., Nieves, J., Formica, C., Henneman, E., Woelfert, L., Shen, V., Dempster, D., Cosman, F., 1997b. Randomised controlled study of effect of parathyroid hormone on vertebral-bone mass and fracture incidence among postmenopausal women on oestrogen with osteoporosis. Lancet 350, 550–555.

Lindsay, R., Cosman, F., Zhou, H., Bostrom, M.P., Shen, V.W., Cruz, J.D., Nieves, J.W., Dempster, D.W., 2006. A novel tetracycline labeling schedule for longitudinal evaluation of the short-term effects of anabolic therapy with a single iliac crest bone biopsy: early actions of teriparatide. J. Bone Miner. Res. 21 (3), 366–373.

Lindsay, R., Zhou, H., Cosman, F., Nieves, J., Dempster, D.W., Hodsman, A.B., 2007. Effects of a one-month treatment with parathyroid hormone (1-34) on bone formation on cancellous, endocortical and periosteal surfaces of the human ilium. J. Bone Miner. Res. 22, 495–502.

Lufkin, E.G., Wahner, H.W., O'Fallon, W.M., Hodgson, S.F., Kotowicz, M.A., Lane, A.W., Judd, H.L., Caplan, R.H., Riggs, B.L., 1992. Treatment of postmenopausal osteoporosis with transdermal estrogen. Ann. Intern. Med. 117 (1), 1–9.

Ma, Y.L., Zeng, Q., Donley, D.W., Ste-Marie, L.G., Gallagher, J.C., Dalsky, G.P., Marcus, R., Eriksen, E.F., 2006. Teriparatide increases bone formation in modeling and remodeling osteons and enhances IGF-II immunoreactivity in postmenopausal women with osteoporosis. J. Bone Miner. Res. 21, 855–864.

Malluche, H.H., Faugere, M.-C., 1987. Atlas of Mineralized Bone Histology. Karger, Basel.

Malluche, H.H., Monier-Faugere, M.C., April 2006. Renal osteodystrophy: what's in a name? Presentation of a clinically useful new model to interpret bone histologic findings. Clin. Nephrol. 65 (4), 235–242.

Malluche, H.H., Ritz, E., Lange, H.P., et al., 1976. Bone histology in incipient and advanced renal failure. Kidney Int. 9, 355–362.

Malluche, H.H., Mawad, H.W., Monier-Faugere, M.C., June 2011. Renal osteodystrophy in the first decade of the new millennium: analysis of 630 bone biopsies in black and white patients. J. Bone Miner. Res. 26 (6), 1368–1376. Erratum, 2011 Erratum in: J. Bone Miner. Res. 2011 Nov; 26 (11), 2793.

Marcus, R., Leary, D., Schneider, D.L., et al., 2000. The contribution of testosterone to skeletal development and maintenance: lessons from the androgen insensitivity syndrome. J. Clin. Endocrinol. Metab. 85, 1032.

Marie, P.J., Caulin, F., 1986. Mechanisms underlying the effects of phosphate and calcitonin on bone histology in postmenopausal osteoporosis. Bone 7 (1), 17–22.

Mashiba, T., Burr, D.B., Turner, C.H., Sato, M., Cain, R.L., HockJ, M., 2001. Effects of human parathyroid hormone (1-34), LY333334, on bone mass, remodeling, and mechanical properties of cortical bone during the first remodeling cycle in rabbits. Bone 28, 538–547.

McClung, M.R., San Martin, J., Miller, P.D., Civitelli, R., Bandeira, F., Omizo, M., Donley, D.W., Dalsky, G.P., Eriksen, E.F., 2005. Opposite bone remodeling effects of teriparatide and alendronate in increasing bone mass. Arch. Intern. Med. 165, 1762–1768.

Melsen, F., Mosekilde, L., Kragstrup, J., 1983. Metabolic bone diseases as evaluated by bone histomorphometry. In: Recker, R.R. (Ed.), Bone Histomorphometry: Techniques and Interpretation. CRC Press, Boca Raton, FL, p. 265.

Meunier, P.J., 1981. Bone biopsy in diagnosis of metabolic bone disease. In: Cohn, T.V., Talmage, R., Matthews, J.L. (Eds.), "Hormonal Control of Calcium Metabolism," Proceedings of the Seventh International Conference on Calcium Regulating Hormones. Excerpta Medica, Amsterdam, p. 109.

Meunier, P.J., 1988. Assessment of bone turnover by histomorphometry in osteoporosis. In: Riggs, B.L., Melton III, L.J. (Eds.), Osteporosis: Etiology, Diagnosis, and Management. Raven Press, New York, p. 317.

Meunier, P.J., Boivin, G., 1997. Bone mineral reflects bone mass but also the degree of mineralization of bone: therapeutic implications. Bone 21, 373–377.

Meunier, P.J., Coindre, J.M., Edouard, C.M., Arlot, M.E., 1980. Bone histomorphometry in Paget's disease quantitative and dynamic analysis of pagetic and non-pagetic bone tissue. Arthritis Rheum. 23, 1095.

Meunier, P.J., Sellami, S., Briancon, D., Edouard, C., 1981. Histological heterogeneity of apparently idiopathic osteoporosis. In: Deluca, H.F., Frost, H.M., Jee, W.S.S., et al. (Eds.), Osteoporosis, Recent Advances in Pathogenesis and Treatment. University Park Press, Baltimore, p. 293.

Miller, P.D., Hattersley, G., Riis, B.J., Williams, G.C., Lau, E., Russo, L.A., Alexandersen, P., Zerbini, C.A., Hu, M.Y., Harris, A.G., Fitzpatrick, L.A., Cosman, F., Christiansen, C., August 16, 2016. ACTIVE study investigators. Effect of abaloparatide vs placebo on new vertebral fractures in postmenopausal women with osteoporosis: a randomized clinical trial. J. Am. Med. Assoc. 316 (7), 722–733.

Misof, B.M., Roschger, P., Cosman, F., Kurland, E.S., Tesch, W., Messmer, P., Dempster, D.W., Nieves, J., Shane, E., Fratzl, P., Klaushofer, K., Bilezikian, J., Lindsay, R., 2003. Effects of intermittent parathyroid hormone administration on bone mineralization density in iliac crest biopsies from patients with osteoporosis: a paired study before and after treatment. J. Clin. Endocrinol. Metab. 88 (3), 1150–1156.

Moreira, C.A., Fitzpatrick, L.A., Wang, Y., Recker, R.R., April 2017. Effects of abaloparatide-SC (BA058) on bone histology and histomorphometry: the ACTIVE phase 3 trial. Bone 97, 314–319.

Moriniere, P., Cohen-Solal, M., Belbrik, S., et al., 1989. Disappearance of aluminic bone disease in a long-term asymptomatic dialysis population restricting Al(OH)3 intake: emergence of an idiopathic adynamic bone disease not related to aluminum. Nephron 53, 975.

Munns, C.F., Rauch, F., Travers, R., Glorieux, F.H., 2005. Effects of intravenous pamidronate treatment in infants with osteogenesis imperfecta: clinical and histomorphometric outcome. J. Bone Miner. Res. 20 (7), 1235–1243.

Neer, R.M., Arnaud, C.D., Zanchetta, J.R., Prince, R., Gaich, G.A., Reginster, J.Y., Hodsman, A.B., Eriksen, E.F., Ish-Shalom, S., Genant, H.K., Wang, O., Mitlak, B.H., 2001. Effect of parathyroid hormone (1-34) on fractures and bone mineral density in postmenopausal women with osteoporosis. N. Engl. J. Med. 344 (19), 1434–1441.

Ott, S.M., Oleksik, A., Lu, Y., Harper, K., Lips, P., 2002. Bone histomorphometric and biochemical marker results of a 2-year placebo-controlled trial of raloxifene in postmenopausal women. J. Bone Miner. Res. 17 (2), 341–348.

Palmieri, G.M., Pitcock, J.A., Brown, P., Karas, J.G., Roen, L.J., 1989. Effect of calcitonin and vitamin D in osteoporosis. Calcif. Tissue Int. 45 (3), 137–141.

Parfitt, A.M., 1983a. Stereological basis of bone histomorphometry: theory of quantitative microscopy and reconstruction of the third dimension. In: Recker, R.R. (Ed.), Bone Histomorphometry: Techniques and Interpretation. CRC Press, Boca Raton, FL, p. 53.

Parfitt, A.M., 1983b. The physiological and clinical significance of bone histomorphometric data. In: Recker, R.R. (Ed.), Bone Histomorphometry: Techniques and Interpretation. CRC Press, Boca Raton, FL, p. 143.

Parfitt, A.M., 1988. Bone remodeling: relationship to the amount and structure of bone, and the pathogenesis and prevention of fractures. In: Riggs, B.L., Melton III, L.J. (Eds.), Osteoporosis: Etiology, Diagnoses, and Management. Raven Press, New York, p. 45.

Parfitt, A.M., Matthews, C.H.E., Villanueva, A.R., et al., 1982. Relationships between surface, volume and thickness of iliac trabecular bone in aging and in osteoporosis. Implications for the microanatomic and cellular mechanisms of bone loss. J. Clin. Invest. 72, 1396.

Parfitt, A.M., Drezner, M.K., Glorieux, F.H., et al., 1987. Bone histomorphometry: standardization of nomenclature, symbols, and units. J. Bone Miner. Res. 2, 595.

Parisien, M., Charhon, S.A., Mainetti, E., et al., 1988. Evidence for a toxic effect of aluminum on ostetoblasts: a histomorphometric study in hemodialysis patients with aplastic bone disorder. J. Bone Miner. Res. 3, 259.

Parisien, M., Silverberg, S.J., Shane, E., et al., 1990. The histomorphometry of bone in primary hyperparathyroidism: preservation of cancellous bone structure. J. Clin. Endocrinol. Metab. 70, 930.

Parisien, M.V., Mellish, R.W.E., Silverberg, S.J., et al., 1992. Maintenance of cancellous bone connectivity in primary hyperparathyroidism: trabecular strut analysis. J. Bone Miner. Res. 7, 913.

Paschalis, E.P., Boskey, A.L., Kassem, M., Eriksen, E.F., 2003. Effect of hormone replacement therapy on bone quality in early postmenopausal women. J. Bone Miner. Res. 18 (6), 955−959.

Paschalis, E.P., Glass, E.V., Donley, D.W., Eriksen, E.F., 2005. Bone mineral and collagen quality in iliac crest biopsies of patients given teriparatide: new results from the fracture prevention trial. J. Clin. Endocrinol. Metab. 90 (8), 4644−4649.

Patel, S., Pazianas, M., Tobias, J., Chambers, T.J., Fox, S., Chow, J., 1999. Early effects of hormone replacement therapy on bone. Bone 24, 245−248.

Pepene, C.E., Seck, T., Diel, I., Minne, H.W., Ziegler, R., Pfeilschifter, J., 2004. Influence of fluor salts, hormone replacement therapy and calcitonin on the concentration of insulin-like growth factor (IGF)-I, IGF-II and transforming growth factor-beta 1 in human iliac crest bone matrix from patients with primary osteoporosis. Eur. J. Endocrinol. 150 (1), 81−91.

Prestwood, K.M., Gunness, M., Muchmore, D.B., Lu, Y., Wong, M., Raisz, L.G., 2000. A comparison of the effects of raloxifene and estrogen on bone in postmenopausal women. J. Clin. Endocrinol. Metab. 85 (6), 2197−2202.

Rao, D.S., 1983. Practical approach to bone biopsy. In: Recker, R.R. (Ed.), Bone Histomorphometry: Techniques and Interpretation. CRC Press, Boca Raton, FL, p. 3.

Recker, R.R., Weinstein, R.S., Chesnut III, C.H., Schimmer, R.C., Mahoney, P., Hughes, C., Bonvoisin, B., Meunier, P.J., 2004. Histomorphometric evaluation of daily and intermittent oral ibandronate in women with postmenopausal osteoporosis: results from the BONE study. Osteoporos. Int. 15 (3), 231−237.

Recker, R., Masarachia, P., Santora, A., Howard, T., Chavassieux, P., Arlot, M., Rodan, G., Wehren, L., Kimmel, D., 2005. Trabecular bone microarchitecture after alendronate treatment of osteoporotic women. Curr. Med. Res. Opin. 21 (2), 185−194.

Recker, R.R., Delmas, P.D., Halse, J., Reid, I.R., Boonen, S., García-Hernandez, P.A., Supronik, J., Lewiecki, E.M., Ochoa, L., Miller, P., Hu, H., Mesenbrink, P., Hartl, F., Gasser, J., Eriksen, E.F., 2008. Effects of intravenous zoledronic acid once yearly on bone remodeling and bone structure. J. Bone Miner. Res. 23, 6−16.

Reeve, J., Meunier, P.J., Parsons, J.A., Bernat, M., Bijvoet, O.L., Courpron, P., Edouard, C., Klenerman, L., Neer, R.M., Renier, J.C., Slovik, D., Vismans, F.J., Potts Jr., J.T., 1980. Anabolic effect of human parathyroid hormone fragment on trabecular bone in involutional osteoporosis: a multicentre trial. Br. Med. J. 280, 1340−1344.

Reeve, J., Bradbeer, J.N., Arlot, M., Davies, U.M., Green, J.R., Hampton, L., Edouard, C., Hesp, R., Hulme, P., Ashby, J.P., Zanelli, J.M., Meunier, P.J., 1991. hPTH 1−34 treatment of osteoporosis with added hormone replacement therapy: biochemical, kinetic and histological responses. Osteoporos. Int. 1 (3), 162−170.

Reid [1], I.R., Miller, P.D., Brown, J.P., Kendler, D.L., Fahrleitner-Pammer, A., Valter, I., Maasalu, K., Bolognese, M.A., Woodson, G., Bone, H., Ding, B., Wagman, R.B., San Martin, J., Ominsky, M.S., Dempster, D.W., October 2010. Effects of denosumab on bone histomorphometry: the FREEDOM and STAND studies. J. Bone Miner. Res. 25 (10), 2256−2265.

Rey, C., Hina, A., Glimcher, M.J., 1995. Maturation of poorly crystalline apatites: chemical and structural aspects in vivo and in vitro. Cells Mater 5, 345−356.

Riggs, B.L., Parfitt, A.M., 2005. Drugs used to treat osteoporosis: the critical need for a uniform nomenclature based on their action on bone remodeling. J. Bone Miner. Res. 20 (2), 177−184.

Roschger, P., Fratzl, P., Klaushofer, K., Rodan, G., 1997. Mineralization of cancellous bone after alendronate and sodium fluoride treatment: a quantitative backscattered electron imaging study on minipig ribs. Bone 20, 393−397.

Roschger, P., Rinnerthaler, S., Yates, J., Rodan, G.A., Fratzl, P., Klaushofer, K., 2001. Alendronate increases degree and uniformity of mineralization in cancellous bone and decreases the porosity in cortical bone of osteoporotic women. Bone 29, 185−191.

Salusky, I.B., Coburn, J.W., Brill, J., et al., 1988. Bone disease in pediatric patients undergoing dialysis with CAPD or CCPD. Kidney Int. 33, 975.

Seeman, E., Delmas, P.D., 2006. Bone quality—the material and structural basis of bone strength and fragility. N. Engl. J. Med. 354, 2250−2261.

Sherrard, D.J., Hercz, G., Pei, Y., et al., 1993. The spectrum of bone disease in end-stage renal failure—an evolving disorder. Kidney Int. 43, 435.

Silverberg, S.J., Shane, E., Dempster, D.W., Bilezikian, J.P., 1990. The effects of vitamin D insufficiency in patients with primary hyperparathyroidism. Am. J. Med. 107, 561.

Silverman, S.L., 2003. Calcitonin. Endocrinol. Metab. Clin. North Am. 32 (1), 273−284.

Siris, E.S., Clemens, T.L., Dempster, D.W., et al., 1987. Tumor-induced osteomalacia. Kinetics of calcium, phosphorus, and vitamin D metabolism and characteristics of bone histomorphometry. Am. J. Med. 82, 307.

Slatopolsky, E., Delmez, J., 2002. Bone disease in chronic renal failure and after renal transplantation. In: Coe, F.L., Favus, M.J. (Eds.), Disorders of Bone and Mineral Metabolism, second ed. Lippincott Williams & Wilkins, New York, p. 865.

Ste-Marie, L.G., Charhon, S.A., Edouard, C., et al., 1984. Iliac bone histomorphometry in adults and children with osteogenesis imperfecta. J. Clin. Pathol. 37, 1801.

Steiniche, T., Hasling, C., Charles, P., Eriksen, E.F., Mosekilde, L., Melsen, F., 1989. A randomized study on the effects of estrogen/gestagen or high dose oral calcium on trabecular bone remodeling in postmenopausal osteoporosis. Bone 10 (5), 313–320.

Steiniche, T., Christiansen, P., Vesterby, A., et al., 1994. Marked changes in iliac crest bone structure in postmenopausal women without any signs of disturbed bone remodeling or balance. Bone 15, 73.

Teitelbaum, S.L., 1980. Pathological manifestations of osteomalacia and rickets. J. Clin. Endocrinol. Metab. 9, 43.

Uusi-Rasi, K., Semanick, L.M., Zanchetta, J.R., Bogado, C.E., Eriksen, E.F., Sato, M., Beck, T.J., 2005. Effects of teriparatide [rhPTH (1-34)] treatment on structural geometry of the proximal femur in elderly osteoporotic women. Bone 36, 948–958.

Valentin-Opran, A., Charhon, S.A., Meunier, P.J., et al., 1982. Quantitative histology of myeloma-induced bone changes. Br. J. Haematol. 52, 601.

Vedi, S., Compston, J.E., 1996. The effects of long-term hormone replacement therapy on bone remodeling in postmenopausal women. Bone 19 (5), 535–539.

Vedi, S., Purdie, D.W., Ballard, P., Bord, S., Cooper, A.C., Compston, J.E., 1999. Bone remodeling and structure in postmenopausal women treated with long-term, high-dose estrogen therapy. Osteoporos. Int. 10 (1), 52–58.

Vedi, S., Bell, K.L., Loveridge, N., Garrahan, N., Purdie, D.W., Compston, J.E., 2003. The effects of hormone replacement therapy on cortical bone in postmenopausal women. A histomorphometric study. Bone 33 (3), 330–334.

Wahab, M., Ballard, P., Purdie, D.W., Cooper, A., Willson, J.C., 1997. The effect of long term oestradiol implantation on bone mineral density in postmenopausal women who have undergone hysterectomy and bilateral oophorectomy. Br. J. Obstet. Gynaecol. 104, 728–731.

Walters, C., Eyre, D.R., 1980. Collagen crosslinks in human dentin: increasing content of hydroxypyridinum residues with age. Calcif. Tissue Int. 35, 401–405.

Weinstein, R.S., 2002. Clinical use of bone biopsy. In: Coe, F.L., Favus, M.J. (Eds.), Disorders of Bone and Mineral Metabolism, second ed. Lippincott Williams & Wilkins, New York, p. 448.

Weinstein, R.S., Parfitt, A.M., Marcus, R., Greenwald, M., Crans, G., Muchmore, D.B., 2003. Effects of raloxifene, hormone replacement therapy, and placebo on bone turnover in postmenopausal women. Osteoporos. Int. 14 (10), 814–822.

Whyte, M.P., Bergfeld, M.A., Murphy, W.A., et al., 1982. Postmenopausal osteoporosis; a heterogeneous disorder as assessed by histomorphometric analysis of iliac crest bone from untreated patients. Am. J. Med. 72, 183.

Wright, C.D., Garrahan, N.J., Stanton, M., Gazet, J.C., Mansell, R.E., Compston, J.E., 1994. Effect of long-term tamoxifen therapy on cancellous bone remodeling and structure in women with breast cancer. J. Bone Miner. Res. 9 (2), 153–159.

Yamauchi, M., 1996. Collagen: the major matrix molecule in mineralized tissues. In: Anderson, J.J.B., Garner, S.C. (Eds.), Calcium and Phosphorus in Health and Disease. CRC Press, New York, pp. 127–141.

Zanchetta, J.R., Bogado, C.E., Ferretti, J.L., Wang, O., Wilson, M.G., Sato, M., Gaich, G.A., Dalsky, G.P., Myers, S.L., 2003. Effects of teriparatide [recombinant human parathyroid hormone (1-34)] on cortical bone in postmenopausal women with osteoporosis. J. Bone Miner. Res. 18, 539–543.

Zoehrer, R., Roschger, P., Paschalis, E.P., Hofstaetter, J.G., Durchschlag, E., Fratzl, P., Phipps, R., Klaushofer, K., 2006. Effects of 3- and 5-year treatment with risedronate on bone mineralization density distribution in triple biopsies of the iliac crest in postmenopausal women. J. Bone Miner. Res. 21, 1106–1119.

Chapter 20

Phosphorus homeostasis and related disorders

Thomas O. Carpenter, Clemens Bergwitz and Karl L. Insogna
Departments of Pediatrics and Internal Medicine, Yale University School of Medicine, New Haven, CT, United States

Chapter outline

Introduction 469
Regulation of phosphate metabolism 470
Overview 470
Hormonal regulators 471
Parathyroid hormone 471
Fibroblast growth factor 23 471
The role of osteocytes 471
Nutritional and gastrointestinal considerations 473
Phosphate and bone mineralization 474
Intracellular/extracellular compartmentalization 475
Mechanisms of phosphate transport 475
Intestinal phosphate transport 476
Renal phosphate transport 477
Ubiquitous metabolic phosphate transporters 478
Primary disorders of phosphate homeostasis 478
Fibroblast growth factor 23–mediated hypophosphatemic disorders 478
X-linked hypophosphatemia (OMIM: 307800) 478
Autosomal dominant hypophosphatemic rickets (OMIM: 193100) 481
Autosomal recessive hypophosphatemic rickets 481
Tumor-induced osteomalacia 484
Other FGF23-mediated hypophosphatemic syndromes 485
Fibroblast growth factor 23–independent hypophosphatemic disorders 486
Hereditary hypophosphatemic rickets with hypercalciuria (OMIM: 241530) 486
Dent's disease (X-linked recessive hypophosphatemic rickets) (OMIM: 300009) 490
Hypophosphatemia with osteoporosis and nephrolithiasis due to SLC34A1 (OMIM: 612286) and NHERF1 mutations (OMIM: 604990) 490
Autosomal recessive Fanconi syndrome (OMIM: 613388) 490
Fanconi–Bickel syndrome (OMIM: 227810) 490
Intestinal malabsorption of phosphate 491
Hyperphosphatemic syndromes 491
Tumoral calcinosis 491
Normophosphatemic disorders of cellular phosphorus metabolism 492
Summary 492
References 494

Introduction

Phosphorus is critical to life as we know it. In mammalian tissues, phosphorus is one of the most abundant components; it has critical roles in growth and development and is a major structural component of bone. Physiologic functions at the organism level include its ability to serve as a buffer in acid–base regulation and multiple roles in cellular metabolism. Inorganic phosphorus exists primarily as phosphate (PO_4), a stable, divalent anionic complex with four oxygen atoms. Phosphate is a requisite component of hydroxyapatite ($Ca_{10}(PO_4)_6(OH)_2$), which crystallizes to form the mineral basis of the vertebrate skeleton. At the molecular level, phosphate provides the molecular backbone of DNA. The chemical bonding of two phosphate molecules creates pyrophosphate (PPi; P_2O_7), a potent inhibitor of bone mineralization, and an intricate balance between local phosphate/PPi balance regulates mineralization. The chemical nature of phosphate–phosphate bonds allows for storage of biological energy as adenosine triphosphate (ATP). In addition, phosphorus influences a variety of enzymatic reactions (e.g., glycolysis) and protein functions (e.g., the oxygen-carrying capacity of hemoglobin by regulation of 2,3-diphosphoglycerate synthesis). Phosphorus influences both the storage and the release of the principal

Principles of Bone Biology. https://doi.org/10.1016/B978-0-12-814841-9.00020-8

carbohydrate energy source, glucose, as glucose is phosphorylated prior to storage as glycogen, and that same phosphorylation step is reversed during glycogenolysis. Finally, bonding (or removal) of phosphorus at specific amino acid residues in proteins provides for highly evolved signaling cascades, as phosphorylation or dephosphorylation of many proteins activates or deactivates a variety of pathways.

Individuals eating a Western diet consume roughly 20 mg/kg of dietary phosphorus daily, varying somewhat by age and sex (Calvo et al., 2014). However, this estimate does not capture the large amount of phosphorus contained in food additives. The majority of phosphorus in the body is contained in mineralized tissue (600–700 g), with the remainder in soft tissues (100–200 g). Less than 1% of the total phosphorus in the body is in extracellular fluids. The plasma contains approximately 12 mg of phosphorus, the majority of which (8 mg) is contained in phospholipids, while only a small amount circulates as inorganic phosphate, either as monohydrogen phosphate or dihydrogen phosphate. The relative concentrations of these two moieties generally favor the monohydrogenated compound (4:1).

The terms "serum phosphorus" and "serum phosphate" are generally used interchangeably when referring to plasma inorganic phosphorus concentrations, because plasma inorganic phosphorus is nearly all in the form of the PO_4 ion. It should be noted that this terminology can be confusing when using mass units (i.e., mg/dL) when the weight of the phosphorus content of the phosphate is reported, yet the term "serum phosphate" is often used in the clinical setting. When using molar units the concentrations of the phosphate and of the phosphorus are equivalent, and less confusion may arise.

In adult humans, most phosphorus is found in bone (600–700 g), whereas the remainder is generally in soft tissue (100–200 g). The plasma contains 11–12 mg/dL of total phosphorus (in both organic and inorganic states) in adults. Inorganic phosphorus is largely in the form of phosphate (PO_4) and is the fraction typically measured in clinical settings. The plasma or serum concentration of phosphorus averages 3.0–4.5 mg/dL in older children and adults, but is greater in infants and younger children. Plasma phosphorus can often be as high as 8 mg/dL in small infants, gradually declining through the first year of life and then further as childhood progresses to adult values.

Rapid shifts in extracellular phosphate can occur during a variety of physiologic processes and pathologic disturbances. For example, the phosphorylation of glucose after it enters the cell explains the rapid depletion of extracellular phosphate with exogenous administration of insulin. Rapid bone remineralization ("hungry bone syndrome") can also cause rapid depletion of extracellular phosphate. Phosphorus deprivation (as occurs with the administration of phosphate-binding resins) can lead to hypophosphatemia. Furthermore, insulin induces shifts of phosphate into cells as is seen in refeeding hypophosphatemia and with the correction of diabetic ketoacidosis. Profound and sustained declines in extracellular phosphate can result in impaired muscle, leukocyte, and platelet function; rhabdomyolysis; and even cardiopulmonary compromise.

Despite the paramount importance of phosphate in mammalian physiology, the local and systemic regulation of phosphate homeostasis is poorly understood. It was not until 2000/2001 that the hormone now recognized as having a critical role in regulating the concentration of extracellular phosphate, fibroblast growth factor 23 (FGF23), was identified. That discovery, while representing a major advance in understanding phosphate homeostatic mechanisms, also highlights a crucial, and as yet unanswered, question: how do osteocytes, the cells that produce FGF23, sense extracellular phosphate and thereby regulate production of FGF23?

Phosphate transporters are present throughout the body and are the portals for entry of phosphate into cells, but their regulation and/or modulation by extracellular phosphate remains to be explored. A better understanding of the regulation of, as well as interactions between, extra- and intracellular phosphate will lead to improved insights into human physiology and better therapies for disorders of phosphate homeostasis.

Regulation of phosphate metabolism

Overview

To provide for these various functions, organisms have developed elaborate mechanisms for regulating supply and intercompartmental transport of phosphate. Various regulatory mechanisms permit modulation based on metabolic phosphorus need and exogenous phosphorus supply. This adaptive homeostatic system primarily adjusts gastrointestinal absorption and renal excretion to control total body stores, while maintaining extracellular concentration of phosphorus within a relatively narrow range. The identification of several hormones (e.g., parathyroid hormone [PTH], FGF23) and transporter proteins (e.g., NPT2, PIT) that are integral to the system has markedly increased our understanding of phosphate homeostasis, but the sensory system underlying regulation of phosphorus balance remains incompletely understood. The majority of ingested phosphate is absorbed in the small intestine, where active transport is mediated by sodium-dependent transporter proteins, and sodium-independent phosphate transport also occurs. The kidney, however, is thought to be the dominant organ involved in the regulation of phosphorus balance, as renal tubular reabsorption of filtered phosphate is adjusted in response to various regulatory factors. Finally, in the setting of severe phosphorus deprivation, the phosphate contained in bone mineral provides a source of phosphorus for the metabolic needs of the organism. The specific roles that the intestine, kidney, and bone play in this complex process are discussed in detail in the following sections.

Hormonal regulators

Parathyroid hormone

PTH was long considered the predominant hormonal effector of phosphate homeostasis. Effects of PTH on proximal tubular phosphate transport can be mediated by both cAMP−protein kinase A (PKA) and phospholipase C−protein kinase C (PKC) signal transduction pathways, although the PKA pathway is the more important pathway in this regard. After interaction with the PTH receptor 1 (PTHR1), PTH effects a rapid removal of the inorganic phosphate transporters NPT2a and NPT2c from proximal tubular apical membranes. Stabilization of NPT2a is mediated by the sodium/hydrogen exchange regulatory cofactor NHERF1, which is phosphorylated by PTH's activation of the PKA and PKC pathways. Overall, PTH appears to serve as a regulator of renal sodium/phosphate cotransport in an acute time frame, and the regulation is determined by changes in the abundance of NPT2 proteins in the renal brush border membrane (Forster et al., 2006). Certain aspects of phosphate homeostasis at the renal level, however, are not explained by actions of PTH. For instance, after removal of the parathyroid glands, regulation of renal phosphate transport by dietary phosphorus content persists, implying that other mediators of this process are at work.

Fibroblast growth factor 23

FGF23 is the most recently identified important physiologic regulator of renal phosphate excretion (Farrow and White, 2010). This novel member of the FGF family is produced by osteocytes and, to some extent, osteoblasts, thus providing a mechanism by which skeletal mineral demands can be communicated to the kidney and thereby influence the phosphate economy of the entire organism. In rodents and humans, after days of dietary phosphate loading, circulating FGF23 levels increase, and similarly, with dietary phosphorus deprivation, FGF23 levels decrease (Antoniucci et al., 2006). FGF23 activates FGF receptors (FGFRs) on the basolateral membrane of renal tubules, resulting in removal of type II sodium-dependent phosphate transporters from the apical surface of the tubular cell by a NHERF1-dependent process, similar to the previously described mechanism for PTH. However, in contrast to PTH, FGF23 actions are mediated by activation of extracellular signal-regulated kinase 1/2 (ERK1/2) rather than the PTH-driven PKA-dependent pathway. Evidence also exists for decreased expression of type II sodium-dependent phosphate transporters via genomic mechanisms. FGF23 interacts with its receptor via a mechanism now identified as characteristic of the endocrine FGFs. FGF23 recognizes its cognate FGFR only in the presence of the coreceptor α-klotho (Urakawa et al., 2006). Activation of this complex results in downstream ERK phosphorylation and subsequently reduced expression of NPT2a and NPT2c, and *CYP27B1* (encoding the vitamin D 1α-hydroxylase), with an increase in expression of *CYP24A1* (encoding the vitamin D 24-hydroxylase). This mechanism of signaling is apparent for the endocrine FGFs, FGF19 and FGF21, which require a separate member of the klotho family (β-klotho) for tissue-specific activation of FGFRs (for a detailed review, see Belov and Mohammadi, 2013).

FGF23 contains a unique C-terminal domain, thought to be the site of the interaction with klotho. The crystal structure of the FGF23/FGFR1c/α-klotho complex has been solved (Chen et al., 2018), confirming this prediction (Fig. 20.1). The C-terminal tail of FGF23 binds both the klotho 1 and the klotho 2 domains of α-klotho, and the FGF-like domain, which is N-terminal to the RXXR furin/subtilisin protease recognition site, binds to the FGFR. α-Klotho is able to associate with "c" isoforms of FGFR1 and FGFR3, and also FGFR4 (Urakawa et al., 2006); renal signaling is thought to primarily occur via FGFR1c. The actions of FGF23 and other related proteins as mediators of disease are discussed in detail later in the section on X-linked hypophosphatemia (XLH).

The role of osteocytes

Osteocytes are the most abundant cell in cortical bone, distributed in an organized array with interconnecting canaliculi (for review, see Dallas et al., 2013). Cellular processes extending from the cell body of the osteocyte tunnel through these canaliculi and communicate with other cells and bony surfaces. The osteocyte produces several proteins involved in phosphate regulation, including: (1) PHEX (phosphate-regulating gene with homologies to endopeptidases on the X chromosome), which regulates FGF23 secretion, with loss of function resulting in elevated circulating FGF23; (2) DMP1 (dentin matrix protein 1), a SIBLING (small integrin-binding ligand N-linked glycoprotein) protein, of which loss of

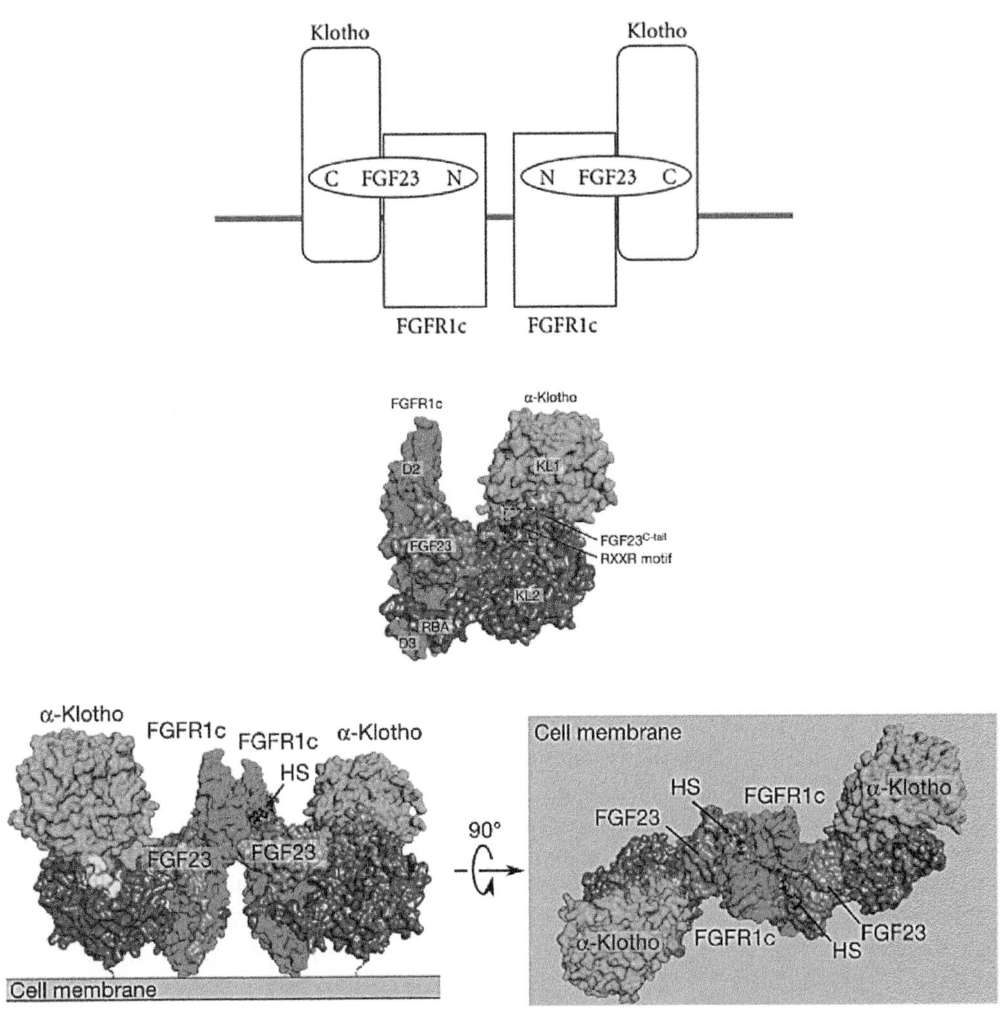

FIGURE 20.1 Schematic and crystal structure of the FGF23-FGFR1c-α-klotho complex. (Top) Schematic diagram of the membrane orientation of fibroblast growth factor 23 (FGF23)—receptor complex. The N terminus (N) binds to the extracellular domain of FGF receptor 1c (FGFR1c) and the C terminus (C) to α-klotho, in a dimerized configuration. (Middle) Surface topology of the ternary structure of this complex: α-klotho is shown in *cyan* (KL1 domain) and *purple* (KL2 domain), with FGF23 in *brown*, with the C terminus (*FGF23*$^{C\text{-}tail}$) bound to α-klotho and the N terminus to FGFR1c, shown in *green*. FGF23 is inactivated by cleavage at the protease recognition site (RXXR). (Bottom) Views of the complex in its dimer formation, stabilized by heparan sulfate (HS), parallel to the cell membrane (left) and looking down on the membrane (right). Colors are as noted for the middle; in *yellow* is the linker between the KL1 and KL2 domains of α-klotho. *Lower and middle images from Chen et al., 2018, with permission.*

function also results in elevated circulating FGF23; (3) FGF23 itself; and (4) FGFR1, which appears to be activated in osteocytes, resulting in elevated FGF23 expression. These observations have led to the consideration that osteocytes may directly respond to phosphate nutritional status, and ultimately relay, via production of FGF23, the mineral demands for bone maintenance to the kidney, where phosphate conservation is regulated. The osteocyte's response to phosphate status does not appear to be an acute process, as is observed with the extracellular calcium-sensing receptor system that regulates PTH secretion in parathyroid glands. It follows that genetic disruption of this pathway may result in the profound systemic disturbances observed in the diseases discussed herein.

In summary, the body's phosphorus status is largely adjusted at the kidney. Phosphate reabsorption is increased under conditions of greater need (rapid growth, pregnancy, lactation, and dietary restriction) and decreased during slow growth, chronic renal failure, or dietary excess. Changes in NPT2 protein abundance parallel these changes and are mediated by FGF23 and other possible factors. Removal of NPT2 cotransporters from the apical membrane of renal tubular cells is an acute process, mediated by PTH. The interaction of PTH and FGF23 in the overall process may also be important. Indeed, ablation of PTH in a murine model of excess FGF23 abrogates hypophosphatemia. Likewise, suppression of PTH may reduce phosphate losses even with persistence of high FGF23 (Bai et al., 2007; Carpenter et al., 2014), suggesting an interaction between the two pathways (Lanske and Razzaque, 2014).

Nutritional and gastrointestinal considerations

Whole-body phosphorus economy is summarized in Fig. 20.2. Dietary phosphorus is estimated at approximately 20 mg/kg/day, when not considering the phosphorus content of food additives. Based on NHANES data, dietary phosphorus intake in adults between the ages of 19 and 70 ranges from 1500 to 1700 mg/day for men and 1000 to 1200 mg/day for women (Calvo et al., 2014). Of the ingested phosphorus there is a net absorption of approximately 13 mg/kg; 16 mg/kg is unidirectionally absorbed, while 3 mg/kg is lost in endogenous gut secretions. The steady-state transit of phosphorus in and out of the skeletal compartment is balanced at roughly 3 mg/kg/day such that the daily urine phosphorus excretion largely reflects net gut absorption. The majority of the body's phosphorus (approximately 85%) is stored as hydroxyapatite in mineralized tissue. Roughly 14% is in soft tissues and only 1% is in the circulation or in the extracellular space. Eight-five percent of circulating phosphorus is ultrafilterable, the remainder is bound to serum proteins. The average woman in the United States has approximately 400 g of phosphorus stored in skeletal tissue, and men have approximately 500 g (Aloia et al., 1984).

The Western diet is awash in phosphorus, and dietary phosphorus estimates do not adequately capture the large amount of phosphorus in food additives. Estimates indicate that, at all ages, the typical phosphorus intake in the United States considerably exceeds the recommended dietary allowance by 1.5- to 2-fold (Calvo et al., 2014). This large phosphorus load may have deleterious consequences that are only now being studied. In a small clinical study, feeding healthy individuals a diet containing 1000 mg of phosphorus to which were added foods containing phosphorus additives resulted in increased circulating levels of FGF23, osteopontin, and osteocalcin, compared with the same diet with no food additives, suggesting the possibility that food additives in the context of the usual dietary phosphorus intakes could alter skeletal homeostasis (Gutierrez et al., 2015). Using NHANES data, a multiply adjusted association between high dietary phosphorus intake and all-cause mortality has been observed, with an inflection point at approximately 1400 mg/day of ingested phosphorus (Chang et al., 2014). It is noteworthy that men in the United States are estimated to have phosphorus intakes that often exceed 1400 mg/day (Calvo et al., 2014).

Intestinal phosphate absorption is highest in infancy and childhood and declines with age, but remains robust at approximately 50%–70% of bioavailable phosphorus. The majority of phosphate is absorbed in the small bowel by two pathways: the first is a passive paracellular pathway, in which phosphate moves through tight junction complexes formed by molecules, such as occludins and claudins. The second pathway is sodium-dependent phosphate cotransport, mediated by the type II sodium-dependent transporter NPT2b (encoded by *SLC34A2*) and, to a lesser extent, by the type III sodium-dependent transporter PIT1. The important role of NPT2b in intestinal phosphate absorption is highlighted by the impaired intestinal phosphate absorption in mice with genetic disruption of the gene (Sabbagh et al., 2009). The contributions of these two pathways to overall intestinal phosphate absorption vary, depending on the type and amount of dietary phosphate

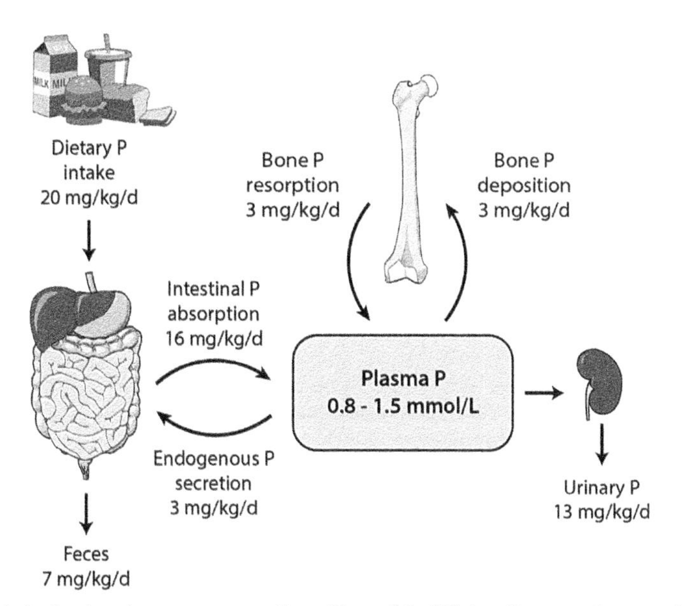

FIGURE 20.2 Summary of whole-body phosphorus economy. *From Figure 8.1, O'Brien, Kerstetter, Insogna. Chapter 8: Phosphorus; in Modern Nutrition in Health and Disease, 11th Edition. A. Ross, B. Caballero, R. Cousin, K. Tucker, T. Ziegler, eds. Lippincott, Williams and Williams. Baltimore, MD. pp 150–158. Reproduced with permission.*

presented to the gut. In experimental animals, it has been estimated that these two pathways contribute roughly equally to intestinal phosphate absorption (Lee and Marks, 2015). However, earlier estimates suggest that the paracellular pathway is the major mechanism for phosphate absorption. Both transcriptional and posttranscriptional regulation of Npt2b has been reported. For example, 1,25-dihydroxyvitamin D ($1,25(OH)_2D$) induces expression of Npt2b via a transcriptional mechanism, while low dietary phosphorus increases the activity of this transporter by posttranscriptional mechanisms. Whether changes in dietary phosphate absorption can lead to rapid changes in renal phosphate handling via as yet uncharacterized pathways that do not involve FGF23 or vitamin D remains an intriguing, but unproven, hypothesis. An early report suggested that this might be the case and, in particular, experimentally induced changes in duodenal phosphate absorption in rats led to rapid changes in renal phosphate handling without changes in systemic mineral homeostasis, suggesting novel "direct" gut/bone communication (Berndt et al., 2007). However, no similar observations have been reported nor have there been any further studies to characterize this pathway.

Finally, a novel interaction between the gut microbiota and FGF23 has been reported in an experimental model. Germ-free mice were found to have elevated circulating levels of FGF23 in the setting of low levels of 25-hydroxyvitamin D (25-OHD) and $1,25(OH)_2D$. Conventionalization of the germ-free mice (with the microbiome of normal mice) resulted in a fall in FGF23 levels and normalization in circulating levels of 25-OHD and $1,25(OH)_2D$. The authors posited that an inflammatory pathway, mediated by gut microbiota, leads to suppression of FGF23 expression. Whether changes in gut microbiota influence systemic phosphorus homeostasis through this pathway in humans remains unclear (Bora et al., 2018).

Phosphate and bone mineralization

The principal form of mineral in bone, hydroxyapatite ($Ca_{10}(PO_4)_6(OH)_2$), contains six atoms of phosphorus per crystal unit (each composed of two molecules). When the extracellular phosphorus supply is chronically deficient, mineralization is disrupted, resulting in the clinical conditions of rickets in children and osteomalacia in adults. Mineralization is essential during growth, when new bone is formed (modeling), as well as after epiphyseal closure, when replacing old bone and repairing damaged bone requires an orchestrated series of cellular events in which discrete packets of bone are removed and replaced (remodeling). Normal mineralization of bone requires the action of osteoblasts and hypertrophic chondrocytes, which secrete matrix vesicles (MVs). MVs have the ability to rapidly induce hydroxyapatite crystal formation in the presence of sufficient concentrations of extracellular calcium and phosphate (Bottini et al., 2018). Phosphate, liberated by the phosphatase PHOSPHO1 from phosphocholine, is then imported into the vesicle via the PIT1 transporter. The ambient extracellular phosphate concentration in bone is regulated by tissue nonspecific alkaline phosphatase (TNAP), which is abundant in MVs. TNAP cleaves PPi generating 2 phosphate molecules, resulting in a phosphate/PPi (Pi/PPi) ratio that favors mineralization. The process of mineralization is critically dependent on this ratio. Hydroxyapatite crystals initially form within MVs but are subsequently propagated onto the collagen matrix through as yet unclear mechanisms. The protein matrix that is eventually mineralized is composed primarily of collagen, but includes many other proteins, such as fibronectin, osteopontin, and osteocalcin.

There is currently intense interest in PPi as a central regulator of mineralization (Bottini et al., 2018; Bonucci, 2012; Zhou et al., 2012; Kim et al., 2010). In addition to TNAP, the extracellular concentration of PPi is regulated by several other proteins, including ENPP1 (ectonucleotide pyrophosphatase/phosphodiesterase 1), an exoenzyme that cleaves ATP to AMP and PPi. With genetic loss of function of ENPP1, extracellular concentrations of PPi are low, leading to rapid and aberrant mineralization of tissues, including the vasculature, a rare but devastating syndrome termed generalized arterial calcification of infancy (GACI) (Ferreira et al., 1993). Conversely, in the genetic absence of TNAP function, high PPi concentrations in the bone microenvironment prevent skeletal mineralization, resulting in the syndrome of hypophosphatasia, which in its most severe form is lethal around the time of birth (Whyte, 2017). ANKH (progressive ankylosis protein homolog) exports intracellular PPi and also plays a role in regulating the Pi/PPi ratio. Mutations in ANKH are associated with rare familial syndromes of calcium PPi deposition disease as well as autosomal dominant craniometaphyseal dysplasia (Pendleton et al., 2002; Nurnberg et al., 2001). Adding further complexity to this process is the fact that the hepatic-specific transporter ABCC6 (ATP binding cassette subfamily C member 6) also supports the extracellular concentration of PPi, since loss-of-function mutations are the basis for the syndrome pseudoxanthoma elasticum, in which serum PPi levels are low and heterotopic calcification, including in vascular tissue, can occur (Kranenburg et al., 2017). The fact that vascular calcification occurs in the genetic absence of both ENPP1 and ABCC6 points to the critical role of PPi in regulating heterotopic mineralization, particularly in the vascular tree. How these proteins modulate tissue mineralization in sites other than bone and vascular tissue remains to be fully explored.

Phosphorus also plays a critical role in chondrocyte maturation. The growth plate is a tightly regulated structure in which cells progress from resting to proliferating chondrocytes and then become prehypertrophic and finally hypertrophic chondrocytes that ultimately apoptose and are replaced by mineralized tissue. The concentration of extracellular phosphate is a key apoptotic signal for hypertrophic chondrocytes (Sabbagh et al., 2005). In the setting of low extracellular phosphate, chondrocytes fail to apoptose, which is part of the pathogenesis of the abnormal growth plate seen in hypophosphatemic rachitic disorders such as XLH. Since hypophosphatemia is also seen with nutritional rickets, the same pathology applies to that condition. A detailed molecular pathway involving ERK1/2–Raf signaling has been described as mediating the proapoptotic effect of phosphate in chondrocytes that, interestingly, is opposed by PTH-related protein, a cytokine that slows chondrocyte apoptosis (Papaioannou et al., 2017; Liu et al., 2014).

A fourth major participant in the regulation of mineralization is the hormone FGF23, which controls extracellular phosphate concentrations. Excess FGF23 levels cause systemic hypophosphatemia and impaired mineralization as described elsewhere in this chapter. In addition, it has been reported that FGF23 directly suppresses TNAP in osteocytes, allowing local concentrations of PPi to rise (Murali et al., 2016). This is another potential mechanism by which FGF23 could impair bone mineralization.

Finally, while the control of circulating levels of phosphorus by systemic hormones like FGF23, PTH, and $1,25(OH)_2D$, coupled with metabolism of PPi by TNAP, ENPP1, ANKH, and ABCC6, seems to be the major regulator of mineralization, other potential mechanisms that interact with phosphorus metabolism have been reported. Thus, the SIBLING proteins, such as osteopontin, as well as peptides derived from their incomplete metabolism in the absence of PHEX, called ASARM (acidic serine aspartate-rich MEPE-associated motif) peptides, have been reported to directly regulate phosphate metabolism and local bone mineralization (David et al., 2011; Boukpessi et al., 2017; Yuan et al., 2014). For this, the reader is referred to a discussion of the SIBLING proteins and ASARM peptides in the section of this chapter devoted to XLH.

Intracellular/extracellular compartmentalization

After absorption from the diet in the gut, phosphate is stored together with calcium as hydroxyapatite in the skeleton, and it is excreted in urine (for reviews see Forster et al., 2006; Prie et al., 2004; Liu and Quarles, 2007; Miyamoto et al., 2007; Shaikh et al., 2008; Strom and Juppner, 2008; White et al., 2006; Kurosu & Kuro-o, 2008) (Fig. 20.3). Phosphate typically moves between the intracellular and the extracellular compartments under the control of transporter proteins. The type III transporters PIT1 (*SLC20A1*) and PIT2 (*SLC20A2*) are ubiquitously expressed and are considered housekeeping transporters that mediate cellular uptake of phosphate to meet the needs of cell metabolism in many tissues. This metabolic uptake of phosphate into muscle, bone, and other tissues is regulated by various factors, such as phosphate itself (Chien et al., 1997; Wang et al., 2001), epinephrine (Suzuki et al., 2001), platelet-derived growth factor (PDGF) (Zhen et al., 1997), insulin and insulin-like growth factor 1 (Polgreen et al., 1994), basic FGF (Suzuki et al., 2000), and transforming growth factor-β (Palmer et al., 2000).

Upregulation of phosphate transporters by insulin, causing sequestration of phosphate into cells and hypophosphatemia, is encountered during refeeding in nutritionally deprived individuals following intravenous glucose infusion or oral carbohydrate intake (Petersen et al., 2005; Price et al., 1996; Butterworth and Younus, 1993; Kemp et al., 1993). Hypophosphatemia can also occur in the setting of high cellular demand for phosphate, for example, during rapid cell growth in hematological malignancies (Liamis and Elisaf, 2000; Milionis et al., 1999) or following hepatic resection (Keushkerian and Wade, 1984; Nafidi et al., 2007). In general, the circulating phosphate concentration is determined by phosphorus absorption from the intestine, excretion in urine, and balance between the intracellular and the extracellular compartments.

Mechanisms of phosphate transport

Transporter-mediated cellular absorption of dietary phosphate (Hilfiker et al., 1998), which is regulated in a $1,25(OH)_2D$-dependent manner (Wilz et al., 1979), accounts for 30% of intestinal phosphate absorption. The remaining 70% is absorbed through a poorly defined paracellular mechanism. The type II cotransporters NPT2a and NPT2c are expressed exclusively at the renal brush border membrane of the proximal tubules, where the bulk of filtered phosphate is reabsorbed (Segawa et al., 2002; Murer et al., 2004; Miyamoto et al., 2004; Shimada et al., 2004a). Based on data obtained in mice and rats, Npt2c is predicted to contribute about 15%–20% to total proximal tubular reabsorption of phosphate (TRP) (Segawa et al., 2002; Ohkido et al., 2003). Npt2c is thought to be rate-limiting only during weaning, with minimal contribution to total sodium/phosphate cotransport in adult rats (Segawa et al., 2002). The type III transporters, Pit1 and Pit2, are ubiquitously

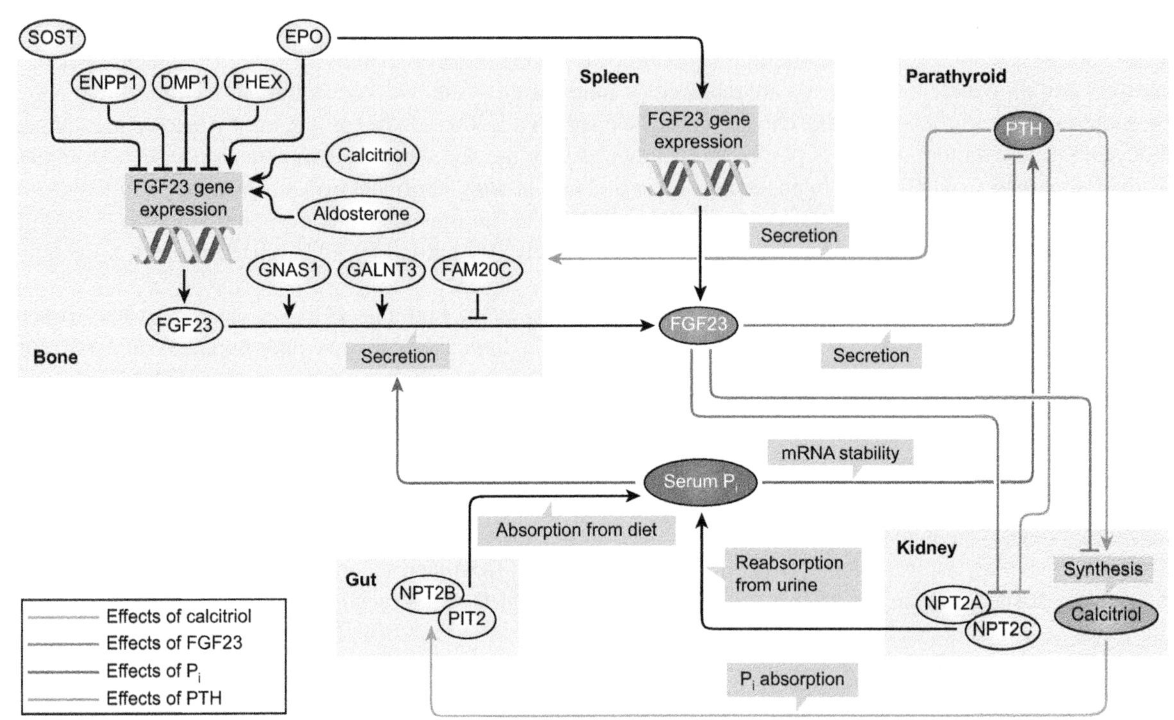

FIGURE 20.3 Regulation of phosphate homeostasis. Endocrine regulation of Pi homeostasis (originally published in Chande and Bergwitz, 2018). Serum phosphate (Pi) stimulates secretion of bioactive fibroblast growth factor 23 (FGF23) in osteoblasts and osteocytes (*blue arrow* (dark grey in print versions)), which directly or indirectly acts at the proximal tubule of the kidneys to inhibit synthesis of calcitriol and the function of NPT2A and NPT2C (*red arrows* (light grey in print versions)). Inhibition of calcitriol reduces absorption of Pi from the diet in the gut (*gray arrow*) and mobilization of Pi from bone mineral. Downregulation of NPT2A and NPT2C reduces renal phosphate reabsorption (*black arrows*). The net effect of FGF23 action is to lower blood levels of Pi. Similar to FGF23, parathyroid hormone (*PTH*) downregulates NPT2A and NPT2B and reduces renal phosphate reabsorption (*green arrows*(grey in print versions)). However, different from FGF23, PTH induces calcitriol and bone turnover, which increase blood Pi (*green arrows*). However, the net effect of PTH is to lower blood levels of Pi. Although not completely understood, golgi associated secretory pathway kinase (FAM20C), DMP1, ENPP1, and PHEX reduce FGF23 expression or secretion, whereas phosphate, iron deficiency, erythropoietin (EPO), GALNT3, and guanine nucleotide binding protein (G protein), alpha stimulating activity polypeptide 1 (GNAS1) stimulate it (*black arrows*). In addition, sclerostin (SOST) seems to negatively regulate FGF23 (*black arrows*). Furthermore, EPO might directly upregulate FGF23 gene expression in myeloid lineage stem cells of the spleen, providing a link to iron homeostasis (*black arrows*). PTH is suppressed by FGF23 in rodents but not in humans (*red arrow*). DMP1, dentin matrix acidic phosphoprotein 1; ENPP1, ectonucleotide pyrophosphatase/phosphodiesterase 1; GALNT3, UDP-*N*-acetyl-α-D-galactosamine:polypeptide *N*-acetylgalactosaminyltransferase, isoform 3; NPT2, solute carrier family 34 (type II) sodium/phosphate cotransporters (with "A", "B", and "C" species); PHEX, phosphate-regulating gene with homologies to endopeptidases on the X chromosome; PIT2, solute carrier family 20 (type III) sodium/phosphate cotransporter ("2" species).

expressed yet account for less than 1% of the mRNAs encoding the various renal sodium/phosphate cotransporters (Tenenhouse et al., 1998). All of these transporter classes utilize the transepithelial sodium gradient to transport phosphate against its electrochemical gradient into cells.

Intestinal phosphate transport

Intestinal uptake of phosphate occurs primarily by passive paracellular diffusion, while only 30% occurs via Na^+-dependent, carrier-mediated transcellular transport (Murer et al., 2000, 2004; Eto et al., 2006; Berndt & Kumar, 2009). The molecular mechanism of the former is unclear; the latter is mediated by the type IIb sodium/phosphate cotransporter NPT2b and the type III sodium/phosphate cotransporter PIT2 (Sabbagh et al., 2011; Giral et al., 2009; Xu et al., 2002, 2003, 2005; Yeh and Aloia, 1987). $1,25(OH)_2D$ and dietary phosphorus depletion are thought to be the most important physiological stimuli of intestinal phosphate absorption and act by increasing the abundance of NPT2b protein (Hattenhauer et al., 1999). Glucocorticoids (Yeh and Aloia, 1987) and estrogen (Xu et al., 2003) increase, and nicotinamide (Eto et al., 2005) decreases, intestinal phosphate transport, but the mechanism is unknown. There is no known direct effect of FGF23 or PTH on intestinal phosphate transporters (Tomoe et al., 2010), but intestine may release a factor(s) that signals dietary phosphate availability to the kidney (Berndt et al., 2007). Evaluation of *Npt2b* (Sabbagh et al., 2009; Knöpfel et al., 2017) conditional knockout (KO) mice suggests that this transporter is probably the primary transepithelial intestinal phosphate transporter. Ablation of *Npt2b* attenuates the rise in blood phosphate seen in the setting of adenine-induced chronic kidney disease (CKD) (Schiavi et al., 2012; Sabbagh and Schiavi, 2014; Ohi et al., 2011). Furthermore, chronic adaptation to a low-phosphate diet in wild-type mice appears to go along with upregulation of *Npt2b* in the duodenum and upregulation of *Npt2a* in the proximal tubules (Giral et al., 2009). However, chronic adaptation in $Npt2b^{+/-}$ (Ohi et al., 2011), conditional $Npt2a^{-/-}$ mice (Schiavi et al., 2012), and, in our experience, $Npt2a^{-/-}$ mice (Pesta et al., 2016) suggests additional transport mechanisms.

Renal phosphate transport

Three classes of sodium/phosphate cotransporters are expressed in the kidney: the type I cotransporter, NPT1 (*SLC17A1*); the type II cotransporters, NPT2a (*SLC34A1*) and NPT2c (*SLC34A3*); and the type III cotransporter PIT2 (*SLC20A2*). Reabsorption of phosphate from the urine in the renal proximal tubules via NPT2a and NPT2c plays a key role in maintaining phosphate homeostasis, while other tubule segments are believed to be of less importance, and there are no known excretory mechanisms for phosphate. The NPT2a and NPT2c transporters preferentially transport divalent phosphate (HPO_4^{2-}) (Biber et al., 2013). Immunohistochemical analysis shows the localization of NPT2a to the apical membrane of renal proximal tubule cells (S1−S3 segments of the proximal tubule) in rodents (Murer et al., 2000, 2004; Berndt et al., 2005). Conversely, NPT2c in rodents appears to be expressed along with PIT2 only in segment 1 of the proximal tubules (Wagner et al., 2017). The NPT2c transporter, which shares approximately 54% amino acid identity with NPT2a, is not electrogenic and mediates cotransport at a ratio of 2 Na^+/1 HPO_4^{2-}. Renal expression of NPT2c in human kidneys was found to resemble the distribution seen in mice (Ohkido et al., 2007) and the response of these individuals to oral phosphorus therapy based on cross-sectional evaluation appears to resemble that described in Npt2a-KO and Npt2a;Npt2c-double-KO (DKO) mice.

Constitutive mouse KO models of Npt2a and Npt2c suggest that Npt2a mediates the bulk of renal reabsorption of phosphate, whereas the contribution of Npt2c to this process is minor and probably restricted to young mice (Miyamoto et al., 2007; Murer et al., 2000, 2004; Berndt et al., 2005; Segawa et al., 2009). Furthermore, rickets or osteomalacia is absent in both mouse models. However, Npt2a;Npt2c DKO suggest that both transporters have nonredundant roles, since these mice exhibit reduced body weight and expansion of the late hypertrophic zone in the bone, which is a characteristic feature of rickets, along with severe hypophosphatemia, hypercalciuria, and renal calcification, which can be reversed by a high-phosphorus diet (Segawa et al., 2009; Tomoe et al., 2009). In contrast to Npt2c-KO mice, humans with homozygous or compound-heterozygous loss-of-function mutations in NPT2c develop hypophosphatemic rickets in the first decade of life. These observations suggest that there may be important differences in renal phosphate handling between mice and humans.

FGF23 reduces reabsorption of phosphate by reducing expression of NPT2a and NPT2c (Yan et al., 2005). FGF23 binds to FGFR1c and its coreceptor klotho (Urakawa et al., 2006; Kuro-o, 2006) with activation of ERK1/2 at the distal convoluted tubules (Farrow et al., 2009). Alternatively, FGF23 may bind to FGFRs in the proximal tubule, leading to phosphorylation of NHERF1 via Sgk1 (Andrukhova et al., 2012). Also, a direct regulation of NPT2a activity by klotho without FGF23 has been demonstrated, whereby klotho causes cleavage and inactivation of NPT2a in the brush border membrane and subsequent internalization (Hu et al., 2010). Internalization of NPT2a from the apical membrane occurs as with receptor-mediated endocytosis (Bacic et al., 2006; Bachmann et al., 2004). Consequently, NPT2a is routed to lysosomes for degradation (Keusch et al., 1998; Pfister et al., 1998; Inoue et al., 2004). The regulation of NPT2c has been studied in less detail: in response to high phosphate, NPT2c downregulation occurs but is slower than for NPT2a (Miyamoto et al., 2007; Picard et al., 2010). Proximal TRP, to a lesser extent, is regulated by insulin, hormones of the somatotropic pituitary axis (Bringhurst and Leder, 2006), FGF7 (Carpenter et al., 2005), and possibly matrix extracellular phosphoglycoprotein (MEPE) and secreted Frizzled-related protein 4 (sFRP4) (White et al., 2006). The signaling mechanisms whereby insulin, growth hormone/IGF-1, FGF7, MEPE, and sFRP4 regulate NPT2a and NPT2c are unknown. The original description of Npt2c suggests that this transporter is exclusively expressed in mouse and rat kidney (Segawa et al., 2002). However, public RNA-sequencing expression databases (GTEx Consortium, 2013) and a qRT-PCR tissue expression analysis (Nishimura and Naito, 2008) suggest low-level expression of NPT2c in many tissues. Furthermore, data from a kidney-specific and inducible Npt2c-deficient mouse Npt2c (CKO) model (Myakala et al., 2014) suggest a role of extrarenal Npt2c in calcium homeostasis. However, the significance of widespread expression of Npt2c outside of the kidneys for human phosphorus and calcium metabolism requires further investigation.

Ubiquitous metabolic phosphate transporters

The PIT transporters serve as "housekeeping" transporters in most tissues (Kavanaugh and Kabat, 1996). Furthermore phosphate activates ERK1/2 in most cell types, and is blocked by pharmacological or genetic ablation of type III transporters (Wittrant et al., 2009). Since PIT1 is able to compensate for the loss of PIT2, intracellular phosphate may be what is "sensed" to activate ERK1/2. However, PIT transporters also bind and signal extracellular phosphate independent of phosphate transport, as shown in HeLa and vascular smooth muscle cells expressing a transporter-deficient mutation, [E70K]PIT1 (Chavkin et al., 2015; Beck et al., 2009). Furthermore, functional cooperation between PIT1 and FGFR1 appears to exist based on in vitro (Yamazaki et al., 2010) and in vivo (Xiao et al., 2014) reports. In addition to ERK1/2, the yeast second messenger inositol heptakisphosphate (IP7) (Azevedo and Saiardi, 2017; Saiardi, 2012) was found to be synthesized by inositol hexakisphosphate kinase 2 in various human cell lines, including HCT116 (colon) and U2OS (bone) cells (Koldobskiy et al., 2010). Furthermore, the highly conserved IP7 binding domain SPX is found in the xenotropic retroviral receptor 1 (XPR1) (Wild et al., 2016). Loss of function of XPR1 causes renal Fanconi syndrome in mice (Ansermet et al., 2017) and primary familial brain calcification (PFBC) in humans (Azevedo and Saiardi, 2017). These findings suggest a role for ERK1/2 and IP7 in metabolic phosphate effects, although details as to the mechanism and whether these second messengers regulate PTH, FGF23, or $1,25(OH)_2D$ remain unknown (Chande and Bergwitz, 2018).

As noted earlier intracellular phosphate may be sequestered into subcellular compartments. PIT1 can localize to the endoplasmic reticulum (ER) and contribute to regulation of ER stress in growth plate chondrocytes (Couasnay et al., 2019). The mitochondrial phosphate carrier PIC (SLC25A3) is part of a multiprotein complex, the mitochondrial permeability transition pore (MPTP), which regulates mitochondrial membrane potential and mitochondrial apoptosis (Pauleau et al., 2008), as well as being important in skeletal and cardiac muscle function (Seifert et al., 2015). The related dicarboxylate carrier (SLC25A10) (Fiermonte et al., 1999) catalyzes transport of dicarboxylates such as malate and succinate across the mitochondrial membrane in exchange for phosphate, sulfate, and thiosulfate.

Surprisingly mild bone and mineral metabolism phenotypes of the individual Pit1- and Pit2-KO mice suggest a high degree of redundance of these generally coexpressed transporters, and bone-selective ablation of Pit1 and Pit2 individually and in combination will likely shed a better light on their metabolic and endocrine functions. For example, acute chondrocyte-specific deletion of Pit1 in mice using a tamoxifen-inducible annexin-Cre in the first 2 days postnatally results in pronounced cell death, possibly as a result of phosphate-transport-independent ER stress (Couasnay et al., 2019), which in hindsight is consistent with earlier reported mildly reduced femur length in Pit1 hypomorphic mice (Bourgine et al., 2013) and the MV mineral defect observed in *collagen2a1*-Cre in Pit1;PHOSPHO1-DKO mice (Yadav et al., 2016).

Primary disorders of phosphate homeostasis

Fibroblast growth factor 23–mediated hypophosphatemic disorders

X-linked hypophosphatemia (OMIM: 307800)

In 1936, Fuller Albright reported a case of rickets unresponsive to vitamin D treatment, and thus termed the condition "vitamin D-resistant rickets "(Albright et al., 1936). In the late 1950s the cardinal feature of renal phosphate wasting, together with an X-linked dominant inheritance pattern, was recognized (Hsia et al., 1959). Subsequent studies in the *Hyp* mouse, a murine model of XLH, demonstrated defective renal tubular phosphate transport (Glorieux and Scriver, 1972; Bell et al., 1988), and decreased expression of renal sodium/phosphate cotransporters in the renal tubules (Tenenhouse et al., 1994; Collins and Ghishan, 1994). In 1974 the observation of persistent hypophosphatemia following renal transplantation in a man with classical XLH raised the consideration that the disease was due to a humoral factor rather than an intrinsic renal defect (Morgan et al., 1974). This concept was substantiated by Meyer and Meyer, who demonstrated that a normal mouse joined in parabiosis to a *Hyp* mouse became hypophosphatemic (Meyer Jr. et al., 1989).

In 1995, loss-of-function mutations in PHEX, a neutral endopeptidase, were shown to be responsible for XLH (HYP-Consortium, 1995). In 2000/2001, a circulating factor responsible for tumor-induced osteomalacia (TIO) and autosomal dominant hypophosphatemic rickets (ADHR) was identified as FGF23, a novel member of the FGF family (ADHR Consortium T, 2000; Shimada et al., 2001). It was quickly realized that FGF23 was overproduced in XLH, establishing that disorder as a classic endocrine syndrome due to overproduction of a hormone (Jonsson et al., 2003; Yamazaki et al., 2002). In XLH, FGF23 is overproduced by osteocytes (Feng et al., 2006), where PHEX is primarily expressed, but how PHEX regulates the production of FGF23 remains unclear.

Pathophysiology

FGF23 directs the kidney to excrete phosphate upon exposure to an excessive phosphate load. Its production is stimulated by increases in extracellular phosphate (and also by $1,25(OH)_2D$) but this axis is perturbed in XLH, such that FGF23 is overproduced, despite the ambient hypophosphatemia. Consequent suppression of transcription of the renal sodium/phosphate cotransporters NPT2a and NPT2c (Shimada et al., 2004b, 2004a; Erben, 2016), as well as their removal from the apical surface of the proximal renal tubule cell (Murali et al., 2016), results in reduced phosphate reclamation. FGF23 also decreases circulating $1,25(OH)_2D$, resulting in impaired intestinal calcium and phosphate absorption.

Abnormal metabolism of bone-derived extracellular matrix proteins may also contribute to the pathophysiology of XLH. Substrates for PHEX may include SIBLING proteins (osteopontin, bone sialoprotein, DMP1, dentin sialophosphoprotein, and MEPE). In vitro evidence suggests that these proteins can be metabolized by PHEX, and in the absence of normal metabolism, protein fragments—ASARM peptides—accumulate, resulting in inhibition of bone mineralization in vitro and renal phosphate wasting in vivo (David et al., 2011; Barros et al., 2013; Addison et al., 2010). Moreover, work has identified increased osteopontin in the osteomalacic osteoid of Hyp mice, and less extensive osteomalacia in the osteopontin null/Hyp (DKO mouse) compared with Hyp littermates (Hoac et al., 2018).

Prevalence

The exact prevalence of XLH is estimated as between 1 in 20,000 live births (United States and Denmark) and 1 in 60,000 (Norway) (Ruppe; Beck-Nielsen et al., 2009; Rafaelsen et al., 2016). As many as 25% of cases appear to be sporadic (Beck-Nielsen et al., 2012; Whyte et al., 1996).

Clinical manifestations

XLH causes rickets in children manifest as bowed lower extremities, short stature, and radiographically abnormal epiphyses. Both children and adults with XLH suffer from the early appearance and persistent recurrence of dental abscesses. These are associated with structural abnormalities in the tooth, including abnormal mineralization of dentin, less abundant cementum, impaired tooth attachment, and increased risk of periodontal disease (Antoniucci et al., 2006; Coyac, 2017). Pulp chambers appear to be enlarged while enamel is largely unaffected (Turan et al., 2010). How these changes relate to the pathogenesis of dental abscesses is not clear. Tooth fragility is often also reported by patients, including fracturing of teeth with normal mastication. Finally, craniosynostosis with associated Chiari malformations are not infrequent occurrences in XLH (Rothenbuhler et al., 2018).

Skeletal disease in XLH was once thought to quiesce following epiphyseal closure, but in fact, adults experience some of the worst complications of the disease. Nearly all affected adults have osteomalacia and poor bone quality leading to increased risks of clinical fracture and insufficiency fractures with little or no trauma, occurring most often in the lower extremities. Abnormal biomechanics (persisting from deformities acquired in childhood) contribute to the high incidence and early onset of osteoarthritis. Osteophytes and calcification of tendons and ligaments at insertion sites (entheses) occur, and can severely limit of range of motion, particularly of the hips, elbows, and shoulders (Liang et al., 2009; Connor et al., 2015). Degenerative changes in the spine may lead to restricted range of motion and chronic low back pain. Calcification of the spinal longitudinal ligaments can occur (Hirao et al., 2016; Forrest et al., 2016; Shiba et al., 2015), as well as spinal stenosis, result in neurologic deficits requiring surgical intervention (Chesher et al., 2018). Scoliosis is common, and hearing loss occurs (Fishman et al., 2004).

Both children and adults with XLH complain of weakness and diminished endurance, which may in part relate to impaired muscle function. Indeed, hypophosphatemia is associated with impaired ATP flux in skeletal muscle, which improves with correction of the hypophosphatemia (Pesta et al., 2016).

Therapy

Medical treatment for XLH historically included phosphate supplements with vitamin D used as a cotherapy. Limited efficacy, secondary hyperparathyroidism, and vitamin D intoxication were not infrequent. In 1980, Glorieux et al. published a landmark study comparing phosphate alone with combined therapy with ergocalciferol and phosphate and with calcitriol and phosphate (Glorieux et al., 1980). Histomorphometric improvement in osteomalacia with calcitriol and phosphate was superior to that of the other two regimens, serum phosphorus levels were improved to a greater extent, and radiographic healing of rickets was superior as well. This work led to a fundamental change in the treatment for XLH, and until 2018, calcitriol and phosphate therapy was the standard medical treatment for children. The therapy, however, does require careful monitoring to avoid hypercalcemia, hypercalciuria, and secondary hyperparathyroidism. Frequent dose adjustments are required as the skeleton grows, necessitating careful monitoring, usually every 3–6 months during rapid growth (Carpenter et al., 2011). Adherence is particularly difficult, as multiple doses of phosphate and two doses of calcitriol daily are usually employed. Phosphate supplementation has frequent gastrointestinal side effects, such as nausea and diarrhea, which further complicate compliance. With good adherence and careful monitoring, and when initiated early in life (i.e., less than 2 years of age), linear growth is improved, and there are fewer rachitic deformities.

Despite these effects, *surgical treatment* is often additionally required in XLH. Correction of genu valgus or genu varum of the lower extremities has traditionally required osteotomies and rodding of the long bones of the lower extremities (Eyres et al., 1993). More recently, when identified early in childhood, and when the bowing deformities are mild–moderate, epiphysiodesis can be used as a less invasive approach to guide growth at the distal femur or tibial growth plates (Ghanem et al., 2011). Children with cranial stenosis sometimes have severe headaches that require surgical treatment.

Growth hormone has been used as an adjunct to standard therapy in children with XLH, with mixed results. Some studies suggest improved linear growth, while others report little long-term benefit and potential worsening of disproportion between upper and lower body segments (Baroncelli et al., 2001; Haffner et al., 2004; Zivicnjak et al., 2011).

In contrast, therapy for adults has remained less widely employed; however, calcitriol and phosphate have been shown to have efficacy in adults: calcitriol and phosphate led to a 50% healing in osteomalacia (over 4 years) and a subjective improvement in pain in an open-label trial (Sullivan et al., 1992). In that study the daily dose of calcitriol and phosphorous ranged between 1.0 and 2.0 μg/day and 1.0 and 1.5 g/day, respectively. Other cross-sectional studies have demonstrated that therapy with calcitriol and phosphate during adult life is associated with less severe dental disease (Connor et al., 2015).

Conventional therapy does prevent enthesopathy, or hearing loss. Even when initiated early in childhood the therapy does not result in a normal adult height. Male sex and obesity portend worse enthesopathy and, importantly, in a cross-sectional study, persistent secondary hyperparathyroidism tended to be associated with worse enthesopathy (Connor et al., 2015). The latter point is highlighted by a study in which paricalcitol was used to suppress PTH in patients with XLH and secondary hyperparathyroidism (Carpenter et al., 2014). Despite the expected rise in FGF23 when paricalcitol was added to conventional therapy, the renal phosphate threshold improved, and bone-specific alkaline phosphatase fell. These data suggest that PTH contributes to the renal pathophysiology in XLH and is a major driver of skeletal turnover. Whether an increase in skeletal turnover contributes to enthesopathy remains unclear, and requires further study.

In 2018, a fully humanized monoclonal blocking antibody to FGF23, burosumab, was approved for treatment of XLH in children and adults in the United States and Canada and in children in the European Union. The drug binds to FGF23, and thereby inhibits the activity of the elevated circulating levels, essentially reversing the major pathophysiologic element of the disease. Renal phosphate reclamation improves and $1,25(OH)_2D$ production is restored, limiting the tendency for PTH oversecretion. In open-label clinical trials in 5- to 12-year-old children (90% of whom were treated with conventional therapy for ~7 years) the serum phosphorus levels corrected and were maintained in the low normal range in more than 90% of the study subjects (Wolfgang et al., 2018), with accompanying improvements in rickets (Carpenter et al., 2018). More recently, in a head-to-head comparison with conventional therapy, burosumab administered every 2 weeks showed more impressive improvements in serum phosphorus, alkaline phosphatase, and radiographic healing of rickets (Imel et al., 2018). Complications in children have been thus far limited to local injection site reactions and rare occurrences of urticaria. In adults, a double-blind, placebo-controlled, 24-week trial showed that monthly administration of burosumab (1 mg/kg) improved serum phosphorus and $1,25(OH)_2D$ levels, with over 90% of patients normalizing serum phosphorus at the dose cycle midpoint (2 weeks after dosing) (Insogna et al., 2018). Stiffness was significantly improved. Comprehensive imaging identified numerous extant fractures and pseudofractures at baseline; subjects treated with burosumab had a 17-fold greater likelihood of healing fractures and pseudofractures than those receiving placebo (Insogna et al., 2018). Preliminary data from a second year of treatment support continued efficacy of the drug in adults (Portale et al., 2018). Treatment of adults with burosumab demonstrated greater than 50% healing of osteomalacia (Insogna et al., 2018) after 1 year of therapy, comparable to that seen after 4½ years in the previously mentioned open-label trial of calcitriol and phosphate. The safety profile of burosumab was good, with restless leg syndrome and local injection site reactions being the only adverse events of note. Importantly, there has not been any evidence of development of nephrocalcinosis or worsening of existing nephrocalcinosis with burosumab, and no evidence for heterotopic cardiac calcification.

Early treatment of XLH in children has been advantageous, and conventional therapy has generally been applied by 6 months of age. It is recommended that treatment of affected children with burosumab be initiated as soon as possible after the first birthday until safety and dosing parameters in infants less than one year of age is established. It is unclear which children should discontinue currently applied conventional therapy and begin treatment with burosumab. Longer term follow-up may provide important guidance in that regard, but thus far, it appears that burosumab is likely to be safer, more convenient, and more effective than conventional therapy. Importantly, adherence is likely to be far better with the parenteral, less frequent dosing regimen. One barrier to the use of burosumab is its high expense, although the manufacturer's program to assist patients while negotiating insurance coverage for the drug has been operational in many settings. In contrast to children, there is no consensus on indications for treatment of any kind for adults with XLH. Suggested indications for treating adults with conventional therapy have been published, although these are not strictly evidenced based and largely represent recommendations from a group of experienced clinicians (Carpenter et al., 2011). Nonetheless they could serve as initial indications for burosumab. They include (1) spontaneous insufficiency fractures, (2) pending orthopedic procedures, (3) biochemical evidence of significant osteomalacia (specifically a significantly elevated bone-specific alkaline phosphatase), and (4) disabling skeletal pain.

Autosomal dominant hypophosphatemic rickets (OMIM: 193100)

Although occurring with far less frequency, ADHR is a disorder with a clinical presentation essentially identical to that of XLH, but occurring in the setting of autosomal dominant inheritance. Patients with an XLH-like phenotype in families with male-to-male transmission should be suspected as having this disorder (Harrison and Harrison, 1979; Econs and McEnery, 1997). Like XLH, hypophosphatemia secondary to renal phosphate wasting occurs, with lower extremity deformities, and rickets/osteomalacia. Affected patients also demonstrate normal serum 25-OHD levels, with inappropriately normal serum concentrations of $1{,}25(OH)_2D$, all hallmarks of XLH (Table 20.1). PTH levels are normal. Long-term studies indicate that a few of the affected female patients demonstrate delayed penetrance of clinically apparent disease and an increased tendency to fracture, features that appear to be less common in XLH. In addition, among patients with the expected biochemical features documented in childhood, some patients, albeit rarely, have been reported to lose the renal phosphate wasting defect after puberty. Specific missense mutations in FGF23 that result in the substitution of an arginine residue at amino acid residue 176 or 179 are present in patients with ADHR (ADHR Consortium, 2000). These mutations disrupt an RXXR subtilisin/furin protease recognition site, and the resultant mutant molecule is thereby protected from proteolysis, and resultant accumulation of unprocessed FGF23 results in elevated circulating intact FGF23 levels.

Circulating FGF23 levels can vary and reflect the activity of disease status (Imel et al., 2007). Exploration of the waxing/waning severity of disease in ADHR has indicated that iron may play a significant role in the regulation of circulating FGF23 (Imel et al., 2011). It is now evident that iron deficiency is able to upregulate FGF23 expression, and in normal individuals the unrestricted processing of the intact protein to its inactive N- and C-terminal fragments is able to compensate for the increased intact FGF23 production observed in the setting of iron deficiency. Therefore normal individuals who become iron deficient maintain normal circulating levels of intact FGF23 despite an increase in production. However, in ADHR, inefficient processing of FGF23, due to the impaired recognition of the protease site, may not be able to compensate for increased FGF23 synthesis during periods of iron deficiency, resulting in elevated active FGF23 accumulation in the circulation with the subsequent renal effects on phosphate excretion. It appears that the waxing and waning clinical severity observed in some cases of ADHR may be amenable to iron supplementation and provide a straightforward approach to therapy. Correction of serum iron levels to high normal levels allowed for discontinuation of conventional rickets medications in one report (Kapelari et al., 2015).

Autosomal recessive hypophosphatemic rickets

Autosomal recessive hypophosphatemic rickets type 1 (ARHR1; OMIM: 241520). Another gene identified as causal to FGF23-mediated hypophosphatemia is *DMP1*, which encodes the SIBLING protein dentin matrix protein 1, yet another product of the osteocyte. Biallelic loss of function mutations in *DMP1* can result in autosomal recessive phosphate wasting rickets due to excess circulating FGF23 levels. The same constellation of progressive rachitic deformities seen in both XLH and ADHR (Feng et al., 2006; Lorenz-Depiereux et al., 2006a) occurs, along with the features of hypophosphatemia, excess urinary phosphate loss and aberrant vitamin D metabolism (normal circulating 25-OHD and $1{,}25(OH)_2D$ levels, despite ambient hypophosphatemia), observed in XLH and ADHR. In addition to the expected phenotypic features consequential to excess FGF23, and in contrast to XLH, spinal radiographs of patients with ARHR reveal noticeably sclerotic vertebral bodies and sclerosis of the skull. The clinical diagnosis of a novel sclerosing dysplasia in adulthood led to the investigation of the causes of osteopetrosis in one recently reported family, but genetic investigation revealed *DMP1* mutations (Gannage-Yared et al., 2014). Mild hypophosphatemia was present. In addition to the enlarged pulp chamber characteristic of teeth in individuals with XLH, enamel hypoplasia can be evident in heterozygotes. Experience with long-term follow-up is not widespread in ARHR and therapeutic response or guidelines have not been definitively established, although severe adult complications of disease such as enthesopathy have been documented (Gannage-Yared et al., 2014; Karaplis et al., 2012). One 2016 study in a *DMP1*-KO animal model of the disease found that an anti-sclerostin antibody improved osteomalacia with little effect on the serum phosphorus or FGF23 levels (Ren et al., 2016).

TABLE 20.1 Human genetic disorders of phosphate homeostasis.

Disorder	Abbreviation	Inheritance	Gene	Mechanism	OMIM No.	Reference
Hyperphosphatemic disorders						
Hyperphosphatemic familial tumoral calcinosis type 1 and the allelic variant Hyperostosis–hyperphosphatemia syndrome	HFTC HSS	AR AR	GALNT3	FGF23 deficiency	211900 610233	(Topaz et al., 2004; Frishberg et al., 2004)
Hyperphosphatemic familial tumoral calcinosis type 2	HFTC	AR	FGF23	FGF23 deficiency	211900	(Ichikawa et al., 2006; Benet-Pages et al., 2005)
Hyperphosphatemic familial tumoral calcinosis type 3	HFTC	AR	KL	FGF23 resistance	211900	Ichikawa et al. (2007)
Pseudohypoparathyroidism	PHP1A PHP1B	AD AD	GNAS GNAS or up-stream regulatory region	PTH resistance FGF23 independent	103580 603233	(Weinstein et al., 1992; Juppner et al., 1998)
Familial isolated hypoparathyroidism	FIH	AD or AR	CaR GCMB PTH	PTH deficiency; FGF23 independent	146200	(Arnold et al., 1990; Pollak et al., 1994; Ding et al., 2001)
Blomstrand disease	BOCD	AR	PTHR1	PTH resistance; FGF23 independent	215045	(Zhang et al., 1998; Karperien et al., 1999)
Hypophosphatemic disorders						
X-linked hypophosphatemia	XLH	X-linked	PHEX	FGF23 dependent	307800	HYP-Consortium (1995)
Autosomal dominant hypophosphatemic rickets	ADHR	AD	FGF23	FGF23 dependent	193100	ADHR Consortium T (2000)
Autosomal dominant hypophosphatemic rickets with hyperparathyroidism	ADHR	AD	KL	FGF23 dependent	612089	Brownstein et al. (2008).
Autosomal recessive hypophosphatemia	ARHP	AR	DMP1	FGF23 dependent	241520	Lorenz-Depiereux et al. (2006)
Hereditary hypophosphatemic rickets with hypercalciuria	HHRH	AR	SLC34A3	Proximal tubular Pi wasting, FGF23 independent	241530	(Bergwitz et al., 2006; Lorenz-Depiereux et al., 2006b)

Vitamin-resistant rickets type 1	VDDR1	AR	CYP27B1	$1,25(OH)_2D$ deficiency, FGF23 independent	264700	Kitanaka et al. (1998)
Vitamin-resistant rickets type 2	VDDR2	AR	VDR	$1,25(OH)_2D$ resistance, FGF23 independent	277440	Hughes et al. (1988)
Familial hypocalciuric hypercalcemia/ neonatal severe hyperparathyroidism	FHH NSHPT	AD/AR	CaR	PTH excess, FGF23 independent	145980 239200	Pollak et al. (1993)
Jansen disease		AD	PTHR1	Constitutively active PTHR1; FGF23 dependent	156400	(Schipani et al., 1995), (Brown et al., 2009)
Normophosphatemic disorders						
Pulmonary alveolar microlithiasis	PAM	AR	SLC34A2	Reduced alveolar epithelial Pi uptake	265100	Sabbagh et al. (2009)
Normophosphatemic familial tumoral calcinosis	NFTC	AR	SAMD9	Unknown	610455	Topaz et al. (2006)
Muscular dystrophy and cardiomyopathy	MDC	AR	SLC25A3	Reduced mitochondrial Pi uptake	610773	(Bhoj et al., 2015; Mayr et al., 2011)
Primary familial basal ganglial calcification type 1	PFBC1	AR	PIT2	Reduced microglial Pi uptake	213600	Lemos et al. (2015)
Primary familial basal ganglial calcification type 6	PFBC6	AR	XPR1	Reduced vascular Pi export	616413	Legati et al. (2015)
Primary familial basal ganglial calcification type 3	PFBC4	AR	PDGFBR	Reduced PIT2 expression	615007	Keller et al. (2013)
Primary familial basal ganglial calcification type 4	PFBC5	AR	PDGFB	Reduced PIT2 expression	615483	Keller et al. (2013)

$1,25(OH)_2D$, 1,25-dihydroxyvitamin D; *AD*, autosomal dominant; *AR*, autosomal recessive; *FGF23*, fibroblast growth factor 23; *Pi*, phosphate; *PIT2*, solute carrier family 20 (type III) sodium/phosphate cotransporter 2; *PTH*, parathyroid hormone.

ARHR2 (OMIM: 613312). Another rare variant of hypophosphatemic rickets with renal phosphate wasting, ARHR2, occurs in association with GACI (Levy-Litan et al., 2010; Lorenz-Depiereux et al., 2010; Rutsch et al., 2003). ARHR2 and GACI are attributed to homozygous loss-of-function mutations of ENPP1. Loss of function of ENPP1 results in the inability to generate the mineralization inhibitor PPi, thereby disrupting the restriction of heterotopic (e.g., vascular) mineralization. GACI is often fatal, but hypophosphatemia, identified in the setting of elevated FGF23 levels in an adult with a homozygous ENPP1 mutation, raised this consideration of rickets in survivors of GACI (Lorenz-Depiereux et al., 2010). Moreover the patient's son was affected with both GACI and hypophosphatemia. The mechanism by which this enzyme influences renal tubular phosphate wasting is not evident, and further study is necessary to understand this intriguing problem. One speculated mechanism may reflect a bone cell response to a relatively hypermineralized (or high phosphate/low PPi) milieu, which results in a compensatory, prolonged secretion of FGF23. Such a mechanism may effectively signal the kidney to reduce the body's mineral load, but apparently cannot be downregulated to protect against excessive phosphate losses. Although there has been concern that the treatment of rickets in patients affected with GACI may promote worsening of vascular calcification (Rutsch et al., 2008), no evidence to sustain this concern has emerged, and one long-term observational report suggests that treatment does not worsen this finding (Ferreira et al., 2016).

Tumor-induced osteomalacia

TIO is a rare paraneoplastic syndrome caused by difficult-to-localize, usually benign tumors that secrete a circulating factor causing renal phosphate wasting. Many of the biochemical features of TIO mimic those seen in XLH, as the secreted factor in most tumors is FGF23, as occurs with XLH. TIO has been reviewed in detail and the reader is referred to that summary for additional details (Chong et al., 2011).

The differential diagnosis of TIO includes other FGF-dependent phosphate wasting disorders, which are often genetic in origin, such that a family history and documentation of time of onset of disease are useful in differentiating these conditions from TIO. Genetic testing may be useful. Among the most important clinical findings in TIO is profound weakness. Despite equivalent degrees of fasting hypophosphatemia, and often a similar extent of histologic osteomalacia, patients with lifelong hypophosphatemia, as in XLH, seem to function much better than patients with TIO, who usually acquire hypophosphatemia later in life. Patients with TIO are not infrequently crippled by their disease, with severe skeletal pain and obvious muscle weakness, manifested by difficulty with ambulation or rising from the floor or chair. Bone pain in TIO is due to the osteomalacia and resulting pseudofractures, which result from chronic phosphate wasting. Fractures occur most often in ribs and in weight-bearing bones such the spine and pelvis, femurs (included the femoral neck), and feet. The muscle weakness is presumably due to a hypophosphatemic myopathy. Correction of the hypophosphatemia often results in improvement in muscle function considerably before healing of the osteomalacia. Since the correct diagnosis is often missed for years, patients with TIO frequently suffer with poor quality of life for extended periods of time (Sanders, 2018).

TIO-causing tumors, categorized as phosphaturic mesenchymal tumors—mixed connective tissue variant (PMTs—MCT) (Folpe et al., 2004), have a varied histologic appearance, frequently with bland, spindle-shaped cells in a myxoid matrix and occasionally osteoclast-like giant cells. A microvascular component is often present. PMTs—MCT can occur in any tissue but frequently arise in soft tissues such as muscle or fat or in skeletal tissues, including bone, cartilage, and tendons. Rarely, tumors other than PMTs—MCT can cause TIO, including myeloma and prostate cancer (Narvaez et al., 2005; Nakahama et al., 1995).

As with XLH, hypophosphatemia occurs with renal phosphate wasting, circulating $1{,}25(OH)_2D$ levels are low, and 25-OHD levels are normal or low normal, excluding vitamin D deficiency as the cause of the patient's symptoms. PTH levels are usually normal, as is serum calcium. If the clinical index of suspicion is high and biochemical features are consistent, then measuring circulating FGF23 is an appropriate confirmatory study. Rarely, other circulating factors have been associated with TIO, including FGF7, FRP4, and MEPE (Carpenter et al., 2005; Chong et al., 2011; Jan de Beur et al., 2002). Interestingly, nearly 40% of TIO tumors have a chromosome rearrangement resulting in expression of a fusion protein in which a fibronectin molecule is fused with the FGFR1 (Lee et al., 2015). Although no direct experimental evidence has yet been reported, it has been speculated that TIO results, in part, from a feed-forward circuit in which tumor-secreted FGF23 activates this hybrid molecule and induces further FGF23 production.

Localizing the offending tumor can be challenging. A careful physical examination, focused on the head, neck, and oral pharynx, with palpation all soft tissues, can sometimes identify a PMT—MCT. Newer imaging technologies have greatly improved tumor localization. Among these is indium-111-labeled octreotide scintigraphy combined with single-photon-emission computed tomography (SPECT/CT) scanning, which allows for 3D visualization as well as taking advantage of the fact that TIO tumors often express somatostatin receptor subtype 2. Fluorodeoxyglucose positron emission tomography (FDG—PET)/CT is increasingly used to localized tumors in patients with TIO, particularly those that are not octreotide avid or are very small. The extreme sensitivity of FDG–PET/CT, however, can lead to false positive results. 1,4,7,10-Tetraazacyclododecane-1,4,7,10-tetraacetic acid (DOTA) (and its derivatives DOTANOC and DOTATATE) PET/CT addresses the "nonspecificity" of conventional FDG–PET/CT by employing a bifunctional chelator that binds gallium and the type 2 somatostatin receptor with high affinity (DOTA NaI^3-octreotide) (Wild et al., 2005; Hesse et al., 2007a; El-Maouche et al., 2016). Reports indicate that this approach may be more sensitive than conventional octreotide SPECT/CT (von Falck et al., 2008). Since the cranial/facial area, as well as distal extremities, can harbor TIO-inducing tumors, it is important to include the entire body in these scan images. In the past, venous sampling has been used in an effort to identify tumors, but the aforementioned imaging studies have largely supplanted the need for that more invasive diagnostic approach.

Effective treatment for TIO is complete surgical resection of the tumor when possible. Generally, these tumors are benign and slow growing with a low mitotic index. Occasionally TIO tumors can arise in a surgically inaccessible site or can be locally invasive, making complete resection impossible. We encountered such a patient, with a cervical vertebral tumor and extensive involvement of paraspinal soft tissues. Rarely TIO tumors can be frankly malignant and metastasize. In patients in whom the tumor cannot be completely resected or identified, medical management can provide significant symptomatic improvement. At this writing, treatment with calcitriol and supplemental phosphorus represents the best medical therapy for TIO. Conventional doses of calcitriol and phosphorus, in the range of 1–3 μg/day calcitriol and 1–3 g/day phosphorus in divided doses, are usually effective. Unlike in XLH, secondary hyperparathyroidism is less often encountered in TIO when using calcitriol/phosphorus, but hypercalciuria and even hypercalcemia can occur, so blood and urine calcium, in addition to serum phosphorus, needs to be monitored during treatment to avoid overreplacement. Unlike in XLH, where FGF23 levels are usually modestly elevated, in TIO FGF23 levels are also often very elevated, despite which conventional therapy seems to provide symptomatic improvement.

The use of a blocking antibody to FGF23 (burosumab) has shown promising results in the treatment of TIO. In an open-label trial, patients with TIO had significant improvement in their hypophosphatemia and marked symptomatic improvement with monthly administration of the drug at doses up to 1 mg/kg (Jan de Beur et al., 2018). A selective FGFR inhibitor (NVP-BGJ398) has, in preliminary studies, been reported to cause dramatic symptomatic improvement in a patient with metastatic TIO. In this patient the drug also caused tumor regression (Collins et al., 2015). Finally, radiofrequency ablation has been used to treat TIO tumors where a non-invasive approach was preferred (Hesse et al., 2007b).

Other FGF23-mediated hypophosphatemic syndromes

In widespread fibrous dysplasia of bone (due to mosaic-activating mutations in GNAS, as part of the McCune–Albright syndrome) (OMIM: 174800), hypophosphatemic osteomalacia/rickets can be a result of elevated circulating FGF23 levels. Indeed, variable degrees of decreased renal tubular phosphate reabsorption, as assessed by tubular maximum reabsorption of phosphate/glomerular filtration rate (TmP/GFR), occur in patients with fibrous dysplasia of bone. Of note, given that activation of GNAS in the renal proximal tubule would be expected to result in phosphaturia, it had long been considered that the mechanism of reduced renal TRP was due to a surmised renal distribution of the mosaic-activating GNAS mutation. The studies of Rimunucci et al. identified elevated circulating FGF23 levels in this condition and that the greater burden of skeletal disease is associated with more impressive hypophosphatemia. Whether the bone affected with fibrous dysplasia or normal bone under other systemic influences accounts for the increased FGF23 secretion has not been clearly settled.

Somatic mutations in HRAS and NRAS appear to generate a mechanism by which mutated skin or bone can increase secretion of FGF23, a condition known as epidermal nevus syndrome (OMIM: 162900), or more recently cutaneous skeletal hypophosphatemia syndrome (Lim et al., 2014; Ovejero et al., 2016).

Other primary skeletal disorders in which elevated FGF23 levels have been reported include osteoglophonic dysplasia (due to mutations in FGFR1) (OMIM: 166250) (White et al., 2005), Jansen metaphyseal chondrodysplasia (due to activating mutations of the PTH1 receptor) (OMIM: 156400) (Brown et al., 2009), neurofibromatosis (OMIM: 162200), and FAM20C mutations (e.g., Raine syndrome, OMIM: 259775) (Rafaelsen et al., 2013). This condition has also been referred to as autosomal recessive hypophosphatemic rickets type 3 (ARHR3). The mechanism(s) by which elevations in FGF23 occur in these settings is not certain at this time. There is evidence to support the finding that FAM20C alters phosphorylation of FGF23, thereby affecting O-glycosylation and allowing for diminished degradation (Tagliabracci et al., 2014). Alternatively, data have shown direct increases in FGF23 secretion with loss of function of FAM20C (Liu et al., 2018). Likewise, loss of function of Nuclear Factor 1 may play a role in the skeletal disease seen in neurofibromatosis by increasing FGF23 production (Kamiya et al., 2017).

Fibroblast growth factor 23–independent hypophosphatemic disorders

Hereditary hypophosphatemic rickets with hypercalciuria (OMIM: 241530)

Hereditary hypophosphatemic rickets with hypercalciuria (HHRH) is a rare autosomal recessive disorder first described in 1985 in a large consanguineous Bedouin kindred (Tieder et al., 1985). HHRH is distinct from the more common XLH (HYP-Consortium, 1995; Holm et al., 1997) with respect to mode of inheritance, and distinct biochemical features associated with increased synthesis and serum levels of 1,25$(OH)_2$D, particularly hypercalciuria (Tieder et al., 1985, 1987; Gazit et al., 1991).

Epidemiology

HHRH has been described in all races, although most reports describe cases of Caucasian and Middle Eastern origin. More than 40 different mutations in *SLC34A3* affecting over 35 kindreds have been reported (Rafaelsen et al., 2016; Daga et al., 2018; Pronicka et al., 2017; Schlingmann et al., 2016; Dhir et al., 2017; Dasgupta et al., 2014; Chi et al., 2014; Ichikawa et al., 2014). Based on the allele frequency of 0.002 for proven pathogenic *SLC34A3* mutations, HHRH has a predicted prevalence of 1:250,000 (Wagner et al., 2017), although the true prevalence is not known. Therefore, HHRH is approximately 10-fold less frequent than the most common inherited disorder of phosphate homeostasis, XLH (1:20,000; Burnett et al., 1964). The predicted prevalence of idiopathic hypercalciuria associated with heterozygous mutation of *SLC34A* is 1:500.

Cloning and identification of human mutations in NPT2c

The genetic defect in HHRH was identified in a combined genome-wide search for linkage and homozygosity mapping using genomic DNA from the original Bedouin kindred (Tieder et al., 1985, 1987). A homozygous deletion SLC34A3.c.228delC was found in all affected individuals, which is predicted to result in complete loss of function of NPT2c (Bergwitz et al., 2006). Different genetic mutations in the same gene were subsequently reported (Rafaelsen et al., 2016; Daga et al., 2018; Pronicka et al., 2017; Schlingmann et al., 2016; Dhir et al., 2017; Dasgupta et al., 2014; Ichikawa et al., 2014; Bhoj et al., 2015) and include missense mutations, mutations affecting potential splice sites, and smaller and larger deletions causing premature stop codons with expression of a truncated transporter protein (Fig. 20.4). Some of these mutations have been functionally characterized in vitro using opossum kidney cells and *Xenopus* oocyte expression systems (Haito-Sugino et al., 2012; Shiozaki et al., 2015; Jaureguiberry et al., 2008).

Pathophysiology

The consequence for human carriers of *SLC34A3*/NPT2c loss-of-function mutations on one or two alleles is isolated proximal tubular phosphate wasting. Renal Fanconi syndrome has not been described for carriers of *SLC34A3*/NPT2c mutations, as occurs with *SLC34A1*/NPT2a mutations. To date no genotype–phenotype correlation has been described, and there is no evidence for dominant negative effects of mutant transporters. Of interest would also be to determine whether loss of NPT2c has consequences for other proximal tubular functions such as CYP24A1 activity, sodium excretion, urine anion gap, and osteopontin and PPi levels (Tenenhouse et al., 2001; Caballero et al., 2017a,b; Li et al., 2017), all of which, together with the characteristic hypercalciuria, may contribute to formation of renal mineral deposits in HHRH and idiopathic hypercalciuria (IH). Haploinsufficiency of NPT2c can cause mild hypophosphatemia, reduced TmP/GFR, and elevations in 1,25$(OH)_2$D and/or urinary calcium excretion (Tieder et al., 1985, 1987; Bergwitz et al., 2006), but, with the exception of one report (Yamamoto et al., 2007), does not generally cause bone disease.

Clinical presentation and diagnostic evaluation

Laboratory findings and genetic testing The clinical assessment of phosphate homeostasis can be challenging: serum phosphorus concentrations are influenced by the time of day, meals, and age. Clinical methods for determination of tubular reabsorption are imprecise. To determine the cause of abnormal serum phosphorus levels in a patient with normal parathyroid and renal function, we first assess TRP (Fig. 20.5) based on the measurement of phosphorus and creatinine in a 2- to 3-h urine collection along with the corresponding serum parameters. TRP, TmP/GFR (Walton and Bijvoet, 1975), or TP/GFR, which provides a more accurate assessment of renal phosphate handling in children (Alon and Hellerstein, 1994), is then calculated. Inappropriately low %TRP in the setting of hypophosphatemia is suggestive of a proximal renal tubular defect. With respect to HHRH, further confirmation can be aided by measurement of circulating 1,25$(OH)_2$D levels. The combined reduction in both TRP and 1,25$(OH)_2$D levels suggests excess FGF23 action, whereas elevated 1,25$(OH)_2$D

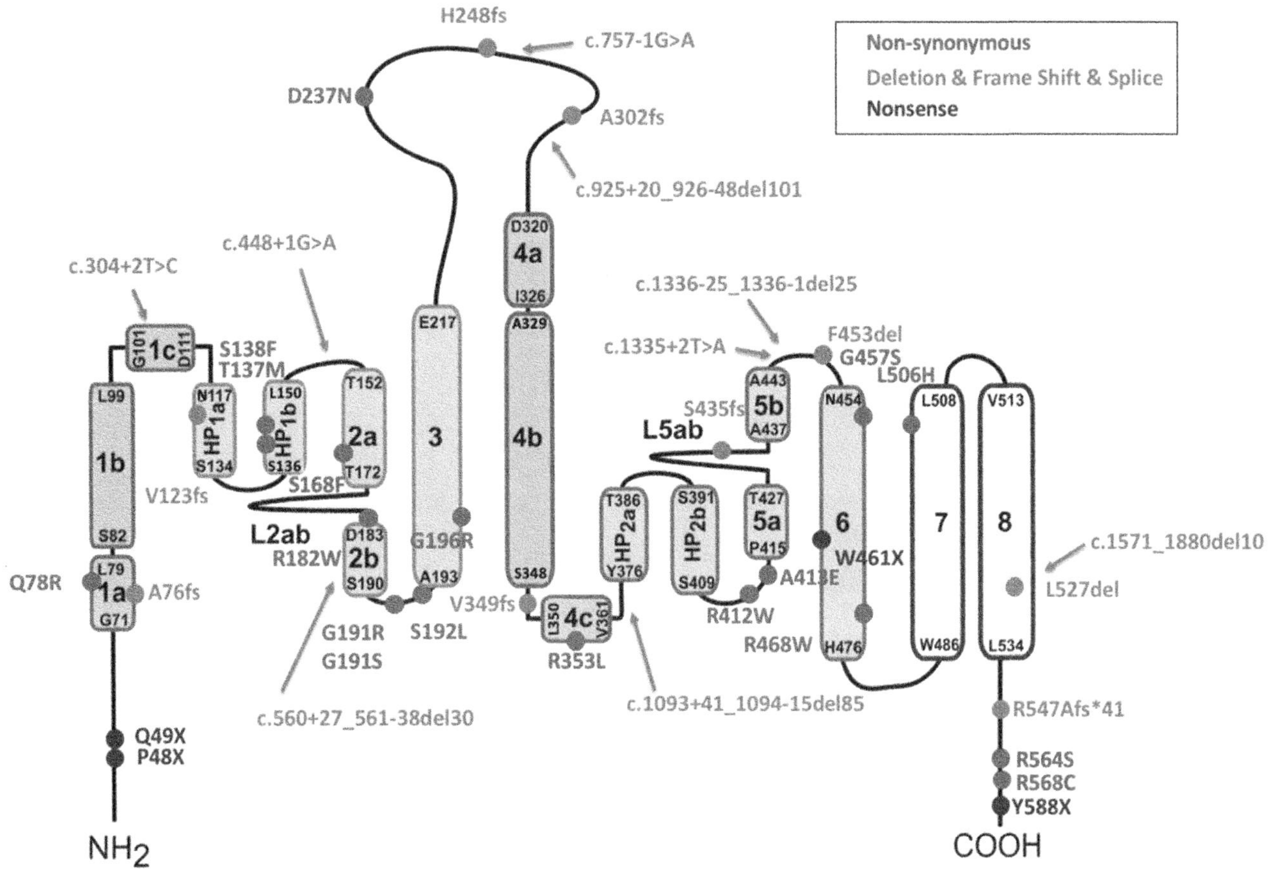

FIGURE 20.4 Reported human mutations in NPT2c. Predicted models of *SLC34A3*/NPT2c (Wagner et al., 2017) and localization of mutations identified in patients. (*Red*, missense mutations; *green*, frameshift, deletion, and splice-site mutations; *purple*, nonsense mutations). The scaffold domain is represented in lavender and the transport domain in orange as in. (Forster and Wagner, 2018; Bergwitz and Miyamoto, 2019). *For details on mutations and references see Rafaelsen, S., Johansson, S., Raeder, H., Bjerknes, R., 2016. Hereditary hypophosphatemia in Norway: a retrospective population-based study of genotypes, phenotypes, and treatment complications. Eur. J. Endocrinol. 174, 125–136 DOI 10.1530/EJE-15-0515; Daga, A., Majmundar, A.J., Braun, D.A., Gee, H.Y., Lawson, J.A., Shril, S., Jobst-Schwan, T., Vivante, A., Schapiro, D., Tan, W., Warejko, J.K., Widmeier, E., Nelson, C.P., Fathy, H.M., Gucev, Z., Soliman, N.A., Hashmi, S., Halbritter, J., Halty, M., Kari, J.A., El-Desoky, S., Ferguson, M.A., Somers, M.J.G., Traum, A.Z., Stein, D.R., Daouk, G.H., Rodig, N.M., Katz, A., Hanna, C., Schwaderer, A.L., Sayer, J.A., Wassner, A.J., Mane, S., Lifton, R.P., Milosevic, D., Tasic, V., Baum, M.A., Hildebrandt, F., 2018. Whole exome sequencing frequently detects a monogenic cause in early onset nephrolithiasis and nephrocalcinosis. Kidney Int. 93, 204–213. DOI S0085-2538(17)30494-5 [pii] 10.1016/j.kint.2017.06.025; Pronicka, E., Ciara, E., Halat, P., Janiec, A., Wojcik, M., Rowinska, E., Rokicki, D., Pludowski, P., Wojciechowska, E., Wierzbicka, A., Ksiazyk, J.B., Jacoszek, A., Konrad, M., Schlingmann, K.P., Litwin, M. 2017. Biallelic mutations in CYP24A1 or SLC34A1 as a cause of infantile idiopathic hypercalcemia (IIH) with vitamin D hypersensitivity: molecular study of 11 historical IIH cases. J. Appl. Genet. 58, 349–353. DOI 10.1007/s13353-017-0397-2 10.1007/s13353-017-0397-2; Schlingmann, K.P., Ruminska, J., Kaufmann, M., Dursun, I., Patti, M., Kranz, B., Pronicka, E., Ciara, E., Akcay, T., Bulus, D., Cornelissen, EAM., Gawlik, A., Sikora, P., Patzer, L., Galiano, M., Boyadzhiev, V., Dumic, M., Vivante, A., Kleta, R., Dekel, B., Levtchenko, E., Bindels, R.J., Rust, S., Forster, I.C., Hernando, N., Jones, G., Wagner, C.A., Konrad, M. 2016. Autosomal-recessive mutations in SLC34A1 encoding sodium-phosphate cotransporter 2A cause idiopathic infantile hypercalcemia. J. Am. Soc. Nephrol. 27, 604–614. DOI 10.1681/asn.2014101025; Dhir, G., Li, D., Hakonarson, H., Levine, M.A. 2017. Late-onset hereditary hypophosphatemic rickets with hypercalciuria (HHRH) due to mutation of SLC34A3/NPT2c. Bone 97, 15–19 DOI S8756-3282(16)30363-5 [pii] 10.1016/j.bone.2016.12.001); Dasgupta, D., Wee, M.J., Reyes, M., Li, Y., Simm, P.J., Sharma, A., Schlingmann, K.P., Janner, M., Biggin, A., Lazier, J., Gessner, M., Chrysis, D., Tuchman, S., Baluarte, H.J., Levine, M.A., Tiosano, D., Insogna, K., Hanley, D.A., Carpenter, T.O., Ichikawa, S., Hoppe, B., Konrad, M., Savendahl, L., Munns, C.F., Lee, H., Juppner, H., Bergwitz, C. 2014. Mutations in SLC34A3/NPT2c are associated with kidney stones and nephrocalcinosis. J. Am. Soc. Nephrol. 25, 2366–2375. DOI 10.1681/ASN.2013101085 ASN, 2013101085 ASN.2013101085; Chi, Y., Zhao, Z., He, X., Sun, Y., Jiang, Y., Li, M., Wang, O., Xing, X., Sun, A.Y., Zhou, X., Meng, X., Xia, W., 2014. A compound heterozygous mutation in SLC34A3 causes hereditary hypophosphatemic rickets with hypercalciuria in a Chinese patient. Bone 59, 114–121. DOI 10.1016/j.bone.2013.11.008 S8756-3282(13)00444-4; Ichikawa, S., Tuchman, S., Padgett, L.R., Gray, A.K., Baluarte, H.J., Econs, M.J., 2014. Intronic deletions in the SLC34A3 gene: a cautionary tale for mutation analysis of hereditary hypophosphatemic rickets with hypercalciuria. Bone 59, 53–56 DOI 10.1016/j.bone.2013.10.018 S8756-3282(13)00428-6; Acar, S., BinEssa, H.A., Demir, K., Al-Rijjal, R.A., Zou, M., Catli, G., Anik, A., Al-Enezi, A.F., Ozisik, S., Al-Faham, M.S.A., Abaci, A., Dundar, B., Kattan, W.E., Alsagob, M., Kavukcu, S., Tamimi, H.E., Meyer, B.F., Bober, E., Shi, Y. 2018. Clinical and genetic characteristics of 15 families with hereditary hypophosphatemia: novel Mutations in PHEX and SLC34A3. PLoS One 13, e0193388. DOI 10.1371/journal.pone.0193388 PONE-D-17-39524.*

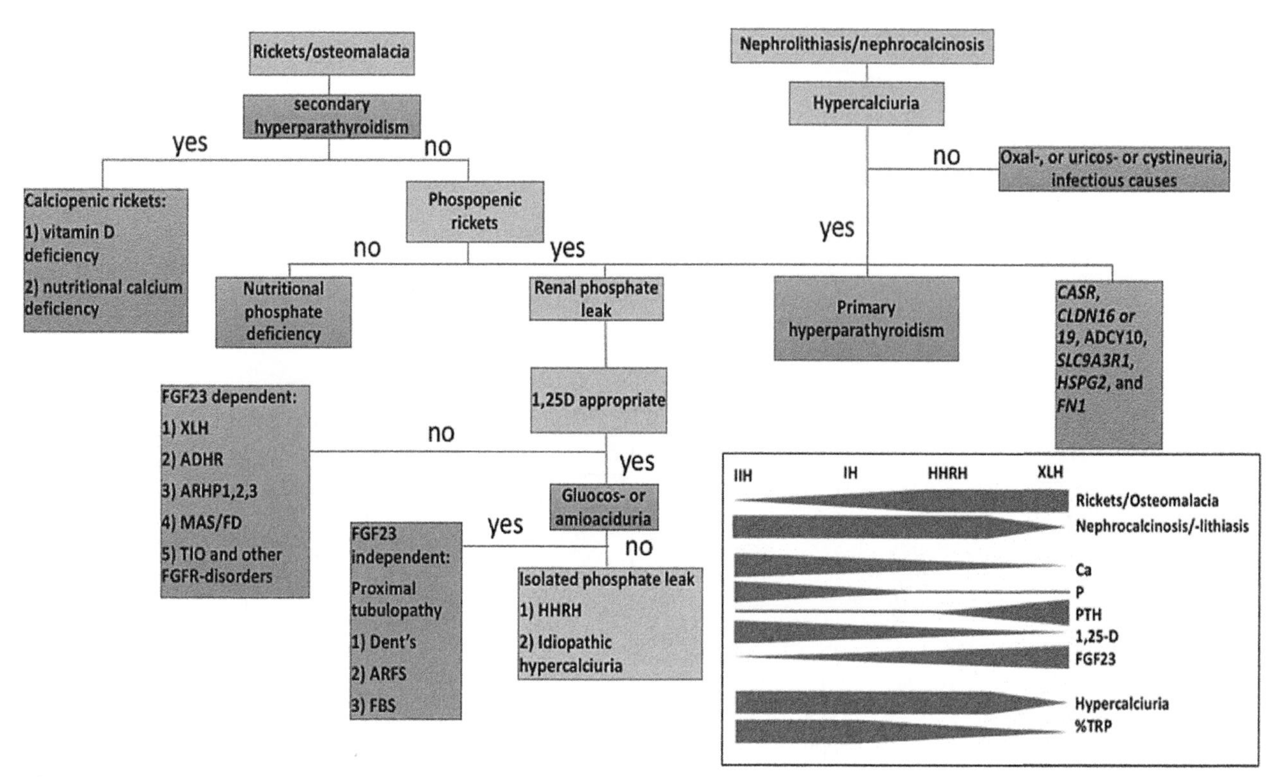

FIGURE 20.5 Differential diagnosis of hypophosphatemic disorders with and without nephrocalcinosis/nephrolithiasis. HHRH may present initially with bone disease or renal calcifications and bone disease is generally missing in heterozygous carriers of *SLC34A3*/NPT2c mutations. Presence of additional symptoms in *orange* further supports the diagnosis, while symptoms in *blue* argue against it. *ADHR*, autosomal dominant hypophosphatemic rickets; *ARFS*, autosomal recessive Fanconi syndrome; *ARHP1,2,3*, autosomal recessive hypophosphatemia type 1, 2, or 3; *FBS*, Fanconi–Bickel syndrome; *FGF23*, fibroblast growth factor 23; *FGFR*, FGF receptor; *HHRH*, hereditary hypophosphatemic rickets with hypercalciuria; *MAS/FD*, McCune–Albright syndrome/fibrous dysplasia; *PTH*, parathyroid hormone; *TIO*, tumor-induced osteomalacia; *%TRP*, percentage tubular reabsorption of phosphate; *XLH*, X-linked dominant hypophosphatemia. For gene names see text. *Originally published in Bergwitz, C., Miyamoto, K.I., 2019. Hereditary hypophosphatemic rickets with hypercalciuria: pathophysiology, clinical presentation, diagnosis and therapy. Pflügers Archiv 471, 149–163. https://doi.org/10.1007/s00424-018-2184-2.*

levels in the context of a low TRP suggest an FGF23-independent process, as would be seen in HHRH. Excess production of $1,25(OH)_2D$ may lead to increased absorption of calcium in the gut, resulting in hypercalciuria and some degree of suppression of PTH production (Tieder et al., 1985). Vitamin D deficiency may mask these findings and needs to be corrected before the aforementioned testing (Kremke et al., 2009). Circulating FGF23 levels can be determined using several commercially available enzyme-linked immunometric assays (Yamazaki et al., 2002). The C-terminal FGF23 assay (Immutopics, Inc., San Clemente, CA, USA) uses antibodies directed against two distinct epitopes within the C-terminal region of FGF23 (Larsson et al., 2005) and, as of this writing, is the only CLIA-certified assay in the United States. This assay in our experience returns FGF23 levels in the low normal range in FGF23-independent hypophosphatemic disorders such as HHRH or renal Fanconi syndrome.

Genetic testing employing whole-exome sequence analysis of leukocyte DNA is increasingly performed, allowing for screening of a panel of genes. Some centers (including the Yale Clinical Genome Research Center) employ techniques with sufficient coverage of intronic sequence to permit detection of the known intronic deletions (Daga et al., 2018; Ma et al., 2015). Single-gene sequencing may be required to detect deletions in 5′ and 3′ UTRs or in the larger intron 12, which, at this writing, is offered only by research laboratories. Fig. 20.4 indicates the locations of known mutations. More detailed information on the type and location of specific mutations is provided at several freely accessible websites: Online Mendelian Inheritance in Man (OMIM) (http://www.ncbi.nlm.nih.gov/omim/), the Exome Variant Server (http://evs.gs.washington.edu/EVS/), and the National Center for Biotechnology Information (NCBI) SNP (http://www.ncbi.nlm.nih.gov/projects/SNP/). Since compound heterozygous loss-of-function mutations in *SLC34A3*/NPT2c have been identified in HHRH (Bergwitz et al., 2006; Lorenz-Depiereux et al., 2006b), genetic evaluation of one parent may be useful to permit allele assignment when heterozygous mutations are discovered.

Musculoskeletal findings Rickets, bowing, and short stature in children and osteopenia/osteoporosis in adults are the most commonly observed features in individuals with classic HHRH. Bone pain is often present (Narchi et al., 2001; Francis and Selby, 1997). The radiological findings of rickets are evident at growth plates, and osteomalacia (under-mineralization of the osteoid) may blur the trabecular architecture. Clinical consequences of these lesions include bone pain and impaired biomechanical properties, which can result in bowing, and insufficiency fractures. Swollen wrists and knees may occur due to growth plate abnormality (Donohue & Demay, 2002). The primary biochemical marker for rickets and osteomalacia is elevated total or bone-specific alkaline phosphatase (Narchi et al., 2001; Francis and Selby, 1997). Bone biopsy has shown increased accumulation of osteoid, compatible with osteomalacia, responsive to oral phosphorus supplementation (Yamamoto et al., 2007; Chen et al., 1989). Interestingly, hypomineralized periosteocytic lesions typical of XLH were absent (Yamamoto et al., 2007). Furthermore, osteoblast surface was increased, while osteoclast number was decreased (Chen et al., 1989). Rickets or osteomalacia occur in many hypophosphatemic disorders, but subtle differences can guide diagnostic and therapeutic decisions. Enthesopathy (painful or indolent mineral deposition of tendon and ligament insertion sites) occurs with XLH, ADHR, and ARHR1, but has not been reported in HHRH (Econs et al., 1994). Likewise, dental cysts, craniosynostosis, midfacial hypoplasia, frontal bossing, scaphocephaly, and Chiari I malformation, all reported in XLH (Jones et al., 2001; Tenenhouse and Econs, 2001), are not present in HHRH, possibly because of the absence of elevated FGF23 in HHRH (Econs et al., 1994; DiMeglio and Econs, 2001). Whether HHRH predisposes to accelerated bone loss in adulthood as may occur with NPT2a or NHERF1 mutations is not known (Prie et al., 2002; Karim et al., 2008).

The mechanism of muscle weakness caused by hypophosphatemia is not well understood and may be related to the role of phosphorus in intracellular signal transduction, phosphocreatine recovery (Clarke et al., 1990; Sinha et al., 2013), and/or ATP synthesis (Pesta et al., 2016; Sinha et al., 2013; Smith et al., 1984; Aono et al., 2011). Sarcopenia may develop in HHRH due to chronic hypophosphatemia. In contrast to the hypophosphatemic myopathy seen in XLH or vitamin D deficiency, HHRH presents with increased calcium, increased $1,25(OH)_2D$, and low FGF23, making this disorder uniquely suited to study of the effects of hypophosphatemia per se on muscle metabolism.

Renal findings None of the originally described HHRH patients were reported to have renal calcifications and kidney stones (Tieder et al., 1985, 1987), but subsequent investigation has revealed that renal complications are common with compound heterozygous or homozygous NPT2c mutations (Kremke et al., 2009; Ichikawa et al., 2006; Page et al., 2008; Phulwani et al., 2011; Mejia-Gaviria et al., 2010; Tencza et al., 2009). Renal calcifications are the only presenting sign in 16% of individuals with HHRH, where concomitant bone disease is not described (Romero et al., 2010; Schissel and Johnson, 2011). Even heterozygous carriers of *SLC34A3*/NPT2c mutations show an approximately threefold higher incidence of renal calcifications, probably related to their intermediate biochemical profile. Therefore, all affected individuals and their first-degree relatives should be examined for renal calcifications. It is unknown whether loss of NPT2c can cause additional proximal tubular phenotypes, such as Fanconi syndrome, as described in two patients with homozygous NPT2a mutations (Magen et al., 2004) who developed CKD later in life. Genome-wide association studies for uric acid nephrolithiasis (Tore et al., 2011), serum phosphorus levels (Kestenbaum et al., 2010), and CKD (Gudbjartsson et al., 2010) have not supported an association of hypercalciuric stone disease with the *SLC34A3*/NPT2c locus, whereas renal calcifications and impaired renal function are so associated (Arcidiacono et al., 2014; Taguchi et al., 2017; Boger et al., 2011; Pattaro et al., 2016). Our meta-analysis clearly suggests that *SLC34A3* should be added to the list of hypercalciuric stone disease genes (Dasgupta et al., 2014).

Therapy and resources

Standard therapy It is important to establish the correct diagnosis of HHRH, as patients require long-term medical therapy with phosphorus supplements (Tieder et al., 1985, 1987), and *not* supplementation with vitamin D, as in nutritional rickets (Reginato and Coquia, 2003). When HHRH is mistaken for XLH, active vitamin D analogs (i.e., calcitriol) are used with oral phosphorus (White et al., 2011), often leading to hypercalcemia, hypercalciuria, nephrocalcinosis, and potentially renal insufficiency (Jaureguiberry et al., 2008; Kremke et al., 2009). When correctly treated with oral phosphorus supplementation only, the rachitic bone disease in HHRH improves quickly. However, the long-term safety of this therapy is unknown with respect to renal calcifications, or the development of hyperparathyroidism or enthesopathy, as occurs in XLH (HYP-Consortium, 1995). Finally, genetic and biochemical data may predict risk for renal calcifications and inform management of oral phosphorus therapy. Serum phosphorus levels, urinary phosphate excretion, and serum $1,25(OH)_2D$ merit further evaluation as predictors of renal calcifications. Our studies in NPT2a-KO mice (Li et al., 2017) suggest that phosphorus can be harmful under certain conditions, suggesting that patients may need to be carefully monitored to avoid renal calcifications despite resolution of hypercalciuria while receiving supplemental phosphorus therapy. Serum $1{,}25(OH)_2D$ may remain elevated long after initiation of phosphorus supplementation and thus it is not clear how suitable this measure is for assessing compliance with oral phosphorus therapy in the short term... (Yu et al., 2012).

Dent's disease (X-linked recessive hypophosphatemic rickets) (OMIM: 300009)

The initial description of X-linked recessive hypophosphatemic rickets involved a family in which males presented with rickets or osteomalacia, hypophosphatemia, and a reduced threshold for renal phosphate reabsorption (Wrong et al., 1994). In contrast to XLH, affected subjects usually exhibit hypercalciuria, elevated serum $1{,}25(OH)_2D$ levels, and proteinuria of up to 3 g/day. Nephrolithiasis and nephrocalcinosis with progressive renal failure occur in early adulthood. Female (heterozygous) carriers are normophosphatemic and lack biochemical abnormalities other than hypercalciuria. Three other independently reported syndromes, X-linked recessive nephrolithiasis with renal failure, Dent's disease, and low-molecular-weight proteinuria with hypercalciuria and nephrocalcinosis, have an overall similar phenotype, but manifest differences in degree of proximal tubular reabsorptive defects, nephrolithiasis, nephrocalcinosis, progressive renal insufficiency, and, in some cases, rickets or osteomalacia. Identification of mutations in the voltage-gated chloride channel gene *CLCN5*, in all four syndromes, has established that these disorders are phenotypic variants of a single disease (Lloyd et al., 1996; Thakker, 2000). The varied manifestations associated with mutations in *CLCN5*, particularly hypophosphatemia and rickets/osteomalacia, emphasize the role of environmental differences, diet, and/or modifying genetic backgrounds on phenotypic expression.

Hypophosphatemia with osteoporosis and nephrolithiasis due to SLC34A1 (OMIM: 612286) and NHERF1 mutations (OMIM: 604990)

Prie et al. investigated a heterogeneous group of patients with idiopathic hypercalciuria, osteoporosis, and renal stones. Using a candidate gene approach they found 2/20 individuals heterozygous for nonsynonymous single-nucleotide polymorphisms (SNPs) in *SLC34A1*/NPT2a (Prie et al., 2002) and 7/94 individuals heterozygous for nonsynonymous SNPs in *SLC9A3R1*/NHERF1 (Karim et al., 2008). Potential dominant negative effects of the *SLC34A1*/NPT2a mutations on proximal renal tubular phosphate reabsorption remain controversial (Virkki et al., 2003), and some of the identified NHERF1 alterations are listed in the NCBI dbSNP database as low-frequency polymorphisms (Bergwitz and Bastepe, 2008). Further study is thus required to prove that these NPT2a or NHERF1 mutations are disease-causing ones. In contrast to HHRH there is generally no history of childhood rickets in individuals with NPT2a mutations.

Autosomal recessive Fanconi syndrome (OMIM: 613388)

In 1988, Tieder et al. described a consanguineous Arab kindred with childhood rickets and defective proximal tubular handing of phosphate, amino acids, and glucose consistent with renal Fanconi syndrome (Tieder et al., 1988). Distinct from other forms of Fanconi, their patients also had elevated $1{,}25(OH)_2D$ levels and absorptive hypercalciuria. In 2010, homozygosity mapping of this kindred showed linkage of the disease to chromosome 5q35.1−q35.3, and subsequent sequence analysis of the *SLC34A1*/NPT2a gene in the linked interval revealed a homozygous duplication, g.2061_2081dup (p.I154V160dup) (Magen et al., 2010). Expression of the mutant NPT2a protein in *Xenopus* oocytes and opossum kidney cells showed complete loss of function and lack of membrane insertion, respectively. The two patients described in the 1988 report, who at this writing are 39 and 43 years of age, continue to have low TmP/GFR, and their FGF23 and PTH levels were recently shown to be low normal (despite impaired renal function), suggesting that their hypophosphatemia is FGF23 and PTH independent. However, their previously documented absorptive hypercalciuria due to increased $1{,}25(OH)_2D$ levels had normalized in the setting of vitamin D deficiency. Although symptoms of rickets were present in childhood, both patients have been relatively asymptomatic during adulthood and discontinued phosphorus supplementation. Both developed CKD stage 2−3 renal failure in their 30s, which is in contrast to the other *SLC34A1*- and *SLC34A3*-related disorders described earlier. Heterozygous carriers had normal renal function and no evidence of proximal tubulopathy, arguing against dominant negative effects of the mutant NPT2a.

Fanconi−Bickel syndrome (OMIM: 227810)

Fanconi−Bickel syndrome is a rare autosomal recessive disorder of hepatorenal glycogen accumulation, proximal renal tubular dysfunction, and impaired utilization of glucose and galactose (Santer et al., 1997, 2002), first described in 1949 and caused by defects in the facilitative glucose transporter 2 (*SLC2A2*/GLUT2) (Santer et al., 1997). Patients present early in life with rickets and hepatomegaly. By 2 years of age enlarged kidneys are evident. Fasting hypoglycemia and postprandial hyperglycemia and hypergalactosemia, as well as hyperlipidemia, may be present. Some cases have presented with isolated but variable increases in urinary phosphate excretion (Mannstadt et al., 2012).

Intestinal malabsorption of phosphate

Primary disorders of intestinal phosphate absorption have not been reported to occur in widespread fashion. However, we have encountered a worrisome pattern of phosphate malabsorption in children with complex disorders associated with intestinal compromise. These children were fed amino acid–based elemental formulas, particularly certain Neocate products (Gonzalez Ballesteros et al., 2017). Associated tube-feeding and use of antacid medications appear to be risk factors for the development of this syndrome, and the phenomenon does not appear to occur when the formulas are used for the labeled indication of milk protein allergy in children who are otherwise healthy (Harvey et al., 2017). Without suspicion of this finding, routine monitoring of biochemical predictors of the condition are generally not monitored until overt skeletal consequences are evident, particularly rickets and fractures. Hypophosphatemia, low-to-undetectable urinary phosphate excretion, and elevated serum alkaline phosphatase levels represent the usually observed biochemical findings. We have recommended that serum phosphorus levels be monitored periodically with the use of such formulas. The disorder can be managed with either phosphate supplementation or transition to alternative formulas. We caution that the transition to an alternative formula should be done gradually, using slowly decreasing percentages of the offending formula (i.e., over 1–2 weeks), as the physiologic adaptation to phosphate deprivation will have probably resulted in marked upregulation of intestinal and renal transporters. Thus hyperabsorption of phosphate, as well as renal retention, is likely to occur initially as abundant phosphate is provided, resulting in hyperphosphatemia and a reciprocal acute hypocalcemia. The problem generally resolves in several days or a few weeks as the system is restored to a normal phosphate-handling equilibrium.

Hyperphosphatemic syndromes

Hyperphosphatemia may result from a variety of physiologic perturbations, such as phosphate loading, tumor lysis syndrome, and CKD, as discussed earlier. This section is focused on FGF23-dependent hyperphosphatemia, of interest as it provides a physiologic example of a symmetrical contrast to the hypophosphatemia seen with excess FGF23 activity.

Tumoral calcinosis

The converse pathophysiology of excess FGF23 can arise with the disorder familial hyperphosphatemic tumoral calcinosis (FHTC) (Folsom and Imel, 2015). This rare disorder is most frequently inherited in an autosomal recessive manner, presenting with findings of ectopic calcifications, often at extensor surfaces and in the pelvis, and in other locations, which appear to be amorphous calcium phosphate precipitates. The precipitates may generate the findings of chronic inflammation at their anatomic location as well (Ramnitz et al., 2016). The biochemical findings in these syndromes are the converse of those seen in XLH, with elevated TmP/GFR, consequent hyperphosphatemia, and normal or elevated $1,25(OH)_2D$ levels.

The disorder is due to deficient FGF23 activity and inherited in an autosomal recessive manner. Most reported cases are due to loss-of-function mutations in *GALNT3* (encoding UDP-*N*-acetyl-α-D-galactosamine:polypeptide *N*-acetylgalactosaminyltransferase 3) (HTC1, OMIM: 211900), resulting in impaired O-glycosylation of FGF23. The consequences of the mutation are increased proteolysis, occurring in the Golgi apparatus, before secretion into the circulation (Topaz et al., 2004). Therefore fragments are the dominant species found in the circulation, such that FGF23 levels will vary dependent on the nature of the immunoassay employed: FGF23 measured using an intact FGF23 assay is usually low, yet high values are detected when employing assays that recognize only the C terminus of the protein. FGF23 mRNA is overexpressed, indicating an intact feedback loop, presumably as the osteocyte responds to the ambient hyperphosphatemia. The disorder has variable effects on the skeleton and has been also described as a hyperostosis/hyperphosphatemia syndrome (Ramnitz et al., 2016). Dental findings may include short, blunt roots and pulpal obliteration (Vieira et al., 2015). We have identified a patient with back pain and intervertebral disc calcifications who has the biochemical phenotype of FHTC, with biallelic *GALNT3* mutations (Lee et al., 2018).

Recessive loss-of-function mutations in FGF23 itself may cause FHTC (HTC2, OMIM: 617993), with mutations evident in conserved residues of the molecule. Presumably, proteolysis is also affected, resulting in the discordance in FGF23 levels depending on the use of intact or C-terminal assays. Finally, FHTC has been reported in one patient who has loss-of-function mutations in α-klotho, the coreceptor for FGF23 (HTC3, OMIM: 617994) (Ichikawa et al., 2007). This individual demonstrated increased levels of intact FGF23 indicating the expected resistance in the setting of the deficient coreceptor. One case of autoimmune HTC has been described, in which high titers of antibodies directed toward FGF23 resulted in resistance to FGF23 action and elevated levels of circulating FGF23 in both intact and C-terminal assays (Roberts et al., 2018). Finally, tumoral calcinosis in the absence of hyperphosphatemia has been described as secondary to CKD and in the setting of loss-of-function mutations in *SAMD9* (OMIM: 610455) (Topaz et al., 2006).

Treatment of the condition has been variably successful with low-phosphorus diet, sevelamer (or other phosphate binders), and acetazolamide. Combinations of these approaches are sometimes attempted. A 2016 report suggests the potential of topical sodium thiosulfate for this difficult to manage clinical situation (Jost et al., 2016).

Other non-FGF23-mediated hyperphosphatemic syndromes, related to hypoparathyroidism, and various forms of resistance to PTH are listed in Table 20.1. These disorders are discussed at length elsewhere (Weinstein et al., 1992; Juppner et al., 1998; Arnold et al., 1990; Pollak et al., 1994; Ding et al., 2001; Zhang et al., 1998; Karperien et al., 1999). Finally, the hyperphosphatemia is well known to increase through increasing stages of CKD, in association with reduction in serum $1,25(OH)_2D$ levels and markedly elevated circulating levels of FGF23.

Normophosphatemic disorders of cellular phosphorus metabolism

Given the widespread role of phosphate in energy metabolism, signaling, protein function, and bone matrix, disorders of phosphate homeostasis impair the function of many organ systems. The majority of disorders of phosphate homeostasis primarily alter extracellular phosphate; however, disorders that primarily change intracellular phosphate have been described. Since extracellular phosphate and intracellular phosphate are intimately related, symptoms of these disorders may overlap. For example, excess phosphate uptake into cells may result in hypophosphatemia and rickets, while reduced phosphate uptake into cells may result in hyperphosphatemia and matrix calcifications. In addition, some transporters have cell-autonomous and systemic functions. For example, loss-of-function mutations in NPT2a may reduce intracellular phosphate and stimulate renal synthesis of $1,25(OH)_2D$ thereby causing hyperabsorption of calcium and phosphate in the gut. However, the net effect of NPT2a loss-of-function mutations can be infantile hypercalcemia with hypercalciuric nephrocalcinosis, while hypophosphatemic rickets are more commonly encountered in the setting of NPT2c loss-of-function mutations.

Individuals with hypertrophic cardiomyopathy muscular dystrophy and lactic acidosis (Bhoj et al., 2015; Mayr et al., 2011) were found to carry loss-of-function mutations in PIC (*SLC25A3*), which mediates uptake of phosphate by the mitochondria. These findings suggest that phosphate is important for the function of the mitochondrial respiratory chain and ATP synthesis as has been reported by us (Pesta et al., 2016) and others (Sinha et al., 2013; Smith et al., 1984; Brown et al., 1985; Choi et al., 2008; Maldonado and Lemasters, 2014). PIC is also a component of the MPTP, which requires the presence of phosphate to permit influx of calcium into the mitochondrial matrix (Seifert et al., 2015). Detailed evaluation of the hearts in a mouse model with an inducible and cardiac-specific deletion of the *SLC25A3* gene (Kwong et al., 2014) showed reduced MPTP opening in response to calcium challenge. PIC therefore also has phosphate transport—independent roles in mitochondrial function.

Loss-of-function mutations in PIT2 (Lemos et al., 2015) and XPR1 (Legati et al., 2015) were reported in individuals suffering from PFBC or Fahr syndrome (OMIM: 213600). These individuals develop vascular calcifications in the basal ganglia of their brain, leading to seizures. Inhibition of phosphate uptake into microglia due to loss of function in PIT2 or inhibition of phosphate export from vascular smooth muscle cells due to loss of function in XPR1 may stimulate formation of calcium phosphate deposits inside these cells (Legati et al., 2015; Anheim et al., 2016). A similar phenotype was observed in human individuals and mouse models with loss-of-function mutations in the PDGFB receptor (PDGFBR) and PDGFB (Keller et al., 2013). Along with reports of a physical interaction of XPR1 with PDGFBR in mice (Yao et al., 2017), these reports suggest that PDGFB, PDGFBR, and phosphate transporters functionally interact. Interestingly, Pit2-KO mice have increased cerebrospinal fluid phosphate levels, which may suggest a role for this transporter in phosphate homeostasis of cerebrospinal fluid (Jensen et al., 2016).

Summary

Advances since the end of the 20th century have led to the identification of intricate mechanisms by which phosphorus is regulated in the whole organism. An improved understanding of P transport processes, the distribution and function of transporters, and the overall hormonal control of this process has emerged. In particular, the FGF23-mediated bone—renal phosphate axis has markedly expanded our conceptual framework for mineral homeostasis. The related disorders of phosphate homeostasis are now better understood, allowing for new molecular targets for treatment of disease, yet raising a multitude of further questions waiting to be explored.

Patient information/web resources:

Orphanet: http://www.orpha.net
OMIM: http://omim.org
NORD: https://rarediseases.org
The XLH Network: http://www.xlhnetwork.org

Key term/acronym	Definition
ATP	Adenosine triphosphate
PPi	Pyrophosphate
PTH	Parathyroid hormone
NPT2	Solute carrier family 34 (type II) sodium/phosphate cotransporters (with "a", "b", and "c" species)
PIT	Solute carrier family 20 (type III) sodium/phosphate cotransporters (with "1" and "2" species)
PKA	Protein kinase A
PKC	Protein kinase C
NHERF1	Sodium/hydrogen exchange regulatory cofactor
GALNT3	UDP-*N*-acetyl-α-D-galactosamine:polypeptide *N*-acetylgalactosaminyltransferase, isoform 3
FGF23	Fibroblast growth factor 23
KL	α-Klotho
PHEX	Phosphate-regulating gene with homologies to endopeptidases on the X chromosome
MV	Matrix vesicle
DMP1	Dentin matrix protein 1
PXE	Pseudoxanthoma elasticum
MEPE	Matrix extracellular phosphoglycoprotein
SIBLING	Small integrin-binding ligand N-linked glycoprotein
FGFR	Fibroblast growth factor receptor
ENPP1	Ectonucleotide pyrophosphatase/phosphodiesterase 1
ANKH	Progressive ankylosis protein homolog
ABCC6	ATP binding cassette subfamily C member 6
ARHR	Autosomal recessive hypophosphatemic rickets
ASARM	Acidic serine aspartate-rich MEPE-associated motif
HHRH	Hereditary hypophosphatemic rickets with hypercalciuria
PTHR1	PTH/PTHrP receptor
VDR	Vitamin D receptor, forms heterodimer with RXR
HFTC	Hyperphosphatemic familial tumoral calcinosis
TNAP	Tissue nonspecific alkaline phosphatase
XLH	X-linked dominant hypophosphatemia
ADHR	Autosomal dominant hypophosphatemic rickets
SLC34	Solute carrier family 34 (sodium/phosphate cotransporter); members 1 and 3 are expressed in the proximal renal tubule, member 2 is expressed in the intestine
CYP27B1	Vitamin D 1-α-hydroxylase
CYP24A1	Vitamin D 24-hydroxylase
GACI	Generalized arterial calcification of infancy
TRPV	Transient receptor potential cation channel, subfamily V; members 5 and 6 are calcium selective
PMCA	Plasma membrane Ca^{2+} ATPase
VDDR1	Vitamin D–dependent rickets type 1

VDDR2	Vitamin D–dependent rickets type 2
FRP4	Frizzled-related protein 4
PFBC	Primary familial brain calcification
MPTP	Mitochondrial permeability transition pore
PIC	Mitochondrial phosphate carrier
PMT–MCT	Phosphaturic mesenchymal tumor–mixed connective tissue variant
TIO	Tumor-induced osteomalacia
PDGF	Platelet-derived growth factor

References

Addison, W.N., Masica, D.L., Gray, J.J., McKee, M.D., 2010. Phosphorylation-dependent inhibition of mineralization by osteopontin ASARM peptides is regulated by PHEX cleavage. J. Bone Miner. Res. 25, 695–705. https://doi.org/10.1359/jbmr.090832.

Acar, S., BinEssa, H.A., Demir, K., Al-Rijjal, R.A., Zou, M., Catli, G., Anik, A., Al-Enezi, A.F., Ozisik, S., Al-Faham, M.S.A., Abaci, A., Dundar, B., Kattan, W.E., Alsagob, M., Kavukcu, S., Tamimi, H.E., Meyer, B.F., Bober, E., Shi, Y., 2018. Clinical and genetic characteristics of 15 families with hereditary hypophosphatemia: novel Mutations in PHEX and SLC34A3. PLoS One 13, e0193388. https://doi.org/10.1371/journal.pone.0193388 pii:PONE-D-17-39524.

ADHR Consortium T, 2000. Autosomal dominant hypophosphataemic rickets is associated with mutations in FGF23. Nat. Genet. 26, 345–348. https://doi.org/10.1038/81664.

Albright, F.B., Allan, M., Bloomberg, E., 1936. Rickets resistant to vitamin D therapy. Am. J. Dis. Child. 9, 529–547.

Aloia, J.F., Vaswani, A., Yeh, J.K., Ellis, K., Cohn, S.H., 1984. Total body phosphorus in postmenopausal women. Miner. Electrolyte Metab. 10, 73–76.

Alon, U., Hellerstein, S., 1994. Assessment and interpretation of the tubular threshold for phosphate in infants and children. Pediatr. Nephrol. 8, 250–251.

Andrukhova, O., Zeitz, U., Goetz, R., Mohammadi, M., Lanske, B., Erben, R.G., 2012. FGF23 acts directly on renal proximal tubules to induce phosphaturia through activation of the ERK1/2-SGK1 signaling pathway. Bone 51, 621–628. https://doi.org/10.1016/j.bone.2012.05.015.

Anheim, M., López-Sánchez, U., Giovannini, D., Richard, A.-C., Touhami, J., N'Guyen, L., Rudolf, G., Thibault-Stoll, A., Frebourg, T., Hannequin, D., Campion, D., Battini, J.-L., Sitbon, M., Nicolas, G., 2016. XPR1 mutations are a rare cause of primary familial brain calcification. J. Neurol. 263, 1559–1564. https://doi.org/10.1007/s00415-016-8166-4.

Ansermet, C., Moor, M.B., Centeno, G., Auberson, M., Hu, D.Z., Baron, R., Nikolaeva, S., Haenzi, B., Katanaeva, N., Gautschi, I., Katanaev, V., Rotman, S., Koesters, R., Schild, L., Pradervand, S., Bonny, O., Firsov, D., 2017. Renal Fanconi syndrome and hypophosphatemic rickets in the absence of xenotropic and polytropic retroviral receptor in the nephron. J. Am. Soc. Nephrol. 28, 1073–1078. https://doi.org/10.1681/ASN.2016070726.

Antoniucci, D.M., Yamashita, T., Portale, A.A., 2006. Dietary phosphorus regulates serum fibroblast growth factor-23 concentrations in healthy men. J. Clin. Endocrinol. Metab. 91, 3144–3149. https://doi.org/10.1210/jc.2006-0021.

Aono, Y., Hasegawa, H., Yamazaki, Y., Shimada, T., Fujita, T., Yamashita, T., Fukumoto, S., 2011. Anti-FGF-23 neutralizing antibodies ameliorate muscle weakness and decreased spontaneous movement of Hyp mice. J. Bone Miner. Res. 26, 803–810. https://doi.org/10.1002/jbmr.275.

Arcidiacono, T., Mingione, A., Macrina, L., Pivari, F., Soldati, L., Vezzoli, G., 2014. Idiopathic calcium nephrolithiasis: a review of pathogenic mechanisms in the light of genetic studies. Am. J. Nephrol. 40, 499–506. https://doi.org/10.1159/000369833.

Arnold, A., Horst, S.A., Gardella, T.J., Baba, H., Levine, M.A., Kronenberg, H.M., 1990. Mutation of the signal peptide-encoding region of the pre-proparathyroid hormone gene in familial isolated hypoparathyroidism. J. Clin. Investig. 86, 1084–1087.

Azevedo, C., Saiardi, A., 2017. Eukaryotic phosphate homeostasis: the inositol pyrophosphate perspective. Trends Biochem. Sci. 42, 219–231. https://doi.org/10.1016/j.tibs.2016.10.008.

Bachmann, S., Schlichting, U., Geist, B., Mutig, K., Petsch, T., Bacic, D., Wagner, C.A., Kaissling, B., Biber, J., Murer, H., Willnow, T.E., 2004. Kidney-specific inactivation of the megalin gene impairs trafficking of renal inorganic sodium phosphate cotransporter (NaPi-IIa). J. Am. Soc. Nephrol. 15, 892–900.

Bacic, D., Lehir, M., Biber, J., Kaissling, B., Murer, H., Wagner, C.A., 2006. The renal Na+/phosphate cotransporter NaPi-IIa is internalized via the receptor-mediated endocytic route in response to parathyroid hormone. Kidney Int. 69, 495–503. https://doi.org/10.1038/sj.ki.5000148 pii:S0085-2538(15)51522-6.

Bai, X., Miao, D., Goltzman, D., Karaplis, A.C., 2007. Early lethality in Hyp mice with targeted deletion of Pth gene. Endocrinology 148, 4974–4983. https://doi.org/10.1210/en.2007-0243.

Baroncelli, G.I., Bertelloni, S., Ceccarelli, C., Saggese, G., 2001. Effect of growth hormone treatment on final height, phosphate metabolism, and bone mineral density in children with X-linked hypophosphatemic rickets. J. Pediatr. 138, 236–243. https://doi.org/10.1067/mpd.2001.108955.

Barros, N.M., Hoac, B., Neves, R.L., Addison, W.N., Assis, D.M., Murshed, M., Carmona, A.K., McKee, M.D., 2013. Proteolytic processing of osteopontin by PHEX and accumulation of osteopontin fragments in Hyp mouse bone, the murine model of X-linked hypophosphatemia. J. Bone Miner. Res. 28, 688–699. https://doi.org/10.1002/jbmr.1766.

Beck, L., Leroy, C., Salaun, C., Margall-Ducos, G., Desdouets, C., Friedlander, G., 2009. Identification of a novel function of PiT1 critical for cell proliferation and independent of its phosphate transport activity. J. Biol. Chem. 284, e99959.

Beck-Nielsen, S.S., Brock-Jacobsen, B., Gram, J., Brixen, K., Jensen, T.K., 2009. Incidence and prevalence of nutritional and hereditary rickets in southern Denmark. Eur. J. Endocrinol. 160, 491–497. https://doi.org/10.1530/EJE-08-0818.

Beck-Nielsen, S.S., Brixen, K., Gram, J., Brusgaard, K., 2012. Mutational analysis of PHEX, FGF23, DMP1, SLC34A3 and CLCN5 in patients with hypophosphatemic rickets. J. Hum. Genet. 57, 453–458. https://doi.org/10.1038/jhg.2012.56.

Bell, C.L., Tenenhouse, H.S., Scriver, C.R., 1988. Primary cultures of renal epithelial cells from X-linked hypophosphatemic (Hyp) mice express defects in phosphate transport and vitamin D metabolism. Am. J. Hum. Genet. 43, 293–303.

Belov, A.A., Mohammadi, M., 2013. Molecular mechanisms of fibroblast growth factor signaling in physiology and pathology. Cold Spring Harb Perspect Biol 5 (6). https://doi.org/10.1101/cshperspect.a015958. Online publication: https://cshperspectives.cshlp.org/content/5/6/a015958.long.

Benet-Pages, A., Orlik, P., Strom, T.M., Lorenz-Depiereux, B., 2005. An FGF23 missense mutation causes familial tumoral calcinosis with hyperphosphatemia. Hum. Mol. Genet. 14, 385–390.

Bergwitz, C., Bastepe, M., 2008. NHERF1 mutations and responsiveness of renal parathyroid hormone. NEJM 359, 2615–2617.

Bergwitz, C., Miyamoto, K.I., 2019. Hereditary hypophosphatemic rickets with hypercalciuria: pathophysiology, clinical presentation, diagnosis and therapy. Pflügers Archiv 471, 149–163. https://doi.org/10.1007/s00424-018-2184-2.

Bergwitz, C., Roslin, N.M., Tieder, M., Loredo-Osti, J.C., Bastepe, M., Abu-Zahra, H., Frappier, D., Burkett, K., Carpenter, T.O., Anderson, D., Garabedian, M., Sermet, I., Fujiwara, T.M., Morgan, K., Tenenhouse, H.S., Juppner, H., 2006. SLC34A3 mutations in patients with hereditary hypophosphatemic rickets with hypercalciuria predict a key role for the sodium-phosphate cotransporter NaPi-IIc in maintaining phosphate homeostasis. Am. J. Hum. Genet. 78, 179–192. https://doi.org/10.1086/499409 pii:S0002-9297(07)62351-9.

Berndt, T., Kumar, R., 2009. Novel mechanisms in the regulation of phosphorus homeostasis. Physiology 24, 17–25.

Berndt, T.J., Schiavi, S., Kumar, R., 2005. "Phosphatonins" and the regulation of phosphorus homeostasis. Am. J. Physiol. Renal. Physiol. 289, F1170–F1182.

Berndt, T., Thomas, L.F., Craig, T.A., Sommer, S., Li, X., Bergstralh, E.J., Kumar, R., 2007. Evidence for a signaling axis by which intestinal phosphate rapidly modulates renal phosphate reabsorption. Proc. Natl. Acad. Sci. U. S. A. 104, 11085–11090. https://doi.org/10.1073/pnas.0704446104.

Bhoj, E.J., Li, M., Ahrens-Nicklas, R., Pyle, L.C., Wang, J., Zhang, V.W., Clarke, C., Wong, L.J., Sondheimer, N., Ficicioglu, C., Yudkoff, M., 2015. Pathologic variants of the mitochondrial phosphate carrier SLC25A3: two new patients and expansion of the cardiomyopathy/skeletal myopathy phenotype with and without lactic acidosis. JIMD Rep 19, 59–66. https://doi.org/10.1007/8904_2014_364.

Biber, J., Hernando, N., Forster, I., 2013. Phosphate transporters and their function. Annu. Rev. Physiol. 75, 535–550. https://doi.org/10.1146/annurev-physiol-030212-183748.

Boger, C.A., Gorski, M., Li, M., Hoffmann, M.M., Huang, C., Yang, Q., Teumer, A., Krane, V., O'Seaghdha, C.M., Kutalik, Z., Wichmann, H.E., Haak, T., Boes, E., Coassin, S., Coresh, J., Kollerits, B., Haun, M., Paulweber, B., Kottgen, A., Li, G., Shlipak, M.G., Powe, N., Hwang, S.J., Dehghan, A., Rivadeneira, F., Uitterlinden, A., Hofman, A., Beckmann, J.S., Kramer, B.K., Witteman, J., Bochud, M., Siscovick, D., Rettig, R., Kronenberg, F., Wanner, C., Thadhani, R.I., Heid, I.M., Fox, C.S., Kao, W.H., Consortium, C.K., 2011. Association of eGFR-related loci identified by GWAS with incident CKD and ESRD. PLoS Genet. 7, e1002292. https://doi.org/10.1371/journal.pgen.1002292 pii:PGENETICS-D-11-00374.

Bonucci, E., 2012. Bone mineralization. Front. Biosci. 17, 100–128.

Bora, S.A., Kennett, M.J., Smith, P.B., Patterson, A.D., Cantorna, M.T., 2018. The gut microbiota regulates endocrine vitamin D metabolism through fibroblast growth factor 23. Front. Immunol. 9, 408. https://doi.org/10.3389/fimmu.2018.00408.

Bottini, M., Mebarek, S., Anderson, K.L., Strzelecka-Kiliszek, A., Bozycki, L., Simao, A.M.S., Bolean, M., Ciancaglini, P., Pikula, J.B., Pikula, S., Magne, D., Volkmann, N., Hanein, D., Millan, J.L., Buchet, R., 2018. Matrix vesicles from chondrocytes and osteoblasts: their biogenesis, properties, functions and biomimetic models. Biochim. Biophys. Acta Gen. Subj. 1862, 532–546. https://doi.org/10.1016/j.bbagen.2017.11.005.

Boukpessi, T., Hoac, B., Coyac, B.R., Leger, T., Garcia, C., Wicart, P., Whyte, M.P., Glorieux, F.H., Linglart, A., Chaussain, C., McKee, M.D., 2017. Osteopontin and the dento-osseous pathobiology of X-linked hypophosphatemia. Bone 95, 151–161. https://doi.org/10.1016/j.bone.2016.11.019.

Bourgine, A., Pilet, P., Diouani, S., Sourice, S., Lesoeur, J., Beck-Cormier, S., Khoshniat, S., Weiss, P., Friedlander, G., Guicheux, J., Beck, L., 2013. Mice with hypomorphic expression of the sodium-phosphate cotransporter PiT1/Slc20a1 have an unexpected normal bone mineralization. PLoS One 8, e65979. https://doi.org/10.1371/journal.pone.0065979 pii:PONE-D-12-23346.

Bringhurst, F.R., Leder, B.Z., 2006. Regulation of calcium and phosphate homeostasis. In: DeGroot, L.J., Jameson, J.L. (Eds.), Endocrinology. W.B. Saunders Co., Philadelphia, pp. 805–843.

Brown, B.J., Robinson, A.E., Brogdon, B.G., 1985. Radiographic findings in hydantoin toxicity. Ala. J. Med. Sci. 22, 428–430.

Brown, W.W., Juppner, H., Langman, C.B., Price, H., Farrow, E.G., White, K.E., McCormick, K.L., 2009. Hypophosphatemia with elevations in serum fibroblast growth factor 23 in a child with Jansen's metaphyseal chondrodysplasia. J. Clin. Endocrinol. Metab. 94, 17–20. https://doi.org/10.1210/jcem.94.2.9988, 10.1210/jc.2008-0220.

Brownstein, C.A., Adler, F., Nelson-Williams, C., Iijima, J., Li, P., Imura, A., Nabeshima, Y., Reyes-Mugica, M., Carpenter, T.O., Lifton, R.P., 2008. A translocation causing increased alpha-klotho level results in hypophosphatemic rickets and hyperparathyroidism. Proc. Natl. Acad. Sci. U. S. A. 105, 3455–3460. https://doi.org/10.1073/pnas.0712361105.

Burnett, C.H., Dent, C.E., Harper, C., Warland, B.J., 1964. Vitamin D-resistant rickets. Analysis of twenty-four pedigrees with hereditary and sporadic cases. Am. J. Med. 36, 222–232.

Butterworth, P.J., Younus, M.J., 1993. Uptake of phosphate by rat hepatocytes in primary culture: a sodium-dependent system that is stimulated by insulin. Biochim. Biophys. Acta 1148, 117–122.

Caballero, D., Li, Y., Fetene, J., Ponsetto, J., Chen, A., Zhu, C., Braddock, D.T., Bergwitz, C., 2017a. Intraperitoneal pyrophosphate treatment reduces renal calcifications in Npt2a null mice. PLoS One 12, e0180098. https://doi.org/10.1371/journal.pone.0180098.

Caballero, D., Li, Y., Ponsetto, J., Zhu, C., Bergwitz, C., 2017b. Impaired urinary osteopontin excretion in Npt2a−/− mice. Am. J. Physiol. Renal. Physiol. 312, F77−F83. https://doi.org/10.1152/ajprenal.00367.2016.

Calvo, M.S., Moshfegh, A.J., Tucker, K.L., 2014. Assessing the health impact of phosphorus in the food supply: issues and considerations. Adv Nutr 5, 104−113. https://doi.org/10.3945/an.113.004861.

Carpenter, T.O., Ellis, B.K., Insogna, K.L., Philbrick, W.M., Sterpka, J., Shimkets, R., 2005. Fibroblast growth factor 7: an inhibitor of phosphate transport derived from oncogenic osteomalacia-causing tumors. J. Clin. Endocrinol. Metab. 90, 1012−1020. https://doi.org/10.1210/jc.2004-0357.

Carpenter, T.O., Imel, E.A., Holm, I.A., Jan de Beur, S.M., Insogna, K.L., 2011. A clinician's guide to X-linked hypophosphatemia. J. Bone Miner. Res. 26, 1381−1388. https://doi.org/10.1002/jbmr.340.

Carpenter, T.O., Olear, E.A., Zhang, J.H., Ellis, B.K., Simpson, C.A., Cheng, D., Gundberg, C.M., Insogna, K.L., 2014. Effect of paricalcitol on circulating parathyroid hormone in X-linked hypophosphatemia: a randomized, double-blind, placebo-controlled study. J. Clin. Endocrinol. Metab. 99, 3103−3111. https://doi.org/10.1210/jc.2014-2017.

Carpenter, T.O., Whyte, M.P., Imel, E.A., Boot, A.M., Hogler, W., Linglart, A., Padidela, R., Van't Hoff, W., Mao, M., Chen, C.Y., Skrinar, A., Kakkis, E., San Martin, J., Portale, A.A., 2018. Burosumab therapy in children with X-linked hypophosphatemia. N. Engl. J. Med. 378, 1987−1998. https://doi.org/10.1056/NEJMoa1714641.

Chande, S., Bergwitz, C., 2018. Role of phosphate sensing in bone and mineral metabolism. Nat. Rev. Endocrinol. 14, 637−655. https://doi.org/10.1038/s41574-018-0076-3.

Chang, A.R., Lazo, M., Appel, L.J., Gutierrez, O.M., Grams, M.E., 2014. High dietary phosphorus intake is associated with all-cause mortality: results from NHANES III. Am. J. Clin. Nutr. 99, 320−327. https://doi.org/10.3945/ajcn.113.073148.

Chavkin, N.W., Chia, J.J., Crouthamel, M.H., Giachelli, C.M., 2015. Phosphate uptake-independent signaling functions of the type III sodium-dependent phosphate transporter, PiT-1, in vascular smooth muscle cells. Exp. Cell Res. 333, 39−48. https://doi.org/10.1016/j.yexcr.2015.02.002.

Chen, C., Carpenter, T., Steg, N., Baron, R., Anast, C., 1989. Hypercalciuric hypophosphatemic rickets, mineral balance, bone histomorphometry, and therapeutic implications of hypercalciuria. Pediatrics 84, 276−280.

Chen, G., Liu, Y., Goetz, R., Fu, L., Jayaraman, S., Hu, M.C., Moe, O.W., Liang, G., Li, X., Mohammadi, M., 2018. alpha-Klotho is a non-enzymatic molecular scaffold for FGF23 hormone signalling. Nature 553, 461−466. https://doi.org/10.1038/nature25451.

Chesher, D., Oddy, M., Darbar, U., Sayal, P., Casey, A., Ryan, A., Sechi, A., Simister, C., Waters, A., Wedatilake, Y., Lachmann, R.H., Murphy, E., 2018. Outcome of adult patients with X-linked hypophosphatemia caused by PHEX gene mutations. J. Inherit. Metab. Dis. 41, 865−876. https://doi.org/10.1007/s10545-018-0147-6.

Chi, Y., Zhao, Z., He, X., Sun, Y., Jiang, Y., Li, M., Wang, O., Xing, X., Sun, A.Y., Zhou, X., Meng, X., Xia, W., 2014. A compound heterozygous mutation in SLC34A3 causes hereditary hypophosphatemic rickets with hypercalciuria in a Chinese patient. Bone 59, 114−121. https://doi.org/10.1016/j.bone.2013.11.008 pii:S8756-3282(13)00444-4.

Chien, M.L., Foster, J.L., Douglas, J.L., Garcia, J.V., 1997. The amphotropic murine leukemia virus receptor gene encodes a 71-kilodalton protein that is induced by phosphate depletion. J. Virol. 71, 4564−4570.

Choi, C.S., Befroy, D.E., Codella, R., Kim, S., Reznick, R.M., Hwang, Y.J., Liu, Z.X., Lee, H.Y., Distefano, A., Samuel, V.T., Zhang, D., Cline, G.W., Handschin, C., Lin, J., Petersen, K.F., Spiegelman, B.M., Shulman, G.I., 2008. Paradoxical effects of increased expression of PGC-1alpha on muscle mitochondrial function and insulin-stimulated muscle glucose metabolism. Proc. Natl. Acad. Sci. U. S. A. 105, 19926−19931. https://doi.org/10.1073/pnas.0810339105.

Chong, W.H., Molinolo, A.A., Chen, C.C., Collins, M.T., 2011. Tumor-induced osteomalacia. Endocr. Relat. Cancer 18, R53−R77. https://doi.org/10.1530/ERC-11-0006.

Clarke, G.D., Kainer, G., Conway, W.F., Chan, J.C., 1990. Intramyocellular phosphate metabolism in X-linked hypophosphatemic rickets. J. Pediatr. 116, 288−292.

Collins, J.F., Ghishan, F.K., 1994. Molecular cloning, functional expression, tissue distribution, and in situ hybridization of the renal sodium phosphate (Na+/Pi) transporter in the control and hypophosphatemic mouse. FASBJ 8, 862−868.

Collins, M., Bergwitz, C., Aitcheson, G., Blau, J., Boyce, A., Gafni, R., Guthrie, L., Miranda, F., Slosberg, E., Graus-Porta, D., Hopmann, C., Welaya, K., Isaacs, R., Miller, C., 2015. Striking response of tumor-induced osteomalacia to the FGFR inhibitor NVP-BGJ398. J. Bone Miner. Res. 30 (Suppl. 1) available at: http://www.asbmr.org/education/AbstractDetail?aid=c5464be6-d873-49f3-bb71-719e2198867e.

Connor, J., Olear, E.A., Insogna, K.L., Katz, L., Baker, S., Kaur, R., Simpson, C.A., Sterpka, J., Dubrow, R., Zhang, J.H., Carpenter, T.O., 2015. Conventional therapy in adults with X-linked hypophosphatemia: effects on enthesopathy and dental disease. J. Clin. Endocrinol. Metab. 100, 3625−3632. https://doi.org/10.1210/JC.2015-2199.

Couasnay, G., Bon, N., Devignes, C.S., Sourice, S., Bianchi, A., Veziers, J., Weiss, P., Elefteriou, F., Provot, S., Guicheux, J., Beck-Cormier, S., Beck, L., 2019. PiT1/Slc20a1 is required for endoplasmic reticulum homeostasis, chondrocyte survival, and skeletal development. J. Bone Miner. Res. 34 (2), 387−398. https://doi.org/10.1002/jbmr.3609 (Epub 2018).

Coyac, B.R., 2017. Tissue-specific mineralization defects in the periodontium of the Hyp mouse model of X-linked hypophosphatemia. Bone 103, 334−346.

Daga, A., Majmundar, A.J., Braun, D.A., Gee, H.Y., Lawson, J.A., Shril, S., Jobst-Schwan, T., Vivante, A., Schapiro, D., Tan, W., Warejko, J.K., Widmeier, E., Nelson, C.P., Fathy, H.M., Gucev, Z., Soliman, N.A., Hashmi, S., Halbritter, J., Halty, M., Kari, J.A., El-Desoky, S., Ferguson, M.A., Somers, M.J.G., Traum, A.Z., Stein, D.R., Daouk, G.H., Rodig, N.M., Katz, A., Hanna, C., Schwaderer, A.L., Sayer, J.A., Wassner, A.J., Mane, S.,

Lifton, R.P., Milosevic, D., Tasic, V., Baum, M.A., Hildebrandt, F., 2018. Whole exome sequencing frequently detects a monogenic cause in early onset nephrolithiasis and nephrocalcinosis. Kidney Int. 93, 204–213. https://doi.org/10.1016/j.kint.2017.06.025 pii:S0085-2538(17)30494-5.

Dallas, S.L., Prideaux, M., Bonewald, L.F., 2013. The osteocyte: an endocrine cell ... and more. Endocr. Rev. 34, 658–690. https://doi.org/10.1210/er.2012-1026.

Dasgupta, D., Wee, M.J., Reyes, M., Li, Y., Simm, P.J., Sharma, A., Schlingmann, K.P., Janner, M., Biggin, A., Lazier, J., Gessner, M., Chrysis, D., Tuchman, S., Baluarte, H.J., Levine, M.A., Tiosano, D., Insogna, K., Hanley, D.A., Carpenter, T.O., Ichikawa, S., Hoppe, B., Konrad, M., Savendahl, L., Munns, C.F., Lee, H., Juppner, H., Bergwitz, C., 2014. Mutations in SLC34A3/NPT2c are associated with kidney stones and nephrocalcinosis. J. Am. Soc. Nephrol. 25, 2366–2375. https://doi.org/10.1681/ASN.2013101085.

David, V., Martin, A., Hedge, A.M., Drezner, M.K., Rowe, P.S., 2011. ASARM peptides: PHEX-dependent and -independent regulation of serum phosphate. Am. J. Physiol. Renal. Physiol. 300, F783–F791. https://doi.org/10.1152/ajprenal.00304.2010.

Dhir, G., Li, D., Hakonarson, H., Levine, M.A., 2017. Late-onset hereditary hypophosphatemic rickets with hypercalciuria (HHRH) due to mutation of SLC34A3/NPT2c. Bone 97, 15–19. https://doi.org/10.1016/j.bone.2016.12.001 pii:S8756-3282(16)30363-5.

DiMeglio, L.A., Econs, M.J., 2001. Hypophosphatemic rickets. Rev. Endocr. Metab. Disord. 2, 165–173.

Ding, C., Buckingham, B., Levine, M.A., 2001. Familial isolated hypoparathyroidism caused by a mutation in the gene for the transcription factor GCMB. J. Clin. Investig. 108, 1215–1220.

Donohue, M.M., Demay, M.B., 2002. Rickets in VDR null mice is secondary to decreased apoptosis of hypertrophic chondrocytes. Endocrinology 143, 3691–3694.

Econs, M.J., McEnery, P.T., 1997. Autosomal dominant hypophosphatemic rickets/osteomalacia: clinical characterization of a novel renal phosphate-wasting disorder. J. Clin. Endocrinol. Metab. 82, 674–681. https://doi.org/10.1210/jcem.82.2.3765.

Econs, M.J., Samsa, G.P., Monger, M., Drezner, M.K., Feussner, J.R., 1994. X-Linked hypophosphatemic rickets: a disease often unknown to affected patients. Bone Miner. 24, 17–24.

El-Maouche, D., Sadowski, S.M., Papadakis, G.Z., Guthrie, L., Cottle-Dellisle, C., Merkel, R., Milo, C., Chen, C.C., Kebebew, E., Collins, M.T., 2016. 68Ga-DOTATATE for tumor localization in tumor-induced osteomalacia. J. Clin. Endocrinol. Metab. 101, 3575–3581.

Erben, R.G., 2016. Update on FGF23 and klotho signaling. Mol. Cell. Endocrinol. 432, 56–65. https://doi.org/10.1016/j.mce.2016.05.008.

Eto, N., Miyata, Y., Ohno, H., Yamashita, T., 2005. Nicotinamide prevents the development of hyperphosphataemia by suppressing intestinal sodium-dependent phosphate transporter in rats with adenine-induced renal failure. Nephrol. Dial. Transplant. 20, 1378–1384. https://doi.org/10.1093/ndt/gfh781 pii:gfh781.

Eto, N., Tomita, M., Hayashi, M., 2006. NaPi-mediated transcellular permeation is the dominant route in intestinal inorganic phosphate absorption in rats. Drug Metab. Pharmacokinet. 21, 217–221.

Eyres, K.S., Brown, J., Douglas, D.L., 1993. Osteotomy and intramedullary nailing for the correction of progressive deformity in vitamin D-resistant hypophosphataemic rickets. J. R. Coll. Surg. Edinb. 38, 50–54.

Farrow, E.G., White, K.E., 2010. Recent advances in renal phosphate handling. Nat. Rev. Nephrol. 6, 207–217. https://doi.org/10.1038/nrneph.2010.17.

Farrow, E.G., Davis, S.I., Ward, L.M., Summers, L.J., Bubbear, J.S., Keen, R., Stamp, T.C., Baker, L.R., Bonewald, L.F., White, K.E., 2009. Molecular analysis of DMP1 mutants causing autosomal recessive hypophosphatemic rickets. Bone 44, 287–294.

Feng, J.Q., Ward, L.M., Liu, S., Lu, Y., Xie, Y., Yuan, B., Yu, X., Rauch, F., Davis, S.I., Zhang, S., Rios, H., Drezner, M.K., Quarles, L.D., Bonewald, L.F., White, K.E., 2006. Loss of DMP1 causes rickets and osteomalacia and identifies a role for osteocytes in mineral metabolism. Nat. Genet. 38, 1310–1315. https://doi.org/10.1038/ng1905.

Ferreira, C., Ziegler, S., Gahl, W.A., 1993. Generalized arterial calcification of infancy. In: Adam, M.P., Ardinger, H.H., Pagon, R.A., Wallace, S.E., Bean, L.J.H., Stephens, K., Amemiya, A. (Eds.), GeneReviews((R)). Seattle (WA).

Ferreira, C.R., Ziegler, S.G., Gupta, A., Groden, C., Hsu, K.S., Gahl, W.A., 2016. Treatment of hypophosphatemic rickets in generalized arterial calcification of infancy (GACI) without worsening of vascular calcification. Am. J. Med. Genet. A 170 (5), 1308–1311. https://doi.org/10.1002/ajmg.a.37574.

Fiermonte, G., Dolce, V., Arrigoni, R., Runswick, M.J., Walker, J.E., Palmieri, F., 1999. Organization and sequence of the gene for the human mitochondrial dicarboxylate carrier: evolution of the carrier family. Biochem. J. 344 (Pt 3), 953–960.

Fishman, G., Miller-Hansen, D., Jacobsen, C., Singhal, V.K., Alon, U.S., 2004. Hearing impairment in familial X-linked hypophosphatemic rickets. Eur. J. Pediatr. 163, 622–623. https://doi.org/10.1007/s00431-004-1504-z.

Folpe, A.L., Fanburg-Smith, J.C., Billings, S.D., Bisceglia, M., Bertoni, F., Cho, J.Y., Econs, M.J., Inwards, C.Y., Jan de Beur, S.M., Mentzel, T., Montgomery, E., Michal, M., Miettinen, M., Mills, S.E., Reith, J.D., O'Connell, J.X., Rosenberg, A.E., Rubin, B.P., Sweet, D.E., Vinh, T.N., Wold, L.E., Wehrli, B.M., White, K.E., Zaino, R.J., Weiss, S.W., 2004. Most osteomalacia-associated mesenchymal tumors are a single histopathologic entity: an analysis of 32 cases and a comprehensive review of the literature. Am. J. Surg. Pathol. 28, 1–30.

Folsom, L.J., Imel, E.A., 2015. Hyperphosphatemic familial tumoral calcinosis: genetic models of deficient FGF23 action. Curr. Osteoporos. Rep. 13, 78–87. https://doi.org/10.1007/s11914-015-0254-3.

Forrest, G., German, J., Giuffrida, A., Luidens, M., Dowling, J., 2016. Hereditary x-linked hypophosphatemia and thoracic myelopathy. Aace Clin. Case Rep. 2, e244–e246.

Forster, I., Wagner, A., 2018. SLC34. In: Choi, S. (Ed.), Encyclopedia of Signalling Molecules. Springer, pp. 5013–5022.

Forster, I.C., Hernando, N., Biber, J., Murer, H., 2006. Proximal tubular handling of phosphate: a molecular perspective. Kidney Int. 70, 1548–1559. https://doi.org/10.1038/sj.ki.5001813.

Francis, R.M., Selby, P.L., 1997. Osteomalacia. Baillieres Clin Endocrinol Metab 11, 145–163.

Frishberg, Y., Araya, K., Rinat, C., Topaz, O., Yamazaki, Y., Feinstein, Y., Navon-Elkan, P., Becker-Cohen, R., Yamashita, T., Igarashi, T., Sprecher, E., 2004. Hyperostosis-hyperphosphatemia syndrome caused by mutations in GALNT3 and associated with augmented processing of FGF-23. Philadelphia, pp. F-P0937 Am. Soc. Nephrol.

Gannage-Yared, M.H., Makrythanasis, P., Chouery, E., Sobacchi, C., Mehawej, C., Santoni, F.A., Guipponi, M., Antonarakis, S.E., Hamamy, H., Megarbane, A., 2014. Exome sequencing reveals a mutation in DMP1 in a family with familial sclerosing bone dysplasia. Bone 68, 142–145. https://doi.org/10.1016/j.bone.2014.08.014.

Gazit, D., Tieder, M., Liberman, U.A., Passi-Even, L., Bab, I.A., 1991. Osteomalacia in hereditary hypophosphatemic rickets with hypercalciuria: a correlative clinical-histomorphometric study. J. Clin. Endocrinol. Metab. 72, 229–235.

Ghanem, I., Karam, J.A., Widmann, R.F., 2011. Surgical epiphysiodesis indications and techniques: update. Curr. Opin. Pediatr. 23, 53–59. https://doi.org/10.1097/MOP.0b013e32834231b3.

Giral, H., Caldas, Y., Sutherland, E., Wilson, P., Breusegem, S., Barry, N., Blaine, J., Jiang, T., Wang, X.X., Levi, M., 2009. Regulation of rat intestinal Na-dependent phosphate transporters by dietary phosphate. Am. J. Physiol. Renal. Physiol. 297, F1466–F1475. https://doi.org/10.1152/ajprenal.00279.2009.

Glorieux, F., Scriver, C.R., 1972. Loss of a parathyroid hormone-sensitive component of phosphate transport in X-linked hypophosphatemia. Science 175, 997–1000.

Glorieux, F.H., Marie, P.J., Pettifor, J.M., Delvin, E.E., 1980. Bone response to phosphate salts, ergocalciferol, and calcitriol in hypophosphatemic vitamin D-resistant rickets. N. Engl. J. Med. 303, 1023–1031. https://doi.org/10.1056/NEJM198010303031802.

Gonzalez Ballesteros, L.F., Ma, N.S., Gordon, R.J., Ward, L., Backeljauw, P., Wasserman, H., Weber, D.R., DiMeglio, L.A., Gagne, J., Stein, R., Cody, D., Simmons, K., Zimakas, P., Topor, L.S., Agrawal, S., Calabria, A., Tebben, P., Faircloth, R., Imel, E.A., Casey, L., Carpenter, T.O., 2017. Unexpected widespread hypophosphatemia and bone disease associated with elemental formula use in infants and children. Bone 97, 287–292. https://doi.org/10.1016/j.bone.2017.02.003.

GTEx Consortium, T., 2013. The genotype-tissue expression (GTEx) project. Nat. Genet. 45, 580–585. https://doi.org/10.1038/ng.2653.

Gudbjartsson, D.F., Holm, H., Indridason, O.S., Thorleifsson, G., Edvardsson, V., Sulem, P., de Vegt, F., d'Ancona, F.C., den Heijer, M., Wetzels, J.F., Franzson, L., Rafnar, T., Kristjansson, K., Bjornsdottir, U.S., Eyjolfsson, G.I., Kiemeney, L.A., Kong, A., Palsson, R., Thorsteinsdottir, U., Stefansson, K., 2010. Association of variants at UMOD with chronic kidney disease and kidney stones-role of age and comorbid diseases. PLoS Genet. 6, e1001039. https://doi.org/10.1371/journal.pgen.1001039.

Gutierrez, O.M., Luzuriaga-McPherson, A., Lin, Y., Gilbert, L.C., Ha, S.W., Beck Jr., G.R., 2015. Impact of phosphorus-based food additives on bone and mineral metabolism. J. Clin. Endocrinol. Metab. 100, 4264–4271. https://doi.org/10.1210/jc.2015-2279.

Haffner, D., Nissel, R., Wuhl, E., Mehls, O., 2004. Effects of growth hormone treatment on body proportions and final height among small children with X-linked hypophosphatemic rickets. Pediatrics 113, e593–e596.

Haito-Sugino, S., Ito, M., Ohi, A., Shiozaki, Y., Kangawa, N., Nishiyama, T., Aranami, F., Sasaki, S., Mori, A., Kido, S., Tatsumi, S., Segawa, H., Miyamoto, K., 2012. Processing and stability of type IIc sodium-dependent phosphate cotransporter mutations in patients with hereditary hypophosphatemic rickets with hypercalciuria. Am. J. Physiol. Cell Physiol. 302, C1316–C1330. https://doi.org/10.1152/ajpcell.00314.2011.

Harrison, H.E., Harrison, H.C., 1979. Disorders of calcium and phosphate metabolism in childhood and adolescence. Major Probl. Clin. Pediatr. 20, 1–314.

Harvey, B.M., Eussen, S., Harthoorn, L.F., Burks, A.W., 2017. Mineral intake and status of cow's milk allergic infants consuming an amino acid-based formula. J. Pediatr. Gastroenterol. Nutr. 65, 346–349. https://doi.org/10.1097/MPG.0000000000001655.

Hattenhauer, O., Traebert, M., Murer, H., Biber, J., 1999. Regulation of small intestinal Na-P(i) type IIb cotransporter by dietary phosphate intake. Am. J. Physiol. 277, G756–G762.

Hesse, E., Moessinger, E., Rosenthal, H., Laenger, F., Brabant, G., Petrich, T., Gratz, K.F., Bastian, L., 2007a. Oncogenic osteomalacia: exact tumor localization by co-registration of positron emission and computed tomography. J. Bone Miner. Res. 22, 158–162. https://doi.org/10.1359/jbmr.060909.

Hesse, E., Rosenthal, H., Bastian, L., 2007b. Radiofrequency ablation of a tumor causing oncogenic osteomalacia. N. Engl. J. Med. 357, 422–424. https://doi.org/10.1056/NEJMc070347.

Hilfiker, H., Kvietikova, I., Hartmann, C.M., Stange, G., Murer, H., 1998. Characterization of the human type II Na/Pi-cotransporter promoter. Pflügers Archiv 436, 591–598.

Hirao, Y., Chikuda, H., Oshima, Y., Matsubayashi, Y., Tanaka, S., 2016. Extensive ossification of the paraspinal ligaments in a patient with vitamin D-resistant rickets: case report with literature review. Int. J. Surg. Case Rep. 27, 125–128.

Hoac, B., Boukpessi, T., Buss, D.J., Chaussain, C., Murshed, M., McKee, M.D., 2018. Ablation of osteopontin in osteomalacic Hyp mice partially rescues the deficient mineralization without correcting hypophosphatemia. In: American Society for Bone and Mineral Research Annual Meeting. Montreal.

Holm, I.A., Huang, X., Kunkel, L.M., 1997. Mutational analysis of the PEX gene in patients with X-linked hypophosphatemic rickets. Am. J. Hum. Genet. 60, 790–797.

Hsia, D.Y., Kraus, M., Samuels, J., 1959. Genetic studies on vitamin D resistant rickets (familial hypophosphatemia). Am. J. Hum. Genet. 11, 156–165.

Hu, M.C., Shi, M., Zhang, J., Pastor, J., Nakatani, T., Lanske, B., Shawkat Razzaque, M., Rosenblatt, K.P., Baum, M.G., Kuro, O.M., Moe, O.W., 2010. Klotho: a novel phosphaturic substance acting as an autocrine enzyme in the renal proximal tubule. FASEB J. 24, 3438–3450.

Hughes, M.R., Malloy, P.J., Kieback, D.G., Kesterson, R.A., Pike, J.W., Feldman, D., O'Malley, B.W., 1988. Point mutations in the human vitamin D receptor gene associated with hypocalcemic rickets. Science 242, 1702–1705.

HYP-Consortium, 1995. A gene (PEX) with homologies to endopeptidases is mutated in patients with X-linked hypophosphatemic rickets. Nat. Genet. 11, 130–136.

Ichikawa, S., Sorenson, A.H., Imel, E.A., Friedman, N.E., Gertner, J.M., Econs, M.J., 2006. Intronic deletions in the SLC34A3 gene cause hereditary hypophosphatemic rickets with hypercalciuria. J. Clin. Endocrinol. Metab. 91, 4022–4027. https://doi.org/10.1210/jc.2005-2840.

Ichikawa, S., Imel, E.A., Kreiter, M.L., Yu, X., Mackenzie, D.S., Sorenson, A.H., Goetz, R., Mohammadi, M., White, K.E., Econs, M.J., 2007. A homozygous missense mutation in human KLOTHO causes severe tumoral calcinosis. J. Clin. Investig. 117, 2684–2691. https://doi.org/10.1172/JCI31330.

Ichikawa, S., Tuchman, S., Padgett, L.R., Gray, A.K., Baluarte, H.J., Econs, M.J., 2014. Intronic deletions in the SLC34A3 gene: a cautionary tale for mutation analysis of hereditary hypophosphatemic rickets with hypercalciuria. Bone 59, 53–56. https://doi.org/10.1016/j.bone.2013.10.018 pii:S8756-3282(13)00428-6.

Imel, E.A., Hui, S.L., Econs, M.J., 2007. FGF23 concentrations vary with disease status in autosomal dominant hypophosphatemic rickets. J. Bone Miner. Res. 22, 520–526. https://doi.org/10.1359/jbmr.070107.

Imel, E.A., Peacock, M., Gray, A.K., Padgett, L.R., Hui, S.L., Econs, M.J., 2011. Iron modifies plasma FGF23 differently in autosomal dominant hypophosphatemic rickets and healthy humans. J. Clin. Endocrinol. Metab. 96, 3541–3549. https://doi.org/10.1210/jc.2011-1239.

Imel, E., Whyte, M., Munns, C., Portale, A., Ward, L., Nilsson, O., Simmons, J., Padidela, R., Namba, N., Cheong, H., Mao, M., Chen, C.-Y., Skrinar, A., San Martin, J., Glorieux, F., 2018. Burosumab improved rickets, phosphate metabolism, and clinical outcomes compared to conventional therapy in children with XLH. J. Bone Miner. Res. 33 (Suppl. 1).

Inoue, M., Digman, M.A., Cheng, M., Breusegem, S.Y., Halaihel, N., Sorribas, V., Mantulin, W.W., Gratton, E., Barry, N.P., Levi, M., 2004. Partitioning of NaPi cotransporter in cholesterol-, sphingomyelin-, and glycosphingolipid-enriched membrane domains modulates NaPi protein diffusion, clustering, and activity. J. Biol. Chem. 279, 49160–49171. https://doi.org/10.1074/jbc.M408942200.

Insogna, K., Rauch, F., Kamenický, P., Ito, N., Takuo Kubota, T., Nakamura, A., Zhang, L., Mealiffe, M., San Martin, J., Portale, A., 2018. The effect of burosumab (KRN23), a fully human anti-FGF23 monoclonal antibody, on osteomalacia in adults with X-linked hypophosphatemia (XLH). J. Bone Miner. Res. 33 (Suppl. 1).

Insogna, K.L., Briot, K., Imel, E.A., Kamenicky, P., Ruppe, M.D., Portale, A.A., Weber, T., Pitukcheewanont, P., Cheong, H.I., Jan de Beur, S., Imanishi, Y., Ito, N., Lachmann, R.H., Tanaka, H., Perwad, F., Zhang, L., Chen, C.Y., Theodore-Oklota, C., Mealiffe, M., San Martin, J., Carpenter, T.O., Investigators, A., 2018. A randomized, double-blind, placebo-controlled, phase 3 trial evaluating the efficacy of burosumab, an anti-FGF23 antibody, in adults with X-linked hypophosphatemia: week 24 primary analysis. J. Bone Miner. Res. 33, 1383–1393. https://doi.org/10.1002/jbmr.3475.

Jan de Beur, S.M., Finnegan, R.B., Vassiliadis, J., Cook, B., Barberio, D., Estes, S., Manavalan, P., Petroziello, J., Madden, S.L., Cho, J.Y., Kumar, R., Levine, M.A., Schiavi, S.C., 2002. Tumors associated with oncogenic osteomalacia express genes important in bone and mineral metabolism. J. Bone Miner. Res. 17, 1102–1110. https://doi.org/10.1359/jbmr.2002.17.6.1102.

Jan de Beur, S., Miller, P.D., Weber, T.J., Peacock, M., Insogna, K.L., Kumar, R., Rauch, F., Luca, D., Theodore-Oklota, C., Lampl, K., San Martin, J., Carpenter, T.O., 2018. Burosumab improved serum phosphorus, osteomalacia, mobility, and fatigue in the 48-week, phase 2 study in adults with tumor-induced osteomalacia syndrome, 2018 J Bone Min Res J. Bone Miner. Res. 33 (Suppl. 1). available at: http://www.asbmr.org/ItineraryBuilder/PresentationDetail.aspx?pid=d311f659-4f63-4aad-9cd4-6f548bc700d3&ptag=WebItinerarySearch.

Jaureguiberry, G., Carpenter, T.O., Forman, S., Juppner, H., Bergwitz, C., 2008. A novel missense mutation in SLC34A3 that causes hereditary hypophosphatemic rickets with hypercalciuria in humans identifies threonine 137 as an important determinant of sodium-phosphate cotransport in NaPi-IIc. Am. J. Physiol. Renal. Physiol. 295, F371–F379. https://doi.org/10.1152/ajprenal.00090.2008.

Jensen, N., Autzen, J.K., Pedersen, L., 2016. Slc20a2 is critical for maintaining a physiologic inorganic phosphate level in cerebrospinal fluid. Neurogenetics 17, 125–130. https://doi.org/10.1007/s10048-015-0469-6.

Jones, A., Tzenova, J., Frappier, D., Crumley, M., Roslin, N., Kos, C., Tieder, M., Langman, C., Proesmans, W., Carpenter, T., Rice, A., Anderson, D., Morgan, K., Fujiwara, T., Tenenhouse, H., 2001. Hereditary hypophosphatemic rickets with hypercalciuria is not caused by mutations in the Na/Pi cotransporter NPT2 gene. J. Am. Soc. Nephrol. 12, 507–514.

Jonsson, K.B., Zahradnik, R., Larsson, T., White, K.E., Sugimoto, T., Imanishi, Y., Yamamoto, T., Hampson, G., Koshiyama, H., Ljunggren, O., Oba, K., Yang, I.M., Miyauchi, A., Econs, M.J., Lavigne, J., Juppner, H., 2003. Fibroblast growth factor 23 in oncogenic osteomalacia and X-linked hypophosphatemia. N. Engl. J. Med. 348, 1656–1663. https://doi.org/10.1056/NEJMoa020881.

Jost, J., Bahans, C., Courbebaisse, M., Tran, T.A., Linglart, A., Benistan, K., Lienhardt, A., Mutar, H., Pfender, E., Ratsimbazafy, V., Guigonis, V., 2016. Topical sodium thiosulfate: a treatment for calcifications in hyperphosphatemic familial tumoral calcinosis? J. Clin. Endocrinol. Metab. 101, 2810–2815. https://doi.org/10.1210/jc.2016-1087.

Juppner, H., Schipani, E., Bastepe, M., Cole, D.E., Lawson, M.L., Mannstadt, M., Hendy, G.N., Plotkin, H., Koshiyama, H., Koh, T., Crawford, J.D., Olsen, B.R., Vikkula, M., 1998. The gene responsible for pseudohypoparathyroidism type Ib is paternally imprinted and maps in four unrelated kindreds to chromosome 20q13.3. Proc. Natl. Acad. Sci. U. S. A. 95, 11798–11803.

Kamiya, N., Yamaguchi, R., Aruwajoye, O., Kim, A.J., Kuroyanagi, G., Phipps, M., Adapala, N.S., Feng, J.Q., Kim, H.K., 2017. Targeted disruption of NF1 in osteocytes increases FGF23 and osteoid with osteomalacia-like bone phenotype. J. Bone Miner. Res. 32, 1716–1726. https://doi.org/10.1002/jbmr.3155.

Kapelari, K., Kohle, J., Kotzot, D., Hogler, W., 2015. Iron supplementation associated with loss of phenotype in autosomal dominant hypophosphatemic rickets. J. Clin. Endocrinol. Metab. 100, 3388–3392. https://doi.org/10.1210/jc.2015-2391.

Karaplis, A.C., Bai, X., Falet, J.P., Macica, C.M., 2012. Mineralizing enthesopathy is a common feature of renal phosphate-wasting disorders attributed to FGF23 and is exacerbated by standard therapy in hyp mice. Endocrinology 153, 5906–5917. https://doi.org/10.1210/en.2012-1551.

Karim, Z., Gerard, B., Bakouh, N., Alili, R., Leroy, C., Beck, L., Silve, C., Planelles, G., Urena-Torres, P., Grandchamp, B., Friedlander, G., Prie, D., 2008. NHERF1 mutations and responsiveness of renal parathyroid hormone. N. Engl. J. Med. 359, 1128–1135.

Karperien, M.C., van der Harten, H.J., van Schooten, R., Farih-Sips, H., den Hollander, N.S., Kneppers, A.L.J., Nijweide, P., Papapoulos, S.E., Löwik, C.W., 1999. A frame-shift mutation in the type I parathyroid hormone/parathyroid hormone-related peptide receptor causing Blomstrand lethal osteochondrodysplasia. J. Clin. Endocrinol. Metab. 84, 3713–3720.

Kavanaugh, M.P., Kabat, D., 1996. Identification and characterization of a widely expressed phosphate transporter/retrovirus receptor family. Kidney Int. 49, 959–963.

Keller, A., Westenberger, A., Sobrido, M.J., García-Murias, M., Domingo, A., Sears, R.L., Lemos, R.R., Ordoñez-Ugalde, A., Nicolas, G., da Cunha, J.E.G., Rushing, E.J., Hugelshofer, M., Wurnig, M.C., Kaech, A., Reimann, R., Lohmann, K., Dobričić, V., Carracedo, A., Petrović, I., Miyasaki, J.M., Abakumova, I., Mäe, M.A., Raschperger, E., Zatz, M., Zschiedrich, K., Klepper, J., Spiteri, E., Prieto, J.M., Navas, I., Preuss, M., Dering, C., Janković, M., Paucar, M., Svenningsson, P., Saliminejad, K., Khorshid, H.R.K., Novaković, I., Aguzzi, A., Boss, A., Le Ber, I., Defer, G., Hannequin, D., Kostić, V.S., Campion, D., Geschwind, D.H., Coppola, G., Betsholtz, C., Klein, C., Oliveira, J.R.M., 2013. Mutations in the gene encoding PDGF-B cause brain calcifications in humans and mice. Nat. Genet. 45, 1077. https://doi.org/10.1038/ng.2723. https://www.nature.com/articles/ng.2723#supplementary-information.

Kemp, G.J., Land, J.M., Coppack, S.W., Frayn, K.N., 1993. Skeletal muscle phosphate uptake during euglycemic-hyperinsulinemic clamp. Clin. Chem. 39, 170–171.

Kestenbaum, B., Glazer, N.L., Kottgen, A., Felix, J.F., Hwang, S.J., Liu, Y., Lohman, K., Kritchevsky, S.B., Hausman, D.B., Petersen, A.K., Gieger, C., Ried, J.S., Meitinger, T., Strom, T.M., Wichmann, H.E., Campbell, H., Hayward, C., Rudan, I., de Boer, I.H., Psaty, B.M., Rice, K.M., Chen, Y.D., Li, M., Arking, D.E., Boerwinkle, E., Coresh, J., Yang, Q., Levy, D., van Rooij, F.J., Dehghan, A., Rivadeneira, F., Uitterlinden, A.G., Hofman, A., van Duijn, C.M., Shlipak, M.G., Kao, W.H., Witteman, J.C., Siscovick, D.S., Fox, C.S., 2010. Common genetic variants associate with serum phosphorus concentration. J. Am. Soc. Nephrol. 21, 1223–1232. https://doi.org/10.1681/ASN.2009111104.

Keusch, I., Traebert, M., Lotscher, M., Kaissling, B., Murer, H., Biber, J., 1998. Parathyroid hormone and dietary phosphate provoke a lysosomal routing of the proximal tubular Na/Pi-cotransporter type II. Kidney Int. 54, 1224–1232. https://doi.org/10.1046/j.1523-1755.1998.00115.x pii:S0085-2538(15)30744-4.

Keushkerian, S., Wade, T., 1984. Hypophosphatemia after major hepatic resection. Curr. Surg. 41, 12–14.

Kim, H.J., Delaney, J.D., Kirsch, T., 2010. The role of pyrophosphate/phosphate homeostasis in terminal differentiation and apoptosis of growth plate chondrocytes. Bone 47, 657–665. https://doi.org/10.1016/j.bone.2010.06.018.

Kitanaka, S., Takeyama, K., Murayama, A., Sato, T., Okumura, K., Nogami, M., Hasegawa, Y., Niimi, H., Yanagisawa, J., Tanaka, T., Kato, S., 1998. Inactivating mutations in the 25-hydroxyvitamin D3 1alpha-hydroxylase gene in patients with pseudovitamin D-deficiency rickets. N. Engl. J. Med. 338, 653–661.

Knöpfel, T., Pastor-Arroyo, E.M., Schnitzbauer, U., Kratschmar, D.V., Odermatt, A., Pellegrini, G., Hernando, N., Wagner, C.A., 2017. The intestinal phosphate transporter NaPi-IIb (Slc34a2) is required to protect bone during dietary phosphate restriction. Sci. Rep. 7, 11018. https://doi.org/10.1038/s41598-017-10390-2.

Koldobskiy, M.A., Chakraborty, A., Werner, J.K., Snowman, A.M., Juluri, K.R., Vandiver, M.S., Kim, S., Heletz, S., Snyder, S.H., 2010. p53-mediated apoptosis requires inositol hexakisphosphate kinase-2. Proc. Natl. Acad. Sci. Unit. States Am. 107, 20947–20951. https://doi.org/10.1073/pnas.1015671107.

Kranenburg, G., de Jong, P.A., Mali, W.P., Attrach, M., Visseren, F.L., Spiering, W., 2017. Prevalence and severity of arterial calcifications in pseudoxanthoma elasticum (PXE) compared to hospital controls. Novel insights into the vascular phenotype of PXE. Atherosclerosis 256, 7–14. https://doi.org/10.1016/j.atherosclerosis.2016.11.012.

Kremke, B., Bergwitz, C., Ahrens, W., Schutt, S., Schumacher, M., Wagner, V., Holterhus, P.M., Juppner, H., Hiort, O., 2009. Hypophosphatemic rickets with hypercalciuria due to mutation in SLC34A3/NaPi-IIc can be masked by vitamin D deficiency and can be associated with renal calcifications. Exp. Clin. Endocrinol. Diabetes 117, 49–56. https://doi.org/10.1055/s-2008-1076716.

Kuro-o, M., 2006. Klotho as a regulator of fibroblast growth factor signaling and phosphate/calcium metabolism. Curr. Opin. Nephrol. Hypertens. 15, 437–441.

Kurosu, H., Kuro-o, M., 2008. The Klotho gene family and the endocrine fibroblast growth factors. Curr. Opin. Nephrol. Hypertens. 17, 368–372.

Kwong, J.Q., Davis, J., Baines, C.P., Sargent, M.A., Karch, J., Wang, X., Huang, T., Molkentin, J.D., 2014. Genetic deletion of the mitochondrial phosphate carrier desensitizes the mitochondrial permeability transition pore and causes cardiomyopathy. Cell Death Differ. 21, 1209–1217. https://doi.org/10.1038/cdd.2014.36.

Lanske, B., Razzaque, M.S., 2014. Molecular interactions of FGF23 and PTH in phosphate regulation. Kidney Int. 86, 1072–1074. https://doi.org/10.1038/ki.2014.316.

Larsson, T., Davis, S.I., Garringer, H.J., Mooney, S.D., Draman, M.S., Cullen, M.J., White, K.E., 2005. Fibroblast growth factor-23 mutants causing familial tumoral calcinosis are differentially processed. Endocrinology 146, 3883–3891.

Lee, G.J., Marks, J., 2015. Intestinal phosphate transport: a therapeutic target in chronic kidney disease and beyond? Pediatr. Nephrol. 30, 363–371. https://doi.org/10.1007/s00467-014-2759-x.

Lee, J.C., Jeng, Y.M., Su, S.Y., Wu, C.T., Tsai, K.S., Lee, C.H., Lin, C.Y., Carter, J.M., Huang, J.W., Chen, S.H., Shih, S.R., Marino-Enriquez, A., Chen, C.C., Folpe, A.L., Chang, Y.L., Liang, C.W., 2015. Identification of a novel FN1-FGFR1 genetic fusion as a frequent event in phosphaturic mesenchymal tumour. J. Pathol. 235, 539–545. https://doi.org/10.1002/path.4465.

Lee, G.S., Brownstein, C.M., Carpenter, T.O., 2018. In: Case Report of a Patient with Hyperphosphatemia and a Novel Mutation in GALNT3Annual Meeting of the Endocrine Society. Chicago.

Legati, A., Giovannini, D., Nicolas, G., Lopez-Sanchez, U., Quintans, B., Oliveira, J.R., Sears, R.L., Ramos, E.M., Spiteri, E., Sobrido, M.J., Carracedo, A., Castro-Fernandez, C., Cubizolle, S., Fogel, B.L., Goizet, C., Jen, J.C., Kirdlarp, S., Lang, A.E., Miedzybrodzka, Z., Mitarnun, W., Paucar, M., Paulson, H., Pariente, J., Richard, A.C., Salins, N.S., Simpson, S.A., Striano, P., Svenningsson, P., Tison, F., Unni, V.K., Vanakker, O., Wessels, M.W., Wetchaphanphesat, S., Yang, M., Boller, F., Campion, D., Hannequin, D., Sitbon, M., Geschwind, D.H., Battini, J.L., Coppola, G., 2015. Mutations in XPR1 cause primary familial brain calcification associated with altered phosphate export. Nat. Genet. 47, 579–581. https://doi.org/10.1038/ng.3289.

Lemos, R.R., Ramos, E.M., Legati, A., Nicolas, G., Jenkinson, E.M., Livingston, J.H., Crow, Y.J., Campion, D., Coppola, G., Oliveira, J.R.M., 2015. Update and mutational analysis of SLC20A2: a major cause of primary familial brain calcification. Hum. Mutat. 36, 489–495. https://doi.org/10.1002/humu.22778.

Levy-Litan, V., Hershkovitz, E., Avizov, L., Leventhal, N., Bercovich, D., Chalifa-Caspi, V., Manor, E., Buriakovsky, S., Hadad, Y., Goding, J., Parvari, R., 2010. Autosomal-recessive hypophosphatemic rickets is associated with an inactivation mutation in the ENPP1 gene. Am. J. Hum. Genet. 86, 273–278. https://doi.org/10.1016/j.ajhg.2010.01.010.

Li, Y., Caballero, D., Ponsetto, J., Chen, A., Zhu, C., Guo, J., Demay, M., Jüppner, H., Bergwitz, C., 2017. Response of Npt2a knockout mice to dietary calcium and phosphorus. PLoS One 12, e0176232. https://doi.org/10.1371/journal.pone.0176232.

Liamis, G., Elisaf, M., 2000. Hypokalemia, hypophosphatemia and hypouricemia due to proximal renal tubular dysfunction in acute myeloid leukemia. Eur. J. Haematol. 64, 277–278.

Liang, G., Katz, L.D., Insogna, K.L., Carpenter, T.O., Macica, C.M., 2009. Survey of the enthesopathy of X-linked hypophosphatemia and its characterization in Hyp mice. Calcif. Tissue Int. 85, 235–246. https://doi.org/10.1007/s00223-009-9270-6.

Lim, Y.H., Ovejero, D., Sugarman, J.S., Deklotz, C.M., Maruri, A., Eichenfield, L.F., Kelley, P.K., Juppner, H., Gottschalk, M., Tifft, C.J., Gafni, R.I., Boyce, A.M., Cowen, E.W., Bhattacharyya, N., Guthrie, L.C., Gahl, W.A., Golas, G., Loring, E.C., Overton, J.D., Mane, S.M., Lifton, R.P., Levy, M.L., Collins, M.T., Choate, K.A., 2014. Multilineage somatic activating mutations in HRAS and NRAS cause mosaic cutaneous and skeletal lesions, elevated FGF23 and hypophosphatemia. Hum. Mol. Genet. 23, 397–407. https://doi.org/10.1093/hmg/ddt429.

Liu, S., Quarles, L.D., 2007. How fibroblast growth factor 23 works. J. Am. Soc. Nephrol. 18, 1637–1647.

Liu, E.S., Zalutskaya, A., Chae, B.T., Zhu, E.D., Gori, F., Demay, M.B., 2014. Phosphate interacts with PTHrP to regulate endochondral bone formation. Endocrinology 155, 3750–3756. https://doi.org/10.1210/en.2014-1315.

Liu, C., Zhou, N., Wang, Y., Zhang, H., Jani, P., Wang, X., Lu, Y., Li, N., Xiao, J., Qin, C., 2018. Abrogation of Fam20c altered cell behaviors and BMP signaling of immortalized dental mesenchymal cells. Exp. Cell Res. 363, 188–195. https://doi.org/10.1016/j.yexcr.2018.01.004.

Lloyd, S.E., Pearce, S.H., Fisher, S.E., Steinmeyer, K., Schwappach, B., Scheinman, S.J., Harding, B., Bolino, A., Devoto, M., Goodyer, P., Rigden, S.P., Wrong, O., Jentsch, T.J., Craig, I.W., Thakker, R.V., 1996. A common molecular basis for three inherited kidney stone diseases. Nature 379, 445–449. https://doi.org/10.1038/379445a0.

Lorenz-Depiereux, B., Bastepe, M., Benet-Pages, A., Amyere, M., Wagenstaller, J., Muller-Barth, U., Badenhoop, K., Kaiser, S.M., Rittmaster, R.S., Shlossberg, A.H., Olivares, J.L., Loris, C., Ramos, F.J., Glorieux, F., Vikkula, M., Juppner, H., Strom, T.M., 2006a. DMP1 mutations in autosomal recessive hypophosphatemia implicate a bone matrix protein in the regulation of phosphate homeostasis. Nat. Genet. 38, 1248–1250. https://doi.org/10.1038/ng1868.

Lorenz-Depiereux, B., Benet-Pages, A., Eckstein, G., Tenenbaum-Rakover, Y., Wagenstaller, J., Tiosano, D., Gershoni-Baruch, R., Albers, N., Lichtner, P., Schnabel, D., Hochberg, Z., Strom, T.M., 2006b. Hereditary hypophosphatemic rickets with hypercalciuria is caused by mutations in the sodium-phosphate cotransporter gene SLC34A3. Am. J. Hum. Genet. 78, 193–201. https://doi.org/10.1086/499410 pii:S0002-9297(07)62352-0.

Lorenz-Depiereux, B., Schnabel, D., Tiosano, D., Hausler, G., Strom, T.M., 2010. Loss-of-function ENPP1 mutations cause both generalized arterial calcification of infancy and autosomal-recessive hypophosphatemic rickets. Am. J. Hum. Genet. 86, 267–272. https://doi.org/10.1016/j.ajhg.2010.01.006.

Ma, S.L., Vega-Warner, V., Gillies, C., Sampson, M.G., Kher, V., Sethi, S.K., Otto, E.A., 2015. Whole exome sequencing reveals novel PHEX splice site mutations in patients with hypophosphatemic rickets. PLoS One 10, e0130729. https://doi.org/10.1371/journal.pone.0130729 pii:PONE-D-15-04157.

Magen, D., Adler, L., Mandel, H., Efrati, E., Zelikovic, I., 2004. Autosomal recessive renal proximal tubulopathy and hypercalciuria: a new syndrome. Am. J. Kidney Dis. 43, 600–606.

Magen, D., Berger, L., Coady, M.J., Ilivitzki, A., Militianu, D., Tieder, M., Selig, S., Lapointe, J.Y., Zelikovic, I., Skorecki, K., 2010. A loss-of-function mutation in NaPi-IIa and renal Fanconi's syndrome. N. Engl. J. Med. 362, 1102–1109.

Maldonado, E.N., Lemasters, J.J., 2014. ATP/ADP ratio, the missed connection between mitochondria and the Warburg effect. Mitochondrion 19 (Pt A), 78–84. https://doi.org/10.1016/j.mito.2014.09.002.

Mannstadt, M., Magen, D., Segawa, H., Stanley, T., Sharma, A., Sasaki, S., Bergwitz, C., Mounien, L., Boepple, P., Thorens, B., Zelikovic, I., Juppner, H., 2012. Fanconi-Bickel syndrome and autosomal recessive proximal tubulopathy with hypercalciuria (ARPTH) are allelic variants caused by GLUT2 mutations. J. Clin. Endocrinol. Metab. 97, E1978–E1986. https://doi.org/10.1210/jc.2012-1279.

Mayr, J.A., Zimmermann, F.A., Horvath, R., Schneider, H.C., Schoser, B., Holinski-Feder, E., Czermin, B., Freisinger, P., Sperl, W., 2011. Deficiency of the mitochondrial phosphate carrier presenting as myopathy and cardiomyopathy in a family with three affected children. Neuromuscul. Disord. 21, 803–808. https://doi.org/10.1016/j.nmd.2011.06.005.

Mejia-Gaviria, N., Gil-Pena, H., Coto, E., Perez-Menendez, T.M., Santos, F., 2010. Genetic and clinical peculiarities in a new family with hereditary hypophosphatemic rickets with hypercalciuria: a case report. Orphanet J. Rare Dis. 5, 1. https://doi.org/10.1186/1750-1172-5-1.

Meyer Jr., R.A., Meyer, M.H., Gray, R.W., 1989. Parabiosis suggests a humoral factor is involved in X-linked hypophosphatemia in mice. J. Bone Miner. Res. 4, 493–500. https://doi.org/10.1002/jbmr.5650040407.

Milionis, H., Pritsivelis, N., Elisaf, M., 1999. Marked hypophosphatemia in a patient with acute leukemia. Nephron 83, 173.

Miyamoto, K., Segawa, H., Ito, M., Kuwahata, M., 2004. Physiological regulation of renal sodium-dependent phosphate cotransporters. Jpn. J. Physiol. 54, 93–102.

Miyamoto, K., Ito, M., Tatsumi, S., Kuwahata, M., Segawa, H., 2007. New aspect of renal phosphate reabsorption: the type IIc sodium-dependent phosphate transporter. Am. J. Nephrol. 27, 503–515. https://doi.org/10.1159/000107069.

Morgan, J.M., Hawley, W.L., Chenoweth, A.I., Retan, W.J., Diethelm, A.G., 1974. Renal transplantation in hypophosphatemia with vitamin D-resistant rickets. Arch. Intern. Med. 134, 549–552.

Murali, S.K., Andrukhova, O., Clinkenbeard, E.L., White, K.E., Erben, R.G., 2016. Excessive osteocytic Fgf23 secretion contributes to pyrophosphate accumulation and mineralization defect in hyp mice. PLoS Biol. 14, e1002427. https://doi.org/10.1371/journal.pbio.1002427.

Murer, H., Hernando, N., Forster, I., Biber, J., 2000. Proximal tubular phosphate reabsorption: molecular mechanisms. Physiol. Rev. 80, 1373–1409. https://doi.org/10.1152/physrev.2000.80.4.1373.

Murer, H., Forster, I., Biber, J., 2004. The sodium phosphate cotransporter family SLC34. Pflügers Archiv 447, 763–767.

Myakala, K., Motta, S., Murer, H., Wagner, C.A., Koesters, R., Biber, J., Hernando, N., 2014. Renal-specific and inducible depletion of NaPi-IIc/Slc34A3, the cotransporter mutated in HHRH, does not affect phosphate or calcium homeostasis in mice. Am. J. Physiol. Renal. Physiol. https://doi.org/10.1152/ajprenal.00133.2013.

Nafidi, O., Lepage, R., Lapointe, R.W., D'Amour, P., 2007. Hepatic resection-related hypophosphatemia is of renal origin as manifested by isolated hyperphosphaturia. Ann. Surg. 245, 1000–1002.

Nakahama, H., Nakanishi, T., Uno, H., Takaoka, T., Taji, N., Uyama, O., Kitada, O., Sugita, M., Miyauchi, A., Sugishita, T., et al., 1995. Prostate cancer-induced oncogenic hypophosphatemic osteomalacia. Urol. Int. 55, 38–40. https://doi.org/10.1159/000282746.

Narchi, H., El Jamil, M., Kulaylat, N., 2001. Symptomatic rickets in adolescence. Arch. Dis. Child. 84, 501–503.

Narvaez, J., Domingo-Domenech, E., Narvaez, J.A., Nolla, J.M., Valverde, J., 2005. Acquired hypophosphatemic osteomalacia associated with multiple myeloma. Joint Bone Spine 72, 424–426. https://doi.org/10.1016/j.jbspin.2004.10.012.

Nishimura, M., Naito, S., 2008. Tissue-specific mRNA expression profiles of human solute carrier transporter superfamilies. Drug Metab. Pharmacokinet. 23, 22–44. JST.JSTAGE/dmpk/23.22.

Nurnberg, P., Thiele, H., Chandler, D., Hohne, W., Cunningham, M.L., Ritter, H., Leschik, G., Uhlmann, K., Mischung, C., Harrop, K., Goldblatt, J., Borochowitz, Z.U., Kotzot, D., Westermann, F., Mundlos, S., Braun, H.S., Laing, N., Tinschert, S., 2001. Heterozygous mutations in ANKH, the human ortholog of the mouse progressive ankylosis gene, result in craniometaphyseal dysplasia. Nat. Genet. 28, 37–41. https://doi.org/10.1038/88236.

Ohi, A., Hanabusa, E., Ueda, O., Segawa, H., Horiba, N., Kaneko, I., Kuwahara, S., Mukai, T., Sasaki, S., Tominaga, R., Furutani, J., Aranami, F., Ohtomo, S., Oikawa, Y., Kawase, Y., Wada, N.A., Tachibe, T., Kakefuda, M., Tateishi, H., Matsumoto, K., Tatsumi, S., Kido, S., Fukushima, N., Jishage, K., Miyamoto, K., 2011. Inorganic phosphate homeostasis in sodium-dependent phosphate cotransporter Npt2b(+)/(-) mice. Am. J. Physiol. Renal. Physiol. 301, F1105–F1113. https://doi.org/10.1152/ajprenal.00663.2010.

Ohkido, I., Segawa, H., Yanagida, R., Nakamura, M., Miyamoto, K., 2003. Cloning, gene structure and dietary regulation of the type-IIc Na/Pi cotransporter in the mouse kidney. Pflügers Archiv 446, 106–115.

Ohkido, I., Hara, S., Segawa, H., Yokoyama, K., Yamamoto, H., Miyamoto, K., Kawaguchi, H., Hosoya, T., 2007. Localization of sodium-phosphate cotransporter NaPi-IIa and -IIc in human proximal renal and distal TubulesRENAL WEEK 2006. J. Am. Soc. Nephrol. San Francisco, pp. abstract SU-PO704.

Ovejero, D., Lim, Y.H., Boyce, A.M., Gafni, R.I., McCarthy, E., Nguyen, T.A., Eichenfield, L.F., DeKlotz, C.M., Guthrie, L.C., Tosi, L.L., Thornton, P.S., Choate, K.A., Collins, M.T., 2016. Cutaneous skeletal hypophosphatemia syndrome: clinical spectrum, natural history, and treatment. Osteoporos. Int. 27, 3615–3626. https://doi.org/10.1007/s00198-016-3702-8.

Page, K., Bergwitz, C., Jaureguiberry, G., Harinarayan, C.V., Insogna, K., 2008. A patient with hypophosphatemia, a femoral fracture, and recurrent kidney stones: report of a novel mutation in SLC34A3. Endocr. Pract. 14, 869–874. https://doi.org/10.4158/EP.14.7.869 pii:K344465V44550341.

Palmer, G., Guicheux, J., Bonjour, J.P., Caverzasio, J., 2000. Transforming growth factor-beta stimulates inorganic phosphate transport and expression of the type III phosphate transporter Glvr-1 in chondrogenic ATDC5 cells. Endocrinology 141, 2236–2243. https://doi.org/10.1210/endo.141.6.7495.

Papaioannou, G., Petit, E.T., Liu, E.S., Baccarini, M., Pritchard, C., Demay, M.B., 2017. Raf kinases are essential for phosphate induction of ERK1/2 phosphorylation in hypertrophic chondrocytes and normal endochondral bone development. J. Biol. Chem. 292, 3164–3171. https://doi.org/10.1074/jbc.M116.763342.

Pattaro, C., Teumer, A., Gorski, M., Chu, A.Y., Li, M., Mijatovic, V., Garnaas, M., Tin, A., Sorice, R., Li, Y., Taliun, D., Olden, M., Foster, M., Yang, Q., Chen, M.H., Pers, T.H., Johnson, A.D., Ko, Y.A., Fuchsberger, C., Tayo, B., Nalls, M., Feitosa, M.F., Isaacs, A., Dehghan, A., d'Adamo, P., Adeyemo, A., Dieffenbach, A.K., Zonderman, A.B., Nolte, I.M., van der Most, P.J., Wright, A.F., Shuldiner, A.R., Morrison, A.C., Hofman, A., Smith, A.V., Dreisbach, A.W., Franke, A., Uitterlinden, A.G., Metspalu, A., Tonjes, A., Lupo, A., Robino, A., Johansson, A., Demirkan, A.,

Kollerits, B., Freedman, B.I., Ponte, B., Oostra, B.A., Paulweber, B., Kramer, B.K., Mitchell, B.D., Buckley, B.M., Peralta, C.A., Hayward, C., Helmer, C., Rotimi, C.N., Shaffer, C.M., Muller, C., Sala, C., van Duijn, C.M., Saint-Pierre, A., Ackermann, D., Shriner, D., Ruggiero, D., Toniolo, D., Lu, Y., Cusi, D., Czamara, D., Ellinghaus, D., Siscovick, D.S., Ruderfer, D., Gieger, C., Grallert, H., Rochtchina, E., Atkinson, E.J., Holliday, E.G., Boerwinkle, E., Salvi, E., Bottinger, E.P., Murgia, F., Rivadeneira, F., Ernst, F., Kronenberg, F., Hu, F.B., Navis, G.J., Curhan, G.C., Ehret, G.B., Homuth, G., Coassin, S., Thun, G.A., Pistis, G., Gambaro, G., Malerba, G., Montgomery, G.W., Eiriksdottir, G., Jacobs, G., Li, G., Wichmann, H.E., Campbell, H., Schmidt, H., Wallaschofski, H., Volzke, H., Brenner, H., Kroemer, H.K., Kramer, H., Lin, H., Mateo Leach, I., Ford, I., Guessous, I., Rudan, I., Prokopenko, I., Borecki, I., Heid, I.M., Kolcic, I., Persico, I., Jukema, J.W., Wilson, J.F., Felix, J.F., Divers, J., Lambert, J.C., Stafford, J.M., Gaspoz, J.M., Smith, J.A., Faul, J.D., Wang, J.J., Ding, J., Hirschhorn, J.N., Attia, J., Whitfield, J.B., Chalmers, J., Viikari, J., Coresh, J., Denny, J.C., Karjalainen, J., Fernandes, J.K., Endlich, K., Butterbach, K., Keene, K.L., Lohman, K., Portas, L., Launer, L.J., Lyytikainen, L.P., Yengo, L., Franke, L., Ferrucci, L., Rose, L.M., Kedenko, L., Rao, M., Struchalin, M., Kleber, M.E., Cavalieri, M., Haun, M., Cornelis, M.C., Ciullo, M., Pirastu, M., de Andrade, M., McEvoy, M.A., Woodward, M., Adam, M., Cocca, M., Nauck, M., Imboden, M., Waldenberger, M., Pruijm, M., Metzger, M., Stumvoll, M., Evans, M.K., Sale, M.M., Kahonen, M., Boban, M., Bochud, M., Rheinberger, M., Verweij, N., Bouatia-Naji, N., Martin, N.G., Hastie, N., Probst-Hensch, N., Soranzo, N., Devuyst, O., Raitakari, O., Gottesman, O., Franco, O.H., Polasek, O., Gasparini, P., Munroe, P.B., Ridker, P.M., Mitchell, P., Muntner, P., Meisinger, C., Smit, J.H., Consortium I, Consortium A, Cardiogram, Group CH-HF, Consortium EC, Kovacs, P., Wild, P.S., Froguel, P., Rettig, R., Magi, R., Biffar, R., Schmidt, R., Middelberg, R.P., Carroll, R.J., Penninx, B.W., Scott, R.J., Katz, R., Sedaghat, S., Wild, S.H., Kardia, S.L., Ulivi, S., Hwang, S.J., Enroth, S., Kloiber, S., Trompet, S., Stengel, B., Hancock, S.J., Turner, S.T., Rosas, S.E., Stracke, S., Harris, T.B., Zeller, T., Zemunik, T., Lehtimaki, T., Illig, T., Aspelund, T., Nikopensius, T., Esko, T., Tanaka, T., Gyllensten, U., Volker, U., Emilsson, V., Vitart, V., Aalto, V., Gudnason, V., Chouraki, V., Chen, W.M., Igl, W., Marz, W., Koenig, W., Lieb, W., Loos, R.J., Liu, Y., Snieder, H., Pramstaller, P.P., Parsa, A., O'Connell, J.R., Susztak, K., Hamet, P., Tremblay, J., de Boer, I.H., Boger, C.A., Goessling, W., Chasman, D.I., Kottgen, A., Kao, W.H., Fox, C.S., 2016. Genetic associations at 53 loci highlight cell types and biological pathways relevant for kidney function. Nat. Commun. 7, 10023. https://doi.org/10.1038/ncomms10023.

Pauleau, A.L., Galluzzi, L., Scholz, S.R., Larochette, N., Kepp, O., Kroemer, G., 2008. Unexpected role of the phosphate carrier in mitochondrial fragmentation. Cell Death Differ. 15, 616. https://doi.org/10.1038/sj.cdd.4402295. https://www.nature.com/articles/4402295#supplementary-information.

Pendleton, A., Johnson, M.D., Hughes, A., Gurley, K.A., Ho, A.M., Doherty, M., Dixey, J., Gillet, P., Loeuille, D., McGrath, R., Reginato, A., Shiang, R., Wright, G., Netter, P., Williams, C., Kingsley, D.M., 2002. Mutations in ANKH cause chondrocalcinosis. Am. J. Hum. Genet. 71, 933–940. https://doi.org/10.1086/343054.

Pesta, D.H., Tsirigotis, D.N., Befroy, D.E., Caballero, D., Jurczak, M.J., Rahimi, Y., Cline, G.W., Dufour, S., Birkenfeld, A.L., Rothman, D.L., Carpenter, T.O., Insogna, K., Petersen, K.F., Bergwitz, C., Shulman, G.I., 2016. Hypophosphatemia promotes lower rates of muscle ATP synthesis. FASEB J. 30, 3378–3387. https://doi.org/10.1096/fj.201600473R.

Petersen, K.F., Dufour, S., Shulman, G.I., 2005. Decreased insulin-stimulated ATP synthesis and phosphate transport in muscle of insulin-resistant offspring of type 2 diabetic parents. PLoS Med. 2, e233. https://doi.org/10.1371/journal.pmed.0020233. 05-PLME-RA-0003R2.

Pfister, M.F., Ruf, I., Stange, G., Ziegler, U., Lederer, E., Biber, J., Murer, H., 1998. Parathyroid hormone leads to the lysosomal degradation of the renal type II Na/Pi cotransporter. Proc. Natl. Acad. Sci. U. S. A. 95, 1909–1914.

Phulwani, P., Bergwitz, C., Jaureguiberry, G., Rasoulpour, M., Estrada, E., 2011. Hereditary hypophosphatemic rickets with hypercalciuria and nephrolithiasis-identification of a novel SLC34A3/NaPi-IIc mutation. Am. J. Med. Genet. 155A, 626–633. https://doi.org/10.1002/ajmg.a.33832.

Picard, N., Capuano, P., Stange, G., Mihailova, M., Kaissling, B., Murer, H., Biber, J., Wagner, C.A., 2010. Acute parathyroid hormone differentially regulates renal brush border membrane phosphate cotransporters. Pflügers Archiv 460, 677–687. https://doi.org/10.1007/s00424-010-0841-1.

Polgreen, K.E., Kemp, G.J., Leighton, B., Radda, G.K., 1994. Modulation of Pi transport in skeletal muscle by insulin and IGF-1. Biochim. Biophys. Acta 1223, 279–284.

Pollak, M.R., Brown, E.M., WuChou, Y.H., Hebert, S.C., Marx, S.J., Steinmann, B., Levi, T., Seidman, C.E., Seidman, J.G., 1993. Mutations in the human Ca^{2+}-sensing receptor gene cause familial hypocalciuric hypercalcemia and neonatal severe hyperparathyroidism. Cell 75, 1297–1303.

Pollak, M.R., Brown, E.M., Estep, H.L., McLaine, P.N., Kifor, O., Park, J., Hebert, S.C., Seidman, C.E., Seidman, J.G., 1994. Autosomal dominant hypocalcaemia caused by a Ca^{2+}-sensing receptor gene mutation. Nat. Genet. 8, 303–307.

Portale, A.A.I., Karl, L., Briot, K., Imel, E., Kamenicky, P., Weber, T., Pitukcheewanont, P., Cheong, H.I., Jan de Beur, S., Imanishi, Y., Ito, N., Lachmann, R., Tanaka, H., Perwad, F., Zhang, L., Theodore-Oklota, C., Mealiffe, M., San Martin, J., Carpenter, T.O., 2018. Continued improvement in clinical outcomes in the phase 3 randomized, double-blind, placebo-controlled study of burosumab, an anti-FGF23 antibody, in adults with X-linked hypophosphatemia (XLH). J. Bone Miner. Res. 33 (Suppl. 1).

Price, T.B., Perseghin, G., Duleba, A., Chen, W., Chase, J., Rothman, D.L., Shulman, R.G., Shulman, G.I., 1996. NMR studies of muscle glycogen synthesis in insulin-resistant offspring of parents with non-insulin-dependent diabetes mellitus immediately after glycogen-depleting exercise. Proc. Natl. Acad. Sci. U. S. A. 93, 5329–5334.

Prie, D., Huart, V., Bakouh, N., Planelles, G., Dellis, O., Gerard, B., Hulin, P., Benque-Blanchet, F., Silve, C., Grandchamp, B., Friedlander, G., 2002. Nephrolithiasis and osteoporosis associated with hypophosphatemia caused by mutations in the type 2a sodium-phosphate cotransporter. N. Engl. J. Med. 347, 983–991.

Prie, D., Beck, L., Friedlander, G., Silve, C., 2004. Sodium-phosphate cotransporters, nephrolithiasis and bone demineralization. Curr. Opin. Nephrol. Hypertens. 13, 675–681.

Pronicka, E., Ciara, E., Halat, P., Janiec, A., Wojcik, M., Rowinska, E., Rokicki, D., Pludowski, P., Wojciechowska, E., Wierzbicka, A., Ksiazyk, J.B., Jacoszek, A., Konrad, M., Schlingmann, K.P., Litwin, M., 2017. Biallelic mutations in CYP24A1 or SLC34A1 as a cause of infantile idiopathic hypercalcemia (IIH) with vitamin D hypersensitivity: molecular study of 11 historical IIH cases. J. Appl. Genet. 58, 349–353. https://doi.org/10.1007/s13353-017-0397-2.

Rafaelsen, S.H., Raeder, H., Fagerheim, A.K., Knappskog, P., Carpenter, T.O., Johansson, S., Bjerknes, R., 2013. Exome sequencing reveals FAM20c mutations associated with fibroblast growth factor 23-related hypophosphatemia, dental anomalies, and ectopic calcification. J. Bone Miner. Res. 28, 1378–1385. https://doi.org/10.1002/jbmr.1850.

Rafaelsen, S., Johansson, S., Raeder, H., Bjerknes, R., 2016. Hereditary hypophosphatemia in Norway: a retrospective population-based study of genotypes, phenotypes, and treatment complications. Eur. J. Endocrinol. 174, 125–136. https://doi.org/10.1530/EJE-15-0515.

Ramnitz, M.S., Gourh, P., Goldbach-Mansky, R., Wodajo, F., Ichikawa, S., Econs, M.J., White, K.E., Molinolo, A., Chen, M.Y., Heller, T., Del Rivero, J., Seo-Mayer, P., Arabshahi, B., Jackson, M.B., Hatab, S., McCarthy, E., Guthrie, L.C., Brillante, B.A., Gafni, R.I., Collins, M.T., 2016. Phenotypic and genotypic characterization and treatment of a cohort with familial tumoral calcinosis/hyperostosis-hyperphosphatemia syndrome. J. Bone Miner. Res. 31, 1845–1854. https://doi.org/10.1002/jbmr.2870.

Reginato, A.J., Coquia, J.A., 2003. Musculoskeletal manifestations of osteomalacia and rickets. Best Pract. Res. Clin. Rheumatol. 17, 1063–1080.

Ren, Y., Han, X., Jing, Y., Yuan, B., Ke, H., Liu, M., Feng, J.Q., 2016. Sclerostin antibody (Scl-Ab) improves osteomalacia phenotype in dentin matrix protein 1(Dmp1) knockout mice with little impact on serum levels of phosphorus and FGF23. Matrix Biol. 52–54, 151–161. https://doi.org/10.1016/j.matbio.2015.12.009.

Roberts, M.S., Burbelo, P.D., Egli-Spichtig, D., Perwad, F., Romero, C.J., Ichikawa, S., Farrow, E., Econs, M.J., Guthrie, L.C., Collins, M.T., Gafni, R.I., 2018. Autoimmune hyperphosphatemic tumoral calcinosis in a patient with FGF23 autoantibodies. J. Clin. Investig. 128, 5368–5373. https://doi.org/10.1172/JCI122004 122004.

Romero, V., Akpinar, H., Assimos, D.G., 2010. Kidney stones: a global picture of prevalence, incidence, and associated risk factors. Rev. Urol. 12, e86–96.

Rothenbuhler, A., Fadel, N., Debza, Y., Bacchetta, J., Diallo, M.T., Adamsbaum, C., Linglart, A., Di Rocco, F., 2018. High incidence of cranial synostosis and Chiari I malformation in children with X-linked hypophosphatemic rickets (XLHR). J. Bone Miner. Res. https://doi.org/10.1002/jbmr.3614.

Ruppe MD X-linked hypophosphatemia. In: Pagon RA, Adam MP, Ardinger HH, al e (eds) Gene Reviews.

Rutsch, F., Ruf, N., Vaingankar, S., Toliat, M.R., Suk, A., Hohne, W., Schauer, G., Lehmann, M., Roscioli, T., Schnabel, D., Epplen, J.T., Knisely, A., Superti-Furga, A., McGill, J., Filippone, M., Sinaiko, A.R., Vallance, H., Hinrichs, B., Smith, W., Ferre, M., Terkeltaub, R., Nurnberg, P., 2003. Mutations in ENPP1 are associated with 'idiopathic' infantile arterial calcification. Nat. Genet. 34, 379–381. https://doi.org/10.1038/ng1221.

Rutsch, F., Boyer, P., Nitschke, Y., Ruf, N., Lorenz-Depierieux, B., Wittkampf, T., Weissen-Plenz, G., Fischer, R.J., Mughal, Z., Gregory, J.W., Davies, J.H., Loirat, C., Strom, T.M., Schnabel, D., Nurnberg, P., Terkeltaub, R., Group, G.S., 2008. Hypophosphatemia, hyperphosphaturia, and bisphosphonate treatment are associated with survival beyond infancy in generalized arterial calcification of infancy. Circ Cardiovasc Genet 1, 133–140. https://doi.org/10.1161/CIRCGENETICS.108.797704.

Sabbagh, Y., Schiavi, S.C., 2014. Role of NPT2b in health and chronic kidney disease. Curr. Opin. Nephrol. Hypertens. 23, 377–384. https://doi.org/10.1097/01.mnh.0000447015.44099.5f.

Sabbagh, Y., Carpenter, T.O., Demay, M.B., 2005. Hypophosphatemia leads to rickets by impairing caspase-mediated apoptosis of hypertrophic chondrocytes. Proc. Natl. Acad. Sci. U. S. A. 102, 9637–9642. https://doi.org/10.1073/pnas.0502249102.

Sabbagh, Y., O'Brien, S.P., Song, W., Boulanger, J.H., Stockmann, A., Arbeeny, C., Schiavi, S.C., 2009. Intestinal npt2b plays a major role in phosphate absorption and homeostasis. J. Am. Soc. Nephrol. 20, 2348–2358. https://doi.org/10.1681/ASN.2009050559.

Sabbagh, Y., Giral, H., Caldas, Y., Levi, M., Schiavi, S.C., 2011. Intestinal phosphate transport. Adv. Chron. Kidney Dis. 18, 85–90. https://doi.org/10.1053/j.ackd.2010.11.004.

Saiardi, A., 2012. How inositol pyrophosphates control cellular phosphate homeostasis? Adv Biol Regul 52, 351–359. https://doi.org/10.1016/j.jbior.2012.03.002.

Sanders, L., 2018. Bone Basics: she was a runner who fractured her foot . It healed but the pain wouldn't subside. Why? N. Y. Times Sunday Magazine, 20.

Santer, R., Schneppenheim, R., Dombrowski, A., Gotze, H., Steinmann, B., Schaub, J., 1997. Mutations in GLUT2, the gene for the liver-type glucose transporter, in patients with Fanconi-Bickel syndrome. Nat. Genet. 17, 324–326. https://doi.org/10.1038/ng1197-324.

Santer, R., Steinmann, B., Schaub, J., 2002. Fanconi-Bickel syndrome–a congenital defect of facilitative glucose transport. Curr. Mol. Med. 2, 213–227.

Schiavi, S.C., Tang, W., Bracken, C., O'Brien, S.P., Song, W., Boulanger, J., Ryan, S., Phillips, L., Liu, S., Arbeeny, C., Ledbetter, S., Sabbagh, Y., 2012. Npt2b deletion attenuates hyperphosphatemia associated with CKD. J. Am. Soc. Nephrol. 23, 1691–1700. https://doi.org/10.1681/ASN.2011121213.

Schipani, E., Kruse, K., Juppner, H., 1995. A constitutively active mutant PTH-PTHrP receptor in Jansen-type metaphyseal chondrodysplasia. Science 268, 98–100.

Schissel, B.L., Johnson, B.K., 2011. Renal stones: evolving epidemiology and management. Pediatr. Emerg. Care 27, 676–681. https://doi.org/10.1097/PEC.0b013e3182228f10 pii:00006565-201107000-00024.

Schlingmann, K.P., Ruminska, J., Kaufmann, M., Dursun, I., Patti, M., Kranz, B., Pronicka, E., Ciara, E., Akcay, T., Bulus, D., Cornelissen, E.A.M., Gawlik, A., Sikora, P., Patzer, L., Galiano, M., Boyadzhiev, V., Dumic, M., Vivante, A., Kleta, R., Dekel, B., Levtchenko, E., Bindels, R.J., Rust, S., Forster, I.C., Hernando, N., Jones, G., Wagner, C.A., Konrad, M., 2016. Autosomal-recessive mutations in SLC34A1 encoding sodium-phosphate cotransporter 2A cause idiopathic infantile hypercalcemia. J. Am. Soc. Nephrol. 27, 604–614. https://doi.org/10.1681/asn.2014101025.

Segawa, H., Kaneko, I., Takahashi, A., Kuwahata, M., Ito, M., Ohkido, I., Tatsumi, S., Miyamoto, K., 2002. Growth-related renal type II Na/Pi cotransporter. J. Biol. Chem. 277, 19665–19672.

Segawa, H., Onitsuka, A., Furutani, J., Kaneko, I., Aranami, F., Matsumoto, N., Tomoe, Y., Kuwahata, M., Ito, M., Matsumoto, M., Li, M., Amizuka, N., Miyamoto, K., 2009. Npt2a and Npt2c in mice play distinct and synergistic roles in inorganic phosphate metabolism and skeletal development. Am. J. Physiol. Renal. Physiol. 297, F671–F678.

Seifert, E.L., Ligeti, E., Mayr, J.A., Sondheimer, N., Hajnoczky, G., 2015. The mitochondrial phosphate carrier: role in oxidative metabolism, calcium handling and mitochondrial disease. Biochem. Biophys. Res. Commun. 464, 369–375. https://doi.org/10.1016/j.bbrc.2015.06.031.

Shaikh, A., Berndt, T., Kumar, R., 2008. Regulation of phosphate homeostasis by the phosphatonins and other novel mediators. Pediatr. Nephrol. 23, 1203–1210.

Shiba, M., Mizuno, M., Kuraishi, K., Suzuki, H., 2015. Cervical ossification of posterior longitudinal ligament in x-linked hypophosphatemic rickets revealing homogeneously increased vertebral bone density. Asian Spine J 9, 106–109. https://doi.org/10.4184/asj.2015.9.1.106.

Shimada, T., Mizutani, S., Muto, T., Yoneya, T., Hino, R., Takeda, S., Takeuchi, Y., Fujita, T., Fukumoto, S., Yamashita, T., 2001. Cloning and characterization of FGF23 as a causative factor of tumor-induced osteomalacia. Proc. Natl. Acad. Sci. U. S. A. 98, 6500–6505. https://doi.org/10.1073/pnas.101545198.

Shimada, T., Hasegawa, H., Yamazaki, Y., Muto, T., Hino, R., Takeuchi, Y., Fujita, T., Nakahara, K., Fukumoto, S., Yamashita, T., 2004a. FGF-23 is a potent regulator of vitamin D metabolism and phosphate homeostasis. J. Bone Miner. Res. 19, 429–435.

Shimada, T., Kakitani, M., Yamazaki, Y., Hasegawa, H., Takeuchi, Y., Fujita, T., Fukumoto, S., Tomizuka, K., Yamashita, T., 2004b. Targeted ablation of Fgf23 demonstrates an essential physiological role of FGF23 in phosphate and vitamin D metabolism. J. Clin. Investig. 113, 561–568. https://doi.org/10.1172/JCI19081.

Shiozaki, Y., Segawa, H., Ohnishi, S., Ohi, A., Ito, M., Kaneko, I., Kido, S., Tatsumi, S., Miyamoto, K., 2015. Relationship between sodium-dependent phosphate transporter (NaPi-IIc) function and cellular vacuole formation in opossum kidney cells. J. Med. Investig. 62, 209–218. https://doi.org/10.2152/jmi.62.209.

Sinha, A., Hollingsworth, K.G., Ball, S., Cheetham, T., 2013. Improving the vitamin D status of vitamin D deficient adults is associated with improved mitochondrial oxidative function in skeletal muscle. J. Clin. Endocrinol. Metab. 98, E509–E513. https://doi.org/10.1210/jc.2012-3592.

Smith, R., Newman, R.J., Radda, G.K., Stokes, M., Young, A., 1984. Hypophosphataemic osteomalacia and myopathy: studies with nuclear magnetic resonance spectroscopy. Clin. Sci. (Lond.) 67, 505–509.

Strom, T.M., Juppner, H., 2008. PHEX, FGF23, DMP1 and beyond. Curr. Opin. Nephrol. Hypertens. 17, 357–362.

Sullivan, W., Carpenter, T., Glorieux, F., Travers, R., Insogna, K., 1992. A prospective trial of phosphate and 1,25-dihydroxyvitamin D3 therapy in symptomatic adults with X-linked hypophosphatemic rickets. J. Clin. Endocrinol. Metab. 75, 879–885. https://doi.org/10.1210/jcem.75.3.1517380.

Suzuki, A., Palmer, G., Bonjour, J.P., Caverzasio, J., 2000. Stimulation of sodium-dependent phosphate transport and signaling mechanisms induced by basic fibroblast growth factor in MC3T3-E1 osteoblast-like cells. J. Bone Miner. Res. 15, 95–102. https://doi.org/10.1359/jbmr.2000.15.1.95.

Suzuki, A., Palmer, G., Bonjour, J.P., Caverzasio, J., 2001. Stimulation of sodium-dependent inorganic phosphate transport by activation of Gi/o-protein-coupled receptors by epinephrine in MC3T3-E1 osteoblast-like cells. Bone 28, 589–594 pii:S8756328201004598.

Tagliabracci, V.S., Engel, J.L., Wiley, S.E., Xiao, J., Gonzalez, D.J., Nidumanda Appaiah, H., Koller, A., Nizet, V., White, K.E., Dixon, J.E., 2014. Dynamic regulation of FGF23 by Fam20C phosphorylation, GalNAc-T3 glycosylation, and furin proteolysis. Proc. Natl. Acad. Sci. U. S. A. 111, 5520–5525. https://doi.org/10.1073/pnas.1402218111.

Taguchi, K., Yasui, T., Milliner, D.S., Hoppe, B., Chi, T., 2017. Genetic risk factors for idiopathic urolithiasis: a systematic review of the literature and causal Network analysis. Eur Urol Focus 3, 72–81. https://doi.org/10.1016/j.euf.2017.04.010 pii:S2405-4569(17)30116-5.

Tencza, A.L., Ichikawa, S., Dang, A., Kenagy, D., McCarthy, E., Econs, M.J., Levine, M.A., 2009. Hypophosphatemic rickets with hypercalciuria due to mutation in SLC34A3/type IIc sodium-phosphate cotransporter: presentation as hypercalciuria and nephrolithiasis. J. Clin. Endocrinol. Metab. 94, 4433–4438. https://doi.org/10.1210/jc.2009-1535.

Tenenhouse, H.S., Econs, M.J., 2001. Mendelian hypophosphatemias. In: Scriver, C.R., Beaudet, A.L., Valle, D., Sly, W.S., Vogelstein, B., Childs, B., Kinzler, K.W. (Eds.), The Metabolic and Molecular Bases of Inherited Diseases. McGraw-Hill, New York, pp. 5039–5067.

Tenenhouse, H.S., Werner, A., Biber, J., Ma, S., Martel, J., Roy, S., Murer, H., 1994. Renal Na(+)-phosphate cotransport in murine X-linked hypophosphatemic rickets. Molecular characterization. J. Clin. Investig. 93, 671–676. https://doi.org/10.1172/JCI117019.

Tenenhouse, H.S., Roy, S., Martel, J., Gauthier, C., 1998. Differential expression, abundance, and regulation of Na+-phosphate cotransporter genes in murine kidney. Am. J. Physiol. 275, F527–F534.

Tenenhouse, H.S., Martel, J., Gauthier, C., Zhang, M.Y., Portale, A.A., 2001. Renal expression of the sodium/phosphate cotransporter gene, Npt2, is not required for regulation of renal 1 alpha-hydroxylase by phosphate. Endocrinology 142, 1124–1129.

Thakker, R.V., 2000. Pathogenesis of Dent's disease and related syndromes of X-linked nephrolithiasis. Kidney Int. 57, 787–793.

Tieder, M., Modai, D., Samuel, R., Arie, R., Halabe, A., Bab, I., Gabizon, D., Liberman, U.A., 1985. Hereditary hypophosphatemic rickets with hypercalciuria. N. Engl. J. Med. 312, 611–617.

Tieder, M., Modai, D., Shaked, U., Samuel, R., Arie, R., Halabe, A., Maor, J., Weissgarten, J., Averbukh, Z., Cohen, N., et al., 1987. Idiopathic" hypercalciuria and hereditary hypophosphatemic rickets. Two phenotypical expressions of a common genetic defect. N. Engl. J. Med. 316, 125–129.

Tieder, M., Arie, R., Modai, D., Samuel, R., Weissgarten, J., Liberman, U.A., 1988. Elevated serum 1,25-dihydroxyvitamin D concentrations in siblings with primary Fanconi's syndrome. N. Engl. J. Med. 319, 845–849.

Tomoe, Y., Segawa, H., Kaneko, I., Furutani, J., Aranami, F., Kuwahara, S., Tominaga, R., Hanabusa, E., Ito, M., Miyamoto, K-i, 2009. Effect of fibroblast growth factor (FGF)23 on Npt2a-/-, Npt2c-/- double-knockout (WKO). Mice J. Am. Soc. Nephrol. Philadelphia, pp. [SA-PO2780].

Tomoe, Y., Segawa, H., Shiozawa, K., Kaneko, I., Tominaga, R., Hanabusa, E., Aranami, F., Furutani, J., Kuwahara, S., Tatsumi, S., Matsumoto, M., Ito, M., Miyamoto, K., 2010. Phosphaturic action of fibroblast growth factor 23 in Npt2 null mice. Am. J. Physiol. Renal. Physiol. 298, F1341−F1350. https://doi.org/10.1152/ajprenal.00375.2009.

Topaz, O., Shurman, D.L., Bergman, R., Indelman, M., Ratajczak, P., Mizrachi, M., Khamaysi, Z., Behar, D., Petronius, D., Friedman, V., Zelikovic, I., Raimer, S., Metzker, A., Richard, G., Sprecher, E., 2004. Mutations in GALNT3, encoding a protein involved in O-linked glycosylation, cause familial tumoral calcinosis. Nat. Genet. 36, 579. https://doi.org/10.1038/ng1358.

Topaz, O., Indelman, M., Chefetz, I., Geiger, D., Metzker, A., Altschuler, Y., Choder, M., Bercovich, D., Uitto, J., Bergman, R., Richard, G., Sprecher, E., 2006. A deleterious mutation in SAMD9 causes normophosphatemic familial tumoral calcinosis. Am. J. Hum. Genet. 79, 759−764.

Tore, S., Casula, S., Casu, G., Concas, M.P., Pistidda, P., Persico, I., Sassu, A., Maestrale, G.B., Mele, C., Caruso, M.R., Bonerba, B., Usai, P., Deiana, I., Thornton, T., Pirastu, M., Forabosco, P., 2011. Application of a new method for GWAS in a related case/control sample with known pedigree structure: identification of new loci for nephrolithiasis. PLoS Genet. 7, e1001281. https://doi.org/10.1371/journal.pgen.1001281.

Turan, S., Aydin, C., Bereket, A., Akcay, T., Guran, T., Yaralioglu, B.A., Bastepe, M., Juppner, H., 2010. Identification of a novel dentin matrix protein-1 (DMP-1) mutation and dental anomalies in a kindred with autosomal recessive hypophosphatemia. Bone 46, 402−409. https://doi.org/10.1016/j.bone.2009.09.016.

Urakawa, I., Yamazaki, Y., Shimada, T., Iijima, K., Hasegawa, H., Okawa, K., Fujita, T., Fukumoto, S., Yamashita, T., 2006. Klotho converts canonical FGF receptor into a specific receptor for FGF23. Nature 444, 770−774. https://doi.org/10.1038/nature05315.

Vieira, A.R., Lee, M., Vairo, F., Loguercio Leite, J.C., Munerato, M.C., Visioli, F., D'Avila, S.R., Wang, S.K., Choi, M., Simmer, J.P., Hu, J.C., 2015. Root anomalies and dentin dysplasia in autosomal recessive hyperphosphatemic familial tumoral calcinosis (HFTC). Oral Surg. Oral Med. Oral Pathol. Oral Radiol. 120, e235−e239. https://doi.org/10.1016/j.oooo.2015.05.006 pii:S2212-4403(15)00918-9.

Virkki, L.V., Forster, I.C., Hernando, N., Biber, J., Murer, H., 2003. Functional characterization of two naturally occurring mutations in the human sodium-phosphate cotransporter type IIa. J. Bone Miner. Res. 18, 2135−2141. https://doi.org/10.1359/jbmr.2003.18.12.2135.

von Falck, C., Rodt, T., Rosenthal, H., Langer, F., Goesling, T., Knapp, W.H., Galanski, M., 2008. (68)Ga-DOTANOC PET/CT for the detection of a mesenchymal tumor causing oncogenic osteomalacia. Eur. J. Nucl. Med. Mol. Imaging 35, 1034. https://doi.org/10.1007/s00259-008-0755-8.

Wagner, C.A., Rubio-Aliaga, I., Hernando, N., 2017. Renal phosphate handling and inherited disorders of phosphate reabsorption: an update. Pediatr. Nephrol. https://doi.org/10.1007/s00467-017-3873-3.

Walton, R.J., Bijvoet, O.L., 1975. Nomogram for derivation of renal threshold phosphate concentration. Lancet 2, 309−310.

Wang, D., Canaff, L., Davidson, D., Corluka, A., Liu, H., Hendy, G.N., Henderson, J.E., 2001. Alterations in the sensing and transport of phosphate and calcium by differentiating chondrocytes. J. Biol. Chem. 276, 33995−34005. https://doi.org/10.1074/jbc.M007757200.

Weinstein, L.S., Gejman, P.V., de Mazancourt, P., American, N., Spiegel, A.M., 1992. A heterozygous 4-bp deletion mutation in the Gs alpha gene (GNAS1) in a patient with Albright hereditary osteodystrophy. Genomics 13, 1319−1321.

White, K.E., Cabral, J.M., Davis, S.I., Fishburn, T., Evans, W.E., Ichikawa, S., Fields, J., Yu, X., Shaw, N.J., McLellan, N.J., McKeown, C., Fitzpatrick, D., Yu, K., Ornitz, D.M., Econs, M.J., 2005. Mutations that cause osteoglophonic dysplasia define novel roles for FGFR1 in bone elongation. Am. J. Hum. Genet. 76, 361−367. https://doi.org/10.1086/427956.

White, K.E., Larsson, T.E., Econs, M.J., 2006. The roles of specific genes implicated as circulating factors involved in normal and disordered phosphate homeostasis: frizzled related protein-4, matrix extracellular phosphoglycoprotein, and fibroblast growth factor 23. Endocr. Rev. 27, 221−241.

White, A.J., Northcutt, M.J., Rohrback, S.E., Carpenter, R.O., Niehaus-Sauter, M.M., Gao, Y., Wheatly, M.G., Gillen, C.M., 2011. Characterization of sarcoplasmic calcium binding protein (SCP) variants from freshwater crayfish *Procambarus clarkii*. Comp. Biochem. Physiol. B Biochem. Mol. Biol. 160 (8−14) https://doi.org/10.1016/j.cbpb.2011.04.003 pii:S1096-4959(11)00080-7.

Whyte, M.P., 2017. Hypophosphatasia: an overview for 2017. Bone 102, 15−25. https://doi.org/10.1016/j.bone.2017.02.011.

Whyte, M.P., Schranck, F.W., Armamento-Villareal, R., 1996. X-linked hypophosphatemia: a search for gender, race, anticipation, or parent of origin effects on disease expression in children. J. Clin. Endocrinol. Metab. 81, 4075−4080. https://doi.org/10.1210/jcem.81.11.8923863.

Wild, D., Macke, H.R., Waser, B., Reubi, J.C., Ginj, M., Rasch, H., Muller-Brand, J., Hofmann, M., 2005. 68Ga-DOTANOC: a first compound for PET imaging with high affinity for somatostatin receptor subtypes 2 and 5. Eur. J. Nucl. Med. Mol. Imaging 32, 724. https://doi.org/10.1007/s00259-004-1697-4.

Wild, R., Gerasimaite, R., Jung, J.Y., Truffault, V., Pavlovic, I., Schmidt, A., Saiardi, A., Jessen, H.J., Poirier, Y., Hothorn, M., Mayer, A., 2016. Control of eukaryotic phosphate homeostasis by inositol polyphosphate sensor domains. Science 352, 986−990. https://doi.org/10.1126/science.aad9858.

Wilz, D.R., Gray, R.W., Dominguez, J.H., Lemann Jr., J., 1979. Plasma 1,25-(OH)2-vitamin D concentrations and net intestinal calcium, phosphate, and magnesium absorption in humans. Am. J. Clin. Nutr. 32, 2052−2060.

Wittrant, Y., Bourgine, A., Khoshniat, S., Alliot-Licht, B., Masson, M., Gatius, M., Rouillon, T., Weiss, P., Beck, L., Guicheux, J., 2009. Inorganic phosphate regulates Glvr-1 and -2 expression: role of calcium and ERK1/2. Biochem. Biophys. Res. Commun. 381, 259−263. https://doi.org/10.1016/j.bbrc.2009.02.034.

Wolfgang, C.T.O., Imel, E., Portale, A.A., Boot, A., Linglart, A., Padidela, R., van't Hoff, W., Gottesman, G.S., Mao, M., Skrinar, A., San Martin, J., Whyte, M.P., 2018. Sustained efficacy and safety of burosumab, an anti-FGF23 monoclonal antibody, for 88 Weeks in children and early adolescents with X-linked hypophosphatemia (XLH). J. Bone Miner. Res. 33 (Suppl. 1).

Wrong, O.M., Norden, A.G., Feest, T.G., 1994. Dent's disease; a familial proximal renal tubular syndrome with low-molecular-weight proteinuria, hypercalciuria, nephrocalcinosis, metabolic bone disease, progressive renal failure and a marked male predominance. QJM 87, 473−493.

Xiao, Z., Huang, J., Cao, L., Liang, Y., Han, X., Quarles, L.D., 2014. Osteocyte-specific deletion of Fgfr1 suppresses FGF23. PLoS One 9, e104154. https://doi.org/10.1371/journal.pone.0104154.

Xu, H., Bai, L., Collins, J.F., Ghishan, F.K., 2002. Age-dependent regulation of rat intestinal type IIb sodium-phosphate cotransporter by 1,25-(OH)(2) vitamin D(3). Am. J. Physiol. Cell Physiol. 282, C487–C493.

Xu, H., Uno, J.K., Inouye, M., Xu, L., Drees, J.B., Collins, J.F., Ghishan, F.K., 2003. Regulation of intestinal NaPi-IIb cotransporter gene expression by estrogen. Am. J. Physiol. Gastrointest. Liver Physiol. 285, G1317–G1324.

Xu, H., Uno, J.K., Inouye, M., Collins, J.F., Ghishan, F.K., 2005. NF1 transcriptional factor(s) is required for basal promoter activation of the human intestinal NaPi-IIb cotransporter gene. Am. J. Physiol. Gastrointest. Liver Physiol. 288, G175–G181.

Yadav, M.C., Bottini, M., Cory, E., Bhattacharya, K., Kuss, P., Narisawa, S., Sah, R.L., Beck, L., Fadeel, B., Farquharson, C., Millan, J.L., 2016. Skeletal mineralization deficits and impaired biogenesis and function of chondrocyte-derived matrix vesicles in Phospho1(-/-) and phospho1/pi t1 double-knockout mice. J. Bone Miner. Res. 31, 1275–1286. https://doi.org/10.1002/jbmr.2790.

Yamamoto, T., Michigami, T., Aranami, F., Segawa, H., Yoh, K., Nakajima, S., Miyamoto, K., Ozono, K., 2007. Hereditary hypophosphatemic rickets with hypercalciuria: a study for the phosphate transporter gene type IIc and osteoblastic function. J. Bone Miner. Metab. 25, 407–413.

Yamazaki, Y., Okazaki, R., Shibata, M., Hasegawa, Y., Satoh, K., Tajima, T., Takeuchi, Y., Fujita, T., Nakahara, K., Yamashita, T., Fukumoto, S., 2002. Increased circulatory level of biologically active full-length FGF-23 in patients with hypophosphatemic rickets/osteomalacia. J. Clin. Endocrinol. Metab. 87, 4957–4960. https://doi.org/10.1210/jc.2002-021105.

Yamazaki, M., Ozono, K., Okada, T., Tachikawa, K., Kondou, H., Ohata, Y., Michigami, T., 2010. Both FGF23 and extracellular phosphate activate Raf/MEK/ERK pathway via FGF receptors in HEK293 cells. J. Cell. Biochem. 111, 1210–1221. https://doi.org/10.1002/jcb.22842.

Yan, X., Yokote, H., Jing, X., Yao, L., Sawada, T., Zhang, Y., Liang, S., Sakaguchi, K., 2005. Fibroblast growth factor 23 reduces expression of type IIa Na+/Pi co-transporter by signaling through a receptor functionally distinct from the known FGFRs in opossum kidney cells. Genes Cells 10, 489–502.

Yao, X.-P., Zhao, M., Wang, C., Guo, X.-X., Su, H.-Z., Dong, E.-L., Chen, H.-T., Lai, J.-H., Liu, Y.-B., Wang, N., Chen, W.-J., 2017. Analysis of gene expression and functional characterization of XPR1: a pathogenic gene for primary familial brain calcification. Cell Tissue Res. 370, 267–273. https://doi.org/10.1007/s00441-017-2663-3.

Yeh, J.K., Aloia, J.F., 1987. Effect of glucocorticoids on the passive transport of phosphate in different segments of the intestine in the rat. Bone Miner. 2, 11–19.

Yu, Y., Sanderson, S.R., Reyes, M., Sharma, A., Dunbar, N., Srivastava, T., Juppner, H., Bergwitz, C., 2012. Novel NaPi-IIc mutations causing HHRH and idiopathic hypercalciuria in several unrelated families: long-term follow-up in one kindred. Bone 50, 1100–1106. https://doi.org/10.1016/j.bone.2012.02.015 pii:S8756-3282(12)00071-3.

Yuan, Q., Jiang, Y., Zhao, X., Sato, T., Densmore, M., Schuler, C., Erben, R.G., McKee, M.D., Lanske, B., 2014. Increased osteopontin contributes to inhibition of bone mineralization in FGF23-deficient mice. J. Bone Miner. Res. 29, 693–704. https://doi.org/10.1002/jbmr.2079.

Zhang, P., Jobert, A.S., Couvineau, A., Silve, C., 1998. A homozygous inactivating mutation in the parathyroid hormone/parathyroid hormone-related peptide receptor causing Blomstrand chondrodysplasia. J. Clin. Endocrinol. Metab. 83, 3365–3368.

Zhen, X., Bonjour, J.P., Caverzasio, J., 1997. Platelet-derived growth factor stimulates sodium-dependent Pi transport in osteoblastic cells via phospholipase Cgamma and phosphatidylinositol 3' -kinase. J. Bone Miner. Res. 12, 36–44. https://doi.org/10.1080/14041049709409113.

Zhou, X., Cui, Y., Zhou, X., Han, J., 2012. Phosphate/pyrophosphate and MV-related proteins in mineralisation: discoveries from mouse models. Int. J. Biol. Sci. 8, 778–790. https://doi.org/10.7150/ijbs.4538.

Zivicnjak, M., Schnabel, D., Staude, H., Even, G., Marx, M., Beetz, R., Holder, M., Billing, H., Fischer, D.C., Rabl, W., Schumacher, M., Hiort, O., Haffner, D., Hypophosphatemic Rickets Study, Group of the Arbeitsgemeinschaft fur Padiatrische E, Gesellschaft fur Padiatrische N, 2011. Three-year growth hormone treatment in short children with X-linked hypophosphatemic rickets: effects on linear growth and body disproportion. J. Clin. Endocrinol. Metab. 96, E2097–E2105. https://doi.org/10.1210/jc.2011-0399.

Chapter 21

Magnesium homeostasis

Karl P. Schlingmann and Martin Konrad
Department of General Pediatrics, University Children's Hospital Münster, Münster, Germany

Chapter outline

Introduction 509
Magnesium physiology 510
Hereditary disorders of magnesium homeostasis 512
Disturbed Mg^{2+} reabsorption in the thick ascending limb 512
Disturbed Mg^{2+} reabsorption in the distal convoluted tubule 514
Acquired hypomagnesemia 518
Cisplatin and carboplatin 518
Aminoglycosides 519
Calcineurin inhibitors 519
Proton pump inhibitors 519
References 520

Introduction

Mg^{2+} is the most prevalent intracellular divalent cation and the fourth most abundant cation in the body. The human body contains approximately 24 g (1000 mmol) of Mg^{2+}, of which 50%–60% is present in bone, while most of the rest is stored in soft tissues. Less than 1% of total body Mg^{2+} is present in blood. Studies on body Mg^{2+} kinetics have already been conducted in the 1960s with the radioactive isotope ^{28}Mg. Avioli and Berman proposed a multicompartmental model of exchangeable Mg^{2+} pools: (1) a Mg^{2+} pool with a relatively fast turnover, comprising ~15% of the estimated body content representing primarily the extracellular fluid, and (2) a slow-turnover intracellular pool comprising >70% of total body Mg^{2+} (Avioli and Berman, 1966).

Therefore, the assessment of Mg^{2+} status represents a difficult task. The most commonly used method for assessing Mg^{2+} status is the measurement of serum Mg^{2+} concentrations. Under physiologic conditions, serum levels are maintained at almost constant values, with a normal serum Mg^{2+} ranging between 0.75 and 1.05 mmol/L. Hypomagnesemia is usually defined as a serum Mg^{2+} level less than 0.75 mmol/L. Unfortunately, there is little correlation with specific tissue concentrations or total body Mg^{2+} stores, and normomagnesemia does not necessarily reflect a sufficient body Mg^{2+} content. Mg^{2+} deficiency and hypomagnesemia often remain asymptomatic. Clinical symptoms are mostly nonspecific and Mg^{2+} deficiency might be associated with additional electrolyte abnormalities, especially hypocalcemia and hypokalemia. Moreover, symptoms do not necessarily correlate with serum Mg^{2+} concentrations. Therefore, additional diagnostic measures have been evaluated to diagnose clinically relevant Mg^{2+} deficiency in the face of normomagnesemia. Examples for such complementary methods, which are usually not available in routine clinical practice, comprise serum ionized Mg^{2+} or erythrocyte Mg^{2+} concentrations.

Mg^{2+} homeostasis primarily depends on the balance between intestinal absorption and renal excretion. Therefore, deficiency can result from reduced dietary intake, intestinal malabsorption or losses, or renal Mg^{2+} wasting. Within physiologic limits, a diminished dietary Mg^{2+} intake is balanced by enhanced Mg^{2+} absorption in the intestine and reduced renal excretion.

In 1964, analyzing a large number of balance studies, Seelig concluded that the intake of Mg^{2+} for a healthy adult required to maintain balance is around 0.25 mmol or 6 mg/kg body weight per day (Seelig, 1964). Actually, the US Food and Nutrition Board recommends a daily intake of Mg^{2+} of 420 mg for men and 320 mg for women (Institutes of Medicine, 1997). However, NHANES data indicate that almost half of the US population consumes considerably lower amounts of Mg^{2+} from daily food (Mosfegh et al., 2006). A reduced nutritional Mg^{2+} intake does not necessarily lead to

Principles of Bone Biology. https://doi.org/10.1016/B978-0-12-814841-9.00021-X

symptomatic Mg^{2+} depletion. However, latent or subclinical hypomagnesemia is relatively frequent, with an estimated prevalence of around 14% in the general population (Schimatschek and Rempis, 2001). This is especially alarming as there is growing evidence of an association of Mg^{2+} deficiency with common chronic diseases such as hypertension, coronary heart disease, metabolic syndrome, or diabetes mellitus (Maier, 2003; Sontia and Touyz, 2007; Ford et al., 2007; Guerrero-Romero and Rodríguez-Morán, 2002; Song et al., 2004; Kieboom et al., 2016).

Magnesium physiology

Under physiologic conditions, around 30%–40% of nutritional Mg^{2+} is absorbed in the small intestine, with smaller amounts taken up in the colon (Quamme, 2008). Intestinal Mg^{2+} absorption occurs via two different transport pathways: (1) a saturable, active transcellular pathway and (2) a nonsaturable, passive paracellular pathway (Kerstan and Quamme., 2002). Whereas at low intraluminal concentrations Mg^{2+} is absorbed primarily via the active transcellular route, the passive paracellular pathway is increasingly used with rising intraluminal concentrations, yielding a curvilinear function for total intestinal Mg^{2+} absorption (Fig. 21.1). The discovery of *TRPM6* mutations in patients with hypomagnesemia with secondary hypocalcemia (HSH) has substantially expanded our understanding of intestinal Mg^{2+} absorption (Quamme, 2008; Schlingmann et al., 2002). The *TRPM6* gene encodes a member of the TRP (transient receptor potential) family of ion channels that is involved in the formation of the apical Mg^{2+}-permeable ion channel in intestinal epithelia as well as in kidney. Therefore, HSH patients, next to an impaired renal Mg^{2+} conservation (see later), exhibit a primary defect in intestinal Mg^{2+} absorption. However, the impaired active transcellular transport can be compensated for by increasing nutritional Mg^{2+} and intraluminal Mg^{2+} concentrations in the intestine, promoting passive paracellular Mg^{2+} uptake. Therefore, HSH patients are able to achieve at least subnormal serum Mg^{2+} levels with high-dose oral Mg^{2+} supplementation. Common acquired causes of hypomagnesemia due to a decreased intestinal absorption include malabsorption syndromes, short bowel syndrome, severe vomiting, diarrhea, or steatorrhea (Hoorn and Zietse, 2013).

The control of body Mg^{2+} homeostasis primarily resides in the kidney. Approximately 80% of total serum Mg^{2+} is filtered in the glomeruli. Thereafter, 95%–97% of filtered Mg^{2+} is reabsorbed along the kidney tubule so that under physiologic, normomagnesemic conditions, 3%–5% of filtered Mg^{2+} is finally excreted in the urine (Fig. 21.2). Fifteen to twenty percent of filtered Mg^{2+} is already reabsorbed in the proximal tubule. The majority of filtered Mg^{2+} (~70%) is reabsorbed in the thick ascending limb (TAL) of Henle's loop. Mg^{2+} reabsorption in the TAL is passive and paracellular in

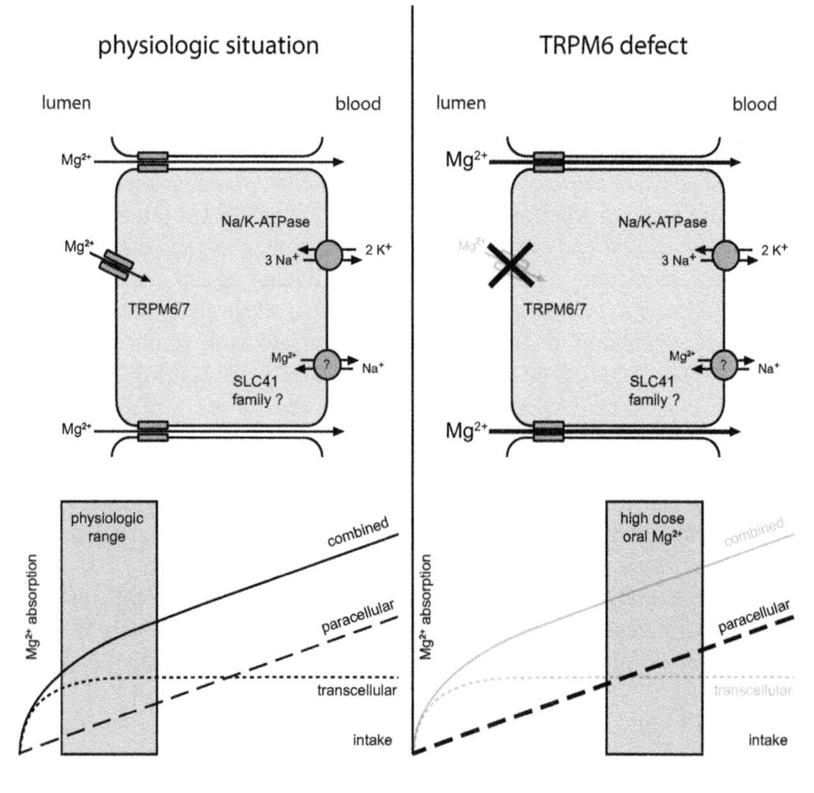

FIGURE 21.1 Intestinal Mg^{2+} absorption in health and disease. Saturable active transcellular transport (*dotted line*), which predominates at low intraluminal Mg^{2+} concentrations, and passive paracellular transport (*dashed line*), which linearly rises with increasing intraluminal Mg^{2+}, result in a curvilinear kinetic profile (left). Transient receptor potential, subfamily M, 6 (*TRPM6*) is a component of the active transcellular pathway. Hypomagnesemia with secondary hypocalcemia patients are able to compensate for TRPM6 deficiency by increasing passive paracellular absorption with high oral Mg^{2+} intake (right). *SLC41*, solute carrier 41 family.

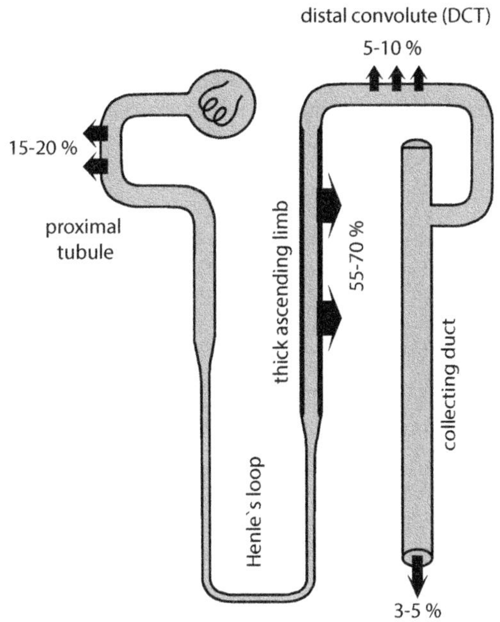

distal convolute (DCT)
5-10 %
15-20 %
proximal tubule
thick ascending limb
55-70 %
collecting duct
Henle´s loop
3-5 %

FIGURE 21.2 Renal tubular magnesium reabsorption. Approximately 80% of total serum Mg^{2+} is filtered in the glomeruli, of which 95%–97% is reabsorbed along the kidney tubule. Fifteen to twenty percent of filtered Mg^{2+} is already reabsorbed in the proximal tubule. The majority of filtered Mg^{2+} (~70%) is reabsorbed in the thick ascending limb of Henle's loop (TAL). Mg^{2+} reabsorption in the TAL is passive and paracellular in nature. Thereafter, 5%–10% of filtered Mg^{2+} is reabsorbed in the distal convoluted tubule (*DCT*) by an active and transcellular process consisting of an apical entry into the DCT cell and a basolateral extrusion into the interstitium. Finally, 3%–5% of the filtered Mg^{2+} is excreted in the urine.

nature and occurs together with calcium (Ca^{2+}) through the intercellular space. Here, specialized tight junctions composed of a specific set of proteins of the claudin family seal the paracellular space for water and electrolytes, but, on the other hand, allow for the selective passage of ions. Paracellular Ca^{2+} and Mg^{2+} transport in the TAL is driven by a lumen-positive transepithelial electric gradient that is generated by active transcellular salt reabsorption (Fig. 21.3). It is negatively regulated by the basolaterally located Ca^{2+}-sensing receptor (CaSR) (Houillier, 2013). The CaSR senses extracellular Ca^{2+} as well as Mg^{2+} concentrations in the distal nephron as well as in other tissues and thereby plays an essential role in Ca^{2+} and Mg^{2+} homeostasis (Brown et al., 1995).

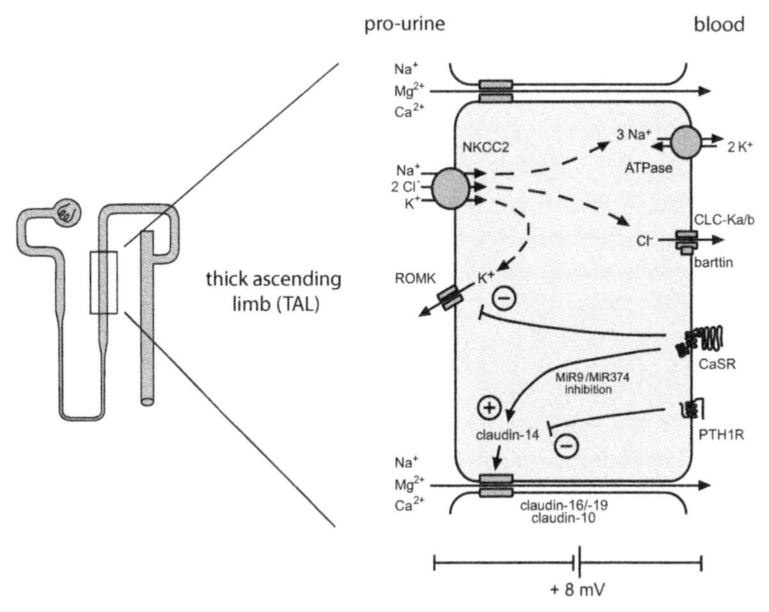

FIGURE 21.3 Mg^{2+} reabsorption in the thick ascending limb of Henle's loop. Mg^{2+} reabsorption is passive and occurs through the paracellular pathway together with Ca^{2+}. The driving force is the lumen-positive transepithelial voltage generated by active transcellular salt reabsorption. Paracellular Ca^{2+} and Mg^{2+} reabsorption occurs through specialized tight junctions. It is negatively regulated by the action of the basolateral Ca^{2+}-sensing receptor (*CaSR*) as well as the parathyroid hormone receptor (*PTH1R*), both influencing the expression of claudin-14, a tight junction protein that seals the paracellular space for both divalent cations.

Finally, 5%−10% of filtered Mg^{2+} is reabsorbed in the distal convoluted tubule (DCT), which defines the final urinary Mg^{2+} excretion as there is no significant reabsorption of Mg^{2+} in the collecting duct. Mg^{2+} reabsorption in the DCT is an active and transcellular process consisting of an apical entry into the DCT cell through a specific and regulated Mg^{2+}-permeable ion channel and a basolateral extrusion into the interstitium. The molecular nature of the apical Mg^{2+} channel was resolved by genetic studies in patients with HSH that identified TRPM6 (see later) (Schlingmann et al., 2002). Unfortunately, the molecular components involved in basolateral Mg^{2+} export are still not completely resolved. Most physiologic studies have favored a Na^{+}-dependent exchange mechanism for basolateral Mg^{2+} extrusion (Quamme, 1997); more recent data point to an involvement of Mg^{2+} transporters of the solute carrier 41 (SLC41) family (Kolisek et al., 2012). Apical Mg^{2+} entry is driven by a favorable transmembrane voltage and represents the rate-limiting step and site of regulation (Dai et al., 2001). Finally, 3%−5% of the filtered Mg^{2+} is excreted in the urine.

For the determination of renal Mg^{2+} conservation and the diagnosis of Mg^{2+} deficiency, the parallel determination of serum and urine Mg^{2+} concentrations is critical. In hypomagnesemic patients, the determination of urinary Mg^{2+} excretions allows for a differentiation between renal Mg^{2+} wasting and extrarenal losses. In the presence of hypomagnesemia, the 24-h Mg^{2+} excretion is expected to be below 1 mmol (Sutton and Domrongkitchaiporn, 1993). In contrast, a decreased 24-h Mg^{2+} excretion might indicate Mg^{2+} deficiency in normomagnesemic individuals, presuming an intact renal Mg^{2+} conservation. Similarly, Mg^{2+}/creatinine ratios and fractional Mg^{2+} excretions calculated from spot urine samples might be useful (Elisaf et al., 1997; Tang et al., 2000). Finally, the parenteral Mg^{2+} loading test (MLT) remains the gold standard for the evaluation of body Mg^{2+} status (Elin, 1994; Hébert et al., 1997). It determines the retention of a defined intravenous Mg^{2+} load as a sensitive index of Mg^{2+} deficiency. Several protocols have been published (Ryzen et al., 1985; Gullestad et al., 1994). A short-term MLT using an infusion of 0.1 mmol Mg^{2+} per kilogram body weight given over 1 h was proposed as a less elaborate alternative to the 8-h standard, yielding comparable results (Rob et al., 1996). Usually, an excretion of less than 70% of the infused Mg^{2+} is considered indicative of Mg^{2+} deficiency (Hébert et al., 1997). Moreover, the MLT can be successfully applied to differentiate between renal and extrarenal Mg^{2+} losses in hypomagnesemic patients to uncover a renal Mg^{2+} leak that becomes evident only at higher serum Mg^{2+} concentrations (Walder et al., 2002).

Hereditary disorders of magnesium homeostasis

Advances in molecular genetics identified a number of genes and their encoded proteins involved in epithelial Mg^{2+} transport (de Baaij et al., 2015b). Knowledge of the underlying genetic defects allows the definition of a growing spectrum of clinical entities underlying hereditary disorders of Mg^{2+} homeostasis. Moreover, the clarification of the molecular structures and of the pathophysiology of hypomagnesemic disorders enables an understanding of disturbances in Mg^{2+} metabolism as a frequent side effect of common drug treatments and raises the alertness of the clinician for the prevention and treatment of acquired hypomagnesemia in their patients.

Disturbed Mg^{2+} reabsorption in the thick ascending limb

In the TAL, the paracellular reabsorption pathway for divalent cations is composed of specialized tight junctions comprising different proteins of the claudin family (i.e., claudin-3, -10, -11, -14, -16, -19). Paracellular transport of divalent cations is passive in nature, driven by the lumen positive transepithelial potential. Patients with genetic defects in *CLDN16* (encoding claudin-16) and *CLDN19* genes (encoding claudin-19) suffer from an autosomal-recessive disease named familial hypomagnesemia with hypercalciuria and nephrocalcinosis (FHHNC; MIM: 248250) (Simon et al., 1999; Konrad et al., 2006). Due to excessive renal Ca^{2+} and Mg^{2+} wasting, patients develop the characteristic triad of hypomagnesemia, hypercalciuria, and nephrocalcinosis. FHHNC patients usually present during childhood with recurrent urinary tract infections, polyuria/polydipsia, nephrolithiasis, and/or failure to thrive. Signs of profound hypomagnesemia such as cerebral seizures or muscular tetany are observed less frequently. An additional ocular involvement (including severe myopia, nystagmus, or chorioretinitis) is common in patients with *CLDN19* mutations (Konrad et al., 2006). Additional laboratory findings include elevated intact parathyroid hormone (iPTH) levels before the onset of renal failure, an incomplete distal tubular acidosis, hypocitraturia, and hyperuricemia present in the majority of patients (Weber et al., 2001). The clinical course of FHHNC patients is often complicated by a continuous decline in renal function leading to end-stage renal disease (ESRD) early in life. A significant number of patients exhibit a marked decline in glomerular filtration rate (GFR) already at the time of diagnosis, and about one-third of patients develop ESRD during adolescence (Konrad et al., 2008). Hypomagnesemia may completely disappear with the decline of GFR. In addition to oral Mg^{2+} supplementation, therapy aims at a reduction in renal Ca^{2+} excretions to prevent the progression of nephrocalcinosis and kidney stone formation, as the extent of renal calcifications has been correlated with the progression of chronic renal failure (Praga et al., 1995). Supportive therapy is critical for the protection of kidney function, including sufficient fluid uptake and an effective treatment of bacterial infections and kidney stone disease. Data on bone mineral density or an additional bone phenotype in patients with FHHNC are sparse. A potential defect in bone mineralization caused by renal Ca^{2+} and Mg^{2+} wasting and early elevations of iPTH might overlap with renal osteodystrophy that develops with the decline in renal function. However, there are recent data on a primary defect in amelogenesis in patients with *CLDN16* mutations (Bardet et al., 2016; Yamaguti et al., 2017). Finally, there is evidence for a heterozygote effect in carriers of *CLDN16* mutations who may present with hypercalciuria, nephrolithiasis, and/or nephrocalcinosis (Weber et al., 2001; Praga et al., 1995). Another study also reported mild hypomagnesemia in family members with heterozygous *CLDN16* mutations (Blanchard et al., 2001). Thus, one might speculate that *CLDN16* mutations could be involved in idiopathic hypercalciuric stone formation.

Interestingly, physiologic data indicate that claudin-16 and claudin-19 do not just facilitate paracellular Mg^{2+} and Ca^{2+} transport (Hou et al., 2008). Rather, the authors suggest that claudin-16 also critically influences paracellular sodium (Na^+) permeability, while claudin-19 seals the paracellular space for chloride (Cl^-). Another study points to the existence of spatially distinct sets of tight junction strands that allow for the permeation of either monovalent or divalent cations (Milatz et al., 2017). In line with these findings are the observations made in mice with a deletion of claudin-10 (*cldn10*), which is thought to form a paracellular Na^+ pore (Breiderhoff et al., 2012). These mice display a decreased paracellular Na^+ permeability but an increase in paracellular Mg^{2+} and Ca^{2+} transport that phenotypically leads to hypermagnesemia and nephrocalcinosis. Similarly, patients with recessive *CLDN10* mutations exhibit a Bartter-like salt wasting phenotype with hypokalemic alkalosis and display serum Ca^{2+} and Mg^{2+} levels that are in the upper normal range or elevated (Bongers et al., 2017; Hadj-Rabia et al., 2018). Finally, the additional knockout of *cldn10* corrects the Mg^{2+} and Ca^{2+} wasting phenotype of *cldn16*-deficient mice (Breiderhoff et al., 2018). Data on bone mineralization have not been reported as of this writing for either patients or mice.

Another claudin that is expressed at tight junctions in the TAL is claudin-14 (*CLDN14*) (Gong et al., 2012). *CLDN14* has been identified in a genome-wide association study (GWAS) as a major risk gene for hypercalciuric nephrolithiasis and a reduced bone mineral density (Thorleifsson et al., 2009). The association of *CLDN14* with bone mineral density was subsequently replicated in two independent studies (Zhang et al., 2014; Tang et al., 2016). Moreover, a variant in the *CLDN14* promotor has been associated with pediatric-onset hypercalciuria and kidney stones (Ure et al., 2017). Claudin-14 interacts with claudin-16 and diminishes paracellular cation permeability in vitro (Gong et al., 2012). Under physiologic conditions, expression of claudin-14 is suppressed via a microRNA pathway (Gong et al., 2015). By activation of the CaSR, extracellular Ca^{2+} is able to relieve this suppression, induce claudin-14 expression, and inhibit paracellular divalent cation reabsorption (Hou et al., 2013). Moreover, claudin-14 expression and therefore paracellular cation transport are critically influenced by parathyroid hormone (PTH) via PTH receptor (PTH1R) signaling (Fig. 21.3). Accordingly, the TAL-specific inactivation of the PTH1R in mice leads to hypercalciuria and diminished serum Ca^{2+} levels, while expression of claudin-14 is significantly increased (Sato et al., 2017). In contrast, the deletion of *cldn14* in mice provokes hypermagnesemia, hypomagnesiuria, and hypocalciuria under a high-Ca^{2+} diet (Gong et al., 2012).

Basolaterally expressed CaSR negatively regulates Mg^{2+} and Ca^{2+} reabsorption in the TAL (Fig. 21.3) (Houillier, 2013). The CaSR senses extracellular Ca^{2+} as well as Mg^{2+} concentrations in the distal nephron (Brown, 2013). In the parathyroid, the CaSR is responsible for adjusting the rate of PTH synthesis and release to the extracellular levels of both divalent cations (Hebert and Brown, 1995). Two different hereditary disorders result from either activating or inactivating CaSR mutations. Heterozygous activating mutations lead to autosomal-dominant hypocalcemia (ADH; MIM: 601198) (Pollak et al., 1994). Patients present with hypocalcemic seizures or muscle spasms. In addition, a significant number of affected patients also exhibit hypomagnesemia (Pearce et al., 1996). Moreover, patients may develop significant renal salt wasting due to parallel inhibition of active transcellular NaCl reabsorption in the TAL (Watanabe et al., 2002). Inappropriately low iPTH levels due to the activation of the CaSR in the parathyroid gland often lead to the diagnosis of primary hypoparathyroidism. In ADH, supplementation with vitamin D and Ca^{2+} should be reserved for symptomatic patients as it may increase calciuria and renal complications (Hannan and Thakker, 2013).

In contrast, inactivating mutations on one or two *CASR* alleles result in familial hypocalciuric hypercalcemia (FHH1; MIM:145980) and neonatal severe hyperparathyroidism (NSHPT; MIM 239200), respectively (Pollak et al., 1993). FHH patients normally present with mild-to-moderate hypercalcemia, accompanied by few if any symptoms, and usually do not require treatment. Urinary excretion rates for Ca^{2+} and Mg^{2+} are markedly reduced, and affected individuals may also exhibit mild hypermagnesemia (Marx et al., 1981). In contrast, NSHPT patients with two mutant *CASR* alleles present in early infancy with severe symptomatic hypercalcemia. If untreated, hyperparathyroidism and hypercalcemia result in diffuse bone demineralization, fractures, and skeletal deformities. Traditionally, partial-to-total parathyroidectomy within the first weeks of life has been propagated and still represents the treatment of choice for most patients (Cole et al., 1997;

Mayr et al., 2016). More recent data indicate that cinacalcet, a calcimimetic drug enhancing CaSR function, may exert positive effects in a subset of patients, while a lack of effect was demonstrated in others (Atay et al., 2014; Gannon et al., 2014). As in symptomatic hypercalcemia of different origin, pamidronate has been shown to effectively lower serum Ca^{2+} levels (Waller et al., 2004). Data on Mg^{2+} metabolism in patients with NSHPT are scarce; however, elevations of serum Mg^{2+} levels to around 50% above the upper normal limit have been reported (Cole et al., 1997). The bone phenotype resulting from inactivating Casr mutations has been extensively studied in mice (Ho et al., 1995; Garner et al., 2001; Liu et al., 2011). $Casr^{-/-}$ mice exhibit markedly elevated iPTH levels, parathyroid hyperplasia, severe hypercalcemia, severe skeletal growth retardation, and premature death (Ho et al., 1995). Garner et al. performed a detailed skeletal analysis in $Casr^{-/-}$ mice and described a rickets phenotype with impaired growth plate calcification and decreased mineralization of metaphyseal bone (Garner et al., 2001).

Disturbed Mg^{2+} reabsorption in the distal convoluted tubule

Genetic studies in patients with inherited hypomagnesemia have identified numerous genes that are involved in active transcellular Mg^{2+} reabsorption in the DCT (Fig. 21.4). Mg^{2+} enters the DCT cell through a specific Mg^{2+}-permeable ion channel. As intraluminal and intracellular Mg^{2+} concentrations are of the same magnitude, the intracellularly negative membrane voltage represents the major driving force for apical Mg^{2+} entry.

Genetic studies in patients with HSH (HOMG1; MIM: 602014) led to the identification of TRPM6 as a critical component of apically located Mg^{2+}-permeable ion channels (Schlingmann et al., 2002; Walder et al., 2002; Chubanov et al., 2004). Recessive loss-of-function mutations in *TRPM6* result in the development of severe hypomagnesemia, cerebral seizures, or other symptoms of increased neuromuscular excitability during infancy as first described in 1968 (Paunier et al., 1968). Delayed diagnosis or noncompliance with treatment can even be fatal or result in permanent neurological damage. In addition to hypomagnesemia, patients exhibit suppressed iPTH levels and consecutive hypocalcemia. The hypocalcemia observed in HSH is resistant to treatment with Ca^{2+} or vitamin D. Relief of clinical symptoms, normocalcemia, and normalization of iPTH levels can be achieved only by administration of high doses of Mg^{2+} (Shalev et al., 1998). The suppression of PTH is thought to result from an inhibition of PTH synthesis and secretion in the presence of profound hypomagnesemia (Anast et al., 1972). The paradoxical block of the parathyroid gland caused by severe Mg^{2+} depletion involves intracellular signaling pathways of the CaSR with an increase in the activity of inhibitory $G\alpha$ subunits (Quitterer et al., 2001). In addition, profound hypomagnesemia might result in an end organ resistance to PTH as intracellular Mg^{2+} is a cofactor for adenylate cyclase (Mori et al., 1992; Mihara et al., 1995). Finally, hypomagnesemia results in a reduction in PTH-induced release of Ca^{2+} from bone, which contributes to the development of secondary hypocalcemia (Freitag et al., 1979). The *TRPM6* gene encodes a member of the TRP family of cation channels. The TRPM6 protein is highly homologous to TRPM7, another Ca^{2+}- and Mg^{2+}-permeable ion channel regulated by Mg-ATP (Nadler et al., 2001). TRPM6 is expressed along the entire gastrointestinal tract as well as in kidney in the DCT, supporting the assumption of HSH being a combined gastrointestinal and renal Mg^{2+} wasting disorder (Voets et al., 2004).

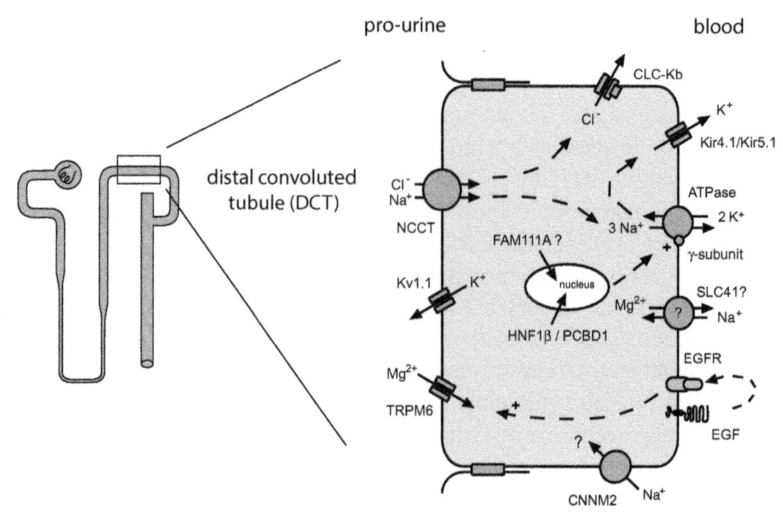

FIGURE 21.4 Mg^{2+} reabsorption in the distal convoluted tubule. Mg^{2+} transport is active and transcellular in nature. Mg^{2+} entry through transient receptor potential, subfamily M, 6 (*TRPM6*) ion channels is dependent on the negative membrane potential. The molecular identity of the basolateral Mg^{2+} extrusion mechanism is still unknown but might involve Na^+/Mg^{2+} exchangers of the solute carrier 41 (*SLC41*) family. Mg^{2+} transport is critically influenced by transcellular salt reabsorption, potassium recycling, and the paracrine action of epidermal growth factor (*EGF*), as well as the transcriptional activity of hepatocyte nuclear factor 1β (*HNF1β*). *CNNM2* might represent a basolaterally expressed Mg^{2+} sensor.

The functional characterization of TRPM6 in different overexpression systems yielded contradictory results. Voets et al. were able to express functional TRPM6 monomers and could demonstrate striking similarities to TRPM7 with respect to gating mechanisms and ion-selectivity profiles, as TRPM6 was shown to be regulated by intracellular Mg^{2+} levels and to be permeable for Mg^{2+} as well as for Ca^{2+} (Voets et al., 2004). Permeation characteristics with currents almost exclusively carried by divalent cations with a higher affinity for Mg^{2+} than for Ca^{2+} supported the role of TRPM6 as the apical Mg^{2+} influx pathway. In contrast, Chubanov and coworkers observed that TRPM6 was present at the cell surface only when associating with TRPM7 (Chubanov et al., 2004). Electrophysiological data in the *Xenopus* oocyte expression system indicated that coexpression of TRPM6 results in a significant amplification of TRPM7-induced currents (Chubanov et al., 2004). The idea of a heteromultimerization of TRPM6 with TRPM7 was also confirmed by another independent group that demonstrated that TRPM6 and TRPM7 are functionally not redundant, but that both proteins can influence each other's biological activity (Schmitz et al., 2005). This group showed that TRPM6 is able to phosphorylate TRPM7, thereby modulating its function.

The most common inherited disorder affecting magnesium conservation in the DCT is the Gitelman syndrome (GS; MIM: 263800), a primary renal salt wasting disorder with an estimated prevalence of approximately 1:40,000. GS is caused by recessive mutations in *SLC12A3* encoding the thiazide-sensitive NaCl cotransporter NCCT (Simon et al., 1996). In contrast to other forms of the Bartter syndrome spectrum, urinary-concentration ability typically is not affected and the extent of renal salt wasting is rather mild. Nevertheless, salt craving is a common clinical feature. Patients usually present at school age or in later life with muscular weakness, cramps, or fatigue. Also, a significant percentage of affected individuals are diagnosed while searching medical consultation because of growth retardation, constipation, or enuresis. Furthermore, many so-called "asymptomatic" patients have been reported, or GS is diagnosed incidentally after the assessment of serum electrolytes for other reasons. Nevertheless, it has been demonstrated that GS should not be considered a mild disorder, since none of the studied patients was truly asymptomatic (Cruz et al., 2001). Salt craving, nycturia, and paresthesia were the most frequent symptoms, all of them significantly affecting the quality of life.

Laboratory examinations in GS patients show the typical constellation of metabolic alkalosis, hypokalemia, and hypomagnesemia; urine analyses reveal hypocalciuria (Bettinelli et al., 1992). The dissociation of renal Ca^{2+} and Mg^{2+} handling was initially considered pathognomonic for GS; however, it has been observed in a number of hypomagnesemic disorders (see later). Mg^{2+} depletion causes neuromuscular irritability and tetany. Decreased renal Ca^{2+} elimination together with Mg^{2+} deficiency favors deposition of mineral Ca^{2+} as demonstrated by increased bone density as well as chondrocalcinosis (Nicolet-Barousse et al., 2005; Ea et al., 2005). Family studies demonstrated that electrolyte imbalances are present from infancy on, although the affected infants remain clinically asymptomatic. Of note, the combination of hypokalemia and hypomagnesemia exerts an exceptionally unfavorable effect on cardiac excitability, which puts these patients at high risk for cardiac arrhythmias (Foglia et al., 2004). Substitution of K^{+} and Mg^{2+} represents the central treatment of this disorder. Avoidance of factors that in addition to hypokalemia and hypomagnesemia might affect cardiac excitability (in particular, QT-time prolonging drugs) is mandatory to prevent life-threatening cardiac arrhythmias (Blanchard et al., 2017). Unfortunately, the exact mechanisms compromising Mg^{2+} reabsorption while favoring reabsorption of Ca^{2+} are not yet completely understood. Studies in *Trpv5*-knockout mice with disrupted Ca^{2+} reabsorption in the DCT that were treated with thiazides to inhibit NCCT pointed to an increase in passive Ca^{2+} reabsorption in the proximal tubule underlying hypocalciuria (Nijenhuis et al., 2005). The same study demonstrated a downregulation of TRPM6 in NCC-knockout mice as well as thiazide-treated mice in parallel to the development of hypomagnesemia.

In 2009, a novel clinical syndrome with autosomal-recessive inheritance combining epilepsy, ataxia, sensorineural deafness, and renal NaCl wasting was described under the acronym EAST or SeSAME syndrome (SESAMES; MIM: 612780) (Bockenhauer et al., 2009; Scholl et al., 2009). While in early life, neurological symptoms predominate, renal salt wasting is typically recognized only later during the course of the disease. As in GS, the renal phenotype includes hypokalemic alkalosis, hypomagnesemia, and hypocalciuria. EAST/SeSAME syndrome is caused by loss-of-function mutations in *KCNJ10* encoding the K^{+} channel Kir4.1 (Reichold et al., 2010). The expression pattern of Kir4.1 fits the disease phenotype with highest expression in brain, the stria vascularis of the inner ear, and the DCT. Here, Kir4.1 is thought to mediate K^{+} recycling over the basolateral membrane and thereby to maintain activity and function of Na/K-ATPase (Fig. 21.4) (Bockenhauer et al., 2009). Accordingly, loss of Kir4.1 function leads to depolarization of the basolateral membrane and therefore to a reduction in the driving force for basolateral Na^{+}-coupled exchangers. By this mechanism, Kir4.1 defects putatively affect the proposed Na^{+}/Mg^{2+} exchanger, which possibly explains the Mg^{2+} wasting observed in EAST/SeSAME syndrome.

An impairment of basolateral Na/K-ATPase function had already been implicated in renal Mg^{2+} wasting when Meij et al. in 1999 discovered a mutation in *FXYD2* in two related families with isolated dominant hypomagnesemia (IDH; MIM: 154020) (Meij et al., 2000). *FXYD2* encodes an accessory γ subunit of Na/K-ATPase that is expressed in a tissue-specific manner (Arystarkhova et al., 2002b). FXYD2 increases the apparent affinity of Na/K-ATPase for ATP while decreasing its Na^+ affinity (Arystarkhova et al., 2002a). Thus, it might provide a mechanism for balancing energy utilization and maintaining appropriate salt gradients. Expression studies of the identified mutant p.G41R γ-subunit in mammalian renal tubule cells revealed a dominant-negative effect of the mutation leading to a retention of the γ-subunit within the cell (Cairo et al., 2008). A possible explanation for the resulting defect in Mg^{2+} reabsorption is based on changes in intracellular Na^+ and K^+ levels: in addition to the presumed effect on basolateral Na^+/Mg^{2+} exchange, Meij et al. suggested that diminished intracellular K^+ might depolarize the apical membrane, resulting in a decrease in Mg^{2+} uptake (Meij et al., 2003). The index patients had presented with cerebral seizures, and laboratory analyses demonstrated profound hypomagnesemia (Geven et al., 1987). Systematic serum Mg^{2+} measurements performed in additional family members revealed hypomagnesemia in numerous additional, apparently healthy individuals. Urine analyses indicated renal Mg^{2+} wasting. Interestingly, urinary Ca^{2+} excretion rates were reduced in all hypomagnesemic individuals, a finding reminiscent of GS patients (mentioned earlier). However, no other associated biochemical abnormalities were reported, especially no hypokalemic alkalosis. Meanwhile, the identical *FXYD2* mutation has been identified in two additional families from the Netherlands and Belgium with dominant hypomagnesemia, substantiating the initial findings (de Baaij et al., 2015a). Several affected family members suffered from muscle cramps and cerebral seizures. Serum Mg^{2+} levels were found to be severely low. Interestingly, two patients exhibited hypokalemia in addition, possibly pointing to a disturbance in cellular Na^+ and K^+ handling with consecutive stimulation of the renin—angiotensin—aldosterone system. Like in the original family, urine analyses demonstrated hypocalciuria in all affected individuals (de Baaij et al., 2015a). This finding without apparent volume depletion potentially contradicts the aforementioned experimental data that favored an increase in proximal tubular Ca^{2+} reabsorption due to volume depletion in GS (Nijenhuis et al., 2005).

Another hereditary defect possibly resulting in defective apical uptake of Mg^{2+} in the DCT was discovered by Glaudemans et al. The authors described a dominant-negative missense mutation in *KCNA1* encoding the voltage-gated potassium (K^+) channel Kv1.1 in a large Brazilian family with hypomagnesemia (EA1; MIM: 176260 - 160120) (Glaudemans et al., 2009). Kv1.1 colocalizes with TRPM6 in the apical DCT cell membrane. These findings for the first time demonstrated a dependency between renal Mg^{2+} and K^+ handling at the molecular level linking Mg^{2+} reabsorption to K^+ secretion. The authors proposed a model in which Kv1.1 contributes to the establishment of a negative apical membrane potential as a prerequisite for TRPM6-mediated Mg^{2+} entry (Fig. 21.4) (Glaudemans et al., 2009). The clinical phenotype of affected patients included muscle cramps, tetanic episodes, tremor, and muscle weakness starting in infancy. Laboratory analyses demonstrated a renal Mg^{2+} leak without alterations in renal Ca^{2+} handling (see later). Interestingly, *KCNA1* mutations had previously been identified in patients with episodic ataxia with myokymia, a neurologic disorder characterized by a periodic appearance of incoordination and imbalance as well as myokymia, an involuntary, spontaneous, and localized trembling of muscles (Browne et al., 1994). Similar clinical findings were also present in members of the kindred studied by Glaudemans et al. with hypomagnesemia (Glaudemans et al., 2009).

Studying a consanguineous family with two affected sisters with isolated recessive hypomagnesemia (HOMG3; MIM: 248250), Groenestege et al. identified a homozygous missense mutation in the *EGF* gene encoding the pro—epidermal growth factor (EGF) protein (Groenestege et al., 2007). The two affected girls initially presented with cerebral seizures during infancy. Unfortunately, a delay in diagnosis resulted in significant neurodevelopmental deficits in both patients. A thorough clinical and laboratory workup at 4 and 8 years of age, respectively, revealed serum Mg^{2+} levels around 0.5—0.6 mmol/L without other associated electrolyte abnormalities. A ^{28}Mg retention study in one patient as well as renal Mg^{2+} excretion rates of 3 to 6 mmol per day in the presence of hypomagnesemia pointed to a primary renal defect, while intestinal Mg^{2+} uptake appeared to be preserved. In contrast, renal Ca^{2+} excretion rates were within the normal range.

Pro-EGF is a small peptide hormone expressed in various tissues, including the DCT in kidney (Groenestege et al., 2007). It has a single transmembrane segment that is inserted in both the luminal and the basolateral membrane of polarized epithelia. After membrane insertion it is processed into active EGF peptide by cleavage from the transmembrane section. EGF activates specialized EGF receptors (EGFRs) that are expressed in the basolateral membrane of DCT cells (Fig. 21.4). This activation was shown to lead to an increase in TRPM6 membrane trafficking and increased Mg^{2+} reabsorption (Thebault et al., 2009). The identified mutation disrupts the basolateral sorting motif in pro-EGF. Hence, the activation of EGFRs in the basolateral membrane is compromised, ultimately leading to reduced Mg^{2+} reabsorption. Despite acting in a paracrine fashion in the DCT, the authors also speculated about a role for EGF as a novel selectively acting magnesiotropic hormone (Groenestege et al., 2007).

Interestingly, renal Mg^{2+} wasting is observed as a common side effect of anti-cancer treatment with EGFR-targeting antibodies (i.e., cetuximab), supporting a critical role of the EGFR signaling pathway in regulating Mg^{2+} reabsorption in the DCT (Tejpar et al., 2007). Inhibitors of EGFR tyrosine kinase (i.e., erlotinib) were shown to have a less pronounced effect on TRPM6 trafficking and function and renal Mg^{2+} conservation in a mouse model (Dimke et al., 2010).

Another form of hereditary hypomagnesemia has been linked to mutations in *CNNM2* (HOMGSMR1; MIM: 613882) (Stuiver et al., 2011). The authors identified heterozygous *CNNM2* mutations in two families in which the affected patients presented with a clinical spectrum ranging from seizures in early childhood to muscle weakness, vertigo, and headache during adolescence. Additional heterozygous mutation carriers from both families even remained asymptomatic. Other than hypomagnesemia, serum electrolytes were within normal ranges. There are no definitive data concerning renal Ca^{2+} conservation and the presence of hypocalciuria. Of note, a neurological phenotype was not reported in any of the four affected family members. Previously, *CNNM2*, next to its homologs *CNNM3* and *CNNM4*, as well as *TRPM6*, had already been associated with serum Mg^{2+} levels in a large GWAS (Meyer et al., 2010). The *CNNM2* gene codes for cyclin M2, a transmembrane protein that is expressed in kidney at the basolateral membranes of TAL and DCT, but also in other organs, especially in brain (Stuiver et al., 2011; de Baaij et al., 2012). Whereas a truncating frameshift mutation was identified in one of the described families, affected individuals of the second family carried a missense mutation leading to a nonconservative amino acid exchange (Stuiver et al., 2011). Contrasting the initial patient cohort, Arjona et al. identified mutations in the *CNNM2* gene in hypomagnesemic patients who presented with seizures and displayed an impaired neurological development despite adequate treatment (Arjona et al., 2014). In addition, patients showed features of autism spectrum disorder, and two affected girls demonstrated severe obesity. *CNNM2* mutations were predominantly identified in the heterozygous state and occurred de novo. Finally, two siblings from a family with suspected parental consanguinity who developed seizures already in the neonatal period and displayed a severe degree of intellectual disability were found to be carriers of a homozygous *CNNM2* mutation. Magnetic resonance imaging (MRI) in these two patients demonstrated an abnormal brain morphology, whereas MRI findings in heterozygous patients were without obvious pathological findings (Arjona et al., 2014). Clearly, the neurological phenotype, i.e., seizure severity and frequency, in these patients appeared to be independent of serum Mg^{2+} levels, suggesting a primary disturbance of brain development or function. In accordance with the human phenotype, heterozygous *Cnnm2*-knockout mice display hypomagnesemia, but in addition show significantly decreased arterial blood pressure (Funato et al., 2017).

Functional studies using overexpression of *CNNM2* in HEK293 cells demonstrated an impaired cellular uptake of Mg^{2+} for mutant in comparison with wild-type CNNM2, pointing to a loss-of-function character of identified *CNNM2* mutations (Arjona et al., 2014). Biotinylation assays demonstrated a lack of cell surface expression at least for a subset of mutated CNNM2 proteins. Yet many questions concerning the cellular localization and function of CNNM2 remain unanswered. Data point to a role for CNNM2 as a potential Mg^{2+} sensor or regulatory factor rather than as a Mg^{2+} transporter, as the authors could not detect CNNM2-mediated Mg^{2+} influx nor efflux (Sponder et al., 2016). Furthermore, the Mg^{2+} uptake stimulated by CNNM2 overexpression is not dependent on extracellular Na^{+} but is abolished by 2-aminoethoxydiphenyl borate, a known inhibitor of TRPM7 ion channels, pointing to a functional coupling of CNNM2 with Mg^{2+}-permeable ion channels (Arjona and de Baaij, 2018). The role of CNNM2 in Mg^{2+} homeostasis and human physiology will be an intriguing topic of future research, as recent GWASs have identified potential links between CNNM2 and body mass index, hypertension, and coronary artery disease, as well as schizophrenia and bipolar disorders (Lv et al., 2017; Takeuchi et al., 2010; Grigoroiu-Serbanescu et al., 2015).

Hepatocyte nuclear factor 1β (*HNF1B*; *TCF2*) is a transcription factor involved in the development of kidney and pancreas. Heterozygous *HNF1B/TCF2* mutations were implicated in a subtype of maturity-onset diabetes of the young (MODY5) before an association with developmental kidney disease was reported (Horikawa et al., 1997). The renal phenotype is highly variable, including enlarged hyperechogenic kidneys, multicystic kidney disease, renal agenesis, renal hypoplasia, and cystic dysplasia, as well as hyperuricemic nephropathy (Heidet et al., 2010). The association with both symptom complexes led to the denomination "renal cysts and diabetes syndrome" (RCAD; MIM: 137920). However, this term was later replaced by the more general description as HNF1B nephropathy because the renal cystic phenotype and the development of diabetes are not uniform clinical findings. Interestingly, approximately half of affected patients present with hypomagnesemia due to renal Mg^{2+} wasting (Adalat et al., 2009). Again, the defect in renal Mg^{2+} conservation is accompanied by hypocalciuria. HNF1B regulates the expression of numerous renal genes, including the *FXYD2* gene, which bears several HNF1B-binding sites in its promoter region (Adalat et al., 2009). Therefore, defective *FXYD2* transcription and diminished expression of the γ-subunit of Na/K-ATPase represent a putative mechanism explaining the renal Mg^{2+} wasting in patients with *HNF1B* mutations.

A similar renal phenotype comprising urinary Mg^{2+} loss has been demonstrated in patients affected with transient neonatal hyperphenylalaninemia due to recessive mutations in *PCBD1* (MIM: 264070) (Ferre et al., 2014). *PCBD1* encodes a bifunctional protein that acts in the salvage pathway for the regeneration of tetrahydrobiopterin, but also as a binding partner of the HNF1B transcription factor. Functional studies demonstrated a dimerization with HNF1B that leads to a stimulation of the *FXYD2* promotor in the DCT (Ferre et al., 2014). Interestingly, a MODY-type diabetes was also observed in two of the described patients.

Kenny−Caffey syndrome type 2 (MIM: 127000) is clinically characterized by severe short stature, small hands and feet, craniofacial anomalies, delayed closure of the anterior fontanelle, and the occurrence of cerebral convulsions. Radiologic studies reveal a cortical thickening and medullary stenosis of tubular bones. Seizures are secondary to severe hypoparathyroidism and resulting hypocalcemia. In addition, patients may develop severe hypomagnesemia. Early manifestations in infancy with profound hypomagnesemia, hypocalcemia, and undetectable serum PTH have been described, resembling the clinical picture of HSH due to *TRPM6* mutations (Isojima et al., 2014). As magnesium repletion was able to correct the hypoparathyroidism, the authors suggested a paradoxical block in PTH synthesis and/or secretion similar to that observed in HSH (see earlier) (Abraham et al., 2017). Kenny−Caffey syndrome type 2 is caused by heterozygous mutations in the *FAM111A* gene that mostly occur de novo (Unger et al., 2013). *FAM111A* codes for an intracellular protein with homology to trypsin-like peptidases with yet undefined function. It is primarily expressed in bone and parathyroid gland and a complex role in the regulation of bone formation as well as parathyroid gland development was suggested (Abraham et al., 2017).

Finally, the clinical and biochemical evaluation of a large Caucasian family in which affected individuals presented with hypomagnesemia, hypercholesterolemia, and hypertension established another entity, termed "metabolic or mitochondrial hypomagnesemia" (MIM: 500005) (Wilson et al., 2004). The transmission of the phenotype exclusively by females suggested mitochondrial inheritance. Genetic studies identified a mutation in the mitochondrial-coded isoleucine tRNA gene, $tRNA^{Ile}$ or MTTI.

The majority of genetically affected family members exhibited at least one of the mentioned symptoms; approximately half showed a combination of two or more symptoms, and around 1/6 had all three features. Serum Mg^{2+} levels of family members in the maternal lineage greatly varied, ranging from ~0.8 to ~2.5 mg/dL (equivalent to ~0.3−~1.0 mmol/L) with approximately 50% of individuals being hypomagnesemic (serum Mg^{2+} <1.8 mg/dL). These hypomagnesemic individuals exhibited higher fractional Mg^{2+} excretions than their normomagnesemic relatives in the maternal lineage, clearly pointing to a renal Mg^{2+} leak. Interestingly, hypomagnesemia was accompanied by decreased urinary Ca^{2+} levels, a finding pointing to the DCT as the affected tubular segment.

The mitochondrial mutation affects the $tRNA^{Ile}$ gene *MTTI* and is located directly adjacent to the anticodon triplet. This position is highly conserved and critical for codon−anticodon recognition. However, the functional link between the tRNA defect and mitochondrial function remains to be elucidated in detail. As ATP consumption along the kidney tubule is highest in the DCT, the authors speculate on an impaired energy metabolism of DCT cells, which in turn could lead to disturbed transcellular Mg^{2+} reabsorption (Wilson et al., 2004).

Acquired hypomagnesemia

Cisplatin and carboplatin

The cytostatic agents cisplatin and carboplatin are widely used in various protocols for anti-cancer treatment of solid tumors. Among various side effects, nephrotoxicity receives the most attention as it represents a major dose-limiting factor. Carboplatin has been reported to have less severe side effects than cisplatin (Boulikas and Vougiouka, 2004; English et al., 1999; Goren, 2003). Hypomagnesemia due to renal Mg^{2+} wasting is regularly observed in patients treated with cisplatin (Goren, 2003; Lajer and Daugaard, 1999). The incidence of Mg^{2+} deficiency is greater than 30% but increases to over 70% with longer usage and greater accumulated doses. Interestingly, cisplatin-induced renal Mg^{2+} wasting is relatively selective (Goren, 2003). Hypocalcemia and hypokalemia may be observed but only with prolonged and severe Mg^{2+} deficiency (Mavichak et al., 1988). The effects of cisplatin may persist for months or years, long after the inorganic platinum has disappeared from the renal tissue (Bianchetti et al., 1991; Markmann et al., 1991).

Studies in cisplatin-treated rats demonstrated a downregulation of TRPM6 as well as EGF expression (van Angelen et al., 2013; Ledeganck et al., 2013). In contrast, the expression of other genes and proteins involved in tubular Mg^{2+} reabsorption remained stable or was even upregulated (*cldn16, cldn19*), potentially reflecting compensatory mechanisms (Ledeganck et al., 2013). Therefore, a defect in EGF-mediated stimulation of TRPM6 in the DCT is suggested to represent the underlying cause of renal Mg^{2+} wasting. Interestingly, cotreatment with Mg^{2+} ameliorated the cisplatin-induced downregulation of TRPM6 and hypomagnesemia and prevented a decline in renal function observed after administration of cisplatin alone (Saito et al., 2017).

Aminoglycosides

Aminoglycosides such as gentamicin induce renal impairment in a significant number of patients, depending on dose and duration of administration. Furthermore, aminoglycosides cause renal Mg^{2+} as well as Ca^{2+} wasting (Shah and Kirschenbaum, 1991). These effects occur soon after onset of therapy; they are dose dependent and reversible upon withdrawal of the drug (Elliott et al., 2000). The combined renal loss of Ca^{2+} and Mg^{2+} suggests that aminoglycosides might affect divalent cation transport in the TAL. As gentamicin is a polyvalent cation it was postulated that it may have its effects on the CaSR (Dai et al., 2001; Ward et al., 2002). As in patients with a constitutive activation of the CaSR (see earlier; Watanabe et al., 2002), a more complex, Bartter-like phenotype consisting of hypokalemic metabolic alkalosis, hypocalcemia, hypomagnesemia, and polyuria has been observed (Singh et al., 2016). In line with the assumption of a TAL defect, studies in mice demonstrated that within hours of gentamicin administration the expression of genes involved in transcellular Ca^{2+} and Mg^{2+} reabsorption in the DCT (Trpv5, Trpv6, Trpm6) is upregulated as a potential compensatory mechanism preventing further electrolyte losses (Lee et al., 2012).

Calcineurin inhibitors

The calcineurin inhibitors cyclosporine and tacrolimus (FK506) are widely used as immunosuppressants in organ transplant recipients and in patients with numerous immunologic disorders. However, patients are at high risk of developing acute and chronic renal injury, including tubular dysfunction with subsequent disturbances in mineral metabolism. Both drugs commonly lead to renal Mg^{2+} wasting and hypomagnesemia (Rob et al., 1996). Calcineurin inhibitor therapy has been shown to be associated with inappropriately high fractional excretion rates for Mg^{2+}, suggesting a distal tubular reabsorption defect (Lote et al., 2000). For tacrolimus, a downregulation of Ca^{2+} and Mg^{2+} transport proteins, including TRPV5 and TRPM6, in the DCT was demonstrated (Nijenhuis et al., 2004). It was speculated that FK506-binding proteins, which are known to regulate Ca^{2+}-permeable TRP-like cation channels, might be involved (Goel et al., 2001). More recent studies in mice demonstrated that cyclosporine downregulates TRPM6 ion channels as well as the NaCl cotransporter NCCT (Ledeganck et al., 2014). The same group also described decreased urinary EGF levels in hypomagnesemic cyclosporine-treated renal transplant patients, pointing to the involvement of the EGF axis in disturbed TRPM6 expression and function (Ledeganck et al., 2014).

Mg^{2+} deficiency has also been implicated as a contributor to calcineurin inhibitor–related nephrotoxicity and arterial hypertension. These adverse effects were shown to depend on dietary salt intake in the rat and may be prevented by oral Mg^{2+} supplementation (Mervaala et al., 1999; Miura et al., 2002).

Proton pump inhibitors

Proton pump inhibitors (PPIs) used for the reduction of gastric acidity have emerged as one of the most widely used classes of drugs worldwide. A small, but significant number of patients receiving PPIs develop moderate to severe hypomagnesemia clinically apparent as muscle cramps, tetany, or even cerebral convulsions (Cundy and Mackay, 2011). In a substantially larger number of patients receiving PPIs, hypomagnesemia or the connection between PPI use and Mg^{2+} deficiency may remain unrecognized. In their review, Cundy and Mackay summarize the data from previous publications revealing severely low serum Mg^{2+} levels of less than 0.4 mmol/L with concomitant hypocalcemia, a finding reminiscent of HSH patients with genetic TRPM6 defects (see earlier) (Cundy and Mackay, 2011). Suppressed PTH levels with consecutive hypocalcemia had already been described before in hypomagnesemic patients receiving PPIs (Epstein et al., 2006). Unfortunately, the mechanism underlying the hypomagnesemia in PPI has not been clarified in detail. As renal Mg^{2+} excretion rates have been reported to be normal or even low, a defective intestinal Mg^{2+} absorption has been suggested (William and Danziger, 2016). However, whether the proposed effects of PPIs on intestinal Mg^{2+} absorption involve the pH-sensitive regulation of TRPM6 ion channels apically located on intestinal brush border cells or other proteins involved in intestinal Mg^{2+} uptake remains to be investigated.

At any rate, it is recommended to closely monitor serum Mg^{2+} levels in patients receiving PPIs, particularly those with concomitant cardiac disease and risk for arrhythmia. The other way around, attention should be drawn to the medication of patients presenting with severe hypomagnesemia and suppressed PTH.

References

Adalat, S., Woolf, A.S., Johnstone, K.A., Wirsing, A., Harries, L.W., Long, D.A., Hennekam, R.C., Ledermann, S.E., Rees, L., van't Hoff, W., Marks, S.D., Trompeter, R.S., Tullus, K., Winyard, P.J., Cansick, J., Mushtaq, I., Dhillon, H.K., Bingham, C., Edghill, E.L., Shroff, R., Stanescu, H., Ryffel, G.U., Ellard, S., Bockenhauer, D., 2009. HNF1B mutations associate with hypomagnesemia and renal magnesium wasting. J. Am. Soc. Nephrol. 20 (5), 1123–1131.

Abraham, M.B., Li, D., Tang, D., O'Connell, S.M., McKenzie, F., Lim, E.M., Hakonarson, H., Levine, M.A., Choong, C.S., 2017. Short stature and hypoparathyroidism in a child with Kenny-Caffey syndrome type 2 due to a novel mutation in FAM111A gene. Int. J. Pediatr. Endocrinol. 2017.

Anast, C.S., Mohs, J.M., Kaplan, S.L., Burns, T.W., 1972. Evidence for parathyroid failure in magnesium deficiency. Science 177 (4049), 606–608.

Arjona, F.J., de Baaij, J.H.F., 2018. CrossTalk opposing view: CNNM proteins are not Na(+)/Mg(2+) exchangers but Mg(2+) transport regulators playing a central role in transepithelial Mg(2+) (re)absorption. J. Physiol. 596 (5), 747–750.

Arjona, F.J., de Baaij, J.H., Schlingmann, K.P., Lameris, A.L., van Wijk, E., Flik, G., Regele, S., Korenke, G.C., Neophytou, B., Rust, S., Reintjes, N., Konrad, M., Bindels, R.J., Hoenderop, J.G., 2014. CNNM2 mutations cause impaired brain development and seizures in patients with hypomagnesemia. PLoS Genet. 10 (4) e1004267.

Arystarkhova, E., Donnet, C., Asinovski, N.K., Sweadner, K.J., 2002a. Differential regulation of renal Na,K-ATPase by splice variants of the gamma subunit. J. Biol. Chem. 277 (12), 10162–10172.

Arystarkhova, E., Wetzel, R.K., Sweadner, K.J., 2002b. Distribution and oligomeric association of splice forms of Na(+)-K(+)-ATPase regulatory gamma-subunit in rat kidney. Am. J. Physiol. Renal. Physiol. 282 (3), F393–F407.

Atay, Z., Bereket, A., Haliloglu, B., Abali, S., Ozdogan, T., Altuncu, E., Canaff, L., Vilaca, T., Wong, B.Y., Cole, D.E., Hendy, G.N., Turan, S., 2014. Novel homozygous inactivating mutation of the calcium-sensing receptor gene (CASR) in neonatal severe hyperparathyroidism-lack of effect of cinacalcet. Bone 64, 102–107.

Avioli, L.V., Berman, M., 1966. Mg28 kinetics in man. J. Appl. Physiol. 21 (6), 1688–1694.

Bardet, C., Courson, F., Wu, Y., Khaddam, M., Salmon, B., Ribes, S., Thumfart, J., Yamaguti, P.M., Rochefort, G.Y., Figueres, M.L., Breiderhoff, T., Garcia-Castano, A., Vallee, B., Le Denmat, D., Baroukh, B., Guilbert, T., Schmitt, A., Masse, J.M., Bazin, D., Lorenz, G., Morawietz, M., Hou, J., Carvalho-Lobato, P., Manzanares, M.C., Fricain, J.C., Talmud, D., Demontis, R., Neves, F., Zenaty, D., Berdal, A., Kiesow, A., Petzold, M., Menashi, S., Linglart, A., Acevedo, A.C., Vargas-Poussou, R., Muller, D., Houillier, P., Chaussain, C., 2016. Claudin-16 deficiency impairs tight junction function in ameloblasts, leading to abnormal enamel formation. J. Bone Miner. Res. 31 (3), 498–513.

Bettinelli, A., Bianchetti, M.G., Girardin, E., Caringella, A., Cecconi, M., Appiani, A.C., Pavanello, L., Gastaldi, R., Isimbaldi, C., Lama, G., et al., 1992. Use of calcium excretion values to distinguish two forms of primary renal tubular hypokalemic alkalosis: Bartter and Gitelman syndromes. J. Pediatr. 120 (1), 38–43.

Bianchetti, M.G., Kanaka, C., Ridolfi-Lüthy, A., Hirt, A., Wagner, H.P., Oetliker, O.H., 1991. Persisting renotubular sequelae after cisplatin in children and adolescents. Am. J. Nephrol. 11 (2), 127–130.

Blanchard, A., Jeunemaitre, X., Coudol, P., Dechaux, M., Froissart, M., May, A., Demontis, R., Fournier, A., Paillard, M., Houillier, P., 2001. Paracellin-1 is critical for magnesium and calcium reabsorption in the human thick ascending limb of Henle. Kidney Int. 59 (6), 2206–2215.

Blanchard, A., Bockenhauer, D., Bolignano, D., Calo, L.A., Cosyns, E., Devuyst, O., Ellison, D.H., Karet Frankl, F.E., Knoers, N.V., Konrad, M., Lin, S.H., Vargas-Poussou, R., 2017. Gitelman syndrome: consensus and guidance from a kidney disease: improving global outcomes (KDIGO) controversies conference. Kidney Int. 91 (1), 24–33.

Bockenhauer, D., Feather, S., Stanescu, H.C., Bandulik, S., Zdebik, A.A., Reichold, M., Tobin, J., Lieberer, E., Sterner, C., Landoure, G., Arora, R., Sirimanna, T., Thompson, D., Cross, J.H., van't Hoff, W., Al Masri, O., Tullus, K., Yeung, S., Anikster, Y., Klootwijk, E., Hubank, M., Dillon, M.J., Heitzmann, D., Arcos-Burgos, M., Knepper, M.A., Dobbie, A., Gahl, W.A., Warth, R., Sheridan, E., Kleta, R., 2009. Epilepsy, ataxia, sensorineural deafness, tubulopathy, and KCNJ10 mutations. N. Engl. J. Med. 360 (19), 1960–1970.

Bongers, E., Shelton, L.M., Milatz, S., Verkaart, S., Bech, A.P., Schoots, J., Cornelissen, E.A.M., Bleich, M., Hoenderop, J.G.J., Wetzels, J.F.M., Lugtenberg, D., Nijenhuis, T., 2017. A novel hypokalemic-alkalotic salt-losing tubulopathy in patients with CLDN10 mutations. J. Am. Soc. Nephrol. 28 (10), 3118–3128.

Boulikas, T., Vougiouka, M., 2004. Recent clinical trials using cisplatin, carboplatin and their combination chemotherapy drugs (review). Oncol. Rep. 11 (3), 559–595.

Breiderhoff, T., Himmerkus, N., Stuiver, M., Mutig, K., Will, C., Meij, I.C., Bachmann, S., Bleich, M., Willnow, T.E., Muller, D., 2012. Deletion of claudin-10 (Cldn10) in the thick ascending limb impairs paracellular sodium permeability and leads to hypermagnesemia and nephrocalcinosis. Proc. Natl. Acad. Sci. U.S.A. 109 (35), 14241–14246.

Breiderhoff, T., Himmerkus, N., Drewell, H., Plain, A., Gunzel, D., Mutig, K., Willnow, T.E., Muller, D., Bleich, M., 2018. Deletion of claudin-10 rescues claudin-16-deficient mice from hypomagnesemia and hypercalciuria. Kidney Int. 93 (3), 580–588.

Brown, E.M., 2013. Role of the calcium-sensing receptor in extracellular calcium homeostasis. Best Pract. Res. Clin. Endocrinol. Metabol. 27 (3), 333–343.

Brown, E.M., Pollak, M., Chou, Y.H., Seidman, C.E., Seidman, J.G., Hebert, S.C., 1995. Cloning and functional characterization of extracellular Ca(2+)-sensing receptors from parathyroid and kidney. Bone 17 (2 Suppl. 1), 7S–11S.

Browne, D.L., Gancher, S.T., Nutt, J.G., Brunt, E.R., Smith, E.A., Kramer, P., Litt, M., 1994. Episodic ataxia/myokymia syndrome is associated with point mutations in the human potassium channel gene, KCNA1. Nat. Genet. 8 (2), 136–140.

Cairo, E.R., Friedrich, T., Swarts, H.G., Knoers, N.V., Bindels, R.J., Monnens, L.A., Willems, P.H., De Pont, J.J., Koenderink, J.B., 2008. Impaired routing of wild type FXYD2 after oligomerisation with FXYD2-G41R might explain the dominant nature of renal hypomagnesemia. Biochim. Biophys. Acta 1778 (2), 398–404.

Chubanov, V., Waldegger, S., Mederos y Schnitzler, M., Vitzthum, H., Sassen, M.C., Seyberth, H.W., Konrad, M., Gudermann, T., 2004. Disruption of TRPM6/TRPM7 complex formation by a mutation in the TRPM6 gene causes hypomagnesemia with secondary hypocalcemia. Proc. Natl. Acad. Sci. U.S.A. 101 (9), 2894–2899.

Cole, D.E., Janicic, N., Salisbury, S.R., Hendy, G.N., 1997. Neonatal severe hyperparathyroidism, secondary hyperparathyroidism, and familial hypocalciuric hypercalcemia: multiple different phenotypes associated with an inactivating Alu insertion mutation of the calcium-sensing receptor gene. Am. J. Med. Genet. 71 (2), 202–210.

Cruz, D.N., Shaer, A.J., Bia, M.J., Lifton, R.P., Simon, D.B., 2001. Gitelman's syndrome revisited: an evaluation of symptoms and health-related quality of life. Kidney Int. 59 (2), 710–717.

Cundy, T., Mackay, J., 2011. Proton pump inhibitors and severe hypomagnesaemia. Curr. Opin. Gastroenterol. 27 (2), 180–185.

Dai, L.J., Ritchie, G., Kerstan, D., Kang, H.S., Cole, D.E., Quamme, G.A., 2001. Magnesium transport in the renal distal convoluted tubule. Physiol. Rev. 81 (1), 51–84.

de Baaij, J.H., Stuiver, M., Meij, I.C., Lainez, S., Kopplin, K., Venselaar, H., Muller, D., Bindels, R.J., Hoenderop, J.G., 2012. Membrane topology and intracellular processing of cyclin M2 (CNNM2). J. Biol. Chem. 287 (17), 13644–13655.

de Baaij, J.H., Dorresteijn, E.M., Hennekam, E.A., Kamsteeg, E.J., Meijer, R., Dahan, K., Muller, M., van den Dorpel, M.A., Bindels, R.J., Hoenderop, J.G., Devuyst, O., Knoers, N.V., 2015a. Recurrent FXYD2 p.Gly41Arg mutation in patients with isolated dominant hypomagnesaemia. Nephrol. Dial. Transplant. 30 (6), 952–957.

de Baaij, J.H., Hoenderop, J.G., Bindels, R.J., 2015b. Magnesium in man: implications for health and disease. Physiol. Rev. 95 (1), 1–46.

Dimke, H., van der Wijst, J., Alexander, T.R., Meijer, I.M., Mulder, G.M., van Goor, H., Tejpar, S., Hoenderop, J.G., Bindels, R.J., 2010. Effects of the EGFR inhibitor erlotinib on magnesium handling. J. Am. Soc. Nephrol. 21 (8), 1309–1316.

Ea, H.K., Blanchard, A., Dougados, M., Roux, C., 2005. Chondrocalcinosis secondary to hypomagnesemia in Gitelman's syndrome. J. Rheumatol. 32 (9), 1840–1842.

Elin, R.J., 1994. Magnesium: the fifth but forgotten electrolyte. Am. J. Clin. Pathol. 102 (5), 616–622.

Elisaf, M., Panteli, K., Theodorou, J., Siamopoulos, K.C., 1997. Fractional excretion of magnesium in normal subjects and in patients with hypomagnesemia. Magnes. Res. 10 (4), 315–320.

Elliott, C., Newman, N., Madan, A., 2000. Gentamicin effects on urinary electrolyte excretion in healthy subjects. Clin. Pharmacol. Ther. 67 (1), 16–21.

English, M.W., Skinner, R., Pearson, A.D., Price, L., Wyllie, R., Craft, A.W., 1999. Dose-related nephrotoxicity of carboplatin in children. Br. J. Canc. 81 (2), 336–341.

Epstein, M., McGrath, S., Law, F., 2006. Proton-pump inhibitors and hypomagnesemic hypoparathyroidism. N. Engl. J. Med. 355 (17), 1834–1836.

Ferre, S., de Baaij, J.H., Ferreira, P., Germann, R., de Klerk, J.B., Lavrijsen, M., van Zeeland, F., Venselaar, H., Kluijtmans, L.A., Hoenderop, J.G., Bindels, R.J., 2014. Mutations in PCBD1 cause hypomagnesemia and renal magnesium wasting. J. Am. Soc. Nephrol. 25 (3), 574–586.

Foglia, P.E., Bettinelli, A., Tosetto, C., Cortesi, C., Crosazzo, L., Edefonti, A., Bianchetti, M.G., 2004. Cardiac work up in primary renal hypokalaemia-hypomagnesaemia (Gitelman syndrome). Nephrol. Dial. Transplant. 19 (6), 1398–1402.

Ford, E.S., Li, C., McGuire, L.C., Mokdad, A.H., Liu, S., 2007. Intake of dietary magnesium and the prevalence of the metabolic syndrome among U.S. adults. Obesity 15 (5), 1139–1146.

Freitag, J.J., Martin, K.J., Conrades, M.B., Bellorin-Font, E., Teitelbaum, S., Klahr, S., Slatopolsky, E., 1979. Evidence for skeletal resistance to parathyroid hormone in magnesium deficiency. Studies in isolated perfused bone. J. Clin. Investig. 64 (5), 1238–1244.

Funato, Y., Yamazaki, D., Miki, H., 2017. Renal function of cyclin M2 Mg2+ transporter maintains blood pressure. J. Hypertens. 35 (3), 585–592.

Gannon, A.W., Monk, H.M., Levine, M.A., 2014. Cinacalcet monotherapy in neonatal severe hyperparathyroidism: a case study and review. J. Clin. Endocrinol. Metab. 99 (1), 7–11.

Garner, S.C., Pi, M., Tu, Q., Quarles, L.D., 2001. Rickets in cation-sensing receptor-deficient mice: an unexpected skeletal phenotype. Endocrinology 142 (9), 3996–4005.

Geven, W.B., Monnens, L.A., Willems, H.L., Buijs, W.C., ter Haar, B.G., 1987. Renal magnesium wasting in two families with autosomal dominant inheritance. Kidney Int. 31 (5), 1140–1144.

Glaudemans, B., van der Wijst, J., Scola, R.H., Lorenzoni, P.J., Heister, A., van der Kemp, A.W., Knoers, N.V., Hoenderop, J.G., Bindels, R.J., 2009. A missense mutation in the Kv1.1 voltage-gated potassium channel-encoding gene KCNA1 is linked to human autosomal dominant hypomagnesemia. J. Clin. Investig. 119 (4), 936–942.

Goel, M., Garcia, R., Estacion, M., Schilling, W.P., 2001. Regulation of Drosophila TRPL channels by immunophilin FKBP59. J. Biol. Chem. 276 (42), 38762–38773.

Gong, Y., Renigunta, V., Himmerkus, N., Zhang, J., Renigunta, A., Bleich, M., Hou, J., 2012. Claudin-14 regulates renal Ca(+)(+) transport in response to CaSR signalling via a novel microRNA pathway. EMBO J. 31 (8), 1999–2012.

Gong, Y., Himmerkus, N., Plain, A., Bleich, M., Hou, J., 2015. Epigenetic regulation of microRNAs controlling CLDN14 expression as a mechanism for renal calcium handling. J. Am. Soc. Nephrol. 26 (3), 663–676.

Goren, M.P., 2003. Cisplatin nephrotoxicity affects magnesium and calcium metabolism. Med. Pediatr. Oncol. 41 (3), 186–189.

Grigoroiu-Serbanescu, M., Diaconu, C.C., Heilmann-Heimbach, S., Neagu, A.I., Becker, T., 2015. Association of age-of-onset groups with GWAS significant schizophrenia and bipolar disorder loci in Romanian bipolar I patients. Psychiatr. Res. 230 (3), 964–967.

Groenestege, W.M., Thébault, S., van der Wijst, J., van den Berg, D., Janssen, R., Tejpar, S., van den Heuvel, L.P., van Cutsem, E., Hoenderop, J.G., Knoers, N.V., Bindels, R.J., 2007. Impaired basolateral sorting of pro-EGF causes isolated recessive renal hypomagnesemia. J. Clin. Investig. 117 (8), 2260–2267.

Guerrero-Romero, F., Rodríguez-Morán, M., 2002. Low serum magnesium levels and metabolic syndrome. Acta Diabetol. 39 (4), 209–213.

Gullestad, L., Midtvedt, K., Dolva, L.O., Norseth, J., Kjekshus, J., 1994. The magnesium loading test: reference values in healthy subjects. Scand. J. Clin. Lab. Invest. 54 (1), 23–31.

Hadj-Rabia, S., Brideau, G., Al-Sarraj, Y., Maroun, R.C., Figueres, M.L., Leclerc-Mercier, S., Olinger, E., Baron, S., Chaussain, C., Nochy, D., Taha, R.Z., Knebelmann, B., Joshi, V., Curmi, P.A., Kambouris, M., Vargas-Poussou, R., Bodemer, C., Devuyst, O., Houillier, P., El-Shanti, H., 2018. Multiplex epithelium dysfunction due to CLDN10 mutation: the HELIX syndrome. Genet. Med. 20 (2), 190–201.

Hannan, F.M., Thakker, R.V., 2013. Calcium-sensing receptor (CaSR) mutations and disorders of calcium, electrolyte and water metabolism. Best Pract. Res. Clin. Endocrinol. Metabol. 27 (3), 359–371.

Hebert, S.C., Brown, E.M., 1995. The extracellular calcium receptor. Curr. Opin. Cell Biol. 7 (4), 484–492.

Hébert, P., Mehta, N., Wang, J., Hindmarsh, T., Jones, G., Cardinal, P., 1997. Functional magnesium deficiency in critically ill patients identified using a magnesium-loading test. Crit. Care Med. 25 (5), 749–755.

Heidet, L., Decramer, S., Pawtowski, A., Morinière, V., Bandin, F., Knebelmann, B., Lebre, A.S., Faguer, S., Guigonis, V., Antignac, C., Salomon, R., 2010. Spectrum of HNF1B mutations in a large cohort of patients who harbor renal diseases. Clin. J. Am. Soc. Nephrol. 5 (6), 1079–1090.

Ho, C., Conner, D.A., Pollak, M.R., Ladd, D.J., Kifor, O., Warren, H.B., Brown, E.M., Seidman, J.G., Seidman, C.E., 1995. A mouse model of human familial hypocalciuric hypercalcemia and neonatal severe hyperparathyroidism. Nat. Genet. 11 (4), 389–394.

Hoorn, E.J., Zietse, R., 2013. Disorders of calcium and magnesium balance: a physiology-based approach. Pediatr. Nephrol. 28 (8), 1195–1206.

Horikawa, Y., Iwasaki, N., Hara, M., Furuta, H., Hinokio, Y., Cockburn, B.N., Lindner, T., Yamagata, K., Ogata, M., Tomonaga, O., Kuroki, H., Kasahara, T., Iwamoto, Y., Bell, G.I., 1997. Mutation in hepatocyte nuclear factor-1 beta gene (TCF2) associated with MODY. Nat. Genet. 17 (4), 384–385.

Hou, J., Renigunta, A., Konrad, M., Gomes, A.S., Schneeberger, E.E., Paul, D.L., Waldegger, S., Goodenough, D.A., 2008. Claudin-16 and claudin-19 interact and form a cation-selective tight junction complex. J. Clin. Investig. 118 (2), 619–628.

Hou, J., Rajagopal, M., Yu, A.S., 2013. Claudins and the kidney. Annu. Rev. Physiol. 75, 479–501.

Houillier, P., 2013. Calcium-sensing in the kidney. Curr. Opin. Nephrol. Hypertens. 22 (5), 566–571.

Institutes of Medicine, 1997. Dietary Reference Intakes for Calcium, Phosphorus, Magnesium, Vitamin D, and Fluoride.

Isojima, T., Doi, K., Mitsui, J., Oda, Y., Tokuhiro, E., Yasoda, A., Yorifuji, T., Horikawa, R., Yoshimura, J., Ishiura, H., Morishita, S., Tsuji, S., Kitanaka, S., 2014. A recurrent de novo FAM111A mutation causes Kenny-Caffey syndrome type 2. J. Bone Miner. Res. 29 (4), 992–998.

Kerstan, D., Quamme, G., 2002. Physiology and pathophysiology of intestinal absorption of magnesium. In: Massry, S.G., M, H., Nishizawa, Y. (Eds.), Calcium in Internal Medicine. Springer-Verlag, Surry, UK, pp. 171–183.

Kieboom, B.C., Niemeijer, M.N., Leening, M.J., van den Berg, M.E., Franco, O.H., Deckers, J.W., Hofman, A., Zietse, R., Stricker, B.H., Hoorn, E.J., 2016. Serum magnesium and the risk of death from coronary heart disease and sudden cardiac death. J Am Heart Assoc 5 (1).

Kolisek, M., Nestler, A., Vormann, J., Schweigel-Rontgen, M., 2012. Human gene SLC41A1 encodes for the Na+/Mg(2)+ exchanger. Am. J. Physiol. Cell Physiol. 302 (1), C318–C326.

Konrad, M., Schaller, A., Seelow, D., Pandey, A.V., Waldegger, S., Lesslauer, A., Vitzthum, H., Suzuki, Y., Luk, J.M., Becker, C., Schlingmann, K.P., Schmid, M., Rodriguez-Soriano, J., Ariceta, G., Cano, F., Enriquez, R., Juppner, H., Bakkaloglu, S.A., Hediger, M.A., Gallati, S., Neuhauss, S.C., Nurnberg, P., Weber, S., 2006. Mutations in the tight-junction gene claudin 19 (CLDN19) are associated with renal magnesium wasting, renal failure, and severe ocular involvement. Am. J. Hum. Genet. 79 (5), 949–957.

Konrad, M., Hou, J., Weber, S., Dötsch, J., Kari, J.A., Seeman, T., Kuwertz-Bröking, E., Peco-Antic, A., Tasic, V., Dittrich, K., Alshaya, H.O., von Vigier, R.O., Gallati, S., Goodenough, D.A., Schaller, A., 2008. CLDN16 genotype predicts renal decline in familial hypomagnesemia with hypercalciuria and nephrocalcinosis. J. Am. Soc. Nephrol. 19 (1), 171–181.

Lajer, H., Daugaard, G., 1999. Cisplatin and hypomagnesemia. Cancer Treat Rev. 25 (1), 47–58.

Ledeganck, K.J., Boulet, G.A., Bogers, J.J., Verpooten, G.A., De Winter, B.Y., 2013. The TRPM6/EGF pathway is downregulated in a rat model of cisplatin nephrotoxicity. PLoS One 8 (2) e57016.

Ledeganck, K.J., De Winter, B.Y., Van den Driessche, A., Jurgens, A., Bosmans, J.L., Couttenye, M.M., Verpooten, G.A., 2014. Magnesium loss in cyclosporine-treated patients is related to renal epidermal growth factor downregulation. Nephrol. Dial. Transplant. 29 (5), 1097–1102.

Lee, C.T., Chen, H.C., Ng, H.Y., Lai, L.W., Lien, Y.H., 2012. Renal adaptation to gentamicin-induced mineral loss. Am. J. Nephrol. 35 (3), 279–286.

Liu, J., Lv, F., Sun, W., Tao, C., Ding, G., Karaplis, A., Brown, E., Goltzman, D., Miao, D., 2011. The abnormal phenotypes of cartilage and bone in calcium-sensing receptor deficient mice are dependent on the actions of calcium, phosphorus, and PTH. PLoS Genet. 7 (9) e1002294.

Lote, C.J., Thewles, A., Wood, J.A., Zafar, T., 2000. The hypomagnesaemic action of FK506: urinary excretion of magnesium and calcium and the role of parathyroid hormone. Clin. Sci. (Lond.) 99 (4), 285–292.

Lv, W.Q., Zhang, X., Zhang, Q., He, J.Y., Liu, H.M., Xia, X., Fan, K., Zhao, Q., Shi, X.Z., Zhang, W.D., Sun, C.Q., Deng, H.W., 2017. Novel common variants associated with body mass index and coronary artery disease detected using a pleiotropic cFDR method. J. Mol. Cell. Cardiol. 112, 1–7.

Maier, J.A., 2003. Low magnesium and atherosclerosis: an evidence-based link. Mol. Aspect. Med. 24 (1–3), 137–146.

Markmann, M., Rothman, R., Reichman, B., Hakes, T., Lewis, J.L., Rubin, S., Jones, W., Almadrones, L., Hoskins, W., 1991. Persistent hypomagnesemia following cisplatin chemotherapy in patients with ovarian cancer. J. Cancer Res. Clin. Oncol. 117 (2), 89–90.

Marx, S.J., Attie, M.F., Levine, M.A., Spiegel, A.M., Downs, R.W., Lasker, R.D., 1981. The hypocalciuric or benign variant of familial hypercalcemia: clinical and biochemical features in fifteen kindreds. Medicine (Baltim.) 60 (6), 397–412.

Mavichak, V., Coppin, C.M., Wong, N.L., Dirks, J.H., Walker, V., Sutton, R.A., 1988. Renal magnesium wasting and hypocalciuria in chronic cis-platinum nephropathy in man. Clin. Sci. (Lond.) 75 (2), 203–207.

Mayr, B., Schnabel, D., Dorr, H.G., Schofl, C., 2016. Genetics in endocrinology: gain and loss of function mutations of the calcium-sensing receptor and associated proteins: current treatment concepts. Eur. J. Endocrinol. 174 (5), R189–R208.

Meij, I.C., Koenderink, J.B., van Bokhoven, H., Assink, K.F., Groenestege, W.T., de Pont, J.J., Bindels, R.J., Monnens, L.A., van den Heuvel, L.P., Knoers, N.V., 2000. Dominant isolated renal magnesium loss is caused by misrouting of the Na(+), K(+)-ATPase gamma-subunit. Nat. Genet. 26 (3), 265–266.

Meij, I.C., Koenderink, J.B., De Jong, J.C., De Pont, J.J., Monnens, L.A., Van Den Heuvel, L.P., Knoers, N.V., 2003. Dominant isolated renal magnesium loss is caused by misrouting of the Na+,K+-ATPase gamma-subunit. Ann. N.Y. Acad. Sci. 986, 437–443.

Mervaala, E.M., Müller, D.N., Park, J.K., Schmidt, F., Löhn, M., Breu, V., Dragun, D., Ganten, D., Haller, H., Luft, F.C., 1999. Monocyte infiltration and adhesion molecules in a rat model of high human renin hypertension. Hypertension 33 (1 Pt 2), 389–395.

Meyer, T.E., Verwoert, G.C., Hwang, S.J., Glazer, N.L., Smith, A.V., van Rooij, F.J., Ehret, G.B., Boerwinkle, E., Felix, J.F., Leak, T.S., Harris, T.B., Yang, Q., Dehghan, A., Aspelund, T., Katz, R., Homuth, G., Kocher, T., Rettig, R., Ried, J.S., Gieger, C., Prucha, H., Pfeufer, A., Meitinger, T., Coresh, J., Hofman, A., Sarnak, M.J., Chen, Y.D., Uitterlinden, A.G., Chakravarti, A., Psaty, B.M., van Duijn, C.M., Kao, W.H., Witteman, J.C., Gudnason, V., Siscovick, D.S., Fox, C.S., Köttgen, A., Consortium, G. F. f. O., Consortium, M. A. o. G. a. I. R. T., 2010. Genome-wide association studies of serum magnesium, potassium, and sodium concentrations identify six Loci influencing serum magnesium levels. PLoS Genet. 6 (8).

Mihara, M., Kamikubo, K., Hiramatsu, K., Itaya, S., Ogawa, T., Sakata, S., 1995. Renal refractoriness to phosphaturic action of parathyroid hormone in a patient with hypomagnesemia. Intern. Med. 34 (7), 666–669.

Milatz, S., Himmerkus, N., Wulfmeyer, V.C., Drewell, H., Mutig, K., Hou, J., Breiderhoff, T., Muller, D., Fromm, M., Bleich, M., Gunzel, D., 2017. Mosaic expression of claudins in thick ascending limbs of Henle results in spatial separation of paracellular Na^{+} and Mg^{2+} transport. Proc. Natl. Acad. Sci. U.S.A. 114 (2), E219–e227.

Miura, K., Nakatani, T., Asai, T., Yamanaka, S., Tamada, S., Tashiro, K., Kim, S., Okamura, M., Iwao, H., 2002. Role of hypomagnesemia in chronic cyclosporine nephropathy. Transplantation 73 (3), 340–347.

Mori, S., Harada, S., Okazaki, R., Inoue, D., Matsumoto, T., Ogata, E., 1992. Hypomagnesemia with increased metabolism of parathyroid hormone and reduced responsiveness to calcitropic hormones. Intern. Med. 31 (6), 820–824.

Mosfegh, A., Goldman, J., Ahuja, J., Rhodes, D., LaComb, R., 2006. What We Eat in America, NHANES 2005–2006: Usual Nutrient Intakes from Food and Water Compared to 1997 Dietary Reference for Vitamin D, Calcium, Phosphorus, and Magnesium, p. 24.

Nadler, M.J., Hermosura, M.C., Inabe, K., Perraud, A.L., Zhu, Q., Stokes, A.J., Kurosaki, T., Kinet, J.P., Penner, R., Scharenberg, A.M., Fleig, A., 2001. LTRPC7 is a Mg.ATP-regulated divalent cation channel required for cell viability. Nature 411 (6837), 590–595.

Nicolet-Barousse, L., Blanchard, A., Roux, C., Pietri, L., Bloch-Faure, M., Kolta, S., Chappard, C., Geoffroy, V., Morieux, C., Jeunemaitre, X., Shull, G.E., Meneton, P., Paillard, M., Houillier, P., De Vernejoul, M.C., 2005. Inactivation of the Na-Cl co-transporter (NCC) gene is associated with high BMD through both renal and bone mechanisms: analysis of patients with Gitelman syndrome and Ncc null mice. J. Bone Miner. Res. 20 (5), 799–808.

Nijenhuis, T., Hoenderop, J.G., Bindels, R.J., 2004. Downregulation of Ca(2+) and Mg(2+) transport proteins in the kidney explains tacrolimus (FK506)-induced hypercalciuria and hypomagnesemia. J. Am. Soc. Nephrol. 15 (3), 549–557.

Nijenhuis, T., Vallon, V., van der Kemp, A.W., Loffing, J., Hoenderop, J.G., Bindels, R.J., 2005. Enhanced passive Ca^{2+} reabsorption and reduced Mg^{2+} channel abundance explains thiazide-induced hypocalciuria and hypomagnesemia. J. Clin. Investig. 115 (6), 1651–1658.

Paunier, L., Radde, I.C., Kooh, S.W., Conen, P.E., Fraser, D., 1968. Primary hypomagnesemia with secondary hypocalcemia in an infant. Pediatrics 41 (2), 385–402.

Pearce, S.H., Williamson, C., Kifor, O., Bai, M., Coulthard, M.G., Davies, M., Lewis-Barned, N., McCredie, D., Powell, H., Kendall-Taylor, P., Brown, E.M., Thakker, R.V., 1996. A familial syndrome of hypocalcemia with hypercalciuria due to mutations in the calcium-sensing receptor. N. Engl. J. Med. 335 (15), 1115–1122.

Pollak, M.R., Brown, E.M., Chou, Y.H., Hebert, S.C., Marx, S.J., Steinmann, B., Levi, T., Seidman, C.E., Seidman, J.G., 1993. Mutations in the human Ca(2+)-sensing receptor gene cause familial hypocalciuric hypercalcemia and neonatal severe hyperparathyroidism. Cell 75 (7), 1297–1303.

Pollak, M.R., Brown, E.M., Estep, H.L., McLaine, P.N., Kifor, O., Park, J., Hebert, S.C., Seidman, C.E., Seidman, J.G., 1994. Autosomal dominant hypocalcaemia caused by a Ca(2+)-sensing receptor gene mutation. Nat. Genet. 8 (3), 303–307.

Praga, M., Vara, J., González-Parra, E., Andrés, A., Alamo, C., Araque, A., Ortiz, A., Rodicio, J.L., 1995. Familial hypomagnesemia with hypercalciuria and nephrocalcinosis. Kidney Int. 47 (5), 1419–1425.

Quamme, G.A., 1997. Renal magnesium handling: new insights in understanding old problems. Kidney Int. 52 (5), 1180–1195.

Quamme, G.A., 2008. Recent developments in intestinal magnesium absorption. Curr. Opin. Gastroenterol. 24 (2), 230–235.

Quitterer, U., Hoffmann, M., Freichel, M., Lohse, M.J., 2001. Paradoxical block of parathormone secretion is mediated by increased activity of G alpha subunits. J. Biol. Chem. 276 (9), 6763–6769.

Reichold, M., Zdebik, A.A., Lieberer, E., Rapedius, M., Schmidt, K., Bandulik, S., Sterner, C., Tegtmeier, I., Penton, D., Baukrowitz, T., Hulton, S.A., Witzgall, R., Ben-Zeev, B., Howie, A.J., Kleta, R., Bockenhauer, D., Warth, R., 2010. KCNJ10 gene mutations causing EAST syndrome (epilepsy, ataxia, sensorineural deafness, and tubulopathy) disrupt channel function. Proc. Natl. Acad. Sci. U.S.A. 107 (32), 14490–14495.

Rob, P.M., Lebeau, A., Nobiling, R., Schmid, H., Bley, N., Dick, K., Weigelt, I., Rohwer, J., Gobel, Y., Sack, K., Classen, H.G., 1996. Magnesium metabolism: basic aspects and implications of ciclosporine toxicity in rats. Nephron 72 (1), 59–66.

Ryzen, E., Elbaum, N., Singer, F.R., Rude, R.K., 1985. Parenteral magnesium tolerance testing in the evaluation of magnesium deficiency. Magnesium 4 (2–3), 137–147.

Saito, Y., Okamoto, K., Kobayashi, M., Narumi, K., Yamada, T., Iseki, K., 2017. Magnesium attenuates cisplatin-induced nephrotoxicity by regulating the expression of renal transporters. Eur. J. Pharmacol. 811, 191–198.

Sato, T., Courbebaisse, M., Ide, N., Fan, Y., Hanai, J.I., Kaludjerovic, J., Densmore, M.J., Yuan, Q., Toka, H.R., Pollak, M.R., Hou, J., Lanske, B., 2017. Parathyroid hormone controls paracellular Ca(2+) transport in the thick ascending limb by regulating the tight-junction protein Claudin14. Proc. Natl. Acad. Sci. U.S.A. 114 (16), E3344–e3353.

Schimatschek, H.F., Rempis, R., 2001. Prevalence of hypomagnesemia in an unselected German population of 16,000 individuals. Magnes. Res. 14 (4), 283–290.

Schlingmann, K.P., Weber, S., Peters, M., Niemann Nejsum, L., Vitzthum, H., Klingel, K., Kratz, M., Haddad, E., Ristoff, E., Dinour, D., Syrrou, M., Nielsen, S., Sassen, M., Waldegger, S., Seyberth, H.W., Konrad, M., 2002. Hypomagnesemia with secondary hypocalcemia is caused by mutations in TRPM6, a new member of the TRPM gene family. Nat. Genet. 31 (2), 166–170.

Schmitz, C., Dorovkov, M.V., Zhao, X., Davenport, B.J., Ryazanov, A.G., Perraud, A.L., 2005. The channel kinases TRPM6 and TRPM7 are functionally nonredundant. J. Biol. Chem. 280 (45), 37763–37771.

Scholl, U.I., Choi, M., Liu, T., Ramaekers, V.T., Häusler, M.G., Grimmer, J., Tobe, S.W., Farhi, A., Nelson-Williams, C., Lifton, R.P., 2009. Seizures, sensorineural deafness, ataxia, mental retardation, and electrolyte imbalance (SeSAME syndrome) caused by mutations in KCNJ10. Proc. Natl. Acad. Sci. U.S.A. 106 (14), 5842–5847.

Seelig, M.S., 1964. The requirement of magnesium by the normal adult. summary and analysis of published data. Am. J. Clin. Nutr. 14, 242–290.

Shah, G.M., Kirschenbaum, M.A., 1991. Renal magnesium wasting associated with therapeutic agents. Miner. Electrolyte Metab. 17 (1), 58–64.

Shalev, H., Phillip, M., Galil, A., Carmi, R., Landau, D., 1998. Clinical presentation and outcome in primary familial hypomagnesaemia. Arch. Dis. Child. 78 (2), 127–130.

Simon, D.B., Nelson-Williams, C., Bia, M.J., Ellison, D., Karet, F.E., Molina, A.M., Vaara, I., Iwata, F., Cushner, H.M., Koolen, M., Gainza, F.J., Gitleman, H.J., Lifton, R.P., 1996. Gitelman's variant of Bartter's syndrome, inherited hypokalaemic alkalosis, is caused by mutations in the thiazide-sensitive Na-Cl cotransporter. Nat. Genet. 12 (1), 24–30.

Simon, D.B., Lu, Y., Choate, K.A., Velazquez, H., Al-Sabban, E., Praga, M., Casari, G., Bettinelli, A., Colussi, G., Rodriguez-Soriano, J., McCredie, D., Milford, D., Sanjad, S., Lifton, R.P., 1999. Paracellin-1, a renal tight junction protein required for paracellular Mg^{2+} resorption. Science 285 (5424), 103–106.

Singh, J., Patel, M.L., Gupta, K.K., Pandey, S., Dinkar, A., 2016. Acquired Bartter syndrome following gentamicin therapy. Indian J. Nephrol. 26 (6), 461–463.

Song, Y., Manson, J.E., Buring, J.E., Liu, S., 2004. Dietary magnesium intake in relation to plasma insulin levels and risk of type 2 diabetes in women. Diabetes Care 27 (1), 59–65.

Sontia, B., Touyz, R.M., 2007. Role of magnesium in hypertension. Arch. Biochem. Biophys. 458 (1), 33–39.

Sponder, G., Mastrototaro, L., Kurth, K., Merolle, L., Zhang, Z., Abdulhanan, N., Smorodchenko, A., Wolf, K., Fleig, A., Penner, R., Iotti, S., Aschenbach, J.R., Vormann, J., Kolisek, M., 2016. Human CNNM2 is not a Mg(2+) transporter per se. Pflügers Archiv 468 (7), 1223–1240.

Stuiver, M., Lainez, S., Will, C., Terryn, S., Günzel, D., Debaix, H., Sommer, K., Kopplin, K., Thumfart, J., Kampik, N.B., Querfeld, U., Willnow, T.E., Němec, V., Wagner, C.A., Hoenderop, J.G., Devuyst, O., Knoers, N.V., Bindels, R.J., Meij, I.C., Müller, D., 2011. CNNM2, encoding a basolateral protein required for renal Mg^{2+} handling, is mutated in dominant hypomagnesemia. Am. J. Hum. Genet. 88 (3), 333–343.

Sutton, R.A., Domrongkitchaiporn, S., 1993. Abnormal renal magnesium handling. Miner. Electrolyte Metab. 19 (4–5), 232–240.

Takeuchi, F., Isono, M., Katsuya, T., Yamamoto, K., Yokota, M., Sugiyama, T., Nabika, T., Fujioka, A., Ohnaka, K., Asano, H., Yamori, Y., Yamaguchi, S., Kobayashi, S., Takayanagi, R., Ogihara, T., Kato, N., 2010. Blood pressure and hypertension are associated with 7 loci in the Japanese population. Circulation 121 (21), 2302–2309.

Tang, N.L., Cran, Y.K., Hui, E., Woo, J., 2000. Application of urine magnesium/creatinine ratio as an indicator for insufficient magnesium intake. Clin. Biochem. 33 (8), 675–678.

Tang, R., Wei, Y., Li, Z., Chen, H., Miao, Q., Bian, Z., Zhang, H., Wang, Q., Wang, Z., Lian, M., Yang, F., Jiang, X., Yang, Y., Li, E., Seldin, M.F., Gershwin, M.E., Liao, W., Shi, Y., Ma, X., 2016. A common variant in CLDN14 is associated with primary biliary cirrhosis and bone mineral density. Sci. Rep. 6, 19877.

Tejpar, S., Piessevaux, H., Claes, K., Piront, P., Hoenderop, J.G., Verslype, C., Van Cutsem, E., 2007. Magnesium wasting associated with epidermal-growth-factor receptor-targeting antibodies in colorectal cancer: a prospective study. Lancet Oncol. 8 (5), 387–394.

Thebault, S., Alexander, R.T., Tiel Groenestege, W.M., Hoenderop, J.G., Bindels, R.J., 2009. EGF increases TRPM6 activity and surface expression. J. Am. Soc. Nephrol. 20 (1), 78–85.

Thorleifsson, G., Holm, H., Edvardsson, V., Walters, G.B., Styrkarsdottir, U., Gudbjartsson, D.F., Sulem, P., Halldorsson, B.V., de Vegt, F., d'Ancona, F.C., den Heijer, M., Franzson, L., Christiansen, C., Alexandersen, P., Rafnar, T., Kristjansson, K., Sigurdsson, G., Kiemeney, L.A., Bodvarsson, M., Indridason, O.S., Palsson, R., Kong, A., Thorsteinsdottir, U., Stefansson, K., 2009. Sequence variants in the CLDN14 gene associate with kidney stones and bone mineral density. Nat. Genet. 41 (8), 926–930.

Unger, S., Gorna, M.W., Le Bechec, A., Do Vale-Pereira, S., Bedeschi, M.F., Geiberger, S., Grigelioniene, G., Horemuzova, E., Lalatta, F., Lausch, E., Magnani, C., Nampoothiri, S., Nishimura, G., Petrella, D., Rojas-Ringeling, F., Utsunomiya, A., Zabel, B., Pradervand, S., Harshman, K., Campos-Xavier, B., Bonafe, L., Superti-Furga, G., Stevenson, B., Superti-Furga, A., 2013. FAM111A mutations result in hypoparathyroidism and impaired skeletal development. Am. J. Hum. Genet. 92 (6), 990–995.

Ure, M.E., Heydari, E., Pan, W., Ramesh, A., Rehman, S., Morgan, C., Pinsk, M., Erickson, R., Herrmann, J.M., Dimke, H., Cordat, E., Lemaire, M., Walter, M., Alexander, R.T., 2017. A variant in a cis-regulatory element enhances claudin-14 expression and is associated with pediatric-onset hypercalciuria and kidney stones. Hum. Mutat. 38 (6), 649–657.

van Angelen, A.A., Glaudemans, B., van der Kemp, A.W., Hoenderop, J.G., Bindels, R.J., 2013. Cisplatin-induced injury of the renal distal convoluted tubule is associated with hypomagnesaemia in mice. Nephrol. Dial. Transplant. 28 (4), 879–889.

Voets, T., Nilius, B., Hoefs, S., van der Kemp, A.W., Droogmans, G., Bindels, R.J., Hoenderop, J.G., 2004. TRPM6 forms the Mg^{2+} influx channel involved in intestinal and renal Mg^{2+} absorption. J. Biol. Chem. 279 (1), 19–25.

Walder, R.Y., Landau, D., Meyer, P., Shalev, H., Tsolia, M., Borochowitz, Z., Boettger, M.B., Beck, G.E., Englehardt, R.K., Carmi, R., Sheffield, V.C., 2002. Mutation of TRPM6 causes familial hypomagnesemia with secondary hypocalcemia. Nat. Genet. 31 (2), 171–174.

Waller, S., Kurzawinski, T., Spitz, L., Thakker, R., Cranston, T., Pearce, S., Cheetham, T., van't Hoff, W.G., 2004. Neonatal severe hyperparathyroidism: genotype/phenotype correlation and the use of pamidronate as rescue therapy. Eur. J. Pediatr. 163 (10), 589–594.

Ward, D.T., McLarnon, S.J., Riccardi, D., 2002. Aminoglycosides increase intracellular calcium levels and ERK activity in proximal tubular OK cells expressing the extracellular calcium-sensing receptor. J. Am. Soc. Nephrol. 13 (6), 1481–1489.

Watanabe, S., Fukumoto, S., Chang, H., Takeuchi, Y., Hasegawa, Y., Okazaki, R., Chikatsu, N., Fujita, T., 2002. Association between activating mutations of calcium-sensing receptor and Bartter's syndrome. Lancet 360 (9334), 692–694.

Weber, S., Schneider, L., Peters, M., Misselwitz, J., Rönnefarth, G., Böswald, M., Bonzel, K.E., Seeman, T., Suláková, T., Kuwertz-Bröking, E., Gregoric, A., Palcoux, J.B., Tasic, V., Manz, F., Schärer, K., Seyberth, H.W., Konrad, M., 2001. Novel paracellin-1 mutations in 25 families with familial hypomagnesemia with hypercalciuria and nephrocalcinosis. J. Am. Soc. Nephrol. 12 (9), 1872–1881.

William, J.H., Danziger, J., 2016. Proton-pump inhibitor-induced hypomagnesemia: current research and proposed mechanisms. World J. Nephrol. 5 (2), 152–157.

Wilson, F.H., Hariri, A., Farhi, A., Zhao, H., Petersen, K.F., Toka, H.R., Nelson-Williams, C., Raja, K.M., Kashgarian, M., Shulman, G.I., Scheinman, S.J., Lifton, R.P., 2004. A cluster of metabolic defects caused by mutation in a mitochondrial tRNA. Science 306 (5699), 1190–1194.

Yamaguti, P.M., Neves, F.A., Hotton, D., Bardet, C., de La Dure-Molla, M., Castro, L.C., Scher, M.D., Barbosa, M.E., Ditsch, C., Fricain, J.C., de La Faille, R., Figueres, M.L., Vargas-Poussou, R., Houillier, P., Chaussain, C., Babajko, S., Berdal, A., Acevedo, A.C., 2017. Amelogenesis imperfecta in familial hypomagnesaemia and hypercalciuria with nephrocalcinosis caused by CLDN19 gene mutations. J. Med. Genet. 54 (1), 26–37.

Zhang, L., Choi, H.J., Estrada, K., Leo, P.J., Li, J., Pei, Y.F., Zhang, Y., Lin, Y., Shen, H., Liu, Y.Z., Liu, Y., Zhao, Y., Zhang, J.G., Tian, Q., Wang, Y.P., Han, Y., Ran, S., Hai, R., Zhu, X.Z., Wu, S., Yan, H., Liu, X., Yang, T.L., Guo, Y., Zhang, F., Guo, Y.F., Chen, Y., Chen, X., Tan, L., Deng, F.Y., Deng, H., Rivadeneira, F., Duncan, E.L., Lee, J.Y., Han, B.G., Cho, N.H., Nicholson, G.C., McCloskey, E., Eastell, R., Prince, R.L., Eisman, J.A., Jones, G., Reid, I.R., Sambrook, P.N., Dennison, E.M., Danoy, P., Yerges-Armstrong, L.M., Streeten, E.A., Hu, T., Xiang, S., Papasian, C.J., Brown, M.A., Shin, C.S., Uitterlinden, A.G., Deng, H.W., 2014. Multistage genome-wide association meta-analyses identified two new loci for bone mineral density. Hum. Mol. Genet. 23 (7), 1923–1933.

Chapter 22

Metal ion toxicity in the skeleton: lead and aluminum

J. Edward Puzas[1] and Brendan F. Boyce[2]
[1]Department of Orthopaedics and Rehabilitation, University of Rochester School of Medicine and Dentistry, Rochester, NY, United States;
[2]Department of Pathology and Laboratory Medicine, University of Rochester School of Medicine and Dentistry, Rochester, NY, United States

Chapter outline

Introduction 527
Research into bone-seeking toxic elements 527
Lead 528
Measurement of lead in bone 529
Lead as an unrecognized risk factor in osteoporosis 529
How does lead cause low bone density? 530
The β-catenin/sclerostin axis: is it the mechanism for lead toxicity in osteoblasts? 531
Mechanism of action of lead: stimulation of sclerostin expression 531
A clinical opportunity 531
Aluminum 531
Summary 534
References 535

Introduction

The skeleton is unique organ system that serves human physiology in many ways. Some of its key functions include mechanical support, protection of vital soft tissue organs, and serving as a reservoir for key minerals such as calcium, phosphorus, protons, and carbonate. The mineral phase of bone is predominantly a crystalline form of hydroxyapatite with the formula $Ca_{10}(PO_4)_6(OH)_2$. However, in actuality, in humans there is extensive substitution of the hydroxyl groups (OH^-) with anions such as fluoride, chloride, and carbonate. About 60% of the volume of bone resides in the mineral phase with about 40% in the organic matrix. And with the skeleton making up about 15% of the total body weight of an average human being, bone can be a very large storage compartment for osseous-seeking elements.

In addition, and with more relevance to this discussion, there can be extensive accumulation of toxic elements that replace calcium in the crystalline structure. These forms of apatite are referred to as "calcium-deficient hydroxyapatites." Approximately 60 cations and anions have been identified as osteotropic elements with avid bone-seeking affinities. About 20 are present in higher quantities with the rest existing in trace amounts. This affinity of charged molecules for hydroxyapatites is a common characteristic of a material with a high charge density. In fact, in the case of hydroxyapatite, its ionic properties have been exploited as a scaffold for the chromatographic partitioning of complex molecules. Thus, it is easy to understand that extending this physical chemistry to the mineral phases in the skeleton provides a large and avid reservoir for an array of molecules and ions.

Research into bone-seeking toxic elements

Surprisingly, it was during the era of the development of the atomic bomb in 1942 that the study of how heavy metals affected bone and other human tissues began in earnest. Literally dozens of elemental by-products were produced by an atomic explosion, but little to nothing was known about how they affected human tissues. The United States dealt with this problem with the launch of a large and sophisticated project known as the Manhattan Project. The project had two major

Principles of Bone Biology. https://doi.org/10.1016/B978-0-12-814841-9.00022-1

scientific directions. The first was centered in the physics laboratories studying the development of the bomb that was carried out in secret atomic energy facilities in Oak Ridge, Tennessee; Los Alamos, New Mexico; and Hanford, Washington. The second direction was devoted to studying how radioactive elements were harbored by specific organs and how they affected tissue function. And as expected, the properties of the skeleton provided a large and high-affinity reservoir for bone-seeking metals such as polonium, plutonium, uranium, thorium, radium, cadmium, cobalt, aluminum, and lead. These investigations were carried out in Rochester, New York, and were the impetus for the growth and development of a world-renowned bone research group in the following decades. The program in Rochester morphed into an Atomic Energy Commission−supported facility in 1947.

After World War II and with the expansion in manufacturing, most of the heavy metal toxins entering the environment were from industrial pollution (Jaishankar, 2014; Nagajyoti, 2010). Heavy metals in wastewater originated from natural sources, soil erosion, natural weathering, and, most importantly, human activities such as mining, industrial effluents, and farming control agents (Morais, 2012).

The most common metals that affect the skeleton are lead and aluminum. Both of these toxic elements contribute to well-recognized bone diseases. The exposure source for lead is generally known to be from the environment and for aluminum is known to be from the environment as well as inadvertent human dosing during clinical procedures such as renal dialysis. Other environmental trace elements that have been identified in bone include cadmium, mercury, and iron, in addition to elements such as cobalt entering humans through injudicious clinical use of orthopedic implants containing mixed metal alloys (Rebolledo, 2011).

Of the key elements that seek the hydroxyapatite of bone, two have been identified as the agents responsible for most of the toxicological pathologies. They are lead and aluminum, with lead being identified as the most dangerous to skeletal metabolism at the present time.

Lead

Much is known regarding the toxicity of lead in neural, renal, and hematopoietic development (reviewed in Wani et al., 2015). Its effects are uniformly devastating to the normal physiology of not only these well-studied tissues but virtually every organ system in the body. However, what has been underappreciated since the 1970's is that bone is also a key target for lead toxicity. This is due to two important features related to how lead interacts with the skeleton; they are the sequestering of lead in bone and the slow turnover and long half-life of lead in the mineral phase of bone.

Virtually all of the lead (94%−97%) in adult humans resides in the skeleton, with the remaining sites being (in order of content) red blood cells, liver, skeletal muscle, skin, kidney, lung, and brain (and other soft tissues) (Barry, 1975). The half-life of lead in bodily tissues is known. It is approximately 1 month in blood, approximately 2 months in soft tissues, and 25−30 years in bone (ATSDR, 2007).

Consider the implications of these facts. Essentially any individual who has been exposed to lead either from the environment or from occupational sources will carry a significant portion of that burden for his or her entire lifetime. In fact, once a person is no longer exposed to a lead source, blood and soft tissue levels may decline to low levels, but skeletal lead can remain unacceptably high and continue to adversely affect cells in the bone compartment.

The reason the skeletal half-life for lead is so long is due to two key properties of bone. First, lead is incorporated into the crystal lattice structure of hydroxyapatite by displacing calcium ions. Its affinity for the mineral phases of bone is very high, such that when lead ions are mobilized through bone resorption, they can redistribute to other mineral sites and not leave the bone compartment. Second, skeletal turnover in humans is slow. Estimates indicate that the entire skeleton turns over at a rate of approximately 10% per year, with cortical bone averaging 4% per year and trabecular bone averaging 28% per year (Manolagas, 2000). Thus, there is little opportunity to decrease the level of lead in osseous tissues through normal metabolism. Also, as might be expected, the accumulation and release of lead levels in cortical and trabecular bone follow the general metabolic activity of their particular compartments. That is, (1) acute lead exposure will lead to a more rapid rise in trabecular bone than cortical bone and (2) lead release from the skeleton will occur first from trabecular bone and later from cortical bone.

Another feature of the long half-life of lead in bone is that the skeleton can continue to dose the blood and soft tissues throughout life. That is, in low-skeletal-turnover states lead remains in the bone compartment, but in high-turnover states it is released into the circulation. This accounts for the long-standing finding that blood lead levels rise in all females during and after the menopause due to enhanced skeletal turnover upon the loss of estrogen (Silbergeld et al., 1988; Potula, 2006). In fact, any acceleration in skeletal turnover (i.e., calcium deficiency, fracture, hyperparathyroidism, pregnancy, lactation, etc.) will lead to elevated blood lead levels (ATSDR, 2007).

Measurement of lead in bone

The measurement of lead levels in a human skeleton can be performed either with invasive or with noninvasive techniques. Invasive measurements are, as one might expect, limited to situations in which an actual sample of bone can be obtained from a living subject undergoing surgery or from a cadaveric specimen. The "pros" to using a specimen of bone are (1) the quantification methods (using atomic absorption spectrometry or mass spectrometry) are highly accurate and (2) different compartments of bone (i.e., cortical vs. trabecular) can be measured separately. The "cons" to using specimens of bone are the invasiveness of the specimen procurement and the unfeasibility of making longitudinal measurements in any one person.

Interestingly, it is possible to measure lead levels in the skeleton in a noninvasive, safe way utilizing an X-ray fluorescence technique. This method was first described in 1976 by Ahlgren et al. (Ahlgren, 1976) and later perfected by the physicists at McMaster University, Hamilton, Ontario (Chettle, 1991).

The principle behind this method depends on the use of an exciting isotope (^{109}Cd) and a sensitive X-ray detection system. ^{109}Cd emits characteristic energies of gamma rays that when directed toward a bone (close to the surface of the skin) can excite an inner shell (i.e., K-shell) electron in lead atoms. Upon deexcitation of the activated lead atom, specific and characteristic X-rays are emitted and detected by the detection sensors. The magnitude of this fluorescence decay is proportional to the number of lead atoms activated by the ^{109}Cd. The system is calibrated with lead-doped samples and after the appropriate calculations are made, a value of micrograms of lead/gram of bone can be made. The "pros" for this method are that (1) it is noninvasive, (2) longitudinal measurements can be made in a single subject, and (3) the radiation dose is extremely low, with little safety concern. The "cons" are (1) accuracy of the measurements in bones with a low lead content is low, (2) the subject must remain in the scanner for 30–45 min for one measurement, and (3) there are no commercially available systems; each system must be constructed from specialty equipment. Nevertheless, the ability to make noninvasive measurements led to a flurry of activity in the 1990s using K-shell X-ray fluorescence to measure many different cohorts of people (reviewed in Hu, 1998).

Lead as an unrecognized risk factor in osteoporosis

An unexpected technical discrepancy in measuring bone mineral density (BMD) by dual-energy X-ray absorptiometry (DXA) prevented clinicians and investigators from recognizing that lead exposure could cause osteopenia and contribute to the genesis of osteoporosis. This technical anomaly occurred in the early years when first-generation DXA scanners were first being used in the diagnosis of osteoporosis. It appears that the presence of lead in bone created an artifactually elevated estimate of bone density. The reason for this is not entirely clear but might be related to the very high atomic mass of lead. Reports in the literature bore out this finding. Escribano and colleagues (1997) showed in an animal study that lead exposure in rats led to a marked decrease in the histomorphometric parameters of trabecular bone volume, number, and thickness. But when these same animals were evaluated with early models of densitometers (i.e., dual-photon absorptiometers) the rat bones showed significant elevations in bone density.

A qualitatively similar result was described by Laraque et al. (1990). In this report, the investigators showed that children with elevated blood lead levels showed a significantly higher bone mineral content (BMC) than a control cohort available at that time. The BMC measurements were made with a single-photon absorptiometer.

To definitively document these discrepancies, it would be best to make a direct comparison of bone density using both the older and the newer-generation technologies. Fortuitously, while upgrading densitometers, our research group had the opportunity to perform a direct comparison between a pencil-beam dual-photon absorptiometer (Lunar DPX-L) and a fan-beam dual X-ray absorptiometer (Lunar Prodigy). In this experiment, bovine bone was doped with known trace amounts of lead and then measured in both scanners. The (unpublished) data are shown in Fig. 22.1. The addition of "micrograms" of lead to "grams" of bone (i.e., parts per million) should not have changed the total mass of the bone sample to any measurable degree. Yet, the DPX-L scanner showed a clear increase in BMD with added lead. If this measurement were being made in a human it would have increased the patient's T score by 0.5–1.0 standard deviations. The Prodigy scanner was also affected by lead in the sample; however, the effect was not evident until a level near 1000 μg Pb/g bone was reached. A level of 1000 μg Pb/g bone is at least 100 times what an exposed human would be expected to have and has never been reported. Discrepancies in the detection and calculations for BMD between dual-photon scanners and dual-X-ray scanners have been documented in the literature (Huffman, 2005).

Two important points can be derived from these observations. First, many years ago lead exposure and its sequestration in the skeleton probably was overlooked as a risk factor for low bone density and osteoporosis. This was due to the apparent interference by lead in the photon-based methods used in bone densitometers. Second, new-generation scanners

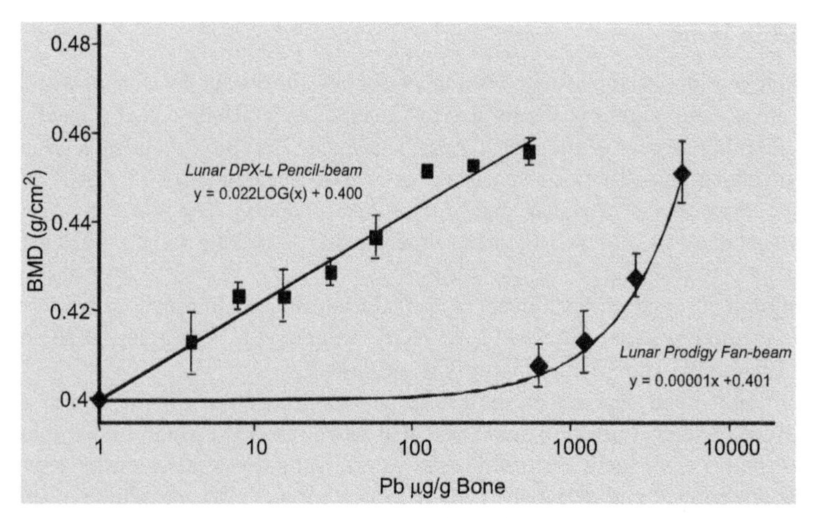

FIGURE 22.1 Bone densitometry in specimens containing measured amounts of lead. Two grams of pulverized bovine bone containing graded amounts of lead salts was scanned with a Lunar Dual-Photon DPX-L densitometer or a GE/Lunar Dual-Energy X-ray absorptiometer. The bone specimens were contained in tissue culture wells (area = 2.0 cm^2). The data are presented as areal density, i.e., "grams of mineralized bone/cm^2 versus micrograms of elemental lead per gram of bone." The bovine bone specimens were not incubated to dryness prior to making the measurements and thus their "weight percent" was approximately 40% mineral and 60% organic matrix + water. Lead (in the parts per million range) affected the bone mineral density (*BMD*) measurement at as little as 5.0 μg/g bone when scanned with the dual-photon densitometer. The dual-energy X-ray densitometer did not detect the added lead until the levels reached supratoxicological levels (approximately 1000 μg/g bone).

with the different mechanisms for generating X-rays and with different calculation algorithms do not appear to be affected by skeletal lead in the range seen in the human population.

How does lead cause low bone density?

In an adult skeleton, bone mass is controlled through regulation of bone-forming activities (by osteoblasts) and bone-resorbing activities (by osteoclasts). When these two cellular processes are in balance with each other, bone mass is maintained at a healthy, steady-state level. However, when these two processes are no longer balanced bone mass can be pathologically increased (i.e., osteopetrosis) or, more commonly, pathologically decreased (i.e., osteoporosis). The skeleton maintains an exquisite series of check and balances to prevent large swings in bone mass. This was recognized many years ago when Harris and Heaney described the kinetically coupled activities of bone-forming and bone-resorbing cells (Harris, 1969). Thus, for an agent such as lead to alter bone mass, it must disrupt the balance between osteoblastic and osteoclastic cellular activities. The question that arises is, which process is the target of lead toxicity in the skeleton?

Given the observation that lead can be easily mobilized in high-turnover states in the skeleton (as mentioned earlier) and that basic science investigations do not show any major effects on osteoclastic bone resorption (Carmouche, 2005), it seems that the important cellular target for lead toxicity is the osteoblast.

Osteoblasts and their progenitor cells residing in the bone compartment can be exposed to locally high concentrations of lead. Assuming conservative values of the lead content of the skeleton (i.e., 5 μg/g bone), the amount of bone resorbed by an osteoclast (i.e., 1 μg), and the diffusion of lead into a radius of 100 μm around a lacuna, an approximate estimate of the lead concentration at a remodeling site can be on the order of 1−10 μM. From basic science investigations, this concentration of lead is sufficient to have profound inhibitory effects on osteoblast function (Beier et al., 2015).

As expected, the in vitro effects of lead manifest themselves in whole-animal models as a decrease in bone quantity and bone quality. Exposing mice (which happen to be an excellent model for lead-induced bone loss) to lead in their drinking water will induce an osteoporotic-like phenotype (Beier et al., 2013). By altering the concentration of lead acetate in the animals' drinking water it is possible to achieve virtually any relevant concentration of the metal ion in blood and bone in 3−6 weeks (Carmouche, 2005; Beier et al., 2013). Interestingly, animals display no difficulty in drinking the water, as the addition of lead confers a pleasant sweet taste. (Unfortunately, this also occurs in human infants around lead-contaminated materials.) After 6 weeks of exposure at 55 ppm lead in water, micro-computed tomographic and histological analyses demonstrate significant changes. They are:

- decrease in skeletal bone volume/total volume
- decrease in volumetric BMD
- decrease in trabecular number
- increase in trabecular separation
- decrease in bone apposition rate
- no change in osteoclast number

The β-catenin/sclerostin axis: is it the mechanism for lead toxicity in osteoblasts?

The *SOST* gene encodes a protein known as sclerostin. The gene is located on chromosome 17q12−q21. Osteocytes were one of the first cells shown to express sclerostin; however, it now appears that the cells from at least four tissues can produce this molecule. They include uterus, kidney, tissues of tooth development, and, most importantly, osteoblasts (Maeda, 2007; Blish, 2008; Murashima-Suginami, 2008; Tanaka, 2008; Rutger, 2005). Sclerostin is a very potent inhibitor of bone formation. It represses osteoblast function and number and can induce apoptotic pathways in these cells. The mechanism of action of sclerostin appears to occur through the Wnt/β-catenin pathway. Data suggest that sclerostin binds to low-density lipoprotein receptor−related protein 5/6, preventing association with the Wnt receptor, Frizzled. This interference blocks Wnt binding and depresses osteoblastic bone formation (van Bezooijen, 2004; Li, 2005).

What is most remarkable regarding this intracellular pathway is that when osteoblasts are exposed to lead there is a manifold upregulation of both the mRNA and the protein for sclerostin (Beier et al., 2015). This stimulation is observed in vitro (with qPCR and western blot assays) and in vivo (with immunohistochemical quantification). Thus, it is likely that during remodeling of lead-containing bone, excess sclerostin expression uncouples osteoblastic bone formation from resorption and an inadequate amount of new bone is deposited. Over many remodeling cycles this would lead to a decrease in skeletal mass.

Mechanism of action of lead: stimulation of sclerostin expression

How can a heavy metal divalent cation specifically affect a key gene in osteoblasts? This is a question that is not yet fully answered; however, the work of Beier (Beier et al. 2013, 2014, 2015) and Holz (Holz, 2012) provides some clues. Lead ion is known to interact with a number of kinase enzymes that utilize ATPase reactions to drive phosphorylation (protein kinase C, calmodulin, Na/K-ATPase, Ca-ATPase) (Flora et al, 2008; Murakami, 1993; Kramer, 1986). One of these kinases is the type 2 transforming growth factor β (TGFβ) receptor, responsible for the phosphorylation of Smad3, and thus it would be expected that lead can block Smad3 reporter activity, decrease phospho-Smad3 intracellular protein levels, and prevent phosphorylation of Smad3 in a cell-free system (Beier et al., 2015). The relevance of these observations lies in the fact that through TGFβ/Smad3 signaling, gene expression of sclerostin is repressed (Beier et al., 2015) and thus, any mechanism that interferes with phospho-Smad3 formation would derepress *SOST* gene activation and stimulate sclerostin levels. A diagrammatic representation of this regulatory pathway is presented in Fig. 22.2.

A clinical opportunity

From a clinical perspective, it may be possible to directly treat lead-induced osteoporosis with a pharmaceutical agent that was recommended in January 2019 for approval by the US FDA Advisory Panel (AMG 785, romosozumab). This humanized blocking antibody inhibits sclerostin activity. If the effect of lead in humans is mediated, even in small part, by upregulation of sclerostin, then this therapeutic agent could be extremely effective in treating the low bone mass. This is a real possibility given that romosozumab has shown very promising results in stimulating bone formation in a large phase III clinical trial (i.e., the ARCH study, Saag, 2017).

Aluminum

Aluminum (Al) toxicity-induced osteomalacia was a common complication of hemodialysis for some patients with chronic renal failure in the 1970s and 1980s, until it was discovered that high levels of aluminum in the tap water used to make the dialysis fluid was the main cause. High serum Al levels (100−800 μg/L vs. normal values of <35 μg/L) from the contaminated water resulted in so-called "dialysis" dementia as well as osteomalacia and microcytic anemia (Alfrey, 1976;

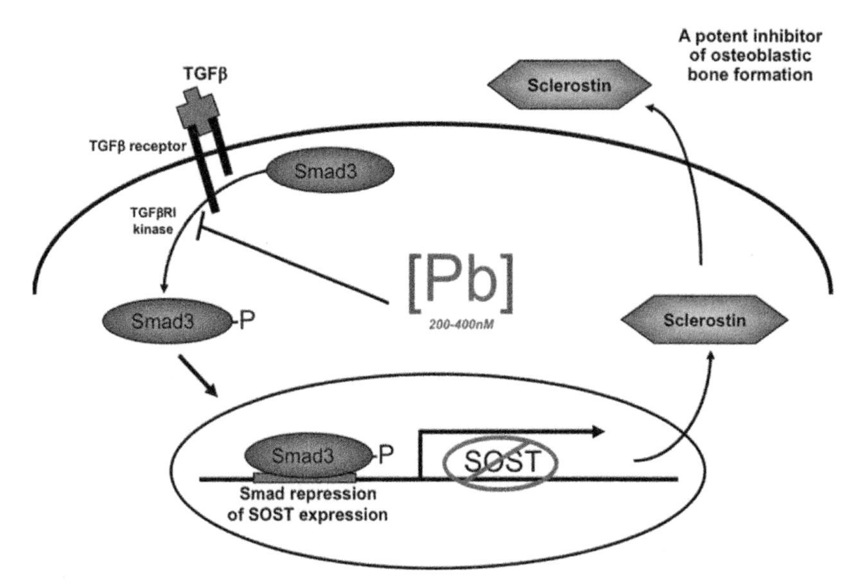

FIGURE 22.2 Intracellular mechanism by which lead induces sclerostin synthesis. Lead, in the 200–400 nM range, can block the phosphorylation of Smad3. Under normal conditions phospho-Smad3 resides in the nucleus and binds to a DNA-response element that represses *SOST* gene expression and sclerostin synthesis. However, in the presence of lead phospho-Smad3 levels are decreased and *SOST* gene activation is "derepressed," evoking a large increase in sclerostin production. *TGFβ*, transforming growth factor β.

Parkinson, 1979; Ellis, 1979; Boyce, 1982), and most patients with this dementia died of it before the cause was discovered. Aluminum is the most common metal in the earth's crust and it can be leached from surface rocks by rain, particularly acid rain (Krewski, 2007). This leaching can lead to high Al levels occurring naturally in water in some reservoirs. Thus, depending upon local conditions, levels of Al in the tap water used in dialysis units or for home dialysis can vary significantly around the world, from <30 μg/L (normal and acceptable) to ~100 μg/L (Boyce, 1982) or >200 μg/L (Krewski, 2007), which is the upper limit of the US Environmental Protection Agency's secondary maximum contaminant level for total Al in water and above which water becomes colored (Willhite, 2012).

Importantly, however, the highest levels of Al were found in water from reservoirs to which Al had been added by water authorities in the form of aluminum sulfate to cause aggregation (flocculation) and subsequent precipitation of suspended particles that discolored the water (Krewski, 2007). This fulfilled legal mandates that tap water must be clear, but it led to Al levels in water from some reservoirs being >1000 μg/L (Boyce, 1982). The Al passed directly from the dialysis fluid into patients' bloodstreams, resulting in cumulative toxicity with each dialysis, typically three per week, because Al is excreted normally by the kidneys, and the lack of renal function in dialysis patients compounded the damage (Alfrey, 1980).

After the cause of the dementia and osteomalacia was identified, Al was removed from contaminated tap water by passing it through reverse osmosis and/or deionization units before being used for dialysis. However, this did not prevent recurrence of the toxicity in all cases until it was fully appreciated that these units needed to be serviced regularly and that Al levels had to be measured in the water flowing from them to confirm that they were working efficiently. This is particularly necessary in home dialysis settings in towns where Al levels are very high as a result of the addition of aluminum sulfate and where monitoring of water Al levels in each home is the responsibility of local authorities.

The identification of aluminum toxicity as a cause of dementia then led to studies to determine if lifelong exposure to aluminum in drinking water and from other sources could be a cause of Alzheimer's disease. Some studies reported Al deposits in brains of Alzheimer's patients (Crapper, 1973; Perl, 1980) and these reports led to numerous epidemiologic studies. Although Al concentrations increase in brain and bone (Hellstrom, 2005) with age in humans, other investigators did not confirm these findings in neuritic plaques in brains from Alzheimer's patients (Landsberg, 1992). Nevertheless, improvement in cognitive function has been reported in some Alzheimer's patients given treatment to reduce blood and tissue levels of aluminum, supporting the "aluminum/dementia hypothesis" (Davenward, 2013). Overall, however, a potential causative role for Al in the pathogenesis of Alzheimer's disease remains controversial (Exley, 2014).

Aluminum toxicity can also occur as a consequence of ingestion of aluminum-containing phosphate-binding drugs given to patients with chronic renal failure to help reduce absorption of phosphate from the intestinal tract (Kaehny et al., 1977a). Serum phosphate levels are high in these patients because they are unable to excrete phosphate through their dysfunctional kidneys. Initially, it was thought that aluminum in phosphate binders, such as aluminum hydroxide, was not absorbed from the gastrointestinal tract. However, this was proven to be wrong by the finding of high serum aluminum levels not only in patients with renal failure not yet on dialysis, but also in individuals with normal renal function taking large quantities of aluminum hydroxide as an antacid for indigestion (Kaehny et al., 1977a). Aluminum absorption from the gut is enhanced by acidic fruits, such as oranges and lemons, and orange juice. Aluminum can be leached from aluminum cooking pots and wrapping foil and this can lead to high Al concentrations, for example, in the juice from chickens wrapped and cooked in aluminum foil along with lemons (Krewski, 2007).

Lead, cadmium, and arsenic can also leach out of cooking utensils during food preparation, which may be a more pressing contemporary issue in developing countries (Weidenhamer, 2017). Aluminum was also found to be a contaminant of casein hydrolysate used as the amino acid source in some intravenous preparations for patients receiving total parenteral nutrition in the 1980s (Klein, 1982), with some patients being reported to receive 2–3 mg Al/day (Kaehny et al., 1977b). Aluminum was also found in some albumin solutions used to expand the intravascular volume in patients with acute hypovolemic states and as the replacement fluid used during plasma exchange (Maharaj, 1987). These problems have largely been solved by removal of Al from medical fluids given to patients.

Patients with end-stage chronic renal failure typically develop hypocalcemia as a consequence of inadequate phosphate excretion and calcium reabsorption in their damaged kidneys, coupled with their inability to 1-hydroxylate 25(OH) vitamin D_3 because their renal tubular cells become deficient in 1α-hydroxylase (Goodman, 2008). Consequently, they develop secondary hyperparathyroidism with associated increased bone remodeling, which can be accompanied by vitamin D deficiency-related osteomalacia unless they are given appropriate treatment to prevent these (Sherrad, 1993). These treatments include dietary phosphate restriction, phosphate binders, calcimimetics, calcitriol, or vitamin D analogs, or a combination of calcimimetics with calcitriol or vitamin D analogs, and addition of calcium to dialysis fluid (KDIGO, 2017; Jamai, 2013; Cocchiara, 2017; Vestergaard, 2011). If patients are exposed to sufficiently high levels of Al, mineralization will be inhibited because the aluminum gets deposited along the mineralization front between osteoid and calcified matrix where it prevents calcification and leads to osteomalacia (Ellis, 1979; Boyce, 1981, 1982, 1992). Al-related osteomalacia became recognized as an entity distinct from vitamin D deficiency-related osteomalacia because of its association with dementia, microcytic anemia, myopathy, fractures, and a tendency to develop hypercalcemia (Boyce, 1982). These were accompanied by a characteristic focal distribution of thickened osteoid seams on trabecular surfaces (Fig. 22.3) (Boyce, 1982) compared with the typical extensive distribution of thickened osteoid seams in vitamin D deficiency-related osteomalacia (Fig. 22.4). The fact that osteoid seams were much thicker than those seen in biopsy specimens from patients with secondary hyperparathyroidism and osteoid seams of normal thickness indicates that aluminum does not inhibit the formation and deposition of osteoid matrix by osteoblasts in vivo in many patients. Some studies have reported aluminum staining along osteoid surfaces associated with normal osteoid thickness in addition to reduced bone formation rates and osteoblast surfaces, suggesting that aluminum may inhibit mineralization and bone formation in some patients

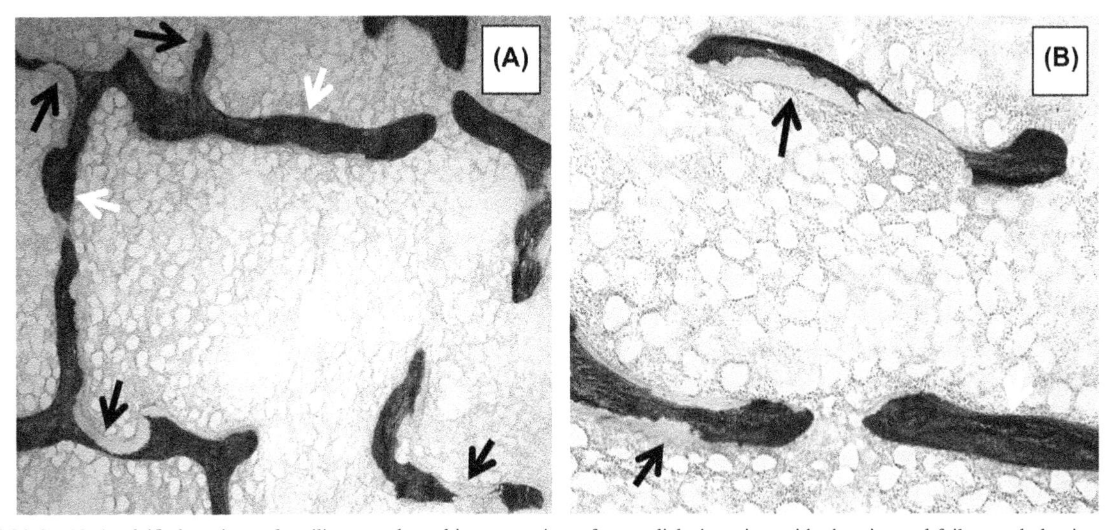

FIGURE 22.3 Undecalcified sections of an iliac crest bone biopsy specimen from a dialysis patient with chronic renal failure and aluminum-induced osteomalacia. Note the characteristic focal distribution of thickened osteoid seams (*black arrows*) on the trabecular surfaces, much of which is fully calcified (*white arrows*). Toluidine blue staining. (A) ×4 original magnification, (B) ×10 original magnification.

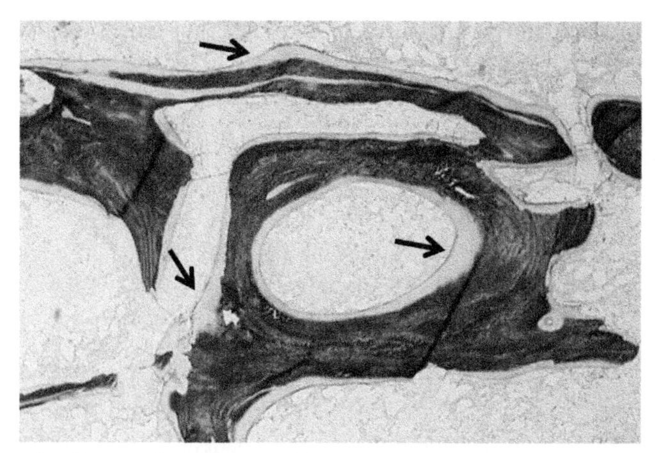

FIGURE 22.4 Undecalcified section of an iliac crest bone biopsy specimen from a patient with alcoholic cirrhosis and vitamin D deficiency–induced osteomalacia. Thickened osteoid seams (*black arrows*) are present on most of the trabecular surfaces. Toluidine blue staining, ×4 original magnification.

(de Vernejoul, 1985; Dunstan, 1984; Parisien, 1988). Indeed, inhibitory effects of aluminum have been reported on osteoblasts and mineralization in vitro (Sprague, 1993; Sun, 2016). Aluminum can be incorporated into mineralizing bone without inhibiting calcification (Chappard, 2016) and interestingly, the thickened osteoid seams can mineralize after Al sources have been removed (Boyce, 1992). The hypercalcemia in aluminum toxicity developed as a consequence of treatment with calcitriol, which enhanced the intestinal absorption of calcium, and the addition of calcium to dialysis fluid (Boyce, 1982). This hypercalcemia sometimes led to unnecessary and potentially fatal parathyroidectomy in patients who were mistakenly diagnosed with tertiary hyperparathyroidism. Aluminum toxicity has also been implicated in the pathogenesis of so-called aplastic or low-turnover renal osteodystrophy (Sherrard, 1996). In this condition, bone turnover is suppressed and osteoid seams are typically thin, and in some of these cases aluminum has been detected along mineralization fronts. It is associated with low bone mass and increased risk of fractures (Cannata-Andia, 2013). Although high concentrations of Al can be toxic to osteoblasts cultured in vitro, the cause of this low-turnover osteodystrophy in renal failure remains poorly understood.

Bone biopsy became the gold standard for the diagnosis and management of Al-induced osteodystrophy in renal failure in many academic nephrology departments in the 1970s through the 1990s. However, as water contamination with Al became more generally recognized as the cause of the osteodystrophy, and water purification measures kept dialysis water Al levels low, and Al-containing phosphate binders were replaced by others containing calcium, bone biopsy became redundant in most dialysis centers. Today, few centers use biopsy to help manage patients with chronic renal failure. Consequently, few nephrologists know how to perform the technique or have the special 6- to 8-mm-wide biopsy needles used typically to get specimens, which were taken from the iliac crest. These biopsy specimens also required special processing and embedding undecalcified in plastic. This allowed osteoid to be distinguished from calcified matrix and bone mineralization and formation rates to be measured in patients given prior double tetracycline labeling. This is not possible when specimens are decalcified and embedded in paraffin, which is how most bone samples are handled in pathology laboratories. Special heavy-duty microtomes are also required to cut sections from the plastic blocks. Most pathology departments do not have these microtomes or technicians who know how to prepare sections using them, nor do they have pathologists trained to interpret sections of undecalcified bone. There are a small number of physicians and scientists in the world today who know how to interpret undecalcified human bone biopsy specimen slides, and most are nearing or have passed retirement age. Thus, it is possible that Al toxicity remains a problem for patients with chronic renal failure and for others exposed to additional sources of Al (Chappard, 2016) but that it is going undetected.

Summary

Metal ion toxicity is an underinvestigated and underappreciated issue in skeletal biology. With the continued accumulation of environmental and industrial agents in our ecosystem that may have an adverse effect on osseous tissues, it might be expected that research on this topic would be an important and well-studied area. However, it seems that among clinicians and scientists in the musculoskeletal field there is little interest in the world of toxicological science, and in the fields of environmental health science there is little interest in bones and joints. Given the fact that human biology must cope with changes in the environment, this lack of interest is unfortunate. As discussed in this treatise, the skeleton can be an avid reservoir for toxins, especially heavy metals. Two of these metals, lead and aluminum, are at the forefront of scientific awareness. It is now time that other skeletal health-modifying environmental compounds be researched.

References

Ahlgren, L., Liden, K., Mattsson, S., Tejning, S., 1976. X-ray fluorescence analysis of lead in human skeleton in vivo. Scand. J. Work. Environ. Health 2, 82–86.

Alfrey, A.C., 1980. Aluminum metabolism in uremia. Neurotoxicology 1, 43–53.

Alfrey, A.C., LeGendre, G.R., Kaehny, W.D., 1976. The dialysis encephalopathy syndrome. Possible aluminum intoxication. N. Engl. J. Med. 294, 184–188.

ATDSR (Agency for Toxic Substance and Disease Registry), 2007. Toxicological Profile for Lead-Update. U.S. Department of Health & Human Services, Public Health Service, Atlanta.

Barry, P.S.I., 1975. A comparison of concentrations of lead in human tissue. Br. J. Ind. Med. 32, 119–139.

Beier, E.E., Maher, J.R., Sheu, T.J., Cory-Slechta, D.A., Berger, A.J., Zuscik, M.J., Puzas, J.E., 2013. Heavy metal lead exposure, osteoporotic-like phenotype in an animal model, and depression of Wnt Signaling. Envion. Health Perspect. 121 (1), 97–104.

Beier, E.E., Sheu, T.J., Buckley, T., Yukata, K., O'Keefe, R.J., Zuscik, M.J., Puzas, J.E., 2014. Inhibition of beta-catenin signaling by Pb leads to incomplete fracture healing. J. Orthoptera Res. 32 (11), 1397–1405.

Beier, E.E., Sheu, T.J., Dang, D., Holz, J.D., Ubayawardena, R., Babij, P., Puzas, J.E., 2015. Heavy metal ion regulation of gene expression: mechanisms by which lead inhibits osteoblastic bone forming activity through modulation of the Wnt/β-catenin signaling pathway. J. Biol. Chem. 290 (29), 18216–18226.

Blish, K.R., Wang, W., Willingham, M.C., Du, W., Birse, C.E., Krishnan, S.R., Brown, J.C., Hawkins, G.A., Garvin, A.J., D'Agostino Jr., R.B., Torti, F.M., Torti, S.V., 2008. A human bone morphogenetic protein antagonist is down-regulated in renal cancer. Mol. Biol. Cell 19 (2), 457–464. PMC2230586.

Boyce, B.F., Elder, H.Y., Fell, G.S., Nicholson, W.A.P., Smith, G.D., Dempster, D.W., Gray, C.C., Boyle, I.T., 1981. Quantitation and localization of aluminium in human cancellous bone in renal osteodystrophy. Scanning Electron. Microsc. 3, 329–337.

Boyce, B.F., Fell, G.S., Elder, H.Y., Junor, B.J., Elliot, H.L., Beastall, G., Fogelman, I., Boyle, I.T., 1982. Hypercalcaemic osteomalacia due to aluminum toxicity. Lancet 2 (8306), 1009–1013.

Boyce, B.F., Byars, J., McWilliams, S., Mocan, M.Z., Elder, H.Y., Boyle, I.T., Junor, B.J., 1992. Histological and electron microprobe studies of mineralization in aluminium-related osteomalacia. J. Clin. Pathol. 45, 502–508.

Cannata-Andía, J.B., Rodriguez-García, M., Gómez-Alonso, C., 2013. Osteoporosis and adynamic bone in chronic kidney disease. J. Nephrol. 26, 73–80.

Carmouche, J.J., Puzas, J.E., Zhang, X., Tiyapatanaputi, P., Cory-Slechta, D.A., Gelein, R., Zuscik, M., Rosier, R.N., Boyce, B.F., O'Keefe, R.J., Schwarz, E.M., 2005. Lead exposure inhibits fracture healing and is associated with increased chondrogenesis, delay in cartilage mineralization, and a decrease in osteoprogenitor frequency. Environ. Health Perspect. 113, 749–755.

Chappard, D., Bizot, P., Mabilleau, G., Hubert, L., 2016. Aluminum and bone: review of new clinical circumstances associated with Al(3+) deposition in the calcified matrix of bone. Morphologie 100, 95–105.

Chettle, D.R., Scott, M.C., Somervaile, L.J., 1991. Lead in bone: sampling and quantitation using K X-rays excited by ^{109}Cd. Environ. Health Perspect. 91, 49–55.

Cocchiara, G., Fazzotta, S., Palumbo, V.D., Damiano, G., Cajozzo, M., Maione, C., Buscemi, S., Spinelli, G., Ficarella, S., Maffongelli, A., Caternicchia, F., Ignazio Lo Monte, A., Buscemi, G., 2017. The medical and surgical treatment in secondary and tertiary hyperparathyroidism. Review. Clin Ter 168, 158–167.

Crapper, D.R., Krishnan, S.S., Dalton, A.J., May 4, 1973. Brain aluminum distribution in Alzheimer's disease and experimental neurofibrillary degeneration. Science 180 (4085), 511–513.

Davenward, S., Bentham, P., Wright, J., et al., 2013. Silicon-rich mineral water as a non-invasive test of the 'aluminium hypothesis' in Alzheimer's disease. J. Alzheimer's Dis. 33 (2), 423–430.

de Vernejoul, M.C., Belenguer, R., Halkidou, H., Buisine, A., Bielakoff, J., Miravet, L., 1985. Histomorphometric evidence of deleterious effect of aluminum on osteoblasts. Bone 6, 15–20.

Dunstan, C.R., Evans, R.A., Hills, E., Wong, S.Y., Alfrey, A.C., 1984. Effect of aluminum and parathyroid hormone on osteoblasts and bone mineralization in chronic renal failure. Calcif. Tissue Int. 36, 133–138.

Ellis, H.A., McCarthy, J.H., Herrington, J., 1979. Bone aluminum in haemodialysed patients and in rats injected with aluminum chloride: relationship to impaired bone metabolism. J. Clin. Pathol. 32, 832–844.

Escribano, A., Revilla, M., Hernandez, E.R., Seco, C., Gonzalez-Riola, J., Villa, L.F., Rico, H., 1997. Effect of lead on bone development and bone mass: a morphometric, densitometric, and histomorphometric study in growing rats. Calcif. Tissue Int. 60, 200–203.

Exley, C., June 2014. What is the risk of aluminium as a neurotoxin? Expert Rev. Neurother. 14 (6), 589–591.

Flora, S.J., Mittal, M., Mehta, A., 2008. Heavy metal induced oxidative stress & its possible reversal by chelation therapy. Indian J. Med. Res. 128 (4), 501–523.

Goodman, W.G., Quarles, L.D., 2008. Development and progression of secondary hyperparathyroidism in chronic kidney disease: lessons from molecular genetics. Kidney Int 74 (3), 276–288.

Harris, W.H., Heaney, R.P., 1969. Skeletal renewal and metabolic bone disease. N. Engl. J. Med. 280 (4), 193–202.

Hellstrom, H.O., Mjoberg, B., Mallmin, H., Michaelsson, K., 2005. The aluminum content of bone increases with age, but is not higher in hip fracture cases with and without dementia compared to controls. Osteoporos. Int. 16, 1982e−1988.

Holz, J.D., Beier, E., Sheu, T.J., Ubayawardena, R., Wang, M., Sampson, E.R., Rosier, R.N., Zuscik, M., Puzas, J.E., 2012. Lead induces an osteoarthritis-like phenotype in articular chondrocytes through disruption of TGF-β signaling. J. Orthop. Res. 30 (11), 1760−1766.

Hu, H., August 1998. Bone lead as a new biologic marker of lead dose: recent findings and implications for public health. Environ. Health Perspect. 106 (Suppl. 4), 961−967.

Huffman, D.M., Niamh, M.A., Landy, M., Potter, E., Nagy, T.R., Gower, B.A., January 1, 2005. Comparison of the Lunar DPX-L and Prodigy dual-energy X-ray absorptiometers for assessing total and regional body composition. Int. J. Body Compos. Res. 3 (1), 25−30.

Jaishankar, M., Mathew, B.B., Shah, M.S., Gowda, K.R.S., 2014. Biosorption of few heavy metal ions using agricultural wastes. J. Environ. Pollution Human Health 2 (1), 1−6.

Jamal, S.A.1, Miller, P.D., 2013 . Secondary and tertiary hyperparathyroidism. J. Clin. Densitom. 16 (1), 64−68.

Kaehny, W.D., Hegg, A.P., Alfrey, A.C., 1977a. Gastrointestinal absorption of aluminum from aluminum-containing antacids. N. Engl. J. Med. 296, 1389−1390.

Kaehny, W.D., Alfrey, A.C., Holman, R.E., et al., 1977b. Aluminum transfer during hemodialysis. Kidney Int. 12, 361−365.

KDIGO, 2017. Clinical practice guideline update for the diagnosis, evaluation, prevention, and treatment of chronic kidney disease−mineral and bone disorder (CKD-MBD). Kidney Int. Suppl. 7, 1−59.

Klein, G.L., Alfrey, A.C., Miller, N.L., et al., 1982. Aluminum loading during total parenteral nutrition. Am. J. Clin. Nutr. 35, 1425−1429.

Kramer, H.J., Gonick, C., Lu, E., 1986. In vitro inhibition of Na-K-ATPase by trace metals: relation to renal and cardiovascular damage. Nephron 44, 329−336.

Krewski, D., et al., 2007. Human health risk assessment for aluminum, aluminum oxide and aluminum hydroxide. J. Toxicol. Environ. Health B Crit. Rev. 10 (Suppl. 1), 1−269.

Landsberg, J.P., McDonald, B., Watt, F., November 5, 1992. Absence of aluminium in neuritic plaque cores in Alzheimer's disease. Nature 360 (6399), 65−68.

Laraque, D., Arena, L., Karp, J., Gruskay, D., 1990. Bone mineral content in black preschoolers: normative data using single photon absorptiometry. Pediatr. Radiol. 20, 461−463.

Li, X., 2005. Sclerostin binds to LRP5/6 and antagonizes canonical Wnt signaling. J. Biol. Chem. 280, 19883−19887.

Maeda, K., Lee, D.S., Yanagimoto Ueta, Y., Suzuki, H., 2007. Expression of uterine sensitization-associated gene-1 (USAG-1) in the mouse uterus during the peri-implantation period. J. Reprod. Dev. 53 (4), 931−936.

Maharaj, D., Fell, G.S., Boyce, B.F., Ng, J.P., Smith, G.D., Boulton-Jones, J.M., Cumming, R.L., Davidson, J.F., 1987. Aluminum bone disease in patients receiving plasma exchange with contaminated albumin. Br. Med. J. 295, 693−696.

Manolagas, S.C., April 2000. Birth and death of bone cells: basic regulatory mechanisms and implications for the pathogenesis and treatment of osteoporosis. Endocr. Rev. 21 (2), 115−137.

Morais, S., Costa, F.G., Pereira, M.L., 2012. Heavy metals and human health. In: Oosthuizen, J. (Ed.), Environmental Health − Emerging Issues and Practice. InTech, pp. 227−246.

Murakami, K., Feng, G., Chen, S.G., February 1993. Inhibition of brain protein kinase C subtypes by lead. J. Pharmacol. Exp. Therap. 264 (2), 757−761.

Murashima-Suginami, A., Takahashi, K., Sakata, T., Tsukamoto, H., Sugai, M., Yanagita, M., Shimizu, A., Sakurai, T., Slavkin, H.C., Bessho, K., 2008. Enhanced BMP signaling results in supernumerary tooth formation in USAG-1 deficient mouse. Biochem. Biophys. Res. Commun. 369 (4), 1012−1016.

Nagajyoti, P.C., Lee, K.D., Sreekanth, T.V.M., 2010. Heavy metals, occurrence and toxicity for plants: a review. Environ. Chem. Lett. 8 (3), 199−216.

Parisien, M., Charhon, S.A., Arlot, M., Mainetti, E., Chavassieux, P., Chapuy, M.C., Meunier, P.J., 1988. Evidence for a toxic effect of aluminum on osteoblasts: a histomorphometric study in hemodialysis patients with aplastic bone disease. J. Bone Miner. Res. 3, 259−267.

Parkinson, I.S., Ward, M.K., Feest, T.G., et al., 1979. Fracturing dialysis osteodystrophy and dialysis encephalopathy. An epidemiological survey. Lancet 1, 406−409.

Perl, D.P., Brody, A.R., April 18, 1980. Alzheimer's disease: X-ray spectrometric evidence of aluminum accumulation in neurofibrillary tangle-bearing neurons. Science 208 (4441), 297−299. PMID: 7367858.

Potula, V.1, Kaye, W., March 2006. The impact of menopause and lifestyle factors on blood and bone lead levels among female former smelter workers: the Bunker Hill Study. Am. J. Ind. Med. 49 (3), 143−152.

Rebolledo, J., Fierens, S., Versporten, A., et al., 2011. Human biomonitoring on heavy metals in Ath: methodological aspects. Arch. Public Health 69, 10.

Rutger, L., van Bezooijena, R.L., 2005. *SOST*/sclerostin, an osteocyte-derived negative regulator of bone formation. Cytokine Growth Factor Rev. 16, 319−327.

Saag, K.G., Petersen, J., Brandi, M.L., Karaplis, A.C., Lorentzon, M., Thomas, T., Maddox, J., Fan, M., Meisner, P.D., Grauer, A., 2017. Romosozumab or alendronate for fracture prevention in women with osteoporosis. N. Engl. J. Med. 377 (15), 1417−1427.

Sherrard, D.J., Hercz, G., Pei, Y., et al., 1993. The spectrum of bone disease in endstage renal failure − an evolving disorder. Kidney Int. 43, 436−442.

Sherrard, D.J., Hercz, G., Pei, Y., Segre, G., 1996. The aplastic form of renal osteodystrophy. Nephrol. Dial. Transplant. 11 (Suppl. 3), 29−31.

Silbergeld, E.K., Schwartz, J., Mahaffey, K., 1988. Lead and osteoporosis: mobilization of lead from bone in postmenopausal women. Environ Res. 47 (1), 79−94.

Sprague, S.M., Krieger, N.S., Bushinsky, D.A., 1993. Aluminum inhibits bone nodule formation and calcification in vitro. Am. J. Physiol. 264, 882−890.

Sun, X., Cao, Z., Zhang, Q., Li, M., Han, L., Li, Y., 2016. Aluminum trichloride inhibits osteoblast mineralization via TGF-β1/Smad signaling pathway. Chem. Biol. Interact. 25 (244), 9–15.

Tanaka, M., Endo, S., Okuda, T., Economides, A.N., Valenzuela, D.M., Murphy, A.J., Robertson, E., Sakurai, T., Fukatsu, A., Yancopoulos, G.D., Kita, T., Yanagita, M., 2008. Expression of BMP-7 and USAG-1 (a BMP antagonist) in kidney development and injury. Kidney Int. 73 (2), 181–191.

van Bezooijen, R.L., 2004. Sclerostin is an osteocyte-expressed negative regulator of bone formation, but not a classical BMP antagonist. J. Exp. Med. 199, 805–814.

Vestergaard, P., Thomsen, S., 2011. Medical treatment of primary, secondary, and tertiary hyperparathyroidism. Curr. Drug Saf. 6, 108–113.

Wani, A.L., Ara, A., Usmani, J.A., June 2015. Lead toxicity: a review. Interdiscip. Toxicol. 8 (2), 55–64.

Weidenhamer, J.D., Fitzpatrick, M.P., Biro, A.M., Kobunski, P.A., Hudson, M.R., Corbin, R.W., Gottesfeld, P., February 1, 2017. Metal exposures from aluminum cookware: an unrecognized public health risk in developing countries. Sci. Total Environ. 579, 805–813.

Willhite, C.C., Ball, G.L., McLellan, C.J., May 2012. Total allowable concentrations of monomeric inorganic aluminum and hydrated aluminum silicates in drinking water. Crit. Rev. Toxicol. 42 (5), 358–442.

Chapter 23

Biology of the extracellular calcium-sensing receptor

Chia-Ling Tu, Wenhan Chang and Dolores M. Shoback
Endocrine Research Unit, Department of Veterans Affairs Medical Center, Department of Medicine, University of California, San Francisco, CA, United States

Chapter outline

Introduction **539**
Structural and biochemical properties of the calcium-sensing receptor **540**
Agonists, antagonists, and modulators of the calcium-sensing receptor **541**
Cationic agonists of the calcium-sensing receptor 541
Allosteric modulators 542
Synthetic modulators 542
Ligand-biased signaling 543
Calcium-sensing receptor intracellular signaling **543**
Calcium-sensing receptor–mediated signaling 543
Calcium-sensing receptor–associated intracellular signaling effectors 545
Calcium-sensing receptor interacting proteins **546**
Regulation of calcium-sensing receptor gene expression **547**
Roles of calcium-sensing receptor in calciotropic tissues **549**
Calcium-sensing receptor in parathyroid glands 549
Calcium-sensing receptor in the kidney 549
Calcium-sensing receptor in bone cells 551
Calcium-sensing receptor in the breast 552
Noncalciotropic roles of the calcium-sensing receptor **553**
Calcium-sensing receptor in the pancreas 553
Calcium-sensing receptor in the gastrointestinal system 554
Calcium-sensing receptor in the peripheral vascular system 555
Calcium-sensing receptor in the lung 556
Calcium-sensing receptor in the epidermis 556
References **557**

Introduction

Extracellular calcium (Ca_o^{2+}) serves as a versatile modulator of numerous physiological processes including hormone secretion, muscle contraction, blood clotting, cell adhesion, and neuronal excitability (Brown, 1991). In mammals, the level of Ca_o^{2+} is maintained within a narrow range (1.1−1.3 mM) by a homeostatic mechanism involving endocrine factors such as the parathyroid hormone (PTH) and the active metabolite of vitamin D, 1,25-dihydroxyvitamin D3 ($1,25(OH)_2D_3$) (Brown, 2013). When circulating $[Ca^{2+}]$ is low, parathyroid chief cells secrete PTH, which acts on bone, kidney, and intestine to mobilize skeletal Ca^{2+}, enhance renal Ca^{2+} reabsorption, and increase intestinal Ca^{2+} absorption, respectively, and to restore normocalcemia; conversely, PTH release decreases with hypercalcemia. In parathyroid cells, there is a very steep inverse sigmoidal relationship between PTH secretion and Ca_o^{2+} concentration (Brent et al., 1988; Rudberg et al., 1982), and the response curves suggest a positive cooperativity by various agonists that could account for the narrow range of Ca_o^{2+} regulating PTH secretion (Brown et al., 1989). In addition, changes in Ca_o^{2+} levels affect several intracellular signaling pathways including cyclic AMP (cAMP)-dependent protein kinase A (PKA) (Chen et al., 1989), phospholipase C (Kifor and Brown, 1988), and inositol phosphate generation (Nemeth and Scarpa, 1987; Shoback et al., 1988) in parathyroid cells. These initial findings supported the hypothesis of an extracellular Ca^{2+}-sensitive surface receptor regulating PTH secretion by coupling to multiple G-protein-dependent pathways (Brown, 1991; Fitzpatrick et al., 1986).

The cloning and characterization of the Ca_o^{2+}-sensing receptor (CaSR), a plasma membrane G-protein-coupled receptor (GPCR), from bovine parathyroid tissue by Brown et al. confirmed the role of Ca^{2+} as an extracellular first messenger in

Principles of Bone Biology. https://doi.org/10.1016/B978-0-12-814841-9.00023-3

controlling parathyroid function and confirmed that other di-, tri- and polyvalent cations mimic the effects of Ca^{2+} (Brown et al., 1993). Functional clones of the CaSR were subsequently isolated from human parathyroid (Garrett et al., 1995a), kidney (Aida et al., 1995b; Riccardi et al., 1995), and thyroid C cells (Garrett et al., 1995b). CaSRs in these tissues function as the "calciostat" to coordinate Ca^{2+} homeostasis by adjusting PTH and calcitonin secretion or renal cation handling. The importance of the CaSR in systemic Ca^{2+} homeostasis is highlighted by the pathological conditions in which inactivating and activating mutations in *CASR* cause hypercalcemia and hypocalcemia, respectively. These disorders include familial hypocalciuric hypercalcemia (FHH) type 1 (FHH1), neonatal severe hyperparathyroidism (NSHPT), and autosomal dominant hypocalcemia (ADH) type 1 (ADH1) (Brown, 2013; Hannan and Thakker, 2013).

Furthermore, CaSRs are widely expressed in cells and tissues not directly involved in regulating systemic mineral ion homeostasis, such as brain (Ruat et al., 1995), liver (Canaff et al., 2001), lung (Finney et al., 2008), vasculature (Alam et al., 2009), and skin (Oda et al., 2000). In addition to Ca^{2+}, the CaSR responds to an assorted array of stimuli that instigate different intracellular signaling pathways regulating a diverse range of biological processes such as hormone secretion (Yano et al., 2004b), gene expression (Yasukawa et al., 2017), ion channel activity (Parkash and Asotra, 2011; Ye et al., 2004), inflammation (Rossol et al., 2012), and cellular proliferation and differentiation (Diez-Fraile et al., 2013; Tu et al., 2011; Zhang X. et al., 2014).

Structural and biochemical properties of the calcium-sensing receptor

The CaSR is a member of a subfamily (family C) of GPCRs that includes eight subtypes of metabotropic glutamate receptors (mGluRs), mGluR1 to mGluR8; two type B γ-aminobutyric acid receptors (GABAB-Rs), GABAB-R1 and GABAB-R2; the promiscuous L-amino acid receptor; three taste receptors; several pheromone receptors; and five orphan receptors (Brauner-Osborne et al., 2007). All these receptors have small molecules as their ligands and share similar structural features: a large extracellular ligand-binding domain (ECD) including a Venus flytrap (VFT) module in the amino-terminal portion of the receptor, a seven-helix transmembrane domain (TMD) characteristic of GPCRs, and a sizable carboxyl-terminal (C-) tail.

The human CaSR cDNA sequence predicts a 120-kDa protein with 11 potential N-linked glycosylation sites in glycosylated ECD (Bai et al., 1996), of which 8 are glycosylated and 3 are not (Ray et al., 1998). While N-glycosylation may not be essential for signal transduction (Brown and MacLeod, 2001), it is critical for normal cell surface expression of the receptor (Ray et al., 1998). An altered CaSR glycosylation pattern has been observed in some forms of FHH (Bai et al., 1996). The CaSR is expressed on the cell surface in the form of disulfide-tethered homodimer (Bai et al., 1998a; Ward et al., 1998). Dimerization occurs in the endoplasmic reticulum during biosynthesis through two disulfide bonds between the cysteine 129 and 131 of each CaSR monomer (Ray et al., 1999), but noncovalent interactions also contribute to receptor dimerization (Jiang et al., 2004). The observation that the coexpression of CaSR monomers carrying inactivating mutations in different functional domains (e.g., the ECD and tail) can reconstitute considerable biological activity suggests intimate functional interactions between the monomeric CaSR subunits in a dimer (Bai et al., 1999). Furthermore, the CaSR can heterodimerize with mGluRs (Gama et al., 2001) and GABAB-Rs (Chang et al., 2007) through disulfide and noncovalent linkages. These heterodimeric interactions alter the protein trafficking, surface expression, and signal transduction properties of the CaSR (Gama et al., 2001) (Chang et al., 2007).

The CaSR ECD is connected to the TMD by a cysteine-rich domain (CRD) that contains nine cysteine residues forming four disulfide bridges within the CRD (Hu et al., 2000). They are critical for maintaining receptor structure, expression, and function (Fan et al., 1998; Hu et al., 2000). The CaSR VFT is formed by two lobes (LB1 and LB2) separated by a cleft representing the ligand-binding site (Silve et al., 2005). It was hypothesized that the binding of ligand in the cleft between the two lobes causes the lobes to close on each other, which directly or indirectly modifies the conformation of the TMD, leading to receptor activation (Hendy et al., 2013). The VFT module is probably evolved from the family of periplasmic binding proteins in bacteria (Felder et al., 1999), which serve as receptors responding to a wide variety of small ligands (e.g., ions, amino acids, and other nutrients) to promote bacterial chemotaxis toward and cellular uptake of environmental nutrients (Tam and Saier, 1993). Studies using CaSR-mGluR chimeric receptor constructs and mutagenesis have predicted five potential Ca^{2+}-binding sites in each monomeric ECD in the hinge region between the lobes of the VFT (site 1) (Huang Y. et al., 2009; Huang Y. et al., 2007; Silve et al., 2005). Functional studies and computational simulation suggested that the interactions between other Ca^{2+}-binding sites in ECD and the ones in site 1 are essential in tuning functionally positive homotropic cooperativity of CaSR activity induced by Ca_o^{2+} (Zhang C. et al., 2014). Other studies showed that L-amino acids could increase the sensitivity of the CaSR toward Ca^{2+} (Conigrave et al., 2007a). A potential L-phenylalanine-binding pocket was identified in the VFT near site 1 (Mun et al., 2004, 2005; Zhang et al., 2002), and mutational studies indicate its importance for positive heterotropic cooperativity between Ca^{2+} and L-amino acids in CaSR-mediated signaling (Zhang C. et al., 2014). Recent X-ray crystallographic study of the CaSR ECD revealed that the interlobe cleft of VFT, the canonical agonist-binding site (Geng et al., 2013; Kunishima et al., 2000; Muto et al., 2007), is empty in the resting state and surprisingly is occupied solely by an L-tryptophan in the active state (Geng et al., 2016). Binding of an L-amino acid closes the interlobe groove in the VFT module, thereby inducing the formation of a novel homodimer interface by LB2 and CRD between subunits (Geng et al., 2016). Four novel Ca^{2+}-binding sites were identified in each protomer in the active structure, and Ca^{2+} stabilizes the active state by enhancing homodimer interactions to fully activate the receptor (Geng et al., 2016). While the presence of Ca_o^{2+} above a threshold level was required for amino acid–mediated CaSR activation, mutations of L-tryptophan-binding residues completely blocked Ca^{2+}-induced intracellular Ca^{2+} mobilization (Conigrave et al., 2007b; Geng et al., 2016), indicating that L-amino acid and Ca^{2+} are coagonists of the CaSR. Furthermore, four anion-binding sites were identified in the interlobe cleft and LB2 in the inactive and active CaSR ECD structures; and anion, such as PO_4^{3-} , may have a negative allosteric effect on receptor activity by stabilizing inactive conformation (Geng et al., 2016).

The structural similarity between CaSR TMD and those of the rhodopsin GPCR family suggests that these receptors share a common mechanism of G protein coupling (Pin et al., 2003) via TMD. The amino acid residues in the second and third intracellular loops are involved in CaSR coupling to G-protein-mediated signaling (Chang et al., 2000). Two cysteine residues, Cys677 and Cys765, within the first and second extracellular loops are critical for maintaining conformation of the CaSR (Ray et al., 2004). A chimeric receptor with the rhodopsin ECD linking with the TMD and tail of CaSR responds to Ca^{2+} in the presence of NPS-R568, a positive CaSR allosteric modulator (Hauache et al., 2000). Computational modeling and mutagenesis studies indicated that extracellular loops 2 and 3 contain additional binding sites for Ca^{2+} and allosteric modulators. Selective positive (calcimimetics) and negative (calcilytics) allosteric modulators bind to distinct but overlapping regions of the TMD (Hu et al., 2002; Petrel et al., 2004). A recent study of specific heterodimers of two loss-of-function CaSR mutants showed that one functional allosteric site per CaSR dimer was sufficient for obtaining the modulatory effects (Jacobsen et al., 2017). In addition to its role in signal transduction, CaSR TMD may also be involved in receptor dimerization through noncovalent interactions (Zhang et al., 2001).

Unlike the ECD and TMD, very few naturally occurring mutations have been identified in the intracellular tail of the CaSR; however, it contains three common polymorphisms, A986S, R990G, and Q1011E, associated with abnormal blood Ca^{2+} levels (Scillitani et al., 2004). A membrane proximal region (residues 863–925) has been shown to be essential to the cell surface expression and biological activity of the receptor (Chang et al., 2001; Goolam et al., 2014; Ray et al., 1997), as a large PEST-like sequence motif may direct lysosomal degradation of the receptor and regulate cell surface receptor level (Zhuang et al., 2012). In addition, a region comprising amino acids 960–984 is involved in binding to accessory proteins (Hjalm et al., 2001). Within its intracellular loops and C tail, the human CaSR harbors five predicted protein kinase C (PKC) and two predicted PKA phosphorylation sites (Garrett et al., 1995a; Hofer and Brown, 2003). Studies utilizing site-directed mutagenesis have demonstrated that phosphorylation of a single key PKC site at Thr888 inhibited most CaSR-mediated stimulation of PLC (Bai et al., 1998b; Davies et al., 2007; Jiang et al., 2002). Phosphorylation of Thr888 was increased by Ca_o^{2+} or calcimimetic and inhibited by calcilytics (Davies et al., 2007). Therefore, PKC-induced phosphorylation of Thr888 at the CaSR tail confers a negative feedback regulation to limit further activation of the PLC pathway. PKA phosphorylation per se plays a minor role in the regulation of CaSR, although one study found that PKA and PKC synergistically inhibited CaSR-mediated activation of PLC (Bosel et al., 2003).

Agonists, antagonists, and modulators of the calcium-sensing receptor

Cationic agonists of the calcium-sensing receptor

The CaSR has a broad spectrum of ligands in addition to Ca^{2+} and other di- and trivalent cations (Be^{2+}, Sr^{2+}, Ni^{2+}, Mn^{2+} Mg^{2+}, Ba^{2+}, Gd^{3+}, and La^{3+}) that bind at the orthosteric site within the interlobe crevice of the VFT module. Organic polycations including polyamines (i.e., spermine, spermidine, and putrescine), aminoglycoside antibiotics (e.g., neomycin, gentamycin), β-amyloid peptides, and basic polypeptides (e.g., polylysine and polyarginine) have been found to bind unidentified sites in VFT (Quinn et al., 1997; Saidak et al., 2009b). Cations and polyamine with higher positive charge density have higher potency as CaSR agonists (Quinn et al., 1997; Riccardi, 2002). All CaSR polycation agonists have been shown to potentiate one another's stimulatory effects on the receptor. In addition to the ECD of the receptor, studies using chimeric receptors indicated that the extracellular loops in TMD may contain binding sites for Ca^{2+} and allosteric modulators (Hammerland et al., 1999; Hauache et al., 2000).

Among these polycation agonists, however, only Ca^{2+}, Mg^{2+}, and spermine are thought to be physiological CaSR agonists. Mg^{2+} behaves as a partial agonist, as it is about twofold less potent than Ca^{2+} on a molar basis in activating the

CaSR (Butters et al., 1997). In the presence of physiological Ca_o^{2+} levels, $Mg^{2+}{}_o$ positively modulates CaSR function (Ruat et al., 1996). Mg^{2+}, in addition to Ca^{2+}, inhibits Ca^{2+} reabsorption in renal cortical thick ascending limb cells (de Rouffignac and Quamme, 1994). Persons with hypercalcemia due to heterozygous or homozygous inactivating mutations of the CaSR (e.g., FHH and NSHPT) manifest varied degrees of hypermagnesemia (Aida et al., 1995a). Conversely, patients with moderate to severe hypomagnesemia also exhibit impaired PTH secretion and develop hypocalcemia (Brown et al., 1999).

Allosteric modulators

Allosteric modulators bind outside of the orthosteric site, most likely changing the three-dimensional receptor conformation and thus affecting receptor affinity and/or ligand-binding efficacy (Hu, 2008). Positive allosteric modulators (PAMs) and negative allosteric modulators respectively increase and decrease CaSR agonist sensitivity. Well-known natural PAMs include aromatic L-amino acids (L-Phe, L-Tyr, L-His, and L-Trp) and short aliphatic amino acids (L-Thr and L-Ala) (Conigrave et al., 2000), which bind at or adjacent to the orthosteric site within the VFT module (Geng et al., 2016; Mun et al., 2005; Zhang et al., 2002). Activation of the CaSR by L-amino acids requires a threshold Ca_o^{2+} level of 1 mM (Conigrave et al., 2000). Acute elevations of amino acid concentrations stimulate Ca_i^{2+} mobilization and suppress PTH secretion in human parathyroid cells in vitro by enhancing the sensitivity of the CaSR to Ca_o^{2+} (Conigrave et al., 2004). These observations support that the CaSR not only is a Ca^{2+}- (and probably an Mg^{2+}-) receptor but also functions as a general "nutrient" sensor. Coordinating mineral ion and protein metabolism might be particularly relevant in the gastrointestinal tract. Indeed, the CaSR has been clearly identified in transgenic mouse studies as an L-amino acid sensor regulating macronutrient-dependent hormone secretion (Brennan et al., 2014).

Polypeptides such as polyarginine, polylysine, protamine (Ruat et al., 1996), and γ-glutamyl peptides (Broadhead et al., 2011; Wang et al., 2006), as well as β-amyloid peptide (Ye et al., 1997), which is excessively produced in the brain of patients with Alzheimer's disease, and polyvalent aminoglycoside antibiotics (McLarnon et al., 2002), are positive modulators that bind in the receptor's VFT domain. Interestingly, the CaSR was found to be activated by glutathione (γ-Glu-Cys-Gly) and its analog γ-Glu-Val-Gly, which can elicit the kokumi taste response and sapid compound denatonium (Maruyama et al., 2012; Rogachevskaja et al., 2011). Coinciding with the fact that these taste-enhancing substances have a positive allosteric effect on the receptor, functional CaSRs have recently been identified in mammalian taste cells (Bystrova et al., 2010; San Gabriel et al., 2009). Glutathione and its analogs may also act as neuromodulators by activating CaSRs on neurons and/or glial cells in the central nervous system (Conigrave and Hampson, 2010).

The CaSR is negatively modulated by pH and ionic strength (e.g., alterations in the concentration of NaCl). Elevated ionic strength suppresses Ca_o^{2+} sensitivity, and reduced ionic strength enhances it (Quinn et al., 1998). On the other hand, raised pH enhances Ca_o^{2+} sensitivity while reduced pH suppresses it (Quinn et al., 2004). Ion strength may have an impact on the binding efficacies of polyvalent cations to CaSR, thereby changing the responsiveness of the receptor to agonists (Quinn et al., 1998). The alterations in CaSR agonist sensitivity by pH are thought to result from changes in the ionization of the charges on CaSR agonists and changes in the conformation of the receptor (Quinn et al., 2004). The impact of changing ionic strength and pH on the responsiveness of the CaSR to divalent cations may be especially relevant in particular microenvironments where ionic strength and pH can vary greatly (Quinn et al., 1998). For instance, Ca_o^{2+}-induced CaSR-mediated acid secretion from intercalated cells in the distal convoluted tubule may be attenuated as local pH falls (Tyler Miller, 2013), and CaSR expression in the subfornical organ may contribute to the ionic strength-dependent regulation of vasopressin secretion and blood pressure (Yano et al., 2004a). Furthermore, anions have a negative allosteric effect on the receptor, as the CaSR ECD contains four anion-binding sites, and the binding of anions, such as SO_4^{2-} and PO_4^{3-}, stabilizes inactive conformation (Geng et al., 2016). SO_4^{2-} decreases Ca_o^{2+}-stimulated IP accumulation in CaSR-expressing HEK-293 cells (Geng et al., 2016). It is assumed that polycation agonists could compete with positively charged residues at anion-binding sites to bind to anions, thereby promoting the dissociation of anions from CaSR ECD and releasing their inhibitory effect on the receptor. This would drive the CaSR toward its active-state conformation (Geng et al., 2016).

Synthetic modulators

Several pharmacological agents that positively modulate CaSR signaling (calcimimetics) or negatively modulate CaSR signaling (calcilytics) (Nemeth, 2002; Nemeth et al., 1998) have been developed. In contrast to cationic agonists that bind to the ECD (Hammerland et al., 1999), selective calcimimetics and calcilytics bind to distinct but overlapping regions in the extracellular loops of the CaSR TMD (Hu, 2008; Petrel et al., 2004). Besides modulating CaSR activity, calcimimetics and calcilytics have been shown to increase and decrease the level of receptor expression by controlling the susceptibility of the receptor to endoplasmic reticulum (ER)-associated degradation (Huang and Breitwieser, 2007).

Calcimimetics such as NPS R-467 and NPS R-568 are small hydrophobic derivatives of phenylalkylamines that activate the CaSR by increasing its affinity for Ca^{2+} (Nemeth et al., 1998). One calcimimetic compound, cinacalcet, has been approved for the treatment of primary hyperparathyroidism in patients with parathyroid carcinoma and the treatment of secondary hyperparathyroidism in patients on dialysis for end-stage renal disease (Nemeth et al., 1998). Cinacalcet targets the parathyroid CaSR and effectively lowers circulating PTH levels, reduces serum phosphorus and FGF23 concentrations, improves bone histopathology, and may diminish skeletal fracture rates and the need for parathyroidectomy (Nemeth and Shoback, 2013). Recent studies demonstrated that a novel therapeutic regimen combining calcimimetics and PTH produced more robust anabolic effects on bones than those resulting from intermittent PTH treatment (Santa Maria et al., 2016). Because CaSRs expressed in chondrocytes, osteoblasts, and osteoclasts have nonredundant roles in modulating the recruitment, proliferation, survival, and differentiation of these cells (Chang et al., 2008), it was hypothesized that calcimimetics in the presence of elevated local $Ca^{2+}{}_{o}$ levels induced by PTH activate endogenous CaSRs in the skeletal tissues, resulting in increased osteoblastic functions and reduced osteoclast-mediated bone resorption (Santa Maria et al., 2016).

Calcilytics, including NPS 2143 and Calhex 231, are CaSR antagonists that stimulate the secretion of PTH and have been tested for the treatment of osteoporosis (Nemeth and Shoback, 2013). Calcilytics might achieve the same osteoanabolic effects as those of intermittent PTH by suppressing parathyroid CaSR activation and producing a "pulse" of endogenous PTH secretion (Gowen et al., 2000; Kumar et al., 2010). Although several calcilytic compounds were evaluated as orally active anabolic therapies for postmenopausal osteoporosis, clinical development has stopped due to lack of efficacy. Calcilytics may be adapted for the treatment of ADH1 or potentially other disorders (Nemeth and Goodman, 2016).

Ligand-biased signaling

"Biased signaling" is the phenomenon by which distinct ligands stabilize distinct receptor conformational states that activate preferred signaling pathways (Leach et al., 2015). The CaSR is subject to biased signaling in response to its endogenous ligands. Aromatic amino acids such as L-Phe have a greater influence on CaSR-mediated mobilization of Ca_i^{2+} (Conigrave et al., 2000, 2004) than on the activation of PI-PLC (Rey et al., 2005), extracellular-signal-regulated kinases (ERKs) 1/2 (Lee et al., 2007), and CREB (Avlani et al., 2013). Similarly, small-molecule drugs such as the calcimimetics NPS-R568 and cinacalcet, and the calcilytic NPS-2143, instigate biased signaling that prefers $Ca^{2+}{}_{i}$ mobilization to ERK1/2 phosphorylation in CaSR-expressing HEK-293 cells (Davey et al., 2012). Certain CaSR mutants underlying disorders of calcium homeostasis also manifest altered biased signaling (Leach et al., 2012). Thus, biased signaling from the CaSR may have important pathophysiological and therapeutic implications.

Calcium-sensing receptor intracellular signaling

Calcium-sensing receptor–mediated signaling

The CaSR activates signaling pathways downstream through three main groups of heterotrimeric G proteins, $G_{q/11}$, $G_{i/o}$, and $G_{12/13}$, and in certain circumstances also activates G_s (see Fig. 23.1). The CaSR's second and third intracellular loops (Chang et al., 2000) and its proximal C-terminus (Chang et al., 2001; Ray et al., 1997) provide key interaction sites with its associated heterotrimeric G proteins—i.e., $G_{q/11}$. CaSR activation in parathyroid cells and other systems elicits $G_{q/11}$-mediated activation of PI-PLCβ, and with the breakdown of membrane phospholipid PIP_2 to IP_3 and diacylglycerol, induces Ca_i^{2+} mobilization and the activation of various isoforms of PKC (Bai et al., 1998b; Brown et al., 1993). The importance of this mechanism for CaSR-mediated inhibition of PTH secretion was demonstrated by the finding that mice with parathyroid-specific deletion of the α-subunit of G_q on a global G_{11} α-subunit-null background developed severe hyperparathyroidism (Wettschureck et al., 2007) similar to that seen in global CaSR exon 5–null mice (Ho et al., 1995) and humans with NSHPT (Pollak et al., 1994).

Besides PLC, the CaSR is able to activate two other phospholipases, phospholipase D (PLD) and phospholipase A2 (PLA2) (Kifor et al., 1997). Studies using specific inhibitors show that PKC and Rho are involved in CaSR-mediated PLD activation (Huang et al., 2004; Kifor et al., 1997). The CaSR primarily modulates PLA2 activity by increasing Ca_i^{2+} levels via activation of the G_q–PLC pathway (Handlogten et al., 2001). The increase in Ca_i^{2+} activates calmodulin and calmodulin-dependent protein kinases, leading to activation of cytosolic PLA2 (Handlogten et al., 2001) and generation of

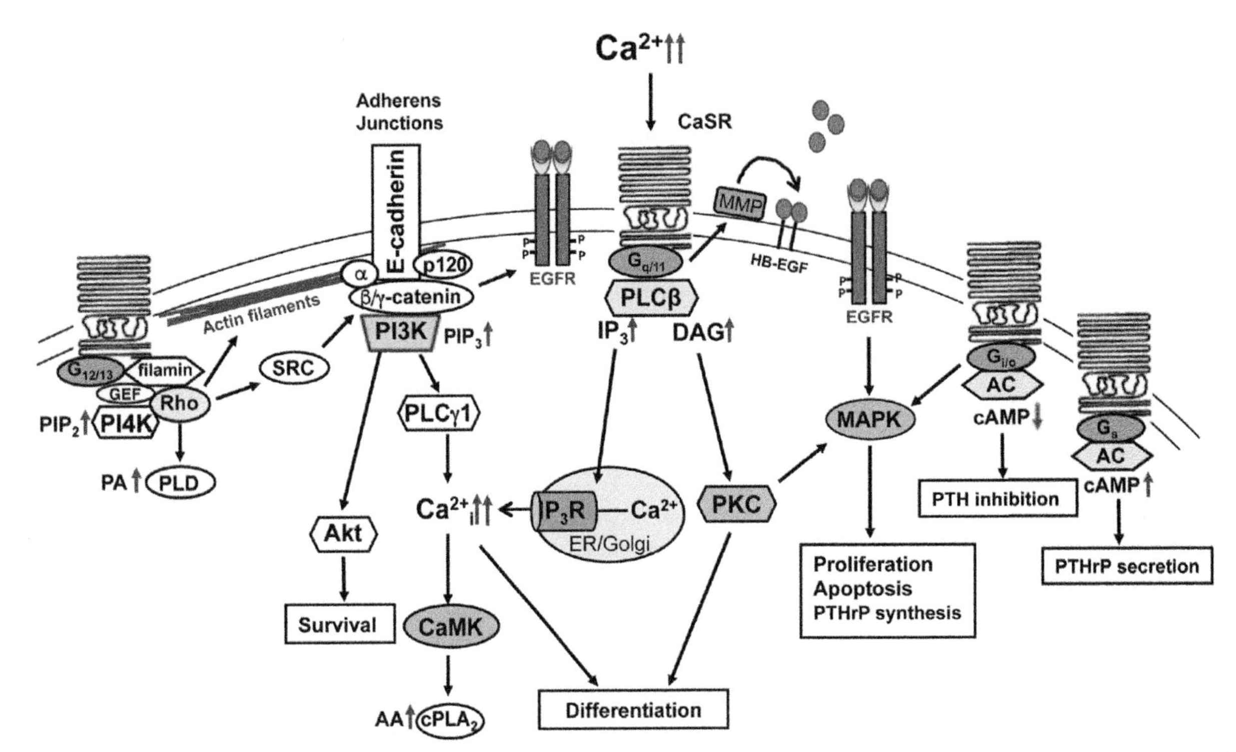

FIGURE 23.1 Schematic representation of calcium-sensing receptor–mediated intracellular signaling. The calcium-sensing receptor (CaSR) typically interacts with three heterotrimeric G protein subfamilies, $G_{q/11}$, $G_{i/o}$, and $G_{12/13}$, and in certain cancer cell types, G_s. Activation of $G_{q/11}$ induces phospholipase C (PLC)-mediated Ca^{2+} mobilization from intracellular stores and protein kinase C (PKC) activation. The increased $Ca^{2+}{}_i$ activates Ca^{2+}/calmodulin-dependent protein kinases (CaMKs) and cytoplasmic phospholipase A2 ($cPLA_2$). Instigating $G_{12/13}$ stimulates Rho-mediated activation of phosphatidylinositol 4-kinase (PI4K) to generate PIP_2, phosphatidylinositol 3-kinase (PI3K) to produce PIP_3 and to activate the downstream effectors Akt and PLCγ1, and cytoskeleton reorganization to promote the formation of stress fibers, cadherin/catenin adhesion complexes, and membrane ruffling. CaSR-mediated activation of phospholipase D (PLD), producing phosphatidic acid, also involves Rho and PKC. Adenylyl cyclase is inhibited by $G_{i/o}$ but activated by G_s in breast cancer cell lines, resulting in decreased or increased cytoplasmic cyclic AMP (cAMP) concentrations, respectively. Activation of the mitogen-activated protein kinase (MAPK) extracellular-signal-related kinases (ERKs) 1/2 involves transactivation of the epidermal growth factor receptor (EGFR) either via liberation of EGF-like ligands by extracellular matrix metalloproteinases (MMPs) or through an E-cadherin-mediated ligand-independent EGFR recruitment and activation. Activation of ERK1/2 also involves PKC and Gi/o-mediated signaling.

second messengers such as arachidonic acid. In kidney, this PLA2 signaling axis mediates the effects of hypercalcemia. However, CaSR-induced activation of PKC and ERK may also contribute to the phosphorylation and activation of PLA2 (Handlogten et al., 2001; Kifor et al., 2001).

In CaSR-expressing HEK-293 cells, elevating Ca^{2+}_o induces sinusoidal oscillations in Ca^{2+}_i (Szekely et al., 2009; Young and Rozengurt, 2002). The repetitive oscillations result from the dynamic phosphorylation and dephosphorylation of a key residue, T888, the primary PKC phosphorylation site of the CaSR located in the proximal intracellular domain (Bai et al., 1998b; Davies et al., 2007; McCormick et al., 2010; Young et al., 2002). Ligand-induced activation of PKC leads to phosphorylation of T888 to uncouple the receptor from $G\alpha_{q/11}$-induced PLC activation and Ca^{2+}_i mobilization (Bai et al., 1998b; Jiang et al., 2002). PKC activation by phorbol esters overcomes the inhibitory effect of high Ca^{2+}_o and increases PTH secretion, whereas disruption of the T888 PKC phosphorylation site increases receptor Ca^{2+}_o sensitivity (Brown et al., 1992; Membreno et al., 1989). The naturally occurring CaSR mutant T888M has been linked to sustained CaSR-mediated suppression of PTH secretion at low Ca^{2+}_o levels causing ADH (Lazarus et al., 2011), demonstrating that the T888 residue and its regulation by PKC are critical for physiological CaSR function in vivo. The phosphorylation state of CaSR T888 depends not only on PKC but also on the dynamic activity of PP2A, a phosphatase that induces dephosphorylation (McCormick et al., 2010). Also, the PLC/Ca^{2+}_i-mediated control of chemotaxis of preosteoblasts to sites of bone resorption may involve a G_{12}-dependent mechanism that sustains Ca^{2+}_i elevations via the activation of PP2A and dephosphorylation of T888 (Godwin and Soltoff, 2002; Zhu et al., 2004, 2007).

Elevated Ca^{2+}_o typically suppresses cAMP levels in CaSR-expressing cells via pertussis toxin-sensitive $G_{i/o}$, which suppresses the activity of various isoforms of adenylyl cyclase. $G_{i/o}$ is responsible for mediating the inhibitory control of

PTH and PTH-related protein (PTHrP) secretion by various CaSR ligands including Ca_o^{2+} (Fitzpatrick et al., 1986; Mamillapalli and Wysolmerski, 2010). Nevertheless, CaSR-mediated activation of PI-PLC provides a $G_{i/o}$-independent mechanism for the inhibition of cAMP synthesis via Ca_i^{2+}-dependent inhibition of adenylyl cyclase isoforms 5, 6, and/or 9 or via phosphodiesterase isoform-1, which breaks down cAMP in response to elevated Ca_i^{2+}/calmodulin (Brown and MacLeod, 2001). This pathway supports Ca_o^{2+}-induced suppression of renin secretion from renal juxtaglomerular cells (Atchison and Beierwaltes, 2013; Beierwaltes, 2010) and vasopressin-induced aquaporin-2 expression in water-reabsorbing principal cells of the collecting ducts (CDs) (Procino et al., 2012). However, the CaSR stimulates rather than inhibits adenylyl cyclase in certain cancers (e.g., breast cancer cells) (Mamillapalli et al., 2008) and in At-T20 pituitary cells, which secrete adrenocorticotropic hormone and PTHrP (Mamillapalli and Wysolmerski, 2010). In normal mammary epithelial cells, CaSR preferably couples to $G_{i/o}$, but the receptor-dependent coupling is switched to G_s in breast cancer cells (Mamillapalli et al., 2008). In contrast, Ca_o^{2+}-induced activation of adenylyl cyclase isoforms 1 and 8 may provide an alternative mechanism for the CaSR-mediated elevation of cAMP (Cooper, 2003).

$G_{12/13}$ modulate receptor-mediated control of shape change, cell migration, and gene expression through the activation of Rho kinase and phosphorylation of serum-response factor (Siehler, 2009). Ca_o^{2+} and the calcimimetic NPS R-467, but not L-amino acids, induced Rho kinase-dependent formation of actin stress fibers in CaSR-expressing HEK-293 cells (Davies et al., 2006); yet G_{12} and Rho are required for L-amino acid-induced Ca_i^{2+} mobilization (Rey et al., 2005). $G_{12/13}$-mediated pathways promote osteoblast differentiation through Wnt3a-β-catenin signaling (Rybchyn et al., 2011) and inhibit osteoclastogenesis by suppressing the expression of receptor activator of nuclear factor kappa-B ligand (RANKL) and promoting the expression of osteoprotegerin (Brennan et al., 2009). $G_{12/13}$ also activates adenylyl cyclase isoform-7 (Jiang et al., 2008), and this pathway may contribute to $Ca^{2+}{}_o$-stimulated cAMP synthesis in primary mouse osteoblasts (Choudhary et al., 2004).

Calcium-sensing receptor–associated intracellular signaling effectors

CaSR activation is linked to proproliferative stimuli in many cell systems, involving the activation of mitogen-activated protein kinases (MAPKs) including ERK1/2, p38, and JNK (Brennan and Conigrave, 2009; Magno et al., 2011b). Ca_o^{2+}-induced ERK1/2 is typically activated through the CaSR–Src–Ras–Raf–MEK–ERK pathway (Hobson et al., 2000). Nevertheless, pertussis toxin, the PLC inhibitor U73122, and PKC inhibitor GF109203X all have been shown to partially inhibit Ca_o^{2+}-induced ERK1/2 phosphorylation in bovine parathyroid and HEK-293 cells expressing the CaSR (Holstein et al., 2004; Kifor et al., 2001). These findings indicate that the CaSR can activate tyrosine kinases and stimulate ERK activity through the $G_{q/11}$–PLC–PKC pathway and by coupling to G_i (Kifor et al., 2001). CaSR-mediated activation of ERK1/2 via G_i requires dynamin/β-arrestin-dependent receptor internalization (Holstein et al., 2004). In addition, CaSR-mediated ERK phosphorylation can occur through a transactivation mechanism in which CaSR stimulates matrix metalloproteinase-mediated release of membrane-bound epidermal growth factor (EGF)-like peptide, which in turn activates EGF receptor (EGFR) and downstream Ras–Raf–MEK–ERK signaling (MacLeod et al., 2004; Yano et al., 2004b). This pathway drives CaSR-mediated cell proliferation in human MCF-7 breast cancer cells (El Hiani et al., 2009b) and PTHrP secretion in human PC-3 prostate cancer cells (Yano et al., 2004b). Furthermore, the engagement of E-cadherin in newly formed cell contacts provides an alternate mechanism for the recruitment and sensitization of EGFR to allow $Ca^{2+}{}_o$ to activate downstream MAPK and Rac1 signaling cascades, together supporting epithelial proliferation (Betson et al., 2002) (Fedor-Chaiken et al., 2003) (Pece and Gutkind, 2000). CaSR-mediated ERK activation in CaSR-expressing HEK-293 cells and ovarian surface epithelial cells may involve the phosphatidylinositol 3-kinase (PI3K) (Hobson et al., 2003). The mouse osteoblastic cell line, MC3T3- E1, responds to CaSR agonists with increased phosphorylation of p38 MAPK (Yamaguchi et al., 2000). CaSR-mediated p38 MAPK signaling modulates the release of PTHrP in CaSR-expressing HEK-293 cells and rat H-500 Leydig cancer cells (MacLeod et al., 2003; Tfelt-Hansen et al., 2003) and regulates vitamin D receptor (VDR) expression in a proximal tubule human kidney epithelial cell line, HK-2G (Maiti et al., 2008). JNK has been shown to control CaSR-mediated proliferation in rat calvarial osteoblasts (Chattopadhyay et al., 2004). JNK and p38 MAPK have also been implicated in other CaSR-mediated pathways regulating proliferation and apoptosis (Corbetta et al., 2000).

In addition to MAPK cascades, CaSR controls cell survival, proliferation, and differentiation through multiple distal effectors. Activation of the CaSR stimulates the expression of Fos, Egr-1, and cyclin-D to promote G1/S cell cycle transition (Chattopadhyay et al., 2004). CaSR-mediated signaling via the PI3K pathway stimulates phosphorylation and activation of the antiapoptotic protein kinase Akt in fetal rat calvarial cells (Dvorak et al., 2004) and opossum kidney cells (Ward et al., 2005). Conversely, CaSR stimulates NF- κB activation and promotes expression of cell death genes such as p53, c-Myc, and Bcl-2 in mature rabbit osteoclasts (Mentaverri et al., 2006). CaSR-mediated apoptosis is induced by $Ca^{2+}{}_i$

overload and activation of the mitochondrial apoptotic pathways and involves the upregulation of caspase-3 and cytochrome *c* as well as Fas/FasL expression (Xing et al., 2011; Zheng et al., 2011). Moreover, CaSR promotes cell differentiation in keratinocytes and colonic epithelial cells (Chakrabarty et al., 2005; Tu et al., 2008, 2011). Stimulating the CaSR increased the expression of E-cadherin in colon carcinoma cells and concomitantly reduced the activation of β-catenin/T-cell factor and suppression of tumor malignancy (Chakrabarty et al., 2005). In keratinocytes, physical interactions between the CaSR, filamin A, and RhoA stimulate E-cadherin-mediated recruitment and the activation of downstream kinases, and also facilitate cell differentiation (Tu et al., 2011). CaSR activators induce the secretion of Wnt5a and expression of Wnt5a receptor Ror2, leading to the inhibition of canonical Wnt/β-catenin signaling and stimulation of cell differentiation in colonic epithelial cells and keratinocytes (Macleod, 2013; Popp et al., 2014).

In kidney, CaSR is coupled to the metabolism of arachidonic acid by cytochrome P450 (Ferreri et al., 2012) and regulates renal calcium excretion in the thick ascending limb independently of PTH by modulating the calcineurin–NFAT1c–microRNA–claudin 14 signaling axis (Toka, 2014). In thyroid C cells, the CaSR controls calcitonin secretion through the activation of voltage-gated Ca^{2+} channels (Freichel et al., 1996). The CaSR has been shown to form signaling complexes with integrins to facilitate cell migration and differentiation in neurons (Tharmalingam et al., 2016). CaSR activation stimulates cyclooxygenase pathways in osteoblasts via a G_i- and PKA-dependent mechanism (Choudhary et al., 2004). Additionally, the CaSR plays a significant role in bone metastasis in renal carcinoma, breast cancer, and prostate cancer (Breuksch et al., 2016). CaSR-mediated activation of ERK1/2, Akt, and PLC signaling determines the tumor cells' ability to proliferate, migrate (Liao et al., 2006; Saidak et al., 2009a), and secrete PTHrP (Mamillapalli and Wysolmerski, 2010), which in turn increases the expression of RANKL in immature osteoblasts and stimulates osteoclastogenesis (Chirgwin and Guise, 2000). These processes promote osteoclast-mediated osteolysis, resulting in the enhanced secretion of growth factors and calcium, again activating tumor cells (Breuksch et al., 2016; Tfelt-Hansen et al., 2003).

Calcium-sensing receptor interacting proteins

The CaSR is a GPCR that operates in the constant presence of agonists, sensing small changes with minimal functional desensitization. Resistance to desensitization requires the maintenance of a functional receptor pool at the cell surface and persistent coupling of the receptor to its heterotrimeric G proteins and downstream signaling pathways. CaSR cell surface expression is maintained by an unusual phenomenon termed agonist-driven insertional signaling (ADIS) (Breitwieser, 2013), in which CaSR activation drives its own trafficking to the plasma membrane. ADIS has potential therapeutic implications for disorders of Ca^{2+} homeostasis in which CaSR expression is impaired due to loss-of-function mutations in the *CASR* gene found in FHH1 and NSHPT (Hannan and Thakker, 2013). However, CaSR-associated disorders can also be caused by different mechanisms. Recent studies revealed that FHH or ADH phenotypes can be associated with mutations on partner proteins associated with CaSR-mediated signaling—e.g., loss-of-function mutations of the sigma subunit of adaptor protein-2 (AP2) and mutations on G protein subunit-α11(GNA11) (Nesbit et al., 2013b; Rogers et al., 2014). A number of molecular binding partners of the CaSR are involved in controlling receptor trafficking to the cell surface and its desensitization, degradation, and signaling (Grant et al., 2011; Huang and Miller, 2007; Ray, 2015).

Once synthesized in the ER, improperly folded CaSR proteins will be moved toward the proteasome for degradation, a process involving osteosarcoma-9 protein, which targets immature CaSRs (Ward et al., 2018). Receptors passing the quality control process will traffic to the plasma membrane or other cellular compartments via interaction with chaperones and small GTP-binding proteins (Breitwieser, 2014). Several proteins interact with the CaSR when the receptor exits the ER. P24A (Strating and Martens, 2009), Sar1 (Zhuang et al., 2010), receptor activity-modifying proteins (Bouschet et al., 2012), and Rab1 (Zhuang et al., 2010) are found to facilitate CaSR trafficking from the ER to the Golgi and increase its plasma membrane expression. On the other hand, the 14-3-3 proteins are predicted to interact with the arginine-rich domain (amino acids 890–898) of the CaSR C-terminal tail and may lead to the retention of the CaSR in the ER (Arulpragasam et al., 2012; Grant et al., 2011). The surface expression of CaSR is also influenced by the interaction of calmodulin (CaM) at residues 871–898 at the C-terminal tail of CaSR, a region involved in phosphorylation and biased signaling (Davey et al., 2012; Leach et al., 2012). CaM binding may stabilize the CaSR on the cell membrane via modulation of anterograde trafficking and thus increase the potency of its functional activity (Bai et al., 1998b; Huang et al., 2010). In addition, dimerization of the CaSR is essential, although not sufficient, for membrane trafficking and full receptor function (Bai et al., 1999).

After CaSRs are inserted into plasma membrane, their interactions with a variety of proteins direct them to different subcellular localizations. Caveolin binds to CaSR intracellular loops 1 and 3 and keeps the receptor highly enriched in invaginations of the plasma membrane, called caveolae, in parathyroid chief cells and cardiac myocytes (Kifor et al., 1998; Sun and Murphy, 2010). Filamin A, an actin-binding scaffold/adaptor protein with the ability to bind various components in MAPK and Rho-GTPase-mediated signaling cascades, interacts with the C-terminal tail of the CaSR (Awata et al., 2001; Hjalm et al., 2001; Pi et al., 2002). Filamin A potentiates CaSR expression in parathyroid cells by protecting the receptor from degradation (Zhang and Breitwieser, 2005). CaSR-dependent activation of downstream ERK1/2 and Rho also relies, in part, on its physical interaction with filamin A (Pi et al., 2002; Rey et al., 2005; Tu et al., 2011). In keratinocytes, disruption of the CaSR–filamin interaction prevents Ca_o^{2+}-induced RhoA activation and the formation of E-cadherin–catenin adhesion complex on the cell membrane, thereby impairing cell–cell adhesion (Tu et al., 2011; Tu and You, 2014). In addition, the CaSR modulates cell adhesion and migration via direct interactions with integrins in developing cerebellum granule cell precursors and medullary thyroid carcinoma cells (Tharmalingam et al., 2011, 2016). Heterodimerization of CaSR with mGluR (Gama et al., 2001) and subunits of GABAB-R (Kim et al., 2014) is found in certain regions of bovine and rat brains. These interactions affect trafficking and surface expression of CaSR as well as receptor signaling to the PLC/IP_3 pathway (Chang et al., 2007; Gama et al., 2001). Testin, a focal adhesion protein, binds to the intracellular tail of the CaSR and positively modulates CaSR-stimulated Rho kinase (Magno et al., 2011a). The binding of hypoxia-induced mitogenic factor to the membrane proximal region in the CaSR tail enhances CaSR-mediated Ca_i^{2+} signaling and hypoxia-evoked proliferation of pulmonary artery smooth cells, leading to the development of pulmonary vascular remodeling and pulmonary hypertension. Additionally, the CaSR's C-terminal tail has been shown to interact directly with two potassium channels, Kir4.1 and Kir4.2, an interaction that inhibits the activities of these channels in the renal thick ascending limb (Huang C. et al., 2007).

Surface-expressed CaSR may undergo endocytosis initiated through phosphorylation by GPCR kinases or PKC (Lorenz et al., 2007; Pi et al., 2005). β-arrestin subsequently binds to the phosphorylated receptor to decrease its capacity to activate G proteins (Lorenz et al., 2007). Endocytosis of CaSR is facilitated by Rab7, Rab11a (Reyes-Ibarra et al., 2007), and AP2 (Nesbit et al., 2013b; Zhuang et al., 2012). The endocytosed receptors are either recycled to the cell membrane, thereby contributing to receptor resensitization, or translocated to lysosomes for degradation (Ray, 2015). Two mechanisms are involved in degradation of the CaSR: (1) CaSR proteins are targeted to the proteasome after being ubiquitinated by the E3 ubiquitin ligase, dorfin, via binding to the C-terminal tail (Huang et al., 2006; Zhuang et al., 2012), or (2) they are degraded in lysosomes, elicited by the binding of AMSH, a deubiquitinating enzyme, to the PEST-like sequence in the C-terminal tail of the CaSR (Herrera-Vigenor et al., 2006; Zhuang et al., 2012). Furthermore, CaSR proteins are sensitive to m-calpain-dependent destruction, adding an additional mechanism for regulating CaSR protein expression (Kifor et al., 2003). These mentioned secretory pathways and endocytosis mechanisms work collaboratively in the continuous presence of agonist stimulation, resulting in a net increase in CaSR expressed on the plasma membrane (Breitwieser, 2013).

Regulation of calcium-sensing receptor gene expression

The human *CASR* gene is located on chromosome 3q13.3–21, and rat and mouse *Casr* genes reside on chromosomes 11 and 16, respectively (Janicic et al., 1995). The *CASR* gene has seven exons (Pearce et al., 1995): Exon 1 contains two alternative promoters for transcription (Chikatsu et al., 2000; Garrett et al., 1995a). Exon 2 encodes a common upstream untranslated region (5′-UTR) and the translation initiation codon ATG. Exons 2 to 6 encode various regions of the ECD, while exon 7 encodes the entire TMD and the C-terminus tail. *CASR* transcripts may have different 3′-untranslated regions (3′-UTRs) derived from two alternative polyadenylation signal sequences in exon 7 (Aida et al., 1995b; Garrett et al., 1995a). Like the human *CASR* gene, mouse and rat genes are organized in a similar manner and have at least two promoters (Hendy et al., 2013). Promoter P1 has TATA and CCAAT boxes upstream to the initiation site, and promoter P2 has Sp1/3 motif at the transcriptional start site; both promoters drive significant levels of basal activity, with P2 more active than P1 (Canaff and Hendy, 2002). Direct analyses of transcripts revealed that transcripts from P2 are expressed at much higher levels than for P1-derived transcripts in human parathyroid, thyroid C cells, and renal proximal tubular cells (Canaff and Hendy, 2002; Mizobuchi et al., 2009). Transcripts from the P1 promoter are reduced in parathyroid adenomas and colorectal carcinomas (Chikatsu et al., 2000; Kallay et al., 2003a). Several factors including active vitamin D, proinflammatory cytokines, and the transcription factors glial cells missing-2 (GCM2) and thyroid transcription factor 1 (TTF1) control *CASR* transcription, and functional cis-elements in *CASR* promoters responsive to these modulators have been characterized (Hendy et al., 2013; Suzuki et al., 1998). Studies of colon carcinomas and neuroblastomas have indicated that epigenetic changes, such as histone deacetylation and promoter hypermethylation of the GC-rich P2 (Casala et al., 2013; Fetahu et al., 2014; Hizaki et al., 2011), contribute to reduced *CASR* expression in these tumors. Although the increased expression of several microRNAs has been proposed as a cause for *CASR* silencing in colorectal tumors (Fetahu et al., 2016; Singh and Chakrabarty, 2013), direct involvement of microRNA in regulating receptor gene expression has not been confirmed.

Alternative spliced *CASR* transcripts have been reported. An exon 3–deleted variant, in which fusion of exons 2 and 4 produces a truncated receptor protein unable to reach the cell surface, is found in placental cytotrophoblast (Bradbury et al., 1998) and parathyroid, thyroid, and kidney tissues (D'Souza-Li et al., 2001). A mouse model of human FHH1 and NSHPT (Ho et al., 1995) in which exon 5 of the *Casr* gene is deleted unexpectedly generates an alternatively spliced *Casr* variant lacking exon 5 in the skin, kidney, and growth plate (Oda et al., 1998; Rodriguez et al., 2005). Although this variant encodes a protein with a 77-amino acid in-frame deletion in the exodomain and exerts a dominant-negative effect on the full-length receptor, reducing its response to Ca_o^{2+} in vitro (Oda et al., 1998), it apparently compensates for the absence of the full-length receptor in bone and cartilage (Rodriguez et al., 2005).

High levels of both Ca_o^{2+} and $1,25(OH)_2D_3$ upregulate CaSR expression in certain cell types. High Ca_o^{2+} increases the expression of the CaSR in mouse pituitary AtT-20 cells (Emanuel et al., 1996) and rat parathyroid glands (Mizobuchi et al., 2004), whereas administration of $1,25(OH)_2D_3$ elevates *CASR* mRNA levels in vivo in kidney and parathyroid gland in rat (Brown et al., 1996; Canaff and Hendy, 2002; Yao et al., 2005). Upregulation of CaSR in the parathyroid gland by $1,25(OH)_2D_3$ increases the responsiveness of the gland to Ca_o^{2+}, thereby reinforcing the negative action of Ca_o^{2+} on PTH synthesis and secretion and parathyroid cell proliferation. Conversely, kidney CaSR expression is downregulated in VDR-null mice and in *Cyp27*$^{-/-}$ mice lacking the 25-hydroxyvitamin D-1α-hydroxylase enzyme (Li et al., 2003). Likewise, reduced CaSR expression in patients with chronic kidney disease could result in part from a concomitant decrease in circulating levels of $1,25(OH)_2D_3$ (Williams et al., 2009). Modulation of the activity of *CASR* promoters by Ca_o^{2+} was shown in vitro by transfecting promoter-reporter constructs in human kidney proximal tubule cells and mouse distal convoluted tubule cells (Hendy and Canaff, 2016). $1,25(OH)_2D_3$ has been shown to potentiate CaSR-mediated antineoplastic effects in differentiated human colon carcinoma cells (Aggarwal et al., 2016). Although functional vitamin D response elements are present in both promoters in the *CASR* gene (Canaff and Hendy, 2002; Klein et al., 2016), Ca_o^{2+} and $1,25(OH)_2D_3$ additively stimulate transcription from promoter P2 but not P1 (Chakrabarty et al., 2005).

GCM2 is a transcription factor essential for the development of the parathyroid gland in terrestrial vertebrates (Okabe and Graham, 2004). The expression of CaSRs in parathyroid cells correlated with GCM2 levels, and the inactivating mutations of *GCM2* gene cause familial isolated hypoparathyroidism (Mannstadt et al., 2008; Okabe and Graham, 2004). GCM2 transactivates the *CASR* gene via GCM2 response elements in promoters P1 and P2 (Canaff et al., 2009). In adenomatous tissue from patients with primary hyperparathyroidism, reduced CaSR expression may be attributed to underexpression of the *GCM2* gene (Correa et al., 2002). TTF1 is present in rat thyroid C cells and parathyroid cells, and it is expressed inversely with the CaSR. TTF1 suppresses the promoter activity of the *Casr* gene via interactions with specific elements on the 5′-flanking regions (Suzuki et al., 1998). However, the regulation of CaSR expression by TTF1 is modulated by $Ca^{2+}{}_i$ levels and may involve other transcription factors (Suzuki et al., 1998).

Expression of the CaSR is regulated physiologically and pathologically at the gene level. CaSR is highly expressed in the developing fetus, with the highest expression levels found in the central and peripheral nervous systems, heart, lung, and cartilage (Riccardi et al., 2009). CaSR plays important roles in neuronal growth (Vizard et al., 2008), lung morphogenesis (Finney et al., 2008), and skeletal development (Chang et al., 2008). In the adult, CaSR expression has been detected in a myriad of cells and tissues. It regulates cellular fate in the parathyroid gland, bone, kidney, blood, and skin as well as in reproductive, cardiovascular, and gastrointestinal tissues (Diez-Fraile et al., 2013). Altered expression of the CaSR has been associated with various disorders in humans. In atherosclerosis, the loss of CaSR expression in vascular smooth muscle cells (VSMCs) leads to vascular calcification (Alam et al., 2009). Animal models of renal insufficiency (Mathias et al., 1998; Toka et al., 2012) show that deficiency of renal CaSR causes PTH-independent hypocalciuria, as the reduction of renal CaSR increases tubular Ca^{2+} reabsorption (Brown and MacLeod, 2001). In genetic forms of hypercalciuria, dysregulation of CaSR expression by vitamin D metabolites may be a critical factor for kidney stone formation. In a rat model of hypercalciuric nephrolithiasis, renal VDR and CaSR levels are elevated and renal Ca^{2+} reabsorption is reduced (Bai and Favus, 2006; Yao et al., 2005). Activation of CaSR inhibits cell proliferation in parathyroid cells, keratinocytes (Tu et al., 2001), and cells of the colon crypts (Rey et al., 2010). CaSR expression is reduced or absent in parathyroid glands of patients with primary or severe uremic secondary hyperparathyroidism (Cetani et al., 2000; Corbetta et al., 2000; Kifor et al., 1996) and in many colon cancers (Bhagavathula et al., 2005; Kallay et al., 2003b). Impaired Ca_o^{2+}-sensing may increase cell proliferation and contribute to parathyroid neoplasia and colorectal tumors. In contrast, the CaSR is overexpressed in breast and prostate cancers with high bone metastatic potential (Liao et al., 2006; Mihai et al., 2006; Saidak et al., 2009a). CaSR activation promotes cancer progression by stimulating secretion of PTHrP, thus contributing to osteolytic bone destruction (Guise et al., 2005; Liao et al., 2006). Moreover, the CaSR is found to be overexpressed in the hippocampus neurons of mice sustaining traumatic or ischemic brain injury (Kim et al., 2011, 2014) and in a model of Alzheimer's disease (Gardenal et al., 2017), leading to calcium overload and neuronal death in brain. The administration of calcilytics, allosteric CaSR inhibitors, renders neuroprotection against the detrimental effects of CaSR overactivation in these pathological conditions (Chiarini et al., 2017; Kim et al., 2013, 2014).

Roles of calcium-sensing receptor in calciotropic tissues

Calcium-sensing receptor in parathyroid glands

CaSRs coevolved with the parathyroid glands in terrestrial vertebrates to maintain systemic Ca^{2+} homeostasis via the close control of the secretion of PTH—the hormone that ultimately controls Ca^{2+} handling by the intestine, kidney, and bone. PTH therefore balances the uptake, storage in bone, and excretion of Ca^{2+} in the body according to systemic needs.

CaSRs enable parathyroid cells to respond to supra- and subphysiological Ca_o^{2+} levels by respectively suppressing and enhancing PTH secretion (Brown, 2013). These critical actions of the parathyroid CaSR are established by the identification of inactivating and activating mutations of the *CASR* in humans with FHH1 and autosomal dominant hypoparathyroidism type 1 (ADH1), respectively (Hannan et al., 2018; Vargas-Poussou et al., 2016). This idea is further supported by the observation that there is a right-shifted Ca_o^{2+}/PTH secretion set point in parathyroid glands cultured from mice with *Casr* deficiency (Cheng et al., 2013). How CaSR activation inversely couples to PTH secretion remains unclear. The ability of homomeric CaSRs to stimulate $G_{q/11}$ proteins and downstream signaling cascades is thought to result from major pathways suppressing PTH secretion at high Ca_o^{2+} concentrations (Conigrave and Ward, 2013). This concept is supported by the development of a phenotype of severe hyperparathyroidism in mice with parathyroid cell-targeted *Gnaq* and *Gna11* double-gene knockouts (Wettschureck et al., 2007) and the association of inactivating and activating *GNA11* mutations in patients afflicted with FHH type 2 (Hannan et al., 2018) and ADH type 2 (Nesbit et al., 2013a) (Howles et al., 2016; Roszko et al., 2016). The linkage of loss-of-function mutations of the sigma subunit of the adapter protein 2 with FHH type 3 (Hovden et al., 2017) further indicates roles for clathrin-mediated endocytosis, receptor trafficking, and receptor reinsertion into the membrane in mediating normal CaSR function in physiological states (Gorvin et al., 2017; Hannan et al., 2016). In contrast, the ability of CaSR to stimulate intracellular Ca^{2+} mobilization can be blunted by the overexpression of regulator of G protein signaling 5 (RGS5) (Balenga et al., 2019), whose expression levels are increased in parathyroid tumors (Koh et al., 2011), indicating an inhibitory action of RGS5 on CaSR signaling. In addition to regulation of acute PTH secretion, CaSR activation suppresses transcription of the PTH gene (Garrett et al., 1995a) and cell growth. The latter is well demonstrated by the presence of significant parathyroid cellular hyperplasia in mice with global knockout of both alleles of *Casr* (Ho et al., 1995) as well as the ability of calcimimetics to prevent parathyroid hyperplasia in rats subjected to the induction of chronic kidney disease (Colloton et al., 2005; Wada et al., 1998). Klotho has also recently been shown to be involved in cross talk with CaSR signaling to suppress PTH synthesis and glandular growth in parathyroid tissue in mice (Fan et al., 2018), implicating a role of FGF23 signaling in mediating parathyroid function.

The ability of parathyroid cells to sense changes in extracellular levels of Ca^{2+} is fundamentally important for skeletal development and maintenance. Deletion of both *Casr* alleles in the parathyroid glands of mice retards perinatal skeletal development in these $^{PTC}Casr^{-/-}$ mice (Chang et al., 2008). While $^{PTC}Casr^{-/-}$ mice recapitulate most biochemical (severe hyperparathyroidism, hypercalcemia, and hypophosphatemia) and skeletal (expansion of growth plate, osteopenia, osteiod accumulation, and excess fractures with poor healing) phenotypes of NSHPT in patients and in mice with global *Casr* knockout (*Casr*$^{-/-}$ mice) (Ho et al., 1995), they manifest hypercalciuria, in contrast to hypocalciuria in *Casr*$^{-/-}$ mice. This is because normal CaSRs are still functioning in the kidney to promote Ca^{2+} excretion in response to the marked hypercalcemia of this conditional knockout mouse ($^{PTC}Casr^{-/-}$ mice).

Given that *Casr* genes are unaltered outside of PTCs in $^{PTC}Casr^{-/-}$ mice, severe PTH excess appears to be sufficient to cause skeletal defects regardless of the status of CaSR function in other tissues. In contrast to severe PTH excess, milder forms of hyperparathyroidism studied in $^{PTC}Casr^{\pm}$ mice with heterozygous *Casr* gene knockout targeted specifically to the parathyroid are anabolic to trabecular bone by protecting the bones from aging-induced bone loss (Cheng et al., 2013). However, these protective effects are at the expense of cortical thinning and demineralization, indicating differential responses of osteoblasts and osteocytes to mild PTH elevations. Given that $^{PTC}Casr^{\pm}$ mice also manifest hypercalcemia, the direct activation of CaSR in osteoblasts, osteocytes, and/or osteoclasts in their skeletons cannot be ruled out as a secondary effect of hyperparathyroidism that contributes to skeletal changes.

Calcium-sensing receptor in the kidney

Molecular mechanisms responsive to changes in the need for Ca^{2+} conservation and the excretion of Ca^{2+} when in excess were postulated to be present in different segments of the kidney long before CaSR cDNA was identified in the parathyroid

(Brown et al., 1993) and later in kidney (Riccardi et al., 1995). This was because evidence had accumulated, from a large body of in vitro and in vivo animal studies, on the many effects that divalent cations (Ca^{2+} and Mg^{2+}) were known to have on renal function, water and solute handling, and ion transport. Aberrant perception of ambient serum Ca^{2+} concentration in the human disorders of FHH and NSHPT lent further support to the critical importance of renal Ca^{2+}- (and Mg^{2+}-) sensing mechanisms operating in human physiology and altered in disease states. Patients with FHH had notable hypocalciuria despite hypercalcemia and normal or even mildly elevated PTH levels (Marx, 2018). This phenotype, when it became clearly evident in the 1970s and 1980s (Attie et al., 1983) (Marx et al., 1978a, 1978b, 1981a, 1981b, 1982a, 1982b, 1985), along with the typical autosomal dominant mode of inheritance of FHH, predicted the condition to be due to a genetic defect with a high degree of penetrance. The phenotypic features further predicted that the responsible gene would be strongly expressed in both parathyroid and renal cells. The fact that many individuals with FHH also had mild hypermagnesemia emphasized that the Ca^{2+}_o-sensing defect of FHH also involved defective Mg^{2+}_o-sensing. The diagnosis of FHH is assessed for clinically by the use of the Ca^{2+}—creatinine clearance ratio, which directly reflects the handling of Ca^{2+} by the kidney in relation to renal function and serum Ca^{2+} levels, and this clearance calculation is widely used (Christensen et al., 2008).

Fluctuations in $[Ca^{2+}]_o$ can exert several different effects on renal function. Increased levels of Ca^{2+}_o promote renal Ca^{2+} excretion, and this response occurs in the absence of PTH. It was observed that patients with FHH who underwent total parathyroidectomy to treat their hypercalcemia continued to demonstrate hypocalciuria, indicating that this was an intrinsic property of the kidney and not a PTH-dependent phenomenon (Attie et al., 1983).

Renal sensitivity to Ca^{2+}_o involves many functions and locations in the kidney (Riccardi and Valenti, 2016). Areas in the nephron and renal functions include (1) Ca^{2+} transport in the thick ascending limb of the loop of Henle (TAL), which is the main site of abnormal Ca^{2+} handling in individuals with FHH (Attie et al., 1983); (2) proximal tubule; (3) CD; (4) distal convoluted tubule; and (5) juxtaglomerular apparatus (JGA).

Where exactly and to what extent CaSR mRNA and protein are expressed within different segments of the nephron have been controversial topics over the years (Riccardi and Valenti, 2016). A lack of reliable reagents for definitively identifying and quantifying the often-low levels of CaSR expression has hindered the development of a clear picture of the sites where CaSRs are expressed. Expression of CaSRs in the TAL is widely accepted and is among the strongest in the kidney. The renal CD is another site where water handling and urinary concentration occur and where CaSRs are prominently expressed. Other sites of CaSR expression include the glomerulus (in podocytes and mesangial cells) and in proximal tubular cells.

Extensive studies have been done in various renal cell systems and mouse models. In cultured mesangial cells, CaSR activation stimulated inositol trisphosphate production and $Ca^{2+}{}_i$ mobilization, similar to the signaling pathways activated by high Ca^{2+}_o and Mg^{2+}_o concentrations and calcimimetics in parathyroid cells (Brown et al., 1993; Riccardi and Valenti, 2016). CaSR activation is also linked to the opening of the transient receptor potential channels (TRPCs) TRPC3 and TRPC6 in these cells (Oh et al., 2011).

The JGA is the principal source of renin release, which is negatively regulated (suppressed) by high Ca^{2+}_o levels but positively regulated by many other important physiological factors. Similar to the signaling pathways activated by high Ca^{2+}_o levels in parathyroid cells, CaSR activation involves Ca^{2+}_i mobilization (via a Gq mechanism) in modulating renin secretion as well as inhibition of cyclic AMP accumulation and adenyl cyclase activity (Beierwaltes, 2010).

Several functions of proximal tubular cells are Ca^{2+}-regulated. One such critical function is phosphate reabsorption. PTH, through its receptor (PTH1R), is a key regulator of phosphate reabsorption. When the PTH1R is stimulated, phosphate excretion increases. CaSR activation by high Ca^{2+}_o or calcimimetic can blunt the ability of PTH to stimulate phosphaturia. This has been used to advantage with the calcimimetic cinacalcet in treating disorders of phosphate wasting such as tumor-induced osteomalacia. High Ca^{2+}_o concentrations also inhibit proximal tubular 1-alpha hydroxylase activity and thereby dampen the production of $1,25(OH)_2D_3$, which will ultimately lower serum Ca^{2+} levels. Acute PTH administration can also reduce CaSR expression in that region of the kidney (Riccardi and Valenti, 2016). Other actions of CaSR signaling in the proximal tubule include luminal acidification and fluid reabsorption (Riccardi and Valenti, 2016). $1,25(OH)_2D_3$ can modulate CaSR expression in the distal tubule, where this hormone promotes Ca^{2+} transport and thereby contributes to the raising of serum Ca^{2+} levels.

A great number of mechanistic studies have been done to clarify the pathways by which CaSRs modulate Ca^{2+} reabsorption in the TAL (Riccardi and Valenti, 2016). PTH stimulates Ca^{2+} reabsorption by increasing Ca^{2+} permeability. CaSRs in the cortical TAL are involved in the actions of hormone-mediated cAMP accumulation such as that due to calcitonin or vasopressin. Decreased cAMP accumulation in these cells will reduce overall Ca^{2+} permeability and thereby Ca^{2+} transport and reabsorption. Ca^{2+} permeability in this portion of the nephron requires NaCl uptake across the lumen of the nephron segment. The transepithelial K^+ gradient that ultimately becomes established in this segment is the driving force of the reabsorption of Ca^{2+}, Mg^{2+}, and Na^{+}. Paracellular Ca^{2+} permeability in the TAL is importantly regulated by various members of the family of claudin proteins. Human disorders of Ca_o^{2+}-sensing also shed light on the role of CaSRs in Ca^{2+} reabsorption by the kidney and in Ca^{2+} excretion. Individuals with germline-activating mutations in the CaSR (ADH1) may have hypercalciuria as a result of their activated renal CaSRs. Renal-specific *Casr* deletion in a mouse model (Toka et al., 2012) produces hypocalciuria along with normal serum Ca^{2+} and PTH levels. This is evidence that renal CaSRs operate to control renal Ca^{2+} excretion independently of changes in PTH in vivo.

The distal convoluted tubule also plays a role in Ca^{2+} reabsorption (Riccardi and Valenti, 2016), and multiple membrane channels and transporters play essential parts in that process. An especially prominent role is played by transient receptor potential cation channel (TRPV) 5, which moves Ca^{2+} across the apical membrane of the distal tubular cell. Other ion transport mechanisms cooperatively participate in Ca^{2+} reabsorption thereafter.

CaSRs are strongly expressed along the CD (Riccardi and Valenti, 2016). In that nephron segment, if high $[Ca^{2+}]_o$ is present, there is urinary acidification due to the activation of an H^{+}-ATPase. CaSRs are also expressed in this nephron segment along with aquaporin 2 water channels. Both CaSRs and aquaporin 2 channels are colocated within the same vesicles in the CD. In this portion of the nephron, the activation of CaSRs is thought to inhibit vasopressin-mediated insertion of aquaporin 2 water channels into the membrane and thereby reduce water reabsorption. This yields water diuresis and "resistance" to the effects of vasopressin when increased levels of CaSR activation are present. The interrelationships between CaSR activity and vasopressin action in this portion of the kidney are thought to underlie the nephrogenic diabetes insipidus commonly seen in individuals with significant hypercalcemia. Thus, there are multiple areas of the kidney where CaSR activation plays a central role in transport and/or permeability to water, Ca^{2+}, Mg^{2+} phosphate, acid, and other ions. CaSR activation also modulates kidney responses to critically important hormones that modulate its function including vasopressin, calcitonin, and others. Human disorders of Ca_o^{2+}-sensing help to underscore the importance of specific physiological pathways and integrate them into in vivo responses.

Calcium-sensing receptor in bone cells

Long before the CaSR cDNA was identified in 1993 by Brown and colleagues (Brown et al., 1993) with an expression cloning strategy using parathyroid gland cRNA in *Xenopus laevis* oocytes, it was clear that changes in the extracellular concentration of Ca^{2+} had significant effects on the function of several different cell types in bone (Dvorak et al., 2004) (Chang et al., 2008; Goltzman and Hendy, 2015). These studies were largely in vitro in cell culture systems of transformed osteosarcoma cell lines and often required higher-than-physiological $[Ca^{2+}]_o$ to achieve effects on signaling pathway activation, Ca_i^{2+} mobilization, or functional or biochemical outcomes (mitogenesis, mineralization, and so forth). The pharmacology of these responses led to controversy in the field as to whether the putative parathyroid Ca_o^{2+}-sensing mechanism responsible for mediating the effects of Ca_o^{2+} on PTH secretion (in a tight physiological range) was the same molecular mechanism operating in different bone cell populations. Once the CaSR cDNA became available, along with highly specific antibodies directed against CaSR epitopes, it became clear, by immunocytochemistry and in situ hybridization, that CaSRs were present in osteoblasts, osteocytes, osteoclasts, marrow mononuclear cells, and macrophages (Chang et al., 1999; Dvorak et al., 2004). Speculations from earlier studies were subsequently supported by strong evidence that CaSRs were playing critical roles in bone and were responsible for the effects of changes in $[Ca^{2+}]_o$ on the activation of signaling responses, alterations in gene expression, and mineralization in bone cell systems (Chang et al., 2008; Dvorak et al., 2004; Goltzman and Hendy, 2015; Rybchyn et al., 2011). Furthermore, the receptor was implicated in regulating bone development and mass in vivo (Chang et al., 2008; Dvorak et al., 2007; Dvorak-Ewell et al., 2011).

The phenotype of global *Casr* knockout mice, $Casr^{-/-}$, generated by Ho et al. (Ho et al., 1995) confirmed the critical role of CaSRs in mediating the control of PTH secretion by Ca_o^{2+}. These mice showed severe hyperparathyroidism, hypercalcemia, and hypophosphatemia and died at 2–3 weeks of age. Dramatic skeletal demineralization was present along with growth retardation and failure to thrive. Given the severe mineral and hormonal disturbances in these mice, it was difficult to ascertain what role bone CaSRs might be playing in the skeletal aspects of their phenotype, because the hyperparathyroidism and serum Ca^{2+} and phosphate derangements in these mice can alone have profound effects on bone growth, development, and mineralization. Knockout mouse models in which the *Pth* gene was deleted on the background of this global $Casr^{-/-}$ mouse (i.e., a double-knockout, $Pth^{-/-}/Casr^{-/-}$) had relatively normal-appearing bone by histology; however, these animals had markedly low PTH levels (Goltzman and Hendy, 2015; Kantham et al., 2009). Similarly, when PTH was eliminated by creating a double-knockout using $Gcm2^{-/-}$ mice in whom parathyroid glands did not develop on the background of this global $Casr^{-/-}$ mouse (i.e., $Gcm2^{-/-}/Casr^{-/-}$), a similarly normal skeletal histology was noted (Tu et al., 2003). This led investigators to surmise that CaSRs were not critical to skeletal development. Thus, it became necessary to generate and assess conditional knockout mice with *Casr* deletion selectively targeted to specific bone cell populations without the complicating features of hypo- or hyperparathyroidism present in the prior models.

The most compelling data for Ca^{2+} as a ligand and CaSR activation as an anabolic pathway in bone emerged from conditional knockout mouse models (Chang et al., 2008). When CaSRs were deleted across the osteoblastic lineage in mice (under control of the 2.3 or 3.6 kb type 1 collagen [Col1] promoter-driven Cre recombinase), the bone phenotypes were dramatic (Chang et al., 2008; Dvorak-Ewell et al., 2011). These bone-specific conditional knockout mice (e.g., $^{2.3Col1}Casr^{-/-}$ and $^{3.6Col1}Casr^{-/-}$ mice) showed growth retardation, spontaneous fractures, low trabecular and cortical bone mass, undermineralized bone, osteoblast and osteocyte apoptosis, and gene expression profiles indicative of arrested osteoblast differentiation (Chang et al., 2008; Dvorak-Ewell et al., 2011). When the conditional knockout of *Casr* was restricted to mature osteoblasts and osteocytes using a *Cre* mouse in which Cre expression was driven by the osteocalcin promoter, markers of osteoblast gene expression were delayed and trabecular bone mass and microarchitecture were reduced (Chang et al., 2008; Dvorak-Ewell et al., 2011), but the phenotype was much milder. These observations in vivo are supported by work in osteoblast cell lines by many labs (Chang et al., 2008; Goltzman and Hendy, 2015; Santa Maria et al., 2016), confirming that CaSR activation by agonists (e.g., high $[Ca^{2+}]_o$) is critical for osteoblast proliferation and differentiation and the maintenance of bone mass.

CaSR activation also modifies the function of mature osteoclasts and other cells within the osteoclastic lineage (monocytes and macrophages) (Goltzman and Hendy, 2015). In osteoclast lineage cells, CaSR activation suppresses resorptive function, expression of markers of osteoclast differentiation, and cell survival (Goltzman and Hendy, 2015; Kameda et al., 1998; Kanatani et al., 1999; Olszak et al., 2000; Santa Maria et al., 2016; Yamaguchi et al., 1998). High $[Ca^{2+}]_o$ also inhibits the resorption of mature osteoclasts. The concentrations of Ca_o^{2+} at which these phenomena are observed are in general much higher than those in the circulation but are ones likely to be present in the bone microenvironment and in the vicinity of resorbing osteoclasts (Datta et al., 1989; Malgaroli et al., 1989; Moonga et al., 1990; Zaidi et al., 1991). Osteoclasts also undergo apoptosis in response to high concentrations of Ca_o^{2+}, and the pathway responsible involves phospholipase C and NF-kappa B (Mentaverri et al., 2006). That a dominant-negative *Casr* construct interferes with these effects is highly supportive of CaSRs as involved in these events in osteoclasts. However, calcimimetics, which stimulate CaSRs and suppress PTH secretion, and calcilytics, which block CaSRs and enhance PTH secretion, do not have the effects predicted of straightforward CaSR activation and inhibition in osteoclasts (Shalhoub et al., 2003) (Gowen et al., 2000).

Observations from certain single-knockout and double-knockout mouse models have shed further light on the interaction of CaSRs with other key molecules involved in Ca^{2+} homeostasis such as PTH and PTH1R and $1,25(OH)_2D_3$. CaSR activation is key in the full actions of PTH on bone compartments. PTH-induced cortical bone demineralization is reduced in $Pth^{-/-}/Casr^{-/-}$ mice compared with $Pth^{-/-}$ mice (Xue et al., 2012), indicating that intact CaSRs are important in the actions of PTH-mediated bone resorption. In addition, the hypercalcemic actions of $1,25(OH)_2D_3$ are at least in part mediated by CaSRs, because deletion of *Casr* in mice on the background of *Cyp27b1* knockout produces less hypercalcemia (Richard et al., 2010). Isolated marrow cells from $Pth^{-/-}/Casr^{-/-}$ mice showed reduced RANKL expression and osteoclastogenesis in response to PTH versus responses in marrow cells from $Pth^{-/-}$ mice (Xue et al., 2012). These studies indicate that in addition to the effects of solitary CaSR activation in bone, there are cooperative effects of CaSR signaling in mediating the actions of PTH and PTH1R and $1,25(OH)_2D_3$, and likely its receptor as well.

Calcium-sensing receptor in the breast

Mammary CaSR plays pivotal roles in modulating maternal Ca^{2+} metabolism and neonatal bone development during lactation. CaSR levels are low during postnatal mammary gland development through late pregnancy but increase to a peak during lactation (VanHouten et al., 2004), when the receptor is expressed primarily on the basolateral surface of ductal epithelial and alveolar cells (Cheng et al., 1998; VanHouten et al., 2004, 2007). During milk production, the mammary gland synthesizes and secretes PTHrP into milk and into the circulation, where it activates bone resorption to mobilize skeletal calcium (Wysolmerski, 2012). Delivery of Ca^{2+} to the mammary gland induces the plasma membrane Ca^{2+}-ATPase 2 (PMCA2) to transport Ca^{2+} from breast epithelial cells across the apical membrane and into the acinar lumen (Reinhardt et al., 2004) (VanHouten and Wysolmerski, 2007). Although the CaSR does not have a dominant role in regulating morphological development or differentiation in the normal mammary gland (Kim and Wysolmerski, 2016), the CaSR adjusts transcellular Ca^{2+} transport, overall milk secretion, and PTHrP production in response to Ca^{2+} availability in the maternal circulation during milk production (Mamillapalli et al., 2013; Vanhouten and Wysolmerski, 2013). The mammary CaSR adjusts milk PTHrP levels in a manner inverse to milk Ca^{2+} content to coordinate maternal and neonatal bone and Ca^{2+} metabolism. In normal mammary epithelial cells, CaSR activation increases milk Ca^{2+} content by stimulating Ca^{2+} pumping and ATPase activity of PMCA2 (VanHouten et al., 2007) while it suppresses PTHrP production via coupling to Gαi to decrease cAMP production by adenylate cyclase and subsequent PKA activation (Mamillapalli et al., 2008). Systemic hypocalcemia reduces overall milk production and Ca^{2+} content and decreases the skeletal Ca^{2+} accrual of suckling pups (Ardeshirpour et al., 2006; VanHouten et al., 2004) but increases PTHrP production, an effect that can be prevented by treatment with a calcimimetic (VanHouten et al., 2004).

The CaSR is overexpressed by many breast cancers; however, CaSR signaling seems to have opposing effects on proliferation and cell death in breast cancer cells. Certain studies have shown that increasing $Ca^{2+}{}_o$ concentration above the physiological range inhibits proliferation, invasion, and anchorage-independent growth of breast cancer cells and that the CaSR mediates growth-inhibitory effects by downregulating the expression of survivin, an antiapoptotic factor (Liu et al., 2009; Promkan et al., 2011). On the contrary, other studies have shown that activation of CaSR promotes cell proliferation and the migration of breast cancer cells via stimulation of PLC, ERK1/2, and EGFR activation, upregulation of phosphocholine and cytokine production, and activation of TRPC1 (El Hiani et al., 2009a; El Hiani et al., 2009b; Hernandez-Bedolla et al., 2015; Huang C. et al., 2009; Saidak et al., 2009a). Knocking out the *Casr* gene or treating breast cancer cells with calcilytic has been shown to inhibit tumor cell proliferation and increase Ca^{2+}-mediated cell death (Kim et al., 2016). Furthermore, CaSR expression was correlated directly with PTHrP protein levels in human breast cancers (Kim et al., 2016; Mu et al., 2012; VanHouten et al., 2010). Unlike normal breast cells, activation of the CaSR in breast cancer cells with Ca_o^{2+}, spermine, aminoglycoside antibiotics, or allosteric calcimimetics upregulates PTHrP secretion through the stimulation of cAMP production (Mamillapalli et al., 2008; Sanders et al., 2000). A switch in G protein usage, by which the CaSR couples to Gαs instead of Gαi, underlies the opposing effects of the CaSR on PTHrP expression in malignant and normal breast cells (Mamillapalli et al., 2008). Nuclear PTHrP stimulates cell proliferation by decreasing the levels of the cell cycle inhibitor p27 kip1 and protects against Ca_o^{2+}-mediated cell death by preventing the nuclear accumulation of apoptosis-inducing factor (Kim et al., 2016). The alteration in PTHrP production has important consequences for the progression of bone metastases (Mihai et al., 2006). Increased PTHrP secretion in response to high Ca_o^{2+} levels acts in a paracrine fashion to accelerate osteolysis (Mamillapalli et al., 2008; Wysolmerski, 2012). As a result of bone resorption, release of growth factors such as TGF-β, IGFs, and FGFs from the bone matrix further stimulates the growth of bone-metastasizing tumors and increases PTHrP production, developing a vicious cycle of osteolysis (Patel et al., 2011; Theriault and Theriault, 2012).

Noncalciotropic roles of the calcium-sensing receptor

Calcium-sensing receptor in the pancreas

CaSR is expressed in the endocrine and exocrine pancreas including β-islet cells, α-cells, duct cells, acinar cells, and various nonexocrine cells such as intrapancreatic nerves and blood vessels (Bruce et al., 1999) (Gray et al., 2006; Racz et al., 2002), whereas somatostatin-secreting δ-cells do not express the CaSR (Squires, 2000). Activation of the CaSR in α-cells stimulates glucagon secretion from human islets (Gray et al., 2006). Within the islet, glucagon can increase insulin secretion despite its counteracting effect on energy homeostasis.

CaSR regulates glucose-evoked insulin secretion by β-cells (Jones et al., 2007; Rybczynska et al., 2017). Elevation of plasma glucose concentration leads to an increase in the ATP/ADP ratio mediated by glucokinase, which closes ATP-dependent K^+ channels, depolarizing the cell membrane and opening voltage-dependent Ca^{2+} channels (VDCCs) to increase Ca_i^{2+} levels and induce insulin release (Squires, 2000). In human islets and insulin-secreting MIN6 cell lines, activation of the CaSR by elevating Ca_o^{2+} induced transient MAPK activation and insulin secretion (Devis et al., 1975; Gray et al., 2006; Squires, 2000). Calcimimetic activation of the CaSR not only enhanced the maximal secretory response to glucose in human and rodent β-cells, but also transiently increased insulin secretion in the absence of an increase in associated nutrient stimulation (Gray et al., 2006). CaSR may modulate insulin secretory responses through the stimulation of β-cell proliferation and cell–cell communication (Hills et al., 2012). Calcimimetic activation of the CaSR in MIN6 increased the expression of E-cadherin and L-type VDCCs and stimulated MAPK-mediated cell proliferation, thereby augmenting cell–cell adhesion and β-cell function to promote the insulin secretion of neighboring cells (Hills et al., 2012; Kitsou-Mylona et al., 2008). Colocalization and spatial interactions between the CaSR and L-type VDCCs may also enhance glucose-induced secretion of insulin (Parkash, 2011). Conversely, cell turnover and glucose-evoked insulin secretion in MIN6 pseudoislets was reduced when CaSR expression was knocked down (Kitsou-Mylona et al., 2008). CaSR-dependent regulation of insulin secretion involves complex signaling cascades including ERK1/2, PLC/IP_3/intracellular Ca^{2+}, and CAMKII (Gray et al., 2006). However, persistent activation of CaSR in human β-cells has been shown to inhibit insulin secretion (Squires et al., 2000). Mice with gain-of-function CaSR mutations manifested hypocalcemia in association with impaired glucose tolerance and insulin secretion (Babinsky et al., 2017; Dong et al., 2015). In one ADH1 mouse model, pancreas islet mass was reduced, and both α-cell and β-cell functions were impaired (Babinsky et al., 2017). The rectification of glucose intolerance and hypoinsulinemia in these mice by a calcilytic highlighted a potential therapeutic application of these compounds for disordered glucose metabolism (Babinsky et al., 2017; Dong et al., 2015).

CaSR is highly expressed in pancreatic duct cells and regulates their proliferation (Racz et al., 2002). It has been reported that the activation of CaSR in the rat pancreas duct luminal membrane increased ductal HCO_3^- secretion (Bruce et al., 1999). These observations suggest that CaSR is able to monitor and regulate Ca^{2+} concentration in pancreatic juice by triggering ductal electrolyte and fluid secretion, and this could reduce the risk of calcium carbonate stone formation and progression to pancreatitis (Racz et al., 2002). Clinical genetic linkage studies have identified that mutations in the CaSR gene and the pancreatic secretory trypsin inhibitor gene in patients with FHH are associated with susceptibility to acute and/or chronic pancreatitis (Felderbauer et al., 2006; Muddana et al., 2008).

Calcium-sensing receptor in the gastrointestinal system

The CaSR is present along the gastrointestinal tract and plays key roles in glandular and fluid secretion and in the occurrence of digestive disease. CaSRs are expressed in the stratified squamous esophageal epithelium (Justinich et al., 2008) and regulate the proliferation, differentiation, cell–cell adhesion, and cytoskeletal organization of epithelial cells (Abdulnour-Nakhoul et al., 2015). Activation of CaSR in the esophageal epithelial cell line HET-1A increases Ca_i^{2+} mobilization and activates MAPK (ERK1/2) signaling pathways to stimulate the secretion of cytokine IL-8 (Justinich et al., 2008). CaSR is implicated in the proliferative response to injury and the pathogenesis of esophagitis, as CaSR expression is increased in eosinophilic esophagitis and Barrett's adenocarcinoma (Abdulnour-Nakhoul et al., 2015). Under the condition of esophagitis, CaSR can be activated by eosinophil-released major basic protein to stimulate FGF9 secretion and activate downstream proliferation-related genes in esophageal epithelial cells (Mulder et al., 2009).

In the stomach, the CaSR is expressed on the basolateral membrane of parietal cells and on both the basolateral and the apical surface on gastrin-releasing G-cells (Cheng et al., 1999). The CaSR is considered a key modulator of gastric acid secretion and gastric luminal pH because it can be activated by the same dietary components (amino acids, amines, calcium) that activate gastrin release from G-cells. Activation of basolateral CaSR by agonists (e.g., Ca^{2+}, Gd^{3+}, and L-type amino acids) stimulates H+ and K+ ATPase and results in the increased production and secretion of gastric acid (Buchan et al., 2001; Busque et al., 2005; Dufner et al., 2005). In contrast, *Casr*$^{-/-}$ animals cannot respond to high intraluminal calcium and L-type amino acid concentrations with gastrin secretion (Feng et al., 2010). The intracellular signal cascades underlying CaSR-mediated H+ and K + ATPase induction include Ca_i^{2+}, PLC, MAPK, and PKC (Kopic et al., 2012; Remy et al., 2007).

Low pH in the gastric lumen produces ionized Ca^{2+} to be absorbed by the small intestine. CaSR participates in the regulation of intestinal secretion and absorption of Ca^{2+} in cooperation with 1,25-dihydroxy vitamin D_3 (Favus et al., 1980, 1981). CaSR is predominantly expressed on both apical and basolateral membranes of colonic crypts in the intestine brush border (Alfadda et al., 2014) and is also localized on the neuronal plexuses, Meissner and Auerbach (Cheng, 2012). Activation of the apical receptor by high luminal Ca^{2+} results in an increase in NaCl, water, and Ca^{2+} absorption. Consequently, elevated levels of blood Ca^{2+} activate intestinal basolateral CaSR to inhibit the absorption, forming a negative feedback loop (Alfadda et al., 2014). CaSR also acts as an important regulator of intestinal enteroendocrine activity in response to nutrient and nonnutrient stimuli. CaSR is involved in the L-amino acid stimulation of gut peptide secretion by K-cell and L-cell in the rat small intestine (Mace et al., 2012) and cholecystokinin secretion in isolated intestinal I cells (Liou et al., 2011). Furthermore, the CaSR plays a central role in intestinal fluid transport. Activation of the colonic CaSR by Ca^{2+}, Gd^{3+}, polyamines, or neomycin induces a rapid increase in $Ca^{2+}{}_i$ levels in both surface and crypt cells and reduces the stimulatory effect of forskolin on net fluid secretion in perfused crypts (Cheng et al., 2002, 2004). While secretagogues such as forskolin and cholera toxin could induce fluid secretion by increasing cyclic nucleotides, activation of either the luminal or the basolateral CaSR by Ca^{2+} and spermine can reverse forskolin-stimulated fluid secretion via facilitation of the destruction of cAMP and cGMP, inhibiting Cl^- secretion and stimulating Na^+ absorption (Geibel et al., 2006). CaSR deficiency in the intestinal epithelium induced a proinflammatory immune response, compromised intestinal barrier function, and altered the composition of the microbiota (Cheng et al., 2014). These findings suggest that CaSR activator and CaSR-based nutrients might provide an effective treatment for disorders such as inflammatory bowel disease, enterocolitis, and secretory diarrhea (Cheng, 2016; Owen et al., 2016).

CaSR negatively modulates colonic epithelial growth and the inactivation of CaSR due to promoter hypermethylation, and defective histone acetylation (Fetahu et al., 2014) plays a key role in colorectal tumorigenesis (Rogers et al., 2012; Singh et al., 2013). High dietary Ca^{2+} intake promotes colonic mucosal epithelial cell differentiation, decreases cell growth, and reduces risk for the development of colorectal cancer (Lamprecht and Lipkin, 2001). Conditional knockout of CaSR in intestinal epithelial cells increases nuclear β-catenin localization, enhances cell proliferation, and diminishes the differentiation of colonic crypts (Rey et al., 2012). Activation of the CaSR by raising Ca^{2+}_o in Caco-2 cells leads to a decrease in c-Myc proto-oncogene expression and abrogates its proproliferative effect (Kallay et al., 1997). Activation of the CaSR in human carcinoma cell lines by elevating Ca^{2+}_o levels decreases cell growth and promotes cell differentiation in correspondence with increased E-cadherin expression and suppressed β-catenin activation (Chakrabarty et al., 2003; Van Aken et al., 2001; Wong and Pignatelli, 2002). Increased E-cadherin stimulated by CaSR can interact with β-catenin to enhance cell–cell and cell–matrix adhesion via remodeling of the actin-cytoskeleton (Brembeck et al., 2006). Meanwhile, the activation of CaSR also prevents nuclear translocation of β-catenin, thereby reducing the formation of β-catenin–TCF4 complex and downregulating c-Myc and cyclin D1 expression (Chakrabarty et al., 2003). In addition, activation of CaSR stimulates the noncanonical Wnt pathway (Wnt5a/Ror2) to inhibit overly active Wnt signaling in colon cancer cell lines (MacLeod et al., 2007) and to decrease the risk of colitis-associated colon cancer by suppressing NFκ-B activity and reducing TNFα secretion and TNFR1 expression (Kelly et al., 2011) (Macleod, 2013). Restoration of CaSR expression in colorectal cancer cells reduces proliferation and increases differentiation and apoptotic potential (Aggarwal et al., 2015) and results in the concurrent reversal of stem cell markers, drug resistance, and epithelial–mesenchymal transition-related transcription factors (Singh and Chakrabarty, 2013). Taken together, these data support a role of CaSR as a tumor suppressor.

Calcium-sensing receptor in the peripheral vascular system

Although arterial blood pressure is regulated by the centrally located circulatory centers in the medulla oblongata, long-term blood pressure is controlled primarily by the volume balance of blood vessels. In addition to the central control mechanisms mediated by vegetative neuronal innervation and hormone secretion, blood vessels are able to independently control their perfusion locally via the actions of the CaSR (Smajilovic et al., 2011). CaSR expression is found in several cell types in vasculature including the endothelium (Ziegelstein et al., 2006), VSMCs (Schepelmann et al., 2016), and the perivascular nerve (Bukoski et al., 1997). The activation of CaSR in rat mesenteric branch arteries induced neuronal release of cannabinoid, causing endothelium-independent hyperpolarization of VSMCs and subsequent relaxation of the arteries (Ishioka and Bukoski, 1999). The CaSR of endothelial cells participates in Ca^{2+}-induced vasorelaxation by two mechanisms, endothelial hyperpolarization and nitric oxide release (Weston et al., 2005) (Greenberg et al., 2016). In mesenteric arteries, activation of CaSR by Ca^{2+}_o or calcimimetic led to the opening of the Ca^{2+}-sensitive K^+ channel, which induced hyperpolarization of endothelium and VSMCs (Greenberg et al., 2016). In the human aorta, activation of CaSR by spermine induced an increase in $Ca^{2+}{}_i$ concentration and nitric oxide (NO) production, resulting in decreased vascular tone (Loot et al., 2013; Ziegelstein et al., 2006). It was suggested that the heteromeric TRPV4/TRPC1 channels in vascular endothelial cells play a role in CaSR-induced NO production and vasorelaxation (Ma et al., 2011) (Greenberg et al., 2017). On the other hand, rises in Ca^{2+}_o induced endothelium-independent contractility in the aorta and mesenteric artery (Schepelmann et al., 2016; Wonneberger et al., 2000). Ablation of CaSR in VSMCs led to lower diastolic and mean arterial blood pressures in *Casr* knockout mice (Schepelmann et al., 2016). Activation of the CaSR of smooth muscle cells induced ERK1/2 phosphorylation and a subsequent increase in proliferation with no elevation in the IP_3 level (Smajilovic et al., 2006). Thus, vascular CaSR plays a dual role in controlling blood vessel tone and heart rate with prorelaxing actions in the endothelium (Lopez-Fernandez et al., 2015) and procontractile effects in vascular smooth muscle (Schepelmann et al., 2016). Nonetheless, a recent study revealed another mechanism for the CaSR to control regional blood flow and blood pressure. Lee et al. emonstrated that activating the CaSR in the olfactory epithelium of rat with Ca^{2+}_o or CaSR agonists increased sympathetic efferent nerve activities and arterial blood pressure and subsequently decreased renal blood flow and renal, hepatic, and enteral microcirculation (Lee et al., 2019). In addition, the suppressive effect on renal perfusion was antagonized by calcilytic NPS-2143 and abolished by mechanic denervation (Lee et al., 2019), supporting a direct coupling between CaSR activation and increased peripheral vascular resistance by sympathetic excitation.

Reduced vasorelaxation by low CaSR expression or function has been linked to the complications of diabetic vasculopathy (Loot et al., 2013). Decreased expression of CaSR is also associated with the development of cardiovascular calcification in individuals with advanced chronic kidney disease (London et al., 2005). In vitro studies show that calcimimetic treatment increases CaSR expression and reduces vascular mineralization (Alam et al., 2009) (Henaut et al., 2014; Molostvov et al., 2015). Additionally, calcimimetic treatments were shown to exert antihypertensive effects in patients with chronic kidney disease (Zitt et al., 2011) and reduced the risk of cardiovascular events in elderly hemodialysis patients (Parfrey et al., 2015; Sumida et al., 2013), possibly because of their ability to restore CaSR expression levels in the vasculature (Henaut et al., 2014).

Calcium-sensing receptor in the lung

CaSR is expressed in both epithelium and smooth muscle in developing fetal lungs (Riccardi et al., 2013). The CaSR has been identified as an important regulator of intraluminal pressure and lung development. A crucial part of lung development is branching morphogenesis during the pseudoglandular stage. Exposure of isolated fetal lung buds to increased Ca_o^{2+} or calcimimetic induced elevations in Ca_i^{2+} and had a suppressive effect on branching because of the inhibition of cell proliferation (Finney et al., 2008). This CaSR-mediated suppression of the lung branching program requires PLC-dependent Ca_i^{2+} release and the activation of PI3K (Finney et al., 2008). Furthermore, higher Ca_o^{2+} or calcimimetic induced lumen distension by increasing luminal fluid volume and stimulating transepithelial fluid transport (Finney et al., 2008). CaSR is also found in the neuroepithelial bodies of the postnatal mouse lung and is implicated in coordinating intercellular communication through Ca_i^{2+} signaling in this intrapulmonary airway stem cell niche (Lembrechts et al., 2013).

Greater expression and function of CaSR in pulmonary arterial smooth muscle cells has been linked with the onset of idiopathic pulmonary arterial hypertension (PAH) (Yamamura et al., 2012) and the development of asthma (Yarova et al., 2015). Unlike the largely protective properties of CaSR in the peripheral vascular system, CaSR is involved in hypoxic vasoconstriction in the pulmonary vasculature and in vascular remodeling during the development and progression of PAH (Peng et al., 2014; Tang et al., 2016; Zhang et al., 2012). Nonetheless, recent work demonstrated that calcilytics can significantly improve the remodeling process of lung vessels in cases of PAH (Yamamura et al., 2015, 2016).

Asthma is characterized by airway accumulation of a group of polycationic proteins, particularly eosinophilic cationic proteins and major basic proteins, due to chronic inflammation (Kurosawa et al., 1992; North et al., 2013). Activation of CaSR by polycations increases Ca_i^{2+} levels and cell proliferation in airway smooth muscle cells, leading to airway hyperresponsiveness, bronchoconstriction, and inflammation in allergic asthma. It was postulated that CaSR may also control the recruitment and activation of inflammatory cells, which in turn release cytokines that subsequently elevate CaSR expression, thus creating a positive feedback loop in asthma (Yarova et al., 2015). These effects, however, could be ameliorated by CaSR ablation from the airway smooth muscle or prevented by nebulized calcilytics (Yarova et al., 2015). The protective effects of nebulized calcilytics are mediated by preventing Ca_i^{2+} increases and by abrogating the signaling pathways associated with airway contractility, such as ERK1/2, p38 MAPK, and PI3K/Akt (Yarova et al., 2015).

Calcium-sensing receptor in the epidermis

The epidermis consists of multiple layers of keratinocytes that differentiate and produce a permeability barrier that provides protection against desiccation, xenobiotic infection, and other environmental insults. Ca_o^{2+} is essential for initiating keratinocyte differentiation and maintaining epidermal functions. Activating CaSR by raising Ca_o^{2+} levels increases Ca_i^{2+} levels through coupling to the $G_{q/11}$/PLC pathway (Bikle et al., 2012) and induces cell–cell adhesion by stimulating the formation of the E-cadherin–catenin protein complex via the activation of Rho small GTPases and Src family tyrosine kinase signaling pathways (Calautti et al., 1998, 2002). The E-cadherin–catenin adhesion complexes in the cell membrane recruit and activate PI3K, Akt, PLCγ1, and EGFR, important regulators of cell proliferation, survival, and differentiation (Calautti et al., 2005; Tu et al., 2018; Xie and Bikle, 2007).

The CaSR is expressed throughout all keratinocyte layers in the epidermis (Komuves et al., 2002) with predominant intracellular perinuclear localization (Tu et al., 2001). Fluorescence immunostaining and coimmunoprecipitation studies revealed that CaSR colocalized and formed a protein complex with PLCγ1, IP_3R, and SPCA1 in the *trans*-Golgi (Tu et al., 2007). Inactivation of CaSR in keratinocytes profoundly reduced releasable Ca_i^{2+} pools (Tu et al., 2007), suggesting that the CaSR may coordinate Ca^{2+} release and the replenishment of stores and Ca^{2+} entry through membrane channels via direct interactions with $Ca^{2+}{}_i$ modulators such as PLC and Ca^{2+}-ATPase. Besides the initial Ca^{2+} release via PLC activation, the CaSR sustains increases in Ca_i^{2+} levels via E-cadherin-mediated signaling (Tu et al., 2008). Ca_o^{2+} stimulates interactions among the CaSR, filamin A, Rho-GEF Trio, RhoA, and E-cadherin at the cell membrane and facilitates CaSR signaling through the Rho and E-cadherin pathways (Tu et al., 2011; Tu and You, 2014). In Casr-deficient keratinocytes, the ability of Ca_o^{2+} to stimulate Ca_i^{2+} mobilization is severely inhibited, and the ineffectual Rho signaling fails to stabilize the E-cadherin–catenin adhesion complex and activate PI3K and PLCγ1, resulting in decreased basal Ca_i^{2+} level, impaired intercellular adhesion, increased cell apoptosis, and reduced differentiation (Tu et al., 2008, 2011). Additionally, depleting CaSRs suppresses keratinocyte proliferation by downregulating the E-cadherin/EGFR/MAPK signaling axis (Tu et al., 2018). Furthermore, the control of keratinocyte proliferation and differentiation by CaSR may involve activation of the Wnt/β-catenin signaling cascade (Popp et al., 2014; Turksen and Troy, 2003).

The physiological importance of CaSR in epidermal development and maintenance is made clear by animal models in which the CaSR gene has been overexpressed or deleted. The skin of keratinocyte-specific CaSR knockout ($^{Epid}Casr^{-/-}$) mice exhibits the loss of an innate epidermal Ca^{2+} gradient, aberrant keratinocyte differentiation, and retarded epidermal repair after injury due to impaired Ca_i^{2+}- and E-cadherin-mediated signaling (Tu et al., 2012, 2018). CaSR abrogation also delays the formation of the permeability barrier during prenatal development and impairs skin barrier function in adults (Tu et al., 2012). On the other hand, transgenic mice with constitutive expression of CaSR in basal keratinocytes manifest advanced differentiation and epidermal permeability barrier formation during embryologic development as well as accelerated hair growth at birth (Turksen and Troy, 2003). These changes in epidermal differentiation may be attributed to enhanced cross talk between CaSR and the Wnt/β-catenin signaling pathway in the CaSR transgenic epidermis (Turksen and Troy, 2003). Similarly, stimulating endogenous epidermal CaSR with calcimimetic NPS-R568 accelerated wound reepithelialization through enhancing of the epidermal $Ca^{2+}{}_i$ signals and E-cadherin membrane expression (Tu et al., 2018). These findings demonstrated a critical role for the CaSR in epidermal homeostasis and the therapeutic potential of calcimimetics for improving skin wound repair.

References

Abdulnour-Nakhoul, S., Brown, K.L., Rabon, E.C., Al-Tawil, Y., Islam, M.T., Schmieg, J.J., et al., 2015. Cytoskeletal changes induced by allosteric modulators of calcium-sensing receptor in esophageal epithelial cells. Phys. Rep. 3 (11).

Aggarwal, A., Hobaus, J., Tennakoon, S., Prinz-Wohlgenannt, M., Graca, J., Price, S.A., et al., 2016. Active vitamin D potentiates the anti-neoplastic effects of calcium in the colon: a cross talk through the calcium-sensing receptor. J. Steroid Biochem. Mol. Biol. 155 (Pt B), 231–238.

Aggarwal, A., Prinz-Wohlgenannt, M., Tennakoon, S., Hobaus, J., Boudot, C., Mentaverri, R., et al., 2015. The calcium-sensing receptor: a promising target for prevention of colorectal cancer. Biochim. Biophys. Acta 1853 (9), 2158–2167.

Aida, K., Koishi, S., Inoue, M., Nakazato, M., Tawata, M., Onaya, T., 1995a. Familial hypocalciuric hypercalcemia associated with mutation in the human $Ca^{(2+)}$-sensing receptor gene. J. Clin. Endocrinol. Metab. 80 (9), 2594–2598.

Aida, K., Koishi, S., Tawata, M., Onaya, T., 1995b. Molecular cloning of a putative $Ca^{(2+)}$-sensing receptor cDNA from human kidney. Biochem. Biophys. Res. Commun. 214 (2), 524–529.

Alam, M.U., Kirton, J.P., Wilkinson, F.L., Towers, E., Sinha, S., Rouhi, M., et al., 2009. Calcification is associated with loss of functional calcium-sensing receptor in vascular smooth muscle cells. Cardiovasc. Res. 81 (2), 260–268.

Alfadda, T.I., Saleh, A.M., Houillier, P., Geibel, J.P., 2014. Calcium-sensing receptor 20 years later. Am. J. Physiol. Cell Physiol. 307 (3), C221–C231.

Ardeshirpour, L., Dann, P., Pollak, M., Wysolmerski, J., VanHouten, J., 2006. The calcium-sensing receptor regulates PTHrP production and calcium transport in the lactating mammary gland. Bone 38 (6), 787–793.

Arulpragasam, A., Magno, A.L., Ingley, E., Brown, S.J., Conigrave, A.D., Ratajczak, T., et al., 2012. The adaptor protein 14-3-3 binds to the calcium-sensing receptor and attenuates receptor-mediated Rho kinase signalling. Biochem. J. 441 (3), 995–1006.

Atchison, D.K., Beierwaltes, W.H., 2013. The influence of extracellular and intracellular calcium on the secretion of renin. Pflügers Archiv 465 (1), 59–69.

Attie, M.F., Gill Jr., J.R., Stock, J.L., Spiegel, A.M., Downs Jr., R.W., Levine, M.A., et al., 1983. Urinary calcium excretion in familial hypocalciuric hypercalcemia. Persistence of relative hypocalciuria after induction of hypoparathyroidism. J. Clin. Investig. 72 (2), 667–676.

Avlani, V.A., Ma, W., Mun, H.C., Leach, K., Delbridge, L., Christopoulos, A., et al., 2013. Calcium-sensing receptor-dependent activation of CREB phosphorylation in HEK293 cells and human parathyroid cells. Am. J. Physiol. Endocrinol. Metab. 304 (10), E1097–E1104.

Awata, H., Huang, C., Handlogten, M.E., Miller, R.T., 2001. Interaction of the calcium-sensing receptor and filamin, a potential scaffolding protein. J. Biol. Chem. 276 (37), 34871–34879.

Babinsky, V.N., Hannan, F.M., Ramracheya, R.D., Zhang, Q., Nesbit, M.A., Hugill, A., et al., 2017. Mutant mice with calcium-sensing receptor activation have hyperglycemia that is rectified by calcilytic therapy. Endocrinology 158 (8), 2486–2502.

Bai, M., Quinn, S., Trivedi, S., Kifor, O., Pearce, S.H., Pollak, M.R., et al., 1996. Expression and characterization of inactivating and activating mutations in the human $Ca^{2+}o$-sensing receptor. J. Biol. Chem. 271 (32), 19537–19545.

Bai, M., Trivedi, S., Brown, E.M., 1998a. Dimerization of the extracellular calcium-sensing receptor (CaR) on the cell surface of CaR-transfected HEK293 cells. J. Biol. Chem. 273 (36), 23605–23610.

Bai, M., Trivedi, S., Kifor, O., Quinn, S.J., Brown, E.M., 1999. Intermolecular interactions between dimeric calcium-sensing receptor monomers are important for its normal function. Proc. Natl. Acad. Sci. U. S. A. 96 (6), 2834–2839.

Bai, M., Trivedi, S., Lane, C.R., Yang, Y., Quinn, S.J., Brown, E.M., 1998b. Protein kinase C phosphorylation of threonine at position 888 in $Ca^{2+}o$-sensing receptor (CaR) inhibits coupling to Ca^{2+} store release. J. Biol. Chem. 273 (33), 21267–21275.

Bai, S., Favus, M.J., 2006. Vitamin D and calcium receptors: links to hypercalciuria. Curr. Opin. Nephrol. Hypertens. 15 (4), 381–385.

Balenga, N., Koh, J., Azimzadeh, P., Hogue, J., Gabr, M., Stains, J.P., et al., 2019 Jan 28. Parathyroid-targeted overexpression of Regulator of G-Protein Signaling 5 (R GS5) causes hyperparathyroidism in transgenic mice. J. Bone Miner. Res. https://doi.org/10.1002/jbmr.3674 [Epub ahead of print].

Beierwaltes, W.H., 2010. The role of calcium in the regulation of renin secretion. Am. J. Physiol. Renal. Physiol. 298 (1), F1–F11.

Betson, M., Lozano, E., Zhang, J., Braga, V.M., 2002. Rac activation upon cell-cell contact formation is dependent on signaling from the epidermal growth factor receptor. J. Biol. Chem. 277 (40), 36962–36969.

Bhagavathula, N., Kelley, E.A., Reddy, M., Nerusu, K.C., Leonard, C., Fay, K., et al., 2005. Upregulation of calcium-sensing receptor and mitogen-activated protein kinase signalling in the regulation of growth and differentiation in colon carcinoma. Br. J. Canc. 93 (12), 1364–1371.

Bikle, D.D., Xie, Z., Tu, C.L., 2012. Calcium regulation of keratinocyte differentiation. Expert Rev. Endocrinol. Metab. 7 (4), 461–472.

Bosel, J., John, M., Freichel, M., Blind, E., 2003. Signaling of the human calcium-sensing receptor expressed in HEK293-cells is modulated by protein kinases A and C. Exp. Clin. Endocrinol. Diabetes 111 (1), 21–26.

Bouschet, T., Martin, S., Henley, J.M., 2012. Regulation of calcium sensing receptor trafficking by RAMPs. Adv. Exp. Med. Biol. 744, 39–48.

Bradbury, R.A., Sunn, K.L., Crossley, M., Bai, M., Brown, E.M., Delbridge, L., et al., 1998. Expression of the parathyroid $Ca^{(2+)}$-sensing receptor in cytotrophoblasts from human term placenta. J. Endocrinol. 156 (3), 425–430.

Brauner-Osborne, H., Wellendorph, P., Jensen, A.A., 2007. Structure, pharmacology and therapeutic prospects of family C G-protein coupled receptors. Curr. Drug Targets 8 (1), 169–184.

Breitwieser, G.E., 2013. The calcium sensing receptor life cycle: trafficking, cell surface expression, and degradation. Best Pract. Res. Clin. Endocrinol. Metabol. 27 (3), 303–313.

Breitwieser, G.E., 2014. Pharmacoperones and the calcium sensing receptor: exogenous and endogenous regulators. Pharmacol. Res. 83, 30–37.

Brembeck, F.H., Rosario, M., Birchmeier, W., 2006. Balancing cell adhesion and Wnt signaling, the key role of beta-catenin. Curr. Opin. Genet. Dev. 16 (1), 51–59.

Brennan, S.C., Conigrave, A.D., 2009. Regulation of cellular signal transduction pathways by the extracellular calcium-sensing receptor. Curr. Pharmaceut. Biotechnol. 10 (3), 270–281.

Brennan, S.C., Davies, T.S., Schepelmann, M., Riccardi, D., 2014. Emerging roles of the extracellular calcium-sensing receptor in nutrient sensing: control of taste modulation and intestinal hormone secretion. Br. J. Nutr. 111 (Suppl. 1), S16–S22.

Brennan, T.C., Rybchyn, M.S., Green, W., Atwa, S., Conigrave, A.D., Mason, R.S., 2009. Osteoblasts play key roles in the mechanisms of action of strontium ranelate. Br. J. Pharmacol. 157 (7), 1291–1300.

Brent, G.A., LeBoff, M.S., Seely, E.W., Conlin, P.R., Brown, E.M., 1988. Relationship between the concentration and rate of change of calcium and serum intact parathyroid hormone levels in normal humans. J. Clin. Endocrinol. Metab. 67 (5), 944–950.

Breuksch, I., Weinert, M., Brenner, W., 2016. The role of extracellular calcium in bone metastasis. J Bone Oncol. 5 (3), 143–145.

Broadhead, G.K., Mun, H.C., Avlani, V.A., Jourdon, O., Church, W.B., Christopoulos, A., et al., 2011. Allosteric modulation of the calcium-sensing receptor by gamma-glutamyl peptides: inhibition of PTH secretion, suppression of intracellular cAMP levels, and a common mechanism of action with L-amino acids. J. Biol. Chem. 286 (11), 8786–8797.

Brown, A.J., Zhong, M., Finch, J., Ritter, C., McCracken, R., Morrissey, J., et al., 1996. Rat calcium-sensing receptor is regulated by vitamin D but not by calcium. Am. J. Physiol. 270 (3 Pt 2), F454–F460.

Brown, E.M., 1991. Extracellular Ca^{2+} sensing, regulation of parathyroid cell function, and role of Ca^{2+} and other ions as extracellular (first) messengers. Physiol. Rev. 71 (2), 371–411.

Brown, E.M., 2013. Role of the calcium-sensing receptor in extracellular calcium homeostasis. Best Pract. Res. Clin. Endocrinol. Metabol. 27 (3), 333–343.

Brown, E.M., Butters, R., Katz, C., Kifor, O., Fuleihan, G.E., 1992. A comparison of the effects of concanavalin-A and tetradecanoylphorbol acetate on the modulation of parathyroid function by extracellular calcium and neomycin in dispersed bovine parathyroid cells. Endocrinology 130 (6), 3143–3151.

Brown, E.M., Chen, C.J., LeBoff, M.S., Kifor, O., Oetting, M.H., el-Hajj, G., 1989. Mechanisms underlying the inverse control of parathyroid hormone secretion by calcium. Soc. Gen. Physiol. 44, 251–268.

Brown, E.M., Gamba, G., Riccardi, D., Lombardi, M., Butters, R., Kifor, O., et al., 1993. Cloning and characterization of an extracellular $Ca^{(2+)}$-sensing receptor from bovine parathyroid. Nature 366 (6455), 575–580.

Brown, E.M., MacLeod, R.J., 2001. Extracellular calcium sensing and extracellular calcium signaling. Physiol. Rev. 81 (1), 239–297.

Brown, E.M., Vassilev, P.M., Quinn, S., Hebert, S.C., 1999. G-protein-coupled, extracellular $Ca^{(2+)}$-sensing receptor: a versatile regulator of diverse cellular functions. Vitam. Horm. 55, 1–71.

Bruce, J.I., Yang, X., Ferguson, C.J., Elliott, A.C., Steward, M.C., Case, R.M., et al., 1999. Molecular and functional identification of a Ca^{2+} (polyvalent cation)-sensing receptor in rat pancreas. J. Biol. Chem. 274 (29), 20561–20568.

Buchan, A.M., Squires, P.E., Ring, M., Meloche, R.M., 2001. Mechanism of action of the calcium-sensing receptor in human antral gastrin cells. Gastroenterology 120 (5), 1128–1139.

Bukoski, R.D., Bian, K., Wang, Y., Mupanomunda, M., 1997. Perivascular sensory nerve Ca^{2+} receptor and Ca^{2+}-induced relaxation of isolated arteries. Hypertension 30 (6), 1431–1439.

Busque, S.M., Kerstetter, J.E., Geibel, J.P., Insogna, K., 2005. L-type amino acids stimulate gastric acid secretion by activation of the calcium-sensing receptor in parietal cells. Am. J. Physiol. Gastrointest. Liver Physiol. 289 (4), G664–G669.

Butters Jr., R.R., Chattopadhyay, N., Nielsen, P., Smith, C.P., Mithal, A., Kifor, O., et al., 1997. Cloning and characterization of a calcium-sensing receptor from the hypercalcemic New Zealand white rabbit reveals unaltered responsiveness to extracellular calcium. J. Bone Miner. Res. 12 (4), 568–579.

Bystrova, M.F., Romanov, R.A., Rogachevskaja, O.A., Churbanov, G.D., Kolesnikov, S.S., 2010. Functional expression of the extracellular- Ca^{2+}-sensing receptor in mouse taste cells. J. Cell Sci. 123 (Pt 6), 972–982.

Calautti, E., Cabodi, S., Stein, P.L., Hatzfeld, M., Kedersha, N., Paolo Dotto, G., 1998. Tyrosine phosphorylation and src family kinases control keratinocyte cell-cell adhesion. J. Cell Biol. 141 (6), 1449–1465.

Calautti, E., Grossi, M., Mammucari, C., Aoyama, Y., Pirro, M., Ono, Y., et al., 2002. Fyn tyrosine kinase is a downstream mediator of Rho/PRK2 function in keratinocyte cell-cell adhesion. J. Cell Biol. 156 (1), 137–148.

Calautti, E., Li, J., Saoncella, S., Brissette, J.L., Goetinck, P.F., 2005. Phosphoinositide 3-kinase signaling to Akt promotes keratinocyte differentiation versus death. J. Biol. Chem. 280 (38), 32856–32865.

Canaff, L., Hendy, G.N., 2002. Human calcium-sensing receptor gene. Vitamin D response elements in promoters P1 and P2 confer transcriptional responsiveness to 1,25-dihydroxyvitamin D. J. Biol. Chem. 277 (33), 30337–30350.

Canaff, L., Petit, J.L., Kisiel, M., Watson, P.H., Gascon-Barre, M., Hendy, G.N., 2001. Extracellular calcium-sensing receptor is expressed in rat hepatocytes. coupling to intracellular calcium mobilization and stimulation of bile flow. J. Biol. Chem. 276 (6), 4070–4079.

Canaff, L., Zhou, X., Mosesova, I., Cole, D.E., Hendy, G.N., 2009. Glial cells missing-2 (GCM2) transactivates the calcium-sensing receptor gene: effect of a dominant-negative GCM2 mutant associated with autosomal dominant hypoparathyroidism. Hum. Mutat. 30 (1), 85–92.

Casala, C., Gil-Guinon, E., Ordonez, J.L., Miguel-Queralt, S., Rodriguez, E., Galvan, P., et al., 2013. The calcium-sensing receptor is silenced by genetic and epigenetic mechanisms in unfavorable neuroblastomas and its reactivation induces ERK1/2-dependent apoptosis. Carcinogenesis 34 (2), 268–276.

Cetani, F., Picone, A., Cerrai, P., Vignali, E., Borsari, S., Pardi, E., et al., 2000. Parathyroid expression of calcium-sensing receptor protein and in vivo parathyroid hormone-$Ca^{(2+)}$ set-point in patients with primary hyperparathyroidism. J. Clin. Endocrinol. Metab. 85 (12), 4789–4794.

Chakrabarty, S., Radjendirane, V., Appelman, H., Varani, J., 2003. Extracellular calcium and calcium sensing receptor function in human colon carcinomas: promotion of E-cadherin expression and suppression of beta-catenin/TCF activation. Cancer Res. 63 (1), 67–71.

Chakrabarty, S., Wang, H., Canaff, L., Hendy, G.N., Appelman, H., Varani, J., 2005. Calcium sensing receptor in human colon carcinoma: interaction with $Ca^{(2+)}$ and 1,25-dihydroxyvitamin D(3). Cancer Res. 65 (2), 493–498.

Chang, W., Chen, T.H., Pratt, S., Shoback, D., 2000. Amino acids in the second and third intracellular loops of the parathyroid Ca^{2+}-sensing receptor mediate efficient coupling to phospholipase C. J. Biol. Chem. 275 (26), 19955–19963.

Chang, W., Pratt, S., Chen, T.H., Bourguignon, L., Shoback, D., 2001. Amino acids in the cytoplasmic C terminus of the parathyroid Ca^{2+}-sensing receptor mediate efficient cell-surface expression and phospholipase C activation. J. Biol. Chem. 276 (47), 44129–44136.

Chang, W., Tu, C., Chen, T.H., Bikle, D., Shoback, D., 2008. The extracellular calcium-sensing receptor (CaSR) is a critical modulator of skeletal development. Sci. Signal. 1 (35), ra1.

Chang, W., Tu, C., Chen, T.H., Komuves, L., Oda, Y., Pratt, S.A., et al., 1999. Expression and signal transduction of calcium-sensing receptors in cartilage and bone. Endocrinology 140 (12), 5883–5893.

Chang, W., Tu, C., Cheng, Z., Rodriguez, L., Chen, T.H., Gassmann, M., et al., 2007. Complex formation with the Type B gamma-aminobutyric acid receptor affects the expression and signal transduction of the extracellular calcium-sensing receptor. Studies with HEK-293 cells and neurons. J. Biol. Chem. 282 (34), 25030–25040.

Chattopadhyay, N., Yano, S., Tfelt-Hansen, J., Rooney, P., Kanuparthi, D., Bandyopadhyay, S., et al., 2004. Mitogenic action of calcium-sensing receptor on rat calvarial osteoblasts. Endocrinology 145 (7), 3451–3462.

Chen, C.J., Barnett, J.V., Congo, D.A., Brown, E.M., 1989. Divalent cations suppress 3',5'-adenosine monophosphate accumulation by stimulating a pertussis toxin-sensitive guanine nucleotide-binding protein in cultured bovine parathyroid cells. Endocrinology 124 (1), 233–239.

Cheng, I., Klingensmith, M.E., Chattopadhyay, N., Kifor, O., Butters, R.R., Soybel, D.I., et al., 1998. Identification and localization of the extracellular calcium-sensing receptor in human breast. J. Clin. Endocrinol. Metab. 83 (2), 703–707.

Cheng, I., Qureshi, I., Chattopadhyay, N., Qureshi, A., Butters, R.R., Hall, A.E., et al., 1999. Expression of an extracellular calcium-sensing receptor in rat stomach. Gastroenterology 116 (1), 118–126.

Cheng, S.X., 2012. Calcium-sensing receptor inhibits secretagogue-induced electrolyte secretion by intestine via the enteric nervous system. Am. J. Physiol. Gastrointest. Liver Physiol. 303 (1), G60–G70.

Cheng, S.X., 2016. Calcium-sensing receptor: a new target for therapy of diarrhea. World J. Gastroenterol. 22 (9), 2711–2724.

Cheng, S.X., Geibel, J.P., Hebert, S.C., 2004. Extracellular polyamines regulate fluid secretion in rat colonic crypts via the extracellular calcium-sensing receptor. Gastroenterology 126 (1), 148–158.

Cheng, S.X., Lightfoot, Y.L., Yang, T., Zadeh, M., Tang, L., Sahay, B., et al., 2014. Epithelial CaSR deficiency alters intestinal integrity and promotes proinflammatory immune responses. FEBS Lett. 588 (22), 4158–4166.

Cheng, S.X., Okuda, M., Hall, A.E., Geibel, J.P., Hebert, S.C., 2002. Expression of calcium-sensing receptor in rat colonic epithelium: evidence for modulation of fluid secretion. Am. J. Physiol. Gastrointest. Liver Physiol. 283 (1), G240–G250.

Cheng, Z., Liang, N., Chen, T.H., Li, A., Santa Maria, C., You, M., et al., 2013. Sex and age modify biochemical and skeletal manifestations of chronic hyperparathyroidism by altering target organ responses to Ca^{2+} and parathyroid hormone in mice. J. Bone Miner. Res. 28 (5), 1087–1100.

Chiarini, A., Armato, U., Gardenal, E., Gui, L., Dal Pra, I., 2017. Amyloid beta-exposed human astrocytes overproduce phospho-tau and overrelease it within exosomes, effects suppressed by calcilytic NPS 2143-further implications for alzheimer's therapy. Front. Neurosci. 11, 217.

Chikatsu, N., Fukumoto, S., Takeuchi, Y., Suzawa, M., Obara, T., Matsumoto, T., et al., 2000. Cloning and characterization of two promoters for the human calcium-sensing receptor (CaSR) and changes of CaSR expression in parathyroid adenomas. J. Biol. Chem. 275 (11), 7553–7557.

Chirgwin, J.M., Guise, T.A., 2000. Molecular mechanisms of tumor-bone interactions in osteolytic metastases. Crit. Rev. Eukaryot. Gene Expr. 10 (2), 159–178.

Choudhary, S., Kumar, A., Kale, R.K., Raisz, L.G., Pilbeam, C.C., 2004. Extracellular calcium induces COX-2 in osteoblasts via a PKA pathway. Biochem. Biophys. Res. Commun. 322 (2), 395–402.

Christensen, S.E., Nissen, P.H., Vestergaard, P., Heickendorff, L., Brixen, K., Mosekilde, L., 2008. Discriminative power of three indices of renal calcium excretion for the distinction between familial hypocalciuric hypercalcaemia and primary hyperparathyroidism: a follow-up study on methods. Clin. Endocrinol. 69 (5), 713–720.
Colloton, M., Shatzen, E., Miller, G., Stehman-Breen, C., Wada, M., Lacey, D., et al., 2005. Cinacalcet HCl attenuates parathyroid hyperplasia in a rat model of secondary hyperparathyroidism. Kidney Int. 67 (2), 467–476.
Conigrave, A.D., Hampson, D.R., 2010. Broad-spectrum amino acid-sensing class C G-protein coupled receptors: molecular mechanisms, physiological significance and options for drug development. Pharmacol. Ther. 127 (3), 252–260.
Conigrave, A.D., Mun, H.C., Brennan, S.C., 2007a. Physiological significance of L-amino acid sensing by extracellular Ca(2+)-sensing receptors. Biochem. Soc. Trans. 35 (Pt 5), 1195–1198.
Conigrave, A.D., Mun, H.C., Delbridge, L., Quinn, S.J., Wilkinson, M., Brown, E.M., 2004. L-amino acids regulate parathyroid hormone secretion. J. Biol. Chem. 279 (37), 38151–38159.
Conigrave, A.D., Mun, H.C., Lok, H.C., 2007b. Aromatic L-amino acids activate the calcium-sensing receptor. J. Nutr. 137 (6 Suppl. 1), 1524S-7S; discussion 48S.
Conigrave, A.D., Quinn, S.J., Brown, E.M., 2000. L-amino acid sensing by the extracellular Ca2+-sensing receptor. Proc. Natl. Acad. Sci. U. S. A. 97 (9), 4814–4819.
Conigrave, A.D., Ward, D.T., 2013. Calcium-sensing receptor (CaSR): pharmacological properties and signaling pathways. Best Pract. Res. Clin. Endocrinol. Metabol. 27 (3), 315–331.
Cooper, D.M., 2003. Molecular and cellular requirements for the regulation of adenylate cyclases by calcium. Biochem. Soc. Trans. 31 (Pt 5), 912–915.
Corbetta, S., Mantovani, G., Lania, A., Borgato, S., Vicentini, L., Beretta, E., et al., 2000. Calcium-sensing receptor expression and signalling in human parathyroid adenomas and primary hyperplasia. Clin. Endocrinol. 52 (3), 339–348.
Correa, P., Akerstrom, G., Westin, G., 2002. Underexpression of Gcm2, a master regulatory gene of parathyroid gland development, in adenomas of primary hyperparathyroidism. Clin. Endocrinol. 57 (4), 501–505.
D'Souza-Li, L., Canaff, L., Janicic, N., Cole, D.E., Hendy, G.N., 2001. An acceptor splice site mutation in the calcium-sensing receptor (CASR) gene in familial hypocalciuric hypercalcemia and neonatal severe hyperparathyroidism. Hum. Mutat. 18 (5), 411–421.
Datta, H.K., MacIntyre, I., Zaidi, M., 1989. The effect of extracellular calcium elevation on morphology and function of isolated rat osteoclasts. Biosci. Rep. 9 (6), 747–751.
Davey, A.E., Leach, K., Valant, C., Conigrave, A.D., Sexton, P.M., Christopoulos, A., 2012. Positive and negative allosteric modulators promote biased signaling at the calcium-sensing receptor. Endocrinology 153 (3), 1232–1241.
Davies, S.L., Gibbons, C.E., Vizard, T., Ward, D.T., 2006. Ca2+-sensing receptor induces Rho kinase-mediated actin stress fiber assembly and altered cell morphology, but not in response to aromatic amino acids. Am. J. Physiol. Cell Physiol. 290 (6), C1543–C1551.
Davies, S.L., Ozawa, A., McCormick, W.D., Dvorak, M.M., Ward, D.T., 2007. Protein kinase C-mediated phosphorylation of the calcium-sensing receptor is stimulated by receptor activation and attenuated by calyculin-sensitive phosphatase activity. J. Biol. Chem. 282 (20), 15048–15056.
de Rouffignac, C., Quamme, G., 1994. Renal magnesium handling and its hormonal control. Physiol. Rev. 74 (2), 305–322.
Devis, G., Somers, G., Malaisse, W.J., 1975. Stimulation of insulin release by calcium. Biochem. Biophys. Res. Commun. 67 (2), 525–529.
Diez-Fraile, A., Lammens, T., Benoit, Y., D'Herde, K.G., 2013. The calcium-sensing receptor as a regulator of cellular fate in normal and pathological conditions. Curr. Mol. Med. 13 (2), 282–295.
Dong, B., Endo, I., Ohnishi, Y., Kondo, T., Hasegawa, T., Amizuka, N., et al., 2015. Calcilytic ameliorates abnormalities of mutant calcium-sensing receptor (CaSR) knock-in mice mimicking autosomal dominant hypocalcemia (ADH). J. Bone Miner. Res. 30 (11), 1980–1993.
Dufner, M.M., Kirchhoff, P., Remy, C., Hafner, P., Muller, M.K., Cheng, S.X., et al., 2005. The calcium-sensing receptor acts as a modulator of gastric acid secretion in freshly isolated human gastric glands. Am. J. Physiol. Gastrointest. Liver Physiol. 289 (6), G1084–G1090.
Dvorak, M.M., Chen, T.H., Orwoll, B., Garvey, C., Chang, W., Bikle, D.D., et al., 2007. Constitutive activity of the osteoblast Ca2+-sensing receptor promotes loss of cancellous bone. Endocrinology 148 (7), 3156–3163.
Dvorak, M.M., Siddiqua, A., Ward, D.T., Carter, D.H., Dallas, S.L., Nemeth, E.F., et al., 2004. Physiological changes in extracellular calcium concentration directly control osteoblast function in the absence of calciotropic hormones. Proc. Natl. Acad. Sci. U. S. A. 101 (14), 5140–5145.
Dvorak-Ewell, M.M., Chen, T.H., Liang, N., Garvey, C., Liu, B., Tu, C., et al., 2011. Osteoblast extracellular Ca^{2+} -sensing receptor regulates bone development, mineralization, and turnover. J. Bone Miner. Res. 26 (12), 2935–2947.
El Hiani, Y., Ahidouch, A., Lehen'kyi, V., Hague, F., Gouilleux, F., Mentaverri, R., et al., 2009a. Extracellular signal-regulated kinases 1 and 2 and TRPC1 channels are required for calcium-sensing receptor-stimulated MCF-7 breast cancer cell proliferation. Cell. Physiol. Biochem. 23 (4–6), 335–346.
El Hiani, Y., Lehen'kyi, V., Ouadid-Ahidouch, H., Ahidouch, A., 2009b. Activation of the calcium-sensing receptor by high calcium induced breast cancer cell proliferation and TRPC1 cation channel over-expression potentially through EGFR pathways. Arch. Biochem. Biophys. 486 (1), 58–63.
Emanuel, R.L., Adler, G.K., Kifor, O., Quinn, S.J., Fuller, F., Krapcho, K., et al., 1996. Calcium-sensing receptor expression and regulation by extracellular calcium in the AtT-20 pituitary cell line. Mol. Endocrinol. 10 (5), 555–565.
Fan, G.F., Ray, K., Zhao, X.M., Goldsmith, P.K., Spiegel, A.M., 1998. Mutational analysis of the cysteines in the extracellular domain of the human Ca^{2+} receptor: effects on cell surface expression, dimerization and signal transduction. FEBS Lett. 436 (3), 353–356.
Fan, Y., Liu, W., Bi, R., Densmore, M.J., Sato, T., Mannstadt, M., et al., 2018. Interrelated role of Klotho and calcium-sensing receptor in parathyroid hormone synthesis and parathyroid hyperplasia. Proc. Natl. Acad. Sci. U. S. A. 115 (16), E3749–E3758.

Favus, M.J., Kathpalia, S.C., Coe, F.L., 1981. Kinetic characteristics of calcium absorption and secretion by rat colon. Am. J. Physiol. 240 (5), G350–G354.

Favus, M.J., Kathpalia, S.C., Coe, F.L., Mond, A.E., 1980. Effects of diet calcium and 1,25-dihydroxyvitamin D3 on colon calcium active transport. Am. J. Physiol. 238 (2), G75–G78.

Fedor-Chaiken, M., Hein, P.W., Stewart, J.C., Brackenbury, R., Kinch, M.S., 2003. E-cadherin binding modulates EGF receptor activation. Cell Commun. Adhes. 10 (2), 105–118.

Felder, C.B., Graul, R.C., Lee, A.Y., Merkle, H.P., Sadee, W., 1999. The Venus flytrap of periplasmic binding proteins: an ancient protein module present in multiple drug receptors. AAPS PharmSci 1 (2), E2.

Felderbauer, P., Klein, W., Bulut, K., Ansorge, N., Dekomien, G., Werner, I., et al., 2006. Mutations in the calcium-sensing receptor: a new genetic risk factor for chronic pancreatitis? Scand. J. Gastroenterol. 41 (3), 343–348.

Feng, J., Petersen, C.D., Coy, D.H., Jiang, J.K., Thomas, C.J., Pollak, M.R., et al., 2010. Calcium-sensing receptor is a physiologic multimodal chemosensor regulating gastric G-cell growth and gastrin secretion. Proc. Natl. Acad. Sci. U. S. A. 107 (41), 17791–17796.

Ferreri, N.R., Hao, S., Pedraza, P.L., Escalante, B., Vio, C.P., 2012. Eicosanoids and tumor necrosis factor-alpha in the kidney. Prostag. Other Lipid Mediat. 98 (3–4), 101–106.

Fetahu, I.S., Hobaus, J., Aggarwal, A., Hummel, D.M., Tennakoon, S., Mesteri, I., et al., 2014. Calcium-sensing receptor silencing in colorectal cancer is associated with promoter hypermethylation and loss of acetylation on histone 3. Int. J. Cancer 135 (9), 2014–2023.

Fetahu, I.S., Tennakoon, S., Lines, K.E., Groschel, C., Aggarwal, A., Mesteri, I., et al., 2016. miR-135b- and miR-146b-dependent silencing of calcium-sensing receptor expression in colorectal tumors. Int. J. Cancer 138 (1), 137–145.

Finney, B.A., del Moral, P.M., Wilkinson, W.J., Cayzac, S., Cole, M., Warburton, D., et al., 2008. Regulation of mouse lung development by the extracellular calcium-sensing receptor, CaR. J. Physiol. 586 (24), 6007–6019.

Fitzpatrick, L.A., Brandi, M.L., Aurbach, G.D., 1986. Calcium-controlled secretion is effected through a guanine nucleotide regulatory protein in parathyroid cells. Endocrinology 119 (6), 2700–2703.

Freichel, M., Zink-Lorenz, A., Holloschi, A., Hafner, M., Flockerzi, V., Raue, F., 1996. Expression of a calcium-sensing receptor in a human medullary thyroid carcinoma cell line and its contribution to calcitonin secretion. Endocrinology 137 (9), 3842–3848.

Gama, L., Wilt, S.G., Breitwieser, G.E., 2001. Heterodimerization of calcium sensing receptors with metabotropic glutamate receptors in neurons. J. Biol. Chem. 276 (42), 39053–39059.

Gardenal, E., Chiarini, A., Armato, U., Dal Pra, I., Verkhratsky, A., Rodriguez, J.J., 2017. Increased calcium-sensing receptor immunoreactivity in the Hippocampus of a triple transgenic mouse model of alzheimer's disease. Front. Neurosci. 11, 81.

Garrett, J.E., Capuano, I.V., Hammerland, L.G., Hung, B.C., Brown, E.M., Hebert, S.C., et al., 1995a. Molecular cloning and functional expression of human parathyroid calcium receptor cDNAs. J. Biol. Chem. 270 (21), 12919–12925.

Garrett, J.E., Tamir, H., Kifor, O., Simin, R.T., Rogers, K.V., Mithal, A., et al., 1995b. Calcitonin-secreting cells of the thyroid express an extracellular calcium receptor gene. Endocrinology 136 (11), 5202–5211.

Geibel, J., Sritharan, K., Geibel, R., Geibel, P., Persing, J.S., Seeger, A., et al., 2006. Calcium-sensing receptor abrogates secretagogue- induced increases in intestinal net fluid secretion by enhancing cyclic nucleotide destruction. Proc. Natl. Acad. Sci. U. S. A. 103 (25), 9390–9397.

Geng, Y., Bush, M., Mosyak, L., Wang, F., Fan, Q.R., 2013. Structural mechanism of ligand activation in human GABA(B) receptor. Nature 504 (7479), 254–259.

Geng, Y., Mosyak, L., Kurinov, I., Zuo, H., Sturchler, E., Cheng, T.C., et al., 2016. Structural mechanism of ligand activation in human calcium-sensing receptor. Elife 5.

Godwin, S.L., Soltoff, S.P., 2002. Calcium-sensing receptor-mediated activation of phospholipase C-gamma1 is downstream of phospholipase C-beta and protein kinase C in MC3T3-E1 osteoblasts. Bone 30 (4), 559–566.

Goltzman, D., Hendy, G.N., 2015. The calcium-sensing receptor in bone–mechanistic and therapeutic insights. Nat. Rev. Endocrinol. 11 (5), 298–307.

Goolam, M.A., Ward, J.H., Avlani, V.A., Leach, K., Christopoulos, A., Conigrave, A.D., 2014. Roles of intraloops-2 and -3 and the proximal C-terminus in signalling pathway selection from the human calcium-sensing receptor. FEBS Lett. 588 (18), 3340–3346.

Gorvin, C.M., Rogers, A., Stewart, M., Paudyal, A., Hough, T.A., Teboul, L., et al., 2017. N-ethyl-N-nitrosourea-Induced adaptor protein 2 sigma subunit 1 (Ap2s1) mutations establish Ap2s1 loss-of-function mice. JBMR Plus 1 (1), 3–15.

Gowen, M., Stroup, G.B., Dodds, R.A., James, I.E., Votta, B.J., Smith, B.R., et al., 2000. Antagonizing the parathyroid calcium receptor stimulates parathyroid hormone secretion and bone formation in osteopenic rats. J. Clin. Investig. 105 (11), 1595–1604.

Grant, M.P., Stepanchick, A., Cavanaugh, A., Breitwieser, G.E., 2011. Agonist-driven maturation and plasma membrane insertion of calcium-sensing receptors dynamically control signal amplitude. Sci. Signal. 4 (200), ra78.

Gray, E., Muller, D., Squires, P.E., Asare-Anane, H., Huang, G.C., Amiel, S., et al., 2006. Activation of the extracellular calcium-sensing receptor initiates insulin secretion from human islets of Langerhans: involvement of protein kinases. J. Endocrinol. 190 (3), 703–710.

Greenberg, H.Z., Shi, J., Jahan, K.S., Martinucci, M.C., Gilbert, S.J., Vanessa Ho, W.S., et al., 2016. Stimulation of calcium-sensing receptors induces endothelium-dependent vasorelaxations via nitric oxide production and activation of IKCa channels. Vasc. Pharmacol. 80, 75–84.

Greenberg, H.Z.E., Carlton-Carew, S.R.E., Khan, D.M., Zargaran, A.K., Jahan, K.S., Vanessa Ho, W.S., et al., 2017. Heteromeric TRPV4/TRPC1 channels mediate calcium-sensing receptor-induced nitric oxide production and vasorelaxation in rabbit mesenteric arteries. Vasc. Pharmacol. 96–98, 53–62.

Guise, T.A., Kozlow, W.M., Heras-Herzig, A., Padalecki, S.S., Yin, J.J., Chirgwin, J.M., 2005. Molecular mechanisms of breast cancer metastases to bone. Clin. Breast Canc. 5 (Suppl. 1(2)), S46–S53.

Hammerland, L.G., Krapcho, K.J., Garrett, J.E., Alasti, N., Hung, B.C., Simin, R.T., et al., 1999. Domains determining ligand specificity for Ca^{2+} receptors. Mol. Pharmacol. 55 (4), 642–648.
Handlogten, M.E., Huang, C., Shiraishi, N., Awata, H., Miller, R.T., 2001. The Ca^{2+}-sensing receptor activates cytosolic phospholipase A2 via a Gqalpha -dependent ERK-independent pathway. J. Biol. Chem. 276 (17), 13941–13948.
Hannan, F.M., Babinsky, V.N., Thakker, R.V., 2016. Disorders of the calcium-sensing receptor and partner proteins: insights into the molecular basis of calcium homeostasis. J. Mol. Endocrinol. 57 (3), R127–R142.
Hannan, F.M., Kallay, E., Chang, W., Brandi, M.L., Thakker, R.V., 2018. The calcium-sensing receptor in physiology and in calcitropic and noncalcitropic diseases. Nat. Rev. Endocrinol. 15 (1), 33–51.
Hannan, F.M., Thakker, R.V., 2013. Calcium-sensing receptor (CaSR) mutations and disorders of calcium, electrolyte and water metabolism. Best Pract. Res. Clin. Endocrinol. Metabol. 27 (3), 359–371.
Hauache, O.M., Hu, J., Ray, K., Xie, R., Jacobson, K.A., Spiegel, A.M., 2000. Effects of a calcimimetic compound and naturally activating mutations on the human Ca2+ receptor and on Ca2+ receptor/metabotropic glutamate chimeric receptors. Endocrinology 141 (11), 4156–4163.
Henaut, L., Boudot, C., Massy, Z.A., Lopez-Fernandez, I., Dupont, S., Mary, A., et al., 2014. Calcimimetics increase CaSR expression and reduce mineralization in vascular smooth muscle cells: mechanisms of action. Cardiovasc. Res. 101 (2), 256–265.
Hendy, G.N., Canaff, L., 2016. Calcium-sensing receptor gene: regulation of expression. Front. Physiol. 7, 394.
Hendy, G.N., Canaff, L., Cole, D.E., 2013. The CASR gene: alternative splicing and transcriptional control, and calcium-sensing receptor (CaSR) protein: structure and ligand binding sites. Best Pract. Res. Clin. Endocrinol. Metabol. 27 (3), 285–301.
Hernandez-Bedolla, M.A., Carretero-Ortega, J., Valadez-Sanchez, M., Vazquez-Prado, J., Reyes-Cruz, G., 2015. Chemotactic and proangiogenic role of calcium sensing receptor is linked to secretion of multiple cytokines and growth factors in breast cancer MDA-MB-231 cells. Biochim. Biophys. Acta 1853 (1), 166–182.
Herrera-Vigenor, F., Hernandez-Garcia, R., Valadez-Sanchez, M., Vazquez-Prado, J., Reyes-Cruz, G., 2006. AMSH regulates calcium-sensing receptor signaling through direct interactions. Biochem. Biophys. Res. Commun. 347 (4), 924–930.
Hills, C.E., Younis, M.Y., Bennett, J., Siamantouras, E., Liu, K.K., Squires, P.E., 2012. Calcium-sensing receptor activation increases cell-cell adhesion and beta-cell function. Cell. Physiol. Biochem. 30 (3), 575–586.
Hizaki, K., Yamamoto, H., Taniguchi, H., Adachi, Y., Nakazawa, M., Tanuma, T., et al., 2011. Epigenetic inactivation of calcium-sensing receptor in colorectal carcinogenesis. Mod. Pathol. 24 (6), 876–884.
Hjalm, G., MacLeod, R.J., Kifor, O., Chattopadhyay, N., Brown, E.M., 2001. Filamin-A binds to the carboxyl-terminal tail of the calcium-sensing receptor, an interaction that participates in CaR-mediated activation of mitogen-activated protein kinase. J. Biol. Chem. 276 (37), 34880–34887.
Ho, C., Conner, D.A., Pollak, M.R., Ladd, D.J., Kifor, O., Warren, H.B., et al., 1995. A mouse model of human familial hypocalciuric hypercalcemia and neonatal severe hyperparathyroidism. Nat. Genet. 11 (4), 389–394.
Hobson, S.A., McNeil, S.E., Lee, F., Rodland, K.D., 2000. Signal transduction mechanisms linking increased extracellular calcium to proliferation in ovarian surface epithelial cells. Exp. Cell Res. 258 (1), 1–11.
Hobson, S.A., Wright, J., Lee, F., McNeil, S.E., Bilderback, T., Rodland, K.D., 2003. Activation of the MAP kinase cascade by exogenous calcium-sensing receptor. Mol. Cell. Endocrinol. 200 (1–2), 189–198.
Hofer, A.M., Brown, E.M., 2003. Extracellular calcium sensing and signalling. Nat. Rev. Mol. Cell Biol. 4 (7), 530–538.
Holstein, D.M., Berg, K.A., Leeb-Lundberg, L.M., Olson, M.S., Saunders, C., 2004. Calcium-sensing receptor-mediated ERK1/2 activation requires Galphai2 coupling and dynamin-independent receptor internalization. J. Biol. Chem. 279 (11), 10060–10069.
Hovden, S., Rejnmark, L., Ladefoged, S.A., Nissen, P.H., 2017. AP2S1 and GNA11 mutations - not a common cause of familial hypocalciuric hypercalcemia. Eur. J. Endocrinol. 176 (2), 177–185.
Howles, S.A., Hannan, F.M., Babinsky, V.N., Rogers, A., Gorvin, C.M., Rust, N., et al., 2016. Cinacalcet for symptomatic hypercalcemia caused by AP2S1 mutations. N. Engl. J. Med. 374 (14), 1396–1398.
Hu, J., 2008. Allosteric modulators of the human calcium-sensing receptor: structures, sites of action, and therapeutic potentials. Endocr. Metab. Immune Disord. - Drug Targets 8 (3), 192–197.
Hu, J., Hauache, O., Spiegel, A.M., 2000. Human Ca^{2+} receptor cysteine-rich domain. Analysis of function of mutant and chimeric receptors. J. Biol. Chem. 275 (21), 16382–16389.
Hu, J., Reyes-Cruz, G., Chen, W., Jacobson, K.A., Spiegel, A.M., 2002. Identification of acidic residues in the extracellular loops of the seven-transmembrane domain of the human Ca^{2+} receptor critical for response to Ca2+ and a positive allosteric modulator. J. Biol. Chem. 277 (48), 46622–46631.
Huang, C., Hujer, K.M., Wu, Z., Miller, R.T., 2004. The Ca^{2+}-sensing receptor couples to Galpha12/13 to activate phospholipase D in Madin-Darby canine kidney cells. Am. J. Physiol. Cell Physiol. 286 (1), C22–C30.
Huang, C., Hydo, L.M., Liu, S., Miller, R.T., 2009a. Activation of choline kinase by extracellular Ca^{2+} is $Ca(^{2+})$-sensing receptor, Galpha12 and Rho-dependent in breast cancer cells. Cell. Signal. 21 (12), 1894–1900.
Huang, C., Miller, R.T., 2007. The calcium-sensing receptor and its interacting proteins. J. Cell Mol. Med. 11 (5), 923–934.
Huang, C., Sindic, A., Hill, C.E., Hujer, K.M., Chan, K.W., Sassen, M., et al., 2007a. Interaction of the Ca^{2+}-sensing receptor with the inwardly rectifying potassium channels Kir4.1 and Kir4.2 results in inhibition of channel function. Am. J. Physiol. Renal. Physiol. 292 (3), F1073–F1081.
Huang, Y., Breitwieser, G.E., 2007. Rescue of calcium-sensing receptor mutants by allosteric modulators reveals a conformational checkpoint in receptor biogenesis. J. Biol. Chem. 282 (13), 9517–9525.

Huang, Y., Niwa, J., Sobue, G., Breitwieser, G.E., 2006. Calcium-sensing receptor ubiquitination and degradation mediated by the E3 ubiquitin ligase dorfin. J. Biol. Chem. 281 (17), 11610–11617.

Huang, Y., Zhou, Y., Castiblanco, A., Yang, W., Brown, E.M., Yang, J.J., 2009b. Multiple $Ca^{(2+)}$-binding sites in the extracellular domain of the $Ca^{(2+)}$-sensing receptor corresponding to cooperative $Ca^{(2+)}$ response. Biochemistry 48 (2), 388–398.

Huang, Y., Zhou, Y., Wong, H.C., Castiblanco, A., Chen, Y., Brown, E.M., et al., 2010. Calmodulin regulates Ca^{2+}-sensing receptor-mediated Ca^{2+} signaling and its cell surface expression. J. Biol. Chem. 285 (46), 35919–35931.

Huang, Y., Zhou, Y., Yang, W., Butters, R., Lee, H.W., Li, S., et al., 2007b. Identification and dissection of $Ca^{(2+)}$-binding sites in the extracellular domain of $Ca^{(2+)}$-sensing receptor. J. Biol. Chem. 282 (26), 19000–19010.

Ishioka, N., Bukoski, R.D., 1999. A role for N-arachidonylethanolamine (anandamide) as the mediator of sensory nerve-dependent Ca^{2+}-induced relaxation. J. Pharmacol. Exp. Ther. 289 (1), 245–250.

Jacobsen, S.E., Gether, U., Brauner-Osborne, H., 2017. Investigating the molecular mechanism of positive and negative allosteric modulators in the calcium-sensing receptor dimer. Sci. Rep. 7, 46355.

Janicic, N., Soliman, E., Pausova, Z., Seldin, M.F., Riviere, M., Szpirer, J., et al., 1995. Mapping of the calcium-sensing receptor gene (CASR) to human chromosome 3q13.3-21 by fluorescence in situ hybridization, and localization to rat chromosome 11 and mouse chromosome 16. Mamm. Genome 6 (11), 798–801.

Jiang, L.I., Collins, J., Davis, R., Fraser, I.D., Sternweis, P.C., 2008. Regulation of cAMP responses by the G12/13 pathway converges on adenylyl cyclase VII. J. Biol. Chem. 283 (34), 23429–23439.

Jiang, Y., Minet, E., Zhang, Z., Silver, P.A., Bai, M., 2004. Modulation of interprotomer relationships is important for activation of dimeric calcium-sensing receptor. J. Biol. Chem. 279 (14), 14147–14156.

Jiang, Y.F., Zhang, Z., Kifor, O., Lane, C.R., Quinn, S.J., Bai, M., 2002. Protein kinase C (PKC) phosphorylation of the Ca^{2+} o-sensing receptor (CaR) modulates functional interaction of G proteins with the CaR cytoplasmic tail. J. Biol. Chem. 277 (52), 50543–50549.

Jones, P.M., Kitsou-Mylona, I., Gray, E., Squires, P.E., Persaud, S.J., 2007. Expression and function of the extracellular calcium-sensing receptor in pancreatic beta-cells. Arch. Physiol. Biochem. 113 (3), 98–103.

Justinich, C.J., Mak, N., Pacheco, I., Mulder, D., Wells, R.W., Blennerhassett, M.G., et al., 2008. The extracellular calcium-sensing receptor (CaSR) on human esophagus and evidence of expression of the CaSR on the esophageal epithelial cell line (HET-1A). Am. J. Physiol. Gastrointest. Liver Physiol. 294 (1), G120–G129.

Kallay, E., Bonner, E., Wrba, F., Thakker, R.V., Peterlik, M., Cross, H.S., 2003a. Molecular and functional characterization of the extracellular calcium-sensing receptor in human colon cancer cells. Oncol. Res. 13 (12), 551–559.

Kallay, E., Kifor, O., Chattopadhyay, N., Brown, E.M., Bischof, M.G., Peterlik, M., et al., 1997. Calcium-dependent c-myc proto-oncogene expression and proliferation of Caco-2 cells: a role for a luminal extracellular calcium-sensing receptor. Biochem. Biophys. Res. Commun. 232 (1), 80–83.

Kallay, E., Wrba, F., Cross, H.S., 2003b. Dietary calcium and colon cancer prevention. Forum Nutr. 56, 188–190.

Kameda, T., Mano, H., Yamada, Y., Takai, H., Amizuka, N., Kobori, M., et al., 1998. Calcium-sensing receptor in mature osteoclasts, which are bone resorbing cells. Biochem. Biophys. Res. Commun. 245 (2), 419–422.

Kanatani, M., Sugimoto, T., Kanzawa, M., Yano, S., Chihara, K., 1999. High extracellular calcium inhibits osteoclast-like cell formation by directly acting on the calcium-sensing receptor existing in osteoclast precursor cells. Biochem. Biophys. Res. Commun. 261 (1), 144–148.

Kantham, L., Quinn, S.J., Egbuna, O.I., Baxi, K., Butters, R., Pang, J.L., et al., 2009. The calcium-sensing receptor (CaSR) defends against hypercalcemia independently of its regulation of parathyroid hormone secretion. Am. J. Physiol. Endocrinol. Metab. 297 (4), E915–E923.

Kelly, J.C., Lungchukiet, P., Macleod, R.J., 2011. Extracellular calcium-sensing receptor inhibition of intestinal EpithelialTNF signaling requires CaSR-mediated Wnt5a/Ror2 interaction. Front. Physiol. 2, 17.

Kifor, O., Brown, E.M., 1988. Relationship between diacylglycerol levels and extracellular Ca^{2+} in dispersed bovine parathyroid cells. Endocrinology 123 (6), 2723–2729.

Kifor, O., Diaz, R., Butters, R., Brown, E.M., 1997. The Ca^{2+}-sensing receptor (CaR) activates phospholipases C, A2, and D in bovine parathyroid and CaR-transfected, human embryonic kidney (HEK293) cells. J. Bone Miner. Res. 12 (5), 715–725.

Kifor, O., Diaz, R., Butters, R., Kifor, I., Brown, E.M., 1998. The calcium-sensing receptor is localized in caveolin-rich plasma membrane domains of bovine parathyroid cells. J. Biol. Chem. 273 (34), 21708–21713.

Kifor, O., Kifor, I., Moore Jr., F.D., Butters Jr., R.R., Brown, E.M., 2003. m-Calpain colocalizes with the calcium-sensing receptor (CaR) in caveolae in parathyroid cells and participates in degradation of the CaR. J. Biol. Chem. 278 (33), 31167–31176.

Kifor, O., MacLeod, R.J., Diaz, R., Bai, M., Yamaguchi, T., Yao, T., et al., 2001. Regulation of MAP kinase by calcium-sensing receptor in bovine parathyroid and CaR-transfected HEK293 cells. Am. J. Physiol. Renal. Physiol. 280 (2), F291–F302.

Kifor, O., Moore Jr., F.D., Wang, P., Goldstein, M., Vassilev, P., Kifor, I., et al., 1996. Reduced immunostaining for the extracellular Ca^{2+}-sensing receptor in primary and uremic secondary hyperparathyroidism. J. Clin. Endocrinol. Metab. 81 (4), 1598–1606.

Kim, J.Y., Ho, H., Kim, N., Liu, J., Tu, C.L., Yenari, M.A., et al., 2014. Calcium-sensing receptor (CaSR) as a novel target for ischemic neuroprotection. Ann Clin Transl Neurol 1 (11), 851–866.

Kim, J.Y., Kim, N., Yenari, M.A., Chang, W., 2011. Mild hypothermia suppresses calcium-sensing receptor (CaSR) induction following forebrain ischemia while increasing GABA-B receptor 1 (GABA-B-R1) expression. Transl Stroke Res 2 (2), 195–201.

Kim, J.Y., Kim, N., Yenari, M.A., Chang, W., 2013. Hypothermia and pharmacological regimens that prevent overexpression and overactivity of the extracellular calcium-sensing receptor protect neurons against traumatic brain injury. J. Neurotrauma 30 (13), 1170–1176.

Kim, W., Takyar, F.M., Swan, K., Jeong, J., VanHouten, J., Sullivan, C., et al., 2016. Calcium-sensing receptor promotes breast cancer by stimulating intracrine actions of parathyroid hormone-related protein. Cancer Res. 76 (18), 5348–5360.

Kim, W., Wysolmerski, J.J., 2016. Calcium-sensing receptor in breast physiology and cancer. Front. Physiol. 7, 440.

Kitsou-Mylona, I., Burns, C.J., Squires, P.E., Persaud, S.J., Jones, P.M., 2008. A role for the extracellular calcium-sensing receptor in cell-cell communication in pancreatic islets of langerhans. Cell. Physiol. Biochem. 22 (5–6), 557–566.

Klein, G.L., Castro, S.M., Garofalo, R.P., 2016. The calcium-sensing receptor as a mediator of inflammation. Semin. Cell Dev. Biol. 49, 52–56.

Koh, J., Dar, M., Untch, B.R., Dixit, D., Shi, Y., Yang, Z., et al., 2011. Regulator of G protein signaling 5 is highly expressed in parathyroid tumors and inhibits signaling by the calcium-sensing receptor. Mol. Endocrinol. 25 (5), 867–876.

Komuves, L., Oda, Y., Tu, C.L., Chang, W.H., Ho-Pao, C.L., Mauro, T., et al., 2002. Epidermal expression of the full-length extracellular calcium-sensing receptor is required for normal keratinocyte differentiation. J. Cell. Physiol. 192 (1), 45–54.

Kopic, S., Wagner, M.E., Griessenauer, C., Socrates, T., Ritter, M., Geibel, J.P., 2012. Vacuolar-type H+-ATPase-mediated proton transport in the rat parietal cell. Pflügers Archiv 463 (3), 419–427.

Kumar, S., Matheny, C.J., Hoffman, S.J., Marquis, R.W., Schultz, M., Liang, X., et al., 2010. An orally active calcium-sensing receptor antagonist that transiently increases plasma concentrations of PTH and stimulates bone formation. Bone 46 (2), 534–542.

Kunishima, N., Shimada, Y., Tsuji, Y., Sato, T., Yamamoto, M., Kumasaka, T., et al., 2000. Structural basis of glutamate recognition by a dimeric metabotropic glutamate receptor. Nature 407 (6807), 971–977.

Kurosawa, M., Shimizu, Y., Tsukagoshi, H., Ueki, M., 1992. Elevated levels of peripheral-blood, naturally occurring aliphatic polyamines in bronchial asthmatic patients with active symptoms. Allergy 47 (6), 638–643.

Lamprecht, S.A., Lipkin, M., 2001. Cellular mechanisms of calcium and vitamin D in the inhibition of colorectal carcinogenesis. Ann. N. Y. Acad. Sci. 952, 73–87.

Lazarus, S., Pretorius, C.J., Khafagi, F., Campion, K.L., Brennan, S.C., Conigrave, A.D., et al., 2011. A novel mutation of the primary protein kinase C phosphorylation site in the calcium-sensing receptor causes autosomal dominant hypocalcemia. Eur. J. Endocrinol. 164 (3), 429–435.

Leach, K., Conigrave, A.D., Sexton, P.M., Christopoulos, A., 2015. Towards tissue-specific pharmacology: insights from the calcium-sensing receptor as a paradigm for GPCR (patho)physiological bias. Trends Pharmacol. Sci. 36 (4), 215–225.

Leach, K., Wen, A., Davey, A.E., Sexton, P.M., Conigrave, A.D., Christopoulos, A., 2012. Identification of molecular phenotypes and biased signaling induced by naturally occurring mutations of the human calcium-sensing receptor. Endocrinology 153 (9), 4304–4316.

Lee, H.J., Mun, H.C., Lewis, N.C., Crouch, M.F., Culverston, E.L., Mason, R.S., et al., 2007. Allosteric activation of the extracellular Ca^{2+}-sensing receptor by L-amino acids enhances ERK1/2 phosphorylation. Biochem. J. 404 (1), 141–149.

Lee, S.P., Wu, W.Y., Hsiao, J.K., Zhou, J.H., Chang, H.H., Chien, C.T., 2019. Aromatherapy: activating olfactory calcium-sensing receptors impairs renal hemodynamics via sympathetic nerve-mediated vasoconstriction. Acta Physiol. 225 (1), e13157.

Lembrechts, R., Brouns, I., Schnorbusch, K., Pintelon, I., Kemp, P.J., Timmermans, J.P., et al., 2013. Functional expression of the multimodal extracellular calcium-sensing receptor in pulmonary neuroendocrine cells. J. Cell Sci. 126 (Pt 19), 4490–4501.

Li, X., Zheng, W., Li, Y.C., 2003. Altered gene expression profile in the kidney of vitamin D receptor knockout mice. J. Cell. Biochem. 89 (4), 709–719.

Liao, J., Schneider, A., Datta, N.S., McCauley, L.K., 2006. Extracellular calcium as a candidate mediator of prostate cancer skeletal metastasis. Cancer Res. 66 (18), 9065–9073.

Liou, A.P., Sei, Y., Zhao, X., Feng, J., Lu, X., Thomas, C., et al., 2011. The extracellular calcium-sensing receptor is required for cholecystokinin secretion in response to L-phenylalanine in acutely isolated intestinal I cells. Am. J. Physiol. Gastrointest. Liver Physiol. 300 (4), G538–G546.

Liu, G., Hu, X., Chakrabarty, S., 2009. Calcium sensing receptor down-regulates malignant cell behavior and promotes chemosensitivity in human breast cancer cells. Cell Calcium 45 (3), 216–225.

London, G.M., Marchais, S.J., Guerin, A.P., Metivier, F., 2005. Arteriosclerosis, vascular calcifications and cardiovascular disease in uremia. Curr. Opin. Nephrol. Hypertens. 14 (6), 525–531.

Loot, A.E., Pierson, I., Syzonenko, T., Elgheznawy, A., Randriamboavonjy, V., Zivkovic, A., et al., 2013. Ca^{2+}-sensing receptor cleavage by calpain partially accounts for altered vascular reactivity in mice fed a high-fat diet. J. Cardiovasc. Pharmacol. 61 (6), 528–535.

Lopez-Fernandez, I., Schepelmann, M., Brennan, S.C., Yarova, P.L., Riccardi, D., 2015. The calcium-sensing receptor: one of a kind. Exp. Physiol. 100 (12), 1392–1399.

Lorenz, S., Frenzel, R., Paschke, R., Breitwieser, G.E., Miedlich, S.U., 2007. Functional desensitization of the extracellular calcium-sensing receptor is regulated via distinct mechanisms: role of G protein-coupled receptor kinases, protein kinase C and beta-arrestins. Endocrinology 148 (5), 2398–2404.

Ma, X., Cheng, K.T., Wong, C.O., O'Neil, R.G., Birnbaumer, L., Ambudkar, I.S., et al., 2011. Heteromeric TRPV4-C1 channels contribute to store-operated $Ca^{(2+)}$ entry in vascular endothelial cells. Cell Calcium 50 (6), 502–509.

Mace, O.J., Schindler, M., Patel, S., 2012. The regulation of K- and L-cell activity by GLUT2 and the calcium-sensing receptor CasR in rat small intestine. J. Physiol. 590 (12), 2917–2936.

Macleod, R.J., 2013. CaSR function in the intestine: hormone secretion, electrolyte absorption and secretion, paracrine non-canonical Wnt signaling and colonic crypt cell proliferation. Best Pract. Res. Clin. Endocrinol. Metabol. 27 (3), 385–402.

MacLeod, R.J., Chattopadhyay, N., Brown, E.M., 2003. PTHrP stimulated by the calcium-sensing receptor requires MAP kinase activation. Am. J. Physiol. Endocrinol. Metab. 284 (2), E435–E442.

MacLeod, R.J., Hayes, M., Pacheco, I., 2007. Wnt5a secretion stimulated by the extracellular calcium-sensing receptor inhibits defective Wnt signaling in colon cancer cells. Am. J. Physiol. Gastrointest. Liver Physiol. 293 (1), G403–G411.

MacLeod, R.J., Yano, S., Chattopadhyay, N., Brown, E.M., 2004. Extracellular calcium-sensing receptor transactivates the epidermal growth factor receptor by a triple-membrane-spanning signaling mechanism. Biochem. Biophys. Res. Commun. 320 (2), 455–460.

Magno, A.L., Ingley, E., Brown, S.J., Conigrave, A.D., Ratajczak, T., Ward, B.K., 2011a. Testin, a novel binding partner of the calcium-sensing receptor, enhances receptor-mediated Rho-kinase signalling. Biochem. Biophys. Res. Commun. 412 (4), 584–589.

Magno, A.L., Ward, B.K., Ratajczak, T., 2011b. The calcium-sensing receptor: a molecular perspective. Endocr. Rev. 32 (1), 3–30.

Maiti, A., Hait, N.C., Beckman, M.J., 2008. Extracellular calcium-sensing receptor activation induces vitamin D receptor levels in proximal kidney HK-2G cells by a mechanism that requires phosphorylation of p38alpha MAPK. J. Biol. Chem. 283 (1), 175–183.

Malgaroli, A., Meldolesi, J., Zallone, A.Z., Teti, A., 1989. Control of cytosolic free calcium in rat and chicken osteoclasts. The role of extracellular calcium and calcitonin. J. Biol. Chem. 264 (24), 14342–14347.

Mamillapalli, R., VanHouten, J., Dann, P., Bikle, D., Chang, W., Brown, E., et al., 2013. Mammary-specific ablation of the calcium-sensing receptor during lactation alters maternal calcium metabolism, milk calcium transport, and neonatal calcium accrual. Endocrinology 154 (9), 3031–3042.

Mamillapalli, R., VanHouten, J., Zawalich, W., Wysolmerski, J., 2008. Switching of G-protein usage by the calcium-sensing receptor reverses its effect on parathyroid hormone-related protein secretion in normal versus malignant breast cells. J. Biol. Chem. 283 (36), 24435–24447.

Mamillapalli, R., Wysolmerski, J., 2010. The calcium-sensing receptor couples to Galpha(s) and regulates PTHrP and ACTH secretion in pituitary cells. J. Endocrinol. 204 (3), 287–297.

Mannstadt, M., Bertrand, G., Muresan, M., Weryha, G., Leheup, B., Pulusani, S.R., et al., 2008. Dominant-negative GCMB mutations cause an autosomal dominant form of hypoparathyroidism. J. Clin. Endocrinol. Metab. 93 (9), 3568–3576.

Maruyama, Y., Yasuda, R., Kuroda, M., Eto, Y., 2012. Kokumi substances, enhancers of basic tastes, induce responses in calcium-sensing receptor expressing taste cells. PLoS One 7 (4), e34489.

Marx, S.J., 2018. Familial hypocalciuric hypercalcemia as an atypical form of primary hyperparathyroidism. J. Bone Miner. Res. 33 (1), 27–31.

Marx, S.J., Attie, M.F., Levine, M.A., Spiegel, A.M., Downs Jr., R.W., Lasker, R.D., 1981a. The hypocalciuric or benign variant of familial hypercalcemia: clinical and biochemical features in fifteen kindreds. Medicine (Baltim.) 60 (6), 397–412.

Marx, S.J., Attie, M.F., Spiegel, A.M., Levine, M.A., Lasker, R.D., Fox, M., 1982a. An association between neonatal severe primary hyperparathyroidism and familial hypocalciuric hypercalcemia in three kindreds. N. Engl. J. Med. 306 (5), 257–264.

Marx, S.J., Attie, M.F., Stock, J.L., Spiegel, A.M., Levine, M.A., 1981b. Maximal urine-concentrating ability: familial hypocalciuric hypercalcemia versus typical primary hyperparathyroidism. J. Clin. Endocrinol. Metab. 52 (4), 736–740.

Marx, S.J., Fraser, D., Rapoport, A., 1985. Familial hypocalciuric hypercalcemia. Mild expression of the gene in heterozygotes and severe expression in homozygotes. Am. J. Med. 78 (1), 15–22.

Marx, S.J., Spiegel, A.M., Brown, E.M., Koehler, J.O., Gardner, D.G., Brennan, M.F., et al., 1978a. Divalent cation metabolism. Familial hypocalciuric hypercalcemia versus typical primary hyperparathyroidism. Am. J. Med. 65 (2), 235–242.

Marx, S.J., Spiegel, A.M., Brown, E.M., Windeck, R., Gardner, D.G., Downs Jr., R.W., et al., 1978b. Circulating parathyroid hormone activity: familial hypocalciuric hypercalcemia versus typical primary hyperparathyroidism. J. Clin. Endocrinol. Metab. 47 (6), 1190–1197.

Marx, S.J., Spiegel, A.M., Levine, M.A., Rizzoli, R.E., Lasker, R.D., Santora, A.C., et al., 1982b. Familial hypocalciuric hypercalcemia: the relation to primary parathyroid hyperplasia. N. Engl. J. Med. 307 (7), 416–426.

Mathias, R.S., Nguyen, H.T., Zhang, M.Y., Portale, A.A., 1998. Reduced expression of the renal calcium-sensing receptor in rats with experimental chronic renal insufficiency. J. Am. Soc. Nephrol. 9 (11), 2067–2074.

McCormick, W.D., Atkinson-Dell, R., Campion, K.L., Mun, H.C., Conigrave, A.D., Ward, D.T., 2010. Increased receptor stimulation elicits differential calcium-sensing receptor(T888) dephosphorylation. J. Biol. Chem. 285 (19), 14170–14177.

McLarnon, S., Holden, D., Ward, D., Jones, M., Elliott, A., Riccardi, D., 2002. Aminoglycoside antibiotics induce pH-sensitive activation of the calcium-sensing receptor. Biochem. Biophys. Res. Commun. 297 (1), 71–77.

Membreno, L., Chen, T.H., Woodley, S., Gagucas, R., Shoback, D., 1989. The effects of protein kinase-C agonists on parathyroid hormone release and intracellular free Ca2+ in bovine parathyroid cells. Endocrinology 124 (2), 789–797.

Mentaverri, R., Yano, S., Chattopadhyay, N., Petit, L., Kifor, O., Kamel, S., et al., 2006. The calcium sensing receptor is directly involved in both osteoclast differentiation and apoptosis. FASEB J. 20 (14), 2562–2564.

Mihai, R., Stevens, J., McKinney, C., Ibrahim, N.B., 2006. Expression of the calcium receptor in human breast cancer–a potential new marker predicting the risk of bone metastases. Eur. J. Surg. Oncol. 32 (5), 511–515.

Mizobuchi, M., Hatamura, I., Ogata, H., Saji, F., Uda, S., Shiizaki, K., et al., 2004. Calcimimetic compound upregulates decreased calcium-sensing receptor expression level in parathyroid glands of rats with chronic renal insufficiency. J. Am. Soc. Nephrol. 15 (10), 2579–2587.

Mizobuchi, M., Ritter, C.S., Krits, I., Slatopolsky, E., Sicard, G., Brown, A.J., 2009. Calcium-sensing receptor expression is regulated by glial cells missing-2 in human parathyroid cells. J. Bone Miner. Res. 24 (7), 1173–1179.

Molostvov, G., Hiemstra, T.F., Fletcher, S., Bland, R., Zehnder, D., 2015. Arterial expression of the calcium-sensing receptor is maintained by physiological pulsation and protects against calcification. PLoS One 10 (10), e0138833.

Moonga, B.S., Moss, D.W., Patchell, A., Zaidi, M., 1990. Intracellular regulation of enzyme secretion from rat osteoclasts and evidence for a functional role in bone resorption. J. Physiol. 429, 29–45.

Mu, L., Tuck, D., Katsaros, D., Lu, L., Schulz, V., Perincheri, S., et al., 2012. Favorable outcome associated with an IGF-1 ligand signature in breast cancer. Breast Canc. Res. Treat. 133 (1), 321–331.

Muddana, V., Lamb, J., Greer, J.B., Elinoff, B., Hawes, R.H., Cotton, P.B., et al., 2008. Association between calcium sensing receptor gene polymorphisms and chronic pancreatitis in a US population: role of serine protease inhibitor Kazal 1type and alcohol. World J. Gastroenterol. 14 (28), 4486–4491.

Mulder, D.J., Pacheco, I., Hurlbut, D.J., Mak, N., Furuta, G.T., MacLeod, R.J., et al., 2009. FGF9-induced proliferative response to eosinophilic inflammation in oesophagitis. Gut 58 (2), 166–173.

Mun, H.C., Culverston, E.L., Franks, A.H., Collyer, C.A., Clifton-Bligh, R.J., Conigrave, A.D., 2005. A double mutation in the extracellular Ca^{2+}-sensing receptor's venus flytrap domain that selectively disables L-amino acid sensing. J. Biol. Chem. 280 (32), 29067–29072.

Mun, H.C., Franks, A.H., Culverston, E.L., Krapcho, K., Nemeth, E.F., Conigrave, A.D., 2004. The Venus Fly Trap domain of the extracellular Ca^{2+} -sensing receptor is required for L-amino acid sensing. J. Biol. Chem. 279 (50), 51739–51744.

Muto, T., Tsuchiya, D., Morikawa, K., Jingami, H., 2007. Structures of the extracellular regions of the group II/III metabotropic glutamate receptors. Proc. Natl. Acad. Sci. U. S. A. 104 (10), 3759–3764.

Nemeth, E.F., 2002. The search for calcium receptor antagonists (calcilytics). J. Mol. Endocrinol. 29 (1), 15–21.

Nemeth, E.F., Goodman, W.G., 2016. Calcimimetic and calcilytic drugs: feats, flops, and futures. Calcif. Tissue Int. 98 (4), 341–358.

Nemeth, E.F., Scarpa, A., 1987. Rapid mobilization of cellular Ca^{2+} in bovine parathyroid cells evoked by extracellular divalent cations. Evidence for a cell surface calcium receptor. J. Biol. Chem. 262 (11), 5188–5196.

Nemeth, E.F., Shoback, D., 2013. Calcimimetic and calcilytic drugs for treating bone and mineral-related disorders. Best Pract. Res. Clin. Endocrinol. Metabol. 27 (3), 373–384.

Nemeth, E.F., Steffey, M.E., Hammerland, L.G., Hung, B.C., Van Wagenen, B.C., DelMar, E.G., et al., 1998. Calcimimetics with potent and selective activity on the parathyroid calcium receptor. Proc. Natl. Acad. Sci. U. S. A. 95 (7), 4040–4045.

Nesbit, M.A., Hannan, F.M., Howles, S.A., Babinsky, V.N., Head, R.A., Cranston, T., et al., 2013a. Mutations affecting G-protein subunit alpha11 in hypercalcemia and hypocalcemia. N. Engl. J. Med. 368 (26), 2476–2486.

Nesbit, M.A., Hannan, F.M., Howles, S.A., Reed, A.A., Cranston, T., Thakker, C.E., et al., 2013b. Mutations in AP2S1 cause familial hypocalciuric hypercalcemia type 3. Nat. Genet. 45 (1), 93–97.

North, M.L., Grasemann, H., Khanna, N., Inman, M.D., Gauvreau, G.M., Scott, J.A., 2013. Increased ornithine-derived polyamines cause airway hyperresponsiveness in a mouse model of asthma. Am. J. Respir. Cell Mol. Biol. 48 (6), 694–702.

Oda, Y., Tu, C.L., Chang, W., Crumrine, D., Komuves, L., Mauro, T., et al., 2000. The calcium sensing receptor and its alternatively spliced form in murine epidermal differentiation. J. Biol. Chem. 275 (2), 1183–1190.

Oda, Y., Tu, C.L., Pillai, S., Bikle, D.D., 1998. The calcium sensing receptor and its alternatively spliced form in keratinocyte differentiation. J. Biol. Chem. 273 (36), 23344–23352.

Oh, J., Beckmann, J., Bloch, J., Hettgen, V., Mueller, J., Li, L., et al., 2011. Stimulation of the calcium-sensing receptor stabilizes the podocyte cytoskeleton, improves cell survival, and reduces toxin-induced glomerulosclerosis. Kidney Int. 80 (5), 483–492.

Okabe, M., Graham, A., 2004. The origin of the parathyroid gland. Proc. Natl. Acad. Sci. U. S. A. 101 (51), 17716–17719.

Olszak, I.T., Poznansky, M.C., Evans, R.H., Olson, D., Kos, C., Pollak, M.R., et al., 2000. Extracellular calcium elicits a chemokinetic response from monocytes in vitro and in vivo. J. Clin. Investig. 105 (9), 1299–1305.

Owen, J.L., Cheng, S.X., Ge, Y., Sahay, B., Mohamadzadeh, M., 2016. The role of the calcium-sensing receptor in gastrointestinal inflammation. Semin. Cell Dev. Biol. 49, 44–51.

Parfrey, P.S., Drueke, T.B., Block, G.A., Correa-Rotter, R., Floege, J., Herzog, C.A., et al., 2015. The effects of cinacalcet in older and younger patients on hemodialysis: the evaluation of cinacalcet HCl therapy to lower cardiovascular events (EVOLVE) trial. Clin. J. Am. Soc. Nephrol. 10 (5), 791–799.

Parkash, J., 2011. Glucose-mediated spatial interactions of voltage dependent calcium channels and calcium sensing receptor in insulin producing beta-cells. Life Sci. 88 (5–6), 257–264.

Parkash, J., Asotra, K., 2011. L-histidine sensing by calcium sensing receptor inhibits voltage-dependent calcium channel activity and insulin secretion in beta-cells. Life Sci. 88 (9–10), 440–446.

Patel, L.R., Camacho, D.F., Shiozawa, Y., Pienta, K.J., Taichman, R.S., 2011. Mechanisms of cancer cell metastasis to the bone: a multistep process. Future Oncol. 7 (11), 1285–1297.

Pearce, S.H., Trump, D., Wooding, C., Besser, G.M., Chew, S.L., Grant, D.B., et al., 1995. Calcium-sensing receptor mutations in familial benign hypercalcemia and neonatal hyperparathyroidism. J. Clin. Investig. 96 (6), 2683–2692.

Pece, S., Gutkind, J.S., 2000. Signaling from E-cadherins to the MAPK pathway by the recruitment and activation of epidermal growth factor receptors upon cell-cell contact formation. J. Biol. Chem. 275 (52), 41227–41233.

Peng, X., Li, H.X., Shao, H.J., Li, G.W., Sun, J., Xi, Y.H., et al., 2014. Involvement of calcium-sensing receptors in hypoxia-induced vascular remodeling and pulmonary hypertension by promoting phenotypic modulation of small pulmonary arteries. Mol. Cell. Biochem. 396 (1–2), 87–98.

Petrel, C., Kessler, A., Dauban, P., Dodd, R.H., Rognan, D., Ruat, M., 2004. Positive and negative allosteric modulators of the Ca^{2+}-sensing receptor interact within overlapping but not identical binding sites in the transmembrane domain. J. Biol. Chem. 279 (18), 18990–18997.

Pi, M., Oakley, R.H., Gesty-Palmer, D., Cruickshank, R.D., Spurney, R.F., Luttrell, L.M., et al., 2005. Beta-arrestin- and G protein receptor kinase-mediated calcium-sensing receptor desensitization. Mol. Endocrinol. 19 (4), 1078–1087.

Pi, M., Spurney, R.F., Tu, Q., Hinson, T., Quarles, L.D., 2002. Calcium-sensing receptor activation of rho involves filamin and rho-guanine nucleotide exchange factor. Endocrinology 143 (10), 3830–3838.

Pin, J.P., Galvez, T., Prezeau, L., 2003. Evolution, structure, and activation mechanism of family 3/C G-protein-coupled receptors. Pharmacol. Ther. 98 (3), 325–354.

Pollak, M.R., Chou, Y.H., Marx, S.J., Steinmann, B., Cole, D.E., Brandi, M.L., et al., 1994. Familial hypocalciuric hypercalcemia and neonatal severe hyperparathyroidism. Effects of mutant gene dosage on phenotype. J. Clin. Investig. 93 (3), 1108–1112.

Popp, T., Steinritz, D., Breit, A., Deppe, J., Egea, V., Schmidt, A., et al., 2014. Wnt5a/beta-catenin signaling drives calcium-induced differentiation of human primary keratinocytes. J. Investig. Dermatol. 134 (8), 2183–2191.

Procino, G., Mastrofrancesco, L., Tamma, G., Lasorsa, D.R., Ranieri, M., Stringini, G., et al., 2012. Calcium-sensing receptor and aquaporin 2 interplay in hypercalciuria-associated renal concentrating defect in humans. An in vivo and in vitro study. PLoS One 7 (3), e33145.

Promkan, M., Liu, G., Patmasiriwat, P., Chakrabarty, S., 2011. BRCA1 suppresses the expression of survivin and promotes sensitivity to paclitaxel through the calcium sensing receptor (CaSR) in human breast cancer cells. Cell Calcium 49 (2), 79–88.

Quinn, S.J., Bai, M., Brown, E.M., 2004. pH Sensing by the calcium-sensing receptor. J. Biol. Chem. 279 (36), 37241–37249.

Quinn, S.J., Kifor, O., Trivedi, S., Diaz, R., Vassilev, P., Brown, E., 1998. Sodium and ionic strength sensing by the calcium receptor. J. Biol. Chem. 273 (31), 19579–19586.

Quinn, S.J., Ye, C.P., Diaz, R., Kifor, O., Bai, M., Vassilev, P., et al., 1997. The Ca^{2+}-sensing receptor: a target for polyamines. Am. J. Physiol. 273 (4 Pt 1), C1315–C1323.

Racz, G.Z., Kittel, A., Riccardi, D., Case, R.M., Elliott, A.C., Varga, G., 2002. Extracellular calcium sensing receptor in human pancreatic cells. Gut 51 (5), 705–711.

Ray, K., 2015. Calcium-sensing receptor: trafficking, endocytosis, recycling, and importance of interacting proteins. Prog Mol Biol Transl Sci 132, 127–150.

Ray, K., Clapp, P., Goldsmith, P.K., Spiegel, A.M., 1998. Identification of the sites of N-linked glycosylation on the human calcium receptor and assessment of their role in cell surface expression and signal transduction. J. Biol. Chem. 273 (51), 34558–34567.

Ray, K., Fan, G.F., Goldsmith, P.K., Spiegel, A.M., 1997. The carboxyl terminus of the human calcium receptor. Requirements for cell-surface expression and signal transduction. J. Biol. Chem. 272 (50), 31355–31361.

Ray, K., Ghosh, S.P., Northup, J.K., 2004. The role of cysteines and charged amino acids in extracellular loops of the human $Ca^{(2+)}$ receptor in cell surface expression and receptor activation processes. Endocrinology 145 (8), 3892–3903.

Ray, K., Hauschild, B.C., Steinbach, P.J., Goldsmith, P.K., Hauache, O., Spiegel, A.M., 1999. Identification of the cysteine residues in the amino-terminal extracellular domain of the human $Ca^{(2+)}$ receptor critical for dimerization. Implications for function of monomeric $Ca^{(2+)}$ receptor. J. Biol. Chem. 274 (39), 27642–27650.

Reinhardt, T.A., Lippolis, J.D., Shull, G.E., Horst, R.L., 2004. Null mutation in the gene encoding plasma membrane Ca^{2+}-ATPase isoform 2 impairs calcium transport into milk. J. Biol. Chem. 279 (41), 42369–42373.

Remy, C., Kirchhoff, P., Hafner, P., Busque, S.M., Mueller, M.K., Geibel, J.P., et al., 2007. Stimulatory pathways of the Calcium-sensing receptor on acid secretion in freshly isolated human gastric glands. Cell. Physiol. Biochem. 19 (1–4), 33–42.

Rey, O., Chang, W., Bikle, D., Rozengurt, N., Young, S.H., Rozengurt, E., 2012. Negative cross-talk between calcium-sensing receptor and beta-catenin signaling systems in colonic epithelium. J. Biol. Chem. 287 (2), 1158–1167.

Rey, O., Young, S.H., Jacamo, R., Moyer, M.P., Rozengurt, E., 2010. Extracellular calcium sensing receptor stimulation in human colonic epithelial cells induces intracellular calcium oscillations and proliferation inhibition. J. Cell. Physiol. 225 (1), 73–83.

Rey, O., Young, S.H., Yuan, J., Slice, L., Rozengurt, E., 2005. Amino acid-stimulated Ca^{2+} oscillations produced by the Ca^{2+}-sensing receptor are mediated by a phospholipase C/inositol 1,4,5-trisphosphate-independent pathway that requires G12, Rho, filamin-A, and the actin cytoskeleton. J. Biol. Chem. 280 (24), 22875–22882.

Reyes-Ibarra, A.P., Garcia-Regalado, A., Ramirez-Rangel, I., Esparza-Silva, A.L., Valadez-Sanchez, M., Vazquez-Prado, J., et al., 2007. Calcium-sensing receptor endocytosis links extracellular calcium signaling to parathyroid hormone-related peptide secretion via a Rab11a-dependent and AMSH-sensitive mechanism. Mol. Endocrinol. 21 (6), 1394–1407.

Riccardi, D., 2002. Wellcome Prize Lecture. Cell surface, ion-sensing receptors. Exp. Physiol. 87 (4), 403–411.

Riccardi, D., Brennan, S.C., Chang, W., 2013. The extracellular calcium-sensing receptor, CaSR, in fetal development. Best Pract. Res. Clin. Endocrinol. Metabol. 27 (3), 443–453.

Riccardi, D., Finney, B.A., Wilkinson, W.J., Kemp, P.J., 2009. Novel regulatory aspects of the extracellular Ca^{2+}-sensing receptor, CaR. Pflügers Archiv 458 (6), 1007–1022.

Riccardi, D., Park, J., Lee, W.S., Gamba, G., Brown, E.M., Hebert, S.C., 1995. Cloning and functional expression of a rat kidney extracellular calcium/polyvalent cation-sensing receptor. Proc. Natl. Acad. Sci. U. S. A. 92 (1), 131–135.

Riccardi, D., Valenti, G., 2016. Localization and function of the renal calcium-sensing receptor. Nat. Rev. Nephrol. 12 (7), 414–425.

Richard, C., Huo, R., Samadfam, R., Bolivar, I., Miao, D., Brown, E.M., et al., 2010. The calcium-sensing receptor and 25-hydroxyvitamin D-1alpha-hydroxylase interact to modulate skeletal growth and bone turnover. J. Bone Miner. Res. 25 (7), 1627–1636.

Rodriguez, L., Tu, C., Cheng, Z., Chen, T.H., Bikle, D., Shoback, D., et al., 2005. Expression and functional assessment of an alternatively spliced extracellular Ca^{2+}-sensing receptor in growth plate chondrocytes. Endocrinology 146 (12), 5294–5303.

Rogachevskaja, O.A., Churbanov, G.D., Bystrova, M.F., Romanov, R.A., Kolesnikov, S.S., 2011. Stimulation of the extracellular $Ca^{(2+)}$-sensing receptor by denatonium. Biochem. Biophys. Res. Commun. 416 (3–4), 433–436.

Rogers, A., Nesbit, M.A., Hannan, F.M., Howles, S.A., Gorvin, C.M., Cranston, T., et al., 2014. Mutational analysis of the adaptor protein 2 sigma subunit (AP2S1) gene: search for autosomal dominant hypocalcemia type 3 (ADH3). J. Clin. Endocrinol. Metab. 99 (7), E1300–E1305.

Rogers, A.C., Hanly, A.M., Collins, D., Baird, A.W., Winter, D.C., 2012. Review article: loss of the calcium-sensing receptor in colonic epithelium is a key event in the pathogenesis of colon cancer. Clin. Colorectal Cancer 11 (1), 24–30.

Rossol, M., Pierer, M., Raulien, N., Quandt, D., Meusch, U., Rothe, K., et al., 2012. Extracellular Ca^{2+} is a danger signal activating the NLRP3 inflammasome through G protein-coupled calcium sensing receptors. Nat. Commun. 3, 1329.

Roszko, K.L., Bi, R.D., Mannstadt, M., 2016. Autosomal dominant hypocalcemia (hypoparathyroidism) types 1 and 2. Front. Physiol. 7, 458.

Ruat, M., Molliver, M.E., Snowman, A.M., Snyder, S.H., 1995. Calcium sensing receptor: molecular cloning in rat and localization to nerve terminals. Proc. Natl. Acad. Sci. U. S. A. 92 (8), 3161–3165.

Ruat, M., Snowman, A.M., Hester, L.D., Snyder, S.H., 1996. Cloned and expressed rat Ca^{2+}-sensing receptor. Differential cooperative responses to calicum and magnesium. J. Biol. Chem. 271 (11), 5972–5975.

Rudberg, C., Akerstrom, G., Ljunghall, S., Grimelius, L., Johansson, H., Pertoft, H., et al., 1982. Regulation of parathyroid hormone release in primary and secondary hyperparathyroidism – studies in vivo and in vitro. Acta Endocrinol. 101 (3), 408–413.

Rybchyn, M.S., Slater, M., Conigrave, A.D., Mason, R.S., 2011. An Akt-dependent increase in canonical Wnt signaling and a decrease in sclerostin protein levels are involved in strontium ranelate-induced osteogenic effects in human osteoblasts. J. Biol. Chem. 286 (27), 23771–23779.

Rybczynska, A., Marchwinska, A., Dys, A., Boblewski, K., Lehmann, A., Lewko, B., 2017. Activity of the calcium-sensing receptor influences blood glucose and insulin levels in rats. Pharmacol. Rep. 69 (4), 709–713.

Saidak, Z., Boudot, C., Abdoune, R., Petit, L., Brazier, M., Mentaverri, R., et al., 2009a. Extracellular calcium promotes the migration of breast cancer cells through the activation of the calcium sensing receptor. Exp. Cell Res. 315 (12), 2072–2080.

Saidak, Z., Brazier, M., Kamel, S., Mentaverri, R., 2009b. Agonists and allosteric modulators of the calcium-sensing receptor and their therapeutic applications. Mol. Pharmacol. 76 (6), 1131–1144.

San Gabriel, A., Uneyama, H., Maekawa, T., Torii, K., 2009. The calcium-sensing receptor in taste tissue. Biochem. Biophys. Res. Commun. 378 (3), 414–418.

Sanders, J.L., Chattopadhyay, N., Kifor, O., Yamaguchi, T., Butters, R.R., Brown, E.M., 2000. Extracellular calcium-sensing receptor expression and its potential role in regulating parathyroid hormone-related peptide secretion in human breast cancer cell lines. Endocrinology 141 (12), 4357–4364.

Santa Maria, C., Cheng, Z., Li, A., Wang, J., Shoback, D., Tu, C.L., et al., 2016. Interplay between CaSR and PTH1R signaling in skeletal development and osteoanabolism. Semin. Cell Dev. Biol. 49, 11–23.

Schepelmann, M., Yarova, P.L., Lopez-Fernandez, I., Davies, T.S., Brennan, S.C., Edwards, P.J., et al., 2016. The vascular Ca^{2+}-sensing receptor regulates blood vessel tone and blood pressure. Am. J. Physiol. Cell Physiol. 310 (3), C193–C204.

Scillitani, A., Guarnieri, V., De Geronimo, S., Muscarella, L.A., Battista, C., D'Agruma, L., et al., 2004. Blood ionized calcium is associated with clustered polymorphisms in the carboxyl-terminal tail of the calcium-sensing receptor. J. Clin. Endocrinol. Metab. 89 (11), 5634–5638.

Shalhoub, V., Grisanti, M., Padagas, J., Scully, S., Rattan, A., Qi, M., et al., 2003. In vitro studies with the calcimimetic, cinacalcet HCl, on normal human adult osteoblastic and osteoclastic cells. Crit. Rev. Eukaryot. Gene Expr. 13 (2–4), 89–106.

Shoback, D.M., Membreno, L.A., McGhee, J.G., 1988. High calcium and other divalent cations increase inositol trisphosphate in bovine parathyroid cells. Endocrinology 123 (1), 382–389.

Siehler, S., 2009. Regulation of RhoGEF proteins by G12/13-coupled receptors. Br. J. Pharmacol. 158 (1), 41–49.

Silve, C., Petrel, C., Leroy, C., Bruel, H., Mallet, E., Rognan, D., et al., 2005. Delineating a Ca^{2+} binding pocket within the venus flytrap module of the human calcium-sensing receptor. J. Biol. Chem. 280 (45), 37917–37923.

Singh, N., Chakrabarty, S., 2013. Induction of CaSR expression circumvents the molecular features of malignant CaSR null colon cancer cells. Int. J. Cancer 133 (10), 2307–2314.

Singh, N., Promkan, M., Liu, G., Varani, J., Chakrabarty, S., 2013. Role of calcium sensing receptor (CaSR) in tumorigenesis. Best Pract. Res. Clin. Endocrinol. Metabol. 27 (3), 455–463.

Smajilovic, S., Hansen, J.L., Christoffersen, T.E., Lewin, E., Sheikh, S.P., Terwilliger, E.F., et al., 2006. Extracellular calcium sensing in rat aortic vascular smooth muscle cells. Biochem. Biophys. Res. Commun. 348 (4), 1215–1223.

Smajilovic, S., Yano, S., Jabbari, R., Tfelt-Hansen, J., 2011. The calcium-sensing receptor and calcimimetics in blood pressure modulation. Br. J. Pharmacol. 164 (3), 884–893.

Squires, P.E., 2000. Non-Ca^{2+}-homeostatic functions of the extracellular Ca^{2+}-sensing receptor (CaR) in endocrine tissues. J. Endocrinol. 165 (2), 173–177.

Squires, P.E., Harris, T.E., Persaud, S.J., Curtis, S.B., Buchan, A.M., Jones, P.M., 2000. The extracellular calcium-sensing receptor on human beta-cells negatively modulates insulin secretion. Diabetes 49 (3), 409–417.

Strating, J.R., Martens, G.J., 2009. The p24 family and selective transport processes at the ER-Golgi interface. Biol. Cell 101 (9), 495–509.

Sumida, K., Nakamura, M., Ubara, Y., Marui, Y., Tanaka, K., Takaichi, K., et al., 2013. Cinacalcet upregulates calcium-sensing receptors of parathyroid glands in hemodialysis patients. Am. J. Nephrol. 37 (5), 405–412.

Sun, J., Murphy, E., 2010. Calcium-sensing receptor: a sensor and mediator of ischemic preconditioning in the heart. Am. J. Physiol. Heart Circ. Physiol. 299 (5), H1309–H1317.

Suzuki, K., Lavaroni, S., Mori, A., Okajima, F., Kimura, S., Katoh, R., et al., 1998. Thyroid transcription factor 1 is calcium modulated and coordinately regulates genes involved in calcium homeostasis in C cells. Mol. Cell Biol. 18 (12), 7410–7422.

Szekely, D., Brennan, S.C., Mun, H.C., Conigrave, A.D., Kuchel, P.W., 2009. Effectors of the frequency of calcium oscillations in HEK-293 cells: wavelet analysis and a computer model. Eur. Biophys. J. 39 (1), 149–165.

Tam, R., Saier Jr., M.H., 1993. Structural, functional, and evolutionary relationships among extracellular solute-binding receptors of bacteria. Microbiol. Rev. 57 (2), 320–346.

Tang, H., Yamamura, A., Yamamura, H., Song, S., Fraidenburg, D.R., Chen, J., et al., 2016. Pathogenic role of calcium-sensing receptors in the development and progression of pulmonary hypertension. Am. J. Physiol. Lung Cell Mol. Physiol. 310 (9), L846–L859.

Tfelt-Hansen, J., MacLeod, R.J., Chattopadhyay, N., Yano, S., Quinn, S., Ren, X., et al., 2003. Calcium-sensing receptor stimulates PTHrP release by pathways dependent on PKC, p38 MAPK, JNK, and ERK1/2 in H-500 cells. Am. J. Physiol. Endocrinol. Metab. 285 (2), E329–E337.

Tharmalingam, S., Daulat, A.M., Antflick, J.E., Ahmed, S.M., Nemeth, E.F., Angers, S., et al., 2011. Calcium-sensing receptor modulates cell adhesion and migration via integrins. J. Biol. Chem. 286 (47), 40922–40933.

Tharmalingam, S., Wu, C., Hampson, D.R., 2016. The calcium-sensing receptor and integrins modulate cerebellar granule cell precursor differentiation and migration. Dev Neurobiol 76 (4), 375–389.

Theriault, R.L., Theriault, R.L., 2012. Biology of bone metastases. Cancer Control 19 (2), 92–101.

Toka, H.R., 2014. New functional aspects of the extracellular calcium-sensing receptor. Curr. Opin. Nephrol. Hypertens. 23 (4), 352–360.

Toka, H.R., Al-Romaih, K., Koshy, J.M., DiBartolo 3rd, S., Kos, C.H., Quinn, S.J., et al., 2012. Deficiency of the calcium-sensing receptor in the kidney causes parathyroid hormone-independent hypocalciuria. J. Am. Soc. Nephrol. 23 (11), 1879–1890.

Tu, C.L., Celli, A., Mauro, T., Chang, W., 2018. The calcium-sensing receptor regulates epidermal intracellular $Ca^{(2+)}$ signaling and re-epithelialization after wounding. J. Investig. Dermatol.

Tu, C.L., Chang, W., Bikle, D.D., 2001. The extracellular calcium-sensing receptor is required for calcium-induced differentiation in human keratinocytes. J. Biol. Chem. 276 (44), 41079–41085.

Tu, C.L., Chang, W., Bikle, D.D., 2007. The role of the calcium sensing receptor in regulating intracellular calcium handling in human epidermal keratinocytes. J. Investig. Dermatol. 127 (5), 1074–1083.

Tu, C.L., Chang, W., Bikle, D.D., 2011. The calcium-sensing receptor-dependent regulation of cell-cell adhesion and keratinocyte differentiation requires Rho and filamin A. J. Investig. Dermatol. 131 (5), 1119–1128.

Tu, C.L., Chang, W., Xie, Z., Bikle, D.D., 2008. Inactivation of the calcium sensing receptor inhibits E-cadherin-mediated cell-cell adhesion and calcium-induced differentiation in human epidermal keratinocytes. J. Biol. Chem. 283 (6), 3519–3528.

Tu, C.L., Crumrine, D.A., Man, M.Q., Chang, W., Elalieh, H., You, M., et al., 2012. Ablation of the calcium-sensing receptor in keratinocytes impairs epidermal differentiation and barrier function. J. Investig. Dermatol. 132 (10), 2350–2359.

Tu, C.L., You, M., 2014. Obligatory roles of filamin A in E-cadherin-mediated cell-cell adhesion in epidermal keratinocytes. J. Dermatol. Sci. 73 (2), 142–151.

Tu, Q., Pi, M., Karsenty, G., Simpson, L., Liu, S., Quarles, L.D., 2003. Rescue of the skeletal phenotype in CasR-deficient mice by transfer onto the Gcm2 null background. J. Clin. Investig. 111 (7), 1029–1037.

Turksen, K., Troy, T.C., 2003. Overexpression of the calcium sensing receptor accelerates epidermal differentiation and permeability barrier formation in vivo. Mech. Dev. 120 (6), 733–744.

Tyler Miller, R., 2013. Control of renal calcium, phosphate, electrolyte, and water excretion by the calcium-sensing receptor. Best Pract. Res. Clin. Endocrinol. Metabol. 27 (3), 345–358.

Van Aken, E., De Wever, O., Correia da Rocha, A.S., Mareel, M., 2001. Defective E-cadherin/catenin complexes in human cancer. Virchows Arch. 439 (6), 725–751.

VanHouten, J., Dann, P., McGeoch, G., Brown, E.M., Krapcho, K., Neville, M., et al., 2004. The calcium-sensing receptor regulates mammary gland parathyroid hormone-related protein production and calcium transport. J. Clin. Investig. 113 (4), 598–608.

VanHouten, J., Sullivan, C., Bazinet, C., Ryoo, T., Camp, R., Rimm, D.L., et al., 2010. PMCA2 regulates apoptosis during mammary gland involution and predicts outcome in breast cancer. Proc. Natl. Acad. Sci. U. S. A. 107 (25), 11405–11410.

VanHouten, J.N., Neville, M.C., Wysolmerski, J.J., 2007. The calcium-sensing receptor regulates plasma membrane calcium adenosine triphosphatase isoform 2 activity in mammary epithelial cells: a mechanism for calcium-regulated calcium transport into milk. Endocrinology 148 (12), 5943–5954.

VanHouten, J.N., Wysolmerski, J.J., 2007. Transcellular calcium transport in mammary epithelial cells. J. Mammary Gland Biol. Neoplasia 12 (4), 223–235.

Vanhouten, J.N., Wysolmerski, J.J., 2013. The calcium-sensing receptor in the breast. Best Pract. Res. Clin. Endocrinol. Metabol. 27 (3), 403–414.

Vargas-Poussou, R., Mansour-Hendili, L., Baron, S., Bertocchio, J.P., Travers, C., Simian, C., et al., 2016. Familial hypocalciuric hypercalcemia types 1 and 3 and primary hyperparathyroidism: similarities and differences. J. Clin. Endocrinol. Metab. 101 (5), 2185–2195.

Vizard, T.N., O'Keeffe, G.W., Gutierrez, H., Kos, C.H., Riccardi, D., Davies, A.M., 2008. Regulation of axonal and dendritic growth by the extracellular calcium-sensing receptor. Nat. Neurosci. 11 (3), 285–291.

Wada, M., Ishii, H., Furuya, Y., Fox, J., Nemeth, E.F., Nagano, N., 1998. NPS R-568 halts or reverses osteitis fibrosa in uremic rats. Kidney Int. 53 (2), 448–453.

Wang, M., Yao, Y., Kuang, D., Hampson, D.R., 2006. Activation of family C G-protein-coupled receptors by the tripeptide glutathione. J. Biol. Chem. 281 (13), 8864–8870.

Ward, B.K., Rea, S.L., Magno, A.L., Pedersen, B., Brown, S.J., Mullin, S., et al., 2018. The endoplasmic reticulum-associated protein, OS-9, behaves as a lectin in targeting the immature calcium-sensing receptor. J. Cell. Physiol. 233 (1), 38–56.

Ward, D.T., Brown, E.M., Harris, H.W., 1998. Disulfide bonds in the extracellular calcium-polyvalent cation-sensing receptor correlate with dimer formation and its response to divalent cations in vitro. J. Biol. Chem. 273 (23), 14476–14483.

Ward, D.T., Maldonado-Perez, D., Hollins, L., Riccardi, D., 2005. Aminoglycosides induce acute cell signaling and chronic cell death in renal cells that express the calcium-sensing receptor. J. Am. Soc. Nephrol. 16 (5), 1236–1244.

Weston, A.H., Absi, M., Ward, D.T., Ohanian, J., Dodd, R.H., Dauban, P., et al., 2005. Evidence in favor of a calcium-sensing receptor in arterial endothelial cells: studies with calindol and Calhex 231. Circ. Res. 97 (4), 391–398.

Wettschureck, N., Lee, E., Libutti, S.K., Offermanns, S., Robey, P.G., Spiegel, A.M., 2007. Parathyroid-specific double knockout of Gq and G11 alpha-subunits leads to a phenotype resembling germline knockout of the extracellular Ca^{2+}-sensing receptor. Mol. Endocrinol. 21 (1), 274–280.

Williams, S., Malatesta, K., Norris, K., 2009. Vitamin D and chronic kidney disease. Ethn. Dis. 19 (4 Suppl. 5). S5-8-11.

Wong, N.A., Pignatelli, M., 2002. Beta-catenin–a linchpin in colorectal carcinogenesis? Am. J. Pathol. 160 (2), 389–401.

Wonneberger, K., Scofield, M.A., Wangemann, P., 2000. Evidence for a calcium-sensing receptor in the vascular smooth muscle cells of the spiral modiolar artery. J. Membr. Biol. 175 (3), 203–212.

Wysolmerski, J.J., 2012. Parathyroid hormone-related protein: an update. J. Clin. Endocrinol. Metab. 97 (9), 2947–2956.

Xie, Z., Bikle, D.D., 2007. The recruitment of phosphatidylinositol 3-kinase to the E-cadherin-catenin complex at the plasma membrane is required for calcium-induced phospholipase C-gamma1 activation and human keratinocyte differentiation. J. Biol. Chem. 282 (12), 8695–8703.

Xing, W.J., Kong, F.J., Li, G.W., Qiao, K., Zhang, W.H., Zhang, L., et al., 2011. Calcium-sensing receptors induce apoptosis during simulated ischaemia-reperfusion in Buffalo rat liver cells. Clin. Exp. Pharmacol. Physiol. 38 (9), 605–612.

Xue, Y., Xiao, Y., Liu, J., Karaplis, A.C., Pollak, M.R., Brown, E.M., et al., 2012. The calcium-sensing receptor complements parathyroid hormone-induced bone turnover in discrete skeletal compartments in mice. Am. J. Physiol. Endocrinol. Metab. 302 (7), E841–E851.

Yamaguchi, T., Chattopadhyay, N., Kifor, O., Sanders, J.L., Brown, E.M., 2000. Activation of p42/44 and p38 mitogen-activated protein kinases by extracellular calcium-sensing receptor agonists induces mitogenic responses in the mouse osteoblastic MC3T3-E1 cell line. Biochem. Biophys. Res. Commun. 279 (2), 363–368.

Yamaguchi, T., Olozak, I., Chattopadhyay, N., Butters, R.R., Kifor, O., Scadden, D.T., et al., 1998. Expression of extracellular calcium (Ca2+o)-sensing receptor in human peripheral blood monocytes. Biochem. Biophys. Res. Commun. 246 (2), 501–506.

Yamamura, A., Guo, Q., Yamamura, H., Zimnicka, A.M., Pohl, N.M., Smith, K.A., et al., 2012. Enhanced $Ca^{(2+)}$-sensing receptor function in idiopathic pulmonary arterial hypertension. Circ. Res. 111 (4), 469–481.

Yamamura, A., Ohara, N., Tsukamoto, K., 2015. Inhibition of excessive cell proliferation by calcilytics in idiopathic pulmonary arterial hypertension. PLoS One 10 (9), e0138384.

Yamamura, A., Yagi, S., Ohara, N., Tsukamoto, K., 2016. Calcilytics enhance sildenafil-induced antiproliferation in idiopathic pulmonary arterial hypertension. Eur. J. Pharmacol. 784, 15–21.

Yano, S., Brown, E.M., Chattopadhyay, N., 2004a. Calcium-sensing receptor in the brain. Cell Calcium 35 (3), 257–264.

Yano, S., Macleod, R.J., Chattopadhyay, N., Tfelt-Hansen, J., Kifor, O., Butters, R.R., et al., 2004b. Calcium-sensing receptor activation stimulates parathyroid hormone-related protein secretion in prostate cancer cells: role of epidermal growth factor receptor transactivation. Bone 35 (3), 664–672.

Yao, J.J., Bai, S., Karnauskas, A.J., Bushinsky, D.A., Favus, M.J., 2005. Regulation of renal calcium receptor gene expression by 1,25-dihydroxyvitamin D3 in genetic hypercalciuric stone-forming rats. J. Am. Soc. Nephrol. 16 (5), 1300–1308.

Yarova, P.L., Stewart, A.L., Sathish, V., Britt Jr., R.D., Thompson, M.A., Lowe, A.P.P., et al., 2015. Calcium-sensing receptor antagonists abrogate airway hyperresponsiveness and inflammation in allergic asthma. Sci. Transl. Med. 7 (284), 284ra60.

Yasukawa, T., Hayashi, M., Tanabe, N., Tsuda, H., Suzuki, Y., Kawato, T., et al., 2017. Involvement of the calcium-sensing receptor in mineral trioxide aggregate-induced osteogenic gene expression in murine MC3T3-E1 cells. Dent. Mater. J. 36 (4), 469–475.

Ye, C., Ho-Pao, C.L., Kanazirska, M., Quinn, S., Rogers, K., Seidman, C.E., et al., 1997. Amyloid-beta proteins activate $Ca^{(2+)}$-permeable channels through calcium-sensing receptors. J. Neurosci. Res. 47 (5), 547–554.

Ye, C.P., Yano, S., Tfelt-Hansen, J., MacLeod, R.J., Ren, X., Terwilliger, E., et al., 2004. Regulation of a Ca^{2+}-activated K^+ channel by calcium-sensing receptor involves p38 MAP kinase. J. Neurosci. Res. 75 (4), 491–498.

Young, S.H., Rozengurt, E., 2002. Amino acids and Ca^{2+} stimulate different patterns of Ca^{2+} oscillations through the Ca^{2+}-sensing receptor. Am. J. Physiol. Cell Physiol. 282 (6), C1414–C1422.

Young, S.H., Wu, S.V., Rozengurt, E., 2002. Ca^{2+}-stimulated Ca^{2+} oscillations produced by the Ca^{2+}-sensing receptor require negative feedback by protein kinase C. J. Biol. Chem. 277 (49), 46871–46876.

Zaidi, M., Kerby, J., Huang, C.L., Alam, T., Rathod, H., Chambers, T.J., et al., 1991. Divalent cations mimic the inhibitory effect of extracellular ionised calcium on bone resorption by isolated rat osteoclasts: further evidence for a "calcium receptor". J. Cell. Physiol. 149 (3), 422–427.

Zhang, C., Huang, Y., Jiang, Y., Mulpuri, N., Wei, L., Hamelberg, D., et al., 2014a. Identification of an L-phenylalanine binding site enhancing the cooperative responses of the calcium-sensing receptor to calcium. J. Biol. Chem. 289 (8), 5296–5309.

Zhang, J., Zhou, J., Cai, L., Lu, Y., Wang, T., Zhu, L., et al., 2012. Extracellular calcium-sensing receptor is critical in hypoxic pulmonary vasoconstriction. Antioxid. Redox Signal 17 (3), 471–484.

Zhang, M., Breitwieser, G.E., 2005. High affinity interaction with filamin A protects against calcium-sensing receptor degradation. J. Biol. Chem. 280 (12), 11140–11146.

Zhang, X., Zhang, T., Wu, J., Yu, X., Zheng, D., Yang, F., et al., 2014b. Calcium sensing receptor promotes cardiac fibroblast proliferation and extracellular matrix secretion. Cell. Physiol. Biochem. 33 (3), 557–568.

Zhang, Z., Qiu, W., Quinn, S.J., Conigrave, A.D., Brown, E.M., Bai, M., 2002. Three adjacent serines in the extracellular domains of the CaR are required for L-amino acid-mediated potentiation of receptor function. J. Biol. Chem. 277 (37), 33727–33735.

Zhang, Z., Sun, S., Quinn, S.J., Brown, E.M., Bai, M., 2001. The extracellular calcium-sensing receptor dimerizes through multiple types of intermolecular interactions. J. Biol. Chem. 276 (7), 5316–5322.

Zheng, H., Liu, J., Liu, C., Lu, F., Zhao, Y., Jin, Z., et al., 2011. Calcium-sensing receptor activating phosphorylation of PKCdelta translocation on mitochondria to induce cardiomyocyte apoptosis during ischemia/reperfusion. Mol. Cell. Biochem. 358 (1–2), 335–343.

Zhu, D., Kosik, K.S., Meigs, T.E., Yanamadala, V., Denker, B.M., 2004. Galpha12 directly interacts with PP2A: evidence FOR Galpha12-stimulated PP2A phosphatase activity and dephosphorylation of microtubule-associated protein, tau. J. Biol. Chem. 279 (53), 54983–54986.

Zhu, D., Tate, R.I., Ruediger, R., Meigs, T.E., Denker, B.M., 2007. Domains necessary for Galpha12 binding and stimulation of protein phosphatase-2A (PP2A): is Galpha12 a novel regulatory subunit of PP2A? Mol. Pharmacol. 71 (5), 1268–1276.

Zhuang, X., Adipietro, K.A., Datta, S., Northup, J.K., Ray, K., 2010. Rab1 small GTP-binding protein regulates cell surface trafficking of the human calcium-sensing receptor. Endocrinology 151 (11), 5114–5123.

Zhuang, X., Northup, J.K., Ray, K., 2012. Large putative PEST-like sequence motif at the carboxyl tail of human calcium receptor directs lysosomal degradation and regulates cell surface receptor level. J. Biol. Chem. 287 (6), 4165–4176.

Ziegelstein, R.C., Xiong, Y., He, C., Hu, Q., 2006. Expression of a functional extracellular calcium-sensing receptor in human aortic endothelial cells. Biochem. Biophys. Res. Commun. 342 (1), 153–163.

Zitt, E., Woess, E., Mayer, G., Lhotta, K., 2011. Effect of cinacalcet on renal electrolyte handling and systemic arterial blood pressure in kidney transplant patients with persistent hyperparathyroidism. Transplantation 92 (8), 883–889.

Section D

Endocrine and paracrine regulation of bone

Chapter 24

Parathyroid hormone molecular biology

Tally Naveh-Many[1], Justin Silver[1] and Henry M. Kronenberg[2]

[1]*Minerva Center for Calcium and Bone Metabolism, Nephrology Services, Hadassah University Hospital, Hebrew University School of Medicine, Jerusalem, Israel;* [2]*Endocrine Unit, Massachusetts General Hospital, Harvard Medical School, Boston, MA, United States*

Chapter outline

The parathyroid hormone gene **575**
Organization of the parathyroid hormone gene 575
Promoter sequences 576
The parathyroid hormone mRNA 576
Mutations in the parathyroid hormone gene 577
Development of the parathyroid **578**
Regulation of parathyroid hormone gene expression **580**
1,25-Dihydroxyvitamin D_3 580
Calcium 582
In vitro studies 582
In vivo studies 582
Phosphate 583
Protein–PTH mRNA interactions determine the posttranscriptional regulation of PTH gene expression by calcium, phosphate, and uremia 583
A conserved sequence in the PTH mRNA 3′ UTR binds parathyroid cytosolic proteins and determines mRNA stability 584
The PTH mRNA 3′-UTR-binding proteins that determine PTH mRNA stability 584
MicroRNA in the parathyroid 585
Sex steroids 586
Fibroblast growth factor 23 587
Fibroblast growth factor 23 decreases parathyroid hormone expression 587
Resistance of the parathyroid to FGF23 in chronic kidney disease 587
Parathyroid cell proliferation and mammalian target of rapamycin **588**
Phosphorylation of ribosomal protein S6 mediates mammalian target of rapamycin–induced parathyroid cell proliferation in secondary hyperparathyroidism 588
Summary **589**
Acknowledgments **589**
References **589**

The parathyroid hormone gene

Organization of the parathyroid hormone gene

The human parathyroid hormone (PTH) gene is localized on the short arm of chromosome 11 at 11p15 (Antonarakis et al., 1983; Zabel et al., 1985). The human and bovine genes have two functional TATA transcription start sites, and the rat has only one. The two homologous TATA sequences flanking the human PTH gene direct the synthesis of two human PTH gene transcripts, both in normal parathyroid glands and in parathyroid adenomas (Igarashi et al., 1986; Kronenberg et al., 1986). The PTH genes in all species that have been cloned have two introns or intervening sequences and three exons (Kronenberg et al., 1986). Strikingly, even though fish do not have discrete parathyroid glands, they do synthesize PTH using two distinct genes that resemble mammalian PTH and that share the same exon–intron pattern found in tetrapod PTH genes (Danks et al., 2003; Gensure et al., 2004). Fish, as well as other vertebrates except mammals, also express a PTH-like gene (PTH4 or PTH-L) that bears some resemblance to PTH-related protein (PTHrP) and may contribute to calcium homeostasis as well (Suarez-Bregua et al., 2017). The elephant shark, which is a jawed cartilaginous fish, has at least two Pth genes and one Pthrp gene. The expression of several Pth gene family members in the elephant shark suggests that the Pthrp and Pth genes evolved in a common ancestor of jawed vertebrates even before the development of bone (Liu et al., 2010).

Principles of Bone Biology. https://doi.org/10.1016/B978-0-12-814841-9.00024-5

Intron A splits the 5′ untranslated sequence of the mRNA five nucleotides before the initiator methionine codon (Bell et al., 2005a,b,c). Intron B splits the fourth codon of the region that codes for the pro sequence of prepro-PTH. The three exons that result are roughly divided into three functional domains. Exon I contains the 5′ untranslated region (UTR). Exon II codes for the pre sequence, or signal peptide, and most of the pro sequence and exon III codes for PTH as well as the 3′ UTR. The structure of the PTH gene is thus consistent with the proposal that exons represent functional domains of the mRNA (Bell et al., 2005a,b,c; Kemper, 1986). Although the introns are at the same location, the large intron A in human is about twice as large as those in the rat and bovine. It is interesting that the human gene is considerably longer in both intron A and the 3′ UTR of the cDNA compared with bovine, rat, and mouse. Both introns have the characteristic splice-site elements. The second exon, containing 106 and 121 nucleotides in the human and bovine pre-mRNA, is much smaller and more homologous in size among the genes than intron A. Of interest, the elephant shark Pth1 and Pth2 genes both contain introns in the 5′ UTR and in the coding region at identical positions that are totally conserved in Pth genes from other vertebrates (Liu et al., 2010). The genes for human PTH and PTHrP are located at similar positions on sibling chromosomes 11 and 12. It is therefore likely that they arose from a common precursor by chromosomal duplication.

Promoter sequences

Regions upstream of the transcribed structural gene often determine tissue specificity and contain many of the regulatory sequences for the gene. For PTH, analysis of this region has been hampered by the lack of a parathyroid cell line. The 5 kb of DNA upstream of the start site of the human PTH gene was able to direct parathyroid gland-specific expression in transgenic mice (Hosokawa and Leahy, 1997). Computer analysis of the human PTH promoter region identified a number of consensus sequences (Kel et al., 2005). These include a sequence resembling the canonical cAMP-responsive element (CRE) 5′-TGACGTCA-3′ at position −81 with a single residue deviation. This element was fused to a reporter gene (chloramphenicol acetyltransferase, or CAT) and then transfected into different cell lines. Pharmacological agents that increase cAMP led to an increased expression of the CAT gene, suggesting a functional role for the CRE. The role of this putative CRE in the PTH gene in the parathyroid remains to be established. Specificity protein (Sp) and the nuclear factor-Y (NF-Y) complex are ubiquitously expressed transcription factors associated with basal expression of a host of gene products. Sp family members and NF-Y can cooperatively enhance transcription of a target gene. A highly conserved Sp1 DNA element present in mammalian PTH promoters has been identified and characterized (Alimov et al., 2005; Koszewski et al., 2004). Coexpression of Sp proteins and NF-Y complex leads to synergistic transactivation of the human PTH promoter, with alignment of the Sp1 DNA element essential for full activation (Alimov et al., 2005). A similar arrangement of DNA-response elements and synergism is present in the bovine PTH promoter, suggesting that this may be a conserved mechanism to enhance transcription of the PTH gene.

Several groups have identified DNA sequences that may mediate the negative regulation of PTH gene transcription by 1,25-dihydroxyvitamin D_3 (1,25$(OH)_2D_3$). Demay et al. (1992) identified DNA sequences in the human PTH gene that bind the 1,25$(OH)_2D_3$ receptor (VDR). Nuclear extracts containing the VDR were examined for binding to sequences in the 5′ flanking region of the human PTH gene. A 25-bp oligonucleotide containing sequences from −125 to −101 from the start of exon I bound nuclear proteins that were recognized by monoclonal antibodies against the VDR. The sequences in this region contained a single copy of a motif (AGGTTCA) that is homologous to the motifs repeated in the upregulatory 1,25$(OH)_2D_3$-response element of the osteocalcin gene. When placed upstream of a heterologous viral promoter, the sequences contained in this 25-bp oligonucleotide mediated transcriptional repression in response to 1,25$(OH)_2D_3$ in GH4C1 cells but not in ROS 17/2.8 cells. Therefore, this downregulatory element differs from upregulatory elements both in sequence composition and in the requirement for particular cellular factors other than the VDR for repressing PTH transcription (Demay et al., 1992). Russell et al. (1999) have shown that there are two negative vitamin D-response elements (VDREs) in the rat PTH gene, with different binding affinities to a VDR/RXR heterodimer. Transfection studies with VDRE−CAT constructs showed that they had an additive effect. Liu et al. (Alon et al., 2008; Liu et al., 1996) have identified such sequences in the chicken PTH gene and demonstrated their functionality after transfection into an opossum kidney cell line. They converted the negative activity imparted by the PTH VDRE to a positive transcriptional response through selective mutations introduced into the element. They showed that there was a p160 protein that specifically interacted with a heterodimer complex bound to the wild-type VDRE, but was absent from complexes bound to response elements associated with positive transcriptional activity.

The parathyroid hormone mRNA

Complementary DNAs encoding human, bovine, rat, mouse, pig, chicken, dog, cat, horse, macaque, fugu fish, and zebrafish PTH have all been cloned. The PTH gene is a typical eukaryotic gene with consensus sequences for initiation of RNA synthesis, RNA splicing, and polyadenylation. The primary RNA transcript consists of RNA transcribed from both

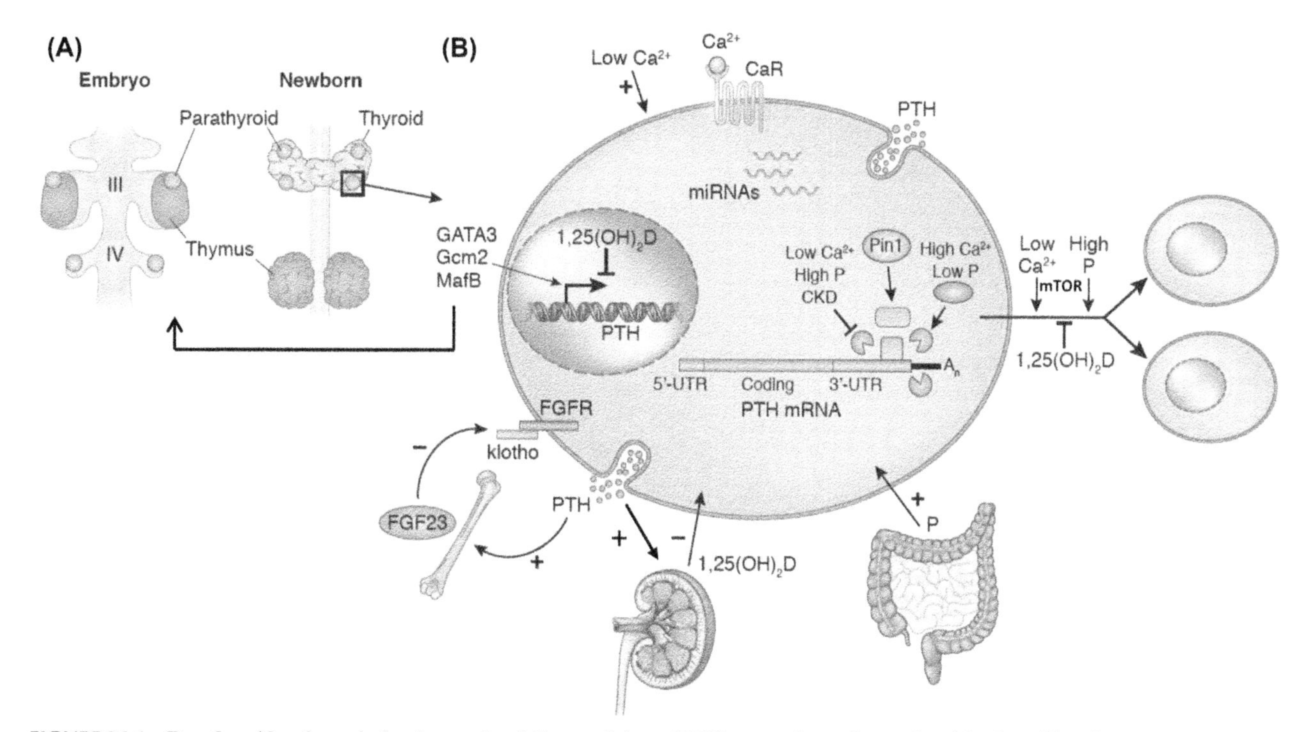

FIGURE 24.1 Parathyroid embryonic development and the regulators of PTH expression and parathyroid cell proliferation. (A) Diagram of the third and fourth pharyngeal pouches showing the embryonic location of the parathyroid- and thymus-fated domains in humans and the postnatal position of the parathyroids adjacent to the thyroid glands. (B) Schematic presentation of a parathyroid cell and daughter cells. The different regulators of PTH gene expression, secretion, and parathyroid cell proliferation are shown. Calcium, $1,25(OH)_2D_3$, the high phosphate of uremia, and FGF23 all regulate PTH secretion and PTH gene expression through transcriptional and posttranscriptional mechanisms. In prolonged hypocalcemia and uremia, there is also parathyroid cell proliferation mediated by mTOR signaling, as shown by the two small cells on the right. miRNAs are also essential for the activation of the parathyroid in secondary hyperparathyroidism. Among the factors affecting parathyroid organogenesis, GATA3, Gcm2, and MafB are all essential for normal gland development and physically interact to activate the PTH gene promoter postnatally. *$1,25(OH)_2D$*, 1,25-dihydroxyvitamin D_3; *CaR*, calcium-sensing receptor; *CKD*, chronic kidney disease; *FGF23*, fibroblast growth factor 23; *FGFR*, fibroblast growth factor receptor; *miRNAs*, microRNAs; *PTH*, parathyroid hormone; *UTR*, untranslated region; *mTOR*, mammalian target of rapamycin. *Modified with permission from* Kidney International *(Naveh-Many, T., Silver, J., 2018. Transcription factors that determine parathyroid development power PTH expression. Kidney Int. 93 (1), 7–9).*

introns and exons, and then RNA sequences derived from the introns are spliced out. The product of this RNA processing, which represents the exons, is the mature PTH mRNA, which will then be translated into prepro-PTH (Figs. 24.1 and 24.2). There is considerable identity among mammalian PTH genes, which is reflected in an 85% identity between human and bovine proteins and 75% identity between human and rat proteins. There is less identity in the 3′ noncoding region. The initiator ATG codons for the human and bovine have been identified by sequencing in vitro translation products of the mRNAs (Habener et al., 1975; Kemper et al., 1976). In the bovine sequence, the first ATG codon is the initiator codon, in accord with many other eukaryotic mRNAs (Kozak, 1991). The human and rat sequences have ATG triplets prior to the probable initiator ATG, which are present 10 nucleotides before the initiation codon and are immediately followed by a termination codon. The termination codon immediately following the codon for glutamine at position 84 of PTH indicates that there are no additional precursors of PTH with peptide extensions at the carboxyl position. A more extensive review of the structure and sequences of the PTH gene has been published elsewhere (Bell et al., 2005a,b,c).

Mutations in the parathyroid hormone gene

Rare patients have been found with abnormal PTH genes that result in hypoparathyroidism (Gafni and Levine, 2005). The PTH gene of a patient with familial isolated hypoparathyroidism (Ahn et al., 1986) has been studied by Arnold et al. (1990), and a point mutation in the hydrophobic core of the signal peptide–encoding region of prepro-PTH was identified. This T-to-C point mutation changed the codon for position 18 of the 31-amino-acid prepro sequence from cysteine to arginine, and in functional studies the mutant protein was processed inefficiently. The mutation impaired interaction of the nascent protein with signal recognition particles and the translocation machinery, and cleavage of the mutant signal sequence by solubilized signal peptidase was slow (Karaplis et al., 1995). A novel mutation of the signal peptide of the

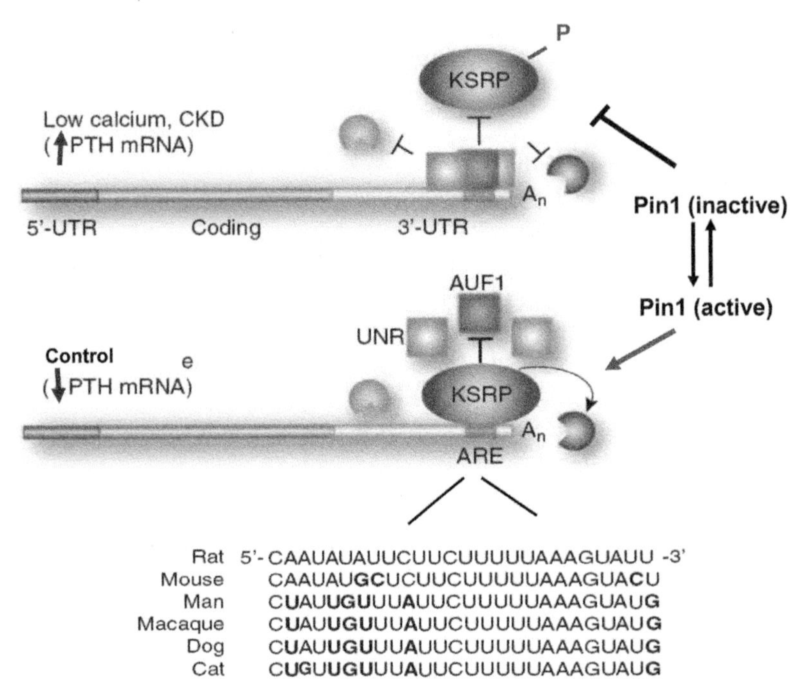

FIGURE 24.2 Model for the regulation of parathyroid hormone (*PTH*) mRNA stability by changes in calcium and kidney failure. A schematic representation of the PTH mRNA and binding proteins and the nucleotide sequence of the 26-nucleotide *cis* element in different species. Nucleotides that differ from the rat sequence are in *bold*. AU-rich binding factor (*AUF1*) and upstream of NRAS (UNR) stabilize and K-homology splicing regulator protein (*KSRP*) destabilizes PTH mRNA. Pin1 is upstream of KSRP and leads to KSRP dephosphorylation and activation. In hypocalcemic and experimental chronic kidney disease (*CKD*) rats, the enzymatic activity of the peptidyl-prolyl *cis/trans* isomerase Pin1 is reduced. As a result, KSRP is phosphorylated and hence less active. The stabilizing proteins AUF1 and UNR then bind the PTH mRNA 3′ UTR AU-rich element (*ARE*) with a greater affinity leading to increased PTH mRNA stability and levels. *Modification of a model shown in* Endocrinology *(Naveh-Many, T., 2010. Minireview: the play of proteins on the parathyroid hormone messenger ribonucleic acid regulates its expression. Endocrinology 151 (4), 1398–1402).*

prepro-PTH gene was associated with autosomal recessive familial isolated hypoparathyroidism (Sunthornthepvarakul et al., 1999). The affected members in this family presented with neonatal hypocalcemic seizures. Their intact PTH levels were undetectable during severe hypocalcemia. A replacement of thymine with a cytosine was found in the first nucleotide of position 23 in the 25-amino-acid signal peptide. This results in the replacement of the normal Ser (TCG) with a Pro (CCG). Only affected family members were homozygous for the mutant allele, whereas the parents were heterozygous, supporting autosomal recessive inheritance. Because this mutation is at the −3 position in the signal peptide of the prepro-PTH gene, the authors hypothesized that the prepro-PTH mutant might not be cleaved by signal peptidase at the normal position and might be degraded in the rough endoplasmic reticulum. The only mutation discovered in the coding region of secreted human PTH is an arginine-to-cysteine mutation at residue 25 of mature PTH(1–84). This ligand has a decreased affinity for the PTH receptor and fails to stimulate hypercalcemia when infused into mice (Lee et al., 2015). Parkinson et al. (Parkinson and Thakker, 1992) studied one kindred with autosomal recessive isolated hypoparathyroidism and identified a G-to-C substitution in the first nucleotide of intron B of the PTH gene. Restriction enzyme cleavage revealed that the patients were homozygous for mutant alleles, unaffected relatives were heterozygous, and unrelated normals were homozygous for the wild-type alleles. Defects in mRNA splicing were investigated by the detection of illegitimate transcription of the PTH gene in lymphoblastoid cells. The mutation resulted in exon skipping with a loss of exon II, which encodes the initiation codon and the signal peptide, thereby causing PTH deficiency. Goswami et al. (2004) studied 51 patients with sporadic idiopathic hypoparathyroidism. Neither the clinical manifestations nor the biochemical indexes of the disease were related to the occurrence of mutations or single-nucleotide polymorphisms in the PTH gene. Autosomal dominant hypocalcemia type 1 and type 2 are genetically distinct disorders that were found to be associated with germ-line gain-of-function mutations of CaR and G11α proteins, respectively (Hannan et al., 2016; Nesbit et al., 2013).

Development of the parathyroid

The parathyroid glands develop from a shared initial organ primordium together with the thymus. Both organs arise from the third pharyngeal pouch endoderm and surrounding neural crest cells. Humans have two additional superior parathyroid glands that develop from the fourth pouch. In mice, the third pouch forms during embryonic days (E) 9–10.5. A transcriptional network involving Hoxa3, Pax1/Pax9, Eya1, Tbx1, Sox3, and Six1/Six4 controls early pouch patterning and formation of the parathyroid/thymus common primordium (Han et al., 2015; Chojnowski et al., 2014). After third pouch formation, it becomes subdivided into two molecularly distinct domains. The parathyroid-destined domain can be distinguished from E10.5 by the expression of the transcription factor Gcm2 (glial cells missing 2), which is required for parathyroid differentiation and survival (Liu et al., 2007). Foxn1 (Forkhead box N1) expression marks the thymus domain. Each primordium separates into a single parathyroid gland and a thymic lobe at E12.5–13.5. The parathyroids are "dragged" along by the migrating thymus lobes until the separation process is complete and then move to their eventual adult locations by about E14.5, adjacent to or embedded in the thyroid glands (Kamitani-Kawamoto et al., 2011; Manley, 2015) (Fig. 24.1). Mice that lack Hoxa3, Pax1/9, Eya1, and Six1/4 have normal initial pouch formation, but then fail to form or have hypoplastic parathyroids and thymus. *Hoxa3*$^{-/-}$ mutants fail to initiate the formation of the parathyroid/thymus primordia. In these mutants, Gcm2 is expressed in its normal domain, but at very low levels, indicating that Hoxa3 upregulates Gcm2 expression. Eya1 and Six1/4 are also required for Gcm2 expression, and mutations result in the loss of parathyroid precursor cells through apoptosis. As loss of Gcm2 itself is sufficient to cause apoptosis, it is possible that the effects of all of these genes, either individually or as a pathway, are mediated by their effect on Gcm2 expression (Manley, 2015).

The DiGeorge syndrome is a human genetic condition resulting from deletion of contiguous genes. The specific gene causing the hypoparathyroidism found in DiGeorge syndrome and the closely related CATCH-22 syndrome (cardiac defects, abnormal facies, thymic hypoplasia, cleft palate, hypocalcemia, associated with chromosome 22 microdeletion) is likely to be Tbx1 (Lindsay et al., 2001). In the DiGeorge syndrome, malformations include absence of the parathyroid glands and thymus as well as the heart outflow tract, and most DiGeorge syndrome patients are hemizygous for a 1.5- to 3.0-Mb region of 22q11. In mice, deletion of one copy of Tbx1 (in the equivalent genetic region in the mouse) affects the development of the fourth pharyngeal arch arteries, whereas homozygous mutation severely disrupts the pharyngeal arch artery system (Lindsay et al., 2001; Merscher et al., 2001). These data in mice suggest that Tbx1 has a major role in the molecular etiology of the DiGeorge syndrome phenotype.

Gata3 (GATA-binding protein 3), Gcm2, and MafB (transcriptional activator v-maf musculoaponeurotic fibrosarcoma oncogene homolog B) transcription factors constitute a genetic cascade regulating parathyroid development and PTH expression (Han et al., 2015). GATA3 haploinsufficiency results in congenital hypoparathyroidism, deafness, and renal dysplasia syndrome in humans. *Gata3*$^{-/-}$ embryos lack Gcm2 expression and exhibit gross defects in the third and fourth pharyngeal pouches, including absent parathyroid–thymus primordia (Grigorieva et al., 2010). *Gata3*$^{+/-}$ mice are viable and have apparently normal parathyroid glands. However, when fed a low calcium and vitamin D diet, plasma PTH levels and calcium concentrations are lower than those in wild-type mice, indicating partial parathyroid dysfunction in the *Gata3*$^{+/-}$ mice (Grigorieva et al., 2010). Gata3 directly binds and regulates Gcm2 expression. Gcm2 is expressed predominately in the pharyngeal pouches and, at later stages, in the developing and mature parathyroid glands. Once the parathyroid domain is established, upregulation of Gcm2 expression is necessary and sufficient for parathyroid differentiation and survival (Manley, 2015). Mutations of the *GCM2* gene in humans lead to familial isolated hypoparathyroidism. Genetic ablation of *Gcm2* in mice results in parathyroid agenesis and hypoparathyroidism. Without Gcm2 function, parathyroid precursor cells fail to differentiate and then undergo apoptosis by E12, resulting in the aparathyroid phenotype (Manley, 2015). Serum PTH levels were low or undetectable in *Gcm2*-knockout mice. In conditional *Gcm2*-knockout mice, immunoreactive PTH was identified within a few thymus cells in wild-type and homozygous knockout mice, consistent with the proposal that expression of PTH in medullary thymic epithelial cells provides a source of self-antigen for negative selection. Work has excluded the thymus as a source of circulating PTH in humans and mice (Yuan et al., 2014). It is possible that the normal process of parathyroid organogenesis in both mice and humans leads to the generation of multiple small parathyroid clusters in addition to the main parathyroid glands (Manley, 2015).

MafB is a basic leucine zipper transcription factor expressed in the developing parathyroid after E11.5 and postnatally. MafB expression is lost in the parathyroid primordium of *Gcm2*-null mice. The parathyroid glands of *MafB*$^{+/-}$ mice are mislocalized between the thymus and the thyroid. In *MafB*$^{-/-}$ mice, the parathyroids do not separate from the thymus during embryological development (Kamitani-Kawamoto et al., 2011). Therefore, MafB regulates later steps of parathyroid development, the separation from the thymus, and migration toward the thyroid.

Among Gata3, Gcm2, and MafB transcription factors, Gata3 is the most upstream, followed by Gcm2, and then MafB, which is the most downstream in parathyroid development (Han et al., 2015). This cascade is critical for parathyroid development. However, expression of Gata3, Gcm2, and MafB persists after parathyroid morphogenesis (Kamitani-Kawamoto et al., 2011), suggesting that they are components of a gene regulatory program that governs parathyroid-specific gene expression and function. Gata3, Gcm2, and MafB form a transcriptional complex that mediates parathyroid-specific PTH expression. Recent in vitro analyses in heterologous cells revealed that Gata3, Gcm2, and MafB physically interact and synergistically activate the *PTH* gene promoter (Han et al., 2015). The in vivo functions of Gata3 and Gcm2 and MafB in parathyroid physiology and disease have been shown. Morito et al. (2018) showed a role for MafB in experimental secondary hyperparathyroidism. Deletion of MafB had no effect on serum PTH levels or mineral metabolism under normal conditions, but decreased serum PTH levels at postnatal day 0 in $MafB^{-/-}$ mice (Kamitani-Kawamoto et al., 2011). Stimulation of the parathyroid by prolonged uremia in $MafB^{+/-}$ mice resulted in an impaired increase in serum PTH, PTH mRNA, and parathyroid cell proliferation. Both PTH and cyclin D2, but not cyclin D1, expression was blunted in $MafB^{+/-}$ uremic mice, suggesting that MafB contributes to the increased PTH and cyclin D2 expression in secondary hyperparathyroidism. Acute hypocalcemia-induced PTH secretion and PTH mRNA were also impaired in $MafB^{+/-}$ and in global *MafB*-knockout mice. Therefore, MafB plays a role in regulation of PTH secretion but this effect may be less important than its role in controlling the development and differentiation of the parathyroid gland. It is intriguing that transcription factors that affect parathyroid morphogenesis are expressed and functional in parathyroid physiology (Fig. 24.1).

Okabe and Graham (2004) have performed elegant studies that demonstrate a role for Gcm2 even in fish, which do not have discrete parathyroid glands. They showed that the parathyroid gland of tetrapods and the gills of fish both express Gcm2 and require this gene for their formation. They also showed that the gill region expresses mRNA encoding the two PTH genes found in fish, as well as mRNA encoding the calcium-sensing receptor (CaR). The conserved role of Gcm2 in forming pharyngeal structures is established, but the relationship between Gcm2 and PTH-producing cells in fish is not clear.

Regulation of parathyroid hormone gene expression

1,25-Dihydroxyvitamin D_3

PTH regulates serum concentrations of calcium and phosphate, which, in turn, regulate the synthesis and secretion of PTH. $1,25(OH)_2D_3$ has independent effects on calcium and phosphate levels and also participates in a well-defined feedback loop between $1,25(OH)_2D_3$ and PTH.

PTH increases the renal synthesis of $1,25(OH)_2D_3$. $1,25(OH)_2D_3$ then increases blood calcium largely by increasing the efficiency of intestinal calcium absorption. $1,25(OH)_2D_3$ also potently decreases transcription of the PTH gene. This action was first demonstrated in vitro in bovine parathyroid cells in primary culture, where $1,25(OH)_2D_3$ led to a marked decrease in PTH mRNA levels (Silver et al., 1985; Russell et al., 1984) and a consequent decrease in PTH secretion (Cantley et al., 1985; Karmali et al., 1989; Chan et al., 1986). The physiological relevance of these findings was established by in vivo studies in rats (Silver et al., 1986). The localization of VDR mRNA to parathyroids was demonstrated by in situ hybridization studies of the thyroparathyroid and duodenum. VDR mRNA was localized to the parathyroids in the same concentration as in the duodenum, the classic target organ of $1,25(OH)_2D_3$ (Naveh-Many and Silver, 1990). Rats injected with amounts of $1,25(OH)_2D_3$ that did not increase serum calcium had marked decreases in PTH mRNA levels, reaching $<4\%$ of control at 48 h (Fig. 24.1). This effect was shown to be transcriptional in both in vivo studies in rats (Silver et al., 1986) and in vitro studies with primary cultures of bovine parathyroid cells (Russell et al., 1986). When 684 bp of the 5′ flanking region of the human PTH gene was linked to a reporter gene and transfected into a rat pituitary cell line (GH4C1), gene expression was lowered by $1,25(OH)_2D_3$ (Okazaki et al., 1988). These studies suggest that $1,25(OH)_2D_3$ decreases PTH transcription by acting on the 5′ flanking region of the PTH gene. The effect of $1,25(OH)_2D_3$ may involve heterodimerization with the retinoic acid receptor. This is because 9-*cis*-retinoic acid, which binds to the retinoic acid receptor, when added to bovine parathyroid cells in primary culture, led to a decrease in PTH mRNA levels. Moreover, combined treatment with retinoic acid and $1,25(OH)_2D_3$ decreased PTH secretion and prepro-PTH mRNA more effectively than either compound alone (Macdonald et al., 1994). Alternatively, retinoic acid receptors might synergize with VDRs through actions on distinct sequences.

A further level at which $1,25(OH)_2D_3$ may regulate the PTH gene would be at the level of the VDR. $1,25(OH)_2D_3$ acts on its target tissues by binding to the VDR, which regulates the transcription of genes with the appropriate recognition sequences. Concentration of the VDR in $1,25(OH)_2D_3$ target sites could allow a modulation of the $1,25(OH)_2D_3$ effect, with an increase in receptor concentration leading to an amplification of its effect and a decrease in receptor concentration dampening the $1,25(OH)_2D_3$ effect. Naveh-Many and Silver (1990) injected $1,25(OH)_2D_3$ into rats and measured the levels of VDR mRNA and PTH mRNA in the thyroparathyroid tissue. They showed that $1,25(OH)_2D_3$ in physiologically relevant doses led to an increase in VDR mRNA levels in the parathyroid glands in contrast to the decrease in PTH mRNA levels. This increase in VDR mRNA occurred after a time lag of 6 h, and a dose response showed a peak at 25 pmol.

Weanling rats fed a diet deficient in calcium were markedly hypocalcemic at 3 weeks and had very high serum $1,25(OH)_2D_3$ levels. Despite the chronically high serum $1,25(OH)_2D_3$ levels, there was no increase in VDR mRNA levels; furthermore, PTH mRNA levels did not fall and were increased markedly. The low calcium in the bloodstream may have prevented the increase in parathyroid VDR levels, which may partially explain PTH mRNA suppression. Whatever the mechanism, the lack of suppression of PTH synthesis in the setting of hypocalcemia and increased serum $1,25(OH)_2D_3$ is crucial physiologically because it allows an increase in both PTH and $1,25(OH)_2D_3$ at a time of chronic hypocalcemic stress. Russell et al. (1993) studied the parathyroids of chicks with vitamin D deficiency and confirmed that $1,25(OH)_2D_3$ regulates PTH and VDR gene expression in the avian parathyroid gland. Brown et al. (1995) studied vitamin D-deficient rats and confirmed that $1,25(OH)_2D_3$ upregulated parathyroid VDR mRNA. Rodriguez et al. (2007) showed that administration of the calcimimetic R568 resulted in increased VDR expression in parathyroid tissue. In vitro studies of the effect of R568 on VDR mRNA and protein were conducted in cultures of whole rat parathyroid glands. Incubation of rat parathyroid glands in vitro with R568 resulted in a dose-dependent decrease in PTH secretion and an increase in VDR expression. Together with previous work on the effect of extracellular calcium to increase parathyroid VDR mRNA in vitro (Garfia et al., 2002), they concluded that activation of the CaR upregulates the parathyroid VDR mRNA.

All these studies show that $1,25(OH)_2D_3$, and calcium in certain circumstances, increases the expression of the VDR gene in the parathyroid gland, which would result in increased VDR protein synthesis and increased binding of $1,25(OH)_2D_3$. This ligand-dependent receptor upregulation would lead to an amplified effect of $1,25(OH)_2D_3$ on the PTH gene and might help explain the dramatic effect of $1,25(OH)_2D_3$ on the PTH gene.

Vitamin D may also amplify its effect on the parathyroid by increasing the activity of the calcium receptor (CaR). Canaff et al. (Canaff and Hendy, 2002) showed that in fact there are VDREs in the human CaR promoter. The CaR, expressed in parathyroid chief cells, thyroid C cells, and cells of the kidney tubule, is essential for maintenance of calcium homeostasis. They showed that parathyroid, thyroid, and kidney CaR mRNA levels increased twofold at 15 h after intraperitoneal injection of $1,25(OH)_2D_3$ in rats. Human thyroid C-cell (TT) and kidney proximal tubule cell (HKC) CaR gene transcription increased approximately twofold at 8 and 12 h after $1,25(OH)_2D_3$ treatment. The human CaR gene has two promoters yielding alternative transcripts containing either exon IA or exon IB 5′ UTR sequences that splice to exon II some 242 bp before the ATG translation start site. $1,25(OH)_2D_3$ stimulated P1 activity 2-fold and P2 activity 2.5-fold. VDREs, in which the half-sites (6 bp) are separated by three nucleotides, were identified in both promoters and shown to confer $1,25(OH)_2D_3$ responsiveness to a heterologous promoter. This responsiveness was lost when the VDREs were mutated. In electrophoretic mobility-shift assays specific protein–DNA complexes were formed in the presence of $1,25(OH)_2D_3$ on oligonucleotides representing the P1 and P2 VDREs. In summary, functional VDREs have been identified in the CaR gene and probably provide the mechanism whereby $1,25(OH)_2D_3$ upregulates parathyroid, thyroid C-cell, and kidney CaR expression.

The use of $1,25(OH)_2D_3$ is limited by its hypercalcemic effect, and therefore a number of $1,25(OH)_2D_3$ analogs have been synthesized that are biologically active but are less hypercalcemic than $1,25(OH)_2D_3$ (Brown, 2005). These analogs usually involve modifications of the $1,25(OH)_2D_3$ side chain, such as 22-oxa-$1,25(OH)_2D_3$, which is the chemical modification in oxacalcitriol (Nishii et al., 1991), or a cyclopropyl group at the end of the side chain in calcipotriol (Kissmeyer and Binderup, 1991). However, detailed in vivo dose–response studies showed that in vivo $1,25(OH)_2D_3$ is the most effective analog for decreasing PTH mRNA levels, even at doses that do not cause hypercalcemia (Naveh-Many and Silver, 1993). The marked activity of $1,25(OH)_2D_3$ analogs in vitro compared with their modest hypercalcemic actions in vivo probably reflects their rapid clearance from the circulation (Bouillon et al., 1991).

The ability of $1,25(OH)_2D_3$ to decrease PTH gene transcription is used therapeutically in the management of patients with chronic renal failure. They are treated with $1,25(OH)_2D_3$, or its prodrug $1\alpha(OH)$-vitamin D_3, to prevent the secondary hyperparathyroidism of chronic renal failure. The poor response in some patients may well result from poor control of serum phosphate, decreased VDR concentration in the patients' parathyroids (Fukuda et al., 1993), an inhibitory effect of a uremic toxin(s) on VDR–VDRE binding (Patel and Rosenthal, 1985), or tertiary hyperparathyroidism with monoclonal parathyroid tumors (Arnold et al., 1995).

Another possible level at which $1,25(OH)_2D_3$ may regulate PTH gene expression involves calreticulin. Calreticulin is a calcium-binding protein present in the endoplasmic reticulum of the cell and may have an additional nuclear function. It regulates gene transcription via its ability to bind a protein motif in the DNA-binding domain of nuclear hormone receptors of sterol hormones. It has been shown to prevent vitamin D's binding and action on the osteocalcin gene in vitro (Wheeler et al., 1995). Sela-Brown et al. (1998) showed that calreticulin inhibits the action of vitamin D on the PTH gene. Hypocalcemic rats had increased levels of calreticulin protein in their parathyroid nuclear fraction, which may explain why hypocalcemia leads to increased PTH gene expression, despite high serum $1,25(OH)_2D_3$ levels.

Calcium

In vitro studies

A remarkable characteristic of the parathyroid is its sensitivity to small changes in serum calcium, which leads to large changes in PTH secretion. This remarkable sensitivity of the parathyroid to increase hormone secretion after small decreases in serum calcium levels is unique to the parathyroid. All other endocrine glands increase hormone secretion after exposure to high extracellular calcium. Calcium sensing also regulates PTH gene expression and parathyroid cell proliferation (Fig. 24.1). In vitro and in vivo data agree that calcium regulates PTH mRNA levels, but data differ in important ways. In vitro studies with bovine parathyroid cells in primary culture showed that calcium regulated PTH mRNA levels (Russell et al., 1983; Brookman et al., 1986), with an effect mainly of high calcium to decrease PTH mRNA. These effects were most pronounced after more prolonged incubations, such as 72 h. The physiologic correlates of these studies in tissue culture are hard to ascertain, as the parathyroid calcium sensor may well have decreased over the time period of the experiment (Mithal et al., 1995). This may explain why the dose response differs from in vivo data, but the dramatic difference in time course suggests that in vivo data reflect something not seen in cultured cells.

In vivo studies

Calcium and phosphate both have marked effects on the levels of PTH mRNA and secretion in vivo. The major effect is for low calcium to increase PTH mRNA levels and low phosphate to decrease PTH mRNA levels. Naveh-Many et al. (1989) showed that a small decrease in serum calcium from 2.6 to 2.1 mmol/L led to large increases in PTH mRNA levels, reaching threefold that of controls at 1 and 6 h. A high serum calcium had no effect on PTH mRNA levels even at concentrations as high as 6.0 mmol/L. Interestingly, in these same thyroparathyroid tissue RNA extracts, calcium had no effect on the expression of the calcitonin gene (Naveh-Many et al., 1989, 1992a,b). Thus, while high calcium is a secretagogue for calcitonin, it does not regulate calcitonin gene expression. Yamamoto et al. (Yamamoto et al., 1989) also studied the in vivo effect of calcium on PTH mRNA levels in rats. They showed that hypocalcemia induced by a calcitonin infusion for 48 h led to a sevenfold increase in PTH mRNA levels. Rats made hypercalcemic (2.9−3.4 mM) for 48 h had the same PTH mRNA levels as controls that had received no infusion (2.5 mM); these levels were modestly lower than those found in rats that had received a calcium-free infusion. In further studies, Naveh-Many et al. (1992a,b) transplanted Walker carcinosarcoma 256 cells into rats. Serum calcium levels increased to 18 mg/dL at day 10 after transplantation. There was no change in PTH mRNA levels in these rats with marked chronic hypercalcemia (Naveh-Many et al., 1992a,b). Differences between in vivo and in vitro results probably reflect the instability of the in vitro system. Nevertheless, the physiological conclusion is that common causes of hypercalcemia in vivo do not importantly decrease PTH mRNA levels; these results emphasize that the gland is geared to respond to hypocalcemia and not hypercalcemia.

The mechanism whereby calcium regulates PTH gene expression is particularly interesting. Changes in extracellular calcium are sensed by a calcium sensor that then regulates PTH secretion (Brown et al., 1993; Yano and Brown, 2005). Signal transduction from the CaR involves activation of phospholipase C, D, and A_2 enzymes (Kifor et al., 1997). It is not known what mechanism transduces the message of changes in extracellular calcium leading to changes in PTH mRNA. However, it has been shown that the response to changes in serum calcium involves the protein phosphatase type 2B, calcineurin (Bell et al., 2005a,b,c). In vivo and in vitro studies demonstrated that inhibition of calcineurin by genetic manipulation or pharmacologic agents affected the response of PTH mRNA levels to changes in extracellular calcium (Bell et al., 2005a,b,c).

Okazaki et al. (1992) identified a negative calcium regulatory element (nCaRE) in the atrial natriuretic peptide gene, with a homologous sequence in the PTH gene. They identified a redox factor protein (ref1), which bound a putative nCaRE, and the level of ref1 mRNA and protein were elevated by an increase in extracellular calcium concentration. They suggested that ref1 had transcription repressor activity in addition to its function as a transcriptional auxiliary protein (Okazaki et al., 1992). Because no parathyroid cell line is available, these studies were performed in nonparathyroid cells, so their relevance to physiologic PTH gene regulation remains to be established.

Moallem et al. (1998) have performed in vivo studies on the effect of hypocalcemia on PTH gene expression. The effect is posttranscriptional in vivo and involves protein−RNA interactions at the 3′ UTR of the PTH mRNA (Moallem et al., 1998). A similar mechanism is involved in the effect of phosphate on PTH gene expression so the mechanisms involved will be discussed later in this chapter.

Phosphate

The demonstration of a direct effect of high phosphate on the parathyroid, independent of calcium and $1,25(OH)_2D_3$, has been difficult. One of the reasons is that the various maneuvers used to increase or decrease serum phosphate invariably lead to a change in the ionized calcium concentration. In moderate renal failure, phosphate clearance decreases and serum phosphate increases; this increase becomes an important problem in severe renal failure. Hyperphosphatemia has always been considered central to the pathogenesis of secondary hyperparathyroidism, but it has been difficult to separate the effects of hyperphosphatemia from those of the attendant hypocalcemia and decrease in serum $1,25(OH)_2D_3$ levels. In the 1970s, Slatopolsky and Bricker (1973) showed in dogs with experimental chronic renal failure that dietary phosphate restriction prevented secondary hyperparathyroidism. Clinical studies demonstrated that phosphate restriction in patients with chronic renal insufficiency is effective in preventing the increase in serum PTH levels (Lucas et al., 1986; Portale et al., 1984; Aparicio et al., 1994; Lafage et al., 1992; Combe and Aparicio, 1994; Combe et al., 1995). The mechanism of this effect was not clear, although at least part of it was considered to be due to changes in serum $1,25(OH)_2D_3$ concentrations. In vitro (Tanaka and DeLuca, 1973; Condamine et al., 1994) and in vivo (Portale et al., 1984, 1989) phosphate directly regulates the production of $1,25(OH)_2D_3$. A raised serum phosphate decreases serum $1,25(OH)_2D_3$ levels, which then leads to decreased calcium absorption from the diet and eventually a low serum calcium. The raised phosphate complexes calcium, which is then deposited in bone and soft tissues and decreases serum calcium. However, a number of careful clinical and experimental studies suggested that the effect of phosphate on serum PTH levels was independent of changes in both serum calcium and $1,25(OH)_2D_3$ levels. In experimental chronic renal failure in dogs, phosphate increased parathyroid cell activity by a mechanism independent of its effect on serum $1,25(OH)_2D_3$ and calcium levels (Lopez-Hilker et al., 1990). Therefore, phosphate plays a central role in the pathogenesis of secondary hyperparathyroidism, by its effects on both serum $1,25(OH)_2D_3$ and serum calcium levels and possibly independently. A raised serum phosphate also stimulates the secretion of fibroblast growth factor 23 (FGF23), which in turn decreases PTH gene expression and serum PTH levels. This effect would act as a counterbalance to the stimulatory effect of phosphate on the parathyroid and is discussed separately in this chapter.

Kilav et al. (1995) were the first to establish in vivo that the effects of serum phosphate on PTH gene expression and serum PTH levels were independent of any changes in serum calcium or $1,25(OH)_2D_3$ (Fig. 24.1). They bred second-generation vitamin D-deficient rats and then placed the weanling vitamin D-deficient rats on a diet with no vitamin D, low calcium, and low phosphate. After one night of this diet, serum phosphate had decreased markedly with no changes in serum calcium or $1,25(OH)_2D_3$. These rats with isolated hypophosphatemia had marked decreases in PTH mRNA levels and serum PTH (Kilav et al., 1995). To establish that the effect of serum phosphate on the parathyroid was indeed a direct effect, in vitro confirmation was needed, which was provided by three groups. Rodriguez et al. showed that increased phosphate levels increased PTH secretion from isolated parathyroid glands in vitro; the effect required maintenance of tissue architecture (Almaden et al., 1996). The effect was found in whole glands or tissue slices but not in isolated cells. This result was confirmed by Slatopolsky et al. (1996). Olgaard's laboratory provided elegant further evidence of the importance of cell–cell communication in mediating the effect of phosphate on PTH secretion (Nielsen et al., 1996). The requirement for intact tissue suggests either that the sensing mechanism for phosphate is damaged during the preparation of isolated cells or that the intact gland structure is important to the phosphate response.

The parathyroid responds to changes in serum phosphate at the level of secretion, gene expression, and cell proliferation, although the signaling involved is unknown. The effect of high phosphate to increase PTH secretion may be mediated by phospholipase A_2-activated signal transduction. Bourdeau et al. (1992, 1994) showed that arachidonic acid and its metabolites inhibit PTH secretion. Almaden et al. (2000) showed in vitro that a high-phosphate medium increased PTH secretion, which was prevented by the addition of arachidonic acid.

Protein–PTH mRNA interactions determine the posttranscriptional regulation of PTH gene expression by calcium, phosphate, and uremia

Diet-induced hypocalcemia and adenine/high phosphorus–induced chronic kidney disease (CKD) increase PTH mRNA levels, and diet-induced hypophosphatemia decreases PTH mRNA levels (Fig. 24.1). In both instances, the effect is posttranscriptional, as shown by nuclear transcript run-on experiments (Moallem et al., 1998; Kilav et al., 2005). Parathyroid cytosolic proteins bind in vitro-transcribed PTH mRNA. Interestingly, this binding was increased with parathyroid proteins from hypocalcemic and CKD rats (with increased PTH mRNA levels) and decreased with parathyroid proteins from hypophosphatemic rats (with decreased PTH mRNA levels). Proteins from other tissues bound to PTH mRNA, but only parathyroid proteins bound PTH mRNA in a way that was regulated by calcium and phosphate. Intriguingly, binding required the presence of the terminal 60 nucleotides of the PTH transcript.

An in vitro degradation assay showed that the effects of hypocalcemic and hypophosphatemic parathyroid proteins on PTH mRNA stability reproduced the differences in PTH mRNA levels observed in vivo (Moallem et al., 1998). Moreover, the difference in RNA stability stimulated by the parathyroid extracts was dependent on an intact 3′ UTR and, in particular, on the terminal 60 nucleotides. Proteins from other tissues in these rats subjected to hypocalcemia, hypophosphatemia, or uremia did not affect PTH mRNA stability in this in vitro assay. Therefore, calcium, phosphate, and CKD change the properties of parathyroid cytosolic proteins, which bind specifically to the PTH mRNA 3′ UTR and determine its stability (Fig. 24.2). What are these proteins?

A conserved sequence in the PTH mRNA 3′ UTR binds parathyroid cytosolic proteins and determines mRNA stability

We have identified the minimal sequence for protein binding in the PTH mRNA 3′ UTR and determined its functionality (Fig. 24.2) (Kilav et al., 2001). A minimum sequence of 26 nucleotides was sufficient for PTH RNA–protein binding. To study the functionality of the sequence in the context of another RNA, a 63-bp cDNA PTH sequence consisting of the 26-nucleotide and flanking regions was fused to growth hormone cDNA. The conserved PTH RNA protein-binding region was necessary and sufficient for responsiveness to calcium and phosphate and determines PTH mRNA stability and levels (Kilav et al., 2001).

The PTH mRNA 3′-UTR-binding element is AU rich and is a type III AU-rich element (ARE). Sequence analysis of the PTH mRNA 3′ UTR of different species revealed a preservation of the 26-nucleotide protein-binding element in rat, murine, human, macaque, feline, and canine 3′ UTRs (Fig. 24.2) (Kilav et al., 2001; Bell et al., 2005a,b,c). In contrast to protein coding sequences that are highly conserved, UTRs are less conserved. The conservation of the protein-binding element in the PTH mRNA 3′ UTR suggests that this element represents a functional unit that has been evolutionarily conserved. The *cis*-acting element is at the 3′ distal end in all species in which it is expressed.

The PTH mRNA 3′-UTR-binding proteins that determine PTH mRNA stability

AU-rich binding factor

Affinity chromatography using the PTH RNA 3′ UTR identified AU-rich-binding factor (AUF1) as a PTH mRNA-binding protein (Sela-Brown et al., 2000; Brewer, 1991). Added recombinant AUF1 stabilized the PTH transcript in the in vitro degradation assay. Therefore, AUF1 is a protein that binds to the PTH mRNA 3′ UTR and stabilizes the PTH transcript. The regulation of protein–PTH mRNA binding involves posttranslational modification of AUF1 (Bell et al., 2005a,b,c; Sela-Brown et al., 2000; Levi et al., 2006). The balance between the stabilizing and the destabilizing proteins determines mRNA levels in response to physiological stimuli (Fig. 24.2) (Brewer, 1991; Naveh-Many, 2010; Nechama et al., 2008; Barreau et al., 2006).

K-homology splicing regulator protein

The mRNA decay-promoting protein K-homology splicing regulator protein (KSRP) binds to PTH mRNA in intact parathyroid glands and in transfected cells (Nechama et al., 2008, 2009; Nechama et al., 2009a,b,c; Gherzi et al., 2004). This binding of KSRP is decreased in glands from calcium-depleted or experimental uremic rats in which PTH mRNA is more stable, compared with parathyroid glands from control and phosphorus-depleted rats in which PTH mRNA is less stable. The differences in KSRP–PTH mRNA binding counter those of AUF1. PTH mRNA decay depends on the KSRP-recruited exosome in parathyroid extracts. In transfected cells, KSRP overexpression and knockdown experiments showed that KSRP decreases PTH mRNA stability and steady-state levels through the PTH mRNA ARE. Overexpression of the PTH mRNA-stabilizing protein AUF1 blocks KSRP–PTH mRNA binding and partially prevents the KSRP-mediated decrease in PTH mRNA levels (Nechama et al., 2008). Therefore, calcium or phosphorus depletion, as well as CKD, regulates the interaction of KSRP and AUF1 with PTH mRNA and its half-life. The balance between the stabilizing and the destabilizing proteins determines PTH mRNA levels in response to physiological stimuli (Fig. 24.2).

Most patients with CKD develop secondary hyperparathyroidism with disabling systemic complications. Calcimimetic agents are effective tools in the management of secondary hyperparathyroidism, acting through allosteric modification of the CaR on the parathyroid gland to decrease PTH secretion and parathyroid cell proliferation. R568 decreased both PTH mRNA and serum PTH levels in adenine/high phosphorus-induced uremia (Nechama et al., 2009a,b,c). The effect of the calcimimetic, similar to that of uremia on PTH gene expression, was posttranscriptional and correlated with differences in protein—RNA binding and posttranslational modifications of AUF1 in the parathyroid. AUF1 modifications were reversed compared with those of normal rats by treatment with R568. Therefore, uremia and activation of the CaR mediated by calcimimetics modify AUF1 posttranslationally. These modifications in AUF1 correlate with changes in protein—PTH mRNA binding and PTH mRNA levels (Nechama et al., 2009a,b,c). In addition, KSRP—PTH mRNA binding was decreased in parathyroids from rats with adenine-induced CKD, a condition in which PTH mRNA is more stable. KSRP—PTH mRNA binding was increased by treatment with R568, correlating with decreased PTH gene expression. This destabilizing effect of R568 was dependent on KSRP and the PTH mRNA 3′ UTR. Therefore, the calcimimetic R568 decreases PTH mRNA levels by altering the balance between KSRP and AUF binding to the PTH mRNA 3′ UTR.

The peptidyl-prolyl isomerase Pin1 determines parathyroid hormone mRNA stability and levels in secondary hyperparathyroidism

Pin1 activity is decreased in parathyroid protein extracts from hypocalcemic and uremic rats. Pharmacologic inhibition of Pin1 increases PTH mRNA levels posttranscriptionally in vivo in the parathyroid and in transfected cells (Nechama et al., 2009a,b,c). Pin1 regulates PTH mRNA stability and levels through the PTH mRNA 3′ UTR *cis*-acting element. Pin1 interacts with the PTH mRNA destabilizing protein, KSRP, and leads to KSRP dephosphorylation and activation. In the parathyroid, Pin1 inhibition in secondary hyperparathyroidism decreases KSRP—PTH mRNA interaction that contributes to the increased PTH gene expression. Furthermore, *Pin1*$^{-/-}$ mice had increased serum PTH and PTH mRNA levels. Therefore, Pin1 determines basal PTH expression in vivo and in vitro, and decreased Pin1 activity correlates with increased PTH mRNA levels in CKD and hypocalcemic rats (Nechama et al., 2009a,b,c). These results demonstrate that Pin1 is a key mediator of PTH mRNA stability and indicate a role for Pin1 in the pathogenesis of the secondary hyperparathyroidism of CKD (Fig. 24.2) (Naveh-Many, 2010; Nechama et al., 2009a,b,c; Kumar, 2009).

Dynein light-chain M_r 8000 binds the PTH mRNA 3′ untranslated region and mediates its association with microtubules

mRNA expression cloning identified dynein light-chain M_r 8000 (LC8) as a 3′-UTR PTH mRNA-binding protein. LC8 is part of the cytoplasmic dynein complexes that function as molecular motors that translocate along microtubules. Recombinant LC8 bound the PTH mRNA 3′ UTR as assessed by RNA mobility-shift assay. PTH mRNA colocalized with polymerized microtubules in the parathyroid gland, as well as with a purified microtubule preparation from calf brain, and this was mediated by LC8 (Epstein et al., 2000). This was the first report of a dynein complex protein binding an mRNA and acting as a motor for the transport and localization of mRNAs in the cytoplasm and the subsequent asymmetric distribution of translated proteins in the cell.

MicroRNA in the parathyroid

Parathyroid-specific deletion of Dicer-dependent microRNA

A further level of posttranscriptional regulation of PTH gene expression is by microRNA (miRNA). miRNAs downregulate gene expression and have vital roles in biology but their functions in the parathyroid have been unexplored. The final step in miRNA maturation is cleavage by Dicer protein in the cytoplasm (Lee et al., 2003). We generated parathyroid-specific Dicer 1-knockout (PT-*Dicer*$^{-/-}$) mice, in which parathyroid miRNA maturation is blocked only in the parathyroid (Shilo et al., 2015). Despite normal basal PTH, under conditions of stress, deletion of Dicer and the subsequent absence of mature miRNA in the mouse parathyroid had surprising effects on parathyroid function. Remarkably, the PT-*Dicer*$^{-/-}$ mice did not increase serum PTH in response to acute hypocalcemia compared with the greater than fivefold increase in controls. PT-*Dicer*$^{-/-}$ glands cultured in low-calcium medium secreted fivefold less PTH at 1.5 h than controls. Chronic hypocalcemia increased serum PTH greater than fourfold less in PT-*Dicer*$^{-/-}$ mice compared with control mice, with no increase in PTH mRNA levels and parathyroid cell proliferation compared with the two- to threefold increase in hypocalcemic controls (Shilo et al., 2015). The importance of our findings is highlighted by the fact that uremic PT-*Dicer*$^{-/-}$ mice with normal serum calcium had an impaired increase in serum PTH similar to their failure to respond to hypocalcemia (Shilo et al., 2015). Therefore, parathyroid miRNAs are necessary for the development of secondary hyperparathyroidism not only of chronic hypocalcemia but also of CKD (Holmes, 2015) (Fig. 24.3). In contrast to the impaired increase in PTH by hypocalcemia and uremia, the PT-*Dicer*$^{-/-}$ mice decreased serum PTH as expected after activation of the parathyroid CaR by both hypercalcemia and a calcimimetic, demonstrating that these processes that suppress PTH secretion are dicer independent. In conclusion, miRNAs are essential for activation of the parathyroid by both acute and chronic hypocalcemia and uremia, the major stimuli for PTH secretion (Shilo et al., 2015, 2016) (Fig. 24.3).

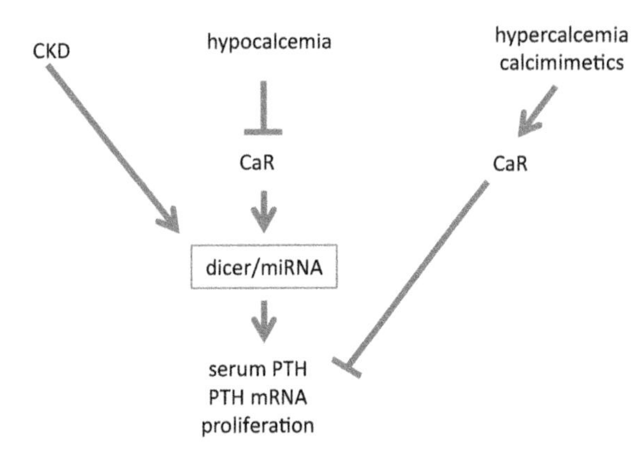

FIGURE 24.3 Model for the role of microRNAs (*miRNA*) in stimulation of the parathyroid by hypocalcemia and chronic kidney disease (*CKD*) but not in suppression of the parathyroid by hypercalcemia or a calcimimetic. Parathyroid miRNAs are essential for the increase in parathyroid gland activity by acute and chronic hypocalcemia and uremia, mediating the increase in parathyroid hormone (*PTH*) mRNA levels, PTH secretion, and parathyroid cell proliferation. Suppression of parathyroid activity by hypercalcemia or a calcimimetic is miRNA independent. *CaR*, calcium-sensing receptor. *With permission from* Current Opinion in Nephrology and Hypertension *(Shilo, V., Silver, J., Naveh-Many, T., 2016. Micro-RNAs in the parathyroid: a new portal in understanding secondary hyperparathyroidism. Curr. Opin. Nephrol. Hypertens. 25 (4), 271–277).*

Let-7 and miRNA-148 regulate parathyroid hormone levels in secondary hyperparathyroidism

miRNA profiling of parathyroid glands from mice and rats with experimental CKD and dialysis patients by miRNA deep sequencing showed that human and rodent parathyroids share similar profiles (Shilo et al., 2017). Parathyroids from uremic and normal rats segregated on the basis of their miRNA expression profiles, and a similar finding was observed in humans. There were several parathyroid miRNAs that were dysregulated in experimental secondary hyperparathyroidism, including miR-29, miR-21, miR-148, miR-30, and miR-141 (upregulated) and miR-10, miR-125, and miR-25 (downregulated). Inhibition of the abundant let-7 family increased PTH secretion in normal and CKD rats, as well as in mouse parathyroid organ cultures. Conversely, inhibition of the upregulated miR-148 family prevented the increase in serum PTH level in CKD rats and decreased levels of secreted PTH in parathyroid cultures. The evolutionary conservation of abundant miRNAs in normal parathyroid glands and the regulation of these miRNAs in secondary hyperparathyroidism indicate their importance for parathyroid function and the development of hyperparathyroidism. Specifically, let-7 and miR-148 antagonism modified PTH secretion in vivo and in vitro, implying roles for these specific miRNAs in the parathyroid (Shilo et al., 2017).

Sex steroids

PTH is anabolic to bone and is an effective means of treating postmenopausal osteoporosis. In postmenopausal women with osteoporosis, time series analysis has shown that there is a loss in the periodicity of PTH secretion (Prank et al., 1995; Fraser et al., 1998). This suggests that estrogens may have an effect on the parathyroid. Estradiol and progesterone both increased the secretion of PTH from bovine parathyroid cells in primary culture (Greenberg et al., 1987). However, transdermal estrogen did not increase serum PTH levels in postmenopausal patients (Prince et al., 1990). Estrogen receptors were not detected in parathyroid tissue by a hormone-binding method (Prince et al., 1991), but were detected by immunohistochemistry and PCR for the estrogen receptor mRNA (Naveh-Many et al., 1992a,b). Moreover, in vivo in ovariectomized rats, both estrogen and progestins regulated PTH gene expression (Naveh-Many et al., 1992a,b).

Further studies were performed on the effects of progestins on PTH gene expression (Epstein et al., 1995). The 19-nor progestin R5020 given to weanling rats or mature ovariectomized rats led to a twofold increase in thyroparathyroid PTH mRNA levels. In addition, in vitro, in primary cultures of bovine parathyroid cells, progesterone increased PTH mRNA levels. The progesterone receptor mRNA was demonstrated in rat parathyroid tissue by in situ hybridization and in human parathyroid adenoma by immunohistochemistry. PTH mRNA levels varied during the rat estrous cycle (Epstein et al., 1995). These results confirm that the parathyroid gland is a target organ for the ovarian sex steroids estrogen and progesterone and they are of physiological relevance, as shown by the changes during estrus. In a rat model of CKD with ovariectomy, estrogen treatment decreased PTH mRNA and serum levels, unlike the increase after estrogen in control rats (Epstein et al., 1995; Carrillo-López et al., 2009). In the CKD rats, estrogens significantly increased FGF23 mRNA and serum levels, suggesting that estrogens may regulate PTH indirectly through FGF23 (Carrillo-López et al., 2009).

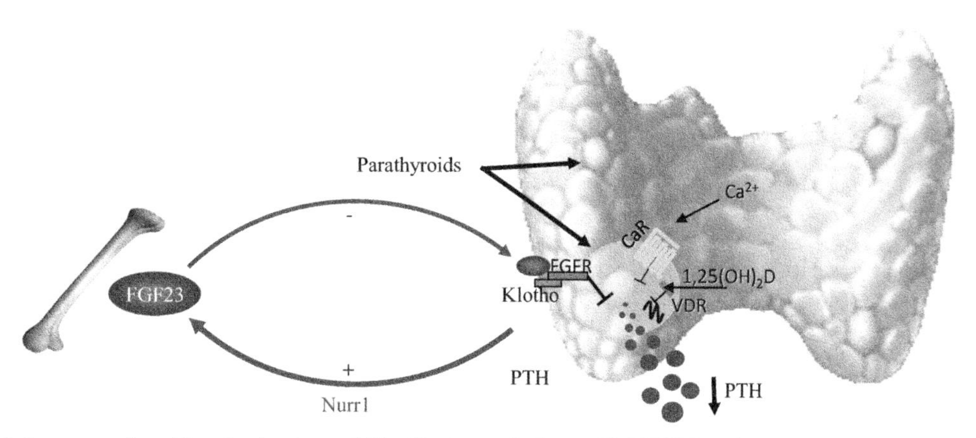

FIGURE 24.4 A bone–parathyroid endocrine loop of fibroblast growth factor 23 (*FGF23*) and parathyroid hormone (*PTH*). Parathyroid function is decreased by calcium through the calcium sensing receptor (*CaR*) and $1,25(OH)_2D_3$ through the vitamin D receptor (*VDR*). FGF23 is secreted by bone and acts on the parathyroid to also decrease PTH synthesis, secretion, and parathyroid cell proliferation by acting on the parathyroid klotho–FGFR1c receptor. PTH in turn increases FGF23 expression by activating the Nurr1 transcription factor in osteoblasts and osteocytes. *Modified from* Kidney International *with permission (Silver, J., Naveh-Many, T., 2009. Phosphate and the parathyroid. Kidney Int. 75 (9), 898–905).*

Fibroblast growth factor 23

Fibroblast growth factor 23 decreases parathyroid hormone expression

Phosphate homeostasis is maintained by a counterbalance between efflux from the kidney and influx from intestine and bone. FGF23 is a bone-derived phosphaturic hormone that acts on the kidney to increase phosphate excretion and suppress biosynthesis of $1,25(OH)_2D_3$. FGF23 signals through FGF receptors (FGFRs) bound by the transmembrane protein klotho (Kurosu et al., 2006; Urakawa et al., 2006). Klotho protein is expressed not only in the kidney but also in the parathyroid, pituitary, and sinoatrial node (Takeshita et al., 2004). Ben Dov et al. (2007) identified the parathyroid as a target organ for FGF23 in rats (Fig. 24.4). The parathyroid gland expressed klotho and FGFRs. The administration of recombinant FGF23 led to an increase in parathyroid klotho levels. In addition, FGF23 activated the mitogen-activated protein kinase (MAPK) pathway in the parathyroid through extracellular signal-regulated kinase 1/2 phosphorylation and increased Egr1 (early growth response) mRNA levels. FGF23 suppressed PTH secretion and PTH gene expression both in vivo in rats and in vitro in parathyroid cultures (Ben Dov et al., 2007). These data indicate that FGF23 acts directly on the parathyroid through the MAPK pathway to decrease serum PTH. Krajisnik et al. (2007) showed similar results using bovine parathyroid cells in primary culture. Interestingly, they also showed that FGF23 led to a dose-dependent increase in the expression of the 1α-hydroxylase enzyme in the parathyroid (Krajisnik et al., 2007). The increased $1,25(OH)_2D_3$ may then act in an autocrine manner to decrease PTH gene transcription. In addition, PTH increases FGF23 expression in bone (Lavi-Moshayoff et al., 2010). PTH increases the levels of the orphan nuclear receptor, Nurr1, in osteocytes and osteoblasts in vivo and in osteoblast-like UMR106 cells, which then acts on the FGF23 promoter to increase FGF23 transcription (Meir et al., 2014). The bone–parathyroid endocrine feedback loop of PTH and FGF23 adds a new dimension to the understanding of mineral homeostasis (Fig. 24.4).

Resistance of the parathyroid to FGF23 in chronic kidney disease

In CKD both serum FGF23 and serum PTH levels are increased. Several studies have shown that a decrease in klotho–FGFR1 expression and signal transduction may explain the resistance of the parathyroid to FGF23 in CKD. In experimental CKD, quantitative immunohistochemistry and qRT-PCR using laser capture microscopy showed that klotho and FGFR1 protein and mRNA levels were decreased in parathyroid sections of rats with adenine diet–induced advanced CKD (Galitzer et al., 2010). Similar results have been shown using the parathyroids from patients with advanced CKD (Komaba et al., 2010). Moreover, in parathyroids of rats with advanced CKD, recombinant FGF23 failed to decrease serum PTH or activate the MAPK pathway. In rat parathyroid organ culture, FGF23 decreased secreted PTH and PTH mRNA levels in control or early CKD rats but not in rats with advanced CKD (Galitzer et al., 2010). Therefore, in advanced experimental CKD, there is a decrease in parathyroid klotho and FGFR1 mRNA and protein levels in rats and in patients with CKD. This decrease corresponds with the resistance of the parathyroid to FGF23 in vivo, which is sustained in parathyroid organ culture in vitro. The increased levels of FGF23 do not decrease PTH levels in established CKD because of a downregulation of its receptor heterodimer complex klotho–FGFR1c. Olauson et al. (2013) showed by genetic and functional studies that a klotho-independent, calcineurin-mediated FGF23 signaling pathway in parathyroid glands can mediate suppression of PTH in the absence of klotho. The presence of klotho-independent FGF23 effects in a klotho-expressing target organ represents a paradigm shift in the conceptualization of FGF23 endocrine action and suggests the centrality of the suppression of FGFR1 more than that of klotho in the resistance to FGF23 in renal failure (Olauson et al., 2013).

Parathyroid cell proliferation and mammalian target of rapamycin

Phosphorylation of ribosomal protein S6 mediates mammalian target of rapamycin–induced parathyroid cell proliferation in secondary hyperparathyroidism

Secondary hyperparathyroidism is characterized by increases in PTH expression and parathyroid cell proliferation (Naveh-Many et al., 1995; Denda et al., 1996; Cozzolino et al., 2005; Silver and Naveh-Many, 2013). Decreased expression of the calcium, $1,25(OH)_2D_3$, and FGF23 receptors contributes to the increased parathyroid cell proliferation in uremia (Garfia et al., 2002; Galitzer et al., 2010; Lewin et al., 2002). Expression of transforming growth factor-α and its receptor, the epidermal growth factor (EGF) receptor (EGFR), is increased in uremic rats and patients (Cozzolino et al., 2005; Gogusev et al., 1996; Drueke, 2000). A dominant negative EGFR gene expressed specifically in the parathyroid glands prevented the activation of endogenous EGFR and the increase in parathyroid gland enlargement and serum PTH (Arcidiacono et al., 2008, 2015; Dusso et al., 2010). Other cell cycle regulators such as cyclin D1 also induce parathyroid cell proliferation, as shown in transgenic mice overexpressing cyclin D1 only in the parathyroid (Imanishi et al., 2001).

Mammalian target of rapamycin (mTOR) integrates signaling pathways to regulate cell growth and proliferation. mTOR is regulated by binding partners found in two complexes, mTOR complex 1 (mTORC1) and mTORC2. Rapamycin inhibits mTORC1 and proliferation in many cell types (Shimobayashi and Hall, 2014; Xu et al., 2015). Different stimuli activate mTORC1 through Akt phosphorylation. 4E-binding protein 1 and the ribosomal protein S6 kinase 1 (S6K1) are mTORC1 targets (Shahbazian et al., 2006). S6K1 phosphorylates ribosomal protein S6 (rpS6) on a cluster of five serine residues at the carboxy terminus. Knock-in mice encoding a mutant rpS6 harboring alanine substitutions at all five phosphorylation sites ($rpS6^{p-/-}$ mice) have reduced size, glucose intolerance, muscle weakness, and impaired renal hypertrophy after uninephrectomy (Ruvinsky et al., 2005, 2009). Volovelsky et al. (2016) have shown that the mTOR pathway is activated in the parathyroid of rats with secondary hyperparathyroidism induced by either chronic hypocalcemia or CKD, as measured by increased phosphorylation of rpS6, a downstream target of the mTOR pathway (Fig. 24.5). This

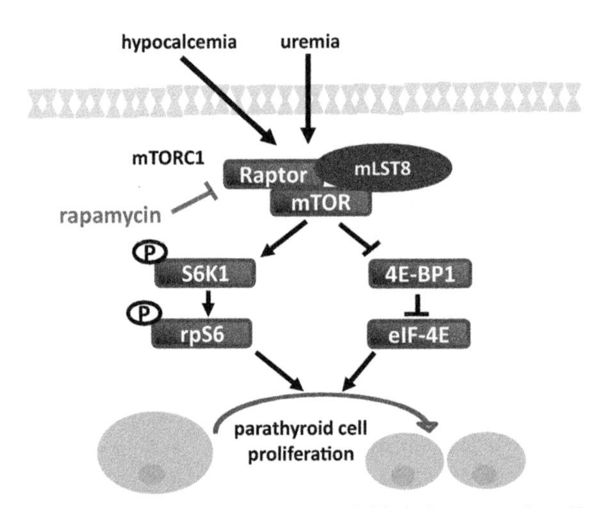

FIGURE 24.5 Mammalian target of rapamycin (*mTOR*) complex 1 (*mTORC1*) induces parathyroid cell proliferation in secondary hyperparathyroidism (SHP) through ribosomal protein S6 (*rpS6*) phosphorylation. The mTOR pathway is activated in the parathyroids of rats and mice with SHP induced by either chronic hypocalcemia or uremia. On activation, mTORC1 phosphorylates and activates S6 kinase 1 (*S6K1*), which then phosphorylates rpS6. mTORC1 also disinhibits eukaryotic translation initiation factor-4E (*eIF-4E*) by inhibiting 4E-binding protein 1 (*4E-BP1*). rpS6 phosphorylation is necessary for the increased parathyroid cell proliferation of SHP. Inhibition of mTORC1 by rapamycin decreases and prevents parathyroid cell proliferation in SHP. *With permission from* Journal of the American Society of Nephrology *(Volovelsky, O., Cohen, G., Kenig, A., Wasserman, G., Dreazen, A., Meyuhas, O., et al., 2016. Phosphorylation of ribosomal protein S6 mediates mammalian target of rapamycin complex 1-induced parathyroid cell proliferation in secondary hyperparathyroidism. J. Am. Soc. Nephrol. 27 (4), 1091–1101).*

activation correlated with increased parathyroid cell proliferation. Inhibition of mTORC1 by rapamycin decreased or prevented parathyroid cell proliferation in rats with secondary hyperparathyroidism and in vitro in uremic rat parathyroid glands in organ culture. Knock-in $rpS6^{p-/-}$ mice, in which rpS6 cannot be phosphorylated, had impaired PTH secretion after experimental uremia. Uremic $rpS6^{p-/-}$ mice had no increase in parathyroid cell proliferation compared with a marked increase in uremic wild-type mice. These results highlight the importance of mTOR activation and rpS6 phosphorylation for the pathogenesis of secondary hyperparathyroidism and indicate that mTORC1 is a significant regulator of parathyroid cell proliferation through rpS6 (Fig. 24.5) (Volovelsky et al., 2016). mTORC1 responds to growth factors, such as insulin, IGF-1, and EGF. The EGFR also activates the Akt–mTORC1 pathway (Huse and Holland, 2010). Therefore, the mTORC1 and EGFR pathways may act together to stimulate parathyroid cell proliferation in secondary hyperparathyroidism.

Summary

The PTH mRNA consists of three exons that are roughly divided into the 5′ UTR, the signal peptide sequence and PTH, and the 3′ UTR. The parathyroid glands develop from a shared initial organ primordium together with the thymus. A transcriptional network determines the embryological development of the parathyroid. $1,25(OH)_2D_3$ decreases PTH gene transcription. The major regulator of the parathyroid is hypocalcemia. Hypocalcemia regulates PTH gene expression in vivo by increasing PTH mRNA levels, and this is mainly posttranscriptional. Phosphate also regulates PTH gene expression, and this effect is independent of the effect of phosphate on serum calcium and $1,25(OH)_2D_3$. The effects of phosphate and uremia are also posttranscriptional. *Trans*-acting parathyroid cytosolic proteins bind to a defined *cis* element in the PTH mRNA 3′ UTR. This binding determines the degradation of PTH mRNA and thereby the PTH mRNA half-life. Changes in the balance of stabilizing and degrading factors on PTH mRNA determine the posttranscriptional effects of calcium, phosphate, and uremia on PTH gene expression. Pin1 regulates protein–PTH mRNA interactions, leading to more rapid PTH mRNA decay. In secondary hyperparathyroidism, Pin1 is less active and is associated with an increase in PTH mRNA stability and levels and thus increased PTH secretion. A further level of the posttranscriptional regulation of PTH gene expression is through miRNAs that are necessary for the increased PTH in acute and chronic hypocalcemia and uremia. mTOR is activated in secondary hyperparathyroidism, correlating with the increase in parathyroid cell proliferation. Inhibition of mTOR decreases parathyroid cell proliferation in both chronic hypocalcemia and CKD. FGF23 acts on its receptor, the klotho–FGFR1c receptor, to decrease PTH mRNA levels and secretion. In advanced CKD there is resistance of the parathyroid to the high FGF23 levels due to downregulation of the FGF23 receptor. $1,25(OH)_2D_3$, hypercalcemia, calcimimetics, hypophosphatemia, and rapamycin all decrease PTH gene expression and parathyroid cell proliferation (Fig. 24.1).

Acknowledgments

This work was supported in part by grants from the National Institutes of Health and the Israel Academy of Sciences (grant 642/16).

References

Alimov, A.P., Park-Sarge, O.K., Sarge, K.D., Malluche, H.H., Koszewski, N.J., 2005. Transactivation of the parathyroid hormone promoter by specificity proteins and the nuclear factor Y complex. Endocrinology 146 (8), 3409–3416.

Ahn, T.G., Antonarakis, S.E., Kronenberg, H.M., Igarashi, T., Levine, M.A., 1986. Familial isolated hypoparathyroidism: a molecular genetic analysis of 8 families with 23 affected persons. Medicine 65, 73–81.

Almaden, Y., Canalejo, A., Hernandez, A., Ballesteros, E., Garcia-Navarro, S., Torres, A., et al., 1996. Direct effect of phosphorus on parathyroid hormone secretion from whole rat parathyroid glands in vitro. J. Bone Miner. Res. 11, 970–976.

Almaden, Y., Canalejo, A., Ballesteros, E., Anon, G., Rodriguez, M., 2000. Effect of high extracellular phosphate concentration on arachidonic acid production by parathyroid tissue in vitro. J. Am. Soc. Nephrol. 11 (9), 1712–1718.

Alon, U.S., Levy-Olomucki, R., Moore, W.V., Stubbs, J., Liu, S., Quarles, L.D., 2008. Calcimimetics as an adjuvant treatment for familial hypophosphatemic rickets. Clin. J. Am. Soc. Nephrol. 3 (3), 658–664.

Antonarakis, S.E., Phillips, J.A., Mallonee, R.L., Kazazian, H.H.J., Fearon, E.R., Waber, P.G., et al., 1983. Beta-globin locus is linked to the parathyroid hormone (PTH) locus and lies between the insulin and PTH loci in man. Proc. Natl. Acad. Sci. U.S.A. 80, 6615–6619.

Aparicio, M., Combe, C., Lafage, M.H., De Precigout, V., Potaux, L., Bouchet, J.L., 1994. In advanced renal failure, dietary phosphorus restriction reverses hyperparathyroidism independent of the levels of calcitriol. Nephron 63, 122–123.

Arcidiacono, M.V., Sato, T., Alvarez-Hernandez, D., Yang, J., Tokurl, M., Gonzalez-Suarez, I., et al., 2008. EGFR activation increases parathyroid hyperplasia and calcitriol resistance in kidney disease. J. Am. Soc. Nephrol. 19 (2), 310–320.

Arcidiacono, M.V., Yang, J., Fernandez, E., Dusso, A., 2015. Parathyroid-specific epidermal growth factor-receptor inactivation prevents uremia-induced parathyroid hyperplasia in mice. Nephrol. Dial. Transplant. 30 (3), 434–440.

Arnold, A., Horst, S.A., Gardella, T.J., Baba, H., Levine, M.A., Kronenberg, H.M., 1990. Mutation of the signal peptide-encoding region of the pre-proparathyroid hormone gene in familial isolated hypoparathyroidism. J. Clin. Investig. 86, 1084–1087.

Arnold, A., Brown, M.F., Urena, P., Gaz, R.D., Sarfati, E., Drueke, T.B., 1995. Monoclonality of parathyroid tumors in chronic renal failure and in primary parathyroid hyperplasia. J. Clin. Investig. 95, 2047–2053.

Barreau, C., Paillard, L., Osborne, H.B., 2006. AU-rich elements and associated factors: are there unifying principles? Nucleic Acids Res. 33 (22), 7138–7150.

Bell, O., Gaberman, E., Kilav, R., Levi, R., Cox, K.B., Molkentin, J.D., et al., 2005a. The protein phosphatase calcineurin determines basal parathyroid hormone gene expression. Mol. Endocrinol. 19, 516–526.

Bell, O., Silver, J., Naveh-Many, T., 2005b. Identification and characterization of *cis*-acting elements in the human and bovine parathyroid hormone mRNA 3'-untranslated region. J. Bone Miner. Res. 20, 858–866.

Bell, O., Silver, J., Naveh-Many, T., 2005c. Parathyroid hormone, from gene to protein. In: Naveh-Many, T. (Ed.), Molecular Biology of the Parathyroid. Molecular Biology Intelligence Unit, first ed. Landes Bioscience and Kluwer Academic/Plenum Publishers, New York, pp. 8–28.

Ben Dov, I.Z., Galitzer, H., Lavi-Moshayoff, V., Goetz, R., Kuro-o, M., Mohammadi, M., et al., 2007. The parathyroid is a target organ for FGF23 in rats. J. Clin. Investig. 117 (12), 4003–4008.

Bouillon, R., Allewaert, K., Xiang, D.Z., Tan, B.K., van-Baelen, H., 1991. Vitamin D analogs with low affinity for the vitamin D binding protein: enhanced in vitro and decreased in vivo activity. J. Bone Miner. Res. 6, 1051–1057.

Bourdeau, A., Souberbielle, J.-C., Bonnet, P., Herviaux, P., Sachs, C., Lieberherr, M., 1992. Phospholipase-A_2 action and arachidonic acid in calcium-mediated parathyroid hormone secretion. Endocrinology 130, 1339–1344.

Bourdeau, A., Moutahir, M., Souberbielle, J.C., Bonnet, P., Herviaux, P., Sachs, C., et al., 1994. Effects of lipoxygenase products of arachidonate metabolism on parathyroid hormone secretion. Endocrinology 135, 1109–1112.

Brewer, G., 1991. An A + U-rich element RNA-binding factor regulates c-myc mRNA stability in vitro. Mol. Cell Biol. 11 (5), 2460–2466.

Brookman, J.J., Farrow, S.M., Nicholson, L., O'Riordan, J.L., Hendy, G.N., 1986. Regulation by calcium of parathyroid hormone mRNA in cultured parathyroid tissue. J. Bone Miner. Res. 1, 529–537.

Brown, A.J., 2005. Vitamin D analogs for the treatment of secondary hyperparathyroidsim in chronic renal failure. In: Naveh-Many, T. (Ed.), Molecular Biology of the Parathyoid. Molecular Biology Intelligence Unit, first ed. Landes Bioscience and Kluwer Academic/Plenum Publishers, New York, pp. 95–112.

Brown, E.M., Gamba, G., Riccardi, D., Lombardi, M., Butters, R., Kifor, O., et al., 1993. Cloning and characterization of an extracellular Ca^{2+} -sensing receptor from bovine parathyroid. Nature 366, 575–580.

Brown, A.J., Zhong, M., Finch, J., Ritter, C., Slatopolsky, E., 1995. The roles of calcium and 1,25-dihydroxyvitamin D3 in the regulation of vitamin D receptor expression by rat parathyroid glands. Endocrinology 136, 1419–1425.

Canaff, L., Hendy, G.N., 2002. Human calcium-sensing receptor gene. Vitamin D response elements in promoters P1 and P2 confer transcriptional responsiveness to 1,25-dihydroxyvitamin D. J. Biol. Chem. 277 (33), 30337–30350.

Cantley, L.K., Ontjes, D.A., Cooper, C.W., Thomas, C.G., Leight, G.S., Wells, S.A.J., 1985. Parathyroid hormone secretion from dispersed human hyperparathyroid cells: increased secretion in cells from hyperplastic glands versus adenomas. J. Clin. Endocrinol. Metab. 60, 1032–1037.

Carrillo-López, N., Román-García, P., Rodríguez-Rebollar, A., Fernández-Martín, J.L., Naves-Díaz, M., Cannata-Andía, J.B., 2009. Indirect regulation of PTH by estrogens may require FGF23. J. Am. Soc. Nephrol. 20 (9), 2009–2017.

Chan, Y.L., McKay, C., Dye, E., Slatopolsky, E., 1986. The effect of 1,25 dihydroxycholecalciferol on parathyroid hormone secretion by monolayer cultures of bovine parathyroid cells. Calcif. Tissue Int. 38, 27–32.

Chojnowski, J.L., Masuda, K., Trau, H.A., Thomas, K., Capecchi, M., Manley, N.R., 2014. Multiple roles for HOXA3 in regulating thymus and parathyroid differentiation and morphogenesis in mouse. Development 141 (19), 3697–3708.

Combe, C., Aparicio, M., 1994. Phosphorus and protein restriction and parathyroid function in chronic renal failure. Kidney Int. 46, 1381–1386.

Combe, C., Morel, D., de-Precigout, V., Blanchetier, V., Bouchet, J.L., Potaux, L., et al., 1995. Long-term control of hyperparathyroidism in advanced renal failure by low-phosphorus low-protein diet supplemented with calcium (without changes in plasma calcitriol). Nephron 70, 287–295.

Condamine, L., Vztovsnik, F., Friedlander, G., Menaa, C., Garabedian, M., 1994. Local action of phosphate depletion and insulin-like growth factor 1 on in vitro production of 1,25-dihydroxyvitamin D by cultured mammalian kidney cells. J. Clin. Investig. 94, 1673–1679.

Cozzolino, M., Lu, Y., Sato, T., Yang, J., Suarez, I.G., Brancaccio, D., et al., 2005. A critical role for enhanced TGF-alpha and EGFR expression in the initiation of parathyroid hyperplasia in experimental kidney disease. Am. J. Physiol. Renal. Physiol. 289 (5), F1096–F1102.

Danks, J.A., Ho, P.M., Notini, A.J., Katsis, F., Hoffmann, P., Kemp, B.E., et al., 2003. Identification of a parathyroid hormone in the fish Fugu rubripes. J. Bone Miner. Res. 18 (7), 1326–1331.

Demay, M.B., Kiernan, M.S., DeLuca, H.F., Kronenberg, H.M., 1992. Sequences in the human parathyroid hormone gene that bind the 1,25-dihydroxyvitamin D-3 receptor and mediate transcriptional repression in response to 1,25-dihydroxyvitamin D-3. Proc. Natl. Acad. Sci. U.S.A. 89, 8097–8101.

Denda, M., Finch, J., Slatopolsky, E., 1996. Phosphorus accelerates the development of parathyroid hyperplasia and secondary hyperparathyroidism in rats with renal failure. Am. J. Kidney Dis. 28 (4), 596–602.

Drueke, T.B., 2000. Cell biology of parathyroid gland hyperplasia in chronic renal failure. J. Am. Soc. Nephrol. 11 (6), 1141–1152.

Dusso, A., Arcidiacono, M.V., Yang, J., Tokumoto, M., 2010. Vitamin D inhibition of TACE and prevention of renal osteodystrophy and cardiovascular mortality. J. Steroid Biochem. Mol. Biol. 121 (1–2), 193–198.

Epstein, E., Silver, J., Almogi, G., Livni, N., Naveh-Many, T., 1995. Parathyroid hormone mRNA levels are increased by progestins and vary during the rat estrous cycle. Am. J. Physiol. 33, E158–E163.

Epstein, E., Sela-Brown, A., Ringel, I., Kilav, R., King, S.M., Benashski, S.E., et al., 2000. Dynein light chain (*M_r 8000*) binds the parathyroid hormone mRNA 3'-untranslated region and mediates its association with microtubules. J. Clin. Investig. 105, 505–512.

Fraser, W.D., Logue, F.C., Christie, J.P., Gallacher, S.J., Cameron, D., OqReilly, D.S., et al., 1998. Alteration of the circadian rhythm of intact parathyroid hormone and serum phosphate in women with established postmenopausal osteoporosis. Osteoporos. Int. 8 (2), 121–126.

Fukuda, N., Tanaka, H., Tominaga, Y., Fukagawa, M., Kurokawa, K., Seino, Y., 1993. Decreased 1,25-dihydroxyvitamin D3 receptor density is associated with a more severe form of parathyroid hyperplasia in chronic uremic patients. J. Clin. Investig. 92, 1436–1443.

Gafni, R.I., Levine, M.A., 2005. Genetic causes of hypoparathyroidism. In: Naveh-Many, T. (Ed.), Molecular Biology of the Parathyoid. Molecular Biology Intelligence Unit, first ed. Landes Bioscience and Kluwer Academic/Plenum Publishers, New York, pp. 159–178.

Galitzer, H., Ben Dov, I.Z., Silver, J., Naveh-Many, T., 2010. Parathyroid cell resistance to fibroblast growth factor 23 in secondary hyperparathyroidism of chronic kidney disease. Kidney Int. 77 (3), 211–218.

Garfia, B., Canadillas, S., Canalejo, A., Luque, F., Siendones, E., Quesada, M., et al., 2002. Regulation of parathyroid vitamin d receptor expression by extracellular calcium. J. Am. Soc. Nephrol. 13 (12), 2945–2952.

Gensure, R.C., Ponugoti, B., Gunes, Y., Papasani, M.R., Lanske, B., Bastepe, M., et al., 2004. Identification and characterization of two parathyroid hormone-like molecules in zebrafish. Endocrinology 145 (4), 1634–1639.

Gherzi, R., Lee, K.Y., Briata, P., Wegmuller, D., Moroni, C., Karin, M., et al., 2004. A KH domain RNA binding protein, KSRP, promotes ARE-directed mRNA turnover by recruiting the degradation machinery. Mol. Cell 14 (5), 571–583.

Gogusev, J., Duchambon, P., Stoermann-Chopard, C., Giovannini, M., Sarfati, E., Drueke, T.B., 1996. De novo expression of transforming growth factor-alpha in parathyroid gland tissue of patients with primary or secondary uraemic hyperparathyroidism. Nephrol. Dial. Transplant. 11 (11), 2155–2162.

Goswami, R., Mohapatra, T., Gupta, N., Rani, R., Tomar, N., Dikshit, A., et al., 2004. Parathyroid hormone gene polymorphism and sporadic idiopathic hypoparathyroidism. J. Clin. Endocrinol. Metab. 89 (10), 4840–4845.

Greenberg, C., Kukreja, S.C., Bowser, E.N., Hargis, G.K., Henderson, W.J., Williams, G.A., 1987. Parathyroid hormone secretion: effect of estradiol and progesterone. Metabolism 36, 151–154.

Grigorieva, I.V., Mirczuk, S., Gaynor, K.U., Nesbit, M.A., Grigorieva, E.F., Wei, Q., et al., 2010. Gata3-deficient mice develop parathyroid abnormalities due to dysregulation of the parathyroid-specific transcription factor Gcm2. J. Clin. Investig. 120 (6), 2144–2155.

Habener, J.F., Kamper, B., Potts, J.T.J., Rich, A., 1975. Preproparathyroid hormone identified by cell-free translation of messenger RNA from hyperplastic human parathyroid tissue. J. Clin. Investig. 56, 1328–1333.

Han, S.I., Tsunekage, Y., Kataoka, K., 2015. Gata3 cooperates with Gcm2 and MafB to activate parathyroid hormone gene expression by interacting with SP1. Mol. Cell. Endocrinol. 411, 113–120.

Hannan, F.M., Babinsky, V.N., Thakker, R.V., 2016. Disorders of the calcium-sensing receptor and partner proteins: insights into the molecular basis of calcium homeostasis. J. Mol. Endocrinol. 57 (3), R127–R142.

Holmes, D., 2015. Parathyroid function: key role for dicer-dependent miRNAs. Nat. Rev. Endocrinol. 11 (8), 445.

Hosokawa, Y.A., Leahy, J.L., 1997. Parallel reduction of pancreas insulin content and insulin secretion in 48-h tolbutamide-infused normoglycemic rats. Diabetes 46 (5), 808–813.

Huse, J.T., Holland, E.C., 2010. Targeting brain cancer: advances in the molecular pathology of malignant glioma and medulloblastoma. Nat. Rev. Canc. 10 (5), 319–331.

Igarashi, T., Okazaki, T., Potter, H., Gaz, R., Kronenberg, H.M., 1986. Cell-specific expression of the human parathyroid hormone gene in rat pituitary cells. Mol. Cell Biol. 6, 1830–1833.

Imanishi, Y., Hosokawa, Y., Yoshimoto, K., Schipani, E., Mallya, S., Papanikolaou, A., et al., 2001. Dual abnormalities in cell proliferation and hormone regulation caused by cyclin D1 in a murine model of hyperparathyroidism. J. Clin. Investig. 107, 1093–1102.

Kamitani-Kawamoto, A., Hamada, M., Moriguchi, T., Miyai, M., Saji, F., Hatamura, I., et al., 2011. MafB interacts with Gcm2 and regulates parathyroid hormone expression and parathyroid development. J. Bone Miner. Res. 26 (10), 2463–2472.

Karaplis, A.C., Lim, S.K., Baba, H., Arnold, A., Kronenberg, H.M., 1995. Inefficient membrane targeting, translocation, and proteolytic processing by signal peptidase of a mutant preproparathyroid hormone protein. J. Biol. Chem. 270, 1629–1635.

Karmali, R., Farrow, S., Hewison, M., Barker, S., O'Riordan, J.L., 1989. Effects of 1,25-dihydroxyvitamin D3 and cortisol on bovine and human parathyroid cells. J. Endocrinol. 123, 137–142.

Kel, A., Scheer, M., Mayer, H., 2005. In silico analysis of regulatory sequences in the human parathyroid hormone gene. In: Naveh-Many, T. (Ed.), Molecular Biology of the Parathyroid. Molecular Biology Intelligence Unit, first ed. Landes Bioscience and Kluwer Academic/Plenum Publishers, New York, pp. 68–83.

Kemper, B., 1986. Molecular biology of parathyroid hormone. CRC Crit. Rev. Biochem. 19, 353–379.

Kemper, B., Habener, J.F., Ernst, M.D., Potts Jr., J.T., Rich, A., 1976. Pre-proparathyroid hormone: analysis of radioactive tryptic peptides and amino acid sequence. Biochemistry 15 (1), 15–19.

Kifor, O., Diaz, R., Butters, R., Brown, E.M., 1997. The Ca2+-sensing receptor (CaR) activates phospholipases C, A2, and D in bovine parathyroid and CaR-transfected, human embryonic kidney (HEK293) cells. J. Bone Miner. Res. 12 (5), 715–725.

Kilav, R., Silver, J., Naveh-Many, T., 1995. Parathyroid hormone gene expression in hypophosphatemic rats. J. Clin. Investig. 96, 327–333.

Kilav, R., Silver, J., Naveh-Many, T., 2001. A conserved cis-acting element in the parathyroid hormone 3'-untranslated region is sufficient for regulation of RNA stability by calcium and phosphate. J. Biol. Chem. 276, 8727–8733.

Kilav, R., Silver, J., Naveh-Many, T., 2005. Regulation of parathyroid hormone mRNA stability by calcium and phosphate. In: Naveh-Many, T. (Ed.), Molecular Biology of the Parathyroid. Molecular Biology Intelligence Unit, first ed. Landes Bioscience and Kluwer Academic/Plenum Publishers, New York, pp. 57–67.

Kissmeyer, A.M., Binderup, L., 1991. Calcipotriol (MC 903): pharmacokinetics in rats and biological activities of metabolites. A comparative study with 1,25(OH)2D3. Biochem. Pharmacol. 41, 1601–1606.

Komaba, H., Goto, S., Fujii, H., Hamada, Y., Kobayashi, A., Shibuya, K., et al., 2010. Depressed expression of Klotho and FGF receptor 1 in hyperplastic parathyroid glands from uremic patients. Kidney Int. 77, 232–238.

Koszewski, N.J., Alimov, A.P., Park-Sarge, O.K., Malluche, H.H., 2004. Suppression of the human parathyroid hormone promoter by vitamin D involves displacement of NF-Y binding to the vitamin D response element. J. Biol. Chem. 279 (41), 42431–42437.

Kozak, M., 1991. Structural features in eukaryotic mRNAs that modulate the initiation of translation. J. Biol. Chem. 266 (30), 19867–19870.

Krajisnik, T., Bjorklund, P., Marsell, R., Ljunggren, O., Akerstrom, G., Jonsson, K.B., et al., 2007. Fibroblast growth factor-23 regulates parathyroid hormone and 1alpha-hydroxylase expression in cultured bovine parathyroid cells. J. Endocrinol. 195 (1), 125–131.

Kronenberg, H.M., Igarashi, T., Freeman, M.W., Okazaki, T., Brand, S.J., Wiren, K.M., et al., 1986. Structure and expression of the human parathyroid hormone gene. Recent Prog. Horm. Res. 42, 641–663.

Kumar, R., 2009. Pin1 regulates parathyroid hormone mRNA stability. J. Clin. Investig. 119 (10), 2887–2891.

Kurosu, H., Ogawa, Y., Miyoshi, M., Yamamoto, M., Nandi, A., Rosenblatt, K.P., et al., 2006. Regulation of fibroblast growth factor-23 signaling by klotho. J. Biol. Chem. 281 (10), 6120–6123.

Lafage, M.H., Combe, C., Fournier, A., Aparicio, M., 1992. Ketodiet, physiological calcium intake and native vitamin D improve renal osteodystrophy. Kidney Int. 42, 1217–1225.

Lavi-Moshayoff, V., Wasserman, G., Meir, T., Silver, J., Naveh-Many, T., 2010. PTH increases FGF23 gene expression and mediates the high-FGF23 levels of experimental kidney failure: a bone parathyroid feedback loop. Am. J. Physiol. Renal. Physiol. 299 (4), F882–F889.

Lee, Y., Ahn, C., Han, J.J., Choi, H., Kim, J., Yim, J., et al., 2003. The nuclear RNase III Drosha initiates microRNA processing. Nature 425 (6956), 415–419.

Lee, S., Mannstadt, M., Guo, J., Kim, S.M., Yi, H.-S., Khatri, A., et al., 2015. A homozygous [Cys25]PTH(1-84) mutation that impairs PTH/PTHrP receptor activation defines a novel form of hypoparathyroidism. J. Bone Miner. Res. 30 (10), 1803–1813.

Levi, R., Ben Dov, I.Z., Lavi-Moshayoff, V., Dinur, M., Martin, D., Naveh-Many, T., et al., 2006. Increased parathyroid hormone gene expression in secondary hyperparathyroidism of experimental uremia is reversed by calcimimetics: correlation with posttranslational modification of the trans acting factor AUF1. J. Am. Soc. Nephrol. 17 (1), 107–112.

Lewin, E., Garfia, B., Recio, F.L., Rodriguez, M., Olgaard, K., 2002. Persistent downregulation of calcium-sensing receptor mRNA in rat parathyroids when severe secondary hyperparathyroidism is reversed by an isogenic kidney transplantation. J. Am. Soc. Nephrol. 13 (8), 2110–2116.

Lindsay, E.A., Vitelli, F., Su, H., Morishima, M., Huynh, T., Pramparo, T., et al., 2001. Tbx1 haploinsufficiency in the DiGeorge syndrome region causes aortic arch defects in mice. Nature 410 (6824), 97–101.

Liu, M., Lee, M.H., Cohen, M., Bommakanti, M., Freedman, L.P., 1996. Transcriptional activation of the Cdk inhibitor p21 by vitamin D3 leads to the induced differentiation of the myelomonocytic cell line U937. Genes Dev. 10, 142–153.

Liu, Z., Yu, S., Manley, N.R., 2007. Gcm2 is required for the differentiation and survival of parathyroid precursor cells in the parathyroid/thymus primordia. Dev. Biol. 305 (1), 333–346.

Liu, Y., Ibrahim, A.S., Tay, B.-H., Richardson, S.J., Bell, J., Walker, T.I., et al., 2010. Parathyroid hormone gene family in a cartilaginous fish, the elephant shark (*Callorhinchus milii*). J. Bone Miner. Res. 25 (12), 2613–2623.

Lopez-Hilker, S., Dusso, A.S., Rapp, N.S., Martin, K.J., Slatopolsky, E., 1990. Phosphorus restriction reverses hyperparathyroidism in uremia independent of changes in calcium and calcitriol. Am. J. Physiol. 259, F432–F437.

Lucas, P.A., Brown, R.C., Woodhead, J.S., Coles, G.A., 1986. 1,25-dihydroxycholecalciferol and parathyroid hormone in advanced chronic renal failure: effects of simultaneous protein and phosphorus restriction. Clin. Nephrol. 25, 7–10.

Macdonald, L.E., Durbin, R.K., Dunn, J.J., McAllister, W.T., 1994. Characterization of two types of termination signal for bacteriophage T7 RNA polymerase. J. Mol. Biol. 238 (2), 145–158.

Manley, N.R., 2015. Embryology of the parathyroid glands. In: Brandi, M.L., Brown, E.M. (Eds.), Hypoparathyroidism. Springer Milan, Milano, pp. 11–18.

Meir, T., Durlacher, K., Pan, Z., Amir, G., Richards, W.G., Silver, J., et al., 2014. Parathyroid hormone activates the orphan nuclear receptor Nurr1 to induce FGF23 transcription. Kidney Int. 86 (6), 1106–1115.

Merscher, S., Funke, B., Epstein, J.A., Heyer, J., Puech, A., Lu, M.M., et al., 2001. TBX1 is responsible for cardiovascular defects in velo-cardio-facial/DiGeorge syndrome. Cell 104 (4), 619–629.

Mithal, A., Kifor, O., Kifor, I., Vassilev, P., Butters, R., Krapcho, K., et al., 1995. The reduced responsiveness of cultured bovine parathyroid cells to extracellular Ca^{2+} is associated with marked reduction in the expression of extracellular Ca^{2+}-sensing receptor messenger ribonucleic acid and protein. Endocrinology 136, 3087–3092.

Moallem, E., Silver, J., Kilav, R., Naveh-Many, T., 1998. RNA protein binding and post-transcriptional regulation of PTH gene expression by calcium and phosphate. J. Biol. Chem. 273, 5253–5259.

Morito, N., Yoh, K., Usui, T., Oishi, H., Ojima, M., Fujita, A., et al., 2018. Transcription factor MafB may play an important role in secondary hyperparathyroidism. Kidney Int. 93 (1), 54–68.

Naveh-Many, T., 2010. Minireview: the play of proteins on the parathyroid hormone messenger ribonucleic acid regulates its expression. Endocrinology 151 (4), 1398–1402.

Naveh-Many, T., Silver, J., 1990. Regulation of parathyroid hormone gene expression by hypocalcemia, hypercalcemia, and vitamin D in the rat. J. Clin. Investig. 86, 1313–1319.

Naveh-Many, T., Silver, J., 1993. Effects of calcitriol, 22-oxacalcitriol and calcipotriol on serum calcium and parathyroid hormone gene expression. Endocrinology 133, 2724–2728.

Naveh-Many, T., Silver, J., 2018. Transcription factors that determine parathyroid development power PTH expression. Kidney Int. 93 (1), 7–9.

Naveh-Many, T., Friedlander, M.M., Mayer, H., Silver, J., 1989. Calcium regulates parathyroid hormone messenger ribonucleic acid (mRNA), but not calcitonin mRNA in vivo in the rat. Endocrinology 125, 275–280.

Naveh-Many, T., Almogi, G., Livni, N., Silver, J., 1992a. Estrogen receptors and biologic response in rat parathyroid tissue and C-cells. J. Clin. Investig. 90, 2434–2438.

Naveh-Many, T., Raue, F., Grauer, A., Silver, J., 1992b. Regulation of calcitonin gene expression by hypocalcemia, hypercalcemia, and vitamin D in the rat. J. Bone Miner. Res. 7, 1233–1237.

Naveh-Many, T., Rahamimov, R., Livni, N., Silver, J., 1995. Parathyroid cell proliferation in normal and chronic renal failure rats: the effects of calcium, phosphate and vitamin D. J. Clin. Investig. 96, 1786–1793.

Nechama, M., Ben Dov, I.Z., Briata, P., Gherzi, R., Naveh-Many, T., 2008. The mRNA decay promoting factor K-homology splicing regulator protein post-transcriptionally determines parathyroid hormone mRNA levels. FASEB J. 22, 3458–3468.

Nechama, M., Ben Dov, I.Z., Silver, J., Naveh-Many, T., 2009a. Regulation of PTH mRNA stability by the calcimimetic R568 and the phosphorus binder lanthanum carbonate in CKD. Am. J. Physiol. Renal. Physiol. 296 (4), F795–F800.

Nechama, M., Peng, Y., Bell, O., Briata, P., Gherzi, R., Schoenberg, D.R., et al., 2009b. KSRP-PMR1-exosome association determines parathyroid hormone mRNA levels and stability in transfected cells. BMC Cell Biol. 10, 70–81.

Nechama, M., Uchida, T., Yosef-Levi, I.M., Silver, J., Naveh-Many, T., 2009c. The peptidyl-prolyl isomerase Pin1 determines parathyroid hormone mRNA levels and stability in rat models of secondary hyperparathyroidism. J. Clin. Investig. 119 (10), 3102–3114.

Nesbit, M.A., Hannan, F.M., Howles, S.A., Babinsky, V.N., Head, R.A., Cranston, T., et al., 2013. Mutations affecting G-protein subunit $\alpha(11)$ in hypercalcemia and hypocalcemia. N. Engl. J. Med. 368 (26), 2476–2486.

Nielsen, P.K., Feldt-Rasmusen, U., Olgaard, K., 1996. A direct effect of phosphate on PTH release from bovine parathyroid tissue slices but not from dispersed parathyroid cells. Nephrol. Dial. Transplant. 11, 1762–1768.

Nishii, Y., Abe, J., Mori, T., Brown, A.J., Dusso, A.S., Finch, J., et al., 1991. The noncalcemic analogue of vitamin D, 22-oxacalcitriol, suppresses parathyroid hormone synthesis and secretion. Contrib. Nephrol. 91, 123–128.

Okabe, M., Graham, A., 2004. The origin of the parathyroid gland. Proc. Natl. Acad. Sci. U.S.A. 101 (51), 17716–17719.

Okazaki, T., Igarashi, T., Kronenberg, H.M., 1988. 5'-flanking region of the parathyroid hormone gene mediates negative regulation by 1,25-(OH)2 vitamin D3. J. Biol. Chem. 263, 2203–2208.

Okazaki, T., Ando, K., Igarashi, T., Ogata, E., Fujita, T., 1992. Conserved mechanism of negative gene regulation by extracellular calcium. J. Clin. Investig. 89, 1268–1273.

Olauson, H., Lindberg, K., Amin, R., Sato, T., Jia, T., Goetz, R., et al., 2013. Parathyroid-specific deletion of klotho unravels a novel calcineurin-dependent FGF23 signaling pathway that regulates PTH secretion. PLoS Genet. 9 (12).

Parkinson, D.B., Thakker, R.V., 1992. A donor splice site mutation in the parathyroid hormone gene is associated with autosomal recessive hypoparathyroidism. Nat. Genet. 1, 149–152.

Patel, S., Rosenthal, J.T., 1985. Hypercalcemia in carcinoma of prostate. Its cure by orchiectomy. Urology 25, 627–629.

Portale, A.A., Booth, B.E., Halloran, B.P., Morris, R.C.J., 1984. Effect of dietary phosphorus on circulating concentrations of 1,25-dihydroxyvitamin D and immunoreactive parathyroid hormone in children with moderate renal insufficiency. J. Clin. Investig. 73, 1580–1589.

Portale, A.A., Halloran, B.P., Curtis Morris, J., 1989. Physiologic regulation of the serum concentration of 1,25-dihydroxyvitamin D by phosphorus in normal men. J. Clin. Investig. 83, 1494–1499.

Prank, K., Nowlan, S.J., Harms, H.M., Kloppstech, M., Brabant, G., Hesch, R.-D., et al., 1995. Time series prediction of plasma hormone concentration. Evidence for differences in predictability of parathyroid hormone secretion between osteoporotic patients and normal controls. J. Clin. Investig. 95, 2910–2919.

Prince, R.L., Dick, I., Garcia-Webb, P., Retallack, R.W., 1990. The effects of the menopause on calcitriol and parathyroid hormone: responses to a low dietary calcium stress test. J. Clin. Endocrinol. Metab. 70, 1119–1123.

Prince, R.L., MacLaughlin, D.T., Gaz, R.D., Neer, R.M., 1991. Lack of evidence for estrogen receptors in human and bovine parathyroid tissue. J. Clin. Endocrinol. Metab. 72, 1226–1228.

Rodriguez, M.E., Almaden, Y., Canadillas, S., Canalejo, A., Siendones, E., Lopez, I., et al., 2007. The calcimimetic R-568 increases vitamin D receptor expression in rat parathyroid glands. AJP Renal Physiol. 292, F1390–F1395.

Russell, J., Lettieri, D., Sherwood, L.M., 1983. Direct regulation by calcium of cytoplasmic messenger ribonucleic acid coding for pre-proparathyroid hormone in isolated bovine parathyroid cells. J. Clin. Investig. 72, 1851–1855.

Russell, J., Silver, J., Sherwood, L.M., 1984. The effects of calcium and vitamin D metabolites on cytoplasmic mRNA coding for pre-proparathyroid hormone in isolated parathyroid cells. Trans. Assoc. Am. Physicians 97, 296–303.

Russell, J., Lettieri, D., Sherwood, L.M., 1986. Suppression by 1,25$(OH)_2D_3$ of transcription of the pre-proparathyroid hormone gene. Endocrinology 119, 2864–2866.

Russell, J., Bar, A., Sherwood, L.M., Hurwitz, S., 1993. Interaction between calcium and 1,25-dihydroxyvitamin D3 in the regulation of preproparathyroid hormone and vitamin D receptor messenger ribonucleic acid in avian parathyroids. Endocrinology 132, 2639–2644.

Russell, J., Ashok, S., Koszewski, N.J., 1999. Vitamin D receptor interactions with the rat parathyroid hormone gene: synergistic effects between two negative vitamin D response elements. J. Bone Miner. Res. 14 (11), 1828–1837.

Ruvinsky, I., Sharon, N., Lerer, T., Cohen, H., Stolovich-Rain, M., Nir, T., et al., 2005. Ribosomal protein S6 phosphorylation is a determinant of cell size and glucose homeostasis. Genes Dev. 19 (18), 2199–2211.

Ruvinsky, I., Katz, M., Dreazen, A., Gielchinsky, Y., Saada, A., Freedman, N., et al., 2009. Mice deficient in ribosomal protein S6 phosphorylation suffer from muscle weakness that reflects a growth defect and energy deficit. PLoS One 4 (5).

Sela-Brown, A., Russell, J., Koszewski, N.J., Michalak, M., Naveh-Many, T., Silver, J., 1998. Calreticulin inhibits vitamin D's action on the PTH gene *in vitro* and may prevent vitamin D's effect *in vivo* in hypocalcemic rats. Mol. Endocrinol. 12, 1193–1200.

Sela-Brown, A., Silver, J., Brewer, G., Naveh-Many, T., 2000. Identification of AUF1 as a parathyroid hormone mRNA 3'-untranslated region binding protein that determines parathyroid hormone mRNA stability. J. Biol. Chem. 275 (10), 7424–7429.

Shahbazian, D., Roux, P.P., Mieulet, V., Cohen, M.S., Raught, B., Taunton, J., et al., 2006. The mTOR/PI3K and MAPK pathways converge on eIF4B to control its phosphorylation and activity. EMBO J. 25 (12), 2781–2791.

Shilo, V., Ben Dov, I.Z., Nechama, M., Silver, J., Naveh-Many, T., 2015. Parathyroid-specific deletion of dicer-dependent microRNAs abrogates the response of the parathyroid to acute and chronic hypocalcemia and uremia. FASEB J. 29 (9), 3964–3976.

Shilo, V., Silver, J., Naveh-Many, T., 2016. Micro-RNAs in the parathyroid: a new portal in understanding secondary hyperparathyroidism. Curr. Opin. Nephrol. Hypertens. 25 (4), 271–277.

Shilo, V., Mor-Yosef Levi, I., Abel, R., Mihailovic, A., Wasserman, G., Naveh-Many, T., et al., 2017. Let-7 and MicroRNA-148 regulate parathyroid hormone levels in secondary hyperparathyroidism. J. Am. Soc. Nephrol. 28 (8), 2353–2363.

Shimobayashi, M., Hall, M.N., 2014. Making new contacts: the mTOR network in metabolism and signalling crosstalk. Nat. Rev. Mol. Cell Biol. 15 (3), 155–162.

Silver, J., Naveh-Many, T., 2009. Phosphate and the parathyroid. Kidney Int. 75 (9), 898–905.

Silver, J., Naveh-Many, T., 2013. FGF-23 and secondary hyperparathyroidism in chronic kidney disease. Nat. Rev. Nephrol. 9 (11), 641–649.

Silver, J., Landau, H., Bab, I., Shvil, Y., Friedlaender, M.M., Rubinger, D., et al., 1985. Vitamin D-dependent rickets types I and II. Diagnosis and response to therapy. Isr. J. Med. Sci. 21, 53–56.

Silver, J., Naveh-Many, T., Mayer, H., Schmelzer, H.J., Popovtzer, M.M., 1986. Regulation by vitamin D metabolites of parathyroid hormone gene transcription in vivo in the rat. J. Clin. Investig. 78, 1296–1301.

Slatopolsky, E., Bricker, N.S., 1973. The role of phosphorus restriction in the prevention of secondary hyperparathyroidism in chronic renal disease. Kidney Int. 4, 141–145.

Slatopolsky, E., Finch, J., Denda, M., Ritter, C., Zhong, A., Dusso, A., et al., 1996. Phosphate restriction prevents parathyroid cell growth in uremic rats. High phosphate directly stimulates PTH secretion in vitro. J. Clin. Investig. 97, 2534–2540.

Suarez-Bregua, P., Cal, L., Cañestro, C., Rotllant, J., 2017. PTH reloaded: a new evolutionary perspective. Front. Physiol. 8, 776.

Sunthornthepvarakul, T., Churesigaew, S., Ngowngarmratana, S., 1999. A novel mutation of the signal peptide of the preproparathyroid hormone gene associated with autosomal recessive familial isolated hypoparathyroidism. J. Clin. Endocrinol. Metab. 84 (10), 3792–3796.

Takeshita, K., Fujimori, T., Kurotaki, Y., Honjo, H., Tsujikawa, H., Yasui, K., et al., 2004. Sinoatrial node dysfunction and early unexpected death of mice with a defect of klotho gene expression. Circulation 109 (14), 1776–1782.

Tanaka, Y., DeLuca, H.F., 1973. The control of vitamin D by inorganic phosphorus. Arch. Biochem. Biophys. 154, 566–570.

Urakawa, I., Yamazaki, Y., Shimada, T., Iijima, K., Hasegawa, H., Okawa, K., et al., 2006. Klotho converts canonical FGF receptor into a specific receptor for FGF23. Nature 444 (7120), 770–774.

Volovelsky, O., Cohen, G., Kenig, A., Wasserman, G., Dreazen, A., Meyuhas, O., et al., 2016. Phosphorylation of ribosomal protein S6 mediates mammalian target of rapamycin complex 1-induced parathyroid cell proliferation in secondary hyperparathyroidism. J. Am. Soc. Nephrol. 27 (4), 1091–1101.

Wheeler, D.G., Horsford, J., Michalak, M., White, J.H., Hendy, G.N., 1995. Calreticulin inhibits vitamin D3 signal transduction. Nucleic Acids Res. 23, 3268–3274.

Xu, J., Chen, J., Dong, Z., Meyuhas, O., Chen, J.K., 2015. Phosphorylation of ribosomal protein S6 mediates compensatory renal hypertrophy. Kidney Int. 87 (3), 543–556.

Yamamoto, M., Igarashi, T., Muramatsu, M., Fukagawa, M., Motokura, T., Ogata, E., 1989. Hypocalcemia increases and hypercalcemia decreases the steady-state level of parathyroid hormone messenger RNA in the rat. J. Clin. Investig. 83, 1053–1056.

Yano, S., Brown, E.M., 2005. The calcium sensing receptor. In: Naveh-Many, T. (Ed.), Molecular Biology of the Parathyroid. Molecular Biology Intelligence Unit, first ed. Landes Bioscience and Kluwer Academic/Plenum Publishers, New York, pp. 44–56.

Yuan, Z., Opas, E.E., Vrikshajanani, C., Libutti, S.K., Levine, M.A., 2014. Generation of mice encoding a conditional null allele of Gcm2. Transgenic Res. 23 (4), 631–641.

Zabel, B.U., Kronenberg, H.M., Bell, G.I., Shows, T.B., 1985. Chromosome mapping of genes on the short arm of human chromosome 11: parathyroid hormone gene is at 11p15 together with the genes for insulin, c-Harvey-ras 1, and beta-hemoglobin. Cytogenet. Cell Genet. 39, 200–205.

Chapter 25

Paracrine parathyroid hormone–related protein in bone: physiology and pharmacology

T. John Martin and Natalie A. Sims

St. Vincent's Institute of Medical Research, Melbourne, Australia; Department of Medicine at St. Vincent's Hospital, The University of Melbourne, Melbourne, Australia

Chapter outline

Introduction: discovery of parathyroid hormone–related protein 595
Primary structure, active domains, processing, and secretion 596
Interaction with parathyroid hormone receptor 1 598
PTHrP tissue distribution and function as a cytokine: the vascular tissue example 599
Nuclear import of parathyroid hormone–related protein 600
Nuclear actions: intracrine and autocrine 600
C-terminal PTHrP and osteostatin 602
PTHrP in fetus: early development and endochondral ossification 602
PTHrP in bone after endochondral ossification 605
Parathyroid hormone–related protein and osteosarcoma 607
Distinct roles of PTH and PTHrP in fetal and postnatal bone 609
PTH and PTHrP in adult bone: PTHrP in physiology and PTH in pharmacology 610
PTHrP analogs in pharmacology: could this change the approach to skeletal anabolic therapy? 612
Conclusion 613
Acknowledgments 613
References 613

Introduction: discovery of parathyroid hormone–related protein

When he carried out 650 autopsies on patients with breast cancer, the English surgeon Stephen Paget noted the remarkable frequency of secondary growths in bone, especially at the ends of the femora and in the skull (Paget, 1889). This led him to write: "in cancer of the breast, the bones suffer in a special way, which cannot be explained by any theory of embolism alone. Some bones suffer more than others, the disease has its seats of election." In that series, he also reported six subjects with breast cancer who had no bone metastases, but who appeared to have a generalized susceptibility of the skeleton to fracture. No explanation of the findings of Paget was forthcoming, but some decades later, the discovery of parathyroid hormone–related protein (PTHrP) as the cause of hypercalcemia in many patients with cancer provided new insights into the pathogenesis of the skeletal complications of malignancy (Suva et al., 1987). Two other chapters (Chapter 54, Goltzman; Chapter 55, Sterling) will consider this topic in more depth, but it is appropriate to introduce the topic of PTHrP in bone with some relevant background.

When Fuller Albright in 1941 was discussing a patient with hypercalcemia accompanying a hypernephroma that had metastasized to the ilium, he suggested that the hypercalcemia might be due to production by the cancer of parathyroid hormone (PTH) (Albright, 1941). This was an influential comment that played a part in the adoption within the next years of the concept of "ectopic" production of PTH by tumors. This became a commonly used explanation through the 1960s and into the 1970s for hypercalcemia in cancer with no or minimal bone metastases (Lafferty, 1966). This seemed at that time to fit with a more general concept of ectopic production of hormones by cancers, for example, with adrenocorticotropic hormone production resulting in hypercorticolism in certain patients with cancer (Meador et al., 1962).

Principles of Bone Biology. https://doi.org/10.1016/B978-0-12-814841-9.00025-7

The first radioimmunoassays for PTH appeared to support this view, with apparently elevated circulating levels of PTH in unselected patients with bronchogenic carcinoma (Berson and Yalow, 1966). When further assays were developed whose specificities were better defined (Riggs et al., 1971), the observations were made that PTH levels in cancer hypercalcemia were lower than those in primary hyperparathyroid subjects with the same degree of elevation of calcium and, furthermore, that immunoreactivity of cancer hypercalcemia samples differed from PTH in that it was nonparallel to PTH standards (Melick et al., 1972; Benson et al., 1974). These findings suggested that the circulating and tumor-derived activity measured in some patients with hypercalcemia in cancer differed immunochemically from authentic PTH itself. Strong support for this was provided by a study in which immunoreactive PTH could not be detected in plasma or tumor extracts from a number of patients whose tumor extracts resorbed bone in vitro (Powell et al., 1973), despite the use of several antibodies of differing specificities. Thus, the term "humoral hypercalcemia of malignancy" (HHM) was used to describe patients with cancer and hypercalcemia without bone metastases (Martin and Atkins, 1979).

A defining step in progress was made with three clinical studies each showing that in patients with HHM the mimicry of PTH action extended beyond the calcium and phosphorus effects to increased excretion of cyclic AMP (cAMP) (Kukreja et al., 1980; Rude et al., 1980; Stewart et al., 1980). This consolidated the view that the tumor activity resembled PTH biologically, resulting in even greater interest in the pathogenesis. By that time rapid, sensitive assays of PTH-like activity that made use of cAMP responses in osteosarcoma (OS) cells had been developed (Majeska et al., 1980; Partridge et al., 1981). These were applied to cancer cell cultures and to tumor extracts from patients with HHM, and resulted in the purification, sequencing, and cloning of what came to be called PTHrP (Moseley et al., 1987; Suva et al., 1987). This newly identified molecule, related to PTH by an evolutionary gene duplication event (Mangin et al., 1990), was recognized as a potent bone-resorbing agent in vivo and in vitro, providing an explanation for the pathogenesis of hypercalcemia in HHM through its actions on bone and kidney. Subsequent to this, PTHrP production by cancer cells was found to favor the establishment and growth of secondary cancer deposits in bone (Powell et al., 1991; Guise et al., 1996), illustrating its behavior as a paracrine agent, analogous with its physiological paracrine/autocrine roles in many normal tissues, including bone.

From its discovery in cancer as a result of its PTH-like actions, PTHrP soon emerged as a paracrine/autocrine mediator of essential processes in endochondral bone formation during skeletal development, and in bone remodeling in the adult skeleton. This chapter will discuss those roles, beginning by considering the unique structural features of PTHrP that equip it to exert biological activities, not only through the common G-protein-coupled receptor that it shares with PTH, but also through other unique recognizable domains within the molecule.

Primary structure, active domains, processing, and secretion

Whereas PTH consists of 84 amino acids, molecular cloning predicted human PTHrP to have three alternatively spliced products of 139, 141, and 173 amino acids (Mangin et al., 1988; Thiede et al., 1988; Yasuda et al., 1989). While the human PTHrP gene undergoes alternative 3′ splicing, the rat and mouse PTHrP genes have a much simpler structure and produce only one form of PTHrP cDNA (Mangin et al., 1990). PTH and PTHrP have identical amino acids in 8 of their first 13 residues, but other similarities within the sequences are no more than would be expected by chance (Martin, 2016). The marked conservation of the PTHrP amino acid sequence in human, rat, mouse, chicken, and canine up to position 111 indicates that important functions are likely to reside in this region.

PTHrP promotes bone resorption, in a manner identical to that of PTH, by acting upon the receptor it shares with PTH (PTHR1). From the time of its discovery, the structural requirements within PTHrP that activate the PTHR1 were recognized to be the same as those required with PTH (Tregear and Potts, 1975; Kemp et al., 1987). Biological assay showed that recombinant PTHrP and synthetic shorter amino-terminal forms were equipotent on a molar basis with one another and with PTH(1−34) in generating cAMP in osteoblast-like cells (Hammonds et al., 1989). They were also recognized equally on a molar basis when measured in a radioimmunoassay directed at the amino terminus of PTHrP (Grill et al., 1991), indicating that the N-terminal region, with high homology to PTH, activates cAMP formation.

Despite their equal ability to activate cAMP via PTHR1, it was clear from the earliest work, even with peptides up to the first 14 residues of PTHrP, that highly specific antibodies that discriminate between PTH and PTHrP could be generated (Moseley et al., 1987). Such antibodies against PTHrP that neutralized its effects in promoting cAMP production completely in vitro, without any detectable neutralizing effect on PTH, were used to prevent and treat hypercalcemia in nude mice bearing xenografts of PTHrP-secreting human cancers (Kukreja et al. 1988, 1990).

PTHrP can be divided into different domains on the basis of its primary amino acid sequence (Fig. 25.1). Intracellular "pre-pro" and "pro" precursors of the mature peptide, essential for intracellular trafficking, are encoded by the first 36 amino acids (−36 to −1); this domain is cleaved from the molecule when it is secreted. The next domain includes the first

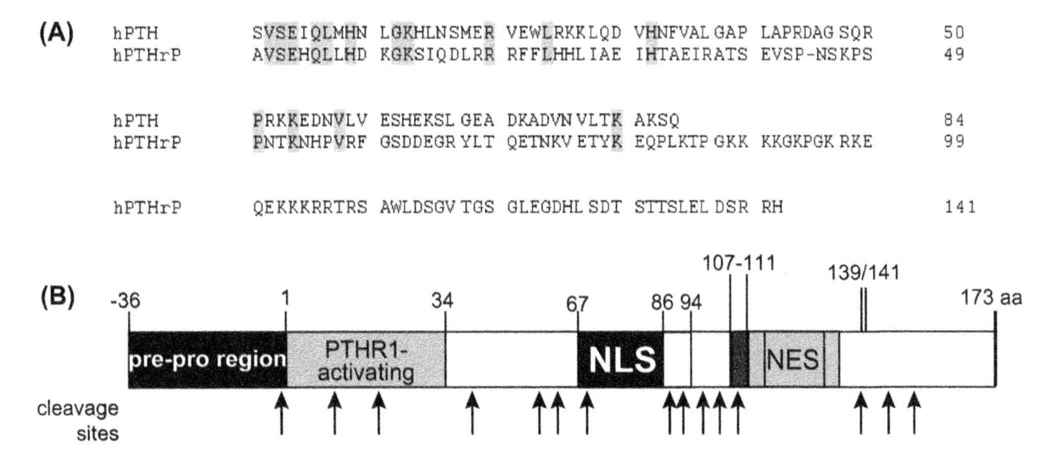

FIGURE 25.1 Protein structure of parathyroid hormone–related protein (PTHrP). (A) Amino acid alignment of the secreted portions of human PTH (*hPTH*) and human PTHrP (*hPTHrP*). Amino acid homologies are marked in *gray*. *Boxes* indicate the common PTH receptor (*PTHR1*)-activating region and the unique nuclear localization sequence (*NLS*) of PTHrP. (B) Protein structure of PTHrP including the amino acid numbers and approximate locations of cleavage sites. Structural domains described in the text are shown, including the pre-pro region, required for PTHrP secretion; the PTHR1-activating region that has high homology with PTH; the NLS; and the nuclear export sequence (*NES*).

13 residues of the mature protein showing the sequence homology with PTH. This domain is critical for most of the agonist effects of PTH and PTHrP on their shared PTHR1 receptor (Kemp et al., 1987). The following residues, PTHrP(14–36), although having almost no homology with PTH, appear to be critical for binding of PTHrP to PTHR1 (Juppner et al., 1991).

Molecular details of the actions of amino acids 36–139/141, including the nuclear localization sequence (NLS), are less well established. The midmolecule portion, between residues 35 and 84, has been shown to be responsible for promoting placental calcium transport, making calcium available for fetal skeletal development (Rodda et al., 1988; Abbas et al., 1989; Kovacs et al., 1996). Residues 107–111 have been reported to inhibit osteoclast activity and bone resorption in vitro (Fenton et al., 1991a; Fenton et al., 1991b) and in vivo (Cornish et al., 1997), and to be mitogenic for osteoblasts (Cornish et al., 1999).

The final tail region of PTHrP, amino acids 142–173, is found only in humans and is encoded by only one of the three human PTHrP mRNA isoforms. Its significance in terms of tissue distribution, processing, or function is unknown, although immunoreactivity for PTHrP(141–173) has been reported in plasma (Burtis et al., 1990) and in amnion (Bruns et al., 1995). It is presumed that the actions ascribed to regions of the molecule beyond the N-terminal 34 amino acids are mediated by non-receptor-mediated actions, or by unique PTHrP receptors, although these have yet to be discovered.

The realization that PTHrP was probably an important paracrine regulator in many tissues directed attention to mechanisms of PTHrP secretion and the nature of the secreted molecule. This topic has increased in interest with the advent of therapeutic approaches for osteoporosis aimed at mimicking PTHrP local action as a skeletal anabolic agent (Neer et al., 2001; Miller et al., 2016).

Two secretory pathways exist in eukaryotic cells: the regulated pathway whereby the proteins are packaged into secretory granules and the constitutive pathway through which the secreted proteins are found in the endoplasmic reticulum (Burgess and Kelly, 1987); PTHrP makes use of both these pathways depending on the cell type. In neuroendocrine cells, PTHrP is secreted via the regulated pathway. PTHrP is a neuroendocrine protein in the sense that it undergoes extensive posttranslational processing in a fashion analogous to that of chromogranin A or somatostatin, and it is a product of a broad variety of neuroendocrine cell types (e.g., pancreatic islet cells, parathyroid cells, adrenal medullary cells, pituitary cells, central nervous system neurons, etc.) (Philbrick et al., 1996; Matsushita et al., 1997; Martin, 2016). In these cells, PTHrP is packaged into secretory granules.

PTHrP is also produced by a broad range of constitutively secreting cell types (e.g., vascular smooth muscle cells, hepatocytes, osteoblasts, osteocytes, keratinocytes, chondrocytes, and renal tubular cells). In these cells, which do not contain the neuroendocrine machinery, PTHrP would be expected to be secreted in a constitutive fashion, analogous to the cytokines and growth factors whose actions it resembles. Indeed, secretion of PTHrP by the constitutive pathway was shown in nonendocrine cell types by Plawner et al. (Plawner et al., 1995), but in that study the nature of the secreted form was not defined, although three region-specific immunoradiometric assays detected secreted products. The first study to define the nature of PTHrP secreted by nonendocrine cells was carried out in osteocytes, of the osteoblast lineage, showing secretion of full-length PTHrP, with no evidence of secretion of any shorter biologically active form (Ansari et al., 2017). This will be discussed later in further detail ("PTH and PTHrP in adult bone: PTHrP in physiology and PTH in pharmacology").

Interaction with parathyroid hormone receptor 1

The discovery of PTHrP was possible because of its very similar biological action compared with that of PTH, in that it promotes adenylyl cyclase activity in cells known to be targets of PTH. The sequencing and cloning data, followed by synthesis of active peptides (Kemp et al., 1987), confirmed this, showing that the structural features within the amino-terminal PTHrP that were required for activation of PTHR1 were closely similar to those in PTH, and that PTH and PTHrP shared actions upon a common receptor, PTHR1 (Juppner et al., 1991), a member of the B class of G-protein-coupled receptors.

Sequence similarity between PTHrP and PTH is contained within the first half of the N-terminal domain (Fig. 25.1), in which 8 of the first 13 residues are identical between the two. Structural studies have helped explain the similar actions of PTHrP and PTH at the receptor, with nuclear magnetic resonance data indicating that there is a relatively stable α-helical portion within amino acids 15–34 of each peptide (Gardella and Vilardaga, 2015). With the use of mutant receptors and altered ligands, a model of interaction of ligand with PTHR1 has been proposed, in which the carboxy-terminal portion of the (1–34) peptide binds to the extracellular domain of the receptor, and then the amino-terminal portion interacts with the receptor's transmembrane domain to induce the conformational change that results in intracellular signaling (Gardella and Vilardaga, 2015). Although the (15–34) sequences of PTH and PTHrP are quite different, in the absence of evidence to the contrary, it must be assumed that the conformation of this portion of PTHrP is sufficiently similar to that in PTH to confer receptor binding capability upon it. Later chapters will consider details of the interaction with PTHR1 (Chapter 28, Gardella and Potts) and postreceptor signaling pathways (Chapter 27, Friedman).

The dominant signaling pathway mediating responses to PTHrP and PTH is the cAMP/protein kinase A (PKA) pathway, with activation by either PTH or PTHrP resulting in coupling of PTHR1 predominantly to Gαs/adenylyl cyclase/cAMP/PKA signaling (Fig. 25.2). Chapter 27 (Bisello and Friedman) considers these activation mechanisms in detail as well as the involvement of other PTHR1-mediated pathways. Downstream effectors include the cAMP-response element binding protein (CREB) and activator protein 1 transcription factors. By using peptide analogs, it has been concluded that cAMP signaling plays the major role in mediating the renal phosphaturic (Nagai et al., 2011) and anabolic responses in bone through the PTHR1 (Yang et al., 2007a).

In light of these very similar actions through the PTHR1, it was not surprising that the effects of pharmacologic PTHrP administration in vitro and in vivo were similar to those of PTH. Thus, addition of exogenous PTHrP(1–34) promoted

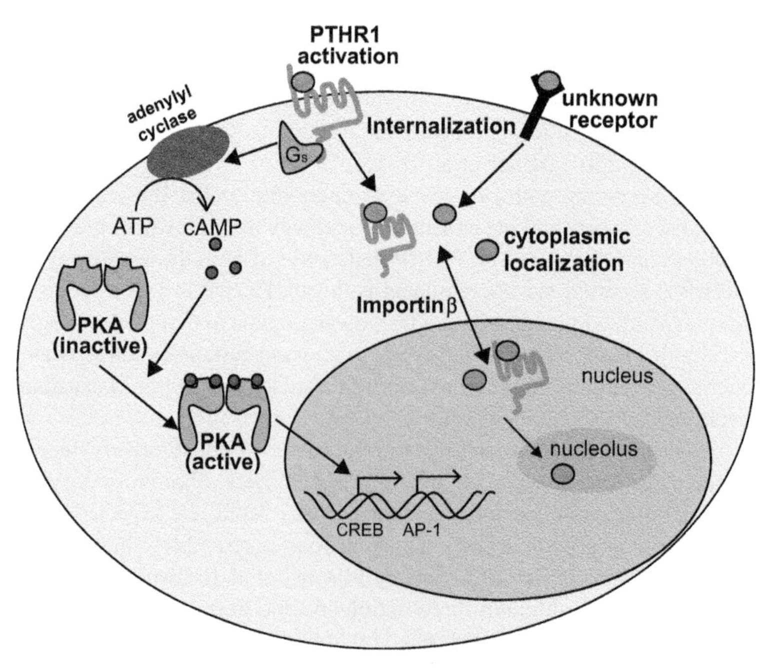

FIGURE 25.2 Multiple known pathways of parathyroid hormone–related protein (PTHrP) action. (1) PTHrP (green (grey in print versions) circle) acts through the G-protein-coupled PTH receptor (*PTHR1*), and via G_s protein, activates adenylyl cyclase to generate cyclic AMP (*cAMP*) from ATP. cAMP activates protein kinase A (*PKA*), which then induces transcription of cAMP-response element binding protein (*CREB*)- and activator protein 1 (*AP-1*)-responsive genes in the nucleus. (2) PTHrP is also internalized by the cell through mechanisms either dependent on the PTHR1 or that make use of an alternative, unknown receptor. PTHrP is also localized within the cytoplasm. PTHrP, both bound to the receptor and unbound, is transported into the nucleus via importin β and can be transported out of the nucleus and into the nucleolus.

bone resorption in organ culture (Kemp et al., 1987), although its lesser potency compared with PTH(1−34) probably reflected its greater susceptibility to proteolytic inactivation in these 48-h cultures (see Fig. 25.1B). Similar effects were observed with amino-terminal PTHrP peptides up to full-length PTHrP (Pilbeam et al., 1993), again with significantly less potency than with PTH(1−34). Such peptides also stimulated osteoclast formation in cocultures of osteoblasts with hemopoietic cells, and both PTHrP(1−34) and the full-length molecule promoted phosphate excretion in the rat when administered intravenously (Kemp et al., 1987; Zhou et al., 1989).

The first demonstration of an anabolic effect of amino-terminal PTHrP in vivo was obtained in the ovariectomized rat, where a dose-dependent anabolic effect of treatment with PTHrP(1−34) was clearly established, but PTHrP(1−34) potency was approximately 25% that of PTH(1−34) in the same experiment (Hock et al., 1989). Lesser potency in vivo of amino-terminal PTHrP peptides was found in virtually all studies (Horwitz et al. 2006, 2013). It might relate to the very marked susceptibility of PTHrP to proteolytic degradation (Diefenbach-Jagger et al., 1995), with the result that when dosage schedules the same as those with PTH are used, less of the injected PTHrP will reach the target cells having avoided degradation. This will be discussed in more detail in a later section ("PTHrP analogs in pharmacology: could this change the approach to skeletal anabolic therapy?"). Comparison of the activities of synthetic and recombinant peptides of various lengths have shown consistently that PTHrP(1−141), (1−84), (1−108), (1−36), and (1−34) are equally potent on a molar basis with one another and with PTH(1−34) in promoting cAMP formation in target cells (Kemp et al., 1987; Hammonds et al., 1989; Li et al., 2012). These assays measure cell responses that are sufficiently rapid that peptide inactivation plays little or no part, unlike prolonged organ culture responses or in vivo effects. These matters need to be taken into consideration in light of the finding that the form of PTHrP released locally, at least by osteocytes, is the full-length molecule (Ansari et al., 2017) (*vide infra*), without any evidence of any shorter forms capable of action through PTHR1. Thus, the form of PTHrP presenting locally as a physiological paracrine/autocrine agent is likely to be the full molecule.

PTHrP tissue distribution and function as a cytokine: the vascular tissue example

It soon became recognized that, apart from its role as a cancer cell product, PTHrP is produced in many tissues throughout the body. The common receptor, PTHR1, is also expressed in these many tissues, often in the same cells as those expressing PTHrP, or closely adjacent to them. Such colocalization was considered consistent with the function of PTHrP as a paracrine/autocrine factor in the developing embryo, particularly in epithelia at many locations and in a number of adult tissues, and raised the possibility that PTHrP is a multifunction cytokine whose actions might influence cell growth and differentiation (Martin et al., 1997; Philbrick, 1998). A body of evidence, particularly in epithelial cells, cancer cells, smooth muscle cells, and osteoblasts, supported this notion, although precise roles are not established in all these cell types.

PTHrP mRNA or protein has been detected in the following human tissues: adrenal, bone, brain, heart, intestine, kidney, liver, lung, mammary gland, ovary, parathyroid, placenta, prostate, skeletal muscle, skin, spleen, stomach, and smooth muscle (Moseley and Gillespie, 1995). PTHrP protein was identified immunologically in normal human and rat fetal bone and cartilage (Moseley et al., 1991; Guenther et al., 1995; Tsukazaki et al., 1995; Suda et al., 1996) and in fetal rat long bones in culture (Nijs-de Wolf et al., 1991), and PTHrP mRNA was detected by RT-PCR in human and rat osteogenic sarcoma cell lines (Rodan et al., 1989; Suda et al., 1996). In situ hybridization analyses localized PTHrP mRNA to active osteoblasts on the bone surface of newborn rat calvarial sections (Suda et al., 1996) and also to spindle-shaped cells of the periosteum, which may represent immature preosteoblasts (Lee et al., 1996; Suda et al., 1996; Rihani-Basharat and Lewinson, 1997). In areas of endochondral bone formation, PTHrP mRNA was detected by in situ hybridization in the perichondrium and maturing chondrocytes in a cell-type and stage-specific manner during fetal rat development (Lee et al., 1996). In a model of intramembranous bone formation in the rabbit, PTHrP mRNA and protein were strongly expressed in osteoblastic cells throughout the bone formation process, including in mature, actively synthetic osteoblasts and in osteocytes (Kartsogiannis et al., 1997).

The vasodilatory effect of PTHrP interacting with PTHR1 on smooth muscle (Roca-Cusachs et al., 1991; Maeda et al., 1996; Massfelder et al., 1998) is instructive in illustrating PTHrP as a physiological paracrine regulator of smooth muscle tone. Decades before the discovery of PTHrP it was known that PTH injection in several animal species resulted in dilatation of vascular beds, increased total blood flow, and decreased blood pressure (Charbon, 1968; Charbon and Hulstaert, 1974; Wang et al., 1984; Mok et al., 1989). These were puzzling observations, not fitting with the accepted role of PTH in regulating blood calcium levels.

The finding that vasoconstrictors such as angiotensin II induced a rapid rise in PTHrP production (Pirola et al., 1993) suggested that vasoconstriction is physiologically limited or reversed through the relaxant action of PTHrP on vascular smooth muscle. For example, in cultured aortic smooth muscle cells subject to stretch, PTHrP responsiveness to angiotensin II was greatly increased and accompanied by increased secretion of PTHrP protein (Noda et al., 1994). In human coronary arteries, the level of PTHrP mRNA expression by smooth muscle cells has been correlated with the degree of coronary atherosclerosis (Nakayama et al., 1994). In coronary arteries restenosing after angioplasty, expression of PTHrP was greatly increased (Ozeki et al., 1996), consistent with a role for PTHrP in the pathogenesis of vascular stenosis. A physiological role of PTHrP was indicated when mice with transgenic overexpression of PTHrP in vascular smooth muscle cells exhibited low blood pressure (Qian et al., 1999). A less conclusive outcome came from conditional knockdown of PTHrP in vascular smooth muscle in mice; no change in blood pressure was found, but renal blood flow was decreased in targeted mice, particularly with saline volume expansion (Raison et al., 2013).

Thus, the discovery of PTHrP and its production and action in vascular smooth muscle provided an explanation for the long-recognized pharmacological response to PTH injection. Despite evidence of a role for PTHrP as a cardiac hormone, a vasodilatory agent, and a local paracrine regulator of smooth muscle tone, PTHrP-null mutant mice develop normal cardiovascular systems (Karaplis et al., 1994). Although this might indicate that PTHrP is not essential for cardiovascular development, any functions for PTHrP in vascular remodeling, local regulation of vascular flow, and vascular response to injury in adult animals might remain undefined, because the animals die at birth of skeletal defects (see "PTHrP in fetus: early development and endochondral ossification").

Nuclear import of parathyroid hormone–related protein

In addition to its actions through the PTHR1, there is evidence that PTHrP also has intracellular signaling capacities (Fig. 25.2). The first evidence of this was the identification of nuclear and nucleolar localization of PTHrP in chondrocytes (Henderson et al., 1995), where its translocation was dependent upon a highly basic region of PTHrP(87–107) homologous to known NLSs within human retroviral regulatory proteins (Truant and Cullen, 1999). Nuclear localization was essential for the ability of PTHrP to confer enhanced survival on the chondrocytes following serum starvation (Henderson et al., 1995). Although the evidence for translocation of PTHrP to the nucleus in both malignant and normal cells is compelling, the molecular mechanisms by which PTHrP acts in the nucleus remain unclear.

The NLS of PTHrP was defined by Lam and colleagues as between residues 67 and 94 (Lam et al. 1999, 2000) (see Fig. 25.1); however there was debate over the precise region required. Others proposed that PTHrP has a bipartite NLS (Massfelder et al., 1998) (Massfelder et al., 1998; Massfelder et al., 1998) and that residues 87–107 include a second NLS that is required for nuclear localization (Henderson et al., 1995). The region 67–94 was found to include the residues necessary for efficient binding to the nuclear transport factor importin β1 (Lam et al. 1999, 2000), rather than to the conventional NLS-binding importin α subunit. In support of this, importin β1 and the monomeric GTP-binding protein Ran were shown to mediate the nuclear import of PTHrP in vitro via the nuclear pore complex in the absence of importin α (Lam et al., 1999). Further, deletion of the basic residues of the NLS resulted in complete cytoplasmic localization of PTHrP. Studies using a series of alanine-mutated PTHrP constructs and truncated importin β1 derivatives showed that PTHrP residues 83–93 are absolutely essential for importin β1 recognition, with residues 71–82 additionally required for high-affinity binding (Lam et al., 2001). The region of 67–94 also includes a candidate nucleus (CcN) motif and a nucleolus-localizing motif. A CcN motif is a type of nuclear localization signal regulated by dual phosphorylation. When its CK2 substrate site is phosphorylated, the rate of NLS-dependent nuclear import is elevated, and when the cdc2 kinase substrate site is phosphorylated, nuclear transport is inhibited. The CcN motif in PTHrP is similar to that described for the archetypal CcN-containing protein, SV40 T-antigen (Jans et al., 1991; Jans, 1995), and for the N-terminal domain of prointerleukin-16 (pro-IL-16) (Wilson et al., 2002) and pro-IL-1α (Stevenson et al., 1997), comprising also in each case consensus protein kinase CK2 and confirmed cyclin-dependent kinase phosphorylation sites. In the case of PTHrP, a threonine residue at amino acid 85 in PTHrP functions as a p34cdc2 kinase phosphorylation site, with its phosphorylation status determining the nuclear localization of PTHrP (Lam et al., 1999).

Most importantly, support for this region of PTHrP as the NLS comes from the crystal structure of importin β bound to PTHrP(67–94), which identified the specific binding sequence within importin β (Cingolani et al., 2002). It remains possible that there may be more than one contributing segment to nuclear translocation. This could even include the triple-arginine motif at position 19–21, which is highly conserved across species (Fig. 25.1A) and which could mediate nuclear trafficking of a recently identified circulating PTHrP(12–48) isoform (Washam et al., 2013).

Nuclear actions: intracrine and autocrine

In studying how PTHrP gains access to the nucleus it was found that when PTHrP(1–141) was transiently expressed in COS-1 cells, it could be detected in cytoplasm and nucleus as well as at the cell membrane (Aarts et al., 1999a,b). This raised the possibility of intracrine actions of PTHrP, for which evidence then came from studies showing that initiation of translation downstream of the methionine initiation site gave rise to forms of PTHrP that could not be secreted, but would be available for nuclear transport (Nguyen et al., 2001). Such a mechanism had been proposed for a second eukaryotic protein, fibroblast growth factor 3 (FGF3), which initiates translation at both a classic AUG codon and an upstream CUG site. For both FGF3 and PTHrP, if the AUG codon is used, the protein may follow the secretory pathway, whereas if the CUG codon is used, the protein may be translocated first to the cytosol and then the nucleus, without secretion (Massfelder et al., 1997; Amizuka et al., 2000; Nguyen et al., 2001). This introduced a new biological concept. Having been discovered as a hormone in cancer, PTHrP was then realized to be an autocrine/paracrine factor in normal physiology, with a nuclear role that could be either autocrine (action by PTHrP is secreted and then internalized) or intracrine (by PTHrP that acts within the cell without requiring secretion).

In vascular smooth muscle cells, nuclear PTHrP localization increased cell proliferation, whereas extracellular PTHrP treatment decreased cell proliferation and enhanced muscle relaxation in the same cells by acting through PTHR1 (Massfelder et al., 1997; de Miguel et al., 2001). This indicates a striking dichotomy of outcome that depends on the mode of delivery of PTHrP to the cell, and needs to be kept in mind when considering any organ in which PTHrP acts in an autocrine/paracrine manner. Perhaps even more remarkably, the increased mitogenesis in vascular smooth muscle cells resulting from PTHrP transfection was found to require not only the NLS, but also the C-terminal 108−139 domain of the molecule (de Miguel et al., 2001), suggesting that additional nonnuclear actions are involved in the intracrine action of PTHrP. The importance of the NLS was illustrated further by the finding that adenoviral delivery at angioplasty of PTHrP lacking the NLS markedly inhibited vascular smooth muscle proliferation and cell cycle progression (Fiaschi-Taesch et al., 2009). The importance of C-terminal intracrine actions has yet to be elucidated.

There are at least two mechanisms by which autocrine PTHrP may gain entry to the nucleus from outside the cell: PTHR1 dependent and PTHR1 independent. Internalization and nuclear localization of fluorescein isothiocyanate-labeled PTHrP(1−108) in keratinocytes required PTHR1 expression (Lam et al., 1999). Similarly, in a study using PTHrP mutant constructs in the MC3T3-E1 osteoblastic cell line, it was concluded that PTHrP needs to be secreted and then internalized via the PTHR1 for NLS-mediated shuttling to the nucleus (Garcia-Martin et al., 2014). In other work though, PTHrP gained entry into chondrocytes independent of PTHR1, with PTHrP(1−141) being transported to the nucleus after binding to the cell surface; this mechanism was dependent on the presence of an intact NLS (Aarts et al., 1999a,b). This effect was abrogated by pretreatment with PTHrP(74−113) alone; since it does not bind to PTHR1, this that indicates endocytosis-dependent nucleolar translocation does not involve PTHR1, but an as-yet undefined cell surface receptor.

In cells in which PTHR1-dependent nuclear translocation of PTHrP is used, if the mechanism is similar to that operating with IL-5, growth hormone (Jans et al., 1997), and vascular endothelial growth factor (Liu et al., 2012), the PTHR1 should be found in the nucleus. There is some evidence for that, with the PTHR1 being found by immunostaining in the nucleus of some cells of all tissues examined in the rat—kidney, liver, small intestine, uterus, and ovary—with PTHrP mRNA protein and mRNA localized in the same or adjacent cells (Watson et al., 2000; Watson et al., 2000). Nuclear localization of PTHR1 was also observed in UMR106, ROS 17.2/8, MC3T3-E1, and SaOS-2 cells (Watson et al., 2000; Watson et al., 2000). Furthermore, the cytoplasmic tail of the PTHR1 at residues 471−488 has a putative NLS similar to that of PTHrP, though whether this is required for nuclear transport has not been shown (Patterson et al., 2007).

In vascular smooth muscle cells (Okano et al., 1995) and keratinocytes (Lam et al., 1997), PTHrP expression was found to be cell cycle dependent. In keratinocytes the highest levels of PTHrP mRNA appeared in response to mitogenic factors only at the G_1 phase of the cell cycle, during which PTHrP localized to the nucleolus (Lam et al., 1997). The possession of a CcN motif and cell cycle regulation of kinases by PTHrP are consistent with a role for PTHrP in cell cycle regulation and/or cell proliferation. This has been the case with a number of other proteins containing CcN motifs, e.g., SV40 T-antigen, nucleoplasmin, interferon-inducible nuclear factor, γ-interferon-inducible protein 16, and p53 (Jans et al., 1991; Robbins et al., 1991; Briggs et al., 2001). Although clearly implicated in delaying apoptosis and promoting proliferation in certain cell types, the precise role of PTHrP in the nucleus/nucleolus remains unclear, and has not been investigated in the osteoblast. RNA has been shown to bind PTHrP (Aarts et al., 1999a,b) through a distinct motif in the nucleolar targeting sequence (Pache et al., 2006), and PTHrP inhibits rRNA synthesis (Aarts et al., 2001). This might indicate a role, perhaps in conjunction with other proteins, as a nuclear export factor for RNA. This would be consistent with the ability of PTHrP to shuttle between nuclear/nucleolar and cytoplasmic compartments. Among the most important tasks will be to identify nuclear binding partners for PTHrP, including the possibility that it could bind DNA directly as a transcription factor.

Several lines of evidence therefore converge to indicate that PTHrP is likely to exert many functions from within the cell: (1) the finding of the CcN motif, (2) the redistribution within the cell as a result of phosphorylation at T85, (3) the striking cell cycle dependence of PTHrP location, (4) the identification of nuclear localization of exogenously supplied PTHrP, (5) the participation of the PTHrP NLS with importin β in a specific, regulated nuclear import process, and (6) the involvement of other sequences within PTHrP in intracrine action. In the many tissues in which PTHrP plays a potent local role, including bone and cartilage, these observations need to be studied further, rather than assuming all local actions of PTHrP can be ascribed to its effects through PTHR1. The fact that PTHrP nuclear localization is integral to its function in any cell studied so far implies that strategies to block PTHrP nuclear localization could have important effects on target cell function.

C-terminal PTHrP and osteostatin

The concept that PTHrP has actions that do not involve signaling through the PTHR1 developed early with the finding that PTHrP promoted the transport of calcium from mother to fetus, making it available for mineralization of the fetal skeleton (Rodda et al., 1988; Kovacs et al., 1996). This action was ascribed to a midmolecule domain of PTHrP when calcium transport could be achieved by placental perfusion with PTHrP(67–86) (Care et al., 1990) or PTHrP(38–94) (Wu et al., 1996). The placental calcium transport data were strongly suggestive of a receptor for PTHrP other than PTHR1. This has not been identified, but a 2014 comprehensive review by Kovacs summarizes the role of PTHrP in the process (Kovacs, 2014).

An alternative receptor for PTHrP has also been long suggested from the outcome of pharmacological experiments using its C-terminal portion. Up to residue 111 a remarkable degree of conservation of PTHrP primary sequence is maintained among human, rat, mouse, dog, cow, and rabbit, with some significant divergence in the remainder of the molecule (Martin, 2016). Interest in the C-terminal domain began with the finding that PTHrP(107–139) inhibited osteoclast activity and bone resorption by isolated rat osteoclasts in vitro. That effect was exerted by the pentapeptide TRSAW (residues 107–111), which was then named "osteostatin" (Fenton et al., 1991; Fenton et al. 1991; Fenton et al. 1993). Similar equipotent inhibitory effects were observed in osteoclasts from embryonic chickens using human or chicken PTHrP(107–139), and the human (TRSAW) and chicken (ARSAW) pentapeptides (Fenton et al., 1994). Although injection of PTHrP(107–139) over the calvariae in mice was found to inhibit bone resorption (Cornish et al., 1997), the antiresorptive effect of TRSAW in organ culture has been controversial, with some investigators not finding this effect in vitro (Sone et al., 1992; Kaji et al., 1995; Murrills et al., 1995).

Although no receptor has yet been identified, both TRSAW and PTHrP(107–139) increased protein kinase C activity in rat splenocytes at low picomolar concentrations (Whitfield et al., 1994), with similar actions in ROS 17.2/8 OS cells (Gagnon et al., 1993). The same group of authors reported protein kinase C activation in OS cells by PTH(28–34) and PTHrP(28–34) (Jouishomme et al., 1992; Gagnon et al., 1993). The actions of these peptides on osteoblasts have been difficult to interpret, since in UMR106 OS cells PTHrP(107–139) and TRSAW inhibited proliferation (Valin et al., 1997), whereas both increased proliferation of fetal rat osteoblasts (Cornish et al., 1999). Others reported that PTHrP(107–139) enhanced human osteoblast differentiation (Alonso et al., 2008) and survival (de Gortazar et al., 2006) through interaction with the vascular endothelial growth factor receptor-2.

In addition to these many in vitro observations, some biological effects of TRSAW and PTHrP(107–139) on bone have been reported in vivo. When neonatal mouse bones were labeled with [^{3}H]tetracycline, TRSAW treatment inhibited PTHrP(1–34)-induced bone resorption, indicated by the release of radioactivity (Rihani-Basharat and Lewinson, 1997). Treatment of streptozotocin-diabetic mice with PTHrP(107–139) promoted bone healing after marrow ablation (Lozano et al., 2011), and perhaps most surprising of all these phenomena and very difficult to explain, high doses of PTHrP(1–34) and PTHrP(107–139) were claimed to be similarly effective in treating mice after initiation of ovariectomy-induced bone loss (de Castro et al., 2012).

Among these reported pharmacologic effects of C-terminal PTHrP(107–139) and TRSAW, many of the in vitro findings were dose dependent and made use of control peptides, but the problems remain that there has still been no receptor identified for any such actions, no non-receptor-mediated internalization mechanism described, nor any signaling pathway that could be held responsible. Furthermore, in most of these studies reported in various species, the C-terminal peptide used was predominantly human PTHrP(107–139). The C-terminal domain is the least conserved among species, with only TRSAW being conserved among mammals. It should be noted that in many of the cited studies, the TRSAW peptide reproduced faithfully the effects of PTHrP(107–139). Thus, this short sequence seems likely to be the most important contributor to the host of pharmacologic effects reported, in which case it could provide a pathway to receptor identification. Although there can be no certainty of any physiological implications, possible roles for the C-terminal domain should continue to be sought, and this would include studies in bone.

PTHrP in fetus: early development and endochondral ossification

Studies using immunohistochemistry, in situ hybridization, or northern blot analyses revealed that PTHrP production took place in the early embryo in mouse (van de Stolpe et al., 1993), chicken (Schermer et al., 1991), rat (Senior et al., 1991), and human (Moseley et al., 1991; Dunne et al., 1994).

In the mouse, the temporal and spatial distribution of PTHrP expression has been described in fetal and adult skeletal and extraskeletal tissues such as brain, kidney, lung, heart, liver, small intestine, skin, and skeletal muscle (Kartsogiannis et al., 1997). Similarly, an extensive range of tissue in the rat expressed PTHrP (Campos et al., 1991; Senior et al., 1991; Lee et al., 1995). Analogous profiles of PTHrP expression were also reported in the human fetus (Moniz et al., 1990; Moseley et al., 1991; Dunne et al., 1994). In the fetus at 7–8 weeks of gestation, ectodermal structures strongly positive for PTHrP expression include the epidermis, the otic placode, and the tooth bud (Dunne et al., 1994). PTHrP-expressing tissues of endodermal origin include the lung, liver, pancreas, stomach, intestine, and hindgut, while those of mesodermal origin include the perichondrium and the developing skeleton. In most of these tissues, PTHrP protein seems to be confined to the epithelial layer, while the mRNA is sometimes seen in mesenchymal components (Moseley et al., 1991; Dunne et al., 1994). Later in embryonic development (18–20 weeks of gestation), PTHrP expression is also apparent in cardiac and skeletal muscle, vascular smooth muscle, neural tissues, and areas of both endochondral and intramembranous osteogenesis in the limb buds and calvariae, respectively (Moniz et al., 1990; Moseley et al., 1991; Dunne et al., 1994). The expression of PTHrP was highest in tissues where active growth and differentiation were occurring, thus favoring the idea that PTHrP may be closely associated with the regulation of fetal, and perhaps neonatal, cellular growth and differentiation.

The spatial and temporal distribution of PTHrP associates strongly with that of the PTHR1 (Karperien et al., 1994). The relative expression levels of PTHrP and its receptor are often inversely correlated within a tissue or in certain locales along a border of apposition. Such a tight inverse coupling of expression would seem to imply either feedback downregulation of the receptor or a precise coordinated regulation of the two genes during the course of fetal development (Lee et al., 1996).

With such evidence of production of PTHrP in many tissues throughout fetal life in a number of species, it was perhaps to be expected that functional evidence of a role for PTHrP and its receptor in mammalian fetal development might come from studies of animals that are transgenic or have gene-targeted knockouts; these models, including later cell-specific models, are summarized in Table 25.1. The first evidence of a physiological role for PTHrP came soon after the discovery of PTHrP, with a dramatic phenotype in mice null for the *Pthlh* gene following homologous recombination (Amizuka et al., 1994; Karaplis et al., 1994; Karaplis and Kronenberg, 1996; Lee et al., 1996). Neonatal mice homozygous for *Pthlh* gene ablation exhibited severe skeletal abnormalities at birth and died within 24 h, with their abnormal ribcage development not allowing adequate respiration and this respiratory distress accentuated by reduced surfactant production and impaired type II alveolar cell development. Whereas the nonskeletal organs and tissues of the *Pthlh* homozygous mutant mice appeared grossly normal, the mice at birth had a domed skull, short snout and mandible, protruding tongue, and disproportionately short limbs. Phenotypic abnormalities were evident throughout the endochondral skeleton (axial as well as appendicular), while, in contrast, no abnormality was noted in skeletal structures that develop entirely by intramembranous ossification.

In the growth plates of $Pthlh^{-/-}$ mutant mice there was a marked reduction in the height of the resting and proliferating chondrocyte zones. This resulted from decreased cell division, and was associated with disorganization of the cartilage columns in the hypertrophic zone and altered deposition of matrix molecules such as type II collagen (Karaplis and Kronenberg, 1996). PTHrP thus proved to be necessary for normal chondrocyte proliferation, and the premature maturation of the skeleton was presumably a consequence of the reduced proliferation and accelerated differentiation/premature hypertrophy and apoptosis of growth plate chondrocytes. These findings were substantiated by a study in which transgenic mice overexpressing PTHrP in chondrocytes (using the mouse collagen type II promoter) were found to have an opposing form of short-limbed dwarfism, in which endochondral ossification was significantly delayed as a result of persistent chondrocyte proliferation and spatially and temporally abnormal chondrocyte hypertrophy (Weir et al., 1996). The delay in endochondral ossification was initially so profound that mice were born with cartilaginous endochondral skeletons. However, by 7 weeks of age in the mice overexpressing *Pthlh* in chondrocytes, this delay in chondrocyte differentiation and ossification was largely corrected, leaving foreshortened and misshapen but histologically near-normal bones.

The alterations in chondrocyte differentiation at the growth plate and ultimate histological healing noted in mice transgenically overexpressing PTHrP in chondrocytes (Weir et al., 1996) are reminiscent of those seen in patients with Jansen's metaphyseal chondrodysplasia, a condition arising from constitutive activation of the PTHR1 (Schipani et al. 1995, 1996, 1997) and in transgenic mice with expression of a constitutively active PTHR1 targeted to the growth plate (Schipani et al., 1997). In the latter study, targeted expression of constitutively active PTHR1 corrected at birth the growth plate abnormalities of $Pthlh^{-/-}$ mice and allowed for their prolonged survival. These "rescued" animals lacked tooth eruption and showed premature epiphyseal closure, indicating the requirement of PTHrP in both processes. Therefore, overexpression of *Pthlh* or constitutive activation of the PTHR1 in the growth plate ultimately result in a similar pattern of abnormalities in endochondral bone formation. Even further evidence for the part played by the PTHrP/PTHR1 signaling pathway comes from the study of Blomstrand's chondrodysplasia, a rare syndrome resulting from loss-of-function

TABLE 25.1 Skeletal phenotypes in mouse models of parathyroid hormone (PTH)–related protein and PTH receptor 1 deficiency.

Targeted gene	Manipulation	Lethality	Bone length	Trabecular bone mass	Osteoblasts	Osteoclasts/ resorption	Marrow adipocytes	Cortical bone	Serum calcium	References
Pthlh	*Global deletion*	Neonatal	Short	Low	Normal	High	—	Normal	—	Amizuka et al. (1994); Karaplis et al. (1994)
	Global haploinsufficiency	Viable	—	Low (12 weeks old)	Low	Low	High	Normal	—	Amizuka et al. (1996); Miao et al. (2005)
	Col1Cre	Viable	—	Low (6 weeks old)	Low	Low	Normal	Normal	Normal	Miao et al. (2005)
	Dmp1(10kb)Cre	Viable	Normal	Low (12 weeks old)	Low	Normal	Normal	Size normal, low strength	Normal	Ansari et al. (2017)
Pth1r	*Global deletion*	Embryonic/neonatal (strain dependent)	Short	Less	More	—	—	Thickened	—	Lanske et al. (1996, 1999)
	Col2Cre	—	Short	—	—	—	—	Thickened	Low	Hirai et al. (2015)
	Prrx1Cre	Viable, impaired survival rate	Short	Less (long bones only)	Less	—	—	Thin, less mineralized	Normal	Fan et al. (2016)
	Osx1Cre	Viable	—	—	—	—	—	—	—	Ono et al. (2016)
	Bglap1Cre	Viable	—	—	—	—	—	—	—	Ono et al. (2016)
	Dmp1(8kb)Cre	Viable	—	High	Normal	Low (variable)	—	Normal	—	Delgado-Calle et al. (2017)
	Dmp1(10kb)Cre	Viable	—	High (12 weeks old)	Low?	Low?	—	Thickened (12 weeks old)	—	Saini et al. (2013)

—, not reported; results with question mark indicate that changes were reported to occur, but were not statistically significant.

mutations within the *PTH1R* gene, in which the human syndrome is recapitulated in mice rendered null for *Pth1r* (Wysolmerski et al., 2001).

It was noted that, just as PTHrP and the PTHR1 localize to the growth plate region of long bones, so too does Indian hedgehog (Ihh) (Lanske et al., 1996; Vortkamp et al., 1996). Ihh belongs to the hedgehog family of genes (McMahon, 2000), which is involved in the regulation of *Drosophila* segment polarity and regulates embryonic patterning during development in many organisms (Goodrich et al., 1996; Iwamoto et al., 1999). Overexpression of *Ihh* in developing chick limbs blocked chondrocyte differentiation in a manner similar to that seen with overexpression of *Pthlh* in chondrocytes. The findings of Lanske et al. (Lanske et al., 1996) and Vortkamp et al. (Vortkamp et al., 1996) showed that PTHrP and the PTHR1 form a negative feedback loop with the Ihh pathway to establish the correct spatial and temporal progression of chondrocyte differentiation and thereby determine the rate and extent of long bone formation. In this instance, the role of PTHrP is a paracrine one, making use of activation of the PTHR1 by locally generated PTHrP. Indeed genetically altered mice with PTHR1 deletion targeted to chondrocytes and their progeny using *Col2Cre* or *Prx1Cre* resulted in a similar shortened bone/accelerated chondrocyte differentiation phenotype compared with the global knockout (Hirai et al., 2015; Fan et al., 2016) (see Table 25.1 for comparison). There has been no investigation of possible actions in chondrocyte differentiation of other functional domains of PTHrP or non-PTHR1-mediated actions, since the comparison of these genetically altered mouse models provides substantial evidence that PTHrP regulates chondrocyte differentiation through the PTHR1.

PTHrP in bone after endochondral ossification

The discovery of the crucial part played by PTHrP in endochondral bone formation in development was a dramatic outcome of the earliest mouse genetic studies. Much more was to come though, when the recognition of widespread production of PTHrP led to evidence of its production in bone (Moseley and Gillespie, 1995) (*vide supra*). In this respect, genetically altered mouse models have again played a critical role in identifying stage-specific roles of PTHrP during osteoblast differentiation in regulating bone formation and resorption (Table 25.1, Fig. 25.3).

Major insights into the role of PTHrP in bone were obtained from further studies carried out in heterozygous $Pthlh^{+/-}$ mice. Although homozygous $Pthlh^{-/-}$ mice died soon after birth (Amizuka et al., 1994), heterozygous $Pthlh^{+/-}$ mice survived and, although phenotypically normal at birth, by 3 months of age exhibited a markedly lower trabecular thickness and connectivity, and histomorphometry revealed low bone formation rate and osteoclast surface (Amizuka et al., 1996). In $Pthlh^{+/-}$ mice the number of apoptotic osteoblasts was greater than in controls, and ex vivo culture of bone marrow cells

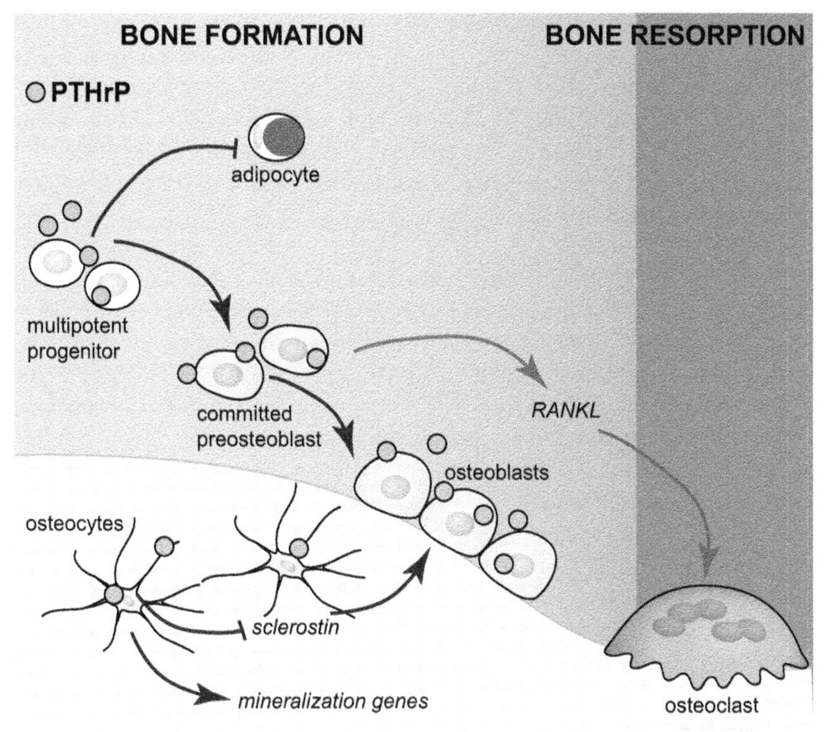

FIGURE 25.3 Stage-specific actions of parathyroid hormone–related protein (*PTHrP*) on the osteoblast and related cell lineages. (1) PTHrP is produced at all stages of osteoblast differentiation. (2) PTHrP acts on multipotent progenitors to inhibit adipogenic differentiation and promote osteoblast differentiation. (3) PTHrP acts on committed osteoblast precursors to promote their continued differentiation to osteoblasts, and promotes their expression of receptor activator of NF-κB ligand (*RANKL*), thereby indirectly stimulating the differentiation of osteoclast precursors to form bone-resorbing osteoclasts. (4) PTHrP acts on mature, bone-forming osteoblasts to promote their bone-forming activity and inhibit their apoptosis. (5) PTHrP acts on osteocytes, resulting in signals that promote bone formation by osteoblasts and signals that maintain mineralization of the bone matrix.

resulted in markedly fewer alkaline phosphatase—positive colonies, indicating impairment of osteogenic cell recruitment and survival. Moreover, *Pthlh*$^{+/-}$ bone marrow contained an abnormally high number of adipocytes. Since the same pluripotent stromal cells in the bone marrow compartment can give rise to adipocytes and osteoprogenitor cells (Owen, 1971), the increased number of adipocytes and osteopenia in these mice could be attributed to preferential development of adipocytes rather than osteoblasts as a consequence of PTHrP haploinsufficiency. Further investigations of the PTHrP mutant mice provided evidence that PTHrP is equally important for the orderly commitment of precursor cells toward the osteogenic lineage and their subsequent maturation and/or function (see Fig. 25.3). In the *Pthlh*$^{+/-}$ mice, osteoblastic progenitor cells (as with chondrocytes) were observed to contain inappropriate accumulations of glycogen, indicative of a defect in cells of the osteogenic lineage arising as a consequence of PTHrP deficiency (Amizuka et al., 1994). The overall conclusion was that PTHrP haploinsufficiency resulted in a low-bone-turnover state with low bone mass resulting from impaired bone formation.

The finding of an osteopenic phenotype in the *Pthlh*$^{+/-}$ mice was sufficiently intriguing to prompt the preparation of mice with osteoblast-specific knockout of *Pthlh* to determine whether it was locally derived PTHrP that was responsible for the low bone mass phenotype. *Col1(2.3kb).Cre* mice were crossed with *Pthlh* floxed mice, resulting in generation of mice that closely reproduced the low-bone-mass phenotype of the earlier *Pthlh*$^{+/-}$ (Miao et al., 2005). Higher levels of osteoblast apoptosis occurred in *Col1(2.3kb).Cre.Pthlh*$^{fl/fl}$ mice than in controls, and histomorphometry showed a low mineral apposition rate, osteoblast number, and osteoid volume, and few osteoclasts. Ex vivo studies showed that precursor cells from these mice were markedly less able to generate osteoblasts. Differences noted between the *Pthlh*$^{+/-}$ and the *Col1(2.3kb).Cre.Pthlh*$^{fl/fl}$ mice were that the latter lacked a marrow adiposity phenotype (Table 25.1). This difference could be explained by the fact that genetic deficiency in the *Pthlh*$^{+/-}$ mice was operative throughout cellular development, whereas in the targeted knockout it came into play at a later stage, after stromal cells had already committed to the osteoblast lineage. The findings in these genetic experiments gave much impetus to the view that locally derived PTHrP, produced by cells in the osteoblast lineage, acts within the lineage to inhibit apoptosis and promote differentiation of committed precursors to become osteoblasts capable of forming bone, and promotes support of osteoclastogenesis by the osteoblast lineage (Fig. 25.3).

The fact that PTHrP protein and mRNA had been noted in osteocytes (Suda et al., 1996; Kartsogiannis et al., 1997), together with the increased awareness of the importance of osteocytes, led to studies that revealed a role for osteocyte-derived PTHrP in bone remodeling, and confirmed suspicions that this might be achieved through autocrine/paracrine mechanisms. First, the effect of disruption of PTHR1 signaling was investigated in *Dmp1Cre.Pth1r*$^{fl/fl}$ mice. This was carried out using two different Cre models: *Dmp1(8kb)Cre* (Delgado-Calle et al., 2017) and *Dmp1(10kb)Cre* (Saini et al., 2013). In contrast to PTHrP deletion, both these PTHR1-deficient models had greater trabecular bone mass than control mice. This phenotype may result from a mild osteopetrosis since both osteoblast and osteoclast surfaces were slightly, but not significantly, lower in the *Dmp1(10kb)Cre.Pth1r* model, and the resorption marker Ctx1 was significantly lower in *Dmp1(8kb)Cre.Pth1r* mice; it does not appear to relate to a high level of bone formation. Furthermore, while both mice showed a blunted anabolic response to intermittent PTH treatment, induction of high circulating PTH levels in the *Dmp1(10kb)Cre.Pth1r* mice by dietary calcium restriction still markedly increased osteoclast formation despite the lack of PTH1R expression in osteocytes. This is consistent with a view that receptor activator of NF-κB ligand (RANKL) production induced through PTHR1 action in less mature osteoblast lineage cells, rather than mature osteoblasts or osteocytes, contributes to osteoclast formation.

The phenotype arising from the osteocyte-specific knockout of PTHrP was strikingly different from the receptor knockout. This was addressed using *Dmp1Cre.Pthlh*$^{f/f}$ mice prepared with the *Dmp1(10kb)Cre*. The mice with PTHrP-deficient osteocytes had low trabecular bone mass with low osteoblast numbers and mineralizing surface, as well as impaired cortical bone strength, indicating a material change in cortical bone matrix (Ansari et al., 2017). This phenotype differed in two important ways from that of *Col1(2.3kb).Cre.Pthlh*$^{fl/fl}$ (Table 25.1). First, the *Dmp1Cre.Pthlh*$^{f/f}$ phenotype was not evident at 6 weeks, but was so at 12 weeks of age, suggesting that while osteoblast-derived PTHrP may play a role in the development of the trabecular network, osteocyte-derived PTHrP is important later in life, when it has a role in remodeling the young adult skeleton. Second, the lack of osteoclast phenotype in the osteocyte-null mice indicated a lesser role for osteocytes in supporting osteoclastogenesis than that for cells earlier in the osteoblast lineage. In addition, the bones from the *Dmp1Cre.Pthlh*$^{f/f}$ mice had impaired bone material strength despite normal size, indicating a role for osteocyte-derived PTHrP in regulating bone composition. PTHrP knockdown, treatment, and overexpression studies in the Ocy454 osteocyte cell line showed that osteocyte-derived PTHrP suppresses sclerostin expression and modifies production of genes known to regulate bone mineralization (*Dmp1*, *Mepe*) (Ansari et al., 2017). This suggests that osteocyte-derived PTHrP not only is physiologically required to promote osteoblast differentiation, but also may regulate the process of bone mineralization, thereby controlling not only bone mass, but also bone strength (Fig. 25.3).

The difference in phenotypes between the high-bone-mass phenotype of the *Dmp1Cre*-driven PTH1R-deficient mice and the low-bone-mass phenotype in *Dmp1Cre*-driven PTHrP ligand–deficient mice suggests that PTHrP could have effects in the osteocyte that are mediated by non-PTH1R-mediated actions. Such noncanonical PTHrP signaling pathways acting in autocrine or paracrine ways have been noted in other cell types, as discussed earlier. The actions of these regions of PTHrP in bone remain poorly defined. Low bone mass, retarded growth, and early senescence (Miao et al., 2008; Toribio et al., 2010) have been reported in knock-in mice expressing PTHrP lacking both its NLS and its C-terminal regions, but retaining the PTHR1-activating N terminus. This provides evidence that these regions are physiologically important. However, the extreme defects in bone growth and development in that model make it difficult to discern their roles in the mature adult skeleton.

Parathyroid hormone–related protein and osteosarcoma

Given the evidence that PTHrP has important roles in osteoblast biology and bone formation, it might have been predicted that it would feature in the pathology of OS. This tumor of the osteoblast lineage is the fifth most common cancer in children and has become a focus of attention, since an improved understanding could lead to new pathways of treatment. We now have evidence from studies of genetically induced OS in the mouse that tumor-derived PTHrP acts in an autocrine/paracrine way in OS to drive its establishment and progression.

Induction of OS in rats with radio-phosphorus injection yielded transplantable tumors that exhibited sensitive dose responsiveness of adenylyl cyclase to PTH activation. PTH-responsive clonal cell lines from those tumors (UMR106 cells) (Partridge et al., 1981) and from a similar rat osteogenic sarcoma (Ros17.2/8 cells) (Majeska et al., 1980) have been used widely in studies of hormone and cytokine actions upon osteoblast-like cells. In that early work, removing the source of PTH by parathyroidectomy had no influence on the growth or on any other aspect of this transplantable OS (Ingleton et al., 1977). These findings came many years before PTHrP was discovered. Subsequent studies in OS cell lines from several species, including human, established that PTHR1 expression is a common, if not universal, feature of OS (e.g., Goerdeladze et al., 1993). Consistent with a pathogenic role for PTHR1 in OS, higher expression of PTHR1 mRNA was detected in metastatic or relapsed human OS samples compared with primary sites, and overexpression of PTHR1 in an OS cell line increased its proliferation and invasion (Yang et al., 2007b).

Although these findings suggested that activation of PTHR1 in OS may promote tumor invasion and proliferation, the role of the receptor signaling or of PTHrP has never been clearly defined. Studies in genetically induced mouse models of OS have pointed to an important pathogenetic role of activation of the cAMP/PKA/CREB pathway in osteoblast lineage cells, driven by autocrine PTHrP acting through the PTHR1 (Ho et al., 2014; Walia et al., 2016). Evidence obtained from human OS is also consistent with this possibility. Although a diverse mutational pattern of single-nucleotide variations has been reported in human OS (Chen et al., 2014), when we compared these proteins known to interact with the cAMP pathway the data suggested that recurrent and enriched changes in the cAMP/CREB pathway occur in osteogenic sarcoma (Walia et al., 2016).

Murine models have been long established for OS, developed from the knockout of genes associated with familial predisposition to OS (such as $p53^{-/-}$; Jacks et al., 1994) or from phenotyping in mutant or transgenic mice (Wang et al., 1995). However, not all cases of murine/rodent OS have proven to be applicable to human OS, with the murine OSs induced by *Nf2* and *Notch1* mutations being key examples (Engin et al., 2008; Mutsaers and Walkley, 2014; Tao et al., 2014; Walia et al., 2018). However, high-fidelity models recapitulating the mutational pattern and phenotypes of human OS have been developed by mesenchymal- and osteoblast-targeted conditional mutant alleles of *p53* and *Rb1,* the key genes implicated in familial OS (Berman et al., 2008; Walkley et al., 2008; Lin et al., 2009; Mutsaers et al., 2013; Quist et al., 2015). These murine models reproduce the cardinal features of the common forms of human OS, where osteoblastic OS is the most common clinical subtype (60%), with fibroblastic and chondroblastic OS comprising about 15% each (Walkley et al., 2008; Mutsaers et al., 2013). The *Osx*-Cre $p53^{fl/fl}pRb^{fl/fl}$ model (osteoblast lineage deletion of *p53*) yields a fibroblastic OS characterized by predominant areas of relatively poorly differentiated (Berman et al., 2008) histology with cells on the surface that appear to be immature osteoblasts (Walkley et al., 2008; Mutsaers et al., 2013). The *Osx*-Cre TRE_shp53.1224$pRb^{fl/fl}$ model (short hairpin RNA [shRNA] knockdown of *p53*) histologically resembles osteoblastic OS, with large mineralized areas and a cell surface phenotype of mature osteoblasts (Mutsaers et al., 2013). The early passage cells from both models have genetic and pharmacological sensitivities comparable to those of primary human patient-derived OS cultures (Baker et al., 2015; Gupte et al., 2015).

In these models of fibroblastic and osteoblastic OS, the primary and metastatic tumors from either subtype all express functional PTHR1. The cultures derived from those models all express PTHrP and PTHR1, and responded to treatment with either PTH or PTHrP with increased adenylyl cyclase activity and consequent increased expression of CREB pathway genes (Mutsaers et al., 2013; Ho et al., 2014; Walia et al., 2016). Conversely, knockdown of PTHR1 in murine OS (fibroblastic) reduced PTHR1 activation, cAMP responses, and tumor cell invasion in vitro (Ho et al., 2014). Although PTHR1 knockdown resulted in greatly reduced proliferation, the tumor was more mineralized in vivo. PTHrP was present intracellularly in OS cells, but was very low as a secreted protein in culture media.

Osteoblastic OS cells expressed high levels of *Pthlh* transcript (Walia et al., 2016), consistent with our previous data identifying substantial levels of intracellular PTHrP in OS cells (Ho et al., 2014). We showed that PTHrP was an OS autocrine ligand by treating cells with phosphodiesterase inhibitor without adding exogenous PTHrP, thus assaying the cAMP induced by an endogenous ligand(s) of the OS cells. This was confirmed when neutralizing anti-PTHrP antibody reduced cAMP formation, and when shRNAs against *Pthlh* reduced cAMP accumulation by >50%. This indicated that OS-derived PTHrP is an endogenous ligand promoting cAMP accumulation in an autocrine/paracrine manner (Fig. 25.4).

An early consequence of p53 deletion in osteoblastic cells was increased cAMP levels, PTHrP production, and autocrine activation of cAMP signaling via PTHrP (Walia et al., 2016). Activation of the PTHrP/cAMP/CREB1 axis was required for the hyperproliferative phenotype of p53-deficient osteoblasts and the maintenance of established OS, identifying this as a tractable pathway for therapeutic inhibition in OS. In support of this, knockdown of *Pthlh* reduced cell proliferation, reduced CREB phosphorylation and transcription of known CREB target genes, and induced apoptosis (Walia et al., 2016). In vivo, two different fibroblastic OS lines infected with sh-*Pthlh* showed reduced tumor growth that

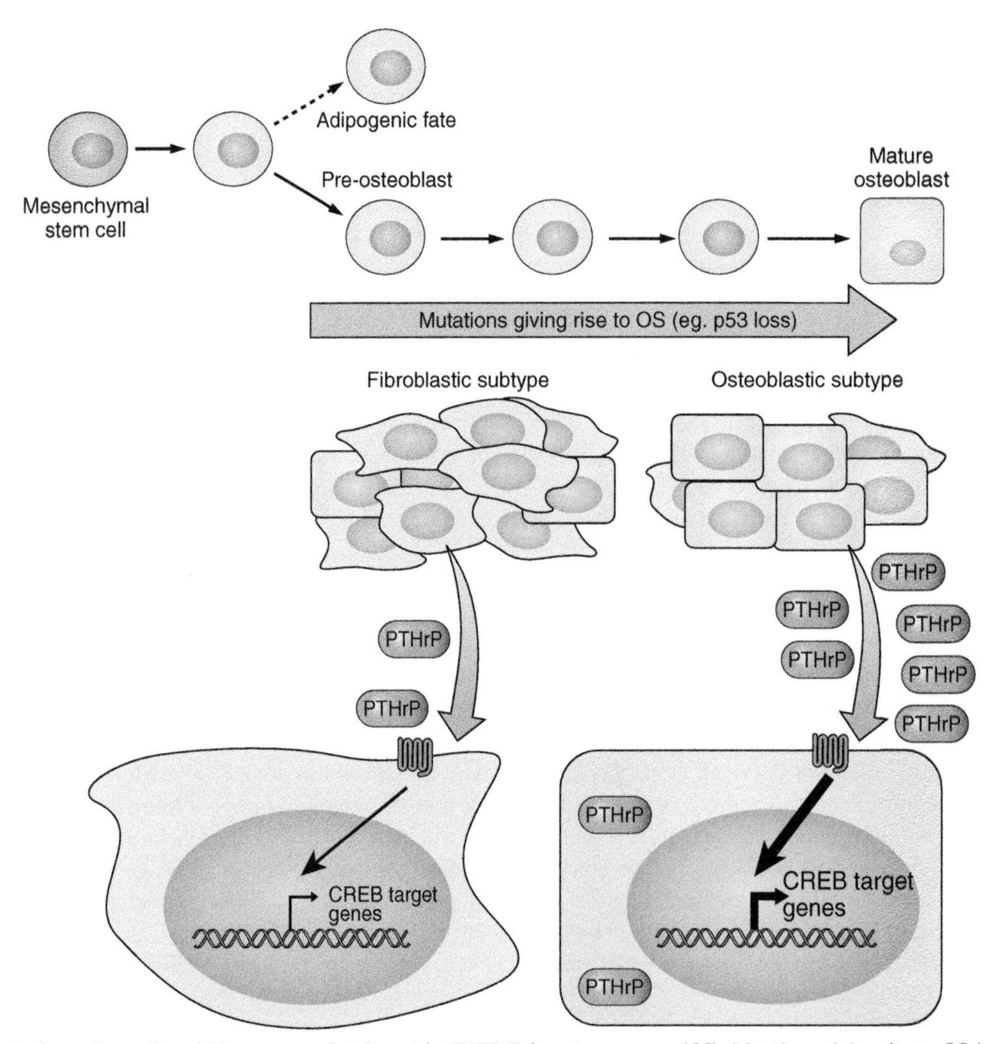

FIGURE 25.4 Actions of parathyroid hormone–related protein (*PTHrP*) in osteosarcoma (*OS*). Mutations giving rise to OS in committed pre-osteoblasts lead to the development of OS of a fibroblastic subtype, which expresses low levels of PTHrP that act through the PTH transmembrane receptor to promote low levels of cAMP-responsive element binding protein (*CREB*) target gene expression. When these mutations occur in more differentiated osteoblasts, this leads to an osteoblastic subtype of OS, which produces high levels of both secreted and intracellular PTHrP, acting in an autocrine/paracrine manner leading to higher levels of CREB target gene expression.

was comparable to the effects of *Pthr1* knockdown in the same OS lines (Ho et al., 2014), providing additional support for the concept that PTHrP is the tumor-derived ligand that promotes PTHR1-dependent OS growth (Fig. 25.4).

The identification of the PTHrP → PTHR1 → CREB-dependent mechanism for OS development may provide an explanation for data showing OS caused by enhanced PKA activity (Molyneux et al., 2010), and suggests that PTHrP → PTHR1 may be an upstream effector of such a pathway (Walia et al., 2016). Such a role extends the involvement of PTHrP in cancer, but in quite a different way compared with that in which it was discovered, as a bone-resorbing factor produced by certain cancers and resulting in hypercalcemia, and possibly also as a paracrine factor contributing to bone metastasis formation.

These data may also have relevance to the high incidence of OS in rats in long-term toxicology studies of any agonist of the PTHR1 (Tashjian and Chabner, 2002). This has resulted in a warning label accompanying these anabolic therapies for bone (Tashjian and Goltzman, 2008). The mechanism of tumor induction in these circumstances is not known. The toxicology study involved daily treatment of Fisher rats from soon after birth to 2 years of age. The high incidence of OS was obtained in rats treated with PTH(1−34) (Vahle et al., 2002), PTH(1−84) (Jolette et al., 2006), and abaloparatide (Jolette et al., 2017). It is conceivable that OS development is related to prolonged stimulation of the PKA/CREB pathway in an animal that might be susceptible to tumor development with such a treatment, but establishing this would require much further work.

Distinct roles of PTH and PTHrP in fetal and postnatal bone

All that has been available in anabolic therapy for the skeleton has been teriparatide (hPTH(1−34)), in use for some years, and now abaloparatide (Tymlos), which is a modified form of PTHrP(1−34) that has been approved by the US FDA. We might make better use of this anabolic pathway if we were to understand better how physiology and pharmacology meet.

A central question of the relative roles of PTH and PTHrP in the regulation of bone remodeling is: to what extent does the hormone PTH contribute to bone remodeling, either directly or indirectly? Further insights into this question came from a series of studies using *Pthlh*, *Pth*, and compound knockout mice.

When *Pth* was ablated in mice, trabecular bone mass in the fetal metaphysis was low compared with controls; in the postnatal state this phenotype reversed, and *Pth*-null mice exhibited high metaphyseal bone mass (Miao et al., 2002). Although this was at first puzzling, a low trabecular bone mass phenotype was then observed when the mice were rendered PTHrP deficient by crossing *Pth*-null mice with $Pthlh^{+/-}$ mice (Miao et al., 2004). In this case PTHrP deficiency prevented the high trabecular bone mass of postnatal PTH-deficient mice, even though the mice were still hypoparathyroid with low serum calcium and high serum phosphate. It was concluded that the PTH-null phenotype of high trabecular bone mass was caused by increased local production of PTHrP; this provided the first indication in vivo for a local role of PTHrP in promoting bone formation.

A comparable series of experiments in older mice provided further insights into the relative roles of PTH and PTHrP in osteogenesis. A model of bone marrow ablation was used (Zhu et al., 2013), in which de novo bone formation precedes resorption during recovery (Suva et al., 1993). When this was carried out in *Pth*-null mice, both osteogenesis and osteoclast formation were delayed and impaired, and *Pthlh* expression in osteoblasts was increased (Zhu et al., 2013). When a similar experiment was carried out in *Pth*-null mice crossed with *Pthlh*-haploinsufficient mice, the response to bone marrow ablation generated fewer osteoblasts, less bone formation, and fewer osteoclasts than in *Pth*-null mice (Zhu et al., 2013). These data suggest a role for PTH in recruiting osteoblast progenitor cells, with locally generated PTHrP being necessary both for further differentiation of osteoblast progenitors to bone-forming osteoblasts and for the generation of osteoclasts.

These genetic experiments point to specific roles for PTH and PTHrP that change between fetal and postnatal environments (Fig. 25.5). In the fetus, ambient calcium is dependent on maternal supply promoted by PTHrP placental action. In the fetus, when PTH promotes osteogenesis, the major function of PTHrP appears to be in directing the sequential controlled proliferation and differentiation of chondrocytes for growth plate development. In the immediate postnatal period, with the major environmental change and the source of nutrient calcium switching from the placenta to the diet, PTH secretion is regulated by extracellular fluid calcium levels, and its action on bone serves primarily to provide calcium to the body by promoting bone resorption. The aforementioned study provided evidence that the other major change in function after birth is that PTHrP becomes the major endogenous stimulus of bone formation, and it does this in a paracrine and possibly an autocrine manner. This is achieved at least partly by action through the PTHR1, but possibly also through actions promoted by other active domains within PTHrP (*vide supra*).

Thus, the anabolic physiological roles of PTH and PTHrP are development and environment dependent. PTH has not been shown to exert a direct physiological anabolic role in the adult animal, where its physiological function rather is the regulation of calcium homeostasis postnatally and in the fetus, and in the latter case is essential in providing calcium for

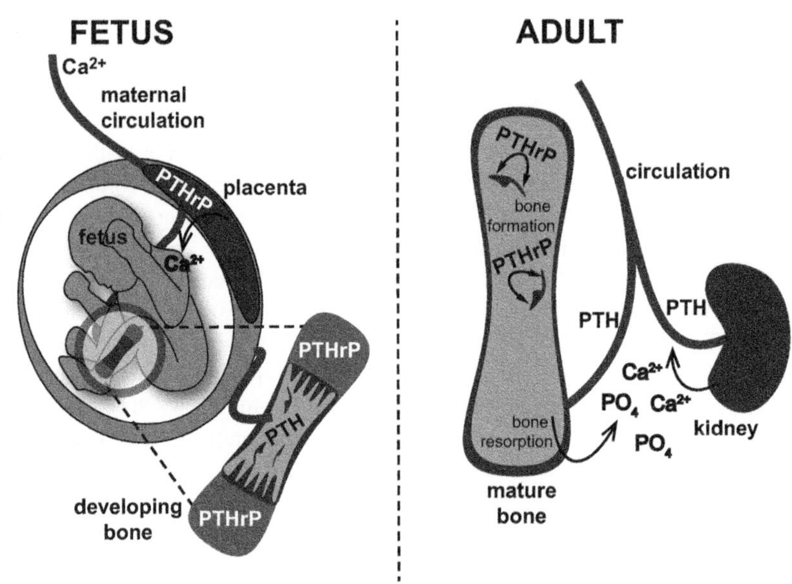

FIGURE 25.5 Roles of parathyroid hormone–related protein (*PTHrP*) and PTH in fetus and adult. Fetal calcium is predominantly provided by the placental action of PTHrP, which is also essential for chondrogenesis. PTH promotes osteogenesis. Postnatally, functions change, with the hormone PTH responsible for calcium homeostasis through actions on the kidney to restrict excretion and through generation of osteoclasts to provide calcium to the extracellular fluid. The bone-forming function is physiologically ascribed to locally generated PTHrP.

mineralization of bone. A hypothesis embracing the relative roles of PTH and PTHrP is that the hormone, PTH, governs osteogenesis through PTHR1 in the fetus and newborn, while PTHrP governs chondrocyte differentiation. The paracrine/autocrine action of PTHrP on bone becomes important when remodeling supervenes sometime after birth, and this role is maintained in the mature skeleton.

PTH and PTHrP in adult bone: PTHrP in physiology and PTH in pharmacology

The observation that physiological functions of PTHrP are carried out by local generation and action made it important to understand the nature of PTHrP released locally in bone and thus available for autocrine and paracrine actions. In discussing processing of PTHrP in an earlier section (*vide supra*) we reviewed early work in neuroendocrine cells that showed PTHrP is subject to the regulated secretory pathway, thereby generating PTHrP(1–36), midterminal peptides, and a C-terminal sequence. Although that might be true of cells that are capable of using the regulated secretory pathway, it would not be expected of mesenchymal cells such as those within the osteoblast lineage.

Accordingly, in studying the role of osteocyte-derived PTHrP, we used knockdown of PTHrP in the immortalized osteocyte cell line Ocy454 to show that PTHrP derived from those cells acts upon PTHR1 in the same cells in an autocrine/paracrine manner to activate adenylyl cyclase and modify gene expression (Ansari et al., 2017). In the same series of experiments full-length and mutant constructs of PTHrP were expressed in the osteocytes, secretion of biologically active PTHrP was established, and a combination of bioassay, radioimmunoassay, and chromatographic and electrophoretic separation methods was used to show that the only protein released from those cells and able to act through PTHR1 was PTHrP of full length. This might not be surprising, given the mesenchymal origin of cells of the osteoblast lineage, which would be expected to use the constitutive pathway of secretion, without the packaging and processing ability inherent in the regulated secretory pathway.

The importance of the observations, though, is that they identify full-length PTHrP as the most likely locally active material released in bone and available to act on the same or nearby cells. Among the questions that this leads to are: (1) does full-length PTHrP interact with the receptor PTHR1 in any unique way that distinguishes it from shorter forms used pharmacologically, (2) do other domains within the PTHrP molecule contribute to actions locally in bone, and (3) what significance might this have for the pharmacological use of peptide analogs that are based to some extent on PTHrP?

The discovery of the role of paracrine/autocrine PTHrP in bone remodeling underlies the thought that has been harbored for some time, that the anabolic action of intermittent (daily) treatment with PTH for osteoporosis mimics pharmacologically the physiological action of PTHrP. That anabolic action is predominantly remodeling based (Lindsay et al., 2006; Ma et al., 2006), with the effect of therapeutic PTH being to promote the recruitment of new basic multicellular units (BMUs) and enhance the activity of those already in progress (Fig. 25.6). Overfilling of BMUs (Compston, 2007), and perhaps also some effect on modeling through activation of lining cells (Kim et al., 2012), results in increased bone

(A) Pharmacology – PTH / Abaloparatide

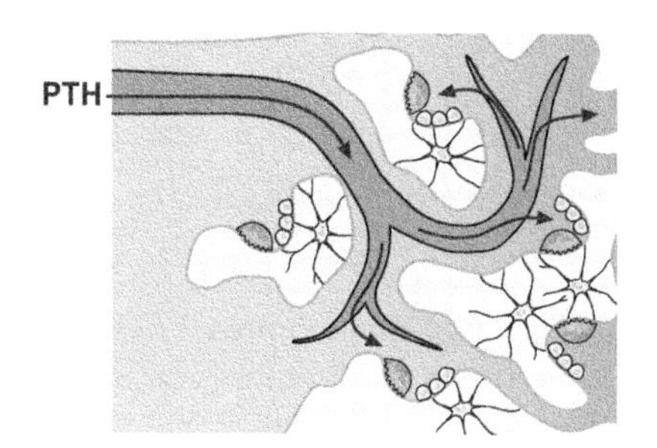

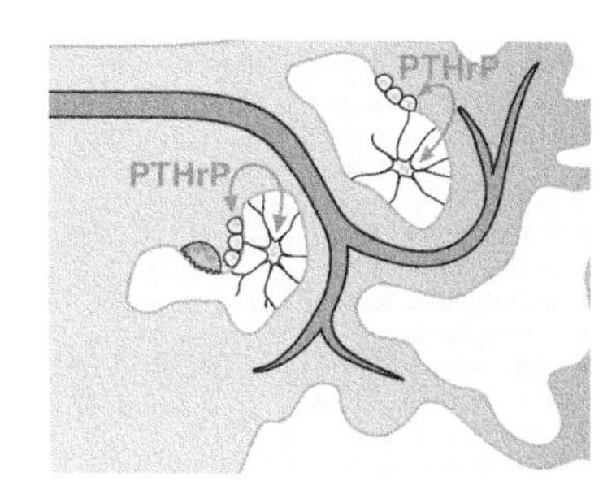

FIGURE 25.6 Physiological actions of parathyroid hormone–related protein (*PTHrP*) versus pharmacologic actions of PTH (teriparatide) and abaloparatide. (A) Pharmacologic administration of intermittent injections of PTH (teriparatide) or abaloparatide provide brief bursts of high systemic levels of these agents that act through the PTH receptor to stimulate bone remodeling at multiple basic multicellular units (BMUs) throughout the skeleton. (B) Physiologically, PTHrP is not provided through the bloodstream, but is provided locally, and stimulates active BMUs at those sites at which they are needed and active.

mass. The cellular actions of exogenous PTH administration are to promote osteoblast differentiation, prevent osteoblast and osteocyte apoptosis, and reduce osteocyte production of the bone formation inhibitor, sclerostin (Jilka, 2007; Martin, 2016). Although PTH promotes bone resorption by stimulating RANKL expression in osteoblasts, in the context of intermittent administration, PTH action on osteoclasts is probably confined to the recruitment of osteoclasts for the initiation of new BMUs, because of the brief exposure of cells to PTH. In contrast, prolonged activation of PTHR1 results in a net greater effect on osteoclast formation and activity. Early experimental evidence for this was obtained in rats using treatment with several different intermittent PTH(1–34) injection schedules and continuous infusion (Frolik et al., 2003). The pharmacokinetic profile and outcome data showed that an anabolic outcome required a short period of elevation of plasma PTH, whereas a catabolic outcome was associated with prolonged elevation of PTH levels.

It is easy to see how PTH given pharmacologically by intermittent injection can mimic the effects of local PTHrP to promote bone formation that take place through the PTH1R, since all these effects of anabolic PTH have been ascribed to locally generated PTHrP (Fig. 25.6). Nevertheless, if the anabolic action of PTHrP is modified in any way, or mediated in part by non-receptor-mediated actions in addition to PTHR1-mediated actions, some differences might be expected. In that case discovering how noncanonical actions might signal through osteoblastic cells could provide ways to optimize PTH-based therapies.

A question related to this concerns the interpretation of experimental data obtained by the treatment of cells in vitro with PTHR1 agonists. Over decades there have been very many published effects of treatment of cells in vitro with PTH, e.g., on gene expression and/or protein production. Although these are often taken to indicate that such pharmacological actions reflect physiological actions of PTH in vivo, evidence for such a physiological role is invariably lacking. In such circumstances, the possibility should be considered that such effects might reflect physiological roles of PTHrP. Some examples in which this might be considered are the PTH promotion of glycolysis in osteoblast lineage cells (Esen et al., 2015), PTH inhibition of salt-induced kinase in osteocytes (Wein et al., 2016), and the finding in osteoblast precursor lineage tracing experiments in mice that PTHR1 is required for the PTH-mediated increase and suppression of apoptosis in early osteoblast precursors without any effect on their rate of proliferation (Balani et al., 2017). If these effects are also induced by physiological PTHrP, this provides further information on how this protein functions in the normal adult skeleton.

Since bone remodeling takes place asynchronously throughout the skeleton at sites where it is needed to replace old or damaged bone and to respond to loading and unloading, it is obvious that the control of remodeling needs to be very local. It might be argued that local PTHrP release would need to be very tightly regulated so that in bone remodeling it is presented only briefly to its target cells so that bone formation is promoted, rather than stimulating both bone formation and resorption. Excessive osteoclast formation in response to PTHrP is unlikely in the context of normal physiological remodeling, where (1) BMUs are active at any one time only in those places in which they are required (Fig. 25.6) and (2) where only osteoblast-derived, and not osteocyte-derived, PTHrP promotes osteoclast formation, as evidenced by the comparison of the cell-specific PTHrP-null mice (see earlier and Table 25.1).

Can changes in the level of the circulating hormone, PTH, exert any fine controlling influence upon these processes that take place asynchronously, scattered throughout the skeleton in virtually a stochastic manner? There have been studies that suggest that circulating PTH levels vary in a circadian manner in human subjects. The changes in circulating levels of PTH in those studies were consistently at a maximum of about 40% and within the normal circulation range of 2–6 pmol/L (Fraser et al., 1998; Rejnmark et al., 2002; Joseph et al., 2007). In one such study (Logue et al., 1989) the late night/early morning increase in plasma PTH rose in parallel with nephrogenous cAMP, suggesting the change in PTH may be sufficient to produce an appropriate response in the kidney target. It remains difficult to envisage how such changes in circulating PTH could bring about initiation of BMUs at sites of need scattered throughout the skeleton. These comments do not apply to situations of PTH excess. Persistent elevation of circulating PTH resulting from primary hyperparathyroidism, from prolonged PTH treatment, or from transgenic overexpression in mice results in a generalized activation of remodeling sites and therefore increased resorption.

PTHrP analogs in pharmacology: could this change the approach to skeletal anabolic therapy?

Studies of the anabolic effect of PTHrP(1−36) in human subjects had their origin in the view that amino-terminal PTHrP is the paracrine ligand for PTHR1 in bone (Wu et al., 1996). The susceptibility of full-length protein to proteolysis (Orloff et al., 1989; Diefenbach-Jagger et al., 1995) precluded its use for this application, so preclinical experiments tested several truncated forms as anabolic agents, including PTHrP(1−34), PTHrP(1−36), and PTHrP(1−74). From the earliest of these studies in the rat (Hock et al., 1989) it was evident that these PTHrP preparations exerted anabolic effects, but were significantly less potent in vivo than PTH, although the two are equally active in promoting cAMP production in vitro. Such a lesser potency in vivo could be anticipated because even these truncated forms of PTHrP have many target sequences that are susceptible to proteolysis (Fig. 25.1). The anabolic action of PTHrP(1−36) in human subjects has been investigated extensively by Stewart and colleagues (Horwitz et al. 2003, 2006, 2010, 2011, 2013; Augustine and Horwitz, 2013). The daily dose of PTHrP(1−36) in human subjects required to increase bone formation markers was manifold higher than the dose of teriparatide needed to achieve these marker levels (Augustine and Horwitz, 2013; Horwitz et al., 2013), probably due to the proteolytic cleavage described earlier. The higher dose requirements might be due to a difference in pharmacokinetics, with PTHrP(1−36) degraded more rapidly following injection, and thus not so widely distributed to activate BMUs. The result would be that lesser amounts of active agonist would be available to the receptor.

Early studies by Stewart and colleagues suggested that PTHrP(1−36) action was relatively free of a resorptive effect as determined by measuring serum biochemical markers of bone formation and resorption. Similar claims have been made for abaloparatide (Leder et al., 2015; Miller et al., 2016), which is a 34-amino-acid peptide in which the first 21 residues are identical with PTHrP. Abaloparatide also contains a number of substitutions that were planned to improve stability (Dong et al., 2001; Hattersley et al., 2016), though the most susceptible of the PTHrP cleavage sites ($-R_{19}R_{20}R_{21}-$) is located within both PTHrP(1−36) and abaloparatide. Abaloparatide (80 μg per day) and teriparatide (20 μg per day) have been reported to have comparable anti-fracture efficacy in the vertebrae, with claims based on biochemical marker data of a greater net bone formation due to a lesser effect on bone resorption marker with abaloparatide than with teriparatide (Dong et al., 2001; Leder et al., 2015; Hattersley et al., 2016). In the same data, the formation marker level was also less with abaloparatide (Dong et al., 2001), so the claim of a lesser effect on resorption might not be soundly based. There remains no information on the cellular basis by which PTHrP(1−36) or abaloparatide could have a lesser effect on bone resorption than formation.

The structural features of abaloparatide have led to it being called a PTHrP analog that is a selective activator of the PTHR1 receptor (Dong et al., 2001; Leder et al., 2015; Hattersley et al., 2016), with adenylyl cyclase assay data showing it to be equipotent with teriparatide. Despite this equivalence, when PTH(1−34) and PTHrP(1−36) were compared for their initial receptor interaction mechanisms, the action of PTHrP(1−36) was restricted to the cell surface, while PTHrP(1−34) was more readily internalized and brought about a somewhat more prolonged increase in cAMP in the target cells through continued endosomal signaling (Ferrandon et al., 2009) (Gardella and Vilardaga, 2015). Similar findings were obtained with abaloparatide (Hattersley et al., 2016), leading to this being suggested as a mechanism by which PTHrP(1−36) and abaloparatide might lead to a lesser resorption response than that with PTH(1−34). These in vitro findings of differing initial interactions of the peptides with receptor have not been taken beyond the stage of cAMP generation. It is difficult to extrapolate them to explain the purportedly different in vivo effects on bone resorption given the lack of evidence of differences in cellular effects beyond cAMP generation in vitro, e.g., on gene expression.

Without information to the contrary it might be assumed that any anabolic action of abaloparatide is through a mechanism shared with teriparatide—predominantly activation of remodeling sites. The possibility of either PTHrP(1−36) or abaloparatide being anabolic and resorption sparing would be an intriguing one, if very difficult to explain. Given that resorption is an inescapable first phase of BMU-based bone remodeling, it is not easy to see how an anabolic effect equivalent to that of teriparatide could be obtained while sparing resorption. The nature of the question is such that it could be resolved only by the use of histomorphometry such as that used to show the predominantly modeling effect of teriparatide when used as an anabolic agent (Lindsay et al., 2006; Ma et al., 2006). A more likely explanation might lie in the pharmacokinetics, with a lesser potency of abaloparatide in vivo because of proteolytic degradation, and less of the agonist being presented to the receptor.

Abaloparatide offers an interesting new approach to anabolic therapy that merits much further understanding of mechanisms. If it can be regarded as a "PTHrP analog," it might be instructive in revealing aspects of the actions of the PTHrP N-terminal domain on target bone cells. The findings indicating full-length PTHrP as the locally generated secreted product highlight the need to consider also the possibility of other PTHrP actions in bone, especially through noncanonical pathways. It has always been assumed that all that is required of PTHrP is that it exerts its activity through acting on the PTHR1. That might not be so, however, in light of the increasing awareness of noncanonical actions of PTHrP.

Conclusion

We have been learning since the 1980s how PTH and PTHrP share a common receptor in many tissues. The current concept is that the hormone PTH regulates calcium homeostasis in development and maturity. PTHrP provides calcium to the developing fetus through its placental calcium transport action, and its local expression directs growth plate development by controlling chondrocyte proliferation and differentiation. Postnatally, PTH assumes the role of supplying calcium by stimulating bone resorption, while PTHrP is the main factor generated locally in bone and controls bone remodeling, without normally contributing to the maintenance of serum calcium. In maturity calcium homeostasis is the sole domain of PTH. Whether PTH physiologically regulates bone formation in maturity remains to be established in human subjects. These views have developed largely as a result of the insights provided from mouse genetic experiments, but much is owed also to pharmacological studies in animals and to the therapeutic efficacy of PTH in the treatment of osteoporosis.

The susceptibility of PTHrP to posttranslational processing is a striking feature of the molecule that results in the generation of a number of peptide components, some of which have been shown to have biological activities that do not require the PTHR1, leading it to be considered as a polyfunctional cytokine. The search for the presumed receptors of these peptides and the identification of the circulating peptide isoforms obviously needs to continue. Such investigation provides attractive challenges, as does the aim of determining the function of nuclear PTHrP. The nuclear role(s) and the susceptibility of PTHrP to proteolysis, together with the lability of PTHrP mRNA and the multiple splice forms, may be the result of evolutionary pressures that equip PTHrP to function as a paracrine agent, as it does in so many tissues.

When used as a skeletal anabolic therapy, PTH appears simply to be reproducing the local physiologic action of PTHrP. The great susceptibility of PTHrP to proteolytic degradation makes it unsuitable as a therapy itself. The possibility is raised in this chapter that domains of PTHrP other than that acting through PTHR1 might contribute to the action on bone. If that were so, the signaling pathways involved could be of interest therapeutically. In the meantime, the use of analogs of PTH and PTHrP that act through PTHR1 are the subject of attention, and much can be learned of the anabolic process through their use.

Acknowledgments

The authors acknowledge research support from the National Health and Medical Research Council (Australia) and the Victorian Government OIS Program.

References

Aarts, M.M., Davidson, D., Corluka, A., Petroulakis, E., Guo, J., Bringhurst, F.R., Galipeau, J., Henderson, J.E., 2001. Parathyroid hormone-related protein promotes quiescence and survival of serum-deprived chondrocytes by inhibiting rRNA synthesis. J. Biol. Chem. 276 (41), 37934–37943.

Aarts, M.M., Levy, D., He, B., Stregger, S., Chen, T., Richard, S., Henderson, J.E., 1999a. Parathyroid hormone-related protein interacts with RNA. J. Biol. Chem. 274 (8), 4832–4838.

Aarts, M.M., Rix, A., Guo, J., Bringhurst, R., Henderson, J.E., 1999b. The nucleolar targeting signal (NTS) of parathyroid hormone related protein mediates endocytosis and nucleolar translocation. J. Bone Miner. Res. 14 (9), 1493–1503.

Abbas, S.K., Pickard, D.W., Rodda, C.P., Heath, J.A., Hammonds, R.G., Wood, W.I., Caple, I.W., Martin, T.J., Care, A.D., 1989. Stimulation of ovine placental calcium transport by purified natural and recombinant parathyroid hormone-related protein (PTHrP) preparations. Q. J. Exp. Physiol. 74 (4), 549–552.

Albright, F., 1941. Case records of the Massachisetts general hospital – case 39061. N. Engl. J. Med. 225, 789–796.

Alonso, V., de Gortazar, A.R., Ardura, J.A., Andrade-Zapata, I., Alvarez-Arroyo, M.V., Esbrit, P., 2008. Parathyroid hormone-related protein (107–139) increases human osteoblastic cell survival by activation of vascular endothelial growth factor receptor-2. J. Cell. Physiol. 217 (3), 717–727.

Amizuka, N., Fukushi-Irie, M., Sasaki, T., Oda, K., Ozawa, H., 2000. Inefficient function of the signal sequence of PTHrP for targeting into the secretory pathway. Biochem. Biophys. Res. Commun. 273 (2), 621–629.

Amizuka, N., Karaplis, A.C., Henderson, J.E., Warshawsky, H., Lipman, M.L., Matsuki, Y., Ejiri, S., Tanaka, M., Izumi, N., Ozawa, H., Goltzman, D., 1996. Haploinsufficiency of parathyroid hormone-related peptide (PTHrP) results in abnormal postnatal bone development. Dev. Biol. 175 (1), 166–176.

Amizuka, N., Warshawsky, H., Henderson, J.E., Goltzman, D., Karaplis, A.C., 1994. Parathyroid hormone-related peptide-depleted mice show abnormal epiphyseal cartilage development and altered endochondral bone formation. J. Cell Biol. 126 (6), 1611–1623.

Ansari, N., Ho, P., Crimeen-Irwin, B., Poulton, I.J., Brunt, A.R., Forwoof, M.R., VDivieti Pajevic, P., Gooi, J., Martin, T.J., Sims, N.A., 2017. Autocrine and paracrine regulation of the murine skeleton by osteocyte -derived parathyroid hormone-related protein. J. Bone Miner. Res. https://doi.org/10.1002/jbmr.3291.

Augustine, M., Horwitz, M.J., 2013. Parathyroid hormone and parathyroid hormone-related protein analogs as therapies for osteoporosis. Curr. Osteoporos. Rep. 11 (4), 400–406.

Baker, E.K., Taylor, S., Gupte, A., Sharp, P.P., Walia, M., Walsh, N.C., Zannettino, A.C., Chalk, A.M., Burns, C.J., Walkley, C.R., 2015. BET inhibitors induce apoptosis through a MYC independent mechanism and synergise with CDK inhibitors to kill osteosarcoma cells. Sci. Rep. 5, 10120.

Balani, D.H., Ono, N., Kronenberg, H.M., 2017. Parathyroid hormone regulates fates of murine osteoblast precursors in vivo. J. Clin. Investig. 127 (9), 3327–3338.

Benson Jr., R.C., Riggs, B.L., Pickard, B.M., Arnaud, C.D., 1974. Immunoreactive forms of circulating parathyroid hormone in primary and ectopic hyperparathyroidism. J. Clin. Investig. 54 (1), 175–181.

Berman, S.D., Calo, E., Landman, A.S., Danielian, P.S., Miller, E.S., West, J.C., Fonhoue, B.D., Caron, A., Bronson, R., Bouxsein, M.L., Mukherjee, S., Lees, J.A., 2008. Metastatic osteosarcoma induced by inactivation of Rb and p53 in the osteoblast lineage. Proc. Natl. Acad. Sci. U.S.A. 105 (33), 11851–11856.

Berson, S.A., Yalow, R.S., 1966. Parathyroid hormone in plasma in adenomatous hyperparathyroidism, uremia, and bronchogenic carcinoma. Science 154 (3751), 907–909.

Briggs, L.J., Johnstone, R.W., Elliot, R.M., Xiao, C.Y., Dawson, M., Trapani, J.A., Jans, D.A., 2001. Novel properties of the protein kinase CK2-site-regulated nuclear- localization sequence of the interferon-induced nuclear factor IFI 16. Biochem. J. 353 (Pt 1), 69–77.

Bruns, M.E., Ferguson 2nd, J.E., Bruns, D.E., Burton, D.W., Brandt, D.W., Juppner, H., Segre, G.V., Deftos, L.J., 1995. Expression of parathyroid hormone-related peptide and its receptor messenger ribonucleic acid in human amnion and chorion-decidua: implications for secretion and function. Am. J. Obstet. Gynecol. 173 (3 Pt 1), 739–746.

Burgess, T.L., Kelly, R.B., 1987. Constitutive and regulated secretion of proteins. Annu. Rev. Cell Biol. 3, 243–293.

Burtis, W.J., Brady, T.G., Orloff, J.J., Ersbak, J.B., Warrell Jr., R.P., Olson, B.R., Wu, T.L., Mitnick, M.E., Broadus, A.E., Stewart, A.F., 1990. Immunochemical characterization of circulating parathyroid hormone-related protein in patients with humoral hypercalcemia of cancer. N. Engl. J. Med. 322 (16), 1106–1112.

Campos, R.V., Asa, S.L., Drucker, D.J., 1991. Immunocytochemical localization of parathyroid hormone-like peptide in the rat fetus. Cancer Res. 51 (23 Pt 1), 6351–6357.

Care, A.D., Abbas, S.K., Pickard, D.W., Barri, M., Drinkhill, M., Findlay, J.B., White, I.R., Caple, I.W., 1990. Stimulation of ovine placental transport of calcium and magnesium by mid-molecule fragments of human parathyroid hormone-related protein. Exp. Physiol. 75 (4), 605–608.

Charbon, G.A., 1968. A rapid and selective vasodialtor effect of parathyroid hormone. Eur. J. Pharmacol. 3 (3), 275–278.

Charbon, G.A., Hulstaert, P.F., 1974. Augmentation of arterial hepatic and renal flow by extracted and synthetic parathyroid hormone. Endocrinology 96 (2), 621–626.

Chen, X., Bahrami, A., Pappo, A., Easton, J., Dalton, J., Hedlund, E., Ellison, D., Shurtleff, S., Wu, G., Wei, L., Parker, M., Rusch, M., Nagahawatte, P., Wu, J., Mao, S., Boggs, K., Mulder, H., Yergeau, D., Lu, C., Ding, L., Edmonson, M., Qu, C., Wang, J., Li, Y., Navid, F., Daw, N.C., Mardis, E.R., Wilson, R.K., Downing, J.R., Zhang, J., Dyer, M.A., P. St Jude Children's Research Hospital-Washington University Pediatric Cancer Genome, 2014. Recurrent somatic structural variations contribute to tumorigenesis in pediatric osteosarcoma. Cell Rep. 7 (1), 104–112.

Cingolani, G., Bednenko, J., Gillespie, M.T., Gerace, L., 2002. Molecular basis for the recognition of a nonclassical nuclear localization signal by importin beta. Mol. Cell 10 (6), 1345–1353.

Compston, J.E., 2007. Skeletal actions of intermittent parathyroid hormone: effects on bone remodelling and structure. Bone 40 (6), 1447–1452.

Cornish, J., Callon, K.E., Lin, C., Xiao, C., Moseley, J.M., Reid, I.R., 1999. Stimulation of osteoblast proliferation by C-terminal fragments of parathyroid hormone-related protein. J. Bone Miner. Res. 14 (6), 915–922.

Cornish, J., Callon, K.E., Nicholson, G.C., Reid, I.R., 1997. Parathyroid hormone-related protein-(107-139) inhibits bone resorption in vivo. Endocrinology 138 (3), 1299–1304.

de Castro, L.F., Lozano, D., Portal-Nunez, S., Maycas, M., De la Fuente, M., Caeiro, J.R., Esbrit, P., 2012. Comparison of the skeletal effects induced by daily administration of PTHrP (1-36) and PTHrP (107-139) to ovariectomized mice. J. Cell. Physiol. 227 (4), 1752–1760.

de Gortazar, A.R., Alonso, V., Alvarez-Arroyo, M.V., Esbrit, P., 2006. Transient exposure to PTHrP (107-139) exerts anabolic effects through vascular endothelial growth factor receptor 2 in human osteoblastic cells in vitro. Calcif. Tissue Int. 79 (5), 360–369.

de Miguel, F., Fiaschi-Taesch, N., Lopez-Talavera, J.C., Takane, K.K., Massfelder, T., Helwig, J.J., Stewart, A.F., 2001. The C-terminal region of PTHrP, in addition to the nuclear localization signal, is essential for the intracrine stimulation of proliferation in vascular smooth muscle cells. Endocrinology 142 (9), 4096–4105.

Delgado-Calle, J., Tu, X., Pacheco-Costa, R., McAndrews, K., Edwards, R., Pellegrini, G.G., Kuhlenschmidt, K., Olivos, N., Robling, A., Peacock, M., Plotkin, L.I., Bellido, T., 2017. Control of bone anabolism in response to mechanical loading and PTH by distinct mechanisms downstream of the PTH receptor. J. Bone Miner. Res. 32 (3), 522–535.

Diefenbach-Jagger, H., Brenner, C., Kemp, B.E., Baron, W., McLean, J., Martin, T.J., Moseley, J.M., 1995. Arg21 is the preferred kexin cleavage site in parathyroid-hormone-related protein. Eur. J. Biochem. 229 (1), 91–98.

Dong, J., Shen, Y., Culler, M., JE, C.-W., Woon, J.-J., Legrand, B., Morgan, M., Chorev, M., Rosenblatt, Nakamoto, C., Moreau, J., 2001. Highly potent analogs of human parathyroid hormone and human parathyroid hormone-related protein. In: Houghten, M.L.a.R. (Ed.), Peptides: The Wave of the Future. American Peptide Society, USA, pp. 668–669.

Dunne, F.P., Ratcliffe, W.A., Mansour, P., Heath, D.A., 1994. Parathyroid hormone related protein (PTHrP) gene expression in fetal and extra-embryonic tissues of early pregnancy. Hum. Reprod. 9 (1), 149–156.

Engin, F., Yao, Z., Yang, T., Zhou, G., Bertin, T., Jiang, M.M., Chen, Y., Wang, L., Zheng, H., Sutton, R.E., Boyce, B.F., Lee, B., 2008. Dimorphic effects of Notch signaling in bone homeostasis. Nat. Med. 14 (3), 299–305.

Esen, E., Lee, S.Y., Wice, B.M., Long, F., 2015. PTH promotes bone anabolism by stimulating aerobic glycolysis via IGF signaling. J. Bone Miner. Res. 30 (11), 1959–1968.

Fan, Y., Bi, R., Densmore, M.J., Sato, T., Kobayashi, T., Yuan, Q., Zhou, X., Erben, R.G., Lanske, B., 2016. Parathyroid hormone 1 receptor is essential to induce FGF23 production and maintain systemic mineral ion homeostasis. FASEB J. 30 (1), 428–440.

Fenton, A.J., Kemp, B.E., Hammonds Jr., R.G., Mitchelhill, K., Moseley, J.M., Martin, T.J., Nicholson, G.C., 1991a. A potent inhibitor of osteoclastic bone resorption within a highly conserved pentapeptide region of parathyroid hormone-related protein; PTHrP[107-111]. Endocrinology 129 (6), 3424–3426.

Fenton, A.J., Kemp, B.E., Kent, G.N., Moseley, J.M., Zheng, M.H., Rowe, D.J., Britto, J.M., Martin, T.J., Nicholson, G.C., 1991b. A carboxyl-terminal peptide from the parathyroid hormone-related protein inhibits bone resorption by osteoclasts. Endocrinology 129 (4), 1762–1768.

Fenton, A.J., Martin, T.J., Nicholson, G.C., 1993. Long-term culture of disaggregated rat osteoclasts: inhibition of bone resorption and reduction of osteoclast-like cell number by calcitonin and PTHrP[107-139]. J. Cell. Physiol. 155 (1), 1–7.

Fenton, A.J., Martin, T.J., Nicholson, G.C., 1994. Carboxyl-terminal parathyroid hormone-related protein inhibits bone resorption by isolated chicken osteoclasts. J. Bone Miner. Res. 9 (4), 515–519.

Ferrandon, S., Feinstein, T.N., Castro, M., Wang, B., Bouley, R., Potts, J.T., Gardella, T.J., Vilardaga, J.P., 2009. Sustained cyclic AMP production by parathyroid hormone receptor endocytosis. Nat. Chem. Biol. 5 (10), 734–742.

Fiaschi-Taesch, N., Sicari, B., Ubriani, K., Cozar-Castellano, I., Takane, K.K., Stewart, A.F., 2009. Mutant parathyroid hormone-related protein, devoid of the nuclear localization signal, markedly inhibits arterial smooth muscle cell cycle and neointima formation by coordinate up-regulation of p15Ink4b and p27kip1. Endocrinology 150 (3), 1429–1439.

Fraser, W.D., Logue, F.C., Christie, J.P., Gallacher, S.J., Cameron, D., O'Reilly, D.S., Beastall, G.H., Boyle, I.T., 1998. Alteration of the circadian rhythm of intact parathyroid hormone and serum phosphate in women with established postmenopausal osteoporosis. Osteoporos. Int. 8 (2), 121–126.

Frolik, C.A., Black, E.C., Cain, R.L., Satterwhite, J.H., Brown-Augsburger, P.L., Sato, M., Hock, J.M., 2003. Anabolic and catabolic bone effects of human parathyroid hormone (1-34) are predicted by duration of hormone exposure. Bone 33 (3), 372–379.

Gagnon, L., Jouishomme, H., Whitfield, J.F., Durkin, J.P., MacLean, S., Neugebauer, W., Willick, G., Rixon, R.H., Chakravarthy, B., 1993. Protein kinase C-activating domains of parathyroid hormone-related protein. J. Bone Miner. Res. 8 (4), 497–503.

Garcia-Martin, A., Ardura, J.A., Maycas, M., Lozano, D., Lopez-Herradon, A., Portal-Nunez, S., Garcia-Ocana, A., Esbrit, P., 2014. Functional roles of the nuclear localization signal of parathyroid hormone-related protein (PTHrP) in osteoblastic cells. Mol. Endocrinol. 28 (6), 925–934.

Gardella, T.J., Vilardaga, J.P., 2015. International Union of Basic and Clinical Pharmacology. XCIII. The parathyroid hormone receptors-family B G protein-coupled receptors. Pharmacol. Rev. 67 (2), 310–337.

Goerdeladze, J., Jablonski, G., Paulssen, R., Mortensen, B., Gutvik, K., Haug, E., Rian, E., Jemtland, R., Friedman, E., Bruland, O., 1993. In: Frontiers in osteosarcoma research, Novak, J.F., McMaster, J.H. (Eds.), G-protein Coupled Signaling in Osteosarcoma Cell Lines. Hogref and Huber, Seattle, Toronto, pp. 297–308.

Goodrich, L.V., Johnson, R.L., Milenkovic, L., McMahon, J.A., Scott, M.P., 1996. Conservation of the hedgehog/patched signaling pathway from flies to mice: induction of a mouse patched gene by Hedgehog. Genes Dev. 10 (3), 301–312.

Grill, V., Ho, P., Body, J.J., Johanson, N., Lee, S.C., Kukreja, S.C., Moseley, J.M., Martin, T.J., 1991. Parathyroid hormone-related protein: elevated levels in both humoral hypercalcemia of malignancy and hypercalcemia complicating metastatic breast cancer. J. Clin. Endocrinol. Metab. 73 (6), 1309–1315.

Guenther, H.L., Hofstetter, W., Moseley, J.M., Gillespie, M.T., Suda, N., Martin, T.J., 1995. Evidence for the synthesis of parathyroid hormone-related protein (PTHrP) by nontransformed clonal rat osteoblastic cells in vitro. Bone 16 (3), 341–347.

Guise, T.A., Yin, J.J., Taylor, S.D., Kumagai, Y., Dallas, M., Boyce, B.F., Yoneda, T., Mundy, G.R., 1996. Evidence for a causal role of parathyroid hormone-related protein in the pathogenesis of human breast cancer-mediated osteolysis. J. Clin. Investig. 98 (7), 1544–1549.

Gupte, A., Baker, E.K., Wan, S.S., Stewart, E., Loh, A., Shelat, A.A., Gould, C.M., Chalk, A.M., Taylor, S., Lackovic, K., Karlstrom, A., Mutsaers, A.J., Desai, J., Madhamshettiwar, P.B., Zannettino, A.C., Burns, C., Huang, D.C., Dyer, M.A., Simpson, K.J., Walkley, C.R., 2015. Systematic screening identifies dual PI3K and mTOR inhibition as a conserved therapeutic vulnerability in osteosarcoma. Clin. Cancer Res. 21 (14), 3216–3229.

Hammonds Jr., R.G., McKay, P., Winslow, G.A., Diefenbach-Jagger, H., Grill, V., Glatz, J., Rodda, C.P., Moseley, J.M., Wood, W.I., Martin, T.J., et al., 1989. Purification and characterization of recombinant human parathyroid hormone-related protein. J. Biol. Chem. 264 (25), 14806–14811.

Hattersley, G., Dean, T., Corbin, B.A., Bahar, H., Gardella, T.J., 2016. Binding selectivity of abaloparatide for PTH-type-1-receptor conformations and effects on downstream signaling. Endocrinology 157 (1), 141–149.

Henderson, J.E., Amizuka, N., Warshawsky, H., Biasotto, D., Lanske, B.M., Goltzman, D., Karaplis, A.C., 1995. Nucleolar localization of parathyroid hormone-related peptide enhances survival of chondrocytes under conditions that promote apoptotic cell death. Mol. Cell Biol. 15 (8), 4064–4075.

Hirai, T., Kobayashi, T., Nishimori, S., Karaplis, A.C., Goltzman, D., Kronenberg, H.M., 2015. Bone is a major target of PTH/PTHrP receptor signaling in regulation of fetal blood calcium homeostasis. Endocrinology 156 (8), 2774–2780.

Ho, P.W., Goradia, A., Russell, M.R., Chalk, A.M., Milley, K.M., Baker, E.K., Danks, J.A., Slavin, J.L., Walia, M., Crimeen-Irwin, B., Dickins, R.A., Martin, T.J., Walkley, C.R., 2014. Knockdown of PTHR1 in osteosarcoma cells decreases invasion and growth and increases tumor differentiation in vivo. Oncogene 34, 2922–2933, 2015.

Hock, J.M., Fonseca, J., Gunness-Hey, M., Kemp, B.E., Martin, T.J., 1989. Comparison of the anabolic effects of synthetic parathyroid hormone-related protein (PTHrP) 1-34 and PTH 1-34 on bone in rats. Endocrinology 125 (4), 2022–2027.

Horwitz, M.J., Augustine, M., Khan, L., Martin, E., Oakley, C.C., Carneiro, R.M., Tedesco, M.B., Laslavic, A., Sereika, S.M., Bisello, A., Garcia-Ocana, A., Gundberg, C.M., Cauley, J.A., Stewart, A.F., 2013. A comparison of parathyroid hormone-related protein (1-36) and parathyroid hormone (1-34) on markers of bone turnover and bone density in postmenopausal women: the PrOP study. J. Bone Miner. Res. 28 (11), 2266–2276.

Horwitz, M.J., Tedesco, M.B., Garcia-Ocana, A., Sereika, S.M., Prebehala, L., Bisello, A., Hollis, B.W., Gundberg, C.M., Stewart, A.F., 2010. Parathyroid hormone-related protein for the treatment of postmenopausal osteoporosis: defining the maximal tolerable dose. J. Clin. Endocrinol. Metab. 95 (3), 1279–1287.

Horwitz, M.J., Tedesco, M.B., Sereika, S.M., Garcia-Ocana, A., Bisello, A., Hollis, B.W., Gundberg, C., Stewart, A.F., 2006. Safety and tolerability of subcutaneous PTHrP(1-36) in healthy human volunteers: a dose escalation study. Osteoporos. Int. 17 (2), 225–230.

Horwitz, M.J., Tedesco, M.B., Sereika, S.M., Hollis, B.W., Garcia-Ocana, A., Stewart, A.F., 2003. Direct comparison of sustained infusion of human parathyroid hormone-related protein-(1-36) [hPTHrP-(1-36)] versus hPTH-(1-34) on serum calcium, plasma 1,25-dihydroxyvitamin D concentrations, and fractional calcium excretion in healthy human volunteers. J. Clin. Endocrinol. Metab. 88 (4), 1603–1609.

Horwitz, M.J., Tedesco, M.B., Sereika, S.M., Prebehala, L., Gundberg, C.M., Hollis, B.W., Bisello, A., Garcia-Ocana, A., Carneiro, R.M., Stewart, A.F., 2011. A 7-day continuous infusion of PTH or PTHrP suppresses bone formation and uncouples bone turnover. J. Bone Miner. Res. 26 (9), 2287–2297.

Ingleton, P.M., Underwood, J.C., Hunt, N.H., Atkins, D., Giles, B., Coulton, L.A., Martin, T.J., 1977. Radiation induced osteogenic sarcoma in the rat as a model of hormone-responsive differentiated cancer. Lab. Anim. Sci. 27 (5 Pt 2), 748–756.

Iwamoto, M., Enomoto-Iwamoto, M., Kurisu, K., 1999. Actions of hedgehog proteins on skeletal cells. Crit. Rev. Oral Biol. Med. 10 (4), 477–486.

Jacks, T., Remington, L., Williams, B.O., Schmitt, E.M., Halachmi, S., Bronson, R.T., Weinberg, R.A., 1994. Tumor spectrum analysis in p53-mutant mice. Curr. Biol. 4 (1), 1–7.

Jans, D.A., 1995. The regulation of protein transport to the nucleus by phosphorylation. Biochem. J. 311 (Pt 3), 705–716.

Jans, D.A., Ackermann, M.J., Bischoff, J.R., Beach, D.H., Peters, R., 1991. p34cdc2-mediated phosphorylation at T124 inhibits nuclear import of SV-40 T antigen proteins. J. Cell Biol. 115 (5), 1203–1212.

Jans, D.A., Briggs, L.J., Gustin, S.E., Jans, P., Ford, S., Young, I.G., 1997. The cytokine interleukin-5 (IL-5) effects cotransport of its receptor subunits to the nucleus in vitro. FEBS Lett. 410 (2–3), 368–372.

Jilka, R.L., 2007. Molecular and cellular mechanisms of the anabolic effect of intermittent PTH. Bone 40 (6), 1434–1446.

Jolette, J., Attalla, B., Varela, A., Long, G.G., Mellal, N., Trimm, S., Smith, S.Y., Ominsky, M.S., Hattersley, G., 2017. Comparing the incidence of bone tumors in rats chronically exposed to the selective PTH type 1 receptor agonist abaloparatide or PTH(1-34). Regul. Toxicol. Pharmacol. 86, 356–365.

Jolette, J., Wilker, C.E., Smith, S.Y., Doyle, N., Hardisty, J.F., Metcalfe, A.J., Marriott, T.B., Fox, J., Wells, D.S., 2006. Defining a noncarcinogenic dose of recombinant human parathyroid hormone 1-84 in a 2-year study in Fischer 344 rats. Toxicol. Pathol. 34 (7), 929–940.

Joseph, F., Chan, B.Y., Durham, B.H., Ahmad, A.M., Vinjamuri, S., Gallagher, J.A., Vora, J.P., Fraser, W.D., 2007. The circadian rhythm of osteoprotegerin and its association with parathyroid hormone secretion. J. Clin. Endocrinol. Metab. 92 (8), 3230–3238.

Jouishomme, H., Whitfield, J.F., Chakravarthy, B., Durkin, J.P., Gagnon, L., Isaacs, R.J., MacLean, S., Neugebauer, W., Willick, G., Rixon, R.H., 1992. The protein kinase-C activation domain of the parathyroid hormone. Endocrinology 130 (1), 53–60.

Juppner, H., Abou-Samra, A.B., Freeman, M., Kong, X.F., Schipani, E., Richards, J., Kolakowski Jr., L.F., Hock, J., Potts Jr., J.T., Kronenberg, H.M., et al., 1991. A G protein-linked receptor for parathyroid hormone and parathyroid hormone-related peptide. Science 254 (5034), 1024–1026.

Kaji, H., Sugimoto, T., Kanatani, M., Fukase, M., Chihara, K., 1995. Carboxyl-terminal peptides from parathyroid hormone-related protein stimulate osteoclast-like cell formation. Endocrinology 136 (3), 842–848.

Karaplis, A.C., Kronenberg, H.M., 1996. Physiological roles for parathyroid hormone-related protein: lessons from gene knockout mice. Vitam. Horm. 52, 177–193.

Karaplis, A.C., Luz, A., Glowacki, J., Bronson, R.T., Tybulewicz, V.L., Kronenberg, H.M., Mulligan, R.C., 1994. Lethal skeletal dysplasia from targeted disruption of the parathyroid hormone-related peptide gene. Genes Dev. 8 (3), 277–289.

Karperien, M., van Dijk, T.B., Hoeijmakers, T., Cremers, F., Abou-Samra, A.B., Boonstra, J., de Laat, S.W., Defize, L.H., 1994. Expression pattern of parathyroid hormone/parathyroid hormone related peptide receptor mRNA in mouse postimplantation embryos indicates involvement in multiple developmental processes. Mech. Dev. 47 (1), 29–42.

Kartsogiannis, V., Moseley, J., McKelvie, B., Chou, S.T., Hards, D.K., Ng, K.W., Martin, T.J., Zhou, H., 1997. Temporal expression of PTHrP during endochondral bone formation in mouse and intramembranous bone formation in an in vivo rabbit model. Bone 21 (5), 385–392.

Kemp, B.E., Moseley, J.M., Rodda, C.P., Ebeling, P.R., Wettenhall, R.E., Stapleton, D., Diefenbach-Jagger, H., Ure, F., Michelangeli, V.P., Simmons, H.A., et al., 1987. Parathyroid hormone-related protein of malignancy: active synthetic fragments. Science 238 (4833), 1568–1570.

Kim, S.W., Pajevic, P.D., Selig, M., Barry, K.J., Yang, J.Y., Shin, C.S., Baek, W.Y., Kim, J.E., Kronenberg, H.M., 2012. Intermittent parathyroid hormone administration converts quiescent lining cells to active osteoblasts. J. Bone Miner. Res. 27 (10), 2075–2084.

Kovacs, C.S., 2014. Bone development and mineral homeostasis in the fetus and neonate: roles of the calciotropic and phosphotropic hormones. Physiol. Rev. 94 (4), 1143–1218.

Kovacs, C.S., Lanske, B., Hunzelman, J.L., Guo, J., Karaplis, A.C., Kronenberg, H.M., 1996. Parathyroid hormone-related peptide (PTHrP) regulates fetal-placental calcium transport through a receptor distinct from the PTH/PTHrP receptor. Proc. Natl. Acad. Sci. U.S.A. 93 (26), 15233–15238.

Kukreja, S.C., Rosol, T.J., Wimbiscus, S.A., Shevrin, D.H., Grill, V., Barengolts, E.I., Martin, T.J., 1990. Tumor resection and antibodies to parathyroid hormone-related protein cause similar changes on bone histomorphometry in hypercalcemia of cancer. Endocrinology 127 (1), 305–310.

Kukreja, S.C., Shemerdiak, W.P., Lad, T.E., Johnson, P.A., 1980. Elevated nephrogenous cyclic AMP with normal serum parathyroid hormone levels in patients with lung cancer. J. Clin. Endocrinol. Metab. 51 (1), 167–169.

Kukreja, S.C., Shevrin, D.H., Wimbiscus, S.A., Ebeling, P.R., Danks, J.A., Rodda, C.P., Wood, W.I., Martin, T.J., 1988. Antibodies to parathyroid hormone-related protein lower serum calcium in athymic mouse models of malignancy-associated hypercalcemia due to human tumors. J. Clin. Investig. 82 (5), 1798–1802.

Lafferty, F.W., 1966. Pseudohyperparathyroidism. Medicine (Baltim.) 45 (3), 247–260.

Lam, M.H., House, C.M., Tiganis, T., Mitchelhill, K.I., Sarcevic, B., Cures, A., Ramsay, R., Kemp, B.E., Martin, T.J., Gillespie, M.T., 1999. Phosphorylation at the cyclin-dependent kinases site (Thr85) of parathyroid hormone-related protein negatively regulates its nuclear localization. J. Biol. Chem. 274 (26), 18559–18566.

Lam, M.H., Hu, W., Xiao, C.Y., Gillespie, M.T., Jans, D.A., 2001. Molecular dissection of the importin beta1-recognized nuclear targeting signal of parathyroid hormone-related protein. Biochem. Biophys. Res. Commun. 282 (2), 629–634.

Lam, M.H., Olsen, S.L., Rankin, W.A., Ho, P.W., Martin, T.J., Gillespie, M.T., Moseley, J.M., 1997. PTHrP and cell division: expression and localization of PTHrP in a keratinocyte cell line (HaCaT) during the cell cycle. J. Cell. Physiol. 173 (3), 433–446.

Lam, M.H., Thomas, R.J., Martin, T.J., Gillespie, M.T., Jans, D.A., 2000. Nuclear and nucleolar localization of parathyroid hormone-related protein. Immunol. Cell Biol. 78 (4), 395–402.

Lanske, B., Amling, M., Neff, L., Guiducci, J., Baron, R., Kronenberg, H.M., 1999. Ablation of the PTHrP gene or the PTH/PTHrP receptor gene leads to distinct abnormalities in bone development. J. Clin. Investig. 104 (4), 399–407.

Lanske, B., Karaplis, A.C., Lee, K., Luz, A., Vortkamp, A., Pirro, A., Karperien, M., Defize, L.H., Ho, C., Mulligan, R.C., Abou-Samra, A.B., Juppner, H., Segre, G.V., Kronenberg, H.M., 1996. PTH/PTHrP receptor in early development and Indian hedgehog-regulated bone growth. Science 273 (5275), 663–666.

Leder, B.Z., O'Dea, L.S., Zanchetta, J.R., Kumar, P., Banks, K., McKay, K., Lyttle, C.R., Hattersley, G., 2015. Effects of abaloparatide, a human parathyroid hormone-related peptide analog, on bone mineral density in postmenopausal women with osteoporosis. J. Clin. Endocrinol. Metab. 100 (2), 697–706.

Lee, K., Deeds, J.D., Segre, G.V., 1995. Expression of parathyroid hormone-related peptide and its receptor messenger ribonucleic acids during fetal development of rats. Endocrinology 136 (2), 453–463.

Lee, K., Lanske, B., Karaplis, A.C., Deeds, J.D., Kohno, H., Nissenson, R.A., Kronenberg, H.M., Segre, G.V., 1996. Parathyroid hormone-related peptide delays terminal differentiation of chondrocytes during endochondral bone development. Endocrinology 137 (11), 5109–5118.

Li, J., Dong, S., Townsend, S.D., Dean, T., Gardella, T.J., Danishefsky, S.J., 2012. Chemistry as an expanding resource in protein science: fully synthetic and fully active human parathyroid hormone-related protein (1-141). Angew Chem. Int. Ed. Engl. 51 (49), 12263–12267.

Lin, P.P., Pandey, M.K., Jin, F., Raymond, A.K., Akiyama, H., Lozano, G., 2009. Targeted mutation of p53 and Rb in mesenchymal cells of the limb bud produces sarcomas in mice. Carcinogenesis 30 (10), 1789–1795.

Lindsay, R., Cosman, F., Zhou, H., Bostrom, M.P., Shen, V.W., Cruz, J.D., Nieves, J.W., Dempster, D.W., 2006. A novel tetracycline labeling schedule for longitudinal evaluation of the short-term effects of anabolic therapy with a single iliac crest bone biopsy: early actions of teriparatide. J. Bone Miner. Res. 21 (3), 366–373.

Liu, Y., Berendsen, A.D., Jia, S., Lotinun, S., Baron, R., Ferrara, N., Olsen, B.R., 2012. Intracellular VEGF regulates the balance between osteoblast and adipocyte differentiation. J. Clin. Investig. 122 (9), 3101–3113.

Logue, F.C., Fraser, W.D., O'Reilly, D.S., Beastall, G.H., 1989. The circadian rhythm of intact parathyroid hormone (1-84) and nephrogenous cyclic adenosine monophosphate in normal men. J. Endocrinol. 121 (1), R1–R3.

Lozano, D., Fernandez-de-Castro, L., Portal-Nunez, S., Lopez-Herradon, A., Dapia, S., Gomez-Barrena, E., Esbrit, P., 2011. The C-terminal fragment of parathyroid hormone-related peptide promotes bone formation in diabetic mice with low-turnover osteopaenia. Br. J. Pharmacol. 162 (6), 1424–1438.

Ma, Y.L., Zeng, Q., Donley, D.W., Ste-Marie, L.G., Gallagher, J.C., Dalsky, G.P., Marcus, R., Eriksen, E.F., 2006. Teriparatide increases bone formation in modeling and remodeling osteons and enhances IGF-II immunoreactivity in postmenopausal women with osteoporosis. J. Bone Miner. Res. 21 (6), 855–864.

Maeda, S., Wu, S., Juppner, H., Green, J., Aragay, A.M., Fagin, J.A., Clemens, T.L., 1996. Cell-specific signal transduction of parathyroid hormone (PTH)-related protein through stably expressed recombinant PTH/PTHrP receptors in vascular smooth muscle cells. Endocrinology 137 (8), 3154–3162.

Majeska, R.J., Rodan, S.B., Rodan, G.A., 1980. Parathyroid hormone-responsive clonal cell lines from rat osteosarcoma. Endocrinology 107 (5), 1494–1503.

Mangin, M., Ikeda, K., Broadus, A.E., 1990. Structure of the mouse gene encoding parathyroid hormone-related peptide. Gene 95 (2), 195–202.

Mangin, M., Ikeda, K., Dreyer, B.E., Milstone, L., Broadus, A.E., 1988. Two distinct tumor-derived, parathyroid hormone-like peptides result from alternative ribonucleic acid splicing. Mol. Endocrinol. 2 (11), 1049–1055.

Martin, T.J., 2016. Parathyroid hormone-related protein, its regulation of cartilage and bone development, and role in treating bone diseases. Physiol. Rev. 96 (3), 831–871.

Martin, T.J., Atkins, D., 1979. Biochemical regulators of bone resorption and their significance in cancer. Essays Med. Biochem. 4, 49–82.

Martin, T.J., Moseley, J.M., Williams, E.D., 1997. Parathyroid hormone-related protein: hormone and cytokine. J. Endocrinol. 154 (Suppl. 1), S23–S37.

Massfelder, T., Dann, P., Wu, T.L., Vasavada, R., Helwig, J.J., Stewart, A.F., 1997. Opposing mitogenic and anti-mitogenic actions of parathyroid hormone-related protein in vascular smooth muscle cells: a critical role for nuclear targeting. Proc. Natl. Acad. Sci. U.S.A. 94 (25), 13630–13635.

Massfelder, T., Fiaschi-Taesch, N., Stewart, A.F., Helwig, J.J., 1998. Parathyroid hormone-related peptide–a smooth muscle tone and proliferation regulatory protein. Curr. Opin. Nephrol. Hypertens. 7 (1), 27–32.

Matsushita, H., Usui, M., Hara, M., Shishiba, Y., Nakazawa, H., Honda, K., Torigoe, K., Kohno, K., Kurimoto, M., 1997. Co-secretion of parathyroid hormone and parathyroid-hormone-related protein via a regulated pathway in human parathyroid adenoma cells. Am. J. Pathol. 150 (3), 861–871.

McMahon, A.P., 2000. More surprises in the Hedgehog signaling pathway. Cell 100 (2), 185–188.

Meador, C.K., Liddle, G.W., Island, D.P., Nicholson, W.E., Lucas, C.P., Nuckton, J.G., Luetscher, J.A., 1962. Cause of Cushing's syndrome in patients with tumors arising from "nonendocrine" tissue. J. Clin. Endocrinol. Metab. 22, 693–703.

Melick, R.A., Martin, T.J., Hicks, J.D., 1972. Parathyroid hormone production and malignancy. Br. Med. J. 2 (5807), 204–205.

Miao, D., He, B., Jiang, Y., Kobayashi, T., Soroceanu, M.A., Zhao, J., Su, H., Tong, X., Amizuka, N., Gupta, A., Genant, H.K., Kronenberg, H.M., Goltzman, D., Karaplis, A.C., 2005. Osteoblast-derived PTHrP is a potent endogenous bone anabolic agent that modifies the therapeutic efficacy of administered PTH 1-34. J. Clin. Investig. 115 (9), 2402–2411.

Miao, D., He, B., Karaplis, A.C., Goltzman, D., 2002. Parathyroid hormone is essential for normal fetal bone formation. J. Clin. Investig. 109 (9), 1173–1182.

Miao, D., Li, J., Xue, Y., Su, H., Karaplis, A.C., Goltzman, D., 2004. Parathyroid hormone-related peptide is required for increased trabecular bone volume in parathyroid hormone-null mice. Endocrinology 145 (8), 3554–3562.

Miao, D., Su, H., He, B., Gao, J., Xia, Q., Zhu, M., Gu, Z., Goltzman, D., Karaplis, A.C., 2008. Severe growth retardation and early lethality in mice lacking the nuclear localization sequence and C-terminus of PTH-related protein. Proc. Natl. Acad. Sci. U.S.A. 105 (51), 20309–20314.

Miller, P.D., Hattersley, G., Riis, B.J., Williams, G.C., Lau, E., Russo, L.A., Alexandersen, P., Zerbini, C.A., Hu, M.Y., Harris, A.G., Fitzpatrick, L.A., Cosman, F., Christiansen, C., Investigators, A.S., 2016. Effect of abaloparatide vs placebo on new vertebral fractures in postmenopausal women with osteoporosis: a randomized clinical trial. J. Am. Med. Assoc. 316 (7), 722–733.

Mok, L.L., Nickols, G.A., Thompson, J.C., Cooper, C.W., 1989. Parathyroid hormone as a smooth muscle relaxant. Endocr. Rev. 10 (4), 420–436.

Molyneux, S.D., Di Grappa, M.A., Beristain, A.G., McKee, T.D., Wai, D.H., Paderova, J., Kashyap, M., Hu, P., Maiuri, T., Narala, S.R., Stambolic, V., Squire, J., Penninger, J., Sanchez, O., Triche, T.J., Wood, G.A., Kirschner, L.S., Khokha, R., 2010. Prkar1a is an osteosarcoma tumor suppressor that defines a molecular subclass in mice. J. Clin. Investig. 120 (9), 3310–3325.

Moniz, C., Burton, P.B., Malik, A.N., Dixit, M., Banga, J.P., Nicolaides, K., Quirke, P., Knight, D.E., McGregor, A.M., 1990. Parathyroid hormone-related peptide in normal human fetal development. J. Mol. Endocrinol. 5 (3), 259–266.

Moseley, J.M., Gillespie, M.T., 1995. Parathyroid hormone-related protein. Crit. Rev. Clin. Lab. Sci. 32 (3), 299–343.

Moseley, J.M., Hayman, J.A., Danks, J.A., Alcorn, D., Grill, V., Southby, J., Horton, M.A., 1991. Immunohistochemical detection of parathyroid hormone-related protein in human fetal epithelia. J. Clin. Endocrinol. Metab. 73 (3), 478–484.

Moseley, J.M., Kubota, M., Diefenbach-Jagger, H., Wettenhall, R.E., Kemp, B.E., Suva, L.J., Rodda, C.P., Ebeling, P.R., Hudson, P.J., Zajac, J.D., et al., 1987. Parathyroid hormone-related protein purified from a human lung cancer cell line. Proc. Natl. Acad. Sci. U.S.A. 84 (14), 5048–5052.

Murrills, R.J., Stein, L.S., Dempster, D.W., 1995. Lack of significant effect of carboxyl-terminal parathyroid hormone-related peptide fragments on isolated rat and chick osteoclasts. Calcif. Tissue Int. 57 (1), 47–51.

Mutsaers, A.J., Ng, A.J., Baker, E.K., Russell, M.R., Chalk, A.M., Wall, M., Liddicoat, B.J., Ho, P.W., Slavin, J.L., Goradia, A., Martin, T.J., Purton, L.E., Dickins, R.A., Walkley, C.R., 2013. Modeling distinct osteosarcoma subtypes in vivo using Cre:lox and lineage-restricted transgenic shRNA. Bone 55 (1), 166–178.

Mutsaers, A.J., Walkley, C.R., 2014. Cells of origin in osteosarcoma: mesenchymal stem cells or osteoblast committed cells? Bone 62, 56–63.

Nagai, S., Okazaki, M., Segawa, H., Bergwitz, C., Dean, T., Potts Jr., J.T., Mahon, M.J., Gardella, T.J., Juppner, H., 2011. Acute down-regulation of sodium-dependent phosphate transporter NPT2a involves predominantly the cAMP/PKA pathway as revealed by signaling-selective parathyroid hormone analogs. J. Biol. Chem. 286 (2), 1618–1626.

Nakayama, T., Ohtsuru, A., Enomoto, H., Namba, H., Ozeki, S., Shibata, Y., Yokota, T., Nobuyoshi, M., Ito, M., Sekine, I., et al., 1994. Coronary atherosclerotic smooth muscle cells overexpress human parathyroid hormone-related peptides. Biochem. Biophys. Res. Commun. 200 (2), 1028–1035.

Neer, R.M., Arnaud, C.D., Zanchetta, J.R., Prince, R., Gaich, G.A., Reginster, J.Y., Hodsman, A.B., Eriksen, E.F., Ish-Shalom, S., Genant, H.K., Wang, O., Mitlak, B.H., 2001. Effect of parathyroid hormone (1-34) on fractures and bone mineral density in postmenopausal women with osteoporosis. N. Engl. J. Med. 344 (19), 1434–1441.

Nguyen, M., He, B., Karaplis, A., 2001. Nuclear forms of parathyroid hormone-related peptide are translated from non-AUG start sites downstream from the initiator methionine. Endocrinology 142 (2), 694–703.

Nijs-de Wolf, N., Pepersack, T., Corvilain, J., Karmali, R., Bergmann, P., 1991. Adenylate cyclase stimulating activity immunologically similar to parathyroid hormone-related peptide can be extracted from fetal rat long bones. J. Bone Miner. Res. 6 (9), 921–927.

Noda, M., Katoh, T., Takuwa, N., Kumada, M., Kurokawa, K., Takuwa, Y., 1994. Synergistic stimulation of parathyroid hormone-related peptide gene expression by mechanical stretch and angiotensin II in rat aortic smooth muscle cells. J. Biol. Chem. 269 (27), 17911–17917.

Okano, K., Pirola, C.J., Wang, H.M., Forrester, J.S., Fagin, J.A., Clemens, T.L., 1995. Involvement of cell cycle and mitogen-activated pathways in induction of parathyroid hormone-related protein gene expression in rat aortic smooth muscle cells. Endocrinology 136 (4), 1782–1789.

Ono, W., Sakagami, N., Nishimori, S., Ono, N., Kronenberg, H.M., 2016. Parathyroid hormone receptor signalling in osterix-expressing mesenchymal progenitors is essential for tooth root formation. Nat. Commun. 7, 11277.

Orloff, J.J., Wu, T.L., Stewart, A.F., 1989. Parathyroid hormone-like proteins: biochemical responses and receptor interactions. Endocr. Rev. 10 (4), 476–495.

Owen, M., 1971. Cellular Dynamics of Bone. Academic Press, New York, N.Y.

Ozeki, S., Ohtsuru, A., Seto, S., Takeshita, S., Yano, H., Nakayama, T., Ito, M., Yokota, T., Nobuyoshi, M., Segre, G.V., Yamashita, S., Yano, K., 1996. Evidence that implicates the parathyroid hormone-related peptide in vascular stenosis. Increased gene expression in the intima of injured carotid arteries and human restenotic coronary lesions. Arterioscler. Thromb. Vasc. Biol. 16 (4), 565–575.

Pache, J.C., Burton, D.W., Deftos, L.J., Hastings, R.H., 2006. A carboxyl leucine-rich region of parathyroid hormone-related protein is critical for nuclear export. Endocrinology 147 (2), 990–998.

Paget, S., 1889. The distribution of secondary growths in cancer of the breast. Lancet 1, 571–573.

Partridge, N.C., Alcorn, D., Michelangeli, V.P., Kemp, B.E., Ryan, G.B., Martin, T.J., 1981. Functional properties of hormonally responsive cultured normal and malignant rat osteoblastic cells. Endocrinology 108 (1), 213–219.

Patterson, E.K., Watson, P.H., Hodsman, A.B., Hendy, G.N., Canaff, L., Bringhurst, F.R., Poschwatta, C.H., Fraher, L.J., 2007. Expression of PTH1R constructs in LLC-PK1 cells: protein nuclear targeting is mediated by the PTH1R NLS. Bone 41 (4), 603–610.

Philbrick, W.M., 1998. Parathyroid hormone-related protein is a developmental regulatory molecule. Eur. J. Oral Sci. 106 (Suppl. 1), 32–37.

Philbrick, W.M., Wysolmerski, J.J., Galbraith, S., Holt, E., Orloff, J.J., Yang, K.H., Vasavada, R.C., Weir, E.C., Broadus, A.E., Stewart, A.F., 1996. Defining the roles of parathyroid hormone-related protein in normal physiology. Physiol. Rev. 76 (1), 127–173.

Pilbeam, C.C., Alander, C.B., Simmons, H.A., Raisz, L.G., 1993. Comparison of the effects of various lengths of synthetic human parathyroid hormone-related peptide (hPTHrP) of malignancy on bone resorption and formation in organ culture. Bone 14 (5), 717–720.

Pirola, C.J., Wang, H.M., Kamyar, A., Wu, S., Enomoto, H., Sharifi, B., Forrester, J.S., Clemens, T.L., Fagin, J.A., 1993. Angiotensin II regulates parathyroid hormone-related protein expression in cultured rat aortic smooth muscle cells through transcriptional and post-transcriptional mechanisms. J. Biol. Chem. 268 (3), 1987–1994.

Plawner, L.L., Philbrick, W.M., Burtis, W.J., Broadus, A.E., Stewart, A.F., 1995. Cell type-specific secretion of parathyroid hormone-related protein via the regulated versus the constitutive secretory pathway. J. Biol. Chem. 270 (23), 14078–14084.

Powell, D., Singer, F.R., Murray, T.M., Minkin, C., Potts Jr., J.T., 1973. Nonparathyroid humoral hypercalcemia in patients with neoplastic diseases. N. Engl. J. Med. 289 (4), 176–181.

Powell, G.J., Southby, J., Danks, J.A., Stillwell, R.G., Hayman, J.A., Henderson, M.A., Bennett, R.C., Martin, T.J., 1991. Localization of parathyroid hormone-related protein in breast cancer metastases: increased incidence in bone compared with other sites. Cancer Res. 51 (11), 3059–3061.

Qian, J., Lorenz, J.N., Maeda, S., Sutliff, R.L., Weber, C., Nakayama, T., Colbert, M.C., Paul, R.J., Fagin, J.A., Clemens, T.L., 1999. Reduced blood pressure and increased sensitivity of the vasculature to parathyroid hormone-related protein (PTHrP) in transgenic mice overexpressing the PTH/PTHrP receptor in vascular smooth muscle. Endocrinology 140 (4), 1826–1833.

Quist, T., Jin, H., Zhu, J.F., Smith-Fry, K., Capecchi, M.R., Jones, K.B., 2015. The impact of osteoblastic differentiation on osteosarcomagenesis in the mouse. Oncogene 34 (32), 4278–4284.

Raison, D., Coquard, C., Hochane, M., Steger, J., Massfelder, T., Moulin, B., Karaplis, A.C., Metzger, D., Chambon, P., Helwig, J.J., Barthelmebs, M., 2013. Knockdown of parathyroid hormone related protein in smooth muscle cells alters renal hemodynamics but not blood pressure. Am. J. Physiol. Renal. Physiol. 305 (3), F333–F342.

Rejnmark, L., Lauridsen, A.L., Vestergaard, P., Heickendorff, L., Andreasen, F., Mosekilde, L., 2002. Diurnal rhythm of plasma 1,25-dihydroxyvitamin D and vitamin D-binding protein in postmenopausal women: relationship to plasma parathyroid hormone and calcium and phosphate metabolism. Eur. J. Endocrinol. 146 (5), 635–642.

Riggs, B.L., Arnaud, C.D., Reynolds, J.C., Smith, L.H., 1971. Immunologic differentiation of primary hyperparathyroidism from hyperparathyroidism due to nonparathyroid cancer. J. Clin. Investig. 50 (10), 2079–2083.

Rihani-Basharat, S., Lewinson, D., 1997. PTHrP(107-111) inhibits in vivo resorption that was stimulated by PTHrP(1-34) when applied intermittently to neonatal mice. Calcif. Tissue Int. 61 (5), 426–428.

Robbins, J., Dilworth, S.M., Laskey, R.A., Dingwall, C., 1991. Two interdependent basic domains in nucleoplasmin nuclear targeting sequence: identification of a class of bipartite nuclear targeting sequence. Cell 64 (3), 615–623.

Roca-Cusachs, A., DiPette, D.J., Nickols, G.A., 1991. Regional and systemic hemodynamic effects of parathyroid hormone-related protein: preservation of cardiac function and coronary and renal flow with reduced blood pressure. J. Pharmacol. Exp. Ther. 256 (1), 110–118.

Rodan, S.B., Wesolowski, G., Ianacone, J., Thiede, M.A., Rodan, G.A., 1989. Production of parathyroid hormone-like peptide in a human osteosarcoma cell line: stimulation by phorbol esters and epidermal growth factor. J. Endocrinol. 122 (1), 219–227.

Rodda, C.P., Kubota, M., Heath, J.A., Ebeling, P.R., Moseley, J.M., Care, A.D., Caple, I.W., Martin, T.J., 1988. Evidence for a novel parathyroid hormone-related protein in fetal lamb parathyroid glands and sheep placenta: comparisons with a similar protein implicated in humoral hypercalcaemia of malignancy. J. Endocrinol. 117 (2), 261–271.

Rude, R.K., Bethune, J.E., Singer, F.R., 1980. Renal tubular maximum for magnesium in normal, hyperparathyroid, and hypoparathyroid man. J. Clin. Endocrinol. Metab. 51 (6), 1425–1431.

Saini, V., Marengi, D.A., Barry, K.J., Fulzele, K.S., Heiden, E., Liu, X., Dedic, C., Maeda, A., Lotinun, S., Baron, R., Pajevic, P.D., 2013. Parathyroid hormone (PTH)/PTH-related peptide type 1 receptor (PPR) signaling in osteocytes regulates anabolic and catabolic skeletal responses to PTH. J. Biol. Chem. 288 (28), 20122–20134.

Schermer, D.T., Chan, S.D., Bruce, R., Nissenson, R.A., Wood, W.I., Strewler, G.J., 1991. Chicken parathyroid hormone-related protein and its expression during embryologic development. J. Bone Miner. Res. 6 (2), 149–155.

Schipani, E., Kruse, K., Juppner, H., 1995. A constitutively active mutant PTH-PTHrP receptor in Jansen-type metaphyseal chondrodysplasia. Science 268 (5207), 98–100.

Schipani, E., Langman, C.B., Parfitt, A.M., Jensen, G.S., Kikuchi, S., Kooh, S.W., Cole, W.G., Juppner, H., 1996. Constitutively activated receptors for parathyroid hormone and parathyroid hormone-related peptide in Jansen's metaphyseal chondrodysplasia. N. Engl. J. Med. 335 (10), 708–714.

Schipani, E., Lanske, B., Hunzelman, J., Luz, A., Kovacs, C.S., Lee, K., Pirro, A., Kronenberg, H.M., Juppner, H., 1997. Targeted expression of constitutively active receptors for parathyroid hormone and parathyroid hormone-related peptide delays endochondral bone formation and rescues mice that lack parathyroid hormone-related peptide. Proc. Natl. Acad. Sci. U.S.A. 94 (25), 13689–13694.

Senior, P.V., Heath, D.A., Beck, F., 1991. Expression of parathyroid hormone-related protein mRNA in the rat before birth: demonstration by hybridization histochemistry. J. Mol. Endocrinol. 6 (3), 281–290.

Sone, T., Kohno, H., Kikuchi, H., Ikeda, T., Kasai, R., Kikuchi, Y., Takeuchi, R., Konishi, J., Shigeno, C., 1992. Human parathyroid hormone-related peptide-(107-111) does not inhibit bone resorption in neonatal mouse calvariae. Endocrinology 131 (6), 2742–2746.

Stevenson, F.T., Turck, J., Locksley, R.M., Lovett, D.H., 1997. The N-terminal propiece of interleukin 1 alpha is a transforming nuclear oncoprotein. Proc. Natl. Acad. Sci. U.S.A. 94 (2), 508–513.

Stewart, A.F., Horst, R., Deftos, L.J., Cadman, E.C., Lang, R., Broadus, A.E., 1980. Biochemical evaluation of patients with cancer-associated hypercalcemia: evidence for humoral and nonhumoral groups. N. Engl. J. Med. 303 (24), 1377–1383.

Suda, N., Gillespie, M.T., Traianedes, K., Zhou, H., Ho, P.W., Hards, D.K., Allan, E.H., Martin, T.J., Moseley, J.M., 1996. Expression of parathyroid hormone-related protein in cells of osteoblast lineage. J. Cell. Physiol. 166 (1), 94–104.

Suva, L.J., Seedor, J.G., Endo, N., Quartuccio, H.A., Thompson, D.D., Bab, I., Rodan, G.A., 1993. Pattern of gene expression following rat tibial marrow ablation. J. Bone Miner. Res. 8 (3), 379–388.

Suva, L.J., Winslow, G.A., Wettenhall, R.E., Hammonds, R.G., Moseley, J.M., Diefenbach-Jagger, H., Rodda, C.P., Kemp, B.E., Rodriguez, H., Chen, E.Y., et al., 1987. A parathyroid hormone-related protein implicated in malignant hypercalcemia: cloning and expression. Science 237 (4817), 893–896.

Tao, J., Jiang, M.M., Jiang, L., Salvo, J.S., Zeng, H.C., Dawson, B., Bertin, T.K., Rao, P.H., Chen, R., Donehower, L.A., Gannon, F., Lee, B.H., 2014. Notch activation as a driver of osteogenic sarcoma. Cancer Cell 26 (3), 390–401.

Tashjian Jr., A.H., Chabner, B.A., 2002. Commentary on clinical safety of recombinant human parathyroid hormone 1-34 in the treatment of osteoporosis in men and postmenopausal women. J. Bone Miner. Res. 17 (7), 1151–1161.

Tashjian Jr., A.H., Goltzman, D., 2008. On the interpretation of rat carcinogenicity studies for human PTH(1-34) and human PTH(1-84). J. Bone Miner. Res. 23 (6), 803–811.

Thiede, M.A., Strewler, G.J., Nissenson, R.A., Rosenblatt, M., Rodan, G.A., 1988. Human renal carcinoma expresses two messages encoding a parathyroid hormone-like peptide: evidence for the alternative splicing of a single-copy gene. Proc. Natl. Acad. Sci. U.S.A. 85 (13), 4605–4609.

Toribio, R.E., Brown, H.A., Novince, C.M., Marlow, B., Hernon, K., Lanigan, L.G., Hildreth 3rd, B.E., Werbeck, J.L., Shu, S.T., Lorch, G., Carlton, M., Foley, J., Boyaka, P., McCauley, L.K., Rosol, T.J., 2010. The midregion, nuclear localization sequence, and C terminus of PTHrP regulate skeletal development, hematopoiesis, and survival in mice. FASEB J. 24 (6), 1947–1957.

Tregear, G.W., Potts Jr., J.T., 1975. Synthetic analogues of residues 1-34 of human parathyroid hormone: influence of residue number 1 on biological potency in vitro. Endocr. Res. Commun. 2 (8), 561–570.

Truant, R., Cullen, B.R., 1999. The arginine-rich domains present in human immunodeficiency virus type 1 Tat and Rev function as direct importin beta-dependent nuclear localization signals. Mol. Cell Biol. 19 (2), 1210–1217.

Tsukazaki, T., Ohtsuru, A., Enomoto, H., Yano, H., Motomura, K., Ito, M., Namba, H., Iwasaki, K., Yamashita, S., 1995. Expression of parathyroid hormone-related protein in rat articular cartilage. Calcif. Tissue Int. 57 (3), 196–200.

Vahle, J.L., Sato, M., Long, G.G., Young, J.K., Francis, P.C., Engelhardt, J.A., Westmore, M.S., Linda, Y., Nold, J.B., 2002. Skeletal changes in rats given daily subcutaneous injections of recombinant human parathyroid hormone (1-34) for 2 years and relevance to human safety. Toxicol. Pathol. 30 (3), 312–321.

Valin, A., Garcia-Ocana, A., De Miguel, F., Sarasa, J.L., Esbrit, P., 1997. Antiproliferative effect of the C-terminal fragments of parathyroid hormone-related protein, PTHrP-(107-111) and (107-139), on osteoblastic osteosarcoma cells. J. Cell. Physiol. 170 (2), 209–215.

van de Stolpe, A., Karperien, M., Lowik, C.W., Juppner, H., Segre, G.V., Abou-Samra, A.B., de Laat, S.W., Defize, L.H., 1993. Parathyroid hormone-related peptide as an endogenous inducer of parietal endoderm differentiation. J. Cell Biol. 120 (1), 235–243.

Vortkamp, A., Lee, K., Lanske, B., Segre, G.V., Kronenberg, H.M., Tabin, C.J., 1996. Regulation of rate of cartilage differentiation by Indian hedgehog and PTH-related protein. Science 273 (5275), 613–622.

Walia, M., Castillo-Tandozo, W., Mutsaers, A.J., Martin, T.J., Walkley, C.R., 2018. Murine models of osteosarcoma: a piece of the translational puzzle. J. Cell. Biochem. in press.

Walia, M.K., Ho, P.M., Taylor, S., Ng, A.J., Gupte, A., Chalk, A.M., Zannettino, A.C., Martin, T.J., Walkley, C.R., 2016. Activation of PTHrP-cAMP-CREB1 signaling following p53 loss is essential for osteosarcoma initiation and maintenance. Elife 5.

Walkley, C.R., Qudsi, R., Sankaran, V.G., Perry, J.A., Gostissa, M., Roth, S.I., Rodda, S.J., Snay, E., Dunning, P., Fahey, F.H., Alt, F.W., McMahon, A.P., Orkin, S.H., 2008. Conditional mouse osteosarcoma, dependent on p53 loss and potentiated by loss of Rb, mimics the human disease. Genes Dev. 22 (12), 1662–1676.

Wang, H.H., Drugge, E.D., Yen, Y.C., Blumenthal, M.R., Pang, P.K., 1984. Effects of synthetic parathyroid hormone on hemodynamics and regional blood flows. Eur. J. Pharmacol. 97 (3–4), 209–215.

Wang, Z.Q., Liang, J., Schellander, K., Wagner, E.F., Grigoriadis, A.E., 1995. c-fos-induced osteosarcoma formation in transgenic mice: cooperativity with c-jun and the role of endogenous c-fos. Cancer Res. 55 (24), 6244–6251.

Washam, C.L., Byrum, S.D., Leitzel, K., Ali, S.M., Tackett, A.J., Gaddy, D., Sundermann, S.E., Lipton, A., Suva, L.J., 2013. Identification of PTHrP(12-48) as a plasma biomarker associated with breast cancer bone metastasis. Cancer Epidemiol. Biomark. Prev. 22 (5), 972–983.

Watson, P.H., Fraher, L.J., Hendy, G.N., Chung, U.I., Kisiel, M., Natale, B.V., Hodsman, A.B., 2000a. Nuclear localization of the type 1 PTH/PTHrP receptor in rat tissues. J. Bone Miner. Res. 15 (6), 1033–1044.

Watson, P.H., Fraher, L.J., Natale, B.V., Kisiel, M., Hendy, G.N., Hodsman, A.B., 2000b. Nuclear localization of the type 1 parathyroid hormone/parathyroid hormone-related peptide receptor in MC3T3-E1 cells: association with serum-induced cell proliferation. Bone 26 (3), 221–225.

Wein, M.N., Liang, Y., Goransson, O., Sundberg, T.B., Wang, J., Williams, E.A., O'Meara, M.J., Govea, N., Beqo, B., Nishimori, S., Nagano, K., Brooks, D.J., Martins, J.S., Corbin, B., Anselmo, A., Sadreyev, R., Wu, J.Y., Sakamoto, K., Foretz, M., Xavier, R.J., Baron, R., Bouxsein, M.L., Gardella, T.J., Divieti-Pajevic, P., Gray, N.S., Kronenberg, H.M., 2016. SIKs control osteocyte responses to parathyroid hormone. Nat. Commun. 7, 13176.

Weir, E.C., Philbrick, W.M., Amling, M., Neff, L.A., Baron, R., Broadus, A.E., 1996. Targeted overexpression of parathyroid hormone-related peptide in chondrocytes causes chondrodysplasia and delayed endochondral bone formation. Proc. Natl. Acad. Sci. U.S.A. 93 (19), 10240–10245.

Whitfield, J.F., Isaacs, R.J., Chakravarthy, B.R., Durkin, J.P., Morley, P., Neugebauer, W., Williams, R.E., Willick, G., Rixon, R.H., 1994. C-terminal fragments of parathyroid hormone-related protein, PTHrP-(107-111) and (107-139), and the N-terminal PTHrP-(1-40) fragment stimulate membrane-associated protein kinase C activity in rat spleen lymphocytes. J. Cell. Physiol. 158 (3), 518–522.

Wilson, K.C., Cruikshank, W.W., Center, D.M., Zhang, Y., 2002. Prointerleukin-16 contains a functional CcN motif that regulates nuclear localization. Biochemistry 41 (48), 14306–14312.

Wu, T.L., Vasavada, R.C., Yang, K., Massfelder, T., Ganz, M., Abbas, S.K., Care, A.D., Stewart, A.F., 1996. Structural and physiologic characterization of the mid-region secretory species of parathyroid hormone-related protein. J. Biol. Chem. 271 (40), 24371–24381.

Wysolmerski, J.J., Cormier, S., Philbrick, W.M., Dann, P., Zhang, J.P., Roume, J., Delezoide, A.L., Silve, C., 2001. Absence of functional type 1 parathyroid hormone (PTH)/PTH-related protein receptors in humans is associated with abnormal breast development and tooth impaction. J. Clin. Endocrinol. Metab. 86 (4), 1788–1794.

Yang, D., Singh, R., Divieti, P., Guo, J., Bouxsein, M.L., Bringhurst, F.R., 2007a. Contributions of parathyroid hormone (PTH)/PTH-related peptide receptor signaling pathways to the anabolic effect of PTH on bone. Bone 40 (6), 1453–1461.

Yang, R., Hoang, B.H., Kubo, T., Kawano, H., Chou, A., Sowers, R., Huvos, A.G., Meyers, P.A., Healey, J.H., Gorlick, R., 2007b. Over-expression of parathyroid hormone Type 1 receptor confers an aggressive phenotype in osteosarcoma. Int. J. Cancer 121 (5), 943–954.

Yasuda, T., Banville, D., Hendy, G.N., Goltzman, D., 1989. Characterization of the human parathyroid hormone-like peptide gene. Functional and evolutionary aspects. J. Biol. Chem. 264 (13), 7720–7725.

Zhou, H., Leaver, D.D., Moseley, J.M., Kemp, B., Ebeling, P.R., Martin, T.J., 1989. Actions of parathyroid hormone-related protein on the rat kidney in vivo. J. Endocrinol. 122 (1), 229–235.

Zhu, Q., Zhou, X., Zhu, M., Wang, Q., Goltzman, D., Karaplis, A., Miao, D., 2013. Endogenous parathyroid hormone-related protein compensates for the absence of parathyroid hormone in promoting bone accrual in vivo in a model of bone marrow ablation. J. Bone Miner. Res. 28 (9), 1898–1911.

Chapter 26

Cardiovascular actions of parathyroid hormone/parathyroid hormone–related protein signaling

Sasan Mirfakhraee and Dwight A. Towler
The University of Texas Southwestern Medical Center, Department of Internal Medicine, Endocrine Division, Dallas, TX, United Sates

Chapter outline

Introduction 623
PTH/PTHrP in cardiovascular development 624
PTH receptor signaling in arterial biology: vascular smooth muscle cell and endothelial responses to PTH and PTHrP 624
PTH2R signaling in vascular pharmacology 630
Parathyroid hormone, hyperparathyroidism, and calcific aortic valve disease 631
Impaired vascular PTH1R signaling and cardiovascular disease: the impact of hyperparathyroidism on cardiovascular mortality, coronary flow reserve, and vascular stiffness 632
Chronic kidney disease–mineral and bone disorder: the metabolic "perfect storm" of cardiovascular risk 633
PTH/PTHrP signaling and the bone–vascular axis 635
PTH1R activation and the renin–angiotensin–aldosterone axis: a feed-forward vicious cycle 636
Summary, conclusions, and future directions 636
Acknowledgments 637
References 637

Introduction

All osteotropic hormones have vasculotropic actions (Thompson and Towler, 2012b; Towler, 2017a). This pithy statement of fact most certainly holds true for the prototypic bone anabolic polypeptides parathyroid hormone (PTH) and parathyroid hormone–related protein (PTHrP) (aka PTHLH). The cardiovascular actions of PTH were first identified in 1925, when Collip and Clark injected anesthetized Dog 164 with intravenous parathyroid extract and documented hypotensive actions (Collip and Clark, 1925; Rambausek et al., 1982). However, only since the end of the 20th century has the fundamental role of PTH/PTHrP receptor (PTH1R) signaling in cardiovascular development, physiology, and disease been fully vetted—but this is not yet widely appreciated (Bilezikian et al., 2015; Tomaschitz et al., 2013). For example, PTH1R signaling is required for normal aortic valve (Gray et al., 2013) and myocardial (Qian et al., 2003) development. Paracrine vascular smooth muscle PTHrP/PTH1R signaling has emerged as important in the homeostatic regulation of renovascular blood flow (Raison et al., 2013), while endothelial PTH1R actions convey the endocrine regulation of blood flow to bone in response to PTH (Benson et al., 2016). In addition to its bone anabolic actions, intermittent pharmacological dosing with PTH(1–34) mitigates the endothelial dysfunction associated with aging (Guers et al., 2017) and suppresses arteriosclerotic responses to dysmetabolic states such as diabetes (Cheng et al., 2010). These observations, among others, provide a second "bookend" that, with Collip and Clark's early observation, serves to brace the first century of PTH/PTHrP research—a century that has *continuously* pointed to cardiovascular actions of the PTH family of ligands. Thus, a fundamental understanding of PTH/PTHrP biology must encompass a better understanding of actions within and upon the cardiovascular system (Cheng et al., 2010).

In this chapter, we build upon our recent reviews (Bilezikian et al., 2015; Thompson and Towler, 2012b; Towler, 2014a) to help highlight once again the importance of PTH/PTHrP to cardiovascular biology and disease. While

Principles of Bone Biology. https://doi.org/10.1016/B978-0-12-814841-9.00026-9

mammalian PTHrP is a dedicated PTH1R ligand, PTH can activate both human PTH1R and human PTH2R (Hoare and Usdin, 2001). Therefore, we also briefly discuss PTH2R and tuberoinfundibular peptide of 39 residues (TIP39), the dedicated PTH2R ligand, as relevant to cardiovascular biology. Finally, we reflect upon chronic kidney disease−mineral and bone disorder (CKD−MBD) as a "perfect storm" of cardiometabolic risk arising in significant part from perturbed PTH1R signaling. We use this discussion as a framework to detail physiologically important interconnections of PTH/PTHrP signaling within the emerging bone−vascular axis (Fadini et al., 2012; Thompson and Towler, 2012a; Towler, 2011). The reader is referred to recent editions of *The Parathyroids* for a more comprehensive historical review (Bilezikian et al., 2015).

PTH/PTHrP in cardiovascular development

Some of the earliest evidence that PTH/PTHrP signaling might contribute to cardiovascular development arose in Buck Strewler's group; they identified expression of PTHrP in the chicken embryonic heart as well as in other mesodermally derived tissues (Schermer et al., 1991). Descriptive studies in the rat revealed that mesenchymal PTH1R expression lay adjacent to epithelial or endothelial sources of PTHrP (Lee et al., 1995). Inductive, paracrine epithelial-to-mesenchymal signals mediated by PTHrP were inferred from the spatial relationship between *PTHrP* ligand and *PTH1R* expression as revealed by these in situ hybridization studies. In dogs, PTH1R was also identified early on as being expressed in heart and aorta, albeit at levels much lower than those observed in kidney and bone (Smock et al., 2001). However, the important role of PTH/PTHrP in cardiovascular development was not truly appreciated until Hank Kronenberg, Tom Clemens, and colleagues began their systematic analyses as to why mice globally lacking the PTH1R exhibit prenatal lethality (Qian et al., 2003). They demonstrated that, as in other vertebrate species, the murine *PTH1R* gene was abundantly expressed in developing mouse cardiomyocytes, but then went on to show that mice genetically lacking PTH1R abruptly die between embryonic day (E) 11.5 and E12.5 due to massive cardiomyocyte apoptosis (Qian et al., 2003). The primary abnormalities in cardiomyocyte mitochondrial morphology and pump function that arise with myocardial PTH1R deficiency were followed by secondary hepatic injury and tissue necrosis (presumptively due to cardiogenic circulatory congestion). The mechanisms whereby PTH1R signaling regulates mitochondrial metabolism to prevent apoptotic cell death in the cardiomyocyte have yet to be fully explored, but based upon recent studies will probably converge on protein kinase A-dependent inhibition of the mitochondrial fission protein Drp1 (Monterisi et al., 2017). Skeletal PTH1R activation has been shown to inhibit apoptosis in several cell types, including in osteoblasts challenged with glucocorticoids (O'Brien et al., 2008) or chondrocytes treated with tumor necrosis factor (Okoumassoun et al., 2007). Thus, PTH1R expression and signaling represent a fundamental component of cardiomyocyte cell physiology. A better understanding of the prosurvival mechanisms afforded by PTH1R signaling in the heart may prove useful in therapeutic approaches to ischemic heart disease as well as cardiotoxicity associated with a subset of chemotherapeutics (Babiker et al., 2018).

Other vertebrate models have confirmed the important role of PTH/PTHrP signaling in cardiovascular development. In zebrafish, a total of three PTH receptors (*pth1r*, *pth2r*, *pth3r*) are expressed (Gray et al., 2013). Knockdown of either *pth1r* or the selective zebrafish *pth1r/pth2r* ligand *pthrp* results in morphants with aortic coarctation defects. Because of the important role for *notch* in mammalian aortic valve development (van den Akker et al., 2012), Chico and colleagues wondered whether *notch* signaling might be perturbed and, indeed, demonstrated that restoration of *notch* signaling in *pth1r*-targeted morphants prevented the aortic defects (Gray et al., 2013). Thus, the variable penetrance for preductal aortic coarctation in lethal Blomstrand chondrodysplasia—a rare disorder due to loss-of-function mutations in the human *PTH1R* (Hoogendam et al., 2007)—may in fact relate to variable compensatory changes in Notch coregulatory pathways that partially compensate for PTH1R deficiency during development. PTH1R and Notch play vital roles in both bone and vascular biology, including vascular endothelial growth factor (VEGF)-dependent processes such as tip-cell selection (Blanco and Gerhardt, 2013). Thus, the cross talk between these morphogenetic signaling systems is likely to help juxtaposition microvascular supply to sites of bone formation in the basic multicellular unit in response to PTH anabolic actions (Prisby et al., 2011).

PTH receptor signaling in arterial biology: vascular smooth muscle cell and endothelial responses to PTH and PTHrP

The acute vasodilatory actions of PTH administration were recognized in the very earliest studies of the hormone's physiological response (Collip and Clark, 1925). Following Collip and Clark's lead, Pang et al. confirmed that bovine PTH(1−34) reduced blood pressure in anesthetized dogs and rats (Pang et al., 1980). Imai and colleagues demonstrated coronary vasodilation in response to PTH that same year (Hashimoto et al., 1981). Ex vivo, human and bovine middle cerebral arteries were identified to undergo vasodilatation as well (Suzuki et al., 1983). These observations prompted additional work that demonstrated vasodilatory PTH responses in multiple vertebrate tissue beds, but responses varied widely in potency and efficacy (Crass et al., 1987; Nickols et al., 1986). Tissue-specific expression of PDZ domain-containing coadapters markedly influences the capacity of the PTH1R to access protein kinase A, phospholipase C, and extracellular signal-regulated kinase (ERK) signaling mediators that shape these responses (Romero et al., 2011). As of this writing, the molecular mechanisms conveying temporal and spatial differences in PTH1R-dependent vasodilation have not been delineated. Of note, in 1986 the salutary actions of PTH in a preclinical model of myocardial ischemia were reported; enhanced collateral blood flow was thought to represent the primary mechanism of benefit (see later) (Feola and Crass, 1986). However, since PTH1R signaling exerts direct antiapoptotic actions within fetal cardiomyocytes (*vide supra*), this response may also contribute to postnatal PTH1R benefits in the setting of ischemia.

Nickols, Barthelmebs, Jean-Jacques Helwig, and colleagues went on to confirm a widespread vasodilatory reaction to PTHrP (Musso et al., 1989; Roca-Cusachs et al., 1991). Endothelial NO production in response to PTH1R activation certainly contributes to vasodilatory PTH/PTHrP actions, as Prisby and colleagues have elegantly demonstrated in bone (Benson et al., 2016) (Fig. 26.1). However, direct PTH1R signaling in arterial vascular smooth muscle cells (VSMC) conveys many beneficial functions in other large conduit vessels. Coronary artery vasodilation is only partly dependent upon endothelial NO release due to direct actions of PTHrP/PTH1R signaling in coronary VSMC that drive vasorelaxation; the latter is dependent upon cAMP-mediated reductions in cytosolic calcium (Ishikawa et al., 1994; Nyby et al., 1995). However, as discussed earlier, this varies with vascular venue. Rhonda Prisby and colleagues identified that the endothelium was vital to PTH actions enhancing blood flow to the skeleton in studies of the femoral principal nutrient artery (Prisby et al., 2013). Whether age-dependent increases in vascular sensitivity to pressors combine with altered nutrient artery endothelial PTH1R signaling to compromise bone health remains to be fully explored.

Bilezikian et al. first established that PTH exerts acute inotropic actions in the isolated perfused heart, dependent upon coronary vasodilatation and independent of direct chronotropy (Ogino et al., 1995). In mechanisms that resemble the epithelial–mesenchymal induction effects of paracrine PTHrP during development (*vide supra*), subsets of endothelial cells appear to provide important sources of PTHrP tone that maintains myocardial function and coronary blood flow. Schlüter determined that coronary endothelial cells produce PTHrP under ischemic conditions to enhance both myocardial inotropy (contraction velocity) and lusitropy (relaxation velocity) (Lutteke et al., 2005; Schluter et al., 2000). Consistent with this, Feola and Crass demonstrated that administration of PTH(1–34) 30 min after the initiation of myocardial ischemia preserved left ventricular function and prevented cardiogenic shock in a canine model of myocardial infarction

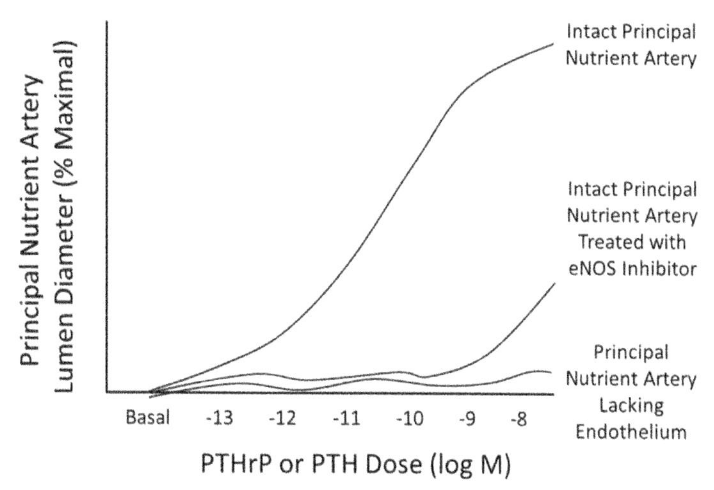

FIGURE 26.1 Parathyroid hormone (*PTH*) and parathyroid hormone–related protein (*PTHrP*) vasodilation of rat femoral nutrient arteries requires intact endothelial nitric oxide synthase (*eNOS*) signaling (Benson et al., 2016). Prisby and colleagues demonstrated that, in addition to aligning the marrow microvasculature closer to the bone multicellular unit as necessary for bone formation (Prisby et al., 2011), PTH/PTHrP signaling also increases blood flow to bone. In elegant ex vivo studies, they demonstrated that PTHrP and PTH induced vasodilation (increased lumen diameter) in the principal nutrient artery supplying rat femurs. This required an intact endothelium (Benson et al., 2016). Inhibition of eNOS with LN^G-nitroarginine methyl ester (L-NAME) also significantly inhibited the vasodilatory responses to either PTH or PTHrP. Importantly, Gohin et al. (2016) have shown that L-NAME inhibits PTH-stimulated increases in bone formation as well. *Graphic representation/summary of results from Benson, T., Menezes, T., Campbell, J., Bice, A., Hood, B., Prisby, R., 2016. Mechanisms of vasodilation to PTH 1-84, PTH 1-34, and PTHrP 1-34 in rat bone resistance arteries. Osteoporos. Int. 27, 1817–1826.*

(Feola and Crass, 1986). However, others have reported direct inotropic effects of PTH(1−34) using isolated cardiomyocytes studied in culture (Tastan et al., 2009). This suggests that pharmacological mimicry of these PTHrP-mediated paracrine endothelial-to-mesenchymal (e.g., coronary smooth muscle, myocardium) signals may hold therapeutic potential in human cardiovascular disease.

Vascular smooth muscle-derived PTHrP has emerged as a stretch-inducible paracrine vasodilator, important for homeostatic roles of PTH1R signaling that control tissue perfusion via regional arterial tone (Noda et al., 1994; Pirola et al., 1994; Takahashi et al., 1995). Clemens's group first established that augmenting local PTHrP/PTH1R signaling has an impact on blood pressure and vascular contractility; sustained reductions in blood pressure and increases in volume-dependent renal tissue perfusion were obtained in novel VSMC-specific transgenic mice expressing either PTHrP or the PTH1R (Noonan et al., 2003). Moreover, pressor responses to angiotensin II were significantly reduced in transgenic animals versus nontransgenic siblings. Wild-type PTH1R activation is ligand dependent; thus, these data strongly suggest that endogenous paracrine PTHrP production—modulated by mechanical, inflammatory, and endocrine cues—locally regulates tissue perfusion (Noonan et al., 2003).

However, it was a seminal study published a decade later by Raison et al. that elegantly confirmed and refined this working model (Raison et al., 2013). Using mice possessing floxed PTHrP alleles and implementing a smooth muscle cell transgene driving tamoxifen-inducible Cre recombinase, postnatal conditional deletion of VSMC PTHrP was achieved in adult mice, thus permitting an assessment of effects on renal blood flow (Raison et al., 2013). Compared with control mice, mice lacking PTHrP in VSMCs (viz., transgenic smooth muscle ERT2-Cre;PTHrP(flox/flox) treated with tamoxifen) exhibited increased renal vascular resistance with concomitant reductions in kidney perfusion and glomerular filtration (Raison et al., 2013). Importantly, renovascular perfusion in response to mechanical stimulation with saline volume expansion (SVE) was also impaired. While normal mice increase renal plasma flow with subsequent diuresis in response to SVE—indices of renovascular vasodilation (see Fig. 26.2)—mice lacking VSMC PTHrP were unable to do so. Interestingly, no change in systemic blood pressure was noted in these VSMC-specific PTHrP-knockout mice even though *pharmacological* dosing with PTHrP simultaneously reduced blood pressure while increasing renal plasma flow (Raison et al., 2013). Thus, endogenous VSMC PTHrP production is required to enable regulated renal perfusion in response to hemodynamic challenge without overt changes in baseline blood pressure.

At first glance, this emerging physiology appears paradoxical when viewed against the backdrop of the cardiovascular disease burden that accrues with primary hyperparathyroidism (HPT) (Yu et al., 2010, 2011, 2013) or the secondary HPT of CKD. Hypertension, arteriosclerosis, valve and vascular calcification, and cardiovascular mortality are all increased with HPT (Carrelli et al., 2013; Fitzpatrick et al., 2008; Iwata et al., 2012; Rubin et al., 2005; Silverberg et al., 2009; Walker

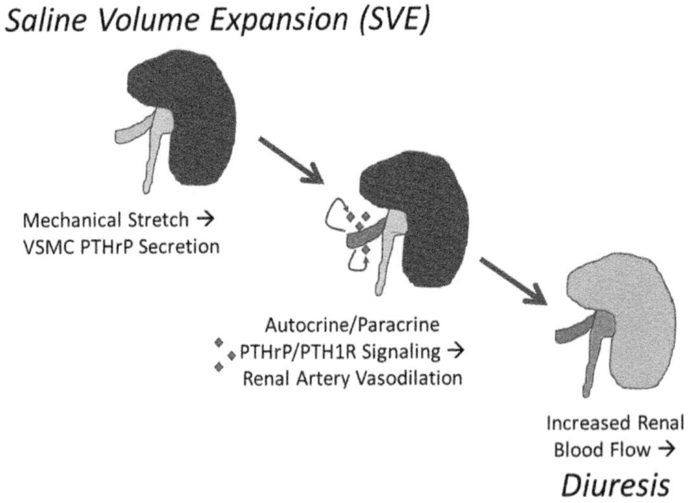

Saline Volume Expansion (SVE)
Mechanical Stretch →
VSMC PTHrP Secretion
Autocrine/Paracrine
PTHrP/PTH1R Signaling →
Renal Artery Vasodilation
Increased Renal
Blood Flow →
Diuresis

FIGURE 26.2 Saline volume expansion (*SVE*) activates vasodilatory vascular smooth muscle cell (*VSMC*) parathyroid hormone−related protein (*PTHrP*) actions that regulate renal blood flow. Barthelmebs and colleagues demonstrated that increases in renal blood flow in response to SVE require VSMC PTHrP expression (Raison et al., 2013). Unlike the model depicted for wild-type mice, conditional deletion of PTHrP in adult VSMCs abrogates physiological increases in renovascular perfusion and diuresis in response to SVE. Interestingly, no change in systemic blood pressure was noted with genetic VSMC PTHrP deficiency, even though *pharmacological* dosing with PTHrP simultaneously reduces blood pressure while increasing renal plasma flow (Raison et al., 2013). *Graphic representation/summary of results from Raison, D., Coquard, C., Hochane, M., Steger, J., Massfelder, T., Moulin, B., Karaplis, A.C., Metzger, D., Chambon, P., Helwig, J.J., et al. 2013. Knockdown of parathyroid hormone related protein in smooth muscle cells alters renal hemodynamics but not blood pressure. Am. J. Physiol. Renal. Physiol. 305, F333−F342.*

et al., 2009, 2010; Walker and Silverberg, 2008; Yu et al., 2010). Vascular desensitization to the aforementioned paracrine PTHrP cues is likely to be one explanation—arising from sustained elevation in circulating PTH with HPT and the inhibitory PTH fragments that accrue in CKD (Friedman and Goodman, 2006; Langub et al., 2003; Nyby et al., 1995). Vascular tissue PTH1R signaling exhibits tachyphylaxis, becoming rapidly refractory to PTH1R-dependent vasorelaxation in response to PTH or PTHrP exposure (Nyby et al., 1995). By studying rat femoral artery segments contracted with norepinephrine ex vivo, Brickman and colleagues demonstrated that vasodilatory responses to either PTHrP or PTH were markedly diminished, by >50%, following 40 min of prior exposure to either of these PTH1R ligands (Nyby et al., 1995) (Fig. 26.3). Similarly, isolated rat VSMCs exhibit tachyphylaxis within 30 min of exposure by using cAMP production as an assay of PTH1R activation (Nyby et al., 1995). Of note, renovascular tachyphylaxis to PTH has been independently documented in the isolated perfused rabbit kidney model (Massfelder et al., 1996). Finally, in otherwise healthy adult humans, while singular administration lowers blood pressure, sustained PTH administration over a period of 12 days actually increases blood pressure (Hulter et al., 1986)—consistent with data from these preclinical disease models (see also "PTH1R activation and the renin–angiotensin–aldosterone axis: a feed-forward vicious cycle"). The precise molecular mechanisms are poorly characterized with respect to the cardiovascular PTH1R; however, as observed in other tissues, modulation of PTH1R actions by PDZ adapter proteins, β-arrestins, and G-protein-receptor kinases certainly contributes to arterial responsivity (Dicker et al., 1999; Peterson and Luttrell, 2017; Romero et al., 2011; Smith and Rajagopal, 2016; Song et al., 2010). Thus, the vasculopathy of primary and secondary HPT will relate in part to arterial desensitization to the important paracrine, PTHrP-dependent regulation of vascular tone and cellular function (Nyby et al., 1995; Raison et al., 2013) (Figs. 26.2 and 26.3). In this model, sustained exposure to PTH and antagonistic PTH degradation fragments can downregulate PTH1R signals regulated by paracrine PTHrP actions that serve homeostatic roles in cardiovascular health (Fig. 26.4) (Friedman and Goodman, 2006; Thompson and Towler, 2012a).

No clinically useful test of *cardiovascular* PTH1R signaling has been developed or validated as of this writing—a clear shortcoming. However, by implementing histomorphometry to quantify the *skeletal* anabolic state (osteoblast surface, mineralizing surfaces) coupled with simultaneous measurements of the prevailing intact PTH level, London and colleagues (London et al., 2015) created an index that reflects an individual's *skeletal* responsiveness to PTH (Towler, 2015) (Fig. 26.5). The investigative team then demonstrated that as a cohort, those dialysis patients with lower ankle–brachial index (ABI)—an index of peripheral arterial disease—exhibit impaired skeletal PTH1R responsiveness (lower osteoblast surface, reduced bone mineralization) compared with dialysis patients with normal ABI (London et al., 2015) (Fig. 26.5). Vascular responses to PTH are important to bone physiology, increasing not only NO-dependent blood flow to bone, but also VEGF-dependent juxtaposition of microvasculature to bone-forming surfaces (Fig. 26.6) (Benson et al., 2016; Guers et al., 2017; Prisby et al., 2011, 2013; Prisby, 2017). Gohin demonstrated that PTH administration increases hindlimb

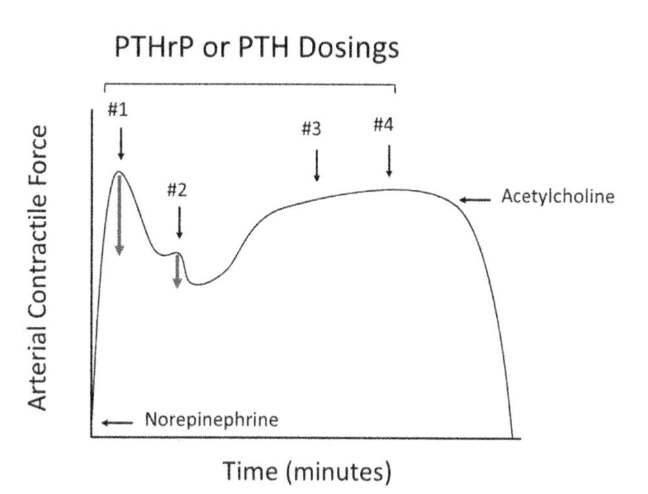

FIGURE 26.3 Arterial desensitization to the vasodilatory actions of parathyroid hormone–related protein (*PTHrP*) or parathyroid hormone (*PTH*). Brickman and colleagues established that following maximal contraction of isolated rat femoral artery segments with the pressor norepinephrine, the initial vasodilatory responses observed with PTH or PTHrP dosing are rapidly lost (Nyby et al., 1995). Compared with the first dose (#1) of either PTH or PTHrP, which resulted in vasorelaxation, subsequent dosing over the next tens of minutes exhibited either blunted (#2) or no (#3, #4) response, indicating tachyphylaxis. However, response to other vasodilators remained intact, as exhibited by preserved vasorelaxation to acetylcholine. *Downward red arrows*, magnitude of vasorelaxation in response to PTHrP or PTH dosing. See text for details. *Graphic representation/summary of results from Nyby, M.D., Hino, T., Berger, M.E., Ormsby, B.L., Golub, M.S., Brickman, A.S., 1995. Desensitization of vascular tissue to parathyroid hormone and parathyroid hormone-related protein. Endocrinology 136, 2497–2504.*

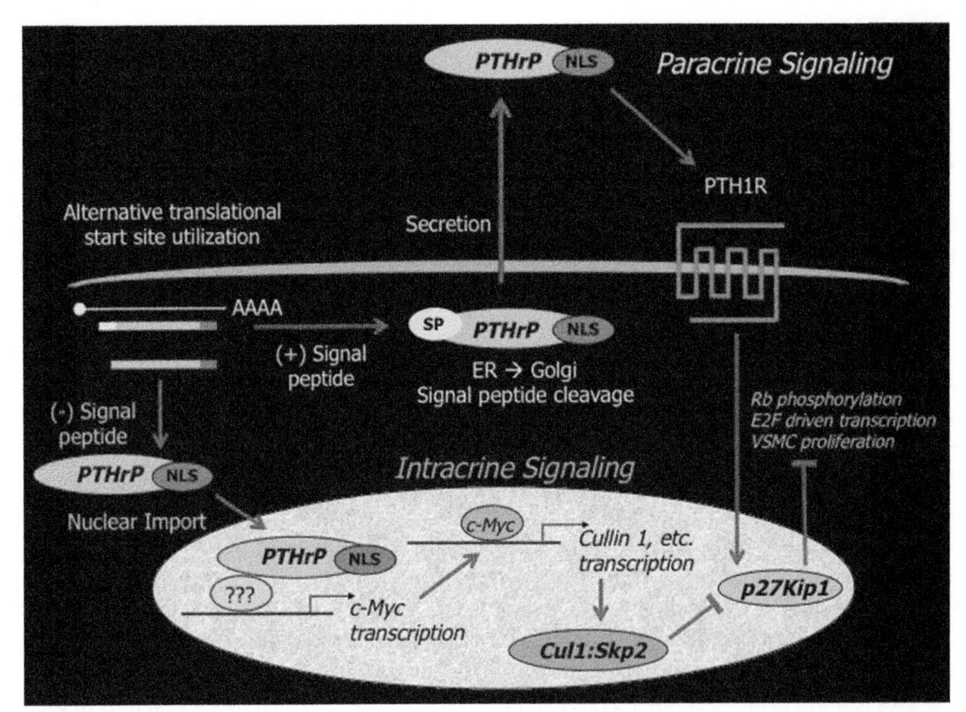

FIGURE 26.4 Paracrine versus intracrine parathyroid hormone–related protein (*PTHrP*) signaling. Translation of the full-length PTHrP protein results in signal peptide (*SP*)-dependent secretion of PTHrP(1–141). Paracrine PTHrP signaling in vascular smooth muscle cells (*VSMCs*), via the G-protein-coupled plasma membrane receptor *PTH1R*, reduces cellular proliferation, promotes vasodilation, and mitigates procalcific signals. Since prior exposure to PTH1R ligands impairs subsequent vasodilatory responses to agonists such as PTHrP (tachyphylaxis; see Fig. 26.3), primary or secondary hyperparathyroidism is predicted to impair the paracrine PTHrP signals important in the hemodynamic regulation of tissue perfusion (e.g., kidney, Fig. 26.2). The nuclear localization signal (*NLS*), although present, is not required for this paracrine activity. Of note, the initiation of PTHrP translation at an alternative internal start site results in a truncated PTHrP protein lacking the N-terminal SP. This nonsecreted PTHrP protein is directed to the nucleus via the NLS encoded at residues 88–106. Nuclear PTHrP activates the c-Myc promoter via this intracrine mechanism, directing expression of programs that enhance Skp2/Cul1-dependent p27Kip1 degradation and promote Cdk2-dependent cellular proliferation. The antagonistic paracrine versus intracrine PTHrP signals differentially regulate p27Kip1— and this emerges as a relevant nodal point in the reciprocal control of VSMC proliferation. Nullification of the paracrine PTHrP actions, as arises with desensitization to chronic PTH exposure (see Fig. 26.3), is predicted to result in unchecked intracrine PTHrP signaling. *ER*, endoplasmic reticulum; *???*, unidentified DNA-binding coregulators of nuclear PTHrP actions. *Reprinted with permission from Towler, D.A., 2014b. The role of PTHrP in vascular smooth muscle. Clin. Rev. Bone Miner. Metabol. 12, 190–196.*

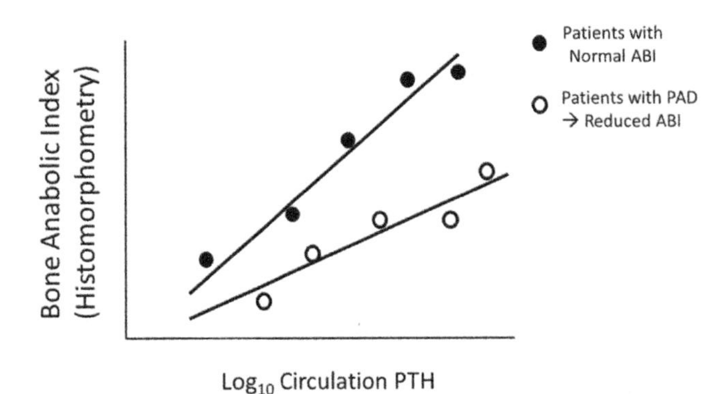

FIGURE 26.5 Histomorphometric evidence of skeletal resistance to parathyroid hormone (*PTH*) in chronic kidney disease patients with peripheral arterial disease (*PAD*). PAD can be identified using a measurement called the ankle–brachial index, or *ABI*. London and colleagues established that, compared with those patients without PAD, patients with PAD exhibit a regression relationship between osteoblast anabolic functions (y axis; double-labeled surface or osteoblast surface with bone surface referent) and intact PTH (*x* axis) that characterizes a reduced PTH sensitivity (London et al., 2015). *Graphic representation of the results of London, G.M., Marchais, S.J., Guerin, A.P., de Vernejoul, M.C. 2015. Ankle-brachial index and bone turnover in patients on dialysis. J. Am. Soc. Nephrol. 26, 476-483; Towler, D.A. 2015. Arteriosclerosis, bone biology, and calciotropic hormone signaling: learning the ABCs of disease in the bone-vascular axis. J. Am. Soc. Nephrol. 26, 243–245.*

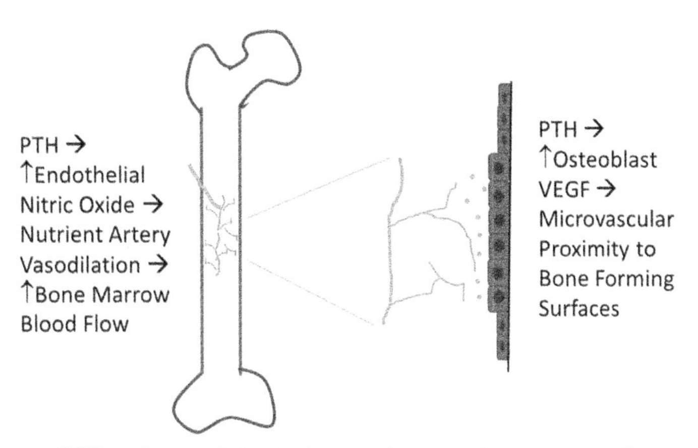

FIGURE 26.6 Parathyroid hormone (*PTH*) actions on skeletal microvasculature are important contributors to anabolic responses. Prisby and colleagues have demonstrated that PTH enhances nutrient artery vasodilation and marrow blood flow in part via endothelial nitric oxide generation by endothelial NO synthase (Prisby et al., 2011) (see also Fig. 26.1). These PTH-induced increases in hindlimb blood flow are important in endocortical bone formation responses (Gohin et al., 2016). Moreover, PTH also increases osteoblast production of vascular endothelial growth factor (*VEGF*; depicted as *red dots*). The paracrine VEGF signaling promotes the juxtaposition of marrow microvasculature to active bone-forming surfaces (Prisby et al., 2011).

blood flow by ca. 30% within 10 min following dosing: a response that is required for PTH-stimulated cortical bone formation (Gohin et al., 2016). Perhaps not surprisingly, then, the vascular dysfunction of arteriosclerosis reduces hip bone mass in humans (Collins et al., 2009) and skeletal perfusion in mice (Shao et al., 2011). However, it remains to be shown that impaired *skeletal* PTH1R signaling is a reliable surrogate for impaired *cardiovascular* PTH1R signaling. Nevertheless, London's important paper (London et al., 2015) highlights that a human metabolic milieu associated with impaired PTH1R signaling exhibits simultaneous reductions in bone and vascular health.

Like CKD, type 2 diabetes and the metabolic syndrome increase the incidence and prevalence of arteriosclerosis, vascular calcification, and conduit vessel stiffness (Stabley and Towler, 2017; Towler, 2017a). This arteriosclerotic vascular stiffness conveys risk for cognitive impairment, stroke, myocardial infarction, heart failure, progression of CKD, and lower extremity amputation (Stabley and Towler, 2017; Thompson and Towler, 2012b; Towler, 2017a). In 2010, our group began to directly assess the potential benefits of cell-autonomous arterial smooth muscle PTH1R signaling in a preclinical model of diabetic arteriosclerosis. We did this by generating a transgenic mouse expressing a constitutively active PTH1R transgene, *PTH1R(H223R)*, in vascular smooth muscle of LDLR$^{-/-}$ mice using the SM22 promoter as a delivery module (Cheng et al., 2010). The PTH1R(H223R) Jansen metaphyseal chondrodysplasia variant is ligand-independent, constitutively active PTH1R that does not undergo homologous desensitization/tachyphylaxis (Schipani et al., 1995). In the LDLR-deficient mouse model of diet-induced diabetic arteriosclerosis, we showed that the VSMC *PTH1R(H223R)* transgene significantly reduced aortic calcification and fibrosis while maintaining arterial compliance (Cheng et al., 2010). This occurred in part via PTH1R-dependent inhibition of arterial prosclerotic Wnt signaling and arterial oxidative stress in VSMCs. Furthermore, intermittent dosing with PTH(1−34) was also shown to reduce vascular osteogenic programs, mineralization, and oxidative stress (Cheng et al., 2010; Shao et al., 2003, 2005). Of note, similar responses have been reported by Morii and colleagues in vitro using cultured primary VSMCs treated with PTHrP (Jono et al., 1997). In vivo, Friedman and coworkers subsequently showed that intermittent PTH(1−34) administration reduces vascular calcification in uremic rats (Sebastian et al., 2008). Likewise, Prisby, Lennon-Edwards, and colleagues showed that intermittent PTH(1−34) administration restores endothelial nitric oxide synthase (NOS) levels and endothelium-dependent arterial vasorelaxation impaired with advanced age in rats (Guers et al., 2017). In collaborative studies that are still ongoing at the time of writing, we have shown that conditional deletion of the PTH1R in VSMCs increases arterial fibrosis in diabetic LDLR$^{-/-}$ mice (Cheng et al., 2016). This confirms and extends our previous results demonstrating that enhancing vascular smooth muscle PTH1R actions reduces arteriosclerosis in this diet-induced murine arteriosclerotic disease model (Cheng et al., 2010). Thus, data from multiple laboratories converge to reveal that maintenance of vascular PTH1R signaling tone helps mitigate the arteriosclerotic responses to diabetes, dyslipidemia, uremia, and aging. However, the precise pharmacokinetic−pharmacodynamic (PK−PD) relationships, including the healthy homeostatic "set point" for cardiovascular PTH1R signaling, remain to be determined.

Of note, a novel intracellular signaling system has been identified for PTHrP actions in both VSMCs and the skeleton that differs dramatically from classical PTH1R activation, and actually antagonizes many of the latter's actions (Fig. 26.4) (Clemens et al., 2001; Miao et al., 2008; Towler, 2014b). This mechanism needs to be recognized in any detailed understanding of PTHrP biology, molecular genetics, and perturbations in circulating PTH levels. The PTHrP gene can encode a transcript wherein translation generates an N-terminal truncation lacking the signal peptide for secretion (Nguyen et al., 2001; Wysolmerski, 2012). A nuclear localization sequence within PTHrP targets the protein to the nucleus where suppression of p27Kip1 expression promotes cell proliferation and neointima formation (Fiaschi-Taesch et al., 2004, 2006, 2009; Fiaschi-Taesch and Stewart, 2003; Massfelder et al., 1997). This intracrine PTHrP/p27Kip1 suppression pathway exerts actions in contradistinction to those elicited by paracrine (secreted) PTHrP/PTH1R signal transduction that suppress proliferation. Thus, when sensitization of the paracrine PTHrP/PTH1R pathway occurs from prior exogenous ligand exposure (Fig. 26.3), the intracrine VSMC PTHrP signaling remains unchecked (Fig. 26.4). The spectrum of nucleoprotein complexes mediating nuclear PTHrP responses involves retinoblastoma-regulated E2F transcriptional pathways (Fiaschi-Taesch et al., 2004; MacLean et al., 2004) and p53 expression (Zhang et al., 2018), but is incompletely characterized.

Thus, strategies that selectively preserve the *paracrine* PTHrP/PTH1R pathway are predicted to exert cardiovascular benefits with respect to enhanced tissue perfusion, reduced arteriosclerotic calcification and vascular stiffness, and restricted neointimal proliferation; these actions, in sum, help to preserve arterial structure, cardiac pump functions, and tissue perfusion. Therefore, in addition to alterations in circulating calcium phosphate levels, the physiology of direct cardiovascular PTH1R signaling and actions deserves full consideration. This concept is very important to embrace as we seek to better define medical and surgical strategies to combat cardiovascular disease in the settings of primary HPT and CKD (Shroff, 2011). As briefly mentioned earlier, no clinically validated metric yet exists to monitor the healthy cardiovascular PTH1R signaling set point. This is all the more difficult since beneficial actions of paracrine PTHrP signaling will be influenced by endocrine PTH tone and circulating inhibitory fragments that accumulate with CKD (Friedman and Goodman, 2006). Moreover, as highlighted (Bilezikian et al., 2015), because variable posttranslational oxidation occurs at residues Met-8 and Met-18 of intact PTH that alters bioactivity, establishing these relationships in certain disease states (CKD, diabetes) may require implementation of an even newer generation of PTH assays (Hocher et al., 2012).

PTH2R signaling in vascular pharmacology

PTH and TIP39 are both agonists for the human PTH2R, while TIP39 is selective for PTH2R and PTHrP is selective for PTH1R activation (Hoare and Usdin, 2001). Compared with PTH1R, far less is known concerning PTH2R signaling and cardiovascular physiology and function. At this early stage of investigation, available data suggest that endogenous TIP39 actions relevant to cardiovascular physiology are likely to be mediated through the central nervous system (CNS). Functional TIP39 protein in peripheral tissues has yet to be unambiguously established; however, since mice transgenic for chondrocyte-specific expression of its cognate receptor PTH2R exhibit skeletal defects, the *TIP39* expression noted in the hypertrophic zone is likely to be functional (Panda et al., 2012). Strategies that rigorously probe *endogenous* TIP39/PTH2R signaling in cardiovascular biology have not yet been pursued. Most relevant studies have pursued pharmacological approaches. In the CNS, intracerebroventricular injection of TIP39 causes a fall in blood pressure in rats (Sugimura et al., 2003). Presynaptic augmentation of glutamatergic autonomic outflow in the hypothalamus (Dimitrov et al., 2011; Dobolyi et al., 2012) and/or arginine vasopressin release may also be important. Of interest, though, TIP39 mRNA has been detected in intrarenal arteries and aorta by RT-PCR (Eichinger et al., 2002).

The *pharmacological* actions of TIP39 administration have been implemented as a strategy to explore contributions of the peripheral PTH2R to cardiovascular physiology. Like the PTH1R, the PTH2R is expressed in endothelial cells, smooth muscle cells, and cardiomyocytes (Potthoff et al., 2011; Ross et al., 2005). Coronary artery flow reserve as modulated by TIP39/PTH2R signaling has been studied in the Langendorff-perfusion isolated rat heart model (Ross et al., 2007). TIP39 treatment exerted little activity in vessels precontracted with phenylephrine. Curiously, precontraction followed by pretreatment with the PTH1R-selective agonist PTHrP enabled vasodilatory response to subsequent TIP39/PTH2R activation (Ross et al., 2007). In vivo, prior cardiac ischemia provided sufficient stimulus for sustained release of endogenous PTHrP from the heart (Monego et al., 2009), and thus potentially enabled the subsequent vasodilatory response to PTH2R ligands. The mechanisms and kinetics of PTH1R–PTH2R cross talk have yet to be fully elucidated, and while TIP39 expression has been detected in the heart and coronary endothelium (Papasani et al., 2004; Ross et al., 2005), endogenous role and regulation therein are virtually unexplored. However, because inducible NOS activity was required for TIP39-mediated dilatation, the coronary endothelium is likely to participate in this complex pharmacology (Ross et al., 2007). Of note, TIP39/PTH2R signaling profoundly reduces the contractility of *isolated* cardiomyocytes in culture (Ross et al., 2005). Thus, as Ross et al. note (Ross et al., 2007), as a dual PTH1R/PTH2R agonist, in some settings human PTH could theoretically exhibit direct negative inotropic actions in vivo via the PTH2R— actions that are countermanded by NOS release via PTH1R. However, this notion contrasts profoundly with in vivo cardiac PTHrP/PTH1R responses; inotropy is augmented with a net positive impact dependent upon endothelial NO release in vivo (Ogino et al., 1995; Schreckenberg et al., 2009). Nevertheless, it remains possible that PTH1R-selective agonists, compared with dual PTH1R/PTH2R agonists, may differ substantially in any putative cardiovascular actions in human subjects.

By deploying newly available murine genetic models in studies of skin biology, TIP39/PTH2R receptor signaling has been shown to support expression of type I collagen and the small proteoglycan decorin as necessary for robust wound healing in the skin (Sato et al., 2017). Since these very same extracellular matrix molecules are also vital to the integrity of large conduit vessels such as aorta (Faarvang et al., 2016; Farb et al., 2004), PTH2R signaling may potentially play a role in cardiovascular tissue matrix remodeling and repair.

In humans, the complex interplay between actions of PTH as an agonist for both PTH1R and PTH2R has yet to be integrated with the kinetics of PTH1R versus PTH2R desensitization (Bisello et al., 2004); this integration will be necessary to provide a unifying model with respect to the PK–PD for PTH1R signaling in cardiovascular health and disease (Thompson and Towler, 2012a). Since PTH binds to both human PTH1R and human PTH2R and can induce tachyphylaxis in PTH1R—and PTH1R downregulation enables cardiovascular PTH2R vasodilation (Ross et al., 2007)—the kinetics of impaired PTH1R versus PTH2R signaling in HPT will matter. Based upon the current data, HPT is predicted to inhibit coronary flow reserve (CFR; discussed later) as supported by either PTH1R or PTH2R, with loss in PTH1R tone exerting the greater impact. Furthermore, the potential relationships between central actions of TIP39 with respect to peripheral PTH/PTHrP biology have yet to be vetted in detail. It is important to note that centrally administered PTHrP (Yamamoto et al., 1998) and TIP39 (Sugimura et al., 2003) differentially regulate arginine vasopressin release (increased by the former, decreased by the latter), which is relevant to cardiovascular pressor/volume homeostasis. Moreover, in contradistinction to peripheral actions, central actions of PTHrP provide a pressor response that is dependent upon CNS-regulated sympathetic outflow (Nagao et al., 1998), while CNS administration of TIP39 reduces blood pressure (Sugimura et al., 2003). Thus, pharmacology targeting the PTH family of ligands must be cognizant that any peripherally administered compounds that cross the blood–brain barrier may exhibit mixed cardiovascular responses.

Parathyroid hormone, hyperparathyroidism, and calcific aortic valve disease

Another vascular structure affected by PTH signaling is the aortic valve (Iwata et al., 2012). As of this writing, the role of PTH/PTHrP signaling in cardiac valve physiology is very poorly understood. As mentioned, knockdown in zebrafish of either *pth1r* or *pthrp* = *pthlh* ligand impairs aortic valve morphogenesis, dependent upon downstream *notch* actions (Gray et al., 2013). In humans, calcific aortic valve disease (CAVD) is prevalent, afflicting approximately 2% of our population over age 60, with up to half of these possessing previously unappreciated bicuspid aortic valves (Yutzey et al., 2014). By age 75 or older, 15% of individuals will have moderate to severe calcific aortic stenosis. Intriguingly, *NOTCH1* mutations in humans convey valve malformations, CAVD, and Notch1 expression inhibits aortic valve interstitial cell calcification in culture (Acharya et al., 2011). For those with valve stenosis and symptoms, surgical intervention is required to mitigate the very high cardiovascular mortality observed in this setting. Several epidemiological studies indicate that primary HPT is associated with CAVD (Iwata et al., 2012; Linhartova et al., 2008; Stefenelli et al., 1997), and genetic variation in the *PTH* gene has been suggested to be a contributor (Gaudreault et al., 2011; Schmitz et al., 2009). Moreover, elevated PTH and hypertension are both risk factors associated with accelerated clinical progression of CAVD in patients with CKD (Iwata et al., 2013). Because the annulus of the heart valve is rich in chondrocytes and chondrocyte progenitors (Lincoln et al., 2006)—cellular targets of PTH1R signaling in the endochondral skeleton (Hirai et al., 2011; Kronenberg, 2006)—the aortic valve response to PTH1R signaling is likely to be quite complex, and in some instances more akin to the tissue response of endochondral bone (Yutzey et al., 2014). With advanced CAVD, true ectopic woven bone formation—replete with marrow elements and osteoclast-dependent bone remodeling—is observed in 13% of specimens (Fuery et al., 2018; Mohler et al., 2001). Medical therapies targeting cholesterol metabolism have failed to have a significant impact on CAVD (Rajamannan et al., 2011), potentially because oxidized phospholipids and lysophosphatidic acid carried by Lp(a) appear much more important (briefly reviewed in Towler, 2017b). Once valve mineralization is advanced, surgical or endovascular strategies for aortic valve replacement are required (Unger et al., 2016). Should surgical treatment of asymptomatic primary HPT (Yu et al., 2010) prove to mitigate risk for future development of clinically significant CAVD (Iwata et al., 2012) in at-risk individuals (e.g., bicuspid valve, high Lp(a) levels, sclerosis without calcinosis on echocardiography), this would (1) provide additional rationale for integrating cardiovascular phenotyping into indications for surgical treatment and (2) potentially alter the age-specific cutoff for recommending surgery (e.g., to age < 60 in lieu of < 50) (Bilezikian, 2018)

Impaired vascular PTH1R signaling and cardiovascular disease: the impact of hyperparathyroidism on cardiovascular mortality, coronary flow reserve, and vascular stiffness

Numerous observational studies have supported the relationship between primary HPT and cardiovascular disease (Walker and Silverberg, 2008). However, one large study of 2.99 million person-years of follow-up in 5735 patients with mild primary HPT presents a clinically compelling epidemiological data set for this pathophysiological relationship (Yu et al., 2010). This study is called PEARS, the Parathyroid Epidemiology and Audits Research Study. PEARS is a retrospective, population-based outcomes study that investigated patients diagnosed with mild primary HPT in Tayside, Scotland, between 1997 and 2006 (Yu et al., 2010). In the 1683 subjects identified with mild primary HPT (2/3rd female), age- and gender-correlated cardiovascular mortality was increased 2.7-fold (Yu et al., 2010). Baseline PTH levels, not calcium, predict long-term outcomes in untreated primary HPT (Yu et al., 2013). The HEALTH ABC cohort, focused on individuals in Memphis and Pittsburgh, confirmed increases in cardiovascular mortality with elevated PTH; elevation conveyed a 1.8-fold increased risk, independent of serum calcium and 25-hydroxyvitamin D levels (Kritchevsky et al., 2012). As discussed before, preclinical models evaluating PTH/PTHrP actions have long indicated the role of the cardiovascular system as a physiologically relevant target, with mechanisms relevant to disease biology rapidly emerging from genetically and pharmacologically manipulated animals (Cheng et al., 2010, 2016; Qian et al., 2003; Raison et al., 2013). However, it is the physiological studies of human subjects that have confirmed and extended these observations in important ways that are likely to alter our recommendations for treatment of HPT in otherwise "asymptomatic" patients (Silverberg et al., 2009).

Macrovascular compliance and conduit functions, known as Windkessel physiology (Safar and Boudier, 2005; Westerhof et al., 2009), are impaired by HPT (Briet et al., 2012; Rubin et al., 2005; Walker et al., 2009). With every heartbeat, part of the energy of each systolic pulse is stored as potential energy within the rubbery elasticity of conduit vessels via systole-induced circumferential stretching of large arteries (Westerhof et al., 2009). During diastole, this energy is released to provide smooth perfusion of the coronary arteries, myocardium, and distal tissues (Thompson and Towler, 2012a). Moreover, with arterial stiffening, retrograde pulse wave reflections arising at vessel branch points travel faster; rather than supporting early diastole, these omnipresent reflections now sum with the orthograde pulse waves of late systole to increase systolic blood pressure (Safar and Boudier, 2005; Soldatos et al., 2011; Westerhof et al., 2009). Thus, with arterial stiffening, not only does perfusion during diastole become more erratically pulsatile, but the workload placed upon the heart increases due to systolic hypertension. Silverberg, Bilezikian, and colleagues established that arterial stiffening is increased in patients with mild HPT. Using the augmentation index (AIx), a measure of the retrograde wave reflection that characterizes vascular stiffening, they uncovered a positive linear relationship between PTH and AIx after adjusting for heart rate, height, gender, blood pressure, age, diabetes mellitus, smoking, and hyperlipidemia (Rubin et al., 2005). Smith and colleagues reported similar responses in their smaller patient cohort (Smith et al., 2000). Importantly, in patients with preexisting cardiovascular abnormalities, parathyroidectomy improves indices of cardiovascular/carotid stiffness (Walker et al., 2012), indicating reversibility. Of note, Prisby and colleagues noted a protective action of intermittent PTH administration on endothelial NOS expression and conduit vessel endothelial function with aging marrow in rodents (Lee et al., 2018). While not studied in great detail, a few small longitudinal studies of primary HPT patients pre− and post−curative parathyroidectomy have shown that the circadian, pulsatile nature of diurnal PTH secretion is normalized as well (Lobaugh et al., 1989).

CFR is a physiological index of epicardial conduit function. CFR assesses not only the significance of any coronary stenosis present, but also the presence of downstream microvascular dysfunction (Bianco and Alpert, 1997). When myocardial oxygen demand downstream of a conduit artery segment cannot be met by vasodilation, CFR is deemed to be compromised (Bianco and Alpert, 1997). Importantly, reduced CFR portends adverse cardiovascular outcomes in men and women (Cortigiani et al., 2012; Pepine et al., 2010). CFR was initially evaluated by angiography following intracoronary administration of vasodilators coupled with Doppler imaging (Pepine et al., 2010); however, transthoracic Doppler echocardiography (Cortigiani et al., 2012) has now been implemented, with peripheral adenosine vasodilator administration, as a mechanism for noninvasively assessing CFR (Caiati et al., 1999; Montisci et al., 2006).

In 2012, Osto and colleagues performed a highly important, longitudinal analysis of CFR in 100 primary HPT patients with solitary adenomas. These patients were studied before and after adenoma resection, comparing CFR results in that population to 50 gender- and age-matched controls (Osto et al., 2012). In this landmark study, they established that PTH, age, and heart rate were the only major variables altering CFR— and that CFR changes with HPT were independent of serum calcium. Moreover, in the 27 primary HPT patients with clearly abnormal CFR ($\leq$2.5 following adenosine infusion), all of these individuals exhibited normal CFR (>2.5) following curative adenoma resection (Osto et al., 2012). Deploying an independent method for CFR measurement—viz., nuclear perfusion imaging—Marini and colleagues also demonstrated improvement of CFR in primary HPT undergoing surgical treatment (Marini et al., 2010). In a clinically important way, these two studies confirmed and extended the previous work of Kosch et al. (Kosch et al., 2000); they had shown that brachial artery flow-mediated vasodilation, an index of endothelial function and NOS activation, is impaired in primary HPT but reversed following surgical treatment (Kosch et al., 2000). The impaired flow-mediated dilation in patients with primary HPT undergoing medical observation has only recently been confirmed by others (Colak et al., 2017). Thus, primary HPT is consistently associated with reversible alterations in coronary and peripheral artery endothelial function that improve following surgical intervention. The improvements in CFR following surgery for HPT have significant implications as the risks versus benefits of parathyroidectomy in otherwise asymptomatic HPT are considered.

Chronic kidney disease–mineral and bone disorder: the metabolic "perfect storm" of cardiovascular risk

CKD is a highly significant cardiovascular risk factor, synergizing with diabetes to increase morbidity and mortality at least 5- to 10-fold (Chang et al., 2013; Debella et al., 2011). While cholesterol-lowering therapies have an impact, reducing nonfatal cardiovascular events by ca. 20% in predialysis CKD, no significant reduction in cardiovascular mortality has been achieved with this strategy in CKD patients on dialysis (Baigent et al., 2011; Baigent and Landry, 2003). This appears to be in large part due to the contributions of perturbed calcium phosphate metabolism to cardiovascular risk (Block et al., 2013) and the bone-derived hormone FGF23 (Moe et al., 2015). Indeed, the CKD–mineral and bone disorder designation was created as an imperfect mechanism meant to capture the implications of this clinical setting (Kidney Disease: Improving Global Outcomes, 2009; Moe et al., 2007).

In the setting of CKD, serum phosphate exerts a stepwise increase in cardiovascular mortality (Eddington et al., 2010). Vascular toxicity is mediated in part via arterial calcification (Shao et al., 2010), and cardiovascular mortality tracks the presence and extent of vascular mineralization (Blacher et al., 2001; London, 2003; London et al., 2003). Phosphate-dependent activation of sodium phosphate cotransporter signaling in VSMCs drives osteogenic mineralization programs (Liu et al., 2013) and proapoptotic responses that perturb mineralizing matrix vesicle clearance, thus augmenting vascular calcium load (Shroff et al., 2008, 2010, 2013). However, at every level of renal function, serum phosphate is a cardiovascular risk factor (Tonelli et al., 2005, 2009) and PTH is a key defense against hyperphosphatemia (Martin and Gonzalez, 2011). PTH directly induces renal phosphate excretion when renal function is intact, recruits osteoblast-derived FGF23 to assist in these phosphaturic actions, and maintains bone formation even in the setting of CKD—a state of variable skeletal resistance to PTH (London et al., 2004; Martin et al., 2012; Quarles, 2013). Thus, actions of PTH play an important role along with FGF23 in mitigating phosphate-dependent vascular toxicity when renal function is intact and normal bone turnover is maintained (London et al., 2004; Martin et al., 2012; Quarles, 2013). Extremes of bone turnover—too high or too low—are likely to negatively influence serum phosphate homeostasis in the setting of CKD (Barreto et al., 2008; Goodman, 2004; London et al., 2004). This may in part explain the bimodal relationship between PTH levels and mortality in dialysis patients (Durup et al., 2012). However, the specific relationship between bone formation (potentially functioning as a "buffer" mitigating vascular phosphate toxicity), changes in PTH tone with declining renal function, and maintaining healthy norms of serum phosphate homeostasis with respect to vascular physiology has not yet been established (Block et al., 2013; Maeda et al., 2007).

Gerard London and colleagues have provided the most important insights into the relationship between PTH, arterial calcification, and bone formation in the setting of CKD (London et al., 2004). Implementing an ultrasound-based method for scoring calcification in the common carotid arteries, the abdominal aorta, the iliofemoral axis, and the lower extremities, his group related the extent of arterial calcification with dynamic histomorphometric assessment of cellular bone functions to circulating PTH. In this important study, they established that patients with the lowest levels of PTH (with or without prior parathyroid surgery) exhibited low-turnover bone disease and the most extensive vascular calcification (London et al., 2004). The relationship between low PTH values and increased coronary artery calcification was subsequently confirmed by others (Kim et al., 2011). Interestingly, similar observations were made by Chertow and colleagues upon comparison of the relationships between bone mineral density, vascular calcium load, and PTH levels in patients treated with calcium-based phosphate binders versus sevelamer (does not contain calcium) (Raggi et al., 2005). Suppression of PTH levels in those subjects given calcium-based binders was associated with lower bone mass and increased vascular calcium load.

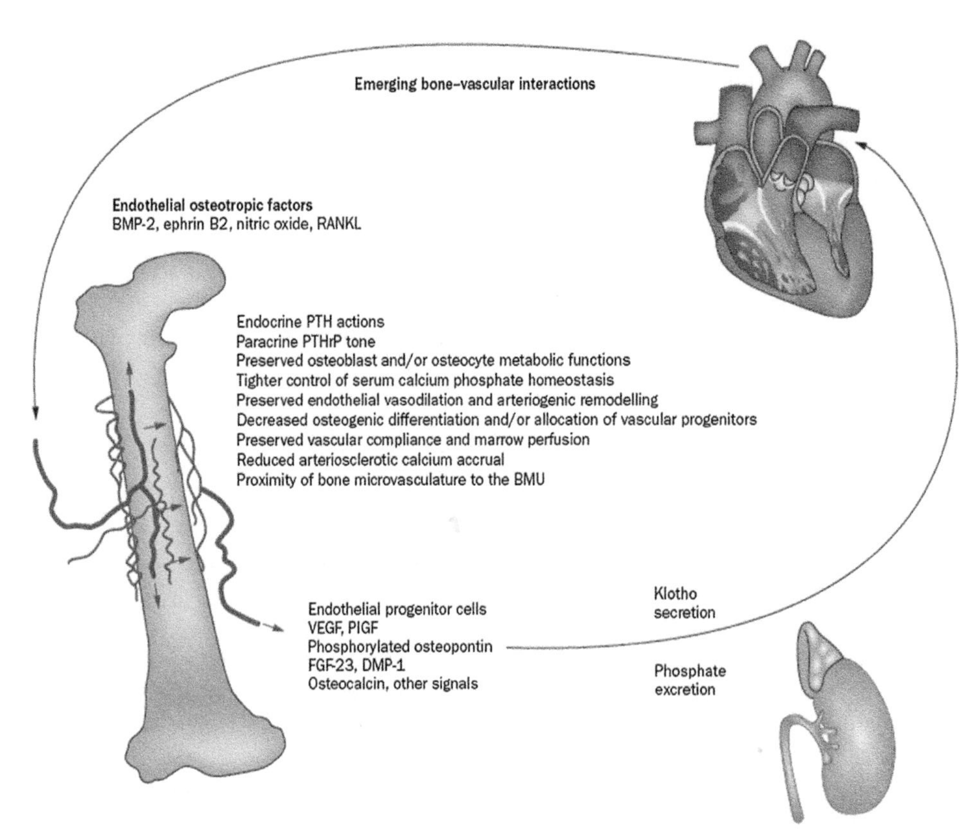

FIGURE 26.7 An emerging bone–vascular axis is connected by calcium, phosphate, hormonal, and cellular signals in cardiovascular health and disease. The skeleton is a parathyroid hormone (*PTH*) receptor 1 (PTH1R)-regulated source/repository of hormones, electrolytes, and cellular elements that profoundly influence cardiovascular health. The kidney plays a pivotal role in modulating this axis by mediating phosphate excretion and by providing additional hormonal relays. The vasculature is also a direct target of PTH actions via both PTH1R and PTH2R in humans. Perturbation of this axis creates the chronic kidney disease–mineral and bone disorder. *BMP-2*, bone morphogenetic protein; *BMU*, basic multicellular unit; *DMP-1*, dentin matrix protein 1; *FGF-23*, fibroblast growth factor 23; *PlGF*, placental growth factor; *PTHrP*, PTH-related protein; *RANKL*, receptor activator of NF-κB ligand; *VEGF*, vascular endothelial growth factor. *See text for details, reprinted with permission from Thompson, B., Towler, D.A. 2012a. Arterial calcification and bone physiology: role of the bone-vascular axis. Nat. Rev. Endocrinol.*

Basal PTH tone most likely exerts beneficial actions via maintenance of bone formation (London et al., 2004). As discussed (Demer and Tintut, 2010; Fadini et al., 2012; Thompson and Towler, 2012a; Zoppellaro et al., 2012), it has become increasingly clear that skeletal maintenance of hematopoiesis, regulation of calcium phosphate exchange, and skeletal production of phosphaturic hormones such as FGF23 play important roles in vascular health (Fig. 26.7; see also discussion of EVOLVE results later). Surgery is the standard of care for severe primary and severe secondary/tertiary HPT (Silverberg et al., 2009). However, surgery is not the first-line approach to the vast majority of patients with significant secondary HPT in CKD because of the clinical need to "titrate" PTH levels to maintain bone formation (London et al., 2004). In the setting of uremia, circulating fragments of PTH and uremic toxins give rise to a skeletal resistance to PTH (Friedman and Goodman, 2006). Based on dynamic histomorphometry performed in patients with CKD5 and on dialysis, PTH levels between 150 and 300 pg/mL were associated with maintenance of normal bone turnover in this setting. However, similar analyses have not been performed in CKD patients prior to the initiation of renal replacement therapy. Moreover, the specific PTH assay used to establish this treatment goal is no longer available, calling into question how one coregisters these prior guidelines with current-generation PTH assays. Hence, current KDIGO (Kidney Disease Improving Global Outcomes) guidelines recommend levels of PTH between two and nine times the upper limit of normal in any given assay, although not strictly coregistered with skeletal or cardiovascular indices of health (London et al., 2010).

Nevertheless, striving to achieve this skeletally defined set point also appears to improve cardiovascular disease risk; and the cardiovascular benefits of pharmacological management of HPT were first established in the setting of CKD5 on dialysis. Treatment with injectable calcitriol or paricalcitol to maintain PTH levels to approximately two to four times the upper limit of normal reduces cardiovascular mortality (Teng et al., 2003, 2005) and points once again to the importance of calciotropic hormones as key modulators of cardiovascular disease in this setting. This biology and pharmacology is likely to be relevant to improving the care of the many aging individuals with CKD3 or worse who do not require renal replacement therapy (30 million individuals estimated in the United States alone). The Systolic Blood Pressure Intervention Trial (SPRINT) noted that predialysis patients with CKD3 or worse and high PTH levels accrued less cardiovascular benefit from intensive blood pressure therapy (Ginsberg et al., 2018). It will be important to determine if normalizing PTH levels in CKD3/CKD4 with novel oral extended-release formulations of calcifediol (Sprague et al., 2017) will restore the cardiovascular benefits of intensive blood pressure control (Ginsberg et al., 2018).

As a type II calcium-sensing receptor mimetic with a short half-life, cinacalcet successfully reduces circulating PTH levels and enables facile titration to goal. Cinacalcet is very effective in treating primary and secondary HPT, but its actions on cardiovascular and fracture risks are still emerging (Verheyen et al., 2013). A meta-analysis of four studies suggested reductions in both hip fracture and cardiovascular hospitalization rates with cinacalcet (Cunningham et al., 2005). In EVOLVE (Evaluation of Cinacalcet Hydrochloride Therapy to Lower Cardiovascular Events), a prospective study to treat secondary HPT in dialysis patients with CKD5, cinacalcet failed to achieve significant reductions in cardiovascular mortality ($P = .11$) in the unadjusted intention-to-treat analysis (Investigators et al., 2012). However, in a second prespecified analysis, older patients on hemodialysis with moderate to severe secondary HPT did accrue cardiovascular benefit with cinacalcet-mediated reductions in PTH (Parfrey et al., 2015). The milieu of CKD5 and dialysis represents a metabolic "perfect storm" for cardiovascular disease; in this setting, secondary HPT is but one important contributor that must be addressed (Towler, 2013). Moving forward, the multifactorial high-risk metabolic and genetic milieu of CKD must be prospectively embraced as strategies targeting PTH biology are evaluated for mitigating cardiovascular risk. Recall that PTH controls FGF23 release from the osteocyte (Rhee et al., 2011). Upon stratification of secondary reductions in FGF23 also achievable by cinacalcet dosing (64% vs. 28% placebo), the EVOLVE trial was able to demonstrate reduced cardiovascular mortality in those patients that exhibited a 30% or greater reduction in FGF23 (Moe et al., 2015). The improvement was achieved via reductions in nonatherosclerotic cardiovascular events, and mechanisms segregating responders from nonresponders have yet to be determined, but may reflect unknown features of the PTH-regulated bone–vascular axis (Fig. 26.7).

Of note, the relationship between bone turnover, PTH, and vascular risk may not hold with respect to the calciphylaxis syndrome, aka calcific uremic arteriolopathy (Shao et al., 2010). This distinct clinical entity involves fibroproliferative calcification of smaller-diameter arterioles in dermal adipose, lung, and mesentery. It is seen most frequently in obese diabetic patients with CKD on warfarin anticoagulation and can be associated with severe HPT (Matsuoka et al., 2005; Rogers et al., 2007). The pathobiology of this important and lethal entity is still poorly understood (Nigwekar et al., 2015).

PTH/PTHrP signaling and the bone–vascular axis

As Scadden, Kronenberg, and colleagues have established, the PTH1R controls the size of the bone marrow hematopoietic niche via its osteotropic actions (Adams et al., 2007; Calvi et al., 2003). This is a critically important concept to consider particularly in the setting of CKD–MBD; circulating hematopoietic progenitors and endothelial progenitor cells derived from the bone marrow niche contribute to both healthy and pathological injury responses in distant venues (Eghbali-Fatourechi et al., 2005, 2007; Fadini et al., 2012). Cognizant of this, Zaruba et al. (Brunner et al., 2008; Zaruba et al., 2008) and Napoli and colleagues (Napoli et al., 2008) examined the impact of PTH(1–34) in models of myocardial and critical limb ischemia, respectively. Zaruba demonstrated that survival following occlusion of the left anterior descending artery in mice was increased from 40% to 60% by postischemia dosing with PTH(1–34). While the relative contributions of vasodilation, angiogenesis, and apoptosis were not examined, survival tracked PTH(1–34)-induced increases in myocardial VEGF expression, neovascularization, and $CD31^+$ and $CD34^+/CD45^+$ cell populations (Zaruba et al., 2008). Border zone cardiomyocyte apoptosis was reduced, with concomitant improvement in cardiac output. Using unilateral femoral artery ligation to model critical limb ischemia, Napoli et al. demonstrated that preligation dosing with PTH(1–34) followed by postischemic granulocyte colony-stimulating factor mobilization of marrow progenitors simultaneously increased limb blood flow and viability and $CD34^+$ stem cell recruitment to and capillary density in ischemic limb skeletal muscle (Napoli et al., 2008).

Thus, an emerging bone–vascular axis has been proposed wherein circulating progenitor cells dependent upon PTH1R bioactivity can favorably influence cardiovascular health (Fig. 26.7). Since circulating endothelial progenitor cell subtypes are differentially altered in patients with peripheral arterial disease, coronary syndromes, or diabetes (Flammer et al., 2012a, 2012b, endothelial progenitor cell quality and character will be extremely important to consider (Gossl et al., 2010). Whether this axis can be productively harnessed with PTH/PTHrP signaling to have an impact on cardiovascular disease burden in humans remains to be explored.

PTH1R activation and the renin–angiotensin–aldosterone axis: a feed-forward vicious cycle

As mentioned before, the relationships between HPT and hypertension are well appreciated (Tomaschitz et al., 2012). In otherwise healthy adults, while singular administration lowers blood pressure, sustained PTH administration over a period of 12 days actually increases blood pressure in humans (Hulter et al., 1986). In addition to the aforementioned PTH1R desensitization mechanisms (Nyby et al., 1995) that impair paracrine PTHrP-mediated vasorelaxation (Maeda et al., 1999; Raison et al., 2013), PTH may have direct effects on aldosterone biosynthesis (Tomaschitz et al., 2012). However, PTH1R-dependent signals in the juxtaglomerular apparatus also support renin production (Saussine et al., 1993a, 1993b). Following surgery for HPT, renin and aldosterone levels fall (Kovacs et al., 1998), albeit with variable improvements in blood pressure and cardiac hypertrophy (Agarwal et al., 2013; Broulik et al., 2011; Ishay et al., 2011; Kovacs et al., 1998; Persson et al., 2011; Vazquez-Diaz et al., 2009). Conversely, it has become clear that angiotensin II augments circulating PTH levels in part via aldosterone/mineralocorticoid receptor signaling pathways in humans (Brown et al., 2013). Thus, a "feed-forward" vicious cycle may exist between dysregulated PTH and the renin–angiotensin–aldosterone (RAA) axis that promotes cardiovascular disease arising from the metabolic derangements of aging, declining renal function, and HPT (Tomaschitz et al., 2013). Of note, the risk for CAVD progression in the Japanese Aortic Stenosis Study tracked with absence of angiotensin receptor blocker therapy in early-stage disease (Yamamoto et al., 2010). While PTH levels were not assessed, it is intriguing to speculate that modulating both arms of the prosclerotic PTH–RAA cycle might help mitigate CAVD risk with aging and uremia (Rajamannan et al., 2011). This potential relationship has yet to be tested in either preclinical or clinical research settings. The use of RAA inhibitors does indeed reduce PTH levels in humans (Brown et al., 2015).

Summary, conclusions, and future directions

The cardiovascular actions of PTH have been known for almost a century (Collip and Clark, 1925). However, the clinical and pharmacological implications of altered PTH/PTHrP signaling with respect to cardiovascular endocrinology have been underappreciated except in the settings of extreme excess with severe hypercalcemia or calciphylaxis. In addition to mild primary HPT, secondary HPT associated with advancing age and/or declining renal function is increasingly prevalent. In addition to the uremic milieu of CKD (Friedman and Goodman, 2006), dyslipidemia also induces a state of PTH/PTHrP resistance (Li et al., 2014; Sage et al., 2011). As such, perturbations in PTH/PTHrP physiology are emerging as one of the most prevalent endocrinopathies affecting human health and health care today. Heart disease remains the No. 1 cause of mortality worldwide and is a growing burden (Gaziano et al., 2006). Given the contributions of PTH/PTHrP signaling in cardiovascular health and disease—and the prevalence of perturbed PTH endocrine physiology—it remains stunning that so very little is known about the fundamentals of PTH/PTHrP biology within the vasculature. The specific protein–protein interactions and signaling cascades conveying PTH1R/PTH2R actions in VSMCs, endothelial cells, and interstitial/adventitial cell populations need to be defined for each relevant vascular bed. Coronary, renal, and iliofemoral vessels may represent the most clinically relevant with respect to cardiovascular and metabolic bone diseases. The mechanisms controlling paracrine PTHrP bioactivities have yet to be identified and fully integrated with the mechanical and neuroendocrine cues that together coordinate tissue perfusion. Most importantly, tissue-specific biomarkers and functional imaging methods are required to quantify cardiovascular PTH/PTHrP actions as a first step toward establishing the healthy set point for signaling tone and dynamics. Because posttranslational modifications and dysmetabolic states induce PTH/PTHrP resistance in the very settings where this biology becomes most clinically significant (Demer and Tintut, 2010; Thompson and Towler, 2012a), titration to a single circulating PTH value range will probably prove to be inadequate. Moreover, whereas surgery is a standard of care in primary HPT that yields reductions in hip fracture risk, renal disease, and hypercalcemic crises (Silverberg et al., 2009; VanderWalde et al., 2006), the impact of parathyroid surgery on cardiovascular morbidity has yet to be firmly established for mild or asymptomatic disease. However, beneficial changes with curative surgery in important cardiovascular parameters, such as vascular stiffness (Walker et al., 2012) and CFR (Osto et al., 2012), are truly compelling. It remains unclear whether surgical and pharmacological interventions for treatment of HPT are functionally equivalent with respect to cardiovascular outcomes. As occurred during the history of the cholesterol controversy, discovery and implementation of clearly effective pharmacotherapy will be important (Steinberg, 2006). Clearly, detailed examination of cardiovascular physiology as regulated by the PTH superfamily will continue to yield important new insights, and will be necessary to devise novel therapeutic strategies that better treat our patients afflicted with cardiometabolic and mineral metabolism disorders.

Acknowledgments

D.A.T. is supported by grants from the National Institutes of Health (HL069229, HL114806) and the American Diabetes Association, the J.D. and Maggie E. Wilson Distinguished Chair in Biomedical Research, and the Louis V. Avioli Professorship in Mineral Metabolism Research.

References

Acharya, A., Hans, C.P., Koenig, S.N., Nichols, H.A., Galindo, C.L., Garner, H.R., Merrill, W.H., Hinton, R.B., Garg, V., 2011. Inhibitory role of Notch1 in calcific aortic valve disease. PLoS One 6, e27743.

Adams, G.B., Martin, R.P., Alley, I.R., Chabner, K.T., Cohen, K.S., Calvi, L.M., Kronenberg, H.M., Scadden, D.T., 2007. Therapeutic targeting of a stem cell niche. Nat. Biotechnol. 25, 238–243.

Agarwal, G., Nanda, G., Kapoor, A., Singh, K.R., Chand, G., Mishra, A., Agarwal, A., Verma, A.K., Mishra, S.K., Syal, S.K., 2013. Cardiovascular dysfunction in symptomatic primary hyperparathyroidism and its reversal after curative parathyroidectomy: results of a prospective case control study. Surgery 154, 1394–1404.

Babiker, H.M., McBride, A., Newton, M., Boehmer, L.M., Drucker, A.G., Gowan, M., Cassagnol, M., Camenisch, T.D., Anwer, F., Hollands, J.M., 2018. Cardiotoxic effects of chemotherapy: a review of both cytotoxic and molecular targeted oncology therapies and their effect on the cardiovascular system. Crit. Rev. Oncol. Hematol. 126, 186–200.

Baigent, C., Landray, M.J., Reith, C., Emberson, J., Wheeler, D.C., Tomson, C., Wanner, C., Krane, V., Cass, A., Craig, J., et al., 2011. The effects of lowering LDL cholesterol with simvastatin plus ezetimibe in patients with chronic kidney disease (Study of Heart and Renal Protection): a randomised placebo-controlled trial. Lancet 377, 2181–2192.

Baigent, C., Landry, M., 2003. Study of heart and renal protection (SHARP). Kidney Int. Suppl. S207–S210.

Barreto, D.V., Barreto Fde, C., Carvalho, A.B., Cuppari, L., Draibe, S.A., Dalboni, M.A., Moyses, R.M., Neves, K.R., Jorgetti, V., Miname, M., et al., 2008. Association of changes in bone remodeling and coronary calcification in hemodialysis patients: a prospective study. Am. J. Kidney Dis. 52, 1139–1150.

Benson, T., Menezes, T., Campbell, J., Bice, A., Hood, B., Prisby, R., 2016. Mechanisms of vasodilation to PTH 1-84, PTH 1-34, and PTHrP 1-34 in rat bone resistance arteries. Osteoporos. Int. 27, 1817–1826.

Bianco, J.A., Alpert, J.S., 1997. Physiologic and clinical significance of myocardial blood flow quantitation: what is expected from these measurements in the clinical ward and in the physiology laboratory? Cardiology 88, 116–126.

Bilezikian, J.P., 2018. Primary hyperparathyroidism. J. Clin. Endocrinol. Metab. 103, 3993–4004.

Bilezikian, J.P., Marcus, R., Levine, M.A., Marococci, C., Silverberg, S.J., Potts, J.T., 2015. The Parathyroids: Basic and Clinical Concepts. Elsevier/AP, Academic Press is an imprint of Elsevier, Amsterdam, Boston.

Bisello, A., Manen, D., Pierroz, D.D., Usdin, T.B., Rizzoli, R., Ferrari, S.L., 2004. Agonist-specific regulation of parathyroid hormone (PTH) receptor type 2 activity: structural and functional analysis of PTH- and tuberoinfundibular peptide (TIP) 39-stimulated desensitization and internalization. Mol. Endocrinol. 18, 1486–1498.

Blacher, J., Guerin, A.P., Pannier, B., Marchais, S.J., London, G.M., 2001. Arterial calcifications, arterial stiffness, and cardiovascular risk in end-stage renal disease. Hypertension 38, 938–942.

Blanco, R., Gerhardt, H., 2013. VEGF and Notch in tip and stalk cell selection. Cold Spring Harb. Perspect. Med. 3, a006569.

Block, G.A., Ix, J.H., Ketteler, M., Martin, K.J., Thadhani, R.I., Tonelli, M., Wolf, M., Juppner, H., Hruska, K., Wheeler, D.C., 2013. Phosphate homeostasis in CKD: report of a scientific symposium sponsored by the National Kidney Foundation. Am. J. Kidney Dis. 62, 457–473.

Briet, M., Pierre, B., Laurent, S., London, G.M., 2012. Arterial stiffness and pulse pressure in CKD and ESRD. Kidney Int. 82, 388–400.

Broulik, P.D., Broulikova, A., Adamek, S., Libansky, P., Tvrdon, J., Broulikova, K., Kubinyi, J., 2011. Improvement of hypertension after parathyroidectomy of patients suffering from primary hyperparathyroidism. Int. J. Endocrinol. 2011, 309068.

Brown, J., de Boer, I.H., Robinson-Cohen, C., Siscovick, D.S., Kestenbaum, B., Allison, M., Vaidya, A., 2015. Aldosterone, parathyroid hormone, and the use of renin-angiotensin-aldosterone system inhibitors: the multi-ethnic study of atherosclerosis. J. Clin. Endocrinol. Metab. 100, 490–499.

Brown, J.M., Williams, J.S., Luther, J.M., Garg, R., Garza, A.E., Pojoga, L.H., Ruan, D.T., Williams, G.H., Adler, G.K., Vaidya, A., 2013. Human interventions to characterize novel relationships between the renin-angiotensin-aldosterone system and parathyroid hormone. Hypertension.

Brunner, S., Zaruba, M.M., Huber, B., David, R., Vallaster, M., Assmann, G., Mueller-Hoecker, J., Franz, W.M., 2008. Parathyroid hormone effectively induces mobilization of progenitor cells without depletion of bone marrow. Exp. Hematol. 36, 1157–1166.

Caiati, C., Montaldo, C., Zedda, N., Bina, A., Iliceto, S., 1999. New noninvasive method for coronary flow reserve assessment: contrast-enhanced transthoracic second harmonic Echo Doppler. Circulation 99, 771–778.

Calvi, L.M., Adams, G.B., Weibrecht, K.W., Weber, J.M., Olson, D.P., Knight, M.C., Martin, R.P., Schipani, E., Divieti, P., Bringhurst, F.R., et al., 2003. Osteoblastic cells regulate the haematopoietic stem cell niche. Nature 425, 841–846.

Carrelli, A.L., Walker, M.D., Di Tullio, M.R., Homma, S., Zhang, C., McMahon, D.J., Silverberg, S.J., 2013. Endothelial function in mild primary hyperparathyroidism. Clin. Endocrinol. 78, 204–209.

Chang, Y.T., Wu, J.L., Hsu, C.C., Wang, J.D., Sung, J.M., 2013. Diabetes and end-stage renal disease synergistically contribute to increased incidence of cardiovascular events: a nation-wide follow-up study during 1998–2009. Diabetes Care.

Cheng, S.L., Behrmann, A.S., Mead, M., Ramachandran, B., Kapoor, K., Perera, R., Kronenberg, H.M., Towler, D.A., 2016. Abstract 219: vascular PTH1R signaling limits aortic collagen deposition in diabetic LDLR-/- mice. Arterioscler. Thromb. Vasc. Biol. 36. Abstract 219.

Cheng, S.L., Shao, J.S., Halstead, L.R., Distelhorst, K., Sierra, O., Towler, D.A., 2010. Activation of vascular smooth muscle parathyroid hormone receptor inhibits Wnt/beta-catenin signaling and aortic fibrosis in diabetic arteriosclerosis. Circ. Res. 107, 271–282.

Clemens, T.L., Cormier, S., Eichinger, A., Endlich, K., Fiaschi-Taesch, N., Fischer, E., Friedman, P.A., Karaplis, A.C., Massfelder, T., Rossert, J., et al., 2001. Parathyroid hormone-related protein and its receptors: nuclear functions and roles in the renal and cardiovascular systems, the placental trophoblasts and the pancreatic islets. Br. J. Pharmacol. 134, 1113–1136.

Colak, S., Aydogan, B.I., Gokcay Canpolat, A., Tulunay Kaya, C., Sahin, M., Corapcioglu, D., Uysal, A.R., Emral, R., 2017. Is primary hyperparathyroidism a cause of endothelial dysfunction? Clin. Endocrinol. 87, 459–465.

Collins, T.C., Ewing, S.K., Diem, S.J., Taylor, B.C., Orwoll, E.S., Cummings, S.R., Strotmeyer, E.S., Ensrud, K.E., 2009. Peripheral arterial disease is associated with higher rates of hip bone loss and increased fracture risk in older men. Circulation 119, 2305–2312.

Collip, J.B., Clark, E.P., 1925. Further studies on the physiological action of a parathyroid hormone. J. Biol. Chem. 64, 485–507.

Cortigiani, L., Rigo, F., Gherardi, S., Bovenzi, F., Molinaro, S., Picano, E., Sicari, R., 2012. Coronary flow reserve during dipyridamole stress echocardiography predicts mortality. JACC Cardiovasc. Imag. 5, 1079–1085.

Crass 3rd, M.F., Jayaseelan, C.L., Darter, T.C., 1987. Effects of parathyroid hormone on blood flow in different regional circulations. Am. J. Physiol. 253, R634–R639.

Cunningham, J., Danese, M., Olson, K., Klassen, P., Chertow, G.M., 2005. Effects of the calcimimetic cinacalcet HCl on cardiovascular disease, fracture, and health-related quality of life in secondary hyperparathyroidism. Kidney Int. 68, 1793–1800.

Debella, Y.T., Giduma, H.D., Light, R.P., Agarwal, R., 2011. Chronic kidney disease as a coronary disease equivalent–a comparison with diabetes over a decade. Clin. J. Am. Soc. Nephrol. 6, 1385–1392.

Demer, L., Tintut, Y., 2010. The bone-vascular axis in chronic kidney disease. Curr. Opin. Nephrol. Hypertens. 19, 349–353.

Dicker, F., Quitterer, U., Winstel, R., Honold, K., Lohse, M.J., 1999. Phosphorylation-independent inhibition of parathyroid hormone receptor signaling by G protein-coupled receptor kinases. Proc. Natl. Acad. Sci. U. S. A. 96, 5476–5481.

Dimitrov, E.L., Kim, Y.Y., Usdin, T.B., 2011. Regulation of hypothalamic signaling by tuberoinfundibular peptide of 39 residues is critical for the response to cold: a novel peptidergic mechanism of thermoregulation. J. Neurosci. 31, 18166–18179.

Dobolyi, A., Dimitrov, E., Palkovits, M., Usdin, T.B., 2012. The neuroendocrine functions of the parathyroid hormone 2 receptor. Front. Endocrinol. 3, 121.

Durup, D., Jorgensen, H.L., Christensen, J., Schwarz, P., Heegaard, A.M., Lind, B., 2012. A reverse J-shaped association of all-cause mortality with serum 25-hydroxyvitamin D in general practice: the CopD study. J. Clin. Endocrinol. Metab. 97, 2644–2652.

Eddington, H., Hoefield, R., Sinha, S., Chrysochou, C., Lane, B., Foley, R.N., Hegarty, J., New, J., O'Donoghue, D.J., Middleton, R.J., et al., 2010. Serum phosphate and mortality in patients with chronic kidney disease. Clin. J. Am. Soc. Nephrol. 5, 2251–2257.

Eghbali-Fatourechi, G.Z., Lamsam, J., Fraser, D., Nagel, D., Riggs, B.L., Khosla, S., 2005. Circulating osteoblast-lineage cells in humans. N. Engl. J. Med. 352, 1959–1966.

Eghbali-Fatourechi, G.Z., Modder, U.I., Charatcharoenwitthaya, N., Sanyal, A., Undale, A.H., Clowes, J.A., Tarara, J.E., Khosla, S., 2007. Characterization of circulating osteoblast lineage cells in humans. Bone 40, 1370–1377.

Eichinger, A., Fiaschi-Taesch, N., Massfelder, T., Fritsch, S., Barthelmebs, M., Helwig, J.J., 2002. Transcript expression of the tuberoinfundibular peptide (TIP)39/PTH2 receptor system and non-PTH1 receptor-mediated tonic effects of TIP39 and other PTH2 receptor ligands in renal vessels. Endocrinology 143, 3036–3043.

Faarvang, A.S., Rordam Preil, S.A., Nielsen, P.S., Beck, H.C., Kristensen, L.P., Rasmussen, L.M., 2016. Smoking is associated with lower amounts of arterial type I collagen and decorin. Atherosclerosis 247, 201–206.

Fadini, G.P., Rattazzi, M., Matsumoto, T., Asahara, T., Khosla, S., 2012. Emerging role of circulating calcifying cells in the bone-vascular axis. Circulation 125, 2772–2781.

Farb, A., Kolodgie, F.D., Hwang, J.Y., Burke, A.P., Tefera, K., Weber, D.K., Wight, T.N., Virmani, R., 2004. Extracellular matrix changes in stented human coronary arteries. Circulation 110, 940–947.

Feola, M., Crass 3rd, M.F., 1986. Parathyroid hormone reduces acute ischemic injury of the myocardium. Surg. Gynecol. Obstet. 163, 523–530.

Fiaschi-Taesch, N., Sicari, B., Ubriani, K., Cozar-Castellano, I., Takane, K.K., Stewart, A.F., 2009. Mutant parathyroid hormone-related protein, devoid of the nuclear localization signal, markedly inhibits arterial smooth muscle cell cycle and neointima formation by coordinate up-regulation of p15Ink4b and p27kip1. Endocrinology 150, 1429–1439.

Fiaschi-Taesch, N., Sicari, B.M., Ubriani, K., Bigatel, T., Takane, K.K., Cozar-Castellano, I., Bisello, A., Law, B., Stewart, A.F., 2006. Cellular mechanism through which parathyroid hormone-related protein induces proliferation in arterial smooth muscle cells: definition of an arterial smooth muscle PTHrP/p27kip1 pathway. Circ. Res. 99, 933–942.

Fiaschi-Taesch, N., Takane, K.K., Masters, S., Lopez-Talavera, J.C., Stewart, A.F., 2004. Parathyroid-hormone-related protein as a regulator of pRb and the cell cycle in arterial smooth muscle. Circulation 110, 177–185.

Fiaschi-Taesch, N.M., Stewart, A.F., 2003. Minireview: parathyroid hormone-related protein as an intracrine factor–trafficking mechanisms and functional consequences. Endocrinology 144, 407–411.

Fitzpatrick, L.A., Bilezikian, J.P., Silverberg, S.J., 2008. Parathyroid hormone and the cardiovascular system. Curr. Osteoporos. Rep. 6, 77–83.

Flammer, A.J., Gossl, M., Li, J., Matsuo, Y., Reriani, M., Loeffler, D., Simari, R.D., Lerman, L.O., Khosla, S., Lerman, A., 2012a. Patients with an HbA1c in the prediabetic and diabetic range have higher numbers of circulating cells with osteogenic and endothelial progenitor cell markers. J. Clin. Endocrinol. Metab. 97, 4761–4768.

Flammer, A.J., Gossl, M., Widmer, R.J., Reriani, M., Lennon, R., Loeffler, D., Shonyo, S., Simari, R.D., Lerman, L.O., Khosla, S., et al., 2012b. Osteocalcin positive CD133+/CD34-/KDR+ progenitor cells as an independent marker for unstable atherosclerosis. Eur. Heart J. 33, 2963–2969.

Friedman, P.A., Goodman, W.G., 2006. PTH(1-84)/PTH(7-84): a balance of power. Am. J. Physiol. Renal. Physiol. 290, F975–F984.

Fuery, M.A., Liang, L., Kaplan, F.S., Mohler 3rd, E.R., 2018. Vascular ossification: pathology, mechanisms, and clinical implications. Bone 109, 28–34.

Gaudreault, N., Ducharme, V., Lamontagne, M., Guauque-Olarte, S., Mathieu, P., Pibarot, P., Bossé, Y., 2011. Replication of genetic association studies in aortic stenosis in adults. Am. J. Cardiol. 108 (9), 1305–1310.

Gaziano, T., Reddy, K.S., Paccaud, F., Horton, S., Chaturvedi, V., 2006. Cardiovascular disease. In: Jamison, D.T., Breman, J.G., Measham, A.R., Alleyne, G., Claeson, M., Evans, D.B., Jha, P., Mills, A., Musgrove, P. (Eds.), Disease Control Priorities in Developing Countries. Washington (DC).

Ginsberg, C., Craven, T.E., Chonchol, M.B., Cheung, A.K., Sarnak, M.J., Ambrosius, W.T., Killeen, A.A., Raphael, K.L., Bhatt, U.Y., Chen, J., et al., 2018. PTH, FGF23, and intensive blood pressure lowering in chronic kidney disease participants in SPRINT. Clin. J. Am. Soc. Nephrol. 13, 1816–1824.

Gohin, S., Carriero, A., Chenu, C., Pitsillides, A.A., Arnett, T.R., Marenzana, M., 2016. The anabolic action of intermittent parathyroid hormone on cortical bone depends partly on its ability to induce nitric oxide-mediated vasorelaxation in BALB/c mice. Cell Biochem. Funct. 34, 52–62.

Goodman, W.G., 2004. The consequences of uncontrolled secondary hyperparathyroidism and its treatment in chronic kidney disease. Semin. Dial. 17, 209–216.

Gossl, M., Modder, U.I., Gulati, R., Rihal, C.S., Prasad, A., Loeffler, D., Lerman, L.O., Khosla, S., Lerman, A., 2010. Coronary endothelial dysfunction in humans is associated with coronary retention of osteogenic endothelial progenitor cells. Eur. Heart J. 31, 2909–2914.

Gray, C., Bratt, D., Lees, J., daCosta, M., Plant, K., Watson, O.J., Solaymani-Kohal, S., Tazzyman, S., Serbanovic-Canic, J., Crossman, D.C., et al., 2013. Loss of function of parathyroid hormone receptor 1 induces Notch-dependent aortic defects during zebrafish vascular development. Arterioscler. Thromb. Vasc. Biol. 33, 1257–1263.

Guers, J.J., Prisby, R.D., Edwards, D.G., Lennon-Edwards, S., 2017. Intermittent parathyroid hormone administration attenuates endothelial dysfunction in old rats. J. Appl. Physiol. 122, 76–81.

Hashimoto, K., Nakagawa, Y., Shibuya, T., Satoh, H., Ushijima, T., Imai, S., 1981. Effects of parathyroid hormone and related polypeptides on the heart and coronary circulation of dogs. J. Cardiovasc. Pharmacol. 3, 668–676.

Hirai, T., Chagin, A.S., Kobayashi, T., Mackem, S., Kronenberg, H.M., 2011. Parathyroid hormone/parathyroid hormone-related protein receptor signaling is required for maintenance of the growth plate in postnatal life. Proc. Natl. Acad. Sci. U. S. A. 108 (1), 191–196.

Hoare, S.R., Usdin, T.B., 2001. Molecular mechanisms of ligand recognition by parathyroid hormone 1 (PTH1) and PTH2 receptors. Curr. Pharmaceut. Des. 7, 689–713.

Hocher, B., Armbruster, F.P., Stoeva, S., Reichetzeder, C., Gron, H.J., Lieker, I., Khadzhynov, D., Slowinski, T., Roth, H.J., 2012. Measuring parathyroid hormone (PTH) in patients with oxidative stress–do we need a fourth generation parathyroid hormone assay? PLoS One 7, e40242.

Hoogendam, J., Farih-Sips, H., Wynaendts, L.C., Lowik, C.W., Wit, J.M., Karperien, M., 2007. Novel mutations in the parathyroid hormone (PTH)/PTH-related peptide receptor type 1 causing Blomstrand osteochondrodysplasia types I and II. J. Clin. Endocrinol. Metab. 92, 1088–1095.

Hulter, H.N., Melby, J.C., Peterson, J.C., Cooke, C.R., 1986. Chronic continuous PTH infusion results in hypertension in normal subjects. J. Clin. Hypertens. 2, 360–370.

Investigators, E.T., Chertow, G.M., Block, G.A., Correa-Rotter, R., Drueke, T.B., Floege, J., Goodman, W.G., Herzog, C.A., Kubo, Y., London, G.M., et al., 2012. Effect of cinacalcet on cardiovascular disease in patients undergoing dialysis. N. Engl. J. Med. 367, 2482–2494.

Ishay, A., Herer, P., Luboshitzky, R., 2011. Effects of successful parathyroidectomy on metabolic cardiovascular risk factors in patients with severe primary hyperparathyroidism. Endocr. Pract. 17, 584–590.

Ishikawa, M., Ouchi, Y., Han, S.Z., Akishita, M., Kozaki, K., Toba, K., Namiki, A., Yamaguchi, T., Orimo, H., 1994. Parathyroid hormone-related protein reduces cytosolic free Ca2+ level and tension in rat aortic smooth muscle. Eur. J. Pharmacol. 269, 311–317.

Iwata, S., Hyodo, E., Yanagi, S., Hayashi, Y., Nishiyama, H., Kamimori, K., Ota, T., Matsumura, Y., Homma, S., Yoshiyama, M., 2013. Parathyroid hormone and systolic blood pressure accelerate the progression of aortic valve stenosis in chronic hemodialysis patients. Int. J. Cardiol. 163 (3), 256–259.

Iwata, S., Walker, M.D., Di Tullio, M.R., Hyodo, E., Jin, Z., Liu, R., Sacco, R.L., Homma, S., Silverberg, S.J., 2012. Aortic valve calcification in mild primary hyperparathyroidism. J. Clin. Endocrinol. Metab. 97, 132–137.

Jono, S., Nishizawa, Y., Shioi, A., Morii, H., 1997. Parathyroid hormone-related peptide as a local regulator of vascular calcification. Its inhibitory action on in vitro calcification by bovine vascular smooth muscle cells. Arterioscler. Thromb. Vasc. Biol. 17, 1135–1142.

Kidney Disease: Improving Global Outcomes, C.K.D.M.B.D.W.G, 2009. KDIGO clinical practice guideline for the diagnosis, evaluation, prevention, and treatment of Chronic Kidney Disease-Mineral and Bone Disorder (CKD-MBD). Kidney Int. Suppl. S1–S130.

Kim, S.C., Kim, H.W., Oh, S.W., Yang, H.N., Kim, M.G., Jo, S.K., Cho, W.Y., Kim, H.K., 2011. Low iPTH can predict vascular and coronary calcifications in patients undergoing peritoneal dialysis. Nephron Clin. Pract. 117, c113–119.

Kosch, M., Hausberg, M., Vormbrock, K., Kisters, K., Gabriels, G., Rahn, K.H., Barenbrock, M., 2000. Impaired flow-mediated vasodilation of the brachial artery in patients with primary hyperparathyroidism improves after parathyroidectomy. Cardiovasc. Res. 47, 813–818.

Kovacs, L., Goth, M.I., Szabolcs, I., Dohan, O., Ferencz, A., Szilagyi, G., 1998. The effect of surgical treatment on secondary hyperaldosteronism and relative hyperinsulinemia in primary hyperparathyroidism. Eur. J. Endocrinol. 138, 543–547.

Kritchevsky, S.B., Tooze, J.A., Neiberg, R.H., Schwartz, G.G., Hausman, D.B., Johnson, M.A., Bauer, D.C., Cauley, J.A., Shea, M.K., Cawthon, P.M., et al., 2012. 25-Hydroxyvitamin D, parathyroid hormone, and mortality in black and white older adults: the health ABC study. J. Clin. Endocrinol. Metab. 97, 4156–4165.

Kronenberg, H.M., 2006. PTHrP and skeletal development. Ann. N Y Acad. Sci. 1068, 1–13.

Langub, M.C., Monier-Faugere, M.C., Wang, G., Williams, J.P., Koszewski, N.J., Malluche, H.H., 2003. Administration of PTH-(7-84) antagonizes the effects of PTH-(1-84) on bone in rats with moderate renal failure. Endocrinology 144, 1135–1138.

Lee, K., Deeds, J.D., Segre, G.V., 1995. Expression of parathyroid hormone-related peptide and its receptor messenger ribonucleic acids during fetal development of rats. Endocrinology 136, 453–463.

Lee, S., Bice, A., Hood, B., Ruiz, J., Kim, J., Prisby, R.D., 2018. Intermittent PTH 1-34 administration improves the marrow microenvironment and endothelium-dependent vasodilation in bone arteries of aged rats. J. Appl. Physiol. 124, 1426–1437.

Li, X., Garcia, J., Lu, J., Iriana, S., Kalajzic, I., Rowe, D., Demer, L.L., Tintut, Y., 2014. Roles of parathyroid hormone (PTH) receptor and reactive oxygen species in hyperlipidemia-induced PTH resistance in preosteoblasts. J. Cell. Biochem. 115, 179–188.

Lincoln, J., Lange, A.W., Yutzey, K.E., 2006. Hearts and bones: shared regulatory mechanisms in heart valve, cartilage, tendon, and bone development. Dev. Biol. 294 (2), 292–302.

Linhartová, K., Veselka, J., Sterbáková, G., Racek, J., Topolcan, O., Cerbák, R., 2008. Parathyroid hormone and vitamin D levels are independently associated with calcific aortic stenosis. Circ. J. 72 (2), 245–250.

Liu, L., Sanchez-Bonilla, M., Crouthamel, M., Giachelli, C., Keel, S., 2013. Mice lacking the sodium-dependent phosphate import protein, PiT1 (SLC20A1), have a severe defect in terminal erythroid differentiation and early B cell development. Exp. Hematol. 41, 432–443 e437.

Lobaugh, B., Neelon, F.A., Oyama, H., Buckley, N., Smith, S., Christy, M., Leight Jr., G.S., 1989. Circadian rhythms for calcium, inorganic phosphorus, and parathyroid hormone in primary hyperparathyroidism: functional and practical considerations. Surgery 106, 1009–1016 discussion 1016-1007.

London, G., Coyne, D., Hruska, K., Malluche, H.H., Martin, K.J., 2010. The new kidney disease: improving global outcomes (KDIGO) guidelines – expert clinical focus on bone and vascular calcification. Clin. Nephrol. 74, 423–432.

London, G.M., 2003. Cardiovascular calcifications in uremic patients: clinical impact on cardiovascular function. J. Am. Soc. Nephrol. 14, S305–S309.

London, G.M., Guerin, A.P., Marchais, S.J., Metivier, F., Pannier, B., Adda, H., 2003. Arterial media calcification in end-stage renal disease: impact on all-cause and cardiovascular mortality. Nephrol. Dial. Transplant. 18, 1731–1740.

London, G.M., Marchais, S.J., Guerin, A.P., de Vernejoul, M.C., 2015. Ankle-brachial index and bone turnover in patients on dialysis. J. Am. Soc. Nephrol. 26, 476–483.

London, G.M., Marty, C., Marchais, S.J., Guerin, A.P., Metivier, F., de Vernejoul, M.C., 2004. Arterial calcifications and bone histomorphometry in end-stage renal disease. J. Am. Soc. Nephrol. 15, 1943–1951.

Lutteke, D., Ross, G., Abdallah, Y., Schafer, C., Piper, H.M., Schluter, K.D., 2005. Parathyroid hormone-related peptide improves contractile responsiveness of adult rat cardiomyocytes with depressed cell function irrespectively of oxidative inhibition. Basic Res. Cardiol. 100, 320–327.

MacLean, H.E., Guo, J., Knight, M.C., Zhang, P., Cobrinik, D., Kronenberg, H.M., 2004. The cyclin-dependent kinase inhibitor p57(Kip2) mediates proliferative actions of PTHrP in chondrocytes. J. Clin. Investig. 113, 1334–1343.

Maeda, H., Tokumoto, M., Yotsueda, H., Taniguchi, M., Tsuruya, K., Hirakata, H., Iida, M., 2007. Two cases of calciphylaxis treated by parathyroidectomy: importance of increased bone formation. Clin. Nephrol. 67, 397–402.

Maeda, S., Sutliff, R.L., Qian, J., Lorenz, J.N., Wang, J., Tang, H., Nakayama, T., Weber, C., Witte, D., Strauch, A.R., et al., 1999. Targeted overexpression of parathyroid hormone-related protein (PTHrP) to vascular smooth muscle in transgenic mice lowers blood pressure and alters vascular contractility. Endocrinology 140, 1815–1825.

Marini, C., Giusti, M., Armonino, R., Ghigliotti, G., Bezante, G., Vera, L., Morbelli, S., Pomposelli, E., Massollo, M., Gandolfo, P., et al., 2010. Reduced coronary flow reserve in patients with primary hyperparathyroidism: a study by G-SPECT myocardial perfusion imaging. Eur. J. Nucl. Med. Mol. Imaging 37, 2256–2263.

Martin, A., David, V., Quarles, L.D., 2012. Regulation and function of the FGF23/klotho endocrine pathways. Physiol. Rev. 92, 131–155.

Martin, K.J., Gonzalez, E.A., 2011. Prevention and control of phosphate retention/hyperphosphatemia in CKD-MBD: what is normal, when to start, and how to treat? Clin. J. Am. Soc. Nephrol. 6, 440–446.

Massfelder, T., Dann, P., Wu, T.L., Vasavada, R., Helwig, J.J., Stewart, A.F., 1997. Opposing mitogenic and anti-mitogenic actions of parathyroid hormone-related protein in vascular smooth muscle cells: a critical role for nuclear targeting. Proc. Natl. Acad. Sci. U. S. A. 94, 13630–13635.

Massfelder, T., Stewart, A.F., Endlich, K., Soifer, N., Judes, C., Helwig, J.J., 1996. Parathyroid hormone-related protein detection and interaction with NO and cyclic AMP in the renovascular system. Kidney Int. 50, 1591–1603.

Matsuoka, S., Tominaga, Y., Uno, N., Goto, N., Sato, T., Katayama, A., Haba, T., Uchida, K., Kobayashi, K., Nakao, A., 2005. Calciphylaxis: a rare complication of patients who required parathyroidectomy for advanced renal hyperparathyroidism. World J. Surg. 29, 632–635.

Miao, D., Su, H., He, B., Gao, J., Xia, Q., Zhu, M., Gu, Z., Goltzman, D., Karaplis, A.C., 2008. Severe growth retardation and early lethality in mice lacking the nuclear localization sequence and C-terminus of PTH-related protein. Proc. Natl. Acad. Sci. U. S. A. 105, 20309–20314.

Moe, S.M., Chertow, G.M., Parfrey, P.S., Kubo, Y., Block, G.A., Correa-Rotter, R., Drueke, T.B., Herzog, C.A., London, G.M., Mahaffey, K.W., et al., 2015. Cinacalcet, fibroblast growth factor-23, and cardiovascular disease in hemodialysis: the evaluation of cinacalcet HCl therapy to lower cardiovascular events (EVOLVE) trial. Circulation 132, 27–39.

Moe, S.M., Drueke, T., Lameire, N., Eknoyan, G., 2007. Chronic kidney disease-mineral-bone disorder: a new paradigm. Adv. Chron. Kidney Dis. 14, 3–12.

Mohler, E.R., 3rd., Gannon, F., Reynolds, C., Zimmerman, R., Keane, M.G., Kaplan, F.S., 2001. Bone formation and inflammation in cardiac valves. Circulation 103 (11), 1522–1528.

Monego, G., Arena, V., Pasquini, S., Stigliano, E., Fiaccavento, R., Leone, O., Arpesella, G., Potena, L., Ranelletti, F.O., Di Nardo, P., et al., 2009. Ischemic injury activates PTHrP and PTH1R expression in human ventricular cardiomyocytes. Basic Res. Cardiol. 104, 427–434.

Monterisi, S., Lobo, M.J., Livie, C., Castle, J.C., Weinberger, M., Baillie, G., Surdo, N.C., Musheshe, N., Stangherlin, A., Gottlieb, E., et al., 2017. PDE2A2 regulates mitochondria morphology and apoptotic cell death via local modulation of cAMP/PKA signalling. Elife 6.

Montisci, R., Chen, L., Ruscazio, M., Colonna, P., Cadeddu, C., Caiati, C., Montisci, M., Meloni, L., Iliceto, S., 2006. Non-invasive coronary flow reserve is correlated with microvascular integrity and myocardial viability after primary angioplasty in acute myocardial infarction. Heart 92, 1113–1118.

Musso, M.J., Plante, M., Judes, C., Barthelmebs, M., Helwig, J.J., 1989. Renal vasodilatation and microvessel adenylate cyclase stimulation by synthetic parathyroid hormone-like protein fragments. Eur. J. Pharmacol. 174, 139–151.

Nagao, S., Seto, S., Kitamura, S., Akahoshi, M., Kiriyama, T., Yano, K., 1998. Central pressor effect of parathyroid hormone-related protein in conscious rats. Brain Res. 785, 75–79.

Napoli, C., William-Ignarro, S., Byrns, R., Balestrieri, M.L., Crimi, E., Farzati, B., Mancini, F.P., de Nigris, F., Matarazzo, A., D'Amora, M., et al., 2008. Therapeutic targeting of the stem cell niche in experimental hindlimb ischemia. Nat. Clin. Pract. Cardiovasc. Med. 5, 571–579.

Nguyen, M., He, B., Karaplis, A., 2001. Nuclear forms of parathyroid hormone-related peptide are translated from non-AUG start sites downstream from the initiator methionine. Endocrinology 142, 694–703.

Nickols, G.A., Metz, M.A., Cline Jr., W.H., 1986. Vasodilation of the rat mesenteric vasculature by parathyroid hormone. J. Pharmacol. Exp. Ther. 236, 419–423.

Nigwekar, S.U., Kroshinsky, D., Nazarian, R.M., Goverman, J., Malhotra, R., Jackson, V.A., Kamdar, M.M., Steele, D.J., Thadhani, R.I., 2015. Calciphylaxis: risk factors, diagnosis, and treatment. Am. J. Kidney Dis. 66, 133–146.

Noda, M., Katoh, T., Takuwa, N., Kumada, M., Kurokawa, K., Takuwa, Y., 1994. Synergistic stimulation of parathyroid hormone-related peptide gene expression by mechanical stretch and angiotensin II in rat aortic smooth muscle cells. J. Biol. Chem. 269, 17911–17917.

Noonan, W.T., Qian, J., Stuart, W.D., Clemens, T.L., Lorenz, J.N., 2003. Altered renal hemodynamics in mice overexpressing the parathyroid hormone (PTH)/PTH-related peptide type 1 receptor in smooth muscle. Endocrinology 144, 4931–4938.

Nyby, M.D., Hino, T., Berger, M.E., Ormsby, B.L., Golub, M.S., Brickman, A.S., 1995. Desensitization of vascular tissue to parathyroid hormone and parathyroid hormone-related protein. Endocrinology 136, 2497–2504.

O'Brien, C.A., Plotkin, L.I., Galli, C., Goellner, J.J., Gortazar, A.R., Allen, M.R., Robling, A.G., Bouxsein, M., Schipani, E., Turner, C.H., et al., 2008. Control of bone mass and remodeling by PTH receptor signaling in osteocytes. PLoS One 3, e2942.

Ogino, K., Burkhoff, D., Bilezikian, J.P., 1995. The hemodynamic basis for the cardiac effects of parathyroid hormone (PTH) and PTH-related protein. Endocrinology 136, 3024–3030.

Okoumassoun, L., Averill-Bates, D., Denizeau, F., Henderson, J.E., 2007. Parathyroid hormone related protein (PTHrP) inhibits TNFalpha-induced apoptosis by blocking the extrinsic and intrinsic pathways. J. Cell. Physiol. 210, 507–516.

Osto, E., Fallo, F., Pelizzo, M.R., Maddalozzo, A., Sorgato, N., Corbetti, F., Montisci, R., Famoso, G., Bellu, R., Luscher, T.F., et al., 2012. Coronary microvascular dysfunction induced by primary hyperparathyroidism is restored after parathyroidectomy. Circulation 126, 1031–1039.

Panda, D.K., Goltzman, D., Karaplis, A.C., 2012. Defective postnatal endochondral bone development by chondrocyte-specific targeted expression of parathyroid hormone type 2 receptor. Am. J. Physiol. Endocrinol. Metab. 303, E1489–E1501.

Pang, P.K., Tenner Jr., T.E., Yee, J.A., Yang, M., Janssen, H.F., 1980. Hypotensive action of parathyroid hormone preparations on rats and dogs. Proc. Natl. Acad. Sci. U. S. A. 77, 675–678.

Papasani, M.R., Gensure, R.C., Yan, Y.L., Gunes, Y., Postlethwait, J.H., Ponugoti, B., John, M.R., Juppner, H., Rubin, D.A., 2004. Identification and characterization of the zebrafish and fugu genes encoding tuberoinfundibular peptide 39. Endocrinology 145, 5294–5304.

Parfrey, P.S., Drueke, T.B., Block, G.A., Correa-Rotter, R., Floege, J., Herzog, C.A., London, G.M., Mahaffey, K.W., Moe, S.M., Wheeler, D.C., et al., 2015. The effects of cinacalcet in older and younger patients on hemodialysis: the evaluation of cinacalcet HCl therapy to lower cardiovascular events (EVOLVE) trial. Clin. J. Am. Soc. Nephrol. 10, 791–799.

Pepine, C.J., Anderson, R.D., Sharaf, B.L., Reis, S.E., Smith, K.M., Handberg, E.M., Johnson, B.D., Sopko, G., Bairey Merz, C.N., 2010. Coronary microvascular reactivity to adenosine predicts adverse outcome in women evaluated for suspected ischemia results from the National Heart, Lung and Blood Institute WISE (Women's Ischemia Syndrome Evaluation) study. J. Am. Coll. Cardiol. 55, 2825–2832.

Persson, A., Bollerslev, J., Rosen, T., Mollerup, C.L., Franco, C., Isaksen, G.A., Ueland, T., Jansson, S., Caidahl, K., Group, S.S., 2011. Effect of surgery on cardiac structure and function in mild primary hyperparathyroidism. Clin. Endocrinol. 74, 174–180.

Peterson, Y.K., Luttrell, L.M., 2017. The diverse roles of arrestin scaffolds in G protein-coupled receptor signaling. Pharmacol. Rev. 69, 256–297.

Pirola, C.J., Wang, H.M., Strgacich, M.I., Kamyar, A., Cercek, B., Forrester, J.S., Clemens, T.L., Fagin, J.A., 1994. Mechanical stimuli induce vascular parathyroid hormone-related protein gene expression in vivo and in vitro. Endocrinology 134, 2230–2236.

Potthoff, S.A., Janus, A., Hoch, H., Frahnert, M., Tossios, P., Reber, D., Giessing, M., Klein, H.M., Schwertfeger, E., Quack, I., et al., 2011. PTH-receptors regulate norepinephrine release in human heart and kidney. Regul. Pept. 171, 35–42.

Prisby, R., Guignandon, A., Vanden-Bossche, A., Mac-Way, F., Linossier, M.T., Thomas, M., Laroche, N., Malaval, L., Langer, M., Peter, Z.A., et al., 2011. Intermittent PTH(1-84) is osteoanabolic but not osteoangiogenic and relocates bone marrow blood vessels closer to bone-forming sites. J. Bone Miner. Res. 26, 2583–2596.

Prisby, R., Menezes, T., Campbell, J., 2013. Vasodilation to PTH (1-84) in bone arteries is dependent upon the vascular endothelium and is mediated partially via VEGF signaling. Bone 54, 68–75.

Prisby, R.D., 2017. Mechanical, hormonal and metabolic influences on blood vessels, blood flow and bone. J. Endocrinol. 235, R77–R100.

Qian, J., Colbert, M.C., Witte, D., Kuan, C.Y., Gruenstein, E., Osinska, H., Lanske, B., Kronenberg, H.M., Clemens, T.L., 2003. Midgestational lethality in mice lacking the parathyroid hormone (PTH)/PTH-related peptide receptor is associated with abrupt cardiomyocyte death. Endocrinology 144, 1053–1061.

Quarles, L.D., 2013. A systems biology preview of the relationships between mineral and metabolic complications in chronic kidney disease. Semin. Nephrol. 33, 130–142.

Raggi, P., James, G., Burke, S.K., Bommer, J., Chasan-Taber, S., Holzer, H., Braun, J., Chertow, G.M., 2005. Decrease in thoracic vertebral bone attenuation with calcium-based phosphate binders in hemodialysis. J. Bone Miner. Res. 20, 764–772.

Raison, D., Coquard, C., Hochane, M., Steger, J., Massfelder, T., Moulin, B., Karaplis, A.C., Metzger, D., Chambon, P., Helwig, J.J., et al., 2013. Knockdown of parathyroid hormone related protein in smooth muscle cells alters renal hemodynamics but not blood pressure. Am. J. Physiol. Renal. Physiol. 305, F333–F342.

Rajamannan, N.M., Evans, F.J., Aikawa, E., Grande-Allen, K.J., Demer, L.L., Heistad, D.D., Simmons, C.A., Masters, K.S., Mathieu, P., O'Brien, K.D., et al., 2011. Calcific aortic valve disease: not simply a degenerative process: a review and agenda for research from the National Heart and Lung and Blood Institute Aortic Stenosis Working Group. Executive summary: calcific aortic valve disease-2011 update. Circulation 124, 1783–1791.

Rambausek, M., Ritz, E., Rascher, W., Kreusser, W., Mann, J.F., Kreye, V.A., Mehls, O., 1982. Vascular effects of parathyroid hormone (PTH). Adv. Exp. Med. Biol. 151, 619–632.

Rhee, Y., Bivi, N., Farrow, E., Lezcano, V., Plotkin, L.I., White, K.E., Bellido, T., 2011. Parathyroid hormone receptor signaling in osteocytes increases the expression of fibroblast growth factor-23 in vitro and in vivo. Bone 49, 636–643.

Roca-Cusachs, A., DiPette, D.J., Nickols, G.A., 1991. Regional and systemic hemodynamic effects of parathyroid hormone-related protein: preservation of cardiac function and coronary and renal flow with reduced blood pressure. J. Pharmacol. Exp. Ther. 256, 110–118.

Rogers, N.M., Teubner, D.J., Coates, P.T., 2007. Calcific uremic arteriolopathy: advances in pathogenesis and treatment. Semin. Dial. 20, 150–157.

Romero, G., von Zastrow, M., Friedman, P.A., 2011. Role of PDZ proteins in regulating trafficking, signaling, and function of GPCRs: means, motif, and opportunity. Adv. Pharmacol. 62, 279–314.

Ross, G., Engel, P., Abdallah, Y., Kummer, W., Schluter, K.D., 2005. Tuberoinfundibular peptide of 39 residues: a new mediator of cardiac function via nitric oxide production in the rat heart. Endocrinology 146, 2221–2228.

Ross, G., Heinemann, M.P., Schluter, K.D., 2007. Vasodilatory effect of tuberoinfundibular peptide (TIP39): requirement of receptor desensitization and its beneficial effect in the post-ischemic heart. Peptides 28, 878–886.

Rubin, M.R., Maurer, M.S., McMahon, D.J., Bilezikian, J.P., Silverberg, S.J., 2005. Arterial stiffness in mild primary hyperparathyroidism. J. Clin. Endocrinol. Metab. 90, 3326–3330.

Safar, M.E., Boudier, H.S., 2005. Vascular development, pulse pressure, and the mechanisms of hypertension. Hypertension 46, 205–209.

Sage, A.P., Lu, J., Atti, E., Tetradis, S., Ascenzi, M.G., Adams, D.J., Demer, L.L., Tintut, Y., 2011. Hyperlipidemia induces resistance to PTH bone anabolism in mice via oxidized lipids. J. Bone Miner. Res. 26, 1197–1206.

Sato, E., Zhang, L.J., Dorschner, R.A., Adase, C.A., Choudhury, B.P., Gallo, R.L., 2017. Activation of parathyroid hormone 2 receptor induces decorin expression and promotes wound repair. J. Investig. Dermatol. 137, 1774–1783.

Saussine, C., Judes, C., Massfelder, T., Musso, M.J., Simeoni, U., Hannedouche, T., Helwig, J.J., 1993a. Stimulatory action of parathyroid hormone on renin secretion in vitro: a study using isolated rat kidney, isolated rabbit glomeruli and superfused dispersed rat juxtaglomerular cells. Clin. Sci. (Lond.) 84, 11–19.

Saussine, C., Massfelder, T., Parnin, F., Judes, C., Simeoni, U., Helwig, J.J., 1993b. Renin stimulating properties of parathyroid hormone-related peptide in the isolated perfused rat kidney. Kidney Int. 44, 764–773.

Schermer, D.T., Chan, S.D., Bruce, R., Nissenson, R.A., Wood, W.I., Strewler, G.J., 1991. Chicken parathyroid hormone-related protein and its expression during embryologic development. J. Bone Miner. Res. 6, 149–155.

Schipani, E., Kruse, K., Juppner, H., 1995. A constitutively active mutant PTH-PTHrP receptor in Jansen-type metaphyseal chondrodysplasia. Science 268, 98–100.

Schmitz, F., Ewering, S., Zerres, K., Klomfass, S., Hoffmann, R., 2009. Ortlepp JR Parathyroid hormone gene variant and calcific aortic stenosis. J. Heart Valve Dis. 18 (3), 262–267.

Schluter, K., Katzer, C., Frischkopf, K., Wenzel, S., Taimor, G., Piper, H.M., 2000. Expression, release, and biological activity of parathyroid hormone-related peptide from coronary endothelial cells. Circ. Res. 86, 946–951.

Schreckenberg, R., Wenzel, S., da Costa Rebelo, R.M., Rothig, A., Meyer, R., Schluter, K.D., 2009. Cell-specific effects of nitric oxide deficiency on parathyroid hormone-related peptide (PTHrP) responsiveness and PTH1 receptor expression in cardiovascular cells. Endocrinology 150, 3735–3741.

Sebastian, E.M., Suva, L.J., Friedman, P.A., 2008. Differential effects of intermittent PTH(1-34) and PTH(7-34) on bone microarchitecture and aortic calcification in experimental renal failure. Bone 43, 1022–1030.

Shao, J.S., Cheng, S.L., Charlton-Kachigian, N., Loewy, A.P., Towler, D.A., 2003. Teriparatide (human parathyroid hormone (1-34)) inhibits osteogenic vascular calcification in diabetic low density lipoprotein receptor-deficient mice. J. Biol. Chem. 278, 50195–50202.

Shao, J.S., Cheng, S.L., Pingsterhaus, J.M., Charlton-Kachigian, N., Loewy, A.P., Towler, D.A., 2005. Msx2 promotes cardiovascular calcification by activating paracrine Wnt signals. J. Clin. Investig. 115, 1210–1220.

Shao, J.S., Cheng, S.L., Sadhu, J., Towler, D.A., 2010. Inflammation and the osteogenic regulation of vascular calcification: a review and perspective. Hypertension 55, 579–592.

Shao, J.S., Sierra, O.L., Cohen, R., Mecham, R.P., Kovacs, A., Wang, J., Distelhorst, K., Behrmann, A., Halstead, L.R., Towler, D.A., 2011. Vascular calcification and aortic fibrosis: a bifunctional role for osteopontin in diabetic arteriosclerosis. Arterioscler. Thromb. Vasc. Biol.

Shroff, R., 2011. Dysregulated mineral metabolism in children with chronic kidney disease. Curr. Opin. Nephrol. Hypertens. 20, 233–240.

Shroff, R., Long, D.A., Shanahan, C., 2013. Mechanistic insights into vascular calcification in CKD. J. Am. Soc. Nephrol. 24, 179–189.

Shroff, R.C., McNair, R., Figg, N., Skepper, J.N., Schurgers, L., Gupta, A., Hiorns, M., Donald, A.E., Deanfield, J., Rees, L., et al., 2008. Dialysis accelerates medial vascular calcification in part by triggering smooth muscle cell apoptosis. Circulation 118, 1748–1757.

Shroff, R.C., McNair, R., Skepper, J.N., Figg, N., Schurgers, L.J., Deanfield, J., Rees, L., Shanahan, C.M., 2010. Chronic mineral dysregulation promotes vascular smooth muscle cell adaptation and extracellular matrix calcification. J. Am. Soc. Nephrol. 21, 103–112.

Silverberg, S.J., Lewiecki, E.M., Mosekilde, L., Peacock, M., Rubin, M.R., 2009. Presentation of asymptomatic primary hyperparathyroidism: proceedings of the third international workshop. J. Clin. Endocrinol. Metab. 94, 351–365.

Smith, J.C., Page, M.D., John, R., Wheeler, M.H., Cockcroft, J.R., Scanlon, M.F., Davies, J.S., 2000. Augmentation of central arterial pressure in mild primary hyperparathyroidism. J. Clin. Endocrinol. Metab. 85, 3515–3519.

Smith, J.S., Rajagopal, S., 2016. The beta-arrestins: multifunctional regulators of G protein-coupled receptors. J. Biol. Chem. 291, 8969–8977.

Smock, S.L., Vogt, G.A., Castleberry, T.A., Lu, B., Owen, T.A., 2001. Molecular cloning and functional characterization of the canine parathyroid hormone/parathyroid hormone related peptide receptor (PTH1). Mol. Biol. Rep. 28, 235–243.

Soldatos, G., Jandeleit-Dahm, K., Thomson, H., Formosa, M., D'Orsa, K., Calkin, A.C., Cooper, M.E., Ahimastos, A.A., Kingwell, B.A., 2011. Large artery biomechanics and diastolic dysfunctionin patients with Type 2 diabetes. Diabet. Med. 28, 54–60.

Song, G.J., Barrick, S., Leslie, K.L., Sicari, B., Fiaschi-Taesch, N.M., Bisello, A., 2010. EBP50 inhibits the anti-mitogenic action of the parathyroid hormone type 1 receptor in vascular smooth muscle cells. J. Mol. Cell. Cardiol. 49, 1012–1021.

Sprague, S.M., Strugnell, S.A., Bishop, C.W., 2017. Extended-release calcifediol for secondary hyperparathyroidism in stage 3-4 chronic kidney disease. Expert Rev. Endocrinol. Metab. 12, 289–301.

Stabley, J.N., Towler, D.A., 2017. Arterial calcification in diabetes mellitus: preclinical models and translational implications. Arterioscler. Thromb. Vasc. Biol. 37, 205–217.

Stefenelli, T., Abela, C., Frank, H., Koller-Strametz, J., Niederle, B., 1997. Time course of regression of left ventricular hypertrophy after successful parathyroidectomy. Surgery 121 (2), 157–161.

Steinberg, D., 2006. Thematic review series: the pathogenesis of atherosclerosis. An interpretive history of the cholesterol controversy, Part V: the discovery of the statins and the end of the controversy. J. Lipid Res. 47, 1339–1351.

Sugimura, Y., Murase, T., Ishizaki, S., Tachikawa, K., Arima, H., Miura, Y., Usdin, T.B., Oiso, Y., 2003. Centrally administered tuberoinfundibular peptide of 39 residues inhibits arginine vasopressin release in conscious rats. Endocrinology 144, 2791–2796.

Suzuki, Y., Lederis, K., Huang, M., LeBlanc, F.E., Rorstad, O.P., 1983. Relaxation of bovine, porcine and human brain arteries by parathyroid hormone. Life Sci. 33, 2497–2503.

Takahashi, K., Inoue, D., Ando, K., Matsumoto, T., Ikeda, K., Fujita, T., 1995. Parathyroid hormone-related peptide as a locally produced vasorelaxant: regulation of its mRNA by hypertension in rats. Biochem. Biophys. Res. Commun. 208, 447–455.

Tastan, I., Schreckenberg, R., Mufti, S., Abdallah, Y., Piper, H.M., Schluter, K.D., 2009. Parathyroid hormone improves contractile performance of adult rat ventricular cardiomyocytes at low concentrations in a non-acute way. Cardiovasc. Res. 82, 77–83.

Teng, M., Wolf, M., Lowrie, E., Ofsthun, N., Lazarus, J.M., Thadhani, R., 2003. Survival of patients undergoing hemodialysis with paricalcitol or calcitriol therapy. N. Engl. J. Med. 349, 446–456.

Teng, M., Wolf, M., Ofsthun, M.N., Lazarus, J.M., Hernan, M.A., Camargo Jr., C.A., Thadhani, R., 2005. Activated injectable vitamin D and hemodialysis survival: a historical cohort study. J. Am. Soc. Nephrol. 16, 1115–1125.

Thompson, B., Towler, D.A., 2012a. Arterial calcification and bone physiology: role of the bone-vascular axis. Nat. Rev. Endocrinol.

Thompson, B., Towler, D.A., 2012b. Arterial calcification and bone physiology: role of the bone-vascular axis. Nat. Rev. Endocrinol. 8, 529–543.

Tomaschitz, A., Ritz, E., Pieske, B., Fahrleitner-Pammer, A., Kienreich, K., Horina, J.H., Drechsler, C., Marz, W., Ofner, M., Pieber, T.R., et al., 2012. Aldosterone and parathyroid hormone: a precarious couple for cardiovascular disease. Cardiovasc. Res. 94, 10–19.

Tomaschitz, A., Ritz, E., Pieske, B., Rus-Machan, J., Kienreich, K., Verheyen, N., Gaksch, M., Grubler, M., Fahrleitner-Pammer, A., Mrak, P., et al., 2013. Aldosterone and parathyroid hormone interactions as mediators of metabolic and cardiovascular disease. Metabolism.

Tonelli, M., Curhan, G., Pfeffer, M., Sacks, F., Thadhani, R., Melamed, M.L., Wiebe, N., Muntner, P., 2009. Relation between alkaline phosphatase, serum phosphate, and all-cause or cardiovascular mortality. Circulation 120, 1784–1792.

Tonelli, M., Sacks, F., Pfeffer, M., Gao, Z., Curhan, G., Cholesterol, and Recurrent Events Trial, I., 2005. Relation between serum phosphate level and cardiovascular event rate in people with coronary disease. Circulation 112, 2627–2633.

Towler, D.A., 2011. Skeletal anabolism, PTH, and the bone-vascular axis. J. Bone Miner. Res. 26, 2579–2582.

Towler, D.A., 2013. Chronic kidney disease: the "perfect storm" of cardiometabolic risk illuminates genetic diathesis in cardiovascular disease. J. Am. Coll. Cardiol. 62, 799–801.

Towler, D.A., 2014a. Physiological actions of PTH and PTHrP IV: vascular, cardiovascular, and CNS biology. In: Bilezikian, J.P. (Ed.), The Parathyroids: Basic and Clinical Concepts. Academic Press in press.

Towler, D.A., 2014b. The role of PTHrP in vascular smooth muscle. Clin. Rev. Bone Miner. Metabol. 12, 190–196.

Towler, D.A., 2015. Arteriosclerosis, bone biology, and calciotropic hormone signaling: learning the ABCs of disease in the bone-vascular axis. J. Am. Soc. Nephrol. 26, 243–245.

Towler, D.A., 2017a. Commonalities between vasculature and bone: an osseocentric view of arteriosclerosis. Circulation 135, 320–322.

Towler, D.A., 2017b. Lipoprotein(a): a taxi for autotaxin takes a toll in calcific aortic valve disease. JACC Basic Transl Sci 2, 241–243.

Unger, P., Clavel, M.A., Lindman, B.R., Mathieu, P., Pibarot, P., 2016. Pathophysiology and management of multivalvular disease. Nat. Rev. Cardiol. 13, 429–440.

van den Akker, N.M., Caolo, V., Molin, D.G., 2012. Cellular decisions in cardiac outflow tract and coronary development: an act by VEGF and NOTCH. Differentiation 84, 62–78.

VanderWalde, L.H., Liu, I.L., O'Connell, T.X., Haigh, P.I., 2006. The effect of parathyroidectomy on bone fracture risk in patients with primary hyperparathyroidism. Arch. Surg. 141, 885–889 discussion 889-891.

Vazquez-Diaz, O., Castillo-Martinez, L., Orea-Tejeda, A., Orozco-Gutierrez, J.J., Asensio-Lafuente, E., Reza-Albarran, A., Silva-Tinoco, R., Rebollar-Gonzalez, V., 2009. Reversible changes of electrocardiographic abnormalities after parathyroidectomy in patients with primary hyperparathyroidism. Cardiol. J. 16, 241–245.

Verheyen, N., Pilz, S., Eller, K., Kienreich, K., Fahrleitner-Pammer, A., Pieske, B., Ritz, E., Tomaschitz, A., 2013. Cinacalcet hydrochloride for the treatment of hyperparathyroidism. Expert Opin. Pharmacother. 14, 793–806.

Walker, M.D., Fleischer, J., Rundek, T., McMahon, D.J., Homma, S., Sacco, R., Silverberg, S.J., 2009. Carotid vascular abnormalities in primary hyperparathyroidism. J. Clin. Endocrinol. Metab. 94, 3849–3856.

Walker, M.D., Fleischer, J.B., Di Tullio, M.R., Homma, S., Rundek, T., Stein, E.M., Zhang, C., Taggart, T., McMahon, D.J., Silverberg, S.J., 2010. Cardiac structure and diastolic function in mild primary hyperparathyroidism. J. Clin. Endocrinol. Metab. 95, 2172–2179.

Walker, M.D., Rundek, T., Homma, S., DiTullio, M., Iwata, S., Lee, J.A., Choi, J., Liu, R., Zhang, C., McMahon, D.J., et al., 2012. Effect of parathyroidectomy on subclinical cardiovascular disease in mild primary hyperparathyroidism. Eur. J. Endocrinol. 167, 277–285.

Walker, M.D., Silverberg, S.J., 2008. Cardiovascular aspects of primary hyperparathyroidism. J. Endocrinol. Investig. 31, 925–931.

Westerhof, N., Lankhaar, J.W., Westerhof, B.E., 2009. The arterial Windkessel. Med. Biol. Eng. Comput. 47, 131–141.

Wysolmerski, J.J., 2012. Parathyroid hormone-related protein: an update. J. Clin. Endocrinol. Metab. 97, 2947–2956.

Yamamoto, K., Yamamoto, H., Yoshida, K., Kisanuki, A., Hirano, Y., Ohte, N., Akasaka, T., Takeuchi, M., Nakatani, S., Ohtani, T., et al., 2010. Prognostic factors for progression of early- and late-stage calcific aortic valve disease in Japanese: The Japanese Aortic Stenosis Study (JASS) retrospective analysis. Hypertens. Res. 33, 269–274.

Yamamoto, S., Morimoto, I., Zeki, K., Ueta, Y., Yamashita, H., Kannan, H., Eto, S., 1998. Centrally administered parathyroid hormone (PTH)-related protein(1-34) but not PTH(1-34) stimulates arginine-vasopressin secretion and its messenger ribonucleic acid expression in supraoptic nucleus of the conscious rats. Endocrinology 139, 383–388.

Yu, N., Donnan, P.T., Flynn, R.W., Murphy, M.J., Smith, D., Rudman, A., Leese, G.P., 2010. Increased mortality and morbidity in mild primary hyperparathyroid patients. The Parathyroid Epidemiology and Audit Research Study (PEARS). Clin. Endocrinol. 73, 30–34.

Yu, N., Donnan, P.T., Leese, G.P., 2011. A record linkage study of outcomes in patients with mild primary hyperparathyroidism: the Parathyroid Epidemiology and Audit Research Study (PEARS). Clin. Endocrinol. 75, 169–176.

Yu, N., Leese, G.P., Donnan, P.T., 2013. What predicts adverse outcomes in untreated primary hyperparathyroidism? The Parathyroid Epidemiology and Audit Research Study (PEARS). Clin. Endocrinol. 79, 27–34.

Yutzey, K.E., Demer, L.L., Body, S.C., Huggins, G.S., Towler, D.A., Giachelli, C.M., Hofmann-Bowman, M.A., Mortlock, D.P., Rogers, M.B., Sadeghi, M.M., et al., 2014. Calcific aortic valve disease: a consensus summary from the alliance of investigators on calcific aortic valve disease. Arterioscler. Thromb. Vasc. Biol. 34, 2387–2393.

Zaruba, M.M., Huber, B.C., Brunner, S., Deindl, E., David, R., Fischer, R., Assmann, G., Herbach, N., Grundmann, S., Wanke, R., et al., 2008. Parathyroid hormone treatment after myocardial infarction promotes cardiac repair by enhanced neovascularization and cell survival. Cardiovasc. Res. 77, 722–731.

Zhang, Y., Chen, G., Gu, Z., Sun, H., Karaplis, A., Goltzman, D., Miao, D., 2018. DNA damage checkpoint pathway modulates the regulation of skeletal growth and osteoblastic bone formation by parathyroid hormone-related peptide. Int. J. Biol. Sci. 14, 508–517.

Zoppellaro, G., Faggin, E., Puato, M., Pauletto, P., Rattazzi, M., 2012. Fibroblast growth factor 23 and the bone-vascular axis: lessons learned from animal studies. Am. J. Kidney Dis. 59, 135–144.

Chapter 27

Parathyroid hormone and parathyroid hormone–related protein actions on bone and kidney

Alessandro Bisello and Peter A. Friedman

Department of Pharmacology and Chemical Biology, Laboratory for GPCR Biology, University of Pittsburgh School of Medicine, Pittsburgh, PA, United States

Chapter outline

Introduction **646**
Receptors and second-messenger systems for parathyroid hormone and parathyroid hormone–related protein **646**
Expression and actions of parathyroid hormone receptor in bone **647**
Effects of parathyroid hormone and parathyroid hormone–related protein on bone cells 648
Molecular mechanisms of action in osteoblasts 648
Adaptor proteins 651
Effects on gap junctions 651
Effects on bone matrix proteins and alkaline phosphatase 651
Effects on bone proteases 652
Effects of parathyroid hormone and parathyroid hormone–related protein on bone cell proliferation 652
Effects of parathyroid hormone and parathyroid hormone–related protein on bone cell differentiation 653
Effects of parathyroid hormone and parathyroid hormone–related protein on bone cells 654
Survival 654
Effects of parathyroid hormone and parathyroid hormone–related protein on bone 654
Bone resorption 654
Effects of parathyroid hormone and parathyroid hormone–related protein on bone 655
Bone formation 655
Parathyroid hormone actions on kidney **655**
Calcium and phosphate homeostasis 656
Calcium chemistry 656
Serum calcium 657
Phosphate chemistry 659
Serum phosphate 659
Parathyroid hormone actions on mineral-ion homeostasis 659
Parathyroid hormone receptor expression, signaling, and regulation in the kidney 660
Parathyroid hormone receptor expression 660
Parathyroid hormone receptor signal transduction in kidney tubular cells 660
Regulation of parathyroid hormone receptor signaling in tubular epithelial cells 661
Calcium absorption and excretion 662
Renal calcium absorption 662
Parathyroid hormone regulation of renal calcium absorption 663
Parathyroid hormone effects on proximal tubule calcium transport 663
Parathyroid hormone effects on distal tubule calcium transport 664
Phosphate excretion **664**
Mechanisms of proximal tubular phosphate absorption 664
Parathyroid hormone regulation of renal phosphate absorption 665
Parathyroid hormone receptor signal transduction in the regulation of calcium and phosphate excretion 665
Parathyroid hormone signaling of renal phosphate transport 666
Sodium and hydrogen excretion **667**
Parathyroid hormone regulation of proximal tubular sodium and hydrogen excretion 668
Vitamin D metabolism **669**
Other renal effects of parathyroid hormone **669**
Renal expression and actions of parathyroid hormone–related protein **670**
Acknowledgments **671**
References **671**

Principles of Bone Biology. https://doi.org/10.1016/B978-0-12-814841-9.00027-0

Introduction

Bone and kidney form a mineral-ion storage depot and regulatory axis that assures normal skeletal growth and development. Parathyroid hormone–related protein (PTHrP) is a major hormonal regulator of bone formation, while PTH (parathyroid hormone) contributes to postnatal skeletal integrity and extracellular mineral-ion homeostasis. PTHrP effects on the kidney are associated with controlling vascular tone, whereas PTH regulates phosphate and calcium absorption and the biosynthesis of vitamin D. Despite these distinct actions and tissue effects, PTH and the type I PTH/PTHrP receptor (PTHR) operate through a single canonical PTH/PTHrP receptor. This chapter reviews both well-accepted and recent advances in our understanding of PTH and PTHrP actions and considers controversial and unsettled elements of their effects.

Receptors and second-messenger systems for parathyroid hormone and parathyroid hormone–related protein

Most PTH and PTHrP effects in bone are mediated by the PTHR (Abou-Samra et al., 1992; Juppner et al., 1991). This family B G-protein-coupled receptor (GPCR) recognizes PTH and PTHrP as well as their biologically active N-terminal peptides PTH(1–34) and (1–36) and PTHrP(1–36). The PTHR binds PTH and PTHrP with equal affinity (although some differences in the affinity for G-protein-coupled and uncoupled states have been reported (Dean et al., 2008)), and in response, ligand binding signals through various cellular effector systems. Signal transduction by the PTHR is primarily mediated by G_s, $G_{q/11}$, G_i, and G_{12}/G_{13} heterotrimeric G proteins (Gardella and Vilardaga, 2015) (see alsoChapters 24, 25, and 28). The particular coupling mechanism for distinct G proteins depends on the cell type in a manner that remains incompletely understood (Abou-Samra et al., 1992; Civitelli et al., 1989, 1990; Juppner et al., 1991; Pines et al., 1996). Studies of the PDZ adapter protein Na^+/H^+ exchange-regulatory factor 1 (NHERF1) revealed that it serves as a switch to control signaling by G_s, $G_{q/11}$, and G_i (Mahon et al., 2002; Wang et al., 2010).

Important effects of the PTHR on mitogen-activated protein kinases (MAPKs), especially the extracellular signaling-related kinases (ERKs) ERK1 and ERK2, have been described (Cole, 1999; Miao et al., 2001; Sneddon et al., 2000, 2007; Swarthout et al., 2001; Syme et al., 2005). The mechanisms of activation of MAPK signaling by the PTHR are complex and involve Src. This nonreceptor tyrosine kinase interacts with the PTHR and its activity contributes to PTH-mediated ERK activation (Rey et al., 2006). In addition, ERK activation may arise from EGFR transactivation (Syme et al., 2005). Polyubiquitination of specific lysine residues in the C-terminal tail of the PTHR also regulates MAPK signaling, specifically activation of ERK1/2 and p38 (Zhang et al., 2018).

More recently, a newer paradigm of PTHR signaling via Gs/cAMP has emerged that differentiates the signaling activities of PTH and PTHrP (and related analogs) (Cheloha et al., 2015). Studies in cell systems show that in response to PTHrP, the PTHR at the cell membrane generates short-lived cAMP signals. PTH also elicits membrane signals via Gs. However, PTH also stimulates the recruitment of β-arrestins by the PTHR followed by its internalization in endosomes. Here, the PTHR assembles a functional complex with β-arrestin and Gs to produce sustained cAMP signaling (Ferrandon et al., 2009). In endosomes, β-arrestin promotes cAMP signaling, contrary to its typical function at the cell membrane, which terminates cAMP signaling. This action involves stimulation of ERK1/2 leading to inhibition of phosphodiesterases, in particular PDE4, and diminished cAMP degradation (Wehbi et al., 2013). Subsequent studies in animals show that this temporally discrete pattern of signaling by PTH and PTHrP results in markedly distinct calcemic responses (Shimizu et al., 2016). Thus, the induction of hypercalcemia by PTHrP is significantly lower than that elicited by PTH. These observations are consistent with studies in humans showing that PTHrP and its modified analog abaloparatide elicit significantly less hypercalcemia than does PTH (Horwitz et al., 2005, 2006; Leder et al., 2015).

The magnitude of physiological responses mediated by the PTHR is tightly linked to the balance between signal generation and termination. The receptor normally behaves in a cyclical pattern of activation and inactivation, where PTHR desensitization guards cells against excessive stimulation, and resensitization protects cells against prolonged hormone resistance. Rapid attenuation of PTHR signaling is mediated by receptor desensitization and internalization, while protracted reductions in responsiveness are due to downregulation and diminished receptor biosynthesis. Desensitization and internalization of the PTHR, as with other G-protein-coupled receptors, is generally thought to be regulated primarily by phosphorylation and arrestins. Ligand binding to the PTHR promotes receptor phosphorylation both by G protein receptor kinases (GRKs) and by two second-messenger-dependent protein kinases, protein kinase A (PKA) and protein kinase C (PKC) (Blind et al., 1996; Dicker et al., 1999; Flannery and Spurney, 2001). GRK2, and to a lesser extent by PKC, mediates phosphorylation of serine residues in the PTHR (Castro et al., 2002; Dicker et al., 1999; Malecz et al., 1998). GRK2 preferentially phosphorylates the distal sites of the intracellular PTHR tail, whereas PKC phosphorylates more upstream residues (Blind et al., 1996). Mice harboring a phosphorylation-resistant PTHR, where serine residues at positions 489, 491, 492, 493, 495, 501, and 504 were mutated to alanine, exhibit essentially normal anabolic responses (Datta et al., 2012), suggesting that PTHR phosphorylation does not importantly affect PTHR internalization or bone anabolism.

Following activation, the PTHR rapidly recruits β-arrestins at the plasma membrane, an event that initiates dynamin-dependent endocytosis. Interestingly, PTHR phosphorylation is not required for the interaction of the receptor with β-arrestins or for receptor internalization (Dicker et al., 1999; Ferrari et al., 1999; Malecz et al., 1998; Sneddon et al., 2003). However, receptor phosphorylation may stabilize the receptor–arrestin complex. As described previously, although the interaction with β-arrestins dampens acute cAMP generation at the plasma membrane, the PTHR, in complex with β-arrestins, remains active in endosomes and induces prolonged cAMP signaling. Although the PTHR lacks a canonical NPXXY internalization motif, other endocytotic signals have been identified within the carboxy terminus. Detailed analysis of the intracellular tail of the PTHR revealed bipartite sequences that negatively or positively regulate receptor endocytosis (Huang et al., 1995). An endocytic signal was detected within residues 475–494 of the opossum kidney PTHR (corresponding to D482-S501 of the human PTHR). Mutations or deletions within this region result in diminished PTH-induced receptor endocytosis (Huang et al., 1995).

In addition to governing PTHR signaling, NHERF1 is importantly involved in determining receptor endocytosis. The interaction of NHERF1 with the PTHR is disrupted by mutating its carboxy-terminal PDZ-binding motif; mutating NHERF1; or depolymerizing the actin cytoskeleton, all of which cause important alterations in PTHR endocytosis (Sneddon et al., 2003). Notably, in the absence of NHERF1, both PTH(1–34) and PTH(7–34) (along with their corresponding full-length peptides) promoted efficient PTHR sequestration. In the presence of NHERF1, however, PTH(1–34)-induced receptor internalization was unaffected, whereas PTH(7–34)-initiated endocytosis was largely inhibited (Sneddon et al., 2003). NHERF1 contains two tandem PDZ domains and an ezrin-binding domain (EBD). PDZ core-binding domains and the NHERF1 EBD domain are required for the inhibition of endocytosis (Wang et al., 2007a).

Expression and actions of parathyroid hormone receptor in bone

The PTHR is expressed widely in cells of the osteoblast lineage. Receptor expression is greater in more differentiated cells such as mature osteoblasts on the trabecular, endosteal, and periosteal surfaces (Fermor and Skerry, 1995; Lee et al., 1993) and osteocytes (Fermor and Skerry, 1995; van der Plas et al., 1994). PTHRs are also expressed in marrow stromal cells near the bone surface (Amizuka et al., 1996), a putatively preosteoblast cell population shown to bind radiolabeled PTH (Rouleau et al., 1988, 1990). However, PTHR is virtually absent in STRO-1–positive alkaline phosphatase-negative marrow stromal cells (Gronthos et al., 1999; Stewart et al., 1999), which are thought to represent relatively early osteoblast precursors. Indeed, PTHR expression can be induced by the differentiation of stromal cells, MC3T3 cells, or C3H10T1/2 cells with dexamethasone, ascorbic acid, or bone morphogenetic proteins (Feuerbach et al., 1997; Hicok et al., 1998; Liang et al., 1999; Stewart et al., 1999; Wang et al., 1999; Yamaguchi et al., 1987). Other data suggest that PTH receptors are limited to a relatively mature population of osteoprogenitor cells that express the osteocalcin gene (Bos et al., 1996).

Whether receptors for PTH or PTHrP are expressed on the osteoclast is controversial. Initial studies using receptor autoradiography failed to demonstrate them (Rouleau et al., 1990; Silve et al., 1982), and other studies have not identified PTHR mRNA or protein on mature osteoclasts (Amizuka et al., 1996; Lee et al., 1993, 1995). However, PTHRs are reportedly present on normal human osteoclasts (Dempster et al., 2005) and osteoclasts from patients with renal failure (Langub et al., 2001). Also, relatively low-affinity binding of radiolabeled PTH peptides to osteoclasts or preosteoclasts has been reported (Teti et al., 1991). The functional importance of such putative receptors is unclear.

PTH and PTHrP have additional receptors besides the PTHR. The PTH2R, which recognizes the amino terminal domain of PTH but not of PTHrP, is a GPCR closely related to the PTHR (Usdin et al., 1995) (Chapter 28). The endogenous ligand for the PTH2R is likely to be tuberoinfundibular peptide of 39 amino acids (TIP39). The PTH2R and TIP39 are expressed predominantly in brain, vasculature, and pancreas (Dobolyi et al., 2003; Eichinger et al., 2002; Usdin et al., 1996, 1999). TIP39 and PTH2R are expressed in distinct areas of the growth plate of newborn mice. Whereas the PTH2R localizes to the resting zone, TIP39 is restricted to prehypertrophic and hypertrophic chondrocytes. In chondrocytes, the TIP39/PTH2R system inhibits proliferation and differentiation (Panda et al., 2009). These findings were expanded to postnatal mice with chondrocyte-specific expression of the PTH2R (Panda et al., 2012). Targeted PTH2R expression in chondrocytes resulted in delayed formation of the secondary ossification center and reduced trabecular bone. Consistent with the finding in cell systems, these mice have attenuated expression of numerous markers of chondrocyte proliferation and differentiation including Sox9, Gdf5, Wdr5, and β-catenin.

A large body of evidence exists for the presence of specific receptor(s) for carboxyl-terminal PTH peptides on osteoblasts (Divieti et al., 2005; Inomata et al., 1995; Nguyen-Yamamoto et al., 2001) and osteocytes (Divieti et al., 2001), and evidence for actions of carboxyl-terminal PTH peptides on bone has been presented (Murray et al., 1989, 1991; Nakamoto et al., 1993; Sutherland et al., 1994). Such a carboxyl-terminal PTHR has thus far eluded molecular cloning.

PTHrP is expressed and secreted by osteoblast-like osteosarcoma cells (Rodan et al., 1989; Suda et al., 1996) and by rat long-bone explants in vitro (Bergmann et al., 1990). Messenger RNA for PTHrP is detected in periosteal cells of fetal rat bones (Karmali et al., 1992). *In situ* hybridization and immunohistochemistry localized PTHrP mRNA and protein to mature osteoblasts on the bone surface of fetal and adult bones from mice and rats (Amizuka et al., 1996; Lee et al., 1995) and to flattened bone-lining cells and some superficial osteocytes (Amizuka et al., 1996) in postnatal mice. In addition, the PTHrP gene is expressed in preosteoblast cells in culture, and in some studies its expression is reduced as preosteoblasts undergo differentiation (Kartsogiannis et al., 1997; Oyajobi et al., 1999; Suda et al., 1996).

As discussed in Chapter 25, PTHrP is cleaved to produce a set of peptides: those that contain the amino terminus (such as PTHrP(1−36)) activate the PTHR, and additional peptides representing the midregion and carboxyl terminus of PTHrP appear to have distinct biological actions mediated by their own receptors (Philbrick et al., 1996; Wysolmerski and Stewart, 1998). Receptors that are specific for amino-terminal PTHrP and do not recognize PTH have been identified in brain (Yamamoto et al., 1997) and other tissues (Gaich et al., 1993; Orloff et al., 1992; Valin et al., 2001), and midregion peptides of PTHrP have actions on placental calcium transport that imply a distinct receptor (Care et al., 1990; Kovacs et al., 1996), but there is presently no evidence for either receptor in bone. Carboxyl-terminal PTHrP fragments [e.g., PTHrP(107−139)] reportedly inhibit bone resorption (Cornish et al., 1997; Fenton et al., 1991a, 1991b, 1993) and stimulate (Goltzman and Mitchell, 1985) or inhibit (Martinez et al., 1997) osteoblast growth and function (Esbrit et al., 2000; Gray et al., 1999), and it is thus likely that a specific receptor for this peptide is present on osteoblasts and conceivably also on osteoclasts.

Effects of parathyroid hormone and parathyroid hormone−related protein on bone cells

Molecular mechanisms of action in osteoblasts

Transcription factors

A major effect of PTH and PTHrP in osteoblasts is directed at modulating the expression and/or function of a number of transcription factors important in bone metabolism. Among these, the most prominent are the cAMP response element binding protein (CREB) (Brindle and Montminy, 1992; Papavassiliou, 1994), the immediate early gene of activator protein-1 (AP-1) family members *c-fos* (c-fos, fra-1, fra-2) and *c-jun* (c-jun, junD) (Clohisy et al., 1992; Lee et al., 1994; McCauley et al., 1997, 2001; Stanislaus et al., 2000a), and Runx2 (Komori, 2002).

PTH induces *c-fos* transcription in a fashion that does not require protein synthesis and is mediated by the transcription factor CREB. Binding and phosphorylation of CREB to cAMP responsive elements within the *c-fos* promoter (Evans et al., 1996; Pearman et al., 1996; Tyson et al., 1999) are required to activate transcription. These events are stimulated by PTH through its ability to activate PKA, whereas the PKC signaling pathway is not involved in this response (Evans et al., 1996; McCauley et al., 1997). Activation of AP-1 transcription factors is important for osteoblast function because the *collagenase-3* (MMP-13) promoter contains AP-1 binding sites (Pendas et al., 1997; Selvamurugan et al., 1998).

One member of the runt-domain transcription factor family, viz. Runx2 (also called cbfa1 and OSF2), is a specific osteoblast transcriptional activator and is required for determination of the osteoblast phenotype (Ducy et al., 1997; Komori, 2002). Runx2 stimulates transcription of a number of key osteoblastic genes such as osteocalcin, osteopontin, collagenase-3, and collagens $\alpha1$ and $\alpha2$ (Banerjee et al., 2001; Porte et al., 1999; Selvamurugan et al., 1998). The importance of Runx2 for osteoblast formation and bone metabolism has been established with in vitro and in vivo models (Ducy et al., 1997, 1999; Komori et al., 1997; Otto et al., 1997). Brief PTH treatment stimulates rapid and transient increases of Runx2 mRNA and protein both in osteoblastic cell cultures and in mice (Krishnan et al., 2003). In addition, PTH, likely via cAMP/PKA activation, stimulates Runx2 activity (Selvamurugan et al., 2000; Winchester et al., 2000). In contrast, long-term treatment with PTH reduces Runx2 levels (Bellido et al., 2003). The dual effect of PTH on Runx2 levels and activity is particularly interesting. As will be discussed later, PTH promotes osteoblast survival in part by increasing Runx2 activity and consequently inducing expression of the survival gene *Bcl2*. However, PTH also induces proteasomal degradation of Runx2 through a mechanism involving Smurf2. Therefore, the ability of PTH to either increase or decrease Runx2 levels in osteoblasts in a fashion dependent on the administration schedule may provide one of the possible (and likely many) mechanisms to explain the complex effect of PTH on bone formation and resorption.

Osterix is a zinc finger domain transcription factor expressed specifically in osteoblasts (Nakashima et al., 2002). Its function is necessary for full osteoblastic differentiation and maturation (Nakashima et al., 2002), and studies in cells (Nishio et al., 2006) and genetically modified mice (Nakashima et al., 2002) have established that osterix lies downstream of Runx2. Similar to Runx2, PTH rapidly stimulates expression of osterix both in cell cultures (Wang et al., 2006) and in vivo (Tanaka et al., 2004).

Finally, PTH injection stimulates rapid and transient increases in the expression of at least three members of the nerve growth factor-inducible factor B (NR4A/NGFI-B) family of orphan nuclear receptors: Nurr1, Nur77, and NOR-1 (Pirih et al., 2005). Nurr1 mediates PTH-stimulated activation of the *osteopontin* (Lammi et al., 2004) and *osteocalcin* (Nervina et al., 2006; Pirih et al., 2004) promoters.

Clearly, PTH has profound effects on gene expression during bone remodeling. The complexity of this regulation is further illustrated by studies showing that PTH induces epigenetic modifications via histone deacetylating enzymes (HDACs). These epigenetic events profoundly impact PTH-mediated regulation of important genes including MMP13, Runx2, RANKL, and sclerostin. In osteoblastic cells, PTH signaling induces PKA-dependent phosphorylation of HDAC4 that in turn regulates expression of MMP13 via Runx2 (Shimizu et al., 2014). In contrast, in calvarial cells PTH elicits polyubiquitylation and degradation of HDAC4, an event that promotes RANKL (Obri et al., 2014). Further understanding of the complex mechanisms by which PTH regulates HDACs comes from studies in osteocytes (Wein et al., 2016, 2017). In these cells, the salt-inducible kinases (SIKs) maintain HDAC4/5 phosphorylated and outside the nucleus. Activation of the PTHR induces PKA-mediated phosphorylation of SIK2, which inhibits its activity. This causes nuclear translocation of HDAC4/5, inhibition of sclerostin transcription, and bone formation. By a similar mechanism, PTH controls the localization of the cAMP-regulated transcriptional coactivators CRTC2. When dephosphorylated, CRTC2 translocates to the nucleus, where it potentiates CREB-dependent transcription of RANKL, thus inducing bone resorption. Collectively, these observations highlight the critical contribution of epigenetic modification to the dual actions of PTH on bone formation and resorption.

The HDAC Sirt1 inhibits PTH-induced MMP13 transcription (Fei et al., 2015) in osteoblastic cells. Other studies show that PTH induces nuclear translocation of HDAC5 with attendant reduction in sclerostin expression, whereas HDAC2 and HDAC3 appear to induce constitutive *SOST* transcription in osteoblastic cells (Baertschi et al., 2014).

Growth factors and cytokines

Insulin-like growth factors. PTH stimulates synthesis and secretion of insulin-like growth factors (IGFs) IGF-I and IGF-II in bone. IGF-I and IGF-II are expressed in the rat (McCarthy et al., 1989) and mouse (Linkhart and Mohan, 1989; Watson et al., 1995). A PTH-dependent increase of IGF requires activation of the cAMP/PKA pathway because its effects are mimicked by cAMP analogs or agents that increase cAMP but not by stimulation of calcium signals or PKC (McCarthy et al., 1990).

IGF-I exerts proliferative and antiapoptotic actions in osteoblasts (Gray et al., 2003), and a number of observations support the requirement of IGF-I as a mediator of the anabolic action of PTH. First, treatment with PTH under conditions where it has an anabolic effect on bone leads to an increase in IGF-I mRNA (Watson et al., 1995) and the bone matrix content of IGF-I. Second, daily PTH injection failed to stimulate bone formation in mice lacking IGF-I (Bikle et al., 2002). Third, a similar absence of an anabolic effect of PTH was observed in mice lacking insulin receptor substrate-1 (Yamaguchi et al., 2005), the key intracellular mediator of IGF-I receptor signaling.

Additionally, both PTH and PTHrP affect the secretion in bone of IGF-binding proteins (IGFBPs), in particular IGFBP-4 (LaTour et al., 1990) and IGFBP-5 (Conover et al., 1993). The role of IGFBPs on IGF function is complex, since they can exert both inhibitory (IGFBP-4) and stimulatory (IGFBP-5) effects on IGF action in a cell- and context-specific manner. Moreover, some IGFBPs have been proposed to have effects independent of IGF (Conover, 2008). Thus, the role of IGFBPs on PTH action remains mostly to be defined.

Fibroblast growth factor-2 (FGF-2). Like IGF-I, FGF-2 is a potent mitogen (Ling et al., 2006), induces differentiation (Woei Ng et al., 2007), and exerts antiapoptotic actions on osteoblasts (Chaudhary and Hruska, 2001; Debiais et al., 2004). PTH rapidly increases FGF-2 and FGF receptor expression in both primary and clonal osteoblastic cells (Hurley et al., 1999). Stimulation of bone formation by PTH is impaired in mice null for FGF-2 (Hurley et al., 2006), and bone marrow cultures from these mice produce fewer osteoclasts in response to PTH (Okada et al., 2003). Collectively, these observations provide evidence that FGF-2 participates in the skeletal actions of PTH in vivo.

Amphiregulin. Recent studies demonstrated an important role for amphiregulin (AR), a member of the epidermal growth factor (EGF) family, as a mediator of PTH action. PTH stimulates AR expression in osteoblastic cells in a cAMP/PKA-dependent manner (Qin and Partridge, 2005). The increase in AR mRNA is mediated by phosphorylation of CREB and binding to a conserved CRE site in the AR promoter. AR appears to be important for bone metabolism because mice lacking AR have decreased trabecular bone (Qin et al., 2005b). In addition, AR stimulates rapid increases in *c-fos* and *c-jun* expression (Qin et al., 2005b). Treatment of primary calvarial cultures in vitro with AR stimulated cell proliferation and concomitant inhibition of osteoblastic differentiation, suggesting a role for AR on preosteoblastic mesenchymal stem cells (Qin and Partridge, 2005; Qin et al., 2005b). It is also notable that PTH stimulates the release of heparin-bound EGF, which belongs to the same family of membrane-bound and releasable EGF receptor agonists as AR, by activation of ADAM proteins in osteoblasts (Ahmed et al., 2003). This process leads to the activation of EGF receptor in an autocrine and paracrine fashion. It is therefore likely that PTH may induce not only the expression of AR in osteoblasts but also its release via a similar mechanism.

Transforming growth factor-β. All three TGF-β isoforms are detected in bone, with TGF-β1 being the most abundant (Hering et al., 2001; Seyedin et al., 1985). TGF-β1 is secreted by bone cells and stored in the extracellular matrix. TGF-β is a potent mitogen for osteoprogenitors (Hock et al., 1990) and has varied effects on differentiation and mineralization. Intermittent PTH treatment of rats increases the bone matrix content of TGF-β1 (Pfeilschifter et al., 1995; Watson et al., 1995) and the expression of TGF-β in osteoblasts (Oursler et al., 1991; Pfeilschifter and Mundy, 1987; Wu and Kumar, 2000). Moreover, PTH and PTHrP increase the secretion of TGF-β by osteoblast-like bone cells (Wu and Kumar, 2000), TGF-β1 activity (Pfeilschifter and Mundy, 1987; Sowa et al., 2003) and the release of TGF-β from bone matrix. Collectively, these observations suggest that TGF-β may contribute to the anabolic effects of intermittent PTH administration.

RANK ligand and osteoprotegerin. The regulation of these molecules by PTH and PTHrP is now recognized as the most prominent mechanism linking PTHR-mediated actions in osteoblasts (and possibly stromal cells) and osteoclastogenesis. RANK ligand (RANKL)[1] is a member of the TNF family of cytokines that is expressed on the osteoblast and stromal cell surface. Its receptor, called receptor activator of NFkB (RANK), is expressed in osteoclast precursors and mature osteoclasts, and its activation by RANKL stimulates osteoclast formation, activity, and survival (Burgess et al., 1999; Fuller et al., 1998; Hofbauer et al., 2000; Lacey et al., 1998). A second component of this system, expressed and secreted by osteoblasts, is osteoprotegerin (OPG) (Simonet et al., 1997; Yasuda et al., 1998), a released protein that binds and sequesters RANKL, preventing its actions on RANK. Therefore, it is the balance between RANKL and OPG (in other words the ratio RANKL:OPG) that ultimately determines the extent of osteoclast formation and function.

PTH and PTHrP regulate the expression of RANKL and OPG. It is interesting to note that the frequency of administration of PTH affects the RANKL:OPG ratio differently. Thus, prolonged exposure to PTH in rats causes sustained stimulation of RANKL expression while inhibiting OPG expression (Huang et al., 2004; Kondo et al., 2002; Lee and Lorenzo, 1999; Ma et al., 2001), with the final effect being an overall stimulation of osteoclast number and resorptive capacity. In contrast, intermittent PTH treatment affected RANKL and OPG only transiently (Ma et al., 2001). Exposure to PTH increases the expression of RANKL in murine bone marrow cultures, cultured osteoblasts, and mouse calvariae (Hofbauer et al., 2000; Lee and Lorenzo, 1999; Tsukii et al., 1998) as well as simultaneously decreasing the expression of OPG (Lee and Lorenzo, 1999). In addition, studies using primary cultures indicate that the effect of PTH on RANKL expression is more pronounced in differentiated osteoblasts, whereas the inhibition of OPG occurs at all differentiation stages (Huang et al., 2004). Notably, the stimulation of osteoclastic differentiation by RANKL requires the presence of macrophage-colony stimulating factor M-CSF, which is also upregulated by PTH (Horowitz et al., 1989; Weir et al., 1989).

Strong evidence supports the key role of the RANK/RANKL/OPG system in the activation of bone resorption upon prolonged exposure to PTH and PTHrP. Stimulation of osteoclastogenesis by PTH is blocked by antibodies to RANKL (Tsukii et al., 1998) or by OPG (Lacey et al., 1998; Yasuda et al., 1998), and OPG inhibits the hypercalcemic response to PTH and PTHrP (Morony et al., 1999; Oyajobi et al., 2001).

Wnt/β-catenin/sclerostin. The Wnt/β-catenin signaling pathway is recognized as an important regulator of bone mass. This is a complex signaling system comprising a number of members. Canonical Wnt signaling is mediated by a receptor complex formed by a Frizzled (a seven-transmembrane domain GPCR) and a lipoprotein-receptor-related protein 5 or 6 (LRP5/6) coreceptor. Upon engagement by Wnt, this complex activates the cytoplasmic protein disheveled (Dvl) followed by the accumulation of unphosphorylated β-catenin. As a result, β-catenin translocates to the nucleus and regulates gene transcription. Moreover, cells secrete proteins such as Dickkopf-1 that interact with the coreceptors LRP5/6 and inhibit Wnt signaling (Baron and Rawadi, 2007). All the key elements of this pathway are expressed in bone cells and osteoblastic cultures and regulated by PTH (Kulkarni et al., 2005). Treatment of rats with PTH increased Frizzled-1 and β-catenin levels and decreased DKK-1, with resulting activation of Wnt responses (Kulkarni et al., 2005). Sclerostin, the product of the *Sost* gene, is also a secreted Wnt inhibitor that binds LRP5 and LRP6 (Li et al., 2005; Semenov et al., 2005), and its

1. RANKL is also called osteoprotegerin ligand (OPGL), osteoclast differentiation factor (ODF), and TNF-related activation-induced cytokine (TRANCE).

level in osteocytes is dramatically reduced by continuous PTH treatment (Bellido, 2006; Bellido et al., 2005). It is interesting that sclerostin appears to inhibit osteoblast differentiation (van Bezooijen et al., 2004, 2007), thereby providing a functional link between osteocytes and osteoblasts. The profound effects of the Wnt signaling complex on bone metabolism are well described by a number of genetic studies in both human and animal models (Baron and Rawadi, 2007), and the observation that PTH engages this pathway in bone cells is indeed of great interest.

Adaptor proteins

G-protein-coupled receptor kinase 2 and β-arrestins. These two ubiquitously expressed proteins play important roles in the regulation of several G-protein-coupled receptors including the PTHR. It is not surprising that alterations in their expression or activity impact the function of osteoblasts and their responsiveness to PTH. G-protein-coupled receptor kinase 2 (GRK2) phosphorylates agonist-occupied PTHR and promotes the binding of β-arrestins (Dicker et al., 1999; Ferrari et al., 1999; Flannery and Spurney, 2001; Vilardaga et al., 2001). These combined actions result in decreased signaling at the cell membrane (Bisello et al., 2002; Dicker et al., 1999; Ferrari and Bisello, 2001). As mentioned earlier, prolonged PTHR signaling in endosomes is promoted by β-arrestins through the inhibition of PDE4 and reduction in cAMP degradation (Gardella, Vilardaga) (Cheloha et al., 2015).

Targeted overexpression of GRK2 in osteoblasts promotes bone loss (Wang et al., 2005a). In contrast, inhibition of GRK2 activity by expression of a dominant negative mutant increases PTH-stimulated cAMP, and mice overexpressing Grk2 have increased bone remodeling with a net gain in bone content (Spurney et al., 2002). Similarly, intermittent PTH increased the number of osteoblasts in mice null for β-arrestin2 (Ferrari et al., 2005). However, no net increase in bone mass was observed, likely due to the intense stimulation of osteoclastogenesis. Another regulator of G-protein-coupled receptor signaling, the regulator of G protein signaling-2 (RGS2), which increases the rate of hydrolysis of GTP bound to G proteins, thereby terminating signaling, has also been implicated in PTHR actions on bone cells. RGS2 mRNA is rapidly and transiently increased by PTH in rat bones as well as osteoblast cultures (Miles et al., 2000b), and its expression in bone cells resulted in decreased PTH-stimulated cAMP production (Thirunavukkarasu et al., 2002). Interestingly, RGS2 upregulation was also observed in cells overexpressing Runx2 (Thirunavukkarasu et al., 2002), suggesting the possibility that mechanisms limiting PTHR signaling by G proteins may be activated upon differentiation of cells along the osteoblastic lineage.

Na^+/H^+ exchange-regulatory factor 1 (NHERF1). NHERF1 is a PDZ-domain scaffolding protein that interacts with the PTHR and regulates various functions including preferential G-protein coupling, membrane retention, and trafficking. Little information exists concerning the role of NHERF1 in regulating the PTH-mediated effect in bone cells. NHERF1 is expressed in active mineralizing osteoblasts, where it regulates PTHR expression during differentiation (Liu et al., 2012). Indeed, when cultured in differentiating medium, mesenchymal stem cells from NHERF1$^{-/-}$ mice show a dramatic reduction in PTHR mRNA accompanied by preferential differentiation to adipocyte. NHERF1 is required for osteoblast differentiation and matrix synthesis, but whether this is directly due to alterations in PTHR signaling is yet unknown. NHERF1-null mice and humans with NHERF1 polymorphisms display osteopenia and increased skeletal fractures (Karim et al., 2008; Morales et al., 2004; Shenolikar et al., 2002). Proliferating osteoblasts express Npt2a and Npt2b, PTHR, and NHERF1 (Wang et al., 2013). In cells from wild-type mice, PTH inhibits phosphate uptake, but this effect was not observed in cells from NHERF1$^{-/-}$ mice. In contrast, PTH increases phosphate uptake in differentiated osteoblasts, and this effect depends on NHERF1 expression.

Effects on gap junctions

PTH increases intercellular communication of bone cells by increasing connexin-43 gene expression (Schiller et al., 1992) and opening gap junctions (Donahue et al., 1995; Schiller et al., 1992). The reduction of connexin-43 levels by transfection of antisense cDNA markedly inhibited the cAMP response to PTH (Vander Molen et al., 1996) and blocked the effect of PTH on mineralization by osteoblast-like cells (Schiller et al., 2001). These effects appear to significantly impair the anabolic action of PTH because increases in bone mineral content in response to intermittent PTH administration were significantly decreased in mice with targeted deletion of connexin43 in osteoblasts (Chung et al., 2006).

Effects on bone matrix proteins and alkaline phosphatase

PTH regulates the expression of a number of bone matrix proteins including type I collagen, osteocalcin, osteopontin, bone sialoprotein, osteonectin, and alkaline phosphatase. In most cases, these genes (with the exception of osteocalcin) are downregulated by continuous exposure to PTH (Wang et al., 2005b), whereas intermittent administration of PTH has inhibitory and stimulatory effects on different genes.

Type I collagen is the most abundant bone matrix protein. PTH and PTHrP acutely inhibit collagen synthesis in vitro (Kream et al., 1980, 1986; Partridge et al., 1989; Pines et al., 1990). Similar inhibition is observed upon infusion of PTH in humans (Simon et al., 1988). However, anabolic PTH treatment can stimulate type I collagen expression (Canalis et al., 1990; Opas et al., 2000), an effect that is likely mediated by increases in IGF-I (Canalis et al., 1989). Intermittent treatment of mouse bone marrow cells with PTH modestly increased collagen expression (Locklin et al., 2003). Similarly, PTH treatment inhibits osteopontin (Noda et al., 1988) and osteonectin (Termine et al., 1981) expression.

The osteocalcin gene is also importantly regulated by PTH (Noda et al., 1988; Towler and Rodan, 1995; Yu and Chandrasekhar, 1997). However, in contrast to the effect on type I collagen, expression of osteocalcin is stimulated by chronic administration of PTH or PTHrP, whereas the acute effect of these hormones is inhibitory (Gundberg et al., 1995).

Finally, the effect of PTH on the expression of bone sialoprotein can be either stimulatory (Ogata et al., 2000; Yang and Gerstenfeld, 1997) or inhibitory (Ma et al., 2001; Wang et al., 2000).

The reported actions of PTH on the expression of alkaline phosphatase are inconsistent. PTH can either stimulate or inhibit secretion of alkaline phosphatase from bone cells (Jongen et al., 1993; Kano et al., 1994; Majeska and Rodan, 1982; McPartlin et al., 1978; Yee, 1985) and may not be particularly indicative of PTH-specific actions. Indeed, although anabolic therapy with PTH generally increases the circulating levels of alkaline phosphatase (Finkelstein et al., 1998), this effect is likely due to an increase in osteoblast number rather than an increase in protein expression.

Effects on bone proteases

PTH stimulates the secretion of a number of matrix metalloproteases (MMPs) in bone cells (see Chapter 16) that are involved in bone remodeling. These include collagenase-3 (MMP-13) (Partridge et al., 1987; Quinn et al., 1990; Scott et al., 1992; Winchester et al., 1999, 2000), stromelysin-1 (Meikle et al., 1992), gelatinase B (Meikle et al., 1992), and the disintegrin and metalloprotease with thrombospondin repeats (Miles et al., 2000a). Bone proteases, in particular collagenase-3 and gelatinase B, partially mediate the stimulation of bone resorption by PTH (Witty et al., 1996). As described above, stimulation of the MMP-13 promoter by PTH requires the combined action of AP-1 transcription factors and Runx2, effects that are mediated by cAMP-dependent activation of CREB (Porte et al., 1999; Selvamurugan et al., 1998, 2000). All of these events are stimulated by PTH. PTH treatment also increases secretion of the tissue inhibitor of matrix metalloproteins (TIMP-1) by osteoblasts (Meikle et al., 1992). This is relevant to the action of PTH because mice overexpressing TIMP-1 in osteoblasts responded to intermittent PTH with increases in bone mineral density higher than those in normal mice (Merciris et al., 2007). This was also accompanied by decreased osteoclastic differentiation (Geoffroy et al., 2004; Merciris et al., 2007).

Effects of parathyroid hormone and parathyroid hormone–related protein on bone cell proliferation

Continuous exposure to relatively high concentrations of PTH(1−34) or PTHrP(1−34) inhibits proliferation of virtually every osteoblastic cell line including UMR 106−01, MC3T3-E1, SaOS-2, and calvarial primary cultures (Civitelli et al., 1990; Kano et al., 1991; Onishi et al., 1997; Qin et al., 2005a). This effect is mediated by changes in the expression levels of several components of the cell cycle, ultimately resulting in arresting cells in the G1 phase. PTH treatment decreases expression of cyclin D1 while increasing the levels of $p21^{Cip1}$ and $p27^{Kip1}$ (Onishi et al., 1997; Qin et al., 2005a), with evident cell cycle arrest.

In contrast, some in vitro studies demonstrated that in certain circumstances PTH stimulates the proliferation of osteoblastic cells (Finkelman et al., 1992; Onishi et al., 1997; Somjen et al., 1990). In particular, very low concentrations of PTH (Swarthout et al., 2001) and brief exposure to PTH (Scutt et al., 1994) resulted in increased cell proliferation. In the preosteoblast cell line TE-85, PTH stimulated proliferation by increasing expression of the cyclin-dependent kinase cdc2 (Onishi et al., 1997).

The effects of PTH and PTHrP on the cell cycle in osteoblastic cells have important consequences in vivo. Several studies indicate that intermittent injections of PTH increase the number of osteoblasts (Kostenuik et al., 1999; Nishida et al., 1994), and this contributes to the stimulation of bone formation. However, this effect does not appear to be related to direct stimulation of osteoblast mitogenesis. Indeed, although intermittent PTH administration in rats greatly increased osteoblast number and function, osteoblast proliferation was not detected (Dobnig and Turner, 1995; Onyia et al., 1995, 1997). It is possible that the increase in osteoblast number produced by intermittent treatment with PTH is due to the activation of bone-lining cells to osteoblasts (Dobnig and Turner, 1995), a process that does not require mitosis. These findings are compatible with the conclusion that PTH inhibits cell cycle progression of committed osteoprogenitors, thereby permitting their maturation.

Effects of parathyroid hormone and parathyroid hormone–related protein on bone cell differentiation

PTH and PTHrP profoundly influence the differentiation program of bone marrow cells to form osteoblasts (Fig. 27.1). Several studies indicate that anabolic administration of PTH stimulates rapid changes in histomorphometry and gene expression, which have been interpreted as resulting from cell differentiation (Hodsman and Steer, 1993; Onyia et al., 1995). The effects of PTH and PTHrP on cell differentiation in culture are also well documented and appear to depend on both the duration and the frequency of PTH exposure. Early, transient PTH treatment enhances the commitment of progenitor cells and increases osteoblast differentiation (Wang et al., 2007b). In vitro, primary osteoblasts briefly (1 h every 48 h) exposed to PTH showed inhibited expression of alkaline phosphatase activity and bone nodule formation (Ishizuya et al., 1997). In contrast, intermittent PTH treatment for 6 h every 2 days stimulated osteoblastic differentiation and formation of mineralized nodules (Ishizuya et al., 1997). Similarly, transient PTH treatment of calvarial osteoblasts inhibited initial osteoblast differentiation but ultimately resulted in increased mineralized nodules and osteoblastic differentiation (Wang et al., 2007b). In cultured murine marrow cells, intermittent PTH treatment increases the expression of osteoblast differentiation markers (such as Runx2, alkaline phosphatase, and type I collagen) (Locklin et al., 2003). In all cases, however, continuous exposure to PTH strongly inhibits osteoblast differentiation (Ishizuya et al., 1997; Wang et al., 2007b). In this respect, it is interesting to note that in humoral hypercalcemia of malignancy (HHM), PTHrP is continuously secreted by tumors and results in the virtual absence of mature osteoblasts (Stewart et al., 1982). In contrast, elevated PTH levels observed in hyperparathyroidism do not have the same effect: indeed, hyperparathyroidism is characterized by increased bone remodeling (i.e., higher formation and resorption) with complex effects on bone (Bilezikian, 2012) (Chapter 54). Although the mechanisms of this difference remain to be fully elucidated, one possible underlying basis is the pulsatile secretion of PTH by the parathyroids in hyperparathyroidism versus the continuous, unregulated secretion of PTHrP in HHM.

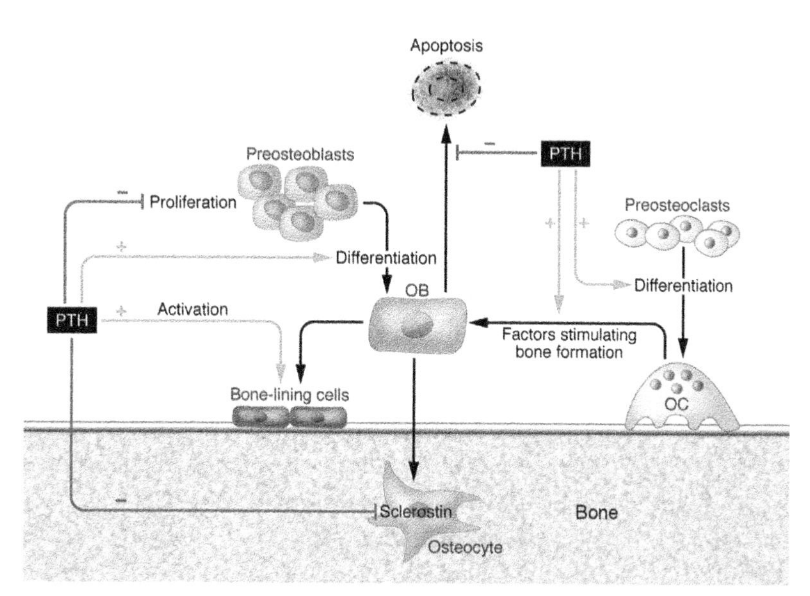

FIGURE 27.1 Cellular effects of PTH on bone. *Bone resorption*. PTH stimulates differentiation and activation of osteoclasts by increasing RANKL and possibly decreasing OPG production by stromal cells and osteoblasts. *Bone formation*. PTH decreases proliferation of preosteoblasts and stimulates their differentiation. PTH decreases osteoblast apoptosis and can activate bone-lining cells into functioning osteoblasts. These combined actions result in increased numbers of osteoblast. Moreover, PTH suppresses the production of the Wnt inhibitor sclerostin by osteocytes, thereby increasing β-catenin signaling in osteoblasts. *From Khosla, S., Westendorf, J.J., Oursler, M.J. 2008. Building bone to reverse osteoporosis and repair fractures. J. Clin. Investig. 118, 421–428.*

Effects of parathyroid hormone and parathyroid hormone–related protein on bone cells

Survival

Cell culture studies show that PTH rapidly stimulates transcription of the prosurvival gene *Bcl-2* while increasing the inactivation of the apoptotic protein Bad (Bellido et al., 2003; Jilka et al., 1998), suggesting that activation of the PTHR in osteoblasts may inhibit apoptosis. Indeed, various studies showed that PTH reduces the apoptotic effects of etoposide, dexamethasone, and serum deprivation in osteoblastic cultures (Bellido et al., 2003; Chen et al., 2002; Jilka et al., 1998). The effect of PTH on the expression of survival proteins appears to require the action of CREB and Runx2 (Bellido et al., 2003). Considering the disparate effects of PTH on Runx2 stability discussed previously, it is therefore possible that the timing and frequency of PTH stimulation may be key determinants for its prosurvival action. Thus, although some studies in mice demonstrated a significant decrease in apoptotic osteoblasts following daily PTH administration (Bellido et al., 2003; Jilka et al., 1999), a study in rats reported an actual increase in apoptotic osteoblasts by intermittent PTH (Stanislaus et al., 2000b). An additional complexity results from the observation that daily administration of PTH in post-menopausal women resulted in increased osteoblast apoptosis (Lindsay et al., 2007). Finally, it should be noted that differences have been observed in the effect of PTH on osteoblast apoptosis at different skeletal sites (cortical and trabecular) and possibly between primary and secondary spongiosa (Jilka et al., 1999; Stanislaus et al., 2000b).

Effects of parathyroid hormone and parathyroid hormone–related protein on bone

Bone resorption

Cellular basis of parathyroid hormone action

PTH and PTHrP increase bone resorption by stimulating osteoclastogenesis and activating the mature osteoclast. These effects require the participation of classical PTH target cells including stromal cells and osteoblasts (Akatsu et al., 1989; McSheehy and Chambers, 1986). Moreover, as described earlier, osteoclasts seemingly do not express high-affinity PTHR (Amizuka et al., 1996; Rouleau et al., 1990; Silve et al., 1982). The identification of the central role of the RANK/RANKL system in osteoclast formation and function, and the appreciation that PTH and PTHrP affect the expression of RANKL and OPG in stromal cells and osteoblasts, provide a molecular basis for understanding PTH-stimulated bone resorption. The precise target cell in the osteoblast lineage responsible for mediating the bone-resorbing effects of PTH and PTHrP, if any, has not been identified, but various marrow stromal cell lines suffice in vitro (Aubin and Bonnelye, 2000), and bone resorption is still active when mature osteoblasts have been ablated (Corral et al., 1998). By binding to its cognate receptor, RANK, on osteoclast precursors and mature osteoclasts, RANKL stimulates osteoclastogenesis and the activity of mature osteoclasts. Therefore, increased RANKL expression mediated the acute and chronic actions of PTH on bone resorption. The major difference in the two effects is likely related to the stronger suppression of the decoy receptor OPG upon prolonged exposure to PTH (Huang et al., 2004). Osteoclast activation by RANKL is apparently responsible for bone resorption at the cellular level and for hypercalcemia, as both are blocked by the decoy receptor OPG (Morony et al., 1999; Yamamoto et al., 1998).

The bone-resorbing effects of amino-terminal PTH and PTHrP are essentially indistinguishable when studied using isolated osteoclasts (Evely et al., 1991; Murrills et al., 1990), bone explant systems (Raisz et al., 1990; Yates et al., 1988), or infusion into the intact animal (Kitazawa et al., 1991; Thompson et al., 1988). In contrast, PTHrP is considerably less potent than PTH in inducing hypercalcemia in humans (Horwitz et al., 2006). This difference is most likely due to lower induction of renal $1.25(OH)_2D_3$ by PTHrP compared with PTH (Horwitz et al., 2003b), though recent findings regarding the persistent noncanonical signaling of cAMP by PTH but not PTHrP described earlier provide a compelling alternative explanation (Cheloha et al., 2015; Ferrandon et al., 2009). Notably, studies in humans showed that intermittent administration of PTHrP(1–36) over 3 months stimulated bone formation without attendant increases in markers of bone resorption (Horwitz et al., 2003b).

As discussed in Chapter 3, PTHrP is a polyhormone, the precursor of multiple biologically active peptides. Carboxy-terminal peptides predicted to arise from cleavage of PTHrP in the polybasic region PTHrP(102–106) have been synthesized and shown to inhibit bone resorption in several explant systems (Fenton et al., 1991a, 1993) and in vivo (Cornish et al., 1997). On this basis, the minimal peptide that inhibits bone resorption, PTHrP(107–111), has been identified and called osteostatin.

Effects of parathyroid hormone and parathyroid hormone–related protein on bone

Bone formation

The mechanisms by which PTH increases bone formation are complex. As described before, PTH and PTHrP exert a variety of effects on osteoblasts. The increase in bone formation in response to intermittent administration of PTH and PTHrP correlates with marked increases in the number of active osteoblasts (Boyce et al., 1996; Dempster et al., 1999; Shen et al., 1993).

It is evident from the previous discussion that every aspect of the osteoblast existence is affected by PTH, and all of these cellular actions may contribute to the increase in osteoblast number observed in the stimulation of bone formation. First, activation of the PTHR produces various actions on the cell cycle of osteoblasts and their precursors. It is clear that in most circumstances PTH and PTHrP cause cell cycle arrest in osteoblasts and preosteoblasts, and this may be a prerequisite to induce further differentiation and activation. Most in vivo evidence does not support a direct proliferative effect of PTH on mature osteoblasts (Dobnig and Turner, 1995; Onyia et al., 1995, 1997). However, PTH and PTHrP increase the expression and release of a number of potent mitogens including IGF-I, TGFβ, and amphiregulin, which may act in a paracrine fashion to expand the pool of osteoprogenitors (Gray et al., 2003; Hock et al., 1990; Qin and Partridge, 2005). Second, intermittent PTH administration stimulates transcription factors, such as Runx2 and osterix, which in turn stimulate differentiation. This effect is accompanied by increases in osteoblastic differentiation markers (Hodsman and Steer, 1993; Onyia et al., 1995; Wang et al., 2007b). Third, some rapid histomorphometric changes observed upon anabolic administration of PTH may derive from stimulation of bone-lining cells to become active osteoblasts (Dobnig and Turner, 1995). Fourth, intermittent PTH administration exerts antiapoptotic actions in osteoblasts and osteocytes (Bellido et al., 2003; Jilka et al., 1999). Obviously, prolonging the life span of osteoblasts would enhance the number of mature osteoblasts. It is interesting that Runx2, a key molecule mediating PTH effects on osteoblast survival, is also involved in the stimulation of osteoblastic differentiation.

The relative contribution of each of these mechanisms to the anabolic action of PTH has not been completely established. It seems likely that the remarkable increases of bone formation in response to intermittent administration of PTH and PTHrP arise from a combination of these effects. Moreover, it is possible that PTH and PTHrP may not have the same effect under all circumstances and in the presence of other treatments affecting bone metabolism.

Finally, it has long been thought that the anabolic actions of PTH and PTHrP are substantially equivalent because most of their cellular effects in vitro and activities in animal models are quite similar. However, it has recently become apparent that some basic differences exist in the action of these two hormones in humans. Intermittent administration of PTHrP(1–36) for 2 weeks (Plotkin et al., 1998) and 3 months (Horwitz et al., 2003a), for instance, leads to increases in biochemical markers of bone formation without changing markers of bone resorption, suggesting that PTHrP(1–36) may uncouple bone formation and resorption.

Parathyroid hormone actions on kidney

PTH regulates renal tubular absorption of phosphate and calcium and synthesis of $1.25(OH)_2D_3$, thereby controlling plasma levels and urinary excretion of these mineral ions and vitamin D (calcitriol) Fig. 27.2). Renal tubular responses to PTH deficiency, PTH or PTHrP excess, and defects in function of the PTHR lead to alterations in blood calcium, phosphate, or vitamin D that are the hallmarks of numerous clinical disorders described later in this volume. Here we review the mechanisms of PTH and PTHrP control of renal tubular calcium and phosphate absorption. The narrative focuses principally on PTH because distinct or unique PTHrP actions on tubular ion transport are not well delineated despite their displaying distinct signaling dynamics (Ferrandon et al., 2009). Because the amino termini of both ligands are similarly recognized by the PTHR, it is likely that PTHrP shares the effects of PTH, though the differences in signaling noted and described earlier may translate to different patterns of dynamic actions. Expression and specific renal actions of PTHrP are associated with regulation of renovascular hemodynamics and may have important effects on hypertension or preeclampsia (Massfelder et al., 1998; Massfelder and Helwig, 1999; Yadav et al., 2014).

Although biologically important, serum magnesium levels do not appear to be significantly regulated by PTH and PTHrP and therefore are not discussed here. Reviews of renal magnesium transport can be found elsewhere (Blaine et al., 2015; Houillier, 2014; Schaffers et al., 2018).

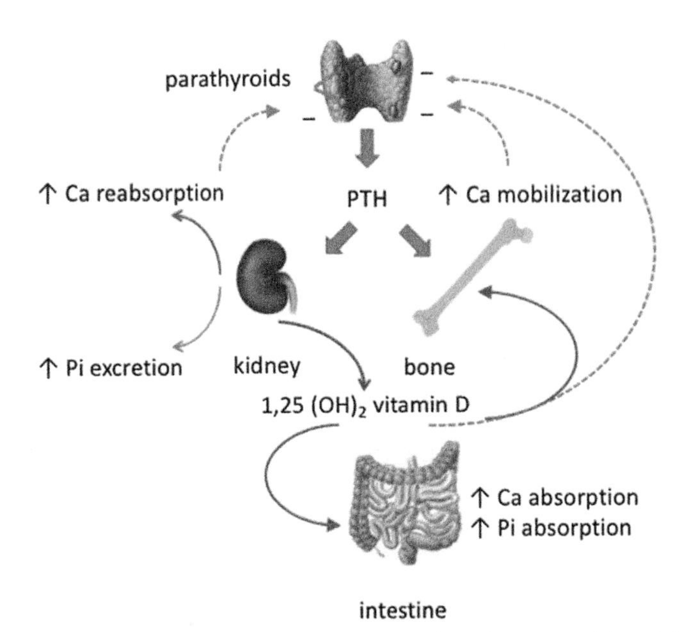

FIGURE 27.2 Parathyroid hormone (PTH) regulation of calcium homeostasis. PTH regulates extracellular calcium homeostasis through a negative feedback scheme. Decreases of extracellular calcium trigger PTH release. PTH in turn stimulates renal calcium reabsorption and mobilizes surface calcium from bone to inhibit further PTH release (dashed lines). PTH also promotes phosphate excretion and stimulates 1α-hydroxylase activity in kidney mitochondria leading to the production of 1,25-dihydroxyvitamin D (calcitriol). Calcitriol, the biologically active metabolite of vitamin D, increases intestinal calcium and phosphate absorption, and calcium mobilization in bone.

Calcium and phosphate homeostasis

Calcium chemistry

Calcium participates in a variety of structural and functional roles in animal cells and tissues. The physical qualities of the calcium ion are especially well suited to the tasks it performs. At first glance, calcium would seem to be an unlikely choice to subserve such diverse functions as an integral macromolecular constituent of bone and at the same time be a primary element in micromolecular signaling. Extracellular concentrations of unbound calcium are roughly 1 mM, whereas its intracellular free concentration under resting conditions is about 0.1 μM—i.e., four orders of magnitude lower. Such extreme differences between intra- and extracellular concentrations place severe constraints and demands on plasma membrane transport proteins to safeguard the integrity of the intracellular milieu. To extrude calcium from the cell, for instance, a prodigious adverse electrochemical gradient must be overcome. Simultaneously, extracellular calcium must be fastidiously regulated. What makes calcium so biologically apt that a variety of evolutionarily taxing developments emerged to accommodate its superior aspects? Some of these structural virtues have been summarized by Williams (1976). For example, calcium, in contrast to magnesium, exhibits a particularly adaptable coordination sphere that facilitates binding to the irregular geometry of proteins. The ability to cross-link two proteins requires an ion with a high coordination number (which dictates the number of electron pairs that can be formed) and is generally six to eight for calcium (Williams, 1976). Such cross-linking of osseous structural proteins is facilitated at the high calcium concentrations found in extracellular fluid. At the same time, the variable bond length of calcium permits formation of the more extensive cross-linking involved in membrane stabilization by facilitating lipid polymorphism and formation of hexagonal arrays. Moreover, unlike disulfide or sugar-peptide cross-links, calcium linking is readily reversible. Despite these virtues, if intracellular free calcium ($[Ca^{2+}]_i$) was similar to its extracellular concentration, the proper functioning of a variety of proteins and macromolecules would be impaired. Thus, there seems to be some evolutionary rationale for maintaining the intracellular concentration of calcium at rather low levels. What benefits, then, result from maintaining low intracellular calcium?

The corollary of the benefits of the physical characteristics of calcium at high extracellular concentrations defines a nearly ideal set of attributes that are desirable at submicromolar intracellular concentrations. By virtue of its low cytoplasmic levels, changes of calcium activity can function as first or second messengers to activate effector targets. The fact that calcium can be rapidly bound and released, together with the high affinity and selectivity of many proteins for calcium, would seem to enhance its ability to serve as a trigger or rapid on/off signaling switch. Another advantage of low free intracellular calcium concentrations is that microcrystallization and precipitation of calcium phosphate is avoided. Circumventing these processes, it has been speculated (396), favors the evolution of high-energy phosphate compounds, which serve as the energy source for a host of biological reactions.

Serum calcium

Calcium may be measured in serum or plasma. As discussed later, this makes little practical difference. Clinical laboratories typically analyze and report results for total calcium; free calcium requires an additional order. Insofar as much circulating calcium is bound either to proteins or as small chemical complexes, the determination of free, or ionized, calcium is useful. Moreover, because pathological conditions including liver disease or kidney failure may change protein levels. Thus, discrepancies between total and ionized calcium may arise but can go undetected without knowing both total and free calcium.

Total calcium concentrations fluctuate with age (Fig. 27.3), averaging 9.5 mg/dL (2.4 mmol/L) (Table 27.1) in adults. Calcium in plasma is present to varying extents in protein-bound, complexed, and ionized forms. Approximately 45% percent of calcium is bound to plasma proteins, mostly to albumin. Smaller amounts of calcium are bound to globulins and a negligible portion to fibrin. Therefore, serum and plasma calcium concentrations are generally indistinguishable (Miles et al., 2004). Another 45% of calcium is ionized (or "free"). The remaining 10% of calcium is associated with small polyvalent anions such as bicarbonate, phosphate, and sulfate. Such ion pairs, e.g., calcium bicarbonate, that arise by electrostatic forces are called "calcium complexes." Together, ionized and complexed calcium are referred to as "diffusible" because only these forms are (1) filtered at the glomerulus and (2) able to cross cell membranes.

Ultrafilterable and ionized fractions of calcium are affected by changes in the total serum calcium concentration, blood pH, plasma protein concentration, and the abundance of complexing anions. Increases in total serum calcium are usually

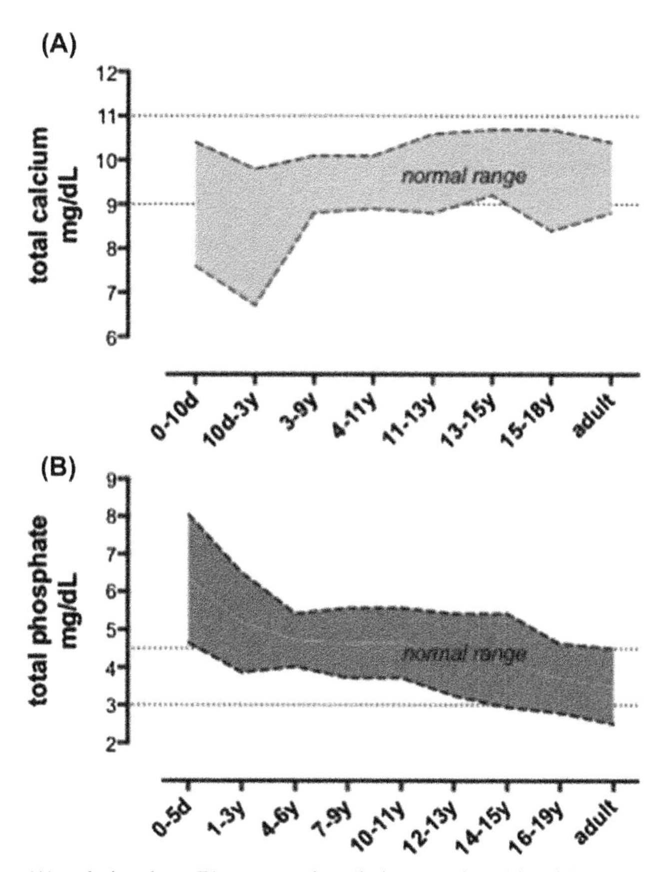

FIGURE 27.3 Total serum calcium (A) and phosphate (B) concentrations during growth and in adults. The normal ranges are highlighted. *Values compiled from various sources Fischbach, F.T., Dunning, M.B. 2015. A manual of laboratory and diagnostic tests. ninth ed. Wolters Kluwer Health, Philadelphia; Meites, S. 1989. Pediatric clinical chemistry: reference (Normal) values. third ed. AACC Press, Washington, DC.*

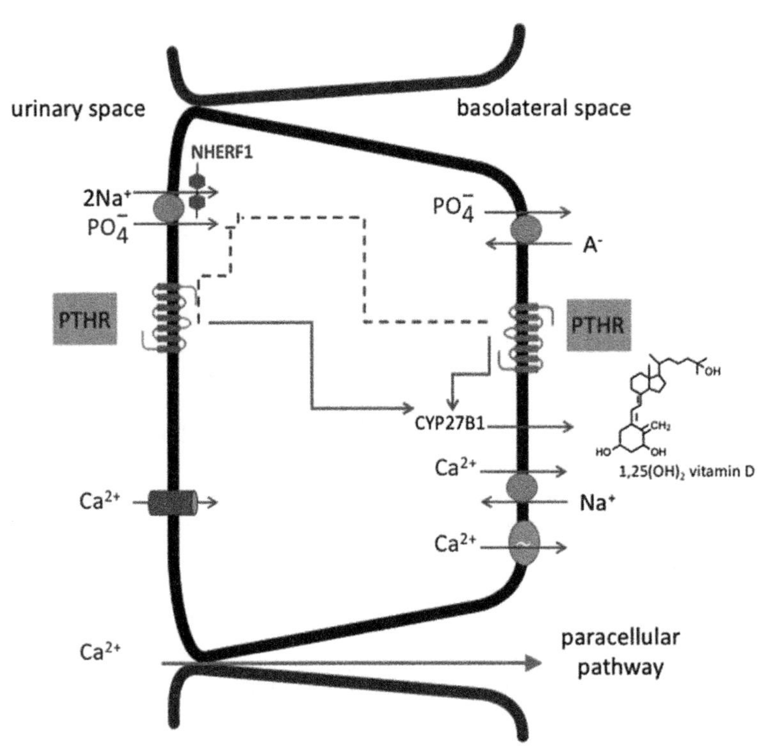

FIGURE 27.4 Phosphate and calcium transport in proximal tubules. Proximal tubule Na-Pi cotransport is mediated principally by apical membrane Npt2a. Two sodium ions are taken up together with a single phosphate molecule. Npt2a is inhibited by PTH as described in the accompanying text. NHERF1 is required for this activity. The mechanism of basolateral PO_4 extrusion is uncertain and is illustrated here as proceeding by a nonspecified anion (A^-) exchanger. Calcium absorption by proximal tubules is largely, if not entirely, a passive process that proceeds through the paracellular pathway of the lateral intercellular spaces (dashed line). Evidence for apical membrane Ca^{2+} entry and basolateral efflux has been described but it is uncertain whether these single processes contribute to transcellular calcium absorption or subserve intracellular calcium homeostasis. PTHR is shown on apical and basolateral cell membranes. Emerging evidence along with published findings suggest the preferential regulation of Npt2a by apical PTHR and induction of CYP27B1 and vitamin D formation by basolateral PTHR.

accompanied by concomitant elevations of ultrafilterable calcium to a total of about 4 mM (Edwards et al., 1974; Le Grimellec et al., 1974). This upper limit of calcium ultrafiltration with hypercalcemia has been postulated to result from the formation of insoluble $[Ca_3(PO_4)_2^-$ −protein] complexes. This idea is supported by the finding that the ultrafilterable phosphate concentration also declines (Cuche et al., 1976). Conversely, hypocalcemia is generally associated with a decline of calcium ultrafiltration (Terepka et al., 1957).

Changes in the concentration of serum proteins are accompanied by parallel alterations of total serum calcium concentration. However, the ultrafilterable fraction generally remains constant. In severe hypoproteinemia, however, the ultrafilterable fraction increases.

TABLE 27.1 Serum calcium (Fischbach and Dunning, 2015).

	mg/dL	mEq/L	mM
Adults			
Total calcium	8.8–10.4	4.4–5.2	2.2–2.6
Ionized calcium	4.6–5.3	2.3–2.6	1.16–1.32
Children			
Total calcium (10d–3yr)	6.7–9.8	3.4–4.9	4.80–5.52
Ionized calcium (1–18yr)	2.24–2.75	1.1–1.4	1.2–1.4

Normal ranges vary between laboratories and depend on applied measurement technology and assay conditions.

In contrast to the parallel relation between total calcium and serum protein, ionized Ca^{2+} levels vary inversely with blood pH. Acidosis increases ionized Ca^{2+} because H^+ ions displace Ca^{2+} from serum proteins. Conversely and for the same reason, alkalosis decreases Ca^{2+} (Hopkins et al., 1952; Loeb, 1926; Peterson et al., 1961). Thus, in these settings the fraction of ionized calcium changes without affecting total calcium. Ionized Ca^{2+} concentrations also change inversely with variations of serum anions. For instance, elevation of phosphate, citrate, sulfate, or bicarbonate increases the serum-free Ca^{2+} secondary to augmented formation of calcium complexes (Walser, 1973).

Symptoms of hypocalcemia vary in relation to the ionized serum calcium concentration. Mild reductions of plasma calcium are associated with paresthesia and muscle cramps; more severe decreases of calcium may induce seizures. Hypercalcemia, on the other hand, has been implicated in the attenuation of the renal effects of PTH, the antidiuretic action of vasopressin, and reduced renal concentrating capacity (Gill and Bartter, 1961; Takaichi and Kurokawa, 1986). The calcium concentration in the extracellular fluid represents a dynamic balance between intestinal absorption, renal reabsorption, and skeletal resorption. Assuming a daily dietary calcium intake of 1000 mg, net intestinal absorption amounts to about 200 mg with the remaining 800 mg excreted in the feces. When in balance, net intestinal absorption is matched by urinary excretion, while calcium accretion and loss from bone are equal. Thus, approximately 200 mg of calcium are excreted daily. In adults, net calcium balance is effectively zero, suggesting that in the absence of a calcium challenge such as lactation, the kidneys represent the dominant regulatory site of calcium metabolism (Peacock et al., 1969).

Phosphate chemistry

Terrestrial mammals are characterized by their avidity for calcium and equally keen mechanisms to eliminate phosphorous. Elemental phosphorous (P) exists in organic and inorganic (Pi) forms. Organic forms include phospholipids and various organic esters. The bulk of phosphate in extracellular fluid exists as the inorganic forms, Na_2HPO_4 and NaH_2PO_4. The Henderson relation determines the ratio of the two:

$$\frac{HPO_4^{2-}}{H_2PO_4^-} = 10^{pH-pKa} \tag{27.1}$$

The dissociation constant, pKa, for phosphate is 6.8. Thus, at pH 7.4, the ratio is essentially 4:1, and the plasma phosphate thus has an intermediate valence of 1.8.[2]

Serum phosphate

Serum phosphate is generally expressed as milligrams per deciliter because concentration in millimolar units can vary with acid–base status as outlined previously. Serum Pi averages between 3 and 4.5 mg/dL and decreases with age (Fig. 27.3B). As with calcium, serum Pi circulates in both free and protein-bound forms. However, some 90% of Pi, whether ionized or complexed with Ca, Mg, or Na, is filtered at the glomerulus. Although it is commonly stated that Pi is freely and completely filtered at the glomerulus, this is incorrect. Ten percent of plasma Pi is bound to protein and not filtered. However, the reduction in ultrafilterable Pi is counterbalanced by an opposite Gibbs–Donnan effect that raises the Pi concentration in the ultrafiltrate. Correcting the volume occupied by plasma proteins (7%) raises the ultrafiltrate Pi concentration by an additional 7.5%, mitigating the reduction of protein-bound Pi. Indeed, both in vivo and in vitro measurements show that the ratio of ultrafilterable Pi to total plasma Pi is close to unity (Harris et al., 1974). Ultrafilterable Pi decreases with the elevation of Ca or Pi, presumably because of formation of high-molecular-weight protein complexes.

Parathyroid hormone actions on mineral-ion homeostasis

By regulating renal tubular absorption of calcium and phosphate and the synthesis of $1.25(OH)_2D_3$, PTH controls the urinary excretion and intestinal absorption of these mineral ions and thereby plays a prominent role in setting their blood levels. Renal tubular responses to PTH deficiency, PTH or PTHrP excess, and defects in PTHR function lead to alterations in blood calcium, phosphate, or $1.25(OH)_2D_3$ that are the hallmarks of numerous clinical disorders, described later in this volume. This chapter reviews current understanding of the mechanisms whereby PTH and PTHrP control renal tubular epithelial function. Because PTH and PTHrP actions in kidney and bone are mediated by the PTHR, its expression, signaling, and trafficking are presented first as a foundation for understanding hormone actions. Although PTH and PTHrP

2. The net valence is calculated from the total number of negative charges, $2 \times 4 + 1$, divided by the number of molecules, $4 + 1$, as determined by the Henderson relation, viz., $9/5 = 1.8$. Thus, at pH 7.4, 1 mmol of phosphorous = 1.8 mEq.

receptors distinct from the PTHR have been described (see Chapter 28), their role, if any, in normal renal physiology is unknown. While not unequivocally proven in each case, it is likely that the effects of PTH and PTHrP described here are mediated by the canonical PTHR.

Parathyroid hormone receptor expression, signaling, and regulation in the kidney

Renal PTHR in is expressed on tubular epithelial cells and vascular endothelial cells. PTHR expression along the nephron is associated with regulatory actions on mineral-ion homeostasis, whereas expression on capillary endothelial cells and glomerular podocytes mediates PTH effects on vascular tone and GFR.

The response to PTHR activation observed in individual renal cells depends upon several factors including (1) receptor location and abundance; (2) the expression of cell-specific adapter proteins that modify PTHR signaling; (3) the array of PTHR-inducible genes; (4) effector proteins including enzymes, ion channels, and transporters; (5) the local concentration of PTH or PTHrP; (6) exposure to other agents that heterologously regulate PTHR function; and (7) the history of recent exposure to PTH or PTHrP, which affects the state of receptor desensitization.

Parathyroid hormone receptor expression

The prominent actions of infused PTH on phosphate and calcium absorption pointed to proximal and distal tubules, respectively, as the principal sites of PTHR expression. These deductions were confirmed as molecular biological tools and antibodies became available. We now recognize PTHR expression at the glomerulus, proximal convoluted and straight tubules, cortical ascending limbs, and distal convoluted tubules. Unexpectedly, detailed examination revealed PTHR localization in proximal tubules on both basolateral and apical cell membranes (Amizuka et al., 1997; Ba et al., 2003; Kaufmann et al., 1994). Possible roles of apical membrane PTHRs are discussed later.

Considerable evidence supports the bilateral expression of PTHR on apical and basolateral surfaces of proximal tubules. Moreover, PTH actions are asymmetrical (Quamme et al., 1989; Reshkin et al., 1990, 1991). Such biased actions suggest differential PTHR coupling to G proteins on apical and basolateral membranes; the presence of adapter proteins such as the PDZ proteins NHERF1 and SCRIBBLE; or A-kinase-anchoring proteins that modify or specify second messenger signaling. The possibility of differential G protein expression seems unlikely because G_s is abundant in brush border membrane vesicles (Brunskill et al., 1991; Stow et al., 1991; Zhou et al., 1990). Interestingly, inhibitory G_i isoforms are found only on apical and not basolateral proximal tubule cells (Stow et al., 1991). Nonetheless, PTH triggers greater cAMP formation when added to the serosal than to the mucosal compartment of cells grown on filter barriers (Reshkin et al., 1991). This finding is compatible with other results showing that apical membrane receptors appear not to be coupled tightly, if at all, to adenylyl cyclase (Kaufmann et al., 1994; Shlatz et al., 1975).

Apical membrane expression of PTHR in proximal tubules was unexpected. The conspicuous presence of apical PTHR raises the question of the nature and function of PTH peptides in the urine. Full-length PTH(1−84) has a molecular weight of 9.4 kDa, is filtered at the glomerulus, and is found in urine (Bethune and Turpin, 1968; Norden et al., 2001). The role of urinary PTH fragments has largely been ignored because of the absence of a conceptual framework meriting their examination, the attendant problems of bioassay and immunometric PTH determinations that have been applied to these assays, and the overriding view that PTHR functionality stems from its expression at basolateral surfaces facing the vasculature. The demonstrable effects of PTH peptides applied to the apical surface of isolated kidney tubules or cultured proximal tubule cells, however, supports the view that PTH may differentially regulate phosphate transport from apical and basolateral surfaces (Kaufmann et al., 1994; Reshkin et al., 1990; Traebert et al., 2000)}.

The *PTHR* gene harbors multiple promoters and 5′-untranslated exons and is thereby capable of generating various transcripts by alternative promoter usage and different RNA splicing patterns (Amizuka et al., 1997; Bettoun et al., 1998; Jobert et al., 1996; Joun et al., 1997; McCuaig et al., 1995). P1 promoters in mouse and P3 in human seem to be used exclusively in kidney cells, whereas the P2 promoter is employed to generate PTHR mRNAs that are widely expressed in extrarenal tissues and organs (Amizuka et al., 1997; Bettoun et al., 1998; Joun et al., 1997). It is presently unknown whether these differences simply reflect opportunities for tissue-specific gene regulation or lead to expression of structurally different forms of the PTHR (Jobert et al., 1996; Joun et al., 1997).

Parathyroid hormone receptor signal transduction in kidney tubular cells

PTH administration promptly increases urinary excretion of cAMP, referred to as nephrogenous cAMP. This effect has been employed clinically to distinguish between primary and secondary hyperparathyroidism, wherein nephrogenous cAMP is elevated in primary but not secondary hyperparathyroidism due to impaired kidney function in the latter condition (Broadus, 1979; Llach and Massry, 1985). Practically speaking, cAMP formation in the nephron originates exclusively from proximal tubules because of the extensive mass of this portion of the nephron compared with other nephron segments. Further, although PTH-induced nephrogenous cAMP is a robust index of the phosphaturic action of PTH, it fails to disclose the effects of PTH on calcium absorption by distal tubules. This too stems from the profusion of proximal tubules compared with distal tubules, possible differences in receptor abundance on the two cell types, the presence of modifying proteins that alter the coupling of the PTHR to G_s, and the signaling array employed.

PTH stimulates adenylyl cyclase with attendant formation of cAMP and activation of PKA in kidney epithelial and vascular cells. PLC may also or alternatively be activated with consequent generation of inositol phosphates and diacylglycerol, release of intracellular calcium, and activation of PKCs, phospholipase A_2, and phospholipase D (PLD). Other signaling mechanisms may participate in mediating PTHR actions in renal cells. For example, PTH-induced stimulation of MAPKs in renal epithelial cells may be triggered by GPCRs, proceeding by activation of nonreceptor tyrosine kinases and transactivation of EGF receptors. MAPK, through PTHR and the FGFR1 fibroblast growth factor receptor, participates in mediating PTH and FGF23 effects, respectively, on phosphate transport (Sneddon et al., 2016). In heterologous cell expression systems, PTHR activation of MAPK occurs by transactivation and PTHR internalization (Syme et al., 2005).

Different patterns of PTHR signaling are present along the nephron. For example, PTH provokes rapid and transient elevations of Ca_i^{2+} in proximal tubule cells (Filburn and Harrison, 1990; Friedman et al., 1999; Hruska et al., 1986, 1987; Tanaka et al., 1995). This response is characteristic of PLC activation, as opposed to calcium influx that produces a more sustained increase of intracellular calcium, which results from the formation of inositol trisphosphate and calcium release from endoplasmic reticulum. Distal tubule cells, in contrast, exhibit a delayed and sustained elevation of Ca_i^{2+} in response to PTH. This effect is due to apical membrane Ca^{2+} entry (Bacskai and Friedman, 1990; Hoenderop et al., 1999). Furthermore, PTH-stimulated calcium transport in distal tubule cells involves PLC-independent PKC activation (Friedman et al., 1996), which is mediated by phospholipase D (Garrido et al., 2009).

More recent characterization of PTHR signaling offers additional and alternative interpretations of cAMP signaling and function. Refined analysis of PTHR activation in single cells using FRET revealed that PTH but not PTHrP promoted sustained elevation of cAMP. Similar findings subsequently were described for other GPCRs (Calebiro et al., 2009; Feinstein et al., 2013; Inda et al., 2016; Kuna et al., 2013; Merriam et al., 2013; Pavlos and Friedman, 2017). In the case of PTHR, sustained cAMP signaling originating at basolateral surfaces may induce 25-hydroxyvitamin D-hydroxylase (*CYP27B1*) transcription, CREB phosphorylation, and 1.25$(OH)_2$vitamin D_3 formation.

Regulation of parathyroid hormone receptor signaling in tubular epithelial cells

As described before, the expression of PTHRs on the surface of kidney cells is controlled by the interaction with adaptor proteins. However, PTHR expression is also controlled by the rate of PTHR gene transcription, though current understanding of this process is incomplete. Hypoparathyroidism, induced by parathyroidectomy or dietary phosphate depletion, upregulates PTHR mRNA levels in rat renal cortex (Kilav et al., 1995a). However, high concentrations of PTH have no detectable effect on PTHR mRNA (Kilav et al., 1995b; Ureña et al., 1994). Renal PTHR mRNA expression is reduced in rats with renal failure, but this apparently is due to some aspect of uremia or renal disease (Disthabanchong et al., 2004) other than secondary hyperparathyroidism per se, as it is not prevented by parathyroidectomy (Largo et al., 1999; Urena et al., 1994; Ureña et al., 1995). In rats with secondary hyperparathyroidism due to vitamin D deficiency, renal cortical PTHR mRNA levels were found to be twice as high as normal and could not be corrected by normalizing serum calcium (Turner et al., 1995). These results suggest that vitamin D impairs PTHR gene transcription in proximal tubules. However, in immortalized distal convoluted tubule cells, PTHR expression was upregulated severalfold by 1.25$(OH)_2D_3$ (Sneddon et al., 1998). In opossum kidney (OK) cells, TGFβ1 diminished PTHR mRNA expression (Law et al., 1994). The physiologic significance of this effect has not been clarified. PTHR mRNA expression was not affected by the mild secondary hypoparathyroidism induced by ovariectomy in rats nor by subsequent estrogen treatment (Cros et al., 1998).

A critical insight to PTHR regulation was achieved with the discovery that the receptor interacts with the cytoplasmic scaffolding proteins Na/H exchanger regulatory factors NHERF1 NHERF2 (NHERF1/2; known also as EBP50 and E3KARP, respectively), which govern certain aspects of signaling and trafficking (Mahon et al., 2002). NHERF1/2 consist of two tandem PSD95/Discs Large/ZO-1 (PDZ) domains and an EBD. The PTHR binds to either PDZ domain through a recognition sequence at its carboxy-terminus. This Glu-Thr-Val-Met sequence corresponds to a canonical PDZ ligand that takes the form D/E–S/T-X-Φ, where X is any amino acid and Φ is a hydrophobic residue—generally L/I/V, but it can also be M (Broadbent et al., 2017). Mutation of any residue of the PDZ sequence of the PTHR, other than the permissive position, abrogates interaction with NHERF1/2 (Mahon et al., 2002; Sneddon et al., 2003). In pivotal work, Mahon and Segre found that NHERF1/2 switched PTHR signaling from adenylyl cyclase to PLC. NHERF1 also regulates mitogen-activated ERK signaling (Wang et al., 2008). NHERF1 is abundantly expressed on proximal tubule apical membranes (Wade et al., 2001). Mouse proximal tubules expresses both NHERF1 and NHERF2 (Wade et al., 2003). NHERF1 is strongly expressed in microvilli, whereas NHERF2 is detected only weakly in microvilli. However, it is expressed predominantly at the base of the microvilli in the vesicle-rich domain. Notably, neither NHERF1 nor NHERF2 is found in distal tubules. Human and mouse kidneys exhibit comparable NHERF1 localization (Shenolikar et al., 2002; Wade et al., 2001, 2003; Weinman et al., 2002). As will be discussed in greater detail later, NHERF1 is expressed in the proximal nephron, where it importantly regulates PTH-dependent phosphate absorption.

Calcium absorption and excretion

The kidneys are responsible for controlling extracellular calcium balance by controlling the amount of calcium retained by the nephron or excreted in the urine. Renal calcium regulation occurs by a series of sequential events as the incipient urine passes through the nephron. The bulk of the filtered calcium is absorbed by proximal tubules, with progressively smaller fractions recovered as the urine passes through downstream tubule segments. Most calcium absorption is not subject to hormone regulation because it proceeds in upstream sites through passive mechanisms that are not regulated by hormones. Hormonal and pharmacological regulation of calcium absorption is achieved by fine-tuning the final few percent of calcium transport in distal nephron segments.

PTH participates importantly in maintaining extracellular calcium homeostasis. Early observations in animals and patients with hypo- or hyperparathyroidism clearly implicated renal calcium handling in PTH abnormalities (Carney, 1996; Gley, 1893; Hackett and Kauffman, 2004; Sandström, 1879–1880).

Renal calcium absorption

PTH reduces renal calcium excretion. Although calcium is absorbed throughout the nephron, the calcium-sparing action of PTH occurs primarily in distal tubule (Friedman, 2008; Sutton and Dirks, 1975). The sites of renal calcium absorption and mechanism of PTH action are summarized here. Extensive descriptions of this older work are available elsewhere (Friedman, 2008; Ko, 2017).

Approximately 20%–25% of the calcium filtered at the glomerulus is absorbed by medullary and cortical thick ascending limbs. In contrast to proximal tubules, calcium movement in thick ascending limbs occurs by parallel transcellular and paracellular calcium transport pathways. Passive paracellular calcium absorption is driven by the favorable lumen-positive transepithelial voltage. In this setting, the rate and magnitude of calcium transport are parallel and proportional to those of sodium. Physiological responses and interventions that enhance sodium absorption increase voltage, thereby augmenting calcium absorption (Hoover et al., 2015). Conversely, maneuvers that decrease sodium absorption concomitantly reduce calcium absorption. Such interventions typically involve the use of diuretics such as furosemide or bumetanide that inhibit sodium absorption by thick ascending limbs and increase calcium excretion. Less commonly, mutations of the Na–K–2Cl cotransporter (*SLC12A1*), as well as the ROMK apical K^+ channel and basolateral ClC–K2 basolateral Cl^- channel associated with the different forms of Bartter's syndrome, are accompanied by hypercalciuria, underscoring the parallel nature of sodium and calcium absorption by thick ascending limbs. PTH stimulates active calcium absorption by cortical thick limbs (Friedman, 2000).

Distal convoluted tubules display a third pattern of transepithelial calcium recovery. Two important features characterize this process (Fig. 27.5). First, calcium transport occurs only by a transcellular mechanism and second, calcium and sodium movement are inversely related. Here, decreased sodium absorption enhances calcium recovery; likewise, increased sodium absorption is accompanied by diminished calcium transport. This inverse relationship has important consequences for calcium economy. For example, thiazide diuretics, which increase sodium excretion, diminish that of calcium. Patients with Gitelman's syndrome further exemplify the inverse association of sodium and calcium movement. This inherited disorder is due to inactivating mutations of the Na–Cl cotransporter (NCC; *SLC12A3*) that mediates sodium absorption by distal convoluted tubules. Patients (Bettinelli et al., 1992; Gitelman et al., 1966) and experimental animals (Schultheis et al., 1998) with Na–Cl cotransporter mutations exhibit characteristic natriuresis accompanied by hypocalciuria, thus exemplifying the inverse relations between sodium and calcium absorption in distal tubules.

Cortical and medullary collecting tubules contribute minimally if at all to renal calcium conservation, and PTH exerts no detectable action on calcium movement by these nephron sites (Bernardo and Friedman, 2013; Friedman, 2014).

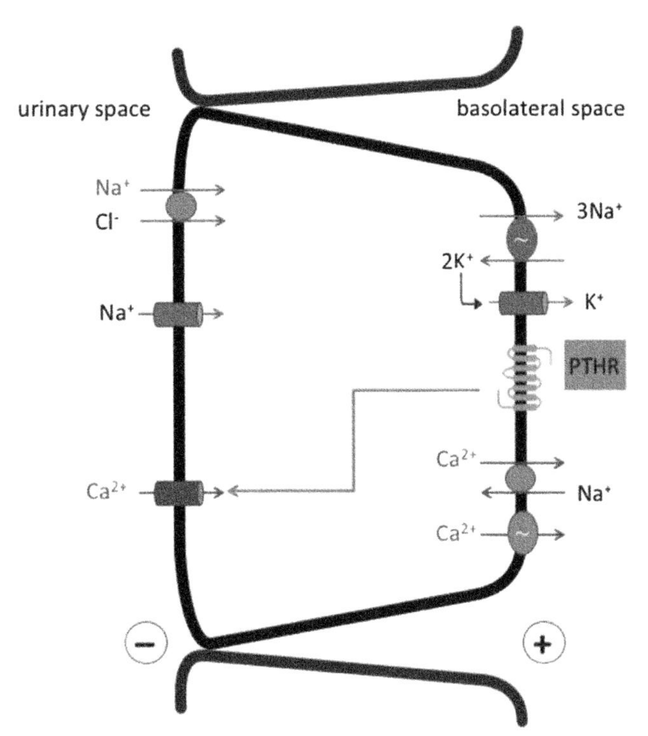

FIGURE 27.5 Calcium absorption by distal convoluted tubules. Two features distinguish distal tubule calcium absorption. First, calcium transport proceeds exclusively through a cellular pathway. Because of the substantial adverse electrochemical barrier, calcium movement is negligible in the absence of PTH (Costanzo and Windhager, 1980). Second, calcium and sodium movement in distal convoluted tubules are inversely related. Thus, thiazide diuretics, which block sodium transport, augment calcium absorption (Gesek and Friedman, 1992). The cell model shows two apical membrane sodium entry mechanisms, the thiazide-diuretic-inhibitable NaCl cotransporter and the amiloride-blockable epithelial Na channel, ENaC. Amiloride also stimulates calcium transport by distal tubule cells (Friedman and Gesek, 1995). Calcium enters the cell across apical membranes through dihydropyridine-sensitive calcium channels. PTHR expressed on basolateral membranes stimulates apical membrane calcium entry that is accompanied by efflux across basolateral membranes. Basolateral calcium efflux is mediated by the plasma membrane Ca^{2+}-ATPase (shown as the energy-dependent [~] process), and the Na^+/Ca^{2+} exchanger as discussed in detail in the text and elsewhere (Ko, 2017; Moe, 2016).

Parathyroid hormone regulation of renal calcium absorption

PTH is the most important hormone regulating renal tubular calcium transport. PTH increases renal calcium reabsorption and lowers urinary calcium excretion. In the complete absence of the hormone or in PTH deficiency states, diminished tubular reabsorption is accompanied by frank hypercalciuria. The stimulatory effect of PTH on calcium transport is limited to distal nephron segments. Paradoxically, PTH decreases calcium absorption by proximal tubules; the calcium-sparing action is due to its effects on distal convoluted tubules, and to a limited extent actions on cortical thick ascending limbs. In some species, the calcium-sparing action proceeds in connecting tubules.

Parathyroid hormone effects on proximal tubule calcium transport

The primary action of PTH on proximal tubule mineral-ion transport is to inhibit Na^+-dependent phosphate (Na^+-Pi) absorption (see below). It is uncertain whether PTH exerts significant or even specific actions on proximal calcium transport. PTH has been reported to increase (234), decrease (5), or not change (150, 251) proximal tubule calcium absorption. Studies showing that PTH alters intracellular free calcium, $[Ca^{2+}]_i$, irrespective of whether it increased or decreased (123, 146, 179, 213, 214, 368), should not be construed as evidence for an effect of PTH on calcium entry, much less on net calcium absorption. Although the aforementioned findings may be associated with net calcium absorption, they may also represent the aggregate of intracellular calcium release, and perhaps individual calcium entry and exit steps that support constitutive cellular activities and intracellular homeostasis. The reason for this is that PTH exerts a host of biochemical effects on proximal tubules such as activation of 25-hydroxyvitamin D-1α-hydroxylase, gluconeogenesis, or ammoniagenesis as well as transport effects on Na^+-Pi absorption, Na^+/H^+ exchange, and Na^+/Ca^{2+} exchange that may involve calcium signaling but do not entail effects on net calcium absorption.

Parathyroid hormone effects on distal tubule calcium transport

Substantial evidence supports the view that PTH increases active transcellular calcium absorption by distal convoluted tubules (Ko, 2017; Moe, 2016). The distal action of PTH may be accompanied by decreased sodium absorption and diuresis (Ko et al., 2011).

PTH potently induces apical calcium entry in distal convoluted tubule cells (165). PTH-stimulated calcium entry in distal convoluted tubules is mediated by the TRPV5/Trpv5 calcium channels (van Goor et al., 2017b). This action is mediated by PKA (de Groot et al., 2009). PKA-induced phosphorylation of TrpV5 Thr^{709} increases open-channel probability (de Groot et al., 2009). Other studies point to a role (Hoenderop et al., 1999) or requirement for combined PKA and PKC action (Friedman et al., 1994).

PTH may further dissociate calcium and sodium absorption by distal convoluted tubules. PTH depresses Slc12a3, the electroneutral NaCl cotransporter, and its surface expression (Ko et al., 2011). Thus, reducing Na^+ entry while enhancing calcium absorption supports the inverse relation between basal and hormone-stimulated calcium transport.

Phosphate excretion

Extracellular phosphate homeostasis is regulated primarily by controlling its renal excretion, where it is absorbed mostly by proximal tubules. The prompt phosphaturia caused by PTH was one of the earliest recognized and best studied actions of PTH (Albright et al., 1929; Greenwald and Gross, 1925). Fibroblast growth factor 23 (FGF23) plays an equally important role in phosphate balance. FGF23 is discussed in Chapter 65 and its renal actions reviewed elsewhere (Erben, 2017; Hu et al., 2013). One point worth emphasizing here is that although PTH and FGF23 exert parallel inhibitory actions on renal phosphate absorption, they display opposing effects on vitamin D metabolism, with PTH increasing vitamin D levels and FGF23 lowering them. The underlying basis for these diverse actions is not known.

NPT2a (SLC9A3R1) and NPT2c (SLC9A3R3) mediate renal Na-Pi cotransport[3] (Table 27.2). Npt2a is found throughout the proximal tubule, with expression decreasing from S1 to S2 and then to S3 segments (Picard et al., 2010; Segawa et al., 2007). Npt2c is found only in the S1 portion of proximal tubules. NPT2b is expressed by intestines, where it participates in phosphate absorption.

Mechanisms of proximal tubular phosphate absorption

Apical phosphate entry is rate-limiting for cellular phosphate absorption and is the major site of its regulation. Phosphate enters the cell across apical brush border membranes against a steep electrochemical gradient established primarily by the strongly negative intracellular voltage. This electrochemical barrier is overcome by NaP_i cotransporters that couple phosphate influx to the favorable dissipative entry of Na^+ (Fig. 27.4).

Type I and type IIa NaP_i cotransporters are localized on apical brush border membranes of proximal convoluted tubule cells. Nonspecific type I transporters exert a variety of functions unrelated to phosphate absorption and are not regulated by PTH or FGF23. Type II cotransporters are structurally and genetically distinct from type I transporters, sharing only 20% identity. As alluded to above, NPT2a and NPT2c are expressed in kidney proximal tubules, whereas NPT2b, an isoform of Npt2a, is expressed exclusively in intestine (Hilfiker et al., 1998). Type III NaP_i cotransporters, originally identified as cell surface virus receptors Glvr-1 and Ram-1, are widely expressed within and outside the kidney (Kavanaugh and Kabat, 1996). Type III cotransporters, found on basolateral cell membranes, are regulated by extracellular phosphate deprivation and PTH (Ohkido et al., 2003; Segawa et al., 2007). Type II cotransporters may also be expressed by distal convoluted tubule cells and may thus be involved in phosphate absorption by both proximal and distal nephrons (Collins et al., 2004; Tenenhouse et al., 1998). Npt1, Npt2a, and Npt3 account for 15%, 84%, and 0.5% of Na-Pi-cotransporter mRNAs (Murer et al., 2003). Type II cotransporters are 80–90-kDa glycoproteins predicted to span the membrane eight times with both their amino and carboxyl termini in the cytoplasm (Murer and Biber, 1997). Recent work shows two variant transcripts of the mouse Npt2a gene, Npt2a-v1 and Npt2a-v2 (Yamamoto et al., 2005). These are characterized by the presence of alternative first exons (either exon 1A or exon 1B). Npt2a-v2 is important for $1.25(OH)_2D_3$-dependent renal cell-specific activation. Npt2a cotransporters are electrogenic and transport Na^+ and $H_2PO^-_4$ at a molar ratio of 3:1 (Murer and Biber, 1997). Expression and activity of NPT2a/Npt2a are strongly regulated by parathyroid status and dietary phosphate (and FGF23) (Keusch et al., 1998; Lotscher et al., 1999; Pfister et al., 1997; Ritthaler et al., 1999; Takahashi et al., 1998). PTH regulation of serum phosphate and tubular NaP_i absorption is lost in mice lacking Npt2a (Zhao and Tenenhouse, 2000).

3. Human genes and gene products are designated by uppercase italics, whereas nonhuman forms are shown in lowercase.

TABLE 27.2 Sodium phosphate transporter family.

Type	GenBank	Function	Reference
NPT1	*SLC17A1*	Modulator of organic and inorganic anion transport	Busch et al. (1996)
NPT2A	*SLC34A1*	3:1 electrogenic BBM proximal tubule Pi uptake	Murer et al. (2000)
NPT2B	*SLC34A2*	Intestinal Pi transport	Hilfiker et al. (1998)
NPT2C	*SLC34A3*	2:1 electroneutral BBM proximal tubule Pi uptake	Bacconi et al. (2005); Bergwitz et al. (2006); Madjdpour et al. (2004); Segawa et al. (2002); Tenenhouse et al. (2003)
NPT3	*SLC20A1*, *SLC20A2*	Pi transporter in bone. Broadly expressed.	Suzuki et al. (2001, 2006)

Npt2c/NPT2c is expressed in rat and human kidneys (Ohkido et al., 2003). Npt2c accounts for approximately 30% of Na/Pi cotransport in kidneys of phosphate-deprived adult mice (Ohkido et al., 2003). Npt2a activity is the principal mechanism whereby PTH controls phosphate absorption in proximal convoluted tubules, at least in rodents. In humans, however, NPT2c plays an important role in phosphate economy. NPT2c mutations are associated with hereditary hypophosphatemic rickets with hypercalciuria, kidney stones, and nephrocalcinosis (Christov and Juppner, 2013; Dasgupta et al., 2014).

Parathyroid hormone regulation of renal phosphate absorption

PTH promptly elevates urinary phosphate elimination. Early studies revealed that this action is involved reduction of Na-Pi cotransport and that restoration of maximal phosphate absorption required microtubules and new protein synthesis (Dousa et al., 1976; Malmström and Murer, 1987).

Functional and immunohistochemical analyses of Npt2 in intact kidneys and cultured cells revealed that PTH induces rapid (15 min) removal of Npt2 from the apical membrane into the subapical endocytic apparatus, followed by microtubule-dependent delivery to lysosomes and proteolytic degradation (Bacic et al., 2006; Kempson et al., 1995; Keusch et al., 1998; Lötscher et al., 1999; Pfister et al., 1997; Zhang et al., 1999). It had been assumed that PTH inhibited Na-Pi cotransport through phosphorylation or other action that would control the kinetics or driving force. This turned out not to be the case. Instead, PTH and FGF23 inhibit Na-Pi cotransport by initiating internalization and downregulation of Npt2a. Though Npt2a phosphorylation apparently is not involved, PTH and FGF23 lead to phosphorylation of NHERF1 (Andrukhova et al., 2012; Deliot et al., 2005; Weinman et al., 2007). It is also worth noting that PTH and FGF23 maximally inhibit only 50%–60% of phosphate transport. It is not known if the balance is subject to hormonal or other regulatory control.

Hypoparathyroidism and hypophosphatemia increase PTHR, Npt2a mRNA, and protein, showing the proteins mediating PTH effects on renal phosphate transport that are coordinately regulated (Kilav et al., 1995a). PTH does not acutely reduce Npt2 mRNA expression, although parathyroidectomy leads to a severalfold increase of apical Npt2a protein and mRNA (Kilav et al., 1995a; Saxena et al., 1995; Takahashi et al., 1998).

Parathyroid hormone receptor signal transduction in the regulation of calcium and phosphate excretion

As described earlier and discussed in Chapter 28, the PTHR mediates its actions through multiple signaling pathways. Signaling is both ligand- and cell type-specific. PTH exerts physiologically distinct and temporospatially separated effects on the kidney. In proximal tubules, PTH inhibits Na-Pi cotransport and Na/H exchange, increases Na^+/Ca^{2+} exchange (Fig. 27.4), and activates CYP27B1 while suppressing 24-hydroxylase activity to increase circulating vitamin D levels. PTH also inhibits Na^+,K^+-ATPase activity, stimulates gluconeogenesis and ammoniagenesis. Complex but precise individual control over these multiple physiological activities assures that all are not simultaneously engaged in response to changing levels of PTH. Just how this is accomplished is far from understood. However, important insights have been gained into PTH-dependent regulation of renal calcium and phosphate transport by identifying cytoplasmic scaffolding proteins that specifically modulate individual PTH actions (Mahon, 2012; Pavlos and Friedman, 2017).

Parathyroid hormone signaling of renal calcium transport

The primary action of PTH is to stimulate renal calcium conservation. This related process proceeds principally in distal tubules. The portion of the distal tubule where this occurs differs somewhat between species. In most species including humans, the rat, the mouse, and the dog, distal convoluted tubules are the primary site of PTH-dependent calcium absorption. This has been demonstrated directly or can be inferred from the localization of the PTHR or the presence of PTH-stimulated adenylyl cyclase activity (Chabardès et al., 1980; Friedman and Gesek, 1993). In the rabbit, connecting tubules are the site of PTH-stimulated calcium transport (Shimizu et al., 1990).

Ca^{2+} entry across apical cell membranes of distal convoluted tubule cells is mediated by TrpV5/TRPV5 calcium channels (Mensenkamp et al., 2007; Na and Peng, 2014). Although activating PKA purportedly accounts for PTH actions on calcium absorption (de Groot et al., 2009), we think it likely that this action involves both PKA and PKC insofar as activation of omnipresent β_2-adrenergic receptors, which are well expressed by distal tubules, fails to augment calcium recovery (Gesek and White, 1997). This suggests that PTH-sensitive calcium transport requires activating both PKA and PKC. Such effects have been reported (Cha et al., 2008; de Groot et al., 2009; Hoover et al., 2015). Some interspecies differences of PKA-sensitive PTH activation of TRPV5 have been noted between rabbit and rodent and human, which behave similarly (van Goor et al., 2017a).

Cellular calcium efflux is mediated by the NCX1 Na^+/Ca^{2+} exchanger and PMCA1 and PMCA4 plasma membrane Ca^{2+}-ATPases (van der Hagen et al., 2014; van Loon et al., 2016), which are expressed on distal tubule cell basolateral membranes (Magyar et al., 2002; White et al., 1997). Although opinions differ regarding the relative contribution of NCX1 and PMCA1/4 to net calcium absorption, we think it likely that Ca^{2+}-ATPase mediates the bulk of Ca^{2+} efflux because the electrochemical driving force for Ca^{2+} extrusion by Na^+/Ca^{2+} exchange is insufficient to account for more than basal calcium transport levels (Friedman, 1998). How these two processes are regulated and synchronized so as to coordinate apical calcium entry with basolateral extrusion is unknown.

Parathyroid hormone signaling of renal phosphate transport

Early experiments in vivo or with isolated renal membranes indicated that regulation of phosphate absorption was mediated by cAMP. The evidence marshaled in support of this conclusion was based, inter alia, on the observation that cAMP analogs and phosphodiesterase inhibitors mimicked the phosphaturic action of PTH (Agus et al., 1973; Coulson and Scheinman, 1989; Gmaj and Murer, 1986; Hammerman, 1986). Indeed, the excretion of urinary cAMP in response to PTH administration, so-called nephrogenic cAMP, reflects virtually exclusively the action of PTH on proximal tubule phosphate absorption (and not distal tubule Ca absorption) and is a direct index of PTHR sensitivity or refractoriness (Besarab and Swanson, 1984). Although these conclusions are solid, the mechanisms by which PTH inhibits renal phosphate transport are complex and incompletely resolved. The intricacy stems from the recognition that the PTHR activates multiple signaling paths. The possibility that differentially expressed PTHR types (Pun et al., 1988) or isoforms account for the heterogeneity of action can be excluded. The first studies profiling the effect of PTH on inositol phosphate and Ca^{2+}-signaling, in fact, preceded the cloning of the PTHR (Bidot-López et al., 1981; Meltzer et al., 1982). PTH stimulation of inositol phosphate and diacylglycerol accumulation is independent of cAMP formation (Hruska et al., 1987) and occurs in parallel (Friedman et al., 1996).

The involvement of specific PTHR-generated signals in regulating Na-P_i cotransport has been pursued extensively in vitro using immortalized OK cells, which exhibit a proximal tubule-like phenotype that includes expression of the Npt2a NaP_i cotransporter and the PTHR (Jüppner et al., 1991; Rabito, 1986). Considerable evidence supports cAMP/PKA-mediated signaling, while other observations point to an important contribution of inositol phosphates/PKC. The extensive body of work on this subject has been reviewed elsewhere (Muff et al., 1992; Murer et al., 2000; Pfister et al., 1999) and described in the previous edition of this text (Bringhurst and Strewler, 2002). Recent incisive work and the discovery of cytoplasmic binding partners for the PTHR and Npt2a have largely reconciled the apparently disparate views. The salient features of PTHR signaling of proximal tubule phosphate transport can be summarized as follows.

PTHRs, as noted earlier, are expressed at both apical and basolateral surfaces of proximal tubule cell membranes (Amizuka et al., 1997; Ba et al., 2003; Traebert et al., 2000). Both PTH(1−34) and PTH(3−34), thought to be a relatively PKC-selective agonist, induced rapid removal of apical NaP_i proteins when selectively applied to apical surfaces of perfused murine proximal tubules or proximal-like OK cells (Pfister et al., 1999; Traebert et al., 2000). Only PTH(1−34) promoted Npt2a endocytosis when added at basolateral surfaces. Two conclusions can be drawn from these findings. First, they are consistent with the expression of PTHRs at both membrane domains. Second, they are compatible with the view that apical PTHR activates both PKA and PKC, whereas basolateral PTHRs couple primarily to PKA. The biological significance of these findings may relate to the dynamics of PTHR signaling and the duration of the phosphaturic action of PTH. Short-term, acute regulation of phosphate absorption is mediated mainly by the PKA pathway, whereas long-term, persistent actions are signaled by PKC (Guo et al., 2013; Nagai et al., 2011).

Independent studies uncovered a likely molecular basis for the polarized PTH action and asymmetric effects and signaling of PKC and PKA. The explanation stems from the localized expression of the adapter protein NHERF1. NHERF1 is a PDZ protein that binds to both the PTHR and Npt2a (Hernando et al., 2002; Mahon et al., 2002; Wade et al., 2003). In the absence of NHERF1, mice exhibit a phosphate-wasting phenotype (Shenolikar et al., 2002). NHERF1 is abundantly expressed at proximal tubule brush border membranes in the rat, and is also detected in the cytoplasm and on basolateral membranes (Wade et al., 2001). In mice, NHERF1 is found exclusively on proximal tubule brush border microvilli (Wade et al., 2003). In OK/H cells, which lack NHERF1, PTH fails to inhibit Na-Pi cotransport, and this defect can be surmounted by overexpressing NHERF1 (Mahon et al., 2003).

According to presently accepted views, PTHR signals through PKA in the absence of NHERF1 and by means of PKC in its presence (Mahon et al., 2002). Thus, at basolateral membranes, where NHERF1 is absent, PTH would be expected to signal through adenylyl cyclase and PKA. At apical (i.e., luminal or mucosal) cell membranes, signaling would be predicted to be mediated by PLC and PKC because of the coupling between the PTHR and PLC (Mahon and Segre, 2004). These predictions are compatible with the observations that PTH(1–34) evoked Npt2a sequestration when added to either cell surface, whereas PTH(3–34), which activates PKC but not PKA, induced Npt2a internalization only when applied to the luminal surface (Traebert et al., 2000). Consistent with this scheme, more recent studies reveal that PTH(3–34) was unable to promote Npt2a internalization from proximal tubule brush border membranes prepared from NHERF1-null mice (Capuano et al., 2006). Activation of PLC by PTH was impaired; bypassing this defect by directly activating PKC with a diacylglycerol analog caused Npt2a internalization in proximal tubules of both wild-type and NHERF1-null mice. Although these are elegant and compelling observations, their relevance to the regulation of phosphate transport under physiological and pathophysiological conditions is unsettled. Complicating this picture, metabolism of cAMP involves nucleotide degradation to adenosine by brush border membrane ectoenzymes such as ecto-5′-nucleotidase. PTH(1–34) augmented ecto-5′-nucleotidase activity in apical membranes of OK cells (Siegfried et al., 1995). This effect was mimicked by PTH(3–34) but not by forskolin. Both PTH fragments increased cytoplasmic Ca^{2+} and stimulated PKC activity. Conversely, stimulation of ecto-5′-nucleotidase activity was blocked by inhibitors of PKC. Thus, PKC-mediated stimulation of 5′-nucleotidase thereby bolsters the effects transduced by PKA.

PTHRs present on proximal tubule luminal membranes are presumably activated by PTH peptides that are filtered at the glomerulus and appear in the luminal fluid. Most of these peptides or smaller fragments are absorbed (Kau and Maack, 1977), internalized (Hilpert et al., 1999), and metabolized (Brown et al., 1991; Daugaard et al., 1994; Yamaguchi et al., 1994) during their passage through the nephron. However, given the increased generation and accumulation of amino-truncated PTH peptides during kidney disease and other pathological settings, it is likely that significant amounts of biologically active PTH reach luminal brush border membranes and would be expected to elicit Npt2a/NPT2a internalization with attendant elevation of urinary phosphate excretion. Such actions might represent an adaptive response to the elevation of plasma phosphate in chronic kidney disease. By contrast, in distal tubules, which lack NHERF1, no such effect would be anticipated.

Not covered here and only partially understood are the convergent actions of PTH and FGF23 on NPT2A and phosphate transport. PTHR and FGFR1, through distinct signaling pathways, both require NHERF1 for their inhibitory effects (Sneddon et al., 2016). The details of how a G-protein-coupled receptor and a receptor tyrosine kinase produce convergent effects to phosphorylate the same critical residue on NHERF1 remains to be defined.

Sodium and hydrogen excretion

Beyond its effects on mineral-ion metabolism, PTH is involved in acid–base balance through its actions on Na/H exchange in proximal tubules. Acute (Ellsworth and Nicholson, 1935; Hellman et al., 1965; Kleeman and Cooke, 1951; Nordin, 1960) or chronic (Hulter and Peterson, 1985) increases in PTH exert a prompt alkalotic response attended by elevated bicarbonate excretion. This action stems from PKA-dependent acute inhibition of NHE3. Chronic exposure to elevated PTH downregulates NHE3 expression (Lee et al., 2017). PTH actions on NHE3 involve and require NHERF1 (Cunningham et al., 2010). Sporadic case reports notwithstanding, primary hyperparathyroidism is generally not attended by consistent or remarkable changes in serum pH and is largely asymptomatic (Bilezikian, 2012; Coe, 1974).

PTH inhibits acidification by suppressing the NHE3, the Na^+/H^+ exchanger. This involves activating adenylyl cyclase, NHERF1, which serves as a scaffold to assemble PKA and NHE3, followed by phosphorylation and inhibition of NHE3

(Moe et al., 1995; Weinman et al., 1998, 2005). Notably, the inhibitory effect of PTH on proximal tubules does not necessarily result in increased bicarbonate excretion (Bank and Aynedjian, 1976; Puschett et al., 1976) and is not accompanied by appreciable reductions in serum bicarbonate (Hulter, 1985). These results suggest that distal tubule compensatory bicarbonate absorption (Bichara et al., 1986) and acidification by downstream tubule segments in response to PTH (Paillard and Bichara, 1989) mitigate the upstream actions of PTH. Secondary actions of PTH to mobilize bicarbonate from bone also contribute to the absence of an effect on systemic pH (Hulter and Peterson, 1985). More important than its modest effects on net bicarbonate balance or pH homeostasis, however, is the permissive action of bicarbonate to facilitate PTH-dependent distal tubule calcium absorption (Marone et al., 1983; Mori et al., 1992; Sutton et al., 1979), a primary function of PTH.

Na^+/H^+ exchange mediates most proximal tubule sodium absorption (Palmer and Schnermann, 2015). Although PTH inhibits Na^+/H^+ exchange and proximal sodium absorption, PTH has a negligible effect on net sodium excretion because of compensatory absorption by downstream tubule segments (Harris et al., 1979). Natriuretic actions of PTH, however, have been described in experimental animals (Bichara et al., 1986; Hellman et al., 1965) and humans (Jespersen et al., 1997).

The fundamental mechanism whereby PTH inhibits Na^+/H^+ exchange differs from its action on Na-Pi cotransport. Whereas PTH decreases Na-Pi cotransport by sequestering Npt2a from brush border membranes, PTH acutely regulates Na^+/H^+ exchange by PKA-dependent NHE3 phosphorylation (Hamm et al., 2008). Decreased NHE3 activity is followed several hours later by reduced brush border membrane NHE3 abundance (Fan et al., 1999). Details of this process and the involvement of NHERF1 are detailed in the next section.

Na^+/H^+ exchange is a dissipative—i.e., secondary active—transport process that depends on the continuous extrusion of Na across basolateral cell membranes. This is accomplished by the ATP-dependent Na^+,K^+-ATPase. Data showing that PTH inhibits the Na^+,K^+-ATPase have been published (Ribeiro and Mandel, 1992; Zhang et al., 1999). An inhibitory action of PTH on Na^+,K^+-ATPase would certainly contribute to, or explain, reduced Na^+/H^+ exchange. However, such an action would be expected to have profound effects on proximal tubule ion transport and all Na-coupled transport processes such as those mediating glucose (Na-glucose) and amino acid (Na-amino acid) cotransport, and perhaps Na-Pi cotransport. Moreover, the inhibitory action of PTH should extend to virtually all cells expressing PTHR including the distal tubule. In this setting, blockade of the Na^+,K^+-ATPase would be expected to reduce Na^+/Ca^{2+} exchange and decrease net calcium absorption, when in fact just the opposite occurs. The described actions of PTH seem remarkably specific for the proximal tubule Na^+,K^+-ATPase and for Na^+/H^+ exchange. A rational explication of these findings awaits further clarification.

Parathyroid hormone regulation of proximal tubular sodium and hydrogen excretion

Efficient bicarbonate absorption by proximal tubules requires NHERF1. Targeted disruption of NHERF1 virtually eliminates PTH- and forskolin-inhibited Na^+/H^+ exchange in proximal tubule cells (Cunningham et al., 2004). This defect could be restored by infecting the cells with an adenovirus containing NHERF1. Interestingly, cAMP generation and PKC activation were equivalent in proximal tubule cells from wild-type mice and NHERF1-null animals. This result seemingly contrasts with the paradigm of the PTH "signaling switch" advanced by Mahon and Segre (Mahon et al., 2002), where PTHR signaling proceeds through cAMP in the presence of NHERF1 but by PKC in its absence. However, mouse proximal tubules express both NHERF1 and NHERF2, and NHERF2 expression is normal in NHERF1-null mice (Cunningham et al., 2008; Shenolikar et al., 2002). Alternate PDZ proteins such as PDZK1 could also compensate for the lack of NHERF1 (Ardura and Friedman, 2011; Yeruva et al., 2015). Thus, it is entirely possible that NHERF2 compensates for the absence of NHERF1, binds the PTH receptor and directs appropriate signaling and cell functions. Notably, OK cells express NHERF1 but not NHERF2 (Wade et al., 2001). Rats too express only NHERF1 on proximal tubule brush border membranes. Thus, OK cells are more of a model for the rat proximal tubule than for that of humans or mice, where both NHERF1 and NHERF2 are expressed (Wade et al., 2003). Thus, results of studies conducted with OK cells demonstrating an absolute role of NHERF1 in PTHR mediated inhibition of Na-Pi cotransport (Mahon et al., 2003) may have fortuitously benefited from the absence of NHERF2.

The majority of PTH effects appear to be mediated by PKA, and to a lesser extent by PKC. However, recent evidence uncovered an alternative pathway that involves cAMP but utilizes exchange protein directly activated by cAMP (EPAC) rather than PKA as an effector (Fujita et al., 2002). Studies employing proximal tubule cells derived from wild-type and NHERF1-null mice revealed a role for EPAC in mediating Na^+/H^+ exchange (Murtazina et al., 2007).

Following secretion of H^+ in exchange for one Na^+ ion by the apical cell membrane Na^+/H^+ exchanger, OH^- is generated within the cell. This is converted to HCO_3^- which is extruded across basolateral cell membranes. This is accomplished by Na^+-dependent-HCO_3^- (NBC1-4, SLC4A4) cotransporters (Boron, 2006; Romero et al., 2004; Soleimani and Burnham, 2001). The possibility that PTH inhibits bicarbonate absorption by an action on basolateral HCO_3^- exit was uncertain and this action was attributed to an effect on the Na^+,K^+-ATPase. More recent studies reveal direct effects of PTH on NBC activity through multiple signaling pathways (Ruiz et al., 1996). This action requires NHERF1 (Bernardo et al., 1999; Weinman et al., 2001). NHE3 is a target for PKA-mediated phosphorylation (Zhao et al., 1999). A role for PKC on PTH-regulated NHE3 is uncertain. Apical PTHR stimulation preferentially activates PKC. PTH induces phosphorylation of multiple serine residues within the cytoplasmic tail of NHE3 (Collazo et al., 2000). Phosphorylation is maximal within 5 min and associated with a reduction in NHE3 activity but not surface expression.

By way of summary, PTH regulates Na^+/H^+ exchange primarily by an immediate inhibitory action that depends upon phosphorylation and requires the presence of NHERF1. PKA and PKC phosphorylate NHE3 (Lee et al., 2017). Following the initial inhibition of membrane-associated NHE3, the protein is withdrawn from the membrane but unlike NPT2A is not degraded or downregulated. Thus, following dephosphorylation the transporter likely is recycled to the plasma membrane.

Vitamin D metabolism

The biologically active form of vitamin D, $1.25(OH)_2D_3$, is synthesized by a cascade of multiorgan processes, the penultimate step of which is 1α hydroxylation that proceeds in proximal renal tubules and is regulated by PTH and the CaSR calcium-sensing receptor (Nolin and Friedman, 2018).

Parathyroidectomy reduces $1.25(OH)_2D_3$ synthesis, whereas PTH administration increases renal 1α-hydroxylase activity (Armbrecht et al., 2003; Fraser and Kodicek, 1973; Walker et al., 1990). Increased synthesis of *CYP27B1*, the $25(OH)D_3$ 1α-hydroxylase gene, results from kidney-specific regulated expression, where the promoter is rapidly induced in response to PTH (Christakos, 2017; Meyer et al., 2017)}. Positive actions of PTH on 1α-hydroxylase expression and $1.25(OH)_2D_3$ biosynthesis are counterregulated by the CaSR (Maiti and Beckman, 2007; Maiti et al., 2008).

PTH not only promotes $1.25(OH)_2D_3$ synthesis but also regulates its degradation. 24-hydroxylation of $1.25(OH)_2D_3$ is the first step in the catabolic inactivation of $1.25(OH)_2D_3$ (Carpenter, 2017; Zierold et al., 2003). It was thought that only proximal renal tubules expressed 24-hydroxylase. Distal tubules, however, take up $1.25(OH)_2D_3$, express the vitamin D receptor, and possess vitamin D-dependent calcium-binding proteins. 24-hydroxylase message, protein, and activity have been demonstrated in distal tubule cells (Yang et al., 1999) and human distal tubules (Blomberg Jensen et al., 2010). Whereas PTH reduces 24-hydroxylase activity in proximal tubule cells, PTH and cAMP augment 24-hydroxylase activity in murine distal tubule cells (Yang et al., 1999). Oppositely oriented actions of PTH on 24-hydroxylase activity suggest differential regulation of 24-hydroxylase expression in proximal and distal tubules.

The signaling pathways employed by the PTHR to increase $1.25(OH)_2D_3$ synthesis have been examined extensively in vivo and in vitro. Involvement of cAMP is suggested by the fact that the PTH effect can be mimicked by cAMP analogs, forskolin, and phosphodiesterase inhibitors (Armbrecht et al., 1984; Henry, 1985; Horiuchi et al., 1977; Korkor et al., 1987; Larkins et al., 1974; Rost et al., 1981; Shigematsu et al., 1986). PTH- or forskolin-stimulated transcriptional induction of 1α-hydroxylase occurred and was blocked by the PKA inhibitor H89 (Murayama et al., 1999). A role for PKC in regulating 1α-hydroxylation has been identified from studies of the effects of PTH on isolated proximal tubules (Janulis et al., 1992, 1993). PKC activation produced rapid (30–60 min) increases of $1.25(OH)_2D_3$ synthesis (PKC and $1.25(OH)_2D_3$ synthesis correlated with PTH concentrations 100-fold to 1000-fold lower than those required for the activation of PKA). Moreover, inhibition of PKC blocked $1.25(OH)_2D_3$ synthesis, and amino-truncated PTH fragments that stimulate PKC but not PKA likewise enhanced $1.25(OH)_2D_3$ formation. Taken together, the results suggest a predominant effect of PKA on transcriptional regulation of *CYP27B1* gene expression and a more rapid posttranscriptional action of PKC on 1α-hydroxylase enzymatic activity.

Other renal effects of parathyroid hormone

PTH exerts a variety of additional described, but far less well understood, effects on ion transport, such as activation of apical Cl^- channels (Suzuki et al., 1991) and metabolism.

The kidney contributes importantly to systemic gluconeogenesis (Friedman and Torretti, 1978; Schoolwerth et al., 1988), although this is not normally apparent because of balancing by renal medullary glycolysis. PTH-dependent renal gluconeogenesis is attended by segmental and internephron heterogeneity. PTH increases gluconeogenesis (Chobanian and Hammerman, 1988), ammoniagenesis (Wang and Kurokawa, 1984), and phosphoenolpyruvate carboxykinase (Watford and Mapes, 1990) mRNA expression in proximal tubules. PTH primarily stimulates gluconeogenesis in cortical S1 and S2 proximal tubules (Wang and Kurokawa, 1984). Although juxtamedullary S1 proximal tubules exhibit the highest rate of gluconeogenesis, it is unaffected by PTH.

Gluconeogenesis is linked to ammoniagenesis because both are stimulated by acidosis and by PTH. Moreover, L-glutamine, which is the major gluconeogenic precursor, is also a substrate for ammoniagenesis. Thus, there may be some metabolic interdependence of PTH actions on gluconeogenesis and ammoniagenesis.

PTH also exerts conspicuous morphological effects causing rapid microvillus shortening (Hruska et al., 1986). This action appears to be specific for proximal tubule cells. Emerging observations suggest that this cell-specific effect may be due to prominent brush border expression of NHERF1 in proximal tubule cells and its absence from distal tubule cells.

It is not known if PTH regulation of Na-Pi cotransport, Na^+/H^+ exchange or Na^+/Ca^{2+} exchange, or its metabolic actions on 25-hydroxyvitamin D-1α-hydroxylase, gluconeogenesis, and ammoniagenesis, involves more complicated second messenger signaling. Notably, PTH receptors activate additional signaling pathways involving phospholipase A_2 (Mandel and Derrickson, 1997; Ribeiro et al., 1994), PLD (Garrido et al., 2009; Somermeyer et al., 1983), and MAPK (Quamme et al., 1994; Swarthout et al., 1997; Verheijen and Defize, 1997). Thus, subtle differences in regulatory control or changes in signaling may arise in pathological settings.

Renal expression and actions of parathyroid hormone–related protein

PTHrP is expressed by glomeruli, distal tubules, and collecting ducts of fetal kidneys and by proximal convoluted tubules, distal convoluted tubules, and glomeruli of adult kidneys (Aya et al., 1999; Philbrick et al., 1996). In rat kidneys, PTHrP mRNA is found in glomeruli, proximal convoluted tubules, and macula densa but not in cortical ascending limbs, medullary ascending limbs, distal convoluted tubules, or collecting ducts (Yang et al., 1997), whereas other studies reported PTHrP transcripts in glomerular mesangial cells and proximal and distal tubules (Soifer et al., 1993). Although PTHrP is critical for normal cardiovascular and bone development, the kidneys of PTHrP-null mice appear histologically normal (Karmali et al., 1992) and a physiological role for PTHrP on the kidney is uncertain (Mundy and Edwards, 2008).

Amino-terminal fragments of PTHrP exhibit renal actions including stimulation of cAMP production, regulation of Na-Pi cotransport, and calcium excretion that are largely indistinguishable from those of PTH (Everhart-Caye et al., 1996; Horiuchi et al., 1987; Horwitz et al., 2003b; Pizurki et al., 1988; Scheinman et al., 1990; Yates et al., 1988). Some differences may be discerned. In distal tubule cells, for instance, PTHrP(1–34) and PTHrP(1–74) were more potent than equimolar concentrations of PTH(1–34) and PTH(1–84) in stimulating adenylyl cyclase (Friedman et al., 1989). In contrast, PTHrP(1–36) has a markedly lower effect at raising serum $1.25(OH)_2D_3$ levels (a consequence of its action on proximal tubules) in human volunteers than does comparable administration of PTH(1–34) (Horwitz et al., 2003b).

PTHrP possesses mid- and carboxy-terminal regions that have distinct biological actions (Clemens et al., 2001; Philbrick et al., 1996). Longer PTHrP fragments may exhibit unique renal properties. For example, bicarbonate excretion in perfused rat kidneys was enhanced equivalently by PTHrP(1–34) and PTH(1–34), whereas PTHrP(1–84), PTHrP(1–108), and PTHrP(1–141) were each less active than PTH(1–34) (Ellis et al., 1990). As discussed in Chapter 25, the PTHrP gene can generate multiple transcripts and protein products, some of which may undergo unique nuclear localization. It is therefore possible that locally expressed PTHrP exerts actions in the kidney that are not shared with PTH.

A possible role for locally produced PTHrP in the renal response to insufficient vascular perfusion has been suggested by findings that PTHrP expression is induced by ischemia or following recovery from ATP depletion (García-Ocaña et al., 1999; Largo et al., 1999; Soifer et al., 1993). PTHrP is expressed by the intima and media of human renal microvessels and in the macula densa (Massfelder et al., 1996). PTHrP (like PTH) increases renin release from the juxtaglomerular apparatus and also stimulates cAMP in renal afferent and efferent arterioles, leading to vasodilation and enhanced renal blood flow (Endlich et al., 1995; Helwig et al., 1991; Musso et al., 1989; Saussine et al., 1993). Both cAMP and nitric oxide have been implicated in PTHrP-induced in vitro vasorelaxation (Massfelder et al., 1996). Thus, enhanced local PTHrP production induced by inadequate renal perfusion or ischemia may be involved in local and systemic autoregulatory mechanisms. According to this view, direct local vasodilatory actions are supplemented by the systemic activation of angiotensinogen, which increases arterial pressure and further sustains renal blood flow.

PTHrP promotes fibrogenesis and is upregulated in experimental nephropathy (Funk, 2001; Ortega et al., 2005, 2006). Locally produced PTHrP may contribute to the proinflammatory effect by activating NF-KB and ERKs (Ramila et al., 2008). The incidence of renal hypertrophy is greater in diabetic mice overexpressing PTHrP than in control animals, suggesting that constitutive PTHrP overexpression may elicit adaptive responses such as nitric oxide production to mitigate against renal damage (Izquierdo et al., 2006).

ACKNOWLEDGMENTS

Original cited work was supported by National Institutes of Health (NIH) awards R01 DK105811-A1, R01 DK111427-A1 (PAF), and R01 HL136382 (AB).

References

Abou-Samra, A.B., Juppner, H., Force, T., Freeman, M.W., Kong, X.F., Schipani, E., Urena, P., Richards, J., Bonventre, J.V., Potts Jr., J.T., et al., 1992. Expression cloning of a common receptor for parathyroid hormone and parathyroid hormone-related peptide from rat osteoblast-like cells: a single receptor stimulates intracellular accumulation of both cAMP and inositol trisphosphates and increases intracellular free calcium. Proc. Natl. Acad. Sci. U. S. A. 89, 2732–2736.

Agus, Z.S., Gardner, L.B., Beck, L.H., Goldberg, M., 1973. Effects of parathyroid hormone on renal tubular reabsorption of calcium, sodium, and phosphate. Am. J. Physiol. 224, 1143–1148.

Ahmed, I., Gesty-Palmer, D., Drezner, M.K., Luttrell, L.M., 2003. Transactivation of the epidermal growth factor receptor mediates parathyroid hormone and prostaglandin F2 alpha-stimulated mitogen-activated protein kinase activation in cultured transgenic murine osteoblasts. Mol. Endocrinol. 17, 1607–1621.

Akatsu, T., Takahashi, N., Udagawa, N., Sato, K., Nagata, N., Moseley, J.M., Martin, T.J., Suda, T., 1989. Parathyroid hormone (PTH)-related protein is a potent stimulator of osteoclast-like multinucleated cell formation to the same extent as PTH in mouse marrow cultures. Endocrinology 125, 20–27.

Albright, F., Bauer, W., Ropes, M., Aub, J.C., 1929. Studies of calcium and phosphorus metabolism: IV. The effect of the parathyroid hormone. J. Clin. Investig. 7, 139–181.

Amizuka, N., Karaplis, A.C., Henderson, J.E., Warshawsky, H., Lipman, M.L., Matsuki, Y., Ejiri, S., Tanaka, M., Izumi, N., Ozawa, H., Goltzman, D., 1996. Haploinsufficiency of parathyroid hormone-related peptide (PTHrP) results in abnormal postnatal bone development. Dev. Biol. 175, 166–176.

Amizuka, N., Lee, H.S., Kwan, M.Y., Arazani, A., Warshawsky, H., Hendy, G.N., Ozawa, H., White, J.H., Goltzman, D., 1997. Cell-specific expression of the parathyroid hormone (PTH)/PTH-related peptide receptor gene in kidney from kidney-specific and ubiquitous promoters. Endocrinology 138, 469–481.

Andrukhova, O., Zeitz, U., Goetz, R., Mohammadi, M., Lanske, B., Erben, R.G., 2012. FGF23 acts directly on renal proximal tubules to induce phosphaturia through activation of the ERK1/2-SGK1 signaling pathway. Bone 51, 621–628.

Ardura, J.A., Friedman, P.A., 2011. Regulation of G protein-coupled receptor function by Na^+/H^+ exchange regulatory factors. Pharmacol. Rev. 63, 882–900.

Armbrecht, H.J., Boltz, M.A., Hodam, T.L., 2003. PTH increases renal 25(OH)D_3-1α-hydroxylase (CYP1α) mRNA but not renal 1,25$(OH)_2D_3$ production in adult rats. Am. J. Physiol. Renal. Physiol. 284, F1032–F1036.

Armbrecht, H.J., Wongsurawat, N., Zenser, T.V., Davis, B.B., 1984. Effect of PTH and 1,25$(OH)_2D_3$ on renal 25(OH)D_3 metabolism, adenylate cyclase, and protein kinase. Am. J. Physiol. Endocrinol. Metab. 246, E102–E107.

Aubin, J.E., Bonnelye, E., 2000. Osteoprotegerin and its ligand: a new paradigm for regulation of osteoclastogenesis and bone resorption. Osteoporos. Int. 11, 905–913.

Aya, K., Tanaka, H., Ichinose, Y., Kobayashi, M., Seino, Y., 1999. Expression of parathyroid hormone-related peptide messenger ribonucleic acid in developing kidney. Kidney Int. 55, 1696–1703.

Ba, J., Brown, D., Friedman, P.A., 2003. Calcium-sensing receptor regulation of PTH-inhibitable proximal tubule phosphate transport. Am. J. Physiol. Renal. Physiol. 285, F1233–F1243.

Bacconi, A., Virkki, L.V., Biber, J., Murer, H., Forster, I.C., 2005. Renouncing electroneutrality is not free of charge: switching on electrogenicity in a Na^+-coupled phosphate cotransporter. Proc. Natl. Acad. Sci. U. S. A. 102, 12606–12611.

Bacic, D., Lehir, M., Biber, J., Kaissling, B., Murer, H., Wagner, C.A., 2006. The renal Na^+/phosphate cotransporter NaPi-IIa is internalized via the receptor-mediated endocytic route in response to parathyroid hormone. Kidney Int. 69, 495–503.

Bacskai, B.J., Friedman, P.A., 1990. Activation of latent Ca^{2+} channels in renal epithelial cells by parathyroid hormone. Nature 347, 388–391.

Baertschi, S., Baur, N., Lueders-Lefevre, V., Voshol, J., Keller, H., 2014. Class I and IIa histone deacetylases have opposite effects on sclerostin gene regulation. J. Biol. Chem. 289, 24995–25009.

Banerjee, C., Javed, A., Choi, J.Y., Green, J., Rosen, V., van Wijnen, A.J., Stein, J.L., Lian, J.B., Stein, G.S., 2001. Differential regulation of the two principal Runx2/Cbfa1 n-terminal isoforms in response to bone morphogenetic protein-2 during development of the osteoblast phenotype. Endocrinology 142, 4026–4039.

Bank, N., Aynedjian, H.S., 1976. A micropuncture study of the effect of parathyroid hormone on renal bicarbonate reabsorption. J. Clin. Investig. 58, 336–344.

Baron, R., Rawadi, G., 2007. Targeting the Wnt/beta-catenin pathway to regulate bone formation in the adult skeleton. Endocrinology 148, 2635–2643.

Bellido, T., 2006. Downregulation of SOST/sclerostin by PTH: a novel mechanism of hormonal control of bone formation mediated by osteocytes. J. Musculoskelet. Neuronal Interact. 6, 358–359.

Bellido, T., Ali, A.A., Gubrij, I., Plotkin, L.I., Fu, Q., O'Brien, C.A., Manolagas, S.C., Jilka, R.L., 2005. Chronic elevation of parathyroid hormone in mice reduces expression of sclerostin by osteocytes: a novel mechanism for hormonal control of osteoblastogenesis. Endocrinology 146, 4577–4583.

Bellido, T., Ali, A.A., Plotkin, L.I., Fu, Q., Gubrij, I., Roberson, P.K., Weinstein, R.S., O'Brien, C.A., Manolagas, S.C., Jilka, R.L., 2003. Proteasomal degradation of Runx2 shortens parathyroid hormone-induced anti-apoptotic signaling in osteoblasts. A putative explanation for why intermittent administration is needed for bone anabolism. J. Biol. Chem. 278, 50259–50272.

Bergmann, P., Nijs-De Wolf, N., Pepersack, T., Corvilain, J., 1990. Release of parathyroid hormonelike peptides by fetal rat long bones in culture. J. Bone Miner. Res. 5, 741–753.

Bergwitz, C., Roslin, N.M., Tieder, M., Loredo-Osti, J.C., Bastepe, M., Abu-Zahra, H., Frappier, D., Burkett, K., Carpenter, T.O., Anderson, D., Garabedian, M., Sermet, I., Fujiwara, T.M., Morgan, K., Tenenhouse, H.S., Juppner, H., 2006. *SLC34A3* mutations in patients with hereditary hypophosphatemic rickets with hypercalciuria predict a key role for the sodium-phosphate Cotransporter NaPi-IIc in maintaining phosphate homeostasis. Am. J. Hum. Genet. 78, 179–192.

Bernardo, A.A., Kear, F.T., Santos, A.V., Ma, J., Steplock, D., Robey, R.B., Weinman, E.J., 1999. Basolateral Na^+/HCO_3^- cotransport activity is regulated by the dissociable Na^+/H^+ exchanger regulatory factor. J. Clin. Investig. 104, 195–201.

Bernardo, J.F., Friedman, P.A., 2013. Renal calcium metabolism. In: Alpern, R.J., Caplan, M.J., Moe, O.W. (Eds.), Seldin and Giebisch's the Kidney: Physiology and Pathophysiology. Academic Press, San Diego, pp. 2225–2247.

Besarab, A., Swanson, J.W., 1984. Tachyphylaxis to PTH in the isolated perfused rat kidney: resistance of anticalciuria. Am. J. Physiol. Renal. Physiol. 247, F240–F245.

Bethune, J.E., Turpin, R.A., 1968. A study of urinary excretion of parathyroid hormone in man. J. Clin. Investig. 47, 1583–1589.

Bettinelli, A., Bianchetti, M.G., Girardin, E., Caringella, A., Cecconi, M., Appiani, A.C., Pavanello, L., Gastaldi, R., Isimbaldi, C., Lama, G., et al., 1992. Use of calcium excretion values to distinguish two forms of primary renal tubular hypokalemic alkalosis: Bartter and Gitelman syndromes. J. Pediatr. 120, 38–43.

Bettoun, J.D., Minagawa, M., Hendy, G.N., Alpert, L.C., Goodyer, C.G., Goltzman, D., White, J.H., 1998. Developmental upregulation of human parathyroid hormone (PTH)/PTH-related peptide receptor gene expression from conserved and human-specific promoters. J. Clin. Investig. 102, 958–967.

Bichara, M., Mercier, O., Paillard, M., Leviel, F., 1986. Effects of parathyroid hormone on urinary acidification. Am. J. Physiol. Renal. Physiol. 251, F444–F453.

Bidot-López, P., Farese, R.V., Sabir, M.A., 1981. Parathyroid hormone and adenosine-3',5'-monophosphate acutely increase phospholipids of the phosphatidate-polyphosphoinositide pathway in rabbit kidney cortex tubules *in vitro* by a cycloheximide-sensitive process. Endocrinology 108, 2078–2081.

Bikle, D.D., Sakata, T., Leary, C., Elalieh, H., Ginzinger, D., Rosen, C.J., Beamer, W., Majumdar, S., Halloran, B.P., 2002. Insulin-like growth factor I is required for the anabolic actions of parathyroid hormone on mouse bone. J. Bone Miner. Res. 17, 1570–1578.

Bilezikian, J.P., 2012. Primary hyperparathyroidism. Endocr. Pract. 18, 781–790.

Bisello, A., Chorev, M., Rosenblatt, M., Monticelli, L., Mierke, D.F., Ferrari, S.L., 2002. Selective ligand-induced stabilization of active and desensitized parathyroid hormone type 1 receptor conformations. J. Biol. Chem. 277, 38524–38530.

Blaine, J., Chonchol, M., Levi, M., 2015. Renal control of calcium, phosphate, and magnesium homeostasis. Clin. J. Am. Soc. Nephrol. 10, 1257–1272.

Blind, E., Bambino, T., Huang, Z., Bliziotes, M., Nissenson, R.A., 1996. Phosphorylation of the cytoplasmic tail of the PTH/PTHrP receptor. J. Bone Miner. Res. 11, 578–586.

Blomberg Jensen, M., Andersen, C.B., Nielsen, J.E., Bagi, P., Jorgensen, A., Juul, A., Leffers, H., 2010. Expression of the vitamin D receptor, 25-hydroxylases, 1alpha-hydroxylase and 24-hydroxylase in the human kidney and renal clear cell cancer. J. Steroid Biochem. Mol. Biol. 121, 376–382.

Boron, W.F., 2006. Acid-base transport by the renal proximal tubule. J. Am. Soc. Nephrol. 17, 2368–2382.

Bos, M.P., van der Meer, J.M., Feyen, J.H., Herrmann-Erlee, M.P., 1996. Expression of the parathyroid hormone receptor and correlation with other osteoblastic parameters in fetal rat osteoblasts. Calcif. Tissue Int. 58, 95–100.

Boyce, R.W., Paddock, C.L., Franks, A.F., Jankowsky, M.L., Eriksen, E.F., 1996. Effects of intermittent hPTH(1-34) alone and in combination with $1,25(OH)_2D_3$ or risedronate on endosteal bone remodeling in canine cancellous and cortical bone. J. Bone Miner. Res. 11, 600–613.

Brindle, P.K., Montminy, M.R., 1992. The CREB family of transcription activators. Curr. Opin. Genet. Dev. 2, 199–204.

Bringhurst, F.R., Strewler, G.J., 2002. Renal and skeletal actions of parathyroid hormone (PTH) and PTH-related protein. In: Bilezikian, J.P., Raisz, L.G., Rodan, G.A. (Eds.), Principles of Bone Biology, 2 ed. Academic Press, San Diego, pp. 483–514.

Broadbent, D., Ahmadzai, M.M., Kammala, A.K., Yang, C., Occhiuto, C., Das, R., Subramanian, H., 2017. Roles of NHERF family of PDZ-binding proteins in regulating GPCR functions. Adv. Immunol. 136, 353–385.

Broadus, A.E., 1979. Nephrogenous cyclic AMP as a parathyroid function test. Nephron 23, 136–141.

Brown, R.C., Silver, A.C., Woodhead, J.S., 1991. Binding and degradation of NH_2-terminal parathyroid hormone by opossum kidney cells. Am. J. Physiol. Endocrinol. Metab. 260, E544–E552.

Brunskill, N., Bastani, B., Hayes, C., Morrissey, J., Klahr, S., 1991. Localization and polar distribution of several G-protein subunits along nephron segments. Kidney Int. 40, 997–1006.

Burgess, T.L., Qian, Y., Kaufman, S., Ring, B.D., Van, G., Capparelli, C., Kelley, M., Hsu, H., Boyle, W.J., Dunstan, C.R., Hu, S., Lacey, D.L., 1999. The ligand for osteoprotegerin (OPGL) directly activates mature osteoclasts. J. Cell Biol. 145, 527–538.

Busch, A.E., Schuster, A., Waldegger, S., Wagner, C.A., Zempel, G., Broer, S., Biber, J., Murer, H., Lang, F., 1996. Expression of a renal type I sodium/phosphate transporter (NaPi-1) induces a conductance in Xenopus oocytes permeable for organic and inorganic anions. Proc. Natl. Acad. Sci. U. S. A. 93, 5347–5351.

Calebiro, D., Nikolaev, V.O., Gagliani, M.C., de Filippis, T., Dees, C., Tacchetti, C., Persani, L., Lohse, M.J., 2009. Persistent cAMP-signals triggered by internalized G-protein-coupled receptors. PLoS Biol. 7, e1000172.

Canalis, E., Centrella, M., Burch, W., McCarthy, T.L., 1989. Insulin-like growth factor I mediates selective anabolic effects of parathyroid hormone in bone cultures. J. Clin. Investig. 83, 60–65.

Canalis, E., McCarthy, T.L., Centrella, M., 1990. Differential effects of continuous and transient treatment with parathyroid hormone related peptide (PTHrp) on bone collagen synthesis. Endocrinology 126, 1806–1812.

Capuano, P., Bacic, D., Roos, M., Gisler, S.M., Stange, G., Biber, J., Kaissling, B., Weinman, E.J., Shenolikar, S., Wagner, C.A., Murer, H., 2006. Defective coupling of apical PTH receptors to phospholipase C prevents internalization of the Na^+/phosphate cotransporter NaPi-IIa in NHERF1 deficient mice. Am. J. Physiol. Cell Physiol. 292, C927–C934.

Care, A.D., Abbas, S.K., Pickard, D.W., Barri, M., Drinkhill, M., Findlay, J.B., White, I.R., Caple, I.W., 1990. Stimulation of ovine placental transport of calcium and magnesium by mid-molecule fragments of human parathyroid hormone-related protein. Exp. Physiol. 75, 605–608.

Carney, J.A., 1996. The glandulae parathyroideae of Ivar Sandstrom. Contributions from two continents. Am. J. Surg. Pathol. 20, 1123–1144.

Carpenter, T.O., 2017. CYP24A1 loss of function: clinical phenotype of monoallelic and biallelic mutations. J. Steroid Biochem. Mol. Biol. 173, 337–340.

Castro, M., Dicker, F., Vilardaga, J.P., Krasel, C., Bernhardt, M., Lohse, M.J., 2002. Dual regulation of the parathyroid hormone (PTH)/PTH-related peptide receptor signaling by protein kinase C and β-arrestins. Endocrinology 143, 3854–3865.

Cha, S.K., Wu, T., Huang, C.L., 2008. Protein kinase C inhibits caveolae-mediated endocytosis of TRPV5. Am. J. Physiol. Renal. Physiol. 294, F1212–F1221.

Chabardès, D., Gagnan-Brunette, M., Imbert-Teboul, M., Gontcharevskaia, O., Montégut, M., Clique, A., Morel, F., 1980. Adenylate cyclase responsiveness to hormones in various portions of the human nephron. J. Clin. Investig. 65, 439–448.

Chaudhary, L.R., Hruska, K.A., 2001. The cell survival signal Akt is differentially activated by PDGF-BB, EGF, and FGF-2 in osteoblastic cells. J. Cell. Biochem. 81, 304–311.

Cheloha, R.W., Gellman, S.H., Vilardaga, J.P., Gardella, T.J., 2015. PTH receptor-1 signalling-mechanistic insights and therapeutic prospects. Nat. Rev. Endocrinol. 11, 712–724.

Chen, H.L., Demiralp, B., Schneider, A., Koh, A.J., Silve, C., Wang, C.Y., McCauley, L.K., 2002. Parathyroid hormone and parathyroid hormone-related protein exert both pro- and anti-apoptotic effects in mesenchymal cells. J. Biol. Chem. 277, 19374–19381.

Chobanian, M.C., Hammerman, M.R., 1988. Parathyroid hormone stimulates ammoniagenesis in canine renal proximal tubular segments. Am. J. Physiol. Renal. Physiol. 255, F847–F852.

Christakos, S., 2017. In search of regulatory circuits that control the biological activity of vitamin D. J. Biol. Chem. 292, 17559–17560.

Christov, M., Juppner, H., 2013. Insights from genetic disorders of phosphate homeostasis. Semin. Nephrol. 33, 143–157.

Chung, D.J., Castro, C.H., Watkins, M., Stains, J.P., Chung, M.Y., Szejnfeld, V.L., Willecke, K., Theis, M., Civitelli, R., 2006. Low peak bone mass and attenuated anabolic response to parathyroid hormone in mice with an osteoblast-specific deletion of connexin43. J. Cell Sci. 119, 4187–4198.

Civitelli, R., Hruska, K.A., Shen, V., Avioli, L.V., 1990. Cyclic AMP-dependent and calcium-dependent signals in parathyroid hormone function. Exp. Gerontol. 25, 223–231.

Civitelli, R., Martin, T.J., Fausto, A., Gunsten, S.L., Hruska, K.A., Avioli, L.V., 1989. Parathyroid hormone-related peptide transiently increases cytosolic calcium in osteoblast-like cells: comparison with parathyroid hormone. Endocrinology 125, 1204–1210.

Clemens, T.L., Cormier, S., Eichinger, A., Endlich, K., Fiaschi-Taesch, N., Fischer, E., Friedman, P.A., Karaplis, A.C., Massfelder, T., Rossert, J., Schluter, K.D., Silve, C., Stewart, A.F., Takane, K., Helwig, J.J., 2001. Parathyroid hormone-related protein and its receptors: nuclear functions and roles in the renal and cardiovascular systems, the placental trophoblasts and the pancreatic islets. Br. J. Pharmacol. 134, 1113–1136.

Clohisy, J.C., Scott, D.K., Brakenhoff, K.D., Quinn, C.O., Partridge, N.C., 1992. Parathyroid hormone induces c-fos and c-jun messenger RNA in rat osteoblastic cells. Mol. Endocrinol. 6, 1834–1842.

Coe, F.L., 1974. Magnitude of metabolic acidosis in primary hyperparathyroidism. Arch. Intern. Med. 134, 262–265.

Cole, J.A., 1999. Parathyroid hormone activates mitogen-activated protein kinase in opossum kidney cells. Endocrinology 140, 5771–5779.

Collazo, R., Fan, L.Z., Hu, M.C., Zhao, H., Wiederkehr, M.R., Moe, O.W., 2000. Acute regulation of Na^+/H^+ exchanger NHE3 by parathyroid hormone via NHE3 phosphorylation and dynamin-dependent endocytosis. J. Biol. Chem. 275, 31601–31608.

Collins, J.F., Bai, L., Ghishan, F.K., 2004. The SLC20 family of proteins: dual functions as sodium-phosphate cotransporters and viral receptors. Pflügers Archiv 447, 647–652.

Conover, C.A., 2008. Insulin-like growth factor-binding proteins and bone metabolism. Am. J. Physiol. Endocrinol. Metab. 294, E10–E14.

Conover, C.A., Bale, L.K., Clarkson, J.T., Torring, O., 1993. Regulation of insulin-like growth factor binding protein-5 messenger ribonucleic acid expression and protein availability in rat osteoblast-like cells. Endocrinology 132, 2525–2530.

Cornish, J., Callon, K.E., Nicholson, G.C., Reid, I.R., 1997. Parathyroid hormone-related protein-(107-139) inhibits bone resorption in vivo. Endocrinology 138, 1299–1304.

Corral, D.A., Amling, M., Priemel, M., Loyer, E., Fuchs, S., Ducy, P., Baron, R., Karsenty, G., 1998. Dissociation between bone resorption and bone formation in osteopenic transgenic mice. Proc. Natl. Acad. Sci. U. S. A. 95, 13835–13840.

Costanzo, L.S., Windhager, E.E., 1980. Effects of PTH, ADH, and cyclic AMP on distal tubular Ca and Na reabsorption. Am. J. Physiol. 239, F478–F485.

Coulson, R., Scheinman, S.J., 1989. Xanthine effects on renal proximal tubular function and cyclic AMP metabolism. J. Pharmacol. Exp. Ther. 248, 589–595.

Cros, M., Silve, C., Graulet, A.M., Morieux, C., Ureña, P., De Vernejoul, M.C., Bouizar, Z., 1998. Estrogen stimulates PTHrP but not PTH/PTHrP receptor gene expression in the kidney of ovariectomized rat. J. Cell. Biochem. 70, 84–93.

Cuche, J.L., Ott, C.E., Marchand, G.R., Diaz-Buxo, J.A., Knox, F.G., 1976. Intrarenal calcium in phosphate handling. Am. J. Physiol. 230, 790–796.

Cunningham, R., Biswas, R., Steplock, D., Shenolikar, S., Weinman, E.J., 2010. Role of NHERF and scaffolding proteins in proximal tubule transport. Urol. Res. 38, 257–262.

Cunningham, R., Steplock, D., Wang, F., Huang, H., E, X., Shenolikar, S., Weinman, E.J., 2004. Defective PTH regulation of NHE3 activity and phosphate adaptation in cultured NHERF-1$^{-/-}$ renal proximal tubule cells. J. Biol. Chem. 279, 37815–37821.

Cunningham, R.M., Esmaili, A., Brown, E., Biswas, R.S., Murtazina, R., Donowitz, M., Dijkman, H.B., van der Vlag, J., Hogema, B.M., De Jonge, H.R., Shenolikar, S., Wade, J.B., Weinman, E.J., 2008. Urine electrolyte, mineral, and protein excretion in NHERF-2 and NHERF-1 null mice. Am. J. Physiol. Renal. Physiol. 294, F1001–F1007.

Dasgupta, D., Wee, M.J., Reyes, M., Li, Y., Simm, P.J., Sharma, A., Schlingmann, K.P., Janner, M., Biggin, A., Lazier, J., Gessner, M., Chrysis, D., Tuchman, S., Baluarte, H.J., Levine, M.A., Tiosano, D., Insogna, K., Hanley, D.A., Carpenter, T.O., Ichikawa, S., Hoppe, B., Konrad, M., Savendahl, L., Munns, C.F., Lee, H., Juppner, H., Bergwitz, C., 2014. Mutations in SLC34A3/NPT2c are associated with kidney stones and nephrocalcinosis. J. Am. Soc. Nephrol. 25, 2366–2375.

Datta, N.S., Samra, T.A., Abou-Samra, A.B., 2012. Parathyroid hormone induces bone formation in phosphorylation-deficient PTHR1 knockin mice. Am. J. Physiol. Endocrinol. Metab. 302, E1183–E1188.

Daugaard, H., Egfjord, M., Lewin, E., Olgaard, K., 1994. Metabolism of N-terminal and C-terminal parathyroid hormone fragments by isolated perfused rat kidney and liver. Endocrinology 134, 1373–1381.

de Groot, T., Lee, K., Langeslag, M., Xi, Q., Jalink, K., Bindels, R.J., Hoenderop, J.G., 2009. Parathyroid hormone activates TRPV5 via PKA-dependent phosphorylation. J. Am. Soc. Nephrol. 20, 1693–1704.

Dean, T., Vilardaga, J.P., Potts Jr., J.T., Gardella, T.J., 2008. Altered selectivity of parathyroid hormone (PTH) and PTH-related protein (PTHrP) for distinct conformations of the PTH/PTHrP receptor. Mol. Endocrinol. 22, 156–166.

Debiais, F., Lefevre, G., Lemonnier, J., Le Mee, S., Lasmoles, F., Mascarelli, F., Marie, P.J., 2004. Fibroblast growth factor-2 induces osteoblast survival through a phosphatidylinositol 3-kinase-dependent, -β-catenin-independent signaling pathway. Exp. Cell Res. 297, 235–246.

Deliot, N., Hernando, N., Horst-Liu, Z., Gisler, S.M., Capuano, P., Wagner, C.A., Bacic, D., O'Brien, S., Biber, J., Murer, H., 2005. Parathyroid hormone treatment induces dissociation of type IIa Na^+-P_i cotransporter-Na^+/H^+ exchanger regulatory factor-1 complexes. Am. J. Physiol. Cell Physiol. 289, C159–C167.

Dempster, D.W., Hughes-Begos, C.E., Plavetic-Chee, K., Brandao-Burch, A., Cosman, F., Nieves, J., Neubort, S., Lu, S.S., Iida-Klein, A., Arnett, T., Lindsay, R., 2005. Normal human osteoclasts formed from peripheral blood monocytes express PTH type 1 receptors and are stimulated by PTH in the absence of osteoblasts. J. Cell. Biochem. 95, 139–148.

Dempster, D.W., Parisien, M., Silverberg, S.J., Liang, X.G., Schnitzer, M., Shen, V., Shane, E., Kimmel, D.B., Recker, R., Lindsay, R., Bilezikian, J.P., 1999. On the mechanism of cancellous bone preservation in postmenopausal women with mild primary hyperparathyroidism. J. Clin. Endocrinol. Metab. 84, 1562–1566.

Dicker, F., Quitterer, U., Winstel, R., Honold, K., Lohse, M.J., 1999. Phosphorylation-independent inhibition of parathyroid hormone receptor signaling by G protein-coupled receptor kinases. Proc. Natl. Acad. Sci. U. S. A. 96, 5476–5481.

Disthabanchong, S., Hassan, H., McConkey, C.L., Martin, K.J., Gonzalez, E.A., 2004. Regulation of PTH1 receptor expression by uremic ultrafiltrate in UMR 106-01 osteoblast-like cells. Kidney Int. 65, 897–903.

Divieti, P., Geller, A.I., Suliman, G., Juppner, H., Bringhurst, F.R., 2005. Receptors specific for the carboxyl-terminal region of parathyroid hormone on bone-derived cells: determinants of ligand binding and bioactivity. Endocrinology 146, 1863–1870.

Divieti, P., Inomata, N., Chapin, K., Singh, R., Juppner, H., Bringhurst, F.R., 2001. Receptors for the carboxyl-terminal region of PTH(1-84) are highly expressed in osteocytic cells. Endocrinology 142, 916–925.

Dobnig, H., Turner, R.T., 1995. Evidence that intermittent treatment with parathyroid hormone increases bone formation in adult rats by activation of bone lining cells. Endocrinology 136, 3632–3638.

Dobolyi, A., Palkovits, M., Usdin, T.B., 2003. Expression and distribution of tuberoinfundibular peptide of 39 residues in the rat central nervous system. J. Comp. Neurol. 455, 547–566.

Donahue, H.J., McLeod, K.J., Rubin, C.T., Andersen, J., Grine, E.A., Hertzberg, E.L., Brink, P.R., 1995. Cell-to-cell communication in osteoblastic networks: cell line-dependent hormonal regulation of gap junction function. J. Bone Miner. Res. 10, 881–889.

Dousa, T.P., Duarte, C.G., Knox, F.G., 1976. Effect of colchicine on urinary phosphate and regulation by parathyroid hormone. Am. J. Physiol. 231, 61–65.

Ducy, P., Starbuck, M., Priemel, M., Shen, J., Pinero, G., Geoffroy, V., Amling, M., Karsenty, G., 1999. A Cbfa1-dependent genetic pathway controls bone formation beyond embryonic development. Genes Dev. 13, 1025–1036.

Ducy, P., Zhang, R., Geoffroy, V., Ridall, A.L., Karsenty, G., 1997. Osf2/Cbfa1: a transcriptional activator of osteoblast differentiation. Cell 89, 747–754.

Edwards, B.R., Sutton, R.A.L., Dirks, J.H., 1974. Effect of calcium infusion on renal tubular reabsorption in the dog. Am. J. Physiol. 227, 13–18.

Eichinger, A., Fiaschi-Taesch, N., Massfelder, T., Fritsch, S., Barthelmebs, M., Helwig, J.J., 2002. Transcript expression of the tuberoinfundibular peptide (TIP)39/PTH2 receptor system and non-PTH1 receptor-mediated tonic effects of TIP39 and other PTH2 receptor ligands in renal vessels. Endocrinology 143, 3036–3043.

Ellis, A.G., Adam, W.R., Martin, T.J., 1990. Comparison of the effects of parathyroid hormone (PTH) and recombinant PTH-related protein on bicarbonate excretion by the isolated perfused rat kidney. J. Endocrinol. 126, 403–408.

Ellsworth, R., Nicholson, W.M., 1935. Further observations upon the changes in the electrolytes of the urine following the injection of parathyroid extract. J. Clin. Investig. 14, 823–827.

Endlich, K., Massfelder, T., Helwig, J.J., Steinhausen, M., 1995. Vascular effects of parathyroid hormone and parathyroid hormone-related protein in the split hydronephrotic rat kidney. J. Physiol. 483, 481–490.

Erben, R.G., 2017. Pleiotropic actions of FGF23. Toxicol. Pathol. https://doi.org/10.1177/0192623317737469.

Esbrit, P., Alvarez-Arroyo, M.V., De Miguel, F., Martin, O., Martinez, M.E., Caramelo, C., 2000. C-terminal parathyroid hormone-related protein increases vascular endothelial growth factor in human osteoblastic cells. J. Am. Soc. Nephrol. 11, 1085–1092.

Evans, D.B., Hipskind, R.A., Bilbe, G., 1996. Analysis of signaling pathways used by parathyroid hormone to activate the c-fos gene in human SaOS2 osteoblast-like cells. J. Bone Miner. Res. 11, 1066–1074.

Evely, R.S., Bonomo, A., Schneider, H.G., Moseley, J.M., Gallagher, J., Martin, T.J., 1991. Structural requirements for the action of parathyroid hormone-related protein (PTHrP) on bone resorption by isolated osteoclasts. J. Bone Miner. Res. 6, 85–93.

Everhart-Caye, M., Inzucchi, S.E., Guinness-Henry, J., Mitnick, M.A., Stewart, A.F., 1996. Parathyroid hormone (PTH)-related protein(1-36) is equipotent to PTH(1-34) in humans. J. Clin. Endocrinol. Metab. 81, 199–208.

Fan, L., Wiederkehr, M.R., Collazo, R., Wang, H., Crowder, L.A., Moe, O.W., 1999. Dual mechanisms of regulation of Na/H exchanger NHE-3 by parathyroid hormone in rat kidney. J. Biol. Chem. 274, 11289–11295.

Fei, Y., Shimizu, E., McBurney, M.W., Partridge, N.C., 2015. Sirtuin 1 is a negative regulator of parathyroid hormone stimulation of matrix metalloproteinase 13 expression in osteoblastic cells: role of sirtuin 1 in the action of PTH on osteoblasts. J. Biol. Chem. 290, 8373–8382.

Feinstein, T.N., Yui, N., Webber, M.J., Wehbi, V.L., Stevenson, H.P., King Jr., J.D., Hallows, K.R., Brown, D., Bouley, R., Vilardaga, J.P., 2013. Noncanonical control of vasopressin receptor type 2 signaling by retromer and arrestin. J. Biol. Chem. 288, 27849–27860.

Fenton, A.J., Kemp, B.E., Hammonds Jr., R.G., Mitchelhill, K., Moseley, J.M., Martin, T.J., Nicholson, G.C., 1991a. A potent inhibitor of osteoclastic bone resorption within a highly conserved pentapeptide region of parathyroid hormone-related protein; PTHrP[107-111]. Endocrinology 129, 3424–3426.

Fenton, A.J., Kemp, B.E., Kent, G.N., Moseley, J.M., Zheng, M.H., Rowe, D.J., Britto, J.M., Martin, T.J., Nicholson, G.C., 1991b. A carboxyl-terminal peptide from the parathyroid hormone-related protein inhibits bone resorption by osteoclasts. Endocrinology 129, 1762–1768.

Fenton, A.J., Martin, T.J., Nicholson, G.C., 1993. Long-term culture of disaggregated rat osteoclasts: inhibition of bone resorption and reduction of osteoclast-like cell number by calcitonin and PTHrP[107-139]. J. Cell. Physiol. 155, 1–7.

Fermor, B., Skerry, T.M., 1995. PTH/PTHrP receptor expression on osteoblasts and osteocytes but not resorbing bone surfaces in growing rats. J. Bone Miner. Res. 10, 1935–1943.

Ferrandon, S., Feinstein, T.N., Castro, M., Wang, B., Bouley, R., Potts, J.T., Gardella, T.J., Vilardaga, J.P., 2009. Sustained cyclic AMP production by parathyroid hormone receptor endocytosis. Nat. Chem. Biol. 5, 734–742.

Ferrari, S.L., Behar, V., Chorev, M., Rosenblatt, M., Bisello, A., 1999. Endocytosis of ligand-human parathyroid hormone receptor 1 complexes is protein kinase C-dependent and involves ß-arrestin2. Real-time monitoring by fluorescence microscopy. J. Biol. Chem. 274, 29968–29975.

Ferrari, S.L., Bisello, A., 2001. Cellular distribution of constitutively active mutant parathyroid hormone (PTH)/PTH-related protein receptors and regulation of cyclic adenosine 3',5'-monophosphate signaling by β-arrestin2. Mol. Endocrinol. 15, 149–163.

Ferrari, S.L., Pierroz, D.D., Glatt, V., Goddard, D.S., Bianchi, E.N., Lin, F.T., Manen, D., Bouxsein, M.L., 2005. Bone response to intermittent parathyroid hormone is altered in mice null for β-arrestin2. Endocrinology 146, 1854–1862.

Feuerbach, D., Loetscher, E., Buerki, K., Sampath, T.K., Feyen, J.H., 1997. Establishment and characterization of conditionally immortalized stromal cell lines from a temperature-sensitive T-Ag transgenic mouse. J. Bone Miner. Res. 12, 179–190.

Filburn, C.R., Harrison, S., 1990. Parathyroid hormone regulation of cytosolic Ca^{2+} in rat proximal tubules. Am. J. Physiol. Renal. Physiol. 258, F545–F552.

Finkelman, R.D., Mohan, S., Linkhart, T.A., Abraham, S.M., Boussy, J.P., Baylink, D.J., 1992. PTH stimulates the proliferation of TE-85 human osteosarcoma cells by a mechanism not involving either increased cAMP or increased secretion of IGF-I, IGF-II or TGFβ. Bone Miner. 16, 89–100.

Finkelstein, J.S., Klibanski, A., Arnold, A.L., Toth, T.L., Hornstein, M.D., Neer, R.M., 1998. Prevention of estrogen deficiency-related bone loss with human parathyroid hormone-(1-34): a randomized controlled trial. J. Am. Med. Assoc. 280, 1067–1073.

Fischbach, F.T., Dunning, M.B., 2015. A Manual of Laboratory and Diagnostic Tests, ninth ed. Wolters Kluwer Health, Philadelphia.

Flannery, P.J., Spurney, R.F., 2001. Domains of the parathyroid hormone (PTH) receptor required for regulation by G protein-coupled receptor kinases (GRKs). Biochem. Pharmacol. 62, 1047–1058.

Fraser, D.R., Kodicek, E., 1973. Regulation of 25-hydroxycholecalciferol-1-hydroxylase activity in kidney by parathyroid hormone. Nat. New Biol. 241, 163–166.

Friedman, P.A., 1998. Codependence of renal calcium and sodium transport. Annu. Rev. Physiol. 60, 179–197.

Friedman, P.A., 2000. Mechanisms of calcium transport. Exp. Nephrol. 8, 343–350.

Friedman, P.A., 2008. Renal calcium metabolism. In: Alpern, R.J., Hebert, S.C. (Eds.), Seldin and Giebisch's the Kidney: Physiology and Pathophysiology, 4 ed. Elsevier, San Diego, pp. 1851–1890.

Friedman, P.A., 2014. Physiological actions of PTH II: renal actions. In: Bilezikian, J.P., Marcus, R., Silverberg, S.J., Marcocci, C., Levine, M.A., Potts Jr., J.T. (Eds.), The Parathyroids, third ed. Elsevier, San Diego, pp. 153–164.

Friedman, P.A., Coutermarsh, B.A., Kennedy, S.M., Gesek, F.A., 1996. Parathyroid hormone stimulation of calcium transport is mediated by dual signaling mechanisms involving PKA and PKC. Endocrinology 137, 13–20.

Friedman, P.A., Coutermarsh, B.A., Kennedy, S.M., Pizzonia, J.H., 1989. Differential stimulation of cAMP formation in renal distal convoluted tubule and cortical thick ascending limb cells by PTH and by PTH-like peptides. J. Bone Miner. Res. 4, S346.

Friedman, P.A., Gesek, F.A., 1993. Calcium transport in renal epithelial cells. Am. J. Physiol. Renal. Physiol. 264, F181–F198.

Friedman, P.A., Gesek, F.A., 1995. Stimulation of calcium transport by amiloride in mouse distal convoluted tubule cells. Kidney Int. 48, 1427–1434.
Friedman, P.A., Gesek, F.A., Coutermarsh, B.A., Kennedy, S.M., 1994. PKA and PKC activation is required for PTH-stimulated calcium uptake by distal convoluted tubule cells. J. Am. Soc. Nephrol. 5, 715.
Friedman, P.A., Gesek, F.A., Morley, P., Whitfield, J.F., Willick, G.E., 1999. Cell-specific signaling and structure-activity relations of parathyroid hormone analogs in mouse kidney cells. Endocrinology 140, 301–309.
Friedman, P.A., Torretti, J., 1978. Regional glucose metabolism in the cat kidney in vivo. Am. J. Physiol. Renal. Physiol. 234, F415–F423.
Fujita, T., Meguro, T., Fukuyama, R., Nakamuta, H., Koida, M., 2002. New signaling pathway for parathyroid hormone and cyclic AMP action on extracellular-regulated kinase and cell proliferation in bone cells. Checkpoint of modulation by cyclic AMP. J. Biol. Chem. 277, 22191–22200.
Fuller, K., Wong, B., Fox, S., Choi, Y., Chambers, T.J., 1998. TRANCE is necessary and sufficient for osteoblast-mediated activation of bone resorption in osteoclasts. J. Exp. Med. 188, 997–1001.
Funk, J.L., 2001. A role for parathyroid hormone-related protein in the pathogenesis of inflammatory/autoimmune diseases. Int. Immunopharmacol. 1, 1101–1121.
Gaich, G., Orloff, J.J., Atillasoy, E.J., Burtis, W.J., Ganz, M.B., Stewart, A.F., 1993. Amino-terminal parathyroid hormone-related protein: specific binding and cytosolic calcium responses in rat insulinoma cells. Endocrinology 132, 1402–1409.
García-Ocaña, A., Galbraith, S.C., Van Why, S.K., Yang, K., Golovyan, L., Dann, P., Zager, R.A., Stewart, A.F., Siegel, N.J., Orloff, J.J., 1999. Expression and role of parathyroid hormone-related protein in human renal proximal tubule cells during recovery from ATP depletion. J. Am. Soc. Nephrol. 10, 238–244.
Gardella, T.J., Vilardaga, J.-P., 2015. International Union of Basic and Clinical Pharmacology. XCIII. The parathyroid hormone receptors—family B G Protein–coupled receptors. Pharmacol. Rev. 67, 310–337.
Garrido, J.L., Wheeler, D., Vega, L.L., Friedman, P.A., Romero, G., 2009. Role of phospholipase D in parathyroid hormone receptor type 1 signaling and trafficking. Mol. Endocrinol. 23, 2048–2059.
Geoffroy, V., Marty-Morieux, C., Le Goupil, N., Clement-Lacroix, P., Terraz, C., Frain, M., Roux, S., Rossert, J., de Vernejoul, M.C., 2004. In vivo inhibition of osteoblastic metalloproteinases leads to increased trabecular bone mass. J. Bone Miner. Res. 19, 811–822.
Gesek, F.A., Friedman, P.A., 1992. Mechanism of calcium transport stimulated by chlorothiazide in mouse distal convoluted tubule cells. J. Clin. Investig. 90, 429–438.
Gesek, F.A., White, K.E., 1997. Molecular and functional identification of β-adrenergic receptors in distal convoluted tubule cells. Am. J. Physiol. Renal. Physiol. 272, F712–F720.
Gill Jr., J.R., Bartter, F.C., 1961. On the impairment of renal concentrating ability in prolonged hypercalcemia and hypercalciuria in man. J. Clin. Investig. 40, 716–722.
Gitelman, H.J., Graham, J.B., Welt, L.G., 1966. A new familial disorder characterized by hypokalemia and hypomagnesemia. Trans. Assoc. Am. Phys. 79, 221–235.
Gley, E., 1893. Glande et glandules thyroides du chien. C. R. Seances Soc. Biol. Fil. 45, 217–219.
Gmaj, P., Murer, H., 1986. Cellular mechanisms of inorganic phosphate transport in kidney. Physiol. Rev. 66, 36–70.
Goltzman, D., Mitchell, J., 1985. Interaction of calcitonin and calcitonin gene-related peptide at receptor sites in target tissues. Science 227, 1343–1345.
Greenwald, I., Gross, J.B., 1925. The effect of the administration of a potent parathyroid extract upon the excretion of nitrogen, phosphorus, calcium, and magnesium, with some remarks on the solubility of calcium phosphate in serum and on the pathogenesis of tetany. J. Biol. Chem. 66, 217–227.
Grey, A., Chen, Q., Xu, X., Callon, K., Cornish, J., 2003. Parallel phosphatidylinositol-3 kinase and p42/44 mitogen-activated protein kinase signaling pathways subserve the mitogenic and antiapoptotic actions of insulin-like growth factor I in osteoblastic cells. Endocrinology 144, 4886–4893.
Grey, A., Mitnick, M.A., Masiukiewicz, U., Sun, B.H., Rudikoff, S., Jilka, R.L., Manolagas, S.C., Insogna, K., 1999. A role for interleukin-6 in parathyroid hormone-induced bone resorption in vivo. Endocrinology 140, 4683–4690.
Gronthos, S., Zannettino, A.C., Graves, S.E., Ohta, S., Hay, S.J., Simmons, P.J., 1999. Differential cell surface expression of the STRO-1 and alkaline phosphatase antigens on discrete developmental stages in primary cultures of human bone cells. J. Bone Miner. Res. 14, 47–56.
Gundberg, C.M., Fawzi, M.I., Clough, M.E., Calvo, M.S., 1995. A comparison of the effects of parathyroid hormone and parathyroid hormone-related protein on osteocalcin in the rat. J. Bone Miner. Res. 10, 903–909.
Guo, J., Song, L., Liu, M., Segawa, H., Miyamoto, K.I., Bringhurst, F.R., Kronenberg, H.M., Juppner, H., 2013. Activation of a non-cAMP/PKA signaling pathway downstream of the PTH/PTHrP receptor is essential for a sustained hypophosphatemic response to PTH infusion in male mice. Endocrinology 154, 1680–1689.
Hackett, D.A., Kauffman Jr., G.L., 2004. Historical perspective of parathyroid disease. Otolaryngol. Clin. 37, 689–700.
Hamm, L.L., Alpern, R.J., Preisig, P.A., 2008. Cellular mechanisms of renal tubular acidification. In: Alpern, R.J., Hebert, S.C. (Eds.), Seldin and Giebisch's the Kindey. Physiology and Pathophysiology. Academic Press, San Diego, pp. 1539–1585.
Hammerman, M.R., 1986. Phosphate transport across renal proximal tubular cell membranes. Am. J. Physiol. Renal. Physiol. 251, F385–F398.
Harris, C.A., Baer, P.G., Chirito, E., Dirks, J.H., 1974. Composition of mammalian glomerular filtrate. Am. J. Physiol. 227, 972–976.
Harris, C.A., Burnatowska, M.A., Seely, J.F., Sutton, R.A.L., Quamme, G.A., Dirks, J.H., 1979. Effects of parathyroid hormone on electrolyte transport in the hamster nephron. Am. J. Physiol. Renal. Physiol. 236, F342–F348.
Hellman, D.E., Au, W.Y., Bartter, F.C., 1965. Evidence for a direct effect of parathyroid hormone on urinary acidification. Am. J. Physiol. 209, 643–650.
Helwig, J.-J., Musso, M.-J., Judes, C., Nickols, G.A., 1991. Parathyroid hormone and calcium: interactions in the control of renin secretion in the isolated, nonfiltering rat kidney. Endocrinology 129, 1233–1242.

Henry, H.L., 1985. Parathyroid hormone modulation of 25-hydroxyvitamin D_3 metabolism by cultured chick kidney cells is mimicked and enhanced by forskolin. Endocrinology 116, 503–510.

Hering, S., Isken, E., Knabbe, C., Janott, J., Jost, C., Pommer, A., Muhr, G., Schatz, H., Pfeiffer, A.F., 2001. TGFβ1 and TGFβ2 mRNA and protein expression in human bone samples. Exp. Clin. Endocrinol. Diabetes 109, 217–226.

Hernando, N., Deliot, N., Gisler, S.M., Lederer, E., Weinman, E.J., Biber, J., Murer, H., 2002. PDZ-domain interactions and apical expression of type IIa Na/P_i cotransporters. Proc. Natl. Acad. Sci. U. S. A. 99, 11957–11962.

Hicok, K.C., Thomas, T., Gori, F., Rickard, D.J., Spelsberg, T.C., Riggs, B.L., 1998. Development and characterization of conditionally immortalized osteoblast precursor cell lines from human bone marrow stroma. J. Bone Miner. Res. 13, 205–217.

Hilfiker, H., Hattenhauer, O., Traebert, M., Forster, I., Murer, H., Biber, J., 1998. Characterization of a murine type II sodium-phosphate cotransporter expressed in mammalian small intestine. Proc. Natl. Acad. Sci. U. S. A. 95, 14564–14569.

Hilpert, J., Nykjaer, A., Jacobsen, C., Wallukat, G., Nielsen, R., Moestrup, S.K., Haller, H., Luft, F.C., Christensen, E.I., Willnow, T.E., 1999. Megalin antagonizes activation of the parathyroid hormone receptor. J. Biol. Chem. 274, 5620–5625.

Hock, J.M., Canalis, E., Centrella, M., 1990. Transforming growth factor-β stimulates bone matrix apposition and bone cell replication in cultured fetal rat calvariae. Endocrinology 126, 421–426.

Hodsman, A.B., Steer, B.M., 1993. Early histomorphometric changes in response to parathyroid hormone therapy in osteoporosis: evidence for de novo bone formation on quiescent cancellous surfaces. Bone 14, 523–527.

Hoenderop, J.G., De Pont, J.J., Bindels, R.J., Willems, P.H., 1999. Hormone-stimulated Ca^{2+} reabsorption in rabbit kidney cortical collecting system is cAMP-independent and involves a phorbol ester-insensitive PKC isotype. Kidney Int. 55, 225–233.

Hofbauer, L.C., Khosla, S., Dunstan, C.R., Lacey, D.L., Boyle, W.J., Riggs, B.L., 2000. The roles of osteoprotegerin and osteoprotegerin ligand in the paracrine regulation of bone resorption. J. Bone Miner. Res. 15, 2–12.

Hoover, R.S., Tomilin, V., Hanson, L.N., Pochynyuk, O., Ko, B., 2015. PTH modulation of NCC activity regulates TRPV5 calcium reabsorption. Am. J. Physiol. Renal. Physiol. 310, F144–F151.

Hopkins, T., Howard, J.E., Eisenberg, H., 1952. Ultrafiltration studies on calcium and phosphorous in human serum. Bull. Johns Hopkins Hosp. 91, 1–21.

Horiuchi, N., Caulfield, M.P., Fisher, J.E., Goldman, M.E., McKee, R.L., Reagan, J.E., Levy, J.J., Nutt, R.F., Rodan, S.B., Schofield, T.L., 1987. Similarity of synthetic peptide from human tumor to parathyroid hormone in vivo and in vitro. Science 238, 1566–1568.

Horiuchi, N., Suda, T., Takahashi, H., Shimazawa, E., Ogata, E., 1977. In vivo evidence for the intermediary role of 3',5'-cyclic AMP in parathyroid hormone-induced stimulation of 1α,25-dihydroxyvitamin D_3 synthesis in rats. Endocrinology 101, 969–974.

Horowitz, M.C., Coleman, D.L., Flood, P.M., Kupper, T.S., Jilka, R.L., 1989. Parathyroid hormone and lipopolysaccharide induce murine osteoblast-like cells to secrete a cytokine indistinguishable from granulocyte-macrophage colony-stimulating factor. J. Clin. Investig. 83, 149–157.

Horwitz, M.J., Tedesco, M.B., Gundberg, C., Garcia-Ocana, A., Stewart, A.F., 2003a. Short-term, high-dose parathyroid hormone-related protein as a skeletal anabolic agent for the treatment of postmenopausal osteoporosis. J. Clin. Endocrinol. Metab. 88, 569–575.

Horwitz, M.J., Tedesco, M.B., Sereika, S.M., Garcia-Ocana, A., Bisello, A., Hollis, B.W., Gundberg, C., Stewart, A.F., 2006. Safety and tolerability of subcutaneous PTHrP(1-36) in healthy human volunteers: a dose escalation study. Osteoporos. Int. 17, 225–230.

Horwitz, M.J., Tedesco, M.B., Sereika, S.M., Hollis, B.W., Garcia-Ocana, A., Stewart, A.F., 2003b. Direct comparison of sustained infusion of human parathyroid hormone-related protein-(1-36) [hPTHrP-(1-36)] versus hPTH-(1-34) on serum calcium, plasma 1,25-dihydroxyvitamin D concentrations, and fractional calcium excretion in healthy human volunteers. J. Clin. Endocrinol. Metab. 88, 1603–1609.

Horwitz, M.J., Tedesco, M.B., Sereika, S.M., Syed, M.A., Garcia-Ocana, A., Bisello, A., Hollis, B.W., Rosen, C.J., Wysolmerski, J.J., Dann, P., Gundberg, C., Stewart, A.F., 2005. Continuous PTH and PTHrP infusion causes suppression of bone formation and discordant effects on 1,25(OH)2 vitamin D. J. Bone Miner. Res. 20, 1792–1803.

Houillier, P., 2014. Mechanisms and regulation of renal magnesium transport. Annu. Rev. Physiol. 76, 411–430.

Hruska, K.A., Goligorsky, M.S., Scoble, J., Tsutsumi, M., Westbrook, S., Moskowitz, D., 1986. Effects of parathyroid hormone on cytosolic calcium in renal proximal tubular primary cultures. Am. J. Physiol. Renal. Physiol. 251, F188–F198.

Hruska, K.A., Moskowitz, D., Esbrit, P., Civitelli, R., Westbrook, S., Huskey, M., 1987. Stimulation of inositol trisphosphate and diacylglycerol production in renal tubular cells by parathyroid hormone. J. Clin. Investig. 79, 230–239.

Hu, M.C., Shiizaki, K., Kuro, O.M., Moe, O.W., 2013. Fibroblast growth factor 23 and klotho: physiology and pathophysiology of an endocrine network of mineral metabolism. Annu. Rev. Physiol. 75, 503–533.

Huang, J.C., Sakata, T., Pfleger, L.L., Bencsik, M., Halloran, B.P., Bikle, D.D., Nissenson, R.A., 2004. PTH differentially regulates expression of RANKL and OPG. J. Bone Miner. Res. 19, 235–244.

Huang, Z., Chen, Y., Nissenson, R.A., 1995. The cytoplasmic tail of the G-protein-coupled receptor for parathyroid hormone and parathyroid hormone-related protein contains positive and negative signals for endocytosis. J. Biol. Chem. 270, 151–156.

Hulter, H.N., 1985. Effects and interrelationships of PTH, Ca^{2+}, vitamin D, and Pi in acid-base homeostasis. Am. J. Physiol. Renal. Physiol. 248, F739–F752.

Hulter, H.N., Peterson, J.C., 1985. Acid-base homeostasis during chronic PTH excess in humans. Kidney Int. 28, 187–192.

Hurley, M.M., Okada, Y., Xiao, L., Tanaka, Y., Ito, M., Okimoto, N., Nakamura, T., Rosen, C.J., Doetschman, T., Coffin, J.D., 2006. Impaired bone anabolic response to parathyroid hormone in Fgf2-/- and Fgf2+/- mice. Biochem. Biophys. Res. Commun. 341, 989–994.

Hurley, M.M., Tetradis, S., Huang, Y.F., Hock, J., Kream, B.E., Raisz, L.G., Sabbieti, M.G., 1999. Parathyroid hormone regulates the expression of fibroblast growth factor-2 mRNA and fibroblast growth factor receptor mRNA in osteoblastic cells. J. Bone Miner. Res. 14, 776–783.

Inda, C., Dos Santos Claro, P.A., Bonfiglio, J.J., Senin, S.A., Maccarrone, G., Turck, C.W., Silberstein, S., 2016. Different cAMP sources are critically involved in G protein-coupled receptor CRHR1 signaling. J. Cell Biol. 214, 181–195.

Inomata, N., Akiyama, M., Kubota, N., Juppner, H., 1995. Characterization of a novel parathyroid hormone (PTH) receptor with specificity for the carboxyl-terminal region of PTH-(1-84). Endocrinology 136, 4732–4740.

Ishizuya, T., Yokose, S., Hori, M., Noda, T., Suda, T., Yoshiki, S., Yamaguchi, A., 1997. Parathyroid hormone exerts disparate effects on osteoblast differentiation depending on exposure time in rat osteoblastic cells. J. Clin. Investig. 99, 2961–2970.

Izquierdo, A., Lopez-Luna, P., Ortega, A., Romero, M., Guitierrez-Tarres, M.A., Arribas, I., Alvarez, M.J., Esbrit, P., Bosch, R.J., 2006. The parathyroid hormone-related protein system and diabetic nephropathy outcome in streptozotocin-induced diabetes. Kidney Int. 69, 2171–2177.

Janulis, M., Tembe, V., Favus, M.J., 1992. Role of PKC in parathyroid hormone stimulation of renal 1,25-dihydroxyvitamin D_3 secretion. J. Clin. Investig. 90, 2278–2283.

Janulis, M., Wong, M.S., Favus, M.J., 1993. Structure-function requirements of parathyroid hormone for stimulation of 1,25-dihydroxyvitamin D_3 production by rat renal proximal tubules. Endocrinology 133, 713–719.

Jespersen, B., Randlov, A., Abrahamsen, J., Fogh-Andersen, N., Kanstrup, I.L., 1997. Effects of PTH(1-34) on blood pressure, renal function, and hormones in essential hypertension: the altered pattern of reactivity may counteract raised blood pressure. Am. J. Hypertens. 10, 1356–1367.

Jilka, R.L., Weinstein, R.S., Bellido, T., Parfitt, A.M., Manolagas, S.C., 1998. Osteoblast programmed cell death (apoptosis): modulation by growth factors and cytokines. J. Bone Miner. Res. 13, 793–802.

Jilka, R.L., Weinstein, R.S., Bellido, T., Roberson, P., Parfitt, A.M., Manolagas, S.C., 1999. Increased bone formation by prevention of osteoblast apoptosis with parathyroid hormone. J. Clin. Investig. 104, 439–446.

Jobert, A.S., Fernandes, I., Turner, G., Coureau, C., Prie, D., Nissenson, R.A., Friedlander, G., Silve, C., 1996. Expression of alternatively spliced isoforms of the parathyroid hormone (PTH)/PTH-related peptide receptor messenger RNA in human kidney and bone cells. Mol. Endocrinol. 10, 1066–1076.

Jongen, J.W., Bos, M.P., van der Meer, J.M., Herrmann-Erlee, M.P., 1993. Parathyroid hormone-induced changes in alkaline phosphatase expression in fetal calvarial osteoblasts: differences between rat and mouse. J. Cell. Physiol. 155, 36–43.

Joun, H., Lanske, B., Karperien, M., Qian, F., Defize, L., Abou-Samra, A., 1997. Tissue-specific transcription start sites and alternative splicing of the parathyroid hormone (PTH)/PTH-related peptide (PTHrP) receptor gene: a new PTH/PTHrP receptor splice variant that lacks the signal peptide. Endocrinology 138, 1742–1749.

Juppner, H., Abou-Samra, A.B., Freeman, M., Kong, X.F., Schipani, E., Richards, J., Kolakowski Jr., L.F., Hock, J., Potts Jr., J.T., Kronenberg, H.M., et al., 1991. A G protein-linked receptor for parathyroid hormone and parathyroid hormone-related peptide. Science 254, 1024–1026.

Kano, J., Sugimoto, T., Fukase, M., Chihara, K., 1994. Direct involvement of cAMP-dependent protein kinase in the regulation of alkaline phosphatase activity by parathyroid hormone (PTH) and PTH-related peptide in osteoblastic UMR-106 cells. Biochem. Biophys. Res. Commun. 199, 271–276.

Kano, J., Sugimoto, T., Fukase, M., Fujita, T., 1991. The activation of cAMP-dependent protein kinase is directly linked to the inhibition of osteoblast proliferation (UMR-106) by parathyroid hormone-related protein. Biochem. Biophys. Res. Commun. 179, 97–101.

Karim, Z., Gerard, B., Bakouh, N., Alili, R., Leroy, C., Beck, L., Silve, C., Planelles, G., Urena-Torres, P., Grandchamp, B., Friedlander, G., Prie, D., 2008. NHERF1 mutations and responsiveness of renal parathyroid hormone. N. Engl. J. Med. 359, 1128–1135.

Karmali, R., Schiffmann, S.N., Vanderwinden, J.M., Hendy, G.N., Nys-DeWolf, N., Corvilain, J., Bergmann, P., Vanderhaeghen, J.J., 1992. Expression of mRNA of parathyroid hormone-related peptide in fetal bones of the rat. Cell Tissue Res. 270, 597–600.

Kartsogiannis, V., Moseley, J., McKelvie, B., Chou, S.T., Hards, D.K., Ng, K.W., Martin, T.J., Zhou, H., 1997. Temporal expression of PTHrP during endochondral bone formation in mouse and intramembranous bone formation in an in vivo rabbit model. Bone 21, 385–392.

Kau, S.T., Maack, T., 1977. Transport and catabolism of parathyroid hormone in isolated rat kidney. Am. J. Physiol. Renal. Physiol. 233, F445–F454.

Kaufmann, M., Muff, R., Stieger, B., Biber, J., Murer, H., Fischer, J.A., 1994. Apical and basolateral parathyroid hormone receptors in rat renal cortical membranes. Endocrinology 134, 1173–1178.

Kavanaugh, M.P., Kabat, D., 1996. Identification and characterization of a widely expressed phosphate transporter/retrovirus receptor family. Kidney Int. 49, 959–963.

Kempson, S.A., Lötscher, M., Kaissling, B., Biber, J., Murer, H., Levi, M., 1995. Parathyroid hormone action on phosphate transporter mRNA and protein in rat renal proximal tubules. Am. J. Physiol. Renal. Physiol. 268, F784–F791.

Keusch, I., Traebert, M., Lotscher, M., Kaissling, B., Murer, H., Biber, J., 1998. Parathyroid hormone and dietary phosphate provoke a lysosomal routing of the proximal tubular Na/Pi-cotransporter type II. Kidney Int. 54, 1224–1232.

Khosla, S., Westendorf, J.J., Oursler, M.J., 2008. Building bone to reverse osteoporosis and repair fractures. J. Clin. Investig. 118, 421–428.

Kilav, R., Silver, J., Biber, J., Murer, H., Naveh-Many, T., 1995a. Coordinate regulation of rat renal parathyroid hormone receptor mRNA and Na-Pi cotransporter mRNA and protein. Am. J. Physiol. Renal. Physiol. 268, F1017–F1022.

Kilav, R., Silver, J., Naveh-Many, T., 1995b. Parathyroid hormone gene expression in hypophosphatemic rats. J. Clin. Investig. 96, 327–333.

Kitazawa, R., Imai, Y., Fukase, M., Fujita, T., 1991. Effects of continuous infusion of parathyroid hormone and parathyroid hormone-related peptide on rat bone in vivo: comparative study by histomorphometry. Bone Miner. 12, 157–166.

Kleeman, C.R., Cooke, R.E., 1951. The acute effects of parathyroid hormone on the metabolism of endogenous phosphate. J. Lab. Clin. Med. 38, 112–127.

Ko, B., 2017. Parathyroid hormone and the regulation of renal tubular calcium transport. Curr. Opin. Nephrol. Hypertens. 26, 405–410.

Ko, B., Cooke, L.L., Hoover, R.S., 2011. Parathyroid hormone (PTH) regulates the sodium chloride cotransporter via Ras guanyl releasing protein 1 (Ras-GRP1) and extracellular signal-regulated kinase (ERK)1/2 mitogen-activated protein kinase (MAPK) pathway. Transl. Res. 158, 282–289.

Komori, T., 2002. Runx2, a multifunctional transcription factor in skeletal development. J. Cell. Biochem. 87, 1–8.

Komori, T., Yagi, H., Nomura, S., Yamaguchi, A., Sasaki, K., Deguchi, K., Shimizu, Y., Bronson, R.T., Gao, Y.H., Inada, M., Sato, M., Okamoto, R., Kitamura, Y., Yoshiki, S., Kishimoto, T., 1997. Targeted disruption of Cbfa1 results in a complete lack of bone formation owing to maturational arrest of osteoblasts. Cell 89, 755–764.

Kondo, H., Guo, J., Bringhurst, F.R., 2002. Cyclic adenosine monophosphate/protein kinase A mediates parathyroid hormone/parathyroid hormone-related protein receptor regulation of osteoclastogenesis and expression of RANKL and osteoprotegerin mRNAs by marrow stromal cells. J. Bone Miner. Res. 17, 1667–1679.

Korkor, A.B., Gray, R.W., Henry, H.L., Kleinman, J.G., Blumenthal, S.S., Garancis, J.C., 1987. Evidence that stimulation of 1,25(OH)$_2$D$_3$ production in primary cultures of mouse kidney cells by cyclic AMP requires new protein synthesis. J. Bone Miner. Res. 2, 517–524.

Kostenuik, P.J., Harris, J., Halloran, B.P., Turner, R.T., Morey-Holton, E.R., Bikle, D.D., 1999. Skeletal unloading causes resistance of osteoprogenitor cells to parathyroid hormone and to insulin-like growth factor-I. J. Bone Miner. Res. 14, 21–31.

Kovacs, C.S., Lanske, B., Hunzelman, J.L., Guo, J., Karaplis, A.C., Kronenberg, H.M., 1996. Parathyroid hormone-related peptide (PTHrP) regulates fetal-placental calcium transport through a receptor distinct from the PTH/PTHrP receptor. Proc. Natl. Acad. Sci. U. S. A. 93, 15233–15238.

Kream, B.E., Rowe, D., Smith, M.D., Maher, V., Majeska, R., 1986. Hormonal regulation of collagen synthesis in a clonal rat osteosarcoma cell line. Endocrinology 119, 1922–1928.

Kream, B.E., Rowe, D.W., Gworek, S.C., Raisz, L.G., 1980. Parathyroid hormone alters collagen synthesis and procollagen mRNA levels in fetal rat calvaria. Proc. Natl. Acad. Sci. U. S. A. 77, 5654–5658.

Krishnan, V., Moore, T.L., Ma, Y.L., Helvering, L.M., Frolik, C.A., Valasek, K.M., Ducy, P., Geiser, A.G., 2003. Parathyroid hormone bone anabolic action requires Cbfa1/Runx2-dependent signaling. Mol. Endocrinol. 17, 423–435.

Kulkarni, N.H., Halladay, D.L., Miles, R.R., Gilbert, L.M., Frolik, C.A., Galvin, R.J., Martin, T.J., Gillespie, M.T., Onyia, J.E., 2005. Effects of parathyroid hormone on Wnt signaling pathway in bone. J. Cell. Biochem. 95, 1178–1190.

Kuna, R.S., Girada, S.B., Asalla, S., Vallentyne, J., Maddika, S., Patterson, J.T., Smiley, D.L., DiMarchi, R.D., Mitra, P., 2013. Glucagon-like peptide-1 receptor-mediated endosomal cAMP generation promotes glucose-stimulated insulin secretion in pancreatic β-cells. Am. J. Physiol. Endocrinol. Metab. 305, E161–E170.

Lacey, D.L., Timms, E., Tan, H.L., Kelley, M.J., Dunstan, C.R., Burgess, T., Elliott, R., Colombero, A., Elliott, G., Scully, S., Hsu, H., Sullivan, J., Hawkins, N., Davy, E., Capparelli, C., Eli, A., Qian, Y.X., Kaufman, S., Sarosi, I., Shalhoub, V., Senaldi, G., Guo, J., Delaney, J., Boyle, W.J., 1998. Osteoprotegerin ligand is a cytokine that regulates osteoclast differentiation and activation. Cell 93, 165–176.

Lammi, J., Huppunen, J., Aarnisalo, P., 2004. Regulation of the osteopontin gene by the orphan nuclear receptor NURR1 in osteoblasts. Mol. Endocrinol. 18, 1546–1557.

Langub, M.C., Monier-Faugere, M.C., Qi, Q., Geng, Z., Koszewski, N.J., Malluche, H.H., 2001. Parathyroid hormone/parathyroid hormone-related peptide type 1 receptor in human bone. J. Bone Miner. Res. 16, 448–456.

Largo, R., Gomez-Garre, D., Santos, S., Penaranda, C., Blanco, J., Esbrit, P., Egido, J., 1999. Renal expression of parathyroid hormone-related protein (PTHrP) and PTH/PTHrP receptor in a rat model of tubulointerstitial damage. Kidney Int. 55, 82–90.

Larkins, R.G., MacAuley, S.J., Rapoport, A., Martin, T.J., Tulloch, B.R., Byfield, P.G., Matthews, E.W., MacIntyre, I., 1974. Effects of nucleotides, hormones, ions, and 1,25-dihydroxycholecalciferon on 1,25-dihydroxycholecalciferol production in isolated chick renal tubules. Clin. Sci. Mol. Med. 46, 569–582.

LaTour, D., Mohan, S., Linkhart, T.A., Baylink, D.J., Strong, D.D., 1990. Inhibitory insulin-like growth factor-binding protein: cloning, complete sequence, and physiological regulation. Mol. Endocrinol. 4, 1806–1814.

Law, F., Bonjour, J.P., Rizzoli, R., 1994. Transforming growth factor-beta: a down-regulator of the parathyroid hormone-related protein receptor in renal epithelial cells. Endocrinology 134, 2037–2043.

Le Grimellec, C., Roinel, N., Morel, F., 1974. Simultaneous Mg, Ca, P, K, Na and Cl analysis in rat tubular fluid. III. During acute Ca plasma loading. Pflügers Archiv 346, 171–188.

Leder, B.Z., O'Dea, L.S., Zanchetta, J.R., Kumar, P., Banks, K., McKay, K., Lyttle, C.R., Hattersley, G., 2015. Effects of abaloparatide, a human parathyroid hormone-related peptide analog, on bone mineral density in postmenopausal women with osteoporosis. J. Clin. Endocrinol. Metab. 100, 697–706.

Lee, J.J., Plain, A., Beggs, M.R., Dimke, H., Alexander, R.T., 2017. Effects of phospho- and calciotropic hormones on electrolyte transport in the proximal tubule. F1000Res 6, 1797.

Lee, K., Deeds, J.D., Bond, A.T., Juppner, H., Abou-Samra, A.B., Segre, G.V., 1993. In situ localization of PTH/PTHrP receptor mRNA in the bone of fetal and young rats. Bone 14, 341–345.

Lee, K., Deeds, J.D., Chiba, S., Un-No, M., Bond, A.T., Segre, G.V., 1994. Parathyroid hormone induces sequential c-fos expression in bone cells in vivo: in situ localization of its receptor and c-fos messenger ribonucleic acids. Endocrinology 134, 441–450.

Lee, K., Deeds, J.D., Segre, G.V., 1995. Expression of parathyroid hormone-related peptide and its receptor messenger ribonucleic acids during fetal development of rats. Endocrinology 136, 453–463.

Lee, S.K., Lorenzo, J.A., 1999. Parathyroid hormone stimulates TRANCE and inhibits osteoprotegerin messenger ribonucleic acid expression in murine bone marrow cultures: correlation with osteoclast-like cell formation. Endocrinology 140, 3552–3561.

Li, X., Zhang, Y., Kang, H., Liu, W., Liu, P., Zhang, J., Harris, S.E., Wu, D., 2005. Sclerostin binds to LRP5/6 and antagonizes canonical Wnt signaling. J. Biol. Chem. 280, 19883–19887.

Liang, J.D., Hock, J.M., Sandusky, G.E., Santerre, R.F., Onyia, J.E., 1999. Immunohistochemical localization of selected early response genes expressed in trabecular bone of young rats given hPTH 1-34. Calcif. Tissue Int. 65, 369–373.

Lindsay, R., Zhou, H., Cosman, F., Nieves, J., Dempster, D.W., Hodsman, A.B., 2007. Effects of a one-month treatment with PTH(1-34) on bone formation on cancellous, endocortical, and periosteal surfaces of the human ilium. J. Bone Miner. Res. 22, 495–502.

Ling, L., Murali, S., Dombrowski, C., Haupt, L.M., Stein, G.S., van Wijnen, A.J., Nurcombe, V., Cool, S.M., 2006. Sulfated glycosaminoglycans mediate the effects of FGF2 on the osteogenic potential of rat calvarial osteoprogenitor cells. J. Cell. Physiol. 209, 811–825.

Linkhart, T.A., Mohan, S., 1989. Parathyroid hormone stimulates release of insulin-like growth factor-I (IGF-I) and IGF-II from neonatal mouse calvaria in organ culture. Endocrinology 125, 1484–1491.

Liu, L., Alonso, V., Guo, L., Tourkova, I., Henderson, S.E., Almarza, A.J., Friedman, P.A., Blair, H.C., 2012. Na+/H+ exchanger regulatory factor 1 (NHERF1) directly regulates osteogenesis. J. Biol. Chem. 287, 43312–43321.

Llach, F., Massry, S.G., 1985. On the mechanism of secondary hyperparathyroidism in moderate renal insufficiency. J. Clin. Endocrinol. Metab. 61, 601–606.

Locklin, R.M., Khosla, S., Turner, R.T., Riggs, B.L., 2003. Mediators of the biphasic responses of bone to intermittent and continuously administered parathyroid hormone. J. Cell. Biochem. 89, 180–190.

Loeb, R.F., 1926. The effect of pure protein solutions and of blood serum on the diffusibility of calcium. J. Gen. Physiol. 8, 451–461.

Lötscher, M., Scarpetta, Y., Levi, M., Halaihel, N., Wang, H.M., Zajicek, H.K., Biber, J., Murer, H., Kaissling, B., 1999. Rapid downregulation of rat renal Na/P_i cotransporter in response to parathyroid hormone involves microtubule rearrangement. J. Clin. Investig. 104, 483–494.

Ma, Y.L., Cain, R.L., Halladay, D.L., Yang, X., Zeng, Q., Miles, R.R., Chandrasekhar, S., Martin, T.J., Onyia, J.E., 2001. Catabolic effects of continuous human PTH (1–38) in vivo is associated with sustained stimulation of RANKL and inhibition of osteoprotegerin and gene-associated bone formation. Endocrinology 142, 4047–4054.

Madjdpour, C., Bacic, D., Kaissling, B., Murer, H., Biber, J., 2004. Segment-specific expression of sodium-phosphate cotransporters NaPi-IIa and -IIc and interacting proteins in mouse renal proximal tubules. Pflügers Archiv 448, 402–410.

Magyar, C.E., White, K.E., Rojas, R., Apodaca, G., Friedman, P.A., 2002. Plasma membrane Ca^{2+}-ATPase and NCX1 Na^+/Ca^{2+} exchanger expression in distal convoluted tubule cells. Am. J. Physiol. Renal. Physiol. 283, F29–F40.

Mahon, M.J., 2012. The parathyroid hormone receptorsome and the potential for therapeutic intervention. Curr. Drug Targets 13, 116–128.

Mahon, M.J., Cole, J.A., Lederer, E.D., Segre, G.V., 2003. Na^+/H^+ exchanger-regulatory factor 1 mediates Inhibition of phosphate transport by parathyroid hormone and second messengers by acting at multiple sites in opossum kidney cells. Mol. Endocrinol. 17, 2355–2364.

Mahon, M.J., Donowitz, M., Yun, C.C., Segre, G.V., 2002. Na^+/H^+ exchanger regulatory factor 2 directs parathyroid hormone 1 receptor signalling. Nature 417, 858–861.

Mahon, M.J., Segre, G.V., 2004. Stimulation by parathyroid hormone of a NHERF-1-assembled complex consisting of the parathyroid hormone I receptor, PLCß, and actin increases intracellular calcium in opossum kidney cells. J. Biol. Chem. 279, 23550–23558.

Maiti, A., Beckman, M.J., 2007. Extracellular calcium is a direct effecter of VDR levels in proximal tubule epithelial cells that counter-balances effects of PTH on renal Vitamin D metabolism. J. Steroid Biochem. Mol. Biol. 103, 504–508.

Maiti, A., Hait, N.C., Beckman, M.J., 2008. Extracellular calcium sensing receptor activation induces vitamin D receptor levels in proximal kidney HK-2G cells by a mechanism that requires phosphorylation of p38α MAPK. J. Biol. Chem. 283, 175–183.

Majeska, R.J., Rodan, G.A., 1982. Alkaline phosphatase inhibition by parathyroid hormone and isoproterenol in a clonal rat osteosarcoma cell line. Possible mediation by cyclic AMP. Calcif. Tissue Int. 34, 59–66.

Malecz, N., Bambino, T., Bencsik, M., Nissenson, R.A., 1998. Identification of phosphorylation sites in the G protein-coupled receptor for parathyroid hormone. Receptor phosphorylation is not required for agonist-induced internalization. Mol. Endocrinol. 12, 1846–1856.

Malmström, K., Murer, H., 1987. Parathyroid hormone regulates phosphate transport in OK cells via an irreversible inactivation of a membrane protein. FEBS Lett. 216, 257–260.

Mandel, L.J., Derrickson, B.H., 1997. Parathyroid hormone inhibits Na^+-K^+ ATPase through G_q/G_{11} and the calcium-independent phospholipase A_2. Am. J. Physiol. Renal. Physiol. 272, F781–F788.

Marone, C.C., Wong, N.L., Sutton, R.A., Dirks, J.H., 1983. Effects of metabolic alkalosis on calcium excretion in the conscious dog. J. Lab. Clin. Med. 101, 264–273.

Martinez, M.E., Garcia-Ocana, A., Sanchez, M., Medina, S., del Campo, T., Valin, A., Sanchez-Cabezudo, M.J., Esbrit, P., 1997. C-terminal parathyroid hormone-related protein inhibits proliferation and differentiation of human osteoblast-like cells. J. Bone Miner. Res. 12, 778–785.

Massfelder, T., Fiaschi-Taesch, N., Stewart, A.F., Helwig, J.J., 1998. Parathyroid hormone-related peptide–a smooth muscle tone and proliferation regulatory protein. Curr. Opin. Nephrol. Hypertens. 7, 27–32.

Massfelder, T., Helwig, J.J., 1999. Parathyroid hormone-related protein in cardiovascular development and blood pressure regulation. Endocrinology 140, 1507–1510.

Massfelder, T., Stewart, A.F., Endlich, K., Soifer, N., Judes, C., Helwig, J.J., 1996. Parathyroid hormone-related protein detection and interaction with NO and cyclic AMP in the renovascular system. Kidney Int. 50, 1591–1603.

McCarthy, T.L., Centrella, M., Canalis, E., 1989. Parathyroid hormone enhances the transcript and polypeptide levels of insulin-like growth factor I in osteoblast-enriched cultures from fetal rat bone. Endocrinology 124, 1247–1253.

McCarthy, T.L., Centrella, M., Canalis, E., 1990. Cyclic AMP induces insulin-like growth factor I synthesis in osteoblast-enriched cultures. J. Biol. Chem. 265, 15353–15356.

McCauley, L.K., Koh, A.J., Beecher, C.A., Rosol, T.J., 1997. Proto-oncogene c-fos is transcriptionally regulated by parathyroid hormone (PTH) and PTH-related protein in a cyclic adenosine monophosphate-dependent manner in osteoblastic cells. Endocrinology 138, 5427–5433.

McCauley, L.K., Koh-Paige, A.J., Chen, H., Chen, C., Ontiveros, C., Irwin, R., McCabe, L.R., 2001. Parathyroid hormone stimulates fra-2 expression in osteoblastic cells in vitro and in vivo. Endocrinology 142, 1975–1981.

McCuaig, K.A., Lee, H.S., Clarke, J.C., Assar, H., Horsford, J., White, J.H., 1995. Parathyroid hormone/parathyroid hormone related peptide receptor gene transcripts are expressed from tissue-specific and ubiquitous promoters. Nucleic Acids Res. 23, 1948–1955.

McPartlin, J., Skrabanek, P., Powell, D., 1978. Early effects of parathyroid hormone on rat calvarian bone alkaline phosphatase. Endocrinology 103, 1573–1578.

McSheehy, P.M., Chambers, T.J., 1986. Osteoblastic cells mediate osteoclastic responsiveness to parathyroid hormone. Endocrinology 118, 824–828.

Meikle, M.C., Bord, S., Hembry, R.M., Compston, J., Croucher, P.I., Reynolds, J.J., 1992. Human osteoblasts in culture synthesize collagenase and other matrix metalloproteinases in response to osteotropic hormones and cytokines. J. Cell Sci. 103 (Pt 4), 1093–1099.

Meites, S., 1989. Pediatric Clinical Chemistry: Reference (Normal) Values, third ed. AACC Press, Washington, DC.

Meltzer, V., Weinreb, S., Bellorin-Font, E., Hruska, K.A., 1982. Parathyroid hormone stimulation of renal phosphoinositide metabolism is a cyclic nucleotide-independent effect. Biochim. Biophys. Acta 712, 258–267.

Mensenkamp, A.R., Hoenderop, J.G., Bindels, R.J., 2007. TRPV5, the gateway to Ca^{2+} homeostasis. Handb. Exp. Pharmacol. 207–220.

Merciris, D., Schiltz, C., Legoupil, N., Marty-Morieux, C., de Vernejoul, M.C., Geoffroy, V., 2007. Over-expression of TIMP-1 in osteoblasts increases the anabolic response to PTH. Bone 40, 75–83.

Merriam, L.A., Baran, C.N., Girard, B.M., Hardwick, J.C., May, V., Parsons, R.L., 2013. Pituitary adenylate cyclase 1 receptor internalization and endosomal signaling mediate the pituitary adenylate cyclase activating polypeptide-induced increase in Guinea pig cardiac neuron excitability. J. Neurosci. 33, 4614–4622.

Meyer, M.B., Benkusky, N.A., Kaufmann, M., Lee, S.M., Onal, M., Jones, G., Pike, J.W., 2017. A kidney-specific genetic control module in mice governs endocrine regulation of the cytochrome P450 gene Cyp27b1 essential for vitamin D3 activation. J. Biol. Chem. 292, 17541–17558.

Miao, D., Tong, X.K., Chan, G.K., Panda, D., McPherson, P.S., Goltzman, D., 2001. Parathyroid hormone-related peptide stimulates osteogenic cell proliferation through protein kinase C activation of the Ras/mitogen-activated protein kinase signaling pathway. J. Biol. Chem. 276, 32204–32213.

Miles, R.R., Roberts, R.F., Putnam, A.R., Roberts, W.L., 2004. Comparison of serum and heparinized plasma samples for measurement of chemistry analytes. Clin. Chem. 50, 1704–1706.

Miles, R.R., Sluka, J.P., Halladay, D.L., Santerre, R.F., Hale, L.V., Bloem, L., Thirunavukkarasu, K., Galvin, R.J., Hock, J.M., Onyia, J.E., 2000a. ADAMTS-1: a cellular disintegrin and metalloprotease with thrombospondin motifs is a target for parathyroid hormone in bone. Endocrinology 141, 4533–4542.

Miles, R.R., Sluka, J.P., Santerre, R.F., Hale, L.V., Bloem, L., Boguslawski, G., Thirunavukkarasu, K., Hock, J.M., Onyia, J.E., 2000b. Dynamic regulation of RGS2 in bone: potential new insights into parathyroid hormone signaling mechanisms. Endocrinology 141, 28–36.

Moe, O.W., Amemiya, M., Yamaji, Y., 1995. Activation of protein kinase A acutely inhibits and phosphorylates Na/H exchanger NHE-3. J. Clin. Investig. 96, 2187–2194.

Moe, S.M., 2016. Calcium homeostasis in health and in kidney disease. Comp. Physiol. 6, 1781–1800.

Morales, F.C., Takahashi, Y., Kreimann, E.L., Georgescu, M.M., 2004. Ezrin-radixin-moesin (ERM)-binding phosphoprotein 50 organizes ERM proteins at the apical membrane of polarized epithelia. Proc. Natl. Acad. Sci. U. S. A. 101, 17705–17710.

Mori, Y., Machida, T., Miyakawa, S., Bomsztyk, K., 1992. Effects of amiloride on distal renal tubule sodium and calcium absorption: dependence on luminal pH. Pharmacol. Toxicol. 70, 201–204.

Morony, S., Capparelli, C., Lee, R., Shimamoto, G., Boone, T., Lacey, D.L., Dunstan, C.R., 1999. A chimeric form of osteoprotegerin inhibits hypercalcemia and bone resorption induced by IL-β, TNF-α, PTH, PTHrP, and 1, $25(OH)_2D_3$. J. Bone Miner. Res. 14, 1478–1485.

Muff, R., Fischer, J.A., Biber, J., Murer, H., 1992. Parathyroid hormone receptors in control of proximal tubule function. Annu. Rev. Physiol. 54, 67–79.

Mundy, G.R., Edwards, J.R., 2008. PTH-related peptide (PTHrP) in hypercalcemia. J. Am. Soc. Nephrol. 19, 672–675.

Murayama, A., Takeyama, K., Kitanaka, S., Kodera, Y., Kawaguchi, Y., Hosoya, T., Kato, S., 1999. Positive and negative regulations of the renal 25-hydroxyvitamin D_3 1α-hydroxylase gene by parathyroid hormone, calcitonin, and $1\alpha,25(OH)_2D_3$ in intact animals. Endocrinology 140, 2224–2231.

Murer, H., Biber, J., 1997. A molecular view of proximal tubular inorganic phosphate (P_i) reabsorption and of its regulation. Pflügers Archiv 433, 379–389.

Murer, H., Hernando, N., Forster, I., Biber, J., 2000. Proximal tubular phosphate reabsorption: molecular mechanisms. Physiol. Rev. 80, 1373–1409.

Murer, H., Hernando, N., Forster, I., Biber, J., 2003. Regulation of Na/Pi transporter in the proximal tubule. Annu. Rev. Physiol. 65, 531–542.

Murray, T.M., Rao, L.G., Muzaffar, S.A., 1991. Dexamethasone-treated ROS 17/2.8 rat osteosarcoma cells are responsive to human carboxylterminal parathyroid hormone peptide hPTH (53-84): stimulation of alkaline phosphatase. Calcif. Tissue Int. 49, 120–123.

Murray, T.M., Rao, L.G., Muzaffar, S.A., Ly, H., 1989. Human parathyroid hormone carboxyterminal peptide (53-84) stimulates alkaline phosphatase activity in dexamethasone-treated rat osteosarcoma cells in vitro. Endocrinology 124, 1097–1099.

Murrills, R.J., Stein, L.S., Fey, C.P., Dempster, D.W., 1990. The effects of parathyroid hormone (PTH) and PTH-related peptide on osteoclast resorption of bone slices in vitro: an analysis of pit size and the resorption focus. Endocrinology 127, 2648–2653.

Murtazina, R., Kovbasnjuk, O., Zachos, N.C., Li, X., Chen, Y., Hubbard, A., Hogema, B.M., Steplock, D., Seidler, U., Hoque, K.M., Tse, C.M., De Jonge, H.R., Weinman, E.J., Donowitz, M., 2007. Tissue-specific regulation of sodium/proton exchanger isoform 3 activity in Na^+/H^+ exchanger regulatory factor 1 (NHERF1) null mice. cAMP inhibition is differentially dependent on NHERF1 and exchange protein directly activated by cAMP in ileum versus proximal tubule. J. Biol. Chem. 282, 25141–25151.

Musso, M.J., Barthelmebs, M., Imbs, J.L., Plante, M., Bollack, C., Helwig, J.J., 1989. The vasodilator action of parathyroid hormone fragments on isolated perfused rat kidney. Naunyn-Schmiedeberg's Arch. Pharmacol. 340, 246–251.

Na, T., Peng, J.B., 2014. TRPV5: a Ca^{2+} channel for the fine-tuning of Ca^{2+} reabsorption. Handb. Exp. Pharmacol. 222, 321–357.

Nagai, S., Okazaki, M., Segawa, H., Bergwitz, C., Dean, T., Potts Jr., J.T., Mahon, M.J., Gardella, T.J., Juppner, H., 2011. Acute down-regulation of sodium-dependent phosphate transporter NPT2a involves predominantly the cAMP/PKA pathway as revealed by signaling-selective parathyroid hormone analogs. J. Biol. Chem. 286, 1618–1626.

Nakamoto, C., Baba, H., Fukase, M., Nakajima, K., Kimura, T., Sakakibara, S., Fujita, T., Chihara, K., 1993. Individual and combined effects of intact PTH, amino-terminal, and a series of truncated carboxyl-terminal PTH fragments on alkaline phosphatase activity in dexamethasone-treated rat osteoblastic osteosarcoma cells, ROS 17/2.8. Acta Endocrinol. 128, 367–372.

Nakashima, K., Zhou, X., Kunkel, G., Zhang, Z., Deng, J.M., Behringer, R.R., de Crombrugghe, B., 2002. The novel zinc finger-containing transcription factor osterix is required for osteoblast differentiation and bone formation. Cell 108, 17–29.

Nervina, J.M., Magyar, C.E., Pirih, F.Q., Tetradis, S., 2006. PGC-1alpha is induced by parathyroid hormone and coactivates Nurr1-mediated promoter activity in osteoblasts. Bone 39, 1018–1025.

Nguyen-Yamamoto, L., Rousseau, L., Brossard, J.H., Lepage, R., D'Amour, P., 2001. Synthetic carboxyl-terminal fragments of parathyroid hormone (PTH) decrease ionized calcium concentration in rats by acting on a receptor different from the PTH/PTH-related peptide receptor. Endocrinology 142, 1386–1392.

Nishida, S., Yamaguchi, A., Tanizawa, T., Endo, N., Mashiba, T., Uchiyama, Y., Suda, T., Yoshiki, S., Takahashi, H.E., 1994. Increased bone formation by intermittent parathyroid hormone administration is due to the stimulation of proliferation and differentiation of osteoprogenitor cells in bone marrow. Bone 15, 717–723.

Nishio, Y., Dong, Y., Paris, M., O'Keefe, R.J., Schwarz, E.M., Drissi, H., 2006. Runx2-mediated regulation of the zinc finger Osterix/Sp7 gene. Gene 372, 62–70.

Noda, M., Yoon, K., Rodan, G.A., 1988. Cyclic AMP-mediated stabilization of osteocalcin mRNA in rat osteoblast-like cells treated with parathyroid hormone. J. Biol. Chem. 263, 18574–18577.

Nolin, T.D., Friedman, P.A., 2018. Agents affecting mineral ion homeostasis and bone turnover. In: Brunton, L.L., Knollman, B., Hilal-Dandan, R. (Eds.), Goodman & Gilman's the Pharmacological Basis of Therapeutics, 13 ed. McGraw Hill, New York, pp. 887–906.

Norden, A.G., Lapsley, M., Lee, P.J., Pusey, C.D., Scheinman, S.J., Tam, F.W., Thakker, R.V., Unwin, R.J., Wrong, O., 2001. Glomerular protein sieving and implications for renal failure in Fanconi syndrome. Kidney Int. 60, 1885–1892.

Nordin, B.E., 1960. The effect of intravenous parathyroid extract on urinary pH, bicarbonate and electrolyte excretion. Clin. Sci. 19, 311–319.

Obri, A., Makinistoglu, M.P., Zhang, H., Karsenty, G., 2014. HDAC4 integrates PTH and sympathetic signaling in osteoblasts. J. Cell Biol. 205, 771–780.

Ogata, Y., Nakao, S., Kim, R.H., Li, J.J., Furuyama, S., Sugiya, H., Sodek, J., 2000. Parathyroid hormone regulation of bone sialoprotein (BSP) gene transcription is mediated through a pituitary-specific transcription factor-1 (Pit-1) motif in the rat BSP gene promoter. Matrix Biol. 19, 395–407.

Ohkido, I., Segawa, H., Yanagida, R., Nakamura, M., Miyamoto, K., 2003. Cloning, gene structure and dietary regulation of the type-IIc Na/Pi cotransporter in the mouse kidney. Pflügers Archiv 446, 106–115.

Okada, Y., Montero, A., Zhang, X., Sobue, T., Lorenzo, J., Doetschman, T., Coffin, J.D., Hurley, M.M., 2003. Impaired osteoclast formation in bone marrow cultures of Fgf2 null mice in response to parathyroid hormone. J. Biol. Chem. 278, 21258–21266.

Onishi, T., Zhang, W., Cao, X., Hruska, K., 1997. The mitogenic effect of parathyroid hormone is associated with E2F-dependent activation of cyclin-dependent kinase 1 (cdc2) in osteoblast precursors. J. Bone Miner. Res. 12, 1596–1605.

Onyia, J.E., Bidwell, J., Herring, J., Hulman, J., Hock, J.M., 1995. In vivo, human parathyroid hormone fragment (hPTH 1-34) transiently stimulates immediate early response gene expression, but not proliferation, in trabecular bone cells of young rats. Bone 17, 479–484.

Onyia, J.E., Miller, B., Hulman, J., Liang, J., Galvin, R., Frolik, C., Chandrasekhar, S., Harvey, A.K., Bidwell, J., Herring, J., Hock, J.M., 1997. Proliferating cells in the primary spongiosa express osteoblastic phenotype in vitro. Bone 20, 93–100.

Opas, E.E., Gentile, M.A., Rossert, J.A., de Crombrugghe, B., Rodan, G.A., Schmidt, A., 2000. Parathyroid hormone and prostaglandin E2 preferentially increase luciferase levels in bone of mice harboring a luciferase transgene controlled by elements of the pro-alpha1(I) collagen promoter. Bone 26, 27–32.

Orloff, J.J., Ganz, M.B., Ribaudo, A.E., Burtis, W.J., Reiss, M., Milstone, L.M., Stewart, A.F., 1992. Analysis of PTHRP binding and signal transduction mechanisms in benign and malignant squamous cells. Am. J. Physiol. Endocrinol. Metab. 262, E599–E607.

Ortega, A., Ramila, D., Ardura, J.A., Esteban, V., Ruiz-Ortega, M., Barat, A., Gazapo, R., Bosch, R.J., Esbrit, P., 2006. Role of parathyroid hormone-related protein in tubulointerstitial apoptosis and fibrosis after folic acid-induced nephrotoxicity. J. Am. Soc. Nephrol. 17, 1594–1603.

Ortega, A., Ramila, D., Izquierdo, A., Gonzalez, L., Barat, A., Gazapo, R., Bosch, R.J., Esbrit, P., 2005. Role of the renin-angiotensin system on the parathyroid hormone-related protein overexpression induced by nephrotoxic acute renal failure in the rat. J. Am. Soc. Nephrol. 16, 939–949.

Otto, F., Thornell, A.P., Crompton, T., Denzel, A., Gilmour, K.C., Rosewell, I.R., Stamp, G.W., Beddington, R.S., Mundlos, S., Olsen, B.R., Selby, P.B., Owen, M.J., 1997. Cbfa1, a candidate gene for cleidocranial dysplasia syndrome, is essential for osteoblast differentiation and bone development. Cell 89, 765–771.

Oursler, M.J., Cortese, C., Keeting, P., Anderson, M.A., Bonde, S.K., Riggs, B.L., Spelsberg, T.C., 1991. Modulation of transforming growth factor-β production in normal human osteoblast-like cells by 17 β-estradiol and parathyroid hormone. Endocrinology 129, 3313–3320.

Oyajobi, B.O., Anderson, D.M., Traianedes, K., Williams, P.J., Yoneda, T., Mundy, G.R., 2001. Therapeutic efficacy of a soluble receptor activator of nuclear factor kappaB-IgG Fc fusion protein in suppressing bone resorption and hypercalcemia in a model of humoral hypercalcemia of malignancy. Cancer Res. 61, 2572–2578.

Oyajobi, B.O., Lomri, A., Hott, M., Marie, P.J., 1999. Isolation and characterization of human clonogenic osteoblast progenitors immunoselected from fetal bone marrow stroma using STRO-1 monoclonal antibody. J. Bone Miner. Res. 14, 351–361.

Paillard, M., Bichara, M., 1989. Peptide hormone effects on urinary acidification and acid-base balance: PTH, ADH, and glucagon. Am. J. Physiol. Renal. Physiol. 256, F973–F985.

Palmer, L.G., Schnermann, J., 2015. Integrated control of Na transport along the nephron. Clin. J. Am. Soc. Nephrol. 10, 676–687.

Panda, D., Goltzman, D., Juppner, H., Karaplis, A.C., 2009. TIP39/parathyroid hormone type 2 receptor signaling is a potent inhibitor of chondrocyte proliferation and differentiation. Am. J. Physiol. Endocrinol. Metab. 297, E1125–E1136.

Panda, D.K., Goltzman, D., Karaplis, A.C., 2012. Defective postnatal endochondral bone development by chondrocyte-specific targeted expression of parathyroid hormone type 2 receptor. Am. J. Physiol. Endocrinol. Metab. 303, E1489–E1501.

Papavassiliou, A.G., 1994. The CREB/ATF family of transcription factors: modulation by reversible phosphorylation. Anticancer Res. 14, 1801–1805.

Partridge, N.C., Dickson, C.A., Kopp, K., Teitelbaum, S.L., Crouch, E.C., Kahn, A.J., 1989. Parathyroid hormone inhibits collagen synthesis at both ribonucleic acid and protein levels in rat osteogenic sarcoma cells. Mol. Endocrinol. 3, 232–239.

Partridge, N.C., Jeffrey, J.J., Ehlich, L.S., Teitelbaum, S.L., Fliszar, C., Welgus, H.G., Kahn, A.J., 1987. Hormonal regulation of the production of collagenase and a collagenase inhibitor activity by rat osteogenic sarcoma cells. Endocrinology 120, 1956–1962.

Pavlos, N.J., Friedman, P.A., 2017. GPCR Signaling and Trafficking: the long and short of it. Trends Endocrinol. Metabol. 28, 213–226.

Peacock, M., Robertson, W.G., Nordin, B.E.C., 1969. Relation between serum and urinary calcium with particular reference to parathyroid activity. Lancet 1, 384–386.

Pearman, A.T., Chou, W.Y., Bergman, K.D., Pulumati, M.R., Partridge, N.C., 1996. Parathyroid hormone induces c-fos promoter activity in osteoblastic cells through phosphorylated cAMP response element (CRE)-binding protein binding to the major CRE. J. Biol. Chem. 271, 25715–25721.

Pendas, A.M., Balbin, M., Llano, E., Jimenez, M.G., Lopez-Otin, C., 1997. Structural analysis and promoter characterization of the human collagenase-3 gene (MMP13). Genomics 40, 222–233.

Peterson, N.A., Feigen, G.A., Crimson, J.M., 1961. Effect of pH on interaction of calcium ions with serum proteins. Am. J. Physiol. 201, 386–392.

Pfeilschifter, J., Laukhuf, F., Muller-Beckmann, B., Blum, W.F., Pfister, T., Ziegler, R., 1995. Parathyroid hormone increases the concentration of insulin-like growth factor-I and transforming growth factor beta 1 in rat bone. J. Clin. Investig. 96, 767–774.

Pfeilschifter, J., Mundy, G.R., 1987. Modulation of type β transforming growth factor activity in bone cultures by osteotropic hormones. Proc. Natl. Acad. Sci. U. S. A. 84, 2024–2028.

Pfister, M.F., Forgo, J., Ziegler, U., Biber, J., Murer, H., 1999. cAMP-dependent and -independent downregulation of type IINa-P_i cotransporters by PTH. Am. J. Physiol. Renal. Physiol. 276, F720–F725.

Pfister, M.F., Lederer, E., Forgo, J., Ziegler, U., Lötscher, M., Quabius, E.S., Biber, J., Murer, H., 1997. Parathyroid hormone-dependent degradation of type II Na^+/P_i cotransporters. J. Biol. Chem. 272, 20125–20130.

Philbrick, W.M., Wysolmerski, J.J., Galbraith, S., Holt, E., Orloff, J.J., Yang, K.H., Vasavada, R.C., Weir, E.C., Broadus, A.E., Stewart, A.F., 1996. Defining the roles of parathyroid hormone-related protein in normal physiology. Physiol. Rev. 76, 127–173.

Picard, N., Capuano, P., Stange, G., Mihailova, M., Kaissling, B., Murer, H., Biber, J., Wagner, C.A., 2010. Acute parathyroid hormone differentially regulates renal brush border membrane phosphate cotransporters. Pflügers Archiv 460, 677–687.

Pines, M., Fukayama, S., Costas, K., Meurer, E., Goldsmith, P.K., Xu, X., Muallem, S., Behar, V., Chorev, M., Rosenblatt, M., Tashjian Jr., A.H., Suva, L.J., 1996. Inositol 1-,4-,5-trisphosphate-dependent Ca2+ signaling by the recombinant human PTH/PTHrP receptor stably expressed in a human kidney cell line. Bone 18, 381–389.

Pines, M., Granot, I., Hurwitz, S., 1990. Cyclic AMP-dependent inhibition of collagen synthesis in avian epiphyseal cartilage cells: effect of chicken and human parathyroid hormone and parathyroid hormone-related peptide. Bone Miner. 9, 23–33.

Pirih, F.Q., Aghaloo, T.L., Bezouglaia, O., Nervina, J.M., Tetradis, S., 2005. Parathyroid hormone induces the NR4A family of nuclear orphan receptors in vivo. Biochem. Biophys. Res. Commun. 332, 494–503.

Pirih, F.Q., Tang, A., Ozkurt, I.C., Nervina, J.M., Tetradis, S., 2004. Nuclear orphan receptor Nurr1 directly transactivates the osteocalcin gene in osteoblasts. J. Biol. Chem. 279, 53167–53174.

Pizurki, L., Rizzoli, R., Moseley, J., Martin, T.J., Caverzasio, J., Bonjour, J.P., 1988. Effect of synthetic tumoral PTH-related peptide on cAMP production and Na-dependent Pi transport. Am. J. Physiol. Renal. Physiol. 255, F957–F961.

Plotkin, H., Gundberg, C., Mitnick, M., Stewart, A.F., 1998. Dissociation of bone formation from resorption during 2-week treatment with human parathyroid hormone-related peptide-(1-36) in humans: potential as an anabolic therapy for osteoporosis. J. Clin. Endocrinol. Metab. 83, 2786–2791.

Porte, D., Tuckermann, J., Becker, M., Baumann, B., Teurich, S., Higgins, T., Owen, M.J., Schorpp-Kistner, M., Angel, P., 1999. Both AP-1 and Cbfa1-like factors are required for the induction of interstitial collagenase by parathyroid hormone. Oncogene 18, 667–678.

Pun, K.K., Arnaud, C.D., Nissenson, R.A., 1988. Parathyroid hormone receptors in human dermal fibroblasts: structural and functional characterization. J. Bone Miner. Res. 3, 453–460.

Puschett, J.B., Zurbach, P., Sylk, D., 1976. Acute effects of parathyroid hormone on proximal bicarbonate transport in the dog. Kidney Int. 9, 501–510.

Qin, L., Li, X., Ko, J.K., Partridge, N.C., 2005a. Parathyroid hormone uses multiple mechanisms to arrest the cell cycle progression of osteoblastic cells from G1 to S phase. J. Biol. Chem. 280, 3104–3111.

Qin, L., Partridge, N.C., 2005. Stimulation of amphiregulin expression in osteoblastic cells by parathyroid hormone requires the protein kinase A and cAMP response element-binding protein signaling pathway. J. Cell. Biochem. 96, 632–640.

Qin, L., Tamasi, J., Raggatt, L., Li, X., Feyen, J.H., Lee, D.C., Dicicco-Bloom, E., Partridge, N.C., 2005b. Amphiregulin is a novel growth factor involved in normal bone development and in the cellular response to parathyroid hormone stimulation. J. Biol. Chem. 280, 3974–3981.

Quamme, G., Pelech, S., Biber, J., Murer, H., 1994. Abnormalities of parathyroid hormone-mediated signal transduction mechanisms in opossum kidney cells. Biochim. Biophys. Acta 1223, 107–116.

Quamme, G., Pfeilschifter, J., Murer, H., 1989. Parathyroid hormone inhibition of Na^+/phosphate cotransport in OK cells: generation of second messengers in the regulatory cascade. Biochem. Biophys. Res. Commun. 158, 951–957.

Quinn, C.O., Scott, D.K., Brinckerhoff, C.E., Matrisian, L.M., Jeffrey, J.J., Partridge, N.C., 1990. Rat collagenase. Cloning, amino acid sequence comparison, and parathyroid hormone regulation in osteoblastic cells. J. Biol. Chem. 265, 22342–22347.

Rabito, C.A., 1986. Sodium cotransport processes in renal epithelial cell lines. Miner. Electrolyte Metab. 12, 32–41.

Raisz, L.G., Simmons, H.A., Vargas, S.J., Kemp, B.E., Martin, T.J., 1990. Comparison of the effects of amino-terminal synthetic parathyroid hormone-related peptide (PTHrP) of malignancy and parathyroid hormone on resorption of cultured fetal rat long bones. Calcif. Tissue Int. 46, 233–238.

Ramila, D., Ardura, J.A., Esteban, V., Ortega, A., Ruiz-Ortega, M., Bosch, R.J., Esbrit, P., 2008. Parathyroid hormone-related protein promotes inflammation in the kidney with an obstructed ureter. Kidney Int. 73, 835–847.

Reshkin, S.J., Forgo, J., Murer, H., 1990. Functional asymmetry in phosphate transport and its regulation in opossum kidney cells: parathyroid hormone inhibition. Pflueg. Arch. Eur. J. Physiol. 416, 624–631.

Reshkin, S.J., Forgo, J., Murer, H., 1991. Apical and basolateral effects of PTH in OK cells: transport inhibition, messenger production, effects of pertussis toxin, and interaction with a PTH analog. J. Membr. Biol. 124, 227–237.

Rey, A., Manen, D., Rizzoli, R., Caverzasio, J., Ferrari, S.L., 2006. Proline-rich motifs in the parathyroid hormone (PTH)/PTH-related protein receptor C terminus mediate scaffolding of c-Src with β-arrestin2 for ERK1/2 activation. J. Biol. Chem. 281, 38181–38188.

Ribeiro, C.P., Dubay, G.R., Falck, J.R., Mandel, L.J., 1994. Parathyroid hormone inhibits Na^+-K^+-ATPase through a cytochrome *P*-450 pathway. Am. J. Physiol. Renal. Physiol. 266, F497–F505.

Ribeiro, C.P., Mandel, L.J., 1992. Parathyroid hormone inhibits proximal tubule Na^+-K^+-ATPase activity. Am. J. Physiol. Renal. Physiol. 262, F209–F216.

Ritthaler, T., Traebert, M., Lötscher, M., Biber, J., Murer, H., Kaissling, B., 1999. Effects of phosphate intake on distribution of type II Na/Pi cotransporter mRNA in rat kidney. Kidney Int. 55, 976–983.

Rodan, S.B., Wesolowski, G., Ianacone, J., Thiede, M.A., Rodan, G.A., 1989. Production of parathyroid hormone-like peptide in a human osteosarcoma cell line: stimulation by phorbol esters and epidermal growth factor. J. Endocrinol. 122, 219–227.

Romero, M.F., Fulton, C.M., Boron, W.F., 2004. The SLC4 family of HCO_3^- transporters. Pflueg. Arch. Eur. J. Physiol. 447, 495–509.

Rost, C.R., Bikle, D.D., Kaplan, R.A., 1981. In vitro stimulation of 25-hydroxycholecalciferol 1α-hydroxylation by parathyroid hormone in chick kidney slices: evidence for a role for adenosine 3',5'-monophosphate. Endocrinology 108, 1002–1006.

Rouleau, M.F., Mitchell, J., Goltzman, D., 1988. In vivo distribution of parathyroid hormone receptors in bone: evidence that a predominant osseous target cell is not the mature osteoblast. Endocrinology 123, 187–191.

Rouleau, M.F., Mitchell, J., Goltzman, D., 1990. Characterization of the major parathyroid hormone target cell in the endosteal metaphysis of rat long bones. J. Bone Miner. Res. 5, 1043–1053.

Ruiz, O.S., Qiu, Y.Y., Wang, L.J., Arruda, J.A.L., 1996. Regulation of the renal Na-HCO_3 cotransporter: V. Mechanism of the inhibitory effect of parathyroid hormone. Kidney Int. 49, 396–402.

Sandström, I., 1879-1880. Om en ny körtel hos menniskan och àtskilliga däggdjur. Upsala Läkareförening Forhandlingar 15, 441–471.

Saussine, C., Massfelder, T., Parnin, F., Judes, C., Simeoni, U., Helwig, J.-J., 1993. Renin stimulating properties of parathyroid hormone-related peptide in the isolated perfused rat kidney. Kidney Int. 44, 764–773.

Saxena, S., Dansby, L., Allon, M., 1995. Adaptation to phosphate depletion in opossum kidney cells. Biochem. Biophys. Res. Commun. 216, 141–147.

Schaffers, O.J.M., Hoenderop, J.G.J., Bindels, R.J.M., de Baaij, J.H.F., 2018. The rise and fall of novel renal magnesium transporters. Am. J. Physiol. Renal. Physiol. 314, F1027–F1033.

Scheinman, S.J., Mitnick, M.E., Stewart, A.F., 1990. Quantitative evaluation of anticalciuretic effects of synthetic parathyroid hormone like peptides. J. Bone Miner. Res. 5, 653–658.

Schiller, P.C., D'Ippolito, G., Balkan, W., Roos, B.A., Howard, G.A., 2001. Gap-junctional communication mediates parathyroid hormone stimulation of mineralization in osteoblastic cultures. Bone 28, 38–44.

Schiller, P.C., Mehta, P.P., Roos, B.A., Howard, G.A., 1992. Hormonal regulation of intercellular communication: parathyroid hormone increases connexin 43 gene expression and gap-junctional communication in osteoblastic cells. Mol. Endocrinol. 6, 1433–1440.

Schoolwerth, A.C., Smith, B.C., Culpepper, R.M., 1988. Renal gluconeogenesis. Miner. Electrolyte Metab. 14, 347–361.

Schultheis, P.J., Lorenz, J.N., Meneton, P., Nieman, M.L., Riddle, T.M., Flagella, M., Duffy, J.J., Doetschman, T., Miller, M.L., Shull, G.E., 1998. Phenotype resembling Gitelman's syndrome in mice lacking the apical Na^+- Cl^- cotransporter of the distal convoluted tubule. J. Biol. Chem. 273, 29150–29155.

Scott, D.K., Brakenhoff, K.D., Clohisy, J.C., Quinn, C.O., Partridge, N.C., 1992. Parathyroid hormone induces transcription of collagenase in rat osteoblastic cells by a mechanism using cyclic adenosine 3',5'-monophosphate and requiring protein synthesis. Mol. Endocrinol. 6, 2153–2159.

Scutt, A., Duvos, C., Lauber, J., Mayer, H., 1994. Time-dependent effects of parathyroid hormone and prostaglandin E2 on DNA synthesis by periosteal cells from embryonic chick calvaria. Calcif. Tissue Int. 55, 208–215.

Segawa, H., Kaneko, I., Takahashi, A., Kuwahata, M., Ito, M., Ohkido, I., Tatsumi, S., Miyamoto, K., 2002. Growth-related renal type II Na/Pi cotransporter. J. Biol. Chem. 277, 19665–19672.

Segawa, H., Yamanaka, S., Onitsuka, A., Tomoe, Y., Kuwahata, M., Ito, M., Taketani, Y., Miyamoto, K.I., 2007. Parathyroid hormone dependent endocytosis of renal type IIc Na/Pi cotransporter. Am. J. Physiol. Renal. Physiol. 292, F395–F403.

Selvamurugan, N., Chou, W.Y., Pearman, A.T., Pulumati, M.R., Partridge, N.C., 1998. Parathyroid hormone regulates the rat collagenase-3 promoter in osteoblastic cells through the cooperative interaction of the activator protein-1 site and the runt domain binding sequence. J. Biol. Chem. 273, 10647–10657.

Selvamurugan, N., Pulumati, M.R., Tyson, D.R., Partridge, N.C., 2000. Parathyroid hormone regulation of the rat collagenase-3 promoter by protein kinase A-dependent transactivation of core binding factor alpha1. J. Biol. Chem. 275, 5037–5042.

Semenov, M., Tamai, K., He, X., 2005. SOST is a ligand for LRP5/LRP6 and a Wnt signaling inhibitor. J. Biol. Chem. 280, 26770–26775.

Seyedin, S.M., Thomas, T.C., Thompson, A.Y., Rosen, D.M., Piez, K.A., 1985. Purification and characterization of two cartilage-inducing factors from bovine demineralized bone. Proc. Natl. Acad. Sci. U. S. A. 82, 2267–2271.

Shen, V., Dempster, D.W., Birchman, R., Xu, R., Lindsay, R., 1993. Loss of cancellous bone mass and connectivity in ovariectomized rats can be restored by combined treatment with parathyroid hormone and estradiol. J. Clin. Investig. 91, 2479–2487.

Shenolikar, S., Voltz, J.W., Minkoff, C.M., Wade, J.B., Weinman, E.J., 2002. Targeted disruption of the mouse NHERF-1 gene promotes internalization of proximal tubule sodium-phosphate cotransporter type IIa and renal phosphate wasting. Proc. Natl. Acad. Sci. U. S. A. 99, 11470–11475.

Shigematsu, T., Horiuchi, N., Ogura, Y., Miyahara, T., Suda, T., 1986. Human parathyroid hormone inhibits renal 24-hydroxylase activity of 25-hydroxyvitamin D_3 by a mechanism involving adenosine 3',5'-monophosphate in rats. Endocrinology 118, 1583–1589.

Shimizu, E., Nakatani, T., He, Z., Partridge, N.C., 2014. Parathyroid hormone regulates histone deacetylase (HDAC) 4 through protein kinase A-mediated phosphorylation and dephosphorylation in osteoblastic cells. J. Biol. Chem. 289, 21340–21350.

Shimizu, M., Joyashiki, E., Noda, H., Watanabe, T., Okazaki, M., Nagayasu, M., Adachi, K., Tamura, T., Potts Jr., J.T., Gardella, T.J., Kawabe, Y., 2016. Pharmacodynamic actions of a long-acting PTH analog (LA-PTH) in thyroparathyroidectomized (TPTX) rats and normal monkeys. J. Bone Miner. Res. 31, 1405–1412.

Shimizu, T., Yoshitomi, K., Nakamura, M., Imai, M., 1990. Effects of PTH, calcitonin, and cAMP on calcium transport in rabbit distal nephron segments. Am. J. Physiol. Renal. Physiol. 259, F408–F414.

Shlatz, L.J., Schwartz, I.L., Kinne-Saffran, E., Kinne, R., 1975. Distribution of parathyroid hormone-stimulated adenylate cyclase in plasma membranes of cells of the kidney cortex. J. Membr. Biol. 24, 131–144.

Siegfried, G., Vrtovsnik, F., Prie, D., Amiel, C., Friedlander, G., 1995. Parathyroid hormone stimulates ecto-5'-nucleotidase activity in renal epithelial cells: role of protein kinase-C. Endocrinology 136, 1267–1275.

Silve, C.M., Hradek, G.T., Jones, A.L., Arnaud, C.D., 1982. Parathyroid hormone receptor in intact embryonic chicken bone: characterization and cellular localization. J. Cell Biol. 94, 379–386.

Simon, L.S., Slovik, D.M., Neer, R.M., Krane, S.M., 1988. Changes in serum levels of type I and III procollagen extension peptides during infusion of human parathyroid hormone fragment (1-34). J. Bone Miner. Res. 3, 241–246.

Simonet, W.S., Lacey, D.L., Dunstan, C.R., Kelley, M., Chang, M.S., Luthy, R., Nguyen, H.Q., Wooden, S., Bennett, L., Boone, T., Shimamoto, G., DeRose, M., Elliott, R., Colombero, A., Tan, H.L., Trail, G., Sullivan, J., Davy, E., Bucay, N., Renshaw-Gegg, L., Hughes, T.M., Hill, D., Pattison, W., Campbell, P., Sander, S., Van, G., Tarpley, J., Derby, P., Lee, R., Boyle, W.J., 1997. Osteoprotegerin: a novel secreted protein involved in the regulation of bone density. Cell 89, 309–319.

Sneddon, W.B., Barry, E.L.R., Coutermarsh, B.A., Gesek, F.A., Liu, F., Friedman, P.A., 1998. Regulation of renal parathyroid hormone receptor expression by 1,25-dihydroxyvitamin D_3 and retinoic acid. Cell. Physiol. Biochem. 8, 261–277.

Sneddon, W.B., Liu, F., Gesek, F.A., Friedman, P.A., 2000. Obligate mitogen-activated protein kinase activation in parathyroid hormone stimulation of calcium transport but not calcium signaling. Endocrinology 141, 4185–4193.

Sneddon, W.B., Ruiz, G.W., Gallo, L.I., Xiao, K., Zhang, Q., Rbaibi, Y., Weisz, O.A., Apodaca, G.L., Friedman, P.A., 2016. Convergent signaling pathways regulate parathyroid hormone and fibroblast growth factor-23 action on NPT2A-mediated phosphate transport. J. Biol. Chem. 291, 18632–18642.

Sneddon, W.B., Syme, C.A., Bisello, A., Magyar, C.E., Weinman, E.J., Rochdi, M.D., Parent, J.L., Abou-Samra, A.B., Friedman, P.A., 2003. Activation-independent parathyroid hormone receptor internalization is regulated by NHERF1 (EBP50). J. Biol. Chem. 278, 43787–43796.

Sneddon, W.B., Yang, Y., Ba, J., Harinstein, L.M., Friedman, P.A., 2007. Extracellular signal-regulated kinase activation by parathyroid hormone in distal tubule cells. Am. J. Physiol. Renal. Physiol. 292, F1028–F1034.

Soifer, N.E., Van Why, S.K., Ganz, M.B., Kashgarian, M., Siegel, N.J., Stewart, A.F., 1993. Expression of parathyroid hormone-related protein in the rat glomerulus and tubule during recovery from renal ischemia. J. Clin. Investig. 92, 2850–2857.

Soleimani, M., Burnham, C.E., 2001. Na^+:HCO_3^- cotransporters (NBC): cloning and characterization. J. Membr. Biol. 183, 71–84.

Somermeyer, M.G., Knauss, T.C., Weinberg, J.M., Humes, H.D., 1983. Characterization of Ca^{2+} transport in rat renal brush-border membranes and its modulation by phosphatidic acid. Biochem. J. 214, 37–46.

Somjen, D., Binderman, I., Schluter, K.D., Wingender, E., Mayer, H., Kaye, A.M., 1990. Stimulation by defined parathyroid hormone fragments of cell proliferation in skeletal-derived cell cultures. Biochem. J. 272, 781–785.

Sowa, H., Kaji, H., Iu, M.F., Tsukamoto, T., Sugimoto, T., Chihara, K., 2003. Parathyroid hormone-Smad3 axis exerts anti-apoptotic action and augments anabolic action of transforming growth factor β in osteoblasts. J. Biol. Chem. 278, 52240–52252.

Spurney, R.F., Flannery, P.J., Garner, S.C., Athirakul, K., Liu, S., Guilak, F., Quarles, L.D., 2002. Anabolic effects of a G protein-coupled receptor kinase inhibitor expressed in osteoblasts. J. Clin. Investig. 109, 1361–1371.

Stanislaus, D., Devanarayan, V., Hock, J.M., 2000a. In vivo comparison of activated protein-1 gene activation in response to human parathyroid hormone (hPTH)(1-34) and hPTH(1-84) in the distal femur metaphyses of young mice. Bone 27, 819–826.

Stanislaus, D., Yang, X., Liang, J.D., Wolfe, J., Cain, R.L., Onyia, J.E., Falla, N., Marder, P., Bidwell, J.P., Queener, S.W., Hock, J.M., 2000b. In vivo regulation of apoptosis in metaphyseal trabecular bone of young rats by synthetic human parathyroid hormone (1-34) fragment. Bone 27, 209–218.

Stewart, A.F., Vignery, A., Silverglate, A., Ravin, N.D., LiVolsi, V., Broadus, A.E., Baron, R., 1982. Quantitative bone histomorphometry in humoral hypercalcemia of malignancy: uncoupling of bone cell activity. J. Clin. Endocrinol. Metab. 55, 219–227.

Stewart, K., Walsh, S., Screen, J., Jefferiss, C.M., Chainey, J., Jordan, G.R., Beresford, J.N., 1999. Further characterization of cells expressing STRO-1 in cultures of adult human bone marrow stromal cells. J. Bone Miner. Res. 14, 1345–1356.

Stow, J.L., Sabolic, I., Brown, D., 1991. Heterogeneous localization of G protein α-subunits in rat kidney. Am. J. Physiol. Renal. Physiol. 261, F831–F840.

Suda, N., Gillespie, M.T., Traianedes, K., Zhou, H., Ho, P.W., Hards, D.K., Allan, E.H., Martin, T.J., Moseley, J.M., 1996. Expression of parathyroid hormone-related protein in cells of osteoblast lineage. J. Cell. Physiol. 166, 94–104.

Sutherland, M.K., Rao, L.G., Wylie, J.N., Gupta, A., Ly, H., Sodek, J., Murray, T.M., 1994. Carboxyl-terminal parathyroid hormone peptide (53-84) elevates alkaline phosphatase and osteocalcin mRNA levels in SaOS-2 cells. J. Bone Miner. Res. 9, 453–458.

Sutton, R.A.L., Dirks, J.H., 1975. The renal excretion of calcium: a review of micropuncture data. Can. J. Physiol. Pharmacol. 53, 979–988.

Sutton, R.A.L., Wong, N.L.M., Dirks, J.H., 1979. Effects of metabolic acidosis and alkalosis on sodium and calcium transport in the dog kidney. Kidney Int. 15, 520–533.

Suzuki, A., Ghayor, C., Guicheux, J., Magne, D., Quillard, S., Kakita, A., Ono, Y., Miura, Y., Oiso, Y., Itoh, M., Caverzasio, J., 2006. Enhanced expression of the inorganic phosphate transporter Pit-1 is involved in BMP-2-induced matrix mineralization in osteoblast-like cells. J. Bone Miner. Res. 21, 674–683.

Suzuki, A., Palmer, G., Bonjour, J.P., Caverzasio, J., 2001. Stimulation of sodium-dependent inorganic phosphate transport by activation of Gi/o-protein-coupled receptors by epinephrine in MC3T3-E1 osteoblast-like cells. Bone 28, 589–594.

Suzuki, M., Morita, T., Hanaoka, K., Kawaguchi, Y., Sakai, O., 1991. A Cl^- channel activated by parathyroid hormone in rabbit renal proximal tubule cells. J. Clin. Investig. 88, 735–742.

Swarthout, J.T., Doggett, T.A., Lemker, J.L., Partridge, N.C., 2001. Stimulation of extracellular signal-regulated kinases and proliferation in rat osteoblastic cells by parathyroid hormone is protein kinase C-dependent. J. Biol. Chem. 276, 7586–7592.

Swarthout, J.T., Lemker, J.F., Wilhelm, D., Dieckmann, A., Angel, P., Partridge, N.C., 1997. Parathyroid hormone regulation of mitogen activated protein kinases in osteoblastic cells. J. Bone Miner. Res. 12, S162.

Syme, C.A., Friedman, P.A., Bisello, A., 2005. Parathyroid hormone receptor trafficking contributes to the activation of extracellular signal-regulated kinases but is not required for regulation of cAMP signaling. J. Biol. Chem. 280, 11281–11288.

Takahashi, F., Morita, K., Katai, K., Segawa, H., Fujioka, A., Kouda, T., Tatsumi, S., Nii, T., Taketani, Y., Haga, H., Hisano, S., Fukui, Y., Miyamoto, K.I., Takeda, E., 1998. Effects of dietary Pi on the renal Na^+-dependent Pi transporter NaPi-2 in thyroparathyroidectomized rats. Biochem. J. 333 (Pt 1), 175–181.

Takaichi, K., Kurokawa, K., 1986. High Ca^{2+} inhibits peptide hormone-dependent cAMP production specifically in thick ascending limbs of Henle. Miner. Electrolyte Metab. 12, 342–346.

Tanaka, H., Smogorzewski, M., Koss, M., Massry, S.G., 1995. Pathways involved in PTH-induced rise in cytosolic Ca^{2+} concentration of rat renal proximal tubule. Am. J. Physiol. Renal. Physiol. 268, F330–F337.

Tanaka, S., Sakai, A., Tanaka, M., Otomo, H., Okimoto, N., Sakata, T., Nakamura, T., 2004. Skeletal unloading alleviates the anabolic action of intermittent PTH(1-34) in mouse tibia in association with inhibition of PTH-induced increase in c-fos mRNA in bone marrow cells. J. Bone Miner. Res. 19, 1813–1820.

Tenenhouse, H.S., Gauthier, C., Martel, J., Gesek, F.A., Coutermarsh, B.A., Friedman, P.A., 1998. Na^+-phosphate cotransport in mouse distal convoluted tubule cells: evidence for *Glvr-1* and *Ram-1* Gene Expression. J. Bone Miner. Res. 13, 590–597.

Tenenhouse, H.S., Martel, J., Gauthier, C., Segawa, H., Miyamoto, K.I., 2003. Differential effects of Npt2a gene ablation and the X-linked Hyp mutation on renal expression of type IIc Na/Pi cotransporter. Am. J. Physiol. Renal. Physiol. 285, F1271–F1278.

Terepka, A.R., Dewey, P.A., Toribara, T.Y., 1957. The ultrafiltrable calcium of human serum. II. Variations is disease states and under experimental conditions. J. Clin. Investig. 37, 87–98.

Termine, J.D., Kleinman, H.K., Whitson, S.W., Conn, K.M., McGarvey, M.L., Martin, G.R., 1981. Osteonectin, a bone-specific protein linking mineral to collagen. Cell 26, 99–105.

Teti, A., Rizzoli, R., Zambonin Zallone, A., 1991. Parathyroid hormone binding to cultured avian osteoclasts. Biochem. Biophys. Res. Commun. 174, 1217–1222.

Thirunavukkarasu, K., Halladay, D.L., Miles, R.R., Geringer, C.D., Onyia, J.E., 2002. Analysis of regulator of G-protein signaling-2 (RGS-2) expression and function in osteoblastic cells. J. Cell. Biochem. 85, 837–850.

Thompson, D.D., Seedor, J.G., Fisher, J.E., Rosenblatt, M., Rodan, G.A., 1988. Direct action of the parathyroid hormone-like human hypercalcemic factor on bone. Proc. Natl. Acad. Sci. U. S. A. 85, 5673–5677.

Towler, D.A., Rodan, G.A., 1995. Identification of a rat osteocalcin promoter 3',5'-cyclic adenosine monophosphate response region containing two PuGGTCA steroid hormone receptor binding motifs. Endocrinology 136, 1089–1096.

Traebert, M., Völkl, H., Biber, J., Murer, H., Kaissling, B., 2000. Luminal and contraluminal action of 1-34 and 3-34 PTH peptides on renal type IIa Na-P_i cotransporter. Am. J. Physiol. Renal. Physiol. 278, F792–F798.

Tsukii, K., Shima, N., Mochizuki, S., Yamaguchi, K., Kinosaki, M., Yano, K., Shibata, O., Udagawa, N., Yasuda, H., Suda, T., Higashio, K., 1998. Osteoclast differentiation factor mediates an essential signal for bone resorption induced by 1 alpha,25-dihydroxyvitamin D3, prostaglandin E2, or parathyroid hormone in the microenvironment of bone. Biochem. Biophys. Res. Commun. 246, 337–341.

Turner, G., Coureau, C., Rabin, M.R., Escoubet, B., Hruby, M., Walrant, O., Silve, C., 1995. Parathyroid hormone (PTH)/PTH-related protein receptor messenger ribonucleic acid expression and PTH response in a rat model of secondary hyperparathyroidism associated with vitamin D deficiency. Endocrinology 136, 3751–3758.

Tyson, D.R., Swarthout, J.T., Partridge, N.C., 1999. Increased osteoblastic c-fos expression by parathyroid hormone requires protein kinase A phosphorylation of the cyclic adenosine 3',5'-monophosphate response element-binding protein at serine 133. Endocrinology 140, 1255–1261.

Ureña, P., Iida-Klein, A., Kong, X.-F., Jüppner, H., Kronenberg, H.M., Abou-Samra, A.B., Segre, G.V., 1994. Regulation of parathyroid hormone (PTH)/PTH-related peptide receptor messenger ribonucleic acid by glucocorticoids and PTH in ROS 17/2.8 and OK cells. Endocrinology 134, 451–456.

Urena, P., Kubrusly, M., Mannstadt, M., Hruby, M., Trinh, M.M., Silve, C., Lacour, B., Abou-Samra, A.B., Segre, G.V., Drueke, T., 1994. The renal PTH/PTHrP receptor is down-regulated in rats with chronic renal failure. Kidney Int. 45, 605–611.

Ureña, P., Mannstadt, M., Hruby, M., Ferreira, A., Schmitt, F., Silve, C., Ardaillou, R., Lacour, B., Abou-Samra, A.B., Segre, G.V., Drüeke, T., 1995. Parathyroidectomy does not prevent the renal PTH/PTHrP receptor down-regulation in uremic rats. Kidney Int. 47, 1797–1805.

Usdin, T.B., Bonner, T.I., Harta, G., Mezey, E., 1996. Distribution of parathyroid hormone-2 receptor messenger ribonucleic acid in rat. Endocrinology 137, 4285–4297.

Usdin, T.B., Gruber, C., Bonner, T.I., 1995. Identification and functional expression of a receptor selectively recognizing parathyroid hormone, the PTH2 receptor. J. Biol. Chem. 270, 15455–15458.

Usdin, T.B., Hilton, J., Vertesi, T., Harta, G., Segre, G., Mezey, E., 1999. Distribution of the parathyroid hormone 2 receptor in rat: immunolocalization reveals expression by several endocrine cells. Endocrinology 140, 3363–3371.

Valin, A., Guillen, C., Esbrit, P., 2001. C-terminal parathyroid hormone-related protein (PTHrP) (107-139) stimulates intracellular Ca(2+) through a receptor different from the type 1 PTH/PTHrP receptor in osteoblastic osteosarcoma UMR 106 cells. Endocrinology 142, 2752–2759.

van Bezooijen, R.L., Roelen, B.A., Visser, A., van der Wee-Pals, L., de Wilt, E., Karperien, M., Hamersma, H., Papapoulos, S.E., ten Dijke, P., Lowik, C.W., 2004. Sclerostin is an osteocyte-expressed negative regulator of bone formation, but not a classical BMP antagonist. J. Exp. Med. 199, 805–814.

van Bezooijen, R.L., Svensson, J.P., Eefting, D., Visser, A., van der Horst, G., Karperien, M., Quax, P.H., Vrieling, H., Papapoulos, S.E., ten Dijke, P., Lowik, C.W., 2007. Wnt but not BMP signaling is involved in the inhibitory action of sclerostin on BMP-stimulated bone formation. J. Bone Miner. Res. 22, 19–28.

van der Hagen, E.A., Lavrijsen, M., van Zeeland, F., Praetorius, J., Bonny, O., Bindels, R.J., Hoenderop, J.G., 2014. Coordinated regulation of TRPV5-mediated Ca^{2+} transport in primary distal convolution cultures. Pflügers Archiv 466, 2077–2087.

van der Plas, A., Aarden, E.M., Feijen, J.H., de Boer, A.H., Wiltink, A., Alblas, M.J., de Leij, L., Nijweide, P.J., 1994. Characteristics and properties of osteocytes in culture. J. Bone Miner. Res. 9, 1697–1704.

van Goor, M.K., Verkaart, S., van Dam, T.J., Huynen, M.A., van der Wijst, J., 2017a. Interspecies differences in PTH-mediated PKA phosphorylation of the epithelial calcium channel TRPV5. Pflügers Archiv.

van Goor, M.K.C., Hoenderop, J.G.J., van der Wijst, J., 2017b. TRP channels in calcium homeostasis: from hormonal control to structure-function relationship of TRPV5 and TRPV6. Biochim. Biophys. Acta 1864, 883–893.

van Loon, E.P., Little, R., Prehar, S., Bindels, R.J., Cartwright, E.J., Hoenderop, J.G., 2016. Calcium extrusion pump PMCA4: a new player in renal calcium handling? PLoS One 11, e0153483.

Vander Molen, M.A., Rubin, C.T., McLeod, K.J., McCauley, L.K., Donahue, H.J., 1996. Gap junctional intercellular communication contributes to hormonal responsiveness in osteoblastic networks. J. Biol. Chem. 271, 12165–12171.

Verheijen, M.H.G., Defize, L.H.K., 1997. Parathyroid hormone activates mitogen-activated protein kinase via a cAMP-mediated pathway independent of Ras. J. Biol. Chem. 272, 3423–3429.

Vilardaga, J.P., Frank, M., Krasel, C., Dees, C., Nissenson, R.A., Lohse, M.J., 2001. Differential conformational requirements for activation of G proteins and the regulatory proteins arrestin and G protein-coupled receptor kinase in the G protein-coupled receptor for parathyroid hormone (PTH)/PTH-related protein. J. Biol. Chem. 276, 33435–33443.

Wade, J.B., Liu, J., Coleman, R.A., Cunningham, R., Steplock, D.A., Lee-Kwon, W., Pallone, T.L., Shenolikar, S., Weinman, E.J., 2003. Localization and interaction of NHERF isoforms in the renal proximal tubule of the mouse. Am. J. Physiol. Cell Physiol. 285, C1494–C1503.

Wade, J.B., Welling, P.A., Donowitz, M., Shenolikar, S., Weinman, E.J., 2001. Differential renal distribution of NHERF isoforms and their colocalization with NHE3, ezrin, and ROMK. Am. J. Physiol. Cell Physiol. 280, C192–C198.

Walker, A.T., Stewart, A.F., Korn, E.A., Shiratori, T., Mitnick, M.A., Carpenter, T.O., 1990. Effect of parathyroid hormone-like peptides on 25-hydroxyvitamin D-1a-hydroxylase activity in rodents. Am. J. Physiol. Endocrinol. Metab. 258, E297–E303.

Walser, M., 1973. Divalent cations: physicochemical state in glomerular filtrate and urine and renal excretion. In: Orloff, J., Berliner, R.W. (Eds.), Handbook of Physiology, Section 8: Renal Physiology, first ed. American Physiological Society, Washington, D.C., pp. 555–586

Wang, A., Martin, J.A., Lembke, L.A., Midura, R.J., 2000. Reversible suppression of in vitro biomineralization by activation of protein kinase A. J. Biol. Chem. 275, 11082–11091.

Wang, B., Ardura, J.A., Romero, G., Yang, Y., Hall, R.A., Friedman, P.A., 2010. Na/H exchanger regulatory factors control PTH receptor signaling by differential activation of Gα protein subunits. J. Biol. Chem. 285, 26976–26986.

Wang, B., Bisello, A., Yang, Y., Romero, G.G., Friedman, P.A., 2007a. NHERF1 regulates parathyroid hormone receptor membrane retention without affecting recycling. J. Biol. Chem. 282, 36214–36222.

Wang, B., Yang, Y., Friedman, P.A., 2008. Na/H Exchange regulator factor 1, a novel Akt-associating protein, regulates extracellular signal-related signaling through a B-Raf-mediated pathway. Mol. Biol. Cell 19, 1637–1645.

Wang, B., Yang, Y., Liu, L., Blair, H.C., Friedman, P.A., 2013. NHERF1 regulation of PTH-dependent bimodal Pi transport in osteoblasts. Bone 52, 268–277.

Wang, B.L., Dai, C.L., Quan, J.X., Zhu, Z.F., Zheng, F., Zhang, H.X., Guo, S.Y., Guo, G., Zhang, J.Y., Qiu, M.C., 2006. Parathyroid hormone regulates osterix and Runx2 mRNA expression predominantly through protein kinase A signaling in osteoblast-like cells. J. Endocrinol. Investig. 29, 101–108.

Wang, D., Christensen, K., Chawla, K., Xiao, G., Krebsbach, P.H., Franceschi, R.T., 1999. Isolation and characterization of MC3T3-E1 preosteoblast subclones with distinct in vitro and in vivo differentiation/mineralization potential. J. Bone Miner. Res. 14, 893–903.

Wang, L., Liu, S., Quarles, L.D., Spurney, R.F., 2005a. Targeted overexpression of G protein-coupled receptor kinase-2 in osteoblasts promotes bone loss. Am. J. Physiol. Endocrinol. Metab. 288, E826–E834.

Wang, M.-S., Kurokawa, K., 1984. Renal gluconeogenesis: axial and internephron heterogeneity and the effect of parathyroid hormone. Am. J. Physiol. Renal. Physiol. 246, F59–F66.

Wang, Y.H., Liu, Y., Buhl, K., Rowe, D.W., 2005b. Comparison of the action of transient and continuous PTH on primary osteoblast cultures expressing differentiation stage-specific GFP. J. Bone Miner. Res. 20, 5–14.

Wang, Y.H., Liu, Y., Rowe, D.W., 2007b. Effects of transient PTH on early proliferation, apoptosis, and subsequent differentiation of osteoblast in primary osteoblast cultures. Am. J. Physiol. Endocrinol. Metab. 292, E594–E603.

Watford, M., Mapes, R.E., 1990. Hormonal and acid-base regulation of phosphoenolpyruvate carboxykinase mRNA levels in rat kidney. Arch. Biochem. Biophys. 282, 399–403.

Watson, P., Lazowski, D., Han, V., Fraher, L., Steer, B., Hodsman, A., 1995. Parathyroid hormone restores bone mass and enhances osteoblast insulin-like growth factor I gene expression in ovariectomized rats. Bone 16, 357–365.

Wehbi, V.L., Stevenson, H.P., Feinstein, T.N., Calero, G., Romero, G., Vilardaga, J.P., 2013. Noncanonical GPCR signaling arising from a PTH receptor-arrestin-Gβγ complex. Proc. Natl. Acad. Sci. U. S. A. 110, 1530–1535.

Wein, M.N., Liang, Y., Goransson, O., Sundberg, T.B., Wang, J., Williams, E.A., O'Meara, M.J., Govea, N., Beqo, B., Nishimori, S., Nagano, K., Brooks, D.J., Martins, J.S., Corbin, B., Anselmo, A., Sadreyev, R., Wu, J.Y., Sakamoto, K., Foretz, M., Xavier, R.J., Baron, R., Bouxsein, M.L., Gardella, T.J., Divieti-Pajevic, P., Gray, N.S., Kronenberg, H.M., 2016. SIKs control osteocyte responses to parathyroid hormone. Nat. Commun. 7, 13176.

Wein, M.N., Liang, Y., Goransson, O., Sundberg, T.B., Wang, J., Williams, E.A., O'Meara, M.J., Govea, N., Beqo, B., Nishimori, S., Nagano, K., Brooks, D.J., Martins, J.S., Corbin, B., Anselmo, A., Sadreyev, R., Wu, J.Y., Sakamoto, K., Foretz, M., Xavier, R.J., Baron, R., Bouxsein, M.L., Gardella, T.J., Divieti-Pajevic, P., Gray, N.S., Kronenberg, H.M., 2017. Corrigendum: SIKs control osteocyte responses to parathyroid hormone. Nat. Commun. 8, 14745.

Weinman, E.J., Biswas, R.S., Peng, G., Shen, L., Turner, C.L., E, X., Steplock, D., Shenolikar, S., Cunningham, R., 2007. Parathyroid hormone inhibits renal phosphate transport by phosphorylation of serine 77 of sodium-hydrogen exchanger regulatory factor-1. J. Clin. Investig. 117, 3412–3420.

Weinman, E.J., Cunningham, R., Wade, J.B., Shenolikar, S., 2005. The role of NHERF-1 in the regulation of renal proximal tubule sodium-hydrogen exchanger 3 and sodium-dependent phosphate cotransporter 2a. J. Physiol. 567, 27–32.

Weinman, E.J., Evangelista, C.M., Steplock, D., Liu, M.Z., Shenolikar, S., Bernardo, A., 2001. Essential role for NHERF in cAMP-mediated inhibition of the Na^+-HCO^-_3 co-transporter in BSC-1 cells. J. Biol. Chem. 276, 42339–42346.

Weinman, E.J., Lakkis, J., Akom, M., Wali, R.K., Drachenberg, C.B., Coleman, R.A., Wade, J.B., 2002. Expression of NHERF-1, NHERF-2, PDGFR-a, and PDGFR-ß in normal human kidneys and in renal transplant rejection. Pathobiology 70, 314–323.

Weinman, E.J., Steplock, D., Tate, K., Hall, R.A., Spurney, R.F., Shenolikar, S., 1998. Structure-function of recombinant Na/H exchanger regulatory factor (NHE-RF). J. Clin. Investig. 101, 2199–2206.

Weir, E.C., Insogna, K.L., Horowitz, M.C., 1989. Osteoblast-like cells secrete granulocyte-macrophage colony-stimulating factor in response to parathyroid hormone and lipopolysaccharide. Endocrinology 124, 899–904.

White, K.E., Gesek, F.A., Nesbitt, T., Drezner, M.K., Friedman, P.A., 1997. Molecular dissection of Ca^{2+} efflux in immortalized proximal tubule cells. J. Gen. Physiol. 109, 217–228.

Williams, R.J.P., 1976. Calcium chemistry and its relation to biological function. Symp. Soc. Exp. Biol. 30, 1–17.

Winchester, S.K., Bloch, S.R., Fiacco, G.J., Partridge, N.C., 1999. Regulation of expression of collagenase-3 in normal, differentiating rat osteoblasts. J. Cell. Physiol. 181, 479–488.

Winchester, S.K., Selvamurugan, N., D'Alonzo, R.C., Partridge, N.C., 2000. Developmental regulation of collagenase-3 mRNA in normal, differentiating osteoblasts through the activator protein-1 and the runt domain binding sites. J. Biol. Chem. 275, 23310–23318.

Witty, J.P., Foster, S.A., Stricklin, G.P., Matrisian, L.M., Stern, P.H., 1996. Parathyroid hormone-induced resorption in fetal rat limb bones is associated with production of the metalloproteinases collagenase and gelatinase B. J. Bone Miner. Res. 11, 72–78.

Woei Ng, K., Speicher, T., Dombrowski, C., Helledie, T., Haupt, L.M., Nurcombe, V., Cool, S.M., 2007. Osteogenic differentiation of murine embryonic stem cells is mediated by fibroblast growth factor receptors. Stem Cell. Dev. 16, 305–318.

Wu, Y., Kumar, R., 2000. Parathyroid hormone regulates transforming growth factor β and β synthesis in osteoblasts via divergent signaling pathways. J. Bone Miner. Res. 15, 879–884.

Wysolmerski, J.J., Stewart, A.F., 1998. The physiology of parathyroid hormone-related protein: an emerging role as a developmental factor. Annu. Rev. Physiol. 60, 431–460.

Yadav, S., Yadav, Y.S., Goel, M.M., Singh, U., Natu, S.M., Negi, M.P., 2014. Calcitonin gene- and parathyroid hormone-related peptides in normotensive and preeclamptic pregnancies: a nested case-control study. Arch. Gynecol. Obstet. 290, 897–903.

Yamaguchi, D.T., Hahn, T.J., Iida-Klein, A., Kleeman, C.R., Muallem, S., 1987. Parathyroid hormone-activated calcium channels in an osteoblast-like clonal osteosarcoma cell line. cAMP-dependent and cAMP-independent calcium channels. J. Biol. Chem. 262, 7711–7718.

Yamaguchi, M., Ogata, N., Shinoda, Y., Akune, T., Kamekura, S., Terauchi, Y., Kadowaki, T., Hoshi, K., Chung, U.I., Nakamura, K., Kawaguchi, H., 2005. Insulin receptor substrate-1 is required for bone anabolic function of parathyroid hormone in mice. Endocrinology 146, 2620–2628.

Yamaguchi, T., Fukase, M., Kido, H., Sugimoto, T., Katunuma, N., Chihara, K., 1994. Meprin is predominantly involved in parathyroid hormone degradation by the microvillar membranes of rat kidney. Life Sci. 54, 381–386.

Yamamoto, H., Tani, Y., Kobayashi, K., Taketani, Y., Sato, T., Arai, H., Morita, K., Miyamoto, K., Pike, J.W., Kato, S., Takeda, E., 2005. Alternative promoters and renal cell-specific regulation of the mouse type IIa sodium-dependent phosphate cotransporter gene. Biochim. Biophys. Acta 1732, 43–52.

Yamamoto, M., Murakami, T., Nishikawa, M., Tsuda, E., Mochizuki, S., Higashio, K., Akatsu, T., Motoyoshi, K., Nagata, N., 1998. Hypocalcemic effect of osteoclastogenesis inhibitory factor/osteoprotegerin in the thyroparathyroidectomized rat. Endocrinology 139, 4012–4015.

Yamamoto, S., Morimoto, I., Yanagihara, N., Zeki, K., Fujihira, T., Izumi, F., Yamashita, H., Eto, S., 1997. Parathyroid hormone-related peptide-(1-34) [PTHrP-(1-34)] induces vasopressin release from the rat supraoptic nucleus in vitro through a novel receptor distinct from a type I or type II PTH/PTHrP receptor. Endocrinology 138, 2066–2072.

Yang, R., Gerstenfeld, L.C., 1997. Structural analysis and characterization of tissue and hormonal responsive expression of the avian bone sialoprotein (BSP) gene. J. Cell. Biochem. 64, 77–93.

Yang, T.X., Hassan, S., Huang, Y.N.G., Smart, A.M., Briggs, J.P., Schnermann, J.B., 1997. Expression of PTHrP, PTH/PTHrP receptor, and Ca^{2+}-sensing receptor mRNAs along the rat nephron. Am. J. Physiol. Renal. Physiol. 272, F751–F758.

Yang, W., Friedman, P.A., Siu-Caldera, M.-L., Reddy, G.S., Kumar, R., Christakos, S., 1999. Expression of 25(OH)D_3 24-hydroxylase in the distal nephron: coordinate regulation by 1,25(OH)$_2D_3$ and or PTH. Am. J. Physiol. Endocrinol. Metab. 276, E793–E805.

Yasuda, H., Shima, N., Nakagawa, N., Mochizuki, S.I., Yano, K., Fujise, N., Sato, Y., Goto, M., Yamaguchi, K., Kuriyama, M., Kanno, T., Murakami, A., Tsuda, E., Morinaga, T., Higashio, K., 1998. Identity of osteoclastogenesis inhibitory factor (OCIF) and osteoprotegerin (OPG): a mechanism by which OPG/OCIF inhibits osteoclastogenesis in vitro. Endocrinology 139, 1329–1337.

Yates, A.J.P., Guttierrez, G.E., Smolens, P., Travis, P.S., Katz, M.S., Aufdemorte, T.B., Boyce, B.F., Hymer, T.K., Poser, J.W., Mundy, G.R., 1988. Effects of a synthetic peptide of a parathyroid hormone-related protein on calcium homeostasis, renal tubular calcium reabsorption, and bone metabolism in vivo and in vitro in rodents. J. Clin. Investig. 81, 932–938.

Yee, J.A., 1985. Stimulation of alkaline phosphatase activity in cultured neonatal mouse calvarial bone cells by parathyroid hormone. Calcif. Tissue Int. 37, 530–538.

Yeruva, S., Chodisetti, G., Luo, M., Chen, M., Cinar, A., Ludolph, L., Lunnemann, M., Goldstein, J., Singh, A.K., Riederer, B., Bachmann, O., Bleich, A., Gereke, M., Bruder, D., Hagen, S., He, P., Yun, C., Seidler, U., 2015. Evidence for a causal link between adaptor protein PDZK1 downregulation and Na^+/H^+ exchanger NHE3 dysfunction in human and murine colitis. Pflügers Archiv 467, 1795–1807.

Yu, X.P., Chandrasekhar, S., 1997. Parathyroid hormone (PTH 1-34) regulation of rat osteocalcin gene transcription. Endocrinology 138, 3085–3092.

Zhang, Q., Xiao, K., Liu, H., Song, L., McGarvey, J.C., Sneddon, W.B., Bisello, A., Friedman, P.A., 2018. Site-specific polyubiquitination differentially regulates parathyroid hormone receptor-initiated MAPK signaling and cell proliferation. J. Biol. Chem. 293, 5556–5571.

Zhang, Y.B., Norian, J.M., Magyar, C.E., Holstein-Rathlou, N.H., Mircheff, A.K., McDonough, A.A., 1999. In vivo PTH provokes apical NHE3 and NaPi2 redistribution and Na-K-ATPase inhibition. Am. J. Physiol. Renal. Physiol. 276, F711–F719.

Zhao, N, Tenenhouse, H.S., 2000. Npt2 gene disruption confers resistance to the inhibitory action of parathyroid hormone on renal sodium-phosphate cotransport. Endocrinology 141, 2159–2165.

Zhao, H., Wiederkehr, M.R., Fan, L., Collazo, R.L., Crowder, L.A., Moe, O.W., 1999. Acute inhibition of Na/H exchanger NHE-3 by cAMP. Role of protein kinase a and NHE-3 phosphoserines 552 and 605. J. Biol. Chem. 274, 3978–3987.

Zhou, J., Sims, C., Chang, C.H., Berti-Mattera, L., Hopfer, U., Douglas, J., 1990. Proximal tubular epithelial cells possess a novel 42-kilodalton guanine nucleotide-binding regulatory protein. Proc. Natl. Acad. Sci. U. S. A. 87, 7532–7535.

Zierold, C., Mings, J.A., DeLuca, H.F., 2003. Regulation of 25-hydroxyvitamin D3-24-hydroxylase mRNA by 1,25-dihydroxyvitamin D3 and parathyroid hormone. J. Cell. Biochem. 88, 234–237.

Chapter 28

Receptors for parathyroid hormone and parathyroid hormone–related protein

Thomas J. Gardella, Harald Jüppner and John T. Potts, Jr.
Endocrine Unit, Department of Medicine and Pediatric Nephrology, MassGeneral Hospital for Children, Massachusetts General Hospital and Harvard Medical School, Boston, MA, United States

Chapter outline

Introduction 691
The PTHR1 is a class B G-protein-coupled receptor 692
Parathyroid hormone receptor gene structure and evolution 694
Structure of the PTHR1 gene 694
Evolution of the parathyroid hormone receptor–ligand system 694
Mechanisms of ligand recognition and activation by parathyroid hormone receptors 695
Basic structural properties of the PTHR1 695
Two-site model of ligand binding to the PTHR1 696
Mechanism of ligand-induced activation at the PTHR1 698
Conformational selectivity and temporal bias at the PTHR1 699
Two high-affinity PTH receptor conformational states, R^0 and RG 699
Conformation-based differences in signaling responses to PTH and PTHrP ligands 699
Endosomal signaling and signal termination at the PTHR1 700
Ligand-directed temporal bias and therapeutic implications 701
LA-PTH, a long-acting PTH/PTHrP analog for hypoparathyroidism 702
Abaloparatide: a PTHrP analog for osteoporosis 702
PTHR1 mutations in disease 703
Jansen's metaphyseal chondrodysplasia 704
Other diseases linked to PTHR1 mutations 704
Nonpeptide mimetic ligands for the PTHR1 704
Other receptors for parathyroid hormone and related ligands 705
PTHR2 and PTHR3 subtypes 705
Possible receptors for C-terminal PTH and PTHrP 706
Conclusions 706
References 707

Introduction

Parathyroid hormone (PTH) and PTH-related protein (PTHrP) mediate their principal biological actions by acting on a single cell-surface receptor, the type-1 PTH/PTHrP receptor, or PTHR1. The PTHR1 thus stands as a key regulatory molecule that is essential for normal physiology at all stages of life. The biological situation is somewhat unique in that the one receptor responds to two endogenous ligands to thereby control two disparate physiological processes: the maintenance of blood calcium and phosphate homeostasis via PTH and the timing of cell differentiation events in the developing skeleton and other tissues via PTHrP. The PTHR1 also holds interest as a prospective drug target, as pharmacologic agents that specifically modulate its activity can potentially be used to treat disturbances of bone and calcium metabolism that occur in a variety of pathologic conditions, such as osteoporosis.

The endogenous PTH and PTHrP ligands are polypeptide chains of 84 and 141 amino acids, respectively, with the first 34 amino acids containing sufficient information for high-affinity binding to the PTHR1 and potent activation of downstream signaling responses. Thus synthetic PTH(1–34) and PTHrP(1–34) peptides can generally mimic most of the biological actions of the full-length molecules. Nevertheless, certain functional activities have been reported for peptides derived from the C-terminal portions of PTH and PTHrP, which has led to the notion that other receptors, distinct from the PTHR1, might exist that bind the C-terminal portion of the ligands, although no such receptor has so far been identified. For the PTHR1, much effort has been directed at elucidating the mechanisms by which the ligands PTH and PTHrP engage the receptor and stimulate signal transduction. These studies generally reveal that PTH(1–34) and PTHrP(1–34) interact

Principles of Bone Biology. https://doi.org/10.1016/B978-0-12-814841-9.00028-2

with the PTHR1 in highly similar, yet not identical, fashions. Consistent with this, the primary structures of the ligands exhibit considerable amino acid sequence homology in the amino-terminal region, particularly at the N-terminal 1–14 portion, which is critical for receptor activation, with eight amino acid identities, while the 15–34 portion, which contributes to binding affinity, shows only a moderate level of homology, with three identities. As discussed later in this chapter, studies suggest that differences can indeed be discerned in the binding modes utilized by PTH(1–34) and PTHrP(1–34) and that such differences can lead to differences in functional effects, particularly in the duration of the signaling responses induced. Such findings further support the notion that the capacity of the PTHR1 to regulate the aforementioned two disparate types of physiological processes—the endocrine control of mineral ion homeostasis and the paracrine control of tissue development—is at least partly based on differences in the mechanisms by which the two respective ligands engage the receptor.

Early work conducted prior to the cloning of the PTHR1 in 1991 and which used canine or bovine renal membrane preparations or cell lines derived from bone or kidney gave initial clues as to the basic pharmacology of PTH ligand binding to endogenous receptor binding sites and the downstream signaling responses that could be activated. This work was made possible in large part by the development of peptide synthesis technology and its early application to PTH (Potts et al., 1971), which yielded most notably the PTH(1–34) peptide, as well as radioiodinated derivatives that enabled direct characterization of the binding sites used (Segre et al., 1979). Such studies further revealed that residues in both the N- and the C-terminal region were important for overall binding affinity (Nussbaum et al., 1980). Parallel studies revealed that PTH peptides induced rapid and robust increases in intracellular cAMP, and that the N-terminal residues of the ligand were critical for activating this response. That this response was mediated by a cell-surface G-protein-coupled receptor (GPCR) was indeed borne out in 1991 by the cloning of the cDNA encoding the PTHR1 from kidney- and bone-derived cell lines (Juppner et al., 1991), which revealed the seven membrane-spanning-domain protein architecture that defines all 800 or so members of the GPCR superfamily. A key goal now is to elucidate the specific mechanisms by which the PTHR1 uniquely recognizes its two specific ligands, PTH and PTHrP, and activates selected intracellular effectors in different target cells. Progress toward this goal has been made via the application of a number of molecular and pharmacological approaches, such as the use of receptor mutagenesis coupled with peptide analog design strategies to reveal sites of specific intermolecular contact and clues about conformational changes involved in activation. The use of genetically modified mice engineered so as to have a specific alteration in a selected component of the ligand–receptor response system have also helped make possible a level of analysis at the whole-animal physiological level. These experimental systems ultimately help reveal the involvement of the PTHR1 and its ligands in skeletal and mineral ion physiology as well in diseases, such that they could lead to the development of new forms of therapy targeted to the PTHR1. Broader views of the roles that the PTHR1 and its ligands play in the systemic control of metabolic processes in bone and kidney, and developing tissue, as well as the downstream processes that the receptor controls within target cells, are discussed in detail in other chapters (see Chapters 24–27, 32, and 52). This chapter then focuses on the molecular properties of the PTHR1 per se, and principally on the mechanisms by which PTH and PTHrP ligands, via their N-terminal portions, interact with this receptor and induce signal transduction processes. It will also discuss findings on the so-called type-2 PTH receptor, or PTHR2, and its endogenous ligand, tuberoinfundibular peptide of 39 residues (TIP39), as well as the possibility that other, as yet unidentified, receptors exist that can bind and potentially respond to the carboxy-terminal portion of the endogenous PTH and PTHrP ligands, which appear not to be contributing to interactions with the PTHR1.

The PTHR1 is a class B G-protein-coupled receptor

The PTHR1 exhibits the general protein architecture seen in each of the class B GPCRs and thus is a single-chain integral membrane protein containing a relatively long amino-terminal extracellular domain (ECD) of ~170 amino acids (after removal of residues 1–22, which act as the signal sequence), a transmembrane domain (TMD) or core region comprising the seven membrane-spanning α helices (TMs 1–7) and their interconnecting extracellular and intracellular loops (ECLs 1–3 and ICLs 1–3), and a C-terminal tail extending from the base of TM7 (~Cys452) to the C terminus (Met593) (Fig. 28.1A). As a class B GPCR, the PTHR1 contains none of the hallmark amino acid residues and sequence motifs that define the three other main classes of GPCRs: the class A receptors represented by the β_2Ar, the class C GPCRs represented by the calcium-sensing receptor, and class F receptors, as represented by the Frizzled receptors (Fredriksson et al., 2003). Instead, the PTHR1 exhibits a distinct pattern of amino acids that is conserved in each of the 14 or so other class B GPCRs, which include the receptors for calcitonin, secretin, glucagon, glucagon-like peptide-1 (GLP1), corticotropin-releasing factor (CRF), and several other peptide hormone ligands (Segre and Goldring, 1993). Nevertheless, comparison of the three-dimensional structures obtained for GPCR representatives from each of the main GPCR families reveals

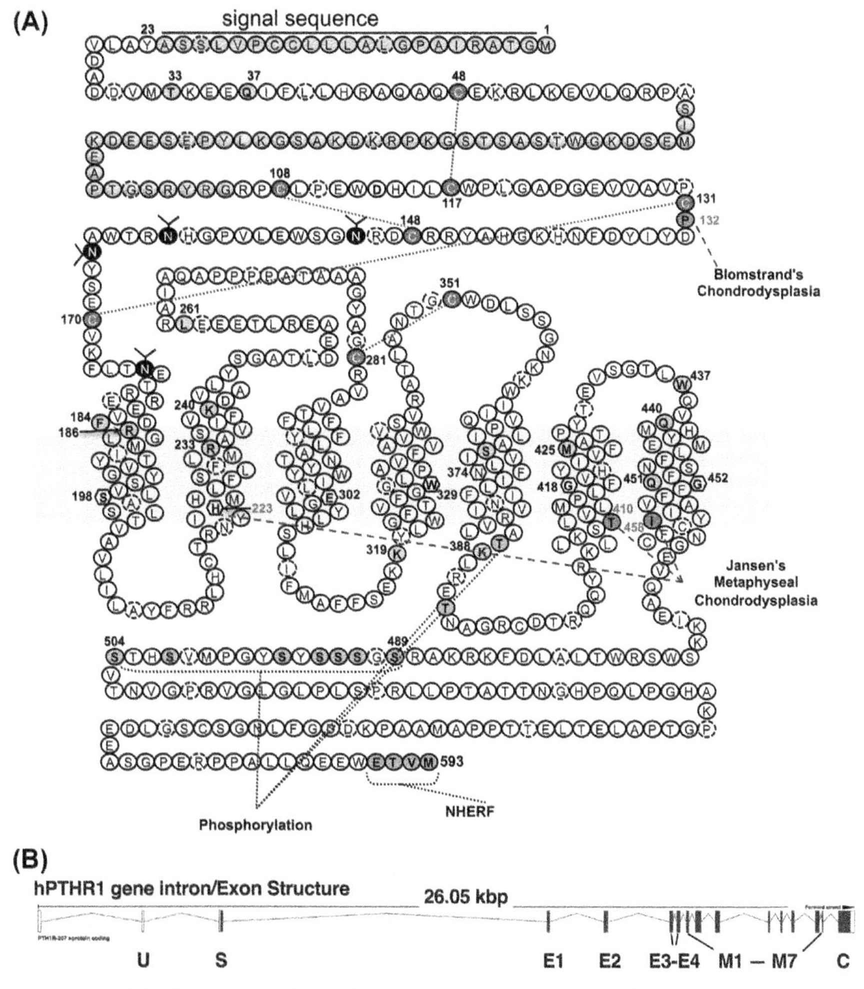

FIGURE 28.1 Primary structure of the human parathyroid hormone receptor type 1 protein and corresponding gene organization. (A) The human parathyroid hormone receptor type 1 (*hPTHR1*) protein is displayed to illustrate the domain organization and locations of selected key residues, including the eight extracellular cysteines involved in a disulfide bond network (*connecting dotted lines*); the four glycosylated extracellular asparagines; Pro132, at which loss-of-function mutations occur in Blomstrand's chondrodysplasia (compound homozygous) and in failed tooth eruption (heterozygous); His223, Thr410, and Ile458, at which activating mutations occur in Jansen's metaphyseal chondrodysplasia; cytoplasmic sites of serine and threonine phosphorylation; the four C-terminal residues involved in PDZ-domain interactions with Na^+/H^+ exchange regulatory factor 1 (*NHERF*) proteins; and a number of residues shown by mutagenesis and/or cross-linking studies to be involved in ligand interaction (*filled shaded circles* with position numbers). Also shown is the residue in each transmembrane domain helix that is the most conserved among the class B G-protein-coupled receptors (*filled hexagons*). (B) The intron/exon structure of the hPTHR1 gene with coding and noncoding exons indicated as *filled* and *open boxes*, respectively.

similarities not only in the basic seven-TMD protein fold that is shared across all GPCR classes, but also in some of the key interhelical contact points that occur within the hepta-helical bundle and are thought to mediate key mechanistic steps of activation (Cvicek et al., 2016). At least some aspects of function are thus likely preserved, at least topologically, between the PTHR1 and perhaps most of the 800 or so other GPCRs encoded in the human genome.

While each class B GPCR responds to a peptide ligand of moderate size—about 30–40 amino acids in length—the PTHR1 is the only class B GPCR for which the endogenous ligands extend more C-terminally, as PTH is 84 amino acids in length and PTHrP is 141 amino acids. Such C-terminal extensions are absent in the ligands identified in the genomes of fish and other nonmammalian species (Rubin and Jüppner, 1999; Mirabeau and Joly, 2013), and thus appear to be modifications that occurred later in vertebrate evolution. In any event, most if not all of the available data indicate that only the first 34 amino acids or so of PTH and PTHrP participate in binding to the PTHR1, such that PTH(1–34) and PTHrP(1–34) peptides exhibit nearly the same affinities and signaling potencies on the PTHR1 as the corresponding full-length polypeptides (Li et al., 2012; Dong et al., 2012).

Parathyroid hormone receptor gene structure and evolution

Structure of the PTHR1 gene

The gene encoding the human PTHR1 resides on chromosome 3 (locus 3p22−p21.1) and spans ~26 kb of DNA (Fig. 28.1B) (McCuaig et al., 1994; Kong et al., 1994). The predicted transcript consists of 14 coding exons and 2 noncoding exons, ranging in size from 42 bp for exon M7 encoding a portion of transmembrane helix 7 to more than 400 bp for exon T encoding the C-terminal tail. The introns vary in size from 81 bp between exons M6 and M6/7 to more than 10 kb between exon S encoding the signal sequence and exon E1 encoding the N-terminal portion of the mature receptor. Three promoter regions have been identified for the gene and shown to be active at different levels in different tissues (Minagawa et al., 2001; Bettoun et al., 1997; Amizuka et al., 1997). Expression of PTHR1 mRNA can be detected in a variety of fetal and adult tissues, including the adrenals, placenta, fat, brain, spleen, liver, lung, and cardiac muscle, with strongest signals observed in adult bone and kidney (Urena et al., 1993; Tian et al., 1993; Uhlen et al., 2015). The genes for the other class B GPCRs generally exhibit a similar intron/exon organization, which is consistent with their evolution from a common ancestor gene and utilization of a similar protein design for engaging their cognate peptide ligands (Hwang et al., 2013).

Evolution of the parathyroid hormone receptor−ligand system

Bioinformatics-based investigations of PTH receptor as well as PTH and PTHrP ligand coding sequences present in the genomes of species representing various stages of evolution have yielded clues about how the PTH ligand and receptor system first emerged and evolved over time to provide the complex developmental and adaptive capacities established in the mammalian species. These studies suggest that PTH receptors and ligands emerged before the evolution of the first vertebrates, as apparent orthologs of the receptor as well as PTH-like ligands have been detected in the genomes of early vertebrates, such as the elephant shark and sea lamprey, as well as some invertebrates, such as the tunicate *Ciona intestinalis* and the amphioxus *Branchiostoma floridae*, which are thought to be representative of early chordates (Fig. 28.2) (Kamesh et al., 2008; Pinheiro et al., 2012; Mirabeau and Joly, 2013; On et al., 2015; Cardoso et al., 2006; Hwang et al., 2013). Sequences with at least some homology (~69% amino acid similarity) have also been identified in the genomes of several insects, such as the red flour beetle (*Tribolium castaneum*) and honeybee (*Apis mellifera*), but not in the fruit fly (*Drosophila melanogaster*) (Li et al., 2013; Cardoso et al., 2014). The biological function of any such invertebrate PTH receptor−like sequence remains unknown.

Genome studies also suggest that chromosomal rearrangements contributed to the diversification of the PTH ligand−receptor system. At least two copies of a PTH receptor gene family member are thus found in the haploid genomes of most vertebrate species, and these can be attributed, in part, to two rounds of whole-genome duplication that are thought to have occurred during the early phases of vertebrate evolution (Fig. 28.2) (Hwang et al., 2013; On et al., 2015; Pinheiro et al., 2012; Cardoso et al., 2006, 2014). These early duplications were followed by other gene duplications as well as deletion events that occurred at different points along the evolutionary paths leading to the divergent animal groups. The genomes of teleost fish, as represented by the zebrafish, *Danio rerio*, thus encode three receptors, the PTHR1, PTHR2, and PTHR3, but lack a PTHR4 gene as a putative partner to the PTHR2, apparently due to a deletion event that happened in early fish evolution (Rubin and Juppner, 1999; Hwang et al., 2013). Bird genomes encode PTHR1 and PTHR3, and lack PTHR2 and PTHR4, while mammals encode PTHR1 and PTHR2, and lack PTHR3 and PTHR4. The absence of a PTHR2 in birds and a PTHR3 in primates apparently reflects separate gene deletion events that occurred at distinct times during the evolution of these two vertebrate groups. The peptide ligands, as represented in humans by PTH, PTHrP, and TIP39, the last a ligand for the PTHR2 as discussed further in a later section, are presumed to have evolved in parallel with their receptors, and thus to have stemmed from some precursor peptide ligand that emerged at or about the time that the first chordates appeared, as indeed, PTH- and/or TIP39-like coding sequences are found in the genomes of early fish, as well as amphioxus and *C. intestinalis* (Rubin and Juppner, 1999, Yan et al., 2012; Pinheiro et al., 2012; Trivett et al., 2005; Liu et al., 2010). These findings are consistent with the notion that the PTH ligand−receptor system is an evolutionarily ancient adaption that emerged in some precursor form before the appearance of the first vertebrates and evolved over time so as to contribute importantly not only to the successful adaption of the animal to changing environments but also to the development of higher-order organ systems, such as the skeleton (Hogan et al., 2005; Suzuki et al., 2011; Trivett et al., 2005).

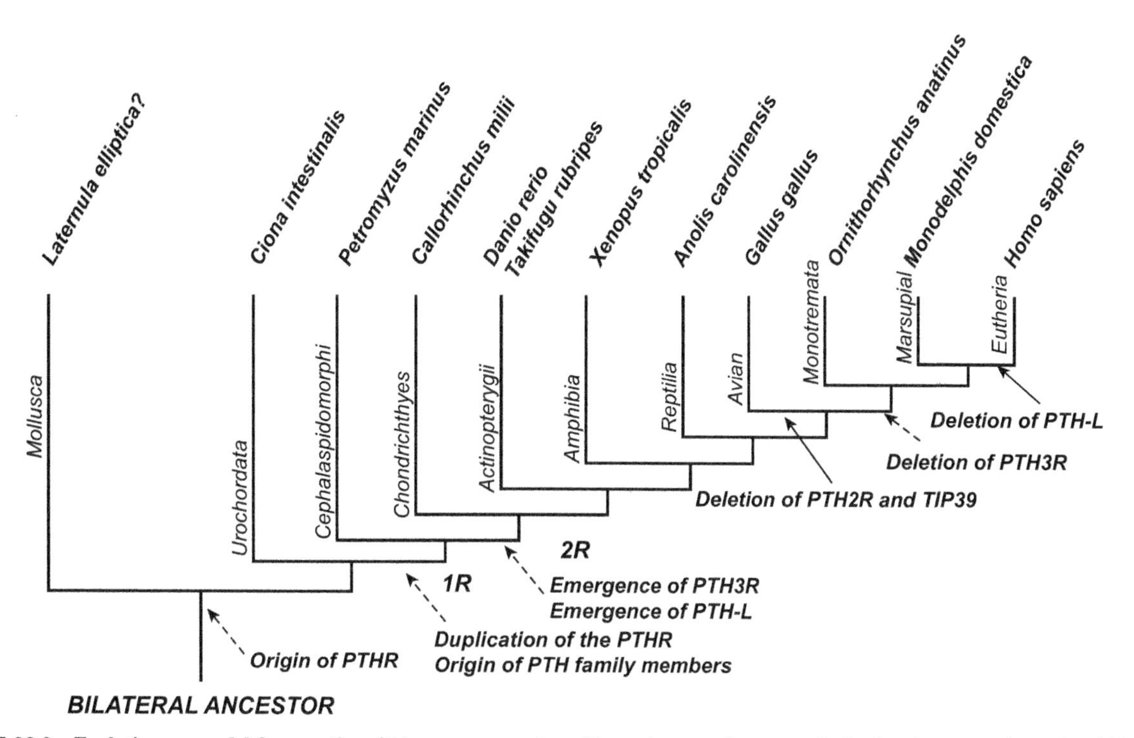

FIGURE 28.2 Evolutionary model for parathyroid hormone receptors. Shown is a tree diagram to depict that the ancestral parathyroid hormone receptor (*PTHR*) sequence probably originated before the appearance of the first chordates, as represented by the tunicate *Ciona intestinalis*, and that two rounds of whole-genome duplication (*1R* and *2R*) occurred before the emergence of the terrestrial vertebrates to result in three receptor subtypes—PTHR1, PTHR2, and PTHR3—with the expected fourth subtype lost by an early gene deletion event. The PTHR2 and PTHR3 subtypes were probably lost by later gene deletion events occurring with the radiations leading to birds and mammals, respectively. *Reproduced pending permission from Pinheiro, P. L., Cardoso, J. C., Power, D. M., Canario, A. V. 2012. Functional characterization and evolution of PTH/PTHrP receptors: insights from the chicken. BMC Evol. Biol. 12, 110.*

Mechanisms of ligand recognition and activation by parathyroid hormone receptors

Basic structural properties of the PTHR1

As in all class B GPCRs, the ECD contains six highly conserved cysteine residues that form a disulfide network (Fig. 28.1A) that maintains the tertiary fold of the ECD, which follows a tripartite helix–β sheet–helix pattern (Pioszak et al., 2008). The ECD contains four asparagines that are glycosylated during intracellular processing and transported to the plasma membrane. A feature of the ECD that is seen only in the mammalian PTHR1s, and not in the mammalian PTHR2s or any other class B GPCR, is a 44-amino-acid segment—Ser61–Gly105 in the human PTHR1—located between the first and the second cysteine. This segment is encoded by a separate exon, called E2, and appears not to contribute to function, as assessed by site-directed mutagenesis strategies and pharmacological analyses in transfected cells (Lee et al., 1994).

Other than the general seven membrane-spanning helical architecture, the class B receptors share no obvious homology with receptors from the other main GPCR subgroups, including the class A GPCRs, which comprise the largest GPCR subgroup and include the β-adrenergic receptor and rhodopsin as well-studied representatives. Nevertheless, the high-resolution structures now available for a number of GPCRs from each main class, including several class B GPCRs, suggest that certain relationships of structure and function might be preserved in all GPCRs, particularly in the intramolecular interaction networks that are located within the TMD bundle and mediate processes of receptor activation and G protein coupling (Bortolato et al., 2014; Cvicek et al., 2016). Within the class B GPCRs there are about 45 amino acid residues that are strongly conserved. These conserved residues are dispersed in the seven TM helices and in the ECD and probably help define the basic three-dimensional scaffold structure of the receptor and as well may participate more directly in basic mechanisms of activation, while other more divergent residues are likely to provide the specific recognition determinants used for selective binding of the appropriate peptide ligand.

Binding of a PTH agonist peptide ligand, such as PTH(1–34), to the extracellular surface of the PTHR1 induces a conformational change in the receptor that leads to G protein coupling and activation of downstream signaling responses (see Chapters 32 and 52 for an in-depth discussion of downstream signaling responses). In most target cells the PTHR1 couples strongly to G proteins containing the GαS subunit, and thus activates the adenylyl cyclase/cAMP/protein kinase A (PKA) signaling system. The agonist-stimulated PTHR1 can also activate other signaling systems, including the Gαq/phospholipase C (PLC)/IP3/iCa^{2+} pathway, the Gα12/DAG/PLD pathway, and the arrestin/ERK1/2 pathway, depending potentially on the type of target cell and PTH or PTHrP ligand analog used (Gesty-Palmer et al., 2006, 2013). PTHR1 signaling is generally thought to be a highly regulated process and thus to terminate relatively soon after initial binding via mechanisms involving receptor phosphorylation and internalization (Bisello et al., 2002; Malecz et al., 1998; Qian et al., 1998), but variations on this theme now seem possible, as discussed in a later section. It is now becoming clear that ligands with different structures can bind to a receptor in different modalities so as to induce different biological outcomes in target cells, in terms of type of pathway activated as well as the duration of the response, a concept generally referred to as biased agonism (Luttrell et al., 2015). This concept has particular relevance for the PTHR1, as it may help explain, at least in part, certain differences in biological actions that have been observed for PTH and PTHrP peptides (Horwitz et al., 2003) and, moreover, could provide a strategic basis for developing new ligand analogs for the PTHR1 that have therapeutic utility.

Two-site model of ligand binding to the PTHR1

Clues regarding the overall mechanism of ligand binding and agonist-induced activation used by the PTHR1 were initially obtained from receptor mutagenesis and photoaffinity cross-linking studies, but more recently have come from the successful application of X-ray crystallography and cryo-electron microscopy (cryo-EM) approaches to the analysis of several other class B GPCRs, either as intact receptors or as the isolated ECD or TMD. Such structural information for the PTHR1, however, is so far reported only for the isolated ECD in complex with PTH(15−34) or PTHrP(12−34) peptides (Pioszak et al., 2008, 2009). The emerging structural data combined are providing valuable insights into the basic mechanisms by which the class B GPCRs engage their cognate peptide ligand and activate signal transduction. The earlier studies on binding mechanisms for the PTHR1 performed in parallel by several groups led to the so-called "two-site" model of ligand−receptor interaction. According to the model, the ligand, such as PTH(1−34), first docks to the receptor via the binding of the C-terminal domain of the ligand, approximately residues 15−34, to the ECD of the receptor. Then the amino-terminal portion of the ligand, approximately residues 1−14, engages the extracellular surface of the TMD portion of the receptor to yield the intact ligand−receptor complex (Fig. 28.3). The ECD component of the interaction contributes predominantly to the overall binding affinity of the complex, while the TMD component mediates the process of receptor activation and G protein coupling. The experimental findings that led to this model came from studies employing, often in parallel, the two complementary approaches of receptor mutagenesis and photoaffinity cross-linking. By the former approach, receptors altered at specific residues or defined regions were generated and analyzed in pharmacologic binding and signaling assays for interaction with various peptide ligand analogs with defined structural modifications such that they could be used as functional probes of specific contact sites or regions. For example, a receptor chimera strategy based on the weaker binding of PTH(7−34) to the rat versus the human PTH receptor was used to establish that the ECD of the receptor was the major site of binding for the C-terminal portion of the ligand (Juppner et al., 1994). A similar chimera strategy based on the capacity of the analog Arg2-PTH(1−34), in which the highly conserved valine-2 is replaced by a bulky arginine, to function as an antagonist on the rat PTHR1 and an agonist on the opossum PTHR1 led to the finding that residue 370 at the extracellular end of TM5, alanine and serine in the opossum and rat receptors, respectively, is a key determinant of ligand-induced activation (Goltzman et al., 1975). Other such mutagenesis studies that followed, extending also to the PTHR2, further supported the two-domain model (Bergwitz et al., 1996, 1997; Turner et al., 1998). The second approach involving the use of photoaffinity cross-linking strategies provided more direct biophysical analyses of ligand and receptor sites in close intermolecular proximity. The method involved ligands modified at a selected position with an amino acid analog containing a photolabile side chain, typically benzophenylalanine (BPA) (Zhou et al., 1997). After binding to the receptor, the complex is UV irradiated, causing the BPA group to covalently link to a site in the target protein chain within a radius of a few angstroms. Initially BPA was thought to react rather nonselectively, but was later discovered to have a propensity to react with methionines (Wittelsberger et al., 2006b). In any event, the strategy resulted in the identification of a number of cross-links that could be mapped to specific receptor residues, including between positions 1 and 2 in the ligand and Met425 at the extracellular end of TM6 (Bisello et al., 1998; Gensure et al., 2001a), between ligand residue 13 and Arg186 at the extracellular end of TM1 (Adams et al., 1998), and between ligand position 23 and Thr33 and/or Gln27 at the N-terminal end of the receptor (Mannstadt et al., 1998). A number of other contacts involving residues in the middle or C-terminal region of PTH(1−34) or PTHrP(1−36) and various sites in the receptor were also mapped, although not always to specific receptor residues (Gensure et al., 2001b, 2003; Greenberg et al., 2000; Wittelsberger et al., 2006a). The information gained from these cross-linking studies provided distance constraints that enabled development of initial molecular models of the PTHR1·PTH ligand complex (Greenberg et al., 2000; Gensure

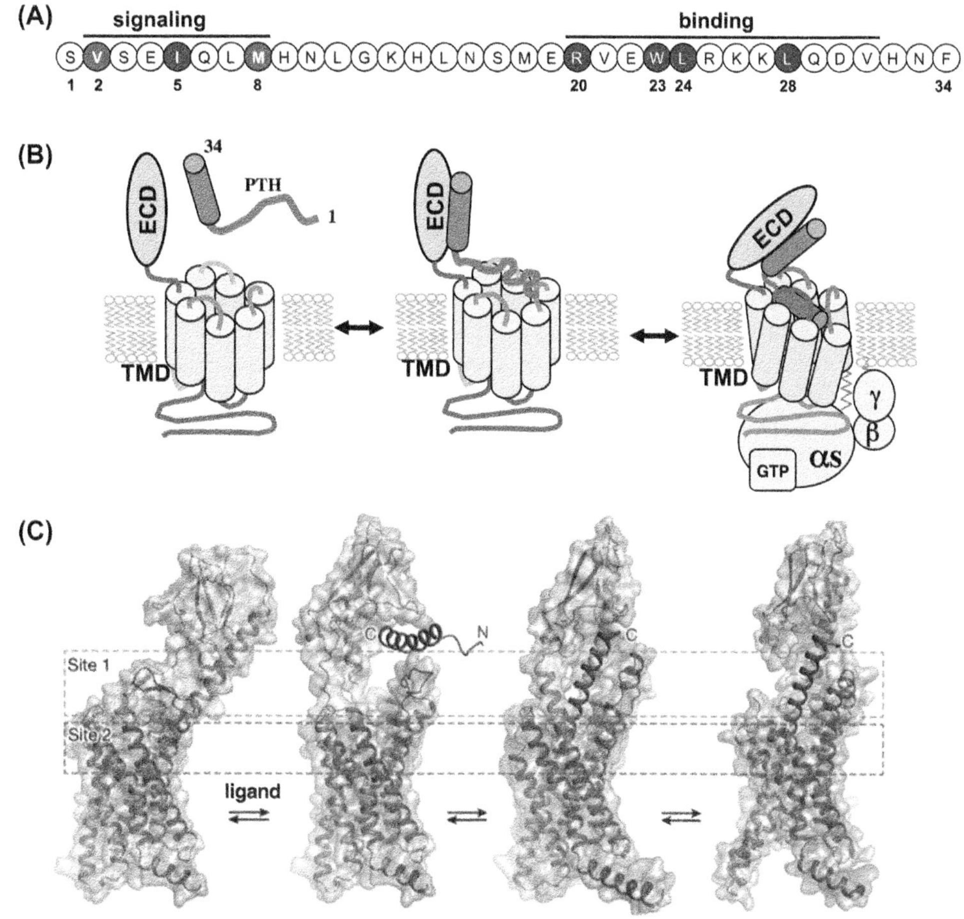

FIGURE 28.3 Ligand binding mechanisms at the parathyroid hormone receptor 1. (A) Sequence of PTH(1–34) highlighting the N-terminal residues Val2, Ile5, and Met8, which play critical roles in receptor activation, and the C-terminal residues Arg20, Trp23, Leu24, and Leu28, which play critical roles in receptor binding. (B) The two-domain model of ligand binding and activation at the PTHR1, as developed by cross-linking and mutagenesis data obtained for the parathyroid hormone receptor 1 (PTHR1): the C-terminal portion of PTH(1–34) first docks to the amino-terminal extracellular domain (*ECD*) of the receptor to provide affinity interactions, and then the amino-terminal portion of the ligand engages the transmembrane domain (*TMD*) of the receptor to induce conformational changes that enable G protein coupling. (C) Refinement of the two-domain model based on high-resolution X-ray crystal and cryo-electron microscopy structures of the glucagon receptor and GLP1 receptor, which are class B GPCRs structurally related to the PTHR1. The ligand, shown in *red*, binds as a nearly linear α helix and makes extensive contacts with exposed extracellular surfaces in both the ECD and the TMD receptor regions. Ligand binding results in an outward movement of the cytoplasmic termini of several of the TM helices, particularly TM6, to thus open a cavity that will accommodate the G protein. *(C) Reproduced pending permission from Zhang, H., Qiao, A., Yang, L., Van Eps, N., Frederiksen, K. S., Yang, D., Dai, A., Cai, X., Zhang, H., Yi, C., Cao, C., He, L., Yang, H., Lau, J., Ernst, O. P., Hanson, M. A., Stevens, R. C., Wang, M. W., Reedtz-Runge, S., Jiang, H., Zhao, Q., Wu, B. 2018a. Structure of the glucagon receptor in complex with a glucagon analogue. Nature 553, 106–110.*

et al., 2003; Wittelsberger et al., 2006a; Piserchio et al., 2002), although the incomplete nature of the input data left a fair amount of uncertainty.

The two-site model was also found to be relevant for most if not all of the other class B GPCRs and their cognate peptide ligands, as elucidated using similar functional and cross-linking-based approaches by other groups (Parthier et al., 2009). This basic model has now been confirmed and refined by high-resolution structural data that have been obtained using X-ray crystallography or cryo-EM approaches for several of the other class B GPCRs, including the CRFR1, the glucagon receptor, and the GLP1 receptor, although not yet for the PTHR1 (de Graaf et al., 2017). A key modification used in the crystallography work, and a potential limitation in terms of data interpretation, is the introduction of thermostabilizing point mutations and insertions of heterologous protein segments known to promote crystallization at dispersed positions in the receptor TMD, which generally tends to be conformationally dynamic. One set of findings from the functional studies that helped define the TMD component of the two-domain model for the PTHR1 was that short amino-terminal PTH fragment peptides, such as PTH(1–14) and PTH(1–11), which as unmodified peptides are inert for binding and signaling due to a lack of stabilizing interactions with the ECD, could act as potent agonists when modified with certain helix-stabilizing and affinity enhancing substitutions. Furthermore, they could stimulate signaling via a PTHR1 construct that lacks most of the ECD (up to Glu182), about as potently as PTH(1−34) does on the intact PTHR1 (Shimizu et al., 2001). These findings highlight the functional autonomy of the N-terminal region of the ligand and the TMD of the receptor, and also suggested that ligands much smaller than PTH(1−34) that target the functionally critical TMD and thus act as potent PTH mimetics could be developed. Native PTH and PTHrP peptides are thought to bind to the PTHR1 via a sequential process that involves initial ligand interactions with the ECD followed by interactions with the TMD and the conformational rearrangements involved in activation (Fig. 28.3) (Hoare and Usdin, 2001). The nature of these conformational changes and the extent to which they may involve a higher-order folding of the complex, for example, a movement of the ECD relative to the TMD, remains an area of uncertainty, but the new class B GPCR structures now available provide clues as to the types of changes that might be possible for the PTHR1−PTH complex.

The crystal structures of the isolated ECD protein in complex with the PTH(15−34) or PTHrP(12−34) fragment suggest that the two ligands bind to the ECD via similar, though not identical, mechanisms (Pioszak et al., 2009). Each peptide thus is bound to the ECD as an α helix and fits into a groove that runs along the center of the ECD. Key hydrophobic interactions occur between Trp23, Leu24, and Leu 28 in the ligand, which align along an apolar face of the helix, and complementary apolar surfaces in the receptor-binding groove. The side chain of Arg20 in the ligand makes extensive interactions with a ring of polar residues located at one end of the ECD. Residues Arg20, Trp23, Leu24, and Leu28 are well conserved in PTH and PTHrP ligands, while residues on the opposite helix face are less well conserved, are mostly polar in character, and can be mutated with relatively little effect on binding. Compared with the PTH(15−34)· ECD structure, the PTHrP helix bends modestly at about Leu24 such that the C-terminal portions of the two helices make distinct contacts, and the C-terminal residue is offset by a few angstroms in the two structures (Pioszak et al., 2009).

Interactions that occur between the N-terminal portion of the ligand, as contained in the PTH(1−14) segment, and the receptor's TMD are of key interest as they are critically involved in receptor activation. There is no direct structural data available for the PTHR1 TMD. The aforementioned mutagenesis and cross-linking data are consistent with the ligand making multiple contacts with an extensive, solvent-exposed surface area that involves the extracellular ends of the TMD helices and the extracellular loops. In the crystal and cryo-EM structures of the TMDs of the other class B GPCRs, the extracellularly exposed surface of the TMD bundle forms a relatively large pocket that accommodates the N-terminal portion of the ligand (Liang et al., 2017; de Graaf et al., 2017). In the intact GLP1R−GLP1 (Zhang et al., 2017, 2018a) and CTR−calcitonin structures, the full-length peptides are bound as linear α helices that extend along the ECD, which is positioned above the TMD, and enter into the TMD bundle such that the N-terminal residues of the ligand helix project into the bundle such that they can trigger the conformational rearrangements involved in receptor activation (Zhang et al., 2018a) (Fig. 28.3C). It is reasonable to suggest that PTH(1−34) binds to the PTHR1 in a fashion similar to that seen for glucagon and calcitonin, but direct confirmation of any such interaction mode for PTH and the PTHR1 remains to be established by direct high-resolution methods.

Mechanism of ligand-induced activation at the PTHR1

Previous mutagenesis and biophysical cross-linking data suggest that substantial conformational changes occur within the TMD bundle of the PTHR1 during agonist-induced activation (Gardella et al., 1996; Gensure et al., 2001a; Thomas et al., 2008). Insights into the types of specific changes that might occur can be inferred by comparing the active- versus inactive-state structures obtained for the several other class B GPCRs. Such comparisons suggest that during the agonist binding and receptor activation process, the ECD of the complex rotates in a counterclockwise direction, relative to the vertical axis of the TMD bundle, by nearly 90 degrees (Zhang et al., 2018a). In addition, two segments at the extracellular ends of the TM1 and TM2 helices isomerize from β-strand conformations to α-helical conformations. Interestingly, whereas the two β strands in the inactive state receptor pair together to form a cover over the ligand-binding pocket of the TMD bundle, the induced α-helical segments extend upward and contact the midregion of the bound ligand helix to thus help stabilize the ligand and guide the N-terminal segment into the TMD pocket (Zhang et al., 2018a). The structural studies on the class B GPCRs also reveal a cluster of highly conserved polar residues within the core of the TMD bundle that form a network of H-bond and electrostatic interactions and thus act to constrain the bundle in a closed, inactive conformation. Agonist interactions within the TMD core trigger a rearrangement of this network to release the constraints and promote the outward movement of the cytoplasmic ends of several of the TM helices, particularly TM6. These movements result in a cavity on the cytoplasmic surface of the receptor that accommodates the G protein (Fig. 28.3C) (Zhang et al., 2017; Yin et al., 2017).

While any such mechanism remains to be assessed for the PTHR1, the studies provide clues as to the types of conformational changes that are likely to occur during the PTHR1 activation process. It is worth noting that several PTHR1 residues that are located at least in the vicinity of the predicted cytoplasmic cavity used by G proteins have been implicated by mutagenesis methods to be involved in G protein interaction. These include Lys319, Val384, and Leu385 for Gαq coupling; Thr387 for Gαs coupling; and Lys388 for Gαs and Gαq coupling (Iida-Klein et al., 1997; Huang et al., 1996). Of further note, these studies led to the generation of the "DSEL" knock-in mouse in which the normal PTH receptor allele is replaced by a PTHR1 mutant having the Glu–Lys–Lys–Tyr sequence at positions 317–320 replaced by Asp–Ser–Glu–Leu and which is thus selectively impaired for Gαq/PLC signaling; the relatively modest effect of the mutant allele on the skeletons of these mice suggests that the Gαq/PLC signaling pathway does not play a critical role in PTH receptor–mediated control of endochondral bone formation (Guo et al., 2002).

Conformational selectivity and temporal bias at the PTHR1

Two high-affinity PTH receptor conformational states, R^0 and RG

The aforementioned structural models are consistent with the notion that structurally distinct ligands for the PTHR1 can stabilize or induce different receptor conformations so as to result in different types of signal transduction responses. The notion that variation in receptor conformation could play an important role in determining the affinity with which certain ligands bind to the PTHR1 emerged from a series of PTH radioligand binding studies that utilized membranes from cells transfected to express the PTHR1 and were performed under conditions to favor either the G-protein-uncoupled receptor conformation or the G-protein-coupled conformation (Hoare et al., 1999b). To promote the G-protein-uncoupled state, the binding reactions were conducted in the presence of GTPγS, which binds to the G protein α subunit and causes it to dissociate from the receptor. To promote the G-protein-coupled state, the binding reactions were conducted using membranes from cells transfected to express excess Gαs, typically a mutant form that binds the receptor with high affinity. The conceptual basis came from early pharmacologic studies performed on the β_2-adrenergic receptor and other such prototypic GPCRs that bind small catecholamine ligands. These studies led to the classical ternary complex model of GPCR action that posits that G protein coupling promotes a high-affinity receptor conformation, while G protein uncoupling, which occurs upon GDP–GTP exchange or GTPγS binding, causes the receptor to relax to a low-affinity state and thus release the bound agonist (De Lean et al., 1980). It was thus surprisingly found that certain ligands for the PTHR1, including PTH(1–34), bound with high affinity even in the presence of GTPγS, while others, such as PTHrP(1–36) and the modified N-terminal PTH fragment M-PTH(1–14), behaved in a fashion more consistent with the classical model and dissociated rapidly upon addition of GTPγS (Hoare et al., 2001; Dean et al., 2006, 2008). The findings thus suggested that the PTHR1 could exist in two distinct high-affinity conformations, depending on the type of ligand bound: one high-affinity conformation, as bound by PTH(1–34), could exist even in the absence of a bound G protein, while the other conformation, as bound by PTHrP(1–36) and M-PTH(1–14), required G protein coupling. Consequently, the two high-affinity states, G protein uncoupled and G protein coupled, were termed R^0 and RG, respectively.

Conformation-based differences in signaling responses to PTH and PTHrP ligands

While the initial findings on distinct affinity PTHR1 conformations came from biochemical studies performed in cell membranes, subsequent studies in intact cells as well as animals supported a biological significance. Thus, cAMP time-course studies performed in PTHR1-expressing cells, in which various ligands were allowed to bind to the receptor typically for 15–30 min and then all unbound ligand was removed by washout, revealed that Gαs/cAMP signaling as induced by R^0-selective ligands such as PTH(1–34) persisted for extended durations after washout. In contrast, cAMP signaling induced by RG-selective ligands such as PTHrP(1–36) or M-PTH(1–14) terminated more rapidly after washout (Okazaki et al., 2007; Dean et al., 2008). The results thus revealed a difference in the mode of binding and hence biological action for the structurally distinct ligands, including for PTH(1–34) and PTHrP(1–36), which until then were thought to bind to and activate the PTHR1 in similar if not identical fashions (Nissenson et al., 1988; Jüppner et al., 1988). However, an intriguing deficiency for stimulation of vitamin D synthesis was suggested for PTHrP(1–36) relative to PTH(1–34) in a human infusion study (Horwitz et al., 2003). The kinetic washout approaches thus enabled differences in modes of binding and signaling to be revealed for different PTH and PTHrP ligands. The kinetic assays initially used were performed in multiwell plate formats and involved measurement of cAMP either by radioimmunoassay of cell lysates or by a luciferase-based Glo sensor reporter that enables the nearly continuous detection of intracellular cAMP in live cells over time (Okazaki et al., 2007; Dean et al., 2008; Maeda et al., 2013). Such studies were subsequently complemented by highly sensitive optical methods of fluorescence resonance energy transfer (FRET) microscopy, which enabled assessment of near-instantaneous changes in intracellular cAMP production at the single-cell level (Ferrandon et al., 2009; Feinstein et al., 2011). The findings also suggest a possible mechanistic basis for the distinct biological roles that PTH and PTHrP play in biology. Thus, a sustained mode of action, as occurs with PTH, would provide a suitable means to maintain blood calcium at the levels needed for normal physiological processes, while a more transient mode of signaling, as occurs with PTHrP, would provide a more suitable means to achieve the precise temporal control of cell differentiation that is required during tissue development, as for example in the skeletal growth plates, in which a precisely timed program of chondrocyte differentiation is required for proper elongation and shape formation of the long bones (Chagin and Kronenberg, 2014; Kronenberg, 2003).

The structural basis for the different modes of binding and hence signaling observed for PTH and PTHrP ligands in these studies is not completely known, but several amino acids that differ in the two ligands have been implicated. Most notably is the divergence at position 5, which is Ile in PTH and His in PTHrP, as it was thus found that the Ile5-PTHrP(1−36) analog binds with higher affinity to the R^0 conformation and exhibits more prolonged cAMP responses after washout than does PTHrP(1−36) (Dean et al., 2008). The divergences in the C-terminal (15−34) regions of the two ligands, which probably result in moderately altered modes of binding to the ECD (Pioszak et al., 2009), could also contribute to the differences in binding modes detected for the two ligands in the functional assays. The PTHrP(12−34) fragment was in fact shown to bind to the isolated ECD of the PTHR1 with an affinity about twofold higher than that of PTH(15−34), which supports an altered mode of binding, although the difference in affinity is opposite to that seen for the N-terminally intact ligands and the intact receptor (Dean et al., 2008). In any case, the findings on conformational selectivity for the PTHR1 provide insights into the mechanisms controlling the duration of ligand-induced signaling at the PTHR1, and furthermore have potential implications for the development of new PTHR1-based therapeutics. These two aspects are discussed in the following sections.

Endosomal signaling and signal termination at the PTHR1

Termination of G protein signaling at most GPCRs has generally been thought to occur relatively soon after initial binding of the agonist ligand and coincident with internalization of the receptor into endosomes (Drake et al., 2006). Findings on the PTHR1, as well as several other GPCRs, however, indicate that G-protein-mediated signaling can continue after most if not all of the receptor has moved into the endosomal domain. According to classical GPCR models, the process of signal termination involves a sequence of biochemical steps that include the phosphorylation of the receptor at serine and threonine residues located on exposed portions of the receptor's C-terminal tail and intracellular loops, the recruitment of β-arrestin proteins to the phosphorylated receptor with a coincident displacement of the G protein, the assembly of the β-arrestin-associated receptor into clathrin-coated pits that then pinch off from the membrane to form vesicles, and the trafficking of the vesicles containing the receptor and ligand as cargo through the endosomal sorting system. The last step directs the cargo to pathways of degradation as mediated by lysosomes and/or the proteasome system or recycles the receptor back to the cell surface (Drake et al., 2006). Many studies provide data generally consistent with this general scenario for the PTHR1. Thus, the agonist-activated PTHR1 is indeed rapidly phosphorylated on at least seven serine and two threonine residues, as well as being ubiquitinated on at least two lysines in the C-terminal tail and intracellular loops; β-arrestin proteins are recruited, the receptor moves into endosomal vesicles and internalizes, and it can be at least partially recycled back to the cell surface (Rey et al., 2006; Tawfeek et al., 2002; Vilardaga et al., 2002; Chauvin et al., 2002; Zhang et al., 2018b). The PTHR1 also engages with certain intracellular scaffolding proteins, particularly members of the Na^+/H^+ exchange regulatory factor family of proteins via PDZ-domain-based anchoring to the receptor's C-terminal tail (Mahon et al., 2003; Mamonova et al., 2012), which further are thought to contribute to the trafficking and signal termination events that occur following agonist binding and activation (see Chapter 27 for further information on PTHR1 trafficking events related to actions in the kidney).

The initial evidence suggesting that PTHR1 could deviate from the classical model of GPCR signal termination came from studies on M-PTH(1−34) and M-PTH(1−28) analogs that contained the same set of affinity- and potency-enhancing modifications as contained in the M-PTH(1−14) analogs mentioned earlier (Okazaki et al., 2008). These M-PTH(1−34) analogs were thus found to mediate cAMP responses in PTHR1-expressing cells that lasted for as long as several hours after washout, and thus for much longer times than those induced by PTH(1−34). Microscopy studies using fluorescent PTH analogs, typically modified with tetramethylrhodamine attached to the ε-amino function of lysine-13, revealed that within 15 min of initial binding, most, if not all, of the ligand was located in cytoplasmic vessels and not on the cell surface. There was thus a spatiotemporal correlation between persistent signaling and location in vesicles, which raised the novel possibility that the agonist-activated PTHR1, at least when bound by certain modified ligands, could mediate prolonged Gαs-mediated cAMP responses from the endosomal compartment. It was further observed that upon injection into mice, such M-PTH(1–34) and related analogs induced elevations in blood calcium that persisted for much longer durations than those induced by PTH(1–34). Importantly, the prolonged calcium responses could not be explained by a simple plasma pharmacokinetic effect, as the modified ligands disappeared from the blood, if anything, more rapidly than did PTH(1–34). The results thus suggested that the modified ligands were rapidly sequestered by binding to receptors in target cells of bone and kidney and thus continued to signal from the endosomal compartment (Okazaki et al., 2008).

Parallel studies by Vilardaga and colleagues using fluorescence microscopy and FRET-based kinetic approaches to track ligand–receptor complexes in real-time in live HEK293 cells supported a novel mechanism, and also identified some of the key cytoplasmic effector proteins involved. These data thus confirmed that the ligand–receptor complexes were translocated rapidly from the plasma membrane to internalized endosomes where they remained associated with the G protein as well as adenylyl cyclase (Ferrandon et al., 2009). Prolonged signaling at the PTHR1 thus clearly correlated not only with binding to a distinct high-affinity state of the receptor, R^0, but also with the formation of signaling complexes that remained stable and presumably active in endosomes. Evidence of such endosomal signaling has now been reported for a number of other GPCRs, including the calcitonin receptor (Andreassen et al., 2014), the calcium-sensing receptor, a class C GPCR (Gorvin et al., 2018), and a number of class A receptors (Godbole et al., 2017; Thomsen et al., 2016).

The concept of endosomal signaling could have implications for understanding the basic processes for how these receptors function. PTHrP(1–36) was found to bind more selectively to the RG conformation of the PTHR1 and thus induces more transient cAMP responses compared with PTH(1–34) or the more strongly R^0-selective analogs, such as M-PTH(1–34) (Dean et al., 2008). Live-cell fluorescence tracking studies in HEK293 cells revealed that PTHrP(1–36) did not colocalize with the PTHR1 in intracellular vesicles (Ferrandon et al., 2009). The findings thus suggested that PTHrP and presumably other RG-selective ligands follow a subcellular trafficking path distinct from that used by PTH(1–34) and the longer-acting analogs and which thus involves differences in processes of signal termination. A key step in the signal termination process for the PTHR1, as induced by PTH(1–34), was thus found to be the association of the internalized receptor with a macromolecular assembly called the retromer complex (Feinstein et al., 2011), which acts at the later stages of endosomal maturation to sort vesicle cargo along pathways of either recycling or degradation. Retromer-mediated signal termination for the PTHR1 was found to be dependent on the constituent proteins, VPS26, VPS35, and SNX27, the last of which binds to the PTHR1 C tail via a PDZ-domain-directed interaction (McGarvey et al., 2016; Chan et al., 2016; Feinstein et al., 2011). The biological relevance for this PTHR1–retromer interaction, initially revealed in HEK293 cells, is supported by studies showing that targeted reduction of VP35 expression in osteoblastic MC3T3 cells prolongs the cAMP signaling response to PTH(1–34), and that mice genetically ablated for VPS35 in osteoblasts exhibit a mild osteoporotic phenotype and an enhanced anabolic response to intermittent PTH(1–34) (Xiong et al., 2016). These studies also revealed that PPP1R14C, which acts to inhibit the PP1 phosphatase, can associate with VPS35 or the internalized PTHR1 to thus result in changes in the phosphorylation status of downstream effectors, including phosphorylated cAMP-responsive element binding protein (CREB). It is thus intriguing to consider that endosomal signaling provides a means for the PTHR1 to activate specific target effectors in defined subcellular compartments so as to achieve efficient control of downstream responses; for example, the phosphorylation of a transcriptional activator such as CREB in the immediate vicinity of the nucleus to thus facilitate its access to the genome.

A second mechanistic aspect of subcellular trafficking and signal termination for the PTHR1 revealed in these studies concerns the role of endosomal acidification. In general, the endosomal interior progressively acidifies from an initial pH of ~7.4 to a low pH of ~4.5, at which most biological ligand–receptor complexes dissociate. This acidification is mediated by the vacuolar ATPase proton pumps, which are activated by cAMP/PKA-dependent phosphorylation. PTH(1–34)-induced cAMP/PKA signaling was thus found to promote vesicle acidification and hence the dissociation of the PTH·PTHR1 complexes, leading to signal termination. Thus PTHR1 signaling appears to be regulated at the level of the endosome by a negative feedback system of acidification (Gidon et al., 2014). These findings further suggest the possibility that structurally distinct ligands might bind to the PTHR1 in different fashions so as to form complexes that have differential sensitivities to endosomal acidification and hence can signal for different durations.

Ligand-directed temporal bias and therapeutic implications

The profound differences in the duration of signaling now seen to be clearly possible for structurally distinct ligand analogs via binding to distinct PTHR1 conformations are consistent with the emerging GPCR concept of ligand-directed temporal bias (Grundmann and Kostenis, 2017). Thus, for the PTHR1, ligands that bind mainly to the G-protein-dependent conformation, RG, mediate transient signaling responses because the ligand–receptor complexes dissociate soon after the first round of G protein activation, hastened by endosomal acidification, whereas ligands that bind with high affinity to the

G-protein-independent conformation, R^0, mediate prolonged signaling responses as the ligand−receptor complexes remain intact upon G protein uncoupling and endosomal acidification and hence can activate multiple G protein coupling cycles (Okazaki et al., 2008). Such temporal bias for the PTHR1, as it relates to distinct PTH and PTHrP analog ligands, has relevance to strategies aimed at the development of new PTHR1-based therapeutics, as the duration of signaling at the PTHR1 can have a profound impact on physiological outcomes, particularly those relating to the balanced processed of bone anabolism and catabolism (Martin et al., 2006).

Ligand-dependent temporal bias at the PTHR1 holds relevance for two key diseases that are treatable with PTHR1-targeted ligands: osteoporosis and hypoparathyroidism. Because of the distinct pathophysiological mechanisms underlying these two diseases—an imbalance in the coupled processes of bone formation bone and resorption leading to an inadequate bone structure in osteoporosis, and a deficiency of PTH production by the parathyroid glands leading to a condition of chronic hypocalcemia in hypoparathyroidism—distinct PTH ligand modes of action and hence pharmacodynamic profiles are needed to achieve effective treatment. Thus, a pulsatile mode of action is needed for osteoporosis, while a more sustained mode of action is needed for hypoparathyroidism (see Chapters 10, 23, and 70 for more detailed information on underlying mechanisms and modes of treatment for these diseases). The concept of temporal bias thus suggests that a short-acting RG-selective ligand, such as PTHrP(1−36), could have utility for osteoporosis, whereas a longer-acting R^0-selective ligand, such as those in the M-PTH(1−34) series of peptides and the long-acting analog LA-PTH, discussed in the next section, could have utility for hypoparathyroidism (Okazaki et al., 2008).

LA-PTH, a long-acting PTH/PTHrP analog for hypoparathyroidism

Based on the initial findings with M-PTH(1−34), efforts were made to develop even longer-acting analogs that could potentially be useful for hypoparathyroidism. One new long-acting analog derived, called LA-PTH, is a hybrid peptide in which the M-PTH(1−14) portion also present in the M-PTH(1−34) is joined to a modified C-terminal 15−36 portion of PTHrP(1−36). This unique structure results in high-affinity binding to the PTHR1 R^0 conformation, and hence markedly prolonged calcemic responses (Fig. 28.4A−D) (Maeda et al., 2013). The functional properties of LA-PTH suggested that it could have therapeutic utility for hypoparathyroidism, a disease for which PTH(1−34) as delivered especially by pump (Winer et al., 2014) and PTH(1−84) by subcutaneous injection by which it exhibits a prolonged pharmacokinetic profile (Mannstadt et al., 2013) have proven to be efficacious, with the latter now available as the drug Natpara. As a potentially improved form of therapy, LA-PTH was tested in several rodent models of the disease in which the parathyroid glands were either surgically removed or, in mice, ablated by diphtheria toxin treatment (Fig. 28.4D), and the results revealed that indeed the analog could normalize blood calcium levels more effectively and for longer periods of time after a single injection than could at least severalfold higher doses of PTH(1−34) or PTH(1−84) (Shimizu et al., 2016; Bi et al., 2016). The observed prolonged pharmacodynamic effects were again not explained by a prolonged pharmacokinetic profile, as the analog was cleared from the circulation if anything more rapidly than PTH(1−34). LA-PTH thus stands as a promising preclinical candidate as a new therapeutic option for hypoparathyroidism.

It should be noted that several other PTH(1−34)-based analogs have been reported to induce sustained pharmacodynamic actions in animals via distinct mechanisms. These mechanisms include the prolongation of the pharmacokinetic profile, as is achieved via fusion of the C terminus of the PTH(1−34) peptide to either the fragment crystallizable (Fc) portion of IgG (Kostenuik et al., 2007) or a 20-kDa polyethylene glycol (PEG) chain (Guo et al., 2017), each of which acts to impede the rate of glomerular filtration, or, in a third analog, via a combination of modifications (Nle8, Lys27, and a C-terminal 2-kDa PEG group) that results apparently in a prolonged retention of the peptide on the cell surface due to impairment of β-arrestin-mediated receptor internalization (Krishnan et al., 2018). Such analogs could also lead to new modes of treatment for hypoparathyroidism. Whether such analogs designed to induce prolonged actions in vivo via distinct mechanisms elicit different types of downstream responses, as, for example, in the specific gene sets regulated, remains to be determined.

Abaloparatide: a PTHrP analog for osteoporosis

Based on the dogma that a pulsatile administration and hence action of PTH is required to achieve an optimal increase in bone structure and strength while avoiding an excess of bone resorption, which typically occurs with a more continuous administration and action of the peptide, it was postulated that a PTHR1 ligand analog that binds selectively to the RG PTHR1 conformation, and thus induces signaling responses of short duration, would be more effective at building new

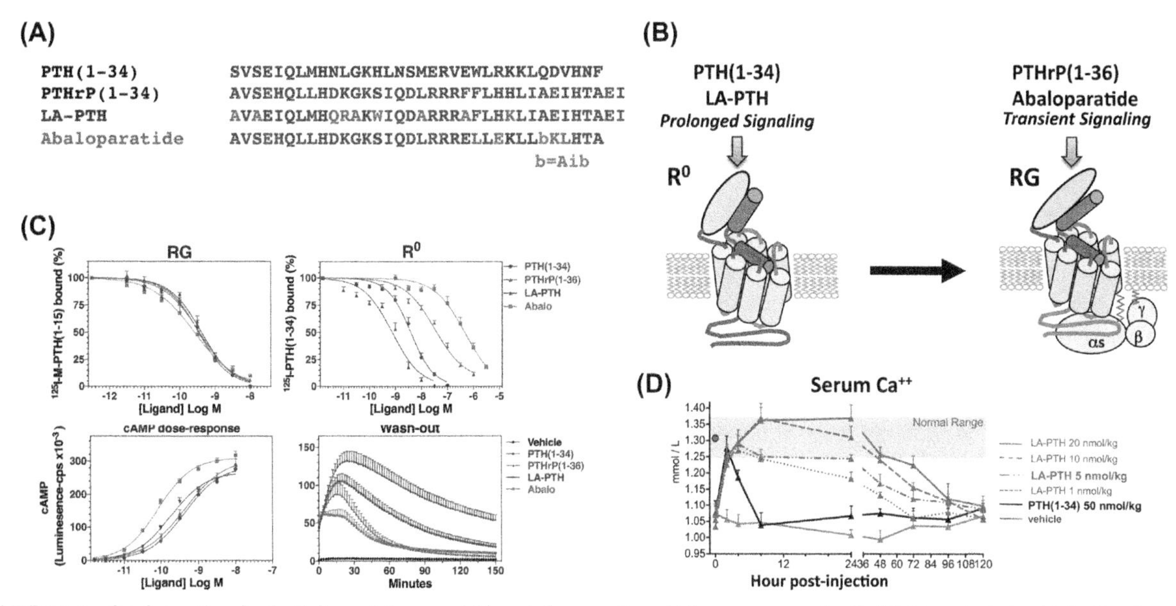

FIGURE 28.4 Conformational selectivity and temporal bias at the parathyroid hormone receptor 1. (A) Ligands that exhibit altered modes of conformational selectivity and temporal bias at the parathyroid hormone receptor 1 (PTHR1). (B) Proposed mechanism of temporal bias. A ligand that binds with high affinity to the G-protein-independent conformation, R^0, mediates prolonged signaling since the complex is stable and can interconvert over time to the biologically active G-protein-coupled conformation, RG; a ligand that binds selectively to RG induces only transient signaling since the ligand dissociates after G protein activation. (C) In HEK293 membranes and cells expressing the human PTHR1, LA-PTH binds with high affinity to R^0 and thus mediates prolonged cAMP signaling responses after washout, whereas abaloparatide binds preferentially to RG and induces more transient cAMP responses. PTH(1−34) and PTHrP(1−36) exhibit intermediate selectivity profiles. (D) In a parathyroidectomized mouse model of hypoparathyroidism, a single injection of LA-PTH, assessed at several doses, results in an extended elevation in serum calcium, whereas PTH(1−34) at an even higher dose results in a more transient elevation. *(C) Reproduced pending permission from Hattersley, G., Dean, T., Corbin, B. A., Bahar, H. & Gardella, T. J. 2016. Binding selectivity of abaloparatide for PTH-type-1-receptor conformations and effects on downstream signaling. Endocrinology 157, 141−149; (D) Reproduced pending permission from Bi, R., Fan, Y., Lauter, K., Hu, J., Watanabe, T., Cradock, J., Yuan, Q., Gardella, T. & Mannstadt, M. 2016. Diphtheria toxin- and GFP-based mouse models of acquired hypoparathyroidism and treatment with a long-acting parathyroid hormone analog. J. Bone Miner. Res. 31, 975−984.*

bone than would an analog that binds selectively to the R^0 conformation and thus induces prolonged responses. Moreover, because excess bone resorption, as occurs with continuous PTHR1 signaling, gives rise to hypercalcemia, an RG-selective analog might provide a reduced risk of this adverse event, which probably was a factor that limited the final dose approved by the US FDA to 20 μg per day rather than 40 μg, even though the latter was more effective at reducing fracture risk (Neer et al., 2001). The newer PTHrP(1−34) analog, called abaloparatide, formerly called BA058, which was approved by the FDA in 2016 as an alternative anabolic treatment for osteoporosis, is thus of interest as data from preclinical and human clinical studies suggest that it increases bone mass at least as effectively as PTH(1−34) but causes less bone resorption and hence hypercalcemia (Miller et al., 2016). Abaloparatide is administered by subcutaneous injection at a dose of 80 μg per day. As both peptide drugs act by binding to the same target PTHR1, the differences in clinical outcomes suggested the possibility that their modes of action on the receptor might not be equivalent. Abaloparatide was found to bind with relatively high selectively to the RG PTHR1 conformation and thus induce more transient cAMP signaling responses in PTHR1-expressing HEK293 cells than PTH(1−34) (Hattersley et al., 2016) (Fig. 28.4A−C). These studies thus revealed a conformation-based mode of temporal bias for abaloparatide that was distinct from that of PTH(1−34), and hence provided at least a plausible mechanism to help explain the distinct effects that the two ligands have on bone metabolism, as seen in the clinical and preclinical testing (Makino et al., 2018).

PTHR1 mutations in disease

The full spectrum of diseases associated with PTHR1 mutations and their biological and clinical aspects are discussed in Chapter 58. The following sections focus on how the disease-associated mutations relate to the structure and functional properties of the receptor, particularly as they are assessed in cell and mouse model systems.

Jansen's metaphyseal chondrodysplasia

Jansen's metaphyseal chondrodysplasia (JMC) is a rare disease associated with skeletal abnormalities, dwarfism, and hypercalcemia, and is caused by heterozygous dominant activating mutations in the PTHR1 (Ohishi et al., 2012; Schipani et al., 1995). Five different PTHR1-activating mutations have been identified in patients with JMC. These mutations occur at the cytoplasmic termini of TM2 (Arg233 → His), TM6 (Thr410 → Pro/Arg), and TM7 (Ile458 → Arg/Lys) (Fig. 28.1A), and each results in agonist-independent cAMP signaling. Strikingly, the mutations each occur at or, in the case of the Ile458 mutations, adjacent to a residue that is involved in the conserved polar network that is present in all class B GPCRs and operates to control receptor activation and deactivation processes and particularly the outward movements of the TMD helices that allow G protein coupling (Yin et al., 2017). Several PTH and PTHrP antagonist ligand analogs, such as [Leu^{11}, D-Trp^{12}]PTHrP(7−34), behave as inverse agonists on the mutant receptors and thus depress their basal signaling activities (Carter et al., 2001, 2015). The mechanism by which these ligands achieve their inverse effect on the receptor's conformational status is unknown, although the Gly12 → D-Trp substitution is required for the effect. In any case, the functional properties of the analogs suggest possible paths toward therapy for JMC, which potentially can be tested in the transgenic models of JMC that have been developed to express the PTHR1-H223R allele in osteoblasts or osteocytes, which leads to marked increases in bone mass (O'Brien et al., 2008; Calvi et al., 2001).

Other diseases linked to PTHR1 mutations

Enchondromatosis (Ollier disease/Maffucci syndrome) is a rare disease characterized by cartilage tumors of the bone, and has been associated with four PTHR1 mutations, Gly121 → Glu, Ala122 → Thr, Arg150 → Cys, and Arg255 → His, each located in the ECD or ECL1 portion of the receptor. One study on the Arg150 → His mutant expressed in COS-7 cells intriguingly suggested a moderately elevated rate of basal cAMP signaling when corrected for a reduced cell-surface expression level (Hopyan et al., 2002), while another study suggested that each of the four mutations causes a loss-of-function phenotype due to effects on expression and/or binding affinity (Couvineau et al., 2008). The mechanism by which these mutations result in cartilage tumors in bone thus remains unclear, but certainly effects on the extracellular scaffold structure seem possible. Blomstrand's chondrodysplasia is a neonatal lethal condition of markedly accelerated bone mineralization that is caused by homozygous loss-of-function mutations in the PTHR1. One coding mutation has been identified, Pro132 → Leu, which affects a conserved site in the core ECD scaffold (Zhang et al., 1998; Karaplis et al., 1998). Eiken syndrome is a very rare skeletal dysplasia that has been associated with a homozygous recessive nonsense mutation in the PTHR1, Arg485 → Stop, that leads to a shortened C-terminal tail (Duchatelet et al., 2005). The phenotype is markedly delayed ossification of the skeleton, which is opposite that of Blomstrand's disease and seems consistent with a gain-of-function effect, albeit a mild one, since the heterozygous mutation is not associated with disease and the homozygous phenotype is distinct from that of JMC. In any case, it seems possible that the mutation leads to altered interactions with cytoplasmic effectors or scaffolding proteins, and hence diversions in subcellular trafficking and signaling, but this remains to be determined. A number of heterozygous loss-of-function mutations have been identified in individuals with defects in tooth eruption (Roth et al., 2014), which is consistent with studies in mice that show that PTHR1 signaling, as induced by PTHrP, is required for proper tooth formation (Ono et al., 2016). Among the mutations found in patients with tooth eruption defects is the Pro132 → Leu also found in Blomstrand's disease, but in that disease the mutation is in a compound-heterozygous arrangement.

Nonpeptide mimetic ligands for the PTHR1

The capacity for PTH agonists to effectively treat osteoporosis and hypoparathyroidism raises interest in the goal of developing orally available nonpeptide mimetics for this receptor. So far two such small-molecule PTHR1 agonists have been reported. The first, AH3960, was tested only in cells and shown to stimulate cAMP formation at doses of about 10 μM or higher (Rickard et al., 2006). More recently, PCO371 was reported and shown to stimulate cAMP formation in cells at concentrations in the low-micromolar range and, moreover, to effectively raise blood calcium levels in TPTX rats (Tamura et al., 2016). Due to an unexpectedly prolonged calcium response, possibly attributable to a prolonged pharmacokinetic profile, PCO371 is under development for hypoparathyroidism, rather than osteoporosis, as treatment of the latter disease requires a transient pharmacodynamic response to avoid excess bone catabolism, whereas treatment of the former disease requires a more sustained pharmacodynamic profile to mimic the effects of the missing hormone. Several nonpeptide antagonist compounds have also been identified for the PTHR1. SW106, discovered by screening a compound library for agents that could inhibit the binding of a radiolabeled PTH(1−14) peptide analog, binds to the PTHR1 with

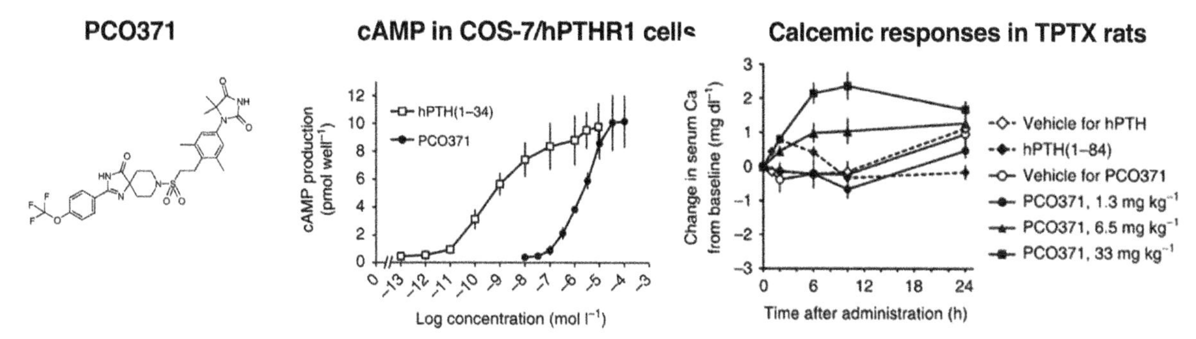

FIGURE 28.5 Small-molecule agonist for the PTHR1. The compound PCO371 was developed from a lead compound identified in a high-throughput screen for parathyroid hormone receptor 1 (*PTHR1*) agonist ligands. In COS-7 cells expressing the human PTHR1, PCO371 is a full agonist for cAMP signaling, albeit its potency is about 3 log-orders weaker than that of PTH(1−34). In TPTX rats, PCO371 induces elevations in serum calcium that are sustained, relative to the effects of PTH(1−34), presumably because of a prolonged pharmacokinetic profile. Consequently, PCO371 is reported to be in development for hypoparathyroidism. *Reproduced pending permission from Tamura, T., Noda, H., Joyashiki, E., Hoshino, M., Watanabe, T., Kinosaki, M., Nishimura, Y., Esaki, T., Ogawa, K., Miyake, T., Arai, S., Shimizu, M., Kitamura, H., Sato, H. & Kawabe, Y. 2016. Identification of an orally active small-molecule PTHR1 agonist for the treatment of hypoparathyroidism. Nat. Commun. 7, 13384.*

micromolar affinity and behaves as a competitive antagonist (Carter et al., 2007, 2015). A broad set of antagonist compounds was reported by a different group, and the most effective of these antagonized the cAMP-stimulating actions of PTH(1−34) in HEK293 cells transfected with the human PTHR1, with inhibitory constants in the 10 nM range, although studies in vivo were not reported (McDonald et al., 2007). Mechanistic studies performed on SW106 and AH3960 (Carter et al., 2015), as well as on PCO371 (Tamura et al., 2016), show that each of these compounds interacts with the TMD region of the receptor, as they exhibit the same effectiveness on the PTHR1 construct that lacks the ECD as they do on the intact PTHR1. Moreover, PCO371 was found to be inactive on the human PTHR2, specifically due to a single divergent residue corresponding to proline-415 in the PTHR1, which is replaced by leucine in the PTHR2. Pro415 is otherwise highly conserved in the class B GPCRs and is located in the middle of TM6, where it is predicted to play a pivotal role in receptor activation (Zhang et al., 2017). Whether PCO371 directly binds to Pro415 or to a site within the extracellularly exposed orthosteric pocket used by the peptide ligand remains to be determined. In any event, it is now clear that small-molecule ligands, both agonists and antagonists, can be developed for the PTHR1, and ultimately might lead to more effective therapies for PTHR1-related diseases (Fig. 28.5).

Other receptors for parathyroid hormone and related ligands

PTHR2 and PTHR3 subtypes

There continues to be appropriate interest in other receptors for the PTH family of ligands. Two apparent receptor subtypes distinct from the PTHR1 have been identified, as discussed in the earlier section on the evolution of this receptor/ligand family. The PTHR3 is present in vertebrate evolution as late as the avian radiation, but its function is not yet clarified or studied extensively, partly because it has been lost from the genome of higher mammals, including humans. Some cell-based tests suggest that it can function in a fashion similar to that of the PTHR1, at least in terms of ligand recognition and response properties (Rubin and Juppner, 1999; Pinheiro et al., 2012). The apparent absence of the PTHR3 in the higher vertebrates leaves open the question of biological importance. On the other hand, the PTHR2 is present in humans and other mammals, and has been characterized sufficiently to suggest a distinct ligand and biological role profile (Dimitrov et al., 2013), although it has been apparently lost in the avian lineage (Pinheiro et al., 2012).

The PTHR2 was initially identified through hybridization-based screening of a human brain cDNA library for PTHR1-related sequences. The identified receptor was thus found to share 51% amino acid identity with the human PTHR1 (Usdin et al., 1995), but to respond weakly to PTH and not at all to PTHrP, while the rat PTHR2 responded to neither (Hoare et al., 1999a). The search for a cognate ligand in bovine hypothalamic extracts yielded the new bioactive peptide called TIP39 (Usdin et al., 1999). TIP39 binds only weakly to the PTHR1 and lacks agonist activity. Nevertheless, TIP39 exhibits some structural homology with PTH and PTHrP (Piserchio et al., 2000), and N-terminally truncated analogs such as TIP9(7−39) bind with improved affinity such that they act as PTHR1 inhibitors (Hoare and Usdin, 2000; Jonsson et al., 2001). Continuing studies with the PTHR2 and TIP39 system since the initial discovery have emphasized its principal role in the central nervous system, with evidence suggesting that it functions to modify behavior, such as combating excessive fear reactions, modifying pain sensations, and playing a role in positive maternal behavior, as studied in suckling rodents, perhaps by stimulating oxytocin release (Cservenak et al., 2017; Usdin et al., 2003). Distinctive other functions include a critical role in spermatid production and fertility through receptors in testicular tissue (Usdin et al., 2008). Other studies indicate the presence of TIP39 and the PTHR2 in skin with a role for keratinocyte differentiation and regulation of extracellular matrix formation and wound repair (Sato et al., 2016). There is no/little evidence to suggest a role in calcium or bone metabolism.

Possible receptors for C-terminal PTH and PTHrP

Other receptors with actions on bone and calcium have been postulated to interact with the carboxyl regions of PTH and PTHrP that lie beyond the amino-terminal (1–34) segment. Indirect evidence for such a possibility is suggested by the moderate degree of sequence homology maintained in the middle and carboxy-terminal regions of ligands from human and related mammalian species and even in chicken for the PTHrP molecule and also, but to a slightly lesser, extent for PTH. Added to this is the awareness that the overall peptide length of these two molecules is considerably greater than that noted for the ligands that activate other members of the class B receptor class, raising the possibility that the C-terminal extensions of PTH and PTHrP were acquired or preserved biologically during evolution. Experimental support for a C-terminal receptor for PTH comes from the capacity of the PTH(39–84) fragment to bind and cross-link to a 90-kDa protein on the surface of ROS17-2.8 cells (Inomata et al., 1995; Takasu et al., 1996), and that the fragment also induced apoptosis in a mouse osteocytic cell line ablated for the PTHR1 (Divieti et al., 2001). For PTHrP, there is evidence from in vivo studies that the mid- or C-terminal region contributes to effects on placental calcium transport as well as tissue development, the latter involving a nuclear localization mechanism (Kovacs et al., 1996; Wu et al., 1996; Gu et al., 2012; Toribio et al., 2010; Lam et al., 1999). On the other hand, genetic data support the view that the actions of PTH and PTHrP on bone are largely attributable to actions through the PTHR1. Thus, homozygous ablation of the PTHR1 in mice gives rise to a neonatal lethal skeletal dysplasia similar to that seen with homozygous ablation of PTHrP (Lanske et al., 1996). In addition, in the very rare human disorder Blomstrand's chondrodysplasia with homozygous loss-of-function PTHR1 alleles there is a similar lethal skeletal chondrodysplasia. Heterozygous loss-of-function mutations of PTHrP in humans give rise to brachydactyly type E (Bae et al., 2018), and such mutations of the PTHR1 give rise to failures in tooth eruption (Ono et al., 2016; Roth et al., 2014), both of which are consistent with a defect in bone development as controlled by the PTHR1. For PTH, conditions of hypoparathyroidism can be sufficiently corrected with PTH(1–34) administration (Winer et al., 2012), while the related condition of pseudohypoparathyroidism results from deficiencies in Gαs (Juppner, 2015), the primary mediator of PTHR1 signaling. In any event, there remains considerable interest in the possibility of such C-terminal receptors, as potential roles in calcium and bone biology have been supported by biological data in studies extending back several decades. However, despite much effort, such putative receptors have not been cloned and therefore the molecular identities not characterized.

Conclusions

Mechanisms of ligand binding and signal transduction at the PTHR1 are becoming increasingly better understood, with the high-resolution X-ray crystal and cryo-EM structures obtained for several related class B GPCRs providing particularly valuable new information. Ligand binding is thus seen to occur via the basic two-site model initially revealed by mutagenesis and cross-linking studies, but to further involve significant conformational rearrangements and molecular movements that ultimately lead to G protein coupling and activation. Structurally distinct PTH and PTHrP ligand analogs can induce or stabilize different PTHR1 conformations to thereby mediate different modes of signaling, with some ligands inducing particularly prolonged cAMP signaling responses, probably from within endosomes, while others induce more transient responses, presumably from the cell surface. Such conformation-based temporal bias for the PTHR1 suggests new approaches for the development of therapies for diseases such as hypoparathyroidism and osteoporosis. Insights into key sites in the receptor involved in function are also provided by PTHR1 mutations identified in certain skeletal disorders, including loss-of-function mutations in cases of failed tooth eruption and gain-of-function mutations in JMC. Mutations of the latter disease reveal the importance of a conserved polar network located at the base of the TMD helical bundle in controlling receptor activation and deactivation. As the views provided by such insights continue to deepen and refine, so should the capacity to design new ligand analogs for the PTHR1 that offer improved efficacy as treatments for diseases of bone and mineral metabolism.

References

Adams, A., Bisello, A., Chorev, M., Rosenblatt, M., Suva, L., 1998. Arginine 186 in the extracellular N-terminal region of the human parathyroid hormone 1 receptor is essential for contact with position 13 of the hormone. Mol. Endocrinol. 12, 1673–1683.

Amizuka, N., Lee, H.S., Kwan, M.Y., Arazani, A., Warshawsky, H., Hendy, G.N., Ozawa, H., White, J.H., Goltzman, D., 1997. Cell-specific expression of the parathyroid hormone (PTH)/PTH-related peptide receptor gene in kidney from kidney-specific and ubiquitous promoters. Endocrinology 138, 469–481.

Andreassen, K.V., Hjuler, S.T., Furness, S.G., Sexton, P.M., Christopoulos, A., Nosjean, O., Karsdal, M.A., Henriksen, K., 2014. Prolonged calcitonin receptor signaling by salmon, but not human calcitonin, reveals ligand bias. PLoS One 9, e92042.

Bae, J., Choi, H., Park, S., Lee, D.-E., Lee, S., 2018. Novel mutation in PTHLH related to brachydactyly type E2 initially confused with unclassical Pseudopseudohypoparathyroidism. Endocrinol. Metab. On-line, 1-8.

Bergwitz, C., Gardella, T.J., Flannery, M.R., Potts, J.T., Kronenberg, H.M., Goldring, S.R., Juppner, H., 1996. Full activation of chimeric receptors by hybrids between parathyroid hormone and calcitonin – evidence for a common pattern of ligand-receptor interaction. J. Biol. Chem. 271, 26469–26472.

Bergwitz, C., Jusseaume, S., Luck, M., Jüppner, H., Gardella, T., 1997. Residues in the membrane-spanning and extracellular loop regions of the PTH-2 receptor determine signaling selectivity for PTH and PTH-related peptide. J. Biol. Chem. 272, 28861–28868.

Bettoun, J.D., Minagawa, M., Kwan, M.Y., Lee, H.S., Yasuda, T., Hendy, G.N., Goltzman, D., White, J.H., 1997. Cloning and characterization of the promoter regions of the human parathyroid hormone (PTH)/PTH-related peptide receptor gene: analysis of deoxyribonucleic acid from normal subjects and patients with pseudohypoparathyroidism type 1b. J. Clin. Endocrinol. Metab. 82, 1031–1040.

Bi, R., Fan, Y., Lauter, K., Hu, J., Watanabe, T., Cradock, J., Yuan, Q., Gardella, T., Mannstadt, M., 2016. Diphtheria toxin- and GFP-based mouse models of acquired hypoparathyroidism and treatment with a long-acting parathyroid hormone analog. J. Bone Miner. Res. 31, 975–984.

Bisello, A., Adams, A.E., Mierke, D., Pellegrini, M., Rosenblatt, M., Suva, L., Chorev, M., 1998. Parathyroid hormone-receptor interactions identified directly by photocross-linking and molecular modeling studies. J. Biol. Chem. 273, 22498–22505.

Bisello, A., Chorev, M., Rosenblatt, M., Monticelli, L., Mierke, D.F., Ferrari, S.L., 2002. Selective ligand-induced stabilization of active and desensitized parathyroid hormone type 1 receptor conformations. J. Biol. Chem. 277, 38524–38530.

Bortolato, A., Dore, A.S., Hollenstein, K., Tehan, B.G., Mason, J.S., Marshall, F.H., 2014. Structure of Class B GPCRs: new horizons for drug discovery. Br. J. Pharmacol. 171, 3132–3145.

Calvi, L.M., Sims, N.A., Hunzelman, J.L., Knight, M.C., Giovannetti, A., Saxton, J.M., Kronenberg, H.M., Baron, R., Schipani, E., 2001. Activated parathyroid hormone/parathyroid hormone-related protein receptor in osteoblastic cells differentially affects cortical and trabecular bone. J. Clin. Investig. 107, 277–286.

Cardoso, J.C., Felix, R.C., Power, D.M., 2014. Nematode and arthropod genomes provide new insights into the evolution of class 2 B1 GPCRs. PLoS One 9, e92220.

Cardoso, J.C., Pinto, V.C., Vieira, F.A., Clark, M.S., Power, D.M., 2006. Evolution of secretin family GPCR members in the metazoa. BMC Evol. Biol. 6, 1–16.

Carter, P.H., Dean, T., Bhayana, B., Khatri, A., Rajur, R., Gardella, T.J., 2015. Actions of the small molecule ligands SW106 and AH-3960 on the type-1 parathyroid hormone receptor. Mol. Endocrinol. 29, 307–321.

Carter, P.H., Liu, R.Q., Foster, W.R., Tamasi, J.A., Tebben, A.J., Favata, M., Staal, A., Cvijic, M.E., French, M.H., Dell, V., Apanovitch, D., Lei, M., Zhao, Q., Cunningham, M., Decicco, C.P., Trzaskos, J.M., Feyen, J.H., 2007. Discovery of a small molecule antagonist of the parathyroid hormone receptor by using an N-terminal parathyroid hormone peptide probe. Proc. Natl. Acad. Sci. U. S. A. 104, 6846–6851.

Carter, P.H., Petroni, B.D., Gensure, R.C., Schipani, E., Potts Jr., J.T., Gardella, T.J., 2001. Selective and nonselective inverse agonists for constitutively active type-1 parathyroid hormone receptors: evidence for altered receptor conformations. Endocrinology 142, 1534–1545.

Chagin, A.S., Kronenberg, H.M., 2014. Role of G-proteins in the differentiation of epiphyseal chondrocytes. J. Mol. Endocrinol. 53, R39–R45.

Chan, A.S., Clairfeuille, T., LANDAO-Bassonga, E., Kinna, G., Ng, P.Y., Loo, L.S., Cheng, T.S., Zheng, M., Hong, W., Teasdale, R.D., Collins, B.M., Pavlos, N.J., 2016. Sorting nexin 27 couples PTHR trafficking to retromer for signal regulation in osteoblasts during bone growth. Mol. Biol. Cell 27, 1367–1382.

Chauvin, S., Bencsik, M., Bambino, T., Nissenson, R.A., 2002. Parathyroid hormone receptor recycling: role of receptor dephosphorylation and beta-arrestin. Mol. Endocrinol. 16, 2720–2732.

Couvineau, A., Wouters, V., Bertrand, G., Rouyer, C., Gerard, B., Boon, L.M., Grandchamp, B., Vikkula, M., Silve, C., 2008. PTHR1 mutations associated with Ollier disease result in receptor loss of function. Hum. Mol. Genet. 17, 2766–2775.

Cservenak, M., Keller, D., Kis, V., Fazekas, E.A., Ollos, H., Leko, A.H., Szabo, E.R., Renner, E., Usdin, T.B., Palkovits, M., Dobolyi, A., 2017. A thalamo-hypothalamic pathway that activates oxytocin neurons in social contexts in female rats. Endocrinology 158, 335–348.

Cvicek, V., Goddard 3rd, W.A., Abrol, R., 2016. Structure-based sequence alignment of the transmembrane domains of all human GPCRs: Phylogenetic, structural and functional implications. PLoS Comput. Biol. 12, e1004805.

De Graaf, C., Song, G., Cao, C., Zhao, Q., Wang, M.W., Wu, B., Stevens, R.C., 2017. Extending the structural view of class B GPCRs. Trends Biochem. Sci. 42, 946–960.

De Lean, A., Stadel, J., Lefkowitz, R., 1980. A ternary complex model explains the agonist-specific binding properties of the adenylate cyclase-coupled b-adrenergic receptor. J. Biol. Chem. 255, 7108–7117.

Dean, T., Linglart, A., Mahon, M.J., Bastepe, M., Juppner, H., Potts Jr., J.T., Gardella, T.J., 2006. Mechanisms of ligand binding to the parathyroid hormone (PTH)/PTH-related protein receptor: selectivity of a modified PTH(1-15) radioligand for GalphaS-coupled receptor conformations. Mol. Endocrinol. 20, 931–943.

Dean, T., Vilardaga, J.P., Potts Jr., J.T., Gardella, T.J., 2008. Altered selectivity of parathyroid hormone (PTH) and PTH-related protein (PTHrP) for distinct conformations of the PTH/PTHrP receptor. Mol. Endocrinol. 22, 156–166.

Dimitrov, E.L., Kuo, J., Kohno, K., Usdin, T.B., 2013. Neuropathic and inflammatory pain are modulated by tuberoinfundibular peptide of 39 residues. Proc. Natl. Acad. Sci. U. S. A. 110, 13156–13161.

Divieti, P., Inomata, N., Chapin, K., Singh, R., Juppner, H., Bringhurst, F.R., 2001. Receptors for the carboxyl-terminal region of pth(1-84) are highly expressed in osteocytic cells. Endocrinology 142, 916–925.

Dong, S., Shang, S., Li, J., Tan, Z., Dean, T., Maeda, A., Gardella, T.J., Danishefsky, S.J., 2012. Engineering of therapeutic polypeptides through chemical synthesis: early lessons from human parathyroid hormone and analogues. J. Am. Chem. Soc. 134, 15122–15129.

Drake, M.T., Shenoy, S.K., Lefkowitz, R.J., 2006. Trafficking of G protein-coupled receptors. Circ. Res. 99, 570–582.

Duchatelet, S., Ostergaard, E., Cortes, D., Lemainque, A., Julier, C., 2005. Recessive mutations in PTHR1 cause contrasting skeletal dysplasias in Eiken and Blomstrand syndromes. Hum. Mol. Genet. 14, 1–5.

Feinstein, T.N., Wehbi, V.L., Ardura, J.A., Wheeler, D.S., Ferrandon, S., Gardella, T.J., Vilardaga, J.P., 2011. Retromer terminates the generation of cAMP by internalized PTH receptors. Nat. Chem. Biol. 7, 278–284.

Ferrandon, S., Feinstein, T.N., Castro, M., Wang, B., Bouley, R., Potts, J.T., Gardella, T.J., Vilardaga, J.P., 2009. Sustained cyclic AMP production by parathyroid hormone receptor endocytosis. Nat. Chem. Biol. 5, 734–742.

Fredriksson, R., Lagerstrom, M.C., Lundin, L.G., Schioth, H.B., 2003. The G-protein-coupled receptors in the human genome form five main families. Phylogenetic analysis, paralogon groups, and fingerprints. Mol. Pharmacol. 63, 1256–1272.

Gardella, T.J., Luck, M.D., Fan, M.H., Lee, C., 1996. Transmembrane residues of the parathyroid hormone (PTH)/PTH-related peptide receptor that specifically affect binding and signaling by agonist ligands. J. Biol. Chem. 271, 12820–12825.

Gensure, R., Carter, P., Petroni, B., Juppner, H., Gardella, T., 2001a. Identification of determinants of inverse agonism in a constitutively active parathyroid hormone/parathyroid hormone related peptide receptor by photoaffinity cross linking and mutational analysis. J. Biol. Chem. 276, 42692–42699.

Gensure, R., Gardella, T., Juppner, H., 2001b. Multiple sites of contact between the carboxyl terminal binding domain of PTHrP (1 36) analogs and the amino terminal extracellular domain of the PTH/PTHrP receptor identified by photoaffinity cross linking. J. Biol. Chem. 276, 28650–28658.

Gensure, R.C., Shimizu, N., Tsang, J., Gardella, T.J., 2003. Identification of a contact site for residue 19 of parathyroid hormone (PTH) and PTH-related protein analogs in transmembrane domain two of the type 1 PTH receptor. Mol. Endocrinol. 17, 2647–2658.

Gesty-Palmer, D., Chen, M., Reiter, E., Ahn, S., Nelson, C.D., Wang, S., Eckhardt, A.E., Cowan, C.L., Spurney, R.F., Luttrell, L.M., Lefkowitz, R.J., 2006. Distinct beta-arrestin- and G protein-dependent pathways for parathyroid hormone receptor-stimulated ERK1/2 activation. J. Biol. Chem. 281, 10856–10864.

Gesty-Palmer, D., Yuan, L., Martin, B., Wood 3rd, W.H., Lee, M.H., Janech, M.G., Tsoi, L.C., Zheng, W.J., Luttrell, L.M., Maudsley, S., 2013. Beta-arrestin-selective G protein-coupled receptor agonists engender unique biological efficacy in vivo. Mol. Endocrinol. 27, 296–314.

Gidon, A., AL-Bataineh, M.M., Jean-Alphonse, F.G., Stevenson, H.P., Watanabe, T., Louet, C., Khatri, A., Calero, G., Pastor-Soler, N.M., Gardella, T.J., Vilardaga, J.P., 2014. Endosomal GPCR signaling turned off by negative feedback actions of PKA and v-ATPase. Nat. Chem. Biol. 10, 707–709.

Godbole, A., Lyga, S., Lohse, M.J., Calebiro, D., 2017. Internalized TSH receptors en route to the TGN induce local Gs-protein signaling and gene transcription. Nat. Commun. 8, 443.

Goltzman, D., Peytremann, A., Callahan, E., Tregear, G.W., Potts Jr., J.T., 1975. Analysis of the requirements for parathyroid hormone action in renal membranes with the use of inhibiting analogues. J. Biol. Chem. 250, 3199–3203.

Gorvin, C.M., Rogers, A., Hastoy, B., Tarasov, A.I., Frost, M., Sposini, S., Inoue, A., Whyte, M.P., Rorsman, P., Hanyaloglu, A.C., Breitwieser, G.E., Thakker, R.V., 2018. AP2? Mutations impair calcium-sensing receptor trafficking and signaling, and show an endosomal pathway to spatially direct G-protein selectivity. Cell Rep. 22, 1054–1066.

Greenberg, Z., Bisello, A., Mierke, D., Rosenblatt, M., Chorev, M., 2000. Mapping the bimolecular interface of the parathyroid hormone (PTH) PTH1 receptor complex: spatial proximity between Lys(27) (of the hormone principal binding domain) and Leu(261) (of the first extracellular loop) of the human PTH1 receptor. Biochemistry 39, 8142–8152.

Grundmann, M., Kostenis, E., 2017. Temporal bias: time-encoded dynamic GPCR signaling. Trends Pharmacol. Sci. 38, 1110–1124.

Gu, Z., Liu, Y., Zhang, Y., Jin, S., Chen, Q., Goltzman, D., Karaplis, A., Miao, D., 2012. Absence of PTHrP nuclear localization and carboxyl terminus sequences leads to abnormal brain development and function. PLoS One 7, e41542.

Guo, J., Chung, U.I., Kondo, H., Bringhurst, F.R., Kronenberg, H.M., 2002. The PTH/PTHrP receptor can delay chondrocyte hypertrophy in vivo without activating phospholipase C. Dev. Cell 3, 183–194.

Guo, J., Khatri, A., Maeda, A., Potts Jr., J.T., Juppner, H., Gardella, T.J., 2017. Prolonged pharmacokinetic and pharmacodynamic actions of a Pegylated parathyroid hormone (1-34) peptide fragment. J. Bone Miner. Res. 32, 86–98.

Hattersley, G., Dean, T., Corbin, B.A., Bahar, H., Gardella, T.J., 2016. Binding selectivity of abaloparatide for PTH-type-1-receptor conformations and effects on downstream signaling. Endocrinology 157, 141–149.

Hoare, S., Usdin, T., 2001. Molecular mechanisms of ligand-recognition by parathyroid hormone 1 (PTH1) and PTH2 receptors. Curr. Pharmaceut. Des. 7, 689–713.

Hoare, S.R., Bonner, T.I., Usdin, T.B., 1999a. Comparison of rat and human parathyroid hormone 2 (PTH2) receptor activation: PTH is a low potency partial agonist at the rat PTH2 receptor. Endocrinology 140, 4419–4425.

Hoare, S.R., DE Vries, G., Usdin, T.B., 1999b. Measurement of agonist and antagonist ligand-binding parameters at the human parathyroid hormone type 1 receptor: evaluation of receptor states and modulation by guanine nucleotide. J. Pharmacol. Exp. Ther. 289, 1323–1333.

Hoare, S.R., Gardella, T.J., Usdin, T.B., 2001. Evaluating the signal transduction mechanism of the parathyroid hormone 1 receptor. Effect of receptor-G-protein interaction on the ligand binding mechanism and receptor conformation. J. Biol. Chem. 276, 7741–7753.

Hoare, S.R., Usdin, T.B., 2000. Tuberoinfundibular peptide (7-39) [TIP(7-39)], a novel, selective, high-affinity antagonist for the parathyroid hormone-1 receptor with no detectable agonist activity. J. Pharmacol. Exp. Ther. 295, 761–770.

Hogan, B.M., Danks, J.A., Layton, J.E., Hall, N.E., Heath, J.K., Lieschke, G.J., 2005. Duplicate zebrafish pth genes are expressed along the lateral line and in the central nervous system during embryogenesis. Endocrinology 146, 547–551.

Hopyan, S., Gokgoz, N., Poon, R., Gensure, R.C., Yu, C., Cole, W.G., Bell, R.S., Juppner, H., Andrulis, I.L., Wunder, J.S., Alman, B.A., 2002. A mutant PTH/PTHrP type I receptor in enchondromatosis. Nat. Genet. 30, 306–310.

Horwitz, M.J., Tedesco, M.B., Sereika, S.M., Hollis, B.W., Garcia-Ocana, A., Stewart, A.F., 2003. Direct comparison of sustained infusion of human parathyroid hormone-related protein-(1-36) [hPTHrP-(1-36)] versus hPTH-(1-34) on serum calcium, plasma 1,25-dihydroxyvitamin D concentrations, and fractional calcium excretion in healthy human volunteers. J. Clin. Endocrinol. Metab. 88, 1603–1609.

Huang, Z., Chen, Y., Pratt, S., Chen, T.-H., Bambino, T., Nissenson, R., Shoback, D., 1996. The N-terminal region of the third intracellular loop of the parathyroid hormone (PTH)/PTH-related peptide receptor is critical for coupling to cAMP and inositol phosphate/Ca^{2++} signal transduction pathways. J. Biol. Chem. 271, 33382–33389.

Hwang, J.I., Moon, M.J., Park, S., Kim, D.K., Cho, E.B., Ha, N., Son, G.H., Kim, K., Vaudry, H., Seong, J.Y., 2013. Expansion of secretin-like G protein-coupled receptors and their peptide ligands via local duplications before and after two rounds of whole-genome duplication. Mol. Biol. Evol. 30, 1119–1130.

Iida-Klein, A., Guo, J., Takemura, M., Drake, M.T., Potts Jt, J.R., Abou-Samra, A., Bringhurst, F.R., Segre, G.V., 1997. Mutations in the second cytoplasmic loop of the rat parathyroid hormone (PTH)/PTH-related protein receptor result in selective loss of PTH-stimulated phospholipase C activity. J. Biol. Chem. 272, 6882–6889.

Inomata, N., Akiyama, M., Kubota, N., Juppner, H., 1995. Characterization of a novel parathyroid hormone (PTH) receptor with specificity for the carboxyl terminal region of PTH (1-84). Endocrinology 136, 4732–4740.

Jonsson, K.B., John, M.R., Gensure, R.C., Gardella, T.J., Juppner, H., 2001. Tuberoinfundibular peptide 39 binds to the parathyroid hormone (PTH)/PTH-related peptide receptor, but functions as an antagonist. Endocrinology 142, 704–709.

Juppner, H., 2015. Genetic and epigenetic defects at the GNAS locus cause different forms of pseudohypoparathyroidism. Ann. Endocrinol. 76, 92–97.

Juppner, H., ABOU-Samra, A.B., Freeman, M., Kong, X.F., Schipani, E., Richards, J., Kolakowski Jr., L.F., Hock, J., Potts Jr., J.T., Kronenberg, H.M., et al., 1991. A G protein-linked receptor for parathyroid hormone and parathyroid hormone-related peptide. Science 254, 1024–1026.

Jüppner, H., Abou-Samra, A.B., Uneno, S., Gu, W.X., Potts Jr., J.T., Segre, G.V., 1988. The parathyroid hormone-like peptide associated with humoral hypercalcemia of malignancy and parathyroid hormone bind to the same receptor on the plasma membrane of ROS 17/2.8 cells. J. Biol. Chem. 263, 8557–8560.

Juppner, H., Schipani, E., Bringhurst, F.R., Mcclure, I., Keutmann, H.T., Potts, J.T., Kronenberg, H.M., Abousamra, A.B., Segre, G.V., Gardella, T.J., 1994. The extracellular amino-terminal region of the parathyroid-hormone (PTH)/PTH-Related peptide receptor determines the binding-affinity for carboxyl-terminal fragments of PTH-(1-34). Endocrinology 134, 879–884.

Kamesh, N., Aradhyam, G.K., Manoj, N., 2008. The repertoire of G protein-coupled receptors in the sea squirt *Ciona intestinalis*. BMC Evol. Biol. 8, 129.

Karaplis, A.C., He, B., Nguyen, M.T., Young, I.D., Semeraro, D., Ozawa, H., Amizuka, N., 1998. Inactivating mutation in the human parathyroid hormone receptor type 1 gene in Blomstrand chondrodysplasia. Endocrinology 139, 5255–5258.

Kong, X.F., Schipani, E., Lanske, B., Joun, H., Karperien, M., Defize, L.H., Juppner, H., Potts Jr., J.T., Segre, G.V., Kronenberg, H.M., et al., 1994. The rat, mouse and human genes encoding the receptor for parathyroid hormone and parathyroid hormone-related peptide are highly homologous. Biochem. Biophys. Res. Commun. 200, 1290–1299.

Kostenuik, P.J., Ferrari, S., Pierroz, D., Bouxsein, M., Morony, S., Warmington, K.S., Adamu, S., Geng, Z., Grisanti, M., Shalhoub, V., Martin, S., Biddlecome, G., Shimamoto, G., Boone, T., Shen, V., Lacey, D., 2007. Infrequent delivery of a long-acting PTH-Fc fusion protein has potent anabolic effects on cortical and cancellous bone. J. Bone Miner. Res. 22, 1534–1547.

Kovacs, C.S., Lanske, B., Hunzelman, J.L., Guo, J., Karaplis, A.C., Kronenberg, H.M., 1996. Parathyroid hormone-related peptide (PTHrP) regulates fetal placental calcium transport through a receptor distinct from the PTH/PTHrP receptor. Proc. Natl. Acad. Sci. U.S.A. 93, 15233–15238.

Krishnan, V., Ma, Y.L., Chen, C.Z., Thorne, N., Bullock, H., Tawa, G., Javella-Cauley, C., Chu, S., Li, W., Kohn, W., Adrian, M.D., Benson, C., Liu, L., Sato, M., Zheng, W., Pilon, A.M., Yang, N.N., Bryant, H.U., 2018. Repurposing a novel parathyroid hormone analogue to treat hypoparathyroidism. Br. J. Pharmacol. 175, 262–271.

Kronenberg, H.M., 2003. Developmental regulation of the growth plate. Nature 423, 332–336.

Lam, M., Briggs, L., Hu, W., Martin, T., Gillespie, M., Jans, D., 1999. Importin beta recognizes parathyroid hormone related protein with high affinity and mediates its nuclear import in the absence of importin alpha. J. Biol. Chem. 274, 7391–7398.

Lanske, B., Karaplis, A., Lee, K., Luz, A., Vortkamp, A., Pirro, A., Karperien, M., Defize, L., Ho, C., Mulligan, R., Abou-Samra, A., Jüppner, H., Segre, G., Kronenberg, H., 1996. PTH/PTHrP receptor in early development and indian hedgehog-regulated bone growth. Science 273, 663–666.

Lee, C.W., Gardella, T.J., Abousamra, A.B., Nussbaum, S.R., Segre, G.V., Potts, J.T., Kronenberg, H.M., Juppner, H., 1994. Role of the extracellular regions of the parathyroid-hormone (PTH) PTH-related peptide receptor in hormone-binding. Endocrinology 135, 1488–1495.

Li, C., Chen, M., Sang, M., Liu, X., Wu, W., Li, B., 2013. Comparative genomic analysis and evolution of family-B G protein-coupled receptors from six model insect species. Gene 519, 1−12.

Li, J., Dong, S., Townsend, S.D., Dean, T., Gardella, T.J., Danishefsky, S.J., 2012. Chemistry as an expanding resource in protein science: fully synthetic and fully active human parathyroid hormone-related protein (1-141). Angew Chem. Int. Ed. Engl. 51, 12263−12267.

Liang, Y.L., Khoshouei, M., Radjainia, M., Zhang, Y., Glukhova, A., Tarrasch, J., Thal, D.M., Furness, S.G.B., Christopoulos, G., Coudrat, T., Danev, R., Baumeister, W., Miller, L.J., Christopoulos, A., Kobilka, B.K., Wootten, D., Skiniotis, G., Sexton, P.M., 2017. Phase-plate cryo-EM structure of a class B GPCR-G-protein complex. Nature 546, 118−123.

Liu, Y., Ibrahim, A.S., Tay, B.H., Richardson, S.J., Bell, J., Walker, T.I., Brenner, S., Venkatesh, B., Danks, J.A., 2010. Parathyroid hormone gene family in a cartilaginous fish, the elephant shark (*Callorhinchus milii*). J. Bone Miner. Res. 25, 2613−2623.

Luttrell, L.M., Maudsley, S., Bohn, L.M., 2015. Fulfilling the promise of "biased" G protein-coupled receptor agonism. Mol. Pharmacol. 88, 579−588.

Maeda, A., Okazaki, M., Baron, D.M., Dean, T., Khatri, A., Mahon, M., Segawa, H., ABOU-Samra, A.B., Jueppner, H., Bloch, K.D., Potts Jr., J.T., Gardella, T.J., 2013. Critical role of parathyroid hormone (PTH) receptor-1 phosphorylation in regulating acute responses to PTH. Proc. Natl. Acad. Sci. U.S.A. 110, 5864−5869.

Mahon, M.J., Cole, J.A., Lederer, E.D., Segre, G.V., 2003. Na+/H+ exchanger-regulatory factor 1 mediates inhibition of phosphate transport by parathyroid hormone and second messengers by acting at multiple sites in opossum kidney cells. Mol. Endocrinol. 17, 2355−2364.

Makino, A., Takagi, H., Takahashi, Y., Hase, N., Sugiyama, H., Yamana, K., Kobayashi, T., 2018. Abaloparatide exerts bone anabolic effects with less stimulation of bone resorption-related factors: a comparison with teriparatide. Calcif. Tissue Int. 103, 289−297.

Malecz, N., Bambino, T., Bencsik, M., Nissenson, R., 1998. Identification of phosphorylation sites in the G protein-coupled receptor for parathyroid hormone. receptor phosphorylation is not required for agonist-induced internalization. Mol. Endocrinol. 12, 1846−1856.

Mamonova, T., Kurnikova, M., Friedman, P.A., 2012. Structural basis for NHERF1 PDZ domain binding. Biochemistry 51, 3110−3120.

Mannstadt, M., Clarke, B.L., Vokes, T., Brandi, M.L., Ranganath, L., Fraser, W.D., Lakatos, P., Bajnok, L., Garceau, R., Mosekilde, L., Lagast, H., Shoback, D., Bilezikian, J.P., 2013. Efficacy and safety of recombinant human parathyroid hormone (1-84) in hypoparathyroidism (REPLACE): a double-blind, placebo-controlled, randomised, phase 3 study. Lancet Diabetes Endocrinol. 1, 275−283.

Mannstadt, M., Luck, M.D., Gardella, T.J., Juppner, H., 1998. Evidence for a ligand interaction site at the amino-terminus of the parathyroid hormone (PTH)/PTH-related protein receptor from cross-linking and mutational studies. J. Biol. Chem. 273, 16890−16896.

Martin, T.J., Quinn, J.M., Gillespie, M.T., Ng, K.W., Karsdal, M.A., Sims, N.A., 2006. Mechanisms involved in skeletal anabolic therapies. Ann. N. Y. Acad. Sci. 1068, 458−470.

Mccuaig, K.A., Clarke, J.C., White, J.H., 1994. Molecular cloning of the gene encoding the mouse parathyroid hormone/parathyroid hormone-related peptide receptor. Proc. Natl. Acad. Sci. U. S. A. 91, 5051−5055.

Mcdonald, I.M., Austin, C., Buck, I.M., Dunstone, D.J., Gaffen, J., Griffin, E., Harper, E.A., Hull, R.A., Kalindjian, S.B., Linney, I.D., Low, C.M., Patel, D., Pether, M.J., Raynor, M., Roberts, S.P., Shaxted, M.E., Spencer, J., Steel, K.I., Sykes, D.A., Wright, P.T., Xun, W., 2007. Discovery and characterization of novel, potent, non-peptide parathyroid hormone-1 receptor antagonists. J. Med. Chem. 50, 4789−4792.

Mcgarvey, J.C., Xiao, K., Bowman, S.L., Mamonova, T., Zhang, Q., Bisello, A., Sneddon, W.B., Ardura, J.A., Jean-Alphonse, F., Vilardaga, J.P., Puthenveedu, M.A., Friedman, P.A., 2016. Actin-sorting nexin 27 (SNX27)-Retromer complex mediates rapid parathyroid hormone receptor recycling. J. Biol. Chem. 291, 10986−11002.

Miller, P.D., Hattersley, G., Riis, B.J., Williams, G.C., Lau, E., Russo, L.A., Alexandersen, P., Zerbini, C.A., Hu, M.Y., Harris, A.G., Fitzpatrick, L.A., Cosman, F., Christiansen, C., Investigators, A.S., 2016. Effect of abaloparatide vs placebo on new vertebral fractures in postmenopausal women with osteoporosis: a randomized clinical trial. J. Am. Med. Assoc. 316, 722−733.

Minagawa, M., Watanabe, T., Kohno, Y., Mochizuki, H., Hendy, G.N., Goltzman, D., White, J.H., Yasuda, T., 2001. Analysis of the P3 promoter of the human parathyroid hormone (PTH)/PTH-related peptide receptor gene in pseudohypoparathyroidism type 1b. J. Clin. Endocrinol. Metab. 86, 1394−1397.

Mirabeau, O., Joly, J.S., 2013. Molecular evolution of peptidergic signaling systems in bilaterians. Proc. Natl. Acad. Sci. U. S. A. 110, E2028−E2037.

Neer, R.M., Arnaud, C.D., Zanchetta, J.R., Prince, R., Gaich, G.A., Reginster, J.Y., Hodsman, A.B., Eriksen, E.F., ISH-Shalom, S., Genant, H.K., Wang, O.H., Mitlak, B.H., 2001. Effect of parathyroid hormone (1-34) on fractures and bone mineral density in postmenopausal women with osteoporosis. N. Engl. J. Med. 344, 1434−1441.

Nissenson, R.A., Diep, D., Strewler, G.J., 1988. Synthetic peptides comprising the amino-terminal sequence of a parathyroid hormone-like protein from human malignancies. Binding to parathyroid hormone receptors and activation of adenylate cyclase in bone cells and kidney. J. Biol. Chem. 263, 12866−12871.

Nussbaum, S.R., Rosenblatt, M., Potts Jr., J.T., 1980. Parathyroid hormone/renal receptor interactions: demonstration of two receptor-binding domains. J. Biol. Chem. 255, 10183−10187.

O'brien, C.A., Plotkin, L.I., Galli, C., Goellner, J.J., Gortazar, A.R., Allen, M.R., Robling, A.G., Bouxsein, M., Schipani, E., Turner, C.H., Jilka, R.L., Weinstein, R.S., Manolagas, S.C., Bellido, T., 2008. Control of bone mass and remodeling by PTH receptor signaling in osteocytes. PLoS One 3, e2942.

Ohishi, M., Ono, W., Ono, N., Khatri, R., Marzia, M., Baker, E.K., Root, S.H., Wilson, T.L., Iwamoto, Y., Kronenberg, H.M., Aguila, H.L., Purton, L.E., Schipani, E., 2012. A novel population of cells expressing both hematopoietic and mesenchymal markers is present in the normal adult bone marrow and is augmented in a murine model of marrow fibrosis. Am. J. Pathol. 180, 811−818.

Okazaki, M., Ferrandon, S., Vilardaga, J.P., Bouxsein, M.L., Potts Jr., J.T., Gardella, T.J., 2008. Prolonged signaling at the parathyroid hormone receptor by peptide ligands targeted to a specific receptor conformation. Proc. Natl. Acad. Sci. U. S. A. 105, 16525−16530.

Okazaki, M., Nagai, S., Dean, T., Potts, J.J., Gardella, T., 2007. Analysis of PTH-PTH receptor interaction mechanisms using a new, long-acting PTH(1-28) analog reveals selective binding to distinct PTH receptor conformations and biological consequences in vivo. J. Bone Miner. Res. 22 (Suppl. 1). Abstract 1190.

On, J.S., Chow, B.K., Lee, L.T., 2015. Evolution of parathyroid hormone receptor family and their ligands in vertebrate. Front. Endocrinol. 6, 28.

Ono, W., Sakagami, N., Nishimori, S., Ono, N., Kronenberg, H.M., 2016. Parathyroid hormone receptor signalling in osterix-expressing mesenchymal progenitors is essential for tooth root formation. Nat. Commun. 7, 11277.

Parthier, C., Reedtz-Runge, S., Rudolph, R., Stubbs, M.T., 2009. Passing the baton in class B GPCRs: peptide hormone activation via helix induction? Trends Biochem. Sci. 34, 303–310.

Pinheiro, P.L., Cardoso, J.C., Power, D.M., Canario, A.V., 2012. Functional characterization and evolution of PTH/PTHrP receptors: insights from the chicken. BMC Evol. Biol. 12, 110.

Pioszak, A.A., Parker, N.R., Gardella, T.J., Xu, H.E., 2009. Structural basis for parathyroid hormone-related protein binding to the parathyroid hormone receptor and design of conformation-selective peptides. J. Biol. Chem. 284, 28382–28391.

Pioszak, A.A., Parker, N.R., Suino-Powell, K., Xu, H.E., 2008. Molecular recognition of corticotropin-releasing factor by its G-protein-coupled receptor CRFR1. J. Biol. Chem. 283, 32900–32912.

Piserchio, A., Shimizu, N., Gardella, T.J., Mierke, D.F., 2002. Residue 19 of the parathyroid hormone: structural consequences. Biochemistry 41, 13217–13223.

Piserchio, A., Usdin, T., Mierke, D., 2000. Structure of tuberoinfundibular peptide of 39 residues. J. Biol. Chem. 275, 27284–27290.

Potts Jr., J.T., Tregear, G.W., Keutmann, H.T., Niall, H.D., Sauer, R., Deftos, L.J., Dawson, B.F., Hogan, M.L., Aurbach, G.D., 1971. Synthesis of a biologically active N-terminal tetratriacontapeptide of parathyroid hormone. Proc. Natl. Acad. Sci. U.S.A. 68, 63–67.

Qian, F., Leung, A., Abou-Samra, A., 1998. Agonist-dependent phosphorylation of the parathyroid hormone/parathyroid hormone-related peptide receptor. Biochemistry 37, 6240–6246.

Rey, A., Manen, D., Rizzoli, R., Caverzasio, J., Ferrari, S.L., 2006. Proline-rich motifs in the parathyroid hormone (PTH)/PTH-related protein receptor C terminus mediate scaffolding of c-Src with beta-arrestin2 for ERK1/2 activation. J. Biol. Chem. 281, 38181–38188.

Rickard, D.J., Wang, F.L., Rodriguez-Rojas, A.M., Wu, Z., Trice, W.J., Hoffman, S.J., Votta, B., Stroup, G.B., Kumar, S., Nuttall, M.E., 2006. Intermittent treatment with parathyroid hormone (PTH) as well as a non-peptide small molecule agonist of the PTH1 receptor inhibits adipocyte differentiation in human bone marrow stromal cells. Bone 39, 1361–1372.

Roth, H., Fritsche, L.G., Meier, C., Pilz, P., Eigenthaler, M., Meyer-Marcotty, P., Stellzig-Eisenhauer, A., Proff, P., Kanno, C.M., Weber, B.H., 2014. Expanding the spectrum of PTH1R mutations in patients with primary failure of tooth eruption. Clin. Oral Investig. 18, 377–384.

Rubin, D.A., Juppner, H., 1999. Zebrafish express the common parathyroid hormone/parathyroid hormone-related peptide receptor (PTH1R) and a novel receptor (PTH3R) that is preferentially activated by mammalian and fugufish parathyroid hormone-related peptide. J. Biol. Chem. 274, 28185–28190.

Sato, E., Muto, J., Zhang, L.J., Adase, C.A., Sanford, J.A., Takahashi, T., Nakatsuji, T., Usdin, T.B., Gallo, R.L., 2016. The parathyroid hormone second receptor PTH2R and its ligand tuberoinfundibular peptide of 39 residues TIP39 regulate intracellular calcium and influence keratinocyte differentiation. J. Investig. Dermatol. 136, 1449–1459.

Schipani, E., Kruse, K., Jüppner, H., 1995. A constitutively active mutant PTH-PTHrP receptor in Jansen-type metaphyseal chondrodysplasia. Science 268, 98–100.

Segre, G.V., Goldring, S.R., 1993. Receptors for secretin, calcitonin, parathyroid hormone (PTH)/PTH-related peptide, vasoactive intestinal peptide, glucagonlike peptide 1, growth hormone-releasing hormone, and glucagon belong to a newly discovered G-protein-linked receptor family. Trends Endocrinol. Metabol. 4, 309–314.

Segre, G.V., Rosenblatt, M., Reiner, B.L., Mahaffey, J.E., Potts Jr., J.T., 1979. Characterization of parathyroid hormone receptors in canine renal cortical plasma membranes using a radioiodinated sulfur-free hormone analogue. correlation of binding with adenylate cyclase activity. J. Biol. Chem. 254, 6980–6986.

Shimizu, M., Carter, P., Khatri, A., Potts, J., Gardella, T., 2001. Enhanced activity in parathyroid hormone (1–14) and (1–11): novel peptides for probing the ligand-receptor interaction. Endocrinology 142, 3068–3074.

Shimizu, M., Joyashiki, E., Noda, H., Watanabe, T., Okazaki, M., Nagayasu, M., Adachi, K., Tamura, T., Potts Jr., J.T., Gardella, T.J., Kawabe, Y., 2016. Pharmacodynamic actions of a long-acting PTH analog (LA-PTH) in thyroparathyroidectomized (TPTX) rats and normal monkeys. J. Bone Miner. Res. 7, 1405–1412.

Suzuki, N., Danks, J.A., Maruyama, Y., Ikegame, M., Sasayama, Y., Hattori, A., Nakamura, M., Tabata, M.J., Yamamoto, T., Furuya, R., Saijoh, K., Mishima, H., Srivastav, A.K., Furusawa, Y., Kondo, T., Tabuchi, Y., Takasaki, I., Chowdhury, V.S., Hayakawa, K., Martin, T.J., 2011. Parathyroid hormone 1 (1–34) acts on the scales and involves calcium metabolism in goldfish. Bone 48, 1186–1193.

Takasu, H., Baba, H., Inomata, N., Uchiyama, Y., Kubota, N., Kumaki, K., Matsumoto, A., Nakajima, K., Kimura, T., Sakakibara, S., Fujita, T., Chihara, K., Nagai, I., 1996. The 69-84 amino acid region of the parathyroid hormone molecule is essential for the interaction of the hormone with the binding sites with carboxyl terminal specificity. Endocrinology 137, 5537–5543.

Tamura, T., Noda, H., Joyashiki, E., Hoshino, M., Watanabe, T., Kinosaki, M., Nishimura, Y., Esaki, T., Ogawa, K., Miyake, T., Arai, S., Shimizu, M., Kitamura, H., Sato, H., Kawabe, Y., 2016. Identification of an orally active small-molecule PTHR1 agonist for the treatment of hypoparathyroidism. Nat. Commun. 7, 13384.

Tawfeek, H.A.W., Qian, F., Abou-Samra, A.B., 2002. Phosphorylation of the receptor for PTH and PTHrP is required for internalization and regulates receptor signaling. Mol. Endocrinol. 16, 1–13.

Thomas, B.E., Sharma, S., Mierke, D.F., Rosenblatt, M., 2008. Parathyroid hormone (PTH) and PTH antagonist induce different conformational changes in the PTHR1 receptor. J. Bone Miner. Res. 5, 925–934.

Thomsen, A.R., Plouffe, B., Cahill 3rd, T.J., Shukla, A.K., Tarrasch, J.T., Dosey, A.M., Kahsai, A.W., Strachan, R.T., Pani, B., Mahoney, J.P., Huang, L., Breton, B., Heydenreich, F.M., Sunahara, R.K., Skiniotis, G., Bouvier, M., Lefkowitz, R.J., 2016. GPCR-G protein-beta-arrestin super-complex mediates sustained G protein signaling. Cell 166, 907–919.

Tian, J., Smogorzewski, M., Kedes, L., Massry, S.G., 1993. Parathyroid hormone-parathyroid hormone related protein receptor messenger RNA is present in many tissues besides the kidney. Am. J. Nephrol. 13, 210–213.

Toribio, R.E., Brown, H.A., Novince, C.M., Marlow, B., Hernon, K., Lanigan, L.G., Hildreth 3rd, B.E., Werbeck, J.L., Shu, S.T., Lorch, G., Carlton, M., Foley, J., Boyaka, P., Mccauley, L.K., Rosol, T.J., 2010. The midregion, nuclear localization sequence, and C terminus of PTHrP regulate skeletal development, hematopoiesis, and survival in mice. FASEB J. 24, 1947–1957.

Trivett, M.K., Potter, I.C., Power, G., Zhou, H., Macmillan, D.L., Martin, T.J., Danks, J.A., 2005. Parathyroid hormone-related protein production in the lamprey *Geotria australis*: developmental and evolutionary perspectives. Dev. Gene. Evol. 215, 553–563.

Turner, P.R., Mefford, S., Bambino, T., Nissenson, R.A., 1998. Transmembrane residues together with the amino terminus limit the response of the parathyroid hormone (PTH) 2 receptor to PTH-related peptide. J. Biol. Chem. 273, 3830–3837.

Uhlen, M., Fagerberg, L., Hallstrom, B.M., Lindskog, C., Oksvold, P., Mardinoglu, A., Sivertsson, A., Kampf, C., Sjostedt, E., Asplund, A., Olsson, I., Edlund, K., Lundberg, E., Navani, S., Szigyarto, C.A., Odeberg, J., Djureinovic, D., Takanen, J.O., Hober, S., Alm, T., Edqvist, P.H., Berling, H., Tegel, H., Mulder, J., Rockberg, J., Nilsson, P., Schwenk, J.M., Hamsten, M., VON Feilitzen, K., Forsberg, M., Persson, L., Johansson, F., Zwahlen, M., VON Heijne, G., Nielsen, J., Ponten, F., 2015. Proteomics. Tissue-based map of the human proteome. Science 347, 1260419.

Urena, P., Kong, X.F., ABOU-Samra, A.B., Juppner, H., Kronenberg, H.M., Potts Jr., J.T., Segre, G.V., 1993. Parathyroid hormone (PTH)/PTH-related peptide receptor messenger ribonucleic acids are widely distributed in rat tissues. Endocrinology 133, 617–623.

Usdin, T., Gruber, C., Bonner, T., 1995. Identification and functional expression of a receptor selectively recognizing parathyroid hormone, the PTH2 receptor. J. Biol. Chem. 270, 15455–15458.

Usdin, T.B., Dobolyi, A., Ueda, H., Palkovits, M., 2003. Emerging functions for tuberoinfundibular peptide of 39 residues. Trends Endocrinol. Metabol. 14, 14–19.

Usdin, T.B., Hoare, S.R., Wang, T., Mezey, E., Kowalak, J.A., 1999. TIP39: a new neuropeptide and PTH2-receptor agonist from hypothalamus. Nat. Neurosci. 2, 941–943.

Usdin, T.B., Paciga, M., Riordan, T., Kuo, J., Parmelee, A., Petukova, G., Camerini-Otero, R.D., Mezey, E., 2008. Tuberoinfundibular Peptide of 39 residues is required for germ cell development. Endocrinology 149, 4292–4300.

Vilardaga, J.P., Krasel, C., Chauvin, S., Bambino, T., Lohse, M.J., Nissenson, R.A., 2002. Internalization determinants of the parathyroid hormone receptor differentially regulate beta-arrestin/receptor association. J. Biol. Chem. 277, 8121–8129.

Winer, K.K., Fulton, K.A., Albert, P.S., Cutler Jr., G.B., 2014. Effects of pump versus twice-daily injection delivery of synthetic parathyroid hormone 1-34 in children with severe congenital hypoparathyroidism. J. Pediatr. 165, 556-563 e1.

Winer, K.K., Zhang, B., Shrader, J.A., Peterson, D., Smith, M., Albert, P.S., Cutler Jr., G.B., 2012. Synthetic human parathyroid hormone 1-34 replacement therapy: a randomized crossover trial comparing pump versus injections in the treatment of chronic hypoparathyroidism. J. Clin. Endocrinol. Metab. 97, 391–399.

Wittelsberger, A., Corich, M., Thomas, B.E., Lee, B.K., Barazza, A., Czodrowski, P., Mierke, D.F., Chorev, M., Rosenblatt, M., 2006a. The mid-region of parathyroid hormone (1-34) serves as a functional docking domain in receptor activation. Biochemistry 45, 2027–2034.

Wittelsberger, A., Thomas, B.E., Mierke, D.F., Rosenblatt, M., 2006b. Methionine acts as a "magnet" in photoaffinity crosslinking experiments. FEBS Lett. 580, 1872–1876.

Wu, T.L., Vasavada, R.C., Yang, K., Massfelder, T., Ganz, M., Abbas, S.K., Care, A.D., Stewart, A.F., 1996. Structural and physiologic characterization of the mid-region secretory species of parathyroid hormone-related protein. J. Biol. Chem. 271, 24371–24381.

Xiong, L., Xia, W.F., Tang, F.L., Pan, J.X., Mei, L., Xiong, W.C., 2016. Retromer in osteoblasts interacts with protein phosphatase 1 regulator subunit 14c, terminates parathyroid hormone's signaling, and promotes its catabolic response. EBioMedicine 9, 45–60.

Yan, Y.L., Bhattacharya, P., He, X.J., Ponugoti, B., Marquardt, B., Layman, J., Grunloh, M., Postlethwait, J.H., Rubin, D.A., 2012. Duplicated zebrafish co-orthologs of parathyroid hormone-related peptide (PTHrp, Pthlh) play different roles in craniofacial skeletogenesis. J. Endocrinol. 214, 421–435.

Yin, Y., DE Waal, P.W., He, Y., Zhao, L.H., Yang, D., Cai, X., Jiang, Y., Melcher, K., Wang, M.W., Xu, H.E., 2017. Rearrangement of a polar core provides a conserved mechanism for constitutive activation of class B G protein-coupled receptors. J. Biol. Chem. 292, 9865–9881.

Zhang, H., Qiao, A., Yang, L., Van Eps, N., Frederiksen, K.S., Yang, D., Dai, A., Cai, X., Zhang, H., Yi, C., Cao, C., He, L., Yang, H., Lau, J., Ernst, O.P., Hanson, M.A., Stevens, R.C., Wang, M.W., Reedtz-Runge, S., Jiang, H., Zhao, Q., Wu, B., 2018a. Structure of the glucagon receptor in complex with a glucagon analogue. Nature 553, 106–110.

Zhang, P., Jobert, A.S., Couvineau, A., Silve, C., 1998. A homozygous inactivating mutation in the parathyroid hormone/parathyroid hormone-related peptide receptor causing Blomstrand chondrodysplasia. J. Clin. Endocrinol. Metab. 83, 3365–3368.

Zhang, Q., Xiao, K., Liu, H., Song, L., Mcgarvey, J.C., Sneddon, W.B., Bisello, A., Friedman, P.A., 2018b. Site-specific polyubiquitination differentially regulates parathyroid hormone receptor-initiated MAPK signaling and cell proliferation. J. Biol. Chem. 293, 5556–5571.

Zhang, Y., Sun, B., Feng, D., Hu, H., Chu, M., Qu, Q., Tarrasch, J.T., Li, S., Sun Kobilka, T., Kobilka, B.K., Skiniotis, G., 2017. Cryo-EM structure of the activated GLP-1 receptor in complex with a G protein. Nature 546, 248–253.

Zhou, A., Bessalle, R., Bisello, A., Nakamoto, C., Rosenblatt, M., Suva, L.J., Chorev, M., 1997. Direct mapping of an agonist-binding domain within the parathyroid hormone/parathyroid hormone-related protein receptor by photoaffinity crosslinking. Proc. Natl. Acad. Sci. U.S.A. 94, 3644–3649.

Chapter 29

Structure and function of the vitamin D-binding proteins

Daniel D. Bikle
VA Medical Center and University of California San Francisco, San Francisco, California, United States

Chapter outline

Introduction 713
Vitamin D-binding protein 714
Genomic regulation 714
Structure and polymorphisms 715
Biologic function 715
Binding to and transport of vitamin D metabolites 715
Actin scavenging 719
Neutrophil recruitment and migration with complement 5a binding 719
Fatty acid binding 719
Formation of vitamin D-binding protein–macrophage-activating factor and its functions 720
Intracellular trafficking of vitamin D metabolites: role of heat shock protein 70 and hnRNPC1/C2 720
The vitamin D receptor 720
Genomic location, protein structure, and regulation 721
Vitamin D receptor mechanism of action: genomic 722
Vitamin D binding sites in the genome 722
Coregulators and epigenetic changes regulating VDR function 724
Negative vitamin D response elements 725
Interaction of VDR with β-catenin signaling 725
Vitamin D receptor mechanism of action: nongenomic 726
Conclusions 727
References 727

Introduction

Vitamin D is obtained by the body either from its production in the skin or through absorption from the intestine. In either case vitamin D must be transported to tissues such as the liver where it is metabolized to its major circulating form, 25-hydroxyvitamin D (25(OH)D), by a number of 25-hydroxylases, the most important of which is Cyp2r1. 25(OH)D is then transported to tissues such as the kidney where it undergoes further metabolism to its major hormonal form 1,25-dihydroxyvitamin D (1,25$(OH)_2$D) by the mitochondrial-based Cyp27b1. Vitamin D-binding protein (DBP) is the key transport protein and, at least in some cells such as the kidney, participates in the transport of 25(OH)D into the cell. However, as will be discussed, DBP has a number of functions independent of its role as a vitamin D transport protein. Within at least some cells the 25(OH)D entering from the blood appears to be shuttled to mitochondria for this enzymatic reaction by what was originally called intracellular vitamin D-binding protein, later identified as the heat shock protein Hsp70, a known chaperone. Hsp70 in some cells may also facilitate the movement of 1,25$(OH)_2$D into the nucleus where it provides the activating ligand for the vitamin D receptor (VDR). The VDR is responsible for all the known genomic actions of vitamin D, most of which require 1,25$(OH)_2$D and its heterodimer partner RXR. However, VDR also has a nongenomic role in the membrane where it responds to a different conformation of ligands in activating a different set of signaling pathways. The membrane VDR shares this role with another protein initially called membrane-associated rapid response steroid (MARRS)-binding protein, more recently known as thioredoxin-like protein (GRP58), endoplasmic reticulum protein 57/60 (ERp57 or 60), or protein disulfide isomerase family A, member 3 (Pdia3). Thus, there is a series of vitamin D-binding proteins that sequentially provide for the transport of vitamin D from its point of origin (skin or gut) to tissues involved with its metabolism to its hormonal form, which mediate the handling of these metabolites within the cells

Principles of Bone Biology. https://doi.org/10.1016/B978-0-12-814841-9.00029-4

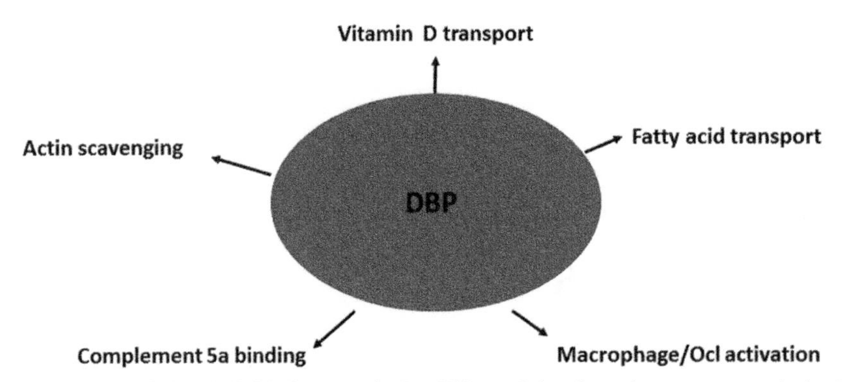

FIGURE 29.1 The multiple functions of vitamin D-binding protein. In addition to being the major transport protein for the vitamin D metabolites, vitamin D-binding protein (*DBP*) along with gelsolin participates in actin scavenging, fatty acid transport, and binding of complement 5a to neutrophils regulating their chemotaxis, and when the Gc1F and Gc1S variants are deglycosylated to form macrophage-activating factor, *DBP*—MAF can activate macrophages and osteoclasts (*Ocl*).

and ultimately to the proteins that initiate the biologic response, genomic and nongenomic. In this chapter aspects of the structures of these proteins relevant to their function and the mechanisms by which these proteins function will be reviewed. Given that vitamin D signaling affects nearly all if not all cells in the body, specific cellular responses will not be reviewed except as they illustrate these mechanisms.

Vitamin D-binding protein

Although the best known function of DBP is the transport of vitamin D metabolites from their source of production to their target tissues, DBP has a number of other functions, including actin scavenging, fatty acid binding, complement 5a (C5a)-binding regulation of neutrophil migration and chemotaxis, and, by way of its conversion to DBP—macrophage-activating factor (DBP—MAF), the regulation of osteoclast activity, immune function, and cancer growth (Fig. 29.1). These functions will be discussed in turn, including their clinical significance and the influence of the different polymorphisms on these functions.

Genomic regulation

The human *DBP* gene is located on chromosome 4q12—q13. It is 35 kb in length and comprises 13 exons encoding 474 amino acids, including a 16-amino-acid leader sequence, which is cleaved before release. The region of the chromosome where *DBP* is located also contains the homologous genes for albumin, α-fetoprotein, and afamin, with which it shares some properties, leading to speculation that this gene cluster arose from gene duplications. However, the transcription of *DBP* goes in the direction opposite that of the other genes in this cluster and it is separated by about 1500 kb (White and Cooke, 2000; Song et al., 1999). Expression of DBP is widespread, but the vast majority originates from the liver (Cooke et al., 1991). The proximal promoter of the *DBP* gene contains three hepatocyte nuclear factor 1 (HNF1) binding sites within 2 kb of the transcription start site (TSS). HNF1 is expressed as the α or β isoform and binds to its response element as a dimer. HNF1α is an activator, whereas HNF1β is inhibitory. In the liver, levels of HNF1α are high, resulting in increased expression of DBP, whereas in the kidney the β form exists, thus suppressing expression, and in the brain, little HNF1 is present (White and Cooke, 2000). The expression of DBP is increased by estrogen (Hagenfeldt et al., 1991) as is well illustrated with the rise in DBP during pregnancy (Moller et al., 2012; Zhang et al., 2014) and with oral contraceptive utilization (Moller et al., 2013). However, whether this is a direct action on the *DBP* gene via the estrogen receptor is not clear. Androgens, on the other hand, do not appear to affect DBP expression (Hagenfeldt et al., 1991). Dexamethasone and certain cytokines such as interleukin-6 (IL-6) also increase DBP production, whereas transforming growth factor β is inhibitory (Guha et al., 1995). As for estrogen, the mechanism underlying such regulation is unclear. However, from a clinical perspective these factors are likely to contribute to the increase in DBP production following trauma (Dahl et al., 2001a) and acute liver failure (Schiodt, 2008), which will be discussed subsequently. Primary hyperparathyroidism, on the other hand, is associated with a reduction in DBP levels, probably contributing to the lower 25(OH)D levels in these patients, as the free 25(OH)D is not reduced (Wang et al., 2017). Vitamin D itself or any of its metabolites does not regulate DBP production.

Structure and polymorphisms

The mature human DBP is 58 kDa in size, although differences in glycosylation of the protein alter the actual size. DBP is composed of three structurally similar domains. The first domain has an α-helical arrangement and is the binding site for the vitamin D metabolites (aa 35–49). Fatty acid binding demonstrates a single high-affinity site for both palmitic acid and arachidonic acid, but only arachidonic acid competes with 25(OH)D for binding, suggesting that the binding of fatty acids is not competitive for the vitamin D binding site in domain 1, but is due to alterations in DBP conformation (Calvo and Ena, 1989; Bouillon et al., 1992). The actin binding site is located in aa 373–403, spanning parts of domains 2 and 3, but part of domain 1 is also involved (Haddad et al., 1992; Head et al., 2002). The C5a/C5a des Arg binding site is to aa 130–149 (Zhang and Kew, 2004). Membrane binding sites have been identified in aa 150–172 and 379–402 (Wilson et al., 2015). DBP is the most polymorphic gene known. Over 120 variants have been described based on electrophoretic properties (Cleve and Constans, 1988), with 1242 polymorphisms listed in the NCBI database as of this writing (Chun, 2012). Of these variants, the Gc1F and Gc1S (rs7041) and Gc2 (rs4588) are the most common (Fig. 29.2). Gc1F and Gc1S involve two polymorphisms, one at aa 432 (416 in the mature DBP) and one at aa 436 (420 in the mature DBP). The 1F allele encodes the sequence of amino acids between 432 and 436 as **D**ATPT; the 1S allele encodes the sequence **E**ATPT. This subtle difference in charge makes GcF run faster and GcS slower during electrophoresis. The Gc2 allele encodes DATP**K**, which runs slower still. Glycosylation further distinguishes the Gc1 variants from the Gc2 variant. The threonine (T) in Gc1 binds *N*-acetylgalactosamine to which galactose and sialic acid bind in tandem. The lysine (K) in the comparable position in Gc2 does not bind these residues (Malik et al., 2013; Nagasawa et al., 2005). This affects the conversion of DBP to DBP–MAF, which involves a partial deglycosylation, removing the galactose and sialic acid by the sequential action of sialidase and β-galactosidase, by T and B cells (Uto et al., 2012).

Biologic function

Binding to and transport of vitamin D metabolites

DBP was discovered by Hirschfeld in 1959 (Hirschfeld, 1959), and was originally called group-specific component (Gc-globulin), and not until 1975 was its function as a carrier of vitamin D metabolites realized (Daiger et al., 1975). This is the function that gives DBP its name, although only about 5% of DBP is required for this purpose. In serum samples from normal individuals, ~85% of circulating vitamin D metabolites are bound to DBP, whereas albumin with its substantially lower binding affinity binds only ~15% of these metabolites despite its 10-fold higher concentration compared with DBP. Approximately 0.4% of total $1,25(OH)_2D_3$ and 0.03% of total $25(OH)D_3$ is free in serum from normal nonpregnant individuals. The affinity of DBP for the vitamin D_2 metabolites is lower (Armas et al., 2004), such that the free fraction would be expected to be higher, but this has not been directly determined. The "bioavailable" vitamin D metabolite comprises the free vitamin D metabolite and the amount bound to albumin, thus measuring around 15% in normal individuals (reviewed in Bikle et al., 2017a). At this point there is little evidence that the albumin fraction is truly bioavailable, and several studies indicate that albumin-bound vitamin D metabolites are not available to cells, at least under static conditions (Bikle and Gee, 1989). However, because the albumin–hormone complexes generally dissociate rapidly this fraction may be more bioavailable in a dynamically perfused tissue (Mendel, 1990). The free hormone hypothesis postulates that only the unbound fraction (the free fraction) of hormones that otherwise circulate in blood bound to their carrier proteins is able to enter cells and exert its biologic effects (Fig. 29.3). Examples include the vitamin D metabolites, about which we are concerned in this chapter; sex steroids; cortisol; and thyroid hormone. These are lipophilic hormones assumed to cross the plasma membrane by diffusion and not by an active transport mechanism. However, at least for some tissues, a transport system has been identified that takes up the 25(OH)D (and presumably other vitamin D metabolites) attached to DBP. That system involves megalin and cubilin. The importance of megalin for vitamin D metabolism was first

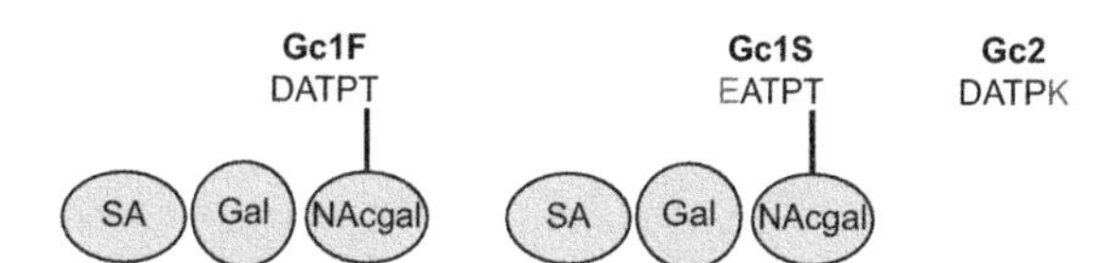

FIGURE 29.2 Major vitamin D-binding protein variants. Gc1F, Gc1S, and Gc2 are the major vitamin D-binding protein (DBP) variants. They differ in the sequence of amino acids from 432 to 436 as shown. DATPT of Gc1F represents Asp–Ala–Thr–Pro–Thr. In Gc1S the aspartate is mutated to glutamate. In Gc2, the threonine is mutated to lysine. The glycosylation at the threonine in Gc1F and Gc1S enables these forms to be converted to DBP–macrophage-activating factor by removal of the *N*-acetylgalactosamine (*NAcgal*) and galactose (*Gal*), leaving sialic acid (*SA*).

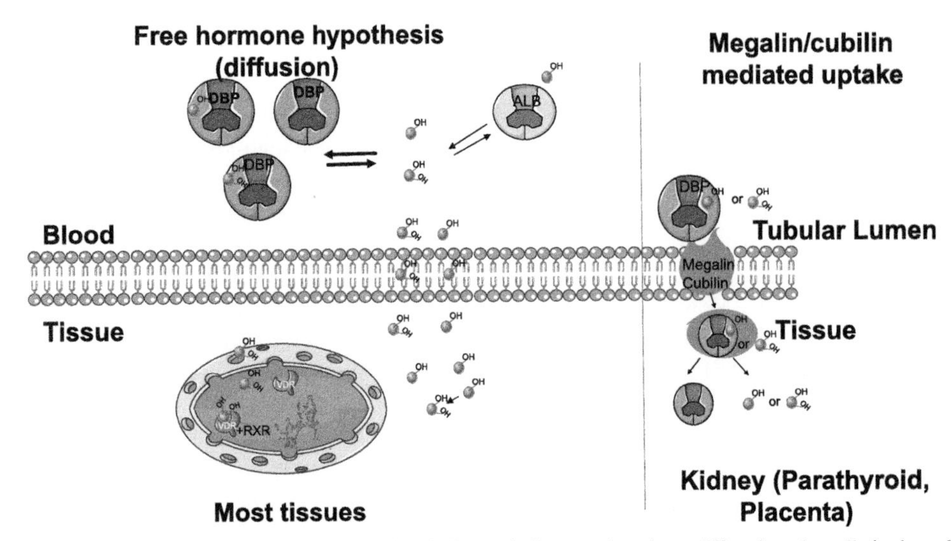

FIGURE 29.3 Free hormone hypothesis. For most cells the vitamin D metabolites are thought to diffuse into the cells in the unbound or free form. However, for some cells, such as the kidney, megalin and cubilin transport the vitamin D into the cell bound to vitamin D-binding protein (*DBP*).

noted by Nykjaer et al. (Nykjaer et al., 1999), who found in the megalin-knockout mouse extensive loss of DBP and 25(OH)D in the urine with the development of bone disease. These mice have very poor survival, but subsequently a kidney-specific knockout of megalin was developed with good survival, enabling the clear demonstration of decreased levels of the vitamin D metabolites, hypocalcemia, and osteomalacia in these mice (Leheste et al., 2003). While megalin has a transcellular domain that is necessary for its endocytosis, cubilin does not, but associates directly with megalin to enable the uptake of DBP and other ligands into the cell (reviewed in Christensen and Birn, 2002). A substantial number of tissues express the megalin/cubilin complex, including the parathyroid gland, choroid plexus, placenta, thyroid, type II pneumocytes, and endometrium, but the role of this complex with respect to its importance for vitamin D metabolism has been best studied in the kidney (Christensen and Birn, 2002), and its role in other tissues with respect to vitamin D metabolite transport into cells is unclear. Although the megalin/cubilin transport system is the best understood mechanism by which DBP-bound 25(OH)D is transported into cells, activated monocytes and the Raji cell line (human B cell lymphoma cell line) have also been reported to accumulate DBP by a mechanism apparently independent of megalin (Chun et al., 2010; Esteban et al., 1992). This mechanism has received little further study.

The knockout of megalin is fruitfully compared with the knockout of DBP in attempting to distinguish the respective roles of DBP and megalin/cubilin with respect to vitamin D entry into cells. In DBP-knockout mice the vitamin D metabolites are presumably all free and/or bioavailable. Unlike the megalin-knockout mice, mice lacking DBP do not show evidence of vitamin D deficiency unless placed on a vitamin D-deficient diet, despite having very low levels of serum 25(OH)D and $1,25(OH)_2D$ and losses of these metabolites in the urine (Safadi et al., 1999). However, on a vitamin D deficient-diet they rapidly develop hyperparathyroidism indicative of vitamin D deficiency. On the other hand, the DBP-knockout mice resisted the development of vitamin D toxicity (renal calcifications) when large doses of vitamin D were administered. In subsequent studies tissue levels of $1,25(OH)_2D$ were found to be normal in the DBP-knockout mice, as was the expression of markers of vitamin D action, such as intestinal transient receptor potential cation channel subfamily V member 6 (TRPV6), calbindin 9k, PMCA1b, and renal TRPV5. Moreover, injection of $1,25(OH)_2D$ into these DBP knockouts showed a more rapid increase in the expression of cytochrome P450 family 24 subfamily A member 1 (Cyp24A1), TRPV5, and TRPV6 (Zella et al., 2008). These studies support the concept that although the megalin/cubilin complex is critical for transporting 25(OH)D—DBP into some tissues such as the kidney, DBP is not necessary for getting the vitamin D metabolites into other cells, thus supporting the free hormone hypothesis for those cells. However, DBP clearly serves as a critical reservoir for the vitamin D metabolites, reducing the risk of vitamin D deficiency when intake or epidermal production is transiently reduced. Further support for this concept comes from studies in cell culture. In keratinocytes, only the free fraction of $1,25(OH)_2D$ was capable of inducing Cyp24A1 under varying conditions of DBP and albumin (Bikle and Gee, 1989). These results were subsequently supported by the observation that serum from DBP-knockout mice enabled the induction of Cyp24A1 by $1,25(OH_2)D$ in MC3T3 cells (Zella et al., 2008) and the induction by 25(OH)D and $1,25(OH)_2D$ of both Cyp24A1 and cathelicidin in monocytes (Chun et al., 2010) better than serum from control mice. These results indicate that DBP and albumin restricted the entry of the vitamin D metabolites into these cells.

To address the free hormone hypothesis, a method to measure the free concentration needed to be developed. This was originally performed by centrifugal ultrafiltration developed by the author to directly determine the free levels of 25(OH)D and $1,25(OH)_2D$ (Bikle et al., 1984, 1986a) in various clinical situations. However, this method is quite labor intensive and has been replaced, at least for free 25(OH)D, by a two-step enzyme-linked immunosorbent assay (ELISA) that directly measures free 25(OH)D (Future Diagnostics Solutions B.V., Wijchen, The Netherlands) using monoclonal antibodies from DIAsource ImmunoAssays (Louvain-la-Neuve, Belgium). This newer assay is dependent on the quality of the antibody used to bind the free 25(OH)D. The antibody in the current assay does not recognize $25(OH)D_2$ as well as $25(OH)D_3$ (77% of the $25(OH)D_3$ value) so it underestimates the free $25(OH)D_2$. However, in most situations where the predominant vitamin D metabolite is $25(OH)D_3$, the data compare quite well with those obtained from similar populations using the centrifugal ultrafiltration assay (Bikle et al., 2017b; Schwartz et al., 2014a).

The centrifugal ultrafiltration method was utilized for the determination of the affinity constants for DBP and albumin binding to 25(OH)D and $1,25(OH)_2D$. These affinity constants were measured in serum from a healthy young adult (the author) and demonstrated an affinity constant (K_a) of DBP for $1,25(OH)_2D$ of $3.7-4.2 \times 10^7\ M^{-1}$ and for 25(OH)D of $7-9 \times 10^8\ M^{-1}$, with a K_a of albumin for $1,25(OH)_2D$ of $5.4 \times 10^4\ M^{-1}$ and for 25(OH)D of $6 \times 10^5\ M^{-1}$ (Bikle et al., 1985, 1986b). Using these affinity constants along with measurements of DBP, albumin, and total vitamin D metabolite of interest permitted the free concentration to be calculated according to the following formula:

$$\text{free vitamin D metabolite} = \frac{\text{total vitamin D metabolite}}{1 + (Ka_{alb} * \text{albumin}) + (Ka_{DBP} * \text{DBP})}.$$

In sera from normal healthy younger individuals, the calculated values using DBP measured with polyclonal antibodies and the calculated values of free 25(OH)D and $1,25(OH)_2D$ correlate well with the directly measured free levels.

When applied to clinical populations with altered DBP levels during either physiologic (e.g., pregnancy) or pathologic (e.g., liver disease) conditions, however, the calculated values no longer are consistent with those measured directly by either centrifugal ultrafiltration or the newly developed ELISA (Schwartz et al., 2014b). There are a number of reasons for this. As noted earlier, calculation of free vitamin D metabolite levels depends on the accurate measurement of total vitamin D metabolite, DBP, and albumin and the affinity constants between DBP and albumin for the vitamin D metabolite whose free concentration is being measured. Although measurement of albumin is considered well standardized, that of the vitamin D metabolites is less so (Fuleihan Gel et al., 2015); although with major efforts to encourage laboratories to use well-defined standards and more advanced technology such as mass spectroscopy, this is improving (Muller and Volmer, 2015; Binkley and Carter, 2017). However, the measurement of DBP has not been standardized, and the results vary substantially between laboratories. One widely used monoclonal assay underestimates the Gc1F allele of DBP, leading to the conclusion that individuals expressing this allele, primarily African Americans, had lower DBP levels than Caucasians (Powe et al., 2013). On the other hand, Nielson et al. (Nielson et al., 2016a,b) subsequently reported their results measuring DBP levels with four different assays, three of which involved polyclonal assays and one the aforementioned monoclonal assay used by Powe et al. (Powe et al., 2013). The monoclonal antibody-based assay resulted in a 54% lower concentration of DBP in African Americans expressing the Gc1F allele, compared with Caucasians, with minimal differences in DBP levels among these groups using the three polyclonal assays. Similarly, mass spectrometry measurements of DBP by Hoofnagle et al. (Hoofnagle et al., 2015) failed to show a major difference in DBP levels with the various alleles, although the results with mass spectroscopy tended to be lower than those with the polyclonal assays. That said, assays using polyclonal assays or mass spectroscopy have consistently demonstrated a modest reduction in the Gc2 variant of DBP compared with the Gc1 variants (Hoofnagle et al., 2015; Lauridsen et al., 2001; Carpenter et al., 2013; Santos et al., 2013). A 2010 genome-wide association study demonstrated that rs2282679, an intronic polymorphism in the DBP gene, was likewise associated with lower 25(OH)D and DBP levels (Wang et al., 2010). This observation has been confirmed in several other populations (Leong et al., 2014; Cheung et al., 2013).

The clinical significance of these allelic differences is unclear. Individuals with the Gc2 variant have been shown to respond to vitamin D supplementation with a more robust increase in 25(OH)D (Fu et al., 2009).

Nevertheless, differences in these alleles were not found to contribute to a difference in fracture rate in a large study including African Americans and Caucasians (Takiar et al., 2015) or in other calcemic and cardiometabolic diseases in the Canadian Multicentre Osteoporosis Study (Leong et al., 2014). However, as reviewed by Malik et al. (Malik et al., 2013) and Speeckaert et al (Speeckaert et al., 2006), a large number of chronic diseases, including types 1 and 2 diabetes (Hirai et al., 1998; Baier et al., 1998; Ye et al., 2001), osteoporosis (Lauridsen et al., 2004; Papiha et al., 1999; Ezura et al., 2003), chronic obstructive lung disease (Chishimba et al., 2010), endometriosis (Faserl et al., 2011), inflammatory bowel disease (Eloranta et al., 2011), some cancers (Abbas et al., 2008; Dimopoulos et al., 1984; Zhou et al., 2012; Poynter et al., 2010) (although see Poynter et al., 2010; McCullough et al., 2007; Ahn et al., 2009), and tuberculosis (Martineau et al., 2010) have been associated with various DBP variants.

A third major problem in attempting to calculate the free fraction of vitamin D metabolites is the assumption that all DBP alleles have the same affinity for the vitamin D metabolites, and that this is invariant under varying clinical conditions. Although differences in the affinity constants for the different DBP alleles have been reported, with Gc1F having the highest affinity and Gc2 the lowest among the common alleles (Arnaud and Constans, 1993), results from other laboratories have not confirmed these differences (Bouillon et al., 1980; Boutin et al., 1989). However, in a more recent study evaluating the half-life of 25(OH)D in serum, subjects homozygous for the Gc1F allele were found to have the shortest half-life, indicating reduced affinity (Jones et al., 2014). On the other hand, serum containing the Gc1F variant of DBP reduced the ability of 25(OH)D and $1,25(OH)_2D$ to induce cathelicidin in monocytes more than serum with the Gc2 allele, suggesting the opposite order of affinity (Chun et al., 2010). Regardless, these potential differences in measured affinity do not begin to explain the large differences between the calculated and the directly measured free metabolite levels in various disease states. Although there are statistically significant correlations between calculated and directly measured free 25(OH)D, the relationship accounts for only 13% of the variation. Calculated free 25(OH)D concentrations are consistently higher than directly measured concentrations in a variety of studies, such as those performed during the third trimester of pregnancy and in patients with liver disease and cystic fibrosis (Schwartz et al., 2014b; Nielson et al., 2016a,b; Lee et al., 2015; Sollid et al., 2016). These studies suggest changes in the affinity of 25(OH)D to DBP independent of allelic variations in at least these clinical conditions.

Studies of normal populations using direct methods of determining free 25(OH)D concentrations have shown good correlations with total 25(OH)D concentrations. The free levels have been reported to be between 0.02% and 0.09% of total 25(OH)D concentrations. Concentrations generally range from 1.2 to 7.9 pg/mL. However, in a study of normal individuals in which DBP alleles and total and free 25(OH)D were measured, those with the Gc2/2 variant had lower DBP and lower total 25(OH)D levels than those with Gc1 alleles, but the differences in directly measured free 25(OH)D levels were nearly the same, unlike the calculated free 25(OH)D levels (Sollid et al., 2016). Parathyroid hormone (PTH) is generally found to be negatively correlated with free 25(OH)D as well as total 25(OH)D, whereas serum C-terminal telopeptide of type I collagen has been reported to have a moderate positive correlation with total and free 25(OH)D (Aloia et al., 2015).

Thus in normal individuals, total 25(OH)D provides a reasonable assessment of vitamin D status. However, this is not the case in conditions in which DBP levels and/or affinities for the vitamin D metabolites are altered. Obesity is associated with reductions in total and free 25(OH)D but not DBP or half-life measurements of 25(OH)D (Walsh et al., 2016). On the other hand, bioavailable 25(OH)D correlated with bone mineral density but total 25(OH)D did not in one study (Powe et al., 2011). Similarly, in hemodialysis patients bioavailable 25(OH)D correlated negatively with PTH and positively with serum calcium, unlike total 25(OH)D. In the nephrotic syndrome, DBP is lost in the urine along with 25(OH)D, reducing the levels of both total and bioavailable 25(OH)D. However, like in the hemodialysis patients, bone mineral density correlated positively and PTH negatively with bioavailable 25(OH)D but not with total 25(OH)D (Aggarwal et al., 2016). In a study of patients with acute kidney injury, bioavailable 25(OH)D but not total 25(OH)D correlated inversely with mortality (Leaf et al., 2013). In acromegaly, DBP levels were observed to be increased, whereas 25(OH)D levels were not, resulting in a reduction in the calculated free 25(OH)D (Altinova et al., 2016) that may contribute to some of the alterations in bone and mineral metabolism in this condition. In a study of HIV^+ patients on triple antiviral therapy, the DBP levels were increased but not the total 25(OH)D, and these changes were associated with an increase in bone turnover markers and PTH, suggesting a fall in free 25(OH)D (which was not measured) (Hsieh et al., 2016). With vitamin D or 25(OH)D supplementation, free 25(OH)D concentrations rise in concert with total 25(OH)D concentrations (Sollid et al., 2016; Aloia et al., 2015; Alzaman et al., 2016; Schwartz et al., 2016; Liu et al., 2006), rising more steeply with D_3 supplementation compared with D_2 (Shieh et al., 2016) and even faster with $25(OH)D_3$ (Shieh et al., 2017). With high-dose D supplementation the changes in PTH were significantly related to changes in free 25(OH)D but not to changes in total 25(OH)D or changes in total $1,25(OH)_2D$ (Shieh et al., 2016), suggesting that free 25(OH)D might be a better marker of the biologically available fraction. In the third trimester of pregnancy, directly measured free 25(OH)D tends to be higher, and free $1,25(OH)_2D$ is substantially higher in pregnant women versus comparator groups of women (Bikle et al., 1984; Schwartz et al., 2014b). These results suggest that the affinity of DBP for the vitamin D metabolites is decreased during pregnancy, perhaps compensating for increased DBP concentrations and the needs of both the mother and the fetus for calcium. Directly measured free 25(OH)D and $1,25(OH)_2D$ tend to be higher in outpatients with cirrhosis compared with other groups (Bikle et al., 1986a; Schwartz et al., 2014b), despite lower total vitamin D metabolite concentrations. The relationship between free 25(OH)D and total 25(OH)D is both steeper and more variable in patients with liver disease than in healthy people, indicating altered affinity of DBP for 25(OH)D in these patients. Those with the most severe cirrhosis and protein synthesis dysfunction have a higher percentage of free 25(OH)D compared with cirrhotics without protein synthesis dysfunction, but free 25(OH)D concentrations are similar due to the presence of both lower total 25(OH)D concentrations and lower DBP. In a vitamin D dose-titration study (Schwartz et al., 2016) of nursing home residents, who are older and likely to have more medical problems and receive more medications than younger people or community-dwelling elderly, free 25(OH)D levels rose along with increases in total 25(OH)D, but responses appeared to be steeper than those of normal subjects, younger outpatients, diabetics, or HIV-infected patients, suggesting altered affinity of 25(OH)D for DBP in this group of individuals. Additional support for the free hormone hypothesis comes from animal studies comparing the effects of vitamin D_2 with vitamin D_3 on calcium metabolism. In one such study mice were raised on a diet containing only vitamin D_2 or vitamin D_3 postweaning. These mice had comparable 25(OH)D levels 8 weeks later, but the D_2-raised mice had higher free 25(OH)D levels with increased numbers of osteoclasts and increased indices of both bone resorption and formation (Chun et al., 2016). These studies, clinical and translational, support the free hormone hypothesis and argue that measurement of the free vitamin D metabolite concentration provides information concerning the vitamin D status of the individual over and above that of the total vitamin D metabolite measurements.

Actin scavenging

A major function of DBP that has received considerably less study than that of vitamin D metabolite binding is its role in actin scavenging. Following trauma (Dahl et al., 2001a), sepsis (Wang et al., 2008; Dahl et al., 2003; Kempker et al., 2012), liver trauma (Schiodt, 2008; Gressner et al., 2009; Schiodt et al., 1997), acute lung injury (Lind et al., 1988), preeclampsia (Tannetta et al., 2014), surgery (Speeckaert et al., 2010; Dahl et al., 2001b), and burn injuries (Koike et al., 2002), large amounts of actin are released from the damaged cells, forming polymerized filamentous F-actin that in combination with coagulation factor Va can lead to disseminated intravascular coagulation and multiorgan failure unless cleared (Meier et al., 2006). The actin scavenging system consists of gelsolin and DBP. Gelsolin depolymerizes F-actin to G (globular)-actin. DBP with its high affinity for G-actin ($K_d = 10$ nM) prevents repolymerization and clears it from the blood (Vasconcellos and Lind, 1993; Mc Leod et al., 1989). There does not appear to be a difference among the major DBP variants for binding to G-actin (Speeckaert et al., 2006). The DBP−actin complexes are rapidly cleared (half-life in blood approximately 30 min) (Dahl et al., 2001b), primarily by the liver, lungs, and spleen, which express receptors for the DBP−actin complexes (Dueland et al., 1991). The immediate result is a drop in DBP levels, potentially altering the bioavailability of the vitamin D metabolites (Schiodt, 2008; Madden et al., 2015; Waldron et al., 2013), but a rise in the DBP−actin complexes (Dahl et al., 2001a; Wang et al., 2008; Schiodt et al., 1997; Lind et al., 1988). The ability of the organism to respond to the insult by increasing DBP production is correlated to survival (Dahl et al., 2001a; Schiodt, 2008; Leaf et al., 2013), and has led to consideration of the use of DBP therapeutically (Pihl et al., 2010; Gomme and Bertolini, 2004).

Neutrophil recruitment and migration with complement 5a binding

Neutrophil activation during inflammation increases their binding sites for DBP (DiMartino et al., 2007), and DBP binding to these sites facilitates C5a-induced chemotaxis (Binder et al., 1999). This role of DBP is not limited to C5a but includes other chemoattractants such as CXCL1 during inflammation (Trujillo et al., 2013). DBP by itself does not promote chemotaxis, nor does the DBP−actin complex discussed before (Ge et al., 2014). The interaction with C5a involves residues 130−149 of DBP, a region that is common to all major DBP alleles (Zhang and Kew, 2004), and no difference in these alleles has been found with respect to their promotion of C5a-mediated chemotaxis (Binder et al., 1999). Binding of $1,25(OH)_2D$ blocks the promotion by DBP of the C5a activity, although 25(OH)D does not (Shah et al., 2006). CD44 and annexin A2 are thought to be part of the DBP cell surface binding site, which also appears to involve cell surface ligands such as chondroitin sulfate proteoglycans, and are involved in C5a chemotaxis (DiMartino et al., 2001; DiMartino and Kew, 1999; McVoy and Kew, 2005). The binding of DBP to its receptor is not altered by and does not alter the affinity of C5a for its receptor (Zhang et al., 2010; Perez, 1994), but does increase the amount of binding of C5a des Arg to the C5a receptor (Perez, 1994).

Fatty acid binding

Like albumin, DBP binds fatty acids but with lower affinity ($K_a = 10^5-10^6 M^{-1}$) and a single binding site (Calvo and Ena, 1989; Swamy and Ray, 2008). Most of the fatty acids binding to DBP are monounsaturated or saturated, with only 5% polyunsaturated. However, only polyunsaturated fatty acids such as arachidonic acid and linoleic acid compete with vitamin D metabolites for DBP binding (Bouillon et al., 1992; Ena et al., 1989), suggesting that the different fatty acids alter the configuration of DBP, which influences binding of the vitamin D metabolites, rather than directly competing with the vitamin D metabolites for their binding site. The role of DBP in fatty acid transport is not clear given its relatively minor participation in fatty acid transport in blood.

Formation of vitamin D-binding protein—macrophage-activating factor and its functions

DBP—MAF is formed from certain alleles of DBP by removal of galactose by membrane-bound β-galactosidase induced in B cells during inflammation by lysophosphatidylcholine, followed by removal of sialic acid by membrane-bound sialidase on T cells (Yamamoto and Homma, 1991). These deglycosylation steps are required for its role in macrophage activation (Uto et al., 2012), but further removal of the *N*-acetylgalactosamine (NAcgal) reduces this activity (Yamamoto et al., 1991). Swamy et al. (Swamy et al., 2001) demonstrated the ability of DBP—MAF to activate osteoclasts, and this activity was independent of its 25(OH)D-binding function. The presence of the NAcgal moiety on DBP—MAF is essential for its ability to activate osteoclasts. DBP—MAF has been shown to stimulate bone resorption in two models of osteopetrosis (Schneider et al., 1995): the osteopetrosis (OP) and the incisor absent (IA) rat. The OP rat lacks osteoclasts, whereas the IA rat has nonfunctional osteoclasts. When given DBP—MAF the OP rats showed increased numbers of osteoclasts, whereas the IA rats showed improved osteoclast function. DBP—MAF has shown efficacy in a number of tumor models. The mouse SCCVII tumor model of squamous cell carcinoma was used to demonstrate that DBP—MAF enhanced the curative effect of photodynamic treatment (Korbelik et al., 1997). The survival time of mice bearing Ehrlich ascites tumors was prolonged by DBP—MAF (Koga et al., 1999). Kister et al. (Kisker et al., 2003) showed that DBP—MAF had antiangiogenic activity and inhibited the growth of pancreatic cancer in SCID mice, which on histologic exam also showed an increase in macrophages with decreased vascularity. Removal of NAcgal by α-NAcgalase blocks DBP—MAF formation, contributing to the loss of immunosuppression in cancer patients (Yamamoto et al., 1996). α-NAcgalase is produced in the liver, and appears to be directly related to tumor burden (Yamamoto et al., 1997). Preparations of DBP—MAF may have therapeutic potential (Nagasawa et al., 2005).

Intracellular trafficking of vitamin D metabolites: role of heat shock protein 70 and hnRNPC1/C2

The role of DBP is to get the vitamin D metabolites to and, for some cells, into the cell. However, a protein originally called intracellular DBP, identified subsequently as Hsp70 (Gacad et al., 1997), was found to promote the intracellular uptake and distribution of the vitamin D metabolites within the cell (Wu et al., 2000). This observation was originally made in New World monkeys, which were resistant to $1,25(OH)_2D$ (Gacad and Adams, 1993), but expressed a lot of this protein compared with Old World monkeys. Overexpression of this protein in cells from Old World monkeys increased the amount of 25(OH)D in these cells, promoted synthesis of $1,25(OH)_2D$, and increased the genomic actions of $1,25(OH)_2D$, such as the induction of Cyp24A1 (Adams et al., 2004). Antisense constructs to Hsp70 blocked these actions. The concept is that Hsp70 functions to transport 25(OH)D to the mitochondria where Cyp27B1 (the 1-hydroxylase) is located, and then transport the product of Cyp27B1, $1,25(OH)_2D$, to the nucleus where it binds the VDR for its genomic actions. The affinities of these proteins favor this trafficking, with higher affinity of Hsp70 for 25(OH)D than DBP and lower affinity of Hsp70 for $1,25(OH)_2D$ than VDR. However, there is an additional intracellular protein that was discovered in New World monkeys that better explains their hormone resistance (Arbelle et al., 1996), and that was subsequently found in a human with vitamin D resistance (Hewison et al., 1993; Chen et al., 2003). This protein, originally called VDRE-BP, is now known to be hnRPC1/C2 protein (Chen et al., 2006). Its overexpression blocks $1,25(OH)_2D$ transcriptional activity. This protein normally sits on the VDRE (vitamin D response element), but is displaced by $1,25(OH)_2D$, allowing the liganded VDR to initiate transcription (Chen et al., 2006).

The vitamin D receptor

The VDR was first discovered as a chromatin-associated $1,25(OH)_2D$-binding protein in 1974 (Brumbaugh & Haussler, 1974, 1975). Although initially identified in the intestine, the site of the best studied actions of vitamin D at the time and the tissue from which it was originally cloned and sequenced (McDonnell et al., 1987), the VDR has subsequently been found in essentially all tissues in which it has been sought (Stumpf et al., 1979; Berger et al., 1988). Even in tissues such as the liver and muscle, low levels of the VDR have been identified. In the liver VDR is primarily expressed in the stellate cells, and not in the more abundant parenchymal cells (Ding et al., 2013). In skeletal muscle the VDR is more highly expressed during development such that in adult tissue the levels are quite low, making it difficult to detect by standard methods (Girgis et al., 2014). Moreover, vitamin D has an impact on many cellular processes. In a 2015 ontology analysis 11,031 putative VDR target genes were identified, of which 43% were involved with metabolism, 19% with cell and tissue morphology, 10% with cell junction and adhesion, 10% with differentiation and development, 9% with angiogenesis, and 5% with epithelial-to-mesenchymal transition (Saccone et al., 2015). Furthermore, VDR can regulate various microRNAs and long noncoding RNAs, regulating the expression of numerous other proteins indirectly (Khanim et al., 2004; Wang et al., 2009; Jiang and Bikle, 2014). As a result of the appreciation that the VDR has such universal distribution and that vitamin D-metabolizing enzymes such as Cyp27B1 (that produces the active metabolite $1,25(OH)_2D$) and Cyp24A1 (that catabolizes 25(OH)D and $1,25(OH)_2D$ to less active forms) are also widely distributed (Bikle, 2014), interest in understanding the role of vitamin D and the VDR in tissues not obviously participating in the classic target tissues regulating calcium and phosphate homeostasis has been substantial. Over 4000 publications regarding vitamin D have appeared each year since 2014, most focused on the nonclassical actions of vitamin D. The remarkable aspect of this widespread distribution of the VDR is how many and yet how specific the actions of VDR are for any given cell. Although we have clues on how this specificity comes about, and these clues will be discussed in this chapter, we have much more to discover. Moreover, although most actions of VDR involve its role as a transcription factor within the nucleus (Carlberg, 2017; Pike et al., 2017), the VDR has also been shown to have nongenomic actions via its location in the membrane (Haussler et al., 2011) and perhaps even in mitochondria (Silvagno and Pescarmona, 2017). In this chapter I will not be dealing with actions of VDR signaling in specific tissues except as they illustrate more general points regarding VDR function, as tissue-specific aspects of VDR function have been well covered in several reviews (Bikle, 2014, 2016; Christakos et al., 2016).

Genomic location, protein structure, and regulation

The human *VDR* is located on chromosome 12. It spans 75 kb of DNA, and comprises nine exons, although the 5′ end is complex, with a number of potential minor TSSs 5′ of the major start site immediately upstream of exon 1a, resulting in alternatively spliced forms (Baker et al., 1988; Crofts et al., 1998; Sunn et al., 2001). Exon 2 contains the translation start site and the nucleotide sequence encoding the short A/B domain (24 aa), to which transcription factors such as TFIIB bind, and the first zinc finger of the DNA-binding domain (DBD) (65 aa). Exon 3 encodes the second zinc finger of the DBD. Exons 4–6 encode the hinge region (143 aa). Exons 7–9 encode part of the hinge region, the entire ligand-binding domain (LBD; E/F) (195 aa) including the AF2 domain, and the extensive 3′ untranslated region (UTR) (Fig. 29.4). A polymorphism (FokI restriction site) is found in the human *VDR*, changing an ATG to an ACG and shifting the translation start site by 3 aa, shifting the total length from 427 to 424 aa. The shorter form may be more transcriptionally active (Jurutka et al., 2000). Numerous other polymorphisms have been described in untranslated regions of the *VDR* gene, to which associations with various diseases have been made, but generally such studies suffer from small numbers and/or lack of reproducibility (Iqbal and Khan, 2017; Bizzaro et al., 2017; Rai et al., 2017) and will not be discussed further. Amino acids 49 and 50 between the two zinc fingers and 102–104 C terminal to the zinc fingers appear to provide the nuclear localization signal (Hsieh et al., 1998; Luo et al., 1994).

The DBD is composed of two zinc fingers held in a tetrahedral configuration by four cysteine residues. The first zinc finger directs DNA binding in the major groove of the DNA binding site. The second zinc finger provides a dimerization interface for its primary partner RXR. A short C-terminal extension from the second zinc finger participates in this dimerization interface (Wan et al., 2015a). The LBD is composed of 12 α helices (H1–H12) as shown in the original crystal structures (Rochel et al., 2000). The terminal H12 provides the essence of the mousetrap model for binding of

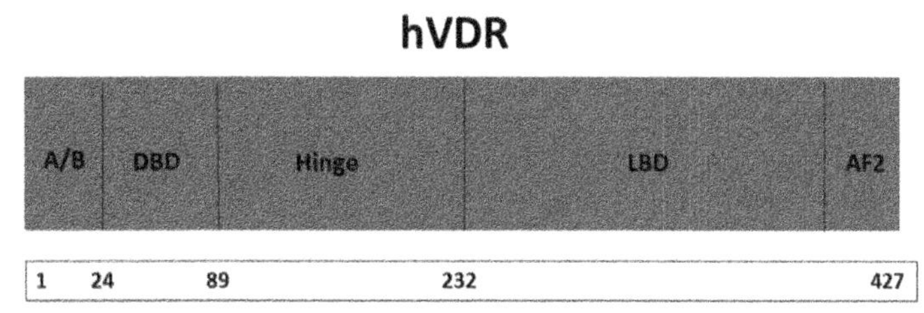

FIGURE 29.4 The structure of the human vitamin D receptor. The major domains of the human vitamin D receptor (*hVDR*) are depicted. The N-terminal A/B or AF2 domain is short in the VDR. It plays a role in TFII binding. The DNA-binding domain (*DBD*) comes next and is the site of the two zinc fingers that directly bind to DNA. The second zinc finger participates in binding to RXR, the major heterodimer partner for VDR. The hinge region also participates in RXR binding. The ligand-binding domain (*LBD*) is the site of binding for $1,25(OH)_2D$ and other ligands, as well as corepressors. When VDR binds to its ligand, AF2 on the C-terminal end undergoes a major conformational change to both enclose the ligand and provide docking for coactivators such as steroid receptor coactivators and Med1 of the Mediator complex.

$1,25(OH)_2D$ to the VDR in that in the nonliganded VDR H12 is open, but with ligand binding H12 moves over the ligand, enclosing it in the ligand-binding pocket (Rochel et al., 2000). This results in very tight binding (K_d approximately 10^{-10} M) (Mellon and DeLuca, 1979). In the closed position H12 along with H3 and H4 provides the interface for coactivators such as the steroid receptor coactivators (SRCs) and mediator complexes, which bind to H12 via their NR boxes containing the LxxLL sequence of amino acids (L for leucine, x any amino acid) (McInerney et al., 1998). The ligand-binding pocket is actually quite large and accommodates a number of ligands, including lithocholic acid. $1,25(OH)_2D$ fits into this pocket in an extended configuration, with its A ring in the β-chair configuration and the 1α group in an equatorial position (6-S *trans* conformation). Thirty-six residues line the ligand pocket, of which six make direct hydrogen bonds with the ligand (Wan et al., 2015b). H9 and H10 along with the second zinc finger provide the heterodimerization site for RXR (Whitfield et al., 1996). The structure of the full human RXR/VDR heterodimer complex with its DR3 DNA binding site has been revealed by cryo-electron microscopy (Orlov et al., 2012). This structure confirms the binding of the DBDs of both RXR and VDR to its major binding site on DNA (a direct repeat element with two hexanucleotide DNA sequences separated by a three-nucleotide spacer) with the RXR DBD binding to the 5′ half-site and the VDR to the 3′ half-site. Their LBDs project perpendicular to the DNA in an open confirmation, enabling binding by coregulators. The membrane VDR accepts a different configuration (6-S *cis* conformation), which led Mizwicki et al. (Mizwicki et al., 2004) to propose an alternative pocket for ligand binding, but this structure has not been confirmed by crystallography.

Amino acid residues critical for the structure/function of the VDR have been well demonstrated by mutations in the VDR in either the human or the mouse gene (reviewed in Malloy and Feldman, 2012). Approximately 45 different mutations in 100 patients have been described as of this writing. Mutations in the DBD block all functions of the VDR. Mutations at aa 274 in the LBD alter the affinity of the VDR for $1,25(OH)_2D$ by affecting the amino acid forming a hydrogen bond with the 1α group. Similarly, mutation at aa 305 alters the hydrogen bond with the 25(OH) group, likewise reducing affinity for the ligand. Mutation of 391 affects the dimerization domain at H9 and H10, reducing transactivation. Mutations of aa 420 in H12 alter coactivator binding. Mutations in the DBD and dimerization domain lead to alopecia as well as rickets, whereas those in the LBD that affect only the affinity for $1,25(OH)_2D$ or the coactivators result in rickets, but not alopecia, indicating that ligand binding is not essential for hair follicle cycling.

The regulation of VDR expression is quite cell specific. For example, $1,25(OH)_2D$ autoregulates VDR expression in bone cells but not in the intestine (Lee et al., 2014; Wood et al., 1998). Moreover, ligand binding to VDR stabilizes it, increasing its levels (Wiese et al., 1992). Many factors, including $1,25(OH)_2D$ itself, have been shown to regulate VDR expression. These include growth factors such as fibroblast growth factor, epidermal growth factor, insulin-like growth factor, and insulin, as well as PTH, glucocorticoids, estrogen, and retinoic acid. A variety of transcription factors have been identified that may mediate their regulation of VDR expression, including activator protein-1 (AP-1), specificity protein-1, CCAAT/enhancer binding protein (C/EBP), and caudal-type homeobox 2 (CDX2). For example, several enhancer elements have been found in the *VDR* gene that bind VDR. These enhancer elements are associated with binding sites for other transcription factors such as C/EBPβ, Runx2, cyclic AMP response element binding protein (CREB), retinoic acid receptor (RAR), and glucocorticoid receptor (Zella et al., 2010). Similarly, calcium upregulates VDR expression in the parathyroid gland presumably through its calcium-sensing receptor (Canadillas et al., 2010). On the other hand, SNAIL1 and SNAIL2 (SLUG) downregulate VDR expression in a number of cancer cell lines (Palmer et al., 2004; Mittal et al., 2008). MicroRNAs have also been shown to regulate VDR levels. miR-125b binds to the 3′ UTR to decrease VDR levels (Mohri et al., 2009; Gu et al., 2011), in the latter study in response to UVB treatment of psoriatic epidermis (Gu et al., 2011). Similarly miR-298 and miR-27b have been shown to bind to the 3′ UTR of VDR to reduce its expression (Pan et al., 2009). In addition the *VDR* promoter may also be hypermethylated by various methylases and methyltransferases, reducing its expression (Chandel et al., 2013).

Vitamin D receptor mechanism of action: genomic

Vitamin D binding sites in the genome

Combining data from chromatin immunoprecipitation sequencing studies of the entire genome from several different cell lines, Carlberg (Carlberg, 2017) reported that the human genome contained over 23,000 VDR binding sites, 70% of which occurred in only one cell type (Tuoresmaki et al., 2014). Only a small percentage of these binding sites can be clearly identified with a target gene (Ramagopalan et al., 2010). Moreover, within a given genome, the number of binding sites varied with the duration of exposure to ligand. Most of these binding sites were ligand dependent, but a substantial number of sites did not require $1,25(OH)_2D$ for VDR binding to occur, although the addition of ligand altered the location of these sites (Heikkinen et al., 2011). For example, in LS180 colon cancer cells 262 VDR binding sites were identified without ligand, but following 1,25(OH)$_2$D 2209 binding sites were found, 71% in intergenic regions, 27% in introns, and only a small number near the TSS (Meyer et al., 2012). Similarly the addition of 1,25(OH)$_2$D also markedly enhances the number of RXR binding sites, a substantial percentage of which colocalized with the VDR binding sites (Meyer et al., 2010). Although the DR3 was the most common binding sequence, it occurred in only a minority of binding sites, albeit the ones with the highest affinity for VDR. An ER9 VDRE (everted repeats with nine spacing nucleotides) has been documented in a number of genes (Schrader et al., 1995, 1997), and a DR6, which incorporated RAR rather than RXR as the heterodimer partner with VDR, was identified in the phospholipase Cγ (PLCγ) promoter (Xie and Bikle, 1997). However, no obvious consensus DNA sequences have emerged for most other binding sites. The DR3, as noted earlier, refers to a DNA sequence with two hexanucleotide half-sites separated by a three-nucleotide spacer. Other nuclear hormone receptors have a similar DNA binding site, but the half-sites are separated by different numbers of nucleotides (Kliewer et al., 1992; Perlmann et al., 1993). The consensus sequence of the half sites is A/G G G/T T C/G A.

The VDR binding sites can be thousands of bases away from the TSS of the genes they regulate, and genes generally have multiple VDR binding sites, the activity of which may vary in different cells and different species. The likely explanation is that the DNA can be looped to bring the relevant regulatory elements adjacent to the TSS. This function is performed by CCCTC-binding factors (CTCFs), which define genomic insulator regions, although not all CTCF sites are involved in insulator actions. Genomic loops of hundreds to thousands of bases can be looped into topologically associating domains (TADs), organizing the genome into several thousand such domains (Dixon et al., 2012). In human monocytes 1,25(OH)$_2$D stimulated CTCF binding to approximately 1300 sites, 50% of which marked the anchors for TADs containing at least one VDR binding site and a 1,25(OH)$_2$D target gene (Neme et al., 2016). An informative example of how this might work in different cells is the regulation of the receptor activator of NF-κB ligand (RANKL) gene (*Tnfsf11*) (Fig. 29.5). This gene is regulated by PTH and 1,25(OH)$_2$D in osteoblasts, but by activators of c-fos in activated T cells. The Pike laboratory identified seven VDR binding sites in this gene up to 88 kb upstream of the TSS, of which the −75 kb site proved most active in the mouse gene (Kim et al., 2006; Martowicz et al., 2011), whereas the proximal site was most active in the human gene (Nerenz et al., 2008). However, in activated T cells, three additional sites even farther

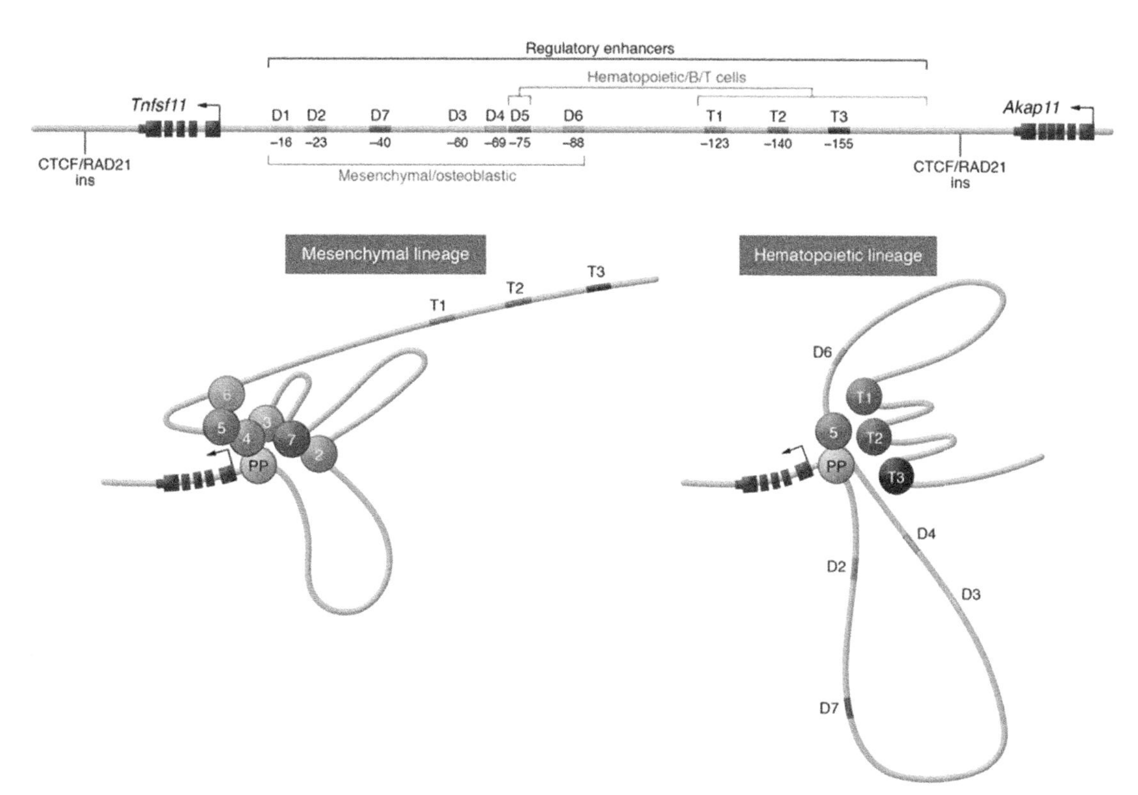

FIGURE 29.5 Theoretical schematic of the *Tnfsf11* (RANKL) gene as it might be configured in osteoblasts or T cells. These two configurations of the gene bring different response elements to the transcription site by alternate forms of looping, illustrating the potential for different means of regulating the same gene in different cellular contexts. The CCCTC-binding factor (*CTCF*) sites demarcate the gene. *Reproduced from Pike et al. Fig. 29.2, J Clin Invest 127:1148, 2017 permission pending.*

upstream of the TSS have been identified as sites of RANKL induction by c-fos (Bishop et al., 2011). CTCF binding sites separate the RANKL gene from adjacent genes, but also separate the VDR binding sites from the more upstream c-fos binding site active in T cells, potentially enabling different looping in these two types of cells (Pike et al., 2017; Bishop et al., 2011). A similar example can be found for Cyp27B1, the enzyme that produces 1,25$(OH)_2$D. This gene is negatively regulated by its product in the kidney, as will be discussed later, but not in other tissues (Bikle, 2016b). Pike et al. demonstrated tissue-specific binding of VDR to response elements for this gene, showing differences between that in the kidney and that in other tissues (Meyer et al., 2017).

The VDR binding sites are generally situated in a region with other transcription factors that may share regulation of that gene. The RANKL gene is again a good example, as the −75 kb VDR binding region contains several CREB sites responsible for PTH regulation of this gene (Fu et al., 2006; Kim et al., 2007). Many genes regulated by 1,25$(OH)_2$D in bone cells also have Runx2 and C/EBPβ binding sites in the same regions as the VDR binding sites and influence VDR activity (Meyer et al., 2014). In intestinal cells the VDR binding sites have been colocalized with C/EBPβ, AP-1, CDX2, and T cell factor-4 (TCF4) (Meyer et al., 2012). Other transcription factors found in the vicinity of VDR binding sites that affect its transcriptional activity include ras-activated Ets transcription factors (Dwivedi et al., 2000) and YY1 (Guo et al., 1997; Raval-Pandya et al., 2001). Pu.1 has been hypothesized to serve as a pioneer factor preceding VDR binding, and has been found associated with a number of VDR binding sites in monocytes (Seuter et al., 2017). Changes in VDR binding sites as can occur during differentiation of the cell can affect the binding of these other transcription factors in similar fashion (Meyer et al., 2015, 2016). In osteoblasts such changes probably account for the ability of 1,25$(OH)_2$D to inhibit the early stages of osteoblast differentiation while promoting the later stages (Bikle, 2012a).

Coregulators and epigenetic changes regulating VDR function

The sites of active transcription are marked by epigenetic changes in both the gene itself and the histones that regulate access of the transcriptional machinery to the gene (reviewed in Long et al., 2015). For example, H3K4me3 (histone 3 in which lysine 4 is triply methylated) is found in sites of active transcription, whereas H3K9 is associated with silent promoter regions. H3K27me is associated with active enhancers, whereas H3K27me is associated with gene suppression. 1,25$(OH)_2$D can increase or decrease the expression of these methylases (Pereira et al., 2012). Sites of epigenetic gene suppression are also generally marked by methylation of cytosine in CpG islands (reviewed in Fetahu et al., 2014) in association with the epigenetic changes in the aforementioned histones (Esteller, 2008). The *VDR* promoter has three CpG islands in its promoter in exon 1a, which in breast cancer are hypermethylated, reducing VDR expression (Marik et al., 2010). Similar effects are seen in HIV-infected T cells (Chandel et al., 2013). 1,25$(OH)_2$D reduces the methylation of the E-cadherin promoter (Lopes et al., 2012), increasing its expression, but may promote methylation of other promoters (Doig et al., 2013). Methylases (amine oxidases or jumonji C domain-containing proteins JmJC) can be activators or suppressors. 1,25$(OH)_2$D regulates these epigenetic changes by affecting the binding of coregulators to the VDR, whether coactivators with histone acetyltransferase activity (HATs) or cosuppressors with histone deacetylase activity (HDACs).

There are over 250 published coregulators interacting with nuclear hormone receptors (McKenna et al., 2009). The best studied coactivators with respect to the VDR are the steroid hormone receptor coactivators (SRCs 1−3) (Xu and O'Malley, 2002) and the Mediator complex (Rachez and Freedman, 2001). Phosphorylation of serine 208 in the VDR following dimerization with RXR enhances their binding (Barletta et al., 2002; Arriagada et al., 2007) The SRCs have three NR boxes containing the LxxLL motif described earlier to which they bind the nuclear hormone receptors (Heery et al., 1997), but VDR binding is primarily to the third NR box (Teichert et al., 2009). Med1, the principal component of the Mediator complex binding to VDR and other nuclear hormone receptors, has two NR boxes, the second NR box being the one binding to VDR (Teichert et al., 2009). SRCs recruit HATs to the VDR, in particular CREB binding protein, CBP/p300 (Ogryzko et al., 1996) and the CBP/p300-associated factor p/CAF (Yang et al., 1996). The Med complex does not contain HAT activity but binds directly to RNA polymerase II to help form the preinitiation complex along with basal transcription factors such as TFIIB and several TAT-binding proteins (Yin and Wang, 2014). The peroxisome proliferator-activated γ coactivator-1α interacts with VDR to regulate genes involved in muscle development and energy metabolism (Narvaez et al., 2009). The VDR also binds histone methyltransferases that can synergize with SRCs in terms of regulation (Koh et al., 2001). These coactivators all bind to the AF2 domain (H12). NCoA62/SKIP does not bind to the AF2 domain but binds to the VDR/RXR heterodimer to increase transcription with SRC, but could also coordinate with corepressors NCoR/SMRT to suppress transcription (Baudino et al., 1998; Leong et al., 2004). WINAC is a member of the SWI/SNF subfamily which is an ATP-dependent chromatin remodeling complex involved in DNA replication and transcriptional elongation (Kitagawa et al., 2003). This complex includes the Williams syndrome transcription factor (WSTF) that when mutated results in Williams syndrome, a developmental disorder manifesting as hypercalcemia and disordered vitamin D metabolism (Morris and Mervis, 2000). This complex plays a central role in the suppression by $1,25(OH)_2D$ of its own production by Cyp27B1, at least in the kidney as will be discussed further (Kim et al., 2007). Meningioma-1 is expressed in osteoblasts, where it enhances $1,25(OH)_2D$/VDR transcription (Sutton et al., 2005).

The best studied corepressors are the SMRT and NCoR complexes (Perissi et al., 1999; Tagami et al., 1998). They have HDAC activity. These corepressors bind to nuclear hormone receptors via their CoRNR boxes, analogous to the NR boxes of the SRC family but with the sequence LxxH/IxxxI/L (Hu and Lazar, 1999). These corepressors do not bind to the AF2 domain (H12) but to H3–H5 in the absence of ligand. In the presence of $1,25(OH)_2D$ and the conformational change with H12, these corepressors are displaced, enabling the coactivators to bind to their sites on H12. Hairless is a corepressor of VDR expressed primarily in the brain and skin (Xie et al., 2006). It binds to the central region of the LBD of VDR, as do NCoR/SMRT to two $\phi xx\phi\phi$ hydrophobic sites (ϕ = hydrophobic amino acid) (Hsieh et al., 2003), but also to H12 via its LxxLL motif, similar to SRC and Med1 (Teichert et al., 2009). The role of Hairless is complex in that it represses ligand-dependent VDR functions with respect to epidermal differentiation (Xie et al., 2006) but appears to be required for ligand-independent VDR regulation of hair follicle cycling (Wang et al., 2007). Alien binds to VDR and inhibits transcription by increasing HDAC activity and nucleosome assembly (Polly et al., 2000; Eckey et al., 2007). SUG1 is a proteasomal component that binds to the AF2 domain of VDR and may be involved in VDR degradation (Masuyama and MacDonald, 1998).

Negative vitamin D response elements

The general (and oversimplified) cycling model for the mechanism by which these coregulators control VDR transcription is that $1,25(OH)_2D$, by enhancing the conformational change in VDR, enhances its dimerization to RXR, strengthening the affinity of the complex for specific sites (e.g., DR3) in the DNA, displacing corepressors, and enabling the binding of the VDR/RXR heterodimer to coactivators. SRCs with their HAT activity are hypothesized to be the first to bind to the VDR, enzymatically altering the histones to open up the site for access to transcription factors. Med1 binding follows, which, by recruiting RNA polymerase II to the gene, facilitates the transcription process (Kim et al., 2005). This cycling model (Carlberg, 2010) does not explain situations such as the differentiation process in the epidermis where Med1 predominates in the early stages of differentiation, whereas SRCs 2 and 3 predominate in the more differentiated keratinocytes, each regulating a different profile of genes (Bikle, 2012b; Oda et al., 2003, 2009; Hawker et al., 2007). Moreover, this model does not explain the role of $1,25(OH)_2D$/VDR as an inhibitor of transcription. In microarray analyses genes downregulated by $1,25(OH)_2D$ are nearly as numerous as upregulated genes (White, 2004), but the mechanisms by which such inhibition takes place are unclear, as the VDR binding sites generally demonstrate no clear nucleotide sequence and may not involve direct binding of the VDR to the DNA. In the PTH and PTH-related protein genes inhibition has been reported to involve either half-sites of a DR3 (Demay et al., 1992) or a full DR3-like sequence (Falzon, 1996), but may also involve the binding of VDR to a vitamin D-interacting repressor (VDIR) that binds to DNA at E-box-like motifs (CANNTG, where N can be any nucleotide) (Kim et al., 2007; Murayama et al., 2004). The Kato lab (Kitagawa et al., 2003; Kato et al., 2004) has studied how $1,25(OH)_2D$ represses its own production in kidney cells through this mechanism. Such suppression appears to be restricted to kidney cells as Cyp27B1 expression in other cells does not show this suppression by $1,25(OH)_2D$ (Xie et al., 2002)). Their studies led them to propose that WSTF in the WINAC complex recruits the ligand-bound VDR to the VDIR sitting on the E-box-like negative VDRE, recruiting corepressors with HDAC activity as well as DNA methyltransferases and a methyl-CpG-binding protein (MBD4) that silences the Cyp27B1 promoter (Kim et al., 2009). PTH stimulated the phosphorylation of MBD4, which promoted demethylation by the glycosylase activity of MBD4, relieving the suppression, thus activating gene expression.

Interaction of VDR with β-catenin signaling

The interaction between VDR and β-catenin in terms of their complementary and opposing transcriptional activities has been well studied in the skin, bone, and intestine. β-Catenin binds to VDR in the AF2 domain (Shah et al., 2006) in a $1,25(OH)_2D$-dependent fashion. Such binding can promote ligand-dependent VDR transcriptional activity and promote cell fate determination (Shah et al., 2006; Palmer et al., 2008), but can also result in inhibition of β-catenin's transcriptional activity with respect to cell proliferation by preventing its interaction with T cell factors/lymphoid enhancer factors (TCFs/LEFs) in the nucleus and increasing its binding to the E-cadherin complex in the membrane (Palmer et al., 2001). VDR and calcium are essential for the formation of the E-cadherin complex (Bikle et al., 2012; Bikle, 2011). Palmer et al. (Palmer et al., 2008) evaluated the interaction between VDR and β-catenin in transcriptional regulation, and identified putative response elements for VDR and β-catenin/LEF in a number of genes, including *Shh*, *Ptch1* and *Ptch2*, and *Gli1* and *Gli2*, members of the Hedgehog signaling pathway. For some genes the interaction is positive. An example of this was shown in TE-85 osteosarcoma cells in a ligand-independent fashion (Haussler et al., 2010). For other genes such as c-myc (where the region showing colocalization of VDR/RXR and β-cateninTCF4 is 335 kb away from the TSS) (Meyer et al., 2012) and cyclin D, 1,25$(OH)_2$D/VDR suppresses β-catenin transcriptional activity (Sancho et al., 2004; Salehi-Tabar et al., 2012). The type of β-catenin transcriptional complex also plays a role in that β-catenin/LEF1 increases VDR induction of osteopontin, whereas β-catenin/TCF4 reduces the induction (Xu et al., 2009). Furthermore, the ability of β-catenin overexpression to induce trichofolliculomas (a benign tumor of the hair follicle) was blocked by an analog of 1,25$(OH)_2$D, but in the absence of VDR, basal cell carcinomas were induced rather than trichofolliculomas. In the hair follicle and intestinal crypt cells, β-catenin signaling is essential for stem cell activation and their proliferation through a process that, at least in hair follicle stem cells, is dependent on the presence of the VDR (Sancho et al., 2004; Lisse et al., 2014; Cianferotti et al., 2007). In the epidermis β-catenin plays a critical role in cell migration and differentiation during wound repair as part of the E-cadherin/catenin complex in the membrane, which as noted is regulated by vitamin D signaling.

Vitamin D receptor mechanism of action: nongenomic

1,25$(OH)_2$D exerts rapid effects on cells that cannot readily be explained by its better understood genomic actions. The first description of such rapid actions was that of rapid increases in intestinal calcium transport following 1,25$(OH)_2$D administration, called transcaltachia (Nemere et al., 1984). However, these rapid actions have subsequently been identified in a number of other cells, including chondrocytes, osteoblasts, keratinocytes, and fibroblasts as well as intestinal cells. A variety of signaling pathways are regulated by these rapid effects of 1,25$(OH)_2$D, including PLC, phospholipase A_2 ,the tyrosine kinase Src, phosphatidylinositol-3-kinase (PI3K), Wnt5a, and p21ras, opening up calcium and chloride channels, with the generation of second messengers such as calcium, cyclic AMP, fatty acids, phosphatidylinositol 3,4,5-triphosphate activating downstream protein kinases A and C, src, mitogen-activated protein kinases, and calmodulin kinase II (reviewed in Hii and Ferrante, 2016; Doroudi et al., 2015). Two receptors for 1,25$(OH)_2$D, both in the membrane, have been identified as mediating these rapid actions. The first receptor isolated from the chick intestinal membrane was initially called MARRS-binding protein (Nemere et al., 1994), which was subsequently shown to be the same protein as GRP58, ERp57 or 60, and Pdia3 (Khanal and Nemere, 2007). The second receptor is the VDR itself, but in its membrane location it is activated by 1,25$(OH)_2$D analogs with a different configuration (6-S *cis*) compared with those that activate the genomic actions of VDR (6-S *trans*), as mentioned earlier (Dormanen et al., 1994; Norman, 2006). In osteoblasts and epidermal fibroblasts both receptors have been shown to be involved in selected nongenomic actions of 1,25$(OH)_2$D (Chen et al., 2013; Sequeira et al., 2012). Moreover, these receptors can be found together with caveolin-1 in caveolae (Chen et al., 2013), where at least in skin fibroblasts they physically interact (Sequeira et al., 2012). Caveolin can provide a scaffold for other signaling molecules with which these nongenomic receptors may interact (Doroudi et al., 2015).

The interaction between the membrane VDR and the nuclear VDR has been suggested, but the mechanism is not totally clear. In kidney cells and the intestinal cell line Caco2 the rapid activation of extracellular signal-regulated kinase 1/2 (ERK1/2), JNK1/2, PKC, PI3K, and p21ras all has been shown to regulate 1,25$(OH)_2$D induction of Cyp24A1, a genomic action of VDR (Dwivedi et al., 2002; Nutchey et al., 2005; Cui et al., 2009). These studies indicated that the rapid activation of ERK1/2 and ERK5 affected 1,25$(OH)_2$D induction of Cyp24A1 by phosphorylating RXRα and Ets-1, respectively, thus modulating the environment of the VDREs in the Cyp24A1 promoter (Dwivedi et al., 2002). In Caco2 cells ERK1/2 activation did not alter VDR binding to its VDRE in the Cyp24A1 promoter but enhanced its transcription by helping recruit the coregulator Med1. On the other hand ERK1/2 activation did not influence another 1,25$(OH)_2$D-induced gene, TRPV6, and did not alter the recruitment of Med1 to the VDR in the TRPV6 promoter (Cui et al., 2009). In human embryonic kidney 293T cells, JNK but not ERK1/2 activation facilitated 1,25$(OH)_2$D induction of Cyp24A1 (Nutchey et al., 2005). Rapid activation of PI3K by 1,25$(OH)_2$D also appears to enhance 1,25$(OH)_2$D induction of Cyp24A1 by mediating 1,25$(OH)_2$D activation of PKCζ and its phosphorylation of SP1, influencing its binding to the GC box in the Cyp24A1 promoter region (Dwivedi et al., 2010). In keratinocytes, activation of ERK1/2 promotes the ability of liganded VDR to induce the antimicrobial peptide cathelicidin (Peric et al., 2009).

VDR also alters the actions of other transcription factors in a nongenomic fashion. For example, the liganded VDR interacts directly with IκB kinase to block NF-κB signaling (Chen et al., 2013), and the liganded MARRS/ERp57/PIA3 has been shown to interact directly with NF-κB, with translocation of the complex into the nucleus where it appears to participate in the differentiation of NB4 leukemia cells (Wu et al., 2010). In T cells VDR/RXR prevents the formation of the NFATc1/AP-1 complex, suppressing the induction of IL-2 and IL-17A (reviewed in Wei and Christakos, 2015).VDR in the absence of 1,25$(OH)_2$D bound Stat1, inhibiting its action (Lange et al., 2014). 1,25$(OH)_2$D plus interferon-α (IFN-α) released this binding, enabling the binding of phosphorylated Stat1 to IFN-α DNA target genes, increasing their expression

(Lange et al., 2014). In some cancer cells nonliganded VDR can block arsenite-induced cell death by binding to and blocking c-jun (Li et al., 2007).

Conclusions

In this chapter I have reviewed three vitamin D-binding proteins: DBP, the protein that transports the vitamin D metabolites between and, in some cases, into cells; the intracellular DBP, otherwise known as Hsp70, that ferries the vitamin D metabolites between the relevant intracellular organelles; and VDR, which mediates the actions of the active vitamin D metabolite, $1,25(OH)_2D$, both in the nucleus for its genomic actions and in the membrane for its nongenomic actions. However, these proteins serve other functions that are not dependent on their binding to the vitamin D metabolites. DBP is an important actin scavenger, binds fatty acids, fixes C5a to neutrophils regulating chemotaxis, and when deglycosylated forms the DBP–MAF, which plays a role in osteoclast activation, macrophage function, and anticancer activity. Hsp70 serves as a chaperone role for other proteins. VDR has functions both genomic and nongenomic that do not require and may be inhibited by its binding to $1,25(OH)_2D$. Regulation of hair follicle cycling is one clear example of this ligand-independent action. Although we have learned much about the structure and function of these vitamin D-binding proteins, there is much more to learn. That said, the huge surge in vitamin D-related publications since 2008 indicates that research in this area will continue to produce new knowledge by which the roles of these proteins in both physiology and pathophysiology will be better understood.

References

Abbas, S., Linseisen, J., Slanger, T., Kropp, S., Mutschelknauss, E.J., Flesch-Janys, D., et al., 2008. The Gc2 allele of the vitamin D binding protein is associated with a decreased postmenopausal breast cancer risk, independent of the vitamin D status. Cancer Epidemiol. Biomark. Prev. 17 (6), 1339–1343.

Adams, J.S., Chen, H., Chun, R., Gacad, M.A., Encinas, C., Ren, S., et al., 2004. Response element binding proteins and intracellular vitamin D binding proteins: novel regulators of vitamin D trafficking, action and metabolism. J. Steroid Biochem. Mol. Biol. 89–90 (1–5), 461–465.

Aggarwal, A., Yadav, A.K., Ramachandran, R., Kumar, V., Kumar, V., Sachdeva, N., et al., 2016. Bioavailable vitamin D levels are reduced and correlate with bone mineral density and markers of mineral metabolism in adults with nephrotic syndrome. Nephrology 21 (6), 483–489.

Ahn, J., Albanes, D., Berndt, S.I., Peters, U., Chatterjee, N., Freedman, N.D., et al., 2009. Vitamin D-related genes, serum vitamin D concentrations and prostate cancer risk. Carcinogenesis 30 (5), 769–776.

Aloia, J., Dhaliwal, R., Mikhail, M., Shieh, A., Stolberg, A., Ragolia, L., et al., 2015. Free 25(OH)D and calcium absorption, PTH, and markers of bone turnover. J. Clin. Endocrinol. Metab. 100 (11), 4140–4145.

Altinova, A.E., Ozkan, C., Akturk, M., Gulbahar, O., Yalcin, M., Cakir, N., et al., 2016. Vitamin D-binding protein and free vitamin D concentrations in acromegaly. Endocrine 52 (2), 374–379.

Alzaman, N.S., Dawson-Hughes, B., Nelson, J., D'Alessio, D., Pittas, A.G., 2016. Vitamin D status of black and white Americans and changes in vitamin D metabolites after varied doses of vitamin D supplementation. Am. J. Clin. Nutr. 104 (1), 205–214.

Arbelle, J.E., Chen, H., Gaead, M.A., Allegretto, E.A., Pike, J.W., Adams, J.S., 1996. Inhibition of vitamin D receptor-retinoid X receptor-vitamin D response element complex formation by nuclear extracts of vitamin D-resistant New World primate cells. Endocrinology 137 (2), 786–789.

Armas, L.A., Hollis, B.W., Heaney, R.P., 2004. Vitamin D2 is much less effective than vitamin D3 in humans. J. Clin. Endocrinol. Metab. 89 (11), 5387–5391.

Arnaud, J., Constans, J., 1993. Affinity differences for vitamin D metabolites associated with the genetic isoforms of the human serum carrier protein (DBP). Hum. Genet. 92 (2), 183–188.

Arriagada, G., Paredes, R., Olate, J., van Wijnen, A., Lian, J.B., Stein, G.S., et al., 2007. Phosphorylation at serine 208 of the 1alpha,25-dihydroxy Vitamin D3 receptor modulates the interaction with transcriptional coactivators. J. Steroid Biochem. Mol. Biol. 103 (3–5), 425–429.

Baier, L.J., Dobberfuhl, A.M., Pratley, R.E., Hanson, R.L., Bogardus, C., 1998. Variations in the vitamin D-binding protein (Gc locus) are associated with oral glucose tolerance in nondiabetic Pima Indians. J. Clin. Endocrinol. Metab. 83 (8), 2993–2996.

Baker, A.R., McDonnell, D.P., Hughes, M., Crisp, T.M., Mangelsdorf, D.J., Haussler, M.R., et al., 1988. Cloning and expression of full-length cDNA encoding human vitamin D receptor. Proc. Natl. Acad. Sci. U.S.A. 85 (10), 3294–3298.

Barletta, F., Freedman, L.P., Christakos, S., 2002. Enhancement of VDR-mediated transcription by phosphorylation: correlation with increased interaction between the VDR and DRIP205, a subunit of the VDR-interacting protein coactivator complex. Mol. Endocrinol. 16 (2), 301–314.

Baudino, T.A., Kraichely, D.M., Jefcoat Jr., S.C., Winchester, S.K., Partridge, N.C., MacDonald, P.N., 1998. Isolation and characterization of a novel coactivator protein, NCoA-62, involved in vitamin D-mediated transcription. J. Biol. Chem. 273 (26), 16434–16441.

Berger, U., Wilson, P., McClelland, R.A., Colston, K., Haussler, M.R., Pike, J.W., et al., 1988. Immunocytochemical detection of 1,25-dihydroxyvitamin D receptors in normal human tissues. J. Clin. Endocrinol. Metab. 67 (3), 607–613.

Bikle, D.D., Gee, E., 1989. Free, and not total, 1,25-dihydroxyvitamin D regulates 25-hydroxyvitamin D metabolism by keratinocytes. Endocrinology 124 (2), 649–654.

Bikle, D.D., Gee, E., Halloran, B., Haddad, J.G., 1984. Free 1,25-dihydroxyvitamin D levels in serum from normal subjects, pregnant subjects, and subjects with liver disease. J. Clin. Investig. 74 (6), 1966–1971.

Bikle, D.D., Siiteri, P.K., Ryzen, E., Haddad, J.G., 1985. Serum protein binding of 1,25-dihydroxyvitamin D: a reevaluation by direct measurement of free metabolite levels. J. Clin. Endocrinol. Metab. 61 (5), 969–975.

Bikle, D.D., Halloran, B.P., Gee, E., Ryzen, E., Haddad, J.G., 1986a. Free 25-hydroxyvitamin D levels are normal in subjects with liver disease and reduced total 25-hydroxyvitamin D levels. J. Clin. Investig. 78 (3), 748–752.

Bikle, D.D., Gee, E., Halloran, B., Kowalski, M.A., Ryzen, E., Haddad, J.G., 1986b. Assessment of the free fraction of 25-hydroxyvitamin D in serum and its regulation by albumin and the vitamin D-binding protein. J. Clin. Endocrinol. Metab. 63 (4), 954–959.

Bikle, D.D., Xie, Z., Tu, C.L., 2012. Calcium regulation of keratinocyte differentiation. Expert Rev. Endocrinol. Metabol. 7 (4), 461–472.

Bikle, D.D., Malmstroem, S., Schwartz, J., 2017a. Current controversies: are free vitamin metabolite levels a more accurate assessment of vitamin D status than total levels? Endocrinol Metab. Clin. N. Am. 46 (4), 901–918.

Bikle, D., Bouillon, R., Thadhani, R., Schoenmakers, I., 2017b. Vitamin D metabolites in captivity? Should we measure free or total 25(OH)D to assess vitamin D status? J. Steroid Biochem. Mol. Biol. 173, 105–116.

Bikle, D.D., 2011. The vitamin D receptor: a tumor suppressor in skin. Discov. Med. 11 (56), 7–17.

Bikle, D.D., 2012a. Vitamin D and bone. Curr. Osteoporos. Rep. 10 (2), 151–159.

Bikle, D.D., 2012b. Vitamin D and the skin: physiology and pathophysiology. Rev. Endocr. Metab. Disord. 13 (1), 3–19.

Bikle, D.D., 2014. Vitamin D metabolism, mechanism of action, and clinical applications. Chem. Biol. 21 (3), 319–329.

Bikle, D.D., 2016a. Extraskeletal actions of vitamin D. Ann. N. Y. Acad. Sci. 1376 (1), 29–52.

Bikle, D.D., 2016b. The endocrine society centennial: extrarenal production of 1,25 dihyroxyvitamin D is now proven. Endocrinology 157 (5), 1717–1718.

Binder, R., Kress, A., Kan, G., Herrmann, K., Kirschfink, M., 1999. Neutrophil priming by cytokines and vitamin D binding protein (Gc-globulin): impact on C5a-mediated chemotaxis, degranulation and respiratory burst. Mol. Immunol. 36 (13–14), 885–892.

Binkley, N., Carter, G.D., 2017. Toward clarity in clinical vitamin D status assessment: 25(OH)D assay standardization. Endocrinol Metab. Clin. N. Am. 46 (4), 885–899.

Bishop, K.A., Coy, H.M., Nerenz, R.D., Meyer, M.B., Pike, J.W., 2011. Mouse Rankl expression is regulated in T cells by c-Fos through a cluster of distal regulatory enhancers designated the T cell control region. J. Biol. Chem. 286 (23), 20880–20891.

Bizzaro, G., Antico, A., Fortunato, A., Bizzaro, N., 2017. Vitamin D and autoimmune diseases: is vitamin D receptor (VDR) polymorphism the culprit? Isr. Med. Assoc. J. 19 (7), 438–443.

Bouillon, R., van Baelen, H., de Moor, P., 1980. Comparative study of the affinity of the serum vitamin D-binding protein. J. Steroid Biochem. 13 (9), 1029–1034.

Bouillon, R., Xiang, D.Z., Convents, R., Van Baelen, H., 1992. Polyunsaturated fatty acids decrease the apparent affinity of vitamin D metabolites for human vitamin D-binding protein. J. Steroid Biochem. Mol. Biol. 42 (8), 855–861.

Boutin, B., Galbraith, R.M., Arnaud, P., 1989. Comparative affinity of the major genetic variants of human group-specific component (vitamin D-binding protein) for 25-(OH) vitamin D. J. Steroid Biochem. 32 (1A), 59–63.

Brumbaugh, P.F., Haussler, M.R., 1974. 1 Alpha,25-dihydroxycholecalciferol receptors in intestine. I. Association of 1 alpha,25-dihydroxycholecalciferol with intestinal mucosa chromatin. J. Biol. Chem. 249 (4), 1251–1257.

Brumbaugh, P.F., Haussler, M.R., 1975. Specific binding of 1alpha,25-dihydroxycholecalciferol to nuclear components of chick intestine. J. Biol. Chem. 250 (4), 1588–1594.

Calvo, M., Ena, J.M., 1989. Relations between vitamin D and fatty acid binding properties of vitamin D-binding protein. Biochem. Biophys. Res. Commun. 163 (1), 14–17.

Canadillas, S., Canalejo, R., Rodriguez-Ortiz, M.E., Martinez-Moreno, J.M., Estepa, J.C., Zafra, R., et al., 2010. Upregulation of parathyroid VDR expression by extracellular calcium is mediated by ERK1/2-MAPK signaling pathway. Am. J. Physiol. Renal. Physiol. 298 (5), F1197–F1204.

Carlberg, C., 2010. The impact of transcriptional cycling on gene regulation. Transcription 1 (1), 13–16.

Carlberg, C., 2017. Molecular endocrinology of vitamin D on the epigenome level. Mol. Cell. Endocrinol. 453, 14–21.

Carpenter, T.O., Zhang, J.H., Parra, E., Ellis, B.K., Simpson, C., Lee, W.M., et al., 2013. Vitamin D binding protein is a key determinant of 25-hydroxyvitamin D levels in infants and toddlers. J. Bone Miner. Res. 28 (1), 213–221.

Chandel, N., Husain, M., Goel, H., Salhan, D., Lan, X., Malhotra, A., et al., 2013. VDR hypermethylation and HIV-induced T cell loss. J. Leukoc. Biol. 93 (4), 623–631.

Chen, H., Hewison, M., Hu, B., Adams, J.S., 2003. Heterogeneous nuclear ribonucleoprotein (hnRNP) binding to hormone response elements: a cause of vitamin D resistance. Proc. Natl. Acad. Sci. U.S.A. 100 (10), 6109–6114.

Chen, H., Hewison, M., Adams, J.S., 2006. Functional characterization of heterogeneous nuclear ribonuclear protein C1/C2 in vitamin D resistance: a novel response element-binding protein. J. Biol. Chem. 281 (51), 39114–39120.

Chen, J., Doroudi, M., Cheung, J., Grozier, A.L., Schwartz, Z., Boyan, B.D., 2013. Plasma membrane Pdia3 and VDR interact to elicit rapid responses to 1alpha,25(OH)(2)D(3). Cell. Signal. 25 (12), 2362–2373.

Chen, Y., Zhang, J., Ge, X., Du, J., Deb, D.K., Li, Y.C., 2013. Vitamin D receptor inhibits nuclear factor kappaB activation by interacting with IkappaB kinase beta protein. J. Biol. Chem. 288 (27), 19450–19458.

Cheung, C.L., Lau, K.S., Sham, P.C., Tan, K.C., Kung, A.W., 2013. Genetic variant in vitamin D binding protein is associated with serum 25-hydroxyvitamin D and vitamin D insufficiency in southern Chinese. J. Hum. Genet. 58 (11), 749–751.

Chishimba, L., Thickett, D.R., Stockley, R.A., Wood, A.M., 2010. The vitamin D axis in the lung: a key role for vitamin D-binding protein. Thorax 65 (5), 456–462.

Christakos, S., Dhawan, P., Verstuyf, A., Verlinden, L., Carmeliet, G., 2016. Vitamin D: metabolism, molecular mechanism of action, and pleiotropic effects. Physiol. Rev. 96 (1), 365–408.

Christensen, E.I., Birn, H., 2002. Megalin and cubilin: multifunctional endocytic receptors. Nat. Rev. Mol. Cell Biol. 3 (4), 256–266.

Chun, R.F., Lauridsen, A.L., Suon, L., Zella, L.A., Pike, J.W., Modlin, R.L., et al., 2010. Vitamin D-binding protein directs monocyte responses to 25-hydroxy- and 1,25-dihydroxyvitamin D. J. Clin. Endocrinol. Metab. 95 (7), 3368–3376.

Chun, R.F., Hernandez, I., Pereira, R., Swinkles, L., Huijs, T., Zhou, R., et al., 2016. Differential responses to vitamin D2 and vitamin D3 are associated with variations in free 25-hydroxyvitamin D. Endocrinology 157 (9), 3420–3430.

Chun, R.F., 2012. New perspectives on the vitamin D binding protein. Cell Biochem. Funct. 30 (6), 445–456.

Cianferotti, L., Cox, M., Skorija, K., Demay, M.B., 2007. Vitamin D receptor is essential for normal keratinocyte stem cell function. Proc. Natl. Acad. Sci. U.S.A. 104 (22), 9428–9433.

Cleve, H., Constans, J., 1988. The mutants of the vitamin-D-binding protein: more than 120 variants of the GC/DBP system. Vox Sang. 54 (4), 215–225.

Cooke, N.E., McLeod, J.F., Wang, X.K., Ray, K., 1991. Vitamin D binding protein: genomic structure, functional domains, and mRNA expression in tissues. J. Steroid Biochem. Mol. Biol. 40 (4–6), 787–793.

Crofts, L.A., Hancock, M.S., Morrison, N.A., Eisman, J.A., 1998. Multiple promoters direct the tissue-specific expression of novel N-terminal variant human vitamin D receptor gene transcripts. Proc. Natl. Acad. Sci. U.S.A. 95 (18), 10529–10534.

Cui, M., Zhao, Y., Hance, K.W., Shao, A., Wood, R.J., Fleet, J.C., 2009. Effects of MAPK signaling on 1,25-dihydroxyvitamin D-mediated CYP24 gene expression in the enterocyte-like cell line, Caco-2. J. Cell. Physiol. 219 (1), 132–142.

Dahl, B., Schiodt, F.V., Gehrchen, P.M., Ramlau, J., Kiaer, T., Ott, P., 2001a. Gc-globulin is an acute phase reactant and an indicator of muscle injury after spinal surgery. Inflamm. Res. 50 (1), 39–43.

Dahl, B., Schiodt, F.V., Rudolph, S., Ott, P., Kiaer, T., Heslet, L., 2001b. Trauma stimulates the synthesis of Gc-globulin. Intensive Care Med. 27 (2), 394–399.

Dahl, B., Schiodt, F.V., Ott, P., Wians, F., Lee, W.M., Balko, J., et al., 2003. Plasma concentration of Gc-globulin is associated with organ dysfunction and sepsis after injury. Crit. Care Med. 31 (1), 152–156.

Daiger, S.P., Schanfield, M.S., Cavalli-Sforza, L.L., 1975. Group-specific component (Gc) proteins bind vitamin D and 25-hydroxyvitamin D. Proc. Natl. Acad. Sci. U.S.A. 72 (6), 2076–2080.

Demay, M.B., Kiernan, M.S., DeLuca, H.F., Kronenberg, H.M., 1992. Sequences in the human parathyroid hormone gene that bind the 1,25-dihydroxyvitamin D3 receptor and mediate transcriptional repression in response to 1,25-dihydroxyvitamin D3. Proc. Natl. Acad. Sci. U.S.A. 89 (17), 8097–8101.

DiMartino, S.J., Kew, R.R., 1999. Initial characterization of the vitamin D binding protein (Gc-globulin) binding site on the neutrophil plasma membrane: evidence for a chondroitin sulfate proteoglycan. J. Immunol. 163 (4), 2135–2142.

DiMartino, S.J., Shah, A.B., Trujillo, G., Kew, R.R., 2001. Elastase controls the binding of the vitamin D-binding protein (Gc-globulin) to neutrophils: a potential role in the regulation of C5a co-chemotactic activity. J. Immunol. 166 (4), 2688–2694.

DiMartino, S.J., Trujillo, G., McVoy, L.A., Zhang, J., Kew, R.R., 2007. Upregulation of vitamin D binding protein (Gc-globulin) binding sites during neutrophil activation from a latent reservoir in azurophil granules. Mol. Immunol. 44 (9), 2370–2377.

Dimopoulos, M.A., Germenis, A., Savides, P., Karayanis, A., Fertakis, A., Dimopoulos, C., 1984. Genetic markers in carcinoma of the prostate. Eur. Urol. 10 (5), 315–316.

Ding, N., Yu, R.T., Subramaniam, N., Sherman, M.H., Wilson, C., Rao, R., et al., 2013. A vitamin D receptor/SMAD genomic circuit gates hepatic fibrotic response. Cell 153 (3), 601–613.

Dixon, J.R., Selvaraj, S., Yue, F., Kim, A., Li, Y., Shen, Y., et al., 2012. Topological domains in mammalian genomes identified by analysis of chromatin interactions. Nature 485 (7398), 376–380.

Doig, C.L., Singh, P.K., Dhiman, V.K., Thorne, J.L., Battaglia, S., Sobolewski, M., et al., 2013. Recruitment of NCOR1 to VDR target genes is enhanced in prostate cancer cells and associates with altered DNA methylation patterns. Carcinogenesis 34 (2), 248–256.

Dormanen, M.C., Bishop, J.E., Hammond, M.W., Okamura, W.H., Nemere, I., Norman, A.W., 1994. Nonnuclear effects of the steroid hormone 1 alpha,25(OH)2-vitamin D3: analogs are able to functionally differentiate between nuclear and membrane receptors. Biochem. Biophys. Res. Commun. 201 (1), 394–401.

Doroudi, M., Olivares-Navarrete, R., Boyan, B.D., Schwartz, Z., 2015. A review of 1alpha,25(OH)2D3 dependent Pdia3 receptor complex components in Wnt5a non-canonical pathway signaling. J. Steroid Biochem. Mol. Biol. 152, 84–88.

Dueland, S., Nenseter, M.S., Drevon, C.A., 1991. Uptake and degradation of filamentous actin and vitamin D-binding protein in the rat. Biochem. J. 274 (Pt 1), 237–241.

Dwivedi, P.P., Omdahl, J.L., Kola, I., Hume, D.A., May, B.K., 2000. Regulation of rat cytochrome P450C24 (CYP24) gene expression. Evidence for functional cooperation of Ras-activated Ets transcription factors with the vitamin D receptor in 1,25-dihydroxyvitamin D(3)-mediated induction. J. Biol. Chem. 275 (1), 47–55.

Dwivedi, P.P., Hii, C.S., Ferrante, A., Tan, J., Der, C.J., Omdahl, J.L., et al., 2002. Role of MAP kinases in the 1,25-dihydroxyvitamin D3-induced transactivation of the rat cytochrome P450C24 (CYP24) promoter. Specific functions for ERK1/ERK2 and ERK5. J. Biol. Chem. 277 (33), 29643–29653.

Dwivedi, P.P., Gao, X.H., Tan, J.C., Evdokiou, A., Ferrante, A., Morris, H.A., et al., 2010. A role for the phosphatidylinositol 3-kinase–protein kinase C zeta–Sp1 pathway in the 1,25-dihydroxyvitamin D3 induction of the 25-hydroxyvitamin D3 24-hydroxylase gene in human kidney cells. Cell. Signal. 22 (3), 543–552.

Eckey, M., Hong, W., Papaioannou, M., Baniahmad, A., 2007. The nucleosome assembly activity of NAP1 is enhanced by Alien. Mol. Cell Biol. 27 (10), 3557–3568.

Eloranta, J.J., Wenger, C., Mwinyi, J., Hiller, C., Gubler, C., Vavricka, S.R., et al., 2011. Association of a common vitamin D-binding protein polymorphism with inflammatory bowel disease. Pharmacogenetics Genom. 21 (9), 559–564.

Ena, J.M., Esteban, C., Perez, M.D., Uriel, J., Calvo, M., 1989. Fatty acids bound to vitamin D-binding protein (DBP) from human and bovine sera. Biochem. Int. 19 (1), 1–7.

Esteban, C., Geuskens, M., Ena, J.M., Mishal, Z., Macho, A., Torres, J.M., et al., 1992. Receptor-mediated uptake and processing of vitamin D-binding protein in human B-lymphoid cells. J. Biol. Chem. 267 (14), 10177–10183.

Esteller, M., 2008. Epigenetics in cancer. N. Engl. J. Med. 358 (11), 1148–1159.

Ezura, Y., Nakajima, T., Kajita, M., Ishida, R., Inoue, S., Yoshida, H., et al., 2003. Association of molecular variants, haplotypes, and linkage disequilibrium within the human vitamin D-binding protein (DBP) gene with postmenopausal bone mineral density. J. Bone Miner. Res. 18 (9), 1642–1649.

Falzon, M., 1996. DNA sequences in the rat parathyroid hormone-related peptide gene responsible for 1,25-dihydroxyvitamin D3-mediated transcriptional repression. Mol. Endocrinol. 10 (6), 672–681.

Faserl, K., Golderer, G., Kremser, L., Lindner, H., Sarg, B., Wildt, L., et al., 2011. Polymorphism in vitamin D-binding protein as a genetic risk factor in the pathogenesis of endometriosis. J. Clin. Endocrinol. Metab. 96 (1), E233–E241.

Fetahu, I.S., Hobaus, J., Kallay, E., 2014. Vitamin D and the epigenome. Front. Physiol. 5, 164.

Fu, Q., Manolagas, S.C., O'Brien, C.A., 2006. Parathyroid hormone controls receptor activator of NF-kappaB ligand gene expression via a distant transcriptional enhancer. Mol. Cell Biol. 26 (17), 6453–6468.

Fu, L., Yun, F., Oczak, M., Wong, B.Y., Vieth, R., Cole, D.E., 2009. Common genetic variants of the vitamin D binding protein (DBP) predict differences in response of serum 25-hydroxyvitamin D [25(OH)D] to vitamin D supplementation. Clin. Biochem. 42 (10–11), 1174–1177.

Fuleihan Gel, H., Bouillon, R., Clarke, B., Chakhtoura, M., Cooper, C., McClung, M., et al., 2015. Serum 25-hydroxyvitamin D levels: variability, knowledge gaps, and the concept of a desirable range. J. Bone Miner. Res. 30 (7), 1119–1133.

Gacad, M.A., Adams, J.S., 1993. Identification of a competitive binding component in vitamin D-resistant New World primate cells with a low affinity but high capacity for 1,25-dihydroxyvitamin D3. J. Bone Miner. Res. 8 (1), 27–35.

Gacad, M.A., Chen, H., Arbelle, J.E., LeBon, T., Adams, J.S., 1997. Functional characterization and purification of an intracellular vitamin D-binding protein in vitamin D-resistant new world primate cells. Amino acid sequence homology with proteins in the hsp-70 family. J. Biol. Chem. 272 (13), 8433–8440.

Ge, L., Trujillo, G., Miller, E.J., Kew, R.R., 2014. Circulating complexes of the vitamin D binding protein with G-actin induce lung inflammation by targeting endothelial cells. Immunobiology 219 (3), 198–207.

Girgis, C.M., Mokbel, N., Cha, K.M., Houweling, P.J., Abboud, M., Fraser, D.R., et al., 2014. The vitamin D receptor (VDR) is expressed in skeletal muscle of male mice and modulates 25-hydroxyvitamin D (25OHD) uptake in myofibers. Endocrinology 155 (9), 3227–3237.

Gomme, P.T., Bertolini, J., 2004. Therapeutic potential of vitamin D-binding protein. Trends Biotechnol. 22 (7), 340–345.

Gressner, O.A., Gao, C., Siluschek, M., Kim, P., Gressner, A.M., 2009. Inverse association between serum concentrations of actin-free vitamin D-binding protein and the histopathological extent of fibrogenic liver disease or hepatocellular carcinoma. Eur. J. Gastroenterol. Hepatol. 21 (9), 990–995.

Gu, X., Nylander, E., Coates, P.J., Nylander, K., 2011. Effect of narrow-band ultraviolet B phototherapy on p63 and microRNA (miR-21 and miR-125b) expression in psoriatic epidermis. Acta Dermato-Venereologica. 91 (4), 392–397.

Guha, C., Osawa, M., Werner, P.A., Galbraith, R.M., Paddock, G.V., 1995. Regulation of human Gc (vitamin D–binding) protein levels: hormonal and cytokine control of gene expression in vitro. Hepatology 21 (6), 1675–1681.

Guo, B., Aslam, F., van Wijnen, A.J., Roberts, S.G., Frenkel, B., Green, M.R., et al., 1997. YY1 regulates vitamin D receptor/retinoid X receptor mediated transactivation of the vitamin D responsive osteocalcin gene. Proc. Natl. Acad. Sci. U.S.A. 94 (1), 121–126.

Haddad, J.G., Hu, Y.Z., Kowalski, M.A., Laramore, C., Ray, K., Robzyk, P., et al., 1992. Identification of the sterol- and actin-binding domains of plasma vitamin D binding protein (Gc-globulin). Biochemistry 31 (31), 7174–7181.

Hagenfeldt, Y., Carlstrom, K., Berlin, T., Stege, R., 1991. Effects of orchidectomy and different modes of high dose estrogen treatment on circulating "free" and total 1,25-dihydroxyvitamin D in patients with prostatic cancer. J. Steroid Biochem. Mol. Biol. 39 (2), 155–159.

Haussler, M.R., Haussler, C.A., Whitfield, G.K., Hsieh, J.C., Thompson, P.D., Barthel, T.K., et al., 2010. The nuclear vitamin D receptor controls the expression of genes encoding factors which feed the "Fountain of Youth" to mediate healthful aging. J. Steroid Biochem. Mol. Biol. 121 (1–2), 88–97.

Haussler, M.R., Jurutka, P.W., Mizwicki, M., Norman, A.W., 2011. Vitamin D receptor (VDR)-mediated actions of 1alpha,25(OH)(2)vitamin D(3): genomic and non-genomic mechanisms. Best Prac. Res. Clin. Endocrinol. Metabol. 25 (4), 543–559.

Hawker, N.P., Pennypacker, S.D., Chang, S.M., Bikle, D.D., 2007. Regulation of human epidermal keratinocyte differentiation by the vitamin D receptor and its coactivators DRIP205, SRC2, and SRC3. J. Investig. Dermatol. 127 (4), 874–880.

Head, J.F., Swamy, N., Ray, R., 2002. Crystal structure of the complex between actin and human vitamin D-binding protein at 2.5 A resolution. Biochemistry 41 (29), 9015–9020.

Heery, D.M., Kalkhoven, E., Hoare, S., Parker, M.G., 1997. A signature motif in transcriptional co-activators mediates binding to nuclear receptors. Nature 387 (6634), 733–736.

Heikkinen, S., Vaisanen, S., Pehkonen, P., Seuter, S., Benes, V., Carlberg, C., 2011. Nuclear hormone 1alpha,25-dihydroxyvitamin D3 elicits a genome-wide shift in the locations of VDR chromatin occupancy. Nucleic Acids Res. 39 (21), 9181–9193.

Hewison, M., Rut, A.R., Kristjansson, K., Walker, R.E., Dillon, M.J., Hughes, M.R., et al., 1993. Tissue resistance to 1,25-dihydroxyvitamin D without a mutation of the vitamin D receptor gene. Clin. Endocrinol. 39 (6), 663–670.

Hii, C.S., Ferrante, A., 2016. The non-genomic actions of vitamin D. Nutrients 8 (3), 135.

Hirai, M., Suzuki, S., Hinokio, Y., Chiba, M., Kasuga, S., Hirai, A., et al., 1998. Group specific component protein genotype is associated with NIDDM in Japan. Diabetologia 41 (6), 742–743.

Hirschfeld, J., 1959. Immune-electrophoretic demonstration of qualitative differences in human sera and their relation to the haptoglobins. Acta Pathol. Microbiol. Scand. 47, 160–168.

Hoofnagle, A.N., Eckfeldt, J.H., Lutsey, P.L., 2015. Vitamin D-binding protein concentrations quantified by mass spectrometry. N. Engl. J. Med. 373 (15), 1480–1482.

Hsieh, J.C., Shimizu, Y., Minoshima, S., Shimizu, N., Haussler, C.A., Jurutka, P.W., et al., 1998. Novel nuclear localization signal between the two DNA-binding zinc fingers in the human vitamin D receptor. J. Cell. Biochem. 70 (1), 94–109.

Hsieh, J.C., Sisk, J.M., Jurutka, P.W., Haussler, C.A., Slater, S.A., Haussler, M.R., et al., 2003. Physical and functional interaction between the vitamin D receptor and hairless corepressor, two proteins required for hair cycling. J. Biol. Chem. 278 (40), 38665–38674.

Hsieh, E., Fraenkel, L., Han, Y., Xia, W., Insogna, K.L., Yin, M.T., et al., 2016. Longitudinal increase in vitamin D binding protein levels after initiation of tenofovir/lamivudine/efavirenz among individuals with HIV. AIDS 30 (12), 1935–1942.

Hu, X., Lazar, M.A., 1999. The CoRNR motif controls the recruitment of corepressors by nuclear hormone receptors. Nature 402 (6757), 93–96.

Iqbal, M.U.N., Khan, T.A., 2017. Association between vitamin D receptor (Cdx2, Fok1, Bsm1, Apa1, Bgl1, Taq1, and poly (a)) gene polymorphism and breast cancer: a systematic review and meta-analysis. Tumour Biol 39 (10), 1010428317731280.

Jiang, Y.J., Bikle, D.D., 2014. LncRNA: a new player in 1alpha, 25(OH)(2) vitamin D(3)/VDR protection against skin cancer formation. Exp. Dermatol. 23 (3), 147–150.

Jones, K.S., Assar, S., Harnpanich, D., Bouillon, R., Lambrechts, D., Prentice, A., et al., 2014. 25(OH)D2 half-life is shorter than 25(OH)D3 half-life and is influenced by DBP concentration and genotype. J. Clin. Endocrinol. Metab. 99 (9), 3373–3381.

Jurutka, P.W., Remus, L.S., Whitfield, G.K., Thompson, P.D., Hsieh, J.C., Zitzer, H., et al., 2000. The polymorphic N terminus in human vitamin D receptor isoforms influences transcriptional activity by modulating interaction with transcription factor IIB. Mol. Endocrinol. 14 (3), 401–420.

Kato, S., Fujiki, R., Kitagawa, H., 2004. Vitamin D receptor (VDR) promoter targeting through a novel chromatin remodeling complex. J. Steroid Biochem. Mol. Biol. 89–90 (1–5), 173–178.

Kempker, J.A., Tangpricha, V., Ziegler, T.R., Martin, G.S., 2012. Vitamin D in sepsis: from basic science to clinical impact. Crit. Care 16 (4), 316.

Khanal, R.C., Nemere, I., 2007. The ERp57/GRp58/1,25D3-MARRS receptor: multiple functional roles in diverse cell systems. Curr. Med. Chem. 14 (10), 1087–1093.

Khanim, F.L., Gommersall, L.M., Wood, V.H., Smith, K.L., Montalvo, L., O'Neill, L.P., et al., 2004. Altered SMRT levels disrupt vitamin D3 receptor signalling in prostate cancer cells. Oncogene 23 (40), 6712–6725.

Kim, S., Shevde, N.K., Pike, J.W., 2005. 1,25-Dihydroxyvitamin D3 stimulates cyclic vitamin D receptor/retinoid X receptor DNA-binding, co-activator recruitment, and histone acetylation in intact osteoblasts. J. Bone Miner. Res. 20 (2), 305–317.

Kim, S., Yamazaki, M., Zella, L.A., Shevde, N.K., Pike, J.W., 2006. Activation of receptor activator of NF-kappaB ligand gene expression by 1,25-dihydroxyvitamin D3 is mediated through multiple long-range enhancers. Mol. Cell Biol. 26 (17), 6469–6486.

Kim, S., Yamazaki, M., Shevde, N.K., Pike, J.W., 2007. Transcriptional control of receptor activator of nuclear factor-kappaB ligand by the protein kinase A activator forskolin and the transmembrane glycoprotein 130-activating cytokine, oncostatin M, is exerted through multiple distal enhancers. Mol. Endocrinol. 21 (1), 197–214.

Kim, M.S., Fujiki, R., Murayama, A., Kitagawa, H., Yamaoka, K., Yamamoto, Y., et al., 2007. 1Alpha,25(OH)2D3-induced transrepression by vitamin D receptor through E-box-type elements in the human parathyroid hormone gene promoter. Mol. Endocrinol. 21 (2), 334–342.

Kim, M.S., Kondo, T., Takada, I., Youn, M.Y., Yamamoto, Y., Takahashi, S., et al., 2009. DNA demethylation in hormone-induced transcriptional derepression. Nature 461 (7266), 1007–1012.

Kisker, O., Onizuka, S., Becker, C.M., Fannon, M., Flynn, E., D'Amato, R., et al., 2003. Vitamin D binding protein-macrophage activating factor (DBP-maf) inhibits angiogenesis and tumor growth in mice. Neoplasia 5 (1), 32–40.

Kitagawa, H., Fujiki, R., Yoshimura, K., Mezaki, Y., Uematsu, Y., Matsui, D., et al., 2003. The chromatin-remodeling complex WINAC targets a nuclear receptor to promoters and is impaired in Williams syndrome. Cell 113 (7), 905–917.

Kliewer, S.A., Umesono, K., Mangelsdorf, D.J., Evans, R.M., 1992. Retinoid X receptor interacts with nuclear receptors in retinoic acid, thyroid hormone and vitamin D3 signalling. Nature 355 (6359), 446–449.

Koga, Y., Naraparaju, V.R., Yamamoto, N., 1999. Antitumor effect of vitamin D-binding protein-derived macrophage activating factor on Ehrlich ascites tumor-bearing mice. Proc. Soc. Exp. Biol. Med. 220 (1), 20–26.

Koh, S.S., Chen, D., Lee, Y.H., Stallcup, M.R., 2001. Synergistic enhancement of nuclear receptor function by p160 coactivators and two coactivators with protein methyltransferase activities. J. Biol. Chem. 276 (2), 1089–1098.

Koike, K., Shinozawa, Y., Yamazaki, M., Endo, T., Nomura, R., Aiboshi, J., et al., 2002. Recombinant human interleukin-1alpha increases serum albumin, Gc-globulin, and alpha1-antitrypsin levels in burned mice. Tohoku J. Exp. Med. 198 (1), 23–29.

Korbelik, M., Naraparaju, V.R., Yamamoto, N., 1997. Macrophage-directed immunotherapy as adjuvant to photodynamic therapy of cancer. Br. J. Canc. 75 (2), 202–207.

Lange, C.M., Gouttenoire, J., Duong, F.H., Morikawa, K., Heim, M.H., Moradpour, D., 2014. Vitamin D receptor and Jak-STAT signaling crosstalk results in calcitriol-mediated increase of hepatocellular response to IFN-alpha. J. Immunol. 192 (12), 6037–6044.

Lauridsen, A.L., Vestergaard, P., Nexo, E., 2001. Mean serum concentration of vitamin D-binding protein (Gc globulin) is related to the Gc phenotype in women. Clin. Chem. 47 (4), 753–756.

Lauridsen, A.L., Vestergaard, P., Hermann, A.P., Moller, H.J., Mosekilde, L., Nexo, E., 2004. Female premenopausal fracture risk is associated with gc phenotype. J. Bone Miner. Res. 19 (6), 875–881.

Leaf, D.E., Waikar, S.S., Wolf, M., Cremers, S., Bhan, I., Stern, L., 2013. Dysregulated mineral metabolism in patients with acute kidney injury and risk of adverse outcomes. Clin. Endocrinol. 79 (4), 491–498.

Lee, S.M., Bishop, K.A., Goellner, J.J., O'Brien, C.A., Pike, J.W., 2014. Mouse and human BAC transgenes recapitulate tissue-specific expression of the vitamin D receptor in mice and rescue the VDR-null phenotype. Endocrinology 155 (6), 2064–2076.

Lee, M.J., Kearns, M.D., Smith, E.M., Hao, L., Ziegler, T.R., Alvarez, J.A., et al., 2015. Free 25-hydroxyvitamin D concentrations in cystic fibrosis. Am. J. Med. Sci. 350 (5), 374–379.

Leheste, J.R., Melsen, F., Wellner, M., Jansen, P., Schlichting, U., Renner-Muller, I., et al., 2003. Hypocalcemia and osteopathy in mice with kidney-specific megalin gene defect. FASEB J. 17 (2), 247–249.

Leong, G.M., Subramaniam, N., Issa, L.L., Barry, J.B., Kino, T., Driggers, P.H., et al., 2004. Ski-interacting protein, a bifunctional nuclear receptor coregulator that interacts with N-CoR/SMRT and p300. Biochem. Biophys. Res. Commun. 315 (4), 1070–1076.

Leong, A., Rehman, W., Dastani, Z., Greenwood, C., Timpson, N., Langsetmo, L., et al., 2014. The causal effect of vitamin D binding protein (DBP) levels on calcemic and cardiometabolic diseases: a Mendelian randomization study. PLoS Med. 11 (10), e1001751.

Li, Q.P., Qi, X., Pramanik, R., Pohl, N.M., Loesch, M., Chen, G., 2007. Stress-induced c-Jun-dependent Vitamin D receptor (VDR) activation dissects the non-classical VDR pathway from the classical VDR activity. J. Biol. Chem. 282 (3), 1544–1551.

Lind, S.E., Smith, D.B., Janmey, P.A., Stossel, T.P., 1988. Depression of gelsolin levels and detection of gelsolin-actin complexes in plasma of patients with acute lung injury. Am. Rev. Respir. Dis. 138 (2), 429–434.

Lisse, T.S., Saini, V., Zhao, H., Luderer, H.F., Gori, F., Demay, M.B., 2014. The vitamin D receptor is required for activation of cWnt and hedgehog signaling in keratinocytes. Mol. Endocrinol. 28 (10), 1698–1706.

Liu, P.T., Stenger, S., Li, H., Wenzel, L., Tan, B.H., Krutzik, S.R., et al., 2006. Toll-like receptor triggering of a vitamin D-mediated human antimicrobial response. Science 311 (5768), 1770–1773.

Long, M.D., Sucheston-Campbell, L.E., Campbell, M.J., 2015. Vitamin D receptor and RXR in the post-genomic era. J. Cell. Physiol. 230 (4), 758–766.

Lopes, N., Carvalho, J., Duraes, C., Sousa, B., Gomes, M., Costa, J.L., et al., 2012. 1Alpha,25-dihydroxyvitamin D3 induces de novo E-cadherin expression in triple-negative breast cancer cells by CDH1-promoter demethylation. Anticancer Res. 32 (1), 249–257.

Luo, Z., Rouvinen, J., Maenpaa, P.H., 1994. A peptide C-terminal to the second Zn finger of human vitamin D receptor is able to specify nuclear localization. Eur. J. Biochem. 223 (2), 381–387.

Madden, K., Feldman, H.A., Chun, R.F., Smith, E.M., Sullivan, R.M., Agan, A.A., et al., 2015. Critically ill children have low vitamin D-binding protein, influencing bioavailability of vitamin D. Ann. Am. Thorac. Soc. 12 (11), 1654–1661.

Malik, S., Fu, L., Juras, D.J., Karmali, M., Wong, B.Y., Gozdzik, A., et al., 2013. Common variants of the vitamin D binding protein gene and adverse health outcomes. Crit. Rev. Clin. Lab. Sci. 50 (1), 1–22.

Malloy, P.J., Feldman, D., 2012. Genetic disorders and defects in vitamin D action. Rheum. Dis. Clin. N. Am. 38 (1), 93–106.

Marik, R., Fackler, M., Gabrielson, E., Zeiger, M.A., Sukumar, S., Stearns, V., et al., 2010. DNA methylation-related vitamin D receptor insensitivity in breast cancer. Cancer Biol. Ther. 10 (1), 44–53.

Martineau, A.R., Leandro, A.C., Anderson, S.T., Newton, S.M., Wilkinson, K.A., Nicol, M.P., et al., 2010. Association between Gc genotype and susceptibility to TB is dependent on vitamin D status. Eur. Respir. J. 35 (5), 1106–1112.

Martowicz, M.L., Meyer, M.B., Pike, J.W., 2011. The mouse RANKL gene locus is defined by a broad pattern of histone H4 acetylation and regulated through distinct distal enhancers. J. Cell. Biochem. 112 (8), 2030–2045.

Masuyama, H., MacDonald, P.N., 1998. Proteasome-mediated degradation of the vitamin D receptor (VDR) and a putative role for SUG1 interaction with the AF-2 domain of VDR. J. Cell. Biochem. 71 (3), 429–440.

Mc Leod, J.F., Kowalski, M.A., Haddad Jr., J.G., 1989. Interactions among serum vitamin D binding protein, monomeric actin, profilin, and profilactin. J. Biol. Chem. 264 (2), 1260–1267.

McCullough, M.L., Stevens, V.L., Diver, W.R., Feigelson, H.S., Rodriguez, C., Bostick, R.M., et al., 2007. Vitamin D pathway gene polymorphisms, diet, and risk of postmenopausal breast cancer: a nested case-control study. Breast Cancer Res. 9 (1), R9.

McDonnell, D.P., Mangelsdorf, D.J., Pike, J.W., Haussler, M.R., O'Malley, B.W., 1987. Molecular cloning of complementary DNA encoding the avian receptor for vitamin D. Science 235 (4793), 1214–1217.

McInerney, E.M., Rose, D.W., Flynn, S.E., Westin, S., Mullen, T.M., Krones, A., et al., 1998. Determinants of coactivator LXXLL motif specificity in nuclear receptor transcriptional activation. Genes Dev. 12 (21), 3357–3368.

McKenna, N.J., Cooney, A.J., DeMayo, F.J., Downes, M., Glass, C.K., Lanz, R.B., et al., 2009. Minireview: evolution of NURSA, the nuclear receptor signaling atlas. Mol. Endocrinol. 23 (6), 740–746.

McVoy, L.A., Kew, R.R., 2005. CD44 and annexin A2 mediate the C5a chemotactic cofactor function of the vitamin D binding protein. J. Immunol. 175 (7), 4754–4760.

Meier, U., Gressner, O., Lammert, F., Gressner, A.M., 2006. Gc-globulin: roles in response to injury. Clin. Chem. 52 (7), 1247–1253.

Mellon, W.S., DeLuca, H.F., 1979. An equilibrium and kinetic study of 1,25-dihydroxyvitamin D3 binding to chicken intestinal cytosol employing high specific activity 1,25-dehydroxy[3H-26, 27] vitamin D3. Arch. Biochem. Biophys. 197 (1), 90–95.

Mendel, C.M., 1990. Rates of dissociation of sex steroid hormones from human sex hormone-binding globulin: a reassessment. J. Steroid Biochem. Mol. Biol. 37 (2), 251–255.

Meyer, M.B., Goetsch, P.D., Pike, J.W., 2010. Genome-wide analysis of the VDR/RXR cistrome in osteoblast cells provides new mechanistic insight into the actions of the vitamin D hormone. J. Steroid Biochem. Mol. Biol. 121 (1–2), 136–141.

Meyer, M.B., Goetsch, P.D., Pike, J.W., 2012. VDR/RXR and TCF4/beta-catenin cistromes in colonic cells of colorectal tumor origin: impact on c-FOS and c-MYC gene expression. Mol. Endocrinol. 26 (1), 37–51.

Meyer, M.B., Benkusky, N.A., Lee, C.H., Pike, J.W., 2014. Genomic determinants of gene regulation by 1,25-dihydroxyvitamin D3 during osteoblast-lineage cell differentiation. J. Biol. Chem. 289 (28), 19539–19554.

Meyer, M.B., Benkusky, N.A., Pike, J.W., 2015. Selective distal enhancer control of the Mmp13 gene identified through clustered regularly interspaced short palindromic repeat (CRISPR) genomic deletions. J. Biol. Chem. 290 (17), 11093–11107.

Meyer, M.B., Benkusky, N.A., Sen, B., Rubin, J., Pike, J.W., 2016. Epigenetic plasticity drives adipogenic and osteogenic differentiation of marrow-derived mesenchymal stem cells. J. Biol. Chem. 291 (34), 17829–17847.

Meyer, M.B., Benkusky, N.A., Kaufmann, M., Lee, S.M., Onal, M., Jones, G., et al., 2017. A kidney-specific genetic control module in mice governs endocrine regulation of the cytochrome P450 gene Cyp27b1 essential for vitamin D3 activation. J. Biol. Chem. 292 (42), 17541–17558.

Mittal, M.K., Myers, J.N., Misra, S., Bailey, C.K., Chaudhuri, G., 2008. In vivo binding to and functional repression of the VDR gene promoter by SLUG in human breast cells. Biochem. Biophys. Res. Commun. 372 (1), 30–34.

Mizwicki, M.T., Keidel, D., Bula, C.M., Bishop, J.E., Zanello, L.P., Wurtz, J.M., et al., 2004. Identification of an alternative ligand-binding pocket in the nuclear vitamin D receptor and its functional importance in 1alpha,25(OH)2-vitamin D3 signaling. Proc. Natl. Acad. Sci. U.S.A. 101 (35), 12876–12881.

Mohri, T., Nakajima, M., Takagi, S., Komagata, S., Yokoi, T., 2009. MicroRNA regulates human vitamin D receptor. Int. J. Cancer 125 (6), 1328–1333.

Moller, U.K., Streym, S., Heickendorff, L., Mosekilde, L., Rejnmark, L., 2012. Effects of 25OHD concentrations on chances of pregnancy and pregnancy outcomes: a cohort study in healthy Danish women. Eur. J. Clin. Nutr. 66 (7), 862–868.

Moller, U.K., Streym, S., Jensen, L.T., Mosekilde, L., Schoenmakers, I., Nigdikar, S., et al., 2013. Increased plasma concentrations of vitamin D metabolites and vitamin D binding protein in women using hormonal contraceptives: a cross-sectional study. Nutrients 5 (9), 3470–3480.

Morris, C.A., Mervis, C.B., 2000. Williams syndrome and related disorders. Annu. Rev. Genom. Hum. Genet. 1, 461–484.

Muller, M.J., Volmer, D.A., 2015. Mass spectrometric profiling of vitamin D metabolites beyond 25-hydroxyvitamin D. Clin. Chem. 61 (8), 1033–1048.

Murayama, A., Kim, M.S., Yanagisawa, J., Takeyama, K., Kato, S., 2004. Transrepression by a liganded nuclear receptor via a bHLH activator through co-regulator switching. EMBO J. 23 (7), 1598–1608.

Nagasawa, H., Uto, Y., Sasaki, H., Okamura, N., Murakami, A., Kubo, S., et al., 2005. Gc protein (vitamin D-binding protein): gc genotyping and GcMAF precursor activity. Anticancer Res. 25 (6A), 3689–3695.

Narvaez, C.J., Matthews, D., Broun, E., Chan, M., Welsh, J., 2009. Lean phenotype and resistance to diet-induced obesity in vitamin D receptor knockout mice correlates with induction of uncoupling protein-1 in white adipose tissue. Endocrinology 150 (2), 651–661.

Neme, A., Seuter, S., Carlberg, C., 2016. Vitamin D-dependent chromatin association of CTCF in human monocytes. Biochim. Biophys. Acta 1859 (11), 1380–1388.

Nemere, I., Yoshimoto, Y., Norman, A.W., 1984. Calcium transport in perfused duodena from normal chicks: enhancement within fourteen minutes of exposure to 1,25-dihydroxyvitamin D3. Endocrinology 115 (4), 1476–1483.

Nemere, I., Dormanen, M.C., Hammond, M.W., Okamura, W.H., Norman, A.W., 1994. Identification of a specific binding protein for 1 alpha,25-dihydroxyvitamin D3 in basal-lateral membranes of chick intestinal epithelium and relationship to transcaltachia. J. Biol. Chem. 269 (38), 23750–23756.

Nerenz, R.D., Martowicz, M.L., Pike, J.W., 2008. An enhancer 20 kilobases upstream of the human receptor activator of nuclear factor-kappaB ligand gene mediates dominant activation by 1,25-dihydroxyvitamin D3. Mol. Endocrinol. 22 (5), 1044–1056.

Nielson, C.M., Jones, K.S., Bouillon, R., Osteoporotic Fractures in Men Research, G., Chun, R.F., Jacobs, J., et al., 2016a. Role of assay type in determining free 25-hydroxyvitamin D levels in diverse populations. N. Engl. J. Med. 374 (17), 1695–1696.

Nielson, C.M., Jones, K.S., Chun, R.F., Jacobs, J.M., Wang, Y., Hewison, M., et al., 2016b. Free 25-hydroxyvitamin D: impact of vitamin D binding protein assays on racial-genotypic associations. J. Clin. Endocrinol. Metab. 101 (5), 2226–2234.

Norman, A.W., 2006. Minireview: vitamin D receptor: new assignments for an already busy receptor. Endocrinology 147 (12), 5542–5548.

Nutchey, B.K., Kaplan, J.S., Dwivedi, P.P., Omdahl, J.L., Ferrante, A., May, B.K., et al., 2005. Molecular action of 1,25-dihydroxyvitamin D3 and phorbol ester on the activation of the rat cytochrome P450C24 (CYP24) promoter: role of MAP kinase activities and identification of an important transcription factor binding site. Biochem. J. 389 (Pt 3), 753–762.

Nykjaer, A., Dragun, D., Walther, D., Vorum, H., Jacobsen, C., Herz, J., et al., 1999. An endocytic pathway essential for renal uptake and activation of the steroid 25-(OH) vitamin D3. Cell 96 (4), 507–515.

Oda, Y., Sihlbom, C., Chalkley, R.J., Huang, L., Rachez, C., Chang, C.P., et al., 2003. Two distinct coactivators, DRIP/mediator and SRC/p160, are differentially involved in vitamin D receptor transactivation during keratinocyte differentiation. Mol. Endocrinol. 17 (11), 2329–2339.

Oda, Y., Uchida, Y., Moradian, S., Crumrine, D., Elias, P.M., Bikle, D.D., 2009. Vitamin D receptor and coactivators SRC2 and 3 regulate epidermis-specific sphingolipid production and permeability barrier formation. J. Investig. Dermatol. 129 (6), 1367–1378.

Ogryzko, V.V., Schiltz, R.L., Russanova, V., Howard, B.H., Nakatani, Y., 1996. The transcriptional coactivators p300 and CBP are histone acetyltransferases. Cell 87 (5), 953–959.

Orlov, I., Rochel, N., Moras, D., Klaholz, B.P., 2012. Structure of the full human RXR/VDR nuclear receptor heterodimer complex with its DR3 target DNA. EMBO J. 31 (2), 291–300.

Palmer, H.G., Gonzalez-Sancho, J.M., Espada, J., Berciano, M.T., Puig, I., Baulida, J., et al., 2001. Vitamin D(3) promotes the differentiation of colon carcinoma cells by the induction of E-cadherin and the inhibition of beta-catenin signaling. J. Cell Biol. 154 (2), 369–387.

Palmer, H.G., Larriba, M.J., Garcia, J.M., Ordonez-Moran, P., Pena, C., Peiro, S., et al., 2004. The transcription factor SNAIL represses vitamin D receptor expression and responsiveness in human colon cancer. Nat. Med. 10 (9), 917–919.

Palmer, H.G., Anjos-Afonso, F., Carmeliet, G., Takeda, H., Watt, F.M., 2008. The vitamin D receptor is a wnt effector that controls hair follicle differentiation and specifies tumor type in adult epidermis. PLoS One 3 (1), e1483.

Pan, Y.Z., Gao, W., Yu, A.M., 2009. MicroRNAs regulate CYP3A4 expression via direct and indirect targeting. Drug Metab. Dispos. 37 (10), 2112–2117.

Papiha, S.S., Allcroft, L.C., Kanan, R.M., Francis, R.M., Datta, H.K., 1999. Vitamin D binding protein gene in male osteoporosis: association of plasma DBP and bone mineral density with (TAAA)(n)-Alu polymorphism in DBP. Calcif. Tissue Int. 65 (4), 262–266.

Pereira, F., Barbachano, A., Singh, P.K., Campbell, M.J., Munoz, A., Larriba, M.J., 2012. Vitamin D has wide regulatory effects on histone demethylase genes. Cell Cycle 11 (6), 1081–1089.

Perez, H.D., 1994. Gc globulin (vitamin D-binding protein) increases binding of low concentrations of C5a des Arg to human polymorphonuclear leukocytes: an explanation for its cochemotaxin activity. Inflammation 18 (2), 215–220.

Peric, M., Koglin, S., Dombrowski, Y., Gross, K., Bradac, E., Ruzicka, T., et al., 2009. VDR and MEK-ERK dependent induction of the antimicrobial peptide cathelicidin in keratinocytes by lithocholic acid. Mol. Immunol. 46 (16), 3183–3187.

Perissi, V., Staszewski, L.M., McInerney, E.M., Kurokawa, R., Krones, A., Rose, D.W., et al., 1999. Molecular determinants of nuclear receptor-corepressor interaction. Genes Dev. 13 (24), 3198–3208.

Perlmann, T., Rangarajan, P.N., Umesono, K., Evans, R.M., 1993. Determinants for selective RAR and TR recognition of direct repeat HREs. Genes Dev. 7 (7B), 1411–1422.

Pihl, T.H., Jorgensen, C.S., Santoni-Rugiu, E., Leifsson, P.S., Hansen, E.W., Laursen, I., et al., 2010. Safety pharmacology, toxicology and pharmacokinetic assessment of human Gc globulin (vitamin D binding protein). Basic Clin. Pharmacol. Toxicol. 107 (5), 853–860.

Pike, J.W., Meyer, M.B., Lee, S.M., Onal, M., Benkusky, N.A., 2017. The vitamin D receptor: contemporary genomic approaches reveal new basic and translational insights. J. Clin. Investig. 127 (4), 1146–1154.

Polly, P., Herdick, M., Moehren, U., Baniahmad, A., Heinzel, T., Carlberg, C., 2000. VDR-Alien: a novel, DNA-selective vitamin D(3) receptor-corepressor partnership. FASEB J. 14 (10), 1455–1463.

Powe, C.E., Ricciardi, C., Berg, A.H., Erdenesanaa, D., Collerone, G., Ankers, E., et al., 2011. Vitamin D-binding protein modifies the vitamin D-bone mineral density relationship. J. Bone Miner. Res. 26 (7), 1609–1616.

Powe, C.E., Evans, M.K., Wenger, J., Zonderman, A.B., Berg, A.H., Nalls, M., et al., 2013. Vitamin D-binding protein and vitamin D status of black Americans and white Americans. N. Engl. J. Med. 369 (21), 1991–2000.

Poynter, J.N., Jacobs, E.T., Figueiredo, J.C., Lee, W.H., Conti, D.V., Campbell, P.T., et al., 2010. Genetic variation in the vitamin D receptor (VDR) and the vitamin D-binding protein (GC) and risk for colorectal cancer: results from the Colon Cancer Family Registry. Cancer Epidemiol. Biomark. Prev. 19 (2), 525–536.

Rachez, C., Freedman, L.P., 2001. Mediator complexes and transcription. Curr. Opin. Cell Biol. 13 (3), 274–280.

Rai, V., Abdo, J., Agrawal, S., Agrawal, D.K., 2017. Vitamin D receptor polymorphism and cancer: an update. Anticancer Res. 37 (8), 3991–4003.

Ramagopalan, S.V., Heger, A., Berlanga, A.J., Maugeri, N.J., Lincoln, M.R., Burrell, A., et al., 2010. A ChIP-seq defined genome-wide map of vitamin D receptor binding: associations with disease and evolution. Genome Res. 20 (10), 1352–1360.

Raval-Pandya, M., Dhawan, P., Barletta, F., Christakos, S., 2001. YY1 represses vitamin D receptor-mediated 25-hydroxyvitamin D(3)24-hydroxylase transcription: relief of repression by CREB-binding protein. Mol. Endocrinol. 15 (6), 1035–1046.

Rochel, N., Wurtz, J.M., Mitschler, A., Klaholz, B., Moras, D., 2000. The crystal structure of the nuclear receptor for vitamin D bound to its natural ligand. Mol. Cell 5 (1), 173–179.

Saccone, D., Asani, F., Bornman, L., 2015. Regulation of the vitamin D receptor gene by environment, genetics and epigenetics. Gene 561 (2), 171–180.

Safadi, F.F., Thornton, P., Magiera, H., Hollis, B.W., Gentile, M., Haddad, J.G., et al., 1999. Osteopathy and resistance to vitamin D toxicity in mice null for vitamin D binding protein. J. Clin. Investig. 103 (2), 239–251.

Salehi-Tabar, R., Nguyen-Yamamoto, L., Tavera-Mendoza, L.E., Quail, T., Dimitrov, V., An, B.S., et al., 2012. Vitamin D receptor as a master regulator of the c-MYC/MXD1 network. Proc. Natl. Acad. Sci. U.S.A. 109 (46), 18827–18832.

Sancho, E., Batlle, E., Clevers, H., 2004. Signaling pathways in intestinal development and cancer. Annu. Rev. Cell Dev. Biol. 20, 695–723.

Santos, B.R., Mascarenhas, L.P., Boguszewski, M.C., Spritzer, P.M., 2013. Variations in the vitamin D-binding protein (DBP) gene are related to lower 25-hydroxyvitamin D levels in healthy girls: a cross-sectional study. Hormone Res. Paediatr. 79, 162–168.

Schiodt, F.V., Ott, P., Bondesen, S., Tygstrup, N., 1997. Reduced serum Gc-globulin concentrations in patients with fulminant hepatic failure: association with multiple organ failure. Crit. Care Med. 25 (8), 1366–1370.

Schiodt, F.V., 2008. Gc-globulin in liver disease. Dan. Med. Bull. 55 (3), 131–146.

Schneider, G.B., Benis, K.A., Flay, N.W., Ireland, R.A., Popoff, S.N., 1995. Effects of vitamin D binding protein-macrophage activating factor (DBP-MAF) infusion on bone resorption in two osteopetrotic mutations. Bone 16 (6), 657–662.

Schrader, M., Nayeri, S., Kahlen, J.P., Muller, K.M., Carlberg, C., 1995. Natural vitamin D3 response elements formed by inverted palindromes: polarity-directed ligand sensitivity of vitamin D3 receptor-retinoid X receptor heterodimer-mediated transactivation. Mol. Cell Biol. 15 (3), 1154–1161.

Schrader, M., Kahlen, J.P., Carlberg, C., 1997. Functional characterization of a novel type of 1 alpha,25-dihydroxyvitamin D3 response element identified in the mouse c-fos promoter. Biochem. Biophys. Res. Commun. 230 (3), 646–651.

Schwartz, J.B., Lai, J., Lizaola, B., Kane, L., Weyland, P., Terrault, N.A., et al., 2014a. Variability in free 25(OH) vitamin D levels in clinical populations. J. Steroid Biochem. Mol. Biol. 144 (Pt A), 156–158.

Schwartz, J.B., Lai, J., Lizaola, B., Kane, L., Markova, S., Weyland, P., et al., 2014b. A comparison of measured and calculated free 25(OH) vitamin D levels in clinical populations. J. Clin. Endocrinol. Metab. 99 (5), 1631–1637.

Schwartz, J.B., Kane, L., Bikle, D., 2016. Response of vitamin D concentration to vitamin D3 administration in older adults without sun exposure: a randomized double-blind trial. J. Am. Geriatr. Soc. 64 (1), 65–72.

Sequeira, V.B., Rybchyn, M.S., Tongkao-On, W., Gordon-Thomson, C., Malloy, P.J., Nemere, I., et al., 2012. The role of the vitamin D receptor and ERp57 in photoprotection by 1alpha,25-dihydroxyvitamin D3. Mol. Endocrinol. 26 (4), 574–582.

Seuter, S., Neme, A., Carlberg, C., Epigenomic, P.U., 2017. 1-VDR crosstalk modulates vitamin D signaling. Biochim. Biophys. Acta 1860 (4), 405–415.

Shah, A.B., DiMartino, S.J., Trujillo, G., Kew, R.R., 2006. Selective inhibition of the C5a chemotactic cofactor function of the vitamin D binding protein by 1,25(OH)2 vitamin D3. Mol. Immunol. 43 (8), 1109–1115.

Shah, S., Islam, M.N., Dakshanamurthy, S., Rizvi, I., Rao, M., Herrell, R., et al., 2006. The molecular basis of vitamin D receptor and beta-catenin crossregulation. Mol. Cell 21 (6), 799–809.

Shieh, A., Chun, R.F., Ma, C., Witzel, S., Meyer, B., Rafison, B., et al., 2016. Effects of high-dose vitamin D2 versus D3 on total and free 25-hydroxyvitamin D and markers of calcium balance. J. Clin. Endocrinol. Metab. 101 (8), 3070–3078.

Shieh, A., Ma, C., Chun, R.F., Witzel, S., Rafison, B., Contreras, H.T.M., et al., 2017. Effects of cholecalciferol vs calcifediol on total and free 25-hydroxyvitamin D and parathyroid hormone. J. Clin. Endocrinol. Metab. 102 (4), 1133–1140.

Silvagno, F., Pescarmona, G., 2017. Spotlight on vitamin D receptor, lipid metabolism and mitochondria: some preliminary emerging issues. Mol. Cell. Endocrinol. 450, 24–31.

Sollid, S.T., Hutchinson, M.Y., Berg, V., Fuskevag, O.M., Figenschau, Y., Thorsby, P.M., et al., 2016. Effects of vitamin D binding protein phenotypes and vitamin D supplementation on serum total 25(OH)D and directly measured free 25(OH)D. Eur. J. Endocrinol. 174 (4), 445–452.

Song, Y.H., Naumova, A.K., Liebhaber, S.A., Cooke, N.E., 1999. Physical and meiotic mapping of the region of human chromosome 4q11-q13 encompassing the vitamin D binding protein DBP/Gc-globulin and albumin multigene cluster. Genome Res. 9 (6), 581–587.

Speeckaert, M., Huang, G., Delanghe, J.R., Taes, Y.E., 2006. Biological and clinical aspects of the vitamin D binding protein (Gc-globulin) and its polymorphism. Clin. Chim. Acta 372 (1–2), 33–42.

Speeckaert, M.M., Wehlou, C., De Somer, F., Speeckaert, R., Van Nooten, G.J., Delanghe, J.R., 2010. Evolution of vitamin D binding protein concentration in sera from cardiac surgery patients is determined by triglyceridemia. Clin. Chem. Lab. Med. 48 (9), 1345–1350.

Stumpf, W.E., Sar, M., Reid, F.A., Tanaka, Y., DeLuca, H.F., 1979. Target cells for 1,25-dihydroxyvitamin D3 in intestinal tract, stomach, kidney, skin, pituitary, and parathyroid. Science 206 (4423), 1188–1190.

Sunn, K.L., Cock, T.A., Crofts, L.A., Eisman, J.A., Gardiner, E.M., 2001. Novel N-terminal variant of human VDR. Mol. Endocrinol. 15 (9), 1599–1609.

Sutton, A.L., Zhang, X., Ellison, T.I., Macdonald, P.N., 2005. The 1,25(OH)2D3-regulated transcription factor MN1 stimulates vitamin D receptor-mediated transcription and inhibits osteoblastic cell proliferation. Mol. Endocrinol. 19 (9), 2234–2244.

Swamy, N., Ray, R., 2008. Fatty acid-binding site environments of serum vitamin D-binding protein and albumin are different. Bioorg. Chem. 36 (3), 165–168.

Swamy, N., Ghosh, S., Schneider, G.B., Ray, R., 2001. Baculovirus-expressed vitamin D-binding protein-macrophage activating factor (DBP-maf) activates osteoclasts and binding of 25-hydroxyvitamin D(3) does not influence this activity. J. Cell. Biochem. 81 (3), 535–546.

Tagami, T., Lutz, W.H., Kumar, R., Jameson, J.L., 1998. The interaction of the vitamin D receptor with nuclear receptor corepressors and coactivators. Biochem. Biophys. Res. Commun. 253 (2), 358–363.

Takiar, R., Lutsey, P.L., Zhao, D., Guallar, E., Schneider, A.L., Grams, M.E., et al., 2015. The associations of 25-hydroxyvitamin D levels, vitamin D binding protein gene polymorphisms, and race with risk of incident fracture-related hospitalization: twenty-year follow-up in a bi-ethnic cohort (the ARIC Study). Bone 78, 94–101.

Tannetta, D.S., Redman, C.W., Sargent, I.L., 2014. Investigation of the actin scavenging system in pre-eclampsia. Eur. J. Obstet. Gynecol. Reprod. Biol. 172, 32–35.

Teichert, A., Arnold, L.A., Otieno, S., Oda, Y., Augustinaite, I., Geistlinger, T.R., et al., 2009. Quantification of the vitamin D receptor-coregulator interaction. Biochemistry 48 (7), 1454–1461.

Trujillo, G., Habiel, D.M., Ge, L., Ramadass, M., Cooke, N.E., Kew, R.R., 2013. Neutrophil recruitment to the lung in both C5a- and CXCL1-induced alveolitis is impaired in vitamin D-binding protein-deficient mice. J. Immunol. 191 (2), 848–856.

Tuoresmaki, P., Vaisanen, S., Neme, A., Heikkinen, S., Carlberg, C., 2014. Patterns of genome-wide VDR locations. PLoS One 9 (4), e96105.

Uto, Y., Yamamoto, S., Mukai, H., Ishiyama, N., Takeuchi, R., Nakagawa, Y., et al., 2012. beta-Galactosidase treatment is a common first-stage modification of the three major subtypes of Gc protein to GcMAF. Anticancer Res. 32 (6), 2359–2364.

Vasconcellos, C.A., Lind, S.E., 1993. Coordinated inhibition of actin-induced platelet aggregation by plasma gelsolin and vitamin D-binding protein. Blood 82 (12), 3648–3657.

Waldron, J.L., Ashby, H.L., Cornes, M.P., Bechervaise, J., Razavi, C., Thomas, O.L., et al., 2013. Vitamin D: a negative acute phase reactant. J. Clin. Pathol. 66 (7), 620–622.

Walsh, J.S., Evans, A.L., Bowles, S., Naylor, K.E., Jones, K.S., Schoenmakers, I., et al., 2016. Free 25-hydroxyvitamin D is low in obesity, but there are no adverse associations with bone health. Am. J. Clin. Nutr. 103 (6), 1465–1471.

Wan, L.Y., Zhang, Y.Q., Chen, M.D., Liu, C.B., Wu, J.F., 2015a. Relationship of structure and function of DNA-binding domain in vitamin D receptor. Molecules 20 (7), 12389–12399.

Wan, L.Y., Zhang, Y.Q., Chen, M.D., Du, Y.Q., Liu, C.B., Wu, J.F., 2015b. Relationship between structure and conformational change of the vitamin D receptor ligand binding domain in 1alpha,25-dihydroxyvitamin D3 signaling. Molecules 20 (11), 20473–20486.

Wang, J., Malloy, P.J., Feldman, D., 2007. Interactions of the vitamin D receptor with the corepressor hairless: analysis of hairless mutants in atrichia with papular lesions. J. Biol. Chem. 282 (35), 25231–25239.

Wang, H., Cheng, B., Chen, Q., Wu, S., Lv, C., Xie, G., et al., 2008. Time course of plasma gelsolin concentrations during severe sepsis in critically ill surgical patients. Crit. Care 12 (4), R106.

Wang, X., Gocek, E., Liu, C.G., Studzinski, G.P., 2009. MicroRNAs181 regulate the expression of p27Kip1 in human myeloid leukemia cells induced to differentiate by 1,25-dihydroxyvitamin D3. Cell Cycle 8 (5), 736–741.

Wang, T.J., Zhang, F., Richards, J.B., Kestenbaum, B., van Meurs, J.B., Berry, D., et al., 2010. Common genetic determinants of vitamin D insufficiency: a genome-wide association study. Lancet 376 (9736), 180–188.

Wang, X., Shapses, S.A., Al-Hraishawi, H., 2017. Free and bioavailable 25-hydroxyvitamin D levels in patients with primary hyperparathyroidism. Endocr. Pract. 23 (1), 66–71.

Wei, R., Christakos, S., 2015. Mechanisms underlying the regulation of innate and adaptive immunity by vitamin D. Nutrients 7 (10), 8251–8260.

White, P., Cooke, N., 2000. The multifunctional properties and characteristics of vitamin D-binding protein. Trends Endocrinol. Metabol. 11 (8), 320–327.

White, J.H., 2004. Profiling 1,25-dihydroxyvitamin D3-regulated gene expression by microarray analysis. J. Steroid Biochem. Mol. Biol. 89–90 (1–5), 239–244.

Whitfield, G.K., Selznick, S.H., Haussler, C.A., Hsieh, J.C., Galligan, M.A., Jurutka, P.W., et al., 1996. Vitamin D receptors from patients with resistance to 1,25-dihydroxyvitamin D3: point mutations confer reduced transactivation in response to ligand and impaired interaction with the retinoid X receptor heterodimeric partner. Mol. Endocrinol. 10 (12), 1617–1631.

Wiese, R.J., Uhland-Smith, A., Ross, T.K., Prahl, J.M., DeLuca, H.F., 1992. Up-regulation of the vitamin D receptor in response to 1,25-dihydroxyvitamin D3 results from ligand-induced stabilization. J. Biol. Chem. 267 (28), 20082–20086.

Wilson, R.T., Bortner Jr., J.D., Roff, A., Das, A., Battaglioli, E.J., Richie Jr., J.P., et al., 2015. Genetic and environmental influences on plasma vitamin D binding protein concentrations. Transl. Res. 165 (6), 667–676.

Wood, R.J., Fleet, J.C., Cashman, K., Bruns, M.E., Deluca, H.F., 1998. Intestinal calcium absorption in the aged rat: evidence of intestinal resistance to 1,25(OH)2 vitamin D. Endocrinology 139 (9), 3843–3848.

Wu, S., Ren, S., Chen, H., Chun, R.F., Gacad, M.A., Adams, J.S., 2000. Intracellular vitamin D binding proteins: novel facilitators of vitamin D-directed transactivation. Mol. Endocrinol. 14 (9), 1387–1397.

Wu, W., Beilhartz, G., Roy, Y., Richard, C.L., Curtin, M., Brown, L., et al., 2010. Nuclear translocation of the 1,25D3-MARRS (membrane associated rapid response to steroids) receptor protein and NFkappaB in differentiating NB4 leukemia cells. Exp. Cell Res. 316 (7), 1101–1108.

Xie, Z., Bikle, D.D., 1997. Cloning of the human phospholipase C-gamma1 promoter and identification of a DR6-type vitamin D-responsive element. J. Biol. Chem. 272 (10), 6573–6577.

Xie, Z., Munson, S.J., Huang, N., Portale, A.A., Miller, W.L., Bikle, D.D., 2002. The mechanism of 1,25-dihydroxyvitamin D(3) autoregulation in keratinocytes. J. Biol. Chem. 277 (40), 36987–36990.

Xie, Z., Chang, S., Oda, Y., Bikle, D.D., 2006. Hairless suppresses vitamin D receptor transactivation in human keratinocytes. Endocrinology 147 (1), 314–323.

Xu, J., O'Malley, B.W., 2002. Molecular mechanisms and cellular biology of the steroid receptor coactivator (SRC) family in steroid receptor function. Rev. Endocr. Metab. Disord. 3 (3), 185–192.

Xu, H., McCann, M., Zhang, Z., Posner, G.H., Bingham, V., El-Tanani, M., et al., 2009. Vitamin D receptor modulates the neoplastic phenotype through antagonistic growth regulatory signals. Mol. Carcinog. 48 (8), 758–772.

Yamamoto, N., Homma, S., 1991. Vitamin D3 binding protein (group-specific component) is a precursor for the macrophage-activating signal factor from lysophosphatidylcholine-treated lymphocytes. Proc. Natl. Acad. Sci. U.S.A. 88 (19), 8539–8543.

Yamamoto, N., Homma, S., Millman, I., 1991. Identification of the serum factor required for in vitro activation of macrophages. Role of vitamin D3-binding protein (group specific component, Gc) in lysophospholipid activation of mouse peritoneal macrophages. J. Immunol. 147 (1), 273–280.

Yamamoto, N., Naraparaju, V.R., Asbell, S.O., 1996. Deglycosylation of serum vitamin D3-binding protein leads to immunosuppression in cancer patients. Cancer Res. 56 (12), 2827–2831.

Yamamoto, N., Naraparaju, V.R., Urade, M., 1997. Prognostic utility of serum alpha-N-acetylgalactosaminidase and immunosuppression resulted from deglycosylation of serum Gc protein in oral cancer patients. Cancer Res. 57 (2), 295–299.

Yang, X.J., Ogryzko, V.V., Nishikawa, J., Howard, B.H., Nakatani, Y., 1996. A p300/CBP-associated factor that competes with the adenoviral oncoprotein E1A. Nature 382 (6589), 319–324.

Ye, W.Z., Dubois-Laforgue, D., Bellanne-Chantelot, C., Timsit, J., Velho, G., 2001. Variations in the vitamin D-binding protein (Gc locus) and risk of type 2 diabetes mellitus in French Caucasians. Metab. Clin. Exp. 50 (3), 366–369.

Yin, J.W., Wang, G., 2014. The Mediator complex: a master coordinator of transcription and cell lineage development. Development 141 (5), 977–987.

Zella, L.A., Shevde, N.K., Hollis, B.W., Cooke, N.E., Pike, J.W., 2008. Vitamin D-binding protein influences total circulating levels of 1,25-dihydroxyvitamin D3 but does not directly modulate the bioactive levels of the hormone in vivo. Endocrinology 149 (7), 3656−3667.

Zella, L.A., Meyer, M.B., Nerenz, R.D., Lee, S.M., Martowicz, M.L., Pike, J.W., 2010. Multifunctional enhancers regulate mouse and human vitamin D receptor gene transcription. Mol. Endocrinol. 24 (1), 128−147.

Zhang, J., Kew, R.R., 2004. Identification of a region in the vitamin D-binding protein that mediates its C5a chemotactic cofactor function. J. Biol. Chem. 279 (51), 53282−53287.

Zhang, J., Habiel, D.M., Ramadass, M., Kew, R.R., 2010. Identification of two distinct cell binding sequences in the vitamin D binding protein. Biochim. Biophys. Acta 1803 (5), 623−629.

Zhang, J.Y., Lucey, A.J., Horgan, R., Kenny, L.C., Kiely, M., 2014. Impact of pregnancy on vitamin D status: a longitudinal study. Br. J. Nutr. 112 (7), 1081−1087.

Zhou, L., Zhang, X., Chen, X., Liu, L., Lu, C., Tang, X., et al., 2012. GC Glu416Asp and Thr420Lys polymorphisms contribute to gastrointestinal cancer susceptibility in a Chinese population. Int. J. Clin. Exp. Med. 5 (1), 72−79.

Chapter 30

Vitamin D gene regulation

Sylvia Christakos[1] and J. Wesley Pike[2]

[1]Department of Microbiology, Biochemistry and Molecular Genetics, Rutgers, New Jersey Medical School, Newark, NJ, United States; [2]Department of Biochemistry, University of Wisconsin–Madison, Madison, WI, United States

Chapter outline

Vitamin D metabolism 739
Role of $1,25(OH)_2D_3$ in classical target tissues 741
Bone 741
Intestine 742
Kidney 742
Parathyroid glands 743
Nonclassical actions of $1,25(OH)_2D_3$ 743
Transcriptional regulation by $1,25(OH)_2D_3$ 743
The vitamin D receptor 743
General features of VDR action 743
Sites of DNA binding 743
Heterodimer formation with retinoid X receptors 744
The vitamin D receptor functions to recruit coregulatory complexes that mediate gene regulation 744
Applying emerging methodologies to study vitamin D receptor action on a genome-wide scale 745
Overarching principles of VDR interaction at target cell genomes in bone cells 745
Genome-wide coregulatory recruitment to target genes via the vitamin D receptor 747
Identifying underlying early mechanistic outcomes in response to VDR/RXR binding 747
The dynamic impact of cellular differentiation and disease on vitamin D receptor cistromes and transcriptional outcomes 748
New approaches to the study of vitamin D-mediated gene regulation in vitro and in vivo 748
Defining the regulatory sites of action of $1,25(OH)_2D_3$, PTH, and FGF23 in the *Cyp27b1* and *Cyp24a1* genes in the kidney 749
Future directions 750
References 750
Further Reading 756

Vitamin D metabolism

Vitamin D is a principal factor required for the development and maintenance of bone as well as for maintaining normal calcium and phosphorus homeostasis. In addition, evidence has indicated the involvement of vitamin D in a number of diverse cellular processes, including effects on differentiation and cell proliferation, cancer progression, and the immune system (DeLuca, 2016; Christakos et al., 2016). For vitamin D to affect mineral metabolism as well as numerous other systems, it must first be metabolized to its active form. Vitamin D, which is taken in from the diet or is synthesized in the skin from 7-dehydrocholesterol in a reaction catalyzed by ultraviolet irradiation, is transported in the blood by the vitamin D-binding protein (DBP) to the liver. In the liver, vitamin D is hydroxylated at the carbon 25 position, resulting in the formation of 25-hydroxyvitamin D_3 ($25(OH)D_3$), the major circulating form of vitamin D, which is one of the most reliable biomarkers of vitamin D status (Bikle et al., 2013). The synthesis of $25(OH)D_3$ has not been reported to be highly regulated (DeLuca, 2008). Cytochrome P450 2R1 (CYP2R1) is probably the key vitamin D 25-hydroxylase responsible for the conversion of vitamin D to $25(OH)D_3$ based in part on genetic evidence that patients with a mutation in *CYP2R1* have vitamin D dysfunction (Cheng et al., 2004). However, in studies in *Cyp2r1*-null mice, $25(OH)D_3$ levels were markedly reduced but not abolished, suggesting that other hydroxylases may also be involved (Zhu et al., 2013). 25-Hydroxyvitamin D proceeds to the kidney via the serum DBP. Megalin, a member of the low-density lipoprotein receptor superfamily, plays an essential role in the renal uptake of $25(OH)D_3$ (Nykjaer et al., 1999). Cubilin and disabled 2 also work in conjunction with megalin for the cellular uptake of DBP/$25(OH)D_3$ (Nykjaer et al., 2001; Morris et al., 2002). In the proximal convoluted and straight tubules of the kidney nephron, $25(OH)D_3$ is hydroxylated at the position of carbon 1 of the A ring

Principles of Bone Biology. https://doi.org/10.1016/B978-0-12-814841-9.00030-0

through the action of mitochondrial 25(OH)D 1α-hydroxylase (CYP27B1), resulting in the formation of the hormonally active form of vitamin D, 1,25-dihydroxyvitamin D_3 (1,25$(OH)_2D_3$) (Fig. 30.1). Mutations in the *CY27B1* gene result in vitamin D-dependent rickets (VDDR) type I, indicating the importance of the *CYP27B1* enzyme (Kitanaka et al., 1998). *Cyp27b1*-null mice have provided a model of VDDR type I (undetectable 1,25$(OH)_2D_3$, low serum calcium, and secondary hyperparathyroidism) (Panda et al., 2001; Dardenne et al., 2001). The kidney can also produce 24,25-dihydroxyvitamin D_3 (24,25$(OH)_2D_3$). CYP24A1 has been reported to be capable of hydroxylating the 24 position of both 25$(OH)D_3$ and 1,25$(OH)_2D_3$ (Omdahl et al., 2003; Jones et al., 2014; Veldurthy et al., 2016) (see Fig. 30.1). It has been suggested that the preferred substrate for CYP24A1 in vivo is 1,25$(OH)_2D_3$ rather than 25$(OH)D_3$ (Shinki et al., 1992). Studies using mice with a targeted inactivating mutation of the *Cyp24a1* gene (*Cyp24a1*-null-mutant mice) provided the first direct in vivo evidence for a role for CYP24A1 in the catabolism of 1,25$(OH)_2D_3$ (St-Arnaud et al., 2000). Both chronic and acute treatment with 1,25$(OH)_2D_3$ resulted in an inability of *Cyp24a1*-deficient mice to clear 1,25$(OH)_2D_3$ from their bloodstream. CYP24A1 limits the amount of 1,25$(OH)_2D_3$ by catalyzing the conversion of 1,25$(OH)_2D_3$ into 24-hydroxylated products targeted for excretion. In addition, production of 24,25$(OH)_2D_3$ by CYP24A1 decreases the pool of 25$(OH)D_3$ available for 1-hydroxylation. CYP24A1 can also catalyze the C23 oxidation pathway, resulting in the formation of 1,25$(OH)_2D_3$ 26,23-lactone from 1,25$(OH)_2D_3$ and 25$(OH)D_3$ 26,23-lactone from 25$(OH)D_3$ (Jones et al., 2014). CYP24A1 is present not only in kidney but also in all cells that contain the vitamin D receptor (VDR). Thus CYP24A1 not only regulates circulating 1,25$(OH)_2D_3$ concentrations, but may also modulate the amount of 1,25$(OH)_2D_3$ in target tissues, resulting in an appropriate cellular response. Inactivating mutations in CYP24A1 have been found in young children with idiopathic infantile hypercalcemia and in adults (Schlingmann et al., 2011; Dinour et al., 2013). In adults, patients were characterized by hypercalcemia, hypercalciuria, and recurring kidney stones (Dinour et al., 2013). These findings provide evidence for a critical role of CYP24A1 in the regulation of 1,25$(OH)_2D_3$ in humans.

The production of 1,25$(OH)_2D_3$ and 24,25$(OH)_2D_3$ is under stringent control. Calcium and phosphorus deprivation results in enhanced production of 1,25$(OH)_2D_3$ (Omdahl et al., 2003; Jones et al., 2014). Elevated parathyroid hormone (PTH) resulting from calcium deprivation may be the primary signal mediating the calcium regulation of 1,25$(OH)_2D_3$ synthesis (Boyle et al., 1971; Henry, 1985). PTH has been reported to stimulate *CYP27B1* (Brezna and DeLuca, 2000; Meyer et al., 2017). 1,25$(OH)_2D_3$ also regulates its own production by inhibiting *CYP27B1* (Brenza and DeLuca, 2000; Meyer et al., 2017). The synthesis of 24,25$(OH)_2D_3$ has been reported to be reciprocally regulated compared with the synthesis of 1,25$(OH)_2D_3$ (stimulated by 1,25$(OH)_2D_3$, and inhibited by low calcium and PTH) (Shinki et al., 1992; Omdahl et al., 2003; Jones et al., 2014) (Fig. 30.1). Fibroblast growth factor 23 (FGF23), which promotes renal phosphate excretion and requires klotho, a transmembrane protein, has also been identified as a physiological regulator of vitamin D metabolism, which suppresses the expression of *CYP27B1* (Shimada et al., 2004; Hu et al., 2013) (Fig. 30.1). Low dietary phosphate intake results in decreased FGF23 levels and a corresponding increase in *CYP27B1* activity (Perwad et al.,

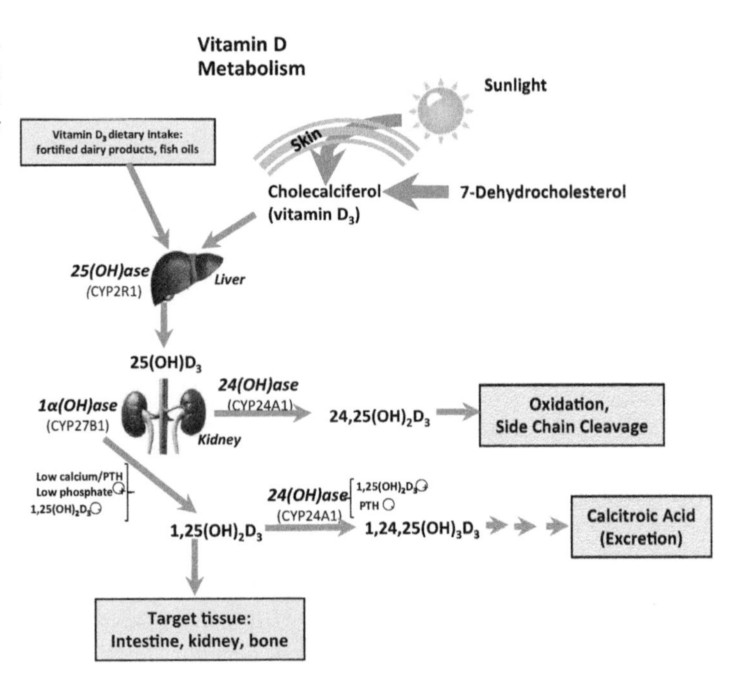

FIGURE 30.1 The pathway of vitamin D metabolism. Cytochrome P450 2R1 (*CYP2R1*) has been identified as a key vitamin D 25-hydroxylase. Parathyroid hormone (*PTH*), fibroblast growth factor 23/α-klotho, and serum calcium and phosphate act together to maintain optimal 1,25$(OH)_2D_3$ levels (Dhawan et al., 2017).

2005). FGF23 requires the transmembrane receptor α-klotho (a multifunctional protein involved in phosphate and calcium homeostasis and in aging that binds to FGF receptors) to activate FGF signaling (Hu et al., 2013). Studies by Meyer et al. (2017) identified a kidney-specific control module generated by a renal cell–specific chromatin structure located distal to *Cyp27b1* that mediates PTH, FGF23, and 1,25$(OH)_2D_3$ regulation of *Cyp27b1* expression, which will be discussed later in this chapter. In addition to the kidney, *CYP27B1* is also expressed during pregnancy in placenta (Zehnder et al., 2002). Extrarenal production by macrophages of CYP27B1 has also been convincingly demonstrated in patients with sarcoidosis (Barbour et al., 1981; Adams et al., 1983). Other immune cells also can produce 1,25$(OH)_2D_3$ (Ooi et al., 2014). In addition, cancer cells have been shown to express *CYP27B1* (Hobaus et al., 2013). Although *CYP27B1* expression has been noted in a number of tissues (Bikle, 2009), whether there is a functional impact of *CYP27B1* activity in vivo in tissues other than the kidney and placenta under normal physiological conditions remains to be determined.

Serum 1,25$(OH)_2D_3$ levels and the capacity of the kidney to hydroxylate 25$(OH)D_3$ to 1,25$(OH)_2D_3$ have been reported to decline with age (Armbrecht et al., 1980). In addition, an increase in renal *CYP24A1* gene expression and an increase in the clearance of 1,25$(OH)_2D_3$ with aging have been reported (Matkovits and Christakos, 1995; Wada et al., 1992). These findings have implications concerning the etiology of osteoporosis and suggest that the combined effect of a decline in the ability of the kidney to synthesize 1,25$(OH)_2D_3$ and an increase in the renal metabolism of 1,25$(OH)_2D_3$ may contribute to age-related bone loss. Whether there is an interrelationship between the decline of sex steroids with age and age-related changes in *CYP27B1* and *CYP24A1* remains to be determined.

Role of 1,25$(OH)_2D_3$ in classical target tissues

Bone

Exactly how 1,25$(OH)_2D_3$ affects mineral homeostasis is a subject of continuing investigation. It has been suggested that the antirachitic action of 1,25$(OH)_2D_3$ is indirect and the result of increased intestinal absorption of calcium and phosphorus by 1,25$(OH)_2D_3$, thus resulting in their increased availability for incorporation into bone (Underwood and DeLuca, 1984; Weinstein et al., 1984). Studies using VDR-ablated mice (VDR-null mice) also suggest that a principal role of the VDR in skeletal homeostasis is its role in intestinal calcium absorption (Li et al., 1997; Yoshizawa et al., 1997; Amling et al., 1999). When VDR-null mice were fed a calcium/phosphorus/lactose-enriched diet, serum-ionized calcium levels were normalized, the development of hyperparathyroidism was prevented, and the animals did not develop rickets or osteomalacia, although alopecia was still observed. In addition, transgenic expression of VDR in the intestine of VDR-null mice results in normalization of serum calcium, bone density, and bone volume (Xue and Fleet, 2009). Thus, it was suggested that skeletal consequences of VDR ablation are due primarily to impaired intestinal calcium absorption. In vitro studies, however, have shown that 1,25$(OH)_2D_3$ can resorb bone (Raisz et al., 1972). Although 1,25$(OH)_2D_3$ stimulates the formation of bone-resorbing osteoclasts, receptors for 1,25$(OH)_2D_3$ are not present in osteoclasts but rather in osteoblasts (Wang et al., 2014). Stimulation of osteoclast formation by 1,25$(OH)_2D_3$ requires cell-to-cell contact between osteoblastic cells and osteoclast precursors and involves upregulation by 1,25$(OH)_2D_3$ in osteoblastic cells of osteoprotegerin ligand or receptor activator of nuclear factor-κB ligand (RANKL; Takeda et al., 1999; Yasuda et al., 1998a, 1998b). RANKL is induced by 1,25$(OH)_2D_3$ (Kim et al., 2007; Nerenz et al., 2008) as well as by PTH in osteoblasts/stromal cells. In *Cyp27a1*-null mice and in VDR-null mice, although PTH levels are markedly elevated, osteoclast numbers are not increased, indicating that both 1,25$(OH)_2D_3$ and VDR are necessary for PTH production of osteoclasts (Amling et al., 1999; Panda et al., 2001). RANKL is a member of the membrane-associated tumor necrosis factor ligand family that enhances osteoclast formation by mediating direct interactions between osteoblasts/stromal cells and osteoclast precursor cells. Osteoclastogenesis-inhibitory factor/osteoprotegerin, a member of the tumor necrosis factor receptor family, is a soluble decoy receptor for RANKL that antagonizes RANKL function, thus blocking osteoclastogenesis. Osteoprotegerin is downregulated by 1,25$(OH)_2D_3$ (Yasuda et al., 1998a).

In addition to increasing the availability of calcium and phosphorus for incorporation into bone and stimulating osteoclast formation, direct effects of 1,25$(OH)_2D_3$ on bone cells have also been demonstrated (Bikle, 2012). For example, 1,25$(OH)_2D_3$ has been reported to stimulate the synthesis of the calcium-binding proteins osteocalcin (OC) (Price and Baukol, 1980; Demay et al., 1990; Kerner et al., 1989) and osteopontin (OPN) (Prince and Butler, 1987; Noda et al., 1990; Shen and Christakos, 2005) in osteoblastic cells. OPN, a major noncollagenous bone protein, has been reported to inhibit bone matrix mineralization (Boskey et al., 1993). 1,25$(OH)_2D_3$ suppresses mineralization by increasing the expression of mineralization inhibitors in osteoblasts in addition to OPN, including *Ennp1* and *Ennp3* (ectonucleotide pyrophosphastase phosphodiesterase 1 and 3) (Lieben et al., 2012). Thus during a negative calcium balance 1,25$(OH)_2D_3$ can promote increased bone resorption and reduced matrix mineralization to maintain normal serum calcium levels at the expense of skeletal integrity (Lieben et al., 2012). Direct effects of $1,25(OH)_2D_3$ on bone cells have also been demonstrated in osteocytes. $1,25(OH)_2D_3$ stimulates FGF23 production from osteocytes, resulting in renal modulation of phosphate levels (Lanske et al., 2014). Collectively these studies support a direct effect of $1,25(OH)_2D_3$ on bone as well as an indirect role through stimulation of intestinal calcium absorption.

Intestine

The major defect in VDR-null mice is in intestinal calcium absorption indicating that a principal action of $1,25(OH)_2D_3$ is to maintain calcium homeostasis via increasing efficiency of intestinal calcium absorption (Li et al., 1997; Yoshizawa et al., 1997; Amling et al., 1999). In addition, when patients with hereditary $1,25(OH)_2D_3$-resistant rickets (HVDRR) are treated with intravenous or high-dose oral calcium the skeletal phenotype of the patients is reversed (Hochberg et al., 1992). Although these studies indicate the importance of the intestine in $1,25(OH)_2D_3$ regulation of intestinal calcium absorption, the mechanisms involved are still incompletely understood. When the demand for calcium increases due to a diet deficient in calcium, demand for skeletal growth, or pregnancy or lactation, synthesis of $1,25(OH)_2D_3$ increases and intestinal calcium absorption occurs predominantly by an active transcellular process (Wasserman and Fullmer, 1995). In the transcellular process $1,25(OH)_2D_3$ has been reported to act by regulating (1) calcium entry through the apical calcium channel transient receptor potential cation channel subfamily V member 6 (TRPV6), (2) transcellular movement of calcium by binding to the calcium-binding protein calbindin, and (3) extrusion of calcium from the cell by the plasma membrane calcium ATPase (PMCA1b). Studies in TRPV6-null and calbindin-D-null mice have challenged this traditional view. There are no phenotypic differences between calbindin-D_{9k}-null mice and TRPV6-null mice and wild-type mice when dietary calcium is normal (Benn et al., 2008; Kutuzova et al., 2006; Kutuzova et al., 2008). However, studies in TRPV6/calbindin-D_{9k} double-null mice under conditions of low dietary calcium have shown that intestinal calcium absorption is least efficient in the absence of both proteins (compared with single-null mice and wild-type mice), suggesting that TRPV6 and calbindin can act together in certain aspects of the absorptive process (Benn et al., 2008). Findings in the single-null mice under adequate calcium conditions suggest that calbindin-D_{9k} and TRPV6 are redundant for intestinal calcium absorption and suggest compensation by other channels or proteins yet to be identified (Benn et al., 2008; Kutuzova et al, 2006, 2008; Christakos, 2012). Although other apical membrane transporters may compensate for the loss of TRPV6, intestine-specific transgenic expression of TRPV6 has been shown to result in a marked increase in intestinal calcium absorption and bone density in VDR-null mice, indicating a direct role for TRPV6 in the calcium absorptive process and that a primary defect in VDR-null mice is low apical membrane calcium uptake (Cui et al., 2012). It should also be noted that although the duodenum has been the focus of research related to $1,25(OH)_2D_3$ regulation of calcium absorption, it is the distal intestine where most of the ingested calcium is absorbed. Studies have shown that transgenic expression of VDR specifically in ileum, cecum, and colon can prevent abnormal calcium homeostasis and rickets in VDR-null mice (Dhawan et al., 2017a, 2017b). These findings indicate that, although calcium is absorbed more rapidly in the duodenum, the distal segments of the intestine contribute significantly to vitamin D regulation of calcium homeostasis. Studies related to mechanisms involved in $1,25(OH)_2D_3$ regulation of calcium absorption in the distal intestine may suggest new strategies to increase the efficiency of calcium absorption in individuals at risk for bone loss due to aging, bariatric surgery, or inflammatory bowel disease.

Kidney

In addition to bone and intestine, a third target tissue involved in the regulation by $1,25(OH)_2D_3$ of mineral homeostasis is the kidney. Although most of the filtered calcium is reabsorbed by a passive, paracellular path in the proximal renal tubule that is independent of $1,25(OH)_2D_3$, 10%−15% of the filtered calcium is reabsorbed in the distal convoluted tubule and connecting tubule and is regulated by PTH and $1,25(OH)_2D_3$ (Boros et al., 2009). $1,25(OH)_2D_3$ has been reported to enhance the stimulatory effect of PTH on calcium transport in the distal nephron in part by increasing PTH receptor mRNA and binding activity in distal tubule cells (Sneddon et al., 1998). Similar to studies in the intestine, $1,25(OH)_2D_3$ regulates an active transcellular process in the distal portion of the nephron by inducing the apical calcium channel TRPV5 (which shows 73.4% sequence homology with TPRV6) and by inducing the calbindins (both calbindin-D_{9k} [9000 M_r] and calbindin-D_{28k} [28,000 M_r] are present in mouse kidney and only calbindin-D_{28k} is present in rat and human kidney) (Boros et al., 2009; Rhoten et al., 1985; Christakos et al., 1989; Christakos, 1995). Calcium is extruded via PMCA1b and the Na^+/Ca^{2+} exchanger (Boros et al., 2009). Mice lacking TRPV5 display diminished calcium reabsorption in the distal tubule, hypercalciuria, and disturbances in bone structure, supporting the suggested role for TRPV5 in renal calcium handling (Hoenderop et al., 2003). Another important effect of $1,25(OH)_2D_3$ in the kidney is inhibition of *CYP27B1* and stimulation of *CYP24A1* (Ohyama et al., 1996; Shinki et al., 1992; Kerry et al., 1996; Zierold et al., 1995; Meyer et al., 2010a, 2010b). In addition to effects on calcium transport in the distal nephron and modulation of the hydroxylases, effects of 1,25$(OH)_2D_3$ on renal phosphate reabsorption have also been suggested. 1,25$(OH)_2D_3$ may regulate phosphate homeostasis by increasing FGF23 expression in osteocytes and by inducing klotho expression in the distal tubule (Lanske et al., 2014; Forster et al., 2011).

Parathyroid glands

1,25$(OH)_2D_3$ and its analogs markedly decrease parathyroid hormone expression (Zella et al., 2014). 1,25$(OH)_2D_3$ upregulates the calcium receptor, which is a major regulator of PTH (Canaff and Hendy, 2002). Parathyroid-specific VDR-null mice showed only a modest increase in PTH levels (Mier et al., 2009). In contrast, deletion of the calcium receptor in parathyroid resulted in severe hyperparathyroidism and the condition was lethal (Wettschureck et al., 2007). These findings suggest that a principal function of 1,25$(OH)_2D_3$ is to sensitize the parathyroid gland to calcium inhibition by upregulating the calcium receptor. Direct action of 1,25$(OH)_2D_3$ on the prepro-PTH gene has also been reported (Demay et al., 1992).

Nonclassical actions of 1,25$(OH)_2D_3$

In addition to the regulation of calcium homeostasis, many additional biological processes are regulated by 1,25$(OH)_2D_3$, including cellular proliferation and differentiation, inhibition of cancer progression, regulation of adaptive and innate immunity, and effects on cardiovascular function (for review see Christakos et al., 2016; Christakos and DeLuca, 2011). Convincing evidence for extraskeletal effects of 1,25$(OH)_2D_3$ has been obtained through studies in cells and animal models. Understanding 1,25$(OH)_2D_3$ extraskeletal biological responses may suggest similar pathways in humans that could lead to the development of new therapies to prevent and treat disease.

Transcriptional regulation by 1,25$(OH)_2D_3$

The vitamin D receptor

The VDR mediates the genomic mechanism of action of 1,25$(OH)_2D_3$ and was first discovered in the chicken intestine in 1974 and shortly thereafter in other tissues, including the parathyroid glands, kidney, and bone (Brumbaugh and Haussler, 1974a, 1974b; Brumbaugh et al., 1975). The protein's biochemical features, including its retention in chromatin (Haussler et al., 1968) and its ability to bind to DNA (Pike and Haussler, 1979), suggested that it was similar to other receptors for known steroid hormones and that it probably played a role in transcriptional regulation. However, it was the cloning of the chicken *VDR* gene (McDonnell et al., 1987) and then the human (Baker et al., 1988) and rat (Burmester et al., 1988) versions that ushered in a new era of defining the mechanisms through which vitamin D operated to control gene expression. Aside from the development of unique experimental probes, the cloning of the VDR enabled studies of the domain structure of the receptor that confirmed that it was a bona fide member of the steroid receptor gene family (Evans, 1988; Jin et al., 1996; McDonnell et al., 1989). Equally important, the cloning and structural analysis of the VDR's human chromosomal gene that followed (Hughes et al., 1988; Miyamoto et al., 1997) led ultimately to the identification of a series of mutations within the gene itself that were responsible for the syndrome of HVDRR (Brooks et al., 1978; Hughes et al., 1991; Malloy et al., 1990; Ritchie et al., 1989; Sone et al., 1989). This syndrome had been identified in 1978 (Brooks et al., 1978) and its etiology suggested at the time that it could be due to a defect(s) in the mechanism through which vitamin D exerted its actions (Eil et al., 1981; Lin et al., 1996; Pike et al., 1984; Sone et al., 1990), a hypothesis that was extended during the intervening years (Malloy et al., 2014). The discovery of mutations in the *VDR* gene solidified the essential role of the VDR as the sole mediator of the activities of the vitamin D hormone. This conclusion was eventually confirmed and extended through studies that recapitulated features of the disease phenotype through deletion of the VDR from the mouse genome.

General features of VDR action

Sites of DNA binding

Initial studies suggested that 1,25$(OH)_2D_3$ activated gene expression programs in a wide variety of cell types and identified numerous gene candidates for further investigation. Most prominent among these were tissues that expressed the vitamin D-dependent calcium-binding proteins (calbindins) (Christakos et al., 1989; Gill and Christakos, 1993), OC (Lian et al.,

1989; Price and Baukol, 1980), OPN (Prince and Butler, 1987), and CYP24A1 (Haussler et al., 1980), although numerous others emerged during the following several decades as well. The cloning of many of the genes for these proteins and identification of their structural organization also prompted exploration of the mechanisms through which $1,25(OH)_2D_3$ and its receptor promoted their regulation. These studies, first with human *BGLP* (OC) (Kerner et al., 1989; Ozono et al., 1990) and subsequently with *Spp1* (OPN) (Noda et al., 1990), *Cyp24a1* (Ohyama et al., 1994, 1996; Zierold et al., 1994, 1995) and others (Carlberg, 2003), suggested that the VDR bound to a 15-bp vitamin D-responsive DNA element (VDRE) comprising two directly repeated consensus AGGTCA hexanucleotide half-sites separated by 3 bp that was generally located within a kilobase or so of the promoters for these genes (Ozono et al., 1990). These features were similar, but not identical, to those for other nuclear receptors. $1,25(OH)_2D_3$ also strongly suppressed the expression of numerous genes, most notably that of *PTH* and *CYP27B1*, but also in more recent studies that of the interleukin-17 (*IL-17*) gene. Repression of IL-17 transcription by $1,25(OH)_2D_3$ is thought to involve, in part, dissociation of histone acetylase activity, recruitment of deacetylase, and VDR/retinoid X receptor (RXR) binding to NFAT sites (Joshi et al., 2011). While some progress has been made, additional details of the mechanisms of suppression for these and other downregulated genes have yet to be fully understood but are almost certain to be highly diverse. Nevertheless, some progress has been made relative to the suppression of *Cyp27b1* by $1,25(OH)_2D_3$, and will be considered in specific detail later in this chapter. Although the presence of unique "negative VDREs" has been suggested at negatively regulated genes such as the PTH gene, this type of mechanism has yet to be substantiated (Demay et al., 1992).

Heterodimer formation with retinoid X receptors

The discovery of the first VDREs enabled the important subsequent finding that VDR binding to these specific DNA sequences was dependent upon an unknown nuclear factor (Liao et al., 1990; Sone et al., 1991a, 1991b). The identity of this protein was revealed shortly thereafter when it was found that RXRs, also members of the steroid receptor family, were capable of forming heterodimeric complexes with the VDR as well as other members of the steroid receptor class (Mangelsdorf and Evans, 1995). Importantly, $1,25(OH)_2D_3$ was found to promote heterodimer formation between VDR and RXR, although the precise cellular location where this interaction occurs between the two remains unclear (Kliewer et al., 1992). It has been suggested that in addition to its contribution to DNA binding, RXR may participate in the transcriptional activation process (Pathrose et al., 2002; Thompson et al., 2001); structural studies suggest otherwise (Orlov et al., 2012).

The vitamin D receptor functions to recruit coregulatory complexes that mediate gene regulation

Early studies revealed that transcription factor binding near promoters leads to an interaction with basal transcriptional machinery that enhances gene output. It is now known, however, that the activity of most DNA-binding transactivators involves the recruitment of additional coregulatory complexes by this protein class that function to modulate chromatin in a highly specific fashion. Overcoming and/or restoring the inherent repressive state of chromatin requires the presence of this complex regulatory machinery, which is able to shift and/or displace nucleosomes (chromatin remodeling), alter the condensation state and therefore the local architecture of chromatin (histone modifications), create or restrict novel binding sites for additional coregulatory complexes (epigenetic sites), and/or facilitate the entry of RNA polymerase II (RNA pol II) at appropriate times and sites. Three complexes that participate in these activities that are known to be recruited by the VDR include (1) vertebrate ATPase-containing homologs of the yeast SWI/SNF complex that utilize the energy of ATP to remodel and reposition nucleosomes (Carlson and Laurent, 1994); (2) complexes that contain either histone acetyltransferases (HATs) or methyltransferases and deacetyltransferases (HDACs) or histone demethylases, which function to modify the lysine- or arginine-containing tails of histone 3 and/or histone 4 at specific locations (Rachez and Freedman, 2000; Smith and O'Malley, 2004); and (3) the Mediator complex, believed to facilitate the entry of RNA pol II into the general transcriptional apparatus and perhaps to play a role in transcriptional reinitiation (Lewis and Reinberg, 2003) (Fig. 30.2). The activities of each of these complexes may be essential for the downstream activity of almost all DNA-binding proteins. Several classes of regulatory transferase complexes comprise components of dynamic and highly active mechanisms that are epigenetic in nature, and involve the coordinated integration of expression of gene networks (Arrowsmith et al., 2012). These programs are widely responsible for development, differentiation, and mature cell function (Gifford et al., 2013; Xie et al., 2013). HATs and their reciprocal HDACs, for example, regulate the level of epigenetic histone H3 and H4 marks, controlling the degree to which chromatin is condensed and therefore the DNA accessible for transcription factor binding (Ho and Crabtree, 2010). The recruitment of these large chromatin regulatory complexes is frequently coordinated by factors such as the p160 family steroid receptor coactivator 1 (SRC1), SRC2, and

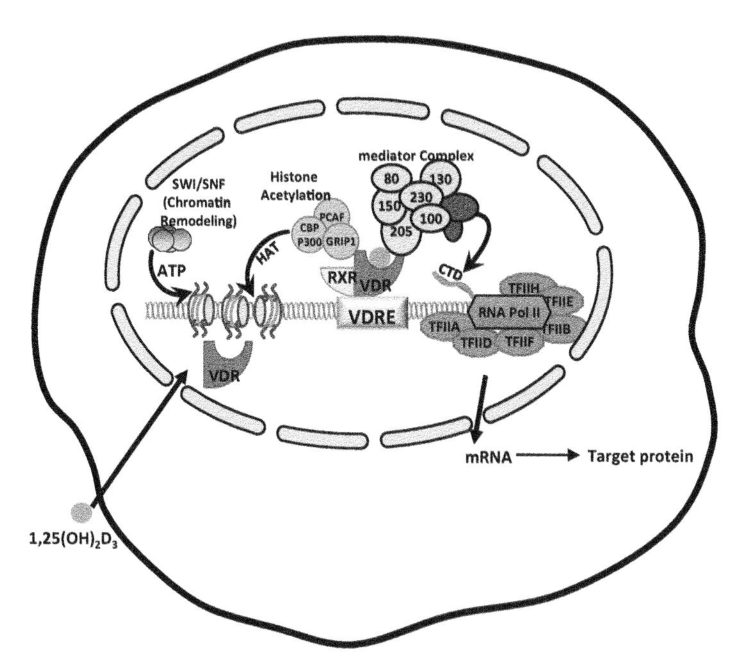

FIGURE 30.2 Mechanism of action of $1,25(OH)_2D_3$ in target cells. $1,25(OH)_2D_3$ regulates gene transcription in target cells by binding to its nuclear receptor, *VDR*. Activated VDR heterodimerizes with retinoid X receptor (*RXR*) and binds to vitamin D-response elements (*VDREs*) in and around target genes. In addition to basal transcription factors, VDR interacts with steroid receptor coactivator 2 (SRC2; also known as *GRIP1*), which has histone acetylase (HAT) activity as a primary coactivator. SRC2 can recruit proteins as secondary coactivators, such as CBP/P300 (which also have HAT activity). VDR also interacts with the Mediator complex, which facilitates the activation of the RNA polymerase II (*RNA Pol II*) holoenzyme through its C-terminal domain (*CTD*), thus promoting the formation of the preinitiation complex. The SWI/SNF complex, which remodels chromatin using the energy of ATP hydrolysis, also contributes to activation by VDR (Dhawan et al., 2017).

SRC3; the HATs CBP and p300; and the corepressor SMRT or NCoR, as well as any one of the many HDACs that interact directly with transcription factors such as the VDR to suppress activity (Smith and O'Malley, 2004). Importantly, activation of the VDR with $1,25(OH)_2D_3$ leads to the creation of a binding site on the VDR protein that mediates the link between the receptor and these coregulatory complexes (McInerney et al., 1998; Perissi et al., 1999; Westin et al., 1998). It is important to note, however, that many of these coregulatory factors are frequently prebound to sites of regulatory action by $1,25(OH)_2D_3$ as a result of their prior recruitment through additional DNA-binding proteins that co-interact at sites of VDR binding. Studies show that the ability of the VDR to recruit several of these coregulatory factors results in striking changes in epigenetic histone marks that facilitate altered gene output (Martowicz et al., 2011; Meyer et al., 2007; Zella et al., 2010). Thus, it is clear that, like other DNA-binding proteins, the function of the VDR is simply to focus and/or modulate the recruitment of transcriptionally active complexes to gene subsets that are then responsible for vitamin D hormone response. Considerable crystallographic information has now accrued to support not only the structural organization of the VDR/RXR heterodimer, but also its association with DNA and its recruitment of coregulators (Orlov et al., 2012). Many of these interactions are prompted by $1,25(OH)_2D_3$, and may well be differentially affected in the present of analogs of the vitamin D hormone. These differential interactions are proposed to account for the concept of analog selectivity in vivo, although this remains unproven and is highly controversial at this writing.

Applying emerging methodologies to study vitamin D receptor action on a genome-wide scale

The study of transcription has been revitalized since 2010, in part as a result of the coupling of chromatin immunoprecipitation (ChIP) to next-generation DNA sequence (ChIP-seq) analysis, providing new methodologies that are now capable of revealing unprecedented mechanistic detail on a genome-wide scale (Ernst and Kellis, 2010; Ernst et al., 2011; Hoffman et al., 2013). Using these methodologies many new principles have emerged, particularly when paired with transcriptomic measurements using RNA-sequencing analysis (Wang et al., 2009). In the following we will highlight the results of some of these collective studies, which have provided important new insights into the transcriptional mechanism of action of $1,25(OH)_2D_3$.

Overarching principles of VDR interaction at target cell genomes in bone cells

A compelling advantage of ChIP-seq analysis is the ability of this approach to identify VDR binding sites and other features of gene regulation in an unbiased manner on a genome-wide scale. Thus, although studies using traditional methods correctly pinpointed the regulatory regions of a few vitamin D target genes located near promoters, including *OC*,

OPN, *Cyp24a1*, and others (Kim et al., 2005), subsequent studies using ChIP-chip and ChIP-seq analyses not only confirmed many of these findings but also provided a broader overarching, genome-wide perspective on binding sites for the VDR in cells in general (see Table 30.1). In osteoblasts, we could detect approximately 1000 residual binding sites for the VDR in the absence of $1,25(OH)_2D_3$ using ChIP-chip analysis, which was increased to approximately 7000–8000 in the presence of $1,25(OH)_2D_3$ (Meyer et al., 2010b). This collection of binding sites has been termed a cellular cistrome and supports the idea that VDR DNA binding is largely hormone dependent (Meyer et al., 2014a). Importantly, similar analyses of binding sites for RXR revealed a more extensive collection of sites that was only modestly increased through $1,25(OH)_2D_3$ activation, emphasizing that RXR is a heterodimer partner not only for the VDR but also for several additional nuclear receptors (Mangelsdorf and Evans, 1995). Interestingly, RXR was frequently prebound at potential VDR binding sites in the absence of $1,25(OH)_2D_3$, suggesting that it might mark potential sites of VDR interaction on DNA. Each of these findings was confirmed in additional studies in mesenchymal stem cells, osteoblasts, osteocytes, and adipocytes, and revealed that while overlap was present, DNA sites for the VDR differed significantly across the genome depending on the cell type and stage examined (Meyer et al., 2014b, 2016; St John et al., 2014). Studies of VDR binding in EB-immortalized human B cells, primary B cells and monocytes (Ramagopalan et al., 2010), and THP-1 monocytes (Heikkinen et al., 2011) confirmed these conclusions. Importantly, although genome-wide ChIP-seq analyses demonstrate binding sites for the VDR near the promoters of genes such as OC, OPN, *Cyp24a1*, and a few others (Meyer et al., 2014b), this technique was unable to confirm a promoter-proximal VDR element for many genes, suggesting that regulatory elements were likely to be located elsewhere. Despite this, however, de novo motif finding analyses of the most common DNA sequence elements found in thousands of identified VDR binding sites confirm that most contain the originally postulated VDRE motif composed of AGGTCA xxg AGGTCA (Meyer et al., 2012, 2014b). Thus, the consensus VDRE initially identified in the human OC gene (Kerner et al., 1989; Ozono et al., 1990) remains the most representative of the DNA sequences with which the VDR interacts in mammalian genomes.

Interestingly, the unbiased and distance-unrestricted nature of ChIP-seq analysis also provided a number of surprising and unexpected insights of major significance (Ong and Corces, 2011). Perhaps most important, although traditional studies of $1,25(OH)_2D_3$ action identified regions immediately upstream of transcriptional start sites as indicated earlier, unbiased ChIP-seq analyses revealed that regulatory regions for the vitamin D hormone and its receptor are more commonly located in clusters within introns or in intergenic regions tens if not hundreds of kilobases upstream or downstream of regulated genes (Meyer et al., 2012, 2014b, 2016). Examples of such distal elements for the VDR abound, but can be found in many of the genes whose putative promoter proximal elements were undetectable by ChIP-seq analysis. They include the mouse *Tnfsf11* (RANKL) gene in bone cells, where at least five intergenic regulatory regions for the VDR are located (Kim et al., 2006); the *Cyp24a1* gene in numerous cell types, including bone cells, where in addition to the well-known promoter proximal element discussed before, a complex downstream cluster of regulatory elements exists in both the mouse and the human genes (Meyer et al., 2010a); the *Vdr* gene in bone cells, where both upstream regulatory regions and several intronic elements are present, a topic to be considered later (Lee et al., 2015a; Zella

TABLE 30.1 Overarching principles of vitamin D action in target cells.

Active transcription unit for induction: the VDR/RXR heterodimer
VDR binding sites (the VDR cistrome): 2000–8000 $1,25(OH)_2D_3$-sensitive binding sites per genome whose number and location are chromatin dependent and a function of cell type
Mode of DNA binding: predominantly, but not exclusively, $1,25(OH)_2D_3$ dependent
VDR/RXR binding site sequence (VDRE): induction mediated by classic hexameric half-sites (AGGTCA) separated by 3 bp; repression mediated by divergent sites
Distal binding site locations: dispersed across the genome in *cis*-regulatory modules (CRMs, or enhancers); located in a cell-type-specific manner near promoters, but preferentially within introns and distal intergenic regions; frequently located in clusters of elements
Epigenetic CRM signatures: defined by dynamically regulated posttranslational histone H3 and H4 modifications
Modular features of CRMs: contain binding sites for multiple transcription factors that facilitate either independent or synergistic interaction and mediate integration
VDR cistromes: dynamic alterations in the cellular epigenome during differentiation, maturation, and disease provoke changes to the VDR cistrome that qualitatively and quantitatively affect the vitamin D-regulated transcriptome

RXR, retinoid X receptor; *VDR*, vitamin D receptor.

et al., 2010); the *TRPV6* gene in intestinal cells, which contains multiple upstream elements (Lee et al., 2015b; Meyer et al., 2006); the *S100g* and PMCA2b (*Atp2b1*) genes in the intestine, which also contain multiple upstream elements (Lee et al., 2015b); and many target genes such as *c-FOS* and *c-MYC* and others in human colorectal cancer cells as well (Meyer et al., 2012). Indeed, enhancers for other transcription factors have been identified more than a megabase from the genes they are known to regulate, although at present the most distal VDR binding site is located 335 kb upstream of the human *c-MYC* promoter. It is important to clarify, however, that the linear/distal nature of regulatory elements for genes is illusionary, as distances do not take into account DNA packaging and looping events imposed on DNA that bring key regulatory segments into the proximity of a gene's promoter region (Deng and Blobel, 2010, 2014; 2017; Deng et al., 2012; Whalen et al., 2016). It is also worth indicating that of the tens of thousands of "putative" VDRE-like sequences that occur across the genome based purely on in silico analyses alone, only a small proportion of these elements mediate vitamin D action due to chromatin restriction (Kellis et al., 2014). An additional observation, as indicated earlier, is the finding through ChIP-seq analysis that most genes are regulated by more than one distal enhancer and, in some cases, by multiple regulatory regions. Recent ENCODE (Encyclopedia of DNA Elements) estimates suggest that genes on a genome-wide scale are regulated by an average of 10 independent enhancers (Kellis et al., 2014).

Modularity comprises an additional important regulatory feature of genes. Thus, individual enhancers often contain organized arrays of linear DNA sequences capable of assembling distinct, nonrandom DNA-binding transcription factor complexes that can function uniquely to regulate the gene with which they are linked. Therefore, it is interesting to note that over 42% of VDR-binding sites in bone cells are located in enhancers that contain prebound CCAAT/enhancer binding protein β (C/EBPβ) and the master regulator RUNX2 (Meyer et al., 2014a, 2014b). Indeed, these factors assemble in a highly organized fashion relative to each other, a nucleoprotein structure we have termed an osteoblast enhancer complex. Not surprisingly, both RUNX2 and C/EBPβ in this configuration can positively and perhaps negatively influence the overall regulatory activity of $1{,}25(OH)_2D_3$ and its receptor. An alternative and dispersed arrangement has also been identified in bone cells as highlighted in the *Mmp13* gene. Here, binding sites for the VDR, C/EBPβ, and RUNX2 are located across three separate upstream enhancers (Meyer et al., 2015a, 2015b). The activities of these three regions do not function independently, however, but rather influence one another's activity in an overall hierarchical manner to modulate the expression of *Mmp13*. It is known that many other genes contain this or similar arrangements as well. A collective summary of many of the newly acquired features of vitamin D-mediated gene regulation obtained via genome-wide analyses is provided in Table 30.1.

Genome-wide coregulatory recruitment to target genes via the vitamin D receptor

The primary function of the VDR is to recruit chromatin-active coregulatory complexes that facilitate in turn the modulation of gene output. Numerous studies at single-gene levels support the capacity of the VDR to recruit these complexes, and ChIP-seq analyses support the presence of these complexes on a genome-wide scale as well. Thus, for example, the VDR was found to recruit coactivators such as SRC1, CBP, and MED1 as well as the corepressors NCoR and SMRT in colorectal LS180 cells (Meyer and Pike, 2013). This recruitment was correlated with VDR binding sites linked to genes that are modulated directly by $1{,}25(OH)_2D_3$, although not preferentially to either induced or suppressed genes, as might be expected, suggesting that the roles of coregulators are not limited specifically to activation or repression and that their activities may be gene-context driven. On the other hand, studies in liver stellate cells suggest that $1{,}25(OH)_2D_3$-mediated repression of a profibrotic gene expression program induced by transforming growth factor β (TGFβ) does not involve the apparent recruitment of corepressors SMRT and NCoR (Ding et al., 2013). In addition to SRC1, CBP, and MED1, Brahma-related gene 1 (BRG1), an ATPase that is a component of the SWI/SNF chromatin remodeling complex, also plays a fundamental role in $1{,}25(OH)_2D_3$-induced transcription (Seth-Vollenweider et al., 2014). In previous studies cooperative effects between the C/EBP family of transcription factors and the VDR in $1{,}25(OH)_2D_3$-induced *CYP24A1* transcription were noted (Dhawan et al., 2005). C/EBP and BRG1 were subsequently found to be components of the same complex that are recruited to the C/EBP site of the *CYP24A1* gene by $1{,}25(OH)_2D_3$, resulting in activation of *CYP24A1* transcription (Seth-Vollenweider et al., 2014). PRMT5, a type II protein arginine methyltransferase that interacts with BRG1, represses $1{,}25(OH)_2D_3$-induced *CYP24A1* transcription via its methylation of H3R8 and H4R3 Seth-Vollenweider et al. (2014). Thus, the SWI/SNF complex can play a role in the silencing as well as in the activation of VDR mediated transcription.

Identifying underlying early mechanistic outcomes in response to VDR/RXR binding

The ability of the VDR to recruit epigenetically active coregulatory complexes such as HATs, HDACs, and a variety of histone methyltransferases that regulate chromatin structure suggests that $1{,}25(OH)_2D_3$ may influence the levels of distinct epigenetic marks imposed by these chromatin modifiers as a means of regulating gene output. Importantly, many such epigenetic marks on histones H3 and H4 are enriched in regions within gene loci that are uniquely active (Ernst and Kellis, 2010; Ernst et al., 2011; Meyer et al., 2015c; Pike et al., 2015). Perhaps of most importance are changes in the levels of acetylation at H4K5 (H4K5ac), H3K9 (H3K9ac), and H3K27 (H3K27ac) that reflect alterations in the transcriptional activity of the genes with which they are linked; these modifications generally occur within enhancers that regulate these genes, although they can also occur at locations within genes as well. Regulatory regions that are marked by both genetic and epigenetic information at gene loci are frequently termed variable chromatin modulators (Deplancke et al., 2016). An increase in several of these acetylation marks occurs at specific sites of VDR binding in genes such as *Opn* and *Cyp24a1*, *Lrp5*, *Tnfsf11*, and *Vdr* following $1,25(OH)_2D_3$ stimulation (Pike et al., 2014, 2015, 2016) and can be used to define sites of action of $1,25(OH)_2D_3$ even in the absence of evidence of VDR occupancy. Overall, histone modification analyses suggest that $1,25(OH)_2D_3$ promotes VDR/RXR binding at sites on cellular genomes that are marked by acetylated H3K9, H3K27, and H4K5, and that these interactions frequently result in an upregulation of acetylation that facilitates enhanced levels of gene expression. These studies provide a global perspective on the actions of vitamin D in several cell types, indicating that the primary role of the VDR is to facilitate the recruitment of chromatin modifiers such as acetyltransferases and deacetyltransferases that function to impose epigenetic histone changes within the enhancers of some but not all vitamin D-sensitive target genes.

The dynamic impact of cellular differentiation and disease on vitamin D receptor cistromes and transcriptional outcomes

Perhaps the most important observation made on a genome-wide scale has been the discovery that cellular differentiation exerts a dramatic quantitative and qualitative impact on sites of genomic VDR binding, an effect that correlates directly with the hormone's ability to regulate the differentiating cell's transcriptome in a highly dynamic manner (Meyer et al., 2014b; St John et al., 2014). This process is probably responsible for the cell-type-specific nature of VDR binding and thus transcriptional outcomes at diverse sets of genes that can be measured in different tissues. A general change in the cellular RNA profile in response to $1,25(OH)_2D_3$ is perhaps not surprising, given the fact that the overall effects of $1,25(OH)_2D_3$ on osteoblast lineage cells are known to differ significantly depending upon the state of bone cell differentiation. This concept of differentiation-induced changes in VDR binding and transcriptional integrity is aptly illustrated through a detailed examination of the differential expression of several genes, including *Mmp13*, in osteoblast precursors and mature mineralizing osteoblasts (Meyer et al., 2015a).

With respect to disease processes, Evans and colleagues have demonstrated that these can affect VDR cistromes as well (Ding et al., 2013). The activation of hepatic stellate cells via the upregulation of TGFβ in the liver induces the expression of a collagen program that causes hepatic fibrosis and can induce cirrhosis of the liver. This disease progression can be ameliorated by simultaneous treatment in vivo with an analog of vitamin D. The authors show that the VDR cistrome, which functions normally to suppress the program of collagen expression, is altered as a result of TGFβ action, redirecting VDR binding to alternative sites of action away from collagen genes, thereby blunting opposing sites of vitamin D action. Interestingly, while these findings identify an important action of the VDR to prevent liver fibrosis, they also highlight the role of the VDR in the disease-potentiating activation of stellate cells, a process that could be considered analogous to that of differentiation. Further studies of this system identify the role of the chromatin regulator BRD4 in this activity, and suggest that direct inhibition of this downstream factor by a small-molecule epigenetic regulator can bypass the positive effects of a vitamin D analog (Ding et al., 2015).

New approaches to the study of vitamin D-mediated gene regulation in vitro and in vivo

Although ChIP-seq analyses are able to identify transcription factor sites of occupancy and the epigenetic landscape that characterizes these regions, the frequently distal nature of these regulatory sites prevents direct identification of the genes with which they are functionally linked (Ong and Corces, 2008; Whalen et al., 2016). It has therefore become a requirement to study genes in the endogenous genomic environment in which they are located, whether in cells in culture or in tissues in vivo. One approach has been to create large minigenes that not only span the transcription unit of interest, but also contain all the putative regulatory regions identified surrounding the gene of interest (Meyer et al., 2010a). A second and perhaps more robust strategy is to create individual enhancer deletions within the context of the genome itself, a homologous recombination approach that is highly amenable to studies in the mouse in vivo (Galli et al., 2008; Onal et al., 2015, 2016a, 2016b). A third approach represents a modification of the second: introducing mutations into genomes using RNA-directed CRISPR/Cas9 nuclease methods, which create precise genomic deletions, insertions, and/or mutations at specific gene locations both in vitro and in vivo (Meyer et al., 2015a, 2015b). In the last two approaches, the phenotypic consequences of specific enhancer deletion can be assessed directly in cells or tissues at the level of the putative target gene's expression as well as at the levels of the gene product's biological function. This loss-of-function approach enables a determination of the functional relationship between a distal enhancer and its target gene and, as importantly, facilitates assessment of the biological function of the enhancer in question through phenotyping. Studies of this nature have been conducted for a number of genes, including *Tnfsf11* (RANKL), *Mmp13*, *Lrp5*, *Sost*, *Cdon*, *Boc*, *TRPV6*, and *VDR* itself. The location and distribution of enhancers that have been identified as responsible for the regulation of VDR gene expression by 1,25(OH)$_2$D$_3$ and other hormones in bone and other cell types are described in Lee et al. (2015a, 2017). As can be seen, these sites of regulation are located near the promoter, within several introns, and at multiple intergenic locations many kilobases upstream of the VDR gene itself. Most importantly, these sites have been validated functionally not only in cell-based studies, but in target tissues in the mouse as well, thereby illustrating many of the overarching principles documented in Table 30.1.

Defining the regulatory sites of action of 1,25(OH)$_2$D$_3$, PTH, and FGF23 in the *Cyp27b1* and *Cyp24a1* genes in the kidney

As discussed in an earlier section of this review, vitamin D is metabolized sequentially in the liver by *Cyp2R1* to 25(OH)D$_3$ and then in the kidney by *Cyp27b1* to 1,25(OH)$_2$D$_3$, the active hormonal form of vitamin D$_3$. Its circulating levels are similarly regulated through degradation in the kidney by *Cyp24a1*. Ironically, while it has long been known that 1,25(OH)$_2$D$_3$ suppresses the expression of the *Cyp27b1* gene in the kidney and that PTH induces the gene's expression in this tissue as well, neither the sites of action of these hormones at the *Cyp27b1* gene nor their mechanisms have emerged. A similar commentary can be made for the observed suppression of *Cyp27b1* in the kidney by FGF23, the more recently discovered phosphaturic hormone. Interestingly, these same hormones reciprocally regulate *Cyp24a1* as well, although aside from 1,25(OH)$_2$D$_3$, the mechanisms underlying this regulation also remain obscure (Meyer et al., 2010a). However, a series of experiments conducted using ChIP-seq and other genomic analyses of kidney tissue and loss-of-function enhancer deletion studies in the mouse in vivo have revealed these sites of hormonal regulatory action (Meyer et al., 2017). Indeed, ChIP-seq analysis identified sites of VDR binding within the *Cyp27b1* gene locus in response to 1,25(OH)$_2$D$_3$, while cAMP-responsive element–binding protein (CREB) binding sites were identified in response to PTH. That these sites represented bona fide regulatory regions was confirmed through the presence of regulated epigenetic histone signatures as well as the presence of DNase I sites in these regions reflective of true enhancers. Although not surprising, given the earlier narrative with regard to the frequent distal locations of enhancers relative to the genes they regulate, a single colocalized VDR and CREB site was located within an intron of the upstream neighboring gene *Mettl1*, and three sites for both factors were located within a large intron of *Mettl21b* located even farther upstream and adjacent to *Mettl1*, while no evidence was found for regulatory activity immediately proximal to the *Cyp27b1* promoter.

Deletion of the *Mettl1* intronic site in the mouse in vivo (loss-of-function analysis) resulted in a striking decrease in *Cyp27b1* expression in the kidney due to loss of sensitivity to PTH, the consequence of which was hypocalcemia, hypophosphatemia, elevated blood PTH, absent FGF23, highly reduced levels of 1,25(OH)$_2$D$_3$, and severe skeletal and other aberrations. Sensitivity to FGF23 and 1,25(OH)$_2$D$_3$ suppression was unaltered. On the other hand, although deletion of the *Mettl21b* intron in the mouse also led to a significant decrease in basal expression of *Cyp27b1*, PTH response was retained, while FGF23 response was lost. In these animals, 1,25(OH)$_2$D$_3$ levels were generally normal and the systemic and skeletal phenotype was near normal. Coincident with the reduction in *Cyp27b1* levels in both enhancer-deleted mouse strains was a coordinated reduction in renal *Cyp24a1* expression levels due to aberrant PTH and FGF23 levels, which limited the turnover of 1,25(OH)$_2$D$_3$ in the blood, thereby enabling higher sustained levels of blood 1,25(OH)$_2$D$_3$ than would ordinarily have been suspected. Interestingly, the expression of *Cyp27b1* in nonrenal target tissues such as skin, immune cells, and bone, expression that is largely insensitive to PTH, FGF23, or 1,25(OH)$_2$D$_3$ and has been suggested to account for local production of 1,25(OH)$_2$D$_3$ as a function of 25(OH)D$_3$, was unaffected by these enhancer deletions and at the same time retained broad sensitivity to the inflammatory modulator lipopolysaccharide. These results suggest that key genomic elements within the *Cyp27b1* locus, but not immediately upstream of the gene, have been identified that mediate basal and PTH, FGF23, and 1,25(OH)$_2$D$_3$ regulation in the kidney in vivo, and that this collective regulatory module is structurally and functionally unique to the kidney (Fig. 30.3). Similar features are resident at the *Cyp24a1* gene, where PTH suppresses and both FGF23 and 1,25(OH)$_2$D$_3$ induce this gene in the kidney. Further studies will be necessary to define the mechanisms through which PTH/CREB, 1,25(OH)$_2$D$_3$/VDR, and FGF23/(transcription factor unknown) actually modulate the reciprocal expression of *Cyp27b1* and *Cyp24a1* in the kidney. Regardless of this, however, these studies of the two key genes involved in the activation and degradation of vitamin D continue to highlight the utility of

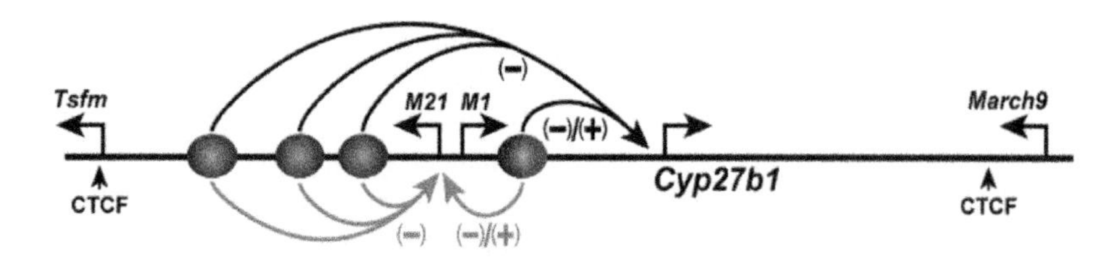

Renal Endocrine Regulatory Module

FIGURE 30.3 Schematic diagram of the kidney-specific endocrine regulatory module that controls renal *Cyp27b1* expression. The single enhancer submodule in the *Mettl1* intron and the triple components of the enhancer submodule in the *Mettl21b* intron are shown as *circles*. *Arrows* from each enhancer component are directed to the *Cyp27b1* gene as well as to the *Mettl21b* gene; both are regulated by parathyroid hormone (PTH), fibroblast growth factor 23 (FGF23), and $1,25(OH)_2D_3$. *Green* indicates a positive regulator (PTH) and *red* indicates negative regulators ($1,25(OH)_2D_3$ and FGF23). The active sites of the two hormonal suppressors have not been resolved functionally within the *Mettl21b* intron. *Arrows* above the line indicate transcriptional start sites for the *Tsfm*, *Mettl21b* (*M21*), *Mettl1* (*M1*), *Cyp27b1*, and *March9* genes. March9 is a member of the March family of membrane-bound E3 ubiquitin protein ligases. The *March9* gene is immediately downstream of *Cyp27b1*. Active CCCTC-binding factor (CTCF) sites are indicated below the line.

emerging genomic principles that are consistent with the mechanisms of action of $1,25(OH)_2D_3$ (and other hormones) at many of its target genes.

Future directions

New target genes and new factors involved in vitamin D-mediated transcription will undoubtedly be identified in numerous different systems that are known to be affected by $1,25(OH)_2D_3$. The new methodology that has enabled the study of these transcriptional mechanisms in unprecedented detail in cells and animal models will lead to new insights into mechanisms through which $1,25(OH)_2D_3$ controls calcium homeostasis and also modulates cell-specific biological processes in numerous extraskeletal systems. In addition to studies concerning the mechanisms involved in mediating the genomic actions of $1,25(OH)_2D_3$, further studies related to the physiological significance of target proteins using null-mutant mice and transgenic mice are needed. Although the principal function of vitamin D is intestinal calcium absorption, the mechanisms involved are still incompletely understood. To identify new therapeutic approaches to sustain calcium balance, mechanisms by which $1,25(OH)_2D_3$ acts at both the proximal and the distal segments of the intestine need to be defined. In vivo studies addressing physiological significance combined with studies related to the molecular mechanisms involved in $1,25(OH)_2D_3$ action are needed to provide new insight into the role of $1,25(OH)_2D_3$ in calcium homoeostasis and extraskeletal health, which could lead to the development of vitamin D therapeutics with selective modulation of responses in bone and other target tissues.

References

Adams, J.S., Sharma, O.P., Gacad, M.A., Singer, F.R., 1983. Metabolism of 25-hydroxyvitamin D3 by cultured pulmonary alveolar macrophages in sarcoidosis. J. Clin. Invest. 72 (5), 1856–1860.

Amling, M., Priemel, M., Holzmann, T., Chapin, K., Rueger, J.M., Baron, R., Demay, M.B., 1999. Rescue of the skeletal phenotype of vitamin D receptor-ablated mice in the setting of normal mineral ion homeostasis: formal histomorphometric and biomechanical analyses. Endocrinology 140, 4982–4987.

Armbrecht, H.J., Zenser, T.V., Davis, B.B., 1980. Effect of age on the conversion of 25-hydroxyvitamin D3 to 1,25-dihydroxyvitamin D3 by kidney of rat. J. Clin. Invest. 66, 1118–1123.

Arrowsmith, C.H., Bountra, C., Fish, P.V., Lee, K., Schapira, M., 2012. Epigenetic protein families: a new frontier for drug discovery. Nat. Rev. Drug Discov. 11 (5), 384–400.

Baker, A.R., McDonnell, D.P., Hughes, M., Crisp, T.M., Mangelsdorf, D.J., Haussler, M.R., et al., 1988. Cloning and expression of full-length cDNA encoding human vitamin D receptor. Proc. Natl. Acad. Sci. U.S.A. 85 (10), 3294–3298.

Barbour, G.L., Coburn, J.W., Slatopolsky, E., Norman, A.W., Horst, R.L., 1981. Hypercalcemia in an anephric patient with sarcoidosis: evidence for extrarenal generation of 1,25-dihydroxyvitamin D. N. Engl. J. Med. 305 (8), 440–443.

Benn, B.S., Ajibade, D., Porta, A., Dhawan, P., Hediger, M., Peng, J.B., Jiang, Y., Oh, G.T., Jeung, E.B., Lieben, L., Bouillon, R., Carmeliet, G., Christakos, S., 2008. Active intestinal calcium transport in the absence of transient receptor vanilloid type 6 and calbindin-D9k. Endocrinology 149, 2196–3205.

Bikle, D.D., 2009. Extra renal synthesis of 1, 25-dihydroxyvitamin D and it's health implications. Clin. Rev. Bone Miner. Metab. 7, 114–125.

Bikle, D.D., Adams, J., Christakos, S., 2013. Vitamin D: production, metabolism and clinical requirements. In: Rosen, C. (Ed.), Primer Metab Bone Dis. Wiley, Hoboken, NJ, pp. 235–245.

Bikle, D.D., 2012. Vitamin D and bone. Curr. Osteoporos. Rep. 10 (2), 151–159.

Boros, S., Bindels, R.J., Hoenderop, J.G., 2009. Active Ca(2+) reabsorption in the connecting tubule. Pflugers Arch. 458 (1), 99–109.

Boskey, A.L., Maresca, M., Ullrich, W., Doty, S.B., Butler, W.T., Prince, C.W., 1993. Osteopontin-hydroxyapatite interactions in vitro: inhibition of hydroxyapatite formation and growth in a gelatin-gel. Bone Miner. 22 (2), 147–159.

Boyle, I.T., Gray, R.W., DeLuca, H.F., 1971. Regulation by calcium of in vitro synthesis of 1,25-dihdroxycholecalciferol and 24,25-dihdroxycholecalciferol. Proc. Natl. Acad. Sci. U.S.A. 68, 2131–2135.

Brenza, H.L., DeLuca, H.F., 2000. Regulation of 25-hydroxyvitamin D3 1alpha-hydroxylase gene expression by parathyroid hormone and 1,25-dihydroxyvitamin D3. Arch. Biochem. Biophys. 381, 143–152.

Brooks, M.H., Bell, N.H., Love, L., Stern, P.H., Orfei, E., Queener, S.F., et al., 1978. Vitamin-D-dependent rickets type II. Resistance of target organs to 1,25-dihydroxyvitamin D. N. Engl. J. Med. 298 (18), 996–999.

Brumbaugh, P., Haussler, M., 1974a. 1 Alpha,25-dihydroxycholecalciferol receptors in intestine. I. Association of 1 alpha,25-dihydroxycholecalciferol with intestinal mucosa chromatin. J. Biol. Chem. 249 (4), 1251–1257.

Brumbaugh, P.F., Haussler, M.R., 1974b. 1a,25-dihydroxycholecalciferol receptors in intestine. II. Temperature-dependent transfer of the hormone to chromatin via a specific cytosol receptor. J. Biol. Chem. 249 (4), 1258–1262.

Brumbaugh, P.F., Hughes, M.R., Haussler, M.R., 1975. Cytoplasmic and nuclear binding components for 1a25-dihydroxyvitamin D3 in chick parathyroid glands. Proc. Natl. Acad. Sci. U.S.A. 72 (12), 4871–4875.

Burmester, J.K., Maeda, N., DeLuca, H.F., 1988. Isolation and expression of rat 1,25-dihydroxyvitamin D3 receptor cDNA. Proc. Natl. Acad. Sci. U.S.A. 85 (4), 1005–1009.

Canaff, L., Hendy, G.N., 2002. Human calcium-sensing receptor gene. Vitamin D response elements in promoters P1 and P2 confer transcriptional responsiveness to 1,25-dihydroxyvitamin D. J. Biol. Chem. 277 (33), 30337–30350.

Carlberg, C., 2003. Molecular basis of the selective activity of vitamin D analogues. J. Cell. Biochem. 88 (2), 274–281.

Carlson, M., Laurent, B.C., 1994. The SNF/SWI family of global transcriptional activators. Curr. Opin. Cell Biol. 6 (3), 396–402.

Cheng, J.B., Levine, M.A., Bell, N.H., Mangelsdorf, D.J., Russell, D.W., 2004. Genetic evidence that the human CYP2R1 enzyme is a key vitamin D 25-hydroxylase. Proc. Natl. Acad. Sci. U.S.A. 101 (20), 7711–7715.

Christakos, S., 1995. Vitamin D-dependent calcium-binding proteins: chemistry distribution, functional considerations, and molecular biology: update 1995. Endocr. Rev. Monogr. 4, 108–110.

Christakos, S., 2012. Recent advances in our understanding of 1,25-dihydroxyvitamin D(3) regulation of intestinal calcium absorption. Arch. Biochem. Biophys. 523 (1), 73–76.

Christakos, S., DeLuca, H.F., 2011. Minireview: vitamin D: is there a role in extraskeletal health? Endocrinology 152 (8), 2930–2936.

Christakos, S., Dhawan, P., Verstuyf, A., Verlinden, L., Carmeliet, G., 2016. Vitamin D: metabolism, molecular mechanism of action, and pleiotropic effects. Physiol. Rev. 96 (1), 365–408.

Christakos, S., Gabrielides, C., Rhoten, W.B., 1989. Vitamin D-dependent calcium binding proteins: chemistry, distribution, functional considerations, and molecular biology. Endocr. Rev. 10 (1), 3–26.

Cui, M., Li, Q., Johnson, R., Fleet, J.C., 2012. Villin promoter-mediated transgenic expression of transient receptor potential cation channel, subfamily V, member 6 (TRPV6) increases intestinal calcium absorption in wild-type and vitamin D receptor knockout mice. J. Bone Miner. Res. 27 (10), 2097–2107.

Dardenne, O., Prud'homme, J., Arabian, A., Glorieux, F.H., St-Arnaud, R., 2001. Targeted inactivation of the 25-hydroxyvitamin D(3)-1(alpha)-hydroxylase gene (CYP27B1) creates an animal model of pseudovitamin D-deficiency rickets. Endocrinology 142 (7), 3135–3141.

DeLuca, H.F., 2008. Evolution of our understanding of vitamin D. Nutr. Rev. 66 (10 Suppl. 2), S73–S87.

DeLuca, H.F., 2016. Vitamin D: historical overview. Vitam. Horm. 100, 1–20.

Demay, M.B., Gerardi, J.M., DeLuca, H.F., Kronenberg, H.M., 1990. DNA sequences in the rat osteocalcin gene that bind the 1,25-dihydroxyvitamin D3 receptor and confer responsiveness to 1,25-dihydroxyvitamin D3. Proc. Natl. Acad. Sci. U.S.A. 87, 369–373.

Demay, M.B., Kiernan, M.S., DeLuca, H.F., Kronenberg, H.M., 1992. Sequences in the human parathyroid hormone gene that bind the 1,25-dihydroxyvitamin D3 receptor and mediate transcriptional repression in response to 1,25-dihydroxyvitamin D3. Proc. Natl. Acad. Sci. U.S.A. 89 (17), 8097–8101.

Deng, W., Blobel, G.A., 2010. Do chromatin loops provide epigenetic gene expression states? Curr. Opin. Genet. Dev. 20 (5), 548–554.

Deng, W., Blobel, G.A., 2014. Manipulating nuclear architecture. Curr. Opin. Genet. Dev. 25, 1–7.

Deng, W., Blobel, G.A., 2017. Detecting long-range enhancer-promoter interactions by quantitative chromosome conformation capture. Methods Mol. Biol. 1468, 51–62.

Deng, W., Lee, J., Wang, H., Miller, J., Reik, A., Gregory, P.D., et al., 2012. Controlling long-range genomic interactions at a native locus by targeted tethering of a looping factor. Cell 149 (6), 1233–1244.

Deplancke, B., Alpern, D., Gardeux, V., 2016. The genetics of transcription factor DNA binding variation. Cell 166 (3), 538–554.

Dhawan, P., Peng, X., Sutton, A.L., MacDonald, P.N., Croniger, C.M., Trautwein, C., Centrella, M., McCarthy, T.L., Christakos, S., 2005. Functional cooperation between CCAAT/enhancer-binding proteins and the vitamin D receptor in regulation of 25-hydroxyvitamin D3 24-hydroxylase. Mol. Cell Biol. 25, 472–487.

Dhawan, P., Veldurthy, V., Yehia, G., Hsaio, C., Porta, A., Kim, K.I., Patel, N., Lieben, L., Verlinden, L., Carmeliet, G., Christakos, S., 2017a. Transgenic expression of the vitamin D receptor restricted to the ileum, cecum, and colon of vitamin D receptor knockout mice rescues vitamin D receptor-dependent rickets. Endocrinology 158 (11), 3792–3804.

Dhawan, P., Wei, R., Veldurthy, V., Christakos, S., 2017b. New developments in our understanding of the regulation of calcium homeostasis by vitamin D. In: Collins, J. (Ed.), Molecular, Genetic and Nutritional Aspects of Major and Trace Minerals. Elsevier Inc., pp. 27–34.

Ding, N., Hah, N., Yu, R.T., Sherman, M.H., Benner, C., Leblanc, M., et al., 2015. BRD4 is a novel therapeutic target for liver fibrosis. Proc. Natl. Acad. Sci. U.S.A. 112 (51), 15713–15718.

Ding, N., Yu, R.T., Subramaniam, N., Sherman, M.H., Wilson, C., Rao, R., et al., 2013. A vitamin D receptor/SMAD genomic circuit gates hepatic fibrotic response. Cell 153 (3), 601–613.

Dinour, D., Beckerman, P., Ganon, L., Tordjman, K., Eisenstein, Z., Holtzman, E.J., 2013. Loss-of-function mutations of CYP24A1, the vitamin D 24-hydroxylase gene, cause long-standing hypercalciuric nephrolithiasis and nephrocalcinosis. J. Urol. 190 (2), 552–557.

Eil, C., Liberman, U.A., Rosen, J.F., Marx, S.J., 1981. A cellular defect in hereditary vitamin-D-dependent rickets type II: defective nuclear uptake of 1,25-dihydroxyvitamin D in cultured skin fibroblasts. N. Engl. J. Med. 304 (26), 1588–1591.

Ernst, J., Kellis, M., 2010. Discovery and characterization of chromatin states for systematic annotation of the human genome. Nat. Biotechnol. 28 (8), 817–825.

Ernst, J., Kheradpour, P., Mikkelsen, T.S., Shoresh, N., Ward, L.D., Epstein, C.B., et al., 2011. Mapping and analysis of chromatin state dynamics in nine human cell types. Nature 473 (7345), 43–49.

Evans, R.M., 1988. The steroid and thyroid hormone receptor superfamily. Science 240 (4854), 889–895.

Forster, R.E., Jurutka, P.W., Hsieh, J.C., Haussler, C.A., Lowmiller, C.L., Kaneko, I., et al., 2011. Vitamin D receptor controls expression of the anti-aging klotho gene in mouse and human renal cells. Biochem. Biophys. Res. Commun. 414 (3), 557–562.

Galli, C., Zella, L.A., Fretz, J.A., Fu, Q., Pike, J.W., Weinstein, R.S., et al., 2008. Targeted deletion of a distant transcriptional enhancer of the receptor activator of nuclear factor-kappaB ligand gene reduces bone remodeling and increases bone mass. Endocrinology 149 (1), 146–153.

Gifford, C.A., Ziller, M.J., Gu, H., Trapnell, C., Donaghey, J., Tsankov, A., et al., 2013. Transcriptional and epigenetic dynamics during specification of human embryonic stem cells. Cell 153 (5), 1149–1163.

Gill, R.K., Christakos, S., 1993. Identification of sequence elements in mouse calbindin-D28k gene that confer 1,25-dihydroxyvitamin D3- and butyrate-inducible responses. Proc. Natl. Acad. Sci. U.S.A. 90 (7), 2984–2988.

Haussler, M.R., Chandler, J.S., Pike, J.W., Brumbaugh, P.F., Speer, D.P., Pitt, M.J., 1980. Physiological importance of vitamin D metabolism. Prog. Biochem. Pharmacol. 17, 134–142.

Haussler, M.R., Myrtle, J.F., Norman, A.W., 1968. The association of a metabolite of vitamin D3 with intestinal mucosa chromatin in vivo. J. Biol. Chem. 243 (15), 4055–4064.

Heikkinen, S., Väisänen, S., Pehkonen, P., Seuter, S., Benes, V., Carlberg, C., 2011. Nuclear hormone 1α,25-dihydroxyvitamin D3 elicits a genome-wide shift in the locations of VDR chromatin occupancy. Nucleic Acids Res. 39 (21), 9181–9193.

Henry, H., 1985. Parathyroid modulation of 25-hydroxyvitamin D3 metabolism by cultured chick kidney cells is mimicked and enhanced by forskolin. Endocrinology 116, 503–510.

Ho, L., Crabtree, G.R., 2010. Chromatin remodelling during development. Nature 463 (7280), 474–484.

Hobaus, J., Thiem, U., Hummel, D.M., Kallay, E., 2013. Role of calcium, vitamin D, and the extrarenal vitamin D hydroxylases in carcinogenesis. Anti Cancer Agents Med. Chem. 13 (1), 20–35.

Hochberg, Z., Tiosano, D., Even, L., 1992. Calcium therapy for calcitriol-resistant rickets. J. Pediatr. 121 (5 Pt 1), 803–808.

Hoenderop, J.G., van Leeuwen, J.P., van der Eerden, B.C., Kersten, F.F., van der Kemp, A.W., Merillat, A.M., Waarsing, J.H., Rossier, B.C., Vallon, V., Hummler, E., Bindels, R.J., 2003. Renal Ca21 wasting, hyperabsorption, and reduced bone thickness in mice lacking TRPV5. J. Clin. Invest. 112, 1906–1914.

Hoffman, M.M., Ernst, J., Wilder, S.P., Kundaje, A., Harris, R.S., Libbrecht, M., et al., 2013. Integrative annotation of chromatin elements from ENCODE data. Nucleic Acids Res. 41 (2), 827–841.

Hu, M.C., Shiizaki, K., Kuro-o, M., Moe, O.W., 2013. Fibroblast growth factor 23 and Klotho: physiology and pathophysiology of an endocrine network of mineral metabolism. Annu. Rev. Physiol. 75, 503–533.

Hughes, M.R., Malloy, P.J., Kieback, D.G., Kesterson, R.A., Pike, J.W., Feldman, D., O'Malley, B.W., 1988. Point mutations in the human vitamin D receptor gene associated with hypocalcemic rickets. Science 242 (4886), 1702–1705.

Hughes, M., Malloy, P., O'Malley, B., Pike, J., Feldman, D., 1991. Genetic defects of the 1,25-dihydroxyvitamin D3 receptor. J. Recept. Res. 11 (1–4), 699–716.

Jin, C.H., Kerner, S.A., Hong, M.H., Pike, J.W., 1996. Transcriptional activation and dimerization functions in the human vitamin D receptor. Mol. Endocrinol. 10 (8), 945–957.

Jones, G., Prosser, D.E., Kaufmann, M., 2014. Cytochrome P450-mediated metabolism of vitamin D. J. Lipid Res. 55 (1), 13–31.

Joshi, S., Pantalena, L.C., Liu, X.K., Gaffen, S.L., Liu, H., Rohowsky-Kochan, C., Ichiyama, K., Yoshimura, A., Steinman, L., Christakos, S., Youssef, S., 2011. 1,25-Dihydroxyvitamin D(3) ameliorates Th17 autoimmunity via transcriptional modulation of interleukin-17A. Mol. Cell Biol. 31 (17), 3653–3669.

Kellis, M., Wold, B., Snyder, M.P., Bernstein, B.E., Kundaje, A., Marinov, G.K., Ward, L.D., Birney, E., Crawford, G.E., Dekker, J., Dunham, I., Elnitski, L.L., Farnham, P.J., Feingold, E.A., Gerstein, M., Giddings, M.C., Gilbert, D.M., Gineras, T.R., Green, E.D., Guigo, R., Hubbard, T., Kent, J., Lieb, J.D., Myers, R.M., Pazin, M.J., Ren, B., Stamatoyannopoulos, J.A., Weng, Z., White, K.P., Hardison, R.C., 2014. Defining functional DNA elements in the human genome. Proc. Natl. Acad. Sci. U.S.A. 111 (17), 6131–6138.

Kerner, S.A., Scott, R.A., Pike, J.W., 1989. Sequence elements in the human osteocalcin gene confer basal activation and inducible response to hormonal vitamin D3. Proc. Natl. Acad. Sci. U.S.A. 86 (12), 4455–4459.

Kerry, D.M., Dwivedi, P.P., Hahn, C.N., Morris, H.A., Omdahl, J.L., May, B.K., 1996. Transcriptional synergism between vitamin D-responsive elements in the rat 25-hydroxyvitamin D3 24-hydroxylase (CYP24) promoter. J. Biol. Chem. 271, 29715–29721.

Kim, S., Shevde, N., Pike, J., 2005. 1,25-Dihydroxyvitamin D3 stimulates cyclic vitamin D receptor/retinoid X receptor DNA-binding, co-activator recruitment, and histone acetylation in intact osteoblasts. J. Bone Miner. Res. 20 (2), 305–317.

Kim, S., Yamazaki, M., Shevde, N.K., Pike, J.W., 2007. Transcriptional control of receptor activator of nuclear factor-kappaB ligand by the protein kinase A activator forskolin and the transmembrane glycoprotein 130-activating cytokine, oncostatin M, is exerted through multiple distal enhancers. Mol. Endocrinol. 21, 197–214.

Kim, S., Yamazaki, M., Zella, L., Shevde, N., Pike, J., 2006. Activation of receptor activator of NF-kappaB ligand gene expression by 1,25-dihydroxyvitamin D3 is mediated through multiple long-range enhancers. Mol. Cell Biol. 26 (17), 6469–6486.

Kitanaka, S., Takeyama, K., Murayama, A., Sato, T., Okumura, K., Nogami, M., et al., 1998. Inactivating mutations in the 25-hydroxyvitamin D3 1alpha-hydroxylase gene in patients with pseudovitamin D-deficiency rickets. N. Engl. J. Med. 338 (10), 653–661.

Kliewer, S.A., Umesono, K., Mangelsdorf, D.J., Evans, R.M., 1992. Retinoid X receptor interacts with nuclear receptors in retinoic acid, thyroid hormone and vitamin D3 signalling. Nature 355 (6359), 446–449.

Kutuzova, G.D., Akhter, S., Christakos, S., Vanhooke, J., Kimmel-Jehan, C., Deluca, H.F., 2006. Calbindin D(9k) knockout mice are indistinguishable from wild-type mice in phenotype and serum calcium level. Proc. Natl. Acad. Sci. U.S.A. 103 (33), 12377–12381.

Kutuzova, G.D., Sundersingh, F., Vaughan, J., Tadi, B.P., Ansay, S.E., Christakos, S., Deluca, H.F., 2008. TRPV6 is not required for 1alpha,25-dihydroxyvitamin D3-induced intestinal calcium absorption in vivo. Proc. Natl. Acad. Sci. U.S.A. 105 (50), 19655–19659.

Lanske, B., Densmore, M.J., Erben, R.G., 2014. Vitamin D endocrine system and osteocytes. Bonekey Rep. 3, 494.

Lee, S.M., Meyer, M.B., Benkusky, N.A., O'Brien, C.A., Pike, J.W., 2015a. Mechanisms of enhancer-mediated hormonal control of vitamin D receptor gene expression in target cells. J. Biol. Chem. 290 (51), 30573–30586.

Lee, S.M., Meyer, M.B., Benkusky, N.A., O'Brien, C.A., Pike, J.W., 2017. The impact of VDR expression and regulation in vivo. J. Steroid Biochem. Mol. Biol. 177, 36–45.

Lee, S.M., Riley, E.M., Meyer, M.B., Benkusky, N.A., Plum, L.A., DeLuca, H.F., Pike, J.W., 2015b. 1,25-Dihydroxyvitamin D3 controls a cohort of vitamin D receptor target genes in the proximal intestine that is enriched for calcium-regulating components. J. Biol. Chem. 290 (29), 18199–18215.

Lewis, B.A., Reinberg, D., 2003. The mediator coactivator complex: functional and physical roles in transcriptional regulation. J. Cell Sci. 116 (Pt 18), 3667–3675.

Li, Y.C., Pirro, A.E., Amling, M., Delling, G., Baron, R., Bronson, R., Demay, M.B., 1997. Targeted ablation of the vitamin D receptor: an animal model of vitamin D-dependent rickets type II with alopecia. Proc. Natl. Acad. Sci. U.S.A. 94, 9831–9835.

Lian, J.B., Stein, G.S., Stewart, C., Puchacz, E., Mackowiak, S., Aronow, M., et al., 1989. Osteocalcin: characterization and regulated expression of the rat gene. Connect. Tissue Res. 21 (1–4), 61–68 discussion 69.

Liao, J., Ozono, K., Sone, T., McDonnell, D., Pike, J., 1990. Vitamin D receptor interaction with specific DNA requires a nuclear protein and 1,25-dihydroxyvitamin D3. Proc. Natl. Acad. Sci. U.S.A. 87 (24), 9751–9755.

Lieben, L., Masuyama, R., Torrekens, S., Van Looveren, R., Schrooten, J., Baatsen, P., et al., 2012. Normocalcemia is maintained in mice under conditions of calcium malabsorption by vitamin D-induced inhibition of bone mineralization. J. Clin. Invest. 122 (5), 1803–1815.

Lin, N., Malloy, P., Sakati, N., al-Ashwal, A., Feldman, D., 1996. A novel mutation in the deoxyribonucleic acid-binding domain of the vitamin D receptor causes hereditary 1,25-dihydroxyvitamin D-resistant rickets. J. Clin. Endocrinol. Metab. 81 (7), 2564–2569.

Malloy, P.J., Tasic, V., Taha, D., Tótóncóler, F., Ying, G.S., Yin, L.K., et al., 2014. Vitamin D receptor mutations in patients with hereditary 1,25-dihydroxyvitamin D-resistant rickets. Mol. Genet. Metab. 111 (1), 33–40.

Malloy, P., Hochberg, Z., Tiosano, D., Pike, J., Hughes, M., Feldman, D., 1990. The molecular basis of hereditary 1,25-dihydroxyvitamin D3 resistant rickets in seven related families. J. Clin. Invest. 86 (6), 2071–2079.

Mangelsdorf, D.J., Evans, R.M., 1995. The RXR heterodimers and orphan receptors. Cell 83 (6), 841–850.

Martowicz, M.L., Meyer, M.B., Pike, J.W., 2011. The mouse RANKL gene locus is defined by a broad pattern of histone H4 acetylation and regulated through distinct distal enhancers. J. Cell. Biochem. 112 (8), 2030–2045.

Matkovits, T., Christakos, S., 1995. Variable in vivo regulation of rat vitamin D dependent genes (osteopontin, Ca,Mg-Adenosine Triphosphatase, and 25-hydroxyvitamin D3 24-hydroxylase): implications for differing mechanisms of regulation and involvement of multiple factors. Endocrinology 136, 3971–3982.

McDonnell, D.P., Mangelsdorf, D.J., Pike, J.W., Haussler, M.R., O'Malley, B.W., 1987. Molecular cloning of complementary DNA encoding the avian receptor for vitamin D. Science 235 (4793), 1214–1217.

McDonnell, D., Scott, R., Kerner, S., O'Malley, B., Pike, J., 1989. Functional domains of the human vitamin D3 receptor regulate osteocalcin gene expression. Mol. Endocrinol. 3 (4), 635–644.

McInerney, E.M., Rose, D.W., Flynn, S.E., Westin, S., Mullen, T.M., Krones, A., et al., 1998. Determinants of coactivator LXXLL motif specificity in nuclear receptor transcriptional activation. Genes Dev. 12 (21), 3357–3368.

Meir, T., Levi, R., Lieben, L., Libutti, S., Carmeliet, G., Bouillon, R., et al., 2009. Deletion of the vitamin D receptor specifically in the parathyroid demonstrates a limited role for the receptor in parathyroid physiology. Am. J. Physiol. Renal. Physiol. 297 (5), F1192–F1198.

Meyer, M.B., Pike, J.W., 2013. Corepressors (NCoR and SMRT) as well as coactivators are recruited to positively regulated 1α,25-dihydroxyvitamin D3-responsive genes. J. Steroid Biochem. Mol. Biol. 136, 120–124.

Meyer, M.B., Benkusky, N.A., Pike, J.W., 2014a. The RUNX2 cistrome in osteoblasts: characterization, down-regulation following differentiation, and relationship to gene expression. J. Biol. Chem. 289 (23), 16016–16031.

Meyer, M.B., Benkusky, N.A., Pike, J.W., 2015c. Profiling histone modifications by chromatin immunoprecipitation coupled to deep sequencing in skeletal cells. Methods Mol. Biol. 1226, 61–70.

Meyer, M.B., Benkusky, N.A., Pike, J.W., 2015a. Selective distal enhancer control of the Mmp13 gene identified through clustered regularly interspaced short palindromic repeat (CRISPR) genomic deletions. J. Biol. Chem. 290 (17), 11093–11107.

Meyer, M.B., Benkusky, N.A., Kaufmann, M., Lee, S.M., Onal, M., Jones, G., Pike, J.W., 2017. A kidney-specific genetic control module in mice governs endocrine regulation of the cytochrome P450 gene Cyp27b1 essential for vitamin D3 activation. J. Biol. Chem. 292 (42), 17541–17558.

Meyer, M.B., Benkusky, N.A., Lee, C.H., Pike, J.W., 2014b. Genomic determinants of gene regulation by 1,25-dihydroxyvitamin D3 during osteoblast-lineage cell differentiation. J. Biol. Chem. 289 (28), 19539–19554.

Meyer, M.B., Benkusky, N.A., Onal, M., Pike, J.W., 2015b. Selective regulation of Mmp13 by 1,25(OH)2D3, PTH, and Osterix through distal enhancers. J. Steroid Biochem. Mol. Biol. 164, 258–264.

Meyer, M.B., Benkusky, N.A., Sen, B., Rubin, J., Pike, J.W., 2016. Epigenetic plasticity drives adipogenic and osteogenic differentiation of marrow-derived mesenchymal stem cells. J. Biol. Chem. 291 (34), 17829–17847.

Meyer, M.B., Goetsch, P.D., Pike, J.W., 2010a. A downstream intergenic cluster of regulatory enhancers contributes to the induction of CYP24A1 expression by 1alpha,25-dihydroxyvitamin D3. J. Biol. Chem. 285 (20), 15599–15610.

Meyer, M.B., Goetsch, P.D., Pike, J.W., 2010b. Genome-wide analysis of the VDR/RXR cistrome in osteoblast cells provides new mechanistic insight into the actions of the vitamin D hormone. J. Steroid Biochem. Mol. Biol. 121 (1–2), 136–141.

Meyer, M.B., Goetsch, P.D., Pike, J.W., 2012. VDR/RXR and TCF4/β-catenin cistromes in colonic cells of colorectal tumor origin: impact on c-FOS and c-MYC gene expression. Mol. Endocrinol. 26 (1), 37–51.

Meyer, M.B., Watanuki, M., Kim, S., Shevde, N.K., Pike, J.W., 2006. The human transient receptor potential vanilloid type 6 distal promoter contains multiple vitamin D receptor binding sites that mediate activation by 1,25-dihydroxyvitamin D3 in intestinal cells. Mol. Endocrinol. 20 (6), 1447–1461.

Meyer, M.B., Zella, L.A., Nerenz, R.D., Pike, J.W., 2007. Characterizing early events associated with the activation of target genes by 1,25-dihydroxyvitamin D3 in mouse kidney and intestine in vivo. J. Biol. Chem. 282 (31), 22344–22352.

Miyamoto, K., Kesterson, R., Yamamoto, H., Taketani, Y., Nishiwaki, E., Tatsumi, S., et al., 1997. Structural organization of the human vitamin D receptor chromosomal gene and its promoter. Mol. Endocrinol. 11 (8), 1165–1179.

Morris, S.M., Tallquist, M.D., Rock, C.O., Cooper, J.A., 2002. Dual roles for the Dab2 adaptor protein in embryonic development and kidney transport. EMBO J. 21 (7), 1555–1564.

Nerenz, R.D., Martowicz, M.L., Pike, J.W., 2008. An enhancer 20 kilobases upstream of the human receptor activator of nuclear factor-kappaB ligand gene mediates dominant activation by 1,25-dihydroxyvitamin D3. Mol. Endocrinol. 22 (5), 1044–1056.

Noda, M., Vogel, R.L., Craig, A.M., Prahl, J., DeLuca, H.F., Denhardt, D.T., 1990. Identification of a DNA sequence responsible for binding of the 1,25-dihydroxyvitamin D3 receptor and 1,25-dihydroxyvitamin D3 enhancement of mouse secreted phosphoprotein 1 (SPP-1 or osteopontin) gene expression. Proc. Natl. Acad. Sci. U.S.A. 87 (24), 9995–9999.

Nykjaer, A., Dragun, D., Walther, D., Vorum, H., Jacobsen, C., Herz, J., Melsen, F., Christensen, E.I., Willnow, T.E., 1999. An endocytic pathway essential for renal uptake and activation of the steroid 25-(OH) vitamin D3. Cell 96, 507–515.

Nykjaer, A., Fyfe, J.C., Kozyraki, R., Leheste, J.R., Jacobsen, C., Nielsen, M.S., et al., 2001. Cubilin dysfunction causes abnormal metabolism of the steroid hormone 25(OH) vitamin D(3). Proc. Natl. Acad. Sci. U.S.A. 98 (24), 13895–13900.

Ohyama, Y., Ozono, K., Uchida, M., Shinki, T., Kato, S., Suda, T., et al., 1994. Identification of a vitamin D-responsive element in the 5′-flanking region of the rat 25-hydroxyvitamin D3 24-hydroxylase gene. J. Biol. Chem. 269 (14), 10545–10550.

Ohyama, Y., Ozono, K., Uchida, M., Yoshimura, M., Shinki, T., Suda, T., Yamamoto, O., 1996. Functional assessment of two vitamin D-responsive elements in the rat 25-hydroxyvitamin D3 24-hydroxylase gene. J. Biol. Chem. 271 (48), 30381–30385.

Omdahl, J.L., Bobrovnikova, E.V., Annalora, A., Chen, P., Serda, R., 2003. Expression, structure-function, and molecular modeling of vitamin D P450s. J. Cell. Biochem. 88, 356–362.

Onal, M., Bishop, K.A., St John, H.C., Danielson, A.L., Riley, E.M., Piemontese, M., et al., 2015. A DNA segment spanning the mouse Tnfsf11 transcription unit and its upstream regulatory domain rescues the pleiotropic biologic phenotype of the RANKL null mouse. J. Bone Miner. Res. 30 (5), 855–868.

Onal, M., St John, H.C., Danielson, A.L., Pike, J.W., 2016a. Deletion of the distal Tnfsf11 RL-D2 enhancer that contributes to PTH-mediated RANKL expression in osteoblast lineage cells results in a high bone mass phenotype in mice. J. Bone Miner. Res. 31 (2), 416–429.

Onal, M., St John, H.C., Danielson, A.L., Markert, J.W., Riley, E.M., Pike, J.W., 2016b. Unique distal enhancers linked to the mouse Tnfsf11 gene direct tissue-specific and inflammation-induced expression of RANKL. Endocrinology 157 (2), 482–496.

Ong, C.T., Corces, V.G., 2008. Modulation of CTCF insulator function by transcription of a noncoding RNA. Dev. Cell 15 (4), 489–490.

Ong, C.T., Corces, V.G., 2011. Enhancer function: new insights into the regulation of tissue-specific gene expression. Nat. Rev. Genet. 12 (4), 283–293.

Ooi, J.H., McDaniel, K.L., Weaver, V., Cantorna, M.T., 2014. Murine $CD8^+$ T cells but not macrophages express the vitamin D 1alpha-hydroxylase. J. Nutr. Biochem. 25 (1), 58–65.

Orlov, I., Rochel, N., Moras, D., Klaholz, B.P., 2012. Structure of the full human RXR/VDR nuclear receptor heterodimer complex with its DR3 target DNA. EMBO J. 31 (2), 291–300.

Ozono, K., Liao, J., Kerner, S.A., Scott, R.A., Pike, J.W., 1990. The vitamin D-responsive element in the human osteocalcin gene. Association with a nuclear proto-oncogene enhancer. J. Biol. Chem. 265 (35), 21881–21888.

Panda, D.K., Miao, D., Tremblay, M.L., Sirois, J., Farookhi, R., Hendy, G.N., Goltzman, D., 2001. Targeted ablation of the 25-hydroxyvitamin D 1alpha-hydroxylase enzyme: evidence for skeletal, reproductive, and immune dysfunction. Proc. Natl. Acad. Sci. U.S.A. 98, 7498–7503.

Pathrose, P., Barmina, O., Chang, C.Y., McDonnell, D.P., Shevde, N.K., Pike, J.W., 2002. Inhibition of 1,25-dihydroxyvitamin D3-dependent transcription by synthetic LXXLL peptide antagonists that target the activation domains of the vitamin D and retinoid X receptors. J. Bone Miner. Res. 17 (12), 2196–2205.

Perissi, V., Staszewski, L.M., McInerney, E.M., Kurokawa, R., Krones, A., Rose, D.W., et al., 1999. Molecular determinants of nuclear receptor-corepressor interaction. Genes Dev. 13 (24), 3198–3208.

Perwad, F., Azam, N., Zhang, M.Y., Yamashita, T., Tenenhouse, H.S., Portale, A.A., 2005. Dietary and serum phosphorus regulate fibroblast growth factor 23 expression and 1,25-dihydroxyvitamin D metabolism in mice. Endocrinology 146, 5358–5364.

Pike, J.W., Haussler, M.R., 1979. Purification of chicken intestinal receptor for 1,25-dihydroxyvitamin D. Proc. Natl. Acad. Sci. U.S.A. 76 (11), 5485–5489.

Pike, J.W., Lee, S.M., Meyer, M.B., 2014. Regulation of gene expression by 1,25-dihydroxyvitamin D3 in bone cells: exploiting new approaches and defining new mechanisms. Bonekey Rep. 3, 482.

Pike, J.W., Meyer, M.B., Benkusky, N.A., Lee, S.M., St John, H., Carlson, A., et al., 2016. Genomic determinants of vitamin D-regulated gene expression. Vitam. Horm. 100, 21–44.

Pike, J.W., Meyer, M.B., St John, H.C., Benkusky, N.A., 2015. Epigenetic histone modifications and master regulators as determinants of context dependent nuclear receptor activity in bone cells. Bone 81, 757–764.

Pike, J., Dokoh, S., Haussler, M., Liberman, U., Marx, S., Eil, C., 1984. Vitamin D3-resistant fibroblasts have immunoassayable 1,25-dihydroxyvitamin D3 receptors. Science 224 (4651), 879–881.

Price, P.A., Baukol, S.A., 1980. 1,25-Dihydroxyvitamin D3 increases synthesis of the vitamin K-dependent bone protein by osteosarcoma cells. J. Biol. Chem. 255 (24), 11660–11663.

Prince, C.W., Butler, W.T., 1987. 1,25-Dihydroxyvitamin D3 regulates the biosynthesis of osteopontin, a bone-derived cell attachment protein, in clonal osteoblast-like osteosarcoma cells. Coll. Relat. Res. 7 (4), 305–313.

Rachez, C., Freedman, L.P., 2000. Mechanisms of gene regulation by vitamin D(3) receptor: a network of coactivator interactions. Gene 246 (1–2), 9–21.

Raisz, L.G., Trammel, C.L., Holick, M.F., DeLuca, H.F., 1972. 1,25-Dihydroxyvitamin D3: a potent stimulator of bone resorption in tissue culture. Science 175, 768–769.

Ramagopalan, S.V., Heger, A., Berlanga, A.J., Maugeri, N.J., Lincoln, M.R., Burrell, A., et al., 2010. A ChIP-seq defined genome-wide map of vitamin D receptor binding: associations with disease and evolution. Genome Res. 20 (10), 1352–1360.

Rhoten, W.B., Bruns, M.E., Christakos, S., 1985. Presence and localization of two vitamin D-dependent calcium-binding proteins in kidneys of higher vertebrates. Endocrinology 117, 674–683.

Ritchie, H., Hughes, M., Thompson, E., Malloy, P., Hochberg, Z., Feldman, D., et al., 1989. An ochre mutation in the vitamin D receptor gene causes hereditary 1,25-dihydroxyvitamin D3-resistant rickets in three families. Proc. Natl. Acad. Sci. U.S.A. 86 (24), 9783–9787.

Schlingmann, K.P., Kaufmann, M., Weber, S., Irwin, A., Goos, C., John, U., et al., 2011. Mutations in CYP24A1 and idiopathic infantile hypercalcemia. N. Engl. J. Med. 365 (5), 410–421.

Seth-Vollenweider, T., Joshi, S., Dhawan, P., Sif, S., Christakos, S., 2014. Novel mechanism of negative regulation of 1,25-dihydroxyvitamin D3-induced 25-hydroxyvitamin D3 24-hydroxylase (Cyp24a1) Transcription: epigenetic modification involving cross-talk between protein-arginine methyltransferase 5 and the SWI/SNF complex. J. Biol. Chem. 289 (49), 33958–33970.

Shen, Q., Christakos, S., 2005. The vitamin D receptor, Runx2, and the Notch signaling pathway cooperate in the transcriptional regulation of osteopontin. J. Biol. Chem. 280, 40589–40598.

Shimada, T., Kakitani, M., Yamazaki, Y., Hasegawa, H., Takeuchi, Y., Fujita, T., Fukumoto, S., Tomizuka, K., Yamashita, T., 2004. Targeted ablation of Fgf23 demonstrates an essential physiological role of FGF23 in phosphate and vitamin D metabolism. J. Clin. Invest. 113, 561–568.

Shinki, T., Jin, C.H., Nishimura, A., Nagai, Y., Ohyama, Y., Noshiro, M., Okuda, K., Suda, T., 1992. Parathyroid hormone inhibits 25-hydroxyvitamin D3-24-hydroxylase mRNA expression stimulated by 125-dihydroxyvitamin D3 in rat kidney but not in intestine. J. Biol. Chem. 267, 13757–13762.

Smith, C.L., O'Malley, B.W., 2004. Coregulator function: a key to understanding tissue specificity of selective receptor modulators. Endocr. Rev. 25 (1), 45–71.

Sneddon, W.B., Barry, E.L., Coutermarsh, B.A., Gesek, F.A., Liu, F., Friedman, P.A., 1998. Regulation of renal parathyroid hormone receptor expression by 1,25-dihydroxyvitamin D3 and retinoic acid. Cell. Physiol. Biochem. 8, 261–277.

Sone, T., Kerner, S., Pike, J.W., 1991a. Vitamin D receptor interaction with specific DNA. Association as a 1,25-dihydroxyvitamin D3-modulated heterodimer. J. Biol. Chem. 266 (34), 23296–23305.

Sone, T., Marx, S., Liberman, U., Pike, J., 1990. A unique point mutation in the human vitamin D receptor chromosomal gene confers hereditary resistance to 1,25-dihydroxyvitamin D3. Mol. Endocrinol. 4 (4), 623–631.

Sone, T., Ozono, K., Pike, J.W., 1991b. A 55-kilodalton accessory factor facilitates vitamin D receptor DNA binding. Mol. Endocrinol. 5 (11), 1578–1586.

Sone, T., Scott, R., Hughes, M., Malloy, P., Feldman, D., O'Malley, B., Pike, J., 1989. Mutant vitamin D receptors which confer hereditary resistance to 1,25-dihydroxyvitamin D3 in humans are transcriptionally inactive in vitro. J. Biol. Chem. 264 (34), 20230–20234.

St John, H.C., Bishop, K.A., Meyer, M.B., Benkusky, N.A., Leng, N., Kendziorski, C., et al., 2014. The osteoblast to osteocyte transition: epigenetic changes and response to the vitamin D3 hormone. Mol. Endocrinol. 28 (7), 1150–1165.

St-Arnaud, R., Arabian, A., Travers, R., Barletta, F., Raval-Pandya, M., Chapin, K., Depovere, J., Mathieu, C., Christakos, S., Demay, M.B., Glorieux, F.H., 2000. Deficient mineralization of intramembranous bone in vitamin D-24-hydroxylase-ablated mice is due to elevated 1,25-dihydroxyvitamin D and not to the absence of 24, 25-dihydroxyvitamin D. Endocrinology 141, 2658–2666.

Takeda, S., Yoshizawa, T., Nagai, Y., Yumato, H., Fukumoto, S., Sekine, K., Kato, S., Matsumoto, T., Fujita, T., 1999. Stimulation of osteoclast formation by 1,25−dihydroxyvitamin D requires its binding to vitamin D receptor (VDR) in osteoblastic cells: studies using VDR knockout mice. Endocrinology 140, 1005−1008.

Thompson, P.D., Remus, L.S., Hsieh, J.C., Jurutka, P.W., Whitfield, G.K., Galligan, M.A., et al., 2001. Distinct retinoid X receptor activation function-2 residues mediate transactivation in homodimeric and vitamin D receptor heterodimeric contexts. J. Mol. Endocrinol. 27 (2), 211−227.

Underwood, J.L., DeLuca, H.F., 1984. Vitamin D is not directly necessary for bone growth and mineralization. Am. J. Physiol. 246, E493−eE498.

Veldurthy, V., Wei, R., Campbell, M., Lupicki, K., Dhawan, P., Christakos, S., 2016. 25-Hydroxyvitamin D(3) 24-hydroxylase: a key regulator of 1,25(OH)(2)D(3) catabolism and calcium homeostasis. Vitam. Horm. 100, 137−150.

Wada, L., Daly, R., Kern, D., Halloran, B., 1992. Kinetics of 1,25-dihydroxyvitamin D metabolism in the aging rat. Am. J. Physiol. 262, E906−eE910.

Wang, Y., Zhu, J., DeLuca, H.F., 2014. Identification of the vitamin D receptor in osteoblasts and chondrocytes but not osteoclasts in mouse bone. J. Bone Miner. Res. 29 (3), 685−692.

Wang, Z., Gerstein, M., Snyder, M., 2009. RNA-Seq: a revolutionary tool for transcriptomics. Nat. Rev. Genet. 10 (1), 57−63.

Wasserman, R.H., Fullmer, C.S., 1995. Vitamin D and intestinal calcium transport: facts, speculations, and hypotheses. J. Nutr. 125, 1971S−1979S.

Weinstein, R.S., Underwood, J.L., Hutson, M.S., DeLuca, H.F., 1984. Bone histomorphometry in vitamin D-deficient rats infused with calcium and phosphorus. Am. J. Physiol. 246, E499−E505.

Westin, S., Kurokawa, R., Nolte, R.T., Wisely, G.B., McInerney, E.M., Rose, D.W., et al., 1998. Interactions controlling the assembly of nuclear-receptor heterodimers and co-activators. Nature 395 (6698), 199−202.

Wettschureck, N., Lee, E., Libutti, S.K., Offermanns, S., Robey, P.G., Spiegel, A.M., 2007. Parathyroid specific double knockout of Gq and G11 a subunits leads to a pheyotype resembling germline knockout of the extracellular Ca^{2+} sending receptor. Mol. Endocrinol. 21, 274−280.

Whalen, S., Truty, R.M., Pollard, K.S., 2016. Enhancer-promoter interactions are encoded by complex genomic signatures on looping chromatin. Nat. Genet. 48 (5), 488−496.

Xie, W., Schultz, M.D., Lister, R., Hou, Z., Rajagopal, N., Ray, P., et al., 2013. Epigenomic analysis of multilineage differentiation of human embryonic stem cells. Cell 153 (5), 1134−1148.

Xue, Y., Fleet, J.C., 2009. Intestinal vitamin D receptor is required for normal calcium and bone metabolism in mice. Gastroenterology 136 (4), 1317−1327 e1311−1312.

Yasuda, H., Shima, N., Nakagawa, N., Mochizuki, S.I., Yano, K., Fujise, N., Sato, Y., Goto, M., Yamaguchi, K., Kuriyama, M., Kanno, T., Murakami, A., Tsuda, E., Morinaga, T., Higashio, K., 1998a. Identity of osteoclastogenesis inhibitory factor (OCIF) and osteoprotegerin (OPG): a mechanism by which OPG/OCIF inhibits osteoclastogenesis in vitro. Endocrinology 139, 1329−1337.

Yasuda, H., Shima, N., Nakagawa, N., Yamaguchi, K., Kinosaki, M., Mochizuki, S.I., Tomoyasu, A., Yano, K., Goto, M., Murakami, A., Tsuda, E., Morinaga, T., Higashio, K., Udagawa, N., Takahashi, N., Suda, T., 1998b. Osteoclast differentiation factor is a ligand for osteoprotegerin/ osteoclastogenesis-inhibitory factor and is identical to TRANCE/RANKL. Proc. Natl. Acad. Sci. U.S.A. 95, 3597−3602.

Yoshizawa, T., Handa, Y., Uematsu, Y., Takeda, S., Sekine, K., Yoshihara, Y., Kawakami, T., Arioka, K., Sato, H., Uchiyama, Y., Masushige, S., Fukamizu, A., Matsumoto, T., Kato, S., 1997. Mice lacking the vitamin D receptor exhibit impaired bone formation, uterine hypoplasia and growth retardation after weaning. Nat. Genet. 16, 391−396.

Zehnder, D., Evans, K.N., Kilby, M.D., Bulmer, J.N., Innes, B.A., Stewart, P.M., Hewison, M., 2002. The ontogeny of 25-hydroxyvitamin D(3) 1alpha-hydroxylase expression in human placenta and decidua. Am. J. Pathol. 161 (1), 105−114.

Zella, J.B., Plum, L.A., Plowchalk, D.R., Potochoiba, M., Clagett-Dame, M., DeLuca, H.F., 2014. Novel, selective vitamin D analog suppresses parathyroid hormone in uremic animals and postmenopausal women. Am. J. Nephrol. 39 (6), 476−483.

Zella, L.A., Meyer, M.B., Nerenz, R.D., Lee, S.M., Martowicz, M.L., Pike, J.W., 2010. Multifunctional enhancers regulate mouse and human vitamin D receptor gene transcription. Mol. Endocrinol. 24 (1), 128−147.

Zhu, J.G., Ochalek, J.T., Kaufmann, M., Jones, G., Deluca, H.F., 2013. CYP2R1 is a major, but not exclusive, contributor to 25-hydroxyvitamin D production in vivo. Proc. Natl. Acad. Sci. U.S.A. 110 (39), 15650−15655.

Zierold, C., Darwish, H.M., DeLuca, H.F., 1994. Identification of a vitamin D-response element in the rat calcidiol (25-hydroxyvitamin D3) 24-hydroxylase gene. Proc. Natl. Acad. Sci. U.S.A. 91 (3), 900−902.

Zierold, C., Darwish, H.M., DeLuca, H.F., 1995. Two vitamin D response elements function in the rat 1,25-dihydroxyvitamin D 24-hydroxylase promoter. J. Biol. Chem. 270 (4), 1675−1678.

Further Reading

Hoenderop, J.G., van der Kemp, A.W., Hartog, A., van de Graaf, S.F., Van Os, C.H., Willems, P.H., Bindels, R.J., 1999. Molecular identification of the apical Ca21 channel in 1,25-dihydroxyvitamin D-responsive epithelia. J. Biol. Chem. 274, 8375−8378.

Maurano, M.T., Wang, H., John, S., Shafer, A., Canfield, T., Lee, K., Stamatoyannopoulos, J.A., 2015. Role of DNA methylation in modulating transcription factor occupancy. Cell Rep. 12 (7), 1184−1195.

Chapter 31

Nonskeletal effects of vitamin D: current status and potential paths forward

Neil Binkley[1], Daniel D. Bikle[2], Bess Dawson-Hughes[3], Lori Plum[4], Chris Sempos[5] and Hector F. DeLuca[4]

[1]*University of Wisconsin School of Medicine and Public Health, Madison, Wisconsin, United States;* [2]*VA Medical Center and University of California San Francisco, San Francisco, California, United States;* [3]*Jean Mayer USDA Human Nutrition Research Center on Aging at Tufts University, Boston, Massachusetts, United States;* [4]*Department of Biochemistry, University of Wisconsin–Madison, Madison, Wisconsin, United States;* [5]*Vitamin D Standardization Program, Havre de Grace, MD, United States*

Chapter outline

Introduction 757
Vitamin D and immunity 758
Vitamin D and muscle performance, balance, and falls 759
Physiology 759
Clinical studies of muscle performance and balance 759
Vitamin D and falls 760
Vitamin D and cancer 761
Cellular mechanisms 761
Vitamin D metabolism 761
MicroRNA 761
Cell cycle regulation and proliferation 761
Apoptosis 761
Animal studies 762
Colorectal cancer 762
Breast cancer 762
Prostate cancer 762
Skin 762
Clinical studies 762
Colorectal cancer 762
Breast cancer 763
Prostate cancer 763
Skin cancer 763
Vitamin D and cardiovascular disease 764
Issues in existing data and paths forward to resolve the role of vitamin D deficiency in nonskeletal disease 764
Vitamin D status assessment 765
Is 25(OH)D measurement enough? 765
Clinical and preclinical studies 766
Conclusion and paths forward 768
References 768

Introduction

It is widely assumed that hypovitaminosis D is highly prevalent worldwide (Palacios and Gonzalez, 2014). This is reasonable as humans historically obtained vitamin D via cutaneous production upon UVB exposure, and sun avoidance is now common and widely advocated (USDDH, 2014). Vitamin D deficiency could potentially have widespread adverse health consequences (Holick, 2017). As many tissues possess both the vitamin D receptor and the 1-hydroxylase, it can be hypothesized that active vitamin D (i.e., $1,25(OH)_2D$) is produced locally, i.e., that vitamin D has autocrine/paracrine effects in addition to the classic endocrine effects on calcium homeostasis (Holick, 2008). The effects of vitamin D deficiency on the musculoskeletal system are well established. However, it is reasonable that vitamin D inadequacy could, potentially, compromise the functions of multiple organs and systems (in addition to the musculoskeletal system) and thereby, ultimately, lead to dysfunction leading to acute and chronic disease (Holick, 2011). Consistent with this, hypovitaminosis D, generally defined as low circulating 25-hydroxyvitamin D (25(OH)D), has been associated with a multitude of human diseases, e.g., muscular dysfunction, cardiovascular disease (CVD), cancer, immune function, hypertension, obesity, diabetes mellitus, and many, many others (Holick, 2017; Wang et al., 2017). If these associational studies are reflective of causality, i.e., if vitamin D inadequacy is contributing in a causal way to a multitude of human diseases with

Principles of Bone Biology. https://doi.org/10.1016/B978-0-12-814841-9.00031-2

immense impact upon quality of life, mortality, and health care expense, then widespread screening for this deficiency is needed and societal efforts such as food supplementation are needed. Indeed, vitamin D inadequacy (using the Institute of Medicine guidance) is present in over 30% of Americans (Looker et al., 2011). This prevalence is comparable to those of obesity, hypertension, and hyperlipidemia, other widely recognized contributors to chronic diseases for which widespread public health efforts to manage exist. Alternatively, it could be hypothesized that the association of low vitamin D status with the multitude of human conditions is nothing more than the use of circulating 25(OH)D to identify those with poor nutritional status and/or suboptimal health, thereby reducing the likelihood of sun exposure. If this is true, widespread 25(OH)D measurement and supplementation is an unnecessary waste of limited health care resources and must cease. Clearly, it is essential to determine whether vitamin D inadequacy does, or does not, contribute to nonskeletal disease.

As such, it is not surprising that a plethora of observational and prospective studies evaluating the association of vitamin D status with many diseases has been and is being performed and published. As systematic reviews of randomized trials are considered the highest level of evidence-based medicine (Sackett et al., 1996), it is expected that meta-analyses relating vitamin D to nonskeletal disease are often being performed. Indeed, a PubMed search in March 2018 found that 747 vitamin D-related meta-analyses were performed from 2013 through 2017, an average of approximately three every week. Moreover, a number of authoritative reviews have evaluated the potential role of vitamin D in nonskeletal conditions (Autier et al., 2017; Rejnmark et al., 2017; Rosen et al., 2012). To succinctly summarize these reviews, it remains unclear what role, if any, vitamin D inadequacy plays in causing nonskeletal diseases. We believe this state of uncertainty reflects deficiencies in assessment of vitamin D status and clinical trial design as is discussed later. As such, this chapter will not replicate recent reviews. However, as examples of existing data suggesting a potential importance in human disease, we will briefly overview data relating vitamin D status to cancer, immune function, falls, and CVD. These conditions were selected as they cause immense morbidity, mortality, and health care cost worldwide.

Vitamin D and immunity

There are many extensive reviews that explore the possible role of vitamin D in the immune system (Colotta et al., 2017; Hayes et al., 2015; Peelen et al., 2011). Certainly, the vitamin D receptor (VDR) is found in activated T cells and macrophages (Wang et al., 2012c). In vitamin D deficiency the VDR is either low or absent from T cells but is present in macrophages. The receptor quickly appears in T cells following vitamin D administration. B cells lack the VDR in our hands, while others claim its presence in these cells. By autoradiography, Stumpf et al. demonstrated that titrated $1,25(OH)_2D_3$ is found in the nuclei of thymus, spleen, lymph nodes, and macrophages, but interestingly not in lymphocytes (Stumpf et al., 1990; Stumpf and Downs, 1987). There is little doubt vitamin D plays a role in immunity; however, there is much debate and little agreement on how vitamin D is involved in immunity.

A great deal of information has been generated, primarily by in vitro methods, on cytokine levels in T cells, with concentrations of $1,25(OH)_2D_3$ higher than found in vivo. Although arguments that these levels can be achieved locally have been made, no such in vivo demonstration has appeared. Further, rarely has true vitamin D deficiency through dietary means been achieved, and most studies examine the effect of supplemental vitamin D above normal. In this brief review an attempt will be made to present the effects of vitamin D deficiency versus sufficiency in animal models when available. Unfortunately, consistent information on whether vitamin D deficiency has any impact on innate or acquired immunity in animals or humans is unavailable.

On the other hand, some information is available in situations in which the adaptive immune system fails, or autoimmunity, in animals. One example is in type 1 diabetes. Vitamin D deficiency accelerated both the onset and the severity of the disease in the NOD mouse (Zella et al., 2003). In this circumstance a major consideration is that the insulin-producing islet cells have a high concentration of the VDR (Wang et al., 2012b). While vitamin D through its active form suppresses the disease (Zella et al., 2003), the suppression could be a result of $1,25(OH)_2D_3$ action on the islet cell rather than an action on cells of the immune system. A combination of cellular targets is also possible, as specific T cell populations show small alterations in NOD mice when the VDR is eliminated (Gysemans et al., 2008). However, the consequences of these changes are not clear.

The opposite impact of vitamin D status is found in the case of experimental autoimmune encephalomyelitis (EAE), a model of multiple sclerosis (MS). Quite unexpected is the finding from our group (DeLuca and Plum, 2011), and independently by a group in France (Fernandes de Abreu et al., 2010), that vitamin D deficiency suppresses EAE instead of exacerbating disease. Thus, vitamin D is required and functions in some manner to program T cells and/or macrophages to attack the spinal cord, precipitating EAE, a widely accepted model of MS. Originally Lemire and Archer reported that

$1,25(OH)_2D_3$ could suppress EAE, a finding that was repeated in other laboratories (Lemire and Archer, 1991; VanAmerongen et al., 2004). In our group the suppression of EAE by $1,25(OH)_2D_3$ is accompanied by hypercalcemia (Cantorna et al., 1996). When dietary calcium was reduced to 0.02%, the effectiveness of $1,25(OH)_2D_3$ to suppress EAE was severely limited (Cantorna et al., 1999). Further hypercalcemia itself suppresses EAE (Meehan et al., 2005). Therefore, the suppression of EAE by $1,25(OH)_2D_3$ is linked to dietary and perhaps serum calcium. Consistent with the removal of vitamin D from the diet, the ablation of VDR also blocks the development of EAE (Wang et al., 2012b). Thus, in contrast to type 1 diabetes, vitamin D is required for the development of the disease, reducing the likelihood vitamin D might be used to suppress MS. Results from meta-analyses of the clinical trials done to test the use of vitamin D to suppress MS seem to confirm this idea (Cashman et al., 2016b; Hempel et al., 2017; Zheng et al., 2018).

Distinct from this finding, a narrow band of UV light (295–315 nm) suppresses EAE by a mechanism not involving vitamin D. Although this may provide an explanation of the low incidence of MS in high-sunlight areas, its mechanism is largely unknown.

Another disease that involves the immune system is inflammatory bowel disease, which has been extensively studied by Cantorna et al. (Bora and Cantorna, 2017). Their results are similar to those observed in animal models of type 1 diabetes, i.e., that VDR and ligand deficiency both exacerbate the disease. The intestinal epithelium possesses large amounts of the VDR and it is unclear whether the effect of the vitamin D is directly on immune cells or through its action on the intestinal villi. It is clear that subsets of T cells, i.e., invariant natural killer T cells as well as CD8aa populations, are significantly reduced in the gastrointestinal tract of mice lacking VDR (Bora and Cantorna, 2017). Whether the decrease in these populations can explain the reduction of disease is unclear.

In conclusion, the presence of the VDR and its ligand in cells of the immune system strongly suggests a role for vitamin D in immunity. However, little in vivo evidence is available on the role of vitamin D in the immune system. Vitamin D deficiency and/or a lack of the VDR does suppress the animal model of MS (EAE). Quite differently, 1,25-dihydroxyvitamin D, the active form of vitamin D, suppresses type 1 diabetes in NOD mice, while vitamin D deficiency exacerbates the disease. A similar relationship is found in the case of inflammatory bowel disease. Although these effects of vitamin D are clear, the underlying mechanisms remain unknown.

Vitamin D and muscle performance, balance, and falls

Physiology

VDRs are present in many tissues, including muscle, but their role in muscle is not entirely clear. The number of VDRs in muscle (Bischoff-Ferrari et al., 2004) and the number of type 2 muscle fibers (Vandervoort, 2002) decline with aging, and preliminary evidence indicates that similar changes may occur in vitamin D deficiency (Ceglia et al., 2013). In a small study of 21 older women with a low mean serum 25(OH)D level of 18.4 ng/mL, daily supplementation with 4000 IU of vitamin D_3 for 4 months increased the intramyonuclear VDR number and the cross-sectional area of type 2 muscle fibers compared with placebo (Ceglia et al., 2013). No significant changes in muscle performance were documented.

With aging, the number of motor neurons enervating muscle declines (Brown, 1972), as does balance. There is limited evidence that vitamin D deficiency may adversely affect motor neurons. Skaria documented reduced nerve conduction velocities as well as myopathy in patients with osteomalacia (Skaria et al., 1975). However, treatment with vitamin D reversed the myopathy but did not improve nerve conduction velocity in these patients, leaving open the possibility that the patients had unrelated peripheral neuropathies (Skaria et al., 1975). Irani (1976) observed no nerve conduction abnormalities in patients with osteomalacia.

Clinical studies of muscle performance and balance

Based on the histomorphometric muscle changes seen in severe vitamin D deficiency, many investigators have hypothesized that supplementation with vitamin D may reduce muscle wasting and improve muscle performance in older adults with milder degrees of vitamin D deficiency. A 2014 meta-analysis of 30 randomized controlled clinical trials (RCTs) involving 5615 older adults examined the impact of vitamin D supplementation on muscle strength, mass, and power (Beaudart et al., 2014). In 29 of these trials, supplemental vitamin D significantly increased a global index of strength comprising mainly grip and leg strength. Subset analyses indicated that benefit was greater for participants with 25(OH)D levels <12 ng/mL, and that leg strength but not grip strength improved significantly with supplementation. In the trials that measured muscle mass (N = 6) and power (N = 5), no significant effect of vitamin D was observed (Beaudart et al., 2014). Several subsequent trials have found no impact of supplemental vitamin D on muscle function in older adults (Hansen et al., 2015; Levis and Gomez-Marin, 2017; Uusi-Rasi et al., 2015), possibly because the study participants were not vitamin D deficient (see "Clinical studies"). In contrast, a study by Cangussu et al. reported improved chair rise performance with vitamin D supplementation over 9 months in postmenopausal women with a mean serum 25(OH)D level of 15 ng/mL (Cangussu et al., 2015).

The role of vitamin D in muscle mass and performance in young adults is less extensively studied. In a large clinical trial in school girls with a mean serum 25(OH)D level of 14 ng/mL, vitamin D supplementation for 1 year increased lean tissue mass (and bone mass) but had no effect on grip strength (El-Hajj Fuleihan et al., 2006). Similarly, in a study of young women, mean age 21 years, with a low mean 25(OH)D level of 9.3 ng/mL, treatment over 6 months increased serum 25(OH)D to near 30 ng/mL but had no significant effect on grip strength (Goswami et al., 2012). Thus current evidence indicates that supplementation modestly increases muscle mass and bone mass, but does not improve muscle strength.

Several trials conducted in older adults have reported vitamin D treatment to improve balance, measured under static conditions by body sway, compared with placebo (Cangussu et al., 2016; Pfeifer et al., 2009). The impact of vitamin D on dynamic balance (a complex process involving sensory information, selection of the appropriate response to maintain balance, and efficient activation of the muscles that respond to a balance disturbance) is unknown, but it may be inferred by the effect of vitamin D on risk of falling (see the next section).

Vitamin D and falls

Prevention of falls is a public health priority because 1%–2% of falls lead to hip fractures, falls are independent determinants of functional decline, and falls lead to 40% of all nursing home admissions (Tinetti and Williams, 1997). Moreover, falls are a substantial and increasing cause of mortality in the very old (Hartholt et al., 2018).

The role of vitamin D in preventing falls has been studied extensively and clinical trials have produced mixed findings. Two meta-analyses published in 2009 and 2011 found that vitamin D supplementation had a modest beneficial effect on fall risk in older adults (Bischoff-Ferrari et al., 2009; Murad et al., 2011). One of these papers suggested that fall risk reduction with supplementation may be more effective in individuals with low 25(OH)D levels (Murad et al., 2011). In 2014, a meta-analysis concluded that the effect estimate for vitamin D with or without calcium on falls did not alter the relative risk by 15% or more (Bolland et al., 2014). A 2015 meta-analysis found vitamin D supplementation to be more effective in elders with low 25(OH)D levels (LeBlanc and Chou, 2015). Since this most recent meta-analysis (LeBlanc and Chou, 2015), several large, placebo-controlled trials have been published that collectively appear to fit the pattern that only vitamin D-insufficient elders are responsive to supplementation with vitamin D. Hansen et al. treated postmenopausal women with a mean serum 25(OH)D level of 21 ng/mL with 50,000 IU of vitamin D_3 twice monthly or 800 IU daily for 1 year and observed no effect on falls (Hansen et al., 2015). Similarly, the Vitamin D Assessment (ViDA) Study found no effect of treatment for 3.4 years with 100,000 IU of vitamin D_3 per month on fall risk in older adults with a mean serum 25(OH)D level of 25 ng/mL (Khaw et al., 2017). In contrast, in postmenopausal women with a very low mean 25(OH)D level of 15 ng/mL, treatment with 1000 IU of vitamin D_3 per day for 9 months reduced fall risk by about 50% (Cangussu et al., 2016).

In summary, available evidence indicates that elders with low 25(OH)D levels are most likely to benefit from supplementation with vitamin D and the degree of fall risk reduction may be proportional to the degree of insufficiency. The results of a large ongoing trial (as of this writing) testing several doses of vitamin D in elders with 25(OH)D levels in the range of 10–29 ng/mL should provide a more precise estimate of the segment of the elderly population that is likely to benefit from supplementation and the optimal replacement dose(s) (NCT02166333).

Several studies have indicated that supplementation under certain circumstances can increase risk of falling. This was first noted by Sanders et al., who found that a large annual oral dose of 500,000 IU of vitamin D increased risk of falls (and fractures) in elders (Sanders et al., 2010). An increase in fall risk was also seen in a trial administering an oral dose of 60,000 IU of vitamin D_3 monthly (Bischoff-Ferrari et al., 2016). In long-term care residents, treatment with 100,000 IU of vitamin D_3 per month, compared with standard dosing, significantly increased fall risk (secondary end point), while significantly lowering the risk of developing an acute respiratory infection (primary outcome) (Ginde et al., 2017). A common feature of these trials is that they employed high intermittent doses of vitamin D. But treatment with 100,000 IU per month does not always increase risk of falls; this dose administered in the ViDA study had no significant effect on the fall rate (Khaw et al., 2017). Nonetheless, there is sufficient evidence for risk that daily or weekly dosing is preferred to higher intermittent dosing.

Vitamin D and cancer

The antiproliferative, prodifferentiating effects of vitamin D signaling on many, if not all, cell types have raised the hope that vitamin D, 1,25(OH)$_2$D, or one or more of its analogs would prove useful in the prevention and/or treatment of cancer. Although the cellular and animal studies have been promising, and at least for some cancers epidemiologic evidence is suggestive, to date the RCTs are less compelling. That said, a number of mechanisms by which 1,25(OH)$_2$D and its analogs might alter tumor development and progression have been documented, at least in the cellular and animal studies as reviewed in 2016 (Bikle, 2016).

Cellular mechanisms

Vitamin D metabolism

Most tumors express the VDR and often express cytochrome P450 27B1 (CYP27B1), the enzyme that converts 25(OH)D to 1,25(OH)$_2$D, thus producing the ligand for VDR. Their expression is often lost as the tumor undergoes progressive dedifferentiation (Brozek et al., 2012; Matusiak and Benya, 2007; Santagata et al., 2014). On the other hand CYP24A1 expression is also often increased in tumors, and is associated with resistance to 1,25(OH)$_2$D (Brozek et al., 2012; Tannour-Louet et al., 2014). These changes in vitamin D metabolism and responsiveness reduce the ability of 1,25(OH)$_2$D to control the proliferation and differentiation of these tumors.

MicroRNA

A number of microRNAs that alter tumor proliferation have been identified as being regulated by 1,25(OH)$_2$D (Christakos et al., 2016). This includes increased expression of miR-145, which blocks the expression of E2F3, a key regulator of proliferation (Chang et al., 2015), and miR-32, which blocks the proapoptotic protein Bim, protecting the cell (human myeloid leukemia) from AraC-induced apoptosis (Gocek et al., 2011).

Cell cycle regulation and proliferation

1,25(OH)$_2$D typically causes arrest at the G0/G1 and/or G1/S transition in the cell cycle, associated with a decrease in cyclins and an increase in the inhibitors of the cyclin-dependent kinases such as p21cip1 and p27kip1 (Hager et al., 2001; Palmer et al., 2003). The family of Forkhead box O proteins blocks proliferation. 1,25(OH)$_2$D promotes their interaction with the VDR as well as their regulation by Sirt1 and protein phosphatase 1, maintaining these proteins in the transcriptionally active dephosphorylated state (An et al., 2010). On the other hand 1,25(OH)$_2$D reduces the expression of proproliferative genes such Myc, Fos, and Jun (Meyer et al., 2012), while stimulating the expression of insulin-like growth factor (IGF)-binding protein 3 (IGFBP3) in prostate and breast cancer (BCa) cells. The increased expression of IGFBP3 would bind the IGFs, blocking their ability to stimulate proliferation (Colston et al., 1998; Huynh et al., 1998). In epithelial cells 1,25(OH)$_2$D stimulates the expression of transforming growth factor β2, which is antiproliferative in these cells (Peehl et al., 2004; Swami et al., 2003; Yang et al., 2001) and suppresses components of the Hedgehog pathway, which when overexpressed result in basal cell carcinomas (BCCs) (Aszterbaum et al., 1998; Teichert et al., 2011). 1,25(OH)$_2$D inhibits the proliferative effects of epidermal growth factor by inhibiting the expression of its receptor in breast cell lines (McGaffin and Chrysogelos, 2005). Constitutive activation of the Wnt/β-catenin pathway is the cause of most colorectal cancers (CRCs). When activated, β-catenin enters the nucleus, where it binds to T cell factor/lymphoid enhancer-binding factor (TCF/LEF) sites in genes promoting proliferation (e.g., cyclin D1). VDR, on binding to its ligand, blocks this pathway both by binding to β-catenin, limiting its access to the TCF/LEF sites in the nucleus, and by stimulating the formation of the E-cadherin/catenin complex in the cell membrane to which β-catenin is bound, limiting its translocation to the nucleus (Byers et al., 2012). Moreover, 1,25(OH)$_2$D can increase the expression of the Wnt inhibitor dickkopf (Dkk)-1 (Aguilera et al., 2007) while inhibiting that of the Wnt activator Dkk-4 (Pendas-Franco et al., 2008) in colon cancer cells.

Apoptosis

1,25(OH)$_2$D stimulates the expression of a number of proapoptotic genes, such as GOS2 (G0/G1 switch gene 2) (Palmer et al., 2003), Bax (Kizildag et al., 2010), DAP (death-associated protein)-3, CFKAR (caspase 8 apoptosis-related cysteine peptidase), FADD (Fas-associated death domain), and caspases (e.g., caspases 3, 4, 6, and 8) (Swami et al., 2003) in a variety of cell lines, while suppressing the expression of proapoptotic genes such as Bcl2 and Bcl-XL in these and other cell lines (Kizildag et al., 2010; Weitsman et al, 2003, 2005).

Animal studies

Animal studies demonstrating the efficacy of $1,25(OH)_2D$ in preventing or slowing the progression of various tumors are numerous, with those of the colorectum (CRC), breast, prostate, and skin being most studied. Below are selected examples.

Colorectal cancer

Diets that are low in calcium and vitamin D (Western diet) fed to mice increase the risk of CRC for these mice. This risk can be reversed by diets supplemented with calcium and vitamin D (Newmark et al., 2009). The administration of vitamin D metabolites to mice in which CRC was induced by azoxymethane and dextran sulfate incorporated into the diet is at least partially protective (Murillo et al., 2010). Activation of the Wnt/β-catenin pathway caused by mutations in adenomatous polyposis coli (APCmin) results in tumor formation much faster when the mice are placed on a low-calcium, low-vitamin-D Western diet (Yang et al., 2008) or on a vitamin D-deficient diet (Xu et al., 2010). Similarly, these tumors develop faster when these mice are bred with VDR-null mice (Zheng et al., 2012).

Breast cancer

BCa induced by dimethylbenzanthracene (DMBA) are increased in number when rats are fed the low-calcium, low-vitamin-D Western diet (Lipkin and Newmark, 1999) or when VDR-null mice are administered DMBA (Zinser and Welsh, 2004). The growth of BCa xenografts can be prevented by vitamin D analogs (VanWeelden et al., 1998).

Prostate cancer

Vitamin D analogs can also inhibit the growth of prostate cancer (PCa) regardless of androgen receptor status (Bhatia et al., 2009). A vitamin D-deficient diet will promote the growth of PC3 PCa cells in bone (Zheng et al., 2011). Breeding the transgenic prostate tumor model LPB-Tag with VDR-null mice stimulates the growth of these tumors (Mordan-McCombs et al., 2010). The development of tumors in the TRAMP (transgenic adenocarcinoma of the mouse prostate) model of PCa can be prevented with high doses of $1,25(OH)_2D$ (Krishnan et al., 2010).

Skin

The most common skin cancers are squamous cell carcinomas (SCCs) and BCCs. VDR-null mice develop skin tumors when exposed to DMBA (with or without topical application of phorbol myristate acetate (TPA)) or after chronic exposure to UVB at a rate far greater than that of their wild-type controls (Ellison et al., 2008; Teichert et al., 2011). Topical $1,25(OH)_2D$ is protective at least from the early effects of UVB on markers of DNA damage such as cyclobutane pyrimidine dimers (Gupta et al., 2007), although extended studies to determine whether topical application of $1,25(OH)_2D$ is protected in UVB-induced tumor formation have not been done.

Clinical studies

The inverse relationship between solar exposure and cancer mortality in North America was first noted by Apperly (1941) in 1941, and then popularized and linked to vitamin D by the Garland brothers (Garland and Garland, 1980) in 1980 in their epidemiologic studies with colon cancer. With the exception of skin cancer, this inverse relationship between solar exposure and cancer has been reported for many types of cancer (van der Rhee et al., 2006). More recent studies have examined the association of vitamin D intake or serum levels of 25(OH)D, generally using case–control and cohort studies. RCTs have generally been smaller, and often do not support the positive associations between sunlight, dietary vitamin D, or 25(OH)D levels and cancer prevention found in the epidemiologic studies. As for animal studies, most data come from studies of colorectal, breast, prostate, and skin cancer.

Colorectal cancer

In meta-analyses of nine studies (eight cohort, one nested case–control; 6466 subjects) evaluating the relationship of vitamin D intake and CRC and nine studies (seven cohort, two case–control; 2764 cases, 3948 controls) evaluating serum levels of 25(OH)D and CRC, Ma et al. (2011) found an overall relative risk (RR) of 0.88 (CI 0.8–0.96) comparing the highest with the lowest categories of vitamin D intake and an RR of 0.67 (CI 0.54–0.80) for the highest to lowest 25(OH) D levels. When dietary calcium was taken into account (higher calcium is better) the risk reduction was enhanced. These meta-analyses confirm previous results from the American Cancer Society cohort study (120,000 men and women) (Cho et al., 2004) and the National Institutes of Health (NIH; 16,000 participants) (Carroll et al., 2010). Thus, at least for CRC, there is a reasonably consistent set of epidemiologic data supporting the protective effect of vitamin D (and calcium) with respect to CRC incidence. However, RCTs have been disappointing. The Women's Health Initiative (WHI), in which 400 IU of vitamin D daily was assessed, failed to show protection against CRC (Wactawski-Wende et al., 2006). The low dose of vitamin D used in this trial may have limited the likelihood of observing an effect. However, a study of 1000 IU vitamin D daily for 3–5 years also failed to show a benefit in the reduction of colon adenomas and colon cancer even in a high-risk population (Baron et al., 2015). Perhaps vitamin D supplementation over short time spans relative to the time required for cancer to manifest is a limiting factor.

Breast cancer

In a large case–control study from Italy (Rossi et al., 2009) a 34% decrease in BCa was found in those ingesting the highest level of vitamin D (>194 IU/day) versus those ingesting the lowest level (<60 IU/day). The largest cohort studies (Lin et al., 2007; Shin et al., 2002) (Nurses Health Study with 88,891 participants and WHI with 31,487 participants) showed a reduction in risk (RR 0.72, CI 0.55–0.94, and RR 0.65, CI 0.42–1.00, respectively) in those ingesting the highest amount of vitamin D, but only in the premenopausal women. Chlebowski (2011) reviewed 10 case–control and 10 cohort studies with respect to vitamin D intake and BCa and four case–control and six nested case–control studies with respect to 25(OH)D levels and BCa. Only when premenopausal/perimenopausal women were included in the analysis was a significant negative association between increased vitamin D intake and BCa incidence found (RR 0.83, CI 0.73–0.95) (Chen et al., 2010). Of the six nested case–control studies assessing the relationship of serum 25(OH)D and BCa, only one showed a significant negative association between high 25(OH)D levels and incidence of BCa. In a separate meta-analysis by Gandini et al. (2011) an RR of 0.89 (0.82–0.98) for a 10 ng/mL increase in 25(OH)D was found when all studies were included and an RR of 0.83 (0.79–0.87) when only case–control studies were pooled. As for CRC, the WHI did not show a protective role for vitamin D in BCa. Thus the epidemiologic data tend to support a protective role for vitamin D against BCa, but the data are not as consistent as for CRC.

Prostate cancer

In contrast to CRC, the role of vitamin D in PCa is decidedly mixed. In a summary of 14 prospective studies (not RCTs) examining the association between 25(OH)D levels and the development of PCa, van der Rhee et al. (2006) found that 11 studies showed no association, and one study (Ahn et al., 2008) showed a positive association with PCa aggressiveness, although this was not seen in other studies (Gilbert et al., 2011). In particular, a meta-analysis of six cohort/nested case–control studies (8722 cases) examining the association of dietary vitamin D intake with PCa found an RR of 1.14 (CI 0.99–1.31) for an increase in dietary vitamin D of 1000 IU (Gilbert et al., 2011). Similarly, a meta-analysis of 14 cohort/nest case–control studies, including 4353 cases, examining the association of serum 25(OH)D and PCa found an RR of 1.04 (CI 0.99–1.1) for a 10 ng/mL increase in 25(OH)D for all PCas (Gilbert et al., 2011). Although an initial phase I clinical trial (ASCENT I) with high dose of $1,25(OH)_2D$ and docetaxol seemed to show promise in the treatment of castration-resistant PCa, this initial success could not be repeated in a larger phase II trial (ASCENT II) potentially flawed by differences in the dose of docetaxol between the calcitriol plus docetaxol arm and that with docetaxol alone (Scher et al., 2011). These trials did establish the safety of using high doses of calcitriol in the treatment of cancer when administered under controlled conditions. Nevertheless, the clinical evidence weighs against vitamin D being beneficial in the prevention or treatment of PCa.

Skin cancer

Skin cancer, in particular nonmelanoma skin cancer (NMSC), is by far the most common cancer, with over 1 million skin cancers occurring annually in the United States. Of these, 96% are NMSC (80% BCC, 16% SCC), with melanomas making up the remaining 4%. Because UVB is the major etiologic agent for NMSC, but is also the major means by which the body makes vitamin D, it has been difficult to separate out the beneficial effects of vitamin D on NMSC from the deleterious effects of UVB exposure. Moreover, most national registries do not include NMSC, leaving no easily accessible database with which to track the development of these cancers. That said, a few epidemiologic studies have been published. An earlier study on vitamin D intake found no association between vitamin D and BCC risk (Hunter et al., 1992). However, some clinical studies of BCC patients show a potential beneficial role for vitamin D. In a nested case–control study of 178 elderly men with NMSC compared with 930 without skin cancer enrolled in the Osteoporotic Fractures in Men (MrOS) study, men with the highest baseline serum 25(OH)D levels (30 ng/mL) had 47% lower odds of NMSC (CI 0.3–0.93) compared with those with the lowest baseline 25(OH)D levels (Tang et al., 2010). However, this observation has not been confirmed in other studies, which have shown a positive correlation between 25(OH)D levels and BCC (Asgari et al., 2010; Eide et al., 2011). In an analysis of 36,282 postmenopausal women in the WHI, no difference was found in the incidence of melanoma between women receiving 400 IU vitamin D plus calcium (1000 mg) supplementation compared with placebo over a follow-up period of 7 years, although a subgroup analysis of women at high risk based on a previous history of NMSC showed a reduction in the incidence of melanoma in those receiving calcium and vitamin D supplementation (Tang et al., 2011).

Vitamin D and cardiovascular disease

In observational studies, low circulating 25(OH)D concentrations have been associated with increased risk of CVD. Although these associations are not seen in all studies, there is no evidence that higher 25(OH)D levels are associated with greater CVD risk (Nemerovski et al., 2009). As such, it is plausible that vitamin D is indeed causally related to CVD. Importantly, both blood vessels and cardiac myocytes have been reported to possess VDRs and CYP27B1, the 1α-hydroxylase enzyme (Tishkoff et al., 2008).

Low vitamin D status has been hypothesized to potentially cause CVD by multiple, not exclusive, mechanisms, including regulation of calcium flux altering vascular and myocardial calcification, mediation of inflammation/cell proliferation, and interactions producing antiatherosclerotic effects (among others) (Norman and Powell, 2014; Cashman, 2014; Pilz et al., 2010). For example, it is plausible that the relationship of 25(OH)D with parathyroid hormone (PTH) might be involved in CVD, as elevated PTH is associated with increased blood pressure, which, if sustained, could lead to cardiac hypertrophy and ultimately to heart failure (Rostand and Drueke, 1999). In addition, chronically high inflammation is positively associated with CVD, and $1,25(OH)_2D$ may downregulate proinflammatory and upregulate antiinflammatory cytokines (Cardus et al., 2009; Chen et al., 2013; Helming et al., 2005). Consistent with a potential effect in CVD, a 2018 meta-analysis of six vitamin D supplementation RCTs in adults with heart failure found vitamin D supplementation to lower tumor necrosis factor α (Rodriguez et al., 2018). Other potential linkages of vitamin D status with CVD include potential effects on the functions of endothelial cells (Dong et al., 2012; Merke et al., 1989) and cardiac myocytes (Pilz et al., 2010; Tishkoff et al., 2008).

Epidemiologic evidence supports a potential relationship of low vitamin D status with CVD. For example, over ~8 years of follow-up, the hazard ratios for heart failure death and sudden cardiac death for those whose 25(OH)D was <10 ng/mL were 2.8 and 5.1, respectively, compared with those with a 25(OH)D of ≥30 ng/mL (Pilz et al., 2008). Similarly, in an NHANES cohort, a composite CVD index (stroke, myocardial infarction, and angina) found that the prevalence of circulating 25(OH)D at <20 ng/mL was higher (29%) in those with CVD than those without (21%) (Kendrick et al., 2009). Moreover, a Framingham Offspring cohort study found circulating 25(OH)D levels below 15 ng/mL to be associated with increased risk of cardiovascular events, with the 5-year rate being double in those with vitamin D deficiency compared with those with higher concentrations (Wang et al., 2008). Importantly, low vitamin D status is associated with increased risk of death due to heart failure, sudden cardiac event, and CVD all-cause mortality. This was demonstrated in the Ludwigshafen Risk and Cardiovascular Health (LURIC) cohort after a median follow-up of 7.7 years. Heart failure and sudden cardiac events were increased in those with 25(OH)D of <10 ng/mL (Pilz et al., 2008) and CVD all-cause mortality was increased in subjects with lower 25(OH)D (Dobnig et al., 2008) even after adjustment for many traditional CVD risk factors. Multiple other observational studies with generally similar results exist and will not be reviewed here; overall, a 2012 meta-analysis of prospective studies relating 25(OH)D with CVD risk found an inverse relationship between 25(OH)D and CVD risk (Wang et al., 2012a).

Issues in existing data and paths forward to resolve the role of vitamin D deficiency in nonskeletal disease

As noted in the introduction, immense interest, clinical trials, publications, and meta-analyses related to vitamin D have been and are currently being published without adequate clarification of the potential role of vitamin D in nonskeletal diseases. As an oversimplification, we believe this reflects two major issues in vitamin D research: inadequate measurement of vitamin D status and limitations in clinical trial design.

Vitamin D status assessment

It is widely accepted that measurement of the circulating total 25-hydroxyvitamin D (25(OH)D) concentration—i.e., the sum of 25(OH)D_3 and 25(OH)D_2—is the best way to define vitamin D status (Scientific Advisory Committee on Nutrition (SACN), 2016; Holick et al. 2011; Institute of Medicine, 2010). However, substantial variability in "25(OH)D" results has existed, and continues to exist (Binkley et al., 2004; Carter et al., 2004; de la Hunty et al., 2010; Le Goff et al., 2015; Wallace et al., 2010). This variability confounds in a major way all efforts to clarify a potential role of vitamin D in nonskeletal disease and moreover to develop clinical and public health guidelines. It seems self-evident that use of different "25(OH)D" values could lead to differences in the definition of vitamin D inadequacy. Indeed, the use of differing 25(OH) D assays yields differing estimates of the prevalence of vitamin D inadequacy; retrospective standardization documents that prevalence can increase or decrease based on whether the initial assay was negatively or positively biased (Cashman et al, 2013, 2015a, 2016a; Sarafin et al., 2015; Schleicher et al., 2016). As such, failure to standardize 25(OH)D measurement standardization could be considered the major issue contributing to the "chaos" that continues to surround vitamin D (Binkley et al., 2017b).

Historically, no reference measurement procedure or reference materials existed to allow 25(OH)D measurement standardization. As such, 25(OH)D assay standardization is essential for clinical, research, and public health recommendation clarity. The Vitamin D Standardization Program (VDSP) was developed in an attempt to rectify this situation (Sempos et al., 2012).

Detailed review of the VDSP is beyond the scope of this chapter. Briefly, the VDSP coordinates the international effort to standardize serum 25(OH)D measurement to gold standard reference assays that have been developed (Mineva et al., 2015; Phinney et al., 2017; Stockl et al., 2009; Tai et al., 2017). In addition, guidance for retrospective standardization of previously published research for which stored specimens exist has been published (Durazo-Arvizu et al., 2017). Thus, VDSP approaches exist to allow performance and reporting of standardized 25(OH)D results (Sempos et al., 2017). Such an approach is essential if 25(OH)D data are to be pooled in meta-analyses.

Moreover, we believe that the NIH and other grant-funding agencies around the world should require grant applicants to demonstrate the procedures they will use to standardize 25(OH)D measurement before funding of a vitamin D-related study is approved. In addition, as certain automated immunoassays may not function properly in some patient populations due to altered vitamin D-binding protein levels, e.g., pregnant women and intensive care unit patients, researchers need to demonstrate that the assay can function in the patient populations to be studied (Apperly, 1941). Finally, we suggest that standardization of 25(OH)D data be a requirement for research study publication. Indeed, the Endocrine Society journals have updated their author instructions to require detailed description of hormone assay requirements. Appropriately, these instructions specifically note "...a requirement for traceability of an assay to a certified standard is not yet available for all steroid hormones, but is an important goal" (Wierman et al., 2014). The VDSP has published guidance that allows researchers to meet this goal (Sempos et al., 2017). Such a requirement for 25(OH)D data would allow pooling of results in meta-analyses; failure to implement this requirement will facilitate continuation of chaos regarding the role of vitamin D in health and disease.

Is 25(OH)D measurement enough?

A multitude of vitamin D metabolites exist, some of which may possess physiologic effects. As such, it is appropriate to consider whether the current approach of measuring only 25(OH)D is adequate to define an individual's vitamin D status. Indeed, it is plausible that considering a compilation of all vitamin D metabolites that possess a physiologic effect should be done to define vitamin D deficiency (Jones and Kaufmann, 2016). Such metabolites might include free 25(OH)D, $1,25(OH)_2D_3$, $24,25(OH)_2D_3$, and cholecalciferol, among others (Herrmann et al., 2017; Hollis and Wagner, 2013). Ultimately, low vitamin D status must be defined based on measurement of vitamin D-related metabolites that best predict health outcomes. As of this writing, the importance of these other metabolites, if any, in defining vitamin D status remains to be determined (Scientific Advisory Committee on Nutrition (SACN), 2016; Bouillon, 2017; Holick et al., 2011; Institute on Medicine, 2010). However, the ratio of $24,25(OH)_2D_3$ to 25(OH)D appears to be a promising potential advance as it may be predictive of response to vitamin D supplementation and provide additional insights into vitamin D status (Cashman et al., 2015b; Graeff-Armas et al., 2018; Jones and Schlingmann, 2018; Kaufmann et al., 2014; Molin et al., 2015; Wagner et al., 2011). Further work is clearly needed.

As part of the VDSP effort, the National Institute of Standards and Technology (NIST), with support from the NIH Office of Dietary Supplements, has developed reference methods and materials for standardizing the measurement of 3-epi-25(OH)D_3 and 24,25$(OH)_2D_3$. NIST is also working on a reference method for vitamin D-binding protein, PTH, and 1,25$(OH)_2D_3$. Efforts such as these are essential; it must be emphasized that if and when it becomes established that additional vitamin D metabolite measurements contribute to the definition of hypovitaminosis D, standardization or harmonization of such assays (where tools to do so exist) will be essential to avoid repeating the history of chaos seen with 25(OH)D measurement (Krishnan et al., 2010). Moreover, efforts to develop assays that simultaneously measure multiple vitamin D metabolites should, as a requirement, include steps to standardize measurements of these metabolites (Jones and Kaufmann, 2016).

Clinical and preclinical studies

Animal research and small clinical studies are essential precursors to successful large RCTs. Appropriate animal models are essential in studying vitamin D's role in the pathophysiology of complex human disease. We need to ask: are we using the appropriate animal models? Mouse models are superb for evaluating vitamin D metabolism using targeted knockout mouse models; are they the best and most appropriate surrogates of human physiology? Should additional studies be performed in Old World primates?

Small clinical studies can play a key role in moving toward large RCTs, as they are useful in evaluating potential treatment regimens, including evaluating possible doses and treatment regimens. Finally, has the vitamin D field moved too quickly to large RCTs? In the past, large clinical trials were not begun, generally, unless there was solid evidence from in vivo research, animal studies, and small clinical studies that a large trial was likely to be successful. The rationale was that we should not subject humans to clinical trials unless and until there were good data to support a result consistent with the hypothesis. Not knowing the answer to a question should not be sufficient criteria to expose humans to potential risks associated with large clinical trials without extensive data from in vivo, animal, and small clinical research indicating that the trial is likely to support the hypothesis being tested. It could be argued that the vitamin D field has not adequately followed this logical approach before embarking on large-scale RCTs.

Nonetheless, if vitamin D supplementation has beneficial effects on nonskeletal disease, it is reasonable to expect that RCTs should demonstrate such effects. However, despite multiple RCTs being conducted, with various vitamin D doses, regimens, durations, and end points, clarity regarding nonskeletal effects remains elusive. Why this is the case was noted by Heaney, who emphasized RCT failure to consider the "dose—response relation that vitamin D shares with most nutrients" (Heaney, 2012). To state this slightly differently, nutrients are not drugs. This concept is depicted in Fig. 31.1,

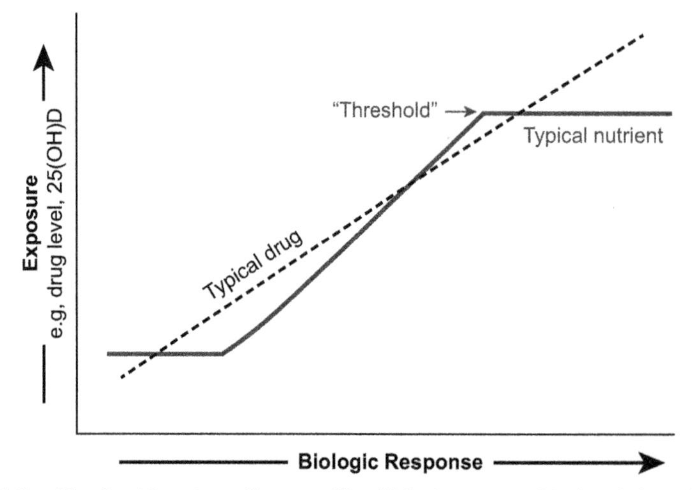

FIGURE 31.1 The sigmoidal relationship of nutrient status ("exposure") to biologic response. Nutrients behave different from drugs; with nutrients a threshold is reached when the individual has an adequate amount of the nutrient. Going above this level does not produce additional beneficial biologic response and could only, potentially, produce toxicity. Failure to consider baseline vitamin D status, specifically, failure to require it to be low as an inclusion criterion for vitamin D supplementation studies, biases such research to a negative conclusion.

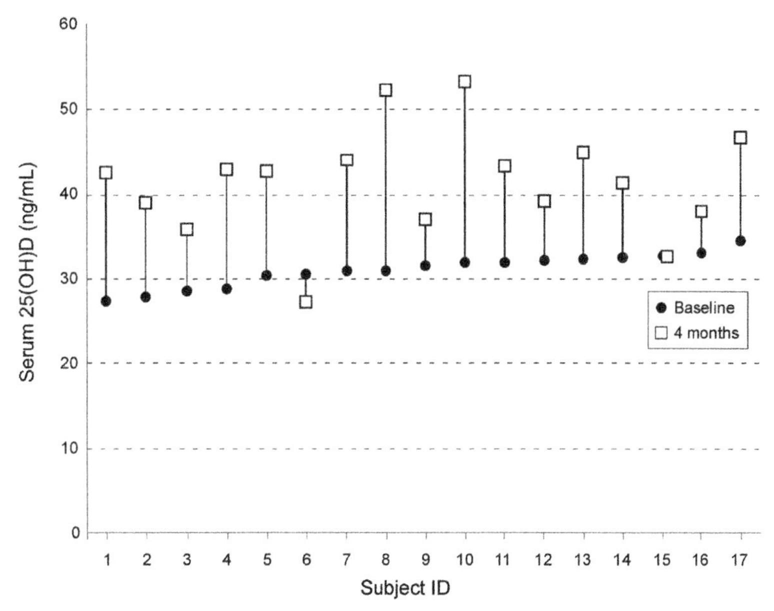

FIGURE 31.2 Variability in 25(OH)D response to a fixed daily supplementation dose. These data demonstrate a substantial variability in the 25(OH) D increase following provision of a fixed supplementation dose. Presented here are 25(OH)D data from postmenopausal women who received 1800 IU of vitamin D_3 daily for 4 months (Binkley et al., 2017a). Among these women, the serum 25(OH)D change ranged from −3.3 ng/mL to +21.4 ng/mL. This variability does not appear to reflect compliance; in this cohort the mean compliance with the daily supplement was 100.3% and ranged from 96% to 105%. Failure to consider such variable response confounds virtually all randomized trials of vitamin D supplementation.

upon review of which it is clear that an adequate dose must be given to individuals who are deficient in the nutrient in question (in this case vitamin D).

Failure to provide a dose that substantially increases 25(OH)D could not be expected to have any beneficial effect. Nonetheless, the vast majority of RCTs simply randomize to a predetermined vitamin D dose without consideration of whether the study subject achieves adequacy. Importantly, between-individual variation in 25(OH)D response to a fixed dose of oral vitamin D supplementation is dramatic (see Fig. 31.2).

Thus, even if a study included only vitamin D-deficient subjects, providing a single dose may not achieve adequacy in all. Perhaps a "treat to target" strategy should be considered in vitamin D supplementation studies (Binkley et al., 2015). Finally, vitamin D supplementation studies have employed various dosing approaches, e.g., daily, weekly, monthly, or less frequently. It has been suggested that daily dosing more closely replicates cutaneous production in circulating cholecalciferol levels (Heaney and Armas, 2015). Indeed, it is plausible that high cholecalciferol concentrations achieved with bolus dosing may induce 24-hydroxylation (Vieth, 2009); a concept supported by work demonstrating that a single vitamin D_3 dose of 150,000 IU led to greater $24,25(OH)_2D$ than daily dosing of 5000 IU for 1 month (Ketha et al., 2018). At a minimum, the possibility that daily and intermittent dosing may have different effects on vitamin D metabolism raises the question of whether these supplementation approaches should be considered equivalent in RCTs.

Of at least equal importance is that nutrients reach a "threshold"; once adequacy is achieved the provision of additional vitamin D could have no beneficial effect and only potentially lead to toxicity (Heaney, 2014). Studies including only those who are deficient in vitamin D could be expected to have a beneficial effect (if one indeed exists), whereas studies of those who are already replete would show no effect. It would seem self-evident that giving "more" vitamin D to a cohort of study subjects who already have "enough" would not inform a potential role of vitamin D in nonskeletal diseases. Nonetheless, the vast majority of vitamin D supplementation RCTs have not required documentation of inadequacy at study baseline. Performing meta-analyses of RCTs conducted without consideration of baseline status or individual response to supplementation should not be expected to clarify whether vitamin D is, or is not, important in nonskeletal disease. Indeed, as Heaney stated, "The question of how much vitamin D is enough is likely to remain muddled as long as meta-analyses focus on trial methodology rather than on biology" (Heaney, 2012). Guidance for the conduct of nutrient supplementation trials have been published and include measurement and use of basal nutrient status as an inclusion criterion, use of a supplementation dose adequate to produce an effect, and documentation of a change in nutrient status (Heaney, 2014). Such guidance must be followed to define what constitutes vitamin D inadequacy.

Conclusion and paths forward

Physiologic rationale combined with in vitro/in vivo preclinical studies and abundant observational data suggest that vitamin D deficiency could contribute to a multitude of human diseases. Unfortunately, despite immense amounts of clinical research and multiple meta-analyses of existing data, we do not know if, or for which nonskeletal diseases, vitamin D deficiency is relevant. This must be resolved. Improvements in clinical trial design and in laboratory measurements to define vitamin D status are needed to clarify what constitutes vitamin D inadequacy and the role, if any, that it plays in nonskeletal disease.

References

Aguilera, O., Pena, C., Garcia, J.M., Larriba, M.J., Ordonez-Moran, P., Navarro, D., Barbachano, A., Lopez de Silanes, I., Ballestar, E., Fraga, M.F., Esteller, M., Gamallo, C., Bonilla, F., Gonzalez-Sancho, J.M., Munoz, A., 2007. The Wnt antagonist DICKKOPF-1 gene is induced by 1alpha,25-dihydroxyvitamin D3 associated to the differentiation of human colon cancer cells. Carcinogenesis 28, 1877–1884.

Ahn, J., Peters, U., Albanes, D., Purdue, M.P., Abnet, C.C., Chatterjee, N., Horst, R.L., Hollis, B.W., Huang, W.Y., Shikany, J.M., Hayes, R.B., 2008. Serum vitamin D concentration and prostate cancer risk: a nested case-control study. J. Natl. Cancer Inst. 100, 796–804.

An, B.S., Tavera-Mendoza, L.E., Dimitrov, V., Wang, X., Calderon, M.R., Wang, H.J., White, J.H., 2010. Stimulation of Sirt1-regulated FoxO protein function by the ligand-bound vitamin D receptor. Mol. Cell. Biol. 30, 4890–4900.

Apperly, F.L., 1941. The relation of solar radiation to cancer mortality in North America. Cancer Res. 1, 191–195.

Asgari, M.M., Tang, J., Warton, M.E., Chren, M.M., Quesenberry Jr., C.P., Bikle, D., Horst, R.L., Orentreich, N., Vogelman, J.H., Friedman, G.D., 2010. Association of prediagnostic serum vitamin D levels with the development of basal cell carcinoma. J. Invest. Dermatol. 130, 1438–1443.

Aszterbaum, M., Rothman, A., Johnson, R.L., Fisher, M., Xie, J., Bonifas, J.M., Zhang, X., Scott, M.P., Epstein Jr., E.H., 1998. Identification of mutations in the human PATCHED gene in sporadic basal cell carcinomas and in patients with the basal cell nevus syndrome. J. Invest. Dermatol. 110, 885–888.

Autier, P., Mullie, P., Macacu, A., Dragomir, M., Boniol, M., Coppens, K., Pizot, C., Boniol, M., 2017. Effect of vitamin D supplementation on non-skeletal disorders: a systematic review of meta-analyses and randomised trials. Lancet Diabetes Endocrinol 5, 986–1004.

Baron, J.A., Barry, E.L., Mott, L.A., Rees, J.R., Sandler, R.S., Snover, D.C., Bostick, R.M., Ivanova, A., Cole, B.F., Ahnen, D.J., Beck, G.J., Bresalier, R.S., Burke, C.A., Church, T.R., Cruz-Correa, M., Figueiredo, J.C., Goodman, M., Kim, A.S., Robertson, D.J., Rothstein, R., Shaukat, A., Seabrook, M.E., Summers, R.W., 2015. A trial of calcium and vitamin D for the prevention of colorectal adenomas. N. Engl. J. Med. 373, 1519–1530.

Beaudart, C., Buckinx, F., Rabenda, V., Gillain, S., Cavalier, E., Slomian, J., Petermans, J., Reginster, J.Y., Bruyere, O., 2014. The effects of vitamin D on skeletal muscle strength, muscle mass, and muscle power: a systematic review and meta-analysis of randomized controlled trials. J. Clin. Endocrinol. Metab. 99, 4336–4345.

Bhatia, V., Saini, M.K., Shen, X., Bi, L.X., Qiu, S., Weigel, N.L., Falzon, M., 2009. EB1089 inhibits the parathyroid hormone-related protein-enhanced bone metastasis and xenograft growth of human prostate cancer cells. Mol. Cancer Ther. 8, 1787–1798.

Bikle, D.D., 2016. Extraskeletal actions of vitamin D. Ann. N.Y. Acad. Sci. 1376, 29–52.

Binkley, N., Borchardt, G., Siglinsky, E., Krueger, D., 2017a. Does vitamin D metabolite measurement help predict 25(OH)D change following vitamin D supplementation? Endocr. Pract. 23, 432–441.

Binkley, N., Dawson-Hughes, B., Durazo-Arvizu, R., Thamm, M., Tian, L., Merkel, J.M., Jones, J.C., Carter, G.D., Sempos, C.T., 2017b. Vitamin D measurement standardization: the way out of the chaos. J. Steroid Biochem. Mol. Biol. 173, 117–121.

Binkley, N., Krueger, D., Cowgill, C.S., Plum, L., Lake, E., Hansen, K.E., DeLuca, H.F., Drezner, M.K., 2004. Assay variation confounds the diagnosis of hypovitaminosis D: a call for standardization. J. Clin. Endocrinol. Metab. 89, 3152–3157.

Binkley, N., Lappe, J., Singh, R.J., Khosla, S., Krueger, D., Drezner, M.K., Blank, R.D., 2015. Can vitamin D metabolite measurements facilitate a "treat-to-target" paradigm to guide vitamin D supplementation? Osteoporos. Int. 26, 1655–1660.

Bischoff-Ferrari, H.A., Borchers, M., Gudat, F., Durmuller, U., Stahelin, H.B., Dick, W., 2004. Vitamin D receptor expression in human muscle tissue decreases with age. J. Bone Miner. Res. 19, 265–269.

Bischoff-Ferrari, H.A., Dawson-Hughes, B., Orav, E.J., Staehelin, H.B., Meyer, O.W., Theiler, R., Dick, W., Willett, W.C., Egli, A., 2016. Monthly high-dose vitamin D treatment for the prevention of functional decline: a randomized clinical trial. JAMA Intern. Med. 176, 175–183.

Bischoff-Ferrari, H.A., Dawson-Hughes, B., Staehelin, H.B., Orav, J.E., Stuck, A.E., Theiler, R., Wong, J.B., Egli, A., Kiel, D.P., Henschkowski, J., 2009. Fall prevention with supplemental and active forms of vitamin D: a meta-analysis of randomised controlled trials. Br. Med. J. 339, b3692.

Bolland, M.J., Grey, A., Gamble, G.D., Reid, I.R., 2014. Vitamin D supplementation and falls: a trial sequential meta-analysis. Lancet Diabetes Endocrinol. 2, 573–580.

Bora, S., Cantorna, M.T., 2017. The role of UVR and vitamin D on T cells and inflammatory bowel disease. Photochem. Photobiol. Sci. 16, 347–353.

Bouillon, R., 2017. Comparative analysis of nutritional guidelines for vitamin D. Nat. Rev. Endocrinol. 13, 466–479.

Brown, W.F., 1972. A method for estimating the number of motor units in thenar muscles and the changes in motor unit count with ageing. J. Neurol. Neurosurg. Psychiatry 35, 845–852.

Brozek, W., Manhardt, T., Kallay, E., Peterlik, M., Cross, H.S., 2012. Relative expression of vitamin D hydroxylases, CYP27B1 and CYP24A1, and of cyclooxygenase-2 and heterogeneity of human colorectal cancer in relation to age, gender, tumor location, and malignancy: results from factor and cluster analysis. Cancers 4, 763–776.

Byers, S.W., Rowlands, T., Beildeck, M., Bong, Y.S., 2012. Mechanism of action of vitamin D and the vitamin D receptor in colorectal cancer prevention and treatment. Rev. Endocr. Metab. Disord. 13, 31–38.

Cangussu, L.M., Nahas-Neto, J., Orsatti, C.L., Bueloni-Dias, F.N., Nahas, E.A., 2015. Effect of vitamin D supplementation alone on muscle function in postmenopausal women: a randomized, double-blind, placebo-controlled clinical trial. Osteoporos. Int. 26, 2413–2421.

Cangussu, L.M., Nahas-Neto, J., Orsatti, C.L., Poloni, P.F., Schmitt, E.B., Almeida-Filho, B., Nahas, E.A., 2016. Effect of isolated vitamin D supplementation on the rate of falls and postural balance in postmenopausal women fallers: a randomized, double-blind, placebo-controlled trial. Menopause 23, 267–274.

Cantorna, M.T., Hayes, C.E., DeLuca, H.F., 1996. 1, 25-dihydroxyvitamin D3 reversibly blocks the progression of relapsing encephalomyelitis, a model of multiple sclerosis. Proc. Natl. Acad. Sci. U.S.A. 93, 7861–7864.

Cantorna, M.T., Humpal-Winter, J., DeLuca, H.F., 1999. Dietary calcium is a major factor in 1,25-dihydroxycholecalciferol suppression of experimental autoimmune encephalomyelitis in mice. J. Nutr. 129, 1966–1971.

Cardus, A., Panizo, S., Encinas, M., Dolcet, X., Gallego, C., Aldea, M., Fernandez, E., Valdivielso, J.M., 2009. 1,25-dihydroxyvitamin D3 regulates VEGF production through a vitamin D response element in the VEGF promoter. Atherosclerosis 204, 85–89.

Carroll, C., Cooper, K., Papaioannou, D., Hind, D., Pilgrim, H., Tappenden, P., 2010. Supplemental calcium in the chemoprevention of colorectal cancer: a systematic review and meta-analysis. Clin. Ther. 32, 789–803.

Carter, G.D., Carter, R., Jones, J., Berry, J., 2004. How accurate are assays for 25-hydroxyvitamin D? Date from the international vitamin D external quality assessment scheme. Clin. Chem. 50, 2195–2197.

Cashman, K.D., 2014. A review of vitamin D status and CVD. Proc. Nutr. Soc. 73, 65–72.

Cashman, K.D., Dowling, K.G., Skrabakova, Z., Gonzalez-Gross, M., Valtuena, J., De Henauw, S., Moreno, L., Damsgaard, C.T., Michaelsen, K.F., Molgaard, C., Jorde, R., Grimnes, G., Moschonis, G., Mavrogianni, C., Manios, Y., Thamm, M., Mensink, G.B., Rabenberg, M., Busch, M.A., Cox, L., Meadows, S., Goldberg, G., Prentice, A., Dekker, J.M., Nijpels, G., Pilz, S., Swart, K.M., van Schoor, N.M., Lips, P., Eiriksdottir, G., Gudnason, V., Cotch, M.F., Koskinen, S., Lamberg-Allardt, C., Durazo-Arvizu, R.A., Sempos, C.T., Kiely, M., 2016a. Vitamin D deficiency in Europe: pandemic? Am. J. Clin. Nutr. 103, 1033–1044.

Cashman, K.D., Dowling, K.G., Skrabakova, Z., Kiely, M., Lamberg-Allardt, C., Durazo-Arvizu, R.A., Sempos, C.T., Koskinen, S., Lundqvist, A., Sundvall, J., Linneberg, A., Thuesen, B., Husemoen, L.L., Meyer, H.E., Holvik, K., Gronborg, I.M., Tetens, I., Andersen, R., 2015a. Standardizing serum 25-hydroxyvitamin D data from four Nordic population samples using the vitamin D standardization program protocols: shedding new light on vitamin D status in Nordic individuals. Scand. J. Clin. Lab. Invest. 75, 549–561.

Cashman, K.D., Hayes, A., Galvin, K., Merkel, J., Jones, G., Kaufmann, M., Hoofnagle, A.N., Carter, G.D., Durazo-Arvizu, R.A., Sempos, C.T., 2015b. Significance of serum 24,25-dihydroxyvitamin D in the assessment of vitamin D status: a double-edged sword? Clin. Chem. 61, 636–645.

Cashman, K.D., Kiely, M., Kinsella, M., Durazo-Arvizu, R.A., Tian, L., Zhang, Y., Lucey, A., Flynn, A., Gibney, M.J., Vesper, H.W., Phinney, K.W., Coates, P.M., Picciano, M.F., Sempos, C.T., 2013. Evaluation of vitamin D standardization program protocols for standardizing serum 25-hydroxyvitamin D data: a case study of the program's potential for national nutrition and health surveys. Am. J. Clin. Nutr. 97, 1235–1242.

Cashman, K.D., Kiely, M., Seamans, K.M., Urbain, P., 2016b. Effect of ultraviolet light-exposed mushrooms on vitamin D status: liquid chromatography-tandem mass spectrometry reanalysis of biobanked sera from a randomized controlled trial and a systematic review plus meta-analysis. J. Nutr. 146, 565–575.

Ceglia, L., Niramitmahapanya, S., da Silva Morais, M., Rivas, D.A., Harris, S.S., Bischoff-Ferrari, H., Fielding, R.A., Dawson-Hughes, B., 2013. A randomized study on the effect of vitamin D(3) supplementation on skeletal muscle morphology and vitamin D receptor concentration in older women. J. Clin. Endocrinol. Metab. 98, E1927–E1935.

Chang, S., Gao, L., Yang, Y., Tong, D., Guo, B., Liu, L., Li, Z., Song, T., Huang, C., 2015. miR-145 mediates the antiproliferative and gene regulatory effects of vitamin D3 by directly targeting E2F3 in gastric cancer cells. Oncotarget 6, 7675–7685.

Chen, P., Hu, P., Xie, D., Qin, Y., Wang, F., Wang, H., 2010. Meta-analysis of vitamin D, calcium and the prevention of breast cancer. Breast Canc. Res. Treat. 121, 469–477.

Chen, Y., Liu, W., Sun, T., Huang, Y., Wang, Y., Deb, D.K., Yoon, D., Kong, J., Thadhani, R., Li, Y.C., 2013. 1,25-Dihydroxyvitamin D promotes negative feedback regulation of TLR signaling via targeting microRNA-155-SOCS1 in macrophages. J. Immunol. 190, 3687–3695.

Chlebowski, R.T., 2011. Vitamin D and breast cancer: interpreting current evidence. Breast Cancer Res. 13, 217.

Cho, E., Smith-Warner, S.A., Spiegelman, D., Beeson, W.L., van den Brandt, P.A., Colditz, G.A., Folsom, A.R., Fraser, G.E., Freudenheim, J.L., Giovannucci, E., Goldbohm, R.A., Graham, S., Miller, A.B., Pietinen, P., Potter, J.D., Rohan, T.E., Terry, P., Toniolo, P., Virtanen, M.J., Willett, W.C., Wolk, A., Wu, K., Yaun, S.S., Zeleniuch-Jacquotte, A., Hunter, D.J., 2004. Dairy foods, calcium, and colorectal cancer: a pooled analysis of 10 cohort studies. J. Natl. Cancer Inst. 96, 1015–1022.

Christakos, S., Dhawan, P., Verstuyf, A., Verlinden, L., Carmeliet, G., 2016. Vitamin D: metabolism, molecular mechanism of action, and pleiotropic effects. Physiol. Rev. 96, 365–408.

Colotta, F., Jansson, B., Bonelli, F., 2017. Modulation of inflammatory and immune responses by vitamin D. J. Autoimmun. 85, 78–97.

Colston, K.W., Perks, C.M., Xie, S.P., Holly, J.M., 1998. Growth inhibition of both MCF-7 and Hs578T human breast cancer cell lines by vitamin D analogues is associated with increased expression of insulin-like growth factor binding protein-3. J. Mol. Endocrinol. 20, 157–162.

de la Hunty, A., Wallace, A.M., Gibson, S., Viljakainen, H., Lamberg-Allardt, C., Ashwell, M., 2010. UK foods standards agency workshop consensus report: the choice of method for measuring 25-hydroxyvitamin D to estimate vitamin D status for the UK National Diet and Nutrition Survey. Br. J. Nutr. 104, 612–619.

DeLuca, H.F., Plum, L.A., 2011. Vitamin D deficiency diminishes the severity and delays onset of experimental autoimmune encephalomyelitis. Arch. Biochem. Biophys. 513, 140–143.

Dobnig, H., Pilz, S., Scharnagl, H., Renner, W., Seelhorst, U., Wellnitz, B., Kinkeldei, J., Boehm, B.O., Weihrauch, G., Maerz, W., 2008. Independent association of low serum 25-hydroxyvitamin D and 1, 25-dihydroxyvitamin D levels with all-cause and cardiovascular mortality. Arch. Intern. Med. 168, 1340–1349.

Dong, J., Wong, S.L., Lau, C.W., Lee, H.K., Ng, C.F., Zhang, L., Yao, X., Chen, Z.Y., Vanhoutte, P.M., Huang, Y., 2012. Calcitriol protects renovascular function in hypertension by down-regulating angiotensin II type 1 receptors and reducing oxidative stress. Eur. Heart J. 33, 2980–2990.

Durazo-Arvizu, R.A., Tian, L., Brooks, S.P.J., Sarafin, K., Cashman, K.D., Kiely, M., Merkel, J., Myers, G.L., Coates, P.M., Sempos, C.T., 2017. The vitamin D standardization program (VDSP) manual for retrospective laboratory standardization of serum 25-hydroxyvitamin D data. J. AOAC Int. 100, 1234–1243.

Eide, M.J., Johnson, D.A., Jacobsen, G.R., Krajenta, R.J., Rao, D.S., Lim, H.W., Johnson, C.C., 2011. Vitamin D and nonmelanoma skin cancer in a health maintenance organization cohort. Arch. Dermatol. 147, 1379–1384.

El-Hajj Fuleihan, G., Nabulsi, M., Tamim, H., Maalouf, J., Salamoun, M., Khalife, H., Choucair, M., Arabi, A., Vieth, R., 2006. Effect of vitamin D replacement on musculoskeletal parameters in school children: a randomized controlled trial. J. Clin. Endocrinol. Metab. 91, 405–412.

Ellison, T.I., Smith, M.K., Gilliam, A.C., MacDonald, P.N., 2008. Inactivation of the vitamin D receptor enhances susceptibility of murine skin to UV-induced tumorigenesis. J. Invest. Dermatol. 128, 2508–2517.

Fernandes de Abreu, D.A., Ibrahim, E.C., Boucraut, J., Khrestchatisky, M., Feron, F., 2010. Severity of experimental autoimmune encephalomyelitis is unexpectedly reduced in mice born to vitamin D-deficient mothers. J. Steroid Biochem. Mol. Biol. 121, 250–253.

Gandini, S., Boniol, M., Haukka, J., Byrnes, G., Cox, B., Sneyd, M.J., Mullie, P., Autier, P., 2011. Meta-analysis of observational studies of serum 25-hydroxyvitamin D levels and colorectal, breast and prostate cancer and colorectal adenoma. Int. J. Cancer 128, 1414–1424.

Garland, C.F., Garland, F.C., 1980. Do sunlight and vitamin D reduce the likelihood of colon cancer? Int. J. Epidemiol. 9, 227–231.

Gilbert, R., Martin, R.M., Beynon, R., Harris, R., Savovic, J., Zuccolo, L., Bekkering, G.E., Fraser, W.D., Sterne, J.A., Metcalfe, C., 2011. Associations of circulating and dietary vitamin D with prostate cancer risk: a systematic review and dose-response meta-analysis. Cancer Causes Control 22, 319–340.

Ginde, A.A., Blatchford, P., Breese, K., Zarrabi, L., Linnebur, S.A., Wallace, J.I., Schwartz, R.S., 2017. High-dose monthly vitamin D for prevention of acute respiratory infection in older long-term care residents: a randomized clinical trial. J. Am. Geriatr. Soc. 65, 496–503.

Gocek, E., Wang, X., Liu, X., Liu, C.G., Studzinski, G.P., 2011. MicroRNA-32 upregulation by 1,25-dihydroxyvitamin D3 in human myeloid leukemia cells leads to Bim targeting and inhibition of AraC-induced apoptosis. Cancer Res. 71, 6230–6239.

Goswami, R., Vatsa, M., Sreenivas, V., Singh, U., Gupta, N., Lakshmy, R., Aggarwal, S., Ganapathy, A., Joshi, P., Bhatia, H., 2012. Skeletal muscle strength in young Asian Indian females after vitamin D and calcium supplementation: a double-blind randomized controlled clinical trial. J. Clin. Endocrinol. Metab. 97, 4709–4716.

Graeff-Armas, L.A., Kaufmann, M., Lyden, E., Jones, G., 2018. Serum 24,25-dihydroxyvitamin D3 response to native vitamin D2 and D3 Supplementation in patients with chronic kidney disease on hemodialysis. Clin. Nutr. 37, 1041–1045.

Gupta, R., Dixon, K.M., Deo, S.S., Holliday, C.J., Slater, M., Halliday, G.M., Reeve, V.E., Mason, R.S., 2007. Photoprotection by 1,25 dihydroxyvitamin D3 is associated with an increase in p53 and a decrease in nitric oxide products. J. Invest. Dermatol. 127, 707–715.

Gysemans, C., van Etten, E., Overbergh, L., Giulietti, A., Eelen, G., Waer, M., Verstuyf, A., Bouillon, R., Mathieu, C., 2008. Unaltered diabetes presentation in NOD mice lacking the vitamin D receptor. Diabetes 57, 269–275.

Hager, G., Formanek, M., Gedlicka, C., Thurnher, D., Knerer, B., Kornfehl, J., 2001. 1,25(OH)2 vitamin D3 induces elevated expression of the cell cycle-regulating genes P21 and P27 in squamous carcinoma cell lines of the head and neck. Acta Otolaryngol. 121, 103–109.

Hansen, K.E., Johnson, R.E., Chambers, K.R., Johnson, M.G., Lemon, C.C., Vo, T.N., Marvdashti, S., 2015. Treatment of vitamin D insufficiency in postmenopausal women: a randomized clinical trial. JAMA Intern. Med. 175, 1612–1621.

Hartholt, K.A., van Beeck, E.F., van der Cammen, T.J.M., 2018. Mortality from falls in Dutch adults 80 Years and older, 2000-2016. J. Am. Med. Assoc. 319, 1380–1382.

Hayes, C.E., Hubler, S.L., Moore, J.R., Barta, L.E., Praska, C.E., Nashold, F.E., 2015. Vitamin D actions on CD4(+) T cells in autoimmune disease. Front. Immunol. 6, 100.

Heaney, R.P., 2012. Vitamin D–baseline status and effective dose. N. Engl. J. Med. 367, 77–78.

Heaney, R.P., 2014. Guidelines for optimizing design and analysis of clinical studies of nutrient effects. Nutr. Rev. 72, 48–54.

Heaney, R.P., Armas, L.A., 2015. Quantifying the vitamin D economy. Nutr. Rev. 73, 51–67.

Helming, L., Bose, J., Ehrchen, J., Schiebe, S., Frahm, T., Geffers, R., Probst-Kepper, M., Balling, R., Lengeling, A., 2005. 1alpha,25-Dihydroxyvitamin D3 is a potent suppressor of interferon gamma-mediated macrophage activation. Blood 106, 4351–4358.

Hempel, S., Graham, G.D., Fu, N., Estrada, E., Chen, A.Y., Miake-Lye, I., Miles, J.N., Shanman, R., Shekelle, P.G., Beroes, J.M., Wallin, M.T., 2017. A systematic review of the effects of modifiable risk factor interventions on the progression of multiple sclerosis. Mult. Scler. 23, 513–524.

Herrmann, M., Farrell, C.L., Pusceddu, I., Fabregat-Cabello, N., Cavalier, E., 2017. Assessment of vitamin D status – a changing landscape. Clin. Chem. Lab. Med. 55, 3–26.

Holick, M.F., 2008. The vitamin D deficiency pandemic and consequences for nonskeletal health: mechanisms of action. Mol. Aspects Med. 29, 361–368.

Holick, M.F., 2011. Vitamin D: evolutionary, physiological and health perspectives. Curr. Drug Targets 12, 4–18.

Holick, M.F., 2017. The vitamin D deficiency pandemic: approaches for diagnosis, treatment and prevention. Rev. Endocr. Metab. Disord. 18, 153–165.

Holick, M.F., Binkley, N., Bischoff-Ferrari, H.A., Gordon, C.M., Hanley, D.A., Heaney, R.P., Murad, M.H., Weaver, C.M., 2011. Evaluation, treatment, and prevention of vitamin D deficiency: an endocrine society clinical practice guideline. J. Clin. Endocrinol. Metab. 96, 1911–1930.

Hollis, B.W., Wagner, C.L., 2013. Clinical review: the role of the parent compound vitamin D with respect to metabolism and function: why clinical dose intervals can affect clinical outcomes. J. Clin. Endocrinol. Metab. 98, 4619–4628.

Hunter, D.J., Colditz, G.A., Stampfer, M.J., Rosner, B., Willett, W.C., Speizer, F.E., 1992. Diet and risk of basal cell carcinoma of the skin in a prospective cohort of women. Ann. Epidemiol. 2, 231–239.

Huynh, H., Pollak, M., Zhang, J.C., 1998. Regulation of insulin-like growth factor (IGF) II and IGF binding protein 3 autocrine loop in human PC-3 prostate cancer cells by vitamin D metabolite 1,25(OH)2D3 and its analog EB1089. Int. J. Oncol. 13, 137–143.

Institute of Medicine, 2010. Dietary Reference Intakes for Calcium and Vitamin D. Available online: http://www.iom.edu/Reports/2010/Dietary-Reference-Intakes-for-Calcium-and-Vitamin-D/Report-Brief.aspx?page=1.

Irani, P.F., 1976. Electromyography in nutritional osteomalacic myopathy. J. Neurol. Neurosurg. Psychiatry 39, 686–693.

Jones, G., Kaufmann, M., 2016. Vitamin D metabolite profiling using liquid chromatography-tandem mass spectrometry (LC-MS/MS). J. Steroid Biochem. Mol. Biol. 164, 110–114.

Jones, G., Schlingmann, K.-P., 2018. Hypercalcemic states associated with abnormalities in vitamin D metabolism. In: Giustina, A., Bilezikian, J.P. (Eds.), Vitamin D in Clinical Medicine, Frontiers of Hormone Research. Karger, Basal, Switzerland, pp. 89–113.

Kaufmann, M., Gallagher, C., Peacock, M., Schlingmann, K.P., Konrad, M., Deluca, H.F., Sigueiro, R., Lopez, B., Mourino, A., Maestro, M., St-Arnaud, R., Finkelstein, J., Cooper, D.P., Jones, G., 2014. Clinical utility of simultaneous quantitation of 25-hydroxyvitamin D & 24,25-dihydroxyvitamin D by LC-MS/MS involving derivatization with DMEQ-TAD. J. Clin. Endocrinol. Metab. 99, 2567–2574.

Kendrick, J., Targher, G., Smits, G., Chonchol, M., 2009. 25-Hydroxyvitamin D deficiency is independently associated with cardiovascular disease in the Third National Health and Nutrition Examination Survey. Atherosclerosis 205, 255–260.

Ketha, H., Thacher, T.D., Oberhelman, S.S., Fischer, P.R., Singh, R.J., Kumar, R., 2018. Comparison of the effect of daily versus bolus dose maternal vitamin D3 supplementation on the 24,25-dihydroxyvitamin D3 to 25-hydroxyvitamin D3 ratio. Bone 110, 321–325.

Khaw, K.T., Stewart, A.W., Waayer, D., Lawes, C.M.M., Toop, L., Camargo Jr., C.A., Scragg, R., 2017. Effect of monthly high-dose vitamin D supplementation on falls and non-vertebral fractures: secondary and post-hoc outcomes from the randomised, double-blind, placebo-controlled ViDA trial. Lancet Diabetes Endocrinol 5, 438–447.

Kizildag, S., Ates, H., Kizildag, S., 2010. Treatment of K562 cells with 1,25-dihydroxyvitamin D3 induces distinct alterations in the expression of apoptosis-related genes BCL2, BAX, BCLXL, and p21. Ann. Hematol. 89, 1–7.

Krishnan, A.V., Trump, D.L., Johnson, C.S., Feldman, D., 2010. The role of vitamin D in cancer prevention and treatment. Endocrinol. Metab. Clin. North Am. 39, 401–418 (table of contents).

Le Goff, C., Cavalier, E., Souberbielle, J.-C., Gonzalez-Antuna, A., Delvin, E., 2015. Measurement of circulating 25-hydroxyvitamin D: a historical review. Pract. Lab. Med. 2, 1–14.

LeBlanc, E.S., Chou, R., 2015. Vitamin D and falls-fitting new data with current guidelines. JAMA Intern. Med. 175, 712–713.

Lemire, J.M., Archer, D.C., 1991. 1,25-dihydroxyvitamin D3 prevents the in vivo induction of murine experimental autoimmune encephalomyelitis. J. Clin. Invest. 87, 1103–1107.

Levis, S., Gomez-Marin, O., 2017. Vitamin D and physical function in sedentary older men. J. Am. Geriatr. Soc. 65, 323–331.

Lin, J., Manson, J.E., Lee, I.M., Cook, N.R., Buring, J.E., Zhang, S.M., 2007. Intakes of calcium and vitamin D and breast cancer risk in women. Arch. Intern. Med. 167, 1050–1059.

Lipkin, M., Newmark, H.L., 1999. Vitamin D, calcium and prevention of breast cancer: a review. J. Am. Coll. Nutr. 18, 392S–397S.

Looker, A.C., Johnson, C.L., Lacher, D.A., Pfeiffer, C.M., Schleicher, R.L., Sempos, C.T., 2011. Vitamin D status: United States, 2001–2006. NCHS Data Brief 1–8.

Ma, Y., Zhang, P., Wang, F., Yang, J., Liu, Z., Qin, H., 2011. Association between vitamin D and risk of colorectal cancer: a systematic review of prospective studies. J. Clin. Oncol. 29, 3775–3782.

Matusiak, D., Benya, R.V., 2007. CYP27A1 and CYP24 expression as a function of malignant transformation in the colon. J. Histochem. Cytochem. 55, 1257–1264.

McGaffin, K.R., Chrysogelos, S.A., 2005. Identification and characterization of a response element in the EGFR promoter that mediates transcriptional repression by 1,25-dihydroxyvitamin D3 in breast cancer cells. J. Mol. Endocrinol. 35, 117–133.

Meehan, T.F., Vanhooke, J., Prahl, J., Deluca, H.F., 2005. Hypercalcemia produced by parathyroid hormone suppresses experimental autoimmune encephalomyelitis in female but not male mice. Arch. Biochem. Biophys. 442, 214–221.

Merke, J., Milde, P., Lewicka, S., Hugel, U., Klaus, G., Mangelsdorf, D.J., Haussler, M.R., Rauteerberg, E.W., Ritz, E., 1989. Identification and regulation of 1, 25-dihydroxyvitamin D3 receptor activity and biosynthesis of 1, 25-dihydroxyvitamin D3: studies in cultured bovine aortic endothelial cells and human dermal capillaries. J. Clin. Invest. 83, 1903–1915.

Meyer, M.B., Goetsch, P.D., Pike, J.W., 2012. VDR/RXR and TCF4/beta-catenin cistromes in colonic cells of colorectal tumor origin: impact on c-FOS and c-MYC gene expression. Mol. Endocrinol. 26, 37–51.

Mineva, E.M., Schleicher, R.L., Chaudhary-Webb, M., Maw, K.L., Botelho, J.C., Vesper, H.W., et al., 2015. A candidate reference measurement procedure for quantifying serum concentrations of 25-hydroxyvitamin D(3) and 25-hydroxyvitamin D(2) using isotope-dilution liquid chromatography-tandem mass spectrometry. Anal. Bioanal. Chem. 407 (19), 5615—5624.

Molin, A., Baudoin, R., Kaufmann, M., Souberbielle, J.C., Ryckewaert, A., Vantyghem, M.C., Eckart, P., Bacchetta, J., Deschenes, G., Kesler-Roussey, G., Coudray, N., Richard, N., Wraich, M., Bonafiglia, Q., Tiulpakov, A., Jones, G., Kottler, M.L., 2015. CYP24A1 mutations in a cohort of hypercalcemic patients: evidence for a recessive trait. J. Clin. Endocrinol. Metab. 100, E1343—E1352.

Mordan-McCombs, S., Brown, T., Wang, W.L., Gaupel, A.C., Welsh, J., Tenniswood, M., 2010. Tumor progression in the LPB-Tag transgenic model of prostate cancer is altered by vitamin D receptor and serum testosterone status. J. Steroid Biochem. Mol. Biol. 121, 368—371.

Murad, M.H., Elamin, K.B., Elnour, N.O.A., Elamin, M.B., Alkatib, A.A., Fatourechi, M.M., Almandoz, J.P., Mullan, R.J., Lane, M.A., Liu, H., Erwin, P.J., Hensrud, D.D., Montori, V.M., 2011. The effect of vitamin D on falls: a systematic review and meta-analysis. J. Clin. Endocrinol. Metab. 96, 2997—3006.

Murillo, G., Nagpal, V., Tiwari, N., Benya, R.V., Mehta, R.G., 2010. Actions of vitamin D are mediated by the TLR4 pathway in inflammation-induced colon cancer. J. Steroid Biochem. Mol. Biol. 121, 403—407.

Nemerovski, C.W., Dorsch, M.P., Simpson, R.U., Bone, H.G., Aaronson, K.D., Bleske, B.E., 2009. Vitamin D and cardiovascular disease. Pharmacotherapy 29, 691—708.

Newmark, H.L., Yang, K., Kurihara, N., Fan, K., Augenlicht, L.H., Lipkin, M., 2009. Western-style diet-induced colonic tumors and their modulation by calcium and vitamin D in C57Bl/6 mice: a preclinical model for human sporadic colon cancer. Carcinogenesis 30, 88—92.

Norman, P.E., Powell, J.T., 2014. Vitamin D and cardiovascular disease. Circ. Res. 114, 379—393.

Palacios, C., Gonzalez, L., 2014. Is vitamin D deficiency a major global public health problem? J. Steroid Biochem. Mol. Biol. 144 (Pt A), 138—145.

Palmer, H.G., Sanchez-Carbayo, M., Ordonez-Moran, P., Larriba, M.J., Cordon-Cardo, C., Munoz, A., 2003. Genetic signatures of differentiation induced by 1alpha,25-dihydroxyvitamin D3 in human colon cancer cells. Cancer Res. 63, 7799—7806.

Peehl, D.M., Shinghal, R., Nonn, L., Seto, E., Krishnan, A.V., Brooks, J.D., Feldman, D., 2004. Molecular activity of 1,25-dihydroxyvitamin D3 in primary cultures of human prostatic epithelial cells revealed by cDNA microarray analysis. J. Steroid Biochem. Mol. Biol. 92, 131—141.

Peelen, E., Knippenberg, S., Muris, A.H., Thewissen, M., Smolders, J., Tervaert, J.W., Hupperts, R., Damoiseaux, J., 2011. Effects of vitamin D on the peripheral adaptive immune system: a review. Autoimmun. Rev. 10, 733—743.

Pendas-Franco, N., Garcia, J.M., Pena, C., Valle, N., Palmer, H.G., Heinaniemi, M., Carlberg, C., Jimenez, B., Bonilla, F., Munoz, A., Gonzalez-Sancho, J.M., 2008. DICKKOPF-4 is induced by TCF/beta-catenin and upregulated in human colon cancer, promotes tumour cell invasion and angiogenesis and is repressed by 1alpha,25-dihydroxyvitamin D3. Oncogene 27, 4467—4477.

Pfeifer, M., Begerow, B., Minne, H.W., Suppan, K., Fahrleitner-Pammer, A., Dobnig, H., 2009. Effects of a long-term vitamin D and calcium supplementation on falls and parameters of muscle function in community-dwelling older individuals. Osteoporos. Int. 20, 315—322.

Phinney, K.W., Tai, S.S., Bedner, M., Camara, J.E., Chia, R.R.C., Sander, L.C., et al., 2017. Development of an Improved Standard Reference Material for Vitamin D Metabolites in Human Serum. Anal. Chem. 89 (9), 4907—4913.

Pilz, S., Marz, W., Wellnitz, B., Seelhorst, U., Fahrleitner-Pammer, A., Dimai, H.P., Boehm, B.O., Dobnig, H., 2008. Association of vitamin D deficiency with heart failure and sudden cardiac death in a large cross-sectional study of patients referred for coronary angiography. J. Clin. Endocrinol. Metab. 93, 3927—3935.

Pilz, S., Tomaschitz, A., Drechsler, C., Dekker, J.M., Marz, W., 2010. Vitamin D deficiency and myocardial diseases. Mol. Nutr. Food Res. 54, 1103—1113.

Rejnmark, L., Bislev, L.S., Cashman, K.D., Eiriksdottir, G., Gaksch, M., Grubler, M., Grimnes, G., Gudnason, V., Lips, P., Pilz, S., van Schoor, N.M., Kiely, M., Jorde, R., 2017. Non-skeletal health effects of vitamin D supplementation: a systematic review on findings from meta-analyses summarizing trial data. PLoS One 12, e0180512.

Rodriguez, A.J., Mousa, A., Ebeling, P.R., Scott, D., de Courten, B., 2018. Effects of vitamin D supplementation on inflammatory markers in heart failure: a systematic review and meta-analysis of randomized controlled trials. Sci. Rep. 8, 1169.

Rosen, C.J., Adams, J.S., Bikle, D.D., Black, D.M., Demay, M.B., Manson, J.E., Murad, M.H., Kovacs, C.S., 2012. The nonskeletal effects of vitamin D: an Endocrine Society scientific statement. Endocr. Rev. 33, 456—492.

Rossi, M., McLaughlin, J.K., Lagiou, P., Bosetti, C., Talamini, R., Lipworth, L., Giacosa, A., Montella, M., Franceschi, S., Negri, E., La Vecchia, C., 2009. Vitamin D intake and breast cancer risk: a case-control study in Italy. Ann. Oncol. 20, 374—378.

Rostand, S.G., Drueke, T.B., 1999. Parathyroid hormone, vitamin D, and cardiovascular disease in chronic renal failure. Kidney Int. 56, 383—392.

Sackett, D.L., Rosenberg, W.M., Gray, J.A., Haynes, R.B., Richardson, W.S., 1996. Evidence based medicine: what it is and what it isn't. BMJ 312, 71—72.

Sanders, K.M., Stuart, A.L., Williamson, E.J., Simpson, J.A., Kotowicz, M.A., Young, D., Nicholson, G.C., 2010. Annual high-dose oral vitamin D and falls and fractures in older women. J. Am. Med. Assoc. 303, 1815—1822.

Santagata, S., Thakkar, A., Ergonul, A., Wang, B., Woo, T., Hu, R., Harrell, J.C., McNamara, G., Schwede, M., Culhane, A.C., Kindelberger, D., Rodig, S., Richardson, A., Schnitt, S.J., Tamimi, R.M., Ince, T.A., 2014. Taxonomy of breast cancer based on normal cell phenotype predicts outcome. J. Clin. Invest. 124, 859—870.

Sarafin, K., Durazo-Arvizu, R., Tian, L., Phinney, K.W., Tai, S., Camara, J.E., Merkel, J., Green, E., Sempos, C.T., Brooks, S.P., 2015. Standardizing 25-hydroxyvitamin D values from the Canadian health measures survey. Am. J. Clin. Nutr. 102, 1044—1050.

Scher, H.I., Jia, X., Chi, K., de Wit, R., Berry, W.R., Albers, P., Henick, B., Waterhouse, D., Ruether, D.J., Rosen, P.J., Meluch, A.A., Nordquist, L.T., Venner, P.M., Heidenreich, A., Chu, L., Heller, G., 2011. Randomized, open-label phase III trial of docetaxel plus high-dose calcitriol versus docetaxel plus prednisone for patients with castration-resistant prostate cancer. J. Clin. Oncol. 29, 2191–2198.

Schleicher, R.L., Sternberg, M.R., Lacher, D.A., Sempos, C.T., Looker, A.C., Durazo-Arvizu, R.A., Yetley, E.A., Chaudhary-Webb, M., Maw, K.L., Pfeiffer, C.M., Johnson, C.L., 2016. The vitamin D status of the US population from 1988 to 2010 using standardized serum concentrations of 25-hydroxyvitamin D shows recent modest increases. Am. J. Clin. Nutr. 104, 454–461.

Scientific Advisory Committee on Nutrition (SACN), 2016. SACN vitamin D and health report. In: England, P.H. (Ed.), SACN: Reports and Position Statements. Public Health England.

Sempos, C.T., Betz, J.M., Camara, J.E., Carter, G.D., Cavalier, E., Clarke, M.W., Dowling, K.G., Durazo-Arvizu, R.A., Hoofnagle, A.N., Liu, A., Phinney, K.W., Sarafin, K., Wise, S.A., Coates, P.M., 2017. General steps to standardize the laboratory measurement of serum total 25-hydroxyvitamin D. J. AOAC Int. 100, 1230–1233.

Sempos, C.T., Vesper, H.W., Phinney, K.W., Thienpont, L.M., Coates, P.M., 2012. Vitamin D status as an international issue: national surveys and the problem of standardization. Scand. J. Clin. Lab. Invest. 72, 32–40.

Shin, M.H., Holmes, M.D., Hankinson, S.E., Wu, K., Colditz, G.A., Willett, W.C., 2002. Intake of dairy products, calcium, and vitamin d and risk of breast cancer. J. Natl. Cancer Inst. 94, 1301–1311.

Skaria, J., Katiyar, B.C., Srivastava, T.P., Dube, B., 1975. Myopathy and neuropathy associated with osteomalacia. Acta Neurol. Scand. 51, 37–58.

Stockl, D., Sluss, P.M., Thienpont, L.M., 2009. Specifications for trueness and precision of a reference measurement system for serum/plasma 25-hydroxyvitamin D analysis. Clin. Chim. Acta. 408 (1-2), 8–13.

Stumpf, W.E., Bidmon, H.J., Murakami, R., Heiss, C., Mayerhofer, A., Bartke, A., 1990. Sites of action of soltriol (vitamin D) in hamster spleen, thymus, and lymph node, studied by autoradiography. Histochemistry 94, 121–125.

Stumpf, W.E., Downs, T.W., 1987. Nuclear receptors for 1,25(OH)2 vitamin D3 in thymus reticular cells studied by autoradiography. Histochemistry 87, 367–369.

Swami, S., Raghavachari, N., Muller, U.R., Bao, Y.P., Feldman, D., 2003. Vitamin D growth inhibition of breast cancer cells: gene expression patterns assessed by cDNA microarray. Breast Canc. Res. Treat. 80, 49–62.

Tai, S., Nelson, M., Bedner, M., Lang, B., Phinney, K., Sander, L., et al., 2017. Development of Standard Reference Material (SRM) 2973 Vitamin D Metabolites in Frozen Human Serum (High Level). J AOAC Int. 100 (5), 1294–1303.

Tang, J.Y., Fu, T., Leblanc, E., Manson, J.E., Feldman, D., Linos, E., Vitolins, M.Z., Zeitouni, N.C., Larson, J., Stefanick, M.L., 2011. Calcium plus vitamin D supplementation and the risk of nonmelanoma and melanoma skin cancer: post hoc analyses of the women's health initiative randomized controlled trial. J. Clin. Oncol. 29, 3078–3084.

Tang, J.Y., Parimi, N., Wu, A., Boscardin, W.J., Shikany, J.M., Chren, M.M., Cummings, S.R., Epstein Jr., E.H., Bauer, D.C., 2010. Inverse association between serum 25(OH) vitamin D levels and non-melanoma skin cancer in elderly men. Cancer Causes Control 21, 387–391.

Tannour-Louet, M., Lewis, S.K., Louet, J.F., Stewart, J., Addai, J.B., Sahin, A., Vangapandu, H.V., Lewis, A.L., Dittmar, K., Pautler, R.G., Zhang, L., Smith, R.G., Lamb, D.J., 2014. Increased expression of CYP24A1 correlates with advanced stages of prostate cancer and can cause resistance to vitamin D3-based therapies. FASEB J. 28, 364–372.

Teichert, A.E., Elalieh, H., Elias, P.M., Welsh, J., Bikle, D.D., 2011. Overexpression of hedgehog signaling is associated with epidermal tumor formation in vitamin D receptor-null mice. J. Invest. Dermatol. 131, 2289–2297.

Tinetti, M.E., Williams, C.S., 1997. Falls, injuries due to falls, and the risk of admission to a nursing home. N. Engl. J. Med. 337, 1279–1284.

Tishkoff, D.X., Nibbelink, K.A., Holmberg, K.H., Dandu, L., Simpson, R.U., 2008. Functional vitamin D receptor (VDR) in the t-tubules of cardiac myocytes: VDR knockout cardiomyocyte contractility. Endocrinology 149, 558–564.

Uusi-Rasi, K., Patil, R., Karinkanta, S., Kannus, P., Tokola, K., Lamberg-Allardt, C., Sievanen, H., 2015. Exercise and vitamin D in fall prevention among older women: a randomized clinical trial. JAMA Intern. Med. 175, 703–711.

USDHHS, 2014. The Surgeon General's Call to Action to Prevent Skin Cancer (Washington, DC).

van der Rhee, H.J., de Vries, E., Coebergh, J.W., 2006. Does sunlight prevent cancer? A systematic review. Eur. J. Cancer 42, 2222–2232.

VanAmerongen, B.M., Dijkstra, C.D., Lips, P., Polman, C.H., 2004. Multiple sclerosis and vitamin D: an update. Eur. J. Clin. Nutr. 58, 1095–1109.

Vandervoort, A.A., 2002. Aging of the human neuromuscular system. Muscle Nerve 25, 17–25.

VanWeelden, K., Flanagan, L., Binderup, L., Tenniswood, M., Welsh, J., 1998. Apoptotic regression of MCF-7 xenografts in nude mice treated with the vitamin D3 analog, EB1089. Endocrinology 139, 2102–2110.

Vieth, R., 2009. How to optimize vitamin D supplementation to prevent cancer, based on cellular adaptation and hydroxylase enzymology. Anticancer Res. 29, 3675–3684.

Wactawski-Wende, J., Kotchen, J.M., Anderson, G.L., Assaf, A.R., Brunner, R.L., O'Sullivan, M.J., Margolis, K.L., Ockene, J.K., Phillips, L., Pottern, L., Prentice, R.L., Robbins, J., Rohan, T.E., Sarto, G.E., Sharma, S., Stefanick, M.L., Van Horn, L., Wallace, R.B., Whitlock, E., Bassford, T., Beresford, S.A., Black, H.R., Bonds, D.E., Brzyski, R.G., Caan, B., Chelebowski, R.T., Cochrane, B., Garland, C., Gass, M., Hays, J., Heiss, G., Hendrix, S.L., Howard, B.V., Hsia, J., Hubbell, F.A., Jackson, R.D., Johnson, K.C., Judd, H., Kooperberg, C.L., Kuller, L.H., LaCroix, A.Z., Lane, D.S., Langer, R.D., Lasser, N.L., Lewis, C.E., Limacher, M.C., Manson, J.E., 2006. Calcium plus vitamin D supplementation and the risk of colorectal cancer. N. Engl. J. Med. 354, 684–696.

Wagner, D., Hanwell, H.E., Schnabl, K., Yazdanpanah, M., Kimball, S., Fu, L., Sidhom, G., Rousseau, D., Cole, D.E.C., Vieth, R., 2011. The ratio of serum 24, 25-dihydroxyvitamin D3 to serum 25-hydroxyvitamin D3 is predictive of 25-hydroxyvitamin D3 response to vitamin D3 supplementation. J. Steroid Biochem. Mol. Biol. 126, 72–77.

Wallace, A.M., Gibson, S., de la Hunty, A., Lamberg-Allardt, C., Ashwell, M., 2010. Measurement of 25-hydroxyvitamin D in the clinical laboratory: current procedures, performance characteristics and limitations. Steroids 75, 477–488.

Wang, H., Chen, W., Li, D., Yin, X., Zhang, X., Olsen, N., Zheng, S.G., 2017. Vitamin D and chronic diseases. Aging Dis 8, 346–353.

Wang, L., Song, Y., Manson, J.E., Pilz, S., Marz, W., Michaelsson, K., Lundqvist, A., Jassal, S.K., Barrett-Connor, E., Zhang, C., Eaton, C.B., May, H.T., Anderson, J.L., Sesso, H.D., 2012a. Circulating 25-hydroxy-vitamin D and risk of cardiovascular disease: a meta-analysis of prospective studies. Circ. Cardiovasc. Qual. Outcomes 5, 819–829.

Wang, T.J., Pencina, M.J., Booth, S.L., Jacques, P.F., Ingelsson, E., Lanier, K., Benjamin, E.J., D'Agostino, R.B., Wolf, M., Vasan, R.S., 2008. Vitamin D deficiency and risk of cardiovascular disease. Circulation 117, 503–511.

Wang, Y., Marling, S.J., Zhu, J.G., Severson, K.S., DeLuca, H.F., 2012b. Development of experimental autoimmune encephalomyelitis (EAE) in mice requires vitamin D and the vitamin D receptor. Proc. Natl. Acad. Sci. U.S.A. 109, 8501–8504.

Wang, Y., Zhu, J., DeLuca, H.F., 2012c. Where is the vitamin D receptor? Arch. Biochem. Biophys. 523, 123–133.

Weitsman, G.E., Koren, R., Zuck, E., Rotem, C., Liberman, U.A., Ravid, A., 2005. Vitamin D sensitizes breast cancer cells to the action of H2O2: mitochondria as a convergence point in the death pathway. Free Radic. Biol. Med. 39, 266–278.

Weitsman, G.E., Ravid, A., Liberman, U.A., Koren, R., 2003. Vitamin D enhances caspase-dependent and independent TNF-induced breast cancer cell death: the role of reactive oxygen species. Ann. N.Y. Acad. Sci. 1010, 437–440.

Wierman, M.E., Auchus, R.J., Haisenleder, D.J., Hall, J.E., Handelsman, D., Hankinson, S., Rosner, W., Singh, R.J., Sluss, P.M., Stanczyk, F.Z., 2014. Editorial: the new instructions to authors for the reporting of steroid hormone measurements. J. Clin. Endocrinol. Metab. 99, 4375.

Xu, H., Posner, G.H., Stevenson, M., Campbell, F.C., 2010. Apc(MIN) modulation of vitamin D secosteroid growth control. Carcinogenesis 31, 1434–1441.

Yang, K., Lamprecht, S.A., Shinozaki, H., Fan, K., Yang, W., Newmark, H.L., Kopelovich, L., Edelmann, W., Jin, B., Gravaghi, C., Augenlicht, L., Kucherlapati, R., Lipkin, M., 2008. Dietary calcium and cholecalciferol modulate cyclin D1 expression, apoptosis, and tumorigenesis in intestine of adenomatous polyposis coli1638N/+ mice. J. Nutr. 138, 1658–1663.

Yang, L., Yang, J., Venkateswarlu, S., Ko, T., Brattain, M.G., 2001. Autocrine TGFbeta signaling mediates vitamin D3 analog-induced growth inhibition in breast cells. J. Cell. Physiol. 188, 383–393.

Zella, J.B., McCary, L.C., DeLuca, H.F., 2003. Oral administration of 1,25-dihydroxyvitamin D3 completely protects NOD mice from insulin-dependent diabetes mellitus. Arch. Biochem. Biophys. 417, 77–80.

Zheng, C., He, L., Liu, L., Zhu, J., Jin, T., 2018. The efficacy of vitamin D in multiple sclerosis: a meta-analysis. Mult. Scler. Relat. Disord. 23, 56–61.

Zheng, W., Wong, K.E., Zhang, Z., Dougherty, U., Mustafi, R., Kong, J., Deb, D.K., Zheng, H., Bissonnette, M., Li, Y.C., 2012. Inactivation of the vitamin D receptor in APC(min/+) mice reveals a critical role for the vitamin D receptor in intestinal tumor growth. Int. J. Cancer 130, 10–19.

Zheng, Y., Zhou, H., Ooi, L.L., Snir, A.D., Dunstan, C.R., Seibel, M.J., 2011. Vitamin D deficiency promotes prostate cancer growth in bone. Prostate 71, 1012–1021.

Zinser, G.M., Welsh, J., 2004. Effect of Vitamin D3 receptor ablation on murine mammary gland development and tumorigenesis. J. Steroid Biochem. Mol. Biol. 89–90, 433–436.

Chapter 32

Cellular actions of parathyroid hormone on bone

Elena Ambrogini[1,2] and Robert L. Jilka[1,2]

[1]Center for Osteoporosis and Metabolic Bone Diseases, University of Arkansas for Medical Sciences Division of Endocrinology and Metabolism, Little Rock, AR, United States; [2]Central Arkansas Veterans Healthcare System, Little Rock, AR, United States

Chapter outline

Introduction 775
The regulation of bone remodeling by parathyroid hormone 776
The generation and Maintenance of basic multicellular units Is governed by parathyroid hormone 777
Parathyroid hormone regulates factors that govern the assembly and maintenance of basic multicellular units 777
Osteoclast differentiation and life span 777
Osteoblast differentiation and the coupling of bone formation to bone resorption 778
The bone anabolic effects of intermittent parathyroid hormone 779
Stimulation of anabolic remodeling and modeling 779
Mechanisms underlying overfill of resorption cavities in response to injections of parathyroid hormone 781
Stimulation of bone modeling by osteoblasts in response to injections of parathyroid hormone 782
Unresolved issues 782
References 782

Introduction

Parathyroid hormone (PTH) maintains serum Ca homeostasis by direct actions on the kidney and on bone. A fall in serum Ca enhances the activity of the Ca-sensing receptor of parathyroid glands, which immediately increases the secretion of PTH into the circulation. The rise in PTH enhances tubular reabsorption of Ca and the synthesis of 1,25-dihydroxyvitamin D3 in the kidney, which in turn stimulates intestinal Ca absorption. PTH also enhances the release of Ca from bone by increasing osteoclastic bone resorption. The cumulative impact is a rise in serum Ca which then lowers the secretion of PTH, leading to reduced tubular Ca reabsorption, Ca absorption from the gut, and osteoclastic bone resorption.

In adults, bone resorption only takes place at sites of bone remodeling (Parfitt, 1996). This process is carried out by anatomically discrete teams of osteoclasts and osteoblasts, which along with an associated capillary are called basic multicellular units (BMUs). PTH drives the assembly and maintenance of BMUs. Remodeling replaces old or damaged bone with new. In the ilium of postmenopausal women, BMUs remodel 18% of the trabecular bone surface per year and 34% of the endocortical surface. BMUs arise within Haversian canals and tunnel through the cortex to remodel 6% of the cortical bone volume per year (Parfitt, 2002). Trabecular and endocortical surfaces undergo remodeling in laboratory rodents, but in these small mammals the Haversian system is absent (Piemontese et al., 2017). PTH may also stimulate osteocytes to degrade their surrounding lacunar bone (Baud and Boivin, 1978; Tazawa et al., 2004). The high demand for Ca during lactation may be met by PTHrP-stimulated osteocytic osteolysis in rodents (Kovacs, 2017), but the role of this phenomenon in day-to-day Ca homeostasis has not been established. In contrast to the role of PTH in Ca homeostasis, daily injections of exogenous PTH (aka intermittent administration) increase bone mass. This latter feature of PTH function has been employed for the therapy of osteoporosis (Hodsman et al., 2005). This chapter focuses on the cellular mechanisms underlying the diverse effects of PTH on bone.

Principles of Bone Biology. https://doi.org/10.1016/B978-0-12-814841-9.00032-4

The regulation of bone remodeling by parathyroid hormone

Osteoclasts initiate bone remodeling by adhering to the bone surface and forming a tightly sealed pocket into which they pump HCl and secrete lysosomes that dissolve the hydroxyapatite and degrade the collagenous and other proteinaceous components of the matrix. The degraded proteins are packaged into endosomes, transcytosed to the apical surface, and released into the extracellular space (Salo et al., 1997). Bone formation is temporally and spatially linked to this resorptive activity within the BMU in a process called coupling, which ensures that sufficient bone matrix is made to replace the one removed by osteoclasts (Sims and Martin, 2014). Mesenchymal precursors of the osteoblast lineage are recruited to the previously resorbed surface where they intermingle with some of the osteoclasts to form a "reversal-resorption" surface (Lassen et al., 2017). The mesenchymal cells deposit a thin layer of glycosaminoglycan-containing matrix that marks the limit of bone resorption. These mesenchymal cells and/or other recruited progenitors differentiate into osteoblasts and elaborate a matrix consisting of type I collagen, a variety of noncollagenous proteins, and growth factors. The matrix is then calcified. During matrix deposition, some osteoblasts differentiate into osteocytes. Osteocytes are embedded into the matrix and form connections with each other and with the bone surface. Osteoblasts eventually cease matrix production and become the very thin lining cells that cover the quiescent surface of bone.

The dependence of bone formation on bone resorption is the "sine qua non" feature of remodeling. If coupling is defective, the thickness and connectivity of trabecular bone declines. Cortical bone becomes thinner because of endosteal bone loss and more porous because the Haversian canals become wider. If excessive matrix is deposited during coupling, trabecular bone mass increases, and cortical bone becomes thicker.

In addition to remodeling, bone also undergoes modeling. As opposed to bone remodeling, bone modeling is a process carried out by osteoclasts and osteoblasts that operate independently of each other. The purpose of modeling is to sculpt the skeleton during growth and adapt it to the prevailing mechanical strains at different sites.

The function of the BMU can be measured using histologic approaches that quantify the rate of resorption and formation (Dempster et al., 2013) (see Box 32.1). The life span of BMUs in humans is 6–9 months, but the life span of osteoclasts is 1 week or so, and the life span of osteoblasts is a few months (Manolagas, 2000). All osteoclasts and the majority of osteoblasts die by apoptosis (Jilka et al., 2007, 2008). Thus, the function of the BMU requires a continuous supply of new osteoclasts and osteoblasts derived from local or circulating progenitors and the timely death of these cells (Manolagas, 2000). Recent advances in the genetic manipulation of mice has permitted more rigorous examination of the origination and fate of cells of the osteoblast lineage (Ono and Kronenberg, 2015). The important role of the timely death of osteoblasts and perhaps osteoblast progenitors by apoptosis has been illustrated by the increased trabecular bone in mice with osteoblast lineage cells lacking Bax and Bak, two proteins critical for apoptosis (Jilka et al., 2014).

BOX 32.1 Histologic measurements of bone modeling and remodeling.

Term	Description
Mineralizing surface	Bone surface undergoing mineralization as visualized by fluorochromes labeling.
Mineral apposition rate	Distance between fluorochromes labels divided by the labeling interval.
Bone formation rate	Product of mineralizing surface and mineral apposition rate.
Activation frequency	Probability that a remodeling event will be initiated on the bone surface.
Cement line	Thin line of toluidine blue staining that marks the boundary between resorption and formation in remodeled bone, or new bone formation in modeled bone.
Wall width	Amount of bone formed during remodeling by a team of osteoblasts. Measured by the average distance between scalloped cement lines and quiescent bone surface.
Eroded surface	Osteoclast surface plus the reversal-osteoclast surface.
Erosion depth	Average depth of active resorption lacunae.

The generation and Maintenance of basic multicellular units Is governed by parathyroid hormone

Bone turnover is primarily influenced by circulating levels of PTH. In hypoparathyroidism, bone remodeling and the number of BMUs is severely reduced (Rubin et al., 2008; Clarke, 2014; Langdahl et al., 1996). The length of the resorption and formation period of each BMU is prolonged, and the resorption depth and the wall thickness is reduced, as is osteoid width, mineral apposition rate, and bone formation rate (see Box 32.1). Nevertheless, cancellous bone volume is increased due to increased trabecular thickness, likely due to reduced erosion depth combined with a reduced proportion of the surface undergoing remodeling. In mice, deletion of the PTH gene has similar effects (Miao et al., 2004).

Hyperparathyroidism increases bone remodeling. In mild primary hyperparathyroidism there is a 50% increase in the activation frequency of the BMUs and increased formation and resorption surfaces (Christiansen et al., 1992; Dempster et al., 1999; Parisien et al., 2001). In cortical bone, increased remodeling is seen at both the endocortical and intracortical surfaces. The result of the increased intracortical remodeling is larger Haversian canals accounting for increased cortical porosity, and "trabecularization" of the endocortical surface (Vu et al., 2013; Stein et al., 2013). Periosteal surfaces are usually unaffected. In severe primary hyperparathyroidism, bone remodeling is dramatically increased. This is manifested by skeletal deformities, pathological fractures, subperiosteal bone erosion, "brown tumors," and peritrabecular bone marrow fibrosis, also called osteitis fibrosa.

Parathyroid hormone regulates factors that govern the assembly and maintenance of basic multicellular units

Osteoclast differentiation and life span

Under normal physiologic conditions, PTH drives the development of osteoclasts. Osteoclastogenesis requires extravasation of monocytic progenitors from capillaries near the bone targeted for renewal or from locally produced progenitors in the bone marrow. Osteoclast differentiation requires macrophage-colony stimulating factor (M-CSF) which stimulates the replication of these progenitors, and receptor activator of NF-kB ligand (RANKL) which promotes differentiation into multinucleated bone resorptive cells (Boyle et al., 2003). Mice lacking either of these two cytokines are osteopetrotic and practically devoid of osteoclasts. RANKL also maintains the viability of osteoclasts. Administration of osteoprotegerin (OPG), a soluble receptor antagonist of RANKL, causes a decline in osteoclast number coincident with the appearance of apoptotic fragments of osteoclasts as early as 6h after administration (Lacey et al., 2000). OPG also reduces osteoblast number and bone formation rate, directly demonstrating the coupling of bone formation to bone resorption (Jilka et al., 2010; Piemontese et al., 2017).

RANKL is produced by many cell types including cells of the osteoblast lineage, T and B lymphocytes, synovial fibroblasts, adipocytes, and hypertrophic chondrocytes (O'Brien et al., 2013). PTH stimulates the production of RANKL by osteoblastic cells (Rodan and Martin, 1981; Lee and Lorenzo, 1999; Yasuda et al., 1998; Fu et al., 2002; McSheehy and Chambers, 1986). These effects are mediated by activation of a distal enhancer located 76 kb from the transcriptional start site of the RANKL gene (Galli et al., 2008; Fu et al., 2006). The osteopetrotic phenotype of mice lacking a functional RANKL gene can be rescued by a 220-Kb transgene containing the RANKL gene and its upstream regulatory domains (Onal et al., 2015). Targeted deletion of RANKL in osteoblasts/osteocytes using DMP1-Cre, or in osteocytes using Sost-Cre, causes a reduction in osteoclast number and an increase in trabecular bone mass (Xiong et al., 2011, 2014, 2015; Nakashima et al., 2011). These changes are apparent as early as 5 weeks of age, indicating that osteocytes represent the main source of RANKL when skeletal growth and modeling begins to slow, and osteoclasts are needed only for remodeling. In contrast, global deletion of RANKL causes osteopetrosis that is evident from birth (Kong et al., 1999). Thus, other cell types produce the RANKL required for the formation of osteoclasts needed for skeletal development, for example resorption of calcified cartilage in the growth plate (Gebhard et al., 2008).

Conditional deletion of RANKL in osteoblasts/osteocytes with DMP1-Cre attenuates the increase in osteoclast number and the bone loss caused by dietary Ca deficiency in adult mice (Xiong et al., 2014). Osteocytes vastly outnumber osteoblasts, making the former cell type the most likely source of RANKL. Moreover, mice lacking PTHR1 in osteoblast/osteocyte fail to exhibit the increase in RANKL and the loss of bone caused by chronic elevation of PTH (Saini et al., 2013). These findings show that the stimulatory effects of chronic elevation of PTH on osteoclast number and bone remodeling are primarily mediated by direct actions of the hormone on RANKL production by osteocytes. Consistent with this contention, expression of a constitutively active PTHR1 mutant in osteocytes (using the DMP1 promoter) increases RANKL and bone remodeling (O'Brien et al., 2008; Ben-Awadh et al., 2014). RANKL is initially expressed as a transmembrane protein on the cell surface. Following cleavage by extracellular endopeptidases, it is released in soluble form (sRANKL). Mice expressing endopeptidase[HYPHEN]resistant RANKL exhibit normal bone development. However, osteoclasts are reduced and trabecular bone is increased in adult mice, indicating that sRANKL contributes to physiological bone remodeling (Xiong et al., 2018). The circulating levels of sRANKL increase in patients with mild primary hyperparathyroidism and correlate with the level of circulating markers of bone resorption and the rate of femoral bone loss (Nakchbandi et al., 2008). However, the level of circulating sRANKL is too low to be biologically relevant, indicating that local production of the cytokine is responsible for the induction and maintenance of BMUs.

PTH exerts a negative effect on the production of OPG (Fu et al., 2002; Onyia et al., 2000). Since bone mass is severely reduced and osteoclast number is dramatically increased in OPG null mice (Bucay et al., 1998), this cytokine must play a fundamental role in the regulation of osteoclast differentiation and survival in physiological circumstances. In parathyroidectomized patients, bone exhibits a decrease in the ratio of RANKL to OPG transcripts (Stilgren et al., 2004), demonstrating that PTH increases the RANKL/OPG ratio. Surprisingly, however, little is known about the underlying genetic regulatory mechanism or the cellular source(s) of OPG.

PTH affects several other factors with proosteoclastogenic properties. The expression of monocyte chemoattractant protein 1 (MCP-1), a cytokine that promotes osteoclast multinucleation, is strongly induced by PTH in cultured osteoblastic cells (Li et al., 2007). Mice with global deletion of MCP-1 have normal bone mass and osteoclast number, but infusion of PTH fails to cause the expected cortical or trabecular bone loss (Siddiqui et al., 2017). PTH also stimulates the production of IL-11 (Walker et al., 2012), which stimulates osteoclastogenesis in ex-vivo cultures of bone marrow cells (Girasole et al., 1994), by stimulating RANKL synthesis (Horwood et al., 1998; Palmqvist et al., 2002). Moreover, global deletion of the IL-11R reduces osteoclast number (Sims et al., 2005). In vitro and in vivo studies show that PTH also stimulates the synthesis of IL-6, which can increase osteoclast number by promoting the replication of monocytic progenitors (Greenfield et al., 1996; Walker et al., 2012; O'Brien et al., 2005). However, mice lacking IL-6 have normal bone (Fattori et al., 1994; Sims et al., 2005); and studies with these mice show that IL-6 is dispensable for the increased RANKL, osteoclast number, and bone loss caused by secondary hyperparathyroidism (O'Brien et al., 2005).

Matrix metalloproteinase 13 (MMP-13) is involved in collagen turnover in many tissues including bone (Stickens et al., 2004). PTH stimulates the synthesis of MMP-13 in osteoblastic cells (Winchester et al., 2000). Moreover, PTH-induced osteoclastogenesis in calvaria is attenuated in mice bearing a mutation that causes resistance of type I collagen to MMP-13 (Zhao et al., 1999). Unexpectedly, MMP-13 may also act as a cytokine, as indicated by the finding that formation of multi-nucleated osteoclasts is dramatically accelerated in vitro by a catalytically inactive mutant of MMP-13 (Fu et al., 2016).

T cells bear PTH receptors, and conditional deletion of PTHR1 in these cells. Using Lck-Cre has no effect on bone mass or osteoclast number (Tawfeek et al., 2010). Infusion of PTH, however, increases T cell production of tumor necrosis factor (TNF), which may potentiate RANKL synthesis. Moreover, TNF stimulates proinflammatory T helper cells (Th17 cells) to produce IL-17A, which may also stimulate RANKL expression (Li et al., 2015). Work with an IL-17A neutralizing antibody indicates that an IL17-/RANKL cascade contributes to the bone effects of hyperparathyroidism (Li et al., 2015).

Osteoblast differentiation and the coupling of bone formation to bone resorption

Osteoblasts originate from multipotential progenitors in the bone marrow, many of which are associated with the vasculature (Bianco et al., 2013). The commitment of these progenitors to the osteoblast lineage and their development into matrix synthesizing cells, is orchestrated by locally produced growth factors like insulin-like growth factor-I (IGF-1), bone morphogenetic proteins (BMPs), transforming growth factor-β (TGFβ), and wingless-related integration site (Wnt) ligands. Osteoblastogenesis critically depends on a variety of transcription regulators including β-catenin, runt related transcription factor 2 (Runx2), and Osterix 1(Osx1). Wnt ligands bind to frizzled (FZD)\lipoprotein related protein 5 (LRP5) or FZD\LRP6 co-receptor complexes to increase proosteogenic β-catenin mediated transcription (Baron and Kneissel, 2013). Wnt signaling is suppressed by sclerostin, which is produced exclusively by osteocytes, as well as by Dkk1 and both block activation of the Wnt receptors LRP5/6. Sost null mice exhibit high bone mass, illustrating the critical role of sclerostin as a brake on bone formation under physiologic circumstances.

Release of proosteoblastogenic factors from osteoclasts, or from the bone matrix during resorption, contribute to the temporal and spatial linkage of bone formation to bone resorption within the BMU (Fuller et al., 2008; Sims and Martin, 2014). Matrix-derived TGFβ (Tang et al., 2009) and IGF-1 (Crane and Cao, 2014) are released during bone resorption and recruit osteoblast progenitors to sites of bone remodeling (Tang et al., 2009; Xian et al., 2012). Osteoclasts themselves also secrete a variety of proosteogenic factors (Sims, 2016). Among these, sphingosine-1-phosphate (S1P) (Lotinun et al., 2013; Sartawi et al., 2017; Keller et al., 2014), and leukemia inhibitory factor LIF (Koide et al., 2017) have received attention.

Importantly, osteoclast-derived LIF may suppress the expression of sclerostin by osteocytes, thus removing a brake on Wnt signaling in the microenvironment of the BMU (Koide et al., 2017). Consistent with this evidence, deletion of gp130, a subunit of the LIF receptor, in osteoblasts/osteocytes reduces osteoblast number and bone mass without altering osteoclast number, consistent with diminished coupling (Johnson et al., 2014).

Osteoclast-independent factors also contribute to coupling. In a study comparing infused RANKL and infused PTH on trabecular bone remodeling in mice (Jilka et al., 2010), both treatments increased osteoclast and osteoblast number and bone formation rate. However, osteogenic indices and bone mass were lower in RANKL-treated mice compared with PTH-treated mice. The increase in osteoblasts in mice infused with RANKL provides functional evidence for the osteoclast-derived proosteogenic signals mentioned above. However, these signals are evidently insufficient for full replacement of the previously resorbed matrix. PTH-stimulated Wnt signaling may play an additive role. This Wnt signaling is independent of osteoclasts since OPG did not affect PTH-stimulated expression of several established Wnt target genes.

PTH stimulates proosteoblastogenic Wnt signaling via several mechanisms. One is suppression of the constitutive synthesis of the Wnt signaling inhibitor sclerostin (Bellido et al., 2005; Keller and Kneissel, 2005). Indeed, the level of circulating sclerostin is decreased or increased in hyperparathyroidism and hypoparathyroidism, respectively (Costa et al., 2011; van Lierop et al., 2010). PTH also suppresses the expression of the Wnt antagonist Dkk1; and mice overexpressing Dkk1 have low bone mass (Guo et al., 2010). Importantly, continuous elevation of PTH does not increase bone formation or cause peritrabecular fibrosis in Dkk-overexpressing mice, presumably because PTH cannot reduce the Dkk transgene in this model. Mice expressing a constitutively active PTHR1 in osteoblasts and osteocytes, on the other hand, have increased bone mass and high remodeling. These changes are associated with suppression of sclerostin and increased Wnt signaling, which were attenuated by deletion of LRP5 (O'Brien et al., 2008). Finally, in vitro evidence indicates that PTH may stimulate Wnt signaling directly by inducing the association of PTH1R with LRP6, and thereby increasing β-catenin (Wan et al., 2008).

Despite increased Wnt signaling and production of other growth factors, bone mass decreases in hyperparathyroidism. This is probably due to excessive genesis and prolongation of the life span of osteoclasts secondary to increased RANKL and other PTH-regulated cytokines. Osteoblast apoptosis is unaffected in hyperparathyroidism (Bellido et al., 2003), but high levels of PTH have a negative impact on osteoblastogenesis. Indeed, the peritrabecular fibroblasts that characterize severe hyperparathyroidism likely represent osteoblastic cells arrested at a late stage of differentiation. Thus, experiments in mice receiving PTH demonstrate rapid development of these cells, and cessation of PTH administration results in their conversion to bone-forming osteoblasts (Lotinun et al., 2005). Excessive levels of PDGFa may be responsible (Lowry et al., 2008). Another cause of arrested osteoblast differentiation may be accelerated proteasomal degradation of essential transcription factors like Runx2 (Bellido et al., 2003).

The elucidation of the cellular mechanisms responsible for PTH-regulated BMU function helps explain the "hungry bone" syndrome that can be seen after parathyroidectomy (Mallette, 1994). In these patients, because of the rapid decline in PTH, RANKL levels fall, leading to premature osteoclast death, thereby reducing Ca release from bone. However, serum Ca continues to be deposited into bone by osteoblasts already assembled in the BMUs that had been generated prior to surgery and by. New osteoblasts generated upon release of progenitors from the direct or indirect inhibitory effects of PTH.

The bone anabolic effects of intermittent parathyroid hormone

Stimulation of anabolic remodeling and modeling

In contrast to hyperparathyroidism, intermittent (usually daily) injections of PTH (iPTH) increase bone mass in animals and humans. Modified amino terminal versions of hPTH(1−34) (Forteo) and hPTHrP(1−36) (Abaloparatide) increase BMD and reduce the incidence of fractures (Jiang et al., 2003; Hodsman et al., 2005; Miller et al., 2016; Doyle et al., 2018; Finkelstein et al., 2003; Orwoll et al., 2003). Anabolism is confined to trabecular, endocortical, and periosteal surfaces. In contrast, iPTH tends to increase cortical porosity (Zebaze et al., 2017; Moreira et al., 2017; Sato et al., 2004; Burr et al., 2001; Doyle et al., 2018; Fox et al., 2007a). The magnitude of the increase in bone mass declines with time with modest if any benefit after 2 years of treatment (Finkelstein et al., 2009; Dempster et al., 2018a).

Injection of a therapeutic dose of 20 μg of hPTH(1−34) increases levels of this peptide to ~150 pg/mL after 30 min, which is 2- to 10-fold higher than the normal circulating level of 15−65 pg/mL of endogenous PTH. The exogenous peptide has a half-life of ~1 h (Satterwhite et al., 2010). The magnitude of the change in circulating PTH required for anabolism is far beyond the ~20%−30% variation that occurs in a diurnal manner (El Hajj et al., 1997). In rodents the minimum effective interval between bouts of PTH administration is 6 h, with smaller intervals resulting in the histologic changes and bone loss characteristic of hyperparathyroidism. Thus, both a substantial rise and rapid decline in the level of PTH are required for the anabolic effect of the hormone.

Intermittent PTH increases bone remodeling secondary to a transient increase of RANKL. However, the increase in osteoblast number is much higher and occurs rapidly (Jilka et al., 1999; Iida-Klein et al., 2002; Bellido et al., 2003; Lindsay et al., 2006; Ma et al., 2006). Administration of fluorochromes to identify sites of bone formation prior to and following daily PTH injections in postmenopausal women have allowed for more detailed investigation of the cellular events involved. (Lindsay et al., 2006). As illustrated in Fig. 32.1, the location of each label with respect to the shape of the underlying cement lines was used to distinguish new bone made during remodeling, new bone made by overflow of osteoblasts outside the boundaries of the BMU, and new bone made by modeling. Intermittent iPTH clearly elevated remodeling on cancellous, endocortical, and intracortical surfaces after 1 month of therapy (Lindsay et al., 2006). More importantly, iPTH induced overflow labeling and modeling, both of which were rare or absent at baseline. These responses occurred in trabecular and endosteal bone but not in Haversian canals. Increased modeling was reported in another study, based on similar labeling criteria (Ma et al., 2006). iPTH also stimulated modeling-based bone formation on the periosteal surface in about half of the subjects. Increased remodeling, overfill, and modeling were also observed after 6 months of treatment, but after 24 months, overfill- and modeling-based bone formation had practically ceased (Lindsay et al., 2007;

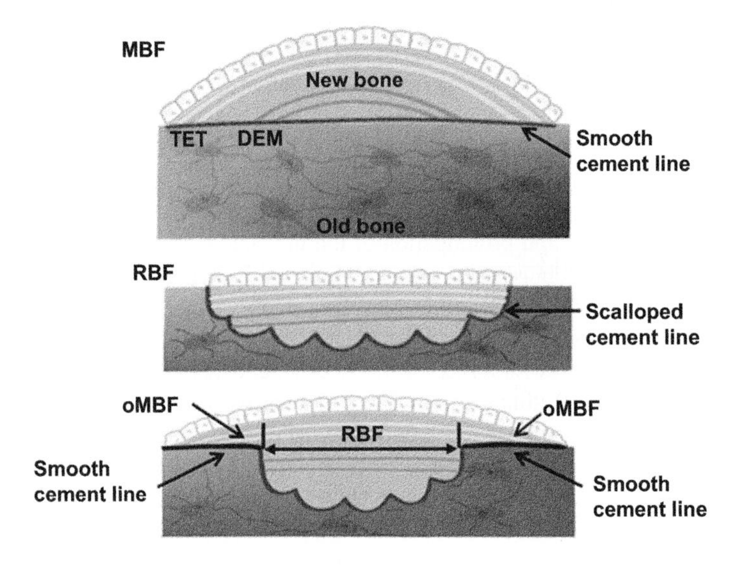

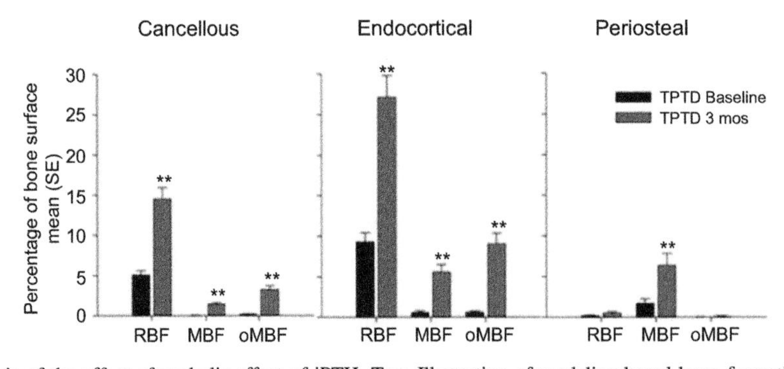

FIGURE 32.1 Cellular basis of the effect of anabolic effect of iPTH. Top. Illustration of modeling-based bone formation (MBF) marked by fluorochrome labeling overlying smooth cement lines; remodeling-based bone formation (RBF) marked by labeling on top of scalloped cement lines that mark the extent of previous osteoclastic bone resorption; and overflow-modeling-based bone formation (oMBF) taking place on smooth cement lines adjacent to scalloped cement lines. Bottom: Effect of 3 months of iPTH on different types of bone formation on cancellous, endocortical, and periosteal surface measured in the same patient using quadrupole fluorochrome labeling. Note absence of MBF and oMBF in cancellous and endocortical bone at baseline. *Reproduced from Dempster, D.W., Zhou, H., Recker, R.R., Brown, J.P., Recknor, C.P., Lewiecki, E.M., Miller, P.D., Rao, S.D., Kendler, D.L., Lindsay, R., Krege, J.H., Alam, J., Taylor, K.A., Melby, T.E., Ruff, V.A. 2018a. Remodeling- and modeling-based bone formation with teriparatide versus denosumab: a longitudinal analysis from baseline to 3 Months in the AVA study. J. Bone Miner. Res. 33, 298–306, with permission.*

Dempster et al., 2018a, 2018b). This coincides with the decline in the stimulatory effect of iPTH on BMD. Moreover, after 24 months, the magnitude of remodeling-based bone formation on cancellous and endocortical surfaces declined two- to threefold from its level at 6 months (Dempster et al., 2018b).

The increased Haversian remodeling in response to intermittent PTH is associated with increased porosity (Dempster et al., 2016; Cohen et al., 2013; Lindsay et al., 2006; Cosman et al., 2016; Ma et al., 2014). This effect is also evident following administration of PTH for 4 months to hypoparathyroid patients (Sikjaer et al., 2012). Increased porosity may be partly due to the formation of new BMUs—i.e., an increase in intracortical remodeling space. Moreover, iPTH fails to arrest ovariectomy-induced development of cortical porosity in non-human primates (Fox et al., 2007b). At a higher dose, iPTH increases cortical porosity even further in this model; yet the anabolic effect in trabecular bone is still present.

Mechanisms underlying overfill of resorption cavities in response to injections of parathyroid hormone

Improved coupling of bone formation to bone resorption at sites of remodeling, resulting from increased osteoblast number within each BMU, most likely accounts for the overfilling of resorption cavities with iPTH. In mice, the anabolic effect of iPTH is attenuated by conditional deletion of the PTHR1 in multipotent osteoblast progenitors (Balani et al., 2017), osteoblasts and osteocytes (Saini et al., 2013; Delgado-Calle et al., 2017) and in T cells (Bedi et al., 2012), suggesting that several interdependent mechanisms are responsible.

Early studies indicated that iPTH increases the number of osteoblast progenitors in the bone marrow (Nishida et al., 1994). Lineage tracing studies in mice have elucidated the underlying mechanisms. In this work the authors utilize mice bearing a floxed gene encoding a fluorescent protein, as well as a transgene consisting of a tamoxifen-inducible Sox9 promoter (known to be expressed in early multipotent mesenchymal cells) driving Cre recombinase (Balani et al., 2017). Administration of tamoxifen for 1 week caused appearance of fluorescence-labeled Sox9+ progenitors within the bone marrow. These progenitors were the source of osteoblasts needed for remodeling, because fluorescent osteoblasts and osteocytes were evident for several months after tamoxifen injection. Intermittent PTH increased the number of fluorescent Sox9+ progenitors as well as the number of fluorescent osteoblasts and osteocytes. FACS analysis of the Sox9+ cells showed that iPTH did not stimulate their replication. This finding is consistent with earlier nucleotide labeling studies (Dobnig and Turner, 1995; Jilka et al., 2009) and evidence that PTH modulates the expression of cell cycle-regulating proteins in cultured osteoblastic cells, resulting in their exit from the cell cycle (summarized in Jilka, 2007). Instead, iPTH increased the number of osteoblast progenitors by attenuating their apoptosis. Moreover, deletion of PTHR1 in Sox9+ cells abrogated the increase in progenitors, as well as the anabolic effect, suggesting that this response depends on direct actions of the hormone on Sox9+ progenitors and/or their progeny. Cessation of iPTH caused the expected loss of anabolism, but interestingly also resulted in an increase in adipocytes that were descended from labeled Sox9+ progenitors.

Besides osteoblast progenitors, intermittent PTH reduces the number of apoptotic osteoblasts in trabecular bone (Jilka et al., 1999; Bellido et al., 2003). PTH-induced survival signaling depends on CREB- and Runx2-mediated transcription, increased expression of the antiapoptotic genes like Bcl-2, and phosphorylation and thereby inactivation of the antiapoptotic protein Bad. The duration of PTH-stimulated antiapoptotic signals is about 6 h in vitro. This is probably because of a separate effect of PTH on stimulating Smurf1-mediated proteosomal proteolysis of the Runx2 required for suppression of apoptosis. This may be the reason that osteoblast apoptosis is unaffected in hyperparathyroidism.

Intermittent PTH modulates the synthesis of autocrine/paracrine factors that accelerate osteoblast differentiation. LRP6 but not LRP5 (Iwaniec et al., 2007) is required for iPTH anabolism, consistent with direct activation of LPR6 by the hormone (Wan et al., 2008). Following injection of PTH, the expression of the Wnt signaling inhibitor sclerostin in osteocytes is transiently suppressed, coinciding with the rise and fall in the circulating level of the injected hormone (Bellido et al., 2005; Keller and Kneissel, 2005). A critical role of the decrease in sclerostin in the effects of iPTH has been suggested by the attenuation of the anabolic effect of iPTH in mice with global deletion of sclerostin (Robling et al., 2011). Moreover, PTH-induced bone formation is attenuated by increasing the dosage of the Sost gene (Kramer et al., 2010). Mice over-expressing sclerostin under the control of the DMP-1 promoter, however, still respond to iPTH (Delgado-Calle et al., 2017). This seeming discrepancy could be explained by the fact that, in the latter studies, sclerostin was overexpressed in both osteocytes and osteoblasts; as opposed to normal mice in which sclerostin is expressed only in osteocytes. Intermittent PTH-stimulated production of Wnt10b by T cells may also contribute to Wnt signaling and bone anabolism (Bedi et al., 2012; Terauchi et al., 2009), but for reasons that are unclear, this mechanism is only relevant to trabecular bone.

Of the many proosteogenic cytokines and growth factors stimulated by PTH, IGF-1 (Bikle et al., 2002; Miyakoshi et al., 2001), FGF2 (Hurley et al., 2006), or ephrinB2 (Tonna et al., 2014) are required for the full anabolic response to iPTH. However, it is unclear whether the IGF-I- and FGF2-dependent anabolism by iPTH reflects increased release from the bone matrix or locally increased biosynthetic actions of the hormone. IGF-I increases glucose uptake and oxidative phosphorylation during osteoblast differentiation, which is critical for the development of the high biosynthetic and secretory capacity of these cells (Esen et al., 2015). Deletion of gp130 − the common co-receptor for IL-6 type cytokines - in osteoblasts/osteocytes attenuates the anabolic effect of iPTH (Standal et al., 2014). This might be explained by a dependence on osteoclast-derived LIF, which can suppress sclerostin as discussed above. Notch signaling normally suppresses osteoblast differentiation, and PTH antagonizes this pathway (Zanotti and Canalis, 2017). Accordingly, mice with conditional deletion of the Notch ligand Jagged 1 in mesenchymal stem cells exhibit increased femoral cancellous bone mass. Moreover, iPTH increases trabecular bone in these mice, perhaps because of loss of Jagged1-dependent limits on osteoblast differentiation (Lawal et al., 2017).

Stimulation of bone modeling by osteoblasts in response to injections of parathyroid hormone

In adults, the periosteum contains a reservoir of osteoblast progenitors that facilitate fracture repair and the slow deposition of periosteal bone that occurs via modeling. Lineage tracing studies in growing mice using tamoxifen-inducible DMP1-Cre transgene showed that some periosteal cells are former osteoblasts (Kim et al., 2012). More important, iPTH reactivated these marked lining cells to resume their matrix-synthesizing function.

Trabecular bone surfaces have lining cells descended from osteoblasts (Kim et al., 2012). Previous morphologic studies had suggested that iPTH rapidly converts these cells into osteoblasts (Leaffer et al., 1995; Dobnig and Turner, 1995), but this has not yet been confirmed using the lineage tracing approach. Nevertheless, administration of an antisclerostin antibody did reactivate trabecular lining cells marked with either DMP1-Cre or OCN-Cre on trabecular and periosteal surfaces (Kim et al., 2012). This anabolic response was short lived, probably because most of the reactivated postmitotic osteoblasts were incorporated into the bone matrix as osteocytes or died; only a minority returned to their former life as lining cells. Thus, the constitutive production of sclerostin by osteocytes may serve an antiproosteoblastogenic role on most skeletal surfaces by suppressing Wnt signaling that would otherwise be induced by locally produced Wnt ligands. Sequestration of sclerostin by the antibody unleashes Wnt signaling throughout the skeleton. Similarly, lining cells may be reactivated by periodic suppression of sclerostin by iPTH.

Unresolved issues

Why does iPTH fail to overfill of BMUs and to reactivate lining cells in Haversian canals? There may be an inhibitory effect of the capillary, since anabolism in Haversian canals has the potential to closing the canal and blocking cortical circulation. Fundamental differences in BMU and lining cell characteristics in Haversian surfaces versus other sites is also possible.

Why does PTH-stimulated BMU overfill and PTH-stimulated modeling wane with time? It is possible that activation of negative feedback pathways on osteoblastogenesis and on reactivation of lining cells is responsible. The new BMUs generated during the later stages of therapy may also have excessive bone resorption. Given the complexity of the cellular response and the absence of anabolism at sites of intracortical remodeling, as reviewed herein, studies of changes in circulating markers of osteoclast and osteoblast function are not very informative (Seeman and Martin, 2015).

References

Balani, D.H., Ono, N., Kronenberg, H.M., 2017. Parathyroid hormone regulates fates of murine osteoblast precursors in vivo. J. Clin. Investig. 127, 3327−3338.

Baron, R., Kneissel, M., 2013. WNT signaling in bone homeostasis and disease: from human mutations to treatments. Nat. Med. 19, 179−192.

Baud, C.A., Boivin, G., 1978. Effects of hormones on osteocyte function and perilacunar wall structure. Clin. Orthop. Relat. Res. 270−281.

Bedi, B., Li, J.Y., Tawfeek, H., Baek, K.H., Adams, J., Vangara, S.S., Chang, M.K., Kneissel, M., Weitzmann, M.N., Pacifici, R., 2012. Silencing of parathyroid hormone (PTH) receptor 1 in T cells blunts the bone anabolic activity of PTH. Proc. Natl. Acad. Sci. Unit. States Am. 109, E725−E733.

Bellido, T., Ali, A.A., Gubrij, I., Plotkin, L.I., Fu, Q., O'brien, C.A., Manolagas, S.C., Jilka, R.L., 2005. Chronic elevation of parathyroid hormone in mice reduces expression of sclerostin by osteocytes: a novel mechanism for hormonal control of osteoblastogenesis. Endocrinology 146, 4577–4583.

Bellido, T., Ali, A.A., Plotkin, L.I., Fu, Q., Gubrij, I., Roberson, P.K., Weinstein, R.S., O'brien, C.A., Manolagas, S.C., Jilka, R.L., 2003. Proteasomal degradation of Runx2 shortens parathyroid hormone-induced anti-apoptotic signaling in osteoblasts: a putative explanation for why intermittent administration is needed for bone anabolism. J. Biol. Chem. 278, 50259–50272.

Ben-Awadh, A.N., Delgado-Calle, J., Tu, X., Kuhlenschmidt, K., Allen, M.R., Plotkin, L.I., Bellido, T., 2014. Parathyroid hormone receptor signaling induces bone resorption in the adult skeleton by directly regulating the RANKL gene in osteocytes. Endocrinology 155, 2797–2809.

Bianco, P., Cao, X., Frenette, P.S., Mao, J.J., Robey, P.G., Simmons, P.J., Wang, C.Y., 2013. The meaning, the sense and the significance: translating the science of mesenchymal stem cells into medicine. Nat. Med. 19, 35–42.

Bikle, D.D., Sakata, T., Leary, C., Elalieh, H., Ginzinger, D., Rosen, C.J., Beamer, W., Majumdar, S., Halloran, B.P., 2002. Insulin-like growth factor I is required for the anabolic actions of parathyroid hormone on mouse bone. J. Bone Miner. Res. 17, 1570–1578.

Boyle, W.J., Simonet, W.S., Lacey, D.L., 2003. Osteoclast differentiation and activation. Nature 423, 337–342.

Bucay, N., Sarosi, I., Dunstan, C.R., Morony, S., Tarpley, J., Capparelli, C., Scully, S., Tan, H.L., Xu, W.L., Lacey, D.L., Boyle, W.J., Simonet, W.S., 1998. *Osteoprotegerin*-deficient mice develop early onset osteoporosis and arterial calcification. Genes Dev. 12, 1260–1268.

Burr, D.B., Hirano, T., Turner, C.H., Hotchkiss, C., Brommage, R., Hock, J.M., 2001. Intermittently administered human parathyroid hormone (1-34) treatment increases intracortical bone turnover and porosity without reducing bone strength in the humerus of ovariectomized cynomolgus monkeys. J. Bone Miner. Res. 16, 157–165.

Christiansen, P., Steiniche, T., Vesterby, A., Mosekilde, L., Hessov, I., Melsen, F., 1992. Primary hyperparathyroidism: iliac crest trabecular bone volume, structure, remodeling, and balance evaluated by histomorphometric methods. Bone 13, 41–49.

Clarke, B.L., 2014. Bone disease in hypoparathyroidism. Arq. Bras. Endocrinol. Metabol. 58, 545–552.

Cohen, A., Stein, E.M., Recker, R.R., Lappe, J.M., Dempster, D.W., Zhou, H., Cremers, S., Mcmahon, D.J., Nickolas, T.L., Muller, R., Zwahlen, A., Young, P., Stubby, J., Shane, E., 2013. Teriparatide for idiopathic osteoporosis in premenopausal women: a pilot study. J. Clin. Endocrinol. Metab. 98, 1971–1981.

Cosman, F., Dempster, D.W., Nieves, J.W., Zhou, H., Zion, M., Roimisher, C., Houle, Y., Lindsay, R., Bostrom, M., 2016. Effect of teriparatide on bone formation in the human femoral neck. J. Clin. Endocrinol. Metab. 101, 1498–1505.

Costa, A.G., Cremers, S., Rubin, M.R., Mcmahon, D.J., Sliney, J., Lazaretti-Castro, M., Silverberg, S.J., Bilezikian, J.P., 2011. Circulating sclerostin in disorders of parathyroid gland function. J. Clin. Endocrinol. Metab. 96, 3804–3810.

Crane, J.L., Cao, X., 2014. Function of matrix IGF-1 in coupling bone resorption and formation. J. Mol. Med. 92, 107–115.

delgado-Calle, J., Tu, X., Pacheco-Costa, R., Mcandrews, K., Edwards, R., Pellegrini, G.G., Kuhlenschmidt, K., Olivos, N., Robling, A., Peacock, M., Plotkin, L.I., Bellido, T., 2017. Control of bone anabolism in response to mechanical loading and PTH by distinct mechanisms downstream of the PTH receptor. J. Bone Miner. Res. 32, 522–535.

Dempster, D.W., Compston, J.E., Drezner, M.K., Glorieux, F.H., Kanis, J.A., Malluche, H., Meunier, P.J., Ott, S.M., Recker, R.R., Parfitt, A.M., 2013. Standardized nomenclature, symbols, and units for bone histomorphometry: a 2012 update of the report of the ASBMR Histomorphometry Nomenclature Committee. J. Bone Miner. Res. 28, 2–17.

Dempster, D.W., Parisien, M., Silverberg, S.J., Liang, X.G., Schnitzer, M., Shen, V., Shane, E., Kimmel, D.B., Recker, R., Lindsay, R., Bilezikian, J.P., 1999. On the mechanism of cancellous bone preservation in postmenopausal women with mild primary hyperparathyroidism. J. Clin. Endocrinol. Metab. 84, 1562–1566.

Dempster, D.W., Zhou, H., Recker, R.R., Brown, J.P., Bolognese, M.A., Recknor, C.P., Kendler, D.L., Lewiecki, E.M., Hanley, D.A., Rao, S.D., Miller, P.D., Woodson 3rd, G.C., Lindsay, R., Binkley, N., Alam, J., Ruff, V.A., Gallagher, E.R., Taylor, K.A., 2016. A longitudinal study of skeletal histomorphometry at 6 and 24 months across four bone envelopes in postmenopausal women with osteoporosis receiving teriparatide or zoledronic acid in the SHOTZ trial. J. Bone Miner. Res. 31, 1429–1439.

Dempster, D.W., Zhou, H., Recker, R.R., Brown, J.P., Recknor, C.P., Lewiecki, E.M., Miller, P.D., Rao, S.D., Kendler, D.L., Lindsay, R., Krege, J.H., Alam, J., Taylor, K.A., Melby, T.E., Ruff, V.A., 2018a. Remodeling- and modeling-based bone formation with teriparatide versus denosumab: a longitudinal analysis from baseline to 3 Months in the AVA study. J. Bone Miner. Res. 33, 298–306.

Dempster, D.W., Zhou, H., Ruff, V.A., Melby, T.E., Alam, J., Taylor, K.A., 2018b. Longitudinal effects of teriparatide or zoledronic acid on bone modeling- and remodeling-based formation in the SHOTZ study. J. Bone Miner. Res. 33, 627–633.

Dobnig, H., Turner, R.T., 1995. Evidence that intermittent treatment with parathyroid hormone increases bone formation in adult rats by activation of bone lining cells. Endocrinology 136, 3632–3638.

Doyle, N., Varela, A., Haile, S., Guldberg, R., Kostenuik, P.J., Ominsky, M.S., Smith, S.Y., Hattersley, G., 2018. Abaloparatide, a novel PTH receptor agonist, increased bone mass and strength in ovariectomized cynomolgus monkeys by increasing bone formation without increasing bone resorption. Osteoporos. Int. 29, 685–697.

EL Hajj, F.G., Klerman, E.B., Brown, E.N., Choe, Y., Brown, E.M., Czeisler, C.A., 1997. The parathyroid hormone circadian rhythm is truly endogenous–a general clinical research center study. J. Clin. Endocrinol. Metab. 82, 281–286.

Esen, E., Lee, S.Y., Wice, B.M., Long, F., 2015. PTH promotes bone anabolism by stimulating aerobic glycolysis via IGF signaling. J. Bone Miner. Res. 30, 1959–1968.

Fattori, E., Cappelletti, M., Costa, P., Sellitto, C., Cantoni, L., Carelli, M., Faggioni, R., Fantuzzi, G., Ghezzi, P., Poli, V., 1994. Defective inflammatory response in interleukin 6-deficient mice. J. Exp. Med. 180, 1243–1250.

Finkelstein, J.S., Hayes, A., Hunzelman, J.L., Wyland, J.J., Lee, H., Neer, R.M., 2003. The effects of parathyroid hormone, alendronate, or both in men with osteoporosis. N. Engl. J. Med. 349, 1216–1226.

Finkelstein, J.S., Wyland, J.J., Leder, B.Z., Burnett-Bowie, S.A., Lee, H., Juppner, H., Neer, R.M., 2009. Effects of teriparatide retreatment in osteoporotic men and women. J. Clin. Endocrinol. Metab. 94, 2495–2501.

Fox, J., Miller, M.A., Newman, M.K., Recker, R.R., Turner, C.H., Smith, S.Y., 2007a. Effects of daily treatment with parathyroid hormone 1–84 for 16 months on density, architecture and biomechanical properties of cortical bone in adult ovariectomized rhesus monkeys. Bone 41, 321–330.

Fox, J., Miller, M.A., Recker, R.R., Turner, C.H., Smith, S.Y., 2007b. Effects of treatment of ovariectomized adult rhesus monkeys with parathyroid hormone 1–84 for 16 months on trabecular and cortical bone structure and biomechanical properties of the proximal femur. Calcif. Tissue Int. 81, 53–63.

Fu, J., Li, S., Feng, R., Ma, H., Sabeh, F., Roodman, G.D., Wang, J., Robinson, S., Guo, X.E., Lund, T., Normolle, D., Mapara, M.Y., Weiss, S.J., Lentzsch, S., 2016. Multiple myeloma-derived MMP-13 mediates osteoclast fusogenesis and osteolytic disease. J. Clin. Investig. 126, 1759–1772.

Fu, Q., Jilka, R.L., Manolagas, S.C., O'brien, C.A., 2002. Parathyroid hormone stimulates receptor activator of NFkappa B ligand and inhibits osteoprotegerin expression via protein kinase A activation of cAMP-response element-binding protein. J. Biol. Chem. 277, 48868–48875.

Fu, Q., Manolagas, S.C., O'brien, C.A., 2006. Parathyroid hormone controls receptor activator of NF-κB ligand gene expression via a distant transcriptional enhancer. Mol. Cell Biol. 26, 6453–6468.

Fuller, K., Lawrence, K.M., Ross, J.L., Grabowska, U.B., Shiroo, M., Samuelsson, B., Chambers, T.J., 2008. Cathepsin K inhibitors prevent matrix-derived growth factor degradation by human osteoclasts. Bone 42, 200–211.

Galli, C., Zella, L.A., Fretz, J.A., Fu, Q., Pike, J.W., Weinstein, R.S., Manolagas, S.C., O'brien, C.A., 2008. Targeted deletion of a distant transcriptional enhancer of the receptor activator of nuclear factor-kappaB ligand gene reduces bone remodeling and increases bone mass. Endocrinology 149, 146–153.

Gebhard, S., Hattori, T., Bauer, E., Schlund, B., Bosl, M.R., de Crombrugghe, B., von der Mark, K., 2008. Specific expression of Cre recombinase in hypertrophic cartilage under the control of a BAC-Col10a1 promoter. Matrix Biol. 27, 693–699.

Girasole, G., Passeri, G., Jilka, R.L., Manolagas, S.C., 1994. Interleukin-11: a new cytokine critical for osteoclast development. J. Clin. Investig. 93, 1516–1524.

Greenfield, E.M., Horowitz, M.C., Lavish, S.A., 1996. Stimulation by parathyroid hormone of interleukin-6 and leukemia inhibitory factor expression in osteoblasts is an immediate-early gene response induced by cAMP signal transduction. J. Biol. Chem. 271, 10984–10989.

Guo, J., Liu, M., Yang, D., Bouxsein, M.L., Saito, H., Galvin, S., Kuhstoss, S.A., Thomas, C.C., Schipani, E., Baron, R., Bringhurst, F.R., Kronenberg, H.M., 2010. Suppression of Wnt signaling by Dkk1 attenuates PTH-mediated stromal cell response and new bone formation. Cell Metabol. 11, 161–171.

Hodsman, A.B., Bauer, D.C., Dempster, D., Dian, L., Hanley, D.A., Harris, S.T., Kendler, D., Mcclung, M.R., Miller, P.D., Olszynski, W.P., Orwoll, E., Yuen, C.K., 2005. Parathyroid hormone and teriparatide for the treatment of osteoporosis: a review of the evidence and suggested guidelines for its use. Endocr. Rev. 26, 688–703.

Horwood, N.J., Elliott, J., Martin, T.J., Gillespie, M.T., 1998. Osteotropic agents regulate the expression of osteoclast differentiation factor and osteoprotegerin in osteoblastic stromal cells. Endocrinology 139, 4743–4746.

Hurley, M.M., Okada, Y., Xiao, L., Tanaka, Y., Ito, M., Okimoto, N., Nakamura, T., Rosen, C.J., Doetschman, T., Coffin, J.D., 2006. Impaired bone anabolic response to parathyroid hormone in Fgf2−/− and Fgf2+/− mice. Biochem. Biophys. Res. Commun. 341, 989–994.

IIDA-Klein, A., Zhou, H., Lu, S.S., Levine, L.R., Ducayen-Knowles, M., Dempster, D.W., Nieves, J., Lindsay, R., 2002. Anabolic action of parathyroid hormone is skeletal site specific at the tissue and cellular levels in mice. J. Bone Miner. Res. 17, 808–816.

Iwaniec, U.T., Wronski, T.J., Liu, J., Rivera, M.F., Arzaga, R.R., Hansen, G., Brommage, R., 2007. PTH stimulates bone formation in mice deficient in Lrp5. J. Bone Miner. Res. 22, 394–402.

Jiang, Y., Zhao, J.J., Mitlak, B.H., Wang, O., Genant, H.K., Eriksen, E.F., 2003. Recombinant human parathyroid hormone (1–34) [teriparatide] improves both cortical and cancellous bone structure. J. Bone Miner. Res. 18, 1932–1941.

Jilka, R.L., 2007. Molecular and cellular mechanisms of the anabolic effect of intermittent PTH. Bone 40, 1434–1446.

Jilka, R.L., Bellido, T., Almeida, M., Plotkin, L.I., O'brien, C., Weinstein, R.S., Manolagas, S.C., 2008. Apoptosis of bone cells. In: Bilezikian, J., Raisz, L., Martin, T. (Eds.), Principles of Bone Biology, third ed. Academic Press.

Jilka, R.L., O'brien, C.A., Ali, A.A., Roberson, P.K., Weinstein, R.S., Manolagas, S.C., 2009. Intermittent PTH stimulates periosteal bone formation by actions on post-mitotic preosteoblasts. Bone 44, 275–286.

Jilka, R.L., O'brien, C.A., Bartell, S.M., Weinstein, R.S., Manolagas, S.C., 2010. Continuous elevation of PTH increases the number of osteoblasts via both osteoclast-dependent and - independent mechanisms. J. Bone Miner. Res. 25, 2427–2437.

Jilka, R.L., O'brien, C.A., Roberson, P.K., Bonewald, L.F., Weinstein, R.S., Manolagas, S.C., 2014. Dysapoptosis of osteoblasts and osteocytes increases cancellous bone formation but exaggerates bone porosity with age. J. Bone Miner. Res. 29, 103–117.

Jilka, R.L., Weinstein, R.S., Bellido, T., Roberson, P., Parfitt, A.M., Manolagas, S.C., 1999. Increased bone formation by prevention of osteoblast apoptosis with parathyroid hormone. J. Clin. Investig. 104, 439–446.

Jilka, R.L., Weinstein, R.S., Parfitt, A.M., Manolagas, S.C., 2007. Quantifying osteoblast and osteocyte apoptosis: challenges and rewards. J. Bone Miner. Res. 22, 1492–1501.

Johnson, R.W., Brennan, H.J., Vrahnas, C., Poulton, I.J., Mcgregor, N.E., Standal, T., Walker, E.C., Koh, T.T., Nguyen, H., Walsh, N.C., Forwood, M.R., Martin, T.J., Sims, N.A., 2014. The primary function of gp130 signaling in osteoblasts is to maintain bone formation and strength, rather than promote osteoclast formation. J. Bone Miner. Res. 29, 1492–1505.

Keller, H., Kneissel, M., 2005. SOST is a target gene for PTH in bone. Bone 37, 148–158.

Keller, J., Catala-Lehnen, P., Huebner, A.K., Jeschke, A., Heckt, T., Lueth, A., Krause, M., Koehne, T., Albers, J., Schulze, J., Schilling, S., Haberland, M., Denninger, H., Neven, M., Hermans-Borgmeyer, I., Streichert, T., Breer, S., Barvencik, F., Levkau, B., Rathkolb, B., Wolf, E., Calzada-Wack, J., Neff, F., Gailus-Durner, V., Fuchs, H., De Angelis, M.H., Klutmann, S., Tsourdi, E., Hofbauer, L.C., Kleuser, B., Chun, J., Schinke, T., Amling, M., 2014. Calcitonin controls bone formation by inhibiting the release of sphingosine 1-phosphate from osteoclasts. Nat. Commun. 5.

Kim, S.W., Pajevic, P.D., Selig, M., Barry, K.J., Yang, J.Y., Shin, C.S., Baek, W.Y., Kim, J.E., Kronenberg, H.M., 2012. Intermittent parathyroid hormone administration converts quiescent lining cells to active osteoblasts. J. Bone Miner. Res. 27, 2075–2084.

Koide, M., Kobayashi, Y., Yamashita, T., Uehara, S., Nakamura, M., Hiraoka, B.Y., Ozaki, Y., Iimura, T., Yasuda, H., Takahashi, N., Udagawa, N., 2017. Bone formation is coupled to resorption via suppression of sclerostin expression by osteoclasts. J. Bone Miner. Res. 32, 2074–2086.

Kong, Y.Y., Yoshida, H., Sarosi, I., Tan, H.L., Timms, E., Capparelli, C., Morony, S., Oliveira, D.S., Van, G., Itie, A., Khoo, W., Wakeham, A., Dunstan, C.R., Lacey, D.L., Mak, T.W., Boyle, W.J., Penninger, J.M., 1999. OPGL is a key regulator of osteoclastogenesis, lymphocyte development and lymph-node organogenesis. Nature 397, 315–323.

Kovacs, C.S., 2017. The skeleton is a storehouse of mineral that is plundered during lactation and (fully?) replenished afterwards. J. Bone Miner. Res. 32, 676–680.

Kramer, I., Loots, G.G., Studer, A., Keller, H., Kneissel, M., 2010. Parathyroid hormone (PTH) induced bone gain is blunted in SOST overexpressing and deficient mice. J. Bone Miner. Res. 25, 178–189.

Lacey, D.L., Tan, H.L., Lu, J., Kaufman, S., Van, G., Qiu, W., Rattan, A., Scully, S., Fletcher, F., Juan, T., Kelley, M., Burgess, T.L., Boyle, W.J., Polverino, A.J., 2000. Osteoprotegerin ligand modulates murine osteoclast survival in vitro and in vivo. Am. J. Pathol. 157, 435–448.

Langdahl, B.L., Mortensen, L., Vesterby, A., Eriksen, E.F., Charles, P., 1996. Bone histomorphometry in hypoparathyroid patients treated with vitamin D. Bone 18, 103–108.

Lassen, N.E., Andersen, T.L., Ploen, G.G., Soe, K., Hauge, E.M., Harving, S., Eschen, G.E.T., Delaisse, J.M., 2017. Coupling of bone resorption and formation in real time: new knowledge gained from human Haversian BMUs. J. Bone Miner. Res. 32, 1395–1405.

Lawal, R.A., Zhou, X., Batey, K., Hoffman, C.M., Georger, M.A., Radtke, F., Hilton, M.J., Xing, L., Frisch, B.J., Calvi, L.M., 2017. The Notch ligand Jagged1 regulates the osteoblastic lineage by maintaining the osteoprogenitor pool. J. Bone Miner. Res. 32, 1320–1331.

Leaffer, D., Sweeney, M., Kellerman, L.A., Avnur, Z., Krstenansky, J.L., Vickery, B.H., Caulfield, J.P., 1995. Modulation of osteogenic cell ultrastructure by RS-23581, an analog of human parathyroid hormone (PTH)-related peptide- (1–34), and bovine PTH-(1–34). Endocrinology 136, 3624–3631.

Lee, S.K., Lorenzo, J.A., 1999. Parathyroid hormone stimulates TRANCE and inhibits osteoprotegerin messenger ribonucleic acid expression in murine bone marrow cultures: correlation with osteoclast-like cell formation. Endocrinology 140, 3552–3561.

Li, J.Y., D'amelio, P., Robinson, J., Walker, L.D., Vaccaro, C., Luo, T., Tyagi, A.M., Yu, M., Reott, M., Sassi, F., Buondonno, I., Adams, J., Weitzmann, M.N., Isaia, G.C., Pacifici, R., 2015. IL-17A is increased in humans with primary hyperparathyroidism and mediates PTH-induced bone loss in mice. Cell Metabol. 22, 799–810.

Li, X., Qin, L., Bergenstock, M., Bevelock, L.M., Novack, D.V., Partridge, N.C., 2007. Parathyroid hormone stimulates osteoblastic expression of MCP-1 to recruit and increase the fusion of pre/osteoclasts. J. Biol. Chem. 282, 33098–33106.

Lindsay, R., Cosman, F., Zhou, H., Bostrom, M.P., Shen, V.W., Cruz, J.D., Nieves, J.W., Dempster, D.W., 2006. A novel tetracycline labeling schedule for longitudinal evaluation of the short-term effects of anabolic therapy with a single iliac crest bone biopsy: early actions of teriparatide. J. Bone Miner. Res. 21, 366–373.

Lindsay, R., Zhou, H., Cosman, F., Nieves, J., Dempster, D.W., Hodsman, A.B., 2007. Effects of a one-month treatment with parathyroid hormone (1–34) on bone formation on cancellous, endocortical and periosteal surfaces of the human ilium. J. Bone Miner. Res. 22, 495–502.

Lotinun, S., Kiviranta, R., Matsubara, T., Alzate, J.A., Neff, L., Luth, A., Koskivirta, I., Kleuser, B., Vacher, J., Vuorio, E., Horne, W.C., Baron, R., 2013. Osteoclast-specific cathepsin K deletion stimulates S1P-dependent bone formation. J. Clin. Investig. 123, 666–681.

Lotinun, S., Sibonga, J.D., Turner, R.T., 2005. Evidence that the cells responsible for marrow fibrosis in a rat model for hyperparathyroidism are preosteoblasts. Endocrinology 146, 4074–4081.

Lowry, M.B., Lotinun, S., Leontovich, A.A., Zhang, M., Maran, A., Shogren, K.L., Palama, B.K., Marley, K., Iwaniec, U.T., Turner, R.T., 2008. Osteitis fibrosa is mediated by platelet-derived growth factor-A via a phosphoinositide 3-kinase-dependent signaling pathway in a rat model for chronic hyperparathyroidism. Endocrinology 149, 5735–5746.

Ma, Y.L., Zeng, Q., Donley, D.W., Ste-Marie, L.G., Gallagher, J.C., Dalsky, G.P., Marcus, R., Eriksen, E.F., 2006. Teriparatide increases bone formation in modeling and remodeling osteons and enhances IGF-II immunoreactivity in postmenopausal women with osteoporosis. J. Bone Miner. Res. 21, 855–864.

Ma, Y.L., Zeng, Q.Q., Chiang, A.Y., Burr, D., Li, J., Dobnig, H., Fahrleitner-Pammer, A., Michalska, D., Marin, F., Pavo, I., Stepan, J.J., 2014. Effects of teriparatide on cortical histomorphometric variables in postmenopausal women with or without prior alendronate treatment. Bone 59, 139–147.

Mallette, L.E., 1994. The functional and pathological spectrum of parathryoid abnormalities in hyperparathyroidism (Chapter 25). In: Bilezikian, J.P., Levine, M., Marcus, R. (Eds.), The Parathyroids. Raven Press Ltd. New York, pp. 423–455.

Manolagas, S.C., 2000. Birth and death of bone cells: basic regulatory mechanisms and implications for the pathogenesis and treatment of osteoporosis. Endocr. Rev. 21, 115–137.

Mcsheehy, P.M., Chambers, T.J., 1986. Osteoblast-like cells in the presence of parathyroid hormone release soluble factor that stimulates osteoclastic bone resorption. Endocrinology 119, 1654–1659.

Miao, D., Li, J., Xue, Y., Su, H., Karaplis, A.C., Goltzman, D., 2004. Parathyroid hormone-related peptide is required for increased trabecular bone volume in parathyroid hormone-null mice. Endocrinology 145, 3554–3562.

Miller, P.D., Hattersley, G., Riis, B.J., Williams, G.C., Lau, E., Russo, L.A., Alexandersen, P., Zerbini, C.A., Hu, M.Y., Harris, A.G., Fitzpatrick, L.A., Cosman, F., Christiansen, C., Investigators, A.S., 2016. Effect of abaloparatide vs placebo on new vertebral fractures in postmenopausal women with osteoporosis: a randomized clinical trial. J. Am. Med. Assoc. 316, 722–733.

Miyakoshi, N., Kasukawa, Y., Linkhart, T.A., Baylink, D.J., Mohan, S., 2001. Evidence that anabolic effects of PTH on bone require IGF-I in growing mice. Endocrinology 142, 4349–4356.

Moreira, C.A., Fitzpatrick, L.A., Wang, Y., Recker, R.R., 2017. Effects of abaloparatide-SC (BA058) on bone histology and histomorphometry: the ACTIVE phase 3 trial. Bone 97, 314–319.

Nakashima, T., Hayashi, M., Fukunaga, T., Kurata, K., OH-Hora, M., Feng, J.Q., Bonewald, L.F., Kodama, T., Wutz, A., Wagner, E.F., Penninger, J.M., Takayanagi, H., 2011. Evidence for osteocyte regulation of bone homeostasis through RANKL expression. Nat. Med. 17, 1231–1234.

Nakchbandi, I.A., Lang, R., Kinder, B., Insogna, K.L., 2008. The role of the receptor activator of nuclear factor-κb ligand/osteoprotegerin cytokine system in primary hyperparathyroidism. J. Clin. Endocrinol. Metab. 93, 967–973.

Nishida, S., Yamaguchi, A., Tanizawa, T., Endo, N., Mashiba, T., Uchiyama, Y., Suda, T., Yoshiki, S., Takahashi, H.E., 1994. Increased bone formation by intermittent parathyroid hormone administration is due to the stimulation of proliferation and differentiation of osteoprogenitor cells in bone marrow. Bone 15, 717–723.

O'brien, C.A., Jilka, R.L., Fu, Q., Stewart, S., Weinstein, R.S., Manolagas, S.C., 2005. IL-6 is not required for parathyroid hormone stimulation of RANKL expression, osteoclast formation, and bone loss in mice. Am. J. Physiol. Endocrinol. Metab. 289, E784–E793.

O'brien, C.A., Nakashima, T., Takayanagi, H., 2013. Osteocyte control of osteoclastogenesis. Bone 54, 258–263.

O'brien, C.A., Plotkin, L.I., Galli, C., Goellner, J.J., Gortazar, A.R., Allen, M.R., Robling, A., Bouxsein, M., Schipani, E., Turner, C.H., Jilka, R.L., Weinstein, R.S., Manolagas, S.C., Bellido, T., 2008. Control of bone mass and remodeling by PTH receptor signaling in osteocytes. PLoS One 3, e2942. https://doi.org/10.1371/journal.pone.0002942.

Onal, M., Bishop, K.A., ST John, H.C., Danielson, A.L., Riley, E.M., Piemontese, M., Xiong, J., Goellner, J.J., O'brien, C.A., Pike, J.W., 2015. A DNA segment spanning the mouse Tnfsf11 transcription unit and its upstream regulatory domain rescues the pleiotropic biologic phenotype of the RANKL null mouse. J. Bone Miner. Res. 30, 855–868.

Ono, N., Kronenberg, H.M., 2015. Mesenchymal progenitor cells for the osteogenic lineage. Curr. Mol. Biol. Rep. 1, 95–100.

Onyia, J.E., Miles, R.R., Yang, X., Halladay, D.L., Hale, J., Glasebrook, A., Mcclure, D., Seno, G., Churgay, L., Chandrasekhar, S., Martin, T.J., 2000. In vivo demonstration that human parathyroid hormone 1-38 inhibits the expression of osteoprotegerin in bone with the kinetics of an immediate early gene. J. Bone Miner. Res. 15, 863–871.

Orwoll, E.S., Scheele, W.H., Paul, S., Adami, S., Syversen, U., Diez-Perez, A., Kaufman, J.M., Clancy, A.D., Gaich, G.A., 2003. The effect of teriparatide [human parathyroid hormone (1–34)] therapy on bone density in men with osteoporosis. J. Bone Miner. Res. 18, 9–17.

Palmqvist, P., Persson, E., Conaway, H.H., Lerner, U.H., 2002. IL-6, leukemia inhibitory factor, and oncostatin M stimulate bone resorption and regulate the expression of receptor activator of NF-kappa B ligand, osteoprotegerin, and receptor activator of NF-kappa B in mouse calvariae. J. Immunol. 169, 3353–3362.

Parfitt, A.M., 1996. Skeletal heterogeneity and the purposes of bone remodeling/Implications for the understanding of osteoporosis. In: Marcus, R., Feldman, D., Kelsey, J. (Eds.), Osteoporosis. Academic Press, San Diego, CA.

Parfitt, A.M., 2002. Misconceptions (2): turnover is always higher in cancellous than in cortical bone. Bone 30, 807–809.

Parisien, M., Dempster, D.W., Shane, E., Bilezikian, J.P., 2001. Histomorphometric analysis of bone in primary hyperparathyroidism. In: Bilezikian, J.P., Marcus, R., Levine, M.A. (Eds.), The Parathyroids. Basic and Clinical Concepts. Academic Press, San Diego.

Piemontese, M., Almeida, M., Robling, A.G., Kim, H.N., Xiong, J., Thostenson, J.D., Weinstein, R.S., Manolagas, S.C., O'brien, C.A., Jilka, R.L., 2017. Old age causes de novo intracortical bone remodeling and porosity in mice. JCI Insight 2.

Robling, A.G., Kedlaya, R., Ellis, S.N., Childress, P.J., Bidwell, J.P., Bellido, T., Turner, C.H., 2011. Anabolic and catabolic regimens of human parathyroid hormone 1–34 elicit bone- and envelope-specific attenuation of skeletal effects in sost-deficient mice. Endocrinology 152, 2963–2975.

Rodan, G.A., Martin, T.J., 1981. Role of osteoblasts in hormonal control of bone resorption - a hypothesis. Calcif. Tissue Int. 33, 349–351.

Rubin, M.R., Dempster, D.W., Zhou, H., Shane, E., Nickolas, T., Sliney, J., Silverberg, S.J., Bilezikian, J.P., 2008. Dynamic and structural properties of the skeleton in hypoparathyroidism. J. Bone Miner. Res. 23, 2018–2024.

Saini, V., Marengi, D.J., Barry, K.J., Fulzele, K.S., Heiden, E., Liu, X., Dedic, C., Maeda, A., Lotinun, S., Baron, R., Pajevic, P.D., 2013. Parathyroid hormone (PTH)/PTH-related peptide type 1 receptor (PPR) signaling in osteocytes regulates anabolic and catabolic skeletal responses to PTH. J. Biol. Chem. 288, 20122–20134.

Salo, J., Lehenkari, P., Mulari, M., Metsikkö, K., Väänänen, H.K., 1997. Removal of osteoclast bone resorption products by transcytosis. Science 276, 270–273.

Sartawi, Z., Schipani, E., Ryan, K.B., Waeber, C., 2017. Sphingosine 1-phosphate (S1P) signalling: role in bone biology and potential therapeutic target for bone repair. Pharmacol. Res. 125, 232–245.

Sato, M., Westmore, M., Ma, Y.L., Schmidt, A., Zeng, Q.Q., Glass, E.V., Vahle, J., Brommage, R., Jerome, C.P., Turner, C.H., 2004. Teriparatide [PTH(1–34)] strengthens the proximal femur of ovariectomized nonhuman primates despite increasing porosity. J. Bone Miner. Res. 19, 623–629.

Satterwhite, J., Heathman, M., Miller, P.D., Marin, F., Glass, E.V., Dobnig, H., 2010. Pharmacokinetics of teriparatide (rhPTH[1–34]) and calcium pharmacodynamics in postmenopausal women with osteoporosis. Calcif. Tissue Int. 87, 485–492.

Seeman, E., Martin, T.J., 2015. Co-administration of antiresorptive and anabolic agents: a missed opportunity. J. Bone Miner. Res. 30, 753–764.

Siddiqui, J.A., Johnson, J., LE Henaff, C., Bitel, C.L., Tamasi, J.A., Partridge, N.C., 2017. Catabolic effects of human PTH (1−34) on bone: requirement of monocyte chemoattractant protein-1 in murine model of hyperparathyroidism. Sci. Rep. 7, 15300.

Sikjaer, T., Rejnmark, L., Thomsen, J.S., Tietze, A., Bruel, A., Andersen, G., Mosekilde, L., 2012. Changes in 3-dimensional bone structure indices in hypoparathyroid patients treated with PTH(1−84): a randomized controlled study. J. Bone Miner. Res. 27, 781−788.

Sims, N.A., 2016. Cell-specific paracrine actions of IL-6 family cytokines from bone, marrow and muscle that control bone formation and resorption. Int. J. Biochem. Cell Biol. 79, 14−23.

Sims, N.A., Jenkins, B.J., Nakamura, A., Quinn, J.M., Li, R., Gillespie, M.T., Ernst, M., Robb, L., Martin, T.J., 2005. Interleukin-11 receptor signaling is required for normal bone remodeling. J. Bone Miner. Res. 20, 1093−1102.

Sims, N.A., Martin, T., 2014. Coupling the activities of bone formation and resorption: a multitude of signals within the basic multicellular unit. BoneKEy Rep. 3.

Standal, T., Johnson, R.W., Mcgregor, N.E., Poulton, I.J., Ho, P.W., Martin, T.J., Sims, N.A., 2014. gp130 in late osteoblasts and osteocytes is required for PTH-induced osteoblast differentiation. J. Endocrinol. 223, 181−190.

Stein, E.M., Silva, B.C., Boutroy, S., Zhou, B., Wang, J., Udesky, J., Zhang, C., Mcmahon, D.J., Romano, M., Dworakowski, E., Costa, A.G., Cusano, N., Irani, D., Cremers, S., Shane, E., Guo, X.E., Bilezikian, J.P., 2013. Primary hyperparathyroidism is associated with abnormal cortical and trabecular microstructure and reduced bone stiffness in postmenopausal women. J. Bone Miner. Res. 28, 1029−1040.

Stickens, D., Behonick, D.J., Ortega, N., Heyer, B., Hartenstein, B., Yu, Y., Fosang, A.J., Schorpp-Kistner, M., Angel, P., Werb, Z., 2004. Altered endochondral bone development in matrix metalloproteinase 13-deficient mice. Development 131, 5883−5895.

Stilgren, L.S., Rettmer, E., Eriksen, E.F., Hegedüs, L., Beck-Nielsen, H., Abrahamsen, B., 2004. Skeletal changes in osteoprotegerin and receptor activator of nuclear factor-κb ligand mRNA levels in primary hyperparathyroidism: effect of parathyroidectomy and association with bone metabolism. Bone 35, 256−265.

Tang, Y., Wu, X., Lei, W., Pang, L., Wan, C., Shi, Z., Zhao, L., Nagy, T.R., Peng, X., Hu, J., Feng, X., Van Hul, W., Wan, M., Cao, X., 2009. TGF-β1-induced migration of bone mesenchymal stem cells couples bone resorption with formation. Nat. Med. 15, 757−765.

Tawfeek, H., Bedi, B., Li, J.Y., Adams, J., Kobayashi, T., Weitzmann, M.N., Kronenberg, H.M., Pacifici, R., 2010. Disruption of PTH receptor 1 in T cells protects against PTH-induced bone loss. PLoS One 5, e12290.

Tazawa, K., Hoshi, K., Kawamoto, S., Tanaka, M., Ejiri, S., Ozawa, H., 2004. Osteocytic osteolysis observed in rats to which parathyroid hormone was continuously administered. J. Bone Miner. Metab. 22, 524−529.

Terauchi, M., Li, J.Y., Bedi, B., Baek, K.H., Tawfeek, H., Galley, S., Gilbert, L., Nanes, M.S., Zayzafoon, M., Guldberg, R., Lamar, D.L., Singer, M.A., Lane, T.F., Kronenberg, H.M., Weitzmann, M.N., Pacifici, R., 2009. T lymphocytes amplify the anabolic activity of parathyroid hormone through Wnt10b signaling. Cell Metabol. 10, 229−240.

Tonna, S., Takyar, F.M., Vrahnas, C., Crimeen-Irwin, B., Ho, P.W.M., Poulton, I.J., Brennan, H.J., Mcgregor, N.E., Allan, E.H., Nguyen, H., Forwood, M.R., Tatarczuch, L., Mackie, E.J., Martin, T.J., Sims, N.A., 2014. EphrinB2 signaling in osteoblasts promotes bone mineralization by preventing apoptosis. FASEB J. 28, 4482−4496.

Van Lierop, A.H., Witteveen, J.E., Hamdy, N.A., Papapoulos, S.E., 2010. Patients with primary hyperparathyroidism have lower circulating sclerostin levels than euparathyroid controls. Eur. J. Endocrinol. 163, 833−837.

Vu, T.D., Wang, X.F., Wang, Q., Cusano, N.E., Irani, D., Silva, B.C., Ghasem-Zadeh, A., Udesky, J., Romano, M.E., Zebaze, R., Jerums, G., Boutroy, S., Bilezikian, J.P., Seeman, E., 2013. New insights into the effects of primary hyperparathyroidism on the cortical and trabecular compartments of bone. Bone 55, 57−63.

Walker, E.C., Poulton, I.J., Mcgregor, N.E., Ho, P.W., Allan, E.H., Quach, J.M., Martin, T.J., Sims, N.A., 2012. Sustained RANKL response to parathyroid hormone in oncostatin M receptor-deficient osteoblasts converts anabolic treatment to a catabolic effect in vivo. J. Bone Miner. Res. 27, 902−912.

Wan, M., Yang, C., Li, J., Wu, X., Yuan, H., Ma, H., He, X., Nie, S., Chang, C., Cao, X., 2008. Parathyroid hormone signaling through low-density lipoprotein-related protein 6. Genes Dev. 22, 2968−2979.

Winchester, S.K., Selvamurugan, N., D'alonzo, R.C., Partridge, N.C., 2000. Developmental regulation of collagenase-3 mRNA in normal, differentiating osteoblasts through the activator protein-1 and the runt domain binding sites. J. Biol. Chem. 275, 23310−23318.

Xian, L., Wu, X., Pang, L., Lou, M., Rosen, C.J., Qiu, T., Crane, J., Frassica, F., Zhang, L., Rodriguez, J.P., Jia, X., Yakar, S., Xuan, S., Efstratiadis, A., Wan, M., Cao, X., 2012. Matrix IGF-1 maintains bone mass by activation of mTOR in mesenchymal stem cells. Nat. Med. 18, 1095−1101.

Xiong, J., Cawley, K., Piemontese, M., Fujiwara, Y., Zhao, H., Goellner, J.J., O'Brien, C.A., 2018 Jul 25. Soluble RANKL contributes to osteoclast formation in adult mice but not ovariectomy-induced bone loss. Nat Comm 9 (1), 2909.

Xiong, J., Onal, M., Jilka, R.L., Weinstein, R.S., Manolagas, S.C., O'brien, C.A., 2011. Matrix-embedded cells control osteoclast formation. Nat. Med. 17, 1235−1241.

Xiong, J., Piemontese, M., Onal, M., Campbell, J., Goellner, J.J., Dusevich, V., Bonewald, L., Manolagas, S.C., O'brien, C.A., 2015. Osteocytes, not osteoblasts or lining cells, are the main source of the RANKL required for osteoclast formation in remodeling bone. PLoS One 10, e0138189.

Xiong, J., Piemontese, M., Thostenson, J.D., Weinstein, R.S., Manolagas, S.C., O'brien, C.A., 2014. Osteocyte-derived RANKL is a critical mediator of the increased bone resorption caused by dietary calcium deficiency. Bone 66, 146−154.

Yasuda, H., Shima, N., Nakagawa, N., Yamaguchi, K., Kinoshaki, M., Mochizuki, S., Tomoyasu, A., Yano, K., Goto, M., Murakami, A., Tsuda, E., Morinaga, T., Higashio, K., Udagawa, N., Takahashi, N., Suda, T., 1998. Osteoclast differentiation factor is a ligand for osteoprotegerin/osteoclastogenesis-inhibitory factor and is identical to TRANCE/RANKL. Proc. Natl. Acad. Sci. Unit. States Am. 95, 3597−3602.

Zanotti, S., Canalis, E., 2017. Parathyroid hormone inhibits Notch signaling in osteoblasts and osteocytes. Bone 103, 159−167.

Zebaze, R., Takao-Kawabata, R., Peng, Y., Zadeh, A.G., Hirano, K., Yamane, H., Takakura, A., Isogai, Y., Ishizuya, T., Seeman, E., 2017. Increased cortical porosity is associated with daily, not weekly, administration of equivalent doses of teriparatide. Bone 99, 80–84.

Zhao, W., Byrne, M.H., Boyce, B.F., Krane, S.M., 1999. Bone resorption induced by parathyroid hormone is strikingly diminished in collagenase-resistant mutant mice. J. Clin. Investig. 103, 517–524.

Chapter 33

Calcitonin peptides

Dorit Naot*, David S. Musson and Jillian Cornish
Department of Medicine, University of Auckland, Auckland, New Zealand

Chapter outline

Introduction **789**
Calcitonin-family gene and peptide structure **789**
Extraskeletal actions of calcitonin-family peptides **790**
Calcitonin 790
Calcitonin gene-related peptide 791
Amylin 792
Receptors for calcitonin-family peptides **792**
Peptide access to the bone microenvironment **793**
Effects on osteoclasts **793**
Calcitonin 794
Calcitonin gene-related peptide 795
Amylin 795
Effects on osteoblasts **796**
Calcitonin 796
Calcitonin gene-related peptide 796
Amylin 797
Effects of local and systemic peptide administration into laboratory animals **798**
Skeletal effects 798
Effects on calcium metabolism 799
The skeletal effects of calcitonin, calcitonin gene-related peptide, and amylin: lessons from genetically modified mice **800**
Calcitonin and calcitonin gene-related peptide 800
Amylin 800
Calcitonin receptor 801
The role of calcitonin and calcitonin receptor in situations of calcium stress 801
Calcitonin, calcitonin gene-related peptide, and amylin—relevance to human bone physiology **802**
References **802**

Introduction

The peptide hormone calcitonin was discovered as the active component in extracts from the thyroid gland that acutely lowered the concentration of circulating calcium (Copp et al., 1962; Copp and Cheney, 1962). Over the years, several peptides with structural similarity to calcitonin have been discovered and became recognized as the "calcitonin family," which includes the calcitonin gene-related peptides (αCGRP and βCGRP), amylin, adrenomedullin, and adrenomedullin 2. The peptides of the calcitonin family signal through related receptors; calcitonin itself binds to the calcitonin receptor, a seven-transmembrane domain G-protein coupled receptor (GPCR), while all other family members require receptor heterodimers that include a GPCR and receptor-activity-modifying protein (RAMP). The calcitonin family peptides have been investigated extensively; in addition to studies of the biochemistry, pharmacology, and physiological activities of the peptides, a large body of literature describes the development of these peptides, their analogues and their antagonists for clinical use. This chapter provides a general introduction to the calcitonin family and summarizes the current knowledge of the skeletal activities of calcitonin, CGRP, and amylin.

Calcitonin-family gene and peptide structure

Calcitonin and αCGRP are generated by alternative splicing of mRNA transcribed from the *CALCA* gene, which in humans is located on the short arm of chromosome 11. The *CALCA* gene contains six exons: exons I–IV are included in the mRNA for the precursor of calcitonin, preprocalcitonin, whereas preproCGRP is encoded by exons I–III and V–IV (Fig. 33.1A). The alternative splicing of *CALCA* mRNA is tissue specific—calcitonin mRNA is produced mainly in parafollicular

* Contributor retains copyright for images.

Principles of Bone Biology. https://doi.org/10.1016/B978-0-12-814841-9.00033-6

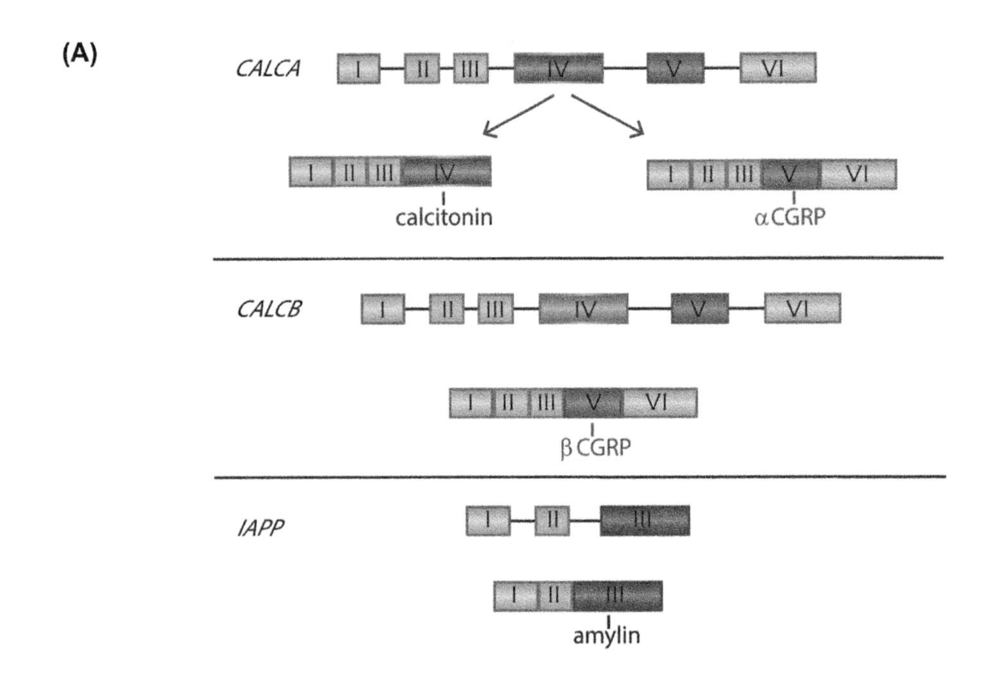

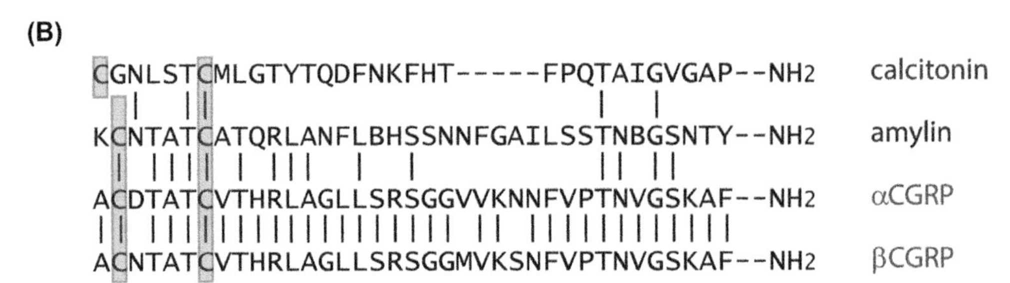

FIGURE 33.1 Genes, mRNA, and peptides of four members of the human calcitonin family: calcitonin, αCGRP, βCGRP, and amylin. (A) The *CALCA* primary transcript is processed through alternative splicing. Calcitonin mRNA contains exons I, II, III, and IV, whereas αCGRP mRNA contains exons I, II, III, V, and VI. Each mRNA is translated into a precursor protein that undergoes proteolytic cleavage to produce the mature peptides. Exon IV encodes the mature calcitonin peptide and exon V encodes the mature αCGRP. βCGRP is the only mature peptide encoded by *CALCB*. The *IAPP* gene contains only three exons, with exon III encoding the mature amylin peptide. (B) Amino acid sequences of the human peptides. All four peptides are amidated at the C terminal. The cysteine residues connected by a disulfide bond in each peptide are shaded, and identical amino acids are indicated by vertical lines.

cells of the thyroid, whereas in the nervous system, αCGRP mRNA is the predominant form (Breimer et al., 1988). In nonmammalian species—birds, fish and reptiles—calcitonin is derived from the ultimobranchial body. Salmon calcitonin (sCT), which shares only 50% sequence identity with the human peptide, has a much higher biological potency in humans than human calcitonin itself (Niall et al., 1969). A second CGRP peptide, βCGRP, which in humans differs from αCGRP by three amino acids, is encoded by *CALCB*, a separate gene located on the short arm of chromosome 11. Human amylin (also called island amyloid polypeptide) is encoded by *IAPP*, a gene that includes three exons and is located on the short arm of chromosome 12. Like calcitonin and CGRP, amylin is synthesized as a prepropeptide, which is processed by proteolytic cleavage to the mature amylin peptide. Amylin is produced mainly in β islet cells of the pancreas but has also been detected in the gastrointestinal tract as well as lung and neuronal tissues. The amino acid sequences of the mature calcitonin, αCGRP, βCGRP, and amylin in humans are presented in Fig. 33.1B. Mature calcitonin is a 32 amino-acid peptide, whereas CGRPs and amylin contain 37 amino acids. All the calcitonin family peptides have an amino acid ring structure at the N terminal created by a disulfide bond between cysteine residues at positions 2 and 7 in CGRP and amylin, and positions 1 and 7 in calcitonin, and all are amidated at the C terminal. These structural elements are shared by the additional members of the calcitonin family, adrenomedullin and adrenomedullin 2, which are not discussed further in this chapter.

Extraskeletal actions of calcitonin-family peptides

Calcitonin

Calcitonin was initially identified as a hormone that induces hypocalcemia through a rapid and potent inhibition of bone resorption by osteoclasts. The great majority of studies of calcitonin have focused on its skeletal activity and its effect on calcium homeostasis. The scientific literature includes a large number of clinical studies of the skeletal effects of sCT, as sCT has been in wide clinical use for a number of years, mainly for patients with Paget's bone disease and osteoporosis (Chesnut 3rd et al., 2008; Henriksen et al., 2016). Although recent studies (discussed later in the chapter) indicate that calcitonin regulates bone formation during development, the physiological role of calcitonin has not yet been clearly established, and it has been suggested that calcitonin might not have a major role in mammals under normal physiological conditions (Davey and Findlay, 2013). Calcium metabolism and bone mineral density (BMD) are unaffected in patients with medullary thyroid carcinoma, who have long-term excess of endogenous calcitonin levels, or who have undergone thyroidectomy and have undetectable circulating calcitonin (Hurley et al., 1987; Wuster et al., 1992). The current understanding is that the main significance of calcitonin appears to be the conservation of body calcium stores in situations of calcium stress, including rapid growth, pregnancy, and lactation.

Outside the skeleton, calcitonin transiently enhances calcium excretion through the inhibition of tubular calcium resorption. In addition, calcitonin was found to increase the urinary excretion rate of sodium, potassium, phosphorus, chloride, and magnesium (Findlay and Sexton, 2004). Calcitonin is also involved in the regulation of vitamin D processing, enhancing the 1-hydroxylation of 25-hydroxyvitamin D in the proximal straight tubule of the kidney. Another target tissue of calcitonin is the central nervous system. Calcitonin receptors are highly expressed in the central nervous system, and calcitonin administration has been shown to affect the pituitary and the hypothalamus. In rats, intracerebral injection of calcitonin reduced food and water intake and changed the pattern of secretion of growth hormone. In humans, high concentrations of sCT reduced the circulating levels of hormones including LH, FSH, and testosterone. Calcitonin has an analgesic effect on chronic bone pain and pain associated with osteoporotic fracture (Silverman and Azria, 2002).

Calcitonin gene-related peptide

CGRP is mainly expressed in the central and peripheral nervous systems. It is stored in large, dense-core vesicles within sensory nerve terminals and is commonly colocalized with substance P. The circulating levels of CGRP are low, and it is suggested that the major effects of CGRP are exerted locally. CGRP is recognized as a potent microvascular vasodilator, and its physiological activity is largely dependent on its release from perivascular nerves (Russell et al., 2014). Interestingly, injections of CGRP antagonists to healthy individuals have no significant effects on blood pressure, suggesting that CGRP has no major role in the control of physiological systemic blood pressure. CGRP-containing nerve fibers are also present in the heart vasculature, where CGRP release produces both positive inotropic and positive chronotropic effects and is thought to play a cardioprotective role (Russell et al., 2014).

CGRP also affects energy metabolism. In skeletal muscle, CGRP inhibits glycogen synthesis and stimulates glycogenolysis, while in the liver it increases gluconeogenesis and glucose output (Lima et al., 2017). A number of animal studies have found that CGRP administration reduces food intake, with CGRP inducing anorexia through activation of the cAMP/protein kinase A pathway. Mice deficient of αCGRP were protected from obesity induced by a high-fat diet. In comparison with wild-type controls, the αCGRP knockout mice had higher metabolic rate, increased energy expenditure, and raised body temperature (Russell et al., 2014).

CGRP is involved in neuronal regulation of the immune system. It has been shown to mediate a host of immune regulatory responses, and in different experimental systems CGRP was found to have either proinflammatory or antiinflammatory effects. CGRP promotes neurogenic inflammation, and its release from nerve terminals causes local vasodilation, edema formation, increased blood flow, and recruitment of inflammatory cells. On the other hand, in murine dendritic cells, CGRP inhibited TLR-stimulated production of the inflammatory mediators tumor necrosis factor (TNFα) and C−C motif chemokine ligand 4 (CCL4), an effect that was dependent on the cAMP/protein kinase A signaling pathway (Harzenetter et al., 2007).

A large number of investigations of CGRP in recent years have focused on its involvement in the pathophysiology of migraines (Iyengar et al., 2017). Raised levels of CGRP are observed both peripherally and centrally in migraine patients. CGRP is released from trigeminal afferent nerve fibers during a migraine attack, causing vasodilatation and neurogenic inflammation. The maintenance of a sensitized, hyperresponsive neuronal state is considered a central underlying mechanism in migraine. CGRP facilitates nociceptive transmission and contributes to the development and maintenance of a sensitized state in peripheral sensory neurons as well as neurons within the central nervous system. Four monoclonal antibodies are currently in development for migraine prevention, three against CGRP itself and one against the CGRP receptor. Results from phase III trials suggest comparable efficacy, safety, and tolerability of all four antibodies (Paemeleire & MaassenVanDenBrink, 2018).

Amylin

Amylin was purified from pancreatic deposits of patients with type 2 diabetes and from human insulinoma and was found to be predominantly expressed in β cells of the islets of Langerhans in the pancreas (Cooper et al., 1987; Westermark et al., 1987). In these cells, amylin is stored in secretory granules together with insulin, and it is cosecreted with insulin in response to changes in circulating glucose levels; hyperglycemia stimulates amylin secretion, whereas hypoglycemia inhibits it. Human amylin monomers are soluble, but aggregate to form pancreatic amyloid in patients with type 2 diabetes as well as spontaneously in vitro, in a concentration-dependent manner (Konarkowska et al., 2006). The role of amylin aggregates in the development of β-cell lesions in type 2 diabetes is not entirely clear, although there is evidence that they contribute to the progressive failure and eventually to the death of islet β-cells (Lorenzo et al., 1994; Zhang et al., 2016).

The main physiological role of amylin is the regulation of energy homeostasis (Hay et al., 2015). Eating leads to a rapid increase in circulating levels of amylin, which produces satiation signals and thus controls ingested meal size. Administration of exogenous amylin into animals reduces eating within minutes, an effect that could be reproduced by amylin analogues and blocked by amylin antagonists. Mechanisms involved in the activity of amylin as a satiety signal include delayed gastric emptying, which leads to satiation and prevents further food intake, and central signaling of amylin in specific areas of the brain stem. In addition to the short-term effect on satiety, amylin is also one of a group of hormones secreted in proportion to body adiposity, and is considered an adiposity signal that has a long-term effect to reduce eating by enhancing satiety. In obese rats, basal plasma levels of amylin were higher in comparison with lean controls, and peripheral or central administration of amylin led to decreased body weight as well as fat gain. In addition to short- and long-term effects on food intake, central administration of amylin increases energy expenditure (Hay et al., 2015).

In recent years, studies have demonstrated the importance of interactions between amylin and leptin, an adipokine secreted predominantly from white fat tissue, in the control of energy homeostasis (Levin and Lutz, 2017; Lutz, 2012). Similar to amylin and insulin, leptin provides negative, catabolic feedback to the brain when body adiposity is increased. Circulating leptin levels are generally elevated in obesity, and obese individuals can become leptin-resistant. Amylin was shown to act as an endogenous leptin sensitizer, and cross talk between leptin and amylin was suggested by the observation that leptin-deficient mice have reduced expression of amylin in the hypothalamus, which could be normalized by exogenous leptin (Li et al., 2015). Moreover, infusion of the amylin antagonist AC187 acutely reduced the anorectic effect of leptin.

The synthetic analogue of human amylin, pramlintide, is currently in clinical use for diabetes as an adjunctive therapy to mealtime insulin (Qiao et al., 2017; Singh-Franco et al., 2007). Multiple clinical studies in patients with type 1 and type 2 diabetes have shown that pramlintide in combination with insulin improved overall glycemic control in comparison with insulin alone, possibly by providing a more physiologically balanced therapeutic approach. Pramlintide was found to reduce body weight in patients with insulin-treated diabetes, suggesting a potential use as an antiobesity agent (Hay et al., 2015; Qiao et al., 2017; Singh-Franco et al., 2007). Clinical studies of pramlintide for weight reduction in obese subjects have so far produced promising results (Boyle et al., 2017).

Receptors for calcitonin-family peptides

Calcitonin receptor (CTR) belongs to the type II seven-transmembrane GPCRs and was cloned initially from porcine (Lin et al., 1991) and subsequently from rat, human, and other species. In most species, calcitonin receptor RNA is alternatively spliced, resulting in the expression of several isoforms (Gorn et al., 1995; Sexton et al., 1993). The various isoforms differ from each other structurally, as well as in their tissue distribution, affinity for ligands and the downstream signaling pathways activated by these ligands. The second GPCR that binds calcitonin-family ligands is calcitonin receptor-like receptor (CRLR), which was identified as a protein that shares about 55% amino acid sequence homology with CTR (Chang et al., 1993; Njuki et al., 1993). In contrast to CTR, CRLR itself is not transported to the cell membrane, and in order to form a functional receptor it requires the coexpression of one of three of the receptor activity-modifying proteins (RAMP1—RAMP3) (McLatchie et al., 1998). These single-transmembrane domain proteins interact with the seven-transmembrane domain GPCRs, and regulate their activity through a number of mechanisms: (1) RAMPs modify the binding specificities of the GPCRs to their ligands. (2) Receptor trafficking—RAMPs act as chaperones, directing CRLR to the cell membrane. (3) Receptor desensitization—following ligand binding, the complexes of CTR-RAMP are internalized by the cell. The RAMPs appear to determine whether the complex will be directed to degradative pathways or recycled back to the cell surface (Bomberger et al., 2005; Klein et al., 2016). (4) Signaling—studies of amylin receptors (AMY1-3) found that downstream signaling pathways of intracellular calcium production depend on the specific interacting RAMP (Morfis et al., 2008). A recent study of the pharmacology of CRLR-RAMP receptor complex demonstrated that the receptors display both ligand- and RAMP-dependent signaling bias among the downstream Gα subunit activation (Weston et al., 2016).

TABLE 33.1 Receptors for the calcitonin-family peptides

Receptor name	Subunits		Ligands
	GPCR	RAMP	
CTR	CTR	-	Calcitonin
AMY_1	CTR	RAMP1	Amylin, CGRP
AMY_2	CTR	RAMP2	Amylin
AMY_3	CTR	RAMP3	Amylin
CGRP	CRLR	RAMP1	CGRP
AM_1	CRLR	RAMP2	Adrenomedullin
AM_2	CRLR	RAMP3	Adrenomedullin, Adrenomedullin2

GPCR; G-protein coupled receptor, RAMP; receptor activity-modifying protein,

It has been well-established that specific combinations of CTR and CRLR with RAMPs produce the receptors for all peptides of the calcitonin family. Thus, while calcitonin is the ligand of CTR, heterodimerization of the CTR with either of the three RAMPs produces receptors with high affinity for amylin, the combination of CRLR with RAMP1 creates a CGRP receptor, and CRLR with either RAMP2 or RAMP3 acts as an adrenomedullin receptor (Lerner, 2006; Poyner et al., 2002) (Table 33.1). Pharmacological studies have shown that each of the receptor combinations typically binds one or two of the calcitonin-family peptides with high affinity, while other members of the family can bind to the same receptor dimer with lower affinities (Bower and Hay, 2016). The cross-reactivity of members of the calcitonin family with the various receptor combinations presents a challenge when interpreting experimental results, as deficiency in one specific component can be masked by interactions of the remaining members of the peptide and receptor families.

Peptide access to the bone microenvironment

The access of calcitonin and amylin to the bone environment is mainly through the circulation. Amylin is secreted from the islets of Langerhans of the pancreas and can be found in the circulation at 5 pmol/L in fasting state, rising to 15–25 pmol/L following a meal (Hay et al., 2015). Amylin secretion is pulsatile, with peaks occurring at about 5-min intervals (Juhl et al., 2000). Increased levels of circulating amylin are found in individuals with obesity or hypertension. In early stages of type 2 diabetes, along with the development of insulin resistance, circulating levels of insulin and amylin are high, whereas at later stages the secretion of both amylin and insulin becomes deficient (Zhang et al., 2016). Calcitonin, secreted predominantly from the C cells of the thyroid, circulates at low concentrations of about 3 pmol/L (Findlay and Sexton, 2004). The release of calcitonin is stimulated by a fall in blood calcium and inhibited by elevated calcium levels. Circulating levels of CT are increased in some pathological states, for example, in patients with medullary thyroid carcinoma.

Different studies determined the concentration of CGRP in circulation within the range of 1–40 pmol/L (Born et al., 1991; Schifter, 1991). Circulating concentrations are increased by sex hormone replacement therapy in postmenopausal women (Spinetti et al., 1997). However, local nerve terminals are the most important source of CGRP in the bone environment, and in certain situations local concentrations are likely to be much higher than those found in the circulation. Sensory nerve fibers containing CGRP are widely distributed in bone and bone marrow, with the richest innervation found at the epiphysis and periosteum (Bjurholm, 1991; Hill and Elde, 1991; Irie et al., 2002). Immuno-staining of CGRP-containing nerve fibers in bone tissue demonstrated changes in distribution during bone development and regeneration (Irie et al., 2002). When bone defects are created surgically, the development of CGRP-containing nerves is noted several days later, often in association with new blood vessels (Aoki et al., 1994), suggesting a role in callus formation and bone healing. Similar responses are seen following fractures (Hukkanen et al., 1993).

Effects on osteoclasts

The osteoclast has been recognized as a main target cell for calcitonin, central to its activity in lowering circulating calcium levels. Following the discovery of other members of the calcitonin family, their activities in osteoclasts have also been studied in detail. Mature osteoclasts express CTR, CRLR, and RAMP1-3 and are therefore a potential target for direct activity by all peptides of the calcitonin family.

Calcitonin

Bone-resorbing osteoclasts become polarized and form specialized structures that attach to the bone surface: the ruffled border, where protons and proteases are secreted to demineralize and degrade the bone matrix; and the surrounding sealing zone, produced by a dense ring of actin-rich podosomes. These structure are highly dynamic and allow continuous resorption while the osteoclast is moving along the bone surface. Early studies have shown that mature osteoclasts express high-affinity calcitonin receptors, and that calcitonin binding causes the immediate arrest of bone resorption (Chambers et al., 1984; Chambers and Magnus, 1982). Within 1 min of binding, calcitonin causes arrest of cell motility, which is then followed by the disruption of the podosomes and ruffled border and detachment of the osteoclast from the bone surface. Interestingly, the effects of calcitonin on osteoclast motility and attachment are mediated via two different signaling pathways: arrest of motility is cAMP-dependent, whereas retraction and disruption of the ruffled border and sealing zone are mediated through intracellular calcium signaling (Zaidi et al., 2002) (Fig. 33.2). Detachment of the sealing zone is mediated through the phosphorylation and changes in the intracellular distribution of Src and by the tyrosine kinase Pyk2, which is highly expressed in osteoclasts and localized mainly in the sealing zone (Shyu et al., 2007; Zhang et al., 2002). In addition to the key effects on mature osteoclast motility and attachment, the antiresorptive activity of calcitonin appears to be mediated by inhibition of early stages of osteoclast differentiation. In murine bone marrow cultures, calcitonin decreased the number of tartrate-resistant acid phosphatase (TRAP)-positive multinucleated cells and reduced the ratio of TRAP-positive multinucleated to mono- and binucleated cells, indicating an inhibitory effect on the fusion of osteoclast precursors (Cornish et al., 2001). Similar results were found in cultures of mouse spleen cells and in bone marrow macrophages induced to differentiate into osteoclasts, where sCT inhibited the formation of TRAP-positive multinucleated cells and the number of resorption pits formed on bone slices (Granholm et al., 2007). The inhibitory effect was reproduced by activation of protein kinase A and cAMP. The presence of a large number of TRAP-positive mono nucleated cells suggested that expression of early osteoclastogenesis markers was not disrupted, but further differentiation and cell fusion were inhibited.

The effect of calcitonin on osteoclast activity is transient, and within 48 h of continuous or repetitive calcitonin treatment the inhibitory effect is lost (Wener et al., 1972). This desensitization of osteoclasts was named the "escape phenomenon," and further investigations have shown that the cells become unresponsive due to a ligand-induced internalization of the calcitonin receptor as well as inhibition of receptor synthesis (Takahashi et al., 1995; Wada et al., 1996). More recent studies examined whether the escape phenomenon is unique to calcitonin or shared by other members of the calcitonin family. Initially, amylin was shown to produce a transient inhibitory effect on bone resorption in neonatal mouse calvaria, and subsequently all members of the calcitonin family were found to downregulate the expression of CTR mRNA and therefore to induce desensitization of osteoclasts to further treatment (Granholm et al., 2011).

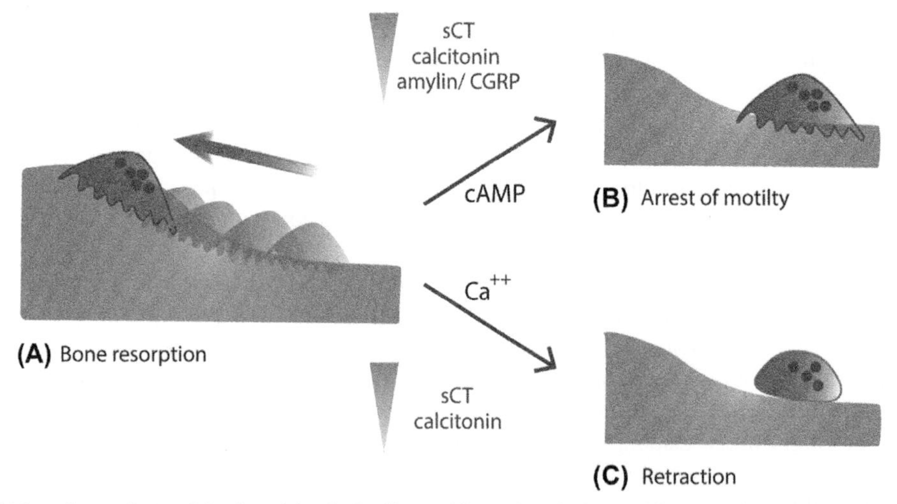

FIGURE 33.2 Inhibition of osteoclast activity by calcitonin-family peptides. (A) Actively resorbing osteoclasts attach to the bone surface through a ruffled border and a surrounding sealing zone, two highly dynamic structures that allow continuous resorption while the osteoclast is moving along the bone surface. (B) Calcitonin, CGRP. and amylin arrest the movement of osteoclast along the bone surface through a cAMP-dependent mechanism. (C) Calcitonin induces the retraction of osteoclasts from the bone surface, an effect that is mediated through intracellular calcium signaling and is not shared by the other members of the calcitonin family. This additional mechanism used by calcitonin may explain its greater potency in inhibiting osteoclast activity in comparison with the other peptides. The peptides are listed in order of potency, which is also indicated by the triangle color gradient. sCT, salmon calcitonin; CGRP, calcitonin gene-related peptide.

Interestingly, CGRP and adrenomedullin, which signal through CRLR-RAMP dimers and not through CTR, were also shown to downregulate the expression of CTR. In contrast to CTR, CRLR is not internalized following the binding of its ligands to the receptor dimers, and its levels remain unchanged.

Calcitonin gene-related peptide

Shortly after the discovery of CGRP, its injection into rats and rabbits was found to produce a calcitonin-like effect, lowering circulating calcium concentrations (Tippins et al., 1984). In rabbits, CGRP was approximately equipotent with calcitonin, whereas concentrations of two to three orders of magnitude higher than those of calcitonin were required to produce hypocalcemia in rats. A number of groups used calvariae organ cultures to study the effect of CGRP in greater detail. In this experimental system, the release of ^{45}Ca from prelabeled neonatal mouse or rat calvariae is used as a quantitative measure of bone resorption. Yamamoto et al. (Yamamoto et al., 1986) have shown that human CGRP produces a comparable level of inhibition of both basal and parathyroid hormone-stimulated resorption, but the half-maximal concentration of CGRP was 500-fold higher than that of human calcitonin. Similar studies in bone organ cultures have confirmed the antiresorptive activity of CGRP, which blocked the stimulation of bone resorption produced by different osteolytic factors (D'Souza et al., 1986; Roos et al., 1986). In cell cultures of neonatal rat osteoclasts, both αCGRP and βCGRP were found to directly inhibit bone resorption, although their potency was three orders of magnitude lower than that of human calcitonin (Zaidi et al., 1987a,b).

Further characterization of CGRP activity in osteoclasts found that it inhibits cell motility but not cell retraction. The inhibition of cell motility is mediated by increase in cAMP levels and could be blocked by the inhibitory analog CGRP(8-37), whereas CGRP had no effect on intracellular calcium levels (Alam et al., 1993a,b; Alam et al., 1991). cAMP production was also demonstrated in osteoclast-like multinucleated cells formed in cocultures of mouse osteoblasts and bone marrow cells in the presence of calcitriol (Tamura et al., 1992). It appears that the consistent lower potency of CGRP in inhibiting bone resorption, compared with calcitonin, results from its selective activation of cAMP and the downstream inhibition of motility and its lack of effect on intracellular calcium, which is linked to the downstream inhibitory effect on cell retraction (Fig. 33.2).

In addition to its activity on mature osteoclasts, evidence suggests that CGRP also affects osteoclast precursors. Specific binding of CGRP to mouse bone marrow cells and macrophages has been demonstrated, and CGRP was found to inhibit the development of osteoclasts in macrophage-osteoblast cocultures (Mullins et al., 1993; Owan and Ibaraki, 1994). In cultures of mouse bone marrow, CGRP inhibited the formation of TRAP-positive mononuclear cells as well as the subsequent fusion of these cells to form multinucleated osteoclasts (Cornish et al., 2001). Studies of the effect of CGRP on the differentiation and activity of osteoclasts in mouse bone marrow cultures found that a 100-fold higher concentration of CGRP is required for an inhibitory effect on differentiation than is required to inhibit pit formation, suggesting that the main effect of CGRP is in inhibiting osteoclast activity rather than formation (Wang et al., 2010).

In recent years, new aspects of CGRP as an inhibitor of bone resorption have come into focus. With the recognition that CGRP is released locally from sensory nerves that innervate the skeleton, a putative protective role for CGRP in the process of particle-induced osteolysis has been suggested. One of the common complications of joint arthroplasty is the formation of wear particles that induce inflammation and the release of cytokines that activate local bone resorption and can lead to aseptic loosening and eventually to early implant failure. The observation that CGRP nerve fibers are present adjacent to sites of periprosthetic osteolysis led to the hypothesis that CGRP acts locally to inhibit osteolysis. Ultra-high-molecular-weight polyethylene (UHMWPE) particles are common prosthetic wear debris and were studied in vitro in primary human osteoblasts and in the human osteosarcoma cell line MG-63 (Kauther et al., 2011; Xu et al., 2010). UHMWPE particles induced RANKL expression and inhibited the expression of OPG, while CGRP reduced UHMWPE-induced RANKL expression, suggesting an indirect inhibition of particle-induced bone resorption. CGRP was also shown to inhibit the secretion of osteolysis-associated proinflammatory cytokines and therefore could have additional beneficial effects on prevention of prosthesis loosening (Jablonski et al., 2015).

Amylin

Injection of amylin into laboratory animals strongly induced hypocalcemia with a potency that was either equal to or lower than that of human calcitonin (Datta et al., 1989; Wimalawansa, 1997; Zaidi et al., 1990). Amylin inhibits bone resorption by mature osteoclasts as determined by the reduced number of resorptive pits per TRAP-positive multinucleated cell in bone marrow cells cultured on bone slices (Cornish et al., 2001). Like CGRP, amylin inhibits the motility of mature osteoclasts in a cAMP-dependent mechanism but has no effect on cell retraction (Fig. 33.2). In organ cultures of neonatal mouse calvariae, where bone resorption was stimulated by 1,25(OH)2D3, amylin increased cAMP levels and reduced both basal and PTH-stimulated resorption (Cornish et al., 1998b; Pietschmann et al., 1993). Fragments of amylin tested in this experimental system had no effect, and an intact amylin molecule was required in order to inhibit bone resorption. This is in contrast to the osteoblast activity of amylin tested in vitro, where a peptide fragment of the eight amino acids of the N-terminal (amylin(1-8)) acts as an agonist, and the fragment amylin(8-37) acts as an antagonist (Cornish et al., 1998a,b). Amylin was also shown to inhibit bone resorption in organ cultures of fetal mouse long bones, where its potency was similar to that of CGRP but 60-fold less than that of human calcitonin (Tamura et al., 1992). In addition to inhibiting the activity of osteoclasts, amylin has been shown to inhibit their differentiation. In mouse bone marrow cultures that were stimulated to generate osteoclasts by 1,25(OH)2D3, amylin inhibited the formation of TRAP-positive mono- and bi-nuclear cells as well as the fusion of these cells and the formation of multinucleated osteoclast-like cells (Cornish et al., 2001). Amylin inhibition of osteoclast differentiation was mediated via ERK1/2 signaling, and the expression of a negative dominant form of ERK1/2 blocked amylin's inhibitory effect (Dacquin et al., 2004). Amylin had no effect on bone marrow cell proliferation, and in the presence of amylin osteoclasts were smaller and contained fewer nuclei. Results from in vitro studies of amylin should be interpreted in light of its marked propensity to adhere to the surfaces of laboratory plasticware (Young et al., 1992), suggesting that the actual concentrations of amylin in all in vitro experiments may be one to two orders of magnitude less than the amount added to the media. Thus, both osteoclast and calvariae studies imply that amylin may regulate bone resorption at physiological concentrations.

Effects on osteoblasts

Calcitonin

A large number of studies have shown that calcitonin receptor is not expressed in osteoblasts, strongly indicating that calcitonin cannot affect these cells directly (Naot and Cornish, 2008). Consistent with this observation, calcitonin had no effect on the proliferation of osteoblasts in vitro, and when injected locally over the calvaria of adult mice, indices of bone formation were not different from those of vehicle-injected control mice (Cornish et al., 1995, 2000). However, a small number of studies found evidence for calcitonin activity in osteoblasts. These studies have shown that calcitonin stimulates the proliferation of primary osteoblasts and osteoblastic cell lines in vitro and increases bone matrix synthesis ex vivo in an organ culture system (Farley et al., 1988; Villa et al., 2003). An earlier study, using subcutaneously implanted demineralized bone matrix in rats, found that administration of calcitonin soon after the implantation stimulated osteoblast proliferation and bone formation, whereas administration of calcitonin after the initiation of bone formation had the reverse, inhibitory effect (Weiss et al., 1981). One possible explanation for the observed activity of calcitonin in osteoblasts in these experiments is that the effects were in fact indirect, mediated via osteoclasts, in a mechanism that has been discovered more recently and will be discussed in detail below (Keller et al., 2014). In addition, given the cross-reactivity of members of the calcitonin family and their receptors, a low-affinity interaction of calcitonin with receptors other than CTR might have contributed to the observed activity.

Calcitonin gene-related peptide

The activity of CGRP in osteoblasts is likely to be mediated by the high-affinity receptor complex CRLR-RAMP1, as genes encoding for these two receptor components are highly expressed in osteoblasts (Naot and Cornish, 2008). Early studies demonstrated binding of CGRP to cells of rat calvaria, and increased cAMP levels in the UMR-106 rat osteosarcoma cell line following CGRP treatment, while the cells were not responsive to calcitonin (Michelangeli et al., 1986). Subsequent studies in bone cell cultures obtained by sequential digestion of neonatal chicken, rat, and mouse calvariae again demonstrated a cAMP response to CGRP with no response to calcitonin (Michelangeli et al., 1989). Several other intracellular signaling pathways are activated by CGRP in osteoblast-like cells. In UMR-106 cells, CGRP stimulated Na+/H+ exchange and induced a transient twofold increase in intracellular calcium concentrations by mobilization of calcium from intracellular stores (Gupta and Schwiening, 1994; Kawase et al., 1995). Detailed analysis of the changes in intracellular calcium in the human osteosarcoma cell line MG-63 found a two-phase response to CGRP: a rapid transient calcium discharge from intracellular stores in a cAMP-independent mechanism followed by a secondary sustained calcium influx into the cell in a cAMP-dependent manner (Burns et al., 2004). In the osteosarcoma cell line OHS-4, CGRP increased intracellular calcium concentrations but had no detectable effect on cAMP, suggesting that in these cells the initial cAMP-independent stage predominates (Drissi et al., 1999). CGRP was also shown to stimulate potassium efflux, inducing membrane hyperpolarization that results in rapid changes in cell morphology in UMR-106 cells (Kawase and Burns, 1998). Structure–function studies of CGRP in preosteoblast KS-4 cells found that cAMP response was greatly reduced with truncation of either the C-terminal amino acids or the disulfide bridge at the N-terminal (Thiebaud et al., 1991).

Studying the effect of CGRP on osteoblast proliferation, Cornish et al. (Cornish et al., 1995) showed a small increase in proliferation in primary rat osteoblast-like cells in response to CGRP, although the potency of CGRP was much lower than that of amylin. CGRP also stimulated proliferation in primary cultures of human osteoblast-like cells, and neither CGRP(8-37) nor amylin (8-37) inhibited this effect (Villa et al., 2000).

CGRP can activate the differentiation of precursor cells into osteoblasts. This has been demonstrated in bone marrow stromal cells derived from either healthy or ovariectomized (OVX) rats, where CGRP induced the formation of mineralizing osteoblasts in vitro (Liang et al., 2015). The effect of CGRP on bone marrow stromal cell differentiation was mediated via the Wnt/β-catenin pathway, and could be inhibited by either CGRP(8-37) or the Wnt pathway antagonist secreted frizzled-related protein (Zhou et al., 2016). The Wnt/β-catenin pathway was also found to mediate the activity of CGRP as a survival factor in human osteoblast-like primary cultures, where CGRP inhibited the apoptosis induced by serum depravation or dexamethasone (Mrak et al., 2010). Another pathway that may be involved in mediating the effect of CGRP in osteoblasts is the BMP-signaling pathway. BMP2 expression was induced by CGRP in MG-63 osteosarcoma cells and contributed to the activity of CGRP to promote osteogenic differentiation, while the proliferative effect of CGRP in these cells appeared to be BMP-independent (Tian et al., 2013).

The positive effect of CGRP on osteoblasts and its potential to enhance bone formation have been studied in the context of orthopedic implants and tissue engineering. In a recent study, Zhang et al. (Zhang et al., 2016) investigated the mechanisms underlying the effect of implants containing biodegradable magnesium to improve fracture healing. Using a fracture repair model in rats, they have shown that implant-derived magnesium increases neuronal CGRP in both the peripheral cortex of the femur and the ipsilateral dorsal root ganglia. CGRP released from sensory neurons was shown to induce local osteogenic differentiation of periosteum-derived stem cells and accelerate healing (Zhang et al., 2016). In other studies, CGRP was added to bone graft substitutes and scaffold materials, and its potential to promote local bone formation and healing was tested. Bio-Oss is a scaffold material prepared by removal of the organic component form bovine bone, and is used as bone graft that allows osteogenic cell migration (Li et al., 2017). CGRP loaded onto Bio-Oss surface was shown to enhance proliferation and differentiation of primary osteoblasts in vitro. CGRP was also shown to promote proliferation and osteogenic differentiation of rabbit adipose-derived stem cells cultured within calcium alginate gel, and a potential use of these constructs in tissue engineering has been suggested (Huang et al., 2015).

While early studies investigated the potential general role of CGRP in regulation of bone remodeling and found that it promotes osteoblast differentiation and only weakly induces proliferation, in later years the focus has shifted, and the activity of CGRP in osteoblasts is mainly studied in the context of bone healing. Thus, the role of CGRP released from nerve endings at sites of fracture or bone lesions, as well as the potential of CGRP as a bone anabolic factor added to bone scaffold materials, are areas of research that are likely to generate important novel findings in the near future.

Amylin

Shortly after its discovery, amylin was shown to stimulate cAMP production in a preosteoblastic cell line, in primary fetal rat osteoblasts and primary human osteoblasts (Tamura et al., 1992; Cornish et al., 1995; Millet and Vignery, 1997; Villa et al., 1997). These studies have shown that amylin stimulated the proliferation of primary osteoblasts from rats and humans at concentrations as low as 10 pmol/L. In rat primary osteoblastic cells, the mitogenic effect of amylin was mediated via the phosphorylation of ERK1/2, most likely through activation of Gi proteins, and was blocked by the specific inhibitor of this pathway, PD-98059 (Cornish et al., 2004). It is not clear which receptor mediates the activity of amylin in osteoblasts, as the three receptors AMY_{1-3} include CTR, which is not expressed in these cells. It is possible that amylin affects osteoblasts through interaction with one of the other receptors for the calcitonin-family peptides. Surprisingly, the proliferative effect of amylin in osteoblasts required the presence of IGF1 receptor, despite the fact there was no direct binding of amylin to IGF1 receptor, nor was there a paracrine effect of osteoblast-derived IGF1. Stimulation of osteoblast proliferation in vitro was also produced by the N-terminal octapeptide fragment amylin(1-8), although its half-maximal effective concentration was 10-fold higher than that of the intact peptide (Cornish et al., 1998a,b).

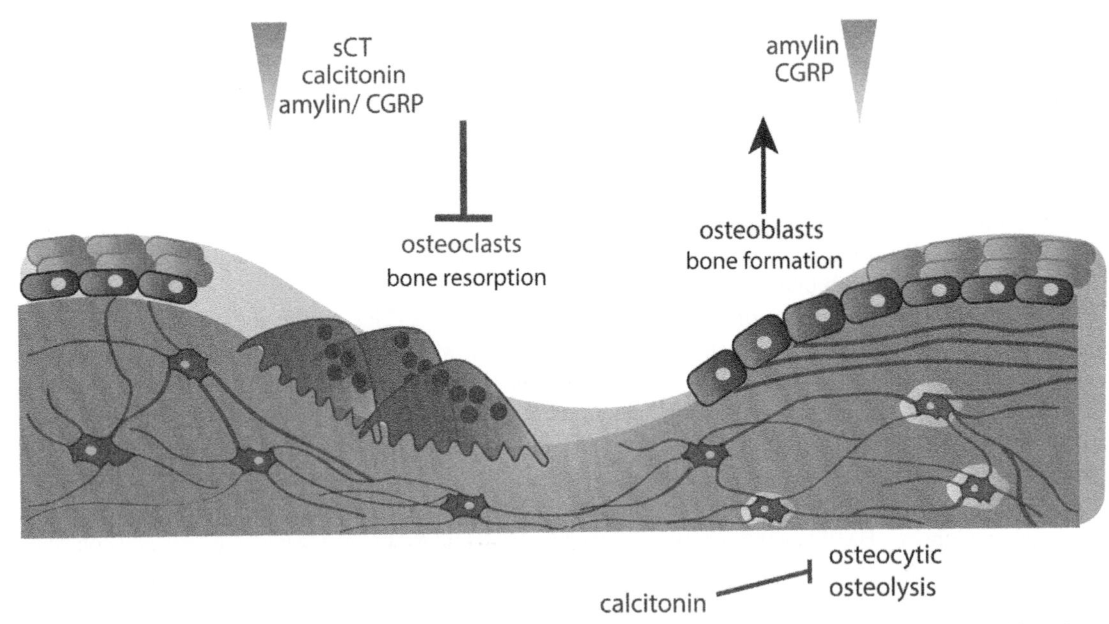

FIGURE 33.3 **The effects of calcitonin-family peptides on bone cells.** In vitro studies have established an overall positive effect of the calcitonin-family peptides in bone. Calcitonin, CGRP, and amylin are inhibitory to osteoclasts, and CGRP and amylin also promote osteoblast formation and activity. The osteocyte appears to be a target cell for calcitonin, as studies in genetically modified mice suggested a protective role for calcitonin in osteocytic osteolysis. The peptides are listed in order of potency, which is also indicated by the color gradient. sCT, salmon calcitonin; CGRP, calcitonin gene-related peptide.

The activities of calcitonin, CGRP and amylin in bone cells, as determined by in vitro studies, are summarized in Fig. 33.3.

Effects of local and systemic peptide administration into laboratory animals

Skeletal effects

OVX rats that were given daily subcutaneous injections of CGRP for 28 days had decreased bone resorption and a modest reduction in postovariectomy bone loss from 60% to 46% (Valentijn et al., 1997). In contrast, injection of CGRP over the calvariae of adult mice detected no significant effects on either bone resorption or formation (Cornish et al., 1995; Cornish and Reid, 1999). A different approach to determining the effect of CGRP on bone mass was used by Hill et al. (Hill et al., 1991). Reasoning that most of the CGRP that gains access to bone does so via sensory nerves, they studied the effect of sensory denervation using capsaicin treatment in rats. This intervention produced no change in cortical or medullary area, or periosteal apposition rate in tibiae, but the osteoclast surface in the mandible was decreased. In a more recent study, Offley et al. showed that capsaicin treatment destroyed the unmyelinated sensory axons containing CGRP, and the reduced CGRP signaling was associated with a decrease in bone mass (Offley et al., 2005).

Local injection of amylin over the calvariae of adult mice produced a substantial decrease in bone resorption, an increase in bone formation, and in mineralized bone area after only five daily injections (Cornish et al., 1995). Systemic administration of amylin to adult mice over 4 weeks produced a 70% increase in total bone volume in the proximal tibia (Fig. 33.4) as well as increased cortical width, tibial growth plate width, and tibial length along with increased body weight and fat mass (Cornish et al., 1998a,b). Structure–function studies identified amylin(1-8) as potentially retaining the beneficial effects in bone but devoid of amylin's activity on energy and carbohydrate metabolism. Thus, administration of amylin(1-8) by local injection over the calvariae of female mice produced a positive effect on bone formation, which was greater than that of an equimolar dose of hPTH-(1–34). Daily systemic administration of amylin (1-8) to sexually mature male mice for 4 weeks produced a near twofold increase in histomorphometric indices of osteoblast activity and resulted in increased bone strength as determined by three-point bending (Cornish et al., 2000). In contrast, amylin(1-8) administration to OVX rats had no effect on parameters of bone formation (Ellegaard et al., 2010). In rats, systemic administration of amylin for 18 days increased trabecular bone volume of the proximal tibia by 25%, although there were no changes in histomorphometric indices of formation or resorption (Romero et al., 1995). In a study of OVX rats, systemic administration of amylin increased bone formation and inhibited bone resorption as determined by analysis of the bone

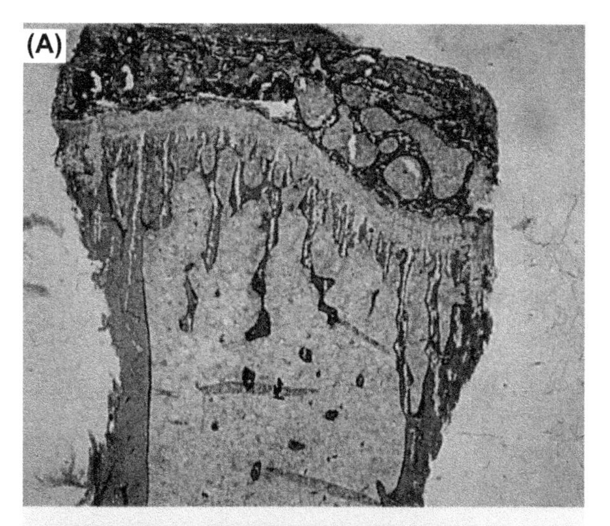

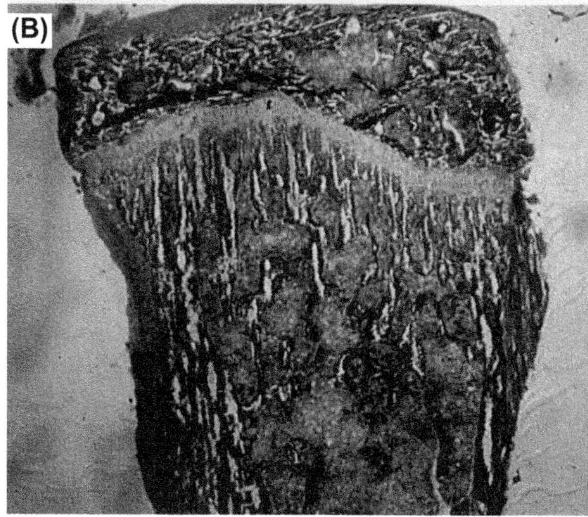

FIGURE 33.4 Increases bone volume induced by systemic administration of amylin. Photomicrographs of proximal tibiae of mice treated systemically with (A) vehicle or (B) amylin (10 μg/day) for 4 weeks. Trabecular bone volume is increased by 70% in amylin-treated animals. *Reprinted from Cornish et al., 1998a, with permission.*

turnover markers: plasma osteocalcin and urinary deoxypyridinoline. Distal metaphyseal and total femoral bone densities were higher in the OVX animals treated with amylin in comparison with untreated controls (Horcajada-Molteni et al., 2000). Interestingly, amylin infusion has also produced osteogenic effects in diabetic rat models, although its efficacy varied depending on the model (Gutierrez-Rojas et al., 2013).

In summary, the great majority of studies found that amylin administration increases bone formation and inhibits bone resorption, producing an overall positive skeletal effects. The investigations of amylin(1-8) produced less consistent results, with increased bone formation seen in a local and systemic administration models but no protective effect found against OVX-induced bone loss. In light of the recent findings regarding the mechanisms of action of calcitonin in bone, as discussed below (Keller et al., 2014), it is possible that the increase in bone formation measured in the animal models results from indirect activity of amylin through cells other than osteoblasts.

Effects on calcium metabolism

The most apparent effect of the injection of amylin or CGRP to laboratory animal is the induction of hypocalcemia. In fasted rats that were given a single intravenous bolus injections of amylin or CGRP, the decrease in total plasma calcium concentration was similar for both peptides (Young et al., 1993). Although the hypocalcemic effect is largely attributable to inhibition of osteoclastic bone resorption, amylin, like calcitonin, may have a direct calciuretic effect on the kidney. Rat amylin bound to receptors in porcine kidney with an affinity comparable to that of porcine calcitonin itself, and induced cAMP production (Sexton et al., 1994). In rat renal tubular membranes, CGRP and amylin stimulated cAMP production with comparable half-maximal concentrations, and the effects of both peptides were blocked by CGRP(8-37) (Osajima et al., 1995). Consistent with these findings, amylin infusion doubled urinary calcium excretion in dogs (Miles et al., 1994). However, the increased urinary loss of calcium accounted for less than 10% of the fall in serum calcium, suggesting that reduced osteoclastic resorption was the principal contributor to the hypocalcemic effect. These changes were accompanied by a doubling of PTH levels, whereas circulating calcitonin concentrations remained unchanged.

The skeletal effects of calcitonin, calcitonin gene-related peptide, and amylin: lessons from genetically modified mice

Calcitonin and calcitonin gene-related peptide

In one of the early studies that examined the skeletal effects of the calcitonin-family peptides in genetically modified mice, CGRP was introduced under the osteocalcin promoter, directing its overexpression specifically to the osteoblast lineage (Ballica et al., 1999). The transgenic mice had increased bone formation rate, as well as higher trabecular bone density and volume in comparison with wild-type (WT) littermate controls, suggesting a positive effect of CGRP in osteoblasts. Later studies used mostly knockout mice to investigate the skeletal effects of the calcitonin family peptides. Mice deficient of the *Calca* gene, missing both calcitonin and αCGRP, were the first knockout model created (Hoff et al., 2002). Based on previous in vitro and in vivo studies of calcitonin activity in bone, the *Calca*-KO mice were expected to display an osteoporotic phenotype due to accelerated bone resorption in the absence of calcitonin. However, the mice had no changes in osteoclast number or bone resorption markers in comparison with WT controls, and surprisingly, had higher bone volume and trabecular number with decreased trabecular spacing. Double calcein-labeling showed the high bone volume was a result of a significant increase in bone formation (Hoff et al., 2002). *Calca*-KO mice were less sensitive to ovariectomy; while WT mice lost about one-third of their bone mass over 2 months after ovariectomy, the *Calca*-KO mice had no change in bone mass. Overall, the unexpected skeletal phenotype of the *Calca*-KO mice suggested that calcitonin or αCGRP inhibited bone formation. A second knockout mouse model, which was only deficient of αCGRP, was generated in order to differentiate between the contributions of the deficiency in calcitonin and that of αCGRP to the increased bone formation seen in the *Calca*-KO mice (Schinke et al., 2004). αCGRP-KO mice had normal levels of circulating calcitonin and unlike the *Calca*-KO, had reduced bone formation rate and developed osteopenia. This phenotype was consistent with the earlier study that demonstrated increased bone formation in mice over-expressing αCGRP (Ballica et al., 1999). A subsequent long-term study of the bone phenotype of the knockout animals showed that the osteopenia displayed by the αCGRP-KO mice was also evident at 6, 12, and 18 months of age (Huebner et al., 2006). In the *Calca*-KO mice the bone phenotype developed with age, and at 12 and 18 months, along with the increased bone formation, there was evidence for increased bone resorption. The high bone turnover resulted in a substantial increase of trabecular structures that lead to trabecularization of cortical bone, and in addition, 20% of the *Calca*-deficient mice had hyperostotic lesions. A number of common potential secondary mechanisms that cause increased bone resorption have been investigated and excluded, as the *Calca*-KO mice had normal histologic appearance of several organs and the absence of hyperparathyroidism and hypogonadism (Huebner et al., 2006). The analyses of the two mice strains, *Calca*-KO and αCGRP-KO, are confounded by the presence of an intact *Calcb* gene that encodes for βCGRP. However, a study of the bone phenotype of *Calcb* -KO mice found no differences between these mice and control WT mice, suggesting that βCGRP does not play an important role in the regulation of bone remodeling (Huebner et al., 2008).

Taken together, the studies of genetically modified mice strains imply that in vivo, αCGRP is a physiological inducer of bone formation, whereas calcitonin inhibits osteoblast activity and bone formation, at least in young animals. Our current understanding of the physiological activity of calcitonin in bone is still incomplete, and a clear understanding of the unusual histological appearance of cortical trabecularization and hyperostotic lesions in the older *Calca*-KO mice is still missing. It is possible that some unanswered questions could be resolved by examining the bone phenotype of mice deficient of calcitonin only, with functional αCGRP. However, this animal model has not yet been described.

Amylin

The phenotype of amylin-deficient mice was described by Dacquin et al (Dacquin et al., 2004). The mice had normal food intake, body weight, and glucose metabolism. At the age of 24 weeks, amylin-deficient mice had a typical osteoporotic phenotype, showing decreased bone density of long bones, low bone mass, and a 50% decrease in trabecular bone volume.

The number of osteoblasts as well as the bone formation rate, assessed by double calcein injection, were similar between amylin-deficient mice and WT controls, suggesting that the osteoporotic phenotype was not a result of a defect in bone formation. In contrast, the amylin-deficient mice had an increased number of osteoclasts and an increase in degradation products of collagen in the urine, suggestive of accelerated bone resorption. Further studies of amylin-deficient mice demonstrated that the bone-effects were sex-dependent. Amylin-deficient males showed increased trabecular thickness at 4 and 6 weeks of age and increased femoral length at 26 weeks, whereas female mice were no different than the wild type (Davey et al., 2006). The role of CTR in mediating the bone-effects of amylin was investigated in a compound hemizygous mouse model: CTR (*Calcr* +/−) and amylin (*Iapp* +/−) (Dacquin et al., 2004). These mice displayed a combination of abnormalities identified in each of the individual hemizygous knockout, leading the authors to conclude that CTR is unlikely to be mediating amylin bone effects.

Calcitonin receptor

The contribution of CTR to skeletal physiology has been studied in three different strains of genetically modified mice. Since global deletion of CTR was embryonic lethal, the skeletal phenotype was initially studied in hemizygote CTR-KO mice (*Calcr*+/-), whose level of expression of CTR in osteoclasts was half that found in the WT mice, providing a model for the study of CTR haploinsufficiency (Dacquin et al., 2004). *Calcr* +/- mice had high bone mass due to increased bone formation, whereas no changes were identified in measures of bone resorption. These observations were consistent with the phenotype of the *Calca*-KO mice, providing further evidence for a physiological role of calcitonin as an inhibitor of bone formation, and indicating its activity is mediated via CTR. A second viable animal model was generated by incomplete deletion of the CTR gene, leaving a low residual level of expression (Davey et al., 2008). In this animal model only male mice had a small increase in bone formation rate, indicating that CTR plays a minor role in the physiological regulation of bone and calcium homeostasis in the basal state. More recently, employing a modified technique to knockout the expression of CTR, viable mice with global deletion of *Calcr* have been generated (Keller et al., 2014). The *Calcr*-KO mice had high bone mass due to increased bone formation, a phenotype that was then reproduced in mice that had the gene deletion restricted to the osteoclast lineage. The study further investigated the underlying mechanisms of the activity of calcitonin, signaling via CTR, to inhibit bone formation. The loss of CTR in osteoclasts has been shown to increase the levels of SPINSTER2, an exporter protein required for the secretion of sphingosine-1-phosphate (S1P), a potent inducer of bone formation. Therefore, the study suggests that calcitonin inhibits bone formation indirectly, through interaction with the osteoclast-expressed CTR (Keller et al., 2014).

Taken together, while pharmacological studies found that calcitonin inhibits of bone resorption, studies of genetically modified animals determined that calcitonin's physiological role is to inhibit bone formation. A similar discrepancy was identified in studies of the effect of PTH in bone; as PTH is a physiological stimulator of bone resorption that stimulates bone formation in pharmacological use (Martin and Sims, 2015).

The role of calcitonin and calcitonin receptor in situations of calcium stress

The roles of calcitonin and calcitonin receptor in situations of calcium stress were investigated in the genetically modified mice models. When hypercalcemia was induced in *Calcr*-KO mice by 1,25$(OH)_2D_3$, the peak levels of total serum calcium were significantly higher in the KO mice in comparison with the WT controls, suggesting that in these conditions CTR is important for calcium homeostasis (Davey et al., 2008). The underlying mechanism for calcitonin activity appears to be the inhibition of osteoclast activity, as a similar impaired response to hypercalcemia was later identified in a mouse model in which CTR was specifically missing in osteoclasts (Turner et al., 2011). A physiological condition of calcium stress is also present during lactation, as the maternal skeleton rapidly demineralizes in order to supply calcium to the milk. In the *Calca*-KO model, deficient in both calcitonin and CGRP, spine BMC dropped during lactation by over 50%, whereas in the WT controls the drop was only of 23.6% (Woodrow et al., 2006). After weaning, spine BMC was slower to return to baseline values in the *Calca*-KO in comparison with WT mice. Injection of sCT normalized the bone parameters, whereas CGRP was without effect, indicating that calcitonin is the peptide that plays an important role in calcium balance and preservation of the maternal skeleton during lactation and its recovery after weaning (Woodrow et al., 2006). Another mechanism that mobilizes calcium from skeletal stores during lactation is osteolytic osteolysis. Interestingly, osteocytes have been shown to express a number of genes in common with osteoclasts and to resorb bone locally. Histomorphometric analysis of the femurs at the end of lactation found that osteocyte lacunar area in *Calcr*-KO mice were larger than in WT, suggesting a role for calcitonin in inhibition of osteocytic osteolysis and protection of the maternal skeleton during lactation (Clarke et al., 2015).

Calcitonin, calcitonin gene-related peptide, and amylin—relevance to human bone physiology

The roles of the three peptides—calcitonin, CGRP, and amylin—in human physiology and pathophysiology have been studied extensively. The current understating is that calcitonin plays an important role in situations of calcium stress, regulating calcium homeostasis mainly through skeletal effects and perhaps through activity in the kidney during rapid growth, pathological hypercalcemia, pregnancy, and lactation. In contrast, the main physiological effects of CGRP and amylin appear not to be related directly to bone, with CGRP being a most potent microvascular vasodilator and contributing to pain especially in migraine, while amylin appears to be primarily important for energy homeostasis, inducing satiety, and increasing energy expenditure. Nevertheless, the peptides are likely to play at least a secondary role in bone metabolism.

It has been hypothesized that amylin secretion following a meal directs the absorbed calcium and protein from the meal into new bone synthesis by increasing bone growth at a time when the substrates are available (MacIntyre, 1989; Zaidi et al., 1993). Amylin may also contribute to the relationship between body mass and bone density. Body mass, and particularly fat mass, is the principal determinant of bone density in women (Reid et al., 1992a,b). A number of factors contribute to the fat—bone relationship; including the effect of weight on skeletal load bearing and adipocyte production of estrogen and adipokines (Naot and Cornish, 2014; Reid, 2002). In addition, the increased circulating levels of amylin and insulin in obesity are likely to have positive effects in bone and thus contribute to the fat mass—bone mass relationship. In fact, insulin levels were found to directly relate to bone density in normal postmenopausal women, and because amylin is cosecreted with insulin, it would seem likely that a similar relationship for this peptide exists (Reid et al., 1993). Pramlintide is a synthetic analog of amylin that is currently approved for the treatment of type 1 and type 2 diabetes, and might be in wider use in the future due to its effect on fat reduction. However, so far the effect of pramlintide on bone metabolism in humans has not been studied extensively. In a study of 23 patients with type 1 diabetes, 12 months of pramlintide treatment had no significant effects on markers on bone turnover or BMD (Chandran, 2017). As in vitro and animal studies determined an overall positive effect of amylin in bone, it would be of great interest to assess the impact of pramlintide treatment on long-term bone health and fracture risk.

Because CGRP access to bone is mainly via sensory innervation, the current understanding is that the bone activity of CGRP is predominantly localized. Evidence suggests that CGRP plays a role in the local bone response to different stimuli, including exercise, injury and the presence of metal implants. Given that monoclonal antibodies against CGRP are likely to be approved for clinical use for migraine prevention in the near future, it will be important to monitor the effects of their long-term use on bone physiology and healing.

References

Alam, A.S., Moonga, B.S., Bevis, P.J., Huang, C.L., Zaidi, M., 1991. Selective antagonism of calcitonin-induced osteoclastic quiescence (Q effect) by human calcitonin gene-related peptide-(Val8Phe37). Biochem. Biophys. Res. Commun. 179 (1), 134–139.

Alam, A.S., Bax, C.M., Shankar, V.S., Bax, B.E., Bevis, P.J., Huang, C.L., et al., 1993a. Further studies on the mode of action of calcitonin on isolated rat osteoclasts: pharmacological evidence for a second site mediating intracellular $Ca2^{+}$ mobilization and cell retraction. J. Endocrinol. 136 (1), 7–15.

Alam, A.S., Moonga, B.S., Bevis, P.J., Huang, C.L., Zaidi, M., 1993b. Amylin inhibits bone resorption by a direct effect on the motility of rat osteoclasts. Exp. Physiol. 78 (2), 183–196.

Aoki, M., Tamai, K., Saotome, K., 1994. Substance P- and calcitonin gene-related peptide-immunofluorescent nerves in the repair of experimental bone defects. Int. Orthop. 18 (5), 317–324.

Ballica, R., Valentijn, K., Khachatryan, A., Guerder, S., Kapadia, S., Gundberg, C., et al., 1999. Targeted expression of calcitonin gene-related peptide to osteoblasts increases bone density in mice. J. Bone Miner. Res. 14 (7), 1067–1074.

Bjurholm, A., 1991. Neuroendocrine peptides in bone. Int. Orthop. 15 (4), 325–329.

Bomberger, J.M., Parameswaran, N., Hall, C.S., Aiyar, N., Spielman, W.S., 2005. Novel function for receptor activity-modifying proteins (RAMPs) in post-endocytic receptor trafficking. J. Biol. Chem. 280 (10), 9297–9307.

Born, W., Beglinger, C., Fischer, J.A., 1991. Diagnostic relevance of the amino-terminal cleavage peptide of procalcitonin (PAS-57), calcitonin and calcitonin gene-related peptide in medullary thyroid carcinoma patients. Regul. Pept. 32 (3), 311–319.

Bower, R.L., Hay, D.L., 2016. Amylin structure-function relationships and receptor pharmacology: implications for amylin mimetic drug development. Br. J. Pharmacol. 173 (12), 1883–1898.

Boyle, C.N., Lutz, T.A., Le Foll, C., 2017. Amylin – its role in the homeostatic and hedonic control of eating and recent developments of amylin analogs to treat obesity. Mol Metab 8, 203–210.

Breimer, L.H., MacIntyre, I., Zaidi, M., 1988. Peptides from the calcitonin genes: molecular genetics, structure and function. Biochem. J. 255 (2), 377–390.

Burns, D.M., Stehno-Bittel, L., Kawase, T., 2004. Calcitonin gene-related peptide elevates calcium and polarizes membrane potential in MG-63 cells by both cAMP-independent and -dependent mechanisms. Am. J. Physiol. Cell Physiol. 287 (2), C457–C467.

Chambers, T.J., Magnus, C.J., 1982. Calcitonin alters behaviour of isolated osteoclasts. J. Pathol. 136 (1), 27–39.

Chambers, T.J., Athanasou, N.A., Fuller, K., 1984. Effect of parathyroid hormone and calcitonin on the cytoplasmic spreading of isolated osteoclasts. J. Endocrinol. 102 (3), 281–286.

Chandran, M., 2017. Diabetes drug effects on the skeleton. Calcif. Tissue Int. 100 (2), 133–149.

Chang, C.P., Pearse 2nd, R.V., O'Connell, S., Rosenfeld, M.G., 1993. Identification of a seven transmembrane helix receptor for corticotropin-releasing factor and sauvagine in mammalian brain. Neuron 11 (6), 1187–1195.

Chesnut 3rd, C.H., Azria, M., Silverman, S., Engelhardt, M., Olson, M., Mindeholm, L., 2008. Salmon calcitonin: a review of current and future therapeutic indications. Osteoporos. Int. 19 (4), 479–491.

Clarke, M.V., Russell, P.K., Findlay, D.M., Sastra, S., Anderson, P.H., Skinner, J.P., et al., 2015. A role for the calcitonin receptor to limit bone loss during lactation in female mice by inhibiting osteocytic osteolysis. Endocrinology 156 (9), 3203–3214.

Cooper, G.J., Willis, A.C., Clark, A., Turner, R.C., Sim, R.B., Reid, K.B., 1987. Purification and characterization of a peptide from amyloid-rich pancreases of type 2 diabetic patients. Proc. Natl. Acad. Sci. U. S. A. 84 (23), 8628–8632.

Copp, D.H., Cheney, B., 1962. Calcitonin-a hormone from the parathyroid which lowers the calcium-level of the blood. Nature 193, 381–382.

Copp, D.H., Cameron, E.C., Cheney, B.A., Davidson, A.G., Henze, K.G., 1962. Evidence for calcitonin–a new hormone from the parathyroid that lowers blood calcium. Endocrinology 70, 638–649.

Cornish, J., Reid, I.R., 1999. Skeletal effects of amylin and related peptides. Endocrinololgist 9, 183–189.

Cornish, J., Callon, K.E., Cooper, G.J., Reid, I.R., 1995. Amylin stimulates osteoblast proliferation and increases mineralized bone volume in adult mice. Biochem. Biophys. Res. Commun. 207 (1), 133–139.

Cornish, J., Callon, K.E., King, A.R., Cooper, G.J., Reid, I.R., 1998a. Systemic administration of amylin increases bone mass, linear growth, and adiposity in adult male mice. Am. J. Physiol. 275 (4 Pt 1), E694–E699.

Cornish, J., Callon, K.E., Lin, C.Q., Xiao, C.L., Mulvey, T.B., Coy, D.H., et al., 1998b. Dissociation of the effects of amylin on osteoblast proliferation and bone resorption. Am. J. Physiol. 274 (5 Pt 1), E827–E833.

Cornish, J., Callon, K.E., Gasser, J.A., Bava, U., Gardiner, E.M., Coy, D.H., et al., 2000. Systemic administration of a novel octapeptide, amylin-(1-8), increases bone volume in male mice. Am. J. Physiol. Endocrinol. Metab. 279 (4), E730–E735.

Cornish, J., Callon, K.E., Bava, U., Kamona, S.A., Cooper, G.J., Reid, I.R., 2001. Effects of calcitonin, amylin, and calcitonin gene-related peptide on osteoclast development. Bone 29 (2), 162–168.

Cornish, J., Grey, A., Callon, K.E., Naot, D., Hill, B.L., Lin, C.Q., et al., 2004. Shared pathways of osteoblast mitogenesis induced by amylin, adrenomedullin, and IGF-1. Biochem. Biophys. Res. Commun. 318 (1), 240–246.

Dacquin, R., Davey, R.A., Laplace, C., Levasseur, R., Morris, H.A., Goldring, S.R., et al., 2004. Amylin inhibits bone resorption while the calcitonin receptor controls bone formation in vivo. J. Cell Biol. 164 (4), 509–514.

Datta, H.K., Zaidi, M., Wimalawansa, S.J., Ghatei, M.A., Beacham, J.L., Bloom, S.R., et al., 1989. In vivo and in vitro effects of amylin and amylin-amide on calcium metabolism in the rat and rabbit. Biochem. Biophys. Res. Commun. 162 (2), 876–881.

Davey, R.A., Findlay, D.M., 2013. Calcitonin: physiology or fantasy? J. Bone Miner. Res. 28 (5), 973–979.

Davey, R.A., Moore, A.J., Chiu, M.W., Notini, A.J., Morris, H.A., Zajac, J.D., 2006. Effects of amylin deficiency on trabecular bone in young mice are sex-dependent. Calcif. Tissue Int. 78 (6), 398–403.

Davey, R.A., Turner, A.G., McManus, J.F., Chiu, W.S., Tjahyono, F., Moore, A.J., et al., 2008. Calcitonin receptor plays a physiological role to protect against hypercalcemia in mice. J. Bone Miner. Res. 23 (8), 1182–1193.

Drissi, H., Lieberherr, M., Hott, M., Marie, P.J., Lasmoles, F., 1999. Calcitonin gene-related peptide (CGRP) increases intracellular free $Ca2^{+}$ concentrations but not cyclic AMP formation in CGRP receptor-positive osteosarcoma cells (OHS-4). Cytokine 11 (3), 200–207.

D'Souza, S.M., MacIntyre, I., Girgis, S.I., Mundy, G.R., 1986. Human synthetic calcitonin gene-related peptide inhibits bone resorption in vitro. Endocrinology 119 (1), 58–61.

Ellegaard, M., Thorkildsen, C., Petersen, S., Petersen, J.S., Jorgensen, N.R., Just, R., et al., 2010. Amylin(1-8) is devoid of anabolic activity in bone. Calcif. Tissue Int. 86 (3), 249–260.

Farley, J.R., Tarbaux, N.M., Hall, S.L., Linkhart, T.A., Baylink, D.J., 1988. The anti-bone-resorptive agent calcitonin also acts in vitro to directly increase bone formation and bone cell proliferation. Endocrinology 123 (1), 159–167.

Findlay, D.M., Sexton, P.M., 2004. Calcitonin. Growth Factors. 22 (4), 217–224.

Gorn, A.H., Rudolph, S.M., Flannery, M.R., Morton, C.C., Weremowicz, S., Wang, T.Z., et al., 1995. Expression of two human skeletal calcitonin receptor isoforms cloned from a giant cell tumor of bone. The first intracellular domain modulates ligand binding and signal transduction. J. Clin. Investig. 95 (6), 2680–2691.

Granholm, S., Lundberg, P., Lerner, U.H., 2007. Calcitonin inhibits osteoclast formation in mouse haematopoetic cells independently of transcriptional regulation by receptor activator of NF-{kappa}B and c-Fms. J. Endocrinol. 195 (3), 415–427.

Granholm, S., Henning, P., Lerner, U.H., 2011. Comparisons between the effects of calcitonin receptor-stimulating peptide and intermedin and other peptides in the calcitonin family on bone resorption and osteoclastogenesis. J. Cell. Biochem. 112 (11), 3300–3312.

Gupta, A., Schwiening, C.J., 1994. Boron WF. Effects of CGRP, forskolin, PMA, and ionomycin on pHi dependence of Na-H exchange in UMR-106 cells. Am. J. Physiol. 266 (4 Pt 1), C1088–C1092.

Gutierrez-Rojas, I., Lozano, D., Nuche-Berenguer, B., Moreno, P., Acitores, A., Ramos-Alvarez, I., et al., 2013. Amylin exerts osteogenic actions with different efficacy depending on the diabetic status. Mol. Cell. Endocrinol. 365 (2), 309–315.

Harzenetter, M.D., Novotny, A.R., Gais, P., Molina, C.A., Altmayr, F., Holzmann, B., 2007. Negative regulation of TLR responses by the neuropeptide CGRP is mediated by the transcriptional repressor ICER. J. Immunol. 179 (1), 607–615.

Hay, D.L., Chen, S., Lutz, T.A., Parkes, D.G., Roth, J.D., 2015. Amylin: pharmacology, physiology, and clinical potential. Pharmacol. Rev. 67 (3), 564–600.

Henriksen, K., Byrjalsen, I., Andersen, J.R., Bihlet, A.R., Russo, L.A., Alexandersen, P., et al., 2016. A randomized, double-blind, multicenter, placebo-controlled study to evaluate the efficacy and safety of oral salmon calcitonin in the treatment of osteoporosis in postmenopausal women taking calcium and vitamin D. Bone 91, 122–129.

Hill, E.L., Elde, R., 1991. Distribution of CGRP-, VIP-, D beta H-, SP-, and NPY-immunoreactive nerves in the periosteum of the rat. Cell Tissue Res. 264 (3), 469–480.

Hill, E.L., Turner, R., Elde, R., 1991. Effects of neonatal sympathectomy and capsaicin treatment on bone remodeling in rats. Neuroscience 44 (3), 747–755.

Hoff, A.O., Catala-Lehnen, P., Thomas, P.M., Priemel, M., Rueger, J.M., Nasonkin, I., et al., 2002. Increased bone mass is an unexpected phenotype associated with deletion of the calcitonin gene. J. Clin. Investig. 110 (12), 1849–1857.

Horcajada-Molteni, M.N., Davicco, M.J., Lebecque, P., Coxam, V., Young, A.A., Barlet, J.P., 2000. Amylin inhibits ovariectomy-induced bone loss in rats. J. Endocrinol. 165 (3), 663–668.

Huang, C.Z., Yang, X.N., Liu, D.C., Sun, Y.G., Dai, X.M., 2015. Calcitonin gene-related peptide-induced calcium alginate gel combined with adipose-derived stem cells differentiating to osteoblasts. Cell Biochem. Biophys. 73 (3), 609–617.

Huebner, A.K., Schinke, T., Priemel, M., Schilling, S., Schilling, A.F., Emeson, R.B., et al., 2006. Calcitonin deficiency in mice progressively results in high bone turnover. J. Bone Miner. Res. 21 (12), 1924–1934.

Huebner, A.K., Keller, J., Catala-Lehnen, P., Perkovic, S., Streichert, T., Emeson, R.B., et al., 2008. The role of calcitonin and alpha-calcitonin gene-related peptide in bone formation. Arch. Biochem. Biophys. 473 (2), 210–217.

Hukkanen, M., Konttinen, Y.T., Santavirta, S., Paavolainen, P., Gu, X.H., Terenghi, G., et al., 1993. Rapid proliferation of calcitonin gene-related peptide-immunoreactive nerves during healing of rat tibial fracture suggests neural involvement in bone growth and remodelling. Neuroscience 54 (4), 969–979.

Hurley, D.L., Tiegs, R.D., Wahner, H.W., Heath 3rd, H., 1987. Axial and appendicular bone mineral density in patients with long-term deficiency or excess of calcitonin. N. Engl. J. Med. 317 (9), 537–541.

Irie, K., Hara-Irie, F., Ozawa, H., Yajima, T., 2002. Calcitonin gene-related peptide (CGRP)-containing nerve fibers in bone tissue and their involvement in bone remodeling. Microsc. Res. Tech. 58 (2), 85–90.

Iyengar, S., Ossipov, M.H., Johnson, K.W., 2017. The role of calcitonin gene-related peptide in peripheral and central pain mechanisms including migraine. Pain 158 (4), 543–559.

Jablonski, H., Kauther, M.D., Bachmann, H.S., Jager, M., Wedemeyer, C., 2015. Calcitonin gene-related peptide modulates the production of pro-inflammatory cytokines associated with periprosthetic osteolysis by THP-1 macrophage-like cells. Neuroimmunomodulation 22 (3), 152–165.

Juhl, C.B., Porksen, N., Sturis, J., Hansen, A.P., Veldhuis, J.D., Pincus, S., et al., 2000. High-frequency oscillations in circulating amylin concentrations in healthy humans. Am. J. Physiol. Endocrinol. Metab. 278 (3), E484–E490.

Kauther, M.D., Bachmann, H.S., Neuerburg, L., Broecker-Preuss, M., Hilken, G., Grabellus, F., et al., 2011. Calcitonin substitution in calcitonin deficiency reduces particle-induced osteolysis. BMC Muscoskelet. Disord. 12, 186.

Kawase, T., Burns, D.M., 1998. Calcitonin gene-related peptide stimulates potassium efflux through adenosine triphosphate-sensitive potassium channels and produces membrane hyperpolarization in osteoblastic UMR106 cells. Endocrinology 139 (8), 3492–3502.

Kawase, T., Howard, G.A., Roos, B.A., Burns, D.M., 1995. Diverse actions of calcitonin gene-related peptide on intracellular free $Ca2^{+}$ concentrations in UMR 106 osteoblastic cells. Bone 16 (4 Suppl. 1), 379S–384S.

Keller, J., Catala-Lehnen, P., Huebner, A.K., Jeschke, A., Heckt, T., Lueth, A., et al., 2014. Calcitonin controls bone formation by inhibiting the release of sphingosine 1-phosphate from osteoclasts. Nat. Commun. 5, 5215.

Klein, K.R., Matson, B.C., Caron, K.M., 2016. The expanding repertoire of receptor activity modifying protein (RAMP) function. Crit. Rev. Biochem. Mol. Biol. 51 (1), 65–71.

Konarkowska, B., Aitken, J.F., Kistler, J., Zhang, S., Cooper, G.J., 2006. The aggregation potential of human amylin determines its cytotoxicity towards islet beta-cells. FEBS J. 273 (15), 3614–3624.

Lerner, U.H., 2006. Deletions of genes encoding calcitonin/alpha-CGRP, amylin and calcitonin receptor have given new and unexpected insights into the function of calcitonin receptors and calcitonin receptor-like receptors in bone. J. Musculoskelet. Neuronal Interact. 6 (1), 87–95.

Levin, B.E., Lutz, T.A., 2017. Amylin and leptin: Co-regulators of energy homeostasis and neuronal development. Trends Endocrinol. Metabol. 28 (2), 153–164.

Li, Z., Kelly, L., Gergi, I., Vieweg, P., Heiman, M., Greengard, P., et al., 2015. Hypothalamic amylin acts in concert with leptin to regulate food intake. Cell Metabol. 22 (6), 1059–1067.

Li, Y., Yang, L., Zheng, Z., Li, Z., Deng, T., Ren, W., et al., 2017. Bio-Oss((R)) modified by calcitonin gene-related peptide promotes osteogenesis in vitro. Exp Ther Med 14 (5), 4001–4008.

Liang, W., Zhuo, X., Tang, Z., Wei, X., Li, B., 2015. Calcitonin gene-related peptide stimulates proliferation and osteogenic differentiation of osteoporotic rat-derived bone mesenchymal stem cells. Mol. Cell. Biochem. 402 (1–2), 101–110.

Lima, W.G., Marques-Oliveira, G.H., da Silva, T.M., Chaves, V.E., 2017. Role of calcitonin gene-related peptide in energy metabolism. Endocrine 58 (1), 3–13.

Lin, H.Y., Harris, T.L., Flannery, M.S., Aruffo, A., Kaji, E.H., Gorn, A., et al., 1991. Expression cloning and characterization of a porcine renal calcitonin receptor. Trans. Assoc. Am. Phys. 104, 265–272.

Lorenzo, A., Razzaboni, B., Weir, G.C., Yankner, B.A., 1994. Pancreatic islet cell toxicity of amylin associated with type-2 diabetes mellitus. Nature 368 (6473), 756–760.

Lutz, T.A., 2012. Control of energy homeostasis by amylin. Cell. Mol. Life Sci. 69 (12), 1947–1965.

MacIntyre, I., 1989. Amylinamide, bone conservation, and pancreatic beta cells. Lancet 2 (8670), 1026–1027.

Martin, T.J., Sims, N.A., 2015. Calcitonin physiology, saved by a lysophospholipid. J. Bone Miner. Res. 30 (2), 212–215.

McLatchie, L.M., Fraser, N.J., Main, M.J., Wise, A., Brown, J., Thompson, N., et al., 1998. RAMPs regulate the transport and ligand specificity of the calcitonin-receptor-like receptor. Nature 393 (6683), 333–339.

Michelangeli, V.P., Findlay, D.M., Fletcher, A., Martin, T.J., 1986. Calcitonin gene-related peptide (CGRP) acts independently of calcitonin on cyclic AMP formation in clonal osteogenic sarcoma cells (UMR 106-01). Calcif. Tissue Int. 39 (1), 44–48.

Michelangeli, V.P., Fletcher, A.E., Allan, E.H., Nicholson, G.C., Martin, T.J., 1989. Effects of calcitonin gene-related peptide on cyclic AMP formation in chicken, rat, and mouse bone cells. J. Bone Miner. Res. 4 (2), 269–272.

Miles, P.D., Deftos, L.J., Moossa, A.R., Olefsky, J.M., 1994. Islet amyloid polypeptide (amylin) increases the renal excretion of calcium in the conscious dog. Calcif. Tissue Int. 55 (4), 269–273.

Millet, I., Vignery, A., 1997. The neuropeptide calcitonin gene-related peptide inhibits TNF-alpha but poorly induces IL-6 production by fetal rat osteoblasts. Cytokine 9 (12), 999–1007.

Morfis, M., Tilakaratne, N., Furness, S.G., Christopoulos, G., Werry, T.D., Christopoulos, A., et al., 2008. Receptor activity-modifying proteins differentially modulate the G protein-coupling efficiency of amylin receptors. Endocrinology 149 (11), 5423–5431.

Mrak, E., Guidobono, F., Moro, G., Fraschini, G., Rubinacci, A., Villa, I., 2010. Calcitonin gene-related peptide (CGRP) inhibits apoptosis in human osteoblasts by beta-catenin stabilization. J. Cell. Physiol. 225 (3), 701–708.

Mullins, M.W., Ciallella, J., Rangnekar, V., McGillis, J.P., 1993. Characterization of a calcitonin gene-related peptide (CGRP) receptor on mouse bone marrow cells. Regul. Pept. 49 (1), 65–72.

Naot, D., Cornish, J., 2008. The role of peptides and receptors of the calcitonin family in the regulation of bone metabolism. Bone 43 (5), 813–818.

Naot, D., Cornish, J., 2014. Cytokines and hormones that contribute to the positive association between fat and bone. Front. Endocrinol. 5, 70.

Niall, H.D., Keutmann, H.T., Copp, D.H., Potts Jr., J.T., 1969. Amino acid sequence of salmon ultimobranchial calcitonin. Proc. Natl. Acad. Sci. U. S. A. 64 (2), 771–778.

Njuki, F., Nicholl, C.G., Howard, A., Mak, J.C., Barnes, P.J., Girgis, S.I., et al., 1993. A new calcitonin-receptor-like sequence in rat pulmonary blood vessels. Clin. Sci. 85 (4), 385–388.

Offley, S.C., Guo, T.-Z., Wei, T., Clark, J.D., Vogel, H., Lindsey, D.P., et al., 2005. Capsaicin-sensitive sensory neurons contribute to the maintenance of trabecular bone integrity. J. Bone Miner. Res. 20 (2), 257–267.

Osajima, A., Mutoh, Y., Uezono, Y., Kawamura, M., Izumi, F., Takasugi, M., et al., 1995. Adrenomedullin increases cyclic AMP more potently than CGRP and amylin in rat renal tubular basolateral membranes. Life Sci. 57 (5), 457–462.

Owan, I., Ibaraki, K., 1994. The role of calcitonin gene-related peptide (CGRP) in macrophages: the presence of functional receptors and effects on proliferation and differentiation into osteoclast-like cells. Bone Miner. 24 (2), 151–164.

Paemeleire, K., MaassenVanDenBrink, A., 2018. Calcitonin-gene-related peptide pathway mAbs and migraine prevention. Curr. Opin. Neurol. 31 (3), 274–280.

Pietschmann, P., Farsoudi, K.H., Hoffmann, O., Klaushofer, K., Horandner, H., Peterlik, M., 1993. Inhibitory effect of amylin on basal and parathyroid hormone-stimulated bone resorption in cultured neonatal mouse calvaria. Bone 14 (2), 167–172.

Poyner, D.R., Sexton, P.M., Marshall, I., Smith, D.M., Quirion, R., Born, W., et al., 2002. International Union of Pharmacology. XXXII. The mammalian calcitonin gene-related peptides, adrenomedullin, amylin, and calcitonin receptors. Pharmacol. Rev. 54 (2), 233–246.

Qiao, Y.C., Ling, W., Pan, Y.H., Chen, Y.L., Zhou, D., Huang, Y.M., et al., 2017. Efficacy and safety of pramlintide injection adjunct to insulin therapy in patients with type 1 diabetes mellitus: a systematic review and meta-analysis. Oncotarget 8 (39), 66504–66515.

Reid, I.R., Ames, R., Evans, M.C., Sharpe, S., Gamble, G., France, J.T., et al., 1992a. Determinants of total body and regional bone mineral density in normal postmenopausal women–a key role for fat mass. J. Clin. Endocrinol. Metab. 75 (1), 45–51.

Reid, I.R., Plank, L.D., Evans, M.C., 1992b. Fat mass is an important determinant of whole body bone density in premenopausal women but not in men. J. Clin. Endocrinol. Metab. 75 (3), 779–782.

Reid, I.R., Evans, M.C., Cooper, G.J., Ames, R.W., Stapleton, J., 1993. Circulating insulin levels are related to bone density in normal postmenopausal women. Am. J. Physiol. 265 (4 Pt 1), E655–E659.

Reid, I.R., 2002. Relationships among body mass, its components, and bone. Bone 31 (5), 547–555.

Romero, D.F., Bryer, H.P., Rucinski, B., Isserow, J.A., Buchinsky, F.J., Cvetkovic, M., et al., 1995. Amylin increases bone volume but cannot ameliorate diabetic osteopenia. Calcif. Tissue Int. 56 (1), 54–61.

Roos, B.A., Fischer, J.A., Pignat, W., Alander, C.B., Raisz, L.G., 1986. Evaluation of the in vivo and in vitro calcium-regulating actions of noncalcitonin peptides produced via calcitonin gene expression. Endocrinology 118 (1), 46–51.

Russell, F.A., King, R., Smillie, S., Kodji, X., Brain, S.D., 2014. Calcitonin gene-related peptide: physiology and pathophysiology. Physiol. Rev. 94 (4), 1099–1142.

Schifter, S., 1991. Circulating concentrations of calcitonin gene-related peptide (CGRP) in normal man determined with a new, highly sensitive radioimmunoassay. Peptides 12 (2), 365–369.

Schinke, T., Liese, S., Priemel, M., Haberland, M., Schilling, A.F., Catala-Lehnen, P., et al., 2004. Decreased bone formation and osteopenia in mice lacking alpha-calcitonin gene-related peptide. J. Bone Miner. Res. 19 (12), 2049–2056.

Sexton, P.M., Houssami, S., Hilton, J.M., O'Keeffe, L.M., Center, R.J., Gillespie, M.T., et al., 1993. Identification of brain isoforms of the rat calcitonin receptor. Mol. Endocrinol. 7 (6), 815–821.

Sexton, P.M., Houssami, S., Brady, C.L., Myers, D.E., Findlay, D.M., 1994. Amylin is an agonist of the renal porcine calcitonin receptor. Endocrinology 134 (5), 2103–2107.

Shyu, J.F., Shih, C., Tseng, C.Y., Lin, C.H., Sun, D.T., Liu, H.T., et al., 2007. Calcitonin induces podosome disassembly and detachment of osteoclasts by modulating Pyk2 and Src activities. Bone 40 (5), 1329–1342.

Silverman, S.L., Azria, M., 2002. The analgesic role of calcitonin following osteoporotic fracture. Osteoporos. Int. 13 (11), 858–867.

Singh-Franco, D., Robles, G., Gazze, D., 2007. Pramlintide acetate injection for the treatment of type 1 and type 2 diabetes mellitus. Clin. Ther. 29 (4), 535–562.

Spinetti, A., Margutti, A., Bertolini, S., Bernardi, F., BiFulco, G., degli Uberti, E.C., et al., 1997. Hormonal replacement therapy affects calcitonin gene-related peptide and atrial natriuretic peptide secretion in postmenopausal women. Eur. J. Endocrinol. 137 (6), 664–669.

Takahashi, S., Goldring, S., Katz, M., Hilsenbeck, S., Williams, R., Roodman, G.D., 1995. Downregulation of calcitonin receptor mRNA expression by calcitonin during human osteoclast-like cell differentiation. J. Clin. Investig. 95 (1), 167–171.

Tamura, T., Miyaura, C., Owan, I., Suda, T., 1992. Mechanism of action of amylin in bone. J. Cell. Physiol. 153 (1), 6–14.

Thiebaud, D., Akatsu, T., Yamashita, T., Suda, T., Noda, T., Martin, R.E., et al., 1991. Structure-activity relationships in calcitonin gene-related peptide: cyclic AMP response in a preosteoblast cell line (KS-4). J. Bone Miner. Res. 6 (10), 1137–1142.

Tian, G., Zhang, G., Tan, Y.H., 2013. Calcitonin gene-related peptide stimulates BMP-2 expression and the differentiation of human osteoblast-like cells in vitro. Acta Pharmacol. Sin. 34 (11), 1467–1474.

Tippins, J.R., Morris, H.R., Panico, M., Etienne, T., Bevis, P., Girgis, S., et al., 1984. The myotropic and plasma-calcium modulating effects of calcitonin gene-related peptide (CGRP). Neuropeptides 4 (5), 425–434.

Turner, A.G., Tjahyono, F., Chiu, W.S., Skinner, J., Sawyer, R., Moore, A.J., et al., 2011. The role of the calcitonin receptor in protecting against induced hypercalcemia is mediated via its actions in osteoclasts to inhibit bone resorption. Bone 48 (2), 354–361.

Valentijn, K., Gutow, A.P., Troiano, N., Gundberg, C., Gilligan, J.P., Vignery, A., 1997. Effects of calcitonin gene-related peptide on bone turnover in ovariectomized rats. Bone 21 (3), 269–274.

Villa, I., Rubinacci, A., Ravasi, F., Ferrara, A.F., Guidobono, F., 1997. Effects of amylin on human osteoblast-like cells. Peptides 18 (4), 537–540.

Villa, I., Melzi, R., Pagani, F., Ravasi, F., Rubinacci, A., Guidobono, F., 2000. Effects of calcitonin gene-related peptide and amylin on human osteoblast-like cells proliferation. Eur. J. Pharmacol. 409 (3), 273–278.

Villa, I., Dal Fiume, C., Maestroni, A., Rubinacci, A., Ravasi, F., Guidobono, F., 2003. Human osteoblast-like cell proliferation induced by calcitonin-related peptides involves PKC activity. Am. J. Physiol. Endocrinol. Metab. 284 (3), E627–E633.

Wada, S., Udagawa, N., Nagata, N., Martin, T.J., Findlay, D.M., 1996. Calcitonin receptor down-regulation relates to calcitonin resistance in mature mouse osteoclasts. Endocrinology 137 (3), 1042–1048.

Wang, L., Shi, X., Zhao, R., Halloran, B.P., Clark, D.J., Jacobs, C.R., et al., 2010. Calcitonin-gene-related peptide stimulates stromal cell osteogenic differentiation and inhibits RANKL induced NF-kappaB activation, osteoclastogenesis and bone resorption. Bone 46 (5), 1369–1379.

Weiss, R.E., Singer, F.R., Gorn, A.H., Hofer, D.P., Nimni, M.E., 1981. Calcitonin stimulates bone formation when administered prior to initiation of osteogenesis. J. Clin. Investig. 68 (3), 815–818.

Wener, J.A., Gorton, S.J., Raisz, L.G., 1972. Escape from inhibition or resorption in cultures of fetal bone treated with calcitoninand parathyroid hromone. Endocrinology 90 (3), 752–759.

Westermark, P., Wernstedt, C., Wilander, E., Hayden, D.W., O'Brien, T.D., Johnson, K.H., 1987. Amyloid fibrils in human insulinoma and islets of Langerhans of the diabetic cat are derived from a neuropeptide-like protein also present in normal islet cells. Proc. Natl. Acad. Sci. U. S. A. 84 (11), 3881–3885.

Weston, C., Winfield, I., Harris, M., Hodgson, R., Shah, A., Dowell, S.J., et al., 2016. Receptor activity-modifying protein-directed G protein signaling specificity for the calcitonin gene-related peptide family of receptors. J. Biol. Chem. 291 (42), 21925–21944.

Wimalawansa, S.J., 1997. Amylin, calcitonin gene-related peptide, calcitonin, and adrenomedullin: a peptide superfamily. Crit. Rev. Neurobiol. 11 (2–3), 167–239.

Woodrow, J.P., Sharpe, C.J., Fudge, N.J., Hoff, A.O., Gagel, R.F., Kovacs, C.S., 2006. Calcitonin plays a critical role in regulating skeletal mineral metabolism during lactation. Endocrinology 147 (9), 4010–4021.

Wuster, C., Raue, F., Meyer, C., Bergmann, M., Ziegler, R., 1992. Long-term excess of endogenous calcitonin in patients with medullary thyroid carcinoma does not affect bone mineral density. J. Endocrinol. 134 (1), 141–147.

Xu, J., Kauther, M.D., Hartl, J., Wedemeyer, C., 2010. Effects of alpha-calcitonin gene-related peptide on osteoprotegerin and receptor activator of nuclear factor-kappaB ligand expression in MG-63 osteoblast-like cells exposed to polyethylene particles. J. Orthop. Surg. Res. 5, 83.

Yamamoto, I., Kitamura, N., Aoki, J., Shigeno, C., Hino, M., Asonuma, K., et al., 1986. Human calcitonin gene-related peptide possesses weak inhibitory potency of bone resorption in vitro. Calcif. Tissue Int. 38 (6), 339–341.

Young, A.A., Gedulin, B., Wolfe-Lopez, D., Greene, H.E., Rink, T.J., Cooper, G.J., 1992. Amylin and insulin in rat soleus muscle: dose responses for cosecreted noncompetitive antagonists. Am. J. Physiol. 263 (2 Pt 1), E274–E281.

Young, A.A., Rink, T.J., Wang, M.W., 1993. Dose response characteristics for the hyperglycemic, hyperlactemic, hypotensive and hypocalcemic actions of amylin and calcitonin gene-related peptide-I (CGRP alpha) in the fasted, anaesthetized rat. Life Sci. 52 (21), 1717–1726.

Zaidi, M., Chambers, T.J., Gaines Das, R.E., Morris, H.R., MacIntyre, I., 1987a. A direct action of human calcitonin gene-related peptide on isolated osteoclasts. J. Endocrinol. 115 (3), 511–518.

Zaidi, M., Fuller, K., Bevis, P.J., GainesDas, R.E., Chambers, T.J., MacIntyre, I., 1987b. Calcitonin gene-related peptide inhibits osteoclastic bone resorption: a comparative study. Calcif. Tissue Int. 40 (3), 149–154.

Zaidi, M., Datta, H.K., Bevis, P.J., Wimalawansa, S.J., MacIntyre, I., 1990. Amylin-amide: a new bone-conserving peptide from the pancreas. Exp. Physiol. 75 (4), 529–536.

Zaidi, M., Shankar, V.S., Huang, C.L.H., Pazianas, M., Bloom, S.R., 1993. Amylin in bone conservation: current evidence and hypothetical considerations. Trends Endocrinol. Metabol. 4, 255–259.

Zaidi, M., Inzerillo, A.M., Moonga, B.S., Bevis, P.J., Huang, C.L., 2002. Forty years of calcitonin–where are we now? A tribute to the work of Iain Macintyre. FRS. Bone. 30 (5), 655–663.

Zhang, Z., Neff, L., Bothwell, A.L., Baron, R., Horne, W.C., 2002. Calcitonin induces dephosphorylation of Pyk2 and phosphorylation of focal adhesion kinase in osteoclasts. Bone 31 (3), 359–365.

Zhang, X.X., Pan, Y.H., Huang, Y.M., Zhao, H.L., 2016. Neuroendocrine hormone amylin in diabetes. World J. Diabetes 7 (9), 189–197.

Zhang, Y., Xu, J., Ruan, Y.C., Yu, M.K., O'Laughlin, M., Wise, H., et al., 2016. Implant-derived magnesium induces local neuronal production of CGRP to improve bone-fracture healing in rats. Nat. Med. 22 (10), 1160–1169.

Zhou, R., Yuan, Z., Liu, J., Liu, J., 2016. Calcitonin gene-related peptide promotes the expression of osteoblastic genes and activates the WNT signal transduction pathway in bone marrow stromal stem cells. Mol. Med. Rep. 13 (6), 4689–4696.

Chapter 34

Regulation of bone remodeling by central and peripheral nervous signals

Patricia Ducy

Department of Pathology & Cell Biology, Columbia University, College of Physicians & Surgeons, New York, NY, United States

Chapter outline

Introduction **809**
Afferent signals regulating bone remodeling via the central nervous system **810**
Negative regulation of bone remodeling by leptin 810
Dual action of adiponectin on bone remodeling 810
Central and efferent regulators of bone remodeling **811**
Leptin's action on bone remodeling is mediated by brain serotonin signaling and the sympathetic nervous system 811
Counterregulation of sympathetic nervous system control of bone remodeling by the parasympathetic nervous system and adiponectin 812
Other regulators of sympathetic nervous system control of bone remodeling 813
NeuromedinU 813
Endocannabinoid signaling 813
Orexin signaling 813
Regulation of bone resorption by melanocortin receptor 4 and cocaine- and amphetamine-regulated transcript 814
Y receptor signaling 814
Brain-derived neurotrophic factor 815
Interleukin-1 815
Evidence of central/neuronal regulations of bone mass in human **815**
Leptin 816
Adrenergic signaling 816
Brain serotonin and neuromedinU 817
Melanocortin receptor 4 and cocaine- and amphetamine-regulated transcript 818
Neuropeptide Y, brain-derived neurotrophic factor, and cannabinoid receptor 2 **818**
Conclusions and perspective **818**
References **818**

Introduction

Hints of a connection between the nervous system and the regulation of bone mass have long been present in the literature. For instance, osteoporosis is a known complication of spinal cord injury and experimental models of sensory or sympathetic denervation have shown that these two neuronal systems could be involved in bone development and remodeling (Chenu, 2004; Jiang et al., 2006). Likewise, major depression is associated with low bone mass and increased incidence of osteoporotic fractures, as is the use of several types of central nervous system-active drugs such as anticonvulsants and opioids (Kinjo et al., 2005). These examples all point at a regulation of bone mass by the nervous system—i.e., at signals efferent from the brain (or nerves) to bone cells. This might explain why most of the earlier studies have focused on peripherally produced neuromediators (Chenu, 2002; Spencer et al., 2007). More recently, however, afferent signals that influence the brain's control of bone mass have been identified, and a comprehensive understanding of their central mode of action has emerged. Overall, the identification of these pathways illustrates the investigative power of mouse genetics and can be viewed as an archetype of the current effort of integrative biology. More importantly, this rapidly evolving field of bone neurobiology has shed light on perplexing clinical observations and may therefore become critical in developing novel concepts in drug development.

Principles of Bone Biology. https://doi.org/10.1016/B978-0-12-814841-9.00034-8

Afferent signals regulating bone remodeling via the central nervous system

At the present time there are two signals known, in vivo, to affect bone mass via a central nervous system (CNS) relay. Remarkably, they both originate from adipocytes, suggesting a preferential relationship between the regulation of energy metabolism and bone homeostasis. One signal is leptin, a 16 kDa peptide hormone that was originally identified by positional cloning of the mutation present in *ob/ob* mice, a natural mutant presenting morbid obesity and sterility (Zhang et al., 1994). The other one is adiponectin, a secreted molecule previously known for its insulin-sensitizing ability in animals fed a high-fat diet (Maeda et al., 2002; Kadowaki and Yamauchi, 2005). Subsequent analysis of their respective central mode of action has identified a significant overlap in their ability to influence bone remodeling.

Negative regulation of bone remodeling by leptin

Mice harboring an inactivating mutation in the leptin-encoding gene (*ob/ob* mice) or in the gene encoding Lepr, its only known receptor (*db/db* mice), are obese and hypogonadic (Zhang et al., 1994; Tartaglia et al., 1995; Ahima, 2004). Given the negative effect of sex steroid hormone depletion on bone mass, they should therefore exhibit a low-bone-mass phenotype. Instead, these two mutant strains display a high-bone-mass phenotype when analyzed by histomorphometry, which given this mice obesity, allows the most objective evaluation of bone parameters as its results do not need to be adjusted for differences of lean/fat mass (Ducy et al., 2000; Baldock et al., 2005). This increased bone mass, which appears limited to cancellous bone, is also observed in genetically engineered *Lepr* knockout rats (Vaira et al., 2012; Solomon et al., 2014) and in transgenic lipodystrophic mice (b-ZIP mice) (Ducy et al., 2000). The latter observation is consistent with the advanced bone age observed in patients with lipodystrophy, a condition associated with a severe deficiency in serum leptin (Westvik, 1996; Elefteriou et al., 2004). Conversely, expression of a leptin transgene and intracerebroventricular infusion (ICV) of leptin can correct the high-bone-mass phenotype of respectively lipodystrophic and *ob/ob* mice, and mice expressing a gain-of-function of the leptin receptor (*l/l* mice) have decreased bone mass (Elefteriou et al., 2004; Shi et al., 2008).

The high-bone-mass phenotype of leptin-signaling deficient mice is due to an increase in bone formation rate that is already present in fat-restricted 1-month-old *ob/ob* mice and heterozygote *ob/+* mice, which are not obese, indicating that it is the absence of leptin not an increase in fat mass that is responsible for this phenotype (Ducy et al., 2000). Consistent with their hypogonadism, *ob/ob* and *db/db* mice have an increase in bone resorption (Ducy et al., 2000). It is, however, milder than their innate and permanent sex steroid depletion would have predicted, because by itself leptin deficiency has a negative effect on bone resorption through its action on the sympathetic tone (see below).

Studies using peripheral injections of pharmacological doses of leptin (over 40 μg/day; i.e., 700-fold the amount used in ICV studies) (Steppan et al., 2000; Cornish et al., 2002; Turner et al., 2013) have reported a beneficial effect on bone mass accrual, and proposed that leptin could act directly on osteoblasts. However, several arguments argue against a direct action on bone cells. First, gene expression in bone tissue or isolated osteoblasts as well as lineage-tracing studies using a Leptin receptor reporter transgene (*Lepr-YFP*) have failed to detect it in osteoblasts (Ducy et al., 2000; Ding et al., 2012). Second, osteoblasts derived from *db/db* mice proliferate and differentiate normally when cultured ex vivo (Ducy et al., 2000). Third, neither transgenic expression of leptin in osteoblasts nor the conditional inactivation of *Lepr* specifically in osteoblasts causes a bone phenotype in mice (Takeda et al., 2002; Shi et al., 2008). In contrast, inactivating *Lepr* in neurons induces the same high-bone-mass phenotype than the one observed in *db/db* mice (Shi et al., 2008; Yadav et al., 2009). The positive effect on bone mass observed upon leptin injections at high doses could in fact be explained by the well-known leptin resistance that centrally occurs when its serum concentration becomes too high, for example due to obesity (Könner and Brüning, 2012; Motyl and Rosen, 2012; Mark, 2013). Since leptin centrally affects bone mass accrual at a lower threshold than it affects body weight, the effect of leptin on bone remodeling is likely to be more quickly impacted by a leptin-resistance mechanism (Elefteriou et al., 2004; Shi et al., 2008). Hence, supraphysiological doses of this hormone injected peripherally could still cause weight loss, as some of the above mentioned studies reported, while already eliciting a central leptin resistance that will cause a gain of bone mass similar to the one observed in the absence of leptin signaling.

Dual action of adiponectin on bone remodeling

Analysis of mice deficient in adiponectin has identified both central and peripheral effects of this hormone on bone mass accrual. Young adult *Adiponectin−/−* mice display a high-bone-mass phenotype that is due to an increase in the number of osteoblasts and bone formation rate (Kajimura et al., 2013). Genetic and in vitro studies attributed this effect to adiponectin's ability to regulate proliferation and apoptosis of osteoblasts in a FoxO1-dependent manner via PI3 kinase activation.

This early high-bone-mass phenotype, however, cannot be observed in older mice that, instead, show a decreased bone mass characterized by a lesser number and proliferation ability of osteoblasts and an increase in bone resorption parameters (Kajimura et al., 2013; Wu et al., 2014). This phenotype is consistent with the observation that injection of an adiponectin-expressing adenovirus increases bone mass by suppressing osteoclastogenesis while activating osteoblastogenesis (Oshima et al., 2005).

The observations mentioned above indicate that the adipocyte secretes two hormones exerting opposite influences on the same physiological functions in adult mice. The fact that b-ZIP lipodystrophic mice have a high-bone-mass phenotype despite lacking both these hormones (Ducy et al., 2000), however, suggests that lack of leptin is dominant over the absence of adiponectin.

Central and efferent regulators of bone remodeling

Leptin's action on bone remodeling is mediated by brain serotonin signaling and the sympathetic nervous system

Leptin's action on appetite and reproduction depends on its binding to a specific receptor expressed by brain neurons (Ahima and Flier, 2000; Ahima, 2004). In agreement with this central mode of action, ICV infusion of low doses of leptin in *ob/ob* mice, at a rate that does not result in any detectable leak of leptin in the general circulation, corrects their high-bone-mass phenotype (Ducy et al., 2000). This rescue is complete and occurs even at minimal doses that do not influence body weight (Elefteriou et al., 2004). Likewise, low doses of leptin administered by ICV infusion in wild-type animals induce a potent antiosteogenic effect resulting in severe bone loss within 1 month (Ducy et al., 2000).

Chemical lesioning experiments identified the ventromedial hypothalamus (VMH) as the relay of the leptin anti-osteogenic effect in the brain (Takeda et al., 2002). Yet deletion of the leptin receptor in these same neurons does not cause a bone phenotype (Balthasar et al., 2004; Yadav et al., 2009). This apparent contradiction is explained by the fact that leptin uses a serotonin relay to act on the VMH neurons of the hypothalamus (Yadav et al., 2009; Ducy and Karsenty, 2010; Oury and Karsenty, 2011). Indeed, leptin signals to neurons of the dorsal and median raphe nuclei in the brain stem to blunt the production of brain serotonin that itself is a positive regulator of bone mass accrual (Yadav et al., 2009) (Fig. 34.1). In support of this mechanism, *Tph2*-deficient mice, which lack brain serotonin, show a low-bone-mass phenotype mirroring the one observed in *ob/ob* mice, and normalizing the brain content of the *ob/ob* mice by inactivating one allele of *Tph2* corrects their bone phenotype (Yadav et al., 2009). Conversely, inactivation of *LepR* in the *Tph2*-expressing neurons of the brain stem causes a bone phenotype similar to the one presented by *db/db* mice (Yadav et al., 2009). The negative influence of a deficit in brain serotonin signaling on bone mass explains the bone loss observed upon chronic use of selective serotonin reuptake inhibitors (SSRIs) as well as in a mouse model of Alzheimer's disease (Dengler-Crish et al., 2016; Ortuno et al., 2016).

Consistent with the lesioning experiments that identified the VMH nuclei as mediating leptin antiosteogenic action, dextran anterograde and retrograde neuron tracing studies showed that serotonergic neurons project from the brain stem to the VMH nuclei (Yadav et al., 2009). Serotonin then signals to VMH neurons via the Htr2c receptor to regulate CREB phosphorylation on Ser133 via Ca^{2+} as a second messenger and a CaMKKβ/CaMKIV cascade (Oury et al., 2010). This osteogenic pathway was demonstrated in vivo as mice with an inactivation of *Creb* in VMH neurons (*Creb* $_{VHM}$*−/−*) or heterozygous compound for *Creb* and *Htr2c* (*Creb* $_{VHM}$*−/−; Htr2c+/−*) or for *Creb* and *CaMKIV* (*Creb* $_{VHM}$*+/−; CaMKIV* $_{VHM}$ *+/−*) in VMH neurons all show a low-bone-mass phenotype characterized by decreased bone formation and increased bone resorption (Oury et al., 2010). Thus, leptin negatively regulates bone mass by inhibiting brain stem serotonergic neurons, which normally signal to VMH neurons via an Htr2c→CaMKKβ→CaMKIV→CREB cascade to increased bone formation and decrease bone resorption (Fig. 34.1).

A recurrent feature of the mouse models mentioned above is an increase in sympathetic tone, as evidenced by high *Ucp1* expression in brown adipose tissue and urinary levels of epinephrine (Oury et al., 2010). In contrast, *ob/ob* mice are known to exhibit low sympathetic activity (Bray and York, 1998) that, when corrected by treatment with an adrenergic agonist, leads to a dramatic decrease in bone mass (Takeda et al., 2002). The role of the sympathetic nervous system (SNS) as an antiosteogenic activity is supported by the observation that mice deficient in dopamine β-hydroxylase (DBH), an enzyme necessary to produce norepinephrine and epinephrine (the catecholamine ligands produced by the SNS), have a high-bone-mass phenotype (Takeda et al., 2002). Also consistent with this notion is the existence of a high-bone-mass phenotype associated with both an increase in bone formation and a decrease in bone resorption in mice deficient globally or only in osteoblasts for the β2 adrenergic receptor, the main adrenergic receptor expressed in these cells (Takeda et al., 2002; Kajimura et al., 2011). Importantly, the bone phenotype observed in Adrβ2-deficient mice cannot be rescued by

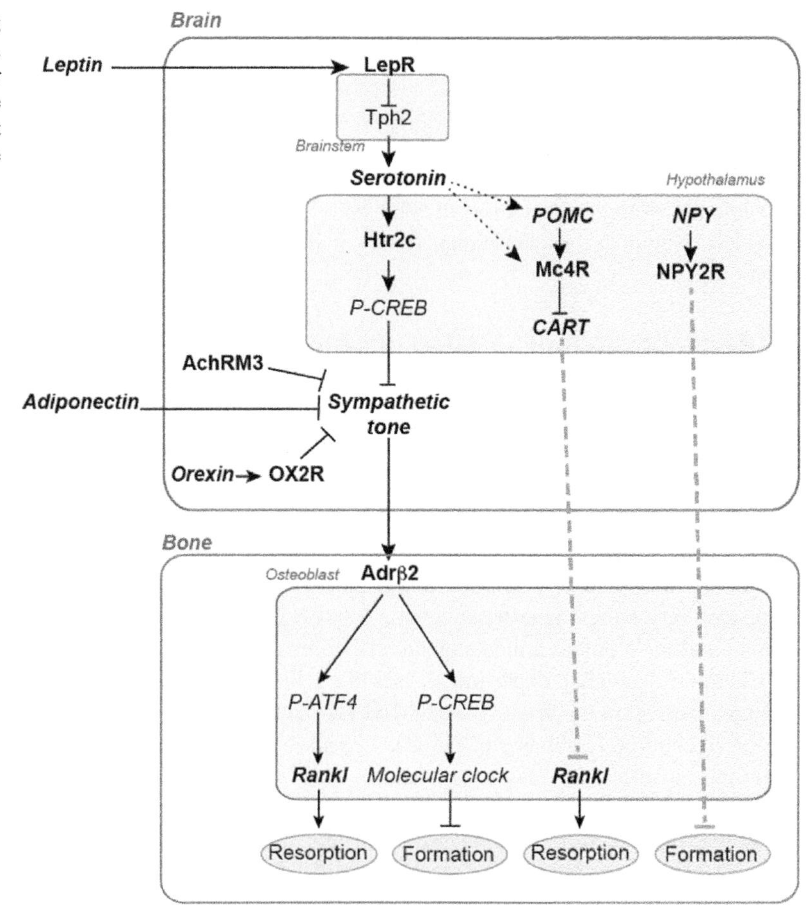

FIGURE 34.1 Central regulation of bone remodeling in mice. Secreted molecules are shown in bold Italics, receptors are shown in bold letters and intracellular effectors are shown in italics. *Black dotted lines* indicate regulation that are most likely to exist but have not yet been demonstrated formally. *Dashed gray lines* indicate mode of actions that have not yet been identified.

leptin ICV infusion demonstrating that the sympathetic nervous system, via Adrβ2 mediates leptin regulation of bone mass (Takeda et al., 2002; Elefteriou et al., 2005). Pharmacologic experiments have confirmed the genetic demonstration of an effect of the SNS on bone mass. Treatment of ovariectomized mice or rats with low doses of propranolol, a nonselective β-blocker, inhibits bone loss and preserves mechanical properties (Takeda et al., 2002; Bonnet et al., 2008; Khajuria et al., 2013). Similar improvements have been observed in a rat model of hypertension associated with bone fragility using low doses of butoxamine, a selective Adrβ2 antagonist (Arai et al., 2013).

At the molecular level, sympathetic signaling acts in osteoblasts through a dual mechanism (Fig. 34.1). It impairs osteoblast proliferation through phosphorylation of CREB, the molecular clock, type D cyclins, and *AP-1* gene expression (Fu et al., 2005; Kajimura et al., 2011). It also promotes *Rankl* expression in osteoblast progenitor cells following protein kinase A phosphorylation of ATF4, a cell-specific CREB-related transcription factor essential (Yang et al., 2004; Elefteriou et al., 2005; Kajimura et al., 2011).

Counterregulation of sympathetic nervous system control of bone remodeling by the parasympathetic nervous system and adiponectin

The low-bone-mass phenotype observed in aged adiponectin-deficient mice, which is due to the conjunction of a decrease in bone formation and an increase in bone resorption parameters is caused by an increase in sympathetic activity. Indeed, infusing adiponectin centrally or removing one allele of *Dbh* from *Adiponectin−/−* mice normalizes *Ucp1* expression in brown fat, norepinephrine content in the brain and urinary epinephrine elimination. and triggers an increase in bone mass secondary to an increase in osteoblast numbers and bone formation rate (Kajimura et al., 2013; Wu et al., 2014). The regulation of SNS output by adiponectin is mediated by its activity on the neurons of the locus coeruleus (LC), the site of *Dbh* expression in the brain stem. Indeed, *Adiponectin−/−* mice lacking only in neurons of the LC one allele of *FoxO1,* which encodes an intracellular effector of adiponectin signaling, have normal sympathetic activity as measured by *Ucp1* expression in brown fat and a significant increase in bone mass, osteoblast number, and bone formation rate while osteoclast number, *Rankl* expression, and serum CTx levels are decreased compared with *Adiponectin−/−* mice (Kajimura et al., 2013).

Signaling through the muscarinic receptor 3 (AchRM3) is another mechanism counteracting the SNS control of bone remodeling. Muscarinic receptors (AchRM1-4) bind acetylcholine, the main neurotransmitter used by the PNS. While inactivation of *AchRM1*, *AchRM2*, or *AchRM4* does not significantly affect bone mass, mice lacking AchRM3 either globally or only in neurons have a low-bone-mass phenotype associated with increased bone resorption and decreased bone formation (Shi et al., 2010). This phenotype can be attributed to the negative action of the SNS, since it is associated with elevated levels of epinephrine in the urine and is corrected when catecholamine signaling is normalized through coinactivation of one allele of *Adrβ2* (*AchRM3−/−; Adrβ2+/−* mice) (Shi et al., 2010). *AchRM3* being expressed in the noradrenergic neurons of the LC but not in the VMH of the hypothalamus or in sympathetic chain ganglia (Shi et al., 2010) suggests a direct effect on central sympathetic output similar to the one described for adiponectin signaling (Fig. 34.1). Whether these two effectors indeed mechanistically overlap still needs to be defined.

Other regulators of sympathetic nervous system control of bone remodeling

NeuromedinU

NeuromedinU (NMU) is a neuropeptide produced by nerve cells in the small intestine and in the brain (Brighton et al., 2004). It is generally assumed that NMU regulates various aspects of physiology including appetite, stress response, and SNS activation (Brighton et al., 2004). NMU-deficient mice present a high-bone-mass phenotype associated with an increase in bone formation similar to the one displayed by *ob/ob* mice (Sato et al., 2007). This phenotype is not cell autonomous, as NMU-deficient osteoblasts in culture are indistinguishable from wild-type osteoblasts and treatment of wild-type osteoblasts with NMU do not affect their proliferation or differentiation. In contrast, ICV infusion of NMU in wild-type, NMU−/− and *ob/ob* mice decreases bone formation and bone volume (Sato et al., 2007). However, ICV infusion of leptin in NMU−/− mice cannot decrease bone mass, indicating that NMU is a mediator for leptin's action on bone formation (Sato et al., 2007). Accordingly, NMU affects the leptin-dependent negative regulation of bone formation by the molecular clock and it was thus proposed that the high-bone-mass phenotype of the NMU−/− mice is caused by their resistance to the antiosteogenic activity of the SNS (Sato et al., 2007). Yet the basis of this mechanism of action still remains incompletely understood.

Endocannabinoid signaling

Endocannabinoids are lipid mediators that regulate analgesia, energy balance, and appetite by binding to the G protein-coupled receptors CB1 and CB2. CB1 is predominantly expressed in the brain and peripheral neurons, where it is responsible for the psychotropic action of cannabinoids, but it is also expressed in peripheral tissues including immune cells as well as the reproductive system, gastrointestinal tract, and lungs (Di Marzo et al., 2004). Global and conditional inactivation of *CB1* in noradrenergic and adrenergic neurons of the central and peripheral nervous system up-regulates bone formation and bone mass while decreasing bone resorption in adult male mice (Idris et al., 2005; Tam et al., 2006; Deis et al., 2018). Although skeletal levels of norepinephrine are not changed in the conditional knockout mice, expression of *Adrβ2* is increased (Deis et al., 2018). Whether this increase explains their bone phenotype and the molecular basis of this observation remain to be defined. In contrast, *CB2* is not expressed in the brain but is mainly expressed by immune cells (Di Marzo et al., 2004). It is also expressed at lower levels in osteoblasts and osteoclasts and CB2-deficient mice have a low-bone-mass phenotype most likely of peripheral origin (Ofek et al., 2006).

Orexin signaling

Orexins are neuropeptides expressed in the lateral hypothalamus, stimulating multiple behaviors including wakefulness and feeding by binding to two receptors OX1R and OX2R. In the absence of orexin, mice develop a low-bone-mass phenotype caused by a decrease in bone formation (Wei et al., 2014). Analysis of mice lacking the orexin receptors identified an underlying complexity to this phenotype (Wei et al., 2014). Indeed, inactivation of OX1R is associated with increased bone formation and bone mass while bone resorption and *Rankl* expression are decreased. This phenotype appears to be caused by a peripheral effect involving signaling in osteoblasts by Ghrelin, a peptide predominantly expressed in stomach that binds to receptors expressed in osteoblasts to regulate their differentiation and function (Delhanty et al., 2014; Ma et al., 2015). In contrast, absence of OX2R is associated with decreased bone formation and bone mass, and this phenotype is centrally mediated since it can be reproduced by ICV injection of an OX2R-selective antagonist (Wei et al., 2014). Orexin's central effect could be achieved through modulation of the sympathetic tone, since *Ucp1* expression in brown adipose tissue is increased in OX2R−/− mice (Wei et al., 2014), and the densest extrahypothalamic projection of orexin-positive neurons extends to the noradrenergic LC (Peyron et al., 1998) (Fig. 34.1). Such a mechanism, however, stills needs to be formally demonstrated.

Regulation of bone resorption by melanocortin receptor 4 and cocaine- and amphetamine-regulated transcript

In addition to the SNS-related pathway mentioned above, the absence of brain serotonin causes a sharp increase in the expression of the melanocortin receptor 4 gene (*Mc4R*) (Yadav et al., 2009). Melanocortins, a family of peptides produced by posttranslational processing of proopiomelanocortin (POMC), regulate food intake and energy expenditure via binding to two melanocortin receptors expressed in the central nervous system (Mc3R and Mc4R). While these receptors show a widespread presence in the rodent brain (Mountjoy et al., 1994; Kishi et al., 2003), POMC has a limited distribution, being described in only two neuronal populations: one in the arcuate nucleus of the hypothalamus (ARC) and the other in the nucleus of the tractus solitarius (NTS) of the brain stem (Palkovits et al., 1987; Bronstein et al., 1992). The POMC neurons of the ARC are known to be responsive to leptin and MC4R has thus been implicated in leptin's control of appetite (Cheung et al., 1997; Cone, 1999). It was first reported that patients lacking Mc4R have high bone mineral density (BMD) and advanced bone age (Farooqi et al., 2000). Subsequently, *Mc4r−/−* mice were shown to have an increase in bone mass caused by an isolated decrease in osteoclast number and function consistent with decreased *Rankl* expression (Elefteriou et al., 2005). The same decrease in bone resorption activity was also noted in MC4R-deficient patients (Ahn et al., 2006).

This high-bone-mass/low-resorption phenotype has been explained by an increase in CART signaling. CART is a neuropeptide encoded by a gene expressed in hypothalamic neurons among other parts of the nervous system and in peripheral organs such as the pancreas and the adrenal glands but not in bone cells (Couceyro et al., 1997; Elias et al., 1998; Wierup et al., 2004; Elefteriou et al., 2005). That *Cart* expression is virtually undetectable in hypothalamic neurons of *ob/ob* mice suggested that it is positively regulated by leptin and could therefore act as a mediator of its functions (Douglass et al., 1995; Kristensen et al., 1998). Yet Cart-deficient mice do not present a body weight or reproduction phenotype (Asnicar et al., 2001). Instead they display a late-onset low-bone-mass phenotype (Elefteriou et al., 2005). Osteoblast numbers and bone formation rates are normal in these mice, but the osteoclast surface and number are nearly doubled, leading to a significant increase in urinary Dpd elimination. Given the absence of *Cart* central expression in *ob/ob* mice, this negative regulation of bone resorption by CART explains, at least in part, the increase in bone resorption observed in these mice.

The action of CART on bone resorption is peripheral, as ICV infusion of recombinant CART in *Cart−/−* mice does not correct their low-bone-mass phenotype while transgenic mice harboring a twofold increase in CART circulating level display high bone mass due to an isolated decrease in osteoclast number. Moreover, this same transgene can rescue the low-bone-mass phenotype of the *Cart−/−* mice (Singh et al., 2008). The *Cart−/−* phenotype is not bone cells autonomous as Cart-deficient bone marrow macrophages differentiate normally into osteoclasts and Cart-deficient bone marrow stromal cells can normally support osteoclastogenesis in coculture experiments (Elefteriou et al., 2005). *Rankl* expression is increased in *Cart−/−* bones, suggesting that Cart regulates bone resorption by ultimately modulating Rankl signaling (Elefteriou et al., 2005; Ahn et al., 2006). In the absence of any identified CART receptor, however, one can only speculate whether this factor acts directly on osteoblasts or uses one or several relays to signal to bone cells.

Mc4r inactivation in mice causes an increase in hypothalamic *Cart* expression, and serum CART levels are significantly elevated in patients heterozygous for inactivating mutations of Mc4R (Elefteriou et al., 2005; Ahn et al., 2006). Moreover, inactivation of one or two copies of the *Cart* gene in Mc4r-deficient mice normalizes their bone resorption parameters, *Rankl* expression, and thereby their bone mass (Ahn et al., 2006). Thus, in addition to the SNS pathway, leptin also controls bone resorption via an MC4R/CART pathway (Fig. 34.1). Although likely, whether this regulation also occurs via a brain serotonin relay remains to be genetically confirmed.

Y receptor signaling

The Y signaling system is complex, consisting of at least five receptors (NPY1R, NPY2R, NPY4R, NPY5R, and NPY6R in mouse) with different binding profiles and sites of expression in the central nervous system and the periphery as well as multiple endogenous ligands: neuropeptide Y (NPY), peptide YY (PYY), and pancreatic polypeptide (PP) (Lin et al., 2004). NPY is widely expressed in the central and peripheral nervous system, and NPY fibers project from the ARC nuclei, which are known to participate in the control of appetite (Hokfelt et al., 1998). The related family members PYY and PP are produced in the small intestine and colon or in the pancreas, respectively (Hazelwood, 1993), where they affect gut motility in addition to pancreatic and gall bladder secretion. PYY is also expressed in a subset of brain stem neurons (Glavas et al., 2008). PYY and PP can signal to specific Y receptors in the hypothalamus and the brain stem to further influence pancreatic and gastric secretion. NPY and PYY have identical affinity for all known Y receptors, and PP binds preferentially to the Y4 receptor (Larhammar 1996).

While *PP−/−* mice do not show a bone phenotype, both PYY-deficient and NPY-deficient mice show an increase in bone mass (Wortley et al., 2007; Baldock et al., 2009; Wong et al., 2012), suggesting that the latter two peptides are negative regulators of bone remodeling. Evidence from overexpressing mice and studies of genetically engineered models suggests that only NPY may exert such activity via central signaling (in addition to direct effects on bone cells) (Matic et al., 2012; Wong et al., 2012) (Fig. 34.1). Indeed, overexpression of NPY in the hypothalamus decreases bone mass by inhibiting bone formation, and this activity can, even if partly, normalize bone mass in the *NPY−/−* mice (Baldock et al., 2005; Baldock et al., 2009). This central activity is most likely mediated by the NPY2R receptor, as its hypothalamus-specific depletion causes an increase in cancellous bone volume by increasing bone formation (Baldock et al., 2002; Baldock et al., 2006; Shi et al., 2010), while the peripheral activity of NPY depends on NPY1R (Baldock et al., 2007). An opposite role has been recently proposed for the Y6 receptor, which is expressed in brain but not in bone and whose global inactivation is associated with reduced bone mass, decreased bone formation, and increased osteoclast number (Khor et al., 2016). This central role, however, awaits to be confirmed via brain-specific inactivation experiments.

Brain-derived neurotrophic factor

Brain-derived neurotrophic factor (BDNF) is a member of the nerve growth factor family of proteins that binds to the tyrosine kinase receptor type B (Trkb). It is expressed in the CNS as well as the peripheral nervous system and is known to affect memory, cognition, and behavior including feeding. Specific inactivation of *Bdnf* in the brain of mice causes a high-bone-mass phenotype associated with decreased osteoclast number, at least in females (Camerino et al., 2012). These mice do not show changes in sympathetic activity or serotonin levels suggesting another, yet unidentified, mode of action.

Interleukin-1

IL-1 is a polypeptide product that mediates several components of acute-phase response to infection and injury. Its main sites of expression are the peripheral immune system and bone cells as well as glia and neuron cells in the CNS (Lorenzo et al., 1990; Dinarello, 1997). When injected subcutaneously, IL-1 is a potent stimulator of bone resorption, and IL-1R-deficient mice do not lose bone after ovariectomy (Sabatini et al., 1988; Lorenzo et al., 1998). Targeted overexpression in the CNS of mice of human IL-1Ra, a natural IL-1 receptor antagonist, results in a low-bone-mass phenotype caused by a doubling of osteoclast numbers (Bajayo et al., 2005). However, the cellular and molecular bases of this central activity are still unknown.

Evidence of central/neuronal regulations of bone mass in human

While genetic engineering in mice has been a powerful tool in identifying many central and neuron-related factors controlling bone remodeling, human genetic evidence confirming these findings is still limited. One possible explanation for this paucity of data stems from the fact that most of the factors identified have major roles regulating other key functions in the body such as feeding, energy metabolism, reproduction, ad behavior, and it is therefore difficult to assess effects on bone mass independently of these other consequences in patients. For example, correcting for body weight or body mass index (BMI) may weaken correlations, since they could be inherent to the trait analyzed and therefore legitimately associated with it but also may simply reflect lifestyle differences. Another issue could come from the exclusion (or absence of exclusion) of patients using medications (yet) considered unrelated to bone health but that are designed to address other activities of these pleiotropic factors. Given these limitations, as much as a positive genetic evidence can be considered a validation of the findings made in rodents, an absence of evidence should be interpreted cautiously.

In addition to the limitations mentioned above, association studies based on the quantification of these factors' blood levels or on therapeutic interventions related to their role in other diseases can be biased by the technical difficulty to accurately extrapolate neuron-produced levels of neuromediators from blood to bone and by the specificity (or lack thereof), dose, effect, and side effects of the drugs evaluated. In that respect, their conclusions should be taken with even more reservation than genetic associations.

Leptin

An association with increased circulating levels of leptin and incidence of osteoporosis has been reported in patients harboring specific alleles at two polymorphism sites of the leptin receptor gene (rs1137100 and rs1137101, both G variants) (Ye and Lu, 2013). Although these variants were not associated with obesity in this particular population, suggesting that they do not cause a receptor loss-of-function, their impact on leptin signaling has not been elucidated, and these data should therefore be interpreted carefully. In contrast, a patient harboring a known obesity-inducing inactivating mutation of the leptin gene was shown to display high bone mass, more definitely indicating that leptin fulfills a role in humans similar to the one it plays in rodents (Elefteriou et al., 2004). Other, although indirect, evidence of such a role is the advanced bone age and presence of osteosclerotic lesions in lipodystrophic patients, whose serum leptin levels are extremely low due to the absence of adipocytes (Westvik, 1996; Elefteriou et al., 2004).

Multiple studies have also assessed a correlation between leptin serum levels and bone mass. However, especially in the case of leptin, one must be cautious when reporting the results of clinical studies including obese participants since their blood leptin levels are expected to be high, but they may be experiencing central leptin resistance. The mere fact that obese individuals are overweight despite high levels of serum leptin most likely demonstrates such a leptin loss-of-function effect. In this case, finding a positive association of leptin levels with BMD would be an expected outcome. In contrast, when leptin levels are moderately elevated and body weight remains within the normal range, negative correlations between serum leptin levels and bone mass have been observed. For instance, an early study noted an inverse association between serum leptin levels adjusted for fat mass and BMD in a group of 221 Japanese men (Sato et al., 2001). Likewise, in a large cohort of young men (the Gothenburg Osteoporosis and Obesity Determinants cohort), leptin was found to be a negative independent predictor of areal BMD at several measured sites, and of cortical bone size at both non—weight-loaded bones (radius) and weight-loaded bones (tibia) (Lorentzon et al., 2006). Importantly, this study was performed on a primarily healthy population with normal body mass indexes sparing the need to adjust BMD readings for differences in lean and fat mass. In another population of healthy subjects without difference on body weight or body mass index, serum levels of leptin were significantly associated with reduced total body, hip, femoral neck, femoral shaft, and total femur BMD (Wu et al., 2009). Other cross-sectional studies have failed to show such a negative association between serum leptin levels and areal BMD (Goulding and Taylor, 1998; Martini et al., 2001; Thomas et al., 2001; Papadopoulou et al., 2004). Most likely, some of the differences between these and other studies can be attributed to the way data are evaluated and are presented, either adjusted or unadjusted for body weight. For instance, in a large North American population-based study including a high representation of the elderly, non-Hispanic blacks, and Mexican Americans, BMD increases with increasing leptin concentration in men. However, after adjustment for BMI and other bone-related factors, an inverse association emerged, being most evident in men younger than 60 years (Ruhl and Everhart, 2002). Similarly, in a few recent studies in men, leptin was inversely correlated to aBMD, an association that became apparent only after adjustment of aBMD for body weight (Sato et al., 2001; Morberg et al., 2003). Lastly, in a small population of middle-age (non-pre-menopausal) Japanese subjects, significant negative correlations were observed between cortical bone thickness and both blood leptin levels and sympathetic activity, as assessed by quantification of heart rate variability, while a positive correlation existed between leptin levels and sympathetic activity (Kuriyama et al., 2017).

Adrenergic signaling

Multiple nongenomic evidence indicates that the SNS function on bone mass is conserved in humans. For instance, patients with reflex sympathetic dystrophy, a disease characterized by localized high sympathetic activity, develop a severe and localized osteoporosis that can be treated by β-blockers (Kurvers, 1998). Likewise, short sleep, which causes an increase in SNS activity, is associated with bone loss; pheochromocytomas, which cause an increase in catecholamine levels, are associated with an increase in bone resorption that can be normalized by adrenalectomy; and sympathetic activity, quantified by microneurography, was found inversely associated with bone volume, trabecular thickness, and bone strength at the distal radius in women (Farr et al., 2012; Veldhuis-Vlug et al., 2012; Kuriyama et al., 2017). Genetic evidence, however, has not been conclusive. While the risk of osteoporosis at the femoral neck was associated with a polymorphism in the *ADRβ2* gene (rs1042713, AG and GG genotypes) in a medium-sized population of postmenopausal Korean women, no correlation could be found with BMD and fracture risk in case-control and meta-analysis studies of populations of European origin (Lee et al., 2014; Veldhuis-Vlug et al., 2015). Additional studies are needed that analyze polymorphisms at this locus in different populations but also in genes encoding other adrenergic receptors, as in humans those may contribute to the effect of the SNS on osteoblasts.

From a pharmacological standpoint, beneficial effects of β-blocker administration on BMD and/or fracture risk in women have been reported by multiple epidemiologic studies. In postmenopausal women (Geelong Osteoporosis Study), higher BMD at the total hip and ultradistal forearm as well as a decreased fracture risk and serum CTx levels were associated with β-blocker use (Pasco et al. 2004, 2005). Likewise, in a large cohort of elderly men (Concord Health and Aging in Men Project, CHAMP), age-related loss of BMD was found attenuated by the use of β-blockers (Bleicher et al., 2013). Analysis of the Dubbo Osteoporosis Epidemiology Study (DOES) cohort also showed that both men and women using β-blockers had higher BMD at the femoral neck and spine as well as lower fracture risk than those not treated (Yang et al., 2011) Case-control studies also showed that the use of β-blockers decreases the odds ratio for facture and was associated with a higher BMD at the spine, femoral neck, and proximal femur (Schlienger et al., 2004; Bonnet et al., 2007). Two prospective case-control studies respectively in elderly and severely burned patients showed that BMD was significantly improved in β-blocker users (Turker et al., 2006; Herndon et al., 2012). More generally, meta-analyses have identified an association between use of β-blockers (but not α-blockers) and a statistically significant protection against all fractures, with protection against hip fractures being stronger (Wiens et al., 2006; Yang et al., 2012).

A few studies also reported less conclusive evidence of a beneficial effect of β-blockers on bone mass in human patients. For instance, in US women enrolled in the Study of Osteoporotic Fractures, total hip BMD was greater among β-blocker users, but adjustment for body weight or other parameters eliminated the difference. Nevertheless, there was a protective effect of β-blockers against hip fracture (hazard ratio for hip fracture associated with β-blocker use was 0.76 [95% CI 0.58−0.99]) (Reid et al., 2005). Analysis of a large cohort of Medicare beneficiaries treated with antihypertensive agents also did not observe a protective effect of β-blockers on the incidence of bone fractures identified based on diagnosis and procedure codes, but in the absence of radiographs and BMD data (Solomon et al., 2011). The discrepancy between the results of these and the studies cited above could have many origins. One of them could relate to the absence of specificity of the β-blocker used, a parameter that is usually not taken into account. As a matter of fact, many studies have been performed on populations using β-blockers following a diagnosis of hypertension and therefore most likely to be treated with β_1-blockers, which are cardioselective. Although a lower risk of fracture was associated with usage of this class of drugs in the Korean Health Insurance Review and Assessment Service database (Song et al., 2012), their biological effect on bone cells is not clear, since studies in mice have suggested that the Adrβ1 receptor could counteract the effect of Adrβ2 on bone remodeling (Elefteriou et al., 2005; Pierroz et al., 2012). Perhaps even more critical, the length of treatment and dose of β-blockers could be other significant factors of variability within and between studies. At least in ovariectomized rats the best preventive effect against bone loss was obtained with the lowest dose of propranolol, the highest dose being ineffective, while the opposite was observed for cardiac hemodynamic functions (Bonnet et al., 2006). Given that most patients analyzed in studies using β-blockers were prescribed these drugs to treat hypertension, it is thus possible that a significant proportion of them may be taking doses inappropriate to achieve a positive effect on bone mass.

Brain serotonin and neuromedinU

Variants of human *TPH2* linked to serotonin deficiency have been reported as associated with a spectrum of neuropsychiatric disorders such as major depression or bipolar disorder, and these conditions are also associated with bone loss and/or increased risk of fractures (Misra et al., 2004; McKinney et al., 2009; Hsu et al., 2016). Although this observation is suggestive of a link between these variants and bone health, specific studies are needed to confirm this hypothesis. Another indication that brain serotonin plays a role in the control of bone mass in human stems from the bone loss observed in patients chronically using SSRIs (Wu et al., 2009; Haney et al., 2010; Rizzoli et al., 2012; Zhou et al., 2018). This same effect can be reproduced in rodents, and molecular studies have shown that it is mediated by a decrease of brain serotonin signaling leading to increased sympathetic output (Warden et al., 2005; Bonnet et al., 2007; Warden et al., 2008; Ortuno et al., 2016). Accordingly, it could be prevented by cotreatment with a β-blocker (Ortuno et al., 2016). This latter observation also awaits confirmation by clinical studies.

In contrast, the contribution of NeuromedinU to the central regulation of bone mass has already been assessed in humans. Analysis of a large population of European children (IDEFICS cohort) has associated variants of the *NMU* gene with bone strength (C variant of both rs6827359 and rs12500837 as well as homozygosity of the C variant of rs9999653) (Gianfagna et al., 2013). Interestingly, this study also reported a synergistic effect between NMU and ADRB2 variants (*NMU* rs6827359, CC variant and *ADRB2* rs1042713, GG variant), confirming the functional link observed in mouse models (Sato et al., 2007).

Melanocortin receptor 4 and cocaine- and amphetamine-regulated transcript

As mentioned above, inactivating mutations in one allele of MC4R are linked to advanced bone age, assessed by radiography of the wrist, in children and a high-bone-mass phenotype in adults (Farooqi et al., 2000). This phenotype is associated with an increase in circulating CART levels and a decrease in the resorption biomarker CTx (Ahn et al., 2006). In addition, a polymorphism in the CART gene (rs2239670, AG variant) was found associated with a higher BMD at the lumbar spine (Chun et al., 2015) in Korean postmenopausal women while a significant association between another one (rs7379701, TC and CC variants) and forearm BMD was reported in Caucasian postmenopausal women (Guerardel et al., 2006).

Neuropeptide Y, brain-derived neurotrophic factor, and cannabinoid receptor 2

A small longitudinal substudy of early postmenopausal women (Kuopio Osteoporosis Risk Factor and Prevention population) has shown that a gain-of-function polymorphism in the *NPY* gene (rs16139, variant C) associates with a better maintenance of femoral neck BMD after 5 years (Heikkinen et al., 2004, Ding et al., 2005) and analysis of a population of Afro-Caribbean men (Tobago Bone Health Study) has identified two single nucleotide polymorphisms (SNPs) in the *NPY* gene associated with higher total-hip BMD (rs16135 and rs16123, G variants) (Goodrich et al., 2009). However, the other variant of the latter polymorphism (rs16123, CC variant), along with another NPY SNP (rs17149106, GG variant), was associated with increased odds ratio for osteoporosis at the lumbar spine (although there was no significant difference in BMD) in postmenopausal Korean women (Chun et al., 2015). This discrepancy may reflect sex-based differences, a trend that was also observed in mouse models (Baldock et al., 2009). Lastly, in this same population of women, the combination of two polymorphisms in the *NPY2R* gene (TT variants of both rs2880415 and rs6857715) was also associated with a higher rate of osteoporosis at the lumber spine (Chun et al., 2015). Altogether, these results support an involvement of NPY signaling in the regulation of bone mass in human.

While the analysis of four SNPs in human *CB1* did not find a statistically significant correlation with bone health, in a small population of postmenopausal women, several *CB2* SNPs, alone or in combination, show a significant increase in the odds ratio for osteoporosis (Karsak et al., 2005). These results are consistent with an involvement of at least the *CB2* locus in human osteoporosis.

Lastly, two studies have shown a link between BDNF and bone mass. A genome-wide search for phospho-SNP associated with BMD at the hip and spine identified a BDNF variant (rs6265, V66M) in both Caucasian and Chinese populations (Deng et al., 2013). This variant, which impairs the phosphorylation of BDNF by the CHEK2 kinase at amino-acid T62, also associates with multiple mental disorders (Gratacos et al., 2007; Deng et al., 2013). In addition, an integrative analysis that combined four genome-wide association study datasets identified two *BDNF* SNPs (rs7124442, 3′UTR variant and rs11030119, intron variant) with spine BMD and osteoporotic fractures (Guo et al., 2016).

Conclusions and perspective

Since the identification of leptin's central regulation of bone remodeling in mice, the past 2 decades have seen a large number of studies revealing the complexity and potential biomedical importance of the neuronal control of bone mass. The pleiotropic nature of these pathways as well as the technical difficulty in measuring central or even peripheral neuromediator levels or activity in patients have somewhat hampered their validation in human physiology, but many elements have also emerged that confirm findings made using animal models. Whether some or most of these regulations take place in humans, they represent potential targets for designing new diagnostic and therapeutic tools that should be further explored to predict or treat bone health disorders.

References

Ahima, R.S., 2004. Body fat, leptin, and hypothalamic amenorrhea. N. Engl. J. Med. 351 (10), 959–962.

Ahima, R.S., Flier, J.S., 2000. Leptin. Annu. Rev. Physiol. 62, 413–437.

Ahn, J.D., Dubern, B., Lubrano-Berthelier, C., Clement, K., Karsenty, G., 2006. Cart overexpression is the only identifiable cause of high bone mass in melanocortin 4 receptor deficiency. Endocrinology 147 (7), 3196–3202.

Arai, M., Sato, T., Takeuchi, S., Goto, S., Togari, A., 2013. Dose effects of butoxamine, a selective beta2-adrenoceptor antagonist, on bone metabolism in spontaneously hypertensive rat. Eur. J. Pharmacol. 701 (1–3), 7–13.

Asnicar, M.A., Smith, D.P., Yang, D.D., Heiman, M.L., Fox, N., Chen, Y.F., Hsiung, H.M., Koster, A., 2001. Absence of cocaine- and amphetamine-regulated transcript results in obesity in mice fed a high caloric diet. Endocrinology 142 (10), 4394–4400.

Bajayo, A., Goshen, I., Feldman, S., Csernus, V., Iverfeldt, K., Shohami, E., Yirmiya, R., Bab, I., 2005. Central IL-1 receptor signaling regulates bone growth and mass. Proc. Natl. Acad. Sci. U. S. A. 102 (36), 12956–12961.

Baldock, P.A., Allison, S., McDonald, M.M., Sainsbury, A., Enriquez, R.F., Little, D.G., Eisman, J.A., Gardiner, E.M., Herzog, H., 2006. Hypothalamic regulation of cortical bone mass: opposing activity of Y2 receptor and leptin pathways. J. Bone Miner. Res. 21 (10), 1600–1607.

Baldock, P.A., Allison, S.J., Lundberg, P., Lee, N.J., Slack, K., Lin, E.J., Enriquez, R.F., McDonald, M.M., Zhang, L., During, M.J., Little, D.G., Eisman, J.A., Gardiner, E.M., Yulyaningsih, E., Lin, S., Sainsbury, A., Herzog, H., 2007. Novel role of Y1 receptors in the coordinated regulation of bone and energy homeostasis. J. Biol. Chem. 282 (26), 19092–19102.

Baldock, P.A., Lee, N.J., Driessler, F., Lin, S., Allison, S., Stehrer, B., Lin, E.J., Zhang, L., Enriquez, R.F., Wong, I.P., McDonald, M.M., During, M., Pierroz, D.D., Slack, K., Shi, Y.C., Yulyaningsih, E., Aljanova, A., Little, D.G., Ferrari, S.L., Sainsbury, A., Eisman, J.A., Herzog, H., 2009. Neuropeptide Y knockout mice reveal a central role of NPY in the coordination of bone mass to body weight. PLoS One 4 (12), e8415.

Baldock, P.A., Sainsbury, A., Allison, S., Lin, E.J., Couzens, M., Boey, D., Enriquez, R., During, M., Herzog, H., Gardiner, E.M., 2005. Hypothalamic control of bone formation: distinct actions of leptin and y2 receptor pathways. J. Bone Miner. Res. 20 (10), 1851–1857.

Baldock, P.A., Sainsbury, A., Couzens, M., Enriquez, R.F., Thomas, G.P., Gardiner, E.M., Herzog, H., 2002. Hypothalamic Y2 receptors regulate bone formation. J. Clin. Investig. 109 (7), 915–921.

Balthasar, N., Coppari, R., McMinn, J., Liu, S.M., Lee, C.E., Tang, V., Kenny, C.D., McGovern, R.A., Chua Jr., S.C., Elmquist, J.K., Lowell, B.B., 2004. Leptin receptor signaling in POMC neurons is required for normal body weight homeostasis. Neuron 42 (6), 983–991.

Bleicher, K., Cumming, R.G., Naganathan, V., Seibel, M.J., Blyth, F.M., Le Couteur, D.G., Handelsman, D.J., Creasey, H.M., Waite, L.M., 2013. Predictors of the rate of BMD loss in older men: findings from the CHAMP study. Osteoporos. Int. 24 (7), 1951–1963.

Bonnet, N., Benhamou, C.L., Malaval, L., Goncalves, C., Vico, L., Eder, V., Pichon, C., Courteix, D., 2008. Low dose beta-blocker prevents ovariectomy-induced bone loss in rats without affecting heart functions. J. Cell. Physiol. 217 (3), 819–827.

Bonnet, N., Bernard, P., Beaupied, H., Bizot, J.C., Trovero, F., Courteix, D., Benhamou, C.L., 2007a. Various effects of antidepressant drugs on bone microarchitecture, mechanical properties and bone remodeling. Toxicol. Appl. Pharmacol. 221 (1), 111–118.

Bonnet, N., Gadois, C., McCloskey, E., Lemineur, G., Lespessailles, E., Courteix, D., Benhamou, C.L., 2007b. Protective effect of beta blockers in postmenopausal women: influence on fractures, bone density, micro and macroarchitecture. Bone 40 (5), 1209–1216.

Bonnet, N., Laroche, N., Vico, L., Dolleans, E., Benhamou, C.L., Courteix, D., 2006. Dose effects of propranolol on cancellous and cortical bone in ovariectomized adult rats. J. Pharmacol. Exp. Ther. 318 (3), 1118–1127.

Bray, G.A., York, D.A., 1998. The MONA LISA hypothesis in the time of leptin. Recent Prog. Horm. Res. 53, 95–117.

Brighton, P.J., Szekeres, P.G., Willars, G.B., 2004. Neuromedin U and its receptors: structure, function, and physiological roles. Pharmacol. Rev. 56 (2), 231–248.

Bronstein, D.M., Schafer, M.K., Watson, S.J., Akil, H., 1992. Evidence that beta-endorphin is synthesized in cells in the nucleus tractus solitarius: detection of POMC mRNA. Brain Res. 587 (2), 269–275.

Camerino, C., Zayzafoon, M., Rymaszewski, M., Heiny, J., Rios, M., Hauschka, P.V., 2012. Central depletion of brain-derived neurotrophic factor in mice results in high bone mass and metabolic phenotype. Endocrinology 153 (11), 5394–5405.

Chenu, C., 2002. Glutamatergic regulation of bone remodeling. J. Musculoskelet. Neuronal Interact. 2 (3), 282–284.

Chenu, C., 2004. Role of innervation in the control of bone remodeling. J. Musculoskelet. Neuronal Interact. 4 (2), 132–134.

Cheung, C.C., Clifton, D.K., Steiner, R.A., 1997. Proopiomelanocortin neurons are direct targets for leptin in the hypothalamus. Endocrinology 138 (10), 4489–4492.

Chun, E.H., Kim, H., Suh, C.S., Kim, J.H., Kim, D.Y., Kim, J.G., 2015. Polymorphisms in neuropeptide genes and bone mineral density in Korean postmenopausal women. Menopause 22 (11), 1256–1263.

Cone, R.D., 1999. The central melanocortin system and energy homeostasis. Trends Endocrinol. Metabol. 10 (6), 211–216.

Cornish, J., Callon, K.E., Bava, U., Lin, C., Naot, D., Hill, B.L., Grey, A.B., Broom, N., Myers, D.E., Nicholson, G.C., Reid, I.R., 2002. Leptin directly regulates bone cell function in vitro and reduces bone fragility in vivo. J. Endocrinol. 175 (2), 405–415.

Couceyro, P.R., Koylu, E.O., Kuhar, M.J., 1997. Further studies on the anatomical distribution of CART by in situ hybridization. J. Chem. Neuroanat. 12 (4), 229–241.

Deis, S., Srivastava, R.K., Ruiz de Azua, I., Bindila, L., Baraghithy, S., Lutz, B., Bab, I., Tam, J., 2018. Age-related regulation of bone formation by the sympathetic cannabinoid CB1 receptor. Bone 108, 34–42.

Delhanty, P.J., van der Velde, M., van der Eerden, B.C., Sun, Y., Geminn, J.M., van der Lely, A.J., Smith, R.G., van Leeuwen, J.P., 2014. Genetic manipulation of the ghrelin signaling system in male mice reveals bone compartment specificity of acylated and unacylated ghrelin in the regulation of bone remodeling. Endocrinology 155 (11), 4287–4295.

Deng, F.Y., Tan, L.J., Shen, H., Liu, Y.J., Liu, Y.Z., Li, J., Zhu, X.Z., Chen, X.D., Tian, Q., Zhao, M., Deng, H.W., 2013. SNP rs6265 regulates protein phosphorylation and osteoblast differentiation and influences BMD in humans. J. Bone Miner. Res. 28 (12), 2498–2507.

Dengler-Crish, C.M., Smith, M.A., Wilson, G.N., 2016. Early evidence of low bone density and decreased serotonergic synthesis in the dorsal raphe of a tauopathy model of Alzheimer's disease. J. Alzheimer's Dis. 55 (4), 1605–1619.

Di Marzo, V., Bifulco, M., De Petrocellis, L., 2004. The endocannabinoid system and its therapeutic exploitation. Nat. Rev. Drug Discov. 3 (9), 771–784.

Dinarello, C.A., 1997. Interleukin-1. Cytokine Growth Factor Rev. 8 (4), 253–265.

Ding, B., Kull, B., Liu, Z., Mottagui-Tabar, S., Thonberg, H., Gu, H.F., Brookes, A.J., Grundemar, L., Karlsson, C., Hamsten, A., Arner, P., Ostenson, C.G., Efendic, S., Monne, M., von Heijne, G., Eriksson, P., Wahlestedt, C., 2005. Human neuropeptide Y signal peptide gain-of-function polymorphism is associated with increased body mass index: possible mode of function. Regul. Pept. 127 (1–3), 45–53.

Ding, L., Saunders, T.L., Enikolopov, G., Morrison, S.J., 2012. Endothelial and perivascular cells maintain haematopoietic stem cells. Nature 481 (7382), 457–462.

Douglass, J., McKinzie, A.A., Couceyro, P., 1995. PCR differential display identifies a rat brain mRNA that is transcriptionally regulated by cocaine and amphetamine. J. Neurosci. 15 (3 Pt 2), 2471–2481.

Ducy, P., Amling, M., Takeda, S., Priemel, M., Schilling, A.F., Beil, F.T., Shen, J., Vinson, C., Rueger, J.M., Karsenty, G., 2000. Leptin inhibits bone formation through a hypothalamic relay: a central control of bone mass. Cell 100 (2), 197–207.

Ducy, P., Karsenty, G., 2010. The two faces of serotonin in bone biology. J. Cell Biol. 191 (1), 7–13.

Elefteriou, F., Ahn, J.D., Takeda, S., Starbuck, M., Yang, X., Liu, X., Kondo, H., Richards, W.G., Bannon, T.W., Noda, M., Clement, K., Vaisse, C., Karsenty, G., 2005. Leptin regulation of bone resorption by the sympathetic nervous system and CART. Nature 434 (7032), 514–520.

Elefteriou, F., Takeda, S., Ebihara, K., Magre, J., Patano, N., Kim, C.A., Ogawa, Y., Liu, X., Ware, S.M., Craigen, W.J., Robert, J.J., Vinson, C., Nakao, K., Capeau, J., Karsenty, G., 2004. Serum leptin level is a regulator of bone mass. Proc. Natl. Acad. Sci. U. S. A. 101 (9), 3258–3263.

Elias, C.F., Lee, C., Kelly, J., Aschkenasi, C., Ahima, R.S., Couceyro, P.R., Kuhar, M.J., Saper, C.B., Elmquist, J.K., 1998. Leptin activates hypothalamic CART neurons projecting to the spinal cord. Neuron 21 (6), 1375–1385.

Farooqi, I.S., Yeo, G.S., Keogh, J.M., Aminian, S., Jebb, S.A., Butler, G., Cheetham, T., O'Rahilly, S., 2000. Dominant and recessive inheritance of morbid obesity associated with melanocortin 4 receptor deficiency. J. Clin. Investig. 106 (2), 271–279.

Farr, J.N., Charkoudian, N., Barnes, J.N., Monroe, D.G., McCready, L.K., Atkinson, E.J., Amin, S., Melton 3rd, L.J., Joyner, M.J., Khosla, S., 2012. Relationship of sympathetic activity to bone microstructure, turnover, and plasma osteopontin levels in women. J. Clin. Endocrinol. Metab. 97 (11), 4219–4227.

Fu, L., Patel, M.S., Bradley, A., Wagner, E.F., Karsenty, G., 2005. The molecular clock mediates leptin-regulated bone formation. Cell 122 (5), 803–815.

Gianfagna, F., Cugino, D., Ahrens, W., Bailey, M.E.S., Bammann, K., Herrmann, D., Koni, A.C., Kourides, Y., Marild, S., Molnár, D., Moreno, L.A., Pitsiladis, Y.P., Russo, P., Siani, A., Sieri, S., Sioen, I., Veidebaum, T., Iacoviello, L., I. c. on behalf of the, 2013. Understanding the links among neuromedin U gene, beta2-adrenoceptor gene and bone health: an observational study in European children. PLoS One 8 (8), e70632.

Glavas, M.M., Grayson, B.E., Allen, S.E., Copp, D.R., Smith, M.S., Cowley, M.A., Grove, K.L., 2008. Characterization of brainstem peptide YY (PYY) neurons. J. Comp. Neurol. 506 (2), 194–210.

Goodrich, L.J., Yerges-Armstrong, L.M., Miljkovic, I., Nestlerode, C.S., Kuipers, A.L., Bunker, C.H., Patrick, A.L., Wheeler, V.W., Zmuda, J.M., 2009. Molecular variation in neuropeptide Y and bone mineral density among men of African Ancestry. Calcif. Tissue Int. 85 (6), 507.

Goulding, A., Taylor, R.W., 1998. Plasma leptin values in relation to bone mass and density and to dynamic biochemical markers of bone resorption and formation in postmenopausal women. Calcif. Tissue Int. 63 (6), 456–458.

Gratacos, M., Gonzalez, J.R., Mercader, J.M., de Cid, R., Urretavizcaya, M., Estivill, X., 2007. Brain-derived neurotrophic factor Val66Met and psychiatric disorders: meta-analysis of case-control studies confirm association to substance-related disorders, eating disorders, and schizophrenia. Biol. Psychiatry 61 (7), 911–922.

Guerardel, A., Tanko, L.B., Boutin, P., Christiansen, C., Froguel, P., 2006. Obesity susceptibility CART gene polymorphism contributes to bone remodeling in postmenopausal women. Osteoporos. Int. 17 (1), 156–157.

Guo, Y., Dong, S.S., Chen, X.F., Jing, Y.A., Yang, M., Yan, H., Shen, H., Chen, X.D., Tan, L.J., Tian, Q., Deng, H.W., Yang, T.L., 2016. Integrating Epigenomic elements and GWASs identifies BDNF gene affecting bone mineral density and osteoporotic fracture risk. Sci. Rep. 6, 30558.

Haney, E.M., Warden, S.J., Bliziotes, M.M., 2010. Effects of selective serotonin reuptake inhibitors on bone health in adults: time for recommendations about screening, prevention and management? Bone 46 (1), 13–17.

Hazelwood, R.L., 1993. The pancreatic polypeptide (PP-fold) family: gastrointestinal, vascular, and feeding behavioral implications. Proc. Soc. Exp. Biol. Med. 202 (1), 44–63.

Heikkinen, A.-M., Niskanen, L.K., Salmi, J.A., Koulu, M., Pesonen, U., Uusitupa, M.I.J., Komulainen, M.H., Tuppurainen, M.T., Kröger, H., Jurvelin, J., Saarikoski, S., 2004. Leucine7 to proline7 polymorphism in prepro-NPY gene and femoral neck bone mineral density in postmenopausal women. Bone 35 (3), 589–594.

Herndon, D.N., Rodriguez, N.A., Diaz, E.C., Hegde, S., Jennings, K., Mlcak, R.P., Suri, J.S., Lee, J.O., Williams, F.N., Meyer, W., Suman, O.E., Barrow, R.E., Jeschke, M.G., Finnerty, C.C., 2012. Long-term propranolol use in severely burned pediatric patients: a randomized controlled study. Ann. Surg. 256 (3), 402–411.

Hokfelt, T., Broberger, C., Zhang, X., Diez, M., Kopp, J., Xu, Z., Landry, M., Bao, L., Schalling, M., Koistinaho, J., DeArmond, S.J., Prusiner, S., Gong, J., Walsh, J.H., 1998. Neuropeptide Y: some viewpoints on a multifaceted peptide in the normal and diseased nervous system. Brain Res. Brain Res. Rev. 26 (2–3), 154–166.

Hsu, C.C., Hsu, Y.C., Chang, K.H., Lee, C.Y., Chong, L.W., Wang, Y.C., Hsu, C.Y., Kao, C.H., 2016. Increased risk of fracture in patients with bipolar disorder: a nationwide cohort study. Soc. Psychiatr. Psychiatr. Epidemiol. 51 (9), 1331–1338.

Idris, A.I., van 't Hof, R.J., Greig, I.R., Ridge, S.A., Baker, D., Ross, R.A., Ralston, S.H., 2005. Regulation of bone mass, bone loss and osteoclast activity by cannabinoid receptors. Nat. Med. 11 (7), 774–779.

Jiang, S.D., Jiang, L.S., Dai, L.Y., 2006. Mechanisms of osteoporosis in spinal cord injury. Clin. Endocrinol. 65 (5), 555–565.

Kadowaki, T., Yamauchi, T., 2005. Adiponectin and adiponectin receptors. Endocr. Rev. 26 (3), 439–451.

Kajimura, D., Hinoi, E., Ferron, M., Kode, A., Riley, K.J., Zhou, B., Guo, X.E., Karsenty, G., 2011. Genetic determination of the cellular basis of the sympathetic regulation of bone mass accrual. J. Exp. Med. 208 (4), 841–851.

Kajimura, D., Lee, H.W., Riley, K.J., Arteaga-Solis, E., Ferron, M., Zhou, B., Clarke, C.J., Hannun, Y.A., DePinho, R.A., Guo, E.X., Mann, J.J., Karsenty, G., 2013. Adiponectin regulates bone mass via opposite central and peripheral mechanisms through FoxO1. Cell Metabol. 17 (6), 901–915.

Karsak, M., Cohen-Solal, M., Freudenberg, J., Ostertag, A., Morieux, C., Kornak, U., Essig, J., Erxlebe, E., Bab, I., Kubisch, C., de Vernejoul, M.C., Zimmer, A., 2005. Cannabinoid receptor type 2 gene is associated with human osteoporosis. Hum. Mol. Genet. 14 (22), 3389–3396.
Khajuria, D.K., Razdan, R., Mahapatra, D.R., Bhat, M.R., 2013. Osteoprotective effect of propranolol in ovariectomized rats: a comparison with zoledronic acid and alfacalcidol. J. Orthop. Sci. 18 (5), 832–842.
Khor, E.C., Yulyaningsih, E., Driessler, F., Kovacic, N., Wee, N.K.Y., Kulkarni, R.N., Lee, N.J., Enriquez, R.F., Xu, J., Zhang, L., Herzog, H., Baldock, P.A., 2016. The y6 receptor suppresses bone resorption and stimulates bone formation in mice via a suprachiasmatic nucleus relay. Bone 84, 139–147.
Kinjo, M., Setoguchi, S., Schneeweiss, S., Solomon, D.H., 2005. Bone mineral density in subjects using central nervous system-active medications. Am. J. Med. 118 (12), 1414.
Kishi, T., Aschkenasi, C.J., Lee, C.E., Mountjoy, K.G., Saper, C.B., Elmquist, J.K., 2003. Expression of melanocortin 4 receptor mRNA in the central nervous system of the rat. J. Comp. Neurol. 457 (3), 213–235.
Könner, A.C., Brüning, J.C., 2012. Selective insulin and leptin resistance in metabolic disorders. Cell Metabol. 16 (2), 144–152.
Kristensen, P., Judge, M.E., Thim, L., Ribel, U., Christjansen, K.N., Wulff, B.S., Clausen, J.T., Jensen, P.B., Madsen, O.D., Vrang, N., Larsen, P.J., Hastrup, S., 1998. Hypothalamic CART is a new anorectic peptide regulated by leptin. Nature 393 (6680), 72–76.
Kuriyama, N., Inaba, M., Ozaki, E., Yoneda, Y., Matsui, D., Hashiguchi, K., Koyama, T., Iwai, K., Watanabe, I., Tanaka, R., Omichi, C., Mizuno, S., Kurokawa, M., Horii, M., Niwa, F., Iwasa, K., Yamada, S., Watanabe, Y., 2017. Association between loss of bone mass due to short sleep and leptin-sympathetic nervous system activity. Arch. Gerontol. Geriatr. 70, 201–208.
Kurvers, H.A., 1998. Reflex sympathetic dystrophy: facts and hypotheses. Vasc. Med. 3 (3), 207–214.
Larhammar, D., 1996. Structural diversity of receptors for neuropeptide Y, peptide YY and pancreatic polypeptide. Regul. Pept. 65 (3), 165–174.
Lee, H.J., Kim, H., Ku, S.Y., Choi, Y.M., Kim, J.H., Kim, J.G., 2014. Association between polymorphisms in leptin, leptin receptor, and beta-adrenergic receptor genes and bone mineral density in postmenopausal Korean women. Menopause 21 (1), 67–73.
Lin, S., Boey, D., Herzog, H., 2004. NPY and Y receptors: lessons from transgenic and knockout models. Neuropeptides 38 (4), 189–200.
Lorentzon, M., Landin, K., Mellstrom, D., Ohlsson, C., 2006. Leptin is a negative independent predictor of areal BMD and cortical bone size in young adult Swedish men. J. Bone Miner. Res. 21 (12), 1871–1878.
Lorenzo, J.A., Naprta, A., Rao, Y., Alander, C., Glaccum, M., Widmer, M., Gronowicz, G., Kalinowski, J., Pilbeam, C.C., 1998. Mice lacking the type I interleukin-1 receptor do not lose bone mass after ovariectomy. Endocrinology 139 (6), 3022–3025.
Lorenzo, J.A., Sousa, S.L., Van den Brink-Webb, S.E., Korn, J.H., 1990. Production of both interleukin-1 alpha and beta by newborn mouse calvarial cultures. J. Bone Miner. Res. 5 (1), 77–83.
Ma, C., Fukuda, T., Ochi, H., Sunamura, S., Xu, C., Xu, R., Okawa, A., Takeda, S., 2015. Genetic determination of the cellular basis of the ghrelin-dependent bone remodeling. Mol. Metabol. 4 (3), 175–185.
Maeda, N., Shimomura, I., Kishida, K., Nishizawa, H., Matsuda, M., Nagaretani, H., Furuyama, N., Kondo, H., Takahashi, M., Arita, Y., Komuro, R., Ouchi, N., Kihara, S., Tochino, Y., Okutomi, K., Horie, M., Takeda, S., Aoyama, T., Funahashi, T., Matsuzawa, Y., 2002. Diet-induced insulin resistance in mice lacking adiponectin/ACRP30. Nat. Med. 8 (7), 731–737.
Mark, A.L., 2013. Selective leptin resistance revisited. Am. J. Physiol. Regul. Integr. Comp. Physiol. 305 (6), R566–R581.
Martini, G., Valenti, R., Giovani, S., Franci, B., Campagna, S., Nuti, R., 2001. Influence of insulin-like growth factor-1 and leptin on bone mass in healthy postmenopausal women. Bone 28 (1), 113–117.
Matic, I., Matthews, B.G., Kizivat, T., Igwe, J.C., Marijanovic, I., Ruohonen, S.T., Savontaus, E., Adams, D.J., Kalajzic, I., 2012. Bone-specific overexpression of NPY modulates osteogenesis. J. Musculoskelet. Neuronal Interact. 12 (4), 209–218.
McKinney, J.A., Turel, B., Winge, I., Knappskog, P.M., Haavik, J., 2009. Functional properties of missense variants of human tryptophan hydroxylase 2. Hum. Mutat. 30 (5), 787–794.
Misra, M., Papakostas, G.I., Klibanski, A., 2004. Effects of psychiatric disorders and psychotropic medications on prolactin and bone metabolism. J. Clin. Psychiatry 65 (12), 1607–1618 quiz 1590, 1760–1601.
Morberg, C.M., Tetens, I., Black, E., Toubro, S., Soerensen, T.I., Pedersen, O., Astrup, A., 2003. Leptin and bone mineral density: a cross-sectional study in obese and nonobese men. J. Clin. Endocrinol. Metab. 88 (12), 5795–5800.
Motyl, K.J., Rosen, C.J., 2012. Understanding leptin-dependent regulation of skeletal homeostasis. Biochimie 94 (10), 2089–2096.
Mountjoy, K.G., Mortrud, M.T., Low, M.J., Simerly, R.B., Cone, R.D., 1994. Localization of the melanocortin-4 receptor (MC4-R) in neuroendocrine and autonomic control circuits in the brain. Mol. Endocrinol. 8 (10), 1298–1308.
Ofek, O., Karsak, M., Leclerc, N., Fogel, M., Frenkel, B., Wright, K., Tam, J., Attar-Namdar, M., Kram, V., Shohami, E., Mechoulam, R., Zimmer, A., Bab, I., 2006. Peripheral cannabinoid receptor, CB2, regulates bone mass. Proc. Natl. Acad. Sci. U. S. A. 103 (3), 696–701.
Ortuno, M.J., Robinson, S.T., Subramanyam, P., Paone, R., Huang, Y.Y., Guo, X.E., Colecraft, H.M., Mann, J.J., Ducy, P., 2016. Serotonin-reuptake inhibitors act centrally to cause bone loss in mice by counteracting a local anti-resorptive effect. Nat. Med. 22 (10), 1170–1179.
Oshima, K., Nampei, A., Matsuda, M., Iwaki, M., Fukuhara, A., Hashimoto, J., Yoshikawa, H., Shimomura, I., 2005. Adiponectin increases bone mass by suppressing osteoclast and activating osteoblast. Biochem. Biophys. Res. Commun. 331 (2), 520–526.
Oury, F., Karsenty, G., 2011. Towards a serotonin-dependent leptin roadmap in the brain. Trends Endocrinol. Metabol. 22 (9), 382–387.
Oury, F., Yadav, V.K., Wang, Y., Zhou, B., Liu, X.S., Guo, X.E., Tecott, L.H., Schutz, G., Means, A.R., Karsenty, G., 2010. CREB mediates brain serotonin regulation of bone mass through its expression in ventromedial hypothalamic neurons. Genes Dev. 24 (20), 2330–2342.
Palkovits, M., Mezey, E., Eskay, R.L., 1987. Pro-opiomelanocortin-derived peptides (ACTH/beta-endorphin/alpha-MSH) in brainstem baroreceptor areas of the rat. Brain Res. 436 (2), 323–338.

Papadopoulou, F., Krassas, G.E., Kalothetou, C., Koliakos, G., Constantinidis, T.C., 2004. Serum leptin values in relation to bone density and growth hormone-insulin like growth factors axis in healthy men. Arch. Androl. 50 (2), 97–103.

Pasco, J.A., Henry, M.J., Nicholson, G.C., Schneider, H.G., Kotowicz, M.A., 2005. Beta-blockers reduce bone resorption marker in early postmenopausal women. Ann. Hum. Biol. 32 (6), 738–745.

Pasco, J.A., Henry, M.J., Sanders, K.M., Kotowicz, M.A., Seeman, E., Nicholson, G.C., 2004. Beta-adrenergic blockers reduce the risk of fracture partly by increasing bone mineral density: geelong osteoporosis study. J. Bone Miner. Res. 19 (1), 19–24.

Peyron, C., Tighe, D.K., van den Pol, A.N., de Lecea, L., Heller, H.C., Sutcliffe, J.G., Kilduff, T.S., 1998. Neurons containing hypocretin (orexin) project to multiple neuronal systems. J. Neurosci. 18 (23), 9996–10015.

Pierroz, D.D., Bonnet, N., Bianchi, E.N., Bouxsein, M.L., Baldock, P.A., Rizzoli, R., Ferrari, S.L., 2012. Deletion of β-adrenergic receptor 1, 2, or both leads to different bone phenotypes and response to mechanical stimulation. J. Bone Miner. Res. 27 (6), 1252–1262.

Reid, I.R., Gamble, G.D., Grey, A.B., Black, D.M., Ensrud, K.E., Browner, W.S., Bauer, D.C., 2005. Beta-blocker use, BMD, and fractures in the study of osteoporotic fractures. J. Bone Miner. Res. 20 (4), 613–618.

Rizzoli, R., Cooper, C., Reginster, J.Y., Abrahamsen, B., Adachi, J.D., Brandi, M.L., Bruyère, O., Compston, J., Ducy, P., Ferrari, S., Harvey, N.C., Kanis, J.A., Karsenty, G., Laslop, A., Rabenda, V., Vestergaard, P., 2012. Antidepressant medications and osteoporosis. Bone 51 (3), 606–613.

Ruhl, C.E., Everhart, J.E., 2002. Relationship of serum leptin concentration with bone mineral density in the United States population. J. Bone Miner. Res. 17 (10), 1896–1903.

Sabatini, M., Boyce, B., Aufdemorte, T., Bonewald, L., Mundy, G.R., 1988. Infusions of recombinant human interleukins 1 alpha and 1 beta cause hypercalcemia in normal mice. Proc. Natl. Acad. Sci. U. S. A. 85 (14), 5235–5239.

Sato, M., Takeda, N., Sarui, H., Takami, R., Takami, K., Hayashi, M., Sasaki, A., Kawachi, S., Yoshino, K., Yasuda, K., 2001. Association between serum leptin concentrations and bone mineral density, and biochemical markers of bone turnover in adult men. J. Clin. Endocrinol. Metab. 86 (11), 5273–5276.

Sato, S., Hanada, R., Kimura, A., Abe, T., Matsumoto, T., Iwasaki, M., Inose, H., Ida, T., Mieda, M., Takeuchi, Y., Fukumoto, S., Fujita, T., Kato, S., Kangawa, K., Kojima, M., Shinomiya, K.I., Takeda, S., 2007. Central control of bone remodeling by neuromedin U. Nat. Med. 10, 1234–1240.

Schlienger, R.G., Kraenzlin, M.E., Jick, S.S., Meier, C.R., 2004. Use of beta-blockers and risk of fractures. JAMA 292 (11), 1326–1332.

Shi, Y., Oury, F., Yadav, V.K., Wess, J., Liu, X.S., Guo, X.E., Murshed, M., Karsenty, G., 2010. Signaling through the M(3) muscarinic receptor favors bone mass accrual by decreasing sympathetic activity. Cell Metabol. 11 (3), 231–238.

Shi, Y., Yadav, V.K., Suda, N., Liu, X.S., Guo, X.E., Myers Jr., M.G., Karsenty, G., 2008. Dissociation of the neuronal regulation of bone mass and energy metabolism by leptin in vivo. Proc. Natl. Acad. Sci. U. S. A 105 (51), 20529–20533.

Shi, Y.C., Lin, S., Wong, I.P., Baldock, P.A., Aljanova, A., Enriquez, R.F., Castillo, L., Mitchell, N.F., Ye, J.M., Zhang, L., Macia, L., Yulyaningsih, E., Nguyen, A.D., Riepler, S.J., Herzog, H., Sainsbury, A., 2010. NPY neuron-specific Y2 receptors regulate adipose tissue and trabecular bone but not cortical bone homeostasis in mice. PLoS One 5 (6), e11361.

Singh, M.K., Elefteriou, F., Karsenty, G., 2008. Cocaine and amphetamine-regulated transcript may regulate bone remodeling as a circulating molecule. Endocrinology 149 (8), 3933–3941.

Solomon, D.H., Mogun, H., Garneau, K., Fischer, M.A., 2011. Risk of fractures in older adults using antihypertensive medications. J. Bone Miner. Res. 26 (7), 1561–1567.

Solomon, G., Atkins, A., Shahar, R., Gertler, A., Monsonego-Ornan, E., 2014. Effect of peripherally administered leptin antagonist on whole body metabolism and bone microarchitecture and biomechanical properties in the mouse. Am. J. Physiol. Endocrinol. Metab. 306 (1), E14–E27.

Song, H.J., Lee, J., Kim, Y.-J., Jung, S.-Y., Kim, H.J., Choi, N.-K., Park, B.-J., 2012. β1 selectivity of β-blockers and reduced risk of fractures in elderly hypertension patients. Bone 51 (6), 1008–1015.

Spencer, G.J., McGrath, C.J., Genever, P.G., 2007. Current perspectives on NMDA-type glutamate signalling in bone. Int. J. Biochem. Cell Biol. 39 (6), 1089–1104.

Steppan, C.M., Crawford, D.T., Chidsey-Frink, K.L., Ke, H., Swick, A.G., 2000. Leptin is a potent stimulator of bone growth in ob/ob mice. Regul. Pept. 92 (1–3), 73–78.

Takeda, S., Elefteriou, F., Levasseur, R., Liu, X., Zhao, L., Parker, K.L., Armstrong, D., Ducy, P., Karsenty, G., 2002. Leptin regulates bone formation via the sympathetic nervous system. Cell 111 (3), 305–317.

Tam, J., Ofek, O., Fride, E., Ledent, C., Gabet, Y., Muller, R., Zimmer, A., Mackie, K., Mechoulam, R., Shohami, E., Bab, I., 2006. Involvement of neuronal cannabinoid receptor CB1 in regulation of bone mass and bone remodeling. Mol. Pharmacol. 70 (3), 786–792.

Tartaglia, L.A., Dembski, M., Weng, X., Deng, N., Culpepper, J., Devos, R., Richards, G.J., Campfield, L.A., Clark, F.T., Deeds, J., et al., 1995. Identification and expression cloning of a leptin receptor, OB-R. Cell 83 (7), 1263–1271.

Thomas, T., Burguera, B., Melton 3rd, L.J., Atkinson, E.J., O'Fallon, W.M., Riggs, B.L., Khosla, S., 2001. Role of serum leptin, insulin, and estrogen levels as potential mediators of the relationship between fat mass and bone mineral density in men versus women. Bone 29 (2), 114–120.

Turker, S., Karatosun, V., Gunal, I., 2006. Beta-blockers increase bone mineral density. Clin. Orthop. Relat. Res. 443, 73–74.

Turner, R.T., Kalra, S.P., Wong, C.P., Philbrick, K.A., Lindenmaier, L.B., Boghossian, S., Iwaniec, U.T., 2013. Peripheral leptin regulates bone formation. J. Bone Miner. Res. 28 (1), 22–34.

Vaira, S., Yang, C., McCoy, A., Keys, K., Xue, S., Weinstein, E.J., Novack, D.V., Cui, X., 2012. Creation and preliminary characterization of a leptin knockout rat. Endocrinology 153 (11), 5622–5628.

Veldhuis-Vlug, A.G., El Mahdiui, M., Endert, E., Heijboer, A.C., Fliers, E., Bisschop, P.H., 2012. Bone resorption is increased in pheochromocytoma patients and normalizes following adrenalectomy. J. Clin. Endocrinol. Metab. 97 (11), E2093–E2097.

Veldhuis-Vlug, A.G., Oei, L., Souverein, P.C., Tanck, M.W.T., Rivadeneira, F., Zillikens, M.C., Kamphuisen, P.W., Maitland-van der Zee, A.H., de Groot, M.C.H., Hofman, A., Uitterlinden, A.G., Fliers, E., de Boer, A., Bisschop, P.H., 2015. Association of polymorphisms in the beta-2 adrenergic receptor gene with fracture risk and bone mineral density. Osteoporos. Int. 26 (7), 2019–2027.

Warden, S.J., Nelson, I.R., Fuchs, R.K., Bliziotes, M.M., Turner, C.H., 2008. Serotonin (5-hydroxytryptamine) transporter inhibition causes bone loss in adult mice independently of estrogen deficiency. Menopause 15 (6), 1176–1183.

Warden, S.J., Robling, A.G., Sanders, M.S., Bliziotes, M.M., Turner, C.H., 2005. Inhibition of the serotonin (5-hydroxytryptamine) transporter reduces bone accrual during growth. Endocrinology 146 (2), 685–693.

Wei, W., Motoike, T., Krzeszinski, J.Y., Jin, Z., Xie, X.J., Dechow, P.C., Yanagisawa, M., Wan, Y., 2014. Orexin regulates bone remodeling via a dominant positive central action and a subordinate negative peripheral action. Cell Metabol. 19 (6), 927–940.

Westvik, J., 1996. Radiological features in generalized lipodystrophy. Acta Paediatr. Suppl. 13, 44–51.

Wiens, M., Etminan, M., Gill, S.S., Takkouche, B., 2006. Effects of antihypertensive drug treatments on fracture outcomes: a meta-analysis of observational studies. J. Intern. Med. 260 (4), 350–362.

Wierup, N., Kuhar, M., Nilsson, B.O., Mulder, H., Ekblad, E., Sundler, F., 2004. Cocaine- and amphetamine-regulated transcript (CART) is expressed in several islet cell types during rat development. J. Histochem. Cytochem. 52 (2), 169–177.

Wong, I.P.L., Driessler, F., Khor, E.C., Shi, Y.-C., Hörmer, B., Nguyen, A.D., Enriquez, R.F., Eisman, J.A., Sainsbury, A., Herzog, H., Baldock, P.A., 2012. Peptide YY regulates bone remodeling in mice: a link between gut and skeletal biology. PLoS One 7 (7), e40038.

Wortley, K.E., Garcia, K., Okamoto, H., Thabet, K., Anderson, K.D., Shen, V., Herman, J.P., Valenzuela, D., Yancopoulos, G.D., Tschöp, M.H., Murphy, A., Sleeman, M.W., 2007. Peptide YY regulates bone turnover in rodents. Gastroenterology 133 (5), 1534–1543.

Wu, J.Y., Scadden, D.T., Kronenberg, H.M., 2009. Role of the osteoblast lineage in the bone marrow hematopoietic niches. J. Bone Miner. Res. 24 (5), 759–764.

Wu, Q., Magnus, J.H., Liu, J., Bencaz, A.F., Hentz, J.G., 2009. Depression and low bone mineral density: a meta-analysis of epidemiologic studies. Osteoporos. Int. 20 (8), 1309–1320.

Wu, Y., Tu, Q., Valverde, P., Zhang, J., Murray, D., Dong, L.Q., Cheng, J., Jiang, H., Rios, M., Morgan, E., Tang, Z., Chen, J., 2014. Central adiponectin administration reveals new regulatory mechanisms of bone metabolism in mice. Am. J. Physiol. Endocrinol. Metabol. 306 (12), E1418–E1430.

Yadav, V.K., Oury, F., Suda, N., Liu, Z.W., Gao, X.B., Confavreux, C., Klemenhagen, K.C., Tanaka, K.F., Gingrich, J.A., Guo, X.E., Tecott, L.H., Mann, J.J., Hen, R., Horvath, T.L., Karsenty, G., 2009. A serotonin-dependent mechanism explains the leptin regulation of bone mass, appetite, and energy expenditure. Cell 138 (5), 976–989.

Yang, S., Nguyen, N.D., Center, J.R., Eisman, J.A., Nguyen, T.V., 2011. Association between beta-blocker use and fracture risk: the Dubbo osteoporosis Epidemiology study. Bone 48 (3), 451–455.

Yang, S., Nguyen, N.D., Eisman, J.A., Nguyen, T.V., 2012. Association between beta-blockers and fracture risk: a Bayesian meta-analysis. Bone 51 (5), 969–974.

Yang, X., Matsuda, K., Bialek, P., Jacquot, S., Masuoka, H.C., Schinke, T., Li, L., Brancorsini, S., Sassone-Corsi, P., Townes, T.M., Hanauer, A., Karsenty, G., 2004. ATF4 is a substrate of RSK2 and an essential regulator of osteoblast biology; implication for Coffin-Lowry Syndrome. Cell 117 (3), 387–398.

Ye, X.L., Lu, C.F., 2013. Association of polymorphisms in the leptin and leptin receptor genes with inflammatory mediators in patients with osteoporosis. Endocrine 44 (2), 481–488.

Zhang, Y., Proenca, R., Maffei, M., Barone, M., Leopold, L., Friedman, J.M., 1994. Positional cloning of the mouse obese gene and its human homologue. Nature 372, 425–432.

Zhou, C., Fang, L., Chen, Y., Zhong, J., Wang, H., Xie, P., 2018. Effect of selective serotonin reuptake inhibitors on bone mineral density: a systematic review and meta-analysis. Osteoporos. Int. 29 (6), 1243–1251.

Section E

Other systemic hormones that influence bone metabolism

Chapter 35

Estrogens and progestins

David G. Monroe and Sundeep Khosla

Department of Medicine, Division of Endocrinology, Mayo Clinic College of Medicine, Rochester, MN, United States; The Robert and Arlene Kogod Center on Aging, Rochester, MN, United States

Chapter outline

Estrogens and estrogen receptors 827
Estrogen receptor mouse models 828
Estrogens—from a clinical perspective 829
Progestins and progesterone receptors in bone biology 832
Summary and conclusions 833
Conflict of Interest 834
References 834

Estrogens and estrogen receptors

Osteoporosis, characterized by loss of bone mass and microarchitecture, is an enormous and growing public health concern worldwide. Indeed, one in three women and one in five men will suffer an osteoporotic fracture in their lifetime (Melton 3rd et al., 1998; Kanis et al., 2000). Moreover, this burden in women exceeds the incidence of breast cancer, stroke, and myocardial infarction combined in a given year (Cauley et al., 2008). Although aging is the greatest risk factor for osteoporotic bone loss, estrogen deficiency in women and men is an important determinant for bone loss later in life (Riggs et al., 2002). Interestingly, although menopause is the major life event leading to the reduction in gonadal function in women, increasing evidence demonstrates that gradually declining levels of biologically available estrogens in men also lead to age-related bone loss (Khosla et al., 2008). Therefore, the understanding of how estrogens modulate bone physiology is critical not only from a scientific perspective, but also from a clinical perspective due to the widespread prevalence of osteoporosis.

The general importance of estrogen in the regulation of bone physiology was first recognized by the studies of Fuller Albright in the 1940s, who demonstrated that estrogen deficiency, through either ovariectomy (OVX) or following menopause, caused bone loss that could be effectively reversed through the administration of estradiol (Albright, 1940). Following this seminal discovery, investigators hypothesized that a protein, or receptor, must exist which can bind estrogens and mediate their effects on cellular and/or tissue physiology. The first evidence for an "estrogen receptor" (ER) came from experiments conducted by Gorski and Jensen in the 1960s, where they showed that rat uterus protein extract homogenate bound labeled-17β-estradiol (Gorski et al., 1968; Jensen et al., 1968; Shyamala and Gorski, 1969).

The molecular evidence for a direct effect of estrogens on bone did not occur for another 20 years, when a number of investigators demonstrated that bone was indeed an estrogen-responsive tissue (Chow et al., 1992; Ernst et al., 1989; Spelsberg et al., 1999; Tobias and Chambers, 1991; Tobias et al., 1991; Turner et al., 1993). Two distinct ERs are now recognized as mediating the effects of estrogen. ERα (*ESR1*) was cloned in the 1980s (Greene et al., 1986; Walter et al., 1985), and ERβ (*ESR2*) was cloned about a decade later (Kraus et al., 1995; Mosselman et al., 1996; Onoe et al., 1997). Both receptors share a structural homology with each other as well as with all members of the nuclear hormone superfamily of receptors (Mangelsdorf et al., 1995): (1) a central DNA binding (DBD) and hinge domain exists that mediates the interaction with specific DNA sequence elements and dimerization, (2) a C-terminal ligand binding domain (LBD), which also contains a transcriptional activation function (AF2) that allows specific ligands to bind causing a conformational change in the 12 α-helices necessary for activation of the ERs, and (3) a variable N-terminal domain with a secondary activation function (AF1). The simplified domain structure of human ERα and ERβ, with percent homologies, is shown in

Principles of Bone Biology. https://doi.org/10.1016/B978-0-12-814841-9.00035-X

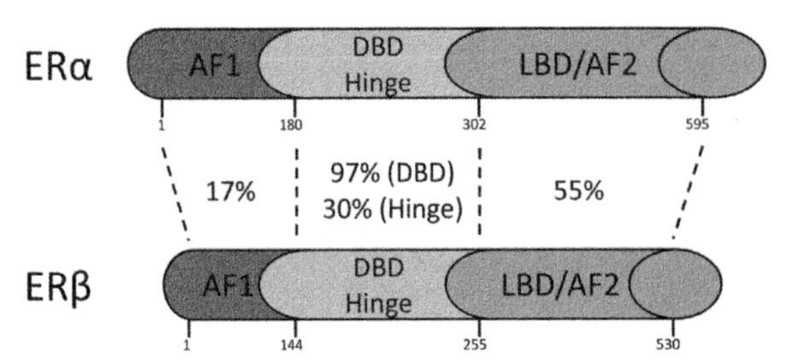

FIGURE 35.1 Structure and amino acid identity between human ERα and ERβ.

Fig. 35.1. ERα is expressed in cells of the osteoblastic (Eriksen et al., 1988; Ikegami et al., 1993) and osteoclastic (Oursler et al., 1991, 1994) lineages. ERβ is also often also coexpressed in these same lineages (Braidman et al., 2001; Vidal et al., 1999; Windahl et al., 2000; Arts et al., 1997; Monroe et al., 2003a).

Since estrogens are steroid molecules and largely hydrophobic, they can easily diffuse into cells from the circulation, where they encounter and bind to the ERs. This complex can directly regulate the transcription of genes under the control of estrogen response elements, called primary response genes, which typically occurs within 30–60 min of estrogen binding. These proteins then can modulate the transcription of further downstream genes (24–48 h later), in a process termed the "cascade model" of steroid hormone action (Landers and Spelsberg, 1992). ERs typically function as dimers, with both homodimers and heterodimers observed in osteoblastic cells. Although ERα and ERβ can regulate some of the same genes, their patterns of expression have been shown to be largely unique in bone cells (Monroe et al., 2003a, 2005). There is also considerable evidence that nongenomic ER pathways exist that are mediated by membrane-bound ERs and activate kinase cascades to elicit very rapid (within minutes) cellular effects (Razandi et al., 1999). These functions have been extensively reviewed (Levin, 2002; Saczko et al., 2017; Kelly and Levin, 2001) and thus will not be discussed in further detail here.

Estrogen receptor mouse models

As described above, the effects of estrogen on bone physiology are largely mediated through two related yet distinct soluble proteins belonging to the nuclear receptor superfamily of transcription factors, termed ERα and ERβ (Mangelsdorf et al., 1995; Monroe et al., 2003b; Almeida et al., 2017). Although more information exists on the role of ERα in bone physiology compared with ERβ, our understanding of the full effects of ERα on bone during growth and aging is far from complete. The first studies conducted on understanding the role of ERα in bone physiology were performed by Sims and colleagues, who examined mice harboring a global deletion (e.g., in all tissues) of ERα (Sims et al., 2002). They found in both males and females that global ERα deletion triggered a complex phenotype, including a decrease in bone turnover, decreased cortical thickness, and an increase in trabecular bone mass. However, the effects were confounded by high circulating estrogen levels in female mice and high testosterone levels in both sexes. Because ER expression has been shown to exhibit a compartment-specific expression pattern (with ERα expression in both cortical and trabecular sites, whereas ERβ expression is only evident in trabecular sites) (Onoe et al., 1997; Bord et al., 2001), these data suggest that high estrogen levels activate ERβ in trabecular bone, leading to a compensatory response at trabecular sites. On the contrary, global deletion of ERβ did not affect circulated steroid levels in either males or females and largely had no effect on bone in males. In females, trabecular bone, periosteal, and endosteal circumference were increased (Windahl et al., 1999, 2000; Sims et al., 2002). Even with the confounding steroid levels in the ERα global knockout mouse models, these studies demonstrated that loss of ERα decreased cortical thickness, that ERβ could compensate for loss of ERα in trabecular bone, and in the setting of intact ERα signaling, ERβ seemed to antagonize ERα action in bone.

Although the data garnered from these global ER knockout mouse models provided some insights into ER function in bone physiology, it was necessary to characterize the functions of these receptors in the various cell types responsible for bone remodeling (i.e., osteoblasts, osteoclasts, osteocytes). Therefore, more conclusive data on the functions of ERs in bone came from utilizing the Cre/LoxP system to achieve cell-specific deletion of ERs. In these systems, a crucial exon in either ERα or ERβ is flanked by loxP recombination sites and crossed into a "driver" mouse model whereby a tissue-specific promoter drives expression of Cre recombinase, thereby triggering a recombination event in which the gene is effectively knocked out in a cell-specific manner. Recently, this system has been used to demonstrate that an important target cell for ERα action is the osteoclast. ERα deletion in the early myeloid lineage using the LysM-Cre model (Martin-Millan et al., 2010) and in the more differentiated osteoclast lineage cells using the Ctsk-Cre model

(Nakamura et al., 2007) led to reduced trabecular bone mass due to increased osteoclast number and bone resorption in female mice. These data suggest that ERα is crucial for the control of osteoclast number and resorption in female mice.

The Cre/LoxP system has also been utilized to examine the role of ERα and ERβ in osteoblasts, using osteoblast lineage-specific Cre mouse drivers. Using Cre drivers that are active in early mesenchymal progenitor cells (Prx1-Cre) (Logan et al., 2002) or in osteoprogenitors (Osx-Cre) (Rodda and McMahon, 2006) to delete ERα resulted in impaired Wnt signaling and impaired optimal cortical bone mass acquisition that was ligand independent (Almeida et al., 2013). No trabecular phenotype was observed in either model. Surprisingly, deletion of ERα using the Col1a1-Cre driver, which is active in fully differentiated osteoblasts, elicited no bone phenotype in either sex up to 26 weeks of age (Almeida et al., 2013). In contrast with these findings, deletion of ERα using an osteocalcin-Cre (Ocn-Cre) model, which also is active in differentiated osteoblasts (although possibly somewhat more mature than the Col1a1 promoter), led to deficits in bone mass at both trabecular and cortical sites in female mice (Maatta et al., 2013; Melville et al., 2014). There was also a deficit in trabecular bone in males, although this phenotype was not apparent until ~6 months of age (whereas the phenotype in females was evident by 3 months of age).

Since the osteocyte has recently been identified as an important controller of bone metabolism (Bonewald, 2017), studies have been conducted in which ERα was conditionally deleted in osteocytes using the osteocyte-dominant Dmp1 promoter-driven Cre (Dmp1-Cre) (Kalajzic et al., 2004). This promoter may slightly overlap with the Ocn-Cre and has yielded somewhat inconsistent results. Windahl and colleagues (Windahl et al., 2013) reported that only male mice exhibited a lowered trabecular bone mass, whereas Kondoh and colleagues (Kondoh et al., 2014) reported the same phenotype but in only female mice. The two studies were consistent in that no cortical phenotype was observed in either sex. Although the true nature of ERα function in osteocytes cannot be cleanly gleaned from these studies, it is clear that ERα signaling is somehow important in late osteoblasts/osteocytes. Future studies will need to be designed to address these discrepant findings.

Studies examining the role of cell-specific deletion of ERβ have begun to shed insight into this molecule as well. Deletion of ERβ in osteoprogenitor (Prx1-Cre) or osteoblasts (Col1a1-Cre) elicited opposite effects to that of ERα deletion, which is an increase in trabecular bone mass with no effect on cortical bone mass in female mice (Nicks et al., 2016). In experiments examining the effects of ERβ deletion using bone marrow stromal cells, an increase in the ratio of colony-forming unit (CFU)-osteoblasts to CFU-fibroblasts was observed, suggesting an increased differentiation capacity of ERβ-negative progenitor cells. Consistent with this, ERβ has been shown to antagonize ERα in other systems (Hall and McDonnell, 1999); it is likely that ERβ plays a similar antagonistic role in bone.

The understanding of how ERα and ERβ function in bone physiology is now becoming more complete and Table 35.1 summarizes the current published knowledge from multiple laboratories concerning the phenotypes of the ER conditional-deletion mouse models. Although certainly discrepancies exist, a general consensus is becoming evident: ERα deletion in osteoclasts leads to reduced trabecular bone but not cortical bone in female (but not male) mice, ERα deletion in later osteoblastic lineages generally leads to reduction in both trabecular and cortical bone mass (although this appears to be somewhat sex-specific in some cases), and that ERβ deletion in osteoblastic lineages leads to increased trabecular bone mass in female mice. However, these data were all collected in mice where the conditional deletion was present from conception onwards and therefore may involve developmental effects of the deletion that may persist into adulthood. In order to understand the specific roles of the receptors in a cell-specific manner in adulthood without the confounding effects of developmental deletion, inducible mouse models in which the deletion can be induced later in life are necessary to further understand and clarify the role of both ERα and ERβ in the adult skeleton.

Estrogens—from a clinical perspective

Over 70 years ago, Fuller Albright demonstrated that estrogen deficiency, achieved through either OVX or menopause, caused a rapid bone loss that estrogen replacement could mitigate (Albright, 1940). This also occurs in men, as it was shown that following the removal of the testes, men also experienced a similar loss of bone (Stepan et al., 1989). Since the major sex steroid in males is testosterone, this at the time led to the generalization that estrogen regulates bone mass in women, and testosterone regulates bone mass in men. This, however, proved to be an oversimplification, and as we will discuss shortly, estrogens prove to be the major sex steroid regulating bone physiology in both sexes. In this section, we will review our understanding of how estrogen regulates bone metabolism in humans in both sexes.

The first crack in the notion that testosterone is the dominant regulator of bone mass in men was the identification of a 28-year old male with homozygous null mutations in the ERα gene (Smith et al., 1994). This man exhibited osteopenia with unfused epiphyses and had a spine bone mineral density (BMD) of 2 standard deviations below the average for normal 15 year-old males. This was followed by two independent reports describing a two males with a complete

TABLE 35.1 Summary of phenotypes for ER and PR conditional mouse knockouts.

Conditional KOs		Female		Male		
Genotype	Cell type	Trabecular bone	Cortical bone	Trabecular bone	Cortical bone	References
ERα/LysM-Cre	Osteoclast	↓	~	~	~	Martin-Millan et al., 2010
ERα/Catk-Cre	Osteoclast	↓	~			Nakamura et al., 2007
ERα/Prx1-Cre	Mesen-Prog	~	↓			Almeida et al., 2013
ERα/Osx1-Cre	Osteo-Prog	~	↓			Almeida et al., 2013
ERα/Col1a1-Cre	Osteoblast	~	~	~	~	Almeida et al., 2013
ERα/Ocn-Cre	Osteoblast	↓	↓	~ then↓	~	Maatta et al., 2013; Melville et al., 2014
ERα/Dmp1-Cre	Osteocyte			↓	~	Windahl et al., 2013
ERα/Dmp1-Cre	Osteocyte	↓	~			Kondoh et al., 2014
ERβ/Prx1-Cre	Mesen-Prog	↑	~			Nicks et al., 2016
ERβ/Col1a1-Cre	Mesen-Prog	↑	~			Nicks et al., 2016
PR/Prx1-Cre	Mesen-Prog	↑	~	↑	~	Zhong et al., 2017
PR/Ocn-Cre	Osteoblast	~	~	~	~	Zhong et al., 2017
PR/Dmp1-Cre	Osteocyte	↑	~	↑	~	Zhong et al., 2017

~ parameter is unchanged; ↓ parameter is decreased; ↑ parameter is increased.

deficiency in the aromatase enzyme, the molecule responsible for the conversion of testosterone to estrogens (Carani et al., 1997; Bilezikian et al., 1998). Interestingly, these males exhibited a virtually indistinguishable bone phenotype for the original ERα-deficient man. These reports demonstrated that estrogens indeed had an important role in both epiphyseal closure and the attainment of peak bone mass that occurs in the late puberty of males. The original ERα-deficient male (Smith et al., 1994) was reanalyzed later in life and a reduction was found in the bone parameters of both trabecular and cortical compartments (Smith et al., 2008), demonstrating that his osteopenia tracked into later life. While the typical forms of osteoporosis, such as in hypogonadal men or estrogen-deficient women, exhibit increased bone turnover (with indices of resorption and formation both increasing), this male had a reduction in both resorption and formation indices. These "experiments of nature" logically lead to an interesting insight into molecular mechanism—that a loss of estrogen (e.g., the ligand) leads to reduced bone mass in a fundamentally different manner than the loss of the receptor. This notion is supported by the presence of many estrogen-independent effects that have been observed in numerous other tissues (Ciana et al., 2003). Interestingly, cultured primary bone cells from the ERα-deficient male exhibited an increase in the ERβ protein, suggesting a compensatory response (Smith et al., 2008). However, when this data is considered in context with the high circulating estrogen levels present in this individual, it suggests that the ERβ pathway cannot adequately compensate for ERα in the regulation of global bone mass. Whether this is due to the loss of ERα itself, the compensatory increase in ERβ protein, changes in transcriptional coregulatory dynamics, or all of the above, is unclear and would warrant further investigation.

In women, a number of observational studies have also substantiated the original observation made by Fuller Albright and further confirm the essential and key role of estrogens in the regulation of bone physiology. The decline in serum estrogens that occurs throughout menopause is tightly associated with an increased in the bone resorption parameter N-telopeptide of type I collagen (Sowers et al., 2013). Since androgen production, albeit at a low level, still continues in the ovaries following menopause, androgen levels are unchanged and argue against their role in estrogen deficiency-mediated bone loss (Handelsman et al., 2016). The radical idea that estrogen regulates BMD in both women and men has been documented in other human observational studies, where serum estradiol levels appear to be more closely correlated to BMD and turnover indices than androgen levels (Khosla et al., 1998).

A number of human interventional studies from various groups have also concluded that estrogen is indeed the dominant sex steroid that regulates bone physiology. The Women's Health Initiative study, which investigated the effects of estrogen/progesterone (vs. placebo) in a large cohort of postmenopausal women, demonstrated that steroid treatment increased BMD at multiple sites and reduced the risk of hip fracture by one-third (Cauley et al., 2003). Another key study compared the effects of estrogen and testosterone treatment in older men (mean age 68 years) where endogenous sex steroid production was pharmacologically eliminated, and exogenously treated with either sex steroid alone or in combination using patches for 3 weeks (Falahati-Nini et al., 2000). Increases in bone resorption indices were prevented by treatment with steroids as well as by the estrogen-replaced group alone. Importantly, the testosterone-replaced group, in which aromatization to estrogens in inhibited, was ineffective in altering bone resorption makers. Using a statistical model, the investigators estimated that 70% of the effect was due to estrogen alone. This notion was confirmed by a similar study that found that testosterone replacement in hypogonadal men (±an aromatase inhibitor) was ineffective in altering serum turnover markers and that estrogen was indeed the major sex steroid involved in the regulation of bone mass in men (Finkelstein et al., 2016). Collectively, these studies provide strong evidence that in both sexes estrogen is the major sex steroid involved in the regulation bone physiology (Khosla, 2015). The function of testosterone in the regulation of bone mass in men appears to be as a precursor for the aromatization to estrogens, although testosterone also likely plays other important roles in periosteal apposition during growth, leading to larger bone size in men versus women (Almeida et al., 2017).

Another important question when considering the effects of estrogens and ERs (ERα and ERβ) is what might be the potential mechanisms and/or mediators of estrogen action in human bone. Based largely on rodent OVX models, genes that encode proinflammatory cytokines have emerged as one class of potential mediators of estrogen action in bone, especially on osteoclast action and bone resorption. Of particular interest, TNFα, IL1α/β, and IL6 have been the most consistent (Weitzmann and Pacifici, 2006). In humans, several interventional studies have addressed the influence of these proinflammatory cytokines on bone health. One study examined a cohort of normal women ranging from 24 to 87 years old and showed that IL6 concentrations correlated highly with age; however, using sophisticated multiple-regression statistical models, no correlation with estrogen status could be established (McKane et al., 1994). To further clarify this issue, a study was designed in which estrogen-withdrawn early postmenopausal women were placed on either an IL1 or TNFα blocker for 3 weeks, and bone resorption indices were analyzed (Charatcharoenwitthaya et al., 2007). They found a reduction in serum resorption markers suggesting that these proinflammatory cytokines mediate in part bone resorption during estrogen deficiency in humans.

Receptor activator of nuclear factor kappa-B ligand (RANKL), which is produced by both osteoblasts and osteocytes in the bone microenvironment, is another key regulator of bone remodeling by influencing the process of osteoclastic differentiation (Xiong and O'Brien, 2012). In one study designed to examine the role of estrogenic regulation of RANKL in bone, RANKL + cells were isolated from premenopausal women (control) and postmenopausal women treated with either vehicle or estradiol (Eghbali-Fatourechi et al., 2003). They found that the number of RANKL + cells increases with age and estrogenic status, and that such increases were reversible with estradiol treatment. These data suggest that loss of estrogens at menopause may contribute to increased bone resorption through the increase in the proosteoclastic factor RANKL. The increased resorption in estrogen-deficient women can be abrogated clinically through treatment with denosumab, a humanized monoclonal antibody to RANKL (Cummings et al., 2009).

Increasing evidence also exists that demonstrates that the secreted Wnt inhibitor, sclerostin, is regulated by estrogen (Drake and Khosla, 2017). Treatment of estrogen-deficient women with estrogen or the select estrogen receptor modulator raloxifene negatively regulates circulating serum sclerostin protein levels (Drake and Khosla, 2017; Chung et al., 2012; Fujita et al., 2014; Modder et al., 2011a,b). Sclerostin mRNA expression, as assessed by RNA sequencing and QPCR analyses, was also decreased in primary bone needle biopsies isolated from postmenopausal women treated with estradiol (Farr et al., 2017). The new therapeutic agent romosozumab (Cosman et al., 2016), which is a humanized monoclonal antibody directed against sclerostin, results in increases in bone formation with a concomitant decrease in bone resorption in postmenopausal women, similar to those changes observed with estrogen treatment (Drake and Khosla, 2017).

Progestins and progesterone receptors in bone biology

Clinically, progesterone has been used in addition to estradiol to treat postmenopausal osteoporosis to minimize the adverse effects of estradiol on the uterus and breast, however the effects of progestins on bone biology have been significantly less studied (Prior et al., 1990). As with estrogen and estrogen receptors, progestins exert their effects via the progesterone receptor (PR), a member of the nuclear receptor superfamily of transcription factors (Mangelsdorf et al., 1995). Activated PR can acts a progesterone response elements to alter gene expression of specific PR target genes, however nongenomic effects have also been observed (Grosse et al., 2000).

In mammals, PRs exists in two related protein isoforms, which arise from the regulated activation of alternative promoters, resulting in the PR-A and PR-B isoforms (Kastner et al., 1990; Giangrande and McDonnell, 1999). Since they are a member of the nuclear receptor superfamily of transcription factors along with ERs (Mangelsdorf et al., 1995), they share the same modular protein structure including a DBD, LBD, and AF domains (Fig. 35.2) (Hill et al., 2012). The two isoforms are identical except for an additional 164 amino acid residues on the N-terminus of PR-B, which contains an additional and unique activation function (AF-3). Interestingly, both PR isoforms are estrogen inducible (Jung-Testas et al., 1991; Harris et al., 1995; Rickard et al., 2002), suggesting cross-talk between the estrogenic and progestin pathways. In general, the PR-B isoform has a more dominant effect on transcriptional regulation, and under certain conditions, the PR-A isoform can act as an antagonist to PR-B function, as can other nuclear hormone receptors such as ERα (Kraus et al., 1995). Expression of both mRNA and protein for PRs has been demonstrated in multiple osteoblastic cell lines (Rickard et al., 2002; Wei et al., 1993; MacNamara et al., 1995, 1998) and in an age-dependent manner in mouse bone tissue (Zhong et al., 2017).

There is considerable in vitro evidence that progestins can affect osteoblast and bone marrow–derived osteoblastic progenitor cells through the regulation of both proliferation and differentiation, although the data are somewhat conflicting. Various reports have demonstrated that progesterone treatment of some osteoblast cell models has shown either an inhibition of proliferation (Canalis and Raisz, 1978), no effect on proliferation when treated alone (Slootweg et al., 1992), or a stimulation of proliferation (Tremollieres et al., 1992; Scheven et al., 1992; Manzi et al., 1994; Verhaar et al., 1994). Although these data are variable, the overall weight of the evidence suggests that progestins indeed stimulate proliferation in certain mouse and rat osteoblast cell models. Using primary rat bone explants, it was shown the progesterone stimulated

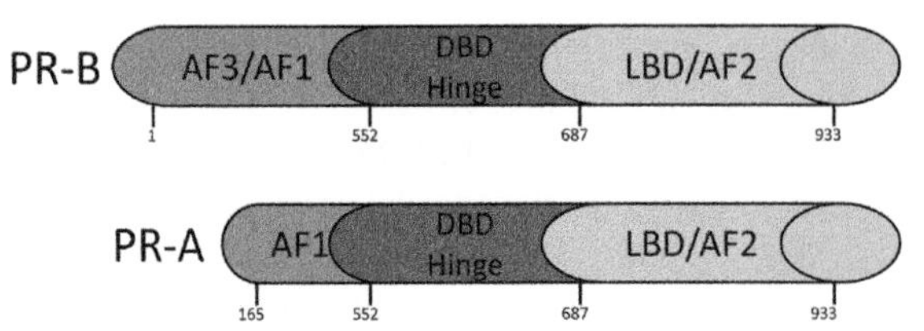

FIGURE 35.2 Structure of human PR-A and PR-B.

the formation of alkaline phosphatase positive (AP+) colonies (Ishida & Heersche, 1997, 1999), suggesting that progesterone may positively influence osteoblast differentiation. In human primary osteoblast cultures and in rat osteosarcoma cells, progesterone treatment stimulated IGF2 secretion (Tremollieres et al., 1992). This regulation of the IGF system was demonstrated to be via activation of IGFBP5, which is known to enhance IGF1/2 activity in osteoblast cultures (Boonyaratanakornkit et al., 1999). In summary, these data support the notion that progesterone positively influences the proliferation and bone-forming activities of osteoblasts.

Initial studies in the 1980s using animal models showed that progesterone has positive effects on bone metabolism in both rats (Burnett and Reddi, 1983) and dogs (Snow and Anderson, 1985). Using an OVX-induced model of bone loss in rats, it was also shown that progesterone replacement had a significant effect on the preservation of bone during sex steroid deficiency (Barengolts et al., 1990, 1996; Bowman and Miller, 1996; Schmidt et al., 2000). However other studies have failed to show an effect on trabecular bone when either treated alone or coadministered with estradiol (Fujimaki et al., 1995; Kalu et al., 1991), creating some uncertainty in the field in regard to the effects of progestins on bone metabolism. Interestingly, the PR antagonist mifepristone (RU486) was also shown to reduce OVX-induced bone loss in rats (Barengolts et al., 1995) and in a patient with Cushing's disease and severe osteoporosis (Newfield et al., 2001), although the latter effects could be due to RU486-mediated antagonism of the glucocorticoid receptor.

Due to these conflicting reports on the effects of progestins on bone metabolism in both animal and human models of osteoporosis, mutant mice have been generated where both PR-A and PR-B have been deleted (the PRKO mouse), in hopes of understanding the true nature of the effects of progestins and PRs in bone physiology. These mice exhibit severe reproductive abnormalities, and the females are infertile (Lydon et al., 1995; Conneely et al., 2001), and also exhibit an increase in trabecular bone mass and bone formation at 12 and 26 weeks of age in female mice (Rickard et al., 2008). Treatment with the PR antagonist RU486 also leads to an increase in trabecular bone mass in young wild-type mice (Yao et al., 2010). Recently, cell-type specific deletions of both PR isoforms using the Cre/LoxP system (the PRcKO mouse) have been made (Zhong et al., 2017). This report confirmed the data from the PRKO knockout mouse and RU486 antagonist data and found a significant increase in trabecular bone mass when PRs were deleted in early mesenchymal progenitor cells (using the Prx1-Cre driver), but not in mature osteoblasts (using the Ocn-Cre driver), and a modest increased trabecular bone mass phenotype following deletion in osteocytes (using the Dmp1-Cre driver). These phenotypes were observed in both sexes (see Table 35.1). Cortical bone mass was unaffected in all models. Taken together, these findings suggest that PR may play a negative role in bone homeostasis. Interestingly, in a specific PR-A mouse knockout model, progesterone had a significantly higher effect on reproductive tissues (Mulac-Jericevic et al., 2000), suggesting that PR-A plays a negative role on the actions of PR-B, although these data need to be confirmed in the skeleton.

Although the data on the effects or progestins in human physiology are somewhat controversial (Waller et al., 1996; De Souza et al., 1997), the bulk of the evidence suggests a role for progestins, albeit modest, in bone physiology. Women who have ovulatory dysfunction due to luteal phase defects have lower bone mass over 1 year compared with normal women (Prior et al., 1990; Prior, 1990), which can be reversed with progesterone replacement therapy (Prior et al., 1994). Additionally, progesterone treatment of women with postmenopausal osteoporosis could protect against further cortical bone loss (Gallagher et al., 1991; Gray et al., 1996; McNeeley Jr. et al., 1991). Although some reports have demonstrated that progesterone treatment of postmenopausal women is as effective as estradiol treatment in preventing further bone loss (Gray et al., 1996; McNeeley Jr. et al., 1991), others have shown a lesser effect when compared to estradiol alone (Prior et al., 1997; PEPI, 1996).

Summary and conclusions

In recent years, many discoveries have advanced our knowledge regarding estrogen action in the bone microenvironment. Although some of the phenotypes of the ERα and ERβ conditional knockout mouse models have been described, it is still unclear the precise mechanism(s) that ERs utilize to regulate bone metabolism, which is especially evident in the role of ERβ itself and ERα/ERβ interactions that may influence bone processes. Given how important estrogen action is for bone, it will be important to continue to define the mechanism(s) that these ERs play, including the role of nonnuclear ERs, to develop novel clinical therapies for aging-related bone diseases such as osteoporosis. It is of interest that two of the new therapies developed for osteoporosis (i.e., denosumab and sclerostin) appear to lie in the estrogen pathway in humans. Therefore, future investigation into how these mechanisms function during aging in bone will be especially important and will help in the improvement of current therapies and in the development of novel therapies for osteoporosis.

CONFLICT OF INTEREST

The authors have declared that no financial conflict of interest exists.

References

Albright, F., 1940. Post-menopausal osteoporosis. Trans. Assoc. Am. Phys. 55, 298–305.
Almeida, M., et al., 2013. Estrogen receptor-alpha signaling in osteoblast progenitors stimulates coritcal bone accrual. J. Clin. Investig. 123, 394–404.
Almeida, M., et al., 2017. Estrogens and androgens in skeletal physiology and pathophysiology. Physiol. Rev. 97, 135–187.
Arts, J., et al., 1997. Differential expression of estrogen receptors alpha and beta mRNA during differentiation of human osteoblast SV-HFO cells. Endocrinology 138, 5067–5070. https://doi.org/10.1210/endo.138.11.5652.
Barengolts, E.I., et al., 1990. Effects of progesterone on postovariectomy bone loss in aged rats. J. Bone Miner. Res. 5, 1143–1147. https://doi.org/10.1002/jbmr.5650051109.
Barengolts, E.I., et al., 1996. Effects of progesterone on serum levels of IGF-1 and on femur IGF-1 mRNA in ovariectomized rats. J. Bone Miner. Res. 11, 1406–1412. https://doi.org/10.1002/jbmr.5650111006.
Barengolts, E.I., Lathon, P.V., Lindh, F.G., 1995. Progesterone antagonist RU 486 has bone-sparing effects in ovariectomized rats. Bone 17, 21–25.
Bilezikian, J.P., Morishima, A., Bell, J., Grumbach, M.M., 1998. Increased bone mass as a result of estrogen therapy in a man with aromatase deficiency. N. Engl. J. Med. 339, 599–603.
Bonewald, L.F., 2017. The role of the osteocyte in bone and nonbone disease. Endocrinol Metab. Clin. N. Am. 46, 1–18. https://doi.org/10.1016/j.ecl.2016.09.003.
Boonyaratanakornkit, V., et al., 1999. Progesterone stimulation of human insulin-like growth factor-binding protein-5 gene transcription in human osteoblasts is mediated by a CACCC sequence in the proximal promoter. J. Biol. Chem. 274, 26431–26438.
Bord, S., Horner, A., Beavan, S., Compston, J., 2001. Estrogen receptors alpha and beta are differentially expressed in developing human bone. J. Clin. Endocrinol. Metab. 86, 2309–2314. https://doi.org/10.1210/jcem.86.5.7513.
Bowman, B.M., Miller, S.C., 1996. Elevated progesterone during pseudopregnancy may prevent bone loss associated with low estrogen. J. Bone Miner. Res. 11, 15–21. https://doi.org/10.1002/jbmr.5650110104.
Braidman, I.P., et al., 2001. Localization of estrogen receptor beta protein expression in adult human bone. J. Bone Miner. Res. 16, 214–220. https://doi.org/10.1359/jbmr.2001.16.2.214.
Burnett, C.C., Reddi, A.H., 1983. Influence of estrogen and progesterone on matrix-induced endochondral bone formation. Calcif. Tissue Int. 35, 609–614.
Canalis, E., Raisz, L.G., 1978. Effect of sex steroids on bone collagen synthesis in vitro. Calcif. Tissue Res. 25, 105–110.
Carani, C., et al., 1997. Effect of testosterone and estradiol in a man with aromatase deficiency. N. Engl. J. Med. 337, 91–95.
Cauley, J.A., et al., 2003. Effects of estrogen plus progestin on risk of fracture and bone mineral density. J. Am. Med. Assoc. 290, 1729–1738.
Cauley, J.A., et al., 2008. Incidence of fractures compared to cardiovascular disease and breast cancer: the Women's Health Initiative Observational Study. Osteoporos. Int. 19, 1717–1723.
Charatcharoenwitthaya, N., Khosla, S., Atkinson, E.J., McCready, L.K., Riggs, B.L., 2007. Effect of blockade of TNF-a and interleukin-1 action on bone resorption in early postmenopausal women. J. Bone Miner. Res. 22, 724–729.
Chow, J.W., Lean, J.M., Chambers, T.J., 1992. 17 beta-estradiol stimulates cancellous bone formation in female rats. Endocrinology 130, 3025–3032. https://doi.org/10.1210/endo.130.5.1572310.
Chung, Y.E., et al., 2012. Long-term treatment with raloxifene, but not bisphosphonates reduces circulating sclerostin levels in postmenopausal women. Osteoporos. Int. 23, 1235–1243.
Ciana, P., et al., 2003. In vivo imaging of transcriptionally active estrogen receptors. Nat. Med. 9, 82–86.
Conneely, O.M., Mulac-Jericevic, B., Lydon, J.P., De Mayo, F.J., 2001. Reproductive functions of the progesterone receptor isoforms: lessons from knock-out mice. Mol. Cell. Endocrinol. 179, 97–103.
Cosman, F., et al., 2016. Romosozumab treatment in postmenopausal women with osteoporosis. N. Engl. J. Med. 375, 1532–1543.
Cummings, S.R., et al., 2009. Denosumab for prevention of fractures in postmenopausal women with osteoporosis. N. Engl. J. Med. 361, 756–765.
De Souza, M.J., et al., 1997. Bone health is not affected by luteal phase abnormalities and decreased ovarian progesterone production in female runners. J. Clin. Endocrinol. Metab. 82, 2867–2876. https://doi.org/10.1210/jcem.82.9.4201.
Drake, M.T., Khosla, S., 2017. Hormomal and systemic regulation of sclerostin. Bone 96, 8–17.
Eghbali-Fatourechi, G., et al., 2003. Role of RANK ligand in mediating increased bone resorption in early postmenopausal women. J. Clin. Investig. 111, 1221–1230.
Eriksen, E.F., et al., 1988. Evidence of estrogen receptors in normal human osteoblast-like cells. Science 241, 84–86.
Ernst, M., Heath, J.K., Schmid, C., Froesch, R.E., Rodan, G.A., 1989. Evidence for a direct effect of estrogen on bone cells in vitro. J. Steroid Biochem. 34, 279–284.
Falahati-Nini, A., et al., 2000. Relative contributions of testosterone and estrogen in regulating bone resorption and formation in normal elderly men. J. Clin. Investig. 106, 1553–1560.
Farr, J.N., et al., 2017. Targeting cellular senescence prevents age-related bone loss in mice. Nat. Med. 23, 1072–1079. https://doi.org/10.1038/nm.4385.
Finkelstein, J.S., et al., 2016. Gonadal steroid-dependent effects on bone turnover and bone mineral density in men. J. Clin. Investig. 126, 1114–1125.
Fujimaki, T., et al., 1995. Effects of progesterone on the metabolism of cancellous bone in young oophorectomized rats. J. Obstet. Gynaecol. 21, 31–36.

Fujita, K., et al., 2014. Effects of estrogen on bone mRNA levels of sclerostin and other genes relevant to bone metabolism in postmenopausal women. J. Clin. Endocrinol. Metab. 99, E81–E88.

Gallagher, J.C., Kable, W.T., Goldgar, D., 1991. Effect of progestin therapy on cortical and trabecular bone: comparison with estrogen. Am. J. Med. 90, 171–178.

Giangrande, P.H., McDonnell, D.P., 1999. The A and B isoforms of the human progesterone receptor: two functionally different transcription factors encoded by a single gene. Recent Prog. Horm. Res. 54, 291–313 discussion 313-294.

Gorski, J., Toft, D., Shyamala, G., Smith, D., Notides, A., 1968. Hormone receptors: studies on the interaction of estrogen with the uterus. Recent Prog. Horm. Res. 24, 45–80.

Greene, G.L., et al., 1986. Sequence and expression of human estrogen receptor complementary DNA. Science 231, 1150–1154.

Grey, A., Cundy, T., Evans, M., Reid, I., 1996. Medroxyprogesterone acetate enhances the spinal bone mineral density response to oestrogen in late postmenopausal women. Clin. Endocrinol. 44, 293–296.

Grosse, B., Kachkache, M., Le Mellay, V., Lieberherr, M., 2000. Membrane signalling and progesterone in female and male osteoblasts. I. Involvement of intracellular Ca(2+), inositol trisphosphate, and diacylglycerol, but not cAMP. J. Cell. Biochem. 79, 334–345.

Hall, J.M., McDonnell, D.P., 1999. The estrogen receptor beta-isoform (ER-beta) of the human estrogen receptor modulates ER-alpha transcriptional activity and is a key regulator of the cellular response to estrogens and antiestrogens. Endocrinology 140, 5566–5578.

Handelsman, D.J., Sikaris, K., Ly, L.P., 2016. Estimating age-specific trends in circulating testosterone and sex hormone-binding globulin in males and females across the lifespan. Ann. Clin. Biochem. 53, 377–384.

Harris, S.A., Enger, R.J., Riggs, B.L., Spelsberg, T.C., 1995. Development and characterization of a conditionally immortalized human fetal osteoblastic cell line. J. Bone Miner. Res. 10, 178–186. https://doi.org/10.1002/jbmr.5650100203.

Hill, K.K., Roemer, S.C., Churchill, M.E., Edwards, D.P., 2012. Structural and functional analysis of domains of the progesterone receptor. Mol. Cell. Endocrinol. 348, 418–429. https://doi.org/10.1016/j.mce.2011.07.017.

Ikegami, A., et al., 1993. Immunohistochemical detection and northern blot analysis of estrogen receptor in osteoblastic cells. J. Bone Miner. Res. 8, 1103–1109. https://doi.org/10.1002/jbmr.5650080911.

Ishida, Y., Heersche, J.N., 1997. Progesterone stimulates proliferation and differentiation of osteoprogenitor cells in bone cell populations derived from adult female but not from adult male rats. Bone 20, 17–25.

Ishida, Y., Heersche, J.N., 1999. Progesterone- and dexamethasone-dependent osteoprogenitors in bone cell populations derived from rat vertebrae are different and distinct. Endocrinology 140, 3210–3218. https://doi.org/10.1210/endo.140.7.6850.

Jensen, E.V., et al., 1968. A two-step mechanism for the interaction of estradiol with rat uterus. Proc. Natl. Acad. Sci. U.S.A. 59, 632–638.

Jung-Testas, I., Renoir, J.M., Gasc, J.M., Baulieu, E.E., 1991. Estrogen-inducible progesterone receptor in primary cultures of rat glial cells. Exp. Cell Res. 193, 12–19.

Kalajzic, I., et al., 2004. Dentin matrix protein 1 expression during osteoblastic differentiation, generation of an osteocyte GFP-transgene. Bone 35, 74–82. https://doi.org/10.1016/j.bone.2004.03.006.

Kalu, D.N., et al., 1991. A comparative study of the actions of tamoxifen, estrogen and progesterone in the ovariectomized rat. Bone Miner. 15, 109–123.

Kanis, J.A., et al., 2000. Long-term risk of osteoporotic fracture in Malmo. Osteoporos. Int. 11, 669–674.

Kastner, P., et al., 1990. Two distinct estrogen-regulated promoters generate transcripts encoding the two functionally different human progesterone receptor forms A and B. EMBO J. 9, 1603–1614.

Kelly, M.J., Levin, E.R., 2001. Rapid actions of plasma membrane estrogen receptors. Trends Endocrinol. Metabol. 12, 152–156.

Khosla, S., et al., 1998. Relationship of serum sex steroid levels and bone turnover markers with bone mineral density in men and women: a key role for bioavailable estrogen. J. Clin. Endocrinol. Metab. 83, 2266–2274.

Khosla, S., Amin, S., Orwoll, E., 2008. Osteoporosis in men. Endocr. Rev. 29, 441–464.

Khosla, S., 2015. New insights into androgen and estrogen receptor regulation of the male skeleton. J. Bone Miner. Res. 30, 1134–1137.

Kondoh, S., et al., 2014. Estrogen receptor α in osteocytes regulates trabecular bone formation in female mice. Bone 60, 68–77.

Kraus, W.L., Weis, K.E., Katzenellenbogen, B.S., 1995. Inhibitory cross-talk between steroid hormone receptors: differential targeting of estrogen receptor in the repression of its transcriptional activity by agonist- and antagonist-occupied progestin receptors. Mol. Cell Biol. 15, 1847–1857.

Landers, J.P., Spelsberg, T.C., 1992. New concepts in steroid hormone action: transcription factors, proto-oncogenes, and the cascade model for steroid regulation of gene expression. Crit. Rev. Eukaryot. Gene Expr. 2, 19–63.

Levin, E.R., 2002. Cellular functions of plasma membrane estrogen receptors. Steroids 67, 471–475.

Logan, M., et al., 2002. Expression of Cre Recombinase in the developing mouse limb bud driven by a Prxl enhancer. Genesis 33, 77–80. https://doi.org/10.1002/gene.10092.

Lydon, J.P., et al., 1995. Mice lacking progesterone receptor exhibit pleiotropic reproductive abnormalities. Genes Dev. 9, 2266–2278.

Maatta, J.A., et al., 2013. Inactivation of estrogen receptor α in bone-forming cells induces bone loss in female mice. FASEB J. 27, 478–488.

MacNamara, P., O'Shaughnessy, C., Manduca, P., Loughrey, H.C., 1995. Progesterone receptors are expressed in human osteoblast-like cell lines and in primary human osteoblast cultures. Calcif. Tissue Int. 57, 436–441.

MacNamara, P., Skillington, J., Loughrey, H.C., 1998. Studies on progesterone receptor expression in human osteoblast cells. Biochem. Soc. Trans. 26, S1.

Mangelsdorf, D.J., et al., 1995. The nuclear receptor superfamily: the second decade. Cell 83, 835–839.

Manzi, D.L., Pilbeam, C.C., Raisz, L.G., 1994. The anabolic effects of progesterone on fetal rat calvaria in tissue culture. J. Soc. Gynecol. Investig. 1, 302–309.

Martin-Millan, M., et al., 2010. The estrogen receptor-alpha in osteoclasts mediates the protective effects of estrogens on cancellous but not cortical bone. Mol. Endocrinol. 24, 323–334.

McKane, W.R., Khosla, S., Peterson, J.M., Egan, K., Riggs, B.L., 1994. Circulating levels of cytokines that modulate bone resorption: effects of age and menopause in women. J. Bone Miner. Res. 9, 1313–1318. https://doi.org/10.1002/jbmr.5650090821.

McNeeley Jr., S.G., Schinfeld, J.S., Stovall, T.G., Ling, F.W., Buxton, B.H., 1991. Prevention of osteoporosis by medroxyprogesterone acetate in postmenopausal women. Int. J. Gynaecol. Obstet. 34, 253–256.

Melton 3rd, L.J., Atkinson, E.J., O'Connor, M.K., O'Fallon, W.M., Riggs, B.L., 1998. Bone density and fracture risk in men. J. Bone Miner. Res. 13, 1915–1923. https://doi.org/10.1359/jbmr.1998.13.12.1915.

Melville, K.M., et al., 2014. Female mice lacking estrogen receptor-alpha in osteoblast have compromised bone mass and strength. J. Bone Miner. Res. 29, 370–379.

Modder, U.I., et al., 2011a. Regulation of circulating sclerostin levels by sex steroids in women and in men. J. Bone Miner. Res. 26, 27–34. https://doi.org/10.1002/jbmr.128.

Modder, U.I., et al., 2011b. Effects of estrogen on osteoprogenitor cells and cytokines/bone-regulatory factors in postmenopausal women. Bone 49, 202–207. https://doi.org/10.1016/j.bone.2011.04.015.

Monroe, D.G., et al., 2003a. Mutual antagonism of estrogen receptors alpha and beta and their preferred interactions with steroid receptor coactivators in human osteoblastic cell lines. J. Endocrinol. 176, 349–357.

Monroe, D.G., et al., 2005. Estrogen receptor alpha and beta heterodimers exert unique effects on estrogen- and tamoxifen-dependent gene expression in human U2OS osteosarcoma cells. Mol. Endocrinol. 19, 1555–1568. https://doi.org/10.1210/me.2004-0381.

Monroe, D.G., Secreto, F.J., Spelsberg, T.C., 2003b. Overview of estrogen action in osteoblasts: role of the ligand, the receptor, and the co-regulators. J. Musculoskelet. Neuronal Interact. 3, 357–362 discussion 381.

Mosselman, S., Polman, J., Dijkema, R., 1996. ER beta: identification and characterization of a novel human estrogen receptor. FEBS Lett. 392, 49–53.

Mulac-Jericevic, B., Mullinax, R.A., DeMayo, F.J., Lydon, J.P., Conneely, O.M., 2000. Subgroup of reproductive functions of progesterone mediated by progesterone receptor-B isoform. Science 289, 1751–1754.

Nakamura, T., et al., 2007. Estrogen prevents bone loss via estrogen receptor alpha and induction of fas ligand in osteoclasts. Cell 130, 811–823.

Newfield, R.S., Spitz, I.M., Isacson, C., New, M.I., 2001. Long-term mifepristone (RU486) therapy resulting in massive benign endometrial hyperplasia. Clin. Endocrinol. 54, 399–404.

Nicks, K.M., et al., 2016. Deletion of estrogen receptor beta in osteoprogenitor cells increases trabecular but not cortical bone mass in female mice. J. Bone Miner. Res. 31, 606–614.

Onoe, Y., Miyaura, C., Ohta, H., Nozawa, S., Suda, T., 1997. Expression of estrogen receptor beta in rat bone. Endocrinology 138, 4509–4512. https://doi.org/10.1210/endo.138.10.5575.

Oursler, M.J., Osdoby, P., Pyfferoen, J., Riggs, B.L., Spelsberg, T.C., 1991. Avian osteoclasts as estrogen target cells. Proc. Natl. Acad. Sci. U.S.A. 88, 6613–6617.

Oursler, M.J., Pederson, L., Fitzpatrick, L., Riggs, B.L., 1994. Human giant cell tumors of the bone (osteoclastomas) are estrogen target cells. Proc. Natl. Acad. Sci. U.S.A. 91, 5227–5231.

PEPI, 1996. Effects of hormone therapy on bone mineral density: results from the postmenopausal estrogen/progestin interventions (PEPI) trial. The Writing Group for the PEPI. J. Am. Med. Assoc. 276, 1389–1396.

Prior, J.C., et al., 1997. Premenopausal ovariectomy-related bone loss: a randomized, double-blind, one-year trial of conjugated estrogen or medroxyprogesterone acetate. J. Bone Miner. Res. 12, 1851–1863. https://doi.org/10.1359/jbmr.1997.12.11.1851.

Prior, J.C., Vigna, Y.M., Schechter, M.T., Burgess, A.E., 1990. Spinal bone loss and ovulatory disturbances. N. Engl. J. Med. 323, 1221–1227. https://doi.org/10.1056/NEJM199011013231801.

Prior, J.C., Vigna, Y.M., Barr, S.I., Rexworthy, C., Lentle, B.C., 1994. Cyclic medroxyprogesterone treatment increases bone density: a controlled trial in active women with menstrual cycle disturbances. Am. J. Med. 96, 521–530.

Prior, J.C., 1990. Progesterone as a bone-trophic hormone. Endocr. Rev. 11, 386–398.

Razandi, M., Pedram, A., Greene, G.L., Levin, E.R., 1999. Cell membrane and nuclear estrogen receptors (ERs) originate from a single transcript: studies of ERalpha and ERbeta expressed in Chinese hamster ovary cells. Mol. Endocrinol. 13, 307–319. https://doi.org/10.1210/mend.13.2.0239.

Rickard, D.J., et al., 2002. Estrogen receptor isoform-specific induction of progesterone receptors in human osteoblasts. J. Bone Miner. Res. 17, 580–592. https://doi.org/10.1359/jbmr.2002.17.4.580.

Rickard, D.J., et al., 2008. Bone growth and turnover in progesterone receptor knockout mice. Endocrinology 149, 2383–2390. https://doi.org/10.1210/en.2007-1247.

Riggs, B.L., Khosla, S., Melton 3rd, L.J., 2002. Sex steroids and the construction and conservation of the adult skeleton. Endocr. Rev. 23, 279–302.

Rodda, S.J., McMahon, A.P., 2006. Distinct roles for Hedgehog and canonical Wnt signaling in specification, differentiation and maintenance of osteoblast progenitors. Development 133, 3231–3244. https://doi.org/10.1242/dev.02480.

Saczko, J., et al., 2017. Estrogen receptors in cell membranes: regulation and signaling. Adv. Anat. Embryol. Cell Biol. 227, 93–105. https://doi.org/10.1007/978-3-319-56895-9_6.

Scheven, B.A., Damen, C.A., Hamilton, N.J., Verhaar, H.J., Duursma, S.A., 1992. Stimulatory effects of estrogen and progesterone on proliferation and differentiation of normal human osteoblast-like cells in vitro. Biochem. Biophys. Res. Commun. 186, 54–60.

Schmidt, I.U., Wakley, G.K., Turner, R.T., 2000. Effects of estrogen and progesterone on tibia histomorphometry in growing rats. Calcif. Tissue Int. 67, 47–52.

Shyamala, G., Gorski, J., 1969. Estrogen receptors in the rat uterus. Studies on the interaction of cytosol and nuclear binding sites. J. Biol. Chem. 244, 1097–1103.

Sims, N.A., et al., 2002. Deletion of estrogen receptors reveals a regulatory role for estrogen receptors beta in bone remodeling in females but not in males. Bone 30, 18–25.

Slootweg, M.C., Ederveen, A.G., Schot, L.P., Schoonen, W.G., Kloosterboer, H.J., 1992. Oestrogen and progestogen synergistically stimulate human and rat osteoblast proliferation. J. Endocrinol. 133, R5–R8.

Smith, E.P., et al., 1994. Estrogen resistance caused by a mutation in the estrogen-receptor gene in a man. N. Engl. J. Med. 331, 1056–1061.

Smith, E.P., et al., 2008. Impact on bone of an estrogen receptor-alpha gene loss of function mutation. J. Clin. Endocrinol. Metab. 93, 3088–3096.

Snow, G.R., Anderson, C., 1985. The effects of continuous progestogen treatment on cortical bone remodeling activity in beagles. Calcif. Tissue Int. 37, 282–286.

Sowers, M.R., et al., 2013. Changes in bone resorption across the menopause transition: effects of reproductive hormones, body size, and ethnicity. J. Clin. Endocrinol. Metab. 98, 2854–2863.

Spelsberg, T.C., Subramaniam, M., Riggs, B.L., Khosla, S., 1999. The actions and interactions of sex steroids and growth factors/cytokines on the skeleton. Mol. Endocrinol. 13, 819–828. https://doi.org/10.1210/mend.13.6.0299.

Stepan, J.J., Lachman, M., Zverina, J., Pacovsky, V., 1989. Castrated men exhibit bone loss: effect of calcitonin treatment on biochemical indices of bone remodeling. J. Clin. Endocrinol. Metab. 69, 523–527.

Tobias, J.H., Chambers, T.J., 1991. The effect of sex hormones on bone resorption by rat osteoclasts. Acta Endocrinol. 124, 121–127.

Tobias, J.H., Chow, J., Colston, K.W., Chambers, T.J., 1991. High concentrations of 17 beta-estradiol stimulate trabecular bone formation in adult female rats. Endocrinology 128, 408–412. https://doi.org/10.1210/endo-128-1-408.

Tremollieres, F.A., Strong, D.D., Baylink, D.J., Mohan, S., 1992. Progesterone and promegestone stimulate human bone cell proliferation and insulin-like growth factor-2 production. Acta Endocrinol. 126, 329–337.

Turner, R.T., Bell, N.H., Gay, C.V., 1993. Evidence that estrogen binding sites are present in bone cells and mediate medullary bone formation in Japanese quail. Poultry Sci. 72, 728–740.

Verhaar, H.J., Damen, C.A., Duursma, S.A., Scheven, B.A., 1994. A comparison of the action of progestins and estrogen on the growth and differentiation of normal adult human osteoblast-like cells in vitro. Bone 15, 307–311.

Vidal, O., Kindblom, L.G., Ohlsson, C., 1999. Expression and localization of estrogen receptor-beta in murine and human bone. J. Bone Miner. Res. 14, 923–929. https://doi.org/10.1359/jbmr.1999.14.6.923.

Waller, K., et al., 1996. Bone mass and subtle abnormalities in ovulatory function in healthy women. J. Clin. Endocrinol. Metab. 81, 663–668. https://doi.org/10.1210/jcem.81.2.8636286.

Walter, P., et al., 1985. Cloning of the human estrogen receptor cDNA. Proc. Natl. Acad. Sci. U.S.A. 82, 7889–7893.

Wei, L.L., Leach, M.W., Miner, R.S., Demers, L.M., 1993. Evidence for progesterone receptors in human osteoblast-like cells. Biochem. Biophys. Res. Commun. 195, 525–532. https://doi.org/10.1006/bbrc.1993.2077.

Weitzmann, M.N., Pacifici, R., 2006. Estrogen deficiency and bone loss: an inflammatory tale. J. Clin. Investig. 116, 1186–1194.

Windahl, S.H., et al., 2013. Estrogen receptor-α in osteocytes is important for trabecular bone formation in male mice. Proc. Natl. Acad. Sci. U.S.A. 110, 2294–2299.

Windahl, S.H., Vidal, O., Andersson, G., Gustafsson, J.A., Ohlsson, C., 1999. Increased cortical bone mineral content but unchanged trabecular bone mineral density in female ERbeta(-/-) mice. J. Clin. Investig. 104, 895–901. https://doi.org/10.1172/JCI6730.

Windahl, S.H., Norgard, M., Kuiper, G.G., Gustafsson, J.A., Andersson, G., 2000. Cellular distribution of estrogen receptor beta in neonatal rat bone. Bone 26, 117–121.

Xiong, J., O'Brien, C.A., 2012. Osteocyte RANKL: new insights into the control of bone remodeling. J. Bone Miner. Res. 27, 499–505. https://doi.org/10.1002/jbmr.1547.

Yao, W., et al., 2010. Inhibition of the progesterone nuclear receptor during the bone linear growth phase increases peak bone mass in female mice. PLoS One 5, e11410. https://doi.org/10.1371/journal.pone.0011410.

Zhong, Z.A., et al., 2017. Sex-dependent, osteoblast stage-specific effects of progesterone receptor on bone acquisition. J. Bone Miner. Res. 32, 1841–1852. https://doi.org/10.1002/jbmr.3186.

Chapter 36

Physiological actions of parathyroid hormone-related protein in epidermal, mammary, reproductive, and pancreatic tissues

Christopher S. Kovacs
Faculty of Medicine, Memorial University of Newfoundland, St. John's, NL, Canada

Chapter outline

Introduction **839**
Skin **840**
Parathyroid hormone-related protein and its receptor expression 840
Biochemistry of parathyroid hormone-related protein 840
Function of parathyroid hormone-related protein 841
Pathophysiology of parathyroid hormone-related protein 841
Mammary gland **842**
Embryonic mammary development 842
Adolescent mammary development 844
Pregnancy and lactation 844
Pathophysiology of parathyroid hormone-related protein in the mammary gland 848
Reproductive tissues **849**
Parathyroid hormone-related protein and placental calcium transport 849
Uterus and extraembryonic tissues 851
Placenta and fetal membranes 851
Implantation and early pregnancy 852
Pathophysiology of parathyroid hormone-related protein in the placenta 853
Summary **853**
Endocrine pancreas **853**
Parathyroid hormone-related protein and its receptors 853
Regulation of parathyroid hormone-related protein and its receptors 854
Biochemistry of parathyroid hormone-related protein 854
Function of parathyroid hormone-related protein 854
Pathophysiology of parathyroid hormone-related protein 855
Conclusions **856**
Acknowledgments **856**
References **856**

Introduction

A substantial breakthrough in the study of parathyroid hormone-related protein (PTHrP) was the development of genetic models that ablated the gene for PTHrP or its receptor, or overexpressed PTHrP within bone cells. These models revealed that PTHrP plays an important role in regulating the terminal differentiation of chondrocytes and the development of the endochondral skeleton; the major findings are reviewed in Chapters 1 and 25. These same models provided insight into PTHrP's previously unknown roles in the development and function of organs and tissues outside the skeleton. The actions of PTHrP in the vascular system and central nervous system are reviewed in Chapter 26, and in bone in Chapter 25. In this chapter we review the known and putative roles of PTHrP in skin, mammary gland, placenta, and other reproductive tissues as well as the islet cells of the pancreas.

Principles of Bone Biology. https://doi.org/10.1016/B978-0-12-814841-9.00036-1

Skin

Parathyroid hormone-related protein and its receptor expression

Normal human keratinocytes were the first nonmalignant cells shown to produce PTHrP (Merendino et al., 1986), and multiple studies have confirmed that rodent and human keratinocytes in tissue culture express the PTHrP gene and secrete bioactive PTHrP (reviewed in Philbrick et al., 1996). PTHrP expression has also been examined in skin in vivo using both immunohistochemistry and in situ hybridization. During fetal development in rats and mice, PTHrP is expressed principally within the epithelial cells of developing hair follicles (Karmali et al., 1992; Lee et al., 1995). In mature skin, the PTHrP gene is also expressed most prominently within hair follicles, and mRNA levels appear to vary with the hair cycle. Cho et al. (2003) found that PTHrP expression increased within the outer root sheath and isthmus of late anagen hair follicles in mice. During catagen and telogen, PTHrP transcripts were abundant within the isthmus, but expression was downregulated during early anagen (Cho et al., 2003). In addition to hair follicles, low levels of PTHrP expression may also be found throughout the interfollicular epidermis from the basal to granular layer. Some studies have suggested that PTHrP is more highly expressed in superbasal keratinocytes (Danks et al., 1989; Hayman et al., 1989), although not all studies have reported this pattern (Atillasoy et al., 1991; Grone et al., 1994). A variety of factors have been reported to regulate PTHrP production by cultured keratinocytes (see Philbrick et al., 1996 for review). For example, glucocorticoids and 1,25-(OH)2D have been shown to suppress PTHrP production, whereas fetal bovine serum, matrigel, and an as-yet-unidentified factor(s) secreted from cultured fibroblasts have been shown to enhance PTHrP production. The upregulation of PTHrP production by fibroblast-conditioned media is particularly interesting, as PTHrP in turn acts back on dermal fibroblasts, suggesting that it may function in a short regulatory loop between keratinocytes and dermal fibroblasts (Shin et al., 1997; Blomme et al., 1999a). Finally, in vivo, PTHrP expression has been shown to be upregulated at the margins of healing wounds in guinea pigs (Blomme et al., 1999b). Interestingly, in this study, PTHrP was also detected in myofibroblasts and macrophages, suggesting that keratinocytes may not be the only source of PTHrP in skin.

The general consensus has been that keratinocytes do not express the type I PTH/PTHrP receptor (PTH1R), but dermal fibroblasts do (Hanafin et al., 1995; Orloff et al., 1995). PTHrP has been shown to bind to skin fibroblasts and elicit biochemical and biological responses in these cells (Shin et al., 1997; Blomme et al., 1999a; Wu et al., 1987). In addition, studies utilizing in situ hybridization in fetal skin have demonstrated that PTH1R mRNA is absent from the epidermis yet abundant in the dermis, especially in those cells adjacent to the keratinocytes (Karmali et al., 1992; Lee et al., 1995; Dunbar et al., 1999). There are fewer data concerning the expression patterns of the PTH1R in more mature skin, but in mice, it appears that the relative amount of PTH1R mRNA in dermal fibroblasts is reduced in adult compared with fetal skin (Cho et al., 2003). As with PTHrP, there also appears to be a hair-cycle-dependent variation in the expression of the PTH1R gene in the connective tissue adjacent to the isthmus of the hair follicles. However, unlike PTHrP mRNA, PTH1R mRNA appears to be most plentiful during early anagen (Cho et al., 2003). In addition to fibroblasts, studies using sensitive PCR-based detection methods have also reported expression of the PTH1R in keratinocytes, although these studies have used cultured cells (Errazahi et al., 2003, 2004). Furthermore, studies have shown that cultured keratinocytes bind and respond to PTHrP by inducing calcium transients, suggesting the presence of PTHrP receptors, either classical PTH1R or nonclassical, alternative PTHrP receptors (Orloff et al., 1992, 1995). However, to date no such alternative receptors have been isolated, so their existence remains uncertain. Furthermore, it is not clear whether PTH1R is expressed on keratinocytes in vivo. Therefore, although the possibility of both paracrine and autocrine signaling exists, no functional studies address the relative importance of either pathway to the biology of PTHrP in skin.

Biochemistry of parathyroid hormone-related protein

As described in Chapter 25, during transcription, the PTHrP gene undergoes alternative splicing to generate multiple mRNAs, which in human cells give rise to three main protein isoforms. In addition, each isoform is subject to posttranslational processing to generate a variety of peptides of varying length. Human keratinocytes have been shown to contain mRNA encoding for each of the three main isoforms, although as in other systems, no clearly defined or unique role has yet emerged for any of the three individual isoforms (Philbrick et al., 1996). Keratinocytes have also been shown to process full-length PTHrP into a variety of smaller peptides, including PTHrP(1−36) and a midregion fragment beginning at amino acid 38 (Soifer et al., 1992). These cells have also been shown to secrete a large (<10 kDa) amino-terminal form that is glycosylated (Wu et al., 1991). There is currently no specific information regarding the secretion of COOH-terminal peptides of PTHrP in skin, but keratinocytes are also likely to produce these peptides.

Function of parathyroid hormone-related protein

Several studies suggest that PTHrP is involved in the regulation of hair growth. As noted earlier, the PTHrP gene in embryonic skin is expressed most prominently in developing hair follicles, and overexpression of PTHrP in the basal keratinocytes of skin in transgenic mice leads to a severe inhibition of ventral hair follicle morphogenesis during fetal development (Wysolmerski et al., 1994). This appears to be the result of an interaction between PTHrP and BMP signaling that normally is involved in the regulation of Msx2 expression, patterning of the mammary mesenchyme, and lateral inhibition of hair follicle development around the nipple (Hens et al., 2007) (see "Mammary Gland"). However, it is unlikely that PTHrP is critical to hair follicle morphogenesis elsewhere, because disruption of the PTHrP or PTH1R genes does not seem to affect hair follicle formation or patterning in mice except for the vicinity of the nipple (Karaplis et al., 1994; Lanske et al., 1996; Foley et al., 1998).

It has also been suggested that PTHrP may participate in the regulation of the hair cycle. Systemic or topical administration of PTH1R antagonists to mice appears to perturb the hair cycle by prematurely terminating telogen, prolonging anagen growth, and inhibiting catagen (Schilli et al., 1997; Safer et al., 2007). In a reciprocal fashion, mice overexpressing PTHrP in their skin (K14-PTHrP mice) demonstrate a delayed emergence of dorsal hair and have shorter hairs, and their hair follicles enter catagen approximately 2 days early (Cho et al., 2003; Diamond et al., 2006). These findings were associated with a lower proliferation rate in the hair matrix and a less well-developed perifollicular vasculature during anagen. Together, these studies imply that PTHrP may regulate the anagen-to-catagen transition, acting to inhibit hair follicle growth by pushing growing hair follicles into the growth-arrested or catagen/telogen phase of the hair cycle. If this hypothesis were correct, one would expect PTHrP knockout mice to exhibit findings similar to PTH1R antagonist-treated mice. However, this does not appear to be the case. In mice that lack PTHrP in their skin, the hair cycle appears to be normal (Foley et al., 1998). In fact, rather than a promotion of hair growth, these mice demonstrate a thinning of their coats over time. These conflicting results are difficult to rationalize at this point, but suggest that although PTHrP may contribute to the regulation of the hair cycle, it is unlikely to be necessary for this process to unfold normally.

PTHrP has also been implicated in the regulation of keratinocyte proliferation and/or differentiation. Data from studies in vitro have suggested that PTHrP promotes the differentiation of keratinocytes (reviewed in Philbrick et al., 1996), but studies in vivo have suggested that PTHrP inhibits keratinocyte differentiation (Foley et al., 1998). A careful comparison of the histology of PTHrP-null and PTHrP-overexpressing skin demonstrated reciprocal changes. In the absence of PTHrP, it appeared that keratinocyte differentiation was accelerated, whereas in skin exposed to PTHrP overexpression, keratinocyte differentiation appeared to be retarded (Foley et al., 1998). Therefore, in a physiological context, PTHrP appears to slow the rate of keratinocyte differentiation and to preserve the proliferative basal compartment. Remarkably, these changes in the rate of keratinocyte differentiation are exactly analogous to those noted for chondrocyte differentiation in the growth plates of mice overexpressing PTHrP compared with PTHrP- and PTH1R-null mice (Philbrick et al., 1996). Again, at present it is difficult to rationalize conflicting data regarding the effects of PTHrP on keratinocyte differentiation, but studies in genetically altered mice indicate that PTHrP participates in the complex regulation of these processes in vivo. Further research will be needed to understand its exact role.

As alluded to previously, an important but still unresolved question is whether the effects of PTHrP on keratinocyte proliferation, differentiation, and hair follicle growth are the result of its effects on keratinocytes directly or via its effects on dermal fibroblasts. At present, more data support the paracrine possibility. PTH1R is expressed on dermal fibroblasts in vivo and in culture (Lee et al., 1995; Hanafin et al., 1995). Dermal fibroblasts have been demonstrated to show biochemical and biological responses to PTHrP (Shin et al., 1997; Blomme et al., 1999a; Wu et al., 1987; Thomson et al., 2003). Furthermore, PTHrP has been shown to induce changes in growth factor and extracellular matrix production that could in turn lead to changes in keratinocyte proliferation and/or differentiation and hair follicle growth (Shin et al., 1997; Blomme et al., 1999a; Insogna et al., 1989). Of course, the autocrine and paracrine signaling pathways are not mutually exclusive, but any direct autocrine effects of PTHrP on keratinocytes would require the presence of PTHrP receptors. An alternative possibility by which PTHrP might have cell autonomous effects on keratinocytes is via an intracrine pathway involving its translocation to the nucleus (Philbrick et al., 1996). Clearly, much research is needed to define the receptors and signaling pathways by which PTHrP acts in skin. Only when this information is available will we be able to understand the mechanisms leading to the skin phenotypes that have been observed in the various transgenic models discussed earlier.

Pathophysiology of parathyroid hormone-related protein

To date, PTHrP has not been clearly implicated in any diseases of the skin. It has been suggested that skin and skin appendage findings in rescued PTHrP-null mice are reminiscent of a series of disorders collectively known as ectodermal dysplasias (Foley et al., 1998), but PTHrP has not been formally linked to any of these diseases. It has also been noted that psoriatic skin may downregulate PTHrP expression, and a small trial of topically administered PTH suggested that stimulation of the PTH1R might improve the histological abnormalities in psoriatic plaques (Hollick et al., 2003). However, these potentially exciting results have not yet been verified in larger trials. Topical application of PTH1R antagonists may be useful for stimulating hair growth, although no data yet suggest that PTHrP is involved in the pathogenesis of alopecia(s) (Safer et al., 2007; Skrok et al., 2015). The most common tumors causing humoral hypercalcemia of malignancy (HHM) are those of squamous histology, but these tumors rarely arise from skin keratinocytes. In fact, the most common skin tumors, basal cell carcinomas, do not overexpress PTHrP and are not associated with hypercalcemia (Philbrick et al., 1996). Primary hyperparathyroidism, hypoparathyroidism, and pseudohypoparathyroidism have each been sporadically reported to have cutaneous manifestations, but whether such findings are directly the result of abnormal PTH1R signaling, abnormal calcitriol concentrations, or a consequence of hypocalcemia or hypercalcemia remains unknown (Skrok et al., 2015). Although PTHrP appears to participate in the normal physiology of the skin, it is not clear at this juncture whether it is involved in skin pathophysiology.

Mammary gland

Very soon after its discovery, PTHrP was reported to be expressed in mammary tissue and secreted into milk at high concentrations (Thiede and Rodan, 1988; Budayr et al., 1989). In retrospect, it would have been much easier to extract PTHrP from human or cow milk than from the tiny amounts expressed in the tumors from which it was first isolated. It is now appreciated that PTHrP is critically important for the proper development and functioning of the mammary gland throughout life. In addition, it has been implicated as an important modulator of the biological behavior of breast cancer. The mammary gland develops in several discrete stages and only reaches its fully differentiated state during pregnancy and lactation. PTHrP appears to serve different functions during the principle stages of mammary development: embryonic development, adolescent growth, and pregnancy and lactation. For each stage, we will first outline the pertinent developmental events in rodents, as data regarding the function(s) of PTHrP largely come from studies in mice and rats. Next, we will discuss the localization of PTHrP and PTHrP receptors and the regulation of the expression of PTHrP and its receptors. Finally, we will address the function of PTHrP.

Embryonic mammary development

In mice, there are two phases of embryonic mammary development. The first involves the formation of five pairs of mammary buds, each of which consists of a bulb-shaped collection of epithelial cells surrounded by several layers of fibroblasts known as the mammary mesenchyme (Sakakura, 1987). After the formation of these buds, mouse mammary development displays a characteristic pattern of sexual dimorphism. In male embryos, in response to androgens, the mammary mesenchyme destroys the epithelial bud and male mice are left without mammary glands or nipples (Sakakura, 1987). In female embryos, however, the mammary buds remain quiescent until embryonic day 16 (E16), when they undergo a transition into the second step of embryonic development, the formation of the rudimentary ductal tree. This process involves the elongation of the mammary bud, its penetration out of the dermis and into a specialized stromal compartment known as the mammary fat pad, and the initiation of ductal branching morphogenesis. At the time of birth, the gland consists of a simple epithelial ductal tree consisting of 15–20 branched tubes within a fatty stroma (Sakakura, 1987). This initial pattern persists until puberty, at which time the mature virgin gland is formed through a second round of branching morphogenesis regulated by circulating hormones (discussed later).

The PTHrP gene is expressed exclusively within epithelial cells of the mammary bud soon after it begins to form. PTHrP mRNA continues to be localized to mammary epithelial cells during the initial round of branching morphogenesis as the bud grows out into the presumptive mammary fat pad and begins to branch (Dunbar et al., 1998, 1999; Wysolmerski et al., 1998). At some point after birth, PTHrP gene expression is downregulated, and in the adult virgin gland, PTHrP mRNA is found only within specific portions of the duct system (Dunbar et al., 1998). In contrast to the PTHrP gene, the PTH1R gene appears to be expressed within the mesenchyme, but its expression is widespread and is not limited to the developing mammary structures. Transcripts for the PTH1R gene are found within the mammary mesenchyme but also throughout the developing dermis (Dunbar et al., 1999; Wysolmerski et al., 1998). It is not clear when the receptor gene is first expressed within the subepidermal mesenchyme. However, it is already present when the mammary bud begins to form, and it continues to be expressed within fibroblasts surrounding the mammary ducts as they begin to extend and branch (Wysolmerski et al., 1998; Dunbar et al., 1998).

Epithelial expression of PTHrP and mesenchymal expression of the PTH1R are not unique to the developing mammary gland, and this pattern has long led to speculation that PTHrP and its receptor might contribute to the regulation of epithelial–mesenchymal interactions during organogenesis. There is now solid evidence that this is the case during embryonic mammary development, where PTHrP acts as an epithelial signal that influences cell fate decisions within the developing mammary mesenchyme. Data supporting this notion come from studies in several genetically altered mouse models. First, in PTHrP or PTH1R knockout mice, there is a failure of the normal androgen-mediated destruction of the mammary bud owing to the failure of the mammary mesenchyme to differentiate properly and to express androgen receptors (Dunbar et al., 1999). Second, in PTHrP and PTH1R knockout mice, the mammary buds fail to grow out into the fat pad and initiate branching morphogenesis, again due to defects in the mammary mesenchyme (Wysolmerski et al., 1998; Dunbar et al., 1998). Finally, in keratin 14 (K14)-PTHrP transgenic mice that ectopically overexpress PTHrP within all the basal keratinocytes of the developing embryo, subepidermal mesenchymal cells, which should acquire a dermal fate, instead react to the excess PTHrP by becoming mammary mesenchyme (Dunbar et al., 1999).

As demonstrated by these studies, PTHrP signaling is essential for mammary gland formation in rodents. When the mammary gland begins to form, PTH1R is expressed in all of the mesenchymal cells underneath the epidermis, but PTHrP is expressed only within the mammary epithelial buds and not within the epidermis itself (Karmali et al., 1992; Thiede and Rodan, 1988; Wysolmerski et al., 1998). As the mammary bud grows down into the mesenchyme, PTHrP, produced by mammary epithelial cells, interacts over short distances with PTH1R on the immature mesenchymal cells closest to the epithelial bud and triggers these cells to differentiate into mammary mesenchyme. PTHrP accomplishes this, at least in part, by upregulating expression of BMP receptors on the mesenchymal cells and sensitizing them to respond in an autocrine fashion to BMP4 expressed in the ventral surface of the embryo (Hens et al., 2007). In this way, PTHrP acts as a patterning molecule contributing to the formation of small patches of mammary-specific stroma around the mammary buds and within the surrounding sea of presumptive dermis (Fig. 36.1; Foley et al., 2001). The process of differentiation set in motion by PTHrP signaling is critical to the ability of the mammary-specific stroma to direct further morphogenesis of the epithelium and to inhibit hair follicle development in the vicinity of the developing nipple. In the absence of this signaling, the mesenchyme can neither destroy the epithelial bud in response to androgens nor trigger the outgrowth of the bud and the initiation of branching morphogenesis (Dunbar et al., 1998, 1999a; Wysolmerski et al., 1998).

Although the model described earlier was developed from studies in mice, PTHrP is also critical to the formation of breast tissues in human fetuses. Blomstrand's chondrodysplasia is a fatal form of dwarfism caused by null mutations of the PTH1R gene (Jobert et al., 1998) (see Chapter 44). Affected fetuses have skeletal abnormalities similar to those caused by deletion of the PTHrP and PTH1R genes in mice (see Chapter 15) and in addition lack breast tissue and nipples (Wysolmerski et al., 1999). In normal human fetuses, the PTHrP gene is expressed within the mammary epithelial bud, and

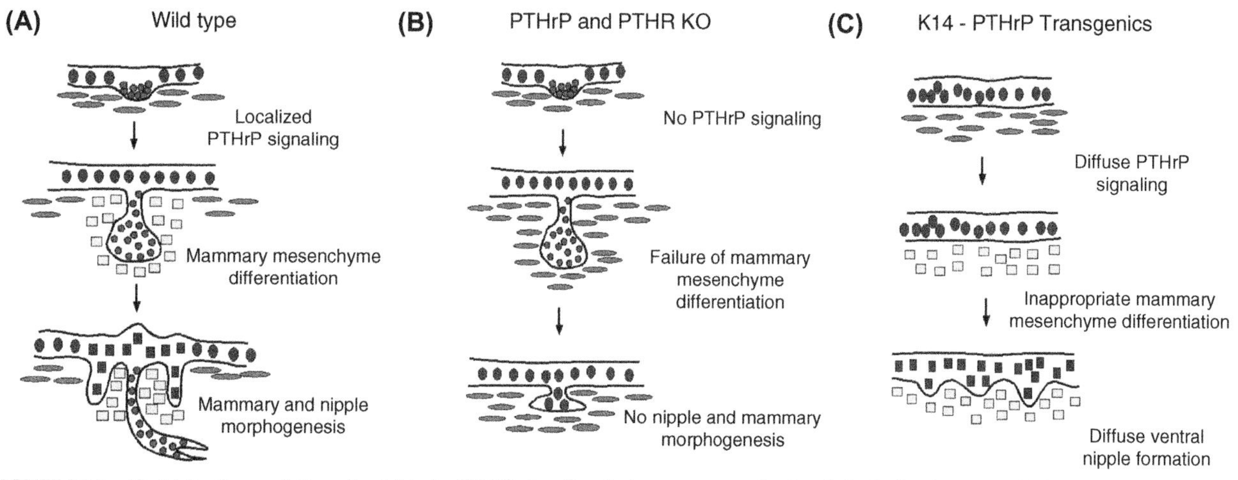

FIGURE 36.1 Model for the regulation of cell fate by PTHrP signaling during mammary gland and nipple development. (A) Normally, the mammary epithelial cells (small circles) express PTHrP after the bud starts to form. PTHrP signals to the dermal mesenchyme (ovals) near the developing bud, and as a result, these cells become mammary mesenchyme (light squares). The mammary mesenchyme maintains the mammary fate of the epithelial cells, triggers their morphogenesis, and induces the overlying epidermis (upright ovals) to become the nipple (dark squares). (B) In the absence of PTHrP signaling, no mammary mesenchyme is formed. Therefore, the mammary epithelial cells revert to an epidermal fate, no morphogenesis occurs, and the nipple does not form. (C) In the presence of diffuse PTHrP signaling, the entire ventral dermis becomes mammary mesenchyme, and the ventral epidermis becomes nipple sheath. *Reproduced from Foley, J., Dann, P., Hong, J., Cosgrove, J., Dreyer, B., Rimm, D., Dunbar, M. E., Philbrick, W., Wysolmerski, J.J., 2001. Parathyroid hormone-related protein maintains mammary epithelial fate and triggers nipple skin differentiation during embryonic mammary development. Development 128, 513–525, with permission.*

the PTH1R gene is expressed in surrounding mesenchyme (Wysolmerski et al., 1999; Cormier et al., 2003). Therefore, in humans as in mice, epithelial-to-mesenchymal PTHrP-to-PTH1R signaling is essential to the formation of the embryonic mammary gland.

Adolescent mammary development

After birth, the murine mammary gland undergoes little development until the onset of puberty. At that point, in response to hormonal changes, the distal ends of the mammary ducts form specialized structures called terminal end buds. These structures serve as sites of cellular proliferation and differentiation during a period of active growth that gives rise to the typical branched duct system of the mature virgin gland (Daniel and Silberstein, 1987). Once formed, the ductal tree remains relatively unchanged until pregnancy when another round of hormonal stimulation induces the formation of the alveolar units that produce milk.

Similar to findings in the embryonic mammary gland, during puberty PTHrP is a product of mammary epithelial cells, and PTH1R is expressed in stromal cells (Dunbar et al., 1998). However, the structure of the pubertal gland is more complex than that of the embryonic gland, and here there are conflicting data regarding the localization of PTHrP and the PTH receptor. Although there is general agreement that PTHrP is expressed in epithelial cells in the postnatal gland, there is some disagreement regarding the specific epithelial compartments in which PTHrP is found. Studies employing in situ hybridization in mice have suggested that after birth, the overall levels of PTHrP gene expression in mammary ducts are reduced except in the terminal end buds during puberty (Dunbar et al., 1998). In these structures, appreciable amounts of PTHrP mRNA were detected in the peripherally located cap cells. In other parts of the gland there was little, if any, specific hybridization for PTHrP. In contrast, studies looking at mature human and canine mammary glands using immunohistochemical techniques have suggested that PTHrP can be found in both luminal epithelial and myoepithelial cells throughout the ducts (Grone et al., 1994; Liapis et al., 1993). Furthermore, studies using cultured cells have suggested that PTHrP is produced by luminal and myoepithelial cells isolated from normal glands (Ferrari et al., 1992; Seitz et al., 1993; Wojcik et al., 1999). Fewer reports have looked at the localization of PTH1R expression in the postnatal mammary gland, but as in embryological development, it is expressed in the mammary stroma (Dunbar et al., 1998). In situ hybridization studies have found the highest concentration of PTH1R mRNA in the stroma immediately surrounding terminal end buds during puberty (Dunbar et al., 1998). This same study found lower levels of PTH1R mRNA distributed generally within the fat pad stroma but very little expression in the dense stroma surrounding the more mature ducts. In addition, these investigators found no evidence of PTH1R mRNA in freshly isolated epithelial cells (Dunbar et al., 1998). In contrast, other studies have suggested that PTH1R is expressed in cultured luminal epithelial and myoepithelial cells (Seitz et al., 1993; Wojcik et al., 1999) as well as in cultured breast cancer cell lines (Birch et al., 1995). In summary, during puberty PTHrP and its receptor are found predominantly within the terminal end buds, with PTHrP localized to the epithelium and PTH1R localized in the stroma. It remains an open and interesting question whether, at some time during mammary ductal development, epithelial cells express low levels of PTH1R.

Studies in transgenic mice have suggested that PTHrP regulates mammary morphogenesis during puberty. Overexpression of PTHrP in mammary epithelial cells using the K14 promoter results in an impairment of ductal branching morphogenesis (Wysolmerski et al., 1995). There are two aspects to the defect. First, the terminal end buds advance through the mammary fat pad at a significantly slower rate than normal. Second, there is a severe reduction in the branching complexity of the ductal tree. As seen in Fig. 36.2, this results in a spare and stunted epithelial duct system. Experiments altering the timing and duration of PTHrP overexpression in the mammary gland using a tetracycline-regulated K14-PTHrP transgene have demonstrated that the two aspects of this pubertal phenotype appear to represent separate functions of PTHrP. The branching (or patterning) defect results from embryonic overexpression of PTHrP, whereas the ductal elongation defect is a function of overexpression of PTHrP during puberty (Dunbar et al., 2001). These effects on ductal patterning provide further evidence of the importance of PTHrP as a regulator of embryonic mammary development. In addition, the localization patterns for PTHrP and PTH1R during puberty, combined with the effects of pubertal overexpression of PTHrP on ductal growth, suggest that PTHrP also functions later in mammary development. During puberty it appears to modulate the epithelial–mesenchymal interactions that govern ductal elongation.

Pregnancy and lactation

Mammary epithelial cells only reach their fully differentiated state during lactation. Under hormonal stimulation during pregnancy, there is a massive wave of epithelial proliferation and morphogenesis that gives rise to multiple terminal ductules and alveolar units. During the later stages of pregnancy, the epithelial cells fully differentiate and then begin to

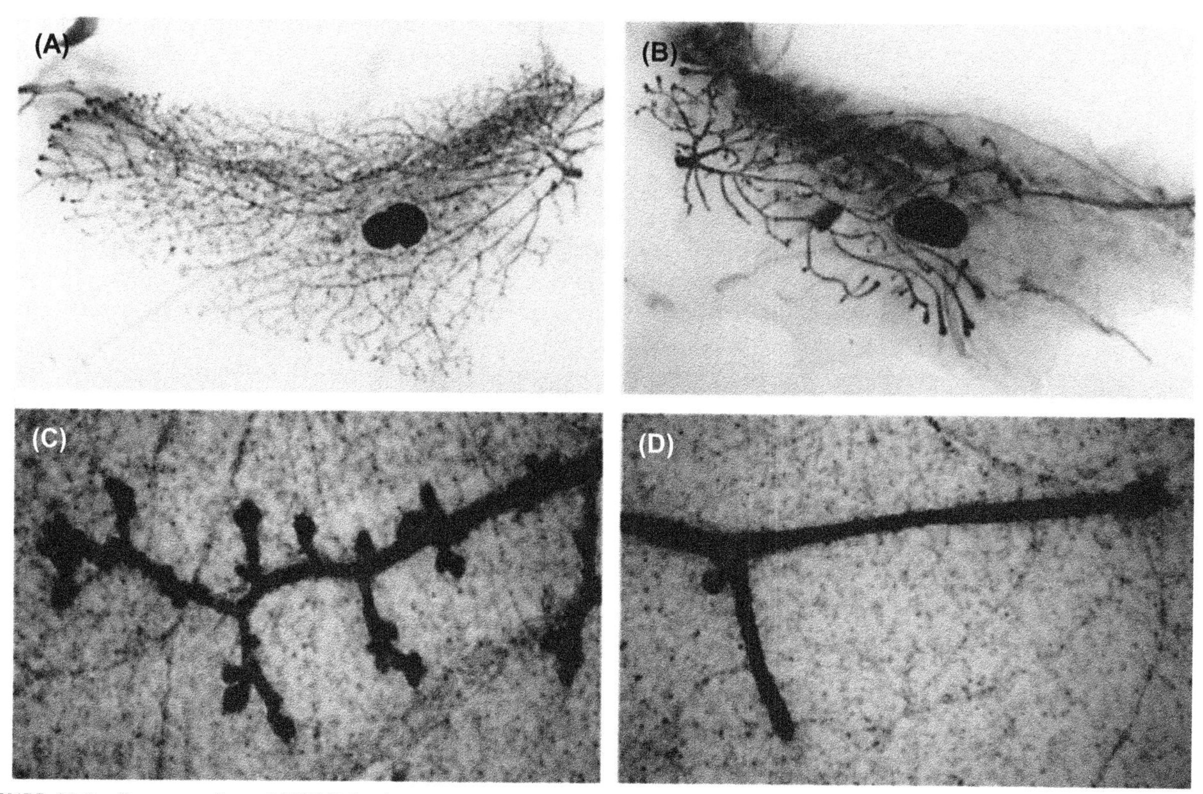

FIGURE 36.2 Overexpression of PTHrP in the mammary gland of K14-PTHrP transgenic mice antagonizes ductal elongation and branching morphogenesis during puberty. **A** and **B** represent typical whole-mount analyses of the fourth inguinal mammary glands from wild-type (A) and K14-PTHrP transgenic mice (B) at 6 weeks of age. The dark oval in the center of each gland is a lymph node. Growth of the ducts during puberty is directional, and each gland is arranged so that the primary duct (the origin of the duct system) is toward the center of the figure. Note that overexpression of PTHrP results in an impairment of the elongation of the ducts through the fat pad as well as a dramatic reduction of the branching complexity of the ductal tree. **C** and **D** represent higher magnifications of a portion of the ducts from the wild-type (C) and transgenic (D) glands demonstrating the reduction in side branching caused by overexpression of PTHrP. *Modified from Wysolmerski, J.J., McCaughern-Carucci, J.F., Daifotis, A.G., Broadus, A.E., Philbrick, W.M., 1995. Overexpression of parathyroid hormone-related protein or parathyroid hormone in transgenic mice impairs branching morphogenesis during mammary gland development. Development 121, 3539–3547 with permission.*

secrete milk during lactation. By the time lactation commences, the fatty stroma of the mammary gland is almost completely replaced by actively secreting alveoli. Upon completion of lactation, widespread apoptosis of the differentiated epithelial cells occurs, and the gland remodels itself into a duct system similar to that of the virgin animal (Daniel and Silberstein, 1987).

Localization studies in humans, rodents, and cows have all noted epithelial cells to be the source of PTHrP in the mammary gland during pregnancy and lactation (Liapis et al., 1993; Wojcik et al., 1998, 1999; Rakopoulos et al., 1992). Based on the assessment of whole-gland RNA, PTHrP expression appears to be upregulated at the start of lactation under the control of both local and systemic factors (Philbrick et al., 1996; Thiede and Rodan, 1988; Thiede, 1989; Thompson et al., 1994; Buch et al., 1992). Thiede and Rodan (1988) and Thiede (1989) originally reported that PTHrP expression in rats is dependent on suckling and on serum prolactin concentrations. However, prolactin must serve only as a permissive factor, for additional studies have shown that the suckling response is a local one and that PTHrP only rises in the milked gland (Thompson et al., 1994). Furthermore, overall PTHrP expression increases gradually over the course of lactation, and in later stages its expression becomes independent of serum prolactin levels (Bucht et al., 1992). Serotonin is now recognized as playing an important role in local control of PTHrP production by mammary tissue (Hernandez et al., 2012; Marshall et al., 2014). It is clear that much of the PTHrP made during lactation ends up in milk, in which levels of PTHrP are up to 10,000-fold higher than in the circulation of normal individuals and 1000-fold higher than in patients suffering from HHM (Philbrick et al., 1996). PTHrP concentrations in milk have generally been found to mirror RNA levels in the gland, increasing over the duration of lactation and rising acutely with suckling (Thompson et al., 1994; Yamamoto et al., 1991; Law et al., 1991; Goff et al., 1991). In addition, evidence shows that PTHrP levels may vary with the calcium content of milk (Yamamoto et al., 1992a; Law et al., 1991; Goff et al., 1991; Uemura et al., 1997; Kovacs and Kronenberg, 1997; Kovacs, 2016). In mice, the calcium-sensing receptor (CaR) is expressed on the basolateral surface of mammary epithelial cells during lactation and regulates PTHrP production such that increased delivery of calcium to the mammary gland decreases PTHrP production and secretion into milk (VanHouten et al., 2004; Ardeshirpour et al., 2006). Selective ablation of CaR within mammary tissue results in locally increased PTHrP production and increased milk PTHrP content (Mamillapalli et al., 2013). Finally, in mice, PTHrP mRNA levels are promptly downregulated during the early stages of involution and then increase to prelactation levels about a week into the remodeling process (Wysolmerksi et al., 2008).

In contrast to PTHrP, little study has been made of the expression and regulation of PTHrP receptors during pregnancy and lactation. In early pregnancy, the PTH/PTHrP receptor is expressed at low levels in the stroma surrounding the developing alveolar units (Dunbar et al., 1998). Studies using whole-gland RNA demonstrate a reciprocal relationship between PTH1R and PTHrP mRNA levels. That is, as PTHrP expression rises during lactation, PTH1R mRNA levels decrease, and as PTHrP mRNA levels fall during early involution, PTH1R expression increases to its former level (Wysolmerski et al., 2008). This may represent active downregulation of the receptor by PTHrP or may simply reflect the changing amount of stroma within the gland at these different stages. However, in a study of cells isolated from lactating rats, it was suggested that epithelial cells, as well as stromal cells, express this receptor (Wojcik et al., 1999), so the regulation of PTH1R expression during pregnancy and lactation may be complex.

Initial reports of the presence of PTHrP in the mammary gland and milk prompted a great deal of speculation regarding its functions in breast tissue during lactation. These proposals revolved around four general hypotheses that PTHrP may be involved in (1) maternal calcium homeostasis and the mobilization of calcium from the maternal skeleton; (2) regulating vascular and/or myoepithelial tone in the lactating mammary gland; (3) transepithelial calcium transport into milk; and (4) neonatal calcium homeostasis or neonatal gut physiology. As will be discussed, there is support for all four hypotheses. The function of PTHrP in the lactating mammary gland in mice was addressed by disrupting the PTHrP gene solely in mammary epithelial cells during lactation (VanHouten et al., 2003). This experiment has supplied evidence to support the role of PTHrP in regulating maternal and neonatal calcium and bone metabolism and has shown that PTHrP plays an indirect role in mediating calcium transport into milk. This point was made clearer in experiments that used calcimimetic drugs or ablated CaR from mammary epithelial cells and found that this increased the mammary production of PTHrP and milk calcium content. It appears that PTHrP's role during lactation is a systemic one to maintain the supply of calcium to the breast tissue that supports milk production, whereas the transport of calcium into milk is locally controlled by CaR and other factors (VanHouten et al., 2004; Ardeshirpour et al., 2006; Mamillapalli et al., 2013; Kovacs, 2016).

Milk production requires a great deal of calcium, and providing an adequate supply to the mammary gland stresses maternal bone and mineral metabolism (Kovacs and Kronenberg, 1997; Kovacs, 2016). Some of the calcium required for milk comes from the diet, as calcium intake is increased owing to the orexigenic effects of suckling (Smith and Grove, 2002). In addition, renal reabsorption of calcium is increased during lactation so that urinary losses are reduced. Finally, a significant proportion of the calcium transported into milk is derived from the maternal skeleton, through the processes of osteoclast-mediated bone resorption and osteocytic osteolysis (Kovacs, 2016). All three processes (intestinal calcium absorption, renal calcium conservation, and skeletal resorption) are stimulated in part by PTHrP. The interplay between intestinal calcium absorption, renal calcium conservation, and skeletal resorption is best illustrated in the rodent, which requires both increased intestinal calcium absorption *and* increased skeletal resorption in order to provide the calcium needed to milk. Increasing the calcium content of the diet causes increased renal calcium excretion and blunts skeletal resorption, whereas a low-calcium diet causes a marked increase in skeletal resorption and reduced renal calcium excretion (Kovacs, 2016). The combination of a low-calcium diet and the use of an antiresorptive medication to block skeletal resorption leads to reduced milk calcium content, tetany, and sudden death of lactating rodents (Kovacs, 2016). In contrast to rodents, intestinal calcium absorption is normal during lactation in women, and low and high intakes of calcium do not influence the magnitude of skeletal resorption during lactation (Kovacs, 2016). Instead, during human lactation, the extent of skeletal resorption appears to be programmed by the intensity of lactation, which in turn determines the production of PTHrP. Breast milk output predicts the rise in PTHrP, and both parameters predict the magnitude of bone mineral density lost during lactation (Kovacs, 2016).

Overall rates of bone turnover are elevated during lactation, but bone resorption and formation are uncoupled, such that bone resorption outstrips bone formation to cause a rapid decline in bone mass. The average nursing woman loses between 5% and 10% of her bone mass over 6 months, especially from the trabecular spine (Kovacs, 2016), whereas rodents lose up to one-third of their skeletal mass over 21 days of lactation (VanHouten and Wysolmerski, 2003). Mobilization of skeletal calcium during lactation does not rely on any of the established calcium-regulating hormones, PTH, 1,25(OH)2D, or calcitonin (Kovacs and Kronenberg, 1997; Kovacs, 2016). Multiple investigations have confirmed that bone resorption is increased because of the combination of decreased systemic estradiol concentrations and elevated circulating PTHrP levels. Suckling directly inhibits hypothalamic gonadotropin-releasing hormone (GnRH) secretion and induces hypogonadotropic hypogonadism (Smith and Grove, 2002). Bone loss correlates with the duration of amenorrhea in nursing women; however, the intensity of lactation predicts bone loss and the duration of amenorrhea, and bone loss has been shown to continue in women whose menses resume despite ongoing intense lactation (Kovacs, 2016). Pharmacological treatment with estradiol in lactating mice reduced bone loss by 60% (VanHouten and Wysolmerski, 2003); however, the resulting estradiol levels were fivefold higher than normal and may therefore indicate suppressed bone loss from the known pharmacological effects of estradiol rather than implicating that estradiol deficiency accounts for 60% of the bone loss that normally occurs during lactation.

Many studies have now documented elevated circulating PTHrP levels in lactating humans and rodents (Kovacs and Kronenberg, 1997; Kovacs, 2016), and PTHrP levels correlate directly with biochemical markers of bone resorption and inversely with bone mass in lactating mice (VanHouten and Wysolmerski, 2003). In addition, circulating PTHrP levels have been shown to correlate with bone density changes in lactating humans (Sowers et al., 1996). Furthermore, there have been multiple case reports of pseudohyperparathyroidism during pregnancy and lactation, in which PTHrP secretion by breasts was determined to have caused hypercalcemia with high plasma concentrations of PTHrP (see below) (Kovacs, 2016). In normal lactating women who have been followed longitudinally, the albumin-corrected serum calcium and ionized calcium rise significantly but stay within the normal range, whereas serum phosphorus rises above normal, with these effects implying the release of calcium and phosphorus from bone (Kovacs, 2016). Therefore, the condition of pseudohyperparathyroidism may represent one extreme of the normal physiology of lactation, in which PTHrP's effects to increase bone resorption lead to an increase in the circulating calcium concentration.

When the PTHrP gene was disrupted specifically in the lactating mammary gland in mice, circulating PTHrP levels declined, milk PTHrP became unmeasurable, rates of bone resorption decreased, and bone loss was reduced by 50% (VanHouten et al., 2003). The circulating PTHrP level remained above nonpregnant values; moreover, the mice developed significant secondary hyperparathyroidism, which would have supported ongoing bone resorption despite lower circulating PTHrP. Similarly, loss of the gene encoding calcitonin led to a doubling of bone loss during lactation, increased milk calcium content, increased PTHrP, and marked secondary hyperparathyroidism (Woodrow et al., 2006; Collins et al., 2013; Kovacs, 2016). Thus, it is now clear that the lactating mammary gland secretes PTHrP into both milk and the circulation, and that secondary hyperparathyroidism can contribute to increasing skeletal resorption during lactation when the demand for skeletal calcium is particularly high or PTHrP is absent. Systemic PTHrP from the breast acts together with estrogen deficiency to stimulate bone resorption and cause bone loss during lactation. As noted previously, the production of PTHrP by the lactating breast is decreased by stimulation of the CaR on mammary epithelial cells (VanHouten et al., 2004).

Fig. 36.3 displays the "breast–brain–bone" circuit, a classic endocrine feedback loop during lactation in which the mammary gland functions as an "accessory" parathyroid gland. It uses PTHrP instead of PTH to ensure an adequate flow of calcium from the maternal skeleton to make milk. If calcium delivery to the gland falls, mammary epithelial cells produce more PTHrP, which stimulates increased bone resorption. The delivery of more calcium to the mammary gland from the skeleton then stimulates the CaR and decreases PTHrP production. But milk production is controlled mainly by suckling on demand, which in turn can provoke maternal hypocalcemia (milk fever in animals) and secondary hyperparathyroidism.

Genetic removal of PTHrP from the mammary gland does not alter milk calcium levels, and PTHrP-null mammary epithelial cells are able to transport calcium in vitro at the same rate as wild-type mammary epithelial cells (VanHouten et al., 2003). Thus, unlike the placenta (see later), it does not appear that PTHrP directly contributes to the regulation of transepithelial calcium transport in the mammary gland. However, as mentioned earlier, the marked secondary hyperparathyroidism that developed in that model would have supported calcium delivery to the mammary tissues despite the absence of PTHrP's normal effects, and the circulating PTHrP level was still elevated during lactation over its nonpregnancy values. Multiple human studies have found that milk calcium content correlates with the content of PTHrP, which is consistent with PTHrP's function to supply calcium to mammary tissue, even if it does not directly stimulate pumping of calcium into milk from mammary epithelial cells (Kovacs, 2016).

The function of PTHrP in milk remains an open question. Murine pups that receive PTHrP-deleted milk accrete more skeletal mineral content over the first 12 days after birth (Mamillapalli, 2013), which suggests that PTHrP within milk may modulate either intestinal calcium absorption or skeletal mineral accretion in neonatal mice. Whether PTHrP in milk is absorbed into the neonatal circulation is an unresolved question, but some animal studies have suggested that it is absorbed intact, and might therefore have effects on the skeletal and mineral metabolism in the neonate (Kovacs, 2016). Further work is needed to understand the role of PTHrP in milk.

Another potential function of PTHrP during lactation is the regulation of vascular and/or myoepithelial cell tone. As discussed in Chapter 26, PTHrP has been shown to modulate smooth muscle cell tone in a variety of organs, including the vascular tree, where it acts as a vasodilator. Consistent with these effects, two studies have shown that PTHrP increases

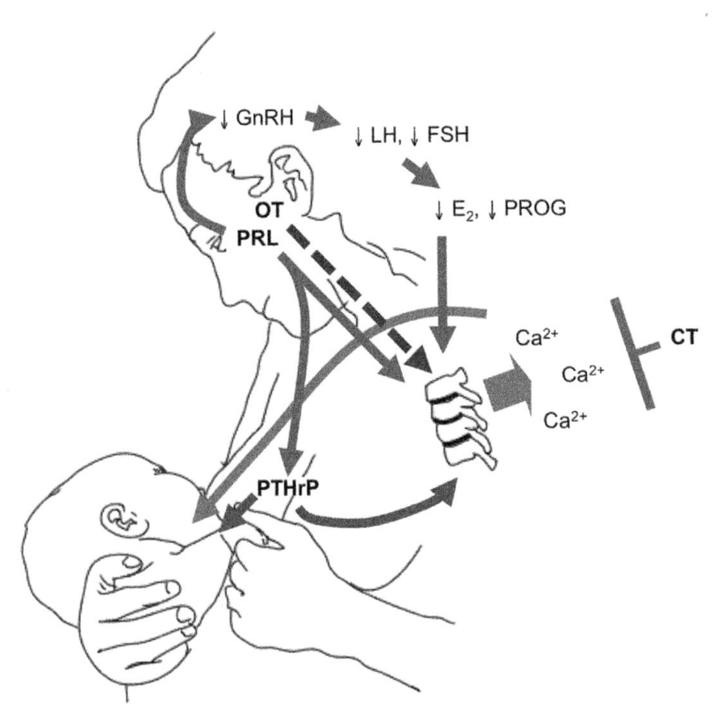

FIGURE 36.3 Breast–Brain–Bone circuit controls lactation. Suckling and prolactin [PRL] both inhibit the hypothalamic GnRH pulse center, which in turn suppresses the gonadotropins (luteinizing hormone [LH] and follicle-stimulating hormone [FSH]), leading to low levels of the ovarian sex steroids (estradiol [E_2] and progesterone [PROG]). Prolactin may also have direct effects on its receptor in bone cells. PTHrP production and release from the breast are stimulated by suckling, prolactin, low estradiol, and the calcium receptor. PTHrP enters the bloodstream and combines with systemically low estradiol levels to markedly upregulate bone resorption and (at least in rodents) osteocytic osteolysis. Increased bone resorption releases calcium and phosphate into the blood stream, which then reaches the breast ducts and is actively pumped into the breast milk. PTHrP also passes into milk at high concentrations, but whether swallowed PTHrP plays a role in regulating calcium physiology of the neonate is uncertain. In addition to stimulating milk ejection, oxytocin may directly affect osteoblast and osteoclast function (dashed line). Calcitonin may inhibit skeletal responsiveness to PTHrP and low estradiol given that mice lacking calcitonin lose twice the amount of bone during lactation as normal mice. Not depicted is that calcitonin may also act on the pituitary to suppress prolactin release, and within breast tissue to reduce PTHrP expression and lower the milk calcium content (see text). *Adapted with kind permission from Kovacs, C.S., 2005a. Calcium and bone metabolism during pregnancy and lactation. J. Mammary Gland Biol. Neoplasia 10 (2), 105–118, © Springer Science and Business Media B.V.*

mammary blood flow during lactation (Davicco et al., 1993; Thiede et al., 1992). The injection of amino-terminal fragments of PTHrP into the mammary artery of dried (nonlactating) ewes was shown to increase mammary blood flow and override the vasoconstrictive effects of endothelin (Davicco et al., 1993). Thiede and colleagues (1992) have demonstrated that the nutrient arteries of the inguinal mammary glands of rats secrete PTHrP and that its production is responsive to suckling and prolactin. Myoepithelial cells in the breast are similar in some ways to vascular smooth muscle cells and are thought to participate in the control of milk ejection by contracting in response to oxytocin (Daniel and Silberstein, 1987). Therefore, it is interesting that myoepithelial cells in culture have been shown to express PTH1R and to respond to PTHrP by elevating intracellular cAMP (Seitz et al., 1993; Wojcik et al., 1999). Furthermore, mirroring the effects of PTHrP on the endothelin-induced contraction of vascular smooth muscle, PTHrP has been shown to block the rise in intracellular calcium normally induced in response to oxytocin in myoepithelial cells (Seitz et al., 1993).

Pathophysiology of parathyroid hormone-related protein in the mammary gland

PTHrP may contribute to pathophysiology in the human breast in several instances. First, as noted previously, fetuses afflicted with Blomstrand's chondrodystrophy lack nipples and breast tissue (Wysolmerski et al., 1999). Second, some cases of PTHrP-mediated hypercalcemia during pregnancy or lactation (pseudohyperparathyroidism) have been reported to be associated with elevations in circulating levels of PTHrP (Khosla et al., 1990; Reid et al., 1992; Kovacs, 2016). Additional cases of pseudohyperparathyroidism have been caused by the placental production of PTHrP; see later. One of these cases was caused by massive breast hyperplasia, and after reduction mammoplasty, the patient's hypercalcemia and elevated PTHrP levels both resolved (Khosla et al., 1990). However, large breasts are not a requirement for pseudohyperparathyroidism to occur. A related aspect is that the production of PTHrP by the lactating mammary gland leads to normalization of mineral and bone metabolism in hypoparathyroid women who breastfeed (Kovacs, 2016). In some cases, hypercalcemia has developed when exogenous use of calcitriol and calcium supplements has not been stopped in advance of lactation in these hypoparathyroid women. Finally, the area with the greatest potential impact on human health is the relationship of PTHrP production to breast cancer. This is evolving into a complicated topic and will be addressed only briefly here. However, it will be reviewed in more depth in Chapters 55 and 56.

It is well documented that PTHrP is produced by a number of primary breast carcinomas and that this sometimes leads to classical HHM (Isales et al., 1987). A potentially more widespread role may be the involvement of PTHrP in the osteotrophism of breast cancer (Guise et al., 1996; Yin et al., 1999). Animal models have suggested that PTHrP production by breast tumor cells is important to their ability to form skeletal metastases (Guise et al., 1996; Yin et al., 1999). However, there is conflicting evidence as to whether PTHrP production by a primary breast tumor is predictive of bone metastases in patients (Bundred et al., 1996; Henderson et al., 2001, 2006). The largest and most carefully controlled study suggested that PTHrP production by the primary tumor is actually a negative rather than positive predictor of skeletal metastases (Henderson et al., 2006). It may be that PTHrP production does not enable a tumor cell to get into the skeleton, but once there, the ability of tumor cells to upregulate PTHrP production within the bone microenvironment becomes important to their ability to grow in the skeleton (Yin et al., 1999). Several studies suggest that TGF-β released from the bone matrix during bone resorption may be particularly important in stimulating PTHrP production (Kakonen et al., 2002). More PTHrP would in turn be expected to increase osteoclast activity and release more TGF-β, setting up a potential feed-forward vicious cycle that may be important in the pathogenesis of bone destruction around the metastatic tumor.

There is a growing literature suggesting that PTHrP may also be important in controlling breast cancer cell proliferation, apoptosis, migration, and invasion. Although it is unclear whether normal mammary epithelial cells express significant levels of the PTH1R, it appears that many breast cancer cell lines do so (Birch et al., 1995). Some of the effects ascribed to PTHrP appear to be mediated by this receptor, but some appear to be mediated by midregion PTHrP fragments acting either through intracrine mechanisms or perhaps via a specific midregion PTHrP cell-surface receptor (Shen et al., 2004; Kumari et al., 2006; Shen and Falzon, 2006; Dittmer et al., 2006; Sirchia et al., 2007). In either case, the effects of midregion PTHrP apparently require that this portion of the peptide enter the nucleus via a distinct bipartite nuclear localization signal. In particular, several studies have suggested that midregion PTHrP fragments may promote cell migration by regulating $\alpha 6\beta 4$ integrin expression through nuclear effects (Shen et al., 2003; Dittmer et al., 2006; Shen and Falzon, 2006). Despite these intriguing findings, at this point it is not clear whether these effects are artifacts of cell lines or will prove to be clinically relevant. Most studies in cell lines suggest that PTHrP expression by a breast tumor would contribute to increased aggressiveness or a tendency to metastasize.

However, animal models have provided seemingly contradictory data regarding PTHrP's potential influence on breast cancer tumor progression. Deletion of PTHrP promoted tumor progression in the *Neu* mouse model of spontaneous tumor development (Fleming et al., 2009), while deletion of PTHrP in the MMTV-PyMT mouse model had the opposing effect of preventing tumor progression (Li et al., 2011). These mouse studies make it unclear as to whether it would be beneficial or harmful to inhibit PTHrP expression in the primary tumors of patients with breast cancer.

On the other hand, human studies have suggested that expression of PTHrP by the primary breast tumor has beneficial effects on multiple outcomes. In the large prospective clinical study of Henderson and colleagues mentioned previously (Henderson et al., 2001, 2006), in which patients were assessed 5 and 10 years after surgery, PTHrP expression by a primary breast tumor was clearly a marker of a less aggressive course correlating with estrogen and progesterone receptor positive status, increased survival, lower rates of recurrence, and lower rates of all metastases (not just bone metastases). An analysis of the METABRIC dataset within The Cancer Genome Atlas breast cancer dataset revealed that expression of the human PTHrP gene, *PTHLH*, was upregulated (not downregulated) at the mRNA level or amplified in breast cancer patients (R.J. Johnson, A.W. Freeman and T.J. Martin, unpublished, 2018). Furthermore, patients with mRNA upregulation of *PTHLH* had significantly increased overall survival. Similar results have been gleaned from other gene expression surveys in breast cancer (www.oncomine.org; Rhodes et al., 2004). Further in vivo studies are clearly needed, but the available human data suggest that inhibiting PTHrP expression in primary breast cancers may be harmful rather than beneficial.

Reproductive tissues

Parathyroid hormone-related protein and placental calcium transport

Nearly all the calcium, and a large proportion of inorganic phosphate (85%) and magnesium (70%), transferred from mother to fetus is associated with development and mineralization of the fetal skeleton (Grace et al., 1986). The bulk of placental calcium transfer occurs rapidly over a short interval late in gestation, such that 80% occurs in the third trimester in humans (Givens and Macy, 1933), and 96% occurs in the last 5 days of gestation in the rat (Comar, 1956). The concentrations of both total and ionized Ca in all mammalian fetuses studied during late gestation have been observed to be higher than maternal levels. As a result of studies in which the sheep was used extensively for the study of fetal calcium control, one of the first suggested physiological roles of PTHrP was that of regulating the transport of calcium from mother to fetus in the mammal, thereby making calcium available to the growing fetal skeleton (Rodda et al., 1988).

Immunoreactive PTH levels were found to be low in fetal lambs, whereas PTH-like biological activity in serum was high (Care et al., 1985), suggesting the presence of another PTH-like substance. Similar findings of low PTH and high PTHrP have been demonstrated in multiple studies of fetal mice and human cord blood (Kovacs, 2003, 2016). Parathyroidectomy in the fetal lamb resulted in loss of the calcium gradient that exists between mother and fetus as well as impairment of bone mineralization, implicating parathyroids as the source of the regulatory agent. Crude, partially purified, or recombinant PTHrP, but neither PTH nor maternal parathyroid extract that contained no immunoreactive PTHrP, restored the gradient (Rodda et al., 1988). Thus, PTHrP appeared to be the active component of the fetal parathyroid glands responsible for maintaining fetal calcium levels and suppressing fetal PTH levels. In support of this hypothesis, immunoreactive PTHrP was found to be readily detectable in sheep fetal parathyroids from the time they form (MacIsaac et al., 1991) and was also found in early placenta, suggesting that the latter tissue may be a source of PTHrP for calcium transport early in gestation.

The portion of PTHrP that appears to be responsible for regulating placental calcium transport lies between residues 67 and 86 (Care et al., 1990), but the responsible receptor has not yet been identified. Although syncytiotrophoblasts are believed to be central in the transport of calcium to the fetus, cytotrophoblasts (which differentiate to form the syncytium) are believed to be the calcium-sensing cells, and raising the extracellular calcium concentration has been shown to inhibit PTHrP release from these cells (Hellman et al., 1992). The CaR has been localized to cytotrophoblasts of human placenta (Bradbury et al., 1997), and the work of Kovacs et al. (1998) has implicated it in placental calcium transport. Specifically, ablation of CaR reduced placental calcium transport, and this was accompanied by a reduction in plasma PTHrP (Kovacs et al., 1998). Furthermore, a calreticulin-like calcium-binding protein has been isolated from trophoblast cells, and its expression is increased by treatment with PTHrP(67–84) but not with N-terminal PTHrP (Hershberger and Tuan, 1998).

Although these observations are strongly suggestive of involvement of PTHrP and the CaR, the mechanisms of placental calcium transport are still not fully understood. Support for the role of PTHrP also comes from the PTHrP gene knockout mouse, in which placental calcium transport is severely impaired (Kovacs et al., 1996). In mice homozygous for deletion of the PTHrP gene, fetal plasma calcium and maternal–fetal calcium gradient were significantly reduced. When fetuses were injected in utero with fragments of PTHrP or PTH, calcium transport was significantly restored only by treatment with a midmolecular region of PTHrP that does not act via PTH1R. Furthermore, in mice rendered null for the PTH1R gene, placental calcium transport was increased, and PTH1R-null fetuses had plasma PTHrP levels more than 10 times higher than controls (Kovacs et al., 2001). The circulating PTHrP in the fetal mice was found to be derived from several tissues, including liver and placenta, but parathyroids were excluded as a source of PTHrP in this setting (Kovacs et al., 2001). Additional studies confirmed that mice lacking parathyroid glands or parathyroid hormone did not have an alteration in circulating PTHrP, nor was placental calcium transport reduced, although the absence of fetal PTH was associated with an even greater reversal of the maternal–fetal calcium gradient than that induced by loss of PTHrP (Kovacs et al., 2001; Simmonds et al., 2010). These findings indicate that fetal parathyroids may not be a source of circulating PTHrP and are compatible with an earlier detailed examination of normal fetal rat parathyroids that found no detectable PTHrP by in situ hybridization, RT-PCR, and immunohistochemistry (Tucci et al., 1996).

Additional support for the role of PTHrP is that placental expression of PTHrP was increased in two different models in which the vitamin D receptor is ablated, and in both studies placental calcium transport was also increased (Kovacs, 2005b; Lieben, 2013).

Study of mice lacking the gene encoding PTH enabled the detection of a low level of expression of PTH mRNA in the murine placenta and alterations in the expression of genes involved in cation transport (Simmonds et al., 2010). Exogenous PTH administration also stimulated placental calcium transport in *Pth*-null fetuses but not in WT fetuses within the same litters (Simmonds et al., 2010). Therefore, the placenta appears to be an ectopic site of low-level PTH expression, but whether this has an important influence on fetal–placental mineral homeostasis in normal mice is unknown. Furthermore, whether PTH is expressed at all in the human placenta has not been examined with modern methods.

Overall, conclusions from the murine studies are similar in many respects to those in sheep, namely that PTHrP contributes to fetal skeleton calcium supply by controlling maternal–fetal calcium transport through actions mediated by a midmolecule portion of the PTHrP molecule. The murine and sheep studies differ in that the parathyroids do not appear to be a dominant source of PTHrP in rodents; the placenta may be the relevant source of PTHrP that controls placental calcium transfer and the fetal–placental gradient. Murine data support reduced placental calcium transport when PTHrP is reduced or absent and increased placental calcium transport when circulating PTHrP concentrations or placental PTHrP expression are increased.

Uterus and extraembryonic tissues

The uterus, both pregnant and nonpregnant, is another of the many sites of production and action of PTHrP. The relaxing effect of PTH on uterine smooth muscle had been long recognized (Shew et al., 1984), and it was not surprising that PTHrP had the same effect (Shew et al., 1991). The finding that expression of mRNA for PTHrP in the myometrium during late gestation in the rat was controlled by intrauterine occupancy by the fetoplacental unit raised the possibility of a role for PTHrP in regulating uterine muscle tone (Thiede et al., 1990).

In studies in rats with or without estrogen treatment, protein and mRNA for PTHrP were localized not only in the myometrium, as had been shown in pregnancy (Thiede et al., 1990), but also in the epithelial cells lining the endometrium and endometrial glands. Indeed, the strongest PTHrP production appeared to be in these sites (Paspaliaris et al., 1992), suggesting that the endometrium and endometrial glands might be the major uterine site of PTHrP production and that PTHrP might be a local regulator of endometrial function and myometrial contractility. Estrogen treatment enhanced uterine production of PTHrP, but most significantly, the relaxing effect of PTHrP on uterine contractility in vitro was enhanced greatly by the pretreatment of noncycling rats with estrogen. In keeping with this observation, uterine horns from cycling rats in proestrous and estrous phases of the cycle showed a greater responsiveness to PTHrP than those from noncycling rats. These findings are consistent with a role for PTHrP as an autocrine and/or paracrine regulator of uterine motility and function. Furthermore, they suggest that PTHrP belongs to a class of other locally acting peptides, such as oxytocin, vasoactive intestinal peptide, and relaxin, for which pretreatment of animals with estrogen increases the response of the uterus (Ottesen et al., 1985; Mercado-Simmen et al., 1982; Fuchs et al., 1982).

Further evidence for a specific and regulated role of PTHrP in the uterus during gestation comes from the observation of a temporal pattern in the relaxation response to PTHrP by longitudinal uterine muscle during pregnancy in the rat, with maximal responses at times when estrogen levels would be high. In contrast, the circular muscle did not respond at any stage during gestation (Williams et al., 1994). The inability of PTHrP to relax uterine muscle in the last stages of gestation does not support a direct role in the onset of parturition. Treatment of pregnant rats with intraperitoneal injections of human PTHrP 1−34 resulted in a significant decrease in the expression of connexin-43 (mRNA and protein) and the oxytocin receptor mRNA in the myometrium, but it did not affect the timing of delivery, progesterone in maternal plasma, or levels of c-fos, fra-2, or PTH1R mRNA on any gestational day. These findings are compatible with the hypothesis that PTHrP may act to keep the myometrium quiescent at a time when progesterone levels are falling, but that the effects of PTHrP signaling are overridden by other factors that dictate the onset of labor (Mitchell et al., 2003). This hypothesis is supported by the demonstration (Thiede et al., 1990) that expression of mRNA was dependent on the presence of the fetus and that levels increased throughout pregnancy and decreased sharply after delivery. It seems likely, therefore, that the observed fall in PTHrP reflects the recontracted state of the uterine muscle, consistent with the observation in the bladder (Yamamoto et al., 1992), and that the level of expression is functionally related to contractility. The temporal expression of PTHrP in endometrial glands and blood vessels (Williams et al., 1994) also supports roles in other regulated functions that might include uterine growth during pregnancy and the regulation of uterine and placental blood flow (Mandsager et al., 1994).

Uterine growth restriction was induced in Wistar−Kyoto rats by ligating uterine vessels, and this resulted in a 15% decrease in fetal weight, a 21% decrease in fetal number, and a 46% decrease in placental PTHrP content, but a 2.5-fold increase in uterine PTHrP content (Wlodek et al., 2005). The increase in uterine PTHrP content may be compensatory to increase uteroplacental blood flow. Other studies have demonstrated that the vasodilatory effect of PTHrP on myometrial blood vessels is dependent upon the presence of functional endothelium, and that the effect is likely mediated by nitric oxide (Meziani et al., 2005). Conversely, treatment of pregnant rats with a PTHrP antagonist (PTHrP 7−34) resulted in evidence of growth restriction (reduced fetal weight and placental weight) and apoptosis within placentas (Thoten et al., 2005). Thus, there is some functional evidence to suggest that PTHrP plays a role in regulating uterine blood flow and growth.

Placenta and fetal membranes

PTHrP mRNA and protein have been detected in rat and human placenta in various cell types (Hellman et al., 1992; Germain et al., 1992; Bowden et al., 1994). In addition, neoplastic cells of placental origin secrete PTHrP, including hydatidiform moles and choriocarcinomas in vitro (Deftos et al., 1994). The presence of PTH/PTHrP receptor mRNA has been demonstrated in rat (Urena et al., 1993), mouse (Kovacs et al., 2002), and human (Curtis et al., 1998) placenta, and infusion of PTHrP(1−34) into isolated human placental lobules stimulates cyclic AMP production (Williams et al., 1991).

Three further sets of observations lend support to the hypothesis that PTHrP is involved in placental/uterine interactions and that its most likely role in the placenta and placental membranes is related to the growth and maintenance of the placenta itself during pregnancy. First, PTHrP production by cultured amniotic cells has been shown to be regulated by prolactin, human placental lactogen, transforming growth factor-β (TGF-β), insulin, insulin-like growth factor, and epidermal growth factor (Dvir et al., 1995). Second, PTHrP has been shown to regulate epidermal growth factor receptor expression in cytotrophoblast cultures (Alsat et al., 1993), an event associated with placental development. Third, studies of vascular reactivity in isolated human placental cotyledons preconstricted with a thromboxane A2 mimetic showed PTHrP to be a very effective vasodilator (Macgill et al., 1997). The narrow concentration range to which the tissue responded, together with the desensitization in response to repeated PTHrP infusions, was consistent with a paracrine and/or an autocrine action of PTHrP in human gestational tissues. Adequacy of the fetoplacental circulation is essential for the nutritional demands of the growing fetus, and both humoral and local factors are likely to be important in its control. It is possible that alterations of the expression, localization, and/or action of PTHrP might contribute to the genesis of conditions such as preeclampsia and intrauterine growth retardation in which placental vascular resistance is increased (Gude et al., 1996). Another related and potentially interacting influence is angiotensin II, known to be a powerful enhancer of PTHrP production in the vasculature and in human placental explants (Li et al., 1998). The ability of angiotensin II to stimulate estradiol production in human placental explants through actions upon its AT1 receptor (Kalenga et al., 1995) provides a further link with PTHrP control.

PTHrP has also been shown to regulate the differentiation of cells explanted from the murine placenta. PTHrP treatment reduced proliferation, inhibited apoptosis, and promoted differentiation into trophoblast giant cells. As well, PTHrP treatment induced the expression of transcription factors known to stimulate giant cell formation (Stra13 and AP-2gamma) and inhibited the formation of other trophoblast cell types by suppressing trophoblast progenitors and spongiotrophoblast-promoting factors (Eomes, Mash-2, and mSNA) (El-Hashashetd et al., 2005). Thus, PTHrP likely plays a role in the differentiation of cells during placentation.

The most likely source of increased amniotic fluid PTHrP concentrations during pregnancy is the amnion itself, because PTHrP mRNA expression is also highest at term and greater in the amnion than in choriodecidua or placenta (Bowden et al., 1994; Ferguson et al., 1992; Wlodek et al., 1996). In tissue from women with full-term pregnancies and not in labor, the concentration of N-terminal PTHrP has been found to be higher in amnion covering the placenta than in the reflected amnion covering the decidua parietalis (Bowden et al., 1994). Nevertheless, the concentration of N-terminal PTHrP in reflected amnion (the layer apposed to the uterus) was inversely related to the interval between rupture of the membranes and delivery. The observation that PTHrP levels in the amnion decrease after rupture of the fetal membranes has led to the proposal that PTHrP derived from the membranes may inhibit uterine contraction and that labor may occur following loss of this inhibition. Plasma levels of PTHrP increase threefold during pregnancy and reach even higher levels postpartum (Bertelloni et al., 1994; Gallacher et al., 1994; Ardawi et al., 1997; Yadav et al., 2014) with the likely sources being placenta and breast, respectively. Human fetal membranes have been shown to inhibit contractions of the rat uterus in vitro (Collins et al., 1993), so this tissue does appear to produce factors that can modulate uterine activity. Furthermore, primary cultures of human amniotic cells secrete PTHrP into the medium (Germain et al., 1992). Thus, although the physiological function(s) of amnion-derived PTHrP is currently unknown, preliminary evidence suggests that it may play a role in regulating the onset of labor. It is also possible that amniotic fluid is a source of PTHrP ingested by the fetus and acts as a growth factor in lung and/or gut development. Consistent with this hypothesis, mice lacking PTHrP have immature lungs associated with arrested type II alveolar cell development and reduced surfactant production in vivo and in vitro (Rubin et al., 2004).

In summary, although many functional studies remain to be completed, potential roles for PTHrP produced by fetal membranes and placenta include transport of calcium across the placenta, accommodation of stretch of membranes, growth and differentiation of fetal and/or maternal tissues, vasoregulation, and regulation of labor.

Implantation and early pregnancy

Some physiological functions other than control of myometrial activity were suggested by findings of Beck et al. (1993), who identified PTHrP mRNA as being limited to epithelial cells of implantation sites. This pregnancy-related expression appeared at day 5.5 in the rat fetus in the antimesometrial uterine epithelium of implantation sites, raising the possibility of a further function of PTHrP playing a part in the localization of implantation or initial decidualization. Decidual cells produced mRNA for PTHrP both in normal gestation and after the induction of deciduomata. Expression of the gene in these cells followed epithelial expression by 48 h. It was concluded from this work that the location of PTHrP gene expression in the uterus, together with the time of its expression, may play a part in implantation of the blastocyst.

Further evidence for a function of PTHrP in the implantation process came from Nowak et al. (1999), who showed that PTHrP and TGF-β were potent stimulators of trophoblast outgrowth by mouse blastocysts cultured in vitro. The TGF-β effect appeared to be mediated by PTHrP, which itself was acting through a mechanism distinct from the PTHR1.

Thus, both the timing and the localization of PTHrP gene expression suggested that it might play a part in the implantation of the blastocyst (Beck et al., 1993). Upon finding substantial levels of immunoreactive PTHrP in the uterine luminal fluid of estrogen-treated immature rats, and because the PTH/PTHrP receptor was known to be expressed in rat uterus (Urena et al., 1993), Williams et al. (1998) investigated the role of PTHrP acting through this receptor in influencing early pregnancy in the rat. Infusion of either a PTHrP antagonist peptide or a monoclonal anti-PTHrP antibody into the uterine lumen during pregnancy resulted in excessive decidualization. The latter appeared to be the result of a decrease in the number of apoptotic decidual cells in the antagonist-infused horn. In pseudopregnant rats, infusion of receptor antagonist into the uterine lumen resulted in increases in wet weight of the infused horn compared with the control side, indicating an effect on deciduoma formation.

These observations suggest that activation of the PTH/PTHrP receptor by locally produced PTHrP might be crucial for normal decidualization during pregnancy in rats, probably not by being involved in the initiation of the decidual reaction but rather in the maintenance of the decidual cell mass.

Pathophysiology of parathyroid hormone-related protein in the placenta

The breasts and placenta contribute PTHrP to maternal circulation during pregnancy, and this occasionally leads to the development of PTHrP-mediated hypercalcemia (pseudohyperparathyroidism) (Kovacs, 2016). This is accompanied by plasma PTHrP concentrations as high as 40 pmol/L and evidence of increased bone resorption. Although uncommon, it suggests that PTHrP production by breasts and placenta may also contribute to the normal regulation of maternal mineral homeostasis during pregnancy, including upregulation of calcitriol production by the maternal kidneys (Kovacs, 2016).

The clinical courses of individual cases of pseudohyperparathyroidism have made evident which of breasts versus placenta were responsible for the excess production of PTHrP. When the breasts (of normal or large size) are the source, the hypercalcemia persists and may worsen postpartum as the onset of lactation induces a further rise in mammary PTHrP production. Breast binders, dopaminergic agents, and weaning usually lead to a prompt resolution of the hypercalcemia, but occasionally it has persisted for months to years or has required a reduction mammoplasty to resolve it (Kovacs, 2016). On the other hand, when the placenta is the source of excess PTHrP, the condition reverses within a few hours of delivery. One pregnant woman with normal-sized breasts developed severe hypercalcemia (5.25 mmol/L) in the third trimester, accompanied by undetectable PTH and a serum PTHrP of 21 pmol/L (Eller-Vainicher et al., 2012). Within 6 hours after an urgent C-section, she was profoundly hypocalcemic, PTHrP had become undetectable, and PTH had risen above the normal range. The rapid reversal from hypercalcemia to hypocalcemia, elevated PTHrP to undetectable PTHrP, and undetectable PTH to high PTH revealed the placenta as the culprit source of excess PTHrP.

Some cases of pseudohyperparathyroidism have presented with vertebral compression fractures during pregnancy or lactation, confirming that the excess production of PTHrP causes marked bone resorption and can temporarily cause skeletal fragility (Kovacs, 2016).

Summary

The multiple roles of PTHrP in the reproductive tissues and cycle, and in the placenta, largely reflect its roles as a paracrine/autocrine/intracrine regulator. Of the many functions it exerts in these systems, probably the only endocrinal one is when PTHrP in the fetal circulation regulates placental calcium transport. There remains much to be learned of the place of PTHrP in reproductive and placental physiology and pathology.

Endocrine pancreas

Parathyroid hormone-related protein and its receptors

The presence of PTHrP in the pancreatic islets became apparent shortly after the identification of PTHrP in 1987. It is expressed as early as day 18 of gestation in the islet cells of fetal rodents (Campos et al., 1991). Asa et al. (1990) demonstrated that PTHrP was present in islet cells and in all 4 cell types, including the α, β, δ, and PP cells. PTHrP mRNA was shown to be present in isolated islet RNA as well (Drucker et al., 1989), demonstrating that the peptide could be produced within the islet. Gaich et al. (1993) confirmed these findings, demonstrating that PTHrP was indeed present in islet cells of all four types and that it was also present in pancreatic ductular epithelial cells. The peptide is not present in adult pancreatic exocrine cells. Plawner et al. (1995) demonstrated that PTHrP is present in individual beta cells in culture and showed that PTHrP colocalized with insulin in the Golgi apparatus as well as in insulin secretory granules. Interestingly, in a perifusion system employing a beta-cell line, PTHrP was shown to be cosecreted with insulin from beta cells following depolarization of the cell (Plawner et al., 1995). The secreted forms of PTHrP included amino-terminal, midregion, and carboxy-terminal forms of PTHrP (see later).

The PTH1R is expressed throughout the gut epithelium, including the pancreas, from day 9 of development in fetal rodents (Karperien et al., 1994). PTH1R mRNA and protein have been confirmed to be present in adult mouse islet cells by RT-PCR and immunostaining (Fujinaka et al., 2004). There is also functional evidence for PTHrP receptors on the pancreatic beta cell, as it is clear that PTHrP(1−36) elicits prompt and vigorous responses in intracellular calcium in cultured beta-cell lines. For example, Gaich et al. (1993) have demonstrated that PTHrP(1−36) in doses as low as 10−12 M stimulates calcium release from intracellular stores. Interestingly, unlike events observed in bone and renal cell types where PTHrP receptor activation is associated with activation of cAMP/PKA, as well as the PKC/intracellular calcium pathways, only the latter is observed in cultured beta cells in response to PTH or PTHrP(1−36) (Gaich et al., 1993). Whether this reflects the presence of a different type of receptor on beta cells or differential coupling of the PTH1R to subsets of specific G-proteins or catalytic subunits in beta cells compared with bone and renal cells has not been studied.

Regulation of parathyroid hormone-related protein and its receptors

There is little information describing how or to what degree PTHrP or the PTH receptor family is regulated in the pancreatic islet. As will become clear from the sections that follow, there are physiological reasons why such regulation might occur under normal circumstances, but this area remains unexplored.

Biochemistry of parathyroid hormone-related protein

PTHrP undergoes extensive posttranslational processing. Much of what is known or inferred regarding PTHrP processing is derived from studies in the rat insulinoma line, RIN-1038 (Soifer et al., 1992; Yang et al., 1994; Wu et al., 1991). These cells model the processing of PTHrP in the regulated secretory pathway, and have been shown to serve as a model for authentic processing of other human neuroendocrine peptides, such as insulin, proopiomelanocortin, glucagon, and calcitonin. Using a combination of untransfected RIN-1038 cells, RIN-1038 cells overexpressing hPTHrP(1−139), hPTHrP(1−141), or hPTHrP(1−173), and a panel of region-specific radioimmunoassays and immunoradiometric assays, RIN cells have been shown to secrete PTHrP(1−36), PTHrP(38−94), PTHrP(38−95), and PTHrP(38−101) (Soifer et al., 1992; Yang et al., 1994; Wu et al., 1991). In addition, RIN 1038 cells have been shown to secrete a form of PTHrP that is recognized by a PTHrP (109−138) radioimmunoassay (Yang et al., 1994) and another form that is recognized by a PTHrP(139−173) radioimmunoassay (Burtis et al., 1992).

As mentioned, these data derive from rat insulinoma cells that use the regulated secretory pathway for the processing of PTHrP. On the other hand, osteoblasts, chondrocytes, and other cells of mesenchymal origin use the constitutive secretory pathway and would not be expected to use the same processing mechanisms as cells that use the regulated secretory pathway. In accord with that, the only isoform of PTHrP found to be secreted by the Ocy454 osteocyte cell line was full-length PTHrP (Ansari et al., 2018).

As described earlier, PTHrP(1−36) stimulates intracellular calcium increments in cultured beta cells (Gaich et al., 1993). PTHrP(38−94) has also been shown to stimulate intracellular calcium release in these cells (Wu et al., 1991). PTHrP(38−94) does not activate adenylyl cyclase in cultured beta cells, and other PTHrP species have not been explored in beta cells in functional terms.

Function of parathyroid hormone-related protein

Pancreas development in rodents begins at approximately day E9−10, and by day E18−19, clusters of beta cells have begun to coalesce and form immature islets (Edlund, 1998). These islet cell clusters continue to increase in number, size, and density of beta cells in the week after delivery and then decline abruptly in number through a wave of beta-cell apoptosis (Finegood et al., 1995).

The role of PTHrP in pancreatic cell development and function is poorly understood at present. The pancreas of PTHrP-null mice (Karaplis et al., 1994) develops normally in anatomic terms (Vasavada et al., 1998), but nothing is known about the function of these islets. PTHrP-null mice die immediately after delivery, so nothing is known of islet function or development following birth in the absence of PTHrP. "Rescued" PTHrP mice do exist (Wysolmerski et al., 1998), and they survive to adulthood. These mice have normal-appearing pancreata and islets (Vasavada et al., 1998), but they have dental abnormalities, are undernourished, and grow poorly. Therefore, it is difficult to characterize their islets in functional terms, as islet mass, proliferation, and function are heavily dependent on fuel availability. Streuker and Drucker (1991) have suggested that PTHrP may play a role in beta-cell differentiation because it is upregulated in beta-cell lines in the presence of the islet-differentiating agent, butyrate.

In vitro studies have demonstrated that, similar to the actions of PTH, PTHrP(1−34) and PTHrP(1−86) can enhance glucose-stimulated insulin release from cultured islet cells, and increased insulin mRNA expression and protein content within cultured islets and islet tumor cell lines (Villanueva-Penacarrillo et al., 1999; Sawada et al., 2001). PTHrP has been shown to increase proliferation of cultured islet cells, with the effects varying by cell passage and the ambient glucose concentration (Sawada et al., 2001; Vasavada et al., 2007). Also within cell lines, there is evidence that PTHrP induces cAMP production, raises intracellular calcium through influx of calcium into the cell, and inhibits JNK1/2 activation (Mozar et al., 2014).

In an effort to understand the role of PTHrP in the pancreatic islet, Vasavada and collaborators have developed transgenic mice that overexpress PTHrP under the control of the rat insulin-II promoter (RIP) (Vasavada et al., 1996; Porter et al., 1998). RIP-PTHrP mice display striking degrees of islet hyperplasia and an increase in islet number as well as the size of individual islets. This increased islet mass is associated with increased function; RIP-PTHrP mice are hyperinsulinemic and hypoglycemic compared with their littermates (Vasavada et al., 1996; Porter et al., 1998). They become profoundly and symptomatically hypoglycemic with fasting. Interestingly, RIP-PTHrP mice are also resistant to the diabetogenic effects of the beta-cell toxin, streptozotocin. Following the administration of streptozotocin, normal mice readily develop diabetes, but RIP-PTHrP mice either fail to become diabetic or develop only mild hyperglycemia (Porter et al., 1998).

The mechanism(s) responsible for the increase in islet mass in the RIP-PTHrP mouse remains undefined. There are two levels at which this question can be addressed: identification of the source of the cells responsible for the increase in islet mass and the signaling mechanisms that are responsible for the increase. With respect to the first, islet mass can, in theory, be increased by three pathways: (1) the recruitment of new islets from the pancreatic duct or its branches distributed throughout the exocrine pancreas in a process referred to as "islet neogenesis"; (2) induction of proliferation of existing beta cells within islets; and (3) prolongation of the life span of existing beta cells. Of these options, there is evidence that PTHrP can drive beta-cell replication (Villanueva et al., 1999; Fujinaka et al., 2004), suggesting that beta-cell proliferation may account for at least part of the phenotype. The RIP promoter is restricted to expression in beta cells and therefore unable to influence pancreatic cells prior to their differentiation into beta cells, suggesting that the neogenesis of beta cells from ductal or other precursors is not a likely contributor (Vasavada et al., 1996). Finally, the bulk of evidence would support a dominant role for PTHrP in enhancing beta-cell survival, as occurs in other cell types (Cebrian et al., 2002).

More recently, daily subcutaneous injections of PTHrP(1−36) in adult mice have been shown to increase the proliferation of beta cells and improve glucose tolerance (Williams et al., 2011). Vasavada's group has shown that after partial pancreatectomy, treatment with PTHrP(1−36) increased beta-cell proliferation over the first 30 days and led to a marked increase in beta cell mass by 90 days (Mozar et al., 2016). Therefore, PTHrP could conceivably be a treatment to improve beta cell mass in patients with diabetes. Further studies are needed.

At the signaling level, little is known regarding the mechanism of action of PTHrP on beta cells. Although it is known that PTHrP can stimulate intracellular calcium in cultured beta-cell lines (Gaich et al., 1993; Wu et al., 1991), it is not known whether this occurs in vivo in normal, nontransformed beta cells within intact islets. Nor is it known whether PTHrP stimulates adenylyl cyclase in normal beta cells in vivo or whether it participates in nuclear or intracrine signaling in beta cells as it appears to in chondrocytes, osteoblasts, vascular smooth muscle cells, or other cell types (Aarts et al., 1999; Massfelder et al., 1997; Lam et al., 1999) (see Chapter 25). These processes, too, are under study.

Pathophysiology of parathyroid hormone-related protein

From the earlier discussion, it is clear that the normal physiological role of PTHrP in the pancreatic islet remains undefined. In contrast, PTHrP plays clear pathophysiological roles in at least some pancreatic islet neoplasms. PTHrP overexpression with resultant development of HHM has been demonstrated on multiple occasions in multiple investigators' hands (Asa et al., 1990; Stewart et al., 1986; Wu et al., 1997; Skrabanek et al., 1980). In the only large series of malignancy-associated hypercalcemia in which tumors have been fully subdivided based on histology (Skrabanek et al., 1980), islet cell carcinomas, which are not particularly common, produce HHM fully as often as pancreatic adenocarcinomas, a very common neoplasm. Historically, islet tumors were among the first in which PTHrP bioactivity was identified (Stewart et al., 1986; Wu et al., 1997). Furthermore, patients with islet carcinomas regularly demonstrate increases in circulating PTHrP as determined by radioimmunoassay or immunoradiometric assays (Lanske et al., 1996). When assessed by immunohistochemistry, these tumors also demonstrate increased staining for PTHrP (Asa et al., 1990; Drucker et al., 1989).

The significance of these findings for islet tumor oncogenesis is not known. Is this simply a random derepression of the PTHrP gene or is it a specific upregulation of the PTHrP gene? Is there a pathological role for PTHrP in the development of pancreatic islet tumors, corresponding to the mass enhancing effects of PTHrP in the islets of the RIP-PTHrP mouse? These questions remain interesting but unanswered at present.

Conclusions

Advances in mouse genetics and transgenic technology have been a boon to the study of physiology. This has certainly been the case for the PTHrP field, where studies in genetically altered mice have provided a starting place for the study of the physiology of a protein that was discovered outside its natural context. This chapter has outlined the current state of knowledge regarding the physiological roles of PTHrP in skin, the mammary gland, placenta, uterus, extraembryonic tissues, and pancreas. Much of this information (although not all) has come from studies performed in a variety of transgenic mice. These studies have shown that PTHrP is important to both the development and physiological functioning of these organs. However, at this point, we continue to have as many questions as answers. Many experiments remain to be done before we comprehend all the nuances of the functions of PTHrP at these sites.

Acknowledgments

This chapter has been updated from a prior version cowritten with Drs. Andrew F. Stewart and John J. Wysolmerski. The work was supported by the Canadian Institutes of Health Research (#133413), the Janeway Research Foundation, and both the Discipline of Medicine and Faculty of Medicine of Memorial University of Newfoundland.

References

Aarts, M.M., Levy, D., He, B., Stregger, S., Chen, T., Richard, S., Henderson, J.E., 1999. Parathyroid hormone-related protein interacts with RNA. J. Biol. Chem. 274, 4832–4838.

Alsat, E., Haziza, J., Scippo, M.L., Frankenne, F., Evain Brion, D., 1993. Increase in epidermal growth factor receptor and its mRNA levels by parathyroid hormone (1–34) and parathyroid hormone-related protein (1–34) during differentiation of human trophoblast cells in culture. J. Cell. Physiol. 53, 32–42.

Ansari, N., Ho, P.W., Crimeen-Irwin, B., Poulton, I.J., Brunt, A.R., Forwood, M.R., Divieti Pajevic, P., Gooi, J.H., Martin, T.J., Sims, N.A., 2018. Autocrine and paracrine regulation of the murine skeleton by osteocyte-derived parathyroid hormone0related protein. J. Bone Miner. Res. 33, 137–153.

Ardawi, M.S., Nasrat, H.A., BA'Aqueel, H.S., 1997. Calcium regulating hormones and parathyroid hormone-related peptide in normal human pregnancy and post-partum: a longitudinal study. Eur. J. Endocrinol. 137 (4), 402–409.

Ardeshirpour, L., Dann, P., Pollak, M., Wysolmerski, J., VanHouten, J., 2006. The calcium-sensing receptor regulates PTHrP production and calcium transport in the lactating mammary gland. Bone 38, 787–793.

Asa, S.L., Henderson, J., Goltzman, D., Drucker, D.J., 1990. Parathyroid hormone-related peptide in normal and neoplastic human endocrine tissues. J. Clin. Endocrinol. Metab. 71, 1112–1118.

Atillasoy, E.J., Burtis, W.J., Milstone, L.M., 1991. Immunohisto-chemical localization of parathyroid hormone-related protein (PTHRP) in normal human skin. J. Investig. Dermatol. 96, 277–280.

Beck, F., Tucci, J., Senior, P.V., 1993. Expression of parathyroid hormone-related protein mRNA by uterine tissues and extraembryonic membranes during gestation in rats. J. Reprod. Fertil. 99, 343–352.

Bertelloni, S., Baroncelli, G.I., Pelletti, A., Battini, R., Saggese, G., 1994. Parathyroid hormone related protein in healthy pregnant women. Calcif. Tissue Int. 54, 195–197.

Birch, M.A., Carron, J.A., Scott, M., Fraser, W.D., Gallagher, J.A., 1995. Parathyroid hormone (PTH)/PTH-related protein (PTHrP) receptor expression and mitogenic responses in human breast cancer cell lines. Br. J. Canc. 72, 90–95.

Blomme, E.A., Sugimoto, Y., Lin, Y.C., Capen, C.C., Rosol, T.J., 1999a. Parathyroid hormone-related protein is a positive regulator of keratinocyte growth factor expression by normal dermal fibroblasts. Mol. Cell. Endocrinol. 152, 189–197.

Blomme, E.A., Zhou, H., Kartsogiannis, V., Capen, C.C., Rosol, T.J., 1999b. Spatial and temporal expression of parathyroid hormone-related protein during wound healing. J. Investig. Dermatol. 112, 788–795.

Bowden, S.J., Emly, J.F., Hughes, S.V., Powell, G., Ahmed, A., Whittle, M.J., Ratcliffe, J.G., Ratcliffe, W.A., 1994. Parathyroid hormone-related protein in human term placenta and membranes. J. Endocrinol. 142, 217–224.

Bradbury, R.A., Sunn, K.L., Crossley, M.C., Bai, M., Brown, F.M., Del-bridge, L., Conigrave, A.D., 1997. Expression of the parathyroid Ca21 sensing receptor in cytotrophoblasts from human term placenta. J. Endocrinol. 156, 425–430.

Bucht, E., Carlqvist, M., Hedlund, B., Bremme, K., Torring, O., 1992. Parathyroid hormone-related peptide in human milk measured by a mid-molecule radioimmunoassay. Metab. Clin. Exp. 41, 11–16.

Budayr, A.A., Halloran, B.R., King, J., Diep, D., Nissenson, R.A., Strewler, G.J., 1989. High levels of parathyroid hormone-related protein in milk. Proc. Natl. Acad. Sci. U. S. A 86, 7183–7185.

Bundred, N.J., Walker, R.A., Ratcliffe, W.A., Warwick, J., Morrison, J.M., 1996. Parathyroid hormone related protein and skeletal morbidity in breast cancer. Eur. J. Cancer 28, 690–692.

Burtis, W.J., Debeyssey, M., Philbrick, W.M., Orloff, J.J., Daifotis, A.G., Soifer, N.E., Milstone, L.M., 1992. Evidence for the presence of an extreme carboxy-terminal parathyroid hormone-related peptide in biological specimens. J. Bone Miner. Res. 7 (Suppl. 1), S225.

Campos, R.V., Asa, S.L., Drucker, D.J., 1991. Immunocytochemical localization of parathyroid hormone-like peptide in the rat fetus. Cancer Res. 51, 6351–6357.

Care, A.D., Caple, I.W., Pickard, D.W., 1985. The roles of the parathyroid and thyroid glands on calcium homeostasis in the ovine fetus. In: Jones, C.T., Nathaniels, P.W. (Eds.), The Physiological Development of the Fetus and Newborn. Academic Press, London, pp. 135–140.

Care, A.D., Abbas, S.K., Pickard, D.W., Barri, M., Drinkhill, M., Findlay, J.B.C., White, I.R., Caple, I.W., 1990. Stimulation of ovine placental transport of calcium and magnesium by mid-molecule fragments of human parathyroid hormone-related protein. Exp. Physiol. 75, 605–608.

Cebrian, A., Garcia-Ocaña, A., Takane, K.K., Sipula, D., Stewart, A.F., Vasavada, R.C., 2002. Overexpression of parathyroid hormone-related protein inhibits pancreatic beta cell death in vivo and in vitro. Diabetes 51, 3003–3013.

Cho, Y.M., Woodard, G.L., Dunbar, M., Gocken, T., Jimenez, J.A., Foley, J., 2003. Hair-cycle-dependent expression of parathyroid hormone-related protein and its type I receptor: evidence for regulation at the anagen to catagen transition. J. Investig. Dermatol. 120, 715–727.

Collins, J.N., Kirby, B.J., Woodrow, J.P., Gagel, R.F., Rosen, C.J., Sims, N.A., Kovacs, C.S., 2013. Lactating *Ctcgrp* nulls lose twice normal bone mineral content due to fewer osteoblasts and more osteoclasts, while bone mass is fully restored post-weaning in association with upregulation of Wnt signaling and other novel genes. Endocrinology 154 (4), 1400–1413.

Collins, P.L., Idriss, E., Moore, J.J., 1993. Human fetal membranes inhibit spontaneous uterine contractions. J. Clin. Endocrinol. Metab. 77, 1479–1484.

Comar, C.L., 1956. Radiocalciums studies in pregnancy. Ann. N. Y. Acad. Sci. 64, 281–298.

Cormier, S., Delezoide, A.L., Silve, C., 2003. Expression patterns of parathyroid hormone-related peptide (PTHrP) and parathyroid hormone receptor type 1 (PTHR1) during human development are suggestive of roles specific for each gene that are not mediated through the PTHrP/PTHR1 paracrine signaling pathway. Gene. Expr. Patterns. 3 (1), 59–63.

Curtis, N.E., King, R.G., Moseley, J.M., Ho, P.W., Rice, G.E., Wlodek, M.E., 1998. Intrauterine expression of parathyroid hormone-related protein in normal and pre-eclamptic pregnancies. Placenta 19, 595–601.

Daniel, C.W., Silberstein, G.B., 1987. Postnatal development of the rodent mammary gland. In: Neville, M.C., Daniel, C.W. (Eds.), The Mammary Gland: Development, Regulation and Function. Plenum, New York, pp. 3–36.

Danks, J.A., Ebeling, P.R., Hayman, J., Chou, S.T., Moseley, J.M., Dunlop, J., Kemp, B.E., Martin, T.J., 1989. PTHRP: immunohisto-chemical localization in cancers and in normal skin. J. Bone Miner. Res. 4, 237–238.

Davicco, M., Rouffet, J., Durand, D., Lefaivre, J., Barlet, J.P., 1993. Parathyroid hormone-related peptide may increase mammary blood flow. J. Bone Miner. Res. 8, 1519–1524.

Deftos, L.J., Burton, D.W., Brant, D.W., Pinar, H., Rubin, L.P., 1994. Neoplastic hormone-producing cells of the placenta produce and secrete parathyroid hormone-related protein. Studies by immuno-histology, immunoassay, and polymerase chain reaction. Lab. Invest. 71, 847–852.

Diamond, A.G., Gonterman, R.M., Anderson, A.L., Menon, K., Offutt, C.D., Weaver, C.H., Philbrick, W.M., Foley, J., 2006. Parathyroid hormone-related protein and the PTH receptor regulate angiogenesis of the skin. J. Investig. Dermatol. 126, 2127–2134.

Dittmer, A., Vetter, M., Schunke, D., Span, P.N., Sweep, F., Thomssen, C., Dittmer, J., 2006. Parathyroid hormone-related protein regulates tumor-relevant genes in breast cancer cells. J. Biol. Chem. 281, 14563–14572.

Drucker, D.J., Asa, S.L., Henderson, J., Goltzman, D., 1989. The PTHrP gene is expressed in the normal and neoplastic human endocrine pancreas. Mol. Endocrinol. 3, 1589–1595.

Dunbar, M.E., Young, P., Zhang, J.P., McCaughern-Carucci, J., Lanske, B., Orloff, J., Karaplis, A., Cunha, G., Wysolmerski, J.J., 1998. Stromal cells are critical targets in the regulation of mammary ductal morphogenesis by parathyroid hormone-related protein (PTHrP). Dev. Biol. 203, 75–89.

Dunbar, M.E., Dann, P.R., Robinson, G.W., Hennighausen, L., Zhang, J.P., Wysolmerski, J.J., 1999. Parathyroid hormone-related protein is necessary for sexual dimorphism during embryonic mammary development. Development 126, 3485–3493.

Dunbar, M.E., Dann, P., Brown, C.W., Dreyer, B., Philbrick, W.P., Wysolmerski, J.J., 2001. Temporally-regulated overexpression of PTHrP in the mammary gland reveals distinct fetal and pubertal phenotypes. J. Endocrinol. 171, 403–416.

Dvir, R., Golander, A., Jaccard, N., Yedwab, G., Otremski, I., Spirer, Z., Weisman, Y., 1995. Amniotic fluid and plasma levels of parathyroid hormone-related protein and hormonal modulation of its secretion by amniotic fluid cells. Eur. J. Endocrinol. 133, 277–282.

Edlund, H., 1998. Transcribing pancreas. Diabetes 47, 1817–1823.

El-Hashash, A.H., Esbrit, P., Kimber, S.J., 2005. PTHrP promotes murine secondary trophoblast giant cell differentiation through induction of endocycle, upregulation of giant-cell-promoting transcription factors and suppression of other trophoblast cell types. Differentiation 73 (4), 154–174.

Eller-Vainicher, C., Ossola, M.W., Beck-Peccoz, P., Chiodini, I., 2012. PTHrP-associated hypercalcemia of pregnancy resolved after delivery: a case report. Eur. J. Endocrinol. 166, 753–756.

Errazahi, A., Bouizar, Z., Lieberherr, M., Souil, E., Rizk-Rabin, M., 2003. Functional type I PTH/PTHrP receptor in freshly isolated newborn rat keratinocytes: identification by RT-PCR and immunohistochemistry. J. Bone. Miner. Res. 18 (4), 737–750.

Errazahi, A., Lieberherr, M., Bouizar, Z., Rizk-Rabin, M., 2004. PTH-1R responses to PTHrP and regulation by vitamin D in keratinocytes and adjacent fibroblasts. J. Steroid Biochem. Mol. Biol. 89–90, 381–385.

Ferguson, J.E., Gorman, J.V., Bruns, D.E., Weir, E.C., Burtis, W.J., Martin, T.J., Bruns, M.E., 1992. Abundant expression of parathyroid hormone-related protein in human amnion and its association with labor. Proc. Natl. Acad. Sci. U.S.A. 89, 8384–8388.

Ferrari, S.L., Rizzoli, R., Bonjour, J.P., 1992. Parathyroid hormone-related protein production by primary cultures of mammary epithelial cells. J. Cell. Physiol. 150, 304–411.

Finegood, D.T., Scaglia, L., Bonner-Weir, S., 1995. Perspectives in diabetes. Dynamics of beta-cell mass in the growing rat pancreas. Estimation with a simple mathematical model. Diabetes 44, 249–256.

Fleming, N.I., Trivett, M.K., George, J., Slavin, J.L., Murray, W.K., Moseley, J.M., Anderson, R.L., Thomas, D.M., 2009. Parathyroid hormone-related protein protects against mammary tumor emergence and is associated with monocyte infiltration in ductal carcinoma in situ. Cancer Res. 69 (18), 7473–7479.

Foley, J., Longely, B.J., Wysolmerski, J.J., Dreyer, B.E., Broadus, A.E., Philbrick, W.M., 1998. Regulation of epidermal differentiation by PTHrP: evidence from PTHrP-null and PTHrP-overexpressing mice. J. Investig. Dermatol. 111, 1122–1128.

Foley, J., Dann, P., Hong, J., Cosgrove, J., Dreyer, B., Rimm, D., Dunbar, M.E., Philbrick, W., Wysolmerski, J.J., 2001. Parathyroid hormone-related protein maintains mammary epithelial fate and triggers nipple skin differentiation during embryonic mammary development. Development 128, 513–525.

Fuchs, A.R., Fuchs, F., Husslein, P., Soloff, M.S., Fernstrom, M., 1982. Oxytocin receptors and human parturition: a dual role of oxytocin in the initiation of labor. Science 215, 1396–1398.

Fujinaka, Y., Sipula, D., Garcia-Ocana, A., Vasavada, R.C., 2004. Characterization of mice doubly transgenic for parathyroid hormone-related protein and murine placental lactogen: a novel role for placental lactogen in pancreatic beta-cell survival. Diabetes 53, 3120–3130.

Gaich, G., Orloff, J.J., Atillasoy, E.J., Burtis, W.J., Ganz, M.B., Stewart, A.F., 1993. Amino-terminal parathyroid hormone-related protein: specific binding and cytosolic calcium responses in rat insulinoma cells. Endocrinology 132, 1402–1409.

Gallacher, S.J., Fraser, W.D., Owens, O.J., Dryburgh, F.J., Logue, F.C., Jenkins, A., Kennedy, J., Boyle, I.T., 1994. Changes in calciotrophic hormones and biochemical markers of bone turnover in normal human pregnancy. Eur. J. Endocrinol. 131 (4), 369–374.

Germain, A.M., Attaroglu, H., MacDonald, P.C., Casey, M.L., 1992. Parathyroid hormone-related protein mRNA in avascular human amnion. J. Clin. Endocrinol. Metab. 75, 1173–1175.

Givens, M.H., Macy, I.C., 1933. The chemical composition of the human fetus. J. Biol. Chem. 102, 7–17.

Goff, J.P., Reinhardt, T.A., Lee, S., Hollis, B.W., 1991. Parathyroid hormone-related peptide content of bovine milk and calf blood as assessed by radioimmunoassay and bioassay. Endocrinology 129, 2815–2819.

Grace, N.D., Atkinson, J.H., Martinson, P.L., 1986. Accumulation of minerals by the foetus(es) and conceptus of single and twin-bearing ewes. N. Z. J. Agric. Res. 29, 207–222.

Grone, A., Werkmeister, J.R., Steinmeyer, C.L., Capen, C.C., Rosol, T.J., 1994. Parathyroid hormone-related protein in normal and neo-plastic canine tissues: immunohistochemical localization and biochemical extraction. Vet. Pathol. 31, 308–315.

Gude, N.M., King, R.G., Brennecke, S.P., 1996. Factors regulating placenta hemodynamics. In: Sastry, B.V.R. (Ed.), Placental Pharmacology. CRC Press, Boca Raton, FL, pp. 23–45.

Guise, T.A., Yin, J.J., Taylor, S.D., Kumagai, Y., Dallas, M., Boyce, B., Yoneda, T., Mundy, G.R., 1996. Evidence for a causal role of parathyroid hormone-related protein in the pathogenesis of human breast cancer-mediated osteolysis. J. Clin. Investig. 98, 1544–1549.

Hanafin, N.M., Chen, T.C., Heinrich, G., Segré, G.V., Holick, M.F., 1995. Cultured human fibroblasts and not cultured human keratinocytes express a PTH/PTHrP receptor mRNA. J. Investig. Dermatol. 105, 133–137.

Hayman, J.A., Danks, J.A., Ebeling, P.R., Moseley, J.M., Kemp, B.E., Martin, T.J., 1989. Expression of PTHRP in normal skin and tumors. J. Pathol. 158, 293–296.

Hellman, P., Ridefelt, P., Juhlin, C., Akerstrom, G., Rastad, J., Gylfe, E., 1992. Parathyroid-like regulation of parathyroid-hormone-related protein release and cytoplasmic calcium in cytotrophoblast cells of human placenta. Arch. Biochem. Biophys. 293, 174–180.

Henderson, M.A., Danks, J.A., Moseley, J.M., Slavin, J.L., Harris, T.L., McKinlay, M.R., Hopper, J.L., Martin, T.J., 2001. Parathyroid hormone-related protein production by breast cancers, improved survival, and reduced bone metastases. J. Natl. Cancer Inst. 93, 234–237.

Henderson, M.A., Danks, J.A., Slavin, J.L., Byrnes, G.B., Choong, P.F., Spillane, J.B., Hopper, J.L., Martin, T.J., 2006. Parathyroid hormone-related protein localization in breast cancers predict improved prognosis. Cancer Res. 66 (4), 2250–2256.

Hens, J., Dann, P., Zhang, J.P., Robinson, G., Wysolmerski, J., 2007. BMP4 and PTHrP interact to stimulate ductal outgrowth and inhibit hair follicle induction during embryonic mammary development. Development 134, 1221–1230.

Hernandez, L.L., Gregerson, K.A., Horseman, N.D., 2012. Mammary gland serotonin regulates parathyroid hormone-related protein and other bone-related signals. Am. J. Physiol. Endocrinol. Metab. 302 (8), E1009–E1015.

Hershberger, M.E., Tuan, R.S., 1998. Placental 57-kDa Ca(2^1)-binding protein: regulation of expression and function in trophoblast calcium transport. Dev. Biol. 199, 80–92.

Holick, M.F., Chimeh, F.N., Ray, S., 2003. Topical PTH (1–34) is a novel, safe and effective treatment for psoriasis: a randomized self-controlled trial and an open trial. Br. J. Dermatol. 149, 370–376.

Insogna, K.L., Stewart, A.F., Morris, C.F., Hough, L.M., Milstone, L.M., Centrella, M., 1989. Native and synthetic analogues of the malignancy-associated parathyroid hormone-like protein have in vivo transforming growth factor-like properties. J. Clin. Investig. 83, 1057–1060.

Isales, C., Carcangiu, M.L., Stewart, A.F., 1987. Hypercalcemia in breast cancer: reassessment of the mechanism. Am. J. Med. 82, 1143.

Jobert, A.S., Zhang, P., Couvineau, A., Bonaventure, J., Roume, J., Le Merer, M., Silve, C., 1998. Absence of functional receptors for parathyroid hormone and parathyroid hormone-related peptide in Blomstrand Chondrodysplasia. J. Clin. Investig. 102, 34–40.

Kalenga, M.K., De Gasparo, M., Thomas, K., De Hertogh, R., 1995. Angiotensin-II stimulates estradiol secretion from human placental explants through A_1 receptor activation. J. Clin. Endocrinol. Metab. 80, 1233–1237.

Karaplis, A.C., Luz, A., Glowacki, J., Bronson, R.T., Tybulewicz, V.L.J., Kronenberg, H.M., Mulligan, R.C., 1994. Lethal skeletal dysplasia from targeted disruption of the parathyroid hormone-related peptide gene. Genes Dev. 8, 277–289.

Karmali, R., Schiffman, S.N., Vanderwinden, J.M., Hendy, G.N., Nys-DeWolf, N., Corvilain, J., Bergmann, P., Vanderhaeghen, J.J., 1992. Expression of mRNA of parathyroid hormone-related peptide in fetal bones of the rat. Cell Tissue Res. 270, 597–600.

Karperien, M., van Dijk, T.B., Hoeijmakers, T., et al., 1994. Expression pattern of parathyroid hormone/parathyroid hormone related peptide receptor mRNA in mouse postimplantation embryos indicates involvement in multiple developmental processes. Mech. Dev. 47, 29–42.

Kakonen, S.M.1, Selander, K.S., Chirgwin, J.M., Yin, J.J., Burns, S., Rankin, W.A., Grubbs, B.G., Dallas, M., Cui, Y., Guise, T.A., 2002. Transforming growth factor-beta stimulates parathyroid hormone-related protein and osteolytic metastases via Smad and mitogen-activated protein kinase signaling pathways. J. Biol. Chem. 277 (27), 24571–24578. Epub 2002 Apr 18.

Khosla, S., van Heerden, J.A., Gharib, H., Jackson, I.T., Danks, J., Hayman, J.A., Martin, T.J., 1990. Parathyroid hormone-related protein and hypercalcemia secondary to massive mammary hyperplasia (letter). N. Engl. J. Med. 322, 1157.

Kovacs, C.S., Kronenberg, H.M., 1997. Maternal-fetal calcium and bone metabolism during pregnancy, puerperium, and lactation. Endocr. Rev. 18, 832–872.

Kovacs, C.S., Lanske, B., Hunzelman, J.L., Guo, J., Karaplis, A.C., Kronenberg, H.M., 1996. Parathyroid hormone-related peptide (PTHrP) regulates fetal-placental calcium transport through a receptor distinct from the PTH/PTHrP receptor. Proc. Natl. Acad. Sci. U.S.A. 93, 15233–15238.

Kovacs, C.S., Ho-Pao, C.I., Hunzelman, J.L., Lanske, B., Fox, J., Seidman, J.G., Seidman, C.E., Kronenberg, H.M., 1998. Regulation of murine fetal-placental calcium metabolism by the calcium-sensing receptor. J. Clin. Investig. 101 (28), 12–20.

Kovacs, C.S., Manley, N.R., Moseley, J.M., Martin, T.J., Kronenberg, H.M., 2001. Fetal parathyroids are not required to maintain placental calcium transport. J. Clin. Investig. 107, 1007–1015.

Kovacs, C.S., Chafe, L.L., Woodland, M.L., McDonald, K.R., Fudge, N.J., Wookey, P.J., 2002. Calcitropic gene expression suggests a role for intraplacental yolk sac in maternal-fetal calcium exchange. Am. J. Physiol. Endocrinol. Metab. 282, E721–E732.

Kovacs, C.S., 2003. Fetal mineral homeostasis. In: Glorieux, F.H., Pettifor, J.M., Jüppner, H. (Eds.), Pediatric Bone: Biology and Diseases. Academic Press, San Diego, pp. 271–302.

Kovacs, C.S., 2005a. Calcium and bone metabolism during pregnancy and lactation. J. Mammary Gland Biol. Neoplasia 10 (2), 105–118.

Kovacs, C.S., Woodland, M.L., Fudge, N.J., Friel, J.K., 2005b. The vitamin D receptor is not required for fetal mineral homeostasis or for the regulation of placental calcium transfer in mice. Am. J. Physiol. Endocrinol. Metab. 289 (1), E133–E144.

Kovacs, C.S., 2016. Maternal mineral and bone metabolism during pregnancy, lactation, and post-weaning recovery. Physiol. Rev. 96, 449–547.

Kumari, R., Robertson, J.F., Watson, S.A., 2006. Nuclear targeting of a midregion PTHrP fragment is necessary for stimulating growth in breast cancer cells. Int. J. Cancer 119, 49–59.

Lam, M.H.C., House, C.M., Tiganis, T., Mitchelhill, K.I., Sarcevic, B., Cures, A., Ramsay, R., Kemp, B.E., Martin, T.J., Gillespie, M.T., 1999. Phosphorylation at the cyclin-dependent kinases site (Thr^{85}) of parathyroid hormone-related protein negatively regulates its nuclear localization. J. Biol. Chem. 274, 18559–18566.

Lanske, B., Karaplis, A., Lee, K., Luz, A., Vortkam, A., Pirro, A., Karperien, M., Defize, L., Ho, C., Mulligan, R., Abou-Samra, A., Jüppner, H., Segré, G., Kronenberg, H., 1996. PTH/PTHrP receptor in early development and Indian hedgehog-regulated bone growth. Science 273, 663–666.

Law, F.M.L., Moate, P.J., Leaver, D.D., Dieffenbach, H., Grill, V., Ho, P.W.M., Martin, T.J., 1991. Parathyroid hormone-related protein in milk and its correlation with bovine milk calcium. J. Endocrinol. 128, 21–26.

Li, J., Karaplis, A.C., Huang, D.C., Siegel, P.M., Camirand, A., Yang, X.F., Muller, W.J., Kremer, R., 2011. PTHrP drives breast tumor initiation, progression, and metastasis in mice and is a potential therapy target. J. Clin. Investig. 121 (12), 4655–4669.

Li, X., Shams, M., Zhu, J., Khalig, A., Wilkes, M., Whittle, M., Barnes, N., Ahmed, A., 1998. Cellular localization of ATI receptor mRNA and protein in normal placenta and its reduced expression in intrauterine growth restriction. Angiotensin II stimulates the release of vasorelaxants. J. Clin. Investig. 101, 442–454.

Liapis, H., Crouch, E.C., Grosso, L.E., Kitazawa, S., Wick, M.R., 1993. Expression of parathyroid like protein in normal, proliferative, and neoplastic human breast tissues. Am. J. Pathol. 143, 1169–1178.

Lieben, L., Stockmans, I., Moermans, K., Carmeliet, G., 2013. Maternal hypervitaminosis D reduces fetal bone mass and mineral acquisition and leads to neonatal lethality. Bone 57, 123–131.

Lee, K., Deeds, J.D., Segre, G.V., 1995. Expression of parathyroid hormone-related peptide and its messenger ribonucleic acids during fetal development of rats. Endocrinology 136, 453–463.

Macgill, K., Mosely, J.M., Martin, T.J., Brennecke, S.P., Rice, G.E., Wlodek, M.E., 1997. Vascular effects of PTHrP (1–34) and PTH (1–34) in the human fetal-placental circulation. Placenta 18, 587–592.

MacIsaac, R.J., Heath, J.A., Rodda, C.P., Mosely, J.M., Care, A.D., Martin, T.J., Caple, I.W., 1991. Role of the fetal parathyroid glands and parathyroid hormone-related protein in the regulation of placental transport of calcium, magnesium and inorganic phosphate. Reprod. Fertil. Dev. 3, 447–457.

Mamillapalli, R., VanHouten, J., Dann, P., Bikle, D., Chang, W., Brown, E., Wysolmerski, J., 2013. Mammary-specific ablation of the calcium-sensing receptor during lactation alters maternal calcium metabolism, milk calcium transport, and neonatal calcium accrual. Endocrinology 154, 3031–3042.

Mandsager, N.T., Brewer, A.S., Myatt, L., 1994. Vasodilator effects of parathyroid hormone, parathyroid hormone-related protein, and calcitonin gene-related peptide in the human fetal-placental circulation. J. Soc. Gynecol. Investig. 1, 19–24.

Marshall, A.M., Hernandez, L.L., Horseman, N.D., 2014. Serotonin and serotonin transport in the regulation of lactation. J. Mammary Gland Biol. Neoplasia 19 (1), 139–146.

Massfelder, T., Dann, P., Wu, T.L., Vasavada, R., Helwig, J.-J., Stewart, A.F., 1997. Opposing mitogenic and anti-mitogenic actions of parathyroid hormone-related protein in vascular smooth muscle cells: a critical role for nuclear targeting. Proc. Natl. Acad. Sci. U.S.A. 94, 13630–13635.

Mercado-Simmen, R., Bryant-Greenwood, G.D., Greenwood, F.C., 1982. Relaxin receptor in the rat myometrium: regulation by estrogen and progesterone. Endocrinology 110, 220–226.

Merendino, J.J., Insogna, K.L., Milstone, L.M., Broadus, A.E., Stewart, A.F., 1986. Cultured human keratinocytes, produce a parathyroid hormone-like protein. Science 231, 388–390.

Meziani, F., Van Overloop, B., Schneider, F., Gairard, A., 2005. Parathyroid hormone-related protein-induced relaxation of rat uterine arteries: influence of the endothelium during gestation. J. Soc. Gynecol. Investig. 12 (1), 14–19.

Mitchell, J.A., Ting, T.C., Wong, S., Mitchell, B.F., Lye, S.J., 2003. Parathyroid hormone-related protein treatment of pregnant rats delays the increase in connexin 43 and oxytocin receptor expression in the myometrium. Biol. Reprod. 69 (2), 556–562.

Mozar, A., Guthalu Kondegowda, N.G., Pollack, I., Fenutria, R., Vasavada, R.C., 2014. The role of PTHrP in pancreatic beta-cells and implications for diabetes pathophysiology and treatment. Clin. Rev. Bone Miner. Metabol. 12, 165–177.

Mozar, A., Lin, H., Williams, K., Chin, C., Li, R., Kondegowda, N.G., Stewart, A.F., Garcia-Ocaña, A., Vasavada, R.C., 2016. Parathyroid hormone-related peptide (1–36) enhances beta cell regeneration and increases beta cell mass in a mouse model of partial pancreatectomy. PLoS One 11 (7), e0158414.

Nowak, R.A., Haimovici, F., Biggers, J.D., Erbach, G.T., 1999. Transforming growth factor-beta stimulates mouse blastocyst out-growth through a mechanism involving parathyroid hormone-related protein. Biol. Reprod. 60, 85–93.

Orloff, J.J., Ganz, M.B., Ribaudo, A.E., Burtis, W.J., Reiss, M., Milstone, L.M., Stewart, A.F., 1992. Analysis of parathyroid hormone-related protein binding and signal transduction mechanisms in benign and malignant squamous cells. Am. J. Physiol. 262, E599–E607.

Orloff, J.J., Kats, J., Urena, P., Schipani, E., Vasavada, R.C., Philbrick, W.M., Behal, A., Abou-Samra, A.B., Segre, G.V., Juppner, H., 1995. Further evidence for a novel receptor for amino-terminal PTHrP on keratinocytes and squamous carcinoma cell lines. Endocrinology 136, 3016–3023.

Ottesen, B., Larsen, J.J., Stau-Olsen, P., Gammeltoft, S., Fahrenkrug, J., 1985. Influence of pregnancy and sex steroids on concentration, motor effect and receptor binding on VIP in the rabbit female genital tract. Regul. Pept. 11, 83–92.

Paspaliaris, V., Vargas, S.J., Gillespie, M.T., Williams, E.D., Danks, J.A., Moseley, J.M., Story, M.E., Pennefather, J.N., Leaver, D.D., Martin, T.J., 1992. Oestrogen enhancement of the myometrial response to exogenous parathyroid hormone-related protein (PTHrP), and tissue localization of endogenous PTHrP and its mRNA in the virgin rat uterus. J. Endocrinol. 134, 415–425.

Philbrick, W.M., Wysolmerski, J.J., Galbraith, S., Holt, E., Orloff, J.J., Yang, K.H., Vasavada, R.C., Weir, E.C., Broadus, A.E., Stewart, A.F., 1996. Defining the roles of parathyroid hormone-related protein in normal physiology. Physiol. Rev. 76, 127–173.

Plawner, L.L., Philbrick, W.M., Burtis, W.J., Broadus, A.E., Stewart, A.F., 1995. Secretion of parathyroid hormone-related protein: cell-specific secretion via the regulated vs. the constitutive secretory pathway. J. Biol. Chem. 270, 14078–14084.

Porter, S.E., Sorenson, R.L., Dann, P., Garcia-Ocana, A., Stewart, A.F., Vasavada, R.C., 1998. Progressive pancreatic islet hyperplasia in the islet-targeted, PTH-related protein-overexpressing mouse. Endocrinology 139, 3743–3745.

Rakopoulos, M., Vargas, S.J., Gillespie, M.T., Ho, P.W.M., Diefenbach-Jagger, H., Leaver, D.D., Grill, V., Moseley, J.M., Danks, J.A., Martin, T.J., 1992. Production of parathyroid hormone-related protein by the rat mammary gland in pregnancy and lactation. Am. J. Physiol. 263, E1077–E1085.

Reid, I.R., Wattie, D.J., Evans, M.C., Budayr, A.A., 1992. Postpregnancy osteoporosis associated with hypercalcemia. Clin. Endocrinol. 37, 298–303.

Rhodes, D.R., Yu, J., Shanker, K., Deshpande, N., Varambally, R., Ghosh, D., Barrette, T., Pandey, A., Chinnaiyan, A.M., 2004. Oncomine: a cancer microarray database and integrated data-mining platform. Neoplasia 6, 1–6.

Rodda, C.P., Kubota, M., Heath, J.A., Ebeling, P.R., Mosely, J.M., Care, A.D., Caple, I.W., Martin, T.J., 1988. Evidence for a novel parathyroid hormone-related protein in fetal lamb parathyroid glands and sheep placenta: comparisons with a similar protein implicated in humoral hypercalcemia of malignancy. J. Endocrinol. 117, 261–271.

Rubin, L.P., Kovacs, C.S., Pinar, H., Tsai, S.W., Torday, J.S., Kronenberg, H.M., 2004. Arrested pulmonary alveolar cytodifferentiation and defective surfactant synthesis in mice missing the gene for parathyroid hormone-related protein. Dev. Dynam. 230 (2), 278–289.

Safer, J.D., Ray, S., Holick, M.F., 2007. A topical parathyroid hormone/parathyroid hormone-related peptide receptor antagonist stimulates hair growth in mice. Endocrinology 148, 1167–1170.

Sakakura, T., 1987. Mammary embryogenesis. In: Neville, M.C., Daniel, C.W. (Eds.), The Mammary Gland: Development, Regulation and Function. Plenum, New York, pp. 37–66.

Sawada, Y., Zhang, B., Okajima, F., Izumi, T., Takeuchi, T., 2001. PTHrP increases pancreatic beta-cell-specific functions in well-differentiated cells. Mol. Cell. Endocrinol. 182, 265–275.

Schilli, M.B., Ray, S., Paus, R., Obi-Tabot, E., Holick, M.F., 1997. Control of hair growth with parathyroid hormone (7–34). J. Investig. Dermatol. 108, 928–932.

Seitz, P.K., Cooper, K.M., Ives, K.L., Ishizuka, J., Townsend, C.M., Rajsraman, S., Cooper, C.W., 1993. Parathyroid hormone-related peptide production and action in a myoepithelial cell line derived from normal human breast. Endocrinology 133, 1116–1124.

Shen, X., Falzon, M., 2003. PTH-related protein modulates PC-3 prostate cancer cell adhesion and integrin subunit profile. Mol. Cell. Endocrinol. 199 (1-2), 165–177. PMID: 12581888.

Shen, X., Falzon, M., 2006. PTH-related protein upregulates integrin a6b4 expression and activates Akt in breast cancer cells. Exp. Cell Res. 312, 3822–3834.

Shen, X., Qian, L., Falzon, M., 2004. PTH-related protein enhances MCF-7 breast cancer cell adhesion, migration, and invasion via an intracrine pathway. Exp. Cell. Res. 294 (2), 420–433.

Shew, R.L., Yee, J.A., Pang, P.K.T., 1984. Direct effect of parathyroid hormone on rat uterine contraction. J. Pharmacol. Exp. Ther. 230, 1–6.

Shew, R.L., Yee, J.A., Kliewer, D.B., Keflemariam, Y.J., McNeill, D.L., 1991. Parathyroid hormone-related peptide inhibits stimulated uterine contraction in vitro. J. Bone Miner. Res. 6, 955–960.

Shin, J.H., Ji, C., Casinghino, S., McCarthy, T.L., Centrella, M., 1997. Parathyroid hormone-related protein enhances insulin-like growth factor-I expression by fetal rat dermal fibroblasts. J. Biol. Chem. 272, 23498–23502.

Simmonds, C.S., Karsenty, G., Karaplis, A.C., Kovacs, C.S., 2010. Parathyroid hormone regulates fetal-placental mineral homeostasis. J. Bone Miner. Res. 25 (3), 594–605.

Sirchia, R., Priulla, M., Sciandrello, G., Caradonna, F., Barbata, G., Luparello, C., 2007. Mid-region parathyroid hormone-related protein (PTHrP) binds chromatin of MDA-MB231 breast cancer cells and isolated oligonucleotides "in vitro". Breast Canc. Res. Treat. 105, 105–116.

Skrabanek, P., McPartlin, J., Powell, D.M., 1980. Tumor hypercalcemia and ectopic hyperparathyroidism. Medicine (Baltim.) 59, 262–282.

Skrok, A., Bednarczuk, T., Skwarek, A., ·Popow, M., Rudnicka, L., Olszewska, M., 2015. The effect of parathyroid hormones on hair follicle physiology: implications for treatment of chemotherapy-induced alopecia. Skin Pharmacol. Physiol. 28, 213–225.

Smith, M.S., Grove, K.L., 2002. Integration of the regulation of reproductive function and energy balance: lactation as a model. Front. Neuroendocrinol. 23 (3), 225–256. In review.

Soifer, N.E., Dee, K.E., Insogna, K.L., Burtis, W.J., Matovcik, L.M., Wu, T.L., Milstone, L.M., Broadus, A.E., Philbrick, W.M., Stewart, A.F., 1992. Secretion of a novel mid-region fragment of parathyroid hormone-related protein by three different cell lines in culture. J. Biol. Chem. 267, 18236–18243.

Sowers, M.F., Hollis, B.W., Shapiro, B., Randolph, J., Janney, C.A., Zhang, D., Schork, A., Crutchfield, M., Stanczyk, F., Russell-Aulet, M., 1996. Elevated parathyroid hormone-related peptide associated with lactation and bone density loss. J. Am. Med. Assoc. 276, 549–554.

Stewart, A.F., Insogna, K.L., Burtis, W.J., Aminiafshar, A., Wu, T., Weir, E.C., Broadus, A.E., 1986. Frequency and partial characterization of adenylate cyclase-stimulating activity in tumors associated with humoral hypercalcemia of malignancy. J. Bone Miner. Res. 1, 267–276.

Streuker, C., Drucker, D.J., 1991. Rapid induction of parathyroid hormone-like peptide gene expression by sodium butyrate in a rat islet cell line. Mol. Endocrinol. 5, 703–708.

Thiede, M.A., 1989. The mRNA encoding a parathyroid hormone-like peptide is produced in mammary tissue in response to elevations in serum prolactin. Mol. Endocrinol. 3, 1443–1447.

Thiede, M.A., Rodan, G.A., 1988. Expression of a calcium-mobilizing parathyroid hormone-like peptide in lactating mammary tissue. Science 242, 278–280.

Thiede, M.A., Daifotis, A.G., Weir, E.C., Brines, M.L., Burtis, W.J., Ikeda, K., Dreyer, B.E., Garfield, R.E., Braodus, A.E., 1990. Intrauterine occupancy controls expression of the parathyroid hormone-related peptide gene in preterm rat myometrium. Proc. Natl. Acad. Sci. U.S.A. 87, 6969–6973.

Thiede, M.A., Grasser, W.A., Peterson, D.N., 1992. Regulation of PTHrP in the mammary blood supply supports a role in mammary gland blood flow. Bone Miner. 17, A8 (Abstract).

Thompson, G.E., Ratcliffe, W.A., Hughes, S., Abbas, S.K., Care, A.D., 1994. Local control of parathyroid hormone-related protein secretion by the mammary gland of the goat. Comp. Biochem. Physiol. A108, 485–490.

Thomson, M., McCarrol, J., Bond, J., Gordon-Thomson, C., Williams, E.D., Moore, G.P.M., 2003. Parathyroid hormone-related peptide modulates signal pathways in skin and hair follicle cells. Exp. Dermatol. 12, 389–395.

Thota, C.S., Reed, L.C., Yallampalli, C., 2005. Effects of parathyroid hormone like hormone (PTHLH) antagonist, PTHLH(7–34), on fetoplacental development and growth during midgestation in rats. Biol. Reprod. 73, 1191–1198.

Tucci, J., Russell, A., Senior, P.V., Fernley, R., Ferraro, T., Beck, F., 1996. The expression of parathyroid hormone and parathyroid hormone-related protein in developing rat parathyroid glands. J. Mol. Endocrinol. 17, 149–157.

Urena, P., Kong, X.F., Abou Samra, A.B., Juppner, H., Kronenberg, H.M., Potts, J.T., Segré, G.V., 1993. Parathyroid hormone (PTH) PTH-related peptide receptor messenger ribonucleic acids are widely distributed in rat tissues. Endocrinology 133, 617–623.

Uemura, H., Yasui, T., Yoneda, N., Irahara, M., Aono, T., 1997. Measurement of N- and C-terminal-region fragments of parathyroid hormone-related peptide in milk from lactating women and investigation of the relationship of their concentrations to calcium in milk. J. Endocrinol. 153, 445–451.

VanHouten, J.N., Dann, P., Stewart, A.F., Watson, C.J., Pollak, M., Karaplis, A.C., Wysolmerski, J.J., 2003. Cre-mediated deletion of PTHrP from the mammary gland reduces bone turnover and preserves bone mass during lactation. J. Clin. Investig. 112, 1429–1436.

VanHouten, J.N., Wysolmerski, J.J., 2003. Low estrogen and high PTHrP levels contribute to accelerated bone resorption and bone loss in lactating mice. Endocrinology 144, 5521–5529.

VanHouten, J., Dann, P., McGeoch, G., Brown, E.M., Krapcho, K., Neville, M., Wysolmerski, J.J., 2004. The calcium sensing receptor regulates mammary gland production of PTHrP and calcium transport into milk. J. Clin. Investig. 113, 598–608.

Vasavada, R., Cavaliere, C., D'Ercole, A.J., Dann, P., Burtis, W.J., Madlener, A.L., Zawalich, K., Zawalich, W., Philbrick, W.M., Stewart, A.F., 1996. Overexpression of PTHrP in the pancreatic islets of transgenic mice causes hypoglycemia, hyperinsulinemia and islet hyperplasia. J. Biol. Chem. 271, 1200–1208.

Vasavada, R.C., Garcia-Ocana, A., Massfelder, T., Dann, P., Stewart, A.F., 1998. Parathyroid hormone-related protein in the pancreatic islet and the cardiovascular system. Recent Prog. Horm. Res. 53, 305–338 discussion 338–340.

Vasavada, R.C., Wang, L., Fujinaka, Y., Takane, K.K., Rosa, T., Mellado-Gil, J.M., Garcia-Ocaña, A., 2007. Protein kinase C-zeta activation markedly enhances beta-cell proliferation: an essential role in growth factor mediated beta-cell mitogenesis. Diabetes 56, 2732–2743.

Villanueva-Penacarrillo, M.L., Cancelas, J., de Miguel, F., Redondo, A., Valin, A., Valverde, I., Esbrit, P., 1999. Parathyroid hormone-related peptide stimulates DNA synthesis and insulin secretion in pancreatic islets. J. Endocrinol. 163, 403–408.

Williams, E.D., Leaver, D.D., Danks, J.A., Moseley, J.M., Martin, T.J., 1994. Effect of parathyroid hormone-related protein (PTHrP) on the contractility of the myometrium and localization of PTHrP in the uterus of pregnant rats. J. Reprod. Fertil. 102, 209–214.

Williams, E.D., Major, B.J., Martin, T.J., Moseley, J.M., Leaver, D.D., 1998. Effect of antagonism of the parathyroid hormone (PTH)/PTH-related protein receptor on decidualization in rat uterus. J. Reprod. Fertil. 112, 59–67.

Williams, J.M.A., Abramovich, D.R., Dacke, C.G., Mayhew, T.M., Page, K.R., 1991. Parathyroid hormone (1–34) peptide activates cyclic AMP in the human placenta. Exp. Physiol. 76, 297–300.

Williams, K., Abanquah, D., Joshi-Gokhale, S., Otero, A., Lin, H., Guthalu, N.K., Zhang, X., Mozar, A., Bisello, A., Stewart, A.F., Garcia-Ocaña, A., Vasavada, R.C., 2011. Systemic and acute administration of parathyroid hormone-related peptide (1–36) stimulates endogenous beta cell proliferation while preserving function in adult mice. Diabetologia 54 (11), 2867–2877.

Wlodek, M.E., Ho, P., Rice, G.E., Moseley, J.M., Martin, T.J., Brennecke, S.P., 1996. Parathyroid hormone-related protein (PTHrP) concentrations in human amniotic fluid during gestation and at the time of labour. Reprod. Fertil. Dev. 7, 1509–1513.

Wlodek, M.E., Westcott, K.T., O'Dowd, R., Serruto, A., Wassef, L., Moritz, K.M., Moseley, J.M., 2005. Uteroplacental restriction in the rat impairs fetal growth in association with alterations in placental growth factors including PTHrP. Am. J. Physiol. 288 (6), R1620–R1627.

Wojcik, S.F., Schanbacher, F.L., McCauley, L.K., Zhou, H., Kartsogiannis, V., Capen, C.C., Rosol, T.J., 1998. Cloning of bovine parathyroid hormone-related protein (PTHrP) cDNA and expression of PTHrP mRNA in the bovine mammary gland. J. Mol. Endocrinol. 20, 271–280.

Wojcik, S.F., Capen, C.C., Rosol, T.J., 1999. Expression of PTHrP and the PTH/PTHrP receptor in purified alveolar epithelial cells, myoepithelial cells and stromal fibroblasts derived from the lactating mammary gland. Exp. Cell Res. 248, 415–422.

Woodrow, J.P., Sharpe, C.J., Fudge, N.J., Hoff, A.O., Gagel, R.F., Kovacs, C.S., 2006. Calcitonin plays a critical role in regulating skeletal mineral metabolism during lactation in mice. Endocrinology 147 (9), 4010–4021.

Wu, T.L., Insogna, K.L., Milstone, L., Stewart, A.F., 1987. Skin-derived fibroblasts respond to human PTH-like adenylate cyclase-stimulating proteins. J. Clin. Endocrinol. Metab. 65, 105–109.

Wu, T.L., Soifer, N.E., Burtis, W.J., Milstone, M., Stewart, A.F., 1991. Glycosylation of parathyroid hormone-related peptide secreted by human epidermal keratinocytes. J. Clin. Endocrinol. Metab. 73, 1002–1007.

Wu, T.-J., Lin, C.-L., Taylor, R.L., Kvols, L.K., Kao, P.C., 1997. Increased parathyroid hormone-related peptide in patients with hypercalcemia associated with islet cell carcinoma. Mayo Clin. Proc. 72, 111–115.

Wysolmerski, J.J., Broadus, A.E., Zhou, J., Fuchs, E., Milstone, L.M., Philbrick, W.P., 1994. Overexpression of parathyroid hormone-related protein in the skin of transgenic mice interferes with hair follicle development. Proc. Natl. Acad. Sci. U.S.A. 91, 1133–1137.

Wysolmerski, J.J., McCaughern-Carucci, J.F., Daifotis, A.G., Broadus, A.E., Philbrick, W.M., 1995. Overexpression of parathyroid hormone-related protein or parathyroid hormone in transgenic mice impairs branching morphogenesis during mammary gland development. Development 121, 3539–3547.

Wysolmerski, J.J., Philbrick, W.M., Dunbar, M.E., Lanske, B., Kronenberg, H., Karaplis, A., Broadus, A.E., 1998. Rescue of the parathyroid hormone-related protein knockout mouse demonstrates that parathyroid hormone-related protein is essential for mammary gland development. Development 125, 1285–1294.

Wysolmerski, J.J., Roume, J., Silve, C., 1999. Absence of functional type I PTH/PTHrP receptors in humans is associated with abnormalities in breast and tooth development. J. Bone Miner. Res. 14, S135.

Wysolmerski, J.J., Stewart, A.F., Kovacs, C.S., 2008. Physiological actions of parathyroid hormone (PTH) and PTH-related protein: epidermal, mammary, reproductive, pancreatic tissues. In: Bilezikian, J.P., Raisz, L.G., Martin, T.J. (Eds.), Principles of Bone Biology, third ed. Academic Press, San Diego, pp. 713–731.

Yadav, S., Yadav, Y.S., Goel, M.M., Singh, U., Natu, S.M., Negi, M.P., 2014. Calcitonin gene- and parathyroid hormone-related peptides in normotensive and preeclamptic pregnancies: a nested case-control study. Arch. Gynecol. Obstet. 290, 897–903.

Yang, K.H., dePapp, A.E., Soifer, N.S., Wu, T.L., Porter, S.E., Bellantoni, M., Burtis, W.J., Broadus, A.E., Philbrick, W.M., Stewart, A.F., 1994. Parathyroid hormone-related protein: evidence for transcript- and tissue-specific post-translational processing. Biochemistry 33, 7460–7469.

Yamamoto, M., Duong, L.T., Fisher, J.E., Thiede, M.A., Caulfield, M.P., Rosenblatt, M., 1991. Suckling-mediated increases in urinary phosphate and 3′-5′-cyclic adenosine monophosphate excretion in lactating rats: possible systemic effects of parathyroid hormone-related protein. Endocrinology 129, 2614–2622.

Yamamoto, M., Harm, S.C., Grasser, W.A., Thiede, M.A., 1992. Parathyroid hormone-related protein in the rat urinary bladder: a smooth muscle relaxant produced locally in response to mechanical stretch. Proc. Natl. Acad. Sci. U.S.A. 89, 5326–5330.

Yin, J.J., Selander, K., Chirgwin, J.M., Dallas, M., Grubbs, B.G., Wieser, R., Massague, J., Mundy, G.R., Guise, T.A., 1999. TGF-β signaling blockade inhibits PTHrP secretion by breast cancer cells and bone metastases development. J. Clin. Investig. 103, 197–206.

Chapter 37

The pharmacology of selective estrogen receptor modulators: past and present*

Jasna Markovac[1] and Robert Marcus[2]

[1]*California Institute of Technology, Pasadena, CA, United States;* [2]*Stanford University, Stanford, CA, United States*

Chapter outline

Introduction 863
Selective estrogen receptor modulator mechanism 865
Selective estrogen receptor modulator chemistry 868
Selective estrogen receptor modulator pharmacology 868
Skeletal system 869
Preclinical studies 869
Clinical studies 870
Reproductive system 871
Uterus 871
Mammary 873
Other 876
Cardiovascular system 877
Cardiovascular safety of selective estrogen receptor modulators 878
Potential cardiovascular benefit of selective estrogen receptor modulators 879
Central nervous system 880
Central nervous system safety of selective estrogen receptor modulators 880
Central nervous system efficacy of selective estrogen receptor modulators 881
General safety profile and other pharmacological considerations 882
Other safety 882
Pharmacokinetics 883
Future directions with selective estrogen receptor modulators 884
Summary 884
References 885

Introduction

In the past decade, essentially no new significant information about the skeletal aspects of selective estrogen receptor modulators (SERMs) has appeared. Ten years ago, several promising candidate molecules were making their way through clinical trials. Some of these showed clinical efficacy in reducing vertebral fractures, particularly lasofoxifene and bazedoxifene. However, except for bazedoxifene, side effects stifled continued development. Moreover, as the efficacy and safety profiles of bazedoxifene closely resembled those of raloxifene (EVISTA, Eli Lilly & Co), which had already been available for patient care for more than 2 decades, no clear path for it to succeed as a stand-alone product was evident. Accordingly, bazedoxifene has been combined with conjugated equine estrogens as a dual preparation (DUAVEE, Pfizer, Inc.) that is FDA-approved for use in postmenopausal women for control of vasomotor symptoms and the prevention (but not treatment) of osteoporosis. Thus, raloxifene remains the only approved SERM for both the treatment and the prevention of osteoporosis.

In recent years, the pharmaceutical industry has shown little or no interest in developing other SERMs for skeletal purposes, although some interest persists in the basic science surrounding the determinants of their tissue specificity. Feng and O'Malley (2014) published some of the more current research dealing with the role of nuclear reception modulators and other coregulators in the action of SERMs. In addition, recent studies gave interesting and potentially clinically useful

* Most of this chapter is an edited version of the chapter from the third edition of this book, written by Henry U. Bryant, Eli Lilly and Company, Indianapolis, IN.

 https://doi.org/10.1016/B978-0-12-814841-9.00037-3

results with SERMs in other therapeutic areas (Mirkin and Pickar, 2015), but discussion of these exceeds the scope of this chapter and of this book.

For this reason, what follows is material little changed from what was published on SERMs in the third edition of this work (2009).

The widespread distribution of estrogen receptors (ERs), and their critical role in normal physiology and various pathophysiological states when estrogen levels decline, indicate the importance of ER-targeted therapies for use in postmenopausal women. Controversy and concern over the use of estrogen replacement therapies created the opportunity to design molecules to selectively modulate estrogen action in those tissues where estrogen agonism is the desired goal while simultaneously producing an estrogen-neutral or estrogen-antagonistic effect in tissues where estrogen-related side effects are a concern. SERMs are in clinical use for the treatment and prevention of osteoporosis (raloxifene), breast cancer prevention (tamoxifen and raloxifene) and treatment (tamoxifen and toremifene), and the induction of ovulation (clomiphene). By 2008, seven different SERMs had reached advanced clinical evaluation for postmenopausal osteoporosis, with three molecules (droloxifene, idoxifene, and levormeloxifene) withdrawn for unfavorable risk/benefit profiles and three additional molecules (lasofoxifene, bazedoxifene, and arzoxifene) in phase 3 status or under regulatory review. Experience with these molecules revealed several key themes for chronic use of SERMs: (1) Each SERM generates a unique complex with the ER that influences cofactor recruitment in estrogen–target tissues responsible for the tissue-selective pharmacological profile, which translates to each SERM generating potentially an entirely unique overall safety and efficacy profile, indicating the need for thorough evaluation of each individual SERM across multiple tissue types for efficacy and safety determination. (2) Uterine safety historically has been the critical safety feature for chronic SERM use in osteoporosis therapy, and careful assessment of the potential for uterine stimulation has been a key element in the consideration of new molecules in this class. (3) The pharmacokinetic and distribution properties of SERMs offered an additional aspect influencing the magnitude of the overall biological response by either improving systemic bioavailability or altering uptake into important estrogen-responsive tissues.

Research in the ER field attracted considerable attention during the 15 years prior to 2009 with significant events ranging from very basic research discoveries, such as resolution of the liganded ER crystal structure and identification of a second ER form (ERβ), to important clinical observations regarding estrogen use in postmenopausal women from the Women's Health Initiative (WHI) trial. Estrogen exhibited a "Jekyll and Hyde" therapeutic profile, as hormone replacement (estrogen 1 progestin) was associated with distinct benefits on the menopausal syndrome including reductions in vasomotor symptoms and fracture incidence as well as other benefits such as a reduction in colon cancer. However, these benefits were offset by significant increases in risk for coronary events (myocardial infarction and stroke) and breast and uterine cancer. In the early 1990s, research around the concept of SERMs—molecules that simultaneously agonized or antagonized estrogen action in different tissue types—offered a new way of looking at ER pharmacology and served to trigger renewed interest in estrogen-related research in the mid-1990s.

Prior to the development of the "SERM-concept," ER ligands were generally thought of as falling into the category of full agonist, partial agonist, or full antagonist across all tissue types (i.e., uterus, mammary, and bone). For example, steroidal hormones such as 17β-estradiol were known to behave as full agonists both in vitro and in vivo across multiple tissue types, whereas compounds such as fulvestrant (ICI-182,780) were known to be complete ER antagonists that bound tightly to the ER but lacked intrinsic activity, and therefore completely blocked the action of full ER agonists like 17β-estradiol. These "pharmacotypes" are depicted in Fig. 37.1A and B, respectively. Conversely, compounds such as tamoxifen were known, in the presence of estrogen, to block estrogen action in estrogen-responsive tissues (i.e., breast cancer cells), but in the absence of estrogen to mimic estrogen in bone and uterus in estrogen-deficient animals and women, thus exhibiting a classical partial-agonist profile (see Fig. 37.1C). Although this profile held some attractive features for use in ER-dependent breast cancer, the profile was prohibitive for chronic use in postmenopausal women for noncancer indications (like osteoporosis), where even the potentially less robust uterine stimulation induced by tamoxifen's partial agonist action produced untenable side effects that created a risk/benefit ratio that was unfavorable for use in diseases such as osteoporosis (Kalu et al., 1991). As a result, virtually no work was being done in the pharmaceutical industry developing novel ER ligands for osteoporosis because the prevailing medical opinion at the time was that any molecule with sufficient estrogen agonism capable of producing a benefit in bone would also generate sufficient agonism (even if a partial agonist) in uterine tissue to create a risk that would unfavorably offset the bone benefit (Feldman et al., 1989).

With the first preclinical and clinical descriptions of the unique profile of raloxifene in estrogen-deficient animals and postmenopausal women (Black et al., 1994; Draper et al., 1997), the SERM concept was born, radically shifting thought around use of ER ligands in postmenopausal women and opening the door for use in chronic diseases such as osteoporosis. Accordingly, the initial goals of SERM-based therapy for osteoporosis required the molecule to have estrogen-like efficacy on bone and concomitant fracture reduction without estrogen-like stimulatory effects on the uterus or mammary tissue. As

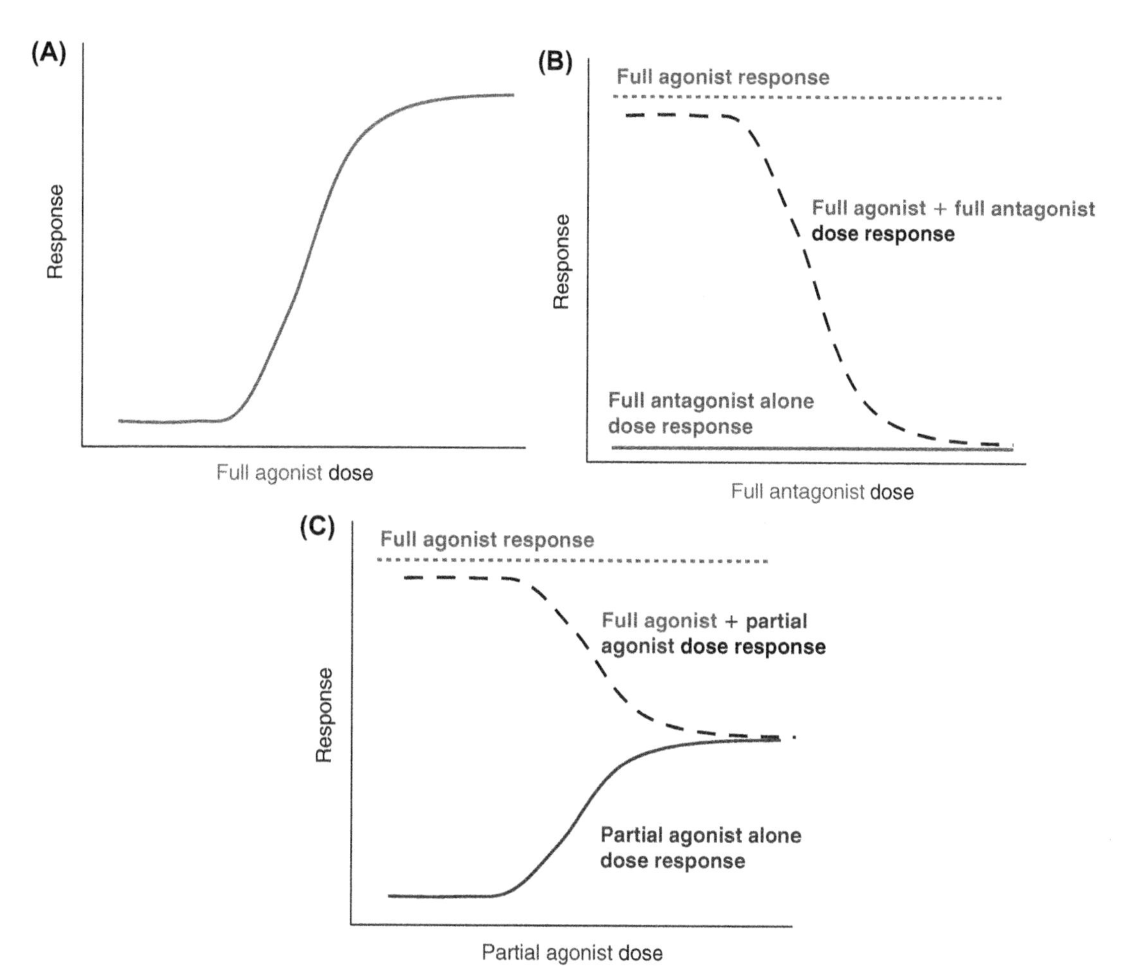

FIGURE 37.1 Potential pharmacotypes resulting from various estrogen receptor–ligand interactions in the form of hypothetical dose–response curves for a full agonist (A), full antagonist (B), and partial agonist (C). (See plate section.)

of the writing of this chapter for the third edition of this book, only four molecules with SERM-like profiles had achieved clinical use (Table 37.1), and only one, raloxifene, had attained approval for use in the treatment and prevention of osteoporosis. However, other molecules had been evaluated clinically for postmenopausal osteoporosis and are reviewed here. In addition, some other SERM applications are presented that could potentially benefit other diseases or disorders in postmenopausal women.

Selective estrogen receptor modulator mechanism

The effects of SERMs on biological systems are predominantly mediated by specific, high-affinity interactions with ERs that are primarily located in target cell nuclei (Nilsson *et al.*, 2001). Their nuclear hormonal action involves the complex interplay of a number of protein and genomic elements that allow SERMs to regulate gene transcription and subsequent protein production by the cell. Three key elements of the SERM mechanism are depicted in Fig. 37.2, which include (1) high-affinity interaction with the ER, (2) ER–ligand dimerization and association with a tissue-specific set of coregulatory proteins, and (3) binding of the ER/adaptor protein complex to specific DNA response elements located in the promoter regions of nuclear target genes and ensuing regulation of gene transcription. Depending on the cellular and promoter context, the DNA-bound receptor can induce or inhibit the transcription of specific genes within the tissue.

The ability to specifically bind to the ER is perhaps the single most important feature of all molecules with a SERM profile. The affinities of several of the more extensively studied SERMs are provided in Table 37.2. An important determinant of the ultimate pharmacological response is the shape of the ligand–ER complex, which is unique with each individual ligand (McDonnell et al., 1995), with different SERMs leading to specific orientation of ER subunits (i.e., raloxifene; Brzozowski et al., 1997). This conformation or shape of the ligand–ER complex then determines the

TABLE 37.1 Selective estrogen receptor modulators currently approved for human use.

SERM	Trade name	Approved indications	Daily dose (mg)
Clomiphene	Clomid	Induction of ovulation	50–100
Raloxifene	EVISTA	Osteoporosis prevention	60
		Osteoporosis treatment	
		Breast cancer risk reduction in postmenopausal osteoporotic women and in high-risk postmenopausal women	
Tamoxifen	Nolvadex	Metastatic breast cancer treatment	20–40
		Adjuvant breast cancer treatment	
		Ductal carcinoma in situ	
		Breast cancer risk reduction in high-risk women	
Toremifene[a]	Fareston	Metastatic breast cancer treatment	60

[a]*Toremifene (Fareston) is currently not approved in the United States but is approved for metastatic breast cancer treatment in Europe.*

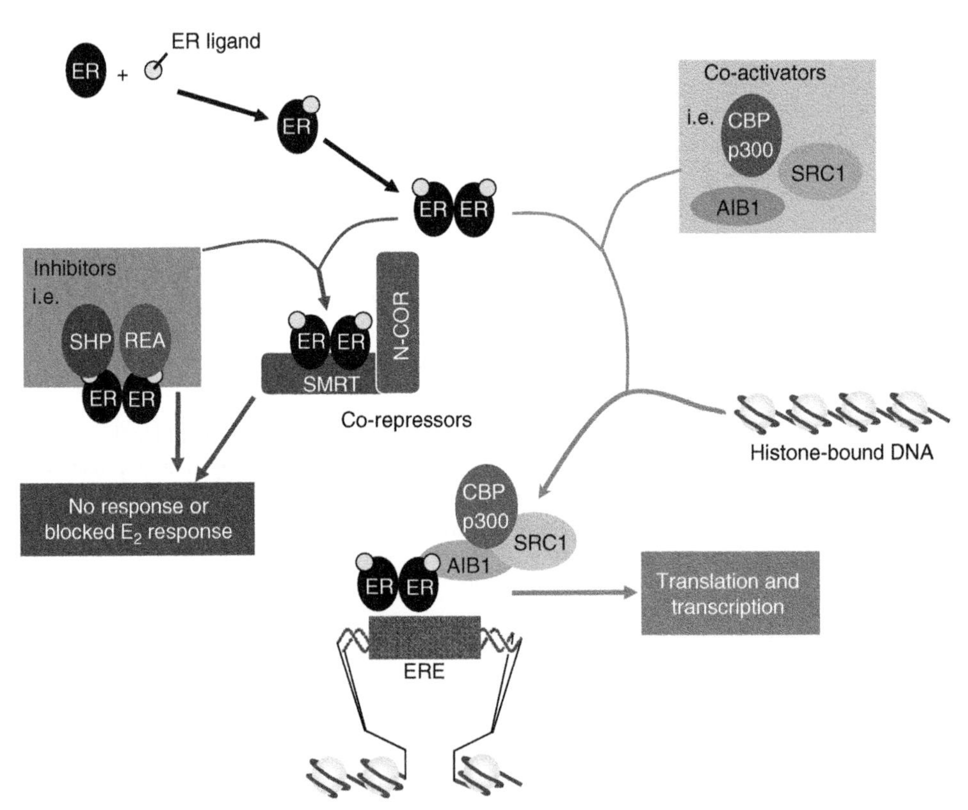

FIGURE 37.2 Selective estrogen receptor (ER) modulator mechanism of action.

subsequent protein–protein interactions that ensue. Herein lies the basis for the wide array of different pharmacological profiles produced by different SERMs, because the confirmation of the ER–SERM complex is distinct for each molecule (McDonnell et al., 1995). It is important to recognize that a second form of ER is known to exist, ERβ (Kuiper et al., 1996). ERα and ERβ display unique patterns of tissue distribution, typically with the expression levels of one subtype dominating (Saunders et al., 1997), although it should be noted that most tissues contain at least small amounts of both subtypes, and with the role of putative α/β heterodimers unknown, it is possible that a low-expression subtype may be a key rate-limiting

TABLE 37.2 Affinities of various SERMs for human estrogen receptors ERα and ERβ.

ER ligand	ERα(IC_{50})	ERβ(IC_{50})
17β-Estradiol	0.3–0.8[a]	0.9–2.5[a]
Tamoxifen	72–138	173–1204
4-Hydroxytamoxifen	0.22–0.98	1.5–2.46
Raloxifene	0.4–1.31	5.6–13.0
Fulvestrant (ICI-182,780)	0.8–1.0	1.12–3.6

[a]*All values are in nM. As binding data often can vary from laboratory to laboratory, and various approaches can be taken to attaining binding affinities, ranges for ERα and ERβ obtained from representative references are presented (Sun et al., 1999; Brady et al., 2001; de Boer et al., 2004; He et al., 2005; Leblanc et al., 2007).*

step in ultimate nuclear activity. All SERMs that have reached advanced clinical evaluation show high affinity for both ERα and ERβ with sufficient circulating and tissue exposure to ensure binding of both subtypes, indicating that for these molecules at least, differential ERα or ERβ activation does not explain the tissue-selective pharmacological effects.

In addition to ERs themselves, a number of other coregulatory proteins, such as coactivators (which enhance transcription) and corepressors (which reduce transcription), play an essential role in determining the ultimate response of an individual cell to liganded ER. A number of coactivators and corepressors that interact with ligand-bound ER have been identified and are reviewed thoroughly elsewhere (Smith, 2006).

The relative tissue expression of the different cofactors and the ability of the ER–ligand complex to interact with those cofactors determine the tissue-selective agonist/antagonist profile of the various SERM molecules. Cofactor tissue expression can also be altered with various pathophysiological states, such as breast cancer (Bautista et al., 1998). The important nature of the tissue-relevant cofactor context was demonstrated by Shang and Brown (2002), who compared the effects of two SERMs, tamoxifen and raloxifene, with estrogen in two tissue contexts, a breast cancer cell line and a uterine endometrial carcinoma cell line. In the mammary cells that proliferate in response to estrogen, 17β-estradiol recruited coactivators leading to increased gene expression. In these same cells, where tamoxifen and raloxifene both display estrogen antagonist pharmacology, the ligand–SERM complex with both molecules recruited corepressors and not the coactivators observed with 17β-estradiol on ER-mediated transcription. However, in a uterine cell line where tamoxifen exhibits estrogen agonist pharmacology and raloxifene is a complete antagonist, tamoxifen was associated with the recruitment of a coactivator protein complex that is expressed at higher levels in uterine cells. The coactivator requirements for estrogen-stimulated gene expression in uterine cells were distinct from those for tamoxifen, indicating multiple signaling mechanisms even for the agonist response. Conversely, raloxifene failed to recruit a coactivator construct and rather induced a corepressor construct in uterine cells (Shang and Brown, 2002). Thus, the relative abundance of ER-associated coactivators and corepressors is an important factor in the tissue-specific pharmacology of SERMs.

In addition to the layers of complexity provided by multiple ER–SERM conformations and tissue-selective cofactor recruitment, the mechanism of tissue selectivity of SERMs is further complicated by the existence of multiple DNA response elements. Many estrogen-responsive genes contain the classical estrogen response element and a number of DNA response elements, such as activator protein-1 and steroidogenic factor-1 response element (Vanacker et al., 1999). The mechanism for SERM activation (or inactivation) of ER-mediated function is further complicated by the presence of novel DNA response elements that are more apparent following formation of the ER–SERM cofactor complex.

Although there have been considerable strides in understanding the molecular biology of SERM action in a general sense, with the critical role of the ER, specific cofactors that are recruited to the transcriptional complex, and the specific DNA response elements activated, it is important to recognize that each SERM has the potential to produce a unique fingerprint of pharmacological activity at the whole-organism level. Contributing to the eventual profile are the molecular mechanisms reviewed earlier as well as other factors, such as absorption, distribution, excretion, and metabolism of the SERM, that add another layer of complexity for ultimate pharmacological response, creating the need to fully characterize the tissue-specific effects of each individual SERM in an in vivo paradigm.

Selective estrogen receptor modulator chemistry

A key element in determining the pharmacological profile of each distinct SERM is influenced heavily by the chemical makeup of the SERM including the basic scaffold, the placement of the hydroxyl moieties, and the positioning and nature of the basic side chain. Crystal structures of various ligands bound to the ER indicate that small molecules can induce a spectrum of receptor conformations. As described previously, the specific SERM–ER conformation has a tremendous impact on cofactor recruitment and ultimate genomic activation or inhibition by the SERM. Chemical scaffolds that have produced SERMs in current clinical use, or at least that have reached phase 3 clinical evaluation in humans, are depicted in Fig. 37.3 and include triphenylethylenes (i.e., tamoxifen, droloxifene, idoxifene, clomiphene, and toremifene), benzothiophenes (raloxifene and arzoxifene), tetrahydronaphthalenes (lasofoxifene and nafoxidine), indoles (bazedoxifene), and benzopyrans (acolbifene and levormeloxifene). Key structural features of these molecules, which are indicated in Fig. 37.4 for raloxifene versus 17β-estradiol, are typical for the entire class, with the most important features being the hydroxyl moieties and the basic side chain, as described thoroughly elsewhere (Grese et al., 1997).

Selective estrogen receptor modulator pharmacology

Given the wide distribution of ER and the pleiotropic nature of estrogen and its multiple metabolites, SERMs should be expected to likewise affect multiple organ systems, and this is indeed the case. Because one area of focus in this volume is skeletal pharmacology, emphasis here will be placed on the pharmacologic effects of SERMs on bone and other tissues that are of relevance to safety in the clinical setting. Accordingly, emphasis is placed on those SERMs where osteoporosis and

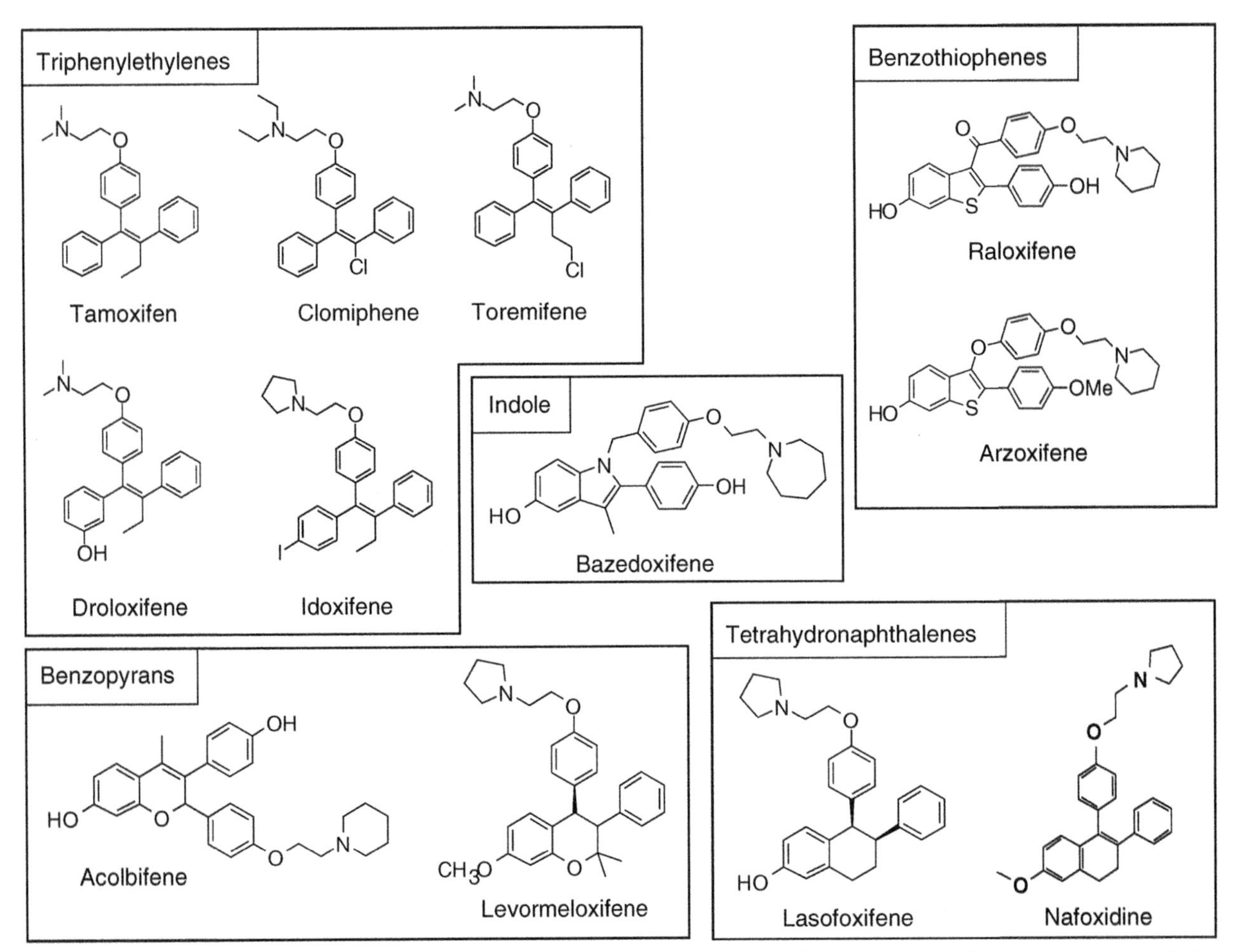

FIGURE 37.3 Structures and chemical classes of selective estrogen receptor modulators currently in clinical use or that have reached phase 3 clinical trials in women.

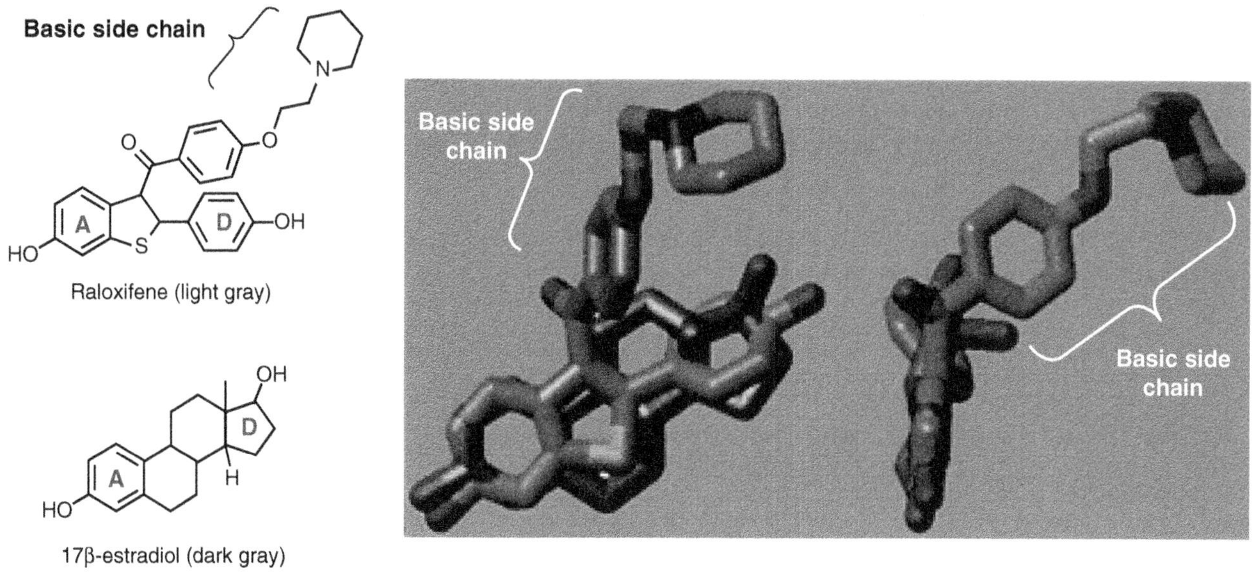

FIGURE 37.4 Key structural features of raloxifene versus 17β-estradiol.

bone have been the primary focus of research. SERMs used primarily for other indications (present or with future potential) will be briefly summarized as well.

Skeletal system

Preclinical studies

Much as in postmenopausal women, estrogen deficiency in ovariectomized (OVX) animals leads to a rapid increase in bone turnover, where excessive osteoclast resorptive activity results in a marked decline in trabecular bone mass and strength with a concomitant increase in fractures. In rats, ovariectomy produces a rapid osteopenic response that can be discerned within 5 weeks. All of the SERMs depicted in Fig. 37.3 have been evaluated in the OVX rat and demonstrate estrogen-like protection from bone loss induced by estrogen deficiency. In the OVX rat model, SERMs like raloxifene (Black et al., 1994), arzoxifene (Sato et al., 1998), tamoxifen (Sato et al., 1996a,b), droloxifene (Ke et al., 1995), idoxifene (Nuttall et al., 1998), clomiphene (Jiminez et al., 1997), bazedoxifene (Komm et al., 2005), lasofoxifene (Ke et al., 1998), levormeloxifene (Galbiati et al., 2002), toremifene (Qu et al., 2000), and acolbifene (Martel et al., 2000) prevent the loss of bone in vertebrae, distal femur, and proximal tibia, all trabecular-rich bone sites. In addition to maintaining bone mass, SERMs also preserve bone strength through improvements in bone microarchitecture (i.e., raloxifene; Turner et al., 1994). For both bone mass and bone strength, the absolute magnitudes of the effects of most SERMs in OVX rats are indistinguishable from those of estrogen and can approach values attained for sham-surgery controls when the SERM is administered in a preventative mode. However, differences in potency for these bone-protective effects can occur, with third-generation SERMs like arzoxifene, bazedoxifene, and lasofoxifene producing efficacy equivalent to that of raloxifene in OVX rat trabecular BMD responses at approximately 10% of the dose (Ke et al., 1998; Sato et al., 1998). There had been preclinical hints of improved efficacy of some of these latter agents as well, for example with arzoxifene, which improves cortical bone strength to a greater degree than observed with maximally effective doses of either estrogen or raloxifene (Sato et al., 1998). Similarly, with bazedoxifene, improved biomechanical properties in trabecular bone were observed relative to estrogen after 1 year of treatment in OVX rats (Komm et al., 2007).

As with estrogen, the primary activity of SERMs that is responsible for the beneficial effect on bone is antiresorptive. Mechanistic studies in vitro demonstrated that raloxifene and estrogen exert their antiresorptive effects primarily as inhibitors of osteoclast differentiation, rather than as direct inhibitors of activated osteoclasts. Raloxifene acts to suppress mediators of osteoclast differentiation, such as the receptor activator of nuclear factor-kappa-β (RANK) and RANK-ligand (Bashir et al., 2005), and increase endogenous antiresorptive factors such as osteoprotegerin (Viereck et al., 2003). Biochemical markers of bone turnover (i.e., serum osteocalcin and urinary collagen cross-links) are suppressed in OVX rats in a manner similar to that observed with estrogen (Frolik et al., 1996). Histomorphometric analysis of bone from raloxifene-treated OVX rats confirmed the antiresorptive mechanism of action for raloxifene (Evans et al., 1994). Similar studies with the other SERMs discussed here indicate the same antiresorptive mechanism for bone protection. Of likely importance with respect to long-term safety in the skeleton is the finding that SERMs such as raloxifene produce their inhibitory action on bone resorption with minimal suppressive effects on bone formation, leaving bone formation rates at levels comparable to those of sham-operated control animals (Evans et al., 1994). The molecular fingerprint of SERMs in estrogen-deficient rat trabecular bone, as assessed by DNA microarray, is unique for each SERM, although it is clear that some SERMs are less suppressive of bone formation. For example, in OVX rats, raloxifene returned a cluster of genes associated with bone formation to ovary-intact control levels, as opposed to a bisphosphonate (alendronate), estrogen, or even another SERM (acolbifene), which exhibited a greater suppressive effect on bone formation–associated genes (Helvering et al., 2005a,b). The overall SERM profile on bone then represents a sharp distinction from the marked suppression of bone formation that occurs with other bone antiresorptives, such as the bisphosphonates. A final bone cell type that may be of relevance to the skeletal-protective effect of SERMs is the osteocyte. Whereas relatively less work has been done with this bone cell owing to technical challenges, work in MLO-Y4 cells (an osteocyte-like cell line) has indicated that raloxifene prevents oxidative stress–induced apoptosis and inhibits the generation of reactive oxygen species with oxidative insults, such as hydrogen peroxide (Mann et al., 2007).

Clinical studies

The abundance of preclinical information on the effects of SERMs on bone was easily matched by a plethora of long-term clinical trials conducted on a number of different SERM molecules, either as the primary element of registration trials for postmenopausal osteoporosis or as part of the safety assessment for use in breast cancer. Certainly the most extensively studied SERM on the human skeleton has been raloxifene, which has been investigated in nearly 40,000 clinical trial subjects enrolled in prospective, randomized trials (placebo or active comparator) that have ranged in duration from 1 to 8 years. In postmenopausal women, raloxifene (60 mg/day) exhibited an antiresorptive action, as evidenced by reductions in accelerated bone turnover as measured by biochemical markers of bone resorption (Ettinger et al., 1999), while only modestly suppressing bone formation. In calcium tracer kinetic studies in postmenopausal women, Heaney and Draper (1997) showed suppression of bone resorption with raloxifene, whereas bone formation was not affected after 31 weeks of treatment. The observation of resorption inhibition with minimal formation suppression by raloxifene was confirmed by histomorphometric analysis of iliac crest bone biopsies (Ott et al., 2002). This antiresorptive activity is associated with an approximately 2.5% increase in vertebral BMD relative to placebo-treated controls. This increase in spine BMD that occurs following raloxifene treatment in postmenopausal women is less marked than observed with a bisphosphonate (alendronate; Johnell et al., 2002). However, this magnitude of BMD improvement in the spine underestimates the mechanical improvement produced by raloxifene as evidenced by the 55% reduction in new vertebral fractures in women with prevalent fractures (Ettinger et al., 1999), a rate comparable to that produced by bisphosphonates. This observation led to increased attentiveness to the potential effects of raloxifene (and putatively other SERMs in the future) on bone quality. The eventual resistance of bone to fracture was the result of both the content, or mass of the material (i.e., BMD), and the quality of that material, which was likely the result of bone microarchitecture and the nature of the mineralized matrix itself. In animal models, increased trabecular thickness and the maintenance of platelike (vs. rodlike) trabecular structures, both of which correlate with improved biomechanical strength of bone, were observed in OVX mice (Cano et al., 2008). Although BMD is a noninvasive, easily quantifiable parameter in clinical trials, bone quality remains more difficult to quantify because it is primarily revealed by the eventual incidence of fracture. A number of efforts targeted better understanding and quantifying of bone quality in instances where raloxifene had shown some benefits over other antiresorptive therapies such as histomorphometric analyses of trabecular bone architecture and microcrack frequency in bone (Allen et al., 2006; Li et al., 2005). One area where some aspect of bone quality was beginning to be elucidated is the proximal femur, where imaging technologies had been applied to postmenopausal clinical trial subjects to show an increase in resistance to axial and bending stresses in raloxifene-treated women (Uusi-Rasi et al., 2006), indicating improved structural components of bone strength and stability with the SERM. Raloxifene produced positive effects on hip BMD, which increased 2.1% versus placebo after 3 years in postmenopausal women (Ettinger et al., 1999), although without a significant effect on nonvertebral fracture rates (Ettinger et al., 1999). However, an interesting trend was noted in a subset of women who entered the trials with severe vertebral fractures. In this subset of more severely osteoporotic women, raloxifene produced a 50% reduction in nonvertebral fractures (Delmas et al., 2003). Finally, in addition to the reduction of vertebral fractures in osteoporotic women, raloxifene provided fracture risk protection to osteopenic women (Kanis et al., 2003).

A number of other SERMs had attempted unsuccessfully to register for an osteoporosis prevention/treatment indication. The failure of those molecules to achieve regulatory approval for osteoporosis was primarily based on safety and risk/

benefit analysis, as each demonstrated some level of improvement on skeletal parameters in earlier clinical trials. Prior to discontinuation of levormeloxifene phase 3 clinical trials because of gynecological-associated adverse events, phase 2 clinical trials demonstrated positive effects of this SERM on BMD and bone turnover (Alexandersen et al., 2001). Idoxifene, a triphenylethylene also discontinued in phase 3 for uterine-adverse events, produced clinically relevant increases in BMD in osteopenic postmenopausal women (Chesnut et al., 1998). The third-generation molecules, lasofoxifene and bazedoxifene, are both very potent SERMs with relatively high bioavailability (Gardner et al., 2006; Patat et al., 2003) and produced advanced clinical testing results for osteoporosis. In a 2-year trial in 410 postmenopausal women, lasofoxifene at 0.25 or 1 mg/day suppressed bone turnover similarly to raloxifene, but lasofoxifene increased lumbar spine BMD by 3.6% and 3.9% (respectively by dosage), which outpaced the increase observed with raloxifene (McClung et al., 2006). A 2-year BMD trial and 3-year fracture prevention trial demonstrated the skeletal protective effects of bazedoxifene relative to raloxifene. In the 3-year trial, nearly 7500 women were treated with 20 or 40 mg/day bazedoxifene, placebo, or 60 mg/day raloxifene. In this trial, bazedoxifene produced a significant reduction in relative risk for new vertebral fractures of 37% for the higher dose and 42% for the lower dose, with raloxifene producing a comparable 42% reduction in relative risk of new vertebral fractures (Silverman et al., 2007). Mean lumbar spine BMD was significantly improved, relative to placebo, by bazedoxifene with a magnitude of response comparable to raloxifene, and biochemical markers of bone turnover were also significantly lowered with bazedoxifene (Miller et al., 2007).

A number of clinical trials have focused on the bone-sparing effects of two triphenylethylene SERMs used for breast cancer treatment, tamoxifen and toremifene. Although most studies demonstrated a skeletal benefit for these two agents, trials have typically been small and not placebo-controlled in design. A consistent benefit was observed with tamoxifen and toremifene primarily at trabecular bone sites, which was consistent with observations made with raloxifene in postmenopausal women. After 3 years of use, tamoxifen or toremifene in breast cancer patients led to a less-than-expected decline in vertebral BMD (Tiitinen et al., 2004). In shorter trials (1 year), similar effects were observed, with the effect of tamoxifen somewhat stronger than that of toremifene (2% higher BMD with tamoxifen vs. toremifene, which basically prevented age-related decline over the 1-year trial; Marttenun et al., 1998). Although many studies had reported similar benefits on BMD in postmenopausal breast cancer patients, particularly with tamoxifen (e.g., Love et al., 1992), there was at least one indication that the use of tamoxifen in normal premenopausal women was associated with a reduction in bone mineral density (Powles et al., 1996).

Reproductive system

Uterus

Atrophy of the uterus accompanies estrogen deficiency in humans and most animal species, and cessation of menses is a hallmark feature of menopause in women. A major side effect of most current ER-based therapies is stimulation of the uterus resulting in part from estrogen-induced proliferation of uterine endometrial tissue. The cancer concern associated with this proliferative effect and the resumption of menses (when combined with progestin regimens as hormonal replacement therapy) have been major limitations to estrogen replacement therapeutic approaches. These uterine side effects of estrogen are often the primary deterrent in the risk/benefit decision for postmenopausal women to utilize or remain compliant with hormonal estrogen therapies for chronic use with diseases such as osteoporosis. For postmenopausal women, a major and significant advantage of SERMs like raloxifene over hormonal estrogen therapies is the lack of uterine stimulation with the SERM. However, not all SERMs share the same degree of uterine safety observed with raloxifene, and thus effects on the uterus also serve as an important distinguishing feature among various SERMs. In this regard, SERMs that have failed in phase 3 clinical trials have done so primarily because of an untenable degree of uterine stimulation or uterus-related adverse events. The accumulated clinical experience with uterine safety for a number of SERMs helps one determine which preclinical models and parameters provide optimal predictive value for uterine safety. The use of in vitro systems, estrogen-depleted animals, and estrogen-replete animals to assess antagonist potential formed a triangulated approach for the uterine safety assessment of SERMs.

Estrogen agonism in the uterus

Initial indication of the uterine estrogenic potential of SERMs was demonstrated using Ishikawa cells, a human endometrial cancer cell line. Ishikawa cell proliferation is stimulated by 17β-estradiol, tamoxifen, and 4-hydroxytamoxifen (active metabolite of tamoxifen), but not by uterine-sparing SERMs such as raloxifene (Koda et al., 2004). More subtle changes among various SERMs were detected by evaluating their effects on ER-mediated alkaline phosphatase production, creatine kinase production, and expression of progesterone receptor. The more uterine-stimulatory SERMs produced greater induction of alkaline phosphatase and progesterone receptor expression and were less effective antagonists of 17β-estradiol-stimulated responses in Ishikawa cells (Bramlett et al., 2003). Of note, whereas uterine-safe SERMs like raloxifene failed to stimulate creatine kinase production in Ishikawa cells, raloxifene induced this activity in cell lines with an osteoblast background, consistent with the "SERM" activity profile (Koda et al., 2004).

Lack of biologically meaningful stimulation of the uterus in the estrogen-depleted state (e.g., in postmenopausal women and OVX animals) was the crux of SERM uterine-safety evaluation. The uterine effects of tamoxifen and raloxifene had been thoroughly evaluated in numerous clinical settings as well as in a variety of preclinical models. Raloxifene did not produce estrogen-like stimulatory effects in the uterus of OVX rats (Black et al., 1994). In the OVX rat model, a slight non-dose-related elevation of uterine weight was frequently observed. However, this phenomenon contrasted markedly with the robust dose-related elevation of uterine weight produced by estrogen in these animals. Raloxifene failed to stimulate other estrogen-sensitive markers in the uterus of OVX rats, such as uterine eosinophilia or uterine epithelial cell height (Black et al., 1994). In large-scale clinical trials, after 8 years of chronic use in postmenopausal women, extensive uterine safety evaluation revealed no significant uterine-stimulatory effects of raloxifene in humans (Delmas et al., 1997). Indeed, a significant reduction in endometrial cancer of the uterus was noted in postmenopausal women using raloxifene (Delmas et al., 1997).

As indicated, not all SERMs exhibit the uterine safety profile demonstrated with raloxifene. This is perhaps most evident in the extensive work done with tamoxifen. Studies in the uterus of OVX rats demonstrated robust dose-related stimulatory effects of tamoxifen on uterine weight, uterine epithelial cell height, and uterine eosinophilia (Adrian et al., 1996). Clinical experience with tamoxifen is consistent with these observations, as uterine bleeding and significant elevation of endometrial cancer has been observed in women exposed to tamoxifen in long-term studies (Fisher et al., 1994).

Thus, raloxifene and tamoxifen serve as bookend profiles for uterine safety for chronic SERM use in postmenopausal women, where molecules can be assessed for either a "raloxifene-like" profile of little or no uterine stimulation or a more estrogenic "tamoxifen-like" profile. Several SERMs have advanced to clinical research and corroborated the preclinical observations for those molecules. For example, droloxifene, idoxifene, and levormeloxifene all produced dose-related elevation of uterine weight, uterine epithelial height, and uterine eosinophilia in the OVX rat (Adrian et al., 1996) and were eventually halted in clinical development on the basis of uterine liabilities such as increased bleeding, increased endometrial thickness, and polyps that developed in phase 3 studies for osteoporosis (Silfen et al., 1999). The other triphenylethylenes used clinically, clomiphene and toremifene, also exhibit an overall uterine stimulatory profile in OVX rats (Turner et al., 1998; Carthew, 1999), although uterine safety is less of a concern with acute therapeutic use of clomiphene, and the risk/benefit profile for toremifene use in breast cancer treatment has a different weighting of risk. Mixed effects have been observed with lasofoxifene, which in some preclinical reports produced significant dose-related stimulation of uterine epithelial cell height and uterine eosinophilia (Cole et al., 1997), but not in others (Ke et al., 1998). The outcome of clinical trials with lasofoxifene is incomplete, although initial observations indicate statistically significant uterine stimulation with an approximately 80% increase in endometrial thickness relative to placebo-treated controls and an increased incidence of leukorrhea (McClung et al., 2006). On the other hand, bazedoxifene, arzoxifene, and acolbifene produced very little stimulation of the uterus in OVX rats (Komm et al., 2005; Sato et al., 1998; Martel et al., 2000), and at least with bazedoxifene, studies in postmenopausal women have corroborated the animal work (Adachi et al., 2007).

A significant uterine-related adverse event that emerged in a number of clinical trials is relaxation of the pelvic floor, typically with associated prolapse of the uterus. Estrogen therapy was traditionally thought to improve the structural integrity of pelvic tissue with expected favorable effects on symptoms such as urinary incontinence. However, data from both the WHI trial and the Heart and Estrogen/progestin Replacement Study (HERS) provided evidence for the increased incidence of urinary incontinence in association with estrogen use (Hendrix et al., 2005). Related to this, pelvic organ prolapse was associated with both levormeloxifene and idoxifene within 1 year of therapy (Fleischer et al., 1999; Goldstein et al., 2002). These observations, taken together, suggest that these uterine wall–associated adverse events with idoxifene and levormeloxifene might be due to their more estrogen-like uterine activity. The mechanism by which estrogen and some SERMs compromise the uterine wall in postmenopausal women may be linked to an increase in collagen-degrading enzymes, such as metalloproteinase-2 (MMP-2), based on OVX rat studies. In OVX rats, uterine MMP-2 levels were reduced relative to those of ovary-intact animals. Estrogen, as well as levormeloxifene, produced marked increases in uterine MMP-2 expression (Helvering et al., 2005a,b). Not all SERMs, however, have been associated with pelvic floor problems, because no increase in this adverse event has been related to the chronic use of tamoxifen, toremifene, bazedoxifene, or lasofoxifene (Fisher et al., 1994; Maenpaa et al., 1997; McClung et al., 2006; Adachi et al., 2007). Consistent with its uterine-safe overall profile, the extensive uterine safety evaluations with raloxifene showed no increase in the incidence of problems associated with pelvic floor relaxation. Rather, in at least one report, raloxifene produced benefit with a significant reduction in the frequency of surgery for pelvic floor relaxation in a population of postmenopausal women (Goldstein et al., 2001).

Estrogen antagonism in the uterus

The final important component in assessing the uterine safety profile of SERMs relied primarily on preclinical data—the ability of SERMs to antagonize the uterine-stimulatory effects of estrogen. Further insight into the uterine activity profile of SERMs was gleaned from effects on the uterus in the presence of estrogen, because this allowed assessment of the uterine estrogen antagonist potential for these compounds. As in the estrogen-deficient state, the various SERMs also produced one of two general activity profiles in estrogen-replete animals: that of either a complete estrogen antagonist or a partial agonist at the ER. In the uterus of either estrogen-treated immature or adult OVX rats, SERMs producing desirable uterine safety profiles in postmenopausal women, such as raloxifene, arzoxifene, and bazedoxifene, produced a complete estrogen antagonistic effect (Adrian et al., 1996; Sato et al., 1998; Komm et al., 2005). This effect is most clearly demonstrated in the estrogen-treated immature rat, a model classically used to determine uterine liability of ligands for the ER. In this model, raloxifene blocks the uterotrophic effects of estrogen with an ED_{50} of 0.3 mg/kg (Silfen et al., 1999), whereas arzoxifene and bazedoxifene completely antagonizes the estrogen response with greater potency (Sato et al., 1998; Komm et al., 2005). A key feature of this antagonistic effect with SERMs like raloxifene in the immature rat uterus is the complete antagonism of the uterotrophic effect of estrogen. That is, the uteri from estrogen-treated immature rats given doses of raloxifene exceeding 3 mg/kg are indistinguishable from those of non-estrogen-treated immature rats (Adrian et al., 1996). This is in dramatic contrast to other SERMs—this is best exemplified by tamoxifen, which behaves as a classical partial agonist in the uterus. That is, tamoxifen does significantly antagonize the effects of estrogen in the immature rat uterus. However, in the case of tamoxifen, the maximal degree of this antagonism is only approximately 50% (Adrian et al., 1996). The primary reason for this incomplete antagonistic effect of tamoxifen is that at higher doses, the inherent uterine stimulatory capacity of tamoxifen limits further suppression of estrogen-induced uterotrophic response. Other triphenylethylene SERMs, as well as levormeloxifene (Adrian et al., 1996), produce partial estrogen agonist profiles in estrogen-stimulated immature rats.

In addition to the usefulness of this parameter for SERM uterine safety assessment, the complete estrogen antagonistic effects of SERMs like raloxifene were utilized successfully in a variety of preclinical models of estrogen-stimulated uterine pathologies with some limited success in clinical studies. For example, in vitro and in vivo models of uterine leiomyoma (fibroids) demonstrated favorable responses to raloxifene (Porter et al., 1998) with some reports of successful use of high-dose raloxifene to prevent the progression of uterine leiomyomas in small groups of premenopausal and postmenopausal women, either alone or in combination with a GnRH agonist (Jirecek et al., 2004; Palomaba et al., 2002b, 2005) via a combined antiproliferative/proapoptotic effect. However, other small clinical trials have failed to detect a benefit of raloxifene on uterine leiomyoma (Palomba et al., 2002a), indicating the need for randomized, well-controlled clinical trials for this indication, which would be necessary to balance the benefit on leiomyoma status versus the additional risk of ovarian stimulation from SERM use in premenopausal women (discussed later; note: raloxifene was not indicated for any uses in premenopausal women). In a rat model of another estrogen-dependent, uterine-related pathology, endometriosis, raloxifene inhibited the estrogen-dependent growth of peritoneal uterine explants (Swisher et al., 1995). In monkeys with spontaneous endometrial lesions, raloxifene eliminated the occurrence of peritoneal endometriosis lesions (Fanning et al., 1996).

The uterine estrogen antagonist effect of SERMs is also of potential benefit in the treatment of endometrial cancer, the most common neoplasm of the female urogenital tract. In many cases, endometrial cancer is susceptible to hormonal-targeted therapies involving both progestin-based and estrogen-depletion strategies. Given the role of estrogens both directly and indirectly via the regulation of progesterone receptor, SERM approaches have been explored. Tamoxifen was evaluated in a small cohort of advanced or recurrent endometrial carcinoma patients with limited success, dissuading the use of tamoxifen as a single agent for this tumor type (Thigpen et al., 2001). Promising efficacy in endometrial cancer treatment was reported with the third-generation SERM, arzoxifene. In early studies in a small number of treatment-refractory endometrial cancer patients, arzoxifene produced a favorable clinical response rate and stabilized the disease in a substantial number of women (Burke and Walker, 2003). In a phase 2 open-label study, ER-positive and progesterone receptor-positive patients with recurrent/advanced endometrial cancer also exhibited a high response rate and duration of response with a favorable side effect profile after administration of arzoxifene (McMeekin et al., 2003).

Mammary

The effects of estrogens on mammary tissue in normal adult animals and women are not typically as clear and robust as those effects observed in the uterus. However, clearly a majority of mammary tumors are ER-positive and respond favorably to estrogen antagonism (Wakeling et al., 1987). The role of estrogen replacement in the risk of breast cancer continues to be a controversial topic, and clearly the concomitant use of progestins in most hormonal replacement therapy paradigms is confounding. In the WHI trial, a significant increase in breast cancer risk in postmenopausal women was a key finding (Writing Group for WHI Investigators, 2002), contributing to dramatic reductions in the use of estrogen replacement strategies for various postmenopausal indications. The role of progestin in this observation was suggested by a follow-up study of hysterectomized women using unopposed conjugated equine estrogens for more than 7 years, where no increase in breast cancer risk was observed (Stefanick et al., 2006). Although some controversy remains over the exact potential for breast cancer risk with estrogen therapies, it is clear that concern over this risk has limited patient compliance with steroidal estrogen therapies. A major advantage of the SERM class of molecules is the lack of this cancer concern with respect to mammary tissue. Indeed, SERMs as a class have demonstrated a benefit in either treating or preventing breast cancer in animal models and women. The effects of SERMs in normal mammary tissue, as breast cancer therapies and preventatives, will be summarized here.

In normal mammary tissue, the effects of SERMs are largely unnoticeable. Extensive use of tamoxifen in both premenopausal and postmenopausal women for the use of breast cancer risk reduction has not been associated with untoward effects on mammary tissue, although in developing mammary glands in mice, tamoxifen impaired the growth of mammary ducts and increased mammary alveolar development (Hovey et al., 2005). A cautionary note in interpreting animal data with respect to SERMs and mammary tissue is worth mentioning. The rat is unique in that male mammary tissue exhibits a layering of epithelial cells with a somewhat hypertrophic appearance, which is an androgen-dependent phenotype in rats. SERMs with highly effective estrogen antagonism in mammary tissue can sufficiently block estrogen influence on mammary epithelia in rats, permitting the low ambient level of androgen in the female rat to predominate and convert the mammary histological phenotype to resemble that of the male rat (Rudmann et al., 2005), a phenomenon known not to exist in humans. Caution should be taken not to confuse this female/male phenotype conversion in the rat with hypertrophy of mammary tissue.

The antitumor effects of SERMs can be demonstrated readily in mammary tumor cell lines (i.e., MCF-7, T-47D, ZR-75-1) in vitro. Each SERM depicted in Fig. 37.3 is an estrogen antagonist in one or all of these cell lines (Short et al., 1996; Simard et al., 1997; Komm et al., 2005). The MCF-7 human mammary tumor cell line is an excellent, estrogen-dependent, in vitro system for determining antagonism of estrogen-induced proliferative activity of compounds. Most SERMs inhibit estrogen-stimulated proliferation in the 0.2−1 nM range in this assay, although for those triphenylethylene SERMs requiring the formation of active metabolites, such as tamoxifen and toremifene, one must evaluate the active metabolite to see this level of potency. For example, the IC_{50} for tamoxifen in MCF-7 cells is 200, vs. 1.2 nM for 4-hydroxytamoxifen (Suh et al., 2001). Des-methylarzoxifene, a likely active metabolite of arzoxifene, is the most potent inhibitor of MCF-7 proliferation, with an IC_{50} value of 0.05 nM (vs. 0.4 nM IC_{50} for arzoxifene in this cell line; Suh et al., 2001). Also of relevance in the MCF-7 tumor cell line, raloxifene fails to induce proliferation of these cells in the absence of exogenous estrogen, contrasting raloxifene with other SERMs, such as tamoxifen, that produce a low level of MCF-7 proliferation in estrogen-deficient cell culture media (Sato et al., 1995).

Consistent with the mammary tumor cell culture work are the estrogen antagonist effects of SERMs in animal models of estrogen-dependent breast cancer. Various animal models have shown the ability of SERMs such as tamoxifen, toremifene, and raloxifene to blunt the growth of established mammary tumors induced by carcinogens such as dimethylbenzanthracene (Clemens et al., 1983; Robinson et al., 1988) or in breast cancer tumor cell line xenografts in athymic mice (Fuchs-Young et al., 1997; Qu et al., 2000). Tamoxifen, raloxifene, and lasofoxifene are also effective in preventing mammary tumors induced by other chemical carcinogens, such as nitrosomethylurea (Gottardis et al., 1987; Anzano et al., 1996; Cohen et al., 2001). Of great interest is the apparent increase in mammary tumor efficacy that has been demonstrated preclinically for some of the more recently developed SERMs. For example, arzoxifene, a third-generation SERM molecule, produced significantly improved efficacy in preventing mammary tumor growth in vivo (Suh et al., 2001).

With respect to treatment of human breast cancer, numerous options have been in use that employ endocrine-based strategies, with some patients considered suitable for estrogen reduction or antagonism strategies alone (i.e., estrogen-positive tumors) and others using endocrine manipulation approaches as an adjunct to traditional tumor chemotherapy. Two SERMs, tamoxifen and toremifene, are approved for chemo/endocrine treatment in the management of postmenopausal women with node-positive breast cancer. With respect to breast cancer, tamoxifen has been in use the longest of any SERM, with greater than 20 years of clinical use. When used as an adjunct, tamoxifen reduces the risk of recurrent cancer and also decreases the risk of new tumors arising in the other breast. Both tamoxifen and toremifene show similar efficacy in terms of 5-year overall survival and disease-free survival rates (the disease-free rate for tamoxifen is 69%, and for toremifene is 72%) for early-stage breast cancer and also demonstrate similar efficacy as first-line therapy for metastatic breast cancer in postmenopausal women (Cuzick et al., 2002 and 2003). As of 2009, the only other SERM in late-stage clinical development with clinical assessment of breast cancer treatment potential was arzoxifene, which was evaluated in several phase 2 trials in advanced breast cancer patients, where some benefit in terms of time to progressive disease and clinical benefit rate were observed (Baselga et al., 2003).

The other significant application of SERMs is in breast cancer prevention. A number of environmental and genetic factors are associated with increased risk of developing breast cancer in women including advanced age, family history of breast cancer, and a greater lifetime estrogen exposure (assessed via surrogate indicators such as estradiol levels, use of estrogen therapy, age at menopause, and body mass index). One of the best tools for overall assessment of breast cancer risk is the Gail model, where a risk factor of greater than or equal to 1.67 defines a woman at risk (Costantino et al., 1989). Tamoxifen was the first SERM to show reduced risk of breast cancer through a number of large, placebo-controlled trials. In the Breast Cancer Prevention Trial, tamoxifen was evaluated in a cohort of 13,388 women at increased risk of breast cancer and produced a 49% reduction in the relative risk of invasive breast cancer, with a 69% reduced risk of ER-positive mammary tumors (Fisher et al., 1998). However, despite this substantial reduction in risk and inclusion of breast cancer risk reduction as an approved use for tamoxifen, the clinical use of tamoxifen for this indication has been rather lackluster—primarily owing to a side effects profile that tilts the risk/benefit ratio in a negative direction in the minds of most physicians and women. Side effects profiles will be reviewed in subsequent sections, but for tamoxifen, side effects include endometrial cancer, uterine sarcoma, stroke, venous thrombus events, and cataracts. The increase in endometrial cancer in postmenopausal women likely stems from the uterine stimulatory properties of tamoxifen and represents one area for improvement in other SERMs. In this regard, raloxifene received approval for reducing the risk of invasive breast cancer in postmenopausal women with osteoporosis and in postmenopausal women at high risk for invasive breast cancer. After 8 years of monitoring 4011 postmenopausal women with osteoporosis, a 66% reduction in the incidence of invasive breast cancer was observed with raloxifene use (Martino et al., 2004). In the Study of Tamoxifen and Raloxifene (STAR) trial, a head-to-head comparison of the two SERMs was conducted in 19,000 postmenopausal women at high risk of breast cancer, where tamoxifen and raloxifene were found to produce similar reductions in the incidence of invasive breast cancer (Vogel et al., 2006), with the primary benefit being a reduced risk of ER-positive invasive breast cancers (Barrett-Conner et al., 2006). The most significant differences between raloxifene and tamoxifen in the STAR trial were significantly fewer uterine-associated adverse events with raloxifene (most notably the lack of endometrial cancer), whereas tamoxifen appeared to have a greater effect on noninvasive breast cancer incidence than raloxifene (Vogel et al., 2006). These differences between tamoxifen and raloxifene, although subtle, indicated a difference from preclinical and even early clinical indicators, and as such demonstrated the need for thorough clinical evaluation before accurate therapeutic risk/benefit assessment and approval of indications can be made for human use. In this regard, several SERMs in development, such as acolbifene and bazedoxifene (Labrie et al., 2004; Adachi et al., 2007), had preclinical and early clinical profiles that were promising for potential use in reduction of the risk for breast cancer, but clinical evaluation was needed to predict the ultimate utility of these molecules in this regard.

The mechanism by which SERMs such as tamoxifen, raloxifene, and toremifene inhibit breast cancer development or progression is likely the result of multiple beneficial effects. Direct antagonism of estrogen action at the ER in target cells in breast tissue, as previously described, is a key component of the action of the SERMs, given the strong positive linkage of estrogen exposure to relative risk for developing breast cancer and the fact that SERMs are much more effective versus ER-positive breast cancers. However, it is also likely that antitumor effects that reduce estrogen bioavailability as well as effects independent of estrogen contribute to the ultimate anti−breast cancer effect. Raloxifene is known to elevate levels of sex hormone-binding globulin (Reindollar et al., 2002), which would be expected to reduce bioavailable estrogen levels and thereby further reduce the risk of estrogen-associated breast cancer. Other beneficial indirect SERM effects include (1) modification of signaling proteins with a role in tumor cell biology, as is observed with tamoxifen on protein kinase C, TGFβ, calmodulin, ceramide, and MAP-kinases (Mandlekar and Kong, 2001); (2) induction of apoptosis in mammary tumor cell lines, as with tamoxifen, raloxifene, and toremifene (Mandlekar and Kong, 2001; Diel et al., 2002; Houvinen et al., 1993); and (3) dampening of growth factor systems known to play a role in tumor progression. The effect of growth factors in the pathogenesis of breast cancer continues to attract considerable attention. The insulin-like growth factor (IGF) system, inclusive of signaling factors, IGF receptors, and IGF-binding proteins, has strong connections to the malignant transformation of normal breast epithelium and thus is implicated in the development and progression of breast cancer (Ward et al., 1994). The IGF system protects cancer cells from apoptosis, thus promoting their survival. Beneficial effects of SERMs on the IGF system were demonstrated in postmenopausal women after 2 years of raloxifene treatment. Raloxifene reduced circulating IGF-1 levels and increased IGF-binding protein-3 levels, which would be expected to be associated with reductions in bioavailable IGF-1 (Lasco et al., 2006).

Mammographic density of the breast depends on the contributions of the predominant cell types in the breast: stromal, epithelial (both of which are higher-density tissue types), and fat tissue (which is relatively radiolucent by standard mammography). Although the role of breast density in the prediction of relative breast cancer risk remains a topic of debate, most investigators agree that breast cancer risk is higher in women with higher-density breast tissue (Boyd et al., 1995). Whether this is related to a protective effect of increased abundance of fatty tissue or simply impedance in detecting small amounts of cancerous tissue in higher-density breast is not clear. However, it is clear that estrogen use is associated with a significant increase in mammographic density (Breendale *et al.*, 1999). Conversely, SERMs, specifically raloxifene and arzoxifene, were associated with a reduction of breast density in postmenopausal women (Lasko et al., 2006; Kimler et al., 2006).

Other

In addition to the uterus and mammary tissue, other components of the female reproductive system are under the direct or indirect control of estrogen and thus are also susceptible to the effects of SERMs. Depending on the patient, those SERM effects may be desirable, undesirable, or neutral.

Ovarian effects

The ovary is an important regulator of cyclicity in the female reproductive system and the predominant source of circulating estrogen, and it is indirectly regulated via estrogen action on the hypothalamic–pituitary–ovarian (HPO) axis. In mice, 15- to 24-month exposure to tamoxifen, toremifene, or raloxifene is associated with ovarian tumors. However, there is no evidence of increased ovarian cancer risk with these agents in women. Careful evaluation of multiple randomized, placebo-controlled studies indicates that raloxifene did not increase ovarian cancer in postmenopausal women compared with placebo (Neven et al., 2002). Thus, the murine observations may be owing to species differences in ovarian responses in terms of tumor development between mice and humans. The endocrine system plays a key role in the production of ovarian tumors in mice, particularly when gonadotropins are sustained at elevated levels for extended periods (Murphy et al., 1973). In this regard, raloxifene produces a sustained, dose-related increase in serum luteinizing hormone (LH) levels and inhibition of ovarian follicle maturation in mice, both effects being reversible upon discontinuation of the SERM (Cohen et al., 2000). In vitro, raloxifene is an antagonist of 17β-estradiol in pituitary gonadotrophs (Ortmann et al., 1988), suggesting that the elevation of LH with SERMs is related to blockade of the feedback-inhibitory properties of estrogen on the HPO axis.

Hormonal effects

In women, the most striking effects of SERMs on ovarian function are primarily through hormonal effects on the HPO axis. Because ovarian function obviously has major differences between the premenopausal and postmenopausal state, the effects of SERMs are likewise different with each state of ovarian function. The HPO axis is a complex endocrine system for which normal physiology requires the interplay of a number of steroid and peptidyl hormones including GnRH from the hypothalamus, follicle-stimulating hormone (FSH), prolactin and LH from the anterior pituitary, and estradiol as a hormone primarily originating from the ovary (although multiple tissue types possess the necessary enzymes to interconvert estrogen and testosterone as well as to generate multiple estrogen metabolites).

In premenopausal women, SERMs predominantly exert a stimulatory effect on the HPO axis and ovulation, with stimulatory effects on GnRH, FSH, and LH and marked increases in serum estradiol levels (Adashi et al., 1996). Depending on the situation, this may be a desired therapeutic goal or an undesirable side effect. This effect has been taken advantage of for more than 40 years with the widespread use of clomiphene for induction of ovulation in women with ovulatory dysfunction who desire pregnancy. For this indication, clomiphene has been used acutely to induce ovulation and increase the number of follicles produced in a given cycle. Use of clomiphene for more than 12 cycles in women who failed to become pregnant has been associated with an increase in ovarian cancer, leading to the recommendation to limit use of this agent to no more than six cycles if a pregnancy does not occur (Rossing et al., 1994). The other two SERMs currently approved for use in premenopausal women are tamoxifen and toremifene, used in the management of breast cancer. In these women, chronic treatment with the SERM results in ovarian-associated adverse effects. Of note are the induction of ovarian cysts and high circulating levels of estradiol that frequently occur with tamoxifen in reproductive-age women (Cook et al., 1995). Although these ovarian cysts can cause discomfort, they rarely require surgical intervention.

By far, the predominant use of SERMs in postmenopausal women has been for osteoporosis treatment or prevention (raloxifene), breast cancer treatment (tamoxifen or toremifene), and breast cancer risk reduction (tamoxifen or raloxifene). After menopause or in estrogen-deficient OVX animals, estrogen levels drop, and the lack of negative feedback provided by estrogen on the hypothalamus-anterior pituitary leads to an increase in FSH and LH levels or pulses. In general, SERMs exhibit a partial agonist effect on the HPO axis in postmenopausal women. Raloxifene, tamoxifen, and toremifene are associated with reductions in LH and FSH levels in postmenopausal women (Cheng et al. 2004; Ellmen et al., 2003). HPO axis-related adverse effects, such as ovarian cysts, have not been typically seen in postmenopausal women and have been reported only occasionally with tamoxifen use (Shushan et al., 1996).

Another important estrogen-regulated hormonal product of the anterior pituitary is prolactin. In premenopausal women, raloxifene failed to alter circulating prolactin levels (Faupel-Badger et al., 2006). In postmenopausal women, raloxifene (Cheng et al., 2004) and clomiphene (Garas et al., 2006) reduced serum prolactin, operating either directly on lactotropes in the anterior pituitary or via increasing opiatergic tone in the hypothalamus (Lasco et al., 2002).

Vaginal effects

The drop in circulating estradiol levels associated with menopause is responsible for various vaginal-related symptoms including itching, dryness, and dyspareunia. Various forms of estrogen and hormone replacement, via both systemic and local delivery routes, have been widely used to provide relief for postmenopausal women who have these symptoms. SERMs demonstrate a range of profiles on vaginal symptoms of menopause. Most reports in postmenopausal women indicated estrogen-like maturation of vaginal epithelial cells with tamoxifen and toremifene (Ellmen et al., 2003; Friedrich et al., 1998). Conversely, there was some indication for antagonism of estrogen influence on vaginal tissue because both toremifene and tamoxifen produced a fourfold increase in vaginal dryness in postmenopausal women (Marttunen et al., 2001) and toremifene partially antagonized the effect of estrogen on vaginal epithelium in postmenopausal women (Homesley et al., 1993). The background estrogen status of the individual may be an important determinant, because in women with a high estrogenic activity (with respect to hormonal cytology), tamoxifen produced no vaginal estrogen-like effects, whereas women with lower estrogen levels saw an increase in vaginal tissue maturation index (Shiota et al., 2002). Even with these modest estrogen-like effects produced by toremifene and tamoxifen on the vagina, no beneficial or untoward urogenital effects of either agent had been observed in postmenopausal women (Marttunen et al., 2001). Raloxifene use in postmenopausal women was associated with a more neutral profile on vaginal tissue in postmenopausal women because it failed to affect vaginal epithelium in this population (Komi et al., 2005) and was not associated with adverse vaginal symptoms (Davies et al., 1999). Raloxifene also did not antagonize the beneficial effect of vaginal estrogen cream or an estradiol-releasing ring on vaginal atrophy or sexual function in postmenopausal women (Kessel et al., 2003). One SERM that was extensively studied for potential beneficial effects on postmenopausal vaginal-associated symptoms in phase 3 clinical trials is lasofoxifene. In clinical trials, lasofoxifene improved the vaginal epithelium maturation index (as was observed with tamoxifen and toremifene) but also reduced vaginal pH and reduced the incidence of dyspareunia (Portman et al., 2004; Bachmann et al., 2004). These unique effects of lasofoxifene might be related to the increased vaginal mucus formation observed in vaginal tissue from OVX rats treated with lasofoxifene (Wang et al., 2006)—effects that were not observed with raloxifene and tamoxifen.

Cardiovascular system

Menopause is associated with a dramatic increase in cardiovascular disease and characteristic changes in a number of heart disease-associated risk factors, such as an increase in serum cholesterol and specifically LDL-cholesterol, a decline in the cardioprotective HDL-cholesterol, elevated levels of lipoprotein(a) [Lp(a)], and increased insulin resistance and fat mass, resulting in a classic "metabolic syndrome" profile for heart disease risk (Spencer et al., 1997). Unfavorable markers of vascular endothelial damage, such as elevations in homocysteine levels and C-reactive protein (CRP), also accompany declining ovarian function during menopause (Hak et al., 2000). The fact that women enjoy a relatively "cardioprotected" status prior to menopause relative to their male counterparts historically presented the impression that estrogen replacement should be beneficial for reducing cardiovascular disease in postmenopausal women. Certainly, in both estrogen-deficient animal models and postmenopausal women, estrogen produces a number of effects that would be associated with an improvement in risk for cardiovascular disease, such as reduction of total circulating cholesterol levels, LDL-cholesterol in particular, and a concomitant increase in HDL-cholesterol. Accordingly, observational clinical trials suggested a significant reduction of cardiovascular disease in women who used hormone replacement therapy after menopause (i.e., Stampfer and Colditz, 1991). Thus, it was an unanticipated result when randomized clinical trials conducted to confirm the beneficial effects of hormone replacement therapy on the cardiovascular system indicated the opposite effect—an increase in cardiovascular disease associated with estrogen/progestin use. The first trial to indicate this was the HERS trial, where no reduction in coronary heart disease was detected, but an increase in cardiovascular-related deaths in the initial year of therapy was observed (Hulley et al., 1998). Similarly, the WHI trial failed to demonstrate any beneficial effects of estrogen use, but again an increased risk of adverse cardiovascular outcomes including increased incidence of myocardial infarction and stroke (Manson et al., 2003). Finally, the Women's Estrogen for Stroke Trial (WEST) indicated an early increase in risk of fatal stroke during the first year of estrogen use in women with preexisting cerebrovascular disease (Viscoli et al., 2001). Both observational and randomized, placebo-controlled studies in postmenopausal women indicated a significant increase in the risk of deep venous thrombosis and pulmonary embolism in association with hormone replacement (Manson et al., 2003).

Thus, it is clear that two important considerations related to SERMs needed to be addressed. First, and foremost, the cardiovascular safety profile of SERMs in postmenopausal women has to be thoroughly evaluated based on the risk ascribed to hormone replacement therapy by trials such as HERS, WHI, and WEST. The second consideration revolves around potential beneficial effects on cardiovascular risk factors, which might be instructive in determining the additional benefit of these molecules. However, if we are to learn a lesson from the hormone replacement field, one needs to be careful to rely on randomized, well-controlled trials to make these assessments. The cardiovascular safety of SERMs in clinical use and their potential cardiovascular benefits are discussed here.

Cardiovascular safety of selective estrogen receptor modulators

The primary adverse event associated with every chronically used ER ligand is the occurrence of venous thrombolic events (VTEs), typically as deep vein thromboses or pulmonary emboli. The relative frequency of VTEs with SERMs is typically two- to threefold greater than for placebo when assessed in randomized clinical trials, a rate that is comparable to that observed with the use of oral estrogen replacement therapy (Cosman and Lindsay, 1999). Such an increase in venous thromboses was described with tamoxifen and toremifene use in breast cancer patients (Cuzick et al., 2003; Harvey et al., 2006) and raloxifene use in osteoporotic patients (Duvernoy et al., 2005). Comparable rates of venous thromboses were also reported in the 2- and 3-year phase 3 clinical trials for lasofoxifene and bazedoxifene (Adachi et al., 2007; McClung et al., 2006). The incident rate of VTEs was elevated in postmenopausal women subjected to prolonged periods of inactivity (i.e., extended bed stay during invasive surgical procedures), and as such, SERMs used chronically carry a recommendation to discontinue drug therapy during periods of anticipated immobility of several hours or more (Cuzick et al., 2002; Seeman, 2001). The mechanism for the increased incidence of VTEs with SERMs is not clear, although a number of clinical trials noted procoagulant changes with estrogen replacement, tamoxifen, or raloxifene along with impairment of anticoagulant factors with these three regimens. Although there was some variability in the results of different studies with respect to specific factor changes, such as fibrinogen that is increased by raloxifene in some reports (Sgarabotto et al., 2006) but decreased in others (Walsh et al., 1998), some clear trends have emerged. In a randomized, placebo-controlled trial in healthy postmenopausal women (Cosman et al., 2005), estrogen replacement increased coagulation factor VII and reduced the anticoagulation factors antithrombin and plasminogen activator inhibitor-1 (PAI-1). Tamoxifen generated an overall procoagulant profile, although via an increase in clotting factor VIII, factor IX, and von Willebrand factor on the coagulant side, and decreases in antithrombin, protein C, and PAI-1 on the anticoagulant side. Raloxifene produced a different pattern of changes with some similarity to that of tamoxifen, although without elevation in clotting factor IX or reduction of protein C (Cosman et al., 2005). In a separate study, Dahm et al. (2006) demonstrated that estrogen replacement, tamoxifen, and raloxifene all acted to reduce human endothelial production of tissue factor pathway inhibitor-1 (TFPI), an anticoagulant factor. Thus, some clear trends have emerged as the most critical determinants for VTE occurrence with estrogen or SERMs, those being the reduction of important factors such as antithrombin, PAI-1 and TFPI, all anticoagulant functional proteins. However, this is more of a correlative hypothesis because no chronically used SERM seemed to avoid the 2%–3% rate of VTE occurrence. No preclinical models or predictors are available to predict the relative likelihood for VTEs in humans with SERMs or estrogens.

With respect to the more severe cardiovascular adverse events observed with hormone replacement in the HERS, WHI, and WEST clinical trials, reports were mixed regarding the incidence of stroke or myocardial infarction with various SERMs. Although a coronary heart disease neutral profile was reported initially with tamoxifen in the breast cancer prevention trial, an increased risk for stroke with tamoxifen was eventually observed (Reis et al., 2001). The long-term clinical trials that supported raloxifene approval for osteoporosis indications found no change in the incidence of myocardial infarction relative to placebo (Martino et al., 2005). Analysis on a per-year basis showed no increase in cardiovascular events in the first year of raloxifene use, which contrasts with the pattern reported for estrogen replacement in the HERS and WHI trials (Keech et al., 2005). Cardiovascular events were the primary outcome of the Raloxifene Use for The Heart, or RUTH, trial, where no increase in coronary events or stroke incidence was observed, but a statistically significant increase in stroke-associated mortality was reported (Barrett-Conner et al., 2006).

Potential cardiovascular benefit of selective estrogen receptor modulators

Although the large, randomized, placebo-controlled trials conducted with SERMs did not show a significant reduction in cardiovascular events, smaller trials and subsets of larger placebo-controlled trials demonstrated some favorable trends, leaving open the possibility that SERMs may offer a degree of cardioprotection. With tamoxifen, one clinical study concluded a slight reduction in cardiac death and reduced risk for myocardial infarction (Rutqvist et al., 1993). In the raloxifene osteoporosis registration clinical trial, a positive effect on the incidence of cardiovascular events was observed in a subset of women at high risk for heart disease (Barrett-Connor et al., 2002). Despite the lack of verifiable cardiovascular outcomes with SERMs, most SERMs evaluated in both clinical and preclinical settings showed largely favorable effects on most cardiovascular risk factors for heart disease including lipid metabolism, clotting factors, and vessel wall factors. However, the "fingerprint" of each SERM on the wide array of cardiovascular surrogates is strikingly distinct, suggesting that each SERM needs careful clinical evaluation before a complete assessment of cardiovascular benefit or risk can be ascribed.

The most clear and robust effect observed with SERMs on cardiovascular-relevant parameters in preclinical models was the reduction of serum cholesterol levels (i.e., raloxifene, Black et al., 1994; tamoxifen, Sato et al., 1996a,b; clomiphene, Turner et al., 1998; toremifene, Qu et al., 2000; acolbifene, Martel et al., 2000; lasofoxifene, Ke et al., 1998; bazedoxifene, Komm et al., 2005; arzoxifene, Palkowitz et al., 1997). As with the skeletal responses in OVX rats, all SERMs depicted in Fig. 37.3 were capable of reducing serum cholesterol by roughly the same magnitude, thus mimicking the response to estrogen in this animal model, although differences in potency are evident. This hypocholesterolemic effect of SERMs is mediated by the ER, as demonstrated in the case of raloxifene by a very close correlation of ER-binding affinity and cholesterol-lowering in vivo for a series of raloxifene analogues (Kauffman et al., 1997).

In clinical trials, most of the SERMs depicted in Fig. 37.3 reduced total serum cholesterol and LDL-cholesterol (Reid et al., 2004; Joensuu et al., 2000; McClung et al., 2006) even in hypertriglyceridemic women (Dayspring et al., 2006), with raloxifene and tamoxifen also reducing Lp(a) (Love et al., 1994; Mijatovic et al., 1999)—all effects that are cardioprotective with respect to cardiovascular disease risk factors. Neither tamoxifen nor raloxifene elevated HDL-cholesterol (Love et al., 1994; Walsh et al., 1998), a cardiovascular-beneficial effect of estrogen replacement. Triglyceride elevation, an undesired effect often associated with estrogen replacement, was increased in most clinical trials with tamoxifen, although a triglyceride-neutral profile was observed with raloxifene and toremifene (Walsh et al., 1998; Kusama et al., 2004).

Preclinical models focused on the vessel wall have produced beneficial effects after SERM administration. In rats, neointimal thickening following aortic denudation injury was reduced by either raloxifene or tamoxifen, with both SERMs shown to regulate vascular smooth muscle cell function, indicating a potential benefit against restenosis following percutaneous transluminal coronary angioplasty (Savolainen-Peltonen et al., 2004). Even though SERMs are linked to increases in VTEs in clinical studies, preclinical work has indicated a beneficial effect of raloxifene in a model of carotid artery thrombosis. In OVX mice, estrogen deficiency amplifies thrombosis following carotid photochemical injury. In this model system, raloxifene as well as estrogen significantly reduces intraarterial thrombosis prolonging time to occlusion, likely via a mechanism that involves reduced platelet adhesion and increased expression of COX-2 (Abu-Fanne et al., 2008). Vascular antiinflammatory effects of raloxifene were also linked to vasorelaxant properties (Pinna et al., 2006).

Cardiovascular disease surrogates associated with vessel wall function and inflammation were also generally improved in clinical studies following raloxifene and tamoxifen. CRP, a marker linked to vascular injury that reflects inflammatory activity in the vascular wall, was elevated with estrogen replacement (Cushman et al., 1999) but reduced with raloxifene and tamoxifen (Cushman et al., 2001; Walsh et al., 2000), as was homocysteine (Anker et al., 1995; De Leo et al., 2001), a circulating factor linked to toxicity of vascular endothelial cells. Improved endothelial function with tamoxifen (Stamatelopoulos et al., 2004) and reduction in endothelin-1 (Saitta et al., 2001), an endogenous vasoconstrictor, have also been described.

Clearly, other factors in addition to circulating lipids influence the ultimate potential cardiovascular benefit provided by compounds such as the SERMs. In this regard, direct effects of SERMs on cardiovascular tissue have been investigated extensively, and agents such as raloxifene produced a number of effects of potential cardiovascular benefit. For example, raloxifene produced an antioxidant effect on serum lipoproteins (Zuckerman and Bryan, 1996), inhibited vascular smooth muscle migration (Wiernicki et al., 1996), and elevated vascular endothelial cell nitric oxide production (Saitta et al., 2001). Preclinical studies in models of atherogenesis have provided mixed results. Raloxifene failed to prevent the reduction of the coronary artery intimal area in cholesterol-fed OVX monkeys (Clarkson et al., 1998). However, 6-month exposure to raloxifene did reduce aortic atherogenesis in cholesterol-fed OVX rabbits (Bjarnason et al., 1997) and improved the coronary artery intimal area in OVX sheep (Gaynor et al., 2000). Given that nearly all direct cardiovascular studies with SERMs in postmenopausal women have been conducted with raloxifene, it is difficult to know whether these particular effects can be generalized across the SERM class or whether, as in the uterus, distinct cardiovascular SERM profiles might emerge. Additional research would be needed.

Central nervous system

The mixed biological results observed with estrogen in the cardiovascular system are paralleled by the profile of estrogen activity in the central nervous system (CNS). Extensive in vitro and animal studies suggest neuroprotective and other beneficial effects of estrogen on central processes ranging from cognition to fine motor control and mood. However, translational work to human neurodegenerative disease has failed to corroborate the preclinical data, and as in the cardiovascular system, suggests potential untoward effects of estrogen on the human CNS.

Clearly, ER is broadly distributed throughout the brain. Original thoughts were that ER was predominantly restricted to the hypothalamus and associated with well-known functions such as regulation of reproductive hormones in the periventricular nucleus and thermoregulation in the lateral hypothalamus. The discovery of the ERβ subtype and improved antibodies for the detection of ERβ, however, led to the reconstruction of ER distribution neuroanatomical charts to include higher brain regions such as the cortex and hippocampus, specifically neurons associated with learning and memory, such as pyramidal cells throughout the CA1 and CA3 regions of the hippocampus and cortex (Shugrue and Merchenthaler, 2001).

Estrogen produces a number of beneficial effects in vitro and in animal models. In neuronal cell culture, 17β-estradiol has neurotrophic effects and inhibits neuronal damage induced by neurotoxicants (O'Neill et al., 2004). In OVX rats, estrogen increases hippocampal choline acetyltransferase activity, which leads to increased levels of acetylcholine, a neurotransmitter associated with cognition (Wu et al., 1999). Estrogen also reduces neural damage after experimental forebrain ischemia (Simpkins et al., 1997) and produces beneficial effects in animal models of Parkinson's disease (Gomez-Mancilla and Bedard, 1992). Estrogen also affects central serotonergic neurotransmission in animal models, leading to increased serotonin production and firing and reduced degradation (Bethea et al., 2002a,b). Dopaminergic neurotransmission in the striatum is influenced by estrogen (Landry et al., 2002), as are glutamate receptors (Cyr et al., 2001).

The preclinical data are consistent in indicating a beneficial effect of estrogen on CNS function. However, clinical data have not supported this conclusion. Certainly, early observational studies suggested up to a 30% reduction in risk of dementia with estrogen use (Yaffe et al., 1998). However, because these trials were not well-controlled prospective studies, there was considerable room for group-related artifacts (e.g., educational status) that may affect data interpretation. In this regard, in large, randomized, placebo-controlled trials, estrogen not only failed to reduce the incidence of dementia but also was associated with increased risk of dementia and stroke (Shumaker et al., 2003). Additional data are clearly needed, because the role of progestin in studies where hormone replacement is employed is a complicating factor—although even with estrogen-only use, a similar profile was observed (Shumaker et al., 2004). Other potential effects of estrogen on the brain (e.g., on mood) are equally controversial. Clearly, in consideration of SERMs, careful attention must be paid not only to potential beneficial effects but also to CNS safety.

Central nervous system safety of selective estrogen receptor modulators

No outwardly neurotoxic effects of tamoxifen, raloxifene, or any SERM depicted in Fig. 37.3 have been reported in neuronal cell culture and animal studies. Some differences among various SERMs have been reported with respect to their ability to antagonize estrogen effects in the brain. For example, tamoxifen blocks the effect of estrogen on serotonin 2A receptor expression in the forebrain or dorsal raphe of rats (Sumner et al., 1999), whereas raloxifene fails to show an antagonist profile on this receptor subtype in these brain regions (Cyr et al., 2000).

In clinical studies, use of tamoxifen for 5 years in women for breast cancer therapy did not alter performance on a series of cognitive tests compared with breast cancer patients who had never used tamoxifen, although an increase in physician visits for memory problems was noted (Paganini-Hill and Clark, 2000b). The confounding variable of ongoing disease state in these women must be considered, though, because studies conducted in elderly women found no differences in mental functional tests or speed of response with tamoxifen use (Ernst et al., 2002). In pilot studies conducted to evaluate the safety of raloxifene on cognitive function in postmenopausal women with osteoporosis, no negative effects on memory function were detected as determined by a number of mental acuity tests (Nickelsen et al., 1999). Follow-up work in more than 7700 postmenopausal women with osteoporosis confirmed the initial observation of no negative effects of raloxifene on cognitive performance (Yaffe et al., 2001). Consistent with the latter observation, no untoward effects of raloxifene were observed on mood, sexual behavior, or sleep in postmenopausal women.

Central nervous system efficacy of selective estrogen receptor modulators

Much as with estrogen, SERMs are largely associated with preclinical profiles that suggest neuroprotection and an overall positive CNS profile, although certain subtle differences can be observed among different SERMs, further indicating the need to thoroughly evaluate each specific SERM molecule. Positive effects of SERMs can be demonstrated in vitro; as in a neural cell line, raloxifene increased neurite outgrowth in culture (Nilsen et al., 1998). Raloxifene and tamoxifen were neuroprotective in a neuroepithelial cell line by conferring resistance to β-amyloid-induced toxicity via elevation of seladin-1 (Benvenuti et al., 2005), a factor known to be downregulated in brain regions affected by Alzheimer's disease (Greeve et al., 2001).

Neurotransmitter-related changes within the CNS by various SERMs can be similar for some transmitters in some brain regions but differ in other neurotransmitter systems. Much as with estrogen, raloxifene and tamoxifen increase hippocampal choline acetyltransferase activity (Wu et al., 1999). Indicating comparable effect of these two SERMs on the neurotransmission of acetylcholine, a neurotransmitter associated with cognition. Raloxifene and tamoxifen, however, produce different overall profiles on serotonin neurotransmission in the brain. In both rats and monkeys, raloxifene produces a spectrum of changes in the forebrain that are favorable for serotonin neurotransmission, such as increased tryptophan hydroxylase, increased serotonin 2A receptor expression, and reduced serotonin transporter (Cyr et al., 2000; Smith et al., 2004). In contrast, tamoxifen does not affect serotonin transporter and reduces tryptophan hydroxylase activity in the forebrain (Sumner et al., 1999), a profile unfavorable for serotonin neurotransmission. Of note, the SERM arzoxifene produces a pattern of effects on forebrain serotonin neurotransmission that parallels that produced by raloxifene (Bethea et al., 2002a,b). Other differences on neurotransmitter profiles can be observed with dopaminergic systems, as with lateral striatum dopamine receptor expression, which is increased in response to raloxifene but not affected by tamoxifen (Landry et al., 2002).

Distinct SERM efficacy profiles can also be detected in various CNS pathology animal models. Hippocampal neurodegeneration can be induced in rats by systemic injection of kainic acid, which leads to induction of inflammatory astroglia and neural loss that is protected by pretreatment with estrogen. In OVX rats injected with kainic acid, hippocampal neural loss was prevented in each case by tamoxifen, raloxifene, and bazedoxifene without affecting the reactive gliosis component (Ciriza et al., 2004), a profile distinct from that of 17β-estradiol, which was both neuroprotective and antiinflammatory in this model. Again, not all SERMs behaved the same in this model system, as lasofoxifene failed to exhibit a neuroprotective profile (Ciriza et al., 2004). Neuroprotective effects of SERMs in other animal models have also been observed. In separate studies, tamoxifen and arzoxifene reduced neural damage following occlusion of the middle cerebral artery in OVX rats, an experimental model for stroke (Mehta et al., 2003; Rossberg et al., 2000). The mechanism for SERM protection from focal cerebral ischemia in these animal models is unclear, because raloxifene induced a relaxation of rat cerebral arteries in vitro via inhibition of L-type calcium channels (Tsang et al., 2004), although direct effects on cerebral blood flow were ruled out in one study (Rossberg, 2000). Attenuation of excitatory amino acid release and putative antioxidant effects have also been demonstrated as potentially contributing to the neuroprotective effects of tamoxifen and raloxifene (Osuka et al., 2001; Siefer et al., 1994). Finally, neuroprotective effects of tamoxifen and raloxifene were observed in 1-methyl-4-phenyl-1,2,3,6-tetrahydropyridine-induced dopamine depletion in a Parkinsonian model (Obata and Kubota, 2001; Grandbois et al., 2000).

A number of metabolic-based imaging and cognitive performance clinical trials suggested potential CNS benefits with some SERMs. Proton magnetic resonance spectroscopy in elderly women who had taken tamoxifen for at least 2 years for breast cancer treatment demonstrated a reduction in myoinositol, a glial marker that reflects glial proliferation in response to brain injury (Ernst et al., 2002), suggesting a neuroprotective effect of tamoxifen. Although the disease status of these women complicated interpretation of the data, it was interesting to note that in a comparable population of women taking estrogen replacement therapy but without breast cancer, a similar effect on myoinositol was observed (Ernst et al., 2002). Assessment of adverse CNS events in the randomized, placebo-controlled osteoporosis registration studies for raloxifene revealed no negative effects during the course of safety assessment. These studies did suggest some interesting trends for raloxifene-related improvement in cognitive performance, specifically higher verbal memory and attention scores; decreases in these areas are putative harbingers of mild cognitive impairment and Alzheimer's disease (Nickelsen et al., 1999; Yaffe et al., 2001). Women over the age of 70 in particular experienced smaller declines in memory and attention on raloxifene (Yaffe et al., 2005). In studies focused on the risk for Alzheimer's disease in postmenopausal women with osteoporosis, Yaffe and colleagues (2005) reported a one-third reduction in the risk of mild cognitive impairment with a trend for reduced risk of Alzheimer's disease. In both studies where cognitive improvement was suggested with raloxifene, it is worth noting that the benefit was primarily observed in women receiving 120 mg/day of raloxifene. These benefits of raloxifene were not observed in similar women taking 60 mg/day of raloxifene, the standard and approved daily dose for osteoporosis prevention and treatment and breast cancer risk reduction. The potential requirement for a greater dose of raloxifene to generate meaningful CNS benefits was consistent with preclinical literature, such as increased hippocampal choline acetyltransferase activity, which also requires higher doses of raloxifene than are necessary for the bone effects in OVX rats (Wu et al., 1999), and the recognition that raloxifene has poor penetration of the blood–brain barrier (Bryant et al., 1997). The relative estrogen background may be a complicating factor because women with undetectable circulating estrogen levels prior to administration of raloxifene showed a greater benefit for mild cognitive impairment risk than women with higher circulating estrogen levels at baseline (Yaffe et al., 2005). The precise process of cognitive function that SERMs affect would require additional research. Although executive function/decision-making is the ultimate output of cognitive networks, these activities depend on more basic processes, such as alertness and arousal, to be operational. In this regard, it is interesting to note that in a study conducted in a small cohort of elderly men using functional magnetic resonance imaging, raloxifene improved memory function via increased arousal during initial encoding of information, likely via a neurogenic effect (Goekoop et al., 2006).

The important principle of drug exposure for SERMs will be reviewed in the next section, but particularly germane to the brain, as a central pharmacological tenant, is that for agents to exert their effect on a receptor, they must be available to the receptor. The limited brain exposure with raloxifene predicted by preclinical models is certainly an important factor. ER-binding studies also suggest reduced exposure of tamoxifen in the brain relative to peripheral tissues, such as the uterus (Bowman et al., 1982). Clearly, the functional studies with various SERMs in animal models with raloxifene and tamoxifen argue strongly that there is sufficient exposure to exert a biological response to these SERMs in the brain. However, critical factors such as dose, duration, and agonist/antagonist potential may be severely or subtly affected by the ability of the molecule to penetrate the brain. As a result, dose–response relationships that are expected based on peripheral tissues may not be monitored for effects within the CNS.

General safety profile and other pharmacological considerations

Other safety

In addition to the adverse events already reviewed (i.e., VTEs observed with all SERMs and uterine stimulation observed with some SERMs), SERMs are associated with other untoward effects that should be considered in the risk/benefit decision for each patient. Of these other adverse events observed in clinical trials, leg cramps and the induction of hot flushes were observed in women, to at least some extent, with all the SERM molecules depicted in Fig. 37.3. Hot flushes, or vasomotor symptoms, are a hallmark indicator of menopausal transition occurring in up to 70% of US women. The incidence of hot flushes is likely reflective of declining or changing estrogen status and typically abates when circulating estrogen levels reach their postmenopausal steady-state concentration. However, in a small percentage of women, vasomotor symptoms can be severe and extend well into menopause. Estrogen replacement clearly is effective in relieving postmenopausal hot flushes. With SERM use, however, it is likely that an estrogen withdrawal-like response is initiated, producing a state similar to that experienced by women who are estrogen depleted and have subsequent hot flushes. Although it is unclear whether this phenomenon is caused by the estrogen antagonist or agonist properties of the SERM at the ER in hypothalamic thermoregulatory centers, SERM-induced hot flushes are transient in nature, because with continued use in most cases, this side effect subsides typically within 6 months, likely as the thermoregulatory set point reestablishes (Tataryn et al., 1980). Consistent with this proposed mechanism, proximity to the climacteric state may influence the incident rate and severity of SERM-induced hot flushes. In postmenopausal women over the age of 55, significantly fewer hot flushes were observed in response to raloxifene compared with those observed in younger postmenopausal women. In a similar context, tamoxifen-induced hot flushes tended to be more severe in premenopausal breast cancer patients than in postmenopausal patients.

One problem in the assessment of the incident rate of hot flush induction following SERM administration is the relatively high placebo response rate in the postmenopausal population. In observational studies performed on randomized, placebo-controlled trials, the rate of reported hot flushes in postmenopausal women receiving placebo was 21% over a 30-month trial period (Cohen and Lu, 2000). In this study, the incidence of hot flushes in postmenopausal women using raloxifene was 28%. Others have confirmed an approximately 7% increase in hot flush incidence as a side effect of raloxifene use (Davies et al., 1999). Of note, the hot flushes induced by raloxifene were in the mild to moderate category in terms of severity, as severe hot flushes in postmenopausal women using raloxifene occurred at a rate indistinguishable from that of placebo controls. Finally, the increase in hot flush incidence with raloxifene was transient, because no differences relative to placebo controls were observed after 6 months (Davies et al., 1999).

The observation of hot flush incidence is typical for other SERM molecules as well. Tamoxifen use in both premenopausal and postmenopausal women for breast cancer treatment has long been associated with hot flushes as a side effect in 10%–20% of patients and is more common in women with higher estrogen levels (Legha, 1988). Toremifene also induces hot flushes at rates equivalent to or slightly greater than that of tamoxifen (Hays et al., 1995). Clomiphene use for induction of ovulation increases hot flushes as well (Derman and Adashi, 1994). Levormeloxifene (Alexanderson et al., 2001), lasofoxifene (McClung et al., 2006), and bazedoxifene (Adachi et al., 2007) also increase hot flush incidence in postmenopausal women.

An interesting note of relevance to hot flushes and SERMs is the potential application of SERMs in combination with estrogen for the treatment of hot flushes. In a small study of postmenopausal women using 17β-estradiol in combination with raloxifene, a significant decrease in hot flushes was observed (compared with raloxifene treatment alone), although signs of endometrial stimulation were detected as well with this combination (Stoval et al., 2007). Large, randomized, placebo-controlled clinical trials have evaluated the combination of bazedoxifene and conjugated equine estrogens as a potential alternative for the treatment of hot flushes (and potentially other menopausal symptoms) without the need to include a progestin for the maintenance of uterine safety. Uterine assessment at 1- and 2-year intervals demonstrated a lack of endometrial hyperplasia for the bazedoxifene/conjugated equine estrogens combination (Pickar et al., 2007), indicating that the SERM had effectively blocked the potent stimulatory action of the estrogenic component of the combination. The combination also produced an increase in lumbar spine BMD in postmenopausal women that was superior to that of both placebo and raloxifene (Lindsay et al., 2007). Full assessment of the risk/benefit ratio of the bazedoxifene/conjugated equine estrogens combination for use in the relief of menopausal symptoms requires full assessment of hot flush and other symptom efficacy as well as effects on other adverse events, such as VTEs.

Other adverse events not already discussed and associated with some triphenylethylene SERMs are those associated with the eye. These ocular-related untoward effects include retinopathy, macular crystal formation, corneal keratopathies, and cataracts. Eye pathologies have been primarily associated with the use of tamoxifen and toremifene, with comparable incidence of 7%–10% (Hays et al., 1995). Visual disturbances were also noted with clomiphene use (Asch and Greenblat, 1976) but were not increased with nontriphenylethylene SERMs, such as with raloxifene use in postmenopausal women (Cohen and Lu, 2000). Length of therapy increased the risk of cataract development with tamoxifen use (Paganini-Hill and Clark, 2000a). One mechanism that has been proposed to address cataract formation with tamoxifen is the blockade of the chloride channels in the lens of the eye, which are important for the maintenance of lens hydration (Zhang et al., 1994). Of importance, this effect of tamoxifen on chloride channels in the lens seems to be independent of interaction with ER, and as such is likely off-target toxicity for certain triphenylethylene SERMs.

Pharmacokinetics

Pharmacokinetic properties of the SERMs are an important consideration in the overall effects of these molecules and served as a focal point for the development of novel and improved agents. For example, the third-generation SERMs arzoxifene, bazedoxifene, and lasofoxifene all have pharmacokinetic properties that represent improvements over raloxifene and tamoxifen. The ultimate advantage of these improvements to patients would require additional research, but as noted earlier in this chapter, much of the research and development has been redirected.

As previously indicated, tamoxifen generates active metabolite(s) with greater affinity and efficacy at the ER (Coezy et al., 1982). After a single oral dose of tamoxifen, maximal plasma levels are reached within several hours with an elimination half-life of 5–7 days (Fromson et al., 1973). Steady-state concentrations are attained within 3–4 weeks of chronic dosing (Adam et al., 1980). The potential for preferential tissue distribution was suggested in animal studies, where greater levels of radioactive tamoxifen were detected in mammary gland, uterus, and liver than in blood (Furr and Jordan, 1984). In humans, elevated uterine levels of tamoxifen with respect to circulating concentrations were observed, with endometrial levels twice those of myometrial or cervical concentrations (Fromson and Sharp, 1974). The primary route of elimination of tamoxifen follows hepatic metabolism, primarily glucuronidation, and subsequent biliary excretion with fairly little of the parent molecule being excreted in the urine and the potential for enterohepatic recirculation suggested in animal studies (Furr and Jordan, 1984).

Toremifene also is a triphenylethylene SERM that generates active metabolites (i.e., deaminohydroxy-toremifene; DeGregorio et al., 2000). Peak plasma concentrations of toremifene occur within 2–4 h after a single oral dose with nearly complete absorption, and plasma levels are linear with dose over a fairly wide dose range (Anttila et al., 1990). The elimination half-life of toremifene is 5 days, and steady-state circulating concentrations are reached within 2–4 weeks of chronic dosing. There is evidence in animal models for relatively increased tissue distribution of toremifene in some tissues, with mammary gland uptake similar to tamoxifen (Kargas et al., 1989). Toremifene is extensively metabolized in the liver via demethylation, hydroxylation, and side-chain oxidation modifications (Anttila et al., 1990), and the primary route of elimination is fecal elimination following enterohepatic recirculation (Anttilla et al., 1990). Hepatic impairment significantly increases the half-life of toremifene, nearly doubling it (Anttila et al., 1995).

Clomiphene is the only triphenylethylene SERM in current clinical use for which there are no known active metabolites. Clomiphene is a racemic mixture of cis- (zuclomifene) and trans- (enclomiphene) isomers that is rapidly absorbed following oral administration. Peak plasma concentrations are achieved in approximately 6 h, with a half-life of approximately 5 days (Dickey and Holtkamp, 1996), although metabolites have been detected up to 6 weeks after a single dose, suggesting likely enterohepatic recirculation (Kausta et al., 1997). Clomiphene is hepatically metabolized and fecally excreted (Adashi, 1996).

The pharmacokinetics of raloxifene, a benzothiophene SERM, have some features similar to those of triphenylethylenes but also some considerable differences. Like tamoxifen and its relatives, raloxifene is rapidly absorbed from the gastrointestinal tract after oral administration, with peak blood levels attained in approximately 6 h and 60% of the oral dose absorbed (Heringa, 2003). Raloxifene is also highly bound to plasma proteins (approximately 95%; Heringa, 2003). However, in contrast to triphenylethylenes, the elimination half-life of raloxifene is considerably shorter at 28 h, and there are no known active metabolites of raloxifene in humans or rodents. Although there is virtually no P450 metabolism of raloxifene in the liver, it is extensively metabolized by first-pass hepatic glucuronidation, yielding an absolute oral bioavailability of only approximately 2% in humans (Snyder et al., 2000). Raloxifene is widely distributed, and as with triphenylethylenes, very little is excreted in the urine, with the bulk of clearance through biliary excretion and loss in the feces (Knadler et al., 1995).

Relatively less information is available on pharmacokinetic profiles of the third-generation SERMs arzoxifene, bazedoxifene, and lasofoxifene. Lasofoxifene demonstrates improved bioavailability because the molecule was designed to resist intestinal wall glucuronidation (Gennari et al., 2006). The elimination half-life of lasofoxifene is 165 h, and there is a linear relationship between plasma concentrations and dose (Gardner et al., 2006). The primary metabolic route for lasofoxifene is hepatic oxidation and subsequent conjugation (Branson et al., 2006). Bazedoxifene produces an absolute bioavailability of 6.2% following an oral dose, which is about threefold greater than that produced by raloxifene. Bazedoxifene demonstrates kinetic linearity at dose levels of 5–40 mg (Ermer et al., 2003). Maximal circulating concentrations of bazedoxifene are achieved within 1–2 h after oral exposure, and the elimination half-life is approximately 28 h with steady-state circulating levels attained in 7 days (Ermer et al., 2003). As with raloxifene, there is very little P450-mediated metabolism of bazedoxifene, and glucuronidation is the major metabolic route. The primary route of elimination is the feces, with evidence for enterohepatic recirculation (Chandrasekaran et al., 2003). Arzoxifene also exhibits pharmacokinetic advantages over raloxifene. Over a dosage range of 10–100 mg, blood levels of arzoxifene increase linearly with respect to dose, with maximal levels attained at 2–6 h and an elimination half-life of 30–35 h over this dose range (Munster et al., 2001). Arzoxifene is metabolized (demethylated) to desmethylarzoxifene, which is an active metabolite with a high binding affinity for the ER (Rash and Knadler, 1997).

Future directions with selective estrogen receptor modulators

In addition to the use of SERMs for osteoporosis, breast cancer, and the induction of ovulation, the ability of ER-activity-modulating agents to favorably impact other diseases and syndromes, in both women and men, is under discovery and clinical development efforts. The overall scope of these drug discovery efforts is broad, changing rapidly, and beyond the focus of this review.

Summary

SERMs are a diverse class of molecules that affect a broad spectrum of biological systems, with potential therapeutic benefit for a variety of diseases. Current concern over long-term use of estrogen-containing regimens has created an opportunity for the application of SERMs to chronic indications such as osteoporosis treatment or prevention. The unique SERM profile also potentially allows their use in other chronic indications of interest to postmenopausal women, most notably breast cancer risk reduction and treatment. However, safety considerations are very important for SERM use in these chronic indications. The pleiotropic nature of the ER and its role in numerous physiological systems raise the importance of considering potential SERM benefits and/or adverse events in the cardiovascular system and other tissues.

References

Abu-Fanne, R., Brzezinski, A., Golomb, M., Grad, E., Foldes, A.J., Shufaro, Y., Varon, D., Brill, A., Lotan, C., Danenberg, H.D., 2008. Effects of estradiol and raloxifene on arterial thrombosis in ovariectomized mice. Menopause 15. Published online 2007 Nov 19 [Epub ahead of print].

Adachi, J.D., Chesnut, C.H., Brown, J.P., Christiansen, C., Russo, L.A., Fernandes, C.E., Menegoci, J.C., King, A., Chines, A.A., Bessac, L., Chakrabarti, D., 2007. Safety and tolerability of bazedoxifene in postmenopausal women with osteoporosis: results from a 3-year, randomized, placebo- and active-controlled clinical trial. J. Bone Miner. Res. 22 (Suppl. 1), S460.

Adam, H.K., Patterson, J.S., Kemp, J.V., 1980. Studies on the mechanism and pharmacokinetics of tamoxifen in normal volunteers. Cancer Treat. 64, 761–764.

Adashi, E.Y., 1996. Ovulation induction: clomiphene citrate. In: Adahsi, E.Y., Rock, J.A., Rosenwalks, Z. (Eds.), Reproductive Endocrinology, Surgery and Technology. Lippincott-Raven, Philadelphia, pp. 1181–1206.

Adrian, M.D., Cole, H.W., Shetler, P.K., Rowley, E.R., Magee, D.E., Pell, T., Zeng, G., Sato, M., Bryant, H.U., 1996. Comparative pharmacology of a series of selective estrogen receptor modulators. J. Bone Miner. Res. 11 (Suppl. 1), S447.

Alexandersen, P., Riss, B.J., Stakkestad, J.A., Delmas, P.D., Christiansen, C., 2001. Efficacy of levormeloxifene in the prevention of postmenopausal bone loss and on the lipid profile compared to low dose hormone replacement therapy. J. Clin. Endocrinol. Metab. 86, 755–760.

Allen, M.R., Iwata, K., Sato, M., Burr, D.B., 2006. Raloxifene enhances vertebral mechanical properties independent of bone density. Bone 39, 1130–1135.

Anker, G., Lonning, P.E., Ueland, P.M., Refsum, H., Lien, E.A., 1995. Plasma levels of the atherogenic amino acid homocysteine in post-menopausal women with breast cancer treated with tamoxifen. Int. J. Cancer 60, 365–368.

Anttila, M., Valavaara, R., Krivinen, S., Maenpaa, J., 1990. Pharmacokinetics of toremifene. J. Steroid Biochem. 36, 249–252.

Anttila, M., Laakso, S., Nylandern, P., Sotaniemi, E.A., 1995. Pharmacokinetics of the novel antiestrogenic agent toremifene in subjects with altered liver and kidney function. Clin. Pharmacol. Ther. 57, 628–635.

Anzano, M.A., Peer, C.W., Smith, J.M., Mullen, L.T., Schrader, W.M., Logsdon, D.L., Driver, C.L., Brown, C.C., Roberts, A.B., Sporn, M.B., 1996. Chemoprevention of mammary carcinogenesis in the rat: combined use of raloxifene and 9-cis-retinoic acid. J. Natl. Cancer Inst. 88, 23–25.

Asch, R.H., Greenblatt, R.B., 1976. Update on the safety and efficacy of clomiphene citrate as a therapeutic agent. J. Reprod. Med. 17, 175–180.

Bachmann, G.A., Gass, Moffett, A., Portman, D., Symons, J., 2004. Lasofoxifene improves symptoms associated with vaginal atrophy. Menopause 11, 669.

Barrett-Connor, E., Grady, D., Sashegyi, A., Anderson, P.W., Cox, D.A., Hoszowski, K., Rautaharju, P., Harper, K.D., 2002. Raloxifene and cardiovascular events in osteoporotic women: four-year results from MORE (Multiple Outcomes of Raloxifene Evaluation) randomized trial. J. Am. Med. Assoc. 287, 847–857.

Barrett-Connor, E., Mosca, L., Collins, P., Geiger, M.J., Grady, D., Kornitzer, M., McNabb, M., Wenger, N., 2006. Effects of raloxifene on cardiovascular events and breast cancer in postmenopausal women. N. Engl. J. Med. 335, 125–137.

Baselga, J., Llombart-Cussa, A., Bellet, M., Guillem-Porta, V., Enas, N., Krejcy, K., Carrasco, E., Kayitalire, L., Kuta, M., Lluch, A., Vodvarka, P., Kerbrat, P., Namer, M., Petruzelka, L., 2003. Randomized, double-blind, multicenter trial comparing two doses of arzoxifene (LY353381) in hormone-sensitive advanced or metastatic breast cancer patients. Ann. Oncol. 14, 1383–1390.

Bashir, A., Mak, Y.T., Sankaralingam, S., Cheung, J., McGowan, N.W.A., Grigoriadis, A.E., Fogelman, I., Hampson, G., 2005. Changes in RANKL/OPG/RANK gene expression in peripheral mononuclear cells following treatment with estrogen or raloxifene. Steroids 70, 847–855.

Bautista, S., Valles, H., Walker, R.L., Anzick, S., Zellinger, R., Meltzer, P., Theillet, C., 1998. In breast cancer, amplification of the steroid receptor coactivator gene AIB1 is correlated with estrogen and progesterone receptor positivity. Clin. Cancer Res. 4, 2925–2929.

Benvenuti, S., Luciani, P., Vannelli, G.B., Gelmini, S., Franceschi, E., Serio, M., Peri, A., 2005. Estrogen and selective estrogen receptor modulators exert neuroprotective effects and stimulate the expression of Selective Alzheimer's Disease Indicator-1, a recently discovered anti-apoptotic gene, in human neuroblast long-term cell cultures. J. Clin. Endocrinol. Metab. 90, 1775–1782.

Bethea, C.L., Lu, N.Z., Gundlah, C., Streicher, J.M., 2002a. Diverse actions of ovarian steroids in the serotonin neural system. Front. Neuroendocrinol. 23, 41–100.

Bethea, C.L., Mirkes, S.J., Su, A., Michelson, D., 2002b. Effects of oral estrogen, raloxifene and arzoxifene on gene expression in serotonin neurons of macaques. Psychoneuroendocrinology 27, 431–445.

Bjarnason, N.H., Haarbo, J., Byrjalsen, I., Kauffman, R.F., Christiansen, C., 1997. Raloxifene inhibits aortic accumulation of cholesterol in ovariectomized, cholesterol-fed, rabbits. Circulation 96, 1964–1969.

Black, L.J., Rowley, E.R., Bekele, A., Sato, M., Magee, D.E., Williams, D.C., Cullinan, G.J., Bendele, R., Kauffman, R.F., Bensch, W., Frolik, C.A., Termine, J.D., Bryant, H.U., 1994. Raloxifene (LY139482 HCl) prevents bone loss and reduces serum cholesterol without causing uterine hypertrophy in ovariectomized rats. J. Clin. Investig. 93, 63–69.

Bowman, S.P., Leake, A., Morris, I.D., 1982. Hypothalamic, pituitary and uterine cytosolic and nuclear oestrogen receptors and their relationship to the serum concentrations of tamoxifen and its metabolite, 4-hydroxytamoxifen, in the ovariectomized rat. J. Endocrinol. 94, 167–175.

Boyd, N.F., Byng, J.W., Jong, R.A., Fishell, E.K., Little, L.E., Miller, A.B., Lockwood, G.A., Tritchler, D.L., Yaffe, M.J., 1995. Quantitative classification of mammographic densities and breast cancer risk: results from the Canadian National Breast Screening Study. J. Natl. Cancer Inst. 87, 670–675.

Brady, H., Doubleday, M., Gayo-Fung, L.M., Hicman, M., Khammungkhune, S., Kois, A., Lipps, S., Pierce, S., Richard, N., Shevlin, G., Sutherland, M.K., Anderson, D.W., Bhagwat, S.S., Stein, B., 2002. Differential response of estrogen receptors alpha and beta to SP500263, a novel potent selective estrogen receptor modulator. Mol. Pharmacol. 61, 562–568.

Bramson, C., Ouellet, D., Roman, D., Randinitis, E., Gardner, M.J., 2006. A single-dose pharmacokinetic study of lasofoxifene in healthy volunteers and subjects with mild and moderate hepatic impairment. J. Clin. Pharmacol. 46, 29–36.

Bramlett, K.S., Burris, T.P., 2003. Target specificity of selective estrogen receptor modulators within human endometrial cancer cells. Steroid Biochem. Mol. Biol. 86, 27–34.

Bryant, H.U., Bales, K.R., Paul, S.M., Yang, H., Cole, H.W., Walker-Daniels, J., McEwen, R.S., Chow, H., Santerre, R.F., 1997. Estrogen agonist effects of selective estrogen receptor modulators in the ovariectomized rat brain. Soc. Neurosci. Abstr. 23, 2377.

Brzozowski, A.M., Pike, A.C., Dauter, Z., Hubbard, R.E., Bonn, T., Engstrom, O., Ohman, L., Green, G.L., Gustafsson, J.A., 1997. Molecular basis of agonism and antagonism in the estrogen receptor. Nature 389, 753–768.

Burke, T.W., Walker, C.L., 2003. Arzoxifene as therapy for endometrial cancer. Gynecol. Oncol. 90 (Pt 2), S40–S46.

Cano, A., Dapia, S., Noguera, I., Pineda, B., hermenegildo, C., del Val, R., Caeiro, J.R., Garcia-Perez, M.A., 2008 Jun. Comparative effects of 17b-estradiol, raloxifene and genistein on bone 3D microarchitecture and volumetric bone mineral density in the ovariectomized mice. Osteoporos. Int. 19 (6), 793–800.

Carthew, P., Edwards, R.E., Nolan, B.M., Tucker, M.J., Smith, L.L., 1999. Compartmentalized uterotrophic effects of tamoxifen, toremifene and estradiol in the ovariectomized Wistar rat. Toxicol. Sci. 48, 197–205.

Chandrasekaran, A., Ermer, J., McKenad, W., Lee, H., DeMaio, W., Kotake, A., Sullivan, P., Orczyk, G., Scantina, J., 2003. Bazedoxifene acetate metabolic disposition in healthy postmenopausal women. J. Clin. Pharm. Ther. 73, 47.

Cheng, W.C., Yen, M.L., Hsu, S.H., Chen, K.H., Tsai, K.S., 2004. Effects of raloxifene, one of the selective estrogen receptor modulators, on pituitary-ovary axis and prolactin in postmenopausal women. Endocrine 23, 215–218.

Chesnut, C., Weiss, S., Mulder, H., Wasnich, R., Greenwald, R., Eastell, R., Fitts, D., Jensen, C., Haines, A., MacDonald, B., 1998. Idoxifene increases bone mineral density in osteopenic postmenopausal women. Bone 23 (Suppl. 1), S389.

Ciriza, I., Carrero, P., Azcoitia, I., Lundeen, S.G., Garcia-Segura, L.M., 2004. Selective estrogen receptor modulators protect hippocampal neurons from kainic acid excitotoxicity: differences with the effect of estradiol. J. Neurobiol. 61, 209–221.

Clarkson, T.B., Anthony, M.S., Jerome, C.P., 1998. Lack of effect of raloxifene on coronary artery atherosclerosis of postmenopausal monkeys. J. Clin. Endocrinol. Metab. 83, 721–726.

Clemens, J.A., Bennett, D.R., Black, L.J., Jones, C.D., 1983. Effects of a new antiestrogen, keoxifene (LY156758), on growth of carcinogen-induced mammary tumors and on LH and prolactin levels. Life Sci. 32, 2869–2875.

Coezy, E., Borgna, J.L., Rochefort, H., 1982. Tamoxifen and metabolites in MCF-7 cells: correlation between binding to estrogen receptor and inhibition of cell growth. Cancer Res. 42, 317–323.

Cohen, F.J., Lu, Y.M., 2000. Characterization of hot flashes reported by healthy postmenopausal women receiving raloxifene or placebo during osteoporosis prevention trials. Maturitas 34, 65–73.

Cohen, I.R., Sims, M.L., Robbins, M.R., Lakshmanan, M.C., Francis, P.C., Long, G.G., 2000. The reversible effects of raloxifene on luteinizing hormone levels and ovarian morphology in mice. Reprod. Toxicol. 14, 37–44.

Cohen, L.A., Pittman, B., Wang, C.X., Aliaga, C., Yu, L., Moyer, J.D., 2001. LAS, a novel selective estrogen receptor modulator with chemopreventative and therapeutic activity in the N-nitroso-N-methylurea-induced rat mammary tumor model. Cancer Res. 61, 8683–8688. I.

Cole, H.W., Adrian, M.D., Shetler, P.K., Sato, M., Rowley, E.R., Glasebrook, A.L., Short, L.L., Grese, T.A., Palkowitz, A.D., Thrasher, K.J., Bryant, H.U., 1997. Comparative pharmacology of high potency selective estrogen receptor modulators: LY353381•HCL and CP-336,156. J. Bone Miner. Res. 12 (Suppl. 1), S349.

Cook, L.S., Weiss, N.S., Schwartz, S.M., White, E., McKnight, B., Moore, D.E., Daling, J.R., 1995. Population-based study of tamoxifen therapy and subsequent ovarian, endometrial, and breast cancers. J. Natl. Cancer Inst. 87, 1259–1364.

Cosman, F., Lindsay, R., 1999. Selective estrogen receptor modulators: clinical spectrum. Endocr. Rev. 20, 418–434.

Cosman, F., Baz-Hecht, M., Cushman, M., Vardy, M.D., Cruz, J.D., Nieves, J.W., Zion, M., Lindsay, R., 2005. Short-term effects of estrogen, tamoxifen and raloxifene on hemostasis: a randomized-controlled study and review of the literature. Thromb. Res. 116, 1–13.

Costantino, J.P., Gail, M.H., Pee, D., Anderson, S., Redmond, C.K., Benichou, J., Wieand, H.S., 1999. Validation studies for models projecting the risk of invasive and total breast cancer incidence. J. Natl. Cancer Inst. 91, 1541–1548.

Cushman, M., Legault, C., Barrett-Connor, E., Stefanick, M.L., Kessler, C., Judd, H.L., Sakkunen, P.A., Tracy, R.P., 1999. Effect of postmenopausal hormones on inflammation-sensitive proteins: the postmenopausal estrogen/progestin interventions (PEPI) study. Circulation 100, 717–722.

Cushman, M., Costantino, J.P., Tracy, R.P., Song, K., Buckle, L., Roberts, J.D., Krag, D.N., 2001. Tamoxifen and cardiac risk factors in healthy women: suggestion of an anti-inflammatory effect. Arterioscler. Thromb. Vasc. Biol. 21, 255–261.

Cuzick, J., Forbes, J., Edwards, R., Baum, M., Cawthorn, S., Contes, A., Hamed, H., Howell, A., Powles, T., 2002. First results from the international breast cancer intervention study (IBIS-I): a randomized prevention trial. Lancet 360, 817–824.

Cuzick, J., Powles, T., Veronesi, U., Forbes, J., Edwards, R., Ashley, S., Boyle, P., 2003. Overview of the main outcomes in breast-cancer prevention trials. Lancet 361, 296–300.

Cyr, M., Landry, M., DiPaolo, T., 2000. Modulation of estrogen receptor directed drugs of 5-hydroxytryptamine-2A receptors in rat brain. Neuropsychopharmacology 23, 69–78.

Cyr, M., Thibault, C., Morissette, M., Landry, M., DiPaolo, T., 2001. Estrogen-like activity of tamoxifen and raloxifene on NMDA receptor binding and expression of its subunits in rat brain. Neuropsychopharmacology 25, 242–257.

Dahm, A.E.A., Iversen, N., Birkenes, B., Ree, A.H., Sandset, P.M., 2006. Estrogens, selective estrogen receptor modulators, and a selective estrogen down-regulator inhibit endothelial production of tissue factor pathway inhibitor I. BMC Cardiovasc. Disord. 6, 40–48.

Davies, G.C., Huster, W.J., Lu, Y., Plouffe, L., Lakshmanan, M., 1999. Adverse events reported by postmenopausal women in controlled trials with raloxifene. Obstet. Gynecol. 93, 558–565.

Dayspring, T., Qu, Y., Keech, C., 2006. Effects of raloxifene on lipid and lipoprotein levels in postmenopausal osteoporotic women with and without hypertriglyceridemia. Metab. Clin. Exp. 55, 972–979.

de Boer, T., Otjens, D., Muntendam, A., Meulman, E., van Oostijen, M., Ensing, K., 2004. Development and validation of fluorescent receptor assays based on the human recombinant estrogen receptor subtypes alpha and beta. J. Pharm. Biomed. Anal. 34, 671–679.

De Leo, V., La Marca, A., Morgante, G., Lanzetta, D., Setaci, C., Petraglia, F., 2001. Randomized control study of the effects of raloxifene on serum lipids and homocysteine in older women. Am. J. Obstet. Gynecol. 184, 350–353.

DeGregoria, M.W., Wurz, G.T., Taras, T.L., Erkkola, R.U., Halonen, K.H., Huupponen, R.K., 2000. Pharmacokinetics of (deaminohydroxy)toremifene in humans: a new, selective estrogen receptor modulator. Eur. J. Clin. Pharmacol. 56, 469–475.

Delmas, P.D., Bjarnason, N.H., Mitlak, B.H., Ravoux, A.-C., Shah, A.S., Huster, W.J., Draper, M., Christiansen, C., 1997. Effects of raloxifene on bone mineral density, serum cholesterol concentrations and uterine endometrium in postemenopausal women. N. Engl. J. Med. 337, 1641–1647.

Delmas, P.D., Genant, H.K., Crans, G.G., Stock, J.L., Wong, M., Siris, E., Adachi, J.C., 2003. Severity of prevalent vertebral fractures and the risk of subsequent vertebral and nonvertebral fractures: results from the MORE trial. Bone 33, 522–532.

Derman, S.G., Adahi, E.Y., 1994. Adverse effects of fertility drugs. Drug Saf. 11, 408–421.

Dickey, R.P., Holtkamp, D.E., 1996. Development, pharmacology and clinical experience with clomiphene citrate. Hum. Reprod. Update 2, 483–506.

Diel, P., Olff, S., Schmidt, S., Michna, H., 2002. Effects of the environmental estrogens bisphenol A, o,p'-DDT, p-tert-octylphenol and coumestrol on apoptosis induction, cell proliferation and the expression of estrogen sensitive molecular parameters in the human breast cancer cell line MCF-7. J. Steroid Biochem. Mol. Biol. 80, 61–70.

Draper, M.W., Flowers, D.E., Huster, W.J., Nield, J.A., Harper, K.D., Arnaud, C., 1997. A controlled trial of raloxifene (LY139481) HCl: impact on bone turnover and serum lipid profile in healthy postmenopausal women. J. Bone Miner. Res. 11, 835–842.

Duvernoy, C.S., Kulkarni, P.M., Dowsett, S.A., Keech, C.A., 2005. Vascular events in the Multiple Outcomes of Raloxifene Evaluation (MORE) trial: incidence, patient characteristics, and effect of raloxifene. Menopause 12, 444–452.

Ellmen, J., Hakulinen, P., Partanen, A., Hayes, D.F., 2003. Estrogenic effects of toremifene and tamoxifen in postmenopausal breast cancer patients. Breast Canc. Res. Treat. 82, 103–111.

Ermer, J., McKeand, W., Sullivan, P., Parker, V., Orczyk, G., 2003. Bazedoxifene acetate dose proportionality in healthy, postmenopausal women. J. Clin. Pharm. Ther. 73, 43.

Ernst, T., Chang, L., Cooray, D., Salvador, C., Jovicich, J., Walot, I., Boone, K., Chlebowski, R., 2002. The effects of tamoxifen and estrogen on brain metabolism in elderly women. J. Natl. Cancer Inst. 94, 592–597.

Ettinger, B., Black, D.M., Mitlak, B.M., Knickerbocker, R.K., Nickelsen, T., Genant, H.K., Christiansen, C., Delmas, P.D., Zanchetta, J.R., Stakkestad, J., Gluer, C.C., Krueger, K., Cohen, F.J., Eckert, S., Ensrud, K.E., Avioli, L.V., Lips, P., Cummings, S.R., 1999. Reduction of vertebral fracture risk in postmenopausal women with osteoporosis treated with raloxifene. J. Am. Med. Assoc. 282, 637–645.

Evans, G., Bryant, H.U., Magee, D., Sato, M., Turner, R.T., 1994. The effects of raloxifene on tibia histomorphometry in ovariectomized rats. Endocrinology 134, 2283–2288.

Fanning, P., Kuehl, T., Lee, R., Pearson, S., Wincek, T., Pliego, J., Spiekeman, A., Bryant, H.U., Rippy, M., 1996. Video mapping to assess efficacy of an antiestrogen (raloxifene) on spontaneous endometriosis in the rhesus monkey, Macaca mulata. In: Juehl, T.J. (Ed.), Bunkley Day Proceedings, vol. 6. Texas A&M University Health Science Centre, Temple TX, pp. 51–61.

Faupel-Badger, J.M., Prindville, S.A., Venzon, D., Vonderhaar, B.K., Zujewski, J.A., Eng-Wong, J., 2006. Effects of raloxifene on circulating prolactin and estradiol levels in premenopausal women at high risk for developing breast cancer. Cancer Epidemiol. Biomark. Prev. 15, 1153–1158.

Feldman, S., Minne, H.W., Parvizi, S., Pfeifer, M., Lempert, U.G., Bauss, F., Ziegler, R., 1989. Antiestrogen and antiandrogen administration reduce bone mass in the rat. Bone Miner. 7, 245–254.

Feng, Q., O'Malley, B.W., 2014. Nuclear receptor modulation-Role of coregulators in selective estrogen receptor (SERM) actions. Steroids 90, 39–43.

Fisher, B., Constantino, J.P., Redmond, C.K., Fisher, E.R., Wickerham, D.L., Cronin, W.M., 1994. Endometrial cancer in tamoxifen treated breast cancer patients: findings from the National surgical adjuvant breast and Bowel Project. J. Natl. Cancer Inst. 86, 527–537.

Fisher, B., Costantino, J.P., Wickerham, D.L., Redmond, C.K., Kavanah, M., Cronin, W.M., Vogel, V., Robidoux, A., Dimitrov, N., Atkins, J., Daly, M., Wieand, S., Tan-Chiu, E., Ford, L., Wolmark, N., 1998. Tamoxifen for prevention of breast cancer: report of the National surgical adjuvant breast and Bowel Project P-1 study. J. Natl. Cancer Inst. 90, 1371–1388.

Fleischer, A.C., Wheeler, J.E., Yeh, I.T., Kravitz, B., Jensen, C., MacDonald, B., 1999. Sonographic assessment of the endometrium in osteopenic postmenopausal women treated with idoxifene. J. Ultrasound Med. 18, 503–512.

Friedrich, M., Mink, D., Villena-Heinsen, C., Woll-Hermann, A., Wagner, S., Schmidt, W., 1998. The influence of tamoxifen on the maturation index of vaginal epithelium. Clin. Exp. Obstet. Gynecol. 25, 121–124.

Frolik, C.A., Bryant, H.U., Black, E.C., Magee, D.E., Chandrasekhar, S., 1996. Time dependent changes in biochemical bone markers and serum cholesterol in ovariectomized rats: effects of raloxifene HCI, tamoxifen, estrogen and alendronate. Bone 18, 621–627.

Fromson, J.M., Sharp, D.S., 1974. The selective uptake of tamoxifen by human uterine tissue. J. Obstet. Gynaecol. Br. Commonw. 81, 321–323.

Fromson, J.M., Pearson, S., Branah, S., 1973. The metabolism of tamoxifen II: in female patients. Xenobiotica 3, 711–714.

Fuchs-Young, R., Iversen, P., Shetler, P., Layman, N., Hale, L., Short, L., Magee, D., Sluka, J., Glasebrook, A., Bryant, H.U., Palkowitz, A., 1997. Preclinical demonstrations of specific and potent inhibition of mammary tumor growth by new selective estrogen receptor modulators. Proc. Am. Assoc. Cancer Res. 38, 573.

Furr, B.J., Jordan, V.C., 1984. The pharmacology and clinical uses of tamoxifen. Pharmacol. Ther. 25, 127–205.

Galbiati, E., Caruso, P.L., Amari, G., Armani, E., Ghirardi, S., Delcanale, M., Civelli, M., 2002. Effects of 3-phenyl-4-[[4-[2-(1-piperidinyl) ethoxy] phenyl]methyl]-2H-1-benzopyran-7-ol (CHF 4056), a novel nonsteroidal estrogen agonist/antagonist, on reproductive and nonreproductive tissue. J. Pharmacol. Exp. Ther. 300, 802–809.

Garas, A., Trypsianis, G., Kallitsaris, G., Milingos, A., Messinis, I.E., 2006. Oestradiol stimulates prolactin secretion in women through oestrogen receptors. Clin. Endocrinol. 65, 638–642.

Gardner, M., Taylor, A., Wei, G., Calcagni, A., Duncan, B., Milton, A., 2006. Clinical pharmacology of multiple doses of lasofoxifene in postmenopausal women. J. Clin. Pharmacol. 46, 52–58.

Gaynor, J.S., Monnet, E., Selzman, C., Parker, D., Kaufman, L., Bryant, H.U., Mallinckrodt, C., Wrigley, R., Whitehill, T., Turner, A.S., 2000. The effect of raloxifene on coronary arteries in aged ovariectomized ewes. J. Vet. Pharmacol. Ther. 23, 175–179.

Gennari, L., Merlotti, D., Martini, G., Nuti, R., 2006. Lasofoxifene: a third-generation selective estrogen receptor modulator for the prevention and treatment of osteoporosis. Expert Opin. Investig. Drugs 15, 1091–1103.

Goldstein, S.R., Nanavati, N., 2002. Adverse events that are associated with the selective estrogen receptor modulator levormeloxifene in an aborted phase III osteoporosis treatment study. Am. J. Obstet. Gynecol. 187, 521–527.

Goekoop, R., Barkhof, F., Duschek, E.J.J., Netlenbos, C., Knol, D.L., Scheltens, P., Rombouts, S.A.R.B., 2006. Raloxifene treatment enhances brain activation during recognition of familiar items: A pharmacologic fMRI study in elderly males. Neuropsychopharmacology 31, 1508–1518.

Goldstein, S.R., Neven, P., Zhou, L., Taylor, Y.L., Ciaccia, A.V., Plouffe, L., 2001. The effect of raloxifene on the frequency of pelvic floor relaxation. Obstet. Gynecol. 98, 91–96.

Gomez-Mancilla, B., Bedard, P.J., 1992. Effect of estrogen and progesterone on L-DOPA induced dyskinesia in MPTP-treated monkeys. Neurosci. Lett. 135, 129–132.

Gottardis, M.M., Jordan, V.C., 1987. Antitumor actions of keoxifene and tamoxifen in the N-nitrosomethylurea induced rat mammary carcinoma model. Cancer Res. 47, 4020–4024.

Grandbois, M., Morissette, M., Callier, S., Di Paolo, T., 2000. Ovarian steroids and raloxifene prevent MPTP-induced dopamine depletion in mice. Neuroreport 11, 343–346.

Greeve, I., Hermans-Borgmeyer, I., Bellinger, C., Kasper, D., Gomez-Isla, T., Behl, C., Levkau, B., Nitsch, R.M., 2001. The human DIMINUTO/DWARF1 homolog seladin-1 confers resistance to Alzheimer's disease-associated neurodegeneration and oxidative stress. J. Neurosci. 20, 7345–7352.

Grese, T.A., Sluka, J.P., Bryant, H.U., Cullinan, G.C., Glasebrook, A.L., Jones, C.D., Matsumoto, K., Palkowitz, A.D., Sato, M., Termine, J.D., Winter, M.A., Yang, N.N., Dodge, J.A., 1997. Molecular determinants of tissue selectivity in estrogen receptor modulators. Proc. Natl. Acad. Sci. U.S.A. 94, 14105–14110.

Hak, A.E., Polderman, K.H., Westendorp, I.C., 2000. Increased plasma homocysteine after menopause. Atherosclerosis 149, 163–168.

Harvey, H.A., Kinura, M., Hajba, A., 2006. Toremifene: an evaluation of its safety profile. Breast 15, 142–157.

Hays, D.F., Van Zyl, J.A., Hacking, A., Goedhals, L., Bezwoda, W.R., Mailliand, J.A., Jones, S.E., Vogel, C.L., Berris, R.F., Shemano, I., 1995. Randomized comparison of tamoxifen and two separate doses of toremifene in postmenopausal patients with metastatic breast cancer. J. Clin. Oncol. 13, 2556–2566.

He, L., Xiang, H., Lu-Yong, Z., Wei-Sheng, T., Hong, H.H., 2005. Novel estrogen receptor ligands and their structure activity relationship evaluated by scintillation proximity assay for high throughput screening. Drug Discov. Res. 64, 203–212.

Heany, R.P., Draper, M.W., 1997. Raloxifene and estrogen: comparative bone-remodelling kinetics. J. Clin. Endocrinol. Metab. 82, 3425–3429.

Helvering, L.M., Adrian, M.D., Geiser, A.G., Estrem, S.T., Wei, T., Huang, S., Chen, P., Dow, E.R., Calley, J.N., Dodge, J.A., Grese, T.A., Jones, S.A., Halladay, D.L., Miles, R.R., Onyia, J.E., Ma, Y.L., Sato, M., Bryant, H.U., 2005a. Differential effects of estrogen and raloxifene on messenger RNA and matrix metalloproteinase 2 activity in the rat uterus. Biol. Reprod. 72, 830–841.

Helvering, L.M., Liu, R., Kulkarni, N.H., Wei, T., Chen, P., Huang, S., Lawrence, F., Halladay, D.L., Miles, R.R., Ambrose, E.M., Sato, M., Ma, Y.L., Frolik, C.A., Dow, E.R., Bryant, H.U., Onyia, J.E., 2005b. Expression profiling of rat femur revealed suppression of bone formation genes by treatment with alendronate and estrogen but not raloxifene. Mol. Pharmacol. 68, 1225–1238.

Hendrix, S.L., Cochrane, B.B., Nygaard, I.E., Handa, V.L., Barnabei, V.M., Iglesia, C., Aragaki, A., Naughton, M.J., Wallace, R.B., McNeely, S.G., 2005. Effects of estrogen with and without progestin on urinary incontinence. J. Am. Med. Assoc. 293, 935–948.

Heringa, M., 2003. Review on raloxifene: profile of a selective estrogen receptor modulator. Int. J. Clin. Pharmacol. Ther. 41, 331–345.

Homesley, H.D., Shemano, I., Gams, R.A., Harry, D.S., Hickox, P.G., Rebar, R.W., Bump, R.C., Mullin, T.J., Wentz, A.C., O'Toole, R.V., Lovelace, J.V., Lyden, C.C.T., 1993. Antiestrogenic potency of toremifene and tamoxifen in postmenopausal women. Am. J. Clin. Oncol. Cancer Clin. Trials 16, 117–122.

Houvinen, R., Warri, A., Collan, Y., 1993. Mitotic activity, apoptosis and TRPM-2 mRNA expression in DMBA-induced rat mammary carcinoma treated with anti-estrogen toremifene. Int. J. Cancer 55, 685–691.

Hovey, R.C., Asai-Sato, M., Warri, A., Terry-Koroma, B., Colyn, N., Ginsburg, E., Vonderhaar, B.K., 2005. Effects of neonatal exposure to diethylstilbesterol, tamoxifen, and toremifene on the BABL/c mouse mammary gland. Biol. Reprod. 72, 423–435.

Hulley, S., Grady, D., Bush, T., Furberf, C., Herrington, D., Riggs, B., Vittinghoff, E., 1998. Randomized trial of estrogen plus progestin for secondary prevention of coronary heart disease in postmenopausal women. Heart and estrogen/progestin replacement study (HERS) research group. J. Am. Med. Assoc. 280, 605–613.

Jimenez, M.A., Magee, D.E., Bryant, H.U., Turner, R.T., 1997. Clomiphene prevents cancellous bone loss from tibia of ovariectomized rats. Endocrinology 138, 1794–1800.

Jirecek, S., Lee, A., Pave, I., Crans, G., Eppel, W., Wenzl, R., 2004. Raloxifene prevents the growth of uterine leiomyomas in premenopausal women. Fertil. Steril. 81, 132–136.

Joensuu, H., Holli, K., Oksanen, H., Valavaara, R., 2000. Serum lipid levels during and after adjuvant toremifene or tamoxifen therapy for breast cancer. Breast Canc. Res. Treat. 63, 225–234.

Johnell, O., Scheele, W.M., Lu, Y., Reginster, J.-Y., Need, A.G., Seeman, E., 2002. Additive effects of raloxifene and alendronate on bone density and biochemical markers of bone remodeling in postmenopausal women with osteoporosis. J. Clin. Endocrinol. Metab. 87, 985–1002.

Kalu, D., Salerno, E., Liu, C.C., Echon, R., Ray, M., Gaza-Zepata, M., Hollis, B.W., 1991. A comparative study of the actions of tamoxifen, estrogen and progesterone in the ovariectomized rat. Bone Miner. 15, 109–124.

Kangas, L., Haaparanta, M., Paul, R., Roeda, D., Sipila, H., 1989. Biodistribution and scintigraphy of 14C-toremifene in rats bearing DMBA-induced mammary carcinoma. Pharmacol. Toxicol. 64, 373–377.

Kauffman, R.F., Bensch, W.R., Roudebush, R.E., Cole, H.W., Bean, J.S., Phillips, D., Monroe, A., Cullinan, G.J., Glasebrook, A.L., Bryant, H.U., 1997. Hypocholesterolemic activity of raloxifene (LY139481): pharmacological characterization as a selective estrogen receptor modulator (SERM). J. Pharmacol. Exp. Ther. 280, 146–153.

Kanis, J.A., Johnell, O., Black, D.M., Downs Jr., R.W., Sarkar, S., Fuerst, T., Secrest, R.J., Pavo, I., 2003. Effect of raloxifene on the risk of new vertebral fracture inpostmenopausal women with osteopenia or osteoporosis: a reanalysis of the Multiple Outcomes of Raloxifene Evaluation trial. Bone 33, 293–300.

Kausta, E., White, D., Franks, S., 1997. Modern use of clomiphene citrate in induction of ovulation. Hum. Reprod. Update 3, 359–365.

Ke, H.Z., Chen, H.K., Qi, H., Pirie, C.M., Simmons, H.A., Ma, Y.F., Jee, W.S.S., Thompson, D.D., 1995. Effects of droloxifene on prevention of cancellous bone loss and bone turnover in the axial skeleton of aged, ovariectomized rats. Bone 17, 491–496.

Ke, H.Z., Paralkar, V.M., Grasser, W.A., Crawford, D.T., Qi, H., Simmons, H.A., Pirie, C.M., Chidsey-Frink, K.L., Owen, T.A., Smock, S.L., Chen, H.K., Jee, W.S., Cameron, K.O., Rosati, R.L., Brown, T.A., Dasilva-Jardine, P., Tompson, D.D., 1998. Effects of CP336,156, a new, non-steroidal estrogen agonist/antagonist on bone, serum cholesterol, uterus and body composition in rat models. Endocrinology 139, 2068–2076.

Keech, C.A., Sashegyi, A., Barrett-Conner, E., 2005. Year-by-year analysis of cardiovascular events in the multiple outcomes of raloxifene evaluation (MORE) trial. Curr. Med. Res. Opin. 21, 135–140.

Kessel, B., Nachtigall, L., Plouffe, L., Siddhanti, S., Rosen, A., Parsons, A., 2003. Effect of raloxifene on sexual function in postmenopausal women. Climacteric 6, 248–256.

Kimler, B.F., Ursin, C., Fabian, J., Anderson, J.R., Chamberlain, C., Mayo, M.S., O'Shaughnessy, J.A., Lynch, H.T., Johnson, K.A., Browne, D., 2006. Effect of the third generation selective estrogen receptor modulator arzoxifene on mammographic breast density. J. Clin. Oncol. 24 (Suppl. 1), 562.

Knadler, M.P., Lantz, R.J., Gillespie, T.A., 1995. The disposition and metabolism of 14C-labelled raloxifene in humans. Pharm. Res. (N. Y.) 12 (Suppl. 1), 372.

Koda, M., Jarzabek, K., Haczynski, J., Knapp, P., Sulkowski, S., Woczynski, S., 2004. Differential effects of raloxifene and tamoxifen on the expression of estrogen receptors and antigen Ki-67 in human endometrial adenocarcinoma cell line. Oncol. Rep. 12, 517–521.

Komi, J., Lankinen, K.S., Harkonen, P., DeGregoria, M.W., Voipio, S., Kivinen, S., Tuimala, R., Vihtamaki, T., Vihko, K., Ylikorkala, O., Erkkola, R., 2005. Effects of ospemifene and raloxifene on hormonal status, lipids, genital tract and tolerability in postmenopausal women. Menopause 12, 202–209.

Komm, B.S., Kharode, Y.P., Bodine, P.V., Harris, H.A., Miller, C.P., Lyttle, C.R., 2005. Bazedoxifene acetate: a selective estrogen receptor modulator with improved selectivity. Endocrinology 146, 3999–4008.

Komm, B.S., Bodine, P.V., Minck, D.R., 2007. Effects of bazedoxifene on bone loss: a 12 month study in ovariectomized rats. J. Bone Miner. Res. 22 (Suppl. 1), S206.

Kuiper, G.G.J.M., Enmark, E., Pelto-Huikko, M., Nilsson, S., Gustafsson, J.A., 1996. Cloning of a novel receptor expressed in rat prostate and ovary. Proc. Natl. Acad. Sci. U.S.A. 93, 5925–5930.

Kusama, M., Miyauchi, K., Aoyama, H., Sano, M., Kimura, M., Mitsuyama, S., Komaki, K., Doihara, H., 2004. Effects of toremifene and tamoxifen on serum lipids in postmenopausal patients with breast cancer. Breast Canc. Res. Treat. 88, 1–8.

Labrie, F., Champagne, P., Labrie, C., Roy, J., Laverdiere, J., Provencher, L., Potvin, M., Drolet, Y., Panasci, L., Esperance, B., Dufresne, J., Latreille, J., Robert, J., Samson, B., Jolivet, J., Yelle, L., Cusan, L., Diamond, P., Candas, B., 2004. Activity and safety of the antiestrogen EM-800, the orally active precursor of acolbifene, in tamoxifen -resistant breast cancer. J. Clin. Oncol. 22, 864–871.

Landry, M., Levesqu, D., DiPaolo, T., 2002. Estrogenic properties of raloxifene, but not tamoxifen, on D2 and D3 dopamine receptors in the rat forebrain. Neuroendocrinology 76, 214–222.

Lasco, A., Cannavo, S., Gaudio, A., Morabito, N., Basile, N., Nicita-Mauro, B., Frisina, N., 2002. Raloxifene and pituitary secretion in post-menopausal women. Eur. J. Endocrinol. 147, 461–465.

Lasco, A., Gaudio, A., Morini, E., Morabito, N., Nicita-Mauro, C., Catalano, A., Denuzzo, G., Sansotta, C., Xourafa, A., Macri, I., Frisina, N., 2006. Effect of long-term treatment with raloxifene on mammary density in postmenopausal women. Menopause 13, 787–792.

Leblanc, K., Sexton, E., Parent, S., Belanger, G., Dery, M.-C., Boucehr, V., Asselin, E., 2007. Effects of 4-hydroxytamoxifen, raloxifene and ICI-182,780 on survival of uterine cancer cell lines in the presence and absence of exogenous estrogens. Int. J. Oncol. 30, 477–487.

Legha, S.S., 1988. Tamoxifen in the treatment of breast cancer. Ann. Intern. Med. 109, 219–228.

Li, J., Sato, M., Jerome, C., Turner, C.H., Fan, Z., Burr, D.B., 2005. Microdamage accumulation in the monkey vertebrae does not occur when bone turnover is suppressed by 50% or less with estrogen or raloxifene. J. Bone Miner. Res. 23, 48–54.

Lindsay, R., Rankin, S., Constantine, G., Olivier, S., Pickar, J., 2007. A double-blind, placebo-controlled, phase III study of bazedoxifene/conjugated equine estrogens in postmenopausal women: effects on BMD. In: Endocrine Society Abstracts 89th Meeting. Abstract 126.

Love, R.R., Mazess, R.B., Barden, H.S., Epstein, S., Newcomb, P.A., Jordan, V.C., Carbone, P.P., DeMets, D.L., 1992. Effects of tamoxifen on bone mineral density in postmenopausal women with breast cancer. N. Engl. J. Med. 326, 852–856.

Love, R.R., Wiebe, D.A., Feyzi, J.M., Newcomb, P.A., Chappell, R.J., 1994. Effects of tamoxifen on cardiovascular risk factors in postmenopausal women after 5 years of treatment. J. Natl. Cancer Inst. 86, 1534–1539.

Maenpaa, J.U., Ala-Fossi, S.L., 1997. Toremefine in postmenopausal breast cancer. Efficacy, safety and cost. Drugs Aging 11, 261–270.

Mandlekar, S., Kong, A.N., 2001. Mechanisms of tamoxifen-induced apoptosis. Apoptosis 6, 469–477.

Mann, V., Huber, C., Kogianni, G., Collins, F., Noble, B., 2007. The antioxidant effect of estrogen and selective estrogen receptor modulators in the inhibition of osteocyte apoptosis in vitro. Bone 40, 674–684.

Manson, J.E., Hsia, J., Johnson, K.C., Rossouw, J.E., Assaf, A.R., Lasser, N.L., Trevisan, M., Black, H.R., Heckbert, S.R., Detrano, R., Strickland, O.L., Wong, N.D., Crouse, J.R., Stein, E., 2003. Estrogen plus progestin and the risk of coronary heart disease. N. Engl. J. Med. 349, 523–534.

Martel, C., Picard, S., Belanger, R.V., Labrie, C., Labrie, F., 2000. Prevention of bone loss by EM-800 and raloxifene in the ovariectomized rat. J. Steroid Biochem. Mol. Biol. 74, 45–56.

Martino, S., Cauley, J.A., Barrett-Connor, E., Powles, T.J., Mershon, J., Disch, D., Secrest, R.J., Cummings, S.R., 2004. Continuing Outcomes Relevant to Evista: breast cancer incidence in postmenopausal osteoporotic women in a randomized trial of raloxifene. J. Natl. Cancer Inst. 96, 1751–1761.

Martino, S., Disch, D., Dowsett, S.A., Keech, C.A., Mershon, J., 2005. Safety assessment of raloxifene over eight years in a clinical trial setting. Curr. Med. Res. Opin. 21, 1441–1452.

Marttunen, M.B., Hietanen, P., Tiitinen, A., Ylikorkala, O., 1998. Comparison of effects of tamoxifen and toremifene on bone biochemistry and bone mineral density in postmenopausal breast cancer patients. J. Clin. Endocrinol. Metab. 83, 1158–1162.

Marttunen, M.B., Cacciatore, B., Hietanen, P., Pyrhonen, S., Tiitinen, A., Wahlstrom, T., Ylikorkala, O., 2001. Prospective study on gynaecological effects of two antiestrogens tamoxifen and toremifene in postmenopausal women. Br. J. Canc. 84, 897–902.

McClung, M.R., Siris, E., Cummings, S., Bolognese, M., Ettinger, M., Moffett, A., Emkey, R., Day, W., Somayaji, V., Lee, A., 2006. Prevention of bone loss in postmenopausal women treated with lasofoxifene compared with raloxifene. Menopause 13, 377–386.

McDonnel, D.P., Clemm, D.L., Hermann, T., Goldman, M.E., Pike, J.W., 1995. Analysis of estrogen receptor function in vitro reveals three distinct classes of anti-estrogens. Mol. Endocrinol. 9, 659–669.

McMeekin, D.S., Gordon, A., Fowler, J., Melemed, A., Buller, R., Burke, T., Bloss, J., Sabbatini, P., 2003. A phase II trial of arzoxifene, a selective estrogen receptor modulator, in patients with recurrent or advanced endometrial cancer. Gynecol. Oncol. 90, 49–64.

Mehta, S.H., Dhandapani, K.M., De Sevilla, L.M., Webb, R.C., Mahesh, V.B., Brann, D.W., 2003. Tamoxifen, a selective estrogen receptor modulator reduces ischemic damage caused by middle cerebral artery occlusion in the ovariectomized female rat. Neuroendocrinology 77, 44–50.

Mijatovic, V., van der Mooren, M.J., Kenemans, P., de Valk-de Roo, G.W., Netlenboss, C., 1999. Raloxifene lowers serum lipoprotein (a) in healthy postmenopausal women: a randomized, double-blind, placebo-controlled comparison with conjugated equine estrogens. Menopause 6, 134–137.

Miller, P.D., Christiansen, C., Hoeck, H.C., Kendler, D.L., Lewiecki, E.M., Woodson, G., Ciesielska, M., Chines, A.A., Constantine, G., Delmas, P.D., 2007. Efficacy of bazedoxifene for prevention of postmenopausal osteoporosis: results of a 2-year, phase III, placebo- and active-controlled study. J. Bone Miner. Res. 22 (Suppl. 1), S59.

Mirkin, S., Pickar, J.H., 2015. Selective estrogen receptor modulators (SERMS): a review of clinical data. Maturitas 80, 52–57.

Munster, P.N., Buzdar, A., Dhingra, K., Enas, N., Ni, L., Major, M., Melemed, A., Seidman, A., Booser, D., Theriault, R., Norton, L., Hudis, C., 2001. Phase I study of a third-generation selective estrogen receptor modulator, LY353381. HCl, in metastatic breast cancer. J. Clin. Oncol. 19, 2002–2009.

Murphy, E.D., Beamer, W.G., 1973. Plasma gonadotropin levels during early stages of ovarian tumorigenesis in mice of the Wx-Wu genotype. Cancer Res. 33, 721–723.

Neven, P., Goldstein, S.R., Ciaccia, A.V., Zhou, L., Silfen, S.L., Muram, D., 2002. The effect of raloxifene on the incidence of ovarian cancer in postmenopausal women. Gynecol. Oncol. 85, 388–390.

Nickelsen, T., Lufkin, E.G., Riggs, B.L., Cox, D.A., Crook, T.H., 1999. Raloxifene hydrochloride, a selective estrogen receptor modulator: safety assessment of effects on cognitive function and mood in postmenopausal women. Psychoneuroendocrinology 24, 115–128.

Nilsen, J., Mor, G., Naftolin, F., 1998. Raloxifene induces neurite outgrowth in estrogen receptor positive PC-12 cells. Menopause 5, 211–216.

Nuttall, M.E., Bradbeer, J.N., Stroup, G.B., Nadeau, D.P., Hoffman, S.J., Zhao, H., Rehm, S., Gowen, M., 1998. Idoxifene: a novel selective estrogen receptor modulator prevents bone loss and lowers cholesterol levels in ovariectomized rats and decreases uterine weight in intact rats. Endocrinology 139, 5224–5234.

Obata, T., Kubota, S., 2001. Protective effect of tamoxifen on 1-methyl-4-phenylpyridine-induced hydroxyl radical generation in the rat striatum. Neurosci. Lett. 308, 87–90.

Ortmann, O., Emons, G., Knuppen, R., Catt, K.J., 1988. Inhibitory actions of keoxifene on luteinizing hormone secretion in pituitary gonadotrophs. Endocrinology 123, 962–968.

Osuka, K., Feustel, P.J., Mongin, A.A., Tranmer, B.I., Kimelberg, H.K., 2001. Tamoxifen inhibits nitrotyrosine formation after reversible middle cerebral artery occlusion in the rat. J. Neurochem. 76, 1842–1850.

Ott, S.M., Oleksik, A., Lu, Y., Harper, K.D., Lips, P., 2002. Bone histomorphometric and biochemical marker results of a two year placebo controlled trial of raloxifene in postmenopausal women. J. Bone Miner. Res. 17, 341–348.

O'Neill, K., Chen, S., Brinton, R.D., 2004. Impact of the selective estrogen receptor modulator, raloxifene, on neuronal survival and out-growth following toxic insults associated with aging and Alzheimer's disease. Exp. Neurol. 185, 63–80.

Paganini-Hill, A., Clark, L.J., 2000a. Eye problems in breast cancer patients treated with tamoxifen. Breast Canc. Res. Treat. 60, 167–172.

Paganini-Hill, A., Clark, L.J., 2000b. Preliminary assessment of cognitive function in breast cancer patients treated with tamoxifen. Breast Canc. Res. Treat. 64, 165–176.

Palkowitz, A.L., Glasebrook, A.L., Thrasher, K.J., Hauser, K.L., Short, L.L., Phillips, D.L., Muehl, B.S., Sato, M., Shetler, P.K., Cullinan, G.J., Zeng, G.Q., Pell, T.R., Bryant, H.U., 1997. Discovery and synthesis of 6-hydroxy-3-[4-(1-piperidinyl)-ethoxy-phenoxy]-2-(4-hydroxyphenyl)benzo [b]-thiophene: a novel, highly potent selective estrogen receptor modulator (SERM). J. Med. Chem. 40, 1407–1416.

Palomba, S., Orio, F., Morelli, M., Russo, T., Pellicano, M., Zupi, E., Lombardi, G., Nappi, C., Panici, P.L.B., Zullo, F., 2002a. Raloxifene administration in premenopausal women with uterine leiomyomas: a pilot study. J. Clin. Endocrinol. Metab. 87, 3603–3608.

Palomba, S., Russo, T., Oria, F., Tauchmanova, K., Supi, E., Panici, P.L.B., Nappi, C., Calao, A., Lombardi, G., Zullo, F., 2002b. Effectiveness of combined GnRH analogue plus raloxifene administration in the treatment of uterine leiomyomas: a prospective, randomized, single-blind, placebo-controlled clinical trial. Hum. Reprod. 17, 3213–3219.

Palomba, S., Orio, F., Russo, T., Falbo, A., Tolino, A., Lombardi, G., Climini, V., Zullo, F., 2005. Antiproliferative and proapoptotic effects of raloxifene on uterine leiomyomas in postmenopausal women. Fertil. Steril. 84, 154–161.

Patat, A., McKeand, W., Baird-Bellaire, S., Ermer, J., LeCoz, F., 2003. Absolute/relative bioavailability of bazedoxifene acetate in healthy post-menopausal women. J. Clin. Pharm. Ther. 73, 43.

Pickar, J.H., Archer, D.F., Constantine, G., Ronkin, S., Speroff, L., 2007. SMART-1: a double-blind, placebo-controlled, phase III study of bazedoxifene/conjugate equine estrogens in postmenopausal women—effects on endometrium. In: Endocrine Society Abstracts 89th Meeting. Abstract 246.

Pinna, C., Bolego, C., Sanvito, P., Pelosi, V., Baetta, R., Corsini, A., Gaion, R.M., Cignarella, A., 2006. Raloxifene elicits combined rapid vasorelaxation and long-term anti-inflammatory actions in rat aorta. J. Pharmacol. Exp. Ther. 319, 1444–1451.

Porter, K.B., Tsibris, J.C., Porter, G.W., Fuchs-Young, R., Nicosia, S.V., O'Brien, W.F., Spellacy, W.N., 1998. Effects of raloxifene in a Guinea pig model for leiomyomas. Am. J. Obstet. Gynecol. 170, 1283–1287.

Portman, D., Moffett, A., Kerber, I., Drossman, S., Somayaji, V., Lee, A., 2004. Lasofoxifene, a selective estrogen receptor modulator, improves objective measure of vaginal atrophy. Menopause 11, 675.

Powles, T.J., Hickish, T., Kanis, J.A., Tidy, A., Ashley, S., 1996. Effect of tamoxifen on bone mineral density measured by dual-energy X-ray absorptiometry in healthy premenopausal and postmenopausal women. J. Clin. Oncol. 14, 78–84.

Rash, T., Knadler, M.P., 1997. The disposition and biotransformation of the selective estrogen receptor modulator, LY353381, in femal Fisher 344 rats following a single oral dose. In: Proceedings of the American Association of Pharmaceutical Science. A1223 [Abstract].

Qu, Q., Zheng, H., Dahlund, J., Laine, A., Cockcroft, N., Peng, Z., Koskinen, M., Hemminki, K., Kangas, L., Vaananen, K., Harkonen, P., 2000. Selective estrogenic effects of a novel triphenylethylene compound, FC-1271a on bone, cholesterol level, and reproductive tissue in intact and ovariectomized rats. Endocrinology 141, 809–820.

Reid, I.R., Eastell, R., Fogelman, I., Adachi, J.D., Rosen, A., Netelenbos, C., Watts, N.B., Seeman, E., Ciaccia, A.V., Draper, M.W., 2004. A comparison of the effects of raloxifene and conjugated equine estrogen on bone and lipids in healthy postmenopausal women. Arch. Intern. Med. 164, 871–879.

Reindollar, R., Koltun, W., Parsons, A., Rosen, A., Siddhanti, S., Plouffe, L., 2002. Effects of oral raloxifene on serum estradiol levels and other markers of estrogenicity. Fertil. Steril. 78, 469–472.

Reis, S.E., Costantino, J.P., Wickerham, D.L., Tan-Chiu, E., Wang, J., Kavanah, M., 2001. Cardiovascular effects of tamoxifen in women with and without heart disease: breast cancer prevention trial. National surgical adjuvant breast and Bowel Project breast cancer prevention trial investigators. J. Natl. Cancer Inst. 93, 16–21.

Robinson, S.P., Mauel, D.P., Jordan, V.C., 1988. Antitumor actions of toremifene in 7,12-dimethylbenzanthrocene (DMBA)-induced rat mammary tumor model. J. Cancer Clin. Oncol. 24, 1817–1821.

Rossberg, M.I., Murphy, S.J., Traystman, R.J., Hurn, P.D., 2000. LY353381.HCl, a selective estrogen receptor modulator, and experimental stroke. Stroke 31, 3041–3046.

Rossing, M.A., Daling, J.R., Weiss, N.S., Moore, D.E., Self, S.G., 1994. Ovarian tumors in a cohort of infertile women. N. Engl. J. Med. 331, 771–776.

Rudmann, D.G., Cohen, I.R., Robbins, M.R., Coutant, D.E., Henck, J.W., 2005. Androgen dependent mammary gland virilism in rats given the selective estrogen receptor modulator LY2066948 hydrochloride. Toxicol. Pathol. 33, 711–719.

Rutqvist, L.E., Mattsson, A., 1993. Cardiac and thromboembolic morbidity among postmenopausal women with early-stage breast cancer in a randomized trial of adjuvant tamoxifen: The Stockholm Breast Cancer Study Group. J. Natl. Cancer Inst. 85, 1398–1406.

Saitta, A., Altavilla, D., Cucinotta, D., Morabito, N., Frisina, N., Corrado, F., D'Anna, R., Lasco, A., Squadrito, G., Caudia, A., Cancellieri, F., Arcoraci, V., Squadrito, F., 2001. Randomized, double-blind, placebo-controlled study on effects of raloxifene and hormone replacement therapy on plasma NO concentrations, endothelin-1 levels, and endothelium-dependent vasodilation in postmenopausal women. Arterioscler. Thromb. Vasc. Biol. 21, 1512–1519.

Sato, M., Glasebrook, A.L., Bryant, H.U., 1995. Raloxifene: a selective estrogen receptor modulator. J. Bone Miner. Res. 12 (Suppl. 2), S9–S20.

Sato, M., Bryant, H.U., Iversen, P., Helterbrand, J., Smietana, F., Bemis, K., Higgs, R., Turner, C., Owan, I., Takano, Y., Burr, D.B., 1996a. Advantages of raloxifene over alendronate or estrogen on non-reproductive and reproductive tissues in the long-term dosing of ovariectomized rats. J. Pharmacol. Exp. Ther. 279, 298–305.

Sato, M., Rippy, M.K., Bryant, H.U., 1996b. Raloxifene, tamoxifen, nafoxidine and estrogen effects on reproductive and nonreproductive tissues in ovariectomized rats. FASEB J. 10, 905–912.

Sato, M., Turner, C.H., Wang, T., Adrian, M.D., Rowley, E., Bryant, H.U., 1998. LY353381.HCl: a novel raloxifene analog with improved SERM potency and efficacy in vivo. J. Pharmacol. Exp. Ther. 287, 1–7.

Saunders, P.T., Maguire, S.M., Gaughan, J., Millar, M.R., 1997. Expression of oestrogen receptor beta (ER beta) in multiple rat tissues visualised by immunohistochemistry. J. Endocrinol. 154, R13–R16.

Savolainen-Peltonone, H., Luoto, N.-M., Kangas, L., Hayry, P., 2004. Selective estrogen receptor modulators prevent neointima formation after vascular injury. Mol. Cell. Endocrinol. 227, 9–20.

Seeman, E., 2001. Raloxifene. J. Bone Miner. Res. 19, 65–75.

Sgarabotto, M., Baldini, M., Dei Cas, A., Manotti, C., Barilli, A.L., Rinaldi, M., Benassi, L., Modena, A.B., 2006. Effects of raloxifene and continuous combined hormone therapy on haemostatis variables: a multicenter, randomized, double-blind study. Thromb. Res. 119, 85–91.

Shang, Y., Brown, M., 2002. Molecular determinants for the tissue specificity of SERMs. Science 295, 2465–2468.

Shita, A., Igarashi, T., Kurose, T., Ohno, M., Hando, T., 2002. Reciprocal effects of tamoxifen on hormonal cytology in postmenopausal women. Acta Cytol. 46, 499–506.

Short, L.L., Glasebrook, A.L., Adrian, M.D., Cole, H., Shetler, P., Rowley, E.R., Magee, D.E., Pell, T., Zeng, G., Sato, M., Bryant, H.U., 1996. Distinct effects of selective estrogen receptor modulators on estrogen dependent and estrogen independent human breast cancer cell proliferation. J. Bone Miner. Res. 11 (Suppl. 1), S482.

Shugrue, P.J., Merchenthaler, I., 2001. Distribution of estrogen receptor immunoreactivity in the central nervous system. J. Comp. Neurol. 43, 64–81.

Shumaker, S.A., Legault, C., Rapp, S.R., Thai, L., Wallace, R.B., Ockene, J.K., Hendrix, S.L., Jones, B.N., Assaf, A.R., Jackson, R.D., Kotchen, J.M., Wassertheil-Smoller, S., Wactawski-Wende, J., 2003. Estrogen plus progestin and the incidence of dementia and mild cognitive impairment in postmenopausal women: a randomized controlled trial. J. Am. Med. Assoc. 289, 2651–2662.

Shumaker, S.A., Legault, C., Kuller, L., Rapp, S.R., Thai, L., Lane, D.S., Fillet, H., Stefanick, M.L., Hendrix, S.L., Lewis, C.E., Masaki, K., Coker, L.H., 2004. Conjugated equine estrogens and incidence of probable dementia and mild cognitive impairment in postmenopausal women: Women's health initiative memory study. J. Am. Med. Assoc. 291, 2947–2958.

Shushan, A., Peretz, T., Uziely, B., Lewin, A., Mor-Yosef, S., 1996. Ovarian cysts in premenopausal and postmenopausal tamoxifen-treated women with breast cancer. Am. J. Obstet. Gynecol. 175 (Pt 1), 752–753.

Siefer, D.B., Roa-Pena, L., Keefe, D.L., Zhang, H., Goodman, S., Jones, E.E., Naftolin, F., 1994. Increasing hypothalamic arcuate nucleus glial peroxidase activity in aging female rats is reduced by an antiestrogen and a gonadotropin-releasing hormone agonist. Menopause 1, 83–90.

Silfen, S.L., Ciaccia, A.V., Bryant, H.U., 1999. Selective estrogen receptor modulators: tissue specificity and differential uterine effects. Climacteric 2, 268–283.

Silverman, S.L., Christiansen, K., Genant, H.K., Zanchetta, J.R., Valter, L., de Villiers, T.J., Constantine, G., Chines, A.A., 2007. Efficacy of bazedoxifene in reducing new vertebral fracture risk in postmenopausal women with osteoporosis from a 3–year randomized, placebo- and active-controlled trial. J. Bone Miner. Res. 22 (Suppl. 1), S58.

Simard, J., Labrie, C., Belanger, A., Ganther, S., Singh, S.M., Merand, Y., Labrie, F., 1997. Characterization of the effects of the novel non-steroidal antiestrogen EM-800 on basal and estrogen-induced proliferation. Int. J. Cancer 73, 104–112.

Simpkins, J.W., Rajakumar, G., Zhang, Y.Q., Simpkins, C.E., Greenwald, D., Yu, C.J., Bodor, N., Day, A.L., 1997. Estrogens may reduce mortality and ischemic damage caused by middle cerebral artery occlusion in the female rat. J. Neurosurg. 87, 724–730.

Smith, M.R., 2006. Treatment related osteoporosis in men with prostate cancer. Clin. Cancer Res. 12, 6315S–6319S.

Smith, L.J., Henderson, J.A., Abell, C.W., Bethea, C.L., 2004. Effects of ovarian steroids and raloxifene on proteins that synthesize, transport and degrade serotonin in the raphe region of macaques. Neuropsychopharmacology 29, 2035–2045.

Snyder, K.R., Sparano, N., Malinowksi, J.M., 2000. Raloxifene hydrochloride. Am. J. Health Syst. Pharm. 57, 1669–1678.

Spencer, C.P., Godsland, I.F., Stevenson, J.C., 1997. Is there a postmenopausal metabolic syndrome? Gynecol. Endocrinol. 11, 341–355.

Stamatelopoulos, K.S., Lekakis, J.P., Poulakaki, N.A., Papamichael, C.M., Venetsanou, K., Aznaouridis, K., Protogerou, A.D., Papaioannou, T.G., Kumar, S., Stamatelopoulos, S.F., 2004. Tamoxifen improves endothelial function and reduces carotid intima-media thickness in postmenopausal women. Am. Heart J. 147, 1093–1099.

Stampfer, M.J., Colditz, G.A., 1991. Estrogen replacement therapy and coronary disease: a quantitative assessment of the epidemiological evidence. Prev. Med. 20, 47–63.

Stefanick, M.L., Anderson, G.L., Margolis, K.L., Hendriz, S.L., Rodabough, R.J., Paskett, E.D., Lane, D.S., Hubbell, F.A., Assaf, A.R., Sarto, G.E., Schenken, R.S., Yasmeen, S., Lessin, L., Shlebowski, R.T., 2006. Effects of conjugated equine estrogens on breast cancer and mammography screening in postmenopausal women with hysterectomy. J. Am. Med. Assoc. 295, 1647–1657.

Stovall, D.W., Utian, W.H., Gass, M.L.S., Qu, Y., Muram, D., Wong, M., Plouffe, L., 2007. The effects of combined raloxifene and oral estrogen on vasomotor symptoms and endometrial safety. Menopause 14, 510–517.

Suh, N., Glasebrook, A.G., Palkowitz, A.D., Bryant, H.U., Burris, L.L., Starling, J.J., Pearce, H.L., Williams, C., Peer, C., Wang, Y., Sporn, M.B., 2001. Arzoxifene, a new selective estrogen receptor modulator for chemoprevention of experimental breast cancer. Cancer Res. 61, 8412–8415.

Sumner, B.E.H., Grant, K.E., Rosie, R., Hegele-Hartung, C., Fritzemeier, K.H., Fink, G., 1999. Effects of tamoxifen on serotonin transporter and 5-hydroxytryptamine 2 A receptor binding sites and mRNA levels in the brain of ovariectomized rats with or without acute estradiol replacement. Mol. Brain Res. 73, 119–128.

Sun, J., Meyers, M.J., Fink, B., Rajendran, R., Katzenellenbogen, J.A., Katzenellenbogen, B.S., 1999. Novel ligands that function as selective estrogens or antiestrogens for estrogen receptor-alpha or estrogen receptor-beta. Endocrinology 140, 800–804.

Swisher, D.K., Tague, R.M., Seyler, D.E., 1995. Effect of the selective estrogen receptor modulator raloxifene on explanted uterine growth in rats. Drug Dev. Res. 36, 43–45.

Thigpen, T., Brady, M.F., Homesley, H.D., Soper, J.T., Bell, J., 2001. Tamoxifen in the treatment of advanced or recurrent endometrial carcinoma: a gynecologic oncology group study. J. Clin. Oncol. 19, 364–367.

Tiitinen, A., Nikander, E., Hietanen, P., Metsa-Heikkila, M., Ylikorkala, O., 2004. Changes in bone mineral density during and after 3 years use of tamoxifen or toremifene. Maturitas 48, 321–327.

Tsang, S.Y., Yao, X., Essin, K., Wone, C.M., Chan, F.L., Gollasch, M., Juang, Y., 2004. Raloxifene relaxes rat cerebral arteries in vitro and inhibits L-type voltage-sensitive Ca1 1channels. Stroke 35, 1709–1714.

Turner, C.H., Sato, M., Bryant, H.U., 1994. Raloxifene preserves bone strength and bone mass in ovariectomized rats. Endocrinology 135, 2001–2005.

Turner, R.T., Evans, G.L., Sluka, J.P., Adrian, M.D., Bryant, H.U., Turner, C.H., Sato, M., 1998. Differential responses of estrogen target tissues in rats including bone to clomiphene, enclomiphene, and zuclomiphene. Endocrinology 139, 3712–3720.

Uusi-Rasi, K., Beck, T.J., Semanick, L.M., Daphtary, M.M., Crans, G.G., Desaiah, D., Harper, K.D., 2006. Structural effects of raloxifene on the proximal femur: results from the multiple outcomes of raloxifene evaluation trial. Osteoporos. Int. 17, 575–586.

Vanacker, J., Pettersson, K., Gustafsson, J.A., Ladet, V., 1999. Transcriptional targets shared by ERRs and ER alpha but not ER beta. EMBO J. 18, 4270–4279.

Viereck, V., Grundker, C., Blaschke, S., Niederkleine, B., Siggelkow, H., Frosck, K.-H., Raddatz, D., Emons, G., Hofbauer, L.C., 2003. Raloxifene concurrently stimulates osteoprotogerin and inhibits interleukin-6 production by human trabecular osteoblasts. J. Clin. Endocrinol. Metab. 88, 4206–4213.

Viscoli, C.M., Brass, L.M., Kernan, W.N., Sarrel, P.M., Suissa, S., Horwitz, R.I., 2001. A clinical trial of estrogen-replacement therapy after ischemic stroke. N. Engl. J. Med. 345, 1243–1249.

Vogel, V.G., Costantino, J.P., Wickerham, D.L., Cronin, W.M., Cecchini, R.S., Atkins, J.N., Bevers, T.B., Fehrenbacher, L., Pajon, E.R., Wade, J.L., Robidoux, A., Margolese, R.G., James, J., Lippman, S.M., Runowicz, C.D., Ganz, P.A., Reis, S.E., McCaskill-Stevens, W., Ford, L.G., Jordan, V.C., Wolmark, N., 2006. Effects of tamoxifen vs raloxifene on the risk of developing invasive breast cancer and other disease outcomes: the NSABP study of tamoxifen and raloxifene (STAR) P-2 trial. J. Am. Med. Assoc. 295, 2727–2741.

Wakeling, A.E., Valcaccia, B., 1987. Antiestrogenic and antitumor activities of a series of non-steroidal antiestrogens. J. Endocrinol. 99, 455–464.

Walsh, B.W., Kuller, L.H., Wild, R.A., Paul, S., Farmer, M., Lawrence, J.B., Shah, A.S., Anderson, P.W., 1998. Effects of raloxifene on serum lipids and coagulation factors in healthy postmenopausal women. J. Am. Med. Assoc. 279, 1445–1451.

Walsh, B.W., Paul, S., Wild, R.A., Dean, R.A., Tracy, R.P., Cox, D.A., Anderson, P.W., 2000. The effects of hormone replacement therapy and raloxifene on C-reactive protein and homocysteine in healthy postmenopausal women: a randomized-controlled trial. J. Clin. Endocrinol. Metab. 85, 214–218.

Wang, X.N., Simmons, H.A., Salatto, C.T., Cosgrove, P.G., Thompson, D.D., 2006. Lasofoxifene enhances vaginal mucus formation without causing hypertrophy and increases estrogen receptor beta and androgen receptor in rats. Menopause 13, 609–620.

Ward, A., Bates, P., Fisher, R., Richardson, L., Graham, C.F., 1994. Disproportionate growth in mice with IGF-2 transgenes. Proc. Natl. Acad. Sci. U.S.A. 91, 10365–10369.

Wiernicki, T., Glasebrook, A.L., Phillips, D.L., Singh, J.P., 1996. Estrogen and a novel tissue selective estrogen receptor modulator raloxifene directly modulate vascular smooth muscle cell functions: Implications in the cardioprotective mechanism of estrogen. Circulation 94 (8 Suppl. I), I278.

Writing Group for Women's Health Initiative Investigators, 2002. Risks and benefits of estrogen plus progestin in healthy postmenopausal women. J. Am. Med. Assoc. 288, 321–333.

Wu, X., Glinn, M.A., Ostrowski, N.L., Su, Y., Ni, B., Cole, H.W., Bryant, H.U., Paul, S.M., 1999. Raloxifene and estradiol benzoate both fully restore hippocampal choline acetyltransferase activity in ovariectomized rats. Brain Res. 847, 98–104.

Yaffe, K., Sawaya, G., Liebergurg, I., Grady, D., 1998. Estrogen therapy in postmenopausal women: effects on cognitive function and dementia. J. Am. Med. Assoc. 279, 688–695.

Yaffe, K., Krueger, K., Sarkar, S., Grady, D., Barrett-Connor, E., Cox, D.A., Nickelsen, T., 2001. Cognitive function in postmenopausal women treated with raloxifene. N. Engl. J. Med. 344, 1207–1213.

Yaffe, K., Krueger, K., Cummings, S.R., Blackwell, T., Henderson, V.W., Sarkar, S., Ensrud, K., Grady, D., 2005. Effect of raloxifene on prevention of dementia and cognitive impairment in older women: The Multiple Outcomes of Raloxifene Evaluation (MORE) randomized trial. Am. J. Psychiatry 162, 683–690.

Zhang, J.J., Jacob, T.J.C., Valverde, M.A., Hardy, S.P., Mintenig, G.M., Sepulveda, F.V., Gill, D.R., Hyde, S.C., Trezise, A.E.O., Higgins, C.F., 1994. Tamoxifen blocks chloride channels: a possible mechanism for cataract formation. J. Clin. Investig. 94, 1690–1697.

Zuckerman, S.H., Bryan, N., 1996. Inhibition of LDL oxidation and myeloperoxidase dependent tyrosyl radical formation by the selective estrogen receptor modulator raloxifene. Atherosclerosis 126, 65–75.

Chapter 38

Thyroid hormone and bone

Peter A. Lakatos[1,a], Bence Bakos[1,a], Istvan Takacs[1] and Paula H. Stern[2]

[1]*1st Department of Medicine, Semmelweis University Medical School, Budapest, Hungary;* [2]*Department of Pharmacology, Northwestern University Feinberg School of Medicine, Chicago, IL, United States*

Chapter outline

Introduction **895**
Intracellular mechanism of thyroid hormone action **895**
Nuclear actions of thyroid hormones 895
Nongenomic actions of thyroid hormones 898
Thyrotropin as an independent agent of bone metabolism 898
Cellular effects of thyroid hormones on the bone **899**
Osteoblasts 900
Osteoclasts 900
Remodeling 901
Chondrocytes 901
In vivo responses of the skeleton to thyroid hormones: animal studies **901**
Hypothyroidism 901
Hyperthyroidism 902
Pathophysiological effects of altered thyroid hormone status in humans **902**
Hypothyroidism 902
Subclinical hypothyroidism 903
Hyperthyroidism 903
Subclinical hyperthyroidism 904
Overview and future directions **904**
References **905**

Introduction

Thyroid hormone is essential for skeletal development and also affects mature bone. Depending on the hormone concentration and stage of life, the effects of thyroid hormone can be either beneficial or deleterious to the skeleton. This chapter focuses on the intracellular and cellular effects of thyroid hormone on bone cells as well as its relationship to the observed effects of the hormone on the skeleton in vivo in both experimental animals and humans. Studies published since the previous edition provide information on thyroid hormone effects on additional genes of interest in osteoblasts and further assessment of the skeletal risks of excess thyroid hormone.

New data regarding nuclear and nongenomic actions of thyroid hormone are discussed. These aspects of hormone effects are accompanied by an expansion of the section on in vivo responses of the skeleton to thyroid function alterations including subclinical conditions.

Intracellular mechanism of thyroid hormone action

Nuclear actions of thyroid hormones

Thyroid hormone receptors (TRs) are members of the nuclear receptor (NR) superfamily (Germain et al., 2006). All of these receptors share a common modular structure with a centrally located DNA-binding domain composed of two zinc fingers, an amino-terminal A/B domain involved in transcription modulation, and a carboxy-terminal ligand-binding domain that is also involved in receptor dimerization and interactions with coactivators and corepressors. TRs are

---

[a] participated equally in the work.

Principles of Bone Biology. https://doi.org/10.1016/B978-0-12-814841-9.00038-5

nuclear proteins capable of binding to cognate DNA elements in the absence of their ligands. Binding of the ligand to the receptor alters the receptor conformation and subsequently enables the activation or repression of specific genes.

Thyroid hormone response elements (TREs) in the promoters of T3 target genes share a common hexanucleotide "half-site" sequence of (A/G)GGT(C/A/G)A. Pairs of this structure may occur as direct, everted, or inverted repeat. Promoters of certain genes have clusters of TREs with differing structures (Yen, 2001).

Activated TRs bind to TREs in the target genes' promoter either as monomer, homodimer, or as heterodimer formed with other members of the NR superfamily. These latter include the vitamin D receptor, retinoid X receptor, and retinoic acid receptors. Heterodimerization is an important modulatory mechanism of TR function and also a means of cross talk with other signaling pathways. In bone (Williams et al., 1994, 1995), as in other tissues (Glass, 1994; Brent et al., 1991), DNA binding and transcriptional activation are enhanced when the thyroid hormone receptor isoforms are present as heterodimers with retinoid or vitamin D receptors. In osteoblast cell lines, interactions among the retinoid, vitamin D, and thyroid hormone ligands appeared to mediate specific responses (Williams et al., 1994, 1995). Studies on the effects of treatment combinations on the expression of osteoblast phenotypic genes in the cell lines revealed complex responses that indicated the importance of dose, treatment duration, and degree of confluence in dictating the magnitude of response (Williams et al., 1995). In primary rat osteoblastic cells, alteration of the ligand combinations did not influence the responses (Bland et al., 1997).

In addition to differences between TREs of specific genes, variations of TR dimerization patterns and tissue-specific and developmental stage-dependent expression of different TR isoforms (discussed later), as well as cellular response to thyroid hormones, are also altered by a number of different nuclear coregulator proteins.

Coactivator proteins that bind to the liganded TR are members of the SRC/p160 family (McKenna et al., 1999; McKenna and O'Malley, 2002) or the TR-associated protein (TRAP) complex (Sharma and Fondell, 2002; Burakov et al., 2002). Besides enhancing transcriptional activity, SRC/p160 proteins such as steroid hormone receptor CoA are characterized by histone acetyltransferase activity.

Thyroid hormone receptors have a dual functionality and may act as repressors on the TRE in the absence of T3. This process is mediated by the interaction of the TR with corepressor proteins including nuclear receptor corepressor, silencing mediator of retinoid and thyroid hormone receptor, Sin3, and histone deacetylases (Torchia et al., 1998; Yen et al., 2006). This results in the condensation of chromatin structure and repression of transcription through decreased access of transcription factors (Koenig, 1998; Wu and Koenig, 2000).

TRs are products of two genes located on two different chromosomes. The TRα gene found on chromosome 17 encodes one isoform (TRα1) with T3 binding capacity, and two C-terminal splice variants (TRα2, TRα3) lacking it (Izumo and Mahdavi, 1988; Chassande et al., 1997). Truncated variants of the TRα1 isoform (TR$\Delta\alpha$1, TR$\Delta\alpha$2), retaining T3 binding capacity but missing the A/B domain, also exist (Plateroti et al., 2001).

Three T3 binding isoforms (TRβ1, TRβ2, TRβ3) varying only in the A/B domain are encoded by the TRβ gene situated on chromosome 3 (Williams, 2000). TR$\Delta\beta$3 has T3 binding activity but lacks both the A/B and DNA binding domains.

TR isoforms are expressed in a complex, tissue, and developmental stage-dependent manner, suggesting functional differences between subtypes. TRα1 is constitutionally expressed during embryonic development and is subsequently the dominant TR isoform in the brain, heart, and bone (Vella and Hollenberg, 2017). TRβ is identified as the major isoform in the liver, kidneys, ears, retinas, the hypothalamus, pituitary, and thyroid. Differences in function between TR subtypes are confirmed by studies of thyroid hormone resistance in humans and genetically engineered mice models.

In skeletal tissues, mRNAs for TRα1, TRα2, and TRβ are found in MG63, ROS 17/2.8, and UMR-106 cell lines (Williams et al., 1994; Allain et al., 1996). TRβ2 mRNA has been found in osteoblasts (Abu et al., 2000). mRNA for TRα1 was 12 times higher than TRβ1 mRNA in tibia and femur of 7-week-old male mice (O'Shea et al., 2003). TRα1, TRα2, TRβ1, and TRβ2 mRNAs were expressed in chondrocytes at all stages of differentiation. TRα1, TRα2, and TRβ1 mRNAs were highly expressed in osteoblasts at bone-remodeling sites; and mRNAs for all isoforms were present and highly expressed in multinucleated osteoclastic cells from an osteoclastoma (Abu et al., 1997). TRα1, TRα2, and TRβ1 mRNA have also been detected in rat femurs and vertebrae (Milne et al., 1999). Immunohistochemical staining with antibodies recognizing a TRα epitope or specific TRα2 and TRβ revealed the presence of receptor protein in osteoblast cell lines and in osteoclasts in tissue smears from a human osteoclastoma (Allain et al., 1996). In contrast to mRNA expression, TRα1 protein expression was not seen in the osteoclastoma cells and was limited to osteoblasts at sites of remodeling and undifferentiated chondrocytes (Abu et al., 2000).

The syndrome of resistance to thyroid hormones (RTH) caused by dominant-negative mutations of the TRβ gene was recognized as early as 1967. Over 3000 cases have occurred, and since the identification of the first causal genetic abnormality in 1989, over 120 different mutations have been documented. In contrast, similar mutations in the TRα gene, and corresponding conditions, have only been recognized since 2012 (Bochukova et al., 2012; van Mullem et al., 2012; Moran et al., 2013).

RTHβ is characterized by an autosomal dominant inheritance, elevated thyroid hormone, and unsuppressible TSH levels. Symptoms manifest with variable expressivity and include short stature, decreased weight, goiter, cognitive impairment, tachycardia, and hypacusis. All reported cases except one are genetically heterozygous. The mutations are clustered and largely located within domains in the carboxy-terminal region. They are mainly nucleotide substitutions that result in single amino acid changes (Refetoff et al., 1993). The mutant alleles act by a dominant-negative mechanism to inhibit the ability of the normal allele to elicit normal receptor function (Chatterjee et al., 1991; Sakurai et al., 1990). The dominant-negative action appears to be at the level of DNA binding (Kopp et al., 1996).

Skeletal alterations include retarded bone age and stippled epiphyses, similar to characteristics of hypothyroidism, with a resulting short stature. In other patients there is accelerated bone age, accelerated chondrocyte maturation, and early epiphyseal closure, again resulting in short stature (Behr et al., 1997). The target sites at which resistance occurs (pituitary or peripheral) may determine the phenotype.

Patients with RTHα present with normal levels of TSH accompanied by low/normal T4 and high/normal T3 concentrations and a decreased FT4/FT3 ratio. While cognitive and motor abnormalities often vary, skeletal development is markedly delayed. Disproportionate growth retardation, delayed bone age, patent skull sutures, macrocephaly, flattened nasal bridge, short stature, epiphyseal dysgenesis, and defective bone mineralization are reported (Tylki-Szymańska et al., 2015). Radiographic features are suggestive of hypothyroidism. Phenotype is largely dependent on the severity of the underlying genetic abnormality, the resulting hormone binding potential, and dominant-negative activity of the TRα1 receptor. Consequently, the response to T4 treatment also varies between patients. A recent report detailed the case of a patient with a mutation affecting both TRα1 and TRα2 leading to a severe, atypical skeletal phenotype with intrauterine growth retardation, macrocephaly, hypertelorism, micrognathia, short and broad nose, clavicular and 12th rib agenesis, elongated thorax, ovoid vertebrae, scoliosis, congenital hip dislocation, short limbs, humeroradial synostosis, and syndactyly (Espiard et al., 2015).

A syndrome resulting in advanced bone age was associated with a mutation in the MCT8 thyroid hormone transporter gene (Herzovich et al., 2007). A single nucleotide change (Q261X) in exon 3 on the X chromosome resulted in low serum T4 and free T4, elevated serum T3 and free T3, slightly elevated TSH. The child had severe neurological abnormalities and normal growth, which had been previously noted in other patients with mutations in this gene (Dumitrescu et al., 2004).

Knockin mice models of RTH have helped shed more light on the roles of different TR isoforms in bone and other tissues (Kaneshige et al., 2001; O'Shea et al., 2006). Carrying a C-terminal frame-shift mutation derived from a patient with severe RTH, the TRβPV mouse has TRβ receptors that are dominant-negative antagonists incapable of transactivation and T3 binding. The resulting phenotype is consistent with human RTHβ, showing serum lipid abnormalities, growth retardation, hearing loss, neurological dysfunctions, and markedly elevated TSH and thyroid hormone levels.

The bone phenotype in these mice is reminiscent of skeletal thyrotoxicosis, with advanced endochondral and intramembranous ossification, premature closure of the growth plates, and shortened body length along with increased mineralization and craniosynostosis. The narrower growth plate appears to be a consequence of faster transition through the proliferative zone. Trabecular bone mass is also decreased with more resorption and a greater number of TRAP-positive cells. Expression of fibroblast growth factor receptor-1 (FGFR1), a skeletal T3-target gene, is increased.

Mice carrying PV and other types of dominant-negative mutations in the TRα1 gene were also created (Kaneshige et al., 2001; O'Shea et al., 2006; Quignodon et al., 2007; Bassett et al., 2014). With the TRα1PV protein being a dominant-negative antagonist of both TRα1 and TRβ, the mutation was lethal in homozygotes. Heterozygotes showed mild elevation in TSH levels, accompanied with normal FT3 and FT4. Bone phenotype was hypothyroid with growth retardation, wider growth plates, delayed intramembranous and endochondral ossification, transiently decreased bone calcification, decreased resorption, and TRAP-positive cells. FGFR1 expression and adult bone remodeling were also reduced.

These findings support TRβ being the predominant regulator of the negative feedback loop of the HPT axis. Changes in the skeleton of RTH patients and PV mice models suggest TRα1 to be the dominant TR isoform in bone.

Studies done with knockout mice also illustrate the complex roles of different TR isoforms. Hormone levels in TRα1−/− mice show mild central hypothyroidism with no gross impairments of skeletal development (Wikstrom et al., 1998; Gauthier et al., 1999; Gothe et al., 1999; Gloss et at., 2001). TRα2−/− animals have mild thyroid dysfunction with decreased peripheral hormone levels accompanied by normal TSH. Growth was not impaired, but adult bone density was decreased (Saltó et al., 2001).

TRα−/−, TRα1−/− TRβ−/−, and TRα−/− TRβ−/− animals exhibit different degrees of similarly impaired bone development with growth retardation, disorganized bone plates, and delayed endochondral ossification (Fraichard et al., 1997; Gauthier et al., 1999; Göthe et al., 1999). TRα−/− mice are hormonally hypothyroid while double-knockout mice lacking TRβ have increased TSH, T3, and T4 levels. TRα0/0 mice missing not just TRα1 and TRα2 but also TRΔα1 and

TRΔα2 transcripts are hormonally euthyroid. Similarly to TRα−/− mice, they exhibit growth retardation and delayed endochondral ossification but have increased adult BMD due to defective remodeling (Bassett et al., 2007).

TRβ−/− animals exhibit elevated TSH and thyroid hormone levels, and signs of skeletal thyrotoxicosis not unlike those seen with TRβPV mice. Accelerated endochondral and intramembranous ossification, advanced bone age, increased mineral deposition, persistent short stature, and later on progressive osteoporosis, are reported (Bassett et al., 2007).

Mice lacking DIO2, the activating deiodinase catalyzing the intracellular synthesis of T3 from T4, show pituitary resistance to T4 resulting in increased TSH, slightly elevated T4, and normal T3 levels. Though postnatal skeletal development and linear growth are normal in DIO2−/− mice, osteoblast function and bone formation in adulthood were shown to be severely impaired (Bassett et al., 2010).

Differential expression and only partial overlap in functions of TR isoforms allow for the development of thyroid hormone analogs that have tissue specificity owing to their preferential interaction with one receptor isoform. As previously outlined, TRβ1 is the predominant isoform in the liver and most peripheral tissues, while heart rate and bone metabolism are influenced mostly via TRα1. Potential benefits of selective TRβ1 agonists would be lowering of body weight, serum lipid, and cholesterol levels without deleterious cardiac or skeletal side effects. Selective and partial TRα1 agonists are hoped to produce positive inotropic cardiac effects while lowering peripheral resistance in heart failure patients. Though no selective TR agonist is in clinical use at the time of writing, several compounds are under different stages of development and investigation. Selective TRβ1 agonists of interest include kb-141 (Grover et al., 2003), GC1 (Chiellini et al., 2002), eprotirome (Ladenson et al., 2010), and tiratricol (Sherman et al., 1997). DITPA is a thyroid hormone analogue tested in the treatment of chronic heart failure (Goldman et al., 2009).

Nongenomic actions of thyroid hormones

Although high-affinity binding sites for thyroid hormones on the plasma membrane were identified as early as the 1980s, specific receptors on the cell membrane, in the cytoplasm, and in mitochondria have been established and their modes of action increasingly elucidated in recent years (see reviews by Davis et al., 2008, 2011, 2013, 2016). The effects are mediated through rapid activation of signaling pathways, leading in various tissues to increased nitric oxide (NO) synthase (Hiroi et al.), mitogen-activated protein kinase (MAPK) (Davis et al., 2000; Lei et al., 2008), and the Akt activator phosphatidylinositol-3-kinase (Hiroi et al., 2006; Moeller et al., 2006). Subsequent cellular responses can result from activation of nuclear transactivator proteins or the direct effects of signaling intermediates on the plasma membrane (Davis et al., 2016). The integrin $\alpha_V\beta_3$ acts as a low-affinity membrane receptor for thyroid hormone (Bergh et al., 2005), mediating angiogenic actions and effects on platelets and neurons as well as on bone cells (Davis et al., 2011). In human osteoblastic MG63 and SaOS2 cells, an $\alpha_V\beta_3$-blocking antibody inhibited thyroid hormone stimulation of the MAPK pathway and thymidine incorporation (Scarlett et al., 2008), suggesting a role of the pathway in bone formation. The resorptive effects of thyroid hormone may also involve $\alpha_V\beta_3$, based on the finding that a vitronectin receptor antagonist inhibited thyroxin-induced bone resorption (Hoffman et al., 2002). A novel high-affinity plasma membrane TRα associated with caveolin domains in human primary osteoblasts mediated the rapid effects of T3 on NO, cyclic guanosine monophosphate (cGMP), Src phosphorylation, and activation of MAPK and Akt, leading to cell proliferation and survival (Kalyanaraman et al., 2014). Hypothyroid mice exhibited cGMP deficiency with osteocyte apoptosis and impaired bone formation (Kalyanaraman et al., 2014). Treatment with T3 rapidly decreased Src Y416 autophosphorylation, thereby stimulating osteocalcin expression in primary calvarial osteoblasts from neonatal mice (Asai et al., 2009).

Rapid (within 30 s) increases in inositol mono-, bis-, and triphosphates were elicited by treatment of fetal rat limb bones with 100 nM and 1 μM T3 (Lakatos and Stern, 1991). The inactive analogs diiodothyronine and rT3 did not increase inositol phosphates. This effect of T3 was inhibited by indomethacin and could represent an initiation pathway for the prostaglandin-dependent effects of thyroid hormones on bone resorption, discussed later. Thyroid hormones at high doses inhibit cyclic AMP phosphodiesterase (Marcus, 1975). T3 at 0.1 and 1 nM increased ornithine decarboxylase and potentiated the responses of this enzyme to parathyroid hormone (PTH) (Schmid et al., 1986).

Thyrotropin as an independent agent of bone metabolism

In the past 10−15 years, thyrotropin (TSH) has been implicated as playing a direct regulatory role in many extrathyroidal tissues. TSH receptor (TSHR) expression has been established in the brain, the pituitary, orbital preadipocytes and fibroblasts, the kidney, ovary and testis, skin and hair follicles, heart, adipose tissue, hematopoietic and immune cells, and bone. In vitro and in vivo studies of TSH action suggest multifaceted regulatory functions as well as a role in disease states such as endocrine ophthalmopathy.

TSHR is expressed on chondrocytes, osteoblasts, and osteoclasts (Abe et al., 2003; Endo and Kobayashi, 2013). Besides systemic TSH, other potential local ligands have been implied for these receptors, such as Tshb, a splice variant expressed by bone marrow derived macrophages (Vincent et al., 2009), and thyrostimulin, a glycoprotein TSH-activating hormone expressed by both osteoblasts and osteoclasts (Bassett et al., 2015).

The physiologic role of TSHR in bone cells, the pathway downstream from the receptor, and the potential effects of TSH in bone metabolism are still contradictory. In certain studies, osteoblastogenesis, type I collagen-, bone sialoprotein-, and osteocalcin expression were inhibited in osteoblasts treated with TSH, presumably via the Wnt signal pathway (Abe et al., 2003). Others found a stimulatory effect on osteoblast differentiation and function on various cell lines and animal models (Sampath et al., 2007; Baliram et al., 2011; Boutin et al., 2014). In other studies the presence and physiologic function of TSHR on osteoblasts were proposed to be insignificant (Tsai et al., 2004; Bassett and Williams, 2008). Most findings point toward a potential inhibitory role of TSH on osteoclast function, an effect mediated primarily via TNFα (Hase et al., 2006; Ma et al., 2011; Sun et al., 2013; Zhang et al., 2014).

Intermittent low-dose TSH treatment of ovariectomized rats resulted in a decrease in resorption markers, an increase in formation markers, and an improvement in BMD and overall bone strength (Sampath et al., 2007; Sun et al., 2008; Dumic-Cule et al., 2014).

Genetically modified mice models investigating the roles of TSH on bone metabolism also reported somewhat conflicting results. TSHR knockout mice that have severe congenital hypothyroidism with extremely elevated TSH levels show growth retardation and BMD reduction with histomorphometric signs of increased bone turnover; though when supplemented with thyroid extract, these animals regain normal weight following a "catch-up" growth, while bone density and calvarial thickness remain reduced. Compared with the wild genotype, a 3-week treatment with supraphysiologic thyroxine resulted in increased bone loss in TSHR knockout animals. Furthermore, while Tshr ± mice are hormonally euthyroid, have normal formation and resorption marker levels, and normal calvarial thickness, BMD was still found to be slightly reduced at sites. These findings suggest that TSH has an inhibitory effect on bone remodeling (Abe et al., 2003; Baliram et al., 2012). Comparison of the skeletal phenotype of Pax8−/− and hyt/hyt mice, however, implies that TSH has no or a minimal effect on bone metabolism. The former have impaired thyroid development due to the lack of a critical transcription factor with resulting hypothyroidism, accompanied by a 2000-fold elevation in TSH levels but an intact TSHR (Mansouri et al., 1998). Hyt/hyt mice are characterized by a loss of function mutation of TSHR also resulting in hypothyroidism and extreme elevation of TSH levels (Beamer et al., 1981). Despite the fundamental difference in TSHR function, both mice models have similar defects of skeletal metabolism typical of severe hypothyroidism. Chondrocyte, osteoblast, and osteoclast activities are decreased while impaired linear growth, delayed endochondral ossification, reduced cortical bone mass, defective trabecular bone remodeling, and reduced bone mineralization are evident (Bassett and Williams, 2008).

Thyrostimulin deficient Gpb5−/− mice show an increase in infantile bone production that is resolved by adulthood (Bassett et al., 2015). However, somewhat contrary to this finding, in vitro treatment of osteoblasts thyrostimulin resulted in no notable change in differentiation or function.

Cellular effects of thyroid hormones on the bone

Thyroid hormones exert a complex, developmental stage, and dose-dependent influence on bone formation, growth, and remodeling, affecting several critical constituents of skeletal metabolism. As we have previously discussed, the dominant TR isoform in the bone is TRα1. While compared with TRβ it is present in bone cells at 10-fold higher levels (O'Shea et al., 2003), recent studies also suggest a role for the latter (Monfoulet et al., 2011). Thyroid receptors are most notably present on reserve zone and proliferating chondrocytes (Ballock et al., 1999; Robson et al., 2000; Stevens et al., 2000), osteoblasts, and osteoblastic bone marrow stromal cells (Rizzoli et al., 1986; Allain et al., 1996; Milne et al., 1999; Siddiqi et al., 2002). Their presence and role on osteocytes and osteoclasts are still unclear (Bassett and Williams, 2003). Differentiated chondrocytes have no TR expression.

The activating deiodinase DIO2, a catalyzing intracellular formation of T3 from T4, is expressed primarily in osteoblasts (Bassett et al., 2010), and in animal models has been demonstrated to be present from early embryonal stages of skeletal development (Gouveia et al., 2005; Capelo et al., 2008). The inactivating DIO3 deiodinase has been found in all skeletal cell lines, with the highest activity in growth plate chondrocytes.

The presence of several thyroid hormone transport proteins such as MCT8, MCT10, LAT1, and LAT2 also has been demonstrated in growth plate chondrocytes, osteoblasts, and osteoclasts (Williams et al., 2008; Capelo et al., 2009; Abe et al., 2012).

Wide concentration ranges of thyroid hormones have been used in experimental studies, especially in vitro, and often markedly different dosages are required to obtain the same response in a different cell line, model system, or laboratory. The differentiation state and the production of modulating factors are potential variables that can affect the response in a given system. In addition, the presence of thyroid hormone in the added sera or the presence of binding sites in stripped sera can dramatically influence the free hormone available to the cells or tissue. Several studies have estimated the amount of free hormone available under the experimental conditions used (Sato et al., 1987; Allain et al., 1992). In one study, an equilibrium dialysis method was used to determine free T4 and T3 after treating fetal calf serum with AG1-X8 resin (Sato et al., 1987). T4 and T3 concentrations in the fetal calf serum prior to extraction were 11.1 μg/dL and 157 ng/dL, respectively. It was determined that the addition of 10 nM T4 to the stripped serum provided 80 p.m. free T4, and that addition of 1 nM T3 provided 40 p.m. free T3. In the other study, in which 10% neonatal calf serum was used, the free T3 was measured by radioimmunoassay (Allain et al., 1992). It was determined that the addition of 10 p.m. T3 yielded a free T3 concentration of 2.1 p.m., that 0.1 nM yielded 4 p.m., that 1 nM yielded 2.1 p.m., and that 10 nM yielded 0.39 p.m. (Allain et al., 1992). Although the type and percentage of serum would influence the final values, these measurements and calculations are of value in comparing studies and in relating in vitro concentrations to the concentrations of thyroid hormones in normal serum.

Osteoblasts

T3 may produce varying responses in osteoblasts in vitro depending on species, anatomical origin of cells, stage of differentiation, passage number, cell confluence, dose of T3, and duration of treatment (Rizzoli et al., 1986; Williams et al., 1994; Cray et al., 2013). Most results suggest a stimulating role of T3 on osteoblast proliferation (Ernst and Froesch, 1987; LeBron et al., 1989; Luegmayr et al., 1996), differentiation (Ohishi et al., 1994; Klaushofer et al., 1995), and function (Fratzl-Zelman et al., 1997).

T3 can increase proliferation of rodent and human osteoblastic cells (Ernst and Froesch, 1987; Kassem et al., 1993). In the rodent cell cultures, 0.01 and 1 nM were stimulatory, and 10 nM was inhibitory in longer term cultures (Ernst and Froesch, 1987). Cell number was decreased after 8 days of incubation with T4 in MC3T3-E1 cells; inhibition was observed with 10 nM T3 and was maximal at 1 μM (Kasono et al., 1988). In other investigations, T3 did not significantly affect the growth of ROS 25/1, UMR-106, and ROS 17/2.8 cells (Sato et al., 1987; LeBron et al., 1989; Williams et al., 1994).

In response to T3 treatment, an increase is seen in the expression of osteoblastic phenotypic markers, such as osteocalcin (Gouveia et al., 2001; Varga et al., 2003), osteoprotegerin (Varga et al., 2004), osteopontin, type I collagen (Kawaguchi et al., 1994a; Varga et al., 2010), and alkaline phosphatase (Sato et al., 1987; Kasono et al., 1988; Banovac and Koren, 2000). Effects are seen at concentrations in the physiologic range; however, the dose-dependence of the response is quite variable and may be dependent on cell type and culture conditions. Levels of IL-6, IL -8, MMP9, MMP13 (Varga et al., 2009), and tissue inhibitor of metalloproteinase-1 are also elevated, as are IGF-1, IGFBP-2, -3, and -4, and FGFR1, which are thought to be major secondary mediators of T3-mediated osteoblast activation (Schmid et al., 1992; Varga et al., 1994; Glantschnig et al., 1996; Huang et al., 2000; Stevens et al., 2003). IGF-I has significant anabolic effects on bone, increasing cell replication and both collagen and noncollagen protein synthesis (Canalis, 1980; Hock et al., 1986; McCarthy et al., 1989; Centrella et al., 1990; Pirskanen et al., 1993). T3 increased IGF-I expression more markedly in cells from vertebral marrow than in cells from femoral marrow (Milne et al., 1998). Interference with IGF-I action by decreasing expression or function of the IGF-I receptor by the use of antisense oligonucleotides, antibodies, and antagonist peptide decreased the anabolic effects of T3 on MC3T3-E1 cells and primary mouse calvarial osteoblasts, including effects on alkaline phosphatase, osteocalcin, and collagen synthesis (Huang et al., 2000). The effects of thyroid hormones on IGFs may be modulated by changes in IGF-binding proteins (IGFBPs). The physiological role of IGFBPs is not fully understood; however, they can influence the cellular uptake and turnover of IGF-I.

Treatment of osteoblasts with T3 stimulated expression of FGFR1 mRNA and protein (Stevens et al., 2003). PTH and PTHrp receptors are also upregulated in response to T3 treatment in osteoblastic cell lines (Schmid et al., 1986; Gu et al., 2001). Osteoblastic cell morphology, cytoskeleton formation, and cell adhesion molecules also respond to T3 stimulus in vitro (Luegmayr et al., 1996, 2000; Fratzl-Zelman et al., 1997).

Osteoclasts

T3 stimulates resorption in bone organ cultures. Fetal rat limb bones (Mundy et al., 1976; Hoffmann et al., 1986; Lakatos and Stern, 1992) and neonatal mouse calvaria (Krieger et al., 1988; Klaushofer et al., 1989; Kawaguchi et al., 1994) are the models that have been studied most extensively. Compared with the effects of PTH, T3 responses are slower to develop, with the dose–response curves being generally shallow (Mundy et al., 1976; Krieger et al., 1988; Klaushofer et al., 1989; Kawaguchi et al., 1994b), and maximal effects are lower. Further evidence for the differing actions of T3 and PTH is supplied by the contrast in their interaction with TGFβ (Lakatos and Stern, 1992). TGFβ enhanced the early responses to PTH and inhibited later effects, whereas interaction with T3 displayed a somewhat reversed time course.

It is still unclear whether the increased osteoclastogenesis and bone resorption seen in vitro in response to T3 and in vivo in thyrotoxicosis is mediated by a direct effect of T3 on osteoclasts or is exclusively secondary to changes in the function of osteoblast, stromal osteocytes, or other bone marrow cells (Mundy et al., 1976; Klaushofer et al., 1989). T3 failed to activate isolated osteoclasts; however, when mixed bone cells were added to the cultures, a significant response was observed with 1 μM T3, although not with lower concentrations (Allain et al., 1992).

Various osteoblast-derived secondary mediators, growth factors, and cytokines have been implicated. In neonatal mouse calvaria, resorption was inhibited by indomethacin, suggesting a prostaglandin-dependent pathway (Krieger et al., 1988; Klaushofer et al., 1989; Kawaguchi et al., 1994b). Other studies have shown prostaglandin-independent effects (Conaway et al., 1998) involving other mediators, such as the RANKL/OPG pathway (Varga et al., 2004), IGF-1 (Stracke et al., 1986; Lakatos et al., 1993),TGFβ (Lakatos and Stern, 1992; Klaushofer et al., 1995), interleukins (Tarjan et al., 1995; Siddiqi et al., 1998; Schiller et al., 1998), and interferon-γ.

Remodeling

Most in vitro studies have focused on either anabolic or catabolic effects of thyroid hormone under conditions designed to optimize the study of the particular response. However, because there are dose-dependent biphasic effects on formation parameters and delayed (Klaushofer et al., 1989) and submaximal (Mundy et al., 1976; Krieger et al., 1988; Lakatos and Stern, 1992) effects on resorption, it may be that neither effect can be studied to the exclusion of the other, and the net effects on bone remodeling may be accessible to in vitro investigation. A model system designed to study growth, mineralization, and resorption in radii and ulnae of 16-day fetal mice (Soskolne et al., 1990) revealed interesting differences between effects of T3 and PTH. Effects of T3 were studied over a 0.1nM–10μM dose range. T3 concentrations in the 10nM–0.3 μM range resulted in increases in diaphyseal length, increased calcium, phosphate, and hydroxyproline content, and decreased 45Ca release. At higher concentrations (1 and 10 μM), T3 stimulated 45Ca release. In contrast, when PTH was studied over a 1pM–0.1 μM range, only resorptive effects were observed, these being at concentrations of 1 nM and higher.

Chondrocytes

Thyroid hormones exert a complex regulatory effect on chondrocyte function including proliferation, matrix synthesis, mineralization, chondrocyte maturation, endochondral ossification, and linear bone growth. In different cultures, T3 is mostly shown to inhibit chondrocyte proliferation, induce hypertrophic differentiation, and help sustain the structure and ossification of epiphyseal cartilage (Burch et al., 1982a, 1982b; Böhme et al., 1992; Quarto et al., 1992; Ballock et al., 1994; Alini et al., 1996; Ishikawa et al., 1998; Rosenthal et al., 1999; Miura et al., 2002). T3 was approximately 50 times more potent than T4 in promoting expression of the hypertrophic markers in prehypertrophic chondrocytes in cells cultured with insulin/transferrin/selenium (Alini et al., 1996). There was a biphasic dose-dependency of the effects of T3 and T4 to stimulate the synthesis of type II collagen and chondroitin sulfate-rich proteoglycans in cultured rabbit articular chondrocytes (Glade et al., 1994). The shift toward hypertrophic phenotype is mainly mediated via the upregulation of cyclin-dependent kinase inhibitors (Ballock et al., 2000). T3 also stimulates the synthesis of extracellular matrix proteins and enzymes involved in mineralization and matrix degradation including collagen X, matrix proteoglycans, ALP, MMP13, and aggrecanase-2.

In vivo responses of the skeleton to thyroid hormones: animal studies

Hypothyroidism

Animal models of hypothyroidism include the use of the antithyroid agents propylthiouracil and methimazole to block the synthesis of thyroid hormones. Treatment of young rats with methimazole for 7 weeks resulted in a marked increase in trabecular bone volume of the subchondral spongiosa of the mandibular condyles and a decrease in cartilage cellularity (Lewinson et al., 1994). IGF-I was present in the condyles of control rats but lacking in hypothyroid rats. Replacement of T4 during the last 2 weeks of treatment restored the parameters to normal (Lewinson et al., 1994). Histomorphometric studies in iliac crest biopsies of young rats made hypothyroid by a 12-week treatment with propylthiouracil showed that both osteoid surfaces and eroded surfaces were reduced, and cancellous bone volume was increased (Allain et al., 1995). In a study in which 21-day rats were made hypothyroid by administration of methimazole, T4 given daily at doses of 2–64 μg/kg/day for 21 days elicited biphasic effects on epiphyseal growth plate width and longitudinal growth rate (Ren et al., 1990). The dose–response curve paralleled that of serum IGF-I concentrations, which were postulated to contribute to the growth responses (Ren et al., 1990). An interesting animal model for hypothyroidism utilizes transgenic mice (line TG66-19) in which the bovine thyroglobulin promoter drives the expression of the herpes simplex type I virus thymidine kinase gene in thyrocytes. This enzyme converts ganciclovir to ganciclovir-59-phosphate, which inhibits DNA replication, resulting in loss of thyrocytes, loss of follicles, and undetectable T3 and T4; levels of PTH and CT are unaffected (Wallace et al., 1991, 1995). In this transgenic mouse model, administration of 15 or 50 μg of ganciclovir to mouse dams during days 14–18 of gestation resulted in growth delay in pups carrying the transgene (Wallace et al., 1995). The authors point out that the reason their effects were more dramatic than those obtained with the hyt/hyt mouse, a strain that has an inactivating mutation in the TSH receptor, is that in the latter model, circulating T4 is still 10%–20% of normal (Adams et al., 1989). Effects of mutations in thyroid hormone receptors in mouse models were discussed earlier.

Hyperthyroidism

A range of T4 regimens has been used to elicit hyperthyroidism in animal models. The duration of treatment is generally at least 3 weeks and dosages range from 200 μg to 1 g per day. Lower concentrations have been used in animals previously made hypothyroid with antithyroid drugs (Lewinson et al., 1994). When thyroid hormones are administered to young rats, bone growth is enhanced (Glasscock and Nicoll, 1981). This response is not seen in older rats, suggesting that the stage of cellular differentiation or the environment in terms of other hormones and local factors can influence the manifestation of thyroid hormone responses. T3 treatment of neonatal rats elicited a narrowing of the sagittal suture and increased mineral apposition rates at the osseous edges of the sutures (Akita et al., 1994). Histomorphometric analysis was consistent with the conclusion that T3 is critical for bone remodeling (Allain et al., 1995). When the animals were rendered hyperthyroid by treatment with T4 (200 μg/day for 12 weeks), the mineral apposition and formation rates were increased markedly, with a smaller increase in eroded surfaces (Allain et al., 1995). A greater sensitivity of cortical bone (femur) than trabecular bone (spine) to thyroid hormone-induced bone loss has been noted in animal models of hyperthyroidism (Ongphiphadhanakul et al., 1993; Suwanwalaikorn et al., 1996, 1997; Gouveia et al., 1997; Zeni et al., 2000). Tooth movement was greater in T3-treated rats undergoing orthodontic procedures than in control untreated animals, probably reflecting greater root resorption (Shirazi et al., 1999). Rats at 10 days old, treated with 100 μg/kg/day for up to 60 days, displayed altered parameters of cranial width, narrowing of the suture gap of the sagittal suture, and intense immunohistochemical staining for IGF-I along the suture margins, consistent with the possibility that local IGF-I is involved in the effect of thyroid hormone to cause premature suture closure (Akita et al., 1996). Ovariectomized rats treated with a low dose of T4 (30 μg/kg/day for 12 weeks) showed increased bone turnover and decreased bone density compared with controls; however, in the presence of 17β-estradiol, their bone mass and mineral apposition rate were greater than those of controls (Yamaura et al., 1994). T4 (250 μg/kg/day for 5 weeks) increased serum osteocalcin and urinary pyridinoline and produced a greater loss of bone mineral compared with either ovariectomy or T4 alone (Zeni et al., 2000). In contrast to the effects of these high doses of T4, administration of a more physiological concentration (10 μg/kg/day) to ovariectomized rats resulted in a generalized increase in bone mineral density at both lumbar and vertebral sites (Gouveia et al., 1997). Estradiol prevented T3-stimulated decreases in bone mineral density in ovariectomized thyroidectomized rats, but had no effect in animals that were not treated with T3 (DiPippo et al., 1995). These results raise the possibility of cross talk at the level of binding of estradiol and T3 receptors to DNA target sites.

Pathophysiological effects of altered thyroid hormone status in humans

Hypothyroidism

With an incidence of 1:1800, hypothyroidism is the most common congenital endocrine disorder, while primary hypothyroidism caused mostly by chronic autoimmune thyroiditis is the leading cause of hypothyroidism in childhood and adolescence. Approximately 2% of US adolescents have elevated TSH. Bone turnover is decreased in hypothyroidism (Mosekilde and Melsen, 1978), which in juvenile cases leads to abnormal endochondral ossification, delayed skeletal maturation and bone age, epiphyseal dysgenesis and short stature. Due to screening efforts in the Western world, the skeletal changes of untreated congenital hypothyroidism are rarely seen today. These might include complete cessation of bone maturation, growth arrest and skeletal dysplasia with a broad face, broad flat nasal bridge, hypertelorism, persistently patent sutures, scoliosis, vertebral immaturity and absence of ossification centers, and congenital hip dislocation. In a study of children with congenital hypothyroidism treated with T4, the bone age at 1.5 years was correlated positively with the dose of T4 administered during the first year and with the concentrations of serum T4 (Heyerdahl et al., 1994).

Adult-onset hypothyroidism leads to reduced remodeling rates with both the activity of osteoblasts and osteoclasts decreased. To present date, there are only a few studies reporting on changes of bone turnover markers in adult hypothyroidism. Urinary pyridinium cross-links (Nakamura et al., 1996) and serum IGF-I (Lakatos et al., 2000) were shown to be reduced in these patients. Changes in bone volume and density are hard to assess, as no patient remains untreated for a significant enough time. Most data come from patients with restored euthyroidism. Nevertheless, low turnover seems to result only in no change or even a slight increase in bone volume and density (Paul et al., 1988; Stamato et al., 2000; Vestergaard et al., 2002; González-Rodríguez et al., 2013).

Consistent with the lack of significant changes in bone density, the few studies that are available suggest that hypothyroidism or elevated TSH does not increase fracture risk in itself. However, as detailed below, larger doses of T4 supplementation or frank overtreatment of hypothyroidism is shown to be an independent risk factor for fracture (Cummings et al., 1995; Melton et al., 2000; Van Den Eeden et al., 2003; Ahmed et al., 2006; Flynn et al., 2010).

Subclinical hypothyroidism

Subclinical hypothyroidism—that is, elevated TSH accompanied with thyroid hormone concentrations in the normal range—is extremely prevalent, especially among the elderly. While some degree of decline in bone turnover is demonstrable in these patients (Meier et al., 2004), no association was shown with changes in BMD or fracture risk (Waring et al., 2013; Garin et al., 2014; Blum et al., 2015). Current treatment recommendations do not include skeletal considerations.

Hyperthyroidism

With Graves' disease being the leading cause, thyrotoxicosis is much less common in infants and adolescents than hypothyroidism. It results in accelerated growth and skeletal development. Advanced bone age due to the increase in chondrocyte differentiation in the epiphyseal growth plates leads to early cessation of growth and persistent short stature (Schlesinger and Fisher, 1951; Saggese et al., 1990). In severe early onset cases, craniosynostosis might occur.

In adults, hyperthyroidism shortens the bone turnover cycle, leading to high turnover rates. Histomorphometric analyses show increased osteoclast numbers and resorbing surfaces, with loss of trabecular bone volume (Mosekilde and Melsen, 1978). Histomorphometric data yield a kinetic model demonstrating accelerated bone remodeling, with a disproportionately greater increase in resorption and a net loss of bone with each cycle of remodeling (Eriksen, 1986). Since the initial description of bone loss and the "worm eaten" appearance of long bones in thyrotoxicosis by von Recklinghausen (1891) more than a century ago, substantial additional evidence has shown that excessive thyroid hormone production can lead to bone loss. In patients with hyperthyroidism, markers of bone turnover are increased in correlation with disease severity. Pyridinoline and hydroxypyridinoline cross-link excretion are elevated (Harvey et al., 1991; Garnero et al., 1994; Nagasaka et al., 1997; Engler et al., 1999), as are urinary N-terminal telopeptide of type I collagen (Mora et al., 1999; Pantazi et al., 2000) and serum carboxyterminal-1-telopeptide (Loviselli et al., 1997; Miyakawa et al., 1996; Nagasaka et al., 1997). Evidence of activation of osteoblasts in hyperthyroidism is the elevation of alkaline phosphatase (Mosekilde and Christesen, 1977; Cooper et al., 1979; Martinez et al., 1986; Nagasaka et al., 1997; Pantazi et al., 2000), osteocalcin (Martinez et al., 1986; Lee et al., 1990; Mosekilde et al., 1990; Nagasaka et al., 1997; Loviselli et al., 1997; Pantazi et al., 2000), and carboxyterminal propeptide of type I procollagen (Nagasaka et al., 1997). Consistent with the molecular studies (Varga et al., 2004) showing that T3 increases OPG in osteoblastic cells, elevated OPG has been reported in patients with Graves' disease (Amato et al., 2004; Mochizuki et al., 2006). Greater increases in the resorption markers than the formation markers verify the imbalance between resorption and formation, leading to bone loss (Garnero et al., 1994; Miyakawa et al., 1996).

Decreases in bone mineral content are established by several studies (Fraser et al., 1971; Krolner et al., 1983; Toh et al., 1985; Guo et al., 1997; Udayakumar et al., 2006; Majima et al., 2006), and a concomitant increase in fracture risk (Fraser et al., 1971; Wejda et al., 1995; Bauer et al., 2001; Vestergaard et al., 2000, 2005; Vestergaard and Mosekilde, 2003 and mortality (Franklyn et al., 1998; Patel et al., 2014) is also well documented. These latter changes are suggested by some to be only partly a result of thyrotoxicosis induced bone loss and partly a result of other aspects of the disease (Cummings et al., 1995). The skeletal effects of hyperthyroidism seem to be especially pronounced in postmenopausal women.

Recovery of bone loss in hyperthyroid patients following antithyroid treatment has been inconsistent (Fraser et al., 1971; Toh et al., 1985; Saggese et al., 1990; Diamond et al., 1994; Mudde et al., 1994; Olkawa et al., 1999; Kumeda et al., 2000; Barsal et al., 2004) but may be achieved more readily in younger individuals (Fraser et al., 1971; Saggese et al., 1990). Studies have documented protective effects of methimazole (Langdahl et al., 1996a; Nagasaka et al., 1997; Mora et al., 1999). Surgery and radioactive iodine also prevented bone loss in hyperthyroid patients (Langdahl et al., 1996b; Arata et al., 1997; Karunakaran et al., 2016) but were less protective than methimazole (Vestergaard et al., 2000a). Calcium supplementation with or without calcitonin (Kung and Yeung, 1996), estrogen (Lakatos et al., 1989), and bisphosphonates (Ongphiphadhanakul et al., 1993; Rosen et al., 1993a; Yamamoto et al., 1993; Rosen et al., 1993b; Kung and Ng., 1994; Lupoli et al., 1996) were also investigated and shown to be protective against ongoing bone loss in hyperthyroidism.

Additional to the treatment of the underlying condition, the therapy for hyperthyroidism-related osteoporosis does not differ from that of other secondary causes. Where data are available, novel antiporotic medications also seem to retain their efficacy in this clinical setting (Mirza and Canalis, 2015; Mana et al., 2017).

Subclinical hyperthyroidism

Subclinical hyperthyroidism is suggested to increase bone turnover rates. However data on whether this is actually the case and if so, whether this translates to an actual loss of bone mass and an increase in fracture risk is still controversial.

Data is derived from multiple sources including patients with endogenous subclinical hyperthyroidism and patients treated with suppressive thyroxine doses primarily for thyroid cancer and in rarer cases for euthyroid goiter. The question whether endogenous subclinical hyperthyroidism or larger amounts of exogenously administered thyroid hormones increase the risk of bone loss, especially among individuals already at risk for osteoporotic fractures, is of particular interest.

Bone turnover markers are shown to be elevated in certain studies while remaining physiologic in others (Nystrom et al., 1989; Karner et al., 2005; Reverter et al., 2005; El Hadidy et al., 2011; Lee et al., 2014). The situation regarding BMD and fracture risk is quite similar. A number of heterogeneous studies conducted in different age, gender, and disease groups show disparate degrees of TSH suppression. Densitometry was performed at varying anatomical locations using different methodologies. There are also large differences in the duration of follow-up. Effects of subclinical hyperthyroidism are thus reported to range from insignificant (Franklyn et al., 1992; Lee et al., 2014; Waring et al., 2013; Garin et al., 2014) to major (Kung and Yeung, 1996, Jódar et al., 1998; Tauchmanovà et al., 2004; Sugitani and Fujimoto, 2011; Kim et al., 2015). Where fracture risk was shown to be increased at all (Bauer et al., 2001; Turner et al., 2011; Abrahamsen et al., 2014, 2015), values of relative risk were reported between 1.25 (Vadiveloo et al., 2011) and 5 (Lee et al., 2010).

While many papers report on some degree of bone loss and increase in fracture risk, the dose of thyroxine treatment, the degree and duration of TSH suppression, sex, and postmenopausal status in women seem to be major determinants. The protective effect of adequate calcium intake was also suggested in one paper (Kung and Yeung, 1996). Meta-analyses seem to confirm these differences (Faber et al., 1994; Uzzan et al., 1996; Quan et al., 2002; Heemstra et al., 2006; Wirth et al., 2014; Blum et al., 2015). The available evidence regarding decreased BMD and increased fracture risk in subclinical hyperthyroidism of both endogenous and exogenous origin is most robust in postmenopausal women.

A few studies have also been conducted on the potential skeletal effects of variations in thyroid markers within the physiologic range. With lower TSH levels and thyroid function at the upper end of the normal range, osteoporosis and fracture risk were found to be increased again, primarily in postmenopausal women (Kim et al., 2006; Morris et al., 2007; Murphy et al., 2010; Lin et al., 2011; van Rijn et al., 2014; Noh et al., 2015; Hwangbo et al., 2016). In the case of premenopausal women and men, changes in bone metabolism related to subclinical hyperthyroidism or normal variations in TSH are much less convincing. For further assessment, possibly more prospective studies are needed.

Overview and future directions

Thyroid hormones play a major role in skeletal physiology and pathology both during development and in adult bone maintenance. Their lack results in cessation of bone maturation, growth, and remodeling, while their excess is associated with accelerated bone maturation and accelerated bone loss in adulthood. Thyrotoxicosis is a prevalent cause of secondary osteoporosis, while subclinical hyperthyroidism, suppressive doses of thyroxine treatment, and possibly higher thyroid activity within the normal range are all associated with increased bone loss and fracture risk, especially in postmenopausal women.

Osteoblasts and chondrocytes are the primary target cells of thyroid hormones, while secondary paracrine mediators are implied in the effects of T3 on osteoclasts. Different isoforms of thyroid hormone receptors, mediating their genomic action, are well characterized, with TRα1 being the main isoform in bone. RTHα, a condition presenting with marked skeletal changes, is caused by the dominant negative mutations of the TRα gene. The disease has only been recognized in the past few years, and additional studies might bring further insights into the role of thyroid hormones on skeletal physiology. Selective thyroid receptor analogues may present a therapeutic option for metabolic syndrome or chronic heart failure in the future.

Nongenomic actions of thyroid hormones mediated via membrane and cytoplasmic receptors are increasingly recognized and pose an interesting subject for further research. The presence of a pituitary–bone axis and the potential role of TSH as a quasi-independent regulator of bone development and remodeling have also recently been posited. Though still contradictory, results are promising in this challenging topic.

References

Abe, E., Marians, R.C., Yu, W., Wu, X.B., Ando, T., Li, Y., Iqbal, J., Eldeiry, L., Rajendren, G., Blair, H.C., Davies, T.F., Zaidi, M., 2003. TSH is a negative regulator of skeletal remodeling. Cell 115 (2), 151–162.

Abe, S., Namba, N., Abe, M., Fujiwara, M., Aikawa, T., Kogo, M., Ozono, K., 2012. Monocarboxylate transporter 10 functions as a thyroid hormone transporter in chondrocytes. Endocrinology 153 (8), 4049–4058.

Abrahamsen, B., Jørgensen, H.L., Laulund, A.S., Nybo, M., Bauer, D.C., Brix, T.H., Hegedüs, L., 2015. The excess risk of major osteoporotic fractures in hypothyroidism is driven by cumulative hyperthyroid as opposed to hypothyroid time: an observational register-based time-resolved cohort analysis. J. Bone Miner. Res. 30 (5), 898–905.

Abrahamsen, B., Jørgensen, H.L., Laulund, A.S., Nybo, M., Brix, T.H., Hegedüs, L., 2014. Low serum thyrotropin level and duration of suppression as a predictor of major osteoporotic fractures-the OPENTHYRO register cohort. J. Bone Miner. Res. 29 (9), 2040–2050.

Abu, E.O., Bord, S., Horner, A., Chatterjee, V.K., Compston, J.E., 1997. The expression of thyroid hormone receptors in human bone. Bone 21 (2), 137–142.

Abu, E.O., Horner, A., Teti, A., Chatterjee, V.K., Compston, J.E., 2000. The localization of thyroid hormone receptor mRNAs in human bone. Thyroid 10 (4), 287–293.

Adams, P.M., Stein, S.A., Palnitkar, M., Anthony, A., Gerrity, L., Shanklin, D.R., 1989. Evaluation and characterization of the hypothyroid hyt/hyt mouse. I: somatic and behavioral studies. Neuroendocrinology 49 (2), 138–143.

Ahmed, L.A., Schirmer, H., Berntsen, G.K., Fønnebø, V., Joakimsen, R.M., 2006. Self-reported diseases and the risk of non-vertebral fractures: the Tromsø study. Osteoporos. Int. 17 (1), 46–53.

Akita, S., Hirano, A., Fujii, T., 1996. Identification of IGF-I in the calvarial suture of young rats: histochemical analysis of the cranial sagittal sutures in a hyperthyroid rat model. Plast. Reconstr. Surg. 97 (1), 1–12.

Akita, S., Nakamura, T., Hirano, A., Fujii, T., Yamashita, S., 1994. Thyroid hormone action on rat calvarial sutures. Thyroid 4 (1), 99–106.

Alini, M., Kofsky, Y., Wu, W., Pidoux, I., Poole, A.R., 1996. In serum-free culture thyroid hormones can induce full expression of chondrocyte hypertrophy leading to matrix calcification. J. Bone Miner. Res. 11 (1), 105–113.

Allain, T.J., Chambers, T.J., Flanagan, A.M., McGregor, A.M., 1992. Tri-iodothyronine stimulates rat osteoclastic bone resorption by an indirect effect. J. Endocrinol. 133 (3), 327–331.

Allain, T.J., Thomas, M.R., McGregor, A.M., Salisbury, J.R., 1995. A histomorphometric study of bone changes in thyroid dysfunction in rats. Bone 16 (5), 505–509.

Allain, T.J., Yen, P.M., Flanagan, A.M., McGregor, A.M., 1996. The isoform-specific expression of the tri-iodothyronine receptor in osteoblasts and osteoclasts. Eur. J. Clin. Investig. 26 (5), 418–425.

Amato, G., Mazziotti, G., Sorvillo, F., Piscopo, M., Lalli, E., Biondi, B., Iorio, S., Molinari, A., Giustina, A., Carella, C., 2004. High serum osteoprotegerin levels in patients with hyperthyroidism: effect of medical treatment. Bone 35 (3), 785–791.

Arata, N., Momotani, N., Maruyama, H., Saruta, T., Tsukatani, K., Kubo, A., Ikemoto, K., Ito, K., 1997. Bone mineral density after surgical treatment for Graves' disease. Thyroid 7 (4), 547–554.

Asai, S., Cao, X., Yamauchi, M., Funahashi, K., Ishiguro, N., Kambe, F., 2009. Thyroid hormone non-genomically suppresses Src thereby stimulating osteocalcin expression in primary mouse calvarial osteoblasts. Biochem. Biophys. Res. Commun. 387 (1), 92–96.

Baliram, R., Latif, R., Berkowitz, J., Frid, S., Colaianni, G., Sun, L., Zaidi, M., Davies, T.F., 2011. Thyroid-stimulating hormone induces a Wnt-dependent, feed-forward loop for osteoblastogenesis in embryonic stem cell cultures. Proc. Natl. Acad. Sci. U. S. A. 108 (39), 16277–16282.

Baliram, R., Sun, L., Cao, J., Li, J., Latif, R., Huber, A.K., Yuen, T., Blair, H.C., Zaidi, M., Davies, T.F., 2012. Hyperthyroid-associated osteoporosis is exacerbated by the loss of TSH signaling. J. Clin. Investig. 122 (10), 3737–3741.

Ballock, R., Mita, B.C., Zhou, X., Chen, D.H., Mink, L.M., 1999. Expression of thyroid hormone receptor isoforms in rat growth plate cartilage in vivo. J. Bone Miner. Res. 14 (9), 1550–1556.

Ballock, R.T., Reddi, A.H., 1994. Thyroxine is the serum factor that regulates morphogenesis of columnar cartilage from isolated chondrocytes in chemically defined medium. J. Cell Biol. 126 (5), 1311–1318.

Ballock, R.T., Zhou, X., Mink, L.M., Chen, D.H., Mita, B.C., Stewart, M.C., 2000. Expression of cyclin-dependent kinase inhibitors in epiphyseal chondrocytes induced to terminally differentiate with thyroid hormone. Endocrinology 141 (12), 4552–4557.

Banovac, K., Koren, E., 2000. Triiodothyronine stimulates the release of membrane-bound alkaline phosphatase in osteoblastic cells. Calcif. Tissue Int. 67 (6), 460–465.

Barsal, G., Taneli, F., Atay, A., Hekimsoy, Z., Erciyas, F., 2004. Serum osteocalcin levels in hyperthyroidism before and after antithyroid therapy. Tohoku J. Exp. Med. 203 (3), 183–188.

Bassett, J.H., Boyde, A., Howell, P.G., Bassett, R.H., Galliford, T.M., Archanco, M., Evans, H., Lawson, M.A., Croucher, P., St Germain, D.L., Galton, V.A., Williams, G.R., 2010. Optimal bone strength and mineralization requires the type 2 iodothyronine deiodinase in osteoblasts. Proc. Natl. Acad. Sci. U. S. A. 107 (16), 7604–7609.

Bassett, J.H., Boyde, A., Zikmund, T., Evans, H., Croucher, P.I., Zhu, X., Park, J.W., Cheng, S.Y., Williams, G.R., 2014. Thyroid hormone receptor α mutation causes a severe and thyroxine-resistant skeletal dysplasia in female mice. Endocrinology 155 (9), 3699–3712.

Bassett, J.H., Nordström, K., Boyde, A., Howell, P.G., Kelly, S., Vennström, B., Williams, G.R., 2007a. Thyroid status during skeletal development determines adult bone structure and mineralization. Mol. Endocrinol. 21 (8), 1893–1904.

Bassett, J.H., O'Shea, P.J., Sriskantharajah, S., Rabier, B., Boyde, A., Howell, P.G., Weiss, R.E., Roux, J.P., Malaval, L., Clement-Lacroix, P., Samarut, J., Chassande, O., Williams, G.R., 2007b. Thyroid hormone excess rather than thyrotropin deficiency induces osteoporosis in hyperthyroidism. Mol. Endocrinol. 21 (5), 1095–1107.

Bassett, J.H., van der Spek, A., Logan, J.G., Gogakos, A., Bagchi-Chakraborty, J., Murphy, E., van Zeijl, C., Down, J., Croucher, P.I., Boyde, A., Boelen, A., Williams, G.R., 2015. Thyrostimulin regulates osteoblastic bone formation during early skeletal development. Endocrinology 156 (9), 3098–3113.

Bassett, J.H., Williams, G.R., 2003. The molecular actions of thyroid hormone in bone. Trends Endocrinol. Metabol. 14 (8), 356–364.

Bassett, J.H., Williams, G.R., 2008. Critical role of the hypothalamic-pituitary-thyroid axis in bone. Bone 43 (3), 418–426.

Bauer, D.C., Ettinger, B., Nevitt, M.C., Stone, K.L., Study of Osteoporotic Fractures Research Group, 2001. Risk for fracture in women with low serum levels of thyroid-stimulating hormone. Ann. Intern. Med. 134 (7), 561–568.

Beamer, W.J., Eicher, E.M., Maltais, L.J., Southard, J.L., 1981. Inherited primary hypothyroidism in mice. Science 212 (4490), 61–63.

Behr, M., Ramsden, D.B., Loos, U., 1997. Deoxyribonucleic acid binding and transcriptional silencing by a truncated c-erbA beta 1 thyroid hormone receptor identified in a severely retarded patient with resistance to thyroid hormone. J. Clin. Endocrinol. Metab. 82 (4), 1081–1087.

Bergh, J.J., Lin, H.Y., Lansing, L., Mohamed, S.N., Davis, F.B., Mousa, S., Davis, P.J., 2005. Integrin alphaVbeta3 contains a cell surface receptor site for thyroid hormone that is linked to activation of mitogen-activated protein kinase and induction of angiogenesis. Endocrinology 146 (7), 2864–2871.

Bland, R., Sammons, R.L., Sheppard, M.C., Williams, G.R., 1997. Thyroid hormone, vitamin D and retinoid receptor expression and signalling in primary cultures of rat osteoblastic and immortalised osteosarcoma cells. J. Endocrinol. 154 (1), 63–74.

Blum, M.R., Bauer, D.C., Collet, T.H., Fink, H.A., Cappola, A.R., da Costa, B.R., Wirth, C.D., Peeters, R.P., Åsvold, B.O., den Elzen, W.P., Luben, R.N., Imaizumi, M., Bremner, A.P., Gogakos, A., Eastell, R., Kearney, P.M., Strotmeyer, E.S., Wallace, E.R., Hoff, M., Ceresini, G., Rivadeneira, F., Uitterlinden, A.G., Stott, D.J., Westendorp, R.G., Khaw, K.T., Langhammer, A., Ferrucci, L., Gussekloo, J., Williams, G.R., Walsh, J.P., Jüni, P., Aujesky, D., Rodondi, N., Collaboration, T.S., 2015. Subclinical thyroid dysfunction and fracture risk: a meta-analysis. J. Am. Med. Assoc. 313 (20), 2055–2065.

Bochukova, E., Schoenmakers, N., Agostini, M., Schoenmakers, E., Rajanayagam, O., Keogh, J.M., Henning, E., Reinemund, J., Gevers, E., Sarri, M., Downes, K., Offiah, A., Albanese, A., Halsall, D., Schwabe, J.W., Bain, M., Lindley, K., Muntoni, F., Vargha-Khadem, F., Khadem, F.V., Dattani, M., Farooqi, I.S., Gurnell, M., Chatterjee, K., 2012. A mutation in the thyroid hormone receptor alpha gene. N. Engl. J. Med. 366 (3), 243–249.

Boutin, A., Eliseeva, E., Gershengorn, M.C., Neumann, S., 2014. β-Arrestin-1 mediates thyrotropin-enhanced osteoblast differentiation. FASEB J. 28 (8), 3446–3455.

Brent, G.A., Moore, D.D., Larsen, P.R., 1991. Thyroid hormone regulation of gene expression. Annu. Rev. Physiol. 53, 17–35.

Burakov, D., Crofts, L.A., Chang, C.P., Freedman, L.P., 2002. Reciprocal recruitment of DRIP/mediator and p160 coactivator complexes in vivo by estrogen receptor. J. Biol. Chem. 277 (17), 14359–14362.

Burch, W.M., Lebovitz, H.E., 1982a. Triiodothyronine stimulates maturation of porcine growth-plate cartilage in vitro. J. Clin. Investig. 70 (3), 496–504.

Burch, W.M., Lebovitz, H.E., 1982b. Triiodothyronine stimulation of in vitro growth and maturation of embryonic chick cartilage. Endocrinology 111 (2), 462–468.

Böhme, K., Conscience-Egli, M., Tschan, T., Winterhalter, K.H., Bruckner, P., 1992. Induction of proliferation or hypertrophy of chondrocytes in serum-free culture: the role of insulin-like growth factor-I, insulin, or thyroxine. J. Cell Biol. 116 (4), 1035–1042.

Canalis, E., 1980. Effect of insulinlike growth factor I on DNA and protein synthesis in cultured rat calvaria. J. Clin. Investig. 66 (4), 709–719.

Capelo, L.P., Beber, E.H., Fonseca, T.L., Gouveia, C.H., 2009. The monocarboxylate transporter 8 and L-type amino acid transporters 1 and 2 are expressed in mouse skeletons and in osteoblastic MC3T3-E1 cells. Thyroid 19 (2), 171–180.

Capelo, L.P., Beber, E.H., Huang, S.A., Zorn, T.M., Bianco, A.C., Gouveia, C.H., 2008. Deiodinase-mediated thyroid hormone inactivation minimizes thyroid hormone signaling in the early development of fetal skeleton. Bone 43 (5), 921–930.

Centrella, M., McCarthy, T.L., Canalis, E., 1990. Receptors for insulin-like growth factors-I and -II in osteoblast-enriched cultures from fetal rat bone. Endocrinology 126 (1), 39–44.

Chassande, O., Fraichard, A., Gauthier, K., Flamant, F., Legrand, C., Savatier, P., Laudet, V., Samarut, J., 1997. Identification of transcripts initiated from an internal promoter in the c-erbA alpha locus that encode inhibitors of retinoic acid receptor-alpha and triiodothyronine receptor activities. Mol. Endocrinol. 11 (9), 1278–1290.

Chatterjee, V.K., Nagaya, T., Madison, L.D., Datta, S., Rentoumis, A., Jameson, J.L., 1991. Thyroid hormone resistance syndrome. Inhibition of normal receptor function by mutant thyroid hormone receptors. J. Clin. Investig. 87 (6), 1977–1984.

Chiellini, G., Nguyen, N.H., Apriletti, J.W., Baxter, J.D., Scanlan, T.S., 2002. Synthesis and biological activity of novel thyroid hormone analogues: 5'-aryl substituted GC-1 derivatives. Bioorg. Med. Chem. 10 (2), 333–346.

Conaway, H.H., Ransjö, M., Lerner, U.H., 1998. Prostaglandin-independent stimulation of bone resorption in mouse calvariae and in isolated rat osteoclasts by thyroid hormones (T4, and T3). Proc. Soc. Exp. Biol. Med. 217 (2), 153–161.

Cooper, D.S., Kaplan, M.M., Ridgway, E.C., Maloof, F., Daniels, G.H., 1979. Alkaline phosphatase isoenzyme patterns in hyperthyroidism. Ann. Intern. Med. 90 (2), 164–168.

Cray, J.J., Khaksarfard, K., Weinberg, S.M., Elsalanty, M., Yu, J.C., 2013. Effects of thyroxine exposure on osteogenesis in mouse calvarial preosteoblasts. PLoS One 8 (7), e69067.

Cummings, S.R., Nevitt, M.C., Browner, W.S., Stone, K., Fox, K.M., Ensrud, K.E., Cauley, J., Black, D., Vogt, T.M., 1995. Risk factors for hip fracture in white women. Study of Osteoporotic Fractures Research Group. N. Engl. J. Med. 332 (12), 767–773.

Davis, P.J., Davis, F.B., Mousa, S.A., Luidens, M.K., Lin, H.Y., 2011. Membrane receptor for thyroid hormone: physiologic and pharmacologic implications. Annu. Rev. Pharmacol. Toxicol. 51, 99–115.

Davis, P.J., Goglia, F., Leonard, J.L., 2016. Nongenomic actions of thyroid hormone. Nat. Rev. Endocrinol. 12 (2), 111–121.

Davis, P.J., Leonard, J.L., Davis, F.B., 2008. Mechanisms of nongenomic actions of thyroid hormone. Front. Neuroendocrinol. 29 (2), 211–218.

Davis, P.J., Lin, H.Y., Tang, H.Y., Davis, F.B., Mousa, S.A., 2013. Adjunctive input to the nuclear thyroid hormone receptor from the cell surface receptor for the hormone. Thyroid 23 (12), 1503–1509.

Davis, P.J., Shih, A., Lin, H.Y., Martino, L.J., Davis, F.B., 2000. Thyroxine promotes association of mitogen-activated protein kinase and nuclear thyroid hormone receptor (TR) and causes serine phosphorylation of TR. J. Biol. Chem. 275 (48), 38032–38039.

Diamond, T., Vine, J., Smart, R., Butler, P., 1994. Thyrotoxic bone disease in women: a potentially reversible disorder. Ann. Intern. Med. 120 (1), 8–11.

DiPippo, V.A., Lindsay, R., Powers, C.A., 1995. Estradiol and tamoxifen interactions with thyroid hormone in the ovariectomized-thyroidectomized rat. Endocrinology 136 (3), 1020–1033.

Dumic-Cule, I., Draca, N., Luetic, A.T., Jezek, D., Rogic, D., Grgurevic, L., Vukicevic, S., 2014. TSH prevents bone resorption and with calcitriol synergistically stimulates bone formation in rats with low levels of calciotropic hormones. Horm. Metab. Res. 46 (5), 305–312.

Dumitrescu, A.M., Liao, X.H., Best, T.B., Brockmann, K., Refetoff, S., 2004. A novel syndrome combining thyroid and neurological abnormalities is associated with mutations in a monocarboxylate transporter gene. Am. J. Hum. Genet. 74 (1), 168–175.

El Hadidy, e. H., Ghonaim, M., El Gawad, S. S. h., El Atta, M.A., 2011. Impact of severity, duration, and etiology of hyperthyroidism on bone turnover markers and bone mineral density in men. BMC Endocr. Disord. 11, 15.

Endo, T., Kobayashi, T., 2013. Excess TSH causes abnormal skeletal development in young mice with hypothyroidism via suppressive effects on the growth plate. Am. J. Physiol. Endocrinol. Metab. 305 (5), E660–E666.

Engler, H., Oettli, R.E., Riesen, W.F., 1999. Biochemical markers of bone turnover in patients with thyroid dysfunctions and in euthyroid controls: a cross-sectional study. Clin. Chim. Acta 289 (1–2), 159–172.

Eriksen, E.F., 1986. Normal and pathological remodeling of human trabecular bone: three dimensional reconstruction of the remodeling sequence in normals and in metabolic bone disease. Endocr. Rev. 7 (4), 379–408.

Ernst, M., Froesch, E.R., 1987. Triiodothyronine stimulates proliferation of osteoblast-like cells in serum-free culture. FEBS Lett. 220 (1), 163–166.

Espiard, S., Savagner, F., Flamant, F., Vlaeminck-Guillem, V., Guyot, R., Munier, M., d'Herbomez, M., Bourguet, W., Pinto, G., Rose, C., Rodien, P., Wémeau, J.L., 2015. A novel mutation in THRA gene associated with an atypical phenotype of resistance to thyroid hormone. J. Clin. Endocrinol. Metab. 100 (8), 2841–2848.

Faber, J., Galløe, A.M., 1994. Changes in bone mass during prolonged subclinical hyperthyroidism due to L-thyroxine treatment: a meta-analysis. Eur. J. Endocrinol. 130 (4), 350–356.

Flynn, R.W., Bonellie, S.R., Jung, R.T., MacDonald, T.M., Morris, A.D., Leese, G.P., 2010. Serum thyroid-stimulating hormone concentration and morbidity from cardiovascular disease and fractures in patients on long-term thyroxine therapy. J. Clin. Endocrinol. Metab. 95 (1), 186–193.

Fraichard, A., Chassande, O., Plateroti, M., Roux, J.P., Trouillas, J., Dehay, C., Legrand, C., Gauthier, K., Kedinger, M., Malaval, L., Rousset, B., Samarut, J., 1997. The T3R alpha gene encoding a thyroid hormone receptor is essential for post-natal development and thyroid hormone production. EMBO J. 16 (14), 4412–4420.

Franklyn, J.A., Betteridge, J., Daykin, J., Holder, R., Oates, G.D., Parle, J.V., Lilley, J., Heath, D.A., Sheppard, M.C., 1992. Long-term thyroxine treatment and bone mineral density. Lancet 340 (8810), 9–13.

Franklyn, J.A., Maisonneuve, P., Sheppard, M.C., Betteridge, J., Boyle, P., 1998. Mortality after the treatment of hyperthyroidism with radioactive iodine. N. Engl. J. Med. 338 (11), 712–718.

Fraser, S.A., Wilson, G.M., 1971. Plasma-calcitonin in disorders of thyroid function. Lancet 1 (7702), 725–726.

Fratzl-Zelman, N., Hörandner, H., Luegmayr, E., Varga, F., Ellinger, A., Erlee, M.P., Klaushofer, K., 1997. Effects of triiodothyronine on the morphology of cells and matrix, the localization of alkaline phosphatase, and the frequency of apoptosis in long-term cultures of MC3T3-E1 cells. Bone 20 (3), 225–236.

Garin, M.C., Arnold, A.M., Lee, J.S., Robbins, J., Cappola, A.R., 2014. Subclinical thyroid dysfunction and hip fracture and bone mineral density in older adults: the cardiovascular health study. J. Clin. Endocrinol. Metab. 99 (8), 2657–2664.

Garnero, P., Vassy, V., Bertholin, A., Riou, J.P., Delmas, P.D., 1994. Markers of bone turnover in hyperthyroidism and the effects of treatment. J. Clin. Endocrinol. Metab. 78 (4), 955–959.

Gauthier, K., Chassande, O., Plateroti, M., Roux, J.P., Legrand, C., Pain, B., Rousset, B., Weiss, R., Trouillas, J., Samarut, J., 1999. Different functions for the thyroid hormone receptors TRalpha and TRbeta in the control of thyroid hormone production and post-natal development. EMBO J. 18 (3), 623–631.

Germain, P., Staels, B., Dacquet, C., Spedding, M., Laudet, V., 2006. Overview of nomenclature of nuclear receptors. Pharmacol. Rev. 58 (4), 685–704.

Glade, M.J., Kanwar, Y.S., Stern, P.H., 1994. Insulin and thyroid hormones stimulate matrix metabolism in primary cultures of articular chondrocytes from young rabbits independently and in combination. Connect. Tissue Res. 31 (1), 37–44.

Glantschnig, H., Varga, F., Klaushofer, K., 1996. Thyroid hormone and retinoic acid induce the synthesis of insulin-like growth factor-binding protein-4 in mouse osteoblastic cells. Endocrinology 137 (1), 281–286.

Glass, C.K., 1994. Differential recognition of target genes by nuclear receptor monomers, dimers, and heterodimers. Endocr. Rev. 15 (3), 391–407.

Glasscock, G.F., Nicoll, C.S., 1981. Hormonal control of growth in the infant rat. Endocrinology 109 (1), 176–184.

Gloss, B., Trost, S., Bluhm, W., Swanson, E., Clark, R., Winkfein, R., Janzen, K., Giles, W., Chassande, O., Samarut, J., Dillmann, W., 2001. Cardiac ion channel expression and contractile function in mice with deletion of thyroid hormone receptor alpha or beta. Endocrinology 142 (2), 544–550.

Goldman, S., McCarren, M., Morkin, E., Ladenson, P.W., Edson, R., Warren, S., Ohm, J., Thai, H., Churby, L., Barnhill, J., O'Brien, T., Anand, I., Warner, A., Hattler, B., Dunlap, M., Erikson, J., Shih, M.C., Lavori, P., 2009. DITPA (3,5-Diiodothyropropionic Acid), a thyroid hormone analog to treat heart failure: phase II trial veterans affairs cooperative study. Circulation 119 (24), 3093–3100.

González-Rodríguez, L.A., Felici-Giovanini, M.E., Haddock, L., 2013. Thyroid dysfunction in an adult female population: a population-based study of Latin American Vertebral Osteoporosis Study (LAVOS) – Puerto Rico site. Puert. Rico Health Sci. J. 32 (2), 57–62.

Gorka, J., Taylor-Gjevre, R.M., Arnason, T., 2013. Metabolic and clinical consequences of hyperthyroidism on bone density. Int. J. Endocrinol. 2013, 638–727.

Gouveia, C.H., Christoffolete, M.A., Zaitune, C.R., Dora, J.M., Harney, J.W., Maia, A.L., Bianco, A.C., 2005. Type 2 iodothyronine selenodeiodinase is expressed throughout the mouse skeleton and in the MC3T3-E1 mouse osteoblastic cell line during differentiation. Endocrinology 146 (1), 195–200.

Gouveia, C.H., Jorgetti, V., Bianco, A.C., 1997. Effects of thyroid hormone administration and estrogen deficiency on bone mass of female rats. J. Bone Miner. Res. 12 (12), 2098–2107.

Gouveia, C.H., Schultz, J.J., Bianco, A.C., Brent, G.A., 2001. Thyroid hormone stimulation of osteocalcin gene expression in ROS 17/2.8 cells is mediated by transcriptional and post-transcriptional mechanisms. J. Endocrinol. 170 (3), 667–675.

Grover, G.J., Mellström, K., Ye, L., Malm, J., Li, Y.L., Bladh, L.G., Sleph, P.G., Smith, M.A., George, R., Vennström, B., Mookhtiar, K., Horvath, R., Speelman, J., Egan, D., Baxter, J.D., 2003. Selective thyroid hormone receptor-beta activation: a strategy for reduction of weight, cholesterol, and lipoprotein (a) with reduced cardiovascular liability. Proc. Natl. Acad. Sci. U. S. A. 100 (17), 10067–10072.

Gu, W.X., Stern, P.H., Madison, L.D., Du, G.G., 2001. Mutual up-regulation of thyroid hormone and parathyroid hormone receptors in rat osteoblastic osteosarcoma 17/2.8 cells. Endocrinology 142 (1), 157–164.

Guo, C.Y., Weetman, A.P., Eastell, R., 1997. Longitudinal changes of bone mineral density and bone turnover in postmenopausal women on thyroxine. Clin. Endocrinol. 46 (3), 301–307.

Göthe, S., Wang, Z., Ng, L., Kindblom, J.M., Barros, A.C., Ohlsson, C., Vennström, B., Forrest, D., 1999. Mice devoid of all known thyroid hormone receptors are viable but exhibit disorders of the pituitary-thyroid axis, growth, and bone maturation. Genes Dev. 13 (10), 1329–1341.

Harvey, R.D., McHardy, K.C., Reid, I.W., Paterson, F., Bewsher, P.D., Duncan, A., Robins, S.P., 1991. Measurement of bone collagen degradation in hyperthyroidism and during thyroxine replacement therapy using pyridinium cross-links as specific urinary markers. J. Clin. Endocrinol. Metab. 72 (6), 1189–1194.

Hase, H., Ando, T., Eldeiry, L., Brebene, A., Peng, Y., Liu, L., Amano, H., Davies, T.F., Sun, L., Zaidi, M., Abe, E., 2006. TNFalpha mediates the skeletal effects of thyroid-stimulating hormone. Proc. Natl. Acad. Sci. U. S. A. 103 (34), 12849–12854.

Heemstra, K.A., Hamdy, N.A., Romijn, J.A., Smit, J.W., 2006. The effects of thyrotropin-suppressive therapy on bone metabolism in patients with well-differentiated thyroid carcinoma. Thyroid 16 (6), 583–591.

Herzovich, V., Vaiani, E., Marino, R., Dratler, G., Lazzati, J.M., Tilitzky, S., Ramirez, P., Iorcansky, S., Rivarola, M.A., Belgorosky, A., 2007. Unexpected peripheral markers of thyroid function in a patient with a novel mutation of the MCT8 thyroid hormone transporter gene. Horm. Res. 67 (1), 1–6.

Heyerdahl, S., Kase, B.F., Stake, G., 1994. Skeletal maturation during thyroxine treatment in children with congenital hypothyroidism. Acta Paediatr. 83 (6), 618–622.

Hiroi, Y., Kim, H.H., Ying, H., Furuya, F., Huang, Z., Simoncini, T., Noma, K., Ueki, K., Nguyen, N.H., Scanlan, T.S., Moskowitz, M.A., Cheng, S.Y., Liao, J.K., 2006. Rapid nongenomic actions of thyroid hormone. Proc. Natl. Acad. Sci. U. S. A. 103 (38), 14104–14109.

Hock, J.M., Gunness-Hey, M., Poser, J., Olson, H., Bell, N.H., Raisz, L.G., 1986. Stimulation of undermineralized matrix formation by 1,25 dihydroxyvitamin D3 in long bones of rats. Calcif. Tissue Int. 38 (2), 79–86.

Hoffman, S.J., Vasko-Moser, J., Miller, W.H., Lark, M.W., Gowen, M., Stroup, G., 2002. Rapid inhibition of thyroxine-induced bone resorption in the rat by an orally active vitronectin receptor antagonist. J. Pharmacol. Exp. Ther. 302 (1), 205–211.

Hoffmann, O., Klaushofer, K., Koller, K., Peterlik, M., Mavreas, T., Stern, P.H., 1986. Indomethacin inhibits thrombin-but not thyroxin-stimulated resorption of fetal rat limb bones. Prostaglandins 31 (4), 601–608.

Huang, B.K., Golden, L.A., Tarjan, G., Madison, L.D., Stern, P.H., 2000. Insulin-like growth factor I production is essential for anabolic effects of thyroid hormone in osteoblasts. J. Bone Miner. Res. 15 (2), 188–197.

Hwangbo, Y., Kim, J.H., Kim, S.W., Park, Y.J., Park, D.J., Kim, S.Y., Shin, C.S., Cho, N.H., 2016. High-normal free thyroxine levels are associated with low trabecular bone scores in euthyroid postmenopausal women. Osteoporos. Int. 27 (2), 457–462.

Ishikawa, Y., Genge, B.R., Wuthier, R.E., Wu, L.N., 1998. Thyroid hormone inhibits growth and stimulates terminal differentiation of epiphyseal growth plate chondrocytes. J. Bone Miner. Res. 13 (9), 1398–1411.

Izumo, S., Mahdavi, V., 1988. Thyroid hormone receptor alpha isoforms generated by alternative splicing differentially activate myosin HC gene transcription. Nature 334 (6182), 539–542.

Jódar, E., Begoña López, M., García, L., Rigopoulou, D., Martínez, G., Hawkins, F., 1998. Bone changes in pre- and postmenopausal women with thyroid cancer on levothyroxine therapy: evolution of axial and appendicular bone mass. Osteoporos. Int. 8 (4), 311–316.

Kalyanaraman, H., Schwappacher, R., Joshua, J., Zhuang, S., Scott, B.T., Klos, M., Casteel, D.E., Frangos, J.A., Dillmann, W., Boss, G.R., Pilz, R.B., 2014. Nongenomic thyroid hormone signaling occurs through a plasma membrane-localized receptor. Sci. Signal. 7 (326), ra48.

Kaneshige, M., Suzuki, H., Kaneshige, K., Cheng, J., Wimbrow, H., Barlow, C., Willingham, M.C., Cheng, S., 2001. A targeted dominant negative mutation of the thyroid hormone alpha 1 receptor causes increased mortality, infertility, and dwarfism in mice. Proc. Natl. Acad. Sci. U. S. A. 98 (26), 15095–15100.

Karner, I., Hrgović, Z., Sijanović, S., Buković, D., Klobucar, A., Usadel, K.H., Fassbender, W.J., 2005. Bone mineral density changes and bone turnover in thyroid carcinoma patients treated with supraphysiologic doses of thyroxine. Eur. J. Med. Res. 10 (11), 480–488.

Karunakaran, P., Maharajan, C., Mohamed, K.N., Rachamadugu, S.V., 2016. Rapid restoration of bone mass after surgical management of hyperthyroidism: a prospective case control study in Southern India. Surgery 159 (3), 771–776.

Kasono, K., Sato, K., Han, D.C., Fujii, Y., Tsushima, T., Shizume, K., 1988. Stimulation of alkaline phosphatase activity by thyroid hormone in mouse osteoblast-like cells (MC3T3-E1): a possible mechanism of hyperalkaline phosphatasia in hyperthyroidism. Bone Miner. 4 (4), 355–363.

Kassem, M., Mosekilde, L., Eriksen, E.F., 1993. Effects of triiodothyronine on DNA synthesis and differentiation markers of normal human osteoblast-like cells in vitro. Biochem. Mol. Biol. Int. 30 (4), 779–788.

Kawaguchi, H., Pilbeam, C.C., Raisz, L.G., 1994a. Anabolic effects of 3,3',5-triiodothyronine and triiodothyroacetic acid in cultured neonatal mouse parietal bones. Endocrinology 135 (3), 971–976.

Kawaguchi, H., Pilbeam, C.C., Woodiel, F.N., Raisz, L.G., 1994b. Comparison of the effects of 3,5,3'-triiodothyroacetic acid and triiodothyronine on bone resorption in cultured fetal rat long bones and neonatal mouse calvariae. J. Bone Miner. Res. 9 (2), 247–253.

Kim, C.W., Hong, S., Oh, S.H., Lee, J.J., Han, J.Y., Kim, S.H., Nam, M., Kim, Y.S., 2015. Change of bone mineral density and biochemical markers of bone turnover in patients on suppressive levothyroxine therapy for differentiated thyroid carcinoma. J. Bone Metab. 22 (3), 135–141.

Kim, D.J., Khang, Y.H., Koh, J.M., Shong, Y.K., Kim, G.S., 2006. Low normal TSH levels are associated with low bone mineral density in healthy postmenopausal women. Clin. Endocrinol. 64 (1), 86–90.

Klaushofer, K., Hoffmann, O., Gleispach, H., Leis, H.J., Czerwenka, E., Koller, K., Peterlik, M., 1989. Bone-resorbing activity of thyroid hormones is related to prostaglandin production in cultured neonatal mouse calvaria. J. Bone Miner. Res. 4 (3), 305–312.

Klaushofer, K., Varga, F., Glantschnig, H., Fratzl-Zelman, N., Czerwenka, E., Leis, H.J., Koller, K., Peterlik, M., 1995. The regulatory role of thyroid hormones in bone cell growth and differentiation. J. Nutr. 125 (7 Suppl. 1), 1996S–2003S.

Koenig, R.J., 1998. Thyroid hormone receptor coactivators and corepressors. Thyroid 8 (8), 703–713.

Kopp, P., Kitajima, K., Jameson, J.L., 1996. Syndrome of resistance to thyroid hormone: insights into thyroid hormone action. Proc. Soc. Exp. Biol. Med. 211 (1), 49–61.

Krieger, N.S., Stappenbeck, T.S., Stern, P.H., 1988. Characterization of specific thyroid hormone receptors in bone. J. Bone Miner. Res. 3 (4), 473–478.

Krølner, B., Jørgensen, J.V., Nielsen, S.P., 1983. Spinal bone mineral content in myxoedema and thyrotoxicosis. Effects of thyroid hormone(s) and antithyroid treatment. Clin. Endocrinol. 18 (5), 439–446.

Kumeda, Y., Inaba, M., Tahara, H., Kurioka, Y., Ishikawa, T., Morii, H., Nishizawa, Y., 2000. Persistent increase in bone turnover in Graves' patients with subclinical hyperthyroidism. J. Clin. Endocrinol. Metab. 85 (11), 4157–4161.

Kung, A.W., Ng, F., 1994. A rat model of thyroid hormone-induced bone loss: effect of antiresorptive agents on regional bone density and osteocalcin gene expression. Thyroid 4 (1), 93–98.

Kung, A.W., Yeung, S.S., 1996. Prevention of bone loss induced by thyroxine suppressive therapy in postmenopausal women: the effect of calcium and calcitonin. J. Clin. Endocrinol. Metab. 81 (3), 1232–1236.

Ladenson, P.W., Kristensen, J.D., Ridgway, E.C., Olsson, A.G., Carlsson, B., Klein, I., Baxter, J.D., Angelin, B., 2010. Use of the thyroid hormone analogue eprotirome in statin-treated dyslipidemia. N. Engl. J. Med. 362 (10), 906–916.

Lakatos, P., Caplice, M.D., Khanna, V., Stern, P.H., 1993. Thyroid hormones increase insulin-like growth factor I content in the medium of rat bone tissue. J. Bone Miner. Res. 8 (12), 1475–1481.

Lakatos, P., Foldes, J., Nagy, Z., Takacs, I., Speer, G., Horvath, C., Mohan, S., Baylink, D.J., Stern, P.H., 2000. Serum insulin-like growth factor-I, insulin-like growth factor binding proteins, and bone mineral content in hyperthyroidism. Thyroid 10 (5), 417–423.

Lakatos, P., Stern, P.H., 1991. Evidence for direct non-genomic effects of triiodothyronine on bone rudiments in rats: stimulation of the inositol phosphate second messenger system. Acta Endocrinol. 125 (5), 603–608.

Lakatos, P., Stern, P.H., 1992. Effects of cyclosporins and transforming growth factor beta 1 on thyroid hormone action in cultured fetal rat limb bones. Calcif. Tissue Int. 50 (2), 123–128.

Lakatos, P., Tarján, G., Mérei, J., Földes, J., Holló, I., 1989. Androgens and bone mineral content in patients with subtotal thyroidectomy for benign nodular disease. Acta Med. Hung. 46 (4), 297–305.

Langdahl, B.L., Loft, A.G., Eriksen, E.F., Mosekilde, L., Charles, P., 1996a. Bone mass, bone turnover, body composition, and calcium homeostasis in former hyperthyroid patients treated by combined medical therapy. Thyroid 6 (3), 161–168.

Langdahl, B.L., Loft, A.G., Eriksen, E.F., Mosekilde, L., Charles, P., 1996b. Bone mass, bone turnover, calcium homeostasis, and body composition in surgically and radioiodine-treated former hyperthyroid patients. Thyroid 6 (3), 169–175.

LeBron, B.A., Pekary, A.E., Mirell, C., Hahn, T.J., Hershman, J.M., 1989. Thyroid hormone 5'-deiodinase activity, nuclear binding, and effects on mitogenesis in UMR-106 osteoblastic osteosarcoma cells. J. Bone Miner. Res. 4 (2), 173–178.

Lee, J.S., Buzková, P., Fink, H.A., Vu, J., Carbone, L., Chen, Z., Cauley, J., Bauer, D.C., Cappola, A.R., Robbins, J., 2010. Subclinical thyroid dysfunction and incident hip fracture in older adults. Arch. Intern. Med. 170 (21), 1876–1883.

Lee, M.S., Kim, S.Y., Lee, M.C., Cho, B.Y., Lee, H.K., Koh, C.S., Min, H.K., 1990. Negative correlation between the change in bone mineral density and serum osteocalcin in patients with hyperthyroidism. J. Clin. Endocrinol. Metab. 70 (3), 766–770.

Lee, M.Y., Park, J.H., Bae, K.S., Jee, Y.G., Ko, A.N., Han, Y.J., Shin, J.Y., Lim, J.S., Chung, C.H., Kang, S.J., 2014. Bone mineral density and bone turnover markers in patients on long-term suppressive levothyroxine therapy for differentiated thyroid cancer. Ann. Surg. Treat. Res. 86 (2), 55–60.

Lei, J., Mariash, C.N., Bhargava, M., Wattenberg, E.V., Ingbar, D.H., 2008. T3 increases Na-K-ATPase activity via a MAPK/ERK1/2-dependent pathway in rat adult alveolar epithelial cells. Am. J. Physiol. Lung Cell Mol. Physiol. 294 (4), L749–L754.

Lewinson, D., Bialik, G.M., Hochberg, Z., 1994. Differential effects of hypothyroidism on the cartilage and the osteogenic process in the mandibular condyle: recovery by growth hormone and thyroxine. Endocrinology 135 (4), 1504–1510.

Lin, J.D., Pei, D., Hsia, T.L., Wu, C.Z., Wang, K., Chang, Y.L., Hsu, C.H., Chen, Y.L., Chen, K.W., Tang, S.H., 2011. The relationship between thyroid function and bone mineral density in euthyroid healthy subjects in Taiwan. Endocr. Res. 36 (1), 1–8.

Loviselli, A., Mastinu, R., Rizzolo, E., Massa, G.M., Velluzzi, F., Sammartano, L., Mela, Q., Mariotti, S., 1997. Circulating telopeptide type I is a peripheral marker of thyroid hormone action in hyperthyroidism and during levothyroxine suppressive therapy. Thyroid 7 (4), 561–566.

Luegmayr, E., Glantschnig, H., Varga, F., Klaushofer, K., 2000. The organization of adherens junctions in mouse osteoblast-like cells (MC3T3-E1) and their modulation by triiodothyronine and 1,25-dihydroxyvitamin D3. Histochem. Cell Biol. 113 (6), 467–478.

Luegmayr, E., Varga, F., Frank, T., Roschger, P., Klaushofer, K., 1996. Effects of triiodothyronine on morphology, growth behavior, and the actin cytoskeleton in mouse osteoblastic cells (MC3T3-E1). Bone 18 (6), 591–599.

Lupoli, G., Nuzzo, V., Di Carlo, C., Affinito, P., Vollery, M., Vitale, G., Cascone, E., Arlotta, F., Nappi, C., 1996. Effects of alendronate on bone loss in pre- and postmenopausal hyperthyroid women treated with methimazole. Gynecol. Endocrinol. 10 (5), 343–348.

Ma, R., Morshed, S., Latif, R., Zaidi, M., Davies, T.F., 2011. The influence of thyroid-stimulating hormone and thyroid-stimulating hormone receptor antibodies on osteoclastogenesis. Thyroid 21 (8), 897–906.

Majima, T., Komatsu, Y., Doi, K., Takagi, C., Shigemoto, M., Fukao, A., Morimoto, T., Corners, J., Nakao, K., 2006. Negative correlation between bone mineral density and TSH receptor antibodies in male patients with untreated Graves' disease. Osteoporos. Int. 17 (7), 1103–1110.

Mana, D.L., Zanchetta, M.B., Zanchetta, J.R., 2017. Retreatment with teriparatide: our experience in three patients with severe secondary osteoporosis. Osteoporos. Int. 28 (4), 1491–1494.

Mansouri, A., Chowdhury, K., Gruss, P., 1998. Follicular cells of the thyroid gland require Pax8 gene function. Nat. Genet. 19 (1), 87–90.

Marcus, R., 1975. Cyclic nucleotide phosphodiesterase from bone: characterization of the enzyme and studies of inhibition by thyroid hormones. Endocrinology 96 (2), 400–408.

Martinez, M.E., Herranz, L., de Pedro, C., Pallardo, L.F., 1986. Osteocalcin levels in patients with hyper- and hypothyroidism. Horm. Metab. Res. 18 (3), 212–214.

McCarthy, T.L., Centrella, M., Canalis, E., 1989. Regulatory effects of insulin-like growth factors I and II on bone collagen synthesis in rat calvarial cultures. Endocrinology 124 (1), 301–309.

McKenna, N.J., O'Malley, B.W., 2002. Combinatorial control of gene expression by nuclear receptors and coregulators. Cell 108 (4), 465–474.

McKenna, N.J., Xu, J., Nawaz, Z., Tsai, S.Y., Tsai, M.J., O'Malley, B.W., 1999. Nuclear receptor coactivators: multiple enzymes, multiple complexes, multiple functions. J. Steroid Biochem. Mol. Biol. 69 (1–6), 3–12.

Meier, C., Beat, M., Guglielmetti, M., Christ-Crain, M., Staub, J.J., Kraenzlin, M., 2004. Restoration of euthyroidism accelerates bone turnover in patients with subclinical hypothyroidism: a randomized controlled trial. Osteoporos. Int. 15 (3), 209–216.

Melton, L.J., Ardila, E., Crowson, C.S., O'Fallon, W.M., Khosla, S., 2000. Fractures following thyroidectomy in women: a population-based cohort study. Bone 27 (5), 695–700.

Milne, M., Kang, M.I., Cardona, G., Quail, J.M., Braverman, L.E., Chin, W.W., Baran, D.T., 1999. Expression of multiple thyroid hormone receptor isoforms in rat femoral and vertebral bone and in bone marrow osteogenic cultures. J. Cell. Biochem. 74 (4), 684–693.

Milne, M., Kang, M.I., Quail, J.M., Baran, D.T., 1998. Thyroid hormone excess increases insulin-like growth factor I transcripts in bone marrow cell cultures: divergent effects on vertebral and femoral cell cultures. Endocrinology 139 (5), 2527–2534.

Mirza, F., Canalis, E., 2015. Management of endocrine disease: secondary osteoporosis: pathophysiology and management. Eur. J. Endocrinol. 173 (3), R131–R151.

Miura, M., Tanaka, K., Komatsu, Y., Suda, M., Yasoda, A., Sakuma, Y., Ozasa, A., Nakao, K., 2002. Thyroid hormones promote chondrocyte differentiation in mouse ATDC5 cells and stimulate endochondral ossification in fetal mouse tibias through iodothyronine deiodinases in the growth plate. J. Bone Miner. Res. 17 (3), 443–454.

Miyakawa, M., Tsushima, T., Demura, H., 1996. Carboxy-terminal propeptide of type 1 procollagen (P1CP) and carboxy-terminal telopeptide of type 1 collagen (1CTP) as sensitive markers of bone metabolism in thyroid disease. Endocr. J. 43 (6), 701–708.

Mochizuki, Y., Banba, N., Hattori, Y., Monden, T., 2006. Correlation between serum osteoprotegerin and biomarkers of bone metabolism during antithyroid treatment in patients with Graves' disease. Horm. Res. 66 (5), 236–239.

Moeller, L.C., Cao, X., Dumitrescu, A.M., Seo, H., Refetoff, S., 2006. Thyroid hormone mediated changes in gene expression can be initiated by cytosolic action of the thyroid hormone receptor beta through the phosphatidylinositol 3-kinase pathway. Nucl. Recept. Signal. 4, e020.

Monfoulet, L.E., Rabier, B., Dacquin, R., Anginot, A., Photsavang, J., Jurdic, P., Vico, L., Malaval, L., Chassande, O., 2011. Thyroid hormone receptor β mediates thyroid hormone effects on bone remodeling and bone mass. J. Bone Miner. Res. 26 (9), 2036–2044.

Mora, S., Weber, G., Marenzi, K., Signorini, E., Rovelli, R., Proverbio, M.C., Chiumello, G., 1999. Longitudinal changes of bone density and bone resorption in hyperthyroid girls during treatment. J. Bone Miner. Res. 14 (11), 1971–1977.

Moran, C., Schoenmakers, N., Agostini, M., Schoenmakers, E., Offiah, A., Kydd, A., Kahaly, G., Mohr-Kahaly, S., Rajanayagam, O., Lyons, G., Wareham, N., Halsall, D., Dattani, M., Hughes, S., Gurnell, M., Park, S.M., Chatterjee, K., 2013. An adult female with resistance to thyroid hormone mediated by defective thyroid hormone receptor α. J. Clin. Endocrinol. Metab. 98 (11), 4254–4261.

Morris, M.S., 2007. The association between serum thyroid-stimulating hormone in its reference range and bone status in postmenopausal American women. Bone 40 (4), 1128–1134.

Mosekilde, L., Christensen, M.S., 1977. Decreased parathyroid function in hyperthyroidism: interrelationships between serum parathyroid hormone, calcium-phosphorus metabolism and thyroid function. Acta Endocrinol. 84 (3), 566–575.

Mosekilde, L., Eriksen, E.F., Charles, P., 1990. Effects of thyroid hormones on bone and mineral metabolism. Endocrinol Metab. Clin. N. Am. 19 (1), 35–63.

Mosekilde, L., Melsen, F., 1978. Morphometric and dynamic studies of bone changes in hypothyroidism. Acta Pathol. Microbiol. Scand. 86 (1), 56–62.

Mudde, A.H., Houben, A.J., Nieuwenhuijzen Kruseman, A.C., 1994. Bone metabolism during anti-thyroid drug treatment of endogenous subclinical hyperthyroidism. Clin. Endocrinol. 41 (4), 421–424.

Mundy, G.R., Shapiro, J.L., Bandelin, J.G., Canalis, E.M., Raisz, L.G., 1976. Direct stimulation of bone resorption by thyroid hormones. J. Clin. Investig. 58 (3), 529–534.

Murphy, E., Glüer, C.C., Reid, D.M., Felsenberg, D., Roux, C., Eastell, R., Williams, G.R., 2010. Thyroid function within the upper normal range is associated with reduced bone mineral density and an increased risk of nonvertebral fractures in healthy euthyroid postmenopausal women. J. Clin. Endocrinol. Metab. 95 (7), 3173–3181.

Nagasaka, S., Sugimoto, H., Nakamura, T., Kusaka, I., Fujisawa, G., Sakuma, N., Tsuboi, Y., Fukuda, S., Honda, K., Okada, K., Ishikawa, S., Saito, T., 1997. Antithyroid therapy improves bony manifestations and bone metabolic markers in patients with Graves' thyrotoxicosis. Clin. Endocrinol. 47 (2), 215–221.

Nakamura, H., Mori, T., Genma, R., Suzuki, Y., Natsume, H., Andoh, S., Kitahara, R., Nagasawa, S., Nishiyama, K., Yoshimi, T., 1996. Urinary excretion of pyridinoline and deoxypyridinoline measured by immunoassay in hypothyroidism. Clin. Endocrinol. 44 (4), 447–451.

Noh, H.M., Park, Y.S., Lee, J., Lee, W., 2015. A cross-sectional study to examine the correlation between serum TSH levels and the osteoporosis of the lumbar spine in healthy women with normal thyroid function. Osteoporos. Int. 26 (3), 997–1003.

Nyström, E., Lundberg, P.A., Petersen, K., Bengtsson, C., Lindstedt, G., 1989. Evidence for a slow tissue adaptation to circulating thyroxine in patients with chronic L-thyroxine treatment. Clin. Endocrinol. 31 (2), 143–150.

O'Shea, P.J., Bassett, J.H., Cheng, S.Y., Williams, G.R., 2006. Characterization of skeletal phenotypes of TRalpha1 and TRbeta mutant mice: implications for tissue thyroid status and T3 target gene expression. Nucl. Recept. Signal. 4, e011.

O'Shea, P.J., Harvey, C.B., Suzuki, H., Kaneshige, M., Kaneshige, K., Cheng, S.Y., Williams, G.R., 2003. A thyrotoxic skeletal phenotype of advanced bone formation in mice with resistance to thyroid hormone. Mol. Endocrinol. 17 (7), 1410–1424.

Ohishi, K., Ishida, H., Nagata, T., Yamauchi, N., Tsurumi, C., Nishikawa, S., Wakano, Y., 1994. Thyroid hormone suppresses the differentiation of osteoprogenitor cells to osteoblasts, but enhances functional activities of mature osteoblasts in cultured rat calvaria cells. J. Cell. Physiol. 161 (3), 544–552.

Olkawa, M., Kushida, K., Takahashi, M., Ohishi, T., Hoshino, H., Suzuki, M., Ogihara, H., Ishigaki, J., Inoue, T., 1999. Bone turnover and cortical bone mineral density in the distal radius in patients with hyperthyroidism being treated with antithyroid drugs for various periods of time. Clin. Endocrinol. 50 (2), 171–176.

Ongphiphadhanakul, B., Jenis, L.G., Braverman, L.E., Alex, S., Stein, G.S., Lian, J.B., Baran, D.T., 1993. Etidronate inhibits the thyroid hormone-induced bone loss in rats assessed by bone mineral density and messenger ribonucleic acid markers of osteoblast and osteoclast function. Endocrinology 133 (6), 2502–2507.

Pantazi, H., Papapetrou, P.D., 2000. Changes in parameters of bone and mineral metabolism during therapy for hyperthyroidism. J. Clin. Endocrinol. Metab. 85 (3), 1099–1106.

Patel, K.V., Brennan, K.L., Brennan, M.L., Jupiter, D.C., Shar, A., Davis, M.L., 2014. Association of a modified frailty index with mortality after femoral neck fracture in patients aged 60 years and older. Clin. Orthop. Relat. Res. 472 (3), 1010–1017.

Paul, T.L., Kerrigan, J., Kelly, A.M., Braverman, L.E., Baran, D.T., 1988. Long-term L-thyroxine therapy is associated with decreased hip bone density in premenopausal women. J. Am. Med. Assoc. 259 (21), 3137–3141.

Pirskanen, A., Jääskeläinen, T., Mäenpää, P.H., 1993. Insulin-like growth factor-1 modulates steroid hormone effects on osteocalcin synthesis in human MG-63 osteosarcoma cells. Eur. J. Biochem. 218 (3), 883–891.

Plateroti, M., Gauthier, K., Domon-Dell, C., Freund, J.N., Samarut, J., Chassande, O., 2001. Functional interference between thyroid hormone receptor alpha (TRalpha) and natural truncated TRDeltaalpha isoforms in the control of intestine development. Mol. Cell Biol. 21 (14), 4761–4772.

Quan, M.L., Pasieka, J.L., Rorstad, O., 2002. Bone mineral density in well-differentiated thyroid cancer patients treated with suppressive thyroxine: a systematic overview of the literature. J. Surg. Oncol. 79 (1), 62–69 discussion 69-70.

Quarto, R., Campanile, G., Cancedda, R., Dozin, B., 1992. Thyroid hormone, insulin, and glucocorticoids are sufficient to support chondrocyte differentiation to hypertrophy: a serum-free analysis. J. Cell Biol. 119 (4), 989–995.

Quignodon, L., Vincent, S., Winter, H., Samarut, J., Flamant, F., 2007. A point mutation in the activation function 2 domain of thyroid hormone receptor alpha1 expressed after CRE-mediated recombination partially recapitulates hypothyroidism. Mol. Endocrinol. 21 (10), 2350–2360.

Refetoff, S., Weiss, R.E., Usala, S.J., 1993. The syndromes of resistance to thyroid hormone. Endocr. Rev. 14 (3), 348–399.

Ren, S.G., Huang, Z., Sweet, D.E., Malozowski, S., Cassorla, F., 1990. Biphasic response of rat tibial growth to thyroxine administration. Acta Endocrinol. 122 (3), 336–340.

Reverter, J.L., Holgado, S., Alonso, N., Salinas, I., Granada, M.L., Sanmartí, A., 2005. Lack of deleterious effect on bone mineral density of long-term thyroxine suppressive therapy for differentiated thyroid carcinoma. Endocr. Relat. Cancer 12 (4), 973–981.

Rizzoli, R., Poser, J., Bürgi, U., 1986. Nuclear thyroid hormone receptors in cultured bone cells. Metabolism 35 (1), 71–74.

Robson, H., Siebler, T., Stevens, D.A., Shalet, S.M., Williams, G.R., 2000. Thyroid hormone acts directly on growth plate chondrocytes to promote hypertrophic differentiation and inhibit clonal expansion and cell proliferation. Endocrinology 141 (10), 3887–3897.

Rosen, H.N., Moses, A.C., Gundberg, C., Kung, V.T., Seyedin, S.M., Chen, T., Holick, M., Greenspan, S.L., 1993a. Therapy with parenteral pamidronate prevents thyroid hormone-induced bone turnover in humans. J. Clin. Endocrinol. Metab. 77 (3), 664–669.

Rosen, H.N., Sullivan, E.K., Middlebrooks, V.L., Zeind, A.J., Gundberg, C., Dresner-Pollak, R., Maitland, L.A., Hock, J.M., Moses, A.C., Greenspan, S.L., 1993b. Parenteral pamidronate prevents thyroid hormone-induced bone loss in rats. J. Bone Miner. Res. 8 (10), 1255–1261.

Rosenthal, A.K., Henry, L.A., 1999. Thyroid hormones induce features of the hypertrophic phenotype and stimulate correlates of CPPD crystal formation in articular chondrocytes. J. Rheumatol. 26 (2), 395–401.

Saggese, G., Bertelloni, S., Baroncelli, G.I., 1990. Bone mineralization and calciotropic hormones in children with hyperthyroidism. Effects of methimazole therapy. J. Endocrinol. Investig. 13 (7), 587–592.

Sakurai, A., Miyamoto, T., Refetoff, S., DeGroot, L.J., 1990. Dominant negative transcriptional regulation by a mutant thyroid hormone receptor-beta in a family with generalized resistance to thyroid hormone. Mol. Endocrinol. 4 (12), 1988–1994.

Saltó, C., Kindblom, J.M., Johansson, C., Wang, Z., Gullberg, H., Nordström, K., Mansén, A., Ohlsson, C., Thorén, P., Forrest, D., Vennström, B., 2001. Ablation of TRalpha2 and a concomitant overexpression of alpha1 yields a mixed hypo- and hyperthyroid phenotype in mice. Mol. Endocrinol. 15 (12), 2115–2128.

Sampath, T.K., Simic, P., Sendak, R., Draca, N., Bowe, A.E., O'Brien, S., Schiavi, S.C., McPherson, J.M., Vukicevic, S., 2007. Thyroid-stimulating hormone restores bone volume, microarchitecture, and strength in aged ovariectomized rats. J. Bone Miner. Res. 22 (6), 849–859.

Sato, K., Han, D.C., Fujii, Y., Tsushima, T., Shizume, K., 1987. Thyroid hormone stimulates alkaline phosphatase activity in cultured rat osteoblastic cells (ROS 17/2.8) through 3,5,3'-triiodo-L-thyronine nuclear receptors. Endocrinology 120 (5), 1873–1881.

Scarlett, A., Parsons, M.P., Hanson, P.L., Sidhu, K.K., Milligan, T.P., Burrin, J.M., 2008. Thyroid hormone stimulation of extracellular signal-regulated kinase and cell proliferationin human osteoblast-like cells is initiated at integrin alphaVbeta3. J. Endocrinol. 196 (3), 509–517.

Schiller, C., Gruber, R., Ho, G.M., Redlich, K., Gober, H.J., Katzgraber, F., Willheim, M., Hoffmann, O., Pietschmann, P., Peterlik, M., 1998. Interaction of triiodothyronine with 1alpha,25-dihydroxyvitamin D3 on interleukin-6-dependent osteoclast-like cell formation in mouse bone marrow cell cultures. Bone 22 (4), 341–346.

Schlesinger, B., Fisher, O.D., 1951. Accelerated skeletal development from thyrotoxicosis and thyroid overdosage in childhood. Lancet 2 (6677), 289–290.

Schmid, C., Schläpfer, I., Futo, E., Waldvogel, M., Schwander, J., Zapf, J., Froesch, E.R., 1992. Triiodothyronine (T3) stimulates insulin-like growth factor (IGF)-1 and IGF binding protein (IGFBP)-2 production by rat osteoblasts in vitro. Acta Endocrinol. 126 (5), 467–473.

Schmid, C., Steiner, T., Froesch, E.R., 1986. Triiodothyronine increases responsiveness of cultured rat bone cells to parathyroid hormone. Acta Endocrinol. 111 (2), 213–216.

Sharma, D., Fondell, J.D., 2002. Ordered recruitment of histone acetyltransferases and the TRAP/Mediator complex to thyroid hormone-responsive promoters in vivo. Proc. Natl. Acad. Sci. U. S. A. 99 (12), 7934–7939.

Sherman, S.I., Ringel, M.D., Smith, M.J., Kopelen, H.A., Zoghbi, W.A., Ladenson, P.W., 1997. Augmented hepatic and skeletal thyromimetic effects of tiratricol in comparison with levothyroxine. J. Clin. Endocrinol. Metab. 82 (7), 2153–2158.

Shirazi, M., Dehpour, A.R., Jafari, F., 1999. The effect of thyroid hormone on orthodontic tooth movement in rats. J. Clin. Pediatr. Dent. 23 (3), 259–264.

Siddiqi, A., Burrin, J.M., Wood, D.F., Monson, J.P., 1998. Tri-iodothyronine regulates the production of interleukin-6 and interleukin-8 in human bone marrow stromal and osteoblast-like cells. J. Endocrinol. 157 (3), 453–461.

Siddiqi, A., Parsons, M.P., Lewis, J.L., Monson, J.P., Williams, G.R., Burrin, J.M., 2002. TR expression and function in human bone marrow stromal and osteoblast-like cells. J. Clin. Endocrinol. Metab. 87 (2), 906–914.

Soskolne, W.A., Schwartz, Z., Goldstein, M., Ornoy, A., 1990. The biphasic effect of triiodothyronine compared to bone resorbing effect of PTH on bone modelling of mouse long bone in vitro. Bone 11 (5), 301–307.

Stamato, F.J., Amarante, E.C., Furlanetto, R.P., 2000. Effect of combined treatment with calcitonin on bone densitometry of patients with treated hypothyroidism. Rev. Assoc. Med. Bras. 46 (2), 177–181.

Stevens, D.A., Harvey, C.B., Scott, A.J., O'Shea, P.J., Barnard, J.C., Williams, A.J., Brady, G., Samarut, J., Chassande, O., Williams, G.R., 2003. Thyroid hormone activates fibroblast growth factor receptor-1 in bone. Mol. Endocrinol. 17 (9), 1751–1766.

Stevens, D.A., Hasserjian, R.P., Robson, H., Siebler, T., Shalet, S.M., Williams, G.R., 2000. Thyroid hormones regulate hypertrophic chondrocyte differentiation and expression of parathyroid hormone-related peptide and its receptor during endochondral bone formation. J. Bone Miner. Res. 15 (12), 2431–2442.

Stracke, H., Rossol, S., Schatz, H., 1986. Alkaline phosphatase and insulin-like growth factor in fetal rat bone under the influence of thyroid hormones. Horm. Metab. Res. 18 (11), 794.

Sugitani, I., Fujimoto, Y., 2011. Effect of postoperative thyrotropin suppressive therapy on bone mineral density in patients with papillary thyroid carcinoma: a prospective controlled study. Surgery 150 (6), 1250–1257.

Sun, L., Vukicevic, S., Baliram, R., Yang, G., Sendak, R., McPherson, J., Zhu, L.L., Iqbal, J., Latif, R., Natrajan, A., Arabi, A., Yamoah, K., Moonga, B.S., Gabet, Y., Davies, T.F., Bab, I., Abe, E., Sampath, K., Zaidi, M., 2008. Intermittent recombinant TSH injections prevent ovariectomy-induced bone loss. Proc. Natl. Acad. Sci. U. S. A. 105 (11), 4289–4294.

Sun, L., Zhu, L.L., Lu, P., Yuen, T., Li, J., Ma, R., Baliram, R., Moonga, S.S., Liu, P., Zallone, A., New, M.I., Davies, T.F., Zaidi, M., 2013. Genetic confirmation for a central role for TNFα in the direct action of thyroid stimulating hormone on the skeleton. Proc. Natl. Acad. Sci. U. S. A. 110 (24), 9891–9896.

Suwanwalaikorn, S., Ongphiphadhanakul, B., Braverman, L.E., Baran, D.T., 1996. Differential responses of femoral and vertebral bones to long-term excessive L-thyroxine administration in adult rats. Eur. J. Endocrinol. 134 (5), 655–659.

Suwanwalaikorn, S., Van Auken, M., Kang, M.I., Alex, S., Braverman, L.E., Baran, D.T., 1997. Site selectivity of osteoblast gene expression response to thyroid hormone localized by in situ hybridization. Am. J. Physiol. 272 (2 Pt 1), E212–E217.

Tarjan, G., Stern, P.H., 1995. Triiodothyronine potentiates the stimulatory effects of interleukin-1 beta on bone resorption and medium interleukin-6 content in fetal rat limb bone cultures. J. Bone Miner. Res. 10 (9), 1321–1326.

Tauchmanovà, L., Nuzzo, V., Del Puente, A., Fonderico, F., Esposito-Del Puente, A., Padulla, S., Rossi, A., Bifulco, G., Lupoli, G., Lombardi, G., 2004. Reduced bone mass detected by bone quantitative ultrasonometry and DEXA in pre- and postmenopausal women with endogenous subclinical hyperthyroidism. Maturitas 48 (3), 299–306.

Toh, S.H., Claunch, B.C., Brown, P.H., 1985. Effect of hyperthyroidism and its treatment on bone mineral content. Arch. Intern. Med. 145 (5), 883–886.

Torchia, J., Glass, C., Rosenfeld, M.G., 1998. Co-activators and co-repressors in the integration of transcriptional responses. Curr. Opin. Cell Biol. 10 (3), 373–383.

Tsai, J.A., Janson, A., Bucht, E., Kindmark, H., Marcus, C., Stark, A., Zemack, H.R., Torring, O., 2004. Weak evidence of thyrotropin receptors in primary cultures of human osteoblast-like cells. Calcif. Tissue Int. 74 (5), 486–491.

Turner, M.R., Camacho, X., Fischer, H.D., Austin, P.C., Anderson, G.M., Rochon, P.A., Lipscombe, L.L., 2011. Levothyroxine dose and risk of fractures in older adults: nested case-control study. BMJ 342, d2238.

Tylki-Szymańska, A., Acuna-Hidalgo, R., Krajewska-Walasek, M., Lecka-Ambroziak, A., Steehouwer, M., Gilissen, C., Brunner, H.G., Jurecka, A., Różdżyńska-Świątkowska, A., Hoischen, A., Chrzanowska, K.H., 2015. Thyroid hormone resistance syndrome due to mutations in the thyroid hormone receptor α gene (THRA). J. Med. Genet. 52 (5), 312–316.

Udayakumar, N., Chandrasekaran, M., Rasheed, M.H., Suresh, R.V., Sivaprakash, S., 2006. Evaluation of bone mineral density in thyrotoxicosis. Singap. Med. J. 47 (11), 947–950.

Uzzan, B., Campos, J., Cucherat, M., Nony, P., Boissel, J.P., Perret, G.Y., 1996. Effects on bone mass of long term treatment with thyroid hormones: a meta-analysis. J. Clin. Endocrinol. Metab. 81 (12), 4278–4289.

Vadiveloo, T., Donnan, P.T., Cochrane, L., Leese, G.P., 2011. The Thyroid Epidemiology, Audit, and Research Study (TEARS): morbidity in patients with endogenous subclinical hyperthyroidism. J. Clin. Endocrinol. Metab. 96 (5), 1344–1351.

Van Den Eeden, S.K., Barzilay, J.I., Ettinger, B., Minkoff, J., 2003. Thyroid hormone use and the risk of hip fracture in women > or = 65 years: a case-control study. J. Womens Health 12 (1), 27–31.

van Mullem, A., van Heerebeek, R., Chrysis, D., Visser, E., Medici, M., Andrikoula, M., Tsatsoulis, A., Peeters, R., Visser, T.J., 2012. Clinical phenotype and mutant TRα1. N. Engl. J. Med. 366 (15), 1451–1453.

van Rijn, L.E., Pop, V.J., Williams, G.R., 2014. Low bone mineral density is related to high physiological levels of free thyroxine in peri-menopausal women. Eur. J. Endocrinol. 170 (3), 461–468.

Varga, F., Rumpler, M., Klaushofer, K., 1994. Thyroid hormones increase insulin-like growth factor mRNA levels in the clonal osteoblastic cell line MC3T3-E1. FEBS Lett. 345 (1), 67–70.

Varga, F., Rumpler, M., Spitzer, S., Karlic, H., Klaushofer, K., 2009. Osteocalcin attenuates T3- and increases vitamin D3-induced expression of MMP-13 in mouse osteoblasts. Endocr. J. 56 (3), 441–450.

Varga, F., Rumpler, M., Zoehrer, R., Turecek, C., Spitzer, S., Thaler, R., Paschalis, E.P., Klaushofer, K., 2010. T3 affects expression of collagen I and collagen cross-linking in bone cell cultures. Biochem. Biophys. Res. Commun. 402 (2), 180–185.

Varga, F., Spitzer, S., Klaushofer, K., 2004. Triiodothyronine (T3) and 1,25-dihydroxyvitamin D3 (1,25D3) inversely regulate OPG gene expression in dependence of the osteoblastic phenotype. Calcif. Tissue Int. 74 (4), 382–387.

Varga, F., Spitzer, S., Rumpler, M., Klaushofer, K., 2003. 1,25-Dihydroxyvitamin D3 inhibits thyroid hormone-induced osteocalcin expression in mouse osteoblast-like cells via a thyroid hormone response element. J. Mol. Endocrinol. 30 (1), 49–57.

Vella, K.R., Hollenberg, A.N., 2017. The actions of thyroid hormone signaling in the nucleus. Mol. Cell. Endocrinol. https://doi.org/10.1016/j.mce.2017.03.001 pii: S0303-7207(17)30169-7. [Epub ahead of print].

Vestergaard, P., Mosekilde, L., 2002. Fractures in patients with hyperthyroidism and hypothyroidism: a nationwide follow-up study in 16,249 patients. Thyroid 12 (5), 411–419.

Vestergaard, P., Mosekilde, L., 2003. Hyperthyroidism, bone mineral, and fracture risk–a meta-analysis. Thyroid 13 (6), 585–593.

Vestergaard, P., Rejnmark, L., Mosekilde, L., 2005. Influence of hyper- and hypothyroidism, and the effects of treatment with antithyroid drugs and levothyroxine on fracture risk. Calcif. Tissue Int. 77 (3), 139–144.

Vestergaard, P., Rejnmark, L., Weeke, J., Mosekilde, L., 2000. Fracture risk in patients treated for hyperthyroidism. Thyroid 10 (4), 341–348.

Vincent, B.H., Montufar-Solis, D., Teng, B.B., Amendt, B.A., Schaefer, J., Klein, J.R., 2009. Bone marrow cells produce a novel TSHbeta splice variant that is upregulated in the thyroid following systemic virus infection. Genes Immun. 10 (1), 18–26.

Wallace, H., Ledent, C., Vassart, G., Bishop, J.O., al-Shawi, R., 1991. Specific ablation of thyroid follicle cells in adult transgenic mice. Endocrinology 129 (6), 3217–3226.

Wallace, H., Pate, A., Bishop, J.O., 1995. Effects of perinatal thyroid hormone deprivation on the growth and behaviour of newborn mice. J. Endocrinol. 145 (2), 251–262.

Waring, A.C., Harrison, S., Fink, H.A., Samuels, M.H., Cawthon, P.M., Zmuda, J.M., Orwoll, E.S., Bauer, D.C., Osteoporotic Fractures in Men (MrOS) Study, 2013. A prospective study of thyroid function, bone loss, and fractures in older men: the MrOS study. J. Bone Miner. Res. 28 (3), 472–479.

Wejda, B., Hintze, G., Katschinski, B., Olbricht, T., Benker, G., 1995. Hip fractures and the thyroid: a case-control study. J. Intern. Med. 237 (3), 241–247.

Wikström, L., Johansson, C., Saltó, C., Barlow, C., Campos Barros, A., Baas, F., Forrest, D., Thorén, P., Vennström, B., 1998. Abnormal heart rate and body temperature in mice lacking thyroid hormone receptor alpha 1. EMBO J. 17 (2), 455–461.

Williams, A.J., Robson, H., Kester, M.H., van Leeuwen, J.P., Shalet, S.M., Visser, T.J., Williams, G.R., 2008. Iodothyronine deiodinase enzyme activities in bone. Bone 43 (1), 126–134.

Williams, G.R., 2000. Cloning and characterization of two novel thyroid hormone receptor beta isoforms. Mol. Cell Biol. 20 (22), 8329–8342.

Williams, G.R., Bland, R., Sheppard, M.C., 1994. Characterization of thyroid hormone (T3) receptors in three osteosarcoma cell lines of distinct osteoblast phenotype: interactions among T3, vitamin D3, and retinoid signaling. Endocrinology 135 (6), 2375–2385.

Williams, G.R., Bland, R., Sheppard, M.C., 1995. Retinoids modify regulation of endogenous gene expression by vitamin D3 and thyroid hormone in three osteosarcoma cell lines. Endocrinology 136 (10), 4304–4314.

Wirth, C.D., Blum, M.R., da Costa, B.R., Baumgartner, C., Collet, T.H., Medici, M., Peeters, R.P., Aujesky, D., Bauer, D.C., Rodondi, N., 2014. Subclinical thyroid dysfunction and the risk for fractures: a systematic review and meta-analysis. Ann. Intern. Med. 161 (3), 189–199.

Wu, Y., Koenig, R.J., 2000. Gene regulation by thyroid hormone. Trends Endocrinol. Metabol. 11 (6), 207–211.

Yamamoto, M., Markatos, A., Seedor, J.G., Masarachia, P., Gentile, M., Rodan, G.A., Balena, R., 1993. The effects of the aminobisphosphonate alendronate on thyroid hormone-induced osteopenia in rats. Calcif. Tissue Int. 53 (4), 278–282.

Yamaura, M., Nakamura, T., Kanou, A., Miura, T., Ohara, H., Suzuki, K., 1994. The effect of 17 beta-estradiol treatment on the mass and the turnover of bone in ovariectomized rats taking a mild dose of thyroxin. Bone Miner. 24 (1), 33–42.

Yen, C.C., Huang, Y.H., Liao, C.Y., Liao, C.J., Cheng, W.L., Chen, W.J., Lin, K.H., 2006. Mediation of the inhibitory effect of thyroid hormone on proliferation of hepatoma cells by transforming growth factor-beta. J. Mol. Endocrinol. 36 (1), 9–21.

Yen, P.M., 2001. Physiological and molecular basis of thyroid hormone action. Physiol. Rev. 81 (3), 1097–1142.

Zeni, S., Gomez-Acotto, C., Di Gregorio, S., Mautalen, C., 2000. Differences in bone turnover and skeletal response to thyroid hormone treatment between estrogen-depleted and repleted rats. Calcif. Tissue Int. 67 (2), 173–177.

Zhang, W., Zhang, Y., Liu, Y., Wang, J., Gao, L., Yu, C., Yan, H., Zhao, J., Xu, J., 2014. Thyroid-stimulating hormone maintains bone mass and strength by suppressing osteoclast differentiation. J. Biomech. 47 (6), 1307–1314.

Chapter 39

Basic and clinical aspects of glucocorticoid action in bone

Hong Zhou[1], Mark S. Cooper[2] and Markus J. Seibel[2]

[1]*The University of Sydney, ANZAC Research Institute, Sydney, NSW, Australia;* [2]*The University of Sydney, ANZAC Research Institute and Department of Endocrinology & Metabolism, Concord Hospital, Sydney, NSW, Australia*

Chapter outline

Introduction 915
The physiological role of glucocorticoids in bone 916
Glucocorticoid signaling and prereceptor metabolism 916
Local glucocorticoid metabolism in bone 917
Novel insights from targeted disruption of glucocorticoid signaling in bone 918
Endogenous glucocorticoids promote osteoblastogenesis 918
Glucocorticoid excess and the skeleton 919
Pathogenesis of glucocorticoid-induced osteoporosis 920
Glucocorticoid excess and its effects on osteoblast differentiation 921
Glucocorticoids prevent osteoblast cell cycle progression 922
Glucocorticoids induce osteoblast apoptosis 923
Glucocorticoid excess and the osteocyte 924
Glucocorticoid excess and the osteoclast 924
Glucocorticoid excess and local glucocorticoid metabolism 925
Indirect mechanisms for glucocorticoid-induced osteoporosis 925
Glucocorticoid excess and systemic fuel metabolism 926
Treatment of glucocorticoid-induced osteoporosis 927
Assessment of the patient with glucocorticoid-induced osteoporosis 927
Management of glucocorticoid-induced osteoporosis 928
Bisphosphonates 929
Denosumab 929
Raloxifene 930
Teriparatide 930
Sex hormone replacement 932
Timing and monitoring of therapy 932
New and emerging therapies 932
Selective glucocorticoid receptor activators 932
Antisclerostin/DKK1 932
Conclusions and future perspectives 933
Acknowledgments 933
References 933

Introduction

More than 80 years ago, Harvey Cushing described 12 patients presenting with either pituitary or adrenal adenomas. Most of these patients were found to have severe osteoporosis and Cushing noted that their "bones were easily cut with a knife" during postmortem examination (Cushing, 1932). The cause of the skeletal pathology of what is known today as "Cushing's disease" or "Cushing's syndrome" was later discovered to be the adrenal steroid hormone, 11β,17α,21-trihydroxypregn-4-ene-3,20-dione, or cortisol. In 1949, Hench and colleagues published on the use of cortisone, a glucocorticoid closely related to cortisol, in patients with rheumatoid arthritis, a report which revolutionized clinical medicine through the discovery of the potent and even today unrivaled antiinflammatory and immunosuppressive properties of glucocorticoids (Hench et al., 1949). Since then, these naturally occurring hormones as well as their pharmaceutical derivatives have been used with great benefit in the treatment of countless conditions including arthritis, inflammatory bowel disease, allergy and hypersensitivity disorders, chronic obstructive pulmonary disease, cancer, and transplant rejection.

While the clinical benefits of glucocorticoid therapy can hardly be overestimated, the use of these drugs is unfortunately associated with numerous unwanted effects such as bone loss and osteoporosis (Reid et al., 2009), insulin resistance and diabetes (Gonzalez- Gonzalez et al., 2013; Niewoehner et al., 1999), proximal myopathy, cataract, skin atrophy, delayed

Principles of Bone Biology. https://doi.org/10.1016/B978-0-12-814841-9.00039-7

wound healing, and increased risk of infection, to name only a few (Kanda et al., 2001; Natsui et al., 2006). These unwanted outcomes create a significant dilemma for clinicians, as improvements in the primary condition often seem to be achievable only by accepting significant adverse effects that are often hard to prevent or treat.

In this chapter, we will review our current understanding of the mechanisms underlying the physiology and pathophysiology of glucocorticoid action on the skeleton and discuss present and future treatments for glucocorticoid-induced osteoporosis (GIO). As a general note, it is important to realize that the actions of glucocorticoids on bone and mineral metabolism are strongly dose and time dependent. Thus, at physiological concentrations, glucocorticoids are key regulators of mesenchymal cell differentiation and therefore bone development, with additional regulatory roles in renal and intestinal calcium handling. However, at supraphysiological—i.e., therapeutic—concentrations, glucocorticoids affect the very same systems in different and often unfavorable ways. Thus, glucocorticoids can act on the skeleton in both anabolic and catabolic ways.

The physiological role of glucocorticoids in bone

Glucocorticoid signaling and prereceptor metabolism

Endogenous glucocorticoids are essential to life, as they control electrolyte and fluid homeostasis, fuel metabolism, and immune and stress responses. These actions are primarily mediated through the glucocorticoid receptor (GR) within target tissues. As such, both GR expression levels and ligand binding affinity influence glucocorticoid actions at the tissue level. The GR resides predominantly in the cytoplasm, complexed with heat-shock protein 90 (Hsp90) and several immunophilins such as FKBP51 (Zhou and Cidlowski, 2005). Upon ligand binding, the GR dissociates from this complex and undergoes a number of conformational changes that result in receptor dimerization and subsequent nuclear translocation (Lu and Cidlowski, 2006). Within the nucleus, the dimerized ligand-receptor complex binds directly to highly conserved glucocorticoid response elements (GREs) in the promoter region of target genes. This process is known as "transactivation" and usually facilitates gene transcription. The ligand-bound homodimer can also bind to negative GREs, thereby inhibiting gene transcription, an effect known as transrepression. On the other hand, the monomeric ligand-receptor complex binds to DNA-bound transcription factors such as STAT5, which enhances gene expression in a process known as tethered transactivation. The monomer can also bind to proinflammatory signaling molecules in the cytosol, preventing interaction with effector transcription factors such as AP-1 or NF-κB. This process is known as indirect or tethered transrepression (Moutsatsou et al., 2012) (Fig. 39.1). The traditional view of glucocorticoid action was that the antiinflammatory effects of glucocorticoids are caused by transrepression, whereas adverse outcomes largely arise from its transactivation capacities (Newton and Holden, 2007). However, it is now clear that this model was oversimplistic, with some antiinflammatory responses being due to transactivation—e.g., transaction induces the expression of the antiinflammatory protein dual specificity phosphatase 1 (DUSP1)—and some adverse outcomes due to transrepression—e.g., glucocorticoid inhibition of bone formation (Cooper et al., 2012; Rauch et al., 2010). Further adding to the diversity of glucocorticoid signaling mechanisms, many of the actions of the glucocorticoid receptors at the level of DNA may in fact by due to monomeric complexes (Meijsing et al., 2009). There is additionally evidence for nongenomic actions of glucocorticoids which typically happen more rapid than traditional genomic responses. These nongenomic mechanisms may involve non-GR-mediated interactions of glucocorticoids with cell membranes, nongenomic effects mediated by cytosolic GRs, or interactions of glucocorticoids with membrane-bound glucocorticoid receptors (Stahn and Buttgereit, 2008).

Endogenous glucocorticoid levels are systemically regulated via the hypothalamic–pituitary–adrenal axis. However, glucocorticoid action depends not only on plasma and interstitial fluid hormone concentrations, but also on cellular glucocorticoid levels. Within specific tissues, the 11β-hydroxysteroid dehydrogenase (11βHSD) enzymes metabolize glucocorticoids at the prereceptor level and control intracellular hormone availability via interconversion of biologically active and inactive ligands (Stewart and Krozowski, 1999). Thus, 11β-HSD type 1 (11βHSD1) predominantly catalyzes formation of active cortisol (in humans) and corticosterone (in rodents) from inactive cortisone and 11-dehydrocorticosterone, respectively, leading to increased intracellular glucocorticoid concentrations. This enzyme also converts inactive prednisone to active prednisolone—individuals who lack this activity are unresponsive to prednisone treatment but remain sensitive to prednisolone. In contrast, 11β-HSD type 2 (11βHSD2) unidirectionally catalyzes conversion of active glucocorticoids to their inactive metabolites (Fig. 39.1). The latter enzyme is primarily expressed in mineralocorticoid target tissues such as the kidney, where it protects the mineralocorticoid receptor from illicit activation by cortisol (Stewart and Krozowski, 1999). Of note, cytokines, growth factors, and certain other enzymes are able to modulate glucocorticoid metabolism via changing 11β-HSD enzyme activities on a predominantly local level (Cooper et al., 2001).

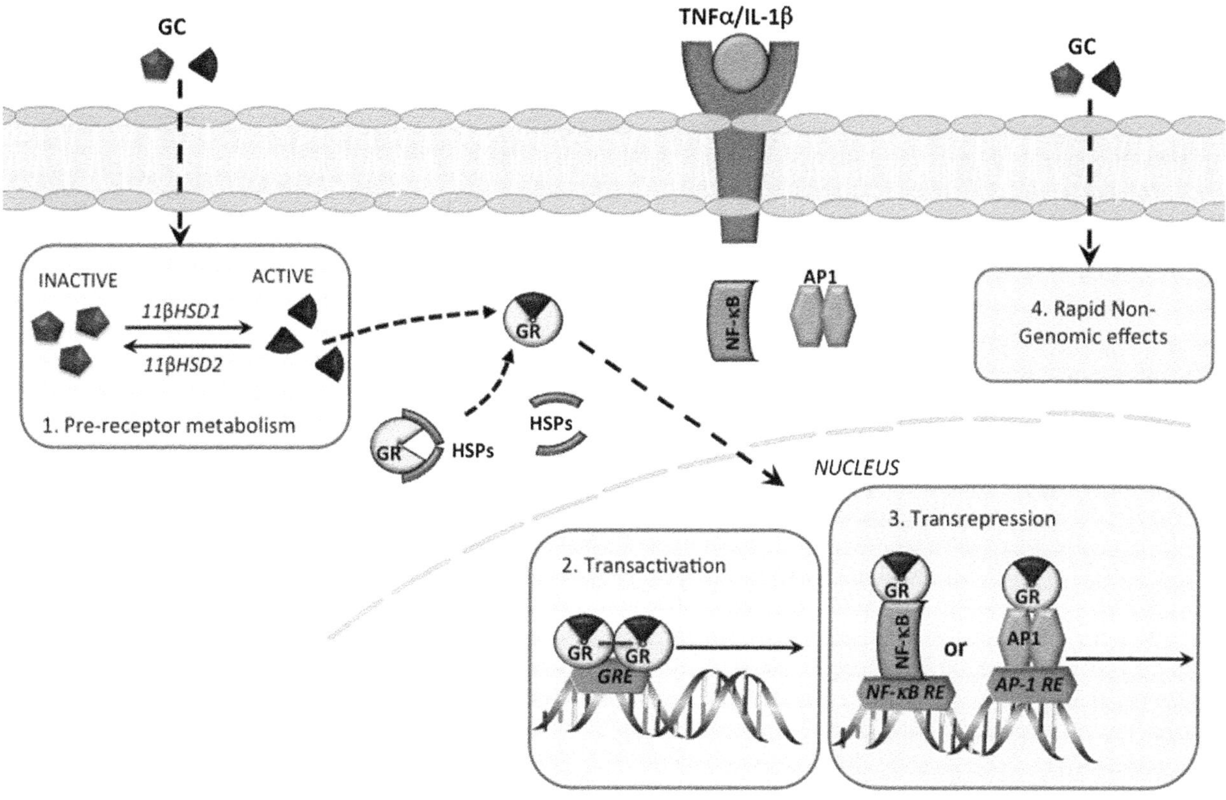

FIGURE 39.1 Glucocorticoid signaling. Glucocorticoids are lipophilic and cross the cell membrane freely. In many tissues, glucocorticoids undergo prereceptor metabolism by the cytosolic 11β-HSD enzymes. Thus, 11β-HSD type 1 converts inactive glucocorticoids (cortisone in humans, dehydrocorticosterone in rodents) to the active form (cortisol and corticosterone, respectively). Conversely, 11β-HSD type 2 converts cortisol and corticosterone into their inactive pendants. Active glucocorticoids bind to the cytoplasmic glucocorticoid receptor to form a complex that is able to translocate to the nucleus, where it either binds to its specific response element or to other transcription factors such as AP-1 (fos and jun) or NF-κβ, resulting in transactivation or transrepression of gene expression. Rapid nongenomic effects have also been described.

Local glucocorticoid metabolism in bone

As discussed above, local metabolism of glucocorticoids impacts their actions within bone. Expression of the 11βHSD1 enzyme has been extensively characterized in bone (Cooper et al., 2000; Weinstein et al., 2010b). Human bone tissue has the ability to interconvert cortisone and cortisol, but the kinetics of the reaction indicate the presence of the 11βHSD1 enzyme rather than 11βHSD2 (Cooper et al., 2000). This is supported by immunohistochemistry and in situ hybridization studies that localize the expression of 11βHSD1 primarily to osteoblasts. In ex vivo studies, the activity of 11βHSD1 in primary human osteoblasts is dependent on donor age, with expression low in young donors but high in osteoblasts isolated from older donors (Cooper et al., 2002). An age-related change in expression is paralleled in mice in vivo (Weinstein et al., 2010b). Mice with global deletion of 11βHSD1 have no overt abnormality in bone structure (Justesen et al., 2004); however, the bone phenotype of aged 11βHSD1 knockout animals has not been examined. In elderly humans, a negative correlation between plasma cortisone levels and osteocalcin has been observed, a relationship independent of plasma cortisol (Cooper et al., 2005). This study suggests that 11βHSD1 activity has functional consequences on osteoblasts in vivo. A similar negative association between cortisone and spine bone density was also seen in this study. Genetic association studies have linked the presence of allelic variants of *HSD11B1* (the gene for 11βHSD1) that predict low activity of 11βHSD1 with an increase in bone density and a reduction in vertebral fracture risk (Hwang et al., 2009). These data suggest that some of the adverse age-related changes in bone could be glucocorticoid mediated, but with exposure restricted to the osteoblast through local production by the 11βHSD1 enzyme. An analogous role for 11βHSD1 and endogenous glucocorticoids in skin aging has recently been demonstrated (Tiganescu et al., 2013).

Novel insights from targeted disruption of glucocorticoid signaling in bone

Investigating the effects of endogenous glucocorticoids on tissue function has been difficult, as global deletion of the GR in mice leads to early postnatal lethality (Cole et al., 1995). Over the past 2 decades, advances in cell-specific disruption of glucocorticoid signaling in mouse models have greatly improved our understanding of the physiological role of glucocorticoids in bone development and structure. Osteoblast-targeted deletion of the GR ($GR^{Runx2Cre}$) in mice results in an overall reduction in bone volume (Rauch et al., 2010). Correspondingly, calvarial osteoblast cultures derived from $GR^{Runx2Cre}$ mice demonstrate a reduction in osteoblast differentiation and the ability to form mineralized nodules (Rauch et al., 2010). Although not expressed naturally in postnatal bone cells, the 11βHSD2 enzyme has been used as a tool to examine the effects of endogenous glucocorticoids on the skeleton by targeted transgenic expression of 11βHSD2 within bone cells. This has the added advantage over GR deletion models that other possible signaling pathways in osteoblasts such as the mineralocorticoid receptor are also targeted. Transgenic mice in which 11βHSD2 expression is driven by a 2.3 kb collagen type I (Col2.3) promoter are characterized by targeted disruption of glucocorticoid signaling exclusively in mature osteoblasts and osteocytes (Sher et al., 2004, 2006). These mice (usually referred to as "Col2.3-11βHSD2 transgenic" mice) exhibit a phenotype of delayed calvarial bone development. Thus, embryonic and neonatal Col2.3-11βHSD2 transgenic mice display calvarial bone hypoplasia, osteopenia, increased suture patency, ectopic differentiation of cartilage in the sagittal suture, and a defect in the postnatal removal of parietal cartilage (Zhou et al., 2009). Similar to $GR^{Runx2Cre}$ mice, Col2.3-11βHSD2 transgenic adult mice of both sexes have reduced bone volume, lower trabecular number, and lower cortical bone mass (Kalak et al., 2009; Sher et al., 2004).

In a different model, the 11βHSD2 transgene is driven by the 3.6 kb collagen type I promoter ("Col3.6-11βHSD2 transgenic mice"). These animals exhibit a phenotype similar to that of Col2.3-11βHSD2 transgenic mice, including defects in calvarial bone development (Yang et al., 2010). Together, these findings strongly support the notion that endogenous glucocorticoids play a critical role in mesenchymal and osteoblast differentiation, accrual of bone mass, and the development of a healthy skeletal structure (Kalak et al., 2009; Rauch et al., 2010; Sher et al., 2004, 2006).

Endogenous glucocorticoids promote osteoblastogenesis

It has long been known from osteoblast cultures that the addition of glucocorticoids to the culture medium is crucial for the formation of mineralized nodules (Bellows et al., 1987), suggesting that these hormones are essential for differentiation of mesenchymal cells into mature osteoblasts (Haynesworth et al., 1992; Herbertson and Aubin, 1995). Evidence obtained from Col2.3-11βHSD2 transgenic mice indicates that this action of glucocorticoids is not through direct effects on mesenchymal progenitor cells but rather mediated indirectly through mature osteoblasts (Sher et al., 2006). Thus, calvarial cell cultures derived from Col2.3-11βHSD2 transgenic mice exhibit greatly reduced osteoblastogenesis and increased adipogenesis compared with cultures from animals with intact osteoblastic glucocorticoid signaling (i.e., wild-type mice). This phenotypic shift in mesenchymal progenitor cell commitment is accompanied by a reduction in Wnt7b and Wnt10b mRNA expression and β-catenin protein levels in transgenic compared with wild-type cultures. In addition, in the calvarial cell culture system, mature osteoblasts appear to be the dominant source of Wnt proteins, in particular Wnt7b, Wnt10b and Wnt9a, which are important stimulators of osteoblastogenesis. Thus, at physiological concentrations, glucocorticoids directly stimulate osteoblasts to produce and release Wnt proteins, which then act in a paracrine manner to activate the canonical Wnt signaling cascade in mesenchymal stem cells (MSCs). This results in the accumulation of β-catenin and the expression of the master regulator of osteoblast differentiation, runt-related transcription factor 2 (RUNX2) in MSCs, which drives the differentiation of these pluripotent cells toward the osteoblastic lineage. The Wnt signaling cascade in MSCs is further enhanced by the glucocorticoid-induced downregulation of potent Wnt suppressors such as sFRP1 (secreted frizzled-related protein 1) (Fig. 39.2) (Mak et al., 2009; Zhou et al., 2008). These pathways act synergistically to direct MSCs away from the adipogenic and toward the osteoblastic lineage. This mechanism is proven to be of importance in cranial bone development in vivo.

The skeletal changes in these mice are due to a lack of paracrine Wnt signaling by Col2.3-11βHSD2 transgenic osteoblasts. In addition to controlling lineage commitment, the glucocorticoid-induced increase in osteoblastic paracrine Wnt signaling affects the surrounding chondrocytes by inducing the expression of matrix metalloproteinase 14 (Mmp14). Mmp14 is central to the breakdown of extracellular matrix during development and tissue remodeling, and vital for cartilage removal during the process of ossification (Zhou et al., 2009). These concurrent and tightly interconnected pathways establish a novel role for both glucocorticoids and osteoblasts in the intricate process of intramembranous bone development.

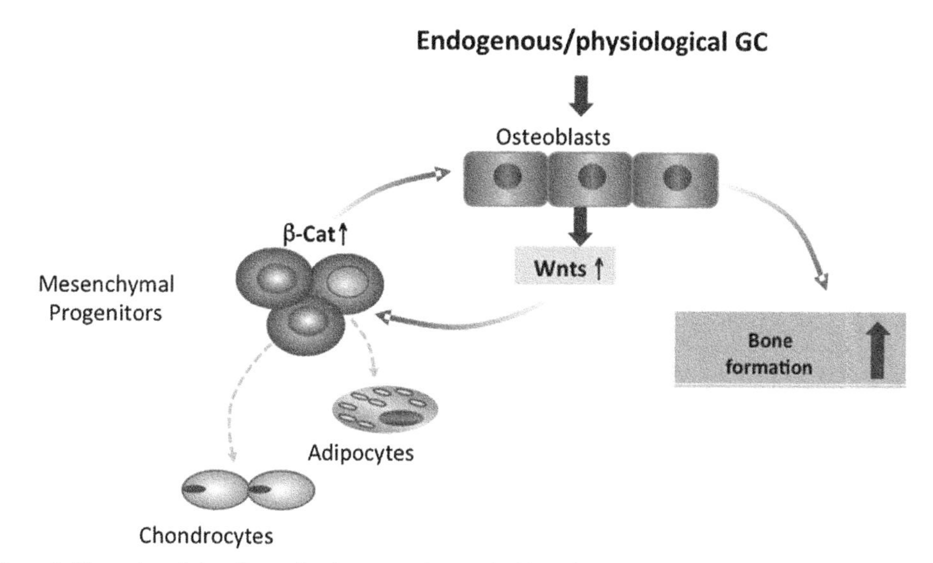

FIGURE 39.2 Schematic illustration of the effects of endogenous glucocorticoids on bone. At physiological concentrations, glucocorticoids stimulate mature osteoblasts to produce canonical Wnt proteins, which activate the β-catenin signaling cascade in mesenchymal progenitor cells. This in turn promotes differentiation of these immature cells toward the osteoblastic cell linage, with negative effects on chondro- and adipocytogenesis. Endogenous glucocorticoid signaling in osteoblasts therefore favors bone formation.

In summary, it is now clear that endogenous glucocorticoids control osteoblast differentiation, skeletal development during normal intramembranous bone formation, and trabecular and cortical bone mass accrual in mice.

Glucocorticoid excess and the skeleton

Nearly 65 years after their introduction into clinical practice by Hench and colleagues (Hench et al., 1949), glucocorticoids are still the most widely and frequently used class of antiinflammatory drugs. Survey data from 1996 estimated that in the UK, 0.5% of the population, 1.75% of women aged over 55 years, and up to 68% of patients with rheumatoid arthritis are on continuous treatment with oral glucocorticoids (Walsh et al., 1996). Later studies, also from the UK, found that up to 2.5% of people aged between 70 and 79 years receive oral glucocorticoids on an ongoing basis (van Staa et al., 2000a). A more recent multinational population-based study indicated that globally, up to 4.6% of postmenopausal women are taking oral glucocorticoids (Diez-Perez et al., 2011). In the US, data from several NHANES survey cycles demonstrate that the prevalence of glucocorticoid use remains above 1% of the general population (Overman et al., 2013) (Fig. 39.3). The major reason for the widespread use of therapeutic glucocorticoids is their unsurpassed antiinflammatory and immunomodulatory efficacy (reviewed in Buttgereit et al., 2011).

Therapeutic glucocorticoids, particularly when used over extended periods, cause significant off-target effects that are similar to the clinical signs and symptoms of endogenous hypercortisolism (Cushing's disease/syndrome). These include but are not limited to musculoskeletal conditions such as osteoporosis, osteonecrosis, myopathy, and sarcopenia, as well as metabolic effects resulting in glucose intolerance or diabetes mellitus, dyslipidemia, and excessive and abnormal fat accrual (Pereira and Freire de Carvalho, 2011; Schakman et al., 2008). Other adverse effects of prolonged glucocorticoid therapy include skin atrophy, hypertension, adrenal suppression, glaucoma, cataract, delayed wound healing, increased risk of infection, and in children, growth retardation.

Recognized as the most common form of secondary osteoporosis, GIO represents an iatrogenic problem of major clinical relevance across numerous specialties. Thus, one in five patients treated with oral glucocorticoids suffers an osteoporotic fracture within the first 12 months of therapy (Adachi et al., 1997; Cohen et al., 1999). This number increases to 50% after 5–10 years (Walsh et al., 1999, 2002), with most fractures occurring in the spine, proximal femur, and ribs. The relative risk of fracture increases as a function of glucocorticoid dose and duration of therapy, while absolute fracture risk is determined by a range of additional clinical risk factors such as age, gender, and comorbidities. Fracture risk is highest during the first few months following commencement of treatment (van Staa et al., 2000b, 2002) and decreases when therapy is ceased. Of note, even daily doses as low as 2.5 mg of prednisolone equivalent have been associated with increased fracture rates (van Staa et al., 2000b) (Fig. 39.4). Long-term inhaled glucocorticoid therapy in children and adults is also associated with significant bone loss (Wong et al., 2000). Despite the clear evidence for the detrimental effects of

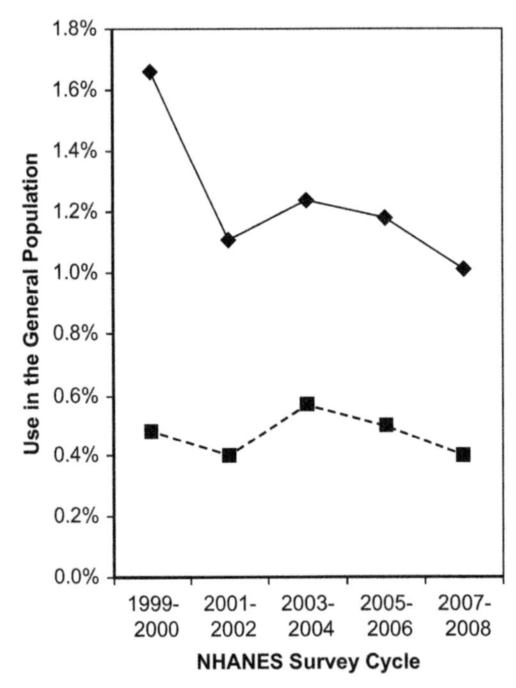

FIGURE 39.3 Prevalence of glucocorticoid use and antiosteoporosis therapy. Data were extracted from five National Health and Nutrition Examination Surveys. Diamonds represent glucocorticoid use. Squares represent concomitant antiosteoporosis therapy (Overman et al., 2013).

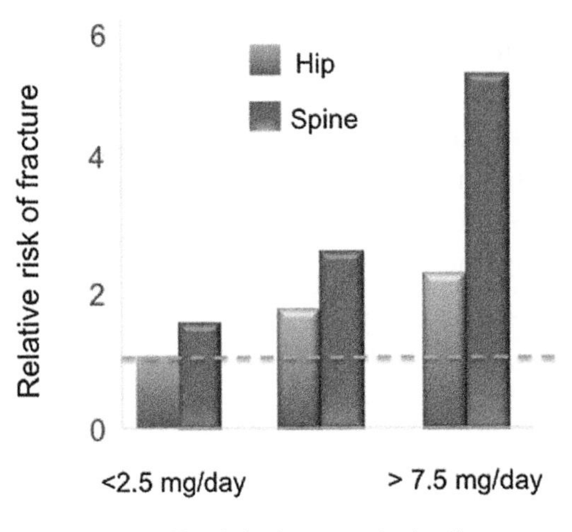

FIGURE 39.4 Relative risk of vertebral and hip fracture according to daily glucocorticoid dose. Note that even glucocorticoid doses considered by most clinicians as "low" (i.e., less than 2.5 mg prednisolone equivalent per day) are associated with a significant increase in fracture risk over time. *Adapted from van Staa, T.P., Leufkens, H.G., Abenhaim, L., Zhang, B., Cooper, C., 2000b. Oral corticosteroids and fracture risk: relationship to daily and cumulative doses. Rheumatology (Oxford) 39, 1383–1389.*

glucocorticoids on bone health, only a small proportion of patients receiving these medications are being treated to prevent bone loss and fractures (Overman et al., 2013) (Fig. 39.3).

Pathogenesis of glucocorticoid-induced osteoporosis

The main mechanisms underlying GIO are the direct action of glucocorticoids on bone cells, in particular on osteoblasts and osteocytes. Other effects are related to mineral metabolism such as decreased intestinal calcium absorption and increased renal calcium clearance (Canalis et al., 2007; Kuroki et al., 2008) as well as decreased growth hormone secretion (Mazziotti and Giustina, 2013) and alteration in sex steroid metabolism (Manolagas, 2013) (Table 39.1).

TABLE 39.1 Extraskeletal pathways contributing to glucocorticoid-induced osteoporosis.

Signaling pathway	Effect	References
Altered calcium balance	Glucocorticoid-induced decrease in calcium resorption and increase in calcium secretion may interfere with bone metabolism	Canalis et al. (2007)
Increased proteolysis in skeletal muscle	May contribute to a higher frequency of falls and fractures	Canalis et al. (2007), Pereira and Freire de Carvalho (2011), Schakman et al. (2008)
Growth hormone (GH) deficiency	Glucocorticoid-induced suppression of GH may contribute to the severity of GIO	Mazziotti and Giustina (2013)
Altered sex hormones	Glucocorticoid-induced suppression of testosterone and/or estrogen may contribute to GIO	Canalis et al. (2007), van Staa (2006)
Altered PTH pulsatility	Glucocorticoids may alter PTH pulsatility, which may contribute to bone loss	Bonadonna et al. (2005), Canalis et al. (2007)

In bone, glucocorticoid excess affects osteoblasts, osteocytes, and osteoclasts. Earlier, mostly histomorphometric studies indicated that the predominant feature of GIO is a suppression of osteoblast activity and hence of bone formation while osteoclast numbers are either unchanged or slightly increased (Weinstein et al., 1998). These data were sufficient to understand that the pathogenesis of GIO was fundamentally different from that of postmenopausal osteoporosis. The mechanisms by which glucocorticoids exert their actions on bone remained elusive, however. With the advent of more sophisticated technologies, we now have a more detailed (although still incomplete) understanding of the cellular and molecular actions of glucocorticoids in bone.

Although the individual effects of glucocorticoid excess on osteoblasts, osteocytes, and osteoclasts have been well characterized in vitro, the high degree of reciprocal interconnectivity between bone cells has made it difficult to dissect the skeletal actions of glucocorticoids in vivo. In this context, the development of genetically modified mouse models in which only one specific subset of bone cells has been shielded from the effects of exogenous glucocorticoids has provided useful information on the main target cells and molecular mechanisms involved in the pathogenesis of GIO. Thus, recent studies in rodents have demonstrated that protecting the *osteoclast* from excessive glucocorticoid signaling partly prevents the increase in osteoclast lifespan and bone resorption normally seen with glucocorticoid treatment (see also below) (Jia et al., 2006; Kim et al., 2006). However, as a consequence of the ongoing suppression of osteoblast and osteocyte function, bone formation remains compromised in most of these models (Jia et al., 2006; Rauch et al., 2010), inevitably leading to reduced bone strength over a prolonged period. Thus, in the long-term, the direct effects of glucocorticoids on osteoclasts appear to contribute little to the loss of bone mass and strength seen in GIO.

Conversely, protecting the osteoblast (and with that the osteocyte) from excessive glucocorticoid exposure and signaling not only prevents osteoblast and osteocyte apoptosis, but in fact preserves osteoblast function and bone formation (O'Brien et al., 2004). Furthermore, in these animal models, bone strength and quality are maintained despite the presence of supraphysiological plasma glucocorticoid levels (Henneicke et al., 2011; O'Brien et al., 2004; Rauch et al., 2010). Abrogating glucocorticoid signaling in osteoblasts/osteocytes also prevents the increase in osteoclast number and activity in addition to changes in bone vascularity and the lacunar–canalicular system normally seen with glucocorticoid excess (Henneicke et al., 2011; Weinstein et al., 2010b). Therefore, some of the effects of exogenous glucocorticoids on the osteoclast are preventable when the osteoblast/osteocyte system is shielded from excessive glucocorticoid signaling, pointing to the cells of the osteoblast lineage as the main target of exogenous glucocorticoids within the skeleton.

Glucocorticoid excess and its effects on osteoblast differentiation

It has been well established that glucocorticoid excess inhibits both osteoblast differentiation and function, while at the same time inducing osteoblast apoptosis, resulting in a rapid (within hours) and profound suppression of bone formation (Kauh et al., 2012; Weinstein et al., 1998). In recent years, a number of molecular targets and mechanisms underlying these osteoblast-specific glucocorticoid actions have been identified. For example, exposure to supraphysiological glucocorticoid concentrations suppresses the very pathways that at physiological levels promote osteoblastogenesis, namely Wnt and bone morphogenetic protein 2 (BMP-2) signaling: At high concentrations, glucocorticoids inhibit the synthesis and release of

Wnt-related transcription factors by mature osteoblasts (Mak et al., 2009). Downregulation of Wnt signaling leads to impaired osteoblast differentiation via a reduction in molecular downstream signals such as β-catenin and RUNX2 transcription factors (see above). Additionally, high intracellular levels of glucocorticoids enhance β-catenin degradation via increased glycogen synthase kinase 3 beta expression, a protein, which drives β-catenin ubiquitinylation and proteasomal degradation (Wang et al., 2009) and promotes the osteoblastic expression of Wnt inhibitors (Hayashi et al., 2009; Kato et al., 2018; Mak et al., 2009). These glucocorticoid-inducible soluble antagonists act on MSCs, inhibiting Wnt/β-catenin signaling and thereby reducing osteoblast differentiation (Yao et al., 2008; Zhou et al., 2008). Moreover, glucocorticoid excess concurrently increases expression of transcription factors critical for adipocyte differentiation such as peroxisome proliferator-activated receptor γ (Yao et al., 2008), suggesting that the commitment of MSC to the adipocyte lineage limits the pool of cells available to differentiate into osteoblasts. However, this hypothesis requires further investigation.

Another key regulator of osteoblast differentiation, BMP-2, is profoundly inhibited in mice receiving glucocorticoids at therapeutic concentrations (Yao et al., 2008). In addition, dexamethasone treatment of osteoblast cultures has been reported to substantially increase the expression of follistatin and Dan, two BMP-2 antagonists (Hayashi et al., 2009), thus exacerbating the inhibition of BMP-2 signaling in osteoblasts. Recent studies have demonstrated that the monomeric glucocorticoid-GR complex is sufficient to significantly impair osteoblast differentiation (Rauch et al., 2010), most likely due to interference with the proinflammatory transcription factor AP-1 and suppression of IL-11 (Rauch et al., 2011). The latter cytokine has previously been shown to promote bone formation (Takeuchi et al., 2002). Indeed, the addition of IL-11 to glucocorticoid-treated osteoblast cultures protects these cells from glucocorticoid-induced reduction in differentiation potential. IL-11 has also been shown to promote BMP-2 signaling (Takeuchi et al., 2002) and more recently demonstrated to suppress the production of Wnt antagonists (Matsumoto et al., 2012). Thus, a reduction in IL-11 activity appears to attenuate two critically important signaling pathways that control differentiation in osteoblasts. Analyzing bone specimens from patients with GIO and prednisolone-treated mice, a recent study (Liu et al., 2017a) documented increased levels of casein kinase-2 interacting protein-1 (CKIP-1), a ubiquitination-related protein facilitating Smurf1-mediated Smad1/5 ubiquitination (Lu et al., 2008). Interestingly, mice with osteoblast-specific CKIP-1 deletion maintained normal BMP signaling and were protected from glucocorticoid-induced bone loss (Liu et al., 2017a).

In recent years, microRNAs (miRs) have been discovered to play an important role in osteoblast differentiation. Excess glucocorticoids modulate the expression pattern of microRNAs, impairing osteoblastogenesis. Thus, miR-29a expression in bone has been demonstrated to decrease with glucocorticoid exposure. Overexpression of miR-29a in transgenic mice mitigates glucocorticoid-induced bone loss by protecting against impaired osteogenic differentiation (Ko et al., 2015). Glucocorticoids promote HDAC4 signaling, subsequently accelerating RUNX2 and β-catenin ubiquitination.

Overexpressing miR-29a significantly attenuates glucocorticoid-mediated RUNX2 deacetylation and β-catenin ubiquitination via inhibiting HDAC4 action (Ko et al., 2013, 2015). Dexamethasone treatment upregulates miR-199a-5p expression, inhibiting osteoblast proliferation through reducing FZD4 and Wnt2 expression. Deletion of miR-199a-5p attenuates dexamethasone-induced inhibition of osteoblast proliferation (Shi et al., 2015). More recently, it has been reported that silencing miR-106b expression provides protection from glucocorticoid-induced bone loss through restoring BMP2 expression, which consequently maintains osteoblast differentiation and bone formation (Liu et al., 2017b). In contrast to these catabolic microRNAs, miR-34a-5p promotes osteoblast differentiation through suppression of Notch signaling. Glucocorticoids reduce the expression of miR-34a-5p, thereby inhibiting osteogenic differentiation (Kang et al., 2016).

The simultaneous reduction in both Wnt/β-catenin and BMP-2 signaling during glucocorticoid excess leads to a significant shift in MSC differentiation, a reduction in osteoblastogenesis, and consequently a loss of osteoblast-generated proteins such as collagen and osteocalcin (Moutsatsou et al., 2012; Yao et al., 2008).

Glucocorticoids prevent osteoblast cell cycle progression

Exposure of osteoblasts to high concentrations of glucocorticoids in vitro has been shown to disrupt physiological cell cycle progression (Chang et al., 2009). A marked decrease in cyclin D2 and an increase in cyclin-dependent kinase inhibitor 1B ($p27^{kip1}$) following dexamethasone treatment of human osteoblasts has been observed, consistent with cell cycle arrest at the G0 and G1 phase (Chang et al., 2009). Dexamethasone has also been reported to increase the expression of DUSP1 in cultured osteoblasts, with a corresponding decrease in the phosphorylation status of mitogen-activated protein kinase 1 (MAPK1). The dephosphorylation of MAPK1 impairs osteoblast response to mitogenic signaling, inhibiting cell proliferation (Horsch et al., 2007). Furthermore, dexamethasone suppresses the osteoblastic expression of cyclin A, a protein crucial for the transition from G1 to S phase during or after commitment to the osteoblast lineage (Gabet et al.,

2011). Cell cycle progression via the canonical Wnt signaling pathway requires the association of β-catenin with LEF/TCF transcription factors, both of which are inhibited by high glucocorticoid levels. In the face of reduced Wnt and β-catenin signaling, the reduction in LEF/TCF further inhibits the G1 to S phase transition (Jia et al., 2006).

Glucocorticoids induce osteoblast apoptosis

Pharmacological doses of glucocorticoids induce osteoblast apoptosis both in vitro and in vivo through a number of proapoptotic pathways. Administration of dexamethasone to osteoblast cultures dose-dependently increases proapoptotic factors of the Bcl-2 family such as Bim (Espina et al., 2008). Correspondingly, knockdown of Bim in osteoblasts significantly reduces glucocorticoid-induced apoptosis (Espina et al., 2008). Furthermore, silencing E4BP4, a basic leucine zipper transcription factor, attenuates glucocorticoid-induced Bim expression and blocks osteoblast apoptosis induced by dexamethasone treatment (Chen et al., 2014). In addition, glucocorticoid-induced increases in Bak, another proapoptotic member of the Bcl-2 family, as well as a decrease in Bcl-XL expression, a prosurvival Bcl-2 protein, further drive osteoblasts toward cell death (Chang et al., 2009). Exposure of osteoblasts to dexamethasone in vitro also results in upregulation of the tumor suppressor and apoptosis regulator, p53, and subsequent Bcl-2 mediated cell apoptosis (Li et al., 2012). Glucocorticoids have also been found to induce osteoblast apoptosis via a reduction in β1-integrin expression, which leads to a loss of cell-matrix adhesion (Naves et al., 2011). High-dose glucocorticoids may induce osteoblast apoptosis through activation of $p66^{shc}$ and an increase in reactive oxygen species (ROS), which in turn leads to constitutive activation of the JNK pathway (Almeida et al., 2011) or induces endoplasmic reticulum (ER) stress (Sato et al., 2015). Opposing ER stress by inhibiting eIF2α dephosphorylation prevents osteoblast and osteocyte apoptosis and attenuates osteoblast dysfunction induced by glucocorticoids in vitro and in vivo (Sato et al., 2015).

In summary, glucocorticoids at supraphysiological or pharmacological concentrations are detrimental to the skeleton through potent suppression of osteoblastogenesis, induction of osteoblast dysfunction, diversion of MSC to the adipocyte lineage, as well as promotion of osteoblastic cell cycle arrest and apoptosis (Fig. 39.5). Overall, exposure to high dose glucocorticoids not only compromises the skeleton's ability to form new bone, but also profoundly disturbs the regulatory balance between bone formation and resorption, which eventually leads to a reduction in bone mass and an increase in fracture risk.

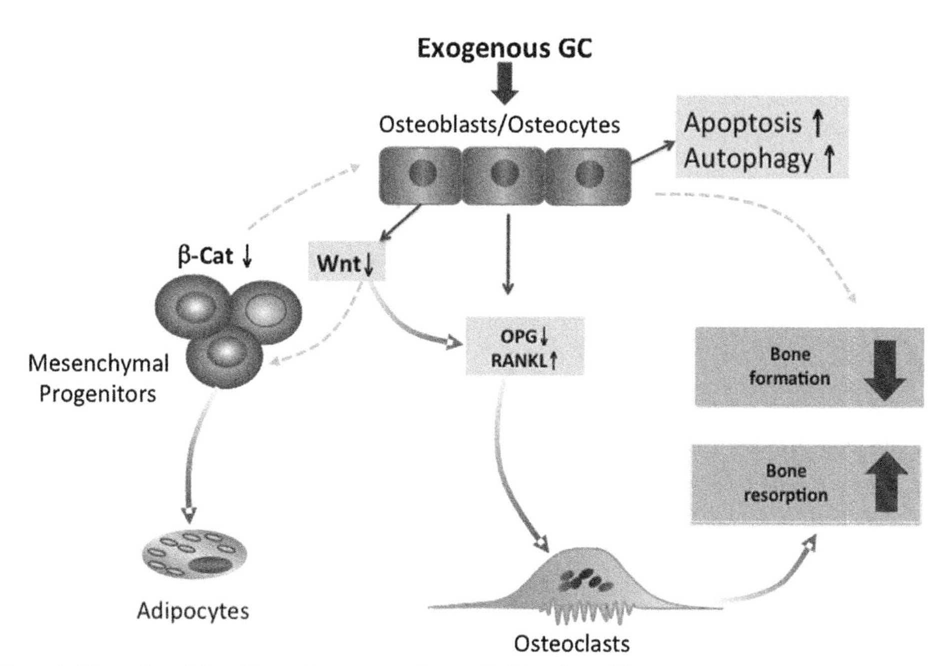

FIGURE 39.5 Schematic illustration of the effects of exogenous glucocorticoid on bone. High-dose glucocorticoids negatively impact osteoblast and osteocyte function through (1) inhibition of Wnt protein expression in mature osteoblasts (which results in mesenchymal progenitor cells differentiating preferentially toward adipocytes and away from osteoblasts); (2) stimulation of RANKL and inhibition of OPG expression (resulting in an increase in the RANKL/OPG ratio which favors bone resorption); and (3) induction of osteoblast and osteocyte apoptosis (which reduces bone formation).

Glucocorticoid excess and the osteocyte

The osteocyte is the most abundant cell type in the skeleton, representing 90%–95% of the entire cell mass (Crockett et al., 2011). The complex osteocyte network functions as a mechanosensor, which maintains bone integrity through the recruitment of osteoblasts and osteoclasts to the sites of active bone remodeling (Temiyasathit and Jacobs, 2010). Furthermore, this system as well as its fluid content has been shown to lend the skeleton a high degree of stiffness (Weinstein et al., 2010b). High glucocorticoid concentrations not only lead to a substantial loss of intraosseous vasculature, but also to a decrease in the solute transport from the circulation to the lacunar–canalicular system (Weinstein et al., 2010a, 2010b). The reduction in overall intraskeletal fluid content is strongly associated with the glucocorticoid-induced decrease in bone strength but not necessarily bone mass (Weinstein et al., 2010a).

Treatment with pharmacological doses of glucocorticoids has been shown to cause intracellular damage and induce osteocyte autophagy both in vitro and in vivo (Jia et al., 2011; Xia et al., 2010). With prolonged exposure to glucocorticoids, pathological buildup of autophagosomes occurs, and as the osteocyte is embedded in the bone matrix, the degraded materials expelled from the autophagosomes create an environment toxic to the cell, leading to cellular demise (Jia et al., 2011). Glucocorticoid-induced autophagy subsequently results in connexin43 degradation, thereby impairing osteocyte cell-cell connectivity (Gao et al., 2016). Additionally, higher rates of osteocyte apoptosis have been linked to activation of proapoptotic kinases, Pyk2 and JNK, in osteocytes (Bellido, 2010). While activation of JNK leads to an increase in ROS and eventually programmed cell death, activation of the protein tyrosine kinase Pyk2 results in the reorganization of the cytoskeleton, cell detachment, and ultimately apoptosis (Bellido, 2010).

In summary, recent evidence has highlighted both the osteocyte and the lacunar–canalicular network as important targets of glucocorticoid-induced damage to bone. The reduction in skeletal fluid content in conjunction with decreased osteocyte viability represents an important novel mechanism by which excessive levels of glucocorticoids may compromise bone stability without affecting actual bone mass.

Glucocorticoid excess and the osteoclast

During the initial stage of glucocorticoid therapy, there is a rapid but transient increase in bone resorption due to an increase in osteoclast number and activity (Dovio et al., 2004; Henneicke et al., 2011; Jia et al., 2006). In humans, biochemical markers of bone resorption increase within 24 h of exposure to therapeutic levels of glucocorticoids (Dovio et al., 2004). Experimental studies have established that glucocorticoids directly extend the lifespan of mature osteoclasts by delaying apoptotic signaling, which over time leads to a significant increase in osteoclast numbers (Jia et al., 2006; Kim et al., 2006). Osteoclast formation is further amplified indirectly through glucocorticoid-induced changes in osteoblast gene expression. For example, high levels of glucocorticoids stimulate the production of RANKL by cells of the osteoblast lineage while simultaneously reducing the expression of the RANKL decoy receptor, OPG (Hofbauer et al., 1999; Sivagurunathan et al., 2005). This imbalance in the RANKL-OPG ratio leads to a profound increase in osteoclast activity and bone resorption.

Furthermore, the effect of glucocorticoids on osteoblastic RANKL expression appears to be mediated by miR-17/20a (Shi et al., 2014).

Under physiological conditions, the major source of RANKL and hence the primary driver of osteoclastogenesis is the osteocyte rather than the osteoblast (Nakashima et al., 2011; Xiong et al., 2011). In vitro studies show that the osteocyte presents membrane-bound RANKL to osteoclast precursors at the extremities of their dendritic processes rather than secreting soluble RANKL (Honma et al., 2013). Whether this process is affected by glucocorticoids is yet to be determined. Apoptosis of osteocytes leads, seemingly paradoxically, to an increase in RANKL concentrations, possibly due to signaling of dying cells to surrounding healthy osteocytes, which then in turn express RANKL (Kennedy et al., 2012). However, another study reported that osteocyte apoptotic bodies strongly induce osteoclastogenesis in vitro, although this occurred independently of RANKL expression (Kogianni et al., 2008). As glucocorticoids are known to induce osteocyte apoptosis, the subsequent induction of RANKL or other signaling pathways may contribute to the increase in osteoclastogenesis observed following glucocorticoid exposure.

Apart from its function as a potent inhibitor of osteoblast function, sclerostin may also have a role as a potential inductor of osteoclast activity and bone resorption (Li et al., 2008). Recent studies provide in vivo evidence that sclerostin-induced bone resorption may be an important contributor to bone loss under conditions of glucocorticoid excess (Henneicke et al., 2017; Sato et al., 2016). When exposed to high concentrations of glucocorticoids, sclerostin-deficient mice are protected from increased bone resorption and bone loss. However, sclerostin deficiency does not prevent glucocorticoid-induced osteoblast/osteocyte apoptosis and impaired bone formation (Sato et al., 2016). A similar mechanism may play a role in relaying the effects of chronic stress on bone resorption. Chronic stress exposure leads to elevated levels of circulating glucocorticoids, resulting in bone loss, increased bone resorption, and enhanced Sost/sclerostin expression in wild-type mice but not in mice where glucocorticoid signaling has been disrupted in osteoblasts and osteocytes (Henneicke et al., 2017). These results indicate that stress-related bone loss is mediated via the effects of glucocorticoids on osteoblasts/osteocytes and subsequent activation of osteoclasts through amplified expression of Sost/sclerostin (Henneicke et al., 2017).

Despite an initial increase in bone resorption, prolonged glucocorticoid excess appears to suppress osteoclast number and function, directly blocking the induction of cytoskeletal changes in the osteoclast required for bone resorption (Kim et al., 2007). There is also evidence that glucocorticoids suppress the proliferation of osteoclast precursors, or more specifically, uncommitted bone marrow macrophages (Kim et al., 2006, 2007).

Overall, exposure to exogenous glucocorticoids causes an initial rise in bone resorption, which contributes to the early and rapid loss of bone mineral density. During long-term therapy, however, the disruptive effects of glucocorticoids on the cytoskeleton and suppression of osteoclastogenesis appear to be the predominant feature as far as the osteoclast is concerned (Fig. 39.5).

Glucocorticoid excess and local glucocorticoid metabolism

Local steroid metabolism through expression of the 11βHSD1 enzyme is also likely to influence glucocorticoid impacts on bone. Prednisone and prednisolone, the main oral glucocorticoids used clinically, are metabolized by this enzyme in a similar fashion to cortisone and cortisol (Cooper et al., 2002). Systemic measures of 11βHSD1 obtained in healthy control subjects strongly predict the extent to which oral glucocorticoids have an impact on bone formation markers such as osteocalcin (Cooper et al., 2003). However, 11βHSD1 expression in osteoblasts is regulated by a range of factors that are likely to be of relevance to people treated with glucocorticoids. Both proinflammatory cytokines and glucocorticoids increase expression and activity of 11βHSD1 (Cooper et al., 2001, 2002; Kaur et al., 2010), potentially increasing the sensitivity of bone to glucocorticoids during inflammatory illness. However, 11βHSD1 expression is also increased in a range of tissues during inflammation (Kaur et al., 2010). Global deletion of 11βHSD1 in mice increases the severity of a range of models of experimental inflammation (Coutinho et al., 2012). It is therefore possible that any increased sensitivity of bone to glucocorticoids through an increase in 11βHSD1 will be matched by increased sensitivity of the underlying disease being treated. No association of systemic 11βHSD1 activity and bone loss was seen in a trial of patients with inflammatory bowel disease treated with high-dose oral prednisolone (Cooper et al., 2011).

Indirect mechanisms for glucocorticoid-induced osteoporosis

The mechanisms described so far concern the direct consequences of glucocorticoid action on bone cells and their molecular signaling. However, particularly when given at higher doses, glucocorticoids have further systemic effects (Table 39.1), not all of which have been clearly delineated. Changes in renal and intestinal calcium handling, as well as sex and growth hormone action, have been well documented. Thus, glucocorticoids reduce intestinal calcium absorption while inhibiting calcium reabsorption in the kidney (Ritz et al., 1984). This leads to a negative net calcium balance, which may adversely affect bone mineralization. It was previously thought that glucocorticoid-mediated alteration of (and in particular increased) PTH secretion was also a common feature to account for bone loss (Fucik et al., 1975). However, many studies have subsequently failed to find direct links to between glucocorticoids and increased PTH signaling, and it is now considered that changes in PTH secretion or action do not play a central role in GIO (Rubin and Bilezikian, 2002). Clearly, however, increased PTH secretion can be a secondary and appropriate response in the context of a prolonged negative net calcium balance induced by glucocorticoids.

When given at higher doses, glucocorticoids can induce hypogonadism through suppression of the HP axis, which in turn may contribute to bone loss and increased fracture risk (Chrousos et al., 1998). Other hormones have also been implicated in indirect effects of glucocorticoids on bone. In particular, glucocorticoids cause a reduction in IGF1, a powerful stimulator of osteoblast function, and thus any reduction in its levels might result in a reduced anabolic action on bone. However, studies involving calvarial bones from mice with deletion of the IGF1 receptor demonstrated that glucocorticoids still have a major negative effect on collagen synthesis, suggesting that changes in IGF1 signaling are not a major part of the actions of glucocorticoids on osteoblasts (Woitge and Kream, 2000). Although a small observational study in children with arthritis treated with glucocorticoids suggested a protective effect of growth hormone (GH) treatment on bone density GH/IGF1, treatment has not been evaluated in the context of a randomized controlled trial in patients taking glucocorticoids.

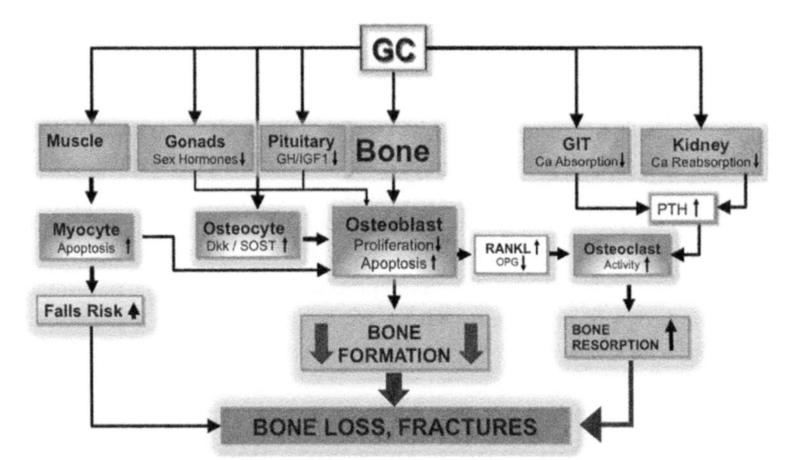

FIGURE 39.6 Schematic summary of the detrimental effects of therapeutic or excess glucocorticoids on skeletal and nonskeletal organ systems. While the skeletal adverse outcomes are mediated mainly through effects on osteoblasts, many other systems contribute to the increase in fracture risk in patients on long-term glucocorticoid therapy.

Proximal myopathy is a well-recognized feature of glucocorticoid excess (Pereira and Freire de Carvalho, 2011) and is thought to affect bone health directly (through its effects on bone strength) and indirectly (through an increase in falls risk) (Fig. 39.6).

Glucocorticoid excess and systemic fuel metabolism

Insulin resistance, dyslipidemia, and abnormal body fat accrual and distribution belong to the most common and problematic adverse effects of glucocorticoid therapy (reviewed in Rafacho et al., 2014). Short-term treatment with glucocorticoids is associated with impaired release of insulin from the pancreas. However, this mode of action does not feature prominently during long-term glucocorticoid use (Beard et al., 1984; Hazlehurst et al., 2013; Nicod et al., 2003; Petersons et al., 2013). Thus, patients with rheumatoid arthritis treated chronically with prednisolone tend to develop insulin resistance (Petersons et al., 2013). For most individuals, blood glucose levels remain in the nondiabetic range, but this is only achieved through a rise in plasma insulin levels. However, a significant proportion of glucocorticoid-treated individuals will develop glucose intolerance or overt diabetes mellitus. These patients tend to be those that already have a degree of insulin resistance due to their age, genetic or ethnic background, or other comorbidities (van Raalte et al., 2009). Many patients exposed to high doses of therapeutic glucocorticoids for prolonged periods also develop changes in body fat accrual and distribution. The basis for the changes in fat redistribution is unclear, but differences in response to glucocorticoids of subcutaneous and visceral adipose tissue have been proposed (Fernandez-Rodriguez et al., 2009).

Until recently, these metabolic changes were thought to be due to the direct effects of glucocorticoids on liver, pancreas, and adipose tissue. However, recent research has raised the possibility that the adverse effects of glucocorticoids on fuel metabolism are mediated, at least in part, through their action on bone. As mentioned above, glucocorticoids suppress osteoblast function and inhibit synthesis of osteocalcin, an osteoblast-derived peptide previously reported to be involved in normal glucose metabolism (Lee et al., 2007). Using transgenic and gene therapy approaches, experimental studies in mice found that the suppression of osteoblast function and associated reduction in osteocalcin signaling may be central to the pathogenesis of glucocorticoid-induced dysmetabolism (Brennan-Speranza et al., 2012). Thus, mice in which glucocorticoid signaling has been disrupted in osteoblasts and osteocytes (Col2.3-11βHSD2 transgenic mice) are not only protected from the bone-wasting effects of glucocorticoid excess but maintain normal insulin sensitivity, glucose tolerance, and body weight despite high-dose glucocorticoid treatment (Brennan-Speranza et al., 2012). Very similar effects were achieved by replacing osteocalcin in the circulation via gene therapy in glucocorticoid-treated animals.

The above studies point toward bone, and specifically the osteoblast and its product, osteocalcin, being significant regulators of energy metabolism in the context of glucocorticoid treatment. This may have clinical implications, since the metabolic adverse effects of glucocorticoids are difficult to counteract clinically. There is only a limited amount of clinical data relating to the relationship between glucocorticoids, osteocalcin, and changes in energy metabolism. The situation is particularly complicated, since most people treated with glucocorticoids will have an underlying disease that itself could influence systemic energy balance and glucose homeostasis (van Staa et al., 2000c). This is compounded by the lack of availability of clinical approaches to selectively raise serum osteocalcin levels. One recent study has overcome some of these limitations by examining patients treated with bisphosphonates, teriparatide (PTH1-34), or calcium/vitamin D alone for GIO (Stockbrugger et al., 2002). Teriparatide is known to stimulate osteocalcin synthesis whereas bisphosphonates achieve the opposite effect. Over 12 months of therapy, bisphosphonates or calcium/vitamin D had no effect on glycemic control, whereas in teriparatide-treated patients, a small but statistically significant improvement in HbA1c was observed. This finding suggests that a treatment that stimulates osteocalcin has the potential to improve glycemic control in the setting of chronic glucocorticoid treatment.

In summary, there is now good evidence that the effects of glucocorticoid excess on fuel metabolism are at least in part mediated through the skeleton. While animal data suggest a relatively strong relationship, clinical studies that fully address this link in humans are lacking. Additionally, the detailed molecular mechanisms by which osteocalcin exerts its action, including the exact receptor(s) and tissues of action, remain uncertain. Although certainly a matter of future research, it is conceivable that the mechanisms related to the actions of osteocalcin and related osteoblast-derived proteins on fuel metabolism may become the basis of future therapies for the prevention and treatment of glucocorticoid-induced dysmetabolism (Ferris and Kahn, 2012).

Treatment of glucocorticoid-induced osteoporosis

While the negative effects of glucocorticoids on bone health have been known to clinicians for many years, GIO remains underdiagnosed and undertreated. Despite increasing awareness of the various forms of osteoporosis among patients and health care providers, recognition of the potentially devastating consequences of glucocorticoid therapy on the skeleton, and the knowledge of how to prevent or treat these outcomes, are still inadequate (Curtis et al., 2005; Feldstein et al., 2005; Overman et al., 2013) (Fig. 39.3). In part, this deficit may be due to the "osteoporosis care gap"—i.e., a lack of adequate medical care that worldwide leaves 70%–80% of patients with osteoporotic fractures undiagnosed and untreated (Eisman et al., 2012). In addition, the complex actions of glucocorticoids in bone and elsewhere, and the uncertainties regarding intervention thresholds for the prevention and treatment of the disease, may contribute to current inadequacies in the recognition and management of patients with GIO.

Assessment of the patient with glucocorticoid-induced osteoporosis

Oral or parental glucocorticoids are almost always used to treat systemic inflammatory conditions, which through increased inflammatory cytokine output affect bone health independent of medication-induced adverse effects (Hardy and Cooper, 2009). This is best demonstrated in rheumatoid arthritis, where inflammatory disease activity can have a substantial negative impact on bone health through the stimulation of osteoclastogenesis and bone erosions and suppression of bone formation (Bultink et al., 2005; Gough et al., 1994; Matzelle et al., 2012). Low-dose glucocorticoid therapy has been reported to suppress inflammation in patients with early arthritis, and despite the possible negative effects of glucocorticoids on bone, results in better bone status. As such, the impact of the underlying disease on bone needs to be balanced against potentially beneficial effects of glucocorticoid therapy that may result in overall beneficial effects on skeletal health if it reduces the inflammatory activity of the underlying inflammatory disease (and hence circulating cytokine levels) (Guler-Yuksel et al., 2008; Landewe et al., 2002; van der Goes et al., 2013). These partly conflicting factors need to be considered when making the clinical decision to intervene. Nevertheless, the use of therapeutic glucocorticoids—particularly at high doses—is associated with a substantial increase in fracture risk (Fig. 39.4) (van Staa et al., 2000b, 2005).

In contrast to postmenopausal osteoporosis, measurement of bone mineral density has limited value in determining fracture risk in the context of glucocorticoid therapy (Van Staa et al., 2003). As a rule, the "fracture threshold" seems to be lower in patients on glucocorticoid therapy—i.e., fractures occur at higher bone mineral density than what has been established for women with postmenopausal osteoporosis. Although new imaging techniques such as high-resolution computed tomography and volumetric quantitative computed tomography have been found to reasonably predict fracture risk even in the context of GIO, their use outside clinical studies currently remains limited (Kalpakcioglu et al., 2011).

A comprehensive review of potential risk factors for bone loss and fracture therefore remains essential to the clinical work-up of patients with GIO. Some guidelines also recommend the use of fracture risk calculators (Grossman et al., 2010; Lekamwasam et al., 2012), although these recommendations are not without controversy. For example, instead of individual or cumulative doses, FRAX uses a fixed ("average") glucocorticoid dose to calculate fracture risk, while the Garvan fracture risk calculator does not account for glucocorticoid treatment at all. It is therefore likely that both algorithms underestimate fracture risk in patients receiving glucocorticoids at therapeutic doses.

While standard diagnostic measures such as bone mineral density testing and laboratory parameters are of limited value in the work-up of the patient with GIO, they can be useful in monitoring treatment efficacy or disease progression. Since bone loss in GIO can vary considerably between individuals, monitoring changes in bone mineral density may be helpful in identifying patients particularly susceptible to the skeletal effects of glucocorticoids. Evaluation of hepatic and renal function as well as determination of serum vitamin D, calcium, phosphate, and parathyroid hormone levels are usually recommended before commencement of treatment (Grossman et al., 2010; Lekamwasam et al., 2012).

Management of glucocorticoid-induced osteoporosis

One of the core principles of treating patients with GIO is to minimize glucocorticoid use. This is easier said than done, as more often than not reductions in glucocorticoid dosing are followed by a flare-up of the underlying condition. In these cases, concomitant treatment with a bone-sparing agent is necessary. Dose, timing, and duration of such "adjuvant therapies" are a matter of ongoing debate and vary across studies and guidelines (Grossman et al., 2010; Hansen et al., 2011; Lekamwasam et al., 2012), but there is some consensus that in patients with low bone mineral density or genuinely increased fracture risk, treatment with a bisphosphonate or teriparatide should coincide with the initiation of glucocorticoid therapy (Table 39.2). In addition, supplementation with calcium and vitamin D as required is recommended (Grossman et al., 2010; Lekamwasam et al., 2012). Thus, the 2017 American College of Rheumatology Guidelines for the Prevention and Treatment of Glucocorticoid-Induced Osteoporosis (Buckley et al., 2017), while noting the "limited evidence regarding the benefits and harms of interventions" in patients treated with glucocorticoids, recommends "treating only with calcium and vitamin D in adults at low fracture risk, treating with calcium and vitamin D plus an additional osteoporosis medication (oral bisphosphonate preferred) in adults at moderate-to-high fracture risk, continuing calcium plus vitamin D but switching from an oral bisphosphonate to another anti-fracture medication in adults in whom oral bisphosphonate treatment is not appropriate, and continuing oral bisphosphonate treatment or switching to another anti-fracture medication in adults who complete a planned oral bisphosphonate regimen but continue to receive glucocorticoid treatment" (Buckley et al., 2017).

TABLE 39.2 Recommended pharmacological interventions for the prevention and treatment of glucocorticoid-induced osteoporosis.

		Evidence grade		
Intervention	Dose	(BMD)	(Fracture)[a]	References[b]
Calcium	Oral: 1000–1500 mg daily	A[c]	–	Amin et al. (1999), Buckley et al. (1996), Sambrook et al. (1993)
Vitamin D	Oral: 800–1000 IU daily	A[d]	–	Amin et al. (1999), Buckley et al. (1996), Sambrook et al. (1993)
Alendronate	Oral: 70 mg once/week	A	B	Adachi et al. (2001), Stoch et al. (2009), Yilmaz et al. (2001)
Risedronate	Oral: 35 mg once/week	A	A	Cohen et al. (1999), Eastell et al. (2000), Reid et al. (2000), (2001), Wallach et al. (2000)
Zoledronic acid	Intravenous: 5 mg once/year	A	–	Kauh et al. (2012), Sambrook et al. (2012)
Teriparatide	Subcutaneous: 20 mg once/day	A	A	Gluer et al. (2013), Karras et al. (2012), Saag et al. (2007), (2009)
Etidronate	Oral: 400 mg daily for 2 weeks every 3 months	A	A	Adachi et al. (1997), Cortet et al. (1999), Roux et al. (1998), Struys et al. (1995)

A, several RCT or meta-analysis; B, one RCT or nonrandomized studies.
[a]*Not a primary end point and vertebral fractures only.*
[b]*Reviewed in Grossman et al. (2010), Lekamwasam et al. (2012), and Rizzoli et al. (2012).*
[c]*In combination with Vitamin D.*
[d]*In combination with calcium.*

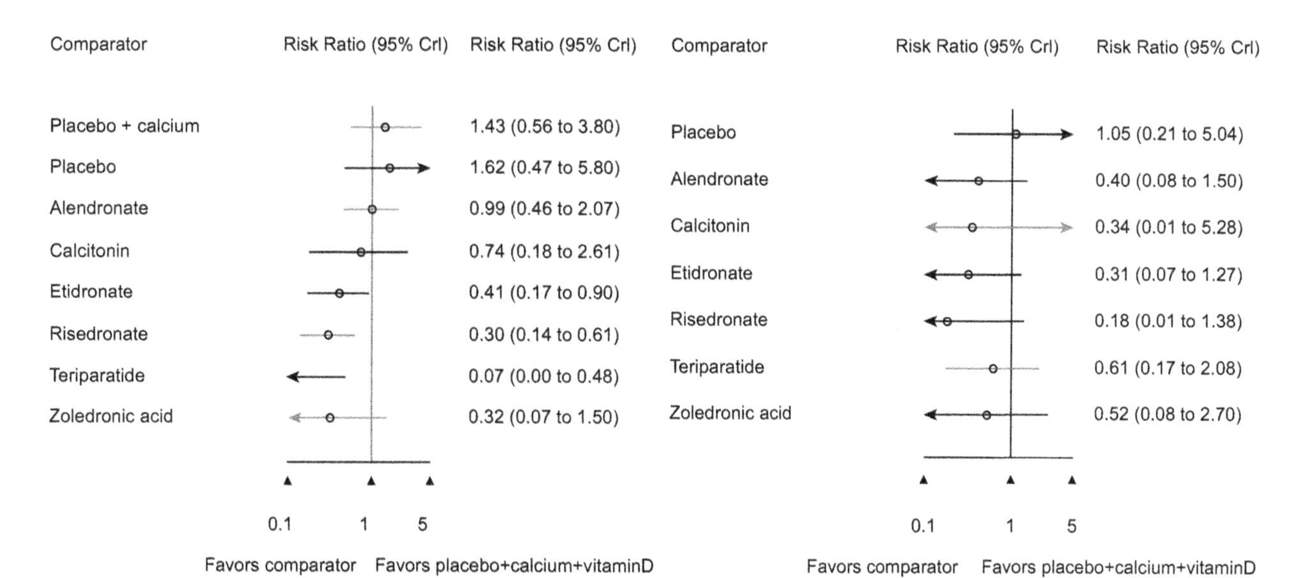

FIGURE 39.7 Efficacy of osteoporosis pharmacotherapies in preventing fracture among oral glucocorticoid users. Forest plots showing risk ratios for fracture for patients on "active" treatment versus placebo. Upper panel: Vertebral fracture. Lower panel: Nonvertebral fracture (Amiche et al., 2016).

Bisphosphonates

Bisphosphonates are the current standard in the prevention and therapy of GIO (Grossman et al., 2010; Lekamwasam et al., 2012). A recent Cochrane review concludes that "there was high-certainty evidence that bisphosphonates are beneficial in reducing the risk of vertebral fractures with data extending to 24 months of use. There was low certainty evidence that bisphosphonates may make little or no difference in preventing non-vertebral fractures" (Allen et al., 2016).

Given orally or intravenously, bisphosphonates generally act by inducing osteoclast apoptosis, thereby reducing the rate of bone resorption. While bisphosphonates prevent rapid loss of bone mineral density during the initial phase of glucocorticoid therapy, their longer-term effect on fracture risk is less well documented (de Nijs et al., 2006; Grossman et al., 2010; Saag et al., 1998). Some reports show significant reductions in vertebral or hip fracture risk (Amiche et al., 2018; Axelsson et al., 2017) but the effect on nonhip, nonvertebral fracture risk is either weak or unknown (Fig. 39.7). Most studies investigating the use of bisphosphonates in the context of GIO are underpowered to detect significant differences in fracture rates and therefore use changes in bone mineral density or other surrogates as the primary outcome measure (Lekamwasam et al., 2012). The absence of reliable fracture data is particularly problematic since glucocorticoids not only affect bone remodeling but also compromise bone strength through a number of different mechanisms that are not targeted by bisphosphonates. As outlined above, glucocorticoid-induced activation of osteoclasts and subsequent loss of bone mineral density is only one of several causal factors in the pathogenesis of GIO. Rodent models suggest that bone strength is closely linked to osteoblast and osteocyte viability, which is readily compromised under conditions of glucocorticoid excess (Weinstein et al., 2010b). Even though bisphosphonates have been shown in animal models to partially prevent glucocorticoid-induced cell apoptosis (Plotkin et al., 1999, 2005), their effects on osteoclast function always leads to a significant reduction in osteoblast activity (Saag et al., 1998). The resulting decline in bone formation comes on top of the glucocorticoid-induced suppression of osteoblast function, which poses a conundrum in regards to the use of bisphosphonates in the setting of GIO.

Denosumab

Denosumab is a potent RANKL inhibitor, and like all antiresorptive agents, acts by inhibiting osteoclast function and bone resorption. Denosumab effectively reduces fracture risk in postmenopausal osteoporosis (Brown et al., 2009), but its use in GIO has thus far not been fully assessed. A subgroup analysis of glucocorticoid-treated patients with rheumatoid arthritis demonstrated a significant increase in lumbar spine and hip bone mineral density after 12 months of treatment with denosumab (60 mg s.c. every 6 months) (Dore et al., 2010).

A recent study of 29 patients on chronic glucocorticoid therapy for a variety of systemic inflammatory conditions found that denosumab was effective in preventing bone resorption and bone loss, with increases in bone density at both the lumbar spine and hip (Sawamura et al., 2017).

In a 12-month study of patients receiving an oral bisphosphonate for chronic glucocorticoid therapy, switching to denosumab resulted in greater gains of lumbar spine bone mineral density and more pronounced suppression of bone turnover markers. The study was not powered for fracture as an end point (Mok et al., 2015). Similar findings were reported in another small study from Japan (Ishiguro et al., 2017). Since denosumab not only reduces osteoclast function but also osteoclast recruitment, activation, and survival together with a significant reduction in osteoblastic bone formation, its use in GIO may compromise bone quality over time. Additional long-term studies of GIO are therefore of paramount importance.

Raloxifene

The selective estrogen receptor modulator raloxifene has been shown to reduce fracture risk in women with postmenopausal osteoporosis (Ettinger et al., 1999). In a recent randomized double-blind placebo-controlled trial, 12 months of raloxifene (60 mg daily) significantly increased spinal and hip bone mineral density in women receiving longer-term glucocorticoids for a number of various conditions. The study was underpowered to measure the effect of raloxifene on fracture risk (Mok et al., 2011).

Teriparatide

Osteoanabolic treatments may have greater beneficial potential in GIO, particularly in patients undergoing long-term treatment. In a mouse model of GIO, intermittent treatment with parathyroid hormone (PTH) not only averted the adverse effects of glucocorticoids on bone turnover but also preserved osteocyte viability (Weinstein et al., 2010a). While the anabolic effects of PTH on bone-forming cells and their molecular pathways are not fully understood, PTH has been shown to induce several molecular pathways that are suppressed by glucocorticoid excess. For example, PTH directly induces canonical Wnt signaling in osteoblasts, a signal profoundly inhibited by supraphysiological levels of glucocorticoids (Keller and Kneissel, 2005; Weinstein et al., 2010a). This anabolic PTH signal is further amplified by the suppression of the Wnt antagonist sclerostin in osteocytes (Anastasilakis et al., 2010; Gatti et al., 2011). Intermittent administration of PTH also induces the expression of Cyclin D1 in osteoblasts and preosteoblasts (Datta et al., 2007) as well as the expression of RUNX2 (Hisa et al., 2011). As mentioned above, both of these factors are key regulators of osteoblast proliferation and differentiation and are suppressed by glucocorticoids at pharmacological doses. Overall, PTH treatment has been shown to induce differentiation of osteoblast precursors and prolong osteoblast and osteocyte survival, thereby increasing osteoblast/osteocyte numbers and overall activity.

On a structural level, PTH treatment increases the amount of trabecular bone (e.g., at the spine), while its anabolic impact is much less pronounced at cortical sites (Dempster et al., 2001; Gluer et al., 2013; Orwoll et al., 2003). This pattern is distinctively different from that of bisphosphonates, which cause smaller increases in bone mineral density across both trabecular and cortical sites (Gluer et al., 2013; Saag et al., 2007, 2009). Clinical studies indicate that the risk of vertebral fractures rises substantially with ongoing glucocorticoid therapy, whereas that of nonvertebral fracture increases only moderately (van Staa et al., 2000c). Given that PTH has its anabolic effects predominantly in trabecular bone (such as the spine), PTH may have greater therapeutic benefits in preventing vertebral fractures than bisphosphonates do. Indeed, an 18-month randomized controlled trial of alendronate versus teriparatide (recombinant human PTH[1−34]) in patients with GIO found teriparatide to be associated with increased markers of bone formation and a greater increase in areal bone mineral density at both the spine and the hip (Saag et al., 2007) (Fig. 39.8). Of note, even though the study was not adequately powered to detect differences in fracture incidence, three incident vertebral fractures occurred among the 173 patients in the teriparatide group and 13 incident vertebral fractures among the 169 patients receiving alendronate (Saag et al., 2007). Similar differences were seen at the 36-month follow-up (Saag et al., 2009). In a recent 18-month RCT of men with GIO, Gluer and colleagues compared the effect of teriparatide versus risedronate on lumbar spine bone mineral density (Gluer et al., 2013). The authors found greater increases in lumbar spine areal density with teriparatide. Interestingly, a subanalysis utilizing high-resolution quantitative CT of the T12 vertebra at 18 months of treatment revealed a greater increase in trabecular bone mineral density for teriparatide compared with risedronate (14.5% vs. 2.0%), while cortical density changed similarly in both groups (8.0% vs. 6.1%). In addition, there was a substantially greater increase in vertebral strength in the teriparatide group compared with the bisphosphonate group (Gluer et al., 2013). These results seem to confirm some predictions from rodent models regarding the effectiveness of osteoanabolic treatments versus bisphosphonates. Specifically, these data and a recent meta-analysis (Amiche et al., 2016) (Fig. 39.7) support the notion that teriparatide has greater efficacy in preserving or even increasing spinal bone strength, and reducing fracture risk in the setting of glucocorticoid treatment.

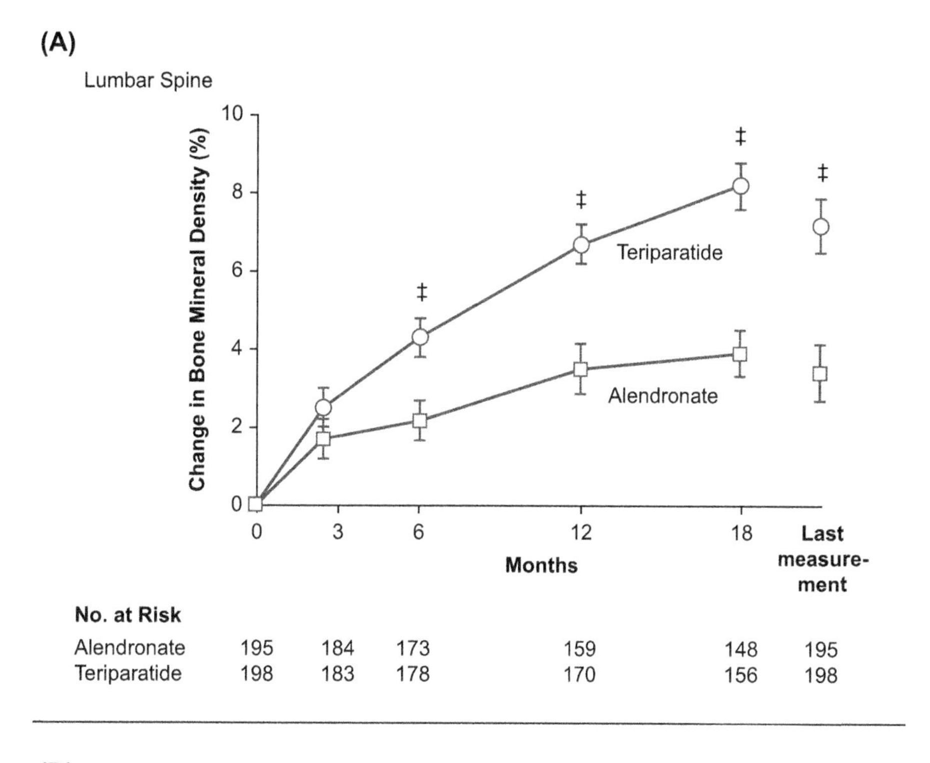

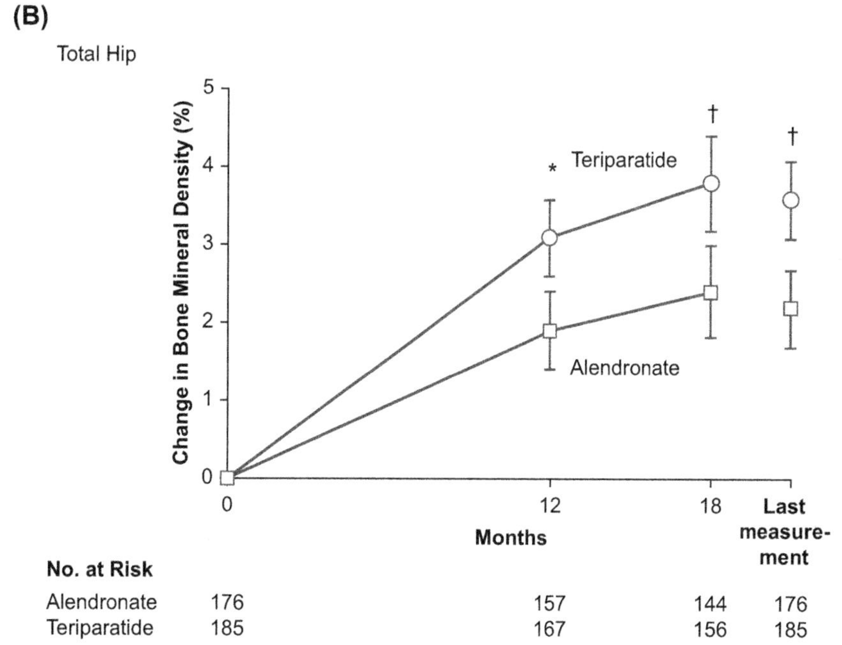

FIGURE 39.8 Effect of teriparatide or alendronate treatment on bone mineral density in patients with glucocorticoid-induced osteoporosis. The graphs show the percent change in mean bone mineral density from baseline to 18 months or last measurement. *P*-values are $* < 0.05$, $† < 0.01$, $‡ < 0.001$. Bars are standard error.

While these results are certainly promising, further clinical studies are required to confirm the effectiveness of osteoanabolic treatments in the management of GIO. In particular, long-term studies large enough to assess fracture rates—not bone mineral density—as a primary outcome are now required. Combinations of antiresorptive agents (such as bisphosphonates) with osteoanabolic agents (such as teriparatide) have recently been studied, with some success in the context of postmenopausal osteoporosis (Cosman et al., 2011). Whether such combination therapies yield beneficial results in GIO remains to be seen.

Sex hormone replacement

Hypogonadism is a known adverse outcome of high-dose, long-term glucocorticoid therapy (Chrousos et al., 1998; Pearce et al., 1998). Many but not all guidelines recommend sex hormone replacement in glucocorticoid-treated patients diagnosed with proven sex hormone deficiency. However, the evidence base for these recommendations is weak, as hormone replacement therapy has been studied in only a few small short-term studies. After due consideration of risks and benefits, hormone replacement might be used as either adjuvant therapy or second-line treatment in patients who cannot be treated with bisphosphonates or teriparatide.

Timing and monitoring of therapy

There is wide agreement that patients should remain on bone-protective treatments for as long as they require therapy with glucocorticoids. Drug holidays, sometimes considered in patients with postmenopausal osteoporosis, are usually not an option in GIO (Grossman et al., 2010; Nawata et al., 2005; Soen, 2011). As adherence with oral bisphosphonates is usually low, continuous surveillance and encouragement of the patient is a necessity. The introduction of intravenous bisphosphonates offers a viable and effective alternative in nonadherent patients.

There is limited data concerning the monitoring of patients with GIO. In patients whose fracture risk is deemed low, baseline and serial measurements of bone mineral density are usually recommended, although there is no clear evidence at what intervals these scans should be performed and what degree of glucocorticoid-induced bone loss should trigger an intervention. Many guidelines do, however, recommend intervening when the T-score at any site is −1.5 or lower. Patients should be carefully monitored for incident minimal trauma fractures, as these are a clear indication for the commencement of therapy. Despite the limitations of bone mineral density as a tool to monitor treatment efficacy, most guidelines recommend that patients on current bone-protective therapy have yearly bone mineral density scans. Serial determinations of bone turnover are usually not recommended.

New and emerging therapies

Despite their limitations, current treatments for GIO have proven to provide a significant degree of protection against bone loss and fracture. New therapies in this area would need to be more efficacious than existing medications, have fewer adverse effects, or have additional nonskeletal benefits.

Selective glucocorticoid receptor activators

In an attempt to develop agents with superior glucocorticoid-like properties but fewer off-target effects, a number of pharmaceutical companies have invested in the development of so-called selective glucocorticoid receptor activators (SEGRAs) (Oakley and Cidlowski, 2013; Schacke et al., 2007; Sundahl et al., 2015). This was on the basis that beneficial antiinflammatory effects of glucocorticoids could be "dissociated" from the adverse effects on bone and carbohydrate metabolism. Unfortunately, few agents able to dissociate these actions have been identified so far. The concept of GR transactivation as the cause of most adverse effects, and transrepression as the basis of most antiinflammatory effects, of glucocorticoids may be overly simplistic. Thus, some important antiinflammatory genes such as DUSP1 are clearly transactivated, and some of the adverse effects, such as the effect of glucocorticoids on bone, are primarily caused by transrepression (Rauch et al., 2010; Tchen et al., 2010). Despite this, there are clearly some agents, such as Compound A, that have useful differential effects on various GR-mediated responses (Lesovaya et al., 2015; Thiele et al., 2012). In a mouse model of GIO, mice were treated with equivalent antiinflammatory doses of prednisolone and Compound A. Prednisolone, as expected, caused a reduction in bone formation and an increased RANKL/OPG ratio (Thiele et al., 2012). These effects were not seen, however, with Compound A, indicating that it has some "bone-sparing" properties. Unfortunately, the therapeutic index for Compound A appears too narrow for use in humans. The attraction of pursuing the development of SEGRAs is their potential to reduce adverse effects on a range of tissues rather than just bone.

Antisclerostin/DKK1

As discussed above, the canonical Wnt signaling pathway is critical to bone formation. Glucocorticoid treatment causes increases in DKK1 and sclerostin, both powerful inhibitors of the Wnt pathway. Antibodies capable of neutralizing these targets are now being evaluated clinically in postmenopausal and cancer-induced bone loss. In contrast to DKK1, sclerostin expression and action appears exclusive to osteoblasts and osteocytes. The antisclerostin antibody romosozumab has been studied in women with postmenopausal osteoporosis, where 12 months of treatment led to remarkable increases in bone mineral density and significant reductions in fracture risk (Cosman et al., 2016; Langdahl et al., 2017; Saag et al., 2017). However, so far no RCT data on the efficacy of romosozumab are available in patients with GIO.

Conclusions and future perspectives

Our understanding of the pleiotropic effects of glucocorticoids on bone cells and their signaling behavior has greatly improved over the past 2 decades, resulting in a clearer and more complete picture of the mechanism behind the developmental anabolic, and the detrimental catabolic, actions of glucocorticoids on the skeleton.

It is now evident that endogenous glucocorticoids at physiological levels are vital for the commitment and differentiation of osteoblasts and play a pivotal role in the control of skeletal development and maintenance of healthy bone. At supraphysiological levels, however, glucocorticoids become largely detrimental to bone and cause bone loss and skeletal fragility. While high levels of glucocorticoids affect all bone cells, rodent models and human studies reliably demonstrate that their key targets in the skeleton are osteoblasts and osteocytes. Newer research indicates that the adverse effects of glucocorticoids on fuel metabolism and body fat accrual are at least in part mediated through their skeletal actions on osteoblasts and osteocytes. These cells are now at the core of new and promising interventional strategies uniquely suited to offset the detrimental effects of glucocorticoids on molecular as well as structural levels.

Although the study was not powered for fracture outcomes, there was a significant difference for vertebral fractures between the alendronate (6.1%) and teriparatide (0.6%) arms ($P < .01$) (Saag et al., 2007). Similar differences were seen at the 36 months follow-up (7.7% vs. 1.7%, $P < .01$) (Saag et al., 2009).

Acknowledgments

The authors thank the National Health and Medical Research Council (NHMRC), Australia, for ongoing research support through project grants #402462, #570946, #612838, #612839, #1086100, #1087271, and #1101879.

References

Adachi, J.D., Bensen, W.G., Brown, J., Hanley, D., Hodsman, A., Josse, R., et al., 1997. Intermittent etidronate therapy to prevent corticosteroid-induced osteoporosis. N. Engl. J. Med. 337, 382–387.

Adachi, J.D., Saag, K.G., Delmas, P.D., Liberman, U.A., Emkey, R.D., Seeman, E., et al., 2001. Two-year effects of alendronate on bone mineral density and vertebral fracture in patients receiving glucocorticoids: a randomized, double-blind, placebo-controlled extension trial. Arthritis Rheum. 44, 202–211.

Allen, C.S., Yeung, J.H., Vandermeer, B., Homik, J., 2016. Bisphosphonates for steroid induced osteoporosis. Cochrane Database Syst. Rev. 10, CD001347.

Almeida, M., Han, L., Ambrogini, E., Weinstein, R.S., Manolagas, S.C., 2011. Glucocorticoids and tumor necrosis factor alpha increase oxidative stress and suppress Wnt protein signaling in osteoblasts. J. Biol. Chem. 286, 44326–44335.

Amiche, M.A., Albaum, J.M., Tadrous, M., Pechlivanoglou, P., Levesque, L.E., Adachi, J.D., Cadarette, S.M., 2016. Efficacy of osteoporosis pharmacotherapies in preventing fracture among oral glucocorticoid users: a network meta-analysis. Osteoporos. Int. 27, 1989–1998.

Amiche, M.A., Levesque, L.E., Gomes, T., Adachi, J.D., Cadarette, S.M., March, 2018. Effectiveness of oral bisphosphonates in reducing fracture risk among oral glucocorticoid users: three matched cohort analyses. J. Bone Miner. Res. 33 (3), 419–429. https://doi.org/10.1002/jbmr.3318.

Amin, S., LaValley, M.P., Simms, R.W., Felson, D.T., 1999. The role of vitamin D in corticosteroid-induced osteoporosis: a meta-analytic approach. Arthritis Rheum. 42, 1740–1751.

Anastasilakis, A.D., Polyzos, S.A., Avramidis, A., Toulis, K.A., Papatheodorou, A., Terpos, E., 2010. The effect of teriparatide on serum Dickkopf-1 levels in postmenopausal women with established osteoporosis. Clin. Endocrinol. 72, 752–757.

Axelsson, K.F., Nilsson, A.G., Wedel, H., Lundh, D., Lorentzon, M., 2017. Association between alendronate use and hip fracture risk in older patients using oral prednisolone. J. Am. Med. Assoc. 318, 146–155.

Beard, J.C., Halter, J.B., Best, J.D., Pfeifer, M.A., Porte Jr., D., 1984. Dexamethasone-induced insulin resistance enhances B cell responsiveness to glucose level in normal men. Am. J. Physiol. 247, E592–E596.

Bellido, T., 2010. Antagonistic interplay between mechanical forces and glucocorticoids in bone: a tale of kinases. J. Cell. Biochem. 111, 1–6.

Bellows, C.G., Aubin, J.E., Heersche, J.N., 1987. Physiological concentrations of glucocorticoids stimulate formation of bone nodules from isolated rat calvaria cells in vitro. Endocrinology 121, 1985–1992.

Bonadonna, S., Burattin, A., Nuzzo, M., Bugari, G., Rosei, E.A., Valle, D., et al., 2005. Chronic glucocorticoid treatment alters spontaneous pulsatile parathyroid hormone secretory dynamics in human subjects. Eur. J. Endocrinol. 152, 199–205.

Brennan-Speranza, T.C., Henneicke, H., Gasparini, S.J., Blankenstein, K.I., Heinevetter, U., Cogger, V.C., et al., 2012. Osteoblasts mediate the adverse effects of glucocorticoids on fuel metabolism. J. Clin. Investig. 122, 4172–4189.

Brown, J.P., Prince, R.L., Deal, C., Recker, R.R., Kiel, D.P., de Gregorio, L.H., et al., 2009. Comparison of the effect of denosumab and alendronate on BMD and biochemical markers of bone turnover in postmenopausal women with low bone mass: a randomized, blinded, phase 3 trial. J. Bone Miner. Res. 24, 153–161.

Buckley, L., Guyatt, G., Fink, H.A., Cannon, M., Grossman, J., Hansen, K.E., et al., 2017. 2017 American College of Rheumatology guideline for the prevention and treatment of glucocorticoid-induced osteoporosis. Arthritis Care Res. 69, 1095–1110.

Buckley, L.M., Leib, E.S., Cartularo, K.S., Vacek, P.M., Cooper, S.M., 1996. Calcium and vitamin D3 supplementation prevents bone loss in the spine secondary to low-dose corticosteroids in patients with rheumatoid arthritis. A randomized, double-blind, placebo-controlled trial. Ann. Intern. Med. 125, 961–968.

Bultink, I.E., Lems, W.F., Kostense, P.J., Dijkmans, B.A., Voskuyl, A.E., 2005. Prevalence of and risk factors for low bone mineral density and vertebral fractures in patients with systemic lupus erythematosus. Arthritis Rheum. 52, 2044–2050.

Buttgereit, F., Burmester, G.R., Straub, R.H., Seibel, M.J., Zhou, H., 2011. Exogenous and endogenous glucocorticoids in rheumatic diseases. Arthritis Rheum. 63, 1–9.

Canalis, E., Mazziotti, G., Giustina, A., Bilezikian, J.P., 2007. Glucocorticoid-induced osteoporosis: pathophysiology and therapy. Osteoporos. Int. 18, 1319–1328.

Chang, J.K., Li, C.J., Liao, H.J., Wang, C.K., Wang, G.J., Ho, M.L., 2009. Anti-inflammatory drugs suppress proliferation and induce apoptosis through altering expressions of cell cycle regulators and pro-apoptotic factors in cultured human osteoblasts. Toxicology 258, 148–156.

Chen, F., Zhang, L., OuYang, Y., Guan, H., Liu, Q., Ni, B., 2014. Glucocorticoid induced osteoblast apoptosis by increasing E4BP4 expression via up-regulation of Bim. Calcif. Tissue Int. 94, 640–647.

Chrousos, G.P., Torpy, D.J., Gold, P.W., 1998. Interactions between the hypothalamicpituitary- adrenal axis and the female reproductive system: clinical implications. Ann. Intern. Med. 129, 229–240.

Cohen, S., Levy, R.M., Keller, M., Boling, E., Emkey, R.D., Greenwald, M., et al., 1999. Risedronate therapy prevents corticosteroid-induced bone loss: a twelvemonth, multicenter, randomized, double-blind, placebo-controlled, parallel-group study. Arthritis Rheum. 42, 2309–2318.

Cole, T.J., Blendy, J.A., Monaghan, A.P., Krieglstein, K., Schmid, W., Aguzzi, A., et al., 1995. Targeted disruption of the glucocorticoid receptor gene blocks adrenergic chromaffin cell development and severely retards lung maturation. Genes Dev. 9, 1608–1621.

Cooper, M.S., Blumsohn, A., Goddard, P.E., Bartlett, W.A., Shackleton, C.H., Eastell, R., et al., 2003. 11beta-hydroxysteroid dehydrogenase type 1 activity predicts the effects of glucocorticoids on bone. J. Clin. Endocrinol. Metab. 88, 3874–3877.

Cooper, M.S., Bujalska, I., Rabbitt, E., Walker, E.A., Bland, R., Sheppard, M.C., et al., 2001. Modulation of 11beta-hydroxysteroid dehydrogenase isozymes by proinflammatory cytokines in osteoblasts: an autocrine switch from glucocorticoid inactivation to activation. J. Bone Miner. Res. 16, 1037–1044.

Cooper, M.S., Kriel, H., Sayers, A., Fraser, W.D., Williams, A.M., Stewart, P.M., et al., 2011. Can 11beta-hydroxysteroid dehydrogenase activity predict the sensitivity of bone to therapeutic glucocorticoids in inflammatory bowel disease? Calcif. Tissue Int. 89, 246–251.

Cooper, M.S., Rabbitt, E.H., Goddard, P.E., Bartlett, W.A., Hewison, M., Stewart, P.M., 2002. Osteoblastic 11beta-hydroxysteroid dehydrogenase type 1 activity increases with age and glucocorticoid exposure. J. Bone Miner. Res. 17, 979–986.

Cooper, M.S., Syddall, H.E., Fall, C.H., Wood, P.J., Stewart, P.M., Cooper, C., Dennison, E.M., 2005. Circulating cortisone levels are associated with biochemical markers of bone formation and lumbar spine BMD: the Hertfordshire Cohort Study. Clin. Endocrinol. 62, 692–697.

Cooper, M.S., Walker, E.A., Bland, R., Fraser, W.D., Hewison, M., Stewart, P.M., 2000. Expression and functional consequences of 11beta-hydroxysteroid dehydrogenase activity in human bone. Bone 27, 375–381.

Cooper, M.S., Zhou, H., Seibel, M.J., 2012. Selective glucocorticoid receptor agonists: glucocorticoid therapy with no regrets? J. Bone Miner. Res. 27, 2238–2241.

Cortet, B., Hachulla, E., Barton, I., Bonvoisin, B., Roux, C., 1999. Evaluation of the efficacy of etidronate therapy in preventing glucocorticoid-induced bone loss in patients with inflammatory rheumatic diseases. A randomized study. Rev. Rhum. Engl. Ed. 66, 214–219.

Cosman, F., Crittenden, D.B., Adachi, J.D., Binkley, N., Czerwinski, E., Ferrari, S., et al., 2016. Romosozumab treatment in postmenopausal women with osteoporosis. N. Engl. J. Med. 375, 1532–1543.

Cosman, F., Eriksen, E.F., Recknor, C., Miller, P.D., Guanabens, N., Kasperk, C., et al., 2011. Effects of intravenous zoledronic acid plus subcutaneous teriparatide [rhPTH(1-34)] in postmenopausal osteoporosis. J. Bone Miner. Res. 26, 503–511.

Coutinho, A.E., Gray, M., Brownstein, D.G., Salter, D.M., Sawatzky, D.A., Clay, S., et al., 2012. 11beta-Hydroxysteroid dehydrogenase type 1, but not type 2, deficiency worsens acute inflammation and experimental arthritis in mice. Endocrinology 153, 234–240.

Crockett, J.C., Rogers, M.J., Coxon, F.P., Hocking, L.J., Helfrich, M.H., 2011. Bone remodelling at a glance. J. Cell Sci. 124, 991–998.

Curtis, J.R., Westfall, A.O., Allison, J.J., Becker, A., Casebeer, L., Freeman, A., et al., 2005. Longitudinal patterns in the prevention of osteoporosis in glucocorticoid-treated patients. Arthritis Rheum. 52, 2485–2494.

Cushing, H., 1932. The basophil adenomas of the pituitary body and their clinical manifestations (pituitary basophilism). Bull. Johns Hopkins Hosp. 50, 137–195.

Datta, N.S., Pettway, G.J., Chen, C., Koh, A.J., McCauley, L.K., 2007. Cyclin D1 as a target for the proliferative effects of PTH and PTHrP in early osteoblastic cells. J. Bone Miner. Res. 22, 951–964.

de Nijs, R.N., Jacobs, J.W., Lems, W.F., Laan, R.F., Algra, A., Huisman, A.M., et al., 2006. Alendronate or alfacalcidol in glucocorticoid-induced osteoporosis. N. Engl. J. Med. 355, 675–684.

Dempster, D.W., Cosman, F., Kurland, E.S., Zhou, H., Nieves, J., Woelfert, L., et al., 2001. Effects of daily treatment with parathyroid hormone on bone microarchitecture and turnover in patients with osteoporosis: a paired biopsy study. J. Bone Miner. Res. 16, 1846–1853.

Diez-Perez, A., Hooven, F.H., Adachi, J.D., Adami, S., Anderson, F.A., Boonen, S., et al., 2011. Regional differences in treatment for osteoporosis. The global longitudinal study of osteoporosis in women (GLOW). Bone 49, 493–498.

Dore, R.K., Cohen, S.B., Lane, N.E., Palmer, W., Shergy, W., Zhou, L., et al., 2010. Effects of denosumab on bone mineral density and bone turnover in patients with rheumatoid arthritis receiving concurrent glucocorticoids or bisphosphonates. Ann. Rheum. Dis. 69, 872–875.

Dovio, A., Perazzolo, L., Osella, G., Ventura, M., Termine, A., Milano, E., et al., 2004. Immediate fall of bone formation and transient increase of bone resorption in the course of high-dose, short-term glucocorticoid therapy in young patients with multiple sclerosis. J. Clin. Endocrinol. Metab. 89, 4923–4928.

Eastell, R., Devogelaer, J.P., Peel, N.F., Chines, A.A., Bax, D.E., Sacco-Gibson, N., et al., 2000. Prevention of bone loss with risedronate in glucocorticoid-treated rheumatoid arthritis patients. Osteoporos. Int. 11, 331–337.

Eisman, J.A., Bogoch, E.R., Dell, R., Harrington, J.T., McKinney Jr., R.E., McLellan, A., et al., 2012. Making the first fracture the last fracture: ASBMR task force report on secondary fracture prevention. J. Bone Miner. Res. 27, 2039–2046.

Espina, B., Liang, M., Russell, R.G., Hulley, P.A., 2008. Regulation of bim in glucocorticoid-mediated osteoblast apoptosis. J. Cell. Physiol. 215, 488–496.

Ettinger, B., Black, D.M., Mitlak, B.H., Knickerbocker, R.K., Nickelsen, T., Genant, H.K., et al., 1999. Reduction of vertebral fracture risk in post-menopausal women with osteoporosis treated with raloxifene: results from a 3-year randomized clinical trial. Multiple Outcomes of Raloxifene Evaluation (MORE) Investigators. J. Am. Med. Assoc. 282, 637–645.

Feldstein, A.C., Elmer, P.J., Nichols, G.A., Herson, M., 2005. Practice patterns in patients at risk for glucocorticoid-induced osteoporosis. Osteoporos. Int. 16, 2168–2174.

Fernandez-Rodriguez, E., Stewart, P.M., Cooper, M.S., 2009. The pituitary-adrenal axis and body composition. Pituitary 12, 105–115.

Ferris, H.A., Kahn, C.R., 2012. New mechanisms of glucocorticoid-induced insulin resistance: make no bones about it. J. Clin. Investig. 122, 3854–3857.

Fucik, R.F., Kukreja, S.C., Hargis, G.K., Bowser, E.N., Henderson, W.J., Williams, G.A., 1975. Effect of glucocorticoids on function of the parathyroid glands in man. J. Clin. Endocrinol. Metab. 40, 152–155.

Gabet, Y., Noh, T., Lee, C., Frenkel, B., 2011. Developmentally regulated inhibition of cell cycle progression by glucocorticoids through repression of cyclin A transcription in primary osteoblast cultures. J. Cell. Physiol. 226, 991–998.

Gao, J., Cheng, T.S., Qin, A., Pavlos, N.J., Wang, T., Song, K., et al., 2016. Glucocorticoid impairs cell-cell communication by autophagy-mediated degradation of connexin 43 in osteocytes. Oncotarget 7, 26966–26978.

Gatti, D., Viapiana, O., Idolazzi, L., Fracassi, E., Rossini, M., Adami, S., 2011. The waning of teriparatide effect on bone formation markers in post-menopausal osteoporosis is associated with increasing serum levels of DKK1. J. Clin. Endocrinol. Metab. 96, 1555–1559.

Gluer, C.C., Marin, F., Ringe, J.D., Hawkins, F., Moricke, R., Papaioannu, N., et al., 2013. Comparative effects of teriparatide and risedronate in glucocorticoid-induced osteoporosis in men: 18-month results of the EuroGIOPs trial. J. Bone Miner. Res. 28, 1355–1368.

Gonzalez-Gonzalez, J.G., Mireles-Zavala, L.G., Rodriguez-Gutierrez, R., Gomez-Almaguer, D., Lavalle-Gonzalez, F.J., Tamez-Perez, H.E., et al., 2013. Hyperglycemia related to high-dose glucocorticoid use in noncritically ill patients. Diabetol. Metab. Syndrome 5, 18.

Gough, A.K., Lilley, J., Eyre, S., Holder, R.L., Emery, P., 1994. Generalised bone loss in patients with early rheumatoid arthritis. Lancet 344, 23–27.

Grossman, J.M., Gordon, R., Ranganath, V.K., Deal, C., Caplan, L., Chen, W., et al., 2010. American College of Rheumatology 2010 recommendations for the prevention and treatment of glucocorticoid-induced osteoporosis. Arthritis Care Res. 62, 1515–1526.

Guler-Yuksel, M., Bijsterbosch, J., Goekoop-Ruiterman, Y.P., de Vries-Bouwstra, J.K., Hulsmans, H.M., de Beus, W.M., et al., 2008. Changes in bone mineral density in patients with recent onset, active rheumatoid arthritis. Ann. Rheu. Dis. 67, 823–828.

Hansen, K.E., Wilson, H.A., Zapalowski, C., Fink, H.A., Minisola, S., Adler, R.A., 2011. Uncertainties in the prevention and treatment of glucocorticoid-induced osteoporosis. J. Bone Miner. Res. 26, 1989–1996.

Hardy, R., Cooper, M.S., 2009. Bone loss in inflammatory disorders. J. Endocrinol. 201, 309–320.

Hayashi, K., Yamaguchi, T., Yano, S., Kanazawa, I., Yamauchi, M., Yamamoto, M., Sugimoto, T., 2009. BMP/Wnt antagonists are upregulated by dexamethasone in osteoblasts and reversed by alendronate and PTH: potential therapeutic targets for glucocorticoid-induced osteoporosis. Biochem. Biophys. Res. Commun. 379, 261–266.

Haynesworth, S.E., Goshima, J., Goldberg, V.M., Caplan, A.I., 1992. Characterization of cells with osteogenic potential from human marrow. Bone 13, 81–88.

Hazlehurst, J.M., Gathercole, L.L., Nasiri, M., Armstrong, M.J., Borrows, S., Yu, J., et al., 2013. Glucocorticoids fail to cause insulin resistance in human subcutaneous adipose tissue in vivo. J. Clin. Endocrinol. Metab. 98, 1631–1640.

Hench, P.S., Kendall, E.C., et al., 1949. The effect of a hormone of the adrenal cortex (17- hydroxy-11-dehydrocorticosterone; compound E) and of pituitary adrenocorticotropic hormone on rheumatoid arthritis. Proc. Staff Meet. Mayo Clin. 24, 181–197.

Henneicke, H., Herrmann, M., Kalak, R., Brennan-Speranza, T.C., Heinevetter, U., Bertollo, N., et al., 2011. Corticosterone selectively targets endo-cortical surfaces by an osteoblast-dependent mechanism. Bone 49, 733–742.

Henneicke, H., Li, J., Kim, S., Gasparini, S.J., Seibel, M.J., Zhou, H., 2017. Chronic mild stress causes bone loss via an osteoblast-specific glucocorticoid-dependent mechanism. Endocrinology 158, 1939–1950.

Herbertson, A., Aubin, J.E., 1995. Dexamethasone alters the subpopulation make-up of rat bone marrow stromal cell cultures. J. Bone Miner. Res. 10, 285–294.

Hisa, I., Inoue, Y., Hendy, G.N., Canaff, L., Kitazawa, R., Kitazawa, S., et al., 2011. Parathyroid hormone-responsive Smad3-related factor, Tmem119, promotes osteoblast differentiation and interacts with the bone morphogenetic protein- Runx2 pathway. J. Biol. Chem. 286, 9787–9796.

Hofbauer, L.C., Gori, F., Riggs, B.L., Lacey, D.L., Dunstan, C.R., Spelsberg, T.C., Khosla, S., 1999. Stimulation of osteoprotegerin ligand and inhibition of osteoprotegerin production by glucocorticoids in human osteoblastic lineage cells: potential paracrine mechanisms of glucocorticoid-induced osteoporosis. Endocrinology 140, 4382–4389.

Honma, M., Ikebuchi, Y., Kariya, Y., Hayashi, M., Hayashi, N., Aoki, S., Suzuki, H., 2013. RANKL subcellular trafficking and regulatory mechanisms in osteocytes. J. Bone Miner. Res. 28, 1936–1949.

Horsch, K., de Wet, H., Schuurmans, M.M., Allie-Reid, F., Cato, A.C., Cunningham, J., et al., 2007. Mitogen-activated protein kinase phosphatase 1/dual specificity phosphatase 1 mediates glucocorticoid inhibition of osteoblast proliferation. Mol. Endocrinol. 21, 2929–2940.

Hwang, J.Y., Lee, S.H., Kim, G.S., Koh, J.M., Go, M.J., Kim, Y.J., et al., 2009. HSD11B1 polymorphisms predicted bone mineral density and fracture risk in postmenopausal women without a clinically apparent hypercortisolemia. Bone 45, 1098–1103.

Ishiguro, S., Ito, K., Nakagawa, S., Hataji, O., Sudo, A., 2017. The clinical benefits of denosumab for prophylaxis of steroid-induced osteoporosis in patients with pulmonary disease. Arch. Osteoporos. 12, 44.

Jia, D., O'Brien, C.A., Stewart, S.A., Manolagas, S.C., Weinstein, R.S., 2006. Glucocorticoids act directly on osteoclasts to increase their life span and reduce bone density. Endocrinology 147, 5592–5599.

Jia, J., Yao, W., Guan, M., Dai, W., Shahnazari, M., Kar, R., et al., 2011. Glucocorticoid dose determines osteocyte cell fate. FASEB J. 25, 3366–3376.

Justesen, J., Mosekilde, L., Holmes, M., Stenderup, K., Gasser, J., Mullins, J.J., et al., 2004. Mice deficient in 11beta-hydroxysteroid dehydrogenase type 1 lack bone marrow adipocytes, but maintain normal bone formation. Endocrinology 145, 1916–1925.

Kalak, R., Zhou, H., Street, J., Day, R.E., Modzelewski, J.R., Spies, C.M., et al., 2009. Endogenous glucocorticoid signalling in osteoblasts is necessary to maintain normal bone structure in mice. Bone 45, 61–67.

Kalpakcioglu, B.B., Engelke, K., Genant, H.K., 2011. Advanced imaging assessment of bone fragility in glucocorticoid-induced osteoporosis. Bone 48, 1221–1231.

Kanda, F., Okuda, S., Matsushita, T., Takatani, K., Kimura, K.I., Chihara, K., 2001. Steroid myopathy: pathogenesis and effects of growth hormone and insulin-like growth factor-I administration. Horm. Res. 56 (Suppl. 1), 24–28.

Kang, H., Chen, H., Huang, P., Qi, J., Qian, N., Deng, L., Guo, L., 2016. Glucocorticoids impair bone formation of bone marrow stromal stem cells by reciprocally regulating microRNA-34a-5p. Osteoporos. Int. 27, 1493–1505.

Karras, D., Stoykov, I., Lems, W.F., Langdahl, B.L., Ljunggren, O., Barrett, A., et al., 2012. Effectiveness of teriparatide in postmenopausal women with osteoporosis and glucocorticoid use: 3-year results from the EFOS study. J. Rheumatol. 39, 600–609.

Kato, T., Khanh, V.C., Sato, K., Kimura, K., Yamashita, T., Sugaya, H., et al., 2018. Elevated expression of Dkk-1 by glucocorticoid treatment impairs bone regenerative capacity of adipose tissue-derived mesenchymal stem cells. Stem Cell. Dev. 27, 85–99.

Kauh, E., Mixson, L., Malice, M.P., Mesens, S., Ramael, S., Burke, J., et al., 2012. Prednisone affects inflammation, glucose tolerance, and bone turnover within hours of treatment in healthy individuals. Eur. J. Endocrinol. 166, 459–467.

Kaur, K., Hardy, R., Ahasan, M.M., Eijken, M., van Leeuwen, J.P., Filer, A., et al., 2010. Synergistic induction of local glucocorticoid generation by inflammatory cytokines and glucocorticoids: implications for inflammation associated bone loss. Ann. Rheum. Dis. 69, 1185–1190.

Keller, H., Kneissel, M., 2005. SOST is a target gene for PTH in bone. Bone 37, 148–158.

Kennedy, O.D., Herman, B.C., Laudier, D.M., Majeska, R.J., Sun, H.B., Schaffler, M.B., 2012. Activation of resorption in fatigue-loaded bone involves both apoptosis and active pro-osteoclastogenic signaling by distinct osteocyte populations. Bone 50, 1115–1122.

Kim, H.J., Zhao, H., Kitaura, H., Bhattacharyya, S., Brewer, J.A., Muglia, L.J., et al., 2007. Glucocorticoids and the osteoclast. Ann. N. Y. Acad. Sci. 1116, 335–339.

Kim, H.J., Zhao, H., Kitaura, H., Bhattacharyya, S., Brewer, J.A., Muglia, L.J., et al., 2006. Glucocorticoids suppress bone formation via the osteoclast. J. Clin. Investig. 116, 2152–2160.

Ko, J.Y., Chuang, P.C., Chen, M.W., Ke, H.C., Wu, S.L., Chang, Y.H., et al., 2013. MicroRNA-29a ameliorates glucocorticoid-induced suppression of osteoblast differentiation by regulating beta-catenin acetylation. Bone 57, 468–475.

Ko, J.Y., Chuang, P.C., Ke, H.J., Chen, Y.S., Sun, Y.C., Wang, F.S., 2015. MicroRNA-29a mitigates glucocorticoid induction of bone loss and fatty marrow by rescuing Runx2 acetylation. Bone 81, 80–88.

Kogianni, G., Mann, V., Noble, B.S., 2008. Apoptotic bodies convey activity capable of initiating osteoclastogenesis and localized bone destruction. J. Bone Miner. Res. 23, 915–927.

Kuroki, Y., Kaji, H., Kawano, S., Kanda, F., Takai, Y., Kajikawa, M., Sugimoto, T., 2008. Short-term effects of glucocorticoid therapy on biochemical markers of bone metabolism in Japanese patients: a prospective study. J. Bone Miner. Metab. 26, 271–278.

Landewe, R.B., Boers, M., Verhoeven, A.C., Westhovens, R., van de Laar, M.A., Markusse, H.M., et al., 2002. COBRA combination therapy in patients with early rheumatoid arthritis: long-term structural benefits of a brief intervention. Arthritis Rheum. 46, 347–356.

Langdahl, B.L., Libanati, C., Crittenden, D.B., Bolognese, M.A., Brown, J.P., Daizadeh, N.S., et al., 2017. Romosozumab (sclerostin monoclonal antibody) versus teriparatide in postmenopausal women with osteoporosis transitioning from oral bisphosphonate therapy: a randomised, open-label, phase 3 trial. Lancet 390, 1585–1594.

Lee, N.K., Sowa, H., Hinoi, E., Ferron, M., Ahn, J.D., Confavreux, C., et al., 2007. Endocrine regulation of energy metabolism by the skeleton. Cell 130, 456–469.

Lekamwasam, S., Adachi, J.D., Agnusdei, D., Bilezikian, J., Boonen, S., Borgstrom, F., et al., 2012. A framework for the development of guidelines for the management of glucocorticoid-induced osteoporosis. Osteoporos. Int. 23, 2257–2276.

Lesovaya, E., Yemelyanov, A., Swart, A.C., Swart, P., Haegeman, G., Budunova, I., 2015. Discovery of Compound A–a selective activator of the glucocorticoid receptor with anti-inflammatory and anti-cancer activity. Oncotarget 6, 30730–30744.

Li, H., Qian, W., Weng, X., Wu, Z., Li, H., Zhuang, Q., et al., 2012. Glucocorticoid receptor and sequential P53 activation by dexamethasone mediates apoptosis and cell cycle arrest of osteoblastic MC3T3-E1 cells. PLoS One 7, e37030.

Li, X., Ominsky, M.S., Niu, Q.T., Sun, N., Daugherty, B., D'Agostin, D., et al., 2008. Targeted deletion of the sclerostin gene in mice results in increased bone formation and bone strength. J. Bone Miner. Res. 23, 860–869.

Liu, J., Lu, C., Wu, X., Zhang, Z., Li, J., Guo, B., et al., 2017a. Targeting osteoblastic casein kinase-2 interacting protein-1 to enhance Smad-dependent BMP signaling and reverse bone formation reduction in glucocorticoid-induced osteoporosis. Sci. Rep. 7, 41295.

Liu, K., Jing, Y., Zhang, W., Fu, X., Zhao, H., Zhou, X., et al., 2017b. Silencing miR- 106b accelerates osteogenesis of mesenchymal stem cells and rescues against glucocorticoid-induced osteoporosis by targeting BMP2. Bone 97, 130–138.

Lu, K., Yin, X., Weng, T., Xi, S., Li, L., Xing, G., et al., 2008. Targeting WW domains linker of HECT-type ubiquitin ligase Smurf1 for activation by CKIP-1. Nat. Cell Biol. 10, 994–1002.

Lu, N.Z., Cidlowski, J.A., 2006. Glucocorticoid receptor isoforms generate transcription specificity. Trends Cell Biol. 16, 301–307.

Mak, W., Shao, X., Dunstan, C.R., Seibel, M.J., Zhou, H., 2009. Biphasic glucocorticoid-dependent regulation of Wnt expression and its inhibitors in mature osteoblastic cells. Calcif. Tissue Int. 85, 538–545.

Manolagas, S.C., 2013. Steroids and osteoporosis: the quest for mechanisms. J. Clin. Investig. 123, 1919–1921.

Matsumoto, T., Kuriwaka-Kido, R., Kondo, T., Endo, I., Kido, S., 2012. Regulation of osteoblast differentiation by interleukin-11 via AP-1 and Smad signaling. Endocr. J. 59, 91–101.

Matzelle, M.M., Gallant, M.A., Condon, K.W., Walsh, N.C., Manning, C.A., Stein, G.S., et al., 2012. Resolution of inflammation induces osteoblast function and regulates the Wnt signaling pathway. Arthritis Rheum. 64, 1540–1550.

Mazziotti, G., Giustina, A., 2013. Glucocorticoids and the regulation of growth hormone secretion. Nat. Rev. Endocrinol. 9, 265–276.

Meijsing, S.H., Pufall, M.A., So, A.Y., Bates, D.L., Chen, L., Yamamoto, K.R., 2009. DNA binding site sequence directs glucocorticoid receptor structure and activity. Science 324, 407–410.

Mok, C.C., Ho, L.Y., Ma, K.M., 2015. Switching of oral bisphosphonates to denosumab in chronic glucocorticoid users: a 12-month randomized controlled trial. Bone 75, 222–228.

Mok, C.C., Ying, K.Y., To, C.H., Ho, L.Y., Yu, K.L., Lee, H.K., Ma, K.M., 2011. Raloxifene for prevention of glucocorticoid-induced bone loss: a 12-month randomised double-blinded placebo-controlled trial. Ann. Rheu. Dis. 70, 778–784.

Moutsatsou, P., Kassi, E., Papavassiliou, A.G., 2012. Glucocorticoid receptor signaling in bone cells. Trends Mol. Med. 18, 348–359.

Nakashima, T., Hayashi, M., Fukunaga, T., Kurata, K., Oh-Hora, M., Feng, J.Q., et al., 2011. Evidence for osteocyte regulation of bone homeostasis through RANKL expression. Nat. Med. 17, 1231–1234.

Natsui, K., Tanaka, K., Suda, M., Yasoda, A., Sakuma, Y., Ozasa, A., et al., 2006. High-dose glucocorticoid treatment induces rapid loss of trabecular bone mineral density and lean body mass. Osteoporos. Int. 17, 105–108.

Naves, M.A., Pereira, R.M., Comodo, A.N., de Alvarenga, E.L., Caparbo, V.F., Teixeira, V.P., 2011. Effect of dexamethasone on human osteoblasts in culture: involvement of beta1 integrin and integrin-linked kinase. Cell Biol. Int. 35, 1147–1151.

Nawata, H., Soen, S., Takayanagi, R., Tanaka, I., Takaoka, K., Fukunaga, M., et al., 2005. Guidelines on the management and treatment of glucocorticoid-induced osteoporosis of the Japanese society for bone and mineral research (2004). J. Bone Miner. Metab. 23, 105–109.

Newton, R., Holden, N.S., 2007. Separating transrepression and transactivation: a distressing divorce for the glucocorticoid receptor? Mol. Pharmacol. 72, 799–809.

Nicod, N., Giusti, V., Besse, C., Tappy, L., 2003. Metabolic adaptations to dexamethasone-induced insulin resistance in healthy volunteers. Obes. Res. 11, 625–631.

Niewoehner, D.E., Erbland, M.L., Deupree, R.H., Collins, D., Gross, N.J., Light, R.W., et al., 1999. Effect of systemic glucocorticoids on exacerbations of chronic obstructive pulmonary disease. Department of Veterans Affairs Cooperative Study Group. N. Engl. J. Med. 340, 1941–1947.

O'Brien, C.A., Jia, D., Plotkin, L.I., Bellido, T., Powers, C.C., Stewart, S.A., et al., 2004. Glucocorticoids act directly on osteoblasts and osteocytes to induce their apoptosis and reduce bone formation and strength. Endocrinology 145, 1835–1841.

Oakley, R.H., Cidlowski, J.A., 2013. The biology of the glucocorticoid receptor: new signaling mechanisms in health and disease. J. Allergy Clin. Immunol. 132, 1033–1044.

Orwoll, E.S., Scheele, W.H., Paul, S., Adami, S., Syversen, U., Diez-Perez, A., et al., 2003. The effect of teriparatide [human parathyroid hormone (1-34)] therapy on bone density in men with osteoporosis. J. Bone Miner. Res. 18, 9–17.

Overman, R.A., Yeh, J.Y., Deal, C.L., 2013. Prevalence of oral glucocorticoid usage in the United States: a general population perspective. Arthritis Care Res. 65, 294–298.

Pearce, G., Tabensky, D.A., Delmas, P.D., Baker, H.W., Seeman, E., 1998. Corticosteroid-induced bone loss in men. J. Clin. Endocrinol. Metab. 83, 801–806.

Pereira, R.M., Freire de Carvalho, J., 2011. Glucocorticoid-induced myopathy. Joint Bone Spine 78, 41–44.

Petersons, C.J., Mangelsdorf, B.L., Jenkins, A.B., Poljak, A., Smith, M.D., Greenfield, J.R., et al., 2013. Effects of low-dose prednisolone on hepatic and peripheral insulin sensitivity, insulin secretion, and abdominal adiposity in patients with inflammatory rheumatologic disease. Diabetes Care 36, 2822–2829.

Plotkin, L.I., Aguirre, J.I., Kousteni, S., Manolagas, S.C., Bellido, T., 2005. Bisphosphonates and estrogens inhibit osteocyte apoptosis via distinct molecular mechanisms downstream of extracellular signal-regulated kinase activation. J. Biol. Chem. 280, 7317–7325.

Plotkin, L.I., Weinstein, R.S., Parfitt, A.M., Roberson, P.K., Manolagas, S.C., Bellido, T., 1999. Prevention of osteocyte and osteoblast apoptosis by bisphosphonates and calcitonin. J. Clin. Investig. 104, 1363–1374.

Rafacho, A., Ortsater, H., Nadal, A., Quesada, I., 2014. Glucocorticoid treatment and endocrine pancreas function: implications for glucose homeostasis, insulin resistance and diabetes. J. Endocrinol. 223, R49–R62.

Rauch, A., Gossye, V., Bracke, D., Gevaert, E., Jacques, P., Van Beneden, K., et al., 2011. An anti-inflammatory selective glucocorticoid receptor modulator preserves osteoblast differentiation. FASEB J. 25, 1323–1332.

Rauch, A., Seitz, S., Baschant, U., Schilling, A.F., Illing, A., Stride, B., et al., 2010. Glucocorticoids suppress bone formation by attenuating osteoblast differentiation via the monomeric glucocorticoid receptor. Cell Metabol. 11, 517–531.

Reid, D.M., Adami, S., Devogelaer, J.P., Chines, A.A., 2001. Risedronate increases bone density and reduces vertebral fracture risk within one year in men on corticosteroid therapy. Calcif. Tissue Int. 69, 242–247.

Reid, D.M., Devogelaer, J.P., Saag, K., Roux, C., Lau, C.S., Reginster, J.Y., et al., 2009. Zoledronic acid and risedronate in the prevention and treatment of glucocorticoid-induced osteoporosis (HORIZON): a multicentre, double-blind, double-dummy, randomised controlled trial. Lancet 373, 1253–1263.

Reid, D.M., Hughes, R.A., Laan, R.F., Sacco-Gibson, N.A., Wenderoth, D.H., Adami, S., et al., 2000. Efficacy and safety of daily risedronate in the treatment of corticosteroid-induced osteoporosis in men and women: a randomized trial. European Corticosteroid-Induced Osteoporosis Treatment Study. J. Bone Miner. Res. 15, 1006–1013.

Ritz, E., Kreusser, W., Rambausek, M., 1984. Effects of glucocorticoids on calcium and phosphate excretion. Adv. Exp. Med. Biol. 171, 381–397.

Rizzoli, R., Adachi, J.D., Cooper, C., Dere, W., Devogelaer, J.P., Diez-Perez, A., et al., 2012. Management of glucocorticoid-induced osteoporosis. Calcif. Tissue Int. 91, 225–243.

Roux, C., Oriente, P., Laan, R., Hughes, R.A., Ittner, J., Goemaere, S., et al., 1998. Randomized trial of effect of cyclical etidronate in the prevention of corticosteroid-induced bone loss. Ciblos Study Group. J. Clin. Endocrinol. Metab. 83, 1128–1133.

Rubin, M.R., Bilezikian, J.P., 2002. Clinical review 151: the role of parathyroid hormone in the pathogenesis of glucocorticoid-induced osteoporosis: a re-examination of the evidence. J. Clin. Endocrinol. Metab. 87, 4033–4041.

Saag, K.G., Emkey, R., Schnitzer, T.J., Brown, J.P., Hawkins, F., Goemaere, S., et al., 1998. Alendronate for the prevention and treatment of glucocorticoid-induced osteoporosis. Glucocorticoid-induced osteoporosis intervention study group. N. Engl. J. Med. 339, 292–299.

Saag, K.G., Petersen, J., Brandi, M.L., Karaplis, A.C., Lorentzon, M., Thomas, T., et al., 2017. Romosozumab or alendronate for fracture prevention in women with osteoporosis. N. Engl. J. Med. 377, 1417–1427.

Saag, K.G., Shane, E., Boonen, S., Marin, F., Donley, D.W., Taylor, K.A., et al., 2007. Teriparatide or alendronate in glucocorticoid-induced osteoporosis. N. Engl. J. Med. 357, 2028–2039.

Saag, K.G., Zanchetta, J.R., Devogelaer, J.P., Adler, R.A., Eastell, R., See, K., et al., 2009. Effects of teriparatide versus alendronate for treating glucocorticoid-induced osteoporosis: thirty-six-month results of a randomized, double-blind, controlled trial. Arthritis Rheum. 60, 3346–3355.

Sambrook, P., Birmingham, J., Kelly, P., Kempler, S., Nguyen, T., Pocock, N., Eisman, J., 1993. Prevention of corticosteroid osteoporosis. A comparison of calcium, calcitriol, and calcitonin. N. Engl. J. Med. 328, 1747–1752.

Sambrook, P.N., Roux, C., Devogelaer, J.P., Saag, K., Lau, C.S., Reginster, J.Y., et al., 2012. Bisphosphonates and glucocorticoid osteoporosis in men: results of a randomized controlled trial comparing zoledronic acid with risedronate. Bone 50, 289–295.

Sato, A.Y., Cregor, M., Delgado-Calle, J., Condon, K.W., Allen, M.R., Peacock, M., et al., 2016. Protection from glucocorticoid-induced osteoporosis by anti- catabolic signaling in the absence of sost/sclerostin. J. Bone Miner. Res. 31, 1791–1802.

Sato, A.Y., Tu, X., McAndrews, K.A., Plotkin, L.I., Bellido, T., 2015. Prevention of glucocorticoid induced-apoptosis of osteoblasts and osteocytes by protecting against endoplasmic reticulum (ER) stress in vitro and in vivo in female mice. Bone 73, 60–68.

Sawamura, M., Komatsuda, A., Togashi, M., Wakui, H., Takahashi, N., 2017. Effects of denosumab on bone metabolic markers and bone mineral density in patients treated with glucocorticoids. Intern. Med. 56, 631–636.

Schacke, H., Berger, M., Rehwinkel, H., Asadullah, K., 2007. Selective glucocorticoid receptor agonists (SEGRAs): novel ligands with an improved therapeutic index. Mol. Cell. Endocrinol. 275, 109–117.

Schakman, O., Gilson, H., Thissen, J.P., 2008. Mechanisms of glucocorticoid-induced myopathy. J. Endocrinol. 197, 1–10.

Sher, L.B., Harrison, J.R., Adams, D.J., Kream, B.E., 2006. Impaired cortical bone acquisition and osteoblast differentiation in mice with osteoblast-targeted disruption of glucocorticoid signaling. Calcif. Tissue Int. 79, 118–125.

Sher, L.B., Woitge, H.W., Adams, D.J., Gronowicz, G.A., Krozowski, Z., Harrison, J.R., Kream, B.E., 2004. Transgenic expression of 11beta-hydroxysteroid dehydrogenase type 2 in osteoblasts reveals an anabolic role for endogenous glucocorticoids in bone. Endocrinology 145, 922–929.

Shi, C., Huang, P., Kang, H., Hu, B., Qi, J., Jiang, M., et al., 2015. Glucocorticoid inhibits cell proliferation in differentiating osteoblasts by microRNA-199a targeting of WNT signaling. J. Mol. Endocrinol. 54, 325–337.

Shi, C., Qi, J., Huang, P., Jiang, M., Zhou, Q., Zhou, H., et al., 2014. MicroRNA- 17/20a inhibits glucocorticoid-induced osteoclast differentiation and function through targeting RANKL expression in osteoblast cells. Bone 68, 67–75.

Sivagurunathan, S., Muir, M.M., Brennan, T.C., Seale, J.P., Mason, R.S., 2005. Influence of glucocorticoids on human osteoclast generation and activity. J. Bone Miner. Res. 20, 390–398.

Soen, S., 2011. Guidelines on the management and treatment of glucocorticoid-induced osteoporosis. Nihon Rinsho. Jpn. J. Clin. Med. 69, 1310–1314.

Stahn, C., Buttgereit, F., 2008. Genomic and nongenomic effects of glucocorticoids. Nat. Clin. Pract. Rheumatol. 4, 525–533.

Stewart, P.M., Krozowski, Z.S., 1999. 11 beta-Hydroxysteroid dehydrogenase. Vitam. Horm. 57, 249–324.

Stoch, S.A., Saag, K.G., Greenwald, M., Sebba, A.I., Cohen, S., Verbruggen, N., et al., 2009. Once-weekly oral alendronate 70 mg in patients with glucocorticoid-induced bone loss: a 12-month randomized, placebo-controlled clinical trial. J. Rheumatol. 36, 1705–1714.

Stockbrugger, R.W., Schoon, E.J., Bollani, S., Mills, P.R., Israeli, E., Landgraf, L., et al., 2002. Discordance between the degree of osteopenia and the prevalence of spontaneous vertebral fractures in Crohn's disease. Aliment Pharmacol. Ther. 16, 1519–1527.

Struys, A., Snelder, A.A., Mulder, H., 1995. Cyclical etidronate reverses bone loss of the spine and proximal femur in patients with established corticosteroid-induced osteoporosis. Am. J. Med. 99, 235–242.

Sundahl, N., Bridelance, J., Libert, C., De Bosscher, K., Beck, I.M., 2015. Selective glucocorticoid receptor modulation: new directions with non-steroidal scaffolds. Pharmacol. Ther. 152, 28–41.

Takeuchi, Y., Watanabe, S., Ishii, G., Takeda, S., Nakayama, K., Fukumoto, S., et al., 2002. Interleukin-11 as a stimulatory factor for bone formation prevents bone loss with advancing age in mice. J. Biol. Chem. 277, 49011–49018.

Tchen, C.R., Martins, J.R., Paktiawal, N., Perelli, R., Saklatvala, J., Clark, A.R., 2010. Glucocorticoid regulation of mouse and human dual specificity phosphatase 1 (DUSP1) genes: unusual cis-acting elements and unexpected evolutionary divergence. J. Biol. Chem. 285, 2642–2652.

Temiyasathit, S., Jacobs, C.R., 2010. Osteocyte primary cilium and its role in bone mechanotransduction. Ann. N. Y. Acad. Sci. 1192, 422–428.

Thiele, S., Ziegler, N., Tsourdi, E., De Bosscher, K., Tuckermann, J.P., Hofbauer, L.C., Rauner, M., 2012. Selective glucocorticoid receptor modulation maintains bone mineral density in mice. J. Bone Miner. Res. 27, 2242–2250.

Tiganescu, A., Tahrani, A.A., Morgan, S.A., Otranto, M., Desmouliere, A., Abrahams, L., et al., July, 2013. 11beta-hydroxysteroid dehydrogenase blockade prevents age-induced skin structure and function defects. J. Clin. Investig. 123 (7), 3051–3060. https://doi.org/10.1172/JCI64162.

van der Goes, M.C., Jacobs, J.W., Jurgens, M.S., Bakker, M.F., van der Veen, M.J., van der Werf, J.H., et al., 2013. Are changes in bone mineral density different between groups of early rheumatoid arthritis patients treated according to a tight control strategy with or without prednisone if osteoporosis prophylaxis is applied? Osteoporos. Int. 24, 1429–1436.

van Raalte, D.H., Ouwens, D.M., Diamant, M., 2009. Novel insights into glucocorticoid-mediated diabetogenic effects: towards expansion of therapeutic options? Eur. J. Clin. Investig. 39, 81–93.

van Staa, T.P., 2006. The pathogenesis, epidemiology and management of glucocorticoid-induced osteoporosis. Calcif. Tissue Int. 79, 129–137.

van Staa, T.P., Geusens, P., Pols, H.A., de Laet, C., Leufkens, H.G., Cooper, C., 2005. A simple score for estimating the long-term risk of fracture in patients using oral glucocorticoids. QJM 98, 191–198.

Van Staa, T.P., Laan, R.F., Barton, I.P., Cohen, S., Reid, D.M., Cooper, C., 2003. Bone density threshold and other predictors of vertebral fracture in patients receiving oral glucocorticoid therapy. Arthritis Rheum. 48, 3224–3229.

van Staa, T.P., Leufkens, H.G., Abenhaim, L., Begaud, B., Zhang, B., Cooper, C., 2000a. Use of oral corticosteroids in the United Kingdom. QJM 93, 105–111.

van Staa, T.P., Leufkens, H.G., Abenhaim, L., Zhang, B., Cooper, C., 2000b. Oral corticosteroids and fracture risk: relationship to daily and cumulative doses. Rheumatology 39, 1383–1389.

van Staa, T.P., Leufkens, H.G., Abenhaim, L., Zhang, B., Cooper, C., 2000c. Use of oral corticosteroids and risk of fractures. J. Bone Miner. Res. 15, 993–1000.

van Staa, T.P., Leufkens, H.G., Cooper, C., 2002. The epidemiology of corticosteroid-induced osteoporosis: a meta-analysis. Osteoporos. Int. 13, 777–787.

Wallach, S., Cohen, S., Reid, D.M., Hughes, R.A., Hosking, D.J., Laan, R.F., et al., 2000. Effects of risedronate treatment on bone density and vertebral fracture in patients on corticosteroid therapy. Calcif. Tissue Int. 67, 277–285.

Walsh, L.J., Lewis, S.A., Wong, C.A., Cooper, S., Oborne, J., Cawte, S.A., et al., 2002. The impact of oral corticosteroid use on bone mineral density and vertebral fracture. Am. J. Respir. Crit. Care Med. 166, 691–695.

Walsh, L.J., Wong, C.A., Cooper, S., Guhan, A.R., Pringle, M., Tattersfield, A.E., 1999. Morbidity from asthma in relation to regular treatment: a community based study. Thorax 54, 296–300.

Walsh, L.J., Wong, C.A., Pringle, M., Tattersfield, A.E., 1996. Use of oral corticosteroids in the community and the prevention of secondary osteoporosis: a cross sectional study. BMJ 313, 344–346.

Wang, F.S., Ko, J.Y., Weng, L.H., Yeh, D.W., Ke, H.J., Wu, S.L., 2009. Inhibition of glycogen synthase kinase-3beta attenuates glucocorticoid-induced bone loss. Life Sci. 85, 685–692.

Weinstein, R.S., Jilka, R.L., Almeida, M., Roberson, P.K., Manolagas, S.C., 2010a. Intermittent parathyroid hormone administration counteracts the adverse effects of glucocorticoids on osteoblast and osteocyte viability, bone formation, and strength in mice. Endocrinology 151, 2641–2649.

Weinstein, R.S., Jilka, R.L., Parfitt, A.M., Manolagas, S.C., 1998. Inhibition of osteoblastogenesis and promotion of apoptosis of osteoblasts and osteocytes by glucocorticoids. Potential mechanisms of their deleterious effects on bone. J. Clin. Investig. 102, 274–282.

Weinstein, R.S., Wan, C., Liu, Q., Wang, Y., Almeida, M., O'Brien, C.A., et al., 2010b. Endogenous glucocorticoids decrease skeletal angiogenesis, vascularity, hydration, and strength in aged mice. Aging Cell 9, 147–161.

Woitge, H.W., Kream, B.E., 2000. Calvariae from fetal mice with a disrupted Igf1 gene have reduced rates of collagen synthesis but maintain responsiveness to glucocorticoids. J. Bone Miner. Res. 15, 1956–1964.

Wong, C.A., Walsh, L.J., Smith, C.J., Wisniewski, A.F., Lewis, S.A., Hubbard, R., et al., 2000. Inhaled corticosteroid use and bone-mineral density in patients with asthma. Lancet 355, 1399–1403.

Xia, X., Kar, R., Gluhak-Heinrich, J., Yao, W., Lane, N.E., Bonewald, L.F., et al., 2010. Glucocorticoid-induced autophagy in osteocytes. J. Bone Miner. Res. 25, 2479–2488.

Xiong, J., Onal, M., Jilka, R.L., Weinstein, R.S., Manolagas, S.C., O'Brien, C.A., 2011. Matrix-embedded cells control osteoclast formation. Nat. Med. 17, 1235–1241.

Yang, M., Trettel, L.B., Adams, D.J., Harrison, J.R., Canalis, E., Kream, B.E., 2010. Col3.6- HSD2 transgenic mice: a glucocorticoid loss-of-function model spanning early and late osteoblast differentiation. Bone 47, 573–582.

Yao, W., Cheng, Z., Busse, C., Pham, A., Nakamura, M.C., Lane, N.E., 2008. Glucocorticoid excess in mice results in early activation of osteoclastogenesis and adipogenesis and prolonged suppression of osteogenesis: a longitudinal study of gene expression in bone tissue from glucocorticoid-treated mice. Arthritis Rheum. 58, 1674–1686.

Yilmaz, L., Ozoran, K., Gunduz, O.H., Ucan, H., Yucel, M., 2001. Alendronate in rheumatoid arthritis patients treated with methotrexate and glucocorticoids. Rheumatol. Int. 20, 65–69.

Zhou, H., Mak, W., Kalak, R., Street, J., Fong-Yee, C., Zheng, Y., et al., 2009. Glucocorticoid-dependent Wnt signaling by mature osteoblasts is a key regulator of cranial skeletal development in mice. Development 136, 427–436.

Zhou, H., Mak, W., Zheng, Y., Dunstan, C.R., Seibel, M.J., 2008. Osteoblasts directly control lineage commitment of mesenchymal progenitor cells through Wnt signaling. J. Biol. Chem. 283, 1936–1945.

Zhou, J., Cidlowski, J.A., 2005. The human glucocorticoid receptor: one gene, multiple proteins and diverse responses. Steroids 70, 407–417.

Chapter 40

Diabetes and bone

Caterina Conte[1,2], Roger Bouillon[3] and Nicola Napoli[4,5]

[1]*Vita-Salute San Raffaele University, Milan, Italy;* [2]*Division of Immunology, Transplantation and Infectious Diseases, IRCCS San Raffaele Scientific Institute, Milan, Italy;* [3]*Laboratory of Clinical and Experimental Endocrinology, Department of Chronic Diseases, Metabolism and Aging, KU Leuven, Belgium;* [4]*Unit of Endocrinology and Diabetes, University Campus Bio-Medico, Rome, Italy;* [5]*Division of Bone and Mineral Diseases, Washington University in St Louis, St Louis, MO, United States*

Chapter outline

Introduction **941**
Effects of diabetes and insulin on endochondral bone growth **942**
Effects of insulin on growth plate cartilage in nondiabetic animal models and in vitro 942
Skeletal growth in T1DM 942
Animal models 942
Children 942
Effects of T1DM on bone 943
T1DM and fracture risk 943
T1DM and bone turnover 943
T1DM, bone density, and bone structure 945
T1DM and bone strength 945
Effects of T2DM and insulin on bone 946
T2DM and fractures 946
T2DM and bone turnover 947
T2DM, bone density, and bone structure 947
T2DM and bone strength 948
Clinical risk factors for fractures in T1DM and T2DM **949**
Effect of diabetes treatments on bone **950**
Bone repair in T1DM and T2DM **951**
Bone and systemic metabolism: a two-way interaction? **951**
Metabolic control of glucose and insulin in bone 951
Bone hormones control systemic metabolism 952
What causes diabetic bone disease? **952**
The diabetic hormonal milieu 952
Lower circulating insulin-like growth factor 1 952
Hypercortisolism 953
Calciotropic hormones 953
Low amylin 954
Impaired vascularization of diabetic bones 954
Altered collagenous bone matrix 954
Increased collagen glycosylation 954
Reduced enzymatic collagen cross-linking 955
Increased bone marrow adiposity 956
Inflammation and oxidative stress 956
Loss of incretin effect 956
Sclerostin 957
Treatment of bone fragility in diabetes **957**
General conclusions **957**
Acknowledgments **958**
References **958**

Introduction

The term "diabetic bone disease" is used to describe the changes in bone growth, remodeling, and density as well as fracture risk imparted by the presence of type 1 diabetes mellitus (T1DM) or type 2 diabetes mellitus (T2DM). T1DM is caused by pancreatic β-cell destruction, usually leading to an absolute insulin deficiency; northern Europeans are among the populations with the highest prevalence of T1DM. However, T2DM is currently by far the most prevalent form of diabetes on all continents. Its pathophysiology is heterogeneous, ranging from predominantly insulin resistance—i.e., at any of the classical insulin-sensitive tissues, liver, skeletal muscles, and adipose tissue—with relative insulin deficiency to a predominantly insulin secretory defect with variable insulin resistance. Hence, insulin levels in T2DM vary widely, anywhere between hyper- and hypoinsulinemia. The majority of patients with T2DM are obese, and obesity itself aggravates insulin resistance. Body weight is, therefore, an important confounding factor when examining the pathophysiology of bone in T2DM individuals.

 https://doi.org/10.1016/B978-0-12-814841-9.00040-3

Effects of diabetes and insulin on endochondral bone growth

Effects of insulin on growth plate cartilage in nondiabetic animal models and in vitro

The classic experiments of Salter and Best (Salter and Best, 1953) demonstrated that insulin administration increased growth plate width in hypophysectomized rats. This could be a local effect, because insulin injection into the proximal tibial growth plate (Heinze et al., 1989) or insulin infusion into one hindlimb (Alarid et al., 1992) produced widening of the treated growth plates only. Such local effect might be mediated by in situ production of insulin-like growth factor 1 (IGF-1), because the trophic effect of insulin was nullified by coinfusion of an IGF-1 antiserum (Alarid et al., 1992). Mice with absent expression of insulin receptors (IRs) in bone showed normal bone length (Irwin et al., 2006). However, they overexpressed the IGF-1 receptor, which could represent a compensatory response.

In vitro studies have documented the presence of IRs in a chondrosarcoma cell line (Foley et al., 1982). Chondrocyte proliferation and $^{35}SO_4$ incorporation were stimulated by insulin in several tissue and cell culture systems (Foley et al., 1982; Heinze et al., 1989; Maor et al., 1993). These effects were obtained at physiological levels of insulin, as low as 1 nM, but equimolar IGF-1 was more potent than insulin.

Collectively, the available data suggest that growth hormone (GH) and IGF-1 are more important than insulin for chondrocyte proliferation and maturation (Yakar and Isaksson, 2016).

Skeletal growth in T1DM

Animal models

The most frequently used animal models are rats or mice with T1DM induced by β-cell-destroying drugs, alloxan or—in the large majority of studies—streptozotocin (STZ), and BB (Bio-Breeding) rats with spontaneous immune-mediated diabetes. Although T1DM can be drug induced at any age, BB rats develop diabetes past the peak growth rate (7 weeks). In both rodent models, insulin levels are very low or undetectable. The growth plate width as well as the endochondral bone growth—assessed by double fluorochrome labeling of the calcifying cartilage—at the proximal tibia was consistently lower in untreated, severely hyperglycemic STZ-induced or spontaneously diabetic BB rats, and was remedied by systemic insulin treatment (Bain et al., 1997; Epstein et al., 1994; Scheiwiller et al., 1986; Verhaeghe et al., 1992). The femur was shorter after 28 days of STZ diabetes (Lucas, 1987).

The effect of insulin on bone growth might be direct and/or indirect, for example, by restoring the depressed hepatic release of IGF-1. Although it was reported that systemic administration of IGF-1 in T1DM rats partly normalized growth plate width (Scheiwiller et al., 1986), this finding was not corroborated in a subsequent study (Verhaeghe et al., 1992). The low $^{35}SO_4$ uptake by rib cartilage explants from T1DM rats was also unresponsive to exogenous IGF-1 (Kelley et al., 1993). These data suggest that diabetic cartilage is, at least in part, resistant to the actions of IGF-1. Blunted stimulatory actions of IGF-1 on osteoblasts due to high concentrations of advanced glycation end products (AGEs) and hyperglycemia may also contribute (McCarthy et al., 2001; Terada et al., 1998).

Children

At the onset of T1DM, there is no difference in height compared with nondiabetic children. In fact, most studies have documented that children are slightly taller at the onset of diabetes compared with reference values (Bonfig et al., 2012; Holl et al., 1998). Increased growth velocity has even been associated with risk of islet autoimmunity in children at risk for T1DM (Beyerlein et al., 2014; Lamb et al., 2009; Yassouridis et al., 2017).

However, growth is affected from the clinical onset onward. In the preinsulin era, prepubertal growth virtually stopped, and stunted growth (Mauriac syndrome) was frequently observed in later decades of irregular insulin treatment. In addition, there was a delay in pubertal development and growth. Retarded growth and pubertal development remain common to this day among African children with T1DM (Elamin et al., 2006). In Western countries, the majority of recent studies have documented a mild reduction in growth from height-for-age (z) charts. This deficit was more pronounced in children with a prepubertal compared with a pubertal onset of T1DM (Holl et al., 1998). Poor glycemic control predicts growth retardation: indeed, glycated hemoglobin A1c (HbA1c) levels (a reflection of glycemic control in the previous 2–3 months) correlated inversely with height velocity (Danne et al., 1997; Holl et al., 1998), and patients with good glycemic control (HbA1c $< 7\%$) had normal near-adult height (Bonfig et al., 2012). In the larger analysis conducted so far, mean near-adult height was reduced by 0.16 SD scores in patients with onset of diabetes before age 10 years (Bonfig et al., 2012).

Considering that mean height at onset of diabetes was +0.25 SD scores (i.e., at diagnosis, children with diabetes were taller than average), 0.41 SD scores were lost since diabetes onset to near-adult age.

At diagnosis, bone maturation—determined by radiographs of hand and wrist—was not different compared with nondiabetic children (Danne et al., 1997; Holl et al., 1998). However, there was a small but significant retardation of bone maturation with increasing T1DM duration (i.e., a 1-year difference between chronological and bone age after 11 years of diabetes) (Danne et al., 1997; Holl et al., 1998). Dysregulation of the GH−IGF-1 axis, which promotes chondrocyte proliferation, differentiation, and hypertrophy, may play a role in the pathogenesis of impaired skeletal growth in T1DM (Raisingani et al., 2017).

Effects of T1DM on bone

T1DM and fracture risk

The available epidemiologic studies concur that the risk of fractures is substantially increased in individuals with T1DM, more than could be explained by the mild bone mineral density (BMD) deficit; in effect, a 1-SD reduction in areal BMD (aBMD) would translate into a twofold increased risk of hip fracture (Johnell et al., 2005). A caveat is that many studies contained few incident fractures in the diabetic group ($n = 1-18$).

In a large meta-analysis that included 7,185,572 participants from 16 prospective and 9 retrospective cohort studies, patients with T1DM had 1.5 times greater risk of total fractures, 4.4 times greater risk of hip fractures, 1.8 times greater risk of upper arm fractures, and 2 times greater risk of ankle fractures compared with nondiabetic individuals (Wang et al., 2019). Compared with T2DM, T1DM was associated with a greater risk of total, hip, and ankle fracture. There was no increase in the risk of forearm or vertebral fractures. Previous meta-analyses reported relative risks ranging from 3.2 to 6.3 for any fracture (Janghorbani et al., 2007; Shah et al., 2015), 3.8 to 6.9 for hip fracture (Fan et al., 2016; Shah et al., 2015; Vestergaard, 2007), and 2.9 for spine fracture (Shah et al., 2015). A meta-analysis that specifically addressed fracture risk in patients with T1DM ages 18−50 years ($n = 35{,}925$) compared with nondiabetic controls ($n = 2{,}455{,}016$) found that T1DM was associated with a nearly twofold increase in the risk of any fracture and a more than fourfold increase in the risk of hip fracture, this increase being greater in women than in men (Thong et al., 2018). Overall, these findings indicate that younger age does not protect against the increase in fracture risk associated with T1DM.

In T1DM, the possible association between BMD and HbA1c is still a matter of debate. One possible explanation of the discordant findings is that the relation between BMD and glycemic control, is, in fact, multifactorial, and therefore may be hardly detectable by traditional statistics. Eller-Vainicher and colleagues (Eller-Vainicher et al., 2011) adopted a special mathematic approach, the artificial neural networks (ANNs), to study 175 eugonadal T1DM patients and 151 age- and body mass index (BMI)-matched controls. The authors found that T1DM subjects had lower spine and femur BMD. Interestingly, spine BMD was independently associated with BMI and daily insulin dose (DID), whereas femur BMD was associated with BMI and creatinine clearance (CrCl). The authors found that a BMI below 23.5 kg/m^2, a DID > 0.67 units/kg, and a CrCl <88.8 mL/min were associated with low BMD. Data were also analyzed using the supervised ANNs and a semantic connectivity map that showed that low BMD at both sites was indirectly connected with HbA1c through the link with chronic diabetes complications. A 2019 large nested case−control analysis corroborated these observations, showing that T1DM patients with poor glycemic control (HbA1c >8%) were 1.4 times more likely to have an incident nonvertebral low-trauma fracture than those with good glycemic control (HbA1c ≤7%) (Vavanikunnel et al., 2019). Furthermore, fracture risk was associated with comorbidities related to micro- and macrovascular complications, such as diabetic retinopathy and ischemic heart disease.

Few studies have investigated the prevalence of asymptomatic vertebral fractures in T1DM. In a study by Zhukouskaya and coauthors, patients with T1DM had a lower BMD at both spine and femur than controls and a higher prevalence of vertebral fractures (24.4% vs. 6.1%) (Zhukouskaya et al., 2013). Age, diabetes duration, age at diabetes onset, HbA1c, BMD, and the prevalence of chronic complications were similar in patients with and without fractures. Interestingly, the elevated prevalence of asymptomatic vertebral fractures was associated with the presence of T1DM independent of BMD, suggesting that impaired bone quality may contribute the increased fracture risk in T1DM.

T1DM and bone turnover

Animal models of T1DM

Several animal models have been used for T1DM (mainly alloxan- or STZ-induced diabetic rats and mice, spontaneously diabetic BB rats, and nonobese diabetic [NOD] mice). Untreated severe T1DM (hyperglycemia >300 mg/dL) was associated with low bone formation as shown by biochemical markers, in particular osteocalcin (OC) concentrations. Plasma OC dropped exponentially after onset of T1DM in BB rats to 25% of nondiabetic levels after 5 weeks (Verhaeghe et al., 1997). Plasma OC gradually returned to the control range with increasing insulin dosage in BB rats (Verhaeghe et al., 1997) and was normalized by pancreas transplantation in STZ-diabetic rats (Ishida et al., 1992).

Histomorphometry consistently showed low bone formation on all bone surfaces (trabecular–endosteal, endocortical, and periosteal) in STZ-diabetic and BB rats. Static morphometry demonstrated a marked decline in osteoblast and osteoid surface or volume, and dynamic morphometry a decrease in both mineralizing surface and mineral apposition rate. These changes were (partly) reversed by insulin (Bain et al., 1997; Epstein et al., 1994; Glajchen et al., 1988; Shires et al., 1981; Verhaeghe et al., 1992). Electron microscopy of the endocortical surface confirmed that active, cuboidal osteoblasts are virtually absent in STZ-diabetic rats and are replaced by inactive bone-lining cells with flattened nuclei and little or no endoplasmic reticulum (Sasaki et al., 1991). Thus, most of the bone surface is in a quiescent state in severely hyperglycemic animals.

OC but not Runx2 or alkaline phosphatase mRNA levels were reduced in tibias from STZ-diabetic mice (Botolin et al., 2005), reinforcing the conclusion that loss of mature osteoblasts is a key finding in T1DM.

Regarding bone resorption, T1DM rats displayed diminished total and creatinine-corrected deoxypyridinoline excretion (Horcajada-Molteni et al., 2001; Verhaeghe et al., 2000). Most but not all histomorphometric data confirmed that the osteoclast surface/number is decreased moderately to severely in T1DM murine models and this is reversed by insulin treatment (Glajchen et al., 1988; Peng et al., 2016; Shires et al., 1981; Verhaeghe et al., 1992). Electron microscopy showed that most osteoclasts in T1DM rats lack a ruffled border–clear zone complex and that acid phosphatase activity is rarely detected (Kaneko et al., 1990). Bacterial inoculation of the scalp in *db/db* mice also generated fewer osteoclasts (He et al., 2004).

Greater marrow adiposity has been a consistent finding in animal models of T1DM compared with nondiabetic controls. Interestingly, bone marrow adiposity was increased in STZ-diabetic mice and NOD mice, with upregulation of adipogenic genes such as peroxisome proliferator-activated receptor γ_2 (PPARγ_2), adipocyte fatty acid–binding protein 2, and resistin (Botolin et al., 2005). STZ-diabetic mice and NOD mice had significant trabecular bone loss, and tended to have cortical bone loss. Despite similar earlier markers of osteoblast differentiation compared with controls, OC mRNA, a marker of mature osteoblasts, was significantly reduced (Botolin and McCabe, 2007). Bone marrow adiposity, but not bone loss, was prevented by treatment with leptin or with a PPARγ antagonist (Botolin and McCabe, 2006; Motyl and McCabe, 2009), indicating that these two manifestations of diabetic bone disease result, at least in part, from different mechanisms.

Clinical data

Only few data on bone histomorphometry are available, and the measurement of biochemical markers of bone formation/mineralization produced variable results. A case–control study of 18 patients with T1DM with varying degrees of glycemic control and a wide range of disease duration showed no significant differences between diabetics and controls in histomorphometric parameters (Armas et al., 2012). However, in the T1DM group, patients with a history of fracture tended to have histomorphometric parameters suggestive of lower bone remodeling compared with nonfractured patients.

No definitive data have been produced regarding biochemical markers of bone resorption in T1DM. In some studies, resorption markers were unchanged, or even slightly increased. A systematic review and meta-analysis of studies that assessed bone turnover markers in diabetes found no difference in procollagen type 1 N-terminal propeptide (P1NP), bone-specific alkaline phosphatase, tartrate-resistant acid phosphatase (TRAP), osteoprotegerin, and N-terminal telopeptide (NTX) (Hygum et al., 2017). Sclerostin, a potent inhibitor of bone formation, was significantly higher, and C-terminal telopeptide of type I collagen (CTX) and OC significantly lower, in patients with T1DM compared with controls. OC levels were affected by plasma glucose levels, whereas HbA1c was a significant effect modifier for sclerostin.

It has also been reported that time-related changes in circulating bone turnover markers in patients with T1DM are not different from those in the general population, with no relevant changes in women and a decrease in the bone formation markers OC and P1NP and a trend toward a decrease in the bone resorption marker CTX (Hamilton et al., 2018).

Although the evidence supports the notion that T1DM is characterized by low bone turnover, what bone turnover markers reflect in T1DM and whether they predict fracture risk have not been fully understood. As an example, an inverse association between sclerostin levels and fracture risk in T1DM has been reported (Starup-Linde et al., 2016a).

In contrast to animal models, most human studies of T1DM did not show a significant increase in bone marrow adiposity, possibly due to the relatively small number of subjects included (Abdalrahaman et al., 2015; Slade et al., 2012).

T1DM, bone density, and bone structure

Animal models of T1DM

Experimental T1DM had a negative effect on bone size (wet and dry weight, diaphyseal width) that was related to at least three factors: earlier age at which diabetes was induced or occurred spontaneously, longer diabetes duration, and more severe hyperglycemia (Dixit and Ekstrom, 1980; Einhorn et al., 1988; Locatto et al., 1993; Lucas, 1987). Thus, the impact on bone size was modest in T1DM models with onset of diabetes past the peak growth rate (Verhaeghe et al., 1994, 2000).

T1DM negatively affects both cortical bone and trabecular bone. The reduction in trabecular bone volume was the result of thinner trabeculae, while the number and separation of trabeculae remained the same (Botolin et al., 2005; Epstein et al., 1994; Suzuki et al., 2003; Thrailkill et al., 2005a). In locations with dense trabecular bone (e.g., femur, mandible), the effect of hyperglycemia was less than that in areas without dense trabecular bone such as the tibial region (Hua et al., 2018).

Clinical data

Reduced trabecular bone density and cortical thickness have been reported in children and adolescents with T1DM (Maratova et al., 2018). Similarly, lower cortical and trabecular volumetric BMD (vBMD) as well as deterioration in bone microarchitecture has been reported in adult patients with T1DM compared with nondiabetic controls (Shanbhogue et al., 2015; Verroken et al., 2017), these reductions being more evident in patients with microvascular disease (Shanbhogue et al., 2015).

Reductions in bone size and bone strength have been reported in prepubertal children with T1DM (Bechtold et al., 2006, 2007; Franceschi et al., 2018). Some (Bechtold et al., 2006, 2007) but not all (Roggen et al., 2013) studies suggest that normalization of bone size occurs after puberty.

Women with long-standing T1DM had higher BMD at the lumbar spine and lower BMD at the femoral neck and higher rates of fragility fractures compared with age- and sex-matched controls, suggesting that factors other than BMD influence fracture risk in long-standing T1DM (Alhuzaim et al., 2019). No difference in BMD was evident between men with long-standing T1DM and age-matched nondiabetic men (Alhuzaim et al., 2019; Maddaloni et al., 2017). Thus, most but not all available data indicate that there is an early but mild BMD deficit in T1DM individuals that does not appear to deteriorate over time (Hamilton et al., 2018; Krakauer et al., 1995; Moyer-Mileur et al., 2004; Shah et al., 2017). Furthermore, BMD in women with T1DM remains similar to that expected for age, BMI, and postmenopausal status (Hamilton et al., 2018). Whether long-term glycemic control is a determinant of the BMD deficit in T1DM remains a matter of controversy (Hough et al., 2016). Inhibition of the bone-anabolic Wnt pathway, as suggested by an increase in circulating sclerostin and dickkopf-1 in T1DM children, may play a role (Faienza et al., 2017; Tsentidis et al., 2017), although not all studies point toward this direction (Tsentidis et al., 2016). The aBMD deficit is more pronounced among T1DM adults with microvascular complications (Campos Pastor et al., 2000; Clausen et al., 1997; Eller-Vainicher et al., 2011; Munoz-Torres et al., 1996; Rozadilla et al., 2000; Strotmeyer et al., 2006).

T1DM and bone strength

Animal models of T1DM

The breaking strength of the femoral or tibial midshaft was assessed in several T1DM models (STZ-diabetic and BB rats, NOD mice) by various methods (perpendicular pressure, torsion, three-point bending) (Dixit and Ekstrom, 1980; Einhorn et al., 1988; Horcajada-Molteni et al., 2001; Silva et al., 2009; Thrailkill et al., 2005a; Verhaeghe et al., 1990, 1994, 2000). Most, though not all, of these studies concluded that strength parameters were reduced after a critical period of diabetes (e.g., 8 weeks in rats), even when correcting for their smaller bone size.

Diabetic bone brittleness may be related to alterations of collagen quality (collagen glycation, integrity, and secondary structure) rather than changes in the mineral component of the bone matrix (Mieczkowska et al., 2015) and appears to increase with increasing duration of T1DM (Nyman et al., 2011). In WBN/Kob rats, a model of spontaneous T1DM occurring at an advanced age (i.e., from 12 months), enzymatic collagen cross-links in femoral bone were reduced from the prediabetic stage (8 months), while nonenzymatic (glycosylated) cross-links increased from 12 months. Interestingly, the collagen cross-link parameters in prediabetic/diabetic WBN bones strongly correlated with biomechanical properties (Saito et al., 2006). Similarly, STZ-induced young diabetic rats had markedly reduced trabecular bone, a relative deficit in cortical bone, and significant deficits in whole bone stiffness and strength (i.e., structural mechanical properties) (Silva et al., 2009). These changes were associated with a significant increase in nonenzymatic cross-links.

Of note, daily subcutaneous injections of insulin nearly restored the diabetes-related loss in peak force endured by the femur midshaft during three-point bending (Rao Sirasanagandla et al., 2014) and peak bending stress of the tibia midshaft (Bortolin et al., 2017) in STZ-induced diabetic rats.

Clinical data

No studies have directly measured bone strength in T1DM. However, bone strength can be estimated using quantitative computed tomography (QCT), which allows three-dimensional noninvasive assessment of bone macro- and micro-architecture (e.g., cortical porosity and trabecular connectivity). In a study that used central QCT to estimate vBMD and strength in femoral bone subfractions of 17 male T1DM patients and 18 sex-matched nondiabetic controls ages 18–49 years, cross-sectional moment of inertia, section modulus (a geometric index of bending stress), and buckling ratio (BR; an index of cortical instability) in the femoral neck and femoral shaft were similar in the two groups (Ishikawa et al., 2015). However, BR in the intertrochanteric region was higher in patients with T1DM. This associated with lower vBMD, cortical cross-sectional area, and cortical thickness in the intertrochanteric region, possibly indicating increased susceptibility to intertrochanteric fractures in young and middle-aged men with T1DM. Another study used high-resolution peripheral QCT (HR-pQCT) to assess peripheral bone microarchitecture and bone strength in T1DM patients, with or without micro-vascular disease, compared with nondiabetic controls (Shanbhogue et al., 2015). Patients with T1DM and microvascular disease had significant deficits in both cortical and trabecular bone microarchitectural parameters and estimated bone strength at the distal radius and tibia compared with matched nondiabetic control subjects, suggesting a role for micro-vascular disease in the pathogenesis of diabetic bone disease.

Effects of T2DM and insulin on bone

T2DM and fractures

Fracture risk is elevated in individuals with T2DM, but much less so than in T1DM. Large cohorts—the Study of Osteoporotic Fractures (Schwartz et al., 2001) and, in particular, the Women's Health Initiative Observational Study (Bonds et al., 2006)—confirmed a mildly increased risk (between one- and twofold). The latter study documented an elevated age-adjusted risk of fracture at all skeletal sites except the forearm/wrist/hand. A meta-analysis of cohort studies confirmed that fracture risk in patients with T2DM is 1.24 times greater than in nondiabetic individuals (Ni and Fan, 2017).

However, comprehensive meta-analyses have confirmed a nearly doubled risk of hip fracture in T2DM (Janghorbani et al., 2007; Wang et al., 2019). An association between T2DM and risk for upper arm and ankle fractures has also been reported (Wang et al., 2019), whereas the risk of spine fractures appears to be similar to that in nondiabetic subjects (Napoli et al., 2018b; Wang et al., 2019). Several mechanisms contribute to bone loss and increased fracture risk in T2DM (Fig. 40.1), as will be detailed in the following paragraphs.

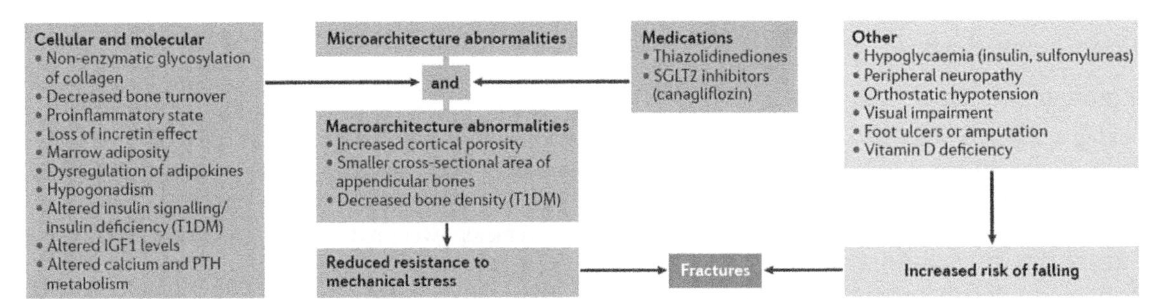

FIGURE 40.1 Mechanisms underlying bone loss and fractures in type 2 diabetes mellitus. Multiple mechanisms can contribute to the increased fracture risk observed in diabetes mellitus. Nonenzymatic glycosylation of collagen, decreased bone turnover, a proinflammatory state, and microvascular disease determine both micro– and macro–bone architecture abnormalities that cause reduced resistance to mechanical stress. Several studies have shown a different trend in terms of bone mineral density, which is generally decreased in type 1 diabetes mellitus (*T1DM*) and normal or increased in type 2 diabetes mellitus (T2DM). Alterations in bone structure in T2DM include increased cortical porosity and reduced cortical density. Insulin treatment is an additional risk factor for falls and fractures, probably because of the increased rate of hypoglycemic episodes in patients treated with insulin. The negative effect of thiazolidinediones on bone health is well known. In patients with T2DM, evidence suggests a potential negative effect also for some sodium/ glucose cotransporter-2 (*SGLT2*) inhibitors. In this clinical scenario characterized by increased bone fragility, typical diabetic complications such as poor balance, diabetic retinopathy, impaired renal function, and neuropathy have been associated with an increased risk of falls and fractures. All these complications have even a greater effect in patients with T1DM. *IGF1*, insulin-like growth factor 1; *PTH*, parathyroid hormone. *Source: Napoli N et al. 2017.* Nat Rev Endocrinol. *13, 208–219.*

Limited evidence indicates that diabetes (T1DM or T2DM) delays the healing of clinical fractures, taking about twice as long (Loder, 1988; Marin et al., 2018). Furthermore, T2DM patients have a 30% increased postfracture mortality compared with nondiabetics, and a significant excess in absolute mortality risk (Martinez-Laguna et al., 2017).

T2DM and bone turnover

Animal models of T2DM

Several rodent models have been used for the study of skeletal fragility in T2DM. These models encompass various T2DM etiologies (spontaneous or diet-induced, single gene or polygenic), body types (obese or lean), and timings of T2DM onset (before or after skeletal maturity). However, current animal models are limited in their ability to mirror key metabolic features of adult-onset, obesity-related T2DM and their impact on bone metabolism, bone mass, and strength (Fajardo et al., 2014). In contrast to observations in humans, in which decreased bone formation and normal or reduced bone resorption are reported, rodent models of T2DM exhibit decreased or normal bone formation and high bone resorption, often resulting in low BMD (Devlin et al., 2014; Ducy et al., 2000; Reinwald et al., 2009; Tanaka et al., 2018; Turner et al., 2013; Zhang et al., 2009). Furthermore, the onset of diabetes is often before skeletal maturation, and inconsistent findings have been reported in some models of T2DM with respect to bone turnover. As an example, both reduced and normal levels of bone formation markers were reported for ZDF rats (Hamann et al., 2011; Liu et al., 2007; Mathey et al., 2002). Similarly, contrasting findings have been reported for bone resorption markers in this model (Hamann et al., 2011, 2013; Mathey et al., 2002).

Clinical data

Like T1DM, T2DM is accompanied by low bone formation and remodeling. In a small but interesting series of eight diabetic subjects (ages 37–67, diabetic for 2–36 years, six with T2DM), a transiliac bone biopsy was performed because of a low BMD; histomorphometry showed a significant decrease in osteoid thickness, mineralizing surface, and mineral apposition rate (Krakauer et al., 1995). Data from Dr. Rubin's group at Columbia University have proved that T2DM patients present significant reductions in mineralizing surface, bone formation rate, and osteoblast surface ($n = 9$) (Manavalan et al., 2012). Moreover, lower RUNX2 expression and an inverse association between glucose levels and mineral apposition rate in T2DM versus nondiabetics was detected (Manavalan et al., 2012).

Biochemical bone remodeling markers are also lower than in nondiabetics.

Most studies showed that the bone formation markers P1NP and OC, as well as the bone resorption marker CTX, are decreased in patients with T2DM compared with nondiabetic controls (Starup-Linde, 2013; Starup-Linde and Vestergaard, 2016). The reduction in markers of bone formation appears to be greater in T2DM than in T1DM (Starup-Linde et al., 2016b). Moreover, Starup-Linde and colleagues demonstrated an inverse relationship between glycemic control (HbA1c) and OC levels, and a similar trend for CTX and P1NP (Starup-Linde et al., 2016b). Interestingly, there appears to be a J- or U-shaped relationship between bone remodeling and fracture risk (Heaney, 2003). It has been postulated that diabetes impairs the healing of microdamage in load-bearing bones because of suppressed bone formation and that accumulated microdamage may predispose diabetic individuals to overt fractures (Krakauer et al., 1995).

Circulating levels of sclerostin are increased in T2DM, possibly accounting for the reduced bone turnover in patients with diabetes (Gaudio et al., 2012; van Lierop et al., 2012). Furthermore, sclerostin levels have been associated with fracture risk in some reports (Heilmeier et al., 2015; Yamamoto et al., 2013). Osteoprotegerin, NTX, and bone-specific alkaline phosphatase are not significantly different between T2DM patients and controls (Hygum et al., 2017).

Impaired fasting glucose was associated with lower OC (Mitchell et al., 2018), CTX, and P1NP (Holloway-Kew et al., 2019; Jiajue et al., 2014) in women, and lower β-CTX and P1NP in men (Holloway-Kew et al., 2019), suggesting that prediabetes is also associated with reduced bone turnover (Starup-Linde et al., 2016b).

T2DM, bone density, and bone structure

Hyperinsulinemia, insulin resistance, and bone density

Hyperinsulinemia is the compensatory answer to peripheral insulin resistance; when inadequate, impaired fasting glucose and/or postload hyperglycemia (impaired glucose tolerance) ensues. Several population studies have demonstrated a correlation between, on one hand, fasting insulin, the homeostatic model of insulin resistance (HOMA-IR), and/or postchallenge glucose and insulin concentrations and, on the other hand, the aBMD at the hip and/or the spine (Conte et al., 2018). The data suggest that aBMD progressively increases with worsening glucose metabolism, i.e., from normal fasting glucose to overt T2DM (Mitchell et al., 2018). Yet adjusting for BMI or fat mass often attenuates or abolishes the correlation between aBMD and glucose metabolism parameters. Higher insulin resistance, as estimated with HOMA-IR, was associated with greater vBMD and generally favorable bone microarchitecture at both non-weight-bearing and weight-bearing skeletal sites, independent of body weight, in postmenopausal nondiabetic women (Shanbhogue et al., 2016). On the other hand, an inverse association between insulin resistance and indices of bone strength has been reported (Ahn et al., 2016; Srikanthan et al., 2014). In older nondiabetic adults participating in the Health ABC Study ($n = 2398$), increasing insulin resistance was associated with lower risk of fracture in unadjusted models (Napoli et al., 2019). However, after adjusting for BMI and BMD this relationship was lost, suggesting that higher BMD and BMI associated with insulin resistance accounted for the inverse association.

Animal models of T2DM

In contrast to T1DM, bone size was increased in T2DM models such as the Goto-Kakizaki rat, particularly at the diaphysis. Whereas cortical BMD, measured by QCT, was only mildly affected, trabecular BMD was markedly (33%–53%) decreased in these rats (Ahmad et al., 2003). T2DM may well have complex effects on bone with divergent responses generated by obesity and hyperglycemia. Disruption of leptin or leptin signaling in mice creates massive obesity with T2DM and increases bone mass via central nervous system pathways, some of which still need to be uncovered (Idelevich and Baron, 2018). As mentioned, however, concerns exist regarding the correlation between human obesity/T2DM and rodent models to study skeletal changes in obesity/T2DM (Lai et al., 2014; Rendina-Ruedy and Smith, 2016), particularly with regard to BMD, which is generally low in rodent models (Devlin et al., 2014; Ducy et al., 2000; Reinwald et al., 2009; Tanaka et al., 2018; Turner et al., 2013; Zhang et al., 2009).

Clinical data

The seminal report by Meema and Meema (Meema and Meema, 1967) that the cortical thickness at the radius was higher in aged (65–101 years) white women with T2DM than in nondiabetics was followed by numerous studies using dual-energy X-ray absorptiometry. The majority of these studies showed that aBMD is significantly increased in T2DM subjects and correlates positively with BMI and insulin levels (Ma et al., 2012). For example, in the Study of Osteoporotic Fractures, T2DM ($n < 500$) was associated with a 4.8% increase in aBMD at the radius and a 3.4% increase in aBMD at the femoral neck in multivariate analyses (Bauer et al., 1993; Orwoll et al., 1996). In the Rotterdam Study, lumbar spine and the femoral neck aBMD were also 3%–4% higher in T2DM individuals ($n = 578$) (van Daele et al., 1995). Similarly, in the Health ABC Study, total hip aBMD was 4%–5% higher in black and white men and women with T2DM ($n = 566$) after adjusting for their altered body composition (i.e., higher fat and lean mass) (Strotmeyer et al., 2004). However, total hip aBMD was inversely related to T2DM duration (Strotmeyer et al., 2004), and white women with T2DM showed increased aBMD loss compared with their nondiabetic counterparts (Schwartz et al., 2005).

Data from diabetic men enrolled in the Osteoporotic Fractures in Men (MrOS) Study have shown significantly higher aBMD (Napoli et al., 2014) and spine vBMD in T2DM subjects compared with nondiabetic ones (Napoli et al., 2018b). Importantly, higher trabecular vBMD was more strongly associated with incident vertebral fractures in nondiabetic men (odds ratio [OR] 0.33; 95% CI 0.24 to 0.45), compared with men with diabetes (OR 0.64; 95% CI 0.36 to 1.13).

Samelson and colleagues investigated the association between T2DM and cortical and trabecular bone microarchitecture and density (HR-pQCT), as well as total bone density and bone area in 1069 male and female members of the Framingham Offspring Cohort (Samelson et al., 2018). After adjustment for age, sex, weight, and height, T2DM subjects had significantly lower cortical vBMD, higher cortical porosity, and smaller cross-sectional area at the tibia, but not the radius. Furthermore, decreased cortical density and increased cortical porosity at the tibia differentiated between T2DM with prior fracture and non-T2DM with prior fracture. Trabecular indices did not differ or were more favorable in patients with T2DM than in nondiabetic subjects.

T2DM and bone strength

Animal models

In general, bone mechanical properties are reduced in murine models of T2DM. In young and adult C57BL/6 mice with high-fat diet (HFD)-induced obesity and hyperglycemia, femoral strength, stiffness, and toughness were dramatically lower than in control mice, possibly due to reduced structural quality (inferior lamellar and osteocyte alignment) (Ionova-Martin et al., 2011). HFD-fed mice demonstrated normal whole-body aBMD and femoral aBMD, but reduced spine aBMD. In Zucker diabetic fatty (ZDF) rats, shear stress testing of the femoral neck demonstrated significantly decreased maximum load and stiffness compared with nondiabetic rats, probably due to lower bone volume (Hamann et al., 2013).

An impairment in bone biomechanical properties is supported by the finding of larger indentation distances in cortical bone of murine models of T2DM (Acevedo et al., 2018; Gallant et al., 2013), possibly due to an increase in AGE content (Acevedo et al., 2018; Devlin et al., 2014).

Humans

Most studies have used HR-pQCT to assess bone microarchitecture and strength indices in humans. In a cross-sectional study of 19 elderly women with T2DM, trabecular microarchitecture was preserved but cortical porosity was higher in the radius and tibia compared with nondiabetic controls (Burghardt et al., 2010). In the radius, compressive bone strength, estimated using finite element analysis (FEA), was significantly lower. In older men with T2DM participating in the MrOS study, total bone cross-sectional area in the radius and tibia was lower in those with diabetes compared with nondiabetics, and bending strength relative to body weight was decreased in cortical regions (Petit et al., 2010). In trabecular regions, bone strength was similar between T2DM and non-T2DM older men. Fractured postmenopausal women with T2DM had significantly greater cortical porosity and lower bone strength parameters (stiffness, failure load, and cortical load fraction) compared with those without fracture (Heilmeier et al., 2016). These indices did not differ between nondiabetic women with or without fracture.

Bone microindentation allows direct assessment of the mechanical characteristics of cortical bone in humans by measuring the distance that a metal probe can enter the tibia for a given applied force (Diez-Perez et al., 2010, 2016; Nilsson et al., 2017). The bone material strength index (BMSi), assessed by bone microindentation, reflects the ability of bone to resist microcrack generation and propagation. Postmenopausal women with T2DM had lower BMSi compared with nondiabetic controls despite similar BMD (Farr et al., 2014; Furst et al., 2016; Nilsson et al., 2017), suggesting a deficit in diabetic bone material properties. In older women (75–80 years) with T2DM, BMSi by reference point indentation was lower than in nondiabetic controls despite generally better measures of trabecular and cortical bone microarchitecture and bone strength estimated by FEA (Nilsson et al., 2017). Furthermore, BMSi was inversely correlated with HbA1c levels over the prior 10 years (Farr et al., 2014) and with duration of diabetes (Furst et al., 2016).

Clinical risk factors for fractures in T1DM and T2DM

Regarding T1DM, a Swedish study (Miao et al., 2005) reported that the age-adjusted risk of hip fracture increased with longer diabetes duration. Poor glycemic control has also been associated with increased fracture risk in T1DM (Vavanikunnel et al., 2019). Longer diabetes duration and poorer glycemic control mean a greater likelihood of micro- and macrovascular complications; indeed, the risk of fracture was particularly high among T1DM patients with eye, nephropathic, neurologic, or cardiovascular complications (Miao et al., 2005; Vavanikunnel et al., 2019).

For T2DM, the elevated risk of fractures was observed in both white and black US women (Bonds et al., 2006). BMI is not a critical factor, because the risk was comparably elevated in nonobese and obese T2DM women, and controlling for BMI did not meaningfully alter the risk ratio (Janghorbani et al., 2006; Nicodemus et al., 2001; Schwartz et al., 2001). In some studies, the risk appeared to be higher among insulin-treated versus non-insulin-treated T2DM individuals (Janghorbani et al., 2006; Napoli et al., 2014; Ottenbacher et al., 2002). This may reflect the association between longer duration of T2DM and fracture: in individuals with >12 years of T2DM, the risk was increased more than twofold (Janghorbani et al., 2006; Nicodemus et al., 2001; Schwartz et al., 2001). Indeed, insulin-treated T2DM individuals showed a fourfold increased risk of falls (Schwartz et al., 2002), and repeated falls in diabetic individuals with a foot ulcer predisposed to fractures (Wallace et al., 2002). Moreover, these patients also have a longer clinical history of other complications like retinopathy or neuropathy. In the Australian Blue Mountains Eye Study, which examined 216 subjects (ages 49–97) with T1DM or T2DM, cortical cataract and especially retinopathy increased the risk of fractures (Ivers et al., 2001). Neuropathy with or without foot ulcers was another important predisposing factor for fractures in general (Strotmeyer et al., 2005), and foot fractures in particular (Cavanagh et al., 1994).

In contrast, studies performed in the Joslin "medalists," long-standing T1DM survivors (at least 50 years) with low cardiovascular risk, showed a very low prevalence of osteoporosis and fractures (Maddaloni et al., 2017). Another possible risk factor is glycemic control, tested in different studies with single measurements of HBA1c. Although limited by the short exposure (3 months), in general there is a progressive increase in risk of fractures in patients with higher HBA1c, mostly in those with levels >9% (Li et al., 2015). However, in a 2019 nested case–control analysis that used the mean of HbA1c values over 3 years to estimate glycemic control, there was no association between poor glycemic control (HbA1c >8%) and risk of nonvertebral low trauma fractures in patients with T2DM (Vavanikunnel et al., 2019). Diabetic individuals with eye, neuropathic, and vascular complications—and poor physical health in general—are more likely to fall.

It is likely, therefore, that falls mediate a significant part of the diabetes-associated risk of fractures. As a corollary, effective fall prevention strategies should be developed to prevent fractures in individuals with long-standing diabetes or diabetes complications. According to guidelines released by the American Diabetes Association/European Association for the Study of Diabetes and by the American College of Physicians (American Diabetes Association, 2019; Qaseem et al., 2018), a less stringent target in glucose control should be achieved in diabetic patients, mostly if affected by diabetes-related complications. This approach will limit the risk of hypoglycemic events, which are a main risk factor for cardiovascular mortality but also for fractures.

Effect of diabetes treatments on bone

Weight loss is recommended for overweight/obese patients with diabetes (American Diabetes Association, 2019). Due to mechanical unloading of the skeleton, weight loss might have a negative impact on bone health. An intensive lifestyle intervention resulting in long-term weight loss was not associated with an increased risk of incident total or hip fracture in overweight/obese adults with T2DM, but the risk of frailty fracture increased following weight loss (Johnson et al., 2017). Body weight loss of $\geq$ 20% from maximum weight is a significant risk factor for fragility fractures in patients with T2DM (Komorita et al., 2018). Bariatric–metabolic surgery has been recognized as an effective approach for the treatment of T2DM in obese patients (Rubino et al., 2016). Patients with T2DM undergoing bariatric surgery (roux-en-Y gastric bypass) exhibit higher bone turnover, lower aBMD, and impaired HR-pQCT-derived measurements of bone mass, structure, and strength, and more than double the likelihood of having osteopenia or osteoporosis despite adequate calcium and vitamin D supplementation compared with nonoperated patients (Madsen et al., 2019). These observations suggest that strict monitoring of bone health should be implemented for patients expected to lose large amounts of weight.

Metformin, unless contraindicated, is recommended as the first-line treatment for patients with T2DM. Most studies reported that metformin has a neutral or even favorable effect of fracture risk. In preclinical studies, metformin was shown to enhance bone-anabolic pathways, stimulating osteoblast differentiation and bone formation (Palermo et al., 2015). Treatment with metformin was associated with a 19% reduction in the risk of any fracture, although no effect on the risk of forearm, spine, or hip fracture was observed in patients with T2DM (Vestergaard et al., 2005). In the same case–control study, diabetic individuals who used sulfonylureas had a lower risk of hip fracture than diabetic individuals in general (Vestergaard et al., 2005). Another study found that treatment with a sulfonylurea was associated with reduced risk of vertebral fracture (Kanazawa et al., 2010). Data from the MrOS cohort indicate that metformin has a neutral effect on fracture risk, whereas there was an increased risk of hip fractures in patients treated with sulfonylureas (Napoli et al., 2014). However, it is likely that sulfonylureas do not exert a direct negative effect on bone but rather are associated with increased risk of hypoglycemia (Monami et al., 2014), which in turn is associated with increased fracture risk (Johnston et al., 2012). Thiazolidinediones (TZDs) or PPARγ agonists (e.g., pioglitazone, rosiglitazone) worsen diabetic bone disease. Indeed, aBMD loss at the spine and the whole body was higher among women—but not men—with T2DM who used a TZD (Schwartz et al., 2006). In line with these data, in the ADOPT trial, which compared metformin, glyburide, and rosiglitazone for 4 years in 4360 T2DM patients, the number of fractures (any site) was higher among women—but not men—treated with the TZD compared with the other two drugs (Kahn et al., 2006). Finally, a meta-analysis of randomized controlled trials of at least 12 months' duration found that TZDs were associated with a 45% increase in fracture, although the increase was significant only in women (Loke et al., 2009). In mice, TZD treatment enhanced bone marrow adiposity and caused trabecular bone loss by uncoupling resorption (activated) and formation (suppressed) (Li et al., 2006), the latter perhaps as a result of osteoblast/osteocyte apoptosis (Sorocéanu et al., 2004). Incretin-based treatments, namely, inhibitors of dipeptidyl peptidase-4 (DPP-4), an enzyme that rapidly inactivates gastric inhibitory polypeptide (GIP) and glucagon-like peptide 1 (GLP1), and GLP1 receptor agonists (GLP1RA), appear to have a neutral effect on bone, although limited data are available. Data on animals indicate that GLP1RAs may improve bone mass and architecture (Lu et al., 2015; Pereira et al., 2015). However, meta-analyses of both randomized clinical trials and population-based real-world data showed no effect of GLP1RAs on fracture risk (Driessen et al., 2017; Mabilleau et al., 2014). Conflicting data exist on DPP-4 inhibitors, which were reported to either reduce (Monami et al., 2011) or have no effect on (Driessen et al., 2017) fracture risk. Sodium/glucose cotransporter-2 (SGLT-2) inhibitors decrease plasma glucose levels by inhibiting glucose reuptake by SGLT-2 in the proximal tubule of the kidney. These drugs may affect calcium and phosphate homeostasis and potentially increase fracture risk (Meier et al., 2016). The SGLT-2 inhibitor canagliflozin was associated with a significant increase in both all fractures and low-trauma fractures in patients with T2DM (Watts et al., 2016). However, no increased risk of fracture was evident in four meta-analyses of placebo-controlled randomized controlled trials of dapagliflozin, canagliflozin, and empagliflozin (Kohler et al., 2018; Ruanpeng et al., 2017; Tang et al., 2016; Wu et al., 2016). In a nested case–control study, the use of SGLT-2 inhibitors was not associated with an increased risk of fractures of the upper or lower limbs compared with use of DPP-4 inhibitors in patients with T2DM (Schmedt et al., 2019). Although these data are reassuring, further evidence is needed to confirm the bone safety of SGLT-2 inhibitors.

Insulin has anabolic effects of bone. Intensive insulin treatment in patients with T1DM was associated with stabilization of BMD over 7 years and a marked decrease in bone resorption, as assessed by circulating bone turnover markers (Campos Pastor et al., 2000). In contrast, a large body of evidence indicates that insulin treatment is associated with increased fracture risk in T2DM (Gilbert and Pratley, 2015). A 2018 population-based study including more than 12,000 patients with T2DM found that insulin monotherapy was associated with a 63% increase in fracture risk compared with metformin monotherapy (Losada et al., 2018). The increased fracture risk associated with insulin treatment in T2DM might be due to more advanced disease, with increased vascular complications, in insulin-treated T2DM patients.

Statins are another frequently used class of drugs in diabetic patients. In the general population, statin use was associated with reduced risk of osteoporosis (Lin et al., 2018), although the effect on fracture risk is uncertain (Toh and Hernandez-Diaz, 2007). A small study suggested that statins increase bone density at the hip in T2DM patients (Chung et al., 2000), and an inverse association of fracture risk with statin use has been reported (Martinez-Laguna et al., 2018).

Bone repair in T1DM and T2DM

Fracture healing in patients with diabetes is delayed by nearly 90%, and the risk of complications such as delayed union, nonunion, re-dislocation, or pseudoarthrosis is more than threefold compared with nondiabetic individuals (Jiao et al., 2015). Several experimental models have demonstrated an impaired bone repair in T1DM with less new bone formed, as assessed by radiographic, micro-computed tomographic, and histomorphometric analyses (Follak et al., 2004; Funk et al., 2000; Gandhi et al., 2006; Kawaguchi et al., 1994; Kayal et al., 2007; Santana et al., 2003; Thrailkill et al., 2005a). Overall, bone formation was approximately 7% lower ($P < .001$) in T1DM models compared with controls (Camargo et al., 2017). Both endochondral and periosteal postfracture repair mechanisms were defective. Both increased cartilage resorption, due to greater osteoclast number (Kayal et al., 2007), and impaired cartilage resorption, due to decreased osteoclast function (Kasahara et al., 2010), have been reported in STZ-diabetic rats. However, some local treatments also (partially) reversed the defective fracture repair, e.g., local administration of insulin (Gandhi et al., 2005; Paglia et al., 2013).

In models of T2DM, bone regeneration was significantly decreased due to impairments in angiogenesis and osteogenesis that could be reversed by local application of fibroblast growth factor 9 and, to a lesser degree, vascular endothelial growth factor (VEGF) A (Wallner et al., 2015). Increased callus adiposity due to the shift of mesenchymal stem cells toward the adipogenic lineage (Brown et al., 2014), delayed periosteal mesenchymal osteogenesis, premature apoptosis of cartilage callus, impaired microvascularization (Roszer et al., 2014), and impairment in osteoblast differentiation (Hamann et al., 2011) may also be responsible for delayed fracture healing in T2DM. Animal studies suggest that DPP-4 inhibitors may enhance fracture healing by altering stem cell lineage allocation (Ambrosi et al., 2017).

Bone and systemic metabolism: a two-way interaction?

Metabolic control of glucose and insulin in bone

Both osteoblasts (Ferron et al., 2010) and osteoclasts (Thomas et al., 1998) express the IR on their surface, insulin signaling through the IR being essential for normal bone acquisition (Fulzele et al., 2010; Thrailkill et al., 2014). In fact, insulin increases osteoblast proliferation and enhances osteoblast differentiation, collagen synthesis, alkaline phosphatase production, and glucose uptake, and inhibits osteoclast activity, leading to increased bone formation (Cornish et al., 1996; Pramojanee et al., 2013; Pun et al., 1989; Thomas et al., 1998; Thrailkill et al., 2005b; Yang et al., 2010). Osteoblasts also express the IGF-1 receptor on their surface (Fulzele et al., 2007). IGF-1 binds both the IGF-1 receptor and, with lower binding affinity, the IR, thus triggering the insulin signaling pathway. Studies investigating the effects of insulin on osteocytes are lacking, as expression of the IR on the osteocyte surface is minimal or absent.

Supraphysiological glucose concentrations and/or AGEs were shown to reduce the expression of proosteogenic markers in human osteoblasts (Ehnert et al., 2015; Miranda et al., 2016), suppress mineralization (Ogawa et al., 2007), and inhibit the Wnt signaling pathway, which is critical for osteogenic differentiation (Lopez-Herradon et al., 2013; Picke et al., 2016). Furthermore, exposure to AGEs induced osteoblast and mesenchymal stem cell apoptosis (Alikhani et al., 2007; Kume et al., 2005). Hyperglycemia inhibits osteoclastogenesis (Wittrant et al., 2008) and impairs osteoclast differentiation and activity, as measured by TRAP in embryonic stem cells (Dienelt and zur Nieden, 2011). Osteocytes exposed to high glucose concentrations increase expression of sclerostin, which inhibits bone formation by antagonizing the Wnt/β-catenin canonical signaling pathway (Tanaka et al., 2015). Osteocyte exposure to AGEs increases sclerostin and decreases expression of receptor activator of nuclear factor κB ligand (RANKL), a member of the tumor necrosis factor (TNF) superfamily that is necessary for osteoclast formation, differentiation, activation, and survival (Tanaka et al., 2015). In addition, hyperglycemia impairs osteocyte response to mechanical stimulation, a mechanism essential for their function (Seref-Ferlengez et al., 2016; Villasenor et al., 2019).

Bone hormones control systemic metabolism

Evidence indicates a reciprocal regulation of glucose and bone metabolism (Conte et al., 2018). The osteoblast-derived bone matrix protein OC acts as a hormone, stimulating β-cell proliferation and insulin secretion. Consistently, OC-knockout mice displayed decreased β-cell proliferation, glucose intolerance, and insulin resistance (Wei et al., 2014a). In vitro, insulin secretion increased when pancreatic β cells were cocultured with osteoblasts or in the presence of supernatants from cultured osteoblasts, suggesting that an osteoblast-derived factor stimulates insulin secretion by β cells (Ferron et al., 2008; Lee et al., 2007) acting upon GPRC6A receptors (Ferron et al., 2008; Wei et al., 2014b). Administration of OC significantly decreased plasma glucose and increased insulin secretion in C57BL/6J wild-type mice (Ferron et al., 2008). In addition to stimulating insulin secretion, OC indirectly improved insulin sensitivity by stimulating the secretion of adiponectin, a white adipocyte-derived insulin-sensitizing adipokine.

Evidence from preclinical studies indicates that only undercarboxylated OC (ucOC) exerts endocrine actions (Lee et al., 2007). OC decarboxylation is a pH-dependent mechanism that occurs during bone resorption, when bone matrix is acidified by osteoclasts. However, treatment with bisphosphonates, which inhibit bone resorption and therefore pH-dependent decarboxylation of OC, is not associated with worsening glucose metabolism (Schwartz et al., 2013a). Treatment with denosumab, a fully human monoclonal antibody that inhibits bone resorption by binding to RANKL, has been associated with reductions in fasting plasma glucose in diabetic subjects (Napoli et al., 2018a). Blockade of nuclear factor κB, a proinflammatory master switch that controls the production of inflammatory markers and mediators, has been postulated as the mechanism responsible for this observation. A number of human studies have explored the relationship between OC and glucose homeostasis. OC levels have been reported to be lower in T2DM compared with healthy subjects (Oz et al., 2006), inversely related to BMI, fat mass, and plasma glucose (Kanazawa et al., 2009, 2011b; Kindblom et al., 2009; Zhou et al., 2009), but also to atherosclerosis and inflammatory parameters such as high sensitive C-reactive protein and interleukin-6 (IL-6) (Pittas et al., 2009). However, OC-stimulated IL-6 release has been postulated as a mechanism to support muscle function during exercise, through enhanced glucose and fatty acid uptake into myofibers (Mera et al., 2016a, 2016b; Tsuka et al., 2015).

Studies investigating the effects of medications or conditions that may affect OC levels yielded conflicting results, questioning the OC–glucose relationship in humans. For example, alendronate therapy, which decreases OC levels, was associated with reduced diabetes risk (Vestergaard, 2011); treatment with vitamin K, which reduces the ucOC/OC ratio, improved insulin resistance (Beulens et al., 2010; Choi et al., 2011; Yoshida et al., 2008); chronic hyperparathyroidism, which is characterized by increased OC release, was associated with increased insulin resistance and impaired glucose regulation (Taylor and Khaleeli, 2001). Less evidence is available for T1DM. No association between OC and β-cell function was found in subjects with new-onset T1DM (Napoli et al., 2013). However, in T1DM, total OC was inversely correlated with BMI and HbA1c, and ucOC was inversely correlated with HbA1c in T1DM patients with mean disease duration of 21 years (Neumann et al., 2016). Despite relevant clinical and preclinical data supporting a positive feedback between osteoblasts and β cells, it is possible that, in a condition of continuous autoimmune damage against β cells such as in T1DM, the ability of OC to modulate β-cell function is lost. On the other hand, Thrailkill et al. reported a positive effect of OC on endogenous insulin production (assessed as C-peptide/glucose ratio) (Thrailkill et al., 2012) in subjects with long-standing disease. The relationship between OC and glucose metabolism requires further investigation.

What causes diabetic bone disease?

The diabetic hormonal milieu

Lower circulating insulin-like growth factor 1

Lower circulating IGF-1 is probably a crucial factor. Downregulation of hepatic IGF-1 release resulted in a 73% drop in serum IGF-1 levels in children with new-onset T1DM (Bereket et al., 1995); circulating IGF-1 remained below control levels in T1DM adolescents and adults receiving insulin treatment (Bouillon et al., 1995; Leger et al., 2006). IGF-1 had powerful effects on osteoblast proliferation and bone matrix formation in vitro (Hock et al., 1988). Also, disruption of the IGF-1 gene resulted in 25%–40% smaller bones in prepubertal mice and prevented periosteal expansion and BMD gain during puberty; serum OC levels in IGF-1-knockout mice were reduced by 24%–54% (Mohan et al., 2003). Depletion of circulating IGF-1 (while skeletal expression of IGF-1 remained normal) equally resulted in decreased growth and periosteal expansion, which was restored by exogenous IGF-1 (Yakar et al., 2002). Insulin and IGF-1 initiate cellular responses by binding to and activating the insulin and IGF-1 (tyrosine kinase) receptors, and phosphorylation of IR substrate (IRS)-1. As would be predicted, IRS-1-knockout mice showed smaller bones with thinner growth plates, lower trabecular and cortical bone volume, thin trabeculae, and low bone formation and resorption (Ogata et al., 2000). These mice also were hyperinsulinemic. Studies in T1DM animals showed a strong correlation between plasma IGF-1 and OC levels (Verhaeghe et al., 1997). Similarly, IGF-1 concentrations were correlated with biochemical bone formation markers in T1DM individuals (Bouillon et al., 1995). In addition, serum IGF-1 was an independent predictor of total body bone mineral content in children and adolescents with T1DM (Leger et al., 2006). Upregulation of the hepatic expression of IGF-binding protein-1 (IGFBP1) by insulin deficiency caused a sevenfold increment in circulating IGFBP1 in children with new-onset T1DM (Bereket et al., 1995). Through the accrued formation of binary IGF-1:IGFBP1 complexes (Frystyk et al., 2002), the decrement in circulating free (bioavailable) IGF-1 in T1DM is more pronounced than that of total IGF-1. It is not surprising, therefore, that IGFBP1 inhibited the osteoblastic effects of IGF-1 (Campbell and Novak, 1991). Elevated IGFBP1 levels might also explain, in part, why bone growth and remodeling in T1DM rats were resistant to the anabolic effects of exogenous IGF-1 (Verhaeghe et al., 1992).

Using the Cre-loxP technique to disrupt IGF-1R in osteoblasts in vitro, Fulzele and colleagues were able to examine insulin signaling in primary osteoblasts exclusively through its cognate receptor (Fulzele et al., 2007). In osteoblasts engineered for conditional disruption of the IGF-1R, insulin signaling was markedly increased and insulin treatment rescued the defective differentiation and mineralization, suggesting that IR signaling can compensate, at least in part, for loss of IGF-1R signaling.

Several in vivo studies have shown that the stimulatory actions of IGF-1 on osteoblasts are blunted by high concentrations of AGEs and that high glucose concentrations or AGEs might induce osteoblast resistance to the actions of IGF-1 (McCarthy et al., 2001; Terada et al., 1998). Serum levels of IGF-1 were found to be inversely associated with the presence of vertebral fractures in postmenopausal women with T2DM independent of age, glycemic control, renal function, insulin secretion, or lumbar spine BMD, and with the number of prevalent vertebral fractures in these women independent of lumbar spine BMD (Kanazawa et al., 2011a).

In 2018, a Japanese group studying a cohort of about 1000 T2DM subjects found that an IGF-1 cutoff value of 127 ng/mL was associated with higher risk of vertebral fracture (Kanazawa et al., 2018). In particular, lower IGF-1 and lower T scores were associated with significantly increased risk of vertebral fractures compared with higher IGF-1 and higher T scores, both in postmenopausal women and in men. Further evidence is needed to validate the predictive value of IGF-1 on fracture risk in individuals with diabetes.

Hypercortisolism

Several studies suggest that T2DM patients with chronic complications have an increased cortisol secretion even though within the normal range (Chiodini et al., 2006, 2007). In a case–control study on 99 T2DM postmenopausal women with good glycemic control without hypercortisolism and 107 matched nondiabetic controls, vertebral fractures were associated with cortisol secretion and sensitivity, as mirrored by the sensitizing polymorphism glucocorticoid receptor gene (Zhukouskaya et al., 2015). Thus, it is possible to hypothesize that in the presence of T2DM complications a slightly enhanced cortisol secretion may contribute to bone damage (Eller-Vainicher et al., 2017). The same authors found that trabecular bone score (TBS) was not different between T2DM postmenopausal women with good glycemic control and matched nondiabetic controls. However, TBS was reduced in T2DM patients with vertebral fractures compared with nonfractured T2DM patients. Interestingly, the presence of a TBS ≤ 1.130 or BMD Z score at the femoral neck below -1.0 had the best diagnostic accuracy for detecting T2DM fractured patients with a sensitivity and negative predictive value above 80% (Zhukouskaya et al., 2016).

Calciotropic hormones

There is a great deal of controversy regarding the effects of T1DM on circulating parathyroid hormone (PTH) and 1,25-dihydroxyvitamin D_3 (1,25$(OH)_2D_3$). Intermittent administration of human PTH(1–34) to T1DM rats improved trabecular bone formation, volume, and strength (Suzuki et al., 2003); however, intermittent PTH is a powerful anabolic agent for trabecular bone in various osteopenic conditions not necessarily associated with hypoparathyroidism (e.g., postmenopausal or glucocorticoid-induced osteopenia). Although T1DM rats displayed a higher metabolic clearance rate of $1,25(OH)_2D_3$ and lower (i.e., total but not free) circulating $1,25(OH)_2D_3$, $1,25(OH)_2D_3$ infusion did not affect the osteoblast and osteoid surface on histomorphometry and barely raised serum OC (Verhaeghe et al., 1993).

Clinical data have shown that high blood glucose may increase urinary calcium excretion and PTH levels (Hofbauer et al., 2007; McNair et al., 1979; Raskin et al., 1978). Conversely, improvement of blood glucose control may reduce calcium and phosphate urinary excretion and $1,25(OH)_2D_3$ levels (Okazaki et al., 1997). Numerous studies have shown that both patients with T1DM and those with T2DM have low vitamin D levels (Scragg et al., 1995). Obesity itself is also associated with low vitamin D levels, although the pathophysiology of this is unclear. One mechanism may be related to the deposition of vitamin D in adipocytes (Wortsman et al., 2000), although altered 25-hydroxylation in the liver of obese individuals is another possibility. It is a matter of debate if vitamin D levels may predict risk of diabetes (Ebeling et al., 2018). Indeed, although higher levels of 25(OH)D have been associated with reduced risk of diabetes, other studies have not found an association between baseline vitamin D and incident diabetes (Napoli et al., 2016; Schafer et al., 2014). Vitamin D supplementation aimed at raising 25(OH)D levels above 30 ng/mL had no effect on insulin secretion, insulin sensitivity, or the development of diabetes compared with placebo (Davidson et al., 2013). Importantly, patients with new-onset T1DM have reduced levels of $1,25(OH)_2D$ and 25(OH)D compared with healthy controls (Pozzilli et al., 2005). A causative role of vitamin D deficiency in diabetes has been suggested although vitamin D supplementation in early diagnosed patients with T1DM has not improved measures of glucose control nor insulin need nor bone turnover markers (Bizzarri et al., 2010; Maddaloni et al., 2018; Napoli et al., 2013).

Low amylin

Amylin is a peptide cosecreted with insulin by pancreatic β cells in response to nutrient stimuli. Amylin treatment of STZ-diabetic rats resulted in some improvement of the diabetic syndrome, and prevented the low-formation osteopenia observed in untreated rats (Horcajada-Molteni et al., 2001). The role of amylin must be clarified in future research.

Impaired vascularization of diabetic bones

It has been postulated by some that diabetic bone disease is another manifestation of microangiopathy, but evidence supporting this hypothesis is lacking. More research on the vascularization of diabetic bones (including the growth plate), and the factors that regulate vascularization (e.g., VEGFs), is sorely needed.

Altered collagenous bone matrix

Increased collagen glycosylation

Nonenzymatic protein glycosylation and oxidation results in the gradual accumulation of AGEs in serum and various tissues. Proteins with a long half-life such as collagen are particularly susceptible to glycosylation. AGE accumulation in bone occurs with normal aging but is much accelerated in diabetes (Katayama et al., 1996). Accumulation of AGEs in aging bone was associated with increased brittleness (Wang et al., 2002), and may also be a determinant of diabetic bone brittleness. Moreover, AGE accumulation in diabetic bones affected bone cell function through inhibition of osteoblastic cell differentiation and function (Katayama et al., 1996) and stimulation of osteoblast apoptosis (Alikhani et al., 2007). RAGE, one of the receptors for AGEs, was overexpressed in healing bone tissue of STZ-diabetic mice, and local delivery of a RAGE ligand delayed bone healing in nondiabetic mice (Santana et al., 2003). These data would indicate that accumulation of AGEs explains in part the impaired fracture repair in diabetes. In a 2018 study, proximal femur specimens from T2DM ($n = 20$) and nondiabetic ($n = 33$) subjects undergoing total hip replacement surgery were imaged by micro-computed tomography, tested by cyclic reference point indentation, and quantified for AGE content (Karim et al., 2018). Bone specimens from patients with T2DM had similar cortical porosity but altered cortical bone biomechanical properties (greater creep indentation distance and indentation distance increase) and tended to have greater AGE content in cortices compared with nondiabetic bones. Conversely, biomechanical properties and AGE content of trabecular bone were similar between T2DM and nondiabetic subjects. The normal trabecular bone and modest cortical bone alterations suggest that other factors than altered bone microstructure could contribute to increased fracture risk in T2DM. A 2019 study by the Donnelly group investigated differences in material properties, microarchitecture, and mechanical performance of trabecular bone in men with and without T2DM undergoing total hip arthroplasty (Hunt et al., 2019).

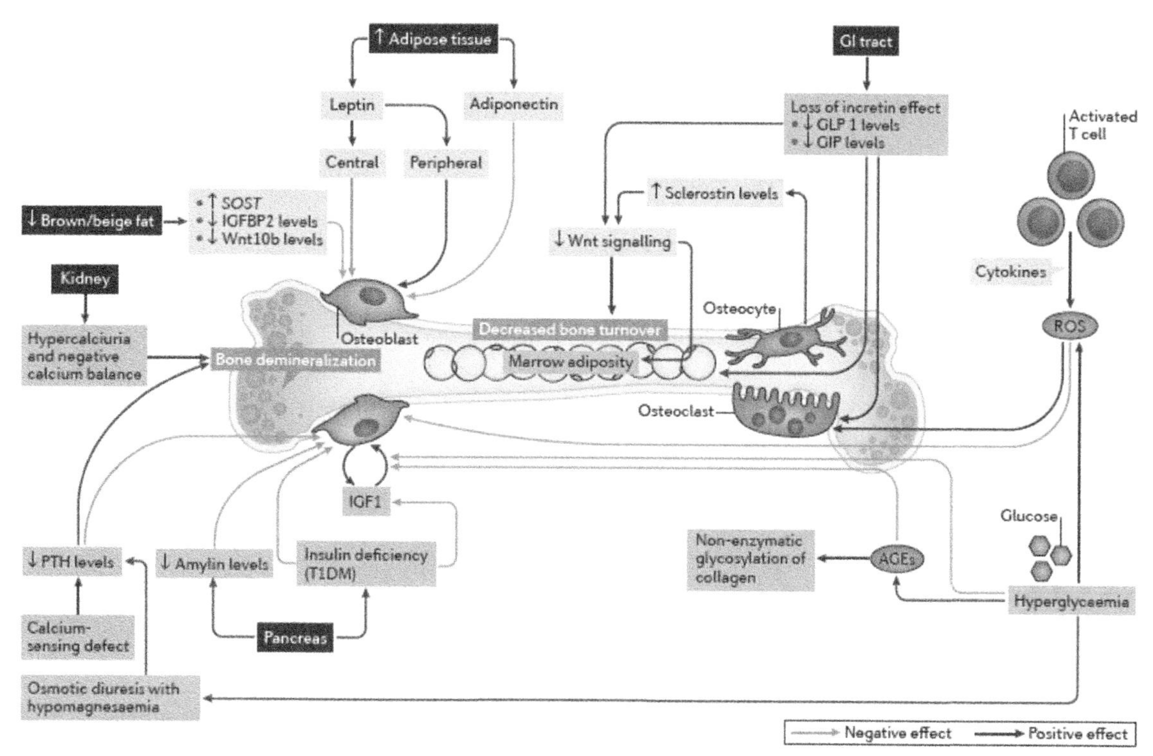

FIGURE 40.2 Cellular and molecular mechanisms of bone diseases in diabetes mellitus. Although several reports consistently indicate an increased risk of fractures in patients with diabetes mellitus, the underlying mechanisms are unclear and there is not enough evidence for a conclusive model of bone fragility in diabetes mellitus; however, some factors should be highly considered. With the decline of β-cell function, chronic hyperglycemia causes oxidative stress, inflammation, and the production of reactive oxygen species (*ROS*) and advanced glycation end products (*AGEs*), causing organ damage and reduced bone strength. In particular, accumulation of AGEs in diabetic bone collagen determines reduced material properties and increased susceptibility to fracture. AGEs and hyperglycemia also directly inhibit bone formation via suppression of osteoblast function. Low bone formation is also caused by disturbances to the Wnt signaling pathway, with increased sclerostin (*SOST*) expression, higher sclerostin levels, and decreased levels of insulin-like growth factor-binding protein 2 (*IGFBP2*) and protein *Wnt10b*. In type 1 diabetes mellitus (*T1DM*) and in the late stages of T2DM, bone formation is also decreased by insulin deficiency through an inhibitory effect on osteoblasts, either directly or through alterations in insulin-like growth factor 1 (*IGF1*) levels. Other factors typically linked to T2DM and obesity interfere with bone health. Dysregulation of adipokines like adiponectin and leptin has a negative effect through complex central and peripheral mechanisms. New evidence also indicates a negative effect on bone health by loss of the incretin effect, with reduced bone formation and increased osteoclastogenesis. Finally, alterations of the calcium–parathyroid hormone (*PTH*) axis result in a negative calcium balance, thereby contributing to bone demineralization in diabetes mellitus. *GI*, gastrointestinal; *GIP*, gastric inhibitory polypeptide; *GLP1*, glucagon-like peptide 1. *Source: Napoli N et al. 2017.* Nat Rev Endocrinol. *13, 208–219.*

Bone specimens from men with T2DM had significantly greater content of the AGE pentosidine (+36%) and sugars bound to the collagen matrix (+42%) compared with specimens from nondiabetic men. Mineral content was significantly higher (+7%) and, consistent with previous reports of increased BMD at the femoral neck and hip in T2DM, bone volume fraction (BV/TV) tended to be greater (+24%) in specimens from diabetic men. These effects favorably influenced bone strength. However, the greater values of pentosidine in T2DM men decreased post-yield strain and toughness. Overall, these findings indicate a beneficial effect of T2DM on trabecular microarchitecture, but an impairment of collagen matrix. High concentrations of AGEs can offset the favorable effect of greater mineral content, thus increasing bone fragility (Fig. 40.2).

Reduced enzymatic collagen cross-linking

Spontaneously diabetic WBN/Kob rats showed reduced serum vitamin B_6 (pyridoxal and pyridoxamine) and immature enzymatic collagen cross-links in femoral bone from the prediabetic stage onward; however, mature cross-links (pyridinoline and deoxypyridinoline) remained normal, while nonenzymatic cross-links increased from the diabetic stage. As mentioned earlier, the altered collagenous matrix composition was correlated with reduced bone strength parameters (Saito et al., 2006).

Increased bone marrow adiposity

In recent years, bone marrow fat (BMF) has received increasing attention as a determinant of bone health that is also involved in whole-body energy regulation, hematopoiesis, and possibly further functions (Veldhuis-Vlug and Rosen, 2017). A positive association between osteocyte-derived sclerostin and BMF has been described in men (Ma et al., 2014), indicating that osteocytes may influence marrow adipogenesis via inhibition of the Wnt signaling pathway, thus favoring adipogenesis at the expense of osteogenesis.

In a cross-sectional analysis of men participating in the MrOS study, a large epidemiological study of nearly 6000 men, those with diabetes ($n = 38$) had significantly higher spine BMF (59% vs. 55%), spine BMD, and total hip than nondiabetic men, despite similar body weight, BMI, and waist circumference (Sheu et al., 2017). BMF was inversely correlated with BMD in diabetic men, but not in controls. Schwartz and colleagues investigated the relationships between vertebral BMF, BMD by QCT, and fracture in 257 older adults (mean age 79 years) (Schwartz et al., 2013b). Mean BMF was 54% in men and 55% in women. Those with prevalent vertebral fracture (21 men, 32 women) had significantly higher mean BMF (57% vs. 54%) after adjustment for age, gender, and trabecular spine vBMD. When men and women were analyzed separately, gender-related differences emerged. The difference in BMF between those with or without a prevalent fracture remained statistically significant only in men. BMF was associated with lower trabecular vBMD at the spine (−10.5% difference for each 1 SD increase in BMF), total hip, and femoral neck, but not with cortical vBMD, in women. In men, BMF was marginally but significantly associated with trabecular spine vBMD (−6.1%). Total hip and spine aBMD were negatively correlated with BMF in women only.

Patients with T1DM appear to have similar marrow adiposity compared with nondiabetics. No differences in BMF were identified between male patients with T1DM and healthy controls (Armas et al., 2012; Carvalho et al., 2018), and neither duration of disease nor glycemic control was related to bone marrow adiposity. This lack of association between BMF and T1DM was confirmed in young women with T1DM compared with healthy controls (Abdalrahaman et al., 2017). Irrespective of the presence of diabetes, in young women BMF was inversely associated with BMD (Abdalrahaman et al., 2017). Carvalho and colleagues showed that BMF quantity and lipid composition (saturated and unsaturated lipids) were similar between T1DM subjects and controls (Carvalho et al., 2018). There was, however, a significant inverse correlation between BMF saturated lipids and BMD.

Inflammation and oxidative stress

Chronic low-grade inflammation underlies several metabolic diseases, including T1DM, T2DM, and associated micro- and macrovascular complications (Clark et al., 2017; Lontchi-Yimagou et al., 2013). Chronic low-grade inflammation could also have a role in diabetic bone disease. Elevated cytokine levels can activate osteoclastogenesis and suppress osteoblast differentiation (Gilbert et al., 2000; Glajchen et al., 1988). Several proinflammatory cytokines that are elevated in both T1DM and T2DM may have detrimental effects on bone health. TNF has been shown to stimulate osteoclastogenesis (Glantschnig et al., 2003) and inhibit osteoblastogenesis (Gilbert et al., 2000). Local production of inflammatory cytokines in bone tissue has been described in murine models of STZ-induced T1DM (Coe et al., 2011; Motyl et al., 2009). Furthermore, systemic exposure to proinflammatory cytokines such as IL-1, IL-6, and TNF results in increased oxidative stress, which affects differentiation and survival of osteoclasts, osteoblasts, and osteocytes (Manolagas, 2010). Mesenchymal stem cells may also be affected, as hyperglycemia-induced oxidative stress was shown to influence mesenchymal stem cell differentiation, with adipogenesis being favored over bone formation (Aguiari et al., 2008). Although these studies point toward a detrimental effect of chronic inflammation and oxidative stress on bone, further studies are needed to better define their contribution to diabetic bone disease.

Loss of incretin effect

Incretins, namely GIP and GLP1, are gut-secreted hormones that are responsible for the "incretin effect," i.e., the greater insulin response to oral versus intravenous glucose administration. Patients with T2DM have a reduced incretin effect, with impaired GLP1 production after a meal (Knop et al., 2007). GLP1 promotes proliferation of mesenchymal stem cells and prevents their differentiation into adipocytes (Sanz et al., 2010). GLP1 receptor knockout mice showed increased osteoclast numbers, bone resorption rate, and bone fragility and low calcitonin levels (Yamada et al., 2008). Increasing doses of exendin-4, a GLP1 mimetic, resulted in proportional increases in BMD, bone strength, and bone formation in rats (Ma et al., 2013). Whether loss of incretin effect contributes to bone fragility in T2DM in humans has not been established. Two categories of incretin-based therapies are available for the treatment of T2DM: inhibitors of DPP-4, and GLP1RA. Observational studies indicate that incretin mimetics have a neutral effect on bone health in humans (Driessen et al., 2017; Mabilleau et al., 2014; Monami et al., 2011), and interventional trials are lacking.

Sclerostin

Osteocytes are the most abundant bone cell type and play a central role in bone modeling and remodeling, regulating the bone biomechanical response to loading and the capacity to repair microcracks (i.e., fracture initiators). The role of osteocytes in diabetic bone disease is often overlooked, and yet to be fully clarified. Bone expression of sclerostin (encoded by *SOST*) and dickkopf-related protein 1, two major inhibitors of bone formation via inhibition of Wnt−β-catenin signaling, were increased in rat models of T1DM and T2DM (Hie et al., 2011; Kim et al., 2013; Nuche-Berenguer et al., 2010). Sclerostin levels were also found to be higher in patients with diabetes, either type 1 or type 2, than in healthy individuals (Gennari et al., 2012; Neumann et al., 2014). In T2DM, this increase was associated with a decrease in markers of bone formation (Gaudio et al., 2012), further suggesting that sclerostin inhibits bone turnover in diabetic states.

While increased circulating levels of sclerostin were associated with vertebral fractures in postmenopausal women with T2DM (Yamamoto et al., 2013), suggesting that sclerostin could be a useful clinical tool for fracture risk assessment in patients with T2DM, T1DM patients with the highest sclerostin levels had an 81% decreased risk of fracture compared with those with the lowest levels (Starup-Linde et al., 2016a).

Treatment of bone fragility in diabetes

Anti-osteoporosis drugs appear to be as effective in diabetic as in nondiabetic individuals, although the number of studies is limited. In a post hoc analysis of the Fracture Intervention Trial, alendronate 5−10 mg for 3 years increased aBMD at the spine and hip and reduced bone remodeling indices comparably in diabetic and nondiabetic postmenopausal women (Keegan et al., 2004). A population-based study that assessed the efficacy of antiresorptive drugs for fracture prevention found no difference in the anti-fracture efficacy between patients with diabetes and nondiabetic controls or between patients with T1DM and those with T2DM (Vestergaard et al., 2011). Efficacy in reducing nonvertebral fractures, increasing BMD, and decreasing back pain was reported with the osteoanabolic agent teriparatide independent of diabetes status in the Direct Assessment of Nonvertebral Fractures in Community Experience (DANCE) observational study (Schwartz et al., 2016). Data from pooled observational studies indicate that treatment with teriparatide for more than 6 months significantly reduces risk of clinical fractures. Interestingly, the efficacy of the medication on fracture risk was superior in diabetics compared with nondiabetics (Langdahl et al., 2018).

General conclusions

- Both T1DM and T2DM are associated with higher incidence of fractures, although the risk is far greater for T1DM. The increased fracture incidence in T1DM is more pronounced than could be explained by the relatively small BMD deficit.
- T2DM is characterized by higher risk of nonvertebral fracture and normal or higher areal and volumetric BMD. The role of cortical porosity, which appears to be increased in T2DM patients, is still controversial. Insulin deficiency may explain, at least in part, bone fragility in T1DM and in the last stages of T2DM. Insulin stimulates growth and bone formation in vitro and in vivo through its receptors on the bone cells. However, absence of IR expression in bone does not cause diabetic bone disease, indicating that the effect of insulin deficiency on bone is largely if not completely indirect.
- Fracture risk increases with diabetes duration and is higher among individuals with diabetes complications and limited physical capabilities; the higher incidence of falls in diabetic individuals is probably an important mediator.
- Glucotoxicity, increased oxidative stress, and inflammation are common findings in diabetics, but their impact on bone fragility is not known.
- Bone strength is clearly impaired in T1DM and T2DM. Diabetes-induced changes in the collagenous matrix are related to its biomechanical properties. These include reduced enzymatic collagen cross-links but increased glycosylated collagen (AGEs).
- Data from histology and bone markers indicate that diabetes is accompanied by low bone remodeling (both formation and resorption), a condition that may affect bone quality. Strong evidence indicates elevated sclerostin levels in T2DM, indicating a possible dysfunction of the Wnt osteoanabolic pathway in T2DM.

- Antidiabetic medications should be used wisely in diabetic patients, avoiding those that may affect bone strength (TZDs) or increase the risk of hypoglycemic events (sulfonylureas). Target HbA1c should not be too stringent, according to the most recent guidelines.
- Antiosteoporotic medications have been proved equally effective and safe in diabetics.

Key points

- Patients with type 1 diabetes mellitus or type 2 diabetes mellitus (T2DM) have an increased risk of fractures; bone mineral density underestimates this risk in individuals with T2DM, making risk assessment challenging.
- Patients with diabetes mellitus with long-term disease, poor glycemic control, β-cell failure and who receive insulin treatment are at the highest risk of fractures.
- Low bone turnover, accumulation of advanced glycation end products, micro- and macroarchitecture alterations, and tissue material damage lead to abnormal biomechanical properties and impair bone strength.
- Other determinants of bone fragility include inflammation, oxidative stress, adipokine alterations, Wnt dysregulation, and increased marrow fat.
- Complications of diabetes mellitus, such as neuropathy, poor balance, sarcopenia, vision impairment, and frequent hypoglycemic events, increase the risk of falls and risk of fracture; preventive measures are advised, especially in patients taking insulin.
- Use of thiazolidinediones, or some SGLT-2 inhibitors might contribute to increased fracture risk; antidiabetic medications with good bone safety profiles such as metformin, glucagon-like peptide 1 analogs, or dipeptidyl peptidase-4 inhibitors are preferred.

Acknowledgments

Portions of the text were reproduced by permission of the Journal of Endocrinology Ltd.

References

Abdalrahaman, N., McComb, C., Foster, J.E., Lindsay, R.S., Drummond, R., McKay, G.A., Perry, C.G., Ahmed, S.F., 2017. The relationship between adiposity, bone density and microarchitecture is maintained in young women irrespective of diabetes status. Clin. Endocrinol. 87, 327.

Abdalrahaman, N., McComb, C., Foster, J.E., McLean, J., Lindsay, R.S., McClure, J., McMillan, M., Drummond, R., Gordon, D., McKay, G.A., Shaikh, M.G., Perry, C.G., Ahmed, S.F., 2015. Deficits in trabecular bone microarchitecture in young women with type 1 diabetes mellitus. J. Bone Miner. Res. 30, 1386.

Acevedo, C., Sylvia, M., Schaible, E., Graham, J.L., Stanhope, K.L., Metz, L.N., Gludovatz, B., Schwartz, A.V., Ritchie, R.O., Alliston, T.N., Havel, P.J., Fields, A.J., 2018. Contributions of material properties and structure to increased bone fragility for a given bone mass in the UCD-T2DM rat model of type 2 diabetes. J. Bone Miner. Res. 33, 1066.

Aguiari, P., Leo, S., Zavan, B., Vindigni, V., Rimessi, A., Bianchi, K., Franzin, C., Cortivo, R., Rossato, M., Vettor, R., Abatangelo, G., Pozzan, T., Pinton, P., Rizzuto, R., 2008. High glucose induces adipogenic differentiation of muscle-derived stem cells. Proc. Natl. Acad. Sci. U. S. A. 105, 1226.

Ahmad, T., Ohlsson, C., Saaf, M., Ostenson, C.G., Kreicbergs, A., 2003. Skeletal changes in type-2 diabetic Goto-Kakizaki rats. J. Endocrinol. 178, 111.

Ahn, S.H., Kim, H., Kim, B.J., Lee, S.H., Koh, J.M., 2016. Insulin resistance and composite indices of femoral neck strength in Asians: the fourth Korea national health and nutrition examination survey (KNHANES IV). Clin. Endocrinol. 84 (2), 185–193.

Alarid, E.T., Schlechter, N.L., Russell, S.M., Nicoll, C.S., 1992. Evidence suggesting that insulin-like growth factor-I is necessary for the trophic effects of insulin on cartilage growth in vivo. Endocrinology 130, 2305.

Alhuzaim, O.N., Lewis, E.J.H., Lovblom, L.E., Cardinez, M., Scarr, D., NBoulet, G., Weisman, A., Lovshin, J.A., Lytvyn, Y., Keenan, H.A., Brent, M.H., Paul, N., Bril, V., Cherney, D.Z.I., Perkins, B.A., 2019. Bone mineral density in patients with longstanding type 1 diabetes: results from the Canadian study of longevity in type 1 diabetes. J. Diabet. Complicat. In press.

Alikhani, M., Alikhani, Z., Boyd, C., MacLellan, C.M., Raptis, M., Liu, R., Pischon, N., Trackman, P.C., Gerstenfeld, L., Graves, D.T., 2007. Advanced glycation end products stimulate osteoblast apoptosis via the MAP kinase and cytosolic apoptotic pathways. Bone 40, 345.

Ambrosi, T.H., Scialdone, A., Graja, A., Gohlke, S., Jank, A.-M., Bocian, C., Woelk, L., Fan, H., Logan, D.W., Schürmann, A., Saraiva, L.R., Schulz, T.J., 2017. Adipocyte accumulation in the bone marrow during obesity and aging impairs stem cell-based hematopoietic and bone regeneration. Cell Stem Cell 20, 771.

American Diabetes Association, 2019. Standards of medical care in diabetes – 2019. Diabetes Care 42 (S1). Available at: http://care.diabetesjournals.org/content/42/Supplement_1.

Armas, L.A.G., Akhter, M.P., Drincic, A., Recker, R.R., 2012. Trabecular bone histomorphometry in humans with type 1 diabetes mellitus. Bone 50, 91.

Bain, S., Ramamurthy, N.S., Impeduglia, T., Scolman, S., Golub, L.M., Rubin, C., 1997. Tetracycline prevents cancellous bone loss and maintains near-normal rates of bone formation in streptozotocin diabetic rats. Bone 21, 147.

Bauer, D.C., Browner, W.S., Cauley, J.A., Orwoll, E.S., Scott, J.C., Black, D.M., Tao, J.L., Cummings, S.R., 1993. Factors associated with appendicular bone mass in older women. The Study of Osteoporotic Fractures Research Group. Ann. Intern. Med. 118, 657.

Bechtold, S., Dirlenbach, I., Raile, K., Noelle, V., Bonfig, W., Schwarz, H.P., 2006. Early manifestation of type 1 diabetes in children is a risk factor for changed bone geometry: data using peripheral quantitative computed tomography. Pediatrics 118, e627.

Bechtold, S., Putzker, S., Bonfig, W., Fuchs, O., Dirlenbach, I., Schwarz, H.P., 2007. Bone size normalizes with age in children and adolescents with type 1 diabetes. Diabetes Care 30, 2046.

Bereket, A., Lang, C.H., Blethen, S.L., Gelato, M.C., Fan, J., Frost, R.A., Wilson, T.A., 1995. Effect of insulin on the insulin-like growth factor system in children with new-onset insulin-dependent diabetes mellitus. J. Clin. Endocrinol. Metab. 80, 1312.

Beulens, J.W., van der, A.D., Grobbee, D.E., Sluijs, I., Spijkerman, A.M., van der Schouw, Y.T., 2010. Dietary phylloquinone and menaquinones intakes and risk of type 2 diabetes. Diabetes Care 33, 1699.

Beyerlein, A., Thiering, E., Pflueger, M., Bidlingmaier, M., Stock, J., Knopff, A., Winkler, C., Heinrich, J., Ziegler, A.G., 2014. Early infant growth is associated with the risk of islet autoimmunity in genetically susceptible children. Pediatr. Diabetes 15, 534.

Bizzarri, C., Pitocco, D., Napoli, N., Di Stasio, E., Maggi, D., Manfrini, S., Suraci, C., Cavallo, M.G., Cappa, M., Ghirlanda, G., Pozzilli, P., Group, I., 2010. No protective effect of calcitriol on beta-cell function in recent-onset type 1 diabetes: the IMDIAB XIII trial. Diabetes Care 33, 1962.

Bonds, D.E., Larson, J.C., Schwartz, A.V., Strotmeyer, E.S., Robbins, J., Rodriguez, B.L., Johnson, K.C., Margolis, K.L., 2006. Risk of fracture in women with type 2 diabetes: the women's health initiative observational study. J. Clin. Endocrinol. Metab. 91, 3404.

Bonfig, W., Kapellen, T., Dost, A., Fritsch, M., Rohrer, T., Wolf, J., Holl, R.W., Diabetes Patienten Verlaufsdokumentationssystem Initiative of the German Working Group for Pediatric, D., the German Bundesministerium fur Bildung und Forschung Competence Net for Diabetes, M, 2012. Growth in children and adolescents with type 1 diabetes. J. Pediatr. 160, 900.

Bortolin, R.H., Freire Neto, F.P., Arcaro, C.A., Bezerra, J.F., da Silva, F.S., Ururahy, M.A., Souza, K.S., Lima, V.M., Luchessi, A.D., Lima, F.P., Lia Fook, M.V., da Silva, B.J., Almeida, M.D., Abreu, B.J., de Rezende, L.A., de Rezende, A.A., 2017. Anabolic effect of insulin therapy on the bone: osteoprotegerin and osteocalcin up-regulation in streptozotocin-induced diabetic rats. Basic Clin. Pharmacol. Toxicol. 120, 227.

Botolin, S., Faugere, M.-C., Malluche, H., Orth, M., Meyer, R., McCabe, L.R., 2005. Increased bone adiposity and peroxisomal proliferator-activated receptor-gamma2 expression in type I diabetic mice. Endocrinology 146, 3622.

Botolin, S., McCabe, L.R., 2006. Inhibition of PPARgamma prevents type I diabetic bone marrow adiposity but not bone loss. J. Cell. Physiol. 209, 967.

Botolin, S., McCabe, L.R., 2007. Bone loss and increased bone adiposity in spontaneous and pharmacologically induced diabetic mice. Endocrinology 148, 198.

Bouillon, R., Bex, M., Van Herck, E., Laureys, J., Dooms, L., Lesaffre, E., Ravussin, E., 1995. Influence of age, sex, and insulin on osteoblast function: osteoblast dysfunction in diabetes mellitus. J. Clin. Endocrinol. Metab. 80, 1194.

Brown, M.L., Yukata, K., Farnsworth, C.W., Chen, D.G., Awad, H., Hilton, M.J., O'Keefe, R.J., Xing, L., Mooney, R.A., Zuscik, M.J., 2014. Delayed fracture healing and increased callus adiposity in a C57BL/6J murine model of obesity-associated type 2 diabetes mellitus. PLoS One 9, e99656.

Burghardt, A.J., Issever, A.S., Schwartz, A.V., Davis, K.A., Masharani, U., Majumdar, S., Link, T.M., 2010. High-resolution peripheral quantitative computed tomographic imaging of cortical and trabecular bone microarchitecture in patients with type 2 diabetes mellitus. J. Clin. Endocrinol. Metab. 95, 5045.

Camargo, W.A., de Vries, R., van Luijk, J., Hoekstra, J.W., Bronkhorst, E.M., Jansen, J.A., van den Beucken, J., 2017. Diabetes mellitus and bone regeneration: a systematic review and meta-analysis of animal studies. Tissue Eng. B Rev. 23, 471.

Campbell, P.G., Novak, J.F., 1991. Insulin-like growth factor binding protein (IGFBP) inhibits IGF action on human osteosarcoma cells. J. Cell. Physiol. 149, 293.

Campos Pastor, M.M., Lopez-Ibarra, P.J., Escobar-Jimenez, F., Serrano Pardo, M.D., Garcia-Cervigon, A.G., 2000. Intensive insulin therapy and bone mineral density in type 1 diabetes mellitus: a prospective study. Osteoporos. Int. 11, 455.

Carvalho, A.L., Massaro, B., Silva, L., Salmon, C.E.G., Fukada, S.Y., Nogueira-Barbosa, M.H., Elias Jr., J., Freitas, M.C.F., Couri, C.E.B., Oliveira, M.C., Simoes, B.P., Rosen, C.J., de Paula, F.J.A., 2018. Emerging aspects of the body composition, bone marrow adipose tissue and skeletal phenotypes in type 1 diabetes mellitus. J. Clin. Densitom. https://doi.org/10.1016/j.jocd.2018.06.007 (In press).

Cavanagh, P.R., Young, M.J., Adams, J.E., Vickers, K.L., Boulton, A.J., 1994. Radiographic abnormalities in the feet of patients with diabetic neuropathy. Diabetes Care 17, 201.

Chiodini, I., Adda, G., Scillitani, A., Coletti, F., Morelli, V., Di Lembo, S., Epaminonda, P., Masserini, B., Beck-Peccoz, P., Orsi, E., Ambrosi, B., Arosio, M., 2007. Cortisol secretion in patients with type 2 diabetes: relationship with chronic complications. Diabetes Care 30, 83.

Chiodini, I., Di Lembo, S., Morelli, V., Epaminonda, P., Coletti, F., Masserini, B., Scillitani, A., Arosio, M., Adda, G., 2006. Hypothalamic-pituitary-adrenal activity in type 2 diabetes mellitus: role of autonomic imbalance. Metabolism 55, 1135.

Choi, H.J., Yu, J., Choi, H., An, J.H., Kim, S.W., Park, K.S., Jang, H.C., Kim, S.Y., Shin, C.S., 2011. Vitamin K2 supplementation improves insulin sensitivity via osteocalcin metabolism: a placebo-controlled trial. Diabetes Care 34, e147.

Chung, Y.S., Lee, M.D., Lee, S.K., Kim, H.M., Fitzpatrick, L.A., 2000. HMG-CoA reductase inhibitors increase BMD in type 2 diabetes mellitus patients. J. Clin. Endocrinol. Metab. 85, 1137.

Clark, M., Kroger, C.J., Tisch, R.M., 2017. Type 1 diabetes: a chronic anti-self-inflammatory response. Front. Immunol. 8, 1898.

Clausen, P., Feldt-Rasmussen, B., Jacobsen, P., Rossing, K., Parving, H.H., Nielsen, P.K., Feldt-Rasmussen, U., Olgaard, K., 1997. Microalbuminuria as an early indicator of osteopenia in male insulin-dependent diabetic patients. Diabet. Med. 14, 1038.

Coe, L.M., Irwin, R., Lippner, D., McCabe, L.R., 2011. The bone marrow microenvironment contributes to type I diabetes induced osteoblast death. J. Cell. Physiol. 226, 477.

Conte, C., Epstein, S., Napoli, N., 2018. Insulin resistance and bone: a biological partnership. Acta Diabetol. 55, 305.

Cornish, J., Callon, K.E., Reid, I.R., 1996. Insulin increases histomorphometric indices of bone formation in vivo. Calcif. Tissue Int. 59, 492.

Danne, T., Kordonouri, O., Enders, I., Weber, B., 1997. Factors influencing height and weight development in children with diabetes. Results of the Berlin Retinopathy Study. Diabetes Care 20, 281.

Davidson, M.B., Duran, P., Lee, M.L., Friedman, T.C., 2013. High-dose vitamin D supplementation in people with prediabetes and hypovitaminosis D. Diabetes Care 36, 260.

Devlin, M.J., Van Vliet, M., Motyl, K., Karim, L., Brooks, D.J., Louis, L., Conlon, C., Rosen, C.J., Bouxsein, M.L., 2014. Early-onset type 2 diabetes impairs skeletal acquisition in the male TALLYHO/JngJ mouse. Endocrinology 155, 3806.

Dienelt, A., zur Nieden, N.I., 2011. Hyperglycemia impairs skeletogenesis from embryonic stem cells by affecting osteoblast and osteoclast differentiation. Stem Cell. Dev. 20, 465.

Diez-Perez, A., Bouxsein, M.L., Eriksen, E.F., Khosla, S., Nyman, J.S., Papapoulos, S., Tang, S.Y., 2016. Technical note: recommendations for a standard procedure to assess cortical bone at the tissue-level in vivo using impact microindentation. Bone Rep. 5, 181.

Diez-Perez, A., Guerri, R., Nogues, X., Caceres, E., Pena, M.J., Mellibovsky, L., Randall, C., Bridges, D., Weaver, J.C., Proctor, A., Brimer, D., Koester, K.J., Ritchie, R.O., Hansma, P.K., 2010. Microindentation for in vivo measurement of bone tissue mechanical properties in humans. J. Bone Miner. Res. 25, 1877.

Dixit, P.K., Ekstrom, R.A., 1980. Decreased breaking strength of diabetic rat bone and its improvement by insulin treatment. Calcif. Tissue Int. 32, 195.

Driessen, J.H.M., de Vries, F., van Onzenoort, H., Harvey, N.C., Neef, C., van den Bergh, J.P.W., Vestergaard, P., Henry, R.M.A., 2017. The use of incretins and fractures – a meta-analysis on population-based real life data. Br. J. Clin. Pharmacol. 83, 923.

Ducy, P., Amling, M., Takeda, S., Priemel, M., Schilling, A.F., Beil, F.T., Shen, J., Vinson, C., Rueger, J.M., Karsenty, G., 2000. Leptin inhibits bone formation through a hypothalamic relay: a central control of bone mass. Cell 100, 197.

Ebeling, P.R., Adler, R.A., Jones, G., Liberman, U.A., Mazziotti, G., Minisola, S., Munns, C.F., Napoli, N., Pittas, A.G., Giustina, A., Bilezikian, J.P., Rizzoli, R., 2018. Management of endocrine disease: Therapeutics of vitamin D. Eur. J. Endocrinol. 179, R239.

Ehnert, S., Freude, T., Ihle, C., Mayer, L., Braun, B., Graeser, J., Flesch, I., Stockle, U., Nussler, A.K., Pscherer, S., 2015. Factors circulating in the blood of type 2 diabetes mellitus patients affect osteoblast maturation – description of a novel in vitro model. Exp. Cell Res. 332, 247.

Einhorn, T.A., Boskey, A.L., Gundberg, C.M., Vigorita, V.J., Devlin, V.J., Beyer, M.M., 1988. The mineral and mechanical properties of bone in chronic experimental diabetes. J. Orthop. Res. 6, 317.

Elamin, A., Hussein, O., Tuvemo, T., 2006. Growth, puberty, and final height in children with Type 1 diabetes. J. Diabet. Complicat. 20, 252.

Eller-Vainicher, C., Scillitani, A., Chiodini, I., 2017. Is the hypothalamic-pituitary-adrenal axis disrupted in type 2 diabetes mellitus and is this relevant for bone health? Endocrine 58, 201.

Eller-Vainicher, C., Zhukouskaya, V.V., Tolkachev, Y.V., Koritko, S.S., Cairoli, E., Grossi, E., Beck-Peccoz, P., Chiodini, I., Shepelkevich, A.P., 2011. Low bone mineral density and its predictors in type 1 diabetic patients evaluated by the classic statistics and artificial neural network analysis. Diabetes Care 34, 2186.

Epstein, S., Takizawa, M., Stein, B., Katz, I.A., Joffe II, Romero, D.F., Liang, X.G., Li, M., Ke, H.Z., Jee, W.S., et al., 1994. Effect of cyclosporin A on bone mineral metabolism in experimental diabetes mellitus in the rat. J. Bone Miner. Res. 9, 557.

Faienza, M.F., Ventura, A., Delvecchio, M., Fusillo, A., Piacente, L., Aceto, G., Colaianni, G., Colucci, S., Cavallo, L., Grano, M., Brunetti, G., 2017. High sclerostin and dickkopf-1 (DKK-1) serum levels in children and adolescents with type 1 diabetes mellitus. J. Clin. Endocrinol. Metab. 102, 1174.

Fajardo, R.J., Karim, L., Calley, V.I., Bouxsein, M.L., 2014. A review of rodent models of type 2 diabetic skeletal fragility. J. Bone Miner. Res. 29, 1025.

Fan, Y., Wei, F., Lang, Y., Liu, Y., 2016. Diabetes mellitus and risk of hip fractures: a meta-analysis. Osteoporos. Int. 27, 219.

Farr, J.N., Drake, M.T., Amin, S., Melton, L.J., McCready, L.K., Khosla, S., 2014. In vivo assessment of bone quality in postmenopausal women with type 2 diabetes. J. Bone Miner. Res. 29, 787.

Ferron, M., Hinoi, E., Karsenty, G., Ducy, P., 2008. Osteocalcin differentially regulates beta cell and adipocyte gene expression and affects the development of metabolic diseases in wild-type mice. Proc. Natl. Acad. Sci. U.S.A. 105, 5266.

Ferron, M., Wei, J., Yoshizawa, T., Del Fattore, A., DePinho, R.A., Teti, A., Ducy, P., Karsenty, G., 2010. Insulin signaling in osteoblasts integrates bone remodeling and energy metabolism. Cell 142, 296.

Foley Jr., T.P., Nissley, S.P., Stevens, R.L., King, G.L., Hascall, V.C., Humbel, R.E., Short, P.A., Rechler, M.M., 1982. Demonstration of receptors for insulin and insulin-like growth factors on Swarm rat chondrosarcoma chondrocytes. Evidence that insulin stimulates proteoglycan synthesis through the insulin receptor. J. Biol. Chem. 257, 663.

Follak, N., Kloting, I., Wolf, E., Merk, H., 2004. Histomorphometric evaluation of the influence of the diabetic metabolic state on bone defect healing depending on the defect size in spontaneously diabetic BB/OK rats. Bone 35, 144.

Franceschi, R., Longhi, S., Cauvin, V., Fassio, A., Gallo, G., Lupi, F., Reinstadler, P., Fanolla, A., Gatti, D., Radetti, G., 2018. Bone geometry, quality, and bone markers in children with type 1 diabetes mellitus. Calcif. Tissue Int. 102, 657.

Frystyk, J., Hojlund, K., Rasmussen, K.N., Jorgensen, S.P., Wildner-Christensen, M., Orskov, H., 2002. Development and clinical evaluation of a novel immunoassay for the binary complex of IGF-I and IGF-binding protein-1 in human serum. J. Clin. Endocrinol. Metab. 87, 260.

Fulzele, K., DiGirolamo, D.J., Liu, Z., Xu, J., Messina, J.L., Clemens, T.L., 2007. Disruption of the insulin-like growth factor type 1 receptor in osteoblasts enhances insulin signaling and action. J. Biol. Chem. 282, 25649.

Fulzele, K., Riddle, R.C., DiGirolamo, D.J., Cao, X., Wan, C., Chen, D., Faugere, M.C., Aja, S., Hussain, M.A., Bruning, J.C., Clemens, T.L., 2010. Insulin receptor signaling in osteoblasts regulates postnatal bone acquisition and body composition. Cell 142, 309.

Funk, J.R., Hale, J.E., Carmines, D., Gooch, H.L., Hurwitz, S.R., 2000. Biomechanical evaluation of early fracture healing in normal and diabetic rats. J. Orthop. Res. 18, 126.

Furst, J.R., Bandeira, L.C., Fan, W.W., Agarwal, S., Nishiyama, K.K., McMahon, D.J., Dworakowski, E., Jiang, H., Silverberg, S.J., Rubin, M.R., 2016. Advanced glycation endproducts and bone material strength in type 2 diabetes. J. Clin. Endocrinol. Metab. 101, 2502.

Gallant, M.A., Brown, D.M., Organ, J.M., Allen, M.R., Burr, D.B., 2013. Reference-point indentation correlates with bone toughness assessed using whole-bone traditional mechanical testing. Bone 53, 301.

Gandhi, A., Beam, H.A., O'Connor, J.P., Parsons, J.R., Lin, S.S., 2005. The effects of local insulin delivery on diabetic fracture healing. Bone 37, 482.

Gandhi, A., Doumas, C., O'Connor, J.P., Parsons, J.R., Lin, S.S., 2006. The effects of local platelet rich plasma delivery on diabetic fracture healing. Bone 38, 540.

Gaudio, A., Privitera, F., Battaglia, K., Torrisi, V., Sidoti, M.H., Pulvirenti, I., Canzonieri, E., Tringali, G., Fiore, C.E., 2012. Sclerostin levels associated with inhibition of the Wnt/β-catenin signaling and reduced bone turnover in type 2 diabetes mellitus. J. Clin. Endocrinol. Metab. 97, 3744.

Gennari, L., Merlotti, D., Valenti, R., Ceccarelli, E., Ruvio, M., Pietrini, M.G., Capodarca, C., Franci, M.B., Campagna, M.S., Calabrò, A., Cataldo, D., Stolakis, K., Dotta, F., Nuti, R., 2012. Circulating sclerostin levels and bone turnover in type 1 and type 2 diabetes. J. Clin. Endocrinol. Metab. 97, 1737.

Gilbert, L., He, X., Farmer, P., Boden, S., Kozlowski, M., Rubin, J., Nanes, M.S., 2000. Inhibition of osteoblast differentiation by tumor necrosis factor-alpha. Endocrinology 141, 3956.

Gilbert, M.P., Pratley, R.E., 2015. The impact of diabetes and diabetes medications on bone health. Endocr. Rev. 36, 194.

Glajchen, N., Epstein, S., Ismail, F., Thomas, S., Fallon, M., Chakrabarti, S., 1988. Bone mineral metabolism in experimental diabetes mellitus: osteocalcin as a measure of bone remodeling. Endocrinology 123, 290.

Glantschnig, H., Fisher, J.E., Wesolowski, G., Rodan, G.A., Reszka, A.A., 2003. M-CSF, TNFalpha and RANK ligand promote osteoclast survival by signaling through mTOR/S6 kinase. Cell Death Differ. 10, 1165.

Hamann, C., Goettsch, C., Mettelsiefen, J., Henkenjohann, V., Rauner, M., Hempel, U., Bernhardt, R., Fratzl-Zelman, N., Roschger, P., Rammelt, S., Günther, K.-P., Hofbauer, L.C., 2011. Delayed bone regeneration and low bone mass in a rat model of insulin-resistant type 2 diabetes mellitus is due to impaired osteoblast function. Am. J. Physiol. Endocrinol. Metabol. 301, E1220.

Hamann, C., Rauner, M., Höhna, Y., Bernhardt, R., Mettelsiefen, J., Goettsch, C., Günther, K.-P., Stolina, M., Han, C.-Y., Asuncion, F.J., Ominsky, M.S., Hofbauer, L.C., 2013. Sclerostin antibody treatment improves bone mass, bone strength, and bone defect regeneration in rats with type 2 diabetes mellitus. J. Bone Miner. Res. 28, 627.

Hamilton, E.J., Drinkwater, J.J., Chubb, S.A.P., Rakic, V., Kamber, N., Zhu, K., Prince, R.L., Davis, W.A., Davis, T.M.E., 2018. A 10-year prospective study of bone mineral density and bone turnover in males and females with type 1 diabetes. J. Clin. Endocrinol. Metab. 103, 3531.

He, H., Liu, R., Desta, T., Leone, C., Gerstenfeld, L.C., Graves, D.T., 2004. Diabetes causes decreased osteoclastogenesis, reduced bone formation, and enhanced apoptosis of osteoblastic cells in bacteria stimulated bone loss. Endocrinology 145, 447.

Heaney, R.P., 2003. Is the paradigm shifting? Bone 33, 457.

Heilmeier, U., Carpenter, D.R., Patsch, J.M., Harnish, R., Joseph, G.B., Burghardt, A.J., Baum, T., Schwartz, A.V., Lang, T.F., Link, T.M., 2015. Volumetric femoral BMD, bone geometry, and serum sclerostin levels differ between type 2 diabetic postmenopausal women with and without fragility fractures. Osteoporos. Int. 26, 1283.

Heilmeier, U., Cheng, K., Pasco, C., Parrish, R., Nirody, J., Patsch, J.M., Zhang, C.A., Joseph, G.B., Burghardt, A.J., Schwartz, A.V., Link, T.M., Kazakia, G., 2016. Cortical bone laminar analysis reveals increased midcortical and periosteal porosity in type 2 diabetic postmenopausal women with history of fragility fractures compared to fracture-free diabetics. Osteoporos. Int. 27, 2791.

Heinze, E., Vetter, U., Voigt, K.H., 1989. Insulin stimulates skeletal growth in vivo and in vitro–comparison with growth hormone in rats. Diabetologia 32, 198.

Hie, M., Iitsuka, N., Otsuka, T., Tsukamoto, I., 2011. Insulin-dependent diabetes mellitus decreases osteoblastogenesis associated with the inhibition of Wnt signaling through increased expression of Sost and Dkk1 and inhibition of Akt activation. Int. J. Mol. Med. 28, 455.

Hock, J.M., Centrella, M., Canalis, E., 1988. Insulin-like growth factor I has independent effects on bone matrix formation and cell replication. Endocrinology 122, 254.

Hofbauer, L.C., Brueck, C.C., Singh, S.K., Dobnig, H., 2007. Osteoporosis in patients with diabetes mellitus. J. Bone Miner. Res. 22, 1317.

Holl, R.W., Grabert, M., Heinze, E., Sorgo, W., Debatin, K.M., 1998. Age at onset and long-term metabolic control affect height in type-1 diabetes mellitus. Eur. J. Pediatr. 157, 972.

Holloway-Kew, K.L., De Abreu, L.L.F., Kotowicz, M.A., Sajjad, M.A., Pasco, J.A., 2019. Bone turnover markers in men and women with impaired fasting glucose and diabetes. Calcif. Tissue Int. https://doi.org/10.1007/s00223-019-00527-y (In press).

Horcajada-Molteni, M.N., Chanteranne, B., Lebecque, P., Davicco, M.J., Coxam, V., Young, A., Barlet, J.P., 2001. Amylin and bone metabolism in streptozotocin-induced diabetic rats. J. Bone Miner. Res. 16, 958.

Hough, F.S., Pierroz, D.D., Cooper, C., Ferrari, S.L., Bone, I.C., Diabetes Working, G., 2016. Mechanisms in endocrinology: mechanisms and evaluation of bone fragility in type 1 diabetes mellitus. Eur. J. Endocrinol. 174, R127.

Hua, Y., Bi, R., Zhang, Y., Xu, L., Guo, J., Li, Y., 2018. Different bone sites-specific response to diabetes rat models: bone density, histology and microarchitecture. PLoS One 13, e0205503.

Hunt, H., Torres, A., Palomino, P., Marty, E., Saiyed, R., Cohn, M., Jo, J., Warner, S., Sroga, G., King, K., Lane, J., Vashishth, D., Hernandez, C., Donnelly, E., 2019. Altered tissue composition, microarchitecture, and mechanical performance in cancellous bone from men with type 2 diabetes mellitus. J. Bone Miner. Res. https://doi.org/10.1002/jbmr.3711 (In press).

Hygum, K., Starup-Linde, J., Harslof, T., Vestergaard, P., Langdahl, B.L., 2017. Mechanisms in endocrinology: diabetes mellitus, a state of low bone turnover – a systematic review and meta-analysis. Eur. J. Endocrinol. 176, R137.

Idelevich, A., Baron, R., 2018. Brain to bone: what is the contribution of the brain to skeletal homeostasis? Bone 115, 31.

Ionova-Martin, S.S., Wade, J.M., Tang, S., Shahnazari, M., Ager 3rd, J.W., Lane, N.E., Yao, W., Alliston, T., Vaisse, C., Ritchie, R.O., 2011. Changes in cortical bone response to high-fat diet from adolescence to adulthood in mice. Osteoporos. Int. 22, 2283.

Irwin, R., Lin, H.V., Motyl, K.J., McCabe, L.R., 2006. Normal bone density obtained in the absence of insulin receptor expression in bone. Endocrinology 147, 5760.

Ishida, H., Seino, Y., Takeshita, N., Kurose, T., Tsuji, K., Okamoto, Y., Someya, Y., Hara, K., Akiyama, Y., Imura, H., et al., 1992. Effect of pancreas transplantation on decreased levels of circulating bone gamma-carboxyglutamic acid-containing protein and osteopenia in rats with streptozotocin-induced diabetes. Acta Endocrinol. 127, 81.

Ishikawa, K., Fukui, T., Nagai, T., Kuroda, T., Hara, N., Yamamoto, T., Inagaki, K., Hirano, T., 2015. Type 1 diabetes patients have lower strength in femoral bone determined by quantitative computed tomography: a cross-sectional study. J Diabetes Investig 6, 726.

Ivers, R.Q., Cumming, R.G., Mitchell, P., Peduto, A.J., Blue Mountains Eye, S., 2001. Diabetes and risk of fracture: the Blue Mountains eye study. Diabetes Care 24, 1198.

Janghorbani, M., Feskanich, D., Willett, W.C., Hu, F., 2006. Prospective study of diabetes and risk of hip fracture: the Nurses' Health Study. Diabetes Care 29, 1573.

Janghorbani, M., Van Dam, R.M., Willett, W.C., Hu, F.B., 2007. Systematic review of type 1 and type 2 diabetes mellitus and risk of fracture. Am. J. Epidemiol. 166, 495.

Jiajue, R., Jiang, Y., Wang, O., Li, M., Xing, X., Cui, L., Yin, J., Xu, L., Xia, W., 2014. Suppressed bone turnover was associated with increased osteoporotic fracture risks in non-obese postmenopausal Chinese women with type 2 diabetes mellitus. Osteoporos. Int. 25, 1999.

Jiao, H., Xiao, E., Graves, D.T., 2015. Diabetes and its effect on bone and fracture healing. Curr. Osteoporos. Rep. 13, 327.

Johnell, O., Kanis, J.A., Oden, A., Johansson, H., De Laet, C., Delmas, P., Eisman, J.A., Fujiwara, S., Kroger, H., Mellstrom, D., Meunier, P.J., Melton 3rd, L.J., O'Neill, T., Pols, H., Reeve, J., Silman, A., Tenenhouse, A., 2005. Predictive value of BMD for hip and other fractures. J. Bone Miner. Res. 20, 1185.

Johnson, K.C., Bray, G.A., Cheskin, L.J., Clark, J.M., Egan, C.M., Foreyt, J.P., Garcia, K.R., Glasser, S., Greenway, F.L., Gregg, E.W., Hazuda, H.P., Hergenroeder, A., Hill, J.O., Horton, E.S., Jakicic, J.M., Jeffery, R.W., Kahn, S.E., Knowler, W.C., Lewis, C.E., Miller, M., Montez, M.G., Nathan, D.M., Patricio, J.L., Peters, A.L., Pi-Sunyer, X., Pownall, H.J., Reboussin, D., Redmon, J.B., Steinberg, H., Wadden, T.A., Wagenknecht, L.E., Wing, R.R., Womack, C.R., Yanovski, S.Z., Zhang, P., Schwartz, A.V., Look, A.S.G., 2017. The effect of intentional weight loss on fracture risk in persons with diabetes: results from the look AHEAD randomized clinical trial. J. Bone Miner. Res. 32, 2278.

Johnston, S.S., Conner, C., Aagren, M., Ruiz, K., Bouchard, J., 2012. Association between hypoglycaemic events and fall-related fractures in Medicare-covered patients with type 2 diabetes. Diabetes Obes. Metab. 14, 634.

Kahn, S.E., Haffner, S.M., Heise, M.A., Herman, W.H., Holman, R.R., Jones, N.P., Kravitz, B.G., Lachin, J.M., O'Neill, M.C., Zinman, B., Viberti, G., Group, A.S., 2006. Glycemic durability of rosiglitazone, metformin, or glyburide monotherapy. N. Engl. J. Med. 355, 2427.

Kanazawa, I., Notsu, M., Miyake, H., Tanaka, K., Sugimoto, T., 2018. Assessment using serum insulin-like growth factor-I and bone mineral density is useful for detecting prevalent vertebral fractures in patients with type 2 diabetes mellitus. Osteoporos. Int. 29, 2527.

Kanazawa, I., Yamaguchi, T., Sugimoto, T., 2011a. Serum insulin-like growth factor-I is a marker for assessing the severity of vertebral fractures in postmenopausal women with type 2 diabetes mellitus. Osteoporos. Int. 22, 1191.

Kanazawa, I., Yamaguchi, T., Yamamoto, M., Sugimoto, T., 2010. Relationship between treatments with insulin and oral hypoglycemic agents versus the presence of vertebral fractures in type 2 diabetes mellitus. J. Bone Miner. Metab. 28, 554.

Kanazawa, I., Yamaguchi, T., Yamamoto, M., Yamauchi, M., Kurioka, S., Yano, S., Sugimoto, T., 2009. Serum osteocalcin level is associated with glucose metabolism and atherosclerosis parameters in type 2 diabetes mellitus. J. Clin. Endocrinol. Metab. 94, 45.

Kanazawa, I., Yamaguchi, T., Yamauchi, M., Yamamoto, M., Kurioka, S., Yano, S., Sugimoto, T., 2011b. Serum undercarboxylated osteocalcin was inversely associated with plasma glucose level and fat mass in type 2 diabetes mellitus. Osteoporos. Int. 22, 187.

Kaneko, H., Sasaki, T., Ramamurthy, N.S., Golub, L.M., 1990. Tetracycline administration normalizes the structure and acid phosphatase activity of osteoclasts in streptozotocin-induced diabetic rats. Anat. Rec. 227, 427.

Karim, L., Moulton, J., Van Vliet, M., Velie, K., Robbins, A., Malekipour, F., Abdeen, A., Ayres, D., Bouxsein, M.L., 2018. Bone microarchitecture, biomechanical properties, and advanced glycation end-products in the proximal femur of adults with type 2 diabetes. Bone 114, 32.

Kasahara, T., Imai, S., Kojima, H., Katagi, M., Kimura, H., Chan, L., Matsusue, Y., 2010. Malfunction of bone marrow-derived osteoclasts and the delay of bone fracture healing in diabetic mice. Bone 47, 617.

Katayama, Y., Akatsu, T., Yamamoto, M., Kugai, N., Nagata, N., 1996. Role of nonenzymatic glycosylation of type I collagen in diabetic osteopenia. J. Bone Miner. Res. 11, 931.

Kawaguchi, H., Kurokawa, T., Hanada, K., Hiyama, Y., Tamura, M., Ogata, E., Matsumoto, T., 1994. Stimulation of fracture repair by recombinant human basic fibroblast growth factor in normal and streptozotocin-diabetic rats. Endocrinology 135, 774.

Kayal, R.A., Tsatsas, D., Bauer, M.A., Allen, B., Al-Sebaei, M.O., Kakar, S., Leone, C.W., Morgan, E.F., Gerstenfeld, L.C., Einhorn, T.A., Graves, D.T., 2007. Diminished bone formation during diabetic fracture healing is related to the premature resorption of cartilage associated with increased osteoclast activity. J. Bone Miner. Res. 22, 560.

Keegan, T.H.M., Schwartz, A.V., Bauer, D.C., Sellmeyer, D.E., Kelsey, J.L., fracture intervention, t., 2004. Effect of alendronate on bone mineral density and biochemical markers of bone turnover in type 2 diabetic women: the fracture intervention trial. Diabetes Care 27, 1547.

Kelley, K.M., Russell, S.M., Matteucci, M.L., Nicoll, C.S., 1993. An insulin-like growth factor I-resistant state in cartilage of diabetic rats is ameliorated by hypophysectomy. Possible role of metabolism. Diabetes 42, 463.

Kim, J.Y., Lee, S.K., Jo, K.J., Song, D.Y., Lim, D.M., Park, K.Y., Bonewald, L.F., Kim, B.J., 2013. Exendin-4 increases bone mineral density in type 2 diabetic OLETF rats potentially through the down-regulation of SOST/sclerostin in osteocytes. Life Sci. 92, 533.

Kindblom, J.M., Ohlsson, C., Ljunggren, O., Karlsson, M.K., Tivesten, A., Smith, U., Mellstrom, D., 2009. Plasma osteocalcin is inversely related to fat mass and plasma glucose in elderly Swedish men. J. Bone Miner. Res. 24, 785.

Knop, F.K., Vilsboll, T., Hojberg, P.V., Larsen, S., Madsbad, S., Volund, A., Holst, J.J., Krarup, T., 2007. Reduced incretin effect in type 2 diabetes: cause or consequence of the diabetic state? Diabetes 56, 1951.

Kohler, S., Kaspers, S., Salsali, A., Zeller, C., Woerle, H.J., 2018. Analysis of fractures in patients with type 2 diabetes treated with empagliflozin in pooled data from placebo-controlled trials and a head-to-head study versus glimepiride. Diabetes Care 41, 1809.

Komorita, Y., Iwase, M., Fujii, H., Ohkuma, T., Ide, H., Jodai-Kitamura, T., Sumi, A., Yoshinari, M., Nakamura, U., Kang, D., Kitazono, T., 2018. Impact of body weight loss from maximum weight on fragility bone fractures in Japanese patients with type 2 diabetes: the Fukuoka diabetes registry. Diabetes Care 41, 1061.

Krakauer, J.C., McKenna, M.J., Buderer, N.F., Rao, D.S., Whitehouse, F.W., Parfitt, A.M., 1995. Bone loss and bone turnover in diabetes. Diabetes 44, 775.

Kume, S., Kato, S., Yamagishi, S., Inagaki, Y., Ueda, S., Arima, N., Okawa, T., Kojiro, M., Nagata, K., 2005. Advanced glycation end-products attenuate human mesenchymal stem cells and prevent cognate differentiation into adipose tissue, cartilage, and bone. J. Bone Miner. Res. 20, 1647.

Lai, M., Chandrasekera, P.C., Barnard, N.D., 2014. You are what you eat, or are you? The challenges of translating high-fat-fed rodents to human obesity and diabetes. Nutr. Diabetes 4, e135.

Lamb, M.M., Yin, X., Zerbe, G.O., Klingensmith, G.J., Dabelea, D., Fingerlin, T.E., Rewers, M., Norris, J.M., 2009. Height growth velocity, islet autoimmunity and type 1 diabetes development: the Diabetes Autoimmunity Study in the Young. Diabetologia 52, 2064.

Langdahl, B.L., Silverman, S., Fujiwara, S., Saag, K., Napoli, N., Soen, S., Enomoto, H., Melby, T.E., Disch, D.P., Marin, F., Krege, J.H., 2018. Real-world effectiveness of teriparatide on fracture reduction in patients with osteoporosis and comorbidities or risk factors for fractures: integrated analysis of 4 prospective observational studies. Bone 116, 58.

Lee, N.K., Sowa, H., Hinoi, E., Ferron, M., Ahn, J.D., Confavreux, C., Dacquin, R., Mee, P.J., McKee, M.D., Jung, D.Y., Zhang, Z., Kim, J.K., Mauvais-Jarvis, F., Ducy, P., Karsenty, G., 2007. Endocrine regulation of energy metabolism by the skeleton. Cell 130, 456.

Leger, J., Marinovic, D., Alberti, C., Dorgeret, S., Chevenne, D., Marchal, C.L., Tubiana-Rufi, N., Sebag, G., Czernichow, P., 2006. Lower bone mineral content in children with type 1 diabetes mellitus is linked to female sex, low insulin-like growth factor type I levels, and high insulin requirement. J. Clin. Endocrinol. Metab. 91, 3947.

Li, C.I., Liu, C.S., Lin, W.Y., Meng, N.H., Chen, C.C., Yang, S.Y., Chen, H.J., Lin, C.C., Li, T.C., 2015. Glycated hemoglobin level and risk of hip fracture in older people with type 2 diabetes: a competing risk analysis of Taiwan diabetes cohort study. J. Bone Miner. Res. 30, 1338.

Li, M., Pan, L.C., Simmons, H.A., Li, Y., Healy, D.R., Robinson, B.S., Ke, H.Z., Brown, T.A., 2006. Surface-specific effects of a PPARgamma agonist, darglitazone, on bone in mice. Bone 39, 796.

Lin, T.K., Chou, P., Lin, C.H., Hung, Y.J., Jong, G.P., 2018. Long-term effect of statins on the risk of new-onset osteoporosis: a nationwide population-based cohort study. PLoS One 13, e0196713.

Liu, Z., Aronson, J., Wahl, E.C., Liu, L., Perrien, D.S., Kern, P.A., Fowlkes, J.L., Thrailkill, K.M., Bunn, R.C., Cockrell, G.E., Skinner, R.A., Lumpkin Jr., C.K., 2007. A novel rat model for the study of deficits in bone formation in type-2 diabetes. Acta Orthop. 78, 46.

Locatto, M.E., Abranzon, H., Caferra, D., Fernandez, M.C., Alloatti, R., Puche, R.C., 1993. Growth and development of bone mass in untreated alloxan diabetic rats. Effects of collagen glycosylation and parathyroid activity on bone turnover. Bone Miner. 23, 129.

Loder, R.T., 1988. The influence of diabetes mellitus on the healing of closed fractures. Clin. Orthop. Relat. Res. 210.

Loke, Y.K., Singh, S., Furberg, C.D., 2009. Long-term use of thiazolidinediones and fractures in type 2 diabetes: a meta-analysis. Can. Med. Assoc. J. 180, 32.

Lontchi-Yimagou, E., Sobngwi, E., Matsha, T.E., Kengne, A.P., 2013. Diabetes mellitus and inflammation. Curr. Diabetes Rep. 13, 435.

Lopez-Herradon, A., Portal-Nunez, S., Garcia-Martin, A., Lozano, D., Perez-Martinez, F.C., Cena, V., Esbrit, P., 2013. Inhibition of the canonical Wnt pathway by high glucose can be reversed by parathyroid hormone-related protein in osteoblastic cells. J. Cell. Biochem. 114, 1908.

Losada, E., Soldevila, B., Ali, M.S., Martinez-Laguna, D., Nogues, X., Puig-Domingo, M., Diez-Perez, A., Mauricio, D., Prieto-Alhambra, D., 2018. Real-world antidiabetic drug use and fracture risk in 12,277 patients with type 2 diabetes mellitus: a nested case-control study. Osteoporos. Int. 29, 2079.

Lu, N., Sun, H., Yu, J., Wang, X., Liu, D., Zhao, L., Sun, L., Zhao, H., Tao, B., Liu, J., 2015. Glucagon-like peptide-1 receptor agonist Liraglutide has anabolic bone effects in ovariectomized rats without diabetes. PLoS One 10, e0132744.

Lucas, P.D., 1987. Reversible reduction in bone blood flow in streptozotocin-diabetic rats. Experientia 43, 894.

Ma, L., Oei, L., Jiang, L., Estrada, K., Chen, H., Wang, Z., Yu, Q., Zillikens, M.C., Gao, X., Rivadeneira, F., 2012. Association between bone mineral density and type 2 diabetes mellitus: a meta-analysis of observational studies. Eur. J. Epidemiol. 27, 319.

Ma, X., Meng, J., Jia, M., Bi, L., Zhou, Y., Wang, Y., Hu, J., He, G., Luo, X., 2013. Exendin-4, a glucagon-like peptide-1 receptor agonist, prevents osteopenia by promoting bone formation and suppressing bone resorption in aged ovariectomized rats. J. Bone Miner. Res. 28, 1641.

Ma, Y.H., Schwartz, A.V., Sigurdsson, S., Hue, T.F., Lang, T.F., Harris, T.B., Rosen, C.J., Vittinghoff, E., Eiriksdottir, G., Hauksdottir, A.M., Siggeirsdottir, K., Sigurdsson, G., Oskarsdottir, D., Napoli, N., Palermo, L., Gudnason, V., Li, X., 2014. Circulating sclerostin associated with vertebral bone marrow fat in older men but not women. J. Clin. Endocrinol. Metab. 99, E2584.

Mabilleau, G., Mieczkowska, A., Chappard, D., 2014. Use of glucagon-like peptide-1 receptor agonists and bone fractures: a meta-analysis of randomized clinical trials. J. Diabetes 6, 260.

Maddaloni, E., Cavallari, I., Napoli, N., Conte, C., 2018. Vitamin D and diabetes mellitus. Front. Horm. Res. 50, 161.

Maddaloni, E., D'Eon, S., Hastings, S., Tinsley, L.J., Napoli, N., Khamaisi, M., Bouxsein, M.L., Fouda, S.M.R., Keenan, H.A., 2017. Bone health in subjects with type 1 diabetes for more than 50 years. Acta Diabetol. 54, 479.

Madsen, L.R., Espersen, R., Ornstrup, M.J., Jorgensen, N.R., Langdahl, B.L., Richelsen, B., 2019. Bone health in patients with type 2 diabetes treated by roux-en-Y gastric bypass and the role of diabetes remission. Obes. Surg. https://doi.org/10.1007/s11695-019-03753-3 (In press).

Manavalan, J.S., Cremers, S., Dempster, D.W., Zhou, H., Dworakowski, E., Kode, A., Kousteni, S., Rubin, M.R., 2012. Circulating osteogenic precursor cells in type 2 diabetes mellitus. J. Clin. Endocrinol. Metab. 97, 3240.

Manolagas, S.C., 2010. From estrogen-centric to aging and oxidative stress: a revised perspective of the pathogenesis of osteoporosis. Endocr. Rev. 31, 266.

Maor, G., Silbermann, M., von der Mark, K., Heingard, D., Laron, Z., 1993. Insulin enhances the growth of cartilage in organ and tissue cultures of mouse neonatal mandibular condyle. Calcif. Tissue Int. 52, 291.

Maratova, K., Soucek, O., Matyskova, J., Hlavka, Z., Petruzelkova, L., Obermannova, B., Pruhova, S., Kolouskova, S., Sumnik, Z., 2018. Muscle functions and bone strength are impaired in adolescents with type 1 diabetes. Bone 106, 22.

Marin, C., Luyten, F.P., Van der Schueren, B., Kerckhofs, G., Vandamme, K., 2018. The impact of type 2 diabetes on bone fracture healing. Front. Endocrinol. 9, 6.

Martinez-Laguna, D., Nogues, X., Abrahamsen, B., Reyes, C., Carbonell-Abella, C., Diez-Perez, A., Prieto-Alhambra, D., 2017. Excess of all-cause mortality after a fracture in type 2 diabetic patients: a population-based cohort study. Osteoporos. Int. 28, 2573.

Martinez-Laguna, D., Tebe, C., Nogues, X., Kassim Javaid, M., Cooper, C., Moreno, V., Diez-Perez, A., Collins, G.S., Prieto-Alhambra, D., 2018. Fracture risk in type 2 diabetic patients: a clinical prediction tool based on a large population-based cohort. PLoS One 13, e0203533.

Mathey, J., Horcajada-Molteni, M.N., Chanteranne, B., Picherit, C., Puel, C., Lebecque, P., Cubizoles, C., Davicco, M.J., Coxam, V., Barlet, J.P., 2002. Bone mass in obese diabetic Zucker rats: influence of treadmill running. Calcif. Tissue Int. 70, 305.

McCarthy, A.D., Etcheverry, S.B., Cortizo, A.M., 2001. Effect of advanced glycation endproducts on the secretion of insulin-like growth factor-I and its binding proteins: role in osteoblast development. Acta Diabetol. 38, 113.

McNair, P., Madsbad, S., Christensen, M.S., Christiansen, C., Faber, O.K., Binder, C., Transbol, I., 1979. Bone mineral loss in insulin-treated diabetes mellitus: studies on pathogenesis. Acta Endocrinol. 90, 463.

Meema, H.E., Meema, S., 1967. The relationship of diabetes mellitus and body weight to osteoporosis in elderly females. Can. Med. Assoc. J. 96, 132.

Meier, C., Schwartz, A.V., Egger, A., Lecka-Czernik, B., 2016. Effects of diabetes drugs on the skeleton. Bone 82, 93.

Mera, P., Laue, K., Ferron, M., Confavreux, C., Wei, J., Galan-Diez, M., Lacampagne, A., Mitchell, S.J., Mattison, J.A., Chen, Y., Bacchetta, J., Szulc, P., Kitsis, R.N., de Cabo, R., Friedman, R.A., Torsitano, C., McGraw, T.E., Puchowicz, M., Kurland, I., Karsenty, G., 2016a. Osteocalcin signaling in myofibers is necessary and sufficient for optimum adaptation to exercise. Cell Metabol. 23, 1078.

Mera, P., Laue, K., Wei, J., Berger, J.M., Karsenty, G., 2016b. Osteocalcin is necessary and sufficient to maintain muscle mass in older mice. Mol Metab 5, 1042.

Miao, J., Brismar, K., Nyren, O., Ugarph-Morawski, A., Ye, W., 2005. Elevated hip fracture risk in type 1 diabetic patients: a population-based cohort study in Sweden. Diabetes Care 28, 2850.

Mieczkowska, A., Mansur, S.A., Irwin, N., Flatt, P.R., Chappard, D., Mabilleau, G., 2015. Alteration of the bone tissue material properties in type 1 diabetes mellitus: a Fourier transform infrared microspectroscopy study. Bone 76, 31.

Miranda, C., Giner, M., Montoya, M.J., Vazquez, M.A., Miranda, M.J., Perez-Cano, R., 2016. Influence of high glucose and advanced glycation end-products (ages) levels in human osteoblast-like cells gene expression. BMC Muscoskelet. Disord. 17, 377.

Mitchell, A., Fall, T., Melhus, H., Wolk, A., Michaëlsson, K., Byberg, L., 2018. Type 2 diabetes in relation to hip bone density, area, and bone turnover in Swedish men and women: a cross-sectional study. Calcif. Tissue Int. 103, 501.

Mohan, S., Richman, C., Guo, R., Amaar, Y., Donahue, L.R., Wergedal, J., Baylink, D.J., 2003. Insulin-like growth factor regulates peak bone mineral density in mice by both growth hormone-dependent and -independent mechanisms. Endocrinology 144, 929.

Monami, M., Dicembrini, I., Antenore, A., Mannucci, E., 2011. Dipeptidyl peptidase-4 inhibitors and bone fractures: a meta-analysis of randomized clinical trials. Diabetes Care 34, 2474.

Monami, M., Dicembrini, I., Kundisova, L., Zannoni, S., Nreu, B., Mannucci, E., 2014. A meta-analysis of the hypoglycaemic risk in randomized controlled trials with sulphonylureas in patients with type 2 diabetes. Diabetes Obes. Metab. 16, 833.

Motyl, K.J., Botolin, S., Irwin, R., Appledorn, D.M., Kadakia, T., Amalfitano, A., Schwartz, R.C., McCabe, L.R., 2009. Bone inflammation and altered gene expression with type I diabetes early onset. J. Cell. Physiol. 218, 575.

Motyl, K.J., McCabe, L.R., 2009. Leptin treatment prevents type I diabetic marrow adiposity but not bone loss in mice. J. Cell. Physiol. 218, 376.

Moyer-Mileur, L.J., Dixon, S.B., Quick, J.L., Askew, E.W., Murray, M.A., 2004. Bone mineral acquisition in adolescents with type 1 diabetes. J. Pediatr. 145, 662.

Munoz-Torres, M., Jodar, E., Escobar-Jimenez, F., Lopez-Ibarra, P.J., Luna, J.D., 1996. Bone mineral density measured by dual X-ray absorptiometry in Spanish patients with insulin-dependent diabetes mellitus. Calcif. Tissue Int. 58, 316.

Napoli, N., Conte, C., Pedone, C., Strotmeyer, E.S., Barbour, K.E., Black, D.M., Samelson, E.J., Schwartz, A.V., 2019. Effect of insulin resistance on BMD and fracture risk in older adults. J. Clin. Endocrinol. Metab. https://doi.org/10.1210/jc.2018-02539 (In press).

Napoli, N., Pannacciulli, N., Vittinghoff, E., Crittenden, D., Yun, J., Wang, A., Wagman, R., Schwartz, A.V., 2018a. Effect of denosumab on fasting glucose in women with diabetes or prediabetes from the freedom trial. Diabetes Metab Res Rev 34, e2991.

Napoli, N., Schafer, A.L., Lui, L.Y., Cauley, J.A., Strotmeyer, E.S., Le Blanc, E.S., Hoffman, A.R., Lee, C.G., Black, D.M., Schwartz, A.V., 2016. Serum 25-hydroxyvitamin D level and incident type 2 diabetes in older men, the Osteoporotic Fractures in Men (MrOS) study. Bone 90, 181.

Napoli, N., Schwartz, A.V., Schafer, A.L., Vittinghoff, E., Cawthon, P.M., Parimi, N., Orwoll, E., Strotmeyer, E.S., Hoffman, A.R., Barrett-Connor, E., Black, D.M., Osteoporotic Fractures in Men Study Research, G., 2018b. Vertebral fracture risk in diabetic elderly men: the MrOS study. J. Bone Miner. Res. 33, 63.

Napoli, N., Strollo, R., Pitocco, D., Bizzarri, C., Maddaloni, E., Maggi, D., Manfrini, S., Schwartz, A., Pozzilli, P., Group, I., 2013. Effect of calcitriol on bone turnover and osteocalcin in recent-onset type 1 diabetes. PLoS One 8, e56488.

Napoli, N., Strotmeyer, E.S., Ensrud, K.E., Sellmeyer, D.E., Bauer, D.C., Hoffman, A.R., Dam, T.-T.L., Barrett-Connor, E., Palermo, L., Orwoll, E.S., Cummings, S.R., Black, D.M., Schwartz, A.V., 2014. Fracture risk in diabetic elderly men: the MrOS study. Diabetologia 57, 2057.

Neumann, T., Lodes, S., Kastner, B., Franke, S., Kiehntopf, M., Lehmann, T., Muller, U.A., Wolf, G., Samann, A., 2014. High serum pentosidine but not esRAGE is associated with prevalent fractures in type 1 diabetes independent of bone mineral density and glycaemic control. Osteoporos. Int. 25, 1527.

Neumann, T., Lodes, S., Kastner, B., Franke, S., Kiehntopf, M., Lehmann, T., Muller, U.A., Wolf, G., Samann, A., 2016. Osteocalcin, adipokines and their associations with glucose metabolism in type 1 diabetes. Bone 82, 50.

Ni, Y., Fan, D., 2017. Diabetes mellitus is a risk factor for low bone mass-related fractures: a meta-analysis of cohort studies. Medicine (Baltim.) 96, e8811.

Nicodemus, K.K., Folsom, A.R., Iowa Women's Health, S., 2001. Type 1 and type 2 diabetes and incident hip fractures in postmenopausal women. Diabetes Care 24, 1192.

Nilsson, A.G., Sundh, D., Johansson, L., Nilsson, M., Mellstrom, D., Rudang, R., Zoulakis, M., Wallander, M., Darelid, A., Lorentzon, M., 2017. Type 2 diabetes mellitus is associated with better bone microarchitecture but lower bone material strength and poorer physical function in elderly women: a population-based study. J. Bone Miner. Res. 32, 1062.

Nuche-Berenguer, B., Moreno, P., Portal-Nunez, S., Dapia, S., Esbrit, P., Villanueva-Penacarrillo, M.L., 2010. Exendin-4 exerts osteogenic actions in insulin-resistant and type 2 diabetic states. Regul. Pept. 159, 61.

Nyman, J.S., Even, J.L., Jo, C.H., Herbert, E.G., Murry, M.R., Cockrell, G.E., Wahl, E.C., Bunn, R.C., Lumpkin Jr., C.K., Fowlkes, J.L., Thrailkill, K.M., 2011. Increasing duration of type 1 diabetes perturbs the strength-structure relationship and increases brittleness of bone. Bone 48, 733.

Ogata, N., Chikazu, D., Kubota, N., Terauchi, Y., Tobe, K., Azuma, Y., Ohta, T., Kadowaki, T., Nakamura, K., Kawaguchi, H., 2000. Insulin receptor substrate-1 in osteoblast is indispensable for maintaining bone turnover. J. Clin. Investig. 105, 935.

Ogawa, N., Yamaguchi, T., Yano, S., Yamauchi, M., Yamamoto, M., Sugimoto, T., 2007. The combination of high glucose and advanced glycation end-products (AGEs) inhibits the mineralization of osteoblastic MC3T3-E1 cells through glucose-induced increase in the receptor for AGEs. Horm. Metab. Res. 39, 871.

Okazaki, R., Totsuka, Y., Hamano, K., Ajima, M., Miura, M., Hirota, Y., Hata, K., Fukumoto, S., Matsumoto, T., 1997. Metabolic improvement of poorly controlled noninsulin-dependent diabetes mellitus decreases bone turnover. J. Clin. Endocrinol. Metab. 82, 2915.

Orwoll, E.S., Bauer, D.C., Vogt, T.M., Fox, K.M., 1996. Axial bone mass in older women. Study of osteoporotic fractures research group. Ann. Intern. Med. 124, 187.

Ottenbacher, K.J., Ostir, G.V., Peek, M.K., Goodwin, J.S., Markides, K.S., 2002. Diabetes mellitus as a risk factor for hip fracture in mexican american older adults. J Gerontol A Biol Sci Med Sci 57, M648.

Oz, S.G., Guven, G.S., Kilicarslan, A., Calik, N., Beyazit, Y., Sozen, T., 2006. Evaluation of bone metabolism and bone mass in patients with type-2 diabetes mellitus. J. Natl. Med. Assoc. 98, 1598.

Paglia, D.N., Wey, A., Breitbart, E.A., Faiwiszewski, J., Mehta, S.K., Al-Zube, L., Vaidya, S., Cottrell, J.A., Graves, D., Benevenia, J., O'Connor, J.P., Lin, S.S., 2013. Effects of local insulin delivery on subperiosteal angiogenesis and mineralized tissue formation during fracture healing. J. Orthop. Res. 31, 783.

Palermo, A., D'Onofrio, L., Eastell, R., Schwartz, A.V., Pozzilli, P., Napoli, N., 2015. Oral anti-diabetic drugs and fracture risk, cut to the bone: safe or dangerous? A narrative review. Osteoporos. Int. 26, 2073.

Peng, J., Hui, K., Hao, C., Peng, Z., Gao, Q.X., Jin, Q., Lei, G., Min, J., Qi, Z., Bo, C., Dong, Q.N., Bing, Z.H., Jia, X.Y., Fu, D.L., 2016. Low bone turnover and reduced angiogenesis in streptozotocin-induced osteoporotic mice. Connect. Tissue Res. 57, 277.

Pereira, M., Jeyabalan, J., Jorgensen, C.S., Hopkinson, M., Al-Jazzar, A., Roux, J.P., Chavassieux, P., Orriss, I.R., Cleasby, M.E., Chenu, C., 2015. Chronic administration of Glucagon-like peptide-1 receptor agonists improves trabecular bone mass and architecture in ovariectomised mice. Bone 81, 459.

Petit, M.A., Paudel, M.L., Taylor, B.C., Hughes, J.M., Strotmeyer, E.S., Schwartz, A.V., Cauley, J.A., Zmuda, J.M., Hoffman, A.R., Ensrud, K.E., Osteoporotic Fractures in Men Study, G., 2010. Bone mass and strength in older men with type 2 diabetes: the Osteoporotic Fractures in Men Study. J. Bone Miner. Res. 25, 285.

Picke, A.K., Salbach-Hirsch, J., Hintze, V., Rother, S., Rauner, M., Kascholke, C., Moller, S., Bernhardt, R., Rammelt, S., Pisabarro, M.T., Ruiz-Gomez, G., Schnabelrauch, M., Schulz-Siegmund, M., Hacker, M.C., Scharnweber, D., Hofbauer, C., Hofbauer, L.C., 2016. Sulfated hyaluronan improves bone regeneration of diabetic rats by binding sclerostin and enhancing osteoblast function. Biomaterials 96, 11.

Pittas, A.G., Harris, S.S., Eliades, M., Stark, P., Dawson-Hughes, B., 2009. Association between serum osteocalcin and markers of metabolic phenotype. J. Clin. Endocrinol. Metab. 94, 827.

Pozzilli, P., Manfrini, S., Crino, A., Picardi, A., Leomanni, C., Cherubini, V., Valente, L., Khazrai, M., Visalli, N., group, I., 2005. Low levels of 25-hydroxyvitamin D3 and 1,25-dihydroxyvitamin D3 in patients with newly diagnosed type 1 diabetes. Horm. Metab. Res. 37, 680.

Pramojanee, S.N., Phimphilai, M., Kumphune, S., Chattipakorn, N., Chattipakorn, S.C., 2013. Decreased jaw bone density and osteoblastic insulin signaling in a model of obesity. J. Dent. Res. 92, 560.

Pun, K.K., Lau, P., Ho, P.W., 1989. The characterization, regulation, and function of insulin receptors on osteoblast-like clonal osteosarcoma cell line. J. Bone Miner. Res. 4, 853.

Qaseem, A., Wilt, T.J., Kansagara, D., Horwitch, C., Barry, M.J., Forciea, M.A., Clinical Guidelines Committee of the American College of, P., 2018. Hemoglobin A1c targets for glycemic control with pharmacologic therapy for nonpregnant adults with type 2 diabetes mellitus: a guidance statement update from the American College of Physicians. Ann. Intern. Med. 168, 569.

Raisingani, M., Preneet, B., Kohn, B., Yakar, S., 2017. Skeletal growth and bone mineral acquisition in type 1 diabetic children; abnormalities of the GH/IGF-1 axis. Growth Hormone IGF Res. 34, 13.

Rao Sirasanagandla, S., Ranganath Pai Karkala, S., Potu, B.K., Bhat, K.M., 2014. Beneficial effect of cissus quadrangularis linn. On osteopenia associated with streptozotocin-induced type 1 diabetes mellitus in male wistar rats. Adv Pharmacol Sci 2014, 483051.

Raskin, P., Stevenson, M.R., Barilla, D.E., Pak, C.Y., 1978. The hypercalciuria of diabetes mellitus: its amelioration with insulin. Clin. Endocrinol. 9, 329.

Reinwald, S., Peterson, R.G., Allen, M.R., Burr, D.B., 2009. Skeletal changes associated with the onset of type 2 diabetes in the ZDF and ZDSD rodent models. Am. J. Physiol. Endocrinol. Metab. 296, E765.

Rendina-Ruedy, E., Smith, B.J., 2016. Methodological considerations when studying the skeletal response to glucose intolerance using the diet-induced obesity model. Bone Rep. 5, 845.

Roggen, I., Gies, I., Vanbesien, J., Louis, O., De Schepper, J., 2013. Trabecular bone mineral density and bone geometry of the distal radius at completion of pubertal growth in childhood type 1 diabetes. Horm. Res. Paediatr 79, 68.

Roszer, T., Jozsa, T., Kiss-Toth, E.D., De Clerck, N., Balogh, L., 2014. Leptin receptor deficient diabetic (db/db) mice are compromised in postnatal bone regeneration. Cell Tissue Res. 356, 195.

Rozadilla, A., Nolla, J.M., Montana, E., Fiter, J., Gomez-Vaquero, C., Soler, J., Roig-Escofet, D., 2000. Bone mineral density in patients with type 1 diabetes mellitus. Joint Bone Spine 67, 215.

Ruanpeng, D., Ungprasert, P., Sangtian, J., Harindhanavudhi, T., 2017. Sodium-glucose cotransporter 2 (SGLT2) inhibitors and fracture risk in patients with type 2 diabetes mellitus: a meta-analysis. Diabetes Metab. Res. Rev. 33.

Rubino, F., Nathan, D.M., Eckel, R.H., Schauer, P.R., Alberti, K.G., Zimmet, P.Z., Del Prato, S., Ji, L., Sadikot, S.M., Herman, W.H., Amiel, S.A., Kaplan, L.M., Taroncher-Oldenburg, G., Cummings, D.E., Delegates of the 2nd Diabetes Surgery, S., 2016. Metabolic surgery in the treatment algorithm for type 2 diabetes: a joint statement by international diabetes organizations. Diabetes Care 39, 861.

Saito, M., Fujii, K., Mori, Y., Marumo, K., 2006. Role of collagen enzymatic and glycation induced cross-links as a determinant of bone quality in spontaneously diabetic WBN/Kob rats. Osteoporos. Int. 17, 1514.

Salter, J., Best, C.H., 1953. Insulin as a growth hormone. Br. Med. J. 2, 353.

Samelson, E.J., Demissie, S., Cupples, L.A., Zhang, X., Xu, H., Liu, C.T., Boyd, S.K., McLean, R.R., Broe, K.E., Kiel, D.P., Bouxsein, M.L., 2018. Diabetes and deficits in cortical bone density, microarchitecture, and bone size: Framingham HR-pQCT study. J. Bone Miner. Res. 33, 54.

Santana, R.B., Xu, L., Chase, H.B., Amar, S., Graves, D.T., Trackman, P.C., 2003. A role for advanced glycation end products in diminished bone healing in type 1 diabetes. Diabetes 52, 1502.

Sanz, C., Vazquez, P., Blazquez, C., Barrio, P.A., Alvarez Mdel, M., Blazquez, E., 2010. Signaling and biological effects of glucagon-like peptide 1 on the differentiation of mesenchymal stem cells from human bone marrow. Am. J. Physiol. Endocrinol. Metab. 298, E634.

Sasaki, T., Kaneko, H., Ramamurthy, N.S., Golub, L.M., 1991. Tetracycline administration restores osteoblast structure and function during experimental diabetes. Anat. Rec. 231, 25.

Schafer, A.L., Napoli, N., Lui, L., Schwartz, A.V., Black, D.M., Study of Osteoporotic, F., 2014. Serum 25-hydroxyvitamin D concentration does not independently predict incident diabetes in older women. Diabet. Med. 31, 564.

Scheiwiller, E., Guler, H.P., Merryweather, J., Scandella, C., Maerki, W., Zapf, J., Froesch, E.R., 1986. Growth restoration of insulin-deficient diabetic rats by recombinant human insulin-like growth factor I. Nature 323, 169.

Schmedt, N., Andersohn, F., Walker, J., Garbe, E., 2019. Sodium-glucose co-transporter-2 inhibitors and the risk of fractures of the upper or lower limbs in patients with type 2 diabetes: a nested case-control study. Diabetes Obes. Metab. 21, 52.

Schwartz, A.V., Hillier, T.A., Sellmeyer, D.E., Resnick, H.E., Gregg, E., Ensrud, K.E., Schreiner, P.J., Margolis, K.L., Cauley, J.A., Nevitt, M.C., Black, D.M., Cummings, S.R., 2002. Older women with diabetes have a higher risk of falls: a prospective study. Diabetes Care 25, 1749.

Schwartz, A.V., Pavo, I., Alam, J., Disch, D.P., Schuster, D., Harris, J.M., Krege, J.H., 2016. Teriparatide in patients with osteoporosis and type 2 diabetes. Bone 91, 152.

Schwartz, A.V., Schafer, A.L., Grey, A., Vittinghoff, E., Palermo, L., Lui, L.Y., Wallace, R.B., Cummings, S.R., Black, D.M., Bauer, D.C., Reid, I.R., 2013a. Effects of antiresorptive therapies on glucose metabolism: results from the FIT, HORIZON-PFT, and FREEDOM trials. J. Bone Miner. Res. 28, 1348.

Schwartz, A.V., Sellmeyer, D.E., Ensrud, K.E., Cauley, J.A., Tabor, H.K., Schreiner, P.J., Jamal, S.A., Black, D.M., Cummings, S.R., Study of Osteoporotic Features Research, G., 2001. Older women with diabetes have an increased risk of fracture: a prospective study. J. Clin. Endocrinol. Metab. 86, 32.

Schwartz, A.V., Sellmeyer, D.E., Strotmeyer, E.S., Tylavsky, F.A., Feingold, K.R., Resnick, H.E., Shorr, R.I., Nevitt, M.C., Black, D.M., Cauley, J.A., Cummings, S.R., Harris, T.B., Health, A.B.C.S., 2005. Diabetes and bone loss at the hip in older black and white adults. J. Bone Miner. Res. 20, 596.

Schwartz, A.V., Sellmeyer, D.E., Vittinghoff, E., Palermo, L., Lecka-Czernik, B., Feingold, K.R., Strotmeyer, E.S., Resnick, H.E., Carbone, L., Beamer, B.A., Park, S.W., Lane, N.E., Harris, T.B., Cummings, S.R., 2006. Thiazolidinedione use and bone loss in older diabetic adults. J. Clin. Endocrinol. Metab. 91, 3349.

Schwartz, A.V., Sigurdsson, S., Hue, T.F., Lang, T.F., Harris, T.B., Rosen, C.J., Vittinghoff, E., Siggeirsdottir, K., Sigurdsson, G., Oskarsdottir, D., Shet, K., Palermo, L., Gudnason, V., Li, X., 2013b. Vertebral bone marrow fat associated with lower trabecular BMD and prevalent vertebral fracture in older adults. J. Clin. Endocrinol. Metab. 98, 2294.

Scragg, R., Holdaway, I., Singh, V., Metcalf, P., Baker, J., Dryson, E., 1995. Serum 25-hydroxyvitamin D3 levels decreased in impaired glucose tolerance and diabetes mellitus. Diabetes Res. Clin. Pract. 27, 181.

Seref-Ferlengez, Z., Maung, S., Schaffler, M.B., Spray, D.C., Suadicani, S.O., Thi, M.M., 2016. P2X7R-Panx1 complex impairs bone mechanosignaling under high glucose levels associated with type-1 diabetes. PLoS One 11, e0155107.

Shah, V.N., Harrall, K.K., Shah, C.S., Gallo, T.L., Joshee, P., Snell-Bergeon, J.K., Kohrt, W.M., 2017. Bone mineral density at femoral neck and lumbar spine in adults with type 1 diabetes: a meta-analysis and review of the literature. Osteoporos. Int. 28, 2601.

Shah, V.N., Shah, C.S., Snell-Bergeon, J.K., 2015. Type 1 diabetes and risk of fracture: meta-analysis and review of the literature. Diabet. Med. 32, 1134.

Shanbhogue, V.V., Finkelstein, J.S., Bouxsein, M.L., Yu, E.W., 2016. Association between insulin resistance and bone structure in nondiabetic postmenopausal women. J. Clin. Endocrinol. Metab. 101, 3114.

Shanbhogue, V.V., Hansen, S., Frost, M., Jørgensen, N.R., Hermann, A.P., Henriksen, J.E., Brixen, K., 2015. Bone geometry, volumetric density, microarchitecture, and estimated bone strength assessed by HR-pQCT in adult patients with type 1 diabetes mellitus. J. Bone Miner. Res. 30, 2188.

Sheu, Y., Amati, F., Schwartz, A.V., Danielson, M.E., Li, X., Boudreau, R., Cauley, J.A., Osteoporotic Fractures in Men Research, G., 2017. Vertebral bone marrow fat, bone mineral density and diabetes: the Osteoporotic Fractures in Men (MrOS) study. Bone 97, 299.

Shires, R., Teitelbaum, S.L., Bergfeld, M.A., Fallon, M.D., Slatopolsky, E., Avioli, L.V., 1981. The effect of streptozotocin-induced chronic diabetes mellitus on bone and mineral homeostasis in the rat. J. Lab. Clin. Med. 97, 231.

Silva, M.J., Brodt, M.D., Lynch, M.A., McKenzie, J.A., Tanouye, K.M., Nyman, J.S., Wang, X., 2009. Type 1 diabetes in young rats leads to progressive trabecular bone loss, cessation of cortical bone growth, and diminished whole bone strength and fatigue life. J. Bone Miner. Res. 24, 1618.

Slade, J.M., Coe, L.M., Meyer, R.A., McCabe, L.R., 2012. Human bone marrow adiposity is linked with serum lipid levels not T1-diabetes. J. Diabetes Complicat. 26, 1.

Sorocéanu, M.A., Miao, D., Bai, X.-Y., Su, H., Goltzman, D., Karaplis, A.C., 2004. Rosiglitazone impacts negatively on bone by promoting osteoblast/osteocyte apoptosis. J. Endocrinol. 183, 203.

Srikanthan, P., Crandall, C.J., Miller-Martinez, D., Seeman, T.E., Greendale, G.A., Binkley, N., Karlamangla, A.S., 2014. Insulin resistance and bone strength: findings from the study of midlife in the United States. J. Bone Miner. Res. 29, 796.

Starup-Linde, J., 2013. Diabetes, biochemical markers of bone turnover, diabetes control, and bone. Front. Endocrinol. 4.

Starup-Linde, J., Lykkeboe, S., Gregersen, S., Hauge, E.-M., Langdahl, B.L., Handberg, A., Vestergaard, P., 2016a. Bone structure and predictors of fracture in type 1 and type 2 diabetes. J. Clin. Endocrinol. Metab. 101, 928.

Starup-Linde, J., Lykkeboe, S., Gregersen, S., Hauge, E.-M., Langdahl, B.L., Handberg, A., Vestergaard, P., 2016b. Differences in biochemical bone markers by diabetes type and the impact of glucose. Bone 83, 149.

Starup-Linde, J., Vestergaard, P., 2016. Biochemical bone turnover markers in diabetes mellitus - a systematic review. Bone 82, 69.

Strotmeyer, E.S., Cauley, J.A., Orchard, T.J., Steenkiste, A.R., Dorman, J.S., 2006. Middle-aged premenopausal women with type 1 diabetes have lower bone mineral density and calcaneal quantitative ultrasound than nondiabetic women. Diabetes Care 29, 306.

Strotmeyer, E.S., Cauley, J.A., Schwartz, A.V., Nevitt, M.C., Resnick, H.E., Bauer, D.C., Tylavsky, F.A., de Rekeneire, N., Harris, T.B., Newman, A.B., 2005. Nontraumatic fracture risk with diabetes mellitus and impaired fasting glucose in older white and black adults: the health, aging, and body composition study. Arch. Intern. Med. 165, 1612.

Strotmeyer, E.S., Cauley, J.A., Schwartz, A.V., Nevitt, M.C., Resnick, H.E., Zmuda, J.M., Bauer, D.C., Tylavsky, F.A., de Rekeneire, N., Harris, T.B., Newman, A.B., Health, A.B.C.S., 2004. Diabetes is associated independently of body composition with BMD and bone volume in older white and black men and women: the Health, Aging, and Body Composition Study. J. Bone Miner. Res. 19, 1084.

Suzuki, K., Miyakoshi, N., Tsuchida, T., Kasukawa, Y., Sato, K., Itoi, E., 2003. Effects of combined treatment of insulin and human parathyroid hormone(1-34) on cancellous bone mass and structure in streptozotocin-induced diabetic rats. Bone 33, 108.

Tanaka, H., Yamashita, T., Yoneda, M., Takagi, S., Miura, T., 2018. Characteristics of bone strength and metabolism in type 2 diabetic model Tsumura, Suzuki, obese diabetes mice. BoneKEy Rep. 9, 74.

Tanaka, K., Yamaguchi, T., Kanazawa, I., Sugimoto, T., 2015. Effects of high glucose and advanced glycation end products on the expressions of sclerostin and RANKL as well as apoptosis in osteocyte-like MLO-Y4-A2 cells. Biochem. Biophys. Res. Commun. 461, 193.

Tang, H.L., Li, D.D., Zhang, J.J., Hsu, Y.H., Wang, T.S., Zhai, S.D., Song, Y.Q., 2016. Lack of evidence for a harmful effect of sodium-glucose cotransporter 2 (SGLT2) inhibitors on fracture risk among type 2 diabetes patients: a network and cumulative meta-analysis of randomized controlled trials. Diabetes Obes. Metab. 18, 1199.

Taylor, W.H., Khaleeli, A.A., 2001. Coincident diabetes mellitus and primary hyperparathyroidism. Diabetes Metab Res Rev 17, 175.

Terada, M., Inaba, M., Yano, Y., Hasuma, T., Nishizawa, Y., Morii, H., Otani, S., 1998. Growth-inhibitory effect of a high glucose concentration on osteoblast-like cells. Bone 22, 17.

Thomas, D.M., Udagawa, N., Hards, D.K., Quinn, J.M., Moseley, J.M., Findlay, D.M., Best, J.D., 1998. Insulin receptor expression in primary and cultured osteoclast-like cells. Bone 23, 181.

Thong, E.P., Herath, M., Weber, D.R., Ranasinha, S., Ebeling, P.R., Milat, F., Teede, H., 2018. Fracture risk in young and middle-aged adults with type 1 diabetes mellitus: a systematic review and meta-analysis. Clin. Endocrinol. 89, 314.

Thrailkill, K., Bunn, R.C., Lumpkin, C., Wahl, E., Cockrell, G., Morris, L., Kahn, C.R., Fowlkes, J., Nyman, J.S., 2014. Loss of insulin receptor in osteoprogenitor cells impairs structural strength of bone. J. Diab. Res. 2014, 703589.

Thrailkill, K.M., Jo, C.H., Cockrell, G.E., Moreau, C.S., Lumpkin, C.K., Fowlkes, J.L., 2012. Determinants of undercarboxylated and carboxylated osteocalcin concentrations in type 1 diabetes. Osteoporos. Int. 23, 1799.

Thrailkill, K.M., Liu, L., Wahl, E.C., Bunn, R.C., Perrien, D.S., Cockrell, G.E., Skinner, R.A., Hogue, W.R., Carver, A.A., Fowlkes, J.L., Aronson, J., Lumpkin, C.K., 2005a. Bone formation is impaired in a model of type 1 diabetes. Diabetes 54, 2875.

Thrailkill, K.M., Lumpkin, C.K., Bunn, R.C., Kemp, S.F., Fowlkes, J.L., 2005b. Is insulin an anabolic agent in bone? Dissecting the diabetic bone for clues. Am. J. Physiol. Endocrinol. Metab. 289, E735.

Toh, S., Hernandez-Diaz, S., 2007. Statins and fracture risk. A systematic review. Pharmacoepidemiol. Drug Saf. 16, 627.

Tsentidis, C., Gourgiotis, D., Kossiva, L., Marmarinos, A., Doulgeraki, A., Karavanaki, K., 2016. Sclerostin distribution in children and adolescents with type 1 diabetes mellitus and correlation with bone metabolism and bone mineral density. Pediatr. Diabetes 17, 289.

Tsentidis, C., Gourgiotis, D., Kossiva, L., Marmarinos, A., Doulgeraki, A., Karavanaki, K., 2017. Increased levels of Dickkopf-1 are indicative of Wnt/beta-catenin downregulation and lower osteoblast signaling in children and adolescents with type 1 diabetes mellitus, contributing to lower bone mineral density. Osteoporos. Int. 28, 945.

Tsuka, S., Aonuma, F., Higashi, S., Ohsumi, T., Nagano, K., Mizokami, A., Kawakubo-Yasukochi, T., Masaki, C., Hosokawa, R., Hirata, M., Takeuchi, H., 2015. Promotion of insulin-induced glucose uptake in C2C12 myotubes by osteocalcin. Biochem. Biophys. Res. Commun. 459, 437.

Turner, R.T., Kalra, S.P., Wong, C.P., Philbrick, K.A., Lindenmaier, L.B., Boghossian, S., Iwaniec, U.T., 2013. Peripheral leptin regulates bone formation. J. Bone Miner. Res. 28, 22.

van Daele, P.L., Stolk, R.P., Burger, H., Algra, D., Grobbee, D.E., Hofman, A., Birkenhager, J.C., Pols, H.A., 1995. Bone density in non-insulin-dependent diabetes mellitus. The Rotterdam Study. Ann. Intern. Med. 122, 409.

van Lierop, A.H., Hamdy, N. a. T., van der Meer, R.W., Jonker, J.T., Lamb, H.J., Rijzewijk, L.J., Diamant, M., Romijn, J.A., Smit, J.W.A., Papapoulos, S.E., 2012. Distinct effects of pioglitazone and metformin on circulating sclerostin and biochemical markers of bone turnover in men with type 2 diabetes mellitus. Eur. J. Endocrinol. 166, 711.

Vavanikunnel, J., Charlier, S., Becker, C., Schneider, C., Jick, S.S., Meier, C.R., Meier, C., 2019. Association between glycemic control and risk of fracture in diabetic patients: a nested case-control study. J. Clin. Endocrinol. Metab. 104, 1645.

Veldhuis-Vlug, A.G., Rosen, C.J., 2017. Mechanisms of marrow adiposity and its implications for skeletal health. Metabolism 67, 106.

Verhaeghe, J., Suiker, A.M., Einhorn, T.A., Geusens, P., Visser, W.J., Van Herck, E., Van Bree, R., Magitsky, S., Bouillon, R., 1994. Brittle bones in spontaneously diabetic female rats cannot be predicted by bone mineral measurements: studies in diabetic and ovariectomized rats. J. Bone Miner. Res. 9, 1657.

Verhaeghe, J., Suiker, A.M., Van Bree, R., Van Herck, E., Jans, I., Visser, W.J., Thomasset, M., Allewaert, K., Bouillon, R., 1993. Increased clearance of 1,25(OH)2D3 and tissue-specific responsiveness to 1,25(OH)2D3 in diabetic rats. Am. J. Physiol. 265, E215.

Verhaeghe, J., Suiker, A.M., Visser, W.J., Van Herck, E., Van Bree, R., Bouillon, R., 1992. The effects of systemic insulin, insulin-like growth factor-I and growth hormone on bone growth and turnover in spontaneously diabetic BB rats. J. Endocrinol. 134, 485.

Verhaeghe, J., Thomsen, J.S., van Bree, R., van Herck, E., Bouillon, R., Mosekilde, L., 2000. Effects of exercise and disuse on bone remodeling, bone mass, and biomechanical competence in spontaneously diabetic female rats. Bone 27, 249.

Verhaeghe, J., Van Herck, E., van Bree, R., Moermans, K., Bouillon, R., 1997. Decreased osteoblast activity in spontaneously diabetic rats. In vivo studies on the pathogenesis. Endocrine 7, 165.

Verhaeghe, J., van Herck, E., Visser, W.J., Suiker, A.M., Thomasset, M., Einhorn, T.A., Faierman, E., Bouillon, R., 1990. Bone and mineral metabolism in BB rats with long-term diabetes. Decreased bone turnover and osteoporosis. Diabetes 39, 477.

Verroken, C., Pieters, W., Beddeleem, L., Goemaere, S., Zmierczak, H.G., Shadid, S., Kaufman, J.M., Lapauw, B., 2017. Cortical bone size deficit in adult patients with type 1 diabetes mellitus. J. Clin. Endocrinol. Metab. 102, 2887.

Vestergaard, P., 2007. Discrepancies in bone mineral density and fracture risk in patients with type 1 and type 2 diabetes–a meta-analysis. Osteoporos. Int. 18, 427.

Vestergaard, P., 2011. Risk of newly diagnosed type 2 diabetes is reduced in users of alendronate. Calcif. Tissue Int. 89, 265.

Vestergaard, P., Rejnmark, L., Mosekilde, L., 2005. Relative fracture risk in patients with diabetes mellitus, and the impact of insulin and oral antidiabetic medication on relative fracture risk. Diabetologia 48, 1292.

Vestergaard, P., Rejnmark, L., Mosekilde, L., 2011. Are antiresorptive drugs effective against fractures in patients with diabetes? Calcif. Tissue Int. 88, 209.

Villasenor, A., Aedo-Martin, D., Obeso, D., Erjavec, I., Rodriguez-Coira, J., Buendia, I., Ardura, J.A., Barbas, C., Gortazar, A.R., 2019. Metabolomics reveals citric acid secretion in mechanically-stimulated osteocytes is inhibited by high glucose. Sci. Rep. 9, 2295.

Wallace, C., Reiber, G.E., LeMaster, J., Smith, D.G., Sullivan, K., Hayes, S., Vath, C., 2002. Incidence of falls, risk factors for falls, and fall-related fractures in individuals with diabetes and a prior foot ulcer. Diabetes Care 25, 1983.

Wallner, C., Schira, J., Wagner, J.M., Schulte, M., Fischer, S., Hirsch, T., Richter, W., Abraham, S., Kneser, U., Lehnhardt, M., Behr, B., 2015. Application of VEGFA and FGF-9 enhances angiogenesis, osteogenesis and bone remodeling in type 2 diabetic long bone regeneration. PLoS One 10, e0118823.

Wang, H., Ba, Y., Xing, Q., Du, J.L., 2019. Diabetes mellitus and the risk of fractures at specific sites: a meta-analysis. BMJ Open 9, e024067.

Wang, X., Shen, X., Li, X., Agrawal, C.M., 2002. Age-related changes in the collagen network and toughness of bone. Bone 31, 1.

Watts, N.B., Bilezikian, J.P., Usiskin, K., Edwards, R., Desai, M., Law, G., Meininger, G., 2016. Effects of canagliflozin on fracture risk in patients with type 2 diabetes mellitus. J. Clin. Endocrinol. Metab. 101, 157.

Wei, J., Ferron, M., Clarke, C.J., Hannun, Y.A., Jiang, H., Blaner, W.S., Karsenty, G., 2014a. Bone-specific insulin resistance disrupts whole-body glucose homeostasis via decreased osteocalcin activation. J. Clin. Investig. 124, 1.

Wei, J., Hanna, T., Suda, N., Karsenty, G., Ducy, P., 2014b. Osteocalcin promotes beta-cell proliferation during development and adulthood through Gprc6a. Diabetes 63, 1021.

Wittrant, Y., Gorin, Y., Woodruff, K., Horn, D., Abboud, H.E., Mohan, S., Abboud-Werner, S.L., 2008. High d(+)glucose concentration inhibits RANKL-induced osteoclastogenesis. Bone 42, 1122.

Wortsman, J., Matsuoka, L.Y., Chen, T.C., Lu, Z., Holick, M.F., 2000. Decreased bioavailability of vitamin D in obesity. Am. J. Clin. Nutr. 72, 690.

Wu, J.H., Foote, C., Blomster, J., Toyama, T., Perkovic, V., Sundstrom, J., Neal, B., 2016. Effects of sodium-glucose cotransporter-2 inhibitors on cardiovascular events, death, and major safety outcomes in adults with type 2 diabetes: a systematic review and meta-analysis. Lancet Diabetes Endocrinol 4, 411.

Yakar, S., Isaksson, O., 2016. Regulation of skeletal growth and mineral acquisition by the GH/IGF-1 axis: lessons from mouse models. Growth Hormone IGF Res. 28, 26.

Yakar, S., Rosen, C.J., Beamer, W.G., Ackert-Bicknell, C.L., Wu, Y., Liu, J.-L., Ooi, G.T., Setser, J., Frystyk, J., Boisclair, Y.R., LeRoith, D., 2002. Circulating levels of IGF-1 directly regulate bone growth and density. J. Clin. Investig. 110, 771.

Yamada, C., Yamada, Y., Tsukiyama, K., Yamada, K., Udagawa, N., Takahashi, N., Tanaka, K., Drucker, D.J., Seino, Y., Inagaki, N., 2008. The murine glucagon-like peptide-1 receptor is essential for control of bone resorption. Endocrinology 149, 574.

Yamamoto, M., Yamauchi, M., Sugimoto, T., 2013. Elevated sclerostin levels are associated with vertebral fractures in patients with type 2 diabetes mellitus. J. Clin. Endocrinol. Metab. 98, 4030.

Yang, J., Zhang, X., Wang, W., Liu, J., 2010. Insulin stimulates osteoblast proliferation and differentiation through ERK and PI3K in MG-63 cells. Cell Biochem. Funct. 28, 334.

Yassouridis, C., Leisch, F., Winkler, C., Ziegler, A.G., Beyerlein, A., 2017. Associations of growth patterns and islet autoimmunity in children with increased risk for type 1 diabetes: a functional analysis approach. Pediatr. Diabetes 18, 103.

Yoshida, M., Jacques, P.F., Meigs, J.B., Saltzman, E., Shea, M.K., Gundberg, C., Dawson-Hughes, B., Dallal, G., Booth, S.L., 2008. Effect of vitamin K supplementation on insulin resistance in older men and women. Diabetes Care 31, 2092.

Zhang, L., Liu, Y., Wang, D., Zhao, X., Qiu, Z., Ji, H., Rong, H., 2009. Bone biomechanical and histomorphometrical investment in type 2 diabetic Goto-Kakizaki rats. Acta Diabetol. 46, 119.

Zhou, M., Ma, X., Li, H., Pan, X., Tang, J., Gao, Y., Hou, X., Lu, H., Bao, Y., Jia, W., 2009. Serum osteocalcin concentrations in relation to glucose and lipid metabolism in Chinese individuals. Eur. J. Endocrinol. 161, 723.

Zhukouskaya, V.V., Eller-Vainicher, C., Gaudio, A., Cairoli, E., Ulivieri, F.M., Palmieri, S., Morelli, V., Orsi, E., Masserini, B., Barbieri, A.M., Polledri, E., Fustinoni, S., Spada, A., Fiore, C.E., Chiodini, I., 2015. In postmenopausal female subjects with type 2 diabetes mellitus, vertebral fractures are independently associated with cortisol secretion and sensitivity. J. Clin. Endocrinol. Metab. 100, 1417.

Zhukouskaya, V.V., Eller-Vainicher, C., Gaudio, A., Privitera, F., Cairoli, E., Ulivieri, F.M., Palmieri, S., Morelli, V., Grancini, V., Orsi, E., Masserini, B., Spada, A.M., Fiore, C.E., Chiodini, I., 2016. The utility of lumbar spine trabecular bone score and femoral neck bone mineral density for identifying asymptomatic vertebral fractures in well-compensated type 2 diabetic patients. Osteoporos. Int. 27, 49.

Zhukouskaya, V.V., Eller-Vainicher, C., Vadzianava, V.V., Shepelkevich, A.P., Zhurava, I.V., Korolenko, G.G., Salko, O.B., Cairoli, E., Beck-Peccoz, P., Chiodini, I., 2013. Prevalence of morphometric vertebral fractures in patients with type 1 diabetes. Diabetes Care 36, 1635.

Chapter 41

Androgen receptor expression and steroid action in bone

Venkatesh Krishnan

Lilly Research Laboratories, Eli Lilly & Company, Lilly Corporate Center, Indianapolis, United States

Chapter outline

Introduction	971
Loss-of-function evidence from rare human variants	972
Gain-of-function evidence from human trials	974
Can we reliably measure androgen gain of function in bone health using end points such as bone mineral density?	975
Gender-specific differences in bone geometry and architecture	975
Gain-of-function studies with testosterone treatment and change in bone architecture	976
Loss of function using genetically modified mice	976
Role of estrogen receptor alpha in bone using genetically modified mice is equivocal	977
Gain-of-function studies using selective androgen receptor modulators in sexually mature animal models	977
The muscle–bone interface and its potential impact on bone strength	978
Future prospects for androgens and male skeletal health	979
References	979

Introduction

Androgens and estrogens belong to the steroid hormone superfamily that binds to nuclear receptors that mediate their canonical actions in specific reproductive and endocrine tissues (Tsai and O'Malley, 1994). Estrogen mediates its action through two cognate receptors termed estrogen receptor alpha (ESR1) and estrogen receptor beta (ESR2)—also known as NR3A1: ERalpha and NR3A2: ERbeta (Nilsson et al., 2001). Androgen signals through the only known cognate receptor, termed androgen receptor (AR, also referred to as NR3C4) (Simental et al., 1991). Previous studies have shown that loss of estrogen following surgical removal of ovaries or an aging-related decrease in estrogen production (defined earlier as menopause) can lead to bone loss in women ((see Fig. 41.2) Albright, 1940). In a prospective study among prostate cancer patients seeking androgen-deprivation therapy (ADT) compared with age-matched patients not seeking ADT, it was shown that the absence of circulating androgens results in an increased rate of bone turnover and a loss in bone mineral density (BMD) over 2 years (Prostate Cancer, 2002). The ADT regimen included an LHRH agonist leuprolide along with concomitant administration of flutamide or bicalutamide, which are specific antagonists of AR. Therefore, both estrogen and androgen deprivation can lead to bone loss in women and men, respectively. Furthermore, surgical gonadectomy of sexually mature men resulted in bone loss as evidenced by a rapid decline in BMD and an increase in bone turnover markers (Stepan et al., 1989). Therefore, this provided a simple model that suggested a prominent role for androgen in maintaining healthy bone mass in men. However, in a landmark mechanistic study conducted by Khosla et al. in 1998, it was shown that estrogen (but not androgen) supplementation was both necessary and sufficient to recover the bone loss initiated by ADT. These results have challenged the dogma that androgens and other AR agonists are exclusive in imparting favorable skeletal effects in men. A subsequent study has corroborated this finding by exploring (Finkelstein et al., 2016) a different study design involving pharmacological induction of hypogonadism in men followed by graded doses of testosterone replacement with or without an aromatase inhibitor. Similar to the earlier study described by Khosla et al., changes in bone resorption and formation markers were predominantly regulated by estrogen and not androgen in these men. Moreover, testosterone, in the absence of aromatization to estrogen (aromatase inhibitor arm), was unable to

Principles of Bone Biology. https://doi.org/10.1016/B978-0-12-814841-9.00041-5

prevent decreases in BMD at multiple sites. Collectively, these two and additional interventional studies (Khosla et al., 2015), using variations of the approaches used in the studies just described, now provide compelling evidence in humans that in both women and men, estrogen is the dominant sex steroid regulating bone metabolism. While these mechanistic studies have employed the use of testosterone and its aromatase enzyme catalyzed conversion to estrogen, in the presence/absence of aromatase inhibitors, they do not address the direct consequences of androgen ligands in bone. These can be primarily achieved by the use of nonaromatizable androgens that may be represented by a wide array of selective androgen–receptor modulators (SARMs) (Narayanan et al., 2018). Unfortunately, most clinical development deploying these SARMs carries unknown long-term risks that have yet to warrant their use as therapeutic options in long-term disease contexts (treatment duration >2 years) other than exigent diseases such as cancer. Therefore, the field lacks access to published data describing carefully designed long-term efficacy studies with SARMs that could allow one to measure the effects of such specific AR ligands in the context of human bone health. In contrast, there is ample evidence that antiandrogens, by way of their use in prostate cancer and other benign prostate diseases, do result in loss of bone health that predisposes patients to a higher risk of fractures (Cheung et al., 2013).

The subsequent sections of the review will now explore the role of AR in affecting male bone health. It will contrast both loss-in-function and gain-in-function paradigms to help ascribe a holistic view on this topic. It is now evident that rodent models that reflect loss of function (LoF), either by way of a whole body knockout of these genes from early development or by way of tissue-specific knockout mice using tissue-specific promoters that drive *Cre* recombinase, provide useful ways to assess the importance of AR in rodent bone health. This review will capture our current understanding of these genetically modified animal models and help reconcile it with the LoF observations in man that may or may not align with these rodent observations. In addition, this article will summarize the numerous gain-in-function paradigms ranging from the use of steroidal testosterone replacement to the emerging discovery and development of SARMs as they impact bone health.

Loss-of-function evidence from rare human variants

The importance of ESR1 in male human skeleton was exemplified by a rare LoF homozygous deletion of ER-alpha in a male individual that rendered him with osteopenia and a spine BMD more than two standard deviations below the age-matched normal for 15-year-old boys (Smith et al., 1994). However, when carefully evaluating the iliac crest biopsy obtained from this rare ER-alpha-null male, it was reported to reflect reduced cancellous bone volume and cortical width at this site. It is conceivable that this is due to the absence of the direct beneficial effect of ER-alpha on strain-dependent bone formation indices, as suggested by Lanyon et al. (Lee et al., 2003). Aligned with this report were two reports of loss of hormone that was realized through a homozygous deletion of the aromatase gene that is crucial for converting circulating testosterone to estrogen, the ligand that targets the receptors ESR1 and ESR2. In both these individuals who lacked adequate circulating estrogen levels, the skeletal phenotype was near identical to that described earlier for the ER-alpha-null human (Carani et al., 1997; Bilezekian et al., 1998). Collectively, these findings, albeit rare, provide a prominent role for estrogens and ESR1 in affecting the peak bone mass in young adolescent males.

The enzyme 5-alpha reductase (type 1 and type 2) catalyzes the reduction of testosterone to the nonaromatizable and potent AR ligand dihydrotestosterone (DHT) (see Fig. 41.1), which can bind AR with greater affinity than testosterone and mediate the biological effects of testosterone as an active metabolite. Patients with mutations of 5-alpha reductase type 2 have been reported to have normal BMD (Sobel et al., 2006). These individuals have low circulating levels of DHT. Treatment of adult men with the 5-alpha reductase inhibitor finasteride does not affect BMD (Matsumoto et al., 2002). However, finasteride is a weak inhibitor of the type 2 isoform of 5-alpha reductase and does not affect the constitutively active prevalent type 1 isoform. Thus, although DHT levels are lower in finasteride-treated men than in healthy untreated controls, they are still in the low normal range in many treated men. More recent studies in dutasteride-treated patients, which is a potent inhibitor of both type 1 and type 2 isoforms of 5-alpha reductase, also have failed to elicit a significant reduction in BMD in men (Amory et al., 2008). These observations of normal bone health in light of reduced conversion of testosterone to its potent metabolite DHT further guide our conclusion for the relative importance of ESR versus AR ligands in maintaining bone mass in males.

The 3-beta hydroxysteroid dehydrogenase 2 (HSD3B2) gene is located on chromosome 1q13.1, and encodes for the human type II (3b-HSD2) isoenzyme, which is only expressed in the adrenal cortex and in steroidogenic cells within the gonads. It is an essential enzyme for the synthesis of testosterone along with cortisol and aldosterone (see Fig. 41.1). An individual who presented at birth with a salt-wasting syndrome as a result of this rare insufficiency variant, which represented an LoF for this enzyme, was rescued with cortisol treatments. This male achieved normal puberty and had the expected developmental defects in his sexual organs associated with androgen insufficiency. However, in spite of low

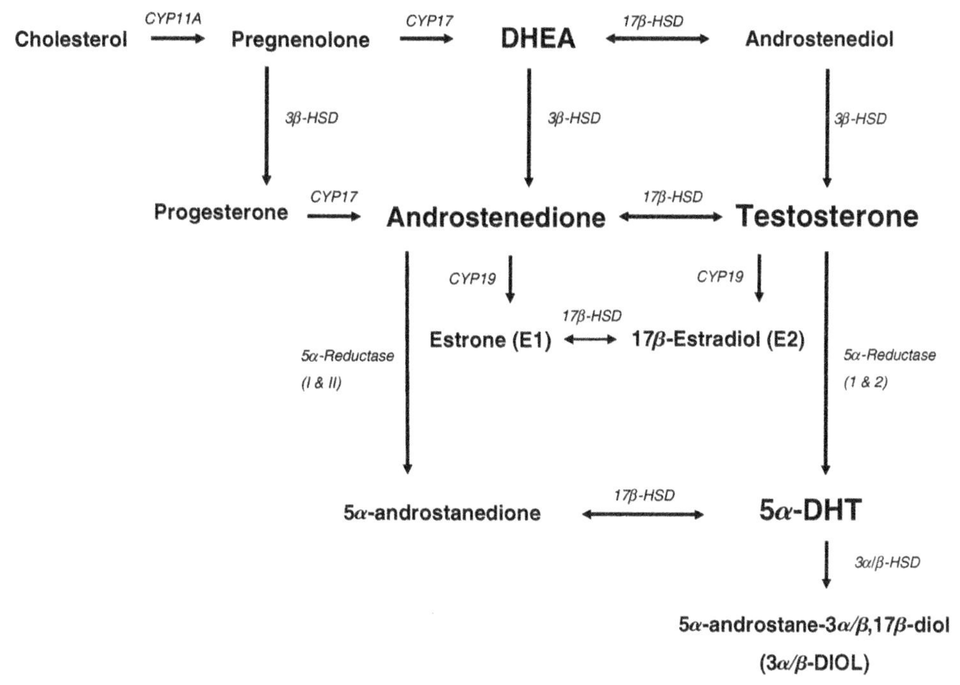

FIGURE 41.1 An outline of the steroidal synthesis of androgen and its metabolism has been provided. Of importance are the roles played by 5alpha-Reductase in converting Testosterone to 5alpha-Dihydrotestosterone, and the conversion of androstenedione to testosterone by 17 Beta HSD.

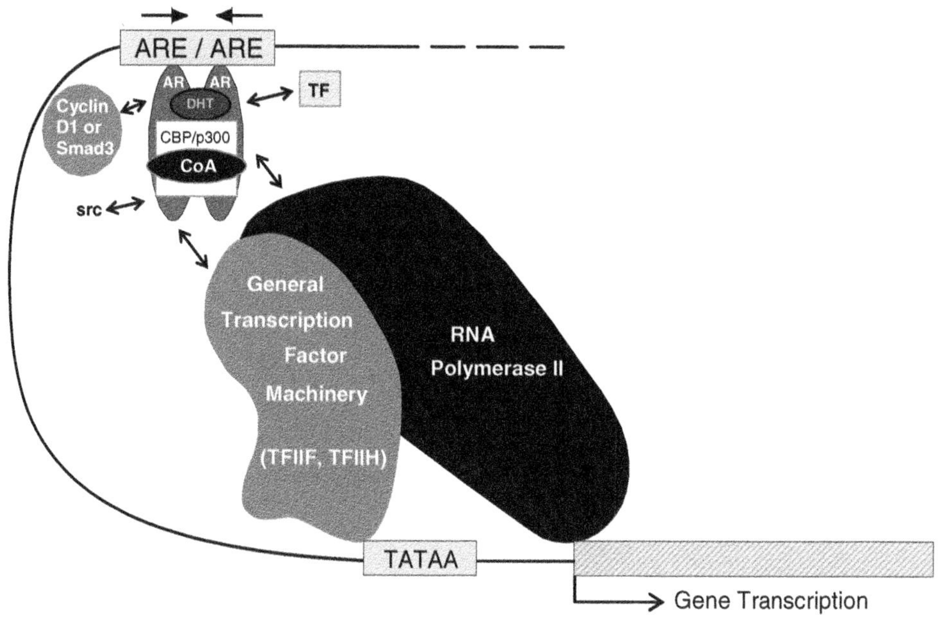

FIGURE 41.2 The molecular action of transcriptional activation by the AR homodimer acting on cis-regulatory elements identified on androgen-regulated genes is shown along with the coactivators that may influence tissue specific androgen-induced gene expression.

circulating testosterone levels (~13.8 nmol/L compared with normal range if 20−40 nmol/L), which is an expected consequence of this rare LoF variant, the patient showed a normal BMD. His bone density measured at approximately 19 years of age was normal with a femoral T-score of (+2.8 s.d.) and vertebral T-score of (+1.9 s.d.), suggestive of an absence of an obvious deleterious effect on BMD in spite of insufficient androgen levels (Bruno et al., 2018). Collectively, the rare occurrences of human LoF variants point to a prominent role for estrogen/ESR1 in maintaining male skeletal health and less so for the hormone testosterone and related androgens such as dihydrotestosterone. However, the absence of well-annotated AR LoF individuals makes it challenging to offer a comparative assessment of androgens (vs. estrogens) and their role in human bone health relying on this approach of rare LoF genetic variants.

Gain-of-function evidence from human trials

Placebo-controlled studies using various approved formulations of testosterone have been limited and often report conflicting results (Snyder et al., 1999; Kenny et al., 2001; Tenover, 1992). Some component of this discrepancy could be attributed to the study design within these trials. In the study conducted by Snyder et al., 108 men over 65 years of age were randomized to receive a testosterone or placebo patch for 36 months (Snyder et al., 1999). The majority of these subjects were neither osteoporotic by current WHO criteria, nor were they hypogonadal (mean testosterone level >350 ng/dL). The primary end point was the measure of vBMD as measured using dual-absorption X-ray absorptiometry (DEXA) measured at 6-month intervals over the 3-year period. The testosterone patch did not show significant improvement in the primary end point of *v*BMD compared with the placebo group, except for the final measure of spine BMD by DEXA at 36 months. In a posthoc analysis using secondary linear regression, it was reported that by stratifying subjects based on their pretreatment serum testosterone concentration, one could clearly show favorable effects of the testosterone patch arm compared with placebo arm on lumbar spine bone density in overtly hypogonadal patients. In contrast to the previous study of nonosteoporotic, eugonadal men, Amory et al. studied hypogonadal men ($T \leq 300$ ng/dL) over age 65, randomly assigned to placebo or testosterone, bimonthly 200 mg intramuscular injections ± finasteride, for 3 years (Amory et al., 2004). Their inclusion criteria for these elderly men used baseline mean T-scores ranging from −0.3 to −0.53, and therefore these hypogonadal men were not deemed osteoporotic. Their peak serum total testosterone levels were at the high normal serum range for the testosterone-only group, with a mean peak value for the group being 35.9 ± 12.1 nmol/L (mean ± sd). In contrast to the Snyder et al. study described above, BMD at a variety of sites including lumbar spine, total hip, and trochanter (femur site) were all increased with the testosterone ± finasteride arms at 12 months and remained significantly increased for the 3-year period compared with the results for placebo injections. It is noteworthy that the magnitude of the increases in lumbar spine BMD seen with intramuscular testosterone at 12 months were comparable to those observed in men with osteoporosis using daily injections of teriparatide, a strong well-known bone anabolic agent (Orwoll et al., 2003). However, this study could not account for these effects as being specific to the increase in serum testosterone, as testosterone could have been aromatized to estrogen in specific tissues. Consistent with this caveat, the authors measured an increase in serum estradiol throughout this 3-year period. Notably, these gains in lumbar BMD over 36 months were positively correlated with the magnitude of increase in serum total testosterone (r^2 0.44; $P \leq .001$), bioavailable testosterone (r^2 0.45 and $P \leq .009$), and serum estradiol (r^2 0.45; $P \leq .0006$). Therefore, in spite of encouraging changes in lumbar BMD comparable to those seen with anabolic agents such as teriparatide, one needs to cautiously ascribe direct beneficial effects in improving BMD for male subjects with androgens. It is important to note, however, that selective modulators of ESR1 and ESR2 ligands such as raloxifene do not provide such a robust change in lumbar spine BMD as seen with intramuscular testosterone after 12 months in patients receiving gonadotropin-releasing hormone (GnRH) agonists for prostate cancer. The mean (BMD ± SE) BMD of the posterior−anterior lumbar spine increased by $1.0 \pm 0.9\%$ in the men treated with raloxifene plus GnRH agonist and decreased by $1.0 \pm 0.6\%$ in the men who did not receive raloxifene but were taking GnRH agonist ($P = .07$) (Smith et al., 2004). It is not clear if the full complement of ESR1- and ESR2-related bone changes ascribed to estrogen-mediated action can be recapitulated by a selective ligand like raloxifene in the context of a GnRH agonist.

The use of intramuscular testosterone enanthate or cypionate, 100 mg/week, in patients with acquired hypogonadism as reported by Katznelson et al. had the most definitive gain seen in vertebral BMD (14% after 18 months) but is confounded by the fact that these were younger adults (average age 53) with lower vertebral BMD as a result of diminished peak bone mass caused by their central hypogonadism, which could have included changes in the growth hormone/IGF-1 axis. Most recently, Snyder et al. published the most extensive 1-year study in elderly (>65 years of age) hypogonadal men (serum $T < 275$ ng/dL), when approximately 790 men were randomized to placebo or a testosterone gel. However, the authors did not measure BMD or biomarkers of turnover such as CTX or P1NP (Snyder et al., 2016). In summary, while a few gain-of-function (GoF) studies have been conducted in elderly hypogonadal men, which included approximately 90 men for a 1-year treatment duration and provided a 90% power to reliably detect vBMD changes based on the average 9% gain in vertebral BMD at 1 year as measured by Katznelson et al. (1996) using individuals with acquired hypogonadism, these results do not provide a compelling reason to ascribe this benefit in men. The primary reason to maintain a cautious position is that while all these studies provide a vignette or two of benefit in increasing BMD, in order to balance any perceived target-related risk, such as prostate and cardiovascular health, one needs larger, well-controlled, and longer-duration studies (>2 years) to provide a balanced view of the benefit–risk profile of androgen replacement to address issues with male skeletal health. In lieu of the robust efficacy of non-AR binding pharmaceuticals such as antiresorptives and anabolic agents in male patients, this data set will be challenging to seek from such well-controlled studies (Snyder et al., 2014).

Can we reliably measure androgen gain of function in bone health using end points such as bone mineral density?

It is now well established that bone quality is a measure that encompasses BMD plus aspects of bone strength that are difficult to measure using objective preestablished criteria such as bone geometry, microarchitecture, and bone tissue properties (Fonseca et al., 2014). The limitations for exclusive reliance on BMD are highlighted by the significant differences in trabecular connectivity with unit areas from subjects matched for bone volume (Recker, 1993). Also, in numerous interventional studies with antiresorptives, significant and clinically relevant reductions in fracture risk are associated with modest changes in BMD (Sarkar et al., 2002; Delmas and Seeman, 2004). Therefore, it is conceivable that clinical end points related to bone microarchitecture and other structural parameters as assessed by high-resolution computer tomography, magnetic resonance imaging, bone mineral strength by microindentation, etc. may provide valuable insight into bone quality that could supplement our experience with BMD to arrive at a holistic view of the benefits of androgen-related effects on bone quality. These alternate end points reflecting the changes in microarchitecture of bone in response to androgens are proposed due to the well-known observations related to striking changes seen in the mass, structure, and geometry of the male versus female skeleton (Bruzek, 2002).

Gender-specific differences in bone geometry and architecture

Higher peak bone mass in males compared with females would posit higher fracture risk for males (compared with females) as they progressively lose bone mass due to aging for a given BMD, because these males have a greater absolute reduction from their peak density. Yet, studies have shown that for a given DXA BMD score, the fracture risk between genders is indistinguishable (Selby et al., 2000). Characteristic differences in bone architecture between males (larger bone area) and females (smaller bone area) perhaps offer a potential advantage to males that may explain their reduced susceptibility to fracture (Seeman, 1993). Therefore, the higher peak bone mass in males compared with age-matched females offers them a distinct advantage (Lambert et al., 2011). It is also well known that the cross-sectional diameters of vertebrae and femoral necks are larger in males and that this is associated with greater bone strength than that of age-matched females (Mosekilde et al., 1990) Decreased axial width along with reduced cross-sectional area, seen in elderly men (MINOS study) with a higher risk of fragility fractures, further underscores the importance of bone geometry as an independent determinant of fracture risk in addition to BMD (Szulc et al., 2006). It is also believed that through aging, women show greater reductions in periosteal thickness compared with men (Christiansen et al., 2011) Interestingly, the type of trabecular changes that occur with aging also differ between the sexes, as males tend to have trabecular thinning and females tend to lose trabecular connectivity (Lambert et al., 2011). Therefore, despite the fact that males will have a larger reduction for any given DXA BMD value, this does not seem to translate into an increased fracture risk, and is perhaps related to the protective benefit of their peak bone characteristics and different types of bone architectural properties compared with females. Consistent with these findings, the prevalence of osteoporosis is significantly greater for women than men when stratified across age groups through deciles (Kanis, 2007). This observation is also substantiated in male rats deficient in androgen and exposed to aromatase inhibitors, which fail to convert testosterone to estrogen and show specific decrement in certain geometric features of the proximal femur, suggesting that androgens may control peak bone mass in the male rat by relying on non-BMD measures (Vanderschueren et al., 1997). The mechanistic underpinnings for the sexually dimorphic nature of skeletal geometry were described recently by Sims et al. Recent studies implicate the regulation of corticalization by SOCS-3. They used a DMP-1–driven Cre to generate an osteocyte-specific knockout of SOCS-3 that allowed them to create a sexually divergent phenotype of corticalization. The nonaromatizable androgen (i.e., DHT) can inhibit the activity of delayed corticalization in Dmp-1–specific SOCS-3-null mice by targeting IL-6 (Cho et al., 2017). This provides us with a mechanistic basis for the increased corticalization seen in the male skeleton, which may effectively contribute to reduced fracture risk. Therefore, androgen ligands that enhance cortical bone parameters may favorably affect fracture risk. However, if these androgens resulted in changes in bone architecture or geometry such as increased cross-sectional area as a result of increased cortical area, even in the context of increasing bone mineral content, it would only reflect as a minimal change in BMD. Therefore, when it comes to androgen-mediated impact on male bone health, it may be more prudent to look beyond BMD.

Gain-of-function studies with testosterone treatment and change in bone architecture

In a small clinical study (32 hypogonadal men) exploring the use of testosterone treatment in hypogonadal men (primary hypogonadism), Leifke et al. showed that intramuscular injection of testosterone enanthate (250 mg every 2 weeks over 3 years) resulted in a 30% gain in trabecular BMD and a 9% increase in cortical BMD at the lumbar spine as measured by quantitative computer tomography (Leifke et al., 1998). These improvements were seen in all patients independent of their leading cause for primary hypogonadism. In contrast to this smaller study, Khosla et al. have shown in 269 men that there is no correlation between microstructural parameters at the wrist bone as measured by high-resolution three-dimensional quantitative computer tomography and those of bioavailable testosterone. Therefore, it is important to recognize that there may be a threshold effect of testosterone on structural properties of bone that may only be evident in men with severe hypogonadism and is less evident in the broad eugonadal population. Also, the anatomical site (wrist vs. spine) may offer distinct responses to androgens. In another very limited study (10 hypogonadal men) explored by Benito et al., it was shown that magnetic resonance microimaging-based end points highlight the integrity of the trabecular network, which is compromised in hypogonadal men and improved with testosterone replacement. These patients showed an increase in the surface-to-curve ratio (+11% vs. baseline), defined earlier (Gomberg et al., 2000), which is the topologic representation of the ratio of trabecular plates to rods and suggests that testosterone replacement partially restored trabecular connectivity. These studies show that testosterone treatment results in improvements typically seen with anabolic agents on trabecular connectivity.

Loss of function using genetically modified mice

The effects of estrogen on bone are mediated by two related but distinct receptors, ESR1 (ER-alpha) and ESR2 (ER-beta) (Nilsson et al., 2001; Almeida et al., 2017). Sims et al. published the global knockout of ESR1, and as a result of alteration in the hypothalamic–pituitary–gonadal axis, these animals had supraphysiological levels of circulating estrogen and androgen levels in females and male knockout mice. Loss of ESR1 from birth resulted in a decrease in bone turnover and an increase in cancellous bone mass accompanied by a decrease in cortical thickness (Sims et al., 2002). A follow-up study reported by Callewaert et al. pursued a strategy of creating both single- and double-knockouts of ESR1 and AR in male mice by crossing these compound heterozygotes. Removal of ESR1 resulted in an unusual gain in trabecular bone volume; in contrast, removal of AR resulted in an opposite and dramatic reduction in trabecular bone volume when assessed using computer tomography of the distal femur. These results would imply that in the absence of hypothalamic–pituitary–gonadal axis-related suppression of circulating hormones mediated by ESR1, systemic increases in serum androgens may result in this surprising increase in bone volume in ESR1-null mice. In contrast, when both ESR1 and AR were knocked out, they observed complete loss of this gain in trabecular bone volume in the single-ESR1-null mice comparable to that of the single-AR-null mice (Callewaert et al., 2009). In addition, cortical area at the femoral site was reduced with the single-ESR1-null mice, which was further reduced with the single-AR-null mice, followed by the greatest reduction in cortical area in the double-null mice. Collectively, AR may be required for male mice to accrue peak trabecular bone mass and peak cortical area. These interpretations from global ESR1 knockout mice are confounded by significant alterations in circulating sex steroid levels. To avoid these dramatic changes in circulating sex steroids that may inadvertently reflect GoF for other steroid hormone receptors that function through these circulating hormones, the field has resorted to a tissue-specific expression of Cre recombinase that allows targeted deletion of ESRs and AR in specific tissues that retain activity off these targeted promoters.

Deletion of ESR1 and AR in osteoprogenitor cells, osteocytes, and osteoclasts (Ucer et al., 2016) reported by Ucer et al. provides the most thorough assessment of the roles of these steroid hormone receptors in developing mice. In these reports, the authors were careful in not realizing systemic supraphysiological increases in circulating hormones by deleting these receptors in specific cell types. They utilized a Prx-Cre mouse to target the osteoprogenitor cells throughout development. Analysis of these mice using computer tomography showed a decrease in cancellous BV/TV and trabecular number, and an increase in trabecular separation, compared with C57BL6 mice that retain AR expression in these cells. The effect of orchidectomy on cancellous bone in the wild-type mice was similar to the effects of AR deletion from Prx-Cre-driven mesenchymal progenitors in androgen-sufficient mice. Therefore, the targeted AR deletion in mesenchymal progenitors is similar to that observed with orchidectomy-induced loss in circulating androgens. Surprisingly, the authors noted an absence of cortical bone changes in these Prx-Cre-driven osteoprogenitor AR-null mice. Furthermore, removal of AR from osteoclast precursors that are reflected in the LysM Cre mice show that AR function in mouse bone is not dependent on osteoclast expression of AR. The LysM gene is specifically expressed in cells of the monocyte/macrophage lineage and their descendants as well as in neutrophils (Clausen et al., 1999).

Role of estrogen receptor alpha in bone using genetically modified mice is equivocal

ESR1 may arguably contribute to changes in both cancellous and cortical bone as described by the use of ESR1-floxed mice crossed to Dmp1-Cre, and ESR1-floxed mice crossed to OCN-Cre mice, both of which result in measurable decreases in cancellous bone (Maatta et al., 2013; Kondoh et al., 2014). In contrast to these reports, others have shown that removal of ESR1 from osteoprogenitor cells or osteocytes does not affect cancellous bone mass. These studies have crossed ESR1-floxed mice to osteoprogenitor-specific Prx1-Cre (Ucer et al., 2016) or osteoblast-expressed Col1-Cre mice (Almeida et al., 2013), and finally osteocyte-specific Dmp1-Cre mice (Windahl et al., 2013). In all these reports, the impact on regulating cancellous bone mass is not very evident for ESR1. However, cortical bone parameters are modestly dependent on ESR1, as reported in findings made by Almeida et al. (2013).

Factors that may contribute to these equivocal observations may be (1) the use of distinct Cre promoters that may vary in tissue specificity that has yet to be determined, (2) the limitations of using the Cre-Lox system, which may yield variable levels of gene deletion (Manolagas et al., 2014), and (3) phenotypes that are quite modest, so that subtle variations over time during development may affect how these results are interpreted. Collectively, one could suggest that ESR1 deletion in osteoblast lineage cells may result in reductions in both cancellous and cortical bone mass in female mice and only cancellous bone mass in male mice. However, an important caveat to all of these studies is that the receptor deletion, although cell-specific, has been present since conception. Therefore, the skeletal findings largely reflect developmental effects of these receptors rather than their exclusive role in the adult skeleton.

Gain-of-function studies using selective androgen receptor modulators in sexually mature animal models

LoF of AR in osteoprogenitor cells (Ucer et al., 2016) or since conception globally (Callewaert et al., 2009) leads to decreases in trabecular bone mass, connectivity, and cortical area as reported earlier. However, the most compelling data set explores GoF paradigms using nonsteroidal, nonaromatizable SARMs and offers the best approach to fully understand the effects of AR ligands in the mature skeleton.

Earlier reports included suggestions of robust activity in male orchidectomized rat models offered by perspective summaries described earlier (Rosen et al., 2002) but these did not provide an opportunity for detailed analysis of the preclinical datasets. However, the most compelling preclinical evidence was offered by Kearbey et al., who showed a dose-dependent gain in trabecular bone mass in female ovariectomized mature rats that were treated with a distinct aryl propionamide SARM (S-4) (Gao et al., 2005a). They showed protection from ovariectomy-induced trabecular bone loss using the S-4 SARM as measured by lumbar spine BMD. This benefit offered by the SARM was dependent on AR signaling, as observed by the reversal of this protection of spine BMD in the arm that combined the S-4 SARM along with the AR antagonist bicalutamide. In addition, they observed enhancements in cortical bone parameters, including cortical width, cortical density, and periosteal circumference at peak doses of 1.0–3.0 mg/kg-day. Collectively, this report provided compelling evidence of the osteoprotective effects of a nonsteroidal SARM that functions as an AR agonist in bone when used immediately after ovariectomy in female SD rats (Kearbey et al., 2007). To further evaluate the potential for an anabolic response, the same group reported deployment of the S-4 SARM in a delayed rat male orchidectomy model. In this model, the orchidectomy induced an expected loss in total whole-body bone mineral content and BMD over the 12-week wasting period. This change in whole-body BMC and BMD was completely recovered after 8 weeks of treatment with S-4 SARM given daily at 10 mg/kg-day. However, there were some modest reductions in osteocalcin levels, which may point to a decrease in bone turnover in these animals (Gao et al., 2005b). It was not clear whether the authors measured favorable changes in markers such as P1NP, which reflect de novo bone formation as reflected by changes in the N-terminus-modified epitopes presented in serum with the synthesis of new collagen molecules placed onto bone (Eastell et al., 2006). Therefore, while these results are encouraging, it is not definitive whether the gains in total bone mineral content and BMD are due to SARM-mediated gains in bone formation. The most compelling report that showcases the direct anabolic action of SARMs was reported by Hanada et al., who utilized a −tetrahydroquinoline derivative to generate

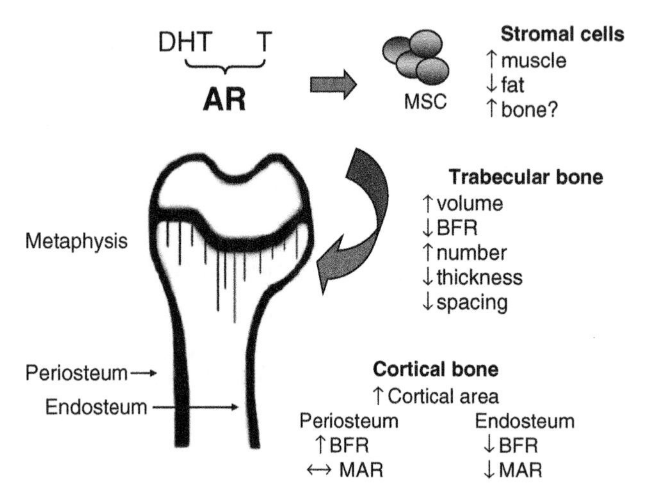

FIGURE 41.3 A summary of the expected changes in response to androgen action, at the various geometric sites associated with appendicular skeletal sites is shown.

an SARM, termed as S-40503 (Hanada et al., 2003). This molecule was shown to be very selective for AR and did not bind ESR1, ESR2, or other nuclear hormone receptors. In this report, they utilized a bilateral orchidectomized male rat that was treated with the S-40503 SARM for 4 weeks 1-day postgonadectomy. They showed increases in femoral BMD that were dose-dependent after the 4-week treatment period. To address direct anabolic action by SARM S-40503, they utilized an ovariectomized female rat model, wherein animals were allowed to lose bone mass for 4 weeks followed by 4 weeks of daily subcutaneous injections with SARM S-40503 (30 mg/kg-day), estrogen (20 μg/kg-day), or DHT (10 mg/kg-day). They measured distal femur cancellous BMD, which was most improved with estrogen followed by SARM S-40503 and DHT to a similar but qualitatively lesser extent. In contrast, they observed a robust increase in cortical BMD that was greater than intact nonovariectomized age-matched animals for both SARM S-40503 and DHT. This gain in cortical bone BMD was not apparent for the estrogen-only arm. Furthermore, they used calcein double-labeling to show direct evidence of an increased mineral apposition rate (MAR μm/day) in response to nonaromatizable DHT and SARM S-40503, which was in contrast to minimal apposition observed with estrogen (no change). Notably, the MAR observed for S-40503-treated animals was higher than for the intact nonovariectomized control, again suggesting direct anabolic action in response to this GoF paradigm using both a nonaromatizable androgen (i.e., DHT) and a nonsteroidal selective SARM, S-40503. Collectively, the use of structurally distinct nonsteroidal SARMs in preclinical models suggests a potential to directly and favorable impact bone mass in animals undergoing gonadectomy-induced bone loss (see Fig. 41.3). However, these preclinical findings have yet to translate to changes in improvements in bone mass or bone biomarkers in clinical studies exploring SARMs. This may be due to the positioning of these SARMs in diseases such as cancer cachexia and sarcopenia, which may require long-term use to determine such bone-related end points.

The muscle–bone interface and its potential impact on bone strength

Androgens in general are well known to be muscle anabolic and form the basis of numerous clinical reports on their ability to increase skeletal muscle mass (Bhasin et al., 1997). Changes in bone and muscle go hand-in-hand to affect the skeletal features of both men and women. Therefore, it is logical to expect the muscle–bone interface to provide favorable strain- or load-induced bone formation cues at these interfaces. The role played by the tendon that presents the collagen-enriched structural bridge between the muscle and bone interface is perhaps less understood for fully appreciating the impact of muscle and mobility on bone formation. The collagen fibrils, considered the fundamental force-transmitting unit of the tendon, are densely arranged within the extracellular matrix of specialized cells called tenocytes and are oriented parallel to the bone–muscle axis (Nourissat et al., 2015). Such an orientation provides a seamless transmission of the mechanical forces transduced by skeletal muscle contractions, which impinge on the enthesis within the bone–tendon interface. Evidence for such an impact was offered by Hamrick et al., who showed that gdf-8 knockout mice that had a dramatic increase in muscle weight and volume exhibited significant changes in cross-sectional area and bone mineral content in the proximal humerus and deltoid crest. These were consistent with areas closely representing sites of attachment of the muscle to the bone and further underscore the impact of muscle volume or mass on bone geometry (Hamrick et al., 2002).

Therefore, the muscle–bone relationship, expressed as the "bone–muscle unit," can be viewed as single coordinate structure that is strongly influenced by mobility-induced mechanical loading. Such differential strains imposed by increases or decreases in mechanical loading, as a result of disability or aging, can affect overall bone geometry and bone strength (Lang, 2011) (see Fig. 41.3). The close coupling between muscle and bone, and the gender differences in the relationship, are often viewed in the context of the mechanostat theory first elaborated by Frost (2000). The independent risk conferred by skeletal muscle was underscored in the Frost mechanostat review that highlighted falls as the largest "risk factor" for traumatic osteoporosis fractures, because without falls, the fractures caused by them would not occur regardless of whether some microscopic fatigue damage in bone increases the bone fragility associated with all osteopenias (Frost, 2003). However, fall risk accumulates as a result of multiple intrinsic and extrinsic risk factors such as changes in balance induced by medications or alcohol, loss in lower limb skeletal power with aging, poor vision, and a variety of environmental conditions. The independent risk factor of lower limb skeletal muscle strength and/or power was addressed by Becker et al. in the first-of-its-kind intervention on elderly weak fallers using an antimyostatin antibody (Becker et al., 2015). Results from this trial emphasized the importance of muscle power a feature of muscle strength over a fixed time period which is essential to how our body reacts to a displacement such as a fall. The intervention provided preliminary evidence that treatment with a muscle mass enhancing antibody that targets gdf-8 might improve functional measures of muscle power in elderly individuals prone to falls. Similar results have been published by Srinivas-Shankar et al., using transdermal testosterone gels for 6 months in elderly frail men with borderline low serum T levels. In this randomized controlled trial, they found that testosterone supplementation showed marked improvements in the primary outcome measure of isometric knee extension peak torque and isokinetic knee extension peak torque. Unfortunately, the study failed to measure markers of bone turnover to help inform the changes that could be affecting the muscle–bone unit. The study also did not measure the impact of testosterone gels to reduce the incidence of falls. In a separate study conducted by Bronwyn et al. in chronic glucocorticoid-treated patients, injectable testosterone esters given intramuscularly every 2 weeks for 12 months showed marked improvement in lumbar spine BMD ($+4.7\% \pm 1.1$) along with improvements in muscle strength as determined by knee flexion and extension using isokinetic dynamometry (Bronwyn et al., 2003).

Collectively, the impact of androgens on various functional parameters of skeletal muscle that impacts mobility may confer added benefit to measures of bone strength and/or geometry through the muscle–bone interface.

Future prospects for androgens and male skeletal health

The seminal findings made by Khosla et al. (1998) have provided some clarity on the importance of the ESR1–estrogen axis-mediated benefit conferred upon the male skeleton in the course of normal physiology. However, clinical interventions that address ESR1- and/or ESR2-mediated pathways to enhance skeletal health are relatively less prominent. The emergence of novel mechanisms of resistance to antagonists or inhibitors of the AR–testosterone axis as it pertains to prostate cancer (Watson et al., 2015) will naturally require better medicines that balance the benefit to the risk profiles of such agents. This will require us to be vigilant about the deleterious effects of androgen ablation as they become more prevalent and mainstream in their targeting of less severe patients suffering from PCa to help prevent the occurrence, or to add as maintenance regimens in patients undergoing remission to prevent the relapse of tumor burden. Recent advances in this approach have been seen with the development of such agents targeting distinct actions of estrogen and androgen action in breast cancer (Yu et al., 2017). Our ability to provide therapeutic options to balance cancer risk while maintaining skeletal health in these individuals would warrant more nuanced classes of agents in the future (Figs. 41.1–41.3).

SARMs offer a variety of chemical scaffolds that allow us to better design the benefit and risk profile of such interventions. However, as noted by recent reviews, the field has progressed slowly in large randomized controlled trials, querying bone-related end points as these novel agents posit risk in terms of their cardiovascular outcomes (Narayanan et al., 2018; Nique et al., 2012; Krishnan et al., 2018; Saeed et al., 2015). It is conceivable, as these agents receive prominence as they begin to discharge such risk in the clinic, that trials exploring their benefit on the muscle–bone interface will become more available for assessing the GoF paradigm with such agents. However, until such time, the use of AR ligands to better define such benefits shall remain elusive to the medical community.

References

Albright, F., 1940. Trans. Assoc. Am. Phys. 55, 298–305.
Almeida, M., et al., 2013. J. Clin. Investig. 123, 394–404.
Almeida, M., et al., 2017. Physiol. Rev. 97, 135–187.
Amory, J.K., et al., 2004. J. Clin. Endocrinol. Metab. 89 (2), 503–510.

Amory, J.K., et al., 2008. J. Urol. 179 (6), 2333–2338.
Becker, C., et al., 2015. Lancet Diabetes Endocrinol. 3, 948–957.
Bhasin, S., et al., 1997. J. Clin. Endocrinol. Metab. 82 (1), 3–8.
Bilezekian, J.P., et al., 1998. N. Engl. J. Med. 339, 599–603.
Bronwyn, A.L.C., et al., 2003. J. Clin. Endocrinol. Metab. 88 (7), 3167–3176.
Bruno, D., et al., 2018. Endocrine Connections 7, 395–402.
Bruzek, J.A., 2002. J. Phys. Anthropol. 117, 157–168.
Callewaert, F., et al., 2009. FASEB J. 23, 232–240.
Carani, C., et al., 1997. N. Engl. J. Med. 337, 91–95.
Cheung, A.S., et al., 2013. Andrology 1, 583–589.
Cho, D.-C., et al., 2017. Nat. Commun. 8 (806), 1–13.
Christiansen, B.A., et al., 2011. J. Bone Miner. Res. 26 (5), 974–983.
Clausen, B.E., et al., 1999. Transgenic Res. 8, 265–277.
Delmas, P.D., Seeman, E., 2004. Bone 34 (4), 599–604.
Eastell, R., et al., 2006. Curr. Med. Res. Opin. 22 (1), 61–66.
Finkelstein, J.S., et al., 2016. J. Clin. Investig. 126, 1114–1125.
Fonseca, H., et al., 2014. Sports Med. 4, 37–53.
Frost, H.M., 2000. Med. Sci. Sports Exerc. 32 (5), 911–917.
Frost, H., 2003. Anat. Rec. A 275A, 1081–1101.
Gao, W., et al., 2005a. Chem. Rev. 105, 3352–3370.
Gao, W., et al., 2005b. Endocrinology 146 (11), 4887–4897.
Gomberg, B.R., et al., 2000. IEEE Trans. Med. Imaging 19, 166.
Hamrick, M.W., et al., 2002. Calcif. Tissue Int. 71, 63–68.
Hanada, K., et al., 2003. Biol. Pharm. Bull. 26 (11), 1563–1569.
Kanis, J.A., 2007. Assessment of Osteoporosis at the Primary Health-Care Level. WHO Collaborating Centre, University of Sheffield, Sheffield, UK.
Katznelson, L., et al., 1996. J. Clin. Endocrinol. Metab. 81, 4358–4365.
Kearbey, J.D., et al., 2007. Pharm. Res. (N. Y.) 24 (2), 328–335.
Kenny, A.,M., et al., 2001. J. Gerontol. A Biol. Sci. Med. Sci. 56 (5), 266–272.
Khosla, S., et al., 1998. J. Clin. Endocrinol. Metab. 83, 2266–2274.
Khosla, S., et al., 2015. J. Bone Miner. Res. 30, 1134–1137.
Kondoh, S., et al., 2014. Bone 60, 68–77.
Krishnan, V., et al., March 2018. Andrology (12).
Lambert, J.K., et al., 2011. Curr. Osteoporos. Rep. 9 (4), 229–236.
Lang, T.F., 2011. J. Osteoporos. 2011, 702735. https://doi.org/10.4061/2011/702735.
Lee, K., et al., 2003. Nature 424, 389.
Leifke, E., et al., 1998. Eur. J. Endocrinol. 138, 51–58.
Maatta, J.A., et al., 2013. FASEB J. 27, 478–488.
Manolagas, S.C., et al., 2014. J. Bone Miner. Res. 29, 2131–2140.
Matsumoto, A.M., et al., 2002. J. Urol. 167 (5), 2105–2108.
Mosekilde, L., et al., 1990. Bone 11 (2), 67–73.
Narayanan, R., et al., 2018. Mol. Cell. Endocrinol. 465, 134–140.
Nilsson, S., et al., 2001. Physiol. Rev. 81, 1535–1565.
Nique, F., et al., 2012. J. Med. Chem. 55, 8225–8235.
Nourissat, G., et al., 2015. Nat. Rev. Rheumatol. 11, 223–233.
Orwoll, E.S., et al., 2003. J. Bone Miner. Res. 18 (1), 9–17.
Prostate Cancer Prostatic Dis. 5, 2002, 304–310.
Recker, R.R., 1993. Calcif. Tissue Int. 53 (1), 139–142.
Rosen, J., et al., 2002. J. Musculoskelet. Neuronal Interact. 2 (3), 222–224.
Saeed, A., et al., December 2015. J. Med. Chem. 59 (2).
Sarkar, S., et al., 2002. J. Bone Miner. Res. 17 (1), 1–10.
Seeman, E., 1993. Am. J. Med. 95 (5), 22–28.
Selby, P.L., et al., 2000. Osteoporos. Int. 11 (2), 153–157.
Simental, J.A., et al., 1991. J. Biol. Chem. 266 (1), 510–518.
Sims, N.A., et al., 2002. Bone 30, 18–25.
Smith, E.P., et al., 1994. N. Engl. J. Med. 331, 1056–1061.
Smith, M.R., et al., 2004. J. Clin. Endocrinol. Metab. 89 (8), 3841–3846.
Snyder, P.J., et al., 1999. J. Clin. Endocrinol. Metab. 84 (6), 1966–1972.
Snyder, P.J., et al., 2014. Clin. Trials 11, 362–375.

Snyder, P.J., et al., 2016. N. Engl. J. Med. 374, 611–624.
Sobel, V., et al., 2006. J. Clin. Endocrinol. Metab. 91 (8), 3017–3023.
Stepan, J.J., et al., 1989. J. Clin. Endocrinol. Metab. 69, 523–527.
Szulc, P., et al., 2006. Bone 38 (4), 595–602.
Tenover, J.S., 1992. J. Clin. Endocrinol. Metab. 75 (4), 1092–1098.
Tsai, M., O'Malley, B.W., 1994. Annu. Rev. Biochem. 63 (1), 451–486.
Ucer, S., et al., 2016. J. Bone Miner. Res. 32, 560–574.
Vanderschueren, D., et al., 1997. Endocrinology 138, 2301–2307.
Watson, P.A., et al., 2015. Nat. Rev. Canc. 15 (12), 701–711.
Windahl, S.H., et al., 2013. Proc. Natl. Acad. Sci. U.S.A. 110, 2294–2299.
Yu, Z., et al., 2017. Clin. Cancer Res. 23 (24), 7608–7620.

Section F

Local regulators

Chapter 42

Growth hormone, insulin-like growth factors, and IGF binding proteins

Clifford J. Rosen[1] and Shoshana Yakar[2]

[1]Maine Medical Center Research Institute, Scarborugh, ME, United States; [2]David B. Kriser Dental Center, Department of Basic Science and Craniofacial Biology, New York University College of Dentistry New York, New York, NY, United States

Chapter outline

Introduction **985**
Physiology of the GH/IGF/IGFBP system **986**
Growth hormone releasing hormone 986
Growth hormone releasing hormone receptor 987
Somatostatin 987
Growth hormone 987
Mechanism of growth hormone secretion 987
Effects of gonadal status on the GH–IGF-I axis 988
Effects of the GH/IGF-I/IGFBP system on the aging skeleton 988
The IGF regulatory system and its relationship to the skeleton **989**
IGF-I, IGF–II, IGFBPs, and IGF receptors 989
Insulin-like growth factor binding proteins 991
Insulin-like growth factor binding protein proteases 993
Growth hormone/IGF actions on the intact skeleton **994**
GH–IGF-I systemic effects on body size and longitudinal growth 994
Growth hormone and IGF-I effects on modeling and remodeling 995
Insulin-like growth factors, other transcription factors, and osteoblasts 996
Insulin-like growth factors and osteoclasts 997
Insulin-like growth factors and osteocytes 997
Energy utilization by skeletal cells and the role of IGF-I 997
Pathogenic role of GH/IGF/IGFBPs in osteoporosis **998**
Effects of growth hormone deficiency on bone metabolism 998
Effects of growth hormone excess on bone mass and bone turnover 998
Changes in the GH–IGF-I axis in patients with osteoporosis 999
GH and IGF-I as treatments for skeletal disorders **1000**
Growth hormone treatment for skeletal disorders 1000
Growth hormone treatment for children with insufficient GH secretion 1000
Growth hormone administration for healthy adults 1001
Growth hormone treatment for adult-onset GH deficiency 1001
Growth hormone administration to elderly men and women 1002
Growth hormone treatment for osteoporotic patients 1003
Insulin-like growth factor I for the treatment of osteoporosis 1004
Overview 1004
Murine studies 1004
Human studies of insulin-like growth factor I and bone mineral density 1005
Limitations to the clinical use of recombinant human insulin-like growth factor I 1007
Summary **1007**
Acknowledgments **1007**
References **1007**
Further reading **1015**

Introduction

I has been 60 years since Salmon and Daughaday first reported the presence of a soluble factor induced by growth hormone (GH) that had insulin-like properties and mediated somatic growth. The subsequent identification of insulin-like growth factors (IGFs) has led to remarkable discoveries and therapeutic advances, particularly in relation to the skeleton. IGFs are 7 kDa peptides that are ubiquitous in nature and regulate cellular proliferation, differentiation, and death (Daughaday et al., 1972). The liver is the main source of circulating IGF-I; fat contributes about 10% and muscle about 5%. Although bone is a very minor contributor to circulating IGFs, it is one of the richest IGF tissues relative to the composition of other peptides

Principles of Bone Biology. https://doi.org/10.1016/B978-0-12-814841-9.00042-7

and growth factors (Finkelman et al., 1990). "Local"/bone tissue IGF pools play a predominant role in skeletal modeling and acquisition. IGFs are also highly conserved across species; hence, the physiology of these peptides can be studied in numerous animal models. Although GH has been recognized as the major determinant of liver-derived IGF-I, the balance between skeletal and serum IGF-I has also been clarified, particularly with the relatively recent introduction of targeted genomic engineering.

The importance of the GH/IGF system in aging has undergone intense scrutiny in the last 3 decades. Biological aging is a normal physiological process, part of the continuum from growth to death. Like other organ systems, skeletal homeostasis is maximized during the second and third decades of life, and IGFs are essential for that process. However, trabecular bone loss also begins by the third decade even though the mechanism is not clear. The hypothalamic–pituitary axis is profoundly affected by aging. GH secretion is reduced, resulting in lower levels of circulating IGF-I and IGF-II but higher levels of IGF binding proteins (IGFBPs) 1, 2, and 4 (Kelijman, 1991; Rudman et al., 1981). Early attempts to link age-related bone loss to a suppressed GH-IGF-I axis, or enhanced IGFBP expression, spawned considerable interest in GH and/or IGF-I as therapeutic tools for osteoporosis and sarcopenia. The advent of recombinant gene technology propelled synthetic growth factors into an ever-expanding therapeutic domain, particularly in regard to short stature. Another therapeutic venue for GH has been in its potential utility for the frail elderly. The paradigm that a "somatopause," or GH deficiency state with aging, produces discrete musculoskeletal changes, has never been firmly defined, nor has it been determined whether these changes can be reversed with GH. On the other hand, GH is indicated for individuals with pituitary insufficiency, and long-term studies suggest a major benefit for GH treatment in respect to bone mineral density (BMD) and quality of life. Regardless, more attention has focused on understanding the IGF regulatory system (i.e., IGFs, IGFBPs, Type I and II IGFRs, and proteases that cleave IGFBPs) in relation to bone acquisition and maintenance (Rudman et al., 1990). In this chapter, we will review the cellular and systemic actions of IGFs in order to more fully understand the functional integration of IGF regulatory components with other skeletal and systemic factors. As such, a thorough overview of the physiology of the IGF system, from GH to receptor signaling, is warranted.

Physiology of the GH/IGF/IGFBP system

Growth hormone releasing hormone

In order to appreciate the complexity and redundancy inherent in the IGF regulatory system in bone, an overview of the physiologic aspects of IGF-I regulation is required. Although IGF-I is regulated by genetic, nutritional, hormonal (insulin and thyroxine), and environmental factors, GH is the major regulator of IGF-I expression in tissues and dominates IGF-I

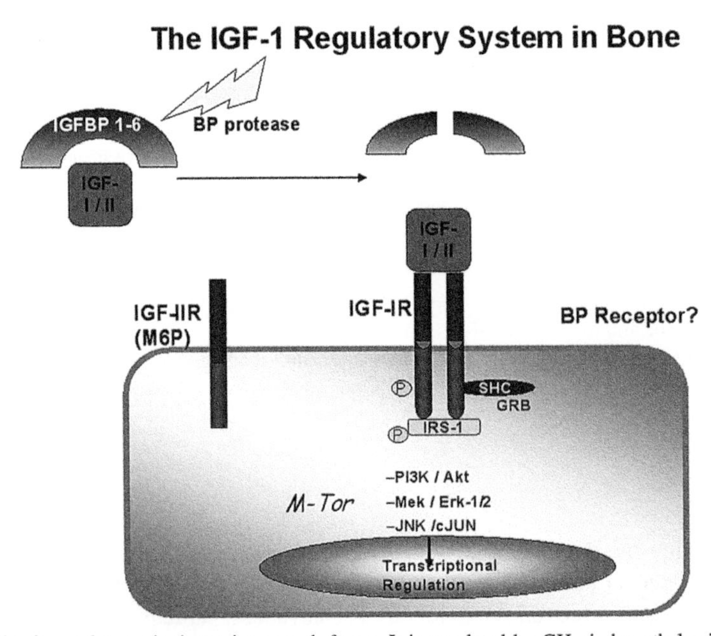

FIGURE 42.1 IGF-I is an endocrine and paracrine/autocrine growth factor. It is regulated by GH via hypothalamic GHRH. The primary source of circulating IGF-I is liver, although bone is a very rich source of IGF-I and IGF-II. Target tissues for IGF endocrine action include fat, muscle, and bone. IGF binding proteins (IGFBPs) circulate in excess of IGFs, but there is a ternary circulating IGF complex that includes the acid labile subunit (ALS) and IGFBP-3. GHBP is a growth hormone binding protein, a circulating GH receptor. The balance between circulating and skeletal IGF-I during growth and maintenance determines the overall effect of this peptide on bone.

levels in the circulation (Fig. 42.1). That process begins in the hypothalamus and ends at the tissue level. The regulation of GH secretion from the pituitary is complex and involves elaboration of discrete neurosecretory peptides from the hypothalamus. Growth hormone releasing hormone (GHRH) is a hypothalamic hormone essential for normal expansion of the somatotrope lineage during pituitary development (Frohman and Kineman, 2002). GHRH, via binding to GHRH receptor (GHRHR), acts on somatotropes to increase GH biosynthesis and secretion and is thought to cause somatotroph proliferation (Petersenn and Schulte, 2000). Release of GHRH is episodic, which accounts for the pulsatile release of GH from the pituitary. A decrease in GHRH is associated with somatotrope hypoplasia, whereas an increase in GHRH is associated with somatotrope hyperplasia (Frohman et al., 2000). Indeed, overexpression of human GHRH (hGHRH) (Mayo et al., 1998) in mice caused increased body size and skeletal gigantism. Excess in hGHRH led to systemic stimulation of endogenous GH and IGF-I, which in turn caused initial increase in bone mass (Tseng and Goldstein, 1998). However, with age, excess GH levels were associated with increases in bone resorption, thinning of the bone cortexes, and compromised mechanical properties. In contrast, a mutation in the mouse GHRH gene (lit/lit) resulted in growth retardation, where mice reached ~50%–60% of normal adult body size (Kasukawa et al., 2000). The blunted GHRH secretion inhibited GH release and IGF-I production in liver and other tissues, leading to significant decreases in BMD and reduced radial bone growth (periosteal circumference) (Kasukawa et al., 2000).

Growth hormone releasing hormone receptor

GHRHR is a G-protein-coupled receptor with seven hydrophobic transmembrane domains. The human GHRHR gene spans 15 kb and comprises 13 exons. The open reading frame was shown to extend 1269 bp and encode a protein of 423 amino acids with a predicted molecular weight of 47-kDa. After release from the hypothalamus, GHRH binds to the GHRHR predominantly located on the pituitary somatotrope (Lin-Su and Wajnrajch, 2002). GHRHR activation leads to the opening of a sodium channel in the somatotrope, which causes its depolarization. The resultant change in the intracellular voltage in turn opens a voltage-gated calcium channel, allowing for calcium influx, which directly causes the release of premade GH stored in secretory granules (Petersenn et al., 2000). The cAMP elevation stimulates protein kinase A, which phosphorylates and activates the transcription factor cAMP response element binding protein, which then stimulates de novo GH production (Petersenn et al., 2000; Muller et al., 1999; Mayo et al., 1995). Autosomal recessive mutations in the GHRHR can result in near total absence of GH and lead to short stature in humans as well as the little phenotype in mice with very low bone mass and reduced fat (Rosen and Donahue, 1998).

Somatostatin

Somatostatin (SMS) is a small (14-amino-acid) but ubiquitous polypeptide that inhibits GH synthesis and release. In concert with GHRH, SMS regulates GH secretion through a dual control system—one stimulatory, the other inhibitory. Several molecular forms of somatostatin, distinct from the native 14-amino-acid peptide, have been isolated. In addition to inhibition of GH release, SMS also inhibits secretion of thyrotropin as well as several pancreatic hormones, including glucagon and insulin. The SMS receptor has been localized to various cell types, especially those of neuroendocrine origin. Localization of this receptor suggests that SMS acts as an endocrine and a paracrine regulator in diverse tissues. A highly potent synthetic analog of SMS, octreotide, has been used therapeutically in acromegaly and diagnostically (in a radiolabeled form) for scintigraphic visualization of neuroendocrine tumors.

Growth hormone

Mechanism of growth hormone secretion

GH and its receptor belong to the cytokine superfamily of receptors and ligands. GH is an anabolic polypeptide hormone produced by the somatotropes of the anterior pituitary gland. GH is secreted under the influence of three hypothalamic hormones: (1) GHRH, which acts via the GHRHR; (2) GHS (also known as Ghrelin), which acts through the GHS receptor; and (3) somatostatin, SMS, which acts on the pituitary to suppress basal and stimulated GH secretion but is not believed to affect GH synthesis (Petersen et al., 2000; Lin-Su et al., 2002). GH secretion is pulsatile (due to the episodic release of GHRH) and circadian with the highest pulse amplitude occurring between 02:00 and 06:00 (Ho and Weissberger, 1990). Puberty has a dramatic effect on the amplitude of GH pulses due to changes in the hypothalamic milieu as a result of rising sex steroid concentrations (Jansson et al., 1985). Through binding directly to growth hormone receptor (GHR), GH has profound effects not only on somatic and bone growth but also on carbohydrate and lipid homeostasis. GH antagonizes insulin action in muscle and fat. It activates lipase in adipose tissue and leads to mobilization of fatty acids from the adipose tissue (Liu et al., 2004; Heffernan et al., 2001). In liver, GH inhibits de novo lipid synthesis (Liu et al., 2016a,b,c).

Deletion of the GHR gene in mice (GHRKO) results in a 50%–60% reduction in body size associated with low serum and tissue IGF-I levels (Liu et al., 2004). Linear skeletal growth of GHRKO mice is severely inhibited and associated with premature growth plate contraction and reduced chondrocyte proliferation. Long bones of GHRKO mice show reductions in trabecular bone volume and all cortical bone traits (Sjogren et al., 2000). Likewise, bone turnover in the GHRKO mice severely reduced and periosteal bone apposition prematurely ceased. Gene inactivation of the GHR mediator, STAT5ab, in mice resulted in a 50% decrease in serum IGF-I levels, reduced linear growth due to abnormalities in the growth plate, and decreased cortical bone area and width.

Effects of gonadal status on the GH–IGF-I axis

Not surprisingly, the pattern of GH secretion in animals and humans depends highly on age and sex (Ho and Weissberger, 1990; Jansson et al., 1985). Both factors strongly influence the frequency and amplitude of GH pulses, which in turn determines GH basal secretory rates, and the levels of serum IGF-I. Characteristic changes during puberty in rats parallel pubertal changes in humans (Jansson et al., 1985). GH secretion in male and female rats is identical after birth but at puberty, a sexually differentiated pattern of secretion appears with male rats displaying high-amplitude, low-frequency pulses, and with female rats displaying pulses of high frequency but low amplitude (Jansson et al., 1985). In humans, sexual differences in GH secretion during puberty are less pronounced, even though administration of gonadal steroids to prepubertal children increases GH pulses and mimics the pubertal milieu of the hypothalamus. Various sampling techniques (profiles vs. stimulatory tests) and assays with different sensitivities have produced disparate findings. However, spontaneous and stimulated GH peaks in humans are enhanced during puberty. Matched for age and body mass index, young girls were found to have higher integrated GH concentrations than boys (Ho and Weissberger, 1990). Other secretory characteristics, including pulse amplitude, frequency, and the fraction of GH secreted as pulses were similar in both sexes of the same age. In one study, black adolescents (males and females) had higher GH secretory rates than age-matched white adolescents (Wright et al., 1995). Higher GH secretion rates in adolescent blacks could lead to a greater acquisition of bone mass.

Clinical and animal studies have shown that sex steroids can modulate GH secretion at the level of the pituitary, as well as GH's action in target tissues such as liver and bone (Meinhardt et al., 2006). GH levels correlate positively with estrogen and estrogen replacement therapy in postmenopausal women independent of serum IGF-I levels (Norman et al., 2013). Mean pulse amplitudes of GH are lower in older women than in premenopausal women. GH secretory indices in postmenopausal women correlate with serum estradiol but not with total serum androgen levels. During menopause, GH secretion is reduced (Ho and Weissberger, 1990). However, oral administration of estradiol (or conjugated equine estrogens) increases GH secretion as a result of reduced hepatic generation of IGF-I (Ho and Weissberger, 1990; Dawson-Hughes et al., 1986). On the other hand, transdermal administration of 17– estradiol increases serum IGF-I concentration, suggesting that suppression of IGF-I by oral estrogens is due to a "first-pass" hepatic effect. Impaired IGF-I generation in the liver removes a key component of negative feedback on the hypothalamus, resulting in increased GH release.

Effects of the GH/IGF-I/IGFBP system on the aging skeleton

GH exerts a multitude of biological effects on various tissues through the GHR. Regulation of GH bioactivity occurs at several pre- and postreceptor levels. Growth hormone binding protein (GHBP) is a plasma binding protein identical to the extracellular domain of the tissue GHR. GHBP binds exclusively to GH, and most if not all serum GH is bound to this carrier protein. Measurements of GHBP in serum are relatively stable and reflect the endogenous status of the GHR in responsive tissues (Baumann et al., 1989). With advanced age, GHBP concentrations increase substantially.

The GH-IGF-I axis undergoes changes over a life span so that elders have lower spontaneous GH secretion rates and serum IGF-I levels than in younger people (Kelijman, 1991; Rudman et al., 1981; Donahue et al., 1990). Most of these age-related differences are a function of an altered hypothalamic–pituitary set point due in part to changes in lifestyle and nutrition. The GH secretory responses to common stimuli, such as GHRH, clonidine, L-dopa, physostigmine, pyridostigmine, hypoglycemia, and met-enkephalin, but not arginine, reduced with aging. Somatotrope responsiveness to GHRH and arginine does not vary with age, implying that the maximal secretory capacity of somatotropic cells is preserved in elderly people (Corpas et al., 1993).

Age-dependent declines in GH/IGF-I signals have been causally linked to frailty in respect to musculoskeletal function. However, large cross-sectional studies have demonstrated only a weak association between diminished serum IGF-I and age-related bone loss, or between serum IGF-I and BMD (Rudman et al., 1981; Nicolas et al., 1994; Donahue et al., 1990; Corpas et al., 1993; Langlois et al., 1998). In one large cohort study, the lowest serum IGF-I quartile was associated with a significantly greater risk of hip fractures (Garnero et al., 2000). On the other hand, skeletal concentrations of IGF-I, IGF-II, and IGFBP-5 in femoral cortical and trabecular bone decline significantly with age, and these have been associated with low BMD (Rudman et al., 1981; Nicolas et al., 1994; Donahue et al., 1990; Corpas et al., 1993; Langlois et al., 1998). In contrast to the multitude of studies linking serum IGF-I to age-related frailty and muscle performance, differences in GH secretion are difficult to determine due to its normal pulsatility. There is one study in older postmenopausal women relating changes in 24-hour GH levels with BMD (Dennison et al., 2003), and a recent large and prospective study showed that GH replacement therapy in postmenopausal women had positive dose-dependent effects on BMD and sustained effects on reducing patient fracture risk, even 7 years after the GH treatment ceased (Krantz et al., 2015). Lastly, a clinical trial performed in which GH treatment was given to elderly patients with hip fractures and resulted in some positive outcomes (Yeo et al., 2003). Overall, clinical data suggest distinct and overlapping effects of the GH/IGF-I axis on the aging skeleton.

The IGF regulatory system and its relationship to the skeleton

IGF-I, IGF–II, IGFBPs, and IGF receptors

The IGFs are single-chain polypeptides. IGF-I consists of 70 amino-acid residues, and IGF-II has 67 amino acids (Fig. 42.2). They have B, C, and A domains similar to proinsulin as well as a D domain that is not found in proinsulin. This D domain may sterically hinder the interaction of IGFs with the insulin receptor (IR), leading to only weak ligand binding of IGFs to the IR. A number of posttranscriptional and posttranslational variants of IGFs have also been described (Slootweg et al., 1990). These IGFs have variable affinities for IGFBPs and IGF receptors (IGFRs). In vitro, these growth factors may have significantly greater activity than that of native IGF-I or IGF-II, especially those that exhibit weak binding to IGFBPs.

IGF-I and IGF-II differ in their abilities to promote tissue growth due in part to the presence of two distinct IGFRs, IGF-IR and IGF-IIR (Le Roith et al., 1997) (see Fig. 42.1). IGF-IR is a tetramer consisting of two identical extracellular

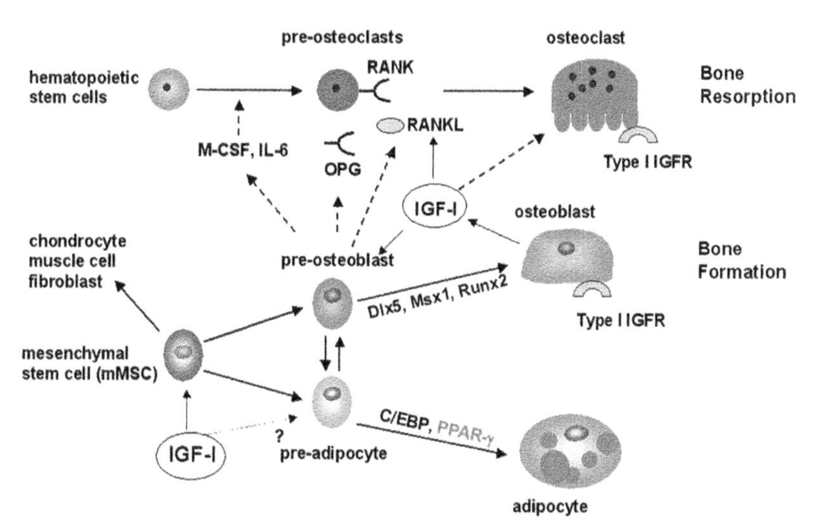

FIGURE 42.2 The IGF regulatory system contributes to bone formation through activation of the IGF type I receptor on osteoblasts. Downstream signaling is via three pathways: 1—PI3 kinase/Akt/mTor, 2—MAP kinase/ERK, and 3—JNK/Jun. PI3K/mTor regulates cell differentiation, while MAP/ERK and JNK/Jun impact cell growth. IGFBPs carry IGFs to the receptor; several proteases can cleave IGFs for binding to the type I receptor. The type II IGF receptor can bind IGF-II and generally is a clearance mechanism for IGF-II; it serves also as a mannose 6 phosphate binding site. Finally, IGFBP receptors have been theorized as possible IGF-independent signaling factors, but none have been proven. However, it is likely that IGFBPs could bind to other receptors (e.g., EGF) and affect cell growth and differentiation.

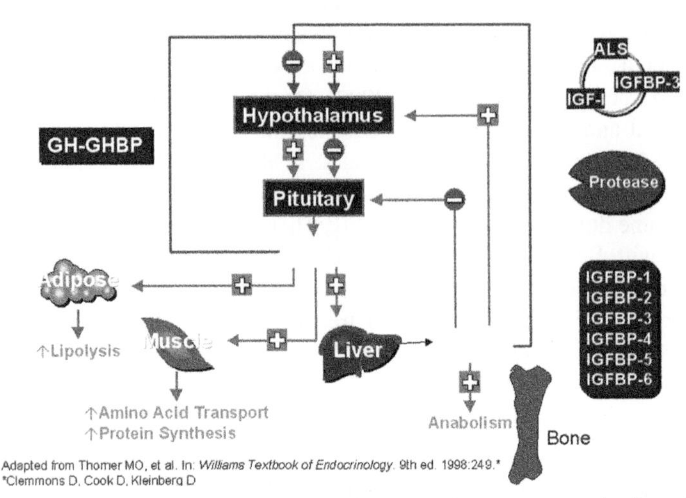

FIGURE 42.3 IGF-I serves as both a mitogen, albeit weak, and a differentiation factor for marrow stromal cells as they enter the osteoblast lineage. In addition, IGF-I can enhance osteoclast differentiation by stimulating RANKL production from stromal cells. There may be a direct effect of IGF-I on osteoclasts, because these cells express the type IGF-1 receptor and signal through IRS-2. IGF-I can also enhance adipocyte differentiation, and its role in promoting marrow adipogenesis is an area of recent investigations. Overall, IGF-I represents a coupling factor for bone remodeling.

αsubunits (conferring ligand-binding specificity) and two identical transmembrane β-subunits (possessing tyrosine kinase activity). IGF-IR resembles the IR and shares amino acid sequence homology (Zeslawski et al., 2001). IGF-II and insulin also bind to IGF-IR but with 2-fold or 15-fold to 1000-fold lower affinities, respectively (D'Ercole, 1996). It has an intrinsic tyrosine kinase activity critical for specific second message generation and indeed, ligand binding to the extracellular domain of IGF-IR results in autophosphorylation and tyrosine phosphorylation of IGF-IR substrates (Fig. 42.2). Tyrosine-phosphorylated insulin-receptor substrate-1 (IRS-1) and SHC bind different effector proteins involved in interconnecting pathways, including Ras/Raf-1/mitogen-activated protein kinase (MAPK, also known as extracellular signal-regulated kinase, or ERK) and phosphatidylinositol 3-kinase (PI3K)/phosphoinositide-dependent kinase-1 (PDK-1)/Akt (Maroni et al., 2004). Activation of the Ras/Raf-1/MAPK pathway is considered critical for cell proliferation, whereas the PI3K/PDK-1/Akt pathway is considered important for cell survival. The protein encoded by the phosphatase and tensin homologue gene deleted on chromosome 10 (PTEN) is a lipid phosphatase that decreases the activation of Akt by dephosphorylating phosphatidylinositol-3,4,5-triphosphate and phosphatidylinositol-3,4-biphosphate. Therefore, PTEN acts as an "off" switch for the PI3K/PDK-1/Akt pathway as well being a tumor suppressor. This tumor-suppressor gene is deleted or mutated in various types of human cancers (Zhao et al., 2004; Sansal and Sellers, 2004). On the other hand, the monomeric IGF type II receptor (IGF-IIR) does not bind insulin but does bind IGF-II, with a 500-fold increased affinity compared with IGF-I (Khandwala et al., 2000).

IGF-IIR exhibits no intrinsic kinase activity but is structurally very similar to the mannose 6-phosphate receptor, which is involved in targeting lysosomal enzymes intracellularly (Le Roith et al., 1997; Sell et al., 1995). Interestingly, it is now clear that in most tissues, except the hepatocytes in the liver, there are hybrid IGF-IR/IR receptors leading to significant cross-utilization between two rather distinct peptides. Hybrid receptors may also explain the growth-promoting activity of insulin, even in skeletal tissue, as well as the hypoglycemic effects of recombinant human IGF-I (rhIGF-I) when administered parenterally. However, there are little data on expression of hybrid receptors in bone.

IGFs stimulate cell proliferation and in some cases late differentiation. However, adequate nutrition is required for the full expression of IGF biologic activity in most tissues, particularly for linear growth. For example, during states of malnutrition, GH production increases but hepatic IGF-I generation is severely impaired. Resistance at the hepatic GHR reduces serum IGF-I and impairs GH bioactivity. The mechanism surrounding these nutrient-sensitive changes has not been clarified, although message stability is clearly reduced by undernutrition. For malnourished children, the result is cessation of linear growth. GH resistance, to lesser degrees, occurs in other conditions such as diabetes mellitus, acute catabolic stresses, and renal insufficiency.

IGFs are produced in virtually every tissue (Rosen et al., 1994). However, with new methods for conditional deletions of IGF-I in various tissues, it is clear that the main source of circulating IGF-I (>75%) is the liver. Other sources include muscle and adipose tissue (Rosen et al., 1994; Mohan and Baylink, 1990). Together these three sites contribute more than 95% to the circulating IGF-I pool. With acute or chronic hepatic insufficiency, both serum IGF-I and IGF-II levels are decreased, and particularly in hepatic osteodystrophy bone formation, and cortical bone mass is impaired due to the low levels of circulating IGF-I (Liu et al., 2017).

In the circulation, IGFs are bound to six identified IGFBPs with a relatively small but detectable amount of "free" IGF-I, which also circulates but has a very short half-life. GH treatment increases total and free IGF-I levels in a dose-dependent manner. However, rhIGF-I increases "free" IGF-I considerably more than rhGH, likely due to inhibition of GH secretion and the GH-dependent production of the acid labile subunit (ALS) and IGFBP-3, the main carriers of IGF-I in serum. The clinical significance of increases in "free" IGF-I versus that bound to IGFBPs is still not clear. IGF-II also circulates in large concentrations bound to IGFBPs, and a small proportion is free. It generally is produced in much greater amounts during prenatal periods. Absence of IGF-II from genetic deletion in the embryo leads to transient but reversible growth retardation in the first 3 weeks of life that is restored by the secondary increases in endogenous IGF-I (Uchimura et al., 2017). In humans in contrast to mice, there is more circulating IGF-II than IGF-I, and this is true for the skeletal composition of IGFs.

The distribution of IGFs in the serum pool is determined by the relative saturation of IGFBPs. This may explain why treatment with IGF-I may have different tissue effects than treatment with GH. Infusions of IGF-I produce a transient rise in free IGF-I and the suppression of IGF-II, insulin, and endogenous GH (Ebeling et al., 1993). During the course of an IGF-I infusion, however, IGF-I is partitioned into several pools. This is due to the unsaturated nature of the lower molecular weight IGFBPs, and the presence of a large (150-kDa) circulating ternary IGF-binding complex. This complex, composed of IGF-I (or II), IGFBP-3 (or IGFBP-5), and an ALS, is the major circulatory reservoir for both IGFs. Normally, the majority of circulating IGF is bound to this saturated intravascular complex. However, with rapid IGF-I infusions, some IGF-I goes into the lower (50-kDa) unsaturated IGFBP fractions where transport into the extravascular space is possible. Partitioning of IGFs into various binding pools is critical to the biologic activity of GH and IGF-I.

Insulin-like growth factor binding proteins

In the circulation and in all tissues, there are six IGFBPs. IGFBPs 1−6 belong to the same gene family, and several features distinguish these known IGFBPs from one other (Rajaram et al., 1997) (Fig. 42.1). Just as IGFBPs serve important regulatory functions within the circulation, their role at the tissue level is also critical for the full biologic expression of IGFs. Binding of IGFBPs to IGFs normally blocks the interaction between IGFs and IGF-IR and consequently suppresses IGF actions (Kelley et al., 1996; Collett-Solberg and Cohen, 1996). Nevertheless, binding of IGFBPs to IGFs can also protect IGFs from proteolytic degradation and consequently enhances IGF actions by augmenting their bioavailability in local tissues (Kelley et al., 1996; Collett-Solberg and Cohen, 1996). Therefore, most IGFBPs function in a bipotential manner, and their impacts on IGFs depend to a large extent on the posttranslational modification of IGFBPs by phosphorylation and proteolysis (Kelley et al., 1996; Collett-Solberg and Cohen, 1996; Coverley and Baxter, 1997; Claussen et al., 1997). Some IGFBPs (IGFBP-3 and IGFBP-5) exhibit IGF-independent or partially independent effects (Kelley et al., 1996; Mohan et al., 1996, Xi et al., 2016).

The predominant binding protein in serum (and bone) is IGFBP-3, a 43-kDa glycosylated peptide. It is present in large concentrations in the serum and is easily measurable by radioimmunoassay (RIA). As noted earlier, IGFBP-3 is part of a larger saturated ternary complex including IGF-I (or II) and an 80-kDa ALS. The association of these three proteins requires the presence of either IGF-I or IGF-II. In turn, this complex prolongs the half-life of IGFs and provides a unique storage site. The levels of circulating IGFBP-3 are principally controlled by GH (Martin and Baxter, 1988). However, IGFBP-3 synthesis outside the liver is regulated by other endocrine and paracrine factors. At a cellular level, IGFBP-3 has stimulatory and inhibitory effects on IGF-I depending on the cell type and the physiologic milieu. IGFBP-3 action at the cell is characterized by its interaction with IGF-I or −II. In vitro, coincubation of IGFBP-3 with IGF-I can block IGF access to IGF-IR (Rosen et al., 1994; Conover, 1991). Conversely, preincubation of IGFBP-3 in certain cell systems facilitates receptor binding of the ligand by attaching to the cell membrane at a site remote from the receptor. In addition, IGFBP-3 may have IGF-independent actions on cell action. Although a putative IGFBP-3 receptor has not been cloned, IGFBP-3 has been shown to downregulate cell proliferation in certain cell lines and to enhance p53 production. Further regulation of IGF-I by IGFBP-3 can occur in the extracellular space if IGFBP-3 undergoes proteolysis. Enzymatic degradation of IGFBP-3 produces low-molecular-weight IGFBP-3 fragments that differ in their affinities for IGFs (Cohen et al., 1992; Mohan, 1993). There are numerous nonspecific IGFBP-3 proteases, which are produced by various cell types, that can be found in the intra- and extravascular space and are regulated by endocrine and paracrine factors. Prostate-specific antigen (PSA) is a serine protease that cleaves IGFBP-3 and may be important in defining skeletal metastases with prostate cancer (Nunn et al., 1997).

IGFBPs 1–6 are also found in the circulation and may play both systemic and local roles in the regulation of IGF bioactivity at the tissue level. In contrast to IGFBP-3, these IGFBPs are not fully saturated, and they easily translocate from the circulation into the extracellular space. IGFBP-1 is a 30-kDa peptide produced primarily in the liver. Serum IGFBP-1 levels correlate inversely with circulating insulin, and in poorly controlled insulin-dependent diabetes mellitus (IDDM), serum IGFBP-1 levels are quite high (Brismar et al., 1988). Hepatic IGFBP-1 production is tightly regulated by insulin and substrate availability. However, unsaturated IGFBP-1 could also serve as a reservoir of binding activity for unbound IGF or as the initial binding site for cell-secreted IGF prior to transfer to the more stable GH-dependent 150-kDa complex. Shifts in the levels of IGFBP-1 may alter the distribution of IGFs among other IGFBPs and thus affect the relative distribution of IGFs between the intra- and extravascular space. This mechanism could be critical in controlling metabolic and mitogenic activities of IGFs (Arany et al., 1994). Indeed, mice overexpressing the hIGFBP-1 in hepatocytes (under the a1 antitrypsin promoter) show severe growth retardation (decreased linear growth) with mineralization defects and reduced BMD effects (Ben Lagha et al., 2006) that were likely mediated by sequestering IGF-I. Similarly, in vitro studies have shown that IGFBP-1 is synthesized by osteoblast and could inhibit IGF actions in bone during states of high IGFBP-1 production, such as starvation, and IDDM. In addition, Wang et al. have demonstrated that endocrine IGFBP-1 from the liver can regulate osteoclastogenesis and mediate the bone resorbing actions of fibroblast growth factor-21 (Wang et al., 2015).

Human IGFBP-2 is a 31-kDa protein that preferentially binds to IGF-II (Christiansen et al., 1991). It is found in high concentrations prenatally, and it is the major IGFBP in cerebrospinal fluid. IGFBP-2 is produced by most cells, but based on tissue-specific conditional genetic deletions, most circulating IGFBP-2 arises from the liver (DeMambro—personal communication). Insulin and dexamethasone have been shown to decrease the production of IGFBP-2 in rat osteoblasts (Schmid et al., 1992). Excess rhIGFBP-2 inhibits IGF-I–stimulated bone cell proliferation, bone collagen synthesis, and bone formation. Skeletal concentrations of IGFBP-2 are not nearly as high as the levels of IGFBPs 3–5. Interestingly, IGFBP-2 concentrations increase in puberty, with GH deficiency (GHD), and with malnutrition, as well as during aging. These changes are likely associated with circulating levels of GH and also IGF-I. Khosla et al. (1998) reported that pro-IGF-II coupled to IGFBP-2 is present in the circulation of patients with osteosclerosis due to hepatitis C infections, suggesting that IGFBP-2 may have a permissive role in enhancing skeletal turnover and in binding, through its heparin-binding domain, to extracellular matrices. In addition, animal studies using IGF-II and IGFBP-2 in a complex have demonstrated that this combination can prevent disuse- and ovariectomy-induced bone loss (Conover et al., 2002). Further support for that tenet comes from studies in the IGFBP-2-null mouse that have shown very low bone mass, particularly in males (Demambro, 2010). Importantly, Clemmons and colleagues reported that in appropriate concentrations, IGFBP-2 synergizes with IGF-I by binding to its own extracellular receptor RPTPb that inhibits PTEN activity and thereby enhances AKT-signaling from IGF-I (Xi et al., 2016). A small recombinant IGFBP-2 peptide fragment (HBD1) has also been shown to stimulate bone formation when pegylated and administered intermittently (Kawai et al., 2011).

IGFBP-4 is a glycosylated 24-kDa-binding protein. It is one IGFBP that is consistently inhibitory for IGFs in numerous cell systems. It was originally isolated from skeletal tissue and was found to inhibit IGF-mediated bone cell proliferation (Mohan et al., 1989b; Scharla et al., 1993). IGFBP-4 may serve as a reservoir for IGF-I and IGF-II in bone, and the relative proportions of IGF-I and IGFBP-4 are critical to define its function. Targeted overexpression of IGFBP-4 in osteoblasts under the osteocalcin promoter results in impaired overall postnatal skeletal growth, and reduced bone formation indices, likely via sequestration of IGF-I by IGFBP-4 and consequent impairment of IGF-I action in skeletal tissue (Zhang et al., 2003). The expression of IGFBP-4 in bone cells is regulated by cyclic adenosine monophosphate (AMP), parathyroid hormone (PTH), and 1,25-dihydroxyvitamin D. In addition, IGF-I stimulates IGFBP-4 proteolysis through the target enzyme, pregnancy-associated plasma protein-A (PAPP-A), thereby providing an autocrine/paracrine loop between the ligand and its binding protein (Durham et al., 1994). PAPP-A–deficient mice show growth retardation phenotype with reduced body size and low bone mass, suggesting a complex role for this autocrine–paracrine loop, particularly in the skeleton. Earlier studies showed that circulating IGFBP-4 was linked to hip fracture risk, although later studies have not validated this binding protein as a risk marker (Rosen et al. (1992). It is clear that there is an age-related increase in this binding protein, and in one case, a relatively strong correlation between PTH and IGFBP-4 (Honda et al., 1996)(Karasik et al., 2003).

More recently, it has been shown that global deletion of *Igfbp4* in mice resulted in sex-specific body composition and bone density phenotypes that have led to a reevaluation of the role of IGFBP-4 in peak bone acquisition. Maridas demonstrated that loss of *Igfbp4* affected mesenchymal stromal cell differentiation, regulated osteoclastogenesis, and influenced skeletal development and adult bone maintenance (Maridas et al., 2017a,b). Female *Igfbp4*-null mice had low trabecular and cortical bone mass, while male *Igfbp4* −/− were protected from changes in bone mass. In addition, fat mass was reduced in both sexes, and IGFBP-4 was found to be essential for adipogenic differentiation and important for depot-specific changes (Maridas et al., 2017a,b).

IGFBP-5 is a nonglycosylated 31-kDa IGFBP produced by osteoblasts and numerous other cell types. It is found in relatively high concentrations in both bone and serum and can be measured by RIA (Nicolas et al., 1994). IGFBP-5 synthesis increased by PTH and other cyclic AMP analogs (Conover et al., 1993). IGFBP-5 is the only IGFBP shown to consistently stimulate osteoblast cell proliferation in vitro, thus increasing the number of osteoblasts. Recent studies suggest that the mitogenic effects of IGFBP-5 may in part be independent of IGFs and mediated through rhIGFBP-5's own signal transduction pathway (Mohan et al., 1995; Andress and Birnbaum, 1991; Schmid et al., 1996; Slootweg et al., 1996; Richman et al., 1999). IGFBP-5 is the most abundant IGFBP stored in the bone, where it is bound to hydroxyapatite and extracellular matrix proteins (Nicolas et al., 1995; Mohan et al., 1995). Intact IGFBP-5's major role in the skeleton may be as a storage component for IGFs, because both IGF-I and IGF-II have very low binding affinity for hydroxyapatite but bind avidly to IGFBP-5 (Rosen et al., 1994). During remodeling, resorption enhances proteolytic cleavage of IGFBP-5. In addition, during formation and mineralization, synthesis and release of IGFBP-5 by bone cells facilitates attachment of IGFs to the newly mineralized matrix (Koutsilieris and Polychronakos, 1992). Intact IGFBP-5 can also be found circulating in the extracellular space. In the circulation, IGFBP-5 can also bind ALS and IGF-I/IGF-II, providing an alternative ternary complex to the main complex ALS–IGFBP-3–IGF-I. In vitro, IGFBP-5 enhances IGF bioactivity, especially in bone. Yet, in vivo, IGFBP-5 action can be either inhibitory or stimulatory depending on the relative concentration of IGF-I and its mode of administration (Salih et al., 2005). Overexpression of IGFBP-5 ubiquitously in mice (under the CMV promoter) associates with reduced BMD in a gender- and an age-dependent manner, with males affected more severely (Salih et al., 2005). Expression of IGFBP-5 in osteoblasts (under the osteocalcin promoter) results in decreases in trabecular bone volume, impaired osteoblastic function, and osteopenia.

IGFBP-6 has been less well studied and slightly differs structurally from the other IGFBPs. Generally, IGFBPs are composed of three domains of approximately equal sizes (Baxter, 2000). The amino- and carboxyl-terminal domains are each internally disulfide-linked and share a high degree of sequence homology across the family (Neumann and Bach, 1999). However, little homology is exhibited among their central L-domains. The disulfide linkages of the amino-domain of IGFBP-6 differ from those of other IGFBPs, whereas the carboxyl-domain disulfides are the same for all IGFBPs so far studied (Neumann and Bach, 1999; Forbes et al., 1998; Chelius et al., 2001). IGFBP-6 differs functionally from other IGFBPs in binding IGF-II, with 20- to 100-fold higher affinity than IGF-I, whereas IGFBPs 1–5 do not have a marked IGF binding preference (Headey et al., 2004). Therefore, IGFBP-6 constitutes a relatively specific IGF-II inhibitor (Bach, 1999).

It is important to note that no mutations have been identified for any IGFBPs in humans. The importance of the ALS in keeping the IGF-I reservoir in serum was not appreciated for many years, until the generation of ALS-null mice (ALSKO) (Uekl et al., 2000) and the identification of several subjects with mutations in the *igfals* gene (Domene et al., 2004). The ALS plays fundamental roles in stabilizing two binary complexes of IGF-I with IGFBPs 3 and 5. Subjects with mutations in the *igfals* gene present with low levels of serum IGF-I, short stature, and reduced BMD (Domene et al., 2010). Similarly, ALSKO mice show ~60% reductions in serum IGF-I levels; they are born at normal weight but show decreases in body size during early postnatal growth (3 weeks) and remain ~20% smaller than controls throughout life (Courtland et al., 2010). Detailed morphology of the femurs by microCT revealed reduced total cross-sectional area, cortical bone area, polar moment of inertia, and reduced mechanical properties (by four-point bending assay).

Insulin-like growth factor binding protein proteases

IGF bioactivity is transcriptionally regulated by hormones and paracrine factors but also by their liberation from their respective binding proteins. Tissue-specific proteases provide another level of regulation of IGFs. Binding-protein-targeted proteases have been identified in serum and in various tissues including bone. These proteases alter the binding capacity of IGFs for IGFBPs, thereby freeing IGFs to bind to their respective IGFRs (Rosen et al., 1994). PSA and PAPP-A represent two of the most important circulating serine proteases that liberate IGFs from their IGFBPs, although these two enzymes are not specific for these binding proteins. Bone is a rich source of binding protein proteases. Proteases that cleave IGFBP play a pivotal role in determining the regulatory effects of IGFBPs on IGF actions. Three categories of known proteases proteolyze IGFBPs: (1) kallikreins; (2) neutral and acid-activated cathepsins; and (3) matrix metalloproteinases (MMPs) (Cohen et al., 1992, 1994; Conover et al., 1995; Fowlkes et al., 1994a). PSA is the first serine kallikrein IGFBP protease found to proteolyze both IGFBP-3 and IGFBP-5 into several lower molecular weight fragments and is regulated at least to some extent by testosterone and other androgens (Collett-Solberg and Cohen, 1996; Cohen et al., 1992; Koutsilieris et al., 1992). Its role in mediating the enhanced bone formation found in the lumbar spine of metastatic prostate cancer patients remains to be determined. However, it is likely one of several mechanisms whereby IGFs stimulate the mitogenic activity of cancer and bone cells.

The concept of "soil and seed" implies that the inherent bioactivity of IGFs, whether bound or free, could stimulate growth of neoplastic tissue after homing of cancer cells to bone. A kallikrein-like IGFBP protease, 7S subunit nerve growth factor, is found to cleave both IGFBP 4 and 6. Cathepsin D is an acid-activated lysosomal protease that can cleave all six IGFBPs. MMPs are able to proteolyze a spectrum of IGFBPs including IGFBPs 2–5 (Collett-Solberg and Cohen, 1996; Fowlkes et al., 1994b; Rajah et al., 1996; Claussen et al., 1997; Marinaro et al., 1999; Braulke et al., 1995; Manes et al., 1999). PAPP-A, which is not specific for IGFBP-4, is activated by IGFs and found within skeletal tissue as well as other organs. Particularly for PAPP-A, it is clear that IGFs can regulate tissue-specific proteases, thereby establishing a complex regulatory loop in which the ligand (i.e., IGF) controls its own bioavailability through transcriptional and nontranscriptional means (Arany et al., 1994). Global deletions of PAPP-A result in smaller mice with reduced bone mass, while targeted transgenic overexpression of PAPP-A in muscle causes hypertrophy and increased body weight (BW). A human inactivating mutation in PAPA-2, a key protease of IGFBP-3 and IGFBP-5 protease, was reported in three subjects. These subjects presented with elevated serum IGF-I levels, elevated levels of IGFBPs 3 and 5, and increased levels of ALS, and showed growth failure and reduced BMD that were attributed to the inability to liberate IGF-I from its major binding proteins (IGFBPs 3 and 5) in the serum and the bone matrix (Dauber, 2016). IGF-I and IGF-II can regulate PAPP-A, but GH does not regulate PAPP-A or other proteases (Angelloz-Nicoud and Binoux, 1995; Skjaerbaek et al., 1998). Insulin and estrogen may also affect IGFBP protease activities (Bereket et al., 1995; Bang et al., 1998; Kudo et al., 1996).

Growth hormone/IGF actions on the intact skeleton

GH–IGF-I systemic effects on body size and longitudinal growth

Liu et al. (1993) were the first to report that newborn mice homozygous for a targeted disruption of Igf1 exhibit a growth deficiency of similar severity to that previously observed in viable Igf2-null mutants (i.e., 60% of normal birth weight). Depending on the genetic background, Liu found that some of the Igf1(−/−) dwarfs died shortly after birth, whereas others survived and reached adulthood (Liu et al., 1993). Conversely, null mutants for the Igf1R gene invariably die at birth of respiratory failure and exhibit a more severe growth deficiency (45% of normal size) compared with wild-type animals. In addition to generalized organ hypoplasia in IgfR(−/−) embryos, including the muscles and developmental delays in ossification, deviations from normalcy were found in the central nervous system and the epidermis. Igf1 −/−/Igf1R(−/−) double mutants did not differ in phenotype from Igf1R (−/−) single mutants, while in Igf 2(−)/Igf1R(−/−) and Igf1(−/−)/Igf 2(−) double mutants, which are phenotypically identical, dwarfism was further exacerbated (i.e., 30% of normal size).

To investigate the role of IGF-I in normal development, Powell-Braxton et al. generated mice with an inactive Igf1 gene. Heterozygous Igf1(±) mice are healthy and fertile; however, they are 10%–20% smaller than wild-type littermates and have lower levels of IGF-I (Powell-Braxton et al., 1993). The size reduction is attributable to a decrease in organ size and bone mass. This was also confirmed by the Kream laboratory, further describing a defect in osteoblast function as the major mechanism for reduced bone mass and size (He et al., 2006).

As previously noted, at birth homozygous mutant Igf1 (−/−) mice are less than 60% the BW of wild-type, and greater than 80% of the pups die perinatally. The survivors are sometimes compromised in terms of homeostatic processes, but the compensatory mechanisms in survivors are interesting. For example, Bikle et al. (2001) analyzed the structural properties of bone from mice rendered IGF-I deficient by homologous recombination. The knockout mice were 24% the size of their wild-type littermates at the time of study (4 months). The knockout tibias were 28%, and L1 vertebrae were 26% the size of wild-type bones. Bone formation rates (BFRs) of knockout tibias were 27% that of wild-type littermates (Bikle et al., 2001). The bones of mutant mice responded normally to GH (1.7-fold increase) and supranormally to IGF-I (5.2-fold increase) with respect to BFR. Cortical thickness of the proximal tibia was reduced 17% in the knockout mouse. However, trabecular bone volume (bone volume/total volume [BV/TV]) in knockout mice compared with wild-type controls was increased 23% in male mice and 88% in female mice as a result of increased connectivity, increased number, and decreased spacing of the trabeculae. Thus, absence of IGF-I causes a smaller bone that may be more compact, almost certainly due to reduced osteoclastic activity as noted earlier. The structural consequences of these bones in respect to fracture have not been studied.

As indicated previously, IGF-I acts in an endocrine (serum) and autocrine/paracrine fashion. Yakar et al. (1999) used the Cre/loxP recombination system whereby mice with loxP-flanked Igf1 gene were mated with albumin-Cre transgenic mice expressing the Cre recombinase exclusively in the liver. Liver-specific Igf1 gene-deletion (LID) mice were macroscopically normal, suggesting that autocrine/paracrine IGF-I could support normal postnatal growth and development (Yakar et al., 1999). Nevertheless, more extensive developmental phenotyping of the LID mouse revealed a marked reduction in bone volume, periosteal circumference, and medial lateral width consistent with the hypothesis that circulating (endocrine) IGF-I had an important role in bone modeling (Yakar et al., 2002). Moreover, recent studies from Yakar et al. have shown that ALSKO mice also have reduced cortical thickness and enhanced trabecular bone, consistent with an endocrine effect of the circulating IGF complex on skeletal acquisition (Yakar et al., 2006). Finally, double-ALS and IGF-I knockout mice have a major growth phenotype as well as markedly reduced BMD despite normal expression of skeletal IGF-I (Yakar et al., 2006). Thus, it seems likely that both local and circulating IGF-I concentrations are essential for peak bone acquisition and maintenance.

GH has distinct IGF-I–independent effects on the skeleton in terms of linear growth and bone remodeling. However, due to the intimate relationship between these two hormones and their feedback loop regulatory system, it has been extremely difficult to ascertain their independent roles. Double-GH/IGF-I knockout mice exhibit significant growth retardation, suggesting that each of these two growth factors contributes distinctly to longitudinal growth (Lupu et al., 2001). Interestingly, in those studies only 17% of the growth in these mice could be attributed to non-GH and non-IGF-I determinants (Lupu et al., 2001). GH may have its own effect on linear growth independent of IGF-I. For example, GH stimulates longitudinal bone growth in normal rats, but rhIGF-I does not (Schmid and Ernst, 1991). Similarly, transgenic mice that overexpress GH grow to twice their normal size, even though administration of IGF-I to normal mice does not provoke a similar growth response. The presence of both GHR and IGFRs on bone cells complicates the interpretation of GH's action.

Longitudinal growth results from the activity of GH on the cartilaginous growth plate. In human bone, proliferating chondrocytes express IGF-IRs and are responsive to paracrine IGFs secreted by differentiated cartilage cells (Isaksson et al., 1987). The target for GH in the growth plate is the differentiated chondrocyte that synthesizes IGF-I in response to GH. Proliferating chondrocytes respond to locally produced IGF-I by differentiation, which in turn leads to cartilage expansion and linear growth. Thus, GH's stimulatory properties on the endochondral growth plate are mediated largely by induction of IGF-I. A detailed analysis of the growth plate of IGF-I–null mice showed that chondrocyte cell number and proliferation were preserved in IGF-I nulls despite significant (~35%) reductions in longitudinal growth, findings attributed to the high levels of GH in those mice (Wang et al., 1999). Further defects were found in chondrocyte hypertrophy, metabolism, and matrix production, suggesting that IGF-I plays a role in chondrocyte hypertrophy. In a separate study, the direct cellular targets for GH in the growth plate were mapped in normal, GH-deficient, and fasting mice and rats (Gevers, 2009). The authors confirmed that GH stimulated prechondrocytes and early chondrocytes but had no effect on hypertrophic chondrocytes. Chondrocyte-specific IGF-IR gene deletion using the type II collagen I (Col2a1)-driven Cre resulted in a more severe phenotype with shorter bones, disorganized chondrocyte columns, delayed ossification, and undermineralized skeletons (Wang et al., 2011). However, chondrocyte-specific IGF-I gene deletion using the same promoter (Col2a1) resulted in decreased body length, areal BMD, and bone mineral content, likely due to the compensatory effects of circulating IGF-I. Collectively, mouse studies strengthen the "dual effector hypothesis," which suggests that GH affects the resting cells of the growth plate, the perichondrium, and the groove of Ranvier to drive prechondrocytes at this region to proliferate, while locally produced IGF-I drives chondrocyte proliferation (clonal expansion at the proliferating zone) and possibly chondrocyte hypertrophy (Green et al., 1985).

Growth hormone and IGF-I effects on modeling and remodeling

Skeletal modeling during growth, and remodeling during maturation and aging, are the sum of several distinct events beginning with activation of lining cells, differentiation of osteoblast and osteocyte formation, and recruitment of osteoclasts. These processes engage multiple signaling pathways. Besides the large quantities of IGFs, there are other growth factors such as transforming growth factor-beta and fibroblast growth factors that are released during active resorption, ensuring that bone formation will be coupled to bone resorption (Mohan et al., 1991). IGFs also stimulate the differentiation and activation of osteoclasts, possibly in concert with cytokines such as receptor activator of nuclear factor Kappa β ligand (RANKL) and macrophage colony-stimulating factor (Slootweg et al., 1992; Mochizuki et al., 1992). Administration of IGF-I enhances bone formation and bone resorption to a similar degree (Ebeling et al., 1993). The acidic pH within the microenvironment of the osteoclast during active resorption permits activation of several proteases, including PAPP-A (Conover et al., 1994), that liberate IGFs from IGFBPs and allow for further recruitment of osteoblasts and osteoclasts to the remodeling site.

In vitro, GH, IGF-I, and IGF-II all have modest mitogenic effects on bone cell growth (Canalis et al., 1989; Mohan et al., 1988). This suggests that GH could act through IGFs to activate skeletal remodeling. Indeed, GH-induced cell proliferation can be blocked by simultaneous addition of a specific monoclonal antibody to IGF-I (Mohan et al., 1991). IGF-II, on the other hand, stimulates mitogenesis independent of GH, and even if high doses of IGF-I are coadministered. This implies that IGF-II could regulate osteoblastic proliferation via IGF-IIR (Mohan et al., 1989a). In vitro, both IGF-I and IGF-II are mitogenic to rodent preosteoblasts, and both rapidly increase mRNA expression of the protooncogene, c-fos, 20- to 40-fold in less than 30 min (Merriman et al., 1990). IGFs also stimulate type I collagen synthesis, alkaline phosphatase activity, and osteocalcin in more differentiated human osteoblast-like cells (Mohan et al., 1989; Schmid et al., 1984). Taken together, IGFs are important for osteoblast activity, particularly in response to intermittent parathyroid hormone, but the effects are dose and time dependent.

Insulin-like growth factors, other transcription factors, and osteoblasts

The exact roles of IGFs in osteoblast and osteoclast differentiation are complex. In part this relates to the developmentally sensitive and finely orchestrated process that drives mesenchymal stromal cells (MSCs) into the osteogenic lineage. Not unexpectedly, this pathway involves multiple transcription factors and cytokines as well as IGFs. Hence, IGFs act directly on MSCs to drive cell growth and differentiation and also synergize with other growth factors (e.g., Wnts) and their signaling pathways (Akt, MAPK, p38, etc.) to promote osteogenesis.

Two critical transcription factors that control osteoblast fate are runt-related transcription factor 2 (Runx2) and osterix (Osx). Runx2, core binding factor 1/polyoma enhancer binding protein 2a, a transcription factor that belongs to the Runx family, is the subunit of a heterodimeric transcription factor, PEBP2/CBF, which is composed of α and β subunits (Ito, 1999). Runx2 is involved in osteoblast differentiation and bone formation. In particular, Runx2 is required for early commitment of mesenchymal stem cells into osteoprogenitors and also functions later in osteoblast differentiation to regulate the formation of the extracellular matrix (Ducy et al., 1999). IGF-I influences Runx2 expression through activation of PI3K, Pak1, and ERK (Xiao et al., 2000; Qiao et al., 2004).

Schnurri-3 (Shn3), a large zinc finger protein that functions as an adapter protein in the immune system (Oukka et al., 2002), has been shown to control the posttranslational protein expression level of Runx2 through promoting Runx2 degradation by recruitment of the WW domain containing E3 ubiquitin ligase1 to Runx2. Mice lacking Shn3 display adult-onset osteosclerosis with increased bone mass due to an augmented osteoblast activity (Jones et al., 2006). Glucose itself may also regulate Runx2 post translationally via Smurf1 that is regulated by AMP-activated protein kinase (AMPK) (Wei et al., 2015).

Osx (Sp7) is a master zinc-finger-containing transcription factor of osteoblast lineage progression that is highly specific to osteoblasts in vivo and acts downstream of bone morphogenetic protein (BMP)-2 Smad signaling (Xiao et al., 2000; Qiao et al., 2004; Nakashima et al., 2002). The Osx amino acid sequence predicts three C2H2-type zinc fingers with a high degree of identity to similar DNA-binding domains in the transcription factors Sp1, Sp3, and Sp4. The expression of Osx is more specific to osteoblasts than Runx2 (Nakashima and de Crombrugghe, 2003). Because no Osx transcripts are detected in the skeletal elements of Runx2-null mice, Osx must be downstream of Runx2 in the pathway of osteoblast differentiation (Nakashima et al., 2002). Both IGF-I and BMP-2 are shown to upregulate Osx expression during early osteoblast differentiation (Celil et al., 2005). In mesenchymal stem cells, it appears that both MAPK and PKD signaling pathways serve as points of convergence for mediating the IGF-I- and BMP-2-induced effects on Osx expression (Celil et al., 2005). IGF-I-mediated Osx expression required all three MAPK components (Erk, p38, and JNK), whereas BMP-2 required p38 and JNK signaling, and the synergistic interactions of BMP-2 and IGF-I were also disrupted by PKD inhibition (Celil et al., 2005).

It should be noted that Runx2-independent pathways of ossification also exist, including (1) the Wnt signaling pathway; (2) the Msx2-dependent vascular ossification pathway; and (3) Osx induction via Dlx5, a homeobox transcription factor, which acts downstream of BMP-2 (Kato et al., 2002; Cheng et al., 2003; Lee et al., 2003). These studies suggest additional pathways may act in parallel to, or independent of, Runx2 to regulate Osx expression during osteogenic lineage progression. Indeed, in the canonical Wnt signaling pathway, there is cross talk with the IGF pathway. For example, β-catenin binds to IRS-1 as well as other factors to enhance its transport from the cytoplasm into the nucleus where it can affect a whole series of downstream target genes.

In sum, IGFs enhance differentiation of mesenchymal stem cells. Targeted ablation of IGF-I in osteoblasts resulted in severe growth retardation and impaired bone mineralization. Reduced bone formation in Col1a2-IGF-IKO mice is attributed to decreased osteoblast function (Kesavan et al., 2011). Conditional deletion of IGF-IR in osteoblasts using the osteocalcin promoter-driven Cre recombinase led to low bone mass (Zhang et al., 2002; Wang et al., 2007). Histomorphometric indices demonstrated sufficient osteoblasts but reduced bone formation and mineralization rates, thereby implying that IGF-I must be important in late differentiation and mineralization (*See IGF actions on the skeleton*).

Moreover, IGF-I is essential for adipogenic differentiation and likely is downstream of peroxisome proliferator-activated receptor gamma (PPARγ) activation. Because mesenchymal stromal cells can enter several distinct lineages, the precise timing of IGF-I action and other factors, active in the marrow niche, some of them undefined, ultimately determine cell fate (Fig. 42.2).

Insulin-like growth factors and osteoclasts

In addition to regulating osteoblast function, several groups have shown that skeletal IGF-I can stimulate osteoclast recruitment and differentiation either directly through IGF-IR or via RANKL expression (Mochizuki et al., 1992; Rubin et al., 2002). This would place IGF-I in the category of a "coupler" for bone remodeling. Indeed, IGF-I-null mice are growth retarded, and most die after birth. Those that survive are very small and have developmental defects in brain, muscle, bone, and lung, and are infertile. However, their skeletal phenotype is particularly striking and is characterized by very little cortical bone but an increased trabecular bone volume fraction (Bikle et al., 2001, 2006). Recently, it has been demonstrated in vivo and in vitro that the absence of IGF-I impairs osteoclast recruitment and activity, although the exact mechanism is unknown (Wang et al., 2006).

Insulin-like growth factors and osteocytes

Osteocytes are terminally differentiated osteoblasts buried in the mineralized matrix and are the most abundant cells in bone. Osteocytes play fundamental roles in bone remodeling and bone response to mechanical loading. These cells are the first responders to crack or fracture and are responsible for sending signals to recruit osteoclasts to resorb and remodel the damaged site. The roles of GH or IGF-I in osteocytes function were partially elucidated recently. Using the DMP-1 promoter-driven cre, the GHR, IGF-IR, or IGF-I genes were inactivated in osteocytes. Ablation of GHR in osteocytes did not affect body weight, body length, body composition, or organ weights but led to slender bones with reduced total cross-sectional area and cortical bone area in both sexes. Likewise, Dmp-1-mediated *Igf-1 receptor (Igf-1r)* gene—deleted mice (DMP-IGF-IRKO) showed normal body and organ weights but exhibited significant reduction in cortical bone area and a reduced cortical bone thickness in both sexes. Notably, cortical bone area decreased by ~20% while marrow area increased by 40%—50% compared with controls, suggesting that endosteal resorption increased following IGF-IR ablation in osteocytes. A similar approach was used to knock down IGF-I in bone (DMP-IGF-IKO mice) (Sheng et al., 2013). Skeletal characterization showed slender bones with reduced total cross-sectional area and reduced cortical bone area in both sexes of DMP-IGF-IKO mice. Collectively, inactivation of the GH/IGF-I axis in osteocytes leads to significant decreases in cortical bone with minimal (~5%) effects on linear bone growth, suggesting that this axis regulates not only longitudinal growth but also radial bone growth via its actions on osteocytes.

As indicated above, osteocytes are the bone mechanosensors that initiate the anabolic response to loading or trigger the resorption process when damage occurs. In vitro and in vivo models of mechanical loading suggest that IGF-I mediates these processes. *Igf1* gene expression increases following mechanical stimulation (Lau, 2006; Reijnders et al., 2007). Mice with *Igf1* gene deletion in osteocytes showed blunted bone response to four-point bending, which was associated with significant decreases in the expression levels of cox2, c-fos, and *igf-1* (the early mechanosensitive genes) in bone and reduced Wnt signaling (Lau, 2013). Skeletal unloading, on the other hand, induced bone loss that was also shown to be dependent on IGF-I. Hind limb suspension in rats inhibited activation of IGF-IR signaling pathways and was associated with a reduced number of osteoblast progenitors, their differentiation, and reduced bone formation (Sakata et al., 2004).

Energy utilization by skeletal cells and the role of IGF-I

Both modeling and remodeling are energy requiring processes. Hence, understanding substrate utilization and ATP production, particularly by osteoblasts, is critical to defining the processes inherent in peak bone mass and maintenance. Glucose has long been known as a major nutrient for osteoblasts and other cells, and IGF-I can modulate glycolysis. Recent work with radiolabeled glucose analogs has confirmed a significant uptake of glucose by bone in the mouse (Wei, 2015; Zoch, 2016). The Glut transporters appear to be mainly responsible for glucose uptake in osteoblast lineage cells. Expression studies detected Glut1 and Glut3 in osteoblastic cell lines (Thomas et al., 1996; Zoidis et al., 2011). More recently, Glut1 was shown to be a major transporter in primary osteoblast cultures and modulates the post-translational modification of Runx2 by suppressing AMPK and blocking ubiquitination of Runx2 (Wei, 2015); selective deletion of Glut1 in osteoblast precursors suppressed osteoblast differentiation in vitro and in vivo. However, others have reported that Glut4 increases in neonatal calvarial osteoblast cultures in response to beta-glycerophosphate and ascorbate, or insulin, even though genetic deletion of Glut4 does not cause an obvious skeletal phenotype (Li et al., 2016).

Recent work has linked the bone anabolic function of Wnts and IGFs with increased aerobic glycolysis in osteoblast-lineage cells (Esen et al., 2013; Lee et al., 2017). Both share common regulatory pathways, the former activating mTORC2 (mammalian target of rapamycin complex 2) and Akt that in turn acutely increase the protein abundance of a number of key glycolytic enzymes (e.g., Hk2, Pfk1, Pfkfb3 [6-phosphofructo-2-kinase/fructose-2,6-bisphosphatase 3], and Ldha), the latter acting through IRS-1 and AKT. But IGF-I is also a major target of PTH's anabolic actions on osteoblasts. PTH has been shown by several investigators to increase glycolysis, and that is presumed to be the source of ATP for enhanced collagen synthesis and bone formation (Zoidis et al., 2011). Studies in MC3T3-E1 cells have uncovered that PTH stimulates aerobic glycolysis via activation of IGF signaling, which in turn activates the PI3K-mTORC2 cascade and upregulates metabolic enzymes such as Hk2, Ldha, and Pdk1 in a manner analogous to the Wnts. Importantly, genetic studies have demonstrated that deletion of either *Igf1* or *Igf1r* in osteoblasts essentially abolishes the bone anabolic effect of PTH in mice (Bikle et al., 2002; Wang et al., 2007). And blocking glycolysis with 2DG or other inhibitors completely prevents osteoblast differentiation. Hence, the mechanisms that allow IGFs to drive osteoblast function are multiple and include several fuel pathways. Undoubtedly, the major driver for osteoblast differentiation is glucose, but fatty acids and glutamine contribute in a temporal-specific manner (Lee et al., 2017).

Pathogenic role of GH/IGF/IGFBPs in osteoporosis

Effects of growth hormone deficiency on bone metabolism

GHD in childhood is associated with growth failure and short stature. However, the effects of GHD on BMD in prepubertal children have been more difficult to quantify, in part because of the mixed hypopituitary syndromes that often accompany GHD. Thus, there is a paucity of studies examining BMD status in GH-deficient children. By single-photon absorptiometry of the wrist, children with GHD have been found to have low BMD (Wuster et al., 1992). Serum concentrations of osteocalcin are also reduced in children with GHD, but the response of osteocalcin to GH administration does not correlate with linear growth (Delmas et al., 1986). In several cross-sectional studies of adults with GHD, lumbar spine BMD was reduced compared with that of age-matched controls (Wuster et al., 1991; Johansson et al., 1992; Hyer et al., 1992; Rosen et al., 1993; Bing-You et al., 1993; DeBoer et al., 1994). In one group of adult GHD patients, the lowest spinal BMD was found in people who were previously treated with rhGH during childhood (Wuster et al., 1991, 1993). This degree of osteopenia was not due to cortisone or thyroxine substitution, because the BMDs of patients on hormonal substitution did not differ from those without hormone replacement (Wuster et al., 1993). In that same study, Wuster et al. (1991) showed an increased prevalence of vertebral osteoporotic fractures among GH-deficient adults. Kaufman et al. (1992) confirmed low BMD in GHD adults with or without hormonal deficiencies. However, Kann et al. (1993) found no difference in the apparent phalangeal ultrasound transmission velocity of GHD patients compared with age- and sex-matched controls. De Boer et al. (1994) noted that low BMD was partly explained by reduced body height, but with correction for body mass index, BMD was still significantly reduced compared with age- and sex-matched controls (Wuster et al., 1992).

The cause of low BMD in adult GHD has been thought to be due to insufficient bone acquisition during the adolescent years (DeBoer et al., 1994). This hypothesis is supported by bone histomorphometry. In 36 men with GHD (primarily of juvenile onset), there were increased eroded surfaces, increased osteoid thickness and increased mineralization lag time, all indicative of delayed mineralization that was probably due to changes in the timing of puberty (DeBoer et al., 1994; Bravenboer et al., 1994). In support of those histomorphometric changes, low serum levels of osteocalcin have been detected in some adult GHD patients (Delmas et al., 1986). This is in sharp contrast to patients with normal GH secretion but multiple pituitary hormone deficiencies, where serum osteocalcin levels are normal but there is markedly increased urinary pyridinoline excretion (de la Piedra et al., 1988). Fracture rates are higher in individuals with GHD, and it is established that chronic GH treatment in GHD prevents further fractures, although interestingly it does not affect quality of life (Krantz, 2015).

Effects of growth hormone excess on bone mass and bone turnover

Chronic GH excess in adults (i.e., acromegaly) has been a surrogate model for studying the effects of GH on the skeleton. Acromegaly is a multisystem progressive disorder caused by uncontrolled secretion of GH, most commonly by a pituitary adenoma (Capatina et al., 2015). A major biochemical characteristic of acromegaly is the increased production of liver-derived IGF-I. In fact, most systemic complications associated with acromegaly can be attributed to the effects of

IGF-I, which mediates the somatic and some of the metabolic effects of GH (Capatina et al., 2015). Subsequently, the objectives of treatment are twofold—inhibition of tumor growth and controlling the secretion of IGF-I.

The effects of excess GH on the skeleton in acromegaly are complicated by GH-mediated changes in vitamin D metabolism in the kidney, secondary effects of GH on PTH production, and gonadotropin secretion (Ezzat et al., 1993; Bouillon, 1991). Increased bone turnover has been reported in acromegaly by biochemical markers and histomorphometric studies (de la Piedra et al., 1988; Ezzat et al., 1993). Studies reporting BMD in patients with acromegaly used different morphometric modalities, and the findings are inconsistent. However, several cross-sectional studies suggested that acromegaly associated with increased skeletal fragility, particularly increased vertebral fracture risk (Wassenaar et al., 2011; Mazziotti et al., 2008; Ueland et al., 2002; Kayath et al., 1997) that does not always correlate with decreased BMD. In one study, computed tomography (CT) measurements of the lumbar spine revealed that trabecular BMD was elevated in only 1 of 14 patients with active acromegaly (Ezzat et al., 1993). This may have been due to hypogonadism in the acromegalics. Wuster et al. (1993) recently studied five patients with active acromegaly treated with octreotide for 5 years. All had achieved normal IGF-I levels during therapy. Spinal BMD was initially decreased in all five patients but normalized in three of them with octreotide. All patients remained eugonadal throughout follow-up.

As noted earlier, biochemical markers of bone turnover are altered in acromegaly. Many of these changes can be related to alterations in gonadal status during disease and its treatment. However, changes in bone turnover with acromegaly reflect persistent coupling of the remodeling cycle with increased resorption and formation. Serum osteocalcin and skeletal alkaline phosphatase are increased in acromegaly, as are urinary calcium and hydroxyproline excretion (de la Piedra et al., 1988; Ezzat et al., 1993). Although serum calcium, total alkaline phosphatase, and phosphorus are usually normal, there may also be increased synthesis of 1,25-dihydroxyvitamin D. This results from significant intracellular phosphate shifts due in part to an increased circulating IGF-I.

Changes in the GH–IGF-I axis in patients with osteoporosis

For several years, attempts have been made to link GH secretory status with low BMD and osteoporosis. As noted previously, efforts to find a relationship between GH secretion and age-related bone loss have been conflicted at best. However, other investigators have examined the relationship of GH to BMD in the immediate menopausal period. These efforts gained prominence in the 1980s when it was reported that GH secretion in patients with osteoporosis was reduced even after stimulation with L-arginine. Low serum IGF-I, IGF-II, and IGFBP-3 levels (by RIA) were noted in 98 females with postmenopausal osteoporosis compared with 59 normal controls and 91 patients with osteoarthritis or degenerative bone disease (Wuster et al., 1993). In a cross-sectional study of a large cohort of older postmenopausal women from Framingham, Langlois et al. (1998) reported very strong correlations between the lowest quintile of IGF-I and BMD at the spine hip and radius. Bauer et al. reported that in the Study of Osteoporotic Fractures, women in the lowest quartile for serum IGF-I had a 60% greater likelihood of hip or spine fractures, even when controlling for BMD (Bauer et al., 1997). Gamero et al. (2000) noted that low serum levels of IGF-I were associated with a significantly greater risk of hip fractures among a large cohort of older postmenopausal women in France. In a study of 61 community-dwelling men over the age of 27, who were randomly selected from the Calgary cohort of 1000 subjects in the Canadian Multicentre Osteoporosis Study, IGF-I was found to be a significant predictor of BMD at the total hip, femoral neck, and femoral trochanter neck (p # 0.001). Szulc et al. (2004) evaluated the correlation of BMD with serum IGF-I in a large cohort of 721 men aged 19–85, considering age, BW, 17 b-estradiol, free testosterone, and PTH. Serum IGF-I decreased with age (r 5 20.44, p 5 0.0001). IGF-I correlated positively with BMD at the whole body and at the third lumbar vertebra. BMD of the total hip was 6% higher in men in the highest quartile of IGF-I than in men in the lowest quartile. However, others have not found a relationship between serum IGF-I and BMD in patients with fractures, in postmenopausal osteoporosis, or in otherwise healthy subjects (Kassem et al., 1994; Lloyd et al., 1996; Rudman and Mattson, 1994).

In male osteoporotic patients, serum IGF-I and IGFBP-3 concentrations are low and correlated with lumbar BMD (Ljunghall et al., 1992; Nakamura et al., 1992). Johannsen et al. (1994a,b) noted that among healthy males, IGFBP-3 was the best predictor of femoral BMD. (Kurland et al., 1997) reported that younger males with idiopathic osteoporosis had low serum levels of IGF-I in relation to age-matched controls. Similar results were noted in premenopausal women (Rubin et al., 2005). Moreover, those men also had low rates of bone turnover by histomorphometry but normal GH dynamics (Kurland et al., 1998), similar to women with idiopathic osteoporosis (Shane). Of potential pathophysiological importance is the observation that patients with osteoarthritis have higher concentrations of IGF-II than those of normal controls (Wuster et al., 1993; Mohan et al., 1991). Other studies have related serum IGFBP-4 and IGFBP-5 due to aging and low bone mass, although causality was not established (Karasik et al., 2002). Also, Dennison and colleagues reported that low

24-hour GH profiles were associated with low BMD of the lumbar spine in older British women (Dennison et al., 2003). Further longitudinal studies will be required to determine the precise relationship among IGFs, GH, and osteoporosis.

GH and IGF-I as treatments for skeletal disorders

Anabolic agents can directly stimulate bone formation and therefore might have great potential to decrease fractures. Therefore, there is an emerging interest in developing anabolic agents including PTH, PTHrp, GH, and IGF-I to treat osteoporosis (Rubin et al., 2002).

Growth hormone treatment for skeletal disorders

GH has direct and indirect effects on bone depending on age and skeletal maturity. Indirectly, GH can enhance bone mass through its effects on muscle mass and calcium transport in the gut through IGF-I–mediated increases in 1 alpha hydroxylase that generates active 1,25 vitamin D, as well as by suppressing adipocyte differentiation in different fat depots including bone marrow (Fleet et al., 1994; Bianda et al., 1998; Menagh et al., 2010). In addition, GH can directly stimulate bone remodeling and increase endochondral growth through its actions on the osteoblast. Overall, GH is considered essential for the growth and maintenance of skeletal mass. Moreover, it is established that for virtually all cohorts of GHD subjects, whether onset is in childhood or adulthood, male or female, there is reduced areal BMD. In some but not all studies, volumetric BMD, measured either by CT or areal-adjusted algorithms, was reduced in children with GHD (Baroncelli et al., 1998). In the largest observational trial of GHD subjects to date—i.e., KIMS—GHD was associated with a marked increase in fracture risk, particularly when compared with age-matched normal controls (Rosen et al., 1997; Wuster et al., 2001). Hence, there is a strong rationale to treat GHD in children.

Substantial differences between the direct and indirect (i.e., via IGF-I) effects of GH on the osteoblast, the marrow stromal cell precursor, and the osteocyte may partially explain changes in skeletal responsiveness to GH and IGF-I. For example, exogenous GH stimulates longitudinal growth in normal rats, but rhIGF-I does not (Schmid et al., 1991). Similarly, transgenic mice who overexpress GH grow to twice their normal size, whereas exogenous administration of IGF-I is far less efficient in stimulating long bone growth. Thus, despite the fact that GH induces IGF-I production in the skeleton and elsewhere, treatments with GH and with IGF-I are not equivalent. In general, skeletal responsiveness to GH and IGF-I depend on the species, GH status of the animal, and mode of administration. Systemic side effects of rhGH and rhIGF-I therapy may differ substantially.

Growth hormone treatment for children with insufficient GH secretion

Intermittent (daily or three times weekly) injections of rhGH to children with GHD resulted in a prolonged and sustained GH profile with resultant catch-up growth evident during the first year of treatment (Rappaport and Czernichow, 1993). This increase in skeletal growth was accompanied by a rise in serum levels of type I procollagen peptide. Although dosage schemes varied between the United States and Europe (0.1 mg/kg/tiw [USA] to 0.7 U/kg/week[Europe]), there was a strong dose-related growth response to rhGH (de Muinck Keizer-Schrama et al., 1992). Indeed, most studies of preadolescent GHD children have shown significant improvements in areal BMD with GH replacement. However, the skeletal response to GH depends on several factors including (1) GH secretory status; 2) pretreatment IGF-I levels; (3) pretreatment height velocity; and (4) GH dosage (de Muinck Keizer-Schrama et al., 1992). The rate of change in serum IGF-I (rather than the absolute level of IGF-I attained by GH treatment) is a relatively nonspecific predictor of growth, as are procollagen-I and osteocalcin concentrations (Delmas et al., 1986; de Muinck Keizer-Schrama et al., 1992). Serum procollagen III levels correlate with growth rates during GH treatment (Delmas et al., 1986).

Linear growth is a measurable response to exogenous GH, but changes in BMD in children are more difficult to quantify. In some studies, bone mineral content (BMC) is increased during GH treatment to a greater extent than expected for change in bone size (Inzucchi and Robbins, 1994). In one of the longest intervention trials to date, 26 GHD children were given rhGH (0.6 IU/kg per week) for 12 months (Saggese et al., 1993). Baseline radial BMC z scores (corrected for their chronological, statural, and bone ages) were significantly reduced, as were serum osteocalcin and procollagen peptide levels. Treatment with rhGH six times per week increased BMC and normalized the z scores of the radius in nearly 50% of the subjects. Serum levels of procollagen peptide during the first week of treatment were positively related to growth velocity at 6 and 12 months and radial BMC at 12 months. In another nonrandomized trial, 32 children with GHD aged 7–16 were treated for nearly 1 year with rhGH and found to have significant improvements in areal BMD and final adult height (Saggese et al., 1996). In adolescent GHD subjects, GH replacement had more variable effects on peak bone acquisition. Even with higher rhGH doses, significant changes in volumetric BMD have not been found, nor has acceleration in skeletal maturation (Baroncelli et al., 2004; Mauras et al., 2000; Kamp et al., 2002). Controversy continues as to whether GH treatment impacts BMD in children with idiopathic short stature or children born small for gestational age (SGA). Arends et al. (2003) demonstrated that prepubertal SGA children given 33 ug/kg/d of rhGH for 3 years had significant increases in height, areal BMD, and areal adjusted spine BMD. However, rhGH administered to children with idiopathic short stature who had low volumetric BMD did not result in further increases in BMD despite significant changes in lean body mass and bone turnover indices (Hogler et al., 2005). In cerebral palsy children (aged 4–15 years), 18 months of rhGH (50 mg/day) in a randomized placebo-controlled trial RPCT, areal BMD increased significantly, as did height, IGF-I, IGFBP-3, and osteocalcin (Ali et al., 2007). In sum, rhGH improves adult height, and areal BMD in prepubertal GHD children treated for at least 1 year. These changes are accompanied by favorable effects on body composition, muscle strength, and overall quality of life. It is still uncertain, however, how beneficial these GH-induced effects are in late adolescent GHD subjects in non-GHD states, or whether true volumetric BMD is significantly improved by long-term GH therapy.

Growth hormone administration for healthy adults

Although there were striking differences between longitudinal growth in children and remodeling in adults, criteria that determine rhGH responsiveness in children still may be relevant for older individuals. It has already been established that biochemical and histomorphometric responses to rhGH in children may differ according to their GH secretory status. The same principle probably holds for adults treated with GH. Three adult populations have been studied before and after GH treatment in order to examine predictors of skeletal responsiveness: (1) healthy adults, 2) GHD adults, and (3) elderly men and women with/without osteoporosis.

Initial studies with rhGH in adults focused primarily on changes in body composition. Short-term treatment with rhGH leads to a decrease in adiposity and an increase in lean body mass (Crist et al., 1988). There is also a marked shift in extracellular water (Holloway et al., 1994). Detailed analysis of skeletal biomarkers during GH treatment (1990) was reported in 20 male volunteers (ages 22–31) given a relatively large dose (0.1 IU/kg) of rhGH twice daily for 7 days. Serum osteocalcin increased after 2 days of treatment and remained elevated for 6 months. Bone alkaline phosphatase decreased initially (during the 7 days of GH treatment) but then increased slightly over 6 months (Brixen et al., 1990). Serum calcium and phosphate increased but only during the 7-day treatment phase. Like bone formation indices, urinary markers of bone resorption (urinary Ca/Cr and hydroxyproline/creatinine) rose during treatment and remained elevated for up to 4 weeks after the discontinuation of therapy.

Treatment with rhGH stimulates bone remodeling. More importantly, the anabolic effect on bone may persist well beyond discontinuation of GH. Early (2-day) and late (2-week) osteocalcin responses imply that GH can stimulate existing osteoblasts and enhance recruitment of new osteoblasts. Still, it is uncertain whether those effects are mediated through IGF-I. For example, Brixen et al. (1990) were unable to find a significant correlation between the rise in serum IGF-I and an increase in osteocalcin or bone alkaline phosphatase. The absence of a significant correlation between bone formation markers and serum IGF-I, however, may be due to the relatively small changes in of serum IGF-I.

Growth hormone treatment for adult-onset GH deficiency

GHD can be diagnosed by insufficient GH secretion in serial measurements in response to provocative stimuli (GHRH, insulin, glucagon) and serial GH measurements. The majority of adult patients treated with rhGH have either idiopathic GHD or a history of previous central nervous system/pituitary-hypothalamic tumors. Early trials with rhGH replacement therapy examined changes in muscle mass, muscle strength, and body fat. Daily administration of subcutaneous rhGH to GHD patients produced a marked rise in serum IGF-I and an increase in muscle mass and basal metabolic rate (Jorgensen et al., 1991). Some of those anabolic changes were noted soon after the initiation of rhGH. For example, mean nitrogen retention during the first 15 days of rhGH treatment was as much as 2.8 g per day (approximately 20 g of muscle mass) (Valk et al., 1994). GH treatment can also increase the total cross-sectional area of thigh muscles and quadriceps as well as improve hip flexors and limb girdle strength (Jorgensen et al., 1991; Valk et al., 1994). At least one group has suggested that rhGH can increase the number of type II muscle fibers. Total fat mass, however, consistently decreases during rhGH treatment (Jorgensen et al., 1989, 1991).

Several biochemical parameters reflect the pharmacologic action of GH on the skeleton. Serum calcium, osteocalcin, and urinary hydroxyproline all increase, while PTH declines slightly during rhGH treatment. Newer and more sensitive markers of bone turnover also reflect changes during rhGH treatment. Urinary deoxypyridinoline increases threefold, and the amino-terminal propeptide of type III procollagen doubles, during 4 months of daily rhGH (Christiansen et al., 1991; Johansen et al., 1990). After cessation of rhGH treatment, deoxypyridinoline excretion decreases, but type III procollagen levels remain higher than controls for several months (Johansen et al., 1990). Serum osteocalcin and procollagen 1 N-terminal propeptide markers of bone formation also increase significantly with rhGH therapy in adults, similar to the change noted in children.

If prolonged GHD in adults results in profound changes in the musculoskeletal system, then GH replacement would be expected to enhance muscle performance and subsequently BMD. When 0.25 IU/kg/week of rhGH was administered to 12 GHD adults for 1 year, there was a marked increase in trabecular BMD (measured by single- and dual-energy QCT of the spine) at 6 and 12 months. At 12 months proximal and distal forearm BMC increased, midthigh muscle area was greater, and fat cross-sectional area decreased. Because the rise in spine BMD was noted with both single- and dual-energy CT measurements of the spine, it is possible that the enhancement in BMD was significant and not related to a reduction in marrow fat.

Several groups have performed longer nonrandomized trials with rhGH in the GHD syndrome. Changes in BMD were not significant at 12 months; however by 24 and 36 months BMD has been reported to increase by as much as 5%−8% in the spine (Janssen et al., 1998; Baum et al., 1996; Papadakis et al., 1996). Krantz et al., in a randomized double-blinded control trial, demonstrated that 10 years of GH could reduce fractures (Krantz et al., 2015). In addition, other investigators have reported a concomitant increase in muscle strength after 2 years of rhGH treatment. It appears from those studies that individuals with earlier onset of GHD, as well as those with the lowest BMD, had the greatest likelihood of showing significant changes in BMD with rhGH. More recently, four randomized placebo-controlled trials have been conducted in adult GHD subjects treated with rhGH for at least 18 months. In one study of men only, BMD increased in the lumbar spine by 5.1% and the femoral neck by 2.4% (Baum et al., 1996). In another study of both men and women, there were no significant differences in BMD after 18 months between the group treated with rhGH and those taking placebo (Sneppen et al., 2002). In the third trial, men but not women showed increases in spine BMD after 24 months of rhGH compared with untreated controls. The most definitive and most recent study was a true randomized placebo-controlled trial using physiologic rather than pharmacologic doses of rhGH in GHD patients. In contrast to the three previous studies, changes in serum IGF-I were titrated within the normal range rather than to the supraphysiologic levels achieved with higher doses of rhGH. Interestingly, in that study of 67 men and women, spine BMD increased after 18 months by nearly 4% in men, which was statistically different from the placebo control but increased much less in women, and those changes were not statistically different from placebo (Snyder et al., 2007). Neither gender showed a significant change in hip BMD in response to rhGH in doses up to 12 ug/kg/day. In sum, there are clear sex- and dose-dependent effects of rhGH on BMD in GHD subjects treated for at least 18 months.

Growth hormone administration to elderly men and women

As previously noted, elderly people have lower GH secretory amplitudes and reduced serum levels of IGF-I and IGFBP-3 compared with younger adults (Kelijman, 1991; Rudman et al., 1981; Donahue et al., 1990). Moreover, the pulse frequency for GH is reduced in the elderly population. Based on these data, it was assumed that skeletal responsiveness to GH in elders would be identical to that seen in GHD patients. In elderly men, one group reported a blunted serum IGF-I response to 0.1 mg/kg GH (36% lower) compared with that in younger men or adults with GHD (Lieberman et al., 1994). However, Rosen et al. (1999) and others have noted that the generation of IGF-I after various doses of rhGH to frail elders was not associated with significant GH resistance. GH replacement for adult GHD or for pharmacologic treatment results in similar IGF-I responses independent of age. Side effects such as fluid retention, gynecomastia, and carpal tunnel appear to be more common in the elderly who are given rhGH compared with young adults who have GHD.

Rudman gave 21 men over age 65 0.03 mg/kg of rhGH three times per week (as a subcutaneous injection), or placebo, in a randomized controlled trial. Twelve men received rhGH while nine men served as observational controls. The men were selected on the basis of a low serum IGF-I concentration (Rudman et al., 1990). The rhGH produced a threefold rise in circulating IGF-I, an increase in lean body mass (as measured by 40K analysis), and a decline in total adipose mass. BMD of the lumbar vertebrae (L1−L4) as measured by dual-energy X-ray absorptiometry (DXA) increased 1.6% after 6 months in the treatment group, while no change was noted in controls. Biochemical markers of bone turnover were not examined, and no changes in BMD were detected in the mid or distal radius or three areas of the hip. Furthermore, the spinal BMD changes at 6 months were not sustained at 1 year.

Marcus et al. studied the effects of rhGH in 16 men and women over age 60 (Marcus et al., 1990). Daily doses of rhGH (0.03, 0.06, or 0.12 mg/kg BW/day) were randomly assigned to each subject and administered once daily for 7 days. Serum IGF-I, osteocalcin, PTH, and calcitriol concentrations all increased during treatment. In this short-term study, there was also a significant rise in urinary hydroxyproline and urinary calcium excretion with a concomitant decline in urinary sodium excretion.

Holloway et al. conducted a longer double-blind RPCT of daily rhGH for 1 year in 27 healthy elderly women, 8 of whom took a stable dose of estrogen throughout the study (Holloway et al., 1994). Thirteen women completed 6 months of treatment, and fourteen women completed 6 months in the placebo group. Side effects prompted a 50% reduction in the original dose of rhGH (from 0.043 mg/kg BW or approximately 0.3 mg rhGH/kg/week to 0.02 mg/kg/day) and led to several dropouts in the treatment group. Fat mass and percentage body fat declined in the treatment group, but there were no changes in BMD at the spine or hip at 6 or 12 months in other groups (Holloway et al., 1994). Although BMD did not change, there were changes in some biochemical parameters. In particular, urinary markers of bone resorption (e.g., hydroxyproline and pyridinoline) increased after 6 months of rhGH treatment. The response of bone formation markers was more variable. Osteocalcin increased, but type I procollagen peptide levels did not change. For women taking estrogen replacement therapy, indices of bone turnover (both formation and resorption) were blunted.

Rosen et al. (1999) reported that there was a dose-dependent decrease in total body BMD after 1 year of rhGH in frail elderly men and women. This occurred despite striking increases in osteocalcin and serum IGF-I with the highest doses of rhGH (0.01 mg/kg/day). In part, the absence of a GH effect on BMD is not surprising, because resorption is coupled to formation and GH activates the entire remodeling sequence. Moreover, the skeletal response was only measured after 1 year of treatment; this was probably inadequate to determine the true effect of rhGH on BMD as noted from earlier studies in GHD. Not surprisingly, in the same trial of 132 frail elderly subjects by Rosen and colleagues, urinary N-telopeptide and osteocalcin rose to the same extent, suggesting that total bone turnover, not just bone formation, was increased by rhGH therapy. The relatively high incidence of acute side effects (weight gain, carpal tunnel syndrome, edema, and glucose intolerance) in GH trials, especially in the frail elderly, has remained particularly troublesome even with titrating doses. Moreover, high serum IGF-I levels for long periods may predispose individuals to certain malignancies. Thus, there is limited enthusiasm for rhGH or rhGHRH treatments in the frail elderly.

Growth hormone treatment for osteoporotic patients

Short nonrandomized clinical trials with GH in osteoporosis were attempted well before GH replacement therapy was approved in pituitary insufficiency. As early as 1975, two patients with osteogenesis imperfecta and one patient with involutional osteoporosis were treated with GH. Histomorphometric parameters of increased bone formation and resorption were noted. Subsequent studies employed GH with and without antiresorptive agents. Aloia et al. (1987) administered between 2 and 6 U/day of GH for 12 months to patients with postmenopausal osteoporosis (the first 6 months of treatment featured low-dose GH; the last 6 months consisted of high-dose GH (6 U/day). Radial BMC dropped slightly, and histomorphometric parameters did not change during treatment. However, severity of back pain decreased considerably in several people. Daily GH injections (4 U/day) combined with alternating doses of calcitonin produced an increase in total body calcium (measured by neutron activation analysis) but a decline in radial bone mass after 16 months. In a separate trial, 14 postmenopausal women were given 2 months of GH, then 3 months of calcitonin, in a modified form of coherence therapy. Total body calcium increased 2.3%/year with few side effects, but there were no changes in BMD or histomorphometric indices. One study administered 16 U of rhGH every other day along with daily sodium fluoride to six women with postmenopausal osteoporosis. On histomorphometric analysis, there was a significant increase in the number of osteoblasts and osteoclasts, but BMD was unchanged.

Johansson et al. (1994) conducted a crossover double-blinded RPCT of rhGH and idiopathic osteoporosis. In this 7-day trial with rhGH 2 IU/m2, procollagen peptide and osteocalcin levels increased after treatment as did urinary markers of bone resorption. The changes in osteocalcin were relatively small and were not sustained after discontinuation of GH treatment. Thus, GH stimulates bone remodeling activity, thereby leaving open the possibility that GH can be coupled to antiresorptive agents. This thesis was tested in a 2-year randomized trial by Holloway and colleagues (Holloway et al., 1997). In that study, rhGH and nasal calcitonin increased spine BMD by approximately 2%. This, however, was not much different than the use of CT alone, and certainly less than what has been seen in very large randomized trials with antiresorptive agents. Once again, there were several side effects that produced limited enthusiasm for rhGH as a primary treatment for osteoporosis. In another combination trial that was larger and longer, 80 osteoporotic postmenopausal women on hormone replacement therapy (HRT) (estrogen with or without progestin) were administered rhGH 10 2.5 U/day or placebo for 18 months, and then open-label rhGH for another 18 months. These women were then followed for an additional 24 months. Women given GH and HRT had a marked increase in total body and spine BMD compared with placebo, which was maintained to year 3 but disappeared by year 5 (Landin-Wilhelmsen et al., 2003). This trial suggested that combination therapy of an anabolic and an antiresorptive could be used in postmenopausal osteoporosis. A similar result was noted in a 7-year follow-up of 30 men and women who received rhGH for 4 years and then were treated with alendronate for an additional 3 years. BMD increased significantly versus a control group after 3 years, especially in the males, and the addition of alendronate further enhanced spine BMD at year 7 (Biermasz et al., 2004). Therefore, it is likely that GH may induce small changes in BMD that over an extended period could translate into fewer spine fractures as noted by Krantz et al.

Insulin-like growth factor I for the treatment of osteoporosis

Overview

In the late 1980s, clinical trials with rhIGF-I for diabetes mellitus were begun. The availability of this recombinant peptide and the absence of other treatments to stimulate bone formation accelerated animal and human studies of rhIGF-I in metabolic bone diseases. Theoretically, there are potential benefits for rhIGF-I compared with rhGH. These include (1) more direct stimulation of bone formation; (2) bypass of skeletal GH resistance; and (3) reduction in GH-induced side effects such as carpal tunnel and diabetes mellitus. There are, however, considerably fewer animal and human studies using rhIGF-I than those using rhGH. Therefore, these advantages have either yet to be fully realized or have not been validated.

IGF-I is not a potent mitogen in most tissues, and bone is no exception (see Chapter 8 by Canalis). There are high-affinity receptors for IGF-I expressed on osteoblasts, and IGF-I can stimulate preosteoblast replication and provoke resting cells to proceed through their growth cycles. IGF-I maintains the differentiated osteoblast phenotype, stimulates collagen synthesis, and prevents collagen degradation. Theoretically, therefore, despite its relatively weak mitogenic properties, IGF-I could potentially have significant anabolic activity on the skeleton.

Murine studies

In hypophysectomized rats, growth can be fully restored by administration of either GH or IGF-I but not IGF-II (Schoenle et al., 1985). A similar growth response occurs after rhIGF-I in streptozotocin-diabetic rats but not in sex-linked dwarf-mutant chickens (Schoenle et al., 1985; Tixier-Boichard et al., 1992). In normal rats, rhIGF-I administrated either systemically or locally (hind limb infusions) does not stimulate longitudinal bone growth. In the spontaneously diabetic BioBreeding (BB) rat, rhIGF-I treatment did not result in changes in epiphyseal width, osteoblast surfaces, or osteocalcin concentration (Verhaeghe et al., 1992).

The skeletal response to rhIGF-I is determined by the GH/IGF-I status of the animal. For example, IGF-I does not increase bone formation in normal rats, whereas it stimulates bone growth and normalizes type I procollagen mRNA levels in hypophysectomized rats (Schmid et al., 1989; Spencer et al., 1991; Tobias et al., 1992). Similarly, in the spontaneous mouse mutant (lit/lit), GHR absence results in very low levels of IGF-I and skeletal dwarfism. IGF-I treatment restores growth and increases total body water but does not enhance BMD in these mice (Donahue et al., 1993). These findings are somewhat similar to the effects of GH on the skeleton in GHD animals. However, rhIGF-I and rhGH differ in their actions on the circulatory IGF regulatory system. GH stimulates hepatic production of IGF-I, IGFBP-3, and ALS, while rhIGF-I administration increases the total circulating pool of IGF-I but suppresses hepatic production of IGFBP-3, primarily through feedback inhibition of GH secretion. It is conceivable that variations in IGF-I biological activity (between direct IGF-I administration and endogenously produced IGF-I as a result of GH treatment) may be due to the relative proportion of free versus bound IGF-I to IGFBP-3.

Several experimental paradigms have been employed to study the effects of IGF-I on bone turnover in animals. These include (1) oophorectomy, (2) diabetes mellitus (spontaneous or induced), and (3) immobilization. In oophorectomized rats, administration of rhIGF-I has variable effects on bone remodeling, BMD, and bone strength. In older oophorectomized rats, rhIGF-I increased midshaft tibial BMD and enhanced periosteal bone apposition (Ammann et al., 1996). Six weeks of rhIGF-I (delivery by mini-osmotic pump) to older rats caused a dose-dependent increase in BMD in the lumbar spine and proximal femur, although bone strength and stiffness did not change. Muller reported that subcutaneous administration of rhIGF-I to adult oophorectomized rats stimulated bone formation as evidenced by increased osteoid surfaces, osteoblast surfaces, and mineral apposition rates. At high doses of rhIGF-I, osteoclast surface and osteoclast number also increased. In contrast, Tobias et al. (1992) found that rhIGF-I (200 mg/kg) administered for 17 days to 15-week-old rats increased longitudinal and periosteal growth of the femur but suppressed trabecular bone formation in oophorectomized and control rats. Bone resorption was also slightly suppressed during rhIGF-I treatment, although not to the extent that bone formation was inhibited.

T1DM (Type I diabetes mellitus) is associated with decreased cortical BMD. Although the pathophysiology of diabetic osteopenia remains unknown, it appears that the duration of diabetes, the extent of diabetic control, and the timing of disease onset are each associated with higher risks of low BMD (McNair, 1988). Numerous clinical studies report abnormalities in the GH/IGF-I axis in TIDM (Rasingani, 2017) including reductions in serum IGF-I due to portal insulinopenia and elevations in GH secretion (Hourd, 1991). Diminished insulin action in TIDM is associated with increased serum IGFBP-1 levels, which can sequester IGF-I, affect its bioavailability to the skeleton, and contribute to impaired bone acquisition. The cellular mechanism leading to low BMD in TIDM relates mainly to defects in osteoblastic function, as serum levels of osteocalcin, a marker of bone formation, are reduced in TIDM (Verhaeghe et al., 1992). Serum markers of bone formation are reduced in type I diabetics, suggesting a possible defect in osteoblastic activity (Verhaeghe et al., 1992). Serum IGF-I levels are either normal or low in IDDM but often are reduced in patients with poor diabetic control. In these same people, serum IGFBP-1 levels are quite high. This has led investigators to believe that changes in the IGF regulatory system during poor metabolic control contribute to impaired growth.

Spontaneously diabetic BB rats exhibit osteopenia and therefore provide a useful model for studying the effects of IGF-I on bone remodeling. Even though bone formation is lower in BB than in control rats (as measured by serum markers), administration of rhIGF-I does not increase bone epiphyseal width, osteoblast surfaces, or serum osteocalcin (Verhaeghe et al., 1992). Thus, despite evidence that circulating levels of IGF-I are reduced in some patients with type I IDDM, preliminary animal studies have failed to show that IGF-I administration can correct any inherent defect in bone formation. However, those studies did not include insulin treatment, so it is unclear whether IGF-I could have anabolic properties on bone in the presence of adequate insulin levels.

Alternate ways of exploiting the anabolic properties of IGF-I in bone have been proposed. IGF-I has been administered by intraarterial infusion or coupled to IGFBP-3. Infusion of rhIGF-I continuously into the arterial supply of the right hind limb of ambulatory rats for 14 days led to a 22% increase in cortical and trabecular bone formation in the infused limb (Spencer et al., 1991). By histomorphometry the number of osteoblasts, but not osteoclasts, was increased. Using an alternative model, Bagi et al. (1994) administered rhIGF-I or a complex of IGF-I-IGFBP-3 to 16-week-old oophorectomized rats. The IGF-I—IGFBP-3 complex (7.5 mg/kg/day) led to greater increases in bone formation than IGF-I alone, though both treatments increased longitudinal bone growth. The highest doses of rhIGF-I and rhIGF-I-GFBP-3 enhanced trabecular thickness in the lumbar vertebrae and femoral epiphyses, and increased bone resorption, but only in the femoral metaphysis. A similar study contrasting IGF-I with IGF-I-GFBP-3 was performed in 22-week-old oophorectomized rats (Brommage et al., 1993). BMD increased in both groups, but fewer than 10% of the rats treated with IGF-I—IGFBP-3 complex developed hypoglycemia, compared with nearly 50% of those treated with rhIGF-I alone.

Human studies of insulin-like growth factor I and bone mineral density

There is one published study of bone markers that employed rhIGF-I to healthy young postmenopausal women. Doses of rhIGF-I from 30 to 180 mg/kg/day were administered daily by subcutaneous injection for 6 days to older postmenopausal women without fractures and with normal BMD (Ebeling, 1993). Very significant dose-dependent increases in serum PICP, osteocalcin, and urinary deoxypyridinoline were reported. Although the rise in PICP was greater than the increase in collagen breakdown (measured by deoxypyridinoline), it is uncertain whether this meant that formation was stimulated more than resorption. For the two highest doses of rhIGF-I (120 and 180 mg/kg/day), orthostasis, weight gain, edema, tachycardia, and parotid discomfort were noted. At lower doses (30 and 60 mg/kg/day), fewer side effects were reported, but less discrete changes in PICP were noted.

One indication for rhIGF-I treatment, which has been approved in the United States, Sweden, and other countries, is the GH-resistant short stature syndrome (Laron dwarf), in which patients lack functional GHRs and thus do not respond to GH. Laron dwarfs show very low levels of IGF-I in serum with high levels of GH (due to a lack of negative feedback on GH by IGF-I) and are growth retarded (Bondy et al., 1994). Underwood treated one such boy (age 9) with 2 weeks of continuous intravenous rhIGF-I (Bondy et al., 1994). Urinary calcium excretion increased while urinary phosphate and sodium decreased. After a 2-week continuous infusion of rhIGF-I, the patient was treated with twice-daily SC rhIGF-I (120 mg/kg) for 2 years. Growth occurred at a rate of 10 cm/year, compared with 5 cm for the 3 years prior to treatment. Subsequently, Underwood and colleagues have treated eight patients in this manner without hypoglycemia, while Laron and his group have treated five children (Bondy et al., 1994; Laron et al., 1992). More recently, a child with an IGF-I deletion mutation in exon 5 has been reported. This patient had very short stature, mental retardation, and other abnormalities along with very low levels of circulating IGF-I (Woods et al., 1996). The rhIGF-I treatment led to a marked increase in linear growth and a huge increase in spinal bone mass. However, when corrected for changes in size of the bone, the incremental changes in volumetric bone mass were much less impressive (Camacho-Hubner et al., 1999). Hypoglycemia was avoided in these cases by having children eat 3 to 4 hours after their IGF-I injection, although several children had selective growth of adenoidal tissue.

Two unique aspects of these IGF-I data challenge previous concepts about the role of GH in skeletal homeostasis. First, IGF-I can act as a classical endocrine hormone stimulating longitudinal growth independent of GH, although not to the same extent as GH; second, GH may not be absolutely essential for statural growth; i.e., the stimulatory effect of GH on chondrocytes that permits skeletal responsiveness to IGF-I may not be as critical as once perceived. In fact, crossing Laron dwarf mice (GHR-null mice, GHRKO) with mice expressing hepatic IGF-I transgene (HIT) restored IGF-I levels in serum (GHRKO-HIT mice) and was able to promote skeletal linear growth and BMD (Wu et al., 2013). However, despite restoration of serum IGF-I in the GHRKO-HIT mice, growth was not normalized due to compromised delivery of IGF-I to the tissues, the inability of endocrine IGF-I to restore bone tissue IGF-I levels, and IGF-I-independent effects of GH on linear and radial bone growth. Caution must be undertaken in examining the effects of rhIGF-I on BMD in children, because most changes in the skeleton relate to linear growth and periosteal enhancement, both of which can contribute to two-dimensional changes in BMD as measured by DXA, but the changes are less when corrected for size (Bachrach et al., 1998).

Idiopathic osteoporosis in men is an ill-defined syndrome of low bone mass and spinal fractures without associated hypogonadism. By histomorphometry, these men often have low bone turnover, suggesting a possible defect in bone formation. Several groups of investigators have suggested that this syndrome is related to low serum IGF-I levels (Ljunghall et al., 1992; Kurland et al., 1997, 1998, Rubin and Bilezikian, 2002). Because the therapeutic options in males with osteoporosis are somewhat limited, and treatment for low bone turnover states is generally frustrating, the therapeutic potential for anabolic agents like IGF-I in this condition should be quite high. In one male with idiopathic osteoporosis and low serum IGF-I, Johansson et al. (1994) administered rhIGF-I (160 mg/kg/day SC) for 7 days. Bone alkaline phosphatase, osteocalcin, and PICP all increased more than 40% over baseline. However, urinary calcium/creatinine and hydroxyproline excretion rose during treatment. In a recent trial of rhIGF-I (at a dose of 80 mg/kg/day) and rhGH (2 IU/m2/day) in 12 men, serum osteocalcin, serum procollagen peptide, and urinary deoxypyridinoline excretion all increased following 7 days of rhIGF-I treatment (Holloway et al., 1997). Although there were slight differences in the response of certain biochemical markers to IGF-I and GH, both forms of therapy produced significant increases in bone resorption.

Anorexia nervosa is a condition characterized by amenorrhea and profoundly low BMD (due to either low peak bone mass or rapid bone loss) as well as reduced body weight, low circulating IGF-I, resistance to GH, and a marked propensity for fractures. Hence, rhIGF-I might be considered an ideal therapeutic option for this group of adolescents and young adults with severe bone disease, particularly because oral contraceptive pills (OCPs) have virtually no effect on BMD in these patients. Grinspoon et al. (2003) studied 60 anorexic women with low BMD in an RPCT of 9-month duration using rhIGF-I 30 ug/kg/d with and without OCPs (Grinspoon, 1993). The group of women receiving rhIGF-I and OCPs had the greatest increase in spine BMD (11.8%); rhIGF-I alone also increased BMD (11.1%), while bone loss occurred in the group receiving placebo or OCPs alone. Interestingly, there were virtually no side effects in the anorexic women, and serum IGFBP-2 was inversely correlated with changes in hip BMD. Although the increase in BMD was relatively modest, considering the lack of other available therapies, these changes are encouraging and suggest that further studies are needed.

Clinical trials provide evidence that IGF-I acts by increasing the generation of new osteons, thereby promoting bone resorption and formation. This action might be ideal for older individuals, because one characteristic of age-related osteoporosis is suppressed bone formation. However, concerns about dosing and side effects have limited enthusiasm for this approach. Yet it is conceivable that low doses of rhIGF-I (,30 mg/kg/day) could differentially stimulate bone formation. In one trial, 16 healthy elderly women were given 60 ug/kg/day (high dose) and 15 mg/kg/day (low dose) of rhIGF-I and tested for 28 days. The high-dose rhIGF-I increased the markers of bone resorption and formation. However, low doses of rhIGF-I caused increases in serum osteocalcin and PICP but had no effect on total pyridinoline excretion (Ghiron et al., 1995). These data would support the thesis that low doses of rhIGF-I may directly increase osteoblastic function with only a minimal increase in bone resorption. However, further studies will be needed to assess the future therapeutic role of low doses of rhIGF-I in osteoporosis.

One strategy is to administer a bone-specific agent that stimulates bone mass, such as PTH. Intermittent hPTH increases trabecular bone by stimulating osteoblasts to synthesize IGF-I and other growth factors (Rosen and Donahue, 1998). Another strategy is to administer IGF-I along with an IGFBP. Bagi et al. (1994) previously reported that the IGF-I/IGFBP-3 complex could enhance bone mass in the metaphysis and epiphysis of rats. One small randomized placebo-controlled trial utilized subcutaneous infusions of IGF-I/IGFBP-3 in 24 older women with hip fractures. Bone loss in the contralateral hip was reduced considerably after 6 months (i.e., from 6% to 1.5%) in those subjects who were given complex

IGF-I/IGFBP-3 versus those receiving saline (Geusens et al., 1998; Boonen et al., 2002). Accompanying that change in BMD was also an increase in grip strength in those that received the active agent, with no significant side effects reported.

Limitations to the clinical use of recombinant human insulin-like growth factor I

IGF-I treatment has its limitations (Zofkova, 2003). The impacts of serious sequelae of long-term administration of IGF-I remain to be evaluated. During IGF-I treatment, undesirable metabolic manifestations may develop hypoglycemia (in particular after large intravenous doses) and hypophosphatemia with subsequent hypotension (Zofkova, 2003). A more frequent incidence of gynecomastia was also observed. The use of IGF-I/IGFBP-3 complex seems to be very useful and safe in women who are older and who have a recent hip fracture (Agnusdei and Gentilella, 2005). Furthermore, the combination of an anabolic agent with an antiresorptive drug (such as calcitonin or alendronate) could be more potent than either agent alone. However, the therapeutic effects of such a combined approach on BMD at different skeletal sites have been controversial (Agnusdei et al., 2005).

Summary

Several lines of evidence point to the importance of local IGF-I in skeletal turnover. Systemic IGF-I, regulated by GH, IGFBPs, and several proteases as well as IGFRs, also must be important for full linear growth and peak skeletal acquisition. Recent work suggests that recombinant growth factors may be anabolic for the skeletal remodeling unit. First, both GH and IGF-I stimulate osteoblastic differentiation. In vivo models using targeted deletion or overexpression of IGFs or IGF-IR support a critical role for this regulatory circuit in peak bone acquisition. In respect to growth hormone, rhGH treatment increases bone mass but must continue for several years, and outcomes should be measured for several years thereafter (as the skeletal response cannot be detected shortly after treatment). Skeletal response to rhGH is greater in men than in women, and the potential long-term risks of elevated IGF-I, as well as the availability of other less expensive therapies (such as bisphosphonates, teriparatide and abaloparatide), precludes more vigorous development of these peptides. Third, in GHD children, both rhGH and rhIGF-I can enhance trabecular and cortical BMD, and certainly recombinant GH is part of current recommendations for this condition. Therefore, unless more favorable responses in properly controlled clinical trials are seen with rhGH or rhIGF-I, these drugs are not recommended for the treatment of postmenopausal osteoporosis.

Acknowledgments

This work was funded through PHS Grants from NIDDK 92790 and NIAMS AR 45433.

References

Agnusdei, D., Gentilella, R., 2005. GH and IGF-I as therapeutic agents for osteoporosis. J. Endocrinol. Investig. 28 (8 Suppl. 1), 32–36.

Ali, O., et al., 2007. Growth hormone therapy improves bone mineral density in children with cerebral palsy: a preliminary pilot study. J. Clin. Endocrinol. Metab. 92 (3), 932–937.

Aloia, J.F., Vaswani, A., Meunier, P.J., Edouard, C.M., Arlot, M.E., Yeh, J.K., Cohn, S.H., 1987. Coherence treatment of postmenopausal osteoporosis with growth hormone and calcitonin. Calcif. Tissue Int. 40 (5), 253–259.

Ammann, P., et al., 1996. Bone density and shape as determinants of bone strength in IGF-I and/or pamidronate-treated ovariectomized rats. Osteoporos. Int. 6 (3), 219–227.

Andress, D.L., Birnbaum, R.S., 1991. A novel human insulin-like growth factor binding protein secreted by osteoblast-like cells. Biochem. Biophys. Res. Commun. 176 (1), 213–218.

Angelloz-Nicoud, P., Binoux, M., 1995. Autocrine regulation of cell proliferation by the insulin-like growth factor (IGF) and IGF binding protein-3 protease system in a human prostate carcinoma cell line (PC-3). Endocrinology 136 (12), 5485–5492.

Arany, E., et al., 1994. Differential cellular synthesis of insulin-like growth factor binding protein-1 (IGFBP-1) and IGFBP-30020sxyeiwithin human liver. J. Clin. Endocrinol. Metab. 79 (6), 1871–1876.

Arends, N.J., et al., 2003. GH treatment and its effect on bone mineral density, bone maturation and growth in short children born small for gestational age: 3-year results of a randomized, controlled GH trial. Clin. Endocrinol. 59 (6), 779–787.

Bach, L.A., 1999. Insulin-like growth factor binding protein-6: the "forgotten" binding protein? Horm. Metab. Res. 31 (2–3), 226–234.

Bachrach, L.K., et al., 1998. Bone mineral, histomorphometry, and body composition in adults with growth hormone receptor deficiency. J. Bone Miner. Res. 13 (3), 415–421.

Bagi, C.M., et al., 1994. Benefit of systemically administered rhIGF-I and rhIGF-I/IGFBP-3 on cancellous bone in ovariectomized rats. J. Bone Miner. Res. 9 (8), 1301–1312.

Bang, P., et al., 1998. Postoperative induction of insulin-like growth factor binding protein-3 proteolytic activity: relation to insulin and insulin sensitivity. J. Clin. Endocrinol. Metab. 83 (7), 2509–2515.

Baroncelli, G.I., et al., 1998. Measurement of volumetric bone mineral density accurately determines degree of lumbar undermineralization in children with growth hormone deficiency. J. Clin. Endocrinol. Metab. 83 (9), 3150–3154.

Baroncelli, G.I., et al., 2004. Longitudinal changes of lumbar bone mineral density (BMD) in patients with GH deficiency after discontinuation of treatment at final height; timing and peak values for lumbar BMD. Clin. Endocrinol. 60 (2), 175–184.

Bauer, D.C., et al., 1998. Low serum IGF-I but not IGFBP-3 predicts hip and spine fracture. The study of osteoporotic fracture. Bone 23, 561.

Baum, H.B., et al., 1996. Effects of physiologic growth hormone therapy on bone density and body composition in patients with adult-onset growth hormone deficiency. A randomized, placebo-controlled trial. Ann. Intern. Med. 125 (11), 883–890.

Baumann, G., et al., 1989. Regulation of plasma growth hormone-binding proteins in health and disease. Metabolism 38 (7), 683–689.

Baxter, R.C., 2000. Insulin-like growth factor (IGF)-binding proteins: interactions with IGFs and intrinsic bioactivities. Am. J. Physiol. Endocrinol. Metab. 278 (6), E967–E976.

Ben Lagha, N., et al., 2006. IGFBP-1 involvement in intrauterine growth retardation. Endocrinology 147, 4730–4737.

Bereket, A., et al., 1995. Insulin-like growth factor binding protein-3 proteolysis in children with insulin-dependent diabetes mellitus: a possible role for insulin in the regulation of IGFBP-3 protease activity. J. Clin. Endocrinol. Metab. 80 (8), 2282–2288.

Biermasz, N.R., et al., 2004. Long-term skeletal effects of recombinant human growth hormone (rhGH) alone and rhGH combined with alendronate in GH-deficient adults: a seven-year follow-up study. Clin. Endocrinol. 60 (5), 568–575.

Bikle, D.D., et al., 2001. The skeletal structure of insulin-like growth factor I-deficient mice. J. Bone Miner. Res. 16 (12), 2320–2329.

Bikle, D.D., et al., 2006. Development and progression of alopecia in the vitamin D receptor null mouse. J. Cell. Physiol. 207 (2), 340–353.

Bikle, D.D., et al., 2002. Insulin-like growth factor I is required for the anabolic actions of parathyroid hormone on mouse bone. J. Bone Miner. Res. 17, 1570–1578.

Bing-You, R.G., et al., 1993. Low bone mineral density in adults with previous hypothalamic-pituitary tumors: correlations with serum growth hormone responses to GH-releasing hormone, insulin-like growth factor I, and IGF binding protein 3. Calcif. Tissue Int. 52 (3), 183–187.

Bianda, T., et al., 1998. Effects of short term IGF-I or GH treatment on bone metabolism and production of 1,25 vitamin D in GH deficient patients. J. Clin. Endocrinol. Metab. 83, 61–67.

Bondy, C.A., et al., 1994. Clinical uses of insulin-like growth factor I. Ann. Intern. Med. 120 (7), 593–601.

Boonen, S., et al., 2002. Musculoskeletal effects of the recombinant human IGF-I/IGF binding protein-3 complex in osteoporotic patients with proximal femoral fracture: a double-blind, placebo-controlled pilot study. J. Clin. Endocrinol. Metab. 87 (4), 1593–1599.

Bouillon, R., 1991. Growth hormone and bone. Horm. Res. 36 (Suppl. 1), 49–55.

Braulke, T., et al., 1995. Proteolysis of IGFBPs by cathepsin D in vitro and in cathepsin D-deficient mice. Prog. Growth Factor Res. 6 (2–4), 265–271.

Bravenboer, N., et al., 1994. The effect of GH on bone mass and bone turnover of GHD men. J. Bone Miner. Res. 9 (S1), B253.

Brismar, K., et al., 1988. Insulin regulates the 35 kDa IGF binding protein in patients with diabetes mellitus. J. Endocrinol. Investig. 11 (8), 599–602.

Brixen, K., et al., 1990. A short course of recombinant human growth hormone treatment stimulates osteoblasts and activates bone remodeling in normal human volunteers. J. Bone Miner. Res. 5 (6), 609–618.

Brommage, R., et al., 1993. Treatment with the rhIGF-I/IGFBP-3 complex increases cortical bone and lean body mass in oophorectomized rats. J. Bone Miner. Res. 9 (Suppl. 1), 1.

Camacho-Hubner, C., et al., 1999. Effects of recombinant human insulin-like growth factor I (IGF-I) therapy on the growth hormone-IGF system of a patient with a partial IGF-I gene deletion. J. Clin. Endocrinol. Metab. 84 (5), 1611–1616.

Canalis, E., et al., 1989. Insulin-like growth factor I mediates selective anabolic effects of parathyroid hormone in bone cultures. J. Clin. Investig. 83 (1), 60–65.

Capatina, C., et al., 2015. Sixty years of neuroendocrinology: Acromegaly. J. Endocrinol. 226, T141–T160.

Celil, A.B., et al., 2005. Osx transcriptional regulation is mediated by additional pathways to BMP2/Smad signaling. J. Cell. Biochem. 95 (3), 518–528.

Chelius, D., et al., 2001. Expression, purification, and characterization of the structure and disulfide linkages of insulin-like growth factor binding protein-4. J. Endocrinol. 168 (2), 283–296.

Cheng, S.L., et al., 2003. MSX2 promotes osteogenesis and suppresses adipogenic differentiation of multipotent mesenchymal progenitors. J. Biol. Chem. 278 (46), 45969–45977.

Christiansen, J.S., et al., 1991. GH-replacement therapy in adults. Horm. Res. 36 (Suppl. 1), 66–72.

Claussen, M., et al., 1997. Proteolysis of insulin-like growth factors (IGF) and IGF binding proteins by cathepsin D. Endocrinology 138 (9), 3797–3803.

Cohen, P., et al., 1992. Prostate-specific antigen (PSA) is an insulin-like growth factor binding protein-3 protease found in seminal plasma. J. Clin. Endocrinol. Metab. 75 (4), 1046–1053.

Cohen, P., et al., 1994. Biological effects of prostate specific antigen as an insulin-like growth factor binding protein-3 protease. J. Endocrinol. 142 (3), 407–415.

Collett-Solberg, P.F., Cohen, P., 1996. The role of the insulin-like growth factor binding proteins and the IGFBP proteases in modulating IGF action. Endocrinol Metab. Clin. N. Am. 25 (3), 591–614.

Conover, C.A., 1991. Glycosylation of insulin-like growth factor binding protein-3 (IGFBP-3) is not required for potentiation of IGF-I action: evidence for processing of cell-bound IGFBP-3. Endocrinology 129 (6), 3259–3268.

Conover, C.A., et al., 1993. Regulation of insulin-like growth factor binding protein-5 messenger ribonucleic acid expression and protein availability in rat osteoblast-like cells. Endocrinology 132 (6), 2525–2530.
Conover, C.A., et al., 1994. Insulin-like growth factor-II enhancement of human fibroblast growth via a nonreceptor-mediated mechanism. Endocrinology 135 (1), 76–82.
Conover, C.A., et al., 1995. Endogenous cathepsin D-mediated hydrolysis of insulin-like growth factor-binding proteins in cultured human prostatic carcinoma cells. J. Clin. Endocrinol. Metab. 80 (3), 987–993.
Conover, C.A., et al., 2002. Subcutaneous administration of insulin-like growth factor (IGF)-II/IGF binding protein-2 complex stimulates bone formation and prevents loss of bone mineral density in a rat model of disuse osteoporosis. Growth Hormone IGF Res. 12 (3), 178–183.
Corpas, E., et al., 1993. Human growth hormone and human aging. Endocr. Rev. 14 (1), 20–39.
Courtland, H.W., et al., 2010. Sex specific regulation of body size by the acid labile subunit. J. Bone Miner. Res. 25, 2059–2068.
Coverley, J.A., Baxter, R.C., 1997. Phosphorylation of insulin-like growth factor binding proteins. Mol. Cell. Endocrinol. 128 (1–2), 1–5.
Crist, D.M., et al., 1988. Body composition response to exogenous GH during training in highly conditioned adults. J. Appl. Physiol. 65 (2), 579–584.
D'Ercole, A.J., 1996. Insulin-like growth factors and their receptors in growth. Endocrinol Metab. Clin. N. Am. 25 (3), 573–590.
Dabuer, A., 2016. Mutations in PAPPA2 causes short stature due to low IGF-1 availability. EMBO Mol. Med. 8, 363–374.
Daughaday, W.H., et al., 1972. Somatomedin: proposed designation for sulphation factor. Nature 235 (5333), 107.
Dawson-Hughes, B., et al., 1986. Regulation of growth hormone and somatomedin-C secretion in postmenopausal women: effect of physiological estrogen replacement. J. Clin. Endocrinol. Metab. 63 (2), 424–432.
de Muinck Keizer-Schrama, S.M., et al., 1992. Dose-response study of biosynthetic human growth hormone (GH) in GH-deficient children: effects on auxological and biochemical parameters. Dutch Growth Hormone Working Group. J. Clin. Endocrinol. Metab. 74 (4), 898–905.
de la Piedra, P.C., et al., 1988. Correlation among plasma osteocalcin, growth hormone, and somatomedin C in acromegaly. Calcif. Tissue Int. 43 (1), 44–45.
DeBoer, H., et al., 1994. Consequences of childhood-onset growth hormone deficiency for adult bone mass. J. Bone Miner. Res. 9, 1319–1326.
Delmas, P.D., et al., 1986. Serum bone GLA-protein in growth hormone deficient children. J. Bone Miner. Res. 1 (4), 333–338.
DeMambro, V.E., Kawai, M., Clemens, T.L., Fulzele, K., Maynard, J.A., de Evsikova, C.M., Johnson, K.R., Canalis, E., Beamer, W.G., Rosen, C.J., Donahue, L.R., 2010. A Novel Spontaneous Mutation of *Irs1* in Mice Results in Hyperinsulinemia, Reduced Growth, Low Bone Mass and Impaired Adipogenesis. J. Endocrinol. 204 (3), 241–253.
Dennison, E.M., et al., 2003. Growth hormone predicts bone density in elderly women. Bone 32 (4), 434–440.
Domene, H.M., et al., 2004. Deficiency of the circulating insulin-like growth factor system associated with inactivation of the acid-labile subunit gene. N. Engl. J. Med. 350, 570–577.
Domene, H.M., et al., 2010. Deficiency of ALS in children with short stature. Pediatr. Endocrinol. Rev. 7 (4), 339–346.
Donahue, L.R., et al., 1990. Age-related changes in serum insulin-like growth factor-binding proteins in women. J. Clin. Endocrinol. Metab. 71 (3), 575–579.
Donahue, L.R., et al., 1993. Regulation of metabolic water and protein compartments by insulin-like growth factor-I and testosterone in growth hormone-deficient lit/lit mice. J. Endocrinol. 139 (3), 431–439.
Ducy, P., et al., 1999. A Cbfa1-dependent genetic pathway controls bone formation beyond embryonic development. Genes Dev. 13 (8), 1025–1036.
Durham, S.K., et al., 1994. The insulin-like growth factor-binding protein-4 (IGFBP-4)-IGFBP-4 protease system in normal human osteoblast-like cells: regulation by transforming growth factor-beta. J. Clin. Endocrinol. Metab. 79 (6), 1752–1758.
Ebeling, P.R., et al., 1993. Short-term effects of recombinant human insulin-like growth factor I on bone turnover in normal women. J. Clin. Endocrinol. Metab. 77 (5), 1384–1387.
Esen, E., et al., 2013. WNT-LRP5 signaling induces Warburg effect through mTORC2 activation during osteoblast differentiation. Cell Metabol. 17, 745–755.
Ezzat, S., et al., 1993. Biochemical assessment of bone formation and resorption in acromegaly. J. Clin. Endocrinol. Metab. 76 (6), 1452–1457.
Finkelman, R.D., Mohan, S., Jennings, J.C., Taylor, A.K., Jepsen, S., Baylink, D.J., 1990. Quantitation of growth factors IGF-I, SGF/IGF-II, and TGF-beta in human dentin. J. Bone Miner. Res. 5 (7), 717–723.
Fleet, J.C., et al., 1994. Growth hormone and parathyroid hormone stimulate intestinal calcium absorption in aged female rats. Endocrinology 134 (4), 1755–1760.
Forbes, B.E., et al., 1998. Localization of an insulin-like growth factor (IGF) binding site of bovine IGF binding protein-2 using disulfide mapping and deletion mutation analysis of the C-terminal domain. J. Biol. Chem. 273 (8), 4647–4652.
Fowlkes, J.L., et al., 1994a. Matrix metalloproteinases degrade insulin-like growth factor-binding protein-3 in dermal fibroblast cultures. J. Biol. Chem. 269 (41), 25742–25746.
Fowlkes, J.L., et al., 1994b. Proteolysis of insulin-like growth factor binding protein-3 during rat pregnancy: a role for matrix metalloproteinases. Endocrinology 135 (6), 2810–2813.
Frohman, L.A., Kineman, R.D., 2002. Growth hormone-releasing hormone and pituitary somatotrope proliferation. Minerva Endocrinol. 27 (4), 277–285.
Frohman, L.A., et al., 2000. Secretagogues and the somatotrope: signaling and proliferation. Recent Prog. Horm. Res. 55, 269–290.
Garnero, P., et al., 2000. Biochemical markers of bone turnover, endogenous hormones, and the risk of fractures in postmenopausal women: the OFELY study. J. Bone Miner. Res. 15 (8), 1526–1536.
Gamero, P., et al., 2000. Low serum IGF-1 and occurrence of osteoporotic fractures in postmenopausal women. Lancet 355 (9207), 898–899.

Geusens, P., et al., 1998. Musculoskeletal effects of rhIGF-I/IGFBP-3 in hip fracture patients: results from double-blind, placebo-controlled phase II study. Bone 23 (Suppl. 1), 157.

Gevers, E.F., 2009. Regulation of rapid signal transducer and activator of transcription 5 phosphorylation in the resting cells of the growth plate. Endocrinology 150, 3627−3636.

Ghiron, L.J., et al., 1995. Effects of recombinant insulin-like growth factor-I and growth hormone on bone turnover in elderly women. J. Bone Miner. Res. 10 (12), 1844−1852.

Green, H., et al., 1985. A dual effector theory of growth hormone action. Differentiation 29, 195−208.

Grinspoon, S., et al., 2003. Effects of recombinant human insulin-like growth factor (IGF)-I and estrogen administration on IGF-I, IGF binding protein (IGFBP)-2, and IGFBP-3 in anorexia nervosa: a randomized-controlled study. J. Clin. Endocrinol. Metab. 88 (3), 1142−1149.

He, J., et al., 2006. Postnatal growth and bone mass in mice with IGF-I haploinsufficiency. Bone 38 (6), 826−835.

Headey, S.J., et al., 2004. C-terminal domain of insulin-like growth factor (IGF) binding protein-6: structure and interaction with IGF-II. Mol. Endocrinol. 18 (11), 2740−2750.

Heffernan, M., et al., 2001. The effects of human GH and its lipolytic fragment (AOD9604) on lipid metabolism following chronic treatment in obese mice and beta(3)-AR knockout mice. Endocrinology 142 (12), 5182−5189.

Ho, K.Y., Weissberger, A.J., 1990. Secretory patterns of growth hormone according to sex and age. Horm. Res. 33 (Suppl. 4), 7−11.

Hogler, W., et al., 2005. Effect of growth hormone therapy and puberty on bone and body composition in children with idiopathic short stature and growth hormone deficiency. Bone 37 (5), 642−650.

Holloway, L., et al., 1994. Effects of recombinant human growth hormone on metabolic indices, body composition, and bone turnover in healthy elderly women. J. Clin. Endocrinol. Metab. 79 (2), 470−479.

Holloway, L., et al., 1997. Skeletal effects of cyclic recombinant human growth hormone and salmon calcitonin in osteopenic postmenopausal women. J. Clin. Endocrinol. Metab. 82 (4), 1111−1117.

Honda, Y., et al., 1996. Recombinant synthesis of insulin-like growth factor-binding protein-4 (IGFBP-4): development, validation, and application of a radioimmunoassay for IGFBP-4 in human serum and other biological fluids. J. Clin. Endocrinol. Metab. 81 (4), 1389−1396.

Hourd, P., et al., 1991. Urinary growth excretion during puberty in type 1 diabetes mellitus. Diabet. Med. 8, 237−242.

Hyer, S.L., et al., 1992. Growth hormone deficiency during puberty reduces adult bone mineral density. Arch. Dis. Child. 67 (12), 1472−1474.

Inzucchi, S.E., Robbins, R.J., 1994. Clinical review 61: effects of growth hormone on human bone biology. J. Clin. Endocrinol. Metab. 79 (3), 691−694.

Isaksson, O.G., et al., 1987. Mechanism of the stimulatory effect of growth hormone on longitudinal bone growth. Endocr. Rev. 8 (4), 426−438.

Ito, Y., 1999. Molecular basis of tissue-specific gene expression mediated by the runt domain transcription factor PEBP2/CBF. Genes Cells 4 (12), 685−696.

Jansson, J.O., et al., 1985. Sexual dimorphism in the control of growth hormone secretion. Endocr. Rev. 6 (2), 128−150.

Janssen, Y.J., et al., 1998. Skeletal effects of two years of treatment with low physiological doses of recombinant human growth hormone (GH) in patients with adult-onset GH deficiency. J. Clin. Endocrinol. Metab. 83 (6), 2143−2148.

Johansen, J.S., et al., 1990. Effects of growth hormone (GH) on plasma bone Gla protein in GH-deficient adults. J. Clin. Endocrinol. Metab. 70 (4), 916−919.

Johansson, A.G., et al., 1992. The bone mineral density in acquired growth hormone deficiency correlates with circulating levels of insulin-like growth factor I. J. Intern. Med. 232 (5), 447−452.

Johansson, A.G., et al., 1994. Effects of short-term treatment with IGF-I and GH on markers of bone metabolism in idiopathic osteoporosis. J. Bone Miner. Res. 9 (Suppl. 1), 328.

Johansson, A.G., et al., 1994. Growth hormone-dependent insulin-like growth factor binding protein is a major determinant of bone mineral density in healthy men. J. Bone Miner. Res. 9 (6), 915−921.

Jones, D.C., et al., 2006. Regulation of adult bone mass by the zinc finger adapter protein Schnurri-3. Science 312 (5777), 1223−1227.

Jorgensen, J.O., et al., 1989. Beneficial effects of growth hormone treatment in GH-deficient adults. Lancet 1 (8649), 1221−1225.

Jorgensen, J.O., et al., 1991. Long-term growth hormone treatment in growth hormone deficient adults. Acta Endocrinol. 125 (5), 449−453.

Kamp, G.A., et al., 2002. High dose growth hormone treatment induces acceleration of skeletal maturation and an earlier onset of puberty in children with idiopathic short stature. Arch. Dis. Child. 87 (3), 215−220.

Kann, P., et al., 1993. Bone quality in growth hormone deficient adults. Acta Endocrinol. 128 (Suppl. 2), 60.

Karasik, D., et al., 2002. Insulin-like growth factor binding proteins 4 and 5 and bone mineral density in elderly men and women. Calcif. Tissue Int. 71 (4), 323−328.

Karasik, D., et al., 2003. Age, gender, and body mass effects on quantitative trait loci for bone mineral density: the Framingham Study. Bone 33 (3), 308−316.

Kassem, M., et al., 1994. No evidence for reduced spontaneous or growth-hormone-stimulated serum levels of insulin-like growth factor (IGF)-I, IGF-II or IGF binding protein 3 in women with spinal osteoporosis. Eur. J. Endocrinol. 131 (2), 150−155.

Kato, M., et al., 2002. Cbfa1-independent decrease in osteoblast proliferation, osteopenia, and persistent embryonic eye vascularization in mice deficient in Lrp5, a Wnt coreceptor. J. Cell Biol. 157 (2), 303−314.

Kaufman, J.M., et al., 1992. Bone mineral status in growth hormone-deficient males with isolated and multiple pituitary deficiencies of childhood onset. J. Clin. Endocrinol. Metab. 74 (1), 118−123.

Kawai, M., et al., 2011. The heparin binding domain of IGFBP-2 has IGF binding independent biologic activity on the growing skeleton. J. Biol. Chem. 286, 14670−14680.

Kayath, M.J., et al., 1997. Osteopenia is present in the minority of patients with acromegaly and is predominant in the spine. Osteoporos. Int. 7, 226–230.
Kelijman, M., 1991. Age-related alterations of the growth hormone/insulin-like-growth-factor I axis. J. Am. Geriatr. Soc. 39 (3), 295–307.
Kelley, K.M., et al., 1996. Insulin-like growth factor-binding proteins (IGFBPs) and their regulatory dynamics. Int. J. Biochem. Cell Biol. 28 (6), 619–637.
Kesavan, C., et al., 2011. Conditional disruption of IGF-I in type I collagen expressing cells shows an essential role of IGF-I in skeletal anabolic response to lading. Am J Physiol Endocrinol 301, E1191–E1197.
Khandwala, H.M., et al., 2000. The effects of insulin-like growth factors on tumorigenesis and neoplastic growth. Endocr. Rev. 21 (3), 215–244.
Khosla, S., et al., 1998. Insulin-like growth factor system abnormalities in hepatitis C-associated osteosclerosis. Potential insights into increasing bone mass in adults. J. Clin. Investig. 101 (10), 2165–2173.
Koutsilieris, M., Polychronakos, C., 1992. Proteinolytic activity against IGF-binding proteins involved in the paracrine interactions between prostate adenocarcinoma cells and osteoblasts. Anticancer Res. 12 (3), 905–910.
Kudo, Y., et al., 1996. Regulation of insulin-like growth factor-binding protein-4 protease activity by estrogen and parathyroid hormone in SaOS-2 cells: implications for the pathogenesis of postmenopausal osteoporosis. J. Endocrinol. 150 (2), 223–229.
Krantz, E., et al., 2015. Effects of GH treatment on fractures and quality of life in postmenopausal women. J. Clin. Endocrinol. Metab. 100, 3251–3259.
Kurland, E.S., et al., 1997. Insulin-like growth factor-I in men with idiopathic osteoporosis. J. Clin. Endocrinol. Metab. 82 (9), 2799–2805.
Kurland, E.S., et al., 1998. Normal growth hormone secretory reserve in men with idiopathic osteoporosis and reduced circulating levels of insulin-like growth factor-I. J. Clin. Endocrinol. Metab. 83 (7), 2576–2579.
Landin-Wilhelmsen, K., et al., 2003. Growth hormone increases bone mineral content in postmenopausal osteoporosis: a randomized placebo-controlled trial. J. Bone Miner. Res. 18 (3), 393–405.
Langlois, J.A., et al., 1998. Association between insulin-like growth factor I and bone mineral density in older women and men: the Framingham Heart Study. J. Clin. Endocrinol. Metab. 83 (12), 4257–4262.
Laron, Z., et al., 1992. Effects of insulin-like growth factor on linear growth, head circumference, and body fat in patients with Laron-type dwarfism. Lancet 339 (8804), 1258–1261.
Lau, K.H., et al., 2006. Upregulation of the Wnt, Estrogen receptor, IGF-I and BMP pathways in C57bl6 mice as opposed to C3H/HeJ mice in part contributes to the differential anabolic response to fluid shear. J. Biol. Chem. 281, 94768–94788.
Lau, K.H., et al., 2013. Osteocyte derived IGF-I is essential for determining bone mechanosensitivity. Am. J. Physiol. Endocrinol. Metab. 305, E271–E281.
Le Roith, D., et al., 1997. The insulin-like growth factor-I receptor and apoptosis. Implications for the aging progress. Endocrine 7 (1), 103–105.
Lee, W.C., et al., 2017. Energy metabolism of the osteoblast:implications for osteoporosis. Endocr. Rev. 38, 255–266.
Lee, M.H., et al., 2003. BMP-2-induced Osterix expression is mediated by Dlx5 but is independent of Runx2. Biochem. Biophys. Res. Commun. 309 (3), 689–694.
Lieberman, S.A., et al., 1994. The insulin-like growth factor I generation test: resistance to growth hormone with aging and estrogen replacement therapy. Horm. Metab. Res. 26 (5), 229–233.
Li, Z., et al., 2016. Glucose transporter-4 facilitates insulin-stimulated glucose uptake in osteoblasts. Endocrinology 157 (11), 4094–4103. Epub 2016 Sep 3.
Lin-Su, K., Wajnrajch, M.P., 2002. Growth hormone releasing hormone (GHRH) and the GHRH receptor. Rev. Endocr. Metab. Disord. 3 (4), 313–323.
Liu, J.L., et al., 2004. Disruption of growth hormone receptor gene causes diminished pancreatic islet size and increased insulin sensitivity in mice. Am. J. Physiol. Endocrinol. Metab. 287 (3), E405–E413.
Liu, J.P., et al., 1993. Mice carrying null mutations of the genes encoding insulin-like growth factor I (Igf-1) and type 1 IGF receptor (Igf1r). Cell 75 (1), 59–72.
Liu, Z., et al., 2016a. GH control of hepatic lipid metabolism. Diabetes 65, 3598–3609.
Liu, Z., et al., 2016b. DMP-1-mediated Ghr gene recombination compromises skeletal development and impairs skeletal response to intermittent PTH. FASEB J. 30, 635–652.
Liu, Z., et al., 2016c. Does the GH/IGF axis contribute to skeletal dimorphism. Growth Hormone IGF Res. 27, 7–17.
Liu, Z., et al., 2017. Reduced serum IGF-I associated with hepatic osteodystrophy is a main determinant of cortical but not trabecular bone mass. J. Bone Miner. Res. Epub ahead of print Sept 2017.
Ljunghall, S., et al., 1992. Low plasma levels of insulin-like growth factor 1 (IGF-1) in male patients with idiopathic osteoporosis. J. Intern. Med. 232 (1), 59–64.
Lloyd, M.E., et al., 1996. Relation between insulin-like growth factor-I concentrations, osteoarthritis, bone density, and fractures in the general population: the Chingford study. Ann. Rheum. Dis. 55 (12), 870–874.
Lupu, F., et al., 2001. Roles of growth hormone and insulin-like growth factor 1 in mouse postnatal growth. Dev. Biol. 229 (1), 141–162.
McNair, P., 1988. Bone mineral metabolism in human type 1 (insulin dependent) diabetes mellitus. Dan. Med. Bull. 35 (2), 109–121.
Manes, S., et al., 1999. The matrix metalloproteinase-9 regulates the insulin-like growth factor-triggered autocrine response in DU-145 carcinoma cells. J. Biol. Chem. 274 (11), 6935–6945.
Marcus, R., et al., 1990. Effects of short-term administration of recombinant human growth hormone to elderly people. J. Clin. Endocrinol. Metab. 70 (2), 519–527.
Maridas, D.E., et al., 2017a. IGFBP-4 regulates skeletal growth in a sex specific manner. J. Endocrinol. 233, 131–144.
Maridas, D.E., et al., 2017b. IGFBP-4 is required for adipogenesis. Endocrinology 158, 3488–3500.

Marinaro, J.A., et al., 1999. HaCaT human keratinocytes express IGF-II, IGFBP-6, and an acid-activated protease with activity against IGFBP-6. Am. J. Physiol. 276 (3 Pt 1), E536–E542.
Maroni, P.D., et al., 2004. Mitogen activated protein kinase signal transduction pathways in the prostate. Cell Commun. Signal. 2 (1), 5.
Martin, J.L., Baxter, R.C., 1988. Insulin-like growth factor-binding proteins (IGF-BPs) produced by human skin fibroblasts: immunological relationship to other human IGF-BPs. Endocrinology 123 (4), 1907–1915.
Mauras, N., et al., 2000. High-dose recombinant human growth hormone (GH) treatment of GH-deficient patients in puberty increases near-final height: a randomized, multicenter trial. Genentech, Inc., Cooperative Study Group. J. Clin. Endocrinol. Metab. 85 (10), 3653–3660.
Mayo, K.E., et al., 1988. Dramatic pituitary hyperplasia in transgenc mice expressing a GH releasing peptide. Mol. Endocrinol. 7, 606–612.
Mayo, K.E., et al., 1995. Growth hormone-releasing hormone: synthesis and signaling. Recent Prog. Horm. Res. 50, 35–73.
Mazziotti, G., et al., 2008. Prevalence of osteoporosis in men with acromegaly. J. Clin. Endocrinol. Metab. 93, 4649–4654.
Meinhardt, U.J., et al., 2006. Modulation of GH secretion by sex steroids. Clin. Endocrinol. 65, 413–422.
Menagh, P.J., et al., 2010. Growth Hormone regulates the balance between bone formation and bone marrow adiposity. J. Bone Miner. Res. 25, 757–768.
Merriman, H.L., et al., 1990. Insulin-like growth factor-I and insulin-like growth factor-II induce c-fos in mouse osteoblastic cells. Calcif. Tissue Int. 46 (4), 258–262.
Mochizuki, H., et al., 1992. Insulin-like growth factor-I supports formation and activation of osteoclasts. Endocrinology 131 (3), 1075–1080.
Mohan, S., 1993. Insulin-like growth factor binding proteins in bone cell regulation. Growth Regul. 3 (1), 67–70.
Mohan, S., Baylink, D.J., 1990. Autocrine-paracrine aspects of bone metabolism. Growth Genet. Horm. 6, 1–9.
Mohan, S., et al., 1988. Primary structure of human skeletal growth factor: homology with human insulin-like growth factor-II. Biochim. Biophys. Acta 966 (1), 44–55.
Mohan, S., et al., 1989a. Characterization of the receptor for insulin-like growth factor II in bone cells. J. Cell. Physiol. 140 (1), 169–176.
Mohan, S., et al., 1989b. Isolation of an inhibitory insulin-like growth factor (IGF) binding protein from bone cell-conditioned medium: a potential local regulator of IGF action. Proc. Natl. Acad. Sci. U. S. A 86 (21), 8338–8342.
Mohan, S., et al., 1991. Increased IGF-I and IGF-II in bone from patients with osteoarthritis. J. Bone Miner. Res. 1 (Suppl. 1), 131.
Mohan, S., et al., 1995. Age-related changes in IGFBP-4 and IGFBP-5 levels in human serum and bone: implications for bone loss with aging. Prog. Growth Factor Res. 6 (2–4), 465–473.
Mohan, S., et al., 1996. Insulin-like growth factor (IGF)-binding proteins in serum–do they have additional roles besides modulating the endocrine IGF actions? J. Clin. Endocrinol. Metab. 81 (11), 3817–3820.
Muller, E.E., et al., 1999. Neuroendocrine control of growth hormone secretion. Physiol. Rev. 79 (2), 511–607.
Nakamura, T., et al., 1992. Clinical significance of serum levels of insulin like growth factors as bone metabolic markers in postmenopausal women. Bone Miner. 17 (Suppl. 1), 170.
Nakashima, K., de Crombrugghe, B., 2003. Transcriptional mechanisms in osteoblast differentiation and bone formation. Trends Genet. 19 (8), 458–466.
Nakashima, K., et al., 2002. The novel zinc finger-containing transcription factor osterix is required for osteoblast differentiation and bone formation. Cell 108 (1), 17–29.
Neumann, G.M., Bach, L.A., 1999. The N-terminal disulfide linkages of human insulin-like growth factor-binding protein-6 (hIGFBP-6) and hIGFBP-1 are different as determined by mass spectrometry. J. Biol. Chem. 274 (21), 14587–14594.
Norman, C., et al., 2013. Estradiol regulates GH releasing peptide interaction with GHRH and somatostain. Eur. J. Endocrinol. 170, 121–129, 2013.
Nicolas, V., et al., 1994. Age-related decreases in insulin-like growth factor-I and transforming growth factor-beta in femoral cortical bone from both men and women: implications for bone loss with aging. J. Clin. Endocrinol. Metab. 78 (5), 1011–1016.
Nicolas, V., et al., 1995. An age-related decrease in the concentration of insulin-like growth factor binding protein-5 in human cortical bone. Calcif. Tissue Int. 57 (3), 206–212.
Nunn, S.E., et al., 1997. Regulation of prostate cell growth by the insulin-like growth factor binding proteins and their proteases. Endocrine 7 (1), 115–118.
Oukka, M., et al., 2002. A mammalian homolog of Drosophila schnurri, KRC, regulates TNF receptor-driven responses and interacts with TRAF2. Mol. Cell 9 (1), 121–131.
Papadakis, M.A., et al., 1996. Growth hormone replacement in healthy older men improves body composition but not functional ability. Ann. Intern. Med. 124 (8), 708–716.
Petersenn, S., Schulte, H.M., 2000. Structure and function of the growth-hormone-releasing hormone receptor. Vitam. Horm. 59, 35–69.
Powell-Braxton, L., et al., 1993. IGF-I is required for normal embryonic growth in mice. Genes Dev. 7 (12B), 2609–2617.
Qiao, M., et al., 2004. Insulin-like growth factor-1 regulates endogenous RUNX2 activity in endothelial cells through a phosphatidylinositol 3-kinase/ERK-dependent and Akt-independent signaling pathway. J. Biol. Chem. 279 (41), 42709–42718.
Raisingani, M., et al., 2017. Skeletal growth and mineral acquisition in Type I diabetic children: abnormalities of the GH/IGF axis. Growth Horm and IGF Res. 34, 13–21.
Rajah, R., et al., 1996. 7S nerve growth factor is an insulin-like growth factor-binding protein protease. Endocrinology 137 (7), 2676–2682.
Rajaram, S., et al., 1997. Insulin-like growth factor-binding proteins in serum and other biological fluids: regulation and functions. Endocr. Rev. 18 (6), 801–831.
Rappaport, R., Czernichow, P., 1993. Disorders of GH and prolactin secretion. In: Bertrand, J., et al. (Eds.), Pediatric Endocrinology. Williams and Wilkins, Baltimore, pp. 220–241.
Reijnders, C.M., et al., 2007. Effect of mechanical loading on IGF-I gene expression in rat tibia. J. Endocrinol. 192, 131–140.

Richman, C., et al., 1999. Recombinant human insulin-like growth factor- binding protein-5 stimulates bone formation parameters in vitro and in vivo. Endocrinology 140 (10), 4699–4705.

Rosen, C.J., et al., 1992. The 24/25-kDa serum insulin-like growth factor- binding protein is increased in elderly women with hip and spine fractures. J. Clin. Endocrinol. Metab. 74 (1), 24–27.

Rosen, C.J., et al., 1994. Insulin-like growth factors and bone: the osteoporosis connection. Proc. Soc. Exp. Biol. Med. 206 (2), 83–102.

Rosen, C.J., Donahue, L.R., 1998. Insulin-like growth factors and bone: the osteoporosis connection revisited. Proc. Soc. Exp. Biol. Med. 219 (1), 1–7.

Rosen, C.J., et al., 1999. The RIGHT Study: a randomized placebo-controlled trial of recombinant human growth hormone in frail elderly: dose response effects on bone mass and bone turnover. J. Bone Miner. Res. 14 (Suppl. 1), 208.

Rosen, T., et al., 1993. Reduced bone mineral content in adult patients with growth hormone deficiency. Acta Endocrinol. 129 (3), 201–206.

Rosen, T., et al., 1997. Increased fracture frequency in adult patients with hypopituitarism and GH deficiency. Eur. J. Endocrinol. 137 (3), 240–245.

Rubin, J., et al., 2002. IGF-I regulates osteoprotegerin (OPG) and receptor activator of nuclear factor-kappaB ligand in vitro and OPG in vivo. J. Clin. Endocrinol. Metab. 87 (9), 4273–4279.

Rubin, M.R., Bilezikian, J.P., 2002. New anabolic therapies in osteoporosis. Curr. Opin. Rheumatol. 14 (4), 433–440.

Rudman, D., Mattson, D.E., 1994. Serum insulin-like growth factor I in healthy older men in relation to physical activity. J. Am. Geriatr. Soc. 42 (1), 71–76.

Rudman, D., et al., 1981. Impaired growth hormone secretion in the adult population: relation to age and adiposity. J. Clin. Investig. 67 (5), 1361–1369.

Rudman, D., et al., 1990. Effects of human growth hormone in men over 60 years old. N. Engl. J. Med. 323 (1), 1–6.

Rubin, M.R., et al., 2005. Idiopathic osteoporosis in premenopausal women. Osteoporos. Int. 16, 526–533.

Saggese, G., et al., 1993. Effects of long-term treatment with growth hormone on bone and mineral metabolism in children with growth hormone deficiency. J. Pediatr. 122 (1), 37–45.

Saggese, G., et al., 1996. The effect of long-term growth hormone (GH) treatment on bone mineral density in children with GH deficiency. Role of GH in the attainment of peak bone mass. J. Clin. Endocrinol. Metab. 81 (8), 3077–3083.

Sakata, T., et al., 2004. Skeletal unloading induces resistance to IGF-I by inhibiting activation of the IGF signaling pathways. J. Bone Miner. Res. 19, 436–446.

Salih, D.A., et al., 2005a. Insulin-like growth factor-binding protein-5 induces a gender-related decrease in bone mineral density in transgenic mice. Endocrinology 146 (2), 931–940.

Sansal, I., Sellers, W.R., 2004. The biology and clinical relevance of the PTEN tumor suppressor pathway. J. Clin. Oncol. 22 (14), 2954–2963.

Scharla, S.H., et al., 1993. 1,25-Dihydroxyvitamin D3 increases secretion of insulin-like growth factor binding protein-4 (IGFBP-4) by human osteoblast-like cells in vitro and elevates IGFBP-4 serum levels in vivo. J. Clin. Endocrinol. Metab. 77 (5), 1190–1197.

Schmid, C., Ernst, M., 1991. IGF, in cytokines and bone metabolism. In: Gowen, M. (Ed.), IGFs. CRC Press, Boca Raton, pp. 229–259.

Schmid, C., et al., 1984. Insulin-like growth factor I supports differentiation of cultured osteoblast-like cells. FEBS Lett. 173 (1), 48–52.

Schmid, C., et al., 1989. Insulin-like growth factor I regulates type I procollagen messenger ribonucleic acid steady state levels in bone of rats. Endocrinology 125 (3), 1575–1580.

Schmid, C., et al., 1992. Differential regulation of insulin-like growth factor binding protein (IGFBP)-2 mRNA in liver and bone cells by insulin and retinoic acid in vitro. FEBS Lett. 303 (2–3), 205–209.

Schmid, C., et al., 1996. Effects and fate of human IGF-binding protein-5 in rat osteoblast cultures. Am. J. Physiol. 271 (6 Pt 1), E1029–E1035.

Schoenle, E., et al., 1985. Comparison of in vivo effects of insulin-like growth factors I and II and of growth hormone in hypophysectomized rats. Acta Endocrinol. 108 (2), 167–174.

Sell, C., et al., 1995. Insulin-like growth factor I (IGF-I) and the IGF-I receptor prevent etoposide-induced apoptosis. Cancer Res. 55 (2), 303–306.

Sheng, M.H., et al., 2013. Disruption of the IGF-1 gene in osteocytes impairs developmental bone growth in mice. Bone 52, 133–144.

Sjogren, K., et al., 2000. Disproportionate skeletal growth and markedly diminished bone mineral content in GH receptor null mice. Biochem. Biophys. Res. Commun. 267, 603–608.

Skjaerbaek, C., et al., 1998. No effect of growth hormone on serum insulin-like growth factor binding protein-3 proteolysis. J. Clin. Endocrinol. Metab. 83 (4), 1206–1210.

Slootweg, M.C., et al., 1990. The presence of classical insulin-like growth factor (IGF) type-I and -II receptors on mouse osteoblasts: autocrine/paracrine growth effect of IGFs? J. Endocrinol. 125 (2), 271–277.

Slootweg, M.C., et al., 1992. Osteoclast formation together with interleukin-6 production in mouse long bones is increased by insulin-like growth factor-I. J. Endocrinol. 132 (3), 433–438.

Slootweg, M.C., et al., 1996. Growth hormone receptor activity is stimulated by insulin-like growth factor binding protein 5 in rat osteosarcoma cells. Growth Regul. 6 (4), 238–246.

Sneppen, S.B., et al., 2002. Bone mineral content and bone metabolism during physiological GH treatment in GH-deficient adults–an 18-month randomized, placebo-controlled, double-blinded trial. Eur. J. Endocrinol. 146 (2), 187–195.

Snyder, P.J., et al., 2007. Effect of growth hormone replacement on bone mineral density in adult-onset growth hormone deficiency. J. Bone Miner. Res. 22 (5), 762–770.

Spencer, E.M., et al., 1991. In vivo actions of insulin-like growth factor-I (IGF-I) on bone formation and resorption in rats. Bone 12 (1), 21–26.

Szulc, P., et al., 2004. Insulin-like growth factor I is a determinant of hip bone mineral density in men less than 60 years of age: MINOS study. Calcif. Tissue Int. 74 (4), 322–329.

Tixier-Boichard, M., et al., 1992. Effects of insulin-like growth factor-I (IGF-I) infusion and dietary tri-iodothyronine (T3) supplementation on growth, body composition, and plasma hormone levels in sex-linked dwarf mutant and normal chickens. J. Endocrinol. 133 (1), 101–110.

Tobias, J.H., et al., 1992. Opposite effects of insulin-like growth factor-I on the formation of trabecular and cortical bone in adult female rats. Endocrinology 131 (5), 2387–2392.

Thomas, D.M., et al., 1996. Dexamethasone modulates insulin receptor expression and subcellular distribution of the glucose transporter GLUT 1 in UMR 106-01, a clonal osteogenic sarcoma cell line. J. Mol. Endocrinol. 17, 7–17, 1996.

Tseng, K.F., Goldstein, S.A., 1998. Systemic over secretion of growth hormone in transgenic mice. J. Bone Miner. Res. 13, 706–715.

Uchimura, T., et al., 2017. An essential role for IGFII in cartilage development and glucose metabolism during postnatal long bone growth. Development 19, 3533–3546.

Uekl, I., et al., 2000. Inactivation of the Acid labile subunit in mice results in mild retardation of growth despite profound disruptions in the IGF regulatory system. Proc. Natl. Acad. Sci. Unit. States Am. 97, 6868–6873.

Ueland, T., et al., 2002. Decreased trabecular bone content, IGFBP-5, apparent density, IGF-II content in acromegaly. Eur. J. Clin. Investig. 32, 122–128.

Valk, N.K., et al., 1994. The effects of human growth hormone (GH) administration in GH-deficient adults: a 20-day metabolic ward study. J. Clin. Endocrinol. Metab. 79 (4), 1070–1076.

Verhaeghe, J., et al., 1992. The effects of systemic insulin, insulin-like growth factor-I, and growth hormone on bone growth and turnover in spontaneously diabetic BB rats. J. Endocrinol. 134 (3), 485–492.

Wang, Y., et al., 2007. IGF-I receptor is essential for the anabolic actions of PTH on bone. J. Bone Miner Res. 22, 1329–1337.

Wang, Y., et al., 2011. IGF-1R signaling in chondrocytes modulates growth plate development by interacting with PTHrp/Ihh. J. Bone Miner. Res. 26, 1437–1446.

Wang, J., et al., 1999. Igf1 promoters long bone growth by IGF augmenting chondrocyte hypertrophy. FASEB J. 13, 1985–1990.

Wang, Y., et al., 2006. Role of IGF-I signaling in regulating osteoclastogenesis. J. Bone Miner. Res. 21 (9), 1350–1358.

Wang, X., et al., 2015. A liver bone Endocrine relay by IGFBP-1 promotes osteoclastogenesis and mediates FGF-21 induced bone resorption, 22, 811–824.

Wassenaar, M.J., et al., 2011. High prevalence of vertebral fractures despite normal bone density in patients with long term controlled acromegaly. Eur. J. Endocrinol. 164, 475–483.

Wei, J., et al., 2015. Glucose uptake and Runx2 synergize to orchestrate osteoblast differentiation. Cell 161, 1576–1591.

Woods, K.A., et al., 1996. Intrauterine growth retardation and postnatal growth failure associated with deletion of the insulin-like growth factor I gene. N. Engl. J. Med. 335 (18), 1363–1367.

Wright, N.M., et al., 1995. Greater secretion of growth hormone in black than in white men: possible factor in greater bone mineral density–a clinical research center study. J. Clin. Endocrinol. Metab. 80 (8), 2291–2297.

Wu, Y., et al., 2013. Serum IGF-I is insufficient to restore skeletal size in total absence of the growth hormone receptor. J. Bone Miner. Res. 28, 1575–1586.

Wuster, C., et al., 1991. Increased prevalence of osteoporosis and arteriosclerosis in conventionally substituted anterior pituitary insufficiency: need for additional growth hormone substitution? German. Klin. Wochenschr. 69 (16), 769–773.

Wuster, C., et al., 1992. Bone mass of spine and forearm in osteoporosis and in German normals: influences of sex, age and anthropometric parameters. Eur. J. Clin. Investig. 22 (5), 336–370.

Wuster, C., et al., 1993. Decreased serum levels of insulin-like growth factors and IGF binding protein 3 in osteoporosis. J. Intern. Med. 234 (3), 249–255.

Wuster, C., et al., 2001. The influence of growth hormone deficiency, growth hormone replacement therapy, and other aspects of hypopituitarism on fracture rate and bone mineral density. J. Bone Miner. Res. 16 (2), 398–405.

Xi, G., et al., 2016. IRS-1 functions as a molecular scaffold to coordinate IGF-I/IGFBP-2 signaling during osteoblast differentiation. J. Bone Miner. Res. 31, 1300–1314.

Xiao, G., et al., 2000. MAPK pathways activate and phosphorylate the osteoblast-specific transcription factor. Cbfa1. J. Biol. Chem. 275 (6), 4453–4459.

Yakar, S., et al., 1999. Normal growth and development in the absence of hepatic insulin-like growth factor I. Proc. Natl. Acad. Sci. U. S. A 96 (13), 7324–7329.

Yakar, S., et al., 2002. Circulating levels of IGF-1 directly regulate bone growth and density. J. Clin. Investig. 110 (6), 771–781.

Yakar, S., et al., 2006. The ternary IGF complex influences postnatal bone acquisition and the skeletal response to intermittent parathyroid hormone. J. Endocrinol. 189 (2), 289–299.

Yeo, A.L., et al., 2003. Frailty and the biochemical effects of recombinant GH in women after surgery for hip fracture. Growth Hormone IGF Res. 13, 361–370.

Zeslawski, W., et al., 2001. The interaction of insulin-like growth factor-I with the N-terminal domain of IGFBP-5. EMBO J. 20 (14), 3638–3644.

Zhang, M., et al., 2002. Osteoblast-specific knockout of the insulin-like growth factor (IGF) receptor gene reveals an essential role of IGF signaling in bone matrix mineralization. J. Biol. Chem. 277 (46), 44005–44012.

Zhang, M., et al., 2003. Paracrine over expression of IGFBP-4 in osteoblasts decreases bone turnover and global growth retardation. J. Bone Miner. Res. 18, 836–843.

Zhao, H., et al., 2004. PTEN inhibits cell proliferation and induces apoptosis by downregulating cell surface IGF-IR expression in prostate cancer cells. Oncogene 23 (3), 786–794.
Zoch, M.L., et al., 2016. In vivo radiometric analysis of glucose uptake and distribution in mouse bone. Bone Res 4, 16004, 2016.
Zofkova, I., 2003. Pathophysiological and clinical importance of insulin-like growth factor-I with respect to bone metabolism. Physiol. Res. 52 (6), 657–679.
Zoidis, E., et al., 2011. Stimulation of glucose transport in osteoblastic cells by parathyroid hormone and insulin-like growth factor I. Mol. Cell. Biochem. 348, 33–42.

Further reading

Andreassen, T.T., Oxlund, H., 2001. The effects of growth hormone on cortical and cancellous bone. J. Musculoskelet. Neuronal Interact. 2 (1), 49–58.
Bikle, D.D., et al., 1994. Skeletal unloading induces resistance to insulin-like growth factor I. J. Bone Miner. Res. 9 (11), 1789–1796.
Canalis, E., Lian, J.B., 1989. Effects of bone associated growth factors on DNA, collagen, and osteocalcin synthesis in cultured fetal rat calvariae. Bone 9 (4), 243–246.
Celil, A.B., Campbell, P.G., 2005. BMP-2 and insulin-like growth factor-I mediate Osterix (Osx) expression in human mesenchymal stem cells via the MAPK and protein kinase D signaling pathways. J. Biol. Chem. 280 (36), 31353–31359.
Daughaday, W.H., 1989. A personal history of the origin of the somatomedin hypothesis and recent challenges to its validity. Perspect. Biol. Med. 32 (2), 194–211.
Diamond, T., et al., 1989. Spinal and peripheral bone mineral densities in acromegaly: the effects of excess growth hormone and hypogonadism. Ann. Intern. Med. 111 (7), 567–573.
Ernst, M., Rodan, G.A., 1990. Increased activity of insulin-like growth factor (IGF) in osteoblastic cells in the presence of growth hormone (GH): positive correlation with the presence of the GH-induced IGF-binding protein BP-3. Endocrinology 127 (2), 807–814.
Heaney, R.P., 1962. Radiocalcium metabolism in disuse osteoporosis in man. Am. J. Med. 33, 188–200.
Kassem, M., et al., 1993. Growth hormone stimulates proliferation and differentiation of normal human osteoblast-like cells in vitro. Calcif. Tissue Int. 52 (3), 222–226.
Kasukawa, et al., 2003. Evidence that sensitivity to GH is growth period and tissue time dependent. Endocrinology 144, 3950–3957.
Madeira, M., et al., 2013. Vertebral fracture assessment in acromegaly. J. Clin. Densitom. 16, 238–243.
Machwate, M., et al., 1994. Insulin-like growth factor-I increases trabecular bone formation and osteoblastic cell proliferation in unloaded rats. Endocrinology 134 (3), 1031–1038.
Maor, G., et al., 1989. Human growth hormone enhances chondrogenesis and osteogenesis in a tissue culture system of chondroprogenitor cells. Endocrinology 125 (3), 1239–1245.
Mathews, L.S., et al., 1988. Expression of insulin-like growth factor I in transgenic mice with elevated levels of growth hormone is correlated with growth. Endocrinology 123 (1), 433–437.
Mohan, S., Baylink, D.J., 1991. Bone growth factors. Clin. Orthop. Relat. Res. 263, 30–48.
Rosen, C.J., 2004. Insulin-like growth factor I and bone mineral density: experience from animal models and human observational studies. Best Pract. Res. Clin. Endocrinol. Metabol. 18 https://doi.org/10.1016/j.beem.2004.02.007.
Salih, D.A., et al., 2005b. IGFBP-5 induces a gender related decrease in bone mineral density in transgenic mice. Endocrinology 146, 931–940.
Seeman, E., et al., 1982. Differential effects of endocrine dysfunction on the axial and the appendicular skeleton. J. Clin. Investig. 69 (6), 1302–1309.
Whitehead, H.M., et al., 1992. Growth hormone treatment of adults with growth hormone deficiency: results of a 13-month placebo controlled cross-over study. Clin. Endocrinol. 36 (1), 45–52.

Further reading

Index for Volumes 1 and 2

'*Note:* Page numbers followed by "f" indicate figures, "t" indicate tables and "b" indicate boxes.'

A

Abaloparatide (ABL), 267, 457–460, 609, 613, 702–703, 1638–1639, 1916
ABCC6. *See* ATP binding cassette subfamily C member 6 (ABCC6)
ABI. *See* Ankle–brachial index (ABI)
ABL. *See* Abaloparatide (ABL)
aBMD. *See* Areal bone mineral density (aBMD)
Abnormal bone pathogenesis, 1472–1475
  abnormalities
    of bone collagen, 1475
    in fibroblast growth factor-23–klotho pathway, 1474
  disordered osteoblast function or differentiation, 1474
  impaired Wnt signaling, 1474–1475
  parathyroid hormone and 1,25$(OH)_2D_3$, 1472–1473
  transforming growth factor β family abnormalities, 1475
AC. *See* Adenylyl cyclase (AC)
Accuracy errors, 1861
Acellular cementoblasts, 1063
N-acetylgalactosamine (NAcgal), 720
AchRM3. *See* Muscarinic receptor 3 (AchRM3)
Acid
  phosphatase, 1570
  solubilization, 295–296
Acid labile subunit (ALS), 991
Acidic FGF, 1114
Acidic serine aspartate-rich MEPE-associated motif (ASARM), 475
ACLT. *See* Anterior cruciate ligament transection (ACLT)
Acolbifene, 868–870, 868f
Acquired FGF23-related hypophosphatemic disease, 1533
Acquired inflammation, SSCs/BMSCs role in, 62–63
Acrodysostosis, 1394
Acromesomelic dysplasia, type Maroteaux (AMDM), 1395–1396
ACTH. *See* Adrenocorticotropic hormone (ACTH)
Actin scavenging, 719
Activated endothelial cells, 1144
Activating transcription factor 4 (ATF4), 166, 1281, 1809
  *Atf*4-deficient mouse model, 168
  as transcriptional regulator of osteoblast functions, 168–169
Activator protein 1 (AP-1), 169, 307–308, 380, 648, 722
Active transcriptional complex formation, 1088–1089
Acute phase reaction (APR), 1675
AD. *See* Alzheimer's disease (AD)
Adams Oliver syndrome (AOS), 1097
Adaptation to exercise, 1935, 1937
Adapter-related protein complex 2 (AP2), 1415
Adaptive homeostatic system, 470–471
Adaptor proteins (APs), 651
  AP2, 546
Adenomatous polyposis coli protein (APC protein), 100, 179, 1412–1413
Adenosine monophosphate (AMP), 992
Adenosine triphosphatase 11C (ATP11C), 1366–1367
Adenosine triphosphate (ATP), 469–470, 1275
Adenylyl cyclase (AC), 1275
Adequate intake (AI), 1646–1647
ADH. *See* Autosomal dominant hypocalcemia (ADH)
Adherens junctions, 423–428, 424f
Adhesion signals, 122–123
Adhesive interactions, 423
ADHR. *See* Autosomal dominant hypophosphatemic rickets (ADHR)
Adipocytes, 75–76, 91–92, 1343–1344
Adipogenesis, 48–49
Adiponectin, 1931–1932
  control counterregulation of bone remodeling by, 812–813
  dual action on bone remodeling, 810–811
ADIS. *See* Agonist-driven insertional signaling (ADIS)
ADO. *See* Autosomal dominant osteopetrosis (ADO)
ADO type 1 (ADO1), 1554
ADO type 2 (ADO2), 1554
Adolescent mammary development, 844
ADOPT trial, 950–951
β2-Adrenergic receptor (β2AR), 1344
Adrenergic signaling, 816–817
Adrenocorticotropic hormone (ACTH), 1436
Adrenocorticotropic hormone-secreting adenoma (ACTHoma), 1300
ADT. *See* Androgen-deprivation therapy (ADT)
Adult bone
  marrow niche, 75–79
  PTH and PTHrP in, 610–612
Adult(s), 479
  adult-onset GH deficiency, 1001–1002
  adult-onset hypothyroidism, 903
  HPP, 1579, 1579f
Adulthood originate before puberty, 251–252
Advanced age, 256–257
  predominance of cortical bone loss, 259–261
Advanced glycation end products (AGEs), 345–347, 346f, 942, 951–952, 954, 1475
  types, 346–347
Adverse immune-mediated perturbations, 1691
Adynamic bone disease, 452, 1474
Afferent signals regulating bone remodeling, 810–811
AFFs. *See* Atypical femoral fractures (AFFs)
Aged skeleton characteristics, 275–276
  human, 275
  rodents, 276
AGEs. *See* Advanced glycation end products (AGEs)
Aggrecan (*Acan*), 1892
Aggrecan and versican (PG-100), 360
*Aggrecan-CreERT*2 mice, 1795
Aging, 827, 1189
  bone cell, 276–278
  GH/IGF-I/IGFBP system on aging skeleton, 988–989
  molecular mechanisms, 278–282
Agonist-driven insertional signaling (ADIS), 546
Agrp, 1931–1932
AHO. *See* Albright hereditary osteodystrophy (AHO)
AI. *See* Adequate intake (AI)
AIRE. *See* Autoimmune regulator (AIRE)
AIx. *See* Augmentation index (AIx)
Akt/PKB, 186–187
Alagille syndrome, 1096–1097
Albers-Schönberg disease. *See* Mild osteopetrosis
Albright hereditary osteodystrophy (AHO), 1393, 1431, 1434–1440, 1435f
  clinical presentation, 1434–1436
  diagnosis and management, 1437–1438
  molecular genetics, 1437

Albright hereditary osteodystrophy (AHO) (*Continued*)
pathogenesis, 1438–1440
Albumin–hormone complexes, 715–716
Aldehyde derivatives, 301
Alendronate, 1496
Alendronate therapy, 952
Alkaline phosphatase (ALP), 363, 1569, 1570f
effects on, 651–652
genomic structure, protein chemistry, and enzymology, 1572–1573
history and physiological roles, 1570–1572
immunoreactivity, 1584
in skeletal mineralization, 1571t
Alkaline phosphatase (ALP), 1802
Alkaline phosphatase positive (AP+), 832–833
Allosteric modulators, 542
Alopecia, 1514
ALP. *See* Alkaline phosphatase (ALP)
Alpha-smooth muscle actin (αSMA), 1892
ALPL gene
gene defects, 1584
structural defects, 1584–1585
Alpl knockout animals, 1591
ALS. *See* Acid labile subunit (ALS); Amyotrophic lateral sclerosis (ALS)
ALS-null mice (ALSKO), 993
Aluminum (Al), 531–534
absorption, 532–533
aluminum/dementia hypothesis, 532
toxicity, 532–533
Alveolar bone, 1064, 1070
Alveolar epithelium, 1199
Alveolar ridge, 1065
Alzheimer's disease (AD), 277–278, 1674
AMDM. *See* Acromesomelic dysplasia, type Maroteaux (AMDM)
Amelotin-null mice, 1075–1076
Amine oxidases, 724
Amino acid residues, 722
155-Amino-acid-long nonglycosylated polypeptide, 1114
Amino-terminal fragment (ATF), 388
Amino-terminal PTH-related protein interaction with cell surface receptors, 1312–1313
Aminoglycosides, 519
Aminopropyltriethoxysilane (APES), 1903
AMP. *See* Adenosine monophosphate (AMP)
AMP-activated protein kinase (AMPK), 996
Amphetamine-regulated transcript, 818
Amphetamine-regulated transcript, bone resorption regulation by, 814–815
Amphiregulin (AR), 649–650
AMPK. *See* AMP-activated protein kinase (AMPK)
Amylin, 789–790, 792, 795–796, 954
amylin–relevance to human bone physiology, 802
calcitonin effects on osteoblasts, 797–798
to human bone physiology, 802
receptors, 792
skeletal effects of, 800–801
Amyotrophic lateral sclerosis (ALS), 1609
Anabolic
agents, 1496, 1821–1822
drug, 460
genes, 1716
to musculoskeletal system, 1769–1770
therapy, 266–268, 1493
Wnt signaling pathway as targeting, 188–189
window, 266
Analgesics, 1562
Analog action, 1746–1748
Analytical factors, 1809
Androgen, 282
deprivation in men with prostate cancer, 1696
gain of function in bone health using end points, 975
and male skeletal health, 979
Androgen receptor (AR), 971–972, 973f, 1340
bone geometry and architecture, 975–976
evidence from human trials, 974–975
loss-of-function
evidence from rare human variants, 972–974
using genetically modified mice, 976–977
muscle–bone interface and potential impact on bone strength, 978–979
studies using selective androgen receptor modulators, 977–978
studies with testosterone treatment and change in bone architecture, 976
Androgen-deprivation therapy (ADT), 971–972, 1340
Angiocrine signaling in bone, 210
Angiogenesis, 1142
Angiotensin II, 599–600
Ankle–brachial index (ABI), 627–629
Ankylosis protein (ANK), 430–431
Annexin A2, 719
Annexin II (ANXA2), 1338
ANNs. *See* Artificial neural networks (ANNs)
Anorexia nervosa, 1006
ANS. *See* Autonomic nervous system (ANS)
Antagonizing BMP signaling, 1192, 1192f
Anterior cruciate ligament transection (ACLT), 1167
Anterior pituitary tumors, 1300–1301
Anterior–posterior axis, 14
Anticatabolic agents
bisphosphonates, 455
calcitonin, 454
denosumab, 455–456
hormone therapy, 454
SERMs, 454–455
SHOTZ, 460
Anticonvulsants, 809
Antiinflammatory genes, 932
Antiproliferative effects, 1740
Antiresorptive
agents, 246, 266, 1496, 1695, 1723, 1916
effect, 1681, 1719
therapy, 263–266
treatment, 1821
Antisclerostin/DKK1, 932–933
Antithrombin, 878
Antitumor effects, 1202, 1673–1674
ANXA2. *See* Annexin II (ANXA2)
AP-1. *See* Activator protein 1 (AP-1)
AP2. *See* Adapter-related protein complex 2 (AP2)
AP2S1 gene, 1415–1416
APC protein. *See* Adenomatous polyposis coli protein (APC protein)
APECED syndrome. *See* Autoimmune polyendocrinopathy-candidiasis-ectodermal dystrophy syndrome (APECED syndrome)
Apis mellifera (Honeybee), 694
Aplastic bone disease, 1474
Apoptotic/apoptosis, 761
cells, 1210
of osteocytes, 137, 254
APs. *See* Adaptor proteins (APs)
Aquaporin 2 channels, 551
AR. *See* Amphiregulin (AR); Androgen receptor (AR)
Arachidonic acid derivatives, 321–322
Arachidonic acid mobilization, 1249
Areal bone mineral density (aBMD), 943, 1618, 1860–1861
fracture prediction using, 1862
Arg–Gly–Asp sequence (RGD sequence), 401–402, 1154–1155
ARHR. *See* Autosomal recessive hypophosphatemic rickets (ARHR)
ARO. *See* Autosomal recessive osteopetrosis (ARO)
Aromatase inhibitors in women with breast cancer, 1696
Aromatic amino acids, 543
β-Arrestins, 646–647, 651, 1277
Arsenic, 533
Arterial biology, PTH receptor signaling in, 624–630
Articular cartilage, 1173, 1194–1195
Articular chondrocytes, 1714
Artificial neural networks (ANNs), 943
Arzoxifene, 868, 868f, 884
ASARM. *See* Acidic serine aspartate-rich MEPE-associated motif (ASARM)
ASCENT II trial, 763
Asfotase alfa treatment for hypophosphatasia, 1591
Aspartic proteinases, 390
Assembly, 297–302
Asthma, 556
At-T20 pituitary cells, 544–545
Ataxia–telangiectasia mutated (ATM), 280, 1674
ATF4. *See* Activating transcription factor 4 (ATF4)

ATG5. *See* Autophagy-related protein 5 (ATG5)
Atherosclerosis, 280, 283–284, 373
ATM. *See* Ataxia–telangiectasia mutated (ATM)
ATP. *See* Adenosine triphosphate (ATP)
ATP binding cassette subfamily C member 6 (ABCC6), 474
ATP11C. *See* Adenosine triphosphatase 11C (ATP11C)
Attachment molecules in bone cells
  CD44, 411
  chondroadherin, 412–413
  glypicans and perlecan, 410–411
  immunoglobulin superfamily members, 411–412
  osteoactivin, 412
  syndecans, 410
Attainment of bone's peak material and structural strength, 246–254
Atypical femoral fractures (AFFs), 1700–1701
AU-rich element (ARE), 584
AU-rich-binding factor (AUF1), 584
Augmentation index (AIx), 632
Autocrine nuclear actions, 600–602
Autoimmune polyendocrinopathy-candidiasis-ectodermal dystrophy syndrome (APECED syndrome), 1362–1363
Autoimmune polyglandular syndrome type 1, 1362–1363
Autoimmune regulator (AIRE), 1362–1363
Autoimmunity, 1202
Autonomic nervous system (ANS), 212–213
Autophagosome, 281
Autophagy, 138
  loss, 281–282
Autophagy-related protein 5 (ATG5), 114
Autophagy–lysosomal system, 281
Autosomal dominant hypocalcemia (ADH), 513, 539–540, 1364, 1368, 1666
  ADH1, 539–540, 549, 1368–1369
    mouse models, 1370
  ADH2, 1369–1370
    mouse models, 1370
  calcium-sensing receptor and Gα11, 1368
Autosomal dominant hypoparathyroidism. *See* Autosomal dominant hypocalcemia (ADH)
Autosomal dominant hypophosphatemic rickets (ADHR), 478, 481, 1114, 1529–1530, 1532
Autosomal dominant osteopetrosis (ADO), 1553
Autosomal dominant syndrome, 1659–1660
Autosomal recessive Fanconi syndrome, 490
Autosomal recessive hypoparathyroidism, 1364–1365
Autosomal recessive hypophosphatemic rickets (ARHR), 481–484, 1533
  ARHR1, 481, 1071
Autosomal recessive osteopetrosis (ARO), 1553–1554
Axial skeleton, 9–11
Axin 1, 182–183
Axin 2, 182–183
Axin protein, 181–183

B

B-lymphocyte-induced maturation protein-1 (BLIMP1), 118, 1206
BAC. *See* Bacterial artificial chromosome (BAC)
BAC recombineering approaches, 1795–1796
Bacterial artificial chromosome (BAC), 1281–1282, 1717, 1795
BAG-75. *See* Bone acidic glyprotein-75 (BAG-75)
Ballooning effect, 1874
Barakat syndrome, 1362
Bariatric–metabolic surgery, 950
Bartter syndrome type V, 1368–1369
Basal cell carcinomas (BCCs), 761
Basic fibroblast growth factor (bFGF), 385
Basic helix–loop–helix transcription factors (bHLH transcription factors), 167, 1089
Basic multicellular units (BMUs), 95, 146, 220–221, 246, 254–256, 276, 610–611, 775
  bone remodeling compartment, 254–255
  coupling occurs locally within, 222
  function, 776
  generation and maintenance of, 777
  multidirectional steps of remodeling cycle, 256
  osteocyte death in signaling bone remodeling, 254
  PTH regulates factors governing assembly and maintenance of, 777–779
  resorption phase of remodeling and cessation in, 223
Bazedoxifene, 868, 868f
BDMR. *See* Brachydactyly mental retardation syndrome (BDMR)
BDNF. *See* Brain-derived neurotrophic factor (BDNF)
Bending tests, 1924–1925
  three-point and four-point, 1924–1925, 1925f
Benign parathyroid tumors, genetic derangements in, 1407–1417
  candidate oncogenes and tumor-suppressor genes, 1412
  CaSR and associated proteins, 1413–1415
  Cyclin D1/PRAD1, 1407–1409, 1408f
  FHH, 1415–1416
  FIHP, 1417
  genetic aspects, 1412–1413
  hyperparathyroidism-jaw tumor syndrome, 1416
  tumor-suppressor genes, 1409–1412
Benign prenatal hypophosphatasia, 1580
Benzopyrans, 868, 868f
Benzothiophenes, 868, 868f
  SERM, 884
3-beta hydroxysteroid dehydrogenase 2 (HSD3B2), 972–974
bFGF. *See* Basic fibroblast growth factor (bFGF)
BFR. *See* Bone formation rate (BFR)
Bglap1 gene, 1932–1933
Bglap2 gene, 1932–1933
bHLH transcription factors. *See* Basic helix–loop–helix transcription factors (bHLH transcription factors)
Biased agonism, 695–696
Biased signaling, 543
Bicalutamide, 971–972
Bidirectional remodeling imbalances, 1644
Biglycan (PG-I), 360–362
Bio-Oss, 797
BioBreeding rat (BB rat), 1004
BioDent, 1927–1928
Biological aging, 986
Biomechanical imaging of bone competence, 1837–1838
Biomedica assay, 1804
Biopsy, 446–447, 446f, 453
Biosynthesis, 297–302
Biphasic effects, 1207–1208
Bisphosphonates (BPs), 137, 455, 929, 1321, 1450, 1496, 1671, 1673f, 1677f, 1679f, 1723, 1819–1820, 1915–1916
  applications of clinical and translational pharmacology of BPs, 1680–1682
  mechanisms of action
    mathematical PD models of BPs, 1677
    side effects, 1676
    therapeutic, 1672–1676
  PK, 1677–1680
BLC. *See* Blomstrand's lethal chondrodysplasia (BLC)
BLIMP1. *See* B-lymphocyte-induced maturation protein-1 (BLIMP1)
β blockers, 213
Blomstrand's lethal chondrodysplasia (BLC), 1386–1392
Blosozumab, 188, 1717–1718
BM40, 365
BMC. *See* Bone mineral content (BMC)
BMD. *See* Bone mineral density (BMD)
BMF. *See* Bone marrow fat (BMF)
BMI. *See* Body mass index (BMI)
BMP receptor II (BMPR-II), 1338–1339
BMP-response element (BRE), 1159
Bmp2 gene function in periodontium, 1065–1070, 1067f
  decreasing trabecular and cortical bone in long bones, 1068f
  example of transcriptome and epigenomic analysis, 1071f
BMPR-II. *See* BMP receptor II (BMPR-II)
BMPs. *See* Bone morphogenetic proteins (BMPs)
BMs. *See* Bone metastases (BMs)
BMSCs. *See* Bone marrow stromal cells (BMSCs)

BMSi. *See* Bone material strength index (BMSi)
BMUs. *See* Basic multicellular units (BMUs)
Body mass index (BMI), 815, 1873
Body size and longitudinal growth, GH–IGF-I systemic effects on, 994–995
Body weight (BW), 941, 994
Bone, 76, 133, 205–206, 296, 646, 985–986, 1172, 1833, 1931
  anabolic effects of intermittent PTH
    mechanisms underlying overfill of resorption cavities, 781–782
    stimulation of anabolic remodeling and modeling, 779–781
  biology, 177
  biopsy, 534, 1475, 1912
    clinical indications, 453
    variables, 450t
  as body's calcium sink and reserve, 1644–1645
  bone metastatic niches, adhesion and invasion into, 1338
  bone-derived extracellular matrix proteins, 479
  bone-lining cells, 94–95, 1718
    express collagenase mRNA, 255
  bone-resorbing osteoclasts, 794
  bone-resorption inhibitors, 1915–1916
  bone-seeking toxic elements, research into, 527–528
  bone–muscle unit, 978–979
  calcium in, 1645–1646
  cellular effects of thyroid hormones on, 899–901
  collagen
    abnormalities, 1475
    cross-linking, 340–342
  cross talk, 1937, 1937f
  culture production, 1252–1253
  density, 947–948, 1475–1476
  development, 1142, 1205–1206
  developmental origins
    germ-layer specifications, 47
    skeletal lineage, 53
  diabetes treatments effect, 950–951
  diseases, 1193–1194
  effects of PTH and PTHrP
    bone cells, 648–652, 654
    proliferation, 652–653
  effects on bone proteases, 652
  expression and actions of parathyroid hormone receptor in, 647–655
  extrinsic mechanisms to skeletal aging
    decreased physical activity, 284
    lipid peroxidation and declining innate immunity, 283–284
    loss of sex steroids, 282–283
  failure, 1837
  fibrous dysplasia of, 58–60
  formation, 655
    patterns and development of pericytes/skeletal stem cells, 47–48
  fragility, 262, 264f, 352–353, 1553
    heterogeneous material and structural basis, 263
    in patients with fractures, 263
    treatment in diabetes, 957
  using genetically modified mice equivocal, 977
  glucocorticoids physiological role
    endogenous glucocorticoids promoting osteoblastogenesis, 918–919
    glucocorticoid signaling and prereceptor metabolism, 916
    local glucocorticoid metabolism in bone, 917
    novel insights from targeted disruption of glucocorticoid signaling, 918
  gp130 contribution in osteoclasts to bone physiology, 1226
  growth, 1205–1206
  health and strength, 1696
  homeostasis, 1201
  hormones control systemic metabolism, 952
  isoenzyme of ALP, 1802
  lead measurement in, 529
  mass, 530
  material composition, 258
  mechanical loading, 1258
  mechanical milieu elicited by physical activity
    locomotion induces nonuniform strain environment, 1761–1762
    muscle on bone's strain environment, 1762–1763
    strains in bone, 1761
  microcrack accumulation in vivo, 1910
  microindentation, 949
  mineral, 1672
    density, 975
  modeling stimulation by osteoblasts in response, 782
  morphogenetic proteins, 19–21
  nerve system of, 210–213
  osteocalcin-or lipocalin-2-independent endocrine functions, 1938–1939
  PA in, 388
  pathologies, 304–305
  peak material and structural strength, 246–254
  perfusion, 1877
  phosphatase, 1570
  quality, 975, 1833–1834
    imaging, 1476–1478
  RANKL inhibitors effects, 1691–1693
  as regulator of appetite, 1938
  resorption, 654
  scans, 1602
  sense, 1770–1771
  sensitivity to mechanical signals, 1760–1761
  strain environment, 1762–1763
  stromal cells, 1773
  tamoxifen effects, 1890–1891
  vasculature of, 206–210
Bone acidic glyprotein-75 (BAG-75), 368
"Bone bruise" condition, 1700
Bone cells, 429–430, 1766
  aging, 276–278
    osteoblast progenitors, 277
    osteocytes, 277–278
  attachment molecules in, 410–413
  bone cell-derived IL-1, 1206
  cadherins in, 406–410
  CaSR in, 551–552
  connexin43
    control mechanisms of function, 434–436, 435f
    function in, 432–433
  integrins in, 401–404
  multiple receptors on, 1191
  Wnt signaling and
    interactions and other pathways, 186–188
    osteoblast differentiation and function, 183–185
    osteoclast function, 185
    osteocyte function, 186
    Wnt signaling pathway as target for anabolic therapy, 188–189
Bone formation, 6, 221, 296–297
  bone-forming response, 1718
  indices, 461
  patterns, 47–48
  and repair
    FGF and FGFR signaling, 1119–1122
    FGF regulation, 1121–1122
Bone formation rate (BFR), 449, 994, 1717, 1765, 1899–1900
Bone histomorphometry, 445, 1899
  cortical, 1907–1909
  cryoembedding and cryosectioning, 1903
  fixation, 1901
  low temperature methylmethacrylate embedding, 1902
  measurement of longitudinal bone growth, 1910
  microdamage measurement technique, 1910–1912
  sample preparation, 1901
  sectioning of plastic-embedded bone specimens, 1902–1903
  staining, 1903–1904
  standard methylmethacrylate embedding, 1901
  studies, 454–455
  in vivo labeling, 1899–1900
Bone loss rodent models, histomorphometry of, 1913–1915
Bone marrow, 1875–1877
  bone marrow–derived MSCs, 277
  bone marrow–derived stromal cells, 280
  cell cultures, 1210
  edema syndrome, 1700
  environment, 254–255
  fibroblasts, 1343
  fibrosis, 1062
  inherited forms of bone marrow failure, 60–61
  mesenchymal stem cells, 1772
  microenvironment, 73
  vasculature, 1773

Bone marrow fat (BMF), 956
Bone marrow stromal cells (BMSCs), 46, 56, 60–61, 64, 1068, 1163
Bone mass, 261–262, 1701
central/neuronal regulations, 815–818
adrenergic signaling, 816–817
amphetamine-regulated transcript, 818
brain serotonin and neuromedinU, 817
cocaine-regulated transcript, 818
leptin, 816
melanocortin receptor 4, 818
NPY, brain-derived neurotrophic factor, and cannabinoid receptor 2, 818
GH excess effects, 998–999
and structure, 1722
Bone material strength index (BMSi), 949
Bone matrix, 1692
bone matrix–derived TGF-β, 1170–1172
glycoproteins, 372–373
proteins, effects on, 651–652
quality, 1722
Bone metabolism, 1161, 1801
biochemical markers, 1802–1803
biological markers, 1803–1810, 1803t
GHD effects, 998
Bone metastases (BMs), 1697, 1802
Bone microenvironment
contribution to bone lesion progression, 1342–1345
adipocytes, 1343–1344
bone marrow fibroblasts, 1343
hypoxia and alteration of cancer cell metabolism, 1344–1345
macrophages, 1342
myeloid cells, 1342
osteoblasts, 1343
osteocytes, 1343
physical microenvironment, 1344
sympathetic and parasympathetic nerve system signaling, 1344
T cells, 1343
peptide access to, 793
Bone mineral content (BMC), 529, 1722, 1861
Bone mineral density (BMD), 259, 304, 365, 529, 790–791, 814, 829–831, 870, 943, 971–972, 986, 1005–1007, 1368–1369, 1473, 1615, 1637–1638, 1649, 1658, 1680, 1692, 1712, 1803, 1833–1834, 1857
Bone mineralization, 352, 363, 474–478
IFITM5, 352
intracellular/extracellular compartmentalization, 475
mechanisms of phosphate transport, 475–478
*SERPINF*1, 352
Bone modeling, 219–220, 246–254, 263–268, 776
and remodeling, 1931–1932
and remodeling during growth, 246–254
bone's material and structural strength, 246–251, 250f
sex and racial differences in bone structure, 252–254
trait variances in adulthood originate before puberty, 251–252
stimulation by osteoblasts, 782
Bone morphogenetic proteins (BMPs), 8, 52, 210, 362, 381, 401–402, 778, 996, 1153–1154, 1189, 1338–1339, 1715
antagonizing BMP signaling, 1192
BMP-1, 297, 301, 351
BMP-2, 184, 227, 425, 921–922, 1341
BMP-7, 1338–1339, 1341
canonical BMP signaling, 1189–1190
combinatorial signals, 1191–1192
diversity of ligand and receptor environment, 1190–1191
fracture repair and periosteum, 1194
multiple receptors on bone cells, 1191
osteoarthritis and articular cartilage maintenance, 1194–1195
osteoporosis, 1194
pathway cross talk in bone, 1192–1193
signaling, 20
therapeutics and bone diseases, 1193–1194
Bone morphology regulation by mechanical stimuli, 1764–1766
cycle number, 1765
differential bone remodeling to distinct components of strain tensor, 1764
fluid flow, 1766
osteogenic parameters of strain milieu, 1764
strain distribution, 1765–1766
strain gradient, 1766
strain magnitude, 1764
strain rate, 1765
Bone proteinases
aspartic proteinases, 390
collagenase-3/MMP-13, 384–387
collagenases, 383
cysteine proteinases, 389–390
MMPs, 379–381
MT-MMPs, 382–383
PA in bone, 388
plasminogen activators, 387–388
stromelysin, 381
tPA, 388
type IV collagenases, 381–382
uPA, 388
Bone remodeling, 136–137, 146, 219–220, 246–254, 263–268, 533–534, 828–829, 1164–1167
activity, 213
afferent signals regulating, 810–811
by BMU, 254–256
central and efferent regulators, 812f
bone resorption regulation, 814–815
counterregulation of SNS control of bone remodeling, 812–813
evidence of central/neuronal regulations of bone mass, 815–818
leptin's action on bone remodeling, 811–812
regulators of SNS control of bone remodeling, 813–814
cycle, 1205–1206
dual action of adiponectin on, 810–811
dynamics, 1907
remodeling-based parameters, 1909t
indices, 456
and microstructure during young adulthood, menopause, and advanced age, 256–257
negative regulation of, 810
osteoblast lineage actions during sequence, 99
parathyroid hormone as endocrine regulator of skeletal TGF-β signaling, 1166–1167
regulation by PTH, 776–777
generation and maintenance of BMUs, 777
histologic measurements of bone modeling and remodeling, 776b
TGF-β as coupler of bone resorption and formation, 1164–1166
unit, 449
Bone remodeling compartment (BRC), 254–255
Bone repair
in T1DM and T2DM, 951
therapeutic potential of VEGF for, 1148–1149
Bone resorption, 221, 775, 1553, 1675, 1699, 1724
activity, 220
bone resorption–formation coupling mechanism, 1915–1916
FGF and FGF receptor signaling in, 1122–1124
and formation, 1164–1166, 1692
mechanism, 113
regulation, 814–815
BDNF, 815
IL-1, 815
Y receptor signaling, 814–815
Bone sialoprotein (BSP), 365–366, 1341
BSP-I, 368–369
BSP-II, 369
Bone strength, 257
and quality, 1722
T1DM and, 945–946
animal models of T1DM, 945–946
clinical data, 946
T2DM and, 948–949
animal models, 948–949
humans, 949
testing in rodents
microscale and nanoscale bone material assessment, 1927–1928
whole-bone mechanical testing, 1924–1927
Bone surface density (BS/TV), 1840
Bone traits in mice, GWAS for, 1622–1624
Bone turnover
denosumab effects, 1701
GH excess effects, 998–999
markers, 461, 1662
T1DM and

Bone turnover (*Continued*)
animal models, 943–944
clinical data, 944
T2DM and, 947
Bone turnover markers (BTMs), 1478–1480, 1680, 1802, 1802t
to assessing skeletal safety of new drugs, 1823–1824
clinical uses of bone markers in osteoporosis, 1812–1816
for development of bone drugs, 1819–1823
diagnosis and monitoring of treatment in Paget's disease of bone, 1817
in metastatic bone disease, 1818–1819
monitoring side effects of osteoporosis therapies, 1816
for rare bone diseases, 1817–1818
reductions, 1694–1695
reference ranges, 1812
treatment holiday monitoring, 1816
variability of biochemical markers of bone turnover, 1810–1812
Bone volume/total volume (BV/TV), 994, 1840
Bone-derived hormones
GPRC6A osteocalcin receptor in β cells, 1934
osteocalcin, 1931–1934
regulation, 1935–1937, 1936f
role in adaptation to exercise, 1935
Bone-specific alkaline phosphatase (BSAP), 480, 1478, 1658, 1720–1722
Bone–vascular axis, 635
BPAG1. *See* Bullous pemphigoid antigen (BPAG1)
BPs. *See* Bisphosphonates (BPs)
Brachydactyly, 1099, 1435
Brachydactyly mental retardation syndrome (BDMR), 1397
Brachydactyly type E (BDE), 1383
Brain serotonin, 817
signaling, 811–812
Brain-derived neurotrophic factor (BDNF), 815, 818
variant, 818
Branchiostoma floridae, 694
BRC. *See* Bone remodeling compartment (BRC)
BRE. *See* BMP-response element (BRE)
Breast, CaSR in, 552–553
Breast cancer (BCa), 762, 1339–1340
in animal studies, 762
aromatase inhibitors in women with, 1696
in clinical studies, 763
development, 1199
Breast Cancer Prevention Trial, 875
Brittle bone disease. *See* OI
Brittle Bone Disorders Consortium (BBDC), 1490
Bronchial carcinoids, 1301
Bronchial neuroendocrine tumors, 1301
Bruck syndrome, 350
BS/TV. *See* Bone surface density (BS/TV)
BSAP. *See* Bone-specific alkaline phosphatase (BSAP)
BSP. *See* Bone sialoprotein (BSP)
BTMs. *See* Bone turnover markers (BTMs)
Bullous pemphigoid antigen (BPAG1), 1074–1075
Burosumab, 1549
BV/TV. *See* Bone volume/total volume (BV/TV)
BW. *See* Body weight (BW)

C

c-Fos gene, 170, 746–747
c-MYC, 746–747
c-natriuretic peptide (CNP), 1118
signaling pathways, 1118
C-reactive protein (CRP), 877–878
C-telopeptide, 1603
histidine residue, 345
C-terminal
LBD, 827–828
PTHrP, 602
receptors for C-terminal PTH, 706
telopeptide, 298
C-terminal telopeptide of type I collagen (CTX), 265, 323–324, 1478, 1493, 1495, 1712, 1802
C-type natriuretic peptide (CNP), 25
C-X-C motif chemokine 12 (CXCL12), 75, 1337
C/EBP. *See* CCAAT/enhancer-binding protein (C/EBP)
C57Bl/6 and 129Sv mice, 1256
C5a binding. *See* Complement 5a binding (C5a binding)
CA β-cat. *See* Constitutively active form of β-catenin (CA β-cat)
Ca/Cr ratio. *See* Calcium-to-creatinine ratio (Ca/Cr ratio)
Ca^{2+}-dependent cytosolic PLA_2s ($cPLA_2s$), 1249
Ca^{2+}-independent PLA_2s ($iPLA_2s$), 1249
Ca^{2+}-sensing receptor (CaSR), 510–511, 513
Cachexia, 1317
N-cadherin modulation of Wnt/β-catenin signaling, 426–428
Cadherin superfamily of cell adhesion molecules, 423–428
Cadherin-11 gene (*Cdh*11 gene), 425
Cadherins, 423, 427f, 429f
in bone cells
chondrocytes, 409–410
osteoblasts and osteocytes, 406–408
osteoclasts, 408–409
in commitment and differentiation of chondro-osteogenic cells, 425–426
in skeletal development, growth, and maintenance, 426
Cadmium, 533
Caenorhabditis elegans, 74, 179, 1164–1165
Caffeine, 1649
CALCA gene, 789–790
Calca-KO mice, 800
Calcific aortic valve disease (CAVD), 631
Calcific periarthritis, 1579
Calcilytics, 543
osteoporosis, 1665
repurposing calcilytics for new indications, 1666–1667
structure, 1666, 1666f
Calcimimetics, 543, 552, 1414, 1662–1663, 1663f
Calcineurin inhibitors, 519
Calcinosis cutis, 1435
Calciostat, 1657
Calciotropic end points, 1661–1662
Calciotropic hormones, 953–954, 1478–1480
Calciotropic tissues, CaSR in, 549–553
Calcitonin, 113, 454, 790–791, 794–795
to human bone physiology, 802
role in situations of calcium stress, 801
skeletal effects, 800
Calcitonin gene-related peptide (CGRP), 789, 791, 795–797
to human bone physiology, 802
skeletal effects, 800
Calcitonin peptides
access to bone microenvironment, 793
calcitonin-family gene and peptide structure, 789–790
extraskeletal actions of calcitonin-family peptides, 790–792
genes, mRNA, and peptides of four members of human calcitonin family, 790f
local and systemic peptide administration effects
effects on calcium metabolism, 799–800
skeletal effects, 798–799
receptors for calcitonin-family peptides, 792–793
skeletal effects of calcitonin, calcitonin gene-related peptide, and amylin, 800–801
Calcitonin receptor (CTR), 792, 801
role in situations of calcium stress, 801
Calcitonin receptor-like receptor (CRLR), 792
Calcitriol, 1507–1508
analogs, 1739–1740
Calcium, 582, 1643, 1774
absorption
by distal convoluted tubules, 663f
and excretion, 662
in bone, 1645–1646
bone as body's calcium sink and reserve, 1644–1645
calcium-deficient hydroxyapatites, 527
chelation, 139
chemistry, 656–657
homeostasis, 656–659, 1161
metabolism, 790–791, 799–800
and osteoporosis treatment, 1650–1653
physiological adaptations to low calcium intake, 1646

recommendations for calcium intake, 1647t
regulation by, 1311
requirement, 1646–1650
dietary sources, 1648
nutritional factors, 1649
toxicity, 1649–1650
signals, 119
supplementation
and bone health, 1651–1652
and cardiovascular risk, 1652–1653
in vitro studies, 582
in vivo studies, 582
Calcium pyrophosphate deposition disease (CPPD), 1579
Calcium receptor (CaR), 581, 1657
Calcium-dependent neutral endoproteinases, 379–380
Calcium-sensing receptor (CaSR), 539–540, 1311, 1363–1364, 1368, 1473, 1659–1660
agonists, antagonists, and modulators, 541–543
allosteric modulators, 542
and associated proteins, 1413–1415
in bone cells, 551–552
in breast, 552–553
in calciotropic tissues, 549–553
CaSR–associated intracellular signaling effectors, 545–546
CaSR–mediated signaling, 543–545, 544f
cationic agonists, 541–542
ECD, 540–541
interacting proteins, 546–547
intracellular signaling, 543–546
in kidney, 549–551
ligand-biased signaling, 543
noncalciotropic roles, 553–557
in parathyroid glands, 549
regulation of CaSR gene expression, 547–549
structural and biochemical properties, 540–541
synthetic modulators, 542–543
Calcium-to-creatinine ratio (Ca/Cr ratio), 1368–1369
Calmodulin (CaM), 546, 581
Calmodulin/Ca^{2+}-ATPase-dependent calcium efflux, 405
Calvarium, 6
CaM. *See* Calmodulin (CaM)
CaMos. *See* Canadian Multicentre Osteoporosis Study (CaMos)
cAMP. *See* Cyclic adenosine monophosphate (cAMP)
cAMP-responsive element (CRE), 576
cAMP/PKA/CREB pathway, 607
Camurati–Engelmann disease (CED), 1163
Canadian Multicentre Osteoporosis Study (CaMos), 1848
Canagliflozin, 950–951
Canalicular fluid flow, 146–149
Cancellous bone volume (Cn-BV), 448
Cancer, 1742
associated with bone matrix–derived TGF-β, 1170–1172
breast, 763, 1339–1340
cancer-induced bone disease, 1339–1342
cell metabolism alteration, 1344–1345
colorectal, 762–763
indications, 1696–1698
denosumab for cancer treatment–induced bone loss, 1696
hypercalcemia of malignancy refractory to BP therapy, 1696–1698
metastatic bone disease, multiple myeloma, and giant cell tumors, 1698
prostate, 763, 1340–1341
skeletal malignancies, 1341–1342
skin, 763–764
treatment–induced bone loss, 1696
androgen deprivation in men with prostate cancer, 1696
aromatase inhibitors in women with breast cancer, 1696
vitamin D and, 761–764
animal studies, 762
cellular mechanisms, 761
Canine distemper virus (CDV), 1603–1604
Cannabinoid receptor 2, 818
Cannabinoids, 813
Canonical BMP signaling, 1189–1190, 1190f
Canonical Wnt signaling pathway, 1064
sclerostin inhibition of, 1715
Canonical Wnt signals, 121–122
Cantilever function, 248–249
Ca_o^{2+} sensing receptor. *See* Calcium-sensing receptor (CaSR)
CaR. *See* Calcium receptor (CaR)
CAR cells. *See* CXCL12-abundant reticular cells (CAR cells)
Carboplatin, 518
Carboxy-terminal collagen crosslinks (s-CTX), 458–459
γ-Carboxyglutamic acid (Gla), 370
Gla-containing proteins, 370–371, 370t
MGP, 370–371
osteocalcin, 371
γ-Carboxylation, 1934–1935
Carcinoids, 1301
Cardiac defects, abnormal facies, thymic hypoplasia, cleft palate, hypocalcemia, associated with chromosome 22 microdeletion (CATCH-22 syndrome), 579
Cardiac excitability, 515
Cardiac muscle function, 478
Cardiotrophin 1 (CT-1), 97, 229, 1222
Cardiotrophin-like cytokine, 1222–1223
Cardiotrophin-like cytokine factor 1 (CLCF1), 1222–1224
Cardiovascular development, PTH and PTHrP in, 624
Cardiovascular disease (CVD), 757–758
vitamin D and, 764
Cardiovascular end points, 1662
Cardiovascular system, 877–880
cardiovascular safety of selective estrogen receptor modulators, 878
potential cardiovascular benefit of SERMs, 879–880
CART signalling, 814
Cartilage, 27, 46, 50, 206
formation, 55, 296–297
matrix production, 409–410
Cartilage-to-bone transition, 1145–1146
Cartilage–associated protein (CRTAP), 348–350
Casein kinase 1 (CK1), 181
Casein kinase-2 interacting protein-1 (CKIP-1), 922
Caspase 8 apoptosis-related cysteine peptidase (CFKAR), 761
Caspases, 761
CaSR. *See* Ca^{2+}-sensing receptor (CaSR); Calcium-sensing receptor (CaSR)
CAT. *See* Chloramphenicol acetyltransferase (CAT)
Catabolic 24-hydroxylase (CYP24A1), 1740
Catabolism, 303–304
β-Catenin, 170, 179, 181–183, 284, 761, 1342, 1637, 1715
β-catenin/sclerostin axis, 531
β-catenin–lymphoid enhancer factor transcriptional activity, 401–402
regulation, 1776
transcriptional regulation by, 183
VDR interaction with signaling, 725–726
WNTs and, 23–24
Cathepsin inhibitors, 389
Cathepsin K, 113, 229–230
cathepsin-K-CD200+ cells, 1062
immunohistochemistry, 1907
inhibitors, 1821
Cationic agonists of calcium-sensing receptor, 541–542
CATSHL syndrome, 1119
Caudal-type homeobox 2 (CDX2), 722
CAVD. *See* Calcific aortic valve disease (CAVD)
Caveolae, 546–547
Caveolar organized membrane, 1774–1775
Caveolin, 546–547
CB-BFs. *See* Cord blood-borne fibroblasts (CB-BFs)
CBBF. *See* Children's Brittle Bone Foundation (CBBF)
CBF. *See* Core-binding factor (CBF)
CBFA1. *See* Runt-related transcription factor 2 (RUNX2)
cbfa1. *See* Runx2
CBP/p300, 724–725
CBP/p300-associated factor p/CAF, 724–725
CBW. *See* Collagen-bound water (CBW)
C–C motif chemokine ligand 4 (CCL4), 791
CCAAT/enhancer-binding protein (C/EBP), 307–308, 722
C/EBPα, 118
C/EBPδ, 52
CCCTC-binding factors (CTCFs), 723–724

CCD. *See* Cleidocranial dysplasia (CCD)
CCL4. *See* C–C motif chemokine ligand 4 (CCL4)
CCN2. *See* Connective tissue growth factor (CTGF)
CD146 cell, 56–57
CD40L. *See* Cluster of differentiation 40 ligand (CD40L)
CD44 cell, 411, 719, 1338
Cdc73 gene, 1416, 1418
*Cdh*11 gene. *See* Cadherin-11 gene (*Cdh*11 gene)
CDKN1B gene, 1411–1412
CDKs. *See* Cyclin dependent kinases (CDKs)
cDNA. *See* Complementary DNA (cDNA)
CDs. *See* Collecting ducts (CDs)
CDV. *See* Canine distemper virus (CDV)
CDX2. *See* Caudal-type homeobox 2 (CDX2)
CED. *See* Camurati–Engelmann disease (CED)
Cell(s)
  adhesion molecules, 401, 423–428
  cell-autonomous programs, 73
  contribution to coupling, 231–234
  culture methods, 98
  cycle
    progression, 1718–1719
    regulation and proliferation, 761
  differentiation process, 163
  divisions, 56–57
  membranes, 113
  reprogramming, 1162
  sense stimulus, 1771–1773
    bone marrow mesenchymal stem cells, 1772
    bone marrow vasculature, 1773
    osteoblasts, 1772
    osteoclasts, 1772
    osteocytes, 1772–1773
  shape, 53–54
  sources, 64
  surface adhesion transmembrane molecules, 401
  therapy, 1497
Cell–cell
  adhesion, 407–408
    molecules, 14–15
  communication, 8, 423
  interactions, 53–54
Cell–matrix adhesion, 142–143
β cells, GPRC6A osteocalcin receptor in, 1934
Cell–substrate interactions, 53–54
Cellular
  aging, 276–277
  cementum, 1070
  defects and clinical features, 1520–1521
  effects of thyroid hormones on bone, 899–901
    osteoblasts, 900
    osteoclasts, 900–901
  machinery, 245
  mechanisms
    apoptosis, 761
    cell cycle regulation and proliferation, 761
    microRNA, 761
    vitamin D metabolism, 761
  metabolism, 469–470
  phosphorus metabolism, 492
  senescence, 280–281
Cement line stain, 1904
Cementocytes, 141
Cementogenesis, 1062, 1065
Cementum, 1064
Central nervous system (CNS), 630, 810, 880–882, 1360. *See also* Sympathetic nervous system (SNS)
  afferent signals regulating bone remodeling via, 810–811
  CNS-active drugs, 809
  efficacy of SERMs, 881–882
  safety of SERMs, 880–881
Centrifugal ultrafiltration method, 717
CFKAR. *See* Caspase 8 apoptosis-related cysteine peptidase (CFKAR)
CFP. *See* Cyan FP (CFP)
CFR. *See* Coronary flow reserve (CFR)
CFU. *See* Colony-forming unit (CFU)
CFU-Fs. *See* Colony-forming units –fibroblasts (CFU-Fs)
CFU-GM. *See* Granulocyte–macrophage colony-forming units (CFU-GM)
CGRP. *See* Calcitonin gene-related peptide (CGRP)
CHAMP. *See* Concord Health and Aging in Men Project (CHAMP)
Channels, 1774
CHARGE syndrome, 1357–1358
CHD7 gene, 1357–1358
Chemiluminescence immunoassay (CLIA), 1810–1812
Chemokine ligand 12. *See* C-X-C motif chemokine 12 (CXCL12)
Chemokine receptor 4 (CXCR4), 1316–1317, 1337
Childhood HPP, 1577–1579, 1578f
Children's Brittle Bone Foundation (CBBF), 1490
ChIP. *See* Chromatin immunoprecipitation (ChIP)
ChIP-seq. *See* Chromatin immunoprecipitation linked to parallel DNA sequencing (ChIP-seq)
Chloramphenicol acetyltransferase (CAT), 308–309, 1519–1520
Chloride channel-7 (ClC-7), 113, 223, 1561
Chondro-osteogenic cells, 425–426
Chondroadherin, 412–413
Chondroblasts, 318
Chondrocytes, 5, 15, 92, 406, 409–410, 901, 1789, 1795, 1892–1893
  chondrocyte-specific genetic tool, 1894
  chondrocyte-specific transgenes, 1894
  maturation, 475
  Notch receptors and ligands in, 1089–1092
    Notch action mechanisms, 1092
    Notch signaling role, 1089
Chondrodysplasia, 384–385, 843–844
  syndromes, 1118–1119
    mutations in FGFR3 and FGF9, 1119
    skeletal overgrowth and CATSHL syndrome, 1119
Chondrogenesis, 26–27, 48–49, 296–297, 1065
  FGF and FGFR signaling
    *Fgfr*1 signaling in hypertrophic chondrocytes, 1118
    FGFR3 signaling in growth plate chondrocytes, 1117–1118
    initiation of chondrogenesis, 1116–1117
    regulation of FGFR3 expression, 1117
Chondrogenic models, 1068
Chondroitin sulfate synthase (CHSY), 1099
Chromatin immunoprecipitation (ChIP), 745
Chromatin immunoprecipitation linked to parallel DNA sequencing (ChIP-seq), 311, 745–747, 1745
Chromatin structure, 171
Chronic hyperparathyroidism, 952
Chronic kidney disease (CKD), 583–584, 1463–1464, 1534–1535, 1660–1661, 1740
  abnormalities in fibroblast growth factor-23–klotho pathway, 1474
  association of bone biomarkers and bone outcomes in, 1479t–1480t
  diagnostic tests for abnormal bone, 1475–1480
  disordered osteoblast function or differentiation, 1474
  hip fracture incidence increases with progressive, 1464f
  overlap between osteoporosis and, 1464f
  pathogenesis of abnormal bone, 1472–1475
Chronic kidney disease–mineral and bone disorder (CKD–MB), 633–635, 1660–1661
  FGF23 and, 1534–1535
Chronic kidney disease–mineral bone disorder (CKD–MBD), 1465–1472
  biochemical abnormalities, 1465–1466
  bone abnormalities, 1467–1472
    historical classification, 1467–1470
    spectrum of renal osteodystrophy, 1471–1472
    TMV classification system, 1470–1471
  KDIGO classification, 1465t
  pathogenesis of arterial calcification, 1468f
  vascular calcification, 1467
Chronic periodontitis, 1064
Chronic renal disease (CKD)
CHSY. *See* Chondroitin sulfate synthase (CHSY)
Ciliary neurotrophic factor (CNTF), 1220, 1223
Ciliary neurotrophic factor receptor (CNTFR), 1220, 1222–1224
Cinacalcet, 543, 635, 1657–1658, 1660–1661

Ciona intestinalis, 694
Circulating tissue-nonspecific alkaline phosphatase, 1590
Circulating tumor cells (CTCs), 1337
Cis-eQTLs, 1619–1620
CISH. *See* Cytokine-inducible SH2-containing protein (CISH)
Cisplatin, 518
CK1. *See* Casein kinase 1 (CK1)
CKD. *See* Chronic kidney disease (CKD)
CKD–MB. *See* Chronic kidney disease–mineral and bone disorder (CKD–MB)
CKD–MBD. *See* Chronic kidney disease–mineral bone disorder (CKD–MBD)
CKIP-1. *See* Casein kinase-2 interacting protein-1 (CKIP-1)
Classical mechanical testing, 1837
Claudin-7 (*CLCN7*), 1558–1559, 1561, 1563–1564
Claudin-10 (*CLDN*10), 513
Claudin-14 (*CLDN*14), 513, 1279
Claudin-16 (*CLDN*16), 512–513
Claudin-19 (*CLDN*19), 512–513
ClC-7. *See* Chloride channel-7 (ClC-7)
CLCF1. *See* Cardiotrophin-like cytokine factor 1 (CLCF1)
CLCN7. *See* Claudin-7 (*CLCN7*)
*CLDN*10. *See* Claudin-10 (*CLDN*10)
*CLDN*14. *See* Claudin-14 (*CLDN*14)
*CLDN*16. *See* Claudin-16 (*CLDN*16)
*CLDN*19. *See* Claudin-19 (*CLDN*19)
Cleidocranial dysplasia (CCD), 7, 164
CLIA. *See* Chemiluminescence immunoassay (CLIA)
Clinical computed tomography, 1845
Clomiphene, 866t, 868, 868f, 884
Clonal analysis, 46, 56–57, 1889–1890
Clonal DNA damage, 1406
Clonality of parathyroid tumors, 1406–1407
Clotting, 73
  and identification of human mutations in NPT2c, 486
Cluster of differentiation 40 ligand (CD40L), 1209
CMD. *See* Craniometaphyseal dysplasia (CMD)
Cn-BV. *See* Cancellous bone volume (Cn-BV)
CNNM2 mutation, 517
CNP. *See* c-natriuretic peptide (CNP); C-type natriuretic peptide (CNP)
CNS. *See* Central nervous system (CNS)
CNTF. *See* Ciliary neurotrophic factor (CNTF)
CNTFR. *See* Ciliary neurotrophic factor receptor (CNTFR)
Coactivators, 867
  coactivators/transcription factors, 1743
  proteins, 896
Coat protein complex II (COPII), 300
Cocaine-regulated transcript, 818
  bone resorption regulation by, 814–815
Codon, 314
Codon 23, 1365
COL1. *See* Collagen type I (COL1)
*Col*1(2. 3*kb*). *Cre* mice, 606
*Col*10*a*1-*Cre* lines, 1795
*Col*11*a*2 promoter, 1789
Col1A1. *See* Collagen type I, alpha 1 (Col1A1)
*Col2a*1. *See Type II collagen alpha*-1 (*Col2a*1)
Collaborative Cross (CC), 1623–1624
Collagen, 296–297, 359
  chaperone, 351
  cross-link analysis
    divalent cross-link analysis, 340
    electrospray mass spectrometry, 340
    mature cross-link analysis, 340
  glycosylation, 954–955
  molecules, 339
  posttranslational modifications, 348–351
    CRTAP, LEPRE1, PPIB, 348–350
    MBPTS2, 351
    PLOD2 and FKBP10, 350
    SC65 and P3H3, 350–351
    TMEM38B, 350
  processing, 351
Collagen type I (COL1), 141–142
  and bone pathologies, 304–305
  degradation and catabolism, 303–304
Collagen type I, alpha 1 (Col1A1), 1491
Collagen type I, alpha 2 (Col1A2), 1495
Collagen type II (COL2), 1145
Collagen-bound water (CBW), 1874
Collagenase digestions, 139
Collagenase-3, 384–387, 648
Collagenases, 383
Collecting ducts (CDs), 544–545
Colony-forming unit (CFU), 829
Colony-forming units–fibroblasts (CFU-Fs), 46, 90
Colony-stimulating factor 1 receptor, 1207
Colorectal cancers (CRCs), 761
  animal studies, 762
  clinical studies, 762–763
Combinatorial signals, 1191–1192
Combined antiresorptive and anabolic therapy, 267–268
Complement 5a binding (C5a binding), 714
  neutrophil recruitment and migration with, 719
Complementary DNA (cDNA), 1308
Component labeling algorithm, 1840
Compression testing, 1926
Computed tomography (CT), 999, 1835–1836, 1863–1871
  cancellous bone, 1858
  components of bone quality, 1865–1866
  cortical bone, 1866–1867
  HR-CT, 1868–1871
  opportunistic screening, 1871
  standard quantitative computed tomography to assessing BMD, 1863–1865
Concord Health and Aging in Men Project (CHAMP), 817
Conditional loss-of-function approaches, 1790–1796
  chondrocytes, 1795
  osteoclasts, 1796
  osteoprogenitors/osteoblasts/osteocytes, 1795–1796
  uncondensed mesenchyme and precartilage condensations, 1791–1795
    Cre lines for skeletal analysis, 1792t–1794t
Conductin. *See* Axin 2
Confocal microscopy, 1910
Conformation-based differences in signaling responses, 699–700
Conformational selectivity at PTHR1, 699–703
Congenital skeletal diseases
  caused by Notch gain of function
    brachydactyly, 1099
    HCS, 1098
  caused by Notch loss of function
    Alagille syndrome, 1096–1097
    AOS, 1097
    spondylocostal and spondylothoracic dysostosis, 1097
Connective tissue growth factor (CTGF), 316–317, 1171
Connexin 43 (Cx43), 14–15, 134–135, 1672–1673
  control mechanisms of bone cell function, 434–436
  function in bone cells, 432–433, 432f
Connexins, 134–135, 428–436, 1774
  diseases affecting skeleton, 429–431
  function of connexin43 in bone cells, 432–433
  mechanisms of connexin43 control of bone cell function, 434–436
  in skeleton across life span, 431–432
Constitutively active form of β-catenin (CA β-cat), 122
Contact-dependent communication, 96–97
Conventional gene deletion, 1790
COPII. *See* Coat protein complex II (COPII)
Cord blood-borne fibroblasts (CB-BFs), 91
Core-binding factor (CBF), 380–381
Corepressors, 867
Coronary flow reserve (CFR), 632
Cortical area (Ct.Ar), 449
Cortical BMC, 1867
Cortical bone, 1866–1867
  histomorphometry, 1907–1909
  loss, 263
    predominance, 259–261
  micromilled cross sections, 1903
Cortical pore water (CPW), 1874
Cortical porosity, 278
Cortical porosity area (Ct.Po.Ar), 449
Cortical porosity number (Ct.Po.N), 449
Cortical thick ascending limb of loop of Henle (CTAL), 1315–1316
Cortical water, 1874–1875
Cortical width (Ct.Wi), 449
Corticosteroid

Corticosteroid (*Continued*)
corticosteroid-treated mice, 254
therapy, 254
Corticotropin-releasing factor (CRF), 692–693
Cortisone, 915
COS-1 cells, 600–601
COS-7 cells, 1390
Coupled remodeling, 221
Coupler of bone resorption and formation, 1164–1166
Coupling, 221
and balance, 222–223
of bone formation to bone resorption, 778–779
cells contribution to, 231–234
coupled remodeling, 221
development, 220
factors, 1218, 1222, 1226
matrix-derived resorption products, 227–228
synthesized and secreted by osteoclasts, 228–230
macrophages, immune cells, and endothelial cells, 233–234
membrane-bound coupling factors synthesized by osteoclasts, 230–231
osteoblast lineage cells–sensing surface and signaling, 232–233
resorption phase of remodeling and cessation in BMUs, 223
reversal phase as, 234–236
COXs. *See* Cyclooxygenases (COXs)
cPLA$_2$s. *See* Ca^{2+}-dependent cytosolic PLA$_2$s (cPLA$_2$s)
CPPD. *See* Calcium pyrophosphate deposition disease (CPPD)
CPW. *See* Cortical pore water (CPW)
Crack
density, 1910–1912
length, 1910–1912
surface density, 1910–1912
Cranial malformations, 7
Cranial nerve compression syndromes, 1553–1554
Craniometaphyseal dysplasia (CMD), 430–431
Craniosynostosis, 7
FGFR signaling in, 1124–1125, 1125f
potential therapeutic approaches, 1125–1126
skeletal phenotype, 1124
syndromes, 8, 24
CrCl. *See* Creatinine clearance (CrCl)
CRCs. *See* Colorectal cancers (CRCs)
CRD. *See* Cysteine-rich domain (CRD)
CRE. *See* cAMP-responsive element (CRE)
Cre lines for skeletal analysis, 1792t–1794t
Cre recombinase, 1788, 1790–1791
Cre-IRES-ALPP, 1795
Cre-Lox system, 977, 1887–1888
Cre-loxP system, 829, 953, 1788, 1790–1791
Cre-mediated lacZ deletion (Z/EG), 1888
Creatinine clearance (CrCl), 943
CREB. *See* Cyclic AMP response element binding protein (CREB)
CreER, 1887–1890
CRF. *See* Corticotropin-releasing factor (CRF)
CRISPR/Cas9 technology, 1787
genome editing technology, 1562
genomic engineering using, 1797
CRLF1. *See* Cytokine receptor-like factor 1 (CRLF1)
CRLR. *See* Calcitonin receptor-like receptor (CRLR)
Cross-link
formation
bone collagen cross-linking, 340–342
glycosylations and glycations, 345–347
glycosylation, 345
lysine-modifying enzymes
LOXs, 348
lysyl hydroxylases, 347
structures
divalent cross-links, 342–343
histidine-containing collagen cross-links and maturation products, 344–345
pyridinium cross-links, 343
pyridinoline and pyrrolic cross-linked peptides in urine, 344
pyrrole cross-links, 343–344
Cross-sectional areas (CSAs), 248, 1859–1860
Cross-sectional moments of inertia (CSMI), 1872
CRP. *See* C-reactive protein (CRP)
cRPI. *See* Cyclic reference point indentation (cRPI)
CRTAP. *See* Cartilage–associated protein (CRTAP)
CRTCs. *See* Cyclic AMP-regulated transcriptional coactivators (CRTCs)
Cryo-electron microscopy (Cryo-EM), 696–697
Cryoembedding, 1903
Cryosectioning, 1903
CSAs. *See* Cross-sectional areas (CSAs)
CSHS. *See* Cutaneous skeletal hypophosphatemia syndrome (CSHS)
CSMI. *See* Cross-sectional moments of inertia (CSMI)
CT. *See* Computed tomography (CT)
CT X-ray absorptiometry (CTXA), 1865
CT-1. *See* Cardiotrophin 1 (CT-1)
Ct.Ar. *See* Cortical area (Ct.Ar)
Ct.Po.Ar. *See* Cortical porosity area (Ct.Po.Ar)
Ct.Po.N. *See* Cortical porosity number (Ct.Po.N)
Ct.Wi. *See* Cortical width (Ct.Wi)
CTAL. *See* Cortical thick ascending limb of loop of Henle (CTAL)
CTCFs. *See* CCCTC-binding factors (CTCFs)
CTCs. *See* Circulating tumor cells (CTCs)
CTGF. *See* Connective tissue growth factor (CTGF)
CTNNB1 mutations, 1412
CTR. *See* Calcitonin receptor (CTR)
Ctsk-Cre model, 828–829
CTX. *See* C-terminal telopeptide of type I collagen (CTX)
CTXA. *See* CT X-ray absorptiometry (CTXA)
Culture shock protein, 365
Cultured human osteoblasts, 1208
Cushing's disease, 915
Cutaneous skeletal hypophosphatemia syndrome (CSHS), 1539–1540
CVD. *See* Cardiovascular disease (CVD)
Cx43. *See* Connexin 43 (Cx43)
CXCL1, 719
CXCL12-abundant reticular cells (CAR cells), 75
CXCL12. *See* C-X-C motif chemokine 12 (CXCL12)
CXCL16 receptor, 1337
CXCL4, 78
CXCR4. *See* Chemokine receptor 4 (CXCR4)
CXCR7, 1337
CY27B1, 739–741
regulatory sites of action, 749–750
Cyan FP (CFP), 1888–1889
Cycle number, 1765
Cyclic adenosine monophosphate (cAMP), 539, 596, 1431, 1635
Cyclic AMP response element binding protein (CREB), 170, 386, 648, 722, 1281, 1312, 1933–1934
Cyclic AMP-regulated transcriptional coactivators (CRTCs), 1281
Cyclic chondroadherin peptide, 413
Cyclic reference point indentation (cRPI), 1927–1928
Cyclin D1/PRAD1, 1407–1409, 1408f
Cyclin dependent kinases (CDKs), 1088–1089, 1314, 1408
inhibitors, 1409–1412
Cyclooxygenases (COXs), 1247–1251, 1255
COX-1, 1250
KO mice, 1256
COX-2, 150, 1247, 1250
KO mice, 1256
Cyclosporine, 519
CYP24. *See* Cytochrome P450 family 24 (CYP24)
Cyp24A1. *See* Cytochrome P450 family 24 subfamily A member 1 (Cyp24A1)
Cyp27B1, 720–721, 723–725, 1510
CYP2R1. *See* Cytochrome P450 2R1 (CYP2R1)
CYP3A4, 1510
Cys677, 541
Cys765, 541
Cysteine proteinases, 389–390
Cysteine-rich domain (CRD), 179
Cytochroma, 1739

Cytochrome P450 2R1 (CYP2R1), 739–740, 1510–1511
Cytochrome P450 family 24 (CYP24), 1733
Cytochrome P450 family 24 subfamily A member 1 (Cyp24A1), 716, 720–721, 739–740
  expression, 761
  regulatory sites of action, 749–750
Cytokine receptor-like factor 1 (CRLF1), 1222–1224
Cytokine-inducible SH2-containing protein (CISH), 1214
Cytokines, 279, 303–304, 314, 319–321, 649–651, 987–988, 1205–1207, 1222–1224, 1322–1323
  CNTF, 1223
  CRLF1and CLCF1, 1223–1224
  IFN-γ, 319–320
  IL-1, 320
  IL-4, 321
  IL-6, 321
  IL-13, 320–321
  interleukins, 321
  NP, 1224
  responsive element, 308
  signals, 124
  TNFα, 319
Cytoplasmic FGF23 staining, 1542
Cytotoxic effects, 1676

D

DANCE study. *See* Direct Assessment of Nonvertebral Fractures in Community Experience study (DANCE study)
Danio rerio (zebrafish), 694
DAP-3. *See* Death-associated protein-3 (DAP-3)
DAP12. *See* DNAX activation protein 12 (DAP12)
DBD. *See* DNA-binding domain (DBD)
DBH. *See* Dopamine β-hydroxylase (DBH)
DBP–macrophage-activating factor (DBP–MAF), 714
DC. *See* Dyskeratosis congenita (DC)
DC-STAMP. *See* Dendritic cell-specific transmembrane protein (DC-STAMP)
DCT. *See* Distal convoluted tubule (DCT)
DDR. *See* DNA damage response (DDR)
Deafness, 1359–1361
Death receptor 5 (DR), 1208
Death-associated protein-3 (DAP-3), 761
Decalcification, 1901
Decapentaplegic gene (dpp gene), 1070
Decorin (PG-II), 360–362
Defect, 1516
  cellular defects and clinical features, 1520–1521
  in DNA-binding region, 1518–1519
  in hormone-binding region, 1516–1518
  in vitro posttranscriptional and transcriptional effects of 1,25(OH)$_2$D$_3$, 1519–1520
15-Dehydroxyprostaglandin dehydrogenase (15-PGDH), 1259
Delta homologues named Delta-like (DLL), 1083
Delta-like homolog (DLK), 1086
Delta-like ligand 4 (DLL-4), 1144
Delta/Notch-like EGF-related receptor (DNER), 1086
Delta/Serrate/Lag2 (DSL), 1086
Demeclocycline, 445–446
Dendritic cell-specific transmembrane protein (DC-STAMP), 113–114
Dendritic processes, 1062
Denosumab, 116, 455–456, 461, 929–930, 1496, 1820–1821
  differential effects, 461
  discontinuation, 1701
  effects, 1694t
    additional denosumab data, 1698–1700
    cancer indications, 1696–1698
    glucocorticoid-induced osteoporosis, 1698
    osteoporosis indications, 1693–1696
    potential applications, 1699–1700
    RA, 1699
  safety, 1700–1702
    AFFs, 1700–1701
    hypersensitivity, serious infections, and musculoskeletal pain, 1701–1702
    hypocalcemia, 1700
    osteonecrosis of jaw, 1700
    theoretical impact of RANKL inhibition, 1702
    use in women of reproductive age, 1702
Dental follicle (DF), 1061–1062
Dentin matrix protein (Dmp), 1095
  DMP1, 89, 137, 145, 365–366, 369, 471–472, 1533, 1772–1773, 1789, 1892
Dentin tubules, 1583
Dentinogenesis, 1583
Dentition, 1583
Dent's disease, 490
Deoxypyridinoline (DPD), 1478, 1603, 1802
Depression, 809
Dermal fibroblasts, 319
Dermal wound healing process, 1156
Destructive process, 220
Developing bone
  innervation of, 210–211
  vascularization of, 206–210
    roles of HIF and VEGF, 207f
DEXA. *See* Dual-absorption X-ray absorptiometry (DEXA)
Dexamethasone, 51, 922–923
DF. *See* Dental follicle (DF)
DGS. *See* DiGeorge syndrome (DGS)
DGS type 1 (DGS1), 1357–1358
DGS type 2 (DGS2), 1357–1358
DHLNL. *See* Dihydroxylysinonorleucine (DHLNL)
DHT. *See* Dihydrotachysterol (DHT); Dihydrotestosterone (DHT)
Diabetes. *See also* Insulin
  bone and systemic metabolism
    bone hormones control systemic metabolism, 952
    metabolic control of glucose and insulin in bone, 951–952
  bone fragility treatment in, 957
  clinical risk factors for fractures in T1DM and T2DM, 949–950
    bone repair in T1DM and T2DM, 951
    diabetes treatments effect on bone, 950–951
  diabetes effects and insulin on endochondral bone growth, 942–949
  diabetic bone disease, 952–957
Diabetic bone disease, 941
  altered collagenous bone matrix
    increased collagen glycosylation, 954–955
    reduced enzymatic collagen cross-linking, 955
  diabetic hormonal milieu, 952–954
  impaired vascularization of diabetic bones, 954
  increased bone marrow adiposity, 956
  inflammation and oxidative stress, 956
  loss of incretin effect, 956–957
  sclerostin, 957
Diabetic hormonal milieu
  calciotropic hormones, 953–954
  hypercortisolism, 953
  low amylin, 954
  lower circulating insulin-like growth factor 1, 952–953
Diacylglycerol (DAG), 1276
"Dialysis" dementia, 531–532
Diaphyseal long bones, 1761
Diazepam, 446–447
DICER, 184
Dickkopf-1 (Dkk-1), 184, 650–651, 1065, 1226, 1338–1340, 1474, 1718–1719, 1805
  antibodies, 189
Dictyostelium discoideum, 425
Differential bone remodeling to distinct components of strain tensor, 1764
Diffusion-weighted MRI, 1877
DiGeorge syndrome (DGS), 579, 1357–1359
  clinical features and genetic abnormalities, 1357–1358
  mouse models developing features, 1358–1359
Dihydrotachysterol (DHT), 1739
Dihydrotestosterone (DHT), 972
Dihydroxylysinonorleucine (DHLNL), 342–343
1,25 Dihydroxyvitamin D (1,25(OH)$_2$D), 473–474, 713–714, 720, 723–724, 761, 1308, 1311, 1323, 1465, 1507–1508, 1515, 1529, 1539
24,25-Dihydroxyvitamin D (24,25(OH)$_2$D), 1507–1508
1,25-Dihydroxyvitamin D$_3$ (1,25(OH)$_2$D$_3$), 98, 143–144, 539, 548, 576, 580–581, 953–954, 1472–1473, 1605
  in classical target tissues, 741–743

1,25-Dihydroxyvitamin D_3 (1,25$(OH)_2D_3$) (*Continued*)
bone, 741–742
intestine, 742
kidney, 742–743
nonclassical actions, 743
parathyroid glands, 743
regulatory sites of action, 749–750
transcriptional regulation, 743–745
in vitro posttranscriptional and transcriptional effects of, 1519–1520
1α,25-Dihydroxyvitamin D_3 (1α,25$(OH)_2D_3$), 1733
3,5-Diiodothyropropionic acid (DITPA), 898
Dimethylbenzanthracene (DMBA), 762
Dipeptidyl peptidase-4 (DPP-4), 950–951
Diphtheria toxin alpha chain (DTA), 1796–1797
Diphtheria toxin receptor (DTR), 1788, 1796–1797
Direct Assessment of Nonvertebral Fractures in Community Experience study (DANCE study), 957
Dishevelled proteins (Dsh proteins), 179, 181–183, 650–651, 1637
Disseminated tumor cells (DTCs), 1338
Distal convoluted tubule (DCT), 512, 662, 1315–1316
disturbed Mg^{2+}reabsorption in, 514–518, 514f
Distal pancreatic resection (DPR), 1297
Distal tubule calcium
parathyroid hormone-mediated control of reabsorption, 1279
PTH effects on transport, 664
Distal-Less Homeobox 5 (*Dlx5*), 1189
Disturbed Mg^{2+}reabsorption
in DCT, 514–518, 514f
in thick ascending limb, 512–514
DITPA. *See* 3,5-Diiodothyropropionic acid (DITPA)
Divalent cross-links, 342–343
analysis, 340
Diversity of ligand and receptor environment, 1190–1191
Diversity Outbred population (DO population), 1623–1624
Dkk-1. *See* Dickkopf-1 (Dkk-1)
DLK. *See* Delta-like homolog (DLK)
DLL. *See* Delta homologues named Delta-like (DLL)
DLL-4. *See* Delta-like ligand 4 (DLL-4)
Dlx5. *See* Distal-Less Homeobox 5 (*Dlx5*)
DMBA. *See* Dimethylbenzanthracene (DMBA)
Dmp. *See* Dentin matrix protein (Dmp)
Dmp1 promoter-driven Cre model (Dmp1-Cre model), 144, 829
DMP1-caPTHR1 animals, 1281
Dmp1-Cre model. *See* Dmp1 promoter-driven Cre model (Dmp1-Cre model)
Dmp1(10kb)Cre.Pth1r model, 606
Dmp1(8kb)Cre.Pth1r mice, 606
DNA damage response (DDR), 280
DNA sequence-specific transcription factors, 1160
DNA-binding domain (DBD), 721, 827–828
defects in DNA-binding region, 1518–1519
DNAX activation protein 12 (DAP12), 119
dependent pathways, 1212
DNER. *See* Delta/Notch-like EGF-related receptor (DNER)
DO population. *See* Diversity Outbred population (DO population)
Dopamine β-hydroxylase (DBH), 811–812
Dorsal–ventral axis, 14
DOTA. *See* 1,4,7,10-Tetraazacyclododecane-1,4,7,10-tetraacetic acid (DOTA)
Double heterozygous mutations, 1660
Doxercalciferol, 1661–1662
Doxycycline (Dox), 1891
DPD. *See* Deoxypyridinoline (DPD)
dpp gene. *See* Decapentaplegic gene (dpp gene)
DPP-4. *See* Dipeptidyl peptidase-4 (DPP-4)
DPR. *See* Distal pancreatic resection (DPR)
DR. *See* Death receptor 5 (DR)
Droloxifene, 868, 868f
Drosophila, 74, 163, 178–179, 1164–1165
Drosophila melanogaster (fruit fly), 300–301, 694, 1083
Dsh proteins. *See* Dishevelled proteins (Dsh proteins)
DSL. *See* Delta/Serrate/Lag2 (DSL)
DTA. *See* Diphtheria toxin alpha chain (DTA)
DTCs. *See* Disseminated tumor cells (DTCs)
DTR. *See* Diphtheria toxin receptor (DTR)
Dual effector hypothesis, 995
Dual specificity phosphatase 1 (DUSP1), 916
Dual-absorption X-ray absorptiometry (DEXA), 974
Dual-action drugs, 1823
Dual-energy X-ray absorptiometry (DXA), 529, 1002, 1475, 1639–1640, 1712, 1769, 1809–1810, 1834, 1845, 1857, 1860–1862
beyond bone mineral density, 1871–1873
Dubbo Osteoporosis Epidemiology Study (DOES), 817
Dubowitz syndrome, 1362
DUSP1. *See* Dual specificity phosphatase 1 (DUSP1)
Dutasteride-treated patients, 972
Dvl protein. *See* Dishevelled proteins (Dsh proteins)
Dwarfism, 25
DXA. *See* Dual-energy X-ray absorptiometry (DXA)
Dynein light-chain M_r 8000, 585
Dyskeratosis congenita (DC), 60
Dyslipidemia, 926

E

E-cadherin, 554–555
E.Pm. *See* Eroded perimeter (E.Pm)
E3KARP, 661–662
EAE. *See* Encephalomyelitis (EAE)
EAR. *See* Estimated average requirement (EAR)
Early growth response factor-1 (EGR-1), 316
EAST syndrome, 515
EB-transformed lymphoblasts. *See* Epstein–Barr-transformed lymphoblasts (EB-transformed lymphoblasts)
EBCT. *See* Electron beam computed tomography (EBCT)
EBD. *See* Ezrin-binding domain (EBD)
eBMD. *See* Estimated BMD (eBMD)
EBP50, 661–662
ECD. *See* Extracellular domain (ECD)
Echo time (TE), 1874
Ectodermal anomalies, 1514
Ectodermal dysplasias, 841–842
Ectonucleotide pyrophosphastase phosphodiesterase 1 and 3 (*Ennp1* and *Ennp3*), 741–742
Ectonucleotide pyrophosphatase phosphodiesterase 1 (ENPP1), 1070, 1533
Ectopic calcifications, 372–373
Ectopic secretion of parathyroid hormone, 1418–1419
EDS. *See* Ehlers–Danlos syndrome (EDS)
EDTA. *See* Ethylenediaminetetraacetic acid (EDTA)
EGF. *See* Epidermal growth factor (EGF)
EGFR. *See* Epidermal growth factor receptor (EGFR)
eGFR. *See* Estimated glomerular filtration rate (eGFR)
EGR-1. *See* Early growth response factor-1 (EGR-1)
Ehlers–Danlos syndrome (EDS), 347, 362
types VIIA and VIIB, 303
Eicosanoids, 1247–1248
Eicosapentaenoic acid (EPA), 1248
Eiken familial skeletal dysplasia, 1392
Eiken syndrome, 704
Elastic modulus and stiffness, 1767
Elderly men and women, GH administration to, 1002–1003
Electron beam computed tomography (EBCT), 1467
Electrophoretic mobility
shift assays, 1520
supershift assays, 319
Electrospray mass spectrometry, 340
ELISA. *See* Enzyme-linked immunosorbent assay (ELISA)
Ellsworth–Howard test, 1437
Embryogenesis, 1714
Embryonic and neonatal heart, 1714
Embryonic development, 5
lineage tracing during, 1892–1893
Embryonic mammary development, 842–844
Embryonic skeletogenic mesenchyme, 48–49
Embryonic stem cells (ESCs), 64, 1787
EMT. *See* Epithelial–mesenchymal transition (EMT)

En bloc basic fuchsin staining technique, 1910
En bloc staining protocol, 1912
EN1. *See* Engrailed homeobox-1 (EN1)
Encephalomyelitis (EAE), 758–759
Enchondromatosis, 704, 1392–1393
Encyclopedia of DNA Elements (ENCODE), 746–747
End-stage kidney disease (ESKD), 1463
End-stage renal disease (ESRD), 512–513
Endocannabinoid signaling, 813
Endochondral bone
  development, 1892–1893
  diabetes and insulin effects on growth
    insulin effects on growth plate cartilage, 942
    skeletal growth in T1DM, 942–943
    T1DM effects on bone, 943–946
    T2DM effects and insulin on bone, 946–949
  formation, 15–17, 384, 1089
    epigenetic factors and microRNAs, 26–27
    growth factor signaling pathways, 19–25
    growth plate, 18–19
    mechanisms of Notch action in, 1092
    mediators of skeleton formation, 19
    Notch signaling role in, 1089
    PTH and PTHrP in, 1382
    transcription factors, 25–26
    vasculature functional roles in, 27
Endochondral ossification, 15, 20–21, 1096, 1144–1146, 1221–1222
  angiogenic-osteogenic coupling during bone development, 1145f
  PTHrP in bone after, 605–607
Endocortical resorption, 250, 262, 262f
Endocortical resorptive modeling, 245
Endocortical surface, 254–255
Endoplasmic reticulum (ER), 299, 345, 478, 1249
  ER-associated degradation, 542–543
Endoplasmic reticulum protein 57/60 (ERp57 or 60), 713–714
Endoscopic ultrasonography (EUS), 1296
Endosomal signaling at PTHR1, 700–701
Endosteal bone loss, 261–263
Endosteal fluorochrome, 276
Endothelial cells, 77, 233–234, 1895–1896
Endothelial NOS, 1775–1777
Endothelial responses to PTH and PTHrP, 624–630
Endothelial signaling in bone, 210
Endothelins (ETs), 1340
Energy utilization by skeletal cells and role of IGF-I, 997–998
Energy-consuming function, 1931–1932
Engrailed homeobox-1 (EN1), 14, 1617
ENPP1. *See* Ectonucleotide pyrophosphatase phosphodiesterase 1 (ENPP1)
Enthesopathy, 480
Enzymatic collagen cross-linking reduction, 955
Enzymatic glycosylation, 345
  potential functions, 345
  tissue-dependent variations in cross-link glycosylation, 345
Enzyme activatipn, 1743
Enzyme replacement therapy, 1586
Enzyme-linked immunosorbent assay (ELISA), 717, 1804
EP1 receptor, 1251
EP2 receptor, 1251
EP3 receptor, 1251
EP4 receptor, 1251
EPA. *See* Eicosapentaenoic acid (EPA)
EPAC. *See* Exchange protein directly activated by cAMP (EPAC)
Eph family, 230–231
EphB4, 1607
EphrinB2, 230–231, 1607
EphrinB2/EphB4 ligand, 96–97
Epidermal dendritic cells, 1202
Epidermal growth factor (EGF), 360, 381, 516, 545, 588, 649–650, 1084, 1310
Epidermal growth factor receptor (EGFR), 516, 545, 588, 1251
Epidermis, CaSR in, 556–557
Epigenetic controls, 52–53
Epigenetic factors, 26–27
Epilepsy, 1587
Epiphyseal closure, 479
Epithelial–mesenchymal transition (EMT), 1100, 1161, 1201
Epstein–Barr-transformed lymphoblasts (EB-transformed lymphoblasts), 1516
eQTL. *See* Expression quantitative trait locus (eQTL)
Equilibrium dialysis method, 900
ER. *See* Endoplasmic reticulum (ER)
ER oxidase 1α enzyme (ERO1α enzyme), 300–301
ERK. *See* Extracellular signal-regulated kinase (ERK)
ERK1/2. *See* Extracellular signal-regulated kinase 1/2 (ERK1/2)
ERO1α enzyme. *See* ER oxidase 1α enzyme (ERO1α; enzyme)
Eroded perimeter (E.Pm), 1907
ERp57 or 60. *See* Endoplasmic reticulum protein 57/60 (ERp57 or 60)
ERs. *See* Estrogen receptors (ERs)
Erythropoietin, 1562
Escherichia coli β-galactosidase, 305
ESCs. *See* Embryonic stem cells (ESCs)
ESKD. *See* End-stage kidney disease (ESKD)
ESR1. *See* Estrogen receptor alpha (ESR1)
ESR2. *See* Estrogen receptor beta (ESR2)
ESRD. *See* End-stage renal disease (ESRD)
Estimated average requirement (EAR), 1646–1647
Estimated BMD (eBMD), 1617
Estimated glomerular filtration rate (eGFR), 1463
17β-Estradiol, 869f, 883, 1890
Estrogen receptor alpha (ESR1), 971–972
  in bone using genetically modified mice equivocal, 977
Estrogen receptor beta (ESR2), 971–972
Estrogen receptor–mediated pathway, 1069–1070
Estrogen receptors (ERs), 827–828, 864, 1790–1791. *See also* Progesterone receptors (PR)
  ERα, 144, 183
  mouse models, 828–829
  structure and amino acid identity between human ERα and ERβ, 828f
Estrogens, 282, 827–828, 971–972, 1069–1070, 1208
  agonism, 864
    in uterus, 871–873
  from clinical perspective, 829–832
  deficiency, 254, 260–261, 829
  replacement, 873–874
    therapies, 864
  treatment, 454
Etelcalcetide, 1663–1665
Ethanol, 1924
Ethylenediaminetetraacetic acid (EDTA), 139
ETs. *See* Endothelins (ETs)
EUS. *See* Endoscopic ultrasonography (EUS)
EUS-guided fine-needle aspiration, 1297
Eutopic PTH-related protein overproduction in malignancy, 1309–1312
  transcriptional and posttranscriptional regulation, 1309–1310
  transcriptional regulators of PTH-related protein, 1310–1311
Evaluation of Cinacalcet Hydrochloride Therapy to Lower Cardiovascular Events (EVOLVE), 635, 1473, 1662
Evidence-based medicine, 758
Evocalcet, 1662–1663
EVs. *See* Extracellular vesicles (EVs)
Exchange protein directly activated by cAMP (EPAC), 668
Exercise
  effects on bone quantity and quality
    biochemical modulation of mechanical signals, 1771–1777
    bone's mechanical milieu elicited by physical activity, 1761–1763
    bone's sensitivity to mechanical signals, 1760–1761
    low-magnitude, high-frequency mechanical signals, 1766–1771
    regulation of bone morphology by mechanical stimuli, 1764–1766
  EXERCISE-induced bone adaptation, 1760
  modulating adaptation to, 1937
  regulation of skeletal muscle energy metabolism during, 1935–1937, 1936f
Exons, 547, 724
Expression quantitative trait locus (eQTL), 1618
  in human bone tissue and cells, 1619–1620
Extracellular domain (ECD), 423–424, 540, 692–693, 1413

Extracellular signal-regulated kinase (ERK), 116–117, 134–135, 624–625, 646, 1672–1673, 1773–1774
Extracellular signal-regulated kinase 1/2 (ERK1/2), 52, 401–402, 434, 543, 726, 1312
Extracellular signal-related 1/2–mitogen-activated protein kinase ((ERK1/2–MAPK), 1276–1277
Extracellular vesicles (EVs), 1714
Extraembryonic tissues, 851
Eya1, 578–579
Ezrin-binding domain (EBD), 647

F

FACS. *See* Fluorescence-activated cell sorting (FACS)
FADD. *See* Fas-associated death domain (FADD)
FAK. *See* Focal adhesion kinase (FAK)
Falls, vitamin D and, 760
False discovery rate (FDR), 1617
FAM111A gene. *See* Family with sequence similarity 111 member A gene (*FAM111A* gene)
FAM20C. *See* Family with sequence similarity 20, member C (FAM20C)
Familial hyperphosphatemic tumoral calcinosis (FHTC), 491
Familial hypocalciuric hypercalcemia (FHH), 513–514, 539–540, 1405, 1415–1416, 1659–1660
  FHH1, 539–540
Familial hypomagnesemia with hypercalciuria and nephrocalcinosis (FHHNC), 512–513
Familial isolated hyperparathyroidism (FIHP), 1409–1410, 1417
Familial syndromes, 1362
Family with sequence similarity 111 member A gene (*FAM111A* gene), 518, 1361–1362
Family with sequence similarity 20, member C (FAM20C), 1533
Family–interacting protein of 200 kDa (FIP200), 281
Fan-beam dual X-ray absorptiometer, 529
Fanconi–Bickel syndrome, 490–491
Farnesyl pyrophosphate pyrophosphate (FPP), 1672
FAs. *See* Fatty acids (FAs)
FAS-1. *See* Fasciclin-1 (FAS-1)
Fas-associated death domain (FADD), 761
Fas-ligand (FasL), 1208
Fasciclin-1 (FAS-1), 1803
Fatty acid binding, 715, 719–720
Fatty acids (FAs), 1935–1936
Fc receptor common γ subunit (FcRγ), 119–120
FcRγ. *See* Fc receptor common γ subunit (FcRγ)
FD. *See* Fibrous dysplasia (FD)
FDA. *See* US Food and Drug Administration (FDA)
FDG–PET. *See* Fluorodeoxyglucose positron emission tomography (FDG–PET)
FDR. *See* False discovery rate (FDR)
FE. *See* Finite element (FE)
FEA. *See* Finite element analysis (FEA)
Feed-forward vicious cycle, 636
Femoral diaphysis, 1723
Femoral neck (FN), 1617
Fetal membranes, 851–852
FGF2. *See* Fibroblast growth factor 2 (FGF2)
FGF23–independent hypophosphatemic disorders, 486–491
  autosomal recessive Fanconi syndrome, 490
  Dent's disease, 490
  Fanconi–Bickel syndrome, 490–491
  HHRH, 486–490
  hypophosphatemia with osteoporosis and nephrolithiasis to SLC34A1, 490
FGF23–mediated hypophosphatemic disorders
  autosomal dominant hypophosphatemic rickets, 481
  autosomal recessive hypophosphatemic rickets, 481–484
  X-linked hypophosphatemia, 478–481
FGF3. *See* Fibroblast growth factor 3 (FGF3)
FGFR. *See* Fibroblast growth factor receptor (FGFR)
FGFR substrate 2α (FRS2α), 1116
FGFs. *See* Fibroblast growth factors (FGFs)
FHH. *See* Familial hypocalciuric hypercalcemia (FHH)
FHHNC. *See* Familial hypomagnesemia with hypercalciuria and nephrocalcinosis (FHHNC)
FHL1 gene. *See* Four-and-a-half LIM domains 1 gene (FHL1 gene)
FHTC. *See* Familial hyperphosphatemic tumoral calcinosis (FHTC)
Fibril-forming collagens, 297
Fibrillar collagens, 296
  family, 296–297
  molecules, 339–340
  type I collagen, 389
Fibrillins, 368
Fibrillogenesis, 301
Fibrinogen, 878
Fibripositors, 301
Fibroblast growth factor 2 (FGF2), 362, 425, 649, 1166
Fibroblast growth factor 23 (FGF23), 89, 137–138, 140, 145, 470–471, 472f, 587–588, 634, 664, 740–741, 1465, 1529–1530, 1539–1540, 1660–1661, 1740, 1807
  abnormalities in FGF23–klotho pathway, 1474
  actions, 1530, 1530f
  decreasing parathyroid hormone expression, 587
  gene, 1529–1530
  hyperphosphatemic diseases caused by impaired actions, 1534–1535
  hypophosphatemic diseases caused by excessive actions, 1531–1534
  level regulation, 1530–1531
  receptor for, 1531
  regulatory sites of action, 749–750
  resistance of parathyroid to FGF23 in chronic kidney disease, 587–588
Fibroblast growth factor 3 (FGF3), 600–601
Fibroblast growth factor receptor (FGFR), 471, 587, 1114
  and chondrodysplasia syndromes, 1118–1119
  and craniosynostosis, 1124–1126
  expression in bone, 1115–1116
  FGFR1, 897, 1533
  FGFR3, 1117–1118
    mutations in FGFR3 and FGF9, 1119
    regulation of FGFR 3 expression, 1117
    signaling in growth plate chondrocytes, 1117–1118
  signaling, 1116
    in bone formation and repair, 1119–1122
    in bone resorption, 1122–1124
    in chondrogenesis, 1116–1118
Fibroblast growth factors (FGFs), 8, 24, 52, 317–318, 410, 995, 1113–1114, 1116, 1142–1143, 1160–1161, 1340–1341
  in bone formation and repair, 1119–1122
  in bone resorption, 1122–1124
  in chondrogenesis, 1116–1118
  FGF-8, 1340–1341
  mutations in FGFR3 and FGF9, 1119
  production and regulation in bone, 1113–1115
  and receptors, 24
Fibroblastic cells, 312
Fibrodysplasia ossificans progressiva (FOP), 1674–1675, 1817–1818, 1895–1896
Fibrogenesis, 1155
Fibrogenic cytokines, 310–311
Fibronectin, 141–142, 367
Fibronectin type III domainecontaining 5 ((FNDC5), 367
Fibronectin–FGF receptor 1 (FN1–FGFR1), 1542
Fibrous dysplasia (FD), 58, 1431, 1443–1451, 1539–1540, 1817
  of bone, 58–60
  classic deformities, 1446f
  clinical features, 1443–1444
  diagnosis and management, 1447–1451
  genetics, 1445
  histopathological appearance, 1447f
  pathogenesis, 1445–1447
  scoliosis in, 1444f
FIHP. *See* Familial isolated hyperparathyroidism (FIHP)
Filamin A, 546–547
Filtration, 1839–1840
Fimbrin, 139
Finite element (FE), 1834

Finite element analysis (FEA), 146, 949, 1841, 1858, 1865, 1872–1873, 1927
FIP200. *See* Family–interacting protein of 200 kDa (FIP200)
*FKBP*10, 350
Fluid shear stress, 146–147, 1774
Fluorescence microscopy, 447
Fluorescence resonance energy transfer microscopy (FRET microscopy), 699–700
Fluorescence-activated cell sorting (FACS), 56–57, 96
Fluoride, 1650–1651
Fluorochrome labels, 445–446
Fluorodeoxyglucose positron emission tomography (FDG–PET), 484–485, 1544–1546
Flutamide, 971–972
Fms-like tyrosine kinase/Flt-1. *See* VEGF receptor 1 (VEGFR-1)
FN. *See* Femoral neck (FN)
FN1–FGFR1. *See* Fibronectin–FGF receptor 1 (FN1–FGFR1)
Focal adhesion kinase (FAK), 401–402, 1773–1774
Focally transient, 257
FOGs. *See* Friends of GATA (FOGs)
FOP. *See* Fibrodysplasia ossificans progressiva (FOP)
Forkhead box N1 (Foxn1), 578–579
Forkhead Box O3 (FOXO3A), 1674
Forkhead-box family member (FoxH1), 1160
Formation period (FP), 1900
Four LTBPs (LTBP1–4), 315
Four-and-a-half LIM domains 1 gene (FHL1 gene), 1368
Four-point bending, 1924–1925, 1925f
Foxa2 (transcription factors), 26
Foxa3 (transcription factors), 26
Foxn1. *See* Forkhead box N1 (Foxn1)
FOXO1, 1938
FOXO3A. *See* Forkhead Box O3 (FOXO3A)
2fp521. *See* Zinc finger protein 521 (2fp521)
FPP. *See* Farnesyl pyrophosphate pyrophosphate (FPP)
Fracture, 1259
  bone fragility in patients with, 263
  healing tissue, 1061–1062
  prediction, 1848–1849
    using areal bone mineral density, 1862
  repair, 1194
    and Notch signaling, 1096
  T2DM and, 946–947
    clinical risk factors for, 949–950
  toughness testing, 1926–1927, 1927f
Fracture Intervention Trial, 957
Fracture risk, 919–920, 1701
  in context of glucocorticoid therapy, 927
  T1DM and, 943–946
    clinical risk factors for, 949–950
Fracture risk assessment tool (FRAX tool), 1475–1476
FRCP2, 367
FREEDOM trial, 1693
FRET microscopy. *See* Fluorescence resonance energy transfer microscopy (FRET microscopy)
Friends of GATA (FOGs), 1360
Frizzled (Fzd), 1715
  family of proteins, 180–181
Frizzled-related proteins (FRPs), 1805
Frost coupling concept, 221
FRPs. *See* Frizzled-related proteins (FRPs)
FRS2α. *See* FGFR substrate 2α (FRS2α)
Functional secretion domain (FSD), 113
Functioning GEP-NETs, 1295
Furin deletion, 1935
*FXYD*2, 515–516
Fzd. *See* Frizzled (Fzd)

G

G-CSF. *See* Granulocyte colony–stimulating factor (G-CSF)
G-protein α-subunit (Gsα), 1393, 1431, 1933–1934
  AHO, 1434–1440
  FD, 1443–1451
  gene *GNAS*, 1433–1434
  G_s GTPase cycle, 1433f
  MAS, 1443–1451
  PHPIB, 1441–1443
  POH, 1440–1441
  structure and function, 1432–1433
G-protein-coupled receptor (GPCR), 539–540, 692, 789, 813, 1251, 1274, 1312, 1363–1364, 1379–1380, 1413, 1657
G-protein-coupled receptor family C group 6 member A (GPRC6A), 1933–1936
  Gprc6a$^{-/-}$mice, 1934
  Gprc6aMck$^{-/-}$mice, 1935–1936
  osteocalcin receptor in β cells, 1934
G-protein-coupled receptor kinases (GRKs), 646–647, 1276
  GRK2, 651
G0/G1 switch gene 2 (GOS2), 761
G5 on dialysis (G5D), 1660–1661
GACI. *See* Generalized arterial calcification of infancy (GACI)
GAGs. *See* Glycosaminoglycans (GAGs)
Gail model, 875
Gain-of-function (GoF), 974–975
  evidence from human trials, 974–975
  studies with testosterone treatment and change in bone architecture, 976
β-Galactosidase, 310, 1796
Galactosyltransferase, 299
Gap junction, 423
  communication, 435–436
  effects on, 651
  intercellular communication, 428–436, 430f
GAS6. *See* Growth arrest specific 6 (GAS6)
Gastric inhibitory polypeptide (GIP), 950–951
Gastrinoma, 1296–1297
Gastroenteropancreatic neuroendocrine tumors (GEP-NETs), 1293, 1295–1300
  gastrinoma, 1296–1297
  glucagonoma, 1299
  insulinoma, 1297–1299
  nonfunctioning neuroendocrine tumors of gastroenteropancreatic tract, 1299–1300
  somatostatinoma, 1299
  VIPoma, 1299
Gastrointestinal (GI), 1661
  CaSR in GI system, 554–555
  consideration, 473–474
  side effects, 479–480
GATA-binding protein 3 (Gata3), 579
  clinical features and role of mutations, 1359–1360
  in developmental pathogenesis, 1361
  Gata3$^{-/-}$mice, 1360
  knockout mouse model phenotype, 1360
Gata3. *See* GATA-binding protein 3 (Gata3)
Gaussian filtration, 1839–1840
Gc-globulin. *See* Group-specific component-globulin (Gc-globulin)
Gc1F allele, 717
Gc1F variant, 715
Gc1S variant, 715
Gc2 variant, 715
GCM2. *See* Glial cells missing-2 (GCM2)
GCs. *See* Glucocorticoids (GCs)
GCTB. *See* Giant cell tumor of bone (GCTB)
GDFs. *See* Growth/differentiation factors (GDFs)
GEFs. *See* Guanine nucleotide exchange factors (GEFs)
Genant semiquantitative scoring method, 1858
Gender-specific differences in bone geometry and architecture, 975–976
Gene clusters, 10–11
Gene expression control, 372
Gene Relationships Among Implicated Loci (GRAIL), 1621
Gene targeting, 1790
  and unique features of regulation, 1746–1748
Gene therapy, 1563
  options, 1497
Generalized arterial calcification of infancy (GACI), 474, 1533
Genes mutation
  modifying synthesis of type I collagen chains, 1492
  regulates maturation of secreted procollagen, 1492–1493
Genetic
  ablation, 254
  lineage tracing, 1894
  manipulation, 1887–1888
    of sclerostin expression in mice, 1716–1717
  models, 75
  mutations
    in TGF-β signaling components, 1168–1170, 1169f

Genetic (*Continued*)
on type I collagen formation, 302–303
testing, 486–488, 1302–1303
Genetically modified animals
advantages and disadvantages of conventional gene deletion, 1790
chondrocytes, 1789
conditional loss-of-function approaches, 1790–1796
gene targeting, 1790
genomic engineering using CRISPR/Cas9, 1797
large-scale phenotyping resources and repositories, 1787–1788
lineage tracing and overexpression tools of *Rosa*26 locus, 1796–1797
osteoblasts/osteocytes, 1789
osteoclasts, 1789
overexpression approaches to assess gene function, 1788–1789
tendon and ligament, 1789
transgenic mouse reporters of signaling pathways, 1789–1790
Genetically modified ES cells, 1787
Genetically modified mice, 976–977, 1914–1915
ESR1in bone using genetically modified mice equivocal, 977
models, 899
Genistein, 1069–1070
Genome-wide association study (GWAS), 513, 1615–1618, 1624
using biological knowledge and networks, 1621
for bone traits in mice, 1622–1624
cell/tissue types, 1620–1621
follow-up of, 1618–1622
future directions, 1625
resources for, 1616t, 1625t
for skeletal traits in mice, 1622–1624
Genome-wide screen, 300
"Genomic desert", 1070
Genomic engineering using CRISPR/Cas9, 1797
Genotoxic effects, 1676
Genotype–phenotype correlation, 486
GEP-NETs. *See* Gastroenteropancreatic neuroendocrine tumors (GEP-NETs)
Geranylgeraniol pyrophosphate (GGPP), 1672
Germ-line, 277
GFD. *See* Growth factor domain (GFD)
GFP. *See* Green fluorescent protein (GFP)
GFR. *See* Glomerular filtration rate (GFR)
GGCX. *See* γ-Glutamyl carboxylase (GGCX)
Ggcx gene, 1932–1933
GGPP. *See* Geranylgeraniol pyrophosphate (GGPP)
GH. *See* Growth hormone (GH)
GH deficiency (GHD), 992
effects on bone metabolism, 998
GH/IGF actions on intact skeleton, 994–998
GH/IGF-I/IGFBP system on aging skeleton, 988–989
GH/IGF/IGFBP system
physiology of
GHRH, 986–987
GHRHR, 987
SMS, 987
GH/IGF/IGFBP system, 986–989
GH/IGF/IGFBPS in osteoporosis, 998–1000
GHBP. *See* Growth hormone binding protein (GHBP)
GHD. *See* GH deficiency (GHD)
GH–IGF-I
axis, 988
systemic effects on body size and longitudinal growth, 994–995
GHR. *See* Growth hormone receptor (GHR)
GHRH. *See* Growth hormone releasing hormone (GHRH)
GHRH receptor (GHRHR), 986–987
GI. *See* Gastrointestinal (GI)
Giant cell tumor of bone (GCTB), 1698
Giant cell tumors, 1698
Gingival inflammation, 1075–1076
GIO. *See* Glucocorticoid-induced osteoporosis (GIO)
GIP. *See* Gastric inhibitory polypeptide (GIP)
Gitelman syndrome (GS), 515
*Gja*4-knockout mice, 431–432
Gla. *See* γ-Carboxyglutamic acid (Gla)
O-GlcNAc. *See* O-linked *N*-acetylglucosamine transferase (O-GlcNAc)
Gli2, 1316–1317, 1339–1340
GLI3 repressor (GLI3R), 14
Glial cells missing-2 (GCM2), 547–548, 578–579, 1358, 1365–1366, 1382–1383, 1413
gene abnormalities, 1365–1366
Glioblastoma-derived T cell suppressor factor, 1153–1154
Global effects, 1224–1227
contributions of gp130
in osteoblast lineage to bone structure, 1225–1226
in osteoclasts to bone physiology, 1226
insights from mice and patient with gp130 signaling mutations, 1225
intracellular negative feedback, 1227
Glomerular filtration rate (GFR), 512–513, 1740, 1807
Glomerulopathy, 302–303
GLP1 receptor agonists (GLP1RA), 950–951
Glucagon-like peptide-1 (GLP1), 692–693, 950–951
Glucagonoma, 1299
Glucocorticoid receptor (GR), 916
Glucocorticoid response elements (GREs), 916
Glucocorticoid-induced osteoporosis (GIO), 916, 919–921, 1681–1682, 1698, 1915
indirect mechanisms for, 925–926
management, 928–932
bisphosphonates, 929
denosumab, 929–930
raloxifene, 930
sex hormone replacement, 932
teriparatide, 930–931
timing and monitoring of therapy, 932
patient assessment with, 927–928
treatment, 927
Glucocorticoids (GCs), 51, 186, 1253, 1562, 1586
excess and skeleton, 919–926
excess and effects on osteoblast differentiation, 921–922
indirect mechanisms for glucocorticoid-induced osteoporosis, 925–926
and local glucocorticoid metabolism, 925
and osteoclast, 924–925
and osteocyte, 924
pathogenesis of GIO, 920–921
excess and systemic fuel metabolism, 926–927
management of glucocorticoid-induced osteoporosis, 928–932
patient assessment with glucocorticoid-induced osteoporosis, 927–928
treatment of glucocorticoid-induced osteoporosis, 927
inducing osteoblast apoptosis, 923
new and emerging therapies, 932–933
antisclerostin/DKK1, 932–933
SEGRAs, 932
physiological role in bone, 916–919
preventing osteoblast cell cycle progression, 922–923
signaling, 917f
therapy, 915–916
treatment, 144
Glucokinase, 553–554
Gluconeogenesis, 670
Glucose, 997–998
glucose-6-phosphate dehydrogenase activity, 1772–1773
metabolic control, 951–952
Glucosepane, 347
GluOCN secretion, 102
γ-Glutamyl carboxylase (GGCX), 1934–1935
Glutaraldehyde, 1924
Glycations, 345–347
advanced glycation end products, 346–347
nonspecific glycations, 345–346
potential consequences, 347
Glycogen synthase kinase-3β (GSK-3β), 179, 181–183, 186, 426–427, 1412–1413, 1637
Glycoprotein 130 (gp130), 1214
contributions
in osteoblast lineage to bone structure, 1225–1226
in osteoclasts to bone physiology, 1226
signaling mutations, 1225
Glycoproteins, 363–365, 364t
alkaline phosphatase, 363

osteonectin, 365
periostin, 365
sclerostin, 363–365
tetranectin, 365
Glycosaminoglycans (GAGs), 360
GAG-bearing perlecan domain I, 410–411
moieties, 302
Glycosylations, 299, 345–347, 540, 1572
enzymatic glycosylation, 345
Glypicans, 410–411
Gly–Xaa–Yaa primary amino acid sequence, 339–340
*GNA*11 mutations, 1382–1383
GNAS gene, 1431
human diseases, 1432t
mutations, 1393–1394
organization and imprinting, 1434f
GNAS1. *See* Guanine nucleotide-binding protein, α stimulating (GNAS1)
GnRH. *See* Gonadotropin-releasing hormone (GnRH)
GoF. *See* Gain-of-function (GoF)
Goldilocks effect, 1069
Gonadal status effects on GHeIGF-I axis, 988
Gonadotropin-releasing hormone (GnRH), 974
GOS2. *See* G0/G1 switch gene 2 (GOS2)
gp130. *See* Glycoprotein 130 (gp130)
GPCR. *See* G-protein-coupled receptor (GPCR)
GPRC6A. *See* G-protein-coupled receptor family C group 6 member A (GPRC6A)
Gq/11/phospholipase C/protein kinase C signaling, 1276
GR. *See* Glucocorticoid receptor (GR)
GRAIL. *See* Gene Relationships Among Implicated Loci (GRAIL)
Granulocyte colony–stimulating factor (G-CSF), 76–77, 1225
Granulocyte/macrophage colony-stimulating factor, 410
Granulocyte–macrophage colony-forming units (CFU-GM), 1604
Graves' disease, 903
GRB2. *See* Growth factor receptor–bound protein 2 (GRB2)
Green fluorescent protein (GFP), 75, 91, 137, 1888
GREs. *See* Glucocorticoid response elements (GREs)
GRKs. *See* G-protein-coupled receptor kinases (GRKs)
gRNAs. *See* Guide RNA sequences (gRNAs)
Group VI enzymes (GVI enzymes), 1249
Group-specific component-globulin (Gc-globulin), 715–716
Growth arrest specific 6 (GAS6), 1338–1339
Growth factor domain (GFD), 388
Growth factor receptor–bound protein 2 (GRB2), 116–117
Growth factors, 314, 649–651
CTGF, 316–317
FGF, 317–318
IGF, 318
signaling pathways
CNP, 25
Notch signaling, 25
parathyroid hormone-related protein and Indian hedgehog, 21–23
transforming growth factor β and bone morphogenetic proteins, 19–21
WNTs and β-catenin, 23–24
TGFβ, 314–316
Growth hormone (GH), 19, 480, 942, 985–986
GH-secreting adenoma, 1300
and IGF-I as treatments for skeletal disorders, 1000–1007
administration for healthy adults, 1001
administration to elderly men and women, 1002–1003
IGF I for treatment of osteoporosis, 1004–1007
treatment for adult-onset GH deficiency, 1001–1002
treatment for children with insufficient GH secretion, 1000–1001
treatment for osteoporotic patients, 1003–1004
treatment for skeletal disorders, 1000
mechanism of growth hormone secretion, 987–988
effects of GH/IGF-I/IGFBP system on aging skeleton, 988–989
effects of gonadal status on GHeIGF-I axis, 988
Growth hormone binding protein (GHBP), 988
Growth hormone receptor (GHR), 987–988
Growth hormone releasing hormone (GHRH), 986–987, 1436
Growth plate cartilage, insulin effects on, 942
Growth plate chondrocytes, FGFR3 signaling in, 1117–1118
Growth-promoting activity, 1069
Growth/differentiation factors (GDFs), 1153–1154
GS. *See* Gitelman syndrome (GS)
GSK-3β. *See* Glycogen synthase kinase-3β (GSK-3β)
Gsα. *See* G-protein α-subunit (Gsα)
GTPγS, 699
Guanine nucleotide exchange factors (GEFs), 1774–1775
Guanine nucleotide-binding protein, α stimulating (GNAS1), 1533
Guide RNA sequences (gRNAs), 1797
Gut-derived serotonin, 170
GVI enzymes. *See* Group VI enzymes (GVI enzymes)
GWAS. *See* Genome-wide association study (GWAS)
Gα11, 1368
Gα12/13-phospholipase-transforming protein RhoA pathway, 1276
Gαi/o pathway, 1276
Gαs/adenylyl cyclase/protein kinase A signaling, 1275

H

H223R mutation, 1386
H3K27me, 724
H3K27me3, 52–53
H3K4me3, 724, 1410
H3K9, 724
H3K9me3, 52–53
HA. *See* Hyaluronan (HA); Hydroxyapatite (HA)
HADHB gene, 1362
Hairy 2 gene (*Hes*2 gene), 15
Hairy and enhancer of split (HES), 1089
Hajdu Cheney syndrome (HCS), 1098
Has2. *See* Hyaluronic acid synthase 2 (Has2)
HATs. *See* Histone acetyltransferases (HATs)
HAV. *See* His-Ala-Val (HAV)
HbA1c. *See* Hemoglobin A1c (HbA1c)
HCCs. *See* Hypertrophic chondrocytes (HCCs)
HCRR. *See* Hereditary calcitriol-resistant rickets (HCRR)
HCS. *See* Hajdu Cheney syndrome (HCS)
HD. *See* Heterodimerization domain (HD)
HDAC. *See* Histone deacetylase (HDAC)
HDMs. *See* Histone demethylases (HDMs)
Health, Aging, and Body Composition Study (Health ABC Study), 632, 1878
Healthy adults, GH administration for, 1001
Heat shock protein 47 (Hsp47), 300, 351
Heat shock protein70 (Hsp70), 713–714, 720
Heat-shock protein 90 (Hsp90), 916
Hedgehog signaling (HH signaling), 8, 1062
in bone and periodontium, 1062–1063
network, 1319–1320
HEK-293 cells, 544
Hematologic progenitor cells, 74
Hematopoiesis, 46, 49, 73
Hematopoietic cells, 77–79, 1205–1206
macrophages, 78
megakaryocytes, 78
neutrophils, 78
osteoclasts, 79
T cells, 79
Hematopoietic markers, 56
Hematopoietic stem cell transplantation (HSCT), 1562
Hematopoietic stem cells (HSCs), 47, 73, 208, 280, 1141, 1163, 1338
microenvironments in embryo and perinatal period, 74
niche, 74
heterogeneity for heterogeneous HSCs, 80–81
hormonal regulation, 79–80
neuronal regulation, 79
Hematopoietic system, 1210

Hemichannels, 136
Hemodialysis, 1662–1663
Hemoglobin A1c (HbA1c), 942–943
Heparan sulfate (HS), 360
Heparan sulfate proteoglycans (HSPGs), 1116, 1117f
Hepatic clearance, 1746
Hepatic IGF-I transgene (HIT), 1006
Hepatic stellate cell (HSC), 300–301
Hepatocyte nuclear factor 1 (HNF1), 714
  HNF1α and HNF1β, 714
Hepatocyte nuclear factor 1β (*HNF1B*), 517
Heptahelical GPCRs, 1382
Hereditary 1,25(OH)$_2$D$_3$-resistant rickets (HVDRR), 742
Hereditary calcitriol-resistant rickets (HCRR), 1508, 1513–1521
  cellular and molecular defects
    methods, 1516
    types of defects, 1516
  clinical and biochemical features, 1514
    ectodermal anomalies, 1514
    mode of inheritance, 1515–1516
    vitamin D metabolism, 1514–1515
Hereditary deficiencies in vitamin D action, 1507
  animal models, 1522
  clinical features of rickets and osteomalacia, 1508–1510
  diagnosis, 1521
  hereditary defects in vitamin D receptor–effector system, 1513–1521
  HVDDR, 1510–1513
  treatment, 1521–1522
Hereditary disorders of magnesium homeostasis, 512–518
  disturbed Mg^{2+}reabsorption
    in DCT, 514–518, 514f
    in thick ascending limb, 512–514
Hereditary hypophosphatemic rickets with hypercalciuria (HHRH), 486–490, 488f
  clinical presentation and diagnostic evaluation
    laboratory findings and genetic testing, 486–488
    musculoskeletal findings, 489
    renal findings, 489
  cloning and identification of human mutations in NPT2c, 486
  epidemiology, 486
  pathophysiology, 486
  therapy and resources
    standard therapy, 489–490
Hereditary multiple exostoses (HME), 362
Hereditary vitamin D–dependent rickets (HVDDR), 1507–1508, 1509t, 1510–1513. *See also* Hereditary calcitriol-resistant rickets (HCRR)
Hereditary vitamin D–dependent rickets type A (HVDDR-A), 1507–1508, 1510–1511
Hereditary vitamin D–dependent rickets type B (HVDDR-B), 1511
Hereditary vitamin D–dependent rickets type C (HVDDR-C), 1510–1512
Heritable disorders, 348
  bone mineralization, 352
  collagen chaperone, 351
  collagen posttranslational modifications, 348–351
  collagen processing, 351
  noncollagen genes, 349t
HES. *See* Hairy and enhancer of split (HES)
Hes-related with YRPW motif (HEY), 1089
Hes2 gene. *See* Hairy 2 gene (*Hes2* gene)
Heterodimerization, 896
Heterodimerization domain (HD), 1084
Heterogeneous nuclear ribonucleoproteins (hnRNPs), 313–314
Heterogenous nuclear ribonucleoproteins (hnRVP), 1520
Heterotopic ossification (HO), 1895–1896
  lineage tracing in, 1895–1896
Heterotypic gap junction, 428–429
Heterozygous $Cdh2^{-/+}$ mice, 426
HETEs. *See* Hydroxyeicosatetraenoic acids (HETEs)
HEY. *See* Hes-related with YRPW motif (HEY)
HFD. *See* High-fat diet (HFD)
hGHRH. *See* Human GHRH (hGHRH)
HGPS. *See* Hutchinson–Gilford progeria syndrome (HGPS)
HH signaling. *See* Hedgehog signaling (HH signaling)
HHM. *See* Humoral hypercalcemia of malignancy (HHM)
HHRH. *See* Hereditary hypophosphatemic rickets with hypercalciuria (HHRH)
Hierarchical functional imaging, 1838
High-fat diet (HFD), 948
  HFD–fed LDL receptor–null mice, 284
High-molecular-weight (HMW), 1115f
High-pass filters, 1839–1840
High-resolution computed tomography (HR-CT), 1865, 1868–1871
High-resolution peripheral quantitative computed tomography (HR-pQCT), 946, 1475–1478, 1618, 1804, 1834, 1846–1847, 1858, 1869–1871
Hind limb immobilization model, 1915
Hip geometry, 1872–1873
Hip structure analysis, 1872–1873
His-Ala-Val (HAV), 423–424
Histidine-containing collagen cross-links and maturation products, 344–345
Histochemical tartrate-resistant acid phosphatase (TRAcP), 1907
  staining, 1904
Histological heterogeneity, 453
Histomorphometric/histomorphometry, 258, 944
  analyses of bone healing, 1913
  of bone healing rodent models, 1912–1913
  of bone loss rodent models, 1913–1915
  data, 1662
  measurement of longitudinal bone growth, 1910
  of pharmacological efficacy in rodents, 1915–1917
  of rodent models of bone healing, 1912–1913
  studies of effects of osteoporosis drugs, 454–461
    anticatabolic agents, 454–456
    AVA study, 461
    comparative studies of anabolic and anticatabolic drugs, 460
    osteoanabolic therapies, 456
    PTH(1–34) and PTH(1–84), 456–460
  techniques, 1899
Histone acetylation, 171
Histone acetyltransferases (HATs), 724, 744–745
  and deacetylation by deacetylases, 52–53
Histone deacetylase (HDAC), 316
  activity, 724
  class I HDACs, 171
  class II HDACs, 171
  enzymes, 649
  HDAC3, 167
  HDAC4, 21–22, 26–27, 387, 1397
Histone deacetyltransferases, 744–745
Histone demethylases (HDMs), 52–53, 744–745
Histone methyltransferases (HMTs), 52–53
Histone-acetyl transferases, 26–27
hnRNPC1/C2, 720
hnRNPs. *See* Heterogeneous nuclear ribonucleoproteins (hnRNPs)
hnRVP. *See* Heterogenous nuclear ribonucleoproteins (hnRVP)
Holoprosencephaly type 3 (HPE3), 1367
HOMA-IR. *See* Homeostatic model of insulin resistance (HOMA-IR)
Homeobox genes (*Hox* genes), 12–13, 1358–1359
  Hox5 paralogs, 10–11
  Hoxa3 gene, 578–579, 1358
Homeodomain, 1358–1359
Homeodomain-containing transcription factor (Hoxa2), 166
Homeostasis, 296, 1141
Homeostatic model of insulin resistance (HOMA-IR), 947–948
Homotrimer biosynthesis, 304–305
Homozygous mutant mice, 299–300
Honeybee. *See Apis mellifera* (Honeybee)
Hormonal osteocalcin, 102
Hormonal regulation, 51
  of hematopoietic stem cell niche, 79–80
Hormonal regulators, 471
  FGF23, 471
  PTH, 471
Hormone replacement therapy (HRT), 1003–1004
Hormone therapy (HT), 454
Hormones, 314, 322–324
  hormone-binding region, defects in, 1516–1518

deficient hormone binding, 1516–1518
deficient nuclear uptake, 1518
hormone-driven mammary tumor models, 1199–1200
hormone–receptor interaction, 1516
receptors in osteocytes, 143–144
Hounsfield units (HU), 1863
Howship's lacunae, 112
Hox genes. *See* Homeobox genes (*Hox* genes)
HP. *See* Hydroxylysyl pyridinoline (HP)
HPE3. *See* Holoprosencephaly type 3 (HPE3)
HPO axis. *See* Hypothalamic–pituitary–ovarian axis (HPO axis)
HPP. *See* Hypophosphatasia (HPP)
HPT. *See* Hyperparathyroidism (HPT)
HPT-JT syndrome. *See* Hyperparathyroidism-jaw tumor syndrome (HPT-JT syndrome)
HR-CT. *See* High-resolution computed tomography (HR-CT)
HR-pQCT. *See* High-resolution peripheral quantitative computed tomography (HR-pQCT)
HREs. *See* Hypoxia response elements (HREs)
HRPT2, 1417
HRT. *See* Hormone replacement therapy (HRT)
HS. *See* Heparan sulfate (HS)
HSC. *See* Hepatic stellate cell (HSC)
HSCs. *See* Hematopoietic stem cells (HSCs)
HSCT. *See* Hematopoietic stem cell transplantation (HSCT)
11βHSD. *See* 11β-Hydroxysteroid dehydrogenase (11βHSD)
HSD3B2. *See* 3-beta hydroxysteroid dehydrogenase 2 (HSD3B2)
HSH. *See* Hypomagnesemia with secondary hypocalcemia (HSH)
Hsp47. *See* Heat shock protein 47 (Hsp47)
HSPGs. *See* Heparan sulfate proteoglycans (HSPGs)
HSs. *See* Hypersensitive sites (HSs)
HT. *See* Hormone therapy (HT)
HTLV-I. *See* Human T-cell leukemia virus type I (HTLV-I)
Htr1B receptor, 170
HU. *See* Hounsfield units (HU)
Human, 275
cancer cells, 1697
collagenase-1, 383
gain-of-function evidence from human trials, 974–975
genetic disorders of phosphate homeostasis, 482t–483t
genome project, 177
human genome-wide association studies, 185
IGFBP-2, 992
monogenic high bone mass conditions, 1712–1714
mutations in NPT2c, 486, 487f
periodontal ligament cells, 1211
pharmacologic inhibition of sclerostin by Scl-Abs in, 1724
RANKL knockin mice, 1699
rhabdomyosarcoma cell line, 1153–1154
skeleton, 1768
studies of IGF-I and BMD, 1005–1007
T2DM and, 949
variants, 972–974
Human GHRH (hGHRH), 986–987
Human or partially humanized RANKL (huRANKL), 1691
Human T-cell leukemia virus type I (HTLV-I), 1310
Humoral hypercalcemia of malignancy (HHM), 596, 653, 841–842, 1308, 1315–1316, 1322–1324
actions in bone, 1316
actions in kidney, 1315–1316
Hungry bone syndrome, 470, 779, 1521
huRANKL. *See* Human or partially humanized RANKL (huRANKL)
Hutchinson–Gilford progeria syndrome (HGPS), 1674
HVDDR. *See* Hereditary vitamin D–dependent rickets (HVDDR)
HVDRR. *See* Hereditary 1, 25$(OH)_2D_3$-resistant rickets (HVDRR)
Hyaluronan (HA), 363, 411
Hyaluronic acid synthase 2 (Has2), 363
Hybrid Mouse Diversity Panel (HMDP), 1623
3-Hydroxy-L-proline, 1808
Hydroxyapatite (HA), 527, 1569, 1645
crystals, 297
Hydroxyeicosatetraenoic acids (HETEs), 1247–1248
1-Hydroxylase, 757–758
24-Hydroxylated successor, 1735–1739
Hydroxylysine (Hyl), 345
residues, 348
Hydroxylysinonorleucine (HLNL), 342–343
Hydroxylysyl pyridinoline (HP), 340
Hydroxyproline, 296, 1603
3-Hydroxyproline (3Hyp), 348
11β-Hydroxysteroid dehydrogenase (11βHSD), 916
11βHSD1, 916
11βHSD2, 322, 916
4-Hydroxytamoxifen (4-OHT), 1887–1888, 1890–1891
25-Hydroxyvitamin D (25(OH)D), 474, 713–714, 718–720, 757–758
25-Hydroxyvitamin D_3 (25$(OH)D_3$), 739–740
Hyl. *See* Hydroxylysine (Hyl)
3Hyp. *See* 3-Hydroxyproline (3Hyp)
Hypercalcemia, 1307, 1576, 1601, 1649–1650, 1659, 1743
characterization in malignancy, 1308
in infantile HPP, 1586
of malignancy refractory to BP therapy, 1696–1698
breast cancer, 1697
giant cell tumors, 1698
multiple myeloma, 1697–1698
prostate cancer, 1697
solid tumors, 1697
systemic and local factors, 1322–1324
Hypercalcemic disorders, 1659–1660
Hypercalciuria, 1385, 1576
Hypercortisolism, 953
Hyperfunctioning glands, 1657–1658
Hyperglycemia, 951–952
hyperglycemia-induced oxidative stress, 956
Hyperinsulinemia, 947–948
Hyperlipidemia, 283
Hyperostotic skeleton, 1712
Hyperparathyroidism (HPT), 1307–1308, 1657, 1666–1667
Hyperparathyroidism, 449, 631, 777, 1405
Hyperparathyroidism-jaw tumor syndrome (HPT-JT syndrome), 1405, 1416
Hyperphosphatemia, 451, 491
Hyperphosphatemic diseases caused by FGF23 impaired actions, 1534–1535
FGF23 and CKD–mineral and bone disorder, 1534–1535
treatment of hyperphosphatemic familial tumoral calcinosis, 1534
Hyperphosphatemic syndromes, 491–492
tumoral calcinosis, 491–492
Hypersensitive sites (HSs), 310
Hypersensitivity, 1701–1702
Hyperthyroidism, 902–904
subclinical, 904
Hypertrophic chondrocytes (HCCs), 15–17, 27, 206, 296–297, 1114, 1145
*Fgfr*1 signaling in, 1118
Hypocalcemia, 1666, 1700
Hypogonadism, 932, 974–975
Hypokalemia, 515
Hypomagnesemia, 512–513, 515
Hypomagnesemia, acquired, 518–519
aminoglycosides, 519
calcineurin inhibitors, 519
cisplatin and carboplatin, 518
PPIs, 519
Hypomagnesemia with secondary hypocalcemia (HSH), 510
Hypomineralized cranial vault, 431
Hypomineralized periosteocytic lesions, 489
Hypomorphic *MIB*1 mutations, 1087
Hypoparathyroidism, 777, 1359–1361
complex syndromes associated with, 1357
inherited forms and chromosomal locations, 1356t
PTH in, 1639–1640
Hypophosphatasia (HPP), 1070, 1570, 1573–1587, 1818
adult, 1579, 1579f
Alpl knockout animals, 1591
Asfotase alfa treatment for, 1591
benign prenatal hypophosphatasia, 1580
biochemical and genetic defects, 1583–1585
childhood, 1577–1579, 1578f
circulating tissue-nonspecific alkaline phosphatase, 1590

Hypophosphatasia (HPP) (*Continued*)
clinical features, 1574–1575, 1575f
fibroblast studies, 1590–1591
history, 1573–1574
hypophosphatasia fibroblast studies, 1590–1591
infantile, 1576–1577, 1577f
inorganic pyrophosphate, 1589–1590
laboratory diagnosis, 1580–1583
biochemical findings, 1580
dentition, 1583
inorganic pyrophosphate, 1582–1583
mineral homeostasis, 1580–1581
phosphoethanolamine, 1581
pyridoxal 5′-phosphate, 1582, 1582f
skeleton, 1583
odontohypophosphatasia, 1580
perinatal, 1575, 1576f
phosphoethanolamine, 1587–1588
physiological role of ALP in, 1587–1591
prenatal diagnosis, 1586–1587
prognosis, 1585
pseudohypophosphatasia, 1580
pyridoxal 5′-phosphate, 1588–1589
serum alkaline phosphatase activity in, 1581f
tissue-nonspecific alkaline phosphatase substrates, 1587
treatment, 1585–1586
medical, 1585–1586
supportive, 1585
Hypophosphatemia, 475, 479, 485, 491, 1508
with osteoporosis and nephrolithiasis to SLC34A1, 490
Hypophosphatemic diseases
acquired FGF23-related hypophosphatemic disease, 1533
ADHR, 1532
ARHR, 1533
caused by FGF23 excessive actions, 1531–1534
hypophosphatemic diseases with known genetic causes, 1533
treatment of FGF23-related hypophosphatemic diseases, 1533–1534
XLH, 1532
Hypophosphatemic myopathy, 484
Hypothalamic hormone, 986–987
Hypothalamic–pituitary axis, 986
Hypothalamic–pituitary–gonadal axis, 976
Hypothalamic–pituitary–ovarian axis (HPO axis), 876
Hypothalamus, 986–987
Hypothyroidism, 897, 901–903
subclinical, 903
Hypovitaminosis D, 757–758, 766
Hypoxia, 1344–1345
Hypoxia response elements (HREs), 1147
Hypoxia-inducible factors (HIFs)), 26, 206–207, 207f, 1147
HIF-1α, 307–308, 1319–1320

I

IDG-SW3
cells, 1283
osteocyte cell line, 141
IDH. *See* Isolated dominant hypomagnesemia (IDH)
Idiopathic aplastic bone disease, 452
Idiopathic hypoparathyroidism, 1355–1357
Idiopathic pulmonary arterial hypertension (IPAH), 1667
Idoxifene, 868, 868f
IFCC. *See* International Federation of Clinical Chemistry and Laboratory Medicine (IFCC)
IFITM5. *See* Interferon-induced transmembrane protein 5 (*IFITM5*)
IGFBPs. *See* Insulin-like growth factor binding proteins (IGFBPs)
IGFRs. *See* Insulin-like growth factor receptors (IGFRs)
IGFs. *See* Insulin-like growth factors (IGFs)
IgG immune complex (IgG IC), 120
IHH. *See* Indian hedgehog (IHH)
IHP. *See* Isolated hypoparathyroidism (IHP)
IkB kinase complex (IKK complex), 1560–1561
IKK complex. *See* IkB kinase complex (IKK complex)
IL. *See* Interleukins (IL)
IL-1 receptor (IL-1R), 124
IL-1 receptor type II, 1206
IL-13 receptor α2 (IL-13Ra2), 320–321
IL-18 binding protein (IL-18BP), 1213
IL-1R. *See* IL-1 receptor (IL-1R)
IL-27Ra-binding cytokines, 1224
IL-6. *See* Interleukin-6 (IL-6)
ILK. *See* Integrin-linked kinase (ILK)
IMKC. *See* International Mouse Knockout Consortium (IMKC)
Immature cells, 74
Immobilization-induced bone loss, 1724
Immune/immunity, 73, 1201–1202
cells, 233–234
response, 423
system development and thermoregulation, 1201–1202
vitamin D and, 758–759
Immunoglobulin superfamily members, 411–412
Immunoreceptor tyrosine-based activation motif costimulatory signals (ITAM costimulatory signals), 119
calcium signals, 119
SIGLEC-15 and FcγR, 119–120
Immunoseparated primary cell populations, 150
IMPC. *See* International Mouse Phenotyping Consortium (IMPC)
In situ hybridization (ISH), 164, 233
In vitro
microimaging, 1834–1838
posttranscriptional effects of 1, 25(OH)$_2$D$_3$, 1519–1520
studies, 582
In vivo
growth factors, 1061
labeling, 1899–1900
mechanisms, 1766
responses of skeleton to thyroid hormone
hyperthyroidism, 902
hypothyroidism, 901–902
studies, 582, 1340–1341
transplantation, 55
In vivo microimaging, 1841–1849
in vivo animal microimaging, 1842–1845
dynamic morphometry, 1844–1845
radiation considerations, 1842–1843
reproducibility, 1843–1844
in vivo human microimaging, 1845–1849
clinical computed tomography, 1845
fracture prediction, 1848–1849
high-resolution peripheral quantitative CT, 1846–1847
normative data, 1848
peripheral quantitative computed tomography, 1846
radiation dose, 1847–1848
Inbred strains, 1623
Incisor absent rat (IA rat), 720
Incretins effect, loss of, 956–957
Indian hedgehog (IHH), 21–23
signaling, 8
Induced pluripotent stem cells (iPSCs), 64, 1162, 1497
technology, 1563
uses as diagnostic tool, 1497–1498
Inducible cyclooxygenase
inducible cyclooxygenase–dependent inhibitor, 1257–1258
knockout mice, 1256
basal skeletal phenotype, 1256
modulation of effects of parathyroid hormone, 1256–1258
INF. *See* Interferons (INF)
Infantile
HPP, 1576–1577, 1577f, 1580–1581
osteopetrosis, 1553–1554
Inflammation, 956, 1202, 1699
in periodontal disease, control, 1073–1074
“Information integration” perspective, 1762
Inheritance, 1584
Inhibitors of MMPs (MMPIs), 380
Inorganic phosphate (P_i), 1071, 1569
Inorganic phosphorus, 470
Inorganic pyrophosphate (PP_i), 1070, 1569, 1582–1583, 1589–1590
histopathological findings, 1582–1583
radiographic findings, 1582
Inositol 1,4,5-trisphosphate receptor type 2, 119
Inositol heptakisphosphate (IP7), 478
Inositol-1, 4, 5-trisphosphate (IP_3), 1276
Insulin, 951, 986–987, 1931–1933
effects on growth plate cartilage, 942
metabolic control of, 951–952
resistance, 926, 941, 947–948, 1702
secretion, 1933–1934
Insulin. *See also* Diabetes

Insulin receptors (IRs), 942
Insulin-dependent diabetes mellitus (IDDM), 992
Insulin-like growth factor binding proteins (IGFBPs), 318, 649, 986, 989–993
IGFBP1, 952–953
IGFBP3, 761
IGFBP-4, 992
IGFBP-5, 993
proteases, 993–994
Insulin-like growth factor receptors (IGFRs), 989–991
Insulin-like growth factors (IGFs), 19, 227, 318, 385, 649, 761, 875, 985–986, 996–997, 1160–1161
IGF-1, 52, 77, 80, 138, 402–403, 649, 778, 942, 952–953, 1283, 1310, 1607, 1637, 1804
as treatments for skeletal disorders, 1000–1007
IGF-II, 649
regulatory system and relationship to skeleton, 989–994
Insulin-receptor substrate-1 (IRS-1), 989–990
Insulinoma, 1297–1299
int-1. *See* Integration site/locus (*int*-1)
Intact parathyroid hormone (iPTH), 461, 512–513
Intact skeleton, GH /IGF actions on
energy utilization by skeletal cells and role of IGF-I, 997–998
GH and IGF-I effects on modeling and remodeling, 995–996
GH–IGF-I systemic effects on body size and longitudinal growth, 994–995
IGFs, 996–997
osteoblasts, 996–997
transcription factors, 996–997
Integration site/locus (*int*-1), 178–179
α4β1 Integrin, 1338
αvβ3 Integrin, 1338
circular "rosette", 404–405
Integrin-linked kinase (ILK), 401–402
β3-Integrin-null osteoclasts, 123
Integrins, 14–15, 401
activation by, 1156–1158, 1157f
in bone cells, 401–406
chondrocytes, 406
osteoblasts and osteocytes, 401–404
osteoclasts, 404–406
and integrin-associated proteins, 1773–1774
Intercellular adhesion molecule (ICAM), 411–412
Interferon regulatory factor (IRF), 313
IRF8, 118
Interferon-induced transmembrane protein 5 (*IFITM5*), 352, 1492
Interferons (INF), 1209
IFN-α, 726–727
IFN-γ, 141, 308, 319–320, 1209
Interleukins (IL), 321, 1171
IL-1, 314, 320, 815, 1206
IL-4, 114, 321, 381, 1213
IL-6, 62–63, 151, 321, 411–412, 714, 1214–1218, 1338–1339, 1450–1451, 1605–1607
cytokines, 97, 1214–1215, 1214f, 1216t
IL-7, 1210
IL-8, 1335–1336
IL-10, 1210–1211
IL-11, 1218–1219, 1319–1320, 1335–1336
IL-12, 1211
IL-13, 320–321, 1213
IL-15, 1211
IL-17, 1211–1212
IL-18, 1212–1213
IL-23, 1212
IL-32, 1213
IL-33, 1212–1213
Intermediate osteopetrosis, 1554
Intermediate recessive osteopetrosis (IRO), 1554
Intermittent parathyroid hormone (iPTH), 407–408, 779, 953–954, 1279–1280, 1637
bone anabolic effects of, 779–782
Internal basal lamina, 1074–1075
International Federation of Clinical Chemistry and Laboratory Medicine (IFCC), 1802
International Mouse Knockout Consortium (IMKC), 1790
International Mouse Phenotyping Consortium (IMPC)
bone phenotype screen, 1788
International Mouse Phenotyping Consortium (IMPC), 1621–1622, 1787–1788
International Osteoporosis Foundation (IOF), 1802
International Society for Clinical Densitometry (ISCD), 1860–1861
Interosteonal (interstitial) matrix, 247
Intestinal malabsorption of phosphate, 491
Intestinal phosphate
absorption, 473–474
transport, 476–477
Intracellular
intracellular/extracellular compartmentalization, 475
mechanism of PTH-related protein action, 1313–1315
mechanism of thyroid hormone action
chondrocytes, 901
nongenomic actions of thyroid hormones, 898
nuclear actions of thyroid hormones, 895–898
remodeling, 901
thyrotropin as independent agent of bone metabolism, 898–899
negative feedback through suppressor-of-cytokine-signaling proteins, 1227
pathways, 401
phosphate, 492
trafficking of vitamin D metabolites, 720
transport, 300–301
Intracerebroventricular infusion (ICV), 810
Intracrine nuclear actions, 600–602
Intraductal papillary mucinous neoplasms (IPMNs), 1447
Intrahelical "sacrificial" bonds, 246–247
Intramembranous bone formation, 1146
Intramembranous ossification, 6–9
axial skeleton, 9–11
cellular composition of suture, 7f
intramembranous cranial bone formation, 6f
limb skeleton, 11–14
sclerotome differentiation, 11
somitogenesis, 9–11
Intravenous meperidine hydrochloride, 446–447
Intronic elements in regulating collagen type I
of pro-COL1A1 gene, 312
of pro-COL1A2 gene, 313
IOF. *See* International Osteoporosis Foundation (IOF)
IP_3. *See* Inositol-1,4,5-trisphosphate (IP_3)
IPAH. *See* Idiopathic pulmonary arterial hypertension (IPAH)
$iPLA_2s$. *See* Ca^{2+}-independent PLA_2s ($iPLA_2s$)
IPMNs. *See* Intraductal papillary mucinous neoplasms (IPMNs)
iPMS. *See* Isopropyl methane sulfonate (iPMS)
IPP. *See* Isopentenyl pyrophosphate (IPP)
iPSCs. *See* Induced pluripotent stem cells (iPSCs)
iPTH. *See* Intact parathyroid hormone (iPTH); Intermittent parathyroid hormone (iPTH)
IR substrate (IRS), 952–953
IRF. *See* Interferon regulatory factor (IRF)
Irisin, 367
IRO. *See* Intermediate recessive osteopetrosis (IRO)
Iron deficiency, 481
IRs. *See* Insulin receptors (IRs)
IRS. *See* IR substrate (IRS)
IRS-1. *See* Insulin-receptor substrate-1 (IRS-1)
ISCD. *See* International Society for Clinical Densitometry (ISCD)
ISH. *See* In situ hybridization (ISH)
Island amyloid polypeptide. *See* Amylin
"Islet neogenesis", 855
Isolated dominant hypomagnesemia (IDH), 515–516
Isolated hypoparathyroidism (IHP), 1355–1357, 1363, 1382–1383
Isopentenyl pyrophosphate (IPP), 1672
Isopropyl methane sulfonate (iPMS), 1370
Isoproterenol, 168
ITAM costimulatory signals. *See* Immunoreceptor tyrosine-based activation motif costimulatory signals (ITAM costimulatory signals)

J

Jackson Laboratories Mouse Genome Informatics website, 1790
Jagged (JAG), 1083
  Jag1, 1319–1320
JAK. *See* Janus kinase (JAK)
Jansen's disease, 1385–1386
Jansen's metaphyseal chondrodysplasia (JMC), 704, 1383–1386
Janus kinase (JAK), 308, 1214, 1215f
JE. *See* Junctional epithelium (JE)
Jellyfish, 295
JGA. *See* Juxtaglomerular apparatus (JGA)
JMC. *See* Jansen's metaphyseal chondrodysplasia (JMC)
JNK. *See* Jun N-terminal kinase (JNK)
Joining exons, 297–298
Jun N-terminal kinase (JNK), 52, 118, 545, 1775–1776
  JNK1, 308
Jun proteins, 169–170
Junctional epithelium (JE), 1074
  in periodontium function, 1074–1076
JunD, 1410
Juxtaglomerular apparatus (JGA), 550

K

K-homology splicing regulator protein (KSRP), 584–585
K-PD model. *See* Kinetic-Pharmacodynamic model (K-PD model)
K-POST, 1804
K14. *See* Keratin 14 (K14)
*KCNA*1 mutation, 516
*KCNJ*10 mutation, 515
KCS. *See* Kenny–Caffey syndrome (KCS)
KDIGO. *See* Kidney Disease Improving Global Outcomes (KDIGO)
Kearns–Sayre syndrome (KSS), 1362
Kenny–Caffey syndrome (KCS), 1361–1362
  type 2, 518
Keratin 14 (K14), 843
Keratinocytes, 840
α-Ketoglutarate, 1392–1393
Kidney, 646
  CaSR in, 549–551
  PTH actions on, 655–664
    calcium absorption and excretion, 662
    calcium and phosphate homeostasis, 656–659
    PTHR expression, signaling, and regulation, 660–662
    regulation of renal calcium absorption, 663–664
  stones, 1650
Kidney Disease Improving Global Outcomes (KDIGO), 634, 1463
Kinase domain region (KDR). *See* VEGF receptor 2 (VEGFR-2)
Kindler syndrome, 1075
Kinetic-Pharmacodynamic model (K-PD model), 1680–1681
Kir4.1, 515
Klotho protein, 587, 1531, 1807
α-Klotho, 1465, 1807
  transmembrane receptor, 740–741
Knockin mice models, 897
Knockout (KO), 406, 1717
  animal models for functional analysis, 1621–1622
  mice, 476–477, 1249
  phenotypes, 302
KSRP. *See* K-homology splicing regulator protein (KSRP)
KSS. *See* Kearns–Sayre syndrome (KSS)

L

La ribonucleoprotein 6 gene (LARP6), 318
LA-PTH, 702
Labile, 352–353, 352f
Lactation, 844–848
  breast–brain–bone circuit controls lactation, 848f
  mammary gland, 1199
Lacunar–canalicular system, 142
LAP. *See* Latency-associated peptide (LAP)
Large latent complex (LLC), 1154
LARP6. *See* La ribonucleoprotein 6 gene (LARP6)
Lasofoxifene, 868, 868f, 874, 877, 884
Latency-associated peptide (LAP), 1154
Latent-TGFβ-binding protein (LTBP), 315, 368, 1154, 1169–1170
  in ECM, 1154
Lateral meningocele syndrome (LMS), 1098–1099
LBD. *See* Ligand-binding domain (LBD)
LC. *See* Locus coeruleus (LC)
LC–MS. *See* Liquid chromatography–mass spectrometry analysis (LC–MS)
LCN2. *See* Lipocalin-2 (LCN2)
LCN2/MC4R interaction, 1938
LD. *See* Linkage disequilibrium (LD)
LDL. *See* Low-density lipoprotein (LDL)
LDLR. *See* Low-density lipoprotein receptor (LDLR)
LDS. *See* Loeys–Dietz syndrome (LDS)
Lead, 528–531, 533
  bone densitometry in specimens, 530f
  β-catenin/sclerostin axis, 531
  clinical opportunity, 531
  low bone density, 530–531
  measurement in bone, 529
  mechanism of action, 531
  as unrecognized risk factor in osteoporosis, 529–530
LEFs. *See* Lymphoid enhancer factors (LEFs)
Lehman syndrome. *See* Lateral meningocele syndrome (LMS)
LepR. *See* Leptin receptor (LepR)
LepR-Cre. *See* Leptin receptor Cre (*LepR-Cre*)
*LEPRE*1 gene, 348–350
Leptin, 816, 1931–1932
  action on bone remodeling, 811–812
  negative regulation of bone remodeling by, 810
Leptin receptor (LepR), 92, 1164, 1893
Leptin receptor Cre (*LepR-Cre*), 1795–1796
Let-7, 586
Leucine, 1511
  leucine-rich proteoglycans, 360
Leucine-rich repeat (LRR), 360
  sequence proteins, 362–363
Leucine–phenylalanine–alanine–asparagine sequence (LFAN sequence), 180
Leukemia inhibitory factor (LIF), 97, 232, 778–779, 1220–1222, 1338–1339
Leukemia inhibitory factor receptor (LIFR), 1220, 1338–1339
  LIFR–binding cytokines, 1220–1224
    CT-1, 1222
    cytokines, 1222–1224
    LIF, 1221–1222
Levormeloxifene, 868, 868f
Levothyroxine, 1438
LFAN sequence. *See* Leucine–phenylalanine–alanine–asparagine sequence (LFAN sequence)
LFNG. *See* Lunatic fringe (LFNG)
LHs. *See* Lysyl hydroxylases (LHs)
LID. *See* Liver-specific Igf1 gene-deletion (LID)
LIF. *See* Leukemia inhibitory factor (LIF)
Life span, osteoclast differentiation and, 777–778
LIFR. *See* Leukemia inhibitory factor receptor (LIFR)
Ligament, 1789
Ligand-binding domain (LBD), 721, 827–828
Ligand(s)
  in chondrocytes, 1089–1092
  diversity and receptor environment, 1190–1191
  effects of ligand exposure, 1634–1635
  ligand-biased signaling, 543
  ligand-directed temporal bias, 701–702
  ligand-induced activation mechanism at PTHR1, 698–699
  ligand–receptor pairing, 1775
  in osteoblasts, 1092–1093
  in osteoclasts, 1095–1096
  in osteocytes, 1095
  recognition and activation by parathyroid hormone receptor, 695–699
  selectivity for receptor conformational state, 1635
Limb skeleton
  anterior–posterior axis, 14
  dorsal–ventral axis, 14
  limb development, 11–12, 12f
  mesenchymal condensation and patterning of skeleton, 14–15
  proximal–distal axis, 12–13
LINC. *See* Linker of nucleoskeleton and cytoskeleton (LINC)
Lineage tracing, 1887
  Cre recombinase, 1887–1888

during embryonic development, 1892–1893
experimental design for, 1889–1890
following injury, 1894–1895
in heterotopic ossification, 1895–1896
intersectional strategies to identify cells, 1891
osteoblast-to-chondrocyte transition, 1894
postnatal lineage tracing, 1893–1894
reporters, 1888–1889
of *Rosa*26 locus, 1796, 1797f
tamoxifen effects on bone, 1890–1891
techniques, 1061–1062
Tet expression systems, 1891
Linear beam theory, 1761–1762
Linear growth, 1000–1001
Linkage disequilibrium (LD), 1616
Linker of nucleoskeleton and cytoskeleton (LINC), 1773
Lipid
lipid-free ALP, 1572
peroxidation and declining innate immunity, 283–284
Lipocalin-2 (LCN2), 1931–1932
anorexigenic function of osteoblast-derived, 1938
regulation of food intake, 1939f
Lipoprotein receptor–related protein (LRP), 407–408, 426–427, 1340, 1637, 1711
LRP-1, 386
LRP4 gene, 186
LRP5 gene, 170, 177–178, 180–181, 279, 650–651, 778, 1561
LRP5^{G171V} mutation, 188
Lrp6 protein, 180–181, 279, 650–651, 1166, 1282–1283
sclerostin interaction with LRP4/5/6, 1714–1715
Lipoxins (LXs), 1073
Liquid chromatography–mass spectrometry analysis (LC–MS), 344, 344f
Liver-specific Igf1 gene-deletion (LID), 994–995
LLC. *See* Large latent complex (LLC)
LMS. *See* Lateral meningocele syndrome (LMS)
Loading cycles, 1765
Local glucocorticoid metabolism, 925
Local osteolysis, treatment of, 1321–1323
Local regulators of bone
additional TNF superfamily members, 1208–1209
effects of global, cell-specific, and pathway-specific gp130 modulation, 1224–1227
IL-1, 1206–1207
IL-4, 1213
IL-6, 1215–1218
cytokine family, 1214–1215
IL-7, 1210
IL-11, 1218–1219
IL-12, 1211
IL-13, 1213
IL-15, 1211
IL-17, 1211–1212
IL-18, 1212–1213
IL-23, 1212
IL-27Ra-binding cytokines, 1224
IL-32, 1213
IL-33, 1212–1213
Interferons, 1209
LIFR–binding cytokines, 1220–1224
MIF, 1213–1214
OSM, 1219–1220
TNF, 1207–1208
Localized osteolysis
adhesion and invasion into bone metastatic niches, 1338
bone microenvironment contribution to bone lesion progression, 1342–1345
cancer-induced bone disease, 1339–1342
osteotropism, 1337
tumor dormancy and awakening, 1338–1339
Locomotion inducing nonuniform strain environment, 1761–1762
Locus coeruleus (LC), 812–813
Loeys–Dietz syndrome (LDS), 1169–1170
Long-coding RNAs, 1070
Loss of function, 1790
evidence from rare human variants, 972–974, 973f
using genetically modified mice, 976–977
Loss of sex steroids, 282–283
Low bone mass, animal models of, 1723–1724
Low temperature methylmethacrylate embedding, 1902
Low-bone-mass phenotype, 281
Low-density lipoprotein (LDL), 279, 386
low-density lipoprotein–related protein 5, 1776
Low-density lipoprotein receptor (LDLR), 177–178
Low-level mechanical signals, 1769
increasing bone quantity and strength, 1767
mitigating bone loss to cancer, 1769
normalizing bone formation, 1767–1768
Low-level vibrations, 1768–1769
Low-magnitude, high-frequency mechanical signals, 1766–1771, 1767f
anabolic to musculoskeletal system, 1769–1770
bone sense, 1770–1771
genetic variations modulate bone's response to mechanical signals, 1768
increasing bone quantity and strength, 1767
inhibition of postmenopausal bone loss by low-level vibrations, 1768–1769
normalizing bone formation, 1767–1768
vibrations, 1768
Low-magnitude whole-body vibration treatment, 1772
Low-mass accelerometer, 1763
Lower circulating insulin-like growth factor 1, 952–953
Lower surface area/matrix volume, 245–246
LOX. *See* Lysyl oxidase (LOX)
LOX-like (LOXL), 348
LoxP sites, 1790
LP. *See* Lysyl pyridinoline (LP)
LRP. *See* Lipoprotein receptor–related protein (LRP)
LRR. *See* Leucine-rich repeat (LRR)
LS. *See* Lumbar spine (LS)
LTBP. *See* Latent-TGFβ-binding protein (LTBP)
LTBP1–4. *See* Four LTBPs (LTBP1–4)
Luciferase transgenes, 310
Ludwigshafen Risk and Cardiovascular Health (LURIC), 764
Lumbar spine (LS), 1617
Lumbar vertebrae, 1722
Lunatic fringe (LFNG), 1085–1086
Lung, CaSR in, 556
LXs. *See* Lipoxins (LXs)
Lymphoid enhancer factors (LEFs), 121, 725–726, 1161
Lysine, 715
LysM-Cre model, 828–829
Lysosomal proteinases, 390
Lysosomes, 112
Lysyl hydroxylases (LHs), 347
consequences of LH gene mutations, 347
Lysyl oxidase (LOX), 339, 341f, 348
Lysyl pyridinoline (LP), 340
Lysyloxidase activity, 316

M

M cells. *See* Microfold cells (M cells)
M-CSF. *See* Macrophage colony-stimulating factor (M-CSF)
mAChRs. *See* Muscarinic acetylcholine receptors (mAChRs)
Macroarchitecture of bone, 1834
Macroimaging
CT, 1863–1871
DXA beyond bone mineral density, 1871–1873
quantitative muscle imaging, 1877–1878
radiography, 1858–1860
standard DXA, 1860–1862
Macrophage colony-stimulating factor (M-CSF), 114–115, 185, 222, 650, 777, 1147
Macrophage inflammatory protein-1α (MIP-1α), 1323
Macrophage migration inhibitory factor (MIF), 1213–1214
Macrophages, 78, 233–234, 1220, 1342
Mad homology 1 (MH1), 1159
Mad homology 2 (MH2), 1154–1155
MAF. *See* Minor allele frequency (MAF)
MafB gene, 579
Maffucci syndrome, 1392–1393
Magnesium (Mg^{2+}), 509
homeostasis, 509
acquired hypomagnesemia, 518–519
hereditary disorders, 512–518
intestinal Mg^{2+} absorption in health and disease, 510f
physiology, 510–512
reabsorption, 511f

Magnesium (Mg^{2+}) (*Continued*)
renal tubular magnesium reabsorption, 511f
Magnetic resonance (MR), 1858
Magnetic resonance imaging (MRI), 517, 1296, 1301, 1858, 1873–1877
MAGP. *See* Microfibril-associated glycoproteins (MAGP)
MAH. *See* Malignancy-associated hypercalcemia (MAH)
Male osteoporosis, 1695
Male skeletal health, 979
Malignancy
characterization of hypercalcemia in, 1308
eutopic PTH-related protein overproduction in, 1309–1312
experimental approaches to controlling overproduction or overactivity, 1321–1322
RANKL in, 1201
resistance to antiresorptive agents in malignancy-associated hypercalcemia, 1320
Malignancy-associated hypercalcemia (MAH), 1307–1308
Malignant osteopetrosis, 1553–1554
MAML. *See* Mastermind-like (MAML)
Mammalian cranium, 6
Mammalian target of rapamycin (mTOR), 187–188, 311–312, 588–589, 1300
Mammary CaSR, 552–553
Mammary ductal, 1199
Mammary gland, 842–849, 873–876. *See also* Parathyroid glands
adolescent mammary development, 844
development, 1199–1201
embryonic mammary development, 842–844
epithelium, 1199–1200
pathophysiology of PTHrP in, 848–849
pregnancy and lactation, 844–848
Mammary mesenchyme, 842
Manhattan Project, 527–528
Manic fringe (MFNG), 1085–1086
MAOA. *See* Monoamine oxidase A (MAOA)
MAPK signaling, 1775–1776
MAPKs. *See* Mitogen-activated protein kinases (MAPKs)
MAR. *See* Mineral apposition rate (MAR)
Marfan syndrome (MFS), 1169–1170
Marfan's disease, 1493–1494
Marker genes, 1788–1789
Markers, 56
Marrow adipocytes, 75–76, 277, 1637
Marrow adipose tissue volume, 1760
Marrow perfusion, 1877
Marrow stromal cells (MSCs), 1219
MARRS. *See* Membrane-associated rapid response steroid (MARRS)
MAS. *See* McCune–Albright syndrome (MAS)
Mass spectrum (MS), 1808
Masson–Goldner stain, 1904
Mastermind-like (MAML), 1088–1089
Mastication, 1064–1065
Mathematical PD models of BPs, 1677
Mathematical PK model, 1680
"Matricellular" protein, 365
Matrilin-1, 15
Matrix
deformation–dependent pathway, 1771
matrix-derived resorption products as coupling factors, 227–228
mineral density, 268
mineralization, 1692–1693
Matrix extracellular phosphoglycoprotein (MEPE), 139, 184, 369–370, 477, 1715–1716
Matrix Gla protein (MGP), 370–371
Matrix metalloproteinases (MMPs), 302, 367, 379–381, 652, 993–994, 1154–1155, 1802
MMP-1, 1319–1320
MMP-9, 113, 206, 379, 1339
MMP-13, 234–235, 384–387, 778
MMP-14, 143–144
Matrix vesicles (MVs), 474, 1571
Maturation products, 344–345
Mature "bone-forming" osteoblasts, 93–94
Mature calcitonin, 789–790
Mature cross-link analysis, 340
Mature osteoblast markers, 1893–1894
Mature osteoclasts, 1226
Mature osteocytes, 134
Mature skeleton
innervation
ANS, 212–213
somatic nervous system, 211–212
vascularization, 208–210
bone cells' control of skeletal vascularization and oxygenation, 208–210
endothelial and angiocrine signaling in bone, 210
Maturity-onset diabetes of the young (MODY5), 517
Mauriac syndrome, 942–943
MBD4. *See* Methyl-CpG-binding protein (MBD4)
MBF. *See* Modeling-based bone formation (MBF)
MBPTS2. *See* Membrane-bound transcription factor peptidase, site 2 (MBPTS2)
Mc4R gene. *See* Melanocortin receptor 4 gene (*Mc4R* gene)
McCune–Albright syndrome (MAS), 58–60, 485, 1431, 1443–1451, 1817
clinical features, 1443–1444
diagnosis and management, 1447–1451
genetics, 1445
pathogenesis, 1445–1447
MCP-1. *See* Monocyte chemotactic protein-1 (MCP-1)
MDSCs. *See* Myeloid-derived suppressor cells (MDSCs)
Measles virus (MV), 1603–1604
Mechanical forces, 53–54
Mechanical signals, biochemical modulation of, 1771–1777
Mechanical stimuli, bone morphology regulation by, 1764–1766
Mechanical testing methods
categories, 1924f
for rodent skeleton, 1923
Mechanically activated intracellular signaling, 1775–1777
activation of Wnt/catenin signals, 1776
MAPK signaling, 1775–1776
nitric oxide signaling, 1776–1777
prostaglandins, 1777
Mechanoreceptors in bone cells, 1773
channels, 1774
connexins, 1774
integrins and integrin-associated proteins, 1773–1774
membrane structure, 1774–1775
nuclear connectivity, 1775
primary cilium, 1775
Mechanoresponsive osteocytes, 149
Mechanosensation, 148
Mechanosensitive channels, 1774
Mechanosensory cells, osteocytes as, 145–146
Mechanotransduction, 151, 1771–1772
Medullary cavity
diameter, 249–250
void volume, 245
MEF2C. *See* Myocyte enhancer factor 2C (MEF2C)
MEFs. *See* Mouse embryonic fibroblasts (MEFs)
Megakaryocytes, 78
Megalin/cubilin transport system, 715–716
Melanocortin receptor 4 gene (*Mc4R* gene), 814, 818
bone resorption regulation by, 814–815
Melanoma, 1342
Melanoma inhibitory activity member 3 (MIA3), 300
MELAS syndrome, 1362
Membrane
deformation, 1773
structure, 1774–1775
Membrane-associated rapid response steroid (MARRS), 713–714
MARRS-binding protein, 726
MARRS/ERp57/PIA3, 726–727
Membrane-bound coupling factors, 230–231
Membrane-bound glutathione-dependent PGES (mPGES-1), 1251
Membrane-bound proteases, 303
Membrane-bound transcription factor peptidase, site 2 (MBPTS2), 351
Membrane-spanning proteins, 1774
Membrane-type matrix metalloproteinases (MT-MMPs), 382–383
MEN. *See* Multiple endocrine neoplasia (MEN)
Men with prostate cancer, androgen deprivation in, 1696
Men1 gene, 1303–1304, 1411–1412

MEN1. *See* Multiple endocrine neoplasia type 1 (MEN1)
MEN2A syndrome, 1412
Menin, 1303, 1410
Meningioma-1, 724–725
Menopause, 256–257, 275, 827, 877–878
  menopausal bone loss, 1652
  reversible and irreversible bone loss and microstructural deterioration, 258–259
MEPE. *See* Matrix extracellular phosphoglycoprotein (MEPE)
Mesenchymal cells, 5, 277, 310, 1212
Mesenchymal condensation, 14–15
Mesenchymal stem/stromal cells (MSCs), 51, 75, 91, 276–277, 401–402, 918, 996, 1062, 1158, 1162–1164, 1165f, 1337, 1474, 1715–1716, 1759–1760
Mesenchymal–epithelial transition (MET), 1162
Meso Scale Discovery (MSD), 1806
Mesoderm-derived progenitor cells, 1062
Messenger RNAs (mRNAs), 1809
MET. *See* Mesenchymal–epithelial transition (MET)
Meta-Analysis Geneset Enrichment of variaNT Associations (MAGENTA), 1621
Metabolic control of glucose and insulin, 951–952
Metabolic diseases, 956
Metabolic or mitochondrial hypomagnesemia (MIM), 518
Metabolic phosphorus, 470–471
Metabolic syndrome, 877–878
Metabolomic signature, 1808
Metabotropic glutamate receptors (mGluRs), 540
Metal ion toxicity in skeleton, 527
  aluminum, 531–534
  lead, 528–531
  research into bone-seeking toxic elements, 527–528
Metastatic bone disease, 1698, 1820
  from solid tumors, 1819
Metastatic carcinoma of breast and prostate, 1099–1100
Metazoans, 1083
Metformin, 950–951
Methyl-CpG-binding protein (MBD4), 725
Methylases, 724
Methylmethacrylate (MMA), 1901
  embedding
    low temperature, 1902
    standard, 1901
MFD. *See* Monostotic fibrous dysplasia (MFD)
MFNG. *See* Manic fringe (MFNG)
MFS. *See* Marfan syndrome (MFS)
Mg^{2+} loading test (MLT), 512
MGI. *See* Mouse genome informatics (MGI)
mGluRs. *See* Metabotropic glutamate receptors (mGluRs)
MGP. *See* Matrix Gla protein (MGP)
MH1. *See* Mad homology 1 (MH1)
MIA3. *See* Melanoma inhibitory activity member 3 (MIA3)
MIB. *See* Mind bomb (MIB)
Mice, 1914–1915
  genetic manipulation of sclerostin expression in, 1716–1717
  and patient with gp130 signaling mutations, 1225
Micro-computed tomography (Micro-CT), 267, 276, 1767, 1834–1836, 1835f–1836f, 1857, 1925
Micro-FE model. *See* Microstructural FE model (Micro-FE model)
Micro-MRI. *See* Micro–magnetic resonance imaging (Micro-MRI)
Microarchitecture of bone, 1834
Microautophagy, 281
Microcracks, 247, 1910–1912
Microdamage, 1910–1912
  measurement technique, 1910–1912
  times for *en bloc* staining of different samples, 1911t
Microfibril-associated glycoproteins (MAGP), 1086
Microfold cells (M cells), 1201
Microimaging
  biomechanical imaging of bone competence, 1837–1838
  hierarchical imaging of bone microarchitecture, 1835–1836
  quantitative image processing, 1838–1841
  in vitro, 1834–1838
  in vivo, 1841–1849
Micro–magnetic resonance imaging (Micro-MRI), 1475–1478, 1842
Micromilled cross sections of cortical bone, 1903
Micropetrosis, 275
MicroRNAs (miRNAs), 26–27, 53, 171–172, 184, 314, 547, 585, 722, 922, 1303–1304, 1809–1810
  in parathyroid, 585–586, 586f
Microscale and nanoscale bone material assessment, 1927–1928
  cRPI, 1927–1928
  nanoindentation, 1928
Microsomes, 1744
Microstructural deterioration, 256–257, 266–267
  reversible and irreversible bone loss and, 258–259
Microstructural FE model (Micro-FE model), 1841
Microtome sectioning, 1902–1903
MIF. *See* Macrophage migration inhibitory factor (MIF)
Mild nondeforming OI, 1491
Mild osteopetrosis, 1554
MIM. *See* Metabolic or mitochondrial hypomagnesemia (MIM)
Mind bomb (MIB), 1087
Mineral apposition rate (MAR), 449, 977–978, 1899–1900
Mineral deposition, 352–353
Mineral homeostasis, 1580–1581
Mineral metabolism and periodontium, 1070–1073, 1072f
Mineral-ion homeostasis, PTH actions on, 659–660
Mineralization, 94, 1722
  density, 246
  inhibitors, 94
  of osteocyte lacunae, 275
Mineralized bone matrix, 447
  volume, 245–246
Mineralized lesions, 1714
Mineralized matrix
  mass, 249
  mineralized matrix-associated cells, 1714
  volume, 263–267
Mineralizing surface (MS), 449, 1767–1768, 1899–1900
Minor allele frequency (MAF), 1617
MIP-1α. *See* Macrophage inflammatory protein-1α (MIP-1α)
miR-124, 171–172
miR-27b, 722
miR-29 expression, 184
miR-298, 722
miRNA-148, 586
miRNAs. *See* MicroRNAs (miRNAs)
miRs. *See* MicroRNAs (miRNAs)
Mitochondrial disorders associated with hypoparathyroidism, 1362
Mitochondrial dysfunction, 278–279
Mitochondrial permeability transition pore (MPTP), 478
Mitochondrial trifunctional protein (MTP), 1362
  deficiency, 1362
Mitogen-activated protein kinases (MAPKs), 118, 401–402, 545, 587, 646, 898, 989–990, 1311, 1773–1774
  MAPK1, 922–923
Mitogenesis, 1718–1719
Mixed lineage leukemia (MLL), 1410
MM. *See* Multiple myeloma (MM)
MMA. *See* Methylmethacrylate (MMA)
MMPIs. *See* Inhibitors of MMPs (MMPIs)
MMPs. *See* Matrix metalloproteinases (MMPs)
Modeling-based bone formation (MBF), 266–267, 1635–1636, 1719–1722
MODY5. *See* Maturity-onset diabetes of the young (MODY5)
Molecular mechanisms, 1689–1690
  of action of vitamin D compound, 1748
  of aging, 278–282
    cellular senescence, 280–281
    loss of autophagy, 281–282
    mitochondrial dysfunction, 278–279
  of vessel remodeling, 312
Molecular oncology, 1405–1406
Molecular sensor in ECM, 1153–1164
Monoamine oxidase A (MAOA), 1345
Monoclonal antibodies, 1711
Monocyte chemotactic protein-1 (MCP-1), 778, 1163

Monocyte chemotactic protein-3 (MCP-3), 137
Monocyte–macrophage precursor cells, 256
Monomeric ligand-receptor complex, 916
Monostotic fibrous dysplasia (MFD), 1443
Mouse embryonic fibroblasts (MEFs), 317
Mouse genome informatics (MGI), 1090t–1092t
Mouse models, 304, 348–352
  for ADH1 and ADH2, 1370
  bone mineralization, 352
  collagen chaperone, 351
  collagen posttranslational modifications, 348–351
  collagen processing, 351
  noncollagen genes, 349t
Mouse pro-Col1a1 proximal promoter, 305–306
Mouse pro-Col1a2 gene, 310–311
  delineating mode of action of tissue-specific elements, 311–312
mPGES-1. *See* Membrane-bound glutathione-dependent PGES (mPGES-1)
MPTP. *See* Mitochondrial permeability transition pore (MPTP)
MR. *See* Magnetic resonance (MR)
MRI. *See* Magnetic resonance imaging (MRI)
mRNAs. *See* Messenger RNAs (mRNAs)
MrOS study. *See* Osteoporotic Fractures in Men study (MrOS study)
MS. *See* Mass spectrum (MS); Mineralizing surface (MS); Multiple sclerosis (MS)
MSCs. *See* Marrow stromal cells (MSCs); Mesenchymal stem/stromal cells (MSCs)
MSD. *See* Meso Scale Discovery (MSD)
MT-MMPs. *See* Membrane-type matrix metalloproteinases (MT-MMPs)
mTOR. *See* Mammalian target of rapamycin (mTOR)
mTOR complex 1 (mTORC1), 588–589, 588f
mTOR complex 2 (mTORC2), 588–589
MTP. *See* Mitochondrial trifunctional protein (MTP)
Multiple endocrine neoplasia (MEN), 1405, 1659
Multiple endocrine neoplasia type 1 (MEN1), 1293, 1409–1412
  anterior pituitary tumors, 1300–1301
  characteristics and frequency of NETs in, 1294t
  diagnosis and clinical surveillance of NETs in, 1303t
  genetics and molecular biology, 1302–1304
  GEP-NETs, 1295–1300
  MEN1-associated angiofibroma, 1295f
  PHPT, 1293–1295
  thymic and bronchial neuroendocrine tumors, 1301
Multiple epidermal growth factor-like domain 7, 1712–1714
Multiple myeloma (MM), 1099, 1323, 1342, 1697–1698, 1805, 1818–1819
Multiple sclerosis (MS), 758–759
Multiple vertebral fractures, 1701
Multiplex
  ligation probe amplification, 1303
  reporter approach, 1888–1889
Murine
  collagenase-2, 383
  models, 607
    of multiple myeloma, 1769
  studies, 1004–1005
Muscarinic acetylcholine receptors (mAChRs), 212
Muscarinic receptor 3 (AchRM3), 813
Muscle
  on bone's strain environment, 1762–1763
  lineage, 1895–1896
  muscle-induced strains, 1762
  muscle–bone interface and potential impact on bone strength, 978–979
  performance and balance, 759–760
Musculoskeletal
  findings, 489
  function, 989
  pain, 1701–1702
  pathologies with aberrant TGF-β signaling, 1167–1172
    disorders associated with genetic mutations in TGF-β signaling components, 1168–1170
    OA associated with aberrant activation of TGF-β signaling, 1167–1168
    skeletal metastases of cancer associated with bone matrix–derived TGF-β, 1170–1172
  system, 1164, 1769–1770
Mutations, 488
  in FGFR3 and FGF9, 1119
MV. *See* Measles virus (MV)
MV nucleocapsid protein (MVNP), 1605–1607
MVs. *See* Matrix vesicles (MVs)
Myeloid cells, 1342
Myeloid-derived suppressor cells (MDSCs), 1342
Myelophthisic anemia, 1575
Myocardial cell hypertrophy, 1740
Myocyte enhancer factor 2C (MEF2C), 387, 1282
Myostatin, 1493–1494

N

N-linked glycosylation, 368–370
N-terminal propeptide, 298–299
N-terminal telopeptide of type I collagen (NTX), 1802
N.Oc./T.Ar. *See* Oc.Ns expressed per tissue area (N.Oc./T.Ar)
Na^+/Ca^{2+} exchanger (NCX1), 1279
αNAC. *See* Nascent polypeptide-associated complex and coregulator alpha (αNAC)
NAcgal. *See* N-acetylgalactosamine (NAcgal)
nAChRs. *See* Nicotinic acetylcholine receptors (nAChRs)
Nafoxidine, 868, 868f
Nano-CT, 1835–1836, 1835f
Nano-technology-based carriers, 1675
Nanoindentation, 1928
Nascent polypeptide-associated complex and coregulator alpha (αNAC), 1282–1283
National Center for Biotechnology Information (NCBI), 488
National Institute of Standards and Technology (NIST), 766
National Institutes of Health (NIH), 762–763
National Organization for Rare Disorders (NORD), 1490
National Osteoporosis Foundation, 1759–1760
Natural killer cells (NK cells), 1211
Natural metabolites, 1734–1735
nCaRE. *See* Negative calcium regulatory element (nCaRE)
NCBI. *See* National Center for Biotechnology Information (NCBI)
NCCs. *See* Neural crest cells (NCCs)
NCoA62/SKIP, 724–725
NCoR complexes, 725
NCX1. *See* Na^+/Ca^{2+} exchanger (NCX1)
Nebulette gene (NEBL gene), 1357–1358
Negative allosteric modulators, 542
Negative calcium regulatory element (nCaRE), 582
Negative regulatory region (NRR), 1084
NEMO. *See* NF-kB essential modulator (*NEMO*)
Neoangiogenesis, 1142–1143
Neonatal severe hyperparathyroidism (NSHPT), 513–514, 539–540, 1405, 1660
Neoplastic cells, 1405–1406
Nephric duct, 1360
Nephroblastoma overexpressed (NOV), 1086
Nephrocalcinosis, 490, 1743
Nephrogenous cAMP, 660–661
Nephrolithiasis, 490
  to SLC34A1, 490
Nerve growth factor (NGF), 210–211
Nerve system of bone
  innervation of developing bone, 210–211
  innervation of mature skeleton, 211–213
Nes-GFP transgene. *See* Nestin-GFP transgene (Nes-GFP transgene)
NESP55, 1434, 1440
Nestin, 47
Nestin-GFP transgene (Nes-GFP transgene), 1164
$Nestin–GFP^+$ cells, 75
NETs. *See* Neuroendocrine tumors (NETs)
Neural crest cells (NCCs), 5, 1146
Neural/glial antigen 2 (NG2), 75
Neurocranium, 6

Neuroectodermal stem cell marker, 1718
Neuroendocrine tumors (NETs), 1293
Neurofibromatosis type 1 (NF1), 1169–1170
Neuroglycopenic syndrome, 1297–1299
Neuromediators, 809
NeuromedinU (NMU), 813, 817
Neuronal regulation of hematopoietic stem cell niche, 79
Neuropeptide Y (NPY), 814–815, 818
Neuropilin (NRP), 1144
Neuropoietin (NP), 1223–1224
Neuropsychiatric disorders, 817
Neutral endopeptidase, 478
Neutrophils, 78
  recruitment and migration with complement 5a binding, 719
Next generation sequencing (NGS), 1624–1625
NF GEP-NETs. *See* Nonfunctioning GEPNETs (NF GEP-NETs)
NF tumors. *See* Nonfunctioning tumors (NF tumors)
NF-kB essential modulator (*NEMO*), 1560–1561
NF-Y. *See* Nuclear factor-Y (NF-Y)
NF-κB. *See* Nuclear factor κB (NF-κB)
NF1. *See* Neurofibromatosis type 1 (NF1)
NFAT. *See* Nuclear factor of activated T cells (NFAT)
NFATc. *See* Nuclear factor of activated T cells (NFAT)
NG2. *See* Neural/glial antigen 2 (NG2)
NGF. *See* Nerve growth factor (NGF)
NGS. *See* Next generation sequencing (NGS)
NHERFs. *See* Sodium/hydrogen exchanger regulatory factors (NHERFs)
NICD. *See* Notch intracellular domain (NICD)
Niche concept, 46, 74
  adipocytes, 75–76
  adult bone marrow niche, 75–79
  endothelial cells, 77
  hematopoietic cells, 77–79
  hematopoietic stem cell niche
    hormonal regulation of, 79–80
    neuronal regulation of, 79
  HSCs microenvironments in embryo and perinatal period, 74
  IGF1, 80
  MSCs, 75
  niche heterogeneity for heterogeneous HSCs and progenitor cells, 80–81
  osteoblastic cells, 76–77
  parathyroid hormone, 79–80
Nicotinic acetylcholine receptors (nAChRs), 212
NIH. *See* National Institutes of Health (NIH)
NIST. *See* National Institute of Standards and Technology (NIST)
Nitric oxide (NO), 142–143, 150–151, 555, 898, 1773
  signaling, 1776–1777
Nitric oxide synthase (NOS), 629, 1774
NK cells. *See* Natural killer cells (NK cells)
NLS. *See* Nuclear localization sequence (NLS)
NMR spectroscopy. *See* Nuclear magnetic resonance spectroscopy (NMR spectroscopy)
NMSC. *See* Nonmelanoma skin cancer (NMSC)
NMU. *See* NeuromedinU (NMU)
NO synthase (Nos2), 412–413
NOD mice. *See* Nonobese diabetic mice (NOD mice)
Non-small-cell lung cancer group, 1697
Non-threshold-based image analysis, 260
Noncaveolar organized membrane, 1774–1775
Noncollagenous matrix proteins, 1645
Noncollagenous proteins, 89
Nondiabetic animal models, 942
Nondrug strategy, 1760
Nonendothelial effects of VEGF, 1146–1147
Nonfunctioning GEPNETs (NF GEP-NETs), 1295
Nonfunctioning neuroendocrine tumors of gastroenteropancreatic tract, 1299–1300
Nonfunctioning tumors (NF tumors), 1293
Nongenomic actions of thyroid hormones, 898, 904–905
Nonmelanoma skin cancer (NMSC), 763–764
Nonobese diabetic mice (NOD mice), 943–944
Nonpeptide mimetic ligands for PTHR1, 704–705
Nonsense-mediated decay, 1491
Nonskeletal "MSCs", 51
Nonskeletal effects of vitamin D
  cancer, 761–764
  cardiovascular disease, 764
  clinical studies of muscle performance and balance, 759–760
  falls, 760
  issues in existing data and paths forward to resolving vitamin D deficiency role, 764–767
    25(OH)D measurement, 765–766
    clinical and preclinical studies, 766–767
    vitamin D status assessment, 765
  physiology, 759
  vitamin D and immunity, 758–759
Nonspecific glycations, 345–346
Nonspecific metabolism, 1746
Nonsteroidal antiinflammatory drugs (NSAIDs), 1248
  skeletal response to, 1259–1260
Nonvertebral fracture risk, 246
19-nor progestin R5020, 586
NORD. *See* National Organization for Rare Disorders (NORD)
Normal B-cell physiology, 1201
Normal bone, 449
Normative data, 1848
Normophosphatemic disorders of cellular phosphorus metabolism, 492
NOS. *See* Nitric oxide synthase (NOS)
Notch action mechanisms
  in endochondral bone formation, 1092
  in osteoblasts, 1093, 1094f
  in osteocytes, 1095
Notch activation mechanisms
  formation of active transcriptional complex, 1088–1089
  generation of notch intracellular domain, 1088
  Notch target genes, 1089
Notch and skeletal malignancies
  metastatic carcinoma of breast and prostate, 1099–1100
  multiple myeloma, 1099
  osteosarcoma, 1099
Notch cognate ligands
  regulatory mechanisms, 1087
  structure and function, 1086
Notch intracellular domain (NICD), 1083, 1088
Notch receptors, 1085f
  function, 1084–1085
  and ligands
    in chondrocytes, 1089–1092
    in osteoblasts, 1092–1093
    in osteoclasts, 1095–1096
    in osteocytes, 1095
  regulatory mechanisms, 1085–1086
  structure, 1084
Notch signaling, 25, 52, 141–142, 624, 1084
  in endochondral bone formation, 1089, 1090t–1092t
  fracture repair and, 1096
  in osteoblasts, 1092–1093
  in osteocytes, 1095
  pathway, 1083
Notch target genes, 1089
*NOTCH*1 mutations, 631
NOV. *See* Nephroblastoma overexpressed (NOV)
Novel image analysis approaches, 1837
NP. *See* Neuropoietin (NP)
NPS-2143, 1370
NPT2a, 664
NPT2c, 664
  human mutations in, 486
NPY. *See* Neuropeptide Y (NPY)
NPY2R gene, 818
NR. *See* Nuclear receptor (NR)
NRP. *See* Neuropilin (NRP)
NRR. *See* Negative regulatory region (NRR)
NS-398 gene, 1253
NSAIDs. *See* Nonsteroidal antiinflammatory drugs (NSAIDs)
NSHPT. *See* Neonatal severe hyperparathyroidism (NSHPT)
NTP-PPi-ase. *See* Nucleoside triphosphate pyrophosphatase (NTP-PPi-ase)
NTS. *See* Nucleus of tractus solitarius (NTS)

NTX. *See* N-terminal telopeptide of type I collagen (NTX)
Nuclear actions, 600–602
of thyroid hormones, 895–898
Nuclear connectivity, 1775
Nuclear factor of activated T cells (NFAT), 24, 310, 1093
inhibitors, 168
NFATc1, 118, 405, 1607
Nuclear factor κB (NF-κB), 118–119, 277
Nuclear factor-Y (NF-Y), 576
Nuclear flecks (Nuf), 1370
Nuclear hormone receptors, 832
Nuclear localization sequence (NLS), 1084, 1303, 1313–1314
Nuclear magnetic resonance spectroscopy (NMR spectroscopy), 1714
Nuclear receptor (NR), 895–896
Nucleoside triphosphate pyrophosphatase (NTP-PPi-ase), 1589
Nucleus of tractus solitarius (NTS), 814

O

O-linked *N*-acetylglucosamine transferase (O-GlcNAc), 1097
O.Th. *See* Osteoid thickness (O.Th)
OA. *See* Osteoarthritis (OA)
OBCD project. *See* Origins of Bone and Cartilage Disease project (OBCD project)
Obesity, 954, 1439–1440
OBs. *See* Osteoblasts (OBs)
OC. *See* Osteocalcin (OC)
OC-STAMP. *See* Osteoclast-specific transmembrane protein (OC-STAMP)
Oc.Ns expressed per tissue area (N.Oc./T.Ar), 1907
OCLs. *See* Osteoclasts (OCLs)
Ocn-Cre model. *See* Osteocalcin-Cre model (Ocn-Cre model)
Ocn$^{-/-}$mice, 1935–1936
Ocn$^{-/-}$mice. *See* Osteocalcin-deficient mice (Ocn$^{-/-}$mice)
OCPs. *See* Oral contraceptive pills (OCPs)
OCT. *See* 22-Oxacalcitriol (OCT)
OctreoScan with single-photon emission CT (OctreoScan SPECT/CT), 1544–1546
Oculodentodigital dysplasia (ODDD), 429–430
Odanacatib, 1821
ODDD. *See* Oculodentodigital dysplasia (ODDD)
ODF. *See* Osteoclast differentiation factor (ODF)
odonto-HPP. *See* Odontohypophosphatasia (odonto-HPP)
Odontohypophosphatasia (odonto-HPP), 1574–1575, 1580
OF45. *See* Osteoblast/osteocyte factor 45 (OF45)
1,25-$(OH)_2$-vitamin D synthesis by CYP27B1 expression, 1277–1278
4-OHT. *See* 4-Hydroxytamoxifen (4-OHT)
OI. *See* Osteogenesis imperfecta (OI)
OIF. *See* Osteogenesis Imperfecta Foundation (OIF)
OK cells. *See* Opossum kidney cells (OK cells)
OL-HED-ID. *See* Osteopetrosis, lymphedema, hypohidrotic ectodermal dysplasia and immunodeficiency (OL-HED-ID)
Ollier's disease, 1392–1393
Oncogenic osteomalacia. *See* Tumor-induced osteomalacia (TIO)
Oncostatin M (OSM), 97, 232–233, 1219–1220, 1338–1339
Oncostatin M receptor (OSMR), 1214
OP. *See* Osteopetrosis (OP)
OPBT8 osteopetrosis, 1559
OPG. *See* Osteoprotegerin (OPG)
OPG ligand (OPGL), 1690
Opioids, 809
OPN. *See* Osteopontin (OPN)
Opossum kidney cells (OK cells), 661
Opportunistic screening, 1857, 1871
OPTB4, 1558–1559
Optical trap device, 143
Optineurin (OPTN), 1607, 1609
Oral contraceptive pills (OCPs), 1006
Orchidectomy (ORX), 1915
Orexins, 1931–1932
signaling, 813–814
Origins of Bone and Cartilage Disease project (OBCD project), 1621–1622, 1788
Orthodontic tooth movement, 1258
ORX. *See* Orchidectomy (ORX)
OS. *See* Osteoid surface (OS); Osteosarcoma (OS)
OSF-1. *See* Osteoblast stimulating factor-1 (OSF-1)
OSM. *See* Oncostatin M (OSM)
OSMR. *See* Oncostatin M receptor (OSMR)
Osteitis fibrosa, 777
Osteo-angiogenic coupling, 213
Osteoactivin, 412
Osteoadherin, 362
Osteoanabolic agents in osteoporosis, 1637–1639
abaloparatide, 1638–1639
teriparatide, 1637–1638
Osteoanabolic responses, 426–428
Osteoanabolic therapies, 456
Osteoarthritis (OA), 280, 1163, 1194–1195, 1675, 1699
associated with aberrant activation of TGF-β signaling, 1167–1168
TGF-β modulation as promising approach to OA management, 1172–1173
Osteoblast lineage, 76–77, 1207–1208, 1718–1719
actions during bone remodeling sequence, 99
to bone structure, 1225–1226
bone-lining cells, 94–95
cells, 1220
osteoblast lineage cells—sensing surface and signaling, 232–233
commitment of osteoblast progenitors, 91–92
communication between different stages of differentiation
IL-6 cytokines, 97
PTHrP/PTHR1, 97
differentiation process, 95–96
example of contact-dependent communication, 96–97
lessons in osteoblast biology from Wnt signaling pathway, 99–101
mature "bone-forming" osteoblasts, 93–94
mesenchymal precursors, 90–91
osteocytes, 95
from paracrinology to endocrinology in bone, 101–102
physical sensing and signaling by osteoblasts and osteocytes, 98
promoting osteoclast formation, 98–99
stages, 90–95, 90f
stages of development, 96–98
Osteoblast stimulating factor-1 (OSF-1), 136
Osteoblast/osteocyte factor 45 (OF45), 139
Osteoblastogenesis, 26–27, 184, 256, 296–297, 426–428
endogenous glucocorticoids promoting, 918–919
Osteoblasts (OBs), 5, 15, 89, 91–92, 320, 401–404, 402f, 406–408, 408f, 776, 900, 996–997, 1213, 1335–1336, 1343, 1602–1603, 1637, 1772, 1795–1796
biology from Wnt signaling pathway, 99–101
bone modeling stimulation by, 782
calcitonin effects, 796
amylin, 797–798
CGRP, 796–797
cells, 6, 76–77, 124, 1069–1070
coculture system, 114
glucocorticoids preventing osteoblast cell cycle progression, 922–923
M-CSF, 114–115
osteoclastogenesis supported by RANKL, 116
osteoprotegerin and RANKL, 115–116
role in osteoclastogenesis, 114–116
commitment of, 91–92
differentiation
additional transcriptional regulators of, 165
and function, 183–185
regulation by transcription factors, 171–172
Runx2 functions during skeletogenesis beyond, 165
Runx2 gene in, 163–165
FGFand FGFR signaling in, 1119–1121
function(s), 402–403
additional transcriptional regulators of, 165

ATF4 as transcriptional regulator of, 168–169
glucocorticoids inducing osteoblast apoptosis, 923
matrix formation, 1900
molecular mechanisms of action in, 648–651
mutations control level of differentiation, 1492
Notch receptors and ligands in
mechanisms of Notch action in, 1093
role of Notch signaling in, 1092–1093
OS cells, 608
osteoblast-like cells, 1698
osteoblast-specific element, 308–309
OSF2, 306
osteoblast-specific transcription factors, 164
osteoblast-to-chondrocyte transition, 1894
osteoblast-to-osteocyte transition, 1714
phase, 1602
production in osteoblastic cultures, 1252–1253
progenitors, 91, 277
transcription factors acting downstream of Wnt signaling in, 170–171
type I collagen, 256
vigor, 1718
Osteocalcin (OC), 89, 101–102, 141–142, 164–166, 168–169, 208, 371, 741–742, 943–944, 1001–1002, 1892, 1931–1934
bone as regulator of appetite, 1938
decarboxylation, 952
and activation during bone resorption, 1935
endocrine function regulation by γ-carboxylation, 1934–1935
gene, 652
GPRC6A osteocalcin receptor in β cells, 1934
modulating adaptation to exercise, 1937
Oc-CreER, 1893–1894
regulation
of activity, 1933f
by proprotein convertase furin, 1935
of skeletal muscle energy metabolism during exercise, 1935–1937, 1936f
role in adaptation to exercise, 1935
Osteocalcin-Cre model (Ocn-Cre model), 829
Osteocalcin-deficient mice (Ocn$^{-/-}$mice), 1932–1933
Osteochondral progenitors, 5–6
Osteoclast differentiation factor (ODF), 114, 1690
Osteoclast-specific transmembrane protein (OC-STAMP), 114
Osteoclastogenesis, 256, 1063, 1211–1212, 1555, 1563, 1689–1690, 1695
osteoblastic cells role in, 114–116
signal transduction in, 116–119
Osteoclasts (OCLs), 79, 111, 255, 318, 404–406, 404f, 408–409, 776, 976–977, 1335–1336, 1602, 1689–1690, 1772, 1796, 1909
activation by, 1158
to bone physiology, 1226
bone resorption, 1171
calcitonin effects, 794–795
amylin, 795–796
CGRP, 795
coupling factors synthesized and secreted by, 228–230
differentiation, 119
and coupling of bone formation to bone resorption, 778–779
and life span, 777–778
function, 112f, 185
bone resorption mechanism, 113
DC-STAMP, 113–114
morphological features, 112–113
ruffled border formation, 114
glucocorticoid excess and, 924–925
hypothetical model of osteoclast differentiation, 125f
induction of osteoclast function
adhesion signals, 122–123
cytokine signals, 124
osteoclast precursors in vivo characteristics, 124
ITAM costimulatory signals, 119–120
lineage, 1719, 1720f
membrane-bound coupling factors synthesized by, 230–231
Notch receptors and ligands in, 1095–1096
osteoblast lineage promoting osteoclast formation, 98–99
osteoblastic cells role in osteoclastogenesis, 114–116
osteoclast-independent factors, 779
osteoclast-inhibiting effects, 1691–1692
osteoclast-mediated bone resorption, 1678–1679
osteoclast-specific genes, 144
resorption, 275
signal transduction in osteoclastogenesis, 116–119
WNT signals, 121–122
Osteocyte(s), 57, 95, 133, 255, 255f, 277–278, 284, 401–404, 406–408, 471–472, 776, 829, 951–952, 957, 976–977, 1258, 1279–1284, 1343, 1493, 1714, 1772–1773, 1795–1796
apoptosis, 254
β-catenin haploinsufficiency, 180–181
cell–matrix adhesion, 142–143
cytoskeleton, 142–143, 1777
death in signaling bone remodeling, 254
formation and death, 136–138
function, 186
blood–calcium/phosphate homeostasis, 144–145
canalicular fluid flow and osteocyte mechanosensing, 146–149
functional adaptation, 145
osteocytes as mechanosensory cells, 145–146
osteocytes response to fluid flow in vitro, 149–151
glucocorticoid excess and, 924
hormone receptors in, 143–144
isolation, 133, 138–139
lacunar density, 275
markers, 139–140
matrix synthesis, 141–142
mechanosensing, 146–149
network, 134–136, 275
isolated osteocytes in culture, 135f
osteon in mature human bone, 134f
notch receptors and ligands in
mechanisms of Notch action in osteocytes, 1095
role of Notch signaling in, 1095
osteocyte-derived PTHrP, 610
osteocyte-like MLO-Y4 cells, 254
osteocytic cell lines, 140–141
physical sensing and signaling by, 98
shape and mechanosensing, 149
viability, 137–138
Osteocytic cell lines, 140–141
Osteogenesis, 48–49, 1142
assay, 91
Osteogenesis imperfecta (OI), 302–303, 345, 348, 1489, 1699, 1711, 1818
clinical classification
mild nondeforming OI, 1491
severe-deforming OI, 1490–1491
molecular classification, 1491–1493
genes mutation modifying synthesis of type I collagen chains, 1492
genes mutation regulates maturation of secreted procollagen, 1492–1493
mutations control level of differentiation of osteoblasts, 1492
primary mutations within type I collagen genes, 1491
pathophysiology of, 1493–1496
OI due to underproduction of a normal type I collagen molecule, 1495–1496
OI secondary to abnormal collagen molecule production, 1493–1495
therapeutic options, 1496–1498
anabolic agents, 1496
anti-TGFß and antiactivin agents, 1496
antiresorptive agents, 1496
cell and gene-therapy options, 1497
iPSCs uses as diagnostic tool, 1497–1498
type V, 1492
type VI, 1492
Osteogenesis Imperfecta Foundation (OIF), 1490
Osteoglophonic dysplasia, 485
Osteoid, 246, 447
islet, 1163
osteoid-osteocytes, 136
Osteoid surface (OS), 448
Osteoid thickness (O.Th), 449
Osteoid volume (OV), 448
Osteoimmunology, 119
Osteoinductive factor, 362

Osteolytic bone disease, 1323
Osteolytic lesions, 1602, 1697
Osteoma, 1435
Osteomalacia, 451, 1522, 1531, 1579
clinical features of, 1508–1510
Osteomodulin. *See* Osteoadherin
Osteonecrosis of jaw, 1700
Osteonectin, 365, 1492–1493
Osteopenia, 1858
Osteopetrosis (OP), 720, 1553
clinical features, 1553–1554, 1555t
current therapies, 1562
future therapeutic scenarios, 1562–1564
genetic features, 1556–1561, 1557t
radiographic features, 1554–1556
Osteopetrosis, lymphedema, hypohidrotic ectodermal dysplasia and immunodeficiency (OL-HED-ID), 1560–1561
Osteopetrosis-transmembrane protein 1 (OSTM1), 113
Osteopontin (OPN), 141–142, 234, 365–366, 368–369, 741–742, 1341
Osteoporosis, 275, 283–284, 452–453, 809, 827, 829–831, 864–865, 1170, 1194, 1202, 1615, 1657, 1665, 1693, 1711, 1722–1724, 1763, 1819–1820, 1857
calcium and osteoporosis treatment, 1650–1653
circumscripta, 1602
clinical uses of bone markers in, 1812–1816
diagnosis of osteoporosis and prognosis of bone loss, 1812
predicting and monitoring treatment efficacy, 1813–1816
prediction of fracture risk, 1813
denosumab in combination/sequence with osteoporosis agents, 1695–1696
drugs, 454–461
GH treatment for osteoporotic patients, 1003–1004
IGF I for, 1004
clinical use of recombinant human insulin-like growth factor I, 1007
human studies of IGF-I and BMD, 1005–1007
murine studies, 1004–1005
indications
denosumab in combination/sequence with osteoporosis agents, 1695–1696
male osteoporosis, 1695
postmenopausal osteoporosis, 1693–1695
lead as unrecognized risk factor in, 529–530
monitoring side effects of osteoporosis therapies, 1816
osteoporosis-associated Sp1 polymorphism, 304
osteoporosis-like phenotype, 315
pathogenic role of GH/IGF/IGFBPS, 998–1000
to SLC34A1, 490
Osteoporotic Fractures in Men study (MrOS study), 763–764, 1873
Osteoprogenitors, 136, 206, 1062, 1795–1796
cells, 149–150, 383, 1121
Osteoprotegerin (OPG), 18, 57, 98, 115–116, 141, 170, 185, 221, 267, 432–433, 650, 777–778, 1123, 1200–1201, 1206, 1319–1320, 1341–1342, 1689–1691, 1719, 1805, 1937
Osteosarcoma (OS), 607–609, 608f, 1099
cells, 596
osteosarcoma-9 protein, 546
Osteosclerotic phase, 1602
Osteostatin, 602
Osteotransmitter, 1215, 1216t–1217t, 1226
Osteotropism, 1337
Osterix (OSX), 8, 74, 76, 167–168, 184, 206–207, 310, 428, 649, 996, 1063, 1142
Osterix-CreERt2 model, 1068
Osterix-GFP-Cre mouse model, 1067
Osx-Cre $p53^{fl/fl}pRb^{fl/fl}$ model, 607
Osx-GFP::Cre mouse, 1891
Osx1, 277, 778
OSTM1. *See* Osteopetrosis-transmembrane protein 1 (OSTM1)
OSX. *See* Osterix (OSX)
OV. *See* Osteoid volume (OV)
Ovarian effects of SERMs, 876
Ovariectomy (OVX), 827, 1718–1719, 1913–1914
animals, 869
rat model, 797–798, 1913–1914
Ovary, 876
OVX. *See* Ovariectomy (OVX)
22-Oxacalcitriol (OCT), 1733
Oxidative stress, 956
Oxidized LDL (OxLDL), 283
Oxidized lipids, 283–284
Oxygen, regulation of VEGF expression by, 1147–1148
Oxygenation, 73
bone cells' control of, 208–210
Oxytetracycline, 445–446

P

p130 CRK-associated substrate ($p130^{CAS}$), 122–123
$p130^{CAS}$. *See* p130 CRK-associated substrate ($p130^{CAS}$)
P1NP. *See* Procollagen type 1 N-terminal propeptide (P1NP)
p38 MAPK, 545
P3H1. *See* Prolyl 3-hydroxylase 1 (P3H1)
P3H3. *See* Prolyl 3-hydroxylase 3 (P3H3)
$p62^{P394L}$-knock-in mice, 1608–1609
4-PA. *See* 4-Pyridoxic acid (4-PA)
Paget's disease of bone, diagnosis and monitoring of treatment in, 1817
Pagetic OCLs, 1604–1605
Paget's disease, 1601
biochemistry, 1603
cellular and molecular biology, 1604–1607
etiology, 1609–1610
evidence for presence of paramyxoviruses, 1603–1604
genetic mutations linked to, 1608–1609
histopathology, 1602–1603
patient, 1601
radiology and nuclear medicine, 1601–1602
treatment, 1603
PAH. *See* Pulmonary arterial hypertension (PAH)
PAI-1. *See* Plasminogen activator inhibitor-1 (PAI-1)
Pain sensation, 212
Paired box genes (Pax genes), 1358–1359
Pax1, 578–579, 1359
Pax3, 1359
Pax7, 1359
Pax9, 578–579, 1359
Paired related homeobox 1 (Prx1), 75
Palmitoylated serine residue, 179
PAMPs. *See* Pathogen-associated molecular patterns (PAMPs)
PAMs. *See* Positive allosteric modulators (PAMs)
Pancreas, CaSR in, 553–554
Pancreatic NETs (pNETs), 1295
Pancreatic polypeptide (PP), 814–815, 1295
Pancreatoduodenectomy (PD), 1297
Pannexin channels, 435–436
PAPP-A. *See* Pregnancy-associated plasma protein-A (PAPP-A)
Parabiosis, 1529
Paracrine PTHrP/PTH1R pathway, 630
Paracrinology to endocrinology in bone, 101–102
Paraffin, 534
Paraformaldehyde (PFA), 1901
Paramyxoviruses in Paget's disease, 1603–1604
Parasympathetic nerve system signaling (PSNS signaling), 1344
Parasympathetic nervous system, control counterregulation of bone remodeling by, 812–813
Parathyroid
carcinoma, 1659
molecular pathogenesis of, 1417–1418
cell proliferation, 588–589
cytosolic proteins, 584
hyperfunction, 1405
hyperplasia, 1406–1407
nuclear fraction, 581
parathyroid-specific deletion of Dicer-dependent microRNA, 585
surgery, 1658
Parathyroid Epidemiology and Audits Research Study (PEARS), 632
Parathyroid glands, 578–580
CaSR in, 549
congenital anomalies of, 1355–1357
transcription factors, 1358f

Parathyroid hormone (PTH), 51, 58, 76–77, 79–80, 92, 114, 137, 186, 189, 228, 266, 317–318, 323, 381, 425, 449, 471, 513, 575–578, 595, 623, 630–631, 646, 691, 718, 740–741, 764, 775, 930, 953–954, 992, 1063, 1114, 1161, 1193, 1209, 1273, 1293, 1307–1308, 1324, 1380–1381, 1405, 1465, 1496, 1529, 1543, 1559, 1580–1581, 1633–1634, 1657, 1695–1696, 1714, 1739, 1803, 1914
- actions in kidney, 1277–1279
  - 1,25-$(OH)_2$-vitamin D synthesis by CYP27B1 expression, 1277–1278
  - parathyroid hormone-mediated control
- of distal tubule calcium reabsorption, 1279
- of proximal tubule phosphate handling, 1278–1279
- actions
  - on kidney, 655–664
  - on mineral-ion homeostasis, 659–660
  - on skeletal microvasculature, 629f
- in adult bone, 610–612
- anabolic action, 1225–1226
- arterial desensitization to vasodilatory actions, 627f
- autosomal dominant hypoparathyroidism, 1364
- autosomal recessive hypoparathyroidism, 1364–1365
- bone anabolic effects of intermittent PTH, 779–782
- bone remodeling regulation by, 776–777
- in cardiovascular development, 624
- cellular basis, 654
- cellular effects of PTH on bone, 653f
- delayed tooth eruption, 1393
- effects
  - on bone cells, 648–652, 654
  - differentiation, 653
  - on distal tubule calcium transport, 664
  - proliferation, 652–653
  - on proximal tubule calcium transport, 663
  - survival, 654
- in endochondral bone formation, 1382
- as endocrine regulator of skeletal TGF-β signaling, 1166–1167
- in fetal and postnatal bone, 609–610, 610f
- GCM2 gene abnormalities, 1365–1366
- gene, 1309
  - abnormalities, 1363–1366
  - structure and function, 1363–1364, 1364f
- histomorphometric evidence of skeletal resistance to, 628f
- human disorders, 1382–1393
- in hypoparathyroidism, 1639–1640
- inducible cyclooxygenase modulation of effects of, 1256–1258
- and ligands, 1274–1277
- mRNA, 576–577, 578f
- mutations in, 577–578
  - in genes downstream, 1393–1397
- organization, 575–576
- in osteocytes, 1282–1284
- parathyroid hormone–like peptide mutations, 1383
- physiology, 1633–1634
- promoter sequences, 576
- PTH mRNA 3' UTR
  - parathyroid cytosolic proteins and mRNA stability, 584
  - PTH mRNA stability, 584–585
- PTH/PTHrP receptor type 1, 1635
- PTH(1–34), 456–460, 457f
- PTH(1–84), 456–460
  - ABL, 457–460
- PTHR2 and PTHR3 subtypes, 705–706
- receptors, 143–144, 705–706
  - mutations, 1383–1393, 1387t–1389t
  - receptors and second-messenger systems, 646–647
  - signaling in arterial biology, 624–630
  - system, 1380–1382
- regulating factors governing assembly and maintenance of BMUs
  - osteoblast differentiation and coupling of bone formation, 778–779
  - osteoclast differentiation and life span, 777–778
- regulation, 580–588
  - of calcium homeostasis, 656f
  - of proximal tubular sodium and hydrogen excretion, 668–669
  - regulatory sites of action, 749–750
  - of renal calcium absorption, 663–664
  - of renal phosphate absorption, 665–666
- renal effects, 669–670
- signaling, 635
  - of renal calcium transport, 666
  - of renal phosphate transport, 666–667
- mechanisms controlling PTH hormone-induced NF-kappa-B ligand, 1280–1281
- unresolved issues, 782
- vascular smooth muscle cell and endothelial responses to, 624–630
- vasodilation of rat femoral nutrient arteries, 625f

Parathyroid hormone receptor 1R (PTH1R), 97, 471, 513, 550, 599, 691–693, 840, 1161, 1273, 1379–1380, 1533
- activation, 636
- conformational selectivity and temporal bias at, 699–703
- endosomal signaling and signal termination at, 700–701
- evolutionary model, 695f
- gene structure and evolution, 694
- interaction with parathyroid hormone receptor 1, 598–599
- ligands, 1274–1277
  - binding mechanisms at, 697f
  - ligand-directed temporal bias and therapeutic implications, 701–702
  - ligand-induced activation mechanism at, 698–699
  - recognition and activation by, 695–699
  - system, 694
- mutations in disease, 703–704
- nonpeptide mimetic ligands for, 704–705
- in osteocytes, 1282–1284
- signaling, 623–624
  - impaired vascular PTH1R signaling and cardiovascular disease, 632–633
- small-molecule agonist for, 705f
- structural properties, 695–696
- structure, 693f, 694
- two-site model of ligand binding to, 696–698

Parathyroid hormone receptor 2R (PTH2R), 623–624, 647
- signaling in vascular pharmacology, 630–631

Parathyroid hormone-related protein (PTHrP), 21–23, 23f, 92, 232, 266, 544–545, 575, 595, 623, 646, 691, 839, 1063, 1219, 1273, 1335–1336, 1379–1382
- in adult bone, 610–612
- analogs in pharmacology, 612–613
- arterial desensitization to vasodilatory actions, 627f
- in bone after endochondral ossification, 605–607
- C-terminal PTHrP and osteostatin, 602
- in cardiovascular development, 624
- characteristics of gene encoding, 1309
- circulating levels, 1318–1319
- delayed tooth eruption, 1393
- detection producing by tumors, 1318–1319
- discovery, 595–596
- effect on tumor progression and survival, 1316–1317
- effects on bone cells, 648–652, 654
  - differentiation, 653
  - proliferation, 652–653
  - survival, 654
- in endochondral bone formation, 1382
- endocrine pancreas, 853–856
- experimental approaches to controlling overproduction or overactivity, 1321–1322
- in fetal and postnatal bone, 609–610, 610f
- in fetus, 602–605
- functions of carboxyl and midregion circulating fragments, 1313
- growth factors and hormone regulation, 1310–1311
- human disorders, 1382–1393
- humoral hypercalcemia manifestations of malignancy, 1315–1316
  - actions in bone, 1316
  - actions in kidney, 1315–1316
- interaction with parathyroid hormone receptor 1, 598–599
- intracellular mechanism of PTH-related protein action, 1313–1315
- in mammary gland, 842–849
- mechanisms of action, 1312–1315
- model of dual action, 1314f
- molecular and cellular biology, 1308–1312
- multiple known pathways, 598f

Parathyroid hormone-related protein (PTHrP) (*Continued*)
mutations in genes downstream of, 1393–1397
nuclear actions, 600–602
nuclear import, 600
and osteosarcoma, 607–609, 608f
paracrine *vs.* intracrine, 628f
physiological actions, 611f
primary structure, active domains, processing, and secretion, 596–598
processing and degradation, 1311–1312
producing tumors with the bone microenvironment, 1319–1320
protein structure, 597f
purification and cloning, 1308
receptors, 706
mutations, 1383–1393, 1387t–1389t
and second-messenger systems, 646–647
system, 1380–1382
renal expression and actions, 670
in reproductive tissues, 849–853
resistance to antiresorptive agents in malignancy-associated hypercalcemia, 1320
signaling, 635
skeletal phenotypes in mouse models, 604t
in skin, 840–842
stage-specific actions, 605f
tissue distribution and function as cytokine, 599–600
transcriptional regulators, 1310–1311
treatment of local osteolysis, 1321–1322
in tumor tissue, 1319
vascular smooth muscle cell and endothelial responses to, 624–630
vasodilation of rat femoral nutrient arteries, 625f
Parathyroid specific Dicer 1-knockout mice (PT-*Dicer*$^{-/-}$ mice), 585
Paraventricular nucleus (PVN), 1938
Paraxial mesoderm, 47
Parfitt's hypothesis, 144
Paricalcitol, 1661–1662
Parkinson's disease, 277–278
PAs. *See* Plasminogen activators (PAs)
Patched homolog 1 (PTCH1), 1345
Pathogen-associated molecular patterns (PAMPs), 283
Pathway cross talk in bone, 1192–1193, 1193f
Pathway-specific gp130 modulation, 1224–1227
Patient-derived xenograft models (PDX models), 1340–1341
Pattern recognition receptors (PRRs), 283
Patterning of skeleton, 14–15
Paucity of bone, 1858
Pax genes. *See* Paired box genes (Pax genes)
PBMCs. *See* Peripheral blood mononuclear cells (PBMCs)
PC. *See* Phosphocholine (PC); Proliferating chondrocyte (PC)
PC1. *See* Polycystin 1 (PC1)
PCa. *See* Prostate cancer (PCa)
*PCBD*1 gene, 517–518
PCL. *See* Poly(ϵ-caprolactone) (PCL)
PD. *See* Pancreatoduodenectomy (PD); Pharmacodynamics (PD)
PDE3A mutations. *See* Phosphodiesterase 3A mutations (PDE3A mutations)
PDE4D. *See* Phosphodiesterase 4D (PDE4D)
PDGF. *See* Platelet-derived growth factor (PDGF)
PDGFB receptor (PDGFBR), 492
PDGFB+ cells. *See* Platelet-derived growth factor subunit B-positive cells (PDGFB+ cells)
PDGFBR. *See* PDGFB receptor (PDGFBR)
Pdia3. *See* Protein disulfide isomerase family A, member 3 (Pdia3)
PDK-1. *See* Phosphoinositide-dependent kinase-1 (PDK-1)
PDLs. *See* Periodontal ligaments (PDLs)
PDX models. *See* Patient-derived xenograft models (PDX models)
PDZ domains. *See* PSD95/Discs Large/ZO-1 domains (PDZ domains)
PEA. *See* Phosphoethanolamine (PEA)
PEARS. *See* Parathyroid Epidemiology and Audits Research Study (PEARS)
PEDF. *See* Pigment epithelium-derived factor (PEDF)
Pencil-beam dual-photon absorptiometer, 529
Peptide, 457–458
access to bone microenvironment, 793
hormone calcitonin, 789
Peptide receptor radionuclide therapy (PRRT), 1300
Peptide YY (PYY), 814–815
Peptidomimetic ligand, 403–404
Peptidylprolyl isomerase B (PPIB), 348–350
Pericytes/skeletal stem cells development, 47–48, 49f
Perinatal HPP, 1575, 1576f
Periodontal bone, 1061
Periodontal disease, methods to control inflammation in, 1073–1074
Periodontal ligaments (PDLs), 1061
Periodontal regenerative therapy, 1073
Periodontal stem/progenitor cells (PSCs), 1061–1062
regeneration, 1073–1074
Periodontitis, 1073
Periodontium, 1061–1062
Bmp2 gene function in, 1065–1070
candidate periodontal stem/progenitor cells, 1062
hh signaling in bone and, 1062–1063
junctional epithelium in periodontium function, 1074–1076
Key regulators of mineral metabolism and, 1070–1073
periodontal stem/progenitor regeneration, 1073–1074
PSCs, 1061–1062
PTH/PTHrP role in long bones and, 1063–1064
Wnt signaling in periodontium and role of Sost gene, 1064–1065
Periosteal cells, 1194
Periosteum, 1194
Periostin (POSTN), 365, 1065, 1803–1804
circulating, 1804
Peripheral blood mononuclear cells (PBMCs), 1516, 1805
Peripheral computed tomography, 251
Peripheral QCT (pQCT), 1834, 1863–1865
Peripheral quantitative computed tomography, 1767, 1846
Peripheral vascular system, CaSR in, 555
Perivascular markers, 1895–1896
Perlecan, 410–411
Peroxisome proliferator-activated receptor gamma (PPARγ), 91–92, 277, 996, 1221
coactivator-1α, 724–725
PPARγ2, 48–49, 944
PET/CT. *See* Positron emission tomographyecomputed tomography (PET/CT)
PFA. *See* Paraformaldehyde (PFA)
PFD. *See* Polyostotic fibrous dysplasia (PFD)
PFE. *See* Primary failure of tooth eruption (PFE)
PFF. *See* Pulsating fluid flow (PFF)
15-PGDH. *See* 15-Dehydroxyprostaglandin dehydrogenase (15-PGDH)
PGE_2. *See* Prostaglandin E_2 (PGE_2)
PGES. *See* Prostaglandin E synthase (PGES)
PGG_2. *See* Prostaglandin G_2 (PGG_2)
PGH_2. *See* Prostaglandin H_2 (PGH_2)
PGHS. *See* Prostaglandin G/H synthase (PGHS)
PGIs. *See* Prostacyclins (PGIs)
PGs. *See* Prostaglandins (PGs)
Pharmacodynamics (PD), 1680
Pharmacokinetic-pharmacodynamic models (PK-PD models), 1680
Pharmacokinetics (PK), 1677–1680
mathematical PK model, 1680
mechanism, 1634–1635
of SERMs, 883–884
studies, 1661
Pharmacologic basis of sclerostin inhibition
pharmacologic inhibition
of sclerostin by Scl-Abs, 1718–1719
of sclerostin by Scl-Abs in vivo, 1719–1722
probing Scl-Ab treatment regimens in preclinical models of osteoporosis, 1722–1724
sclerostin biology and biochemistry, 1712–1717
Pharmacologic inhibition
of sclerostin by Scl-Abs, 1718–1719
effects of Scl-Abs in animal models of postmenopausal osteoporosis, 1719–1722
in humans, 1724
in vivo, 1719–1722

Pharmacological effects of vitamin D compounds, 1742–1746
  activating enzymes, 1743
  hepatic clearance or nonspecific metabolism, 1746
  target cell catabolic enzymes, 1745–1746
  vitamin D receptor/retinoid X receptor/ vitamin D response element interactions, 1744–1745
  vitamin D-binding protein, 1743–1744
Pharmacological mechanisms of therapeutics
  denosumab
    effects, 1693–1700
    safety, 1700–1702
  osteoprotegerin/RANKL-based drug development, 1689–1691
  physiologic mechanisms and effects of RANKL inhibitors in bone, 1691–1693
Pharmacological perturbations, 1926
Pharmacologically important vitamin D compounds, 1734–1740, 1735t
  1α,25$(OH)_2D_3$, 1736t–1737t
  calcitriol analogs, 1739–1740
  miscellaneous vitamin D analogs and associated drugs, 1740
  natural metabolites, 1734–1735
  prodrugs, 1734t, 1735–1739
Pharmacology, PTHrP analogs in, 612–613
Pharyngeal pouches, 1355–1357
PHCs. *See* Prehypertrophic chondrocytes (PHCs)
PHD. *See* Prolhydroxylase (PHD)
PHDs. *See* Prolyl hydroxylase enzymes (PHDs)
Phenotypes, 1790
Phenotyping
  pipeline, 1788
  resources and repositories, 1787–1788
PHEX. *See* Phosphate-regulating gene with homologies to endopeptidases on X chromosome (PHEX)
Phorbol myristate acetate (PMA), 381
Phosphatase and tensin homologue gene deleted on chromosome 10 (PTEN), 989–990
Phosphate (PO_4), 469–470, 474–478, 583–586
  chemistry, 659
  excretion, 664–667
  homeostasis, 144–145, 486–488, 587, 656–659
    primary disorders of, 478–491
  metabolism regulation, 470–471
    hormonal regulators, 471
    nutritional and gastrointestinal considerations, 473–474
    osteocytes, 471–472
  protein–PTH mRNA interactions, 583–584
  transport mechanisms, 475–478
    intestinal phosphate transport, 476–477
    renal phosphate transport, 477
    ubiquitous metabolic phosphate transporters, 478
Phosphate-regulating gene with homologies to endopeptidases on X chromosome (PHEX), 137, 145, 471–472, 1529
Phosphate/PPi (Pi/PPi), 474
Phosphatidylinositol 3-kinase (PI3K), 52, 116–117, 186, 401–402, 427–428, 545, 726, 989–990
Phosphatidylinositol 3-kinase/ phosphoinositide-dependent kinase-1/ Akt pathway (PI3K/PDK-1/Akt pathway), 989–990
Phosphatonins, 1539
Phosphaturic mesenchymal tumors (PMTs), 1539–1542
Phosphaturic mesenchymal tumors–mixed connective tissue variant (PMTs–MCT variant), 484
Phosphocholine (PC), 283
Phosphodiesterase 3A mutations (PDE3A mutations), 1397
Phosphodiesterase 4D (PDE4D), 1394, 1396–1397
Phosphoethanolamine (PEA), 1573–1574, 1581, 1587–1588
Phosphoethanolaminuria, 1581
Phosphoinositide-dependent kinase-1 (PDK-1), 989–990
Phospholipase A_2 (PLA_2), 543–544, 1249
Phospholipase C (PLC), 695–696, 1251
Phospholipase Cγ (PLCγ), 119, 722–723, 1116
  PLCγ2, 1342
Phospholipase D (PLD), 543, 661, 1276
Phosphomonoester phosphohydrolase, 1569
Phosphoproteins, 137
Phosphorus homeostasis
  hyperphosphatemic syndromes, 491–492
  normophosphatemic disorders of cellular phosphorus metabolism, 492
  phosphate and bone mineralization, 474–478
  primary disorders of phosphate homeostasis, 478–491
  regulation of phosphate metabolism, 470–474
PHP. *See* Pseudohypoparathyroidism (PHP)
PHP1A, 1393–1394, 1436
PHP1B. *See* Pseudohypoparathyroidism type Ib (PHP1B)
PHPIA. *See* Pseudohypoparathyroidism type IA (PHPIA)
PHPT. *See* Primary hyperparathyroidism (PHPT)
Physical activity, 284
P_i. *See* Inorganic phosphate (P_i)
Pi/PPi. *See* Phosphate/PPi (Pi/PPi)
PI3K. *See* Phosphatidylinositol 3-kinase (PI3K)
PICP. *See* Procollagen type I C-terminal propeptide (PICP)
Pigment epithelium-derived factor (PEDF), 352, 1492
Pin1, 585
Pituitary adenomas, 1301
Pituitary tumors, 1293
Piwi RNAs, 53
PK. *See* Pharmacokinetics (PK)
PK-PD models. *See* Pharmacokinetic-pharmacodynamic models (PK-PD models)
PKA. *See* Protein kinase A (PKA)
PKC. *See* Protein kinase C (PKC)
PKCα. *See* Protein kinase Cα (PKCα)
PKN3. *See* Protein kinase N3 (PKN3)
PLA_2. *See* Phospholipase A_2 (PLA_2)
Placebo-controlled trials, 760, 974
Placenta, 851–852
Placental ALP, 1573, 1586
Placental calcium transport, PTHrP and, 849–851
Placental growth factor (PlGF), 1142–1143
Plakoglobin, 424–425
Plasma membrane Ca^{2+}-ATPase 2 (PMCA2), 552–553
Plasma membrane calcium ATPase (PMCA1b), 742, 1279
Plasminogen activator inhibitor-1 (PAI-1), 386, 878
Plasminogen activators (PAs), 379, 387–388
  in bone, 388
Plastic-embedded bone specimens
  micromilled cross sections of cortical bone, 1903
  microtome sectioning, 1902–1903
Platelet factor 4. *See* CXCL4
Platelet-derived growth factor (PDGF), 206, 227, 475
  PDGF-bb, 1165–1166
Platelet-derived growth factor subunit B-positive cells (PDGFB+ cells), 51
PLC. *See* Phospholipase C (PLC)
PLCγ. *See* Phospholipase Cγ (PLCγ)
PLD. *See* Phospholipase D (PLD)
PLEKHM1 gene, 1559
Plexins, 231
PLGA. *See* Poly(lactic-coglycolic acid) (PLGA)
PlGF. *See* Placental growth factor (PlGF)
PLOD1–3. *See* Procollagen-lysine, 2-oxoglutarate 5-dioxygenases (PLOD1–3)
PLOD2 gene, 350
PLP. *See* Pyridoxal 5′-phosphate (PLP)
Pluriglandular autoimmune hypoparathyr-oidism, 1362–1363
PMA. *See* Phorbol myristate acetate (PMA)
PMCA1b. *See* Plasma membrane calcium ATPase (PMCA1b)
PMCA2. *See* Plasma membrane Ca^{2+}-ATPase 2 (PMCA2)
PMCA2b gene, 746–747
PMTs. *See* Phosphaturic mesenchymal tumors (PMTs)
PMTs–MCT variant. *See* Phosphaturic mesenchymal tumors–mixed connective tissue variant (PMTs–MCT variant)
pNETs. *See* Pancreatic NETs (pNETs)

POC formation. *See* Primary ossification center formation (POC formation)
Podosomes, 404–405
POGLUT. *See* Protein *O*-glucosyltransferase (POGLUT)
POH. *See* Progressive osseous heteroplasia (POH)
"Point-counting" techniques, 447
Poly(lactic-coglycolic acid) (PLGA), 64
Poly(ϵ-caprolactone) (PCL), 64
Polycationic agonists, 1414
Polycystin 1 (PC1), 148
Polyostotic fibrous dysplasia (PFD), 1443, 1817
Polypeptides, 542, 987
Polyunsaturated fatty acids (PUFAs), 1247–1248
POMC. *See* Proopiomelanocortin (POMC)
Population variance, 252
Porcupine gene, 179
Positive allosteric modulators (PAMs), 542, 1414
Positron emission tomographyecomputed tomography (PET/CT), 1544–1546
Postmenopausal bone loss inhibition by low-level vibrations, 1768–1769
Postmenopausal osteoporosis, 452, 1693–1695
  action of Scl-Abs on MBF and RBF, 1719–1722
  Scl-Abs effects
    on bone mass and structure, 1722
    on bone strength and quality, 1722
POSTN. *See* Periostin (POSTN)
Postnatal bone
  PTH and PTHrP in, 609–610, 610f
  SSCs/BMSCs role in postnatal bone turnover and remodeling, 57–58
Postnatal lineage tracing, 1893–1894
Posttranscriptional regulation of type I collagen, 313–314
PP. *See* Pancreatic polypeptide (PP)
PP2A. *See* Protein phosphatase 2A (PP2A)
PPARγ. *See* Peroxisome proliferator-activated receptor gamma (PPARγ)
PPD. *See* Preaxial polydactyly (PPD)
PPHP. *See* Pseudopseudohypoparathyroidism (PPHP)
PP_i. *See* Inorganic pyrophosphate (PP_i)
PPi-ase. *See* Pyrophosphatase (PPi-ase)
PPIB. *See* Peptidylprolyl isomerase B (PPIB)
PPIs. *See* Proton pump inhibitors (PPIs)
pQCT. *See* Peripheral QCT (pQCT)
PR. *See* Progesterone receptors (PR)
Prdm5, 298
Preaxial polydactyly (PPD), 1367
Precartilage condensations, 1791–1795
Prediabetes, 947
Predominance of cortical bone loss, 259–261
Pregnancy, 844–848
Pregnancy-associated plasma protein-A (PAPP-A), 992
Prehypertrophic chondrocytes (PHCs), 1116–1117
Prereceptor metabolism, glucocorticoid signaling and, 916–919
Presenilin (PSEN), 1088
  presenilin 1/2 proteins, 25
Presomitic mesoderm (PSM), 9
Primary cilium, 1775
Primary disorders of phosphate homeostasis
  FGF23
    independent hypophosphatemic disorders, 486–491
    mediated hypophosphatemic disorders, 478–484
    mediated hypophosphatemic syndromes, 485
  intestinal malabsorption of phosphate, 491
  tumor-induced osteomalacia, 484–485
Primary failure of tooth eruption (PFE), 1386
Primary human breast tumors, 1200
Primary hyperparathyroidism (PHPT), 626–627, 1293–1295, 1405, 1634, 1657–1659
Primary ossification center formation (POC formation), 206, 1142
PRKAR1A. *See* Protein kinase type 1A regulatory subunit protein (PRKAR1A)
PRLoma. *See* Prolactinoma (PRLoma)
Pro-Col1a1 gene, 303, 308–310, 312
Pro-COL1A2
  gene, 303, 313
  proximal promote, 307–308
Pro-EGF, 516
pro-Ocn. *See* Proosteocalcin (pro-Ocn)
Pro-α1(I) proximal promoter, 305–307
Procollagen, 1492–1493
  C-proteinase, 301
  suicide, 302–303
Procollagen type 1 N-terminal propeptide (P1NP), 267, 323–324, 944, 1478, 1603, 1712, 1802
Procollagen type I C-terminal propeptide (PICP), 1813
Procollagen-lysine, 2-oxoglutarate 5-dioxygenases (PLOD1–3), 347
Productive process, 220
Progenitor cells, niche heterogeneity for, 80–81
Progesterone receptors (PR), 832
  in bone biology, 832–833
    structure of human PR-A and PR-B, 832f
  progesterone receptor–positive epithelial cells, 1199
Progestins in bone biology, 832–833
Progressive osseous heteroplasia (POH), 1431, 1440–1441
  clinical features, 1440
  genetics, 1440
  pathogenesis, 1441
Proinflammatory cytokines, 956
Prolactin-secreting adenoma. *See* Prolactinoma (PRLoma)
Prolactinoma (PRLoma), 1293, 1300
Prolhydroxylase (PHD), 1147
Proliferating chondrocyte (PC), 1114
Proline, 1365
Proline-rich tyrosine kinase 2 (PYK2), 122–123, 401–402
Prolyl 3-hydroxylase 1 (P3H1), 348
Prolyl 3-hydroxylase 3 (P3H3), 350–351
Prolyl hydroxylase enzymes (PHDs), 208
Proopiomelanocortin (POMC), 814
Proosteocalcin (pro-Ocn), 1935
Proprotein convertase furin, 1935
Prostacyclins (PGIs), 1247–1248
Prostaglandin E synthase (PGES), 1251
Prostaglandin E_2 (PGE_2), 114, 135–136, 186, 314, 434, 1247, 1253
  and bone formation, 1252–1255
  and bone physiology, 1258–1260
    fracture and wound healing, 1259
    mechanical loading of bone, 1258
    skeletal response to nonsteroidal antiinflammatory drugs, 1259–1260
  and bone resorption, 1252–1255
  receptors, 1251–1252
  synthases, 1251
Prostaglandin G/H synthase (PGHS), 1248
Prostaglandin G_2 (PGG_2), 1248
Prostaglandin H_2 (PGH_2), 1248
Prostaglandins (PGs), 385, 1206, 1247, 1777
  isoforms for prostaglandin G/H synthase, 1249–1251
  production, 1247–1251
  receptors, 144
Prostate cancer (PCa), 762, 1340–1341, 1697
  androgen deprivation in men with, 1696
  in animal studies, 762
  clinical studies, 763
Prostate-specific antigen (PSA), 991–992
Protein disulfide isomerase family A, member 3 (Pdia3), 713–714
Protein kinase A (PKA), 21–22, 181, 428, 471, 539, 598, 646–647, 695–696, 1114, 1252, 1310, 1379–1380, 1635
Protein kinase C (PKC), 52, 381, 471, 541, 646–647, 1251, 1310, 1635
Protein kinase Cα (PKCα), 425
Protein kinase N3 (PKN3), 122
Protein kinase type 1A regulatory subunit protein (PRKAR1A), 1394–1395
  mutations, 1394–1396
Protein *O*-glucosyltransferase (POGLUT), 1086
Protein phosphatase 2A (PP2A), 182, 1117–1118
Proteinases, 379
Proteoglycans, 142, 360–363
  aggrecan and versican (PG-100), 360
  in bone matrix, 361t
  decorin (PG-II) and biglycan (PG-I), 360–362
  leucine-rich repeat sequence proteins and, 362–363
Proteolysis, 481, 1088

Proteomic signature, 1807–1808
Proteostasis, 281
Proton nuclear magnetic resonance (1H NMR), 1808
Proton pump inhibitors (PPIs), 519, 1296–1297
Protooncogenes, 1406
Proximal femur, 1693–1694
Proximal promoters of type I collagen genes
  factors binding to pro-COL1A2 proximal promote, 307–308
  transcription factors binding to pro-α1(I) proximal promoter, 305–307
Proximal tibial metaphysis, 1722
Proximal tubule
  parathyroid hormone-mediated control of proximal tubule phosphate handling, 1278–1279
  proximal tubular phosphate absorption, mechanisms of, 664–665
  PTH effects on proximal tubule calcium transport, 663
Proximal–distal axis, 12–13
PRRs. *See* Pattern recognition receptors (PRRs)
PRRT. *See* Peptide receptor radionuclide therapy (PRRT)
Prrx1-Cre, 1791–1795
Prx1. *See* Paired related homeobox 1 (Prx1)
PSA. *See* Prostate-specific antigen (PSA)
PSCs. *See* Periodontal stem/progenitor cells (PSCs)
PSD95/Discs Large/ZO-1 domains (PDZ domains), 661–662
PSEN. *See* Presenilin (PSEN)
Pseudohypoparathyroidism (PHP), 1275, 1382–1383
Pseudohypoparathyroidism type IA (PHPIA), 1431
Pseudohypoparathyroidism type Ib (PHP1B), 1382–1383, 1431, 1441–1443
  clinical features, 1441
  genetics, 1441–1442
  pathogenesis, 1442–1443
Pseudohypophosphatasia, 1574–1575, 1580
Pseudopseudohypoparathyroidism (PPHP), 1393–1394, 1431
Pseudovitamin D deficiency, 1510
Pseudoxanthoma elasticum (PXE), 1674–1675
PSM. *See* Presomitic mesoderm (PSM)
PSNS signaling. *See* Parasympathetic nerve system signaling (PSNS signaling)
Psoriasis, 1742
PTCH1. *See* Patched homolog 1 (PTCH1)
PTEN. *See* Phosphatase and tensin homologue gene deleted on chromosome 10 (PTEN)
PTH. *See* Parathyroid hormone (PTH)
PTH1R. *See* Parathyroid hormone receptor 1R (PTH1R)
Pthlh gene, 603, 1309
Pthlh$^{+/-}$ mice, 606
PTHR. *See* PTHrP receptor (PTHR)
PTHrP. *See* Parathyroid hormone-related protein (PTHrP)
PTHrP receptor (PTHR), 646–647
  expression, 660
  signal transduction in kidney tubular cells, 660–661
  signal transduction in regulation of calcium and phosphate excretion, 665–666
  signaling in tubular epithelial cells, 661–662
Puberty, 253, 253f
PubMed database, 183–184
PUFAs. *See* Polyunsaturated fatty acids (PUFAs)
Pulmonary arterial hypertension (PAH), 556
Pulmonary indications, 1667
Pulsating fluid flow (PFF), 137, 150–151
pVHL. *See* von Hippel–Lindau tumor-suppressor protein (pVHL)
PVN. *See* Paraventricular nucleus (PVN)
PXE. *See* Pseudoxanthoma elasticum (PXE)
Pycnodysostosis, 1557–1558, 1712
PYD. *See* Pyridinoline (PYD)
PYK2. *See* Proline-rich tyrosine kinase 2 (PYK2)
Pyridinium cross-links, 343
Pyridinoline (PYD), 344, 1478, 1603, 1802
Pyridoxal 5′-phosphate (PLP), 1573–1574, 1582, 1582f, 1588–1589
4-Pyridoxic acid (4-PA), 1588
Pyrophosphatase (PPi-ase), 1571
Pyrophosphate, 47
  arthropathy, 1579
Pyrrole cross-links, 343–344
  peptides in urine, 344
PYY. *See* Peptide YY (PYY)

Q

QCT. *See* Quantitative computed tomography (QCT)
qOPs. *See* Quiescent osteoclast precursors (qOPs)
Quantitative computed tomography (QCT), 946, 1476–1478, 1834, 1857
Quantitative image processing, 1838–1841
  filtration and segmentation, 1839–1840
  finite element analysis, 1841
  quantitative morphometry, 1840–1841
Quiescent osteoclast precursors (qOPs), 124

R

R-Smads. *See* Receptor-activated Smads (R-Smads)
R-SMADs. *See* Regulatory SMADs (R-SMADs)
R26-DTA mice, 1796–1797
RA. *See* Retinoic acid (RA); Rheumatoid arthritis (RA)
RAA axis. *See* Renin–angiotensin–aldosterone axis (RAA axis)
*RAD*51 gene, 1412
*RAD*54 gene, 1412
Radiolabeled bisphosphonates, 1602
Radiological studies, 1384
Raloxifene, 866t, 868, 868f–869f, 872, 874, 877, 881, 884, 930
RAM. *See* RBPJ-association module (RAM)
RAMP. *See* Receptor-activity-modifying protein (RAMP)
Randomized controlled trial (RCT), 1647–1648, 1661, 1769
Randomized placebo-controlled trial, 1002
RANK. *See* Receptor activator of nuclear factor kappa B (RANK)
RANKL. *See* Receptor activator of nuclear factor Kappa β ligand (RANKL)
Rap1, 1432
RAR. *See* Retinoic acid receptor (RAR)
Rare Diseases Clinical Research Network (RDCRN), 1490
Ras-like GTPase domain, 1433
Rat insulin-II promoter (RIP), 855
RBF. *See* Remodeling-based bone formation (RBF)
RBP-J. *See* Recombinant recognition sequence binding protein at Jk site (RBP-J)
RBPJ-association module (RAM), 1084
RBPjk. *See* Recombination signal binding protein for immunoglobulin k J region (RBPjk)
RCAD. *See* Renal cysts and diabetes syndrome (RCAD)
RCT. *See* Randomized controlled trial (RCT)
RD. *See* Runt domain (RD)
RDA. *See* Recommended dietary allowance (RDA)
RDCRN. *See* Rare Diseases Clinical Research Network (RDCRN)
Reactive oxygen species (ROS), 18–19, 278–279, 1154–1155, 1158–1159
  activation by, 1158–1159
Receptor activator of NF-kB ligand (RANKL)
  extraskeletal effects
    functions in immune and thermal regulation, 1201–1202
    in malignancies, 1201
    in mammary gland development and tumorigenesis, 1199–1201
Receptor activator of nuclear factor kappa B (RANK), 113, 118, 650, 869–870, 1690, 1690f
Receptor activator of nuclear factor Kappa β ligand (RANKL), 18, 57, 95, 98–99, 124, 140, 185, 221–222, 255, 277, 362, 411, 546, 606, 650, 723–724, 777, 832, 951–952, 995, 1095, 1199, 1200f, 1255, 1280, 1320, 1335–1336, 1559–1560, 1604–1605, 1635–1636, 1690, 1719, 1772, 1805
  functions in immune and thermal regulation, 1201–1202
    immune system development and thermoregulation, 1201–1202
    inflammation, autoimmunity, and antitumor effects, 1202

Receptor activator of nuclear factor Kappa β ligand (RANKL) (*Continued*)
inhibition, 1702
inhibitors effects in bone, 1691–1693
in malignancies, 1201
in mammary gland development and tumorigenesis, 1199–1201
in breast cancer development and metastasis, 1199
in lactating mammary gland, 1199
osteoclastogenesis supported by, 116
osteoprotegerin and, 115–116
Receptor tyrosine kinases (RTKs), 1310–1311
Receptor-activated Smads (R-Smads), 1159
Receptor-activity-modifying protein (RAMP), 789
Receptors
for calcitonin-family peptides, 792–793
desensitization, 792
environment, 1190–1191
for PTH and PTHrP, 646–647
Recombinant human BMPs (rhBMPs), 1193–1194
Recombinant human IGF-I (rhIGF-I), 990, 1007
Recombinant inbred strains (RI strains), 1623
Recombinant recognition sequence binding protein at Jk site (RBP-J), 118–119
Recombinant sclerostin, 1718
Recombination signal binding protein for immunoglobulin k J region (RBPjk), 25, 1083
Recommended dietary allowance (RDA), 1646–1647
Reference intervals (RIs), 1812
Reference point indentation (RPI), 1927–1928, 1928f
cyclic, 1927–1928
Regenerative medicine, skeletal stem cells and, 65
Regulating collagen type I, intronic elements in, 312–313
Regulator of G protein signaling (RGS), 1432–1433
RGS2, 651
RGS5, 549
Regulatory mechanisms, 1085–1086
of notch ligands, 1087
Regulatory SMADs (R-SMADs), 19–20
Regulatory T cells (Tregs), 1201, 1362–1363
Remodeling-based bone formation (RBF), 266–267, 1719–1722
Renal anomalies syndrome, 1359–1361
Renal cysts and diabetes syndrome (RCAD), 517
Renal Mg^{2+} wasting, 516–517
Renal osteodystrophy, 451–452, 1465, 1467
bone histology, 1469f
contributions to pathogenesis, 1473t
historical classification, 1468t
KDIGO classification of, 1465t
spectrum in CKD, 1471–1472
Renal phosphate absorption, PTH regulation of, 665–666
Renal phosphate transport, 477
PTH signaling of, 666–667
Renal sodium/phosphate cotransporters, 478
Renal toxicity, 1676
Renin–angiotensin–aldosterone axis (RAA axis), 636
Reporters, 1888–1889
Reproductive system
hormonal effects, 876–877
mammary, 873–876
ovarian effects, 876
uterus, 871–873
vaginal effects, 877
Reproductive tissues, 849–853
implantation and early pregnancy, 852–853
pathophysiology of PTHrP in placenta, 853
placenta and fetal membranes, 851–852
PTHrP and placental calcium transport, 849–851
uterus and extraembryonic tissues, 851
Reprogramming
somatic cells, 1563
technology, 1563
Repurposing calcilytics for new indications
HPT, 1666–1667
pulmonary indications, 1667
RER. *See* Rough endoplasmic reticulum (RER)
Resolvin E1 (RvE1), 1073–1074
Resorption
activity, 1768
lacuna, 404–405
lacunae, 113
phase of remodeling and cessation, 223
Respiratory syncytial virus (RSV), 1603–1604
Restriction fragment length polymorphisms (RFLPs), 1363–1364
RET gene, 1412
Retinoic acid (RA), 9–10
Retinoic acid receptor (RAR), 722
Retinoid X receptor (RXR), 323, 1508
Reversal phase as coupling mechanism, 234–236
"Reversal resorption" surface, 776
Reverse tetracycline transactivator (rtTA), 1797, 1891
Reversible bone loss, 256–259
RFLPs. *See* Restriction fragment length polymorphisms (RFLPs)
RGD sequence. *See* Arg–Gly–Asp sequence (RGD sequence)
RGD-containing glycoproteins, 365–371, 366t
BAG-75, 368
γ-carboxyglutamic acid-containing proteins, 370–371
fibrillins, 368
fibronectin, 367
irisin, 367
small integrin-binding ligands with N-linked glycosylation, 368–370
TSPs, 367
vitronectin, 368
RGS. *See* Regulator of G protein signaling (RGS)
rhBMPs. *See* Recombinant human BMPs (rhBMPs)
Rheumatoid arthritis (RA), 1168, 1206, 1212, 1699. *See also* Osteoarthritis (OA)
rhIGF-I. *See* Recombinant human IGF-I (rhIGF-I)
Rho-associated protein kinase (ROCK), 402–403
RI strains. *See* Recombinant inbred strains (RI strains)
Ribosomal protein S6 (rpS6), 588–589
phosphorylation, 588–589
Rickets, 1508–1510, 1531
RIP. *See* Rat insulin-II promoter (RIP)
RIs. *See* Reference intervals (RIs)
Risedronate, 1496
RNA polymerase II (RNA pol II), 744–745
RNA-directed CRISPR/Cas9 nuclease methods, 748–749
Robison's hypothesis, 1570–1571
ROCK. *See* Rho-associated protein kinase (ROCK)
Rodents, 276, 1899
histomorphometry, 1899–1900
histomorphometry of rodents pharmacological efficacy, 1915–1917
models, 1915, 1923
Romosozumab, 188, 460, 832, 1717–1718, 1823
ROR2 conditional knockout mice (ROR2 cKO mice), 122
ROS. *See* Reactive oxygen species (ROS)
*Rosa*26 locus
lineage tracing, 1796
overexpression using, 1796–1797
Rough endoplasmic reticulum (RER), 1365
RPI. *See* Reference point indentation (RPI)
rpS6. *See* Ribosomal protein S6 (rpS6)
rs2282679, 717
RSK2 gene, 168
RSV. *See* Respiratory syncytial virus (RSV)
RTKs. *See* Receptor tyrosine kinases (RTKs)
rtTA. *See* Reverse tetracycline transactivator (rtTA)
"Rugger-jersey spine", 1554–1556
Runt domain (RD), 380–381
Runt-related transcription factor 2 (RUNX2), 8, 25, 52, 163–165, 648, 778, 918, 996, 1341, 1718
accumulation and function, 165–167
functions during skeletogenesis beyond osteoblast differentiation, 165
Osterix, 167–168
RunX2/Cbfa1-binding element, 306
transcription factor, 313
RUNX2. *See* Runt-related transcription factor 2 (RUNX2)

RvE1. *See* Resolvin E1 (RvE1)
RXR. *See* Retinoid X receptor (RXR)

S

S100g, 746–747
S1P. *See* Sphingosine-1-phosphate (S1P)
S6 kinase 1 (S6K1), 588–589
SAA. *See* Serum amyloid A (SAA)
Saccharomyces cerevisiae, 300
Saline volume expansion (SVE), 626, 626f
Salmon calcitonin (sCT), 789–790
Salt craving, 515
Salt-inducible kinase 2 (SIK2), 649, 1282, 1283f
Sandwich assay, 1318
Sanjad–Sakati syndrome, 1361–1362
SARMs. *See* Selective androgen–receptor modulators (SARMs)
SASP. *See* Senescence-associated secretory phenotype (SASP)
Satb2 (nuclear matrix protein), 168–169
SBE. *See* Smad binding element (SBE)
sBMD. *See* Standardized aBMD (sBMD)
SBP. *See* Subchondral bone plate (SBP)
SC65. *See* Synaptonemal complex 65 (SC65)
Scaffolds, 64
Scanning electron microscopy (SEM), 1926
SCF. *See* Stem cell factor (SCF)
Scientific Advisory Committee on Nutrition (SACN), 765
Scl. *See* Sclerostin (SOST)
Scl-Abs. *See* Sclerostin antibodies (Scl-Abs)
Scleraxis (*Scx*), 1789
Sclerosing bone dysplasias, 1712
Sclerosteosis, 1712–1714, 1713t
Sclerostin (SOST), 140, 186, 188, 277, 363–365, 531, 650–651, 957, 1065, 1281–1282, 1474, 1637, 1711, 1716f, 1777, 1803, 1806–1807
  biology and biochemistry
    expression of sclerostin protein, 1714
    genetic manipulation of sclerostin expression in mice, 1716–1717
    human monogenic high bone mass conditions related to sclerostin, 1712–1714
    sclerostin mechanism of action in skeleton, 1714–1716
    structure and functional domains of sclerostin, 1714
  β-catenin/sclerostin axis, 531
  gene, 140, 177–178, 189, 531, 650–651, 1064–1065, 1711
  knockout and overexpression, 1716–1717
  mechanism of action in skeleton, 1714–1716
  mRNA expression, 832
  protein, 1714
  by Scl-Abs in vivo, 1719–1722
  stimulation, 531
  structure and functional domains of, 1714
  transgenic mice, 1715–1716
Sclerostin antibodies (Scl-Abs), 267, 1724, 1725f, 1916–1917
  effects in animal models of postmenopausal osteoporosis, 1719–1722
  stimulated bone formation, 1916–1917
  in vivo, 1719–1722
Sclerotic bone, 1602
Sclerotome differentiation, 11
sCNTFR. *See* Soluble ciliary neurotrophic factor receptor (sCNTFR)
Scout view, 1843
sCT. *See* Salmon calcitonin (sCT)
Scx. *See* Scleraxis (*Scx*)
SDF1. *See* Stromal cell-derived factor 1 (SDF1)
Sea anemones, 295
Second-messenger systems for PTH and PTHrP, 646–647
Secondary hyperparathyroidism (SHPT), 588, 1509, 1660–1665, 1741–1742
  calciotropic end points, 1661–1662
  cardiovascular end points, 1662
  cinacalcet, 1660–1661
  etelcalcetide, 1663–1665
  evocalcet, 1662–1663
  skeletal end points, 1662
Secondary ossification center (SOC), 1142
Secreted Frizzled-related protein 4 (sFRP4), 477
Secreted Frizzled-related protein-1 (sFRP-1), 185
Secreted noncollagenous proteins of bone
  bone matrix glycoproteins and ectopic calcifications, 372–373
  control of gene expression, 372
  glycoproteins, 363–365
  hyaluronan, 363
  proteins, 372
  proteoglycans, 360–363
  RGD-containing glycoproteins, 365–371
  serum proteins, 371–372
Secreted PLA$_2$s (sPLA$_2$s), 1249
Secreted protein acidic and rich in cysteine–osteonectin (SPARC), 303–304, 365, 1338–1339
Sectioning of plastic-embedded bone specimens, 1902–1903
SEF. *See* Similar expression to FGF (SEF)
Segmentation, 1839–1840
  clock, 9–10
SEGRAs. *See* Selective glucocorticoid receptor activators (SEGRAs)
Selective androgen–receptor modulators (SARMs), 972, 977–978
Selective estrogen receptor modulators (SERMs), 454–455, 863, 865–867, 1650
  affinities for human estrogen receptors ERα and ERβ, 867t
  approved for human use, 866t
  chemistry, 868
  future directions with, 884
  general safety profile and other pharmacological considerations
    pharmacokinetics, 883–884
    safety, 882–883
  mechanism of action, 866f
  pharmacology, 868–884
    cardiovascular system, 877–880
    CNS, 880–882
    reproductive system, 871–877
    skeletal system, 869–871
Selective glucocorticoid receptor activators (SEGRAs), 932
Selective intraarterial calcium injection (SACI), 1298–1299
Selective serotonin reuptake inhibitors (SSRIs), 811
Self-renewal determination, skeletal stem cell, 56–57
SEM. *See* Scanning electron microscopy (SEM)
Semaphorin 4D (Sema4D), 99, 231
Semaphorins, 231
Senescence-associated secretory phenotype (SASP), 278
Senescent cells, 280
Senolytics, 280
Sensory organelle, 1775
Sensory organs, 6
Sequestosome 1 (SQSTM1), 1608–1609
SERMs. *See* Selective estrogen receptor modulators (SERMs)
SERPINF1, 352
SERPINH1, 351
Serum, 1252
  biomarkers, 1695
  calcium, 657–659, 658t, 1001–1002, 1700
  phosphate, 470, 476f, 659–660
  phosphorus, 470, 1658–1659
  proteins, 371–372, 371t, 1677
  tartrate-resistant acid phosphatase, 1603
Serum amyloid A (SAA), 1257–1258
  SAA3, 1257–1258
SeSAME syndrome (SESAMES), 515
Setrusumab, 1717–1718
Severe-deforming OI, 1490–1491
Sex
  hormone replacement, 932
  and racial differences in bone structure, 252–254
Sex steroids, 586, 988
  deficiency, 276
  loss of, 282–283
Sexual dimorphism in trabecular and cortical bone loss, 263
sFRP-1. *See* Secreted Frizzled-related protein-1 (sFRP-1)
sFRP4. *See* Secreted Frizzled-related protein 4 (sFRP4)
sFRPs. *See* Soluble FRPs (sFRPs)
SGA. *See* Small for gestational age (SGA)
SGLT-2 inhibitor. *See* Sodium/glucose cotransporter-2 inhibitor (SGLT-2 inhibitor)
SGS. *See* Shprintzen–Goldberg syndrome (SGS)
SHH. *See* Sonic Hedgehog (SHH)
Shn3 protein, 166–167
Short-hairpin RNAs (shRNAs), 53

SHOTZ. *See* Skeletal Histomorphometry in Patients on Teriparatide or Zoledronic Acid Therapy study (SHOTZ)
Shprintzen–Goldberg syndrome (SGS), 1169–1170
SHPT. *See* Secondary hyperparathyroidism (SHPT)
shRNAs. *See* Short-hairpin RNAs (shRNAs)
"Shutter-binding" mechanism, 1714–1715
Sialic acid-binding immunoglobulin-like lectin 15 (SIGLEC-15), 119–120
SIBLING family. *See* Small, integrin-binding ligand, N-linked glycoprotein family (SIBLING family)
Sieverts (Sv), 1847
SIGLEC-15. *See* Sialic acid-binding immunoglobulin-like lectin 15 (SIGLEC-15)
Signal transducer and activator of transcriptions (STATs), 1116
  STAT1, 308
  STAT6, 1213
Signal-to-noise ratio (SNR), 1873
Signal(ing)
  bone remodeling, 254
  molecules, 89
  pathways, 52
    transgenic mouse reporters of, 1789–1790
  peptides, 299
  termination at PTHR1, 700–701
  transduction in osteoclastogenesis
    M-CSF receptor FMS, 116–118
    RANK, 118
    tumor necrosis factor receptors, 118–119
SIK2. *See* Salt-inducible kinase 2 (SIK2)
sIL. *See* Soluble receptor (sIL)
Similar expression to FGF (SEF), 1116
Single-cell
  sequencing methods, 95
  transcriptome analysis, 1064
Single-chain glycosylated 72-kDa polypeptide, 388
Single-ESR1-null mice, 976
Single-nucleotide polymorphisms (SNPs), 490, 818, 1194–1195, 1608–1609, 1615–1616
Single-photonemission computed tomography (SPECT/CT), 484–485
siRNA. *See* Small interfering RNA (siRNA)
siRNA targeting, 1788
Sirtuin (SIRT1), 26–27, 649
Six1/Six4, 578–579
SK-MSCT. *See* Skeletal mesenchymal stem cell transplantation (SK-MSCT)
Skeletal aging, bone extrinsic mechanisms to, 282–284
Skeletal blood vessels, 208, 210, 1142
Skeletal cells and role of IGF-I, 997–998
Skeletal development, 1141
  growth and maintenance, 426
Skeletal disorders
  GH and IGF-I as treatments for, 1000–1007
  GH treatment for, 1000
Skeletal end points, 1662
Skeletal formation, 5
Skeletal fracture, 1923
Skeletal fragility, 1923
Skeletal growth in T1DM, 942–943
Skeletal Histomorphometry in Patients on Teriparatide or Zoledronic Acid Therapy study (SHOTZ), 460
Skeletal lineage, 53
Skeletal malignancies, 1341–1342
Skeletal mesenchymal stem cell transplantation (SK-MSCT), 1563
Skeletal metastases of cancer associated with bone matrix–derived TGF-β, 1170–1172
Skeletal morphology, 1761
Skeletal muscle, 479
  cross talk, 1937, 1937f
  function, 478
Skeletal overgrowth, 1119
Skeletal parathyroid hormone actions, 1279–1284
Skeletal retention, 1678
Skeletal stem cells (SSC), 46, 51, 90
  developmental origins of bone and, 47–48
  in disease, 58–63
    FD of bone and MAS, 58–60
    inherited forms of bone marrow failure, 60–61
    skeletal stem cells in tissue engineering, 63–64
    SSC/BMSC role in acquired inflammation, 62–63
  and regenerative medicine, 65
  self-renewal determination, 56–57
  SSCs/BMSCs
    cell shape, 53–54
    cell–cell and cell–substrate interactions, 53–54
    characterization, 54–57
    epigenetic controls, 52–53
    hormonal regulation, 51
    isolation, 54
    markers, 56
    mechanical forces, 53–54
    microRNAs, 53
    potency, 54–55
    regulation of, 51–54
    role in postnatal bone turnover and remodeling, 57–58
    signaling pathways and transcription factors, 52
    skeletal stem cell self-renewal determination, 56–57
  stem cell and nonestem cell functions of skeletal stem cells, 65–66
  in tissue engineering, 63–64
Skeletal stem cells, 184
Skeletal structure, 219
Skeletal system, 5
  SERMs in
    clinical studies, 870–871
    preclinical studies, 869–870
Skeletal TGF-β signaling, 1166–1167
Skeletal tissues, gene function in, 1788–1789
Skeletal traits in mice, GWAS for, 1622–1624
Skeletal vascular system, 1142
Skeletal vascularization
  bone cells' control of, 208–210
  processes, 1141
Skeletal-related events (SREs), 1698, 1818–1819
Skeletogenesis, 19
  Runx2 functions during, 165
Skeleton, 89
  IGF regulatory system and relationship to, 989f
    IGF-I, 989–991
    IGFBPs, 989–993
    IGF–II, 989–991
    IGFRs, 989–991
    proteases, 993–994
  sclerostin mechanism of action in
    effects on mesenchymal stem cells, 1715–1716
    inhibition of canonical Wnt signaling pathway, 1715
    interaction with LRP4/5/6, 1714–1715
SKI. *See* Sloan-Kettering Institute proto-oncoprotein (SKI)
Skin
  biochemistry of PTHrP, 840
  cancer, 763–764
  pathophysiology of PTHrP, 841–842
  PTHrP and receptor expression, 840–842
SLC. *See* Small latent TGF-β1 complex (SLC)
SLC12A3 mutation, 515
SLC34A3/NPT2c mutations, 489
SLC41. *See* Solute carrier 41 (SLC41)
SLE. *See* Systemic lupus erythematosus (SLE)
Sloan-Kettering Institute proto-oncoprotein (SKI), 1169–1170
SLP. *See* SRC homology 2 domain-containing leukocyte protein (SLP)
SLRP. *See* Small leucine-rich proteoglycan (SLRP)
αSMA. *See Alpha-smooth muscle actin* (αSMA)
SMAA. *See* Smooth muscle α-actin (SMAA)
Smad binding element (SBE), 314–315, 386
Smad-independent signaling pathways, 1161–1162
Smad-mediated signaling, 1155–1156
Small, integrin-binding ligand, N-linked glycoprotein family (SIBLING family), 94, 139–140, 1533
Small for gestational age (SGA), 1000–1001
Small integrin-binding ligands with N-linked glycosylation, 368–370
  BSP-II, 369
  DMP1, 369
  MEPE, 369–370
  osteopontin, 368–369
Small interfering RNA (siRNA), 1563–1564

Small latent TGF-β1 complex (SLC), 1154
Small leucine-rich proteoglycan (SLRP), 298, 360
Small noncoding RNAs, 53
SMD. *See* Spondylometaphyseal dysplasia (SMD)
SMI. *See* Structure model index (SMI)
Smooth muscle α-actin (SMAA), 91
SMRT complexes, 725
SMS. *See* Somatostatin (SMS)
SNPs. *See* Single-nucleotide polymorphisms (SNPs)
SNR. *See* Signal-to-noise ratio (SNR)
SNS. *See* Sympathetic nervous system (SNS)
SOC. *See* Secondary ossification center (SOC)
SOCS. *See* Suppressors of cytokine signaling (SOCS)
Sodium excretion, 667–669
Sodium phosphate transporter family, 665t
Sodium/glucose cotransporter-2 inhibitor (SGLT-2 inhibitor), 950–951
Sodium/hydrogen exchanger regulatory factors (NHERFs), 646–647, 651, 1277
  NHERF1, 471, 667
    mutations, 490
  NHERF1/2, 661–662
Solid tumors, 1697
Solid-phase binding studies, 350
Soluble ciliary neurotrophic factor receptor (sCNTFR), 1222–1223
Soluble form RANKL (sRANKL), 116, 777–778
Soluble FRPs (sFRPs), 1805
Soluble receptor (sIL), 1218–1219
Solute carrier 41 (SLC41), 512
Somatic lateral plate mesoderm, 47
Somatostatin (SMS), 987
Somatostatin analogues (SSAs), 1296–1297
Somatostatin receptors (SSTRs), 1296–1297
Somatostatinoma, 1299
Somatotrope, 987
Sonic Hedgehog (SHH), 9, 14, 1345
Sonography, 1580
SOST. *See* Sclerostin (SOST)
SOX3. *See* Sry-box3 (SOX3)
Sox9. *See* SRY-Box 9 (*Sox9*)
Sp. *See* Specificity protein (Sp)
SPARC. *See* Secreted protein acidic and rich in cysteine–osteonectin (SPARC)
Specialized proresolving lipid mediators (SPMs), 1073
Specificity protein (Sp), 576
SPECT/CT. *See* Single-photonemission computed tomography (SPECT/CT)
Spectrin, 142–143
Sphingosine-1-phosphate (S1P), 229–230, 778–779, 801, 1806
sPLA$_2$s. *See* Secreted PLA$_2$s (sPLA$_2$s)
Splanchnic mesoderm, 47
Spleen tyrosine kinase (SYK), 119
SPMs. *See* Specialized proresolving lipid mediators (SPMs)
Spondylocostal and spondylothoracic dysostosis, 1097
Spondyloepiphyseal dysplasia, 360
Spondylometaphyseal dysplasia (SMD), 367
Spondylothoracic dysostosis, 1097
Sport-induced enhancement in bone, 1765–1766
spp, 368–369
SPRINT. *See* Systolic Blood Pressure Intervention Trial (SPRINT)
SQSTM1. *See* Sequestosome 1 (SQSTM1)
SR. *See* Synchrotron radiation (SR)
sRANKL. *See* Soluble form RANKL (sRANKL)
SRC homology 2 domain-containing leukocyte protein (SLP), 119
SRCs. *See* Steroid receptor coactivators (SRCs)
SREs. *See* Skeletal-related events (SREs)
SRY-Box 9 (*Sox9*), 8, 25–26, 184, 1892
Sry-box3 (SOX3), 578–579, 1366–1368
SSAs. *See* Somatostatin analogues (SSAs)
SSC. *See* Skeletal stem cells (SSC)
SSc. *See* Systemic sclerosis (SSc)
SSRIs. *See* Selective serotonin reuptake inhibitors (SSRIs)
SSTRs. *See* Somatostatin receptors (SSTRs)
Staining
  bone remodeling dynamics, 1907
  cement line stain, 1904
  Masson–Goldner stain, 1904
  toluidine blue stain, 1903
  TRAcP, 1904
  of undecalcified bone sections, 1908f
  von Kossa/MacNeal's stain, 1903
Standard DXA, 1860–1862
  DXA technique, 1860–1862
  fracture prediction using areal bone mineral density, 1862
  monitoring osteoporosis treatment with DXA, 1862
  VFA, 1862
Standard methylmethacrylate embedding, 1901
Standard therapy, 489–490
Standardized aBMD (sBMD), 1861
Staphylococcus aureus, 1211
STAR trial. *See* Study of Tamoxifen and Raloxifene trial (STAR trial)
Static morphometry, 944
Static parameters, 448–449
Statins, 951
  Stat1 protein, 166
  STAT5, 916
STATs. *See* Signal transducer and activator of transcriptions (STATs)
Steady-state locomotion, 1761
Stem cell factor (SCF), 75
Stem cell(s), 45–46, 1889
  functions of skeletal stem cells, 65–66
  niche concept, 74
Steroid hormones, 1311
Steroid receptor coactivators (SRCs), 721–722, 724–725
  SRC1, 744–745
Strain
  in bone, 1761
  distribution, 1765–1766
  gradient, 1766
  magnitude, 1764
  milieu, 1764
  rate, 1765
  strain-derived interstitial fluid flow, 147
  tensor, 1764
Stress fracture healing, 1063
Stromal cell-derived factor 1 (SDF1), 75, 1337
Stromelysin, 381
Structure model index (SMI), 1840
Study of Tamoxifen and Raloxifene trial (STAR trial), 875
Stüve–Wiedemann syndrome (STWS), 1221
Subchondral bone, 1167–1168, 1173
Subchondral bone plate (SBP), 1167–1168
SUG1, 725
Sulfhydryl-donating proteins, 1663–1664
Superoxide dismutase enzymes, 278–279
Suppressors of cytokine signaling (SOCS), 1122–1123
  proteins, 1227
Supraphysiological glucose concentrations, 951–952
Sv. *See* Sieverts (Sv)
SVE. *See* Saline volume expansion (SVE)
SYK. *See* Spleen tyrosine kinase (SYK)
Sympathetic nervous system (SNS), 79, 811–812, 1439–1440. *See also* Central nervous system (CNS)
  control counterregulation of bone remodeling, 812–813
  control regulators of bone remodeling
    endocannabinoid signaling, 813
    NMU, 813
    orexin signaling, 813–814
  leptin's action on bone remodeling mediated by, 811–812
  signaling, 1344
Synaptonemal complex 65 (SC65), 350–351
Synaptotagmin I, 114
Synchrotron radiation (SR), 1836
Syndecans, 410
Synthetic modulators, 542–543
Systemic fuel metabolism, glucocorticoid excess and, 926–927
Systemic inflammation, 151
Systemic lupus erythematosus (SLE), 1202
Systemic sclerosis (SSc), 311–312
Systolic Blood Pressure Intervention Trial (SPRINT), 634–635

T

T cell factor (TCF), 279
T cell factor/lymphoid enhancer-binding factor (TCF/LEF), 179, 761
T cells, 79, 1343
T scores, 1861

T-RIIΔk. *See* Transgenic mice expressing kinase-deficient type II TGF receptor (T-RIIΔk)
T1DM. *See* Type 1 diabetes mellitus (T1DM); Type I diabetes mellitus (T1DM)
T2DM. *See* Type 2 diabetes mellitus (T2DM)
T3 binding isoforms, 896
TADs. *See* Topological-associated domains (TADs)
TAFII-20, 1605
TAK1. *See* TGF-β activated kinase 1 (TAK1)
TAL. *See* Thick ascending limb (TAL)
TALE proteins, 11
Tamoxifen, 866t, 868, 868f, 872, 874–875, 879, 881, 1069–1070, 1887–1889, 1894–1895
  effects on bone, 1890–1891
  Tamoxifen-inducible CreER^{T2} genetic tools, 1892
TAMs. *See* Tumor-associated macrophages (TAMs)
TANGO genes. *See* Transport and Golgi organization genes (TANGO genes)
TANK-binding kinase 1 (TBK1), 1607
TAP. *See* Total alkaline phosphatase (TAP)
Target cell catabolic enzymes, 1745–1746
Tartrate-resistant acid phosphatase (TRAP), 113, 234, 794, 944, 1064, 1122
  immunoreactivity, 234
  TRAcP, 1559
  TRAP 5b, 1478, 1719
  TRAPc histochemistry, 1901
Tartrate-specific acid phosphatase (TRACP), 389
Tax, 1310
Tb.N. *See* Trabecular number (Tb.N)
Tb.Sp. *See* Trabecular separation (Tb.Sp)
Tb.Th. *See* Trabecular thickness (Tb.Th)
TBCE. *See* Tubulin-specific chaperone (TBCE)
TBDs. *See* Telomere biology disorders (TBDs)
TBK1. *See* TANK-binding kinase 1 (TBK1)
TBS. *See* Trabecular bone score (TBS)
TbSp. *See* Trabecular spacing (TbSp)
*TBX*1 gene, 578–579, 1357–1358
TBX6 (T-box transcription factor), 9–10
TCF. *See* T cell factor (TCF)
TCF/LEF. *See* T cell factor/lymphoid enhancer-binding factor (TCF/LEF)
Tcfl gene, 170–171
TE. *See* Echo time (TE)
TECOmedical, 1806
Telomerase, 60
Telomere biology disorders (TBDs), 60
Telopeptide
  hydroxylysines, 343
  lysines, 340–342, 348
Testin, 546–547
Testosterone, 829–831, 972–974
  treatment and change in bone architecture, 976
Tet expression systems, 1891
Tet-off system, 1891
Tet-On system, 1891
1,4,7,10-Tetraazacyclododecane-1,4,7,10-tetraacetic acid (DOTA), 484–485, 1544–1546
Tetracycline labeling and surgical procedure, 445–453
  clinical indications for bone biopsy, 453
  double tetracycline label, 446f
  hyperparathyroidism, 449
  normal bone, 449
  osteomalacia, 451
  osteoporosis, 452–453
  renal osteodystrophy, 451–452
  routine histomorphometric variables, 447–449
  sample preparation and analysis, 447, 448f
Tetracycline-controlled transactivator protein (tTA), 1891
Tetracycline-responsive promoter element (TRE), 1891
Tetrahydronaphthalenes, 868, 868f
Tetrahydroquinoline derivative, 977–978
Tetranectin, 365
TFPI. *See* Tissue factor pathway inhibitor-1 (TFPI)
TGF. *See* Transforming growth factor (TGF)
TGF-β activated kinase 1 (TAK1), 1190
TGFβ. *See* Transforming growth factor β (TGFβ)
TGFβ type II receptors (TGFβRIIs), 321, 1168, 1338
TGFβ-responsive element (TβRE), 307–308
TGFβRIIs. *See* TGFβ type II receptors (TGFβRIIs)
TGN. *See* *Trans*-golgi network (TGN)
TH. *See* Tyrosine hydroxylase (TH)
Therapeutic(s)
  and bone diseases, 1193–1194
  glucocorticoids, 919
  potential of VEGF for bone repair, 1148–1149
Thermal regulation, 1201–1202
Thermoregulation, 1201–1202
Thiazide diuretics, 662
Thiazolidinediones (TZDs), 950–951
Thick ascending limb (TAL), 510–511, 550
  disturbed Mg^{2+}reabsorption in, 512–514
Thioredoxin-like protein, 713–714
Three dimensions (3D), 1834
  DXA, 1873
Three-point bending, 1924–1925, 1925f
3D quantification of trabecular architecture, 1857
Thrombopoietin, 78
Thrombospondin (TSP), 365–367
  TSP-1, 1154–1156, 1338–1339
Thromboxanes (TXAs), 1247–1248
Thymic neuroendocrine tumors, 1301
Thyroid hormone receptors (TRs), 895–896
Thyroid hormone response elements (TREs), 896
Thyroid hormones, 323, 895
  cellular effects of thyroid hormones on bone, 899–901
  future directions, 904–905
  intracellular mechanism of action, 895–899
  pathophysiological effects of altered thyroid hormone status in humans
    hyperthyroidism, 903–904
    hypothyroidism, 902–903
    subclinical hyperthyroidism, 904
    subclinical hypothyroidism, 903
  in vivo responses of skeleton to, 901–902
Thyroid transcription factor 1 (TTF1), 547
Thyroid-stimulating hormone (TSH), 1436
Thyrotoxicosis, 904
Thyrotropin (TSH), 898
  as independent agent of bone metabolism, 898–899
Thyroxine, 986–987
TIBD. *See* Tumor-induced bone disease (TIBD)
TID. *See* Total indentation distance (TID)
Time-dependent accumulation, 278
Time-lapse imaging, 1845
TIMP. *See* Tissue inhibitor of metalloproteinases (TIMP)
TIO. *See* Tumor-induced osteomalacia (TIO)
TIP39. *See* Tuberoinfundibular peptide of 39 residues (TIP39)
TIP39/PTH2R receptor signaling, 630–631, 705–706
Tissue engineering
  cell sources, 64
  scaffolds, 64
  skeletal stem cells in, 63–64
Tissue factor pathway inhibitor-1 (TFPI), 878
Tissue inhibitor of metalloproteinases (TIMP), 303–304, 379–380
  TIMP-1, 652
Tissue nonspecific alkaline phosphatase (TNAP), 363, 474
  ALPL
    gene defects, 1584
    structural defects, 1584–1585
  deficiency, 1583–1585
  inheritance, 1584
  substrates, 1587
Tissue nonspecific isoenzyme of ALP (TNSALP), 1570–1572
Tissue-dependent variations in cross-link glycosylation, 345
Tissue-nonspecific alkaline phosphate (TNAP), 1070
Tissue-specific elements, 311–312
Tissue-type PA (tPA), 386, 388
TissueTek, 1903
TLRs. *See* Toll-like receptors (TLRs)
TMD. *See* Transmembrane domain (TMD)
TMEM38B gene, 350, 1492

TmP/GFR. *See* Tubular maximum reabsorption of phosphate/glomerular filtration rate (TmP/GFR)
TMV. *See* Turnover, mineralization and volume (TMV)
TNAP. *See* Tissue nonspecific alkaline phosphatase (TNAP)
TNBCs. *See* Triple-negative breast cancers (TNBCs)
TNF. *See* Tumor necrosis factor (TNF)
TNF-related activation-induced cytokine (TRANCE), 115–116
TNF-related apoptosis-inducing ligand (TRAIL), 1208
TNFR. *See* Tumor necrosis factor receptors (TNFR)
TNFR type I (TNFRI), 118–119
TNFR type II (TNFRII), 118–119
*TNFS*11 gene, 1559–1560
*Tnfsf*11 gene, 723–724, 723f, 1746, 1749f
TNSALP. *See* Tissue nonspecific isoenzyme of ALP (TNSALP)
Toll-like receptors (TLRs), 1206
Toluidine blue stain, 1903
TOPMed. *See* Trans-Omics for Precision Medicine (TOPMed)
Topological-associated domains (TADs), 723–724, 1070
Toremifene, 866t, 868, 868f, 883–884
Torsion to failure, 1926
Total alkaline phosphatase (TAP), 1478
Total indentation distance (TID), 1928f
"Toulouse-Lautrec" disease, 1557–1558
Toxoplasma gondii, 62–63
tPA. *See* Tissue-type PA (tPA)
Tpr-Met, 1310–1311
TR-associated protein (TRAP), 896
Trabecular bone*See*. CT cancellous bone
Trabecular bone score (TBS), 1475–1478, 1871–1872
Trabecular bone volume, 1890
Trabecular density, 253
Trabecular number (Tb. N), 1840
Trabecular separation (Tb. Sp), 1840
Trabecular spacing (TbSp), 1618
Trabecular structure, 1873–1874
Trabecular surface, 254–255, 259–260
Trabecular thickness (Tb. Th), 1840
TRAcP. *See* Histochemical tartrate-resistant acid phosphatase (TRAcP)
TRACP. *See* Tartrate-specific acid phosphatase (TRACP)
Traditional primary therapeutic approach, 1660
TRAFs. *See* Tumor necrosis factor receptor-associated factors (TRAFs)
TRAIL. *See* TNF-related apoptosis-inducing ligand (TRAIL)
Trait variances in adulthood originate before puberty, 251–252
TRANCE. *See* TNF-related activation-induced cytokine (TRANCE)
Trans-dimerization, 423–424
Trans-golgi network (TGN), 301
Trans-Omics for Precision Medicine (TOPMed), 1625
Transcaltachia, 726
Transcription
  effects of 1, $25(OH)_2D_3$, 1519–1520
  factors, 52, 163, 648–649, 996–997, 1069, 1746–1748
    acting downstream of Wnt signaling, 170–171
    binding to pro-α1(I) proximal promoter, 305–307, 306f
    regulation of osteoblast differentiation by, 171–172
Transcription start site (TSS), 714
Transcriptional regulation, 297–299
  by β-catenin, 183
  of type I collagen genes, 305–308
    proximal promoters, 305–308
Transforming growth factor (TGF), 8
Transforming growth factor β (TGFβ), 52, 79, 223, 232, 302, 314–316, 360, 381, 409, 531, 650, 778, 851–852, 995, 1114, 1142–1143, 1153–1154, 1310, 1335–1336
  activation, 1154–1159
    by integrins, 1156–1158
    by osteoclasts, 1158
    proteolytic activation, 1155
    by ROS, 1158–1159
    by TSP-1, 1155–1156
  activators, 1154
  family abnormalities, 1475
  modulation as promising approach to management of OA, 1172–1173
  as molecular sensor in ECM, 1153–1164
  musculoskeletal pathologies associated with aberrant TGF-β signaling, 1167–1172
  signaling, 1155, 1164–1167
    and bone remodeling, 1164–1167
    canonical signaling pathways, 1155–1156
    and cell reprogramming, 1162
    in mesenchymal stem cells, 1162–1164
    Smad-independent signaling pathways, 1161–1162
  TGF-β1, 78, 650, 1338–1339
  TGF-β2, 1338–1339
  TGF-β3, 1338–1339
Transgenic approaches, 1788
Transgenic mice, 308–309, 362
  models, 133, 1141
  reporters of signaling pathways, 1789–1790
Transgenic mice expressing kinase-deficient type II TGF receptor (T-RIIΔk), 315–316
Transgenic reporter lines, 1789–1790
Transgenic-overexpression approaches, 1788
Transhepatic portal venous sampling, 1297
Transient receptor potential cation channel 5 (TRPV 5), 551, 1279
Transient receptor potential cation channel subfamily V member 6 (TRPV6), 716, 742, 746–747
Transient receptor potential channels (TRPCs), 550
Translation(al)
  control, 299–300
  mechanical information, 1775
  mechanisms, 304–305
  pharmacology of BPs, 1680–1682
Transmembrane domain (TMD), 540, 692–693, 1413
Transmembrane glycoproteins, 406–407, 423–424
Transplant bone marrow cells, 74
Transplantation assay, 56–57
Transport and Golgi organization genes (TANGO genes), 300
Transrepression, 916
Transsphenoidal surgery, 1301
TRAP. *See* Tartrate-resistant acid phosphatase (TRAP); TR-associated protein (TRAP)
Trauma-induced soft tissue damage, 1912
TRE. *See* Tetracycline-responsive promoter element (TRE)
"Treat to target" strategy, 767
Treatment holiday monitoring, 1816
Tregs. *See* Regulatory T cells (Tregs)
TREs. *See* Thyroid hormone response elements (TREs)
Tribolium castaneum (Red flour beetle), 694
Triphenylethylenes, 868
Triple-helical domain, 297–298, 299t
Triple-negative breast cancers (TNBCs), 1200
TrkA. *See* Tyrosine kinase receptor 1 (TrkA)
Trkb. *See* Tyrosine kinase receptor type B (Trkb)
TRP. *See* Tubular reabsorption of phosphate (TRP)
TRPCs. *See* Transient receptor potential channels (TRPCs)
TRPM6 gene, 510, 514–515, 518
TRPM7 gene, 515
TRPV 5. *See* Transient receptor potential cation channel 5 (TRPV 5)
TRPV6. *See* Transient receptor potential cation channel subfamily V member 6 (TRPV6)
TRs. *See* Thyroid hormone receptors (TRs)
TSH. *See* Thyroid-stimulating hormone (TSH); Thyrotropin (TSH)
TSH receptor (TSHR), 898
TSP. *See* Thrombospondin (TSP)
TSRs. *See* Type 1 repeats (TSRs)
TSS. *See* Transcription start site (TSS)
tTA. *See* Tetracycline-controlled transactivator protein (tTA)
TTF1. *See* Thyroid transcription factor 1 (TTF1)
Tuberoinfundibular peptide of 39 residues (TIP39), 623–624
Tubular maximum reabsorption of phosphate/glomerular filtration rate (TmP/GFR), 485, 1542–1543
Tubular reabsorption of phosphate (TRP), 475–476, 1542–1543

Tubulin-specific chaperone (TBCE), 1361–1362
Tumor
cells, 1319–1320
dormancy and awakening, 1338–1339
PTH-related protein in tumor tissue, 1319
tumor-suppressor genes, 1409–1412
Tumor necrosis factor (TNF), 115–116, 778, 951–952, 1206, 1604–1605. *See also* Epidermal growth factor (EGF)
additional TNF superfamily members, 1208–1209
FasL, 1208
TRAIL, 1208
superfamily members, 1208–1209
TNFα, 307, 319, 791
Tumor necrosis factor receptor-associated factors (TRAFs), 111
Tumor necrosis factor receptors (TNFR), 118–119
Tumor-associated macrophages (TAMs), 1342
Tumor-induced bone disease (TIBD), 1335–1336, 1339
key cellular and molecular mechanisms in, 1336f
Tumor-induced osteomalacia (TIO), 478, 484–485, 1529, 1533–1534, 1539. *See also* Osteomalacia
clinical presentation and diagnosis, 1542–1543
diagnostic algorithm for, 1544f
future directions, 1549
PMTs, 1540–1542
treatment
conventional medical treatment, 1547–1549
minimally invasive treatment, 1547
surgical treatment, 1547
tumor localization, 1543–1546
Tumoral calcinosis, 491–492
Tumorigenesis, 1199–1201
Turnover, mineralization and volume (TMV), 452, 1470–1471, 1470t
Twist proteins, 167
TWIST1 gene, 8–9, 167
Two-site model of ligand binding to PTHR1, 696–698
TXAs. *See* Thromboxanes (TXAs)
Type 1 diabetes mellitus (T1DM), 941
bone repair in, 951
effects on bone, 943–946
bone strength, 945–946
bone turnover, 943–944
fracture risk, 943–946
T1DM, bone density, and bone structure, 945
skeletal growth in
animal models, 942
children, 942–943
Type 1 repeats (TSRs), 1155
Type 2 diabetes mellitus (T2DM), 941, 1806–1807
bone repair in, 951
effects and insulin on bone
bone strength, 948–949
bone turnover, 947
fractures, 946–947
T2DM, bone density, and bone structure, 947–948
Type B γ-aminobutyric acid receptors (GABAB-Rs), 540
Type I collagen, 296, 652, 1491
alpha-1, 1892
and bone pathologies, 304–305
degradation and catabolism, 303–304
family of fibrillar collagens, 296–297
gene regulation, 314–324
arachidonic acid derivatives, 321–322
cytokines, 319–321
growth factors, 314–318
hormones and vitamins, 322–324
genetic mutations on type I collagen formation, 302–303
intronic elements in regulating collagen type I, 312–313
posttranscriptional regulation, 313–314
structure, biosynthesis, transport, and assembly, 297–302
structure and functional organization, 308–312
transcriptional regulation, 305–308
type I collagen–rich mineralizing tissues, 1493
Type I diabetes mellitus (T1DM), 1005
Type I NaP_i cotransporters, 664
Type I OI, 1491
Type II collagen alpha-1 (*Col2a1*), 1892
Col2a1-Cre line, 1795
transcription, 311
Type IIa NaP_i cotransporters, 664
Type III NaP_i cotransporters, 664
Type IV collagenases, 381–382
Tyrosine hydroxylase (TH), 212
Tyrosine kinase receptor 1 (TrkA), 210–211
Tyrosine kinase receptor type B (Trkb), 815
TZDs. *See* Thiazolidinediones (TZDs)
TβRE. *See* TGFβ-responsive element (TβRE)

U

U73122 (PLC inhibitor), 545
Ubiquitin–proteasome system, 281
Ubiquitous metabolic phosphate transporters, 478
ucOC. *See* Undercarboxylated OC (ucOC)
Ultrasensitive multiplex two-site immunoassay, 1319
Ultrashort TE (UTE), 1874
Uncondensed mesenchyme, 1791–1795
Undercarboxylated OC (ucOC), 952
Unmineralized bone matrix, 447
3'-Untranslated regions (3'-UTRs), 547, 721
uPA. *See* Urokinase-type PA (uPA)
Upstream segments of type I collagen genes, 308–312
upstream elements
of mouse pro-Col1a2 gene, 310–311
in pro-Col1a1 gene, 308–310
Urinary hydroxyproline, 1001–1002
Urinary N-telopeptide, 1603
Urine, pyrrolic cross-linked peptides in, 344
Urokinase-type PA (uPA), 386, 388
US Food and Drug Administration (FDA), 1637–1638
UTE. *See* Ultrashort TE (UTE)
Uterus, 851, 871–873
estrogen agonism in, 871–873
estrogen antagonism in uterus, 873
3'-UTRs. *See* 3'-Untranslated regions (3'-UTRs)
UV light narrow band, 759

V

Vacuolar proton ATPase (V-ATPase), 113
Vaginal effects of SERMs, 877
Valosin-containing protein, 1608–1609
van Buchem disease, 1282, 1712–1714
Vascular calcification, 1467, 1702
Vascular cell adhesion molecule 1 (VCAM-1), 411–412, 1338
Vascular endothelial growth factor (VEGF), 15–17, 206–207, 209–210, 382, 624, 951, 1120, 1141, 1143f, 1166
bone development and skeletal vascular system, 1142
crucial angiogenic factor, 1142–1144
and endochondral ossification, 1144–1146
homologues, 1147
during intramembranous bone formation, 1146
nonendothelial effects, 1146–1147
regulation of VEGF expression by oxygen, 1147–1148
therapeutic potential of VEGF for bone repair, 1148–1149
VEGF-A, 1474
Vascular pharmacology, PTH2R signaling in, 630–631
Vascular smooth muscle cell (vSMC), 312, 548–549, 625, 1169–1170
to PTH and PTHrP, 624–630
vSMC-derived PTHrP, 626
Vascular system, 1141
Vasculature functional roles in endochondral bone formation, 27
Vasculature of bone
vascularization of developing bone, 206–210
vascularization of mature skeleton, 208–210
Vasoconstrictors, 599–600
Vasodilator-stimulated phosphoprotein (VASP), 412–413
VAV3 exchange factor, 123
vBMD. *See* Volumetric bone mineral density (vBMD)
VCAM-1. *See* Vascular cell adhesion molecule 1 (VCAM-1)

VDCCs. *See* Voltage-dependent Ca^{2+} channels (VDCCs)
VDDR type I. *See* Vitamin D-dependent rickets type I (VDDR type I)
VDIR. *See* Vitamin D-interacting repressor (VDIR)
VDR. *See* Vitamin D receptor (VDR)
VDREs. *See* Vitamin D–response elements (VDREs)
VDR–RXR. *See* Vitamin D receptor-retinoid X receptor (VDR–RXR)
VDSP. *See* Vitamin D Standardization Program (VDSP)
VEGF. *See* Vascular endothelial growth factor (VEGF)
VEGF receptor 1 (VEGFR-1), 1144
VEGF receptor 2 (VEGFR-2), 1144
Venous thrombolic events (VTEs), 878
Ventromedial hypothalamus (VMH), 811
Venus flytrap (VFT), 540, 1368, 1413
Vertebral fracture assessment (VFA), 1860–1862
Vertebral fracture risk, 246
Very late antigen-4 (VLA-4), 1338
VFA. *See* Vertebral fracture assessment (VFA)
VFT. *See* Venus flytrap (VFT)
VHL. *See* Von Hippel–Lindau (VHL)
Vibrations, 1768
"Vicious cycle" model, 1335–1336
ViDA. *See* Vitamin D Assessment (ViDA)
VIPoma, 1299
Viral proteins, 1310
Virtual experiment, 1841
Vitamin B_6 deficiency, 1589
Vitamin B_6–dependent seizures, 1574
Vitamin D, 323–324, 713–714, 1507
  analogs, 1915
    clinical applications, 1741–1742
    gene targets and unique features of regulation, 1746–1748
    molecular mechanisms of action, 1748
    pharmacological effects, 1742–1746
    pharmacologically vitamin D compounds, 1734–1740
  binding sites in genome, 722–724
  binding to and transport of vitamin D metabolites, 715–719
  deficiency, 486–488, 757–759, 1543, 1733
  and immunity, 758–759
  intracellular trafficking of vitamin D metabolites, 720
  loading test, 1513
  metabolism, 669, 739–741, 1514–1515
    pathway, 740f
  negative vitamin D response elements, 725
  prodrugs, 1735–1739
  status assessment, 765
  vitamin D-mediated gene regulation in vitro and in vivo, 748–749
  vitamin D-resistant rickets, 478
Vitamin D Assessment (ViDA), 760
Vitamin D compounds, clinical applications of, 1741–1742
  hyperproliferative conditions, 1742
  secondary hyperparathyroidism, 1741–1742
Vitamin D receptor (VDR), 79, 323, 545, 720–727, 743, 757–759, 1413, 1508, 1733
  with β-catenin signaling, 725–726
  applying emerging methodologies to studying, 745–748
  coregulators and epigenetic changes regulating VDR function, 724–725
  dynamic impact of cellular differentiation and disease on, 748
  functions to recruiting coregulatory complexes, 744–745
  general features, 743–745
  genome-wide coregulatory recruitment, 747
  genomic location, protein structure, and regulation, 721–722, 721f
  hereditary defects in vitamin D receptor–effector system, 1513–1521
  heterodimer formation with retinoid X receptors, 744
  interaction at target cell genomes in bone cells, 745–747, 746t
  knockout mice, 1522
  mechanism of action, 722–727
  mechanistic outcomes in response to VDR/RXR binding, 747–748
  sites of DNA binding, 743–744
  VDR gene, 1412
  vitamin D receptor/retinoid X receptor/VDR element interactions, 1744–1745
Vitamin D receptor-retinoid X receptor (VDR–RXR), 1311
Vitamin D Standardization Program (VDSP), 765
Vitamin D-binding protein (DBP), 713–720, 739–740, 1735–1739, 1743–1744. *See also* Latent-TGFβ-binding protein (LTBP)
  actin scavenging, 719
  biologic function, 715–720
  fatty acid binding, 719–720
  genomic regulation, 714
  multiple functions, 714f
  neutrophil recruitment and migration with complement 5a binding, 719
  structure and polymorphisms, 715
Vitamin D-binding protein–macrophage-activating factor (DBP–MAF), 720
Vitamin D-dependent rickets type I (VDDR type I), 739–740
Vitamin D-interacting repressor (VDIR), 725
Vitamin D–response elements (VDREs), 743–744, 1508, 1739
Vitamin K epoxide, 1934–1935
Vitamins, 314, 322–324
  corticosteroids, 322
  parathyroid hormone, 323
  thyroid hormones, 323
Vitronectin, 368
  receptors, 122–123
VLA-4. *See* Very late antigen-4 (VLA-4)
VMH. *See* Ventromedial hypothalamus (VMH)
Voltage-dependent Ca^{2+} channels (VDCCs), 553–554
Volume repletion, 1321
Volumetric bone mineral density (vBMD), 249, 252f, 1617
Volumetric QCT of spine and hip, 1863–1865
Volumetric spatial decomposition, 1840–1841
Von Hippel–Lindau (VHL), 208
von Hippel–Lindau tumor-suppressor protein (pVHL), 1147
von Kossa/MacNeal's stain, 1903
vSMC. *See* Vascular smooth muscle cell (vSMC)
VTEs. *See* Venous thrombolic events (VTEs)

W

Warburg effect, 1345
Watery diarrhea, hypokalemia, and achlorhydria syndrome (WDHA syndrome), 1299
Wear debris–induced osteoclastic bone resorption, 1699
Weight
  loss, 950
  weight-bearing exercise, 1759–1760
WES. *See* Whole-exome sequencing (WES)
WEST. *See* Women's Estrogen for Stroke Trial (WEST)
Western Ontario and McMaster Universities Osteoarthritis (WOMAC), 1172–1173
Western/southwestern blot analysis, 1520
Wg. See Wingless gene (*Wg*)
WHI. *See* Women's Health Initiative (WHI)
WHO. *See* World Health Organization (WHO)
Whole-body phosphorus economy, 473, 473f
Whole-body vibration intervention, 1768
Whole-bone mechanical testing, 1924–1927
  compression testing, 1926
  finite element analysis, 1927
  fracture toughness testing, 1926–1927, 1927f
  outcomes, 1925t
  in preclinical drug development, 1926
  specimen preparation, 1924
  standard three-point and four-point bending to failure, 1924–1925, 1925f
  torsion to failure, 1926
Whole-exome sequencing (WES), 488, 1624–1625
Wild-type controls (WT controls), 1249
Williams syndrome transcription factor (WSTF), 724–725
WINAC, 724–725
Windkessel physiology, 632
Wingless gene (*Wg*), 178–179
Wingless-related integration site (Wnt), 23–24, 778, 1340

Wingless-related integration site (Wnt) (*Continued*)
and bone cell function, 183–186
Dsh, glycogen synthase kinase-3β, Axin, and β-catenin, 181–183
genes and proteins, 178–179
Lrp5, Lrp6, and Frizzled, 180–181
signaling, 8
osteoblast biology, 99–101
pathway, 650–651, 951–952, 1711, 1776, 1805
in periodontium and Sost gene, 1064–1065, 1066f
transcription factors acting downstream of, 170–171
signals
canonical, 121–122
noncanonical, 122
transcriptional regulation by β-catenin, 183
Wnt/calcium pathway, 1715
Wnt/catenin signals, 1776
Wnt/β-catenin pathway, 177, 178f, 531, 761, 1715
components of, 179–183
signaling, 8, 15, 426–428, 1120–1121
Wnt1, 184
WNT16, 122, 185
mRNA, 121
WNT3A, 121, 184–185
WNT5a, 24, 122
Wnt7a, 184
Wnt7b, 1340
Wnt. *See* Wingless-related integration site (Wnt)
Wnt/catenin signal activation, 1776
Wolff's law, 145
Women
aromatase inhibitors in women with breast cancer, 1696
of reproductive age, 1702
Women's Estrogen for Stroke Trial (WEST), 877–878
Women's Health Initiative (WHI), 762–763, 831, 1652
World Health Organization (WHO), 1857
Wound healing, 1259
WSTF. *See* Williams syndrome transcription factor (WSTF)

X

X-linked hyper IgM (XHIM), 1209
X-linked hypophosphatemia (XLH), 471, 478–481, 1540, 1818. *See also* hypophosphatemia
clinical manifestations, 479
pathophysiology, 479
prevalence, 479
therapy, 479
X-linked hypophosphatemic rickets, 1529, 1532
X-linked osteopetrosis (XLO), 1553–1554
X-linked recessive hypoparathyroidism, 1366–1368
X-linked recessive hypophosphatemic rickets, 490
X-ray
diffraction, 295–296
examination, 1712
fluorescence technique, 529
tomographic microscopes, 1834
x-ray-based technique, 1842
Xenotropic retroviral receptor 1 (XPR1), 478
XHIM. *See* X-linked hyper IgM (XHIM)
XLH. *See* X-linked hypophosphatemia (XLH)
XLO. *See* X-linked osteopetrosis (XLO)
XPR1. *See* Xenotropic retroviral receptor 1 (XPR1)

Y

Y-box binding factor (YB-1), 308
"Yamanaka" factors, 1162
Yellow FP (YFP), 1888–1889
Young adulthood, 256–263
reversible bone loss and microstructural deterioration, 256–257
Young normal reference population (YN reference population), 1861

Z

Z/EG. *See* Cre-mediated lacZ deletion (Z/EG)
ZDF rats. *See* Zucker diabetic fatty rats (ZDF rats)
Zebrafish, 1787
zebrafish. *See* *Danio rerio* (zebrafish)
ZES. *See* Zollinger–Ellison syndrome (ZES)
Zinc finger protein 521 (2fp521), 167
Zinc-dependent neutral endoproteinases, 379–380
"Zipper-like" fashion, 299–300
ZO-1. *See* Zonula occludens-1 (ZO-1)
Zoledronate, 1496, 1675
Zoledronic acid, 460, 1322
Zollinger–Ellison syndrome (ZES), 1296
Zone of polarizing activity (ZPA), 14
Zonula occludens-1 (ZO-1), 424–425
Zucker diabetic fatty rats (ZDF rats), 948